TEACHER'S EDITION

# PRECALCULUS

GRAPHING, NUMERICAL, ALGEBRAIC

**FIFTH EDITION**

**Franklin Demana**
Ohio State University

**Bert K. Waits**
Ohio State University

**Gregory D. Foley**
Appalachian State University

**Daniel Kennedy**
Baylor School

An imprint of Addison Wesley Longman, Inc.

Reading, Massachusetts • San Francisco, California • New York • Harlow, England
Don Mills, Ontario • Sydney • Mexico City • Madrid • Amsterdam

Sponsoring Editor: Bill Poole
Project Manager: Jennifer Kerber
Senior Production Supervisor: Peggy McMahon
Design Supervisor: Barbara Atkinson
Cover Design: Leslie Haimes
Cover Photo: © D. Boone/Corbis
Production Services: Laurel Technical Services
Marketing Manager: Brenda Bravener
Marketing Coordinator: Laura Potter

**Credits**

**Photo Credits xx** ©Lonnie Duka/Index Stock Imagery **60** ©Benelux Press/Index Stock Imagery **154** ©Taste of Europe/Index Stock Imagery **258** ©Catrina Genovese/Index Stock Imagery **336** ©John Wood/Index Stock Imagery **422** ©Wendy Shattil/Index Stock Imagery **478** ©John Warden/Index Stock Imagery **542** ©Jacob Halaska/Index Stock Imagery **602** ©Mark Gibson/Index Stock Imagery **668** ©Carol Werner/Index Stock Imagery **752** ©Tim Haske/Index Stock Imagery©

**505** Exploration is based on a problem originally posed by Neal Koblitz in March 1988 issue of The American Mathematical Monthly.

ISBN 0-201-70318-1

1 2 3 4 5 6 7 8 9 10–RNT–020100

# Preface

Our text combines appropriate use of technology with standard "paper and pencil" analytic techniques to provide a balanced approach to the study and implementation of precalculus. Technology is fully integrated, rather than just added. The text encourages graphical, numerical, and algebraic modeling of functions as well as problem solving, conceptual understanding, and facility with technology.

Our primary objectives are:

- to help students to truly understand the fundamental concepts of algebra, trigonometry, and analytic geometry,

- to foreshadow important ideas of calculus, and

- to show how algebra and trigonometry can be used to model real-life problems.

In writing this edition, we have followed the guidelines and recommendations published by AMATYC, MAA, and NCTM, and also responded to the many helpful suggestions of both students and instructors. As a result, we believe that the changes made in this edition make this the most effective text available to prepare students for calculus, science, and advanced mathematics courses.

## The Rule of Four - A Balanced Approach

A principal feature of this edition is the balance among the algebraic, numerical, graphical, and verbal methods of representing problems: the rule of four. For instance, we obtain solutions algebraically when that is the most appropriate technique to use, and we obtain solutions graphically or numerically when algebra is difficult to use. We urge students to solve problems by one method and then support or confirm their solutions by using another method. We believe that students must learn the value of each of these methods of representation and must learn to choose the one most appropriate for solving the particular problem under consideration. This approach reinforces the idea that to understand a problem fully, students need to understand it algebraically as well as graphically and numerically.

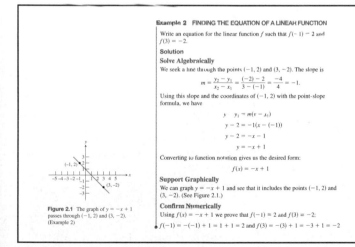

**Example 2**   FINDING THE EQUATION OF A LINEAR FUNCTION

Write an equation for the linear function $f$ such that $f(-1) = 2$ and $f(3) = -2$.

**Solution**

**Solve Algebraically**

We seek a line through the points $(-1, 2)$ and $(3, -2)$. The slope is

$$m = \frac{y_2 - y_1}{x_2 - x_1} = \frac{(-2) - 2}{3 - (-1)} = \frac{-4}{4} = -1.$$

Using this slope and the coordinates of $(-1, 2)$ with the point-slope formula, we have

$$y - y_1 = m(x - x_1)$$
$$y - 2 = -1(x - (-1))$$
$$y - 2 = -x - 1$$
$$y = -x + 1$$

Converting to function notation gives us the desired form:

$$f(x) = -x + 1$$

**Support Graphically**

We can graph $y = -x + 1$ and see that it includes the points $(-1, 2)$ and $(3, -2)$. (See Figure 2.1.)

**Confirm Numerically**

Using $f(x) = -x + 1$ we prove that $f(-1) = 2$ and $f(3) = -2$:
$f(-1) = -(-1) + 1 = 1 + 1 = 2$ and $f(3) = -(3) + 1 = -3 + 1 = -2$

**Figure 2.1** The graph of $y = -x + 1$ passes through $(-1, 2)$ and $(3, -2)$. (Example 2)

For Example 2, see page 157.

## Algebraic Skills

- The Prerequisite Chapter includes a review of solving equations and inequalities algebraically and graphically.

- Appendix A gives instructors the option of including algebraic skills review when necessary.

- Quick Review exercises precede each section exercise set so that students can review the algebraic skills needed to solve the exercises in that section.

## Applications and Real Data

The majority of the applications in the text are based on real data from cited sources, and their presentations are self-contained; students will not need any experience in the fields from which the applications are drawn.

As they work through the applications, students are exposed to functions as mechanisms for modeling data and are motivated to learn about how various functions can help model real-life problems. They learn to analyze and

**Example 6** MODELING U.S. POPULATION USING EXPONENTIAL REGRESSION

Use the 1900–1990 data in Table 3.9 and exponential regression to predict the U.S. population for 1998. Compare the result with the listed value for 1998.

**Solution**

**Model**

Let $P(t)$ be the population (in millions) of United States $t$ years after 1900. Figure 3.16a shows a scatter plot of the data. Using exponential regression, we find a model for the data.

$$P(t) = 80.075 \cdot 1.0131^t$$

Figure 3.16b on the next page shows the scatter plot of the data with a graph of the population model just found. You can see that the curve fits the data fairly well. The coefficient of determination is $r^2 \approx 0.9945$, indicating a close fit and supporting the visual evidence.

**Solve Graphically**

To predict the 1998 U.S. population we substitute $t = 98$ into the regression model. Figure 3.16c reports that $P(98) = 80.075 \cdot 1.0131^{98} \approx 286.7$.

**Interpret**

The model predicts the U.S. population was 286.7 million in 1998. See Figure 3.16c. The actual population was 273.8 million. We overestimated by 12.9 million, slightly less than a 5% error.

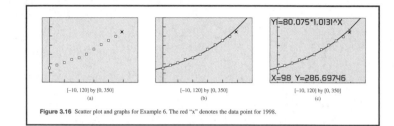

[−10, 120] by [0, 350]     [−10, 120] by [0, 350]     [−10, 120] by [0, 350]
(a)                        (b)                        (c)

**Figure 3.16** Scatter plot and graphs for Example 6. The red "x" denotes the data point for 1998.

For Example 6, see pages 276–277.

| Table 3.9 | U.S. Population (in millions) |
|---|---|
| Year | Population |
| 1900 | 76.2 |
| 1910 | 92.2 |
| 1920 | 106.0 |
| 1930 | 123.2 |
| 1940 | 132.2 |
| 1950 | 151.3 |
| 1960 | 179.3 |
| 1970 | 203.3 |
| 1980 | 226.5 |
| 1990 | 248.7 |
| 1998 | 273.8 |

Source: 1999 New York Times Almanac

model data, represent data graphically, interpret from graphs, and fit curves. Additionally, the tabular representation of data presented in this text highlights the concept that a function is a correspondence between numerical variables. This helps students build the connection between the numbers and graphs and recognize the importance of a full graphical, numerical, and algebraic understanding of the problem. For a complete listing of applications, please see the Applications Index on page 1021.

## Chapter Openers

The answer to the chapter opening application is given at the end of the section in which the appropriate concepts are covered. Frequently, real data is presented in these problems to enable students to explore realistic situations using graphical, numerical and algebraic methods. In other cases, students are asked to model problem situations using the functions studied in the chapter.

For Chapter Opener, see page 60.

# Problem Solving Approach

Systematic problem solving is emphasized in the examples throughout the text, using the following variation of Polya's problem-solving process:

*understand* the problem;

*develop* a mathematical model;

*solve* the mathematical model and support or confirm the solutions;

and *interpret* the solution.

Students are encouraged to use this process throughout the text.

For Example 2, see page 544.

---

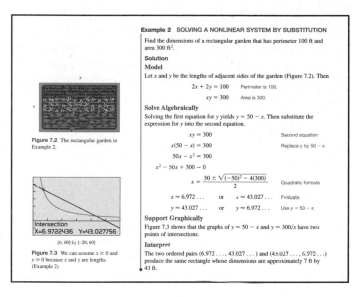

**Example 2**  SOLVING A NONLINEAR SYSTEM BY SUBSTITUTION

Find the dimensions of a rectangular garden that has perimeter 100 ft and area 300 ft$^2$.

**Solution**

**Model**

Let $x$ and $y$ be the lengths of adjacent sides of the garden (Figure 7.2). Then

$$2x + 2y = 100 \qquad \text{Perimeter is 100.}$$
$$xy = 300 \qquad \text{Area is 300.}$$

**Solve Algebraically**

Solving the first equation for $y$ yields $y = 50 - x$. Then substitute the expression for $y$ into the second equation.

$$xy = 300 \qquad \text{Second equation}$$
$$x(50 - x) = 300 \qquad \text{Replace } y \text{ by } 50 - x.$$
$$50x - x^2 = 300$$
$$x^2 - 50x + 300 = 0$$
$$x = \frac{50 \pm \sqrt{(-50)^2 - 4(300)}}{2} \qquad \text{Quadratic formula}$$
$$x \approx 6.972 \ldots \quad \text{or} \quad x \approx 43.027 \ldots \qquad \text{Evaluate.}$$
$$y \approx 43.027 \ldots \quad \text{or} \quad y \approx 6.972 \ldots \qquad \text{Use } y = 50 - x.$$

**Support Graphically**

Figure 7.3 shows that the graphs of $y = 50 - x$ and $y = 300/x$ have two points of intersections.

**Interpret**

The two ordered pairs $(6.972 \ldots, 43.027 \ldots)$ and $(43.027 \ldots, 6.972 \ldots)$ produce the same rectangle whose dimensions are approximately 7 ft by 43 ft.

**Figure 7.2**  The rectangular garden in Example 2.

**Figure 7.3**  We can assume $x \geq 0$ and $y \geq 0$ because $x$ and $y$ are lengths. (Example 2)

Intersection
X=6.9722436   Y=43.027756
[0, 60] by [-20, 60]

---

# Graphing Utilities

Students are expected to use a graphing utility (grapher) to visualize and solve problems. Throughout the text, the student is prompted to *recognize* that a graph is reasonable; *identify* all the important characteristics of a graph, *interpret* those characteristics, and *confirm* them using analytic techniques. By using this method to analyze the numerous figures presented in the examples and exercises, students develop excellent graph-viewing skills.

For Example 9, see page 194.

---

**Figure 2.33**  Cut square corners from a piece of cardboard, and fold the flaps to make a box. (Example 9)

| X | Y1 |
|---|---|
| 1 | 414 |
| 2 | 672 |
| 3 | 798 |
| 4 | 816 |
| 5 | 750 |
| 6 | 624 |
| 7 | 462 |

Y1 ◼ X(20−2X)(25−...

**Figure 2.34**  A table to get a feel for the volume values in Example 9.

[0, 10] by [0, 1000]

**Figure 2.35**  $y_1 = x(25 - 2x)(20 - 2x)$ and $y_2 = 484$. (Example 9)

**Example 9**  DESIGNING A BOX

Dixie Packaging Company has contracted to make boxes with a volume of approximately 484 in$^3$. Squares are to be cut from the corners of a 20-in. by 25-in. piece of cardboard, and the flaps folded up to make an open box. (See Figure 2.33.) What size squares should be cut from the cardboard?

**Solution**

**Model**

We know that the volume $V = \text{height} \times \text{length} \times \text{width}$. So let

$$x = \text{edge of cut-out square (height of box)}$$
$$25 - 2x = \text{length of the box}$$
$$20 - 2x = \text{width of the box}$$
$$V = x(25 - 2x)(20 - 2x)$$

**Solve Numerically and Graphically**

For a volume of 484, we solve the equation $x(25 - 2x)(20 - 2x) = 484$. Because the width of the cardboard is 20 in., $0 \leq x \leq 10$. We use the table in Figure 2.34 to get a sense of the volume values to set the window for the graphs in Figure 2.35. The cubic volume function intersects the constant volume of 484 at $x \approx 1.22$ and $x \approx 6.87$.

**Interpret**

Squares with lengths of approximately 1.22 in. or 6.87 in. should be cut from the cardboard to produce a box with a volume of 484 in.$^3$.

Just as any two points in the Cartesian plane with different $x$ values and different $y$ values determine a unique slant line and its related linear function, any three noncollinear points with different $x$ values determine a quadratic function. In general, $(n + 1)$ points positioned with sufficient generality determine a polynomial function of degree $n$. The process of fitting a polynomial of degree $n$ to $(n + 1)$ points is **polynomial interpolation**. Exploration 2 involves two polynomial interpolation problems.

## Chapter 5 Project

### Modeling the Illumination of the Moon

From the earth, the Moon appears to be a circular disk in the sky that is illuminated to varying degrees by direct sunlight. During each lunar orbit the Moon varies from a status of being a New Moon with no visible illumination to that of a Full Moon which is fully illuminated by direct sunlight. The United States Naval Observatory has developed a mathematical model to find the fraction of the Moon that is illuminated by the Sun. The data in the table below (obtained from the U.S. Naval Observatory web site, http://www.usno.navy.mil/, Astronomical Applications Department) shows the fraction of the Moon illuminated at midnight for each day in January 2000.

| | Fraction of the Moon Illuminated, January 2000 | | | | | | |
|---|---|---|---|---|---|---|---|
| Day # | Fraction illuminated | Day # | Fraction illuminated | Day # | Fraction illuminated | Day # | Fraction illuminated |
| 1 | 0.25 | 9 | 0.05 | 17 | 0.78 | 25 | 0.80 |
| 2 | 0.18 | 10 | 0.11 | 18 | 0.87 | 26 | 0.71 |
| 3 | 0.11 | 11 | 0.18 | 19 | 0.94 | 27 | 0.61 |
| 4 | 0.06 | 12 | 0.26 | 20 | 0.99 | 28 | 0.51 |
| 5 | 0.02 | 13 | 0.36 | 21 | 1.00 | 29 | 0.42 |
| 6 | 0.00 | 14 | 0.46 | 22 | 0.99 | 3 | 0.32 |
| 7 | 0.00 | 15 | 0.57 | 23 | 0.94 | 31 | 0.24 |
| 8 | 0.02 | 16 | 0.68 | 24 | 0.88 | | |

### Explorations

1. Enter the data in the table above into your grapher or computer. Create a scatter plot of the data.

2. Find values for $a$, $b$, $c$, and $d$ so the equation $y = a \cos (b(x - c)) + d$ models the data in the data plot.

3. Verify graphically the cofunction identity $\sin (\pi/2 - \theta) = \cos \theta$ by substituting $(\pi/2 - \theta)$ for $\theta$ in the model above and using sine instead of cosine. (Note $\theta = b(x - c)$.) Observe how well this new model fits the data.

4. Verify graphically the odd-even identity $\cos (\theta) = \cos (-\theta)$ for the model in #2 by substituting $-\theta$ for $\theta$ and observing how well the graph fits the data.

5. Find values for $a$, $b$, $c$, and $d$ so the equation $y = a \sin (b(x - c)) + d$ fits the data in the table.

6. Verify graphically the cofunction identity $\cos (\pi/2 - \theta) = \sin \theta$ by substituting $(\pi/2 - \theta)$ for $\theta$ in the model above and using cosine rather than sine. (Note $\theta = b(x - c)$.) Observe the fit of this model to the data.

7. Verify graphically the odd-even identity $\sin (-\theta) = -\sin (\theta)$ for the model in #5 by substituting $-\theta$ for $\theta$ and graphing $-a \sin (-\theta) + d$. How does this model compare to the original one?

For Chapter Project, see page 477.

## Chapter Projects

Each chapter concludes with a project that requires students to analyze data, and can be assigned as either individual or group work. Each project expands upon concepts and ideas taught in the chapter, and many projects refer students to the Web for further investigation of real data.

## Tips and Historical Notes

**Tips** throughout the text offer practical advice to students on using their grapher to obtain the best, most accurate results. **Margin Notes** include historical information, hints about examples, and provide additional insight to help students avoid common pitfalls and errors.

### Some Words of Warning

In Figure 3.21, notice we used "10^Ans" instead of "10^1.537819095" to check log (34.5). This is because graphers generally store more digits than they display and so we can obtain a more accurate check. Even so, because log (34.5) is an irrational number, a grapher cannot produce its exact value, so checks like those shown in Figure 3.21 may not always work out so perfectly.

For tip, see page 285.

### History of Conic Sections

Parabolas, ellipses, and hyperbolas had been studied for many years when Apollonius (c. 262–190 B.C.) wrote his eight-volume *Conic Sections*. Apollonius, who was born in Perga in northwestern Asia Minor, was the first to unify these three curves as cross sections of a cone and was the first to view the hyperbola as having two branches. Interest in these curves was renewed in the 17th century when Galileo proved that projectiles follow parabolic paths and Kepler discovered that planets travel in elliptical orbits.

For historical note, see page 603.

## Math at Work

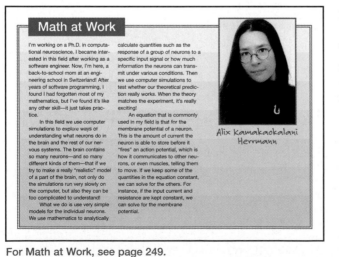

Math at Work

I'm working on a Ph.D. in computational neuroscience. I became interested in this field after working as a software engineer. Now, I'm here, a back-to-school mom at an engineering school in Switzerland! After years of software programming, I found I had forgotten most of my mathematics, but I've found it's like any other skill—it just takes practice.

In this field we use computer simulations to explore ways of understanding what neurons do in the brain and the rest of our nervous systems. The brain contains so many neurons—and so many different kinds of them—that if we try to make a really "realistic" model of a part of the brain, not only do the simulations run very slowly on the computer, but also they can be too complicated to understand! What we do is use very simple models for the individual neurons. We use mathematics to analytically

calculate quantities such as the response of a group of neurons to a specific input signal or how much information the neurons can transmit under various conditions. Then we use computer simulations to test whether our theoretical prediction really works. When the theory matches the experiment, it's really exciting!

An equation that is commonly used in my field is that for the membrane potential of a neuron. This is the amount of current the neuron is able to store before it "fires" an action potential, which is how it communicates to other neurons, or even muscles, telling them to move. If we keep some of the quantities in the equation constant, we can solve for the others. For instance, if the input current and resistance are kept constant, we can solve for the membrane potential.

Alix Kamakaokalani Herrmann

The Math at Work feature introduces the student to individuals who are using mathematics in their jobs. Students see how the ideas they are studying are used in real life.

For Math at Work, see page 249.

For Chapter 10, see page 752.

## Looking Ahead to Calculus

One of the goals of this text is to help build an intuitive foundation for calculus.

- Throughout the text, many examples and topics are marked with an icon, to point out concepts that students will encounter again in calculus. Ideas that foreshadow calculus are highlighted, such as limits, maximum and minimum, asymptotes, and continuity.
- Early in the text, students are introduced to the idea of the limit using an intuitive and conceptual approach. Some calculus notation and language is introduced in the beginning chapters and used throughout the text to establish familiarity.
- Chapter 10, A Preview of Calculus, is new to this edition. The chapter first provides an historical perspective to calculus by presenting the classical studies of motion through the tangent line and area problems. Limits are then investigated further, and the chapter concludes with graphical and numerical examinations of derivatives and integrals.

## Exercise Sets

The exercise sets have been extensively revised for this edition. There are over 6000 exercises including 680 Quick Review Exercises. These Quick Review Exercises appear at the beginning of each exercise set and help students review skills that will be needed in the rest of the exercise set. Following the Quick Review are exercises that allow students to practice the algebraic skills that they have learned in that section. These have been carefully graded from routine to challenging. The following types of skills are tested in each exercise set:

- Algebraic and analytic manipulation
- Connecting algebra to geometry
- Interpretation of graphs
- Graphical and numerical representations of functions
- Data analysis

Also included in the exercise sets are thought-provoking exercises:

- **Explorations** are opportunities for students to discover mathematics on their own or in groups. These exercises often require the use of critical thinking to explore the ideas.
- **Writing to Learn** exercises give students practice at communicating about mathematics and opportunities to demonstrate their understanding of important ideas.
- **Extending the Ideas** exercises go beyond what is presented in the textbook. These exercises are challenging extensions of the book's material. Students may be asked to work on these problems in groups or to solve them as individual or group projects.

This variety provides sufficient flexibility to emphasize the skills most needed for each student or class. Answers to odd numbered problems appear at the back of the student text while all answers appear either on the page or at the end of the teacher's text. Exercises marked with a icon are multi-part Advanced Placement (AP) preparation questions and require more thought and analysis than some of our standard exercises.

## Chapter Review

At the end of each chapter are sections dedicated to helping students review the chapter concepts and material. "Key Ideas" has four parts: Concepts; Properties, Theorems and Formulas; Procedures; and Gallery of Functions. The "Review Exercises" represent the full range of exercises covered in the chapter and give students additional practice with the ideas developed in the chapter. The exercises with colored numbers indicate problems that we feel would make up a good practice test.

## Teacher Edition

The Teacher Edition includes notes written specifically for the teacher. These notes include chapter and section objectives, suggested assignments, and

teaching tips. Various examples are marked with a ▦ icon. Alternates are provided for these examples.

## Changes to this edition

### Chapter P

The new Prerequisites Chapter contains two new sections on solving equations and inequalities graphically and algebraically. The algebra review material has been moved to the new Appendix A. This new appendix will help students review, sharpen, and hone their algebra skills.

### Chapter 1

The basic functions are presented early in this chapter, enabling students to work with a richer variety of functions when learning the basic concepts—a benefit of grapher technology that gives coherence to the rest of the course. The idea of the limit is introduced conceptually along with limit notation to help describe end behavior and asymptotes. Graphical, algebraic, and numerical modeling with functions is emphasized from the beginning.

### Chapter 2

This chapter covers polynomial, rational, and power functions. The treatment of rational functions has been streamlined, but coverage of power functions has been extended. We introduce average rate of change of a function to show how the rate of change of a linear function can be related to slope. Students will encounter average rate of change again in calculus, and its inclusion in this chapter allows for the investigation of more interesting applications and models.

### Chapter 3

The material in this chapter has been reorganized. We include more on mathematical modeling with special emphasis on this technique in Sections 3.2 and 3.5. We include more material on logistic functions and the natural base $e$, and also address the concept of "orders of magnitude."

### Chapter 4

In this chapter we introduce the trigonometric functions using balanced algebraic and graphical approaches. This allows the students to better understand how the functions behave. We have moved the discussion of vectors to Chapter 6 to allow for a more extended treatment of this concept.

## Chapter 5

This is the second of our new trigonometry chapters. Here, we use trigonometric identities to teach mathematical proof. We also use word ladders to illustrate the strategy for proving an identity. The inclusion of more "explorations" allows students to discover trigonometric relationships rather than just memorize them.

## Chapter 6

This chapter is dedicated to the discussion of vectors and parametric and polar equations. Coverage of vectors has been extended to two sections, rather than one in this edition. The material on Parametric Equations has been consolidated into its own section, and is followed by sections on polar coordinates and polar equations.

## Chapter 7

This chapter is a reorganization of Chapter 10 of the previous edition. We have included a new section on matrix algebra, and delayed our discussion of partial fractions until this chapter.

## Chapter 8

The material in this chapter has also been reorganized. Coverage of conics has been expanded to three separate sections on parabolas, ellipses, and hyperbolas for a more thorough and complete treatment. We have revised our approach to translations and rotations and have added a new section on three-dimensional analytic geometry. We feature a variety of applications of conics, including a great deal on reflective properties and applications to celestial mechanics.

## Chapter 9

This is our chapter on Discrete Mathematics. The chapter is organized so that coverage of combinatorics appears before probability and the binomial theorem. The section on probability includes a discussion of conditional probability. Data analysis is emphasized throughout the chapter, especially in the statistics sections where we use real and up-to-date data in our examples and exercises.

## Chapter 10

This introduction to calculus chapter is new to this text. The chapter first provides an historical perspective to calculus by presenting the classical studies of motion through the tangent line and area problems. Limits are then investigated further, and the chapter concludes with graphical and numerical examinations of derivatives and integrals.

## Supplements

### For the Teacher

**Resource Manual** (0-201-70049-2)

- Major Concepts review

- Group Activity Worksheet

- Sample Chapter Tests

- Standardized Test Preparation Questions

- Contest Problems

**Complete Solutions Manual** (0-201-69973-7)

- Complete solutions to all exercises, including chapter review exercises

**Tests and Quizzes** (0-201-70067-0)

- Two parallel test per chapter

- Two quizzes for every 3–4 sections

- Two parallel mid-term tests, covering Chapters P–5

- Two parallel end-of-year tests, covering Chapters 6–10

**Acetates and Transparencies** (0-201-70320-3)

- 1-color acetates of all Quick Review Exercises

- 4-color acetates of 10-15 useful general transparencies

- One Alternate Example per lesson.

**TestGen: Computerized Test Generator** (0-201-69972-9)

- The teacher can easily view, edit, and add questions, transfer questions to tests, and print tests in a variety of fonts and formats.

- Built-in question editor gives the ability to create graphs, import graphics, insert mathematical symbols and templates, and insert variable numbers or text.

- The teacher can create and save tests using TestGen-EQ so students can take them for practice or a grade on a computer network using QuizMaster-EQ.

- The teacher can set preferences for how and when tests are administered with QuizMaster-EQ.

- All tests can be saved as HTML for use on the Web.

## For the Student

**Student's Solution Manual** (0-201-69975-3)

- Contains detailed, worked-out solutions to odd-numbered exercises

**Graphing Calculator Manual** (0-201-70068-9)

- Grapher Workshop provides detailed instruction on important grapher features.

- Features TI-83 and -85, Casio, and HP.

**InterAct Math** (0-201-69976-1)

- **CD version:** Available bundled with the text, InterAct Math on CD-ROM provides extra practice and interactive tutorials with algorithmically generated exercises correlated to this text. Each practice exercise is accompanied by an example and guided solution designed to involve students in the solution process. The software recognizes common student errors and provides appropriate feedback. Student performance history is recorded with each use.

- **Website version:** On the Web, http://www.awl.com/demana, you can use our InterAct Math tutorials correlated to this text free of charge.

- **LAN version:** InterAct Math Plus is available for installation on your school's LAN. InterAct Math Plus enables instructors to create and administer online tests, summarize students' results, and monitor students' progress in the tutorial software, providing an invaluable teaching and tracking resource.

- **Full Web System:** Delivered over the Web, InterAct MathXL issues a diagnostic test correlated to this textbook, individualized study plans based on those test results, targeted practice and tutorial help in InterAct Math, and student tracking. After practicing exercises from their study plan, students can take another test and compare their results. A free InterAct MathXL registration coupon is available for bundle with this text.

## Website

Found at http://www.awl.com/demana, our free dedicated Web site provides dynamic resources for instructors and students. Some of the resources include: a syllabus manager, TI graphing calculator downloads, online quizzing, teaching tips, study tips, Explorations and end-of-chapter projects, InterAct Math tutorials, "Looking Ahead to Calculus" hints, and more.

## Acknowledgments

We would like to express our gratitude to the reviewers of this edition who provided such invaluable insight and comment.

| | |
|---|---|
| Brenda Batten | Thomas Heyward Academy |
| Allan Bellman | Blake High School |
| Charles Brown | Joel E. Ferris High School |
| Ellen S. Hook | Granby High School |
| Cathy Jahr | Westview High School |
| Mike Koehler | Blue Valley North High School |
| Kevin Quattrin | St. Ignatius Prep School |
| Vicki Shirley | Corinth High School |
| Susanna Uhl | Tamalpais Union High School |
| Nancy Wawer | Mark Keppel High School |
| Harlan Weber | Sheboygan Area School District |

We express special thanks to Chris Brueningsen, Linda Antinone, and Bill Bower for their work on the Chapter Projects. Our appreciation extends to the staff at Addison-Wesley for their advice and support in revising this text, particularly Jennifer Kerber, Bill Poole, Greg Tobin, Peggy McMahon, and Karen Guardino.

We would also like to thank the staff at Laurel Technical Services for their help with preparing and proofreading the manuscript, preparing and proofreading the art manuscript, and preparing and checking the print supplements. We are especially grateful to our "Taskmaster," Rochelle Blair, for keeping us in line and on schedule.

# About the Authors

## Franklin D. Demana

Frank Demana received his master's degree in mathematics and his Ph.D. from Michigan State University. Currently, he is Professor Emeritus of Mathematics at The Ohio State University. As an active supporter of the use of technology to teach and learn mathematics, he is cofounder of the national Teachers Teaching with Technology (T³) professional development program. He has been the director and codirector of more than $10 million of National Science Foundation (NSF) and foundational grant activities over the past 15 years. Along with frequent presentations at professional meetings, he has published a variety of articles in the areas of computer and calculator-enhanced mathematics instruction. Dr. Demana is also the cofounder (with Bert Waits) of the annual International Conference on Technology in Collegiate Mathematics (ICTCM). He his corecipient of the 1997 Glenn Gilbert National Leadership Award presented by the National Council of Supervisors of Mathematics.

Dr. Demana coauthored *Essential Algebra: A Calculator Approach*; *Transition to College Mathematics*; *College Algebra and Trigonometry: A Graphing Approach*; *College Algebra: A Graphing Approach*; *and Intermediate Algebra*: *A Graphing Approach.*

## Bert K. Waits

Bert Waits received his Ph.D. from The Ohio State University and is currently Professor Emeritus of Mathematics there. Dr. Waits is cofounder of the national Teachers Teaching with Technology (T³) professional development program, and has been codirector or principal investigator on several large NSF projects. Dr. Waits has published articles in more than 50 nationally recognized professional journals. He frequently gives invited lectures, workshops, and minicourses at national meetings of the MAA and the National Council of Teachers of Mathematics (NCTM) on how to use computer technology to enhance the teaching and learning of mathematics. He has given invited presentations at the International Congress on Mathematical Education (ICME 6, 7, and 8) in Budapest (1998), Quebec (1992), and Seville (1996). Dr. Waits is corecipient of the 1997 Glenn Gilbert National Leadership Award presented by the National Council of Supervisors of Mathematics, and is the cofounder (with Frank Demana) of the ICTCM. Dr Waits was elected to be a member of the National Council of Teachers of Mathematics Board of Directors 2000–2002.

Dr. Waits coauthored *Calculus: Graphical, Numerical, and Algebraic Calculus: A Complete Course*; *College Algebra and Trigonometry: A Graphing Approach*; *College Algebra: A Graphing Approach*; *and Intermediate Algebra: A Graphing Approach.*

## Gregory D. Foley

Gregory D. Foley is Distinguished Professor of Mathematics Education in the Department of Mathematical Sciences at Appalachian State University in Boone, North Carolina. He obtained B.A. and M.A. degrees in mathematics and a Ph.D. in mathematics education from The University of Texas at Austin. Dr. Foley has held faculty positions at North Harris County College, Austin Community College, The Ohio State University, and Sam Houston State University. He is a member of the Academic Coordinator Council for the Teachers Teaching with Technology (T³) program, and the Committee on Research in Undergraduate Mathematics Education of the American

Mathematical Society (AMS) and the Mathematical Association of America (MAA). Dr. Foley has presented numerous talks internationally and has directed projects for high school mathematics teachers. In 19998, Dr. Foley received the biennial American Mathematical Association of Two-Year Colleges (AMATYC) Award for Mathematics Excellence for his lifetime achievements.

## Daniel Kennedy

Dan Kennedy received his undergraduate degree from the College of the Holy Cross and his master's and Ph.D. in mathematics from the University of North Carolina at Chapel Hill. Since 1973 he has taught mathematics at the Baylor School in Chattanooga, Tennessee, where he holds the Cartter Lupton Distinguished professorship. Dr. Kennedy became an Advanced Placement Calculus reader in 1978, which led to an increasing level of involvement with the program as workshop consultant, table leader, and exam leader. He joined the Advanced Placement Calculus Test Development Committee in 1986, then in 1990 became the first high school teacher in 35 years to chair that committee. It was during his tenure as chair that the program moved to require graphing calculators and laid the early groundwork for the recent reform of the Advanced Placement Calculus curriculum. The author of the 1997 *Teacher's Guide—AP® Calculus*, Dr. Kennedy has conducted more than 50 workshops and institutes for high school calculus teachers. His articles on mathematics teaching have appeared in the *Mathematics Teacher* and the *American Mathematical Monthly*, and he is a frequent speaker on calculus curriculum reform at professional and civic meetings. Dr. Kennedy was named a Tandy Technology Scholar in 1992 and a Presidential Award winner in 1995.

Dr. Kennedy coauthored *Calculus: Graphical, Numerical, and Algebraic* and *Calculus: A Complete Course.*

# Contents

# P
## Prerequisites

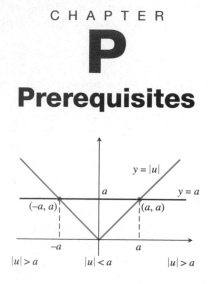

$$y = |u|$$
$$y = a$$
$$(-a, a)$$
$$(a, a)$$
$$-a \qquad a$$
$$|u| > a \qquad |u| < a \qquad |u| > a$$

CHAPTER

# 1
## Functions and Graphs

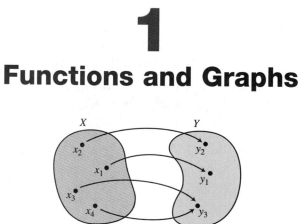

A function

CHAPTER

# 2

# Polynomial, Power and Rational Functions

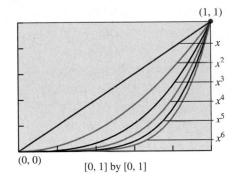

(1, 1)

$x$
$x^2$
$x^3$
$x^4$
$x^5$
$x^6$

(0, 0)

[0, 1] by [0, 1]

CHAPTER

# 3

# Exponential, Logistic, and Logarithmic Functions

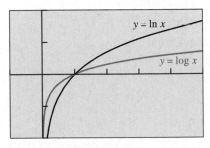

$y = \ln x$

$y = \log x$

[–1, 5] by [–2, 2]

CHAPTER

# 4

# Trigonometric Functions

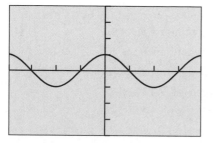

$[-2\pi, 2\pi]$ by $[-4, 4]$

CHAPTER

# 5

# Analytic Trigonometry

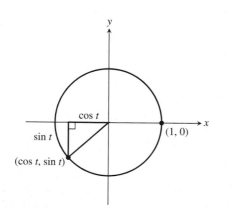

C H A P T E R

# 6

# Vectors, Parametric Equations, and Polar Equations

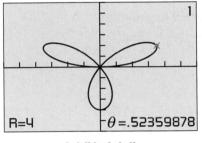

[–6,6] by [–6, 6]

C H A P T E R

# 7

# Systems and Matrices

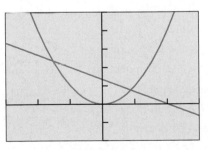

[–3, 3] by [–2, 5]

# 8

# Analytic Geometry in Two and Three Dimensions

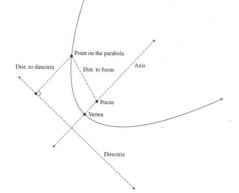

CHAPTER

# 9

# Discrete Mathematics

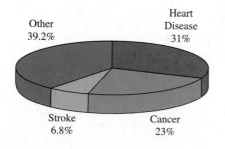

CHAPTER

# 10

# An Introduction to Calculus: Limits, Derivatives, and Integrals

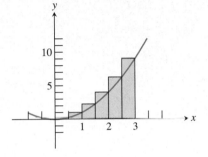

**N**othing can travel faster than the speed of light. Engineers planning space missions use the speed of light to measure large distances in units called light years. Using the speed of light, we can use the length of time it takes light to go from one object in space to another to determine the distance between the objects.

**T**he Moon is about 237,000 miles from Earth and light travels from the Earth to the Moon in approximately 1.27 seconds. The Earth is about 93 million miles from the Sun and light travels from the Earth to the Sun in 8.32 minutes. Given this information, how can we determine the speed of light? What is the distance from the Sun to Pluto if light can travel the distance in 5 hours and 29 minutes? We can answer these questions using methods from Section P.4. The solution for this problem is given on page 34.

## Chapter P Overview

Algebra is a language of representation. Historically, mathematicians and scientists used algebra to represent problems with symbols (algebraic model) and solve them by reducing the solution to algebraic manipulation of the symbols. This technique is still important today and, in this chapter, we review some of the basic algebraic techniques and properties that you used in previous courses.

Graphing calculators were invented in the 1980s, and we are now able to approach problems by representing them with graphs (graphical models) and solve them with numerical and graphical techniques. In this chapter we review the basics of solving equations and inequalities using both algebraic and graphical techniques.

We begin with some standard properties of the real number system, including order. Real numbers are represented as points on a number line, and ordered pairs of real numbers are graphed as points in the coordinate plane. We discuss the relationship among inequalities, interval notation, and intervals of real numbers. We review absolute value of real numbers, distance between points, midpoints of line segments, and equations of circles.

We study slope of a line and the usual basic forms for equations of lines. We write equations for lines that are parallel or perpendicular to a given line.

## Bibliography

**For students:** *Great Jobs for Math Majors*, Stephen Lambert, Ruth J. DeCotis. Mathematical Association of America, 1998.

**For teachers:** *Algebra in a Technological World, Addenda Series, Grades 9–12.* National Council of Teachers of Mathematics, 1995.

*Why Numbers Count—Quantitative Literacy for Tommorrow's America*, Lynn Arthur Steen (Ed.) National Council of Teachers of Mathematics, 1997.

## P.1 Real Numbers

Representing Real Numbers • Order and Interval Notation • Basic Properties of Algebra • Integer Exponents • Scientific Notation

### Representing Real Numbers

A **real number** is any number that can be written as a decimal. Real numbers are used in mathematics, science, and daily life to represent quantities or measurements such as the balance in your checking account, your height, age, distance, temperature, miles per gallon, and so forth. Real numbers are represented by symbols such as $6$, $-8$, $0$, $1.75$, $2.333...$, $8/5$, $\sqrt{3}$, $\sqrt[3]{16}$, $e$, and $\pi$.

The set of real numbers contains several important subsets:

The **natural (or counting) numbers:** $\{1, 2, 3, ...\}$

The **whole numbers:** $\{0, 1, 2, 3, ...\}$

The **integers:** $\{..., -3, -2, -1, 0, 1, 2, 3, ...\}$

The braces { } are used to enclose the **elements**, or **objects** of the set. The rational numbers are another important subset of the real numbers. A **rational number** is any number that can be written as a ratio $a/b$ of two integers, where $b \neq 0$. We can use **set-builder notation** to describe the rational numbers:

$$\left\{ \frac{a}{b} \,\middle|\, a, b \text{ are integers, and } b \neq 0 \right\}$$

The vertical bar that follows $a/b$ is read "such that."

The decimal form of a rational number either **terminates** like $7/4 = 1.75$, or is **infinitely repeating** like $4/11 = 0.363636... = 0.\overline{36}$. The bar over the 36

**Objective**

Students will be able to convert between decimals and fractions, write inequalities, apply the basic properties of algebra, and work with exponents and scientific notation.

**Motivate**

Ask students how real numbers can be classified. Have students discuss ways to display very large or very small numbers without using a lot of zeros.

indicates the block of digits that repeats. A real number is **irrational** if it is *not* rational. The decimal form of an irrational number is infinitely nonrepeating. For example, $\sqrt{3} = 1.7320508\ldots$ and $\pi = 3.14159265\ldots$.

Real numbers can be approximated with calculators. However, calculators give only a few of the digits of an irrational number and we sometimes can find the decimal form of rational numbers with calculators.

### Example 1 EXAMINING DECIMAL FORMS OF RATIONAL NUMBERS

Determine the decimal form of 1/16, 55/27, and 1/17.

**Solution** Figure P.1 suggests that the decimal form of 1/16 terminates and that of 55/27 repeats in blocks of 037.

$$\frac{1}{16} = 0.0625 \quad \text{and} \quad \frac{55}{27} = 2.\overline{037}$$

We can not predict the *exact* decimal form of 1/17 from Figure P.1, however we can say that $1/17 \approx 0.0588235294$. The symbol $\approx$ is read "*is approximately equal to.*" We can use long division (see Exercise 46) to show that

$$\frac{1}{17} = 0.\overline{0588235294117647}.$$

The real numbers and the points of a line can be matched *one-to-one* to form a **real number line**. We start with a horizontal line and match the real number zero with a point $O$, the **origin**. **Positive numbers** are assigned to the right of the origin, and **negative numbers** to the left, as shown in Figure P.2.

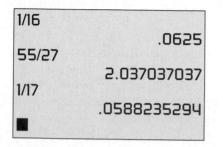

**Figure P.1** Calculator decimal representations of 1/16, 55/27, and 1/17 with the calculator set in floating decimal mode. (Example 1)

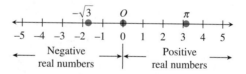

**Figure P.2** The real number line.

Every real number corresponds to one and only one point on the real number line, and every point on the real number line corresponds to one and only one real number. Between every pair of real numbers on the number line there are infinitely many more real numbers.

The number associated with a point is **the coordinate of the point**. As long as the context is clear, we will follow the standard convention of using the real number for both the name of the point and its coordinate.

## Order and Interval Notation

The set of real numbers is **ordered**. This means that we can compare any two real numbers that are not equal using inequalities and say that one is "less than" or "greater than" the other.

### Unordered Systems

Not all number systems are ordered. For example, the complex number system, to be introduced in Chapter 2, cannot be ordered.

---

$a < 0 \Rightarrow -a > 0$
If $a < 0$, then $a$ is to the left of 0 on the real number line, and its opposite, $-a$, is to the right of 0. Thus, $-a > 0$.

---

### Order of Real Numbers

Let $a$ and $b$ be any two real numbers.

| Symbol | Definition | Read |
|--------|-----------|------|
| $a > b$ | $a - b$ is positive | $a$ is greater than $b$ |
| $a < b$ | $a - b$ is negative | $a$ is less than $b$ |
| $a \geq b$ | $a - b$ is positive or zero | $a$ is greater than or equal to $b$ |
| $a \leq b$ | $a - b$ is negative or zero | $a$ is less than or equal to $b$ |

The symbols $>$, $<$, $\geq$, and $\leq$ are **inequality symbols**.

Geometrically, $a > b$ means that $a$ is to the right of $b$ (equivalently $b$ is to the left $a$) on the real number line. For example, since $6 > 3$, 6 is to the right of 3 on the real number line. Note also that $a > 0$ means that $a - 0$, or simply $a$, is positive and $a < 0$ means that $a$ is negative.

We are able to compare any two real numbers because of the following important property of the real numbers.

### Trichotomy Property

Let $a$ and $b$ be any two real numbers. Exactly one of the following is true:

$$a < b, \quad a = b, \quad \text{or} \quad a > b.$$

Inequalities can be used to describe **intervals** of real numbers, as illustrated in Example 2.

### Example 2   INTERPRETING INEQUALITIES

Describe and graph the intervals of real numbers for the inequality

**(a)** $x < 3$        **(b)** $-1 < x \leq 4$

**Solution**

**(a)** The inequality $x < 3$ describes all real numbers less than 3 (Figure P.3i).

**(b)** The *double inequality* $-1 < x \leq 4$ represents all real numbers between $-1$ and 4, excluding $-1$ and including 4 (Figure P.3ii).

### Example 3   WRITING INEQUALITIES

Write an interval of real numbers using an inequality and draw its graph.

**(a)** The real numbers between $-4$ and $-0.5$

**(b)** The real numbers greater than or equal to zero

**Solution**

**(a)** $-4 < x < -0.5$ (see Figure P.3iii)

**(b)** $x \geq 0$ (see Figure P.3iv)

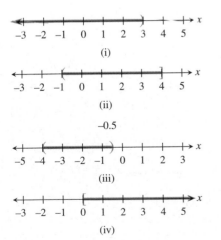

**Figure P.3** In graphs of inequalities parentheses correspond to $<$ and $>$, and brackets to $\leq$ and $\geq$. (Examples 2 and 3)

As shown in Example 2, inequalities define *intervals* on the real number line. We often use [2, 5] to describe the *bounded interval* determined by $2 \leq x \leq 5$. This interval is **closed** because it contains its *endpoints*, 2 and 5. There are four types of **bounded intervals**.

---

**Bounded Intervals of Real Numbers**

Let $a$ and $b$ be real numbers with $a < b$.

| Interval Notation | Interval Type | Inequality Notation | Graph |
|---|---|---|---|
| $[a, b]$ | Closed | $a \leq x \leq b$ | |
| $(a, b)$ | Open | $a < x < b$ | |
| $[a, b)$ | Half-open | $a \leq x < b$ | |
| $(a, b]$ | Half-open | $a < x \leq b$ | |

The numbers $a$ and $b$ are the **endpoints** of each interval.

---

**Interval Notation at ±∞**

Because $-\infty$ is *not* a real number, we use $(-\infty, 2)$ instead of $[-\infty, 2)$ to describe $x < 2$. Similarly, we use $[-1, \infty)$ instead of $[-1, \infty]$ to describe $x \geq -1$.

The interval of real numbers determined by the inequality $x < 2$ can be described by the *unbounded interval* $(-\infty, 2)$. This interval is **open** because it does *not* contain its endpoint 2.

We use the interval notation $(-\infty, \infty)$ to represent the entire set of real numbers. The symbols $-\infty$ (*negative infinity*) and $\infty$ (*positive infinity*) allow us to use interval notation for unbounded intervals and are *not* real numbers. There are four types of **unbounded intervals**.

---

**Unbounded Intervals of Real Numbers**

Let $a$ and $b$ be real numbers.

| Interval Notation | Interval Type | Inequality Notation | Graph |
|---|---|---|---|
| $[a, \infty)$ | Closed | $x \geq a$ | |
| $(a, \infty)$ | Open | $x > a$ | |
| $(-\infty, b]$ | Closed | $x \leq b$ | |
| $(-\infty, b)$ | Open | $x < b$ | |

Each of these intervals has exactly one endpoint, namely $a$ or $b$.

---

**Example 4    CONVERTING BETWEEN INTERVALS AND INEQUALITIES**

Convert interval notation to inequality notation or vice versa. State whether the interval is bounded, its type, and graph the interval.

**(a)** $[-6, 3)$      **(b)** $(-\infty, -1)$      **(c)** $-2 \leq x \leq 3$

## Solution

**(a)** The interval $[-6, 3)$ corresponds to $-6 \le x < 3$, and is bounded and half-open (see Figure P.4a).

**(b)** The interval $(-\infty, -1)$ corresponds to $x < -1$, and is unbounded and open (see Figure P.4b).

**(c)** The inequality $-2 \le x \le 3$ corresponds to the closed, bounded interval $[-2, 3]$ (see Figure P.4c).

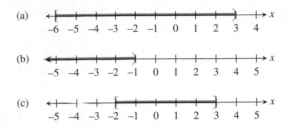

**Figure P.4** Graphs of the intervals of real numbers in Example 4.

## Basic Properties of Algebra

Algebra involves the use of letters and other symbols to represent real numbers. A **variable** is a letter or symbol (for example, $x$, $y$, $t$, $\theta$) that represents an unspecified real number. A **constant** is a letter or symbol (for example, $-2$, $0$, $\sqrt{3}$, $\pi$) that represents a specific real number. An **algebraic expression** is a combination of variables and constants involving addition, subtraction, multiplication, division, powers, and roots.

We state some of the properties of the arithmetic operations of addition, subtraction, multiplication, and division, represented by the symbols $+$, $-$, $\times$ (or $\cdot$) and $\div$ (or $/$), respectively. Addition and multiplication are the primary operations. Subtraction and division are defined in terms of addition and multiplication.

**Subtraction**: $\quad a - b = a + (-b)$

**Division**: $\qquad \dfrac{a}{b} = a\left(\dfrac{1}{b}\right), b \ne 0$

In the above definitions, $-b$ is the **additive inverse** or **opposite** of $b$, and $1/b$ is the **multiplicative inverse** or **reciprocal** of $b$. Perhaps surprisingly, additive inverses are not always negative numbers. The additive inverse of 5 is the negative number $-5$. However, the additive inverse of $-3$ is the positive number 3.

The following properties hold for real numbers, variables, and algebraic expressions.

**Subtraction vs. negative numbers**

On many calculators, there are two "$-$" keys, one for subtraction and one for negative numbers or opposites. Be sure you know how to use both keys correctly. Misuse can lead to incorrect results.

### Properties of Algebra

Let $u$, $v$, and $w$ be real numbers, variables, or algebraic expressions.

**1. Commutative property**
Addition: $u + v = v + u$
Multiplication: $uv = vu$

**2. Associative property**
Addition:
$(u + v) + w = u + (v + w)$
Multiplication: $(uv)w = u(vw)$

**3. Identity property**
Addition: $u + 0 = u$
Multiplication: $u \cdot 1 = u$

**4. Inverse property**
Addition: $u + (-u) = 0$
Multiplication: $u \cdot \dfrac{1}{u} = 1,\ u \neq 0$

**5. Distributive property**
Multiplication over addition
$u(v + w) = uv + uw$
$(u + v)w = uw + vw$

Multiplication over subtraction
$u(v - w) = uv - uw$
$(u - v)w = uw - vw$

The left-hand sides of the equations for the distributive property show the **factored form** of the algebraic expressions, and the right-hand sides show the **expanded form**.

### Example 5 USING THE DISTRIBUTIVE PROPERTY

**(a)** Write the expanded form of $(a + 2)x$.

**(b)** Write the factored form of $3y - by$.

**Solution**

**(a)** $(a + 2)x = ax + 2x$

**(b)** $3y - by = (3 - b)y$

Here are some properties of the additive inverse together with examples that help illustrate their meanings.

**Teaching Note**

Remind students that $-n$ can be either positive or negative, depending on the sign of $n$.

### Properties of the Additive Inverse

Let $u$ and $v$ be real numbers, variables, or algebraic expressions.

| Property | Example |
|---|---|
| 1. $-(-u) = u$ | $-(-3) = 3$ |
| 2. $(-u)v = u(-v) = -(uv)$ | $(-4)3 = 4(-3) = -(4 \cdot 3) = -12$ |
| 3. $(-u)(-v) = uv$ | $(-6)(-7) = 6 \cdot 7 = 42$ |
| 4. $(-1)u = -u$ | $(-1)5 = -5$ |
| 5. $-(u + v) = (-u) + (-v)$ | $-(7 + 9) = (-7) + (-9) = -16$ |

## Integer Exponents

Exponential notation is used to shorten products of factors that repeat. For example,

$$(-3)(-3)(-3)(-3) = (-3)^4 \quad \text{and} \quad (2x + 1)(2x + 1) = (2x + 1)^2.$$

---

**Exponential Notation**

Let $a$ be a real number, variable, or algebraic expression, and $n$ a positive integer. Then

$$a^n = \underbrace{a \cdot a \cdot \ldots \cdot a,}_{n \text{ factors}}$$

where $n$ is the **exponent**, $a$ is the **base**, and $a^n$ is the **$n$th power of $a$**, read as "$a$ to the $n$th power."

---

The two exponential expressions in Example 6 have the same value but have different bases. Be sure you understand the difference.

### Example 6   IDENTIFYING THE BASE

**(a)** In $(-3)^5$, the base is $-3$.

**(b)** In $-3^5$, the base is 3.

Here are the basic properties of exponents together with examples that help illustrate their meanings.

---

**Properties of Exponents**

Let $u$ and $v$ be real numbers, variables, or algebraic expressions, and $m$ and $n$ be integers. All bases are assumed to be nonzero.

| Property | Example |
|---|---|
| **1.** $u^m u^n = u^{m+n}$ | $5^3 \cdot 5^4 = 5^{3+4} = 5^7$ |
| **2.** $\dfrac{u^m}{u^n} = u^{m-n}$ | $\dfrac{x^9}{x^4} = x^{9-4} = x^5$ |
| **3.** $u^0 = 1$ | $8^0 = 1$ |
| **4.** $u^{-n} = \dfrac{1}{u^n}$ | $y^{-3} = \dfrac{1}{y^3}$ |
| **5.** $(uv)^m = u^m v^m$ | $(2z)^5 = 2^5 z^5 = 32z^5$ |
| **6.** $(u^m)^n = u^{mn}$ | $(x^2)^3 = x^{2 \cdot 3} = x^6$ |
| **7.** $\left(\dfrac{u}{v}\right)^m = \dfrac{u^m}{v^m}$ | $\left(\dfrac{a}{b}\right)^7 = \dfrac{a^7}{b^7}$ |

---

To simplify an expression involving powers means to rewrite it so that each factor appears only once, all exponents are positive, and exponents and constants are combined as much as possible.

---

**Calculator Note**

$(-3)^2 = 9$

$-3^2 = -9$

Be careful!

---

**Teaching Note**

Students should get plenty of practice in applying the properties of exponents. Remind them that the absence of an exponent implies the exponent is 1.

---

**Alert**

Some students may attempt to simplify an expression such as $x^4 + x^3$ by adding or multiplying the exponents. Remind them that some expressions cannot be simplified.

## Moving Factors

Be sure you understand how exponent Property 4 permits us to move factors from the numerator to the denominator and vice-versa:

$$\frac{v^{-m}}{u^{-n}} = \frac{u^n}{v^m}$$

## Example 7 SIMPLIFYING EXPRESSIONS INVOLVING POWERS

**(a)** $(2ab^3)(5a^2b^5) = 10(aa^2)(b^3b^5) = 10a^3b^8$

**(b)** $\dfrac{u^2v^{-2}}{u^{-1}v^3} = \dfrac{u^2u^1}{v^2v^3} = \dfrac{u^3}{v^5}$

**(c)** $\left(\dfrac{x^2}{2}\right)^{-3} = \dfrac{(x^2)^{-3}}{2^{-3}} = \dfrac{x^{-6}}{2^{-3}} = \dfrac{2^3}{x^6} = \dfrac{8}{x^6}$

## Scientific Notation

Any positive number can be written in **scientific notation**,

$$c \times 10^m, \text{ where } 1 \le c < 10 \text{ and } m \text{ is an integer.}$$

This notation provides a way to work with very large and very small numbers. For example, the distance between the Earth and the sun is about 93,000,000 miles. In scientific notation,

$$93,000,000 \text{ mi} = 9.3 \times 10^7 \text{ mi.}$$

The *positive exponent* 7 indicates that moving the decimal point in 9.3 to the right 7 places produces the decimal form of the number.

The weight of an oxygen molecule is about

$$0.000\ 000\ 000\ 000\ 000\ 000\ 000\ 053 \text{ grams}$$

In scientific notation,

$$0.000\ 000\ 000\ 000\ 000\ 000\ 000\ 053 \text{ g} = 5.3 \times 10^{-23} \text{ g.}$$

The *negative exponent* $-23$ indicates that moving the decimal point in 5.3 to the left 23 places produces the decimal form of the number.

## Example 8 CONVERTING TO AND FROM SCIENTIFIC NOTATION

**(a)** $2.375 \times 10^8 = 237,500,000$

**(b)** $0.000000349 = 3.49 \times 10^{-7}$

## Example 9 USING SCIENTIFIC NOTATION

Simplify $\dfrac{(370,000)(4,500,000,000)}{18,000}$.

**Solution**

**Using Mental Arithmetic**

$$\frac{(370,000)(4,500,000,000)}{18,000} = \frac{(3.7 \times 10^5)(4.5 \times 10^9)}{1.8 \times 10^4}$$

$$= \frac{(3.7)(4.5)}{1.8} \times 10^{5+9-4}$$

$$= 9.25 \times 10^{10}$$

$$= 92,500,000,000$$

### Using a Calculator

Figure P.5 shows two ways to perform the computation. In the first, the numbers are entered in decimal form. In the second, the numbers are entered in scientific notation. The calculator uses "9.25E10" to stand for $9.25 \times 10^{10}$.

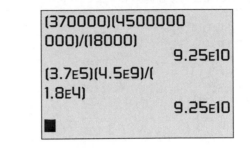

**Figure P.5** Be sure you understand how your calculator displays scientific notation. (Example 9)

# Quick Review P.1

1. List the positive integers between $-3$ and 7.

2. List the integers between $-3$ and 7.

3. List all negative integers greater than $-4$.  $\{-3, -2, -1\}$

4. List all positive integers less than 5.  $\{1, 2, 3, 4\}$

In Exercises 5 and 6, use a calculator to evaluate the expression. Round the value to two decimal places.

5. (a) $4(-3.1)^3 - (-4.2)^5$    (b) $\dfrac{2(-5.5) - 6}{7.4 - 3.8}$

6. (a) $5[3(-1.1)^2 - 4(-0.5)^3]$    (b) $5^{-2} + 2^{-4}$

In Exercises 7 and 8, evaluate the algebraic expression for the given values of the variables.

7. $x^3 - 2x + 1, x = -2, 1.5$  $-3; 1.375$

8. $a^2 + ab + b^2, a = -3, b = 2$  7

9. List the possible remainders when the positive integer $n$ is divided by 7.  0, 1, 2, 3, 4, 5, 6

10. List the possible remainders when the positive integer $n$ is divided by 13.  0, 1, 2, 3, 4, 5, 6, 7, 8, 9, 10, 11, 12

# Section P.1 Exercises

In Exercises 1–4, find the decimal form for the rational number. State whether it repeats or terminates.

1. $-37/8$

2. $15/99$  $0.\overline{15}$ (repeating)

3. $-13/6$  $-2.1\overline{6}$ (repeating)

4. $5/37$  $0.\overline{135}$ (repeating)

In Exercise 5–10, graph the interval of real numbers.

5. $x \le 2$

6. $-2 \le x < 5$

7. $(-\infty, 7)$

8. $[-3, 3]$

9. $x$ is negative

10. $x$ is greater than or equal to 2 and less than or equal to 6.

In Exercises 11–16, use an inequality to describe the interval of real numbers.

11. $[-1, 1)$  $-1 \le x < 1$

12. $(-\infty, 4]$

13.

-5 -4 -3 -2 -1  0  1  2  3  4  5  →  $x$

14.

-5 -4 -3 -2 -1  0  1  2  3  4  5  →  $x$

15. $x$ is between $-1$ and 2.  $-1 < x < 2$

16. $x$ is greater than or equal to 5.  $5 \le x < \infty$, or $x \ge 5$

In Exercises 17–22, use interval notation to describe the interval of real numbers.

17. $x > -3$  $(-3, \infty)$

18. $-7 < x < -2$  $(-7, -2)$

19.

-5 -4 -3 -2 -1  0  1  2  3  4  5  →  $x$   $(-2, -1)$

20.

-5 -4 -3 -2 -1  0  1  2  3  4  5  →  $x$   $[-1, \infty)$

21. $x$ is greater than $-3$ and less than or equal to 4.  $(-3, 4]$

22. $x$ is positive.  $(0, \infty)$

In Exercises 23–28, use words to describe the interval of real numbers.

**23.** $4 < x \le 9$

**24.** $x \ge -1$

**25.** $[-3, \infty)$

**26.** $(-5, 7)$

**27.**

$$\xleftarrow{\qquad} \begin{array}{ccccccccccc} & & & & & & & & & & \\ -5 & -4 & -3 & -2 & -1 & 0 & 1 & 2 & 3 & 4 & 5 \end{array} \xrightarrow{\quad} x$$

**28.**

$$\xleftarrow{\qquad} \begin{array}{ccccccccccc} & & & & & & & & & & \\ -5 & -4 & -3 & -2 & -1 & 0 & 1 & 2 & 3 & 4 & 5 \end{array} \xrightarrow{\quad} x$$

In Exercises 29–32, find the endpoints and state whether the interval is bounded or unbounded, and its type.

**29.** $(-3, 4]$

**30.** $(-3, -1)$

**31.** $(-\infty, 5)$

**32.** $[-6, \infty)$

In Exercises 33–36, use both inequality and interval notation to describe the set of numbers. State the meaning of any variables you use.

**33. Writing to Learn** Bill is at least 29 years old.

**34. Writing to Learn** No item at Sarah's Variety Store costs more than $2.00

**35. Writing to Learn** The price of a gallon of gasoline varies from $1.099 to $1.399.

**36. Writing to Learn** Salary raises at the State University of California at Chico will average between 2% and 6.5%

In Exercises 37 and 38, use the distributive property to write the expanded form of the expression.

**37.** $a(x^2 + b)$   $ax^2 + ab$

**38.** $(y - z^3)c$   $yc - z^3c$

In Exercises 39 and 40, use the distributive property to write the factored form of the expression.

**39.** $ax^2 + dx^2$   $(a + d)x^2$

**40.** $a^3z + a^3w$   $a^3(z + w)$

In Exercises 41 and 42, find the additive inverse of the number.

**41.** $6 - \pi$   $\pi - 6$

**42.** $-7$   $7$

**43. Group Activity** Discuss which algebraic property or properties are illustrated by the equation. Try to reach a consensus.

(a) $(3x)y = 3(xy)$

(b) $a^2b = ba^2$

(c) $a^2b + (-a^2b) = 0$

(d) $(x + 3)^2 + 0 = (x + 3)^2$

(e) $a(x + y) = ax + ay$

**44. Group Activity** Discuss which algebraic property or properties are illustrated by the equation. Try to reach a consensus.

(a) $(x + 2)\dfrac{1}{x + 2} = 1$

(b) $1 \cdot (x + y) = x + y$

(c) $2(x - y) = 2x - 2y$

(d) $2x + (y - z) = 2x + (y + (-z)) = (2x + y) + (-z) = (2x + y) - z$

(e) $\dfrac{1}{a}(ab) = \left(\dfrac{1}{a}a\right)b = 1 \cdot b = b$

## Explorations

**45. Investigating Exponents** For positive integers $m$ and $n$, we can use the definition to show that $a^m a^n = a^{m+n}$.

(a) Examine the equation $a^m a^n = a^{m+n}$ for $n = 0$ and explain why it is reasonable to define $a^0 = 1$ for $a \ne 0$.

(b) Examine the equation $a^m a^n = a^{m+n}$ for $n = -m$ and explain why it is reasonable to define $a^{-m} = 1/a^m$ for $a \ne 0$.

 **46. Decimal Forms of Rational Numbers** Here is the third step when we divide 1 by 17. (The first two steps are not shown, because the quotient is 0 in both cases.)

$$\begin{array}{r} 0.05 \\ 17\overline{)1.00} \\ \underline{85} \\ 15 \end{array}$$

By convention we say that 1 is the first remainder in the long division process, 10 is the second, and 15 is the third remainder.

(a) Continue this long division process until a remainder is repeated, and complete the following table:

| Step | Quotient | Remainder |
|------|----------|-----------|
| 1 | 0 | 1 |
| 2 | 0 | 10 |
| 3 | 5 | 15 |
| ⋮ | ⋮ | ⋮ |

(b) Explain why the digits that occur in the quotient between the pair of repeating remainders determine the infinitely repeating portion of the decimal representation. In this case

$$\frac{1}{17} = 0.\overline{0588235294117647}.$$

(c) Explain why this procedure will always determine the infinitely repeating portion of a rational number whose decimal representation does not terminate. ■

In Exercises 47–52, simplify the expression. Assume that the variables in the denominators are nonzero.

**47.** $\dfrac{x^4y^3}{x^2y^5} \cdot \dfrac{x^2}{y^2}$   $\dfrac{x^4}{y^4}$

**48.** $\dfrac{(3x^2)^2y^4}{3y^2}$   $3x^4y^2$

**49.** $\left(\dfrac{4}{x^2}\right)^2$   $\dfrac{16}{x^4}$

**50.** $\left(\dfrac{2}{xy}\right)^{-3}$   $\dfrac{x^3y^3}{8}$

**51.** $\dfrac{(x^{-3}y^2)^{-4}}{(y^6x^{-4})^{-2}}$   $x^4y^4$

**52.** $\left(\dfrac{4a^3b}{a^2b^3}\right)\left(\dfrac{3b^2}{2a^2b^4}\right)$   $\dfrac{6}{ab^4}$

The data in Table P.1 give details about the year 2000 budget for education in President Clinton's budget proposal.

| Table P.1 Department of Education | |
| --- | --- |
| Budget Item | Amount ($) |
| Eisenhower Professional Development Program | 335 million |
| Goals 2000 | 491 million |
| Title I | 7.9 billion |
| Safe, Drug Free Schools | 591 million |
| Charter Schools | 130 million |
| America Reads Challenge | 286 million |
| Bilingual, Immigrant Education | 415 million |
| Special Education | 5.1 billion |

*Source: National Science Teachers Association, NSTA Reports, April 1999, Vol. 10, No. 5, p 3.*

In Exercises 53–56, write the amount of the budget item in Table P.1 in scientific notation.

**53.** Goals 2000  $4.91 \times 10^8$

**54.** Title I  $7.9 \times 10^9$

**55.** Charter Schools

**56.** America Reads Challenge

In Exercises 57 and 58, write the number in scientific notation.

**57.** The mean distance from Jupiter to the sun is about 483,900,000 miles.  $4.839 \times 10^8$

**58.** The electric charge, in coulombs, of an electron is about $-0.000\ 000\ 000\ 000\ 000\ 000\ 16$.  $-1.6 \times 10^{-19}$

In Exercises 59–62, write the number in decimal form.

**59.** $3.33 \times 10^{-8}$

**60.** $6.73 \times 10^{11}$

**61.** The distance that light travels in 1 year (*one light-year*) is about $5.87 \times 10^{12}$ mi.  5,870,000,000,000

**62.** The mass of a neutron is about $1.6747 \times 10^{-24}$ g.

In Exercises 63 and 64, use scientific notation to simplify.

**63.** $\dfrac{(1.35 \times 10^{-7})(2.41 \times 10^8)}{1.25 \times 10^9}$

**64.** $\dfrac{(3.7 \times 10^{-7})(4.3 \times 10^6)}{2.5 \times 10^7}$

## Extending the Ideas

The **magnitude** of a real number is its distance from the origin.

**65.** List the whole numbers whose magnitudes are less than 7.

**66.** List the natural numbers whose magnitudes are less than 7.

**67.** List the integers whose magnitudes are less than 7.

---

## P.2  Cartesian Coordinate System

Cartesian Plane • Absolute Value of a Real Number • Distance Formulas • Midpoint Formulas • Equations of Circles

### Cartesian Plane

The points in a plane correspond to ordered pairs of real numbers, just as the points on a line can be associated with individual real numbers. This correspondence creates the **Cartesian plane**, or the **rectangular coordinate system** in the plane.

To construct a rectangular coordinate system, or a Cartesian plane, draw a pair of perpendicular real number lines, one horizontal and the other vertical, with the lines intersecting at their respective 0-points (Figure P.6). The horizontal line is usually the **x-axis** and the vertical line is usually the **y-axis**. The positive direction on the *x*-axis is to the right, and the positive direction on the *y*-axis is up. Their point of intersection, *O*, is the **origin of the Cartesian plane**.

Each point *P* of the plane is associated with an **ordered pair (x, y)** of real numbers, the **(Cartesian) coordinates of the point**. The **x-coordinate** represents the intersection of the *x*-axis with the perpendicular from *P*, and the **y-coordinate** represents the intersection of the *y*-axis with the perpendicular from *P*. Figure P.6 shows the points *P* and *Q* with coordinates (4, 2) and (−6, −4), respectively. As with real numbers and a number line, we use the ordered pair (*a*, *b*) for both the name of the point and its coordinates.

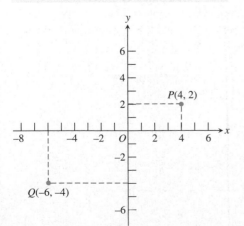

**Figure P.6** The Cartesian coordinate plane.

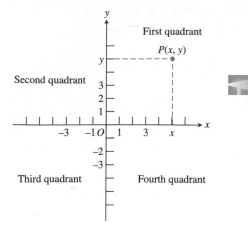

**Figure P.7**   The four quadrants. Points on the *x*- or *y*-axis are *not* in any quadrant.

---

**Alert**

Many students think that | −*a* | is equal to *a*. Show them that $|-(-5)| \neq -5$.

---

| Table P.2  U.S. Exports to Mexico | |
| --- | --- |
| Year | U.S. Exports (billions of dollars) |
| 1991 | 33.3 |
| 1992 | 40.6 |
| 1993 | 41.6 |
| 1994 | 50.8 |
| 1995 | 46.3 |
| 1996 | 56.8 |
| 1997 | 71.4 |
| 1998 | 78.8 |

*Source: Bureau of the Census, Foreign Trade Division, FINAL 1991–1998*

**U.S. Exports to Mexico**

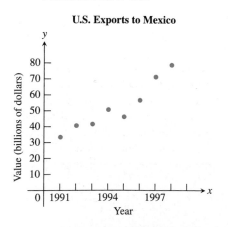

**Figure P.8**   The graph for Example 1.

---

The coordinate axes divide the Cartesian plane into four **quadrants**, as shown in Figure P.7.

### Example 1   PLOTTING DATA ON U.S. EXPORTS TO MEXICO

The value in billions of dollars of U.S. exports to Mexico from 1991 to 1998 is given in Table P.2. Plot the (year, export value) ordered pairs in a rectangular coordinate system.

**Solution**   The points are plotted in Figure P.8 at the bottom of the page.

A **scatter plot** is a plotting of the (*x*, *y*) data pairs on a Cartesian plane. Figure P.8 shows a scatter plot of the data from Table P.2.

## Absolute Value of a Real Number

The *absolute value of a real number* suggests its **magnitude** (size). For example, the absolute value of 3 is 3 and the absolute value of −5 is 5.

---

**Definition   Absolute Value of a Real Number**

The **absolute value of a real number** *a* is

$$|a| = \begin{cases} a, & \text{if } a > 0 \\ -a, & \text{if } a < 0 \\ 0, & \text{if } a = 0 \end{cases}$$

---

### Example 2   USING THE DEFINITION OF ABSOLUTE VALUE

Evaluate:

**(a)** $|-4|$     **(b)** $|\pi - 6|$

**Solution**

**(a)** Because $-4 < 0$, $|-4| = -(-4) = 4$.

**(b)** Because $\pi \approx 3.14$, $\pi - 6$ is negative, so $\pi - 6 < 0$. Thus, $|\pi - 6| = -(\pi - 6) = 6 - \pi \approx 2.858$.

It follows from the above definition that the absolute value of any nonzero real number *a* is positive. And, since $|a| = 0$ when $a = 0$, we conclude that $|a| \geq 0$ for any real number *a*. This and other important properties of absolute value are summarized here.

---

**Properties of Absolute Value**

Let *a* and *b* be real numbers.

**1.** $|a| \geq 0$     **2.** $|-a| = |a|$

**3.** $|ab| = |a||b|$     **4.** $\left|\dfrac{a}{b}\right| = \dfrac{|a|}{|b|}, b \neq 0$

---

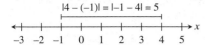

$$|4 - (-1)| = |-1 - 4| = 5$$

**Figure P.9** Finding the distance between $-1$ and 4.

### Absolute Value and Distance

If we let $b = 0$ in the distance formula we see that the distance between $a$ and 0 is $|a|$. Thus, the absolute value of a number is its distance from zero.

### Teaching Note

It is sometimes helpful for students to see the distance between two points by plotting the points on a number line. For example, in Example 2b students can plot a point for 6 on the number line and then plot a point representing $\pi$, somewhere between 3 and 4. If $a$ and $b$ are the two points to be plotted on the number line and $a < b$, then the distance can always be written without absolute value notation as $b - a$. Since 6 is to the right of $\pi$ on the number line, $6 - \pi$ is the correct representation.

## Distance Formulas

The *distance* between $-1$ and 4 on the number line is 5 (see Figure P.9). This distance may be found by subtracting the smaller number from the larger: $4 - (-1) = 5$. If we use absolute value, the order of subtraction does not matter: $|4 - (-1)| = |-1 - 4| = 5$.

---

**Distance Formula (Number Line)**

Let $a$ and $b$ be real numbers. The **distance between $a$ and $b$** is

$$|a - b|.$$

Note that $|a - b| = |b - a|$.

---

**Example 3**  USING AN INEQUALITY TO EXPRESS DISTANCE

We can state that "the distance between $x$ and $-3$ is less than 9" using the inequality

$$|x - (-3)| < 9 \quad \text{or} \quad |x + 3| < 9$$

To find the *distance* between two points that lie on the same horizontal or vertical line in the Cartesian plane, we use the distance formula for points on a number line. For example, the distance between points $x_1$ and $x_2$ on the $x$-axis is $|x_1 - x_2| = |x_2 - x_1|$ and the distance between the points $y_1$ and $y_2$ on the $y$-axis is $|y_1 - y_2| = |y_2 - y_1|$.

To find the distance between two points $P(x_1, y_1)$ and $Q(x_2, y_2)$ that do not line on the same horizontal or vertical line we form the right triangle determined by $P$, $Q$, and $R(x_2, y_1)$ (Figure P.10).

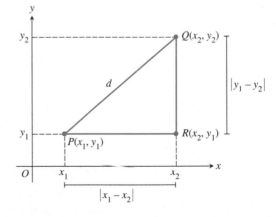

**Figure P.10** Forming a right triangle with hypotenuse $PQ$.

The distance from $P$ to $R$ is $|x_1 - x_2|$, and the distance from $R$ to $Q$ is $|y_1 - y_2|$. By the **Pythagorean Theorem** (see Figure P.11), the distance $d(P, Q)$ between $P$ and $Q$ is

$$d(P, Q) = \sqrt{|x_1 - x_2|^2 + |y_1 - y_2|^2}.$$

Because $|x_1 - x_2|^2 = (x_1 - x_2)^2$ and $|y_1 - y_2|^2 = (y_1 - y_2)^2$, we obtain the following formula.

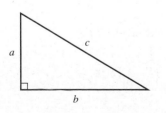

**Figure P.11** The Pythagorean theorem: $c^2 = a^2 + b^2$

> **Distance Formula (Coordinate Plane)**
>
> The **distance $d(P, Q)$ between points $P(x_1, y_1)$ and $Q(x_2, y_2)$** in the coordinate plane is
>
> $$d(P, Q) = \sqrt{(x_1 - x_2)^2 + (y_1 - y_2)^2}$$

### Example 4 FINDING THE DISTANCE BETWEEN TWO POINTS

Find the distance between the points $P(1, 5)$ and $Q(6, 2)$.

**Solution**

$$\begin{aligned} d(P, Q) &= \sqrt{(1 - 6)^2 + (5 - 2)^2} && \text{The distance formula} \\ &= \sqrt{(-5)^2 + 3^2} \\ &= \sqrt{25 + 9} \\ &= \sqrt{34} \approx 5.83 && \text{Using a calculator} \end{aligned}$$

The converse of the Pythagorean theorem is true. That is, if the sum of squares of the lengths of the two sides of a triangle equals the square of the length of the third side, then the triangle is a right triangle.

### Example 5 VERIFYING RIGHT TRIANGLES

Use the converse of the Pythagorean theorem and the distance formula to show that the points $(-3, 4)$, $(1, 0)$, and $(5, 4)$ determine a right triangle.

**Solution** The three points are plotted in Figure P.12. We need to show that the lengths of the sides of the triangle satisfy the Pythagorean relationship $a^2 + b^2 = c^2$. Applying the distance formula we find that

$$\begin{aligned} a &= \sqrt{(-3 - 1)^2 + (4 - 0)^2} = \sqrt{32}, \\ b &= \sqrt{(1 - 5)^2 + (0 - 4)^2} = \sqrt{32}, \\ c &= \sqrt{(-3 - 5)^2 + (4 - 4)^2} = \sqrt{64}. \end{aligned}$$

The triangle is a right triangle because

$$a^2 + b^2 = (\sqrt{32})^2 + (\sqrt{32})^2 = 32 + 32 = 64 = c^2.$$

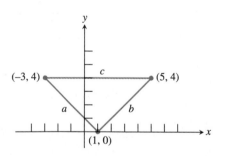

**Figure P.12** The triangle in Example 5.

## Midpoint Formulas

When the endpoints of a segment in a number line are known, we take the average of their coordinates to find the midpoint of the segment.

> **Midpoint Formula (Number Line)**
>
> The **midpoint of the line segment with endpoints $a$ and $b$** is
>
> $$\frac{a + b}{2}.$$

### Example 6 FINDING THE MIDPOINT OF A LINE SEGMENT

The midpoint of the line segment with endpoints $-9$ and $3$ on a number line is

$$\frac{(-9) + 3}{2} = \frac{-6}{2} = -3.$$

Just as with number lines, the midpoint of a line segment in the coordinate plane is determined by its endpoints. Each coordinate of the midpoint is the average of the corresponding coordinates of its endpoints.

> **Midpoint Formula (Coordinate Plane)**
>
> The **midpoint of the line segment with endpoints $(a, b)$ and $(c, d)$ is**
>
> $$\left(\frac{a + c}{2}, \frac{b + d}{2}\right).$$

### Example 7 FINDING THE MIDPOINT OF A LINE SEGMENT

The midpoint of the line segment with endpoints $(-5, 2)$ and $(3, 7)$ is

$$(x, y) = \left(\frac{-5 + 3}{2}, \frac{2 + 7}{2}\right) = (-1, 4.5).$$

Properties of geometric figures can sometimes be confirmed using analytic methods such as the midpoint formulas.

### Example 8 USING THE MIDPOINT FORMULA

It is a fact from geometry that the diagonals of a parallelogram bisect each other. Prove this with a midpoint formula.

**Solution** We can position a parallelogram in the rectangular coordinate plane as shown in Figure P.13. Applying the midpoint formula for the coordinate plane to segments $OB$ and $AC$, we find that

$$\text{midpoint of segment } OB = \left(\frac{0 + a + c}{2}, \frac{0 + b}{2}\right) = \left(\frac{a + c}{2}, \frac{b}{2}\right),$$

$$\text{midpoint of segment } AC = \left(\frac{a + c}{2}, \frac{b + 0}{2}\right) = \left(\frac{a + c}{2}, \frac{b}{2}\right).$$

The midpoints of segments $OA$ and $AC$ are the same, so the diagonals of the parallelogram $OABC$ meet at their midpoints and thus bisect each other.

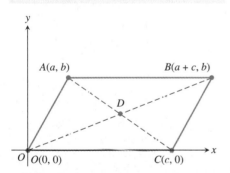

**Figure P.13** The coordinates of $B$ must be $(a + c, b)$ in order for $CB$ to be parallel to $OA$. (Example 8)

## Equations of Circles

A **circle** is the set of points in a plane at a fixed distance (**radius**) from a fixed point (**center**). Figure P.14 shows the circle with center $(h, k)$ and radius $r$. If $(x, y)$ is any point on the circle, the distance formula gives

$$\sqrt{(x - h)^2 + (y - k)^2} = r.$$

Squaring both sides, we obtain the following equation for a circle.

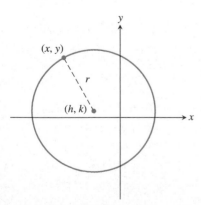

**Figure P.14** The circle with center $(h, k)$ and radius $r$.

**Definition** **Standard Form Equation of a Circle**

The **standard form equation of a circle** with center $(h, k)$ and radius $r$ is

$$(x - h)^2 + (y - k)^2 = r^2.$$

**Example 9** FINDING STANDARD FORM EQUATIONS OF CIRCLES

Find the standard form equation of the circle.

**(a)** Center $(-4, 1)$, radius 8    **(b)** Center $(0, 0)$, radius 5

**Solution**

**(a)**
$$(x - h)^2 + (y - k)^2 = r^2 \qquad \text{Standard form equation}$$
$$(x - (-4))^2 + (y - 1)^2 = 8^2 \qquad \text{Substitute } h = -4, k = 1, r = 8$$
$$(x + 4)^2 + (y - 1)^2 = 64$$

**(b)**
$$(x - h)^2 + (y - k)^2 = r^2 \qquad \text{Standard form equation}$$
$$(x - 0)^2 + (y - 0)^2 = 5^2 \qquad \text{Substitute } h = 0, k = 0, r = 5$$
$$x^2 + y^2 = 25$$

# Quick Review P.2

In Exercises 1 and 2, plot the two numbers on a number line. Then find the distance between them.

**1.** $\sqrt{7}, \sqrt{2}$

**2.** $-5/3, -9/5$

In Exercises 3 and 4, plot the real numbers on a number line.

**3.** $-3, 4, 2.5, 0, -1.5$

**4.** $-\dfrac{5}{2}, -\dfrac{1}{2}, \dfrac{2}{3}, 0, -1$

In Exercises 5 and 6, plot the points.

**5.** $A(3, 5), B(-2, 4), C(3, 0), D(0, -3)$

**6.** $A(-3, -5), B(2, -4), C(0, 5), D(-4, 0)$

In Exercises 7–10, use a calculator to evaluate the expression. Round your answer to two decimal places.

**7.** $\dfrac{-17 + 28}{2}$    5.5

**8.** $\sqrt{13^2 + 17^2}$    21.40

**9.** $\sqrt{6^2 + 8^2}$    10

**10.** $\sqrt{(17 - 3)^2 + (-4 - 8)^2}$

# Section P.2 Exercises

In Exercises 1 and 2, estimate the coordinates of the points.

**1.**

**2.**

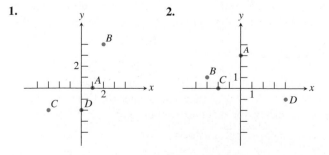

In Exercises 3 and 4, name the quadrants containing the points.

**3.** **(a)** $(2, 4)$    **(b)** $(0, 3)$    **(c)** $(-2, 3)$    **(d)** $(-1, -4)$

**4.** **(a)** $\left(\dfrac{1}{2}, \dfrac{3}{2}\right)$    **(b)** $(-2, 0)$    **(c)** $(-1, -2)$    **(d)** $\left(-\dfrac{3}{2}, -\dfrac{7}{3}\right)$

In Exercises 5–8, evaluate the expression.

**5.** $3 + |-3|$    6

**6.** $2 - |-2|$    0

**7.** $|(-2)3|$    6

**8.** $\dfrac{-2}{|-2|}$    −1

In Exercises 9 and 10, rewrite the expression without using absolute value symbols.

**9.** $|\pi - 4|$   $4 - \pi$      **10.** $|\sqrt{5} - 5/2|$

In Exercises 11–18, find the distance between the points.

**11.** $-9.3, 10.6$   $19.9$      **12.** $-5, -17$   $12$

**13.** $(-3, -1), (5, -1)$   $8$      **14.** $(-4, -3), (1, 1)$

**15.** $(0, 0), (3, 4)$   $5$      **16.** $(-1, 2), (2, -3)$

**17.** $(-2, 0), (5, 0)$   $7$      **18.** $(0, -8), (0, -1)$   $7$

In Exercises 19–22, find the area and perimeter of the figure determined by the points.

**19.** $(-5, 3), (0, -1), (4, 4)$

**20.** $(-2, -2), (-2, 2), (2, 2), (2, -2)$

**21.** $(-3, -1), (-1, 3), (7, 3), (5, -1)$

**22.** $(-2, 1), (-2, 6), (4, 6), (4, 1)$   Perimeter $= 22$; Area $= 30$

In Exercises 23–28, find the midpoint of the line segment with the given endpoints.

**23.** $-9.3, 10.6$   $0.65$

**24.** $-5, -17$   $-11$

**25.** $(-1, 3), (5, 9)$   $(2, 6)$

**26.** $(3, \sqrt{2}), (6, 2)$   $\left(\dfrac{9}{2}, \dfrac{2 + \sqrt{2}}{2}\right)$

**27.** $(-7/3, 3/4), (5/3, -9/4)$   $\left(-\dfrac{1}{3}, -\dfrac{3}{4}\right)$

**28.** $(5, -2), (-1, -4)$   $(2, -3)$

In Exercises 29–34, draw a scatter plot of the data given in the table.

**29. Trends in Nursing Education** The number ($y$) in thousands of first-year fall enrollments in bachelor's-level nursing programs at 326 nursing schools that reported data for all of the five years ($x$). (*Source:* National League for Nursing as reported in *The Chronicle of Higher Education*, September 24, 1999.)

| $x$ | 1994 | 1995 | 1996 | 1997 | 1998 |
|-----|------|------|------|------|------|
| $y$ | 77.1 | 76.5 | 71.9 | 67.2 | 63.7 |

**30. U.N. Written Documents** The number $y$ of pages of printed material (in millions) produced by the United Nations for each year $x$ from 1990 to 1997 is given in the table. (*Source:* United Nations, as reported in the *USA TODAY*, March 5, 1998)

| $x$ | 1990 | 1991 | 1992 | 1993 |
|-----|------|------|------|------|
| $y$ | 733 | 735 | 795 | 749 |

| $x$ | 1994 | 1995 | 1996 | 1997 |
|-----|------|------|------|------|
| $y$ | 775 | 794 | 553 | 550 |

**31. U.S. Imports from Mexico** The total in billions of dollars of U.S. imports from Mexico from 1991 to 1998 is given in Table P.3.

| Table P.3   U.S. Imports from Mexico | |
|---|---|
| Year | U.S. Imports (billions of dollars) |
| 1991 | 31.1 |
| 1992 | 35.2 |
| 1993 | 39.9 |
| 1994 | 49.5 |
| 1995 | 61.7 |
| 1996 | 74.3 |
| 1997 | 85.9 |
| 1998 | 94.6 |

*Source: Bureau of the Census, Foreign Trade Division, FINAL 1991–1998*

**32. U.S. Farm Exports** The total in billions of dollars of U.S. farm exports from 1991 to 1998 is given in Table P.4.

| Table P.4   U.S. Farm Exports | |
|---|---|
| Year | U.S. Farm Exports (billions of dollars) |
| 1991 | 39.4 |
| 1992 | 43.1 |
| 1993 | 42.9 |
| 1994 | 46.3 |
| 1995 | 56.3 |
| 1996 | 60.5 |
| 1997 | 57.1 |
| 1998 | 52.0 |

*Source: U.S. Department of Commerce, as reported in USA TODAY, August 24, 1999*

**33. U.S. Farm Trade Surplus** The total in billions of dollars of U.S. farm trade surplus from 1991 to 1998 is given in Table P.5.

| Table P.5   U.S. Farm Trade Surplus | |
|---|---|
| Year | U.S. Farm Trade Surplus (billions of dollars) |
| 1991 | 17.2 |
| 1992 | 19.6 |
| 1993 | 19.2 |
| 1994 | 20.3 |
| 1995 | 27.1 |
| 1996 | 27.9 |
| 1997 | 21.9 |
| 1998 | 16.2 |

*Source: U.S. Department of Commerce, as reported in USA TODAY, August 24, 1999*

**34. U.S. Exports to Botswana** The total in millions of dollars of U.S. exports to Botswana from 1991 to 1998 is given in Table P.6.

| Table P.6 U.S. Exports to Botswana | |
|---|---|
| Year | U.S. Exports (millions of dollars) |
| 1991 | 30.8 |
| 1992 | 46.5 |
| 1993 | 24.9 |
| 1994 | 22.7 |
| 1995 | 35.8 |
| 1996 | 28.9 |
| 1997 | 43.1 |
| 1998 | 35.6 |

*Source: Bureau of the Census, Foreign Trade Division, FINAL 1991–1998*

In Exercises 35 and 36, use the graph of the stock appreciation value of a $10,000 investment in an index of the S&P 500 as of January 1994, shown below.

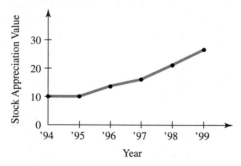

**35. Reading from Graphs** Use the graph to estimate the value of the $10,000 investment in **(a)** January 1997 and **(b)** January 1999. (a) $17,000 (b) $27,000

**36. Percent Increase** Estimate the percent increase in the value of the $10,000 investment in **(a)** January 1997 and **(b)** January 1999. (a) 70% (b) 170%

**37. Group Activity** Prove that the figure determined by the points is an isosceles triangle: (1, 3), (4, 7), (8, 4).

**38. Group Activity** Prove that the diagonals of the figure determined by the points bisect each other.

**(a)** Square $(-7, -1)$, $(-2, 4)$, $(3, -1)$, $(-2, -6)$

**(b)** Parallelogram $(-2, -3)$, $(0, 1)$, $(6, 7)$, $(4, 3)$

**39. (a)** Find the lengths of the sides of the triangle in the figure.

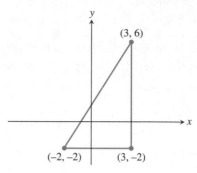

**(b) Writing to Learn** Show that the triangle is a right triangle. $8^2 + 5^2 = 64 + 25 = 89 = (\sqrt{89})^2$

**40. (a)** Find the lengths of the sides of the triangle in the figure.

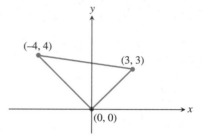

**(b) Writing to Learn** Show that the triangle is a right triangle.

In Exercises 41–44, find the standard form equation for the circle.

**41.** Center (1, 2), radius 5      **42.** Center $(-3, 2)$, radius 1

**43.** Center $(-1, -4)$, radius 3    **44.** Center (0, 0), radius $\sqrt{3}$

In Exercises 45–48, find the center and radius of the circle.

**45.** $(x - 3)^2 + (y - 1)^2 = 36$   center: (3, 1); radius: 6

**46.** $(x + 4)^2 + (y - 2)^2 = 121$   center: $(-4, 2)$; radius: 11

**47.** $x^2 + y^2 = 5$      **48.** $(x - 2)^2 + (y + 6)^2 = 25$

In Exercises 49–52, write the statement using absolute value notation.

**49.** The distance between $x$ and 4 is 3. $|x - 4| = 3$

**50.** The distance between $y$ and $-2$ is greater than or equal to 4.

**51.** The distance between $x$ and $c$ is less than $d$ units.

**52.** $y$ is more than $d$ units from $c$.

## Explorations

**53. Dividing a Line Segment Into Thirds**

**(a)** Find the coordinates of the points one-third and two-thirds of the way from $a = 2$ to $b = 8$ on a number line. 4, 6

**(b)** Repeat (a) for $a = -3$ and $b = 7$.   $\dfrac{1}{3}, \dfrac{11}{3}$

**(c)** Find the coordinates of the points one-third and two-thirds of the way from $a$ to $b$ on a number line.

**(d)** Find the coordinates of the points one-third and two-thirds of the way from the point $(1, 2)$ to the point $(7, 11)$ in the coordinate plane.  $(3, 5); (5, 8)$

**(e)** Find the coordinates of the points one-third and two-thirds of the way from the point $(a, b)$ to the point $(c, d)$ in the coordinate plane.  ■

**54. Determining a Line Segment with Given Midpoint** Let $(4, 4)$ be the midpoint of the line segment determined by the points $(1, 2)$ and $(a, b)$. Determine $a$ and $b$.  $7; 6$

**55. Writing to Learn  Isosceles but Not Equilateral Triangle** Prove that the triangle determined by the points $(3, 0)$, $(-1, 2)$, and $(5, 4)$ is isosceles but not equilateral.

**56. Writing to Learn  Equidistant Point from Vertices of a Right Triangle** Prove that the midpoint of the hypotenuse of the right triangle with vertices $(0, 0)$, $(5, 0)$, and $(0, 7)$ is equidistant from the three vertices.

**57. Writing to Learn** Describe the set of real numbers that satisfy $|x - 2| < 3$.  $-1 < x < 5$

**58. Writing to Learn** Describe the set of real numbers that satisfy $|x + 3| \geq 5$.  $x \leq -8$ or $x \geq 2$

## Extending the Ideas

**59. Writing to Learn  Equidistant Point from Vertices of a Right Triangle** Prove that the midpoint of the hypotenuse of any right triangle is equidistant from the three vertices.

**60. Comparing Areas** Consider the four points $A(0, 0)$, $B(0, a)$, $C(a, a)$, and $D(a, 0)$. Let $P$ be the midpoint of the line segment $CD$ and $Q$ the point one-fourth of the way from $A$ to $D$ on segment $AD$.

**(a)** Find the area of triangle $BPQ$.  $\dfrac{7a^2}{16}$

**(b)** Compare the area of triangle $BPQ$ with the area of $ABCD$.  $\triangle BPQ = \dfrac{7}{16}$. (area of $\square ABCD$)

In Exercises 61–63, let $P(a, b)$ be a point in the first quadrant.

**61.** Find the coordinates of the point $Q$ in the fourth quadrant so that $PQ$ is perpendicular to the $x$-axis.  $Q(a, -b)$

**62.** Find the coordinates of the point $Q$ in the second quadrant so that $PQ$ is perpendicular to the $y$-axis.  $Q(-a, b)$

**63.** Find the coordinates of the point $Q$ in the third quadrant so that the origin is the midpoint of the segment $PQ$.

**64. Writing to Learn** Prove that the distance formula for the number line is a special case of the distance formula for the Cartesian plane.

---

**P.3**
# Linear Equations and Inequalities

Equations • Solving Equations • Linear Equations in One Variable • Linear Inequalities in One Variable

## Equations

An **equation** is a statement of equality between two expressions. Here are some properties of equality that we will use to solve equations algebraically.

**Properties of Equality**

Let $u$, $v$, $w$, and $z$ be real numbers, variables, or algebraic expressions.

| | |
|---|---|
| **1. Reflexive** | $u = u$ |
| **2. Symmetric** | If $u = v$, then $v = u$. |
| **3. Transitive** | If $u = v$, and $v = w$, then $u = w$. |
| **4. Addition** | If $u = v$ and $w = z$, then $u + w = v + z$. |
| **5. Multiplication** | If $u = v$ and $w = z$, then $uw = vz$. |

## Solving Equations

A **solution of an equation in $x$** is a value of $x$ for which the equation is true. To **solve an equation in $x$** means to find all values of $x$ for which the equation is true, that is, to find all solutions of the equation.

**Example 1  CONFIRMING A SOLUTION**

Prove that $x = -2$ is a solution of the equation $x^3 - x + 6 = 0$.

**Solution**

$$(-2)^3 - (-2) + 6 \overset{?}{=} 0$$

$$-8 + 2 + 6 \overset{?}{=} 0$$

$$0 = 0$$

## Linear Equations in One Variable

The most basic equation in algebra is a *linear equation*.

---

**Definition  Linear Equation in $x$**

A **linear equation in $x$** is one that can be written in the form

$$ax + b = 0,$$

where $a$ and $b$ are real numbers with $a \neq 0$.

---

The equation $2z - 4 = 0$ is linear in the variable $z$. The equation $3u^2 - 12 = 0$ is *not* linear in the variable $u$. It turns out that a linear equation in one variable has exactly one solution. We solve such an equation by transforming it into an *equivalent equation* whose solution is obvious. Two or more equations are **equivalent** if they have the same solutions. For example, the equations $2z - 4 = 0$, $2z = 4$, and $z = 2$ are all equivalent. Here are operations that produce equivalent equations.

**Teaching Note**

Compare an equation to a scale. If the scale balances, it is possible to add or subtract equal weights on each side, and the scale will still balance.

**Alert**

Point out that multiplying or dividing both sides of an equation by $x$ does not always result in an equivalent equation.

---

**Operations for Equivalent Equations**

An equivalent equation is obtained if one or more of the following operations are performed.

| Operation | Given Equation | Equivalent Equation |
|---|---|---|
| **1.** Combine like terms, reduce fractions, and remove grouping symbols | $2x + x = \dfrac{3}{9}$ | $3x = \dfrac{1}{3}$ |
| **2.** Perform the same operation on both sides. | | |
|   **(a)** Add $(-3)$. | $x + 3 = 7$ | $x = 4$ |
|   **(b)** Subtract $(2x)$. | $5x = 2x + 4$ | $3x = 4$ |
|   **(c)** Multiply by a nonzero constant $(1/3)$. | $3x = 12$ | $x = 4$ |
|   **(d)** Divide by a nonzero constant $(3)$. | $3x = 12$ | $x = 4$ |

---

The next two examples illustrate how to use equivalent equations to solve linear equations.

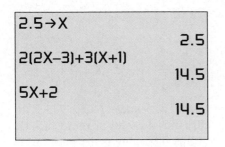

**Figure P.15** The top line stores the number 2.5 into the variable $x$. (Example 2)

**Example 2** SOLVING A LINEAR EQUATION

Solve $2(2x - 3) + 3(x + 1) = 5x + 2$. Support the result with a calculator.

**Solution**

$$2(2x - 3) + 3(x + 1) = 5x + 2$$

| | |
|---|---|
| $4x - 6 + 3x + 3 = 5x + 2$ | Distributive property |
| $7x - 3 = 5x + 2$ | Combine like terms. |
| $2x = 5$ | Add 3, and subtract $5x$ |
| $x = 2.5$ | Divide by 2. |

To support our algebraic work we can use a calculator to evaluate the original equation for $x = 2.5$. Figure P.15 shows that each side of the original equation is equal to 14.5 if $x = 2.5$.

If an equation involves fractions, find the least common denominator (LCD) of the fractions and multiply both sides by the LCD. This is sometimes referred to as *clearing the equation of fractions*. Example 3 illustrates.

**Example 3** SOLVING A LINEAR EQUATION INVOLVING FRACTIONS

Solve

$$\frac{5y - 2}{8} = 2 + \frac{y}{4}$$

**Solution** The denominators are 8, 1, and 4. The LCD of the fractions is 8 (see Appendix A.3 if necessary)

| | |
|---|---|
| $\dfrac{5y - 2}{8} = 2 + \dfrac{y}{4}$ | |
| $8\left(\dfrac{5y - 2}{8}\right) = 8\left(2 + \dfrac{y}{4}\right)$ | Multiply by the LCD 8. |
| $8 \cdot \dfrac{5y - 2}{8} = 8 \cdot 2 + 8 \cdot \dfrac{y}{4}$ | Distributive property |
| $5y - 2 = 16 + 2y$ | Simplify. |
| $5y = 18 + 2y$ | Add 2. |
| $3y = 18$ | Subtract $2y$. |
| $y = 6$ | Divide by 3. |

We leave it to you to check the solution either using paper and pencil or a calculator.

## Linear Inequalities in One Variable

We used inequalities to describe order on the number line in Section P.1. For example, if $x$ is to the left of 2 on the number line, or if $x$ is any real number less than 2, we write $x < 2$. The most basic inequality in algebra is a *linear inequality*.

**Definition** Linear Inequality in $x$

A **linear inequality in $x$** is one that can be written in the form

$$ax + b < 0, \ ax + b \leq 0, \ ax + b > 0, \ \text{or} \ ax + b \geq 0,$$

where $a$ and $b$ are real numbers with $a \neq 0$.

To **solve an inequality in $x$** means to find all values of $x$ for which the inequality is true. A **solution of an inequality in $x$** is a value of $x$ for which the inequality is true. The set of all solutions of an inequality is the **solution set** of the inequality. We **solve an inequality** by finding its solution set. Here is a list of properties we use to solve inequalities.

**Direction of an Inequality**

Multiplying (or dividing) an inequality by a positive number preserves the direction of the inequality. Multiplying (or dividing) an inequality by a negative number reverses the direction.

**Properties of Inequalities**

Let $u$, $v$, $w$, and $z$ be real numbers, variables, or algebraic expressions, and $c$ a real number.

**1. Transitive**      If $u < v$ and $v < w$, then $u < w$.

**2. Addition**        If $u < v$, then $u + w < v + w$.
                       If $u < v$ and $w < z$, then $u + w < v + z$.

**3. Multiplication**  If $u < v$ and $c > 0$, then $uc < vc$.
                       If $u < v$ and $c < 0$, then $uc > vc$.

The above properties are true if $<$ is replaced by $\leq$. There are similar properties for $>$ and $\geq$.

It turns out that the set of solutions of a linear inequality in one variable form an interval of real numbers. Just as with linear equations, we solve a linear inequality by transforming it into an *equivalent inequality* whose solutions are obvious. Two or more inequalities are **equivalent** if they have the same set of solutions. The properties of inequalities listed above describe operations that transform an inequality into an equivalent one.

**Example 4**  SOLVING A LINEAR INEQUALITY

Solve $3(x - 1) + 2 \leq 5x + 6$.

**Solution**

$$3(x - 1) + 2 \leq 5x + 6$$

$$3x - 3 + 2 \leq 5x + 6 \qquad \text{Distributive property}$$

$$3x - 1 \leq 5x + 6 \qquad \text{Simplify.}$$

$$3x \leq 5x + 7 \qquad \text{Add 1.}$$

$$-2x \leq 7 \qquad \text{Subtract } 5x.$$

$$\left(-\frac{1}{2}\right) \cdot -2x \geq \left(-\frac{1}{2}\right) \cdot 7 \qquad \text{Multiply by } -1/2 \text{ (The inequality reverses.)}$$

$$x \geq -3.5$$

The solution set of the inequality is the set of all real numbers greater than or equal to $-3.5$. In interval notation, the solution set is $[-3.5, \infty)$.

Because the solution set of a linear inequality is an interval of real numbers, we can display the solution set with a number line graph as illustrated in Example 5.

### Example 5   SOLVING A LINEAR INEQUALITY INVOLVING FRACTIONS

Solve the inequality and graph its solution set.

$$\frac{x}{3} + \frac{1}{2} > \frac{x}{4} + \frac{1}{3}$$

**Solution**   The LCD of the fractions is 12.

$$\frac{x}{3} + \frac{1}{2} > \frac{x}{4} + \frac{1}{3} \qquad \text{The original inequality}$$

$$12 \cdot \left( \frac{x}{3} + \frac{1}{2} \right) > 12 \cdot \left( \frac{x}{4} + \frac{1}{3} \right) \qquad \text{Multiply by the LCD 12.}$$

$$4x + 6 > 3x + 4 \qquad \text{Simplify.}$$

$$x + 6 > 4 \qquad \text{Subtract } 3x.$$

$$x > -2 \qquad \text{Subtract 6.}$$

The solution set is the interval $(-2, \infty)$. Its graph is shown in Figure P.16.

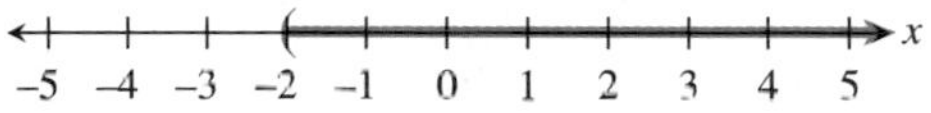

**Figure P.16**   The graph of the solution set of the inequality in Example 5.

Sometimes two inequalities are combined in a **double inequality,** whose solution set is a double inequality with $x$ isolated as the middle term. Example 6 illustrates.

### Example 6   SOLVING A DOUBLE INEQUALITY

Solve the inequality and graph its solution set.

$$-3 < \frac{2x + 5}{3} \le 5$$

**Solution**

$$-3 < \frac{2x + 5}{3} \le 5$$

$$-9 < 2x + 5 \le 15 \qquad \text{Multiply by 3.}$$

$$-14 < 2x \le 10 \qquad \text{Subtract 5.}$$

$$-7 < x \le 5 \qquad \text{Divide by 2.}$$

The solution set is the set of all real numbers greater than $-7$ and less than or equal to 5. In interval notation, the solution is set $(-7, 5]$. Its graph is shown in Figure P.17.

---

**Notes on Examples**

In Example 6, point out that the double inequality $-7 < x \le 5$ means that $x$ must satisfy both inequalities, $-7 < x$ and $x \le 5$.

**Follow-up**

Ask why one should avoid multiplying both sides of an equation or inequality by 0.

**Assignment Guide**

Ex. 3–60, multiples of 3, 62, 65, 66

**Cooperative Learning**

Group Activity: Ex. 61, 62

**Notes on Exercises**

Ex. 1-4 and 31–34 emphasize the definition of a solution of an equation or inequality.
Ex. 1–28 and 35–42 require students to solve equations and inequalities.
Ex 45-48 involve double inequalities.
Ex. 64–67 require students to solve a formula for one of the variables.

**Ongoing Assignment**

Self-Assessment: Ex 23, 53, 61
Embedded Assessment: Ex. 28, 64

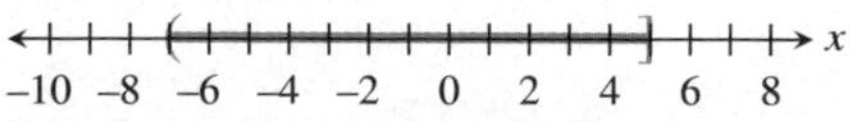

**Figure P.17**   The graph of the solution set of the double inequality in Example 6.

## Quick Review P.3

In Exercises 1 and 2, simplify the expression by combining like terms.

**1.** $2x + 5x + 7 + y - 3x + 4y + 2$  $4x + 5y + 9$

**2.** $4 + 2x - 3z + 5y - x + 2y - z - 2$  $x + 7y - 4z + 2$

In Exercises 3 and 4, use the distributive property to expand the products. Simplify the resulting expression by combining like terms.

**3.** $3(2x - y) + 4(y - x) + x + y$  $3x + 2y$

**4.** $5(2x + y - 1) + 4(y - 3x + 2) + 1$  $-2x + 9y + 4$

In Exercises 5–10, use the LCD to combine the fractions. Simplify the resulting fraction.

**5.** $\dfrac{2}{y} + \dfrac{3}{y}$  $\dfrac{5}{y}$

**6.** $\dfrac{1}{y - 1} + \dfrac{3}{y - 2}$

**7.** $2 + \dfrac{1}{x}$  $\dfrac{2x + 1}{x}$

**8.** $\dfrac{1}{x} + \dfrac{1}{y} - x$  $\dfrac{y + x - x^2y}{xy}$

**9.** $\dfrac{x + 4}{2} + \dfrac{3x - 1}{5}$

**10.** $\dfrac{x}{3} + \dfrac{x}{4}$  $\dfrac{7x}{12}$

## Section P.3 Exercises

In Exercises 1–4, find which values of $x$ are solutions of the equation.

**1.** $2x^2 + 5x = 3$  (a) and (c)

   **(a)** $x = -3$      **(b)** $x = -\dfrac{1}{2}$      **(c)** $x = \dfrac{1}{2}$

**2.** $\dfrac{x}{2} + \dfrac{1}{6} = \dfrac{x}{3}$  (a)

   **(a)** $x = -1$      **(b)** $x = 0$      **(c)** $x = 1$

**3.** $\sqrt{1 - x^2} + 2 = 3$  (b)

   **(a)** $x = -2$      **(b)** $x = 0$      **(c)** $x = 2$

**4.** $(x - 2)^{1/3} = 2$  (c)

   **(a)** $x = -6$      **(b)** $x = 8$      **(c)** $x = 10$

In Exercises 5–10, determine whether the equation is linear in $x$.

**5.** $5 - 3x = 0$  yes

**6.** $5 = 10/2$  no

**7.** $x + 3 = x - 5$  no

**8.** $x - 3 = x^2$  no

**9.** $2\sqrt{x} + 5 = 10$  no

**10.** $x + \dfrac{1}{x} = 1$  no

In Exercises 11–24, solve the equation.

**11.** $3x = 24$  $x = 8$

**12.** $4x = -16$  $x = -4$

**13.** $3t - 4 = 8$  $t = 4$

**14.** $2t - 9 = 3$  $t = 6$

**15.** $2x - 3 = 4x - 5$  $x = 1$

**16.** $4 - 2x = 3x - 6$  $x = 2$

**17.** $4 - 3y = 2(y + 4)$

**18.** $4(y - 2) = 5y$  $y = -8$

**19.** $\dfrac{1}{2}x = \dfrac{7}{8}$  $x = \dfrac{7}{4} = 1.75$

**20.** $\dfrac{2}{3}x = \dfrac{4}{5}$  $x = \dfrac{6}{5} = 1.2$

**21.** $\dfrac{1}{2}x + \dfrac{1}{3} = 1$  $x = \dfrac{4}{3}$

**22.** $\dfrac{1}{3}x + \dfrac{1}{4} = 1$

**23.** $2(3 - 4z) - 5(2z + 3) = z - 17$  $z = \dfrac{8}{19}$

**24.** $3(5z - 3) - 4(2z + 1) = 5z - 2$  $z = \dfrac{11}{2} = 5.5$

In Exercises 25–28, solve the equation. Support your answer with a calculator.

**25.** $\dfrac{2x - 3}{4} + 5 = 3x$

**26.** $2x - 4 = \dfrac{4x - 5}{3}$

**27.** $\dfrac{t + 5}{8} - \dfrac{t - 2}{2} = \dfrac{1}{3}$

**28.** $\dfrac{t - 1}{3} + \dfrac{t + 5}{4} = \dfrac{1}{2}$

**29. Writing to Learn** Write a statement about solutions of equations suggested by the computations in the figure.

   **(a)**

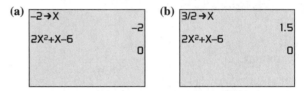

   **(b)**

**30. Writing to Learn** Write a statement about solutions of equations suggested by the computations in the figure.

   **(a)**

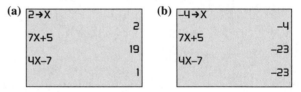

   **(b)**

In Exercises 31–34, find which values of $x$ are solutions of the inequality.

**31.** $2x - 3 < 7$  (a)

   **(a)** $x = 0$      **(b)** $x = 5$      **(c)** $x = 6$

**32.** $3x - 4 \geq 5$  (b) and (c)

   **(a)** $x = 0$      **(b)** $x = 3$      **(c)** $x = 4$

**33.** $-1 < 4x - 1 \leq 11$  (b) and (c)

   **(a)** $x = 0$      **(b)** $x = 2$      **(c)** $x = 3$

**34.** $-3 \leq 1 - 2x \leq 3$  (a), (b), and (c)

   **(a)** $x = -1$      **(b)** $x = 0$      **(c)** $x = 2$

In Exercises 35–42, solve the inequality, and draw a number line graph of the solution set.

**35.** $x - 4 < 2$

**36.** $x + 3 > 5$

**37.** $2x - 1 \leq 4x + 3$

**38.** $3x - 1 \geq 6x + 8$

**39.** $2 \leq x + 6 < 9$

**40.** $-1 \leq 3x - 2 < 7$

**41.** $2(5 - 3x) + 3(2x - 1) \le 2x + 1$

**42.** $4(1 - x) + 5(1 + x) > 3x - 1$

In Exercises 43–54, solve the inequality.

**43.** $\dfrac{5x + 7}{4} \le -3$  $x \le -\dfrac{19}{5}$

**44.** $\dfrac{3x - 2}{5} > -1$  $x > -1$

**45.** $4 \ge \dfrac{2y - 5}{3} \ge -2$

**46.** $1 > \dfrac{3y - 1}{4} > -1$

**47.** $0 \le 2z + 5 < 8$

**48.** $-6 < 5t - 1 < 0$

**49.** $\dfrac{x - 5}{4} + \dfrac{3 - 2x}{3} < -2$

**50.** $\dfrac{3 - x}{2} + \dfrac{5x - 2}{3} < -1$

**51.** $\dfrac{2y - 3}{2} + \dfrac{3y - 1}{5} < y - 1$  $y < \dfrac{7}{6}$

**52.** $\dfrac{3 - 4y}{6} - \dfrac{2y - 3}{8} \ge 2 - y$  $y \ge \dfrac{27}{2}$

**53.** $\dfrac{1}{2}(x - 4) - 2x \le 5(3 - x)$  $x \le \dfrac{34}{7}$

**54.** $\dfrac{1}{2}(x + 3) + 2(x - 4) < \dfrac{1}{3}(x - 3)$  $x < \dfrac{33}{13}$

In Exercises 55–58, find the solutions of the equation or inequality displayed in Figure P.18.

**55.** $x^2 - 2x < 0$  $x = 1$  
**56.** $x^2 - 2x = 0$  $x = 0, 2$

**57.** $x^2 - 2x > 0$  $x = 3, 4, 5, 6$  
**58.** $x^2 - 2x \le 0$  $x = 0, 1, 2$

| X | Y1 |
|---|---|
| 0 | 0 |
| 1 | -1 |
| 2 | 0 |
| 3 | 3 |
| 4 | 8 |
| 5 | 15 |
| 6 | 24 |

Y1 ≡ X²−2X

**Figure P.18** The second column gives values of $y_1 = x^2 - 2x$ for $x = 0, 1, 2, 3, 4, 5,$ and 6.

**59. Writing to Learn** Explain how the second equation was obtained from the first.

$$x - 3 = 2x + 3, \quad 2x - 6 = 4x + 6$$

**60. Writing to Learn** Explain how the second equation was obtained from the first.

$$2x - 1 = 2x - 4, \quad x - \frac{1}{2} = x - 2$$

**61. Group Activity** Determine whether the two equations are equivalent.

**(a)** $3x = 6x + 9, \quad x = 2x + 9$  no

**(b)** $6x + 2 = 4x + 10, \quad 3x + 1 = 2x + 5$  yes

**62. Group Activity** Determine whether the two equations are equivalent.

**(a)** $3x + 2 = 5x - 7, \quad -2x + 2 = -7$  yes

**(b)** $2x + 5 = x - 7, \quad 2x = x - 7$  no

# Explorations

**63. Testing Inequalities on a Calculator**

**(a)** The calculator we use indicates that the statement $2 < 3$ is true by returning the value 1 (for true) when $2 < 3$ is entered. Try it with your calculator.

**(b)** The calculator we use indicates that the statement $2 < 1$ is false by returning the value 0 (for false) when $2 < 1$ is entered. Try it with your calculator.

**(c)** Use your calculator to test which of these two numbers is larger: 799/800, 800/801.  800/801 > 799/800

**(d)** Use your calculator to test which of these two numbers is larger: $-102/101, -103/102$.  $-103/102 > -102/101$

**(e)** If your calculator returns 0 when you enter $2x + 1 < 4$, what can you conclude about the value stored in $x$?  ∎

# Extending the Ideas

**64. Perimeter of a rectangle** The formula for the perimeter $P$ of a rectangle is

$$P = 2(L + W).$$

Solve this equation for $W$.

**65. Area of a Trapezoid** The formula for the area $A$ of a trapezoid is

$$A = \frac{1}{2}h(b_1 + b_2).$$

Solve this equation for $b_1$.

**66. Volume of a Sphere** The formula for the volume $V$ of a sphere is

$$V = \frac{4}{3}\pi r^3. \quad r = \sqrt[3]{\frac{3V}{4\pi}}$$

Solve this equation for $r$.

**67. Celsius and Fahrenheit** The formula for Celsius temperature in terms of Fahrenheit temperature is

$$C = \frac{5}{9}(F - 32). \quad F = \frac{9}{5}C + 32$$

Solve the equation for $F$.

## Lines in the Plane

Slope of a Line • Point-Slope Form Equation of a Line • Slope-Intercept Form Equation of a Line • Graphing Linear Equations in Two Variables • Parallel and Perpendicular Lines • Applying Linear Equations in Two Variables

### Slope of a Line

The slope of a nonvertical line is the ratio of the amount of vertical change to the amount of horizontal change between two points. For the points $(x_1, y_1)$ and $(x_2, y_2)$, the vertical change is $\triangle y = y_2 - y_1$ and the horizontal change is $\triangle x = x_2 - x_1$. $\triangle y$ is read "delta" $y$. See Figure P.19.

---

**Definition  Slope of a Line**

The **slope** of the nonvertical line through the points $(x_1, y_1)$ and $(x_2, y_2)$ is

$$m = \frac{\triangle y}{\triangle x} = \frac{y_2 - y_1}{x_2 - x_1}.$$

If the line is vertical, then $x_1 = x_2$ and the slope is undefined.

---

**Example 1  FINDING THE SLOPE OF A LINE**

Find the slope of the line through the two points. Sketch a graph of the line.

**(a)** $(-1, 2)$ and $(4, -2)$       **(b)** $(1, 1)$, and $(3, 4)$

**Solution**

**(a)** The two points are $(x_1, y_1) = (-1, 2)$ and $(x_2, y_2) = (4, -2)$. Thus,

$$m = \frac{y_2 - y_1}{x_2 - x_1} = \frac{(-2) - 2}{4 - (-1)} = -\frac{4}{5}.$$

**(b)** The two points are $(x_1, y_1) = (1, 1)$ and $(x_2, y_2) = (3, 4)$. Thus,

$$m = \frac{y_2 - y_1}{x_2 - x_1} = \frac{4 - 1}{3 - 1} = \frac{3}{2}.$$

The graphs of these two lines are shown in Figure P.20.

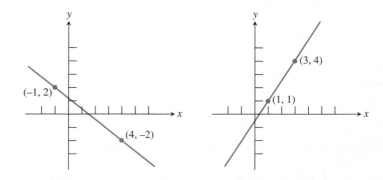

**Figure P.20** The graphs of the two lines in Example 1.

---

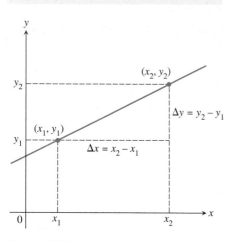

**Figure P.19** The slope of a line can be found from the coordinates of any two points of the line.

Figure P.21 shows a vertical line through the points $(3, 2)$ and $(3, 7)$. If we try to calculate its slope using the slope formula $(y_2 - y_1)/(x_2 - x_1)$, we get zero in the denominator. So, it makes sense to say that a vertical line does not have a slope, or that its slope is undefined.

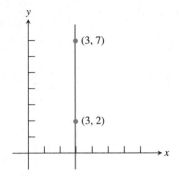

**Figure P.21**   Applying the slope formula to this vertical line gives $m = 5/0$, which is not defined. Thus, the slope of a vertical line is undefined.

## Point-Slope Form Equation of a Line

If we know the coordinates of one point on a line and the slope of the line, then we can find an equation for that line. For example, the line in Figure P.22 passes through the point $(x_1, y_1)$ and has slope $m$. If $(x, y)$ is any other point on this line, the definition of the slope yields the equation

$$m = \frac{y - y_1}{x - x_1} \quad \text{or} \quad y - y_1 = m(x - x_1).$$

An equation written this way is in the *point-slope form*.

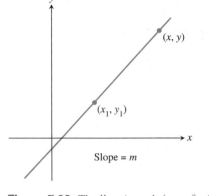

**Figure P.22**   The line through $(x_1, y_1)$ with slope $m$.

> **Definition   Point-Slope Form of an Equation of a Line**
>
> The **point-slope form** of an equation of a line that passes through the point $(x_1, y_1)$ and has slope $m$ is
>
> $$y - y_1 = m(x - x_1).$$

**Example 2   USING THE POINT-SLOPE FORM**

Use the point-slope form to find an equation of the line that passes through the point $(-3, -4)$ and has slope 2.

**Solution**   Substitute $x_1 = -3$, $y_1 = -4$, and $m = 2$ into the point-slope form, and simplify the resulting expression.

$$y - y_1 = m(x - x_1) \qquad \text{Point-slope form}$$
$$y - (-4) = 2(x - (-3)) \qquad x_1 = -3, y_1 = -4, m = 2$$
$$y + 4 = 2x - 2(-3) \qquad \text{Distributive property}$$
$$y + 4 = 2x + 6$$
$$y = 2x + 2 \qquad \text{A common simplified form}$$

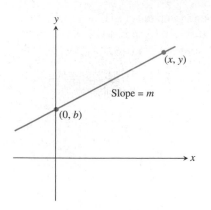

**Figure P.23** The line with slope $m$ and $y$-intercept $(0, b)$.

## Slope-Intercept Form Equation of a Line

The **$y$-intercept** of a nonvertical line is the point where the line intersects the $y$-axis. If we know the $y$-intercept and the slope of the line, we can apply the point-slope form to find an equation of the line.

Figure P.23 shows a line with slope $m$ and $y$-intercept $(0, b)$. A point-slope form equation for this line is $y - b = m(x - 0)$. By rewriting this equation we obtain the form known as the *slope-intercept form.*

> **Definition** **Slope-Intercept Form of an Equation of a Line**
>
> The slope-intercept form of an equation of a line with slope $m$ and $y$-intercept $(0, b)$ is
>
> $$y = mx + b.$$

### Example 3   USING THE SLOPE-INTERCEPT FORM

Write an equation of the line with slope 3 that passes through the point $(-1, 6)$ using the slope-intercept form.

**Solution**

| | |
|---|---|
| $y = mx + b$ | Slope-intercept form |
| $y = 3x + b$ | $m = 3$ |
| $6 = 3(-1) + b$ | $y = 6$ when $x = -1$ |
| $b = 9$ | |

The slope-intercept form of the equation is $y = 3x + 9$.

We cannot use the phrase "*the* equation of a line" because each line has many different equations. Every line has an equation that can be written in the form $Ax + By + C = 0$ where $A$ and $B$ are not both zero. This form is the **general form** for an equation of a line.

If $B \neq 0$, the general form can be changed to the slope-intercept form as follows:

$$Ax + By + C = 0$$

$$By = -Ax - C$$

$$y = \underbrace{-\frac{A}{B}x}_{\text{slope}} + \underbrace{\left(-\frac{C}{B}\right)}_{\text{y-intercept}}$$

> **Forms of Equations of Lines**
>
> | | |
> |---|---|
> | General form: | $Ax + By + C = 0$, $A$ and $B$ not both zero |
> | Slope-intercept form: | $y = mx + b$ |
> | Point-slope form: | $y - y_1 = m(x - x_1)$ |
> | Vertical line: | $x = a$ |
> | Horizontal line: | $y = b$ |

**Teaching Note**

Point out that in the point-slope form, $(x_1, y_1)$ can represent any point on the line. This form is useful for determining the equation if the slope and one or more points are known.

## Graphing Linear Equations in Two Variables

A **linear equation in $x$ and $y$** is one that can be written in the form

$$Ax + By = C,$$

where $A$ and $B$ are not both zero. Rewriting this equation in the form $Ax + By - C = 0$ we see that it is the general form of an equation of a line. If $B = 0$, the line is vertical, and if $A = 0$, the line is horizontal.

The **graph** of an equation in $x$ and $y$ consists of all pairs $(x, y)$ that are solutions of the equation. For example, $(1, 2)$ is a **solution** of the equation $2x + 3y = 8$ because substituting $x = 1$ and $y = 2$ into the equation leads to the true statement $8 = 8$. The pairs $(-2, 4)$ and $(2, 4/3)$ are also solutions.

The *point-plotting method* is a traditional way to sketch the graph of an equation using pencil and paper. Make a table of several solutions, plot these solutions as points in the coordinate plane, and connect the points with a smooth curve.

Because the graph of a linear equation in $x$ and $y$ is a straight line, we really need only to find two solutions and then connect them with a straight line to draw its graph. If a line is neither horizontal nor vertical, then two easy points to find are its $x$-intercept and the $y$-intercept. The **$x$-intercept** is the point $(x', 0)$ where the graph intersects the $x$-axis. Set $y = 0$ and solve for $x$ to find the $x$-intercept. The coordinates of the $y$-intercept are $(0, y')$. Set $x = 0$ and solve for $y$ to find the $y$-intercept.

The point-plotting method of graphing is also the basis for the way a graphing utility, often referred to as a *grapher*, does its curve sketching. We simply enter the equation into the grapher. Then we let the graphing utility compute and plot the points.

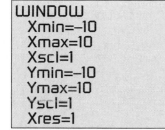

**Figure P.24** The window dimensions for the *standard window*. The notation "$[-10, 10]$ by $[-10, 10]$" is used to represent window dimensions like these.

---

### Graphing with a Graphing Utility

To draw a graph of an equation using a grapher:

**1.** Rewrite the equation in the form $y = $ (an expression in $x$).

**2.** Enter the equation into the grapher.

**3.** Select an appropriate **viewing window** (see Figure P.24).

**4.** Press the "graph" key.

---

### Example 4  USE A GRAPHING UTILITY

Draw the graph of $2x + 3y = 6$.

**Solution**  First we solve for $y$.

$$2x + 3y = 6$$
$$3y = -2x + 6$$
$$y = -\frac{2}{3}x + 2$$

Figure P.25 shows the graph of $y = -(2/3)x + 2$, or equivalently, the graph of the linear equation $2x + 3y = 6$ in the $[-4, 6]$ by $[-3, 5]$ viewing window.

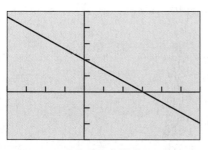

[–4, 6] by [–3, 5]

**Figure P.25** The graph of $2x + 3y = 6$. The points $(0, 2)$ ($y$-intercept) and $(3, 0)$ ($x$-intercept) appear to lie on the graph and, as pairs, are solutions of the equation, providing visual support that the graph is correct. (Example 4)

# Parallel and Perpendicular Lines

**Exploration 1** **Investigating Graphs of Linear Equations**

1. Graph $y = mx$ for $m = -3, -2, -1, 1, 2, 3$ in $[-8, 8]$ by $[-5, 5]$. What do these graphs have in common? How are they different?

2. If $m > 0$, what do the graphs of $y = mx$ and $y = -mx$ have in common? How are they different?

3. Graph $y = 0.3x + b$ for $b = -3, -2, -1, 0, 1, 2, 3$ in $[-8, 8]$ by $[-5, 5]$. What do these graphs have in common? How are they different?

4. What do the graphs of $y = mx + b$ and $y = mx + c$, $b \ne c$, have in common? How are they different?

5. Graph $y = 2x$ and $y = -(1/2)x$ in a *square viewing window* (see margin note). On the calculator we use, the "decimal window" $[-4.7, 4.7]$ by $[-3.1, 3.1]$ is square. Estimate the angle between the two lines.

6. Repeat part 5 for $y = mx$ and $y = -(1/m)x$ with $m = 1, 3, 4$, and 5.

Parallel lines and perpendicular lines were involved in Exploration 1. Using a grapher to decide when lines are parallel or perpendicular is risky. Here is an algebraic test to determine when two lines are parallel or perpendicular.

**Parallel and Perpendicular Lines**

1. Two nonvertical lines are parallel if and only if their slopes are equal.

2. Two nonvertical lines are perpendicular if and only if their slopes $m_1$ and $m_2$ are opposite reciprocals. That is, if and only if
$$m_1 = -\frac{1}{m_2}.$$

### Example 5 FINDING AN EQUATION OF A PARALLEL LINE

Find an equation of the line through $P(1, -2)$ that is parallel to the line $L$ with equation $3x - 2y = 1$.

**Solution** We find the slope of $L$ by writing its equation in slope-intercept form.

$$3x - 2y = 1 \qquad \text{Equation for } L$$
$$-2y = -3x + 1 \qquad \text{Subtract } 3x$$
$$y = \frac{3}{2}x - \frac{1}{2} \qquad \text{Divide by } -2$$

The slope of $L$ is 3/2.

The line whose equation we seek has slope 3/2 and contains the point $(x_1, y_1) = (1, -2)$. Thus, the point-slope form equation for the line we seek is

$$y + 2 = \frac{3}{2}(x - 1)$$

$$y + 2 = \frac{3}{2}x - \frac{3}{2} \qquad \text{Distributive property}$$

$$y = \frac{3}{2}x - \frac{7}{2}$$

### Example 6  FINDING AN EQUATION OF A PERPENDICULAR LINE

Find an equation of the line through $P(2, -3)$ that is perpendicular to the line $L$ with equation $4x + y = 3$. Support the result with a grapher.

**Solution**  We find the slope of $L$ by writing its equation in slope-intercept form.

$$4x + y = 3 \qquad \text{Equation for } L.$$

$$y = -4x + 3 \qquad \text{Subtract } 4x$$

The slope of $L$ is $-4$.

The line whose equation we seek has slope $-1/(-4) = 1/4$ and passes through the point $(x_1, y_1) = (2, -3)$. Thus, the point-slope form equation for the line we seek is

$$y - (-3) = \frac{1}{4}(x - 2)$$

$$y + 3 = \frac{1}{4}x - \frac{2}{4} \qquad \text{Distributive property}$$

$$y = \frac{1}{4}x - \frac{7}{2}$$

Figure P.26 shows the graphs of the two equations in a square viewing window and suggests that the graphs are perpendicular.

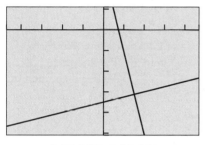

[−4.7, 4.7] by [−5.1, 1.1]

**Figure P.26**  The graphs of $y = -4x + 3$ and $y = (1/4)x - 7/2$ in this square viewing window appear to intersect at a right angle. (Example 6)

## Applying Linear Equations in Two Variables

Linear equations and their graphs occur frequently in applications. Algebraic solutions to these application problems often require finding an equation of a line and solving a linear equation in one variable. Grapher techniques complement algebraic ones.

### Example 7  FINDING THE DEPRECIATION OF REAL ESTATE

Camelot Apartments purchased a $50,000 building and depreciates it $2000 per year over a 25-year period.

**(a)** Write a linear equation giving the value ($y$) of the building in terms of the years ($x$) after the purchase.

**(b)** In how many years will the value of the building be $24,500?

**Solution**

**(a)** We need to determine the value of $m$ and $b$ so that $y = mx + b$, where $0 \le x \le 25$. We know that $y = 50,000$ when $x = 0$, so the line has $y$-

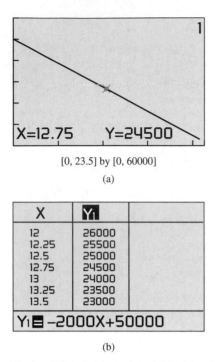

[0, 23.5] by [0, 60000]

(a)

| X | Y₁ | |
|------|-------|--|
| 12 | 26000 | |
| 12.25 | 25500 | |
| 12.5 | 25000 | |
| 12.75 | 24500 | |
| 13 | 24000 | |
| 13.25 | 23500 | |
| 13.5 | 23000 | |

Y₁ ■ −2000X+50000

(b)

**Figure P.27** A (a) graph and (b) table of values for $y = -2000x + 50,000$. (Example 7)

intercept (0, 50 000) and $b = 50,000$. One year after purchase ($x = 1$), the value of the building is $50,000 - 2,000 = 48,000$. So when $x = 1$, $y = 48,000$. Using algebra, we find

$$y = mx + b$$

$$48,000 = m \cdot 1 + 50,000 \qquad \text{y = 48,000 when x = 1}$$

$$-2000 = m$$

The value $y$ of the building $x$ years after its purchase is

$$y = -2000x + 50,000.$$

**(b)** We need to find the value of $x$ when $y = 24,500$.

$$y = -2000x + 50,000$$

Again, using algebra we find

$$24,500 = -2000x + 50,000 \qquad \text{Set y = 24,500}$$

$$-25,500 = -2000x$$

$$12.75 = x$$

The depreciated value of the building will be $24,500 exactly 12.75 years, or 12 years 9 months, after purchase by Camelot Apartments. We can support our algebraic work both graphically and numerically. The trace coordinates in Figure P.27a show graphically that (12.75, 24 500) is a solution of $y = -2000x + 50,000$. This means that $y = 24,500$ when $x = 12.75$.

Figure P.27b is a table of values for $y = -2000x + 50,000$ for a few values of $x$. The fourth line of the table shows numerically that $y = 24,500$ when $x = 12.75$.

Figure P.28 shows Americans' income from August 1998 through July 1999 in trillions of dollars, adjusted for inflation and other factors. In Example 8 we model the data in Figure P.28 with a linear equation.

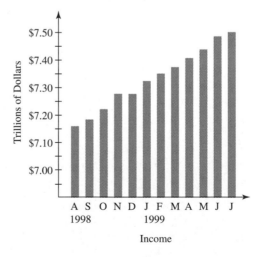

**Figure P.28** Americans' income in July 1998 was 7.13 trillion dollars and in July 1999 was 7.5 trillion dollars. *Source:* AP, Commerce Department as reported in *The Columbus Dispatch* on August 28, 1999. (Example 8)

**Example 8** FINDING A LINEAR MODEL FOR AMERICANS'
PERSONAL INCOME

From July 1998 to July 1999, Americans' income rose from 7.13 trillion
dollars to 7.50 trillion dollars as displayed in Figure P.28.

**(a)** Let $x = 0$ represent July 1998, $x = 1$ represent August 1998, …, and
$x = 12$ represent July 1999. Write a linear equation for Americans' income
$y$ in terms of the month $x$ using the points $(0, 7.13)$ and $(12, 7.50)$.

**(b)** Use the equation in (a) to estimate Americans' income in December
1998.

**(c)** Use the equation in (a) to predict Americans' income in July 2002.

**Solution**

**(a)** Let $y = mx + b$. The value of 7.13 trillion dollars in July 1998 gives
$y = 7.13$ when $x = 0$. So, $b = 7.13$. The value of 7.50 trillion in July 1999
gives $y = 7.5$ when $x = 12$.

$$y = mx + b$$

$$y = mx + 7.13 \qquad \text{\small $y = 7.13$ when $x = 0$}$$

$$7.5 = m \cdot 12 + 7.13 \qquad \text{\small $y = 7.50$ when $x = 12$}$$

$$0.37 = 12m$$

$$m = 0.0308\overline{3}$$

The linear equation we seek is $y = 0.0308\overline{3}x + 7.13$.

**(b)** December 1998 is represented by $x = 5$.

$$y = 0.0308\overline{3}x + 7.13$$

$$y = 0.0308\overline{3} \cdot 5 + 7.13 \qquad \text{\small Set $x = 5$.}$$

$$y \approx 7.28 \qquad \text{\small Simplify}$$

Using the linear model found in (a) we estimate American's income in
December 1998 to bc 7.28 trillion dollars.

**(c)** July 2002 is represented by $x = 48$.

$$y = 0.0308\overline{3}x + 7.13$$

$$y = 0.0308\overline{3} \cdot 48 + 7.13 \qquad \text{\small Set $x = 48$.}$$

$$y \approx 8.61 \qquad \text{\small Simplify.}$$

Using the linear model found in (a) we estimate Americans' income in July
2002 to be 8.61 trillion dollars.

**Problem**

The Moon is about 237,000 miles from Earth and light travels from the Earth to the Moon in approximately 1.27 seconds. The Earth is about 93 million miles from the Sun and light travels from the Earth to the Sun in 8.32 minutes. Given this information, how can we determine the speed of light? What is the distance from the Sun to Pluto if light can travel the distance in 5 hours and 29 minutes?

**Solution**

Solution: We assume that the speed of light is a constant and write a linear equation giving the value ($y$) of distance between the two objects, in terms of the time required to travel between the two objects ($x$):

$$y = mx + b.$$

The slope ($m$) in this equation represents the speed of light because it measures the change in distance divided by the change in time.

Both travel times must be converted to the same unit. We change 8.32 minutes to seconds:

$$8.32 \text{ min} \cdot \frac{60 \text{ sec}}{1 \text{ min}} \approx 499 \text{ sec}$$

So we have two points $(x_1, y_1) = (1.27, 237{,}000)$ and $(x_2, y_2) = (499, 93{,}000{,}000)$.

We can find the slope

$$m = (y_2 - y_1)/(x_2 - x_1) = (93{,}000{,}000 - 237{,}000)/(499 - 1.27)$$
$$= 186{,}372.1295 \approx 186{,}000 \qquad \text{rounding to the nearest thousand}$$

We can also find the $y$-intercept

$$y_1 = mx_1 + b$$
$$237{,}000 = 186{,}000(1.27) + b$$
$$b = 780 \approx 0$$

The slope, $m$, is the speed of light, approximately 186,000 miles per second.

Now, since we know light can travel from the Sun to Pluto in

$$5 \text{ hours and } 29 \text{ minutes} = 5(60) + 29 \text{ minutes}$$
$$= 329 \text{ minutes} = 329(60) \text{ seconds}$$
$$= 19{,}740 \text{ seconds},$$

we can find the distance from the Sun to Pluto:

$$y = 186{,}000x = 186{,}000 \frac{\text{mi}}{\text{sec}} (19{,}740 \text{ seconds})$$
$$= 3{,}671{,}640{,}000$$
$$\approx 3.7 \times 10^9 \text{ miles}.$$

# Quick Review P.4

In Exercises 1–4, solve for $x$.

**1.** $-75x + 25 = 200$   $x = -\dfrac{7}{3}$   **2.** $400 - 50x = 150$   $x = 5$

**3.** $3(1 - 2x) + 4(2x - 5) = 7$   **4.** $2(7x + 1) = 5(1 - 3x)$

In Exercises 5–8, solve for $y$.

**5.** $2x - 5y = 21$   **6.** $\dfrac{1}{3}x + \dfrac{1}{4}y = 2$

**7.** $2x + y = 17 + 2(x - 2y)$   **8.** $x^2 + y = 3x - 2y$

In Exercises 9 and 10, simplify the fraction.

**9.** $\dfrac{9 - 5}{-2 - (-8)}$   $\dfrac{2}{3}$   **10.** $\dfrac{-4 - 6}{-14 - (-2)}$   $\dfrac{5}{6}$

# Section P.4 Exercises

In Exercises 1 and 2, estimate the slope of the line.

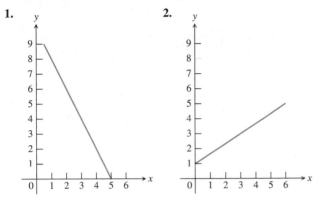

**1.**   **2.**

In Exercises 3–6, graph the linear equation on a grapher. Choose a viewing window that shows the line intersecting both the $x$- and $y$-axes.

**3.** $8x + y = 49$   **4.** $2x + y = 35$

**5.** $123x + 7y = 429$   **6.** $2100x + 12y = 3540$

In Exercises 7–10, find the slope of the line through the pair of points.

**7.** $(-3, 5)$ and $(4, 9)$   $\dfrac{4}{7}$   **8.** $(-2, 1)$ and $(5, -3)$   $-\dfrac{4}{7}$

**9.** $(-2, -5)$ and $(-1, 3)$   8   **10.** $(5, -3)$ and $(-4, 12)$   $-\dfrac{5}{3}$

In Exercises 11–14, find the value of $x$ or $y$ so that the line through the pair of points has the given slope.

| | Points | Slope | |
|---|---|---|---|
| **11.** | $(x, 3)$ and $(5, 9)$ | $m = 2$ | $x = 2$ |
| **12.** | $(-2, 3)$ and $(4, y)$ | $m = -3$ | $y = -15$ |
| **13.** | $(-3, -5)$ and $(4, y)$ | $m = 3$ | $y = 16$ |
| **14.** | $(-8, -2)$ and $(x, 2)$ | $m = 1/2$ | $x = 0$ |

In Exercises 15–18, find a *point-slope form* equation for the line through the point with given slope.

| | Point | Slope | | Point | Slope |
|---|---|---|---|---|---|
| **15.** | $(1, 4)$ | $m = 2$ | **16.** | $(-4, 3)$ | $m = -2/3$ |
| **17.** | $(5, -4)$ | $m = -2$ | **18.** | $(-3, 4)$ | $m = 3$ |

In Exercises 19–24, find a *general form equation* for the line through the pair of points.

**19.** $(-7, -2)$ and $(1, 6)$   **20.** $(-3, -8)$ and $(4, -1)$

**21.** $(1, -3)$ and $(5, -3)$   **22.** $(-1, -5)$ and $(-4, -2)$

**23.** $(-1, 2)$ and $(2, 5)$   **24.** $(4, -1)$ and $(4, 5)$

In Exercises 25–30, find a slope-intercept form equation for the line.

**25.** The line through $(0, 5)$ with slope $m = -3$   $y = -3x + 5$

**26.** The line through $(1, 2)$ with slope $m = 1/2$   $y = \dfrac{1}{2}x + \dfrac{3}{2}$

**27.** The line through the points $(-4, 5)$ and $(4, 3)$

**28.** The line through the points $(4, 2)$ and $(-3, 1)$

**29.** The line $2x + 5y = 12$   $y = -\dfrac{2}{5}x + \dfrac{12}{5}$

**30.** The line $7x - 12y = 96$   $y = \dfrac{7}{12}x - 8$

In Exercises 31 and 32, the line contains the origin and the point in the upper right corner of the grapher screen.

**31. Writing to Learn** Which line shown here has the greater slope? Explain.

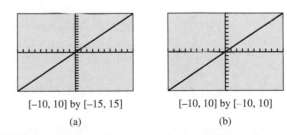

$[-10, 10]$ by $[-15, 15]$     $[-10, 10]$ by $[-10, 10]$
(a)     (b)

**32. Writing to Learn** Which line shown here has the greater slope? Explain.

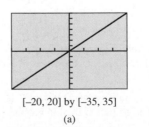 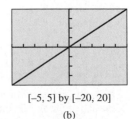

$[-20, 20]$ by $[-35, 35]$     $[-5, 5]$ by $[-20, 20]$
(a)     (b)

In Exercises 33–36, find the value of *x* and the value of *y* for which $(x, 14)$ and $(18, y)$ are points on the graph.

**33.** $y = 0.5x + 12$          **34.** $y = -2x + 18$

**35.** $3x + 4y = 26$          **36.** $3x - 2y = 14$

In Exercises 37–40, find the values for Ymin, Ymax, and Yscl that will make the graph of the line appear in the viewing window as shown here.

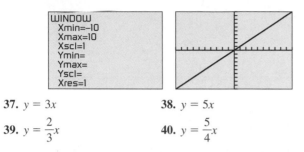

```
WINDOW
Xmin=-10
Xmax=10
Xscl=1
Ymin=
Ymax=
Yscl=
Xres=1
```

**37.** $y = 3x$          **38.** $y = 5x$

**39.** $y = \dfrac{2}{3}x$          **40.** $y = \dfrac{5}{4}x$

In Exercises 41–44, **(a)** find an equation for the line passing through the point and parallel to the given line, and **(b)** find an equation for the line passing through the point and perpendicular to the given line. Support your work graphically.

|       | Point     | Line          |
|-------|-----------|---------------|
| **41.** | $(1, 2)$  | $y = 3x - 2$   |
| **42.** | $(-2, 3)$ | $y = -2x + 4$  |
| **43.** | $(3, 1)$  | $2x + 3y = 12$ |
| **44.** | $(6, 1)$  | $3x - 5y = 15$ |

**45. Real Estate Appreciation** Bob Michaels purchased a house 8 years ago for $42,000. This year it was appraised at $67,500.

**(a)** A linear equation $V = mt + b$, $0 \le t \le 15$, represents the value *V* of the house for 15 years after it was purchased. Determine *m* and *b*.

**(b)** Graph the equation and trace to estimate in how many years after purchase this house will be worth $72,500.

**(c)** Write and solve an equation algebraically to determine how many years after purchase this house will be worth $74,000.

**(d)** Determine how many years after purchase this house will be worth $80,250.

**46. Investment Planning** Mary Ellen plans to invest $18,000, putting part of the money (*x*) into a savings that pays 5% annually and the rest into an account that pays 8% annually.

**(a)** What are the possible values of *x* in this situation?

**(b)** If Mary Ellen invests *x* dollars at 5%, write an equation that describes the total interest *I* received from both accounts at the end of one year.

**(c)** Graph and trace to estimate how much Mary Ellen invested at 5% if she earned $1,020 in total interest at the end of the first year.

**(d)** Use your grapher to generate a table of values for *I* to find out how much Mary Ellen should invest at 5% to earn $1,185 in total interest in one year.

**47. Navigation** A commercial jet airplane climbs at takeoff with slope $m = 3/8$. How far in the horizontal direction will the airplane fly to reach an altitude of 12,000 ft above the takeoff point? $x = 32{,}000$ ft

**48. Grade of a Highway** Interstate 70 west of Denver, Colorado, has a section posted as a 6% grade. This means that for a horizontal change of 100 ft there is a 6-ft vertical change.

6% grade

**(a)** Find the slope of this section of the highway.  0.06

**(b)** On a highway with a 6% grade what is the horizontal distance required to climb 250 ft?  4166.6 ft

**(c)** A sign along the highway says 6% grade for the next 7 mi. Estimate how many feet of vertical change there are along those next 7 mi. (There are 5280 ft in 1 mile.)

**49. Writing to Learn  Building Specifications** Asphalt shingles do not meet code specifications on a roof that has less than a 4-12 pitch. A 4-12 pitch means there are 4 ft of vertical change in 12 ft of horizontal change. A certain roof has slope $m = 3/8$. Could asphalt shingles be used on that roof? Explain.

**50. Revisiting Example 8** Use the linear equation found in Example 8 to estimate Americans' income in the months displayed in Figure P.28.

**51. Americans' Spending** From July 1998 to July 1999, Americans' spending rose from 5.82 trillion dollars to 6.20 trillion dollars as illustrated in Figure P.29.

**(a)** Let $x = 0$ represent July 1998, $x = 1$ represent August 1998, …, and $x = 12$ represent July 1999. Write a linear equation for Americans' spending in terms of the month *x* using the pairs $(0, 5.82)$ and $(12, 6.20)$.

**(b)** Use the equation in (a) to estimate Americans' spending in the months displayed in Figure P.29.

**(c)** Use the equation in (a) to predict Americans' spending in July 2002.  7.34 trillion dollars

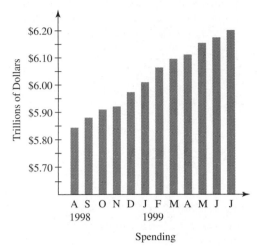

**Figure P.29** Americans' spending in July 1998 was 5.82 trillion dollars and in July 1999 was 6.20 trillion dollars. *Source:* AP, Commerce Department as reported in *The Columbus Dispatch* on August 28, 1999.

**52. U.S. Imports from Mexico**  The total (*y*) in billions of dollars of U.S. imports from Mexico for each year (*x*) from 1991 to 1998 is given in the table. (Source: Bureau of the Census, Foreign Trade Division, FINAL 1991–1998)

| *x* | 1991 | 1992 | 1993 | 1994 | 1995 | 1996 | 1997 | 1998 |
|---|---|---|---|---|---|---|---|---|
| *y* | 31.1 | 35.2 | 39.9 | 49.5 | 61.7 | 74.3 | 85.9 | 94.6 |

**(a)** Use the pairs (1992, 35.2) and (1996, 74.3) to write a linear equation for *x* and *y*.  $y = 9.775x - 19{,}436.6$

**(b)** Superimpose the graph of the linear equation in (a) on a scatter plot of the data.

**(c)** Use the equation in (a) to predict the total U.S. Imports from Mexico in 2001.  123.2

**53. World Population**  The midyear world population for the years 1993 to 1998 (in millions) is shown in Table P.7.

**Table P.7  World Population**

| Year | Population (millions) |
|---|---|
| 1993 | 5523 |
| 1994 | 5603 |
| 1995 | 5682 |
| 1996 | 5761 |
| 1997 | 5840 |
| 1998 | 5919 |

*Source: U.S. Bureau of the Census, International Data Base, Data updated 12-28-98*

**(a)** Let *x* = 0 represent 1990, *x* = 1 represent 1991, and so forth. Draw a scatter plot of the data.

**(b)** Use the 1993 and 1998 data to write a linear equation for the population *y* in terms of the year *x*. Superimpose the graph of the linear equation on the scatter plot in (a).

**(c)** Use the equation in (b) to predict the midyear world population in 2003. Compare it with the Census Bureau estimate of 6301 million.

**54. U.S. Imports from Mexico**  The total in billions of dollars of U.S. imports from Mexico from 1991 to 1998 is given in Table P.8.

**Table P.8  U.S. Imports from Mexico**

| Year | U.S. Imports (billions of dollars) |
|---|---|
| 1991 | 31.1 |
| 1992 | 35.2 |
| 1993 | 39.9 |
| 1994 | 49.5 |
| 1995 | 61.7 |
| 1996 | 74.3 |
| 1997 | 85.9 |
| 1998 | 94.6 |

*Source: Bureau of the Census, Foreign Trade Division, FINAL 1991–1998*

**(a)** Let *x* = 0 represent 1990, *x* = 1 represent 1991, and so forth. Draw a scatter plot of the data.

**(b)** Use the 1992 and 1998 data to write a linear equation for the U.S. Imports from Mexico (*y*) in terms of the year (*x*). Superimpose the graph of the linear equation on the scatter plot in (a).  $y = 9.9(x - 2) + 35.2 = 9.9x + 15.4$

**(c)** Use the equation in (b) to predict the U.S. Imports from Mexico in 2002.  134.2 billion dollars

In Exercises 55 and 56, determine *a* so that the line segments *AB* and *CD* are parallel.

**55.** **56.**

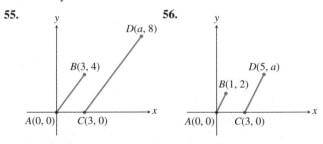

In Exercises 57 and 58, determine *a* and *b* so that figure *ABCD* is a parallelogram.

**57.** **58.**

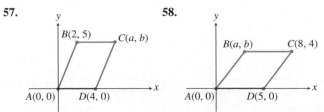

## Explorations

**59. Exploring the Graph of** $\dfrac{x}{a} + \dfrac{y}{b} = c,\, a \neq, 0,\, b \neq 0$

Let $c = 1$.

**(a)** Draw the graph for $a = 3,\, b = -2$.

**(b)** Draw the graph for $a = -2,\, b = -3$.

**(c)** Draw the graph for $a = 5,\, b = 3$.

**(d)** Use your graphs in (a), (b), (c) to conjecture what $a$ and $b$ represent when $c = 1$. Prove your conjecture.

**(e)** Repeat **(a)–(d)** for $c = 2$.

**(f)** If $c = -1$, what do $a$ and $b$ represent? ∎

**60. Writing to Learn Perpendicular Lines**

**(a)** Is it possible for two lines with positive slopes to be perpendicular? Explain.

**(b)** Is it possible for two lines with negative slopes to be perpendicular? Explain.

**61. Group Activity Parallel and Perpendicular Lines**

**(a)** Assume that $c \neq d$ and $a$ and $b$ are not both zero. Show that $ax + by = c$ and $ax + by = d$ are parallel lines. Explain why the restrictions on $a$, $b$, $c$, and $d$ are necessary.

**(b)** Assume that $a$ and $b$ are not both zero. Show that $ax + by = c$ and $bx - ay = d$ are perpendicular lines. Explain why the restrictions on $a$ and $b$ are necessary.

## Extending the Ideas

**62. Connecting Algebra and Geometry** Show that if the midpoints of consecutive sides of any quadrilateral (see figure) are connected, the result is a parallelogram.

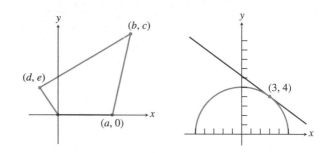

Art for Exercise 62      Art for Exercise 63

**63. Connecting Algebra and Geometry** Consider the semicircle of radius 5 centered at $(0, 0)$ as shown in the figure. Find an equation of the line tangent to the semicircle at the point $(3, 4)$. (*Hint:* A line tangent to a circle is perpendicular to the radius at the point of tangency.)

**64. Connecting Algebra and Geometry** Show that in any triangle (see figure), the line segment joining the midpoints of two sides is parallel to the third side and half as long.

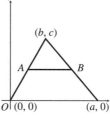

**P.5**

# Solving Equations Graphically, Numerically, and Algebraically

Solving Equations Graphically • Solving Equations Numerically with Tables • Solving Quadratic Equations • Solving Equations by Factoring • Solving Equations Involving Fractions • Solving Equations by Finding Intersections

## Solving Equations Graphically

The graph of the equation $y = 2x - 5$ (in $x$ and $y$) can be used to solve the equation $2x - 5 = 0$ (in $x$). Using the techniques of Section P.3, we can show algebraically that $x = 5/2$ is a solution of $2x - 5 = 0$. Therefore, the ordered pair $(5/2, 0)$ is a solution of $y = 2x - 5$. Figure P.30 suggests that the $x$-intercept of the graph of the line $y = 2x - 5$ is the point $(5/2, 0)$ as it should be.

One way to solve an equation graphically is to find all its $x$-intercepts. There are many graphical techniques that can be used to find $x$-intercepts.

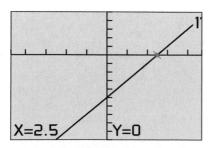

[–4.7, 4.7] by [–10, 5]

**Figure P.30** Using trace we see that $(2.5, 0)$ is an $x$-intercept of the graph of $y = 2x - 5$ and, therefore, $x = 2.5$ is a solution of the equation $2x - 5 = 0$.

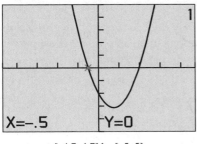

[–4.7, 4.7] by [–5, 5]

**Figure P.31** It appears that $(-0.5, 0)$ and $(2, 0)$ are $x$-intercepts of the graph of $y = 2x^2 - 3x - 2$. (Example 1)

**Alert**

Many students will assume that all solutions are contained in their default viewing window. Remind them that it is frequently necessary to zoom-out in order to see the general behavior of the graph, and then zoom-in to find more exact values of $x$ at the intersection points.

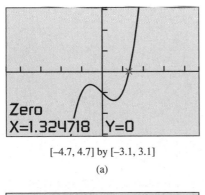

[–4.7, 4.7] by [–3.1, 3.1]

(a)

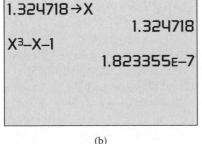

(b)

**Figure P.32** The graph of $y = x^3 - x - 1$. Part (a) shows that $(1.324718, 0)$ is an approximation to the $x$-intercept of the graph. Part (b) supports this conclusion. (Example 2)

**Example 1  SOLVING BY FINDING *X*-INTERCEPTS**

Solve the equation $2x^2 - 3x - 2 = 0$ graphically.

**Solution**  We find the $x$-intercepts of the graph of $y = 2x^2 - 3x - 2$ (Figure P.31). We use trace to see that $(-0.5, 0)$ and $(2, 0)$ are $x$-intercepts of this graph. Thus, the solutions of this equation are $x = -0.5$ and $x = 2$.

We can *only* be sure that the numbers that we read from the graph in Figure P.31 are approximations to the solutions of the equation. In this case, we can use algebra to find exact values.

$$2x^2 - 3x - 2 = 0$$
$$(2x + 1)(x - 2) = 0 \qquad \text{Factor.}$$

We can conclude that

$$2x + 1 = 0 \qquad \text{or} \qquad x - 2 = 0$$
$$x = -1/2 \qquad \text{or} \qquad x = 2.$$

So, $x = -1/2$ and $x = 2$ are the solutions of the original equation.

The algebraic solution procedure used in Example 1 is a special case of the following important property.

> Let $a$ and $b$ be real numbers.
>
> If $ab = 0$, then $a = 0$ or $b = 0$.

A solution of the equation $x^3 - x - 1 = 0$ is a value of $x$ that makes the value of $y = x^3 - x - 1$ equal to zero. Example 2 illustrates a built-in procedure on graphing calculators to find such values of $x$.

**Example 2  SOLVING GRAPHICALLY**

Solve the equation $x^3 - x - 1 = 0$ graphically.

**Solution**  Figure P.32a suggests that $x = 1.324718$ is the solution we seek. Figure P.32b provides numerical support that $x = 1.324718$ is a close approximation to the solution because, when $x = 1.324718$, $x^3 - x - 1 \approx 1.82 \times 10^{-7}$, which is nearly zero.

When solving equations graphically we usually get approximate solutions and not exact solutions. We will use the following agreement about accuracy in this book.

> **Agreement about Approximate Solutions**
>
> Round approximate solutions to hundredths unless directed otherwise. For applications, round to a value that is reasonable for the context of the problem.

With this accuracy agreement, we would report the solution found in Example 2 as 1.32.

## Solving Equations Numerically with Tables

The table feature on graphing calculators provide a numerical *zoom-in procedure* that we can use to find accurate solutions of equations. We illustrate this procedure in Example 3 using the same equation of Example 2.

### Example 3 SOLVING BY USING TABLES

Solve the equation $x^3 - x - 1 = 0$ using grapher tables.

**Solution** From Figure P.32a, we know that the solution we seek is between $x = 1$ and $x = 2$. Figure P.33a sets the starting point of the table (TblStart = 1) at $x = 1$ and increments the numbers in the table ($\triangle$Tbl = 0.1) by 0.1. Figure P.33b shows that the zero of $x^3 - x - 1$ is between $x = 1.3$ and $x = 1.4$.

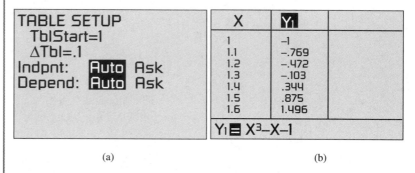

| TABLE SETUP |
| --- |
| TblStart=1 |
| $\triangle$Tbl=.1 |
| Indpnt:  **Auto**  Ask |
| Depend:  **Auto**  Ask |

| X | Y₁ | |
| --- | --- | --- |
| 1 | –1 | |
| 1.1 | –.769 | |
| 1.2 | –.472 | |
| 1.3 | –.103 | |
| 1.4 | .344 | |
| 1.5 | .875 | |
| 1.6 | 1.496 | |
| Y₁ ▆ X³–X–1 | | |

(a)                                    (b)

**Figure P.33** Part (a) gives the setup that produces the table in (b). (Example 3)

The next two steps in this process are shown in Figure P.34.

| X | Y₁ | |
| --- | --- | --- |
| 1.3 | –.103 | |
| 1.31 | –.0619 | |
| 1.32 | –.02 | |
| 1.33 | .02264 | |
| 1.34 | .0661 | |
| 1.35 | .11038 | |
| 1.36 | .15546 | |
| Y₁ ▆ X³–X–1 | | |

| X | Y₁ | |
| --- | --- | --- |
| 1.32 | –.02 | |
| 1.321 | –.0158 | |
| 1.322 | –.0116 | |
| 1.323 | –.0073 | |
| 1.324 | –.0031 | |
| 1.325 | .0012 | |
| 1.326 | .00547 | |
| Y₁ ▆ X³–X–1 | | |

(a)                                    (b)

**Figure P.34** In (a) TblStart = 1.3 and $\triangle$Tbl = 0.01, and in (b) TblStart = 1.32 and $\triangle$Tbl = 0.001. (Example 3)

From Figure P.34a, we can read that the zero is between $x = 1.32$ and $x = 1.33$; and from Figure P.34b, we can read that the zero is between $x = 1.324$ and $x = 1.325$. Because all such numbers round to 1.32, we can report the zero as 1.32 with our accuracy agreement.

> ### Exploration 1 Finding Real Zeros of Equations
>
> Consider the equation $4x^2 - 12x + 7 = 0$.
>
> 1. Use a graph to show that this equation has two real solutions, one between 0 and 1 and the other between 1 and 2.
> 2. Use the numerical zoom-in procedure illustrated in Example 3 to find each zero accurate to two decimal places. 0.79; 2.21
> 3. Use the built-in zero finder (see Example 2) to find the two solutions. Then round them to two decimal places.
> 4. If you are familiar with the graphical zoom-in process, use it to find each solution accurate to two decimal places.
> 5. Compare the numbers obtained in parts 2, 3, and 4.
> 6. Support the results obtained in parts 2, 3, and 4 numerically.
> 7. Use the numerical zoom-in procedure illustrated in Example 3 to find each zero accurate to six decimal places. Compare with the answer found in part 3 with the zero finder. 0.792893; 2.207107

**Exploration Extensions**

Repeat Exploration 1 for another function, such as $y = |x| - |2x - 3|$.

## Solving Quadratic Equations

Linear equations ($ax + b = 0$) and *quadratic equations* are two members of the family of *polynomial equations*, which will be studied in more detail in Chapter 2.

> **Definition** Quadratic Equation in $x$
>
> A **quadratic equation in $x$** is one that can be written in the form
> $$ax^2 + bx + c = 0,$$
> where $a$, $b$, and $c$ are real numbers with $a \neq 0$.

**Square Roots**

If $t^2 = K > 0$, then $t = \sqrt{K}$ or $t = -\sqrt{K}$.

We review some of the basic algebraic techniques for solving quadratic equations. One algebraic technique that we have already used in Example 1 is *factoring*.

Quadratic equations of the form $(ax + b)^2 = c$ are fairly easy to solve as illustrated in Example 4.

### Example 4 SOLVING BY EXTRACTING SQUARE ROOTS

Solve $(2x - 1)^2 = 9$ algebraically.

**Solution**

$$(2x - 1)^2 = 9$$
$$2x - 1 = \pm 3 \qquad \text{Extract square roots.}$$
$$2x = 4 \quad \text{or} \quad 2x = -2$$
$$x = 2 \quad \text{or} \quad x = -1$$

**Alert**

Many students will forget the $\pm$ sign when solving by extracting square roots. Emphasize that these problems usually have *two* solutions and that both are important.

The technique of Example 4 is more general than you might think because every quadratic equation can be written in the form $(x + b)^2 = c$. The procedure we need to accomplish this is *completing the square*.

> **Completing the Square**
>
> To **complete the square** for the expression $x^2 + bx$, add $(b/2)^2$. The completed square is
>
> $$x^2 + bx + \left(\frac{b}{2}\right)^2 = \left(x + \frac{b}{2}\right)^2.$$

To solve a quadratic equation by completing the square, we simply divide both sides of the equation by the coefficient of $x^2$ and then complete the square as illustrated in Example 5.

### Example 5   SOLVING BY COMPLETING THE SQUARE

Solve $4x^2 - 20x + 17 = 0$ by completing the square.

**Solution**

$$4x^2 - 20x + 17 = 0$$

$$x^2 - 5x + \frac{17}{4} = 0 \qquad \text{Divide by 4}$$

$$x^2 - 5x = -\frac{17}{4} \qquad \text{Subtract } \frac{17}{4}.$$

Completing the square on $x^2 - 5x$ we obtain

$$x^2 - 5x + \left(-\frac{5}{2}\right)^2 = -\frac{17}{4} + \left(-\frac{5}{2}\right)^2 \qquad \text{Add } \left(-\frac{5}{2}\right)^2.$$

$$\left(x - \frac{5}{2}\right)^2 = 2 \qquad \text{Factor and simplify.}$$

$$x - \frac{5}{2} = \pm\sqrt{2} \qquad \text{Extract square roots.}$$

$$x = \frac{5}{2} \pm \sqrt{2}$$

$$x = \frac{5}{2} + \sqrt{2} \approx 3.91 \text{ or } x = \frac{5}{2} - \sqrt{2} \approx 1.09$$

The procedure of Example 5 can be applied to the general quadratic equation $ax^2 + bx + c = 0$ to produce the following formula for its solutions (see Exercise 63).

**Alert**

Point out that the quadratic *formula* is used to solve quadratic *equations*. Some students confuse these concepts.

> **Quadratic Formula**
>
> The solutions of the quadratic equation $ax^2 + bx + c = 0$, where $a \neq 0$, are given by the **quadratic formula**
>
> $$x = \frac{-b \pm \sqrt{b^2 - 4ac}}{2a}.$$

### Example 6  SOLVING USING THE QUADRATIC FORMULA

Solve the equation $3x^2 - 6x = 5$.

**Solution**  First we subtract 5 from both sides of the equation to put it in the form $ax^2 + bx + c = 0$: $3x^2 - 6x - 5 = 0$. We can see that $a = 3$, $b = -6$, and $c = -5$.

$$x = \frac{-b \pm \sqrt{b^2 - 4ac}}{2a} \qquad \text{Quadratic formula}$$

$$x = \frac{-(-6) \pm \sqrt{(-6)^2 - 4(3)(-5)}}{2(3)} \qquad a = 3, b = -6, c = -5$$

$$x = \frac{6 \pm \sqrt{96}}{6} \qquad \text{Simplify.}$$

$$x = \frac{6 + \sqrt{96}}{6} \approx 2.63 \quad \text{or} \quad x = \frac{6 - \sqrt{96}}{6} \approx -0.63$$

The graph of $y = 3x^2 - 6x - 5$ in Figure P.35 supports that the $x$-intercepts are approximately $-0.63$ and $2.63$.

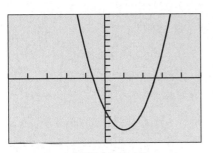

[–5, 5] by [–10, 10]

**Figure P.35**  The graph of $y = 3x^2 - 6x - 5$. (Example 6)

---

**Solving Quadratic Equations Algebraically**

There are four basic ways to solve quadratic equations algebraically.

1. **Factoring** (see Example 1)

2. **Extracting Square Roots** (see Example 4)

3. **Completing the Square** (see Example 5)

4. **Quadratic Formula** (see Example 6)

---

## Solving Equations by Factoring

We can sometimes use factoring to solve polynomial equations other than quadratic equations.

### Example 7  SOLVING A CUBIC EQUATION BY FACTORING

Solve the equation $3x^3 + x^2 - 6x - 2 = 0$.

**Solution**

$$3x^3 + x^2 - 6x - 2 = 0$$

$$x^2(3x + 1) - 2(3x + 1) = 0 \qquad \text{Group terms.}$$

$$(3x + 1)(x^2 - 2) = 0 \qquad \text{Remove common factor.}$$

$$3x + 1 = 0 \quad \text{or} \quad x^2 - 2 = 0 \qquad ab = 0 \Rightarrow a = 0 \text{ or } b = 0$$

$$x = -\frac{1}{3} \quad \text{or} \quad x^2 = 2$$

$$x = -\frac{1}{3} \quad \text{or} \quad x = \pm\sqrt{2}$$

This equation has three solutions: $-1/3$, $\sqrt{2}$, and $-\sqrt{2}$.

Figure P.36 provides graphical support for the solutions found algebraically.

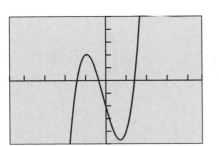

[–4.7, 4.7] by [–5, 5]

**Figure P.36**  The graph of $y = 3x^3 + x^2 - 6x - 2$ supports $x$-intercepts of $(-\sqrt{2}, 0)$, $(-1/3, 0)$, and $(\sqrt{2}, 0)$. (Example 7)

### Example 8  SOLVING A QUARTIC EQUATION BY FACTORING

Solve the equation $x^4 - 7x^2 + 12 = 0$.

**Solution**  The left-hand side of this equation is quadratic in $x^2$.

$$x^4 - 7x^2 + 12 = (x^2)^2 - 7(x^2) + 12$$

We can use this information to factor the expression

$$x^4 - 7x^2 + 12 = 0$$

$$(x^2 - 4)(x^2 - 3) = 0 \qquad \text{Factor}$$

$$x^2 - 4 = 0 \quad \text{or} \quad x^2 - 3 = 0 \qquad ab = 0 \Rightarrow a = 0 \text{ or } b = 0$$

$$x = \pm 2 \quad \text{or} \quad x = \pm\sqrt{3} \qquad \text{Solve by extracting square roots}$$

This equation has four solutions: $-2$, $2$, $-\sqrt{3}$, and $\sqrt{3}$. We leave it to you to either check these solutions in the original equation, or to use a graph to support the results.

## Solving Equations Involving Fractions

To solve an equation involving fractions we begin by finding the LCD (least common denominator) of all the terms of the equation. Then we clear the equation of fractions by multiplying each side of the equation by the LCD, just as we did in Section P.3. Sometimes the LCD contains variables.

When we multiply or divide an equation by an expression containing variables, the resulting equation may have solutions that are *not* solutions of the original equation. These are **extraneous solutions**. For this reason we must check each solution of the resulting equation in the original equation.

### Example 9  SOLVING BY CLEARING FRACTIONS

Solve the equation $\dfrac{x}{x + 1} + \dfrac{3}{x - 4} = \dfrac{15}{x^2 - 3x - 4}$.

**Solution**  Because $x^2 - 3x - 4 = (x + 1)(x - 4)$, the LCD is $(x + 1)(x - 4)$. We multiply each side of the original equation by the LCD.

$$\left[\frac{x}{x + 1} + \frac{3}{x - 4}\right](x + 1)(x - 4) = \left[\frac{15}{x^2 - 3x - 4}\right](x + 1)(x - 4)$$

$$\frac{x(x + 1)(x - 4)}{x + 1} + \frac{3(x + 1)(x - 4)}{x - 4} = \frac{15(x + 1)(x - 4)}{x^2 - 3x - 4} \quad \text{Distributive Property}$$

$$x(x - 4) + 3(x + 1) = 15 \qquad \text{Simplify.}$$

$$x^2 - 4x + 3x + 3 = 15 \qquad \text{Distributive Property}$$

$$x^2 - x - 12 = 0 \qquad \text{Simplify.}$$

$$(x - 4)(x + 3) = 0 \qquad \text{Factor.}$$

$$x - 4 = 0 \quad \text{or} \quad x + 3 = 0 \qquad ab = 0 \Rightarrow a = 0 \text{ or } b = 0$$

$$x = 4 \quad \text{or} \quad x = -3 \qquad \text{Solve.}$$

You can check that $x = -3$ is a solution of the original equation. Setting $x = 4$ in the original leads to zero in two of the denominators. So, $x = 4$ is not a solution, that is, $x = 4$ is extraneous. The original equation has exactly one solution, namely $x = -3$.

## Solving Equations by Finding Intersections

Sometimes we can rewrite an equation and solve it graphically by finding the *points of intersection* of two graphs. A point $(a, b)$ is a **point of intersection** of two graphs if it lies on both graphs.

We illustrate this procedure with the absolute value equation in Example 10.

### Example 10   SOLVING BY FINDING INTERSECTIONS

Solve the equation $|2x - 1| = 6$.

**Solution**  Figure P.37 suggests that the V-shaped graph of $y = |2x - 1|$ intersects the graph of the horizontal line $y = 6$ twice. We can use trace or the intersection feature of our grapher to see that the two points of intersection have coordinates $(-2.5, 6)$ and $(3.5, 6)$. This means that the original equation has two solutions: $-2.5$ and $3.5$.

We can use algebra to find the exact solutions. The only two real numbers with absolute value 6 are 6 itself and $-6$. So, if $|2x - 1| = 6$, then

$$2x - 1 = 6 \quad \text{or} \quad 2x - 1 = -6$$

$$x = \frac{7}{2} = 3.5 \quad \text{or} \quad x = -\frac{5}{2} = -2.5$$

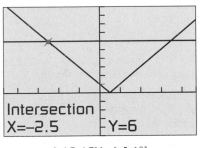

[−4.7, 4.7] by [−5, 10]

**Figure P.37**  The graphs of $y = |2x - 1|$ and $y = 6$ intersect at $(-2.5, 6)$ and $(3.5, 6)$. (Example 10)

## Quick Review P.5

In Exercises 1–4, expand the product.

**1.** $(3x - 4)^2$  $9x^2 - 24x + 16$  **2.** $(2x + 3)^2$  $4x^2 + 12x + 9$

**3.** $(2x + 1)(3x - 5)$    **4.** $(3y - 1)(5y + 4)$

In Exercises 5–8, factor completely.

**5.** $25x^2 - 20x + 4$  $(5x - 2)^2$  **6.** $15x^3 - 22x^2 + 8x$

**7.** $3x^3 + x^2 - 15x - 5$   **8.** $y^4 - 13y^2 + 36$

In Exercises 9 and 10, combine the fractions and reduce the resulting fraction to lowest terms.

**9.** $\dfrac{x}{2x + 1} - \dfrac{2}{x - 3}$    **10.** $\dfrac{x + 1}{x^2 - 5x + 6} - \dfrac{3x + 11}{x^2 - x - 6}$

## Section P.5 Exercises

In Exercises 1–6, solve the equation by extracting square roots.

**1.** $4x^2 = 25$  $x = \pm\dfrac{5}{2}$    **2.** $2(x - 5)^2 = 17$

**3.** $3(x + 4)^2 = 8$    **4.** $4(u + 1)^2 = 18$

**5.** $2y^2 - 8 = 6 - 2y^2$    **6.** $(2x + 3)^2 = 169$

In Exercises 7–12, solve the equation by completing the square.

**7.** $x^2 + 6x = 7$    **8.** $x^2 + 5x - 9 = 0$

**9.** $x^2 - 7x + \dfrac{5}{4} = 0$    **10.** $4 - 6x = x^2$

**11.** $2x^2 - 7x + 9 = (x - 3)(x + 1) + 3x$  $x = 2$ or $x = 6$

**12.** $3x^2 - 6x - 7 = x^2 + 3x - x(x + 1) + 3$

In Exercises 13–18, solve the equation using the quadratic formula.

**13.** $x^2 + 8x - 2 = 0$    **14.** $2x^2 - 3x + 1 = 0$

**15.** $3x + 4 = x^2$    **16.** $x^2 - 5 = \sqrt{3}x$

**17.** $x(x + 5) = 12$  $x = -\dfrac{5}{2} \pm \dfrac{\sqrt{73}}{2} = -6.772\ldots$ or $1.772\ldots$

**18.** $x^2 - 2x + 6 = 2x^2 - 6x - 26$  $x = -4$ or $x = 8$

In Exercises 19–24, solve the equation by factoring. Support your work graphically.

**19.** $x^2 - x - 20 = 0$    **20.** $2x^2 + 5x - 3 = 0$

**21.** $4x^2 - 8x + 3 = 0$    **22.** $x^2 - 8x = -15$

**23.** $x(3x - 7) = 6$    **24.** $x(3x + 11) = 20$

In Exercises 25–28, estimate any *x*- and *y*-intercepts that are shown in the graph.

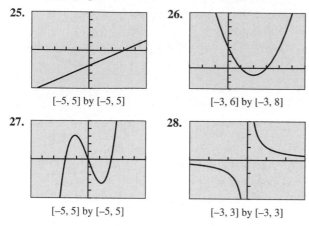

**25.** [−5, 5] by [−5, 5]

**26.** [−3, 6] by [−3, 8]

**27.** [−5, 5] by [−5, 5]

**28.** [−3, 3] by [−3, 3]

In Exercises 29–34, solve the equation graphically by finding *x*-intercepts.

**29.** $x^2 + x - 1 = 0$

**30.** $4x^2 + 20x + 23 = 0$

**31.** $x^3 + x^2 + 2x - 3 = 0$

**32.** $x^3 - 4x + 2 = 0$

**33.** $x^2 + 4 = 4x$

**34.** $x^2 + 2x = -2$

In Exercises 35–40, solve the equation graphically by finding intersections. Confirm your answer algebraically.

**35.** $|t - 8| = 2$

**36.** $|x + 1| = 4$

**37.** $|2x + 5| = 7$

**38.** $|3 - 5x| = 4$

**39.** $|2x - 3| = x^2$

**40.** $|x + 1| = 2x - 3$

In Exercises 41 and 42, the table permits you to estimate a zero of an expression. State the expression and give the zero as accurately as can be read from the table.

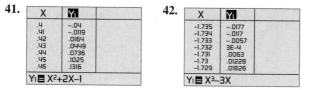

**41.**

| X | Y1 |
|---|---|
| .4 | −.04 |
| .41 | −.0119 |
| .42 | .0164 |
| .43 | .0449 |
| .44 | .0736 |
| .45 | .1025 |
| .46 | .1316 |

Y1 ■ X²+2X−1

**42.**

| X | Y1 |
|---|---|
| −1.735 | −.0177 |
| −1.734 | −.0117 |
| −1.733 | −.0057 |
| −1.732 | 3E−4 |
| −1.731 | .0063 |
| −1.73 | .01228 |
| −1.729 | .01826 |

Y1 ■ X³−3X

In Exercises 43 and 44, use tables to find the indicated number of solutions of the equation accurate to two decimal places.

**43.** Two solutions of $x^2 - x - 1 = 0$  1.62; −0.62

**44.** One solution of $-x^3 + x + 1 = 0$  1.32

**45. Interpreting Graphs** The graphs in the two viewing windows shown here can be used to solve the equation $3\sqrt{x + 4} = x^2 - 1$ graphically.

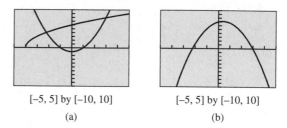

[−5, 5] by [−10, 10]
(a)

[−5, 5] by [−10, 10]
(b)

**(a)** The viewing window in (a) illustrates the intersection method for solving. Identify the two equations that are graphed.

**(b)** The viewing window in (b) illustrates the *x*-intercept method for solving. Identify the equation that is graphed.

**(c) Writing to Learn** How are the intersection points in (a) related to the *x*-intercepts in (b)?

**46. Writing to Learn Revisiting Example 3** Explain why all real numbers *x* that satisfy $1.324 < x < 1.325$ round to 1.32.

In Exercises 47–62, use a method of your choice to solve the equation.

**47.** $x^2 + x - 2 = 0$

**48.** $x^2 - 3x = 12 - 3(x - 2)$

**49.** $|2x - 1| = 5$

**50.** $x + 2 - 2\sqrt{x + 3} = 0$

**51.** $x^3 + 4x^2 - 3x - 2 = 0$

**52.** $\dfrac{x - 1}{x + 2} = 3$

**53.** $x + \dfrac{10}{x} = 7$

**54.** $x + 2 = \dfrac{15}{x}$

**55.** $\dfrac{1}{x} - \dfrac{2}{x - 3} = 4$  $x = \dfrac{11}{8} \pm \dfrac{1}{8}\sqrt{73}$

**56.** $\dfrac{3x}{x + 5} + \dfrac{1}{x - 2} = \dfrac{7}{x^2 + 3x - 10}$  $x = -\dfrac{1}{3}$

**57.** $\dfrac{x - 3}{x} - \dfrac{3}{x + 1} + \dfrac{3}{x^2 + x} = 0$  $x = 5$

**58.** $\dfrac{3}{x + 2} + \dfrac{6}{x^2 + x} = \dfrac{3 - x}{x}$  $x = -3$

**59.** $|x^2 + 4x - 1| = 7$

**60.** $|x + 5| = |x - 3|$  $x = -1$

**61.** $|0.5x + 3| = x^2 - 4$

**62.** $\sqrt{x + 7} = -x^2 + 5$

## Explorations

**63. Deriving the Quadratic Formula** Follow these steps to use completing the square to solve $ax^2 + bx + c = 0$, $a \neq 0$.

**(a)** Subtract *c* from both sides of the original equation and divide both sides of the resulting equation by *a* to obtain

$$x^2 + \frac{b}{a}x = -\frac{c}{a}.$$

**(b)** Add the square of one-half of the coefficient of *x* in (a) to both sides and simplify to obtain

$$\left(x + \frac{b}{2a}\right)^2 = \frac{b^2 - 4ac}{4a^2}.$$

**(c)** Extract square roots in (b) and solve for $x$ to obtain the quadratic formula

$$x = \frac{-b \pm \sqrt{b^2 - 4ac}}{2a}. \qquad \blacksquare$$

**64. Group Activity  Discriminant of a Quadratic**  The radicand $b^2 - 4ac$ in the quadratic formula is called the **discriminant** of the quadratic polynomial $ax^2 + bx + c$ because it can be used to describe the nature of its zeros.

**(a) Writing to Learn**  If $b^2 - 4ac > 0$, what can you say about the zeros of the quadratic polynomial $ax^2 + bx + c$? Explain your answer.

**(b) Writing to Learn**  If $b^2 - 4ac = 0$, what can you say about the zeros of the quadratic polynomial $ax^2 + bx + c$? Explain your answer.

**(c) Writing to Learn**  If $b^2 - 4ac < 0$, what can you say about the zeros of the quadratic polynomial $ax^2 + bx + c$? Explain your answer.

**65. Group Activity  Discriminant of a Quadratic**  Use the information learned in Exercise 64 to create a quadratic polynomial with the following number of real zeros. Support your answer graphically.

**(a)** Two real zeros          **(b)** Exactly one real zero

**(c)** No real zeros.

**66. Size of a Soccer Field**  Several of the World Cup '94 soccer matches were played in Stanford University's stadium in Menlo Park, California. The field is 30 yd longer than it is wide, and the area of the field is 8800 yd². What are the dimensions of this soccer field?  80 yd wide; 110 yd long

**67. Height of a Ladder**  John's paint crew knows from experience that its 18-ft ladder is particularly stable when the distance from the ground to the top of the ladder is 5 ft more than the distance from the building to the base of the ladder as shown in the figure. In this position, how far up the building does the ladder reach?  $\approx$ 14.98 ft up the wall

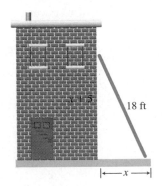

**68. Finding the Dimensions of a Norman Window**  A Norman window has the shape of a square with a semicircle mounted on it. Find the width of the window if the total area of the square and the semicircle is to be 200 ft².

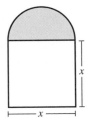

## Extending the Ideas

**69. Finding Number of Solutions**  Consider the equation $|x^2 - 4| = c$

**(a)** Find a value of $c$ for which this equation has four solutions. (There are many such values.)  $c = 2$

**(b)** Find a value of $c$ for which this equation has three solutions. (There is only one such value.)  $c = 4$

**(c)** Find a value of $c$ for which this equation has two solutions. (There are many such values.)  $c = 5$

**(d)** Find a value of $c$ for which this equation has no solutions. (There are many such values.)  $c = -1$

**(e) Writing to Learn**  Are there any other possible number of solutions of this equation? Explain.

**70. Sums and Products of Solutions of** $ax^2 + bx + c = 0$, $a \neq 0$
Suppose that $b^2 - 4ac > 0$.

**(a)** Show that the sum of the two solutions of this equation is $-(b/a)$.

**(b)** Show that the product of the two solutions of this equation is $c/a$.

**71. Exercise 70 Continued**  The equation $2x^2 + bx + c = 0$ has two solutions $x_1$ and $x_2$. If $x_1 + x_2 = 5$ and $x_1 \cdot x_2 = 3$, find the two solutions.

$2.5 \pm \dfrac{1}{2}\sqrt{13}$, or approximately 0.697 and 4.303

# P.6 Solving Inequalities Algebraically and Graphically

Solving Absolute Value Inequalities • Solving Quadratic Inequalities • Solving Inequalities Involving Fractions • Application

## Solving Absolute Value Inequalities

The methods for solving inequalities parallel the methods for solving equations. Here are two basic rules we apply to solve absolute value inequalities.

> **Solving Absolute Value Inequalities**
>
> Let $u$ be an algebraic expression in $x$ and let $a$ be a real number with $a \geq 0$.
>
> **1.** If $|u| < a$, then $u$ is in the interval $(-a, a)$. That is,
>
> $$|u| < a \quad \text{if and only if} \quad -a < u < a$$
>
> **2.** If $|u| > a$, then $u$ is in the interval $(-\infty, -a)$ or $(a, \infty)$, that is,
>
> $$|u| > a \quad \text{if and only if} \quad u < -a \text{ or } u > a$$
>
> The inequalities $<$ and $>$ can be replaced with $\leq$ and $\geq$, respectively. See Figure P.38.

**Objective**

Students will be able to solve inequalities involving absolute value, quadratic polynomials, and expressions involving fractions.

**Motivate**

Ask...
On a graph of $y = x^2 - 4$, what are the values of $x$ where $y = x^2 - 4$ lies below the $x$-axis? **($-2 < x < 2$)**

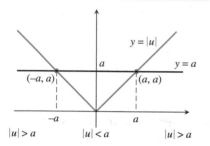

**Figure P.38** The solution of $|u| < a$ is represented by the portion of the number line where the graph of $y = |u|$ is below the graph of $y = a$. The solution of $|u| > a$ is represented by the portion of the number line where the graph of $y = |u|$ is above the graph of $y = a$.

**Alert**

Instead of "$u < a$ or $u > a$," some students may write "$a < u < -a$." Emphasize the important distinction between *and* and *or* in this context.

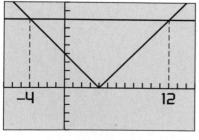

[–7, 15] by [–5, 10]

**Figure P.39** The graphs of $y = |x - 4|$ and $y = 8$. (Example 1)

### Example 1  SOLVING AN ABSOLUTE VALUE INEQUALITY

Solve $|x - 4| < 8$.

**Solution**

$$|x - 4| < 8 \qquad \text{Original inequality}$$
$$-8 < x - 4 < 8 \qquad \text{Equivalent double inequality}$$
$$-4 < x < 12 \qquad \text{Add 4.}$$

As an interval the solution is $(-4, 12)$.

Figure P.39 shows that points on the graph of $y = |x - 4|$ are below the points on the graph of $y = 8$ for values of $x$ between $-4$ and $12$.

### Example 2  SOLVING ANOTHER ABSOLUTE VALUE INEQUALITY

Solve $|3x - 2| \geq 5$.

**Solution**  The solution of this absolute value inequality consists of the solutions of both of these inequalities.

$$3x - 2 \leq -5 \quad \text{or} \quad 3x - 2 \geq 5$$
$$3x \leq -3 \quad \text{or} \quad 3x \geq 7 \qquad \text{Add 2.}$$
$$x \leq -1 \quad \text{or} \quad x \geq \frac{7}{3} \qquad \text{Divide by 3.}$$

## Union of Two Sets

The **union of two sets *A* and *B*,** denoted by $A \cup B$ is the set of all objects that belong to *A* or *B* or both.

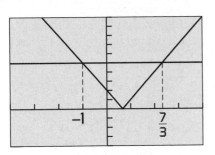

[–4, 4] by [–4, 10]

**Figure P.40**  The graphs of $y = |3x - 2|$ and $y = 5$. (Example 2)

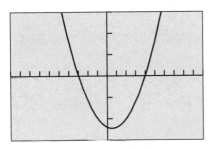

[–10, 10] by [–15, 15]

**Figure P.41**  The graph of $y = x^2 - x - 12$ appears to cross the *x*-axis at $x = -3$ and $x = 4$. (Example 3)

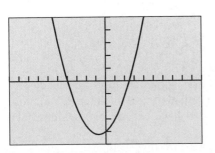

[–10, 10] by [–25, 25]

**Figure P.42**  The graph of $y = 2x^2 + 3x - 20$ appears to be below the *x*-axis for $-4 < x < 2.5$. (Example 4)

The solution consists of all numbers that are in either one of the two intervals $(-\infty, -1]$ or $[7/3, \infty)$, which may be written as $(-\infty, -1] \cup [7/3, \infty)$. The notation "$\cup$" is read as "union."

Figure P.40 shows that points on the graph of $y = |3x - 2|$ are above or on the points on the graph of $y = 5$ for values of *x* to the left of and including $-1$ and to the right of and including $7/3$.

## Solving Quadratic Inequalities

To solve a quadratic inequality such as $x^2 - x - 12 > 0$ we begin by solving the corresponding quadratic equation $x^2 - x - 12 = 0$. Then we determine the values of *x* for which the graph of $y = x^2 - x - 12$ lies above the *x*-axis.

### Example 3  SOLVING A QUADRATIC INEQUALITY

Solve $x^2 - x - 12 > 0$.

**Solution**  First we solve the corresponding equation $x^2 - x - 12 = 0$.

$$x^2 - x - 12 = 0$$

$$(x - 4)(x + 3) = 0 \qquad \text{Factor.}$$

$$x - 4 = 0 \quad \text{or} \quad x + 3 = 0 \qquad ab = 0 \Rightarrow a = 0 \text{ or } b = 0$$

$$x = 4 \quad \text{or} \quad x = -3 \qquad \text{Solve for } x.$$

The solutions of the corresponding quadratic equation are $-3$ and $4$, and they are not solutions of the original inequality because $0 > 0$ is false. Figure P.41 shows that the points on the graph of $y = x^2 - x - 12$ are above the *x*-axis for values of *x* to the left of $-3$ and to the right of $4$.

The solution of the original inequality is $(-\infty, -3) \cup (4, \infty)$.

In Example 4, the quadratic inequality involves the symbol $\leq$. In this case, the solutions of the corresponding quadratic equation are also solutions of the inequality.

### Example 4  SOLVING ANOTHER QUADRATIC INEQUALITY

Solve $2x^2 + 3x \leq 20$.

**Solution**  First we subtract 20 from both sides of the inequality to obtain $2x^2 + 3x - 20 \leq 0$. Next, we solve the corresponding quadratic equation $2x^2 + 3x - 20 = 0$.

$$2x^2 + 3x - 20 = 0$$

$$(x + 4)(2x - 5) = 0 \qquad \text{Factor.}$$

$$x + 4 = 0 \quad \text{or} \quad 2x - 5 = 0 \qquad ab = 0 \Rightarrow a = 0 \text{ or } b = 0$$

$$x = -4 \quad \text{or} \quad x = \frac{5}{2} \qquad \text{Solve for } x.$$

The solutions of the corresponding quadratic equation are $-4$ and $5/2 = 2.5$. You can check that they are also solutions of the inequality.

Figure P.42 shows that the points on the graph of $y = 2x^2 + 3x - 20$ are below the *x*-axis for values of *x* between $-4$ and $2.5$. The solution of the original inequality is $[-4, 2.5]$. We use square brackets because the numbers $-4$ and $2.5$ are also solutions of the inequality.

**Example 5** SOLVING A QUADRATIC INEQUALITY GRAPHICALLY

Solve $x^2 - 4x + 1 \geq 0$ graphically.

**Solution** We can use the graph of $y = x^2 - 4x + 1$ in Figure P.43 to determine that the solutions of the equation $x^2 - 4x + 1 = 0$ are about 0.27 and 3.73. Thus, the solution of the original inequality is $(-\infty, 0.27] \cup [3.73, \infty)$. We use square brackets because the zeros of the quadratic equation are solutions of the inequality even though we only have approximations to their values.

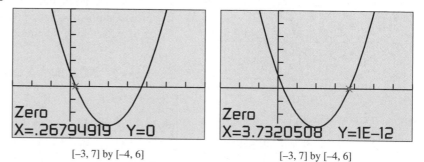

[−3, 7] by [−4, 6]       [−3, 7] by [−4, 6]

**Figure P.43** This figure suggests that $y = x^2 - 4x + 1$ is zero for $x \approx 0.27$ and $x \approx 3.73$.

**Example 6** SHOWING THAT A QUADRATIC INEQUALITY HAS NO SOLUTION

Solve $x^2 + 2x + 2 < 0$.

**Solution** Figure P.44 shows that the graph of $y = x^2 + 2x + 2$ lies above the x-axis for all values for x. Thus, the inequality $x^2 + 2x + 2 < 0$ has *no* solution.

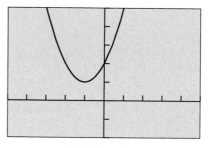

[−5, 5] by [−2, 5]

**Figure P.44** The values of $y = x^2 + 2x + 2$ are never negative. (Example 6)

Figure P.44 also shows that the solutions of the inequality $x^2 + 2x + 2 > 0$ is the set of all real numbers or, in interval notation, $(-\infty, \infty)$. A quadratic inequality can also have exactly one solution (see Exercise 37).

In Example 7, we use algebra and graphs to solve a cubic inequality.

**Example 7   SOLVING A CUBIC INEQUALITY**

Solve $x^3 - 4x \geq 0$.

**Solution**   First we solve the corresponding cubic equation algebraically.

$$x^3 - 4x = 0$$

$$x(x^2 - 4) = 0 \qquad \text{Remove common factor } x.$$

$$x(x - 2)(x + 2) = 0 \qquad \text{Factor.}$$

$$x = 0 \ \text{ or } \ x = 2 \ \text{ or } \ x = -2 \qquad ab = 0 \Rightarrow a = 0 \text{ or } b = 0$$

The solutions of the corresponding cubic equation are $-2$, $0$, and $2$. You can check that they are also solutions of the inequality.

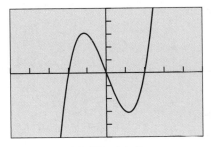

[−5, 5] by [−5, 5]

**Figure P.45**   The graph of $y = x^3 - 4x$ appears to cross the $x$-axis at $x = -2$, $x = 0$, and $x = 2$. (Example 7)

Figure P.45 shows that the points on the graph of $y = x^3 - 4x$ are above the $x$-axis for values of $x$ between $-2$ and $0$, and for values of $x$ to the right of $2$.

The solution of the original inequality is $[-2, 0] \cup [2, \infty)$. We use square brackets because the numbers $-2$, $0$, and $2$ are also solutions of the inequality.

## Solving Inequalities Involving Fractions

We can use what we know about solving quadratic inequalities to solve certain inequalities involving fractions.

**Example 8   SOLVING AN INEQUALITY INVOLVING A FRACTION**

Solve $\dfrac{x - 3}{2x + 3} > 0$.

**Solution**   The fraction $(x - 3)/(2x + 3)$ and the product $(x - 3)(2x + 3)$ have the same sign for all values of $x$ except $x = -3/2$, the value that makes the denominator zero. This means that the two inequalities,

$$\frac{x - 3}{2x + 3} > 0 \quad \text{and} \quad (x - 3)(2x + 3) > 0,$$

have the same solution. The solution of the inequality is $(-\infty, -1.5) \cup (3, \infty)$.

We can rewrite any inequality involving fractions so that its right-hand side is zero, and its left-hand side is a single fraction in reduced form. The values of $x$ that make the denominator zero are never part of the solution of the inequality, but those that make the numerator zero can be, as illustrated in Example 9.

### Example 9 SOLVING ANOTHER INEQUALITY INVOLVING A FRACTION

Solve $\dfrac{x+6}{2x+3} \geq 1$.

**Solution** We rewrite the inequality so that the right-hand side is zero, and the left-hand side is a single fraction.

$$\frac{x+6}{2x+3} \geq 1$$

$$\frac{x+6}{2x+3} - 1 \geq 0 \qquad \text{Subtract 1}$$

$$\frac{x+6}{2x+3} - \frac{2x+3}{2x+3} \geq 0 \qquad 1 = \frac{2x+3}{2x+3}$$

$$\frac{x+6-(2x+3)}{2x+3} \geq 0 \qquad \text{Combine fractions.}$$

$$\frac{-x+3}{2x+3} \geq 0$$

$$\frac{x-3}{2x+3} \leq 0 \qquad \text{Multiply by } -1.$$

This is the same fraction that was involved in Example 8. The fraction is zero for $x = 3$, so $x = 3$ is a solution of the inequality. The fraction is negative for all values of $x$ between $-1.5$ and $3$. The solution of the original inequality is $(-1.5, 3]$.

## Application

The movement of an object that is propelled vertically, but then subject only to the force of gravity, is an example of **projectile motion.**

> ### Projectile Motion
>
> Suppose an object is launched vertically from a point $s_0$ feet above the ground with an initial velocity of $v_0$ feet per second. The vertical position $s$ (in feet) of the object $t$ seconds after it is launched is
>
> $$s = -16t^2 + v_0 t + s_0.$$

### Example 10 FINDING HEIGHT OF A PROJECTILE

A projectile is launched straight up from ground level with an initial velocity of 288 ft/sec.

**(a)** When will the projectile's height above ground be 1152 ft?

**(b)** When will the projectile's height above ground be at least 1152 ft?

**Solution** Here $s_0 = 0$ and $v_0 = 288$. So, the projectile's height is $s = -16t^2 + 288t$.

**(a)** We need to determine when $s = 1152$.

$$s = -16t^2 + 288t$$

$$1152 = -16t^2 + 288t \qquad \text{Substitute } s = 1152.$$

$$16t^2 - 288t + 1152 = 0 \qquad \text{Add } 16t^2 - 288t.$$

$$t^2 - 18t + 72 = 0 \qquad \text{Divide by 16.}$$

$$(t - 6)(t - 12) = 0 \qquad \text{Factor.}$$

$$t = 6 \quad \text{or} \quad t = 12 \qquad \text{Solve for } t.$$

The projectile is 1152 ft above ground twice; the first time at $t = 6$ sec on the way up, and the second time at $t = 12$ sec on the way down (Figure P.46).

**(b)** The projectile will be at least 1152 ft above ground when $s \geq 1152$. We can see from Figure P.46 together with the algebraic work in (a) that the solution is $[6, 12]$. This means that the projectile is at least 1152 ft above ground for times between $t = 6$ sec and $t = 12$ sec, including 6 and 12 sec.
    In Exercise 38 we ask you to use algebra to solve the inequality $s = -16t^2 + 288t \geq 1152$.

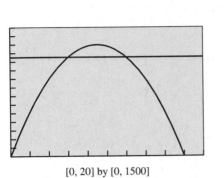

[0, 20] by [0, 1500]

**Figure P.46** The graphs of $s = -16t^2 + 288t$ and $s = 1152$. We know from Example 10a that the two graphs intersect at $(6, 1152)$ and $(12, 1152)$.

## Quick Review P.6

In Exercises 1–3, solve for x.

**1.** $-7 < 2x - 3 < 7$       **2.** $5x - 2 \geq 7x + 4$

**3.** $|x + 2| = 3$  $x = 1$ or $x = -5$

In Exercises 4–6, factor the expression completely.

**4.** $4x^2 - 9$  $(2x - 3)(2x + 3)$     **5.** $x^3 - 4x$  $x(x - 2)(x + 2)$

**6.** $9x^2 - 16y^2$  $(3x - 4y)(3x + 4y)$

In Exercises 7 and 8, reduce the fraction to lowest terms.

**7.** $\dfrac{z^2 - 25}{z^2 - 5z}$  $\dfrac{z + 5}{z}$     **8.** $\dfrac{x^2 + 2x - 35}{x^2 - 10x + 25}$  $\dfrac{x + 7}{x - 5}$

In Exercises 9 and 10, add the fractions and simplify.

**9.** $\dfrac{x}{x - 1} + \dfrac{x + 1}{3x - 4}$     **10.** $\dfrac{2x - 1}{x^2 - x - 2} + \dfrac{x - 3}{x^2 - 3x + 2}$

## Section P.6 Exercises

In Exercises 1–8, solve the inequality algebraically. Write the solution in interval notation and draw its number line graph.

**1.** $|x + 4| \geq 5$       **2.** $|2x - 1| > 3.6$

**3.** $|x - 3| < 2$       **4.** $|x + 3| \leq 5$

**5.** $|4 - 3x| - 2 < 4$       **6.** $|3 - 2x| + 2 > 5$

**7.** $\left|\dfrac{x + 2}{3}\right| \geq 3$       **8.** $\left|\dfrac{x - 5}{4}\right| \leq 6$

In Exercises 9–16, solve the inequality. Use algebra to solve the corresponding equation.

**9.** $2x^2 + 17x + 21 \leq 0$     **10.** $6x^2 - 13x + 6 \geq 0$

**11.** $2x^2 + 7x > 15$     **12.** $4x^2 + 2 < 9x$

**13.** $2 - 5x - 3x^2 < 0$     **14.** $21 + 4x - x^2 > 0$  $(-3, 7)$

**15.** $x^3 - x \geq 0$       **16.** $x^3 - x^2 - 30x \leq 0$

In Exercises 17–20, solve the inequality graphically.

**17.** $x^2 - 4x < 1$  $(-0.24, 4.24)$  **18.** $12x^2 - 25x + 12 \geq 0$

**19.** $8x - 2x^3 < 0$     **20.** $3x^3 - 12x \geq 0$

In Exercises 21–30, use a method of your choice to solve the inequality.

**21.** $6x^2 - 5x - 4 > 0$     **22.** $4x^2 - 1 \leq 0$  $\left[-\dfrac{1}{2}, \dfrac{1}{2}\right]$

**23.** $9x^2 + 12x - 1 \geq 0$     **24.** $4x^2 - 12x + 7 < 0$

**25.** $4x^2 + 1 > 4x$     **26.** $x^2 + 9 \leq 6x$  $x = 3$

**27.** $x^2 - 8x + 16 < 0$     **28.** $9x^2 + 12x + 4 \geq 0$

**29.** $2x^3 > 18x$     **30.** $12x \leq 27x^3$

In Exercises 31–36, solve the inequality algebraically.

**31.** $\dfrac{5}{x+2} \geq 1$  $(-2, 3]$

**32.** $\dfrac{13}{x+5} > 2$  $\left(-5, \dfrac{3}{2}\right)$

**33.** $\dfrac{x+3}{x+5} > 0$

**34.** $\dfrac{x-2}{x-9} \geq 0$

**35.** $\dfrac{x+2}{2x-3} \geq 1$  $\left[\dfrac{3}{2}, 5\right]$

**36.** $\dfrac{2x+1}{3x+2} \leq 1$

**37. Group Activity** Give an example of a quadratic inequality with the indicated solution.

    **(a)** All real numbers     **(b)** No solution

    **(c)** Exactly one solution     **(d)** $[-2, 5]$

    **(e)** $(-\infty, -1) \cup (4, \infty)$     **(f)** $(-\infty, 0] \cup [4, \infty)$

**38. Revisiting Example 10** Solve the inequality $-16t^2 + 288t \geq 1152$ algebraically and compare your answer with the result obtained in Example 10. $[6, 12]$

## Explorations

**39. Constructing a Box with No Top** An open box is formed by cutting squares from the corners of a regular piece of cardboard (see figure) and folding up the flaps.

    **(a)** What size corner squares should be cut to yield a box with a volume of 125 in.$^3$?

    **(b)** What size corner squares should be cut to yield a box with a volume more than 125 in.$^3$? $(0.94, 3.78)$

    **(c)** What size corner squares should be cut to yield a box with a volume of at most 125 in.$^3$?

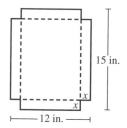

15 in.

*x*

12 in.

**40. Projectile Motion** A projectile is launched straight up from ground level with an initial velocity of 256 ft/sec.

    **(a)** When will the projectile's height above ground be 768 ft? $t = 4$ sec up; $t = 12$ sec down

    **(b)** When will the projectile's height above ground be at least 768 ft? when $t$ is in the interval $[4, 12]$

    **(c)** When will the projectile's height above ground be less than or equal to 768 ft?

**41. Projectile Motion** A projectile is launched straight up from ground level with an initial velocity of 272 ft/sec.

    **(a)** When will the projectile's height above ground be 960 ft? $t = 5$ sec up; $t = 12$ sec down

    **(b)** When will the projectile's height above ground be more than 960 ft? when t is in the interval $(5, 12)$

    **(c)** When will the projectile's height above ground be less than or equal to 960 ft?

**42. Writing to Learn** Explain the role of equation solving in the process of solving an inequality. Give an example.

**43. Travel Planning** Barb wants to drive to a city 105 mi from her home in no more than 2 h. What is the lowest average speed she must maintain on the drive? 52.5 mph

**44. Connecting Algebra and Geometry** Consider the collection of all rectangles that have length 2 in. less than twice their width.

    **(a)** Find the possible widths (in inches) of these rectangles if their perimeters are less than 200 in. 1 in. $< x <$ 34 in.

    **(b)** Find the possible widths (in inches) of these rectangles if their areas are less than or equal to 1200 in.$^2$.

**45. Boyle's Law** For a certain gas, $P = 400/V$, where $P$ is pressure and $V$ is volume. If $20 \leq V \leq 40$, what is the corresponding range for $P$? $10 \leq P \leq 20$

**46. Cash-Flow Planning** A company has current assets (cash, property, inventory, and accounts receivable) of \$200,000 and current liabilities (taxes, loans, and accounts payable) of \$50,000. How much can it borrow if it wants its ratio of assets to liabilities to be no less than 2? Assume the amount borrowed is added to both current assets and current liabilities. no more than \$100,000

## Extending the Ideas

In Exercises 47 and 48, use a combination of algebraic and graphical techniques to solve the inequalities.

**47.** $|2x^2 + 7x - 15| < 10$     **48.** $|2x^2 + 3x - 20| \geq 10$

# Chapter P    Key Ideas

## Concepts

absolute value inequality (p. 48)
absolute value of a real number (p. 12)
additive inverse of a real number (p. 5)
algebraic expression (p. 5)
bounded interval (p. 4)
Cartesian (rectangular) coordinate system
    (p. 11)
Cartesian coordinates of a point (p. 11)
closed interval (p. 4)
constant (p. 5)
coordinate of a point on number line
    (p. 2)
Cartesian plane (p. 11)
coordinates of a point (p. 11)
discriminant of a quadratic polynomial
    (p. 47)
double inequality (p. 23)
element (object) of a set (p. 1)
endpoints of an interval (p. 4)
equation (p. 19)
equivalent equation (p. 20)
equivalent inequality (p. 22)
expanded form (p. 6)
exponential notation (p. 7)
extraneous solution (p. 44)
factored form (p. 6)

graph of an equation (p. 29)
half-open interval (p. 4)
inequality symbols (p. 3)
infinitely repeating decimal (p. 1)
interval of real numbers (p. 4)
irrational number (p. 2)
linear equation in $x$ (p. 20)
linear equation in $x$ and $y$ (p. 29)
linear inequality in $x$ (p. 22)
magnitude of a real number (p. 12)
multiplicative inverse (reciprocal) of a
    real number (p. 5)
open interval (p. 4)
opposite (additive inverse) of a real
    number (p. 5)
ordered pair of real numbers (p. 11)
ordered set (p. 2)
origin (p. 2 & p. 11)
parallel lines (p. 30)
perpendicular lines (p. 30)
point of intersection of two graphs (p. 45)
projectile motion (p. 52)
Pythagorean Theorem (p. 13)
quadrants (p. 12)
quadratic equation in $x$ (p. 41)
rational number (p. 1)

real number (p. 1)
real number line (p. 2)
reciprocal (multiplicative inverse) of a
    real number (p. 5)
scatter plot (p. 12)
scientific notation (p. 8)
set-builder notation (p. 1)
slope of a line (p. 26)
solution of an equation in $x$ (p. 19)
solution of an equation in $x$ and $y$ (p. 29)
solution of an inequality in $x$ (p. 22)
solution set of an inequality (p. 22)
square viewing window (p. 30)
terminating decimal (p. 1)
unbounded interval (p. 4)
union of two sets (p. 49)
variable (p. 5)
viewing window (p. 29)
$x$-axis (p. 11)
$x$-coordinate of a point (p. 11)
$x$-intercept (p. 29)
$y$-axis (p. 11)
$y$-coordinate of a point (p. 11)
$y$-intercept (p. 28)

## Properties, Theorems, and Formulas

### Trichotomy Property

Let $a$ and $b$ be any two real numbers. Exactly one of the following is true:
$a < b$,   $a = b$,   or   $a > b$.

### Properties of Algebra

Let $u$, $v$, and $w$ be real numbers, variables, or algebraic expressions.

**1. Commutative property**
Addition: $u + v = v + u$
Multiplication: $uv = vu$

**2. Associative property**
Addition:
$(u + v) + w = u + (v + w)$
Multiplication: $(uv)w = u(vw)$

**3. Identity property**
Addition: $u + 0 = u$
Multiplication: $u \cdot 1 = u$

**4. Inverse property**
Addition: $u + (-u) = 0$

Multiplication: $u \cdot \dfrac{1}{u} = 1, u \neq 0$

**5. Distributive property**
Multiplication over
addition
$u(v + w) = uv + uw$
$(u + v)w = uw + vw$

Multiplication over
subtraction
$u(v - w) = uv - uw$
$(u - v)w = uw - vw$

### Properties of Equality

Let $u$, $v$, $w$, and $z$ be real numbers, variables, or algebraic expressions.

| | |
|---|---|
| **1. Reflexive** | $u = u$ |
| **2. Symmetric** | If $u = v$, then $v = u$. |
| **3. Transitive** | If $u = v$, and $v = w$, then $u = w$. |
| **4. Addition** | If $u = v$ and $w = z$, then $u + w = v + z$. |
| **5. Multiplication** | If $u = v$ and $w = z$, then $uw = vz$. |

### Properties of Inequalities

Let $u$, $v$, $w$, and $z$ be real numbers, variables, or algebraic expressions, and $c$ a real number.

| | |
|---|---|
| **1. Transitive** | If $u < v$ and $v < w$, then $u < w$. |
| **2. Addition** | If $u < v$, then $u + w < v + w$. |
| | If $u < v$ and $w < z$, then $u + w < v + z$. |
| **3. Multiplication** | If $u < v$ and $c > 0$, then $uc < vc$. |
| | If $u < v$ and $c < 0$, then $uc > vc$. |

The above properties are true if $<$ is replaced by $\leq$. There are similar properties for $>$ and $\geq$.

### Distance Formulas

*Number Line:* Let $a$ and $b$ be real numbers. The **distance between $a$ and $b$** is $|a - b|$.

*Coordinate Plane:* The **distance $d(P, Q)$ between points $P(x_1, y_1)$ and $Q(x_2, y_2)$** in the coordinate plane is

$$d(P, Q) = \sqrt{(x_1 - x_2)^2 + (y_1 - y_2)^2}$$

### Midpoint Formula (Coordinate Plane)

The **midpoint of the line segment with endpoints $(a, b)$ and $(c, d)$ is** $\left( \dfrac{a + c}{2}, \dfrac{b + d}{2} \right)$.

### Quadratic Formula

The solutions of the quadratic equation $ax^2 + bx + c = 0$, where $a \neq 0$, are given by the **quadratic formula**

$$x = \frac{-b \pm \sqrt{b^2 - 4ac}}{2a}.$$

### Equations

| | |
|---|---|
| General form equation of a line: | $Ax + By + C = 0$ |
| Slope-intercept form equation of a line: | $y = mx + b$ |
| Point-slope form equation of a line: | $y - y_1 = m(x - x_1)$ |
| Equation of a vertical line: | $x = a$ |
| Equation of a horizontal line: | $y = b$ |

Standard form equation of a circle with center $(h, k)$ and radius $r$: $(x - h)^2 + (y - k)^2 = r^2$.

## Procedures

### Agreement about Approximate Solutions

Round approximate solutions to hundredths unless directed otherwise. For applications, round to a value that is reasonable for the context of the problem.

### Completing the Square

To complete the square for the expression $x^2 + bx$, add $(b/2)^2$. The completed square is

$$x^2 + bx + \left(\frac{b}{2}\right)^2 = \left(x + \frac{b}{2}\right)^2.$$

### Solving Quadratic Equations Algebraically

1. Factoring
2. Extracting Square Roots
3. Completing the Square
4. Quadratic Formula

## Chapter P  Review Exercises

The collection of exercises marked in red could be used as a chapter test.

In Exercises 1 and 2, find the endpoints and state whether the interval is bounded or unbounded.

**1.** $[0, 5]$  **2.** $(2, \infty)$

**3. Distributive Property** Use the distributive property to write the expanded form of $2(x^2 - x)$.  $2x^2 - 2x$

**4. Distributive Property** Use the distributive property to write the factored form of $2x^3 + 4x^2$.  $2x^2(x + 2)$

In Exercises 5 and 6, simplify the expression. Assume that denominators are not zero.

**5.** $\dfrac{(uv^2)^3}{v^2u^3}$  $v^4$  **6.** $(3x^2y^3)^{-2}$  $\dfrac{1}{9x^4y^6}$

In Exercises 7 and 8, write the number in scientific notation.

**7.** The mean distance from Pluto to the sun is about 3,680,000,000 miles.  $3.68 \times 10^9$

**8.** The diameter of a red blood corpuscle is about 0.000007 meter.  $7 \times 10^{-6}$

In Exercises 9 and 10, write the number in decimal form.

**9.** Our solar system is about $5 \times 10^9$ years old.  5,000,000,000

**10.** The mass of an electron is about $9.1066 \times 10^{-28}$ g (gram).

**11.** The data in Table P.9 give details about the year 2000 budget for education in President Clinton's budget proposal. Write the amount of the budget item in Table P.9 in scientific notation.

**(a)** Eisenhower Professional Development Program

**(b)** Safe, Drug Free Schools  $5.91 \times 10^8$

**(c)** Bilingual, Immigrant Education  $4.15 \times 10^8$

**(d)** Special Education  $5.1 \times 10^9$

### Table P.9  Department of Education

| Budget Item | Amount ($) |
|---|---|
| Eisenhower Professional Development Program | 335 million |
| Goals 2000 | 491 million |
| Title I | 7.9 billion |
| Safe, Drug Free Schools | 591 million |
| Charter Schools | 130 million |
| American Reads Challenge | 286 million |
| Bilingual, Immigrant Education | 415 million |
| Special Education | 5.1 billion |

*Source: National Science Teachers Association, NSTA Reports, April 1999, Vol. 10, No. 5, p 3.*

**12. Decimal Form** Find the decimal form for $-5/11$. State whether it repeats or terminates.  $-0.\overline{45}$ (repeating)

In Exercises 13 and 14, find **(a)** the distance between the points and **(b)** the midpoint of the line segment determined by the points.

**13.** $-5$ and 14  19; 4.5  **14.** $(-4, 3)$ and $(5, -1)$

In Exercises 15 and 16, show that the figure determined by the points is the indicated type.

**15.** Right triangle: $(-2, 1)$, $(3, 11)$, $(7, 9)$  $5\sqrt{5}; 2\sqrt{5}; \sqrt{145}$

**16.** Equilateral triangle: $(0, 1)$, $(4, 1)$, $(2, 1 - 2\sqrt{3})$  4; 4; 4

In Exercises 17 and 18, find the standard form equation for the circle,

**17.** Center $(0, 0)$, radius 2  **18.** Center $(5, -3)$, radius 4

In Exercises 19 and 20, find the center and radius of the circle.

**19.** $(x + 5)^2 + (y + 4)^2 = 9$  **20.** $x^2 + y^2 = 1$

**21. (a)** Find the length of the sides of the triangle in the figure.

**(b) Writing to Learn** Show that the triangle is a right triangle.

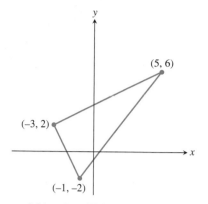

**22. Distance and Absolute Value** Use absolute value notation to write the statement that the distance between $z$ and $-3$ is less than or equal to 1.

**23. Finding a Line Segment with Given Midpoint** Let $(3, 5)$ be the midpoint of the line segment with endpoints $(-1, 1)$ and $(a, b)$. Determine $a$ and $b$.    $a = 7; b = 9$

**24. Finding Slope** Find the slope of the line through the points $(-1, -2)$ and $(4, -5)$.    $-\dfrac{3}{5}$

**25. Finding Point-Slope Form Equation** Find an equation in point-slope form for the line through the point $(2, -1)$ with slope $m = -2/3$.    $y + 1 = -\dfrac{2}{3}(x - 2)$

**26.** Find an equation of the line through the points $(-5, 4)$ and $(2, -5)$ in the general form $Ax + By + C = 0$.

In Exercises 27–32, find an equation in slope-intercept form for the line.

**27.** The line through $(3, -2)$ with slope $m = 4/5$

**28.** The line through the points $(-1, -4)$ and $(3, 2)$

**29.** The line through $(-2, 4)$ with slope $m = 0$    $y = 4$

**30.** The line $3x - 4y = 7$    $y = \dfrac{3}{4}x - \dfrac{7}{4}$

**31.** The line through $(2, -3)$ and parallel to the line $2x + 5y = 3$.    $y = -\dfrac{2}{5}x - \dfrac{11}{5}$

**32.** The line through $(2, -3)$ and perpendicular to the line $2x + 5y = 3$.    $y = \dfrac{5}{2}x - 8$

**33. Americans' Work Time** The data in Table P.10 shows the number of hours a week Americans say they spend at work for several years.

**(a)** Let $x = 0$ represent 1970, $x = 1$ represent 1971, and so forth. Draw a scatter plot of the data.

**(b)** Use the 1973 and 1999 data to write a linear equation for the number of hours $y$ Americans say they work in a week in terms of the year $x$. Superimpose the graph of the linear equation on the scatter plot in (a).

**(c)** Use the equation in (b) to estimate the number of hours Americans say they work in a week in 1990.    46.9

**(d)** Use the equation in (b) to predict the number of hours Americans will say they work in a week in 2002.    51.3

| Table P.10 Americans' Weekly Work Time | |
|---|---|
| Year | Number of hours |
| 1973 | 40.6 |
| 1980 | 46.9 |
| 1987 | 48.8 |
| 1994 | 50.7 |
| 1997 | 50.8 |
| 1998 | 49.9 |
| 1999 | 50.2 |

*Source: Harris Poll, as reported in the USA TODAY, September 9, 1999*

**34.** Consider the point $(-6, 3)$ and Line $L$: $4x - 3y = 5$. Write an equation **(a)** for the line passing through this point and parallel to $L$, and **(b)** for the line passing through this point and perpendicular to $L$. Support your work graphically.

In Exercises 35 and 36, assume that each graph contains the origin and the upper right-hand corner of the viewing window.

**35.** Find the slope of the line in the figure.

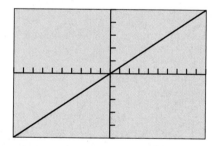

[–10, 10] by [–25, 25]

**36. Writing to Learn** Which line has the greater slope? Explain.

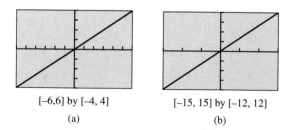

| [–6,6] by [–4, 4] | [–15, 15] by [–12, 12] |
|---|---|
| (a) | (b) |

In Exercises 37–50, solve the equation algebraically.

**37.** $3x - 4 = 6x + 5$    $x = -3$    **38.** $\dfrac{x - 2}{3} + \dfrac{x + 5}{2} = \dfrac{1}{3}$

**39.** $2(5 - 2y) - 3(1 - y) = y + 1$    $y = 3$

**40.** $3(3x - 1)^2 = 21$

**41.** $x^2 - 4x - 3 = 0$      **42.** $16x^2 - 24x + 7 = 0$

**43.** $6x^2 + 7x = 3$      **44.** $\dfrac{2x - 3}{x + 1} = 1$    $x = 4$

**45.** $x(2x + 5) = 4(x + 7)$          **46.** $|4x + 1| = 3$

**47.** $4x^2 - 20x + 25 = 0$ $x = \dfrac{5}{2}$ **48.** $-9x^2 + 12x - 4 = 0$

**49.** $\dfrac{3x}{x + 1} + \dfrac{5}{x - 2} = \dfrac{15}{x^2 - x - 2}$ $x = -\dfrac{5}{3}$

**50.** $\dfrac{3}{x} - x = 2$

**51. Completing the Square**  Use completing the square to solve the equation $2x^2 - 3x - 1 = 0$.

**52. Quadratic Formula**  Use the quadratic formula to solve the equation $3x^2 + 4x - 1 = 0$.

In Exercises 53 and 54, use factoring to solve the equation.

**53.** $3x^3 - 19x^2 - 14x = 0$     **54.** $x^3 + 2x^2 - 4x - 8 = 0$

In Exercises 55 and 56, solve the equation graphically.

**55.** $x^3 - 2x^2 - 2 = 0$        **56.** $|2x - 1| = 4 - x^2$

In Excrcises 57 and 58, solve the inequality and draw a number line graph of the solution.

**57.** $-2 < x + 4 \le 7$         **58.** $5x + 1 \ge 2x - 4$

In Exercises 59–70, solve the inequality.

**59.** $\dfrac{3x - 5}{4} \le -1$ $\left(-\infty, \dfrac{1}{3}\right]$  **60.** $|2x - 5| < 7$ $(-1, 6)$

**61.** $|3x + 4| \ge 2$          **62.** $4x^2 + 3x > 10$

**63.** $x^3 \le 9x$             **64.** $4x^3 - 9x > 0$

**65.** $\dfrac{5}{x - 2} < 1$        **66.** $\dfrac{x - 2}{x + 8} > 0$

**67.** $\left|\dfrac{x + 7}{5}\right| > 2$      **68.** $2x^2 + 3x - 35 < 0$

**69.** $4x^2 + 12x + 9 \ge 0$       **70.** $x^2 - 6x + 9 < 0$

**71. Projectile Motion**  A projectile is launched straight up from ground level with an initial velocity of 320 ft/sec.

   **(a)** When will the projectile's height above ground be 1538 ft? $t \approx 8$ sec up; $t \approx 12$ sec down

   **(b)** When will the projectile's height above ground be at most 1538 ft?

   **(c)** When will the projectile's height above ground be greater than or equal to 1538 ft?

**72. Navigation**  A commercial jet airplane climbs at takeoff with slope $m = 4/9$. How far in the horizontal direction will the airplane fly to reach an altitude of 20,000 ft above the takeoff point? 45,000 ft

**73. Connecting Algebra and Geometry**  Consider the collection of all rectangles that have length 1 cm more than three times their width $w$.

   **(a)** Find the possible widths (in cm) of these rectangles if their perimeters are less than or equal to 150 cm.

   **(b)** Find the possible widths (in cm) of these rectangles if their areas are greater than 1500 cm$^2$.

**A**s the earth's population continues to grow, the solid waste generated by the population grows with it. Government must plan for disposal and recycling of ever growing amounts of solid waste. Planners can use data from the past to predict future waste generation and plan for enough facilities for disposing of and recycling the waste.

**G**iven the following data on the waste generated in Florida from 1990–1994, how can we construct a function to predict the waste that was generated in the years 1995–1999? This solution to this problem is given on page 143 in Section 1.6.

| Year | Tons of Solid Waste Generated (in thousands) |
|------|-----------------------------------|
| 1990 | 19,358 |
| 1991 | 19,484 |
| 1992 | 20,293 |
| 1993 | 21,499 |
| 1994 | 23,561 |

### Bibliography

**For students:** *How to Solve It: A New Aspect of Mathematical Method,* Second Edition, George Pólya. Princeton Science Library, 1991. Available through Dale Seymour Publications

*Functions and Graphs*, I.M. Gelfand, E. G. Glagoleva, E. E. Shnol. Birkauser, 1990.

**For teachers:** *The Language of Functions and Graphs,* Shell Centre for Mathematical Education. Available through Dale Seymour Publications

## Chapter 1 Overview

Historically, geometry developed much more quickly than algebra. One reason is that geometric figures could be *seen*: they gave visible meaning to numerical quantities and visible hints about how they behaved. It took many centuries for mathematicians to develop the algebraic notation we take for granted today, finally giving unknown algebraic quantities a visible life of their own. When calculus entered the picture, the mathematics demanded a more appropriate notation with which to communicate it. That was when functions made their debut.

In this chapter we begin the study of functions that will continue throughout the book. Your previous courses and your graphing calculators have introduced you to some basic functions, the properties of which we will explore at an intuitive level just to get you familiar with the terminology. This familiarity will serve you well in later chapters when we explore the properties of functions in greater depth.

# Modeling and Equation Solving

Numerical Models • Algebraic Models • Graphical Models • The Importance of Zeros • Problem Solving • Grapher Failure and Hidden Behavior • A Word About Proof

## Numerical Models

Scientists and engineers have always used mathematics to model the real world and thereby to unravel its mysteries. A **mathematical model** is a mathematical structure that approximates phenomena for the purpose of studying or predicting their behavior. Thanks to advances in computer technology, the process of devising mathematical models is now a rich field of study itself, **mathematical modeling**.

We will be concerned primarily with three types of mathematical models in this book: *numerical models*, *algebraic models*, and *graphical models*. Each type of model gives insight into real-world problems, but the best insights are often gained by switching from one kind of model to another. Developing the ability to do that will be one of the goals of this course.

Perhaps the most basic kind of mathematical model is the **numerical model**, in which numbers (or *data*) are analyzed to gain insights into phenomena. A numerical model can be as simple as the major league baseball standings or as complicated as the network of interrelated numbers that measure the global economy.

### Example 1 TRACKING THE MINIMUM WAGE

The numbers in Table 1.1 show the growth of the minimum hourly wage from 1938 to 1997. The table also shows the minimum wage adjusted to the purchasing power of 1997 dollars (using the urban consumer price index). Answer the following questions using only the data in the table.

**(a)** In what year did the actual minimum wage increase the most?

**(b)** In what year did a worker earning the minimum wage enjoy the greatest purchasing power?

**(c)** What was the longest period during which the minimum wage did not increase?

**(d)** A worker on minimum wage in 1980 was earning nearly twice as much as a worker on minimum wage in 1970, and yet there was great pressure to raise the minimum wage again. Why?

### Table 1.1  The Minimum Hourly Wage

| Year | Minimum Hourly Wage | Purchasing Power in 1997 Dollars | Year | Minimum Hourly Wage | Purchasing Power in 1997 Dollars |
|------|------|------|------|------|------|
| 1938 | 0.25 | 2.85 | 1978 | 2.65 | 6.52 |
| 1940 | 0.30 | 3.44 | 1979 | 2.90 | 6.41 |
| 1945 | 0.40 | 3.57 | 1980 | 3.10 | 6.04 |
| 1950 | 0.75 | 4.99 | 1981 | 3.35 | 5.92 |
| 1955 | 0.75 | 4.49 | 1985 | 3.35 | 5.00 |
| 1960 | 1.00 | 5.42 | 1989 | 3.35 | 4.34 |
| 1965 | 1.25 | 6.37 | 1990 | 3.80 | 4.67 |
| 1967 | 1.40 | 6.73 | 1991 | 4.25 | 5.01 |
| 1968 | 1.60 | 7.38 | 1992 | 4.25 | 4.86 |
| 1970 | 1.60 | 6.62 | 1993 | 4.25 | 4.72 |
| 1974 | 2.00 | 6.51 | 1994 | 4.25 | 4.60 |
| 1975 | 2.10 | 6.26 | 1995 | 4.25 | 4.48 |
| 1976 | 2.30 | 6.49 | 1996 | 4.75 | 4.86 |
| 1977 | 2.30 | 6.09 | 1997 | 5.15 | 5.15 |

*Source: U.S. Bureau of Labor Statistics*

### Solution

**(a)** In 1996, up $0.50 from 1995. Notice that the minimum wage never goes *down*, so we can tell that there were no other increases of this magnitude even though we do not have data from every year in the table.

**(b)** In 1968 (although we can not rule out 1969 from the data in the table).

**(c)** The nine-year period from 1981 to 1989.

**(d)** Although the minimum wage increased from $1.60 to $3.10 in that period, the purchasing power actually dropped $0.58 (in 1997 dollars). This is one effect of inflation on the economy.

The numbers in Table 1.1 provide a numerical model for one aspect of the U.S. economy by using another numerical model, the urban consumer price index (CPI-U), to adjust the data. Working with large numerical models is standard operating procedure in business and industry, where computers are relied upon to provide fast and accurate data processing.

## Example 2 ANALYZING PRISON POPULATIONS

Table 1.2 shows the growth in the number of prisoners incarcerated in state and federal prisons from 1980 to 1997. Is the proportion of female prisoners over the years increasing?

### Table 1.2 U.S. Prison Population

| Year | Total | Male | Female | Year | Total | Male | Female |
|------|-------|------|--------|------|-------|------|--------|
| 1980 | 315,974 | 303,643 | 12,331 | 1989 | 680,907 | 643,643 | 37,264 |
| 1981 | 353,167 | 338,940 | 14,227 | 1990 | 739,980 | 699,416 | 40,564 |
| 1982 | 394,374 | 378,045 | 16,329 | 1991 | 789,610 | 745,808 | 43,802 |
| 1983 | 419,820 | 402,391 | 17,429 | 1992 | 846,277 | 799,776 | 46,501 |
| 1984 | 443,398 | 424,193 | 19,205 | 1993 | 932,074 | 878,037 | 54,037 |
| 1985 | 480,268 | 458,972 | 21,296 | 1994 | 1,016,691 | 956,566 | 60,125 |
| 1986 | 522,084 | 497,540 | 24,544 | 1995 | 1,085,022 | 1,021,059 | 63,963 |
| 1987 | 560,812 | 533,990 | 26,822 | 1996 | 1,138,984 | 1,069,257 | 69,727 |
| 1988 | 603,732 | 573,587 | 30,145 | 1997 | 1,197,590 | 1,123,478 | 74,112 |

*Source: U.S. Justice Department*

**Solution** The *number* of female prisoners over the years is certainly increasing, but so is the total number of prisoners, so it is difficult to discern from the data whether the *proportion* of female prisoners is increasing. What we need is another column of numbers showing the ratio of female prisoners to total prisoners.

We could compute all the ratios separately, but it is easier to do this kind of repetitive calculation with a single command on a computer spreadsheet. You can also do this on a graphing calculator by manipulating lists (see Exercise 51). Table 1.3 shows the percentage of the total population each year that consists of female prisoners. With this data to extend our numerical model, it is clear that the proportion of female prisoners is increasing.

### Table 1.3 Female Percentage of U.S. Prison Population

| Year | % Female | Year | % Female | Year | % Female |
|------|----------|------|----------|------|----------|
| 1980 | 3.90 | 1986 | 4.70 | 1992 | 5.49 |
| 1981 | 4.03 | 1987 | 4.78 | 1993 | 5.80 |
| 1982 | 4.14 | 1988 | 4.99 | 1994 | 5.91 |
| 1983 | 4.15 | 1989 | 5.47 | 1995 | 5.90 |
| 1984 | 4.33 | 1990 | 5.48 | 1996 | 6.12 |
| 1985 | 4.43 | 1991 | 5.55 | 1997 | 6.19 |

## Algebraic Models

An **algebraic model** uses formulas to relate variable quantities associated with the phenomena being studied. The added power of an algebraic model over a numerical model is that it can be used to generate numerical values of unknown quantities by relating them to known quantities.

### Example 3  COMPARING PIZZAS

A pizzeria sells a rectangular 18" by 24" pizza for the same price as its large round pizza (24" diameter). If both pizzas are of the same thickness, which option gives the most pizza for the money?

**Solution**  We need to compare the *areas* of the pizzas. Fortunately, geometry has provided algebraic models that allow us to compute the areas from the given information.

For the rectangular pizza:

$$Area = l \times w = 18 \times 24 = 432 \text{ square inches.}$$

For the circular pizza:

$$Area = \pi r^2 = \pi \left(\frac{24}{2}\right)^2 = 144\pi \approx 452.4 \text{ square inches.}$$

The round pizza is larger and therefore gives more for the money.

The algebraic models in Example 3 come from geometry, but you have probably encountered algebraic models from many other sources in your algebra and science courses.

---

**Exploration 1**  Designing an Algebraic Model

**Exploration Extensions**

Suppose that after the sale, the merchandise prices are increased by 25%. If *m* represents the marked price before the sale, find an algebraic model for the post-sale price, including tax.

A department store is having a sale in which everything is discounted 25% off the marked price. The discount is taken at the sales counter, and then a state sales tax of 6.5% and a local sales tax of 0.5% are added on.

1. The discount price $d$ is related to the marked price $m$ by the formula $d = km$, where $k$ is a certain constant. What is $k$?  0.75
2. The actual sale price $s$ is related to the discount price $d$ by the formula $s = d + td$, where $t$ is a constant related to the total sales tax. What is $t$?  7% or 0.07
3. Using the answers from steps 1 and 2 you can find a constant $p$ that relates $s$ directly to $m$ by the formula $s = pm$. What is $p$?  0.8025
4. If you only have \$30, can you afford to buy a shirt marked \$36.99?  yes
5. If you have a credit card but are determined to spend no more than \$100, what is the maximum total value of your marked purchases before you present them at the sales counter?  \$124.61

---

The ability to generate numbers from formulas makes an algebraic model far more useful as a predictor of behavior than a numerical model. Indeed, one optimistic goal of scientists and mathematicians when modeling phenomena is to fit an algebraic model to numerical data and then (even more optimistically) to analyze why it works. Not all numerical models can be modeled algebraically. For example, nobody has ever devised a successful formula for predicting the ups and downs of the stock market as a function of time, although that does not stop investors from trying.

If numerical data do behave reasonably enough to suggest that an algebraic model might be found, it is often helpful to look at a picture first. That brings us to graphical models.

## Graphical Models

A **graphical model** is a visible representation of a numerical model or an algebraic model that gives insight into the relationships between variable quantities. Learning to interpret and use graphs is a major goal of this book.

### Example 4   VISUALIZING GALILEO'S GRAVITY EXPERIMENTS

Galileo Galilei (1564–1642) spent a good deal of time rolling balls down inclined planes, carefully recording the distance they traveled as a function of elapsed time. His experiments are commonly repeated in physics classes today, so it is easy to reproduce a typical table of Galilean data.

| Elapsed time (seconds) | 0 | 1 | 2 | 3 | 4 | 5 | 6 | 7 | 8 |
|---|---|---|---|---|---|---|---|---|---|
| Distance traveled (inches) | 0 | 0.75 | 3 | 6.75 | 12 | 18.75 | 27 | 36.75 | 48 |

What graphical model fits the data? Can you find an algebraic model that fits?

**Solution**  A scatter plot of the data is shown in Figure 1.1.

Galileo's experience with quadratic functions suggested to him that this figure was a parabola with its vertex at the origin, he therefore modeled the effect of gravity as a quadratic function:

$$d = kt^2$$

Because the ordered pair (1, 0.75) must satisfy the equation, it follows that $k = 0.75$, yielding the equation

$$d = 0.75t^2.$$

You can verify numerically that this algebraic model correctly predicts the rest of the data points. We will have much more to say about parabolas in Chapter 2.

This insight led Galileo to discover several basic laws of motion that would eventually be named after Isaac Newton. While Galileo had found the algebraic model to describe the path of the ball, it would take Newton's calculus to explain why it worked.

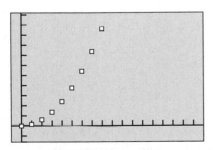

[–1, 18] by [–8, 56]

**Figure 1.1**  A scatter plot of data from a Galileo gravity experiment. (Example 4)

### Example 5   FITTING A CURVE TO DATA

We showed in Example 3 that the percentage of females in the U.S. prison population has been steadily growing over the years. Model this growth graphically and use the graphical model to suggest an algebraic model.

**Solution**  Let $t$ be the number of years after 1980, and let $F$ be the percentage of females in the prison population from year 0 to year 17.

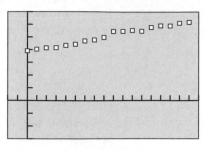

[–2, 18] by [–3, 7]

**Figure 1.2** A scatter plot of the data in Table 1.4. (Example 5)

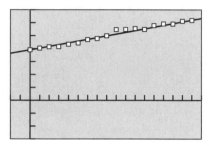

[–2, 18] by [–3, 7]

**Figure 1.3** The line with equation $y = 0.135x + 3.90$ is a good model for the data in Table 1.4. (Example 5)

**Exploration Extensions**

What are the advantages of a linear model over a quadratic model for these data?

From the data in Table 1.3 we get the corresponding data in Table 1.4:

**Table 1.4** Percentage (*F*) of females in the prison population *t* years after 1980

| *t* | 0 | 1 | 2 | 3 | 4 | 5 | 6 | 7 | 8 |
|---|---|---|---|---|---|---|---|---|---|
| *F* | 3.90 | 4.03 | 4.14 | 4.15 | 4.33 | 4.43 | 4.70 | 4.78 | 4.99 |

| *t* | 9 | 10 | 11 | 12 | 13 | 14 | 15 | 16 | 17 |
|---|---|---|---|---|---|---|---|---|---|
| *F* | 5.47 | 5.48 | 5.55 | 5.49 | 5.80 | 5.91 | 5.90 | 6.12 | 6.19 |

A scatter plot of the data is shown in Figure 1.2.

This pattern looks remarkably linear. If we use a line as our graphical model, we can find an algebraic model by finding the equation of the line. We will describe in Chapter 2 how a statistician would find the best line to fit the data, but we can get a pretty good fit for now by finding the line through the points (0, 3.90) and (17, 6.19).

The slope is $(6.19 - 3.90)/(17 - 0) \approx 0.135$ and the *y*-intercept is 3.90. Therefore, the line has equation $y = 0.135x + 3.90$. You can see from Figure 1.3 that this line does a very nice job of modeling the data.

---

**Exploration 2** **Interpreting the Model**

The parabola in Example 4 arose from a law of physics that governs falling objects, which should inspire more confidence than the linear model in Example 5. We can repeat Galileo's experiment many times with differently-sloped ramps, different units of measurement, and even on different planets, and a quadratic model will fit it every time. The purpose of this Exploration is to think more deeply about the linear model in the prison example.

1. The linear model we found will not continue to predict the percentage of female prisoners in the U.S. indefinitely. Why must it eventually fail? Percentages must be ≤ 100

2. Do you think that our linear model will give an accurate estimate of the percentage of female prisoners in the U.S. in 2009? Why or why not? Yes, 2009 is still close to the data we are modeling

3. The linear model is such a good fit that it actually calls our attention to the unusual jump in the percentage of female prisoners in 1989. Statisticians would look for some unusual "confounding" factor in 1989 that might explain the jump. What sort of factors do you think might explain it?

4. Does Table 1.1 suggest a possible factor that might influence female crime statistics?

---

There are other ways of graphing numerical data that are particularly useful for statistical studies. We will treat some of them in Chapter 9. The scatter plot will be our choice of data graph for the time being, as it provides the closest connection to graphs of functions in the Cartesian plane.

# The Importance of Zeros

The main reason for studying algebra through the ages has been to solve equations. We develop algebraic models for phenomena so that we can solve problems, and the solution to the problems usually come down to finding solutions of algebraic equations.

If we are fortunate enough to be solving an equation in a single variable, we might proceed as in the following example.

**Prerequisite Chapter**

In the Prerequisite Chapter we defined solution of an equation, solving an equation, *x*-intercept, and graph of an equation in *x* and *y*.

### Example 6    SOLVING AN EQUATION ALGEBRAICALLY

Find all real numbers $x$ for which $6x^3 = 11x^2 + 10x$.

**Solution**  We begin by changing the form of the equation to $6x^3 - 11x^2 - 10x = 0$.

We can then solve this equation algebraically by factoring:

$$6x^3 - 11x^2 - 10x = 0$$
$$x(6x^2 - 11x - 10) = 0$$
$$x(2x - 5)(3x + 2) = 0$$
$$x = 0 \quad \text{or} \quad 2x - 5 = 0 \quad \text{or} \quad 3x + 2 = 0$$
$$x = 0 \quad \text{or} \quad x = \frac{5}{2} \quad \text{or} \quad x = -\frac{2}{3}$$

In Example 6, we used the important Zero Factor Property of real numbers.

> **The Zero Factor Property**
>
> A product of real numbers is zero if and only if at least one of the factors in the product is zero.

It is this property that enables us to find the three real solutions to the cubic equation in Example 6. Similarly, it is this property that has focused the attention of centuries of algebra students on various tricks (like factoring) designed to solve equations in which an expression is set equal to zero. While these techniques are still well worth knowing, modern problem-solvers are fortunate to have an alternative way to find such solutions.

If we graph the expression, then the *x*-intercepts of the graph of the expression will be the values for which the expression equals 0.

### Example 7    SOLVING AN EQUATION GRAPHICALLY

Solve the equation $x^2 = 10 - 4x$ graphically.

**Solution**  We first find an equivalent equation with 0 on the right hand side: $x^2 + 4x - 10 = 0$. We then graph the equation $y = x^2 + 4x - 10$, as shown in Figure 1.4.

We then use the grapher to locate the *x*-intercepts of the graph:

$$x \approx 1.7416574 \quad \text{and} \quad x \approx -5.741657$$

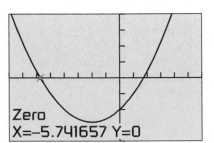

Zero
X=−5.741657 Y=0

[−8, 6] by [−20, 20]

**Figure 1.4**  The graph of $y = x^2 + 4x - 10$. (Example 7)

This quadratic equation has irrational solutions that can be found by the quadratic formula. We can use this technique to *confirm our answer algebraically*:

$$x = \frac{-4 + \sqrt{16 + 40}}{2} \approx 1.7416574$$

and

$$x = \frac{-4 - \sqrt{16 + 40}}{2} \approx -5.7416574$$

While the decimal answers are certainly accurate enough for all practical purposes, it is important to note that only the expressions found by the quadratic formula give the *exact* real number answers. The tidiness of exact answers is a worthy mathematical goal. Realistically, however, exact answers are often impossible to obtain, even with the most sophisticated mathematical tools.

We used the graphing utility of the calculator to **solve graphically** in Example 7. Most calculators also have solvers that would enable us to **solve numerically** for the same decimal approximations without considering the graph. Some calculators have computer algebra systems that will solve numerically to produce exact answers in certain cases. In this book we will distinguish between these two technological methods and the traditional pencil-and-paper methods used to **solve algebraically.**

Every method of solving an equation usually comes down to finding where an expression equals zero. If we use $f(x)$ to denote an algebraic expression in the variable $x$, the connections are as follows.

> **Fundamental Connection**
>
> If $a$ is a real number that solves the equation $f(x) = 0$, then these three statements are equivalent:
>
> 1. The number $a$ is a **root** (or **solution**) of the **equation** $f(x) = 0$.
> 2. The number $a$ is a **zero** of $y = f(x)$.
> 3. The number $a$ is an **x-intercept** of the **graph** of $y = f(x)$. (Sometimes the *point* $(a, 0)$ is referred to as an *x*-intercept.)

## Problem-Solving

George Pólya (1887–1985) is sometimes called the father of modern problem solving, not only because he was good at it (as he certainly was) but also because he published the most famous analysis of the problem-solving process: *How to Solve It: A New Aspect of Mathematical Method.* His "four steps" are well known to most mathematicians:

**Pólya's Four Problem-Solving Steps**

**1.** Understand the problem.

**2.** Devise a plan.

**3.** Carry out the plan.

**4.** Look back.

The problem-solving process that we recommend you use throughout this course will be the following version of Pólya's four steps.

**A Problem-Solving Process**

**Step 1—Understand the problem.**

- Read the problem as stated, several times if necessary.
- Be sure you understand the meaning of each term used.
- Restate the problem in your own words. Discuss the problem with others if you can.
- Identify clearly the information that you need to solve the problem.
- Find the information you need from the given data.

**Step 2—Develop a mathematical model of the problem.**

- Draw a picture to visualize the problem situation. It usually helps.
- Introduce a variable to represent the quantity you seek. (In some cases there may be more than one.)
- Use the statement of the problem to find an equation or inequality that relates the variables you seek to quantities that you know.

**Step 3—Solve the mathematical model and support or confirm the solution**

- **Solve algebraically** using traditional algebraic methods and **support graphically or support numerically** using a graphing utility.
- **Solve graphically or numerically** using a graphing utility and **confirm algebraically** using traditional algebraic methods.
- **Solve graphically or numerically** because there is no other way possible.

**Step 4—Interpret the solution in the problem setting**

- Translate your mathematical result into the problem setting and decide whether the result makes sense.

**Example 8** APPLYING THE PROBLEM-SOLVING PROCESS

The engineers at an auto manufacturer pay students $0.08 per mile plus $25 per day to road test their new vehicles.

**(a)** How much did the auto manufacturer pay Sally to drive 440 miles in one day?

**(b)** John earned $93 test-driving a new car in one day. How far did he drive?

## Solution

### Model

A picture of a car or of Sally or John would not be helpful, so we go directly to designing the model. Both John and Sally earned $25 for one day, plus $0.08 per mile. Multiply dollars/mile by miles to get dollars.

So if $p$ represents the pay for driving $x$ miles in one day, our algebraic model is

$$p = 25 + 0.08x.$$

### Solve Algebraically

**(a)** To get Sally's pay we let $x = 440$ and solve for $p$:

$$p = 25 + 0.08(440)$$

$$= 60.20$$

**(b)** To get John's mileage we let $p = 93$ and solve for $x$:

$$93 = 25 + 0.08x$$

$$68 = 0.08x$$

$$x = \frac{68}{0.08}$$

$$x = 850$$

### Support Graphically

Figure 1.5a shows that the point $(440, 60.20)$ is on the graph of $y = 0.08x + 25$, supporting our answer to a. Figure 1.5b shows that the point $(850, 93)$ is on the graph of $y = 0.08x + 25$, supporting our answer to b. (We could also have **supported** our answer **numerically** by simply substituting in for each $x$ and confirming the value of $p$.)

### Interpret

Sally earned $60.20 for driving 440 miles in one day. John drove 850 miles in one day to earn $93.00.

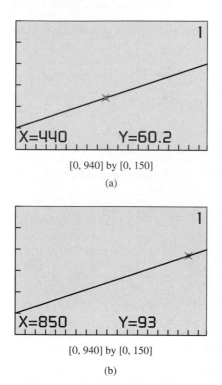

[0, 940] by [0, 150]

(a)

[0, 940] by [0, 150]

(b)

**Figure 1.5** Graphical support for the algebraic solutions in Example 8.

It is not really necessary to *show* written support as part of an algebraic solution, but it is good practice to support answers wherever possible simply to reduce the chance for error. We will often show written support of our solutions in this book in order to highlight the connections among the algebraic, graphical, and numerical models.

## Grapher Failure and Hidden Behavior

While the graphs produced by computers and graphing calculators are wonderful tools for understanding algebraic models and their behavior, it is important to keep in mind that machines have limitations. Occasionally they can produce graphical models that misrepresent the phenomena we wish to study, a problem we call **grapher failure**. Sometimes the viewing window will be too large, obscuring details of the graph which we call **hidden behavior**. We will give an example of each just to illustrate what can happen, but rest assured that these difficulties rarely occur with graphical models that arise from real-world problems.

**Example 9** SEEING GRAPHER FAILURE

Look at the graph of $y = 3/(2x - \pi)$ on a graphing calculator. Is there an $x$-intercept?

**Solution** A graph in the window $[-4, 4]$ by $[-3, 3]$ is shown in Figure 1.6.

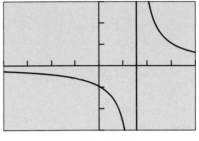

[–4, 4] by [–3, 3]

**Figure 1.6** A graph with a mysterious $x$-intercept. (Example 9)

The graph seems to show an $x$-intercept about halfway between 1 and 2. To confirm this algebraically, we would set $y = 0$ and solve for $x$:

$$0 = \frac{3}{2x - \pi}$$

$$0(2x - \pi) = 3$$

$$0 = 3$$

The statement $0 = 3$ is false for all $x$, so there can be no value that makes $y = 0$, and hence there can be no $x$-intercept for the graph. What went wrong?

The answer is a simple form of grapher failure: the vertical line should not be there! The actual graph of $y = 3/(2x - \pi)$ goes off to $-\infty$ on the left of $\pi/2$ and comes down from $+\infty$ on the right of $\pi/2$ (more on this later), but the grapher can't do that. It plots points at regular increments from left to right, *connecting the points* as it goes. It hits some low point off the screen to the left of $\pi/2$, followed immediately by some high point off the screen to the right of $\pi/2$, and it connects them with that unwanted line.

**Example 10** NOT SEEING HIDDEN BEHAVIOR

Solve graphically: $x^3 - 1.1x^2 - 65.4x + 229.5 = 0$

**Solution** Figure 1.7a on page 12 shows the graph in the standard $[-10, 10]$ by $[-10, 10]$ window, an inadequate choice because too much of the graph is off the screen. Our horizontal dimensions look fine, so we adjust our vertical dimensions to $[-500, 500]$, yielding the graph in Figure 1.7b.

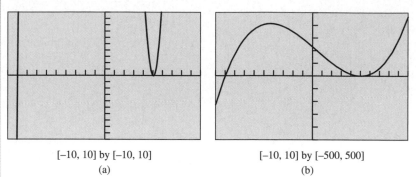

[–10, 10] by [–10, 10]
(a)

[–10, 10] by [–500, 500]
(b)

**Figure 1.7** The graph of $y = x^3 - 1.1x^2 - 65.4x + 229.5$ in two viewing windows. (Example 10)

We use the grapher to locate an $x$-intercept near $-9$ (which we find to be $-9$) and then an $x$-intercept near 5 (which we find to be 5). The graph leads us to believe that we are done. However, if we zoom in closer to observe the behavior near $x = 5$, the graph tells a new story (Figure 1.8).

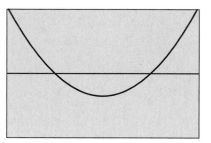

[4.95, 5.15] by [–0.1, 0.1]

**Figure 1.8** A closer look at the graph of $y = x^3 - 1.1x^2 - 65.4x + 229.5$. (Example 10)

In this graph we see that there are actually *two* $x$-intercepts near 5 (which we find to be 5 and 5.1). There are therefore three roots (or zeros) of the equation $x^3 - 1.1x^2 - 65.4x + 229.5 = 0$: $x = -9$, $x = 5$, and $x = 5.1$.

You might wonder if there could be still *more* hidden $x$-intercepts in Example 10! We will learn in Chapter 2 how the *Fundamental Theorem of Algebra* guarantees that there are not.

## A Word About Proof

While Example 10 is still fresh on our minds, let us point out a subtle, but very important, consideration about our solution.

We *solved graphically* to find two solutions, then eventually three solutions, to the given equation. Although we did not show the steps, it is easy to *confirm numerically* that the three numbers found are actually solutions by substituting them into the equation. But the problem asked us to find *all* solutions. While we could explore that equation graphically in a hundred more viewing windows and never find another solution, our failure to find them

---

**Teacher Note**

Sometimes it is impossible to show all of the details of a graph in a single window. For example, in Example 10 the graph in Figure 1.8 reveals minute details of the graph, but it hides the overall shape of the graph.

would not *prove* that they are not out there somewhere. That is why the Fundamental Theorem of Algebra is so important. It tells us that there can be at most three real solutions to *any* cubic equation, so we know for a fact that there are no more.

Exploration is encouraged throughout this book because it is how mathematical progress is made. Mathematicians are never satisfied, however, until they have *proved* their results. We will show you proofs in later chapters and we will ask you to produce proofs occasionally in the exercises. That will be a time for you to set the technology aside, get out a pencil, and show in a logical sequence of algebraic steps that something is undeniably and universally true.

### Example 11  PROVING A PECULIAR NUMBER FACT

Prove that 6 is a factor of $n^3 - n$ for every positive integer $n$.

**Solution**  You can explore this expression for various values of $n$ on your calculator. Table 1.5 shows it for the first 12 values of $n$.

**Table 1.5  The first 12 values of $n^3 - n$**

| $n$ | 1 | 2 | 3 | 4 | 5 | 6 | 7 | 8 | 9 | 10 | 11 | 12 |
|---|---|---|---|---|---|---|---|---|---|---|---|---|
| $n^3 - n$ | 0 | 6 | 24 | 60 | 120 | 210 | 336 | 504 | 720 | 990 | 1320 | 1716 |

All of these numbers are divisible by 6, but that does not prove that they will continue to be divisible by 6 for all values of $n$. In fact, a table with a billion values, all divisible by 6, would not constitute a proof. Here is a proof:

Let $n$ be *any* positive integer.

- We can factor $n^3 - n$ as the product of three numbers: $(n - 1)(n)(n + 1)$.

- The factorization shows that $n^3 - n$ is always the product of three consecutive integers.

- Every set of three consecutive integers must contain a multiple of 3.

- Since 3 divides a factor of $n^3 - n$, it follows that 3 is a factor of $n^3 - n$ itself.

- Every set of three consecutive integers must contain a multiple of 2.

- Since 2 divides a factor of $n^3 - n$, it follows that 2 is a factor of $n^3 - n$ itself.

- Since both 2 and 3 are factors of $n^3 - n$, we know that 6 is a factor of $n^3 - n$.

End of proof!

## Quick Review 1.1

Factor the following expressions completely over the real numbers.

**1.** $x^2 - 16$ $(x + 4)(x - 4)$ **2.** $x^2 + 10x + 25$

**3.** $81y^2 - 4$ $(9y + 2)(9y - 2)$ **4.** $3x^3 - 15x^2 + 18x$

**5.** $16h^4 - 81$ **6.** $x^2 + 2xh + h^2$

**7.** $x^2 + 3x - 4$ $(x + 4)(x - 1)$ **8.** $x^2 - 3x + 4$ $x^2 - 3x + 4$

**9.** $2x^2 - 11x + 5$ **10.** $x^4 + x^2 - 20$

## Section 1.1 Exercises

In Exercises 1–10, match the numerical model to the corresponding graphical model ($a$–$j$) and algebraic model ($k$–$t$).

**1.** (d)(q)

| $x$ | 3 | 5 | 7 | 9 | 12 | 15 |
|---|---|---|---|---|---|---|
| $y$ | 6 | 10 | 14 | 18 | 24 | 30 |

**2.** (f)(r)

| $x$ | 0 | 1 | 2 | 3 | 4 | 5 |
|---|---|---|---|---|---|---|
| $y$ | 2 | 3 | 6 | 11 | 18 | 27 |

**3.** (a)(p)

| $x$ | 2 | 4 | 6 | 8 | 10 | 12 |
|---|---|---|---|---|---|---|
| $y$ | 4 | 10 | 16 | 22 | 28 | 34 |

**4.** (h)(o)

| $x$ | 5 | 10 | 15 | 20 | 25 | 30 |
|---|---|---|---|---|---|---|
| $y$ | 90 | 80 | 70 | 60 | 50 | 40 |

**5.** (e)(l)

| $x$ | 1 | 2 | 3 | 4 | 5 | 6 |
|---|---|---|---|---|---|---|
| $y$ | 39 | 36 | 31 | 24 | 15 | 4 |

**6.** (b)(s)

| $x$ | 1 | 2 | 3 | 4 | 5 | 6 |
|---|---|---|---|---|---|---|
| $y$ | 5 | 7 | 9 | 11 | 13 | 15 |

**7.** (g)(t)

| $x$ | 5 | 7 | 9 | 11 | 13 | 15 |
|---|---|---|---|---|---|---|
| $y$ | 1 | 2 | 3 | 4 | 5 | 6 |

**8.** (j)(k)

| $x$ | 4 | 8 | 12 | 14 | 18 | 24 |
|---|---|---|---|---|---|---|
| $y$ | 20 | 72 | 156 | 210 | 342 | 600 |

**9.** (i)(m)

| $x$ | 3 | 4 | 5 | 6 | 7 | 8 |
|---|---|---|---|---|---|---|
| $y$ | 8 | 15 | 24 | 35 | 48 | 63 |

**10.** (c)(n)

| $x$ | 4 | 7 | 12 | 19 | 28 | 39 |
|---|---|---|---|---|---|---|
| $y$ | 1 | 2 | 3 | 4 | 5 | 6 |

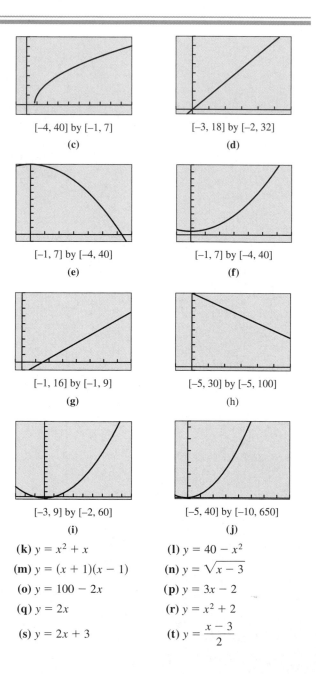

[−4, 40] by [−1, 7]
(c)

[−3, 18] by [−2, 32]
(d)

[−1, 7] by [−4, 40]
(e)

[−1, 7] by [−4, 40]
(f)

[−1, 16] by [−1, 9]
(g)

[−5, 30] by [−5, 100]
(h)

[−3, 9] by [−2, 60]
(i)

[−5, 40] by [−10, 650]
(j)

[−2, 14] by [−4, 36]
(a)

[−1, 6] by [−2, 20]
(b)

**(k)** $y = x^2 + x$ **(l)** $y = 40 - x^2$

**(m)** $y = (x + 1)(x - 1)$ **(n)** $y = \sqrt{x - 3}$

**(o)** $y = 100 - 2x$ **(p)** $y = 3x - 2$

**(q)** $y = 2x$ **(r)** $y = x^2 + 2$

**(s)** $y = 2x + 3$ **(t)** $y = \dfrac{x - 3}{2}$

Exercises 11–18 refer to the data in Table 1.6 below showing the percentage of the female and male populations in the United States employed in the civilian work force in certain years since 1955.

| Table 1.6 Employment Statistics | | |
| --- | --- | --- |
| Year | Women (%) | Men (%) |
| 1955 | 35.7 | 85.4 |
| 1960 | 37.7 | 83.3 |
| 1965 | 39.3 | 80.7 |
| 1970 | 43.3 | 79.7 |
| 1975 | 46.3 | 77.9 |
| 1980 | 51.5 | 77.4 |
| 1985 | 54.5 | 76.3 |
| 1990 | 57.5 | 76.4 |
| 1995 | 58.9 | 75.0 |

*Source: U.S. Bureau of Labor Statistics*

11. According to the numerical model, what has been the trend in females joining the work force over the years since 1955?

12. According to the numerical model, what has been the trend in males joining the work force since 1955? decreasing

13. Model the data graphically with two scatter plots on the same graph, one showing the percentage of women employed as a function of time and the other showing the same for men. Measure time in years since 1955.

14. Are the male percentages falling faster than the female percentages are rising, or vice-versa? vice-versa

15. Model the data algebraically with linear equations of the form $y = mx + b$. Use the 1955 and 1995 data to compute the slopes.

16. If the percentages continue to follow the current linear models, what will be the employment percentages for women and for men in the year 2005? M: 72.4%; W: 64.7%

17. If the percentages continue to follow the current linear models, when will the percentages of women and men in the civilian work force be the same? What percentage will that be? 2014; 70%

18. **Writing to Learn** Explain why the percentages can not continue to follow the current linear models.

Exercises 19–22 refer to the graph below, which shows the *minimum* salaries in major league baseball over a recent 18-year period and the *average* salaries in major league baseball over the same period. Salaries are measured in dollars and time is measured after the starting year (year 0).

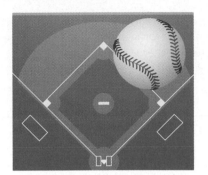

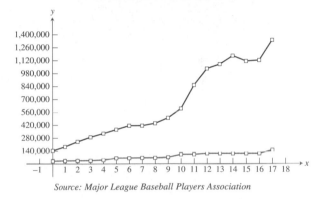

*Source: Major League Baseball Players Association*

19. Which line is which, and how do you know?

20. After Peter Ueberroth's resignation as baseball commissioner in 1988 and his successor's untimely death in 1989, the team owners broke free of previous restrictions and began an era of competitive spending on player salaries. Identify where the 1990 salaries appear in the graph and explain how you can spot them.

21. The owners attempted to halt the uncontrolled spending by proposing a salary cap, which prompted a players' strike in 1994. The strike caused the 1995 season to be shortened and left many fans angry. Identify where the 1995 salaries appear in the graph and explain how you can spot them.

22. **Writing to learn** Analyze the general patterns in the graphical model and give your thoughts about what the long-term implications might be for

(a) the players;

(b) the team owners;

(c) the baseball fans.

In Exercises 23–32, solve the equation algebraically.

23. $v^2 - 5 = 8 - 2v^2$   $\pm\sqrt{\dfrac{13}{3}}$   24. $(x + 11)^2 = 121$   $-22; 0$

25. $x^2 - 7x - \dfrac{3}{4} = 0$   $3.5 \pm \sqrt{13}$

26. $2x^2 - 5x + 2 = (x - 3)(x - 2) + 3x$   $-1; 4$

**27.** $x(2x - 5) = 12$   $-1.5; 4$    **28.** $x(2x - 1) = 10$   $-2; 2.5$

**29.** $x(x + 7) = 14$   $-\dfrac{7}{2} \pm \dfrac{1}{2}\sqrt{105}$

**30.** $x^2 - 3x + 4 = 2x^2 - 7x - 8$   $-2; 6$

**31.** $x + 1 - 2\sqrt{x + 4} = 0$   $5$    **32.** $\sqrt{x} + x = 1$   $\dfrac{3}{2} - \dfrac{1}{2}\sqrt{5}$

In Exercises 33–40, solve the equation graphically by converting it to an equivalent equation with 0 on the right hand side and then finding the *x*-intercepts.

**33.** $2x - 5 = \sqrt{x + 4}$      **34.** $|3x - 2| = 2\sqrt{x + 8}$

**35.** $|2x - 5| = 4 - |x - 3|$     **36.** $\sqrt{x + 6} = 6 - 2\sqrt{5 - x}$

**37.** $2x - 3 = x^3 - 5$       **38.** $x + 1 = x^3 - 2x - 5$

**39.** $(x + 1)^{-1} = x^{-1} + x$    **40.** $x^2 = |x|$   $\{0, 1, -1\}$

**41. Connecting graphs and equations** The curves on the graph below are the graphs of the three curves given by

$$y_1 = 4x + 5$$
$$y_2 = x^3 + 2x^2 - x + 3$$
$$y_3 = -x^3 - 2x^2 + 5x + 2.$$

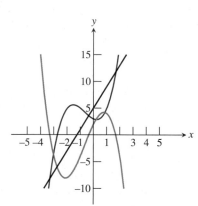

**(a)** Write an equation that can be solved to find the points of intersection of the graphs of $y_1$ and $y_2$.

**(b)** Write an equation that can be solved to find the *x*-intercepts of the graph of $y_3$.

**(c) Writing to Learn** How does the graphical model reflect the fact that the answers to (a) and (b) are equivalent algebraically?

**(d)** Confirm numerically that the *x*-intercepts of $y_3$ give the same values when substituted into the expressions for $y_1$ and $y_2$.

**42. Free Fall of a Smoke Bomb** At the Oshkosh, WI, air show, Jake Trouper drops a smoke bomb to signal the official beginning of the show. Ignoring air resistance, an object in free fall will fall *d* feet in *t* seconds, where *d* and *t* are related by the algebraic model $d = 16t^2$.

**(a)** How long will it take the bomb to fall 180 feet?

**(b)** If the smoke bomb is in free fall for 12.5 seconds after it is dropped, how high was the airplane when the smoke bomb was dropped?   2500 ft

**43. U.S. Air Travel** The number of revenue passengers enplaned in the U.S. over the 11-year period from 1987 to 1997 is shown in the table below.

| Table 1.7  U.S. Air Travel | | | |
|---|---|---|---|
| Year | Passengers (millions) | Year | Passengers (millions) |
| 1987 | 447.7 | 1993 | 488.5 |
| 1988 | 454.6 | 1994 | 528.8 |
| 1989 | 453.7 | 1995 | 547.8 |
| 1990 | 465.6 | 1996 | 581.2 |
| 1991 | 452.3 | 1997 | 598.9 |
| 1992 | 475.1 | | |

*Source: Air Transport Association*

**(a)** Graph a scatter plot of the data.

**(b)** Model the data algebraically with the equation $P = 2t^2 - 4.6t + 452$, where *P* is the number of passengers in millions, and *t* is the number of years after 1987. Superimpose the graph of the model on the scatter plot.

**(c)** According to the algebraic model, when will the number of passengers reach 730 million?

**(d)** Do you think this algebraic model will still be valid in the year 2007? Why or why not?

**44. Connecting Algebra and Geometry** Explain how the algebraic equation $(x + b)^2 = x^2 + 2bx + b^2$ models the areas of the regions in the geometric figure shown below on the left:

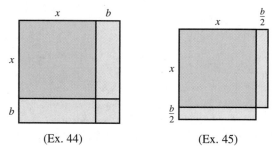

(Ex. 44)              (Ex. 45)

**45. Connecting Algebra and Geometry** The geometric figure shown on the right above is a large square with a small square missing.

**(a)** Find the area of the figure.   $x^2 + bx$

**(b)** What area must be added to complete the large square?

**(c)** Explain how the algebraic formula for completing the square models the completing of the square in (b).

**46. Exploring Hidden Behavior** Solving graphically, find all real solutions to the following equations. Watch out for hidden behavior.

**(a)** $y = 10x^3 + 7.5x^2 - 54.85x + 37.95$   $-3$ or 1.1 or 1.15

**(b)** $y = x^3 + x^2 - 4.99x + 3.03$   $-3$

**47. Exploring Grapher Failure** Let $y = (x^{200})^{1/200}$.

(a) Explain algebraically why $y = x$ for all $x \geq 0$.

(b) Graph the equation $y = (x^{200})^{1/200}$ in the window $[0, 1]$ by $[0, 1]$.

(c) Is the graph different from the graph of $y = x$? Yes

(d) Can you explain why the grapher failed?

**48. Writing to learn** The graph below shows the distance from home against time for a jogger. Using information from the graph, write a paragraph describing the jogger's workout.

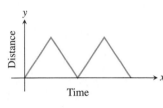

# Explorations

**49. Analyzing the Market** Both Ahmad and LaToya watch the stock market throughout the year for stocks that make significant jumps from one month to another. When they spot one, each buys 100 shares. Ahmad's rule is to sell the stock if it fails to perform well for three months in a row. LaToya's rule is to sell in December if the stock has failed to perform well since its purchase.

The graph below shows the monthly performance (Jan–Dec) of a stock that both Ahmad and LaToya have been watching.

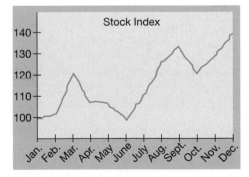

(a) Both Ahmad and LaToya bought the stock early in the year. In which month? March

(b) At approximately what price did they buy the stock?

(c) When did Ahmad sell the stock?

(d) How much did Ahmad lose on the stock? About $2000

(e) **Writing to learn** Explain why LaToya's strategy was better than Ahmad's for this particular stock in this particular year.

(f) Sketch a 12-month graph of a stock's performance that would favor Ahmad's strategy over LaToya's.

**50. Group Activity  Creating Hidden Behavior** You can create your own graphs with hidden behavior. Working in groups of two or three, try this exploration.

(a) Graph the equation $y = (x + 2)(x^2 - 4x + 4)$ in the window $[-4, 4]$ by $[-10, 10]$.

(b) Confirm algebraically that this function has zeros only at $x = -2$ and $x = 2$.

(c) Graph the equation $y = (x + 2)(x^2 - 4x + 4.01)$ in the window $[-4, 4]$ by $[-10, 10]$

(d) Confirm algebraically that this function has only one zero, at $x = -2$. (Use the discriminant.)

(e) Graph the equation $(x + 2)(x^2 - 4x + 3.99)$ in the window $[-4, 4]$ by $[-10, 10]$.

(f) Confirm algebraically that this function has three zeros. (Use the discriminant.)

**51. Doing Arithmetic With Lists** Enter the data from the "Total" column of Table 1.2 of Example 2 into list $L_1$ in your calculator. Enter the data from the "Female" column into list $L_2$. Check a few computations to see that the procedures in (a) and (b) cause the calculator to divide each element of $L_2$ by the corresponding entry in $L_1$, multiply it by 100, and store the resulting list of percentages in $L_3$.

(a) On the home screen, enter the command $100 \times L_2 / L_1 \rightarrow L_3$.

(b) Go to the top of list $L_3$ and enter $L_3 = 100(L_2/L_1)$.  ■

# Extending the Ideas

**52. The Proliferation of Cell Phones** Table 1.8 shows the number of cellular phone subscribers in the U.S. and their average local monthly bill in the years from 1988 to 1997.

| Table 1.8  Cellular Phone Subscribers | | |
|---|---|---|
| Year | Subscribers (number) | Average Local Monthly Bill ($) |
| 1988 | 2,069,411 | 98.02 |
| 1989 | 3,508,944 | 89.30 |
| 1990 | 5,283,055 | 80.90 |
| 1991 | 7,557,148 | 72.74 |
| 1992 | 11,032,753 | 68.68 |
| 1993 | 16,009,461 | 61.48 |
| 1994 | 24,134,421 | 56.21 |
| 1995 | 33,785,661 | 51.00 |
| 1996 | 44,042,992 | 47.70 |
| 1997 | 55,132,293 | 42.78 |

*Source: Cellular Telecommunications Industry Assocation*

(a) Graph the scatter plots of the number of subscribers and the average local monthly bill as functions of time, letting time $t = $ the number of years after 1988.

(b) One of the scatter plots clearly suggests a quadratic model with formula $y = ax^2 + b$. Use the point at $t = 0$ to solve for $b$; then use the point at $t = 9$ to solve for $a$.

**(c)** Superimpose the graph of the quadratic model onto the scatter plot. Does the fit appear to be good?

**(d)** The other scatter plot suggests a linear model with equation $y = mx + b$. Use the points at $t = 0$ and $t = 9$ to find the linear model.

**(e)** Superimpose the graph of the linear model onto the scatter plot. Does the fit appear to be good?

**(f)** Do you think that a quadratic model might be even better than the linear model? Explain.

**(g)** If your calculator does quadratic regression, use it to fit a quadratic curve to the linear-looking data. Superimpose the graph onto the scatter plot. Does the fit appear to be better than the line?

**53. Group activity**  (Continuation of Exercise 52) Discuss the economic forces suggested by the two models in Exercise 52 and speculate about the future by analyzing the graphs.

**54. Proving a Theorem**  Prove that if $n$ is a positive integer, then $n^2 + 2n$ is either odd or a multiple of 4. Compare your proof with those of your classmates.

---

## 1.2  Functions and Their Properties

Definition and Notation • Domain and Range • Continuity •
Increasing and Decreasing • Boundedness • Local and Absolute
Extrema • Symmetry • Asymptotes • End Behavior

### Definition and Notation

Mathematics and its applications abound with examples of formulas by which quantitative variables are related to each other. For many centuries the algebraic emphasis was on how to manipulate formulas to solve for one variable in terms of the others—and that was important—but with calculus the emphasis changed to studying the relationships among the formulas themselves. The language and notation of functions is ideal for that purpose.

A function is actually a simple concept; if it were not, history would have replaced it with a simpler one by now. Here is the definition.

> **Definition**  **Function, Domain, and Range**
>
> A function from a set $D$ to a set $R$ is a rule that assigns to every element in $D$ a unique element in $R$. The set $D$ of all input values is the **domain** of the function, and the set $R$ of all output values is the **range** of the function.

There are many ways to look at functions. One of the most intuitively helpful is the "machine" concept, in which values of the domain ($x$) are fed into the machine (the function $f$) to produce range values ($y$). To indicate that $y$ comes from the function acting on $x$, we use Euler's elegant **function notation** $y = f(x)$ (which we read as "**y equals f of x**" or "**the value of f at x**"). Here $x$ is the **independent variable** and $y$ is **the dependent variable**.

A function can also be viewed as a **mapping** of the elements of the domain onto the elements of the range. Figure 1.10a shows a function that maps elements from the domain $X$ onto elements of the range $Y$. Figure 1.10b shows another such mapping, but *this one is not a function*, since the rule does not assign the element $x_1$ to a *unique* element of $Y$.

**A Bit of History**

The word function in its mathematical sense is generally attributed to Gottfried Leibniz (1646–1716), one of the pioneers in the methods of calculus. His attention to clarity of notation is one of his greatest contributions to scientific progress, which is why we still use his notation in calculus courses today. Ironically, it was not Leibniz but Leonhard Euler (1707–1783) who introduced the familiar notation $f(x)$, a notation so perfectly suited to the subject that it is difficult to imagine how modern mathematicians could communicate without it.

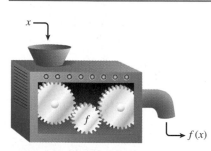

**Figure 1.9**  A "machine" diagram for a function.

**Lesson Guide**

Day 1: Definition and Notation; Domain and Range; Continuity; Increasing and Decreasing

Day 2: Increasing and Decreasing; Local and Absolute Extrema

Day 3: Symmetry; Asymptotes; End Behavior.

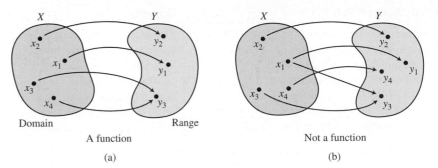

Figure 1.10 The diagram in (a) depicts a maping from $X$ to $Y$ that is a function. The diagram in (b) depicts a maping from $X$ to $Y$ that is not a function.

This uniqueness of the range value is very important to us as we study function behavior. Knowing that $f(2) = 8$ tells us something about $f$, and that understanding would be contradicted if we were to discover later that $f(2) = 4$. That is why you will never see a function defined by an ambiguous formula like $f(x) = 3x \pm 2$.

### Example 1 DEFINING A FUNCTION

Does the $y = x^2$ formula define $y$ as a function of $x$?

**Solution**

Yes, $y$ is a function of $x$. In fact, we can write the formula in function notation: $f(x) = x^2$. When a number $x$ is substituted into the function, the square of $x$ will be the output, and there is no ambiguity about what the square of $x$ is.

Another useful way to look at functions is graphically. The **graph of the function $y = f(x)$** is the set of all points $(x, f(x))$, $x$ in the domain of $f$. We match domain values along the $x$-axis with their range values along the $y$-axis to get the ordered pairs that yield the graph of $y = f(x)$.

### Example 2 SEEING A FUNCTION GRAPHICALLY

Of the three graphs shown in Figure 1.11, which is *not* the graph of a function? How can you tell?

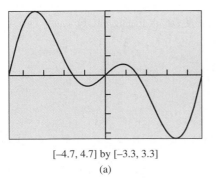

[−4.7, 4.7] by [−3.3, 3.3]

(a)

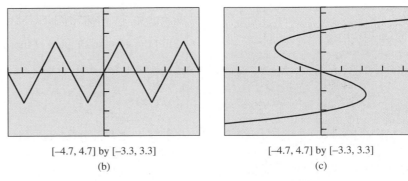

[−4.7, 4.7] by [−3.3, 3.3]　　　　　[−4.7, 4.7] by [−3.3, 3.3]

(b)　　　　　　　　　　　　　　　　(c)

Figure 1.11 One of these is not the graph of a function. (Example 2)

## What about data?

When moving from a numerical model to an algebraic model we will often use a function to approximate data pairs that by themselves violate our definition. In Figure 1.12 we can see that several pairs of data points fail the vertical line test, and yet the linear function approximates the data quite well.

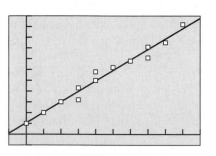

[–1, 10] by [–1, 11]

**Figure 1.12** The data points fail the vertical line test but are nicely approximated by a linear function.

## Solution

The graph in (c) is not the graph of a function. There are three points on the graph with *x*-coordinate 0, so the graph does not assign a *unique* value to 0. (Indeed, we can see that there are plenty of numbers between –2 and 2 to which the graph assigns multiple values.) The other two graphs do not have a comparable problem because no vertical line intersects either of the other graphs in more than one point. Graphs that pass this *vertical line test* are the graphs of functions.

---

### Vertical Line Test

A graph (set of points $(x, y)$) in the *xy*-plane defines *y* as a function of *x* if and only if no vertical line intersects the graph in more than one point.

---

## Domain and Range

We will usually define functions algebraically, giving the rule explicitly in terms of the domain variable. The rule, however, does not tell the complete story without some consideration of what the domain actually is.

For example, we can define the volume of a sphere as a function of its radius by the formula

$$V(r) = \frac{4}{3}\pi r^3 \text{ (Note that this is ``V of r''—not ``V · r'')}$$

This *formula* is defined for all real numbers, but the volume *function* is not defined for negative *r* values. So, if our intention were to study the volume function, we would restrict the domain to be all $r \geq 0$.

---

### Agreement

Unless we are dealing with a model (like volume) that necessitates a restricted domain, we will assume that the domain of a function defined by an algebraic expression is the same as the domain of the algebraic expression, the **implied domain**. For models, we will use a domain that fits the situation, the **relevant domain**.

---

### Example 3   FINDING THE DOMAIN OF A FUNCTION

Find the domain of each of these functions:

**(a)** $f(x) = \sqrt{x + 3}$

**(b)** $g(x) = \dfrac{\sqrt{x}}{x - 5}$

**(c)** $A(s) = (\sqrt{3}/4)s^2$, where $A(s)$ is the area of an equilateral triangle with sides of length *s*.

## Teaching Note

Many students have difficulty with the concepts of domain and range. Provide opportunities to discuss and write down the domains and ranges of many functions. The examples and exercises in this section lend themselves to using cooperative groups in the classroom. Working in groups allows students to help each other and to take advantage of a variety of exploratory activities.

## Solution

### Solve Algebraically

**(a)** The expression under a radical may not be negative. We set $x + 3 \geq 0$ and solve to find $x \geq -3$. The domain of $f$ is the interval $[-3, \infty)$.

**(b)** The expression under a radical may not be negative; therefore $x \geq 0$. Also, the denominator of a fraction may not be zero; therefore $x \neq 5$. The domain of $g$ is the interval $[0, \infty)$ with the number 5 removed, which we can write as the *union* of two intervals: $[0, 5) \cup (5, \infty)$.

**(c)** The algebraic expression has domain all real numbers, but the behavior being modeled restricts $s$ from being negative. The domain of $A$ is the interval $[0, \infty)$.

### Support Graphically

We can support our answers in (a) and (b) graphically, as the calculator should not plot points where the function is undefined.

**(a)** Notice that the graph of $y = \sqrt{x + 3}$ (Figure 1.13a) shows points only for $x \geq -3$, as expected.

**(b)** The graph of $y = \sqrt{x}/(x - 5)$ (Figure 1.13b) shows points only for $x \geq 0$, as expected, but shows an unexpected line through the $x$-axis at $x = 5$. This line, a form of grapher failure described in the previous section, should not be there. Ignoring it, we see that 5, as expected, is not in the domain.

**(c)** The graph of $y = (\sqrt{3}/4)s^2$ (Figure 1.13c) shows the unrestricted domain of the algebraic expression: all real numbers. The calculator has no way of knowing that $s$ is the length of a side of a triangle.

**Teaching Note**

It may be necessary to review the notation for the union of two sets in the context of intervals.

**Note**

The symbol "$\cup$" is read "union." It means that the elements of the two sets are combined to form one set.

**Teaching Note**

The graphing calculator allows us to explore many real-world phenomena that can be represented geometrically using this technology. This is important because students should be exposed to many different algebraic representations of functions

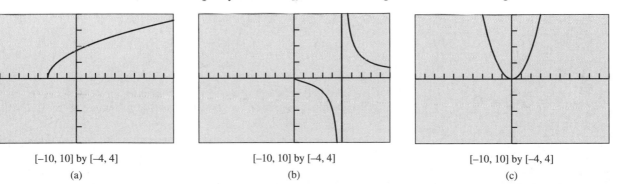

| $[-10, 10]$ by $[-4, 4]$ | $[-10, 10]$ by $[-4, 4]$ | $[-10, 10]$ by $[-4, 4]$ |
|:---:|:---:|:---:|
| (a) | (b) | (c) |

**Figure 1.13** Graphical support of the algebraic solutions in Example 3. The vertical line in (b) should be ignored because it results from grapher failure. The points in (c) with negative $x$-coordinates should be ignored because the calculator does not know that $x$ is a length (but we do).

Finding the range of a function algebraically is often much harder than finding the domain, although graphically the things we look for are similar: to find the *domain* we look for all *x-coordinates* that correspond to points on the graph, and to find the *range* we look for all *y-coordinates* that correspond to points on the graph. A good approach is to use graphical and algebraic approaches simultaneously, as we show in Example 4.

**GRAPHER NOTE**
**Function Notation**

A grapher typically does not use function notation. So the function $f(x) = x^2 + 1$ is entered as $y_1 = x^2 + 1$. On some graphers you can evaluate $f$ at $x = 3$ by entering $y_1(3)$ on the home screen. On the other hand, on other graphers $y_1(3)$ means $y_1 * 3$. (Consult Technology Auxillary.)

### Example 4   FINDING THE RANGE OF A FUNCTION

Find the range of the function $f(x) = \dfrac{2}{x}$.

**Solution**

The graph of $y = \dfrac{2}{x}$ is shown in Figure 1.14.

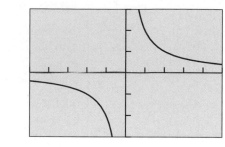

[−5, 5] by [−3, 3]

**Figure 1.14** The graph of $y = 2/x$. Is $y = 0$ in the range?

It appears that $x = 0$ is not in the domain (as expected, because a denominator can not be zero). It also appears that $y = 0$ is not in the range. We confirm this algebraically by trying to solve $2/x = 0$:

$$\frac{2}{x} = 0$$
$$2 = 0 \cdot x$$
$$2 = 0$$

Since the equation $2 = 0$ is never true, $2/x = 0$ has no solutions, and so $y = 0$ is not in the range. But how do we know that all other real numbers *are* in the range? We let $k$ be any other real number and try to solve $2/x = k$:

$$\frac{2}{x} = k$$
$$2 = k \cdot x$$
$$x = \frac{2}{k}$$

As you can see, there was no problem finding an $x$ this time, so 0 is the only number not in the range of $f$. We write the range $(-\infty, 0) \cup (0, \infty)$.

You can see that this is considerably more involved than finding a domain, but we are hampered at this point by not having many tools with which to analyze function behavior. We will revisit the problem of finding ranges in Example 12, after having developed the tools that will simplify the analysis.

## Continuity

One of the most important properties of the majority of functions that model real-world behavior is that they are *continuous*. Graphically speaking, a function is continuous at a point if the graph does not come apart at that point. We can illustrate the concept with a few graphs (Figure 1.15):

**Graphing Activity**

Sketch the graph of a function that has domain $[-5, -1) \cup (2, 4]$ and range $[-4, -1] \cup [1, \infty)$.

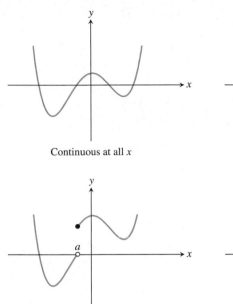

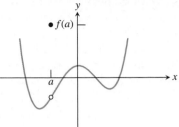

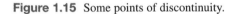

Continuous at all *x*    Removable discontinuity    Removable discontinuity

Jump discontinuity    Infinite discontinuity

**Figure 1.15** Some points of discontinuity.

Let us look at these cases individually.

This graph is continuous everywhere. Notice that the graph has no breaks. This means that if we are studying the behavior of the function *f* for *x* values close to any particular real number *a*, we can be assured that the *f*(*x*) values will be close to *f*(*a*).

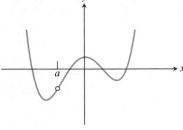

Continuous at all *x*

This graph is continuous everywhere except for the "hole" at *x* = *a*. If we are studying the behavior of this function *f* for *x* values close to *a*, we can *not* be assured that the *f*(*x*) values will be close to *f*(*a*). In this case, *f*(*x*) is smaller than *f*(*a*) for *x* near *a*. This is called a **removable discontinuity** because it can be patched up by re-defining *f*(*a*) so as to plug the hole.

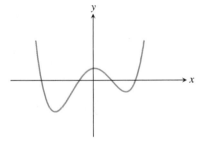

Removable discontinuity

This graph also has a **removable discontinuity** at $x = a$. If we are studying the behavior of this function $f$ for $x$ values close to $a$, we are still not assured that the $f(x)$ values will be close to $f(a)$, because in this case $f(a)$ doesn't even exist. It is removable because we could define $f(a)$ in such a way as to plug the hole and make $f$ continuous at $a$.

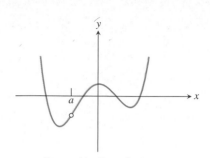

Removable discontinuity

Here is a discontinuity that is *not* removable. It is a **jump discontinuity** because there is more than just a hole at $x = a$; there is a *jump* in function values that makes the gap impossible to plug with a single point $(a, f(a))$, no matter how we try to redefine $f(a)$.

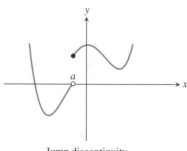

Jump discontinuity

This is a function with an **infinite discontinuity** at $x = a$. It is definitely not removable.

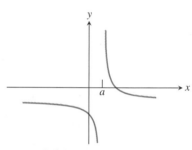

Infinite discontinuity

The simple geometric concept of an unbroken graph at a point is one of those visual notions that is extremely difficult to communicate accurately in the language of algebra. The key concept from the pictures seems to be that we want the point $(x, f(x))$ to slide smoothly onto the point $(a, f(a))$ without missing it, from either direction. This merely requires that $f(x)$ approach $f(a)$ as a *limit* as $x$ approaches $a$. A function $f$ is **continuous at $x = a$** if $\lim_{x \to a} f(x) = f(a)$.

### Example 5   IDENTIFYING POINTS OF DISCONTINUITY

Judging from the graphs, which of the following figures shows functions that are discontinuous at $x = 2$? Are any of the discontinuities removable?

**Choosing Viewing Windows**

Some viewing windows will show a vertical line for the function in Figure 1.16. It is sometimes possible to choose a viewing window in which the vertical line does not appear, as we did in Figure 1.14.

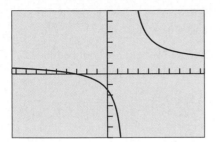

[−9.4, 9.4] by [−6, 6]

**Figure 1.16**  $f(x) = \dfrac{x + 3}{x - 2}$

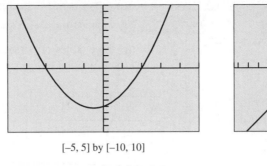

[−5, 5] by [−10, 10]

**Figure 1.17**  $g(x) = (x + 3)(x - 2)$

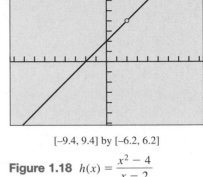

[−9.4, 9.4] by [−6.2, 6.2]

**Figure 1.18**  $h(x) = \dfrac{x^2 - 4}{x - 2}$

### Solution

Figure 1.16 shows a function that is undefined at $x = 2$ and hence not contiuous there. The discontinuity at $x = 2$ is not removable.

The function graphed in Figure 1.17 is a quadratic polynomial whose graph is a parabola, a graph that has no breaks because its domain includes all real numbers. It is continuous for all $x$.

The function graphed in Figure 1.18 is not defined at $x = 2$ and so can not be continuous there. The graph looks like the graph of the line $y = x + 2$, except that there is a hole where the point $(2, 4)$ should be. This is a removable discontinuity.

## Increasing and Decreasing

Another function concept that is easy to understand graphically is the property of being increasing, decreasing, or constant on an interval. We illustrate the concept with a few graphs (Figure 1.19):

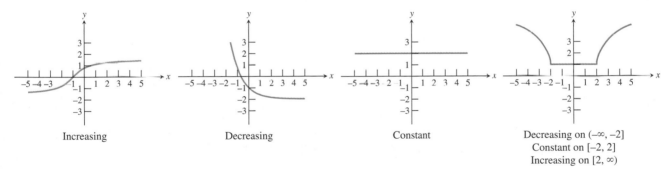

Increasing                Decreasing                Constant           Decreasing on $(-\infty, -2]$
Constant on $[-2, 2]$
Increasing on $[2, \infty)$

**Figure 1.19**  Examples of increasing, decreasing, or constant on an interval.

Once again the idea is easy to communicate graphically, but how can we identify these properties of functions algebraically? Exploration 1 will help to set the stage for the algebraic definition.

**Exploration 1**  Increasing, Decreasing, and Constant Data

1. Of the three tables of numerical data below, which would be modeled by a function that is (a) increasing, (b) decreasing, (c) constant?

| X | Y1 |
|---|----|
| −2 | 12 |
| −1 | 12 |
| 0 | 12 |
| 1 | 12 |
| 3 | 12 |
| 7 | 12 |

| X | Y2 |
|---|----|
| −2 | 3 |
| −1 | 1 |
| 0 | 0 |
| 1 | −2 |
| 3 | −6 |
| 7 | −12 |

| X | Y2 |
|---|----|
| −2 | −5 |
| −1 | −3 |
| 0 | −1 |
| 1 | 1 |
| 3 | 4 |
| 7 | 10 |

2. Make a list of $\triangle Y1$, the *change* in Y1 values as you move down the list. As you move from $Y1 = a$ to $Y1 = b$, the change is $\triangle Y1 = b - a$. Do the same for the values of Y2 and Y3.

| X moves from | $\triangle$X | $\triangle$Y1 | X moves from | $\triangle$X | $\triangle$Y2 | X moves from | $\triangle$X | $\triangle$Y3 |
|---|---|---|---|---|---|---|---|---|
| −2 to −1 | 1 | | −2 to −1 | 1 | | −2 to −1 | 1 | |
| −1 to 0 | 1 | | −1 to 0 | 1 | | −1 to 0 | 1 | |
| 0 to 1 | 1 | | 0 to 1 | 1 | | 0 to 1 | 1 | |
| 1 to 3 | 2 | | 1 to 3 | 2 | | 1 to 3 | 2 | |
| 3 to 7 | 4 | | 3 to 7 | 4 | | 3 to 7 | 4 | |

3. What is true about the quotients $\triangle Y/\triangle X$ for an increasing function? For a decreasing function? For a constant function?

4. Where else have you seen the quotient $\triangle Y/\triangle X$? Does this reinforce your answers in part 3?

Your analysis of the quotients $\triangle Y/\triangle X$ in the exploration should help you to understand the following definition.

**Definition**  Increasing, Decreasing, and Constant Function on an Interval

A function $f$ is **increasing** on an interval if, for any two points in the interval, a positive change in $x$ results in a positive change in $f(x)$.

A function $f$ is **decreasing** on an interval if, for any two points in the interval, a positive change in $x$ results in a negative change in $f(x)$.

A function $f$ is **constant** on an interval if, for any two points in the interval, a positive change in $x$ results in a zero change in $f(x)$.

### Example 6  ANALYZING A FUNCTION FOR INCREASING-DECREASING BEHAVIOR

For each function, tell the intervals on which it is increasing and the intervals on which it is decreasing.

**(a)** $f(x) = (x + 2)^2$

**(b)** $g(x) = \dfrac{x^2}{x^2 - 1}$

### Solution

### Solve Graphically

**(a)** We see from the graph in Figure 1.20 that $f$ is decreasing on $(-\infty, -2]$ and increasing on $[-2, \infty)$. (Notice that we include $-2$ in both intervals. Don't worry that this sets up some contradiction about what happens *at* $-2$, because we only talk about functions increasing or decreasing on *intervals*, and $-2$ is not an interval.)

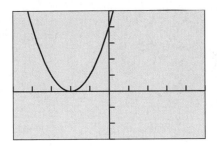

[−5, 5] by [−3, 5]

**Figure 1.20**  The function $f(x) = (x + 2)^2$ decreases on $(-\infty, -2]$ and increases on $[-2, \infty)$. (Example 6)

**(b)** We see from the graph in Figure 1.21 that $g$ is increasing on $(-\infty, -1)$, increasing again on $(-1, 0]$, decreasing on $[0, 1)$, and decreasing again on $(1, \infty)$.

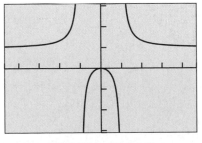

[−4.7, 4.7] by [−3.1, 3.1]

**Figure 1.21**  The function $g(x) = x^2/(x^2 - 1)$ increases on $(-\infty, -1)$ and $(-1, 0]$; the function decreases on $[0, 1)$ and $(1, \infty)$. (Example 6)

You may have noticed that we are making some assumptions about the graphs. How do we know that they don't turn around somewhere off the screen? We will develop some ways to answer that question later in the book, but the most powerful methods will await you when you study calculus.

---

**Notes on Examples**

Explain how the graph of the function is related to the definitions of decreasing on an interval, and increasing on an interval.

---

**Alert**

It is tempting to say that since $g$ increases on $(-\infty, -1)$ and on $(-1, 0]$, then $g$ increases on $(-\infty, 0]$, but this is not true. We can not say that $g(x)$ increases on $(-\infty, 0]$ because we can not choose just any two points in $(-\infty, 0]$ and produce a positive change in $g(x)$ from a positive change in $x$. For example, $g(-2) > g(-1/2)$.

## Boundedness

The concept of *boundedness* is fairly simple to understand both graphically and algebraically. We will move directly to the algebraic definition after motivating the concept with some typical graphs (Figure 1.22).

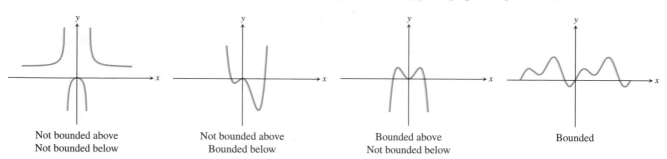

Not bounded above
Not bounded below

Not bounded above
Bounded below

Bounded above
Not bounded below

Bounded

**Figure 1.22** Some examples of graphs bounded and not bounded above and below.

---

**Definition** Lower Bound, Upper Bound, and Bounded

A function $f$ is **bounded below** if there is some number $b$ that is less than or equal to every number in the range of $f$. Any such number $b$ is called a **lower bound** of $f$.

A function $f$ is **bounded above** if there is some number $B$ that is greater than or equal to every number in the range of $f$. Any such number $B$ is called an **upper bound** of $f$.

A function $f$ is **bounded** if it is bounded both above and below.

---

We can extend the above definition to the idea of **bounded on an interval** by restricting the domain of consideration in each part of the definition to the interval we wish to consider. For example, the function $f(x) = 1/x$ is bounded above on the interval $(-\infty, 0)$ and bounded below on the interval $(0, \infty)$.

**Example 7** CHECKING BOUNDEDNESS

Identify each of these functions as bounded below, bounded above, or bounded.

**(a)** $w(x) = 3x^2 - 4$

**(b)** $p(x) = \dfrac{x}{1 + x^2}$

**Solution**

**Solve Graphically**

The two graphs are shown in Figure 1.23.

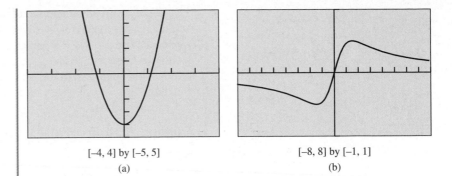

[−4, 4] by [−5, 5]        [−8, 8] by [−1, 1]
(a)                      (b)

**Figure 1.23** The graphs for Example 7. Which are bounded where?

**(a)** The graph of *w* is bounded below. We can confirm our answer algebraically by finding a lower bound, as follows:

$$x^2 \geq 0 \qquad \text{$x^2$ is nonnegative}$$

$$3x^2 \geq 0 \qquad \text{Multiply by 3.}$$

$$3x^2 - 4 \geq 0 - 4 \qquad \text{Subtract 4.}$$

$$3x^2 - 4 \geq -4$$

Thus, $-4$ is a lower bound for $w(x) = 3x^2 - 4$.

**(b)** The graph of *p* is bounded. We leave the verification as an exercise (Exercise 63).

## Local and Absolute Extrema

Many graphs are characterized by peaks and valleys where they change from increasing to decreasing and vice-versa. The extreme values of the function (or *local extrema*) can be characterized as either *local maxima* or *local minima*. The distinction can be easily seen graphically. Figure 1.24 shows a graph with three local extrema: local maxima at points *P* and *R* and a local minimum at *Q*.

This is another function concept that is easier to see graphically than to describe algebraically. Notice that a local maximum does not have to be *the* maximum value of a function; it only needs to be the maximum value of the function on *some* tiny interval.

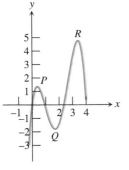

**Figure 1.24** The graph suggests that *f* has a local maximum at *P*, a local minimum at *Q*, and a local maximum at *R*.

> **Definition  Local and Absolute Extrema**
>
> A **local maximum** of a function *f* is a value *f*(*c*) that is greater than or equal to all range values of *f* on some open interval containing *c*. If *f*(*c*) is greater than or equal to all range values of *f*, then *f*(*c*) is the **maximum** (or **absolute maximum**) value of *f*.
>
> A **local minimum** of a function *f* is a value *f*(*c*) that is less than or equal to all range values of *f* on some open interval containing *c*. If *f*(*c*) is less than or equal to all range values of *f*, then *f*(*c*) is the **minimum** (or **absolute minimum**) value of *f*.
>
> Local extrema are also called **relative extrema**.

We have already mentioned that the best method for analyzing increasing and decreasing behavior involves calculus. Not surprisingly, the same is true for local extrema. We will generally be satisfied in this course with approximating local extrema using a graphing calculator, although sometimes an algebraic confirmation will be possible when we learn more about specific functions.

### Example 8  IDENTIFYING LOCAL EXTREMA

Decide whether $f(x) = x^4 - 7x^2 + 6x$ has any local maxima or local minima. If so, find each local maximum value or minimum value and the value of $x$ at which each occurs.

**Solution** The graph of $y = x^4 - 7x^2 + 6x$ (Figure 1.25) suggests that there are two local minimum values and one local maximum value. We use the graphing calculator to approximate local minima as $-24.06$ (which occurs at $x \approx -2.06$) and $-1.77$ (which occurs at $x \approx 1.60$). Similarly, we identify the (approximate) local maximum as $1.32$ (which occurs at $x \approx 0.46$).

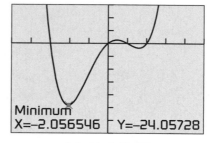

[−5, 5] by [−35, 15]

**Figure 1.25** A graph of $y = x^4 - 7x^2 + 6x$. (Example 8)

### Using a Grapher to Find Local Extrema

Most modern graphers have built in "maximum" and "minimum" finders that identify local extrema by looking for sign changes in $\triangle y$. It is not easy to find local extrema by zooming in on them, as the graphs tend to flatten out and hide the very behavior you are looking for. If you use this method, keep narrowing the vertical window to maintain some curve in the graph.

## Symmetry

In the graphical sense, the word "symmetry" in mathematics carries essentially the same meaning as it does in art: the picture (in this case, the graph) "looks the same" when viewed in more than one way. The interesting thing about mathematical symmetry is that it can be characterized numerically and algebraically as well. We will be looking at three particular types of symmetry, each of which can be spotted easily from a graph, a table of values, or an algebraic formula, once you know what to look for. Since it is the connections among the three models (graphical, numerical, and algebraic) that we need to emphasize in this section, we will illustrate the various symmetries in all three ways, side-by-side.

## Symmetry with respect to the *y*-axis
Example: $f(x) = x^2$

| Graphically | Numerically | Algebraically |

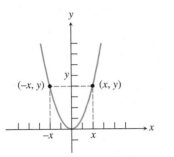

| $x$ | $f(x)$ |
| --- | --- |
| $-3$ | 9 |
| $-2$ | 4 |
| $-1$ | 1 |
| 1 | 1 |
| 2 | 4 |
| 3 | 9 |

For all $x$ in the domain of $f$,

$$f(-x) = f(x)$$

Functions with this property (e.g., $x^n$, $n$ even) are **even** functions.

**Figure 1.26**  The graph looks the same to the left of the *y*-axis as it does to the right of it.

## Symmetry with respect to the *x*-axis
Example: $x = y^2$

| Graphically | Numerically | Algebraically |

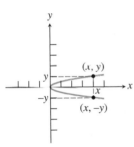

| $x$ | $y$ |
| --- | --- |
| 9 | $-3$ |
| 4 | $-2$ |
| 1 | $-1$ |
| 1 | 1 |
| 4 | 2 |
| 9 | 3 |

Graphs with this kind of symmetry are not usually functions, but we can say that $(x, -y)$ is on the graph whenever $(x, y)$ is on the graph.

**Figure 1.27**  The graph looks the same above the *x*-axis as it does below it.

## Symmetry with respect to the origin
Example: $f(x) = x^3$

| Graphically | Numerically | Algebraically |

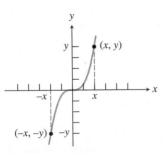

| $x$ | $y$ |
| --- | --- |
| $-3$ | $-27$ |
| $-2$ | $-8$ |
| $-1$ | $-1$ |
| 1 | 1 |
| 2 | 8 |
| 3 | 27 |

For all $x$ in the domain of $f$,

$$f(-x) = -f(x).$$

Functions with this property (e.g., $x^n$, $n$ odd) are **odd** functions.

**Figure 1.28**  The graph looks the same upside-down as it does rightside-up.

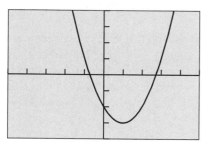

[–5, 5] by [–4, 4]

**Figure 1.30** This graph does not appear to be symmetric with respect to either the $y$-axis or the origin, so we conjecture that $g$ is neither even nor odd.

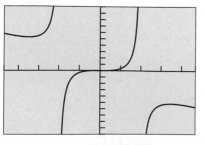

[–4.7, 4.7] by [–10, 10]

**Figure 1.31** This graph appears to be symmetric with respect to the origin, so we conjecture that $h$ is an odd function.

**Example 9   CHECKING FUNCTIONS FOR SYMMETRY**

Tell whether each of the following functions is odd, even, or neither.

**(a)** $f(x) = x^2 - 3$

**(b)** $g(x) = x^2 - 2x - 2$

**(c)** $h(x) = \dfrac{x^3}{4 - x^2}$

**Solution**

**(a) Solve Graphically:**

The graphical solution is shown in Figure 1.29.

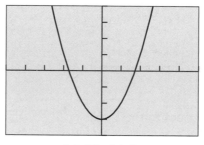

[–5, 5] by [–4, 4]

**Figure 1.29** This graph appears to be symmetric with respect to the $y$-axis, so we conjecture that $f$ is an even function.

**Confirm Algebraically:**

We need to verify that

$$f(-x) = f(x)$$

for all $x$ in the domain of $f$.

$$f(-x) = (-x)^2 - 3 = x^2 - 3$$
$$= f(x)$$

Since this identity is true for all $x$, the function $f$ is indeed even.

**(b) Solve Graphically:**

The graphical solution is shown in Figure 1.30.

**Confirm Algebraically:**

We need to verify that

$$g(-x) \neq g(x) \text{ and } g(-x) \neq -g(x)$$
$$g(-x) = (-x)^2 - 2(-x) - 2$$
$$= x^2 + 2x - 2$$
$$g(x) = x^2 - 2x - 2$$
$$-g(x) = -x^2 + 2x + 2$$

So $g(-x) \neq g(x)$ and $g(-x) \neq -g(x)$

We conclude that $g$ is neither odd nor even.

**(c) Solve Graphically:**

The graphical solution is shown in Figure 1.31.

**Confirm Algebraically:**

We need to verify that

$$h(-x) = -h(x)$$

for all $x$ in the domain of $h$.

$$h(-x) = \frac{(-x)^3}{4 - (-x)^2} = \frac{-x^3}{4 - x^2}$$

$$= -h(x)$$

Since this identity is true for all $x$ except $\pm 2$ (which are not in the domain of $h$), the function $h$ is odd.

## Asymptotes

Consider the graph of the function $f(x) = \dfrac{2x^2}{4 - x^2}$ in Figure 1.32.

The graph appears to flatten out to the right and to the left, getting closer and closer to the horizontal line $y = -2$. We call this line a *horizontal asymptote*. Similarly, the graph appears to flatten out as it goes off the top and bottom of the screen, getting closer and closer to the vertical lines $x = -2$ and $x = 2$. We call these lines *vertical asymptotes*. If we superimpose the asymptotes onto Figure 1.32 as dashed lines, you can see that they form a kind of template that describes the limiting behavior of the graph (Figure 1.33).

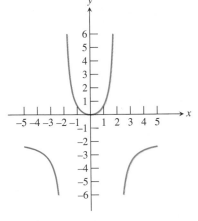

**Figure 1.32** The graph of $f(x) = 2x^2/(4 - x^2)$ has two vertical asymptotes and one horizontal asymptote.

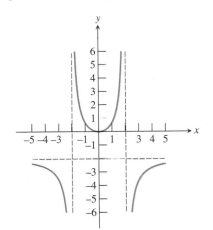

**Figure 1.33** The graph of $f(x) = 2x^2/(4 - x^2)$ with the asymptotes shown as dashed lines.

Since asymptotes describe the behavior of the graph at its horizontal or vertical extremities, the definition of an asymptote can best be stated with limit notation. In this definition, note that $x \to a^-$ means "$x$ approaches $a$ from the left," while $x \to a^+$ means "$x$ approaches $a$ from the right."

> **Definition** Horizontal and Vertical Asymptotes
>
> The line $y = b$ is a **horizontal asymptote** of the graph of a function $y = f(x)$ if $f(x)$ approaches a limit of $b$ as $x$ approaches $+\infty$ or $-\infty$.
>
> In limit notation:
>
> $$\lim_{x \to -\infty} f(x) = b \quad \text{or} \quad \lim_{x \to +\infty} f(x) = b.$$
>
> The line $x = a$ is a **vertical asymptote** of the graph of a function $y = f(x)$ if $f(x)$ approaches a limit of $+\infty$ or $-\infty$ as $x$ approaches $a$ from either direction.
>
> In limit notation:
>
> $$\lim_{x \to a^-} f(x) = \pm\infty \quad \text{or} \quad \lim_{x \to a^+} f(x) = \pm\infty.$$

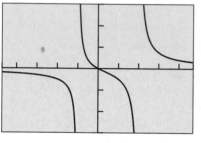

[–4.7, 4.7] by [–3, 3]

**Figure 1.34** The graph of $y = x/(x^2 - x - 2)$ has vertical asymptotes of $x = -1$ and $x = 2$ and a horizontal asymptote of $y = 0$. (Example 10)

### Example 10 IDENTIFYING THE ASYMPTOTES OF A GRAPH

Identify any horizontal or vertical asymptotes of the graph of

$$y = \frac{x}{x^2 - x - 2}.$$

**Solution** The quotient $x/(x^2 - x - 2) = x/((x + 1)(x - 2))$ is undefined at $x = -1$ and $x = -2$, which makes them likely sites for vertical asymptotes. The graph (Figure 1.34) provides support, showing vertical asymptotes of $x = -1$ and $x = 2$.

For large values of $x$, the numerator (a large number) is dwarfed by the denominator (a *product* of *two* large numbers), suggesting that $\lim_{x \to \infty} x/((x + 1)(x - 2)) = 0$. This would indicate a horizontal asymptote of $y = 0$. The graph (Figure 1.34) provides support, showing a horizontal asymptote of $y = 0$ as $x \to \infty$. Similar logic suggests that $\lim_{x \to -\infty} x/((x + 1)(x - 2)) = -0 = 0$, indicating the same horizontal asymptote as $x \to -\infty$. Again, the graph provides support for this.

## End Behavior

A horizontal asymptote gives one kind of end behavior for a function because it shows how the function behaves as it goes off toward either "end" of the $x$-axis. Not all graphs approach lines, but it is helpful to consider what *does* happen "out there." We illustrate with a few examples.

### Example 11 MATCHING FUNCTIONS USING END BEHAVIOR

Match the functions with the graphs in Figure 1.35 by considering end behavior. All graphs are shown in the same viewing window.

**(a)** $y = \dfrac{3x}{x^2 + 1}$ **(b)** $y = \dfrac{3x^2}{x^2 + 1}$ **(c)** $y = \dfrac{3x^3}{x^2 + 1}$ **(d)** $y = \dfrac{3x^4}{x^2 + 1}$

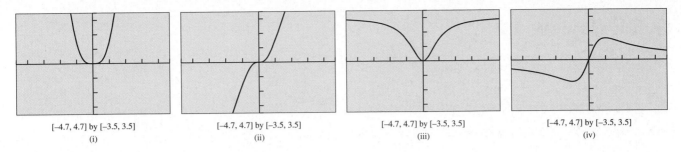

$[-4.7, 4.7]$ by $[-3.5, 3.5]$      $[-4.7, 4.7]$ by $[-3.5, 3.5]$      $[-4.7, 4.7]$ by $[-3.5, 3.5]$      $[-4.7, 4.7]$ by $[-3.5, 3.5]$

(i)          (ii)          (iii)          (iv)

**Figure 1.35** Match the graphs with the functions in Example 11.

**Solution** When $x$ is very large, the denominator $x^2 + 1$ in each of these functions is almost the same number as $x^2$. If we replace $x^2 + 1$ in each denominator by $x^2$ and then reduce the fractions, we get the simpler functions

**(a)** $y = \dfrac{3}{x}$ (close to $y = 0$ for large $x$)   **(b)** $y = 3$   **(c)** $y = 3x$   **(d)** $y = 3x^2$.

So, we look for functions that have end behavior resembling, respectively, the functions

**(a)** $y = 0$   **(b)** $y = 3$   **(c)** $y = 3x$   **(d)** $y = 3x^2$.

Graph (iv) approaches the line $y = 0$. Graph (iii) approaches the line $y = 3$. Graph (ii) approaches the line $y = 3x$. Graph (i) approaches the parabola $y = 3x^2$. So, (a) matches (iv), (b) matches (iii), (c) matches (ii), and (d) matches (i).

For more complicated functions we are often content with knowing whether the end behavior is bounded or unbounded in either direction.

As promised following Example 4 of this section, we will now revisit the task of finding the range of a function.

**Example 12  FINDING THE RANGE OF A FUNCTION (REVISITED)**

Find the range of the function $f(x) = \dfrac{3x^2 - 1}{2x^2 + 1}$.

**Solution**

**Solve Graphically**

Figure 1.36 suggests that $f$ has a minimum value of $-1$ at $x = 0$. It also suggests that the graph of $f$ has a horizontal asymptote of $y = 1.5$. Barring the existence of any hidden behavior, we would conclude that the range of $f$ is the interval $[-1, 1.5)$. In Exercise 72, we ask you to confirm this result algebraically.

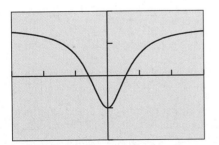

$[-3, 3]$ by $[-2, 2]$

**Figure 1.36** The graph of $f(x) = (3x^2 - 1)/(2x^2 + 1)$. What is its range? (Example 12)

# Quick Review 1.2

In Exercises 1–4, solve the equation or inequality.

**1.** $x^2 - 16 = 0$   $x = \pm 4$     **2.** $9 - x^2 = 0$   $x = \pm 3$

**3.** $x - 10 < 0$   $x < 10$     **4.** $5 - x \le 0$   $x \ge 5$

**7.** $\sqrt{x - 16}$   $x < 16$     **8.** $\dfrac{\sqrt{x^2 + 1}}{x^2 - 1}$   $x = \pm 1$

**9.** $\dfrac{\sqrt{x + 2}}{\sqrt{3 - x}}$   $-2 \le x < 3$     **10.** $\dfrac{x^2 - 2x}{x^2 - 4}$   $x = \pm 2$

In Exercises 5–10, find all values of $x$ algebraically for which the algebraic expression is *not* defined. Support your answer graphically.

**5.** $\dfrac{x}{x - 16}$   $x = 16$     **6.** $\dfrac{x}{x^2 - 16}$   $x = \pm 4$

# Section 1.2 Exercises

In Exercises 1–8, find the domain of the function algebraically and support your answer graphically.

**1.** $f(x) = \sqrt{x^2 + 4}$   $(-\infty, \infty)$   **2.** $h(x) = \dfrac{5}{x - 3}$

**3.** $f(x) = \dfrac{3x - 1}{(x + 3)(x - 1)}$   **4.** $f(x) = \dfrac{1}{x} + \dfrac{5}{x - 3}$

**5.** $g(x) = \dfrac{x}{x^2 - 5x}$   **6.** $h(x) = \dfrac{\sqrt{4 - x^2}}{x - 3}$   $[-2, 2]$

**7.** $h(x) = \dfrac{\sqrt{4 - x}}{(x + 1)(x^2 + 1)}$   **8.** $f(x) = \sqrt{x^4 - 16x^2}$

In Exercises 9–12, use the vertical line test to determine whether the curve is the graph of a function.

**9.**   yes   **10.**   no

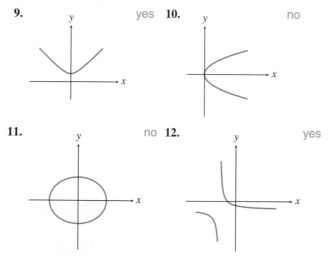

**11.**   no   **12.**   yes

**13. Continuity in the Real World** Tell which of the following functions would be continuous.

  **(a)** Outdoor temperature as a function of time   yes

  **(b)** Price of a stock as a function of time   no

  **(c)** Weight of a baby as a function of time after birth   yes

  **(d)** Cost of U.S. postage as a function of weight of the letter   no

  **(e)** Time of travel between two cities as a function of driving speed   yes

  **(f)** Number of soft drinks sold at a ballpark as a function of outdoor temperature.   no

**14. Increasing and Decreasing in the Real World** Identify each of the following functions as increasing, decreasing, or neither.

  **(a)** Outdoor temperature as a function of time

  **(b)** World population since 1900 as a function of time

  **(c)** Cost of a year at Harvard as a function of time

  **(d)** Time of travel between two cities as a function of driving speed

  **(e)** The Dow Jones Industrial Average as a function of time

  **(f)** Air pressure in earth's atmosphere as a function of altitude

In Exercises 15–18, graph the function and tell whether or not it has a point of discontinuity at $x = 0$. If there is a discontinuity, tell whether it is removable or non-removable.

**15.** $g(x) = \dfrac{3}{x}$     **16.** $h(x) = \dfrac{x^3 + x}{x}$

**17.** $f(x) = \dfrac{|x|}{x}$     **18.** $g(x) = \dfrac{x}{x - 2}$

In Exercises 19–22, state whether each labeled point identifies a local minimum, a local maximum, or neither. Identify intervals on which the function is decreasing and increasing.

**19.**     **20.**

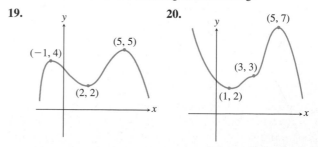

**21.**

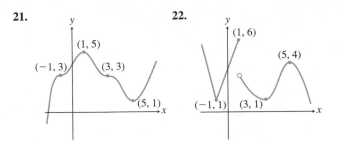

**22.**

**(c)** $h(x) = \dfrac{x^2}{x^3 + 1}$

In Exercises 23–28, use a grapher to find all local maxima and minima and the values of $x$ where they occur. Give values rounded to two decimal places.

**23.** $f(x) = 4 - x + x^2$

**24.** $g(x) = x^3 - 4x + 1$

**25.** $h(x) = -x^3 + 2x - 3$

**26.** $f(x) = (x + 3)(x - 1)^2$

**27.** $h(x) = x^2\sqrt{x + 4}$

**28.** $g(x) = x|2x + 5|$

In Exercises 29–34, graph the function and identify intervals on which the function is increasing, decreasing, or constant.

**29.** $f(x) = |x + 2| - 1$

**30.** $f(x) = |x + 1| + |x - 1| - 3$

**31.** $g(x) = |x + 2| + |x - 1| - 2$

**32.** $h(x) = 0.5(x + 2)^2 - 1$     **33.** $g(x) = 3 - (x - 1)^2$

**34.** $f(x) = x^3 - x^2 - 2x$

In Exercises 35–42, state whether the function is odd, even, or neither. Support graphically and confirm algebraically.

**35.** $f(x) = 2x^4$

**36.** $g(x) = x^3$

**37.** $f(x) = \sqrt{x^2 + 2}$

**38.** $g(x) = \dfrac{3}{1 + x^2}$

**39.** $f(x) = -x^2 + 0.03x + 5$

**40.** $f(x) = x^3 + 0.04x^2 + 3$

**41.** $g(x) = 2x^3 - 3x$

**42.** $h(x) = \dfrac{1}{x}$

In Exercises 43–50, use a method of your choice to find all horizontal and vertical asymptotes of the function.

**43.** $f(x) = \dfrac{x}{x - 1}$  $x = 1; y = 1$  **44.** $q(x) = \dfrac{x - 1}{x}$  $x = 0; y = 1$

**45.** $g(x) = \dfrac{x + 2}{3 - x}$

**46.** $q(x) = 1.5^x$  $y = 0$

**47.** $f(x) = \dfrac{x^2 + 2}{x^2 - 1}$

**48.** $p(x) = \dfrac{4}{x^2 + 1}$

**49.** $g(x) = \dfrac{4x - 4}{x^3 - 8}$

**50.** $h(x) = \dfrac{2x - 4}{x^2 - 4}$

**51. Can a graph cross its own asymptote?** The Greek roots of the word "asymptote" mean "not meeting," since graphs tend to approach, but not meet, their asymptotes. Which of the following functions have graphs that *do* intersect their horizontal asymptotes?

**(a)** $f(x) = \dfrac{x}{x^2 - 1}$

**(b)** $g(x) = \dfrac{x}{x^2 + 1}$

**52. Can a graph have two horizontal asymptotes?** Although most graphs have at most one horizontal asymptote, it is possible for a graph to have more than one. Which of the following functions have graphs with more than one horizontal asymptote?  (a) and (c)

**(a)** $f(x) = \dfrac{|x^3 + 1|}{8 - x^3}$

**(b)** $g(x) = \dfrac{|x - 1|}{x^2 - 4}$

**(c)** $h(x) = \dfrac{x}{\sqrt{x^2 - 4}}$

**53. Can a graph intersect its own vertical asymptote?**

Graph the function $f(x) = \dfrac{x - |x|}{x^2} + 1$.

**(a)** The graph of this function does not intersect its vertical asymptote. Explain why it does not.

**(b)** Show how you can add a single point to the graph of $f$ and get a graph that *does* intersect its vertical asymptote.

**(c)** Is the graph in (b) the graph of a function?

**54. Writing to Learn**  Explain why a graph can not have more than two horizontal asymptotes.

In Exercises 55–60, determine whether the function is bounded above, bounded below, or bounded on its domain

**55.** $y = 32$  bounded     **56.** $y = 2 - x^2$

**57.** $y = 2^x$     **58.** $y = 2^{-x}$

**59.** $y = \sqrt{1 - x^2}$     **60.** $y = x - x^3$

**61.** Find the range of the function $f(x) = \dfrac{x^2}{1 - x^2}$.

**62.** Find the range of the function $g(x) = \dfrac{3 + x^2}{4 - x^2}$.

## Explorations

**63. Bounded Functions**  As promised in Example 7 of this section, we will give you a chance to prove algebraically that $p(x) = x/(1 + x^2)$ is bounded.

**(a)** Graph the function and find the smallest integer $k$ that appears to be an upper bound.  $k = 1$

**(b)** Verify that $x/(1 + x^2) < k$ by proving the equivalent inequality $kx^2 - x + k > 0$. (Use the quadratic formula to show that the quadratic has no real zeros.)

**(c)** From the graph, find the greatest integer $k$ that appears to be a lower bound.  $k = -1$

**(d)** Verify that $x/(1 + x^2) > k$ by proving the equivalent inequality $kx^2 - x + k < 0$.

**64. Baylor School Grade Point Averages**  Baylor School uses a sliding scale to convert the percentage grades on its transcripts to grade point averages (GPAs). Table 1.9 shows the GPA equivalents for selected grades:

| Table 1.9 Converting Grades | |
|---|---|
| Grade ($x$) | GPA ($y$) |
| 60 | 0.00 |
| 65 | 1.00 |
| 70 | 2.05 |
| 75 | 2.57 |
| 80 | 3.00 |
| 85 | 3.36 |
| 90 | 3.69 |
| 95 | 4.00 |
| 100 | 4.28 |

*Source: Baylor School College Counselor*

**(a)** Considering GPA ($y$) as a function of percentage grade ($x$), is it increasing, decreasing, constant, or none of these?

**(b)** Make a table showing the *change* ($\triangle y$) in GPA as you move down the list. (See Exploration 1.)

**(c)** Make a table showing the change in $\triangle y$ as you move down the list. (This is $\triangle\triangle y$.) Considering the *change* ($\triangle y$) in GPA as a function of percentage grade ($x$), is it increasing, decreasing, constant, or none of these?

**(d)** In general, what can you say about the shape of the graph if $y$ is an increasing function of $x$ and $\triangle y$ is a decreasing function of $x$?

**(e)** Sketch the graph of a function $y$ of $x$ such that $y$ is a decreasing function of $x$ and $\triangle y$ is an increasing function of $x$. ◼

**65. Group Activity** Sketch (freehand) a graph of a function with domain all real numbers that satisfies all of the following conditions:

**(a)** $f$ is continuous for all $x$;

**(b)** $f$ is increasing on $(-\infty, 0]$ and on $[3, 5]$;

**(c)** $f$ is decreasing on $[0, 3]$ and on $[5, \infty)$;

**(d)** $f(0) = f(5) = 2$;

**(e)** $f(3) = 0$.

**66. Group Activity** Sketch (freehand) a graph of a function $f$ with domain all real numbers that satisfies all of the following conditions:

**(a)** $f$ is decreasing on $(-\infty, 0)$ and decreasing on $(0, \infty)$;

**(b)** $f$ has a non-removable point of discontinuity at $x = 0$;

**(c)** $f$ has a horizontal asymptote at $y = 1$;

**(d)** $f(0) = 0$;

**(e)** $f$ has a vertical asymptote at $x = 0$.

**67. Group Activity** Sketch (freehand) a graph of a function $f$ with domain all real numbers that satisfies all of the following conditions:

**(a)** $f$ is continuous for all $x$;

**(b)** $f$ is an even function;

**(c)** $f$ is increasing on $[0, 2]$ and decreasing on $[2, \infty)$;

**(d)** $f(2) = 3$.

**68. Group Activity** Get together with your classmates in groups of two or three. Sketch a graph of a function, but do not show it to the other members of your group. Using the language of functions (as in Exercises 65–67), describe your function as completely as you can. Exchange descriptions with the others in your group and see if you can reproduce each other's graphs. Answers vary

## Extending the Ideas

**69.** A function that is bounded above has an infinite number of upper bounds, but there is always a *least upper bound*, i.e., an upper bound that is less than all the others. This least upper bound may or may not be in the range of $f$. For each of the following functions, find the least upper bound and tell whether or not it is in the range of the function.

**(a)** $f(x) = 2 - 0.8x^2$  2. It is in the range.

**(b)** $g(x) = \dfrac{3x^2}{3 + x^2}$  3. It is not in the range.

**(c)** $h(x) = \dfrac{1 - x}{x^2}$  $h(x)$ is not bounded above.

**(d)** $p(x) = 2 \sin(x)$  2. It is in the range.

**(e)** $q(x) = \dfrac{4x}{x^2 + 2x + 1}$  1. It is in the range.

**70.** A continuous function $f$ has domain all real numbers. If $f(-1) = 5$ and $f(1) = -5$, explain why $f$ must have at least one zero in the interval $[-1, 1]$. (This generalizes to a property of continuous functions known as the Intermediate Value Theorem.)

**71. Proving a Theorem** Prove that the graph of every odd function with domain all real numbers must pass through the origin.

**72. Revisiting Example 12** Show algebraically that

$$-1 \le \frac{3x^2 - 1}{2x^2 + 1} \le 1.5$$

for all $x$.

**Ten Basic Functions**

What Graphs Can Tell Us  •  Ten Basic Functions
 •  Analyzing Functions Graphically

## What Graphs Can Tell Us

The preceding section has given us a vocabulary for talking about functions and their properties. We have an entire book ahead of us to study these functions in depth, but in this section we want to set the scene by just *looking* at the graphs of ten "basic" functions that are available on your graphing calculator. You may have seen some of these functions in earlier algebra courses, but that does not matter. Our goal here is to see what can be learned about these functions by just looking at their graphs.

You will find that function attributes like domain, range, continuity, asymptotes, extrema, increasingness, decreasingness, and end behavior are every bit as graphical as they are algebraic. Moreover, the visual cues are often much easier to spot than the algebraic ones.

In future chapters you will learn more about the algebraic properties that make these functions behave as they do. Only then will you able to *prove* what is visually apparent in these graphs. Do not worry about that yet, though; this is intended to be an introduction in which your first impressions can actually be trusted.

## Ten Basic Functions

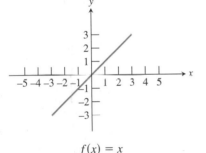

$$f(x) = x$$

Also called: The identity function.
Interesting fact: This is the only function that acts on every real number by leaving it alone.

**Figure 1.37**

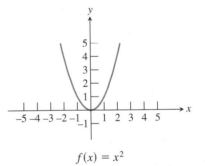

$$f(x) = x^2$$

Interesting fact: The graph of this function, called a parabola, has a reflection property that is useful in making flashlights and satellite dishes.

**Figure 1.38**

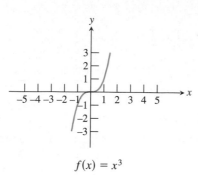

$$f(x) = x^3$$

Interesting fact: The origin is called a "point of inflection" for this curve because the graph changes curvature at that point.

**Figure 1.39**

$$f(x) = \frac{1}{x}$$

Interesting fact: This curve, called a hyperbola, also has a reflection property that is useful in satellite dishes.

**Figure 1.40**

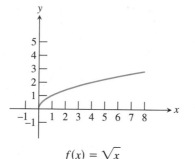

$$f(x) = \sqrt{x}$$

Interesting fact: Put any positive number into your calculator. Take the square root. Then take the square root again. Then take the square root again, and so on. Eventually you will always get 1.

**Figure 1.41**

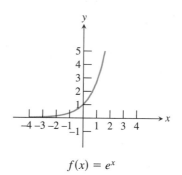

$$f(x) = e^x$$

Interesting fact: The number $e$ is an irrational number (like $\pi$) that shows up in a variety of applications. The symbols $e$ and $\pi$ were both brought into popular use by the great Swiss mathematician Leonhard Euler (1707–1783).

**Figure 1.42**

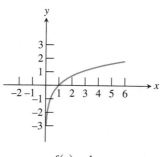

$$f(x) = \ln x$$

Also called: The natural logarithm function.
Interesting fact: This function increases very slowly. If the $x$-axis and $y$-axis were both scaled with unit lengths of one inch, you would have to travel more than two and a half miles along the curve just to get a foot above the $x$-axis.

**Figure 1.43**

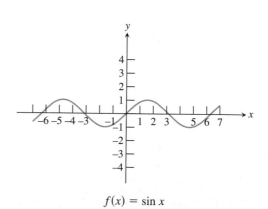

$$f(x) = \sin x$$

Also called: The sine function.
Interesting fact: This function and the sinus cavities in your head derive their names from a common root: the Latin word for "bay." This is due to a 12th-century mistake made by Robert of Chester, who translated a word incorrectly from an Arabic manuscript.

**Figure 1.44**

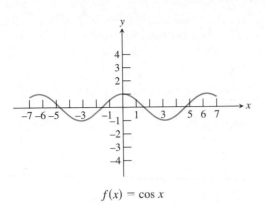

$$f(x) = \cos x$$

Also called: The cosine function.

Interesting fact: The local extrema of the cosine function occur exactly at the zeros of the sine function, and vice-versa.

**Figure 1.45**

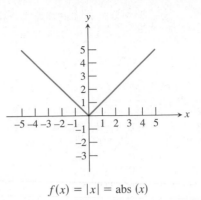

$$f(x) = |x| = \text{abs}\,(x)$$

Also called: The absolute value function.

Interesting fact: This function has an abrupt change of direction (a "corner") at the origin, while our other functions are all "smooth" on their domains.

**Figure 1.46**

### Example 1  LOOKING FOR DOMAINS

**(a)** Seven of the functions have domain the set of all real numbers. Which three do not?

**(b)** One of the functions has domain the set of all reals except 0. Which function is it, and why isn't zero in its domain?

**(c)** Which two functions have no negative numbers in their domains? Of these two, which one is defined at zero?

### Solution

**(a)** Imagine dragging a vertical line along the $x$-axis. If the function has domain the set of all real numbers, then the line will always intersect the graph. The intersection might occur off screen, but the TRACE function on the calculator will show the $y$-coordinate if there is one. Looking at the graphs in Figures 1.40, 1.41, and 1.43, we conjecture that there are vertical lines that do not intersect the curve. A TRACE at the suspected $x$-coordinates confirms our conjecture (Figure 1.47). The functions are $y = 1/x$, $y = \ln x$, and $y = \sqrt{x}$.

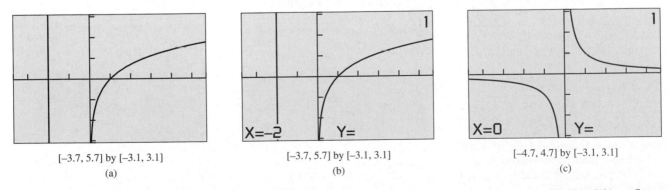

[−3.7, 5.7] by [−3.1, 3.1]
(a)

[−3.7, 5.7] by [−3.1, 3.1]
(b)

[−4.7, 4.7] by [−3.1, 3.1]
(c)

**Figure 1.47** A vertical line through −2 on the $x$-axis (Graph a) appears to miss the graph of $y = \ln x$. A TRACE (Graph b) confirms that −2 is not in the domain. A TRACE at $x = 0$ confirms that 0 is not in the domain of $y = 1/x$ (Graph c). (Example 1)

**(b)** The function $y = 1/x$, with a vertical asymptote at $x = 0$, is defined for all real numbers except 0. This is explained algebraically by the fact that division by zero is undefined.

**(c)** The functions $y = \sqrt{x}$ and $y = \ln x$ have no negative numbers in their domains. (We already knew that about the square root function.) While 0 is in the domain of $y = \sqrt{x}$, we can see by tracing that it is not in the domain of $y = \ln x$. We will see the algebraic reason for this in Chapter 3.

### How Can a Function with a Discontinuity be Continuous?

It might seem strange that a function can have a point of discontinuity and still be called continuous. The same sort of technicality arises when they call you a legal driver. There are things that will make it illegal for you to drive, but it is assumed that those will remain outside your domain.

### Example 2   LOOKING FOR CONTINUITY

Only one of the ten basic functions has a point of discontinuity. Is the point in the domain of the function?

**Solution**  All of the functions have continuous, unbroken graphs except for $y = 1/x$, which clearly has an infinite discontinuity at $x = 0$ (Figure 1.40). We saw in Example 1 that 0 is not in the domain of the function. Since $y = 1/x$ is continuous for every point *in its domain*, it is called a **continuous function.**

### Example 3   LOOKING FOR BOUNDEDNESS

Only two of the ten basic functions are bounded (above and below). Which two?

**Solution**  A function that is bounded must have a graph that lies entirely between two horizontal lines. The graphs of $y = \sin x$ and $y = \cos x$ appear to have this property (Figure 1.48). It looks like the graph of $y = \sqrt{x}$ might also have this property, but we know that the end behavior of the square root function is unbounded: $\lim\limits_{x \to \infty} \sqrt{x} = \infty$, so it is really only bounded below. You will learn in Chapter 4 why the sine and cosine functions are bounded.

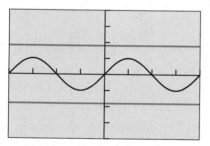

$[-2\pi, 2\pi]$ by $[-4, 4]$

(a)

### Example 4   LOOKING FOR SYMMETRY

Three of the ten basic functions are even. Which are they?

**Solution**  Recall that the graph of an even function is symmetric with respect to the $y$-axis. Three of the functions exhibit the required symmetry: $y = x^2$, $y = \cos x$, and $y = |x|$ (Figure 1.49).

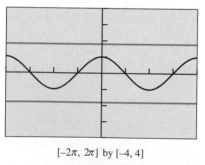

$[-2\pi, 2\pi]$ by $[-4, 4]$

(b)

**Figure 1.48** The graphs of $y = \sin x$ and $y = \cos x$ lie entirely between two horizontal lines and are therefore graphs of bounded functions. (Example 3)

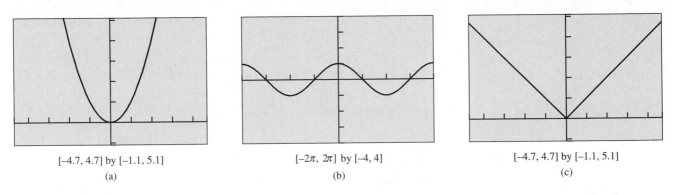

[−4.7, 4.7] by [−1.1, 5.1]
(a)

[−2π, 2π] by [−4, 4]
(b)

[−4.7, 4.7] by [−1.1, 5.1]
(c)

**Figure 1.49** The graphs of $y = x^2$, $y = \cos x$, and $y = |x|$ are symmetric with respect to the $y$-axis, indicating that the functions are even. (Example 4)

### Example 5  IDENTIFYING A PIECEWISE-DEFINED FUNCTION

Which of the ten basic functions has the following **piecewise** definition over separate intervals of its domain?

$$f(x) = \begin{cases} x & \text{if } x \geq 0 \\ -x & \text{if } x < 0 \end{cases}$$

**Solution**  You may recognize this as the definition of the absolute value function (Chapter P). Or, you can reason that the graph of this function must look just like the line $y = x$ to the right of the $y$-axis, but just like the graph of the line $y = -x$ to the left of the $y$-axis. That is a perfect description of the absolute value graph in Figure 1.46. Either way, we recognize this as a piecewise definition of $f(x) = |x|$.

## Analyzing Functions Graphically

We could continue to explore the ten basic functions as in the first four examples, but we also want to make the point that there is no need to restrict ourselves to the basic ten. We can alter the basic functions slightly and see what happens to their graphs, thereby gaining further visual insights into how functions behave.

### Example 6  ANALYZING A FUNCTION GRAPHICALLY

Graph the function $y = (x - 2)^2$. Then answer the following questions:

**(a)** On what interval is the function increasing? On what interval is it decreasing?

**(b)** Is the function odd, even, or neither?

**(c)** Does the function have any extrema?

**(d)** How does the graph relate to the graph of the basic function $y = x^2$?

**Solution**  The graph is shown in Figure 1.50.

**(a)** The function is increasing if its graph is headed upward as it moves from left to right. We see that it is increasing on the interval $[2, \infty)$. The function is decreasing if its graph is headed downward as it moves from left to right. We see that it is decreasing on the interval $(-\infty, 2]$.

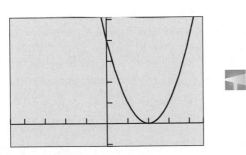

[−4.7, 4.7] by [−1.1, 5.1]

**Figure 1.50** The graph of $y = (x - 2)^2$. (Example 6)

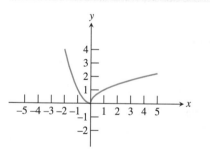

**Figure 1.51** A piecewise-defined function. (Example 7)

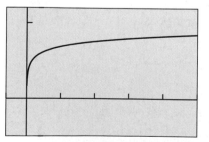

[–600, 5000] by [–5, 12]

**Figure 1.52** The graph of $y = \ln x$ still appears to have a horizontal asymptote, despite the much larger window than in Figure 1.43. (Example 8)

**(b)** The graph is not symmetric with respect to the $y$-axis, nor is it symmetric with respect to the origin. The function is neither.

**(c)** Yes, we see that the function has a minimum value of 0 at $x = 2$. (This is easily confirmed by the algebraic fact that $(x - 2)^2 \geq 0$ for all $x$.)

**(d)** We see that the graph of $y = (x - 2)^2$ is just the graph of $y = x^2$ moved two units to the right.

---

**Exploration 1** Looking for Asymptotes

**1.** Two of the basic functions have vertical asymptotes at $x = 0$. Which two? $f(x) = 1/x$, $f(x) = \ln x$

**2.** Form a new function by adding these functions together. Does the new function have a vertical asymptote at $x = 0$? Yes

**3.** Two of the basic functions have horizontal asymptotes at $y = 0$. Which two? $f(x) = 1/x$, $f(x) = e^x$

**4.** Form a new function by adding these functions together. Does the new function have a horizontal asymptote at $y = 0$? Yes

---

**Example 7** DEFINING A FUNCTION PIECEWISE

Using basic functions from this section, construct a piecewise definition for the function whose graph is shown in Figure 1.51. Is your function continuous?

**Solution** This appears to be the graph of $y = x^2$ to the left of $x = 0$ and the graph of $y = \sqrt{x}$ to the right of $x = 0$. We can therefore define it piecewise as

$$f(x) = \begin{cases} x^2 & \text{if } x \leq 0 \\ \sqrt{x} & \text{if } x > 0 \end{cases}$$

The function is continuous.

You can go a long way toward understanding a function's behavior by looking at its graph. We will continue that theme in the exercises and then revisit it throughout the book. However, you can't go *all* the way toward understanding a function by looking at its graph, as our final example shows.

**Example 8** LOOKING FOR A HORIZONTAL ASYMPTOTE

Does the graph of $y = \ln x$ (Figure 1.43) have a horizontal asymptote?

**Solution** In Figure 1.43 it certainly *looks* like there is a horizontal asymptote that the graph is approaching from below. If we choose a much larger window (Figure 1.52), it still looks that way. In fact, we could zoom out on this function all day long and it would *always* look like it is approaching some horizontal asymptote—but it is not. We will show algebraically in Chapter 3 that the end behavior of this function is $\lim_{x \to \infty} \ln x = \infty$, so its graph must eventually rise above the level of any horizontal line. That rules out any horizontal asymptote, even though there is no *visual* evidence of that fact that we can see by looking at its graph.

# Quick Review 1.3

In Exercises 1–10, evaluate the expression without using a calculator.

**1.** $|-59.34|$  59.34

**2.** $|5 - \pi|$  $5 - \pi$

**3.** $|\pi - 7|$  $7 - \pi$

**4.** $\sqrt{(-3)^2}$  3

**5.** $\ln(1)$  0

**6.** $e^0$  1

**7.** $(\sqrt[3]{3})^3$  3

**8.** $\sqrt[3]{(-15)^3}$  $-15$

**9.** $\sqrt[3]{-8^2}$  $-4$

**10.** $|1 - \pi| - \pi$  $-1$

# Section 1.3 Exercises

In Exercises 1–10, each graph is a slight variation on the graph of one of the ten basic functions described in this section. Match the graph to one of the ten functions (a)–(j) and then support your answer by checking the graph on your calculator. (All graphs are shown in the window $[-4.7, 4.7]$ by $[-3.1, 3.1]$.)

**(a)** $y = -\sin x$
**(b)** $y = \cos x + 1$
**(c)** $y = e^x - 2$

**(d)** $y = (x + 2)^3$
**(e)** $y = x^3 + 1$
**(f)** $y = (x - 1)^2$

**(g)** $y = |x| - 2$
**(h)** $y = -1/x$
**(i)** $y = -x$

**(j)** $y = -\sqrt{x}$

**1.**

$[-4.7, 4.7]$ by $[-3.1, 3.1]$

**2.**

$[-4.7, 4.7]$ by $[-3.1, 3.1]$

**3.**

$[-4.7, 4.7]$ by $[-3.1, 3.1]$

**4.**

$[-4.7, 4.7]$ by $[-3.1, 3.1]$

**5.**

$[-4.7, 4.7]$ by $[-3.1, 3.1]$

**6.**

$[-4.7, 4.7]$ by $[-3.1, 3.1]$

**7.**

$[-4.7, 4.7]$ by $[-3.1, 3.1]$

**8.**

$[-4.7, 4.7]$ by $[-3.1, 3.1]$

**9.**

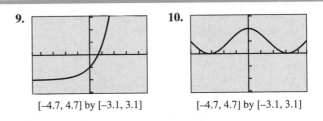

$[-4.7, 4.7]$ by $[-3.1, 3.1]$

**10.**

$[-4.7, 4.7]$ by $[-3.1, 3.1]$

In Exercises 11–20, identify which of the ten basic functions fit the description given.

**11.** The four functions that are odd.

**12.** The five functions that are increasing on their entire domains.  $y = x$, $y = x^3$, $y = \sqrt{x}$, $y = e^x$, $y = \ln x$

**13.** The three functions that are decreasing on the interval $(-\infty, 0)$.  $y = x^2$, $y = 1/x$, $y = |x|$

**14.** The two functions with infinitely many local extrema.

**15.** The two functions with no zeros.  $y = 1/x$, $y = e^x$

**16.** The three functions with range {all real numbers}.

**17.** The three functions that do *not* have end behavior $\lim\limits_{x \to +\infty} f(x) = +\infty$.  $y = 1/x$, $y = \sin x$, $y = \cos x$

**18.** The two functions with end behavior $\lim\limits_{x \to -\infty} f(x) = -\infty$.

**19.** The four functions whose graphs look the same when turned upside-down and flipped about the $y$-axis.

**20.** The two functions whose graphs are identical except for a horizontal shift.  $y = \sin x$, $y = \cos x$

In Exercises 21–30, use your graphing calculator to produce a graph of the function. Then determine the domain and range of the function by looking at its graph.

**21.** $f(x) = x^2 - 5$

**22.** $g(x) = |x - 4|$

**23.** $h(x) = \ln(x + 6)$

**24.** $k(x) = 1/x + 3$

**25.** $p(x) = (x + 3)^2$

**26.** $r(x) = \sqrt{x - 10}$

**27.** $f(x) = \sin(x) + 5$

**28.** $q(x) = e^x + 2$

**29.** $h(x) = |x| - 10$

**30.** $g(x) = 4\cos(x)$

In Exercises 31–36, sketch the graph of the piecewise-defined function. (Try doing it without a calculator.) In each case, state whether the function is continuous or discontinuous at $x = 0$.

**31.** $f(x) = \begin{cases} x & \text{if } x \leq 0 \\ x^2 & \text{if } x > 0 \end{cases}$

**32.** $g(x) = \begin{cases} x^3 & \text{if } x \leq 0 \\ e^x & \text{if } x > 0 \end{cases}$

**33.** $h(x) = \begin{cases} |x| & \text{if } x < 0 \\ \sin x & \text{if } x \geq 0 \end{cases}$ **34.** $w(x) = \begin{cases} 1/x & \text{if } x < 0 \\ \sqrt{x} & \text{if } x \geq 0 \end{cases}$

**35.** $f(x) = \begin{cases} \cos x & \text{if } x \leq 0 \\ e^x & \text{if } x > 0 \end{cases}$ **36.** $g(x) = \begin{cases} |x| & \text{if } x < 0 \\ x^2 & \text{if } x \geq 0 \end{cases}$

**37. Writing to Learn** The function $f(x) = \sqrt{x^2}$ is one of our ten basic functions written in another form.

**(a)** Graph the function and identify which basic function it is.

**(b)** Explain algebraically why the two functions are equal.

**38. Uncovering Hidden Behavior** The function

$g(x) = \sqrt{x^2 + 0.0001} - 0.01$ is *not* one of our ten basic functions written in another form.

**(a)** Graph the function and identify which basic function it appears to be.

**(b)** Verify numerically that it is not the basic function that it appears to be.

**39. Writing to Learn** The function $f(x) = \ln(e^x)$ is one of our ten basic functions written in another form.

**(a)** Graph the function and identify which basic function it is.

**(b)** Explain how the equivalence of the two functions in (a) shows that the natural logarithm function is *not* bounded above (even though it *appears* to be bounded above in Figure 1.43).

## Explorations

**40. Which is Bigger?** For positive values of $x$, we wish to compare the values of the basic functions $x^2$, $x$, and $\sqrt{x}$.

**(a)** How would you order them, from least to greatest?

**(b)** Graph the three functions in the viewing window [0, 30] by [0, 20]. Does the graph confirm your response in (a)?

**(c)** Now graph the three functions in the viewing window [0, 2] by [0, 1.5].

**(d)** Write a careful response to the question in (a) that accounts for all positive values of $x$.

**41. Odds and Evens** There are four odd functions and three even functions in the gallery of Ten Basic Functions. After multiplying these functions together pairwise in different combinations and exploring the graphs of the products, make a conjecture about the symmetry of:

**(a)** a product of two odd functions.

**(b)** a product of two even functions.

**(c)** a product of an odd function and an even function.

**42. Group Activity** Assign to each student in the class the name of one of the ten basic functions, but secretly so that no student knows the "name" of another. (The same function name could be given to several students, but all the functions should be used at least once.) Let each student make a one-sentence self-introduction to the class that reveals something personal "about who I am that really identifies me." The rest

of the students then write down their guess as to the function's identity. Hints should be subtle and cleverly anthropomorphic. (For example, the absolute value function saying "I have a very sharp smile" is subtle and clever, while "I am absolutely valuable" is not very subtle at all.)

**43. Pepperoni pizzas** For a statistics project, a student counted the number of pepperoni slices on pizzas of various sizes at a local pizzeria, compiling the following table:

**Table 1.10**

| Type of Pizza | Radius | Pepperoni count |
|---|---|---|
| Personal | 4" | 12 |
| Medium | 6" | 27 |
| Large | 7" | 37 |
| Extra Large | 8" | 48 |

**(a)** Explain why the pepperoni count ($P$) ought to be proportional to the square of the radius ($r$).

**(b)** Assuming that $P = k \cdot r^2$, use the data pair (4, 12) to find the value of $k$.  0.75

**(c)** Does the algebraic model fit the rest of the data well?

**(d)** Some pizza places have charts showing their kitchen staff how much of each topping should be put on each size of pizza. Do you think this pizzeria uses such a chart? Explain ∎

## Extending the Ideas

**44. Inverse Functions** Two functions are said to be *inverses* of each other if the graph of one can be obtained from the graph of the other by reflecting it across the line $y = x$. For example, the functions with the graphs shown below are inverses of each other:

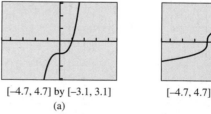

[−4.7, 4.7] by [−3.1, 3.1]    [−4.7, 4.7] by [−3.1, 3.1]
(a)    (b)

**(a)** Two of the Ten Basic Functions in this section are inverses of each other. Which are they? $y = e^x$ and $y = \ln x$

**(b)** Two of the Ten Basic Functions in this section are their own inverses. Which are they? $y = x$ and $y = 1/x$

**(c)** If you restrict the domain of one of the Ten Basic Functions to [0, ∞), it becomes the inverse of another one. Which are they?

**45. Identifying a Function by Its Properties**

**(a)** Six of the Ten Basic Functions have the property that $f(0) = 0$. Which four do not?

**(b)** Only one of the Ten Basic Functions has the property that $f(x + y) = f(x) + f(y)$ for all $x$ and $y$ in its domain. Which one is it? $f(x) = x$

**(c)** One of the Ten Basic Functions has the property that $f(x + y) = f(x)f(y)$ for all $x$ and $y$ in its domain. Which one is it? $f(x) = e^x$

**(d)** One of the Ten Basic Functions has the property that $f(xy) = f(x) + f(y)$ for all $x$ and $y$ in its domain. Which one is it? $f(x) = \ln x$

**(e)** Four of the Ten Basic Functions have the property that $f(x) + f(-x) = 0$ for all $x$ in their domains. Which four are they?

**46. The Greatest Integer Function** Another function available on your calculator is the greatest integer function, $y = \text{int}(x)$. (It is often written as $y = [x]$.) Set your calculator to DOT mode and graph this function in the window $[-4.7, 4.7]$ by $[-3.1, 3.1]$. Then answer the following questions.

**(a)** What appear to be the domain and range of this function? Domain: all reals. Range: all integers

**(b)** Is this function continuous? How can you tell? No. It has jump discontinuities at all integers, which are in the domain of the function.

---

## 1.4 Building Functions from Functions

Combining Functions Algebraically • Composition of Functions • Relations and Implicitly Defined Functions • Relations Defined Parametrically • Inverse Relations and Inverse Functions

## Combining Functions Algebraically

Knowing how a function is "put together" is an important first step when applying the tools of calculus. Functions have their own algebra based on the same operations we apply to real numbers (addition, subtraction, multiplication, and division). One way to build new functions is to apply these operations, using the following definitions.

---

**Definition** Sum, Difference, Product, and Quotient of Functions

Let $f$ and $g$ be two functions with intersecting domains. Then for all values of $x$ in the intersection, the algebraic combinations of $f$ and $g$ are defined by the following rules:

**Sum:** $\qquad (f + g)(x) = f(x) + g(x)$

**Difference:** $\qquad (f - g)(x) = f(x) - g(x)$

**Product:** $\qquad (fg)(x) = f(x)g(x)$

**Quotient:** $\qquad \left(\dfrac{f}{g}\right)(x) = \dfrac{f(x)}{g(x)}$, provided $g(x) \neq 0$

In each case, the domain of the new function consists of all numbers that belong to both the domain of $f$ and the domain of $g$. As noted, the zeros of the denominator are excluded from the domain of the quotient.

---

Euler's function notation works so well in the above definitions that it almost obscures what is really going on. The "+" in the expression "$(f + g)(x)$" stands for a brand new operation called *function addition*. It builds a new function, $f + g$, from the given functions $f$ and $g$. Like any function, $f + g$ is defined by what it does: it takes a domain value $x$ and returns a range value $f(x) + g(x)$. Note that the "+" sign in "$f(x) + g(x)$" *does* stand for the familiar operation of real number addition. So, with the same symbol taking on different roles on either side of the equal sign, there is more to the above definitions than first meets the eye.

Fortunately, the definitions are easy to apply.

### Example 1    DEFINING NEW FUNCTIONS ALGEBRAICALLY

**Lesson Guide**

Day 1:Skills
Day 2: Applications

Let $f(x) = x^2$ and $g(x) = \sqrt{x + 1}$

Find formulas for the functions $f + g, f - g, fg, f/g$, and $gg$. Give the domain of each.

**Solution**  We first determine that $f$ has domain all real numbers and that $g$ has domain $[-1, \infty)$. These domains overlap, the intersection being the interval $[-1, \infty)$. So:

$$(f + g)(x) = f(x) + g(x) = x^2 + \sqrt{x + 1} \quad \text{with domain } [-1, \infty).$$
$$(f - g)(x) = f(x) - g(x) = x^2 - \sqrt{x + 1} \quad \text{with domain } [-1, \infty).$$
$$(fg)(x) = f(x)g(x) = x^2\sqrt{x + 1} \quad \text{with domain } [-1, \infty).$$
$$\left(\frac{f}{g}\right)(x) = \frac{f(x)}{g(x)} = \frac{x^2}{\sqrt{x + 1}} \quad \text{with domain } (-1, \infty).$$
$$(gg)(x) = g(x)g(x) = (\sqrt{x + 1})^2 \quad \text{with domain } [-1, \infty).$$

Note that we could express $(gg)(x)$ more simply as $x + 1$. That would be fine, but the simplification would not change the fact that the domain of $gg$ is (by definition) the interval $[-1, \infty)$. Under other circumstances the function $f(x) = x + 1$ would have domain all real numbers, but under these circumstances it cannot; it is a product of two functions with restricted domains.

## Composition of Functions

**Alert**

Students must understand how the composition notation works. The composition notation $(f \circ g)(x)$ should be compared to the $f(g(x))$ notation so that students can move from one notation to the other with facility. You may wish to read $f \circ g$ as "$f$ following $g$," to emphasize the order in which the functions are applied. Be aware that not all textbooks use the same convention.

It is not hard to see that the function $\sin(x^2)$ is built from the basic functions $\sin x$ and $x^2$, but the functions are not put together by addition, subtraction, multiplication, or division. Instead, the two functions are combined by simply applying them in order—first the squaring function, then the sine function. This operation for combining functions, which has no counterpart in the algebra of real numbers, is called *function composition*.

> **Definition**  **Composition of Functions**
>
> Let $f$ and $g$ be two functions such that the domain of $f$ intersects the range of $g$. The **composition $f$ of $g$**, denoted $f \circ g$, is defined by the rule
>
> $$(f \circ g)(x) = f(g(x)).$$
>
> The domain of $f \circ g$ consists of all $x$-values in the domain of $g$ that map to $g(x)$-values in the domain of $f$. (See Figure 1.53.)

The composition $g$ of $f$, denoted $g \circ f$, is defined similarly. In most cases $g \circ f$ and $f \circ g$ are different functions. (In the language of algebra, "function composition is not a commutative operation.")

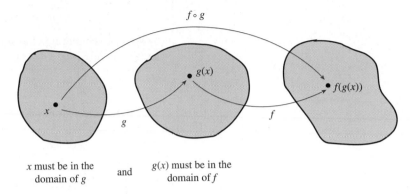

$f \circ g$

$g(x)$

$f(g(x))$

$x$

$g$

$f$

$x$ must be in the domain of $g$    and    $g(x)$ must be in the domain of $f$

**Figure 1.53** In the composition $f \circ g$, the function $g$ is applied first and then $f$. This is the reverse of the order in which we read the symbols.

## Example 2 COMPOSING FUNCTIONS

Let $f(x) = e^x$ and $g(x) = \sqrt{x}$. Find $(f \circ g)(x)$ and $(g \circ f)(x)$ and verify that the functions $f \circ g$ and $g \circ f$ arc not the same.

**Solution**

$$(f \circ g)(x) = f(g(x)) = f(\sqrt{x}) = e^{\sqrt{x}}$$
$$(g \circ f)(x) = g(f(x)) = g(e^x) = \sqrt{e^x}$$

One verification that these functions are not the same is that they have different domains: $f \circ g$ is defined only for $x \geq 0$, while $g \circ f$ is defined for all real numbers. We could also consider their graphs (Figure 1.54), which agree only at $x = 0$ and $x = 4$.

Finally, the graphs suggest a numerical verification: find a single value of $x$ for which $f(g(x))$ and $g(f(x))$ give different values. For example, $f(g(1)) = e$ and $g(f(1)) = \sqrt{e}$. The graph helps us to make a judicious choice of $x$. You do not want to check the functions at $x = 0$ and $x = 4$ and conclude that they are the same!

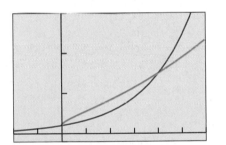

[–2, 6] by [–1, 15]

**Figure 1.54** The graphs of $y = e^{\sqrt{x}}$ and $y = \sqrt{e^x}$ are not the same. (Example 2).

In Example 2 two functions were *composed* to form new functions. There are times in calculus when we need to reverse the process. That is, we may begin with a function $h$ and *decompose* it by finding functions whose composition is $h$.

## Example 3 DECOMPOSING FUNCTIONS

For each function $h$, find functions $f$ and $g$ such that $h(x) = f(g(x))$.

**(a)** $h(x) = (x + 1)^2 - 3(x + 1) + 4$

**(b)** $h(x) = \sqrt{x^3 + 1}$

## Solution

**(a)** We can see that $h$ is quadratic in $x + 1$. Let $f(x) = x^2 - 3x + 4$ and let $g(x) = x + 1$. Then

$$h(x) = f(g(x)) = f(x + 1) = (x + 1)^2 - 3(x + 1) + 4.$$

**(b)** We can see that $h$ is the square root of the function $x^3 + 1$. Let $f(x) = \sqrt{x}$ and let $g(x) = x^3 + 1$. Then

$$h(x) = f(g(x)) - f(x^3 + 1) = \sqrt{x^3 + 1}.$$

There is often more than one way to decompose a function. For example, an alternate way to decompose $h(x) = \sqrt{x^3 + 1}$ in Example 3b is to let $f(x) = \sqrt{x + 1}$ and let $g(x) = x^3$. Then $h(x) = f(g(x)) = f(x^3) = \sqrt{x^3 + 1}$.

### Example 4 MODELING WITH FUNCTION COMPOSITION

In the medical procedure known as angioplasty, doctors insert a catheter into a heart vein (through a large peripheral vein) and inflate a small, spherical balloon on the tip of the catheter. Suppose the balloon is inflated at a constant rate of 44 cubic millimeters per second. (See Figure 1.55.)

**(a)** Find the volume after 5 seconds.

**(b)** Find the radius after 5 seconds.

**(c)** The balloon is ordinarily inflated to a diameter of no more than 9 mm. How long will it take for the balloon to inflate to a diameter of 9 mm?

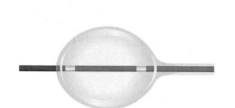

**Figure 1.55** (Example 4)

### Solution

**(a)** The volume will be $44t$ after $t$ seconds. At $t = 5$, the volume will be 220 mm³.

**(b) Solve Algebraically**

$$\frac{4}{3}\pi r^3 = 220$$

$$\pi r^3 = 165$$

$$r = \sqrt[3]{\frac{165}{\pi}} \approx 3.745$$

### Interpret

The radius after 5 seconds will be about 3.75 mm.

**(c)** The volume of a balloon with diameter 9 mm will be

$$\frac{4}{3}\pi(4.5)^3 = \frac{364.5\pi}{3} \approx 381.7 \text{ mm}^3.$$

The time to inflate the balloon to that volume will be

$$\frac{381.7 \text{ mm}^3}{44 \text{ mm}^3 / \text{sec}} \approx 8.7 \text{ seconds}.$$

### Notes on Examples

Example 5b emphasizes that the domain of a composite function cannot always be determined by merely looking at the final function equation.

### Example 5 FINDING THE DOMAIN OF A COMPOSITION

Let $f(x) = x^2 - 1$ and let $g(x) = \sqrt{x}$. Find the domains of the composite functions

**(a)** $g \circ f$                                         **(b)** $f \circ g$

## Solution

**(a)** We compose the functions in the order specified:

$$(g \circ f)(x) = g(f(x))$$
$$= \sqrt{x^2 - 1}$$

For $x$ to be in the domain of $g \circ f$, we must first find $f(x) = x^2 - 1$, which we can do for all real $x$. Then we must take the square root of the result, which we can only do for non-negative values of $x^2 - 1$.

Therefore, the domain of $g \circ f$ consists of all real numbers for which $x^2 - 1 \geq 0$, namely the union $(-\infty, -1] \cup [1, \infty)$.

**(b)** Again, we compose the functions in the order specified:

$$(f \circ g)(x) = f(g(x))$$
$$= (\sqrt{x})^2 - 1$$

For $x$ to be in the domain of $f \circ g$, we must first find $g(x) = \sqrt{x}$, which we can only do for non-negative values of $x$. Then we must square the result and subtract 1, which we can do for all real numbers.

Therefore, the domain of $f \circ g$ consists of the interval $[0, \infty)$.

### Support Graphically

We can graph the composition functions to see if the grapher respects the domain restrictions. The screen to the left of each graph shows the set-up in the "Y=" editor. Figure 1.56b shows the graph of $y = (g \circ f)(x)$, while Figure 1.56d shows the graph of $y = (f \circ g)(x)$. The graphs support our algebraic work quite nicely.

> **Caution**
>
> We might choose to express $(f \circ g)$ more simply as $x - 1$. However, you must remember that the domain of the composition is the same as the domain of $g(x) = \sqrt{x}$ or $[0, \infty)$. The domain of $x - 1$ is all real numbers. It is a good idea to work out the domain of a composition before you simpify the expression for $f(g(x))$. One way to simplify and maintain the restriction on the domain in Example 5 is to write $(f \circ g) = x - 1, x \geq 0$.

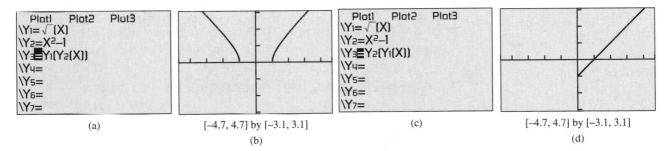

(a)

[−4.7, 4.7] by [−3.1, 3.1]
(b)

(c)

[−4.7, 4.7] by [−3.1, 3.1]
(d)

**Figure 1.56** The functions Y1 and Y2 are composed to get the graphs of $y = (g \circ f)(x)$ and $y = (f \circ g)(x)$, respectively. The graphs support our conclusions about the domains of the two composite functions. (Example 5)

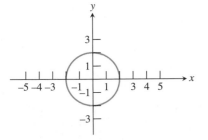

**Figure 1.57** A circle of radius 2 centered at the origin. This set of ordered pairs $(x, y)$ defines a *relation* that is not a function, because the graph fails the vertical line test.

## Relations and Implicitly Defined Functions

There are many useful curves in mathematics that fail the vertical line test and therefore are not graphs of functions. One such curve is the circle in Figure 1.57. While $y$ is not related to $x$ as a function in this instance, there is certainly some sort of relationship going on. In fact, not only does the shape of the graph show a significant *geometric* relationship among the points, but the ordered pairs $(x, y)$ exhibit a significant *algebraic* relationship as well: they consist exactly of the solutions to the equation $x^2 + y^2 = 4$.

The general term for a set of ordered pairs $(x, y)$ is a **relation**. If the relation happens to relate a *single* value of $y$ to each value of $x$, then the relation is also a function and its graph will pass the vertical line test. In the case of the circle with equation $x^2 + y^2 = 4$, both $(0, 2)$ and $(0, -2)$ are in the relation, so $y$ is not a function of $x$.

### Example 6   VERIFYING PAIRS IN A RELATION

Determine which of the ordered pairs $(2, -5)$, $(1, 3)$, and $(2, 1)$ are in the relation defined by $x^2y + y^2 = 5$. Is the relation a function?

**Solution**  We simply substitute the $x$- and $y$-coordinates of the ordered pairs into $x^2y + y^2$ and see if we get 5.

$(2, -5)$:    $(2)^2(-5) + (-5)^2 = 5$    Substitute $x = 2$, $y = -5$.

$(1, 3)$:    $(1)^2(3) + (3)^2 = 12 \neq 5$    Substitute $x = 1$, $y = 3$.

$(2, 1)$:    $(2)^2(1) + (1)^2 = 5$    Substitute $x = 2$, $y = 1$.

So, $(2, -5)$ and $(2, 1)$ are in the relation, but $(1, 3)$ is not.

Since the equation relates two different $y$-values ($-5$ and 1) to the same $x$-value (2), the relation cannot be a function.

Let us revisit the circle $x^2 + y^2 = 4$. While it is not a function itself, we can split it into two equations that *do* define functions, as follows:

$$x^2 + y^2 = 4$$
$$y^2 = 4 - x^2$$
$$y = +\sqrt{4 - x^2} \text{ or } y = -\sqrt{4 - x^2}$$

The graphs of these two functions are, respectively, the upper and lower semicircles of the circle in Figure 1.57. They are shown in Figure 1.58. Since all the ordered pairs in either of these functions satisfy the equation $x^2 + y^2 = 4$, we say that the relation defines the two functions **implicitly**.

### Example 7   USING IMPLICITLY DEFINED FUNCTIONS

Describe the graph of the relation $x^2 + 2xy + y^2 = 1$.

**Solution**  This looks like a difficult task at first, but notice that the expression on the left of the equal sign is a factorable trinomial. This enables us to split the function into two implicitly defined functions as follows:

$$x^2 + 2xy + y^2 = 1$$
$$(x + y)^2 = 1 \qquad \text{Factor.}$$
$$x + y = \pm 1 \qquad \text{Extract square roots.}$$

$x + y = 1$        or    $x + y = -1$

$y = -x + 1$    or        $y = -x - 1$    Solve for $y$.

The graph consists of two parallel lines (Figure 1.59), each the graph of one of the implicitly defined functions.

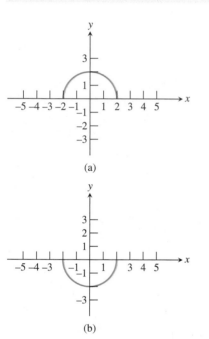

(a)

(b)

**Figure 1.58**  The graphs of
(a) $y = +\sqrt{4 - x^2}$ and
(b) $y = -\sqrt{4 - x^2}$. In each case, $y$ is defined as a function of $x$. These two functions are defined *implicitly* by the relation $x^2 + y^2 = 4$.

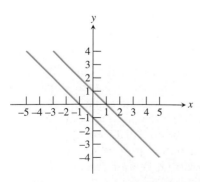

**Figure 1.59**  The graph of the relation $x^2 + 2xy + y^2 = 1$. (Example 7)

## Relations Defined Parametrically

Another natural way to define functions or, more generally, relations, is to define *both* elements of the ordered pair $(x, y)$ in terms of another variable $t$, called a **parameter**. We illustrate with an example.

### Example 8 DEFINING A FUNCTION PARAMETRICALLY

Consider the set of all ordered pairs $(x, y)$ defined by the equations

$$x = t + 1$$
$$y = t^2 + 2t$$

where $t$ is any real number.

**(a)** Find the points determined by $t = -3, -2, -1, 0, 1, 2,$ and 3.

**(b)** Does this relation define $y$ as a function of $x$?

**(c)** Graph the relation in the $(x, y)$ plane.

**(d)** Use the table feature of a grapher set in parametric mode to compute the values of $x$ and $y$ found in (a).

#### Solution

**(a)** Substitute each value of $t$ into the formulas for $x$ and $y$ to find the point that it determines parametrically:

| $t$ | $x = t + 1$ | $y = t^2 + 2t$ | $(x, y)$ |
|---|---|---|---|
| $-3$ | $-2$ | 3 | $(-2, 3)$ |
| $-2$ | $-1$ | 0 | $(-1, 0)$ |
| $-1$ | 0 | $-1$ | $(0, -1)$ |
| 0 | 1 | 0 | $(1, 0)$ |
| 1 | 2 | 3 | $(2, 3)$ |
| 2 | 3 | 8 | $(3, 8)$ |
| 3 | 4 | 15 | $(4, 15)$ |

**(b)** We can find the relationship between $x$ and $y$ algebraically by the method of substitution. First solve for $t$ in terms of $x$ to obtain $t = x - 1$.

| | |
|---|---|
| $y = t^2 + 2t$ | Given |
| $y = (x - 1)^2 + 2(x - 1)$ | $t = x - 1$ |
| $\quad = x^2 - 2x + 1 + 2x - 2$ | Expand. |
| $\quad = x^2 - 1$ | Simplify. |

This is consistent with the ordered pairs we had found in the table. As $t$ varies over all real numbers, we will get all the ordered pairs in the relation $y = x^2 - 1$, which does indeed define $y$ as a function of $x$.

**(c)** We can get the graph by simply graphing the parabola $y = x^2 - 1$, or we can graph the curve in *parametric mode* on the grapher. There is a spot in the "Y=" screen to enter both X1 and Y1 as functions of the parameter T (Figure 1.60a). The "WINDOW" screen has slots for the usual *x*-axis and *y*-axis information, plus slots for "Tmin" and "Tmax" and "Tstep" (Figure 1.60b). Since $t = x - 1$, we set Tmin and Tmax to values one less than those for Xmin and Xmax. The Tstep determines how far the grapher will go from one value of *t* to the next as it computes the ordered pairs. With a window width of 10 units and a Tstep of 0.1, the grapher will compute 100 points, which is sufficient. (The more points, the smoother the graph. See Exploration 1.) Trace along the graph to find the values computed in (a).

Plot1  Plot2  Plot3
\X₁ᴛ◼T+1
Y₁ᴛ◼T²+2T
\X₂ᴛ=
Y₂ᴛ=
\X₃ᴛ=
Y₃ᴛ=
\X₄ᴛ=

(a)

WINDOW
 Tmin=–6
 Tmax=4
 Tstep=.1
 Xmin=–5
 Xmax=5
 Xscl=1
↓Ymin=–3

(b)

[–5, 5] by [–3, 3]

(c)

**Figure 1.60** The graph of a parabola in parametric mode on a graphing calculator. (Example 8)

**(d)** With the grapher set as in (c), use the table setup in Figure 1.61a to obtain the table in Figure 1.61b. These values agree with those found in (a).

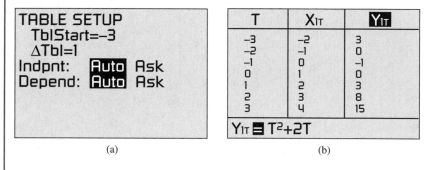

TABLE SETUP
 TblStart=–3
 ΔTbl=1
 Indpnt: **Auto** Ask
 Depend: **Auto** Ask

(a)

| T | X₁ᴛ | Y₁ᴛ |
|---|---|---|
| –3 | –2 | 3 |
| –2 | –1 | 0 |
| –1 | 0 | –1 |
| 0 | 1 | 0 |
| 1 | 2 | 3 |
| 2 | 3 | 8 |
| 3 | 4 | 15 |

Y₁ᴛ ◼ T²+2T

(b)

**Figure 1.61** Using the table feature of a grapher set in parametric mode. (Example 8)

**Teaching Note**

The parametric graphing capabilities of modern technology will greatly enhance a student's ability to visualize the graph of a relation.

**Exploration Extensions**

Make a change to the window that will cause the portion of the parabola above the *x*-axis to disappear.

**Time for T**

Functions defined by parametric equations are frequently encountered in problems of motion, where the *x*- and *y*-coordinates of a moving object are computed as functions of time. This makes time the parameter, which is why you almost always see parameters given as "*t*" in parametric equations.

---

**Exploration 1  Watching your Tstep**

1. Graph the parabola in Example 8 in parametric mode as described in the solution. Press TRACE and observe the values of T, X, and Y. At what value of T does the calculator begin tracing? What point on the parabola results? (It's off the screen.) At what value of T does it stop tracing? What point on the parabola results? How many points are computed as you trace from start to finish?

2. Leave everything else the same and change the Tstep to 0.01. Do you get a smoother graph? Why or why not?

3. Leave everything else the same and change the Tstep to 1. Do you get a smoother graph? Why or why not?

4. What effect does the Tstep have on the speed of the grapher? Is this easily explained?

5. Now change the Tstep to 2. Why does the bottom of the parabola disappear? (It may help to TRACE along the curve.)

6. Change the Tstep back to 0.1 and change the Tmin to $-1$. Why does the left side of the parabola disappear? (Again, it may help to TRACE.)

7. Make a change to the window that will cause the grapher to show the left side of the parabola but not the right.

---

## Inverse Relations and Inverse Functions

What happens when we reverse the coordinates of all the ordered pairs in a relation? We obviously get another relation, as it is another set of ordered pairs, but does it bear any resemblance to the original relation? If the original relation happens to be a function, will the new relation also be a function?

We can get some idea of what happens by switching the coordinates of the parabola in Example 8. This is easy to do in a function defined parametrically; we just switch the parametric equations for *x* and *y*.

### Example 9  SWITCHING COORDINATES OF A RELATION

Consider the set of all ordered pairs $(x, y)$ defined by the equations

$$x = t^2 + 2t$$

$$y = t + 1$$

where *t* is any real number. (These equations will generate the same numbers as in Example 8, but the old *x*'s become the new *y*'s and vice-versa.)

**(a)** Find the points determined by $t = -3, -2, -1, 0, 1, 2,$ and 3.

**(b)** Does this relation define *y* as a function of *x*?

**(c)** Graph the relation in the $(x, y)$ plane.

**Solution**

**(a)** The points are as shown in the table below. (Compare this to the table in Example 8.)

| $t$ | $x = t^2 + 2t$ | $y = t + 1$ | $(x, y)$ |
|-----|----------------|-------------|----------|
| $-3$ | 3 | $-2$ | $(3, -2)$ |
| $-2$ | 0 | $-1$ | $(0, -1)$ |
| $-1$ | $-1$ | 0 | $(-1, 0)$ |
| 0 | 0 | 1 | $(0, 1)$ |
| 1 | 3 | 2 | $(3, 2)$ |
| 2 | 8 | 3 | $(8, 3)$ |
| 3 | 15 | 4 | $(15, 4)$ |

**(b)** We can see from the points $(0, -1)$ and $(0, 1)$ in the table above that this does not define $y$ as a function of $x$.

**(c)** We can use the same algebraic steps as in Example 8 to get the relation in terms of $x$ and $y$: $x = y^2 - 1$. This equation defines two functions implicitly that could be graphed to give the graph of the relation (Figure 1.62).

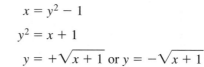

$$x = y^2 - 1$$
$$y^2 = x + 1$$
$$y = +\sqrt{x + 1} \text{ or } y = -\sqrt{x + 1}$$

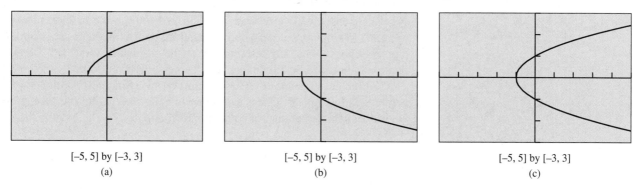

|  |  |  |
|--|--|--|
| [–5, 5] by [–3, 3] | [–5, 5] by [–3, 3] | [–5, 5] by [–3, 3] |
| (a) | (b) | (c) |

**Figure 1.62** The relation $x = y^2 - 1$ can be graphed in (c) by combining the graphs of the two functions defined implicitly by the relation, namely $y = +\sqrt{x + 1}$ in (a) and $y = -\sqrt{x + 1}$ in (b). (Example 9)

Alternatively, we can graph the new function in parametric mode by just switching the parametric equations for *x* and *y* as shown in Figure 1.63.

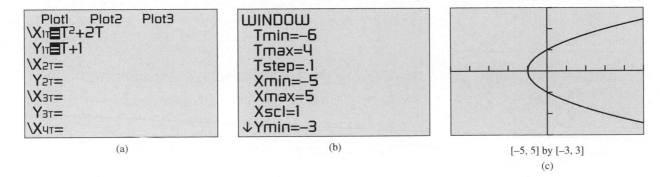

(a)          (b)          [−5, 5] by [−3, 3]
(c)

**Figure 1.63** The same graph as Figure 1.62c is produced here in parametric mode. (Example 9)

The relation in Example 9 is the *inverse relation* of the relation in Example 8. We define an inverse relation by the set of ordered pairs it contains.

> The ordered pair $(a, b)$ is in a relation if and only if the ordered pair $(b, a)$ is in the **inverse relation**.

We will be most interested in inverse relations that happen to be *functions*. Notice that the graph of the inverse relation in Example 9 fails the vertical line test and is therefore not the graph of a function. Can we predict this failure by considering the graph of the original relation in Example 8? Figure 1.64 suggests that we can.

The inverse graph in Figure 1.64b fails the vertical line test because two different *y*-values have been paired with the same *x*-value. This is a direct consequence of the fact that the original relation in Figure 1.64a paired two different *x*-values with the same *y*-value. The inverse graph fails the *vertical* line test precisely because the original graph fails the *horizontal* line test. This gives us a test for relations whose inverses are functions.

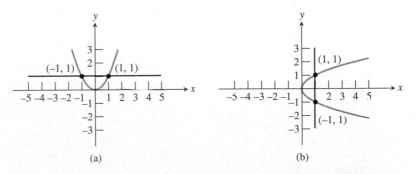

(a)          (b)

**Figure 1.64** The inverse relation in (b) fails the vertical line test because the original relation in (a) fails the horizontal line test.

**Horizontal Line Test**

The inverse of a relation is a function if and only if each horizontal line intersects the graph of the original relation in at most one point.

**Caution about function notation**

The symbol $f^{-1}$ is read "$f$ inverse" and should never be confused with the reciprocal of $f$. If $f$ is a *function*, the symbol $f^{-1}$ can *only* mean $f$ inverse. The reciprocal of $f$ must be written as $1/f$.

### Example 10   APPLYING THE HORIZONTAL LINE TEST

Which of the graphs (1)–(4) in Figure 1.65 are graphs of

(a) relations that are functions?

(b) relations that have inverses that are functions?

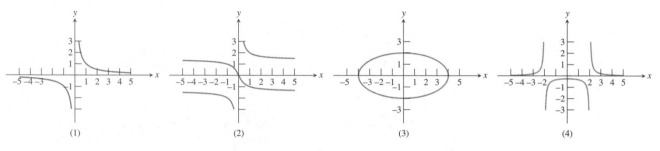

**Figure 1.65**  (Example 10)

**Teaching Note**

Some students may need to be reminded that in considering inverse relations, it is important to consider what happens to points. Thus, examining (a, b) as an element of a relation and (b, a) as an element of the inverse relation is critical to understanding inverse relations. The concept of symmetry drives much of the thinking about inverses.

**Solution**

(a) Graphs (1) and (4) are graphs of functions because these graphs pass the vertical line test. Graphs (2) and (3) are not graphs of functions because these graphs fail the vertical line test.

(b) Graphs (1) and (2) are graphs of relations whose inverses are functions because these graphs pass the horizontal line test. Graphs (3) and (4) fail the horizontal line test so their inverse relations are not functions.

A *function* whose inverse is a function has a graph that passes both the horizontal and vertical line tests (such as graph (1) in Example 10). Such a function is **one-to-one**, since every $x$ is paired with a unique $y$ and every $y$ is paired with a unique $x$.

**Definition   Inverse Function**

If $f$ is a one-to-one function with domain $D$ and range $R$, then the **inverse function of $f$**, denoted $f^{-1}$, is the function with domain $R$ and range $D$ defined by

$$f^{-1}(b) = a \quad \text{if and only if} \quad f(a) = b.$$

### Example 11   FINDING AN INVERSE FUNCTION ALGEBRAICALLY

Find an equation for $f^{-1}(x)$ if $f(x) = x/(x + 1)$.

**Solution**  The graph of $f$ in Figure 1.66 suggests that $f$ is one-to-one. The original function satisfies the equation $y = x/(x + 1)$. If $f$ truly is one-to-

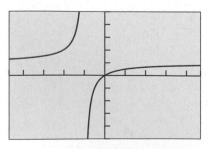

[–4.7, 4.7] by [–5, 5]

**Figure 1.66**  The graph of $f(x) = x/(x + 1)$. (Example 11)

one, the inverse function $f^{-1}$ will satisfy the equation $x = y/(y + 1)$. (Note that we just switch the $x$ and the $y$.)

If we solve this new equation for $y$ we will have a formula for $f^{-1}(x)$:

$$x = \frac{y}{y + 1}$$

$x(y + 1) = y$         Multiply by $y + 1$.

$xy + x = y$         Distributive property

$xy - y = -x$         Isolate the $y$ terms.

$y(x - 1) = -x$         Factor out $y$.

$$y = \frac{-x}{x - 1}$$         Divide by $x - 1$.

$$y = \frac{x}{1 - x}$$         Multiply numerator and denominator by $-1$.

Therefore $f^{-1}(x) = x/(x + 1)$.

Let us candidly admit two things regarding Example 11 before moving on to a graphical model for finding inverses. First, many functions are not one-to-one and so do not have inverse functions to begin with. Second, the algebra involved in finding an inverse function in the manner of Example 11 can be extremely difficult. We will actually find very few inverses this way. As you will learn in future chapters, we will usually rely on our understanding of how $f$ maps $x$ to $y$ to understand how $f^{-1}$ maps $y$ to $x$.

It is possible to use the graph of $f$ to produce a graph of $f^{-1}$ without doing any algebra at all, thanks to the following geometric reflection property:

> **The Inverse Reflection Principle**
>
> The points $(a, b)$ and $(b, a)$ in the coordinate plane are symmetric with respect to the line $y = x$. The points $(a, b)$ and $(b, a)$ are **reflections** of each other in the line $y = x$.

### Example 12   FINDING AN INVERSE FUNCTION GRAPHICALLY

The graph of a function $y = f(x)$ is shown in Figure 1.67. Sketch a graph of the function $y = f^{-1}(x)$.

**Solution** We need not find a formula for $f^{-1}(x)$. All we need to do is to find the reflection of the given graph in the line $y = x$. There are two ways to do this geometrically.

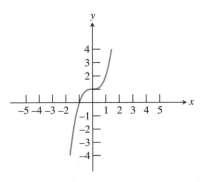

**Figure 1.67** The graph of a one-to-one function. (Example 12)

> **Method 1 (The Mirror Method)** Imagine a mirror along the line $y = x$ and draw the reflection of the given graph in the mirror. (See Figure 1.68.)

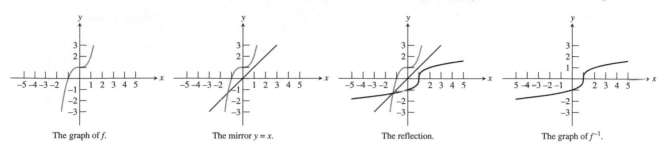

The graph of $f$.     The mirror $y = x$.     The reflection.     The graph of $f^{-1}$.

**Figure 1.68** The Mirror Method. The graph of $f$ is reflected in an imaginary mirror along the line $y = x$ to produce the graph of $f^{-1}$. (Example 12)

> **Method 2 (The Rotation Method)** Imagine the graph to be drawn on a large pane of glass. Rotate the glass around the line $y = x$ so that the *positive* x-axis switches places with the *positive* y-axis. (The back of the glass must be rotated to the front for this to occur.) The graph of $f$ will then become the graph of $f^{-1}$. (See Figure 1.69.)

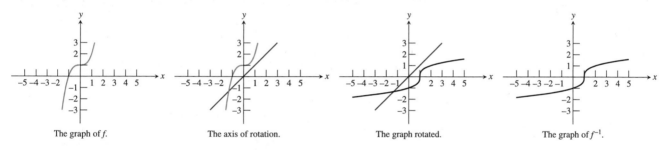

The graph of $f$.     The axis of rotation.     The graph rotated.     The graph of $f^{-1}$.

**Figure 1.69** The Rotation Method. The graph of $f$ is rotated around the line $y = x$ to produce the graph of $f^{-1}$. (Example 12)

There is a natural connection between inverses and function composition that gives further insight into what an inverse actually does: It "undoes" the action of the original function. This leads to the following rule:

---

**The Inverse Composition Rule**

A function $f$ is one-to-one with inverse function $g$ if and only if

$$f(g(x)) = x \text{ for every } x \text{ in the domain of } g, \text{ and}$$
$$g(f(x)) = x \text{ for every } x \text{ in the domain of } f.$$

---

**Example 13** VERIFYING INVERSE FUNCTIONS

Show algebraically that $f(x) = x^3 + 1$ and $g(x) = \sqrt[3]{x - 1}$ are inverse functions.

**Solution** We use the Inverse Composition Rule.

$$f(g(x)) = f(\sqrt[3]{x - 1}) = (\sqrt[3]{x - 1})^3 + 1 = x - 1 + 1 = x$$
$$g(f(x)) = g(x^3 + 1) = \sqrt[3]{(x^3 + 1) - 1} = \sqrt[3]{x^3} = x$$

Since these equations are true for all $x$, the Inverse Composition Rule guarantees that $f$ and $g$ are inverses.

You do not have far to go to find graphical support of this algebraic verification, since these are the functions whose graphs are shown in Example 12!

Some functions are so important that we need to study their inverses even though they are not one-to-one. A good example is the square root function, which is the "inverse" of the square function. It is not the inverse of the *entire* squaring function, because the full parabola fails the horizontal line test. Figure 1.70 shows that the function $y = \sqrt{x}$ is really the inverse of a "restricted-domain" version of $y = x^2$ defined only for $x \geq 0$.

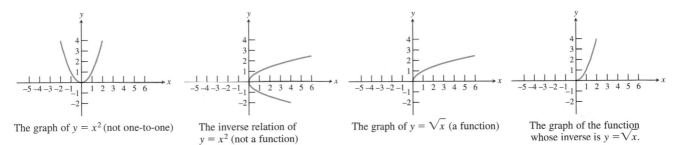

The graph of $y = x^2$ (not one-to-one)

The inverse relation of $y = x^2$ (not a function)

The graph of $y = \sqrt{x}$ (a function)

The graph of the function whose inverse is $y = \sqrt{x}$.

**Figure 1.70** The function $y = x^2$ has no inverse function, but $y = \sqrt{x}$ is the inverse function of $y = x^2$ on the restricted domain $[0, \infty)$.

The consideration of domains adds a refinement to the algebraic inverse-finding method of Example 11, which we now summarize:

---

**How to Find an Inverse Function Algebraically**

Given a formula for a function $f$, proceed as follows to find a formula for $f^{-1}$.

1. Determine that there is a function $f^{-1}$ by checking that $f$ is one-to-one. State any restrictions on the domain of $f$. (Note that it might be necessary to impose some to get a one-to-one version of $f$.)
2. Switch $x$ and $y$ in the formula $y = f(x)$.
3. Solve for $y$ to get the formula $y = f^{-1}(x)$. State any restrictions on the domain of $f^{-1}$.

---

**Example 14** FINDING AN INVERSE FUNCTION

Show that $f(x) = \sqrt{x + 3}$ has an inverse function and find a rule for $f^{-1}(x)$. State any restrictions on the domains of $f$ and $f^{-1}$.

## Solution

### Solve Algebraically

The graph of $f$ passes the horizontal line test, so $f$ has an inverse function (Figure 1.71). Note that $f$ has domain $[-3, \infty)$ and range $[0, \infty)$.

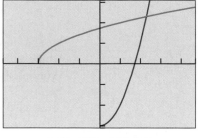

[–4.7, 4.7] by [–3.1, 3.1]

**Figure 1.71** The graph of $f(x) = \sqrt{x + 3}$ and its inverse, a restricted version of $y = x^2 - 3$. (Example 14)

To find $f^{-1}$ we write

$$y = \sqrt{x + 3} \quad \text{where } x \geq -3, y \geq 0$$

$$x = \sqrt{y + 3} \quad \text{where } y \geq -3, x \geq 0 \qquad \text{Interchange } x \text{ and } y.$$

$$x^2 = y + 3 \quad \text{where } y \geq -3, x \geq 0 \qquad \text{Square.}$$

$$y = x^2 - 3 \quad \text{where } y \geq -3, x \geq 0 \qquad \text{Solve for } y.$$

Thus $f^{-1}(x) = x^2 - 3$, with an "inherited" domain restriction of $x \geq 0$. Figure 1.71 shows the two functions. Note the domain restriction of $x \geq 0$ imposed on the parabola $y = x^2 - 3$.

### Support Graphically

Use a grapher in parametric mode and compare the graphs of the two sets of parametric equations with Figure 1.71:

$$x = t \qquad \text{and} \qquad x = \sqrt{t + 3}$$
$$y = \sqrt{t + 3} \qquad\qquad\qquad y = t$$

# Quick Review 1.4

In Exercises 1–10, solve the equation for $y$.

**1.** $x = 3y - 6$  $y = \dfrac{1}{3}x + 2$  **2.** $x = 0.5y + 1$  $y = 2x - 2$

**3.** $x = y^2 + 4$  $y = \pm\sqrt{x - 4}$  **4.** $x = y^2 - 6$  $y = \pm\sqrt{x + 6}$

**5.** $x = \dfrac{y - 2}{y + 3}$  $y = \dfrac{3x + 2}{1 - x}$  **6.** $x = \dfrac{3y - 1}{y + 2}$  $y = \dfrac{2x + 1}{3 - x}$

**7.** $x = \dfrac{2y + 1}{y - 4}$  $y = \dfrac{4x + 1}{x - 2}$  **8.** $x = \dfrac{4y + 3}{3y - 1}$  $y = \dfrac{x + 3}{3x - 4}$

**9.** $x = \sqrt{y + 3}, y \geq -3$  **10.** $x = \sqrt{y - 2}, y \geq 2$

# Section 1.4 Exercises

In Exercises 1–4, find formulas for the functions $f + g, f - g$, and $fg$. Give the domain of each. Also state the domain of each function.

**1.** $f(x) = 2x - 1; g(x) = x^2$

**2.** $f(x) = (x - 1)^2; g(x) = 3 - x$

**3.** $f(x) = \sqrt{x}; g(x) = \sin x$

**4.** $f(x) = \sqrt{x + 5}; g(x) = |x + 3|$

In Exercises 5 and 6, find formulas for $f/g$ and $g/f$. Give the domain of each. Also state the domain of each function.

**5.** $f(x) = \sqrt{x + 3}$; $g(x) = x^2$

**6.** $f(x) = \sqrt{x - 2}$; $g(x) = \sqrt{x + 4}$

**7.** $f(x) = x^2$ and $g(x) = 1/x$ are shown below in the viewing window $[0, 5]$ by $[0, 5]$. Sketch the graph of the sum $(f + g)(x)$ by adding the $y$-coordinates directly from the graphs. Then graph the sum on your calculator and see how close you came.

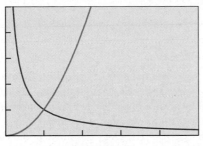

$[0, 5]$ by $[0, 5]$

**8.** The graphs of $f(x) = x^2$ and $g(x) = 4 - 3x$ are shown below in the viewing window $[-5, 5]$ by $[-10, 25]$. Sketch the graph of the difference $(f - g)(x)$ by subtracting the $y$-coordinates directly from the graphs. Then graph the difference on your calculator and see how close you came.

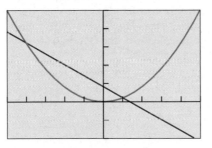

$[-5, 5]$ by $[-10, 25]$

In Exercises 9 and 10, find $(f \circ g)(3)$ and $(g \circ f)(-2)$

**9.** $f(x) = 2x - 3$; $g(x) = x + 1$  5; −6

**10.** $f(x) = x^2 - 1$; $g(x) = 2x - 3$  8; 3

In Exercises 11–14, find $f(g(x))$ and $g(f(x))$. State the domain of each.

**11.** $f(x) = 3x + 2$; $g(x) = x - 1$

**12.** $f(x) = x^2 - 1$; $g(x) = \dfrac{1}{x - 1}$

**13.** $f(x) = x^2 - 2$; $g(x) = \sqrt{x + 1}$

**14.** $f(x) = \dfrac{1}{x - 1}$; $g(x) = \sqrt{x}$

In Exercises 15–20, find $f(x)$ and $g(x)$ so that the function can be described as $y = f(g(x))$.

**15.** $y = \sqrt{x^2 - 5x}$  **16.** $y = (x^3 + 1)^2$

**17.** $y = |3x - 2|$  **18.** $y = \dfrac{1}{x^3 - 5x + 3}$

**19.** $y = (x - 3)^5 + 2$  **20.** $y = e^{\sin x}$

In Exercises 21–24, find the $(x, y)$ pair for the value of the parameter.

**21.** $x = 3t$ and $y = t^2 + 5$ for $t = 2$  $(6, 9)$

**22.** $x = 5t - 7$ and $y = 17 - 3t$ for $t = -2$  $(-17, 23)$

**23.** $x = t^3 - 4t$ and $y = \sqrt{t + 1}$ for $t = 3$  $(15, 2)$

**24.** $x = |t + 3|$ and $y = 1/t$ for $t = -8$  $(5, -1/8)$

In Exercises 25–28, find a direct relationship between $y$ and $x$ and determine whether the parametric equations determine $y$ as a function of $x$.

**25.** $x = 2t$ and $y = 3t - 1$  **26.** $x = t + 1$ and $y = t^2 - 2t$

**27.** $x = t^2$ and $y = t - 2$  **28.** $x = \sqrt{t}$ and $y = 2t - 5$

In Exercises 29–32, find two functions defined implicitly by each given relation.

**29.** $x^2 + y^2 = 25$  **30.** $x + y^2 = 25$

**31.** $x^2 - y^2 = 25$  **32.** $3x^2 - y^2 = 25$

In Exercises 33–38, confirm that $f$ and $g$ are inverses by showing that $f(g(x)) = x$ and $g(f(x)) = x$.

**33.** $f(x) = 3x - 2$ and $g(x) = \dfrac{x + 2}{3}$

**34.** $f(x) = \dfrac{x + 3}{4}$ and $g(x) = 4x - 3$

**35.** $f(x) = x^3 + 1$ and $g(x) = \sqrt[3]{x - 1}$

**36.** $f(x) = \dfrac{7}{x}$ and $g(x) = \dfrac{7}{x}$

**37.** $f(x) = \dfrac{x + 1}{x}$ and $g(x) = \dfrac{1}{x - 1}$

**38.** $f(x) = \dfrac{x + 3}{x - 2}$ and $g(x) = \dfrac{2x + 3}{x - 1}$

In Exercises 39–42, determine whether the function is one-to-one. If it is one-to-one, sketch the graph of the inverse.

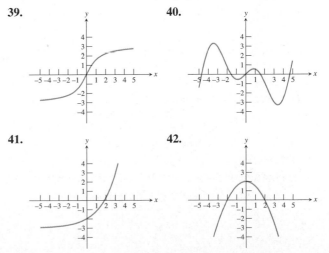

**39.**

**40.**

**41.**

**42.**

In exercises 43–52, find a formula for $f^{-1}(x)$. Give the domain of $f^{-1}$, including any restrictions "inherited" from $f$.

**43.** $f(x) = 3x - 6$

**44.** $f(x) = 2x + 5$

**45.** $f(x) = \dfrac{2x - 3}{x + 1}$

**46.** $f(x) = \dfrac{x + 3}{x - 2}$

**47.** $f(x) = \sqrt{x + 3}$

**48.** $f(x) = \sqrt{x + 2}$

**49.** $f(x) = x^3$

**50.** $f(x) = x^3 + 5$

**51.** $f(x) = \sqrt[3]{x + 5}$

**52.** $f(x) = \sqrt[3]{x - 2}$

**53. Weather Balloons** A high-altitude spherical weather balloon expands as it rises due to the drop in atmospheric pressure. Suppose that the radius $r$ increases at the rate of 0.03 inches per second and that $r = 48$ inches at time $t = 0$. Determine an equation that models the volume $V$ of the balloon at time $t$ and find the volume when $t = 300$ seconds.

**54. A Snowball's Chance** Jake stores a small cache of 4-inch diameter snowballs in the basement freezer, unaware that the freezer's self-defrosting feature will cause each snowball to lose about 1 cubic inch of volume every 40 days. He remembers them a year later (call it 360 days) and goes to retrieve them. What is their diameter then?

**55. Satellite Photography** A satellite camera takes a rectangular-shaped picture. The smallest region that can be photographed is a 5-km by 7-km rectangle. As the camera zooms out, the length $l$ and width $w$ of the rectangle increase at a rate of 2 km/sec. How long does it take for the area $A$ to be at least 5 times its original size?

**56. Computer Imaging** New Age Special Effects, Inc., prepares computer software based on specifications prepared by film directors. To simulate an approaching vehicle, they begin with a computer image of a 5-cm by 7-cm by 3-cm box. The program increases each dimension at a rate of 2 cm/sec. How long does it take for the volume $V$ of the box to be at least 5 times its initial size?    $t \approx 1.62$ seconds

**57. Currency Conversion** In September of 1999 the exchange rate for converting U.S. dollars ($x$) to French francs ($y$) was $y = 6.314x$.

**(a)** How many French francs could you get for \$100 U.S.?

**(b)** What is the inverse function, and what conversion does it represent?

**(c)** A 1999 late-summer tourist had an elegant lunch in Provence ordering from a "fixed price" 125-franc menu. How much was that in U.S. dollars?

**58. Temperature Conversion** The formula for converting Celsius temperature ($x$) to Kelvin temperature is $k(x) = x + 273.16$. The formula for converting Fahrenheit temperature ($x$) to Celsius temperature is $c(x) = (5/9)(x - 32)$

**(a)** Find a formula for $c^{-1}(x)$. What is this formula used for?

**(b)** Find $(k \circ c)(x)$. What is this formula used for?

## Explorations

**59. Function Properties Inherited by Inverses** There are some properties of functions that are automatically shared by inverse functions (when they exist) and some that are not. Suppose that $f$ has an inverse function $f^{-1}$. Give an algebraic or graphical argument (not a rigorous formal proof) to show that each of these properties of $f$ must necessarily be shared by $f^{-1}$.

**(a)** $f$ is continuous.

**(b)** $f$ is one-to-one.

**(c)** $f$ is odd (graphically, symmetric with respect to the origin).

**(d)** $f$ is increasing.

**60. Function Properties not Inherited by Inverses** There are some properties of functions that are not necessarily shared by inverse functions, even if the inverses exist. Suppose that $f$ has an inverse function $f^{-1}$. For each of the following properties, give an example to show that $f$ can have the property while $f^{-1}$ does not.

**(a)** $f$ has a graph with a horizontal asymptote.

**(b)** $f$ has domain all real numbers.

**(c)** $f$ has a graph that is bounded above.

**(d)** $f$ has a removable discontinuity at $x = 5$.    ■

**61. Scaling Algebra Grades** A teacher gives a challenging algebra test to her class. The lowest score is 52, which she decides to scale to 70. The highest score is 88, which she decides to scale to 97.

**(a)** Using the points (52, 70) and (88, 97), find a linear equation that can be used to convert raw scores to scaled grades.    $y = 0.75x + 31$

**(b)** Find the inverse of the function defined by this linear equation. What does the inverse function do?

**62. Writing to Learn** (Continuation of Exercise 61) Explain why it is important for fairness that the scaling function used by the teacher be an *increasing* function. (Caution: It is *not* because "everyone's grade must go up." What would the scaling function in Exercise 61 do for a student who does enough "extra credit" problems to get a raw score of 136?)

## Extending the Ideas

**63. Modeling a Fly Ball Parametrically** A baseball that leaves the bat at an angle of 60° from horizontal traveling 110 feet per second follows a path that can be modeled by the following pair of parametric equations. (You might enjoy verifying this if you have studied motion in physics):

$$x = 110(t) \cos (60°)$$

$$y = 110(t) \sin (60°) - 16t^2$$

You can simulate the flight of the ball on a grapher. Set your grapher to parametric mode and put the functions above in for X2T and Y2T. Set X1T = 325 and Y1T = 5T to draw a 30-foot fence 325 feet from home plate. Set Tmin = 0,

Tmax = 6, Tstep = 0.1, Xmin = 0, Xmax = 350, Xscl = 0, Ymin = 0, Ymax = 300, and Yscl = 0.

**(a)** Now graph the function. Does the fly ball clear the fence?

**(b)** Change the angle to 30° and run the simulation again. Does the ball clear the fence?

**(c)** What angle is optimal for hitting the ball? Does it clear the fence when hit at that angle?

**64. The Baylor GPA Scale Revisited** *(See Problem 64 in Section 1.2.)* The function used to convert Baylor School percentage grades to GPA's on a 4-point scale is

$$y = \left( \frac{3^{1.7}}{30}(x - 65) \right)^{\frac{1}{1.7}} + 1.$$

The function has domain [65, 100]. Anything below 65 is a failure and automatically converts to a GPA of 0.

**(a)** Find the inverse function algebraically. What can the inverse function be used for?

**(b)** Does the inverse function have any domain restrictions?

**(c)** Verify with a graphing calculator that the function found in (a) and the given function are really inverses.

**65. Group Activity** (Continuation of Exercise 64) The number 1.7 that appears in two places in the GPA scaling formula is called the scaling factor ($k$). The value of $k$ can be changed to alter the curvature of the graph while keeping the points (1, 65) and (95, 4) fixed. It was felt that the lowest D (65) needed to be scaled to 1.0, while the middle A (95) needed to be scaled to 4.0. The faculty's Academic Council considered several values of $k$ before settling on 1.7 as the number that gives the "fairest" GPA's for the other percentage grades.

Try changing $k$ to other values between 1 and 2. What kind of scaling curve do you get when $k = 1$? Do you agree with the Baylor decision that $k = 1.7$ gives the fairest GPA's?

**66. Revisiting Example 6** Solve $x^2y + y^2 - 5$ for $y$ using the quadratic formula and graph the pair of implicit functions.

---

**1.5**

# Graphical Transformations

Transformations  •  Vertical and Horizontal Translations  •  Reflections Across Axes  •  Vertical and Horizontal Stretches and Shrinks  •  Combining Transformations

## Transformations

The following functions are all different:

$$y = x^2$$

$$y = (x - 3)^2$$

$$y = 1 - x^2$$

$$y = x^2 - 4x + 5$$

However, a look at their graphs shows that, while no two are exactly the same, all four have the same identical *shape* and *size*. Understanding how algebraic alterations change the shapes, sizes, positions, and orientations of graphs is helpful for understanding the connection between algebraic and graphical models of functions.

In this section we relate graphs using **transformations,** which are functions that map real numbers to real numbers. By acting on the $x$-coordinates and $y$-coordinates of points, transformations change graphs in predictable ways. **Rigid transformations,** which leave the size and shape of a graph unchanged, include horizontal translations, vertical translations, reflections, or any combination of these. **Non-rigid transformations,** which generally distort the shape of a graph, include horizontal or vertical stretches and shrinks.

## Vertical and Horizontal Translations

A **vertical translation** of the graph of $y = f(x)$ is a shift of the graph up or down in the coordinate plane. A **horizontal translation** is a shift of the graph

### Objective

Students will be able to algebraically and graphically represent translations, reflections, stretches, and shrinks of functions and parametric relations.

### Motivate

Ask...
How is the graph of $(x - 2)^2 + (y + 1)^2 = 16$ related to the graph of $x^2 + y^2 = 16$? **(Translated 2 units right and 1 unit down.)**

### Lesson Guide

Day 1: Skills
Day 2: Applications

to the left or the right. The following exploration will give you a good feel for what translations are and how they occur.

---

**Exploration 1** | **Introducing Translations**

Set your viewing window to $[-5, 5]$ by $[-5, 15]$ and your graphing mode to sequential as opposed to simultaneous.

1. Graph the functions

$$y_1 = x^2 \qquad\qquad y_4 = y_1(x) - 2 = x^2 - 2$$
$$y_2 = y_1(x) + 3 = x^2 + 3 \qquad y_5 = y_1(x) - 4 = x^2 - 4$$
$$y_3 = y_1(x) + 1 = x^2 + 1$$

on the same screen. What effect do the $+3$, $+1$, $-2$, and $-4$ seem to have?

2. Graph the functions

$$y_1 = x^2 \qquad\qquad y_4 = y_1(x - 2) = (x - 2)^2$$
$$y_2 = y_1(x + 3) = (x + 3)^2 \qquad y_5 = y_1(x - 4) = (x - 4)^2$$
$$y_3 = y_1(x + 1) = (x + 1)^2$$

on the same screen. What effect do the $+3$, $+1$, $-2$, and $-4$ seem to have?

3. Repeat steps 1 and 2 for the functions $y_1 = x^3$, $y_1 = |x|$, and $y_1 = \sqrt{x}$. Do your observations agree with those you made after steps 1 and 2?

---

In general, *replacing x by x − c* shifts the graph horizontally $c$ units. Similarly, *replacing y by y − c* shifts the graph vertically $c$ units. If $c$ is positive the shift is to the right or up; if $c$ is negative the shift is to the left or down.

This is a nice, consistent rule that unfortunately gets complicated by the fact that the $c$ for a vertical shift rarely shows up being subtracted from $y$. Instead, it usually shows up on the other side of the equal sign being *added* to $f(x)$. That leads us to the following rule, which only *appears* to be different for horizontal and vertical shifts:

---

Let $c$ be a positive real number. Then the following transformations result in translations of the graph of $y = f(x)$:

**Horizontal translations**

$$y = f(x - c) \qquad \text{a translation to the right by } c \text{ units}$$
$$y = f(x + c) \qquad \text{a translation to the left by } c \text{ units}$$

**Vertical translations**

$$y = f(x) + c \qquad \text{a translation up by } c \text{ units}$$
$$y = f(x) - c \qquad \text{a translation down by } c \text{ units}$$

**Example 1** FINDING EQUATIONS FOR TRANSLATIONS

Each view in Figure 1.72 shows the graph of $y_1 = x^3$ and a vertical or horizontal translation $y_2$. Write an equation for $y_2$ as shown in each graph.

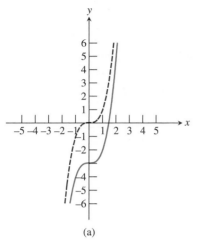
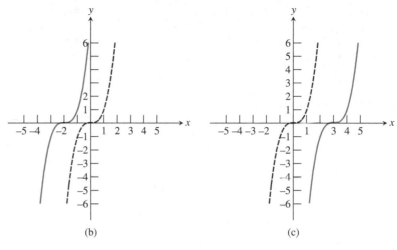

(a)  (b)  (c)

**Figure 1.72** Translations of $y_1 = x^3$. (Example 1)

**Solution**

**(a)** $y_2 = x^3 - 3$   (a vertical translation down by 3 units)

**(b)** $y_2 = (x + 2)^3$   (a horizontal translation left by 2 units)

**(c)** $y_2 = (x - 3)^3$   (a horizontal translation right by 3 units)

## Reflections Across Axes

Points $(x, y)$ and $(x, -y)$ are **reflections** of each other across the $x$-axis. Points $(x, y)$ and $(-x, y)$ are **reflections** of each other across the $y$-axis. (See Figure 1.73.) Two points (or graphs) that are symmetric with respect to a line are **reflections** of each other across that line.

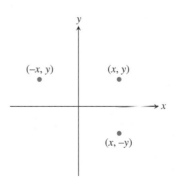

**Figure 1.73** The point $(x, y)$ and its reflections across the $x$- and $y$-axes.

**Exploration 2** **Introducing Reflections**

Set your viewing window to $[-5, 5]$ by $[-4, 4]$ and your graphing mode to sequential as opposed to simultaneous.

1. Graph the functions $y_1 = (x - 2)^2 + 1$ and $y_2 = -y_1(x)$ on the same screen. What is the geometrical relationship between the graphs? They are reflections of each other across the x-axis.

2. Graph the functions $y_1 = (x - 2)^2 + 1$ and $y_2 = y_1(-x)$ on the same screen. What is the geometrical relationship between the graphs? They are reflections of each other across the y-axis.

3. Graph the functions $y_1 = \sqrt{x + 2} + 1$ and $y_2 = -y_1(x)$ on the same screen. What is the geometrical relationship between the graphs? They are reflections of each other across the x-axis.

4. Graph the functions $y_1 = \sqrt{x + 2} + 1$ and $y_2 = y_1(-x)$ on the same screen. What is the geometrical relationship between the graphs? They are reflections of each other across the y-axis.

5. Graph the functions $y_1 = x^3 - 3x^2 + x + 1$ and $y_2 = -y_1(x)$ on the same screen. What is the geometrical relationship between the graphs? They are reflections of each other across the x-axis.

6. Graph the functions $y_1 = x^3 - 3x^2 + x + 1$ and $y_2 = y_1(-x)$ on the same screen. What is the geometrical relationship between the graphs? They are reflections of each other across the y-axis.

Exploration 2 suggests that a reflection across the x-axis results when $y$ is replaced by $-y$, and a reflection across the y-axis results when $x$ is replaced by $-x$. This should make sense to you if you understand Figure 1.73.

The following transformations result in reflections of the graph of $y = f(x)$:

**Across the x-axis**

$$y = -f(x)$$

**Across the y-axis**

$$y = f(-x)$$

**Example 2** FINDING EQUATIONS FOR REFLECTIONS

Find an equation for the reflection of $f(x) = \dfrac{5x - 9}{x^2 + 3}$ across each axis.

**Solution**

**Solve Algebraically**

Across the x-axis: $y = -f(x) = -\dfrac{5x - 9}{x^2 + 3} = \dfrac{9 - 5x}{x^2 + 3}$

Across the y-axis: $y = f(-x) = \dfrac{5(-x) - 9}{(-x)^2 + 3} = \dfrac{-5x - 9}{x^2 + 3}$

### Support Graphically

The graphs in Figure 1.74 support our algebraic work.

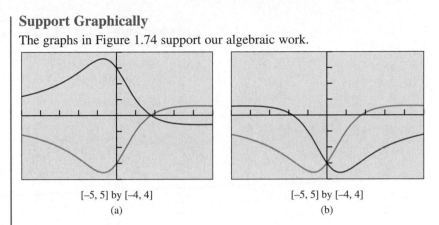

[–5, 5] by [–4, 4]

(a)

[–5, 5] by [–4, 4]

(b)

**Figure 1.74** Reflections of $f(x) = (5x - 9)/(x^2 + 3)$ across (a) the $x$-axis and (b) the $y$-axis. (Example 2)

You might expect that odd and even functions, whose graphs already possess special symmetries, would exhibit special behavior when reflected across the axes. They do, as Example 3 shows.

### Example 3  REFLECTING ODD AND EVEN FUNCTIONS

**(a)** Prove that the graph of an even function remains unchanged when it is reflected across the $y$-axis.

**(b)** Prove that the graph of an odd function is the same when reflected across the $x$-axis as it is when reflected across the $y$-axis.

**Solution**  Note that we can get plenty of graphical support for these statements by reflecting the graphs of various odd and even functions, but what is called for here is **proof**, which will require algebra.

**(a)** Let $f$ be an even function; that is, $f(-x) = f(x)$ for all $x$ in the domain of $f$. To reflect the graph of $y = f(x)$ across the $y$-axis, we make the transformation $y = f(-x)$. But $f(-x) = f(x)$ for all $x$ in the domain of $f$, so this transformation results in $y = f(x)$. The graph of $f$ therefore remains unchanged.

**(b)** Let $f$ be an odd function; that is, $f(-x) = -f(x)$ for all $x$ in the domain of $f$. To reflect the graph of $y = f(x)$ across the $y$-axis, we make the transformation $y = f(-x)$. But $f(-x) = -f(x)$ for all $x$ in the domain of $f$, so this transformation results in $y = -f(x)$. That is exactly the transformation that reflects the graph of $f$ across the $x$-axis, so the two reflections yield the same graph.

## Vertical and Horizontal Stretches and Shrinks

We now investigate what happens when we multiply all the $y$-coordinates (or all the $x$-coordinates) of a graph by a fixed real number.

**Teaching Note**

A drawing program on a computer can be used to illustrate the concepts of vertical and horizontal stretch and shrink.

---

**Exploration 3** | **Introducing Stretches and Shrinks**

Set your viewing window to $[-4.7, 4.7]$ by $[-1.1, 5.1]$ and your graphing mode to sequential as opposed to simultaneous.

**1.** Graph the functions

$$y_1 = \sqrt{4 - x^2}$$
$$y_2 = 1.5y_1(x) = 1.5\sqrt{4 - x^2}$$
$$y_3 = 2y_1(x) = 2\sqrt{4 - x^2}$$
$$y_4 = 0.5y_1(x) = 0.5\sqrt{4 - x^2}$$
$$y_5 = 0.25y_1(x) = 0.25\sqrt{4 - x^2}$$

on the same screen. What effect do the 1.5, 2, 0.5, and 0.25 seem to have?

**2.** Graph the functions

$$y_1 = \sqrt{4 - x^2}$$
$$y_2 = y_1(1.5x) = \sqrt{4 - (1.5x)^2}$$
$$y_3 = y_1(2x) = \sqrt{4 - (2x)^2}$$
$$y_4 = y_1(0.5x) = \sqrt{4 - (0.5x)^2}$$
$$y_5 = y_1(0.25x) = \sqrt{4 - (0.25x)^2}$$

on the same screen. What effect do the 1.5, 2, 0.5, and 0.25 seem to have?

---

**Exploration Extensions**

Graph the functions $y_1 = \sin(\pi x)$, $y_2 = 5 \sin(1/2(\pi x))$, and $y_3 = 1/2 \sin(5\pi x)$. What effect do the multiplied constants 1/2 and 5 seem to have?

---

Exploration 3 suggests that multiplication of $x$ or $y$ by a constant results in a horizontal or vertical stretching or shrinking of the graph.

In general, *replacing $x$ by $x/c$ distorts the graph horizontally by a factor of $c$.* Similarly, *replacing $y$ by $y/c$ distorts the graph vertically by a factor of $c$.* If $c$ is greater than 1 the distortion is a stretch; if $c$ is less than 1 the distortion is a shrink.

Reminiscent of translations, this is a nice, consistent rule that unfortunately gets complicated by the fact that the $c$ for a vertical stretch or shrink rarely shows up as a divisor of $y$. Instead, it usually shows up on the other side of the equal sign as a *factor* multiplied by $f(x)$. That leads us to the following rule:

Let $c$ be a positive real number. Then the following transformations result in stretches or shrinks of the graph of $y = f(x)$:

**Horizontal stretches or shrinks**

$$y = f\left(\frac{x}{c}\right) \quad \begin{cases} \text{a stretch by a factor of } c & \text{if } c > 1 \\ \text{a shrink by a factor of } c & \text{if } c < 1 \end{cases}$$

**Vertical stretches or shrinks**

$$y = c \cdot f(x) \quad \begin{cases} \text{a stretch by a factor of } c & \text{if } c > 1 \\ \text{a shrink by a factor of } c & \text{if } c < 1 \end{cases}$$

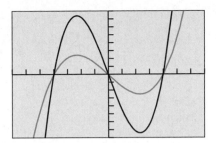

[−7, 7] by [−80, 80]

(a)

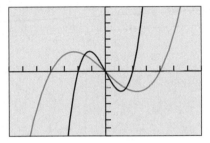

[−7, 7] by [−80, 80]

(b)

**Figure 1.75** The graph of $y_1 = f(x) = x^3 - 16x$, shown with (a) a vertical stretch and (b) a horizontal shrink. (Example 4)

**Example 4**  FINDING EQUATIONS FOR STRETCHES AND SHRINKS

Let $C_1$ be the curve defined by $y_1 = f(x) = x^3 - 16x$. Find equations for the following non-rigid transformations of $C_1$:

**(a)** $C_2$ is a vertical stretch of $C_1$ by a factor of 3.

**(b)** $C_3$ is a horizontal shrink of $C_1$ by a factor of $1/2$.

**Solution**

**Solve Algebraically**

**(a)** Denote the equation for $C_2$ by $y_2$. Then

$$y_2 = 3 \cdot f(x)$$
$$= 3(x^3 - 16x)$$
$$= 3x^3 - 48x$$

**(b)** Denote the equation for $C_3$ by $y_3$. Then

$$y_3 = f\left(\frac{x}{1/2}\right)$$
$$= f(2x)$$
$$= (2x)^3 - 16(2x)$$
$$= 8x^3 - 32x$$

**Support Graphically**

The graphs in Figure 1.75 support our algebraic work.

## Combining Transformations

Transformations may be performed in succession—one after another. If the transformations include stretches, shrinks, or reflections, the order in which the transformations are performed may make a difference. In those cases, be sure to pay particular attention to order.

**Example 5**  COMBINING TRANSFORMATIONS IN ORDER

**(a)** The graph of $y = x^2$ undergoes the following transformations, in order. Find the equation of the graph that results.

- a horizontal shift 2 units to the right

- a vertical stretch by a factor of 3

- a vertical translation 5 units up

**(b)** Apply the transformations in (a) in the opposite order and find the equation of the graph that results.

**Solution**

**(a)** Applying the transformations in order, we have

$$x^2 \Rightarrow (x - 2)^2 \Rightarrow 3(x - 2)^2 \Rightarrow 3(x - 2)^2 + 5$$

Expanding the final expression, we get the function $y = 3x^2 - 12x + 17$.

**(b)** Applying the transformations in the opposite order, we have

$$x^2 \Rightarrow x^2 + 5 \Rightarrow 3(x^2 + 5) \Rightarrow 3((x - 2)^2 + 5)$$

Expanding the final expression, we get the function $y = 3x^2 - 12x + 27$.

The second graph is ten units higher than the first graph because the vertical stretch lengthens the vertical translation when the translation occurs first. Order often matters when stretches, shrinks, or reflections are involved.

### Example 6   TRANSFORMING A GRAPH GEOMETRICALLY

The graph of $y = f(x)$ is shown in Figure 1.76. Determine the graph of the composite function $y = 2f(x + 1) - 3$ by showing the effect of a sequence of transformations on the graph of $y = f(x)$.

**Solution**  The graph of $y = 2f(x + 1) - 3$ can be obtained from the graph of $y = f(x)$ by the following sequence of transformations:

**(a)** a vertical stretch by a factor of 2 to get $y = 2f(x)$ (Figure 1.77a)

**(b)** a horizontal translation 1 unit to the left to get $y = 2f(x + 1)$ (Figure 1.77b)

**(c)** a vertical translation 3 units down to get $y = 2f(x + 1) - 3$ (Figure 1.77c)

(The order of the first two transformations can be reversed without changing the final graph.)

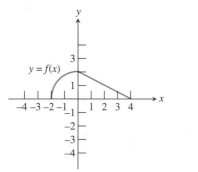

**Figure 1.76**  The graph of the function $y = f(x)$ in Example 6.

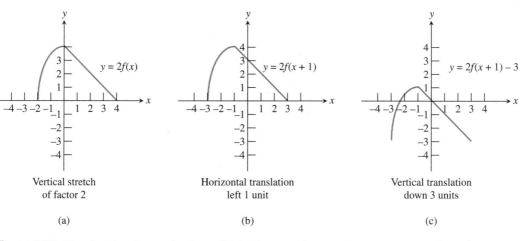

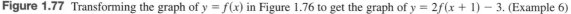

Vertical stretch of factor 2

(a)

Horizontal translation left 1 unit

(b)

Vertical translation down 3 units

(c)

**Figure 1.77**  Transforming the graph of $y = f(x)$ in Figure 1.76 to get the graph of $y = 2f(x + 1) - 3$. (Example 6)

**Alert**

Students sometimes make errors in the order in which they apply transformations to a graph. Warn students of the danger of reversing the order of some of these transformations. Point out that, in general, they are not commutative. Explore which transformations are commutative.

# Quick Review 1.5

In Exercises 1–6, write the expression as a binomial squared.

**1.** $x^2 + 2x + 1$  $(x + 1)^2$    **2.** $x^2 - 6x + 9$  $(x - 3)^2$

**3.** $x^2 + 12x + 36$  $(x + 6)^2$    **4.** $4x^2 + 4x + 1$  $(2x + 1)^2$

**5.** $x^2 - 5x + \dfrac{25}{4}$  $(x - 5/2)^2$    **6.** $4x^2 - 20x + 25$  $(2x - 5)^2$

In Exercises 7–10, perform the indicated operations and simplify.

**7.** $(x - 2)^2 + 3(x - 2) + 4$    **8.** $2(x + 3)^2 - 5(x + 3) - 2$

**9.** $(x - 1)^3 + 3(x - 1)^2 - 3(x - 1)$  $x^3 - 6x + 5$

**10.** $2(x + 1)^3 - 6(x + 1)^2 + 6(x + 1) - 2$  $2x^3$

# Section 1.5 Exercises

In Exercises 1–8, describe how the graph of $y = x^2$ can be transformed to the graph of the given equation.

**1.** $y = x^2 - 3$    **2.** $y = x^2 + 5.2$

**3.** $y = (x + 4)^2$    **4.** $y = (x - 3)^2$

**5.** $y = (100 - x)^2$    **6.** $y = x^2 - 100$

**7.** $y = (x - 1)^2 + 3$    **8.** $y = (x + 50)^2 - 279$

In Exercises 9–12, describe how the graph of $y = \sqrt{x}$ can be transformed to the graph of the given equation.

**9.** $y = -\sqrt{x}$    **10.** $y = \sqrt{x - 5}$

**11.** $y = \sqrt{-x}$    **12.** $y = \sqrt{3 - x}$

In Exercises 13–16, describe how the graph of $y = x^3$ can be transformed to the graph of the given equation.

**13.** $y = 2x^3$    **14.** $y = (2x)^3$

**15.** $y = (0.2x)^3$    **16.** $y = 0.3x^3$

In Exercises 17–20, describe how to transform the graph of $f$ into the graph of $g$.

**17.** $f(x) = \sqrt{x + 2}$ and $g(x) = \sqrt{x - 4}$

**18.** $f(x) = (x - 1)^2$ and $g(x) = -(x + 3)^2$

**19.** $f(x) = (x - 2)^3$ and $g(x) = -(x + 2)^3$

**20.** $f(x) = |2x|$ and $g(x) = 4|x|$

In Exercises 21–24, sketch the graphs of $f$, $g$, and $h$ by hand. Support your answers with a grapher.

**21.** $f(x) = (x + 2)^2$    **22.** $f(x) = x^3 - 2$

   $g(x) = 3x^2 - 2$        $g(x) = (x + 4)^3 - 1$

   $h(x) = -2(x - 3)^2$     $h(x) = 2(x - 1)^3$

**23.** $f(x) = \sqrt[3]{x + 1}$    **24.** $f(x) = -2|x| - 3$

   $g(x) = 2\sqrt[3]{x} - 2$        $g(x) = 3|x + 5| + 4$

   $h(x) = -\sqrt[3]{x - 3}$        $h(x) = |3x|$

In Exercises 25–28, the graph is that of a function $y = f(x)$ that can be obtained by transforming the graph of $y = \sqrt{x}$. Write a formula for the function $f$.

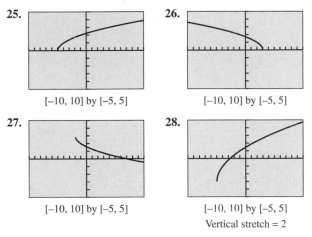

**25.**

[−10, 10] by [−5, 5]

**26.**

[−10, 10] by [−5, 5]

**27.**

[−10, 10] by [−5, 5]

**28.**

[−10, 10] by [−5, 5]

Vertical stretch = 2

In Exercises 29–32, find the equation of the reflection of $f$ across **(a)** the $x$-axis and **(b)** the $y$-axis.

**29.** $f(x) = x^3 - 5x^2 - 3x + 2$    **30.** $f(x) = 2\sqrt{x + 3} - 4$

**31.** $f(x) = \sqrt[3]{8x}$    **32.** $f(x) = 3|x + 5|$

In Exercises 33–36, a graph of the given function can be obtained from $y = x^2$ by both a vertical stretch (or shrink) and by a horizontal shrink (or stretch). In each case, identify the shrink and stretch factors.

**33.** $y = 9x^2$    **34.** $y = \dfrac{1}{4}x^2$

**35.** $y = 16x^2$    **36.** $y = 5x^2$

In Exercises 37–40, transform the given function by **(a)** a vertical stretch by a factor of 2, and **(b)** a horizontal shrink by a factor of 1/3.

**37.** $f(x) = x^3 - 4x$    **38.** $f(x) = |x + 2|$

**39.** $f(x) = x^2 + x - 2$    **40.** $f(x) = \dfrac{1}{x + 2}$

In Exercises 41–44, describe a basic graph and a sequence of transformations that can be used to produce a graph of the given function.

**41.** $y = 2(x - 3)^2 - 4$     **42.** $y = -3\sqrt{x + 1}$

**43.** $y = (3x)^2 - 4$     **44.** $y = -2|x + 4| + 1$

In Exercises 45–48, a graph $G$ is obtained from a graph of $y$ by the sequence of transformations indicated. Write an equation whose graph is $G$.

**45.** $y = x^2$: a vertical stretch by a factor of 3, then a shift right 4 units.   $y = 3(x - 4)^2$

**46.** $y = x^2$: a shift right 4 units, then a vertical stretch by a factor of 3.   $y = 3(x - 4)^2$

**47.** $y = |x|$: a shift left 2 units, then a vertical stretch by a factor of 2, and finally a shift down 4 units.   $y = 2|x + 2| - 4$

**48.** $y = |x|$: a shift left 2 units, then a horizontal shrink by a factor of 1/2, and finally a shift down 4 units.

Exercises 49–52 refer to the function $f$ whose graph is shown below.

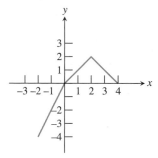

**49.** Sketch the graph of $y = 2 + 3f(x + 1)$.

**50.** Sketch the graph of $y = -f(x + 1) + 1$.

**51.** Sketch the graph of $y = f(2x)$.

**52.** Sketch the graph of $y = 2f(x - 1) + 2$.

**53. Writing to Learn** Graph some examples to convince yourself that a reflection and a translation can have a different effect when combined in one order than when combined in the opposite order. Then explain in your own words why this can happen.

**54. Writing to Learn** Graph some examples to convince yourself that vertical stretches and shrinks do not affect a graph's $x$-intercepts. Then explain in your own words why this is so.

**55. Celsius vs. Fahrenheit** The graph shows the temperature in degrees Celsius in Windsor, Ontario, for one 24-hour period. Describe the transformations that convert this graph to one showing degrees Fahrenheit. (Hint: $F(t) = (9/5)C(t) + 32$.)

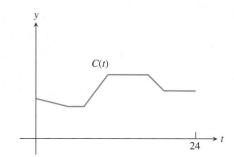

**56. Fahrenheit vs. Celsius** The graph shows the temperature in degrees Fahrenheit in Mt. Clemens, Michigan, for one 24-hour period. Describe the transformations that convert this graph to one showing degrees Celsius. (Hint: $F(t) = (9/5)C(t) + 32$.)

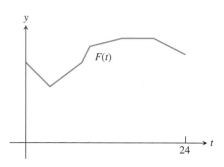

## Explorations

 **57. International Investment** Table 1.11 shows the price of a share of stock in Debeers Consolidated Mines for the first 8 months of 1999:

| Table 1.11 Debeers Consolidated Mines | |
| --- | --- |
| Month | Price ($) |
| 1 | 14.25 |
| 2 | 14.89 |
| 3 | 18.94 |
| 4 | 24.50 |
| 5 | 21.44 |
| 6 | 23.88 |
| 7 | 24.75 |
| 8 | 27.19 |

*Source: Salomon Smith Barney*

**(a)** Graph price ($y$) as a function of month ($x$) as a line graph, connecting the points to make a continuous graph.

**(b)** Explain what transformation you would apply to this graph to produce a graph showing the price of the stock in Japanese yen.

**58. Group Activity** Get with a friend and graph the function $y = x^2$ on both your graphers. Apply a horizontal or vertical stretch or shrink to the function on one of the graphers. Then change the *window* of that grapher to make the two graphs look the same. Can you formulate a general rule for how to find the window?

## Extending the Ideas

**59. The Absolute Value Transformation**  Graph the function $f(x) = x^4 - 5x^3 + 4x^2 + 3x + 2$ in the viewing window $[-5, 5]$ by $[-10, 10]$. (Put the equation in Y1.)

**(a)** Study the graph and try to predict what the graph of $y = |f(x)|$ will look like. Then turn Y1 off and graph Y2 = abs (Y1). Did you predict correctly?

**(b)** Study the original graph again and try to predict what the graph of $y = f(|x|)$ will look like. Then turn Y1 off and graph Y2 = Y1(abs (X)). Did you predict correctly?

**(c)** Given the graph of $y = g(x)$ shown below, sketch a graph of $y = |g(x)|$.

**(d)** Given the graph of $y = g(x)$ shown below, sketch a graph of $y = g(|x|)$.

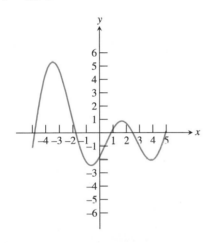

**60. Parametric Circles and Ellipses**  Set your grapher to parametric and radian mode and your window as follows:

Tmin = 0, Tmax = 7, Tstep = 0.1

Xmin = −4.7, Xmax = 4.7, Xscl = 1

Ymin = −3.1, Ymax = 3.1, Yscl = 1

**(a)** Graph the parametric equations $x = \cos t$ and $y = \sin t$. You should get a circle of radius 1.

**(b)** Use a transformation of the parametric function of $x$ to produce the graph of an ellipse that is 4 units wide and 2 units tall.

**(c)** Use a transformation of both parametric functions to produce a circle of radius 3.

**(d)** Use a transformation of both functions to produce an ellipse that is 8 units wide and 4 units tall.

(You will learn more about ellipses in Chapter 8.)

---

## 1.6 Modeling with Functions

Functions from Data  •  Functions from Formulas  •  Functions from Graphs  •  Functions from Verbal Descriptions

### Functions from Data

Now that you have learned more about what functions are and how they behave, we want to return to the modeling theme of Section 1.1. In that section we stressed that one of the goals of this course was to become adept at using numerical, algebraic, and graphical models of the real world in order to solve problems. We now want to focus your attention more precisely on modeling with *functions* so that you can put the subject matter of this chapter to practical use. The more you learn about particular functions later in the book, the more powerful these models will become. With calculus they will become more powerful still.

People who work with data in the real world are grateful when they can find a function that will model one variable under observation in terms of another. It is not always possible (the elusive "stock market price as a function of time" function being an obvious example), but when it *is* possible, the payoff in understanding is considerable. In this course we will use the following 3-step strategy to construct functions from data.

---

**Constructing a Function from Data**

Given a set of data points of the form $(x, y)$, to construct a formula that approximates $y$ as a function of $x$:

1. Make a scatter plot of the data points. The points do not need to pass the vertical line test.

2. Determine from the shape of the plot whether the points seem to follow the graph of a familiar type of function (line, parabola, cubic, sine curve, etc.).

3. Transform a basic function of that type to fit the points as closely as possible.

---

Step 3 might seem like a lot of work, and for earlier generations it certainly was; it required all of the tricks of Section 1.5 and then some. We, however, will gratefully use technology to do this "curve-fitting" step for us, as shown in Example 1.

### Example 1   CURVE-FITTING WITH TECHNOLOGY

Table 1.12 records the low and high daily temperatures observed on 9/9/99 in 20 major American cities. Find a function that approximates the high temperature ($y$) as a function of the low temperature ($x$). Use this function to predict the high temperature that day for Madison, WI, given that the low was 46.

| Table 1.12 Temperature on 9/9/99 | | | | | |
|---|---|---|---|---|---|
| City | Low | High | City | Low | High |
| New York, NY | 70 | 86 | Miami, FL | 76 | 92 |
| Los Angeles, CA | 62 | 80 | Honolulu, HI | 70 | 85 |
| Chicago, IL | 52 | 72 | Seattle, WA | 50 | 70 |
| Houston, TX | 70 | 94 | Jacksonville, FL | 67 | 89 |
| Philadelphia, PA | 68 | 86 | Baltimore, MD | 64 | 88 |
| Albuquerque, NM | 61 | 86 | St. Louis, MO | 57 | 79 |
| Phoenix, AZ | 82 | 106 | El Paso, TX | 62 | 90 |
| Atlanta, GA | 64 | 90 | Memphis, TN | 60 | 86 |
| Dallas, TX | 65 | 87 | Milwaukee, WI | 52 | 68 |
| Detroit, MI | 54 | 76 | Wilmington, DE | 66 | 84 |

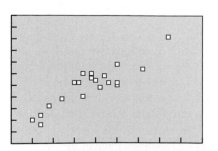

[45, 90] by [60, 115]

**Figure 1.78** The scatter plot of the temperature data in Example 1.

**Solution**  The scatter plot is shown in Figure 1.78.

Notice that the points do not fall neatly along a well-known curve, but they do seem to fall *near* an upwardly-sloping line. We therefore choose to model the data with a function whose graph is a line. We could fit the line by sight (as we did in Example 5 in Section 1.1), but this time we will use the calculator to find the line of "best fit," called the **regression line**. (See your grapher's *Owner's Manual* for how to do this.) The regression line is found to be approximately $y = 0.97x + 23$. As Figure 1.79 shows, the line fits the data as well as can be expected.

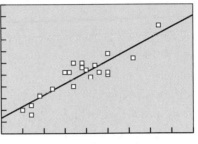

[45, 90] by [60, 115]

**Figure 1.79**  The temperature scatter plot with the regression line shown. (Example 1)

If we use this function to predict the high temperature for the day in Madison, WI, we get $y = 0.97(46) + 23 = 67.62$. (For the record, the high that day was 67.)

Professional statisticians would be quick to point out that this function should not be trusted as a model for all cities, despite the fairly successful prediction for Madison. (For example, the prediction for San Francisco, with a low of 54 and a high of 64, is off by more than 11 degrees.) *The effectiveness of a data-based model is highly dependent on the number of data points and on the way they were selected.* If we had really wanted to find a function to model these temperature spreads nationally, we would have chosen more cities, paying more attention to climatological diversity than to whether the city is "major" or not. While the science of statistical sampling is interesting and important, it is beyond the scope of this course. Consequently, the functions we construct from data in this book should be analyzed for how well they model the data, not for how well they model the larger population from which the data came.

In addition to lines, we can model scatter plots with several other curves by choosing the appropriate regression option on a calculator or computer. The options to which we will refer in this book (and the chapters in which we will study them) are shown on the next page:

| Regression Type | Equation | Graph | Applications |
|---|---|---|---|
| Linear (Chaper 2) | $y = ax + b$ | | Fixed cost plus variable cost, linear growth, free-fall velocity, simple interest, linear depreciation, many others |
| Quadratic (Chapter 2) | $y = ax^2 + bx + c$ (requires at least 3 points) | | Position during free fall, projectile motion, parabolic reflectors, area as a function of lincar dimension, quadratic growth, etc. |
| Cubic (Chapter 2) | $y = ax^3 + bx^2 + cx + d$ (requires at least 4 points) | | Volume as a function of linear dimension, cubic growth, "S"-shaped scatter plots, etc. |
| Quartic (Chapter 2) | $y = ax^4 + bx^3 + cx^2 + dx + e$ (requires at least 5 points) | | "W"-shaped scatter plots, quartic growth, etc. |
| Natural logarithmic (ln) (Chapter 3) | $y = a + b \ln x$ (requires $x > 0$) | | Logarithmic growth, decibels (sound), Richter scale (earthquakes), inverse exponential models |
| Exponential ($b > 1$) (Chapter 3) | $y = a \cdot b^x$ (requires $y > 0$) | | Exponential growth, compound interest, population models, |
| Exponential ($0 < b < 1$) (Chapter 3) | $y = a \cdot b^x$ (requires $y > 0$) | | Exponential decay, depreciation, temperature loss of a cooling body, etc. |
| Power (requires $x, y > 0$)) (Chapter 2) | $y = a \cdot x^b$ | | Modeling slower growth with powers $< 1$ (e.g., square roots), miscellaneous monomial models |
| Logistic (Chapter 3) | $y = \dfrac{c}{1 + a \cdot e^{-bx}}$ (requires $y > 0$) | | Logistic growth: spread of a rumor, population models |
| Sinusoidal (Chapter 4) | $y = a \sin(bx + c) + d$ | | Periodic behavior: harmonic motion, waves, circular motion, etc. |

These graphs are only examples, as they can vary in shape and orientation. (For example, any of the curves could appear upside-down.) The grapher uses various strategies to fit these curves to the data, most of them based on combining function composition with linear regression. Depending on the regression type, the grapher may display a number $r$ called the **correlation coefficient** or a number $r^2$ or $R^2$ called the **coefficient of determination**. In either case, a useful "rule of thumb" is: *the closer the absolute value of this number is to 1, the better the curve fits the data.*

We can use this fact to help choose a regression type, as in Exploration 1.

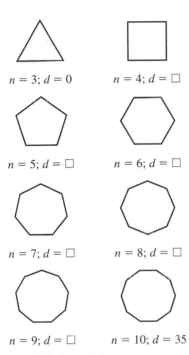

Figure 1.80 Some Polygons. (Exploration 1)

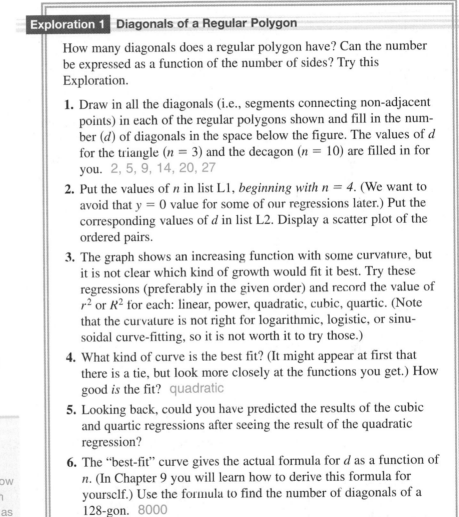

**Exploration 1** **Diagonals of a Regular Polygon**

How many diagonals does a regular polygon have? Can the number be expressed as a function of the number of sides? Try this Exploration.

1. Draw in all the diagonals (i.e., segments connecting non-adjacent points) in each of the regular polygons shown and fill in the number ($d$) of diagonals in the space below the figure. The values of $d$ for the triangle ($n = 3$) and the decagon ($n = 10$) are filled in for you. 2, 5, 9, 14, 20, 27

2. Put the values of $n$ in list L1, *beginning with n = 4*. (We want to avoid that $y = 0$ value for some of our regressions later.) Put the corresponding values of $d$ in list L2. Display a scatter plot of the ordered pairs.

3. The graph shows an increasing function with some curvature, but it is not clear which kind of growth would fit it best. Try these regressions (preferably in the given order) and record the value of $r^2$ or $R^2$ for each: linear, power, quadratic, cubic, quartic. (Note that the curvature is not right for logarithmic, logistic, or sinusoidal curve-fitting, so it is not worth it to try those.)

4. What kind of curve is the best fit? (It might appear at first that there is a tie, but look more closely at the functions you get.) How good *is* the fit? quadratic

5. Looking back, could you have predicted the results of the cubic and quartic regressions after seeing the result of the quadratic regression?

6. The "best-fit" curve gives the actual formula for $d$ as a function of $n$. (In Chapter 9 you will learn how to derive this formula for yourself.) Use the formula to find the number of diagonals of a 128-gon. 8000

We will have more to say about curve fitting as we study the various function types in later chapters.

## Functions from Formulas

You have already seen quite a few formulas in the course of your education. Formulas involving two variable quantities always relate those variables implicitly, and quite often the formulas can be solved to give one variable explicitly as a function of the other. In this book we will use a variety of for-

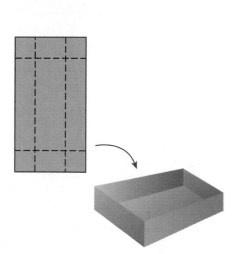

**Figure 1.81** An open-topped box made by cutting the corners from a piece of cardboard and folding up the sides. (Example 3)

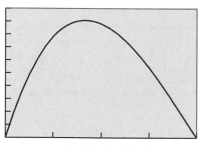

[0, 4] by [0, 100]

(a)

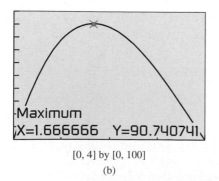

Maximum
X=1.666666   Y=90.740741

[0, 4] by [0, 100]

(b)

**Figure 1.82** The graph of the volume of the box in Example 3.

mulas to pose and solve problems algebraically, although we will not assume prior familiarity with those formulas that we borrow from other subject areas (like physics or economics). We *will* assume familiarity with certain key formulas from mathematics.

### Example 2 FORMING FUNCTIONS FROM FORMULAS

Write the area $A$ of a circle as a function of its

(a) radius $r$.

(b) diameter $d$.

(c) circumference $C$.

**Solution**

(a) The familiar area formula from geometry gives $A$ as a function of $r$:

$$A = \pi r^2.$$

(b) This formula is not so familiar. However, we know that $r = d/2$, so we can substitute that expression for $r$ in the area formula:

$$A = \pi r^2 = \pi(d/2)^2 = (\pi/4)d^2.$$

(c) Since $C = 2\pi r$, we can solve for $r$ to get $r = C/(2\pi)$. Then substitute to get $A$: $A = \pi r^2 = \pi(C/(2\pi))^2 = \pi C^2/(4\pi^2) = C^2/(4\pi)$.

### Example 3 A MAXIMUM VALUE PROBLEM

A square of side $x$ inches is cut out of each corner of an 8 in. by 15 in. piece of cardboard and the sides are folded up to form an open-topped box (Figure 1.81).

(a) Write the volume $V$ of the box as a function of $x$.

(b) Find the domain of $V$ as a function of $x$. (Note that the model imposes restrictions on $x$.)

(c) Graph $V$ as a function of $x$ over the domain found in part (b) and use the maximum finder on your grapher to determine the maximum volume such a box can hold.

(d) How big should the cut-out squares be in order to produce the box of maximum volume?

**Solution**

(a) The box will have a base with sides of width $8 - 2x$ and length $15 - 2x$. The depth of the box will be $x$ when the sides are folded up. Therefore $V = x(8 - 2x)(15 - 2x)$.

(b) The formula for $V$ is a polynomial with domain all reals. However, the depth $x$ must be non-negative, as must the width of the base, $8 - 2x$. Together, these two restrictions yield a domain of $[0, 4]$. (The endpoints give a box with no volume, which is as mathematically feasible as other zero concepts.)

(c) The graph is shown in Figure 1.82. The maximum finder shows that the maximum occurs at the point $(5/3, 90.74)$. The maximum volume is about 90.74 in.$^3$.

(d) Each square should have sides of one-and-two-thirds inches.

## Functions from Graphs

When "thinking graphically" becomes a genuine part of your problem-solving strategy, it is sometimes actually easier to start with the graphical model than it is to go straight to the algebraic formula. The graph provides valuable information about the function.

### Example 4   PROTECTING AN ANTENNA

A small satellite dish is packaged with a cardboard cylinder for protection. The parabolic dish is 24 in. in diameter and 6 in. deep, and the diameter of the cardboard cylinder is 12 in. How tall must the cylinder be to fit in the middle of the dish and be flush with the top of the dish? (See Figure 1.83.)

### Solution

#### Solve Algebraically

The diagram in Figure 1.83a showing the cross-section of this 3-dimensional problem is also a 2-dimensional graph of a quadratic function. We can transform our basic function $y = x^2$ with a vertical shrink so that it goes through the points $(12, 6)$ and $(-12, 6)$, thereby producing a graph of the parabola in the coordinate plane (Figure 1.83b).

$$y = kx^2 \qquad \text{Vertical shrink}$$

$$6 = k(\pm 12)^2 \qquad \text{Substitute } x = \pm 12, y = 6$$

$$k = \frac{6}{144} = \frac{1}{24} \qquad \text{Solve for } k.$$

Thus, $y = \dfrac{1}{24}x^2$

To find the height of the cardboard cylinder, we first find the $y$-coordinate of the parabola 6 inches from the center, that is, when $x = 6$:

$$y = \frac{1}{24}(6)^2 = 1.5$$

From that point to the top of the dish is $6 - 1.5 = 4.5$ in.

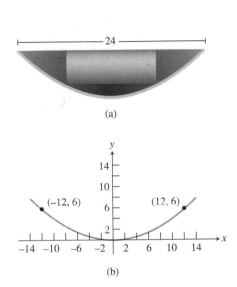

**Figure 1.83** (a) A parabolic satellite dish with a protective cardboard cylinder in the middle for packaging. (b) The parabola in the coordinate plane. (Example 4)

Although Example 4 serves nicely as a "functions from graphs" example, it is also an example of a function that must be constructed by gathering relevant information from a verbal description and putting it together in the right way. People who do mathematics for a living are accustomed to confronting that challenge regularly as a necessary first step in modeling the real world. In honor of its importance, we have saved it until last to close out this chapter in style.

## Functions from Verbal Descriptions

There is no fail-safe way to form a function from a verbal description. It can be hard work, frequently a good deal harder than the mathematics required to solve the problem once the function has been found. The 4-step problem-solving process in Section 1.1 gives you several valuable tips, perhaps the most important of which is to *read* the problem carefully. Understanding what the words say is critical if you hope to model the situation they describe.

**Figure 1.84** A cone with equal height and radius. (Example 5)

**Example 5    FINDING THE MODEL AND SOLVING**

Grain is leaking through a hole in a storage bin at a constant rate of 8 cubic inches per minute. The grain forms a cone-shaped pile on the ground below. As it grows, the height of the cone always remains equal to its radius. If the cone is one foot tall now, how tall will it be in one hour?

**Solution**

Reading the problem carefully, we realize that the formula for the volume of the cone is needed (Figure 1.84). From memory or from the Key Formulas list, we get the formula $V = (1/3)\pi r^2 h$. A careful reading also reveals that the height and the radius are always equal, so we can get volume directly as a function of height: $V = (1/3)\pi h^3$

When $h = 12$ in., the volume is $V = (\pi/3)(12)^3 = 576\pi$ in.$^3$.

One hour later, the volume will have grown by $(60 \text{ min})(8 \text{ in.}^3/\text{min}) = 480 \text{ in.}^3$. The total volume of the pile at that point will be $576\pi + 480$ in.$^3$. Finally, we use the volume formula once again to solve for $h$:

$$\frac{1}{3}\pi h^3 = 576\pi + 480$$

$$h^3 = \frac{3(576\pi + 480)}{\pi}$$

$$h = \sqrt[3]{\frac{3(576\pi + 480)}{\pi}}$$

$$h \approx 13 \text{ inches}$$

**Example 6    LETTING UNITS WORK FOR YOU**

How many rotations does a 15 in. (radius) tire make per second on a sport utility vehicle traveling 70 mph?

**Solution**  It is the perimeter of the tire that comes in contact with the road, so we first find the circumference of the tire:

$$C = 2\pi r = 2\pi(15) = 30\pi \text{ in.}$$

That means that 1 rotation = $30\pi$ in. From this point we proceed by converting "miles per hour" to "rotations per second" by a series of **conversion factors** that are really factors of 1:

$$\frac{70 \text{ miles}}{1 \text{ hour}} \times \frac{1 \text{ hour}}{60 \text{ min}} \times \frac{1 \text{ min}}{60 \text{ sec}} \times \frac{5280 \text{ feet}}{1 \text{ mile}} \times \frac{12 \text{ inches}}{1 \text{ foot}} \times \frac{1 \text{ rotation}}{30\pi \text{ inches}}$$

$$= \frac{70 \times 5280 \times 12 \text{ rotations}}{60 \times 60 \times 30\pi \text{ sec}} \approx 13.07 \text{ rotations per second}$$

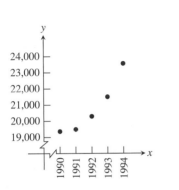

**Figure 1.85**  Scatter plot of the waste data

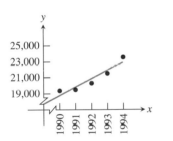

**Figure 1.86**  Scatter plot with the regression line shown.

### Problem

As the earth's population continues to grow, the solid waste generated by the population grows with it. Government must plan for disposal and recycling of ever growing amounts of solid waste. Planners can use data from the past to predict future waste generation and plan for enough facilities for disposing of and recycling the waste.

Given the following data on the waste generated in Florida from 1990–1994, how can we construct a function to predict the waste that was generated in the years 1995–1999? The scatter plot is shown in Figure 1.85.

| Year | Tons of Solid Waste Generated (in thousands) |
|------|-----------------------------------------------|
| 1990 | 19,358 |
| 1991 | 19,484 |
| 1992 | 20,293 |
| 1993 | 21,499 |
| 1994 | 23,561 |

### Solution

Since the data points fall near an upward sloping line, we can use a calculator to compute a regression line to model the data. The regression line is found to be approximately $y = 1042.1x - 2,055,024$.

As Figure 1.86 shows, the line fits the data well.

If we use this function to predict the waste that was generated in 1995, we get $y = 1042.1(1995) - 2,055,024 = 23,965.5$ thousand tons of waste. Similarly, we can predict the waste generated in each of the years 1996-1999.

| Year | Predicted Tons of Solid Waste Generated (in thousands) |
|------|---------------------------------------------------------|
| 1996 | $y = 1042.1(1996) - 2,055,024 = 25,008$ tons |
| 1997 | $y = 1042.1(1997) - 2,055,024 = 26,050$ tons |
| 1998 | $y = 1042.1(1998) - 2,055,024 = 27,092$ tons |
| 1999 | $y = 1042.1(1999) - 2,055,024 = 28,134$ tons |

*Data Source: http://www.fsu.edu/~cpm/FACT/sec_A/total.html*

# Quick Review 1.6

In Exercises 1–10, solve the given formula for the given variable.

**1. Area of a triangle** Solve for $h$: $A = \dfrac{1}{2}bh$   $h = 2(A/b)$

**2. Area of a trapezoid** Solve for $h$: $A = \dfrac{1}{2}(b_1 + b_2)h$

**3. Volume of a right circular cylinder** Solve for $h$:
$V = \pi r^2 h$   $h = V/(\pi r^2)$

**4. Volume of a right circular cone** Solve for $h$:

$V = \dfrac{1}{3}\pi r^2 h$   $h = 3V/(\pi r^2)$

**5. Volume of a sphere** Solve for $r$: $V = \dfrac{4}{3}\pi r^3$

**6. Surface area of a sphere** Solve for $r$: $A = 4\pi r^2$

**7. Surface area of a right circular cylinder** Solve for $h$:
$A = 2\pi rh + 2\pi r^2$

**8. Simple interest** Solve for $t$: $I = Prt$   $t = I/(Pr)$

**9. Compound interest** Solve for $P$: $A = P\left(1 + \dfrac{r}{n}\right)^{nt}$

**10. Free-fall from height $H$** Solve for $t$: $s = H - \dfrac{1}{2}gt^2$
$t = \sqrt{2(H-s)/g}$

# Section 1.6 Exercises

In Exercises 1–10, write a mathematical expression for the quantity described verbally:

**1.** Five more than three times a number $x$.   $3x + 5$

**2.** A number $x$ increased by 5 and then tripled.   $3(x + 5)$

**3.** Seventeen percent of a number $x$.   $0.17x$

**4.** Four more than 5% of a number $x$.   $0.05x + 4$

**5. Area of a rectangle** The area of a rectangle whose length is 12 more than its width $x$.   $(x + 12)(x)$

**6. Area of a triangle** The area of a triangle whose altitude is 2 more than its base length $x$.   $1/2(x)(x + 2)$

**7. Salary increase** A salary after a 4.5% increase, if the original salary is $x$ dollars.   $1.045x$

**8. Income loss** Income after a 3% drop in the current income of $x$ dollars.   $0.97x$

**9. Sale price** Sale price of an item marked $x$ dollars, if 40% is discounted from the marked price.   $0.60x$

**10. Including tax** Actual cost of an item selling for $x$ dollars if the sales tax rate is 8.75%.   $1.0875x$

In Exercises 11–14, choose a variable and write a mathematical expression for the quantity described verbally.

**11. Total cost** The total cost is $34,500 plus $5.75 for each item produced.

**12. Total cost** The total cost is $28,000 increased by 9% plus $19.85 for each item produced.

**13. Revenue** The revenue when each item sells for $3.75.

**14. Profit** The profit consists of a franchise fee of $200,000 plus 12% of all sales.

In Exercises 15–20, write the specified quantity as a function of the specified variable. It will help in each case to draw a picture.

**15.** The height of a right circular cylinder equals its diameter. Write the volume of the cylinder as a function of its radius.

**16.** One leg of a right triangle is twice as long as the other. Write the length of the hypotenuse as a function of the length of the shorter leg.   $a\sqrt{5}$

**17.** The base of an isosceles triangle is half as long as the two equal sides. Write the area of the triangle as a function of the length of the base.   $A = a^2\sqrt{15}/4$

**18.** A square is inscribed in a circle. Write the area of the square as a function of the radius of the circle.   $A = 2r^2$

**19.** A sphere is contained in a cube, tangent to all six faces. Find the surface area of the cube as a function of the radius of the sphere.   $A = 24r^2$

**20.** An isosceles triangle has its base along the $x$-axis with one base vertex at the origin and its vertex in the first quadrant on the graph of $y = 6 - x^2$. Write the area of the triangle as a function of the length of the base.   $A = (24b - b^3)/8$

In Exercises 21–36, write an equation for the problem and solve the problem.

**21.** One positive number is 4 times another positive number. The sum of the two numbers is 620. Find the two numbers.

**22.** When a number is added to its double and its triple, the sum if 714. Find the three numbers.

**23. Salary increase** Mark received a 3.5% salary increase. His salary after the raise was $36,432. What was his salary before the raise?   $1.035x = 36,432; x = 35,200$

**24. Consumer Price Index** The consumer price index for food and beverages in 1997 was 161.3 after a modest 1.7% increase from the previous year. What was the index the previous year? (Source: *U.S. Bureau of Labor Statistics*)

**25. Travel time**  A traveler averaged 52 miles per hour on a 182-mile trip. How many hours were spent on the trip?

**26. Travel time**  On their 560-mile trip, the Bruins basketball team spent two more hours on the interstate highway than they did on local highways. They averaged 45 mph on local highways and 55 mph on the interstate highways. How many hours did they spend driving on local highways?

**27. Sale prices**  At a shirt sale, Jackson sees two shirts that he likes equally well. Which is the better bargain, and why?

$33

40% off

$27

25% off

**28. Job offers**  Ruth is weighing two job offers from the sales departments of two competing companies. One offers a base salary of $25,000 plus 5% of gross sales; the other offers a base salary of $20,000 plus 7% of gross sales. What would Ruth's gross sales total need to be to make the second job offer more attractive than the first?  At least $250,000

**29. Personal computers**  From 1996 to 1997, the worldwide shipments of personal computers grew from 71,065,000 to 82,400,000. What was the percentage increase in worldwide personal computer shipments? (Source: *Dataquest*)  15.95%

**30. Personal computers**  From 1996 to 1997, the U.S. shipments of personal computers grew from 26,650,000 to 30,989,000. What was the percentage increase in U.S. personal computer shipments? (Source: *Dataquest*)  16.28%

**31. Mixing solutions**  How much 10% solution and how much 45% solution should be mixed together to make 100 gallons of 25% solution?

Solution 1

x gallons
10%

+

Solution 2

(100 − x)
gallons
45%

=

Combined
solution

100 gallons
25%

**(a)** Write an equation that models this problem.

**(b)** Solve the equation graphically.

**32. Mixing solutions**  The chemistry lab at the University of Hardwoods keeps two acid solutions on hand. One is 20% acid and the other is 35% acid. How much 20% acid

solution and how much 35% acid solution should be used to prepare 25 liters of a 26% acid solution?

**33. Residential construction**  DDL Construction is building a rectangular house that is 16 feet longer than it is wide. A rain gutter is to be installed in four sections around the 136-foot perimeter of the house. What lengths should be cut for the four sections?

**34. Interior design**  Renée's Decorating Service recommends putting a border around the top of the four walls in a dining room that is 3 feet longer than it is wide. Find the dimensions of the room if the total length of the border is 54 feet.  $2x + 2(x + 3) = 54$; 12 ft × 15 ft

**35. Investment returns**  Reggie invests $12,000, part at 7% annual interest and part at 8.5% annual interest. How much is invested at each rate if Reggie's total annual interest is $900?

**36. Investment returns**  Jackie invests $25,000, part at 5.5% annual interest and the balance at 8.3% annual interest. How much is invested at each rate if Jackie receives a 1-year interest payment of $1571?

## Exploration

**37. Manufacturing**  The Buster Green Shoe Company determines that the annual cost $C$ of making $x$ pairs of one type of shoe is $30 per pair plus $100,000 in fixed overhead costs. Each pair of shoes that is manufactured is sold wholesale for $50.

**(a)** Find an equation that models the cost of producing $x$ pairs of shoes.  $C = 100,000 + 30x$

**(b)** Find an equation that models the revenue produced from selling $x$ pairs of shoes.  $R = 50x$

**(c)** Find how many pairs of shoes must be made and sold in order to break even.  $x = 5000$ pairs of shoes

**(d)** Graph the equations in (a) and (b). What is the graphical interpretation of the answer in (c)?

**38. Employee benefits**  John's company issues employees a contract that identifies salary and the company's contributions to pension, health insurance premiums, and disability insurance. The company uses the following formulas to calculate these values.

| Salary | $x$ (dollars) |
| --- | --- |
| Pension | 12% of salary |
| Health Insurance | 3% of salary |
| Disability Insurance | 0.4% of salary |

If John's total contract with benefits is worth $48,814.20, what is his salary?  $42,300

**39. Manufacturing** Queen, Inc., a tennis racket manufacturer, determines that the annual cost $C$ of making $x$ rackets is $23 per racket plus $125,000 in fixed overhead costs. It costs the company $8 to string a racket.

**(a)** Find a function $y_1 = u(x)$ that models the cost of producing $x$ unstrung rackets. $y_1 = u(x) = 125{,}000 + 23x$

**(b)** Find a function $y_2 = s(x)$ that models the cost of producing $x$ strung rackets. $y_2 = s(x) = 125{,}000 + 31x$

**(c)** Find a function $y_3 = R_u(x)$ that models the revenue generated by selling $x$ unstrung rackets. $y_3 = r_u(x) = 56x$

**(d)** Find a function $y_4 = R_s(x)$ that models the revenue generated by selling $x$ strung rackets. $y_4 = r_s(x) = 79x$

**(e)** Graph $y_1$, $y_2$, $y_3$, and $y_4$ simultaneously in the window [0, 10,000] by [0, 500,000].

**(f) Writing to Learn** Write a report to the company recommending how they should manufacture their rackets, strung or unstrung. Assume that you can include the viewing window in (e) as a graph in the report, and use it to support your recommendation.

**40. Home schooling growth** The estimated number of U.S. children that were home-schooled in the years from 1992 to 1997 were:

| Table 1.13 Home Schooling | |
|---|---|
| Year | Number |
| 1992 | 703,000 |
| 1993 | 808,000 |
| 1994 | 929,000 |
| 1995 | 1,060,000 |
| 1996 | 1,220,000 |
| 1997 | 1,347,000 |

*Source: National Home Education Research Institute*

**(a)** Produce a scatter plot of the number of children home-schooled in thousands ($y$) as a function of years since 1990 ($x$).

**(b)** Find the linear regression equation. (Round the coefficients to the nearest 0.01.) $y = 131.06x + 421.41$

**(c)** Does the value of $r^2$ suggest that the linear model is appropriate? yes

**(d)** Find the quadratic regression equation. (Round the coefficients to the nearest 0.01.)

**(e)** Does the value of $R^2$ suggest that a quadratic model is appropriate? yes

**(f)** Use both curves to predict the number of U.S. children that are home-schooled in the year 2005. How different are the estimates?

**(g) Writing to Learn** Use the results of this exploration to explain why it is risky to use regression equations to predict $y$-values for $x$ values that are not very close to the data points, even when the curves fit the data points very well. ■

## Extending the Ideas

**41. Newton's Law of Cooling** A 190° cup of coffee is placed on a desk in a 72° room. According to Newton's Law of Cooling, the temperature $T$ of the coffee after $t$ minutes will be

$$T = (190 - 72)b^t + 72,$$

where $b$ is a constant that depends on how easily the cooling substance loses heat. The data in Table 1.14 are from a simulated experiment of gathering temperature readings from a cup of coffee in a 72° room at 20 one-minute intervals:

| Table 1.14 Cooling a Cup of Coffee | | | |
|---|---|---|---|
| Time | Temp | Time | Temp |
| 1 | 184.3 | 11 | 140.0 |
| 2 | 178.5 | 12 | 136.1 |
| 3 | 173.5 | 13 | 133.5 |
| 4 | 168.6 | 14 | 130.5 |
| 5 | 164.0 | 15 | 127.9 |
| 6 | 159.2 | 16 | 125.0 |
| 7 | 155.1 | 17 | 122.8 |
| 8 | 151.8 | 18 | 119.9 |
| 9 | 147.0 | 19 | 117.2 |
| 10 | 143.7 | 20 | 115.2 |

**(a)** Make a scatter plot of the data, with the times in list L1 and the temperatures in list L2.

**(b)** Store L2 − 72 in list L3. The values in L3 should now be an exponential function ($y = a \times b^x$) of the values in L1.

**(c)** Find the exponential regression equation for L3 as a function of L1. How well does it fit the data?

**42. Group Activity Newton's Law of Cooling** If you have access to laboratory equipment (such as a CBL or CBR unit for your grapher), gather experimental data such as in Exercise 41 from a cooling cup of coffee. Proceed as follows: Answers will vary in (a)–(e)

**(a)** First, use the temperature probe to record the temperature of the room. It is a good idea to turn off fans and air conditioners that might affect the temperature of the room during the experiment. It should be a constant.

**(b)** Heat the coffee. It need not be boiling, but it should be at least 160°. (It also need not be coffee.)

**(c)** Make a new list consisting of the temperature values minus the room temperature. Make a scatter plot of this list ($y$) against the time values ($x$). It should appear to approach the $x$-axis as an asymptote.

**(d)** Find the equation of the exponential regression curve. How well does it fit the data?

**(e)** What is the equation predicted by Newton's Law of Cooling? (Substitute your initial coffee temperature and the temperature of your room for the 190 and 72 in the equation in Exercise 41.)

**(f) Group Discussion** What sort of factors would affect the value of $b$ in Newton's Law of Cooling? Discuss your ideas with the group.

## Math at Work

When I was a kid, I always liked seeing how things work. I read books to see how things worked and I liked to tinker with things like bicycles, roller skates, or whatever. So it was only natural that I would become a mechanical engineer.

I design methods of controlling turbine speed, so a turbine can be used to generate electric power. The turbine is pushed by steam. A valve is opened to let the steam in, and the speed of the turbine is dictated by how much steam is allowed into the system. Once the turbine is spinning it generates electricity, which runs the pumps, which determine the position of the valves. In this way, the turbine actually controls its own speed.

We use oil pressure to make all of this happen. Oil cannot be compressed, so when we push oil inside a tube, the oil opens the valve, which determines how much steam is pushing the turbine. We need the pumps to push the oil, which controls the opening or closing of the valve. The turbine needs be able to get to speed in given amount of time, as well as shut

down fast enough in case of emergency, and the opening and closing of the valve controls this speed.

While computers do much of the math these days, we use calculus for stress and motion analysis, as well as vibration analysis, fluid flow, and heat analysis. One of the equations we use is:

$$\gamma_1 + \frac{v_1^2}{2} + gh_1 = \gamma_2 + \frac{v_2^2}{2} + gh^2$$

This is a *energy balance equation*. The $\gamma$'s represent density of the oil, the $v$'s represent flow velocity of the oil, the $g$'s represent the force of gravity, and the $h$'s represent the height of the oil above the point where it exerts pressure. The subscript 1's indicate these quantities at one point in the system, and the subscript 2's indicate these quantities at another point in the system.

What this equation means is that the total energy at one point in the system must equal the total energy at another point in the system. Thus, we know how much energy is exerted against the valve controls, wherever they are located in the system.

*John Jay*

## Chapter 1 Key Ideas

### Concepts

absolute maximum (p. 89)
absolute minimum (p. 89)
algebraic model (p. 63)
bounded (p. 88)
bounded above (p. 88)
bounded below (p. 88)
coefficient of determination (p. 139)
composition of functions (p. 108)
constant on an interval (p. 86)
continuous at a point (p. 84)
continuous function (p. 102)
correlation coefficient (p. 139)
decomposition of functions (p. 109)
decreasing on an interval (p. 86)
dependent variable (p. 78)
difference of functions (p. 107)
domain of a function (p. 78)
even function (p. 91)
function (p. 78)
function notation (p. 78)
graph of a function (p. 79)
grapher failure (p. 70)
graphical model (p. 65)
hidden behavior (p. 70)
horizontal asymptote (p. 94)
Horizontal Line Test (p. 118)
horizontal shrink (p. 130)
horizontal stretch (p. 130)
horizontal translation (p. 125)

implicitly-defined function (p. 112)
implied domain (p. 80)
increasing on an interval (p. 86)
independent variable (p. 78)
infinite discontinuity (p. 84)
Inverse Composition Rule (p. 120)
inverse function (p. 118)
Inverse Reflection Principle (p. 119)
inverse relation (p. 117)
jump discontinuity (p. 84)
limit notation (p. 84)
local (relative) minimum (p. 89)
local (relative) maximum (p. 89)
lower bound (p. 88)
mapping (p. 78)
mathematical model (p. 61)
mathematical modeling (p. 61)
Mirror Method (of graphing an inverse) (p. 120)
non-rigid transformations (p. 125)
numerical model (p. 61)
odd function (p. 91)
one-to-one function (p. 118)
parametrically-defined relation (p. 113)
piecewise defined function (p. 103)
product of functions (p. 107)
proof (p. 129)
quotient of functions (p. 107)
range of a function (p. 78)

reflection across $x$-axis (p. 127)
reflection across $y$-axis (p. 127)
reflection through a line (p. 127)
regression line (p. 138)
regression types (p. 138)
relation (p. 111)
relevant domain (p. 80)
removable discontinuity (p. 83)
rigid transformation (p. 125)
root (solution) of an equation (p. 68)
Rotation Method (of graphing an inverse) (p. 120)
solve algebraically (p. 68 and 69)
solve graphically (p. 68 and 69)
solve numerically (p. 68 and 69)
sum of functions (p. 107)
symmetry of graphs (p. 91)
transformation (p. 125)
upper bound (p. 88)
vertical asymptote (p. 94)
Vertical Line Test (p. 80)
vertical shrink (p. 130)
vertical stretch (p. 130)
vertical translation (p. 125)
$x$-intercept of a graph (p. 68)
Zero Factor Property (p. 67)
zero of a function (p. 68)

### Properties, Theorems, and Formulas

#### The Zero Factor Property

A product of real numbers is zero if and only if at least one of the factors in the product is zero.

#### Vertical Line Test

A graph in the $xy$-plane defines $y$ as a function of $x$ if and only if no vertical line intersects the graph in more than one point.

#### Tests for an Even Function

**Graphical:** The graph of $y = f(x)$ is symmetric with respect to the $y$-axis.

**Algebraic:** $f(-x) = f(x)$ for all $x$ in the domain of $f$.

#### Tests for an Odd Function

**Graphical:** The graph of $y = f(x)$ is symmetric with respect to the origin, or, equivalently, looks the same upside down as it does rightside up.

**Algebraic:** $f(-x) = -f(x)$ for all $x$ in the domain of $f$.

## Horizontal Line test

The inverse of a relation is a function if and only if each horizontal line intersects the graph of the original relation in at most one point.

## Inverse Reflection Principle

The points $(a, b)$ and $(b, a)$ are symmetric with respect to the line $y = x$. Equivalently, the points $(a, b)$ and $(b, a)$ are reflections of each other in the line $y = x$.

## Inverse Composition Rule

A function $f$ is one-to-one with inverse function $g$ if and only if

$f(g(x)) = x$ for every $x$ in the domain of $g$, and

$g(f(x)) = x$ for every $x$ in the domain of $f$.

## Translations of Graphs

The graph of $y = f(x)$ is translated:

to the **right** $c$ units to become the graph of $y = f(x - c)$;

to the **left** $c$ units to become the graph of $y = f(x + c)$;

**upward** $c$ units to become the graph of $y = f(x) + c$;

**downward** $c$ units to become the graph of $y = f(x) - c$.

## Reflections of Graphs Across Axes

The graph of $y = f(x)$ is reflected:

across the $x$-axis to become the graph of $y = -f(x)$;

across the $y$-axis to become the graph of $y = f(-x)$.

## Stretches and Shrinks of Graphs

The graph of $y = f(x)$ is:

**stretched vertically** by a factor of $c > 1$ to become the graph of $y = cf(x)$;

**shrunk vertically** by a factor of $0 < c < 1$ to become the graph of $y = cf(x)$;

**stretched horizontally** by a factor of $c > 1$ to become the graph of $y = f\left(\dfrac{x}{c}\right)$;

**shrunk horizontally** by a factor of $0 < c < 1$ to become the graph of $y = f\left(\dfrac{x}{c}\right)$;

# Procedures

## Root, Zero, $x$-intercept

If $f(a) = 0$, then $a$ is:

a **root** (or **solution**) of the **equation** $f(x) = 0$;

a **zero** of the **function** $y = f(x)$;

an **$x$-intercept** of the **graph** of $y = f(x)$.

## Problem-Solving

1. Understand the problem.

2. Develop a mathematical model of the problem.

3. Solve the mathematical model and support or confirm the solution.

4. Interpret the solution in the problem setting.

### Agreement about Domain

We will assume that the **domain** of a function defined by an algebraic expression is the domain of the algebraic expression, the **implied domain**. For models, we will use a domain that fits the situation, the **practical domain**.

### Inverse Notation

If $f$ is a function then $f^{-1}$ denotes the inverse of $f$, not the reciprocal of $f$.

### Finding the Inverse of a Function

**Graphical:** The mirror method (reflection in the line $y = x$) or the rotation method (rotating the graph around the line $y = x$).

**Algebraic:** Follow these three steps:

1. Determine that there is a function $f^{-1}$ by checking that $f$ is one-to-one. State any restrictions on the domain of $f$. (It might be necessary to impose some to get a one-to-one version of $f$.)

2. Switch $x$ and $y$ in the formula $y = f(x)$.

3. Solve for $y$ to get the formula $y = f^{-1}(x)$. State any restrictions on the domain of $f^{-1}$.

## Chapter 1 Review Exercises

The collection of exercises marked in red could be used as a chapter test.

In Exercises 1–10, match the graph with the corresponding function (a)–(j) from the list below. Use your knowledge of function behavior, *not* your grapher.

(a) $f(x) = x^2 - 1$

(b) $f(x) = x^2 + 1$

(c) $f(x) = (x - 2)^2$

(d) $f(x) = (x + 2)^2$

(e) $f(x) = \dfrac{x - 1}{2}$

(f) $f(x) = |x - 2|$

(g) $f(x) = |x + 2|$

(h) $f(x) = -\sin x$

(i) $f(x) = e^x - 1$

(j) $f(x) = 1 + \cos x$

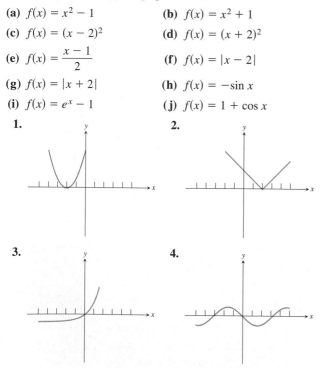

**1.**　**2.**

**3.**　**4.**

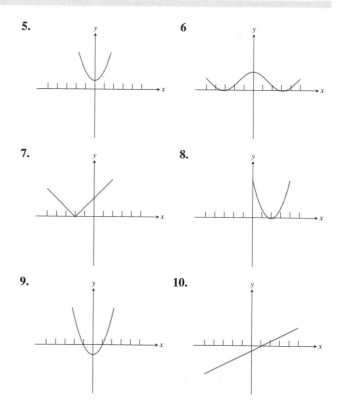

**5.**　**6.**

**7.**　**8.**

**9.**　**10.**

In Exercises 11–18, find **(a)** the domain and **(b)** the range of the function.

**11.** $g(x) = x^3$

**12.** $f(x) = 35x - 602$

**13.** $g(x) = x^2 + 2x + 1$

**14.** $h(x) = (x - 2)^2 + 5$

**15.** $g(x) = 3|x| + 8$

**16.** $k(x) = \sqrt{4 - x^2} - 2$

**17.** $f(x) = \dfrac{x}{x^2 - 2x}$     **18.** $k(x) = \dfrac{1}{\sqrt{9 - x^2}}$

In Exercises 19 and 20, graph the function, and state whether the function is continuous at $x = 0$. If it is discontinuous, state whether the discontinuity is removable or non-removable.

**19.** $f(x) = \dfrac{x^2 - 3}{x + 2}$     **20.** $k(x) = \begin{cases} 2x + 3 & \text{if } x > 0 \\ 3 - x^2 & \text{if } x \le 0 \end{cases}$

In Exercises 21–24, find all **(a)** vertical asymptotes and **(b)** horizontal asymptotes of the graph of the function. Be sure to state your answers as equations of lines.

**21.** $y = \dfrac{5}{x^2 - 5x}$     **22.** $y = \dfrac{3x}{x - 4}$  $x = 4;\ y = 3$

**23.** $y = \dfrac{7x}{\sqrt{x^2 + 10}}$     **24.** $y = \dfrac{|x|}{x + 1}$

In Exercises 25–28, graph the function and state the intervals on which the function is *increasing*.

**25.** $y = \dfrac{x^3}{6}$  $(-\infty, \infty)$     **26.** $y = 2 + |x - 1|$  $[1, \infty)$

**27.** $y = \dfrac{x}{1 - x^2}$     **28.** $y = \dfrac{x^2 - 1}{x^2 - 4}$

In Exercises 29–32, graph the function and tell whether the function is bounded above, bounded below, or bounded.

**29.** $f(x) = x + \sin x$     **30.** $g(x) = \dfrac{6x}{x^2 + 1}$  bounded

**31.** $h(x) = 5 - e^x$     **32.** $k(x) = 1000 + \dfrac{x}{1000}$

In Exercises 33–36, use a grapher to find all **(a)** relative maximum values and **(b)** relative minimum values of the function. Also state the value of $x$ at which each relative extremum occurs.

**33.** $y = (x + 1)^2 - 7$     **34.** $y = x^3 - 3x$

**35.** $y = \dfrac{x^2 + 4}{x^2 - 4}$     **36.** $y = \dfrac{4x}{x^2 + 4}$

In Exercises 37–40, graph the function and state whether the function is odd, even, or neither.

**37.** $y = 3x^2 - 4|x|$  even     **38.** $y = \sin x - x^3$  odd

**39.** $y = \dfrac{x}{e^x}$  neither     **40.** $y = x \cos (x)$  odd

In Exercises 41–44, find a formula for $f^{-1}(x)$.

**41.** $f(x) = 2x + 3$  $(x - 3)/2$     **42.** $f(x) = \sqrt[3]{x - 8}$  $x^3 + 8$

**43.** $f(x) = \dfrac{2}{x}$  $2/x$     **44.** $f(x) = \dfrac{6}{x + 4}$  $6/x - 4$

Exercises 45–52 refer to the function $y = f(x)$ whose graph is given below.

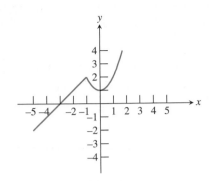

**45.** Sketch the graph of $y = f(x) - 1$.

**46.** Sketch the graph of $y = f(x - 1)$.

**47.** Sketch the graph of $y = f(-x)$.

**48.** Sketch the graph of $y = -f(x)$.

**49.** Sketch a graph of the inverse relation.

**50.** Does the inverse relation define $y$ as a function of $x$?  No

**51.** Sketch a graph of $y = f(|x|)$.

**52.** Define $f$ algebraically as a piecewise function. [Hint: the pieces are translations of two of our "basic" functions.]

In Exercises 53–58, let $f(x) = \sqrt{x}$ and let $g(x) = x^2 - 4$.

**53.** Find an expression for $(f \circ g)(x)$ and give its domain.

**54.** Find an expression for $(g \circ f)(x)$ and give its domain.

**55.** Find an expression for $(f \cdot g)(x)$ and give its domain.

**56.** Find an expression for $\left(\dfrac{f}{g}\right)(x)$ and give its domain.

**57.** Describe the end behavior of the graph of $y = f(x)$.

**58.** Describe the end behavior of the graph of $y = f(g(x))$.

In Exercises 59–64, write the specificd quantity as a function of the specified variable. Remember that drawing a picture will help.

**59. Square inscribed in a circle**  A square of side $s$ is inscribed in a circle. Write the area of the circle as a function of $s$.  $\pi s^2/2$

**60. Circle inscribed in a square**  A circle is inscribed in a square of side $s$. Write the area of the circle as a function of $s$.  $\pi s^2/4$

**61. Volume of a cylindrical tank**  A cylindrical tank with diameter 20 feet is partially filled with oil to a depth of $h$ feet. Write the volume of oil in the tank as a function of $h$.

**62. Draining a cylindrical tank**  A cylindrical tank with diameter 20 feet is filled with oil to a depth of 40 feet. The oil begins draining at a constant rate of 2 cubic feet per second. Write the volume of the oil remaining in the tank $t$ seconds later as a function of $t$.  $4000\pi - 2t$

**63. Draining a cylindrical tank** A cylindrical tank with diameter 20 feet is filled with oil to a depth of 40 feet. The oil begins draining at a constant rate of 2 cubic feet per second. Write the depth of the oil remaining in the tank $t$ seconds later as a function of $t$. $\quad 40 - t/(50\pi)$

**64. Draining a cylindrical tank** A cylindrical tank with diameter 20 feet is filled with oil to a depth of 40 feet. The oil begins draining so that the depth of oil in the tank decreases at a constant rate of 2 feet per hour. Write the volume of oil remaining in the tank $t$ hours later as a function of $t$. $\quad V = 4000\pi - 200\pi t$

**65.** The number of new packaged-goods products introduced into the marketplace each year from 1986 to 1997 is shown in Table 1.15:

| Table 1.15 New Packaged-Goods Products | |
|---|---|
| Year | New Products |
| 1986 | 12,436 |
| 1987 | 14,254 |
| 1988 | 13,421 |
| 1989 | 13,382 |
| 1990 | 15,879 |
| 1991 | 15,401 |
| 1992 | 15,886 |
| 1993 | 17,363 |
| 1994 | 21,986 |
| 1995 | 20,808 |
| 1996 | 24,486 |
| 1997 | 25,261 |

*Source: Marketing Intelligence, Ltd.*

**(a)** Sketch a scatter plot of new products ($y$) as a function of years since 1980 ($x$). (The values of $x$ will run from 6 to 17.)

**(b)** Find the equation of the linear regression line.

**(c)** Based on the regression line, approximately how many new products would be introduced in the year 2000?

**66.** The winning times in the Women's 100-Meter Freestyle event at the Summer Olympic Games since 1948 are shown in Table 1.16:

| Table 1.16 Women's 100-Meter Freestyle | | | |
|---|---|---|---|
| Year | Time | Year | Time |
| 1948 | 66.3 | 1976 | 55.65 |
| 1952 | 66.8 | 1980 | 54.79 |
| 1956 | 62.0 | 1984 | 55.92 |
| 1960 | 61.2 | 1988 | 54.93 |
| 1964 | 59.5 | 1992 | 54.64 |
| 1968 | 60 | 1996 | 54.5 |
| 1972 | 58.59 | | |

*Source: The World Almanac*

**(a)** Sketch a scatter plot of the times ($y$) as a function of the years ($x$) beyond 1900. (The values of $x$ will run from 48 to 96.)

**(b)** Explain why a linear model can not be appropriate for these times over the long term.

**(c)** The points appear to be approaching a horizontal asymptote of $y = 52$. What would this mean about the times in this Olympic event?

**(d)** Subtract 52 from all of the times so that they will approach an asymptote of $y = 0$. Redo the scatter plot with the new $y$-values. Now find the exponential regression curve and superimpose its graph on the vertically-shifted scatter plot.

**(e)** According to the regression curve, what will be the winning time in the Women's 100-Meter Freestyle at the 2000 Summer Olympics? $\quad 53.83$ seconds

**67. Inscribing a cylinder inside a sphere** A right circular cylinder of radius $r$ and height $h$ is inscribed inside a sphere of radius $\sqrt{3}$ inches.

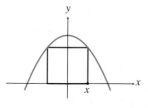

**(a)** Use the Pythagorean Theorem to write $h$ as a function of $r$. $\quad h = 2\sqrt{3 - r^2}$

**(b)** Write the volume $V$ of the cylinder as a function of $r$.

**(c)** What values of $r$ are in the domain of $V$?

**(d)** Sketch a graph of $V(r)$ over the domain $[0, \sqrt{3}]$.

**(e)** Use your grapher to find the maximum volume that such a cylinder can have.

**68. Inscribing a rectangle under a parabola**
A rectangle is inscribed between the $x$-axis and the parabola $y = 36 - x^2$ with one side along the $x$-axis, as shown in the figure below.

**(a)** Let $x$ denote the $x$-coordinate of the point highlighted in the figure. Write the area $A$ of the rectangle as a function of $x$. $\quad 72x - 2x^3$

**(b)** What values of $x$ are in the domain of $A$? $\quad [0, 6]$

**(c)** Sketch a graph of $A(x)$ over the domain.

**(d)** Use your grapher to find the maximum area that such a rectangle can have. $\quad 166.28$ square units

# Chapter 1 Project

## Modeling the Growth of a Business

In 1971, Starbucks Coffee opened its first location in Pike Place Market—Seattle's legendary open-air farmer's market. By 1987, the number of Starbucks stores had grown to 17 and by 1997 there were 1412 locations. The data in the table below (obtained from Starbucks Coffee's web site, www.starbucks.com) summarizes the growth of this company from 1987 through 1997.

| Year | Number of locations |
|------|---------------------|
| 1987 | 17 |
| 1988 | 33 |
| 1989 | 55 |
| 1990 | 84 |
| 1991 | 116 |
| 1992 | 165 |
| 1993 | 272 |
| 1994 | 425 |
| 1995 | 676 |
| 1996 | 1015 |
| 1997 | 1412 |

## Explorations

1. Enter the data in the table above into your grapher or computer. (Let $t = 0$ represent 1987.) Draw a scatter plot for the data.

2. Refer to page 138 in this chapter. Look at the types of graphs displayed and the associated regression types. Notice that the Exponential Regression model with $b > 1$ seems to most closely match the plotted data. Use your grapher or computer to find an exponential regression equation to model this data set (see your grapher's guidebook for instructions on how to do this).

3. Use the exponential model you just found to predict the total number of Starbucks locations for 1998 and 1999. Go to the Starbucks web site (www.starbucks.com) and navigate to the Timeline (found under the "Company Overview" section of the web site). Record the actual number of Starbucks locations for 1998 and 1999. Why is there such a big difference between your predicted values and the actual number of Starbucks locations? What real-world feature of business growth was not accounted for in the exponential growth model?

4. You need to model the data set with an equation that takes into account the fact that a business's growth eventually levels out or reaches a carrying capacity. Refer to page 138 again. Notice that the Logistic Regression modeling graph appears to show exponential growth at first, but eventually levels out. Use your grapher or computer to find the logistic regression equation to model this data set (see your grapher's guidebook for instructions on how to do this).

5. Use the logistic model you just found to predict the total number of Starbucks locations for 1998 and 1999. How do your predictions compare with the actual number of locations for 1998 and 1999 that you found earlier? How many locations do you think there will be in the year 2020?

# CHAPTER

# 2

# Polynomial, Power, and Rational Functions

**H**umidity is a measure of how moist the air is. Humidity affects both our comfort and our health. If humidity is too low, our skin can become dry and cracked, and viruses can live longer. If it is too high, it can make warm temperatures feel even warmer, and mold, fungi, and dust mites can live longer.

**R**elative humidity is a measure used by weather forecasters. It is the amount of water vapor in the air (called vapor pressure), compared with the amount of vapor needed to make the air saturated at the current air temperature (called saturated vapor pressure). The higher the air temperature, the more water it can hold. Relative humidity changes with the temperature and water vapor in the air.

**F**or a constant vapor pressure of 12.26 millibars and varying saturated vapor pressure, *SVP*, (i.e., varying temperatures), we can compute the relative humidity (in percent) by the following formula:

$$RH = 12.26/SVP \times 100$$

**G**iven saturated vapor pressures ranging from 12.26 to 65.43 millibars, and a constant vapor pressure of 12.26 millibars, what range of relative humidity is possible? Does relative humidity increase or decrease with increasing saturated vapor pressure? These questions are answered in Section 2.7 on page 235.

**Bibliography**

**For students:** *Aha! Insight,* Martin Gardner. W. H. Freeman, 1978. Available through Dale Seymour Publications.

*The Imaginary Tale of the Story of i,* Paul J. Nahin. Mathematical Association of America, 1998.

## Chapter 2 Overview

Chapter 1 laid a foundation of the general characteristics of functions, equations, and graphs. In this chapter and the next two, we will explore the theory and applications of specific families of functions. We begin this exploration by studying three interrelated families of functions: polynomial, power, and rational functions.

These three families of functions are used in the social, behavioral, and natural sciences. Linear functions, one of the simplest types of polynomial functions, play a central role in studying rates of change in calculus. Quadratic and higher-degree polynomial functions are used in manufacturing and advanced mathematical modeling. Power functions specify the proportional relationships of geometry, chemistry, and physics.

This chapter includes a thorough study of the theory of polynomial equations. We investigate algebraic methods for finding both real- and complex-number solutions of such equations and relate these methods to the graphical behavior of polynomial and rational functions. The chapter closes by extending these methods to inequalities in one variable.

## 2.1 Linear and Quadratic Functions with Modeling

Polynomial Functions • Linear Functions and Their Graphs • Average Rate of Change • Linear Correlation and Modeling • Quadratic Functions and Their Graphs • Free Fall

### Polynomial Functions

Polynomial functions are among the most familiar of all functions.

> **Definition  Polynomial Function**
>
> Let $n$ be a nonnegative integer and let $a_0, a_1, a_2, \ldots, a_{n-1}, a_n$ be real numbers with $a_n \neq 0$. The function given by
>
> $$f(x) = a_n x^n + a_{n-1} x^{n-1} + \cdots + a_2 x^2 + a_1 x + a_0$$
>
> is a **polynomial function of degree $n$**. The **leading coefficient** is $a_n$.
>
> The zero function $f(x) = 0$ is a polynomial function. It has no degree and no leading coefficient.

Polynomial functions are defined and continuous on all real numbers. It is important to recognize whether a function is a polynomial function.

### Example 1  IDENTIFYING POLYNOMIAL FUNCTIONS

Which of the following are polynomial functions? For those that are polynomial functions, state the degree and leading coefficient. For those that are not, explain why not.

**Bibliography**

**For teachers:** *Explorations in Precalculus Using the TI-82/TI-83, With Apendix Notes for TI-85,* Cochener and Hodge. 1997, Brooks/Cole Publishing Co., 1997.

*Algebra Experiments II, Exploring Nonlinear Functions,* Ronald J. Carlson and Mary Jean Winter. Addison-Wesley, 1993. Available through Dale Seymour Publications.

*Videos: Polynomials, Project Mathematics!,* California Institute of Technology. Available through Dale Seymour Publications.

**(a)** $f(x) = 4x^3 - 5x - \dfrac{1}{2}$  **(b)** $g(x) = 6x^{-4} + 7$

**(c)** $h(x) = \sqrt{9x^4 + 16x^2}$  **(d)** $k(x) = 15x - 2x^4$

**Solution**

**(a)** $f$ is a polynomial function of degree 3 with leading coefficient 4.

**(b)** $g$ is not a polynomial function because of the exponent $-4$.

**(c)** $h$ is not a polynomial function because it cannot be simplified into polynomial form. Notice that $\sqrt{9x^4 + 16x^2} \neq 3x^2 + 4x$.

**(d)** $k$ is a polynomial function of degree 4 with leading coefficient $-2$.

The zero function and all constant functions are polynomial functions. Some other familiar functions are also polynomial functions, as shown below.

| **Polynomial Functions of No and Low Degree** | | |
| :--- | :--- | :--- |
| **Name** | **Form** | **Degree** |
| Zero function | $f(x) = 0$ | Undefined |
| Constant function | $f(x) = a \ (a \neq 0)$ | 0 |
| Linear function | $f(x) = ax + b \ (a \neq 0)$ | 1 |
| Quadratic function | $f(x) = ax^2 + bx + c \ (a \neq 0)$ | 2 |

We study polynomial functions of degree higher than 2 in Section 2.3. For the remainder of this section, we turn our attention to the nature and uses of linear and quadratic polynomial functions.

## Linear Functions and Their Graphs

Linear equations and graphs of lines were reviewed in Sections P.3 and P.4, and many of the examples in Chapter 1 involved linear functions. We now take a closer look at the properties of linear functions.

A **linear function** is a polynomial function of degree 1. In other words, a linear function is defined for all real numbers and has the form

$$f(x) = ax + b, \text{ where } a \text{ and } b \text{ are constants and } a \neq 0.$$

If we use $m$ for the leading coefficient instead of $a$ and let $y = f(x)$, then this equation becomes the familiar slope-intercept form of a line:

$$y = mx + b$$

**Surprising Fact**

Not all lines in the Cartesian plane are graphs of linear functions.

Vertical lines are not graphs of functions because they fail the vertical line test, and horizontal lines are graphs of constant functions. A line in the Cartesian plane is the graph of a linear function if and only if it is a **slant line**, that is, neither horizontal nor vertical. We can apply the formulas and methods of Section P.4 to problems involving linear functions.

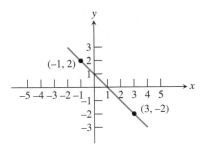

**Figure 2.1** The graph of $y = -x + 1$ passes through $(-1, 2)$ and $(3, -2)$. (Example 2)

### Example 2 FINDING THE EQUATION OF A LINEAR FUNCTION

Write an equation for the linear function $f$ such that $f(-1) = 2$ and $f(3) = -2$.

**Solution**

**Solve Algebraically**

We seek a line through the points $(-1, 2)$ and $(3, -2)$. The slope is

$$m = \frac{y_2 - y_1}{x_2 - x_1} = \frac{(-2) - 2}{3 - (-1)} = \frac{-4}{4} = -1.$$

Using this slope and the coordinates of $(-1, 2)$ with the point-slope formula, we have

$$y - y_1 = m(x - x_1)$$
$$y - 2 = -1(x - (-1))$$
$$y - 2 = -x - 1$$
$$y = -x + 1$$

Converting to function notation gives us the desired form:

$$f(x) = -x + 1$$

**Support Graphically**

We can graph $y = -x + 1$ and see that it includes the points $(-1, 2)$ and $(3, -2)$. (See Figure 2.1.)

**Confirm Numerically**

Using $f(x) = -x + 1$ we prove that $f(-1) = 2$ and $f(3) = -2$:

$f(-1) = -(-1) + 1 = 1 + 1 = 2$ and $f(3) = -(3) + 1 = -3 + 1 = -2$

## Average Rate of Change

Another property that characterizes a linear function is its *rate of change*. The **average rate of change** of a function $y = f(x)$ between $x = a$ and $x = b$, $a \neq b$, is

$$\frac{f(b) - f(a)}{b - a}.$$

The following relationship can be proved using calculus.

> **Constant Rate of Change**
>
> A function defined on all real numbers is a linear function if and only if it has a constant nonzero average rate of change between any two points on its graph.

Because the average rate of change of a linear function is constant, it is called simply the **rate of change** of the linear function. The slope $m$ in the formula $f(x) = mx + b$ is the rate of change of the linear function. In Exploration 1, we revisit Example 7 of Section P.4 in light of the rate of change concept.

---

**Exploration 1** **Modeling Depreciation with a Linear Function**

Camelot Apartments bought a $50,000 building and for tax purposes are depreciating it $2000 per year over a 25-year period using straight-line depreciation.

1. What is the rate of change of the value of the building?
2. Write an equation for the value $v(t)$ of the building as a linear function of the time $t$ since the building was placed in service.
3. Evaluate $v(0)$ and $v(16)$.   50,000; 18,000
4. Solve $v(t) = 39,000$.   $t = 5.5$

---

The rate of change of a linear function is the slope of its graph. Graphically, the slope of a linear function tells us the direction of the associated line. A linear function with a positive slope is increasing (Figures 2.2a and b), while one with a negative slope is decreasing (Figure 2.2c). In a square window, the slope also tells us the steepness of a line: the greater the absolute value of the slope, the steeper the line. So, for example, a line with slope $m = 3$ (Figure 2.2a) is steeper than a line with slope $m = 0.3$ (Figure 2.2b).

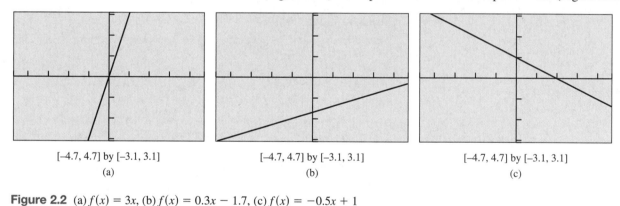

| [−4.7, 4.7] by [−3.1, 3.1] | [−4.7, 4.7] by [−3.1, 3.1] | [−4.7, 4.7] by [−3.1, 3.1] |
| (a) | (b) | (c) |

**Figure 2.2** (a) $f(x) = 3x$, (b) $f(x) = 0.3x - 1.7$, (c) $f(x) = -0.5x + 1$

The rate of change of a linear function is the signed ratio of the corresponding line's rise over run. That is, for a linear function $f(x) = mx + b$,

$$\text{rate of change} = m = \frac{\text{rise}}{\text{run}} = \frac{\text{change in } y}{\text{change in } x} = \frac{\triangle y}{\triangle x}.$$

This formula allows to interpret the slope, or rate of change, of a linear function numerically. For instance, in Exploration 1 the value of the apartment building fell from $50,000 to $0 over a 25-yr period. In Table 2.1 we compute $\triangle y/\triangle x$ for the apartment building's value (in dollars) as a function of time (in years). Because the average rate of change $\triangle y/\triangle x$ is the nonzero constant $-2000$, the building's value is a linear function of time decreasing at a rate of $2000/yr.

| Table 2.1 Rate of Change of the Value of the Apartment Building in Exploration 1: $y = -2000x + 50,000$ | | | | |
|---|---|---|---|---|
| $x$ (time) | $y$ (value) | $\triangle x$ | $\triangle y$ | $\triangle y / \triangle x$ |
| 0 | 50,000 | | | |
| | | 1 | $-2000$ | $-2000$ |
| 1 | 48,000 | | | |
| | | 1 | $-2000$ | $-2000$ |
| 2 | 46,000 | | | |
| | | 1 | $-2000$ | $-2000$ |
| 3 | 44,000 | | | |
| | | 2.5 | $-5000$ | $-2000$ |
| 5.5 | 39,000 | | | |
| | | 10.5 | $-21,000$ | $-2000$ |
| 16 | 18,000 | | | |
| | | 9 | $-18,000$ | $-2000$ |
| 25 | 0 | | | |

In Exploration 1, as in other applications of linear functions, the constant term represents the value of the function for an input of 0. In general, for any function $f$, $f(0)$ is in **initial value of** $f$. So for a linear function $f(x) = mx + b$, the constant term $b$ is the initial value. For any polynomial function $f(x) = a_n x^n + \cdots + a_1 x + a_0$, the **constant term** $f(0) = a_0$ is its initial value. Finally, the initial value of any function—polynomial or otherwise—is the $y$-intercept of its graph.

We now summarize what we have learned about linear functions.

**Characterizing the Nature of a Linear Function**

| Point of View | Characterization |
|---|---|
| Verbal | polynomial of degree 1 |
| Algebraic | $f(x) = mx + b \quad (m \neq 0)$ |
| Graphical | slant line with slope $m$ and $y$-intercept $b$ |
| Analytical | function with constant nonzero rate of change $m$: $f$ is increasing if $m > 0$, decreasing if $m < 0$; initial value of the function $= f(0) = b$ |

## Linear Correlation and Modeling

In Section 1.6 we approached modeling from several points of view. Along the way we learned how to use a grapher to create a scatter plot, compute a regression line for a data set, and overlay a regression line on a scatter plot. We touched on the notion of correlation coefficient. We now go deeper into these modeling and regression concepts.

Figure 2.3 shows five types of scatter plots. When the points of a scatter plot are clustered along a line, we say there is a **linear correlation** between the quantities represented by the data. When an oval is drawn around the points in the scatter plot, generally speaking, the narrower the oval, the stronger the linear correlation.

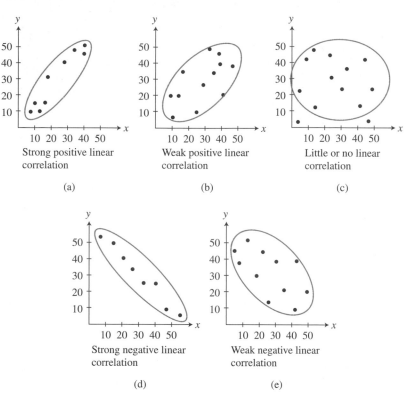

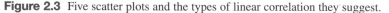

**Figure 2.3** Five scatter plots and the types of linear correlation they suggest.

When the oval tilts like a line with positive slope (as in Figure 2.3a and b), the data have a **positive linear correlation**. On the other hand, when it tilts like a line with negative slope (as in Figure 2.3d and e), the data have a **negative linear correlation**. Some scatter plots exhibit little or no linear correlation (as in Figure 2.3c), or have non-linear patterns.

A number that measures the strength and direction of the linear correlation of a data set is the **(linear) correlation coefficient, $r$**.

**Properties of the Correlation Coefficient, $r$**

1. $-1 \leq r \leq 1$.
2. When $r > 0$, there is a positive linear correlation.
3. When $r < 0$, there is a negative linear correlation.
4. When $|r| \approx 1$, there is a strong linear correlation.
5. When $r \approx 0$, there is weak or no linear correlation.

Correlation informs the modeling process by giving us a measure of goodness of fit. Good modeling practice, however, demands that we have a theoretical reason for selecting a model. In business, for example, fixed cost is modeled by a constant function. (Otherwise, the cost would not be fixed.)

In economics, a linear model is often used for the demand for a product as a function of its price. For instance, suppose Twin Pixie, a large supermarket chain, conducts a market analysis on its store brand of doughnut-shaped oat breakfast cereal. The chain sets various prices for its 15-oz box at its different stores over a period of time. Then using this data the Twin Pixie researchers project the weekly sales at the entire chain of stores for each price and obtain the data shown in Table 2.2.

| Table 2.2  Weekly Sales Data Based on Marketing Research | |
|---|---|
| Price per box | Boxes sold |
| $2.40 | 38,320 |
| $2.60 | 33,710 |
| $2.80 | 28,280 |
| $3.00 | 26,550 |
| $3.20 | 25,530 |
| $3.40 | 22,170 |
| $3.60 | 18,260 |

### Example 3  MODELING AND PREDICTING DEMAND

Use the data in Table 2.2 to write a linear model for demand (in boxes sold per week) as a function of the price per box (in dollars). Describe the strength and direction of the linear correlation. Then use the model to predict weekly cereal sales if the price is dropped to $2.00 or raised to $4.00 per box.

### Solution

#### Model

We enter the data and obtain the scatterplot shown in Figure 2.9a. It appears that the data have a strong negative correlation.

We then find the linear regression model to be approximately

$$y = -15,358.93x + 73,622.50,$$

where $x$ is the price per box of cereal and $y$ the number of boxes sold.

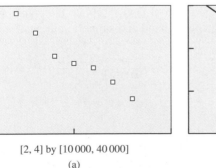

[2, 4] by [10 000, 40 000]

(a)

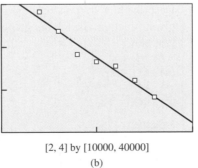

[2, 4] by [10000, 40000]

(b)

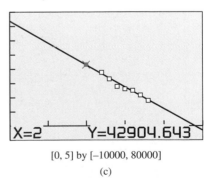

X=2    Y=42904.643

[0, 5] by [−10000, 80000]

(c)

**Figure 2.4**  Scatter plot and regression line graphs for Example 3.

**Teacher's Note**

On some graphers, you need to select Diagnostic On from the Catalog to see the correlation coefficient.

Figure 2.4b shows the scatterplot for Table 2.2 together with a graph of the regression line. You can see that the line fits the data fairly well. The correlation coefficient of $r \approx -0.98$ supports this visual evidence.

### Solve Graphically

Our goal is to predict the weekly sales for prices of $2.00 and $4.00 per box. Using the Trace feature of the grapher, as shown in Figure 2.9b, we see that $y$ is about 42,900 when $x$ is 2. In a similar manner we could find that $y \approx 12,190$ when $x$ is 4.

### Interpret

If Twin Pixie drops the price for its store brand of doughnut-shaped oat breakfast cereal to $2.00 per box, sales will rise to about 42,900 boxes per week. On the other hand, if they raise the price to $4.00 per box, sales will drop to around 12,190 boxes per week.

We summarize for future reference the analysis used in Example 3.

---

**Regression Analysis**

1. Enter and plot the data (scatter plot).
2. Find the regression model that fits the problem situation.
3. Superimposed the graph of the regression model on the scatter plot, and observe the fit.
4. Use the regression model to make the predictions called for in the problem.

---

## Quadratic Functions and Their Graphs

A **quadratic function** is a polynomial function of degree 2. Recall from Section 1.3 that the graph of the squaring function $f(x) = x^2$ is a parabola. We will see that the graph of every quadratic function is an upward or downward opening parabola. This is because the graph of any quadratic function can be obtained from the graph of the squaring function $f(x) = x^2$ by a sequence of translations, reflections, stretches, and shrinks.

**Example 4** TRANSFORMING THE SQUARING FUNCTION

Describe how to transform the graph of $f(x) = x^2$ into the graph of the given function. Sketch the graph by hand and support your answer with a grapher.

**(a)** $g(x) = \dfrac{1}{2}x^2$                  **(b)** $h(x) = -3x^2$

**(c)** $k(x) = -x^2 + 3$            **(d)** $q(x) = (x + 2)^2 + 1$

**Solution**

**(a)** The graph of $g(x) = (1/2)x^2$ is obtained by vertically shrinking the graph of $f(x) = x^2$ by a factor of 1/2. See Figure 2.5a.

**(b)** The graph of $h(x) = -3x^2$ is obtained by reflecting the graph of $f(x) = x^2$ across the $x$-axis and vertically stretching it by a factor of 3. See Figure 2.5b.

**(c)** The graph of $k(x) = -x^2 + 3$ is obtained by reflecting the graph of $f(x) = x^2$ across the $x$-axis and vertically translating it up 3 units. See Figure 2.5c.

**(d)** The graph of $q(x) = (x + 2)^2 + 1$ is obtained by shifting the graph of $f(x) = x^2$ to the left 2 units and up 1 unit. See Figure 2.5d.

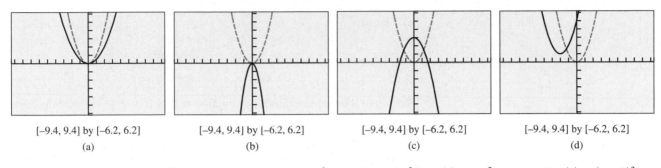

| $[-9.4, 9.4]$ by $[-6.2, 6.2]$ | $[-9.4, 9.4]$ by $[-6.2, 6.2]$ | $[-9.4, 9.4]$ by $[-6.2, 6.2]$ | $[-9.4, 9.4]$ by $[-6.2, 6.2]$ |
| :---: | :---: | :---: | :---: |
| (a) | (b) | (c) | (d) |

**Figure 2.5** The graph $f(x) = x^2$ shown with (a) $g(x) = (1/2)x^2$, (b) $h(x) = -3x^2$, (c) $k(x) = -x^2 + 3$, and (d) $q(x) = (x + 2)^2 + 1$. (Example 4)

The functions in Example 4a and Example 4b are of the form $f(x) = ax^2$. Their orientation is typical of functions of this type. When $a > 0$, $f(x) = ax^2$ graphs as an upward opening parabola. When $a < 0$, its graph is a downward opening parabola. Regardless of the sign of $a$, the $y$-axis is the line of symmetry for the graph of $f(x) = ax^2$. The line of symmetry for a parabola is its **axis of symmetry**, or **axis** for short. The point on the parabola that intersects its axis is the **vertex** of the parabola. Because the graph of a quadratic function is always an upward or downward opening parabola, its vertex is always the lowest or highest point of the parabola. The vertex of $f(x) = ax^2$ is always the origin, as seen in Figure 2.6.

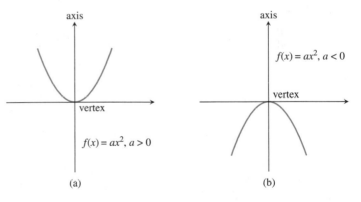

| (a) | (b) |
| :---: | :---: |

**Figure 2.6** The graph $f(x) = ax^2$ for (a) $a > 0$ and (b) $a < 0$.

In Example 4d, $q(x)$ is not written in the **standard quadratic form** $ax^2 + bx + c$, where the powers of $x$ are arranged in descending order. Is $q$ a quadratic function? We can see that it is by expanding the given formula for $q(x)$.

$$q(x) = (x + 2)^2 + 1$$
$$= x^2 + 4x + 4 + 1$$
$$= x^2 + 4x + 5$$

This expansion process can be generalized.

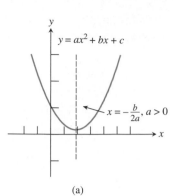

(a)

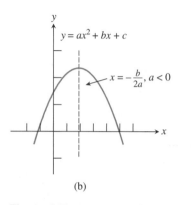

(b)

**Figure 2.7**  The vertex is at $x = -b/(2a)$, which therefore also describes the axis of symmetry. (a) When $a > 0$, the parabola opens upward. (b) When $a < 0$, the parabola opens downward.

---

**Teacher's Note**

Another way of converting a quadratic function from the polynomial form $f(x) = ax^2 + bx + c$ to the vertex form $f(x) = a(x + h)^2 + k$ is completing the square, a method used in Section P.5 to solve quadratic equations. The "confirm" step in Example 5 illustrates this method.

---

$$f(x) = a(x - h)^2 + k$$
$$= a(x^2 - 2hx + h^2) + k$$
$$= ax^2 + (-2ah)x + (ah^2 + k)$$
$$= ax^2 + bx + c \qquad \text{Let } b = -2ah \text{ and } c = ah^2 + k.$$

Because $b = -2ah$ and $c = ah^2 + k$ in the last line above, $h = -b/(2a)$ and $k = c - ah^2$. Using these formulas, any quadratic function $f(x) = ax^2 + bx + c$ can be rewritten in the form

$$f(x) = a(x - h)^2 + k.$$

This **vertex form** for a quadratic function makes it easy to identify the vertex and axis of the graph the function, and to sketch the graph.

---

**Vertex Form of a Quadratic Function**

Any quadratic function $f(x) = ax^2 + bx + c$, $a \neq 0$, can be written in the vertex form

$$f(x) = a(x - h)^2 + k.$$

The graph of $f$ is a parabola with vertex $(h, k)$ and axis $x = h$, where $h = -b/(2a)$ and $k = c - ah^2$. If $a > 0$, the parabola opens upward, and if $a < 0$, it opens downward. (See Figure 2.7.)

---

The formula $h = -b/(2a)$ is useful for locating the vertex and axis of the parabola associated with a quadratic function. To help you remember it, notice that $-b/(2a)$ is part of the quadratic formula

$$x = \frac{-b \pm \sqrt{b^2 - 4ac}}{2a}.$$

(Cover the radical term.) You need not remember $k = c - ah^2$ because you can use $k = f(h)$ instead.

**Example 5    FINDING THE VERTEX AND AXIS OF A QUADRATIC FUNCTION**

Find the vertex and axis of the graph of $f(x) = 6x - 3x^2 - 5$. Rewrite the equation in vertex form.

**Solution**

**Solve Algebraically**

*The standard polynomial form of $f$ is*

$$f(x) = -3x^2 + 6x - 5$$

So $a = -3$, $b = 6$, and $c = -5$, and the coordinates of the vertex are

$$h = \frac{-b}{2a} = \frac{-6}{2(-3)} = 1 \quad \text{and}$$
$$k = f(h) = f(1) = -3(1)^2 + 6(1) - 5 = -2.$$

The equation of the axis is $x = 1$, and the vertex form of $f$ is

$$f(x) = -3(x - 1)^2 + (-2).$$

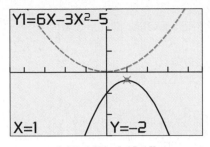

[−4.7, 4.7] by [−15, 15]

**Figure 2.8** The graph of $f(x) = 6x - 3x^2 - 5$ with its vertex highlighted. (Example 5)

## Support Graphically

The graph of $f$ is the downward opening parabola shown in Figure 2.8.

## Confirm Algebraically

*Completing the square provides an alternative way to obtain the vertex form.*

$$f(x) = -3x^2 + 6x - 5$$

$$= -3(x^2 - 2x) - 5 \qquad \text{Factor } -3 \text{ from the } x \text{ terms.}$$

$$= -3(x^2 - 2x + (\ \ ) - (\ \ )) - 5 \qquad \text{Prepare to complete the square.}$$

$$= -3(x^2 - 2x + 1 - 1) - 5 \qquad \text{Complete the square.}$$

$$= -3(x^2 - 2x + 1) - 3(-1) - 5 \qquad \text{Distribute the } -3.$$

$$= -3(x - 1)^2 - 2$$

We now summarize what we know about quadratic functions.

---

### Characterizing the Nature of a Quadratic Function

| Point of View | Characterization |
|---|---|
| Verbal | polynomial of degree 2 |
| Algebraic | $f(x) = ax^2 + bx + c$ or $f(x) = a(x - h)^2 + k \ (a \neq 0)$ |
| Graphical | parabola with vertex $(h, k)$ and axis $x = h$; |
| | opens upward if $a > 0$, opens downward if $a < 0$; |
| | initial value = $y$-intercept = $f(0) = c$; |
| | $x\text{-intercepts} = \dfrac{-b \pm \sqrt{b^2 - 4ac}}{2a}$ |

---

In economics, when demand is linear, revenue is quadratic. Example 6 illustrates this by extending the Twin Pixie model of Example 3.

### Example 6   PREDICTING MAXIMUM REVENUE

Use the model $y = -15{,}358.93x + 73{,}622.50$ from Example 3 to develop a model for the total weekly revenue on doughnut-shaped cereal sales. Determine the maximum revenue and how to achieve it.

#### Solution

#### Model

Revenue can be found by multiplying the price per box, $x$, by the number of boxes sold, $y$. So the total revenue is given by

$$R(x) = x \cdot y = -15{,}358.93x^2 + 73{,}622.50x,$$

a quadratic model.

#### Solve Graphically

In Figure 2.9, we find a maximum of about 88,227 occurs when $x$ is about 2.40.

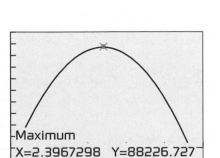

[0, 5] by [−10000, 100000]

**Figure 2.9** The revenue model for Example 6.

## Confirm Algebraically

The maximum value should occur at the vertex of the parabola, whose $x$-coordinate is

$$x = \frac{-b}{2a} = \frac{-73{,}622.50}{2(-15{,}358.93)} \approx 2.40$$

## Interpret

To maximize revenue, Twin Pixie should set the price for its store brand of doughnut-shaped oat breakfast cereal at $2.40 per box. Based on the model, this will yield a total revenue of about $88,227.

Example 6 shows that we sometimes need to extend a model to find the answer we want. By using a linear model for the demand function in Example 3, the revenue function in Example 6 was quadratic. This made it easy to determine the price that would yield maximum revenue.

Recall that the average rate of change of a linear function is constant. If we look at a table of values for the quadratic function in Example 6 (Figure 2.10), we can see that the average rate of change is not constant. In fact, the average rate of change is first positive and then negative. This should seem reasonable because in Example 6 we found the maximum point of the quadratic function, the point where the function switched from increasing to decreasing.

In calculus you will study not only average rate of change but also *instantaneous rate of change*. Such instantaneous rates include velocity and acceleration, which we now begin to investigate.

| PRICE | REV | RTCHG |
|---|---|---|
| .5 | 32972 | 50584 |
| 1 | 58264 | 35225 |
| 1.5 | 75876 | 19866 |
| 2 | 85809 | 4507.3 |
| 2.5 | 88063 | –10852 |
| 3 | 82637 | –26211 |
| 3.5 | 69532 |  |

**Figure 2.10** The average rate of change of a quadratic function is not constant. The list RTCHG in Column 3 can be found using

$$\text{RTCHG} = \frac{\triangle \text{LIST (LREV)}}{\triangle \text{LIST (LPRICE)}}.$$

## Free Fall

Since the time of Galileo Galilei (1564–1642) and Isaac Newton (1642–1727), the vertical motion of a body in free fall has been well understood. The *vertical velocity* and *vertical position (height)* of a free-falling body (as functions of time) are classical applications of linear and quadratic functions.

> ### Vertical Free-Fall Motion
>
> The **height** $s$ and **vertical velocity** $v$ of an object in free fall are given by
>
> $$s(t) = -\frac{1}{2}gt^2 + v_0 t + s_0 \quad \text{and} \quad v(t) = -gt + v_0,$$
>
> where $t$ is time (in seconds), $g \approx 32$ ft/sec$^2$ $\approx 9.8$ m/sec$^2$ is the **acceleration due to gravity**, $v_0$ is the *initial vertical velocity* of the object, and $s_0$ is its *initial height*.

These formulas disregard air resistance, and the two values given for $g$ are valid at sea level. We apply these formulas in Examples 7 and 8, and will use them from time to time throughout the rest of the textbook.

### Example 7 INVESTIGATING FREE-FALL MOTION

As a promotion for the new Houston Astros downtown ballpark, a competition is held to see who can throw a baseball the farthest up in the air from the front row of the upper deck of seats, 83 ft above field level. The winner throws the ball with an initial vertical velocity of 92 ft/sec and it lands on the infield grass. Find

**(a)** the maximum height of the baseball,

**(b)** its time in the air, and

**(c)** its vertical velocity when it hits the ground.

### Solution

### Model

Because the distance units are in feet, we use $g \approx 32$ ft/sec$^2$. It is given that $s_0 = 83$ ft and $v_0 = 92$ ft/sec. So the models are

$$s(t) = -16t^2 + 92t + 83 \quad \text{and} \quad v(t) = -32t + 92.$$

### (a) Solve Algebraically

The maximum height is the maximum value of the quadratic function $s$. This maximum will occur at the vertex, which has coordinates

$$h = \frac{-b}{2a} = \frac{-92}{2(-16)} = 2.875 \quad \text{and}$$

$$k = s(2.875) = -16(2.875)^2 + 92(2.875) + 83 = 215.25$$

### Interpret

The maximum height of the baseball is about 215 ft above field level.

### (b) Solve Graphically

In Figure 2.11, we plot height versus time and calculate the positive-valued zero of the height function, which is $t \approx 6.54$.

### Confirm Algebraically

Using the quadratic formula, we obtain

$$t = \frac{-92 \pm \sqrt{(92)^2 - 4(-16)(83)}}{2(-16)} = \frac{-92 \pm \sqrt{13776}}{-32} \approx -0.79 \text{ or } 6.54.$$

### Interpret

The baseball spends about 6.54 sec in the air.

### (c) Solve Numerically

We now turn to the linear model for vertical velocity.

$$v(6.54\ldots) = -32(6.54\ldots) + 92 \approx -117 \text{ ft/sec.}$$

### Interpret

The baseball's downward rate is about 117 ft/sec when it hits the ground.

Example 7 shows how the theory of free-fall motion can be used in practice. Though this theory was developed long before modern data collection devices arrived upon the scene, devices such as the Calculator-Based Ranger™ (CBR™) can help us appreciate this classical theory.

The data in Table 2.3 were collected on Monday, October 11, 1999, in Boone, North Carolina (about 1 km above sea level), using a CBR and a 15-cm rubber air-filled ball. The CBR was placed on the floor face up. The ball was thrown upward above the CBR, and it landed directly on the face of the device.

---

**Rounding Answers**

Remember to carry all decimals in your calculations until you reach a final answer.

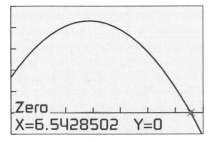

[0, 7] by [−50, 250]

**Figure 2.11** The graph of $y = -16x^2 + 92x + 83$ where $y$ represents the height of the baseball in feet and $x$ represents time in seconds. (Example 7)

| Table 2.3 Rubber Ball Data from CBR | |
| --- | --- |
| Time (sec) | Height (m) |
| 0.0000 | 1.03754 |
| 0.1080 | 1.40205 |
| 0.2150 | 1.63806 |
| 0.3225 | 1.77412 |
| 0.4300 | 1.80392 |
| 0.5375 | 1.71522 |
| 0.6450 | 1.50942 |
| 0.7525 | 1.21410 |
| 0.8600 | 0.83173 |

## Assignment Guide

Day 1: Ex. 1–12 odds
Day 2: Ex. 15–39, multiples of 3
Day 3: Ex. 42, 46, 47, 49, 50, 51, 59, 60, 61, 62, 70, 71

## Cooperative Learning

Group Activity: Ex. 55, 56

## Notes on Exercises

Ex. 35–40 require the student to produce a quadratic function with certain qualities.
Ex. 51–59 offer some classical applications of quadratic functions.
Ex. 60, 62, 63 are applications of quadratic and linear regression.

## Ongoing Assessment

Self-Assessment: Ex. 6, 23, 57, 64
Embedded Assessment: Ex. 36, 59

## Reminder

Recall from Section 1.6 that $R^2$ is the coefficient of determination, which measures goodness of fit.

### Example 8   MODELING VERTICAL FREE-FALL MOTION

Use the data in Table 2.3 to write models for the height and vertical velocity of the rubber ball. Then use these models to predict the maximum height of the ball and its vertical velocity when it hits the face of the CBR.

**Solution**

### Model

First we make a scatter plot of the data, as shown in Figure 2.12a. Using quadratic regression, we find the model for the height of the ball to be about

$$s(t) = -4.676t^2 + 3.758t + 1.045,$$

with $R^2 \approx 0.999$, indicating an excellent fit.

Our free-fall theory says the leading coefficient of $-4.676$ is $-g/2$, giving us a value for $g \approx 9.352$ m/sec$^2$, which is a bit less than the theoretical value of 9.8 m/sec$^2$. We also obtain $v_0 \approx 3.758$ m/sec. So the model for vertical velocity becomes

$$v(t) = -gt + v_0 \approx -9.352t + 3.758.$$

### Solve Graphically and Numerically

The maximum height is the maximum value of $s(t)$, which occurs at the vertex of its graph. We can see from Figure 2.12b that the vertex has coordinates of about (0.402, 1.800).

In Figure 2.12c, to determine when the ball hits the face of the CBR, we calculate the positive-valued zero of the height function, which is $t \approx 1.022$. We turn to our linear model to compute the vertical velocity at impact:

$$v(1.022) = -9.352(1.022) + 3.758 \approx -5.80 \text{ m/sec}$$

### Interpret

The maximum height the ball achieved was about 1.80 m above the face of the CBR. The ball's downward rate is about 5.80 m/sec when it hits the CBR.

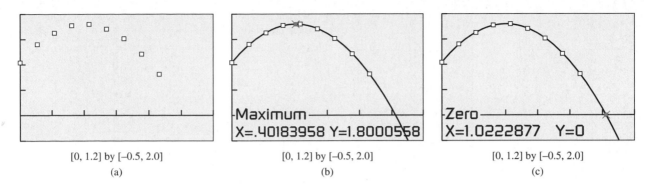

| [0, 1.2] by [−0.5, 2.0] | [0, 1.2] by [−0.5, 2.0] | [0, 1.2] by [−0.5, 2.0] |
| (a) | (b) | (c) |

**Figure 2.12** Scatter plot and graph of height versus time for Example 8.

**Follow-up**

Ask the students to write a quadratic equation that is shifted 3 units to the left and 4 units down from the origin. ($f(x) = x^2 + 6x + 5$)

The curve in Figure 2.12b appears to fit the data extremely well, and $R^2 \approx 0.999$. You may have noticed, however, that Table 2.3 contains the ordered pair (0.4300, 1.80392) and that $1.80392 > 1.800$, which is the maximum shown in Figure 2.12b. So, even though our model is theoretically based and an excellent fit to the data, it is not a perfect model. Despite its imperfections, the model provides accurate and reliable predictions about the CBR experiment.

# Quick Review 2.1

In Exercises 1–2, write an equation in slope-intercept form for a line with the given slope $m$ and $y$-intercept $b$.

**1.** $m = 8$, $b = 3.6$  **2.** $m = -1.8$, $b = -2$

In Exercises 3–4, write an equation for the line containing the given points. Graph the line and points.

**3.** $(-2, 4)$ and $(3, 1)$  **4.** $(1, 5)$ and $(-2, -3)$

In Exercises 5–8, expand the expression.

**5.** $(x + 3)^2$ $x^2 + 6x + 9$  **6.** $(x - 4)^2$ $x^2 - 8x + 16$
**7.** $3(x - 6)^2$ $3x^2 - 36x + 108$ **8.** $-3(x + 7)^2$

In Exercises 9–10, factor the trinomial.

**9.** $2x^2 - 4x + 2$ $2(x - 1)^2$ **10.** $3x^2 + 12x + 12$ $3(x + 2)^2$

# Section 2.1 Exercises

In Exercises 1–6, determine which are polynomial functions. For those that are, state the degree and leading coefficient. For those that are not, explain why not.

**1.** $f(x) = 3x^{-5} + 17$  **2.** $f(x) = -9 + 2x$

**3.** $f(x) = 2x^5 - \dfrac{1}{2}x + 9$  **4.** $f(x) = 13$

**5.** $h(x) = \sqrt[3]{27x^3 + 8x^6}$  **6.** $k(x) = 4x - 5x^2$

In Exercises 7–12, write an equation for the linear function $f$ satisfying the given conditions. Graph $y = f(x)$.

**7.** $f(-5) = -1$  and  $f(2) = 4$
**8.** $f(-3) = 5$  and  $f(6) = -2$
**9.** $f(-4) = 6$  and  $f(-1) = 2$
**10.** $f(1) = 2$  and  $f(5) = 7$
**11.** $f(0) = 3$  and  $f(3) = 0$
**12.** $f(-4) = 0$  and  $f(0) = 2$

In Exercises 13–18, match a graph to the function. Explain your choice.

**13.** $f(x) = 2(x + 1)^2 - 3$  (a)  **14.** $f(x) = 3(x + 2)^2 - 7$  (d)
**15.** $f(x) = 4 - 3(x - 1)^2$  (b)  **16.** $f(x) = 12 - 2(x - 1)^2$  (f)
**17.** $f(x) = 2(x - 1)^2 - 3$  (e)  **18.** $f(x) = 12 - 2(x + 1)^2$  (c)

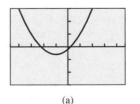
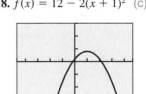

(a)                    (b)

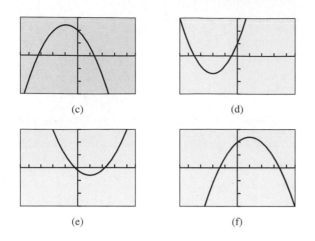

(c)          (d)

(e)          (f)

In Exercises 19–28, find the vertex and axis of the graph of the function. Support your answer graphically.

**19.** $f(x) = 3(x - 1)^2 + 5$     **20.** $g(x) = -3(x + 2)^2 - 1$

**21.** $f(x) = 5(x - 1)^2 - 7$     **22.** $g(x) = 2(x - \sqrt{3})^2 + 4$

**23.** $f(x) = 3x^2 + 5x - 4$     **24.** $f(x) = -2x^2 + 7x - 3$

**25.** $f(x) = 8x - x^2 + 3$     **26.** $f(x) = 6 - 2x + 4x^2$

**27.** $g(x) = 5x^2 + 4 - 6x$     **28.** $h(x) = -2x^2 - 7x - 4$

In Exercises 29–34, use algebraic methods to describe the graph of the function. Sketch the graph by hand. Support your answers graphically.

**29.** $f(x) = x^2 - 4x + 6$     **30.** $g(x) = x^2 - 6x + 12$

**31.** $f(x) = 10 - 16x - x^2$     **32.** $h(x) = 8 + 2x - x^2$

**33.** $f(x) = 2x^2 + 6x + 7$     **34.** $g(x) = 5x^2 - 25x + 12$

In Exercises 35–38, write an equation for the parabola shown, using the fact that one of the given points is the vertex.

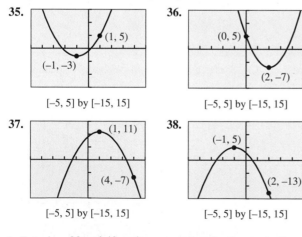

**35.**
$(1, 5)$
$(-1, -3)$
$[-5, 5]$ by $[-15, 15]$

**36.**
$(0, 5)$
$(2, -7)$
$[-5, 5]$ by $[-15, 15]$

**37.**
$(1, 11)$
$(4, -7)$
$[-5, 5]$ by $[-15, 15]$

**38.**
$(-1, 5)$
$(2, -13)$
$[-5, 5]$ by $[-15, 15]$

In Exercises 39 and 40, write an equation for the quadratic function whose graph contains the given vertex and point.

**39.** Vertex $(1, 3)$, point $(0, 5)$

**40.** Vertex $(-2, -5)$, point $(-4, -27)$

In Exercises 41–44, describe the strength and direction of the linear correlation.

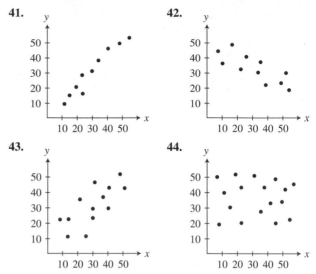

**41.**

**42.**

**43.**

**44.**

**45. Comparing Age and Weight** A group of male children were weighed. Their ages and weights are recorded in Table 2.4.

**(a)** Draw a scatter plot of these data.

**(b) Writing to Learn** Describe the strength and direction of the correlation between age and weight.

| Table 2.4 Children's Age and Weight | |
|---|---|
| Age (months) | Weight (pounds) |
| 18 | 23 |
| 20 | 25 |
| 24 | 24 |
| 26 | 32 |
| 27 | 33 |
| 29 | 29 |
| 34 | 35 |
| 39 | 39 |
| 42 | 44 |

**46. Life Expectancy** Table 2.5 shows the average number of additional years a U.S. citizen is expected to live for various ages.

**(a)** Draw a scatter plot of these data.

**(b) Writing to Learn** Describe the strength and direction of the correlation between age and life expectancy.

**Table 2.5 U.S. Life Expectancy**

| Age (years) | Life Expectancy (years) |
|---|---|
| 10 | 66 |
| 20 | 56 |
| 30 | 47 |
| 40 | 37 |
| 50 | 29 |
| 60 | 20 |
| 70 | 14 |
| 80 | 8 |

*Source: U.S. National Center for Health Statistics, Vital Statistics of the United States.*

**47. Straight-Line Depreciation** MaiLee bought a computer for her home office and depreciated it over 5 years using the straight-line method. If its initial value was $2350, what is its value 3 years later? $940

**48. Costly Doll Making** Patrick's doll-making business has weekly fixed costs of $350. If the cost for materials is $4.70 per doll and his total weekly costs average $500, about how many dolls does Patrick make each week? 32

**49.** Table 2.6 shows the mean annual compensation of construction workers. Let $x = 0$ stand for 1990, $x = 1$ for 1991, and so forth.

**Table 2.6 Construction Worker Average Annual Compensation**

| Year | Annual Compensation (dollars) |
|---|---|
| 1990 | 33,701 |
| 1994 | 37,275 |
| 1995 | 37,417 |
| 1996 | 38,456 |

*Source: U.S. Bureau of Economic Analysis, Statistical Abstract of the United States, 1998.*

**(a)** Find the linear regression model for the data. What does the slope in the regression model represent?

**(b)** Use the linear regression model to predict the construction worker average annual compensation in the year 2002. About $43,168

**50. Total Revenue** The per unit price $p$ (in dollars) of a popular toy when $x$ units (in thousands) are produced is modeled by the function

$$\text{price} = p = 12 - 0.025x.$$

The revenue (in thousands of dollars) is the product of the price per unit and the number of units (in thousands) produced. That is,

$$\text{revenue} = xp = x(12 - 0.025x).$$

**(a)** State the dimensions of a viewing window that shows a graph of the revenue model for producing 0 to 100,000 units.

**(b)** How many units should be produced if the total revenue is to be $1,000,000.

**51. Finding Maximum Area** Among all the rectangles whose perimeters are 100 ft, find the dimensions of the one with maximum area. 25 × 25 ft

**52. Finding the Dimensions of a Painting** A large painting in the style of Rubens is 3 ft longer than it is wide. If the wooden frame is 12 in. wide, the area of the picture and frame is 208 ft², find the dimensions of the painting.

**53. Using Algebra in Landscape Design** Julie Stone designed a rectangular patio that is 25 ft by 40 ft. This patio is surrounded by a terraced strip of uniform width planted with small trees and shrubs. If the area $A$ of this terraced strip is 504 ft², find the width $x$ of the strip. 3.5 ft

**54. Management Planning** The Welcome Home apartment rental company has 1600 units available, of which 800 are currently rented at $300 per month. A market survey indicates that each $5 decrease in monthly rent will result in 20 new leases.

**(a)** Determine a function $R(x)$ that models the total rental income realized by Welcome Home, where $x$ is the number of $5 decreases in monthly rent.

**(b)** Find a graph of $R(x)$ for rent levels between $175 and $300 (that is, $0 \le x \le 25$) that clearly shows a maximum for $R(x)$.

**(c)** What rent will yield Welcome Home the maximum monthly income? $250 per month

**55. Group Activity Beverage Business** The Sweet Drip Beverage Co. sells cans of soda pop in machines. It finds that sales average 26,000 cans per month when the cans sell for 50¢ each. For each nickel increase in the price, the sales per month drop by 1000 cans.

**(a)** Determine a function $R(x)$ that models the total revenue realized by Sweet Drip, where $x$ is the number of $0.05 increases in the price of a can.

**(b)** Find a graph of $R(x)$ that clearly shows a maximum for $R(x)$.

**(c)** How much should Sweet Drip charge per can to realize the maximum revenue? What is the maximum revenue?

**56. Group Activity Sales Manager Planning** Jack was named District Manager of the Month at the Sylvania Wire Co. due to his hiring study. It shows that each of the 30 salespersons he supervises average $50,000 in sales each month, and that for each additional salesperson he would hire, the average sales would decrease $1000 per month. Jack concluded his study by suggesting a number of salespersons that he should hire to maximize sales. What was that number? 10

**57. Baseball Throwing Machine** The Sandusky Little League uses a baseball throwing machine to help train 10-year-old players to catch high pop-ups. It throws the baseball straight up with an initial velocity of 48 ft/sec from a height of 3.5 ft.

**(a)** Find an equation that models the height of the ball $t$ seconds after it is thrown. $h = -16t^2 + 48t + 3.5$

**(b)** What is the maximum height the baseball will reach? How many seconds will it take to reach that height?

**58. Fireworks Planning** At the Bakersville Fourth of July celebration, fireworks are shot by remote control into the air from a pit that is 10 ft below the earth's surface.

**(a)** Find an equation that models the height of an aerial bomb $t$ seconds after it is shot upwards with an initial velocity of 80 ft/sec. Graph the equation in both function mode and parametric mode.

**(b)** What is the maximum height above ground level that the aerial bomb will reach? How many seconds will it take to reach that height? 90 ft, 2.5 sec

**59. Landscape Engineering** In her first project after being employed by Land Scapes International, Becky designs a decorative water fountain that will shoot water to a maximum height of 48 ft. What should be the initial velocity of each drop of water to achieve this maximum height? (*Hint*: Use a grapher and a guess-and-check strategy.)

**60. Patent Applications** Using quadratic regression on the data in Table 2.7, predict the year when the number of patent application will reach 300,000. Let $x = 0$ stand for 1980, $x = 1$ for 1981 and so forth. 2003

| Table 2.7 U.S. Patent Applications | |
| --- | --- |
| Year | Applications (thousands) |
| 1980 | 113.0 |
| 1985 | 127.1 |
| 1990 | 176.7 |
| 1991 | 178.4 |
| 1992 | 187.2 |
| 1993 | 189.4 |
| 1984 | 206.9 |
| 1995 | 226.9 |
| 1996 | 211.6 |

*Source: U.S. Bureau of the Census, Statistical Abstract of the United States, 1998 (Washington, D.C., 1998).*

## Explorations

**61. Highway Engineering** Interstate 70 west of Denver, Colorado, has a section posted as a 6% grade. This means that for a horizontal change of 100 ft there is a 6-ft vertical change.

6% grade

**(a)** Find the slope of this section of the highway. 0.06

**(b)** On a highway with a 6% grade what is the horizontal distance required to climb 250 ft? $\approx$ 4167 ft

**(c)** A sign along the highway says 6% grade for the next 7 mi. Estimate how many feet of vertical change there are along those next 7 mi. (There are 5280 ft in 1 mile.)

**62.** A group of female children were weighed. Their ages and weights are recorded in Table 2.8.

**(a)** Draw a scatter plot of the data.

**(b)** Find the linear regression model. $y \approx 0.68x + 9.01$

**(c)** Interpret the slope of the linear regression equation.

**(d)** Superimpose the regression line on the scatter plot.

**(e)** Use the regression model to predict the weight of a 30-month-old girl. $\approx$ 29.41 lb

| Table 2.8  Children's Ages and Weights | |
|---|---|
| Age (months) | Weight (pounds) |
| 19 | 22 |
| 21 | 23 |
| 24 | 25 |
| 27 | 28 |
| 29 | 31 |
| 31 | 28 |
| 34 | 32 |
| 38 | 34 |
| 43 | 39 |

**63.** Table 2.9 gives the average salaries for NBA players over several seasons.

**(a)** Draw a scatter plot of the data.

**(b)** Find the linear regression model.

**(c)** Interpret the slope of the linear regression equation.

**(d)** Superimpose the regression line on the scatter plot.

**(e)** Use the regression model to predict the average salary for NBA players during the 2001–2002 season.

| Table 2.9    National Basketball Association Average Player Salaries | |
|---|---|
| Season | Salary |
| 1990–1991 | $1,034,000 |
| 1991–1992 | $1,202,000 |
| 1992–1993 | $1,348,000 |
| 1993–1994 | $1,558,000 |
| 1994–1995 | $1,800,000 |
| 1995–1996 | $2,027,261 |
| 1996–1997 | $2,189,442 |

*Source: Paul D. Staudohar, Salary Caps in Team Sports, Compensation and Working Conditions, Spring 1998.*

**64. Writing to Learn   Identifying Graphs of Linear Functions**

**(a)** Which of the lines graphed below are graphs of linear functions? Explain.

**(b)** Which of the lines graphed below are graphs of functions? Explain.

**(c)** Which of the lines graphed below are not graphs of functions? Explain.

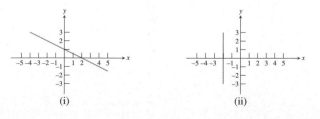

(i)                                (ii)

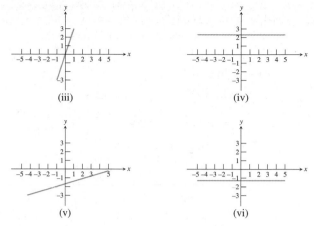

(iii)                              (iv)

(v)                               (vi)

**65. Average Rate of Change**  Let $f(x) = x^2$, $g(x) = 3x + 2$, $h(x) = 7x - 3$, $k(x) = mx + b$, and $l(x) = x^3$.

**(a)** Compute the average rate of change of $f$ from $x = 1$ to $x = 3$.  4

**(b)** Compute the average rate of change of $f$ from $x = 2$ to $x = 5$.  7

**(c)** Compute the average rate of change of $f$ from $x = a$ to $x = c$.  $c + a$

**(d)** Compute the average rate of change of $g$ from $x = 1$ to $x = 3$.  3

**(e)** Compute the average rate of change of $g$ from $x = 1$ to $x = 4$.  3

**(f)** Compute the average rate of change of $g$ from $x = a$ to $x = c$.  3

**(g)** Compute the average rate of change of $h$ from $x = a$ to $x = c$.  7

**(h)** Compute the average rate of change of $k$ from $x = a$ to $x = c$.  $m$

**(i)** Compute the average rate of change of $l$ from $x = a$ to $x = c$.  $c^2 + ac + a^2$

## Extending the Ideas

**66. Minimizing Sums of Squares**  The linear regression line is often called the **least squares line** because it minimizes the sum of the squares of the **residuals**, the differences between actual $y$ values and predicted $y$ values:

$$residual = y_i - (ax_i + b),$$

where $(x_i, y_i)$ are the given data pairs and $y = ax + b$ is the regression equation, as shown in the figure on the next page.

Use these definitions to explain why the regression line obtained from reversing the ordered pairs in Table 2.2 is not the inverse of the function obtained in Example 3.

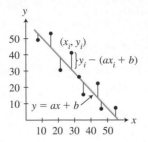

**67. Median-Median Line** Read about the median-median line by going to the Internet, your grapher owner's manual, or a library. Then use the following data set to complete this problem.

$$\{(2, 8), (3, 6), (5, 9), (6, 8), (8, 11), (10, 13), (12, 14), (15, 4)\}$$

**(a)** Draw a scatter plot of the data.

**(b)** Find the linear regression equation and graph it.

**(c)** Find the median-median line equation and graph it.

**(d) Writing to Learn** For this data, which of the two lines appears to be the line of better fit? Why?

**68.** Suppose $b^2 - 4ac > 0$ for the equation $ax^2 + bx + c = 0$.

**(a)** Prove that the sum of the two solutions of this equation is $-b/a$.

**(b)** Prove that the product of the two solutions of this equation is $c/a$.

**69. Connecting Algebra and Geometry** Prove that the axis of the graph of $f(x) = (x - a)(x - b)$ is $x = (a + b)/2$, where $a$ and $b$ are real numbers.

**70. Connecting Algebra and Geometry** Identify the vertex of the graph of $f(x) = (x - a)(x - b)$, where $a$ and $b$ are any real numbers.

**71. Connecting Algebra and Geometry** Prove that if $x_1$ and $x_2$ are real numbers and are zeros of the quadratic function $f(x) = ax^2 + bx + c$, then the axis of the graph of $f$ is $x = (x_1 + x_2)/2$.

---

## 2.2 Power Functions with Modeling

Power Functions and Variation • Monomial Functions and Their Graphs • Graphs of Power Functions • Modeling with Power Functions

### Power Functions and Variation

Five of the basic functions introduced in Section 1.3 were power functions. Power functions are an important family of functions in their own right and are important building blocks for other functions.

> **Definition** Power Function
>
> Any function that can be written in the form
>
> $$f(x) = k \cdot x^a, \text{ where } k \text{ and } a \text{ are nonzero constants,}$$
>
> is a **power function**. The constant $a$ is the **power**, and $k$ is the **constant of variation**, or **constant of proportion**. We say $f(x)$ **varies as** the $a^{\text{th}}$ power of $x$, or $f(x)$ **is proportional to** the $a^{\text{th}}$ power of $x$.

In general, if $y = f(x)$ varies as a constant power of $x$, then $y$ is a power function of $x$. Many of the most common formulas from geometry and science are power functions.

| Name | Formula | Power | Constant of Variation |
|------|---------|-------|----------------------|
| Circumference | $C = 2\pi r$ | 1 | $2\pi$ |
| Area of a circle | $A = \pi r^2$ | 2 | $\pi$ |
| Force of gravity | $F = k/d^2$ | $-2$ | $k$ |
| Boyle's Law | $V = k/P$ | $-1$ | $k$ |

These four power function models involve output-from-input relationships that can be expressed in the language of *variation and proportion*:

- The circumference of a circle varies directly as its radius.

- The area enclosed by a circle is directly proportional to the square of its radius.

- The force of gravity acting on an object is inversely proportional to the square of the distance from the object to the center of the earth.

- Boyle's Law states that the volume of an enclosed gas (at a constant temperature) varies inversely as the applied pressure.

The power function formulas with positive powers are statements of **direct variation** and power function formulas with negative powers are statements of **inverse variation**. Unless the word inversely is included in a variation statement, the variation is assumed to be direct, as in Example 1.

**Example 1** WRITING A POWER FUNCTION FORMULA

From empirical evidence and the laws of physics it has been found that the period of time *T* for the full swing of a pendulum varies as the square root of the pendulum's length *l*, provided that the swing is small relative to the length of the pendulum. Express this relationship as a power function.

**Solution** Because it does not state otherwise, the variation is direct. So the power is positive. The wording tells us that *T* is a function of *l*. Using *k* as the constant of variation gives us

$$T(l) = k\sqrt{l} = k \cdot l^{1/2}.$$

Section 1.3 introduced five basic power functions:

$$x, x^2, x^3, x^{-1} = \frac{1}{x}, \quad \text{and} \quad x^{1/2} = \sqrt{x}.$$

Example 2 tells the mathematical life stories of two other basic power functions: the *cube root function* and the "inverse-square" function.

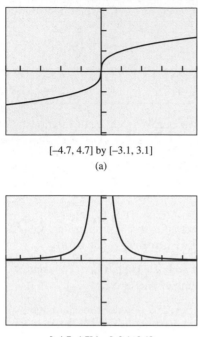

[–4.7, 4.7] by [–3.1, 3.1]

(a)

[–4.7, 4.7] by [–3.1, 3.1]

(b)

**Figure 2.13** The graphs of
(a) $f(x) = \sqrt[3]{x} = x^{1/3}$ and
(b) $g(x) = 1/x^2 = x^{-2}$. (Example 2)

### Example 2 ANALYZING POWER FUNCTIONS

State the power and constant of variation for the function, graph it, and analyze it.

**(a)** $f(x) = \sqrt[3]{x}$          **(b)** $g(x) = \dfrac{1}{x^2}$

#### Solution

**(a)** Because $f(x) = \sqrt[3]{x} = x^{1/3} = 1 \cdot x^{1/3}$, its power is 1/3, and its constant of variation is 1. The graph of $f$ is shown in Figure 2.13a.

Domain: All reals.
Range: All reals.
Continuous.
Increasing for all $x$.
Symmetric with respect to the origin (an odd function).
Not bounded above or below.
No local extrema.
No asymptotes.
End behavior: $\lim\limits_{x \to -\infty} \sqrt[3]{x} = -\infty$ and $\lim\limits_{x \to \infty} \sqrt[3]{x} = \infty$.
Interesting fact: The cube root function $f(x) = \sqrt[3]{x}$ is the inverse of the cubing function.

**(b)** Because $g(x) = 1/x^2 = x^{-2} = 1 \cdot x^{-2}$, its power is $-2$, and its constant of variation is 1. The graph of $g$ is shown in Figure 2.13b.

Domain: $(-\infty, 0) \cup (0, \infty)$.
Range: $(0, \infty)$.
Continuous on its domain. Discontinuous at $x = 0$.
Increasing on $(-\infty, 0)$. Decreasing on $(0, \infty)$.
Symmetric with respect to the $y$-axis (an even function).
Bounded below, but not above.
No local extrema.
Horizontal asymptote: $y = 0$. Vertical asymptote: $x = 0$.
End behavior: $\lim\limits_{x \to -\infty} (1/x^2) = 0$ and $\lim\limits_{x \to \infty} (1/x^2) = 0$.
Interesting fact: $g(x) = 1/x^2$ is the basis of scientific *inverse-square laws*, for example, the inverse-square gravitational principle $F = k/d^2$ mentioned above.

So $g(x) = 1/x^2$ is sometimes called the inverse-square function, but it is *not* the inverse of the squaring function but rather its *multiplicative* inverse.

## Monomial Functions and Their Graphs

An important subclass of power functions is the set of *monomial functions*, or single-term polynomial functions.

> **Definition** Monomial Function
>
> Any function that can be written as
>
> $$f(x) = k \quad \text{or} \quad f(x) = k \cdot x^n,$$
>
> where $k$ is a constant and $n$ is a positive integer, is a **monomial function**.

So the zero function and constant functions are monomial functions, but the more typical monomial function is a power function with a positive integer power, which is the degree of the monomial. The basic functions $x$, $x^2$, and $x^3$ are typical monomial functions. It is important to understand the graphs of monomial functions because every polynomial function is a sum of monomial functions.

In Exploration 1 we take a close look at six basic monomial functions. They have the form $x^n$ for $n = 1, 2, \ldots, 6$. We group them by even and odd powers.

---

**Exploration 1** **Comparing Graphs of Monomial Functions**

Graph the triplets of functions in the stated windows and explain how the graphs are alike and how they are different. Consider the relevant aspects of analysis from Example 2. Which ordered pairs do all three graphs have in common?

1. $f(x) = x$, $g(x) = x^3$, and $h(x) = x^5$ in the window $[-2.35, 2.35]$ by $[-1.5, 1.5]$, then $[-5, 5]$ by $[-15, 15]$, and finally $[-20, 20]$ by $[-200, 200]$.
2. $f(x) = x^2$, $g(x) = x^4$, and $h(x) = x^6$ in the window $[-1.5, 1.5]$ by $[-0.5, 1.5]$, then $[-5, 5]$ by $[-5, 25]$, and finally $[-15, 15]$ by $[-50, 400]$.

---

**Exploration Extensions**

Graph $f(x) = 2x$, $g(x) = 2x^3$, and $h(x) = 2x^5$. How do these graphs differ from the graphs in #1? Do they have any ordered pairs in common? Now graph $f(x) = -2x$, $g(x) = -2x^3$, and $h(x) = -2x^5$. Compare these graphs with the graphs in the first part of the exploration.

From Exploration 1 we see that

$$f(x) = x^n \text{ is even if } n \text{ is even and odd if } n \text{ is odd.}$$

Because of this symmetry, it is enough to know the first quadrant behavior of $f(x) = x^n$. Figure 2.14 shows the graphs of $f(x) = x^n$ for $n = 1, 2, \ldots, 6$ in the first quadrant near the origin.

To understand why the functions in Figure 2.14 graph as they do it is helpful to consider the numerical values displayed in Table 2.10. If $0 < x < 1$, the value of $x^n$ becomes smaller as $n$ becomes larger. So for example, when $0 < x < 1$, the graph of $x^6$ lies below the graphs of $x^n$ for $n = 1, 2, \ldots, 5$.

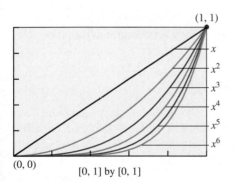

$(1, 1)$

$x$
$x^2$
$x^3$
$x^4$
$x^5$
$x^6$

$(0, 0)$
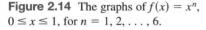
$[0, 1]$ by $[0, 1]$

**Figure 2.14** The graphs of $f(x) = x^n$, $0 \le x \le 1$, for $n = 1, 2, \ldots, 6$.

| Table 2.10 Powers of selected numbers between 0 and 1 | | | | | |
|---|---|---|---|---|---|
| $x$ | 0.2 | 0.4 | 0.6 | 0.8 | 1.0 |
| $x^2$ | 0.04 | 0.16 | 0.36 | 0.64 | 1.0 |
| $x^3$ | 0.008 | 0.064 | 0.216 | 0.512 | 1.0 |
| $x^4$ | 0.0016 | 0.0256 | 0.1296 | 0.4096 | 1.0 |
| $x^5$ | 0.00032 | 0.01024 | 0.07776 | 0.32768 | 1.0 |
| $x^6$ | 0.000064 | 0.004096 | 0.046656 | 0.262144 | 1.0 |

If $x > 1$, the order reverses: The value of $x^n$ becomes greater as $n$ becomes larger. So for example, when $x > 1$, the graph of $x^6$ lies above the graphs of $x^n$ for $n = 1, 2, \ldots, 5$, as shown in Figure 2.15.

So far we have ignored the effect of the constant of variation $k$ on the graph of $f(x) = k \cdot x^n$. From our work with transformations in Section 1.5, we can predict the effect:

- If $k > 0$, the graph of $f(x) = k \cdot x^n$ is a vertical stretch of the graph of $g(x) = x^n$ by a factor of $k$.

- If $k < 0$, the graph of $f(x) = k \cdot x^n$ is a vertical stretch of the graph of $g(x) = x^n$ by a factor of $|k|$, followed by a reflection across the $x$-axis.

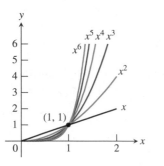

**Figure 2.15** The graphs of $f(x) = x^n$, $0 \le x \le 2$, for $n = 1, 2, \ldots, 6$.

### Example 3 GRAPHING MONOMIAL FUNCTIONS

Describe how to obtain the graph of the given function from the graph of $g(x) = x^n$ with the same power $n$. Sketch the graph by hand and support your answer with a grapher.

**(a)** $f(x) = 2x^3$        **(b)** $f(x) = -\dfrac{2}{3}x^4$

**Solution**

**(a)** We obtain the graph of $f(x) = 2x^3$ by vertically stretching the graph of $g(x) = x^3$ by a factor of 2. Both are odd functions. See Figure 2.16a.

**(b)** We obtain the graph of $f(x) = -(2/3)x^4$ by vertically shrinking the graph of $g(x) = x^4$ by a factor of 2/3 and then reflecting it across the $x$-axis. Both are even functions. See Figure 2.16b.

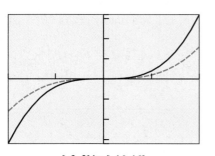

[–2, 2] by [–16, 16]

(a)

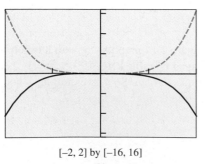

[–2, 2] by [–16, 16]

(b)

**Figure 2.16** The graphs of (a) $f(x) = 2x^3$ with basic monomial $g(x) = x^3$, and (b) $f(x) = -(2/3)x^4$ with basic monomial $g(x) = x^4$. (Example 3)

## Graphs of Power Functions

As with monomial functions, we begin studying the graphs of general power functions by letting the constant of variation $k = 1$ and focusing on the effect of the power $a$ on the graph of the power function $f(x) = x^a$. We have already seen the graphs of several specific power functions:

$$x, x^2, x^3, x^4, x^5, x^6, x^{1/2}, x^{1/3}, x^{-1}, \text{ and } x^{-2}.$$

The first six are monomials, which we have examined in depth. We now explore the graphical behavior of power functions of the form $x^{-n}$ and $x^{1/n}$, where $n$ is a positive integer.

| **Exploration 2** | **Comparing the Graphs of $x^{-1}, x^{-2}, x^{-3}, x^{-4}$** |

Graph the functions $f(x) = x^{-1}$, $g(x) = x^{-2}$, $h(x) = x^{-3}$, and $j(x) = x^{-4}$ in the windows: (a) $[0, 1]$ by $[0, 5]$, (b) $[0, 3]$ by $[0, 3]$, and (c) $[-2, 2]$ by $[-2, 2]$.

Explain how the graphs are alike and how they are different. Consider the relevant aspects of analysis from Example 2. Which ordered pairs do all three graphs have in common?

**Exploration Extensions**

Graph the functions $f(x) = -2x^{-1}$, $g(x) = -2x^{-2}$, $h(x) = -2x^{-3}$ (adjust the windows as necessary). Compare to your other graphs. Explain the results.

| **Exploration 3** | **Comparing the Graphs of $x^{1/2}, x^{1/3}, x^{1/4}, x^{1/5}$** |

Graph the functions $f(x) = x^{1/2}$, $g(x) = x^{1/3}$, $h(x) = x^{1/4}$, and $j(x) = x^{1/5}$ in the windows: (a) $[0, 1]$ by $[0, 1]$, (b) $[0, 3]$ by $[0, 2]$, and (c) $[-3, 3]$ by $[-2, 2]$.

Explain how the graphs are alike and how they are different. Consider the relevant aspects of analysis from Example 2. Which ordered pairs do all three graphs have in common?

**Exploration Extensions**

Graph the functions $f(x) = x^{-1/2}$, $g(x) = x^{-1/3}$, and $h(x) = x^{-1/4}$ in the same window. Compare to your other graphs. Explain the results.

Explorations 1, 2, and 3 examined three specific subclasses of power functions. These are not the only kinds of power functions, but they are among the ones most commonly used in applications, and they provide useful baseline information. We now proceed to the general case.

The graphs in Figure 2.17 represent the four general shapes that are possible for power functions of the form $f(x) = x^a$ for $x \geq 0$. In every case, the graph of $f$ contains $(1, 1)$. Those with positive powers also pass through $(0, 0)$. Those with negative exponents are asymptotic to both axes. Notice the following distinctive features of these four general shapes and the effect of the power $a$:

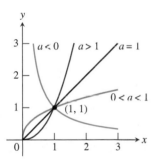

**Figure 2.17** The first-quadrant graphs of $f(x) = x^a$ for various powers $a$.

- If $a = 1$, $f(x) = x$ and the graph of $f$ is a straight line with slope $= 1$ passing through the origin.

- If $a > 1$, the first-quadrant graph of $f(x) = x^a$ is increasing and **concave up**, with its "hollowed out" part facing upward, and the graph of $f$ passes through $(0, 0)$. Examples include $x^2$, $x^3$, $x^4$, $x^5$, and $x^6$.

- If $0 < a < 1$, the first-quadrant graph of $f(x) = x^a$ is increasing and **concave down**, with its "hollowed out" part facing downward, and the graph of $f$ contains the point $(0, 0)$. Examples include $x^{1/2}$ and $x^{1/3}$.

- If $a < 0$, the first-quadrant graph of $f(x) = x^a$ is decreasing and concave up, and the graph of $f$ asymptotically approaches both the $x$-axis and the $y$-axis. The function $f$ is undefined at $x = 0$. Examples include $x^{-1}$ and $x^{-2}$.

We now turn our attention to the effect of the constant of variation $k$ on the graphical behavior of the general power function $f(x) = k \cdot x^a$:

- If $k > 0$, the graph of $f(x) = k \cdot x^a$ is a vertical stretch of the graph of $g(x) = x^a$ by a factor of $k$.

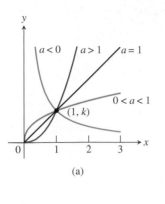

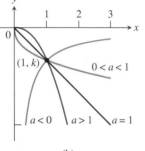

**Figure 2.18** The graphs of $f(x) = k \cdot x^a$ for $x \geq 0$. (a) $k > 0$, (b) $k < 0$

• If $k < 0$, the graph of $f(x) = k \cdot x^a$ is a vertical stretch of the graph of $g(x) = x^a$ by a factor of $|k|$, followed by a reflection across the $x$-axis.

Figure 2.18 illustrates these two cases for $x \geq 0$. When $k > 0$, the graph lies in Quadrant I, but when $k < 0$, the graph is in Quadrant IV. Notice that every power function passes through the point $(1, k)$.

The graphs in Figures 2.17 and 2.18 only tell half the story of the graphical behavior of power functions: They tell what happens for $x \geq 0$. In general, for any power function $f(x) = k \cdot x^a$, one of three following things happens to tell the rest of the story:

• $f$ is undefined for $x < 0$, as is the case for $f(x) = x^{1/2}$ and $f(x) = x^{\pi}$.

• $f$ is an even function, so $f$ is symmetric about the $y$-axis, as is the case for $f(x) = x^{-2}$ and $f(x) = x^{2/3}$.

• $f$ is an odd function, so $f$ is symmetric about the origin, as is the case for $f(x) = x^{-1}$ and $f(x) = x^{7/3}$.

Predicting the general shape of the graph a power function is a two-step process.

---

**Predicting the Graph of a Power Function $f(x) = k \cdot x^a$**

1. **Behavior for $f$ for $x \geq 0$.** Observe the values of the constant of variation $k$ and the power $a$. Are they positive or negative? If $a$ is positive, is it greater than, less than, or equal to 1? The answers to these two questions indicate the behavior for $f$ for $x \geq 0$, which will be one of the eight possibilities shown in Figure 2.18. Describe or sketch the portion of the curve lying in Quadrant I or IV, which will include the point $(1, k)$.

2. **Behavior for $f$ for $x < 0$.** Is $f$ even, is it odd, or is it undefined when $x < 0$? If $f$ is undefined for $x < 0$, there are no points in Quadrants II or III. If $f$ is even or odd, complete the description or sketch of $f$ using symmetry.

---

Example 4 illustrates this two-step approach to approximating the graphical behavior of a power function. Included are the various possibilities for $k$ and $a$ values and for the symmetry of $f(x) = k \cdot x^a$.

**Example 4   GRAPHING POWER FUNCTIONS $f(x) = k \cdot x^a$**

Observe the values of the constants $k$ and $a$. Describe the portion of the curve that lies in Quadrant I or IV. Determine whether $f$ is even, odd, or undefined for $x < 0$. Describe the rest of the curve if any. Graph the function to see whether it matches the description.

**(a)** $f(x) = 2x^{-3}$        **(b)** $f(x) = -0.4x^{1.5}$        **(c)** $f(x) = -x^{0.4}$

**Solution**

**(a)** Because $k = 2$ is positive and $a = -3$ is negative, the graph passes through $(1, 2)$ and is asymptotic to both axes. The function $f$ is odd because

$$f(-x) = 2(-x)^{-3} = \frac{2}{(-x)^3} = -\frac{2}{x^3} = -2x^{-3} = -f(x).$$

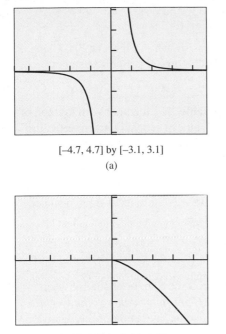

[−4.7, 4.7] by [−3.1, 3.1]

(a)

[−4.7, 4.7] by [−3.1, 3.1]

(b)

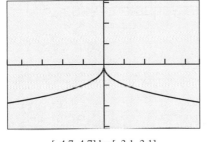

[−4.7, 4.7] by [−3.1, 3.1]

(c)

**Figure 2.19** The graphs of
(a) $f(x) = 2x^{-3}$, (b) $f(x) = -0.4x^{1.5}$, and
(c) $f(x) = -x^{0.4}$. (Example 4)

So its graph is symmetric about the origin. The graph in Figure 2.19a supports all aspects of the description.

**(b)** Because $k = -0.4$ is negative and $a = 1.5 > 1$, the graph contains $(0, 0)$ and passes through $(1, -0.4)$. In the fourth quadrant, it is decreasing and concave down. The function $f$ is undefined for $x < 0$ because

$$f(x) = -0.4x^{1.5} = -\frac{2}{5}x^{3/2} = -\frac{2}{5}(\sqrt{x})^3,$$

and the square-root function is undefined for $x < 0$. So the graph of $f$ has no points in Quadrants II or III. The graph in Figure 2.19b matches the description.

**(c)** Because $k = -1$ is negative and $0 < a < 1$, the graph contains $(0, 0)$ and passes through $(1, -1)$. In the fourth quadrant, it is decreasing and concave up. The function $f$ is even because

$$f(-x) = -(-x)^{0.4} = -(-x)^{2/5} = -(\sqrt[5]{-x})^2 = -(-\sqrt[5]{x})^2$$
$$= -(\sqrt[5]{x})^2 = -x^{0.4} = f(x).$$

So the graph of $f$ is symmetric about the $y$-axis. The graph in Figure 2.19c fits the description.

## Modeling with Power Functions

Noted astronomer Johannes Kepler (1571–1630) developed three laws of planetary motion that are used to this day. Kepler's Third Law states that the square of the period of orbit $T$ (the time required for one full revolution around the Sun) for each planet is proportional to the cube of its average distance $a$ from the Sun. Table 2.11 gives the relevant data for the six planets that were known in Kepler's time. The distances are given in millions of kilometers, or gigameters (Gm).

**Table 2.11  Average Distances and Orbit Periods for the Six Innermost Planets**

| Planet | Average Distance from Sun (Gm) | Period of Orbit (days) |
|--------|-------------------------------|------------------------|
| Mercury | 57.9 | 88 |
| Venus | 108.2 | 225 |
| Earth | 149.6 | 365.2 |
| Mars | 227.9 | 687 |
| Jupiter | 778.3 | 4332 |
| Saturn | 1427 | 10,760 |

*Source: Shupe, Dorr, Payne, Hunsiker, et al., National Geographic Atlas of the World (rev. 6th ed.). Washington, DC: National Geographic Society, 1992, plate 116.*

### Example 5  MODELING PLANETARY DATA WITH A POWER FUNCTION

Use the data in Table 2.11 to obtain a power function model for orbital period as a function of average distance from the Sun. Then use the model to predict the orbital period for Neptune, which is 4497 Gm from the Sun on average.

## Solution

### Model

First we make a scatter plot of the data, as shown in Figure 2.20a. Using power regression, we find the model for the orbital period to be about

$$T(a) = 0.20008a^{1.49963} \approx 0.20a^{1.5} = 0.20a^{3/2} = 0.20\sqrt{a^3}.$$

Figure 2.20b shows the scatter plot for Table 2.11 together with a graph of the power regression model just found. You can see that the curve fits the data quite well. The coefficient of determination is $R^2 \approx 0.999999912$, indicating an amazingly close fit and supporting the visual evidence.

### Solve Numerically

To predict the orbit period for Neptune we substitute its average distance from the Sun in the power regression model:

$$T(4497) = 0.20008(4497)^{1.49963} \approx 60,150$$

### Interpret

It takes Neptune about 60,150 days to orbit the Sun, or about 165 years, which is the value given in the *National Geographic Atlas of the World*. Figure 2.20c reports this result and gives some indication of the relative distances involved. Neptune is much farther from the Sun than the six innermost planets and especially the four closest to the Sun—Mercury, Venus, Earth, and Mars.

**A Bit of History**

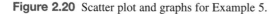

Example 5 shows the predictive power of a well founded model. Exercise 59 asks you to find Kepler's elegant form of the equation, $T^2 = a^3$, which he reported in *The Harmony of the World* in 1619.

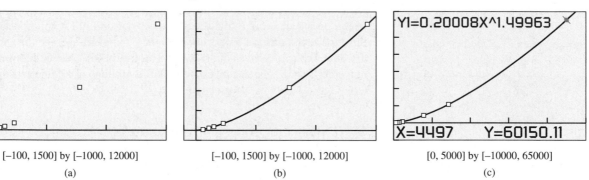

|  |  |  |
|---|---|---|
| [−100, 1500] by [−1000, 12000] | [−100, 1500] by [−1000, 12000] | [0, 5000] by [−10000, 65000] |
| (a) | (b) | (c) |

**Figure 2.20** Scatter plot and graphs for Example 5.

In Example 6, we return to free-fall motion, with a new twist. The data in the table come from the same CBR experiment referenced in Example 8 of Section 2.1. This time we are looking at the downward distance (in meters) the ball has traveled since reaching its peak height and its downward speed (in meters per second). It can be shown (see Exercise 60) that free-fall speed is proportional to a power of the distance traveled.

**Follow-up**

Ask...

What would the graph of $f(x) = -3x^{1/2}$ look like? **(Since $0 < 1/2 < 1$, the graph is concave down in the first quadrant and undefined for $x < 0$; the coefficient $-3$ stretches the graph and reflects it over the $x$-axis.)**

**A Word of Warning**

The regression routine traditionally used
to compute power function models
involves taking logarithms of the data,
and therefore, all of the data must be
strictly positive numbers. So we must
leave out (0, 0) to compute the power
regression equation.

**Why *p*?**

We use *p* for speed to distinguish it from
velocity *v*. Recall that speed is the
*absolute value* of velocity.

**Table 2.12  Rubber Ball Data from CBR Experiment**

| Distance (m) | Speed (m/s) |
|---|---|
| 0.00000 | 0.00000 |
| 0.04298 | 0.82372 |
| 0.16119 | 1.71163 |
| 0.35148 | 2.45860 |
| 0.59394 | 3.05209 |
| 0.89187 | 3.74200 |
| 1.25557 | 4.49558 |

**Example 6  MODELING FREE-FALL SPEED VERSUS DISTANCE**

Use the data in Table 2.12 to obtain a power function model for speed $p$
versus distance traveled $d$. Then use the model to predict the speed of the
ball at impact given that impact occurs when $d \approx 1.80$ m.

**Solution**

**Model**

Figure 2.21a is a scatter plot of the data. Using power regression, we find
the model for speed versus distance to be about

$$p(d) = 4.02708d^{0.49436} \approx 4.03d^{0.5} = 4.03d^{1/2} = 4.03\sqrt{d}.$$

(See margin notes.) Figure 2.21b shows the scatter plot for Table 2.12
together with a graph of the power regression equation just found. You can
see that the curve nicely fits the data. The coefficient of determination is
$r^2 \approx 0.99770$, indicating a close fit and supporting the visual evidence.

**Solve Numerically**

To predict the speed at impact, we substitute $d \approx 1.80$ into the obtained
power regression model:

$$p(1.80) = 4.02708(1.80)^{0.49436} \approx 5.39$$

See Figure 2.21c.

**Interpret**

The speed at impact is about 5.39 m/sec. This is slightly less than the value
obtain in Example 8 of Section 2.1, using a different modeling process for
the same experiment.

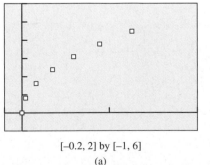

[–0.2, 2] by [–1, 6]
(a)

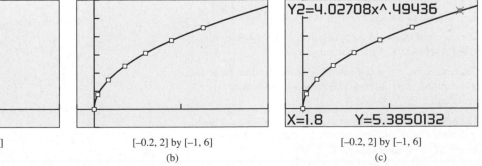

[–0.2, 2] by [–1, 6]
(b)

[–0.2, 2] by [–1, 6]
(c)

**Figure 2.21**  Scatter plot and graphs for Example 6.

# Quick Review 2.2

In Exercises 1–6, write the following expressions using only positive integer powers.

**1.** $x^{2/3}$ $\sqrt[3]{x^2}$

**2.** $p^{5/2}$ $\sqrt{p^5}$

**3.** $d^{-2}$ $1/d^2$

**4.** $x^{-7}$ $1/x^7$

**5.** $q^{-4/5}$ $1/\sqrt[5]{q^4}$

**6.** $m^{-1.5}$ $1/\sqrt{m^3}$

In Exercises 7–10, write the following expressions in the form $k \cdot x^a$ using a single rational number for the power $a$.

**7.** $\sqrt{9x^3}$ $3x^{3/2}$

**8.** $\sqrt[3]{8x^5}$ $2x^{5/3}$

**9.** $\sqrt[3]{\dfrac{5}{x^4}}$ $\approx 1.71x^{-4/3}$

**10.** $\dfrac{4x}{\sqrt{32x^3}}$ $\approx 0.71x^{-1/2}$

# Section 2.2 Exercises

In Exercises 1–10, determine if the function is a power function, given that $c$, $g$, $k$, and $\pi$ represent constants. For those that are, state the power and constant of variation. For those that are not, explain why not. If function notation is not used, name the independent variable.

**1.** $f(x) = -\dfrac{1}{2}x^5$

**2.** $f(x) = 9x^{5/3}$

**3.** $f(x) = 3 \cdot 2^x$

**4.** $f(x) = 13$

**5.** $E(m) = mc^2$

**6.** $KE(v) = \dfrac{1}{2}kv^5$

**7.** $d = \dfrac{1}{2}gt^2$

**8.** $V = \dfrac{4}{3}\pi r^3$

**9.** $I = \dfrac{k}{d^2}$

**10.** $F = m \cdot a$

In Exercises 11–16, determine if the function is a monomial function, given that $l$ and $\pi$ represent constants. For those that are, state the degree and leading coefficient. For those that are not, explain why not. If function notation is not used, name the independent variable.

**11.** $f(x) = -4$

**12.** $f(x) = 3 \cdot x^{-5}$

**13.** $y = -6 \cdot x^7$

**14.** $y = -2 \cdot 5^x$

**15.** $S = 4\pi r^2$

**16.** $A = l \cdot w$

In Exercises 17–22, write the statement as a power function equation.

**17.** The area $A$ of an equilateral triangle varies directly as the square of the length $s$ of its sides. $A = ks^2$

**18.** The volume $V$ of a circular cylinder with fixed height is proportional to the square of its radius $r$. $V = kr^2$

**19.** The current $I$ in an electrical circuit is inversely proportional to the resistance $R$, with constant of variation $V$. $I = V/R$

**20.** Charles's Law states, the volume $V$ of an enclosed ideal gas at a constant pressure varies directly as the absolute temperature $T$. $V = kT$

**21.** The energy $E$ produced in a nuclear reaction is proportional to the mass $m$, with the constant of variation being $c^2$, the square of the speed of light. $E = mc^2$

**22.** The speed $p$ of a free-falling object that has been dropped from rest varies as the square root of the distance traveled $d$, with a constant of variation $k = \sqrt{2g}$. $p = \sqrt{2gd}$

In Exercises 23–26, write a sentence that expresses the relationship in the formula, using the language of variation or proportion.

**23.** $w = mg$, where $w$ and $m$ are the weight and mass of an object and $g$ is the constant acceleration due to gravity.

**24.** $C = \pi D$, where $C$ and $D$ are the circumference and diameter of a circle and $\pi$ is the usual mathematical constant.

**25.** $n = c/v$, where $n$ is the refractive index a of medium, $v$, is the velocity of light in the medium, and $c$ is the constant velocity of light in free space.

**26.** $d = p^2/(2g)$, where $d$ is the distance traveled by a free-falling object dropped from rest, $p$ is the speed of the object, and $g$ is the constant acceleration due to gravity.

In Exercises 27–28, data are given for $y$ as a power function of $x$. Write an equation for the power function, and state its power and constant of variation.

**27.**

| $x$ | 2 | 4 | 6 | 8 | 10 |
|---|---|---|---|---|---|
| $y$ | 2 | 0.5 | 0.222... | 0.125 | 0.08 |

**28.**

| $x$ | 1 | 4 | 9 | 16 | 25 |
|---|---|---|---|---|---|
| $y$ | $-2$ | $-4$ | $-6$ | $-8$ | $-10$ |

In Exercises 29–34, match the equation to one of the curves labeled in the figure.

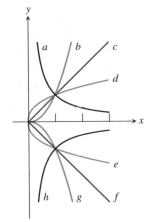

**29.** $f(x) = -\dfrac{2}{3}x^4$ (g)

**30.** $f(x) = \dfrac{1}{2}x^{-5}$ (a)

**31.** $f(x) = 2x^{1/4}$ (d)

**32.** $f(x) = -x^{5/3}$ (g)

**33.** $f(x) = -2x^{-2}$ (h)

**34.** $f(x) = 1.7x^{2/3}$ (d)

In Exercises 35–40, describe how to obtain the graph of the given monomial function from the graph of $g(x) = x^n$ with the same power $n$. State whether $f$ is even or odd. Sketch the graph by hand and support your answer with a grapher.

**35.** $f(x) = \dfrac{2}{3}x^4$

**36.** $f(x) = 5x^3$

**37.** $f(x) = -1.5x^5$

**38.** $f(x) = -2x^6$

**39.** $f(x) = \dfrac{1}{4}x^8$

**40.** $f(x) = \dfrac{1}{8}x^7$

In Exercises 41–44, state the power and constant of variation for the function, graph it, and analyze it in the manner of Example 2 of this section.

**41.** $f(x) = 2x^4$

**42.** $f(x) = -3x^3$

**43.** $f(x) = \dfrac{1}{2}\sqrt[4]{x}$

**44.** $f(x) = -2x^{-3}$

In Exercises 45–50, state the values of the constants $k$ and $a$ for the function $f(x) = k \cdot x^a$. Describe the portion of the curve that lies in Quadrant I or IV. Determine whether $f$ is even, odd, or undefined for $x < 0$. Describe the rest of the curve if any. Graph the function to see whether it matches the description.

**45.** $f(x) = 3x^{1/4}$

**46.** $f(x) = -4x^{2/3}$

**47.** $f(x) = -2x^{4/3}$

**48.** $f(x) = \dfrac{2}{5}x^{5/2}$

**49.** $f(x) = \dfrac{1}{2}x^{-3}$

**50.** $f(x) = -x^{-4}$

**51. Boyle's Law** The volume of an enclosed gas (at a constant temperature) varies inversely as the pressure. If the pressure of a 3.46-L sample of neon gas at 302°K is 0.926 atm, what would the volume be at a pressure of 1.452 atm if the temperature does not change? 2.21 L

**52. Charles's Law** The volume of an enclosed gas (at a constant pressure) varies directly as the absolute temperature. If the pressure of a 3.46-L sample of neon gas at 302°K is 0.926 atm, what would the volume be at a temperature of 338°K if the pressure does not change?

**53. Diamond Refraction** Diamonds have the extremely high refraction index of $n = 2.42$ on average over the range of visible light. Use the formula from Exercise 25 and the fact that $c \approx 3.00 \times 10^8$ m/sec to determine the speed of light through a diamond. $1.24 \times 10^8$ m/sec

**54. Windmill Power** The power $P$ (in watts) produced by a windmill is proportional to the cube of the wind speed $v$ (in mph). If a wind of 10 mph generates 15 watts of power, how much power is generated by winds of 20, 40, and 80 mph? Make a table and explain the pattern.

## Explorations

**55. Keeping Warm** For mammals and other warm-blooded animals to stay warm requires quite a bit of energy. Temperature loss is related to surface area, which is related to body weight, and temperature gain is related to circulation, which is related to pulse rate. In the final analysis, scientists have concluded that the pulse rate $r$ of mammals is a power function of their body weight $w$.

**(a)** Draw a scatter plot of the data in Table 2.13.

**(b)** Find the power regression model.

**(c)** Superimpose the regression curve on the scatter plot.

**(d)** Use the regression model to predict the pulse rate for a 450-kg horse. Is the result close to the 38 beats/min reported by A. J. Clark in 1927?

**Table 2.13 Weight and Pulse Rate of Selected Mammals**

| Mammal | Body weight (kg) | Pulse rate (beats/min) |
|---|---|---|
| Rat | 0.2 | 420 |
| Guinea pig | 0.3 | 300 |
| Rabbit | 2 | 205 |
| Small dog | 5 | 120 |
| Large dog | 30 | 85 |
| Sheep | 50 | 70 |
| Human | 70 | 72 |

*Source: A. J. Clark, Comparative Physiology of the Heart (New York: Macmillan, 1927).*

**56. Light Intensity** Velma and Reggie gathered the data in Table 2.14 using a 100-watt light bulb and a Calculator-Based Laboratory™ (CBL™) with a light-intensity probe.

**(a)** Draw a scatter plot of the data in Table 2.14.

**(b)** Find the power regression model. Is the power close to the theoretical value of $a = -2$?

**(c)** Superimpose the regression curve on the scatter plot.

**(d)** Use the regression model to predict the light intensity at distances of 1.7 m and 3.4 m.

| Table 2.14  Light Intensity Data for a 100-W Light Bulb | |
|---|---|
| Distance (m) | Intensity (W/m²) |
| 1.0 | 7.95 |
| 1.5 | 3.53 |
| 2.0 | 2.01 |
| 2.5 | 1.27 |
| 3.0 | 0.90 |

**57. Group Activity  Rational Powers**  Working in a group of three students, investigate the behavior of power functions of the form $f(x) = k \cdot x^{m/n}$, where $m$ and $n$ are positive with no factors in common.  Have one group member investigate each of the following cases:

- $n$ is even
- $n$ is odd and $m$ is even
- $n$ is odd and $m$ is odd

For each case, decide whether $f$ is even, $f$ is odd, or $f$ is undefined for $x < 0$. Solve graphically and confirm algebraically in a way to convince the rest of your group and your entire class.

## Extending the Ideas

**58. Writing to Learn  Irrational Powers**  A negative number to an irrational power is undefined. Analyze the graphs of $f(x) = x^\pi, x^{1/\pi}, x^{-\pi}, -x^\pi, -x^{1/\pi}$, and $-x^{-\pi}$. Prepare a sketch of all six graphs on one set of axes, labeling each of the curves. Write an explanation for why each graph is positioned and shaped as it is.

**59. Planetary Motion Revisited**  Convert the time and distance units in Table 2.11 to the Earth-based units of years and astronomical units using

$$1 \text{ yr} = 365.2 \text{ days}  \text{ and }  1 \text{ AU} = 149.6 \text{ Gm}.$$

Use this "re-expressed" data to obtain a power function model.  Show algebraically that this model closely approximates Kepler's equation $T^2 = a^3$.

**60. Free Fall Revisited**  The **speed** $p$ of an object is the absolute value of its velocity $v$. The distance traveled $d$ by an object dropped from an initial height $s_0$ with a current height $s$ is given by

$$d = s_0 - s$$

until it hits the ground. Use this information and the free-fall motion formulas from Section 2.1 to prove that

$$d = \frac{1}{2}gt^2, p = gt, \text{ and therefore } p = \sqrt{2gd}.$$

Do the results of Example 6 approximate this last formula?

**61.** Prove that $g(x) = 1/f(x)$ is even if and only if $f(x)$ is even and that $g(x) = 1/f(x)$ is odd if and only if $f(x)$ is odd.

**62.** Use the results in Exercise 61 to prove that $g(x) = k \cdot x^{-a}$ is even if and only if $f(x) = k \cdot x^a$ is even and that $g(x) = k \cdot x^{-a}$ is odd if and only if $f(x) = k \cdot x^a$ is odd.

**63. Joint Variation**  If a variable $z$ varies as the product of the variables $x$ and $y$, we say $z$ **varies jointly** as $x$ and $y$, and we write $z = k \cdot x \cdot y$, where $k$ is the constant of variation. Write a sentence that expresses the relationship in the formula, using the language of joint variation.

**(a)** $F = m \cdot a$, where $F$ and $a$ are the force and acceleration acting on an object of mass $m$.

**(b)** $KE = (1/2)m \cdot v^2$, where $KE$ and $v$ are the kinetic energy and velocity of an object of mass $m$.

**(c)** $F = G \cdot m_1 \cdot m_2/r^2$, where $F$ is the force of gravity acting on objects of masses $m_1$ and $m_2$ with a distance $r$ between their centers and $G$ is the universal gravitational constant.

---

### 2.3

# Polynomial Functions of Higher Degree with Modeling

Graphs of Polynomial Functions • End Behavior of Polynomial Functions • Zeros of Polynomial Functions • Intermediate Value Theorem • Modeling

## Graphs of Polynomial Functions

As we saw in Section 2.1, a polynomial function of degree 0 is a constant function and graphs as a horizontal line. A polynomial function of degree 1 is a linear function; its graph is a slant line. A polynomial function of degree 2 is a quadratic function; its graph is a parabola.

We now consider polynomial functions of higher degree. These include **cubic functions** (polynomials of degree 3) and **quartic functions** (polynomials of degree 4). Every polynomial function is the sum of one or more monomial functions. Recall that a polynomial function of degree $n$ can be written in the form

$$f(x) = a_n x^n + a_{n-1} x^{n-1} + \cdots + a_2 x^2 + a_1 x + a_0, \ a_n \neq 0.$$

Here are some important definitions associated with polynomial functions and this equation.

---

**Definitions  The Vocabulary of Polynomials**

- Each monomial in this sum—$a_n x^n$, $a_{n-1} x^{n-1}$, ..., $a_0$—is a **term** of the polynomial.
- A polynomial function written in this way, with terms in descending degree, is written in **standard form.**
- The constants $a_n, a_{n-1}, \ldots, a_0$ are the **coefficients** of the polynomial.
- The term $a_n x^n$ is the **leading term**, and $a_0$ is the constant term.

---

In Example 1 we use the fact that the constant term $a_0$ of a polynomial function $p$ is the initial value of the function $p(0)$ and the $y$-intercept of the graph to provide a quick and easy check of the transformed graphs.

### Example 1  GRAPHING TRANSFORMATIONS OF MONOMIAL FUNCTIONS

Describe how to transform the graph of an appropriate monomial function $f(x) = a_n x^n$ into the graph of the given function. Sketch the transformed graph by hand and support your answer with a grapher. Compute the location of the $y$-intercept as a check on the transformed graph.

**(a)** $g(x) = 4(x + 1)^3$  **(b)** $h(x) = -(x - 2)^4 + 5$

**Solution**

**(a)** You can obtain the graph of $g(x) = 4(x + 1)^3$ by shifting the graph of $f(x) = 4x^3$ one unit to the left, as shown in Figure 2.22a. The $y$-intercept of the graph of $g$ is $g(0) = 4(0 + 1)^3 = 4$, which appears to agree with the transformed graph.

**(b)** You can obtain the graph of $h(x) = -(x - 2)^4 + 5$ by shifting the graph of $f(x) = -x^4$ two units to the right and five units up, as shown in Figure 2.22b. The $y$-intercept of the graph of $h$ is $h(0) = -(0 - 2)^4 + 5$ $= -16 + 5 = -11$, which appears to agree with the transformed graph.

In Example 1 we were able to determine the value of the constant terms of $g$ and $h$ (their values at $x = 0$) without expanding the functions into standard form. Example 2 shows what can happen when simple monomial functions are combined to obtain polynomial functions. The resulting polynomials are *not* mere translations of monomials.

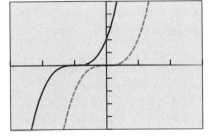

[–3, 3] by [–10, 10]

(a)

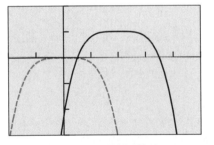

[–2, 5] by [–15, 10]

(b)

**Figure 2.22** (a) The graphs of $g(x) = 4(x + 1)^3$ and $f(x) = 4x^3$. (b) The graphs of $h(x) = -(x - 2)^4 + 5$ and $f(x) = -x^4$. (Example 1)

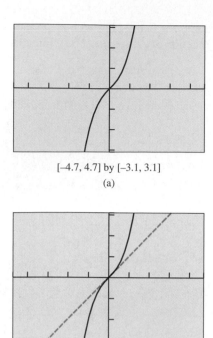

[−4.7, 4.7] by [−3.1, 3.1]

(a)

[−4.7, 4.7] by [−3.1, 3.1]

(b)

**Figure 2.23** The graph of $f(x) = x^3 + x$ (a) by itself and (b) with $y = x$. (Example 2a)

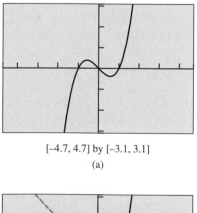

[−4.7, 4.7] by [−3.1, 3.1]

(a)

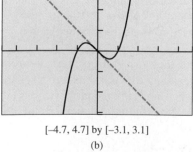

[−4.7, 4.7] by [−3.1, 3.1]

(b)

**Figure 2.24** The graph of $g(x) = x^3 - x$ (a) by itself and (b) with $y = -x$. (Example 2b)

### Example 2   GRAPHING COMBINATIONS OF MONOMIAL FUNCTIONS

Graph the polynomial function, locate its extrema and zeros, and explain how it is related to the monomials from which it is built.

**(a)** $f(x) = x^3 + x$ **(b)** $g(x) = x^3 - x$

**Solution**

**(a)** The graph of $f(x) = x^3 + x$ is shown in Figure 2.23a. The function $f$ is strictly increasing with no extrema and one zero located at $x = 0$.

The general shape of the graph is much like the graph of its leading term $x^3$, but near the origin $f$ behaves much like its other term $x$, as seen in Figure 2.22b. The function $f$ is odd, just like its two building block monomials.

**(b)** The graph of $g(x) = x^3 - x$ is shown in Figure 2.24a. The function $g$ has a local maximum of about 0.38 at $x \approx -0.58$ and a local minimum of about $-0.38$ at $x \approx 0.58$. The function factors as $g(x) = x(x + 1)(x - 1)$ and has zeros located at $x = -1$, $x = 0$, and $x = 1$.

The general shape of the graph is much like the graph of its leading term $x^3$, but near the origin $g$ behaves much like its other term $-x$, as seen in Figure 2.24b. The function $g$ is odd, just like its two building block monomials.

We have now seen a few examples of graphs of polynomial functions, but are these typical? What do graphs of polynomials look like in general?

To begin our answer, let's first recall that every polynomial function is defined and continuous for all real numbers. Not only are graphs of polynomials unbroken without jumps or holes, but they are *smooth*, unbroken lines or curves, with no sharp corners or cusps. Typical graphs of cubic and quartic functions are shown in Figures 2.25 and 2.26.

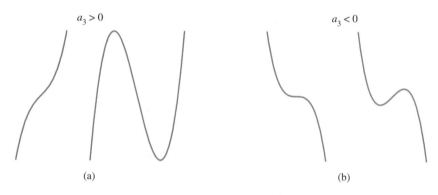

$a_3 > 0$

$a_3 < 0$

(a)

(b)

**Figure 2.25** Graphs of four typical cubic functions: (a) two with positive and (b) two with negative leading coefficients.

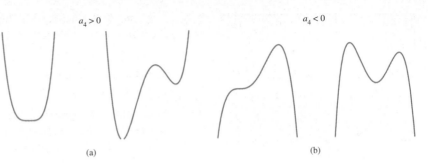

**Figure 2.26**  Graphs of four typical quartic functions: (a) two with positive and (b) two with negative leading coefficients.

Imagine horizontal lines passing through the graphs in Figures 2.25 and 2.26, acting as *x*-axes. Each intersection would be an *x*-intercept that would correspond to a zero of the function. From this mental experiment, we see that cubic functions have at most three zeros and quartic functions have at most four zeros.

Focusing on the high and low points in Figures 2.25 and 2.26, we see that cubic functions have at most two local extrema and quartic functions have at most three local extrema. The following can be proved in calculus:

> **Theorem  Local Extrema and Zeros of Polynomial Functions**
>
> A polynomial function of degree $n$ has at most $n - 1$ local extrema and at most $n$ zeros.

## End Behavior of Polynomial Functions

An important characteristic of polynomial functions is their end behavior. As we shall see, the end behavior of a polynomial is closely related to the end behavior of its leading term. The leading term of a polynomial of degree $n$ is a monomial of degree $n$. Exploration 1 examines the end behavior of monomial functions.

> **Exploration 1  Investigating the End Behavior of $f(x) = a_n x^n$**
>
> Graph each function in the window $[-5, 5]$ by $[-15, 15]$. Describe the end behavior using $\lim_{x \to \infty} f(x)$ and $\lim_{x \to -\infty} f(x)$.
>
> **1. (a)** $f(x) = 2x^3$          **(b)** $f(x) = -x^3$
> **(c)** $f(x) = x^5$              **(d)** $f(x) = -0.5x^7$
>
> **2. (a)** $f(x) = -3x^4$         **(b)** $f(x) = 0.6x^4$
> **(c)** $f(x) = 2x^6$             **(d)** $f(x) = -0.5x^2$
>
> **3. (a)** $f(x) = -0.3x^5$       **(b)** $f(x) = -2x^2$
> **(c)** $f(x) = 3x^4$             **(d)** $f(x) = 2.5x^3$
>
> Describe the patterns you observe. In particular, how do the values of the coefficient $a_n$ and the degree $n$ affect the end behavior of $f(x) = a_n x^n$?

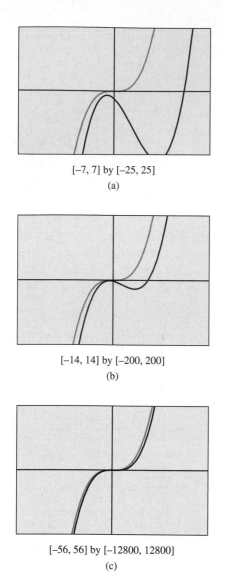

[-7, 7] by [-25, 25]

(a)

[-14, 14] by [-200, 200]

(b)

[-56, 56] by [-12800, 12800]

(c)

**Figure 2.27** As the viewing window gets larger, the graphs of $f(x) = x^3 - 4x^2 - 5x - 3$ and $g(x) = x^3$ look more and more alike. (Example 3)

Example 3 illustrates the link between the end behavior of a polynomial $f(x) = a_n x^n + \cdots + a_1 x + a_0$ and its leading term $a_n x^n$.

### Example 3 COMPARING THE GRAPHS OF A POLYNOMIAL AND ITS LEADING TERM

Superimpose the graphs of $f(x) = x^3 - 4x^2 - 5x - 3$ and $g(x) = x^3$ in successively larger viewing windows, a process called **zoom out**. Continue zooming out until the graphs look nearly identical.

**Solution** Figure 2.27 shows three views of the graphs of $f(x) = x^3 - 4x^2 - 5x - 3$ and $g(x) = x^3$ in progressively larger viewing windows. As the dimensions of the window increase it gets harder to tell them apart. Moreover,

$$\lim_{x \to \infty} f(x) = \lim_{x \to \infty} g(x) = \infty \text{ and } \lim_{x \to -\infty} f(x) = \lim_{x \to -\infty} g(x) = -\infty.$$

Example 3 illustrates something that is true for all polynomials: *In sufficiently large viewing windows, the graph of a polynomial and the graph of its leading term appear to be identical.* Said another way, the leading term *dominates* the behavior of the polynomial as $|x| \to \infty$. Based on this fact and what we have seen in Exploration 1, there are four possible end behavior patterns for a polynomial function. The power and coefficient of the leading term tell us which one of the four patterns occurs.

### Leading Term Test for Polynomial End Behavior

For any polynomial function $f(x) = a_n x^n + \cdots + a_1 x + a_0$, the limits $\lim_{x \to \infty} f(x)$ and $\lim_{x \to -\infty} f(x)$ are determined by the degree $n$ of the polynomial and its leading coefficient $a_n$:

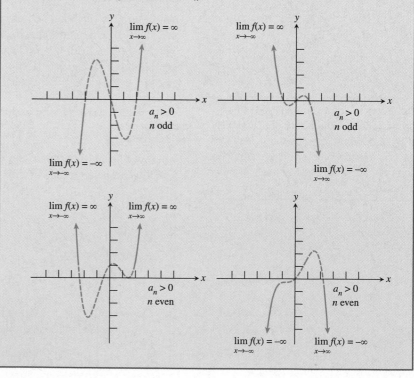

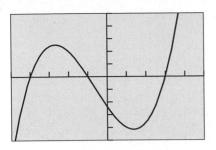

[−5, 5] by [−25, 25]

(a)

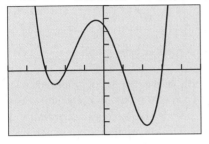

[−5, 5] by [−50, 50]

(b)

**Figure 2.28**
(a) $f(x) = x^3 + 2x^2 - 11x - 12$,
(b) $g(x) = 2x^4 + 2x^3 - 22x^2 - 18x + 35$.
(Example 4)

---

**Teaching Note**

Discuss the terminology and notation associated with zeros of a function: *x*-intercept, root, and solution.

---

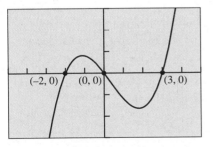

[−5, 5] by [−15, 15]

**Figure 2.29** The graph of
$y = x^3 - x^2 - 6x$ showing the *x*-intercepts.
(Example 5)

---

### Example 4 APPLYING POLYNOMIAL THEORY

Graph the polynomial in a window showing its extrema and zeros and its end behavior. Describe the end behavior using limits.

**(a)** $f(x) = x^3 + 2x^2 - 11x - 12$

**(b)** $g(x) = 2x^4 + 2x^3 - 22x^2 - 18x + 35$

**Solution**

**(a)** The graph of $f(x) = x^3 + 2x^2 - 11x - 12$ is shown in Figure 2.28a. The function $f$ has 2 extrema and 3 zeros, the maximum number possible for a cubic. $\lim_{x \to \infty} f(x) = \infty$ and $\lim_{x \to -\infty} f(x) = -\infty$.

**(b)** The graph of $g(x) = 2x^4 + 2x^3 - 22x^2 - 18x + 35$ is shown in Figure 2.28b. The function $g$ has 3 extrema and 4 zeros, the maximum number possible for a quartic. $\lim_{x \to \infty} g(x) = \infty$ and $\lim_{x \to -\infty} g(x) = \infty$.

## Zeros of Polynomial Functions

Recall that finding the real-number zeros of a function $f$ is equivalent to finding the *x*-intercepts of the graph of $y = f(x)$ or the solutions to the equation $f(x) = 0$. Example 5 illustrates that *factoring* a polynomial function makes solving these three related problems an easy matter.

### Example 5 FINDING THE ZEROS OF A POLYNOMIAL FUNCTION

Find the zeros of $f(x) = x^3 - x^2 - 6x$.

**Solution**

**Solve Algebraically**

We solve the related equation $f(x) = 0$ by factoring:

$$x^3 - x^2 - 6x = 0$$
$$x(x^2 - x - 6) = 0 \qquad \text{Remove common factor } x.$$
$$x(x - 3)(x + 2) = 0 \qquad \text{Factor quadratic.}$$
$$x = 0, \, x - 3 = 0, \, \text{or } x + 2 = 0 \qquad \text{Zero factor property}$$
$$x = 0, \quad x = 3, \quad \text{or} \quad x = -2$$

So the zeros of $f$ are 0, 3, and −2.

**Support Graphically**

Figure 2.29 shows that these same three values are *x*-intercepts of the graph of $y = f(x)$.

From Example 5, we see that if a polynomial function $f$ is presented in factored form, each factor $(x - k)$ corresponds to a zero $x = k$, and if $k$ is a real number, $(k, 0)$ is an *x*-intercept of the graph of $y = f(x)$. We apply this principle in Example 6.

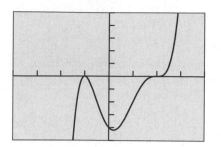

[–4, 4] by [–10, 10]

**Figure 2.30** The graph of $f(x) = (x - 2)^3(x + 1)^2$ showing the $x$-intercepts. (Example 6)

**Example 6**  FINDING THE ZEROS OF A FACTORED POLYNOMIAL FUNCTION

Find the zeros of $f(x) = (x - 2)^3(x + 1)^2$.

**Solution**

**Solve Algebraically**

We solve the related equation $f(x) = 0$:

$$(x - 2)^3(x + 1)^2 = 0$$
$$(x - 2)(x - 2)(x - 2)(x + 1)(x + 1) = 0$$
$$x = 2 \quad \text{or} \quad x = -1$$

So the zeros of $f$ are 2 and $-1$.

**Support Graphically**

Figure 2.30 shows that 2 and $-1$ are $x$-intercepts of the graph of $y = f(x)$.

When a factor is repeated as in Example 6, we say the polynomial function has a *repeated zero*. The function in Example 6 has two repeated zeros. Because the factor $x - 2$ occurs three times, 2 is a zero of *multiplicity* 3. Similarly, $-1$ is a zero of multiplicity 2. The following definition generalizes this concept.

---

**Definition**  Multiplicity of a Zero of a Polynomial Function

If $f$ is a polynomial function and $(x - c)^m$ is a factor of $f$ but $(x - c)^{m+1}$ is not, then $c$ is a zero of **multiplicity $m$** of $f$.

---

A zero of multiplicity $m \geq 2$ is a **repeated zero**. Notice in Figure 2.30 that the graph of $f$ just *kisses* the $x$-axis without crossing it at $(-1, 0)$, but that the graph of $f$ crosses the $x$-axis at $(2, 0)$. This too can be generalized.

---

**Zeros of Odd and Even Multiplicity**

If a polynomial function $f$ has a real zero $c$ of odd multiplicity, then the graph of $f$ crosses the $x$-axis at $(c, 0)$ and the value of $f$ changes sign at $x = c$.

If a polynomial function $f$ has a real zero $c$ of even multiplicity, then the graph of $f$ does not cross the $x$-axis at $(c, 0)$ and the value of $f$ does not change sign at $x = c$.

---

In Example 5 none of the zeros were repeated. Because a nonrepeated zero has multiplicity 1, and 1 is odd, the graph of a polynomial function crosses the $x$-axis and has a sign change at every nonrepeated zero (Figure 2.29). Knowing when a graph crosses the $x$-axis and when it doesn't is important in curve-sketching and in solving inequalities.

## Intermediate Value Theorem

The *Intermediate Value Theorem* tells us that a sign change implies a real zero.

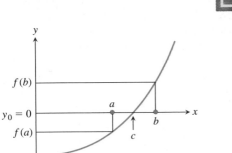

**Figure 2.31**  If $f(a) < 0 < f(b)$, then there is a zero $x = c$ between $a$ and $b$.

> **Theorem  Intermediate Value Theorem**
>
> If $a$ and $b$ are real numbers with $a < b$ and if $f$ is continuous on the interval $[a, b]$, then $f$ takes on every value between $f(a)$ and $f(b)$. In other words, if $y_0$ is between $f(a)$ and $f(b)$, then $y_0 = f(c)$ for some number $c$ in $[a, b]$.
>
> In particular, if $f(a)$ and $f(b)$ have opposite signs (i.e., one is negative and the other is positive), then $f(c) = 0$ for some number $c$ in $[a, b]$ (Figure 2.31).

### Example 7  USING THE INTERMEDIATE VALUE THEOREM

Prove that every polynomial function of odd degree has at least one real zero.

**Solution**  Let $f$ be a polynomial function of odd degree. Because $f$ is odd, the leading term test tells us that $\lim\limits_{x \to \infty} f(x) = - \lim\limits_{x \to -\infty} f(x)$. So there exist real numbers $a$ and $b$ with $a < b$ and such that $f(a)$ and $f(b)$ have opposite signs. Because every polynomial function is defined and continuous for all real numbers, $f$ is continuous on the interval $[a, b]$. Therefore, by the Intermediate Value Theorem, $f(c) = 0$ for some number $c$ in $[a, b]$, and thus $c$ is a real zero of $f$.

In practice the Intermediate Value Theorem is used in combination with our other mathematical knowledge and technological know-how.

> **Alert**
>
> Some students fail to make the distinction between exact and approximate answers. They may think that $\sqrt{3}$ (exact) and 1.732 (approximate) are equivalent.

### Example 8  UNCOVERING HIDDEN BEHAVIOR

Find all of the real zeros of $f(x) = x^4 + 0.1x^3 - 6.5x^2 + 7.9x - 2.4$.

**Solution**

**Solve Graphically**

Because $f$ is of degree 4, there are at most four zeros. The graph in Figure 2.32a suggests a single zero (multiplicity 1) around $x = -3$ and a triple zero (multiplicity 3) around $x = 1$. Closer inspection around $x = 1$ in Figure 2.32b reveals three separate zeros. Using the grapher, we find the four zeros to be $x \approx 1.37$, $x \approx 1.13$, $x \approx 0.50$, and $x \approx -3.10$.

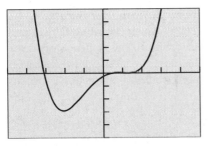

[−5, 5] by [−50, 50]
(a)

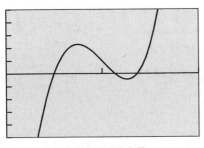

[0, 2] by [−0.5, 0.5]
(b)

**Figure 2.32**  (Example 8)

## Modeling

In the problem-solving process presented in Section 1.1, Step 2 is to develop a mathematical model of the problem. When the model being developed is a polynomial function of higher degree, the algebraic and geometric thinking required can be rather involved. In solving Example 9 you may find it helpful to make a physical model out of paper or cardboard.

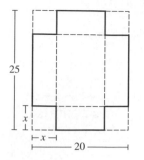

**Figure 2.33** Cut square corners from a piece of cardboard, and fold the flaps to make a box. (Example 9)

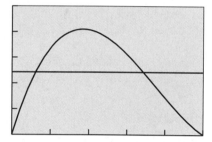

| X | Y₁ | |
|---|---|---|
| 1 | 414 | |
| 2 | 672 | |
| 3 | 798 | |
| 4 | 816 | |
| 5 | 750 | |
| 6 | 624 | |
| 7 | 462 | |

Y₁ ▤ X(20−2X)(25−...

**Figure 2.34** A table to get a feel for the volume values in Example 9.

[0, 10] by [0, 1000]

**Figure 2.35** $y_1 = x(25 - 2x)(20 - 2x)$ and $y_2 = 484$. (Example 9)

### Example 9   DESIGNING A BOX

Dixie Packaging Company has contracted to make boxes with a volume of approximately 484 in.³. Squares are to be cut from the corners of a 20-in. by 25-in. piece of cardboard, and the flaps folded up to make an open box. (See Figure 2.33.) What size squares should be cut from the cardboard?

#### Solution
#### Model
We know that the volume $V$ = height × length × width. So let

$$x = \text{edge of cut-out square (height of box)}$$
$$25 - 2x = \text{length of the box}$$
$$20 - 2x = \text{width of the box}$$
$$V = x(25 - 2x)(20 - 2x)$$

#### Solve Numerically and Graphically
For a volume of 484, we solve the equation $x(25 - 2x)(20 - 2x) = 484$. Because the width of the cardboard is 20 in., $0 \le x \le 10$. We use the table in Figure 2.34 to get a sense of the volume values to set the window for the graphs in Figure 2.35. The cubic volume function intersects the constant volume of 484 at $x \approx 1.22$ and $x \approx 6.87$.

#### Interpret
Squares with lengths of approximately 1.22 in. or 6.87 in. should be cut from the cardboard to produce a box with a volume of 484 in.³.

Just as any two points in the Cartesian plane with different $x$ values and different $y$ values determine a unique slant line and its related linear function, any three noncollinear points with different $x$ values determine a quadratic function. In general, $(n + 1)$ points positioned with sufficient generality determine a polynomial function of degree $n$. The process of fitting a polynomial of degree $n$ to $(n + 1)$ points is **polynomial interpolation**. Exploration 2 involves two polynomial interpolation problems.

---

**Exploration 2** | **Interpolating Points with a Polynomial**

**1.** Use cubic regression to fit a curve through the four points given in the table.

| $x$ | −2 | 1 | 3 | 8 |
|---|---|---|---|---|
| $y$ | 2 | 0.5 | −0.2 | 1.25 |

**2.** Use quartic regression to fit a curve through the five points given in the table.

| $x$ | 3 | 4 | 5 | 6 | 8 |
|---|---|---|---|---|---|
| $y$ | −2 | −4 | −1 | 8 | 3 |

How good is the fit in each case? Why?

Generally we want a reason beyond, "it fits good," to choose a model for genuine data. However, when no theoretical basis exists for picking a model, a balance between goodness of fit and simplicty of model is sought. For polynomials, we try to pick a model with the lowest possible degree that has a reasonably good fit.

| Table 2.15 U.S. Farm Exports | |
|---|---|
| Year | Amount (billions) |
| 1991 | 39.4 |
| 1992 | 43.1 |
| 1993 | 42.9 |
| 1994 | 46.3 |
| 1995 | 56.3 |
| 1996 | 60.5 |
| 1997 | 57.1 |
| 1998 | 52.0 |

*Source: U.S. Department of Commerce as reported in USA Today, August 30, 1999.*

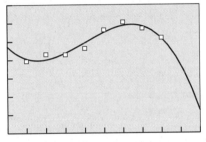

[0, 10] by [0, 70]

**Figure 2.36** Scatter plot and cubic regression graph for Example 10.

**Example 10   Modeling Data—Curve Fitting**

Table 2.15 shows the total U.S. Farm Exports in billions for 1991 through 1998. Try various polynomial regression models (degrees $n = 1, 2, 3, 4$) and decide which is the best model.

**Solution**   Let $x = 0$ represent the year 1990, $x = 1$ the year 1991, and so on. Making a scatter plot of the data and superimposing the polynomial regression functions of degrees $n = 1, 2, 3, 4$ leads us to the conclusion that the cubic model

$$y = -0.3409x^3 + 4.1070x^2 - 10.4687x + 47.3071$$

does the best job of balancing goodness of fit with lowest possible degree. See Figure 2.36.

# Quick Review 2.3

In Exercises 1–6, factor the polynomial into linear factors.

**1.** $x^2 - x - 12$   $(x - 4)(x + 3)$   **2.** $x^2 - 11x + 28$

**3.** $3x^2 - 11x + 6$               **4.** $6x^2 - 5x + 1$

**5.** $3x^3 - 5x^2 + 2x$             **6.** $6x^3 - 22x^2 + 12x$

In Exercises 7–10, solve the equation mentally.

**7.** $x(x - 1) = 0$   $x = 0, x = 1$   **8.** $x(x + 2)(x - 5) = 0$

**9.** $(x + 6)^3(x + 3)(x - 1.5) = 0$   $x = -6, x = -3, x = 1.5$

**10.** $(x + 6)^2(x + 4)^4(x - 5)^3 = 0$   $x = -6, x = -4, x = 5$

# Section 2.3 Exercises

In Exercises 1–6, describe how to transform the graph of an appropriate monomial function $f(x) = a_n x^n$ into the graph of the given polynomial function. Sketch the transformed graph by hand and support your answer with a grapher. Compute the location of the y-intercept as a check on the transformed graph.

**1.** $g(x) = 2(x - 3)^3$          **2.** $g(x) = -(x + 5)^3$

**3.** $g(x) = -\dfrac{1}{2}(x + 1)^3 + 2$          **4.** $g(x) = \dfrac{2}{3}(x - 3)^3 + 1$

**5.** $g(x) = -2(x + 2)^4 - 3$          **6.** $g(x) = 3(x - 1)^4 - 2$

In Exercises 7–10, state the degree and list the zeros of the polynomial function. State the multiplicity of each zero and whether the graph crosses the x-axis at the corresponding x-intercept. Then sketch the graph of the polynomial function by hand and support your answer with a grapher.

**7.** $f(x) = x(x - 3)^2$          **8.** $f(x) = -x^3(x - 2)$

**9.** $f(x) = (x - 2)^3(x + 1)^2$          **10.** $f(x) = 7(x - 3)^2(x + 5)^4$

In Exercises 11 and 12, graph the polynomial function, locate its extrema and zeros, and explain how it is related to the monomials from which it is built.

**11.** $f(x) = -x^4 + 2x$          **12.** $g(x) = 2x^4 - 5x^2$

In Exercises 13–16, match the polynomial function with its graph. Explain your choice.

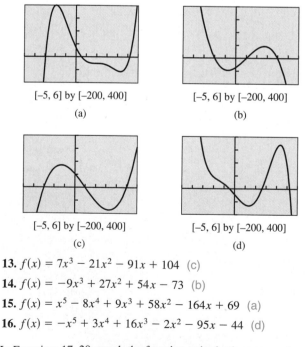

[−5, 6] by [−200, 400]
(a)

[−5, 6] by [−200, 400]
(b)

[−5, 6] by [−200, 400]
(c)

[−5, 6] by [−200, 400]
(d)

**13.** $f(x) = 7x^3 - 21x^2 - 91x + 104$   (c)

**14.** $f(x) = -9x^3 + 27x^2 + 54x - 73$   (b)

**15.** $f(x) = x^5 - 8x^4 + 9x^3 + 58x^2 - 164x + 69$   (a)

**16.** $f(x) = -x^5 + 3x^4 + 16x^3 - 2x^2 - 95x - 44$   (d)

In Exercises 17–20, graph the function pairs in the same series of viewing windows. Zoom out until the two graphs look nearly identical and state your final viewing window.

**17.** $f(x) = x^3 - 4x^2 - 5x - 3$ and $g(x) = x^3$

**18.** $f(x) = x^3 + 2x^2 - x + 5$ and $g(x) = x^3$

**19.** $f(x) = 2x^3 + 3x^2 - 6x - 15$ and $g(x) = 2x^3$

**20.** $f(x) = 3x^3 - 12x + 17$ and $g(x) = 3x^3$

In Exercises 21–28, graph the function in a viewing window that shows all of its extrema and x-intercepts.

**21.** $f(x) = (x - 1)(x + 2)(x + 3)$

**22.** $f(x) = (2x - 3)(4 - x)(x + 1)$

**23.** $f(x) = -x^3 + 4x^2 + 31x - 70$

**24.** $f(x) = x^3 - 2x^2 - 41x + 42$

**25.** $f(x) = (x - 2)^2(x + 1)(x - 3)$

**26.** $f(x) = (2x + 1)(x - 4)^3$

**27.** $f(x) = 2x^4 - 5x^3 - 17x^2 + 14x + 41$

**28.** $f(x) = -3x^4 - 5x^3 + 15x^2 - 5x + 19$

In Exercises 29–32, describe the end behavior of the polynomial function using $\lim\limits_{x \to \infty} f(x)$ and $\lim\limits_{x \to -\infty} f(x)$.

**29.** $f(x) = 3x^4 - 5x^2 + 3$   $\infty$, $\infty$

**30.** $f(x) = -x^3 + 7x^2 - 4x + 3$   $-\infty$, $\infty$

**31.** $f(x) = 7x^2 - x^3 + 3x - 4$   $-\infty$, $\infty$

**32.** $f(x) = x^3 - x^4 + 3x^2 - 2x + 7$   $-\infty$, $-\infty$

In Exercises 33–36, match the polynomial function with its graph. Approximate all of the real zeros of the function.

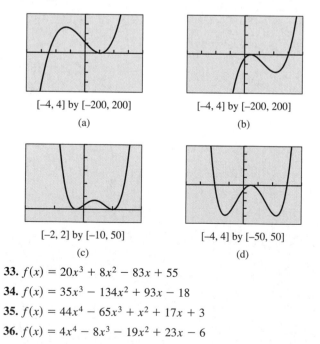

[−4, 4] by [−200, 200]
(a)

[−4, 4] by [−200, 200]
(b)

[−2, 2] by [−10, 50]
(c)

[−4, 4] by [−50, 50]
(d)

**33.** $f(x) = 20x^3 + 8x^2 - 83x + 55$

**34.** $f(x) = 35x^3 - 134x^2 + 93x - 18$

**35.** $f(x) = 44x^4 - 65x^3 + x^2 + 17x + 3$

**36.** $f(x) = 4x^4 - 8x^3 - 19x^2 + 23x - 6$

In Exercises 37–42, find the zeros of the function algebraically.

**37.** $f(x) = x^2 + 2x - 8$          **38.** $f(x) = 3x^2 + 4x - 4$

**39.** $f(x) = 9x^2 - 3x - 2$

**40.** $f(x) = x^3 - 25x$

**41.** $f(x) = 3x^3 - x^2 - 2x$

**42.** $f(x) = 5x^3 - 5x^2 - 10x$

In Exercises 43–48, graph the function in a viewing window that shows all of its $x$-intercepts and approximate all of its zeros.

**43.** $f(x) = 2x^3 + 3x^2 - 7x - 6$

**44.** $f(x) = -x^3 + 3x^2 + 7x - 2$

**45.** $f(x) = x^3 + 2x^2 - 4x - 7$

**46.** $f(x) = -x^4 - 3x^3 + 7x^2 + 2x + 8$

**47.** $f(x) = x^4 + 3x^3 - 9x^2 + 2x + 3$

**48.** $f(x) = 2x^5 - 11x^4 + 4x^3 + 47x^2 - 42x - 8$

In Exercises 49–52, find the zeros of the function algebraically or graphically.

**49.** $f(x) = x^3 - 36x$  0, −6, and 6

**50.** $f(x) = x^3 + 2x^2 - 109x - 110$  −11, −1, and 10

**51.** $f(x) = x^3 - 7x^2 - 49x + 55$  −5, 1, 11

**52.** $f(x) = x^3 - 4x^2 - 44x + 96$  −6, 2, and 8

In Exercises 53–56, using only algebra, find a cubic function with the given zeros. Support by graphing your answer.

**53.** 3, −4, 6

**54.** −2, 3, −5

**55.** $\sqrt{3}, -\sqrt{3}, 4$

**56.** $1, 1 + \sqrt{2}, 1 - \sqrt{2}$

**57.** Use cubic regression to fit a curve through the four points given in the table.

| $x$ | −3 | −1 | 1 | 3 |
|---|---|---|---|---|
| $y$ | 22 | 25 | 12 | −5 |

**58.** Use cubic regression to fit a curve through the four points given in the table.

| $x$ | −2 | 1 | 4 | 7 |
|---|---|---|---|---|
| $y$ | 2 | 5 | 9 | 26 |

**59.** Use quartic regression to fit a curve through the five points given in the table.

| $x$ | 3 | 4 | 5 | 6 | 8 |
|---|---|---|---|---|---|
| $y$ | −7 | −4 | −11 | 8 | 3 |

**60.** Use quartic regression to fit a curve through the five points given in the table.

| $x$ | 0 | 4 | 5 | 7 | 13 |
|---|---|---|---|---|---|
| $y$ | −21 | −19 | −12 | 8 | 3 |

**61. Analyzing Profit** Economists for Smith Brothers, Inc., find the company profit $P$ by using the formula $P = R - C$, where $R$ is the total revenue generated by the business and $C$ is the total cost of operating the business.

**(a)** Using data from past years, the economists determined that $R(x) = 0.0125x^2 + 412x$ models total revenue, and $C(x) = 12{,}225 + 0.00135x^3$ models the total cost of doing business, where $x$ is the number of customers patronizing the

business. How many customers must Smith Bros. have to be profitable each year?

**(b)** How many customers must there be for Smith Bros. to realize an annual profit of $60,000?

**62. Stopping Distance** A state highway patrol safety division collected the data on stopping distances in Table 2.16.

**(a)** Draw a scatter plot of the data.

**(b)** Find the quadratic regression model.

**(c)** Superimpose the regression curve on the scatter plot.

**(d)** Use the regression model to predict the stopping distance for a vehicle traveling at 25 mph.

**(e)** Use the regression model to predict the speed of a car if the stopping distance is 300 ft.

| Table 2.16  Highway Safety Division | |
|---|---|
| Speed (mph) | Stopping Distance (ft) |
| 10 | 15.1 |
| 20 | 39.9 |
| 30 | 75.2 |
| 40 | 120.5 |
| 50 | 175.9 |

**63. Circulation of Blood** Research conducted at a national health research project shows that the speed at which a blood cell travels in an artery depends on its distance from the center of the artery. The function $v = 1.19 - 1.87r^2$ models the velocity (in centimeters per second) of a cell that is $r$ centimeters from the center of an artery.

**(a)** Find a graph of $v$ that reflects values of $v$ appropriate for this problem. Record the viewing-window dimensions.

**(b)** If a blood cell is traveling at 0.975 cm/sec, estimate the distance the blood cell is from the center of the artery.

**64. Volume of a Box** Dixie Packaging Co. has contracted to manufacture a box with no top that is to be made by removing squares of width $x$ from the corners of a 15-in. by 60-in. piece of cardboard.

**(a)** Show that the volume of the box is modeled by $V(x) = x(60 - 2x)(15 - 2x)$.

**(b)** Determine $x$ so that the volume of the box is at least 450 in.$^3$

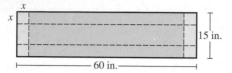

**65. Volume of a Box** Squares of width $x$ are removed from a 10-cm by 25-cm piece of cardboard, and the resulting edges are folded up to form a box with no top. Determine all values of $x$ so that the volume of the resulting box is at most 175 cm$^3$.

**66. Volume of a Box** The function $V = 2666x - 210x^2 + 4x^3$ represents the volume of a box that has been made by removing squares of width $x$ from each corner of a rectangular sheet of material and then folding up the sides. What values are possible for $x$?

# Explorations

In Exercises 67 and 68, two views of the function are given.

**67. Writing to Learn** Describe why each view of the function
$$f(x) = x^5 - 10x^4 + 2x^3 + 64x^2 - 3x - 55,$$
by itself, may be considered inadequate.

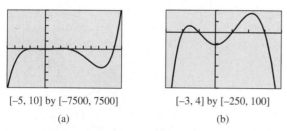

[−5, 10] by [−7500, 7500]    [−3, 4] by [−250, 100]
(a)    (b)

**68. Writing to Learn** Describe why each view of the function
$$f(x) = 10x^4 + 19x^3 - 121x^2 + 143x - 51,$$
by itself, may be considered inadequate.

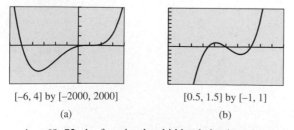

[−6, 4] by [−2000, 2000]    [0.5, 1.5] by [−1, 1]
(a)    (b)

In exercises 69–72, the function has hidden behavior when viewed in the window $[-10, 10]$ by $[-10, 10]$. Describe what behavior is hidden, and state the dimensions of a viewing widnow that reveals the hidden behavior.

**69.** $f(x) = 10x^3 - 40x^2 + 50x - 20$

**70.** $f(x) = 0.5(x^3 - 8x^2 + 12.99x - 5.94)$

**71.** $f(x) = 11x^3 - 10x^2 + 3x + 5$

**72.** $f(x) = 33x^3 - 100x^2 + 101x - 40$ ■

# Extending the Ideas

**73.** Graph the left side of the equation
$$3(x^3 - x) = a(x - b)^3 + c.$$
Then explain why there are no real numbers $a$, $b$, and $c$ that make the equation true. (*Hint*: Use your knowledge of $y = x^3$ and transformations.)

**74.** Graph the left side of the equation
$$x^4 + 3x^3 - 2x - 3 = a(x - b)^4 + c.$$
Then explain why there are no real numbers $a$, $b$, and $c$ that make the equation true.

**75. Looking Ahead to Calculus** The figure shows a graph of both $f(x) = -x^3 + 2x^2 + 9x - 11$ and the line $L$ defined by $y = 5(x - 2) + 7$.

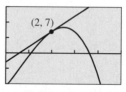

[0, 5] by [−10, 15]

**(a)** Confirm that the point $Q(2, 7)$ is a point of intersection of the two graphs.

**(b)** Zoom in at point $Q$ to develop a visual understanding that $y = 5(x - 2) + 7$ is a *linear approximation* for $y = f(x)$ near $x = 2$.

**(c)** Recall that a line is *tangent* to a circle at a point $P$ if it intersects the circle only at point $P$. View the two graphs in the window $[-5, 5]$ by $[-25, 25]$, and explain why that definition of tangent line is not valid for the graph of $f$.

**76. Looking Ahead to Calculus** Consider the function $f(x) = x^n$ where $n$ is an odd integer.

**(a)** Suppose that $a$ is a positive number. Show that the slope of the line through the points $P(a, f(a))$ and $Q(-a, f(-a))$ is $a^{n-1}$.

**(b)** Let $x_0 = a^{1/(n-1)}$. Find an equation of the line through point $(x_0, f(x_0))$ with the slope $a^{n-1}$.

**(c)** Consider the special case $n = 3$ and $a = 3$. Show both the graph of $f$ and the line from part b in the window $[-5, 5]$ by $[-30, 30]$.

**77. Derive an Algebraic Model of a Problem** Show that the distance $x$ in the figure is a solution of the equation

$x^4 - 16x^3 + 500x^2 - 8000x + 32{,}000 = 0$ and find the value of $D$ by following these steps.

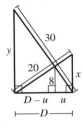

**(a)** Use the similar triangles in the diagram and the properties of proportions learned in geometry to show that

$$\frac{8}{x} = \frac{y-8}{y}.$$

**(b)** Show that $y = \dfrac{8x}{x-8}$.

**(c)** Show that $y^2 - x^2 = 500$. Then substitute for $y$, and simplify to obtain the desired degree 4 equation in $x$.

**(d)** Find the distance $D$.

**78. Group Learning Activity  Work in groups of three**
Consider functions of the form $f(x) = x^3 + bx^2 + x + 1$ where $b$ is a nonzero real number.

**(a)** Discuss as a group how the value of $b$ affects the graph of the function.

**(b)** After completing (a), have each member of the group (individually) predict what the graphs of $f(x) = x^3 + 15x^2 + x + 1$ and $g(x) = x^3 - 15x^2 + x + 1$ will look like.

**(c)** Compare your predictions with each other. Confirm whether they are correct.

---

## 2.4  Real Zeros of Polynomial Functions

Long Division and the Division Algorithm  •  Remainder and Factor Theorems  •  Synthetic Division  •  Rational Zeros Theorem  •  Upper and Lower Bounds

### Long Division and the Division Algorithm

We have seen that a polynomial in factored form reveals its zeros and much about its graph. Understanding the division of polynomials will give us new and better ways to find factors. First we observe that the division of polynomials closely resembles the division of integers. Examine the following side-by-side long divisions. How are they alike?

| | |
|---|---|
| $$\begin{array}{r} 112 \\ 32\overline{)3587} \\ \underline{32\phantom{00}} \\ 387 \\ \underline{32\phantom{0}} \\ 67 \\ \underline{64} \\ 3 \end{array}$$ | $$\begin{array}{r} 1x^2 + 1x + 2 \\ 3x+2\overline{)3x^3 + 5x^2 + 8x + 7} \\ \underline{3x^3 + 2x^2\phantom{+8x+7}} \\ 3x^2 + 8x + 7 \\ \underline{3x^2 + 2x\phantom{+7}} \\ 6x + 7 \\ \underline{6x + 4} \\ 3 \end{array}$$ |

$\leftarrow$ Quotient
$\leftarrow$ Dividend
$\leftarrow$ Multiply: $1x^2 \cdot (3x+2)$
$\leftarrow$ Subtract
$\leftarrow$ Multiply: $1x \cdot (3x+2)$
$\leftarrow$ Subtract
$\leftarrow$ Multiply: $2 \cdot (3x+2)$
$\leftarrow$ Remainder

Every division, whether integer or polynomial, involves a *dividend* divided by a *divisor* to obtain a *quotient* and a *remainder*. We can always check and summarize our result with an equation of the form

$$(\text{Divisor})(\text{Quotient}) + \text{Remainder} = \text{Dividend}$$

For instance, in the example above, to check or summarize these long divisions we could write

$$32 \times 112 + 3 = 3587 \qquad (3x+2)(x^2 + x + 2) + 3 = 3x^3 + 5x^2 + 8x + 7$$

The *division algorithm* contains such a summary *polynomial equation*, but with the dividend written on the left side of the equation.

### Division Algorithm for Polynomials

Let $f(x)$ and $d(x)$ be polynomials with the degree of $f$ greater than or equal to the degree of $d$, and $d(x) \neq 0$. Then there are unique polynomials $q(x)$ and $r(x)$, called the **quotient** and **remainder**, such that

$$f(x) = d(x) \cdot q(x) + r(x) \qquad (1)$$

where either $r(x) = 0$ or the degree of $r$ is less than the degree of $d$.

The function $f(x)$ in the division algorithm is the **dividend**, and $d(x)$ is the **divisor**. If $r(x) = 0$, we say $d(x)$ **divides evenly** into $f(x)$.

The summary statement (1) is sometimes written in *fraction form* as follows.

$$\frac{f(x)}{d(x)} = q(x) + \frac{r(x)}{d(x)} \qquad (2)$$

For instance, in the polynomial division example above we could write

$$\frac{3x^3 + 5x^2 + 8x + 7}{3x + 2} = x^2 + x + 2 + \frac{3}{3x + 2}$$

to summarize the relationships between the polynomials being divided and the resulting quotient and remainder.

### Example 1  USING POLYNOMIAL LONG DIVISION

Use long division to find the quotient and remainder when $2x^4 - x^3 - 2$ is divided by $2x^2 + x + 1$. Write a summary statement in both polynomial and fraction form.

**Solution**

**Solve Algebraically**

$$
\require{enclose}
\begin{array}{r}
x^2 - x \phantom{00000000000} \\
2x^2 + x + 1 \enclose{longdiv}{2x^4 - x^3 + 0x^2 + 0x - 2} \\
\underline{2x^4 + x^3 + x^2 \phantom{00000000}} \\
-2x^3 - x^2 + 0x - 2 \\
\underline{-2x^3 - x^2 - \phantom{0}x \phantom{000}} \\
x - 2
\end{array}
$$

← Quotient

← Remainder

From the division algorithm we know

$$2x^4 - x^3 - 2 = (2x^2 + x + 1)(x^2 - x) + (x - 2).$$

Using equation (2), we obtain the fraction form

$$\frac{2x^4 - x^3 - 2}{2x^2 + x + 1} = x^2 - x + \frac{x - 2}{2x^2 + x + 1}.$$

**Support Graphically**

In Figure 2.37 we graph

$y = 2x^4 - x^3 - 2$ and

$y = (2x^2 + x + 1)(x^2 - x) + (x - 2)$

in the same window and see that the graphs appear to be the same.

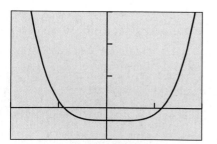

[–2, 2] by [–5, 15]

**Figure 2.37** The graphs of
$y = 2x^4 - x^3 - 2$ and
$y = (2x^2 + x + 1)(x^2 - x) + (x - 2)$.
(Example 1)

## Remainder and Factor Theorems

An important special case of the division algorithm occurs when the divisor is of the form $d(x) = x - k$, where $k$ is a real number. Because the degree of $d(x) = x - k$ is 1, the remainder is a real number. So we obtain the following simplified summary statement for the division algorithm:

$$f(x) = (x - k)q(x) + r \qquad (3)$$

We use this special case of the division algorithm throughout the rest of the section.

Using Equation (3), we evaluate the polynomial $f(x)$ at $x = k$:

$$f(k) = (k - k)q(k) + r$$

$$= 0 \cdot q(k) + r$$

$$= 0 + r$$

$$= r$$

So $f(k) = r$, which is the remainder. This reasoning gives us the following theorem.

> **Theorem**  **Remainder Theorem**
>
> If a polynomial $f(x)$ is divided by $x - k$, then the remainder is $r = f(k)$.

Example 2 shows a clever use of the Remainder Theorem that gives information about the factors, zeros, and $x$-intercepts.

**Example 2**  USING THE REMAINDER THEOREM

Find the remainder when $f(x) = 3x^2 + 7x - 20$ is divided by

**(a)** $x - 2$ **(b)** $x + 1$ **(c)** $x + 4$

**Solution**

**Solve Numerically (by hand)**

**(a)** We can find the remainder without doing long division! Using the Remainder Theorem with $k = 2$ we find that
$r = f(2) = 3(2)^2 + 7(2) - 20 = 12 + 14 - 20 = 6$.

**(b)** $r = f(-1) = 3(-1)^2 + 7(-1) - 20 = 3 - 7 - 20 = -24$.

**(c)** $r = f(-4) = 3(-4)^2 + 7(-4) - 20 = 48 - 28 - 20 = 0$.

**Interpret**

Because the remainder in part (c) is 0, $x + 4$ *divides evenly* into $f(x) = 3x^2 + 7x - 20$. So, $x + 4$ is a factor of $f(x) = 3x^2 + 7x - 20$, $-4$ is a solution of $3x^2 + 7x - 20 = 0$, and $-4$ is an $x$-intercept of the graph of $y = 3x^2 + 7x - 20$. We know all of this without ever dividing, factoring, or graphing!

**Support Numerically (using a grapher)**

We can find the remainders of several division problems at once using the table feature of a grapher. (See Figure 2.38.)

Our interpretation of Example 2c leads us to the following theorem.

| X | Y1 | |
|---|---|---|
| -4 | 0 | |
| -3 | -14 | |
| -2 | -22 | |
| -1 | -24 | |
| 0 | -20 | |
| 1 | -10 | |
| 2 | 6 | |
| Y1 ▤ 3X^2+7X-20 | | |

**Figure 2.38**  Table for $f(x) = 3x^2 + 7x - 20$ showing the remainders obtained when $f(x)$ is divided by $x - k$, for $k = -4, -3, \ldots, 1, 2$.

> ### Theorem   Factor Theorem
>
> A polynomial function $f(x)$ has a factor $x - k$ if and only if $f(k) = 0$.

### Proof

If $f(x)$ has a factor $x - k$, then there is a polynomial $g(x)$ such that

$$f(x) = (x - k)g(x) = (x - k)g(x) + 0.$$

By the uniqueness condition of the Division Algorithm $g(x)$ is the quotient and 0 is the remainder, and by the Remainder Theorem, $f(k) = 0$.

Conversely, if $f(k) = 0$, then the remainder $r = 0$, and $x - k$ divides evenly into $f(x)$ and $x - k$ is a factor of $f(x)$, with $f(x) = (x - k)q(x)$.

Applying the ideas of the Factor Theorem and its proof to Example 2, we can factor $f(x) = 3x^2 + 7x - 20$ by dividing it by the known factor $x + 4$.

$$
\begin{array}{r}
3x - 5 \\
x + 4 \overline{)\,3x^2 + 7x - 20} \\
\underline{3x^2 + 12x\phantom{mmm}} \\
-5x - 20 \\
\underline{-5x - 20} \\
0
\end{array}
$$

So, $f(x) = 3x^2 + 7x - 20 = (x + 4)(3x - 5)$. In this case, there really is no need to use long division, the Remainder Theorem, or the Factor Theorem; traditional factoring methods can do the job. However, when the polynomial function involved has degree 3 or higher, these sophisticated methods can be quite helpful in solving equations and finding factors, zeros, and $x$-intercepts. In fact, the Factor Theorem ties in nicely with earlier connections we have made in the following way.

> ### Fundamental Connections for Polynomial Functions
>
> For a polynomial function $f$ and a real number $k$, the following statements are equivalent:
>
> **1.** $x = k$ is a solution (or root) of the equation $f(x) = 0$
>
> **2.** $k$ is a zero of the function $f$.
>
> **3.** $k$ is an $x$-intercept of the graph of $y = f(x)$.
>
> **4.** $x - k$ is a factor of $f(x)$.

## Synthetic Division

We continue with the important special case of polynomial division that occurs when the divisor is of the form $x - k$. The Remainder Theorem gave us a way to find remainders in this case without long division. We now learn a method for finding both quotients and remainders for division by $x - k$ without long division. This short-cut method for the division of a polynomial by a linear divisor $x - k$ is **synthetic division**.

We illustrate the evolution of this method below in stages, progressing from long division through two intermediate stages to synthetic division.

Moving from stage to stage, focus on the coefficients and their relative positions. Moving from Stage 1 to Stage 2, we suppress the variable $x$ and the powers of $x$, and then from Stage 2 to Stage 3, we eliminate unneeded duplications and collapse vertically.

**Stage 1**
**Long Division**

$$
\begin{array}{r}
2x^2 + 3x + 4 \\
x - 3 \overline{)2x^3 - 3x^2 - 5x - 12} \\
\underline{2x^3 - 6x^2} \\
3x^2 - 5x - 12 \\
\underline{3x^2 - 9x} \\
4x - 12 \\
\underline{4x - 12} \\
0
\end{array}
$$

**Stage 2**
**Variables Suppressed**

$$
\begin{array}{r}
2 \quad 3 \quad 4 \\
-3\overline{)2 \ -3 \ -5 \ -12} \\
\underline{2 \ -6} \\
3 \ -5 \ -12 \\
\underline{3 \ -9} \\
4 \ -12 \\
\underline{4 \ -12} \\
0
\end{array}
$$

**Stage 3**
**Collapsed Vertically**

$$
\begin{array}{r|rrrr}
-3 & 2 & -3 & -5 & -12 \\
& & -6 & -9 & -12 \\
\hline
& 2 & 3 & 4 & 0
\end{array}
$$
Dividend

Quotient, remainder

Finally, moving from Stage 3 to Stage 4, we change the sign of the coefficient representing the divisor and the sign of each of the numbers on the second line of our division scheme. These sign changes yield two advantages:

- The number standing for the divisor $x - k$ is now $k$, its zero.

- Changing the signs in the second line allows us to add rather than subtract.

**Stage 4**
**Synthetic Division**

Zero of divisor →
$$
\begin{array}{r|rrrr}
3 & 2 & -3 & -5 & -12 \\
& & 6 & 9 & 12 \\
\hline
& 2 & 3 & 4 & 0
\end{array}
$$
Dividend

Quotient, remainder

With Stage 4 we have achieved our goal of synthetic division, a highly streamlined version of dividing a polynomial by $x - k$. How does this "bare bones" division work? Example 3 explains the steps.

### Example 3 LEARNING SYNTHETIC DIVISION

Divide $2x^3 - 3x^2 - 5x - 12$ by $x - 3$ using synthetic division.

**Solution**

**Set Up**

The zero of the divisor $x - 3$ is 3. We write 3 in the divisor position. Because the dividend is in standard form, we write its coefficients in order in the dividend position. We then prepare for the division process by leaving space for the line for products and by drawing a horizontal line below the space. (See below.)

**Calculate**

- Because the leading coefficient of the dividend must be the leading coefficient of the quotient, copy the 2 into the first quotient position.

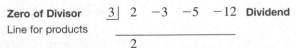

**Zero of Divisor**   $\underline{3}\,\big|\ 2 \quad -3 \quad -5 \quad -12$   **Dividend**
Line for products
$\qquad\qquad\qquad\qquad 2$

- Multiply the zero of the divisor (3) by the most recently determined coefficient of the quotient (2). Write the product above the line and one column to the right.

- Add the next coefficient of the dividend to the product just found and record the sum below the line in the same column.

- Repeat the "multiply" and "add" steps until the last row is completed.

| **Zero of Divisor** | 3⌋ | 2 | −3 | −5 | −12 | **Dividend** |
|---|---|---|---|---|---|---|
| Line for products | | | 6 | 9 | 12 | |
| Line for sums | | 2 | 3 | 4 | 0 | **Remainder** |
| | | | **Quotient** | | | |

### Interpret

The last line of numbers are the coefficients of the quotient polynomial and the remainder. The quotient must be a quadratic function. (Why?) So the quotient is $2x^2 + 3x + 4$ and the remainder is 0. So we conclude that

$$\frac{2x^3 - 3x^2 - 5x - 12}{x - 3} = 2x^2 + 3x + 4, \; x \neq 3.$$

### Example 4  PRACTICING SYNTHETIC DIVISION

Divide $x^4 - 8x^3 + 11x - 6$ by $x + 3$ synthetically, and summarize the results.

**Solution**  The zero of the divisor $x + 3$ is $-3$. The dividend in standard form is

$$x^4 - 8x^3 + 0x^2 + 11x - 6.$$

So our synthetic division scheme becomes:

| **Zero of Divisor** | −3⌋ | 1 | −8 | 0 | 11 | −6 | **Dividend** |
|---|---|---|---|---|---|---|---|
| | | | −3 | 33 | −99 | 264 | |
| | | 1 | −11 | 33 | −88 | 258 | **Remainder** |
| | | | **Quotient** | | | | |

The last line of numbers are the coefficients of the cubic quotient polynomial and the remainder. So we conclude that

$$\frac{x^4 - 8x^3 + 11x - 6}{x + 3} = x^3 - 11x^2 + 33x - 88 + \frac{258}{x + 3}.$$

## Rational Zeros Theorem

Real zeros of polynomial functions are either **rational zeros**—zeros that are rational numbers—or **irrational zeros**—zeros that are irrational numbers. For example,

$$f(x) = 4x^2 - 9 = (2x + 3)(2x - 3)$$

has the rational zeros $-3/2$ and $3/2$, and

$$f(x) = x^2 - 2 = (x + \sqrt{2})(x - \sqrt{2})$$

has the irrational zeros $-\sqrt{2}$ and $\sqrt{2}$.

The Rational Zeros Theorem tells us how to make a list of all potential rational zeros for a polynomial function with integer coefficients.

**Teaching Note**

It is best to see if the coefficients of the terms of $f(x)$ have a common factor that can be factored out before applying the rational zeros test. Factoring out a common factor will reduce the number of extra candidates found.

> ### Theorem  Rational Zeros Theorem
>
> Suppose $f$ is a polynomial function of degree $n \geq 1$ of the form
> $$f(x) = a_n x^n + a_{n-1} x^{n-1} + \cdots + a_0,$$
> with every coefficient an integer and $a_0 \neq 0$. If $x = p/q$ is a rational zero of $f$, where $p$ and $q$ have no common integer factors other than 1, then
>
> - $p$ is an integer factor of the constant coefficient $a_0$, and
> - $q$ is an integer factor of the leading coefficient $a_n$.

### Example 5  SEARCHING FOR RATIONAL ZEROS

Find the rational zeros of $f(x) = x^3 - 3x^2 + 1$.

**Solution**  Because the leading and constant coefficients are both 1, according to the Rational Zeros Theorem, the only potential rational zeros of $f$ are 1 and $-1$. So we check to see if they are in fact zeros of $f$:
$$f(1) = (1)^3 - 3(1)^2 + 1 = -1 \neq 0$$
$$f(-1) = (-1)^3 - 3(-1)^2 + 1 = -3 \neq 0$$

So $f$ has no rational zeros. Figure 2.39 shows that the graph of $f$ has three $x$-intercepts. So $f$ has three real zeros. All three must be irrational numbers.

In Example 5 the Rational Zeros Theorem gave us only two candidates for rational zeros, neither of which "checked out." Often this theorem suggests many candidates, as we see in Example 6. In such a case, we use technology and a variety of algebraic methods to track down the rational zeros.

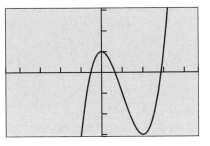

[−4.7, 4.7] by [−3.1, 3.1]

**Figure 2.39**  The function $f(x) = x^3 - 3x^2 + 1$ has three real zeros. (Example 5)

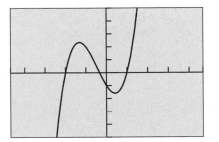

[−4.7, 4.7] by [−10, 10]

**Figure 2.40**  The function $f(x) = 3x^3 - 4x^2 - 5x - 2$ has three real zeros. (Example 6)

### Example 6  SEARCHING FOR AND FINDING RATIONAL ZEROS

Find the rational zeros of $f(x) = 3x^3 + 4x^2 - 5x - 2$.

**Solution**  Because the leading coefficient is 3 and constant coefficient is $-2$, the Rational Zeros Theorem gives us a list of several potential rational zeros of $f$. So we take an organized approach to our solution.

*Potential Rational Zeros*:
$$\frac{\text{Factors of } -2}{\text{Factors of } 3} : \frac{\pm 1, \pm 2}{\pm 1, \pm 3} : \pm 1, \ \pm 2, \ \pm\frac{1}{3}, \ \pm\frac{2}{3}$$

Figure 2.40 suggests that, among our candidates, 1, $-2$, and possibly $-1/3$ or $-2/3$ are the most likely to be rational zeros. We use synthetic division because it tells us whether a number is a zero and, if so, how to factor the polynomial. To see whether 1 is a zero of $f$, we synthetically divide $f(x)$ by $x - 1$:

```
Zero of Divisor   1│  3   4   −5   −2   Dividend
                        3    7    2
                  ─────────────────────
                     3   7    2    0    Remainder
                  Quotient
```

So because the remainder is 0, $x - 1$ is a factor of $f(x)$ and 1 is a zero of $f$. By the Division Algorithm and factoring, we conclude

$$f(x) = 3x^3 + 4x^2 - 5x - 2$$
$$= (x - 1)(3x^2 + 7x + 2)$$
$$= (x - 1)(3x + 1)(x + 2)$$

So the rational zeros of $f$ are 1, $-1/3$, and $-2$.

## Upper and Lower Bounds

We narrow our search for real zeros by using a test that identifies upper and lower bounds for real zeros. A number $k$ is an **upper bound for the real zeros** of $f$ if $f(x)$ is never zero when $x$ is greater than $k$. On the other hand, a number $k$ is a **lower bound for the real zeros** of $f$ if $f(x)$ is never zero when $x$ is less than $k$. So if $c$ is a lower bound and $d$ is an upper bound for the real zeros of a function $f$, all of the real zeros of $f$ must lie in the interval $[c, d]$. Figure 2.41 illustrates this situation.

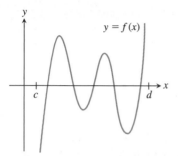

**Figure 2.41** $c$ is a lower bound and $d$ is an upper bound on the real zeros of $f$.

---

**Upper and Lower Bound Tests for Real Zeros**

Let $f$ be a polynomial function of degree $n \geq 1$ with a positive leading coefficient. Suppose $f(x)$ is divided by $x - k$ using synthetic division.

- If $k \geq 0$ and every number in the last line is nonnegative (positive or zero), then $k$ is an *upper bound* for the real zeros of $f$.
- If $k \leq 0$ and the numbers in the last line are alternately nonnegative and nonpositive, then $k$ is a *lower bound* for the real zeros of $f$.

---

**Example 7  ESTABLISHING BOUNDS FOR REAL ZEROS**

Prove that all of the real zeros of $f(x) = 2x^4 - 7x^3 - 8x^2 + 14x + 8$ must lie in the interval $[-2, 5]$.

**Solution**  We must prove that 5 is an upper bound and $-2$ is a lower bound on the real zeros of $f$. The function $f$ has a positive leading coefficient, so we employ the Upper and Lower Bound Test, and use synthetic division:

$$
\begin{array}{r|rrrrr}
5 & 2 & -7 & -8 & 14 & 8 \\
  &   & 10 & 15 & 35 & 245 \\
\hline
  & 2 & 3 & 7 & 49 & 253 \quad \text{Last line}
\end{array}
$$

$$
\begin{array}{r|rrrrr}
-2 & 2 & -7 & -8 & 14 & 8 \\
   &   & -4 & 22 & -28 & 28 \\
\hline
   & 2 & -11 & 14 & -14 & 36 \quad \text{Last line}
\end{array}
$$

Because the last line in the first division scheme consists of all positive numbers, 5 is an upper bound. Because the last line in the second division consists of numbers of alternating signs, $-2$ is a lower bound. All of the real zeros of $f$ must therefore lie in the closed interval $[-2, 5]$.

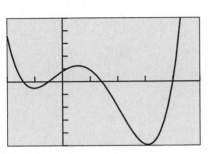

[–2, 5] by [–50, 50]

**Figure 2.42**
$f(x) = 2x^4 - 7x^3 - 8x^2 + 14x + 8$ has all of its real zeros in $[-2, 5]$. (Example 8)

**Example 8**  FINDING THE REAL ZEROS OF
A POLYNOMIAL FUNCTION

Find all of the real zeros of $f(x) = 2x^4 - 7x^3 - 8x^2 + 14x + 8$.

**Solution**  From Example 7 we know that all of the real zeros of $f$ must lie in the closed interval $[-2, 5]$. So in Figure 2.42 we set our Xmin and Xmax accordingly.

Next we use the Rational Zeros Theorem.

*Potential Rational Zeros*:

$$\frac{\text{Factors of 8}}{\text{Factors of 2}} : \frac{\pm 1, \pm 2, \pm 4, \pm 8}{\pm 1, \pm 2} : \pm 1, \pm 2, \pm 4, \pm 8, \pm \frac{1}{2}$$

We compare the $x$-intercepts of the graph in Figure 2.42 and our list of candidates, and decide 4 and $-1/2$ are the only potential rational zeros worth pursuing.

$$
\begin{array}{r|rrrrr}
4 & 2 & -7 & -8 & 14 & 8 \\
  &   & 8 & 4 & -16 & -8 \\
\hline
  & 2 & 1 & -4 & -2 & 0 \\
\end{array}
$$

From this first synthetic division we conclude

$$f(x) = 2x^4 - 7x^3 - 8x^2 + 14x + 8$$
$$= (x - 4)(2x^3 + x^2 - 4x - 2)$$

and we now divide the cubic factor $2x^3 + x^2 - 4x - 2$ by $x + 1/2$:

$$
\begin{array}{r|rrrr}
-1/2 & 2 & 1 & -4 & -2 \\
     &   & -1 & 0 & 2 \\
\hline
     & 2 & 0 & -4 & 0 \\
\end{array}
$$

This second synthetic division allows to complete the factoring of $f(x)$.

$$f(x) = (x - 4)(2x^3 + x^2 - 4x - 2)$$
$$= (x - 4)\left(x + \frac{1}{2}\right)(2x^2 - 4)$$
$$= 2(x - 4)\left(x + \frac{1}{2}\right)(x^2 - 2)$$
$$= (x - 4)(2x + 1)(x + \sqrt{2})(x - \sqrt{2})$$

So the zeros of $f$ are the rational numbers 4 and $-1/2$ and the irrational numbers $-\sqrt{2}$ and $\sqrt{2}$.

A polynomial function cannot have more real zeros than its degree, but it can have fewer. When a polynomial has fewer real zeros than its degree, the Upper and Lower bound Test helps us know that we have found them all, as illustrated by Example 9.

**Example 9**  FINDING THE REAL ZEROS OF
A POLYNOMIAL FUNCTION

Prove that all of the real zeros of $f(x) = 10x^5 - 3x^2 + x - 6$ lie in the interval $[0, 1]$, and find them.

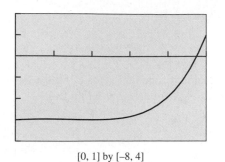

[0, 1] by [−8, 4]

**Figure 2.43** $y = 10x^5 - 3x^2 + x - 6$.

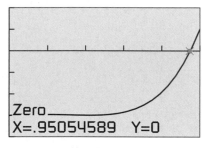

Zero
X=.95054589   Y=0

[0, 1] by [−8, 4]

**Figure 2.44** An approximation for the irrational zero of
$f(x) = 10x^5 - 3x^2 + x - 6$. (Example 9)

**Solution** We first prove that 1 is an upper bound and 0 is a lower bound on the real zeros of $f$. The function $f$ has a positive leading coefficient, so we use synthetic division and the Upper and Lower Bounds Test:

$$
\begin{array}{r|rrrrrr}
1 & 10 & 0 & 0 & -3 & 1 & -6 \\
  &    & 10 & 10 & 10 & 7 & 8 \\
\hline
  & 10 & 10 & 10 & 7 & 8 & 2 \quad \text{Last line}
\end{array}
$$

$$
\begin{array}{r|rrrrrr}
0 & 10 & 0 & 0 & -3 & 1 & -6 \\
  &    & 0 & 0 & 0 & 0 & 0 \\
\hline
  & 10 & 0 & 0 & -3 & 1 & -6 \quad \text{Last line}
\end{array}
$$

Because the last line in the first division scheme consists of all positive numbers, 1 is an upper bound. Because the last line in the second division consists of numbers of alternating signs, 0 is a lower bound. All of the real zeros of $f$ must therefore lie in the closed interval [0, 1]. So in Figure 2.43 we set our Xmin and Xmax accordingly.

Next we use the Rational Zeros Theorem.

*Potential Rational Zeros:*

$$
\frac{\text{Factors of } -6}{\text{Factors of } 10} : \frac{\pm 1, \pm 2, \pm 3, \pm 6}{\pm 1, \pm 2, \pm 5, \pm 10}:
$$

$$
\pm 1, \pm 2, \pm 3, \pm 6, \pm \frac{1}{2}, \pm \frac{3}{2}, \pm \frac{1}{5}, \pm \frac{2}{5}, \pm \frac{3}{5}, \pm \frac{6}{5}, \pm \frac{1}{10}, \pm \frac{3}{10}
$$

We compare the $x$-intercepts of the graph in Figure 2.43 and our list of candidates, and decide $f$ has no rational zeros. From Figure 2.43 we see that $f$ changes sign on the interval [0.8, 1], so by the Intermediate Value Theorem must have a real zero on this interval. Because it is not rational we conclude that it is irrational. Figure 2.44 shows that this lone real zero of $f$ is approximately 0.95.

# Quick Review 2.4

In Exercises 1–4, rewrite the expression as a polynomial in standard form.

**1.** $\dfrac{x^3 - 4x^2 + 7x}{x}$  $x^2 - 4x + 7$  **2.** $\dfrac{2x^3 - 5x^2 - 6x}{2x}$

**3.** $\dfrac{x^4 - 3x^2 + 7x^5}{x^2}$  **4.** $\dfrac{6x^4 - 2x^3 + 7x^2}{3x^2}$

In Exercises 5–10, factor the polynomial into linear factors.

**5.** $x^3 - 4x$  $x(x + 2)(x - 2)$  **6.** $6x^2 - 54$  $6(x + 3)(x - 3)$

**7.** $4x^2 + 8x - 60$  **8.** $15x^3 - 22x^2 + 8x$

**9.** $x^3 + 2x^2 - x - 2$  **10.** $x^4 + x^3 - 9x^2 - 9x$

# Section 2.4 Exercises

In Exercises 1–6, divide $f(x)$ by $d(x)$, and write a summary statement in polynomial form.

**1.** $f(x) = x^2 - 2x + 3; d(x) = x - 1$  $f(x) = (x - 1)^2 + 2$

**2.** $f(x) = x^3 - 1; d(x) = x + 1$

**3.** $f(x) = x^3 + 4x^2 + 7x - 9; d(x) = x + 3$

**4.** $f(x) = 4x^3 - 8x^2 + 2x - 1; d(x) = 2x + 1$

**5.** $f(x) = x^4 - 2x^3 + 3x^2 - 4x + 6; d(x) = x^2 + 2x - 1$

**6.** $f(x) = x^4 - 3x^3 + 6x^2 - 3x + 5; d(x) = x^2 + 1$

In Exercises 7–12, divide using synthetic division, and write a summary statement in fraction form.

**7.** $\dfrac{x^3 - 5x^2 + 3x - 2}{x + 1}$

**8.** $\dfrac{2x^4 - 5x^3 + 7x^2 - 3x + 1}{x - 3}$

**9.** $\dfrac{9x^3 + 7x^2 - 3x}{x - 10}$

**10.** $\dfrac{3x^4 + x^3 - 4x^2 + 9x - 3}{x + 5}$

**11.** $\dfrac{5x^4 - 3x + 1}{4 - x}$

**12.** $\dfrac{x^8 - 1}{x + 2}$

In Exercises 13–18, use the Remainder Theorem to find the remainder when $f(x)$ is divided by $x - k$. Check by using synthetic division.

**13.** $f(x) = 2x^2 - 3x + 1; k = 2$  3

**14.** $f(x) = x^4 - 5; k = 1$  −4

**15.** $f(x) = x^3 - x^2 + 2x - 1; k = -3$  −43

**16.** $f(x) = x^3 - 3x + 4; k = -2$  2

**17.** $f(x) = 2x^3 - 3x^2 + 4x - 7; k = 2$  5

**18.** $f(x) = x^5 - 2x^4 + 3x^2 - 20x + 3; k = -1$  23

In Exercises 19–24, use the Factor Theorem to determine whether the first polynomial is a factor of the second polynomial.

**19.** $x - 1; x^3 - x^2 + x - 1$   **20.** $x - 3; x^3 - x^2 - x - 15$

**21.** $x - 2; x^3 + 3x - 4$   **22.** $x - 2; x^3 - 3x - 2$

**23.** $x + 2; 4x^3 + 9x^2 - 3x - 10$

**24.** $x + 1; 2x^{10} - x^9 + x^8 + x^7 + 2x^6 - 3$

In Exercises 25 and 26, use the graph to guess possible linear factors of $f(x)$. Then completely factor $f(x)$ with the aid of synthetic division.

**25.** $f(x) = 5x^3 - 7x^2 - 49x + 51$

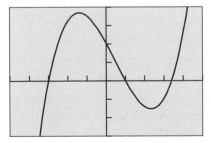

[−5, 5] by [−75, 100]

**26.** $f(x) = 5x^3 - 12x^2 - 23x + 42$

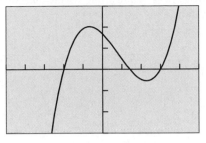

[−5, 5] by [−75, 75]

In Exercises 27–30, find the polynomial function with leading coefficient 2 that has the given degree and zeros.

**27.** Degree 3, with −2, 1, and 4 as zeros

**28.** Degree 3, with −1, 3, and −5 as zeros

**29.** Degree 3, with 2, $\frac{1}{2}$, and $\frac{3}{2}$ as zeros

**30.** Degree 4, with −3, −1, 0, and $\frac{5}{2}$ as zeros

In Exercises 31 and 32, using only algebraic methods, find the cubic function with the given table of values. Check with a grapher.

**31.**

| $x$ | −4 | 0 | 3 | 5 |
|---|---|---|---|---|
| $f(x)$ | 0 | 180 | 0 | 0 |

$f(x) = 3(x + 4)(x - 3)(x - 5)$

**32.**

| $x$ | −2 | −1 | 1 | 5 |
|---|---|---|---|---|
| $f(x)$ | 0 | 24 | 0 | 0 |

$f(x) = 2(x + 2)(x - 1)(x - 5)$

In Exercises 33–36, use the Rational Zeros Theorem to write a list of all potential rational zeros. Then determine which ones, if any, are zeros.

**33.** $f(x) = 6x^3 - 5x - 1$

**34.** $f(x) = 3x^3 - 7x^2 + 6x - 14$

**35.** $f(x) = 2x^3 - x^2 - 9x + 9$

**36.** $f(x) = 6x^4 - x^3 - 6x^2 - x - 12$

In Exercises 37–40, use synthetic division to prove that the number $k$ is an upper bound for the real zeros of the function $f$.

**37.** $k = 3; f(x) = 2x^3 - 4x^2 + x - 2$

**38.** $k = 5; f(x) = 2x^3 - 5x^2 - 5x - 1$

**39.** $k = 2; f(x) = x^4 - x^3 + x^2 + x - 12$

**40.** $k = 3; f(x) = 4x^4 - 6x^3 - 7x^2 + 9x + 2$

In Exercises 41–44, use synthetic division to prove that the number $k$ is a lower bound for the real zeros of the function $f$.

**41.** $k = -1; f(x) = 3x^3 - 4x^2 + x + 3$

**42.** $k = -3; f(x) = x^3 + 2x^2 + 2x + 5$

**43.** $k = 0; f(x) = x^3 - 4x^2 + 7x - 2$

**44.** $k = -4; f(x) = 3x^3 - x^2 - 5x - 3$

In Exercises 45–48, use the upper and lower bound tests to decide whether there is a real zero for the function outside the window shown.

**45.** $f(x) = 6x^4 - 11x^3 - 7x^2 + 8x - 34$

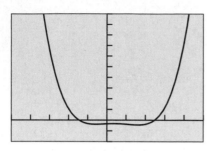

[–5, 5] by [–200, 1000]

**46.** $f(x) = x^5 - x^4 + 21x^2 + 19x - 3$

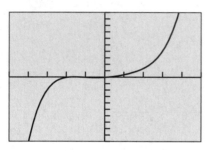

[–5, 5] by [–1000, 1000]

**47.** $f(x) = x^5 - 4x^4 - 129x^3 + 396x^2 - 8x + 3$

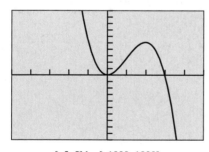

[–5, 5] by [–1000, 1000]

**48.** $f(x) = 2x^5 - 5x^4 - 141x^3 + 216x^2 - 91x + 25$

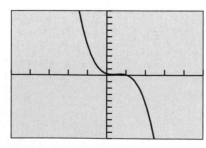

[–5, 5] by [–1000, 1000]

In Exercises 49–56, find all of the real zeros of the function, finding exact values whenever possible. Identify each zero as rational or irrational.

**49.** $f(x) = 2x^3 - 3x^2 - 4x + 6$

**50.** $f(x) = x^3 + 3x^2 - 3x - 9$

**51.** $f(x) = x^3 + x^2 - 8x - 6$

**52.** $f(x) = x^3 - 6x^2 + 7x + 4$

**53.** $f(x) = x^4 - 3x^3 - 6x^2 + 6x + 8$

**54.** $f(x) = x^4 - x^3 - 7x^2 + 5x + 10$

**55.** $f(x) = 2x^4 - 7x^3 - 2x^2 - 7x - 4$

**56.** $f(x) = 3x^4 - 2x^3 + 3x^2 + x - 2$

**57. Setting Production Schedules** The Sunspot Small Appliance Co. determines that the supply function for their EverCurl hair dryer is $S(p) = 6 + 0.001p^3$ and that its demand function is $D(p) = 80 - 0.02p^2$, where $p$ is the price. Determine the price for which the supply equals the demand and the number of hair dryers corresponding to this equilibrium price.

**58. Setting Production Schedules** The Pentkon Camera Co. determines that the supply and demand functions for their 35 mm − 70 mm zoom lens are $S(p) = 200 - p + 0.000007p^4$ and $D(p) = 1500 - 0.0004p^3$, where $p$ is the price. Determine the price for which the supply equals the demand and the number of zoom lenses corresponding to this equilibrium price.

**59.** Find the remainder when $x^{40} - 3$ is divided by $x + 1$. $-2$

**60.** Find the remainder when $x^{63} - 17$ is divided by $x - 1$. $-16$

## Explorations

**61.** Let $f(x) = x^4 + 2x^3 - 11x^2 - 13x + 38$

**(a)** Use the upper and lower bound tests to prove that all of the real zeros of $f$ lie on the interval $[-5, 4]$.

**(b)** Find all of the rational zeros of $f$. 2 is a zero of $f(x)$

**(c)** Factor $f(x)$ using the rational zero(s) found in (b).

**(d)** Approximate all of the irrational zeros of $f$.

**(e)** Use synthetic division and the irrational zero(s) found in (d) to continue the factorization of $f(x)$ begun in (c).

**62. Retail Sales** The amount in billions of retail sales by Fuel Dealers for several years from 1985 to 1997 is given in Table 2.17. Let $x = 0$ stand for 1980, $x = 1$ for 1981, and so forth.

**(a)** Find a quartic regression model, and graph it together with a scatter plot of the data.

**(b)** Find a quadratic regression model, and graph it together with a scatter plot of the data.

**(c)** Use the quartic and quadratic regressions from (a) and (b) to get two estimates of the amount of retail sales for Fuel Dealers in 1999.

**(d) Writing to Learn** Give scenarios to justify each of the estimates in (c).

| Table 2.17 Fuel Dealers Retail Sales | |
| --- | --- |
| Year | Amount (billions) |
| 1985 | 16.8 |
| 1990 | 15.6 |
| 1993 | 15.1 |
| 1994 | 16.0 |
| 1995 | 16.9 |
| 1996 | 19.0 |
| 1997 | 17.7 |

*Source: U.S. Bureau of the Census, Current Busiess Reports, Statistical Abstracts of the United States, 1998.*

**63. Archimedes' Principle** A spherical buoy has a radius of 1 m and a density one-fourth that of seawater. By Archimedes' Principle, the weight of the displaced water will equal the weight of the buoy.

- Let $x$ = the depth to which the buoy sinks.
- Let $d$ = the density of seawater.
- Let $r$ = the radius of the circle formed where buoy, air, and water meet. See the figure below.

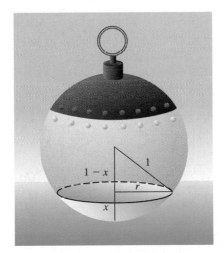

Notice that $0 < x < 1$ and that
$(1 - x)^2 + r^2 = 1,$
$$r^2 = 1 - (1 - x)^2$$
$$= 2x - x^2.$$

**(a)** Verify that the volume of the buoy is $4\pi/3$.

**(b)** Use your result from (a) to establish the weight of the buoy as $\pi d/3$.

**(c)** Prove the weight of the displaced water is $\pi d \cdot x(3r^2 + x^2)/6$.

**(d)** Approximate the depth to which the buoy will sink.

**64. Archimedes' Principle** Using the scenario of Exercise 63, find the depth to which the buoy will sink if its density is one-fifth that of seawater. $\approx 0.57$ m

**65. Biological Research** Stephanie, a biologist who does research for the poultry industry, models the population $P$ of wild turkeys, $t$ days after being left to reproduce, with the function

$$P(t) = -0.00001t^3 + 0.002t^2 + 1.5t + 100.$$

**(a)** Graph the function $y = P(t)$ for appropriate values of $t$.

**(b)** Find what the maximum turkey population is and when it occurs.

**(c)** Assuming that this model continues to be accurate, when will this turkey population become extinct?

**(d) Writing to Learn** Create a scenario that could explain the growth exhibited by this turkey population.

**66. Architectural Engineering** Dave, an engineer at the Trumbauer Group, Inc., an architectural firm, completes structural specifications for a 172-ft-long steel beam, anchored at one end to a piling 20 ft above the ground. He knows that when a 200-lb object is placed $d$ feet from the anchored end, the beam bends $s$ feet where

$$s = (3 \times 10^{-7})d^2(550 - d).$$

**(a)** What is the independent variable in this polynomial function? $d$

**(b)** What are the dimensions of a viewing window that shows a graph for the values that make sense in this problem situation? $[0, 172] \times [0, 5]$

**(c)** How far is the 200-lb object from the anchored end if the vertical deflection is 1.25 ft? $\approx 95.777$ ft

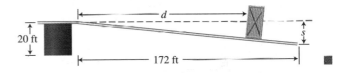

## Extending the Ideas

**67. Writing to Learn** Graph each side of the Example 3 summary equation:

$$f(x) = \frac{2x^3 - 3x^2 - 5x - 12}{x - 3} \quad \text{and}$$

$$g(x) = 2x^2 + 3x + 4, \quad x \neq 3$$

How are these functions related? Include a discussion of the domain and continuity of each function.

**68. Writing to Learn** Graph each side of the Example 4 summary equation:

$$f(x) = \frac{x^4 - 8x^3 + 11x - 6}{x + 3} \quad \text{and}$$

$$g(x) = x^3 - 11x^2 + 33x - 88 + \frac{258}{x + 3}$$

How are these functions related? Include a discussion of the domain and continuity of each function.

**69. Writing to Learn** Explain how to carry out the following division using synthetic division. Work through the steps with complete explanations.

$$\frac{4x^3 - 5x^2 + 3x + 1}{2x - 1}$$

**70. Writing to Learn** The figure shows a graph of $f(x) = x^4 + 0.1x^3 - 6.5x^2 + 7.9x - 2.4$. Explain how to use a grapher to justify the statement.

$$f(x) = x^4 + 0.1x^3 - 6.5x^2 + 7.9x - 2.4$$
$$\approx (x + 3.10)(x - 0.5)(x - 1.13)(x - 1.37)$$

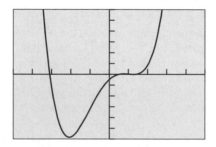

[–5, 5] by [–30, 30]

**71. (a) Writing to Learn** Write a paragraph that describes how the zeros of $f(x) = (1/3)x^3 + x^2 + 2x - 3$ are related to the zeros of $g(x) = x^3 + 3x^2 + 6x - 9$. In what ways does this example illustrate how the Rational Zeros Theorem can be applied to find the zeros of a polynomial with *rational* number coefficients?

**(b)** Find the rational zeros of $f(x) = x^3 - \dfrac{7}{6}x^2 - \dfrac{20}{3}x + \dfrac{7}{2}$.

**(c)** Find the rational zeros of $f(x) = x^3 - \dfrac{5}{2}x^2 - \dfrac{37}{12}x + \dfrac{5}{2}$.

**72.** Use the Rational Zeros Theorem to prove $\sqrt{2}$ is irrational.

**73. Group Activity** *Work in groups of three.* Graph $f(x) = x^4 + x^3 - 8x^2 - 2x + 7$.

**(a)** Use grapher methods to find approximate real number zeros.

**(b)** Identify a list of four linear factors whose product could be called an *approximate factorization of* $f(x)$.

**(c)** Discuss what graphical and numerical methods you could use to show that the factorization from (b) is reasonable.

**74.** A classic theorem, **Descartes' Rule of Signs**, tells us about the number of positive and negative real zeros of a polynomial function, by looking at the polynomial's variations in sign. A *variation in sign* occurs when consecutive coefficients (in standard form) have opposite signs.

If $f(x) = a_n x^n + \cdots + a_0$ is a polynomial of degree $n$, then

- The number of positive real zeros of $f$ is equal to the number of variations in sign of $f(x)$, or that number less some even number.

- The number of negative real zeros of $f$ is equal to the number of variations in sign of $f(-x)$, or that number less some even number.

Use Descartes' Rule of Signs to determine the possible numbers of positive and negative real zeros of the function.

**(a)** $f(x) = x^3 + x^2 - x + 1$

**(b)** $f(x) = x^3 + x^2 + x + 1$

**(c)** $f(x) = 2x^3 + x - 3$

**(d)** $g(x) = 5x^4 + x^2 - 3x - 2$

## 2.5 Complex Numbers

Complex Numbers • Operations with Complex Numbers • Complex Conjugates and Division • Complex Solutions of Quadratic Equations • Complex Plane

### Complex Numbers

Figure 2.45 shows that the function $f(x) = x^2 + 1$ has no real zeros, so $x^2 + 1 = 0$ has no real-number solutions. To remedy this situation, mathematicians in the 17th century extended the definition of $\sqrt{a}$ to include negative real numbers $a$. First the number $i = \sqrt{-1}$ is defined as a solution of the equation $i^2 + 1 = 0$ and is the **imaginary unit**. Then for any negative real number $a$, $\sqrt{a} = \sqrt{|a|} \cdot i$.

The extended system of numbers, called the *complex numbers*, consists of all real numbers and sums of real numbers and real number multiples of $i$. The following are all examples of complex numbers:

$$-6, \quad 5i, \quad \sqrt{5}, \quad -7i, \quad \frac{5}{2}i \quad \frac{2}{3}, \quad -2 + 3i, \quad 5 - 3i, \quad \frac{1}{3} + \frac{4}{5}i$$

[–5, 5] by [–3, 10]

**Figure 2.45** The graph of $f(x) = x^2 + 1$ has no $x$-intercepts.

**Historical Note**

René Descartes (1596–1650) coined the term imaginary in a time when negative solutions to equations were considered *false*. Carl Friedrich Gauss (1777–1855) gave us the term complex number and the symbol *i* for $\sqrt{-1}$. Today practical applications of complex numbers abound.

**Objective**

Students will be able to add, subtract, multiply, and divide complex numbers; and to find complex zeros of quadratic functions.

**Motivate**

Ask...
If *i* is the square root of $-1$, what is the square of $-i$? **($-1$)**

**Teaching Note**

The material in this section is not in the AP Calculus topics syllabi.

> **Definition   Complex Number**
>
> A **complex number** is any number that can be written in the form
>
> $$a + bi,$$
>
> where $a$ and $b$ are real numbers. The real number $a$ is the **real part**, the real number $b$ is the **imaginary part**, and $a + bi$ is the **standard form**.

A real number $a$ is the complex number $a + 0i$, so *all real numbers are also complex numbers*. If $a = 0$ and $b \neq 0$, then $a + bi$ becomes $bi$, and is an **imaginary number**. For instance, $5i$ and $-7i$ are imaginary numbers.

Two complex numbers are **equal** if and only if their real and imaginary parts are equal. For example,

$$x + yi = 2 + 5i \quad \text{if and only if } x = 2 \text{ and } y = 5.$$

## Operations with Complex Numbers

Adding complex numbers is done by adding their real and imaginary parts separately. Subtracting complex numbers is also done by parts.

> **Definition   Addition and Subtraction of Complex Numbers**
>
> If $a + bi$ and $c + di$ are two complex numbers, then
>
> **Sum:**  $(a + bi) + (c + di) = (a + c) + (b + d)i,$
>
> **Difference:**  $(a + bi) - (c + di) = (a - c) + (b - d)i.$

**Example 1   ADDING AND SUBTRACTING COMPLEX NUMBERS**

**(a)** $(7 - 3i) + (4 + 5i) = (7 + 4) + (-3 + 5)i = 11 + 2i$

**(b)** $(2 - i) - (8 + 3i) = (2 - 8) + (-1 - 3)i = -6 - 4i$

The **additive identity** for the complex numbers is $0 = 0 + 0i$. The **additive inverse** of $a + bi$ is $-(a + bi) = -a - bi$ because

$$(a + bi) + (-a - bi) = 0 + 0i = 0.$$

Many of the properties of real numbers also hold for complex numbers. These include:

- Commutative properties of addition and multiplication,

- Associative properties of addition and multiplication, and

- Distributive properties of multiplication over addition and subtraction.

Using these properties and the fact that $i^2 = -1$, complex numbers can be multiplied by treating them as algebraic expressions.

**Teaching Note**

It is useful to compare operations with complex numbers to operations with binomials (i.e., multiply by the FOIL method), but stress that $i$ is not a variable.

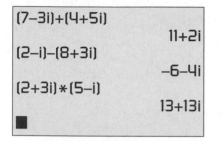

**Figure 2.46** Complex number operations on a grapher. (Examples 1 and 2)

**Teaching Note**

Exercise 55 (Powers of $i$) could be used as an in-class exploration before addressing the powers of a more general complex number, as is done in Example 3.

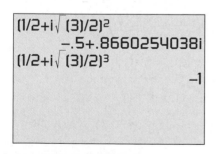

**Figure 2.47** The square and cube of a complex number. (Example 3)

### Example 2 MULTIPLYING COMPLEX NUMBERS

$$(2 + 3i) \cdot (5 - i) = 2(5 - i) + 3i(5 - i)$$
$$= 10 - 2i + 15i - 3i^2$$
$$= 10 + 13i - 3(-1)$$
$$= 13 + 13i$$

We can generalize Example 2 as follows:

$$(a + bi)(c + di) = ac + adi + bci + bdi^2$$
$$= (ac - bd) + (ad + bc)i$$

Many graphers can perform basic calculations on complex numbers. Figure 2.46 shows how the operations of Examples 1 and 2 look on some graphers.

We compute positive integer powers of complex numbers by treating them as algebraic expressions.

### Example 3 RAISING A COMPLEX NUMBER TO A POWER

If $z = \dfrac{1}{2} + \dfrac{\sqrt{3}}{2} i$, find $z^2$ and $z^3$.

**Solution**

$$z^2 = \left(\frac{1}{2} + \frac{\sqrt{3}}{2} i\right)\left(\frac{1}{2} + \frac{\sqrt{3}}{2} i\right)$$
$$= \frac{1}{4} + \frac{\sqrt{3}}{4} i + \frac{\sqrt{3}}{4} i + \frac{3}{4} i^2$$
$$= \frac{1}{4} + \frac{2\sqrt{3}}{4} i + \frac{3}{4} (-1)$$
$$= -\frac{1}{2} + \frac{\sqrt{3}}{2} i$$

$$z^3 = z^2 \cdot z = \left(-\frac{1}{2} + \frac{\sqrt{3}}{2} i\right)\left(\frac{1}{2} + \frac{\sqrt{3}}{2} i\right)$$
$$= -\frac{1}{4} - \frac{\sqrt{3}}{4} i + \frac{\sqrt{3}}{4} i + \frac{3}{4} i^2$$
$$= -\frac{1}{4} + 0i + \frac{3}{4} (-1)$$
$$= -1$$

Figure 2.47 supports these results numerically.

Example 3 demonstrates that $1/2 + (\sqrt{3}/2)i$ is a cube root of $-1$ and a solution of $x^3 + 1 = 0$. In Section 2.6 complex zeros of polynomial functions will be explored in depth.

## Complex Conjugates and Division

The product of the complex numbers $a + bi$ and $a - bi$ is a positive real number:

$$(a + bi) \cdot (a - bi) = a^2 - (bi)^2 = a^2 + b^2.$$

We introduce the following definition to describe this special relationship.

---

**Definition**  **Complex Conjugate**

The **complex conjugate** of the complex number $z = a + bi$ is

$$\bar{z} = \overline{a + bi} = a - bi.$$

---

The **multiplicative identity** for the complex numbers is $1 = 1 + 0i$. The **multiplicative inverse**, or **reciprocal**, of $z = a + bi$ is

$$z^{-1} = \frac{1}{z} = \frac{1}{a + bi} = \frac{1}{a + bi} \cdot \frac{a - bi}{a - bi} = \frac{a}{a^2 + b^2} - \frac{b}{a^2 + b^2}i.$$

In general, a quotient of two complex numbers, written in fraction form, can be simplified as we just simplified $1/z$—by multiplying the numerator and denominator of the fraction by the complex conjugate of the denominator.

**Example 4**  DIVIDING COMPLEX NUMBERS

Write the complex number in standard form.

(a) $\dfrac{2}{3 - i}$  (b) $\dfrac{5 + i}{2 - 3i}$

**Solution**  Multiply the numerator and denominator by the complex conjugate of the denominator.

(a)
$$\frac{2}{3 - i} = \frac{2}{3 - i} \cdot \frac{3 + i}{3 + i}$$
$$= \frac{6 + 2i}{3^2 + 1^2}$$
$$= \frac{6}{10} + \frac{2}{10}i$$
$$= \frac{3}{5} + \frac{1}{5}i$$

(b)
$$\frac{5 + i}{2 - 3i} = \frac{5 + i}{2 - 3i} \cdot \frac{2 + 3i}{2 + 3i}$$
$$= \frac{10 + 15i + 2i + 3i^2}{2^2 + 3^2}$$
$$= \frac{7 + 17i}{13}$$
$$= \frac{7}{13} + \frac{17}{13}i$$

## Complex Solutions of Quadratic Equations

Recall that the solutions of the quadratic equation $ax^2 + bx + c = 0$, where $a$, $b$, and $c$ are real numbers and $a \neq 0$, are given by the quadratic formula

$$x = \frac{-b \pm \sqrt{b^2 - 4ac}}{2a}.$$

The radicand $b^2 - 4ac$ is the **discriminant**, and tells us whether the solutions are real numbers. In particular, if $b^2 - 4ac < 0$, the solutions involve the square root of a negative number and so lead to complex-number solutions. In all, there are three cases, which we now summarize:

> **Discriminant of a Quadratic Equation**
>
> For a quadratic equation $ax^2 + bx + c = 0$, where $a$, $b$, and $c$ are real numbers and $a \neq 0$,
>
> • If $b^2 - 4ac > 0$, there are two distinct real solutions.
>
> • If $b^2 - 4ac = 0$, there is one repeated real solution.
>
> • If $b^2 - 4ac < 0$, there is a complex conjugate pair of solutions.

### Example 5   SOLVING A QUADRATIC EQUATION

Solve $x^2 + x + 1 = 0$.

**Solution**

**Solve Algebraically**

Using the quadratic formula with $a = b = c = 1$, we obtain

$$x = \frac{-(1) \pm \sqrt{(1)^2 - 4(1)(1)}}{2(1)}$$

$$= \frac{-1 \pm \sqrt{-3}}{2}$$

$$= -\frac{1}{2} \pm \frac{\sqrt{3}}{2}i$$

So the solutions are $-1/2 + (\sqrt{3}/2)i$ and $-1/2 - (\sqrt{3}/2)i$, a complex conjugate pair.

**Confirm Numerically**

Substituting $-1/2 + (\sqrt{3}/2)i$ into the original equation, we obtain

$$\left(-\frac{1}{2} + \frac{\sqrt{3}}{2}i\right)^2 + \left(-\frac{1}{2} + \frac{\sqrt{3}}{2}i\right) + 1$$

$$= \left(-\frac{1}{2} - \frac{\sqrt{3}}{2}i\right) + \left(-\frac{1}{2} + \frac{\sqrt{3}}{2}i\right) + 1 = 0$$

By a similar computation we can confirm the second solution.

## Complex Plane

Just as every real number is associated with a point of the real number line, every complex number is associated with a point of the **complex plane**. This idea evolved through the works of Caspar Wessel (1745–1818), Jean-Robert Argand (1768–1822), and Gauss (1777–1855). Real numbers are placed along the horizontal axis and imaginary numbers along the vertical axis. So the axes become the **real axis** and the **imaginary axis**. The complex number $a + bi$ is located at the point $(a, b)$. For example, the complex number $2 + 3i$ is associated with the point 2 units to the right and 3 units up from the origin. (See Figure 2.48.)

**Notes on Examples**

Have the students graph the equation to see that it does not touch or cross the *x*-axis. Also, have the students calculate the value of the discriminant and confirm the prediction that there would be a complex conjugate pair of solutions.

**Teaching Note**

Note that the complex conjugate of 3 is 3, and the complex conjugate of 5*i* is −5*i*.

**Follow-up**

Discuss the similarities and differences between the complex plane and the ordinary Cartesian plane.

**Assignment Guide**

Ex. 3–48, multiples of 3, 51, 53, 56, 57, 59, 60, 63, 64

**Cooperative Learning**

Group Activity: Ex. 55

**Notes on Exercises**

Ex. 1–44 are straightforward computational or graphing exercises relating to powers of *i* and operations on complex numbers.
Ex. 49–54, 56, require the use of the distance and/or midpoint formulas

**Ongoing Assessment**

Self-Assessment: Ex. 17, 28, 49, 54
Embedded Assessment: Ex. 39, 57

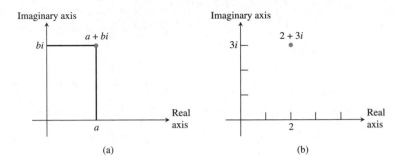

Figure 2.48   Plotting points in the complex plane.

### Example 6   PLOTTING COMPLEX NUMBERS

Plot $u = 1 + 3i$, $v = 2 - i$, and $u + v$ in the complex plane.

**Solution**   First notice that $u + v = (1 + 3i) + (2 - i) = 3 + 2i$. The numbers $u$, $v$, and $u + v$ are plotted in Figure 2.49.

Figure 2.49 suggests that a quadrilateral formed with two complex numbers, their sum, and the origin as the vertices is a parallelogram. You will be asked to prove this in Exercise 59.

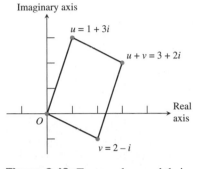

Figure 2.49   Two numbers and their sum in the complex plane. (Example 6)

> **Definition**   **Absolute Value of a Complex Number**
>
> The **absolute value**, or **modulus**, of the complex number $z = a + bi$, where $a$ and $b$ are real numbers, is
>
> $$|z| = |a + bi| = \sqrt{a^2 + b^2}.$$
>
> See Figure 2.50.

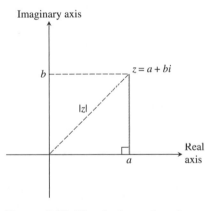

Figure 2.50   The absolute value of a complex number.

Some of geometric formulas we studied in the Cartesian plane actually become simpler in the complex plane. For example, the **distance** between the points $u$ and $v$ in the complex plane is $d = |u - v|$, and the **midpoint** of the line segment connecting $u$ and $v$ is $(u + v)/2$ (see Exercises 53 and 54). These formulas should remind you of the distance formula and the midpoint formula used for points on the (real) number line.

### Example 7   USING THE DISTANCE AND MIDPOINT FORMULAS

Find the distance between $u = -4 + i$ and $v = 2 + 5i$ in the complex plane, and find the midpoint of the segment connecting $u$ and $v$.

**Solution**   The distance between $u$ and $v$ is given by

$$|u - v| = |(-4 + i) - (2 + 5i)| = |-6 - 4i|$$
$$= \sqrt{(-6)^2 + (-4)^2} = 2\sqrt{13} \approx 7.21.$$

The midpoint of the segment connecting $u$ and $v$ is

$$\frac{u + v}{2} = \frac{(-4 + i) + (2 + 5i)}{2} = \frac{-2 + 6i}{2} = -1 + 3i.$$

**Exploration Extensions**

Factor $x^3 - 27$, and then solve $x^3 - 27 = 0$. Plot the solutions in the complex plane. Compute their distance from the origin. Is this the result you expected? **(Distance of all points from origin is 3 units.)**

---

**Exploration 1** **Solving a Cubic Equation**

1. Factor $x^3 - 1$, and then solve $x^3 - 1 = 0$. Compare your results with Example 5. Plot the three solutions in the complex plane. How are the three points related (compute their distance from the origin)?

2. Factor $x^3 + 1$, and then solve $x^3 + 1 = 0$. Compare your results with Example 3. Plot the three solutions in the complex plane. How are the three points related?

---

# Quick Review 2.5

In Exercises 1–4, add or subtract, and simplify.

**1.** $(2x + 3) + (-x + 6)$  $x + 9$   **2.** $(3y - x) + (2x - y)$

**3.** $(2a + 4d) - (a + 2d)$    **4.** $(6z - 1) - (z + 3)$  $5z - 4$

In Exercises 5–10, multiply and simplify.

**5.** $(x - 3)(x + 2)$        **6.** $(2x - 1)(x + 3)$

**7.** $(x - \sqrt{2})(x + \sqrt{2})$  $x^2 - 2$

**8.** $(x + 2\sqrt{3})(x - 2\sqrt{3})$  $x^2 - 12$

**9.** $[x - (1 + \sqrt{2})][x - (1 - \sqrt{2})]$  $x^2 - 2x - 1$

**10.** $[x - (2 + \sqrt{3})][x - (2 - \sqrt{3})]$  $x^2 - 4x + 1$

---

# Section 2.5 Exercises

In Exercises 1–8, write the sum or difference in the standard form $a + bi$.

**1.** $(2 - 3i) + (6 + 5i)$  $8 + 2i$   **2.** $(2 - 3i) + (3 - 4i)$

**3.** $(7 - 3i) + (6 - i)$  $13 - 4i$   **4.** $(2 + i) - (9i - 3)$  $5 - 8i$

**5.** $(2 - i) + (3 - \sqrt{-3})$  $5 - (1 + \sqrt{3})i$

**6.** $(\sqrt{5} - 3i) + (-2 + \sqrt{-9})$  $\sqrt{5} - 2$

**7.** $(i^2 + 3) - (7 + i^3)$  $-5 + i$

**8.** $(\sqrt{7} + i^2) - (6 - \sqrt{-81})$  $(\sqrt{7} - 7) + 9i$

In Exercises 9–16, write the product in standard form.

**9.** $(2 + 3i)(2 - i)$  $7 + 4i$     **10.** $(2 - i)(1 + 3i)$  $5 + 5i$

**11.** $(1 - 4i)(3 - 2i)$  $-5 - 14i$  **12.** $(5i - 3)(2i + 1)$  $-13 - i$

**13.** $(7i - 3)(2 + 6i)$  $-48 - 4i$  **14.** $(\sqrt{-4} + i)(6 - 5i)$

**15.** $(-3 - 4i)(1 + 2i)$  $5 - 10i$  **16.** $(\sqrt{-2} + 2i)(6 + 5i)$

In Exercises 17–20, write the expression in the form $bi$, where $b$ is a real number.

**17.** $\sqrt{-16}$  $4i$          **18.** $\sqrt{-25}$  $5i$

**19.** $\sqrt{-3}$  $\sqrt{3}i$       **20.** $\sqrt{-5}$  $\sqrt{5}i$

In Exercises 21–24, find the real numbers $x$ and $y$ that make the equation true.

**21.** $2 + 3i = x + yi$       **22.** $3 + yi = x - 7i$

**23.** $(5 - 2i) - 7 = x - (3 + yi)$  $x = 1, y = 2$

**24.** $(x + 6i) = (3 - i) + (4 - 2yi)$  $x = 7, y = -7/2$

In Exercises 25–28, write the expression in standard form.

**25.** $(3 + 2i)^2$  $5 + 12i$         **26.** $(1 - i)^3$  $-2 - 2i$

**27.** $\left(\dfrac{\sqrt{2}}{2} + \dfrac{\sqrt{2}}{2}i\right)^4$  $-1$   **28.** $\left(\dfrac{\sqrt{3}}{2} + \dfrac{1}{2}i\right)^3$  $i$

In Exercises 29–32, find the product of the complex number and its conjugate.

**29.** $2 - 3i$  $13$             **30.** $5 - 6i$  $61$

**31.** $-3 + 4i$  $25$            **32.** $-1 - \sqrt{2}i$  $3$

In Exercises 33–40, write the expression in standard form.

**33.** $\dfrac{1}{2 + i}$  $2/5 - 1/5i$        **34.** $\dfrac{i}{2 - i}$  $-1/5 + 2/5i$

**35.** $\dfrac{2 + i}{2 - i}$  $3/5 + 4/5i$      **36.** $\dfrac{2 + i}{3i}$  $1/3 - 2/3i$

**37.** $\dfrac{(2 + i)^2(-i)}{1 + i}$  $1/2 - 7/2i$  **38.** $\dfrac{(2 - i)(1 + 2i)}{5 + 2i}$

**39.** $\dfrac{(1 - i)(2 - i)}{1 - 2i}$  $7/5 - 1/5i$  **40.** $\dfrac{(1 - \sqrt{2}i)(1 + i)}{(1 + \sqrt{2}i)}$

In Exercises 41–44, match the complex number with one of the points labeled a–f.

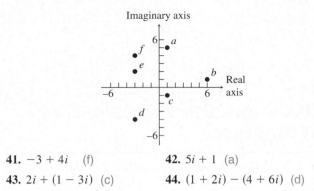

Imaginary axis

**41.** $-3 + 4i$ (f)

**42.** $5i + 1$ (a)

**43.** $2i + (1 - 3i)$ (c)

**44.** $(1 + 2i) - (4 + 6i)$ (d)

In Exercises 45–48, solve the equation.

**45.** $x^2 + 2x + 5 = 0$

**46.** $3x^2 + x + 2 = 0$

**47.** $4x^2 - 6x + 5 = x + 1$

**48.** $x^2 + x + 11 = 5x - 8$

In Exercises 49–52, plot the given points in the complex plane. Draw the line segment connecting them, and find the length and midpoint of the segment.

**49.** $2 + 4i, 2 - 4i$

**50.** $-3 + 7i, -3 - 7i$

**51.** $1 - 4i, -2 - 6i$

**52.** $-5 + i, 3 + 4i$

**53.** Let $u = a + bi$ and $v = c + di$. Prove that the distance between the points $u$ and $v$ is $|u - v|$.

**54.** Let $u = a + bi$ and $v = c + di$. Prove that the midpoint of the line segment determined by $u$ and $v$ is $(u + v)/2$.

## Explorations

**55. Group Activity  The Powers of $i$**

**(a)** Simplify the complex numbers $i, i^2, \ldots, i^8$ by evaluating each one.  $i; -1; -i; 1; i; -1; -i; 1$

**(b)** Simplify the complex numbers $i^{-1}, i^{-2}, \ldots, i^{-8}$ by evaluating each one.  $-i; -1; i; 1; -i; -1; i; 1$

**(c)** Evaluate $i^0$.  1

**(d) Writing to Learn**  Discuss your results from (a)–(c) with the members of your group, and write a summary statement about the integer powers of $i$.

**56. Absolute Value of a Complex Number**

**(a)** Plot $2 + 3i$ in the complex plane.

**(b)** Evaluate $|2 + 3i|$.  $\sqrt{13}$

**(c) Writing to Learn**  Explain how $|2 + 3i|$ is related to the distance between $2 + 3i$ and the origin.

**(d) Writing to Learn**  Repeat (a)–(c) for the complex number $a + bi$, where $a$ and $b$ are any real numbers.

**57. The Powers of a Complex Number**

**(a)** Selected powers of $1 + i$ are displayed in Table 2.18. Try to find a pattern in the data, and use it to predict the real and imaginary parts of $(1 + i)^7$ and $(1 + i)^8$.

**(b)** Plot $(1 + i)^0, (1 + i)^1, \ldots, (1 + i)^6$ in the complex plane. Try to find a pattern in the arrangement of the points, and use it to predict the locations of $(1 + i)^7$ and $(1 + i)^8$.

**(c)** Do your answers from (a) and (b) agree?

**(d)** Compute $(1 + i)^7$ and $(1 + i)^8$, and reconcile your results in (a) and (b), if necessary, with the computed values.

**Table 2.18  Powers of $1 + i$.**

| Power | Real Part | Imaginary Part |
| --- | --- | --- |
| 0 | 1 | 0 |
| 1 | 1 | 1 |
| 2 | 0 | 2 |
| 3 | −2 | 2 |
| 4 | −4 | 0 |
| 5 | −4 | −4 |
| 6 | 0 | −8 |

**58. Writing to Learn**  Describe the nature of the graph of $f(x) = ax^2 + bx + c$ when $a$, $b$, and $c$ are real numbers and the equation $ax^2 + bx + c = 0$ has nonreal complex solutions.

## Extending the Ideas

**59. Connecting Algebra and Geometry**  Verify that 0, $1 + 3i$, $2 - i$, and $3 + 2i$, shown in Figure 2.49, are the vertices of a parallelogram. Then prove that for any distinct, nonzero complex numbers $u$ and $v$, the points 0, $u$, $v$, and $u + v$ are the vertices of a parallelogram.

**60. Writing to Learn**  Describe the graphical relationship between a complex number and its complex conjugate.

**61.** The complex numbers $z$, 1, and $i$ are plotted in the complex plane. Match the complex numbers $1/z, -z, -1/z, \bar{z}, -\bar{z}, |z|, -|z|, |z|i$ with the points labeled a–h.

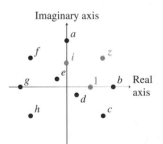

Imaginary axis

**62.** Prove that the difference between a complex number and its conjugate is a complex number whose real part is 0.

**63.** Let $u = a + bi$ and $v = c + di$ be complex numbers in standard form. Determine whether each of the following is always true, and explain why or why not.

**(a)** $|u + v| = |u| + |v|$

**(b)** $|u + v| \le |u| + |v|$

**(c)** $|u + v| \ge |u| + |v|$

**64.** Let $u = a + bi$ and $v = c + di$ be complex numbers in standard form. Determine whether each of the following is always true, and explain why or why not.

(a) $|u \cdot v| = |u| \cdot |v|$

(b) $|u \cdot v| < |u| \cdot |v|$

(c) $|u \cdot v| > |u| \cdot |v|$

# 2.6 Complex Zeros and the Fundamental Theorem of Algebra

Two Major Theorems • Complex Conjugate Zeros • Factoring with Real Number Coefficients

## Two Major Theorems

We have seen that a polynomial function of degree $n$ can have at most $n$ real zeros. We have also seen that a polynomial can have nonreal complex zeros. The *Fundamental Theorem of Algebra* tells us that a polynomial function of degree $n$ has $n$ complex zeros (real and nonreal).

> **Theorem  Fundamental Theorem of Algebra**
>
> A polynomial function of degree $n > 0$ has $n$ complex zeros. Some of these zeros may be repeated.

The Factor Theorem extends to the complex zeros of a polynomial function. Thus, $k$ is a complex zero of a polynomial if and only if $x - k$ is a factor of the polynomial, even if $k$ is not a real number. We combine this fact with the Fundamental Theorem of Algebra to obtain the following theorem.

> **Theorem  Linear Factorization Theorem**
>
> If $f(x)$ is a polynomial function of degree $n > 0$, then $f(x)$ has precisely $n$ linear factors and
> $$f(x) = a(x - z_1)(x - z_2) \cdots (x - z_n)$$
> where $a$ is the leading coefficient of $f(x)$ and $z_1, z_2, \ldots, z_n$ are the complex zeros of $f(x)$. The $z_i$ are not necessarily distinct numbers; some may be repeated.

The Fundamental Theorem of Algebra and the Linear Factorization Theorem are *existence* theorems. They tell us of the existence of zeros and linear factors, but not how to find them.

One connection is lost going from real zeros to complex zeros. If $k$ is a *nonreal* complex zero of a polynomial function $f(x)$, then $k$ is *not* an $x$-intercept of the graph of $f$. The other connections are preserved:

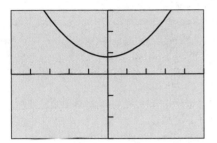

[–5, 5] by [–15, 15]

(a)

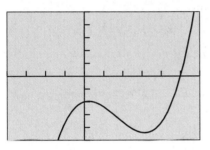

[–4, 6] by [–25, 25]

(b)

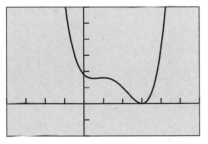

[–4, 6] by [–10, 30]

(c)

**Figure 2.51** The graphs of (a) $y = x^2 + 4$, (b) $y = x^3 - 5x^2 + 2x - 10$, and (c) $y = x^4 - 6x^3 + 10x^2 - 6x + 9$. (Example 1)

> **Fundamental Polynomial Connections in the Complex Case**
>
> The following statements about a polynomial function $f$ are equivalent even if $k$ is a nonreal complex number:
>
> 1. $x = k$ is a solution (or root) of the equation $f(x) = 0$.
> 2. $k$ is a zero of the function $f$.
> 3. $x - k$ is a factor of $f(x)$.

**Example 1** EXPLORING FUNDAMENTAL POLYNOMIAL CONNECTIONS

Write the polynomial function in standard form, and identify the zeros of the function and the $x$-intercepts of its graph.

**(a)** $f(x) = (x - 2i)(x + 2i)$

**(b)** $f(x) = (x - 5)(x - \sqrt{2}i)(x + \sqrt{2}i)$

**(c)** $f(x) = (x - 3)(x - 3)(x - i)(x + i)$

**Solution**

**(a)** The quadratic function $f(x) = (x - 2i)(x + 2i) = x^2 + 4$. The function has two zeros: $x = 2i$ and $x = -2i$. Because the zeros are not real, the graph of $f$ has no $x$-intercepts.

**(b)** The cubic function

$$f(x) = (x - 5)(x - \sqrt{2}i)(x + \sqrt{2}i)$$
$$= (x - 5)(x^2 + 2)$$
$$= x^3 - 5x^2 + 2x - 10$$

has three zeros: $x = 5$, $x = \sqrt{2}i$, and $x = -\sqrt{2}i$. Of the three, only $x = 5$ is an $x$-intercept.

**(c)** The quartic function

$$f(x) = (x - 3)(x - 3)(x - i)(x + i)$$
$$= (x^2 - 6x + 9)(x^2 + 1)$$
$$= x^4 - 6x^3 + 10x^2 - 6x + 9$$

has four zeros: $x = 3$, $x = 3$, $x = i$, and $x = -i$. There are only three distinct zeros. The real zero $x = 3$ is a repeated zero of multiplicity two. Due to this even multiplicity, the graph of $f$ touches but does not cross the $x$-axis at $x = 3$, the only $x$-intercept.

Figure 2.51 supports our conclusions regarding $x$-intercepts.

## Complex Conjugate Zeros

In Section 2.5 we saw that, for quadratic equations $ax^2 + bx + c = 0$ with real coefficients, if the discriminant $b^2 - 4ac$ is negative, the solutions are a conjugate pair of complex numbers. This relationship generalizes to polynomial functions of higher degree in the following way:

### Complex Conjugate Zeros

Suppose that $f(x)$ is a polynomial function with *real coefficients*. If $a$ and $b$ are real numbers with $b \neq 0$ and $a + bi$ is a zero of $f(x)$, then its complex conjugate $a - bi$ is also a zero of $f(x)$.

**Exploration Extensions**

In questions 1 and 2, check to see if the complex conjugates of the given zeros are also zeros. How does the answer relate to the complex conjugate zeros statement above? **(No, they are not zeros.)**

**Exploration 1** **What Can Happen If the Coefficients Are Not Real?**

1. Use substitution to verify that $x = 2i$ and $x = -i$ are zeros of $f(x) = x^2 - ix + 2$.

2. Use substitution to verify that $x = i$ and $x = 1 - i$ are zeros of $f(x) = x^2 - x + (1 + i)$.

3. What conclusions can you draw from parts 1 and 2? Write a summary of your findings.

### Example 2  FINDING A POLYNOMIAL FROM GIVEN ZEROS

Write the standard form of a polynomial function of degree 5 with real coefficients whose zeros include $x = 1$, $x = 1 + 2i$, $x = 1 - i$.

**Solution**  Because the coefficients are real and $1 + 2i$ is a zero, $1 - 2i$ must also be a zero. Therefore, $x - (1 + 2i)$ and $x - (1 - 2i)$ must both be factors of $f(x)$. Likewise, because $1 - i$ is a zero, $1 + i$ must also be a zero. It follows that $x - (1 - i)$ and $x - (1 + i)$ must both be factors of $f(x)$. Therefore,

$$f(x) = (x - 1)[x - (1 + 2i)][x - (1 - 2i)][x - (1 + i)][x - (1 - i)]$$
$$= (x - 1)(x^2 - 2x + 5)(x^2 - 2x + 2)$$
$$= (x^3 - 3x^2 + 7x - 5)(x^2 - 2x + 2)$$
$$= x^5 - 5x^4 + 15x^3 - 25x^2 + 24x - 10$$

is a polynomial of the type we seek. Any nonzero real number multiple of $f(x)$ will also be such a polynomial.

### Example 3  FACTORING A POLYNOMIAL WITH COMPLEX ZEROS

Find all zeros of $f(x) = x^5 - 3x^4 - 5x^3 + 5x^2 - 6x + 8$, and write $f(x)$ in its linear factorization.

**Solution**  Figure 2.52 suggests that the real zeros of $f$ are $x = -2$, $x = 1$, and $x = 4$.

Using synthetic division we can verify these zeros and show that $x^2 + 1$ is also a factor of $f$. So $x = i$ and $x = -i$ are also zeros. Therefore,

$$f(x) = x^5 - 3x^4 - 5x^3 + 5x^2 - 6x + 8$$
$$= (x + 2)(x - 1)(x - 4)(x^2 + 1)$$
$$= (x + 2)(x - 1)(x - 4)(x - i)(x + i).$$

Synthetic division can be used with complex number divisors in the same way it is used with real number divisors.

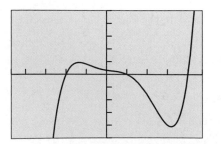

[–4.7, 4.7] by [–125, 125]

**Figure 2.52**
$f(x) = x^5 - 3x^4 - 5x^3 + 5x^2 - 6x + 8$ has three real zeros. (Example 3)

### Example 4  FINDING COMPLEX ZEROS

The complex number $z = 1 - 2i$ is a zero of
$f(x) = 4x^4 + 17x^2 + 14x + 65$. Find the remaining zeros of $f(x)$,
and write it in its linear factorization.

**Solution**  We use a synthetic division to show that $f(1 - 2i) = 0$:

$$
\begin{array}{r|ccccc}
1-2i & 4 & 0 & 17 & 14 & 65 \\
     &   & 4-8i & -12-16i & -27-26i & -65 \\
\hline
     & 4 & 4-8i & 5-16i & -13-26i & 0
\end{array}
$$

Thus $1 - 2i$ is a zero of $f(x)$. The complex conjugate $1 + 2i$ must also be a
zero. We use synthetic division on the quotient found above to find the
remaining quadratic factor:

$$
\begin{array}{r|cccc}
1+2i & 4 & 4-8i & 5-16i & -13-26i \\
     &   & 4+8i & 8+16i & 13+26i \\
\hline
     & 4 & 8 & 13 & 0
\end{array}
$$

Finally, we use the quadratic formula to find the two zeros of
$4x^2 + 8x + 13$:

$$
\begin{aligned}
x &= \frac{-8 \pm \sqrt{64 - 208}}{8} \\[2mm]
  &= \frac{-8 \pm \sqrt{-144}}{8} \\[2mm]
  &= \frac{-8 \pm 12i}{8} \\[2mm]
  &= -1 \pm \frac{3}{2}i
\end{aligned}
$$

Thus the four zeros of $f(x)$ are $1 - 2i$, $1 + 2i$, $-1 + (3/2)i$, and
$-1 - (3/2)i$. Because the leading coefficient of $f(x)$ is 4, we obtain

$$
\begin{aligned}
f(x) = 4&[x - (1 - 2i)][x - (1 + 2i)] \\
&\times \left[x - \left(-1 + \frac{3}{2}i\right)\right]\left[x - \left(-1 - \frac{3}{2}i\right)\right].
\end{aligned}
$$

If we wish to remove fractions in the factors, we can distribute the 4 to get

$$
f(x) = [x - (1 - 2i)][x - (1 + 2i)][2x - (-2 + 3i)][2x - (-2 - 3i)].
$$

## Factoring with Real Number Coefficients

Let $f(x)$ be a polynomial function with real coefficients. The Linear
Factorization Theorem tells us that $f(x)$ can be factored into the form

$$
f(x) = a(x - z_1)(x - z_2) \cdots (x - z_n),
$$

where $z_i$ are complex numbers. Recall, however, that nonreal complex zeros
occur in conjugate pairs. The product of $x - (a + bi)$ and $x - (a - bi)$ is

$$
\begin{aligned}
[x - (a + bi)][x - (a - bi)] &= x^2 - (a - bi)x - (a + bi)x \\
&\quad + (a + bi)(a - bi) \\
&= x^2 - 2ax + (a^2 + b^2).
\end{aligned}
$$

So the quadratic expression $x^2 - 2ax + (a^2 + b^2)$ is a factor of $f(x)$, and its coefficients are real numbers. A quadratic with no real zeros is **irreducible over the reals**. In other words, if we require that the factors of a polynomial have real coefficients, the factorization can be accomplished with linear factors and irreducible quadratic factors.

---

**Factors of a Polynomial with Real Coefficients**

Every polynomial function with real coefficients can be written as a product of linear factors and irreducible quadratic factors, each with real coefficients.

---

### Example 5 FACTORING A POLYNOMIAL

Write $f(x) = 3x^5 - 2x^4 + 6x^3 - 4x^2 - 24x + 16$ as a product of linear and quadratic factors, each with real coefficients.

**Solution** The Rational Zeros Theorem provides a list of candidates for the rational zeros of $f$. The graph of $f$ in Figure 2.53 suggests which candidates to try first. Using synthetic division, we find that $x = 2/3$ is a zero. Thus,

$$f(x) = \left(x - \frac{2}{3}\right)(3x^4 + 6x^2 - 24)$$

$$= \left(x - \frac{2}{3}\right)(3)(x^4 + 2x^2 - 8)$$

$$= (3x - 2)(x^2 - 2)(x^2 + 4)$$

$$= (3x - 2)(x - \sqrt{2})(x + \sqrt{2})(x^2 + 4)$$

Because the zeros of $x^2 + 4$ are complex, any further factorization would introduce nonreal complex coefficients. We have taken the factorization of $f$ as far as possible, subject to the condition that *each factor has real coefficients*.

We have seen that if a polynomial function has real coefficients, then its nonreal complex zeros occur in conjugate pairs. *Because a polynomial of odd degree has an odd number of zeros, it must have one zero that is real.* This confirms Example 7 of Section 2.3 in light of complex numbers.

---

**Polynomial Function of Odd Degree**

Every polynomial function of odd degree with real coefficients has at least one real zero.

---

The function $f(x) = 3x^5 - 2x^4 + 6x^3 - 4x^2 - 24x + 16$ in Example 5 fits the condition of this theorem, so we know right away that we are on the right track in searching for at least one real zero.

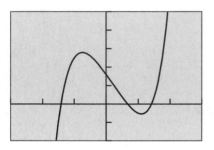

[–3, 3] by [–20, 50]

**Figure 2.53** $f(x) = 3x^5 - 2x^4 + 6x^3 - 4x^2 - 24x + 16$ has three real zeros. (Example 5)

**Follow-up**

Ask students if they can think of another way to justify the statement that an odd-degree polynomial function with real coefficients has at least one real zero. Have them think about end behavior and continuity.

**Assignment Guide**

Ex. 3–36, multiples of 3, 39–44, 46, 47, 48, 49, 52, 54, 55, 56

**Cooperative Learning**

Group Activity: Ex. 50

**Notes on Exercises**

Ex. 1–12, and 45–46 require that students produce a polynomial with desired characteristics.
Ex. 47–49 require students to use cubic or quadratic regression.

**Ongoing Assessment**

Self-Assessment: Ex. 11, 20, 25, 45, 53
Embedded Assessment: Ex. 42, 44

# Quick Review 2.6

In Exercises 1–4, factor the quadratic expression.

**1.** $2x^2 - x - 3$  $(2x - 3)(x + 1)$ **2.** $6x^2 - 13x - 5$

**3.** $10x^2 - 3x - 4$ **4.** $9x^2 - 6x - 8$

In Exercises 5–6, solve the quadratic equation.

**5.** $x^2 - 5x + 11 = 0$ **6.** $2x^2 + 3x + 7 = 0$

In Exercises 7–10, list all potential rational zeros.

**7.** $3x^4 - 5x^3 + 3x^2 - 7x + 2$  $\pm 1, \pm 2, \pm 1/3, \pm 2/3$

**8.** $4x^5 - 7x^2 + x^3 + 13x - 3$

**9.** $5x^3 + 4x^2 + 3x - 4$

**10.** $12x^4 + 8x^3 - 7x + 1$

# Section 2.6 Exercises

In Exercises 1–8, write a polynomial function in standard form with real coefficients whose zeros include those listed.

**1.** $i$ and $-i$  $x^2 + 1$

**2.** $1 - 2i$ and $1 + 2i$

**3.** $1, 3i$, and $-3i$

**4.** $-4, 1 - i$, and $1 + i$

**5.** $2, 3$, and $i$

**6.** $-1, 2$, and $1 - i$

**7.** $5$ and $3 + 2i$

**8.** $-2$ and $1 + 2i$

In Exercises 9–12, write a polynomial function in standard form with real coefficients whose zeros and their multiplicities include those listed.

**9.** $1$ (multiplicity 2), $-2$ (multiplicity 3)

**10.** $-1$ (multiplicity 3), $3$ (multiplicity 1)  $x^4 - 6x^2 - 8x - 3$

**11.** $2$ (multiplicity 2), $3 + i$ (multiplicity 1)

**12.** $-1$ (multiplicity 2), $-2 - i$ (multiplicity 1)

In Exercises 13–16, match the polynomial function graph to the given zeros and multiplicities.

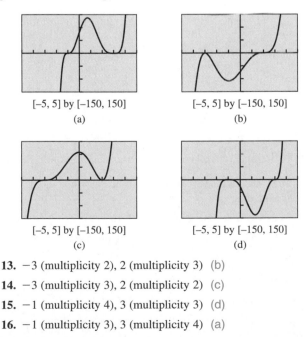

[−5, 5] by [−150, 150]
(a)

[−5, 5] by [−150, 150]
(b)

[−5, 5] by [−150, 150]
(c)

[−5, 5] by [−150, 150]
(d)

**13.** $-3$ (multiplicity 2), $2$ (multiplicity 3)  (b)

**14.** $-3$ (multiplicity 3), $2$ (multiplicity 2)  (c)

**15.** $-1$ (multiplicity 4), $3$ (multiplicity 3)  (d)

**16.** $-1$ (multiplicity 3), $3$ (multiplicity 4)  (a)

In Exercises 17–22, state how many complex and real zeros the function has.

**17.** $f(x) = x^2 - 2x + 7$  2 complex zeros; none real

**18.** $f(x) = x^3 - 3x^2 + x + 1$  3 complex zeros; all 3 real

**19.** $f(x) = x^3 - x + 3$  3 complex zeros; 1 real

**20.** $f(x) = x^4 - 2x^2 + 3x - 4$  4 complex zeros; 2 real

**21.** $f(x) = x^4 - 5x^3 + x^2 - 3x + 6$  4 complex zeros; 2 real

**22.** $f(x) = x^5 - 2x^2 - 3x + 6$  5 complex zeros; 1 real

In Exercises 23–28, find all of the zeros and write a linear factorization of the function.

**23.** $f(x) = x^3 + 4x - 5$

**24.** $f(x) = x^3 - 10x^2 + 44x - 69$

**25.** $f(x) = x^4 + x^3 + 5x^2 - x - 6$

**26.** $f(x) = 3x^4 + 8x^3 + 6x^2 + 3x - 2$

**27.** $f(x) = 6x^4 - 7x^3 - x^2 + 67x - 105$

**28.** $f(x) = 20x^4 - 148x^3 + 269x^2 - 106x - 195$

In Exercises 29–32, using the given zero, find all of the zeros and write a linear factorization of $f(x)$.

**29.** $1 + i$ is a zero of $f(x) = x^4 - 2x^3 - x^2 + 6x - 6$.

**30.** $4i$ is a zero of $f(x) = x^4 + 13x^2 - 48$.

**31.** $3 - 2i$ is a zero of $f(x) = x^4 - 6x^3 + 11x^2 + 12x - 26$.

**32.** $1 + 3i$ is a zero of $f(x) = x^4 - 2x^3 + 5x^2 + 10x - 50$.

In Exercises 33–38, write the function as a product of linear and irreducible quadratic factors all with real coefficients.

**33.** $f(x) = x^3 - x^2 - x - 2$  $(x - 2)(x^2 + x + 1)$

**34.** $f(x) = x^3 - x^2 + x - 6$  $(x - 2)(x^2 + x + 3)$

**35.** $f(x) = 2x^3 - x^2 + 2x - 3$  $(x - 1)(2x^2 + x + 3)$

**36.** $f(x) = 3x^3 - 2x^2 + x - 2$  $(x - 1)(3x^2 + x + 2)$

**37.** $f(x) = x^4 + 3x^3 - 3x^2 + 3x - 4$  $(x - 1)(x + 4)(x^2 + 1)$

**38.** $f(x) = x^4 - 2x^3 + x^2 - 8x - 12$  $(x - 3)(x + 1)(x^2 + 4)$

In Exercises 39 and 40, use *Archimedes' principle*, which states that when a sphere of radius $r$ with density $d_S$ is placed in a liquid of density $d_L = 62.5$ lb/ft$^3$, it will sink to a depth $h$ where

$$\frac{\pi}{3}(3rh^2 - h^3)d_L = \frac{4}{3}\pi r^3 d_S.$$

Find an approximate value for $h$ if:

**39.** $r = 5$ ft and $d_S = 20$ lb/ft$^3$. $h \approx 3.776$ ft

**40.** $r = 5$ ft and $d_S = 45$ lb/ft$^3$. $h \approx 6.513$ ft

In Exercises 41–44, answer yes or no. If yes, include an example. If no, give a reason.

**41. Writing to Learn** Is it possible to find a polynomial of degree 3 with real number coefficients that has $-2$ as its only real zero?

**42. Writing to Learn** Is it possible to find a polynomial of degree 3 with real coefficients that has $2i$ as its only nonreal zero?

**43. Writing to Learn** Is it possible to find a polynomial $f(x)$ of degree 4 with real coefficients that has zeros $-3$, $1 + 2i$, and $1 - i$?

**44. Writing to Learn** Is it possible to find a polynomial $f(x)$ of degree 4 with real coefficients that has zeros $1 + 3i$ and $1 - i$?

In Exercises 45 and 46, find the unique polynomial with real coefficients that meets these conditions.

**45.** Degree 4; zeros at $x = 3$, $x = -1$, and $x = 2 - i$; $f(0) = 30$

**46.** Degree 4; zeros at $x = 1 - 2i$ and $x = 1 + i$; $f(0) = 20$

# Explorations

**47.** Lewis's distance $D$ from a motion detector is given by the data in Table 2.19.

| Table 2.19  Motion Detector Data for Exercise 47 | | | |
|---|---|---|---|
| $t$ (sec) | $D$ (m) | $t$ (sec) | $D$ (m) |
| 0.0 | 1.00 | 4.5 | 0.99 |
| 0.5 | 1.46 | 5.0 | 0.84 |
| 1.0 | 1.99 | 5.5 | 1.28 |
| 1.5 | 2.57 | 6.0 | 1.87 |
| 2.0 | 3.02 | 6.5 | 2.58 |
| 2.5 | 3.34 | 7.0 | 3.23 |
| 3.0 | 2.91 | 7.5 | 3.78 |
| 3.5 | 2.31 | 8.0 | 4.40 |
| 4.0 | 1.57 | | |

**(a)** Find a cubic regression model, and graph it together with a scatter plot of the data.

**(b)** Use the cubic regression model to estimate how far Lewis is from the motion detector initially.

**(c)** Use the cubic regression model to estimate when Lewis changes direction. How far from the motion detector is he when he changes direction?

**48.** Sally's distance $D$ from a motion detector is given by the data in Table 2.20

| Table 2.20  Motion Detector Data for Exercise 48 | | | |
|---|---|---|---|
| $t$ (sec) | $D$ (m) | $t$ (sec) | $D$ (m) |
| 0.0 | 3.36 | 4.5 | 3.59 |
| 0.5 | 2.61 | 5.0 | 4.15 |
| 1.0 | 1.86 | 5.5 | 3.99 |
| 1.5 | 1.27 | 6.0 | 3.37 |
| 2.0 | 0.91 | 6.5 | 2.58 |
| 2.5 | 1.14 | 7.0 | 1.93 |
| 3.0 | 1.69 | 7.5 | 1.25 |
| 3.5 | 2.37 | 8.0 | 0.67 |
| 4.0 | 3.01 | | |

**(a)** Find a cubic regression model, and graph it together with a scatter plot of the data.

**(b)** Describe Sally's motion.

**(c)** Use the cubic regression model to estimate when Sally changes direction. How far is she from the motion detector when she changes direction?

**49.** Jacob's distance $D$ from a motion detector is given by the data in Table 2.21.

| Table 2.21  Motion Detector Data for Exercise 49 | | | |
|---|---|---|---|
| $t$ (sec) | $D$ (m) | $t$ (sec) | $D$ (m) |
| 0.0 | 4.59 | 4.5 | 1.70 |
| 0.5 | 3.92 | 5.0 | 2.25 |
| 1.0 | 3.14 | 5.5 | 2.84 |
| 1.5 | 2.41 | 6.0 | 3.39 |
| 2.0 | 1.73 | 6.5 | 4.02 |
| 2.5 | 1.21 | 7.0 | 4.54 |
| 3.0 | 0.90 | 7.5 | 5.04 |
| 3.5 | 0.99 | 8.0 | 5.59 |
| 4.0 | 1.31 | | |

**(a)** Find a quadratic regression model, and graph it together with a scatter plot of the data.

**(b)** Describe Jacob's motion.

**(c)** Use the quadratic regression model to estimate when Jacob changes direction. How far is he from the motion detector when he changes direction?

**50. Group Activity  The Square Roots of $i$** Let $a$ and $b$ be real numbers such that $(a + bi)^2 = i$.

**(a)** Expand the left-hand side of the given equation.

**(b)** Think of the right-hand side of the equation as $0 + 1i$, and separate the real and imaginary parts of the equation to obtain two equations.  $a^2 - b^2 = 0; \; 2ab = 1$

**(c)** Solve for $a$ and $b$.

**(d)** Check your answer by substituting them in the original equation.

**(e)** What are the two square roots of $i$?

**51.** Verify that the complex number $i$ is a zero of the polynomial $f(x) = x^3 - ix^2 + 2ix + 2$.

**52.** Verify that the complex number $-2i$ is a zero of the polynomial $f(x) = x^3 - (2 - i)x^2 + (2 - 2i)x - 4$.  ■

## Extending the Ideas

In Exercises 53–54, verify that $g(x)$ is a factor of $f(x)$. Then find $h(x)$ so that $f = g \cdot h$.

**53.** $g(x) = x - i; \quad f(x) = x^3 + (3 - i)x^2 - 4ix - 1$

**54.** $g(x) = x - 1 - i; \quad f(x) = x^3 - (1 + i)x^2 + x - 1 - i$

**55. Connecting Algebra and Geometry** In Exercise 59 of Section 2.5 we proved that $0$, $1 + 3i$, $2 - i$, and $3 + 2i$, shown in Figure 2.49, are the vertices of a parallelogram. Which other points in the complex plane complete a parallelogram with $1 + 3i$, $2 - i$, and $3 + 2i$?

**56.** Find the three cube roots of 8 by solving $x^3 = 8$.

**57.** Find the three cube roots of $-64$ by solving $x^3 = -64$.

---

## 2.7  Rational Functions and Equations

Rational Functions  •  Transformations of the Reciprocal Function
•  Limits and Asymptotes  •  Extraneous Solutions  •  Applications

### Rational Functions

Rational functions are ratios (or quotients) of polynomial functions.

> **Definition  Rational Functions**
>
> Let $f$ and $g$ be polynomial functions with $g(x) \neq 0$. Then the function given by
>
> $$r(x) = \frac{f(x)}{g(x)}$$
>
> is a **rational function**.

The domain of a rational function is the set of all real numbers except the zeros of its denominator. Every rational function is continuous on its domain.

### Example 1  IDENTIFYING RATIONAL FUNCTIONS

Which of the following are rational functions? For those that are rational functions, state the domain. For those that are not, explain why not.

**(a)** $f(x) = \dfrac{3x}{x^2 + 1}$  **(b)** $g(x) = \dfrac{x^3 + 2}{x^{1/2}}$  **(c)** $h(x) = \dfrac{\sqrt{25x^4 + 4}}{7x^3 + 6x}$

**(d)** $k(x) = \dfrac{3x - 5}{6x - 2x^3}$  **(e)** $p(x) = \dfrac{-9}{x + 8}$  **(f)** $q(x) = 3x^2 - x + 4$

**Solution**

**(a)** $f$ is a rational function, with domain $(-\infty, \infty)$ because $x^2 + 1$ has no real zeros.

**(b)** $g$ is not a rational function because of the exponent 1/2 in the denominator function.

**(c)** $h$ is not a rational function because the numerator cannot be simplified into polynomial form. Notice that $\sqrt{25x^4 + 4} \neq 5x^2 + 2$.

**(d)** $k$ is a rational function. To determine the domain we factor the denominator:

$$6x - 2x^3 = 2x(3 - x^2)$$
$$= 2x(\sqrt{3} + x)(\sqrt{3} - x)$$

So the zeros of the denominator are $0$, $-\sqrt{3}$, and $\sqrt{3}$, and the domain of $k$ is $(-\infty, -\sqrt{3}) \cup (-\sqrt{3}, 0) \cup (0, \sqrt{3}) \cup (\sqrt{3}, \infty)$.

**(e)** $p$ is a rational function, with domain $(-\infty, -8) \cup (-8, \infty)$ because $x + 8 = 0$ when $x = -8$.

**(f)** $q$ is a rational function because $q(x) = 3x^2 - x + 4 = (3x^2 - x + 4)/1$ and the constant function in the denominator is a polynomial function. The domain of $q$ is $(-\infty, \infty)$.

From Example 1f, we can generalize that every polynomial function is also a rational function.

## Transformations of the Reciprocal Function

One of the simplest rational functions is the reciprocal function

$$f(x) = \frac{1}{x},$$

which is one of the basic functions introduced in Chapter 1. It can be used to generate many other rational functions.

**Exploration Extensions**

Have the students use graphers to graph $f(x) = \dfrac{1}{x} + 1$, $f(x) = \dfrac{1}{x} + 2$, $f(x) = \dfrac{1}{x + 1}$, and $f(x) = \dfrac{1}{x + 2}$. Discuss the results in the shift of the graph of $f(x) = \dfrac{1}{x}$.

---

**Exploration 1**  **Comparing Graphs of Rational Functions**

1. Sketch the graph and find an equation for the function $g$ whose graph is obtained from the reciprocal function $f(x) = 1/x$ by a translation of 2 units to the right.

2. Sketch the graph and find an equation for the function $h$ whose graph is obtained from the reciprocal function $f(x) = 1/x$ by a translation of 5 units to the right, followed by a reflection across the axis.

3. Sketch the graph and find an equation for the function $k$ whose graph is obtained from the reciprocal function $f(x) = 1/x$ by a translation of 4 units to the left, followed by a vertical stretch by a factor of 3, and finally a translation 2 units down.

---

The graph of any rational function of the form

$$g(x) = \frac{ax + b}{cx + d}$$

can be obtained through transformations of the graph of the reciprocal function as illustrated in Example 2. If the degree of the numerator is greater than or equal to the degree of the denominator, we can use polynomial division to rewrite the rational function.

**Teaching Note**

If possible, have students compare the graphs of rational functions using both the dot mode and the connected mode of a grapher. In the connected mode, students may think that the grapher is purposely displaying the vertical asymptotes of the function. The asymptotes that appear on the grapher are actually artifacts of the algorithm used to create the graphical image on the screen. In Example 2a, for example, the asymptote shown might be a result of connecting points such as $(-3.05, -40)$ and $(-2.95, 40)$. If a column of pixels corresponds to exactly $x = -3$, the grapher finds that the function is undefined at that value, so the value is skipped and the asymptote does not appear.

**Example 2   TRANSFORMING THE RECIPROCAL FUNCTION**

Describe how the graph of the given function can be obtained by transforming the graph of the reciprocal function $f(x) = 1/x$. Identify the horizontal and vertical asymptotes.

**(a)** $g(x) = \dfrac{2}{x + 3}$  **(b)** $h(x) = \dfrac{3x - 7}{x - 2}$

**Solution**

**(a)** $g(x) = \dfrac{2}{x + 3}$

$\qquad = 2\left(\dfrac{1}{x + 3}\right)$

$\qquad = 2f(x + 3)$

The graph of $g$ is the graph of the reciprocal function shifted left 3 units and then stretched vertically by a factor of 2. So the lines $x = -3$ and $y = 0$ are vertical and horizontal asymptotes, respectively.

**(b)** We begin with polynomial division:

$$
\begin{array}{r}
3 \phantom{x - 7} \\
x - 2 \overline{)\, 3x - 7} \\
3x - 6 \\
\hline
-1
\end{array}
$$

So, $h(x) = \dfrac{3x - 7}{x - 2} = 3 - \dfrac{1}{x - 2} = -f(x - 2) + 3.$

Thus the graph of $h$ is the graph of the reciprocal function translated 2 units to the right, followed by a reflection across the $x$-axis, and then translated 3 units up. (Note that the reflection must be executed before the vertical translation.) So the lines $x = 2$ and $y = 3$ are vertical and horizontal asymptotes, respectively.

Figure 2.54 supports our conclusions.

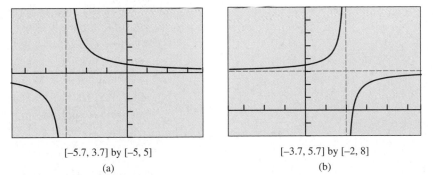

[–5.7, 3.7] by [–5, 5]  [–3.7, 5.7] by [–2, 8]
(a)  (b)

**Figure 2.54**  The graphs of (a) $g(x) = 2/(x + 3)$ and (b) $h(x) = (3x - 7)/(x - 2)$, with their asymptotes overlaid.

## Limits and Asymptotes

In Example 2 we found asymptotes by translating the known asymptotes of the reciprocal function. In Example 3, we use limits to find an asymptote.

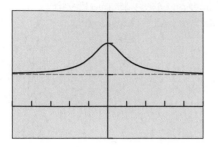

[–5, 5] by [–1, 3]

**Figure 2.55** The graph of $f(x) = (x^2 + 2)/(x^2 + 1)$ with its horizontal asymptote $y = 1$.

### Example 3   FINDING A HORIZONTAL ASYMPTOTE

Describe the end behavior and find the horizontal asymptote of $f(x) = (x^2 + 2)/(x^2 + 1)$.

**Solution**

**Solve Graphically**

The graph of $f$ in Figure 2.55 suggests that $\lim_{x \to \infty} f(x) = \lim_{x \to -\infty} f(x) = 1$ and that the horizontal asymptote is $y = 1$.

**Confirm Algebraically**

Using polynomial long division, we find that

$$f(x) = \frac{x^2 + 2}{x^2 + 1} = 1 + \frac{1}{x^2 + 1}.$$

When the value of $|x|$ is large, the denominator $x^2 + 1$ is a large positive number, and $1/(x^2 + 1)$ is a small positive number, getting closer to zero as $|x|$ increases. Therefore,

$$\lim_{x \to \infty} f(x) = \lim_{x \to -\infty} f(x) = 1,$$

so $y = 1$ is indeed a horizontal asymptote.

Example 3 shows the connection between the end behavior of a rational function and its horizontal asymptote. We now generalize this relationship and summarize other features of the graph of a rational function:

---

**Graphical Features of a Rational Function**
$y = f(x)/d(x) = (a_n x^n + \cdots)/(b_m x^m + \cdots)$

1. **End behavior asymptote:**

   If $n < m$, the end behavior asymptote is the horizontal asymptote $y = 0$.

   If $n = m$, the end behavior asymptote is the horizontal asymptote $y = a_n/b_m$.

   If $n > m$, the end behavior asymptote is the quotient polynomial function $y = q(x)$, where $f(x) = d(x)q(x) + r(x)$.

2. **Vertical asymptotes:** These occur at the zeros of the denominator, provided that the zeros are not also zeros of the numerator of equal or greater multiplicity.

3. **$x$-intercepts:** These occur at the zeros of the numerator, which are not also zeros of the denominator.

4. **$y$-intercept:** This is the value of $f(0)$, if defined.

---

It is a good idea to find all of the asymptotes and intercepts when graphing a rational function. If the end behavior asymptote of a rational function is a slant line, we call it a **slant asymptote**, as illustrated in Example 4.

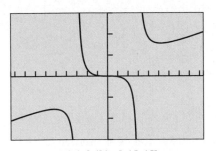

[−9.4, 9.4] by [−15, 15]

(a)

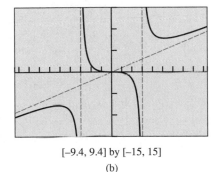

[−9.4, 9.4] by [−15, 15]

(b)

**Figure 2.56** The graph of $f(x) = x^3/(x^2 - 9)$ (a) by itself and (b) with its asymptotes.

**Extraneous Solutions**

Recall that an extraneous solution is a solution of a transformed equation that is not a solution of the original equation being solved.

## Example 4   GRAPHING A RATIONAL FUNCTION

Find the asymptotes and intercepts of the function $f(x) = x^3/(x^2 - 9)$ and graph the function.

**Solution**   The degree of the numerator is greater than the degree of the denominator, so the end behavior asymptote is the quotient polynomial. Using polynomial long division, we obtain

$$f(x) = \frac{x^3}{x^2 - 9} = x + \frac{9x}{x^2 - 9}.$$

So the quotient polynomial is $q(x) = x$, a slant asymptote. Factoring the denominator,

$$x^2 - 9 = (x - 3)(x + 3),$$

shows that the zeros of the denominator are $x = 3$ and $x = -3$. Consequently, $x = 3$ and $x = -3$ are the vertical asymptotes of $f$. The only zero of the numerator is 0, so $f(0) = 0$, and thus we see that the point $(0, 0)$ is the only $x$-intercept and the $y$-intercept of the graph of $f$.

The graph of $f$ in Figure 2.56a passes through $(0, 0)$ and suggests the vertical asymptotes $x = 3$ and $x = -3$ and the slant asymptote $y = q(x) = x$. Figure 2.56b shows the graph of $f$ with its asymptotes overlaid.

## Extraneous Solutions

In Section P.5 we saw that when both sides of an equation in $x$ are multiplied by an expression containing $x$, the resulting equation can have *extraneous solutions*. Extraneous solutions can be identified by checking them in the original equation.

## Example 5   ELIMINATING EXTRANEOUS SOLUTIONS

Solve the equation

$$\frac{2x}{x - 1} + \frac{1}{x - 3} = \frac{2}{x^2 - 4x + 3}$$

**Solution**

**Solve Algebraically**

The denominator of the right-hand side, $x^2 - 4x + 3$, factors into $(x - 1)(x - 3)$. So the least common denominator (LCD) of the equation is $(x - 1)(x - 3)$, and we multiply both sides of the equation by this LCD:

$$(x - 1)(x - 3)\left(\frac{2x}{x - 1} + \frac{1}{x - 3}\right) = (x - 1)(x - 3)\left(\frac{2}{x^2 - 4x + 3}\right)$$

$$2x(x - 3) + (x - 1) = 2$$

$$2x^2 - 5x - 3 = 0$$

$$(2x + 1)(x - 3) = 0$$

$$x = -\frac{1}{2} \quad \text{or} \quad x = 3$$

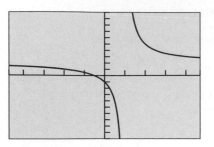

[−4.7, 4.7] by [−10, 10]

**Figure 2.57** The graph of $f(x) = 2x/(x − 1) + 1/(x − 3) − 2/(x^2 − 4x + 3)$. (Example 5)

Pure acid

50 mL of a 35% acid solution

**Figure 2.58** Mixing solutions. (Example 6)

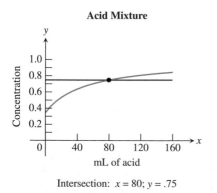

**Acid Mixture**

Intersection: $x = 80$; $y = .75$

**Figure 2.59** The graphs of $f(x) = (x + 17.5)/(x + 50)$ and $g(x) = 0.75$ (Example 6).

## Confirm Numerically

We replace $x$ by $−1/2$ in the original equation:

$$\frac{2(−1/2)}{(−1/2) − 1} + \frac{1}{(−1/2) − 3} \stackrel{?}{=} \frac{2}{(−1/2)^2 − 4(−1/2) + 3}$$

$$\frac{2}{3} − \frac{2}{7} \stackrel{?}{=} \frac{8}{21}$$

The equation is true, so $x = −1/2$ is a valid solution. The original equation is not defined for $x = 3$, so $x = 3$ is an extraneous solution.

## Support Graphically

The graph of

$$f(x) = \frac{2x}{x − 1} + \frac{1}{x − 3} − \frac{2}{x^2 − 4x + 3}$$

in Figure 2.57 suggests that $x = −1/2$ is an $x$-intercept and $x = 3$ is not.

## Interpret

Only $x = −1/2$ is a solution.

## Applications

### Example 6 CALCULATING ACID MIXTURES

How much pure acid must be added to 50 mL of a 35% acid solution to produce a mixture that is at least 75% acid? (See Figure 2.58.)

### Solution

**Model**

*Word statement*:  $\dfrac{\text{mL of pure acid}}{\text{mL of mixture}}$ = concentration of acid

$$0.35 \times 50 \text{ or } 17.5 = \text{mL of pure acid in 35% solution}$$

$$x = \text{mL of acid added}$$

$$x + 17.5 = \text{mL of pure acid in resulting mixture}$$

$$x + 50 = \text{mL of the resulting mixture}$$

$$\frac{x + 17.5}{x + 50} = \text{concentration of acid}$$

**Solve Graphically**

$$\frac{x + 17.5}{x + 50} \geq 0.75 \qquad \text{Inequality to be solved}$$

Figure 2.59 shows graphs of $f(x) = (x + 17.5)/(x + 50)$ and $g(x) = 0.75$. The point of intersection is $(80, 0.75)$. To the right of this point, $f(x) > g(x)$.

**Interpret**

We need to add 80 mL or more of pure acid to the 35% acid solution to make a solution that is at least 75% acid.

### Example 7   FINDING A MINIMUM PERIMETER

Find the dimensions of the rectangle with minimum perimeter if its area is 200 square meters. Find this least perimeter.

**Solution**

**Model**

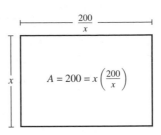

**Figure 2.60**   A rectangle with area 200 m$^2$. (Example 7)

$$\textit{Word statement}: \quad \text{Perimeter} = 2 \times \text{length} + 2 \times \text{width}$$

$$x = \text{width in meters}$$

$$\frac{200}{x} = \frac{\text{area}}{\text{width}} = \text{length in meters}$$

$$\textit{Function to be minimized}: \quad P(x) = 2x + 2\left(\frac{200}{x}\right) = 2x + \frac{400}{x}$$

**Solve Graphically**

The graph of $P$ in Figure 2.61 shows a minimum of approximately 56.57, occurring when $x \approx 14.14$.

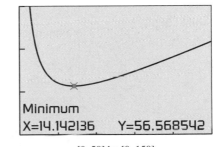

Minimum
X=14.142136      Y=56.568542

[0, 50] by [0, 150]

**Figure 2.61**   A graph of $P(x) = 2x + 400/x$. (Example 7)

**Interpret**

The width is about 14.14 m, and the minimum perimeter is about 56.57 m. Because $200/14.14 \approx 14.14$, the dimensions of the rectangle with minimum perimeter are 14.14 m by 14.14 m, a square.

### Example 8   DESIGNING A JUICE CAN

Stewart Cannery will package tomato juice in 2-liter cylindrical cans. Find the radius and height of the cans if the cans have a surface area of 1000 cm$^2$. (See Figure 2.62.)

**Figure 2.62**   A tomato juice can. (Example 8)

**Solution**

**Model**

$$S = \text{surface area of can in cm}^2$$

$$r = \text{radius of can in centimeters}$$

$$h = \text{height of can in centimeters}$$

Using volume ($V$) and surface area ($S$) formulas and the fact that $1 \text{ L} = 1000 \text{ cm}^3$, we conclude that

$$V = \pi r^2 h = 2000 \quad \text{and} \quad S = 2\pi r^2 + 2\pi rh = 1000.$$

So

$$2\pi r^2 + 2\pi rh = 1000$$

$$2\pi r^2 + 2\pi r\left(\frac{2000}{\pi r^2}\right) = 1000 \qquad \text{Substitute } h = 2000/(\pi r^2)$$

$$2\pi r^2 + \frac{4000}{r} = 1000 \qquad \text{Equation to be solved}$$

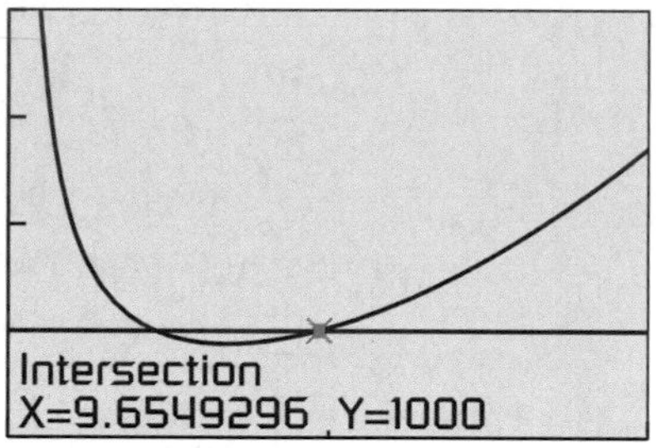

Intersection
X=9.6549296  Y=1000

[0, 20] by [0, 4000]
(a)

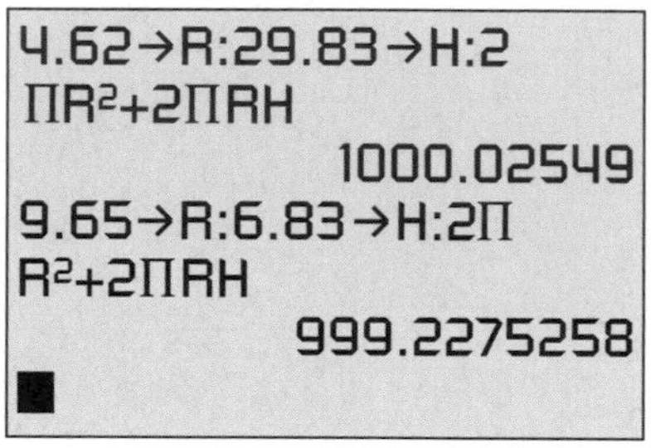

```
4.62→R:29.83→H:2
πR²+2πRH
            1000.02549
9.65→R:6.83→H:2π
R²+2πRH
            999.2275258
■
```
(b)

**Figure 2.63**  (Example 8)

**Solve Graphically**

Figure 2.63a shows the graphs of $f(x) = 2\pi r^2 + 4000/r$ and $g(x) = 1000$. One point of intersection occurs when $r$ is approximately 9.65. A second point of intersection occurs when $r$ is approximately 4.62.

Because $h = 2000/(\pi r^2)$, the corresponding values for $h$ are

$$h = \frac{2000}{\pi(4.619\ldots)^2} \approx 29.83 \quad \text{and} \quad h = \frac{2000}{\pi(9.654\ldots)^2} \approx 6.83.$$

**Support Numerically**

Figure 2.63b shows that the surface area $S$ is about 1000 when $r = 4.62$ and $h = 29.83$, or when $r = 9.65$ and $h = 6.83$.

**Interpret**

With a surface area of 1000 cm$^2$, the cans either have a radius of 4.62 cm and a height of 29.83 cm or have a radius of 9.65 cm and a height of 6.83 cm.

**Problem**

Given saturated vapor pressures ranging from 12.26 to 65.43 millibars, and a constant vapor pressure of 12.26 millibars, what range of relative humidity is possible? Does relative humidity increase or decrease with increasing saturated vapor pressure?

**Solution**

Given a constant vapor pressure of 12.26, and the following saturated vapor pressures, we can compute relative humidity. (Saturated vapor pressure of 12.26 millibars corresponds to air temperature of about 50 degrees Fahrenheit, and saturated vapor pressure of 65.43 millibars corresponds to air temperature of about 100 degrees Fahrenheit.)

| Saturated Vapor Pressure |
| --- |
| 12.26 |
| 17.65 |
| 25.01 |
| 34.94 |
| 48.12 |
| 65.43 |

When the saturated vapor pressure is 12.26, the relative humidity is $RH = 12.26/SVP \times 100 = 12.26/12.26 \times 100 = 100\%$. Similarly, we can compute the relative humidity for the other values of saturated vapor pressure:

| Saturated Vapor Pressure | Relative Humidity (%) |
| --- | --- |
| 12.26 | $12.26/12.26 \times 100 = 100$ |
| 17.65 | $12.26/17.65 \times 100 = 69.5$ |
| 25.01 | $12.26/25.01 \times 100 = 49$ |
| 34.94 | $12.26/34.94 \times 100 = 35.1$ |
| 48.12 | $12.26/48.12 \times 100 = 25.5$ |
| 65.43 | $12.26/65.43 \times 100 = 18.7$ |

So the relative humidity for these saturated vapor pressures ranges from 18.7% to 100%, and we see that relative humidity decreases as saturated vapor pressure (i.e., temperature) increases.

**Follow-up**

Ask students how one can tell whether or not a rational function has a slant asymptote without doing any division. **(There is a slant asymptote if the degree of the numerator minus the degree of the denominator is one.)**

**Assignment Guide**

Day 1: Ex. 3–42, multiples of 3
Day 2: Ex. 45, 48, 49, 51, 52, 55, 56, 58, 59, 60, 62, 63

**Cooperative Learning**

Group Activity: Ex. 53–54

**Notes on Exercises**

Ex. 17–24 give students practice in evaluating limits.

**Ongoing Assessment**

Self-Assessment: Ex. 17, 31, 35, 47, 50
Embedded Assessment: Ex. 56, 58

# Quick Review 2.7

In Exercises 1–6, use factoring to find the real zeros of the function.

**1.** $f(x) = 2x^2 + 5x - 3$

**2.** $f(x) = 3x^2 - 2x - 8$

**3.** $g(x) = x^2 - 4$  $x = \pm 2$

**4.** $g(x) = x^2 - 1$  $x = \pm 1$

**5.** $h(x) = x^3 - 1$  $x = 1$

**6.** $h(x) = x^2 + 1$

In Exercises 7–10, find the quotient and remainder when $f(x)$ is divided by $d(x)$.

**7.** $f(x) = 2x + 1, d(x) = x - 3$  2; 7

**8.** $f(x) = 4x + 3, d(x) = 2x - 1$  2; 5

**9.** $f(x) = 3x - 5, d(x) = x$  3; $-5$

**10.** $f(x) = 5x - 1, d(x) = 2x$  5/2; $-1$

# Section 2.7 Exercises

In Exercises 1–8, which of the following are rational functions? For those that are rational functions, state the domain. For those that are not, explain why not.

**1.** $f(x) = \dfrac{2x^2 - 3x + 4}{x^3 + x + 1}$

**2.** $f(x) = \dfrac{3x - 1}{x^2 + \sqrt{x - 1}}$

**3.** $f(x) = \dfrac{x^2 - 3\sqrt{x} + 1}{2x - 3}$

**4.** $f(x) = \dfrac{1}{3x + 4}$

**5.** $f(x) = \dfrac{1}{x} - 4$

**6.** $f(x) = 2 - \dfrac{1}{x}$

**7.** $f(x) = 4 + \dfrac{1}{x + 1}$

**8.** $f(x) = -2 + \dfrac{2}{x + 3}$

In Exercises 9–16, describe how the graph of the given function can be obtained by transforming the graph of the reciprocal function $f(x) = 1/x$. Identify the horizontal and vertical asymptotes.

**9.** $f(x) = \dfrac{1}{x - 3}$

**10.** $f(x) = \dfrac{1}{x + 3}$

**11.** $f(x) = -\dfrac{2}{x + 5}$

**12.** $f(x) = \dfrac{2}{x + 2}$

**13.** $f(x) = \dfrac{2x - 1}{x + 3}$

**14.** $f(x) = \dfrac{3x - 2}{x - 1}$

**15.** $f(x) = \dfrac{5 - 2x}{x + 4}$

**16.** $f(x) = \dfrac{4 - 3x}{x - 5}$

In Exercises 17–20, evaluate the limit based on the graph of $f$ shown.

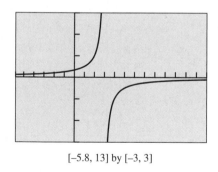

[–5.8, 13] by [–3, 3]

**17.** $\lim\limits_{x \to 3^-} f(x)$   $\infty$

**18.** $\lim\limits_{x \to 3^+} f(x)$   $-\infty$

**19.** $\lim\limits_{x \to \infty} f(x)$   $0$

**20.** $\lim\limits_{x \to -\infty} f(x)$   $0$

In Exercises 21–24, evaluate the limit based on the graph of $f$ shown.

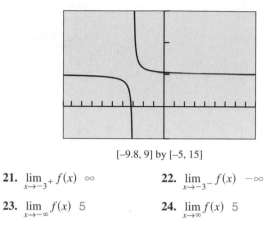

[–9.8, 9] by [–5, 15]

**21.** $\lim\limits_{x \to -3^+} f(x)$   $\infty$

**22.** $\lim\limits_{x \to -3^-} f(x)$   $-\infty$

**23.** $\lim\limits_{x \to -\infty} f(x)$   $5$

**24.** $\lim\limits_{x \to \infty} f(x)$   $5$

In Exercises 25–36, find the asymptotes and intercepts of the function, and graph the function.

**25.** $f(x) = \dfrac{2}{x - 3}$

**26.** $f(x) = \dfrac{-3}{x + 2}$

**27.** $g(x) = \dfrac{x - 2}{x^2 - 2x - 3}$

**28.** $g(x) = \dfrac{x + 2}{x^2 + 2x - 3}$

**29.** $h(x) = \dfrac{2}{x^3 - x}$

**30.** $h(x) = \dfrac{3}{x^3 - 4x}$

**31.** $k(x) = \dfrac{x - 1}{x^2 + 3}$

**32.** $k(x) = \dfrac{x + 3}{x^2 + 1}$

**33.** $f(x) = \dfrac{2x^2 + x - 2}{x^2 - 1}$

**34.** $g(x) = \dfrac{-3x^2 + x + 12}{x^2 - 4}$

**35.** $f(x) = \dfrac{x^2 - 2x + 3}{x + 2}$

**36.** $g(x) = \dfrac{x^2 - 3x - 7}{x + 3}$

In Exercises 37–42, match the rational function with its graph. Identify the viewing window and the scale used on each axis.

**37.** $f(x) = \dfrac{1}{x - 4}$

**38.** $f(x) = -\dfrac{1}{x + 3}$

**39.** $f(x) = 2 + \dfrac{3}{x - 1}$

**40.** $f(x) = 1 + \dfrac{1}{x + 3}$

**41.** $f(x) = -1 + \dfrac{1}{4 - x}$

**42.** $f(x) = 3 - \dfrac{2}{x - 1}$

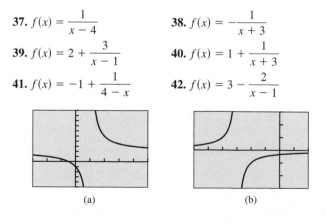

(a)                                        (b)

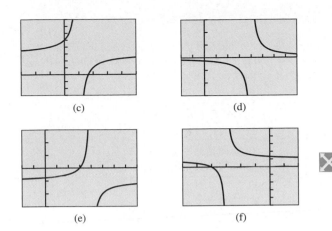

(c)  (d)

(e)  (f)

In Exercises 43–48, solve the equation algebraically. Check for extraneous solutions. Support your answer graphically.

**43.** $\dfrac{3x}{x+5} + \dfrac{1}{x-2} = \dfrac{7}{x^2 + 3x - 10}$

**44.** $\dfrac{4x}{x+4} + \dfrac{3}{x-1} = \dfrac{15}{x^2 + 3x - 4}$

**45.** $\dfrac{x-3}{x} - \dfrac{3}{x+1} + \dfrac{3}{x^2 + x} = 0$

**46.** $\dfrac{x+2}{x} - \dfrac{4}{x-1} + \dfrac{2}{x^2 - x} = 0$

**47.** $\dfrac{3}{x+2} + \dfrac{6}{x^2 + 2x} = \dfrac{3-x}{x}$

**48.** $\dfrac{x+3}{x} - \dfrac{2}{x+3} = \dfrac{6}{x^2 + 3x}$

**49. Acid Mixture** Suppose that $x$ mL of pure acid are added to 125 mL of a 60% acid solution. How many mL of pure acid must be added to obtain a solution of at least 83% acid?

**(a)** Explain why the concentration $C(x)$ of the new mixture is

$$C(x) = \dfrac{x + 0.6(125)}{x + 125}.$$

**(b)** Suppose the viewing window in the figure is used to find a solution to the problem. What is the equation of the horizontal line? $y = 0.83$

**(c)** Write and solve an inequality that answers the question of this problem.

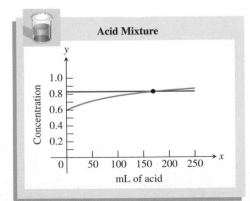

Acid Mixture

**50. Acid Mixture** Suppose that $x$ mL of pure acid are added to 100 mL of a 35% acid solution.

**(a)** Express the concentration $C(x)$ of the new mixture as a function of $x$.

**(b)** Use a graph to determine how much pure acid should be added to the 35% solution to produce a new solution that is less than 75% acid.

**(c)** Solve (b) algebraically.

**51. Breaking Even** Mid Town Sports Apparel, Inc., has found that it needs to sell golf hats for $2.75 each in order to be competitive. It costs $2.12 to produce each hat, and it has weekly overhead costs of $3000.

**(a)** Let $x$ be the number of hats produced each week. Express the average cost (including overhead costs) of producing one hat as a function of $x$.

**(b)** Solve algebraically to find the number of golf hats that must be sold each week to make a profit. Support your answer graphically.

**(c)** How many golf hats must be sold to make a profit of $1000 in 1 week?

**52. Bear Population** The number of bears at any time $t$ (in years) in a federal game reserve is given by

$$P(t) = \dfrac{500 + 250t}{10 + 0.5t}.$$

**(a)** Find the population of bears when the value of $t$ is 10, 40, and 100.

**(b)** Does the graph of the bear population have a horizontal asymptote? If so, what is it? If not, why not?

**(c)** According to this model, what is the largest the bear population can become?

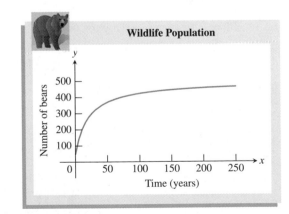

Wildlife Population

## Explorations

**53. Group Activity** *Work in groups of two.* Compare the functions

$$f(x) = \dfrac{x^2 - 9}{x - 3} \quad \text{and} \quad g(x) = x + 3.$$

**(a)** Are the domains equal?

**(b)** Does $f$ have a vertical asymptote? Explain.

**(c)** Explain why the graphs appear to be identical.

**(d)** Are the functions identical?

54. **Group Activity** Explain why the functions are identical or not. Include the graphs and a comparison of the functions' asymptotes, intercepts, and domain.

**(a)** $f(x) = \dfrac{x^2 + x - 2}{x - 1}$ and $g(x) = x + 2$

**(b)** $f(x) = \dfrac{x^2 - 1}{x + 1}$ and $g(x) = x - 1$

**(c)** $f(x) = \dfrac{x^2 - 1}{x^3 - x^2 - x + 1}$ and $g(x) = \dfrac{1}{x - 1}$

**(d)** $f(x) = \dfrac{x - 1}{x^2 + x - 2}$ and $g(x) = \dfrac{1}{x + 2}$

55. **Boyle's Law** This ideal gas law states that the volume of an enclosed gas at a fixed temperature varies inversely as the pressure.

**(a) Writing to Learn** Explain why Boyle's Law yields both a rational function model and a power function model.

**(b)** Which power functions are also rational functions?

**(c)** If the pressure of a 2.59-L sample of nitrogen gas at 291°K is 0.866 atm, what would the volume be at a pressure of 0.532 atm if the temperature does not change?  4.22 L

56. **Light Intensity** Aileen and Malachy gathered the data in Table 2.22 using a 75-watt light bulb and a Calculator-Based Laboratory™ (CBL™) with a light-intensity probe.

**(a)** Draw a scatter plot of the data in Table 2.22.

**(b)** Find an equation for the data assuming it has the form $f(x) = k/x^2$ for some constant $k$. Explain your method for choosing $k$.  $k = 5.81$

**(c)** Superimpose the regression curve on the scatter plot.

**(d)** Use the regression model to predict the light intensity at distances of 2.2 m and 4.4 m.  1.20 W/m²; 0.30 W/m²

**Table 2.22  Light Intensity Data for a 75-W Light Bulb**

| Distance (m) | Intensity (W/m²) |
|---|---|
| 1.0 | 6.09 |
| 1.5 | 2.51 |
| 2.0 | 1.56 |
| 2.5 | 1.08 |
| 3.0 | 0.74 |

## Extending the Ideas

In Exercises 57–60, find the asymptotes and intercepts of the function, and graph the function. Overlay a graph of the end behavior asymptote.

57. $f(x) = \dfrac{x^3 - 2x^2 + x - 1}{2x - 1}$

58. $g(x) = \dfrac{2x^3 - 2x^2 - x + 5}{x - 2}$

59. $f(x) = \dfrac{2x^4 - x^3 - 16x^2 + 17x - 6}{2x - 5}$

60. $g(x) = \dfrac{2x^5 - 3x^3 + 2x - 4}{x - 1}$

In Exercises 61–64, graph the function. Express the function as a piecewise-defined function without absolute value, and use the result to confirm the graph's asymptotes and intercepts algebraically.

61. $h(x) = \dfrac{2x - 3}{|x| + 2}$

62. $h(x) = \dfrac{3x + 5}{|x| + 3}$

63. $f(x) = \dfrac{5 - 3x}{|x| + 4}$

64. $f(x) = \dfrac{2 - 2x}{|x| + 1}$

65. Describe how the graph of

$$f(x) = \frac{ax + b}{cx + d}$$

can be obtained from the graph of $y = 1/x$. (*Hint:* Use long division.)

66. **Writing to Learn** Let $f(x) = 1/(x - (1/x))$ and $g(x) = (x^3 + x^2 - x)/(x^3 - x)$. Does $f = g$? Support your answer by making a comparative analysis of all of the features of $f$ and $g$, including asymptotes, intercepts, and domain.

# 2.8 Solving Inequalities in One Variable

Polynomial Inequalities • Rational Inequalities
• Other Inequalities • Applications

## Polynomial Inequalities

Every polynomial inequality can be written in the form $f(x) > 0$, $f(x) \geq 0$, $f(x) < 0$, or $f(x) \leq 0$, where $f(x)$ is a polynomial. There is a fundamental connection between inequalities and the positive or negative sign of the corresponding expression $f(x)$:

• To solve the inequality $f(x) > 0$ is to find the values of $x$ that make $f(x)$ *positive*.

• To solve the inequality $f(x) < 0$ is to find the values of $x$ that make $f(x)$ *negative*.

If the expression $f(x)$ is a product, we can determine its sign by determining the sign of each of its factors. So the fundamental connection among a polynomial's factors, real zeros, and $x$-intercepts plays a key role in solving polynomial inequalities, just as it does in solving polynomial equations. Example 1 illustrates that a polynomial function changes sign only at its real zeros of odd multiplicity.

**Example 1  FINDING WHERE A POLYNOMIAL IS ZERO, POSITIVE, OR NEGATIVE**

Let $f(x) = (x+3)(x^2+1)(x-4)^2$. Determine the real number values of $x$ that cause $f(x)$ to be **(a)** zero, **(b)** positive, **(c)** negative.

**Solution**  We begin by verbalizing our analysis of the problem:

**(a)** The real zeros of $f(x)$ are $-3$ (with multiplicity 1) and 4 (with multiplicity 2). So $f(x)$ is zero if $x = -3$ or $x = 4$.

**(b)** The factor $x^2 + 1$ is positive for all real numbers $x$. The factor $(x-4)^2$ is positive for all real numbers $x$ except $x = 4$, which makes $(x-4)^2 = 0$. The factor $x + 3$ is positive if and only if $x > -3$. So $f(x)$ is positive if $x > -3$ and $x \neq 4$.

**(c)** By process of elimination, $f(x)$ is negative if $x < -3$.

This verbal reasoning process is aided by the following **sign chart**, which shows the $x$-axis as a number line with the real zeros displayed as the locations of potential sign change and the factors displayed with their sign value in the corresponding interval:

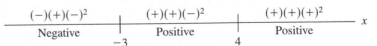

Figure 2.64 supports our findings because the graph of $f$ is above the $x$-axis for $x$ in $(-3, 4)$ or $(4, \infty)$, is on the $x$-axis for $x = -3$ or $x = 4$, and is below the $x$-axis for $x$ in $(-\infty, -3)$.

Our work in Example 1 allows us to report the solutions of four polynomial inequalities:

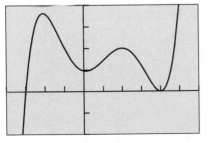

[−4, 6] by [−100, 200]

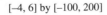

**Figure 2.64**  The graph of $f(x) = (x+3)(x^2+1)(x-4)^2$. (Example 1)

- The solution of $(x + 3)(x^2 + 1)(x - 4)^2 > 0$ is $(-3, 4) \cup (4, \infty)$.
- The solution of $(x + 3)(x^2 + 1)(x - 4)^2 \geq 0$ is $[-3, \infty)$.
- The solution of $(x + 3)(x^2 + 1)(x - 4)^2 < 0$ is $(-\infty, -3)$.
- The solution of $(x + 3)(x^2 + 1)(x - 4)^2 \leq 0$ is $(-\infty, -3]$.

**Teaching Note**

This section presents two types of sign charts: one type is based on the signs of the factors (as shown in Example 1), and the other type is based on end behavior and multiplicity (as shown in Figure 2.65). Using these two methods gives students a deep understanding of several important concepts and a way to check their results, as in Exploration 1.

Example 1 illustrates some important general characteristics of polynomial functions and polynomial inequalities. The polynomial function $f(x) = (x + 3)(x^2 + 1)(x - 4)^2$ in Example 1 and Figure 2.64

- changes sign at its real zero of odd multiplicity $(x = -3)$;
- touches the $x$-axis but does not change sign at its real zero of even multiplicity $(x = 4)$;
- has no $x$-intercepts or sign changes at its nonreal complex zeros associated with the irreducible quadratic factor $(x^2 + 1)$.

This is consistent with what we learned about the relationships between zeros and graphs of polynomial functions in Sections 2.3 and 2.6. The real zeros and their multiplicity together with the end behavior of a polynomial function give us sufficient information about the polynomial to sketch its graph well enough to obtain a correct sign chart without considering the signs of the factors. See Figure 2.65.

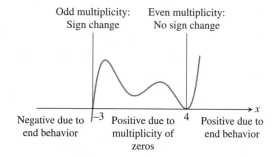

**Figure 2.65** The sign chart and graph of $f(x) = (x + 3)(x^2 + 1)(x - 4)^2$ overlaid.

**Exploration 1** **Sketching a Graph of a Polynomial from Its Sign Chart**

Use your knowledge of end behavior and multiplicity of real zeros to create a sign chart and sketch the graph of the function. Check your sign chart using the factor method of Example 1. Then check your sketch using a grapher.

**1.** $f(x) = 2(x - 2)^3(x + 3)^2$

**2.** $f(x) = -(x + 2)^4(x + 1)(2x^2 + x + 1)$

**3.** $f(x) = 3(x - 2)^2(x + 4)^3(-x^2 - 2)$

**Exploration Extensions**

For your graph of question 1, for what values of $x$ is $f(x) \geq 0$? And for what values of $x$ is $f(x) < 0$?

So far in this section all of the polynomials have been presented in factored form and all of the inequalities have had zero on one of the sides. Examples 2 and 3 show us how to solve polynomial inequalities when they are not given in such a convenient form.

### Example 2  SOLVING A POLYNOMIAL INEQUALITY ANALYTICALLY

Solve $2x^3 - 7x^2 - 10x + 24 > 0$ analytically.

**Solution**  Let $f(x) = 2x^3 - 7x^2 - 10x + 24$. The Rational Zeros Theorem suggests several possible rational zeros of $f$ for factoring purposes:

$$\pm 1, \ \pm 2, \ \pm 3, \ \pm 4, \ \pm 6, \ \pm 8, \ \pm 12, \ \pm 24, \ \pm \frac{1}{2}, \ \pm \frac{3}{2}$$

A table or graph of $f$ can suggest which of these candidates to try. In this case, $x = 4$ is a rational zero of $f$, as the following synthetic division shows:

$$
\begin{array}{r|rrrr}
4 & 2 & -7 & -10 & 24 \\
  &   & 8 & 4 & -24 \\
\hline
  & 2 & 1 & -6 & 0
\end{array}
$$

The synthetic division allows us to start the factoring process, which can then be completed using basic factoring methods:

$$
\begin{aligned}
f(x) &= 2x^3 - 7x^2 - 10x + 24 \\
     &= (x - 4)(2x^2 + x - 6) \\
     &= (x - 4)(2x - 3)(x + 2)
\end{aligned}
$$

So the zeros of $f$ are 4, 3/2, and $-2$. They are all real and all of multiplicity 1, so each will yield a sign change in $f(x)$. Because the degree of $f$ is odd and its leading coefficient is positive, the end behavior of $f$ is given by

$$\lim_{x \to \infty} f(x) = \infty \quad \text{and} \quad \lim_{x \to -\infty} f(x) = -\infty$$

Our analysis yields the following sign chart:

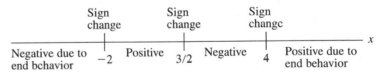

The solution of $2x^3 - 7x^2 - 10x + 24 > 0$ is $(-2, 3/2) \cup (4, \infty)$.

### Example 3  SOLVING A POLYNOMIAL INEQUALITY GRAPHICALLY

Solve $x^3 - 6x^2 \leq 2 - 8x$ graphically.

**Solution**  First we rewrite the inequality as $x^3 - 6x^2 + 8x - 2 \leq 0$. Then we let $f(x) = x^3 - 6x^2 + 8x - 2$ and find the real zeros of $f$ graphically as shown in Figure 2.66. The three real zeros are approximately 0.32, 1.46, and 4.21. The solution consists of the $x$ values for which the graph is on or below the $x$-axis. So the solution of $x^3 - 6x^2 \leq 2 - 8x$ is approximately $(-\infty, 0.32] \cup [1.46, 4.21]$.

The end points of these intervals are only accurate to two decimal places. We use square brackets because the zeros of the polynomial are solutions of the inequality, even though we only have approximations of their values.

When a polynomial function has no sign changes, the solutions of the associated inequalities can look a bit unusual, as illustrated in Example 4.

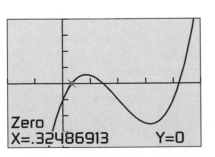

[–2, 5] by [–8, 8]

**Figure 2.66**  The graph of $f(x) = x^3 - 6x^2 + 8x - 2$, with one of three real zeros highlighted. (Example 3)

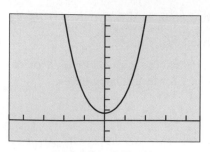

[–4.7, 4.7] by [–20, 100]

(a)

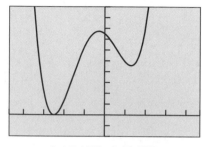

[–4.7, 4.7] by [–20, 100]

(b)

**Figure 2.67** The graphs of
(a) $f(x) = (x^2 + 7)(2x^2 + 1)$ and
(b) $g(x) = (x^2 - 3x + 3)(2x + 5)^2$.
(Example 4)

### Example 4 SOLVING POLYNOMIAL INEQUALITIES WITH UNUSUAL ANSWERS

**(a)** The inequalities associated with the strictly positive polynomial function $f(x) = (x^2 + 7)(2x^2 + 1)$ have unusual solution sets. We use Figure 2.67a as a guide to solving the inequalities:

- The solution of $(x^2 + 7)(2x^2 + 1) > 0$ is $(-\infty, \infty)$, all real numbers.

- The solution of $(x^2 + 7)(2x^2 + 1) \geq 0$ is also $(-\infty, \infty)$.

- The solution set of $(x^2 + 7)(2x^2 + 1) < 0$ is empty. We sometimes say an inequality of this sort has no solution.

- The solution set of $(x^2 + 7)(2x^2 + 1) \leq 0$ is also empty.

**(b)** The inequalities associated with the nonnegative polynomial function $g(x) = (x^2 - 3x + 3)(2x + 5)^2$ also have unusual solution sets. We use Figure 2.67b as a guide to solving the inequalities:

- The solution of $(x^2 - 3x + 3)(2x + 5)^2 > 0$ is $(-\infty, -5/2) \cup (-5/2, \infty)$, all real numbers except $x = -5/2$, the lone real zero of $g$.

- The solution of $(x^2 - 3x + 3)(2x + 5)^2 \geq 0$ is $(-\infty, \infty)$, all real numbers.

- The solution set of $(x^2 - 3x + 3)(2x + 5)^2 < 0$ is empty.

- The solution of $(x^2 - 3x + 3)(2x + 5)^2 \leq 0$ is the single number $x = -5/2$.

## Rational Inequalities

A polynomial function $p(x)$ is positive, negative, or zero for all real numbers $x$, but a rational function $r(x)$ can be positive, negative, zero, or *undefined*. In particular, a rational function is undefined at the zeros of its denominator. So when solving rational inequalities we modify the kind of sign chart used in Example 1 to include the real zeros of both the numerator and the denominator as locations of potential sign change.

### Example 5 CREATING A SIGN CHART FOR A RATIONAL FUNCTION

Let $r(x) = (2x + 1)/((x + 3)(x - 1))$. Determine the values of $x$ that cause $r(x)$ to be **(a)** zero, **(b)** undefined. Then make a sign chart to determine the values of $x$ that cause $r(x)$ to be **(c)** positive, **(d)** negative.

**Solution** Denote the numerator $2x + 1$ as $n(x)$ and the denominator $(x + 3)(x - 1)$ as $d(x)$.

**(a)** The real zeros of $r(x)$ are the real zeros of $n(x) = 2x + 1$. The only zero of $n(x)$ is $-1/2$. So $r(x)$ is zero if $x = -1/2$.

**(b)** $r(x)$ is undefined when $d(x) = (x + 3)(x - 1) = 0$. So $r(x)$ is undefined if $x = -3$ or $x = 1$.

These findings lead to the following sign chart, with three points of potential sign change:

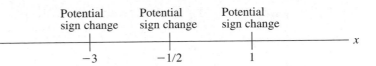

Analyzing the factors of the numerator and denominator yields:

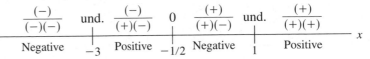

**(c)** So $r(x)$ is positive if $-3 < x < -1/2$ or $x > 1$, and the solution of $(2x + 1)/((x + 3)(x - 1)) > 0$ is $(-3, -1/2) \cup (1, \infty)$.

**(d)** Similarly, $r(x)$ is negative if $x < -3$ or $-1/2 < x < 1$, and the solution of $(2x + 1)/((x + 3)(x - 1)) < 0$ is $(-\infty, -3) \cup (-1/2, 1)$.

Figure 2.68 supports our findings because the graph of $r$ is above the $x$-axis for $x$ in $(-3, -1/2) \cup (1, \infty)$ and is below the $x$-axis for $x$ in $(-\infty, -3) \cup (-1/2, 1)$.

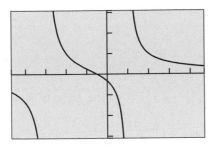

[–4.7, 4.7] by [–3.1, 3.1]

**Figure 2.68** The graph of $r(x) = (2x + 1)/((x + 3)(x - 1))$. (Example 5)

### Example 6   SOLVING A RATIONAL INEQUALITY BY COMBINING FRACTIONS

Solve $5/(x + 3) + 3/(x - 1) < 0$.

**Solution**  We combine the two fractions on the left-hand side of the inequality using the least common denominator $(x + 3)(x - 1)$:

$$\frac{5}{x + 3} + \frac{3}{x - 1} < 0 \qquad \text{Original inequality}$$

$$\frac{5(x - 1)}{(x + 3)(x - 1)} + \frac{3(x + 3)}{(x + 3)(x - 1)} < 0 \qquad \text{Use LCD to rewrite fractions.}$$

$$\frac{5(x - 1) + 3(x + 3)}{(x + 3)(x - 1)} < 0 \qquad \text{Add fractions.}$$

$$\frac{5x - 5 + 3x + 9}{(x + 3)(x - 1)} < 0 \qquad \text{Distributive property}$$

$$\frac{8x + 4}{(x + 3)(x - 1)} < 0 \qquad \text{Simplify.}$$

$$\frac{2x + 1}{(x + 3)(x - 1)} < 0 \qquad \text{Divide both sides by 4.}$$

This is the inequality of Example 5d. So the solution is $(-\infty, -3) \cup (-1/2, 1)$.

**Alert**

When solving a rational inequality some students will attempt to multiply both sides by the LCD without regard to the direction of the inequality sign. They need to realize that since the sign of the LCD (and hence the direction of the inequality sign in the resulting inequality) may depend on $x$, this method will not work.

### Other Inequalities

The sign chart method can be adapted to solve other types of inequalities, and we can support our solutions graphically as needed or desired.

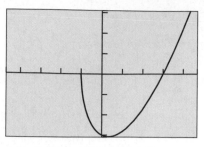

[−4.7, 4.7] by [−3.1, 3.1]

**Figure 2.69** The graph of $f(x) = (x − 3)\sqrt{x + 1}$. (Example 7)

**Example 7 SOLVING AN INEQUALITY INVOLVING A RADICAL**

Solve $(x − 3)\sqrt{x + 1} \geq 0$.

**Solution** Let $f(x) = (x − 3)\sqrt{x + 1}$. Because of the factor $\sqrt{x + 1}$, $f(x)$ is undefined if $x < −1$. The zeros of $f$ are 3 and −1. These findings, along with a sign analysis of the two factors, lead to the following sign chart:

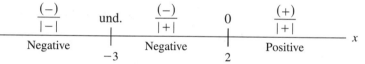

So the solution of $(x − 3)\sqrt{x + 1} \geq 0$ is $\{−1\} \cup [3, \infty)$. The graph of $f$ in Figure 2.69 supports this solution.

**Example 8 SOLVING AN INEQUALITY INVOLVING ABSOLUTE VALUE**

Solve $\dfrac{x − 2}{|x + 3|} \leq 0$.

**Solution** Let $f(x) = (x − 2)/|x + 3|$. Because $|x + 3|$ is in the denominator, $f(x)$ is undefined if $x = −3$. The only zero of $f$ is 2. These findings, along with a sign analysis of the two factors, lead to the following sign chart:

So the solution of $(x − 2)/|x + 3| \leq 0$ is $(−\infty, −3) \cup (−3, 2]$. The graph of $f$ in Figure 2.70 supports this solution.

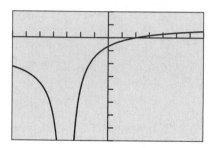

[−7, 7] by [−8, 2]

**Figure 2.70** The graph of $f(x) = (x − 2)/|x + 3|$. (Example 8)

## Applications

**Example 9 DESIGNING A BOX—REVISITED**

Dixie Packaging Company has contracted with another firm to design boxes with a volume of *at least* 600 in.³. Squares are still to be cut from the corners of a 20-in. by 25-in. piece of cardboard, with the flaps folded up to make an open box. What size squares should be cut from the cardboard? (See Example 9 of Section 2.3 and Figure 2.33.)

**Solution**

**Model**

Recall that the volume $V$ of the box is given by

$$V(x) = x(25 − 2x)(20 − 2x),$$

where $x$ represents both the side length of the removed squares and the height of the box. To obtain a volume of at least 600 in.³, we solve the inequality

$$x(25 − 2x)(20 − 2x) \geq 600.$$

**Solve Graphically**

Because the width of the cardboard is 20 in., $0 \leq x \leq 10$, and we set our window accordingly. In Figure 2.71, we find the values of $x$ for which the

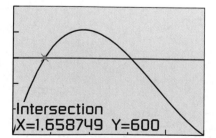

Intersection
X=1.658749  Y=600

[0, 10] by [0, 1000]

**Figure 2.71** The graphs of $y_1 = x(25 − 2x)(20 − 2x)$ and $y_2 = 600$. (Example 9)

cubic function is on or above the horizontal line. The solution is the interval $[1.66, 6.16]$.

### Interpret

Squares with side lengths between 1.66 in. and 6.16 in., inclusive, should be cut from the cardboard to produce a box with a volume of at least 600 in.$^3$.

### Example 10   DESIGNING A JUICE CAN—REVISITED

Stewart Cannery will package tomato juice in 2-liter (2000 cm$^3$) cylindrical cans. Find the radius and height of the cans if the cans have a surface area that is less than 1000 cm$^2$. (See Example 8 of Section 2.7 and Figure 2.62.)

### Solution

#### Model

Recall that the surface area $S$ is given by

$$S(r) = 2\pi r^2 + \frac{4000}{r}.$$

So the inequality to be solved is

$$2\pi r^2 + \frac{4000}{r} < 1000.$$

#### Solve Graphically

Figure 2.72 shows the graphs of $y_1 = S(r) = 2\pi r^2 + 4000/r$ and $y_2 = 1000$. Using grapher methods we find that the two curves intersect at approximately $r \approx 4.619\ldots$ and $r \approx 9.654\ldots$. (We carry all the extra decimal places for greater accuracy in a computation below.)  So the surface area is less than 1000 cm$^3$ if

$$4.62 < r < 9.65.$$

The volume of a cylindrical can is $V = \pi r^2 h$ and $V = 2000$. Using substitution we see that $h = 2000/(\pi r^2)$. To find the values for $h$ we build a double inequality for $2000/(\pi r^2)$.

| | | | |
|---|---|---|---|
| $4.62 <$ | $r$ | $< 9.65$ | Original inequality |
| $4.62^2 <$ | $r^2$ | $< 9.65^2$ | $0 < a < b \Rightarrow a^2 < b^2$. |
| $\pi \cdot 4.62^2 <$ | $\pi r^2$ | $< \pi \cdot 9.65^2$ | Multiply by $\pi$. |
| $\dfrac{1}{\pi \cdot 4.62^2} >$ | $\dfrac{1}{\pi r^2}$ | $> \dfrac{1}{\pi \cdot 9.65^2}$ | $0 < a < b \Rightarrow \dfrac{1}{a} > \dfrac{1}{b}$. |
| $\dfrac{2000}{\pi \cdot 4.62^2} >$ | $\dfrac{2000}{\pi r^2}$ | $> \dfrac{2000}{\pi \cdot 9.65^2}$ | Multiply by 2000. |
| $\dfrac{2000}{\pi (4.619\ldots)^2} >$ | $h$ | $> \dfrac{2000}{\pi (9.654\ldots)^2}$ | Use the extra decimal places now. |
| $29.83 >$ | $h$ | $> 6.83$ | Compute. |

### Interpret

The surface area of the can will be less than 1000 cm$^3$ if its radius is between 4.62 cm and 9.65 cm and its height is between 6.83 cm and 29.83 cm.

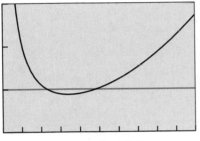

[0, 20] by [0, 3000]

**Figure 2.72** The graphs of $y_1 = 2\pi x^2 + 4000/x$ and $y_2 = 1000$. (Example 10)

---

**Follow-up**

Ask students to discuss why one must find the zeros of both the numerator and denominator to solve a rational inequality of the form $p(x)/h(x) > 0$.

**Assignment Guide**

Day 1: Ex. 3–18, multiples of 3
Day 2: Ex. 18–48, multiples of 3, 51–54, 57, 58, 60, 64

**Notes on Exercises**

Ex. 52–58 include many opportunities for developing student's problem-solving skills.

**Cooperative Learning**

Group Activity: Ex. 63

**Ongoing Assessment**

Self-Assessment: Ex. 5, 11, 16, 34, 47, 55
Embedded Assessment: Ex. 50, 58

# Quick Review 2.8

In Exercises 1–4, use limits to state the end behavior of the function.

**1.** $f(x) = 2x^3 + 3x^2 - 2x + 1$   $\lim\limits_{x \to \infty} f(x) = \infty$; $\lim\limits_{x \to -\infty} f(x) = -\infty$

**2.** $f(x) = -3x^4 - 3x^3 + x^2 - 1$

**3.** $g(x) = \dfrac{x^3 - 2x^2 + 1}{x - 2}$

**4.** $g(x) = \dfrac{2x^2 - 3x + 1}{x + 1}$

In Exercises 5–8, combine the fractions and reduce your answer to lowest terms.

**5.** $x^2 + \dfrac{5}{x}$   $(x^3 + 5)/x$

**6.** $x^2 - \dfrac{3}{x}$   $(x^3 - 3)/x$

**7.** $\dfrac{x}{2x + 1} - \dfrac{2}{x - 3}$

**8.** $\dfrac{x}{x - 1} + \dfrac{x + 1}{3x - 4}$

In Exercises 9 and 10, **(a)** list all the possible rational zeros of the polynomial and **(b)** factor the polynomial completely.

**9.** $2x^3 + x^2 - 4x - 3$

**10.** $3x^3 - x^2 - 10x + 8$

# Section 2.8 Exercises

In Exercises 1–6, determine the $x$ values that cause the polynomial function to be **(a)** zero, **(b)** positive, and **(c)** negative.

**1.** $f(x) = (x + 2)(x + 1)(x - 5)$

**2.** $f(x) = (x - 7)(3x + 1)(x + 4)$

**3.** $f(x) = (x + 7)(x + 4)(x - 6)^2$

**4.** $f(x) = (5x + 3)(x^2 + 6)(x - 1)$

**5.** $f(x) = (2x^2 + 5)(x - 8)^2(x + 1)^3$

**6.** $f(x) = (x + 2)^3(4x^2 + 1)(x - 9)^4$

In Exercises 7–12, complete the factoring if needed, and solve the polynomial inequality using a sign chart. Support graphically.

**7.** $(x + 1)(x - 3)^2 > 0$   $(-1, 3) \cup (3, \infty)$

**8.** $(2x + 1)(x - 2)(3x - 4) \le 0$   $(-\infty, -1/2] \cup [4/3, 2]$

**9.** $(x + 1)(x^2 - 3x + 2) < 0$   **10.** $(2x - 7)(x^2 - 4x + 4) > 0$

**11.** $2x^3 - 3x^2 - 11x + 6 \ge 0$   **12.** $x^3 - 4x^2 + x + 6 \le 0$

In Exercises 13–20, solve the polynomial inequality graphically.

**13.** $x^3 - x^2 - 2x \ge 0$        **14.** $2x^3 - 5x^2 + 3x < 0$

**15.** $2x^3 - 5x^2 - x + 6 > 0$   **16.** $x^3 - 4x^2 - x + 4 \le 0$

**17.** $3x^3 - 2x^2 - x + 6 \ge 0$   **18.** $-x^3 - 3x^2 - 9x + 4 < 0$

**19.** $2x^4 - 3x^3 - 6x^2 + 5x + 6 < 0$   $(3/2, 2)$

**20.** $3x^4 - 5x^3 - 12x^2 + 12x + 16 \ge 0$

In Exercises 21–28, determine the $x$ values that cause the function to be **(a)** zero, **(b)** undefined, **(c)** positive, and **(d)** negative.

**21.** $f(x) = \dfrac{x - 1}{(2x + 3)(x - 4)}$   **22.** $f(x) = \dfrac{(2x - 7)(x + 1)}{x + 5}$

**23.** $f(x) = x\sqrt{x + 3}$   **24.** $f(x) = x^2|2x + 9|$

**25.** $f(x) = \dfrac{\sqrt{x + 5}}{(2x + 1)(x - 1)}$   **26.** $f(x) = \dfrac{x - 1}{(x - 4)\sqrt{x + 2}}$

**27.** $f(x) = \dfrac{(2x + 5)\sqrt{x - 3}}{(x - 4)^2}$   **28.** $f(x) = \dfrac{3x - 1}{(x + 3)\sqrt{x - 5}}$

In Exercises 29–40, solve the inequality using a sign chart. Support graphically.

**29.** $\dfrac{x - 1}{x^2 - 4} < 0$   **30.** $\dfrac{x + 2}{x^2 - 9} < 0$

**31.** $\dfrac{x^2 - 1}{x^2 + 1} \le 0$   $[-1, 1]$   **32.** $\dfrac{x^2 - 4}{x^2 + 4} > 0$

**33.** $\dfrac{x^2 + x - 12}{x^2 - 4x + 4} > 0$   **34.** $\dfrac{x^2 + 3x - 10}{x^2 - 6x + 9} < 0$   $(-5, 2)$

**35.** $\dfrac{x^3 - x}{x^2 + 1} \ge 0$   **36.** $\dfrac{x^3 - 4x}{x^2 + 2} \le 0$

**37.** $x|x - 2| > 0$   **38.** $\dfrac{x - 3}{|x + 2|} < 0$

**39.** $(2x - 1)\sqrt{x + 4} < 0$   **40.** $(3x - 4)\sqrt{2x + 1} \ge 0$

In Exercises 41–50, solve the inequality.

**41.** $\dfrac{x^3(x - 2)}{(x + 3)^2} < 0$   $(0, 2)$   **42.** $\dfrac{(x - 5)^4}{x(x + 3)} \ge 0$

**43.** $x^2 - \dfrac{2}{x} > 0$   **44.** $x^2 + \dfrac{4}{x} \ge 0$

**45.** $\dfrac{1}{x + 1} + \dfrac{1}{x - 3} \le 0$   **46.** $\dfrac{1}{x + 2} - \dfrac{2}{x - 1} > 0$

**47.** $(x + 3)|x - 1| \ge 0$   **48.** $(3x + 5)^2|x - 2| < 0$

**49.** $\dfrac{(x - 5)|x - 2|}{\sqrt{2x - 3}} \ge 0$   **50.** $\dfrac{x^2(x - 4)^3}{\sqrt{x + 1}} < 0$

**51. Writing to Learn**   Write a paragraph that explains two ways to solve the inequality $3(x - 1) + 2 \le 5x + 6$.

**52. Company Wages**   Pederson Electric Co. charges $25 per service call plus $18 per hour for home repair work. How long did an electrician work if the charge was less than $100? Assume the electrician rounds off the time to the nearest quarter hour.

**53. Connecting Algebra and Geometry**   Consider the collection of all rectangles that have lengths 2 in. less than twice their widths. Find the possible widths (in inches) of these rectangles if their perimeters are less than 200 in.

**54. Planning for Profit**   The Grovenor Candy Co. finds that the cost of making a certain candy bar is \$0.13 per bar. Fixed costs amount to \$2000 per week. If each bar sells for \$0.35, find the minimum number of candy bars that will earn the company a profit.   9091 candy bars

**55. Designing a Cardboard Box**   Picaro's Packaging Plant wishes to design boxes with a volume of *not more than* 100 in.$^3$. Squares are to be cut from the corners of a 12-in. by 15-in. piece of cardboard (see figure), with the flaps folded up to make an open box. What size squares should be cut from the cardboard?

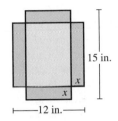

15 in.

*x*

*x*

|———12 in.———|

**56. Cone Problem**   Beginning with a circular piece of paper with a 4-inch radius, as shown in (a), cut out a sector with an arc of length *x*. Join the two radial edges of the remaining portion of the paper to form a cone with radius *r* and height *h*, as shown in (b). What length of arc will produce a cone with a volume greater than 21 in.$^3$?

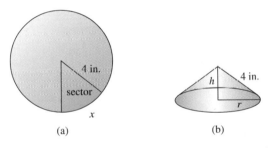

4 in.

sector

*x*

(a)

*h*

4 in.

*r*

(b)

**57. Resistors**   The total electrical resistance *R* of two resistors connected in parallel with resistances $R_1$ and $R_2$ is given by

$$\frac{1}{R} = \frac{1}{R_1} + \frac{1}{R_2}.$$

One resistor has a resistance of 2.3 ohms. Let *x* be the resistance of the second resistor.

**(a)** Express the total resistance *R* as a function of *x*.

**(b)** Find the resistance in the second resistor if the total resistance of the pair is at least 1.7 ohms.

**58. Design a Juice Can**   Flannery Cannery packs peaches in 0.5-L cylindrical cans.

**(a)** Express the surface area *S* of the can as a function of the radius *x* (in cm).   $S = 2\pi x^2 + 1000/x$

**(b)** Find the dimensions of the can if the surface is less than 900 cm$^2$.

**(c)** Find the least possible surface area of the can.

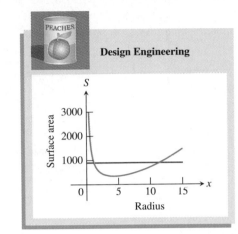

Design Engineering

*S*

Surface area

3000

2000

1000

0   5   10   15

Radius

*x*

**59. Per Capita Income**   The U.S. average per capita income for several years from 1990 to 1997 is given in Table 2.23. Let *x* = 0 represent 1990, *x* = 1 represent 1991, and so forth.

**(a)** Find the linear regression model for the data and superimpose its graph on a scatter plot of the data.

**(b)** Use the model to predict when the per capita income exceeds \$30,000.

| Table 2.23  Per Capita Income | |
| --- | --- |
| Year | Amount (dollars) |
| 1990 | 19,188 |
| 1995 | 23,359 |
| 1996 | 24,436 |
| 1997 | 25,598 |

*Source:  U.S. Bureau of the Census, Survey of Current Business, Statistical Abstract of the United States, 1998*

**60. Annual Housing Cost**   The average annual expenditure for housing for several years from 1989 to 1995 is given in Table 2.24. Let $x = 0$ represent 1980, $x = 1$ represent 1981, and so forth.

**(a)** Find the linear regression model for the data and superimpose its graph on a scatter plot of the data.

**(b)** Use the model to predict when the average annual expenditure for housing exceeds $12,000. in the year 1999

| Table 2.24  Average Annual Housing Cost | |
| --- | --- |
| Year | Amount (dollars) |
| 1989 | 8434 |
| 1990 | 8703 |
| 1991 | 9252 |
| 1992 | 9477 |
| 1993 | 9636 |
| 1994 | 10,106 |
| 1995 | 10,465 |

*Source: U.S. Bureau of the Census, Consumer Expenditures, Statistical Abstract of the United States, 1998*

## Explorations

In Exercises 61 and 62, find the vertical asymptotes and intercepts of the rational function. Then use a sign chart and a table of values to graph the function by hand. Support your result using a grapher. (Hint: You may need to graph the function in more than one window to see different parts of the overall graph.)

**61.** $f(x) = \dfrac{(x - 1)(x + 2)^2}{(x - 3)(x + 1)}$      **62.** $g(x) = \dfrac{(x - 3)^4}{x^2 + 4x}$

## Extending the Ideas

**63. Group Activity  Looking Ahead to Calculus**  Let $f(x) = 3x - 5$.

**(a)** Assume $x$ is in the interval defined by $|x - 3| < 1/3$. Give a convincing argument that $|f(x) - 4| < 1$.

**(b) Writing to Learn**  Explain how (a) is modeled by the figure at the top of the next column.

**(c)** Show how the algebra used in (a) can be modified to show that if $|x - 3| < 0.01$, then $|f(x) - 4| < 0.03$. How would the figure below change to reflect these inequalities?

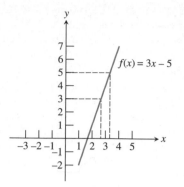

**64. Writing to Learn  Boolean Operators**  The Test menu of many graphers contains inequality symbols that can be used to construct inequality statements, as shown in (a). An answer of 1 indicates the statement is true, and 0 indicates the statement is false. In (b), the graph of $Y_1 = x^2 - 4 \geq 0$ is shown using Dot mode and the window $[-4.7, 4.7]$ by $[-3.1, 3.1]$.  Experiment with the Test menu, and then write a paragraph explaining how to interpret the graph in (b).

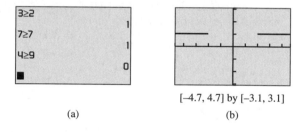

$[-4.7, 4.7]$ by $[-3.1, 3.1]$
(a)                                           (b)

In Exercises 65–66, use the properties of inequality from Chapter P to prove the statement.

**65.** If $0 < a < b$, then $a^2 < b^2$.

**66.** If $0 < a < b$, then $\dfrac{1}{a} > \dfrac{1}{b}$.

# Math at Work

I'm working on a Ph.D. in computational neuroscience. I became interested in this field after working as a software engineer. Now, I'm here, a back-to-school mom at an engineering school in Switzerland! After years of software programming, I found I had forgotten most of my mathematics, but I've found it's like any other skill—it just takes practice.

In this field we use computer simulations to explore ways of understanding what neurons do in the brain and the rest of our nervous systems. The brain contains so many neurons—and so many different kinds of them—that if we try to make a really "realistic" model of a part of the brain, not only do the simulations run very slowly on the computer, but also they can be too complicated to understand!

What we do is use very simple models for the individual neurons. We use mathematics to analytically calculate quantities such as the response of a group of neurons to a specific input signal or how much information the neurons can transmit under various conditions. Then we use computer simulations to test whether our theoretical prediction really works. When the theory matches the experiment, it's really exciting!

An equation that is commonly used in my field is that for the membrane potential of a neuron. This is the amount of current the neuron is able to store before it "fires" an action potential, which is how it communicates to other neurons, or even muscles, telling them to move. If we keep some of the quantities in the equation constant, we can solve for the others. For instance, if the input current and resistance are kept constant, we can solve for the membrane potential.

*Alix Kamakaokalani Herrmann*

---

## Chapter 2 Key Ideas

### Concepts

## Properties, Theorems, and Formulas

### Properties of the Correlation Coefficient, *r*

- $-1 \leq r \leq 1$.

- When $r > 0$, there is a positive linear correlation.

- When $r < 0$, there is a negative linear correlation.

- When $|r| \approx 1$, there is a strong linear correlation.

- When $r \approx 0$, there is weak or no linear correlation.

### Vertex Form of a Quadratic Function

Any quadratic function $f(x) = ax^2 + bx + c$, $a \neq 0$ can be written in vertex form as

$$f(x) = a(x - h)^2 + k.$$

The graph of $f$ is a parabola with vertex $(h, k)$ and axis $x = h$, where $h = -b/2a$ and $k = c - ah^2$. If $a > 0$, the parabola opens upward, and if $a < 0$, it opens downward.

### Vertical Free-Fall Motion

The height $s$ and vertical velocity $v$ of an object in free fall are given by

$$s(t) = -\frac{1}{2}gt^2 + v_0 t + s_0 \text{ and } v(t) = -gt + v_0,$$

where $t$ is time (in seconds), $g \approx 32$ ft/sec$^2 \approx 9.8$ m/sec$^2$ is the **acceleration due to gravity**, $v_0$ is the *initial vertical velocity* of the object, and $s_0$ is its *initial height*.

### Theorem Local Extrema and Zeros of Polynomial Functions

A polynomial function of degree $n$ has at most $n - 1$ local extrema and at most $n$ zeros.

### Leading Term Test for Polynomial End Behavior

For any polynomial function $f(x) = a_x x^n + \cdots + a_1 x + a_0$, $\lim\limits_{x \to \infty} f(x)$ and $\lim\limits_{x \to -\infty} f(x)$ are determined by the degree $n$ of the polynomial and its leading coefficient $a_n$:

### Zeros of Odd and Even Multiplicity

- If a polynomial function $f$ has a real zero $c$ of odd multiplicity, then the graph of $f$ crosses the $x$-axis at $(c, 0)$ and the value of $f$ changes sign at $x = c$.

- If a polynomial function $f$ has a real zero $c$ of even multiplicity, then the graph of $f$ does not cross the $x$-axis at $(c, 0)$ and the value of $f$ does not change sign at $x = c$.

### Intermediate Value Theorem

If $a$ and $b$ are real numbers with $a < b$ and if $f$ is continuous on the interval $[a, b]$, then $f$ takes on every value between $f(a)$ and $f(b)$. In other words, if $y_0$ is between $f(a)$ and $f(b)$, then $y_0 = f(c)$ for some number $c$ in $[a, b]$. In particular, if $f(a)$ and $f(b)$ have opposite signs (i.e., one is negative and the other is positive), then $f(c) = 0$ for some number $c$ in $[a, b]$.

## Division Algorithm for Polynomials

Let $f(x)$ and $d(x)$ be polynomials with the degree of $f$ greater than or equal to the degree of $d$, and $d(x) \neq 0$. Then there are unique polynomials $q(x)$ and $r(x)$, called the **quotient** and **remainder**, such that

$$f(x) = d(x) \cdot q(x) + r(x)$$

where either $r(x) = 0$ or the degree of $r$ is less than the degree of $d$.

## Remainder Theorem

If a polynomial $f(x)$ is divided by $x - k$, then the remainder is $r = f(k)$. **10Factor Theorem**

A polynomial function $f(x)$ has a factor $x - k$ if and only if $f(k) = 0$.

## Factor Theorem

A polynomial function $f(x)$ has a factor $x - k$ if and only if $f(k) = 0$.

## Fundamental Connections for Polynomial Functions

For a polynomial function $f$ and a real number $k$, the following statements are equivalent:

1. $x = k$ is a solution (or root) of the equation $f(x) = 0$.
2. $k$ is a zero of the function $f$.
3. $k$ is an $x$-intercept of the graph of $y = f(x)$.
4. $x - k$ is a factor of $f(x)$.

## Rational Zeros Theorem

Suppose $f$ is a polynomial function of degree $n \geq 1$ of the form

$$f(x) = a_n x^n + a_{n-1} x^{n-1} + \cdots + a_0,$$

with every coefficient an integer and $a_0 \neq 0$. In that case, if $x = p/q$ is a rational zero of $f$, where $p$ and $q$ have no common integer factors greater than 1, then

1. $p$ is an integer factor of the constant coefficient $a_0$, and

2. $q$ is an integer factor of the leading coefficient $a_n$.

## Fundamental Theorem of Algebra

A polynomial function of degree $n > 0$ has $n$ complex zeros. These zeros may be repeated.

## Linear Factorization Theorem

If $f(x)$ is a polynomial function of degree $n > 0$, then $f(x)$ has precisely $n$ linear factors and

$$f(x) = a(x - z_1)(x - z_2) \cdots (x - z_n)$$

where $a$ is the leading coefficient of $f(x)$ and $z_1, z_2, \ldots, z_n$ are the complex zeros of $f(x)$. The $z_i$ are not necessarily distinct numbers; some may be repeated.

## Fundamental Polynomial Connections in the Complex Case

The following statements about a polynomial function $f$ are equivalent even if $k$ is a nonreal complex number:

1. $x = k$ is a solution (or root) of the equation $f(x) = 0$.
2. $k$ is a zero of the function $f$.
3. $x - k$ is a factor of $f(x)$.

## Complex Conjugate Zeros

Suppose that $f(x)$ is a polynomial function with *real coefficients*. If $a$ and $b$ are real numbers with $b \neq 0$ and $a + bi$ is a zero of $f(x)$, then its complex conjugate $a - bi$ is also a zero of $f(x)$.

## Factors of a Polynomial with Real Coefficients

Every polynomial function with real coefficients can be written as a product of linear factors and irreducible quadratic factors, each with real coefficients.

## Polynomial Function of Odd Degree

Every polynomial function of odd degree with real coefficients has at least one real zero.

## Procedures

### Regression Analysis

1. Enter and plot the data (scatter plot).

2. Find the regression model that fits the problem situation.

3. Superimpose the graph of the regression model on the scatter plot, and observe the fit.

4. Use the regression model to make the predictions called for in the problem.

### Completing the Square

The algebraic process of changing a quadratic function from standard form to vertex form.

### Polynomial Long Division

A method of dividing polynomials that closely resembles whole-number long division.

### Synthetic Division

A short-cut method for the division of a polynomial by a linear divisor $x - k$.

### Solving Inequalities Using Sign Charts

Determine the values of $f(x)$ that cause $f(x)$ to be (a) zero, (b) undefined. Then make a sign chart to determine the values of $x$ that cause $f(x)$ to be (c) positive, (d) negative.

## Gallery of Functions

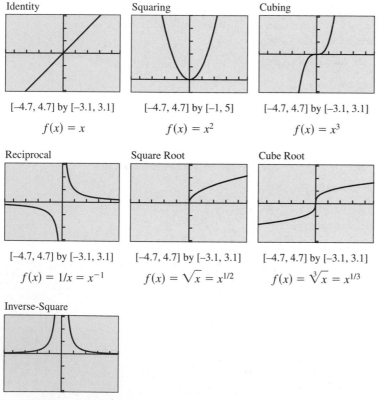

Identity

$[-4.7, 4.7]$ by $[-3.1, 3.1]$

$$f(x) = x$$

Squaring

$[-4.7, 4.7]$ by $[-1, 5]$

$$f(x) = x^2$$

Cubing

$[-4.7, 4.7]$ by $[-3.1, 3.1]$

$$f(x) = x^3$$

Reciprocal

$[-4.7, 4.7]$ by $[-3.1, 3.1]$

$$f(x) = 1/x = x^{-1}$$

Square Root

$[-4.7, 4.7]$ by $[-3.1, 3.1]$

$$f(x) = \sqrt{x} = x^{1/2}$$

Cube Root

$[-4.7, 4.7]$ by $[-3.1, 3.1]$

$$f(x) = \sqrt[3]{x} = x^{1/3}$$

Inverse-Square

$[-4.7, 4.7]$ by $[-3.1, 3.1]$

$$f(x) = 1/x^2 = x^{-2}$$

## Chapter 2    Review Exercises

The collection of exercises marked in red could be used as a chapter test.

In Exercises 1 and 2, write an equation for the linear function $f$ satisfying the given conditions. Graph $y = f(x)$.

**1.** $f(-3) = -2$ and $f(4) = -9$    **2.** $f(-3) = 6$ and $f(1) = -2$

In Exercises 3 and 4, describe how to transform the graph of $f(x) = x^2$ into the graph of the given function. Sketch the graph by hand and support your answer with a grapher.

**3.** $f(x) = 3(x - 2)^2 + 4$    **4.** $g(x) = -(x + 3)^2 + 1$

In Exercises 5–8, find the vertex and axis of the graph of the function. Support your answer graphically.

**5.** $f(x) = -2(x + 3)^2 + 5$    **6.** $g(x) = 4(x - 5)^2 - 7$

**7.** $f(x) = -2x^2 - 16x - 31$    **8.** $g(x) = 3x^2 - 6x + 2$

In Exercises 9 and 10, write an equation for the quadratic function whose graph contains the given vertex and point.

**9.** Vertex $(-2, -3)$, point $(1, 2)$    $y = 5/9(x + 2)^2 - 3$

**10.** Vertex $(-1, 1)$, point $(3, -2)$    $y = -3/16(x + 1)^2 + 1$

In Exercises 11 and 12, write an equation for the quadratic function with graph shown, given one of the labeled points is the vertex of the parabola.

**11.** 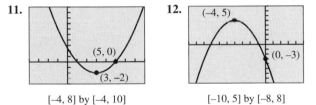    **12.**

[–4, 8] by [–4, 10]            [–10, 5] by [–8, 8]

In Exercises 13–16, graph the function in a viewing window that shows all of its extrema and x-intercepts.

**13.** $f(x) = x^2 + 3x - 40$    **14.** $f(x) = -8x^2 + 16x - 19$

**15.** $f(x) = x^3 + x^2 + x + 5$    **16.** $f(x) = x^3 - x^2 - 20x - 2$

In Exercises 17 and 18, write the statement as a power function equation.

**17.** The surface area $S$ of a sphere varies directly as the square of the radius $r$.    $S = kr^2$

**18.** The force of gravity $F$ acting on an object is inversely proportional to the square of the distance $d$ from the object to the center of the earth.    $F = k/d^2$

In Exercises 19 and 20, write a sentence that expresses the relationship in the formula, using the language of variation or proportion.

**19.** $F = kx$, where $F$ is the force it takes to stretch a spring $x$ units from its unstressed length and $k$ is the spring's force constant.

**20.** $A = \pi \cdot r^2$, where $A$ and $r$ are the area and radius of a circle and $\pi$ is the usual mathematical constant.

In Exercises 21–24, state the values of the constants $k$ and $a$ for the function $f(x) = k \cdot x^a$. Describe the portion of the curve that lies in Quadrant I or IV. Determine whether $f$ is even, odd, or undefined for $x < 0$. Describe the rest of the curve if any. Graph the function to see whether it matches the description.

**21.** $f(x) = 4x^{1/3}$    **22.** $f(x) = -2x^{3/4}$

**23.** $f(x) = -2x^{-3}$    **24.** $f(x) = (2/3)x^{-4}$

In Exercises 25–28, divide $f(x)$ by $d(x)$, and write a summary statement in polynomial form.

**25.** $f(x) = 2x^3 - 7x^2 + 4x - 5$; $d(x) = x - 3$

**26.** $f(x) = x^4 + 3x^3 + x^2 - 3x + 3$; $d(x) = x + 2$

**27.** $f(x) = 2x^4 - 3x^3 + 9x^2 - 14x + 7$; $d(x) = x^2 + 4$

**28.** $f(x) = 3x^4 - 5x^3 - 2x^2 + 3x - 6$; $d(x) = 3x + 1$

In Exercises 29 and 30, use the Remainder Theorem to find the remainder when $f(x)$ is divided by $x - k$. Check by using synthetic division.

**29.** $f(x) = 3x^3 - 2x^2 + x - 5$; $k = -2$

**30.** $f(x) = -x^2 + 4x - 5$; $k = 3$

In Exercises 31 and 32, use the Factor Theorem to determine whether the first polynomial is a factor of the second polynomial.

**31.** $x - 2$; $x^3 - 4x^2 + 8x - 8$    **32.** $x + 3$; $x^3 + 2x^2 - 4x - 2$

In Exercises 33 and 34, use synthetic division to prove that the number $k$ is an upper bound for the real zeros of the function $f$.

**33.** $k = 5$; $f(x) = x^3 - 5x^2 + 3x + 4$

**34.** $k = 4$; $f(x) = 4x^4 - 16x^3 + 8x^2 + 16x - 12$

In Exercises 35 and 36, use synthetic division to prove that the number $k$ is a lower bound for the real zeros of the function $f$.

**35.** $k = -3$; $f(x) = 4x^4 + 4x^3 - 15x^2 - 17x - 2$

**36.** $k = -3$; $f(x) = 2x^3 + 6x^2 + x - 6$

In Exercises 37 and 38, use the Rational Zeros Theorem to write a list of all potential rational zeros. Then determine which ones, if any, are zeros.

**37.** $f(x) = 2x^4 - x^3 - 4x^2 - x - 6$

**38.** $f(x) = 6x^3 - 20x^2 + 11x + 7$

In Exercises 39–46, perform the indicated operation, and write the result in the standard form $a + bi$.

**39.** $(3 - 2i) + (-2 + 5i)$    **40.** $(5 - 7i) - (3 - 2i)$

**41.** $(1 + 2i)(3 - 2i)$  $7 + 4i$    **42.** $(1 + i)^3$  $-2 + 2i$

**43.** $(1 + 2i)^2(1 - 2i)^2$  25

**44.** $i^{29}$  $i$

**45.** $\sqrt{-16}$  $4i$

**46.** $\dfrac{2 + 3i}{1 - 5i}$  $-1/2 + 1/2i$

In Exercises 47 and 48, solve the equation.

**47.** $x^2 - 6x + 13 = 0$  $3 \pm 2i$  **48.** $x^2 - 2x + 4 = 0$

In Exercises 49–52, match the polynomial function with its graph. Explain your choice.

**49.** $f(x) = (x - 2)^2$  (c)

**50.** $f(x) = (x - 2)^3$  (d)

**51.** $f(x) = (x - 2)^4$  (b)

**52.** $f(x) = (x - 2)^5$  (a)

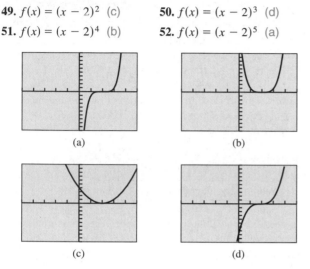

(a)

(b)

(c)

(d)

In Exercises 53–56, find all of the real zeros of the function, finding exact values whenever possible. Identify each zero as rational or irrational. State the number of nonreal complex zeros.

**53.** $f(x) = x^4 - 10x^3 + 23x^2$  **54.** $k(t) = t^4 - 7t^2 + 12$

**55.** $h(x) = x^3 - 2x^2 - 8x + 5$

**56.** $k(x) = x^4 - x^3 - 14x^2 + 24x + 5$

In Exercises 57–60, find all of the zeros and write a linear factorization of the function.

**57.** $f(x) = 2x^3 - 9x^2 + 2x + 30$

**58.** $f(x) = 5x^3 - 24x^2 + x + 12$

**59.** $f(x) = 6x^4 + 11x^3 - 16x^2 - 11x + 10$

**60.** $f(x) = x^4 - 8x^3 + 27x^2 - 50x + 50$, given that $1 + 2i$ is a zero.

In Exercises 61–64, write the function as a product of linear and irreducible quadratic factors all with real coefficients.

**61.** $f(x) = x^3 - x^2 - x - 2$

**62.** $f(x) = 9x^3 - 3x^2 - 13x - 1$

**63.** $f(x) = 2x^4 - 9x^3 + 23x^2 - 31x + 15$

**64.** $f(x) = 3x^4 - 7x^3 - 3x^2 + 17x + 10$

In Exercises 65–70, write a polynomial function with real coefficients whose zeros and their multiplicities include those listed.

**65.** Degree 3; zeros: $\sqrt{5}$, $-\sqrt{5}$, 3  $x^3 - 3x^2 - 5x + 15$

**66.** Degree 2; $-3$ only real zero  $x^2 + 6x + 9$

**67.** Degree 4; zeros: 3, $-2$, 1/3, $-1/2$

**68.** Degree 3; zeros: $1 + i$, 2  $x^3 - 4x^2 + 6x - 4$

**69.** Degree 4; zeros: $-2$(multiplicity 2), 4(multiplicity 2)

**70.** Degree 3; zeros: $2 - i$, $-1$, and $f(2) = 6$

In Exercises 71 and 72, describe how the graph of the given function can be obtained by transforming the graph of the reciprocal function $f(x) = 1/x$. Identify the horizontal and vertical asymptotes.

**71.** $f(x) = \dfrac{-x + 7}{x - 5}$

**72.** $f(x) = \dfrac{3x + 5}{x + 2}$

In Exercises 73–76, find the asymptotes and intercepts of the function, and graph it.

**73.** $f(x) = \dfrac{x^2 + x + 1}{x^2 - 1}$

**74.** $f(x) = \dfrac{2x^2 + 7}{x^2 + x - 6}$

**75.** $f(x) = \dfrac{x^2 - 4x + 5}{x + 3}$

**76.** $g(x) = \dfrac{x^2 - 3x - 7}{x + 3}$

In Exercises 77–84, solve the equation or inequality algebraically, and support graphically.

**77.** $2x + \dfrac{12}{x} = 11$  $x = \dfrac{3}{2}$ or $x = 4$

**78.** $\dfrac{x}{x + 2} + \dfrac{5}{x - 3} = \dfrac{25}{x^2 - x - 6}$  $x = -5$

**79.** $2x^3 + 3x^2 - 17x - 30 < 0$  $(-\infty, 5/2) \cup (-2, 3)$

**80.** $3x^4 + x^3 - 36x^2 + 36x + 16 \geq 0$  $(-\infty, -4] \cup [-1/3, \infty)$

**81.** $\dfrac{x + 3}{x^2 - 4} \geq 0$

**82.** $\dfrac{x^2 - 7}{x^2 - x - 6} < 1$

**83.** $(2x - 1)^2|x + 3| \leq 0$

**84.** $\dfrac{(x - 1)^2|x - 4|}{\sqrt{x + 3}} > 0$

**85.** Plot $-3 - 2i$ in the complex plane.

**86. Writing to Learn** Determine whether

$$f(x) = x^5 - 10x^4 - 3x^3 + 28x^2 + 20x - 2$$

has a zero outside the viewing window. Explain.

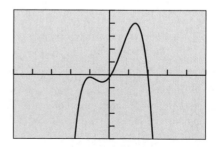

[–5, 5] by [–50, 50]

**87. Launching a Rock** Larry uses a slingshot to launch a rock straight up from a point 6 ft above level ground with an initial velocity of 170 ft/sec.

(a) Find an equation that models the height of the rock $t$ seconds after it is launched and graph the equation. (See Example 7 in Section 2.1.)

(b) What is the maximum height of the rock? When will it reach that height?

(c) When will the rock hit the ground?

**88. Volume of a Box** Edgardo Paper Co. has contracted to manufacture a box with no top that is to be made by removing squares of width $x$ from the corners of a 30-in. by 70-in. piece of cardboard.

(a) Find an equation that models the volume of the box.

(b) Determine $x$ so that the box has a volume of 5800 in.$^3$.

**89. Architectural Engineering** Donoma, an engineer at J. P. Cook, Inc., completes structural specifications for a 255-ft-long steel beam anchored between two pilings 50 ft above ground, as shown in the figure. She knows that when a 250-lb object is placed $d$ feet from the west piling, the beam bends $s$ feet where

$$s = (8.5 \times 10^{-7})d^2(255 - d).$$

(a) Graph the function $s$.

(b) What are the dimensions of a viewing window that shows a graph for the values that make sense in this problem situation?

(c) What is the greatest amount of vertical deflection $s$, and where does it occur?

(d) **Writing to Learn** Give a possible scenario explaining why the solution to (c) does not occur at the halfway point.

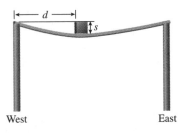

West                    East

**90. Storage Container** A liquid storage container on a truck is in the shape of a cylinder with hemispheres on each end as shown in the figure. The cylinder and hemispheres have the same radius. The total length of the container is 140 ft.

(a) Determine the volume $V$ of the container as a function of the radius $x$.

(b) Graph the function $y = V(x)$.

(c) What is the radius of the container with the largest possible volume? What is the volume?

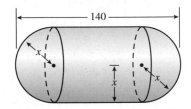

**91. Dow Strategy** A *Dog* stock is one of the 10 Dow stocks with highest dividend yield at the end of a year compared with their stock price. Similarly, a *Star* stock is one of the 10 Dow stocks with lowest dividend yield at the end of a year compared with their stock price. Table 2.25 shows the annual performance for the following year for Dog and Star stocks for several years. Let $x = 0$ represent 1990, $x = 1$ represent 1991, and so forth.

| Table 2.25  Dog and Star Dow Stock Performace | | |
|---|---|---|
| Year | Dog Performance (%) | Star Performance (%) |
| 1996 | 19.8 | 27.9 |
| 1997 | 20.4 | 6.2 |
| 1998 | 9.5 | 17.5 |
| 1999 | −2.4 | 22.4 |
| 2000 | −1.6 | 1.7 |

*Source: www.dogsofthedow.com as reported in the USA Today, February 21, 2000.*

(a) Find a cubic regression model for the Dog stocks, and graph it together with a scatter plot of the Dog data.

(b) Find a cubic regression model for the Star stocks, and graph it together with a scatter plot of the Star data.

(c) **Writing to Learn** Determine the end behavior (lim) of the two regression models. What do the end behavior models say about using the Dog or Star investment strategy.

**92. National Institute of Health Spending** Table 2.26 shows the spending at the National Institute of Health for several years. Let $x = 0$ represent 1990, $x = 1$ represent 1991, and so forth.

**Table 2.26 Spending at the National Institute of Health**

| Year | Amount (billion) |
|------|------------------|
| 1993 | 10.32 |
| 1994 | 10.95 |
| 1995 | 11.30 |
| 1996 | 11.93 |
| 1997 | 12.74 |
| 1998 | 13.64 |
| 1999 | 15.65 |
| 2000 | 17.91 |

*Source: National Institute of Health as reported in The Chronicle of Higher Education November 26, 1999.*

**(a)** Find a linear regression model, and graph it together with a scatter plot of the data.

**(b)** Find a quadratic regression model, and graph it together with a scatter plot of the data.

**(c)** Use the linear and quadratic regression models to estimate when the amount of spending will exceed $20 billion.

**93. Breaking Even** Midtown Sporting Goods has determined that it needs to sell its soccer shinguards for $5.25 a pair in order to be competitive. It costs $4.32 to produce each pair of shinguards, and the weekly overhead cost is $4000.

**(a)** Express the average cost that includes the overhead of producing one shinguard as a function of the number $x$ of shinguards produced each week.

**(b)** Solve algebraically to find the number of shinguards that must be sold each week to make $8000 in profit. Support your work graphically.

**94. Deer Population** The number of deer $P$ at any time $t$ (in years) in a federal game reserve is given by

$$P(t) = \frac{800 + 640t}{20 + 0.8t}.$$

**(a)** Find the number of deer when $t$ is 15, 70, and 100.

**(b)** Find the horizontal asymptote of the graph of $y = P(t)$.

**(c)** According to the model, what is the largest possible deer population?

**95. Resistors** The total electrical resistance $R$ of two resistors connected in parallel with resistances $R_1$ and $R_2$ is given by

$$\frac{1}{R} = \frac{1}{R_1} + \frac{1}{R_2}.$$

The total resistance is 1.2 ohms. Let $x = R_1$.

**(a)** Express the second resistance $R_2$ as a function of $x$.

**(b)** Find $R_2$ if $R_1$ is 3 ohms.

**96. Acid Mixture** Suppose that $x$ ounces of distilled water are added to 50 oz of pure acid.

**(a)** Express the concentration $C(x)$ of the new mixture as a function of $x$.

**(b)** Use a graph to determine how much distilled water should be added to the pure acid to produce a new solution that is less than 60% acid.

**(c)** Solve (b) algebraically.

**97. Industrial Design** Johnson Cannery will pack peaches in 1-L cylindrical cans. Let $x$ be the radius of the base of the can in centimeters.

**(a)** Express the surface area $S$ of the can as a function of $x$.

**(b)** Find the radius and height of the can if the surface area is 900 cm².

**(c)** What dimensions are possible for the can if the surface area is to be less than 900 cm²?

**98. Industrial Design** Gilman Construction is hired to build a rectangular tank with a square base and no top. The tank is to hold 1000 ft³ of water. Let $x$ be a length of the base.

**(a)** Express the outside surface area $S$ of the tank as a function of $x$.

**(b)** Find the length, width, and height of the tank if the outside surface area is 600 ft².

**(c)** What dimensions are possible for the tank if the outside surface area is to be less than 600 ft²?

# Chapter 2 Project

## Modeling the Height of a Bouncing Ball

When a ball is bouncing up and down on a flat surface, its height with respect to time can be modeled using a quadratic equation. One form of a quadratic equation is the vertex form:

$$y = a(x - h)^2 + k$$

In this equation, $y$ represents the height of the ball and $x$ represents the total elapsed time. For this project, you will use a motion detection device to collect distance and time data for a bouncing ball, then find a mathematical model that describes the position of the ball.

## Explorations

| Total elapsed time (seconds) | Height of the ball (meters) | Total elapsed time (seconds) | Height of the ball (meters) |
|---|---|---|---|
| 0.688 | 0 | 1.118 | 0.828 |
| 0.731 | 0.155 | 1.161 | 0.811 |
| 0.774 | 0.309 | 1.204 | 0.776 |
| 0.817 | 0.441 | 1.247 | 0.721 |
| 0.860 | 0.553 | 1.290 | 0.650 |
| 0.903 | 0.643 | 1.333 | 0.563 |
| 0.946 | 0.716 | 1.376 | 0.452 |
| 0.989 | 0.773 | 1.419 | 0.322 |
| 1.032 | 0.809 | 1.462 | 0.169 |
| 1.075 | 0.828 | | |

1. If you collected motion data using a CBL or CBR, a plot of height versus time or distance versus time should be shown on your grapher or computer screen. Either plot will work for this project. If you do not have access to a CBL/CBR, enter the data from the table above into your grapher/computer. Create a scatter plot for the data.

2. Find values for $a$, $h$, and $k$ so that the equation $y = a(x - h)^2 + k$ fits one of the bounces contained in the data plot. Approximate the vertex $(h, k)$ from your data plot and solve for the value of $a$ algebraically.

3. Change the values of $a$, $h$, and $k$ in the model found above and observe how the graph of the function is affected on your grapher or computer. Generalize how each of these changes affects the graph.

4. Expand the equation you found in #2 above so that it is in the standard quadratic form: $y = ax^2 + bx + c$.

5. Use your grapher or computer to select the data from the bounce you modeled above and then use a quadratic regression to find a model for this data set. (See your grapher's guidebook for instructions on how to do this.) How does this model compare with the standard quadratic form found in #4?

6. Complete the square to transform the regression model to the vertex form of a quadratic and compare it to the original vertex model found in #2. (Round the values of $a$, $b$, and $c$ to the nearest 0.001 before completing the square if desired.)

**Definition** Exponential Functions

Let $a$ and $b$ be real number constants. An **exponential function** in $x$ is a function that can be written in the form

$$f(x) = a \cdot b^x,$$

where $a$ is nonzero, $b$ is positive, and $b \neq 1$. The constant $a$ is the *initial value* of $f$ (the value at $x = 0$), and $b$ is the **base**.

Exponential functions are defined and continuous for all real numbers. It is important to recognize whether a function is an exponential function.

**Example 1** IDENTIFYING EXPONENTIAL FUNCTIONS

Which of the following are exponential functions? For those that are exponential functions, state the initial value and the base. For those that are not, explain why not.

(a) $f(x) = 3^x$      (b) $g(x) = 6x^{-4}$      (c) $h(x) = -2 \cdot 1.5^x$

(d) $k(x) = 7 \cdot 2^{-x}$      (e) $p(x) = \sqrt{7^x}$      (f) $q(x) = 5 \cdot 6^\pi$

**Solution**

(a) $f$ is an exponential function, with an initial value of 1 and base of 3.

(b) $g$ is not an exponential function because the base $x$ is a variable and the exponent is a constant; $g$ is a power function.

(c) $h$ is an exponential function, with an initial value of $-2$ and base of 1.5.

(d) $k$ is an exponential function, with an initial value of 7 and base of 1/2 because $2^{-x} = (2^{-1})^x = (1/2)^x$.

(e) $p$ is an exponential function, with an initial value of 1 and base of $\sqrt{7}$ because $\sqrt{7^x} = (7^x)^{1/2} = (7^{1/2})^x = (\sqrt{7})^x$.

(f) $q$ is not an exponential function because the exponent $\pi$ is a constant; $q$ is a constant function.

One way to evaluate an exponential function involving rational number exponents is using the properties of exponents.

**Example 2** COMPUTING EXPONENTIAL FUNCTION VALUES FOR RATIONAL NUMBER INPUTS

For $f(x) = 2^x$,

(a) $f(4) = 2^4 = 2 \cdot 2 \cdot 2 \cdot 2 = 16$.

(b) $f(0) = 2^0 = 1$

(c) $f(-3) = 2^{-3} = \dfrac{1}{2^3} = \dfrac{1}{8} = 0.125$

(d) $f\left(\dfrac{1}{2}\right) = 2^{1/2} = \sqrt{2} = 1.4142\ldots$

(e) $f\left(-\dfrac{3}{2}\right) = 2^{-3/2} = \dfrac{1}{2^{3/2}} = \dfrac{1}{\sqrt{2^3}} = \dfrac{1}{\sqrt{8}} = 0.35355\ldots$

There is no way to use properties of exponents to express an exponential function's value for *irrational* inputs. For example, if $f(x) = 2^x$, $f(\pi) = 2^\pi$, but what does $2^\pi$ mean? Using properties of exponents, $2^3 = 2 \cdot 2 \cdot 2$, $2^{3.1} = 2^{31/10} = \sqrt[10]{2^{31}}$. So we can find meaning for $2^\pi$ by using successively closer *rational* approximations to $\pi$ as shown in Table 3.1

| | | | | | | |
|---|---|---|---|---|---|---|
| **Table 3.1** Values of $f(x) = 2^x$ for Rational Numbers $x$ Approaching $\pi = 3.14159265\ldots$ | | | | | | |
| $x$ | 3 | 3.1 | 3.14 | 3.141 | 3.1415 | 3.14159 |
| $2^x$ | 8 | $8.5\ldots$ | $8.81\ldots$ | $8.821\ldots$ | $8.8244\ldots$ | $8.82496\ldots$ |

We can conclude that $f(\pi) = 2^\pi \approx 8.82$, which could also be found directly using a grapher. The methods of calculus permit a more rigorous definition of exponential functions than we give here. For now, we accept without proof that exponential functions are defined and continuous for all real numbers.

A characteristic of exponential functions that makes them useful in applications is the pattern by which their value changes as their input changes by a fixed amount. This pattern can best be observed in tabular form.

### Example 3 FINDING AN EXPONENTIAL FUNCTION FROM ITS TABLE OF VALUES

Determine formulas for the exponential functions $g$ and $h$ whose values are given in Table 3.2.

| | | |
|---|---|---|
| **Table 3.2** Values for Two Exponential Functions | | |
| $x$ | $g(x)$ | $h(x)$ |
| $-2$ | $4/9$ | 128 |
| $-1$ | $4/3$ | 32 |
| 0 | 4 | 8 |
| 1 | 12 | 2 |
| 2 | 36 | $1/2$ |

**Solution** Because $g$ is exponential, $g(x) = a \cdot b^x$. Because $g(0) = 4$, the initial value $a$ is 4. Because $g(1) = 4 \cdot b^1 = 12$, the base $b$ is 3. So,

$$g(x) = 4 \cdot 3^x.$$

Because $h$ is exponential, $h(x) = a \cdot b^x$. Because $h(0) = 8$, the initial value $a$ is 8. Because $h(1) = 8 \cdot b^1 = 2$, the base $b$ is 1/4. So,

$$h(x) = 8 \cdot \left(\frac{1}{4}\right)^x.$$

Figure 3.2 shows the graphs of these functions pass through the points whose coordinates are given in Table 3.2.

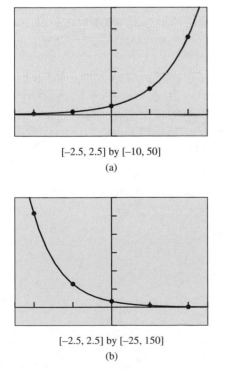

[−2.5, 2.5] by [−10, 50]

(a)

[−2.5, 2.5] by [−25, 150]

(b)

**Figure 3.2** Graphs of (a) $g(x) = 4 \cdot 3^x$ and (b) $h(x) = 8 \cdot (1/4)^x$. (Example 3)

Observe the patterns in the $g(x)$ and $h(x)$ columns of Table 3.2. The $g(x)$ values increase by a factor of 3 and the $h(x)$ values decrease by a factor of 1/4, as we add 1 to $x$ moving from one row of the table to the next. In each case, the change factor is the base of the exponential function. This pattern generalizes to all exponential functions as illustrated in Table 3.3.

**Table 3.3**  Values for a General Exponential Function $f(x) = a \cdot b^x$

| $x$ | $a \cdot b^x$ | |
|---|---|---|
| 0 | $a$ | $\rbrace \times b$ |
| 1 | $ab$ | $\rbrace \times b$ |
| 2 | $ab^2$ | $\rbrace \times b$ |
| 3 | $ab^3$ | $\rbrace \times b$ |
| 4 | $ab^4$ | |

In Table 3.3, as $x$ increases by 1, the function value is multiplied by the base $b$. So the base $b$ determines how an exponential function changes.

**How an Exponential Function Changes**

For any exponential function $f(x) = a \cdot b^x$ and any real number $x$,

$$f(x + 1) = b \cdot f(x).$$

If $a > 0$ and $b > 1$, the function $f$ is increasing and is an **exponential growth function.** The base $b$ is its **growth factor.**

If $a > 0$ and $b < 1$, $f$ is decreasing and is an **exponential decay function.** The base $b$ is its **decay factor.**

In Example 3 and Table 3.2, $g$ is an exponential growth function, and $h$ is an exponential decay function. As $x$ increases by 1, $g(x) = 4 \cdot 3^x$ grows by a factor of 3, and $h(x) = 8 \cdot (1/4)^x$ decays by a factor of 1/4.

The base of an exponential function is analogous to the slope of a linear function. Each tells us whether the function is increasing or decreasing and by how much.

So far, we have focused most of our attention on the algebraic and numerical aspects of exponential functions. We now turn our attention to the graphs of these functions.

> **Exploration 1  Graphs of Exponential Functions**
>
> 1. Graph each function in the viewing window $[-2, 2]$ by $[-1, 6]$.
>
>    **(a)** $y_1 = 2^x$   **(b)** $y_2 = 3^x$   **(c)** $y_3 = 4^x$   **(d)** $y_4 = 5^x$
>
>    • Which point is common to all four graphs?
>    • Analyze the functions for domain, range, continuity, intercepts, increasing or decreasing behavior, symmetry, boundedness, extrema, asymptotes, end behavior, and concavity.
>
> 2. Graph each function in the viewing window $[-2, 2]$ by $[-1, 6]$.
>
>    **(a)** $y_1 = \left(\dfrac{1}{2}\right)^x$   **(b)** $y_2 = \left(\dfrac{1}{3}\right)^x$   **(c)** $y_3 = \left(\dfrac{1}{4}\right)^x$   **(d)** $y_4 = \left(\dfrac{1}{5}\right)^x$
>
>    • Which point is common to all four graphs?
>    • Analyze the functions for domain, range, continuity, intercepts, increasing or decreasing behavior, symmetry, boundedness, extrema, asymptotes, end behavior, and concavity.

We summarize what we have learned about exponential functions with an initial value of 1.

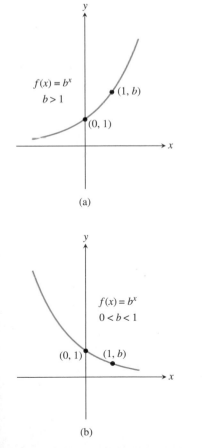

**Characteristics of Basic Exponential Functions**

For any exponential function in the form $f(x) = b^x$,

Domain: All reals.    Range: $(0, \infty)$.
$x$-intercepts: none.    $y$-intercept $= 1$.
Continuous.
No symmetry: neither even nor odd.
Bounded below, but not above.
No local extrema.
Horizontal asymptote: $y = 0$.
No vertical asymptotes.
Passes through $(0, 1)$ and $(1, b)$.
One-to-one.
Concave up.

If $b > 1$ (see Figure 3.3a), then also

• $f$ is an increasing function,
• $\lim\limits_{x \to -\infty} f(x) = 0$ and $\lim\limits_{x \to \infty} f(x) = \infty$.

If $0 < b < 1$ (see Figure 3.3b), then also

• $f$ is a decreasing function,
• $\lim\limits_{x \to -\infty} f(x) = \infty$ and $\lim\limits_{x \to \infty} f(x) = 0$.

The translations, reflections, stretches, and shrinks studied in Section 1.5 together with our knowledge of the graphs of basic exponential functions allow us to predict the graphs of the functions in Example 4.

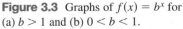

**Figure 3.3** Graphs of $f(x) = b^x$ for (a) $b > 1$ and (b) $0 < b < 1$.

**Example 4    TRANSFORMING EXPONENTIAL FUNCTIONS**

Describe how to transform the graph of $f(x) = 2^x$ into the graph of the given function. Sketch the graph by hand and support your answer with a grapher.

**(a)** $g(x) = 2^{x-1}$          **(b)** $h(x) = 2^{-x}$

**(c)** $k(x) = 3 \cdot 2^x$          **(d)** $q(x) = 2^x - 1$

**Solution**

**(a)** The graph of $g(x) = 2^{x-1}$ is obtained by translating the graph of $f(x) = 2^x$ by 1 unit to the right (Figure 3.4a). Because $g(x) = 2^{x-1} = 2^x \cdot 2^{-1} = 1/2 \cdot 2^x$, it can also be obtained from $f(x) = 2^x$ using a vertical shrink by a factor of $1/2$.

**(b)** We can obtain the graph of $h(x) = 2^{-x}$ by reflecting the graph of $f(x) = 2^x$ across the $y$-axis (Figure 3.4b). Because $2^{-x} = (2^{-1})^x = (1/2)^x$, we can also think of $h$ as an exponential function with an initial value of 1 and a base of $1/2$.

**(c)** We can obtain the graph of $k(x) = 3 \cdot 2^x$ by vertically stretching the graph of $f(x) = 2^x$ by a factor of 3 (Figure 3.4c).

**(d)** The graph of $q(x) = 2^x - 1$ is obtained by shifting the graph of $f(x) = 2^x$ by 1 unit down (Figure 3.4d).

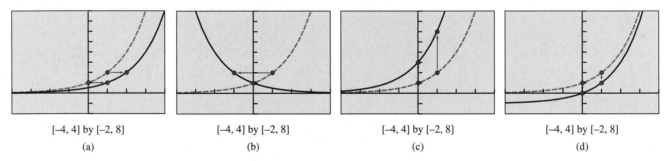

| $[-4, 4]$ by $[-2, 8]$ | $[-4, 4]$ by $[-2, 8]$ | $[-4, 4]$ by $[-2, 8]$ | $[-4, 4]$ by $[-2, 8]$ |
| :---: | :---: | :---: | :---: |
| (a) | (b) | (c) | (d) |

**Figure 3.4** The graph of $f(x) = 2^x$ shown with (a) $g(x) = 2^{x-1}$, (b) $h(x) = 2^{-x}$, (c) $k(x) = 3 \cdot 2^x$, and (d) $q(x) = 2^x - 1$. (Example 4)

## The Natural Base e

The function $f(x) = e^x$ is one of the basic functions introduced in Section 1.3, and is an exponential growth function.

**The Exponential Function**

$f(x) = e^x$

Domain: All reals.
Range: $(0, \infty)$.
Continuous.
Increasing for all $x$.
No symmetry
Bounded below, but not above.
No local extrema.
Horizontal asymptote: $y = 0$.
No vertical asymptotes.
End behavior: $\lim\limits_{x \to -\infty} e^x = 0$ and $\lim\limits_{x \to \infty} e^x = \infty$.

$[-4, 4]$ by $[-1, 5]$

**Figure 3.5** The graph of $f(x) = e^x$.

Because $f(x) = e^x$ is an exponential growth function, it is one-to-one with a concave up graph. But what is $e$, and what makes this exponential function *the* exponential function?

The letter $e$ is the initial of the last name of Leonhard Euler (1707–1783), who introduced the notation. Because $f(x) = e^x$ has special calculus properties that simplify many calculations, it is the *natural base* of exponential functions for calculus purposes.

---

**Definition  The Natural Base $e$**

$$e = \lim_{x \to \infty} \left(1 + \frac{1}{x}\right)^x$$

---

We cannot compute the irrational number $e$ directly, but using this definition we can obtain successively closer approximations to $e$, as shown in Table 3.4. Continuing the process in Table 3.4 with a sufficiently accurate computer, it can be shown that $e \approx 2.718281828459$.

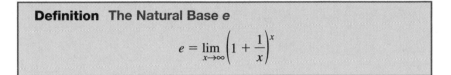

**Table 3.4  Approximations Approaching the Natural Base $e$**

| $x$ | 1 | 10 | 100 | 1000 | 10,000 | 100,000 |
|---|---|---|---|---|---|---|
| $(1 + 1/x)^x$ | 2 | 2.5... | 2.70... | 2.716... | 2.7181... | 2.71826... |

We are usually more interested in the exponential function $f(x) = e^x$ and variations of this function than in the irrational number $e$. In fact, any exponential function can be expressed in terms of the natural base $e$.

---

**Theorem  Exponential Functions and the Base $e$**

Any exponential function $f(x) = a \cdot b^x$ can be rewritten as

$$f(x) = a \cdot e^{kx},$$

for an appropriately chosen real number constant $k$.

If $a > 0$ and $k > 0$, $f(x) = a \cdot e^{kx}$ is an exponential growth function. (See Figure 3.6a.)

If $a > 0$ and $k < 0$, $f(x) = a \cdot e^{kx}$ is an exponential decay function. (See Figure 3.6b.)

---

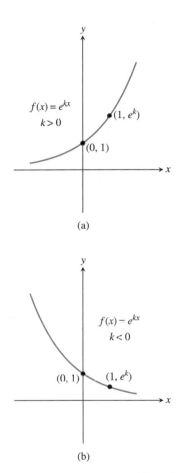

(a)

(b)

**Figure 3.6** Graphs of $f(x) = e^{kx}$ for (a) $k > 0$ and (b) $k < 0$.

In Section 3.3 we will develop some mathematics so that, for any positive number $b \neq 1$, we can easily find the value of $k$ such that $e^{kx} = b^x$. In the meantime, we can use graphical and numerical methods to approximate $k$, as you will discover in Exploration 2.

**Exploration Extensions**

Calculate the values of $e^{0.4}$, $e^{0.5}$, $e^{0.6}$, $e^{0.7}$, and $e^{0.8}$ and discuss how these values relate to the results from Steps 2 and 3.

> **Exploration 2** **Choosing $k$ so that $e^{kx} = 2^x$**
>
> **1.** Graph $f(x) = 2^x$ in the viewing window $[-4, 4]$ by $[-2, 8]$.
> **2.** One at a time, overlay the graphs of $g(x) = e^{kx}$ for $k = 0.4, 0.5, 0.6, 0.7, 0.8$. For which of these values of $k$ does the graph of $g$ most closely match the graph of $f$?
> **3.** Using tables, find the 3-decimal-place value of $k$ for which the values of $g$ most closely approximate the values of $f$.

### Example 5 TRANSFORMING EXPONENTIAL FUNCTIONS

Describe how to transform the graph of $f(x) = e^x$ into the graph of the given function. Sketch the graph by hand and support your answer with a grapher.

**(a)** $g(x) = e^{2x}$                                **(b)** $h(x) = e^{-x}$

**(c)** $k(x) = 3e^x$                             **(d)** $q(x) = 3e^{-2x}$

**Solution**

**(a)** The graph of $g(x) = e^{2x}$ is obtained by horizontally shrinking the graph of $f(x) = e^x$ by a factor of 2 (Figure 3.7a).

**(b)** We can obtain the graph of $h(x) = e^{-x}$ by reflecting the graph of $f(x) = e^x$ across the $y$-axis (Figure 3.7b).

**(c)** We can obtain the graph of $k(x) = 3e^x$ by vertically stretching the graph of $f(x) = e^x$ by a factor of 3 (Figure 3.7c).

**(d)** The graph of $q(x) = 3e^{-2x}$ is obtained by horizontally shrinking the graph of $f(x) = e^x$ by a factor of 2, then reflecting the result across the $y$-axis, and finally vertically stretching the graph by a factor of 3 (Figure 3.7d).

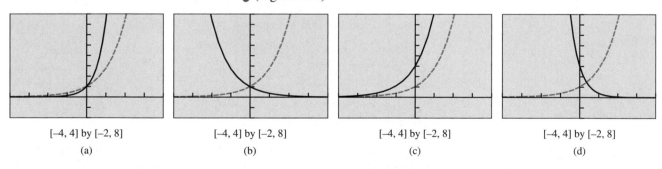

| $[-4, 4]$ by $[-2, 8]$ | $[-4, 4]$ by $[-2, 8]$ | $[-4, 4]$ by $[-2, 8]$ | $[-4, 4]$ by $[-2, 8]$ |
| :---: | :---: | :---: | :---: |
| (a) | (b) | (c) | (d) |

**Figure 3.7** The graph of $f(x) = e^x$ shown with (a) $g(x) = e^{2x}$, (b) $h(x) = e^{-x}$, (c) $k(x) = 3e^x$, and (d) $q(x) = 3e^{-2x}$. (Example 5)

## Logistic Functions and Their Graphs

Exponential growth is unrestricted. An exponential growth function increases at an ever increasing rate and is not bounded above. In many growth situations, there is a limit to the possible growth. A plant can only grow so tall. The number of goldfish in an aquarium is limited by the size of the aquarium. In such situations the growth often begins in an exponential manner, but the growth eventually slows and the graph levels out. The associated growth function is bounded both below and above by horizontal asymptotes.

---

**Definition** Logistic Growth Functions

Let $a$, $b$, $c$, and $k$ be positive constants, with $b < 1$. A **logistic growth function** in $x$ is a function that can be written in the form

$$f(x) = \frac{c}{1 + a \cdot b^x} \quad \text{or} \quad f(x) = \frac{c}{1 + a \cdot e^{-kx}}$$

where the constant $c$ is the **limit to growth**.

---

If $b > 1$ or $k < 0$, these formulas yield **logistic decay functions**. Unless otherwise stated, all logistic functions in this book will be logistic growth functions.

By setting $a = c = k = 1$, we obtain the **basic logistic function**

$$f(x) = \frac{1}{1 + e^{-x}}.$$

This function, though related to the exponential function $e^x$, *cannot* be obtained from $e^x$ by translations, reflections, and horizontal and vertical stretches and shrinks. So we give the basic logistic function a formal introduction:

**Basic Logistic Function**

$$f(x) = \frac{1}{1 + e^{-x}}$$

Domain: All reals.   Range: $(0, 1)$.
$x$-intercepts: none.   $y$-intercept $= 1/2$.
Continuous.
Increasing for all $x$.
Symmetric about $(0, 1/2)$, but neither even nor odd.
Bounded below and above.
No local extrema.
Horizontal asymptotes:   $y = 0$ and $y = 1$.
No vertical asymptotes.
End behavior: $\lim\limits_{x \to -\infty} f(x) = 0$ and $\lim\limits_{x \to \infty} f(x) = 1$.
One-to-one.
Concave up on $(-\infty, 0)$; concave down on $(0, \infty)$.

All logistic growth functions have graphs much like the basic logistic function. Their end behavior is always described by the equations

$$\lim\limits_{x \to -\infty} f(x) = 0 \quad \text{and} \quad \lim\limits_{x \to \infty} f(x) = c,$$

where $c$ is the limit to growth (see Exercise 65). All logistic functions are bounded by their horizontal asymptotes, $y = 0$ and $y = c$, and have a range of $(0, c)$. Although every logistic function is symmetric about the point of its graph with $y$-coordinate $c/2$, this point of symmetry is usually not the $y$-intercept, as we can see in Example 6.

**Aliases for Logistic Growth**

Logistic growth is also known as *restricted*, *inhibited*, or *constrained* exponential growth.

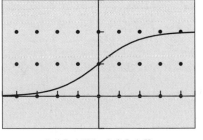

[−4.7, 4.7] by [−0.5, 1.5]
(a)

**Figure 3.8** The graph of $f(x) = 1/(1 + e^{-x})$, with a grid of points to emphasize its symmetry.

## Example 6   GRAPHING LOGISTIC GROWTH FUNCTIONS

Graph the function. Find the $y$-intercept and the horizontal asymptotes.

**(a)** $f(x) = \dfrac{8}{1 + 3 \cdot 0.7^x}$  **(b)** $g(x) = \dfrac{20}{1 + 2e^{-3x}}$

**Solution**

**(a)** The graph of $f(x) = 8/(1 + 3 \cdot 0.7^x)$ is shown in Figure 3.9a. The $y$-intercept is

$$f(0) = \frac{8}{1 + 3 \cdot 0.7^0} = \frac{8}{1 + 3} = 2.$$

Because the limit to growth is 8, the horizontal asymptotes are $y = 0$ and $y = 8$.

**(b)** The graph of $g(x) = 20/(1 + 2e^{-3x})$ is shown in Figure 3.9b. The $y$-intercept is

$$g(0) = \frac{20}{1 + 2e^{-3 \cdot 0}} = \frac{20}{1 + 2} = 20/3 \approx 6.67.$$

Because the limit to growth is 20, the horizontal asymptotes are $y = 0$ and $y = 20$.

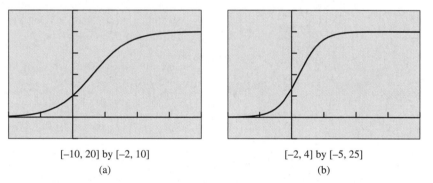

[–10, 20] by [–2, 10]           [–2, 4] by [–5, 25]
(a)                             (b)

**Figure 3.9** The graphs of (a) $f(x) = 8/(1 + 3 \cdot 0.7^x)$ and (b) $g(x) = 20/(1 + 2e^{-3x})$. (Example 6)

## Applications

Exponential and logistic functions have many applications. One area where both types of functions are used is in modeling population. Between 1990 and 1996, both Phoenix and San Antonio passed the 1 million mark. With its Silicon Valley industries, San Jose, California appears to be the next U.S. city destined to surpass 1 million residents. When a city and its population are growing rapidly, as in the case of San Jose, exponential growth is a reasonable model.

| Table 3.5  The Population of San Jose, California | |
|---|---|
| Year | Population |
| 1990 | 782,224 |
| 1996 | 838,744 |

*Source: U.S. Bureau of the Census*

**Notes on Examples**

We have chosen to use graphical solutions here because the algebraic methods needed to solve the exponential equations are presented in Sections 3.5 and 3.6. Examples 7 and 8 set the stage for modeling in Section 3.2.

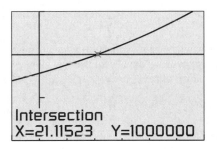

[–10, 60] by [0, 1 500 000]

**Figure 3.10** A population model for San Jose, California. (Example 7)

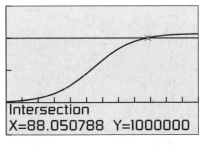

[0, 120] by [–500 000, 1 500 000]

**Figure 3.11** A population model for Dallas, Texas. (Example 8)

### Example 7  MODELING SAN JOSE'S POPULATION

Using the data in Table 3.5 and assuming the growth is exponential, when will the population of San Jose surpass 1 million persons?

**Solution**

**Model**

Let $P(t)$ be the population of San Jose $t$ years after 1990. Because $P$ is exponential, $P(t) = P_0 \cdot b^t$, where $P_0$ is the initial (1990) population of 782,224. From Table 3.5 we see that $P(6) = 782,224b^6 = 838,744$. So,

$$b = \sqrt[6]{\frac{838,744}{782,224}} \approx 1.0117$$

and $P(t) = 782,224 \cdot 1.0117^t$.

**Solve Graphically**

Figure 3.10 shows that this population model intersects $y = 1,000,000$ when the value of the independent variable is about 21.12.

**Interpret**

Because $1990 + 21 = 2011$, if the growth of its population is exponential, San Jose will surpass the 1 million mark in 2011.

While San Jose's population is soaring, in other major cities, such as Dallas, the population growth is slowing. The once sprawling Dallas is now *constrained* by its growing neighbors, such as Garland, Grand Prairie, Irving, and Mesquite, which are becoming major cities in their own right. *A logistic function is often an appropriate model for restricted growth*, such as the growth that Dallas is experiencing.

### Example 8  MODELING DALLAS' POPULATION

Based on recent census data, a logistic model for the population of Dallas, $t$ years after 1900, is

$$P(t) = \frac{1,068,578}{1 + 70.65e^{-0.0789x}}.$$

According to this model, when was the population 1 million?

**Solution**

Figure 3.11 shows that the population model intersects $y = 1,000,000$ when the value of the independent variable is about 88.05. Because $1900 + 88 = 1988$, if the growth of Dallas' population has followed this logistic model, its population was 1 million in 1988.

# Quick Review 3.1

In Exercises 1–4, evaluate the expression without using a calculator.

**1.** $\sqrt[3]{-216}$  $-6$

**2.** $\sqrt[3]{\dfrac{125}{8}}$  $\dfrac{5}{2} = 2.5$

**3.** $27^{2/3}$  $9$

**4.** $4^{5/2}$  $32$

In Exercises 5–8, rewrite the expression using a single positive exponent.

**5.** $(2^{-3})^4$  $1/2^{12}$

**6.** $(3^4)^{-2}$  $1/3^8$

**7.** $(a^{-2})^3$  $1/a^6$

**8.** $(b^{-3})^{-5}$  $b^{15}$

In Exercises 9–10, use a calculator to evaluate the expression.

**9.** $\sqrt[5]{-5.37824}$  $-1.4$

**10.** $\sqrt[4]{92.3521}$  $3.1$

# Section 3.1 Exercises

In Exercises 1–6, which of the following are exponential functions? For those that are exponential functions, state the initial value and the base. For those that are not, explain why not.

**1.** $y = x^8$

**2.** $y = 3^x$

**3.** $y = 5^x$

**4.** $y = 4^2$

**5.** $y = x^{\sqrt{x}}$

**6.** $y = x^{1.3}$

In Exercises 7–10, compute the exact value of the function for the given $x$-value without using a calculator.

**7.** $f(x) = 3 \cdot 5^x$  for  $x = 0$  $3$

**8.** $f(x) = 6 \cdot 3^x$  for  $x = -2$  $2/3$

**9.** $f(x) = -2 \cdot 3^x$  for  $x = 1/3$  $-2\sqrt[3]{3}$

**10.** $f(x) = 8 \cdot 4^x$  for  $x = -3/2$  $1$

In Exercises 11 and 12, determine a formula for the exponential function whose values are given in Table 3.6.

**11.** $f(x)$  $3/2(1/2)^x$

**12.** $g(x)$  $12(1/3)^x$

| Table 3.6 Values for Two Exponential Functions | | |
|---|---|---|
| $x$ | $f(x)$ | $g(x)$ |
| $-2$ | 6 | 108 |
| $-1$ | 3 | 36 |
| 0 | 3/2 | 12 |
| 1 | 3/4 | 4 |
| 2 | 3/8 | 4/3 |

In Exercises 13 and 14, determine a formula for the exponential function whose graph is shown in the figure.

**13.** $f(x)$  $3 \cdot 2^{x/2}$

**14.** $g(x)$  $2e^{-x}$

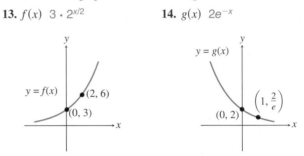

In Exercises 15–24, describe how to transform the graph of $f$ into the graph of $g$. Sketch the graph by hand and support your answer with a grapher.

**15.** $f(x) = 2^x, g(x) = 2^{x-3}$

**16.** $f(x) = 3^x, g(x) = 3^{x+4}$

**17.** $f(x) = 4^x, g(x) = 4^{-x}$

**18.** $f(x) = 2^x, g(x) = 2^{5-x}$

**19.** $f(x) = 0.5^x, g(x) = 3 \cdot 0.5^x + 4$

**20.** $f(x) = 0.6^x, g(x) = 2 \cdot 0.6^{3x}$

**21.** $f(x) = e^x, g(x) = e^{-2x}$

**22.** $f(x) = e^x, g(x) = -e^{-3x}$

**23.** $f(x) = e^x, g(x) = 2e^{3-3x}$

**24.** $f(x) = e^x, g(x) = 3e^{2x} - 1$

In Exercises 25–30, **(a)** match the given function with its graph. **(b) Writing to Learn** Explain how to make the choice without using a grapher.

**25.** $y = 3^x$

**26.** $y = 2^{-x}$

**27.** $y = -2^x$

**28.** $y = -0.5^x$

**29.** $y = 3^{-x} - 2$

**30.** $y = 1.5^x - 2$

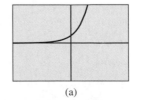

(a)

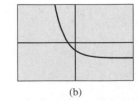

(b)

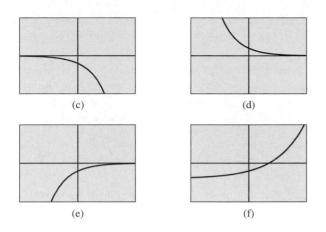

(c)                    (d)

(e)                    (f)

In Exercises 31–34, state whether the function is an exponential growth function or exponential decay function, and describe its end behavior using limits.

**31.** $f(x) = 3^{-2x}$    **32.** $f(x) = \left(\dfrac{1}{e}\right)^x$

**33.** $f(x) = 0.5^x$    **34.** $f(x) = 0.75^{-x}$

In Exercises 35–38, solve the inequality graphically.

**35.** $9^x < 4^x$  $x < 0$    **36.** $6^{-x} > 8^{-x}$  $x > 0$

**37.** $\left(\dfrac{1}{4}\right)^x > \left(\dfrac{1}{3}\right)^x$  $x < 0$    **38.** $\left(\dfrac{1}{3}\right)^x < \left(\dfrac{1}{2}\right)^x$  $x > 0$

In Exercises 39 and 40, use the properties of exponents to prove that two of the given three exponential functions are identical. Support graphically.

**39. Group Activity**

    **(a)** $y_1 = 3^{2x+4}$    **(b)** $y_2 = 3^{2x} + 4$    **(c)** $y_3 = 9^{x+2}$

**40. Group Activity**

    **(a)** $y_1 = 4^{3x-2}$    **(b)** $y_2 = 2(2^{3x-2})$    **(c)** $y_3 = 2^{3x-1}$

In Exercises 41–44, use a grapher to graph the function. Find the y-intercept and the horizontal asymptotes.

**41.** $f(x) = \dfrac{12}{1 + 2 \cdot 0.8^x}$    **42.** $f(x) = \dfrac{18}{1 + 5 \cdot 0.2^x}$

**43.** $f(x) = \dfrac{16}{1 + 3e^{-2x}}$    **44.** $g(x) = \dfrac{9}{1 + 2e^{-x}}$

In Exercises 45–50, graph the function, and analyze it for domain, range, continuity, intercepts, increasing or decreasing behavior, symmetry, boundedness, extrema, asymptotes, end behavior, and concavity.

**45.** $f(x) = 3 \cdot 2^x$    **46.** $f(x) = 4 \cdot 0.5^x$

**47.** $f(x) = 4 \cdot e^{3x}$    **48.** $f(x) = 5 \cdot e^{-x}$

**49.** $f(x) = \dfrac{5}{1 + 4 \cdot e^{-2x}}$    **50.** $f(x) = \dfrac{6}{1 + 2 \cdot e^{-x}}$

**51. Population Growth** Using the data in Table 3.7 and assuming the growth is exponential, when will the population of Columbus surpass 700,000 persons?  In 2006

**52. Population Growth** Using the data in Table 3.7 and assuming the growth is exponential, when will the population of El Paso surpass 800,000 persons?  In 2007

**53. Population Growth** Using the data in Table 3.7 and assuming the growth is exponential, in which year are the populations of Columbus and El Paso equal?  In 2001

| Table 3.7 Populations of Two Major U.S. Cities | | |
|---|---|---|
| City | 1990 Population | 1996 Population |
| Columbus, Ohio | 632,945 | 657,053 |
| El Paso, Texas | 515,342 | 599,865 |

*Source: U.S. Bureau of the Census*

**54. Population Growth** Using 20th century U.S. census data, the population of New York state can be modeled by $P(t) = 19.71/(1 + 61.22e^{-0.03563x})$, where $P$ is the population in millions and $t$ in the number of years since 1800. Based on this model,

    **(a)** What was the population of New York in 1850?

    **(b)** What will New York state's population be in 2010?

    **(c)** What is New York's *maximum sustainable population* (limit to growth)?  19,710,000 people

**55. Bacteria Growth** The number $B$ of bacteria in a petri dish culture after $t$ hours is given by

$$B = 100e^{0.693t}.$$

    **(a)** What was the initial number of bacteria present?  100

    **(b)** How many bacteria are present after 6 hours?  $\approx$ 6394

**56. Carbon Dating** The amount $C$ in grams of carbon-14 present in a certain substance after $t$ years is given by

$$C = 20e^{-0.0001216t}.$$

    **(a)** What was the initial amount of carbon-14 present?

    **(b)** How much is left after 10,400 years? When will the amount left be 10 g?

## Explorations

**57.** Graph each function, and analyze it for domain, range, intercepts, increasing or decreasing behavior, boundedness, extrema, asymptotes, end behavior, and concavity.

    **(a)** $f(x) = x \cdot e^x$    **(b)** $g(x) = \dfrac{e^{-x}}{x}$

**58.** Use the properties of exponents to solve each equation. Support graphically.

    **(a)** $2^x = 4^2$  $x = 4$    **(b)** $3^x = 27$  $x = 3$

    **(c)** $8^{x/2} = 4^{x+1}$  $x = -4$    **(d)** $9^x = 3^{x+1}$  $x = 1$  ∎

## Extending the Ideas

**59. Writing to Learn** Table 3.8 gives function values for $y = f(x)$ and $y = g(x)$. Also, three different graphs are shown.

**Table 3.8 Data for Two Functions**

| $x$ | $f(x)$ | $g(x)$ |
|-----|--------|--------|
| 1.0 | 5.50 | 7.40 |
| 1.5 | 5.35 | 6.97 |
| 2.0 | 5.25 | 6.44 |
| 2.5 | 5.17 | 5.76 |
| 3.0 | 5.13 | 4.90 |
| 3.5 | 5.09 | 3.82 |
| 4.0 | 5.06 | 2.44 |
| 4.5 | 5.05 | 0.71 |

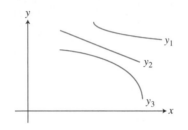

**(a)** Which curve of those shown in the graph most closely resembles the graph of $y = f(x)$? Explain your choice.

**(b)** Which curve most closely resembles the graph of $y = g(x)$? Explain your choice.

**60. Writing to Learn** Let $f(x) = 2^x$. Explain why the graph of $f(ax + b)$ can be obtained by applying one transformation to the graph of $y = c^x$ for an appropriate value of $c$. What is $c$?

Exercises 61–64 refer to the expression $f(a, b, c) = a \cdot b^c$. If $a = 2$, $b = 3$, and $c = x$, the expression is $f(2, 3, x) = 2 \cdot 3^x$, an exponential function.

**61.** If $b = x$, state conditions on $a$ and $c$ under which the expression $f(a, b, c)$ is a quadratic power function.

**62.** If $b = x$, state conditions on $a$ and $c$ under which the expression $f(a, b, c)$ is a decreasing linear function.

**63.** If $c = x$, state conditions on $a$ and $b$ under which the expression $f(a, b, c)$ is an increasing exponential function.

**64.** If $c = x$, state conditions on $a$ and $b$ under which the expression $f(a, b, c)$ is a decreasing exponential function.

**65.** Prove that $\lim\limits_{x \to -\infty} \dfrac{c}{1 + a \cdot b^x} = 0$ and $\lim\limits_{x \to \infty} \dfrac{c}{1 + a \cdot b^x} = c$, for constant $a, b, c > 0$ and $b < 1$.

---

## 3.2 Exponential and Logistic Modeling

Constant Percentage Rate and Exponential Functions • Exponential Growth and Decay Models • Using Regression to Model Population • Other Logistic Models

### Constant Percentage Rate and Exponential Functions

Suppose that a population is *increasing* at a **constant percentage rate** $r > 0$ per year, where $r$ is the percent rate of change expressed in decimal form. Then the population obeys the following pattern:

| | |
|---|---|
| $P_0$ | Initial population |
| $P_0 + P_0 r = P_0(1 + r)$ | Population after 1 year |
| $P_0(1 + r) + P_0(1 + r)r = P_0(1 + r)^2$ | Population after 2 years |
| $\vdots$ | |
| $P_0(1 + r)^{t-1} + P_0(1 + r)^{t-1}r = P_0(1 + r)^t$ | Population after $t$ years |

So the population is the exponential growth function

$$P(t) = P_0(1 + r)^t.$$

The growth factor is the base $1 + r$. In words,

$$\text{Growth Factor} = 1 + \text{Percentage Rate}$$

### Example 1 REVISITING POPULATION OF SAN JOSE

By what annual percentage rate did the population of San Jose grow during 1990–1996?

**Solution** In Example 7 of Section 3.1, we determined the following model for San Jose's population:

$$P(t) = 782,224 \cdot 1.0117^t$$

Setting $1 + r = 1.0117$, we see that the percentage rate $r = 0.0117 = 1.17\%$.

If a population is *decreasing*, then the constant percentage rate $r$ is *negative*, and the population is still given by

$$P(t) = P_0(1 + r)^t.$$

When the rate $r$ is negative, the base $1 + r < 1$, $P(t)$ is an exponential decay function, and $1 + r$ is the decay factor for the population.

### Example 2 MODELING DETROIT'S POPULATION

The population of Detroit has been declining since 1950 at an annual rate of about 1.44%. Assuming the decline has been exponential and given that the 1980 census population was 1,203,000, when did the population of Detroit drop below 1 million persons?

**Solution**

**Model**

Let $P(t)$ be the population of Detroit $t$ years after 1980. Because $P$ is exponential, $P(t) = P_0(1 + r)^t$, where $P_0 = 1,203,000$ and $r = -0.0144$. So, $P(t) = 1,203,000(1 - 0.0144)^t = 1,203,000 \cdot 0.9856^t$.

**Solve Graphically**

Figure 3.12 shows that the population function intersects $y = 1,000,000$ when $t \approx 12.74$.

**Interpret**

Because $1980 + 13 = 1993$, if the decline of Detroit's population has been exponential, its population was 1 million in 1993.

According the U.S. Bureau of the Census, Detroit's population dropped below 1 million in 1996. So, even though our model is not perfect, it is reasonably accurate.

## Exponential Growth and Decay Models

Exponential growth and decay apply to populations of animals, bacteria, and even of radioactive atoms. Exponential growth and decay apply to any situation where the growth is proportional to the current size. Such situations are frequently encountered in biology, chemistry, business, and the social sciences.

Exponential growth models can be developed in terms of the time it takes for the amount to double. On the flip side, exponential decay models can be developed in terms of the time it takes for the population to be halved. Examples 3–5 use these strategies.

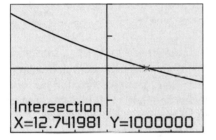

Intersection
X=12.741981  Y=1000000

[–30, 30] by [0, 2 000 000]

**Figure 3.12** A population model for Detroit, Michigan. (Example 2)

### Example 3 MODELING BACTERIA GROWTH

Suppose a culture of 100 bacteria is put into a petri dish and the culture doubles every hour. Predict when the number of bacteria will be 350,000.

#### Solution
#### Model

$$200 = 100 \cdot 2 \qquad \text{Total bacteria after 1 hr}$$

$$400 = 100 \cdot 2^2 \qquad \text{Total bacteria after 2 hr}$$

$$800 = 100 \cdot 2^3 \qquad \text{Total bacteria after 3 hr}$$

$$\vdots$$

$$P(t) = 100 \cdot 2^t \qquad \text{Total bacteria after } t \text{ hr}$$

So the function $P(t) = 100 \cdot 2^t$ represents the bacteria population $t$ hr after it is placed in the petri dish.

#### Solve Graphically

Figure 3.13 shows that the population function intersects $y = 350,000$ when $t \approx 11.77$.

#### Interpret

The population of the bacteria in the petri dish will be 350,000 in about 11 hr and 46 min.

Exponential decay functions model the amount of a radioactive substance present in a sample. The number of atoms of a specific element that change from a radioactive state to a nonradioactive state is a fixed fraction per unit time. The process is called **radioactive decay**, and the time it takes for half of a sample to change its state is the **half-life** of the radioactive substance.

### Example 4 MODELING RADIOACTIVE DECAY

Suppose the half-life of a certain radioactive substance is 20 days and there are 5 g (grams) present initially. Find the time when there will be 1 g of the substance remaining.

#### Solution
#### Model

If $t$ is the time in days, the number of half-lives will be $t/20$.

$$\frac{5}{2} = 5\left(\frac{1}{2}\right)^{20/20} \qquad \text{Grams after 20 days}$$

$$\frac{5}{4} = 5\left(\frac{1}{2}\right)^{40/20} \qquad \text{Grams after 2(20) = 40 days}$$

$$\vdots$$

$$f(t) = 5\left(\frac{1}{2}\right)^{t/20} \qquad \text{Grams after } t \text{ days}$$

Thus the function $f(t) = 5 \cdot 0.5^{t/20}$ models the mass in grams of the radioactive substance at time $t$.

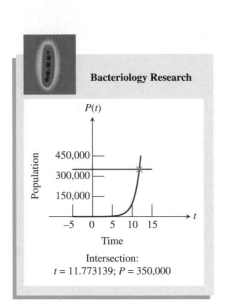

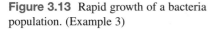

**Bacteriology Research**

P(t)

Intersection:
t = 11.773139; P = 350,000

**Figure 3.13** Rapid growth of a bacteria population. (Example 3)

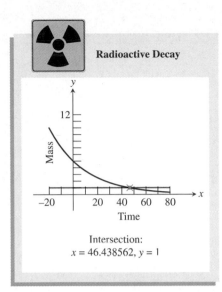

Intersection:
$x = 46.438562, y = 1$

**Figure 3.14** Radioactive decay.
(Example 4)

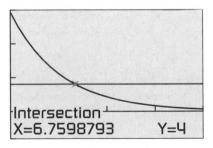

[0, 20] by [−4, 15]

**Figure 3.15** A model for atmospheric
pressure. (Example 5)

### Solve Graphically

Figure 3.14 shows that the graph of $f(t) = 5 \cdot 0.5^{t/20}$ intersects $y = 1$ when $t \approx 46.44$.

### Interpret

There will be 1 g of the radioactive substance left after approximately 46.44 days, or about 46 days 11 hr.

Scientists have established that atmospheric pressure at sea level is 14.7 lb/in.$^2$, and the pressure is reduced by half for each 3.6 mi above sea level. For example, the pressure 3.6 mi above sea level is $(1/2)(14.7) = 7.35$ lb/in.$^2$. This rule for atmospheric pressure holds for altitudes up to 50 mi above sea level. Though the context is different, the mathematics of atmospheric pressure closely resembles the mathematics of radioactive decay.

### Example 5 DETERMINING ALTITUDE FROM ATMOSPHERIC PRESSURE

Find the altitude above sea level at which the atmospheric pressure is 4 lb/in.$^2$.

### Solution
### Model

$$7.35 = 14.7 \cdot 0.5^{3.6/3.6} \qquad \text{Pressure at 3.6 mi}$$

$$3.675 = 14.7 \cdot 0.5^{7.2/3.6} \qquad \text{Pressure at 2(3.6) = 7.2 mi}$$

$$\vdots$$

$$P(h) = 14.7 \cdot 0.5^{h/3.6} \qquad \text{Pressure at } h \text{ mi}$$

So $P(h) = 14.7 \cdot 0.5^{h/3.6}$ models the atmospheric pressure $P$ (in pounds per square inch) as a function of the height $h$ (in miles above sea level). We must find the value of $h$ that satisfies the equation

$$14.7 \cdot 0.5^{h/3.6} = 4$$

### Solve Graphically

Figure 3.15 shows that the graph of $P(h) = 14.7 \cdot 0.5^{h/3.6}$ intersects $y = 4$ when $h \approx 6.76$.

### Interpret

The atmospheric pressure is 4 lb/in.$^2$ at an altitude of approximately 6.76 mi above sea level.

## Using Regression to Model Population

So far, our models have been given to us or developed algebraically. We now use exponential and logistic regression to build models from population data.

Due to the post-World War II baby boom and other factors, exponential growth is not a perfect model for the U.S. population. It does, however provide a means to make approximate predictions, as illustrated in Example 6.

| Table 3.9 U.S. Population (in millions) | |
|---|---|
| Year | Population |
| 1900 | 76.2 |
| 1910 | 92.2 |
| 1920 | 106.0 |
| 1930 | 123.2 |
| 1940 | 132.2 |
| 1950 | 151.3 |
| 1960 | 179.3 |
| 1970 | 203.3 |
| 1980 | 226.5 |
| 1990 | 248.7 |
| 1998 | 273.8 |

*Source: 1999 New York Times Almanac*

### Example 6  MODELING U.S. POPULATION USING EXPONENTIAL REGRESSION

Use the 1900–1990 data in Table 3.9 and exponential regression to predict the U.S. population for 1998. Compare the result with the listed value for 1998.

**Solution**

**Model**

Let $P(t)$ be the population (in millions) of United States $t$ years after 1900. Figure 3.16a shows a scatter plot of the data. Using exponential regression, we find a model for the data:

$$P(t) = 80.075 \cdot 1.0131^t$$

Figure 3.16b on the next page shows the scatter plot of the data with a graph of the population model just found. You can see that the curve fits the data fairly well. The coefficient of determination is $r^2 \approx 0.9945$, indicating a close fit and supporting the visual evidence.

**Solve Graphically**

To predict the 1998 U.S. population we substitute $t = 98$ into the regression model. Figure 3.16c reports that $P(98) = 80.075 \cdot 1.0131^{98} \approx 286.7$.

**Interpret**

The model predicts the U.S. population was 286.7 million in 1998. See Figure 3.16c. The actual population was 273.8 million. We overestimated by 12.9 million, slightly less than a 5% error.

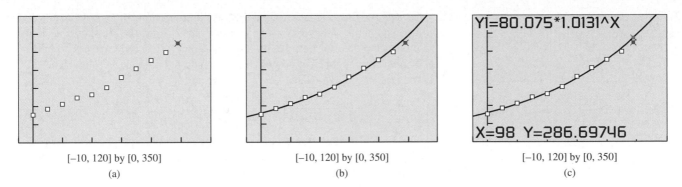

| [−10, 120] by [0, 350] | [−10, 120] by [0, 350] | [−10, 120] by [0, 350] |
|:---:|:---:|:---:|
| (a) | (b) | (c) |

**Figure 3.16** Scatter plot and graphs for Example 6. The red "x" denotes the data point for 1998.

Exponential growth is unrestricted, but population growth often is not. For many populations, the growth begins exponentially, but eventually slows and approaches a limit to growth called the **maximum sustainable population**.

In Section 3.1 we modeled Dallas's population with a logistic function. We now use logistic regression to do the same for the populations of Florida and Pennsylvania. As the data in Table 3.10 suggest, Florida has had rapid growth in the second half of the 20th century, whereas Pennsylvania appears to be approaching its maximum sustainable population.

**Table 3.10 Official Census Populations (in millions), 1900–1990**

| Year | Florida | Pennsylvania |
|:---:|:---:|:---:|
| 1900 | 0.5 | 6.3 |
| 1910 | 0.8 | 7.7 |
| 1920 | 1.0 | 8.7 |
| 1930 | 1.5 | 9.6 |
| 1940 | 1.9 | 9.9 |
| 1950 | 2.8 | 10.5 |
| 1960 | 5.0 | 11.3 |
| 1970 | 6.8 | 11.8 |
| 1980 | 9.7 | 11.9 |
| 1990 | 12.9 | 11.9 |

*Source: U.S. Census Bureau as reported in 1999 New York Times Almanac.*

**Example 7** MODELING TWO STATES' POPULATIONS USING LOGISTIC REGRESSION

Use the data in Table 3.10 and logistic regression to predict the maximum sustainable populations for Florida and Pennsylvania. Graph the logistic models and interpret their significance.

**Solution** Let $F(t)$ and $P(t)$ be the populations (in millions) of Florida and Pennsylvania, respectively, $t$ years after 1800. Figure 3.17a shows a scatter plot of the data for both states. The data for Florida is concave up, and for Pennsylvania, concave down. Using logistic regression, we obtain the models for the two states:

$$F(t) = \frac{33.478}{1 + 7887.9e^{-0.044798t}} \quad \text{and} \quad P(t) = \frac{12.477}{1 + 31.678e^{-0.035307t}}$$

Figure 3.17b shows the scatter plots of the data with graphs of the two population models. You can see that the curves fit the data fairly well. From the numerators of the models we see that

$$\lim_{t \to \infty} F(t) = 33.478 \quad \text{and} \quad \lim_{t \to \infty} P(t) = 12.477.$$

So the maximum sustainable population for Florida is about 33.5 million, and for Pennsylvania is about 12.5 million.

Figure 3.17c shows a three century span for the two states. Pennsylvania had rapid growth in the 19th and 20th centuries, and is approaching its limit to growth in the 21st century. Florida, on the other hand, is currently experiencing extremely rapid growth but should be approaching its maximum sustainable population by the end of the 21st century according to these models.

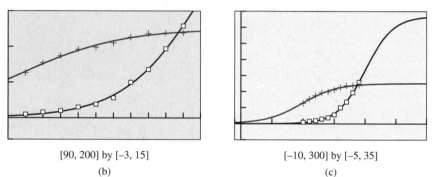

| [90, 200] by [−3, 15] | [90, 200] by [−3, 15] | [−10, 300] by [−5, 35] |
|:---:|:---:|:---:|
| (a) | (b) | (c) |

**Figure 3.17** Scatter plot and graphs for Example 7.

## Other Logistic Models

In Example 3, the bacteria cannot continue to grow exponentially forever because they cannot grow beyond the confines of the petri dish. In Example 7, though Florida's population is booming now, it will eventually level off, just as Pennsylvania's has done. Sunflowers and many other plants grow to a natural height following a logistic pattern. Chemical acid-base titration curves are logistic. Yeast cultures grow logistically. Contagious diseases and even rumors spread according to logistic models.

### Example 8   MODELING A RUMOR

Watauga High School has 1200 students. Bob, Carol, Ted, and Alice start a rumor, which spreads logistically so that $S(t) = 1200/(1 + 39 \cdot e^{-0.9t})$ models the number of students who have heard the rumor by the end of $t$ days.

**(a)** How many students have heard the rumor by the end of Day 0?

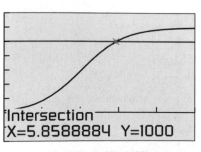

[0, 10] by [−400, 1400]

**Figure 3.18** The spread of a rumor. (Example 8)

**(b)** How long does it take for 1000 students to hear the rumor?

**Solution**

**(a)** $S(0) = \dfrac{1200}{1 + 39 \cdot e^{-0.9 \cdot 0}} = \dfrac{1200}{1 + 39} = 30.$

**(b)** We need to solve $\dfrac{1200}{1 + 39e^{-0.9t}} = 1000.$

Figure 3.18 shows that the graph of $S(t) = 1200/(1 + 39 \cdot e^{-0.9t})$ intersects $y = 1000$ when $t \approx 5.86$. So toward the end of Day 6 the rumor has reached the ears of 1000 students.

# Quick Review 3.2

In Exercises 1 and 2, convert the percent to decimal form or the decimal into a percent.

**1.** 15%  0.15

**2.** 0.04  4%

**3.** Show how to increase 23 by 7% using a single multiplication. 23 · 1.07

**4.** Show how to decrease 52 by 4% using a single multiplication? 52 · 0.96

In Exercises 5 and 6, solve the equation algebraically.

**5.** $40 \cdot b^2 = 160$  ±2

**6.** $243 \cdot b^3 = 9$  1/3

In Exercises 7–10, solve the equation numerically.

**7.** $782b^6 = 838$  1.01

**8.** $93b^5 = 521$  1.41

**9.** $672b^4 = 91$  0.607

**10.** $127b^7 = 56$  0.89

# Section 3.2 Exercises

In Exercises 1–6, tell whether the function is an exponential growth function or exponential decay function, and find the constant percentage rate of growth or decay for the function.

**1.** $f(x) = 3.5 \cdot 1.09^x$

**2.** $f(x) = 4.3 \cdot 1.018^x$

**3.** $f(x) = 78963 \cdot 0.968^x$

**4.** $f(x) = 5607 \cdot 0.9968^x$

**5.** $f(x) = 247 \cdot 2^x$

**6.** $f(x) = 43 \cdot 0.05^x$

In Exercises 7–18, find the exponential function that satisfies the given conditions.

**7.** Initial value = 5, increasing at a rate of 17% per year

**8.** Initial value = 52, increasing at a rate of 2.3% per day

**9.** Initial value = 16, decreasing at a rate of 50% per month

**10.** Initial value = 5, decreasing at a rate of 0.59% per week

**11.** Initial population = 502,000, increasing at a rate of 1.7% per year  502,000 · 1.017^x

**12.** Initial population = 28,900, decreasing at a rate of 2.6% per year  28,900 · 0.974^x

**13.** Initial height = 18 cm, growing at a rate of 5.2% per week

**14.** Initial mass = 15 g, decreasing at a rate of 4.6% per day

**15.** Initial mass = 0.6 g, doubling every 3 days  0.6 · 2^{x/3}

**16.** Initial population = 250, doubling every 7.5 hours

**17.** Initial mass = 592 g, halving once every 6 years

**18.** Initial mass = 17 g, halving once every 32 hours

In Exercises 19 and 20, determine a formula for the exponential function whose values are given in Table 3.11.

**19.** $f(x)$  2.3 · 1.25^x

**20.** $g(x)$  −5.8 · 0.8^x

**Table 3.11  Values for Two Exponential Functions**

| $x$ | $f(x)$ | $g(x)$ |
|---|---|---|
| −2 | 1.472 | −9.0625 |
| −1 | 1.84 | −7.25 |
| 0 | 2.3 | −5.8 |
| 1 | 2.875 | −4.64 |
| 2 | 3.59375 | −3.712 |

In Exercises 21 and 22, determine a formula for the exponential function whose graph is shown in the figure.

**21.**

**22.**

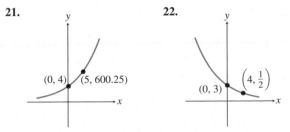

In Exercises 23–26, find the logistic function that satisfies the given conditions.

23. Initial value = 10, limit to growth = 40, passing through (1, 20). $40/[1 + 3 \cdot (1/3)^x]$

24. Initial value = 12, limit to growth = 60, passing through (1, 24). $60/[1 + 4(3/8)^x]$

25. Initial population = 16, maximum sustainable population = 128, passing through (5, 32). $128/(1 + 7 \cdot 0.844^x)$

26. Initial height = 5, limit to growth = 30, passing through (3, 15). $30/(1 + 5 \cdot 0.585^x)$

In Exercises 27 and 28, determine a formula for the logistic function whose graph is shown in the figure.

27.            28.

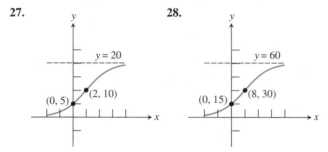

29. **Population Growth** The population of Knoxville is 475,000 and is increasing at the rate of 3.75% each year. Predict when the population will be 1 million.

30. **Population Growth** The population of Glenbrook is 350,000 and is increasing at the rate of 3.25% each year. Predict when the population will be 1 million.

31. **Population Growth** The population of Silver Run in the year 1890 was 6250. Assume the population increased at a rate of 2.75% per year.

    (a) Estimate the population in 1915 and 1940.

    (b) Predict when the population reached 50,000.

32. **Population Growth** The population of Centerville in the year 1910 was 4200. Assume the population increased at a rate of 2.25% per year.

    (a) Estimate the population in 1930 and 1945.

    (b) Predict when the population reached 20,000.

33. **Radioactive Decay** The half-life of a certain radioactive substance is 14 days. There are 6.6 g present initially.

    (a) Express the amount of substance remaining as a function of time $t$.

    (b) When will there be less than 1 g remaining?

34. **Radioactive Decay** The half-life of a certain radioactive substance is 65 days. There are 3.5 g present initially.

    (a) Express the amount of substance remaining as a function of time $t$.

    (b) When will there be less than 1 g remaining?

35. **Writing to Learn** Without using formulas or graphs, compare and contrast exponential functions and linear functions.

36. **Writing to Learn** Without using formulas or graphs, compare and contrast exponential functions and logistic functions.

37. **Writing to Learn** Using the population model that is graphed, explain why the time it takes the population to double (doubling time) is independent of the population size.

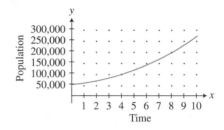

38. **Writing to Learn** Explain why the half-life of a radioactive substance is independent of the initial amount of the substance that is present.

39. **Bacteria Growth** The number $B$ of bacteria in a petri dish culture after $t$ hours is given by

    $$B = 100e^{0.693t}.$$

    When will the number of bacteria be 200? Estimate the doubling time of the bacteria. when $t = 1$; every hour

40. **Carbon Dating** The amount $C$ in grams of carbon-14 present in a certain substance after $t$ years is given by

    $$C = 20e^{-0.0001216t}.$$

    Estimate the half-life of carbon-14. about 5700 years

41. **Atmospheric Pressure** Determine the atmospheric pressure outside an aircraft flying at 52,800 ft (10 mi above sea level). 2.14 lb/in.$^2$

42. **Atmospheric Pressure** Find the altitude above sea level at which the atmospheric pressure is 2.5 lb/in.$^2$.

43. **Spread of Flu** The number of students infected with flu at Springfield High School after $t$ days is modeled by the function

    $$P(t) = \frac{800}{1 + 49e^{-0.2t}}.$$

    (a) What was the initial number of infected students?

    (b) When will the number of infected students be 200?

    (c) The school will close when 300 of the 800-student body are infected. When will the school close? about 17 days

44. **Population of Deer** The population of deer after $t$ years in Cedar State Park is modeled by the function

    $$P(t) = \frac{1001}{1 + 90e^{-0.2t}}.$$

    (a) What was the initial population of deer? 11

    (b) When will the number of deer be 600?

**(c)** What is the maximum number of deer possible in the park?  1001

**45. Population Growth**  Using all of the data in Table 3.9, compute a logistic regression model, and use it to predict the U.S. population in 2010.  ≈ 307,800,000

**46. Population Growth**  Using the data in Table 3.12, confirm the model used in Example 8 of Section 3.1.

**Table 3.12  The Population of Dallas, Texas**

| Year | Population |
|------|-----------|
| 1950 | 434,462 |
| 1960 | 679,684 |
| 1970 | 844,000 |
| 1980 | 905,000 |
| 1990 | 1,007,618 |
| 1996 | 1,053,292 |

*Source: U.S. Bureau of the Census*

**47. Population Growth**  Using the data in Table 3.13, confirm the model used in Exercise 54 of Section 3.1.

**48. Population Growth**  Using the data in Table 3.13, compute a logistic regression model for Arizona's population. Based on your model and the New York population model from Exercise 54 of Section 3.1, will the population of Arizona ever surpass that of New York?

**Table 3.13  Official Census Populations (in millions), 1900–1990**

| Year | Arizona | New York |
|------|---------|----------|
| 1900 | 0.1 | 7.3 |
| 1910 | 0.2 | 9.1 |
| 1920 | 0.3 | 10.3 |
| 1930 | 0.4 | 12.6 |
| 1940 | 0.5 | 13.5 |
| 1950 | 0.7 | 14.8 |
| 1960 | 1.3 | 16.8 |
| 1970 | 1.8 | 18.2 |
| 1980 | 2.7 | 17.6 |
| 1990 | 3.7 | 18.0 |

*Source: U.S. Census Bureau as reported in 1999 New York Times Almanac.*

## Explorations

**49. Population Growth**

**(a)** Use the 1900–1990 data in Table 3.9 and *logistic* regression to predict the U.S. population for 1998.

**(b) Writing to Learn**  Compare the prediction with the value listed in the table for 1998 and with the results of Example 6.

**(c)** Which model—exponential or logistic—makes the better prediction in this case?  logistic model

**50. Population Growth**  Use the data in Tables 3.9 and 3.14.

**(a)** Based on exponential growth models, will Mexico's population surpass that of the United States and if so, when?

**(b)** Based on logistic growth models, will Mexico's population surpass that of the United States and if so, when?  will not exceed

**(c)** What are the maximum sustainable populations for the two countries?

**(d) Writing to Learn**  Which model—exponential or logistic—is more valid in this case? Justify your choice.

**Table 3.14  Population of Mexico Mexico (in millions)**

| Year | Population |
|------|-----------|
| 1900 | 13.6 |
| 1950 | 25.8 |
| 1960 | 34.9 |
| 1970 | 48.2 |
| 1980 | 66.8 |
| 1990 | 88.1 |
| 1998 | 95.8 |
| 2025 | 130.2 |
| 2050 | 154 |

*Source: 1992 Statesman's Yearbook and 1999 New York Times Almanac.*

## Extending the Ideas

**51.** The **hyperbolic sine function** is defined by $\sinh(x) = (e^x - e^{-x})/2$. Prove that sinh is an odd function.

**52.** The **hyperbolic cosine function** is defined by $\cosh(x) = (e^x + e^{-x})/2$. Prove that cosh is an even function.

**53.** The **hyperbolic tangent function** is defined by $\tanh(x) = (e^x - e^{-x})/(e^x + e^{-x})$.

**(a)** Prove that $\tanh(x) = \sinh(x)/\cosh(x)$.

**(b)** Prove that tanh is an odd function.

**(c)** Prove that $f(x) = 1 + \tanh(x)$ is a logistic function.

# 3.3 Logarithmic Functions and Their Graphs

Inverses of Exponential Functions • Common Logarithms—Base 10 • Natural Logarithms—Base $e$ • Graphs of Logarithmic Functions • Applications

## Inverses of Exponential Functions

The graph of an exponential function $f(x) = b^x$, where $b > 0$ and $b \neq 1$, looks like one of the two graphs in Figure 3.19. We see that these graphs, in either of the two forms, satisfy the horizontal line test (every horizontal line intersects the graph at most once). Consequently each exponential function is one-to-one and its inverse is a function. The inverse of an exponential function $f(x) = b^x$ with base $b$ is the **logarithmic function with base $b$**, denoted $\log_b(x)$. That is, if $f(x) = b^x$ with $b > 0$ and $b \neq 1$, then $f^{-1}(x) = \log_b(x)$. See Figure 3.20.

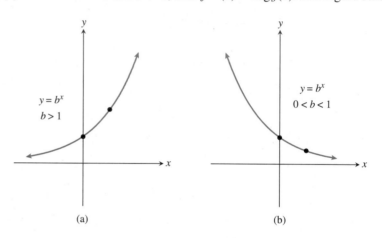

**Figure 3.19** Exponential functions are either (a) increasing or (b) decreasing.

An immediate and useful consequence of this definition is the close connection between an exponential equation and its logarithmic counterpart.

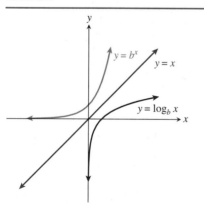
> ### Changing Between Logarithmic and Exponential Form
> If $x > 0$ and $0 < b \neq 1$, then
> $$y = \log_b(x) \quad \text{if and only if} \quad b^y = x.$$

Notice that *a logarithm is an exponent*. To evaluate $\log_3(9)$, we seek the exponent $y$ such that $3^y = 9$. So $y = 2$, and $\log_3(9) = 2$. When there is no possibility for confusion, we often drop the parentheses from a logarithmic expression and write, for example, $\log_3 9 = 2$. Because logarithms are exponents, evaluating simple logarithmic expressions relies on our understanding of exponents.

### Example 1 EVALUATING LOGARITHMS

**(a)** $\log_2 8 = 3$ because $2^3 = 8$.

**(b)** $\log_3 \sqrt{3} = 1/2$ because $3^{1/2} = \sqrt{3}$.

**(c)** $\log_5 \dfrac{1}{25} = -2$ because $5^{-2} = \dfrac{1}{5^2} = \dfrac{1}{25}$.

**(d)** $\log_4 1 = 0$ because $4^0 = 1$.

**(e)** $\log_7 7 = 1$ because $7^1 = 7$.

From our work with logarithms so far, we can generalize some relationships. We state these as properties with justification.

---

### Basic Properties of Logarithms

For $0 < b \neq 1$, $x > 0$, and any real number $y$,

- $\log_b 1 = 0$ because $b^0 = 1$.
- $\log_b b = 1$ because $b^1 = b$.
- $\log_b b^y = y$ because $b^y = b^y$.
- $b^{\log_b x} = x$ because $\log_b x = \log_b x$.

---

These properties give us alternatives for evaluating simple logarithms and a means for evaluating some exponential expressions. The first two parts of Example 2 are the same as the first two parts of Example 1.

### Example 2   EVALUATING LOGARITHMIC AND EXPONENTIAL EXPRESSIONS

**(a)** $\log_2 8 = \log_2 2^3 = 3$.

**(b)** $\log_3 \sqrt{3} = \log_3 3^{1/2} = 1/2$.

**(c)** $6^{\log_6 11} = 11$.

Logarithmic functions are inverses of exponential functions. So the tables and graphs of exponential and logarithmic functions are related in the same way as the tables and graphs of all inverse functions: The inputs and outputs are switched. Table 3.15 illustrates this relationship for $f(x) = 2^x$ and $f^{-1}(x) = \log_2 x$.

**Teaching Note**

Remind students how to find the inverse of $f(x) = x^3$ using the method shown below. Then ask whether they can find the inverse of $g(x) = 3^x$ by the same method. This emphasizes the need for logarithms.

$$f(x) = x^3$$
$$y = x^3$$
$$x = y^3$$
$$\sqrt[3]{x} = y$$
$$f^{-1}(x) = \sqrt[3]{x}$$

#### Table 3.15  An Exponential Function and Its Inverse

| $x$ | $f(x) = 2^x$ | $x$ | $f^{-1}(x) = \log_2 x$ |
|-----|--------------|-----|------------------------|
| $-3$ | $1/8$ | $1/8$ | $-3$ |
| $-2$ | $1/4$ | $1/4$ | $-2$ |
| $-1$ | $1/2$ | $1/2$ | $-1$ |
| $0$ | $1$ | $1$ | $0$ |
| $1$ | $2$ | $2$ | $1$ |
| $2$ | $4$ | $4$ | $2$ |
| $3$ | $8$ | $8$ | $3$ |

This relationship can be used to produce both tables and graphs for logarithmic functions, as you will discover in Exploration 1.

**Exploration Extensions**

Analyze the characteristics of the graphs (domain, range, intercepts, increasing or decreasing behavior, symmetry, asymptotes, end behavior, concavity) and discuss how the two graphs are related to each other visually.

---

**Exploration 1**  Comparing Exponential and Logarithmic Functions

1. Set your grapher to Parametric mode and Simultaneous graphing mode.

   Set $X1T = T$ and $Y1T = 2^{\wedge}T$.

   Set $X2T = 2^{\wedge}T$  and $Y2T = T$.

   *Creating Tables.* Set $TblStart = -3$ and $\Delta Tbl = 1$. Use the Table feature of your grapher to obtain the decimal form of both parts of Table 3.15. Be sure to scroll to the right to see X2T and Y2T.

   *Drawing Graphs.* Set $Tmin = -6$, $Tmax = 6$, and $Tstep = 0.5$. Set the $(x, y)$ window to $[-6, 6]$ by $[-4, 4]$. Use the Graph feature to obtain the simultaneous graphs of $f(x) = 2^x$ and $f^{-1}(x) = \log_2 x$. Use the Trace feature to explore the numerical relationships within the graphs.

2. *Graphing in Function mode.* Graph $y = 2^x$ in the decimal window. Then use the "draw inverse" command to draw the graph of $y = \log_2 x$.

## Common Logarithms—Base 10

Because of their close connection to our base 10 numeration system and to scientific notation, logarithms with base 10 are especially useful. Logarithms with base 10 are **common logarithms**. We often drop the subscript of 10 for the base when using common logarithms. The common logarithmic function $\log_{10} x = \log x$ is the inverse of the exponential function $f(x) = 10^x$. So

$$y = \log x \quad \text{if and only if} \quad 10^y = x.$$

Applying this relationship, we can obtain other fundamental relationships for logarithms with base 10.

---

**Basic Properties of Common Logarithms**

Let $x$ and $y$ be real numbers with $x > 0$.

- $\log 1 = 0$ because $10^0 = 1$.
- $\log 10 = 1$ because $10^1 = 10$.
- $\log 10^y = y$ because $10^y = 10^y$.
- $10^{\log x} = x$ because $\log x = \log x$.

---

Using the definition of common logarithm or these basic properties, we can evaluate expression involving a base of 10.

**Example 3**  EVALUATING LOGARITHMIC AND EXPONENTIAL EXPRESSIONS—BASE 10

(a) $\log 1000 = \log_{10} 1000 = 3$ because $10^3 = 1000$.

(b) $\log \sqrt{10} = \dfrac{1}{2}$ because $10^{1/2} = \sqrt{10}$.

## Some Words of Warning

In Figure 3.21, notice we used "10^Ans" instead of "10^1.537819095" to check log (34.5). This is because graphers generally store more digits than they display and so we can obtain a more accurate check. Even so, because log (34.5) is an irrational number, a grapher cannot produce its exact value, so checks like those shown in Figure 3.21 may not always work out so perfectly.

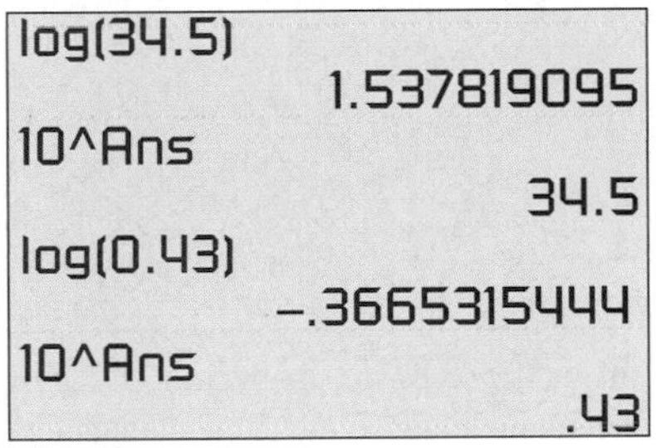

**(c)** $\log \dfrac{1}{10{,}000} = -4$ because $10^{-4} = \dfrac{1}{10^4} = \dfrac{1}{10{,}000}$.

**(d)** $\log \sqrt[5]{10} = \log 10^{1/5} = \dfrac{1}{5}$.

**(e)** $\log \dfrac{1}{1000} = \log \dfrac{1}{10^3} = \log 10^{-3} = -3$.

**(f)** $10^{\log 6} = 6$.

Common logarithms can be evaluated by using the $\boxed{\text{LOG}}$ key on a calculator, as illustrated in Example 4.

### Example 4 EVALUATING COMMON LOGARITHMS WITH A CALCULATOR

Use a calculator to evaluate the logarithmic expression, if it is defined, and check your result by evaluating the corresponding exponential expression.

**(a)** $\log 34.5 = 1.537\ldots$ because $10^{1.537\ldots} = 34.5$.

**(b)** $\log 0.43 = -0.366\ldots$ because $10^{-0.366\ldots} = 0.43$.

See Figure 3.21.

**(c)** $\log(-3)$ is undefined because there is no real number $y$ such that $10^y = -3$.

A grapher will yield either an error message or a complex number answer for entries such as $\log(-3)$. We shall restrict the domain of logarithmic functions to the set of positive real numbers and ignore such complex number answers.

```
log(34.5)
            1.537819095
10^Ans
                   34.5
log(0.43)
            -.3665315444
10^Ans
                    .43
```

**Figure 3.21** Doing and checking common logarithmic computations. (Example 4)

Changing from logarithmic form to exponential form or vice versa sometimes is enough to solve an equation involving exponential and logarithmic functions.

### Notes on Examples

Examples 5 and 8 involve simple equations that can be solved by switching between exponential and logarithmic form. Tougher equations are addressed in Section 3.5.

### Example 5 SOLVING EQUATIONS WITH LOGARITHMS AND EXPONENTS—BASE 10

Solve the equation.

**(a)** $10^x = 4.6$      **(b)** $\log x = -1.3$

**Solution**

**(a)** $10^x = 4.6$

$\qquad x = \log 4.6$          Change to logarithmic form.

$\qquad x = 0.662\ldots \approx 0.66$

**(b)** $\log x = -1.3$

$\qquad x = 10^{-1.3}$          Change to exponential form.

$\qquad x = 0.050\ldots \approx 0.05$

## Natural Logarithms—Base e

Because of their special calculus properties, logarithms with the natural base $e$ are used in many situations. Logarithms with base $e$ are **natural logarithms**. We often use the special abbreviation "ln" (without a subscript) to denote a natural logarithm. Thus, the natural logarithmic function $\log_e x = \ln x$. It is the inverse of the exponential function $f(x) = e^x$. So

$$y = \ln x \quad \text{if and only if} \quad e^y = x.$$

Applying this relationship, we can obtain other fundamental relationships for logarithms with the natural base $e$.

---

### Basic Properties of Natural Logarithms

Let $x$ and $y$ be real numbers with $x > 0$.

- $\ln 1 = 0$ because $e^0 = 1$.
- $\ln e = 1$ because $e^1 = e$.
- $\ln e^y = y$ because $e^y = e^y$.
- $e^{\ln x} = x$ because $\ln x = \ln x$.

---

Using the definition of natural logarithm or these basic properties, we can evaluate expressions involving the natural base $e$.

### Example 6 EVALUATING LOGARITHMIC AND EXPONENTIAL EXPRESSIONS—BASE e

(a) $\ln \sqrt{e} = \log_e \sqrt{e} = 1/2$ because $e^{1/2} = \sqrt{e}$.

(b) $\ln e^5 = \log_e e^5 = 5$.

(c) $e^{\ln 4} = 4$.

Natural logarithms can be evaluated by using the $\boxed{\text{LN}}$ key on a calculator, as illustrated in Example 7.

### Example 7 EVALUATING NATURAL LOGARITHMS WITH A CALCULATOR

Use a calculator to evaluate the logarithmic expression, if it is defined, and check your result by evaluating the corresponding exponential expression.

(a) $\ln 23.5 = 3.157\ldots$ because $e^{3.157\ldots} = 23.5$.

(b) $\ln 0.48 = -0.733\ldots$ because $e^{-0.733\ldots} = 0.48$.

See Figure 3.22.

(c) $\ln(-5)$ is undefined because there is no real number $y$ such that $e^y = -5$.

A grapher will yield either an error message or a complex number answer for entries such as $\ln(-5)$. We will continue to restrict the domain of logarithmic functions to the set of positive real numbers and ignore such complex number answers.

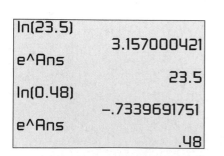

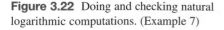

**Figure 3.22** Doing and checking natural logarithmic computations. (Example 7)

**Example 8** SOLVING EQUATIONS WITH LOGARITHMS
AND EXPONENTS—BASE $e$

Solve the equation.

**(a)** $\ln x = 4.62$ **(b)** $e^x = 7.68$

**Solution**

**(a)** $\ln x = 4.62$

$\qquad x = e^{4.62}$ Change to exponential form.

$\qquad x = e^{4.62} = 101.494\ldots \approx 101.49$

**(b)** $e^x = 7.68$

$\qquad x = \ln 7.68$ Change to logarithmic form.

$\qquad x = 2.038\ldots \approx 2.04$

In Exploration 2 of Section 3.1 we used graphical and numerical methods to approximate the value of $k$ such that $e^{kx} = 2^x$. We now can see that $k = \ln 2$ because $2^x = (e^{\ln 2})^x = e^{(\ln 2)x}$.

## Graphs of Logarithmic Functions

The natural logarithmic function $f(x) = \ln x$ is one of the basic functions introduced in Section 1.3. We now list its properties.

### The Natural Logarithmic Function

$f(x) = \ln x$

Domain: $(0, \infty)$. Range: All reals.
$x$-intercepts: $(1, 0)$. $y$-intercepts: nonc.
Continuous on $(0, \infty)$.
Increasing on $(0, \infty)$.
No symmetry.
Not bounded above or below.
No local extrema.
No horizontal asymptotes.
Vertical asymptote: $x = 0$.
End behavior: $\lim\limits_{x \to \infty} \ln x = \infty$

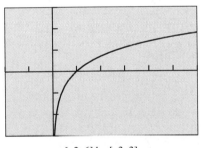

[−2, 6] by [−3, 3]

**Figure 3.23** The graph of $f(x) = \ln x$.

The natural logarithmic function $f(x) = \ln x$ is a one-to-one function and its graph is concave down. Any logarithmic function $g(x) = \log_b x$ with $b > 1$ has the same domain, range, continuity, intercept, increasing behavior, lack of symmetry, and other general behavior as $f(x) = \ln x$. It is rare that we are interested in logarithmic functions $g(x) = \log_b x$ with $0 < b < 1$. So, the graph and behavior of $f(x) = \ln x$ is typical of logarithmic functions.

In this section, we focus on the graphs of the common and natural logarithmic functions, and geometric transformations of these two graphs. To understand the nature of the graphs of $y = \log x$ and $y = \ln x$, we can compare each to the graph of its inverse, $y = 10^x$ and $y = e^x$, respectively. Figure 3.24a shows that the graphs of $y = \ln x$ and $y = e^x$ are reflections of each other across the line $y = x$. Similarly, Figure 3.24b shows that the graphs of $y = \log x$ and $y = 10^x$ are reflections of each other across this same line.

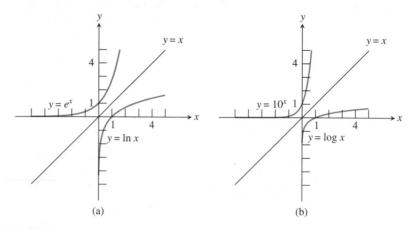

(a)                                                    (b)

**Figure 3.24** Two pairs of inverse functions.

From Figure 3.24 we can see that the graphs of $y = \log x$ and $y = \ln x$ have much in common. Figure 3.25 shows how they differ.

The geometric transformations studied in Section 1.5 together with our knowledge of the graphs of $y = \ln x$ and $y = \log x$ allow us to predict the graphs of the functions in Examples 9 and 10.

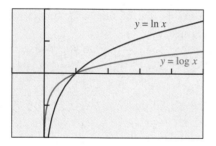

[−1, 5] by [−2, 2]

**Figure 3.25** The graphs of the common and natural logarithmic functions.

### Example 9  TRANSFORMING THE NATURAL LOGARITHMIC FUNCTION

Describe how to transform the graph of $f(x) = \ln x$ into the graph of the given function. Sketch the graph by hand and support your answer with a grapher.

**(a)** $g(x) = \ln (x + 2)$          **(b)** $h(x) = \ln (3 - x)$

**Solution**

**(a)** The graph of $g(x) = \ln (x + 2)$ is obtained by translating the graph of $f(x) = \ln (x)$ 2 units to the left. See Figure 3.26a.

**(b)** $h(x) = \ln (3 - x)$

$\qquad = \ln [-(x - 3)]$

$\qquad = f(-(x - 3))$          A reflection followed by a translation

So we can obtain the graph of $h(x) = \ln (3 - x)$ from the graph of $f(x) = \ln x$ by applying, in order, a reflection across the $y$-axis followed by a translation 3 units to the right. See Figure 3.26b.

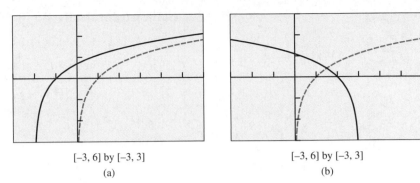

[−3, 6] by [−3, 3]
(a)

[−3, 6] by [−3, 3]
(b)

**Figure 3.26** Transforming $f(x) = \ln x$ to obtain (a) $g(x) = \ln (x + 2)$ and (b) $h(x) = \ln (3 − x)$. (Example 9)

## Example 10 TRANSFORMING THE COMMON LOGARITHMIC FUNCTION

Describe how to transform the graph of $f(x) = \log x$ into the graph of the given function. Sketch the graph by hand and support your answer with a grapher.

**(a)** $g(x) = 3 \log x$        **(b)** $h(x) = 1 + \log x$

**Solution**

**(a)** The graph of $g(x) = 3 \log x$ is obtained by vertically stretching the graph of $f(x) = \log x$ by a factor of 3. See Figure 3.27a.

**(b)** We can obtain the graph of $h(x) = 1 + \log x$ from the graph of $f(x) = \log x$ by a translation 1 unit up. See Figure 3.27b.

## Applications

Logarithmic functions model a variety of scientific and natural phenomena. Example 11 illustrates one such situation.

## Example 11 MODELING DRUG ABSORPTION

A certain drug is administered intravenously for pain. Beginning with 100 cc (cubic centimeters) of the drug, the number of cubic centimeters present in the body after $t$ hours, $f(t)$, is modeled by

$$f(t) = 100 − 43 \ln (1 + t), \quad 0 \le t \le 6.$$

How long will it be until 25 cc of the drug remain?

**Solution**

**Model**

Because $f(t) = 100 − 43 \ln (1 + t)$ models the number of cubic centimeters of the drug in the body at time $t$, we need to solve the equation

$$100 − 43 \ln (1 + t) = 25.$$

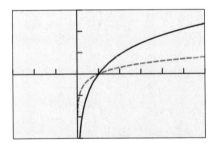

[−3, 6] by [−3, 3]
(a)

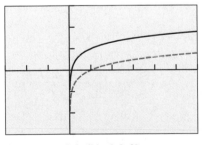

[−3, 6] by [−3, 3]
(b)

**Figure 3.27** Transforming $f(x) = \log x$ to obtain (a) $g(x) = 3 \log x$ and (b) $h(x) = 1 + \log x$. (Example 10)

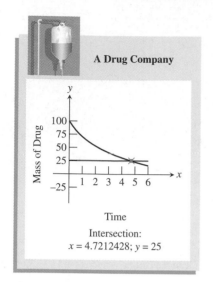

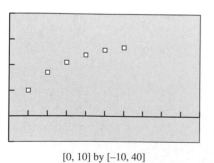

**Figure 3.28** Graph for Example 11.

### Solve Algebraically

$$43 \ln (1 + t) + 25 = 100$$

$$43 \ln (1 + t) = 100 - 25$$

$$\ln (1 + t) = \frac{75}{43}$$

$$1 + t = e^{75/43} \qquad \text{Change to exponential form.}$$

$$t = e^{75/43} - 1$$

$$t = 4.721\ldots$$

### Support Graphically

Figure 3.28 shows that the graphs of $f(t) = 100 - 43 \ln (1 + t)$ and $y = 25$ intersect at $t \approx 4.72$.

### Interpret

After roughly 4 hr 43 min only 25 cc of the drug will remain.

Sometimes the curve of best fit through a set of data is a logarithmic curve, as illustrated in Example 12.

### Example 12   USING LOGARITHMIC REGRESSION

Find a regression equation of the form $f(x) = a + b \ln x$ that fits the data. Then predict the value for $f(9)$.

| $x$ | 1 | 2 | 3 | 4 | 5 | 6 |
|---|---|---|---|---|---|---|
| $f(x)$ | 10 | 17 | 21 | 24 | 26 | 27 |

**Solution**  We enter the data and create a scatter plot. See Figure 3.29a. We then compute the logarithmic regression equation to be

$$f(x) = 10.205 + 9.6929 \ln x.$$

Figure 3.29b shows the scatter plot of the data with a graph of the regression model just found. You can see that the curve fits the data well. The coefficient of determination $r^2 \approx 0.9973$ supports the visual evidence. To predict $f(9)$ we substitute $x = 9$ into the regression model. Figure 3.29c reports that

$$f(9) = 10.205 + 9.6929 \ln 9 \approx 31.50.$$

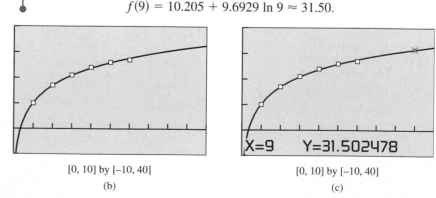

[0, 10] by [−10, 40]

(a)

[0, 10] by [−10, 40]

(b)

[0, 10] by [−10, 40]

(c)

**Figure 3.29** Scatter plot and graphs for Example 12.

**Problem**

If the noise inside the subway is measured at $10^{-3}$ watts/m$^2$, what is the decibel level?

**Solution**

We can compute the decibel intensity of a sound by using the equation

$$\text{Intensity level in dB} = 10 \log (I/I_0)$$

We know that the threshold of hearing is $10^{-12}$ watts/m$^2$ and the magnitude of the subway noise is measured at $10^{-3}$ watts/m$^2$, so we can compute the decibel level as

$$\begin{aligned}
\text{Intensity level} &= 10 \log (I/I_0) \\
&= 10 \log (10^{-3}/10^{-12}) \\
&= 10 \log (10^9) \\
&= 10 (9) \\
&= 90 \text{ dB}
\end{aligned}$$

So the sound intensity inside the subway is 90 dB.

# Quick Review 3.3

In Exercises 1–6, evaluate the expression without using a calculator.

**1.** $5^{-2}$  1/25 = 0.04

**2.** $10^{-3}$  1/1000 = 0.001

**3.** $\dfrac{4^0}{5}$  1/5 = 0.2

**4.** $\dfrac{1^0}{2}$  1/2 = 0.5

**5.** $\dfrac{8^{11}}{2^{28}}$  32

**6.** $\dfrac{9^{13}}{27^8}$  9

In Exercises 7–10, rewrite as a base raised to a rational number exponent.

**7.** $\sqrt{5}$  $5^{1/2}$

**8.** $\sqrt[3]{10}$  $10^{1/3}$

**9.** $\dfrac{1}{\sqrt{e}}$  $e^{-1/2}$

**10.** $\dfrac{1}{\sqrt[3]{e^2}}$  $e^{-2/3}$

# Section 3.3 Exercises

In Exercises 1–18, evaluate the logarithmic expression without using a calculator.

**1.** $\log_4 4$  1

**2.** $\log_6 1$  0

**3.** $\log_2 32$  5

**4.** $\log_3 81$  4

**5.** $\log_5 \sqrt[3]{25}$  2/3

**6.** $\log_6 \dfrac{1}{\sqrt[5]{36}}$  −2/5

**7.** $\log 10^3$  3

**8.** $\log 10{,}000$  4

**9.** $\log 100{,}000$  5

**10.** $\log 10^{-4}$  −4

**11.** $\log \sqrt[3]{10}$  1/3

**12.** $\log \dfrac{1}{\sqrt{1000}}$  −3/2

**13.** $\ln e^3$  3

**14.** $\ln e^{-4}$  −4

**15.** $\ln \dfrac{1}{e}$  −1

**16.** $\ln 1$  0

**17.** $\ln \sqrt[4]{e}$  1/4

**18.** $\ln \dfrac{1}{\sqrt{e^7}}$  −7/2

In Exercises 19–22, evaluate the expression without using a calculator.

**19.** $7^{\log_7 3}$  3

**20.** $5^{\log_5 8}$  8

**21.** $e^{\ln 6}$  6

**22.** $10^{\log 14}$  14

In Exercises 23–30, solve the equation by changing it to exponential form.

**23.** $\log x = 2$  100

**24.** $\log x = 4$  10,000

**25.** $\log x = -1$  0.1

**26.** $\log x = -3$  0.001

**27.** $\ln x = 3$  $e^3 \approx 20.0855$

**28.** $\ln x = 5$  $e^5 \approx 148.4132$

**29.** $\ln x = -2$  $e^{-2} \approx 0.1353$

**30.** $\ln x = -1$  $e^{-1} \approx 0.3679$

In Exercises 31–36, solve the equation by changing it to logarithmic form.

**31.** $10^x = 3$  log 3 ≈ 0.4771    **32.** $10^x = 5.1$

**33.** $e^x = 4.2$  ln 4.2 ≈ 1.4351    **34.** $e^x = 7.3$  ln 7.3 ≈ 1.9879

**35.** $e^{2x} = 5.3$    **36.** $10^{3x} = 9.2$

In Exercises 37–40, solve the equation.

**37.** ln $x = 2.8$  $e^{2.8}$ ≈ 16.4446    **38.** ln $x = 3.1$  $e^{3.1}$ ≈ 22.1980

**39.** log $x = 8.23$    **40.** log $x = 5.25$

In Exercises 41–44, match the function with its graph.

**41.** $f(x) = \log(1 - x)$  (d)    **42.** $f(x) = \log(x + 1)$  (b)

**43.** $f(x) = -\ln(x - 3)$  (a)    **44.** $f(x) = -\ln(4 - x)$  (c)

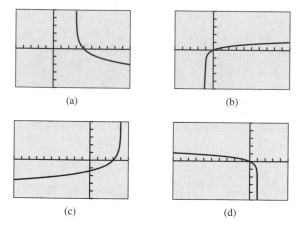

(a)            (b)

(c)            (d)

In Exercises 45–48, describe how to transform the graph of $g(x) = \ln x$ into the graph of the given function. Sketch the graph by hand and support with a grapher.

**45.** $f(x) = \ln(-x) + 3$    **46.** $f(x) = \ln(-x) - 2$

**47.** $f(x) = \ln(2 - x)$    **48.** $f(x) = \ln(5 - x)$

In Exercises 49–52, describe how to transform the graph of $g(x) = \log x$ into the graph of the given function. Sketch the graph by hand and support with a grapher.

**49.** $f(x) = -2 \log(-x)$    **50.** $f(x) = -3 \log(-x)$

**51.** $f(x) = 2 \log(3 - x) - 1$    **52.** $f(x) = -3 \log(1 - x) + 1$

In Exercises 53–56, graph the function, and analyze it for domain, range, continuity, intercepts, increasing or decreasing behavior, symmetry, boundedness, extrema, asymptotes, end behavior, and concavity.

**53.** $f(x) = \log(x - 2)$    **54.** $f(x) = \ln(x + 1)$

**55.** $f(x) = -\ln(x - 1)$    **56.** $f(x) = -\log(x + 2)$

**57. Drug Absorption** A drug is administered intravenously for pain. The function $f(t) = 80 - 23 \ln(1 + t)$, $0 \le t \le 24$, models the amount of the drug in the body after $t$ hours.

**(a)** What was the initial ($t = 0$) number of units of drug administered? 80 units

**(b)** How much is present after 2 h?

**(c)** Draw the graph of $f$.

**58. Light Absorption** The Beer-Lambert law of absorption applied to Lake Erie states that the light intensity $I$ (in lumens), at a depth of $x$ feet, satisfies the equation

$$\log \frac{I}{12} = -0.00235x.$$

Find the intensity of the light at a depth of 30 ft.

**59.** Estimate the $y$-value associated with $x = 25$ as predicted by the logarithmic regression equation for the data  $y \approx 3.44$

| $x$ | 10 | 20 | 30 | 40 |
|---|---|---|---|---|
| $y$ | 1.61 | 2.99 | 3.80 | 4.38 |

**60.** Estimate the $y$-value associated with $x = 13$ as predicted by the logarithmic regression equation for the data.  $y \approx 9.69$

| $x$ | 6 | 12 | 18 | 24 |
|---|---|---|---|---|
| $y$ | 7.38 | 9.45 | 10.67 | 11.53 |

**61. Algerian Oil Production** Find a natural logarithmic regression equation for the data in Table 3.16. Use it to estimate production for the year 1980. Use 1950 as $t = 0$.

| Table 3.16 Algerian Oil Production ||
|---|---|
| Year | Metric Tons |
| 1960 | 8.63 |
| 1970 | 47.25 |
| 1990 | 56.67 |

*Source: The Statesman's Yearbook, 129th ed. (London: The Macmillan Press, Ltd., 1992).*

**62. Canadian Oil Production** Find a natural logarithmic regression equation for the data in Table 3.17. Use it to estimate production for the year 1985. Use 1950 as $t = 0$.

| Table 3.17 Canadian Oil Production ||
|---|---|
| Year | Metric Tons |
| 1960 | 27.48 |
| 1970 | 69.95 |
| 1990 | 92.24 |

*Source: The Statesman's Yearbook, 129th ed. (London: The Macmillan Press, Ltd., 1992).*

## Explorations

**63. Writing to Learn  Parametric Graphing** In the manner of Exploration 1, make tables and graphs for $f(x) = 3^x$ and its inverse $f^{-1}(x) = \log_3 x$. Write a comparative analysis of the two functions regarding domain, range, intercepts, and asymptotes.

**64. Writing to Learn Parametric Graphing** In the manner of Exploration 1, make tables and graphs for $f(x) = 5^x$ and its inverse $f^{-1}(x) = \log_5 x$. Write a comparative analysis of the two functions regarding domain, range, intercepts, and asymptotes.

**65. Group Activity Parametric Graphing** In the manner of Exploration 1, find the number $b > 1$ such that the graphs of $f(x) = b^x$ and its inverse $f^{-1}(x) = \log_b x$ have exactly one point of intersection. What is the one point that is in common to the two graphs? $b = \sqrt[e]{e}$; $(e, e)$ ∎

**66. Writing to Learn** Explain why zero is not in the domain of the logarithmic functions $y = \log_3 x$ and $y = \log_5 x$.

## Extending the Ideas

**67.** Describe how to transform the graph of $f(x) = \ln x$ into the graph of $g(x) = \log_{1/e} x$. reflect across the *x*-axis

**68.** Describe how to transform the graph of $f(x) = \log x$ into the graph of $g(x) = \log_{0.1} x$. reflect across the *x*-axis

---

| **3.4** | **Properties of Logarithmic Functions** |

Properties of Logarithms • Change of Base • Graphs of Logarithmic Functions with Base $b$ • Re-Expressing Data

**Properties of Exponents**

Let $b$, $x$, and $y$ be real numbers with $b > 0$.

**1.** $b^x \cdot b^y = b^{x+y}$.

**2.** $\dfrac{b^x}{b^y} = b^{x-y}$.

**3.** $(b^x)^y = b^{xy}$.

### Properties of Logarithms

Logarithms have distinctive algebraic characteristics that historically made them indispensable for calculations and that still make them valuable in many areas of applications and modeling. In Section 3.3 we learned about the inverse relationship between exponents and logarithms and how to apply some basic properties of logarithms. We now delve deeper into the nature of logarithms to prepare for equation solving and modeling.

**Objective**

Students will be able to apply the properties of logarithms to evaluate expressions and graph functions, and be able to re-express data.

**Motivate**

Have students use a grapher to compare the graphs of $y = \log x$ and $y = \log (10x)$. Discuss the graphs.

**Lesson Guide**

Day 1: Properties of Logarithms; Change of Base; Graphs of Logarithmic Functions with Base $b$.
Day 2: Re-Expressing Data

> **Properties of Logarithms**
>
> Let $b$, $R$, and $S$ be positive real numbers with $b \neq 1$, and $c$ any real number.
>
> - **Product rule:** $\quad \log_b (RS) = \log_b R + \log_b S$
>
> - **Quotient rule:** $\quad \log_b \dfrac{R}{S} = \log_b R - \log_b S$
>
> - **Power rule:** $\quad \log_b R^c = c \log_b R$

The properties of exponents in the margin are the basis for these three properties of logarithms. For instance, the first exponent property listed in the margin is used to verify the product rule. Example 1 uses this property in an approach to proving the product rule that can be adjusted to prove the quotient and power rules for logarithms, which you are asked to do in Exercise 62.

**Example 1 PROVING THE PRODUCT RULE FOR LOGARITHMS**

Prove $\log_b (RS) = \log_b R + \log_b S$.

**Solution** Let $x = \log_b R$ and $y = \log_b S$. The corresponding exponential statements are $b^x = R$ and $b^y = S$. Therefore,

$$RS = b^x \cdot b^y$$

$$= b^{x+y} \qquad \text{First property of exponents.}$$

$$\log_b (RS) = x + y \qquad \text{Change to logarithmic form.}$$

$$= \log_b R + \log_b S \qquad \text{Use the definitions of } x \text{ and } y.$$

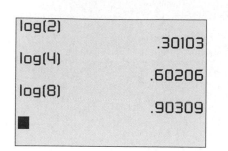

**Figure 3.30**  An arithmetic pattern of logarithms. (Exploration 1)

**Exploration Extensions**

Using the information given in Figure 3.30 and the value of log 5 found in Step 4, evaluate log (16/5), log (5/16), and log (2/5) without using a calculator

---

**Exploration 1**   **Exploring the Arithmetic of Logarithms**

Use the 5-decimal approximations shown in Figure 3.30 to support the properties of logarithms numerically.

**1.** Product    $\log (2 \cdot 4) = \log 2 + \log 4$

**2.** Quotient    $\log \left(\dfrac{8}{2}\right) = \log 8 - \log 2$

**3.** Power    $\log 2^3 = 3 \log 2$

Now evaluate the common logs of other positive integers using the information given in Figure 3.30 and without using your calculator.

**4.** Use the fact that $5 = 10/2$ to evaluate log 5.   0.69897

**5.** Use the fact that 16, 32, and 64 are powers of 2 to evaluate log 16, log 32, and log 64.   1.20412; 1.50515; 1.80618

**6.** Evaluate log 25, log 40, and log 50.   1.39794; 1.60206; 1.69897

List all of the positive integers less than 100 whose common logs can be evaluated knowing only log 2 and the properties of logarithms and without using a calculator.

---

When we solve equations algebraically that involve logarithms, we often have to rewrite expressions using properties of logarithms. Sometimes we need to expand as far as possible, and other times we combine as much as possible. The next three examples illustrate how properties of logarithms can be used to change the form of expressions involving logarithms.

**Example 2   REWRITING THE LOGARITHM OF A PRODUCT**

Assuming $x$ and $y$ are positive, use properties of logarithms to write $\log (8xy^4)$ as a sum of logarithms or multiples of logarithms.

**Solution**

$$\log (8xy^4) = \log 8 + \log x + \log y^4 \qquad \text{Product rule}$$
$$= \log 2^3 + \log x + \log y^4$$
$$= 3 \log 2 + \log x + 4 \log y \qquad \text{Power rule}$$

**Example 3   REWRITING THE LOGARITHM OF A QUOTIENT**

Assuming $x$ is positive, use properties of logarithms to write $\ln (\sqrt{x^2 + 5}/x)$ as a sum or difference of logarithms or multiples of logarithms.

**Solution**

$$\ln \frac{\sqrt{x^2 + 5}}{x} = \ln \frac{(x^2 + 5)^{1/2}}{x}$$
$$= \ln (x^2 + 5)^{1/2} - \ln x \qquad \text{Quotient rule}$$
$$= \frac{1}{2} \ln (x^2 + 5) - \ln x \qquad \text{Power rule}$$

### Example 4   CONDENSING A LOGARITHMIC EXPRESSION

Assuming $x$ and $y$ are positive, use properties of logarithms to write $\ln x^5 - 2 \ln (xy)$ as a single logarithm.

**Solution**

$$\ln x^5 - 2 \ln (xy) = \ln x^5 - \ln (xy)^2 \qquad \text{Power rule}$$

$$= \ln x^5 - \ln (x^2 y^2)$$

$$= \ln \frac{x^5}{x^2 y^2} \qquad \text{Quotient rule}$$

$$= \ln \frac{x^3}{y^2}$$

As we have seen, logarithms have some surprising properties. It is easy to overgeneralize and fall into misconceptions about logarithms. Exploration 2 should help you discern what is true and false about logarithmic relationships.

**Exploration Extensions**

Repeat the Exploration exercise with following relationships:

**9.** $\log_3 5 = (\log_3 x)(\log_x 5)$

**10.** $e^{\ln 6 + \ln x} = 6x$

---

**Exploration 2**   **Discovering Relationships and Non-Relationships**

Of the eight relationships suggested here, four are *true* and four are *false* (using values of $x$ within the domains of both sides of the equations). Thinking about the properties of logarithms, make a prediction about the truth of each statement. Then test each with some specific numerical values for $x$. Finally, compare the graphs of the two sides of the equation.

**1.** $\ln (x + 2) = \ln x + \ln 2$  false     **2.** $\log_3 (7x) = 7 \log_3 x$  false

**3.** $\log_2 (5x) = \log_2 5 + \log_2 x$  true  **4.** $\ln \dfrac{x}{5} = \ln x - \ln 5$  true

**5.** $\log \dfrac{x}{4} = \dfrac{\log x}{\log 4}$  false      **6.** $\log_4 x^3 = 3 \log_4 x$  true

**7.** $\log_5 x^2 = (\log_5 x)(\log_5 x)$  false  **8.** $\log |4x| = \log 4 + \log |x|$  true

Which four are true, and which four are false?

---

## Change of Base

When working with a logarithmic expression with an undesirable base, it is possible to change the expression into a quotient of logarithms with a different base.

### Example 5   CHANGING THE BASE OF A LOGARITHM

Prove that $\log_4 7 = \dfrac{\ln 7}{\ln 4}$.

**Solution**   Let $y = \log_4 7$. Then $4^y = 7$. Taking the natural log of each side of this exponential equation yields

$$\ln 4^y = \ln 7$$

$$y \ln 4 = \ln 7 \qquad \text{Power rule}$$

$$y = \frac{\ln 7}{\ln 4} \qquad \text{Divide by ln 4}$$

In Exercise 63 you are asked to prove the following generalization of Example 5.

> **Change-of-Base Formula for Logarithms**
>
> For positive real numbers $a$, $b$, and $x$ with $a \neq 1$ and $b \neq 1$,
>
> $$\log_b x = \frac{\log_a x}{\log_a b}.$$

Calculators and graphers generally have two logarithm keys—$\boxed{\text{LOG}}$ and $\boxed{\text{LN}}$—which correspond to the bases 10 and $e$, respectively. So we often use the change-of-base formula in one of the following two forms:

$$\log_b x = \frac{\log x}{\log b} \quad \text{or} \quad \log_b x = \frac{\ln x}{\ln b}$$

These two forms are useful in evaluating logarithms and graphing logarithmic functions.

**Example 6  EVALUATING LOGARITHMS BY CHANGING THE BASE**

**(a)** $\log_3 16 = \dfrac{\ln 16}{\ln 3} = 2.523\ldots \approx 2.52$

**(b)** $\log_6 10 = \dfrac{\log 10}{\log 6} = \dfrac{1}{\log 6} = 1.285\ldots \approx 1.29$

**(c)** $\log_{1/2} 2 = \dfrac{\ln 2}{\ln (1/2)} = \dfrac{\ln 2}{\ln 1 - \ln 2} = \dfrac{\ln 2}{-\ln 2} = -1$

## Graphs of Logarithmic Functions with Base $b$

Using the change-of-base formula we can rewrite any logarithmic function $g(x) = \log_b x$ as

$$g(x) = \frac{\ln x}{\ln b} = \frac{1}{\ln b} \ln x.$$

So every logarithmic function is a constant multiple of the natural logarithmic function $f(x) = \ln x$. If the base is $b > 1$, the graph of $g(x) = \log_b x$ is a vertical stretch or shrink of the graph of $f(x) = \ln x$ by the factor $1/\ln b$. If $0 < b < 1$, a reflection across the $x$-axis is required as well.

**Example 7  GRAPHING LOGARITHMIC FUNCTIONS**

Describe how to transform the graph of $f(x) = \ln x$ into the graph of the given function. Sketch the graph by hand and support your answer with a grapher.

**(a)** $g(x) = \log_5 x$          **(b)** $h(x) = \log_{1/4} x$

**Solution**

**(a)** Because $g(x) = \log_5 x = \ln x/\ln 5$, its graph is obtained by vertically shrinking the graph of $f(x) = \ln x$ by a factor of $1/\ln 5 \approx 0.62$. See Figure 3.31a.

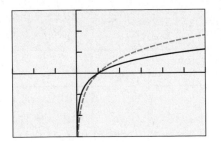

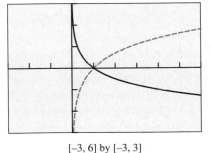

[-3, 6] by [-3, 3]

(a)

[-3, 6] by [-3, 3]

(b)

**Figure 3.31** Transforming $f(x) = \ln x$ to obtain (a) $g(x) = \log_5 x$ and (b) $h(x) = \log_{1/4} x$. (Example 7)

**(b)** $h(x) = \log_{1/4} x = \dfrac{\ln x}{\ln 1/4} = \dfrac{\ln x}{\ln 1 - \ln 4} = \dfrac{\ln x}{-\ln 4} = -\dfrac{1}{\ln 4} \ln x$

So we can obtain the graph of $h$ from the graph of $f(x) = \ln x$ by applying, in either order, a reflection across the $x$-axis and a vertical shrink by a factor of $1/\ln 4 \approx 0.72$. See Figure 3.31b.

We can generalize Example 7b in the following way: If $b > 1$, then $0 < 1/b < 1$ and

$$\log_{1/b} x = -\log_b x.$$

So when given a function like $h(x) = \log_{1/4} x$, with a base between 0 and 1, we can immediately rewrite it as $h(x) = -\log_4 x$. Because we can so readily change the base of logarithms with bases between 0 and 1, such logarithms are rarely encountered or used. Instead, we work with logarithms that have bases $b > 1$, which behave much like natural and common logarithms, as we have seen and now summarize.

---

**Logarithmic Functions with $b > 1$**

Domain: $(0, \infty)$. Range: All reals.
$x$-intercept $= 1$. $y$-intercepts: none.
Continuous.
Increasing on its domain.
No symmetry: neither even nor odd.
Not bounded above or below.
No local extrema.
No horizontal asymptotes.
Vertical asymptote: $x = 0$.
End behavior: $\lim\limits_{x \to \infty} \log_b x = \infty$
Passes through $(1, 0)$ and $(b, 1)$.
One-to-one.     Concave down.

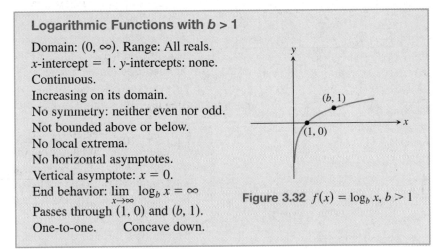

**Figure 3.32** $f(x) = \log_b x$, $b > 1$

---

## Re-Expressing Data

When seeking a model for a set of data it is often helpful to transform the data by applying a function to one or both of the variables in the data set. Such a transformation of a data set is a **re-expression** of the data.

Recall from Section 2.2 that Kepler's Third Law states that the square of the period of orbit $T$ for each planet is proportional to the cube of its average distance $a$ from the sun. If we re-express the Kepler planetary data in Table 2.11 using earth-based units, the constant of proportion becomes 1 and the "is proportional to" in Kepler's Third Law becomes "equals." We can do this by dividing the "average distance" column by 149.6 Gm/AU and the "period of orbit" column by 365.2 days/yr. The re-expressed data are shown in Table 3.18.

**Astronomically speaking**

Recall that an astronomical unit (AU) is the average distance between the Earth and the Sun, about 149.6 million kilometers (149.6 Gm).

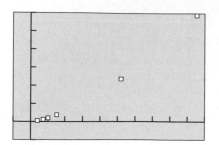

[-1, 10] by [-5, 30]

(a)

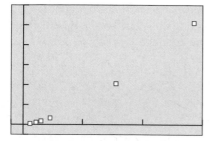

[-100, 1500] by [-1000, 12 000]

(b)

**Figure 3.33** Scatter plots of the planetary data from (a) Table 3.18 and (b) Table 2.11.

**Table 3.18** Average Distances and Orbit Periods for the Six Innermost Planets

| Planet | Average Distance from Sun (AU) | Period of Orbit (yr) |
|--------|-------------------------------|----------------------|
| Mercury | 0.3870 | 0.2410 |
| Venus | 0.7233 | 0.6161 |
| Earth | 1.000 | 1.000 |
| Mars | 1.523 | 1.881 |
| Jupiter | 5.203 | 11.86 |
| Saturn | 9.539 | 29.46 |

*Re-expression of data from: Shupe, et al., National Geographic Atlas of the World (rev. 6th ed.). Washington, DC: National Geographic Society, 1992, plate 116.*

The next step in the modeling process is to make a scatter plot and look for a pattern. Unfortunately, the pattern in the scatter plot of these re-expressed data, shown in Figure 3.33a, is essentially the same as the pattern in the scatter plot of the original data, shown in Figure 3.33b. What we have done is to make the numerical values of the data more convenient and to guarantee that our plot contains the ordered pair (1, 1) for Earth, which could potentially simplify our model. What we have *not* done and what we still wish to do is to clarify the relationship between the variables $a$ (average distance from the Sun) and $T$ (period of orbit).

One of the many uses of logarithms is to re-express data in order to clarify relationships and uncover hidden patterns. In the case of the planetary data, if we plot $(\ln a, \ln T)$ pairs instead of $(a, T)$ pairs, the pattern is much clearer. In Example 8, we carry out this re-expression of the data and then use an algebraic tour de force to obtain Kepler's Third Law.

**Example 8** ESTABLISHING KEPLER'S THIRD LAW USING LOGARITHMIC RE-EXPRESSION

Re-express the $(a, T)$ data pairs in Table 3.18 as $(\ln a, \ln T)$ pairs. Find a linear regression model for the $(\ln a, \ln T)$ pairs. Rewrite the linear regression in terms of $a$ and $T$, and rewrite the equation in a form with no logs or fractional exponents.

**Solution**

**Model**

We use grapher list operations to obtain the $(\ln a, \ln T)$ pairs (see Figure 3.34a). We make a scatter plot of the re-expressed data in Figure 3.34b. The $(\ln a, \ln T)$ pairs appear to lie along a straight line.

We let $y = \ln T$ and $x = \ln a$, and using linear regression, we obtain the following model:

$$y = 1.49950x + 0.00070 \approx 1.5x.$$

Figure 3.31c shows the scatter plot for the $(x, y) = (\ln a, \ln T)$ pairs together with a graph of $y = 1.5x$. You can see that the line fits the re-expressed data quite well.

**Remodel**

Returning to the original variables $a$ and $T$, we obtain:

$$\ln T = 1.5 \cdot \ln a \qquad y = 1.5x$$

$$\frac{\ln T}{\ln a} = 1.5 \qquad \text{Divide by } \ln a$$

$$\log_a T = \frac{3}{2} \qquad \text{Change of base}$$

$$T = a^{3/2} \qquad \text{Switch to exponential form}$$

$$T^2 = a^3 \qquad \text{Square both sides}$$

**Interpret**

This is Kepler's Third Law!

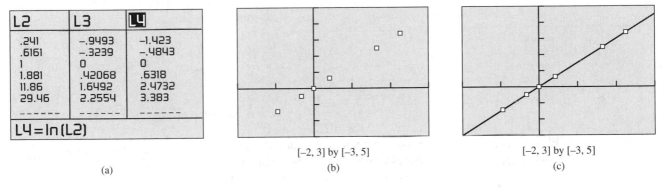

| L2 | L3 | L4 |
|---|---|---|
| .241 | -.9493 | -1.423 |
| .6161 | -.3239 | -.4843 |
| 1 | 0 | 0 |
| 1.881 | .42068 | .6318 |
| 11.86 | 1.6492 | 2.4732 |
| 29.46 | 2.2554 | 3.383 |
| ------ | ------ | ------ |
| L4 = ln(L2) | | |

(a)

[–2, 3] by [–3, 5]

(b)

[–2, 3] by [–3, 5]

(c)

**Figure 3.34**  Scatter plot and graphs for Example 8.

# Quick Review 3.4

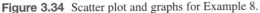

In Exercises 1–4, evaluate the expression without using a calculator.

**1.** $\log 10^2$  2

**2.** $\ln e^3$  3

**3.** $\ln e^{-2}$  −2

**4.** $\log 10^{-3}$  −3

In Exercises 5–10, simplify the expression.

**5.** $\dfrac{x^5 y^{-2}}{x^2 y^{-4}}$  $x^3 y^2$

**6.** $\dfrac{u^{-3} v^7}{u^{-2} v^2}$  $v^5/u$

**7.** $(x^6 y^{-2})^{1/2}$  $|x|^3/|y|$

**8.** $(x^{-8} y^{12})^{3/4}$  $|y|^9/x^6$

**9.** $\dfrac{(u^2 v^{-4})^{1/2}}{(27 u^6 v^{-6})^{1/3}}$  $1/(3|u|)$

**10.** $\dfrac{(x^{-2} y^3)^{-2}}{(x^3 y^{-2})^{-3}}$  $x^{13}/y^{12}$

# Section 3.4 Exercises

In Exercises 1–6, evaluate the logarithm.

**1.** $\log_2 7$  $\ln 7/\ln 2 \approx 2.8074$

**2.** $\log_5 19$

**3.** $\log_8 175$

**4.** $\log_{12} 259$

**5.** $\log_{0.5} 12$

**6.** $\log_{0.2} 29$

In Exercises 7–10, write the expression using only natural logarithms.

**7.** $\log_3 x$  $\ln x/\ln 3$

**8.** $\log_7 x$  $\ln x/\ln 7$

**9.** $\log_2 (a + b)$  $\ln (a + b)/\ln 2$  **10.** $\log_5 (c - d)$  $\ln (c - d)/\ln 5$

In Exercises 11–14, write the expression using only common logarithms.

**11.** $\log_2 x$  $\log x/\log 2$

**12.** $\log_4 x$  $\log x/\log 4$

**13.** $\log_{1/2} (x + y)$

**14.** $\log_{1/3} (x - y)$

In Exercises 15–18, graph the function, and state its domain and range.

**15.** $f(x) = \log_4 x$

**16.** $g(x) = \log_5 x$

**17.** $f(x) = \log_7 (x - 2)$

**18.** $g(x) = \log_3 (2 - x)$

In Exercises 19–22, match the function with its graph. Identify the window dimensions, Xscl, and Yscl of the graph.

**19.** $f(x) = \log_4 (2 - x)$      **20.** $f(x) = \log_6 (x - 3)$

**21.** $f(x) = \log_{0.5} (x - 2)$      **22.** $f(x) = \log_{0.7} (3 - x)$

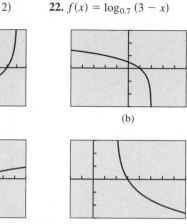

(a)                    (b)

(c)                    (d)

In Exercises 23–34, assuming $x$ and $y$ are positive, use properties of logarithms to write the expression as a sum or difference of logarithms or multiples of logarithms.

**23.** $\ln 8x$    $3 \ln 2 + \ln x$      **24.** $\ln 9y$    $2 \ln 3 + \ln y$

**25.** $\log \dfrac{3}{x}$    $\log 3 - \log x$      **26.** $\log \dfrac{2}{y}$    $\log 2 - \log y$

**27.** $\log_2 y^5$    $5 \log_2 y$      **28.** $\log_2 x^{-2}$    $-2 \log_2 x$

**29.** $\log x^3y^2$            **30.** $\log xy^3$

**31.** $\ln \dfrac{x^2}{y^3}$    $2 \ln x - 3 \ln y$      **32.** $\log 1000x^4$    $3 + 4 \log x$

**33.** $\log \sqrt[4]{\dfrac{x}{y}}$   $\dfrac{1}{4}\log x - \dfrac{1}{4}\log y$      **34.** $\ln \dfrac{\sqrt[3]{x}}{\sqrt[3]{y}}$   $\dfrac{1}{3}\ln x - \dfrac{1}{3}\ln y$

In Exercises 35–44, assuming $x$, $y$, and $z$ are positive, use properties of logarithms to write the expression as a single logarithm.

**35.** $\log x + \log y$    $\log xy$      **36.** $\log x + \log 5$    $\log 5x$

**37.** $\ln y - \ln 3$    $\ln (y/3)$      **38.** $\ln x - \ln y$    $\ln (x/y)$

**39.** $\dfrac{1}{3} \log x$    $\log \sqrt[3]{x}$      **40.** $\dfrac{1}{5} \log z$    $\log \sqrt[5]{z}$

**41.** $2 \ln x + 3 \ln y$    $\ln (x^2y^3)$      **42.** $4 \log y - \log z$    $\log (y^4/z)$

**43.** $4 \log (xy) - 3 \log (yz)$      **44.** $3 \ln (x^3y) + 2 \ln (yz^2)$

In Exercises 45–48, graph the function, and analyze it for domain, range, continuity, intercepts, increasing or decreasing behavior, asymptotes, and end behavior.

**45.** $f(x) = \log_2 (8x)$      **46.** $f(x) = \log_{1/3} (9x)$

**47.** $f(x) = \log (x^2)$      **48.** $f(x) = \ln (x^3)$

**49. Sound Intensity** The level of sound intensity is defined in terms of **decibels** (dB):

$$\beta(I) = 10 \log (I/I_0),$$

where $\beta$ is the sound intensity level in dB, $I$ is the sound intensity in watts/m², and $I_0 = 10^{-12}$ watts/m² is the threshold of human hearing (the quietest audible sound intensity). Compute the sound intensity level in decibels for each sound listed in Table 3.19.

**Table 3.19 Approximate Intensities for Selected Sounds**

| Sound | Intensity (watts/m²) |
|---|---|
| **(a)** Hearing threshold | $10^{-12}$ |
| **(b)** Rustling leaves | $10^{-11}$ |
| **(c)** Conversation | $10^{-6}$ |
| **(d)** School cafeteria | $10^{-4}$ |
| **(e)** Jack hammer | $10^{-2}$ |
| **(f)** Pain threshold | $1$ |

*Sources: J. J. Dwyer, College Physics (Belmont, CA: Wadsworth, 1984), and E. Connally et al., Functions Modeling Change (New York: Wiley, 2000).*

**50. Earthquake Intensity** The **Richter scale** magnitude $R$ of an earthquake is based on the features of the associated seismic wave and is measured by

$$R = \log (a/T) + B,$$

where $a$ is the amplitude in $\mu$m (micrometers), $T$ is the period in seconds, and $B$ accounts for the weakening of the seismic wave due to the distance from the epicenter. Compute the earthquake magnitude $R$ for each set of values.

**(a)** $a = 250$, $T = 2$,   and   $B = 4.25$

**(b)** $a = 300$, $T = 4$,   and   $B = 3.5$

**51. Light Intensity in Lake Erie** The relationship between intensity $I$ of light (in lumens) at a depth of $x$ feet in Lake Erie is given by

$$\log \frac{I}{12} = -0.00235x.$$

What is the intensity at a depth of 40 ft?

**52. Light Intensity in Lake Superior** The relationship between intensity $I$ of light (in lumens) at a depth of $x$ feet in Lake Superior is given by

$$\log \frac{I}{12} = -0.0125x.$$

What is the intensity at a depth of 10 ft?

**53. Writing to Learn** Use the change-of-base formula to explain how we know that the graph of $f(x) = \log_3 x$ can be obtained by applying a transformation to the graph of $g(x) = \ln x$.

**54. Writing to Learn** Use the change-of-base formula to explain how the graph of $f(x) = \log_{0.8} x$ can be obtained by applying transformations to the graph of $g(x) = \log x$.

## Explorations

**55. (a)** Compute the power regression model for the following data. $2.75x^{5.0}$

| $x$ | 4 | 6.5 | 8.5 | 10 |
|---|---|---|---|---|
| $y$ | 2816 | 31,908 | 122,019 | 275,000 |

**(b)** Predict the $y$ value associated with $x = 7.1$ using the power regression model. 49,616

**(c)** Re-express the data in terms of their natural logarithms, and make a scatter plot of $(\ln x, \ln y)$.

**(d)** Compute the linear regression model $(\ln y) = a(\ln x) + b$ for $(\ln x, \ln y)$. $5.00 (\ln x) + 1.01$

**(e)** Confirm that $y = e^b \cdot x^a$ is the power regression model found in (a).

**56. (a)** Compute the power regression model for the following data. $y = 8.095x^{-0.113}$

| $x$ | 2 | 3 | 4.8 | 7.7 |
|---|---|---|---|---|
| $y$ | 7.48 | 7.14 | 6.81 | 6.41 |

**(b)** Predict the $y$ value associated with $x = 9.2$ using the power regression model. 6.30

**(c)** Re-express the data in terms of their natural logarithms, and make a scatter plot of $(\ln x, \ln y)$.

**(d)** Compute the linear regression model $(\ln y) = a(\ln x) + b$ for $(\ln x, \ln y)$. $-0.113(\ln x) + 2.091$

**(e)** Confirm that $y = e^b \cdot x^a$ is the power regression model found in (a).

**57. Keeping Warm—Revisited** Recall from Exercise 55 of Section 2.2 that scientists have found the pulse rate $r$ of mammals is a power function of their body weight $w$.

**(a)** Re-express the data in Table 3.20 in terms of their *common* logarithms, and make a scatter plot of $(\log w, \log r)$.

**(b)** Compute the linear regression model $(\log r) = a(\log w) + b$ for $(\log w, \log r)$.

**(c)** Superimpose the regression curve on the scatter plot.

**(d)** Use the regression equation to predict the pulse rate for a 450-kg horse. Is the result close to the 38 beats/min reported by A. J. Clark in 1927? $r \approx 37.69$; very close

**(e) Writing to Learn** Why can we use either common or natural logarithms to re-express data that fit a power regression model?

**Table 3.20 Weight and Pulse Rate of Selected Mammals**

| Mammal | Body weight (kg) | Pulse rate (beats/min) |
|---|---|---|
| Rat | 0.2 | 420 |
| Guinea pig | 0.3 | 300 |
| Rabbit | 2 | 205 |
| Small dog | 5 | 120 |
| Large dog | 30 | 85 |
| Sheep | 50 | 70 |
| Human | 70 | 72 |

*Source: A. J. Clark, Comparative Physiology of the Heart (New York: Macmillan, 1927).*

**58.** Let $a = \log 2$ and $b = \log 3$. Then, for example $\log 6 = a + b$. List the common logs of all the positive integers less than 100 that can be expressed in terms of $a$ and $b$, writing equations such as $\log 6 = a + b$ for each case.

## Extending the Ideas

**59.** Solve $\ln x > \sqrt[3]{x}$. (6.41, 93.35)

**60.** Solve $1.2^x \leq \log_{1.2} x$. [1.26, 14.77]

**61. Group Activity** Work in groups of three. Have each group member graph and compare the domains for one pair of functions.

- $f(x) = 2 \ln x + \ln (x - 3)$ and $g(x) = \ln x^2(x - 3)$
- $f(x) = \ln (x + 5) - \ln (x - 5)$ and $g(x) = \ln \dfrac{x + 5}{x - 5}$
- $f(x) = \log (x + 3)^2$ and $g(x) = 2 \log (x + 3)$

**Writing to Learn** After discussing your findings, write a brief group report that includes your overall conclusions and insights.

**62.** Prove the quotient and power rules for logarithms.

**63.** Prove the change-of-base formula for logarithms.

**64.** Prove that $f(x) = \log x / \ln x$ is a constant function with restricted domain by finding the exact value of the constant $\log x / \ln x$ expressed as a common logarithm.

**65.** Graph $f(x) = \ln (\ln (x))$, and analyze it for domain, range, continuity, intercepts, increasing or decreasing behavior, symmetry, asymptotes, end behavior, invertibility, and concavity.

**Equation Solving and Modeling**

Solving Exponential Equations • Solving Logarithmic Equations • Orders of Magnitude and Logarithmic Models • Newton's Law of Cooling • Logarithmic Re-Expression

## Solving Exponential Equations

Some exponential and logarithmic equations can be solved by changing between exponential and logarithmic forms, as we saw in Examples 5 and 8 of Section 3.3. For other equations, the properties of exponents or the properties of logarithms are used. A property of both exponential and logarithmic functions that is often helpful for solving equations is that they are one-to-one functions.

### Objective

Students will be able to apply the properties of logarithms to solve exponential and logarithmic equations algebraically and solve applications problems using these equations.

### Motivate

Ask students to use a grapher to graph $y = \log x^2$ and $y = 2 \log x$, and to comment on any differences they see. Do the results contradict the power rule for logarithms? **(No)**

### Lesson Guide

Day 1: Solving Exponential Equations; Solving Logarithmic Equations; Orders of Magnitude and Logarithmic Models
Day 2: Newton's Law of Cooling; Logarithmic Re-Expression

---

**One-to-One Properties**

For any exponential function $f(x) = b^x$,

- If $b^u = b^v$, then $u = v$.

For any logarithmic function $f(x) = \log_b x$,

- If $\log_b u = \log_b v$, then $u = v$.

---

Example 1 shows how the one-to-oneness of exponential functions can be used. Examples 3 and 4 illustrate the use of the one-to-one property of logarithms.

**Example 1** SOLVING AN EXPONENTIAL EQUATION ALGEBRAICALLY

Solve $20(1/2)^{x/3} = 5$.

**Solution**

$$20\left(\frac{1}{2}\right)^{x/3} = 5$$

$$\left(\frac{1}{2}\right)^{x/3} = \frac{1}{4}$$

$$\left(\frac{1}{2}\right)^{x/3} = \left(\frac{1}{2}\right)^{2}$$

$$\frac{x}{3} = 2 \qquad \text{One-to-one property}$$

$$x = 6$$

The equation in Example 2 involves a difference of two exponential functions, which makes it difficult to solve algebraically. So we start with a graphical approach.

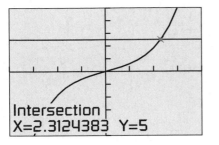

Intersection
X=2.3124383  Y=5

[–4, 4] by [–10, 10]

**Figure 3.35** $y = (e^x - e^{-x})/2$ and $y = 5$.
(Example 2)

## A Cinch?

You may recognize the left-hand side of the equation in Example 2 as the *hyperbolic sine function* that was introduced in Exercise 51 of Section 3.2. This function is often used in calculus. We write $\sinh(x) = (e^x - e^{-x})/2$. "Sinh" is pronounced as if spelled "cinch."

## Teaching Note

Students have probably seen extraneous solutions in previous mathematics courses, but they may not have seen examples where solutions are missed. However, this can also happen with simpler equations. For example, if one attempts to solve $2x = x^2$ by dividing both sides by $x$, then 0 is deleted from the domain and is missed as a solution.

### Example 2  SOLVING AN EXPONENTIAL EQUATION

Solve $(e^x - e^{-x})/2 = 5$.

**Solution**

**Solve Graphically**

Figure 3.35 shows that the graphs of $y = (e^x - e^{-x})/2$ and $y = 5$ intersect when $x \approx 2.31$.

**Confirm Algebraically**

The algebraic approach involves some ingenuity. If we multiply each side of the original equation by $2e^x$ and rearrange the terms we can obtain a quadratic equation in $e^x$:

$$\frac{e^x - e^{-x}}{2} = 5$$

$$e^{2x} - e^0 = 10e^x \qquad \text{Multiply by } 2e^x$$

$$(e^x)^2 - 10(e^x) - 1 = 0 \qquad \text{Subtract } 10e^x \text{ from each side}$$

If we let $w = e^x$, this equation becomes $w^2 - 10w - 1 = 0$, and the quadratic formula gives

$$w = e^x = \frac{10 \pm \sqrt{104}}{2} = 5 \pm \sqrt{26}.$$

Because $e^x$ is always positive, we reject the possibility that $e^x$ has the negative value $5 - \sqrt{26}$. Therefore,

$$e^x = 5 + \sqrt{26}$$

$$x = \ln(5 + \sqrt{26}) \qquad \text{Convert to logarithmic form.}$$

$$x = 2.312\ldots \approx 2.31$$

## Solving Logarithmic Equations

When logarithmic equations are solved algebraically, it is important to keep track of the domain of each expression in the equation as it is being solved. A particular algebraic method may introduce extraneous solutions or worse yet *lose* some valid solutions, as illustrated in Example 3.

### Example 3  SOLVING A LOGARITHMIC EQUATION

Solve $\log x^2 = 2$.

**Solution**

**Method 1**

Use the one-to-one property of logarithms.

$$\log x^2 = 2$$

$$\log x^2 = \log 10^2 \qquad y = \log 10^y$$

$$x^2 = 10^2 \qquad \text{One-to-one property}$$

$$x^2 = 100$$

$$x = 10 \quad \text{or} \quad x = -10$$

### Method 2

Change the equation from logarithmic to exponential form.

$$\log x^2 = 2$$

$$x^2 = 10^2 \qquad \text{Change to exponential form.}$$

$$x^2 = 100$$

$$x = 10 \quad \text{or} \quad x = -10$$

### Method 3 (Incorrect)

Use the power rule of logarithms.

$$\log x^2 = 2$$

$$2 \log x = 2 \qquad \text{Power rule}$$

$$\log x = 1 \qquad \text{Divide by 2}$$

$$x = 10 \qquad \text{Change to exponential form.}$$

### Support Graphically

Figure 3.36 shows that the graphs of $f(x) = \log x^2$ and $y = 2$ intersect when $x = -10$. From the symmetry of the graphs due to $f$ being an even function, we can see that $x = 10$ is also a solution.

### Interpret

Method 1 and 2 are correct. Method 3 fails because the domain of $\log x^2$ is all nonzero real numbers, but the domain of $\log x$ is only the positive real numbers. The correct solution is $x = 10$ or $x = -10$ because both of these $x$-values make the original equation true.

Method 3 of Example 3 violates an easily overlooked but important condition of the power rule $\log_b R^c = c \log_a R$, namely, that the rule holds *if R is positive*. In the expression $\log x^2$, $x$ plays the role of $R$, and $x$ can be $-10$, which is *not* positive. Because algebraic manipulation of a logarithmic equation can produce expressions with different domains, a graphical solution is often less prone to error.

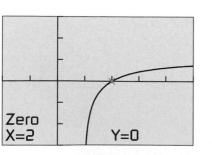

[−15, 15] by [−3, 3]

**Figure 3.36** Graphs of $f(x) = \log x^2$ and $y = 2$. (Example 3)

### Example 4 SOLVING A LOGARITHMIC EQUATION

Solve $\ln (3x - 2) + \ln (x - 1) = 2 \ln x$.

### Solution

### Solve Graphically

To use the $x$-intercept method, we rewrite the equation as

$$\ln (3x - 2) + \ln (x - 1) - 2 \ln x = 0,$$

and graph

$$f(x) = \ln (3x - 2) + \ln (x - 1) - 2 \ln x,$$

as shown in Figure 3.37. The $x$-intercept is $x = 2$, which is the solution to the equation.

[−2, 5] by [−3, 3]

**Figure 3.37** The zero of $f(x) = \ln (3x - 2) + \ln (x - 1) - 2 \ln x$ is $x = 2$. (Example 4)

**Confirm Algebraically**

$$\begin{aligned}
\ln(3x-2)+\ln(x-1) &= 2\ln x && \text{Original equation}\\
\ln[(3x-2)(x-1)] &= \ln x^2 && \text{Product and power rules}\\
(3x-2)(x-1) &= x^2 && \text{One-to-one property}\\
2x^2-5x+2 &= 0\\
(2x-1)(x-2) &= 0\\
x = \frac{1}{2} \quad &\text{or} \quad x = 2
\end{aligned}$$

We recognize $x = 1/2$ is an extraneous solution of the original equation because $\ln(1/2 - 1)$ is not defined. By substituting $x = 2$, we find that

$$\ln 4 + \ln 1 = \ln(4 \cdot 1) = \ln 2^2 = 2\ln 2.$$

Thus, the only solution to the original equation is $x = 2$.

## Orders of Magnitude and Logarithmic Models

When comparing quantities, their sizes sometimes span a wide range. This is why scientific notation was developed.

For instance, the planet Mercury is 57.9 billion meters from the sun; whereas Pluto is 5900 billion meters from the sun, roughly 100 times farther! In scientific notation, Mercury is $5.79 \times 10^{10}$ m from the sun, and Pluto is $5.9 \times 10^{12}$ m from the sun. So Pluto's distance is 2 powers of ten greater than Mercury's distance. From Figure 3.38, we see that the difference in the common logs of these two distances is about 2. The common logarithm of a positive quantity is its **order of magnitude**. So we say, Pluto's distance from the sun is 2 orders of magnitude greater than Mercury's.

Orders of magnitude can be used to compare any like quantities:

- A kilometer is 3 orders of magnitude longer than a meter.
- A dollar is 2 orders of magnitude greater than a penny.
- A horse weighing 400 kg is 4 orders of magnitude heavier than a mouse weighing 40 g.
- New York City with 7 million people is 6 orders of magnitude bigger than Earmuff Junction with a population of 7.

```
log(5.79*10^10)
              10.76267856
log(5.9*10^12)
              12.77085201
log(5.9*10^12)-l
og(5.79*10^10)
              2.008173448
```

**Figure 3.38** Pluto is two orders of magnitude farther from the sun than Mercury.

**Teaching Note**

Orders of magnitude are widely used in science and engineering, but are seldom defined and usually ignored in mathematics. The concept should be new to most students but easy to motivate.

**Exploration 1 Comparing Scientific Notation and Common Logarithms**

Using a calculator compute $\log(4 \cdot 10)$, $\log(4 \cdot 10^2)$, $\log(4 \cdot 10^3), \ldots, \log(4 \cdot 10^{10})$.

What is the pattern in the integer parts of these numbers?

What is the pattern of their decimal parts?

How many orders of magnitude greater is $4 \cdot 10^{10}$ than $4 \cdot 10$?

**Exploration Extensions**

Have the students calculate $10^{1.60205991}$, $10^{2.60205991}$, and $10^{3.60205991}$ and discuss their findings.

Orders of magnitude are used to compare the severity of earthquakes and the acidity of chemical solutions. We now turn our attention to these two applications.

As mentioned in Exercise 50 of Section 3.4, the *Richter scale* magnitude $R$ of an earthquake is based on the features of the associated seismic wave and is measured using the equation

$$R = \log \frac{a}{T} + B,$$

where $a$ is the amplitude in micrometers ($\mu$m) of the vertical ground motion at the receiving station, $T$ is the period of the seismic wave in seconds, and $B$ accounts for the weakening of the seismic wave with increasing distance from the epicenter of the earthquake.

### Example 5  COMPARING EARTHQUAKE INTENSITIES

A particular earthquake has an amplitude $a$ during its early stages, and at later stages the amplitude becomes $100a$ (one hundred times more powerful). By how much does the Richter scale rating $R$ increase as the earthquake progresses?

**Solution**

**Model**

The earthquake intensities can be modeled by these equations:

$$R_1 = \log \frac{a}{T} + B$$

$$R_2 = \log \frac{100a}{T} + B$$

**Solve Algebraically**

We seek the change in Richter scale ratings $R_2 - R_1$:

$$R_2 - R_1 = \left( \log \frac{100a}{T} + B \right) - \left( \log \frac{a}{T} + B \right)$$

$$= \log \frac{100a}{T} - \log \frac{a}{T}$$

$$= \log \frac{100a/T}{a/T}$$

$$= \log 100$$

$$= 2$$

**Interpret**

When the amplitude increases by a factor of $100 = 10^2$, the Richter scale rating increases by 2.

From Example 5 we can see, for instance, that an earthquake that is 8 on the Richter scale is 100 times more severe than an earthquake rated as a 6 on the Richter scale. The earthquake intensities differ by $8 - 6 = 2$ orders of magnitude.

In chemistry, the acidity of a water-based solution is measured by the concentration of hydrogen ions in the solution (in moles per liter). The hydrogen-ion concentration is written $[H^+]$. Because such concentrations usually involve *negative* powers of ten, *negative* orders of magnitude are used to compare acidity levels. The measure of acidity used is **pH**, the opposite of the common log of the hydrogen-ion concentration:

$$pH = -\log[H^+]$$

More acidic solutions have higher hydrogen-ion concentrations and lower pH values.

### Example 6 COMPARING CHEMICAL ACIDITY

Some especially sour vinegar has a pH of 2.4, and a box of Leg and Sickle baking soda has a pH of 8.4.

**(a)** What are their hydrogen-ion concentrations?

**(b)** How many times greater is the hydrogen-ion concentration of the vinegar than that of the baking soda?

**(c)** By how many orders of magnitude do the concentrations differ?

**Solution**

**(a)** Vinegar:
$$-\log[H^+] = 2.4$$
$$\log[H^+] = -2.4$$
$$[H^+] = 10^{-2.4} \approx 3.98 \times 10^{-3} \text{ moles per liter}$$

Baking soda:
$$-\log[H^+] = 8.4$$
$$\log[H^+] = -8.4$$
$$[H^+] = 10^{-8.4} \approx 3.98 \times 10^{-9} \text{ moles per liter}$$

**(b)** $$\frac{[H^+] \text{ of vinegar}}{[H^+] \text{ of baking soda}} = \frac{10^{-2.4}}{10^{-8.4}} = 10^{(-2.4)-(-8.4)} = 10^6$$

**(c)** The hydrogen-ion concentration of the vinegar is 6 orders of magnitude greater than that of the Leg and Sickle baking soda, exactly the difference in their pH values.

## Newton's Law of Cooling

An object that has been heated will cool to the temperature of the medium in which it is placed, such as the surrounding air or water. The temperature $T$ of the object at time $t$ can be modeled by

$$T(t) = T_m + (T_0 - T_m)e^{-kt}$$

for an appropriate value of $k$, where

$$T_m = \text{the temperature of the surrounding medium,}$$

$$T_0 = \text{initial temperature of the object.}$$

This model assumes that the surrounding medium, although taking heat from the object, essentially maintains a constant temperature. In honor of English mathematician and physicist Isaac Newton (1643–1727), this model is called **Newton's law of cooling**.

### Example 7 APPLYING NEWTON'S LAW OF COOLING

A hard-boiled egg at temperature 96° C is placed in 16° C water to cool. Four minutes later the temperature of the egg is 45° C. Use Newton's law of cooling to determine when the egg will be 20° C.

**Solution**

**Model**

We know from the given information that

$T_0 = 96$, the initial temperature of the egg,

$T_m = 16$, the temperature of the medium.

Substituting into $T(t) = T_m + (T_0 - T_m)e^{-kt}$, we have

$$T(t) = 16 + (96 - 16)e^{-kt}.$$

To find the value of $k$ we use the fact that $T = 45$ when $t = 4$.

$$45 = 16 + 80e^{-4k}$$

$$\frac{29}{80} = e^{-4k}$$

$$\ln \frac{29}{80} = -4k \qquad \text{Change to logarithmic form.}$$

$$k = \frac{\ln (29/80)}{-4}$$

$$k \approx 0.254$$

So the Newton's law model for the temperature $T$ of the egg at time $t$ is $T(t) = 16 + 80e^{-0.254t}$. To find $t$ when $T = 20°$ C, we must solve the equation

$$20 = 16 + 80e^{-0.254t}$$

**Solve Algebraically**

$$20 = 16 + 80e^{-0.254t}$$

$$\frac{4}{80} = e^{-0.254t}$$

$$\ln \frac{1}{20} = -0.254t \qquad \text{Change to logarithmic form.}$$

$$t = \frac{\ln (1/20)}{-0.254}$$

$$t \approx 11.794$$

**Interpret**

The temperature of the egg will be 20° C after about 11.8 min (11 min 48 sec).

We can rewrite Newton law of cooling in the following form:

$$T(t) - T_m = (T_0 - T_m)e^{-kt}$$

We use this form of Newton's law when working with temperature data gathered from an actual experiment. Because the difference $T - T_m$ is an exponen-

Section 3.5 *Equation Solving and Modeling* **309**

tial function of time $t$, we can use exponential regression on $T - T_m$ versus $t$ to obtain a model, as illustrated in Example 8.

### Example 8 MODELING WITH NEWTON'S LAW OF COOLING

In an experiment, a temperature probe connected to a *Calculator Based Laboratory*™ device was removed from a cup of hot coffee and placed in a glass of cold water. The first two columns of Table 3.21 show the resulting data for time $t$ (in seconds since the probe was placed in the water) and temperature $T$ (in ° C). In the third column, the temperature data have been re-expressed by subtracting the temperature of the water, which was 4.5° C.

**(a)** Estimate the temperature of the coffee.

**(b)** Estimate the time when the temperature probe reading was 40°C.

### Solution
### Model

Figure 3.39a shows a scatter plot of the re-expressed temperature data. Using exponential regression, we obtain the following model:

$$T(t) - 4.5 = 61.656 \times 0.92770^t$$

Figure 3.39b shows the graph of this modeled with the scatter plot of the data. You can see that the curve fits the data fairly well.

**(a) Solve Algebraically**

From the model we see that $T_0 - T_m \approx 61.656$. So

$$T_0 \approx 61.656 + T_m = 61.656 + 4.5 \approx 66.16$$

**(b) Solve Graphically**

Figure 3.39c shows that the graph of $T(t) - 4.5 = 61.656 \times 0.92770^t$ intersects $y = 40 - 4.5 = 35.5$ when $t \approx 7.36$.

### Interpret

The temperature of the coffee was roughly 66.2° C, and the probe reading was 40° C about 7.4 sec after it was placed in the water.

**Table 3.21** Temperature Data from a *CBL*™ Experiment

| Time $t$ | Temp $T$ | $T - T_m$ |
|----------|----------|-----------|
| 2 | 64.8 | 60.3 |
| 5 | 49.0 | 44.5 |
| 10 | 31.4 | 26.9 |
| 15 | 22.0 | 17.5 |
| 20 | 16.5 | 12.0 |
| 25 | 14.2 | 9.7 |
| 30 | 12.0 | 7.5 |

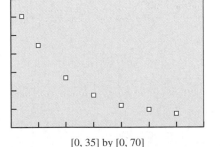

[0, 35] by [0, 70]

(a)

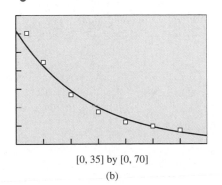

[0, 35] by [0, 70]

(b)

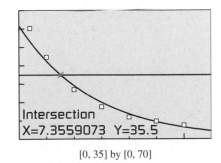

Intersection
X=7.3559073  Y=35.5

[0, 35] by [0, 70]

(c)

**Figure 3.39** Scatter plot and graphs for Example 8.

## Logarithmic Re-Expression

In Section 3.4 we learned that data pairs $(x, y)$ which fit a power model have a linear relationship when re-expressed as $(\ln x, \ln y)$ pairs. We now illustrate that data pairs $(x, y)$ which fit a logarithmic or exponential regression model can be *linearized* through *logarithmic re-expression*. In other words, four of the most commonly encountered models are related by logarithmic re-expression.

---

**Regression Models Related by Logarithmic Re-Expression**

- **Linear regression:**                                  $y = ax + b$
- **Natural logarithmic regression:**    $y = a + b \ln x$
- **Exponential regression:**                   $y = a \cdot b^x$
- **Power regression:**                            $y = a \cdot x^b$

---

When we examine a scatter plot of data pairs $(x, y)$, we should ask whether one of these four regression models could be the best choice for the data. If the data plot appears to be linear, a linear regression may be the best choice. But when it is visually evident that the data plot is not linear, the best choice may be a natural logarithmic, exponential, or power regression.

Knowing the shapes of logarithmic, exponential, and power function graphs helps us choose an appropriate model. In addition, it is often helpful to re-express the $(x, y)$ data pairs as $(\ln x, y)$, $(x, \ln y)$, or $(\ln x, \ln y)$, and create scatter plots of the re-expressed data. If any of the scatter plots appear to be linear, then we have a likely choice for an appropriate model. Figures 3.40–3.42 provide a summary.

**Follow-up**

Ask students to explain how extraneous solutions might be introduced by replacing

$\ln x - \ln (x - 2)$ with $\ln \left( \dfrac{x}{x - 2} \right)$

in an equation.

**Assignment Guide**

Day 1: Ex. 3–39, multiples of 3
Day 2: Ex. 42, 43, 45, 48, 50, 53, 55, 56, 60, 62, 64

**Cooperative Learning**

Group Activity: Ex. 53

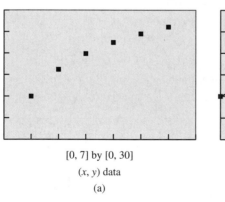

[0, 7] by [0, 30]

$(x, y)$ data

(a)

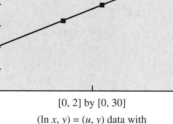

**Conclusion:**

$y = a \ln x + b$ is the logarithmic regression model for the $(x, y)$ data.

[0, 2] by [0, 30]

$(\ln x, y) = (u, y)$ data with linear regression model

$y = au + b$

(b)

**Figure 3.40** Natural logarithmic regression

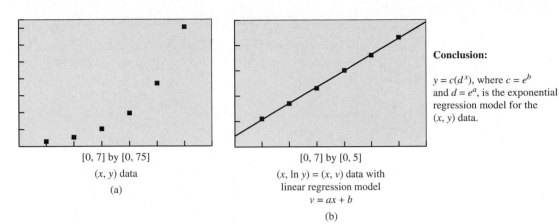

**Conclusion:**

$y = c(d^x)$, where $c = e^b$ and $d = e^a$, is the exponential regression model for the $(x, y)$ data.

[0, 7] by [0, 75]

$(x, y)$ data

(a)

[0, 7] by [0, 5]

$(x, \ln y) = (x, v)$ data with linear regression model $v = ax + b$

(b)

**Figure 3.41**  Exponential regression

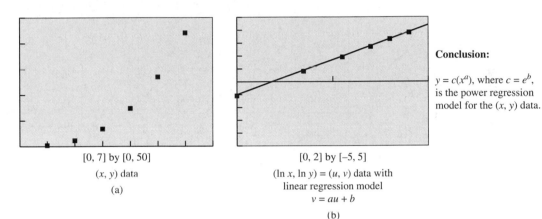

**Conclusion:**

$y = c(x^a)$, where $c = e^b$, is the power regression model for the $(x, y)$ data.

[0, 7] by [0, 50]

$(x, y)$ data

(a)

[0, 2] by [−5, 5]

$(\ln x, \ln y) = (u, v)$ data with linear regression model $v = au + b$

(b)

**Figure 3.42**  Power regression

The three regression models can be justified algebraically. We give the justification for exponential regression, and leave the other two justifications as exercises.

$$v = ax + b$$

$$\ln y = ax + b \qquad v = \ln y.$$

$$y = e^{ax+b} \qquad \text{Change to exponential form.}$$

$$y = e^{ax} \cdot e^b \qquad \text{Use the laws of exponents.}$$

$$y = e^b \cdot (e^a)^x$$

$$y = c \cdot (d^x) \qquad \text{Let } c = e^b \text{ and } d = e^a.$$

Example 9 illustrates how a combination of knowledge about the shapes of logarithmic, exponential, and power function graphs is used in combination with logarithmic re-expression to choose a curve of best fit.

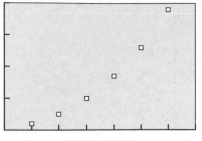

[0, 7] by [0, 40]

**Figure 3.43** A scatter plot of the original data of Example 9.

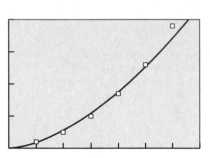

[0, 7] by [0, 40]

**Figure 3.45** A power regression model fits the data of Example 9.

### Example 9    SELECTING A REGRESSION MODEL

Decide whether these data can be best modeled by logarithmic, exponential, or power regression. Find the appropriate regression model.

| $x$ | 1 | 2 | 3 | 4 | 5 | 6 |
|---|---|---|---|---|---|---|
| $y$ | 2 | 5 | 10 | 17 | 26 | 38 |

**Solution**  The shape of the scatter plot of the data in Figure 3.43 suggests that the data might be modeled by an exponential or power function.

Figure 3.44a shows the $(x, \ln y)$ scatter plot, and Figure 3.44b shows the $(\ln x, \ln y)$ scatter plot. Of these two plots, the $(\ln x, \ln y)$ plot appears to be more linear, so we find the power regression model for the original data.

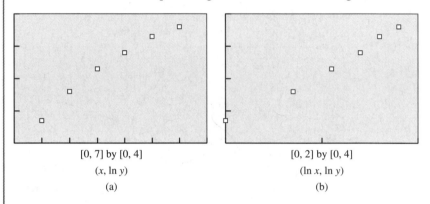

[0, 7] by [0, 4]             [0, 2] by [0, 4]
$(x, \ln y)$                 $(\ln x, \ln y)$
(a)                          (b)

**Figure 3.44** Two logarithmic re-expressions of the data of Example 9.

Figure 3.45 shows the scatter plot of the original $(x, y)$ data with the graph of the power regression model $y = 1.7910x^{1.6472}$ superimposed.

---

# Quick Review 3.5

In Exercises 1–4, prove that each function in the given pair is the inverse of the other.

**1.** $f(x) = e^{2x}$ and $g(x) = \ln (x)^{1/2}$

**2.** $f(x) = 10^{x/2}$ and $g(x) = \log x^2, x > 0$

**3.** $f(x) = (1/3) \ln x$ and $g(x) = e^{3x}$

**4.** $f(x) = 3 \log x^2, x > 0$ and $g(x) = 10^{x/6}$

In Exercises 5 and 6, write the number in scientific notation.

**5.** The mean distance from Jupiter to the Sun is about 778,300,000 km.

**6.** An atomic nucleus has a diameter of about 0.000000000000001 m.

In Exercises 7 and 8, write the number in decimal form.

**7.** Avogadro's number is about $6.02 \times 10^{23}$.

**8.** The atomic mass unit is about $1.66 \times 10^{-27}$ kg.

In Exercises 9 and 10, use scientific notation to simplify the expression; leave your answer in scientific notation.

**9.** $(186,000)(31,000,000)$     **10.** $\dfrac{0.0000008}{0.000005}$   $1.6 \times 10^{-1}$

# Section 3.5 Exercises

In Exercises 1–16, solve each equation algebraically. Obtain a numerical approximation for your solution and check it by substituting into the original equation.

**1.** $1.06^x = 4.1$

**2.** $0.98^x = 1.6$

**3.** $\log x = 4$   $10^4 = 10000$

**4.** $\ln x = -1$   $e^{-1} \approx 0.3679$

**5.** $50e^{0.035x} = 200$

**6.** $80e^{0.045x} = 240$

**7.** $2(10^{-x/3}) = 20$   $-3$

**8.** $3(5^{-x/4}) = 15$   $-4$

**9.** $20\left(\dfrac{1}{2}\right)^{x/4} = 8$

**10.** $40\left(\dfrac{1}{2}\right)^{x/3} = 10$   $6$

**11.** $3 + 2e^{-x} = 6$

**12.** $7 - 3e^{-x} = 2$

**13.** $\log_4 (x - 5) = -1$   $5.25$

**14.** $\log_4 (1 - x) = 1$   $-3$

**15.** $3 \ln (x - 3) + 4 = 5$

**16.** $3 - \log (x + 2) = 5$   $-1.99$

In Exercises 17–20, by how many orders of magnitude do the quantities differ?

**17.** A \$100 bill and a dime.   3

**18.** A canary weighing 20 g and a hen weighing 2 kg.   2

**19.** An earthquake rated 7 on the Richter scale and one rated 5.5.

**20.** Lemon juice with pH = 2.3 and beer with pH = 4.1.   1.8

In Exercises 21–26, state the domain of each function. Then match the function with its graph. (Each graph shown has a window of $[-4.7, 4.7]$ by $[-3.1, 3.1]$.)

**21.** $f(x) = \log x(x + 1)$

**22.** $g(x) = \log x + \log (x + 1)$

**23.** $f(x) = \ln \dfrac{x}{x + 1}$

**24.** $g(x) = \ln x - \ln (x + 1)$

**25.** $f(x) = 2 \ln x$

**26.** $g(x) = \ln x^2$

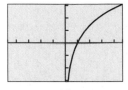

(a)

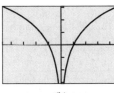

(b)

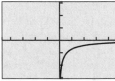

(c)

(d)

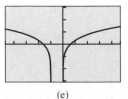

(e)

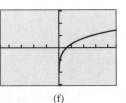

(f)

In Exercises 27–36, solve each equation by the method of your choice. Support your solution by a second method.

**27.** $\dfrac{2^x - 2^{-x}}{3} = 4$

**28.** $\dfrac{2^x + 2^{-x}}{2} = 3$

**29.** $\dfrac{e^x + e^{-x}}{2} = 4$

**30.** $2e^{2x} + 5e^x - 3 = 0$

**31.** $\dfrac{500}{1 + 25e^{0.3x}} = 200$

**32.** $\dfrac{400}{1 + 95e^{-0.6x}} = 150$

**33.** $\dfrac{1}{2} \ln (x + 3) - \ln x = 0$

**34.** $\log x - \dfrac{1}{2} \log (x + 4) = 1$

**35.** $\ln (x - 3) + \ln (x + 4) = 3 \ln 2$   4

**36.** $\log (x - 2) + \log (x + 5) = 2 \log 3$

**37. Comparing Earthquakes** How many times more severe was the 1978 Mexico City earthquake ($R = 7.9$) than the 1994 Los Angeles earthquake ($R = 6.6$)?

**38. Comparing Earthquakes** How many times more severe was the 1995 Kobe, Japan, earthquake ($R = 7.2$) than the 1994 Los Angeles earthquake ($R = 6.6$)?

**39. Chemical Acidity** The pH of carbonated water is 3.9, and the pH of household ammonia is 11.9.

**(a)** What are their hydrogen-ion concentrations?

**(b)** How many times greater is the hydrogen-ion concentration of the carbonated water than that of the ammonia?

**(c)** By how many orders of magnitude do the concentrations differ?   8

**40. Chemical Acidity** Stomach acid has a pH of about 2.0, and blood has a pH of 7.4.

**(a)** What are their hydrogen-ion concentrations?

**(b)** How many times greater is the hydrogen-ion concentration of the stomach acid than that of the blood?

**(c)** By how many orders of magnitude do the concentrations differ?   5.4

**41. Newton's Law of Cooling** A cup of coffee has cooled from 92° C to 50° C after 12 min in a room at 22° C. How long will the cup take to cool to 30° C?   $\approx 28.41$ minutes

**42. Newton's Law of Cooling** A cake is removed from an oven at 350° F and cools to 120° F after 20 min in a room at 65° F. How long will the cake take to cool to 90° F?

**43. Newton's Law of Cooling Experiment** A thermometer is removed from a cup of coffee and placed in water whose temperature $(T_m)$ is 10° C. The data in Table 3.22 were collected over the next 30 sec.

| Table 3.22 Experimental Data | | |
|---|---|---|
| Time $t$ | Temp $T$ | $T - T_m$ |
| 2 | 80.47 | 70.47 |
| 5 | 69.39 | 59.39 |
| 10 | 49.66 | 39.66 |
| 15 | 35.26 | 25.26 |
| 20 | 28.15 | 18.15 |
| 25 | 23.56 | 13.56 |
| 30 | 20.62 | 10.62 |

**(a)** Find a scatter plot of the data $T - T_m$.

**(b)** Find an exponential regression equation for the $T - T_m$ data. Superimpose its graph on the scatter plot.

**(c)** Estimate the thermometer reading when it was removed from the coffee. 89.47° C

**44. Newton's Law of Cooling Experiment** A thermometer was removed from a cup of hot chocolate and placed in water whose temperature $T_m = 0°$ C. The data in Table 3.23 were collected over the next 30 sec.

**(a)** Find a scatter plot of the data $T - T_m$.

**(b)** Find an exponential regression equation for the $T - T_m$ data. Superimpose its graph on the scatter plot.

**(c)** Estimate the thermometer reading when it was removed from the hot chocolate. 79.96° C

| Table 3.23 Experimental Data | | |
|---|---|---|
| Time $t$ | Temp $T$ | $T - T_m$ |
| 2 | 74.68 | 74.68 |
| 5 | 61.99 | 61.99 |
| 10 | 34.89 | 34.89 |
| 15 | 21.95 | 21.95 |
| 20 | 15.36 | 15.36 |
| 25 | 11.89 | 11.89 |
| 30 | 10.02 | 10.02 |

**45. Penicillin Use** The use of penicillin became so widespread in the 1980s in Hungary that it became practically useless against common sinus and ear infections. Now the use of more effective antibiotics has caused a decline in penicillin resistance. The bar graph shows the use of penicillin in Hungary for selected years.

**(a)** From the bar graph we read the data pairs to be approximately $(1, 11)$, $(8, 6)$, $(15, 4.8)$, $(16, 4)$, and $(17, 2.5)$, using $t = 1$ for 1976, $t = 8$ for 1983, and so on. Complete a scatter plot for these data.

**(b) Writing to Learn** Discuss whether the bar graph shown or the scatter plot that you completed best represents the data and why.

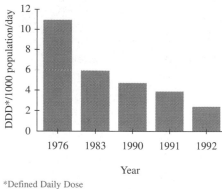

Nationwide Consumption of Penicillin

*Defined Daily Dose

*Source: Science,* Vol. 264, April 15, 1994, American Association for the Advancement of Science.

**46. Writing to Learn** Which regression model would you use for the data in Exercise 45? Discuss various options, and explain why you chose the model you did. Support your writing with tables and graphs as needed.

**Writing to Learn** In Exercises 47–50, tables of $(x, y)$ data pairs are given. Determine whether a linear, logarithmic, exponential, or power regression equation is the best model for the data. Explain your choice. Support your writing with tables and graphs as needed.

**47.**

| $x$ | 1 | 2 | 3 | 4 |
|---|---|---|---|---|
| $y$ | 3 | 4.4 | 5.2 | 5.8 |

**48.**

| $x$ | 1 | 2 | 3 | 4 |
|---|---|---|---|---|
| $y$ | 6 | 18 | 54 | 162 |

**49.**

| $x$ | 1 | 2 | 3 | 4 |
|---|---|---|---|---|
| $y$ | 3 | 6 | 12 | 24 |

**50.**

| $x$ | 1 | 2 | 3 | 4 |
|---|---|---|---|---|
| $y$ | 5 | 7 | 9 | 11 |

## Explorations

In Exercises 51 and 52, use the data in Table 3.24. Determine whether a linear, logarithmic, exponential, power, or logistic regression equation is the best model for the data. Explain your choice. Support your writing with tables and graphs as needed.

**Table 3.24 Official Census Populations (in thousands), 1900–1990**

| Year | Alaska | Hawaii |
|------|--------|--------|
| 1900 | 63.6 | 154 |
| 1910 | 64.4 | 192 |
| 1920 | 55.0 | 256 |
| 1930 | 59.2 | 368 |
| 1940 | 72.5 | 423 |
| 1950 | 128.6 | 500 |
| 1960 | 226.2 | 633 |
| 1970 | 302.6 | 770 |
| 1980 | 401.9 | 965 |
| 1990 | 550.0 | 1108 |

*Source: U.S. Census Bureau as reported in 1999 New York Times Almanac.*

**51. Writing to Learn Modeling Population** Which regression equation is the best model for Alaska's population? logistic regression

**52. Writing to Learn Modeling Population** Which regression equation is the best model for Hawaii's population? logistic regression

**53. Group Activity Normal Distribution** The function

$$f(x) = k \cdot e^{-cx^2},$$

where $c$ and $k$ are positive constants, is a bell-shaped curve that is useful in probability and statistics.

**(a)** Graph $f$ for $c - 1$ and $k = 0.1, 0.5, 1, 2, 10$. Explain the effect of changing $k$.

**(b)** Graph $f$ for $k = 1$ and $c = 0.1, 0.5, 1, 2, 10$. Explain the effect of changing $c$.

## Extending the Ideas

**54. Writing to Learn** Prove, if $u/v = 10^n$ for $u > 0$ and $v > 0$, then $\log u - \log v = n$. Explain how this result relates to powers of ten and orders of magnitude.

**55. Potential Energy** The potential energy $E$ (the energy stored for use at a later time) between two ions in a certain molecular structure is modeled by the function

$$E = -\frac{5.6}{r} + 10e^{-r/3}$$

where $r$ is the distance separating the nuclei.

**(a) Writing to Learn** Graph this function in the window $[-10, 10]$ by $[-10, 30]$, and explain which portion of the graph does not represent this potential energy situation.

**(b)** Identify a viewing window that shows that portion of the graph (with $r \le 10$) which represents this situation, and find the maximum value for $E$.

**56.** In Example 8, the Newton's law of cooling model was

$$T(t) - T_m = (T_0 - T_m)e^{-kt} = 61.656 \times 0.92770^t$$

Determine the value of $k$.

**57.** Justify the conclusion made about natural logarithm regression in Figure 3.40.

**58.** Justify the conclusion made about power regression in Figure 3.42.

In Exercises 59–64, solve the equation or inequality.

**59.** $e^x + x = 5$

**60.** $e^{2x} - 8x + 1 = 0$

**61.** $e^x < 5 + \ln x$

**62.** $\ln |x| - e^{2x} \ge 3$

**63.** $2 \log x - 4 \log 3 > 0$

**64.** $2 \log (x + 1) - 2 \log 6 < 0$

---

## 3.6 Mathematics of Finance

Interest Compounded Annually • Interest Compounded $k$ Times per Year • Interest Compounded Continuously • Annual Percentage Yield • Annuities—Future Value • Loans and Mortgages—Present Value

### Interest Compounded Annually

In business, as the saying goes, "time is money." We must pay for the use of property or money over time. Interest is payment for the use of money over a specified period. When we borrow money, we pay interest, and when we loan money, we receive interest. When we invest in a savings account, we are actually lending money to the bank.

Suppose a principal of $P$ dollars is invested in an account bearing a 5% interest rate, calculated at the end of each year. If $A_1$ represents the total amount in the account after year 1, then

$$A_1 = P + 0.05P = P(1 + 0.05).$$

The amount at the end of year 2 is

$$A_2 = P(1 + 0.05) + 0.05P(1 + 0.05)$$
$$= P(1 + 0.05)(1 + 0.05) \qquad \text{Factor out } P(1 + 0.05)$$
$$= P(1 + 0.05)^2$$

Extending this pattern with an interest rate $r$, we have

$$A_3 = P(1 + r)^3,$$
$$\vdots$$
$$A_n = P(1 + r)^n.$$

So the total value $A$ of the investment at the end of $n$ years is modeled by

$$A = P(1 + r)^n,$$

where $P$ is the principal and $r$ is the constant percentage rate expressed in decimal form.

   **Compound interest** is interest that becomes part of the investment, so that interest is earned on the interest itself, as happened in the situation just described. The formula $A = P(1 + r)^n$ is used when interest is *compounded annually*, that is, compounded once a year.

   Notice that $A = P(1 + r)^n$ is an exponential function with base $1 + r$ and exponent $n$. If we write the function as $A(t) = P(1 + r)^t$, where $t$ represents time in years, we can see how closely this compound interest formula parallels the population model developed at the beginning of Section 3.2.

### Example 1  COMPOUNDING ANNUALLY

Suppose Quan Li invests $500 at 7% interest compounded annually. Find the value of her investment 10 years later.

**Solution**  Letting $P = 500$, $r = 0.07$, and $n = t = 10$,
$A = 500(1 + 0.07)^{10} = 983.575. \ldots$ Rounding to the nearest cent, we see that the value of Quan Li's investment after 10 years is $983.58.

## Interest Compounded *k* Times per Year

Suppose a principal $P$ is invested at an annual interest rate $r$ compounded $k$ times a year for $t$ years. Then $r/k$ is the interest rate per compounding period, and $kt$ is the number of compounding periods. The amount $A$ in the account after $t$ years is

$$A = P\left(1 + \frac{r}{k}\right)^{kt}.$$

### Example 2  COMPOUNDING MONTHLY

Suppose Roberto invests $500 at 9% annual interest *compounded monthly*, that is, compounded 12 times a year. Find the value of his investment 5 years later.

**Solution**  Letting $P = 500$, $r = 0.09$, $k = 12$, and $t = 5$,

$$A = 500\left(1 + \frac{0.09}{12}\right)^{12(5)} = 782.840. \ldots$$

So the value of Roberto's investment after 5 years is $782.84.

The problems in Examples 1 and 2 required that we calculate $A$. Examples 3 and 4 illustrate situations that require us to determine the values of other variables in the compound interest formula.

### Example 3   FINDING THE TIME PERIOD OF AN INVESTMENT

Judy has $500 to invest at 9% annual interest compounded monthly. How long will it take for her investment to grow to $3000?

**Solution**

**Model**

Let $P = 500$, $r = 0.09$, $k = 12$, and $A = 3000$ in the equation

$$A = P\left(1 + \frac{r}{k}\right)^{kt},$$

and solve for $t$.

**Solve Graphically**

For

$$3000 = 500\left(1 + \frac{0.09}{12}\right)^{12t},$$

we let

$$f(t) = 500\left(1 + \frac{0.09}{12}\right)^{12t} \quad \text{and} \quad y = 3000,$$

and then find the point of intersection of the graphs. Figure 3.46 shows that this occurs at $t \approx 19.98$.

**Confirm Algebraically**

$$3000 = 500(1 + 0.09/12)^{12t}$$

$$6 = 1.0075^{12t}$$

$$\ln 6 = \ln (1.0075^{12t}) \qquad \text{Apply ln to each side.}$$

$$\ln 6 = 12t \ln (1.0075) \qquad \text{Power rule}$$

$$t = \frac{\ln 6}{12 \ln 1.0075}$$

$$= 19.983\ldots$$

**Interpret**

So it will take Judy 20 years for the value of the investment to reach (and slightly exceed) $3000.

### Example 4   FINDING AN INTEREST RATE

Stephen has $500 to invest. What annual interest rate *compounded quarterly* (four times per year) is required to double his money in 10 years?

**Solution**

**Model**

Letting $P = 500$, $k = 4$, $t = 10$, and $A = 1000$ yields the equation

$$1000 = 500\left(1 + \frac{r}{4}\right)^{4(10)}$$

that we solve for $r$.

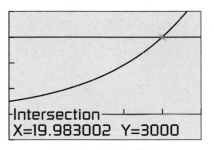

[0, 25] by [–1000, 4000]

**Figure 3.46** Graphs for Example 3.

**Solve Graphically**

Figure 3.47 shows that $f(r) = 500(1 + r/4)^{40}$ and $y = 1000$ intersect at $r \approx 0.07$, or $r = 7\%$.

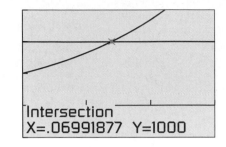

[0, 0.15] by [−500, 1500]

**Figure 3.47** Graphs for Example 4.

**Confirm Algebraically**

$$1000 = 500(1 + r/4)^{40}$$

$$2 = (1 + r/4)^{40}$$

$$\ln 2 = \ln (1 + r/4)^{40}$$

$$\ln 2 = 40 \ln (1 + r/4)$$

$$\frac{\ln 2}{40} = \ln \left(1 + \frac{r}{4}\right)$$

$$e^{(\ln 2)/40} = 1 + r/4$$

$$\frac{r}{4} = e^{(\ln 2)/40} - 1$$

$$r = 4e^{(\ln 2)/40} - 4$$

$$= 0.06991\ldots$$

**Interpret**

Note that we delay computation until the last step for the best possible accuracy. Stephen's investment of $500 will double in 10 years at an annual interest rate of 7% compounded quarterly.

## Interest Compounded Continuously

In Exploration 1, $1000 is invested for 1 year at a 10% interest rate. We investigate the value of the investment at the end of 1 year as the number of compounding periods $k$ increases. In other words, we determine the "limiting" value of the expression $1000(1 + 0.1/k)^k$ as $k$ assumes larger and larger integer values.

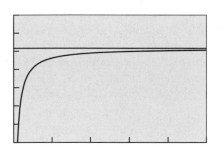

[0, 50] by [1100, 1107]

**Figure 3.48** Graphs for Exploration 1.

**Exploration Extensions**

Repeat the Exploration for an interest rate of 5%. Use an appropriate function for *A*, and use $y = 1000e^{0.05}$.

---

**Exploration 1**   Increasing the Number of Compounding Periods Boundlessly

Let $A = 1000\left(1 + \dfrac{0.1}{k}\right)^{k}$.

1. Complete a table of values of *A* for *k* = 10, 20, . . . , 100. What pattern do you observe?

2. Figure 3.48 shows the graphs of the function $A(k) = 1000(1 + 0.1/k)^{k}$ and the horizontal line $y = 1000e^{0.1}$. Interpret the meanings of these graphs.

---

Recall from Section 3.1 that $e = \lim\limits_{x \to \infty} (1 + 1/x)^{x}$. Therefore, for a fixed interest rate *r*, if we let $x = k/r$,

$$\lim_{k \to \infty} \left(1 + \frac{r}{k}\right)^{k/r} = e.$$

We do not know enough about limits yet, but with some calculus, it can be proved that $\lim\limits_{k \to \infty} P(1 + r/k)^{kt} = Pe^{rt}$. So $A = Pe^{rt}$ is the formula used when interest is **compounded continuously**. In nearly any situation, one of the following two formulas can be used to compute compound interest:

**Compound Interest—Value of an Investment**

Suppose a principal *P* is invested at a fixed annual interest rate *r*. The value of the investment after *t* years is

- $A = P\left(1 + \dfrac{r}{k}\right)^{kt}$   when interest compounds *k* times per year,

- $A = Pe^{rt}$   when interest compounds continuously.

**Example 5   COMPOUNDING CONTINUOUSLY**

Suppose LaTasha invests $100 at 8% annual interest compounded continuously. Find the value of her investment at the end of each of the years 1, 2, . . . , 7.

**Solution**   Substituting into the formula for continuous compounding, we obtain $A(t) = 100e^{0.08t}$. Figure 3.49 shows the values of $y_1 = A(x) = 1000e^{0.08x}$ for *x* = 1, 2, . . . , 7. For example, the value of her investment is $149.18 at the end of 5 years, and $175.07 at the end of 7 years.

| X | Y₁ | |
|---|---|---|
| 1 | 108.33 | |
| 2 | 117.35 | |
| 3 | 127.12 | |
| 4 | 137.71 | |
| 5 | 149.18 | |
| 6 | 161.61 | |
| 7 | 175.07 | |

Y₁ ▤ 100e^(0.08X)

**Figure 3.49** Table of values for Example 5.

## Annual Percentage Yield

With so many different interest rates and methods of compounding it is sometimes difficult for a consumer to compare two different options. For example, would you prefer an investment earning 8.75% annual interest compounded quarterly or one earning 8.7% compounded monthly?

A common basis for comparing investments is the **annual percentage yield (APY)**—the percentage rate that, compounded annually, would yield the same return as the given interest rate with the given compounding period.

### Example 6  COMPUTING ANNUAL PERCENTAGE YIELD (APY)

Ursula invests $2000 with Crab Key Bank at 5.15% annual interest compounded quarterly. What is the equivalent APY?

**Solution**

**Model**

Let $x$ = the equivalent APY. The value of the investment at the end of 1 year using this rate is $A = 2000(1 + x)$. Thus, we have

$$2000(1 + x) = 2000\left(1 + \frac{0.0515}{4}\right)^4.$$

**Solve Algebraically**

$$(1 + x) = \left(1 + \frac{0.0515}{4}\right)^4$$

$$x = \left(1 + \frac{0.0515}{4}\right)^4 - 1$$

$$\approx 0.0525$$

**Interpret**

The annual percentage yield is 5.25%. In other words, Ursula's $2000 invested at 5.15% compounded quarterly for 1 year earns the same interest and yields the same value as $2000 invested elsewhere paying 5.25% interest once at the end of the year.

Example 6 shows that the APY does not depend on the principal $P$ because both sides of the equation were divided by $P = 2000$. So we can assume that $P = 1$ when comparing investments.

### Example 7  COMPARING ANNUAL PERCENTAGE YIELD (APYs)

Which investment is more attractive, one that pays 8.75% compounded quarterly or another than pays 8.7% compounded monthly?

**Solution**

**Model**

Let

$$r_1 = \text{the APY for the 8.75\% rate,}$$

$$r_2 = \text{the APY for the 8.7\% rate.}$$

**Solve Numerically**

$$1 + r_1 = \left(1 + \frac{0.0875}{4}\right)^4 \qquad 1 + r_2 = \left(1 + \frac{0.087}{12}\right)^{12}$$

$$r_1 = \left(1 + \frac{0.0875}{4}\right)^4 - 1 \qquad r_2 = \left(1 + \frac{0.087}{12}\right)^{12} - 1$$

$$\approx 0.09041 \qquad\qquad \approx 0.09055$$

**Interpret**

The 8.7% rate compounded monthly is more attractive because its APY is 9.055% compared with 9.041% for the 8.75% rate compounded quarterly.

## Annuities—Future Value

So far, in all of the investment situations we have considered, the investor has made a single *lump-sum* deposit. But suppose an investor makes regular deposits monthly, quarterly, or yearly—the same amount each time. This is an *annuity* situation.

An **annuity** is a sequence of equal periodic payments. The annuity is **ordinary** if deposits are made at the end of each period at the same time the interest is posted in the account. Figure 3.50 represents this situation graphically. We will consider only ordinary annuities in this textbook.

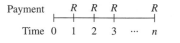

**Figure 3.50** Payments into an ordinary annuity.

### Example 8   COMPUTING AN ORDINARY ANNUITY

Sarah makes quarterly $500 payments at the end of each quarter into a retirement account that pays 8% interest compounded quarterly. How much will be in Sarah's account at the end of the first year?

**Solution**   Because Sarah deposits $500 on the last day of Quarter 1 and $500 on the last day of Quarter 2, the total in her account after two quarters will be $500 + $500(1.02). This is the $500 Quarter 2 payment plus the principal and interest on the Quarter 1 payment. The growth in the value of her account is shown below. Notice the pattern.

End of Quarter 1:

$\quad$ $500 = $500

End of Quarter 2:

$\quad$ $500 + $500(1.02) = $1010

End of Quarter 3:

$\quad$ $500 + $500(1.02) + $500(1.02)^2 = $1530.20

End of the year:

$\quad$ $500 + $500(1.02) + $500(1.02)^2 + $500(1.02)^3 = $2060.80

Thus the total value of the investment returned from an annuity consists of all the periodic payments together with all the interest. This value is called the **future value** of the annuity because it is typically calculated when projecting into the future. We give a mathematical shortcut for computing future value. Because its justification is based on series and mathematical induction, we will revisit this formula in Section 9.4. Note that the interest rate $i$ in the future value formula is the rate for the payment interval, not the annual rate $r$ of previous formulas in this section.

**Teaching Notes**

Show students how an investment in an annuity can accumulate over a 40-year period if they were to begin making monthly deposits of $50 the first month after graduating from high school. Use several different interest rates.

You may wish to point out that in the future and present value formulas, $i = r/k$ and $n = kt$ using the notation of compound interest.

> **Future Value of an Annuity**
>
> The future value $FV$ of an annuity consisting of $n$ equal periodic payments of $R$ dollars at an interest rate $i$ per compounding period (payment interval) is
>
> $$FV = R\frac{(1 + i)^n - 1}{i}.$$

### Example 9  CALCULATING THE VALUE OF AN ANNUITY

At the end of each quarter year, Emily makes a $500 payment into the Lanaghan Mutual Fund. If her investments earn 7.88% annual interest compounded quarterly, what will be the value of Emily's annuity in 20 years?

**Solution**

**Model**

We use $R = 500$, $i = 0.0788/4$, $n = 20(4) = 80$ in the equation

$$FV = R\frac{(1 + i)^n - 1}{i}.$$

**Solve Algebraically**

$$FV = 500 \cdot \frac{(1 + 0.0788/4)^{80} - 1}{0.0788/4}.$$

$$FV = 95{,}483.389\ldots$$

**Interpret**

The value of Emily 's annuity in 20 years will be $95,483.39.

**Notes on Examples**

You can challenge your students using Example 9. Have them keep $FV$ and $R$ constant and try to solve for $i$ (the quarterly interest rate) using an algebraic method. Then have them graph $FV$ as a function of $i$ by letting the variable $x$ replace $i$ in the formula.

## Loans and Mortgages—Present Value

An annuity is a sequence of equal period payments. The net amount of money put into an annuity is its **present value**. The net amount returned from the annuity is its future value. The periodic and equal payments on a loan or mortgage actually constitute an annuity.

How does the bank determine what the periodic payments should be? It considers what would happen to the present value as an investment with interest compounding over the term of the loan and compares the result to the future value of the loan repayment annuity.

We illustrate this reasoning by assuming that a bank lends you a present value $PV = \$50{,}000$ at 6% to purchase a house with the expectation that you will make a mortgage payment each month (at the monthly interest rate of $0.06/12 = 0.005$).

- The future value of an investment at 6% compounded monthly for $n$ months is

$$PV(1 + i)^n = 50{,}000(1 + 0.005)^n.$$

- The future value of an annuity of $R$ dollars (the loan payments) is

$$R\frac{(1 + i)^n - 1}{i} = R\frac{(1 + 0.005)^n - 1}{0.005}.$$

To find $R$, we would solve the equation

$$50{,}000(1 + 0.005)^n = R\frac{(1 + 0.005)^n - 1}{0.005}.$$

In general, the monthly payments of $R$ dollars for a loan of $PV$ dollars must satisfy the equation

$$PV(1 + i)^n = R\frac{(1 + i)^n - 1}{i}.$$

Dividing both sides by $(1 + i)^n$ leads to the following formula for the present value of an annuity.

**Present Value of an Annuity**

The present value $PV$ of an annuity consisting of $n$ equal payments of $R$ dollars earning an interest rate $i$ per period (payment interval) is

$$PV = R\frac{1 - (1 + i)^{-n}}{i}.$$

The annual interest rate charged on consumer loans is the **annual percentage rate (APR)**. The APY for the lender is higher than the APR. See Exercise 56.

**Example 10  CALCULATING LOAN PAYMENTS**

Carlos purchases a new pick-up truck for $18,500. What are the monthly payments for a 4-year loan with a $2000 down payment if the annual interest rate (APR) is 2.9%?

**Solution**

**Model**

The down payment is $2000, so the amount borrowed is $16,500. Since APR = 2.9%, $i = 0.029/12$ and the monthly payment is the solution to

$$16{,}500 = R\frac{1 - (1 + 0.029/12)^{-4(12)}}{0.029/12}.$$

**Solve Algebraically**

$$R\left[1 - \left(1 + \frac{0.029}{12}\right)^{-4(12)}\right] = 16{,}500\left(\frac{0.029}{12}\right)$$

$$R = \frac{16{,}500\,(0.029/12)}{1 - (1 + 0.029/12)^{-48}}$$

$$= 364.487\ldots$$

**Interpret**

Carlos will have to pay $364.49 per month for 47 months, and slightly less the last month (see Exercise 59).

**Teaching Note**

Have students investigate the current rates of interest available for financing a new automobile. Let students determine what auto they want to purchase and how much they would need to finance through a loan. Students should use the formulas from this section to calculate the monthly payment necessary to amortize the loan over a fixed number of months.

Sample loans for houses may also be investigated. It is interesting to calculate the interest paid for a 20 year or 30 year loan on $100,000.

**Follow-up**

Ask students how the interest rate affects the present and future values of an annuity. (**A higher interest rate gives a lower present value and a higher future value.**)

**Assignment Guide**

Day 1: Ex. 1–36, multiples of 3
Day 2: Ex. 39, 42, 43, 45, 46, 48, 50, 52, 53, 55, 58, 59, 60, 61

**Cooperative Learning**

Group Activity: Ex. 57

**Notes on Exercises**

The exercises in this section should be interesting to students because they deal with real-life financial situations. Students can apply these methods to their own financial planning.
Ex. 53, 54 illustrate the results of making accelerated payments on a mortgage.

**Ongoing Assessment**

Self-Assessment: Ex. 48, 8, 14, 18, 24
Embedded Assessment: Ex. 53, 55

# Quick Review 3.6

1. Find 3.5% of 200. 7
2. Find 2.5% of 150. 3.75
3. What is one-fourth of 7.25%? 1.8125%
4. What is one-twelfth of 6.5%? ≈ 0.5417%
5. 78 is what percent of 120? 65%
6. 28 is what percent of 80? 35%

7. 48 is 32% of what number? 150
8. 176.4 is 84% of what number? 210
9. How much does Jane have at the end of 1 year if she invests $300 at 5% simple interest? $315
10. How much does Reggie have at the end of 1 year if he invests $500 at 4.5% simple interest? $522.50

# Section 3.6 Exercises

In Exercises 1–4, find the amount $A$ accumulated after investing a principal $P$ for $t$ years at an interest rate $r$ compounded annually.

1. $P = \$1500$, $r = 7\%$, $t = 6$ $2251.10
2. $P = \$3200$, $r = 8\%$, $t = 4$ $4353.56
3. $P = \$12,000$, $r = 7.5\%$, $t = 7$ $19,908.59
4. $P = \$15,500$, $r = 9.5\%$, $t = 12$ $46,057.58

In Exercises 5–8, find the amount $A$ accumulated after investing a principal $P$ for $t$ years at an interest rate $r$ compounded $k$ times per year.

5. $P = \$1500$, $r = 7\%$, $t = 5$, $k = 4$ $2122.17
6. $P = \$3500$, $r = 5\%$, $t = 10$, $k = 4$ $5752.67
7. $P = \$40,500$, $r = 3.8\%$, $t = 20$, $k = 12$ $86,496.26
8. $P = \$25,300$, $r = 4.5\%$, $t = 25$, $k = 12$ $77,765.69

In Exercises 9 and 10, find the amount $A$ accumulated after investing a principal $P$ for $t$ years at interest rate $r$ compounded continuously.

9. $P = \$1250$, $r = 5.4\%$, $t = 6$ $1728.31
10. $P = \$3350$, $r = 6.2\%$, $t = 8$ $5501.17

In Exercises 11–14, find the future value $FV$ accumulated in an annuity after investing periodic payments $R$ for $t$ years at an annual interest rate $r$, with payments made and interest credited $k$ times per year.

11. $R = \$500$, $r = 7\%$, $t = 6$, $k = 4$ $14,755.51
12. $R = \$300$, $r = 6\%$, $t = 12$, $k = 4$ $20,869.57
13. $R = \$450$, $r = 5.25\%$, $t = 10$, $k = 12$ $70,819.63
14. $R = \$610$, $r = 6.5\%$, $t = 25$, $k = 12$ $456,790.28

In Exercises 15 and 16, find the present value $PV$ of a loan with an annual interest rate $r$ and periodic payments $R$ for a term of $t$ years, with payments made and interest charged 12 times per year.

15. $r = 4.7\%$, $R = \$815.37$, $t = 5$ $43,523.31
16. $r = 6.5\%$, $R = \$1856.82$, $t = 30$ $293,769.01

In Exercises 17 and 18, find the periodic payment $R$ of a loan with present value $PV$ and an annual interest rate $r$ for a term of $t$ years, with payments made and interest charged 12 times per year.

17. $PV = \$18,000$, $r = 5.4\%$, $t = 6$ $293.24
18. $PV = \$154,000$, $r = 7.2\%$, $t = 15$ $1401.47

19. **Finding Time** If John invests $2300 in a savings account with a 9% interest rate compounded quarterly, how long will it take until John's account has a balance of $4150?

20. **Finding Time** If Joelle invests $8000 into a retirement account with a 9% interest rate compounded monthly, how long will it take until this single payment has grown in her account to $16,000?

21. **Trust Officer** Megan is the trust officer for an estate. If she invests $15,000 into an account that carries an interest rate of 8% compounded monthly, how long will it be until the account has a value of $45,000 for Megan's client?

22. **Chief Financial Officer** Willis is the financial officer of a private university with the responsibility for managing an endowment. If he invests $1.5 million at an interest rate of 8% compounded quarterly, how long will it be until the account exceeds $3.75 million?

23. **Finding the Interest Rate** What interest rate compounded daily (365 days/years) is required for a $22,000 investment to grow to $36,500 in 5 years? ≈ 10.13%

24. **Finding the Interest Rate** What interest rate compounded monthly is required for an $8500 investment to triple in 5 years? ≈ 22.17%

25. **Pension Officer** Jack is an actuary working for a corporate pension fund. He needs to have $14.6 million grow to $22 million in 6 years. What interest rate compounded annually does he need for this investment?

26. **Bank President** The president of a bank has $18 million in his bank's investment portfolio that he wants to grow to $25 million in 8 years. What interest rate compounded annually does he need for this investment? ≈ 4.19%

**27. Doubling Your Money** Determine how much time is required for an investment to double in value if interest is earned at the rate of 5.75% compounded quarterly.

**28. Tripling Your Money** Determine how much time is required for an investment to triple in value if interest is earned at the rate of 6.25% compounded monthly.

In Exercises 29–32, complete the table about continuous compounding.

| Continuous Compounding | | | |
|---|---|---|---|
| Initial Investment | APR | Time to Double | Amount in 15 years |
| **29.** $12,500 | 9% | ? | ? |
| **30.** $32,500 | 8% | ? | ? |
| **31.** $ 9500 | ? | 4 years | ? |
| **32.** $16,800 | ? | 6 years | ? |

In Exercises 33–38, complete the table about doubling time of an investment.

| Doubling Time | | |
|---|---|---|
| APR | Compounding Periods | Time to Double |
| **33.** 4% | Quarterly | ? |
| **34.** 8% | Quarterly | ? |
| **35.** 7% | Annually | ? |
| **36.** 7% | Quarterly | ? |
| **37.** 7% | Monthly | ? |
| **38.** 7% | Continuously | ? |

In Exercises 39–42, find the annual percentage yield (APY) for the investment.

**39.** $3000 at 6% compounded quarterly  ≈ 6.14%

**40.** $8000 at 5.75% compounded daily  ≈ 5.92%

**41.** $P$ dollars at 6.3% compounded continuously  ≈ 6.50%

**42.** $P$ dollars at 4.7% compounded monthly  ≈ 4.80%

**43. Comparing Investments** Which investment is more attractive, 5% compounded monthly or 5.1% compounded quarterly?  5.1% quarterly

**44. Comparing Investments** Which investment is more attractive, $5\frac{1}{8}$% compounded annually or 5% compounded continuously?  $5\frac{1}{8}$% annually

In Exercises 45–48, payments are made and interest is credited at the end of each month.

**45. An IRA Account** Amy contributes $50 per month into the Lincoln National Bond Fund that earns 7.26% annual interest. What is the value of Amy's investment after 25 years?  $42,211.46

**46. An IRA Account** Matthew contributes $50 per month into the Hoffbrau Fund that earns 15.5% annual interest. What is the value of Matthew's investment after 20 years?

**47. An Investment Annuity** Betsy contributes to the Celebrity Retirement Puritan Fund that earns 12.4% annual interest. What should her monthly payments be if she wants to accumulate $250,000 in 20 years?  $239.42 per month

**48. An Investment Annuity** Diego contributes to the Commercial National Money Market Account that earns 4.5% annual interest. What should his monthly payments be if he wants to accumulate $120,000 in 30 years?

**49. Car Loan Payment** What is Kim's monthly payment for a 4-year $9000 car loan with an APR of 7.95% from Century Bank?  $219.51 per month

**50. Car Loan Payment** What is Ericka's monthly payment for a 3-year $4500 car loan with an APR of 10.25% from County Savings Bank?  $145.74 per month

**51. House Mortgage Payment** Gendo obtains a 30-year $86,000 house loan with an APR of 8.75% from National City Bank. What is her monthly payment?

**52. House Mortgage Payment** Roberta obtains a 25-year $100,000 house loan with an APR of 9.25% from NBD Bank. What is her monthly payment?  $856.39 per month

**53. Mortgage Payment Planning** An $86,000 mortgage for 30 years at 12% APR requires monthly payments of $884.61. Suppose you decided to make monthly payments of $1050.00.

**(a)** When would the mortgage be completely paid?

**(b)** How much do you save with the greater payments compared with the original plan?  $137,859.60

**54. Mortgage Payment Planning** Suppose you make payments of $884.61 for the $86,000 mortgage in Exercise 53 for 10 years and then make payments of $1050 until the loan is paid.

**(a)** When will the mortgage be completely paid under these circumstances?  22 years 2 months

**(b)** How much do you save with the greater payments compared with the original plan?  $59,006.40

**55. Writing to Learn** Explain why computing the APY for an investment does not depend on the actual amount being invested. Give a formula for the APY on a $1 investment at annual rate $r$ compounded $k$ times a year. How do you extend the result to a $1000 investment?

**56. Writing to Learn** Give reasons why banks might not announce their APY on a loan they would make to you at a given APR. What is the bank's APY on a loan that they make at 4.5% APR?

**57. Group Activity  Work in groups of three or four** Consider population growth of humans or other animals, bacterial growth, radioactive decay, and compounded interest. Explain how these problem situations are similar and how they are different. Give examples to support your point of view.

58. **Simple Interest versus Compounding Annually**
Steve purchases a $1000 certificate of deposit and will earn 6% each year. The interest will be mailed to him, so he will not earn interest on his interest.

(a) Show that after $t$ years, the total amount of interest he receives from his investment plus the original $1000 is given by

$$f(t) = 1000(1 + 0.06t).$$

(b) Steve invests another $1000 at 6% compounded annually. Make a table that compares the value of the two investments for $t = 1, 2, \ldots, 10$ years.

## Explorations

59. **Loan Payoff** Use the information about Carlos's truck loan in Example 10 to make a spreadsheet of the payment schedule. The first few lines of the spreadsheet should look like the following table:

| Month No. | Payment | Interest | Principal | Balance |
|---|---|---|---|---|
| 0 | | | | $16,500.00 |
| 1 | $364.49 | $39.88 | $324.61 | $16,175.39 |
| 2 | $364.49 | $39.09 | $325.40 | $15,849.99 |

To create the spreadsheet successfully, however, you need to use formulas for many of the cells, as shown in boldface type in the following sample:

| Month No. | Payment | Interest | Principal | Balance |
|---|---|---|---|---|
| 0 | | | | $16,500.00 |
| =A2+1 | $364.49 | =round(E2*2.9%/12,2) | =B3–C3 | =E2–D3 |
| =A3+1 | $364.49 | =round(E3*2.9%/12,2) | =B4–C4 | =E3–D4 |

Continue the spreadsheet using copy-and-paste techniques, and determine the amount of the 48th and final payment so that the final balance is $0.00.  $364.38

60. **Writing to Learn  Loan Payoff** Which of the following graphs is an accurate graph of the loan balance as a function of time, based on Carlos's truck loan in Example 10 and Exercise 59? Explain your choice based on increasing or decreasing behavior and other analytical characteristics. Would you expect the graph of loan balance versus time for a 30-year mortgage loan at twice the interest rate to have the same shape or a different shape as the one for the truck loan? Explain.

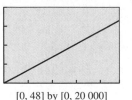

[0, 48] by [0, 20 000]
(a)

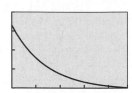

[0, 48] by [0, 20 000]
(b)

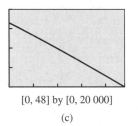

[0, 48] by [0, 20 000]
(c)

## Extending the Ideas

61. The function

$$f(x) = 100\frac{(1 + 0.08/12)^x - 1}{0.08/12}$$

describes the future value of a certain annuity.

(a) What is the annual interest rate?  8%

(b) How many payments per year are there?  12

(c) What is the amount of each payment?  $100

62. The function

$$f(x) = 200\frac{1 - (1 + 0.08/12)^{-x}}{0.08/12}$$

describes the present value of a certain annuity.

(a) What is the annual interest rate?  8%

(b) How many payments per year are there?  12

(c) What is the amount of each payment?  $200

# Chapter 3    Key Ideas

## Concepts

algebraic function (p. 259)
annual percentage rate (APR) (p. 323)
annual percentage yield (APY) (p. 320)
annuity (p. 321)
base of a logarithmic function (p. 282)
base of an exponential function (p. 260)
basic logistic function (p. 267)
common logarithm (p. 284)
compound interest (p. 316, 319)
compounded annually (p. 316)
compounded continuously (p. 319)
constant percentage rate (p. 272)
decay factor (p. 262)
decibel (p. 258)

exponential decay function (p. 262)
exponential function (p. 260)
exponential growth function (p. 262)
future value of an annuity (p. 321–322)
growth factor (p. 262)
half-life (p. 274)
hyperbolic sine function (p. 281)
hyperbolic cosine function (p. 281)
hyperbolic tangent function (p. 281)
limit to growth (p. 267)
logarithmic function with base $b$
   (p. 282, 297)
logistic decay function (p. 267)
logistic growth function (p. 267)

maximum sustainable population (p. 277)
natural base $e$ (p. 265)
natural exponential function (p. 264)
natural logarithm (p. 286)
natural logarithm function (p. 287)
Newton's law of cooling (p. 307)
order of magnitude (p. 305)
ordinary annuity (p. 321)
pH (p. 307)
present value of an annuity (p. 322–323)
radioactive decay (p. 274)
Richter scale (p. 300)
transcendental function (p. 259)

## Properties, Theorems, and Formulas

### Exponential Growth and Decay Functions

For any exponential function $f(x) = a \cdot b^x$ and any real number $x$,

$$f(x + 1) = b \cdot f(x).$$

If $a > 0$ and $b > 1$, the function $f$ is increasing and is an *exponential growth* function. The base $b$ is its *growth factor*.

If $a > 0$ and $b < 1$, $f$ is decreasing and is an *exponential decay function*. The base $b$ is its *decay factor*.

### Characteristics of Basic Exponential Functions

For any exponential function in the form $f(x) = b^x$.

Domain $= (-\infty, \infty)$.    Range $= (0, \infty)$.
$x$-intercepts: none.    $y$-intecept $= 1$.
Continuous.
No symmetry: neither even nor odd.
Bounded below, but not above.
No local extrema.
Horizontal asymptote: $y = 0$.
No vertical asymptotes.
Passes through $(0, 1)$ and $(1, b)$.
One-to-one.
Concave up.

If $b > 1$, then also

• $f$ is an increasing function,

• $\lim_{x \to -\infty} f(x) = 0$ and $\lim_{x \to \infty} f(x) = \infty$.

If $0 < b < 1$, then also

• $f$ is a decreasing function,

• $\lim_{x \to -\infty} f(x) = \infty$ and $\lim_{x \to \infty} f(x) = 0$.

## Theorem Exponential Functions and the base e

Any exponential function $f(x) = a \cdot b^x$ can be rewritten as

$$f(x) = a \cdot e^{kx}$$

for an approximately chosen real real number constant $k$.

If $a > 0$ and $k > 0$, $f(x) = a \cdot e^{kx}$ is an exponential growth function.

If $a > 0$ and $k < 0$, $f(x) = a \cdot e^{kx}$ is an exponential decay function.

## Basic Properties of Logarithms

For $0 < b \neq 1$, $x > 0$, and any real number $y$,

- $\log_b 1 = 0$ because $b^0 = 1$.
- $\log_b b = 1$ because $b^1 = b$.
- $\log_b b^y = y$ because $b^y = b^y$.
- $b^{\log_b x} = x$ because $\log_b x = \log_b x$.

## Basic Properties of Common Logarithms

For $x > 0$, and any real number $y$,

- $\log 1 = 0$ because $10^0 = 1$.
- $\log 10 = 1$ because $10^1 = 10$.
- $\log 10^y = y$ because $10^y = 10^y$.
- $10^{\log x} = x$ because $\log x = \log x$.

## Basic Properties of Natural Logarithms

For $x > 0$, and any real number $y$,

- $\ln 1 = 0$ because $e^0 = 1$.
- $\ln e = 1$ because $e^1 = e$.
- $\ln e^y = y$ because $e^y = e^y$.
- $e^{\ln x} = x$ because $\ln x = \ln x$.

## Properties of Exponents

For $b > 0$ and any real numbers $x$ and $y$,

- $b^x \cdot b^y = b^{x+y}$.
- $\dfrac{b^x}{b^y} = b^{x-y}$.
- $(b^x)^y = b^{xy}$.

## Properties of Logarithms

If $b$, $R$, and $S$ are positive real numbers such that $b \neq 1$, and $c$ is any real number,

- **Product rule:**   $\log_b(RS) = \log_b R + \log_b S$
- **Quotient rule:**   $\log_b \dfrac{R}{S} = \log_b R - \log_b S$
- **Power rule:**   $\log_b R^c = c \log_a R$

## Change-of-Base Formula for Logarithms

For positive real numbers $a$, $b$, and $x$ with $a \neq 1$ and $b \neq 1$,

$$\log_b x = \frac{\log_a x}{\log_a b}$$

## Characterisics of Logarithmic Functions with $b > 1$

Domain: $(0, \infty)$     Range: All reals.
$x$-intercept $= 1$.     $y$-intercepts: none.
Continuous.
Increasing on its domain.
No symmetry: neither even nor odd.
Not bounded above or below.
No local extrema.
No horizontal asymptotes.
Vertical asymptote: $x = 0$.
End behvior: $\lim\limits_{x \to \infty} x = \infty$

Passes through $(1, 0)$ and $(b, 1)$.
One-to-one.     Concave down.

## One-to-One Properties

For any exponential function $f(x) = b^x$,

- If $b^u = b^v$, then $u = v$.

For any logarithmic function $f(x) = \log_b x$,

- If $\log_b u = \log_b v$, then $u = v$.

## Future Value of an Annuity

The future value $FV$ of an annuity consisting of $n$ equal periodic payments of $R$ dollars earning at an interest rate $i$ per compounding period (payment interval) is

$$FV = R\frac{(1 + i)^n - 1}{i}$$

## Present Value of an Annuity

The present value $PV$ of an annuity consisting of $n$ equal payments of $R$ dollars earning at an interest rate $i$ per period (payment interval) is

$$PV = R\frac{1 - (1 + i)^{-n}}{i}.$$

# Procedures

## Changing Between Logarithmic and Exponential Form

If $x > 0$ and $0 < b \neq 1$, then

$$y = \log_b (x) \quad \text{if and only if} \quad b^y = x.$$

## Re-Expression of Data

The process of applying a function to one or both of the variables in an $(x, y)$ data set to put the data in a more convenient form or to discover a relationship between the variables.

## Logarithmic Re-Expression of Data

The process of rewritting $(x, y)$ data pairs as $(\ln x, y)$, $(x, \ln y)$, or $(\ln x, \ln y)$ to discover a relationship between the variables $x$ and $y$.

- If the $(\ln x, y)$ data pairs are linear, then the original $(x, y)$ data pairs have a logarithmic function relationship

- If the $(x, \ln y)$ data pairs are linear, then the original $(x, y)$ data pairs have a exponential function relationship

- If the $(\ln x, \ln y)$ data pairs are linear, then the original $(x, y)$ data pairs have a power function relationship

## Gallery of Functions

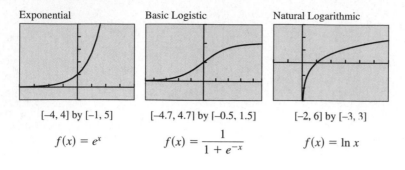

| Exponential | Basic Logistic | Natural Logarithmic |
|---|---|---|
| $[-4, 4]$ by $[-1, 5]$ | $[-4.7, 4.7]$ by $[-0.5, 1.5]$ | $[-2, 6]$ by $[-3, 3]$ |
| $f(x) = e^x$ | $f(x) = \dfrac{1}{1 + e^{-x}}$ | $f(x) = \ln x$ |

## Chapter 3 Review Exercises

The collection of exercises marked in red could be used as a chapter test.

In Exercises 1 and 2, compute the exact value of the function for the given $x$ value without using a calculator.

**1.** $f(x) = -3 \cdot 4^x$ for $x = \dfrac{1}{3}$  $-3\sqrt[3]{4}$

**2.** $f(x) = 6 \cdot 3^x$ for $x = -\dfrac{3}{2}$  $\dfrac{2}{\sqrt{3}}$

In Exercises 3 and 4, determine a formula for the exponential function whose graph is shown in the figure.

**3.**

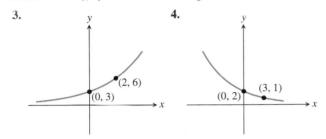

(2, 6)
(0, 3)

**4.**

(0, 2)
(3, 1)

In Exercises 5–10, describe how to transform the graph of $f$ into the graph of $g(x) = 2^x$ or $h(x) = e^x$. Sketch the graph by hand and support your answer with a grapher.

**5.** $f(x) = 4^{-x} + 3$

**6.** $f(x) = -4^{-x}$

**7.** $f(x) = -8^{-x} - 3$

**8.** $f(x) = 8^{-x} + 3$

**9.** $f(x) = e^{2x-3}$

**10.** $f(x) = e^{3x-4}$

In Exercises 11 and 12, graph the function. Find the $y$-intercept and the horizontal asymptotes.

**11.** $f(x) = \dfrac{100}{5 + 3e^{-0.05x}}$

**12.** $f(x) = \dfrac{50}{5 - 2e^{-0.04x}}$

In Exercises 13 and 14, state whether the function is an exponential growth function or an exponential decay function, and describe its end behavior using limits.

**13.** $y = e^{4-x} + 2$

**14.** $y = 2(5^{x-3}) + 1$

In Exercises 15–18, graph the function, and analyze it for domain, range, continuity, intercepts, increasing or decreasing behavior, symmetry, boundedness, extrema, asymptotes, end behavior, and concavity.

**15.** $y = e^{3-x} + 1$

**16.** $y = 3(4^{x+1}) - 2$

**17.** $g(x) = \dfrac{100}{4 + 2e^{-0.01x}}$

**18.** $g(x) = \dfrac{24}{2 - 5e^{-0.01x}}$

In Exercises 19–22, find the exponential function that satisfies the given conditions.

**19.** Initial value = 24, increasing at a rate of 5.3% per day

**20.** Initial population = 67,000, increasing at a rate of 1.67% per year  $f(x) = 67{,}000 \cdot (1.0167)^x$

**21.** Initial height = 18 cm, doubling every 3 weeks

**22.** Initial mass = 117 g, halving once every 262 hours

In Exercises 23 and 24, find the logistic function that satisfies the given conditions.

**23.** Initial value = 12, limit to growth = 30, passing through (2, 20).  $f(x) = 30/(1 + 1.5e^{-0.55x})$

**24.** Initial height = 6, limit to growth = 20, passing through (3, 15).  $f(x) = 20/(1 + 2.33e^{-0.65x})$

In Exercises 25 and 26, determine a formula for the logistic function whose graph is shown in the figure.

**25.**

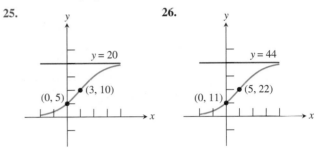

$y = 20$
(0, 5)  (3, 10)

**26.**

$y = 44$
(0, 11)  (5, 22)

In Exercises 27–30, evaluate the logarithmic expression without using a calculator.

**27.** $\log_2 32$  5

**28.** $\log_3 81$  4

**29.** $\log \sqrt[3]{10}$  1/3

**30.** $\ln \dfrac{1}{\sqrt{e^7}}$  $-7/2$

In Exercises 31–34, rewrite the equation in exponential form.

**31.** $\log_3 x = 5$  $3^5 = x$

**32.** $\log_2 x = y$  $2^y = x$

**33.** $\ln \dfrac{x}{y} = -2$  $y = xe^2$

**34.** $\log \dfrac{a}{b} = -3$  $b = 1000a$

In Exercises 35–38, describe how to transform the graph of $y = \log_2 x$ into the graph of the given function. Sketch the graph by hand and support with a grapher.

**35.** $f(x) = \log_2 (x + 4)$

**36.** $g(x) = \log_2 (4 - x)$

**37.** $h(x) = -\log_2 (x - 1) + 2$

**38.** $h(x) = -\log_2 (x + 1) + 4$

In Exercises 39–42, graph the function, and analyze it for domain, range, continuity, intercepts, increasing or decreasing behavior, symmetry, boundedness, extrema, asymptotes, end behavior, and concavity.

**39.** $f(x) = x \ln x$

**40.** $f(x) = x^2 \ln x$

**41.** $f(x) = x^2 \ln |x|$

**42.** $f(x) = \dfrac{\ln x}{x}$

In Exercises 43–54, solve the equation.

**43.** $10^x = 4$  $\log 4 \approx 0.6021$

**44.** $e^x = 0.25$

**45.** $1.05^x = 3$

**46.** $\ln x = 5.4$  $e^{5.4} \approx 221.4064$

**47.** $\log x = -7$

**48.** $3^{x-3} = 5$

**49.** $3 \log_2 x + 1 = 7$   4

**50.** $2 \log_3 x - 3 = 4$

**51.** $\dfrac{3^x - 3^{-x}}{2} = 5$

**52.** $\dfrac{50}{4 + e^{2x}} = 11$

**53.** $\log (x + 2) + \log (x - 1) = 4$

**54.** $\ln (3x + 4) - \ln (2x + 1) = 5$

In Exercises 55 and 56, write the expression using only natural logarithms.

**55.** $\log_2 x$   ln $x$/ln 2

**56.** $\log_{1/6} (6x^2)$

In Exercises 57 and 58, write the expression using only common logarithms.

**57.** $\log_5 x$   log $x$/log 5

**58.** $\log_{1/2} (4x^3)$

In Exercises 59–62, match the function with its graph. All graphs are drawn in the window $[-4.7, 4.7]$ by $[-3.1, 3.1]$.

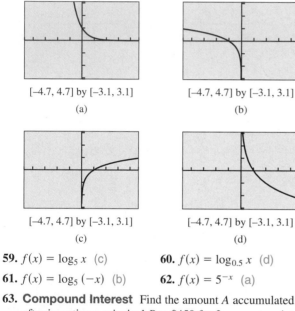

[–4.7, 4.7] by [–3.1, 3.1]
(a)

[–4.7, 4.7] by [–3.1, 3.1]
(b)

[–4.7, 4.7] by [–3.1, 3.1]
(c)

[–4.7, 4.7] by [–3.1, 3.1]
(d)

**59.** $f(x) = \log_5 x$   (c)

**60.** $f(x) = \log_{0.5} x$   (d)

**61.** $f(x) = \log_5 (-x)$   (b)

**62.** $f(x) = 5^{-x}$   (a)

**63. Compound Interest** Find the amount $A$ accumulated after investing a principal $P = \$450$ for 3 years at an interest rate of 4.6% compounded annually.   $515.00

**64. Compound Interest** Find the amount $A$ accumulated after investing a principal $P = \$4800$ for 17 years at an interest rate 6.2% compounded quarterly.   $13,660.81

**65. Compound Interest** Find the amount $A$ accumulated after investing a principal $P$ for $t$ years at interest rate $r$ compounded continuously.   $Pe^{rt}$

**66. Future Value** Find the future value $FV$ accumulated in an annuity after investing periodic payments $R$ for $t$ years at an annual interest rate $r$, with payments made and interest credited $k$ times per year.

**67. Present Value** Find the present value $PV$ of a loan with an annual interest rate $r = 5.5\%$ and periodic payments $R = \$550$ for a term of $t = 5$ years, with payments made and interest charged 12 times per year.   $28,794.06

**68. Present Value** Find the present value $PV$ of a loan with an annual interest rate $r = 7.25\%$ and periodic payments $R = \$953$ for a term of $t = 15$ years, with payments made and interest charged 26 times per year.   $226,396.22

In Exercises 69 and 70, determine the value of $k$ so that the graph of $f$ passes through the given point.

**69.** $f(x) = 20e^{-kx}$,   $(3, 50)$

**70.** $f(x) = 20e^{-kx}$,   $(1, 30)$

In Exercises 71 and 72, use the data in Table 3.25.

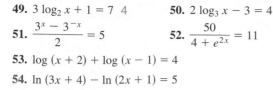

| Table 3.25 Official Census Populations (in millions), 1900–1990 | | |
| --- | --- | --- |
| Year | Georgia | Illinois |
| 1900 | 2.2 | 4.8 |
| 1910 | 2.6 | 5.6 |
| 1920 | 2.9 | 6.5 |
| 1930 | 2.9 | 7.6 |
| 1940 | 3.1 | 7.9 |
| 1950 | 3.4 | 8.7 |
| 1960 | 3.9 | 10.1 |
| 1970 | 4.6 | 11.1 |
| 1980 | 5.5 | 11.4 |
| 1990 | 6.5 | 11.4 |

*Source: U.S. Census Bureau as reported in 1999 New York Times Almanac.*

**71. Modeling Population** Find a exponential regression model for Georgia's population, and use it to predict the population in 2005.

**72. Modeling Population** Find a logistic regression model for Illinois' population, and use it to predict the population in 2010.

**73. Drug Absorption** A drug is administered intravenously for pain. The function $f(t) = 90 - 52 \ln (1 + t)$, where $0 \le t \le 4$, gives the amount of the drug in the body after $t$ hours.

**(a)** What was the initial ($t = 0$) number of units of drug administered?   90 units

**(b)** How much is present after 2 h?   32.8722

**(c)** Draw the graph of $f$.

**74. Population Decrease** The population of Metroville is 123,000 and is decreasing by 2.4% each year.

**(a)** Write a function that models the population as a function of time $t$.   $P(t) = 123{,}000(0.976)^t$

**(b)** Predict when the population will be 90,000.

**75. Population Decrease** The population of Preston is 89,000 and is decreasing by 1.8% each year.

**(a)** Write a function that models the population as a function of time $t$.   $P(t) = 89{,}000(0.982)^t$

**(b)** Predict when the population will be 50,000.

**76. Spread of Flu** The number $P$ of students infected with flu at Northridge High School $t$ days after exposure is modeled by

$$P(t) = \frac{300}{1 + e^{4-t}}.$$

**(a)** What was the initial ($t = 0$) number of students infected with the flu? 5 or 6 students

**(b)** How many students were infected after 3 days?

**(c)** When will 100 students be infected?

**(d)** What would be the maximum number of students infected? 300

**77. Rabbit Population** The number of rabbits in Elkgrove doubles every month. There are 20 rabbits present initially.

**(a)** Express the number of rabbits as a function of the time $t$.

**(b)** How many rabbits were present after 1 year? after 5 years?

**(c)** When will there be 10,000 rabbits? $\approx 8.9658$ months

**78. Guppy Population** The number of guppies in Susan's aquarium doubles every day. There are four guppies initially.

**(a)** Express the number of guppies as a function of time $t$.

**(b)** How many guppies were present after 4 days? after 1 week? 64; 512

**(c)** When will there be 2000 guppies? $\approx 8.9658$ days

**79. Radioactive Decay** The half-life of a certain radioactive substance is 1.5 sec. The initial amount of substance is $S_0$ grams.

**(a)** Express the amount of substance $S$ remaining as a function of time $t$. $S(t) = S_0 \cdot (1/2)^{t/1.5}$

**(b)** How much of the substance is left after 1.5 sec? after 3 sec? $S_0/2$; $S_0/4$

**(c)** Determine $S_0$ if there was 1 g left after 1 min.

**80. Radioactive Decay** The half-life of a certain radioactive substance is 2.5 sec. The initial amount of substance is $S_0$ grams.

**(a)** Express the amount of substance $S$ remaining as a function of time $t$. $S(t) = S_0 \cdot (1/2)^{t/2.5}$

**(b)** How much of the substance is left after 2.5 sec? after 7.5 sec? $S_0/2$; $S_0/8$

**(c)** Determine $S_0$ if there was 1 g left after 1 min.

**81. Richter Scale** Afghanistan suffered two major earthquakes in 1998. The one on February 4 had a Richter magnitude of 6.1, causing about 2300 deaths, and the one on May 30 measured 6.9 on the Richter scale, killing about 4700 people. How many times more powerful was the deadlier quake? 6.31

**82. Chemical Acidity** The pH of seawater is 7.6, and the pH of milk of magnesia is 10.5.

**(a)** What are their hydrogen-ion concentrations?

**(b)** How many times greater is the hydrogen-ion concentration of the seawater than that of milk of magnesia?

**(c)** By how many orders of magnitude do the concentrations differ? 2.9

**83. Annuity Finding Time** If Joenita invests $1500 into a retirement account with an 8% interest rate compounded quarterly, how long will it take this single payment to grow to $3750? 11.75 years

**84. Annuity Finding Time** If Juan invests $12,500 into a retirement account with a 9% interest rate compounded continuously, how long will it take this single payment to triple in value? $\approx 12.2068$ years

**85. Monthly Payments** The time $t$ in months that it takes to pay off a $60,000 loan at 9% annual interest with monthly payments of $x$ dollars is given by

$$t = 133.83 \ln \left( \frac{x}{x - 450} \right).$$

Estimate the length (term) of the $60,000 loan if the monthly payments are $700.

**86. Monthly Payments** Using the equation in Exercise 85, estimate the length (term) of the $60,000 loan if the monthly payments are $500. about 25 years 9 months

**87. Finding APY** Find the annual percentage yield for an investment with an interest rate of 8.25% compounded monthly. $\approx 8.57\%$

**88. Finding APY** Find the annual percentage yield that can be used to advertise an account that pays interest at 7.20% compounded countinuously. $\approx 7.47\%$

**89. Light Absorption** The Beer-Lambert law of absorption applied to Lake Superior states that the light intensity $I$ (in lumens) at a depth of $x$ feet satisfies the equation

$$\log \frac{I}{12} = -0.0125x.$$

Find the light intensity at a depth of 25 ft. $\approx 5.84$ lumens

**90.** For what values of $b$ is $\log_b x$ a vertical stretch of $y = \ln x$? A vertical shrink of $y = \ln x$?

**91.** For what values of $b$ is $\log_b x$ a vertical stretch of $y = \log x$? A vertical shrink of $y = \log x$?

**92.** If $f(x) = ab^x$, $a > 0$, $b > 0$, prove that $g(x) = \ln f(x)$ is a linear function. Find its slope and $y$-intercept.

**93. Spread of Flu** The number of students infected with flu after $t$ days at Springfield High School is modeled by the function

$$P(t) = \frac{1600}{1 + 99e^{-0.4t}}.$$

**(a)** What was the initial number of infected students? 16

**(b)** When will 800 students be infected? about $11\frac{1}{2}$ days

**(c)** The school will close when 400 of the 1600 student body are infected. When would the school close?

**94. Population of Deer** The population $P$ of deer after $t$ years in Briggs State Park is modeled by the function

$$P(t) = \frac{1200}{1 + 99e^{-0.4t}}.$$

**(a)** What was the inital population of deer?  12 deer

**(b)** When will there be 1000 deer?  about $15\frac{1}{2}$ years

**(c)** What is the maximum number of deer planned for the park?  1200

**95. Newton's Law of Cooling** A cup of coffee cooled from $96°$ C to $65°$ C after 8 min in a room at $20°$ C. When will it cool to $25°$ C?  $\approx 41.54$ minutes

**96. Newton's Law of Cooling** A cake is removed from an oven at $220°$ F and cools to $150°$ F after 35 min in a room at $75°$ F. When will it cool to $95°$ F?  $\approx 105.17$ minutes

**97.** The function

$$f(x) = 100\frac{(1 + 0.09/4)^x - 1}{0.09/4}$$

describes the future value of a certain annuity.

**(a)** What is the annual interest rate?  9%

**(b)** How many payments per year are there?  4

**(c)** What is the amount of each payment?  $100

**98.** The function

$$g(x) = 200\frac{1 - (1 + 0.11/4)^{-x}}{0.11/4}$$

describes the present vlaue of a certain annuity.

**(a)** What is the annual interest rate?  11%

**(b)** How many payments per year are there?  4

**(c)** What is the amount of each payment?  $200

**99. Simple Interest versus Compounding Continuously** Grace purchases a $1000 certificate of deposit that will earn 5% each year. The interest will be mailed to her, so she will not earn interest on her interest.

**(a)** Show that after $t$ years, the total amount of interest she receives from her investment plus the original $1000 is given by

$$f(t) = 1000(1 + 0.05t).$$

**(b)** Grace invests another $1000 at 5% compounded continuously. Make a table that compares the values of the two investments for $t = 1, 2, \ldots, 10$ years.

# Chapter 3 Project

## Analyzing a Bouncing Ball

When a ball bounces up and down on a flat surface, the maximum height of the ball decreases with each bounce. Each rebound is a percentage of the previous height. For most balls, the percentage is a constant. In this project, you will use a motion detection device to collect distance and time data for a ball bouncing underneath a motion detector, then find a mathematical model that describes the maximum bounce height as a function of bounce number.

## Collecting the Data

Setup the Calculator Based Laboratory (CBL) system with a motion detector or a Calculator Based Ranger (CBR) system to collect ball bounce data using a ball bounce program for the CBL or the Ball Bounce Application for the CBR. See the CBL/CBR guidebook for specific setup instruction.

Hold the ball at least 2 feet below the detector and release it so that it bounces straight up and down beneath the detector. These programs convert distance versus time data to height from the ground versus time. The graph shows a plot of sample data collected with a racquet ball and CBR. The data table below shows the original height and each maximum height collected.

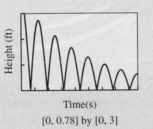

Time(s)
[0, 0.78] by [0, 3]

| Bounce Number | Maximum Height (feet) |
|:-------------:|:---------------------:|
| 0 | 2.7188 |
| 1 | 2.1426 |
| 2 | 1.6565 |
| 3 | 1.2640 |
| 4 | 0.98309 |
| 5 | 0.77783 |

## Explorations

1. If you collected motion data using a CBL or CBR, a plot of height versus time should be shown on your graphing calculator or computer screen. Trace to the maximum height for each bounce and record your data in a table and use other lists in your calculator to enter this data. If you don't have access to a CBL/CBR, enter the data in the table above into your graphing calculator/computer.

2. Bounce height 1 is what percentage of the original height? Calculate the percentage return for each bounce. The numbers should be fairly constant.

3. Create a scatter plot for maximum height versus bounce number.

4. For the first bounce, the height is predicted by multiplying the original height $H$ by the percentage $P$. The second height is predicted by multiplying this height $HP$ by $P$ which gives the $HP^2$. Explain why $Y = HP^x$ is the appropriate model for this data.

5. Enter this equation into your calculator using your values for H and P. How does the model fit your data?

6. Use the statistical features of the calculator to find the exponential regression for this data. Compare it to the equation that you used as a model.

7. How would your data and equation change if you used a different type of ball?

8. What factors would change the $H$ value and what factors affect the $P$ value?

9. Rewrite your equation using base $e$ instead of using $P$ as the base for the exponential equation.

10. What do you predict the graph of ln (bounce height) versus bounce number to look like?

11. Plot ln (bounce height) versus bounce number. Calculate the linear regression and use the concept of linear re-expression to explain how the slope and $y$-intercept are related to $P$ and $H$.

**Bibliography**

**For students:** *16–19 Mathematics: Modelling with Circular Motion, The School Mathematics Project.* Cambridge University Press,1990. Available through Dale Seymour Publications.

**For teachers:** *Trigonometry,* I.M. Gelfand, M. Saul. Birkhauser, 2000.

*A History of Mathematics,* Carl B. Boyer (Second edition revised by Uta C. Merzbach. John Wiley & Sons Inc., 1991.

Videos: *M! Project Mathematics! Sines and Cosines, Part 1,* Tom Apostol. NCTM, 1992.

**Calculator Conversions**

Your calculator probably has built-in functionality to convert degrees to DMS. Consult your *Owner's Manual.* Meanwhile, you should try some conversions the "long way" to get a better feel for how DMS works.

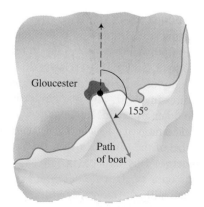

**Figure 4.2** The course of a fishing boat bearing 155° out of Gloucester.

in a circle of radius $h$, and $\theta$ is a **central angle** of the circle intercepting a circular arc of length $a$. In geometry we might call $a$ a "40-degree arc" because of its direct association with the central angle, but notice that $a$ also has a *length* that can be measured in the *same units* as the other lengths in the picture. Over time it became natural to think of the angle being determined by the arc rather than the arc being determined by the angle, and that led to radian measure.

## Degrees and Radians

A **degree**, represented by the symbol °, is a unit of angular measure equal to 1/180th of a straight angle. In the DMS (degree-minute-second) system of angular measure, each degree is subdivided into 60 **minutes** (denoted by ′) and each minute is subdivided into 60 **seconds** (denoted by ″). (Notice that Sumerian influence again.)

Example 1 illustrates how to convert from degrees in decimal form to DMS and vice-versa.

### Example 1  WORKING WITH DMS MEASURE

**(a)** Convert 37.425° to DMS.

**(b)** Convert 42°24′36″ to degrees.

**Solution**

**(a)** We need to convert the fractional part to minutes and seconds. First we convert 0.425° to minutes:

$$0.425° \left( \frac{60'}{1°} \right) = 25.5'$$

Then we convert 0.5 minutes to seconds:

$$0.5' \left( \frac{60''}{1'} \right) = 30''$$

Putting it all together, we find that 37.425° = 37°25′30″.

**(b)** Each minute is 1/60th of a degree, and each second is 1/3600th of a degree. Therefore,

$$42°24'36'' = 42° + \left( \frac{24}{60} \right)^{°} + \left( \frac{36}{3600} \right)^{°} = 42.41°$$

In navigation, the **course** or **bearing** of an object is sometimes given as the angle of the **line of travel** measured clockwise from due north. For example, the line of travel in Figure 4.2 has the bearing of 155°.

In this book we use degrees to measure angles in their familiar geometric contexts, especially when applying trigonometry to real-world problems in surveying, construction, and navigation, where degrees are still the accepted units of measure. When we shift our attention to the trigonometric *functions*, however, we will measure angles in *radians* so that domain and range values can be measured on comparable scales.

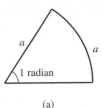

(a)

**Figure 4.3** In a circle, a central angle of 1 radian intercepts an arc of length 1 radius.

> **Definition** Radian
>
> A central angle of a circle has measure 1 **radian** if it intercepts an arc with the same length as the radius. (See Figure 4.3.)

---

**Exploration 1** **Constructing a 1-Radian Angle**

Carefully draw a large circle on a piece of paper, either by tracing around a circlar object or by using a compass. Identify the center of the circle (*O*) and draw a radius horizontally from *O* toward the right, intersecting the circle at point *A*. Then cut a piece of thread or string the same size as the radius. Place one end of the string at *A* and bend it around the circle counterclockwise, marking the point *B* on the circle where the other end of the string ends up. Draw the radius from *O* to *B*.

    The measure of angle *AOB* is one radian.

1. What is the circumference of the circle, in terms of its radius *r*?
2. How many radians must therefore be in a complete circle?
3. If we cut a piece of thread 3 times as big as the radius, would it extend halfway around the circle? Why or why not?
4. How many radians are in a straight angle? $\pi$ radians

---

**Example 2** WORKING WITH RADIAN MEASURE

(a) How many radians are in 90 degrees?

(b) How many degrees are in $\pi/3$ radians?

(c) Find the length of an arc intercepted by a central angle of 1/2 radian in a circle of radius 5 inches.

(d) Find the radian measure of a central angle that intercepts an arc of length *s* in a circle of radius *r*.

**Solution**

(a) Since $\pi$ radians and 180° both measure a straight angle, we can use the conversion factor $(\pi \text{ radians})/(180°) = 1$ to convert radians to degrees:

$$90°\left(\frac{\pi \text{ radians}}{180°}\right) = \frac{90\pi}{180} \text{ radians} = \frac{\pi}{2} \text{ radians}$$

(b) In this case, we use the conversion factor $(180°)/(\pi \text{ radians}) = 1$ to convert radians to degrees:

$$\left(\frac{\pi}{3} \text{ radians}\right)\left(\frac{180°}{\pi \text{ radians}}\right) = \frac{180°}{3} = 60°$$

(c) A central angle of 1 radian intercepts an arc of length 1 radius, which is 5 inches. Therefore, a central angle of 1/2 radian intercepts an arc of length 1/2 radius, which is 2.5 inches.

(d) We can solve this problem with ratios:

$$\frac{x \text{ radians}}{s \text{ units}} = \frac{1 \text{ radian}}{r \text{ units}}$$

$$xr = s$$

$$x = \frac{s}{r}$$

---

**Degree-Radian Conversion**

To convert radians to degrees, multiply by $\dfrac{180°}{\pi \text{ radians}}$.

To convert degrees to radians, multiply by $\dfrac{\pi \text{ radians}}{180°}$.

---

With practice, you can perform these conversions in your head. The key is to think of a straight angle equaling $\pi$ radians as readily as you think of it equaling 180 degrees. If you were asked to draw "by sight" a 60° angle with vertex $A$ in the diagram below, how would you proceed?

$$\underset{A}{\rule{3cm}{0.4pt}\!\bullet\!\rule{3cm}{0.4pt}}$$

Your thinking would probably go something like this: "I need to divide that straight angle (180°) into 3 equal pieces because 60° is one-third of 180°." Then you would draw the line:

Notice that this same challenge is actually *easier* if you are asked to draw an angle of $\pi/3$ radians. The angle measure itself tells you how to divide the straight angle! Similarly, divide a straight angle in two ($\pi/2$ radians) and you get a right angle. Double a straight angle ($2\pi$ radians) and you get a full circular rotation. A 45° angle, which is 1/4th of a straight angle, is $\pi/4$ radians— and so on.

## Circular Arc Length

Since a central angle of 1 radian always intercepts an arc of 1 radius in length, it follows that a central angle of $\theta$ radians in a circle of radius $r$ intercepts an arc of length $\theta r$. This gives us a convenient formula for measuring arc length.

---

**Arc Length Formula (Radian Measure)**

If $\theta$ is a central angle in a circle of radius $r$, and if $\theta$ is measured in radians, then the length $s$ of the intercepted arc is given by

$$s = r\theta$$

---

A somewhat less simple formula (which incorporates the degree-radian conversion formula) applies when $\theta$ is measured in degrees.

### Does a Radian have Units?

The formula $s = r\theta$ implies an interesting fact about radians: as long as $s$ and $r$ are measured in the same units, the radian is unit-neutral. For example, if $r = 5$ inches and $\theta = 2$ radians, then $s = 10$ inches (not 10 "inch-radians"). This unusual situation arises from the fact that the definition of the radian is tied to the length of the radius, units and all.

**Figure 4.4**  A 60° slice of a large pizza. (Example 3)

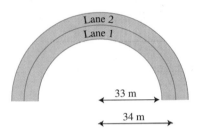
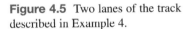

**Figure 4.5**  Two lanes of the track described in Example 4.

---

### Arc Length Formula (Degree Measure)

If $\theta$ is a central angle in a circle of radius $r$, and if $\theta$ is measured in degrees, then the length $s$ of the intercepted arc is given by

$$s = \frac{\pi r \theta}{180}.$$

### Example 3   PERIMETER OF A PIZZA SLICE

Find the perimeter of a 60° slice of a large (7 in. radius) pizza.

**Solution**  The perimeter (Figure 4.4) is 7 in. + 7 in. + $s$ in., where $s$ is the arc length of the pizza's curved edge. By the arc length formula:

$$s = \frac{\pi(7)(60)}{180} = \frac{7\pi}{3} \approx 7.3$$

The perimeter is approximately 21.3 in.

### Example 4   DESIGNING A RUNNING TRACK

The running lanes at the Emery Sears track at Bluffton College are 1 meter wide. The inside radius of lane 1 is 33 meters and the inside radius of lane 2 is 34 meters. How much longer is lane 2 than lane 1 around one turn? (See Figure 4.5.)

**Solution**  We think this solution through in radians. Each lane is a semicircle with central angle $\theta = \pi$ and length $s = r\theta = r\pi$. The difference in their lengths, therefore, is $34\pi - 33\pi = \pi$. Lane 2 is about 3.14 meters longer than lane 1.

## Angular and Linear Motion

In applications it is sometimes necessary to connect *angular speed* (measured in units like revolutions per minute) to *linear speed* (measured in units like miles per hour). The connection is usually provided by one of the arc length formulas or by a conversion factor that equates "1 radian" of angular measure to "1 radius" of arc length.

### Example 5   USING ANGULAR SPEED

Albert Juarez's truck has wheels 36 inches in diameter. If the wheels are rotating at 630 rpm (revolutions per minute), find the truck's speed in miles per hour.

**Solution**  We convert revolutions per minute to miles per hour by a series of unit conversion factors:

$$\frac{630 \text{ rev}}{1 \text{ min}} \times \frac{60 \text{ min}}{1 \text{ hr}} \times \frac{2\pi \text{ radians}}{1 \text{ rev}} \times \frac{18 \text{ in.}}{1 \text{ radian}} \times \frac{1 \text{ ft}}{12 \text{ in.}} \times \frac{1 \text{ mi}}{5280 \text{ ft}}$$

$$\approx 67.47 \, \frac{\text{mi}}{\text{hr}}$$

A **nautical mile** (naut mi) is the length of 1 minute of arc along the earth's equator. Figure 4.6 shows, though not to scale, a central angle *AOB* of the earth

**Notes on Exercises**

Ex. 39 and 49 require students to use arc length to approximate linear distance.

Ex. 40–42, 50, 51, 63 are application problems involving angular speed.

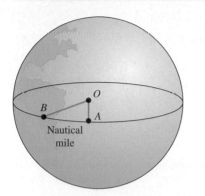

**Figure 4.6** Although the earth is not a perfect sphere, its diameter is, on the average, 7912.18 statute miles. A nautical mile is 1′ of earth's circumference at the equator.

**Ongoing Assessment**

Self Assessment: Ex. 2, 16, 17, 22, 32, 37, 41, 55, 59
Embedded Assessment: Ex. 39, 63

that measures 1/60 of a degree. It intercepts an arc 1 naut mi long. The arc length formula allows us to convert between nautical miles and **statute miles** (stat mi), the familiar "land mile" of 5280 feet.

**Example 6   CONVERTING TO NAUTICAL MILES**

Megan McCarty, a pilot for Western Airlines, frequently pilots flights from Boston to San Francisco, a distance of 2698 stat mi. Captain McCarty's calculations of flight time are based on nautical miles. How many nautical miles is it from Boston to San Francisco?

**Solution**   The radius of the earth at the equator is approximately 3956 stat mi. Convert 1 minute to radians:

$$1' = \left(\frac{1}{60}\right)^{\circ} \times \frac{\pi \text{ rad}}{180^{\circ}} = \frac{\pi}{10{,}800} \text{ radians}$$

Now we can apply the formula $s = r\theta$:

$$1 \text{ naut mi} = (3956)\left(\frac{\pi}{10{,}800}\right) \text{ stat mi}$$

$$\approx 1.15 \text{ stat mi}$$

$$1 \text{ stat mi} = \left(\frac{10{,}800}{3956\pi}\right) \text{ naut mi}$$

$$\approx 0.87 \text{ naut mi}$$

The distance from Boston to San Francisco is

$$2698 \text{ stat mi} = \frac{2698 \cdot 10{,}800}{3956\pi} \approx 2345 \text{ naut mi}$$

# Quick Review 4.1

In Exercises 1 and 2, find the circumference of the circle with the given radius *r*. State the correct unit.

**1.** $r = 2.5$ in.   $5\pi$ in.   **2.** $r = 4.6$ m   $9.2\pi$ m

In Exercises 3 and 4, find the radius of the circle with the given circumference *C*.

**3.** $C = 12$ m   $\dfrac{6}{\pi}$ m   **4.** $C = 8$ ft   $\dfrac{4}{\pi}$ ft

In Exercises 5 and 6, evaluate the expression for the given values of the variables. State the correct unit.

**5.** $s = r\theta$

  **(a)** $r = 9.9$ ft   $\theta = 4.8$ rad   47.52 ft

  **(b)** $r = 4.1$ km   $\theta = 9.7$ rad   39.77 km

**6.** $v = r\omega$

  **(a)** $r = 8.7$ m   $\omega = 3.0$ rad/sec   26.1 m/sec

  **(b)** $r = 6.2$ ft   $\omega = 1.3$ rad/sec   8.06 ft/sec

In Exercises 7–10, convert from miles per hour to feet per second or from feet per second to miles per hour.

**7.** 60 mph   88 ft/sec   **8.** 45 mph   66 ft/sec

**9.** 8.8 ft/sec   6 mph   **10.** 132 ft/sec   90 mph

# Section 4.1 Exercises

In Exercises 1–4, convert from DMS to decimal form.

**1.** 23°12′  23.2°  **2.** 35°24′  35.4°

**3.** 118°44′15″  118.7375°  **4.** 48°30′36″  48.51°

In Exercises 5–8, convert from decimal form to degrees, minutes, seconds (DMS).

**5** 21.2°  21°12′  **6.** 49.7°  49°42′

**7.** 118.32°  118°19′12″  **8.** 99.37°  99°22′12″

In Exercises 9–16, use the appropriate arc length formula to find the missing information.

| $s$ | $r$ | $\theta$ |
|-----|-----|----------|
| **9.** ? | 2 in. | 25 rad  50 in. |
| **10.** ? | 1 cm | 70 rad  70 cm |
| **11.** 1.5 ft | ? | $\pi/4$ rad  $6/\pi$ ft |
| **12.** 2.5 cm | ? | $\pi/3$ rad  $7.5/\pi$ cm |
| **13.** 3 m | 1 m | ?  3 (radians) |
| **14.** 4 in. | 7 in. | ?  $\frac{4}{7}$(radians) |
| **15.** 40 cm | ? | 20°  $360/\pi$ cm |
| **16.** ? | 5 ft | 18°  $\pi/2$ ft |

In Exercises 17 and 18, a central angle $\theta$ intercepts arcs $s_1$ and $s_2$ on two concentric circles with radii $r_1$ and $r_2$ respectively. Find the missing information.

| | $\theta$ | $r_1$ | $s_1$ | $r_2$ | $s_2$ |
|--|----------|-------|-------|-------|-------|
| **17.** | ? | 11 cm | 9 cm | 44 cm | ? |
| **18.** | ? | 8 km | 36 km | ? | 72 km |

In Exercises 19–26, convert from degrees to radians.

**19.** 60°  $\pi/3$  **20.** 90°  $\pi/2$

**21.** 120°  $2\pi/3$  **22.** 150°  $5\pi/6$

**23.** 71.72°  $\approx 1.2518$ rad  **24.** 11.83°  $\approx 0.2065$ rad

**25.** 61°24′  $\approx 1.0716$ rad  **26.** 75°30′  $\approx 1.3177$ rad

In Exercises 27–34, convert from radians to degrees.

**27.** $\pi/6$  30°  **28.** $\pi/4$  45°

**29.** $\pi/10$  18°  **30.** $3\pi/5$  108°

**31.** $7\pi/9$  140°  **32.** $13\pi/20$  117°

**33.** 2  $\approx 114.59°$  **34.** 1.3  $\approx 74.48°$

Exercises 35–38 refer to the 16 compass bearings shown. North corresponds to an angle of 0°, and other angles are measured clockwise from north.

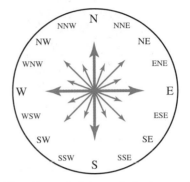

**35. Compass Reading** Find the angle in degrees that describes the compass bearing.

(a) NE (northeast)  45°

(b) NNE (north-northeast)  22.5°

(c) WSW (west-southwest)  247.5°

**36. Compass Reading** Find the angle in degrees that describes the compass bearing.

(a) SSW (south-southwest)  202.5°

(b) WNW (west-northwest)  292.5°

(c) NNW (north-northwest)  337.5°

**37. Compass Reading** Which compass direction is closest to a bearing of 121°?  ESE is closest at 112.5°

**38. Compass Reading** Which compass direction is closest to a bearing of 219°?  SW is closest at 225°

**39. Navigation** Two Coast Guard patrol boats leave Cape May at the same time. One travels with a bearing of 42°30′ and the other with a bearing of 52°12′. If they travel at the same speed, approximately how far apart will they be when they are 25 statute miles from Cape May?

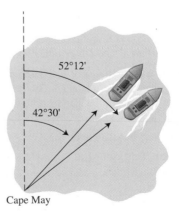

Cape May

**40. Automobile Design** The wheel (including the tire) of a sports car under development by one of the Big Three auto companies has an 11-inch radius. What would be the car's speed in miles per hour if its wheels are turning at 800 rpm?

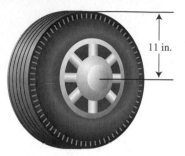

11 in.

**41. Bicycle Racing** Cathy Nguyen races on a bicycle with 13-inch-radius wheels. When she is traveling at a speed of 44 ft/sec, how many revolutions per minute are her wheels making? ≈ 387.85 rpm

**42. Tool Design** A radial arm saw has a circular cutting blade with a diameter of 10 inches. It spins at 2000 rpm. If there are 12 cutting teeth per inch on the cutting blade, how many teeth cross the cutting surface each second? ≈ 12,566.37

**43. Navigation** Sketch a diagram of a ship on the given course.

(a) 35°   (b) 128°   (c) 310°

**44. Navigation** The captain of the tourist boat Julia out of Oak Harbor follows a 38° course for 2 miles and then changes to a 47° course for the next 4 miles. Draw a sketch of this trip.

**45. Navigation** Points *A* and *B* are 257 nautical miles apart. How far apart are *A* and *B* in statute miles?

**46. Navigation** Points *C* and *D* are 895 statute miles apart. How far apart are *C* and *D* in nautical miles?

**47. Designing a Sports Complex** Example 4 describes how lanes 1 and 2 compare in length around one turn of a track. Find the differences in the lengths of these lanes around one turn of the same track.

(a) Lanes 5 and 6   (b) Lanes 1 and 6

**48. Mechanical Engineering** A simple pulley with the given radius *r* used to lift heavy objects is positioned 10 feet above

ground level. Given that the pully roates $\theta°$, determine the height to which the object is lifted.

(a) *r* = 4 in., $\theta$ = 720°   (b) *r* = 2 ft, $\theta$ = 180°

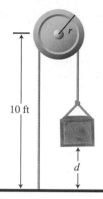

10 ft

*d*

**49. Foucault Pendulum** In 1851 the French physicist Jean Foucault used a pendulum to demonstrate the earth's rotation. There are now over 30 Foucault pendulum displays in the United States. The Foucault pendulum at the Smithsonian Institution in Washington, D.C., consists of a large brass ball suspended by a thin 52-foot cable. If the ball swings through an angle of 1°, how far does it travel?

**50. Group Activity Air Conditioning Belt** The belt on an automobile air conditioner connects metal wheels with radii *r* = 4 cm and *R* = 7 cm. The angular speed of the larger wheel is 120 rpm.

(a) What is the angular speed of the larger wheel in radians per second? $4\pi$ rad/sec

(b) What is the linear speed of the belt in centimeters per second? $28\pi$ cm/sec

(c) What is the angular speed of the smaller wheel in radians per second? $7\pi$ rad/sec

**51. Group Activity Ship's Propeller** The propellers of the *Amazon Paradise* have a radius of 1.2 m. At full throttle the propellers turn at 135 rpm.

(a) What is the angular speed of a propeller blade in radians per second? $4.5\pi$ rad/sec

(b) What is the linear speed of the tip of the propeller blade in meters per second? $5.4\pi$ m/sec

(c) What is the linear speed (in meters per second) of a point on a blade halfway between the center of the propeller and the tip of the blade? $2.7\pi$ m/sec

## Explorations

Table 4.1 shows the latitude-longitude locations of several U.S. cities. Latitude is measured from the equator. Longitude is measured from the Greenwich meridian that passes north-south through London.

| City | Latitude | Longitude |
|------|----------|-----------|

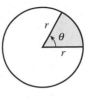

**Table 4.1  Latitude and Longitude Locations of U.S. Cities**

| City | Latitude | Longitude |
|------|----------|-----------|
| Atlanta | 33°45′ | 84°23′ |
| Chicago | 41°49′ | 87°37′ |
| Detroit | 42°22′ | 83°10′ |
| Los Angeles | 35°12′ | 118°02′ |
| Miami | 25°45′ | 80°11′ |
| Minneapolis | 44°58′ | 93°15′ |
| New Orleans | 30°00′ | 90°05′ |
| New York | 40°40′ | 73°58′ |
| San Diego | 32°43′ | 117°10′ |
| San Francisco | 37°45′ | 112°26′ |
| Seattle | 47°36′ | 122°20′ |

In Exercises 52–55, find the difference in longitude between the given cities.

**52.** Atlanta and San Francisco   **53.** New York and San Diego

**54.** Minneapolis and Chicago   **55.** Miami and Seattle ▪

In Exercises 56–59, assume that the two cities have the same longitude (that is, assume that one is directly north of the other), and find the distance between them in nautical miles. Use 7912 miles for the diameter of the earth.

**56.** San Diego and Los Angeles  149 naut mi

**57.** Seattle and San Francisco  591 naut mi

**58.** New Orleans and Minneapolis  898 naut mi

**59.** Detroit and Atlanta  517 naut mi

**60. Group Activity  Area of a Sector**  A *sector of a circle* (shaded in the figure) is a region bounded by a central angle of a circle and its intercepted arc. Use the fact that the areas of sectors are proportional to their central angles to prove that

$$A = \frac{1}{2}r^2\theta$$

where $r$ is the radius and $\theta$ is in radians.

## Extending the Ideas

**61. Area of a Sector**  Use the formula $A = (1/2)r^2\theta$ to determine the area of the sector with given radius $r$ and central angle $\theta$.

   **(a)** $r = 5.9$ ft,  $\theta = \pi/5$       **(b)** $r = 1.6$ km,  $\theta = 3.7$

**62. Navigation**  Control tower $A$ is 60 miles east of control tower $B$. At a certain time an airplane is on bearings of 340° from tower $A$ and 37° from tower $B$. Use a drawing to model the exact location of the airplane.

**63. Bicycle Racing**  Ben Scheltz's bike wheels are 28 inches in diameter, and for high gear the pedal sprocket is 9 inches in diameter and the wheel sprocket is 3 inches in diameter. Find the angular speed in radians per second of the wheel and of both sprockets when Ben reaches his peak racing speed of 66 ft/sec in high gear.

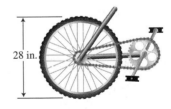

---

**4.2  Trigonometric Functions of Acute Angles**

Right Triangle Trigonometry • Two Famous Triangles • Evaluating Trigonometric Functions with a Calculator • Applications of Right Triangle Trigonometry

### Right Triangle Trigonometry

Recall that geometric figures are **similar** if they have the same shape even though they may have different sizes. Having the same shape means that the angles of one are congruent to the angles of the other and their corresponding sides are proportional. Similarity is the basis for many applications, including scale drawings, maps, and **right triangle trigonometry**, which is the topic of this section.

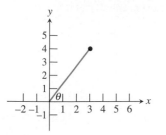

**Figure 4.7** An acute angle $\theta$ in standard position, with one ray along the positive *x*-axis and the other extending into the first quadrant.

It is a theorem of geometry that two triangles are similar if the angles of one are congruent to the angles of the other. For two *right* triangles we need only know that an acute angle of one is equal to an acute angle of the other for the triangles to be similar. Thus one acute angle determines an entire family of similar right triangles. Similarly, a single acute angle $\theta$ of a right triangle determines six distinct ratios of side lengths. Each ratio can therefore be considered a function of $\theta$ as $\theta$ takes on values from 0° to 90° or from 0 radians to $\pi/2$ radians. We wish to study these functions of acute angles more closely.

To bring the power of coordinate geometry into the picture, we will often put our acute angles in **standard position** in the *xy*-plane, with the vertex at the origin, one ray along the positive *x*-axis, and the other ray extending into the first quadrant. (See Figure 4.7.)

The six ratios of side lengths in a right triangle are the six *trigonometric functions* (often abbreviated to "trig functions") of the acute angle $\theta$. We will define them here with reference to the right $\triangle ABC$ as labeled in Figure 4.8. The abbreviations *opp*, *adj*, and *hyp* refer to the lengths of the side opposite $\theta$, the side adjacent to $\theta$, and the hypotenuse, respectively.

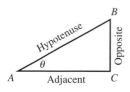

**Figure 4.8** The triangle referenced in our definition of the trigonometric functions.

---

**Definition** Trigonometric Functions

Let $\theta$ be an acute angle in the right $\triangle ABC$ (Figure 4.8). Then

$$\textbf{sine } (\theta) = \sin \theta = \frac{opp}{hyp} \qquad \textbf{cosecant } (\theta) = \csc \theta = \frac{hyp}{opp}$$

$$\textbf{cosine } (\theta) = \cos \theta = \frac{adj}{hyp} \qquad \textbf{secant } (\theta) = \sec \theta = \frac{hyp}{adj}$$

$$\textbf{tangent } (\theta) = \tan \theta = \frac{opp}{adj} \qquad \textbf{cotangent } (\theta) = \cot \theta = \frac{adj}{opp}$$

---

**Function Reminder**

Both $\sin \theta$ and $\sin (\theta)$ represent a function of the variable $\theta$. Neither notation implies multiplication by $\theta$. The notation $\sin (\theta)$ is just like the notation $f(x)$, while the notation $\sin \theta$ is a widely-accepted shorthand. The same note applies to all six trigonometric functions.

**Exploration Extensions**

Have the students calculate the sine, cosine, and tangent of the triangle shown in two ways: (1) use the lengths of the sides and find the values of their ratios; (2) use the trig functions for 30°. Compare results.

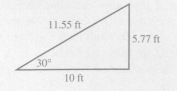

(Remind students to use degree mode.)

---

**Exploration 1** Exploring Trigonometric Functions

There are twice as many trigonometric functions as there are triangle sides which define them, so we can already explore some ways in which the trigonometric functions relate to each other. Doing this Exploration will help you learn which ratios are which.

1. Each of the six trig functions can be paired to another that is its reciprocal. Find the three pairs of reciprocals.
2. Which trig function can be written as the quotient $\sin \theta/\cos \theta$?
3. Which trig function can be written as the quotient $\csc \theta/\cot \theta$?
4. What is the (simplified) product of all six trig functions multiplied together?  1
5. Which two trig functions must be less than 1 for any acute angle $\theta$? [*Hint:* What is always the longest side of a right triangle?]

---

## Two Famous Triangles

Evaluating trigonometric functions of particular angles used to require "trig tables" or slide rules; now it only requires a calculator. There are some angles, however, that appear in right triangles and whose side ratios can be found *geo-*

*metrically.* Every student of trigonometry should be able to find these "special" ratios without a calculator.

### Example 1   EVALUATING TRIGONOMETRIC FUNCTIONS OF 45°

Find the values of all six trigonometric functions for an angle of 45°.

**Solution**  A 45° angle occurs in an *isosceles right triangle*, with angles 45°–45°–90° (see Figure 4.9).

Since the size of the triangle does not matter, we set the length of the two equal legs to 1. The hypotenuse, by the Pythagorean theorem, is $\sqrt{1+1} = \sqrt{2}$. Applying the definitions, we have

$$\sin 45° = \frac{opp}{hyp} = \frac{1}{\sqrt{2}} \approx 0.707 \qquad \csc 45° = \frac{hyp}{opp} = \frac{\sqrt{2}}{1} \approx 1.414$$

$$\cos 45° = \frac{adj}{hyp} = \frac{1}{\sqrt{2}} \approx 0.707 \qquad \sec 45° = \frac{hyp}{adj} = \frac{\sqrt{2}}{1} \approx 1.414$$

$$\tan 45° = \frac{opp}{adj} = \frac{1}{1} = 1 \qquad \cot 45° = \frac{adj}{opp} = \frac{1}{1} = 1$$

Whenever two sides of a right triangle are known, the third side can be found using the Pythagorean theorem. All six trigonometric functions of either acute angle can then be found. We illustrate this in Example 2 with another famous triangle.

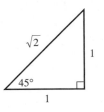

**Figure 4.9**  An isosceles right triangle. (Example 1)

### Example 2   EVALUATING TRIGONOMETRIC FUNCTIONS OF 30°

Find the values of all six trigonometric functions for an angle of 30°.

**Solution**  A 30° angle occurs in a **30°–60°–90°** triangle, which can be constructed from an equilateral (60°–60°–60°) triangle by constructing an altitude to any side. Since size does not matter, start with an equilateral triangle with sides 2 units long. The altitude splits it into two congruent 30°–60°–90° triangles, each with hypotenuse 2 and smaller leg 1. By the Pythagorean theorem, the longer leg has length $\sqrt{2^2 - 1^2} = \sqrt{3}$. (See Figure 4.10.)

We apply the definitions of the trigonometric functions to get:

$$\sin 30° = \frac{opp}{hyp} = \frac{1}{2} \qquad\qquad \csc 30° = \frac{hyp}{opp} = \frac{2}{1} = 2$$

$$\cos 30° = \frac{adj}{hyp} = \frac{\sqrt{3}}{2} \approx 0.866 \qquad \sec 30° = \frac{hyp}{adj} = \frac{2}{\sqrt{3}} \approx 1.155$$

$$\tan 30° = \frac{opp}{adj} = \frac{1}{\sqrt{3}} \approx 0.577 \qquad \cot 30° = \frac{adj}{opp} = \frac{\sqrt{3}}{1} \approx 1.732$$

**Figure 4.10**  An altitude to any side of an equilateral triangle creates two congruent 30°–60°–90° triangles. If each side of the equilateral triangle has length 2, then the two 30°–60°–90° triangles have sides of length 2, 1, and $\sqrt{3}$. (Example 2)

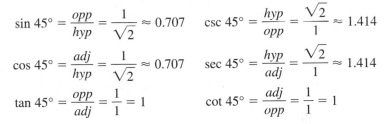

**Exploration 2** **Evaluating Trigonometric Functions of 60°**

1. Find the values of all six trigonometric functions for an angle of 60°. Note that most of the preliminary work has been done in Example 2.

2. Compare the six function values for 60° with the six function values for 30°. What do you notice?

3. We will eventually learn a rule that relates trigonometric functions of any angle with trigonometric functions of the *complementary* angle. (Recall from geometry that 30° and 60° are complementary because they add up to 90°.) Based on this exploration, can you predict what that rule will be? [*Hint:* The "co" in cosine, cotangent, and cosecant actually comes from "complement.")

Example 3 illustrates that knowing one trigonometric ratio in a right triangle is sufficient for finding all the others.

### Example 3 USING ONE TRIGONOMETRIC RATIO TO FIND THEM ALL

Let $\theta$ be an acute angle such that $\sin \theta = 5/6$. Evaluate the other five trigonometric functions of $\theta$.

**Solution** Sketch a triangle showing an acute angle $\theta$. Label the opposite side 5 and the hypotenuse 6. (See Figure 4.11.) Since $\sin \theta = 5/6$, this must be our angle! Now all we need to do is find the other side of the triangle (labeled $x$ in the figure).

From the Pythagorean theorem it follows that $x^2 + 5^2 = 6^2$, so $x = \sqrt{36 - 25} = \sqrt{11}$. Applying the definitions,

$$\sin \theta = \frac{opp}{hyp} = \frac{5}{6} \approx 0.833 \qquad \csc \theta = \frac{hyp}{opp} = \frac{6}{5} = 1.2$$

$$\cos \theta = \frac{adj}{hyp} = \frac{\sqrt{11}}{6} \approx 0.553 \qquad \sec \theta = \frac{hyp}{adj} = \frac{6}{\sqrt{11}} \approx 1.809$$

$$\tan \theta = \frac{opp}{adj} = \frac{5}{\sqrt{11}} \approx 1.508 \qquad \cot \theta = \frac{adj}{opp} = \frac{\sqrt{11}}{5} \approx 0.663$$

**Figure 4.11** How to create an acute angle $\theta$ such that $\sin \theta = 5/6$. (Example 3)

## Evaluating Trigonometric Functions with a Calculator

Using a calculator for the evaluation step enables you to concentrate all your problem-solving skills on the modeling step, which is where the real trigonometry occurs. The danger is that your calculator will try to evaluate what you ask it to evaluate, even if you ask it to evaluate the wrong thing. If you make a mistake, you might be lucky and see an error message. In most cases you will be unlucky and see an answer, which you will assume is correct but is actually wrong. We list the most common calculator errors associated with evaluating trigonometric functions.

## Common Calculator Errors When Evaluating Trig Functions

1. **Using the Calculator in the Wrong Angle Mode (Degrees/Radians)**
   This error is so common that everyone encounters it once in a while.
   You just hope to recognize it when it occurs. For example, suppose we
   are doing a problem in which we need to evaluate the sine of
   10 degrees. Our calculator shows us this (Figure 4.12):

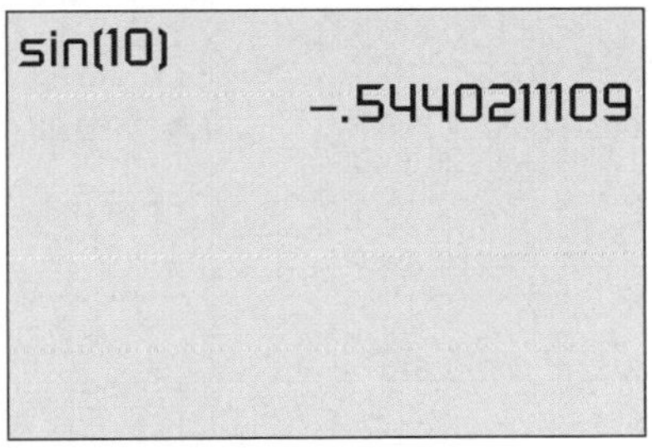

   $$\begin{array}{l} \text{sin(10)} \\ \qquad\qquad -.5440211109 \end{array}$$

   **Figure 4.12** Wrong mode for finding sin (10°).

Why is the answer negative? Our first instinct should be to check the
mode. Sure enough, it is in radian mode. Changing to degrees, we get
sin (10) = 0.1736481777, which is a reasonable answer. (That still
leaves open the question of why the sine of 10 radians is negative, but
that is a topic for the next section.) We will revisit the mode problem
later when we look at trigonometric graphs.

2. **Using the Inverse Trig Keys to Evaluate cot, sec, and csc** There are
   no buttons on most calculators for cotangent, secant, and cosecant. This
   is because they can be easily evaluated by finding reciprocals of tangent,
   cosine, and sine, respectively. Here, for example, is the correct way to
   evaluate the cotangent of 30 degrees (Figure 4.13).

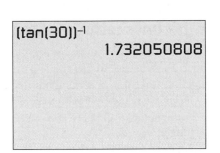

**Figure 4.13** Finding cot (30°).

There is also a key on the calculator for "TAN$^{-1}$"—but this is *not* the
cotangent function! Remember that an exponent of −1 on a *function* is
*never* used to denote a reciprocal; it is always used to denote the *inverse
function*. We will study the inverse trigonometric functions in a later section, but meanwhile you can see that it is a bad way to evaluate cot (30)
(Figure 4.14).

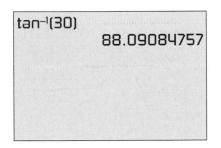

**Figure 4.14** This is not the cot (30°).

3. **Using Function Shorthand that the Calculator Does Not Recognize**
   This error is less dangerous because it usually results in an error message. We will often abbreviate powers of trig functions, writing (for
   example) "$\sin^3 \theta - \cos^3 \theta$" instead of the more cumbersome
   "$(\sin (\theta))^3 - (\cos (\theta))^3$." The calculator does not recognize the shorthand
   notation and interprets it as a syntax error.

4. **Not Closing Parentheses** This is a general algebraic error, but one that
   is easy to make on calculators that automatically open a parenthesis pair
   whenever you type a function key. Check your calculator by pressing the
   SIN key. If the screen displays "sin (" instead of just "sin" then you
   have such a calculator. The danger arises because the calculator will
   automatically *close* the parenthesis pair at the end of a command if you
   have forgotten to do so. That is fine if you *want* the parenthesis at the
   end of the command, but it is bad if you want it somewhere else. For
   example if you want "sin (30)" and you type "sin (30" you will get

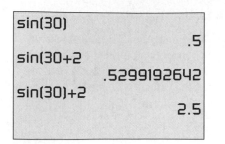

**Figure 4.15** A correct and incorrect way to find sin (30°) + 2.

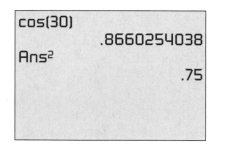

**Figure 4.16** (Example 4)

away with it. But if you want "sin (30) + 2" and you type "sin (30 + 2" you will not (Figure 4.15).

It is usually impossible to find an "exact" answer on a calculator, especially when evaluating trigonometric functions. The actual values are usually irrational numbers with non-terminating, non-repeating decimal expansions. However, you can find some exact answers if you know what you are looking for, as in Example 4.

**Example 4** GETTING AN "EXACT ANSWER" ON A CALCULATOR

Find the exact value of cos 30° on a calculator.

**Solution** As you see in Figure 4.16, the calculator gives the answer 0.8660254038. However, if we recognize 30° as one of our special angles (see Example 2 in this section), we might recall that the exact answer can be written in terms of a square root. We square our answer and get 0.75, which suggests that the exact value of cos 30° is $\sqrt{3/4} = \sqrt{3}/2$.

## Applications of Right Triangle Trigonometry

A triangle has six "parts", three angles and three sides, but you do not need to know all six parts to determine a triangle up to congruence. In fact, three parts are usually sufficient. The trigonometric functions take this observation a step further by giving us the means for actually *finding* the rest of the parts once we have enough parts to establish congruence. Using some of the parts of a triangle to solve for all the others is **solving a triangle**.

We will learn about solving general triangles in Sections 5.5 and 5.6, but we can already do some right triangle solving just by using the trigonometric ratios.

**Example 5** SOLVING A RIGHT TRIANGLE

A right triangle with a hypotenuse of 8 includes a 37° angle (Figure 4.17). Find the measures of the other two angles and the lengths of the other two sides.

**Figure 4.17** (Example 5)

**Solution** Since it is a right triangle, one of the other angles is 90°. That leaves 180° − 90° − 37° = 53° for the third angle.

Referring to the labels in Figure 4.17, we have

$$\sin 37° = \frac{a}{8} \qquad \cos 37° = \frac{b}{8}$$
$$a = 8 \sin 37° \qquad b = 8 \cos 37°$$
$$a \approx 4.81 \qquad b \approx 6.39$$

The real-world applications of triangle-solving are many, reflecting the frequency with which one encounters triangular shapes in everyday life.

### Example 6  FINDING THE HEIGHT OF A BUILDING

From a point 340 feet away from the base of the Peachtree Center Plaza in Atlanta, Georgia, the angle of elevation to the top of the building is 65°. (See Figure 4.18.) Find the height $h$ of the building.

**Solution**  We need a ratio that will relate an angle to its opposite and adjacent sides. The tangent function is the appropriate choice.

$$\tan 65° = \frac{h}{340}$$
$$h = 340 \tan 65°$$
$$h \approx 729 \text{ feet}$$

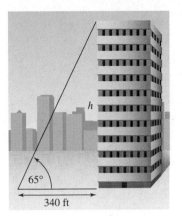

**Figure 4.18**  (Example 6)

10 ft

**Figure 4.19**  A large, helium-filled penguin. (Example 7)

### A Word About Rounding Answers

Notice in Example 6 that we rounded the answer to the nearest integer. In applied problems it is illogical to give answers with more decimal places of accuracy than can be guaranteed for the input values. An answer of 729.132 feet implies razor-sharp accuracy, whereas the reported height of the building (340 feet) implies a much less precise measurement. (So does the angle of 65°.) Indeed, an engineer following specific rounding criteria based on "significant digits" would probably report the answer to Example 6 as 730 feet. We will not get too picky about rounding, but we will try to be sensible.

### Example 7  FINDING HEIGHT ABOVE GROUND

A large, helium-filled penguin is moored at the beginning of a parade route awaiting the start of the parade. Two cables attached to the underside of the penguin make angles of 48° and 40° with the ground and are in the same plane as a perpendicular line from the penguin to the ground. (See Figure 4.19.) If the cables are attached to the ground 10 feet from each other, how high above the ground is the penguin?

**Solution**  We can simplify the drawing to the two right triangles in Figure 4.20 that share the common side $h$.

**Model**

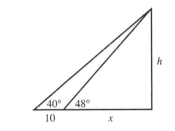

**Figure 4.20**  (Example 7)

By the definition of the tangent function,

$$\frac{h}{x} = \tan 48° \text{ and } \frac{h}{x + 10} = \tan 40°.$$

**Solve Algebraically**

Solving for $h$,

$$h = x \tan 48° \quad \text{and} \quad h = (x + 10) \tan 40°.$$

Set these two expressions for $h$ equal to each other and solve the equation for $x$:

$$x \tan 48° = (x + 10) \tan 40°$$
$$x \tan 48° = x \tan 40° + 10 \tan 40°$$

$$x \tan 48° - x \tan 40° = 10 \tan 40°$$

$$x(\tan 48° - \tan 40°) = 10 \tan 40°$$

$$x = \frac{10 \tan 40°}{\tan 48° - \tan 40°} \approx 30.90459723$$

We retain the full display for $x$ because we are not done yet; we need to solve for $h$:

$$h = x \tan 48° = (30.90459723) \tan 48° \approx 34.32$$

The penguin is approximately 34 feet above ground level.

# Quick Review 4.2

In Exercises 1–4, use the Pythagorean theorem to solve for $x$.

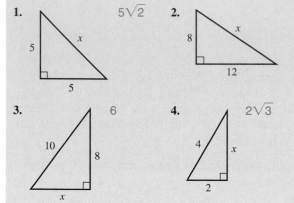

**1.** $5\sqrt{2}$    **2.**

**3.**    **4.** $2\sqrt{3}$

In Exercises 5 and 6, convert units.

**5.** 8.4 ft to inches   100.8 in.    **6.** 940 ft to miles

In Exercises 7–10, solve the equation. State the correct unit.

**7.** $0.388 = \dfrac{a}{20.4 \text{ km}}$    **8.** $1.72 = \dfrac{23.9 \text{ ft}}{b}$

**9.** $\dfrac{2.4 \text{ in.}}{31.6 \text{ in.}} = \dfrac{a}{13.3}$    **10.** $\dfrac{5.9}{\beta} = \dfrac{8.66 \text{ cm}}{6.15 \text{ cm}}$

# Section 4.2 Exercises

In Exercises 1–8, evaluate all six trigometric functions of the angle $\theta$.

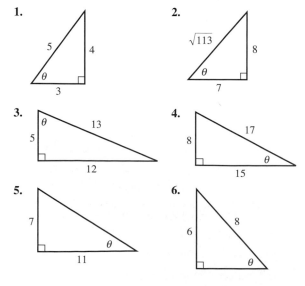

**1.**    **2.** $\sqrt{113}$

**3.**    **4.**

**5.**    **6.**

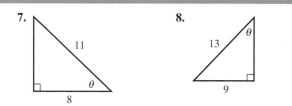

**7.**    **8.**

In Exercises 9–18, assume that $\theta$ is an acute angle in a right triangle satisfying the given conditions. Evaluate the remaining trigonometric functions.

**9.** $\sin \theta = \dfrac{3}{7}$    **10.** $\sin \theta = \dfrac{2}{3}$

**11.** $\cos \theta = \dfrac{5}{11}$    **12.** $\cos \theta = \dfrac{5}{8}$

**13.** $\tan \theta = \dfrac{5}{9}$    **14.** $\tan \theta = \dfrac{12}{13}$

**15.** $\cot \theta = \dfrac{11}{3}$    **16.** $\csc \theta = \dfrac{12}{5}$

**17.** $\csc \theta = \dfrac{23}{9}$    **18.** $\sec \theta = \dfrac{17}{5}$

In Exercises 19–24, evaluate *without* using a calculator.

**19.** $\sin\left(\dfrac{\pi}{3}\right)$   $\dfrac{\sqrt{3}}{2}$

**20.** $\tan\left(\dfrac{\pi}{4}\right)$   1

**21.** $\cot\left(\dfrac{\pi}{6}\right)$   $\sqrt{3}$

**22.** $\sec\left(\dfrac{\pi}{3}\right)$   2

**23.** $\cos\left(\dfrac{\pi}{4}\right)$   $\dfrac{\sqrt{2}}{2}$

**24.** $\csc\left(\dfrac{\pi}{3}\right)$   $2\sqrt{3}$

In Exercises 25–36, evaluate using a calculator. Be sure the calculator is in the correct mode. Give answers correct to three decimal places.

**25.** $\sin 74°$   0.961

**26.** $\tan 8°$   0.141

**27.** $\cos 19°23'$   0.943

**28.** $\tan 23°42'$   0.439

**29.** $\tan\left(\dfrac{\pi}{12}\right)$   0.268

**30.** $\sin\left(\dfrac{\pi}{15}\right)$   0.208

**31.** $\sec 49°$   1.524

**32.** $\csc 19°$   3.072

**33.** $\cot 0.89$   0.810

**34.** $\sec 1.24$   3.079

**35.** $\cot\left(\dfrac{\pi}{8}\right)$   2.414

**36.** $\csc\left(\dfrac{\pi}{10}\right)$   3.236

In Exercises 37–44, find the acute angle $\theta$ that satisfies the given equation. Give $\theta$ in both degrees and radians. You should do these problems without a calculator.

**37.** $\sin\theta = \dfrac{1}{2}$   $30° = \dfrac{\pi}{6}$

**38.** $\sin\theta = \dfrac{\sqrt{3}}{2}$   $60° = \dfrac{\pi}{3}$

**39.** $\cot\theta = \dfrac{1}{\sqrt{3}}$   $60° = \dfrac{\pi}{3}$

**40.** $\cos\theta = \dfrac{\sqrt{2}}{2}$   $45° = \dfrac{\pi}{4}$

**41.** $\sec\theta = 2$   $60° = \dfrac{\pi}{3}$

**42.** $\cot\theta = 1$   $45° = \dfrac{\pi}{4}$

**43.** $\tan\theta = \dfrac{\sqrt{3}}{3}$   $30° = \dfrac{\pi}{6}$

**44.** $\cos\theta = \dfrac{\sqrt{3}}{2}$   $30° = \dfrac{\pi}{6}$

In Exercises 45–50, solve for the variable shown.

**45.**

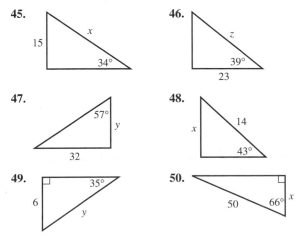

**46.**

**47.**

**48.**

**49.**

**50.**

In Exercises 51–54, solve the right triangle $\triangle ABC$ for all of its unknown parts.

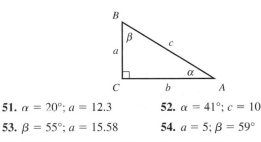

**51.** $\alpha = 20°;\ a = 12.3$

**52.** $\alpha = 41°;\ c = 10$

**53.** $\beta = 55°;\ a = 15.58$

**54.** $a = 5;\ \beta = 59°$

**55. Writing to Learn** What is $\lim\limits_{\theta \to 0} \sin\theta$? Explain your answer in terms of right triangles in which $\theta$ gets smaller and smaller and smaller.

**56. Writing to Learn** What is $\lim\limits_{\theta \to 0} \cos\theta$? Explain your answer in terms of right triangles in which $\theta$ gets smaller and smaller.

**57. Height** A guy wire from the top of the transmission tower at WJBC forms a 75° angle with the ground at a 55-foot distance from the base of the tower. How tall is the tower?

**58. Height** Kirsten places her surveyor's telescope on the top of a tripod 5 feet above the ground. She measures an 8° elevation above the horizontal to the top of a tree that is 120 feet away. How tall is the tree?   $\approx 21.86$ ft

**59. Group Activity Area** For locations between 20° and 60° north latitude a solar collector panel should be mounted so that its angle with the horizontal is 20 greater than the local latitude. Consequently, the solar panel mounted on the roof of Solar Energy, Inc., in Atlanta (latitude 34°) forms a 54° angle with the horizontal. The bottom edge of the 12-ft long panel is resting on the roof, and the high edge is 5 ft above the roof. What is the total area of this rectangular collector panel? ≈ 74.16 ft²

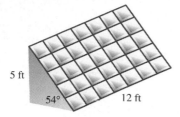

**60. Height** The Chrysler Building in New York City was the tallest building in the world at the time it was built. It casts a shadow approximately 130 feet long on the street when the sun's rays form an 82.9° angle with the earth. How tall is the building? ≈ 1043.70 ft

**61. Distance** DaShanda's team of surveyors had to find the distance *AC* across the lake at Montgomery County Park. Field assistants positioned themselves at points *A* and *C* while DaShanda set up an angle-measuring instrument at point *B*, 100 feet from *C* in a perpendicular direction. DaShanda measured ∠*ABC* as 75°12′42″. What is the distance *AC*? ≈ 378.80 ft

**62. Group Activity Garden Design** Allen's garden is in the shape of a quarter-circle with radius 10 ft. He wishes to plant his garden in four parallel strips, as shown in the diagram on the left below, so that the four arcs along the circular edge of the garden are all of equal length. After measuring four equal arcs, he carefully measures the widths of the four strips and records his data in the table shown at the right below.

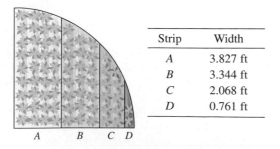

| Strip | Width |
|-------|---------|
| A | 3.827 ft |
| B | 3.344 ft |
| C | 2.068 ft |
| D | 0.761 ft |

Alicia sees Allen's data and realizes that he could have saved himself some work by figuring out the strip widths by trigonometry. By checking his data with a calculator she is able to correct two measurement errors he has made. Find Allen's two errors and correct them.

## Explorations

**63. Mirrors** In the figure, a light ray shining from point *A* to point *P* on the mirror will bounce to point *B* in such a way that the *angle of incidence* α will equal the *angle of reflection* β. This is the *law of reflection* derived from physical experiments. Both angles are measured from the *normal line*, which is perpendicular to the mirror at the point of reflection *P*. If *A* is 2 m farther from the mirror than is *B*, and if α = 30° and *AP* = 5 m, what is the length *PB*?

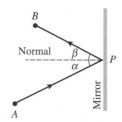

**64. Pool** On the pool table shown in the figure, where along the portion *CD* of the railing should you direct ball *A* so that it will bounce off *CD* and strike ball *B*? Assume that *A* obeys the law of reflection relative to rail *CD*.

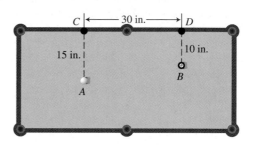

## Extending the Ideas

**65.** Using the labeling of the triangle below, prove that if θ is an acute angle in any right triangle,

$$(\sin \theta)^2 + (\cos \theta)^2 = 1$$

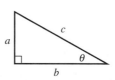

**66.** Using the labeling of the triangle below, prove that the area of the triangle is equal to (1/2) *ab* sin θ. [*Hint:* Start by drawing the altitude to side *b* and finding its length.]

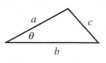

# Trigonometry Extended: The Circular Functions

Trigonometric Functions of Any Angle • Trigonometric Functions of Real Numbers • Periodic Functions

## Trigonometric Functions of Any Angle

We now extend the definitions of the six basic trigonometric functions beyond triangles so that we do not have to restrict our attention to acute angles, or even to positive angles. Our applications of angle measure to the field of navigation made good sense of angles like 270° even though they were greater than straight angles, and we will soon encounter applications that can make use of negative angles and angles greater than 360°.

In geometry we think of an angle as a union of two rays with a common vertex. Trigonometry takes a more dynamic view by thinking of an angle in terms of a rotating ray. The beginning position of the ray, the **initial side**, is rotated about its endpoint, called the **vertex**. The final position is called the **terminal side**. The **measure of an angle** is a number that describes the amount of rotation from the initial side to the terminal side of the angle. **Positive angles** are generated by counterclockwise rotations and **negative angles** are generated by clockwise rotations. Figure 4.21 shows an angle of measure $\alpha$, where $\alpha$ is a positive number.

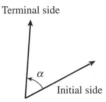

**Figure 4.21**  An angle with positive measure $\alpha$.

To bring the power of coordinate geometry into the picture (literally), we will usually place an angle in **standard position** in the Cartesian plane, with the vertex of the angle at the origin and its initial side lying along the positive *x*-axis. Figure 4.22 shows two angles in standard position, one with positive measure $\alpha$ and the other with negative measure $\beta$.

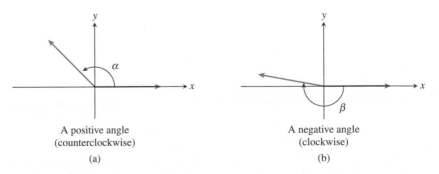

**Figure 4.22**  Two angles in standard position. In (a) the counterclockwise rotation generates an angle with positive measure $\alpha$. In (b) the clockwise rotation generates an angle with negative measure $\beta$.

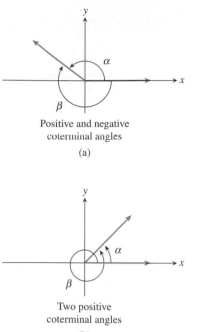

Positive and negative
coterminal angles

(a)

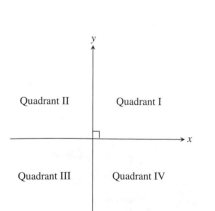

Two positive
coterminal angles

(b)

**Figure 4.23** Coterminal angles. In (a) a positive angle and a negative angle are coterminal, while in (b) both coterminal angles are positive.

Two angles in this expanded angle-measurement system can have the same initial side and the same terminal side, yet have different measures. We call such angles **coterminal angles**. (See Figure 4.23.) For example, angles of 90°, 450°, and −270° are all coterminal, as are angles of $\pi$ radians, $3\pi$ radians, and $-99\pi$ radians. In fact, angles are coterminal whenever they differ by an integer multiple of 360 degrees or by an integer multiple of $2\pi$ radians.

**Example 1** FINDING COTERMINAL ANGLES

Find and draw a positive angle and a negative angle that are coterminal with the given angle.

(a) 30°          (b) −150°          (c) $\dfrac{2\pi}{3}$ radians

**Solution** There are infinitely many possible solutions; we will show two for each angle.

**(a)** Add 360°:   $30° + 360° = 390°$

Subtract 360°:   $30° - 360° = -330°$

Figure 4.24 shows these two angles, which are coterminal with the 30° angle.

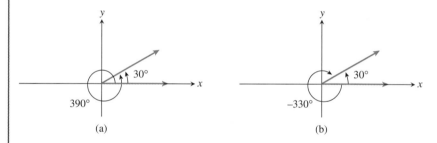

(a)                                    (b)

**Figure 4.24** Two angles coterminal with 30°. (Example 1a)

**(b)** Add 360°:   $-150° + 360° = 210°$

Subtract 720°:   $-150° - 720° = -870°$

We leave it to you to draw the coterminal angles.

**(c)** Add $2\pi$:   $\dfrac{2\pi}{3} + 2\pi = \dfrac{2\pi}{3} + \dfrac{6\pi}{3} = \dfrac{8\pi}{3}$

Subtract $2\pi$:   $\dfrac{2\pi}{3} - 2\pi = \dfrac{2\pi}{3} - \dfrac{6\pi}{3} = -\dfrac{4\pi}{3}$

Again, we leave it to you to draw the coterminal angles.

Extending the definitions of the six basic trigonometric functions so that they can apply to any angle is surprisingly easy, but first you need to see how our current definitions relate to the $(x, y)$ coordinates in the Cartesian plane. We start in the first quadrant (see Figure 4.25), where the angles are all acute. Work through Exploration 1 before moving on.

| Quadrant II | Quadrant I |
|---|---|
| Quadrant III | Quadrant IV |

**Figure 4.25** The four quadrants of the Cartesian plane. Both $x$ and $y$ are positive in QI (Quadrant I). Quadrants, like Super Bowls, are invariably designated by Roman numerals.

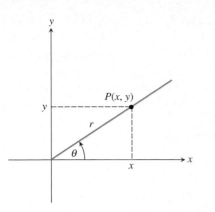

**Figure 4.26** A point P($x$, $y$) in Quadrant I determines an acute angle $\theta$. The number $r$ denotes the distance from $P$ to the origin. (Exploration 1)

---

**Exploration Extensions**

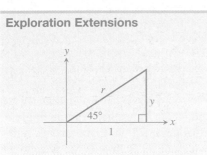

From the diagram shown, calculate the missing values $r$ and $y$, then label the three vertices with their ordered pairs.

**Teaching Note**

Explain that since $r = x^2 + y^2$, its value is always positive regardless of the quadrant of its location.

---

**Exploration 1    Investigating First Quadrant Trigonometry**

Let $P(x, y)$ be any point in the first quadrant (QI), and let $r$ be the distance from $P$ to the origin. (See Figure 4.26)

1. Use the acute angle definition of the sine function (Section 4.2) to prove that $\sin \theta = y/r$.
2. Express $\cos \theta$ in terms of $x$ and $r$.   $\cos \theta = x/r$
3. Express $\tan \theta$ in terms of $x$ and $y$.   $\tan \theta = y/x$
4. Express the remaining three basic trigonometric functions in terms of $x$, $y$, and $r$.

If you have successfully completed Exploration 1 you should have no trouble verifying the solution to Example 2, which we show without the details.

**Example 2    EVALUATING TRIG FUNCTIONS DETERMINED BY A POINT IN QI**

Let $\theta$ be the acute angle in standard position whose terminal side contains the point $(5, 3)$. Find the six trigonometric functions of $\theta$.

**Solution**   The distance from $(5, 3)$ to the origin is $\sqrt{34}$.

So

$$\sin \theta = \frac{3}{\sqrt{34}} \approx 0.514 \qquad \csc \theta = \frac{\sqrt{34}}{3} \approx 1.944$$

$$\cos \theta = \frac{5}{\sqrt{34}} \approx 0.857 \qquad \sec \theta = \frac{\sqrt{34}}{5} \approx 1.166$$

$$\tan \theta = \frac{3}{5} = 0.6 \qquad \cot \theta = \frac{5}{3} \approx 1.667$$

Now we have an easy way to extend the trigonometric functions to any angle: Use the same definitions in terms of $x$, $y$, and $r$—*whether $x$ and $y$ are positive or not.* Compare Example 3 to Example 2.

**Example 3    EVALUATING TRIG FUNCTIONS DETERMINED BY A POINT IN QII**

Let $\theta$ be any angle in standard position whose terminal side contains the point $(-5, 3)$. Find the six trigonometric functions of $\theta$.

**Solution**   The distance from $(-5, 3)$ to the origin is $\sqrt{34}$.

So

$$\sin \theta = \frac{3}{\sqrt{34}} \approx 0.514 \qquad \csc \theta = \frac{\sqrt{34}}{3} \approx 1.944$$

$$\cos \theta = \frac{-5}{\sqrt{34}} \approx -0.857 \qquad \sec \theta = \frac{\sqrt{34}}{-5} \approx -1.166$$

$$\tan \theta = \frac{3}{-5} = -0.6 \qquad \cot \theta = \frac{-5}{3} \approx -1.667$$

Notice in Example 3 that $\theta$ is *any* angle in standard position whose terminal side contains the point $(-5, 3)$. There are infinitely many coterminal angles that could play the role of $\theta$, some of them positive and some of them negative. The values of the six trigonometric functions would be the same for all of them.

We are now ready to state the formal definition.

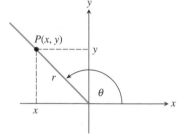

**Figure 4.27** Defining the six trig functions of $\theta$.

---

### Definition  Trigonometric Functions of any Angle

Let $\theta$ be any angle in standard position and let $P(x, y)$ be any point on the terminal side of the angle (except the origin). Let $r$ denote the distance from $P(x, y)$ to the origin, i.e., let $r = \sqrt{x^2 + y^2}$. (See Figure 4.27.) Then

$$\sin \theta = \frac{y}{r} \qquad\qquad \csc \theta = \frac{r}{y} \ (y \neq 0)$$

$$\cos \theta = \frac{x}{r} \qquad\qquad \sec \theta = \frac{r}{x} \ (x \neq 0)$$

$$\tan \theta = \frac{y}{x} \ (x \neq 0) \qquad \cot \theta = \frac{x}{y} \ (y \neq 0)$$

---

Examples 2 and 3 both began with a point $P(x, y)$ rather than an angle $\theta$. Indeed, the point gave us so much information about the trigonometric ratios that we were able to compute them all without ever finding $\theta$. So what do we do if we start with an angle $\theta$ in standard position and we want to evaluate the trigonometric functions? We try to find a point $(x, y)$ on its terminal side. We illustrate this process with Example 4.

### Example 4  EVALUATING THE TRIG FUNCTIONS OF 315°

Find the six trigonometric functions of 315°.

**Solution**  First we draw an angle of 315° in standard position. Without declaring a scale, pick a point $P$ on the terminal side and connect it to the $x$-axis with a perpendicular segment. Notice that the triangle formed (called a **reference triangle**) is a 45°–45°–90° triangle. If we arbitrarily choose the horizontal and vertical sides of the reference triangle to be of length 1, then $P$ has coordinates $(1, -1)$. (See Figure 4.28.)

We can now use the definitions, with $x = 1$, $y = -1$, and $r = \sqrt{2}$.

$$\sin 315° = \frac{-1}{\sqrt{2}} = -\frac{1}{\sqrt{2}} \qquad \csc 315° = \frac{\sqrt{2}}{-1} = -\sqrt{2}$$

$$\cos 315° = \frac{1}{\sqrt{2}} \qquad\qquad \sec 315° = \frac{\sqrt{2}}{1} = \sqrt{2}$$

$$\tan 315° = \frac{-1}{1} = -1 \qquad \cot 315° = \frac{1}{-1} = -1$$

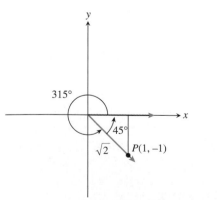

**Figure 4.28** An angle of 315° in standard position determines a 45°–45°–90° *reference triangle*. (Example 4)

The happy fact that the reference triangle in Example 4 was a 45°–45°–90° triangle enabled us to label a point $P$ on the terminal side of the 315° angle and then to find the trigonometric function values. We would also

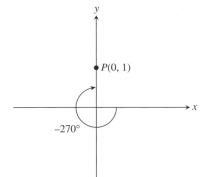

**Figure 4.29** (Example 5a)

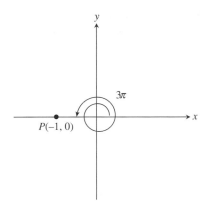

**Figure 4.30** (Example 5b)

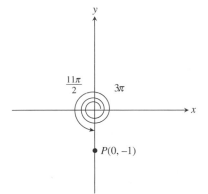

**Figure 4.31** (Example 5c)

be able to find *P* if the given angle were to produce a 30°–60°–90° reference triangle. Angles whose terminal sides lie along one of the coordinate axes are called **quadrantal angles**, and although they do not produce reference triangles at all, it is easy to pick a point *P* along one of the axes.

### Example 5 EVALUATING TRIG FUNCTIONS OF QUADRANTAL ANGLES

Find each of the following, if it exists. If the value does not exist, write DNE.

**(a)** $\sin(-270°)$

**(b)** $\tan 3\pi$

**(c)** $\sec \dfrac{11\pi}{2}$

**Solution**

**(a)** In standard position, the terminal side of an angle of $-270°$ lies along the positive *y*-axis (Figure 4.29). A convenient point *P* along the positive *y*-axis is the point for which $r = 1$, namely (0, 1). Therefore

$$\sin(-270°) = \frac{y}{r} = \frac{1}{1} = 1$$

**(b)** In standard position, the terminal side of an angle of $3\pi$ lies along the negative *x*-axis (Figure 4.30). A convenient point *P* along the negative *x*-axis is the point for which $r = 1$, namely $(-1, 0)$. Therefore

$$\tan 3\pi = \frac{y}{x} = \frac{0}{-1} = 0$$

**(c)** In standard position, the terminal side of an angle of $11\pi/2$ lies along the negative *y*-axis (Figure 4.31). A convenient point *P* along the negative *y*-axis is the point for which $r = 1$, namely $(0, -1)$. Therefore

$$\sec \frac{11\pi}{2} = \frac{r}{x} = \frac{1}{0} \quad \text{DNE.}$$

---

### Evaluating Trig Functions of a Non-quadrantal Angle $\theta$

**1.** Draw the angle $\theta$ in standard position, being careful to place the terminal side in the correct quadrant.

**2.** Without declaring a scale on either axis, label a point *P* (other than the origin) on the terminal side of $\theta$.

**3.** Draw a perpendicular segment from *P* to the *x*-axis, determining the *reference triangle*. If this triangle is one of the triangles whose ratios you know, label the sides accordingly. If it is not, then you will need to use your calculator.

**4.** Use the sides of the triangle to determine the coordinates of point *P*, making them positive or negative according to the signs of *x* and *y* in that particular quadrant.

**5.** Use the coordinates of point *P* and the definitions to determine the six trig functions.

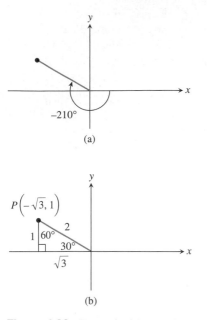

**Figure 4.32** (Example 6a)

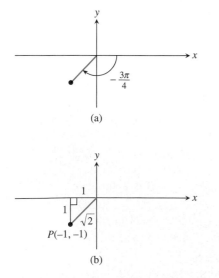

**Figure 4.34** (Example 6c)

## Example 6 EVALUATING MORE TRIG FUNCTIONS

Find the following without a calculator:

**(a)** sin (−210°)

**(b)** tan (5π/3)

**(c)** sec (−3π/4)

**Solution**

**(a)** An angle of −210° in standard position determines a 30°–60°–90° reference triangle in the second quadrant (Figure 4.32). We label the sides accordingly, then use the lengths of the sides to determine the point $P(-\sqrt{3}, 1)$. (Note that the *x*-coordinate is negative in the second quadrant.) The hypotenuse is $r = 2$. Therefore sin (−210°) = $y/r = 1/2$.

**(b)** An angle of 5π/3 radians in standard position determines a 30°–60°–90° reference triangle in the fourth quadrant (Figure 4.33). We label the sides accordingly, then use the lengths of the sides to determine the point $P(1, -\sqrt{3})$. (Note that the *y*-coordinate is negative in the fourth quadrant.) The hypotenuse is $r = 2$. Therefore tan (5π/3) = $y/x = -\sqrt{3}/1 = -\sqrt{3}$.

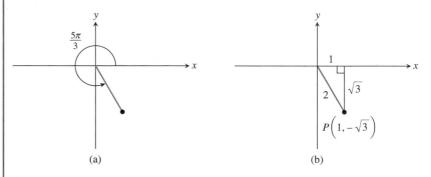

**Figure 4.33** (Example 6b)

**(c)** An angle of −3π/4 radians in standard position determines a 45°–45°–90° reference triangle in the third quadrant (Figure 4.34). We label the sides accordingly, then use the lengths of the sides to determine the point $P(-1, -1)$. (Note that both coordinates are negative in the third quadrant.) The hypotenuse is $r = \sqrt{2}$. Therefore sec (−3π/4) = $r/x = \sqrt{2}/-1 = -\sqrt{2}$.

You might wonder why we would go through this procedure to produce values that could be found so easily with a calculator. The short answer is that we want you to understand how trigonometry *works* in the coordinate plane. Ironically, technology has made these computational exercises more important than ever, since calculators have eliminated the need for the repetitive evaluations that once allowed students to gain their initial insights into the basic trig functions.

Another good exercise is to use information from one trigonometric ratio to produce the other five. We do not need to know the angle θ, although we do need a hint as to the location of its terminal side so that we can sketch a reference triangle in the correct quadrant (or place a quadrantal angle on the correct side of the origin). Example 7 illustrates how this is done.

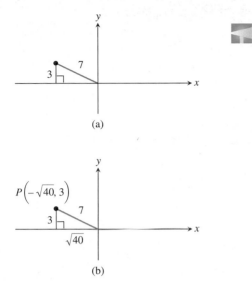

(a)

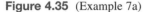

$P\left(-\sqrt{40}, 3\right)$

(b)

**Figure 4.35** (Example 7a)

### Example 7 USING ONE TRIG RATIO TO FIND THE OTHERS

Find $\cos\theta$ and $\tan\theta$ by using the given information to construct a reference triangle.

**(a)** $\sin\theta = \dfrac{3}{7}$ and $\tan\theta < 0$

**(b)** $\sec\theta = 3$ and $\sin\theta > 0$

**(c)** $\cot\theta$ is undefined and $\sec\theta$ is negative

#### Solution

**(a)** Since $\sin\theta$ is positive, the terminal side is either in QI or QII. The added fact that $\tan\theta$ is negative means that the terminal side is in QII. We draw a reference triangle in QII with $r = 7$ and $y = 3$ (Figure 4.35); then we use the Pythagorean theorem to find that $x = -\sqrt{7^2 - 3^2} = -\sqrt{40}$. (Note that $x$ is negative in QII.)

We then use the definitions to get

$$\cos\theta = \frac{-\sqrt{40}}{7} \approx -0.904 \text{ and } \tan\theta = \frac{3}{-\sqrt{40}} \approx -0.474.$$

**(b)** Since $\sec\theta$ is positive, the terminal side is either in QI or QIV. The added fact that $\sin\theta$ is positive means that the terminal side is in QI. We draw a reference triangle in QI with $r = 3$ and $x = 1$ (Figure 4.36); then we use the Pythagorean theorem to find that $y = \sqrt{3^2 - 1^2} = \sqrt{8}$. (Note that $y$ is positive in QI.)

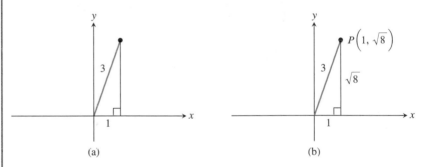

**Figure 4.36** (Example 7b)

We then use the definitions to get

$$\cos\theta = \frac{1}{3} \approx 0.333 \text{ and } \tan\theta - \frac{\sqrt{8}}{1} \approx 2.828.$$

(We could also have found $\cos\theta$ directly as the reciprocal of $\sec\theta$.)

**(c)** Since $\cot\theta$ is undefined, we conclude that $y = 0$ and that $\theta$ is a quadrantal angle on the $x$-axis. The added fact that $\sec\theta$ is negative means that the terminal side is along the negative $x$-axis. We choose the point $(-1, 0)$ on the terminal side and use the definitions to get

$$\cos\theta = -1 \text{ and } \tan\theta = \frac{0}{-1} = 0.$$

### Why Not Degrees?

One could actually develop a consistent theory of trigonometric functions based on a re-scaled $x$-axis with "degrees." For example, your graphing calculator will probably produce reasonable-looking graphs in degree mode. Calculus, however, uses rules that *depend* on radian measure for all trigonometric functions, so it is prudent for precalculus students to become accustomed to that now.

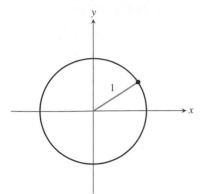

**Figure 4.37** The Unit Circle.

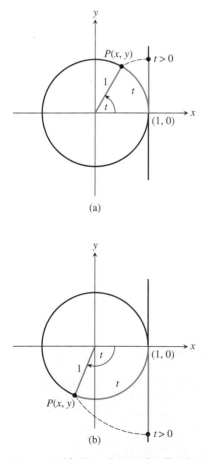

(a)

(b)

**Figure 4.38** How the number line is wrapped onto the unit circle. Note that each number *t* (positive or negative) is "wrapped" to a point *P* that lies on the terminal side of an angle of *t* radians in standard position.

# Trigonometric Functions of Real Numbers

Now that we have extended the six basic trigonometric functions to apply to any angles, we are almost ready to appreciate them as functions of real numbers and to study their behavior. First, for reasons discussed in the first section of this chapter, we must agree to measure $\theta$ in radian mode so that the real number units of the input will match the real number units of the output.

> When considering the trigonometric functions as functions of real numbers, the angles will be measured in radians.

> **Definition** Unit Circle
>
> The unit circle is a circle of radius 1 centered at the origin (Figure 4.37).

The **unit circle** is the ultimate connection between triangle trigonometry and the trigonometric functions. Because arc length along the unit circle corresponds exactly to radian measure, we can use the circle itself as a sort of "number line" for the input values of our functions.

Picture a number line with its origin tangent to the unit circle at the point $(1, 0)$, the point from which we measure angles in standard position. Imagine wrapping this number line around the unit circle in both the positive (counterclockwise) and negative (clockwise) directions. Each point corresponding to a number *t* on the number line will fall on a point *P* that lies on the terminal side of an angle of *t* radians in standard position, as shown in Figure 4.38.

Recall that the point *P* was the key to the definition of the trigonometric functions of any angle. Using its coordinates $(x, y)$ and its distance from the origin $(r)$, we formed all six ratios in terms of those three variables. We are now in a position to do that again, only this time we can begin with a real number on a number line rather than with an angle. Moreover, we can eliminate a variable because $r = 1$.

> **Definition** Trigonometric Functions of Real Numbers
>
> Let *t* be any real number, and let $P(x, y)$ be the point corresponding to *t* when the number line is wrapped onto the unit circle as described above. Then
>
> $$\sin t = y \qquad\qquad \csc t = \frac{1}{y} \ (y \neq 0)$$
>
> $$\cos t = x \qquad\qquad \sec t = \frac{1}{x} \ (x \neq 0)$$
>
> $$\tan t = \frac{y}{x} \ (x \neq 0) \qquad\qquad \cot t = \frac{x}{y} \ (y \neq 0)$$
>
> Therefore, the number *t* on the number line always wraps onto the point $(\cos t, \sin t)$ on the unit circle (Figure 4.39).

Although it is still helpful to draw reference triangles inside the unit circle to see the ratios geometrically, this latest round of definitions does not invoke triangles at all. The real number $t$ determines a point on the unit circle, and the $(x, y)$ coordinates of the point determine the six trigonometric ratios. For this reason, the trigonometric functions when applied to real numbers are usually called the **circular functions**.

---

**Exploration 2**  **Exploring the Unit Circle**

This works well as a group exploration. Get together in groups of two or three and explain to each other why these statements are true. Base your explanations on the unit circle (Figure 4.39). Remember that $-t$ wraps the same distance as $t$, but in the opposite direction.

1. For any $t$, the value of cos $t$ lies between $-1$ and $1$ inclusive.

2. For any $t$, the value of sin $t$ lies between $-1$ and $1$ inclusive.

3. The values of cos $t$ and cos $(-t)$ are always equal to each other. (Recall that this is the check for an *even* function.)

4. The values of sin $t$ and sin $(-t)$ are always opposites of each other. (Recall that this is the check for an *odd* function.)

5. The values of sin $t$ and sin $(t + 2\pi)$ are always equal to each other. In fact, that is true of all six trig functions on their domains, and for the same reason.

6. The values of sin $t$ and sin $(t + \pi)$ are always opposites of each other. The same is true of cos $t$ and cos $(t + \pi)$.

7. The values of tan $t$ and tan $(t + \pi)$ are always equal to each other (unless they are both undefined).

8. The sum $(\cos t)^2 + (\sin t)^2$ always equals 1.

9. (Challenge) Can you discover a similar relationship that is not mentioned in our list of seven? There are some to be found.

---

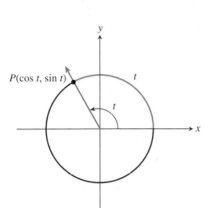

**Figure 4.39** The real number $t$ always wraps onto the point (cos $t$, sin $t$) on the unit circle.

---

**Exploration Extensions**

Put your calculator in radian mode and chose any value for $t$. Verify Steps 1–8 using your chosen value.

---

## Periodic Functions

Statements 5 and 7 in Exploration 2 reveal an important property of the circular functions that we need to define for future reference.

---

**Definition  Periodic Function**

A function $y = f(t)$ is **periodic** if there is a positive number $c$ such that $f(t + c) = f(t)$ for all values of $t$ in the domain of $f$.

The smallest such number $c$ is called the **period** of the function.

---

Exploration 2 suggests that the sine and cosine functions have period $2\pi$ and that the tangent function has period $\pi$. We use this periodicity later to model predictably repetitive behavior in the real world, but meanwhile we can also use it to solve little non-calculator training problems like in some of the previous examples in this section.

**Example 8   USING PERIODICITY**

Find each of the following numbers without a calculator.

(a) $\sin\left(\dfrac{57{,}801\pi}{2}\right)$

(b) $\cos(288.45\pi) - \cos(280.45\pi)$

(c) $\tan\left(\dfrac{\pi}{4} - 99{,}999\pi\right)$

**Solution**

(a) $\sin\left(\dfrac{57{,}801\pi}{2}\right) = \sin\left(\dfrac{\pi}{2} + \dfrac{57{,}800\pi}{2}\right) = \sin\left(\dfrac{\pi}{2} + 28{,}900\pi\right)$

$= \sin\left(\dfrac{\pi}{2}\right) = 1$

Notice that $28{,}900\pi$ is just a large multiple of $2\pi$, so $\pi/2$ and $((\pi/2) + 28{,}900\pi)$ wrap to the same point on the unit circle, namely $(0, 1)$.

(b) $\cos(288.45\pi) - \cos(280.45\pi) =$
$\cos(280.45\pi + 8\pi) - \cos(280.45\pi) = 0.$

Notice that $280.45\pi$ and $(280.45\pi + 8\pi)$ wrap to the same point on the unit circle, so the cosine of one is the same as the cosine of the other.

(c) Since the period of the tangent function is $\pi$ rather than $2\pi$, $99{,}999\pi$ is a large multiple of the period of the tangent function. Therefore,

$$\tan\left(\frac{\pi}{4} - 99{,}999\pi\right) = \tan\left(\frac{\pi}{4}\right) = 1.$$

We take a closer look at the properties of the six circular functions in the next two sections.

# Quick Review 4.3

In Exercises 1–4, give the quadrant in which the terminal side of the angle $\theta$ lies.

**1.** $\theta = \dfrac{-\pi}{6}$  QIV

**2.** $\theta = -\dfrac{5\pi}{6}$  QIII

**3.** $\theta = \dfrac{25\pi}{4}$  QI

**4.** $\theta = \dfrac{16\pi}{3}$  QIII

In Exercises 5–8, use special triangles to evaluate:

**5.** $\tan\dfrac{\pi}{6}$  $\sqrt{3}/3$

**6.** $\cot\dfrac{\pi}{4}$  1

**7.** $\csc\dfrac{\pi}{4}$  $\sqrt{2}$

**8.** $\sec\dfrac{\pi}{3}$  2

In Exercises 9 and 10, use a right triangle to find the other five trigonometric functions of the *acute* angle $\theta$.

**9.** $\sin\theta = \dfrac{5}{13}$

**10.** $\cos\theta = \dfrac{15}{17}$

# Section 4.3 Exercises

In Exercises 1–4, evaluate the six trigonometric functions of the angle $\theta$.

**1.**

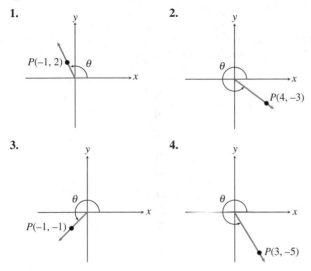

**2.**

**3.**

**4.**

In Exercises 5–10, point $P$ is on the terminal side of angle $\theta$. Evaluate the six trigonometric functions for $\theta$.

**5.** $P(3, 4)$        **6.** $P(-3, 0)$

**7.** $P(0, 5)$        **8.** $P(-4, -6)$

**9.** $P(5, -2)$       **10.** $P(22, -22)$

In Exercises 11–14, state the sign (+ or −) of **(a)** sin $t$, **(b)** cos $t$, and **(c)** tan $t$ for values of $t$ in the interval given.

**11.** $\left(0, \dfrac{\pi}{2}\right)$  +, +, +       **12.** $\left(\dfrac{\pi}{2}, \pi\right)$  +, −, −

**13.** $\left(\pi, \dfrac{3\pi}{2}\right)$  −, −, +      **14.** $\left(\dfrac{3\pi}{2}, 2\pi\right)$  −, +, −

In Exercises 15–20, determine the sign (+ or −) of the given value without the use of a calculator.

**15.** cos 143°  −         **16.** tan 192°  +

**17.** $\cos \dfrac{7\pi}{8}$  −        **18.** $\tan \dfrac{4\pi}{5}$  −

**19.** $\cos \dfrac{-\pi}{12}$  +      **20.** $\sin \dfrac{3\pi}{4}$  +

In Exercises 21–24, choose the point on the terminal side of $\theta$.

**21.** $\theta = 45°$  (a)

    **(a)** $(2, 2)$     **(b)** $(1, \sqrt{3})$     **(c)** $(\sqrt{3}, 1)$

**22.** $\theta = \dfrac{2\pi}{3}$  (b)

    **(a)** $(-1, 1)$    **(b)** $(-1, \sqrt{3})$    **(c)** $(-\sqrt{3}, 1)$

**23.** $\theta = \dfrac{7\pi}{6}$  (a)

    **(a)** $(-\sqrt{3}, -1)$   **(b)** $(-1, \sqrt{3})$   **(c)** $(-\sqrt{3}, 1)$

**24.** $\theta = -60°$  (b)

    **(a)** $(-1, -1)$    **(b)** $(1, -\sqrt{3})$    **(c)** $(-\sqrt{3}, 1)$

In Exercises 25–30, find **(a)** sin $\theta$, **(b)** cos $\theta$, and **(c)** tan $\theta$ for the given quadrantal angle. If the value is undefined, write "undefined."

**25.** $-450°$              **26.** $-270°$

**27.** $7\pi$                **28.** $\dfrac{11\pi}{2}$

**29.** $\dfrac{-7\pi}{2}$            **30.** $-4\pi$

In Exercises 31–42, evaluate without using a calculator by using ratios in a reference triangle.

**31.** cos 120°  $-1/2$      **32.** tan 300°  $-\sqrt{3}$

**33.** $\sec \dfrac{\pi}{3}$  2        **34.** $\csc \dfrac{3\pi}{4}$  $\sqrt{2}$

**35.** $\sin \dfrac{13\pi}{6}$  $\dfrac{1}{2}$      **36.** $\cos \dfrac{7\pi}{3}$  $\dfrac{1}{2}$

**37.** $\tan \dfrac{-15\pi}{4}$  1     **38.** $\cot \dfrac{13\pi}{4}$  1

**39.** $\cos \dfrac{23\pi}{6}$  $\dfrac{\sqrt{3}}{2}$    **40.** $\cos \dfrac{17\pi}{4}$  $\dfrac{\sqrt{2}}{2}$

**41.** $\sin \dfrac{11\pi}{3}$  $-\dfrac{\sqrt{3}}{2}$    **42.** $\cot \dfrac{19\pi}{6}$  $\sqrt{3}$

In Exercises 43–48, evaluate without using a calculator.

**43.** Find sin $\theta$ and tan $\theta$ if $\cos \theta = \dfrac{2}{3}$ and cot $\theta > 0$.

**44.** Find cos $\theta$ and cot $\theta$ if $\sin \theta = \dfrac{1}{4}$ and tan $\theta < 0$.

**45.** Find tan $\theta$ and sec $\theta$ if $\sin \theta = \dfrac{-2}{5}$ and cos $\theta > 0$.

**46.** Find sin $\theta$ and cos $\theta$ if $\cos \theta = \dfrac{3}{7}$ and sec $\theta < 0$.

**47.** Find sec $\theta$ and csc $\theta$ if $\cot \theta = \dfrac{-4}{3}$ and cos $\theta < 0$.

**48.** Find csc $\theta$ and cot $\theta$ if $\tan \theta = \dfrac{-4}{3}$ and sin $\theta > 0$.

In Exercises 49–52, evaluate by using the period of the function.

**49.** $\sin \left(\dfrac{\pi}{6} + 49{,}000\pi\right)$  1/2

**50.** $\tan (1{,}234{,}567\pi) - \tan (7{,}654{,}321\pi)$  0

**51.** $\cos \left(\dfrac{5{,}555{,}555\pi}{2}\right)$  0    **52.** $\tan \left(\dfrac{3\pi - 70{,}000\pi}{2}\right)$

**53. Group Activity**  Use a calculator to evaluate the expressions in Exercises 49–52. Does your calculator give the correct answers? Many calculators miss all four. Give a brief explanation of what probably goes wrong.

54. **Writing to Learn** Give a convincing argument that the period of sin $t$ is $2\pi$. That is, show that there is no smaller positive real number $p$ such that sin $(t + p) =$ sin $t$ for all real numbers $t$.

55. **Refracted Light** Light is *refracted* (bent) as it passes through glass. In the figure below $\theta_1$ is the angle of incidence and $\theta_2$ is the *angle of refraction*. The *index of refraction* is a constant $\mu$ that satisfies the equation

$$\sin \theta_1 = \mu \sin \theta_2.$$

If $\theta_1 = 83°$ and $\theta_2 = 36°$ for a certain piece of flint glass, find the index of refraction.

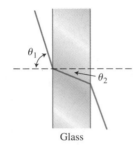

Glass

56. **Refracted Light** A certain piece of crown glass has an index of refraction of 1.52. If a light ray enters the glass at an angle $\theta_1 = 42°$, what is the angle of refraction $\theta_2$?

57. **Damped Harmonic Motion** A weight suspended from a spring is set into motion. Its dispacement $d$ from equilibrium is modeled by the equation

$$d = 0.4e^{-0.2t} \cos 4t.$$

where $d$ is the displacement in inches and $t$ is the time in seconds. Find the displacement at the given time. Use radian mode.

**(a)** $t = 0$  0.4 in.  **(b)** $t = 3$  $\approx 0.1852$ in.

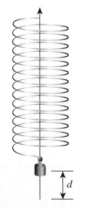

58. **Swinging Pendulum** The Columbus Museum of Science and Industry exhibits a Foucault pendulum 32 ft long that swings back and forth on a cable once in approximately 6 sec. The angle $\theta$ (in radians) between the cable and an imaginary vertical line is modeled by the equation

$$\theta = 0.25 \cos t.$$

Find the measure of angle $\theta$ when $t = 0$ and $t = 2.5$.

59. **Too Close for Comfort** An F-15 aircraft flying at an altitude of 8000 ft passes directly over a group of vacationers hiking at 7400 ft. If $\theta$ is the angle of elevation from the hikers to the F-15, find the distance $d$ from the group to the jet for the given angle.

**(a)** $\theta = 45°$  **(b)** $\theta = 90°$  **(c)** $\theta = 140°$

60. **Manufacturing Swimwear** Get Wet, Inc. manufactures swimwear, a seasonal product. The monthly sales $x$ (in thousands) for Get Wet swimsuits are modeled by the equation

$$x = 72.4 + 61.7 \sin \frac{\pi t}{6},$$

where $t = 1$ represents January, $t = 2$ February, and so on. Estimate the number of Get Wet swimsuits sold in January, April, June, October, and December. For which two of these months are sales the same? Explain why this might be so.

## Explorations

In Exercises 61–64, find the value of the unique real number $\theta$ between 0 and $2\pi$ that satisfies the two given conditions.

61. $\sin \theta = \dfrac{1}{2}$ and $\tan \theta < 0$.  $5\pi/6$

62. $\cos \theta = \dfrac{\sqrt{3}}{2}$ and $\sin \theta < 0$.  $11\pi/6$

63. $\tan \theta = -1$ and $\sin \theta < 0$.  $7\pi/4$

64. $\sin \theta = -\dfrac{\sqrt{2}}{2}$ and $\tan \theta > 0$.  $5\pi/4$

Exercises 65–68 refer to the unit circle in this figure. Point $P$ is on the terminal side of an angle $t$ and point $Q$ is on the terminal side of an angle $t + \pi/2$.

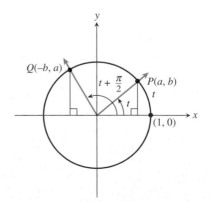

**65. Using Geometry in Trigonometry** Drop perpendiculars from points $P$ and $Q$ to the $x$-axis to form two right triangles. Explain how the right triangles are related.

**66. Using Geometry in Trigonometry** If the coordinates of point $P$ are $(a, b)$, explain why the coordinates of point $Q$ are $(-b, a)$.

**67.** Explain why $\sin\left(t + \dfrac{\pi}{2}\right) = \cos t$.

**68.** Explain why $\cos\left(t + \dfrac{\pi}{2}\right) = -\sin t$. ■

**69. Writing to Learn** In the figure for Exercises 65–68, $t$ is an angle with radian measure $0 < t < \pi/2$. Draw a similar figure for an angle with radian measure $\pi/2 < t < \pi$ and use it to explain why $\sin(t + \pi/2) = \cos t$.

**70. Writing to Learn** Use the accompanying figure to explain each of the following.

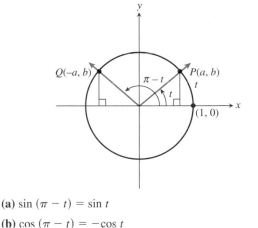

**(a)** $\sin(\pi - t) = \sin t$

**(b)** $\cos(\pi - t) = -\cos t$

## Extending the Ideas

**71. Approximation and Error analysis** Use your grapher to complete the table to show that $\sin \theta \approx \theta$ (in radians) when $|\theta|$ is small. Physicists often use the approximation $\sin \theta \approx \theta$ for small values of $\theta$. For what values of $\theta$ is the *magnitude of the error* in approximating $\sin \theta$ by $\theta$ less than 1% of $\sin \theta$? That is, solve the relation

$$|\sin \theta - \theta| < 0.01\,|\sin \theta|.$$

(*Hint:* Extend the table to include a column for values of $\dfrac{|\sin \theta - \theta|}{|\sin \theta|}$.)

| $\theta$ | $\sin \theta$ | $\sin \theta - \theta$ |
|---|---|---|
| $-0.03$ | | |
| $-0.02$ | | |
| $-0.01$ | | |
| $0$ | | |
| $0.01$ | | |
| $0.02$ | | |
| $0.03$ | | |

**72. Proving a Theorem** If $t$ is any real number, prove that $1 + (\tan t)^2 = (\sec t)^2$.

**Taylor Polynomials** Radian measure allows the trigonometric functions to be approximated by simple polynomial functions. For example, in Exercises 73 and 74, sine and cosine are approximated by *Taylor polynomials*, named after the English mathematician Brook Taylor (1685–1731). Complete each table showing a Taylor polynomial in the third column. Describe the patterns in the table.

**73.**

| $\theta$ | $\sin \theta$ | $\theta - \dfrac{\theta^3}{6}$ | $\sin \theta - \left(\theta - \dfrac{\theta^3}{6}\right)$ |
|---|---|---|---|
| $-0.3$ | $-0.295\ldots$ | | |
| $-0.2$ | $-0.198\ldots$ | | |
| $-0.1$ | $-0.099\ldots$ | | |
| $0$ | $0$ | | |
| $0.1$ | $0.099\ldots$ | | |
| $0.2$ | $0.198\ldots$ | | |
| $0.3$ | $0.295\ldots$ | | |

**74.**

| $\theta$ | $\cos \theta$ | $1 - \dfrac{\theta^2}{2} + \dfrac{\theta^4}{24}$ | $\cos \theta - \left(1 - \dfrac{\theta^2}{2} + \dfrac{\theta^4}{24}\right)$ |
|---|---|---|---|
| $-0.3$ | $0.955\ldots$ | | |
| $-0.2$ | $0.980\ldots$ | | |
| $-0.1$ | $0.995\ldots$ | | |
| $0$ | $1$ | | |
| $0.1$ | $0.995\ldots$ | | |
| $0.2$ | $0.980\ldots$ | | |
| $0.3$ | $0.955\ldots$ | | |

# Graphs of Sine and Cosine: Sinusoids

The Basic Waves Revisited • Sinusoids and Transformations • Modeling Periodic Behavior with Sinusoids

## The Basic Waves Revisited

In the first three sections of this chapter you saw how the trigonometric functions are rooted in the geometry of triangles and circles. It is these connections with geometry that give trigonometric functions their mathematical power and make them widely applicable in many fields.

The unit circle in Section 4.3 was the key to defining the trigonometric functions as functions of real numbers. This makes them available for the same kind of analysis as the other functions introduced in Chapter 1. (Indeed, two of our "Ten Basic Functions" are trigonometric.) We now take a closer look at the algebraic, graphical, and numerical properties of the trigonometric functions, beginning with sine and cosine.

Recall that we can learn quite a bit about the sine function by looking at its graph. Here is a summary of sine facts:

### The Sine Function

$f(x) = \sin x$

Domain: All reals.
Range: $[-1, 1]$.
Continuous.
Alternately increasing and decreasing in periodic waves.
Symmetric with respect to the origin (odd).
Bounded.
Absolute maximum of 1.
Absolute minimum of $-1$.
No horizontal asymptotes.
No vertical asymptotes.
End behavior: $\lim\limits_{x \to -\infty} \sin x$ and $\lim\limits_{x \to \infty} \sin x$ do not exist. (The function values continually oscillate between $-1$ and 1 and approach no limit.)

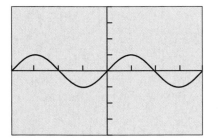

$[-2\pi, 2\pi]$ by $[-4, 4]$

We can add to this list that $y = \sin x$ is *periodic*, with period $2\pi$. We can also add understanding of where the sine function comes from: by definition, $\sin t$ is the $y$-coordinate of the point $P$ on the unit circle to which the real number $t$ gets wrapped (or, equivalently, the point $P$ on the unit circle determined by an angle of $t$ radians in standard position). In fact, now we can see where the wavy graph comes from. Try this Exploration.

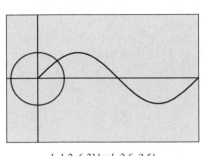

[–1.2, 6.3] by [–2.5, 2.5]

(a)

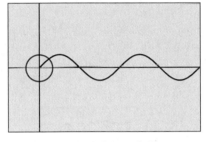

[–2.4, 12.6] by [–5, 5]

(b)

**Figure 4.40** The graph of $y = \sin t$ tracks the $y$-coordinate of the point determined by $t$ as it moves around the unit circle.

---

**Exploration Extensions**

Ask students to describe how to find the sine of 30° on the circle graph and then on the sine wave graph.

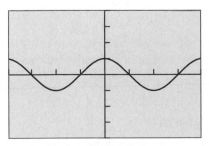

[–2π, 2π] by [–4, 4]

---

| **Exploration 1**  Graphing sin *t* as a Function of *t* |

Set your grapher to radian mode, parametric, and "simultaneous" graphing modes.
Set Tmin = 0, Tmax = 6.3, Tstep = $\pi/24$.
Set the $(x, y)$ window to $[-1.2, 6.3]$ by $[-2.5, 2.5]$.
Set X1T = cos (T) and Y1T = sin (T). This will graph the unit circle.
Set X2T = T and Y2T = sin (T). This will graph sin (T) as a function of T.

Now start the graph and watch the point go counterclockwise around the unit circle as $t$ goes from 0 to $2\pi$ in the positive direction. You will simultaneously see the $y$-coordinate of the point being graphed as a function of $t$ along the horizontal $t$-axis. You can clear the drawing and watch the graph as many times as you need to in order to answer the following questions.

1. Where is the point on the unit circle when the wave is at its highest?
2. Where is the point on the unit circle when the wave is at its lowest?
3. Why do both graphs cross the $x$-axis at the same time?
4. Double the value of Tmax and change the window to $[-2.4, 12.6]$ by $[-5, 5]$. If your grapher can change "style" to show a moving point, choose that style for the unit circle graph. Run the graph and watch how the sine curve tracks the $y$-coordinate of the point as it moves around the unit circle.
5. Explain from what you have seen why the period of the sine function is $2\pi$.
6. Challenge: Can you modify the grapher settings to show dynamically how the cosine function tracks the $x$-coordinate as the point moves around the unit circle?

Although a static picture does not do the dynamic simulation justice, Figure 4.40 shows the final screens for the two graphs in Exploration 1.

Here is a summary of cosine facts:

### The Cosine Function

$f(x) = \cos x$

Domain: All reals.
Range: $[-1, 1]$.
Continuous.
Alternately increasing and decreasing in periodic waves.
Symmetric with respect to the $y$-axis (even).
Bounded.
Absolute maximum of 1.
Absolute minimum of $-1$.
No horizontal asymptotes.
No vertical asymptotes.
End behavior: $\lim_{x \to -\infty} \cos x$ and $\lim_{x \to \infty} \cos x$ do not exist. (The function values continually oscillate between $-1$ and 1 and approach no limit.)

### Exponent Notation

Example 1 introduces a shorthand notation for powers of trigonometric functions: $(\sin\theta)^n$ can be written as $\sin^n\theta$. (Caution: This shorthand notation will probably not be recognized by your calculator.)

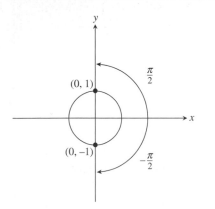

**Figure 4.41** Two solutions to the equation $\sin^2\theta = 1$. (Example 1)

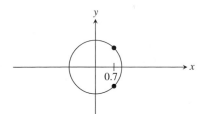

**Figure 4.42** There are two points on the unit circle with $x$-coordinate 0.7. (Example 2)

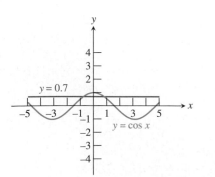

**Figure 4.43** Intersecting the graphs of $y = \cos x$ and $y = 0.7$ gives two solutions to the equation $\cos t = 0.7$. (Example 2)

As with the sine function, we can add the observation that it is periodic, with period $2\pi$.

There are two effective geometric strategies for analyzing the behavior of a sine or cosine function. One involves the unit circle, and the other involves the periodic wave that is the graph of the function itself. (Both are visible in Figure 4.40.) We illustrate how these strategies can both be useful when solving a trigonometric equation.

### Example 1   SOLVING A TRIGONOMETRIC EQUATION

Find all solutions to the equation $\sin^2\theta = 1$.

**Solution** We are looking for all values of $\theta$ such that $\sin\theta = \pm 1$. On the unit circle, these values of $\theta$ must wrap to the points $(0, 1)$ and $(0, -1)$. One solution to $\sin\theta = 1$ is $\pi/2$, and one solution to $\sin\theta = -1$ is $-\pi/2$ (Figure 4.41).

To get all the other solutions, we use the period $(2\pi)$ of the sine wave. The solution $\pi/2$ gives an infinite set of solutions of the form $\pi/2 + n(2\pi)$ for $n = 0, \pm 1, \pm 2, \pm 3, \ldots$, while the solution $-\pi/2$ gives an infinite set of solutions of the form $-\pi/2 + n(2\pi)$ for $n = 0, \pm 1, \pm 2, \pm 3, \ldots$. Putting these together, we get the solution set

$$\{\pm\pi/2 + 2n\pi \mid n = 0, \pm 1, \pm 2, \pm 3, \ldots\}.$$

Alternatively, we can see from the unit circle that all the solutions can be obtained by adding integer multiples of $\pi$ to the single solution $\pi/2$. The solution set

$$\{\pi/2 + n\pi \mid n = 0, \pm 1, \pm 2, \pm 3, \ldots\}$$

is equivalent to the first solution set we obtained.

We can also use periodicity in conjunction with our calculators to solve equations, as shown in Example 2.

### Example 2   SOLVING A TRIG EQUATION WITH A CALCULATOR

Find all solutions to the equation $\cos t = 0.7$.

**Solution** Figure 4.42 shows that there are two points on the unit circle with an $x$-coordinate of 0.7. We do not recognize this value as one of our special triangle ratios, but we can use our calculators to find the smallest positive and negative values for which $\cos x = 0.7$ by intersecting the graphs of $y = \cos x$ and $y = 0.7$ (Figure 4.43).

The two values are predictably opposites of each other: $t \approx \pm 0.7954$. Using the period of cosine (which is $2\pi$), we get the complete solution set:

$$\{\pm 0.7954 + 2n\pi \mid n = 0, \pm 1, \pm 2, \pm 3, \ldots\}.$$

## Sinusoids and Transformations

A comparison of the graphs of $y = \sin x$ and $y = \cos x$ suggests that either one can be obtained from the other by a horizontal translation (Section 1.5). In fact, we will prove later in this section that $\cos x = \sin(x + \pi/2)$. Each graph is an example of a *sinusoid*. In general, any transformation of a sine function (or the graph of such a function) is a sinusoid.

> **Definition  Sinusoid**
>
> A function is a **sinusoid** if it can be written in the form
>
> $$f(x) = a \sin (bx + c) + d$$
>
> where $a$, $b$, $c$, and $d$ are constants and neither $a$ nor $b$ is 0.

Since cosine functions are themselves translations of sine functions, any transformation of a cosine function is also a sinusoid by the above definition.

There is a special vocabulary used to describe some of our usual graphical transformations when we apply them to sinusoids. Horizontal stretches and shrinks affect the *period* and the *frequency*, vertical stretches and shrinks affect the *amplitude*, and horizontal translations bring about *phase shifts*. All of these terms are associated with waves, and waves are quite naturally associated with sinusoids.

> **Definition  Amplitude of a Sinusoid**
>
> The **amplitude** of the sinusoid $f(x) = a \sin (bx + c) + d$ is $|a|$. Similarly, the amplitude of $f(x) = a \cos (bx + c) + d$ is $|a|$.
>
> Graphically, the amplitude is half the height of the wave.

### Example 3  VERTICAL STRETCH OR SHRINK AND AMPLITUDE

Find the amplitude of each function and use the language of transformations to describe how the graphs are related.

**(a)** $y_1 = \cos x$

**(b)** $y_2 = \dfrac{1}{2}\cos x$

**(c)** $y_3 = -3 \cos x$

**Solution**  The amplitudes are (a) 1, (b) 1/2, and (c) $|-3| = 3$.

The graph of $y_2$ is a vertical shrink of the graph of $y_1$ by a factor of 1/2.

The graph of $y_3$ is a vertical stretch of the graph of $y_1$ by a factor of 3, and a reflection across the *x*-axis, performed in either order. (We do not call this a vertical stretch by a factor of $-3$, nor do we say that the amplitude is $-3$.)

The graphs of the three functions are shown in Figure 4.44. You should be able to tell which is which quite easily by checking the amplitudes.

You learned in Section 1.5 that the graph of $y = f(bx)$ when $|b| > 1$ is a horizontal shrink of the graph of $y = f(x)$ by a factor of $1/|b|$. That is exactly what happens with sinusoids, but we can add the observation that the period shrinks by the same factor. When $|b| < 1$, the effect on both the graph and the period is a horizontal stretch by a factor of $1/|b|$, plus a reflection across the *y*-axis if $b < 0$.

**Teaching Note**

Students should be expected to apply previously examined transformations to the graphs of sine and cosine. Give attention to translation, reflection, stretching and shrinking. A brief review on how $f(x) = x^2$ is related to $f(x) = ax^2$ and to $f(x) = (x - h)^2$ may be valuable. Relate this discussion to the trigonometric functions.

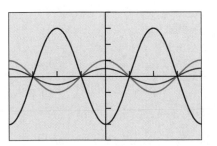

$[-2\pi, 2\pi]$ by $[-4, 4]$

**Figure 4.44**  Sinusoids (in this case, cosine curves) of different amplitudes. (Example 3)

**Period of a Sinusoid**

The period of the sinusoid $f(x) = a \sin(bx + c) + d$ is $2\pi/|b|$.
Similarly, the period of $f(x) = a \cos(bx + c) + d$ is $2\pi/|b|$.
  Graphically, the period is the length of one full cycle of the wave.

**Example 4   HORIZONTAL STRETCH OR SHRINK AND PERIOD**

Find the period of each function and use the language of transformations to describe how the graphs are related.

**(a)** $y_1 = \sin x$

**(b)** $y_2 = -2 \sin\left(\dfrac{x}{3}\right)$

**(c)** $y_3 = 3 \sin(-2x)$

**Solution**

**Solve Algebraically**

The periods are (a) $2\pi$, (b) $2\pi/(1/3) = 6\pi$, and (c) $2\pi/|-2| = \pi$.
  The graph of $y_2$ is a horizontal stretch of the graph of $y_1$ by a factor of 3, a vertical stretch by a factor of 2, and a reflection across the $x$-axis, performed in any order.
  The graph of $y_3$ is a horizontal shrink of the graph of $y_1$ by a factor of 1/2, a vertical stretch by a factor of 3, and a reflection across the $y$-axis, performed in any order. (Note that we do not call this a horizontal shrink by a factor of $-1/2$, nor do we say that the period is $-\pi$.)

**Support Graphically**

  The graphs of the three functions are shown in Figure 4.45. You should be able to tell which is which quite easily by checking the periods or the amplitudes.

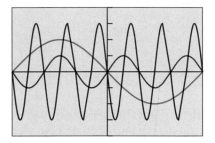

$[-3\pi, 3\pi]$ by $[-4, 4]$

**Figure 4.45** Sinusoids (in this case, sine curves) of different amplitudes and periods. (Example 4)

  In some applications the *frequency* of a sinusoid is an important consideration. The frequency is simply the reciprocal of the period.

**Frequency of a Sinusoid**

The **frequency** of the sinusoid $f(x) = a \sin(bx + c) + d$ is $|b|/2\pi$.
Similarly, the frequency of $f(x) = a \cos(bx + c) + d$ is $|b|/2\pi$.
  Graphically, the frequency is the number of complete cycles the wave completes in a unit interval.

**Example 5   FINDING THE FREQUENCY OF A SINUSOID**

Find the frequency of the function $f(x) = 4 \sin(2\pi x/3)$ and interpret its meaning graphically.

**Solution** The frequency is $(2\pi/3)/2\pi = 1/3$. This is the reciprocal of the period, which is 3.

The graphical interpretation is that the graph completes 1/3 of a full cycle in a unit interval, or 1 full cycle in an interval of length 3. That, of course, is what having period 3 is all about.

Recall from Section 1.5 that the graph of $y = f(x + c)$ is a translation of the graph of $y = f(x)$ by $c$ units to the left when $c > 0$. That is exactly what happens with sinusoids, but using terminology with its roots in electrical engineering, we say that the wave undergoes a **phase shift** of $-c$.

### Example 6   GETTING ONE SINUSOID FROM ANOTHER BY A PHASE SHIFT

**(a)** Write the cosine function as a phase shift of the sine function.

**(b)** Write the sine function as a phase shift of the cosine function.

**Solution**

**(a)** The function $y = \sin x$ has a maximum at $x = \pi/2$, while the function $y = \cos x$ has a maximum at $x = 0$. Therefore, we need to shift the sine curve $\pi/2$ units to the *left* to get the cosine curve:

$$\cos x = \sin (x + \pi/2).$$

**(b)** It follows from the work in (a) that we need to shift the cosine curve $\pi/2$ units to the right to get the sine curve:

$$\sin x = \cos (x - \pi/2).$$

You can support with your grapher that these statements are true. Incidentally, there are many other translations that would have worked just as well. Adding any integral multiple of $2\pi$ to the phase shift would result in the same graph.

One note of caution applies when combining these transformations. A horizontal stretch or shrink affects the variable along the horizontal axis, so it *also affects the phase shift*. Consider the transformation in Example 7.

### Example 7   COMBINING A PHASE SHIFT WITH A PERIOD CHANGE

Construct a sinusoid with period $\pi/5$ and amplitude 6 that goes through (2, 0).

**Solution**   To find the coefficient of $x$, we set $2\pi/|b| = \pi/5$ and solve to find that $b = \pm 10$. We arbitrarily choose $b = 10$. (Either will satisfy the specified conditions.)

For amplitude 6, we have $|a| = 6$. Again, we arbitrarily choose the positive value. The graph of $y = 6 \sin (10x)$ has the required amplitude and period, but it does not go through the point (2, 0). It does, however, go through the point (0, 0), so all that is needed is a phase shift of $+2$ to finish our function. Replacing $x$ by $x - 2$, we get

$$y = 6 \sin (10(x - 2)) = 6 \sin (10x - 20).$$

Notice that we did *not* get the function $y = 6 \sin (10x - 2)$. That function would represent a phase shift of $y = \sin (10x)$, but only by 2/10, not 2.

Parentheses are important when combining phase shifts with horizontal stretches and shrinks.

---

**Characteristics of Sinusoids**

The graphs of $y = a \sin (b(x - h)) + k$ and $y = a \cos (b(x - h)) + k$ (where $a \neq 0$ and $b \neq 0$) have the following characteristics:

amplitude $= |a|$;

$$\text{period} = \frac{2\pi}{|b|};$$

$$\text{frequency} = \frac{|b|}{2\pi}.$$

When compared to the graphs of $y = a \sin bx$ and $y = a \cos bx$ respectively, they also have the following characteristics:

a phase shift of $h$;

a vertical translation of $k$.

---

### Example 8   CONSTRUCTING A SINUSOID BY TRANSFORMATIONS

Construct a sinusoid $y = f(x)$ that rises from a minimum value of $y = 5$ at $x = 0$ to a maximum value of $y = 25$ at $x = 32$ (Figure 4.46).

#### Solution
#### Solve Algebraically

The amplitude of this sinusoid is half the height of the graph: $(25 - 5)/2 = 10$. So $|a| = 10$. The period is 64 (since a full period goes from minimum to maximum and back down to the minimum). So set $2\pi/|b| = 64$. Solving, we get $|b| = \pi/32$.

We need a sinusoid that takes on its minimum value at $x = 0$. We could shift the graph of sine or cosine horizontally, but it is easier to take the cosine curve (which assumes its *maximum* value at $x = 0$) and turn it upside down. This reflection can be obtained by letting $a = -10$ rather than 10.

So far we have:

$$y = -10 \cos \left( \pm \frac{\pi}{32} x \right)$$

$$= -10 \cos \left( \frac{\pi}{32} x \right) \quad \text{(Since cos is an even function)}$$

Finally, we note that this function ranges from a minimum of $-10$ to a maximum of 10. We shift the graph vertically by 15 to obtain a function that ranges from a minimum of 5 to a maximum of 25, as required. Thus

$$y = -10 \cos \left( \frac{\pi}{32} x \right) + 15.$$

#### Support Graphically

We support our answer graphically by graphing the function (Figure 4.47).

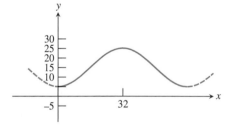

**Figure 4.46** A sinusoid with specifications. (Example 8)

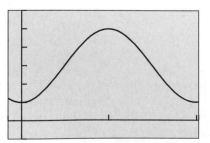

[−5, 65] by [−5, 30]

**Figure 4.47** The graph of the function $y = -10 \cos ((\pi/32)x) + 15$. (Example 8)

## Modeling Periodic Behavior with Sinusoids

Example 8 was intended as more than just an exercise for reviewing the graphical transformations. Constructing a sinusoid with specific properties is often the key step in modeling physical situations that exhibit periodic behavior. We can apply the same technique to the following problem about tides.

### Example 9 CALCULATING THE EBB AND FLOW OF TIDES

One particular July 4th in Galveston, TX, high tide occurred at 9:36 A.M. At that time the water at the end of the 61st Street Pier was 2.7 meters deep. Low tide occurred at 3:48 P.M., at which time the water was only 2.1 meters deep. Assume that the depth of the water is a sinusoidal function of time with a period of half a lunar day (about 12 hours 24 minutes).

**(a)** At what time on the 4th of July did the first low tide occur?

**(b)** What was the approximate depth of the water at 6:00 A.M. and at 3:00 P.M. that day?

**(c)** What was the first time on July 4th when the water was 2.4 meters deep?

### Solution

#### Model

We want to model the depth $D$ as a sinusoidal function of time $t$. The amplitude of the sinusoid is half the height of the graph: $(2.7 - 2.1)/2 = 0.3$. Therefore $a = 0.3$. The period is 12 hours 24 minutes, which converts to 12.4 hours. Solving the equation $2\pi/b = 12.4$, we find that $b = \pi/6.2$.

We need a sinusoid that assumes its maximum value at 9:36 A.M. (which converts to $t = 9.6$ hours after midnight). Cosine assumes its maximum value at $t = 0$, so we can translate a suitable cosine curve 9.6 units to the right.

So far we have

$$y = 0.3 \cos\left(\frac{\pi}{6.2}(t - 9.6)\right).$$

This function ranges from a low of $-0.3$ to a high of $0.3$, so we shift it vertically by 2.4 units to get our function $D$:

$$D(t) = 0.3 \cos\left(\frac{\pi}{6.2}(t - 9.6)\right) + 2.4.$$

#### Solve Graphically

The graph over the 24-hour period of July 4th is shown in Figure 4.48.

We now use the graph to answer the questions posed.

**(a)** The first low tide corresponds to the first local minimum on the graph. We find graphically that this occurs at $t = 3.4$. This translates to $3 + (0.4)(60) = 3:24$ A.M.

**(b)** The depth at 6:00 A.M. is $D(6) \approx 2.32$ meters. The depth at 3:00 P.M. is $D(12 + 3) = D(15) \approx 2.12$ meters.

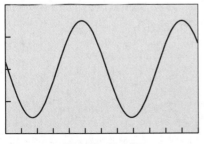

[0, 24] by [2, 2.8]

**Figure 4.48** The Galveston tide graph. (Example 9)

(c) The first time the water is 2.4 meters deep corresponds to the leftmost intersection of the sinusoid with the line $y = 2.4$. We use the grapher to find that $t = 0.3$. This translates to $0 + (0.3)(60) = 00:18$ A.M., which we write as 12:18 A.M.

## Quick Review 4.4

In Exercises 1–3, state the sign (positive or negative) of the function in each quadrant.

**1.** $\sin x$  In order: $+, +, -, -$    **2.** $\cos x$  In order: $+, -, -, +$

**3.** $\tan x$  In order: $+, -, +, -$

In Exercises 4–6, give the radian measure of the angle.

**4.** $135°$  $3\pi/4$

**5.** $-150°$  $-5\pi/6$

**6.** $450°$  $5\pi/2$

In Exercises 7–10, find a transformation that will transform the graph of $y_1$ to the graph of $y_2$.

**7.** $y_1 = \sqrt{x}$ and $y_2 = 3\sqrt{x}$    **8.** $y_1 = e^x$ and $y_2 = e^{-x}$

**9.** $y_1 = \ln x$ and $y_2 = 0.5 \ln x$    **10.** $y_1 = x^3$ and $y_2 = x^3 - 2$

## Section 4.4 Exercises

In Exercises 1–6, graph one period of the function. Use your understanding of transformations, not your graphing calculators. Be sure to show the scale on both axes.

**1.** $y = 2 \sin x$        **2.** $y = 2.5 \sin x$

**3.** $y = 3 \cos x$        **4.** $y = -2 \cos x$

**5.** $y = -0.5 \sin x$     **6.** $y = 4 \cos x$

In Exercises 7–12, graph three periods of the function. Use your understanding of transformations, not your graphing calculators. Be sure to show the scale on both axes.

**7.** $y = 5 \sin 2x$        **8.** $y = 3 \cos \dfrac{x}{2}$

**9.** $y = 0.5 \cos 3x$      **10.** $y = 20 \sin 4x$

**11.** $y = 4 \sin \dfrac{x}{4}$    **12.** $y = 8 \cos 5x$

In Exercises 13–18, specify the period and amplitude of each function. Then give the viewing window in which the graph is shown. Use your understanding of transformations, not your graphing calculators.

**13.** $y = 1.5 \sin 2x$     **14.** $y = 2 \cos 3x$

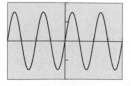 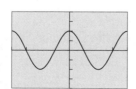

**15.** $y = -3 \cos 2x$      **16.** $y = 5 \sin \dfrac{x}{2}$

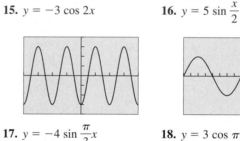

**17.** $y = -4 \sin \dfrac{\pi}{3}x$    **18.** $y = 3 \cos \pi x$

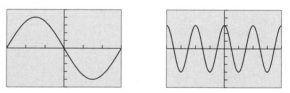

In Exercises 19–24, identify the maximum and minimum values and the zeros of the function in the interval $[-2\pi, 2\pi]$. Use your understanding of transformations, not your graphing calculators.

**19.** $y = 2 \sin x$        **20.** $3 \cos \dfrac{x}{2}$

**21.** $y = \cos 2x$         **22.** $y = \dfrac{1}{2} \sin x$

**23.** $y = -\cos 2x$        **24.** $y = -2 \sin x$

In Exercises 25–30, solve for $x$ in the given interval. You should be able to find these numbers without a calculator, using reference triangles in the proper quadrants.

**25.** $\sin x = \dfrac{1}{2}$,        $0 \le x \le \dfrac{\pi}{2}$  $\dfrac{\pi}{6}$

**26.** $\sin x = \dfrac{1}{2}$,        $\dfrac{\pi}{2} \le x \le \pi$  $\dfrac{5\pi}{6}$

**27.** $\cos x = \dfrac{-\sqrt{2}}{2}, \qquad \dfrac{\pi}{2} \le x \le \pi \quad \dfrac{3\pi}{4}$

**28.** $\cos x = \dfrac{-\sqrt{2}}{2}, \qquad \pi \le x \le \dfrac{3\pi}{2} \quad \dfrac{5\pi}{4}$

**29.** $\tan x = 1, \qquad 2\pi \le x \le \dfrac{5\pi}{2} \quad \dfrac{9\pi}{4}$

**30.** $\tan x = 1, \qquad -\pi \le x \le \dfrac{-\pi}{2} \quad -\dfrac{3\pi}{4}$

In Exercises 31–34, use a calculator to solve for $x$ in the given interval.

**31.** $\sin x = \dfrac{3}{4}, \qquad 0 \le x \le \dfrac{\pi}{2} \approx 0.85$

**32.** $\cos x = \dfrac{2}{3}, \qquad 0 \le x < \dfrac{\pi}{2} \approx 0.84$

**33.** $\cos x = 0.4, \qquad \dfrac{3\pi}{2} \le x < 2\pi \approx 5.12$

**34.** $\sin x = 0.6, \qquad \dfrac{\pi}{2} \le x < \pi \approx 2.50$

In Exercises 35–40, describe the transformations required to obtain the graph of the given function from a basic trigonometric graph.

**35.** $y = 0.5 \sin 3x$

**36.** $y = 1.5 \cos 4x$

**37.** $y = -\dfrac{2}{3} \cos \dfrac{x}{3}$

**38.** $y = \dfrac{3}{4} \sin \dfrac{x}{5}$

**39.** $y = 3 \cos \dfrac{2\pi x}{3}$

**40.** $y = -2 \sin \dfrac{\pi x}{4}$

In Exercises 41–44, describe the transformations required to obtain the graph of $y_2$ from the graph of $y_1$.

**41.** $y_1 = \cos 2x$ and $y_2 = \dfrac{5}{3} \cos 2x$

**42.** $y_1 = 2 \cos \left( x + \dfrac{\pi}{3} \right)$ and $y_2 = \cos \left( x + \dfrac{\pi}{4} \right)$

**43.** $y_1 = 2 \cos \pi x$ and $y_2 = 2 \cos 2\pi x$

**44.** $y_1 = 3 \sin \dfrac{2\pi x}{3}$ and $y_2 = 2 \sin \dfrac{\pi x}{3}$

In Exercises 45–48, select the pair of functions that have identical graphs.

**45.** **(a)** $y = \cos x$      **(b)** $y = \sin \left( x + \dfrac{\pi}{2} \right)$

     **(c)** $y = \cos \left( x + \dfrac{\pi}{2} \right)$ (a) and (b)

**46.** **(a)** $y = \sin x$      **(b)** $y = \cos \left( x - \dfrac{\pi}{2} \right)$

     **(c)** $y = \cos x$ (a) and (b)

**47.** **(a)** $y = \sin \left( x + \dfrac{\pi}{2} \right)$      **(b)** $y = -\cos (x - \pi)$

     **(c)** $y = \cos \left( x - \dfrac{\pi}{2} \right)$ (a) and (b)

**48.** **(a)** $y = \sin \left( 2x + \dfrac{\pi}{4} \right)$      **(b)** $y = \cos \left( 2x - \dfrac{\pi}{2} \right)$

     **(c)** $y = \cos \left( 2x - \dfrac{\pi}{4} \right)$ (a) and (c)

In Exercises 49–56, state the amplitude and period of the sinusoid, and (relative to the basic function) the phase shift and vertical translation.

**49.** $y = -2 \sin \left( x - \dfrac{\pi}{4} \right) + 1$

**50.** $y = -3.5 \sin \left( 2x - \dfrac{\pi}{2} \right) - 1$

**51.** $y = 5 \cos \left( 3x - \dfrac{\pi}{6} \right) + 0.5$    **52.** $y = 3 \cos (x + 3) - 2$

**53.** $y = 2 \cos 2\pi x + 1$        **54.** $y = 4 \cos 3\pi x - 2$

**55.** $y = \dfrac{7}{3} \sin \left( x + \dfrac{5}{2} \right) - 1$    **56.** $y = \dfrac{2}{3} \cos \left( \dfrac{x - 3}{4} \right) + 1$

In Exercises 57 and 58, find values $a$, $b$, $h$, and $k$ so that the graph of the function $y = a \sin (b(x + h)) + k$ is the curve shown.

**57.**                 **58.**

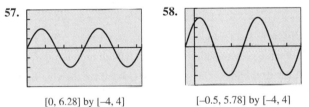

[0, 6.28] by [–4, 4]         [–0.5, 5.78] by [–4, 4]

**59. Points of Intersection** Graph $y = 1.3^{-x}$ and $y = 1.3^{-x} \cos x$ for $x$ in the interval $[-1, 8]$.

     **(a)** How many points of intersection do there appear to be?

     **(b)** Find the coordinates of each point of intersection.

**60. Motion of a Buoy** A signal buoy in the Chesapeake Bay bobs up and down with the height $h$ of its transmitter (in feet) above sea level modeled by $h = a \sin bt + 5$. During a small squall its height varies from 1 ft to 9 ft and there are 3.5 sec from one 9-ft height to the next. What are the values of the constants $a$ and $b$? $a = 4$ and $b = 4\pi/7$

**61. Ferris Wheel** A Ferris wheel 50 ft in diameter makes one revolution every 40 sec. If the center of the wheel is 30 ft above the ground, how long after reaching the low point is a rider 50 ft above the ground?  ≈ 15.90 sec

**62. Tsunami Wave** An earthquake occurred at 9:40 A.M. on Nov. 1, 1755, at Lisbon, Portugal, and started a *tsunami* (often called a tidal wave) in the ocean. It produced waves that traveled more than 540 ft/sec (370 mph) and reached a height of 60 ft. If the period of the waves was 30 min or 1800 sec, estimate the length *L* between the crests.

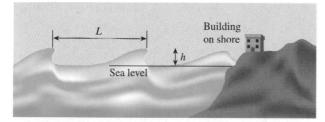

**63. Blood Pressure** The function

$$P = 120 + 30 \sin 2\pi t$$

models the blood pressure (in millimeters of mercury) for a person who has a (high) blood pressure of 150/90; *t* represents seconds.

**(a)** What is the period of this function?  1 second

**(b)** How many heartbeats are there each minute?  60

**(c)** Graph this function to model a 10-sec time interval.

**64. Ebb and Flow** On a particular Labor Day, the high tide in southern California occurs at 7:12 A.M. At that time you measure the water at the end of the Santa Monica Pier to be 11 ft deep. At 1:24 P.M. it is low tide, and you measure the water to be only 7 ft deep. Assume the depth of the water is a sinusoidal function of time with a period of 1/2 a lunar day, which is about 12 h 24 min.

**(a)** At what time on that Labor Day does the first low tide occur?  1:00 A.M.

**(b)** What was the approximate depth of the water at 4:00 A.M. and at 9:00 P.M.?  8.90 ft; 10.52 ft

**(c)** What is the first time on that Labor Day that the water is 9 ft deep?  4:06 A.M.

## Explorations

**65. Approximating Cosine**

**(a)** Draw a scatter plot (*x*, cos *x*) for the 17 special angles *x*, where −π ≤ *x* ≤ π.

**(b)** Find a quartic regression for the data.

**(c)** Compare the approximation to the cosine function given by the quartic regression with the Taylor polynomial approximations given in Exercise 74 of Section 4.3.

**66. Approximating Sine**

**(a)** Draw a scatter plot (*x*, sin *x*) for the 17 special angles *x*, where −π ≤ *x* ≤ π.

**(b)** Find a cubic regression for the data.

**(c)** Compare the approximation to the sine function given by the cubic regression with the Taylor polynomial approximations given in Exercise 73 of Section 4.3.

**67. Visualizing a Musical Note** A piano tuner strikes a tuning fork for the note middle C and creates a sound wave that can be modeled by

$$y = 1.5 \sin 524\pi t,$$

where *t* is the time in seconds.

**(a)** What is the period *p* of this function?  1/262 sec

**(b)** What is the frequency $f = 1/p$ of this note?

**(c)** Graph the function.

**68. Writing to Learn** In a certain video game a cursor bounces back and forth horizontally across the screen at a constant rate. Its distance *d* from the center of the screen varies with time *t* and hence can be described as a function of *t*. Explain why this horizontal distance *d* from the center of the screen *does not vary* according to an equation $d = a \sin bt$, where *t* represents seconds. You may find it helpful to include a graph in your explanation.  ∎

**69. Group Activity** Using only integer values of *a* and *b* between 1 and 9 inclusive, look at graphs of functions of the form

$$y = \sin(ax) \cos(bx) - \cos(ax) \sin(bx)$$

for various values of *a* and *b*. (A group can look at more graphs at a time than one person can.)

**(a)** Some values of *a* and *b* result in the graph of *y* = sin *x*. Find a general rule for such values of *a* and *b*.

**(b)** Some values of *a* and *b* result in the graph of *y* = sin 2*x*. Find a general rule for such values of *a* and *b*.

**(c)** Can you guess which values of *a* and *b* will result in the graph of *y* = sin *kx* for an arbitrary integer *k*?

**70. Group Activity** Using only integer values of *a* and *b* between 1 and 9 inclusive, look at graphs of functions of the form

$$y = \cos(ax) \cos(bx) + \sin(ax) \sin(bx)$$

for various values of *a* and *b*. (A group can look at more graphs at a time than one person can.)

**(a)** Some values of *a* and *b* result in the graph of *y* = cos *x*. Find a general rule for such values of *a* and *b*.

**(b)** Some values of *a* and *b* result in the graph of *y* = cos 2*x*. Find a general rule for such values of *a* and *b*.

**(c)** Can you guess which values of *a* and *b* will result in the graph of *y* = cos *kx* for an arbitrary integer *k*?

## Extending the Ideas

In Exercises 71–74, the graphs of the sine and cosine functions are waveforms like the figure below. By correctly labeling the coordinates of points *A*, *B*, and *C*, you will get the graph of the function given.

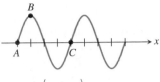

**71.** $y = 3 \cos 2x$ and $A = \left(-\dfrac{\pi}{4}, 0\right)$. Find *B* and *C*.

**72.** $y = 4.5 \sin\left(x - \dfrac{\pi}{4}\right)$ and $A = \left(\dfrac{\pi}{4}, 0\right)$. Find *B* and *C*.

**73.** $y = 2 \sin\left(3x - \dfrac{\pi}{4}\right)$ and $A = \left(\dfrac{\pi}{12}, 0\right)$. Find *B* and *C*.

**74.** $y = 3 \sin(2x - \pi)$, and *A* is the first *x*-intercept on the right of the *y*-axis. Find *A*, *B*, and *C*.

**75. The Ultimate Sinusoidal Equation**

$$y = a \sin[b(x - h)] + k$$

**(a)** Explain why you can assume *b* is positive. (*Hint:* Replace *b* by $-B$ and simplify.)

**(b)** Use one of the horizontal translation identities to prove that the equation

$$y = a \cos[b(x - h)] + k$$

has the same graph as

$$y = a \sin[b(x - H)] + k$$

for a correctly chosen value of *H*. Explain how to choose *H*.

**(c)** Give a unit-circle argument for the identity $\sin(\theta + \pi) = -\sin\theta$. Support your unit-circle argument graphically.

**(d)** Use the identity from (c) to prove that

$$y = -a \sin[b(x - h)] + k, \; a > 0,$$

has the same graph as

$$y = a \sin[b(x - H)] + k, \; a > 0$$

for a correctly chosen value of *H*. Explain how to choose *H*.

**(e)** Combine your results from (a)–(d) to prove that any sinusoid can be represented by the equation

$$y = a \sin[b(x - H)] + k,$$

where *a* and *b* are both positive.

---

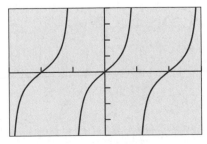

[$-3\pi/2$, $3\pi/2$] by [$-4$, 4]

# 4.5 Graphs of Tangent, Cotangent, Secant, and Cosecant

The Tangent Function • The Cotangent Function • The Secant Function • The Cosecant Function

## The Tangent Function

The graph of the tangent function is shown to the left. As with the sine and cosine graphs, this graph tells us quite a bit about the function's properties. Here is a summary of tangent facts:

### The Tangent Function

$f(x) = \tan x$

Domain: All reals except odd multiples of $\pi/2$.
Range: All reals
Continuous (i.e., continuous on its domain).
Increasing on each interval in its domain.
Symmetric with respect to the origin (odd).
Not bounded above or below.
No local extrema.
No horizontal asymptotes.
Vertical asymptotes: $x = k \cdot (\pi/2)$ for all odd integers *k*.
End behavior: $\lim\limits_{x \to -\infty} \tan x$ and $\lim\limits_{x \to \infty} \tan x$ do not exist. (The function values continually oscillate between $-\infty$ and $\infty$ and approach no limit.)

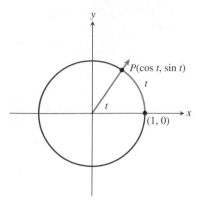

**Figure 4.49** How $t$ determines a point on the unit circle.

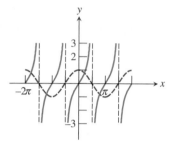

**Figure 4.50** The tangent function has asymptotes at the zeros of cosine.

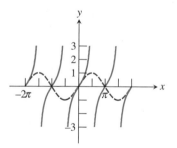

**Figure 4.51** The tangent function has zeros at the zeros of sine.

We now use the unit circle to analyze why the graph of $f(x) = \tan x$ behaves the way it does. To avoid confusion with the $(x, y)$ point on the unit circle, we will temporarily look at $f(t) = \tan t$. The number $t$ wraps to a point $P(x, y)$ on the unit circle (Figure 4.49), from which we use the definition to get

$$\tan t = \frac{y}{x} = \frac{\sin t}{\cos t}.$$

Unlike the sinusoids, the tangent function has a denominator that might be zero, thus making the function undefined. Not only does this actually happen, but it happens an infinite number of times: at all the values of $t$ for which $\cos t = 0$. The points at which $\cos t = 0$ are $(0, 1)$ and $(0, -1)$. We found in Example 1 of Section 4.4 (solving $\sin^2 \theta = 1$) that the values of $t$ that get wrapped to those points are all numbers of the form $\pi/2 + n\pi$ where $n$ is an integer. That is why the tangent function has vertical asymptotes at those values (Figure 4.50). The zeros of the tangent function are the same as the zeros of the sine function: all the integer multiples of $\pi$ (Figure 4.51).

Because $\sin t$ and $\cos t$ are both periodic with period $2\pi$, you might expect the period of the tangent function to be $2\pi$ also. The graph shows, however, that it is $\pi$.

The constants $a, b, h,$ and $k$ influence the behavior of $y = a\tan(b(x - h)) + k$ in much the same way that they do for the graph of $y = a\sin(b(x - h)) + k$. The constant $a$ yields a vertical stretch or shrink, $b$ affects the period, $h$ causes a horizontal translation, and $k$ causes a vertical translation. The terms *amplitude* and *phase shift*, however, are not used, as they apply only to sinusoids.

**Example 1   GRAPHING A TANGENT FUNCTION**

Describe the graph of $y = -\tan 2x$. Graph four periods and locate the vertical asymptotes.

**Solution**   The coefficient ($b$) of $x$ is 2, causing the period (in this case, $\pi$) to be divided by 2. This is the same effect that $b$ has on sinusoids. Four periods would yield an interval of length $4(\pi/2) = 2\pi$, so any viewing window in which Xmax − Xmin = $2\pi$ will show four periods of $y = -\tan 2x$.

The value $a = -1$ causes the graph of $y = \tan 2x$ (Figure 4.52a) to be reflected across the $x$-axis to obtain the graph of $y = -\tan 2x$ (Figure 4.52b).

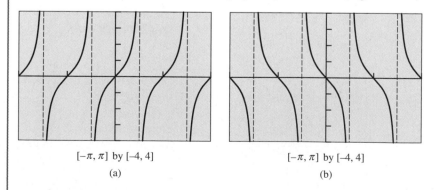

$[-\pi, \pi]$ by $[-4, 4]$          $[-\pi, \pi]$ by $[-4, 4]$
(a)                                    (b)

**Figure 4.52** The graph of (a) $y = \tan 2x$ is reflected across the $x$-axis to produce the graph of (b) $y = -\tan 2x$. (Example 1)

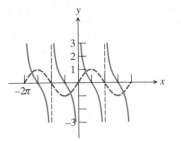

**Figure 4.53** The cotangent has asymptotes at the zeros of the sine function.

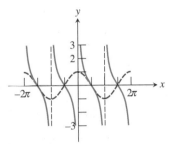

**Figure 4.54** The cotangent has zeros at the zeros of the cosine function.

---

**Cotangent on the Calculator**

If your calculator does not have a "cotan" button, you can use the fact that cotangent and tangent are reciprocals. For example, the function in Example 2 can be entered in a calculator as $y = 3/\tan(x/2) + 1$ or as $y = 3(\tan(x/2))^{-1} + 1$. Remember that it can *not* be entered as $y = 3 \tan^{-1}(x/2) + 1$. (The $-1$ exponent in that position represents a function *inverse*, not a *reciprocal*.)

---

The graph of $y = \tan 2x$ is a horizontal shrink of $\tan x$ by a factor of $1/2$. Thus, the vertical asymptotes of $y = \tan x$ (all odd multiples of $\pi/2$) are shrunk by $1/2$ to become the vertical asymptotes of $y = \tan 2x$ (all odd multiples of $\pi/4$). The asymptotes are unchanged by the reflection across the $x$-axis.

The other three trigonometric functions (cotangent, secant, and cosecant) have been omitted from our list of basic functions because they are reciprocals of tangent, cosine, and sine, respectively. (This is the same reason that you probably do not have buttons for them on your calculators.) As functions they are certainly interesting, but as basic functions they are unnecessary—we can do our trigonometric modeling and equation-solving with the other three. Nonetheless, we give each of them a brief section of its own in this book.

## The Cotangent Function

The cotangent function is the reciprocal of the tangent function. If the number $t$ wraps to a point $P(x, y)$ on the unit circle, then by definition

$$\cot t = \frac{x}{y} = \frac{\cos t}{\sin t}.$$

We analyze this function in virtually the same way that we analyzed the tangent function. The graph of $y = \cot t$ will have asymptotes at the zeros of the sine function (Figure 4.53) and zeros at the zeros of the cosine function (Figure 4.54).

**Example 2   GRAPHING A COTANGENT FUNCTION**

Describe the graph of $f(x) = 3 \cot(x/2) + 1$. Graph two periods.

**Solution**  The period of $f$ is $2\pi$, twice the period of $y = \cot x$, as a result of the horizontal stretch by 2 caused by the constant $b = 1/2$. The value $a = 3$ causes a vertical stretch by a factor of 3. The constant $k = 1$ causes a vertical translation up one unit.

Figure 4.55 shows two periods of the graph of $f$.

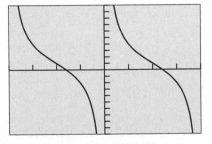

$[-2\pi, 2\pi]$ by $[-10, 10]$

**Figure 4.55** Two periods of $f(x) = 3 \cot(x/2) + 1$. (Example 2)

## The Secant Function

Important characteristics of the secant function can be inferred from the fact that it is the reciprocal of the cosine function.

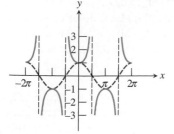

**Figure 4.56** Characteristics of the secant function are inferred from the fact that it is the reciprocal of the cosine function.

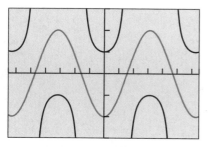

[–6.5, 6.5] by [–3, 3]

**Figure 4.57** The graphs of $y = \sec x$ and $y = -2 \cos x$. (Exploration 1)

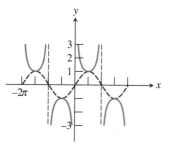

**Figure 4.58** Characteristics of the cosecant function are inferred from the fact that it is the reciprocal of the sine function.

Whenever $\cos x = 1$, its reciprocal $\sec x$ is also 1. The graph of the secant function has asymptotes at the zeros of the cosine function. The period of the secant function is $2\pi$, the same as its reciprocal, the cosine function.

The graph of $y = \sec x$ is shown with the graph of $y = \cos x$ in Figure 4.56. A local maximum of $y = \cos x$ corresponds to a local minimum of $y = \sec x$, while a local minimum of $y = \cos x$ corresponds to a local maximum of $y = \sec x$.

---

**Exploration 1** **Proving a Graphical Hunch**

Figure 4.57 shows that the graphs of $y = \sec x$ and $y = -2 \cos x$ never seem to intersect.

If we stretch the reflected cosine graph vertically by a large enough number, will it continue to miss the secant graph? Or is there a large enough (positive) value of $k$ so that the graph of $y = \sec x$ *does* intersect the graph of $y = -k \cos x$?

1. Try a few other values of $k$ in your calculator. Do the graphs intersect? The graphs do not seem to intersect.

2. Your exploration should lead you to conjecture that the graphs of $y = \sec x$ and $y = -k \cos x$ will *never* intersect for any positive value of $k$. Verify this conjecture by proving **algebraically** that the equation

$$-k \cos x = \sec x$$

has no real solutions when $k$ is a positive number.

---

## The Cosecant Function

Important characteristics of the cosecant function can be inferred from the fact that it is the reciprocal of the sine function.

Whenever $\sin x = 1$, its reciprocal $\csc x$ is also 1. The graph of the cosecant function has asymptotes at the zeros of the sine function. The period of the cosecant function is $2\pi$, the same as its reciprocal, the sine function.

The graph of $y = \csc x$ is shown with the graph of $y = \sin x$ in Figure 4.58. A local maximum of $y = \sin x$ corresponds to a local minimum of $y = \csc x$, while a local minimum of $y = \sin x$ corresponds to a local maximum of $y = \csc x$.

**Example 3** SOLVING A TRIGONOMETRIC EQUATION GRAPHICALLY

Find the smallest positive number $x$ such that $x^2 = \csc x$.

**Solution** There is no algebraic attack that looks hopeful, so we solve this equation graphically. The intersection point of the graphs of $y = x^2$ and $y = \csc x$ that has the smallest positive $x$-coordinate is shown in Figure 4.59. We use the grapher to determine that $x \approx 1.068$.

## Are Cosecant Curves Parabolas?

Figure 4.59 shows a parabola intersecting one of the infinite number of U-shaped curves that make up the graph of the cosecant function. In fact, the parabola intersects *all* of those curves that lie above the *x*-axis, since the parabola must spread out to cover the entire domain of $y = x^2$, which is all real numbers! The cosecant curves do not keep spreading out, as they are hemmed in by asymptotes. That means that the U-shaped curves in the cosecant function are not parabolas.

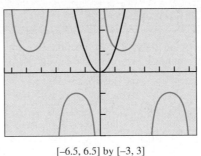

[–6.5, 6.5] by [–3, 3]

**Figure 4.59** A graphical solution of a trigonometric equation. (Example 3)

To close this section, we summarize the properties of the six basic trigonometric functions in tabular form. The "*n*" that appears in several places should be understood as taking on all possible integer values: $0, \pm 1, \pm 2, \pm 3, \ldots$.

| Function | Period | Domain | Range | Asymptotes | Zeros | Even/Odd |
|---|---|---|---|---|---|---|
| $\sin x$ | $2\pi$ | All reals | $[-1, 1]$ | None | $n\pi$ | Odd |
| $\cos x$ | $2\pi$ | All reals | $[-1, 1]$ | None | $\pi/2 + n\pi$ | Even |
| $\tan x$ | $\pi$ | $x \neq \pi/2 + n\pi$ | All reals | $x = \pi/2 + n\pi$ | $n\pi$ | Odd |
| $\cot x$ | $\pi$ | $x \neq n\pi$ | All reals | $x = n\pi$ | $\pi/2 + n\pi$ | Odd |
| $\sec x$ | $2\pi$ | $x \neq \pi/2 + n\pi$ | $(-\infty, -1] \cup [1, \infty)$ | $x = \pi/2 + n\pi$ | None | Even |
| $\csc x$ | $2\pi$ | $x \neq n\pi$ | $(-\infty, -1] \cup [1, \infty)$ | $x = n\pi$ | None | Odd |

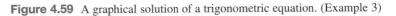

# Quick Review 4.5

In Exercises 1–4, state the period of the function.

**1.** $y = \cos 2x$   $\pi$

**2.** $y = \sin 3x$   $2\pi/3$

**3.** $y = \sin \dfrac{1}{3}x$   $6\pi$

**4.** $y = \cos \dfrac{1}{2}x$   $4\pi$

In Exercises 5–8, find the zeros and vertical asymptotes of the function.

**5.** $y = \dfrac{x - 3}{x + 4}$

**6.** $y = \dfrac{x + 5}{x - 1}$

**7.** $y = \dfrac{x + 1}{(x - 2)(x + 2)}$

**8.** $y = \dfrac{x + 2}{x(x - 3)}$

In Exercises 9 and 10, tell whether the function is odd, even, or neither.

**9.** $y = x^2 + 4$   even

**10.** $y = \dfrac{1}{x}$   odd

## Notes on Exercises

Ex. 31–36 can also be solved numerically or algebraically. The algebraic approach will be covered in section 5.1. Help students make the connection between the different solution methods. Ex. 56 is an advanced regression curve fitting problem.

## Follow-up

Ask what the relationship is between the domain of the tangent and secant functions and the zeros of the cosine function. (**The domain of the tangent and secant functions is the set of all real numbers that are not zeros of the cosine function.**)

## Assignment Guide

Day 1: Ex. 3-36, multiples of 3
Day 2: Ex. 37, 38, 40, 42, 45, 46, 54, 55, 56

**Cooperative Learning**

Group Activity: Ex. 53

# Section 4.5 Exercises

In Exercises 1–4, identify the graph of each function. Use your understanding of transformations, not your graphing calculator.

**1.** Graphs of one period of csc $x$ and 2 csc $x$ are shown.

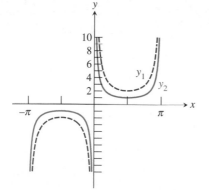

**2.** Graphs of two periods of 0.5 tan $x$ and 5 tan $x$ are shown.

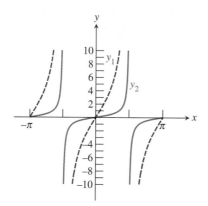

**3.** Graphs of csc $x$ and 3 csc $2x$ are shown.

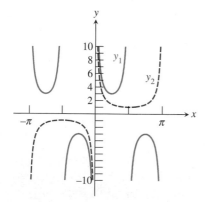

**4.** Graphs of cot $x$ and cot $(x - 0.5) + 3$ are shown.

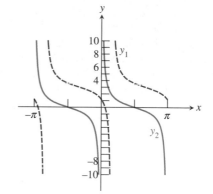

In Exercises 5–12, graph two periods of the function. Use your understanding of transformations, not your graphing calculator. Be sure to show the scale on both axes.

**5.** $y = \tan 2x$        **6.** $y = -\cot 3x$

**7.** $y = \sec 3x$        **8.** $y = \csc 2x$

**9.** $y = 2 \cot 2x$        **10.** $y = 3 \tan (x/2)$

**11.** $y = \csc (x/2)$        **12.** $y = 3 \sec 4x$

In Exercises 13–16, match the trigonometric function with its graph. Then give the Xmin and Xmax values for the viewing window in which the graph is shown. Use your understanding of transformations, not your graphing calculator.

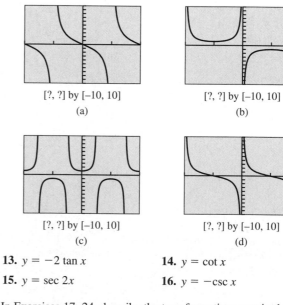

[?, ?] by [−10, 10]
(a)

[?, ?] by [−10, 10]
(b)

[?, ?] by [−10, 10]
(c)

[?, ?] by [−10, 10]
(d)

**13.** $y = -2 \tan x$        **14.** $y = \cot x$

**15.** $y = \sec 2x$        **16.** $y = -\csc x$

In Exercises 17–24, describe the transformations required to obtain the graph of the given function from a basic trigonometric graph.

**17.** $y = 3 \tan x$        **18.** $y = -\tan x$

**19.** $y = 3 \csc x$        **20.** $y = 2 \tan x$

**21.** $y = -3 \cot \frac{1}{2}x$

**22.** $y = -2 \sec \frac{1}{2}x$

**23.** $y = -\tan \frac{\pi}{2}x + 2$

**24.** $y = 2 \tan \pi x - 2$

In Exercises 25–30, solve for $x$ in the given interval. You should be able to find these numbers without a calculator, using reference triangles in the proper quadrants.

**25.** $\sec x = 2$,  $0 \le x \le \pi/2$  $\pi/3$

**26.** $\csc x = 2$,  $\pi/2 \le x \le \pi$  $5\pi/6$

**27.** $\cot x = -\sqrt{3}$,  $\pi/2 \le x \le \pi$  $5\pi/6$

**28.** $\sec x = -\sqrt{2}$,  $\pi \le x \le 3\pi/2$  $5\pi/4$

**29.** $\csc x = 1$,  $2\pi \le x \le 5\pi/2$  $5\pi/2$

**30.** $\cot x = 1$,  $-\pi \le x \le -\pi/2$  $-3\pi/4$

In Exercises 31–36, use a calculator to solve for $x$ in the given interval.

**31.** $\tan x = 1.3$,  $0 \le x < \dfrac{\pi}{2}$  $x \approx 0.92$

**32.** $\sec x = 2.4$,  $0 \le x \le \dfrac{\pi}{2}$  $x \approx 1.14$

**33.** $\cot x = -0.6$,  $\dfrac{3\pi}{2} \le x \le 2\pi$  $x \approx 5.25$

**34.** $\csc x = -1.5$,  $\pi \le x \le \dfrac{3\pi}{2}$  $x \approx 3.87$

**35.** $\csc x = 2$,  $0 \le x \le 2\pi$  $x \approx 0.52$ or $x \approx 2.62$

**36.** $\tan x = 0.3$,  $0 \le x \le 2\pi$  $x \approx 0.29$ or $x \approx 3.43$

**37. Writing to Learn** The figure shows a unit circle and an angle $t$ whose terminal side is in quadrant III.

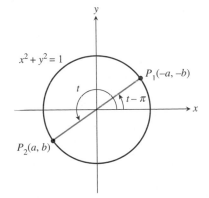

**(a)** If the coordinates of point $P_2$ are $(a, b)$, explain why the coordinates of point $P_1$ on the circle and the terminal side of angle $t - \pi$ are $(-a, -b)$.

**(b)** Explain why $\tan t = \dfrac{b}{a}$.

**(c)** Find $\tan(t - \pi)$, and show that $\tan t = \tan(t - \pi)$.

**(d)** Explain why the period of the tangent function is $\pi$.

**(e)** Explain why the period of the cotangent function is $\pi$.

**38. Writing to Learn** Explain why it is correct to say $y = \tan x$ is the slope of the terminal side of angle $x$ in standard position. $P$ is on the unit circle.

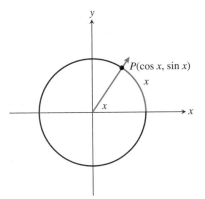

**39. Periodic Functions** Let $f$ be a periodic function with period $p$. That is, $p$ is the smallest positive number such that

$$f(x + p) = f(x)$$

for any value of $x$ in the domain of $f$. Show that the reciprocal $1/f$ is periodic with period $p$.

**40. Identities** Use the unit circle to give a convincing argument for the identity.

**(a)** $\sin(t + \pi) = -\sin t$

**(b)** $\cos(t + \pi) = -\cos t$

**(c)** Use (a) and (b) to show that $\tan(t + \pi) = \tan t$. Explain why this is *not* enough to conclude that the period of tangent is $\pi$.

**41. Lighthouse Coverage** The Bolivar Lighthouse is located on a small island 350 ft from the shore of the mainland as shown in the figure.

**(a)** Express the distance $d$ as a function of the angle $x$.

**(b)** If $x$ is 1.55 rad, what is $d$?  $\approx 16{,}831$ ft

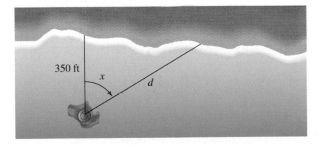

**42. Hot-Air Balloon** A hot-air balloon over Albuquerque, New Mexico, is being blown due east from point $P$ and traveling at a constant height of 800 ft. The angle $y$ is formed by the ground and the line of vision from $P$ to the balloon. This angle changes as the balloon travels.

**(a)** Express the horizontal distance $x$ as a function of the angle $y$. $x = 800 \cot y$

**(b)** When the angle is $\pi/20$ rad, what is its horizontal distance from $P$? $\approx 5051$ ft

**(c)** An angle of $\pi/20$ rad is equivalent to how many degrees?

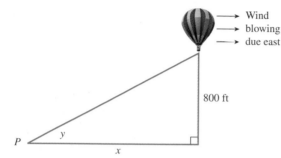

In Exercises 43–46, find approximate solutions for the equation in the interval $-\pi < x < \pi$.

**43.** $\tan x = \csc x$ $\approx \pm 0.905$ **44.** $\sec x = \cot x$

**45.** $\sec x = 5 \cos x$ **46.** $4 \cos x = \tan x$

# Explorations

In Exercises 47 and 48, graph both $f$ and $g$ in the $[-\pi, \pi]$ by $[-10, 10]$ viewing window. Estimate values in the interval $[-\pi, \pi]$ for which $f > g$.

**47.** $f(x) = 5 \sin x$ and $g(x) = \cot x$

**48.** $f(x) = -\tan x$ and $g(x) = \csc x$

**49. Writing to Learn** Graph the function $f(x) = -\cot x$ on the interval $(-\pi, \pi)$. Explain why it is correct to say that $f$ is increasing on the interval $(0, \pi)$, but it is not correct to say that $f$ is increasing on the interval $(-\pi, \pi)$.

**50. Writing to Learn** Graph functions $f(x) = -\sec x$ and

$$g(x) = \frac{1}{x - (\pi/2)}$$

simultaneously in the viewing window $[0, \pi]$ by $[-10, 10]$. Discuss whether you think functions $f$ and $g$ are equivalent.

**51.** Write $\csc x$ as a horizontal translation of $\sec x$.

**52.** Write $\cot x$ as the reflection about the $x$-axis of a horizontal translation of $\tan x$. $\cot x = -\tan (x - \pi/2)$ ∎

# Extending the Ideas

**53. Group Activity Television Coverage** A television camera is on a platform 30 m from the point on High Street where the Worthington Memorial Day Parade will pass. Express the distance $d$ from the camera to a particular parade float as a function of the angle $x$, and graph the function over the interval $-\pi/2 < x < \pi/2$.

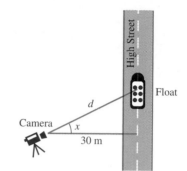

**54. What's In a Name?** The word *sine* comes from the Latin word *sinus*, which means "bay" or "cove." It entered the language through a mistake (variously attributed to Gerardo of Cremona or Robert of Chester) in translating the Arabic word "jiba" (chord) as if it were "jaib" (bay). This was due to the fact that the Arabs abbreviated their technical terms, much as we do today. Imagine someone unfamiliar with the technical term "cosecant" trying to reconstruct the English word that is abbreviated by "csc." It might well enter their language as their word for "cascade."

The names for the other trigonometric functions can all be explained.

**(a)** *Cosine* means "*sine* of the *co*mplement." Explain why this is a logical name for cosine.

**(b)** In the figure on page 387, $BC$ is perpendicular to $OC$, which is a radius of the unit circle. By a familiar geometry theorem, $BC$ is tangent to the circle. $OB$ is part of a secant that intersects the unit circle at $A$. It lies along the terminal side of an angle of $t$ radians in standard position. Write the coordinates of $A$ as functions of $t$. $(\cos t, \sin t)$

**(c)** Use similar triangles to find length $BC$ as a trig function of $t$. $\tan t$

**(d)** Use similar triangles to find length $OB$ as a trig function of $t$. $\sec t$

**(e)** Use the results from (a), (c), and (d) to explain where the names "tangent, cotangent, secant," and "cosecant" came from.

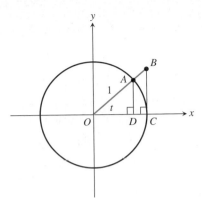

**55. Capillary Action**  A film of liquid in a thin (capillary) tube has surface tension $\gamma$ (gamma) given by

$$\gamma = \frac{1}{2} h\rho g r \sec \phi,$$

where $h$ is the height of the liquid in the tube, $\rho$ (rho) is the density of the liquid, $g = 9.8$ m/sec$^2$ is the acceleration due to gravity, $r$ is the radius of the tube, and $\phi$ (phi) is the angle of contact between the tube and the liquid's surface. Whole blood has a surface tension of 0.058 N/m (newton per meter) and a density of 1050 kg/m$^3$. Suppose that blood rises to a height of 1.5 m in a capillary blood vessel of radius $4.7 \times 10^{-6}$ m. What is the contact angle between the capillary vessel and the blood surface? ($1$ N $= 1$ (kg $\cdot$ m)/sec$^2$) $\approx 0.8952$ radians $\approx 51.29°$

**56. Advanced Curve Fitting**  A researcher has reason to believe that the data in the table below can best be described by an algebraic model involving the secant function:

$$y = a \sec (bx).$$

Unfortunately, her calculator will only do sine regression. She realizes that the following two facts will help her:

$$\frac{1}{y} = \frac{1}{a \sec (bx)} = \frac{1}{a}\cos (bx)$$

and

$$\cos (bx) = \sin\left(bx + \frac{\pi}{2}\right).$$

**(a)** Use these two facts to show that

$$\frac{1}{y} = \frac{1}{a} \sin\left(bx + \frac{\pi}{2}\right).$$

**(b)** Store the $x$ values in the table in L1 in your calculator and the $y$ values in L2. Store the *reciprocals* of the $y$ values in L3. Then do a sine regression for L3 ($1/y$) as a function of L1 ($x$). Write the regression equation.

**(c)** Use the regression equation in (b) to determine the values of $a$ and $b$.

**(d)** Write the secant model: $y = a \sec (bx)$. Does the curve fit the (L1, L2) scatter plot?

| $x$ | 1 | 2 | 3 | 4 |
|---|---|---|---|---|
| $y$ | 5.0703 | 5.2912 | 5.6975 | 6.3622 |

| $x$ | 5 | 6 | 7 | 8 |
|---|---|---|---|---|
| $y$ | 7.4359 | 9.2541 | 12.716 | 21.255 |

**4.6**

# Graphs of Composite Trigonometric Functions

**Objective**

Students will be able to graph sums, differences, and other combinations of trigonometric and algebraic functions.

**Motivate**

Have students use a grapher to graph $f(x)$, $g(x)$, and the sum $(f + g)(x)$, where

$f(x) = x^2$ and $g(x) = \dfrac{x^2}{2} - 3x$.

Point out that the sum of two quadratic functions is another quadratic function (or a line or a constant) and ask whether students think it is possible for the sum of two sinusoids to be a sinusoid. **(Yes)**

Sums and Differences that are Sinusoids  •  Sums and Differences that are Not Sinusoids  •  Combining Trigonometric and Algebraic Functions  •  Damped Oscillation

## Sums and Differences That Are Sinusoids

Section 4.4 introduced you to sinusoids, functions that can be written in the form

$$y = a \sin (b(x - h)) + k$$

and therefore have the shape of a sine curve.

Sinusoids model a variety of physical and social phenomena—such as sound waves, voltage in alternating electrical current, the velocity of air flow during the human respiratory cycle, and many others. Sometimes these phenomena interact in an additive fashion. For example, if $y_1$ models the sound of one tuning fork and $y_2$ models the sound of a second tuning fork, then $y_1 + y_2$

**Lesson Guide**

Day 1: Skills
Day 2: Applications

**Exploration Extensions**

Repeat the exercise for the functions
$y = \sin x + \sin 2x$ and
$y = 2 \sin x - 2 \cos (x - \pi/2)$.

models the sound when they are both struck simultaneously. So we are interested in whether the sums and differences of sinusoids are again sinusoids.

---

**Exploration 1** | **Investigating Sinusoids**

Graph these functions, one at a time, in the viewing window $[-2\pi, 2\pi]$ by $[-6, 6]$. Which ones appear to be sinusoids?

$$y = 3 \sin x + 2 \cos x \qquad\qquad y = 2 \sin x - 3 \cos x$$

$$y = 2 \sin 3x - 4 \cos 2x \qquad\qquad y = 2 \sin (5x + 1) - 5 \cos 5x$$

$$y = \cos \left( \frac{7x - 2}{5} \right) + \sin \left( \frac{7x}{5} \right) \qquad y = 3 \cos 2x + 2 \sin 7x$$

What relationship between the sine and cosine functions ensures that their sum or difference will again be a sinusoid? Check your guess on a graphing calculator by constructing your own examples.

---

The rule turns out to be fairly simple: Sums and differences of sinusoids with the same period are again sinusoids. We state this rule more explicitly as follows.

---

**Sums That Are Sinusoid Functions**

If $y_1 = a_1 \sin (b(x - h_1))$ and $y_2 = a_2 \cos (b(x - h_2))$, then

$$y_1 + y_2 = a_1 \sin (b(x - h_1)) + a_2 \cos (b(x - h_2))$$

is a sinusoid with period $2\pi/|b|$.

---

For the sum to be a sinusoid, the two sinusoids being added together must have the same period, and the sum has the same period as both of them. Also, although the rule is stated in terms of a sine function being added to a cosine function, the fact that every cosine function is a translation of a sine function (and vice-versa) makes the rule equally applicable to the sum of two sine functions or the sum of two cosine functions. If they have the same period, their sum is a sinusoid.

### Example 1 EXPRESSING THE SUM OF SINUSOIDS AS A SINUSOID

Let $f(x) = 2 \sin x + 5 \cos x$. From the discussion above, you should conclude that $f(x)$ is a sinusoid.

**(a)** Find the period of $f$.

**(b)** Find the amplitude and phase shift (to the nearest hundredth).

**(c)** Support your answer graphically.

**Solution**

**(a)** The period of $f$ is the same as the period of $\sin x$ and $\cos x$, namely $2\pi$.

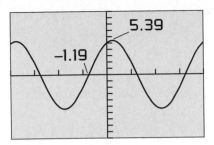

[−2π, 2π] by [−10, 10]

**Figure 4.60** The sum of two sinusoids: $f(x) = 2 \sin x + 5 \cos x$. (Example 1)

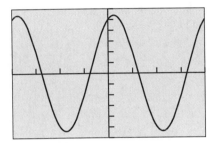

[−2π, 2π] by [−6, 6]

**Figure 4.61** The graphs of $y = 2 \sin x + 5 \cos x$ and $y = 5.39 \sin (x + 1.19)$ appear to be identical. (Example 1)

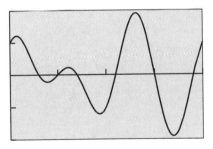

[0, 2π] by [−2, 2]

**Figure 4.62** One period of $f(x) = \sin 2x + \cos 3x$. (Example 2)

---

**Notes on Examples**

Example 2 shows how to find the period of a sum of sinusoids when the sum is not a sinusoid.

**Note**

Recall that $\sin^3 x$ means $(\sin x)^3$

---

**Solve Graphically**

**(b)** We will learn an algebraic way to find the amplitude and phase shift in the next chapter, but for now we will find this information graphically. Figure 4.60 suggests that $f$ is indeed a sinusoid. That is, for some $a$ and $b$,

$$2 \sin x + 5 \cos x = a \sin (x - h).$$

The *maximum value*, rounded to the nearest hundredth, is 5.39, so the amplitude of $f$ is about 5.39. The *x-intercept* closest to $x = 0$, rounded to the nearest hundredth, is −1.19, so the phase shift of the sine function is about −1.19. We conclude that

$$f(x) = a \sin (x + h) \approx 5.39 \sin (x + 1.19).$$

**(c)** We support our answer graphically by showing that the graphs of $y = 2 \sin x + 5 \cos x$ and $y = 5.39 \sin (x + 1.19)$ are virtually identical (Figure 4.61).

## Sums and Differences That Are Not Sinusoids

Example 2 illustrates that if the periods of the individual terms of a sum or difference of sinusoids are unequal, then the sum will not be a sinusoid.

Finding the period of a sum of periodic functions can be tricky. Here is a useful fact to keep in mind. If $f$ is periodic, and if $f(x + s) = f(x)$ for all $x$ in the domain of $f$, then the period of $f$ divides $s$ exactly. In other words, $s$ is either the period or a multiple of the period.

**Example 2 SHOWING A FUNCTION IS PERIODIC BUT NOT A SINUSOID**

Show that $f(x) = \sin 2x + \cos 3x$ is periodic but not a sinusoid. Graph one period.

**Solution** First we show that $2\pi$ is a candidate for the period of $f$, that is, $f(x + 2\pi) = f(x)$ for all $x$.

$$f(x + 2\pi) = \sin (2(x + 2\pi)) + \cos (3(x + 2\pi))$$
$$= \sin (2x + 4\pi) + \cos (3x + 6\pi)$$
$$= \sin 2x + \cos 3x$$
$$= f(x)$$

This means either that $2\pi$ is the period of $f$ or that the period is an exact divisor of $2\pi$. Figure 4.62 suggests that the period is not smaller than $2\pi$, so it must be $2\pi$.

Notice that indeed $f$ is not a sinusoid.

## Combining Trigonometric and Algebraic Functions

A theme of this text has been "families of functions." We have studied polynomial functions, exponential functions, logarithmic functions, and rational functions (to name a few), and in this chapter we have studied trigonometric functions. Now we consider adding, multiplying, or composing trigonometric functions with functions from these other families.

The notable property that distinguishes trigonometric functions from all the others is periodicity. Example 3 shows that when a trigonometric function

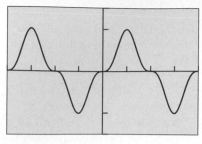

[$-2\pi$, $2\pi$] by [$-1.5$, $1.5$]

**Figure 4.63** The graph of $f(x) = \sin^3 x$. (Example 3)

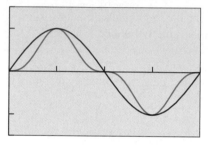

[$0$, $2\pi$] by [$-1.5$, $1.5$]

**Figure 4.64** The graph suggests that $|\sin^3 x| \le |\sin x|$. (Example 3)

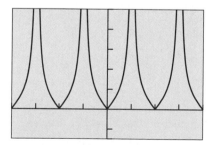

[$-2\pi$, $2\pi$] by [$-1.5$, $5$]

**Figure 4.65** $f(x) = |\tan x|$ has the same period as $y = \tan x$. (Example 4a)

---

**Follow-up**

Ask students whether the sum of two periodic functions is always periodic. **(No, not if the ratio of the periods is irrational.)**

---

is composed with a polynomial, the resulting function may continue to be periodic.

### Example 3   COMPOSING $y = \sin x$ AND $y = x^3$

Graph $f(x) = \sin^3 x$ and prove that it is a periodic function. Then prove that $|\sin^3 x| \le |\sin x|$.

**Solution**

**Solve Graphically**

The graph of $f(x) = \sin^3 x$ in Figure 4.63 suggests that $f$ is periodic with a period no smaller than $2\pi$. To confirm that $f$ is periodic we show that $f(x + 2\pi) = f(x)$ for all $x$.

$$f(x + 2\pi) = \sin^3(x + 2\pi)$$
$$= [\sin (x + 2\pi)]^3$$
$$= (\sin x)^3$$
$$= \sin^3 x = f(x)$$

Figure 4.64 suggests that the graph of $f(x) = \sin^3 x$ is never farther from the $x$-axis than is the graph of $y = \sin x$, or that $|\sin^3 x| \le |\sin x|$.

**Confirm Algebraically**

| | |
|---|---|
| $|\sin x| \le 1$ | Range of sin $x$ is $[-1, 1]$. |
| $|\sin x|^2 \le |\sin x|$ | Multiply by $|\sin x|$. |
| $|\sin x|^3 \le |\sin x|^2$ | Multiply by $|\sin x|$. |
| $|\sin x|^3 \le |\sin x|$ | Transitive property of $\le$. |

The absolute value of a periodic function is also a periodic function. Example 4 illustrates this for two of the basic trigonometric functions.

### Example 4   CREATING NON-NEGATIVE PERIODIC FUNCTIONS

Find the domain, range, and period of each of the following functions:

**(a)** $f(x) = |\tan x|$

**(b)** $g(x) = |\sin x|$

**Solution**

**(a)** Whenever tan $x$ is defined, so is $|\tan x|$. Therefore, the domain of $f$ is the same as the domain of the tangent function, that is, all real numbers except $\pi/2 + n\pi$, $n = 0, \pm 1, \ldots$. Because $f(x) = |\tan x| \ge 0$ and the range of tan $x$ is $(-\infty, \infty)$, the range of $f$ is $[0, \infty)$. The period of $f$, like that of $y = \tan x$, is $\pi$. The graph of $y = f(x)$ is shown in Figure 4.65.

**(b)** Whenever sin $x$ is defined, so is $|\sin x|$. Therefore, the domain of $g$ is the same as the domain of the sine function, that is, all real numbers. Because $g(x) = |\sin x| \ge 0$ and the range of sin $x$ is $[-1, 1]$ the range of $g$ is $[0, 1]$.

The period of $g$ is only half the period of $y = \sin x$, for reasons that are apparent from viewing the graph. The negative sections of the sine curve below the $x$-axis are reflected above the $x$-axis, where they become

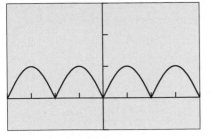

$[-2\pi, 2\pi]$ by $[-1, 3]$

**Figure 4.66**  $g(x) = |\sin x|$ has half the period of $y = \sin x$. (Example 4b)

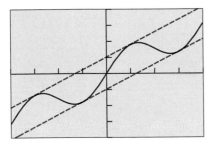

$[-2\pi, 2\pi]$ by $[-4, 4]$

**Figure 4.67**  The graph of $f(x) = 0.5x + \sin x$ oscillates between the lines $y = 0.5x + 1$ and $y = 0.5x - 1$. Although the wave repeats its shape, it is not periodic. (Example 5)

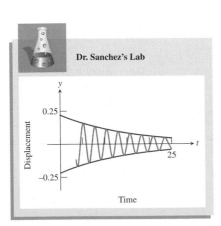

**Figure 4.68**  Damped oscillation in the physics lab. (Example 6)

repetitions of the positive sections. The graph of $y = g(x)$ is shown in Figure 4.66.

### Example 5  ADDING A SINUSOID TO A LINEAR FUNCTION

Graph $f(x) = 0.5x + \sin x$ over the interval $[-2\pi, 2\pi]$ and state its domain and range. Is this function periodic?

**Solution**  The graph is shown in Figure 4.67. It oscillates between the lines $y = 0.5x + 1$ and $y = 0.5x - 1$, taking on an infinite number of local maximum values and an infinite number of local minimum values, each in an increasing pattern. Although the increasing wave repeats its *shape* at regular intervals, the function is not periodic because it does not repeat the same function values. The domain and range are both $(-\infty, \infty)$.

## Damped Oscillation

An oscillating spring under the influence of friction and the amount of electrical charge on a capacitor over time are two examples of quantities that can be modeled by an equation that combines an exponential function and a trigonometric function as a product:

$$y = Ae^{-at} \cos bt.$$

The factor $Ae^{-at}$ is called the **damping factor** because the two functions $y = Ae^{-at}$ and $y = -Ae^{-at}$ provide upper and lower boundaries for the oscillating quantity.

### Example 6  MODELING AN OSCILLATING SPRING

Dr. Sanchez's physics class collected data for an air table glider that oscillates between two springs. The class determined from the data that the equation

$$y = 0.22e^{-0.065t} \cos 2.4t$$

modeled the displacement $y$ of the spring from its original position as a function of time $t$.

**(a) Show algebraically** and **support graphically** that the inequality $-0.22e^{-0.065t} \le y \le 0.22e^{-0.065t}$ is valid for $0 \le t \le 25$.

**(b)** Approximately how long does it take for the spring to be damped so that $-0.1 \le y \le 0.1$?

**Solution**

**(a)** The cosine function oscillates between $-1$ and $1$. Therefore $-1 \le \cos 2.4t \le 1$. If we multiply each expression in this inequality by the (positive) damping factor, we get $-0.22e^{-0.065t} \le 0.22e^{-0.065t} \cos 2.4t \le 0.22e^{-0.065t}$, which, for the students' model, is $-0.22e^{-0.065t} \le y \le 0.22e^{-0.065t}$.

Figure 4.68 supports the inequality graphically, showing the graph of $y = 0.22e^{-0.065t} \cos 2.4t$ oscillating back and forth between the graphs of $y = 0.22e^{-0.065t}$ and $y = -0.22e^{-0.065t}$.

**(b)** We want to find how soon the curve $y = 0.22e^{-0.065t} \cos 2.4t$ falls entirely between the lines $y = -0.1$ and $y = 0.1$. By zooming in on the region indi-

### Assignment Guide

Day 1: Ex. 3–57, multiples of 3
Day 2: Ex. 61, 62, 63, 66, 67, 69, 71, 76, 78

### Notes on Exercises

Ex. 31–34 are similar to Example 5. You may wish to specify that $y_1$ and $y_2$ should actually be tangent to the graph of $y$.

### Cooperative Learning

Group Activity: Ex. 71, 72

### Ongoing Assessment

Self-Assessment: Ex. 1, 9, 15, 20, 27, 35, 39, 45, 51, 55
Embedded Assessment: Ex. 63, 66

cated in Figure 4.69a and using grapher methods, we find that it it takes approximately 11.86 seconds until the graph of $y = 0.22e^{-0.065t} \cos 2.4t$ lies entirely between $y = -0.1$ and $y = 0.1$ (Figure 4.69b).

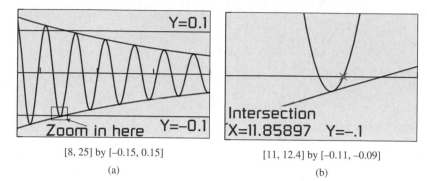

[8, 25] by [–0.15, 0.15]
(a)

[11, 12.4] by [–0.11, –0.09]
(b)

**Figure 4.69** The damped oscillation in Example 6 eventually gets to be less than 0.1 in either direction.

## Quick Review 4.6

In Exercises 1–6, state the domain and range of the function.

**1.** $f(x) = 3 \sin 2x$
**2.** $f(x) = -2 \cos 3x$
**3.** $f(x) = \sqrt{x - 1}$
**4.** $f(x) = \sqrt{x}$
**5.** $f(x) = |x| - 2$
**6.** $f(x) = |x + 2| + 1$

In Exercises 7 and 8, describe the end behavior of the function, that is, the behavior as $|x| \to \infty$.

**7.** $f(x) = 5e^{-2x}$
**8.** $f(x) = -0.2(5^{-0.1x})$

In Exercises 9 and 10, form the compositions $f \circ g$ and $g \circ f$. State the domain of each function.

**9.** $f(x) = x^2 - 4$ and $g(x) = \sqrt{x}$
**10.** $f(x) = x^2$ and $g(x) = \cos x$

## Section 4.6 Exercises

In Exercises 1–6, determine whether $f(x)$ is a sinusoid.

**1.** $f(x) = \sin x - 3 \cos x$  yes
**2.** $f(x) = 4 \cos x + 2 \sin x$
**3.** $f(x) = 2 \cos \pi x + \sin \pi x$
**4.** $f(x) = 2 \sin x - \tan x$  no
**5.** $f(x) = 3 \sin 2x - 5 \cos x$
**6.** $f(x) = \pi \sin 3x - 4\pi \sin 2x$

In Exercises 7–12, find $a$, $b$, and $h$ so that $f(x) \approx a \sin (b(x - h))$.

**7.** $f(x) = 2 \sin 2x - 3 \cos 2x$
**8.** $f(x) = \cos 3x + 2 \sin 3x$
**9.** $f(x) = \sin \pi x - 2 \cos \pi x$
**10.** $f(x) = \cos 2\pi x + 3 \sin 2\pi x$
**11.** $f(x) = 2 \cos x + \sin x$
**12.** $f(x) = 3 \sin 2x - \cos 2x$

In Exercises 13–16, graph one period of the function.

**13.** $y = 2 \cos x + \cos 3x$
**14.** $y = 2 \sin 2x + \cos 3x$
**15.** $y = \cos 3x - 4 \sin 2x$
**16.** $y = \sin 2x + \sin 5x$

In Exercises 17–20, match the function with its graph.

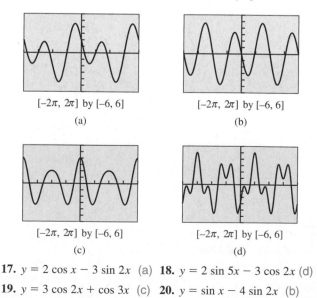

[–2π, 2π] by [–6, 6]
(a)

[–2π, 2π] by [–6, 6]
(b)

[–2π, 2π] by [–6, 6]
(c)

[–2π, 2π] by [–6, 6]
(d)

**17.** $y = 2 \cos x - 3 \sin 2x$  (a)
**18.** $y = 2 \sin 5x - 3 \cos 2x$ (d)
**19.** $y = 3 \cos 2x + \cos 3x$  (c)
**20.** $y = \sin x - 4 \sin 2x$ (b)

In Exercises 21–24, graph two periods of $f$ and compare it to the graph of $y = \cos x$. Show that $f$ is a periodic function and state its period.

**21.** $f(x) = \cos^2 x$        **22.** $f(x) = \cos^3 x$

**23.** $f(x) = \sqrt{\cos^2 x}$        **24.** $f(x) = |\cos^3 x|$

In Exercises 25–30, state the domain and range of the function and graph four periods.

**25.** $y = \cos^2 x$        **26.** $y = |\cos x|$

**27.** $y = |\cot x|$        **28.** $y = \cos|x|$

**29.** $y = -\tan^2 x$        **30.** $y = -\sin^2 x$

In Exercises 31–34, the function is of the form $y = g(x) + \cos bx$. Find two linear functions $y_1$ and $y_2$ related to $g$ such that $y_1 \leq y \leq y_2$. Graph all three functions $y_1$, $y$, and $y_2$ in the same viewing window.

**31.** $y = 2x + \cos x$        **32.** $y = 1 - 0.5x + \cos 2x$

**33.** $y = 2 - 0.3x + \cos x$        **34.** $y = 1 + x + \cos 3x$

In Exercises 35–38, graph $f$ over the interval $[-4\pi, 4\pi]$. Describe how the graphs of $y = x$ and $y = -x$ are related to the graph of $f$.

**35.** $f(x) = x \sin x$        **36.** $f(x) = -x|\cos x|$

**37.** $f(x) = |x \sin x|$        **38.** $f(x) = x \cos |x|$

In Exercises 39–42, graph both $f$ and plus or minus its damping factor in the same viewing window. Describe the behavior of the function $f$ for $x > 0$. What is the end behavior of $f$?

**39.** $f(x) = 1.2^{-x} \cos 2x$        **40.** $f(x) = 2^{-x} \sin 4x$

**41.** $f(x) = x^{-1} \sin 3x$        **42.** $f(x) = e^{-x} \cos 3x$

In Exercises 43–46, find the period and graph the function over two periods.

**43.** $y = \sin 3x + 2 \cos 2x$   $2\pi$

**44.** $y = 4 \cos 2x - 2 \cos (3x - 1)$   $2\pi$

**45.** $y = 2 \sin (3x + 1) - \cos (5x - 1)$   $2\pi$

**46.** $y = 3 \cos (2x - 1) - 4 \sin (3x - 2)$   $2\pi$

In Exercises 47–52, graph $f$ over the interval $[-4\pi, 4\pi]$. Determine whether the function is periodic and, if it is, state the period.

**47.** $f(x) = \left| \sin \frac{1}{2}x \right| + 2$        **48.** $f(x) = 3x + 4 \sin 2x$

**49.** $f(x) = x - \cos x$        **50.** $f(x) = x + \sin 2x$

**51.** $f(x) = \frac{1}{2}x + \cos 2x$        **52.** $f(x) = 3 - x + \sin 3x$

In Exercises 53–60. Find the domain and range of the function.

**53.** $f(x) = 2x + \cos x$        **54.** $f(x) = 2 - x + \sin x$

**55.** $f(x) = |x| + \cos x$        **56.** $f(x) = -2x + |3 \sin x|$

**57.** $f(x) = \sqrt{\sin x}$        **58.** $f(x) = \sin |x|$

**59.** $f(x) = \sqrt{|\sin x|}$        **60.** $f(x) = \sqrt{\cos x}$

**61. Predicting Economic Growth** Stogner Auto Parts, Inc., had total sales of $75 million in 1995. Historically the industry has an annual growth of 4% subject to an adjustment due to its cyclical nature. The company's economists have found that the equation

$$y = 75(1.04^x) + 2 \sin \pi x$$

models the sales (in millions of dollars), where $x = 0$ corresponds to the year 1995.

**(a)** Graph this model to reflect the years from 1995 to 2010.

**(b)** How many years are there in the economic cycle?

**(c)** What does the model predict for sales in 2001?

**62. Oscillating Spring** The oscillations of a spring subject to friction are modeled by the equation $y = 0.43e^{-0.55t} \cos 1.8t$.

**(a)** Graph $y$ and its two damping curves in the same viewing window for $0 \leq t \leq 12$.

**(b)** Approximately how long does it take for the spring to be damped so that $-0.2 \leq y \leq 0.2$?

**63. Writing to Learn** Example 3 shows that the function $y = \sin^3 x$ is periodic. Explain whether you think that $y = \sin x^3$ is periodic and why.

**64. Writing to Learn** Example 4 shows that $y = |\tan x|$ is periodic. Write a convincing argument that $y = \tan |x|$ is not a periodic function.

In Exercises 65 and 66, select the one correct inequality, **(a)** or **(b)**. Give a convincing argument.

**65. (a)** $x - 1 \leq x + \sin x \leq x + 1$ for all $x$   (a)

     **(b)** $x - \sin x \leq x + \sin x$ for all $x$.

**66. (a)** $-x \leq x \sin x \leq x$ for all $x$

     **(b)** $-|x| \leq x \sin x \leq |x|$ for all $x$.   (b)

In Exercises 67–70, match the function with its graph. In each case state the viewing window.

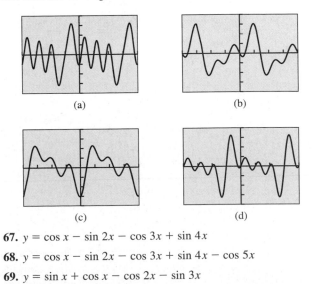

(a)            (b)

(c)            (d)

**67.** $y = \cos x - \sin 2x - \cos 3x + \sin 4x$

**68.** $y = \cos x - \sin 2x - \cos 3x + \sin 4x - \cos 5x$

**69.** $y = \sin x + \cos x - \cos 2x - \sin 3x$

**70.** $y = \sin x - \cos x - \cos 2x - \cos 3x$

## Explorations

**71. Group Activity  Inaccurate or Misleading Graphs**

**(a)** Set $X$min $= 0$ and $X$max $= 2\pi$. Move the cursor along the *x*-axis. What is the distance between one pixel and the next (to the nearest hundredth)?

**(b)** What is the period of $f(x) = \sin 250x$? Consider that the period is the length of one full cycle of the graph. Approximately how many cycles should there be between two adjacent pixels? Can your grapher produce an accurate graph of this function between 0 and $2\pi$?

**72. Group Activity  Length of Days**  The graph shows the number of hours of daylight in Boston as a function of the day of the year, from September 21, 1983, to December 15, 1984. Key points are labeled and other critical information is provided. Write a formula for the sinusoidal function and check it by graphing.

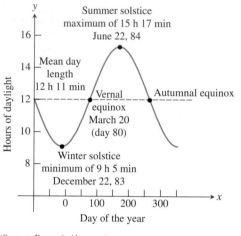

*(Source: Farmer's Almanac.)*

## Extending the Ideas

In Exercises 73–80, first try to predict what the graph will look like (without too much effort, that is, just for fun). Then graph the function in one or more viewing windows to determine the main features of the graph, and draw a summary sketch. Where applicable, name the period, amplitude, domain, range, asymptotes, and zeros.

**73.** $f(x) = \cos e^x$

**74.** $g(x) = e^{\tan x}$

**75.** $f(x) = \sqrt{x} \sin x$

**76.** $g(x) = \sin \pi x + \sqrt{4 - x^2}$

**77.** $f(x) = \dfrac{\sin x}{x}$

**78.** $g(x) = \dfrac{\sin x}{x^2}$

**79.** $f(x) = x \sin \dfrac{1}{x}$

**80.** $g(x) = x^2 \sin \dfrac{1}{x}$

---

## 4.7  Inverse Trigonometric Functions

Inverse Sine Function • Inverse Cosine and Tangent Functions • Composing Trigonometric and Inverse Trigonometric Functions • Applications of Inverse Trigonometric Functions

### Inverse Sine Function

You learned in Section 1.4 that each function has an inverse relation, and that this inverse relation is a function only if the original function is one-to-one. The six basic trigonometric functions, being periodic, fail the horizontal line test for one-to-oneness rather spectacularly. However, you also learned in Section 1.4 that some functions are important enough that we want to study their inverse behavior despite the fact that they are not one-to-one. We do this by restricting the domain of the original function to an interval on which it *is* one-to-one, then finding the inverse of the restricted function. (We did this when defining the square root function, which is the inverse of the function $y = x^2$ restricted to a non-negative domain.)

If you restrict the domain of $y = \sin x$ to the interval $[-\pi/2, \pi/2]$, as shown in Figure 4.70a, the restricted function is one-to-one. The **inverse sine function** $y = \sin^{-1} x$ is the inverse of this restricted portion of the sine function (Figure 4.70b).

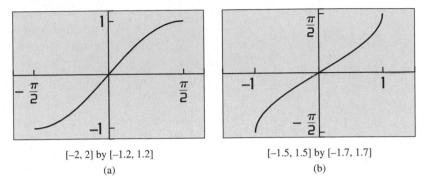

[−2, 2] by [−1.2, 1.2]

(a)

[−1.5, 1.5] by [−1.7, 1.7]

(b)

**Figure 4.70** The (a) restriction of $y = \sin x$ is one-to-one and (b) has an inverse, $y = \sin^{-1} x$.

By the usual inverse relationship, the statements

$$y = \sin^{-1} x \qquad \text{and} \qquad x = \sin y$$

are equivalent for $y$-values in the restricted domain $[-\pi/2, \pi/2]$ and $x$-values in $[-1, 1]$. This means that $\sin^{-1} x$ can be thought of as *the angle between $-\pi/2$ and $\pi/2$ whose sine is $x$*. Since angles and directed arcs on the unit circle have the same measure, the angle $\sin^{-1} x$ is also called the **arcsine of $x$**.

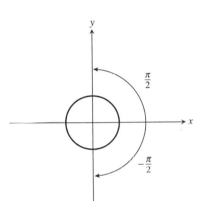

**Figure 4.71** The values of $y = \sin^{-1} x$ will always be found on the right-hand side of the unit circle, between $-\pi/2$ and $\pi/2$.

> **Inverse Sine Function (Arcsine Function)**
>
> The unique angle $y$ in the interval $[-\pi/2, \pi/2]$ such that $\sin y = x$ is the **inverse sine** (or **arcsine**) of $x$, denoted **$\sin^{-1} x$** or **$\arcsin x$**.
>
> The domain of $y = \sin^{-1} x$ is $[-1, 1]$ and the range is $[-\pi/2, \pi/2]$.

It helps to think of the range of $y = \sin^{-1} x$ as being along the right-hand side of the unit circle, which is traced out as angles range from $-\pi/2$ to $\pi/2$ (Figure 4.71).

### Example 1  EVALUATING $\sin^{-1} x$ WITHOUT A CALCULATOR

Find the exact value of each expression without a calculator.

**(a)** $\sin^{-1}\left(\dfrac{1}{2}\right)$   **(b)** $\sin^{-1}\left(-\dfrac{\sqrt{3}}{2}\right)$   **(c)** $\sin^{-1}\left(\dfrac{\pi}{2}\right)$

**(d)** $\sin^{-1}\left(\sin\left(\dfrac{\pi}{9}\right)\right)$   **(e)** $\sin^{-1}\left(\sin\left(\dfrac{5\pi}{6}\right)\right)$

**Solution**

**(a)** Find the point on the right half of the unit circle whose $y$-coordinate is $1/2$ and draw a reference triangle (Figure 4.72). We recognize this as one of our special ratios, and the angle in the interval $[-\pi/2, \pi/2]$ whose sine is $1/2$ is $\pi/6$. Therefore

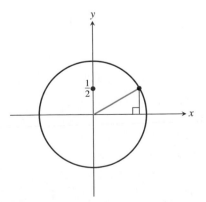

**Figure 4.72** $\sin^{-1}(1/2) = \pi/6$.
(Example 1a)

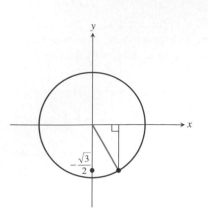

**Figure 4.73** $\sin^{-1}(-\sqrt{3}/2) = -\pi/3$. (Example 1b)

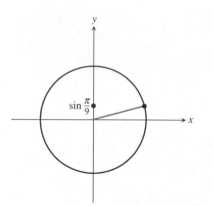

**Figure 4.74** $\sin^{-1}(\sin(\pi/9)) = \pi/9$. (Example 1d)

### What About the Inverse Composition Rule?

Does Example 1e violate the Inverse Composition Rule of Section 1.4? That rule guarantees that $f^{-1}(f(x)) = x$ for every $x$ in the domain of $f$. Keep in mind, however, that the domain of $f$ might need to be restricted in order for $f^{-1}$ to exist. That is certainly the case with the sine function. So Example 1e does not violate the Inverse Composition Rule, because that rule *does not apply* at $x = 5\pi/6$. It lies outside the (restricted) domain of sine.

$$\sin^{-1}\left(\frac{1}{2}\right) = \frac{\pi}{6}.$$

**(b)** Find the point on the right half of the unit circle whose $y$-coordinate is $-\sqrt{3}/2$ and draw a reference triangle (Figure 4.73). We recognize this as one of our special ratios, and the angle in the interval $[-\pi/2, \pi/2]$ whose sine is $-\sqrt{3}/2$ is $-\pi/3$. Therefore

$$\sin^{-1}\left(-\frac{\sqrt{3}}{2}\right) = -\frac{\pi}{3}.$$

**(c)** $\sin^{-1}(\pi/2)$ does not exist, because the domain of $\sin^{-1}$ is $[-1, 1]$ and $\pi/2 > 1$.

**(d)** Draw an angle of $\pi/9$ in standard position and mark its $y$-coordinate on the $y$-axis (Figure 4.74). The angle in the interval $[-\pi/2, \pi/2]$ whose sine is this number is $\pi/9$. Therefore

$$\sin^{-1}\left(\sin\left(\frac{\pi}{9}\right)\right) = \frac{\pi}{9}.$$

**(e)** Draw an angle of $5\pi/6$ in standard position (notice that this angle is *not* in the interval $[-\pi/2, \pi/2]$) and mark its $y$-coordinate on the $y$-axis (Figure 4.75). The angle in the interval $[-\pi/2, \pi/2]$ whose sine is this number is $\pi - 5\pi/6 = \pi/6$. Therefore

$$\sin^{-1}\left(\sin\left(\frac{5\pi}{6}\right)\right) = \frac{\pi}{6}.$$

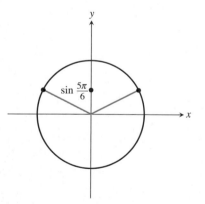

**Figure 4.75** $\sin^{-1}(\sin(5\pi/6)) = \pi/6$. (Example 1e)

### Example 2   EVALUATING sin⁻¹x WITH A CALCULATOR

Use a calculator in radian mode to evaluate these inverse sine values:

**(a)** $\sin^{-1}(-0.81)$

**(b)** $\sin^{-1}(\sin(3.49\pi))$

**Solution**

**(a)** $\sin^{-1}(-0.81) = -0.9441521\ldots \approx -0.944$

**(b)** $\sin^{-1}(\sin(3.49\pi)) = -1.5393804\ldots \approx -1.539$

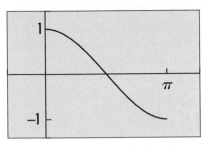

[-1, 4] by [-1.4, 1.4]

(a)

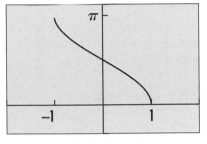

[-2, 2] by [-1, 3.5]

(b)

**Figure 4.76** The (a) restriction of $y = \cos x$ is one-to-one and (b) has an inverse, $y = \cos^{-1} x$.

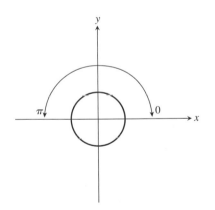

**Figure 4.77** The values of $y = \cos^{-1} x$ will always be found on the top half of the unit circle, between 0 and $\pi$.

Although this is a calculator answer, we can use it to get an exact answer if we are alert enough to expect a multiple of $\pi$. Divide the answer by $\pi$:

$$\text{Ans}/\pi = -0.49.$$

Therefore, we conclude that $\sin^{-1}(\sin(3.49\pi)) = -0.49\pi$.

You should also try to work Example 2 without a calculator. It is possible!

## Inverse Cosine and Tangent Functions

If you restrict the domain of $y = \cos x$ to the interval $[0, \pi]$, as shown in Figure 4.76a, the restricted function is one-to-one. The **inverse cosine function** $y = \cos^{-1} x$ is the inverse of this restricted portion of the cosine function (Figure 4.76b).

By the usual inverse relationship, the statements

$$y = \cos^{-1} x \quad \text{and} \quad x = \cos y$$

are equivalent for $y$-values in the restricted domain $[0, \pi]$ and $x$-values in $[-1, 1]$. This means that $\cos^{-1} x$ can be thought of as *the angle between* 0 *and* $\pi$ *whose cosine is* $x$. The angle $\cos^{-1} x$ is also the **arccosine of** $x$.

> **Inverse Cosine Function (Arccosine Function)**
>
> The unique angle $y$ in the interval $[0, \pi]$ such that $\cos y = x$ is the **inverse cosine** (or **arccosine**) of $x$, denoted $\cos^{-1} x$ or $\arccos x$.
>
> The domain of $y = \cos^{-1} x$ is $[-1, 1]$ and the range is $[0, \pi]$.

It helps to think of the range of $y = \cos^{-1} x$ as being along the top half of the unit circle, which is traced out as angles range from 0 to $\pi$ (Figure 4.77).

If you restrict the domain of $y = \tan x$ to the interval $(-\pi/2, \pi/2)$, as shown in Figure 4.78a, the restricted function is one-to-one. The **inverse tangent function** $y = \tan^{-1} x$ is the inverse of this restricted portion of the tangent function (Figure 4.78b).

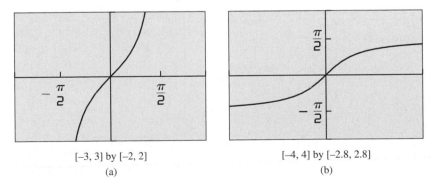

[-3, 3] by [-2, 2]

(a)

[-4, 4] by [-2.8, 2.8]

(b)

**Figure 4.78** The (a) restriction of $y = \tan x$ is one-to-one and (b) has an inverse, $y = \tan^{-1} x$.

By the usual inverse relationship, the statements

$$y = \tan^{-1} x \quad \text{and} \quad x = \tan y$$

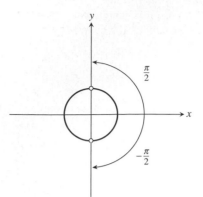

**Figure 4.79** The values of $y = \tan^{-1} x$ will always be found on the right-hand side of the unit circle, between (but not including) $-\pi/2$ and $\pi/2$.

**Alert**

The symbol used to denote inverse functions, $f^{-1}$, frequently causes confusion because it is similar to the symbol $x^{-1}$ used for the reciprocal of a number. Stress that $x^{-1} = 1/x$, but $f^{-1}(x) \neq 1/f(x)$, and $\sin^{-1}(x) \neq 1/\sin x$.

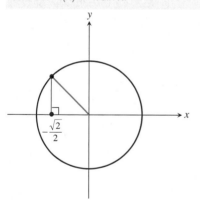

**Figure 4.80** $\cos^{-1}(-\sqrt{2}/2) = 3\pi/4$. (Example 3a)

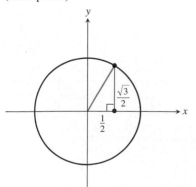

**Figure 4.81** $\tan^{-1}\sqrt{3} = \pi/3$. (Example 3b)

are equivalent for $y$-values in the restricted domain $(-\pi/2, \pi/2)$ and $x$-values in $(-\infty, \infty)$. This means that $\tan^{-1} x$ can be thought of as *the angle between $-\pi/2$ and $\pi/2$ whose tangent is $x$*. The angle $\tan^{-1} x$ is also the **arctangent of $x$**.

---

**Inverse Tangent Function (Arctangent Function)**

The unique angle $y$ in the interval $(-\pi/2, \pi/2)$ such that $\tan y = x$ is the **inverse tangent** (or **arctangent**) of $x$, denoted **$\tan^{-1} x$** or **arctan $x$**.

The domain of $y = \tan^{-1} x$ is $(-\infty, \infty)$ and the range is $(-\pi/2, \pi/2)$.

---

It helps to think of the range of $y = \tan^{-1} x$ as being along the right-hand side of the unit circle (minus the top and bottom points), which is traced out as angles range from $-\pi/2$ to $\pi/2$ (non-inclusive) (Figure 4.79).

**Example 3   EVALUATING INVERSE TRIG FUNCTIONS WITHOUT A CALCULATOR**

Find the exact value of the expression without a calculator.

**(a)** $\cos^{-1}\left(-\dfrac{\sqrt{2}}{2}\right)$

**(b)** $\tan^{-1}\sqrt{3}$

**(c)** $\cos^{-1}(\cos(-1.1))$

**Solution**

**(a)** Find the point on the top half of the unit circle whose $x$-coordinate is $-\sqrt{2}/2$ and draw a reference triangle (Figure 4.80). We recognize this as one of our special ratios, and the angle in the interval $[0, \pi]$ whose cosine is $-\sqrt{2}/2$ is $3\pi/4$. Therefore

$$\cos^{-1}\left(-\frac{\sqrt{2}}{2}\right) = \frac{3\pi}{4}.$$

**(b)** Find the point on the right side of the unit circle whose $y$-coordinate is $\sqrt{3}$ times its $x$-coordinate and draw a reference triangle (Figure 4.81). We recognize this as one of our special ratios, and the angle in the interval $(-\pi/2, \pi/2)$ whose tangent is $\sqrt{3}$ is $\pi/3$. Therefore

$$\tan^{-1}\sqrt{3} = \frac{\pi}{3}.$$

**(c)** Draw an angle of $-1.1$ in standard position (notice that this angle is *not* in the interval $[0, \pi]$) and mark its $x$-coordinate on the $x$-axis (Figure 4.82). The angle in the interval $[0, \pi]$ whose cosine is this number is 1.1. Therefore

$$\cos^{-1}(\cos(-1.1)) = 1.1.$$

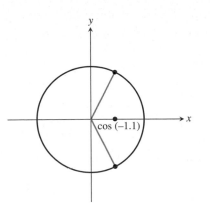

**Figure 4.82** $\cos^{-1}(\cos(-1.1)) = 1.1$.
(Example 3c)

---

**Alert**

Students often expect the inverse sine, inverse cosine, and inverse tangent functions to have the same range. This is not the case because, for example, the sine and cosine functions are not one-to-one on the same interval.

---

**Teaching Note**

Check that in using calculators to produce solutions, students are aware of the restrictions for inverses that affect how many solutions they can produce. Focus students' attention on what happens to domain and range.

**Example 4   DESCRIBING END BEHAVIOR**

Describe the end behavior of the function $y = \tan^{-1} x$.

**Solution**  We can get this information most easily by considering the graph of $y = \tan^{-1} x$, remembering how it relates to the restricted graph of $y = \tan x$. (See Figure 4.83.)

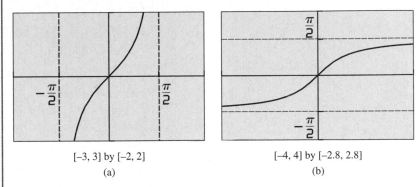

[–3, 3] by [–2, 2]
(a)

[–4, 4] by [–2.8, 2.8]
(b)

**Figure 4.83**  The graphs of (a) $y = \tan x$ (restricted) and (b) $y = \tan^{-1} x$. The vertical asymptotes of $y = \tan x$ are reflected to become the horizontal asymptotes of $y = \tan^{-1} x$. (Example 4)

When we reflect the graph of $y = \tan x$ about the line $y = x$ to get the graph of $y = \tan^{-1} x$, the vertical asymptotes $x = \pm\pi/2$ become horizontal asymptotes $y = \pm\pi/2$. We can state the end behavior accordingly:

$$\lim_{x \to -\infty} \tan^{-1} x = -\frac{\pi}{2} \quad \text{and} \quad \lim_{x \to +\infty} \tan^{-1} x = \frac{\pi}{2}.$$

## Composing Trigonometric and Inverse Trigonometric Functions

We have already seen the need for caution when applying the Inverse Composition Rule to the trigonometric functions and their inverses (Examples 1e and 3c above). The following equations are *always* true whenever they are defined:

$$\sin(\sin^{-1}(x)) = x \qquad \cos(\cos^{-1}(x)) = x \qquad \tan(\tan^{-1}(x)) = x.$$

On the other hand, the following equations are only true for $x$ values in the "restricted" domains of sin, cos, and tan:

$$\sin^{-1}(\sin(x)) = x \qquad \cos^{-1}(\cos(x)) = x \qquad \tan^{-1}(\tan(x)) = x$$

An even more interesting phenomenon occurs when we compose inverse trigonometric functions of one kind with trigonometric functions of another kind, as in $\sin(\tan^{-1} x)$. Surprisingly, these highly trigonometric compositions reduce to algebraic functions that involve no trigonometry at all! This curious situation has profound implications in calculus, where it is sometimes useful to decompose non-trigonometric functions into trigonometric components that seem to come out of nowhere. Try the following Exploration.

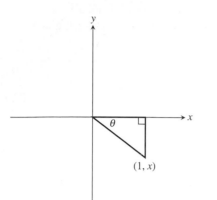

**Figure 4.84** If $x < 0$, then $\theta = \tan^{-1} x$ is an angle in the fourth quadrant. (Exploration 1)

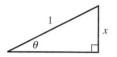

**Figure 4.85** A triangle in which $\theta = \sin^{-1} x$. (Example 5)

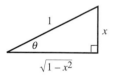

**Figure 4.86** If $\theta = \sin^{-1} x$, then $\cos \theta = \sqrt{1 - x^2}$. Note that $\cos \theta$ will be positive because $\sin^{-1} x$ can only be in Quadrant I or IV. (Example 5)

**Exploration Extensions**

Evaluate: (a) $\sin^{-1} (\sin 7\pi/6)$
(b) $\cos^{-1} (\cos -\pi/4)$
**((a) $-\pi/6$; (b) $\pi/4$)**

**Follow-up**

Ask students to name the values of $x$ for which $x = \cos^{-1} (\cos x)$ and the values for which $x = \cos (\cos^{-1} x)$.
**($0 \le x \le \pi$; $-1 \le x \le 1$)**

**Assignment Guide**

Day 1: Ex. 3–48, multiples of 3
Day 2: Ex. 51, 52, 53, 54, 55, 56, 59

---

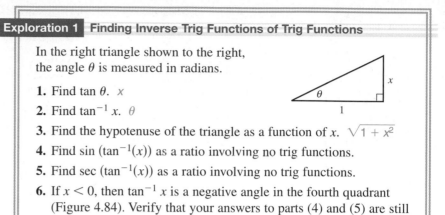

**Exploration 1** Finding Inverse Trig Functions of Trig Functions

In the right triangle shown to the right, the angle $\theta$ is measured in radians.

1. Find $\tan \theta$.  $x$
2. Find $\tan^{-1} x$.  $\theta$
3. Find the hypotenuse of the triangle as a function of $x$.  $\sqrt{1 + x^2}$
4. Find $\sin (\tan^{-1}(x))$ as a ratio involving no trig functions.
5. Find $\sec (\tan^{-1}(x))$ as a ratio involving no trig functions.
6. If $x < 0$, then $\tan^{-1} x$ is a negative angle in the fourth quadrant (Figure 4.84). Verify that your answers to parts (4) and (5) are still valid in this case.

### Example 5   COMPOSING TRIG FUNCTIONS WITH ARCSINE

Compose each of the six basic trig functions with $\sin^{-1} x$ and reduce the composite function to an algebraic expression involving no trig functions.

**Solution**  This time we begin with the triangle shown in Figure 4.85, in which $\theta = \sin^{-1} x$. (This triangle could appear in the fourth quadrant if $x$ were negative, but the trig ratios would be the same.)

The remaining side of the triangle (which is $\cos \theta$) can be found by the Pythagorean theorem. If we denote the unknown side by $s$, we have

$$s^2 + x^2 = 1$$
$$s^2 = 1 - x^2$$
$$s = \pm \sqrt{1 - x^2}$$

Note the ambiguous sign, which requires a further look. Since $\sin^{-1} x$ is always in Quadrant I or IV, the horizontal side of the triangle can only be positive. Therefore, we can actually write $s$ unambiguously as $\sqrt{1 - x^2}$, giving us the triangle in Figure 4.86.

We can now read all the required ratios straight from the triangle:

$$\sin (\sin^{-1}(x)) = x \qquad\qquad \csc (\sin^{-1}(x)) = \frac{1}{x}$$

$$\cos (\sin^{-1}(x)) = \sqrt{1 - x^2} \qquad \sec (\sin^{-1}(x)) = \frac{1}{\sqrt{1 - x^2}}$$

$$\tan (\sin^{-1}(x)) = \frac{x}{\sqrt{1 - x^2}} \qquad \cot (\sin^{-1}(x)) = \frac{\sqrt{1 - x^2}}{x}$$

## Applications of Inverse Trigonometric Functions

When an application involves an angle as a dependent variable, as in $\theta = f(x)$, then to solve for $x$, it is natural to use an inverse trigonometric function and find $x = f^{-1}(\theta)$.

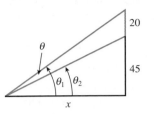

**Figure 4.87** The diagram for the stadium screen. (Example 6)

## Example 6 CALCULATING A VIEWING ANGLE

The bottom of a 20-foot replay screen at Dodger Stadium is 45 feet above the playing field. As you move away from the wall, the angle formed by the screen at your eye changes. There is a distance from the wall at which the angle is the greatest. What is that distance?

### Solution

#### Model

The angle subtended by the screen is represented in Figure 4.87 by $\theta$, and $\theta = \theta_1 - \theta_2$. Since $\tan \theta_1 = 65/x$, it follows that $\theta_1 = \tan^{-1}(65/x)$. Similarly, $\theta_2 = \tan^{-1}(45/x)$. Thus,

$$\theta = \tan^{-1} \frac{65}{x} - \tan^{-1} \frac{45}{x}.$$

#### Solve Graphically

Figure 4.88 shows a graph of $\theta$ that reflects degree mode. The question about distance for maximum viewing angle can be answered by finding the $x$-coordinate of the maximum point of this graph. Using grapher methods we see that this maximum occurs when $x \approx 54$ feet.

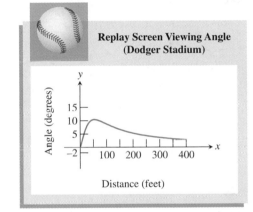

**Figure 4.88** Viewing angle $\theta$ as a function of distance $x$ from the wall. (Example 6)

Therefore the maximum angle subtended by the replay screen occurs about 54 feet from the wall.

# Quick Review 4.7

In Exercises 1–4, state the sign (positive or negative) of the sine, cosine, and tangent in the quadrant.

**1.** Quadrant I  **2.** Quadrant II

**3.** Quadrant III  **4.** Quadrant IV

In Exercises 5–10, find the exact value.

**5.** $\sin (\pi/6)$  1/2  **6.** $\tan (\pi/4)$  1

**7.** $\cos (2\pi/3)$  $-1/2$  **8.** $\sin (2\pi/3)$  $\sqrt{3}/2$

**9.** $\sin (-\pi/6)$  $-1/2$  **10.** $\cos (-\pi/3)$  1/2

# Section 4.7 Exercises

In Exercises 1–12, find the exact value.

**1.** $\sin^{-1}\left(\dfrac{\sqrt{3}}{2}\right)$  $\pi/3$

**2.** $\sin^{-1}\left(-\dfrac{1}{2}\right)$  $-\pi/6$

**3.** $\tan^{-1} 0$  $0$

**4.** $\cos^{-1} 1$  $0$

**5.** $\cos^{-1}\left(\dfrac{1}{2}\right)$  $\pi/3$

**6.** $\tan^{-1} 1$  $\pi/4$

**7.** $\tan^{-1}(-1)$  $-\pi/4$

**8.** $\cos^{-1}\left(-\dfrac{\sqrt{3}}{2}\right)$  $5\pi/6$

**9.** $\sin^{-1}\left(-\dfrac{1}{\sqrt{2}}\right)$  $-\pi/4$

**10.** $\tan^{-1}(-\sqrt{3})$  $-\pi/3$

**11.** $\cos^{-1} 0$  $\pi/2$

**12.** $\sin^{-1} 1$  $\pi/2$

In Exercises 13–18, use a calculator to find the approximate value. Express your answer in degrees.

**13.** $\sin^{-1}(0.362)$  $21.22°$

**14.** $\arcsin 0.67$  $42.07°$

**15.** $\tan^{-1}(-12.5)$  $-85.43°$

**16.** $\cos^{-1}(-0.23)$  $103.30°$

**17.** $\arctan 23.8$  $87.59°$

**18.** $\arccos 0.17$  $80.21°$

In Exercises 19–22, use a calculator to find the approximate value. Express your result in radians.

**19.** $\tan^{-1}(2.37)$  $1.172$

**20.** $\tan^{-1}(22.8)$  $1.527$

**21.** $\sin^{-1}(-0.46)$  $-0.478$

**22.** $\cos^{-1}(-0.853)$  $2.593$

In Exercises 23–32, find the exact value without a calculator.

**23.** $\cos\left[\sin^{-1}(1/2)\right]$  $\sqrt{3}/2$

**24.** $\sin\left[\tan^{-1} 1\right]$  $\sqrt{2}/2$

**25.** $\sin^{-1}\left[\cos(\pi/4)\right]$  $\pi/4$

**26.** $\cos^{-1}\left[\cos(7\pi/4)\right]$  $\pi/4$

**27.** $\cos\left[2\sin^{-1}(1/2)\right]$  $1/2$

**28.** $\sin\left[\tan^{-1}(-1)\right]$  $-\sqrt{2}/2$

**29.** $\arcsin\left[\cos(\pi/3)\right]$  $\pi/6$

**30.** $\arccos\left[\tan(\pi/4)\right]$  $0$

**31.** $\cos(\tan^{-1}\sqrt{3})$  $1/2$

**32.** $\tan^{-1}(\cos\pi)$  $-\pi/4$

In Exercises 33–36, use transformations to describe how the graph of the function is related to a basic inverse trigonometric graph. State the domain and range.

**33.** $f(x) = \sin^{-1}(2x)$

**34.** $g(x) = 3\cos^{-1}(2x)$

**35.** $h(x) = 5\tan^{-1}(x/2)$

**36.** $g(x) = 3\arccos(x/2)$

In Exercises 37–42, find an exact solution to the equation without a calculator.

**37.** $\sin(\sin^{-1} x) = 1$  $1$

**38.** $\cos^{-1}(\cos x) = 1$

**39.** $2\sin^{-1} x = 1$

**40.** $\tan^{-1} x = -1$

**41.** $\cos(\cos^{-1} x) = 1/3$

**42.** $\sin^{-1}(\sin x) = \pi/10$

In Exercises 43–48, find an algebraic expression equivalent to the given expression. (*Hint:* Form a right triangle as done in Example 5.)

**43.** $\sin(\tan^{-1} x)$  $x/\sqrt{1+x^2}$

**44.** $\cos(\tan^{-1} x)$  $x/\sqrt{1+x^2}$

**45.** $\tan(\arcsin x)$  $x/\sqrt{1-x^2}$

**46.** $\cot(\arccos x)$  $x/\sqrt{1-x^2}$

**47.** $\cos(\arctan 2x)$

**48.** $\sin(\arccos 3x)$  $\sqrt{1-9x^2}$

**49. Group Activity  Viewing Angle**  You are standing in an art museum viewing a picture. The bottom of the picture is 2 ft above your eye level, and the picture is 12 ft tall. Angle $\theta$ is formed by the lines of vision to the bottom and to the top of the picture.

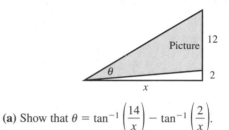

**(a)** Show that $\theta = \tan^{-1}\left(\dfrac{14}{x}\right) - \tan^{-1}\left(\dfrac{2}{x}\right)$.

**(b)** Graph $\theta$ in the $[0, 25]$ by $[0, 55]$ viewing window. Use your grapher to show that the maximum value of $\theta$ occurs approximately 5.3 ft from the picture.

**(c)** How far (to the nearest foot) are you standing from the wall if $\theta = 35°$?

**50. Group Activity  Analysis of a Lighthouse**  A rotating beacon $L$ stands 3 m across the harbor from the nearest point $P$ along a straight shoreline. As the light rotates, it forms an angle $\theta$ as shown in the figure, and illuminates a point $Q$ on the same shoreline as $P$.

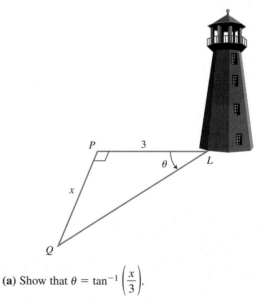

**(a)** Show that $\theta = \tan^{-1}\left(\dfrac{x}{3}\right)$.

**(b)** Graph $\theta$ in the viewing window $[-20, 20]$ by $[-90, 90]$. What do negative values of $x$ represent in the problem? What does a positive angle represent? A negative angle?

**(c)** Find $\theta$ when $x = 15$.  $\approx 78.69°$

**51. Rising Hot-Air Balloon**  The hot-air balloon festival held each year in Phoenix, Arizona, is a popular event for photographers. Jo Silver, an award-winning photographer at the event, watches a balloon rising from ground level from a point 500 ft away on level ground.

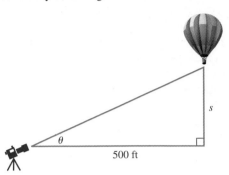

(a) Write $\theta$ as a function of the height $s$ of the balloon.

(b) Is the change in $\theta$ greater as $s$ changes from 10 ft to 20 ft, or as $s$ changes from 200 ft to 210 ft? Explain.

(c) **Writing to Learn**  In the graph of this relationship shown here, do you think that the $x$-axis represents the height $s$ and the $y$-axis angle $\theta$, or does the $x$-axis represent angle $\theta$ and the $y$-axis height $s$? Explain.

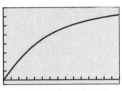

[0, 1500] by [–5, 80]

**52.** Find the domain and range of each of the following functions.

(a) $f(x) = \sin(\sin^{-1} x)$

(b) $g(x) = \sin^{-1}(x) + \cos^{-1}(x)$

(c) $h(x) = \sin^{-1}(\sin x)$

(d) $k(x) = \sin(\cos^{-1} x)$

(e) $q(x) = \cos^{-1}(\sin x)$

## Explorations

**53. Writing to Learn**  Using the format demonstrated in this section for the inverse sine, cosine, and tangent functions, give a careful definition of the inverse cotangent function. (*Hint:* The range of $y = \cot^{-1} x$ is $(0, \pi)$.)

**54. Writing to Learn**  Use an appropriately labeled triangle to explain why $\sin^{-1} x + \cos^{-1} x = \pi/2$. For what values of $x$ is the left-hand side of this equation defined?

**55.** Graph each of the following functions and interpret the graph to find the domain, range, and period of each function. Which of the three functions has points of discontinuity? Are the discontinuities removable or non-removable?

(a) $y = \sin^{-1}(\sin x)$

(b) $y = \cos^{-1}(\cos x)$

(c) $y = \tan^{-1}(\tan x)$  ◾

## Extending the Ideas

**56. Practicing for Calculus**  Express each of the following functions as an algebraic expression involving no trig functions.

(a) $\cos(\sin^{-1} 2x)$   $\sqrt{1 - 4x^2}$   (b) $\sec^2(\tan^{-1} x)$   $1 + x^2$

(c) $\sin(\cos^{-1} \sqrt{x})$   $\sqrt{1 - x}$   (d) $-\csc^2(\cot^{-1} x)$   $-x^2 - 1$

(e) $\tan(\sec^{-1} x^2)$   $\sqrt{x^4 - 1}$

**57. Arccotangent on the Calculator**  Most graphing calculators do not have a button for the inverse cotangent. The graph is shown below. Find an expression that you can put into your calculator to produce a graph of $y = \cot^{-1} x$.

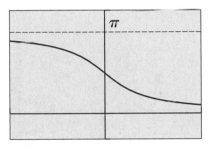

[–3, 3] by [–1, 4]

**58. Advanced Decomposition**  Decompose each of the following algebraic functions by writing it as a trig function of an arctrig function.

(a) $\sqrt{1 - x^2}$   (b) $\dfrac{x}{\sqrt{1 + x^2}}$   (c) $\dfrac{x}{\sqrt{1 - x^2}}$

**59.** Use elementary transformations and the arctangent function to construct a function with domain all real numbers that has horizontal asymptotes at $y = 24$ and $y = 42$.

**60. Avoiding Ambiguities**  When choosing the right triangle in Example 5, we used a hypotenuse of 1. It is sometimes necessary to use a variable quantity for the hypotenuse, in which case it is a good idea to use $x^2$ rather than $x$, just in case $x$ is negative. (All of our definitions of the trig functions have involved triangles in which the hypotenuse is assumed to be positive.)

(a) If we use the triangle below to represent $\theta = \sin^{-1}(1/x)$, explain why side $s$ must be positive regardless of the sign of $x$.

(b) Use the triangle in (a) to find $\tan(\sin^{-1}(1/x))$.

(c) Using an appropriate triangle, find $\sin(\cos^{-1}(1/x))$.

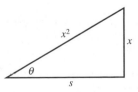

# Solving Problems with Trigonometry

More Right Triangle Problems • Simple Harmonic Motion

## More Right Triangle Problems

We close this first of two trigonometry chapters by revisiting some of the applications of Section 4.2 (right triangle trigonometry) and Section 4.4 (sinusoids).

An **angle of elevation** is the angle through which the eye moves up from horizontal to look at something above, and an **angle of depression** is the angle through which the eye moves down from horizontal to look at something below. For two observers at different elevations looking at each other, the angle of elevation for one equals the angle of depression for the other. The concepts are illustrated in Figure 4.89 as they might apply to observers at Mount Rushmore or the Grand Canyon.

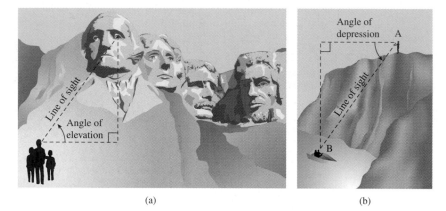

(a)                           (b)

**Figure 4.89** (a) Angle of elevation at Mount Rushmore. (b) Angle of depression at the Grand Canyon.

### Example 1  USING ANGLE OF DEPRESSION

The angle of depression of a buoy from the top of the Barnegat Bay lighthouse 130 feet above the surface of the water is 6°. Find the distance $x$ from the base of the lighthouse to the buoy.

**Solution**  Figure 4.90 **models** the situation.

In the diagram, $\theta = 6°$ because the angle of elevation from the buoy equals the angle of depression from the lighthouse. We **solve algebraically** using the tangent function:

$$\tan \theta = \tan 6° = \frac{130}{x}$$

$$x = \frac{130}{\tan 6°} \approx 1236.9$$

**Interpreting** our answer, we find that the buoy is about 1237 feet from the base of the lighthouse.

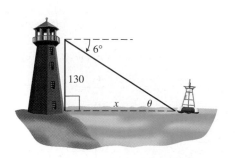

**Figure 4.90** A big lighthouse and a little buoy. (Example 1)

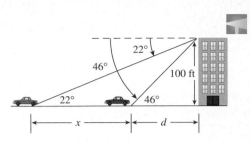

**Figure 4.91** A car approaches Altgelt Hall. (Example 2)

### Example 2 MAKING INDIRECT MEASUREMENTS

From the top of the 100-ft-tall Altgelt Hall a man observes a car moving toward the building. If the angle of depression of the car changes from 22° to 46° during the period of observation, how far does the car travel?

**Solution**

**Solve Algebraically**

Figure 4.91 models the situation. Notice that we have labeled the acute angles at the car's two positions as 22° and 46° (because the angle of elevation from the car equals the angle of depression from the building). Denote the distance the car moves as $x$. Denote its distance from the building at the second observation as $d$.

From the smaller right triangle we conclude:

$$\tan 46° = \frac{100}{d}$$

$$d = \frac{100}{\tan 46°}$$

From the larger right triangle we conclude:

$$\tan 22° = \frac{100}{x + d}$$

$$x + d = \frac{100}{\tan 22°}$$

$$x = \frac{100}{\tan 22°} - d$$

$$x = \frac{100}{\tan 22°} - \frac{100}{\tan 46°}$$

$$x \approx 150.9$$

**Interpreting** our answer, we find that the car travels about 151 feet.

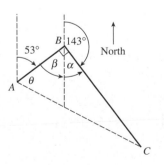

**Figure 4.92** Path of travel for a Coast Guard boat that corners well at 35 knots. (Example 3)

### Example 3 USING TRIGONOMETRY IN NAVIGATION

A U.S. Coast Guard patrol boat leaves Port Cleveland and averages 35 knots (nautical mph) traveling for 2 hours on a course of 53° and then 3 hours on a course of 143°. What is the boat's bearing and distance from Port Cleveland?

**Solution** Figure 4.92 models the situation.

**Solve Algebraically**

In the diagram, line $AB$ is a transversal that cuts a pair of parallel lines. Thus, $\beta = 53°$ because they are alternate interior angles. Angle $\alpha$, as the supplement of a 143° angle, is 37°. Consequently $\angle ABC = 90°$, and $AC$ is the hypotenuse of right $\triangle ABC$.

Use distance = rate × time to determine distances $AB$ and $BC$.

$$AB = (35 \text{ knots})(2 \text{ hours}) = 70 \text{ nautical miles}$$

$$BC = (35 \text{ knots})(3 \text{ hours}) = 105 \text{ nautical miles}.$$

Solve the right triangle for $AC$ and $\theta$.

$$AC = \sqrt{70^2 + 105^2} \qquad \text{Pythagorean theorem}$$

$$AC \approx 126.2$$

$$\theta = \tan^{-1}\left(\frac{105}{70}\right)$$

$$\theta \approx 56.3°$$

**Interpreting** our answers, we find that the boat's bearing from Port Cleveland is $53° + \theta$, or approximately $109.3°$. They are about 126 nautical miles out.

## Simple Harmonic Motion

The sine and cosine functions, because of their periodic nature, are helpful in describing the motion of objects that oscillate, vibrate, or rotate. For example, the linkage in Figure 4.93 converts the rotary motion of a motor to the back-and-forth motion needed for some machines. When the wheel rotates, the piston moves back and forth.

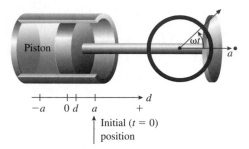

**Figure 4.93** A piston operated by a wheel rotating at a constant rate demonstrates simple harmonic motion.

If the wheel rotates at a constant rate $\omega$ radians per second, the back-and-forth motion of the piston is an example of *simple harmonic motion* and can be modeled by an equation of the form

$$d = a \cos \omega t, \quad \omega > 0,$$

where $a$ is the radius of the wheel and $d$ is the directed distance of the piston from its center of oscillation.

For the sake of simplicity, we will define simple harmonic motion in terms of a point moving along a number line.

### Frequency and Period

Notice that harmonic motion is sinusoidal, with amplitude $|a|$ and period $2\pi/\omega$. The frequency is the reciprocal of the period.

---

**Simple Harmonic Motion**

A point moving on a number line is in **simple harmonic motion** if its directed distance $d$ from the origin is given by either

$$d = a \sin \omega t \quad \text{or} \quad d = a \cos \omega t,$$

where $a$ and $\omega$ are real numbers and $\omega > 0$. The motion has **frequency** $\omega/2\pi$, which is the number of oscillations per unit of time.

**Exploration Extensions**

Change the values of the range to Ymin = −1 and Ymax = 1 and try the problem again. Discuss your observations.

**Exploration 1  Watching Harmonic Motion**

You can watch harmonic motion on your graphing calculator. Set your grapher to parametric mode and set X1T = cos (T) and Y1T = sin (T). Set Tmin = 0, Tmax = 25, Tstep = 0.2, Xmin = −1.5, Xmax = 1.5, Xscl = 1, Ymin = −100, Ymax = 100, Yscl = 0.

If your calculator allows you to change style to graph a moving ball, choose that style. When you graph the function, you will see the ball moving along the x-axis between −1 and 1 in simple harmonic motion. If your grapher does not have the moving ball option, wait for the grapher to finish graphing, then press TRACE and keep your finger pressed on the right arrow key to see the tracer move in simple harmonic motion.

1. For each value of *T*, the parametrization gives the point (cos (*T*), sin (*T*)). What well-known curve should this parametrization produce?  the unit circle

2. Why does the point seem to go back and forth on the *x*-axis when it should be following the curve identified in part (1)? (*Hint:* Check that viewing window again!)

3. Why does the point slow down at the extremes and speed up in the middle? (*Hint:* Remember that the grapher is really following the curve identified in part (1).)

4. How can you tell that this point moves in simple harmonic motion?

### Example 4   CALCULATING HARMONIC MOTION

In a mechanical linkage like the one shown in Figure 4.93, a wheel with an 8-cm radius turns with an angular velocity of $8\pi$ radians/sec.

**(a)** What is the frequency of the piston?

**(b)** What is the distance from the starting position ($t = 0$) exactly 3.45 seconds after starting?

**Solution**  Imagine the wheel to be centered at the origin and let $P(x, y)$ be a point on its perimeter (Figure 4.94). As the wheel rotates and $P$ goes around, the motion of the piston follows the path of the $x$-coordinate of $P$ along the $x$-axis. The angle determined by $P$ at any time $t$ is $8\pi t$, so its $x$-coordinate is $8 \cos 8\pi t$. Therefore, the sinusoid $d = 8 \cos 8\pi t$ models the motion of the piston.

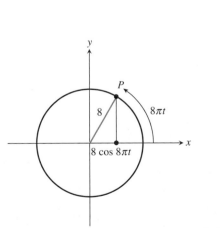

**Figure 4.94**  Modeling the path of a piston by a sinusoid. (Example 4)

**(a)** The frequency of $d = 8 \cos 8\pi t$ is $8\pi/2\pi$, or 4. The piston makes four complete back-and-forth strokes per second. The graph of $d$ as a function of $t$ is shown in Figure 4.95. The four cycles of the sinusoidal graph in the interval $[0, 1]$ model the four cycles of the motor or the four strokes of the piston. Note that the sinusoid has a period of $1/4$, the reciprocal of the frequency.

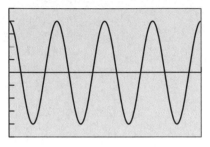

[0, 1] by [−10, 10]

**Figure 4.95** A sinusoid with frequency 4 models the motion of the piston in Example 4.

**(b)** We must find the distance between the positions at $t = 0$ and $t = 3.45$. The initial position at $t = 0$ is

$$d(0) = 8.$$

The position at $t = 3.45$ is

$$d(3.45) = 8 \cos (8\pi \cdot 3.45) \approx 2.47.$$

The distance between the two positions is approximately $8 - 2.47 = 5.53$.

   **Interpreting** our answer, we conclude that the piston is approximately 5.53 cm from its starting position after 3.45 seconds.

### Example 5   CALCULATING HARMONIC MOTION

A mass oscillating up and down on the bottom of a spring (assuming perfect elasticity and no friction or air resistance) can be modeled as harmonic motion. If the weight is displaced a maximum of 5 cm, find the modeling equation if it takes 2 seconds to complete one cycle. (See Figure 4.96.)

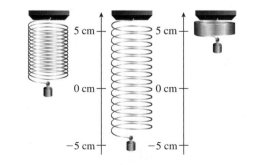

**Figure 4.96** The mass and spring in Example 5.

**Solution** We have our choice between the two equations $d = a \sin \omega t$ or $d = a \cos \omega t$. Assuming that the spring is at the origin of the coordinate system when $t = 0$, we choose the equation $d = a \sin \omega t$.

**Follow-up**

Ask students if they can think of other equations that model the situation in Example 5.

**Assignment Guide**

Day 1: Ex. 1, 2, 5, 6, 7, 10, 12, 13, 14, 16, 17, 19, 20, 21, 25
Day 2: Ex. 29, 31, 33, 34, 36, 37, 38, 40, 42

**Cooperative Learning**

Group Activity: Ex. 35, 43

**Notes on Exercises**

Ex. 27–30 deal with simple harmonic motion.
Ex. 33, 35, and 43 require students to find sinusoidal functions that approximate data. Ex. 37–39 use geometry concepts.

**Ongoing Assessment**

Self-Assessment: Ex. 3, 9, 15, 23, 41
Embedded Assessment: Ex. 34, 36

Because the maximum displacement is 5 cm, we conclude that the amplitude $a = 5$.

Because it takes 2 seconds to complete one cycle, we conclude that the period is 2 and the frequency is 1/2. Therefore,

$$\frac{\omega}{2\pi} = \frac{1}{2},$$

$$\omega = \pi.$$

Putting it all together, our modeling equation is $d = 5 \sin \pi t$.

---

### Problem

If we know that the musical note A above middle C has a pitch of 440 Hertz, how can we model the sound produced by it at 80 decibels?

### Solution

Sound is modeled by simple harmonic motion, with frequency perceived as pitch and measured in cycles per second, and amplitude perceived as loudness and measured in decibels. So for the musical note A with a pitch of 440 Hertz, we have frequency $= \omega/2\pi = 440$ and therefore $\omega = 2\pi 440 = 880\pi$

If this note is played at a loudness of 80 decibels, we have $|a| = 80$. Using the simple harmonic motion model $d = a \sin \omega t$ we have

$$d = 80 \sin 880\pi t.$$

---

# Quick Review 4.8

In Exercises 1–4, find the lengths $a$, $b$, and $c$.

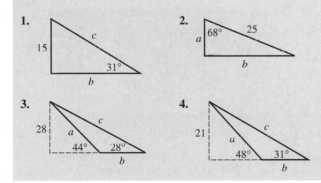

**1.** 15, $c$, $b$, 31°

**2.** 68°, 25, $a$, $b$

**3.** 28, $a$, $c$, 44°, 28°, $b$

**4.** 21, $u$, $c$, 48°, 31°, $b$

In Exercises 5 and 6, find the complement and supplement of the angle.

**5.** 32°      **6.** 73°

In Exercises 7 and 9, state the bearing that describes the direction.

**7.** NE (northeast)  45°      **8.** SSW (south-southwest)

In Exercises 9 and 10, state the amplitude and period of the sinusoid.

**9.** $-3 \sin 2(x - 1)$      **10.** $4 \cos 4(x + 2)$

# Section 4.8 Exercises

In Exercises 1–43, solve the problem using your knowledge of geometry and the techniques of this section. Sketch a figure if one is not provided.

1. **Finding a Cathedral Height** The angle of elevation of the top of the Ulm Cathedral from a point 300 ft away from the base of its steeple on level ground is 60°. Find the height of the cathedral. $300\sqrt{3} \approx 519.62$ ft

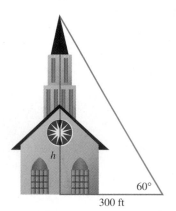

2. **Finding a Monument Height** From a point 100 ft from its base, the angle of elevation of the top of the Arch of Septimus Severus, in Rome, Italy, is 34°13′12″. How tall is this monument? $100 \tan 34°13'12'' \approx 68.01$ ft

3. **Finding a Distance** The angle of depression from the top of the Smoketown Lighthouse 120 ft above the surface of the water to a buoy is 10°. How far is the buoy from the lighthouse? $120 \cot 10° \approx 680.55$ ft

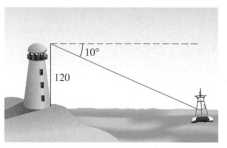

4. **Finding a Baseball Stadium Dimension** The top row of the red seats behind home plate at Cincinnati's Riverfront Stadium is 90 ft above the level of the playing field. The angle of depression to the base of the left field wall is 14°. How far is the base of the left field wall from a point on level ground directly below the top row?

5. **Finding a Guy-Wire Length** A guy wire connects the top of an antenna to a point on level ground 5 ft from the base of the antenna. The angle of elevation formed by this wire is 80°. What are the length of the wire and the height of the antenna?

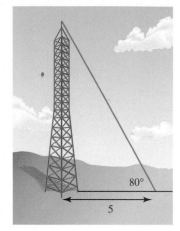

6. **Finding a Length** A wire stretches from the top of a vertical pole to a point on level ground 16 ft from the base of the pole. If the wire makes an angle of 62° with the ground, find the height of the pole and the length of the wire.

7. **Height of Eiffel Tower** The angle of elevation of the top of the TV antenna mounted on top of the Eiffel Tower in Paris is measured to be 80°1′12″ at a point 185 ft from the base of the tower. How tall is the tower plus TV antenna?

8. **Finding the Height of Tallest Chimney** The world's tallest smokestack at the International Nickel Co., Sudbury, Ontario, casts a shadow that is approximately 1580 ft long when the sun's angle of elevation (measured from the horizon) is 38°. How tall is the smokestack?

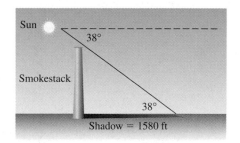

9. **Cloud Height** To measure the height of a cloud, you place a bright searchlight directly below the cloud and shine the beam straight up. From a point 100 ft away from the searchlight, you measure the angle of elevation of the cloud to be 83°12′. How high is the cloud?

10. **Ramping Up** A ramp leading to a freeway overpass is 470 ft long and rises 32 ft. What is the average angle of inclination of the ramp to the nearest tenth of a degree? 3.9°

11. **Antenna Height**  A guy wire attached to the top of the KSAM radio antenna is anchored at a point on the ground 10 m from the antenna's base. If the wire makes an angle of 55° with level ground, how high is the KSAM antenna?

12. **Building Height**  To determine the height of the Louisiana-Pacific (LP) Tower, the tallest building in Conroe, Texas, a surveyor stands at a point on the ground, level with the base of the LP building. He measures the point to be 125 ft from the building's base and the angle of elevation to the top of the building to be 29°48′. Find the height of the building.  125 tan 29°48′ ≈ 71.6 ft

13. **Navigation**  The *Paz Verde*, a whalewatch boat, is located at point *P*, and *L* is the nearest point on the Baja California shore. Point *Q* is located 4.25 mi down the shoreline from *L* and $\overline{PL} \perp \overline{LQ}$. Determine the distance that the *Paz Verde* is from the shore if ∠*PQL* = 35°.  4.25 tan 35° ≈ 2.976 mi

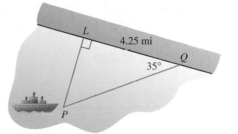

14. **Recreational Hiking**  While hiking on a level path toward Colorado's front range, Otis Evans determines that the angle of elevation to the top of Long's Peak is 30°. Moving 1000 ft closer to the mountain, Otis determines the angle of elevation to be 35°. How much higher is the top of Long's Peak than Otis's elevation?

15. **Civil Engineering**  The angle of elevation from an observer to the bottom edge of the Delaware River drawbridge observation deck located 200 ft from the observer is 30°. The angle of elevation from the observer to the top of the observation deck is 40°. What is the height of the observation deck?

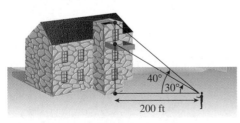

16. **Traveling Car**  From the top of a 100-ft building a man observes a car moving toward him. If the angle of

depression of the car changes from 15° to 33° during the period of observation, how far does the car travel?

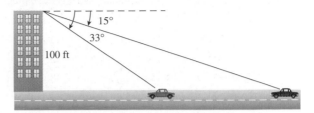

17. **Navigation**  The Coast Guard cutter *Angelica* travels at 30 knots from its home port of Corpus Christi on a course of 95° for 2 h and then changes to a course of 185° for 2 h. Find the distance and the bearing from the Corpus Christi port to the boat.

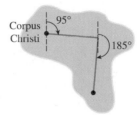

18. **Navigation**  The *Cerrito Lindo* travels at a speed of 40 knots from Fort Lauderdale on a course of 65° for 2 h and then changes to a course of 155° for 4 h. Determine the distance and the bearing from Fort Lauderdale to the boat.

19. **Land Measure**  The angle of depression is 19° from a point 7256 ft above sea level on the north rim of the Grand Canyon level to a point 6159 ft above sea level on the south rim. How wide is the canyon at that point?

20. **Ranger Fire Watch**  A ranger spots a fire from a 73-ft tower in Yellowstone National Park. She measures the angle of depression to be 1°20′. How far is the fire from the tower?  73 cot 1°20′ ≈ 3136.4 ft

21. **Civil Engineering**  The bearing of the line of sight to the east end of the Royal Gorge footbridge from a point 325 ft due north of the west end of the footbridge across the Royal Gorge is 117°. What is the length *l* of the bridge?

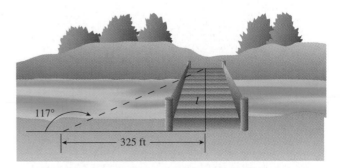

22. **Space Flight** The angle of elevation of the space shuttle *Columbia* from Cape Canaveral is 17° when the shuttle is directly over a ship 12 mi downrange. What is the altitude of the shuttle when it is directly over the ship?

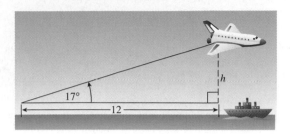

23. **Architectural Design** A barn roof is constructed as shown in the figure. What is the height of the vertical center span? 36.5 tan 15° ≈ 9.8 ft

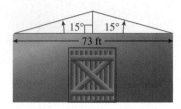

24. **Recreational Flying** A hot-air balloon over Park City, Utah, is 760 ft above the ground. The angle of depression from the balloon to an observer is 5.25°. Assuming the ground is relatively flat, how far is the observer from a point on the ground directly under the balloon?

25. **Navigation** A shoreline runs north-south, and a boat is due east of the shoreline. The bearings of the boat from two points on the shore are 110° and 100°. Assume the two points are 550 ft apart. How 1far is the boat from the shore?

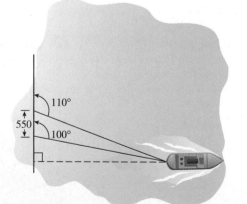

26. **Navigation** Milwaukee, Wisconsin, is directly west of Grand Haven, Michigan, on opposite sides of Lake Michigan. On a foggy night, a law enforcement boat leaves from Milwaukee on a course of 105° at the same time that a small smuggling craft steers a course of 195° from Grand Haven. The law enforcement boat averages 23 knots and collides with the smuggling craft. What was the smuggling boat's average speed? 23 tan 15° ≈ 6.2 knots

27. **Mechanical Design** *Refer to Figure 4.93*. The wheel in a piston linkage like the one shown in the figure has a radius of 6 in. It turns with an angular velocity of $16\pi$ rad/sec. The initial position is the same as that shown in Figure 4.93.

   (a) What is the frequency of the piston? 8 cycles/sec

   (b) What equation models the motion of the piston?

   (c) What is the distance from the initial position 2.85 sec after starting?

28. **Mechanical Design** Suppose the wheel in a piston linkage like the one shown in Figure 4.93 has a radius of 18 cm and turns with an angular velocity of $\pi$ rad/sec.

   (a) What is the frequency of the piston? 1/2 cycle/sec

   (b) What equation models the motion of the piston?

   (c) How many cycles does the piston make in 1 min?

29. **Vibrating Spring** A mass on a spring oscillates back and forth and completes one cycle in 0.5 sec. Its maximum displacement is 3 cm. Write an equation that models this motion. $d = 3 \cos 4\pi t$

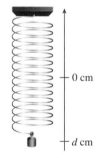

30. **Tuning Fork** A point on the tip of a tuning fork vibrates in harmonic motion described by the equation $d = 14 \sin \omega t$. Find $\omega$ for a tuning fork that has a frequency of 528 vibrations per second. $1056\pi$ radians/sec

31. **Ferris Wheel Motion** The Ferris wheel shown in this figure makes one complete turn every 20 sec. A rider's height, $h$, above the ground can be modeled by the equation $h = a \sin \omega t + k$, where $h$ and $k$ are given in feet and $t$ is given in seconds.

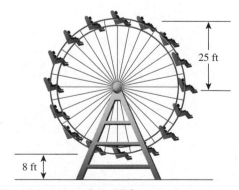

   (a) What is the value of $a$? 25 ft

   (b) What is the value of $k$? 33 ft

   (c) What is the value of $\omega$? $\pi$/10 radians/sec

**32. Ferris Wheel Motion**   Jacob and Emily ride a Ferris wheel at a carnival in Billings, Montana. The wheel has a 16-m diameter and turns at 3 rpm with its lowest point 1 m above the ground. Assume that Jacob and Emily's height $h$ above the ground is a sinusoidal function of time $t$ (in seconds), where $t = 0$ represents the lowest point of the wheel.

(a) Write an equation for $h$.   $h = -8 \cos \pi t/10 + 9$

(b) Draw a graph of $h$ for $0 \le t \le 30$.

(c) Use $h$ to estimate Jacob and Emily's height above the ground at $t = 4$ and $t = 10$.

**33.** Assume that the data shown in Table 4.2 and graphed in the figure repeat each year, and you want to find an equation of the form $y = a \sin [b(t - h)] + k$ that models these data.

| Table 4.2   Temperature Data for St. Louis | |
|---|---|
| Time (months) | Average Temperature |
| 1 | 34° |
| 2 | 30° |
| 3 | 39° |
| 4 | 44° |
| 5 | 58° |
| 6 | 67° |
| 7 | 78° |
| 8 | 80° |
| 9 | 72° |
| 10 | 63° |
| 11 | 51° |
| 12 | 40° |

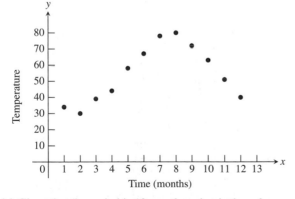

Time (months)

(a) Given that the period is 12 months, what is the value of $b$?   $\pi/6$

(b) How is the amplitude $a$ related to the difference $80° - 30°$?

(c) Use the information in (b) to find $k$.   55°

(d) Find $h$, and write an equation for $y$.

(e) Superimpose a graph of $y$ on a scatter plot of the data in Table 4.2.

(f) Use the modeling equation from (d) (with the assumption that $t = 0$ means January 1) to predict dates when the temperature will be 70°.

**34. Writing to Learn**   For the Ferris wheel in Exercise 31, which equation correctly models the height of a rider who begins the ride at the bottom of the wheel when $t = 0$?

(a) $h = 25 \sin \dfrac{\pi t}{10}$

(b) $h = 25 \sin \dfrac{\pi t}{10} + 8$

(c) $h = 25 \sin \dfrac{\pi t}{10} + 33$

(d) $h = 25 \sin \left( \dfrac{\pi t}{10} + \dfrac{3\pi}{2} \right) + 33$

Explain your thought process, and use of a graphing utility in choosing the correct modeling equation.

## Explorations

**35. Group Activity**   The data for displacement (pressure) versus time on a tuning fork, shown in Table 4.3, were collected using a **CBL** and a microphone.

| Table 4.3   Tuning Fork Data | | | |
|---|---|---|---|
| Time | Displacement | Time | Displacement |
| 0.00091 | −0.080 | 0.00362 | 0.217 |
| 0.00108 | 0.200 | 0.00379 | 0.480 |
| 0.00125 | 0.480 | 0.00398 | 0.681 |
| 0.00144 | 0.693 | 0.00416 | 0.810 |
| 0.00162 | 0.816 | 0.00435 | 0.827 |
| 0.00180 | 0.844 | 0.00453 | 0.749 |
| 0.00198 | 0.771 | 0.00471 | 0.581 |
| 0.00216 | 0.603 | 0.00489 | 0.346 |
| 0.00234 | 0.368 | 0.00507 | 0.077 |
| 0.00253 | 0.099 | 0.00525 | −0.164 |
| 0.00271 | −0.141 | 0.00543 | −0.320 |
| 0.00289 | −0.309 | 0.00562 | −0.354 |
| 0.00307 | −0.348 | 0.00579 | −0.248 |
| 0.00325 | −0.248 | 0.00598 | −0.035 |
| 0.00344 | −0.041 | | |

(a) Graph a scatter plot of the data in the $[0, 0.0062]$ by $[-0.5, 1]$ viewing window.

(b) Select the equation that appears to be the best fit of these data.   The first is the best.

   i. $y = 0.6 \sin (2464x - 2.84) + 0.25$

   ii. $y = 0.6 \sin (1210x - 2) + 0.25$

   iii. $y = 0.6 \sin (2440x - 2.1) + 0.15$

(c) What is the approximate frequency of the tuning fork?

**36. Writing to Learn** Human sleep—awake cycles at three different ages are described by the accompanying graphs. The portions of the graphs above the horizontal lines represent times awake, and the portions below represent times asleep.

**Newborn**

6 PM    12    6 AM    12    6 PM

**Four years**

6 PM    12    6 AM    12    6 PM

**Adult**

6 PM    12    6 AM    12    6 PM

**(a)** What is the period of the sleep—awake cycle of a newborn? Of a four year old? Of an adult?

**(b)** Which of these three sleep—awake cycles is the closest to being modeled by a function $y = a \sin bx$?

**Using Trigonometry in Geometry** In a *regular polygon* all sides have equal length and all angles have equal measure. In Exercises 37 and 38, consider the regular seven-sided polygon whose sides are 5 cm.

5 cm

$r$

$a$

**37.** Find the length of the *apothem*, the segment from the center of the seven-sided polygon to the midpoint of a side.

**38.** Find the radius of the circumscribed circle of the regular seven-sided polygon. $2.5 \csc \pi/7 \approx 5.8$ cm

**39.** A *rhombus* is a quadrilateral with all sides equal in length. Recall that a rhombus is also a parallelogram. Find length $AC$ and length $BD$ in the rhombus shown here.

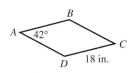

$B$

$A$ $42°$

$C$

$D$ 18 in.

# Extending the Ideas

**40.** A roof has two sections, one with a 50° elevation and the other with a 20° elevation, as shown in the figure.

**(a)** Find the height $BE$. $20 \tan 50° \approx 23.8$ ft

**(b)** Find the height $CD$. $BE + 45 \tan 20° \approx 40.2$ ft

**(c)** Find the length $AE + ED$, and double it to find the length of the cross section of the roof.

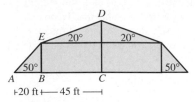

$D$

$E$  20°    20°

$A$  $B$    $C$

$50°$    $50°$

$\vdash$20 ft$\dashv\vdash$—— 45 ft ——$\dashv$

**41. Steep Trucking** The *percentage grade* of a road is its slope expressed as a percentage. A tractor-trailer rig passes a sign that reads, "6% grade next 7 miles." What is the average angle of inclination of the road?

**42. Television Coverage** Many satellites travel in *geosynchronous orbits*, which means that the satellite stays over the same point on the earth. A satellite that broadcasts cable television is in geosynchronous orbit 100 mi above the earth. Assume that the earth is a sphere with radius 4000 mi, and find the arc length of coverage area for the cable television satellite on the earth's surface. $\approx 1771$ mi

**43. Group Activity** A musical note like that produced with a tuning fork or pitch meter is a pressure wave. Typically, frequency is measured in hertz (1 Hz = 1 cycle per second). Table 4.4 gives frequency (in Hz) of several musical notes. The time-vs.-pressure tuning fork data in Table 4.5 was collected using a **CBL** and a microphone.

| Table 4.4    Tuning Fork Data | |
|---|---|
| Note | Frequency (Hz) |
| C | 262 |
| C$^\sharp$ or D$^\flat$ | 277 |
| D | 294 |
| D$^\sharp$ or E$^\flat$ | 311 |
| E | 330 |
| F | 349 |
| F$^\sharp$ or G$^\flat$ | 370 |
| G | 392 |
| G$^\sharp$ or A$^\flat$ | 415 |
| A | 440 |
| A$^\sharp$ or B$^\flat$ | 466 |
| B | 494 |
| C (next octave) | 524 |

| Table 4.5 | Tuning Fork Data | | |
|---|---|---|---|
| Time (sec) | Pressure | Time (sec) | Pressure |
| 0.0002368 | 1.29021 | 0.0049024 | −1.06632 |
| 0.0005664 | 1.50851 | 0.0051520 | 0.09235 |
| 0.0008256 | 1.51971 | 0.0054112 | 1.44694 |
| 0.0010752 | 1.51411 | 0.0056608 | 1.51411 |
| 0.0013344 | 1.47493 | 0.0059200 | 1.51971 |
| 0.0015840 | 0.45619 | 0.0061696 | 1.51411 |
| 0.0018432 | −0.89280 | 0.0064288 | 1.43015 |
| 0.0020928 | −1.51412 | 0.0066784 | 0.19871 |
| 0.0023520 | −1.15588 | 0.0069408 | −1.06072 |
| 0.0026016 | −0.04758 | 0.0071904 | −1.51412 |
| 0.0028640 | 1.36858 | 0.0074496 | −0.97116 |
| 0.0031136 | 1.50851 | 0.0076992 | 0.23229 |
| 0.0033728 | 1.51971 | 0.0079584 | 1.46933 |
| 0.0036224 | 1.51411 | 0.0082080 | 1.51411 |
| 0.0038816 | 1.45813 | 0.0084672 | 1.51971 |
| 0.0041312 | 0.32185 | 0.0087168 | 1.50851 |
| 0.0043904 | −0.97676 | 0.0089792 | 1.36298 |
| 0.0046400 | −1.51971 | | |

**(a)** Graph a scatter plot of the data.

**(b)** Determine $a$, $b$, and $h$ so that the equation $y = a \sin [b(t - h)]$ is a model for the data.

**(c)** Determine the frequency of the sinusoid in (b), and use Table 4.4 to identify the musical note produced by the tuning fork.

**(d)** Identify the musical note produced by the tuning fork used in Exercise 35.  G

| Chapter 4 | Key Ideas |

## Concepts

amplitude (p. 371)
angle of depression (p. 404)
angle of elevation (p. 404)
angular speed (p. 341)
arccosine (inverse cosine) (p. 397)
arcsine (inverse sine) (p. 395)
arctangent (inverse tangent) (p. 397–398)
bearing (p. 338)
central angle (p. 338)
circular functions (p. 363)
cosecant (p. 346, 358, 362, 382)
cosine (p. 346, 358, 362, 369)
cotangent (p. 346, 358, 362, 381)
coterminal angles (p. 356)
course (bearing) (p. 338)
damped oscillation (p. 391)

damping factor (p. 391)
degree (p. 338)
DMS measure (p. 338)
frequency (p. 372, 406)
length of an arc (p. 340–341)
line of travel (p. 338)
linear speed (p. 341)
minute (p. 338)
nautical mile (p. 341)
negative angle (p. 355)
period (p. 363, 372)
periodic function (p. 363)
phase shift (p. 373)
positive angle (p. 355)
quadrantal angle (p. 359)
radian (p. 339)

reference triangle (p. 358)
right triangle trigonometry (p. 345)
secant (p. 346, 358, 362, 381)
second (p. 338)
simple harmonic motion (p. 406)
sine (p. 346, 358, 362, 368)
sinusoid (p. 371, 374)
standard position of an angle
    (p. 346, 355)
statute mile (p. 342)
tangent (p. 346, 358, 362, 379)
trigonometric functions
    (p. 346, 358, 362)
unit circle (p. 362)

## Properties, Theorems, and Formulas

### Arc Length

If $\theta$ is a central angle (measured in radians) in a circle of radius $r$, then the length $s$ of the intercepted arc is $s = r\theta$.

### Right Triangle Trigonometric Ratios

$$\sin \theta = \frac{\text{opposite}}{\text{hypotenuse}} \qquad \tan \theta = \frac{\text{opposite}}{\text{adjacent}} \qquad \sec \theta = \frac{\text{hypotenuse}}{\text{adjacent}}$$

$$\cos \theta = \frac{\text{adjacent}}{\text{hypotenuse}} \qquad \cot \theta = \frac{\text{adjacent}}{\text{opposite}} \qquad \csc \theta = \frac{\text{hypotenuse}}{\text{opposite}}$$

### Trigonometric Functions of Real Numbers

If $(x, y)$ is the point on the unit circle on the terminal side of an angle $\theta$ in standard position, then

$$\sin \theta = y \qquad \tan \theta = \frac{y}{x} \ (x \neq 0) \qquad \sec \theta = \frac{1}{x} \ (x \neq 0)$$

$$\cos \theta = x \qquad \cot \theta = \frac{x}{y} \ (y \neq 0) \qquad \csc \theta = \frac{1}{y} \ (y \neq 0)$$

### Special Angles

|  | degrees | radians | $\sin \theta$ | $\cos \theta$ | $\tan \theta$ | $\cot \theta$ | $\sec \theta$ | $\csc \theta$ |
|---|---|---|---|---|---|---|---|---|
| $\theta$ | 0 | 0 | 0 | 1 | 0 | undef | 1 | undef |
| $\theta$ | 30 | $\pi/6$ | $1/2$ | $\sqrt{3}/2$ | $1/\sqrt{3}$ | $\sqrt{3}$ | $2/\sqrt{3}$ | 2 |
| $\theta$ | 45 | $\pi/4$ | $1/\sqrt{2}$ | $1/\sqrt{2}$ | 1 | 1 | $\sqrt{2}$ | $\sqrt{2}$ |
| $\theta$ | 60 | $\pi/3$ | $\sqrt{3}/2$ | $1/2$ | $\sqrt{3}$ | $1/\sqrt{3}$ | 2 | $2/\sqrt{3}$ |
| $\theta$ | 90 | $\pi/2$ | 1 | 0 | undef | 0 | undef | 1 |
| $\theta$ | 180 | $\pi$ | 0 | $-1$ | 0 | undef | $-1$ | undef |
| $\theta$ | 270 | $3\pi/2$ | $-1$ | 0 | undef | 0 | undef | $-1$ |
| $\theta$ | 360 | $2\pi$ | 0 | 1 | 0 | undef | 1 | undef |

### Sinusoids

The sinusoid $f(x) = a \sin (b(x - h)) + k$ has

amplitude $|a|$;

period $2\pi/|b|$;

frequency $|b|/(2\pi)$;

phase shift $h$;

vertical shift $k$.

If $y_1 = a_1 \sin (b(x - h_1))$ and $y_2 = a_2 \cos (b(x - h))$, then

$$y_1 + y_2 = a_1 \sin (b(x - h_1)) + a_2 \cos (b(x - h_2))$$

is a sinusoid with period $2\pi/|b|$.

## Procedures

### Angle Measure Conversion

To convert radians to degrees, multiply by $\dfrac{180°}{\pi \text{ radians}}$.

To convert degrees to radians, multiply by $\dfrac{\pi \text{ radians}}{180°}$.

## Gallery of Functions

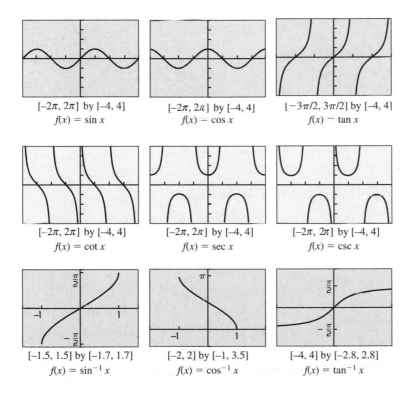

$[-2\pi, 2\pi]$ by $[-4, 4]$
$f(x) = \sin x$

$[-2\pi, 2\pi]$ by $[-4, 4]$
$f(x) = \cos x$

$[-3\pi/2, 3\pi/2]$ by $[-4, 4]$
$f(x) = \tan x$

$[-2\pi, 2\pi]$ by $[-4, 4]$
$f(x) = \cot x$

$[-2\pi, 2\pi]$ by $[-4, 4]$
$f(x) = \sec x$

$[-2\pi, 2\pi]$ by $[-4, 4]$
$f(x) = \csc x$

$[-1.5, 1.5]$ by $[-1.7, 1.7]$
$f(x) = \sin^{-1} x$

$[-2, 2]$ by $[-1, 3.5]$
$f(x) = \cos^{-1} x$

$[-4, 4]$ by $[-2.8, 2.8]$
$f(x) = \tan^{-1} x$

The collection of exercises marked in red could be used as a chapter test.

In Exercises 1–8, determine the quadrant of the terminal side of the angle in standard position. Convert degree measures to radians and radian measures to degrees.

**1.** $\dfrac{5\pi}{2}$ positive y-axis; 450°  **2.** $\dfrac{3\pi}{4}$ QII; 135°

**3.** −135° QIII; −3π/4  **4.** −45° QIV; −π/4

**5.** 78° QI; 13π/30  **6.** 112° QII; 28π/45

**7.** $\dfrac{\pi}{12}$ QI; 15°  **8.** $\dfrac{7\pi}{10}$ QII; 126°

In Exercises 9 and 10, determine the angle measure in both degrees and radians. Draw the angle in standard position if its terminal side is obtained as described.

**9.** A three-quarters counterclockwise rotation

**10.** Two and one-half counterclockwise rotations

In Exercises 11–16, the point is on the terminal side of an angle in standard position. Give the smallest positive angle measure in both degrees and radians.

**11.** $(\sqrt{3}, 1)$ 30° = π/6 rad  **12.** $(-1, 1)$ 135° = 3π/4 rad

**13.** $(-1, \sqrt{3})$120° = 2π/3 rad **14.** $(-3, -3)$225° = 5π/4 rad

**15.** $(6, -12)$  **16.** $(2, 4)$

In Exercises 17–28, evaluate the expression exactly without a calculator.

**17.** sin 30° 1/2  **18.** cos 330° $\sqrt{3}/2$

**19.** tan −135° 1  **20.** sec −135° $-\sqrt{2}$

**21.** $\sin \dfrac{5\pi}{6}$ 1/2  **22.** $\csc \dfrac{2\pi}{3}$ $2/\sqrt{3}$

**23.** $\sec -\dfrac{\pi}{3}$ 2  **24.** $\tan -\dfrac{2\pi}{3}$ $\sqrt{3}$

**25.** csc 270° −1  **26.** sec 180° −1

**27.** cot −90° 0  **28.** tan 360° 0

In Exercises 29–32, evaluate exactly all six trigonometric functions of the angle. Use reference triangles and not your calculator.

**29.** $-\dfrac{\pi}{6}$  **30.** $\dfrac{19\pi}{4}$

**31.** −135°  **32.** 420°

**33.** Find all six trigonometric functions of $\alpha$ in $\triangle ABC$.

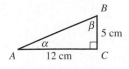

**34.** Use a right triangle to determine the values of all trigonometric functions of $\theta$, where cos $\theta$ = 5/7.

**35.** Use a right triangle to determine the values of all trigonometric functions of $\theta$, where tan $\theta$ = 15/8.

**36.** Use a calculator in degree mode to solve cos $\theta$ = 3/7 if $0° \le \theta \le 90°$. ≈ 64.623°

**37.** Use a calculator in radian mode to solve tan $x$ = 1.35 if $\pi \le x \le 3\pi/2$. ≈ 4.075 radians

**38.** Use a calculator in radian mode to solve sin $x$ = 0.218 if $0 \le x \le 2\pi$.

In Exercises 39–44, solve the right triangle $\triangle ABC$.

**39.** $\alpha = 35°$, $c = 15$  **40.** $b = 8$, $c = 10$

**41.** $\beta = 48°$, $a = 7$  **42.** $\alpha = 28°$, $c = 8$

**43.** $b = 5$, $c = 7$  **44.** $a = 2.5$, $b = 7.3$

In Exercises 45–48, $x$ is an angle in standard position with $0 \le x \le 2\pi$. Determine the quadrant of $x$.

**45.** sin $x$ < 0 and tan $x$ > 0 QIII **46.** cos $x$ < 0 and csc $x$ > 0 QII

**47.** tan $x$ < 0 and sin $x$ > 0  QII **48.** sec $x$ < 0 and csc $x$ > 0 QII

In Exercises 49–52, point $P$ is on the terminal side of angle $\theta$. Evaluate the six trigonometric functions for $\theta$.

**49.** $(-3, 6)$  **50.** $(12, 7)$

**51.** $(-5, -3)$  **52.** $(4, 9)$

In Exercises 53–60, use transformations to describe how the graph of the function is related to a basic trigonometric graph. Graph two periods.

**53.** $y = \sin (x + \pi)$  **54.** $y = 3 + 2 \cos x$

**55.** $y = -\cos (x + \pi/2) + 4$  **56.** $y = -2 - 3 \sin (x - \pi)$

**57.** $y = \tan 2x$  **58.** $y = -2 \cot 3x$

**59.** $y = -2 \sec \dfrac{x}{2}$  **60.** $y = \csc \pi x$

In Exercises 61–66, state the amplitude, period, phase shift, domain, and range for the sinusoid.

**61.** $f(x) = 2 \sin 3x$  **62.** $g(x) = 3 \cos 4x$

**63.** $f(x) = 1.5 \sin (2x - \pi/4)$  **64.** $g(x) = -2 \sin (3x - \pi/3)$

**65.** $y = 4 \cos (2x - 1)$  **66.** $g(x) = -2 \cos (3x + 1)$

In Exercises 67 and 68, graph the function. Then estimate the values of $a$, $b$, and $h$ so that $f(x) \approx a \sin (b(x - h))$.

**67.** $f(x) = 2 \sin x - 4 \cos x$   $a \approx 4.47$, $b = 1$, and $h \approx 1.11$

**68.** $f(x) = 3 \cos 2x - 2 \sin 2x$

In Exercises 69–72, use a calculator to evaluate the expression. Express your answer in both degrees and radians.

**69.** $\sin^{-1}(0.766)$                    **70.** $\cos^{-1}(0.479)$

**71.** $\tan^{-1} 1$   $45° = \pi/4$ rad      **72.** $\sin^{-1} \left( \dfrac{\sqrt{3}}{2} \right)$

In Exercises 73–76, use transformations to describe how the graph of the function is related to a basic inverse trigonometric graph. State the domain and range.

**73.** $y = \sin^{-1} 3x$                     **74.** $y = \tan^{-1} 2x$

**75.** $y = \sin^{-1} (3x - 1) + 2$           **76.** $y = \cos^{-1} (2x + 1) - 3$

In Exercises 77–82, find the exact value of $x$ without using a calculator.

**77.** $\sin x = 0.5$,   $\pi/2 \leq x \leq \pi$   $5\pi/6$

**78.** $\cos x = \sqrt{3}/2$,   $0 \leq x \leq \pi$   $\pi/6$

**79.** $\tan x = -1$,   $0 \leq x \leq \pi$   $3\pi/4$

**80.** $\sec x = 2$,   $\pi \leq x \leq 2\pi$   $5\pi/3$

**81.** $\csc x = -1$,   $0 \leq x \leq 2\pi$   $3\pi/2$

**82.** $\cot x = -\sqrt{3}$,   $0 \leq x \leq \pi$   $5\pi/6$

In Exercises 83 and 84, describe the end behavior of the function.

**83.** $\dfrac{\sin x}{x^2}$          **84.** $\dfrac{3}{5} e^{-x/12} \sin (2x - 3)$

In Exercises 85–88, evaluate the expression without a calculator.

**85.** $\tan (\tan^{-1} 1)$   1         **86.** $\cos^{-1} (\cos \pi/3)$   $\pi/3$

**87.** $\tan (\sin^{-1} 3/5)$   $3/4$    **88.** $\cos^{-1} (\cos -\pi/7)$   $\pi/7$

In Exercises 89–92, determine whether the function is periodic. State the period (if applicable), the domain, and the range.

**89.** $f(x) = |\sec x|$

**90.** $g(x) = \sin |x|$

**91.** $f(x) = 2x + \tan x$

**92.** $g(x) = 2 \cos 2x + 3 \sin 5x$

**93. Arc Length**  Find the length of the arc intercepted by a central angle of $2\pi/3$ rad in a circle with radius 2.   $4\pi/3$

**94. Algebraic Expression**  Find an algebraic expression equivalent to $\tan (\cos^{-1} x)$.   $\sqrt{1 - x^2}/x$

**95. Height of Building**  The angle of elevation of the top of a building from a point 100 m away from the building on level ground is 78°. Find the height of the building.

**96. Height of Tree**  A tree casts a shadow 51 ft long when the angle of elevation of the sun (measured with the horizon) is 25°. How tall is the tree?   $51 \tan 25° \approx 23.8$ ft

**97. Traveling Car**  From the top of a 150-ft building Flora observes a car moving toward her. If the angle of depression of the car changes from 18° to 42° during the observation, how far does the car travel?

**98. Finding Distance**  A lighthouse $L$ stands 4 mi from the closest point $P$ along a straight shore (see figure). Find the distance from $P$ to a point $Q$ along the shore if $\angle PLQ = 22°$.   $4 \tan 22° \approx 1.62$ mi

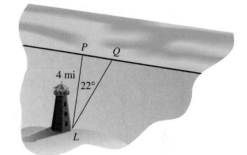

**99. Navigation**  An airplane is flying due east between two signal towers. One tower is due north of the other. The bearing from the plane to the north tower is 23°, and to the south tower is 128°. Use a drawing to show the exact location of the plane.

**100. Finding Distance**  The bearings of two points on the shore from a boat are 115° and 123°. Assume the two points are 855 ft apart. How far is the boat from the nearest point on shore if the shore is straight and runs north-south?

**101. Height of Tree**  Dr. Thom Lawson standing on flat ground 62 ft from the base of a Douglas fir measures the angle of elevation to the top of the tree as 72°24′. What is the height of the tree?   $62 \tan 72°24′ \approx 195.4$ ft

**102. Storing Hay**  A 75-ft-long conveyor is used at the Lovelady Farm to put hay bales up for winter storage. The conveyor is tilted to an angle of elevation of 22°.

(a) To what height can the hay be moved?

(b) If the conveyor is repositioned to an angle of 27°, to what height can the hay be moved?

**103. Swinging Pendulum**  In the Hardy Boys Adventure *While the Clock Ticked*, the pendulum of the grandfather clock at the Purdy place is 44 in. long and swings through an arc of 6°. Find the length of the arc that the pendulum traces.   $22\pi/15 \approx 4.6$ in.

**104. Finding Area**  A windshield wiper on a Plymouth Acclaim is 20 in. long and has a blade 16 in. long. If the wiper sweeps through an angle of 110°, how large an area does the wiper blade clean? (See Exercise 60 in Section 4.1.)

**105. Modeling Mean Temperature** The average daily air temperature (°F) for Fairbanks, Alaska, from 1941 to 1970, can be modeled by the equation

$$T(x) = 37 \sin\left[\frac{2\pi}{365}(x - 101)\right] + 25,$$

where $x$ is time in days with $x = 1$ representing January 1. On what days do you expect the average temperature to be 32°F?

**106. Taming The Beast** The Beast is a featured roller coaster at the King Island's amusement park just north of Cincinnati. On its first and biggest hill, The Beast drops from a height of 52 ft above the ground along a sinusoidal path to a depth 18 ft underground as it enters a frightening tunnel. The mathematical model for this part of track is

$$h(x) = 35 \cos\left(\frac{x}{35}\right) + 17, 0 \le x \le 110,$$

where $x$ is the horizontal distance from the top of the hill and $h(x)$ is the vertical position relative to ground level (both in feet). What is the horizontal distance from the top of the hill to the point where the track reaches ground level?

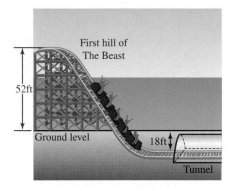

# Chapter 4 Project

## Modeling the Motion of a Pendulum

As a simple pendulum swings back and forth, its displacement can be modeled using a standard sinusoidal equation of the form:

$$y = a \cos (b(x - c)) + d$$

where *y* represents the pendumum's distance from a fixed point and *x* represents total elapsed time. In this project, you will use a motion detection device to collect distance and time data for a swinging pendulum, then find a mathematical model that describes the pendulum's motion.

## Collecting the Data

To start, construct a simple pendulum by fastening about 1 meter of string to the end of a ball. Setup the Calculator Based Laboratory (CBL) system with a motion detector or a Calculator Based Ranger (CBR) system to collect time and distance readings for between 2 and 4 seconds (enough time to capture at least one complete swing of the pendulum). See the CBL/CBR guidebook for specific setup instructions. Start the pendulum swinging in front of the detector, then activate the system. The data table below shows a sample set of data collected as a pendulum swung back and forth in front of a CBR.

| Total elapsed time (seconds) | Distance from the CBR (meters) | Total elapsed time (seconds) | Distance from the CBR (meters) |
|---|---|---|---|
| 0 | 0.665 | 1.1 | 0.525 |
| 0.1 | 0.756 | 1.2 | 0.509 |
| 0.2 | 0.855 | 1.3 | 0.495 |
| 0.3 | 0.903 | 1.4 | 0.521 |
| 0.4 | 0.927 | 1.5 | 0.575 |
| 0.5 | 0.931 | 1.6 | 0.653 |
| 0.6 | 0.897 | 1.7 | 0.741 |
| 0.7 | 0.837 | 1.8 | 0.825 |
| 0.8 | 0.753 | 1.9 | 0.888 |
| 0.9 | 0.663 | 2.0 | 0.921 |
| 1.0 | 0.582 | | |

## Explorations

1. If you collected motion data using a CBL or CBR, a plot of distance versus time should be shown on your graphing calculator or computer screen. If you don't have access to a CBL/CBR, enter the data in the table above into your graphing calculator/computer. Create a scatter plot for the data.

2. Find values for *a*, *b*, *c*, and *d* so that the equation $y = a \cos (b(x - c)) + d$ fits the distance versus time data plot. Refer to the information box on page XXX in this chapter to review sinusoidal graph characteristics.

3. What are the physical meanings of the constants *a* and *d* in the modeling equation $y = a \cos (b(x - c)) + d$? (Hint: What distances do *a* and *d* measure?).

4. Which, if any, of the values of *a*, *b*, *c*, and/or *d* would change if you used the equation $y = a \sin (b(x - c)) + d$ to model the data set?

5. Use your calculator or computer to find a sinusoidal regression equation to model this data set (see your grapher's guidebook for instructions on how to do this). If your calculator/computer uses a different sinusoidal form, compare it to the modeling equation you found earlier, $y = a \cos (b(x - c)) + d$.

**Teaching Note**

The identities and proofs presented in this section are fairly standard. Although the role of identities is different in today's technologically oriented mathematics, you should feel free to explore the fundamental identities using a grapher and to design other approaches to teaching them.

## Basic Trigonometric Identities

**Reciprocal Identities**

$$\csc\theta = \frac{1}{\sin\theta} \qquad \sec\theta = \frac{1}{\cos\theta} \qquad \cot\theta = \frac{1}{\tan\theta}$$

$$\sin\theta = \frac{1}{\csc\theta} \qquad \cos\theta = \frac{1}{\sec\theta} \qquad \tan\theta = \frac{1}{\cot\theta}$$

**Quotient Identities**

$$\tan\theta = \frac{\sin\theta}{\cos\theta} \qquad \cot\theta = \frac{\cos\theta}{\sin\theta}$$

**Exploration 1** Making a Point about Domain of Validity

1. $\theta = 0$ is in the domain of validity of exactly three of the basic identities. Which three?

2. For exactly two of the basic identities, one side of the equation is defined at $\theta = 0$ and the other side is not. Which two?

3. For exactly three of the basic identities, both sides of the equation are undefined at $\theta = 0$. Which three?

**Exploration Extensions**

$\theta = \pi/2$ is in the domain of validity of exactly three of the Basic Identities. Which three?

## Pythagorean Identities

Exploration 2 in Section 4.3 introduced you to the fact that, for any real number $t$, the numbers $(\cos t)^2$ and $(\sin t)^2$ always sum to 1. This is clearly true for the quadrantal angles that wrap to the points $(\pm 1, 0)$ and $(0, \pm 1)$, and it is true for any other $t$ because $\cos t$ and $\sin t$ are the (signed) lengths of legs of a reference triangle with hypotenuse 1 (Figure 5.1). No matter what quadrant the triangle lies in, the Pythagorean theorem guarantees the following identity: $(\cos t)^2 + (\sin t)^2 = 1$.

If we divide each term of the identity by $(\cos t)^2$, we get an identity that involves tangent and secant:

$$\frac{(\cos t)^2}{(\cos t)^2} + \frac{(\sin t)^2}{(\cos t)^2} = \frac{1}{(\cos t)^2}$$

$$1 + (\tan t)^2 = (\sec t)^2$$

If we divide each term of the identity by $(\cos t)^2$, we get an identity that involves cotangent and cosecant:

$$\frac{(\cos t)^2}{(\sin t)^2} + \frac{(\sin t)^2}{(\sin t)^2} = \frac{1}{(\sin t)^2}$$

$$(\cot t)^2 + 1 = (\csc t)^2$$

These three identities are called the *Pythagorean Identities*, which we re-state below using the shorthand notation for powers of trigonometric functions.

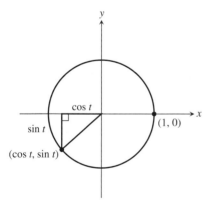

**Figure 5.1** By the Pythagorean theorem, $(\cos t)^2 + (\sin t)^2 = 1$.

> **Pythagorean Identities**
>
> $$\cos^2 \theta + \sin^2 \theta = 1$$
> $$1 + \tan^2 \theta = \sec^2 \theta$$
> $$\cot^2 \theta + 1 = \csc^2 \theta$$

### Example 1   USING IDENTITIES

Find $\sin \theta$ and $\cos \theta$ if $\tan \theta = 5$ and $\cos \theta > 0$.

**Solution**  We could solve this problem by the reference triangle techniques of Section 4.3 (see Example 7 in that section), but we will show an alternate solution here using only identities.

First, we note that $\sec^2 \theta = 1 + \tan^2 \theta = 1 + 5^2 = 26$, so $\sec \theta = \pm\sqrt{26}$.

Since $\sec \theta = \pm\sqrt{26}$, we have $\cos \theta = 1/\sec \theta = 1/\pm\sqrt{26}$.

But $\cos \theta > 0$, so $\cos \theta = 1/\sqrt{26}$.

Finally,

$$\tan \theta = 5$$

$$\frac{\sin \theta}{\cos \theta} = 5$$

$$\sin \theta = 5 \cos \theta = 5\left(\frac{1}{\sqrt{26}}\right) = \frac{5}{\sqrt{26}}.$$

Therefore, $\sin \theta = \dfrac{5}{\sqrt{26}}$ and $\cos \theta = \dfrac{1}{\sqrt{26}}$.

If you find yourself preferring the reference triangle method, that's fine. Remember that combining the power of geometry and algebra to solve problems is one of the themes of this book, and the instinct to do so will serve you well in calculus.

## Cofunction Identities

If $C$ is the right angle in right $\triangle ABC$, then angles $A$ and $B$ are complements. Notice what happens if we use the usual triangle ratios to define the six trigonometric functions of angles $A$ and $B$ (Figure 5.2).

Angle $A$:   $\sin A = \dfrac{y}{r}$   $\tan A = \dfrac{y}{x}$   $\sec A = \dfrac{r}{x}$

$\cos A = \dfrac{x}{r}$   $\cot A = \dfrac{x}{y}$   $\csc A = \dfrac{r}{y}$

Angle $B$:   $\sin B = \dfrac{x}{r}$   $\tan B = \dfrac{x}{y}$   $\sec B = \dfrac{r}{y}$

$\cos B = \dfrac{y}{r}$   $\cot B = \dfrac{y}{x}$   $\csc B = \dfrac{r}{x}$

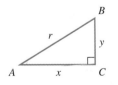

**Figure 5.2** Angles $A$ and $B$ are complements in right $\triangle ABC$.

Do you see what happens? In every case, the value of a function at $A$ is the same as the value of its cofunction at $B$. This always happens with complementary angles; in fact, it is this phenomenon that gives a "co"function its name. The "co" stands for "complement."

---

**Cofunction Identities**

$$\sin\left(\frac{\pi}{2} - \theta\right) = \cos\theta \qquad \cos\left(\frac{\pi}{2} - \theta\right) = \sin\theta$$

$$\tan\left(\frac{\pi}{2} - \theta\right) = \cot\theta \qquad \cot\left(\frac{\pi}{2} - \theta\right) = \tan\theta$$

$$\sec\left(\frac{\pi}{2} - \theta\right) = \csc\theta \qquad \csc\left(\frac{\pi}{2} - \theta\right) = \sec\theta$$

---

Although our argument on behalf of these equations was based on acute angles in a triangle, these equations are genuine identities, valid for all real numbers for which both sides of the equation are defined. We could extend our acute-angle argument to produce a general proof, but it will be easier to wait and use the identities of Section 5.3. We will revisit this particular set of fundamental identities in that section.

## Odd–Even Identities

We have seen that every basic trigonometric function is either odd or even. Either way, the usual function relationship leads to another fundamental identity.

---

**Odd–Even Identities**

$$\sin(-x) = -\sin x \qquad \cos(-x) = \cos x \qquad \tan(-x) = -\tan x$$

$$\csc(-x) = -\csc x \qquad \sec(-x) = \sec x \qquad \cot(-x) = -\cot x$$

---

**Notes on Examples**

For Example 2, it is instructive to make a sketch showing the possible values of $\theta$. Then show that any value of $\theta$ satisfying $\cos\theta = 0.34$ will also satisfy $\sin(\theta - \pi/2) = -0.34$.

### Example 2 USING MORE IDENTITIES

If $\cos\theta = 0.34$, find $\sin(\theta - \pi/2)$.

**Solution** This problem can best be solved using identities.

$$\sin\left(\theta - \frac{\pi}{2}\right) = -\sin\left(\frac{\pi}{2} - \theta\right) \qquad \text{Because sine is odd}$$

$$= -\cos\theta \qquad \text{Cofunction Identity}$$

$$= -0.34$$

## Simplifying Trigonometric Expressions

In calculus it is often necessary to deal with expressions that involve trigonometric functions. Some of those expressions start out looking fairly complicated, but it is often possible to use identities along with algebraic techniques (e.g., factoring or combining fractions over a common denominator) to *simplify* the expressions before dealing with them. In some cases the simplifications can be dramatic.

## Example 3  SIMPLIFYING BY FACTORING AND USING IDENTITIES

Simplify the expression $\sin^3 x + \sin x \cos^2 x$.

### Solution

### Solve Algebraiclly

$$\sin^3 x + \sin x \cos^2 x = \sin x \, (\sin^2 x + \cos^2 x)$$

$$= \sin x \, (1) \qquad \text{Pythagorean Identity}$$

$$= \sin x$$

### Support Graphically

We recognize the graph of $y = \sin^3 x + \sin x \cos^2 x$ as the same as the graph of $y = \sin x$. (Figure 5.3)

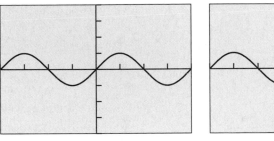

| $[-2\pi, 2\pi]$ by $[-4, 4]$ | $[-2\pi, 2\pi]$ by $[-4, 4]$ |
|:---:|:---:|
| (a) | (b) |

**Figure 5.3**  Graphical support of the identity $\sin^3 x + \sin x \cos^2 x = \sin x$. (Example 3)

## Example 4  SIMPLIFYING BY EXPANDING AND USING IDENTITIES

Simplify the expression $[(\sec x + 1)(\sec x - 1)]/\sin^2 x$.

### Solution

### Solve Algebraically

$$\frac{(\sec x + 1)(\sec x - 1)}{\sin^2 x} = \frac{\sec^2 x - 1}{\sin^2 x} \qquad (a + b)(a - b) = a^2 - b^2$$

$$= \frac{\tan^2 x}{\sin^2 x} \qquad \text{Pythagorean Identity}$$

$$= \frac{\sin^2 x}{\cos^2 x} \cdot \frac{1}{\sin^2 x} \qquad \tan x = \frac{\sin x}{\cos x}$$

$$= \frac{1}{\cos^2 x}$$

$$= \sec^2 x$$

### Support Graphically

The graphs of $y = \dfrac{(\sec x + 1)(\sec x - 1)}{\sin^2 x}$ and $y = \sec^2 x$ appear to be identical, as expected. (Figure 5.4)

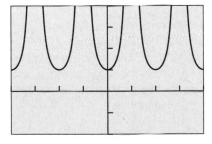

$[-2\pi, 2\pi]$ by $[-2, 4]$

(a)

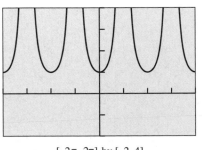

$[-2\pi, 2\pi]$ by $[-2, 4]$

(b)

**Figure 5.4**  Graphical support of the identity $(\sec x + 1)(\sec x - 1)/\sin^2 x = \sec^2 x$. (Example 4)

**Example 5   SIMPLIFYING BY COMBINING FRACTIONS AND USING IDENTITIES**

Simplify the expression $\dfrac{\cos x}{1 - \sin x} - \dfrac{\sin x}{\cos x}$.

**Solution**

$$\frac{\cos x}{1 - \sin x} - \frac{\sin x}{\cos x}$$

$$= \frac{\cos x}{1 - \sin x} \cdot \frac{\cos x}{\cos x} - \frac{\sin x}{\cos x} \cdot \frac{1 - \sin x}{1 - \sin x} \qquad \text{Rewrite using common denominator}$$

$$= \frac{(\cos x)(\cos x) - (\sin x)(1 - \sin x)}{(1 - \sin x)(\cos x)}$$

$$= \frac{\cos^2 x - \sin x + \sin^2 x}{(1 - \sin x)(\cos x)}$$

$$= \frac{1 - \sin x}{(1 - \sin x)(\cos x)} \qquad \text{Pythagorean Identity}$$

$$= \frac{1}{\cos x}$$

$$= \sec x$$

(We leave it to you to provide the graphical support.)

We will use these same simplifying techniques to prove trigonometric identities in Section 5.2.

## Solving Trigonometric Equations

The equation-solving capabilities of calculators have made it possible to solve trigonometric equations without understanding much trigonometry. This is fine, to the extent that solving equations is our goal. However, since understanding trigonometry is also a goal, we will occasionally pause in our development of identities to solve some trigonometric equations with paper and pencil, just to get some practice in using the identities.

**Example 6   SOLVING A TRIGONOMETRIC EQUATION**

Find all values of $x$ in the interval $[0, 2\pi)$ that solve $\cos^3 x / \sin x = \cot x$.

**Solution**

$$\frac{\cos^3 x}{\sin x} = \cot x$$

$$\frac{\cos^3 x}{\sin x} = \frac{\cos x}{\sin x}$$

$$\cos^3 x = \cos x \qquad \text{Multiply both sides by } \sin x$$

$$\cos^3 x - \cos x = 0$$

$$(\cos x)(\cos^2 x - 1) = 0$$

$$(\cos x)(-\sin^2 x) = 0 \qquad \text{Pythagorean Identity}$$

$$\cos x = 0 \quad \text{or} \quad \sin x = 0$$

We reject the possibility that $\sin x = 0$ because it would make both sides of the original equation undefined.

The values in the interval $[0, 2\pi)$ that solve $\cos x = 0$ (and therefore $\cos^3 x/\sin x = \cot x$) are $\pi/2$ and $3\pi/2$.

## Example 7  SOLVING A TRIGONOMETRIC EQUATION BY FACTORING

Find all values of $x$ that solve $2 \sin^2 x + \sin x = 1$.

**Solution**  Let $y = \sin x$. The equation $2y^2 + y = 1$ can be solved by factoring:

$$2y^2 + y = 1$$
$$2y^2 + y - 1 = 0$$
$$(2y - 1)(y + 1) = 0$$
$$2y - 1 = 0 \quad \text{or} \quad y + 1 = 0$$
$$y = \frac{1}{2} \quad \text{or} \quad y = -1$$

So, in the original equation, $\sin x = 1/2$ or $\sin x = -1$. Figure 5.5 shows that the solutions in the interval $[0, 2\pi)$ are $\pi/6$, $5\pi/6$, and $3\pi/2$.

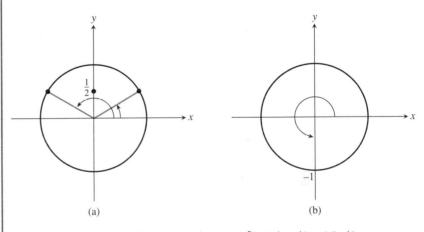

(a)                                      (b)

**Figure 5.5**  (a) $\sin x - 1/2$ has two solutions in $[0, 2\pi)$: $\pi/6$ and $5\pi/6$.
(b) $\sin x = -1$ has one solution in $[0, 2\pi)$: $3\pi/2$. (Example 7)

To get *all* real solutions, we simply add integer multiples of the period, $2\pi$, of the periodic function $\sin x$:

$$x = \frac{\pi}{6} + 2n\pi \quad \text{or} \quad x = \frac{5\pi}{6} + 2n\pi \quad \text{or} \quad x = \frac{3\pi}{2} + 2n\pi \quad (n = 0, \pm 1, \pm 2, \ldots)$$

You might try solving the equation in Example 7 on your grapher for the sake of comparison. Finding *all* real solutions still requires an understanding of periodicity, and finding *exact* solutions requires the savvy to divide your calculator answers by $\pi$. It is likely that anyone who knows that much trigonometry will actually find the algebraic solution to be easier!

# Quick Review 5.1

In Exercises 1–4, evaluate the expression.

**1.** $\sin^{-1}\left(\dfrac{12}{13}\right)$   **2.** $\cos^{-1}\left(\dfrac{3}{5}\right)$

**3.** $\cos^{-1}\left(-\dfrac{4}{5}\right)$   **4.** $\sin^{-1}\left(-\dfrac{5}{13}\right)$

In Exercises 5–8, factor the expression into a product of linear factors

**5.** $a^2 - 2ab + b^2$  $(a-b)^2$   **6.** $4u^2 + 4u + 1$  $(2u+1)^2$

**7.** $2x^2 - 3xy - 2y^2$   **8.** $2v^2 - 5v - 3$

In Exercises 9–12, simplify the expression.

**9.** $\dfrac{1}{x} - \dfrac{2}{y}$  $\dfrac{y-2x}{xy}$   **10.** $\dfrac{a}{x} + \dfrac{b}{y}$  $\dfrac{ay+bx}{xy}$

**11.** $\dfrac{x+y}{(1/x)+(1/y)}$  $xy$   **12.** $\dfrac{x}{x-y} - \dfrac{y}{x+y}$  $\dfrac{x^2+y^2}{x^2-y^2}$

# Section 5.1 Exercises

In Exercises 1–8, use basic identities to simplify the expression.

**1.** $\tan x \cos x$  $\sin x$   **2.** $\cot x \tan x$  $1$

**3.** $\sec y \sin (\pi/2 - y)$  $1$   **4.** $\cot u \sin u$  $\cos u$

**5.** $\dfrac{1 + \tan^2 x}{\csc^2 x}$  $\tan^2 x$   **6.** $\dfrac{1 - \cos^2 \theta}{\sin \theta}$  $\sin \theta$

**7.** $\cos x - \cos^3 x$  $\cos x \sin^2 x$   **8.** $\dfrac{\sin^2 u + \tan^2 u + \cos^2 u}{\sec u}$

In Exercises 9–14, simplify the expression to either 1 or −1.

**9.** $\sin x \csc (-x)$  $-1$   **10.** $\sec (-x) \cos (-x)$  $1$

**11.** $\cot (-x) \cot (\pi/2 - x)$  $-1$   **12.** $\cot (-x) \tan (-x)$  $1$

**13.** $\sin^2 (-x) + \cos^2 (-x)$  $1$   **14.** $\sec^2 (-x) - \tan^2 x$  $1$

In Exercises 15–18, simplify the expression to either a constant or a basic trigonometric function. Support your result graphically.

**15.** $\dfrac{\tan (\pi/2 - x) \csc x}{\csc^2 x}$  $\cos x$   **16.** $\dfrac{1 + \tan x}{1 + \cot x}$  $\tan x$

**17.** $(\sec^2 x + \csc^2 x) - (\tan^2 x + \cot^2 x)$  $2$

**18.** $\dfrac{\sec^2 u - \tan^2 u}{\cos^2 v + \sin^2 v}$  $1$

In Exercises 19–24, use the basic identities to change the expression to one involving only sines and cosines. Then simplify to a basic trigonometric function.

**19.** $(\sin x)(\tan x + \cot x)$  $\sec x$

**20.** $\sin \theta - \tan \theta \cos \theta + \cos (\pi/2 - \theta)$  $\sin \theta$

**21.** $\sin x \cos x \tan x \sec x \csc x$  $\tan x$

**22.** $\dfrac{(\sec y - \tan y)(\sec y + \tan y)}{\sec y}$  $\cos y$

**23.** $\dfrac{\tan x}{\csc^2 x} + \dfrac{\tan x}{\sec^2 x}$  $\tan x$

**24.** $\dfrac{\sec^2 x \csc x}{\sec^2 x + \csc^2 x}$  $\sin x$

In Exercises 25–30, combine the fractions and simplify to a multiple of a power of a basic trigonometric function (e.g., $3 \tan^2 x$).

**25.** $\sin \theta + \tan \theta \cos \theta$   **26.** $\dfrac{1}{1 - \sin x} + \dfrac{1}{1 + \sin x}$

**27.** $\dfrac{\sin x}{\cot^2 x} - \dfrac{\sin x}{\cos^2 x}$  $-\sin x$   **28.** $\dfrac{1}{\sec x - 1} - \dfrac{1}{\sec x + 1}$

**29.** $\dfrac{\sec x}{\sin x} - \dfrac{\sin x}{\cos x}$  $\cot x$   **30.** $\dfrac{\sin x}{1 - \cos x} + \dfrac{1 - \cos x}{\sin x}$

In Exercises 31–38, write each expression in factored form as an algebraic expression of a single trigonometric function (e.g., $(2 \sin x + 3)(\sin x - 1)$).

**31.** $\cos^2 x + 2 \cos x + 1$   **32.** $1 - 2 \sin x + \sin^2 x$

**33.** $1 - 2 \sin x + (1 - \cos^2 x)$   **34.** $\sin x - \cos^2 x - 1$

**35.** $\cos x - 2 \sin^2 x + 1$   **36.** $\sin^2 x + \dfrac{2}{\csc x} + 1$

**37.** $4 \tan^2 x - \dfrac{4}{\cot x} + \sin x \csc x$  $(2 \tan x - 1)^2$

**38.** $\sec^2 x - \sec x + \tan^2 x$  $(2 \sec x + 1)(\sec x - 1)$

In Exercises 39–42, write each expression as an algebraic expression of a single trigonometric function (e.g., $2 \sin x + 3$).

**39.** $\dfrac{1 - \sin^2 x}{1 + \sin x}$  $1 - \sin x$   **40.** $\dfrac{\tan^2 \alpha - 1}{1 + \tan \alpha}$  $\tan \alpha - 1$

**41.** $\dfrac{\sin^2 x}{1 + \cos x}$  $1 - \cos x$   **42.** $\dfrac{\tan^2 x}{\sec x + 1}$  $\sec x - 1$

In Exercises 43–48, find all solutions in the interval $[0, 2\pi)$.

**43.** $2 \cos x \sin x - \cos x = 0$  $\pi/6,\ \pi/2,\ 5\pi/6,\ 3\pi/2$

**44.** $\sqrt{2} \tan x \cos x - \tan x = 0$  $0,\ \pi/4,\ \pi,\ 7\pi/4$

**45.** $\tan x \sin^2 x = \tan x$  $0,\ \pi$   **46.** $\sin x \tan^2 x = \sin x$

**47.** $\tan^2 x = 3$   **48.** $2 \sin^2 x = 1$

In Exerercises 49–54, find all real solutions.

**49.** $4 \cos^2 x - 4 \cos x + 1 = 0$   **50.** $2 \sin^2 x + 3 \sin x + 1 = 0$

**51.** $\sin^2 \theta - 2 \sin \theta = 0$  **52.** $3 \sin t = 2 \cos^2 t$

**53.** $\cos (\sin x) = 1$  **54.** $2 \sin^2 x + 3 \sin x = 2$

In Exercises 55–60, make the suggested trigonometric substitution, and then use Pythagorean Identities to write the resulting function as a multiple of a basic trigonometric function.

**55.** $\sqrt{1 - x^2}$,  $x = \cos \theta$  **56.** $\sqrt{x^2 + 1}$,  $x = \tan \theta$

**57.** $\sqrt{x^2 - 9}$,  $x = 3 \sec \theta$  **58.** $\sqrt{36 - x^2}$,  $x = 6 \sin \theta$

**59.** $\sqrt{x^2 + 81}$,  $x = 9 \tan \theta$  $9 |\sec \theta|$

**60.** $\sqrt{x^2 - 100}$,  $x = 10 \sec \theta$  $10 |\tan \theta|$

# Explorations

**61.** Write all six basic trigonometric functions entirely in terms of $\sin x$.

**62.** Write all six basic trigonometric functions entirely in terms of $\cos x$.

**63. Writing to Learn** Graph the functions $y = \sin^2 x$ and $y = -\cos^2 x$ in the standard trigonometric viewing window. Describe the apparent relationship between these two graphs and verify it with a trigonometric identity.

**64. Writing to Learn** Graph the functions $y = \sec^2 x$ and $y = \tan^2 x$ in the standard trigonometric viewing window. Describe the apparent relationship between these two graphs and verify it with a trigonometric identity.  ■

**65. Orbit of the Moon** Because its orbit is elliptical, the distance from the moon to the earth in miles (measured from the center of the moon to the center of the earth) varies periodically. On Monday, January 11, 1999, the moon was at its apogee (farthest from the earth). The distance of the moon from the earth on the first ten Mondays of 1999 are recorded in Table 5.1:

**Table 5.1 Distance from Earth to Moon**

| Date | Day | Distance |
|---|---|---|
| Jan 11 | 0 | 251,967 |
| Jan 18 | 7 | 238,344 |
| Jan 25 | 14 | 225,785 |
| Feb 1 | 21 | 240,390 |
| Feb 8 | 28 | 251,807 |
| Feb 15 | 35 | 236,309 |
| Feb 22 | 42 | 226,104 |
| Mar 1 | 49 | 242,400 |
| Mar 8 | 56 | 251,331 |
| Mar 15 | 63 | 234,337 |

*Source: The World Almanac and Book of Facts, 1999.*

**(a)** Draw a scatter plot of the data, using "day" as $x$ and "distance" as $y$.

**(b)** Use your calculator to do a sine regression of $y$ on $x$. Find the equation of the best-fit sine curve and superimpose its graph on the scatter plot.

**(c)** What is the approximate number of days from apogee to apogee? Interpret this number in terms of the orbit of the moon.

**(d)** Approximately how far is the moon from the earth at perigee (closest distance)?

**(e)** Since the data begin at apogee, perhaps a cosine curve would be a more appropriate model. Use the sine curve in (b) and a cofunction identity to write a cosine curve that fits the data.

**66. Group Activity** Split the class into six groups. Assign each group one of the basic trigonometric functions and have each group construct a list of five different expressions that can be simplified to its function. Have each group exchange its list with its "cofunction" group to check for accuracy. Later, construct a quiz using two functions from each list, the object of the quiz being to simplify each expression to a basic trigonometric function.  Answers will vary.

# Extending the Ideas

**67.** Prove that $\sin^4 \theta - \cos^4 \theta = \sin^2 \theta - \cos^2 \theta$.

**68.** Find all values of $k$ that result in $\sin^2 x + 1 = k \sin x$ having an infinite solution set.

**69.** Use the cofunction identities and odd–even identities to prove that $\sin (\pi - x) = \sin x$.
[*Hint*: $\sin (\pi - x) = \sin (\pi/2 - (x - \pi/2))$].

**70.** Use the cofunction identities and odd–even identities to prove that $\cos (\pi - x) = -\cos x$.
[*Hint*: $\cos (\pi - x) = \cos (\pi/2 - (x - \pi/2))$].

**71.** Use the identity in Exercise 69 to prove that in any $\triangle ABC$, $\sin (A + B) = \sin C$.

**72.** Use the identities in Exercises 69 and 70 to find an identity for simplifying $\tan (\pi - x)$.

**Proving Trigonometric Identities**

A Proof Strategy • Proving Identities • Disproving Non-Identities • Identities in Calculus

## A Proof Strategy

Proving trigonometric identities can be confusing because you may not be accustomed to having the answer to a problem before you start solving it. Your first instinct is to write down the equation and do something to it, which is not the way to approach an identity. To illustrate the strategy we follow for proving identities, we depart from mathematics for a moment and look at a classic type of puzzle known as a word ladder.

### Example 1  CONSTRUCTING A WORD LADDER

Change the word LOVE to TRIG by changing a single letter at a time. At each step, the letters must form a common English word (no proper nouns allowed).

**Solution**

LOVE

LORE

LORD

LARD

LAID

LAIN

GAIN

GRIN

GRIP

TRIP

TRIG

This took ten steps, but it's a hard one. Can you find a shorter path?

---

**Exploration 1**  Trying Your Hand at a Word Ladder

Change LOVE to MATH under the rules stated in Example 1. (This can be done in far fewer steps than the ladder in Example 1.)

---

The object in a word ladder is to change one word into another as efficiently as possible by taking small steps, but the small steps must follow the important rule that a common English word be formed. We can easily recognize that the following ladder, while shorter than our solution to Example 1, violates the rule:

$$\text{LOVE}$$
$$\text{TOVE}$$
$$\text{TOIE}$$
$$\text{TOIG}$$
$$\text{TRIG}$$

## Proving Identities

Trigonometric identity proofs follow a strategy similar to word ladders. We are told that two expressions are equal, and the object is to prove that they are equal. We do this by changing one expression into the other by a series of intermediate steps that follow the important rule that *every intermediate step yields an expression that is equivalent to the first*.

The changes at every step are accomplished by algebraic manipulations or identities, but the manipulations or identities should be sufficiently obvious as to require no additional justification. Since "obvious" is often in the eye of the beholder, it is usually safer to err on the side of including too many steps than too few.

By working through several examples, we try to give you a sense for what is appropriate as we illustrate some of the algebraic tools that you have at your disposal.

### Example 2 PROVING AN IDENTITY

Prove the identity: $\tan x + \cot x = \sec x \csc x$.

**Solution** We begin by deciding whether to start with the expression on the right or the left. It is usually best to start with the more complicated expression, as it is easier to proceed from the complex toward the simple than to go in the other direction. The expression on the left is slightly more complicated because it involves two terms.

| | |
|---|---|
| $\tan x + \cot x = \dfrac{\sin x}{\cos x} + \dfrac{\cos x}{\sin x}$ | Basic identities |
| $= \dfrac{\sin x}{\cos x} \cdot \dfrac{\sin x}{\sin x} + \dfrac{\cos x}{\sin x} \cdot \dfrac{\cos x}{\cos x}$ | Setting up common denominator |
| $= \dfrac{\sin^2 x + \cos^2 x}{\cos x \cdot \sin x}$ | |
| $= \dfrac{1}{\cos x \cdot \sin x}$ | Pythagorean identity |
| $= \dfrac{1}{\cos x} \cdot \dfrac{1}{\sin x}$ | (A step you could choose to omit) |
| $= \sec x \csc x$ | Basic identities |

The margin notes on the right, called "floaters," are included here for instructional purposes and are not really part of the proof. A good proof should consist of steps for which a knowledgeable reader could readily supply the floaters.

The preceding example illustrates three general strategies that are often useful in proving identities.

---

**Alert**

Some students will test an identity by substituting a number for $x$. Caution them that it is possible for an equation that is not an identity to work for certain values of $x$. For exampe, the equation $\sin x = \cos x$ is true for $x = \pi/4$, but it is not an identity.

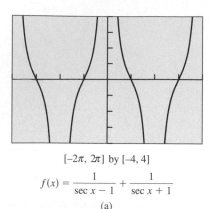

$$[-2\pi, 2\pi] \text{ by } [-4, 4]$$

$$f(x) = \frac{1}{\sec x - 1} + \frac{1}{\sec x + 1}$$

(a)

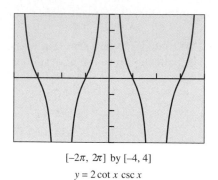

$$[-2\pi, 2\pi] \text{ by } [-4, 4]$$

$$y = 2 \cot x \csc x$$

(b)

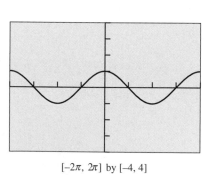

$$[-2\pi, 2\pi] \text{ by } [-4, 4]$$

$$y = \frac{1}{\sec x}$$

(c)

**Figure 5.6** A grapher can be useful for identifying possible identities. (Example 3)

---

**Notes on Examples**

The process used to solve Example 4 can be compared with using a conjugate to rationalize the denominator of $3/(2 - \sqrt{3})$.

---

**General Strategies I**

1. Begin with the more complicated expression and work toward the less complicated expression.

2. If no other move suggests itself, convert the entire expression to one involving sines and cosines.

3. Combine fractions by combining them over a common denominator.

---

**Example 3  IDENTIFYING AND PROVING AN IDENTITY**

Match the function

$$f(x) = \frac{1}{\sec x - 1} + \frac{1}{\sec x + 1}$$

with one of the following. Then confirm the match with a proof.

**(i)** $2 \cot x \csc x$ **(ii)** $\dfrac{1}{\sec x}$

**Solution** Figures 5.6a, b, and c show the graphs of the functions $y = f(x)$, $y = 2 \cot x \csc x$, and $y = 1/\sec x$, respectively. The graphs in (a) and (c) show that $f(x)$ is not equal to the expression in (ii). From the graphs in (a) and (b), it appears that $f(x)$ is equal to the expression in (i). To confirm, we begin with the expression for $f(x)$.

$$\frac{1}{\sec x - 1} + \frac{1}{\sec x + 1}$$

$$= \frac{\sec x + 1}{(\sec x - 1)(\sec x + 1)} + \frac{\sec x - 1}{(\sec x - 1)(\sec x + 1)} \quad \text{Common denominator}$$

$$= \frac{\sec x + 1 + \sec x - 1}{\sec^2 x - 1}$$

$$= \frac{2 \sec x}{\tan^2 x} \quad \text{Pythagorean identity}$$

$$= \frac{2}{\cos x} \cdot \frac{\cos^2 x}{\sin^2 x} \quad \text{Basic identities}$$

$$= \frac{2 \cos x}{\sin x} \cdot \frac{1}{\sin x}$$

$$= 2 \cot x \csc x$$

The next example illustrates how the algebraic identity $(a + b)(a - b) = a^2 - b^2$ can be used to set up a Pythagorean substitution.

**Example 4  SETTING UP A DIFFERENCE OF SQUARES**

Prove the identity: $\cos t/(1 - \sin t) = (1 + \sin t)/\cos t$.

**Solution** The left-hand expression is slightly more complicated, as we can handle extra terms in a numerator more easily than in a denominator. So we begin with the left.

$$\frac{\cos t}{1 - \sin t} = \frac{\cos t}{1 - \sin t} \cdot \frac{1 + \sin t}{1 + \sin t} \quad \text{Setting up a difference of squares}$$

$$= \frac{(\cos t)(1 + \sin t)}{1 - \sin^2 t}$$

$$= \frac{(\cos t)(1 + \sin t)}{\cos^2 t} \qquad \text{Pythagorean Identity}$$

$$= \frac{1 + \sin t}{\cos t}$$

Notice that we kept $(\cos t)(1 + \sin t)$ in factored form in the hope that we could eventually eliminate the factor $\cos t$ and be left with the numerator we need. It is always a good idea to keep an eye on the "target" expression toward which your proof is aimed.

---

**General Strategies II**

1. Use the algebraic identity $(a + b)(a - b) = a^2 - b^2$ to set up applications of the Pythagorean identities.

2. Always be mindful of the "target" expression, and favor manipulations that bring you closer to your goal.

---

In more complicated identities (and word ladders) it is sometimes helpful to see if both sides can be manipulated toward a common intermediate expression. The proof can then be reconstructed in a single path.

**Example 5  WORKING FROM BOTH SIDES**

Prove the identity: $\cot^2 u/(1 + \csc u) = (\cot u)(\sec u - \tan u)$.

**Solution** Both sides are fairly complicated, but the left-hand side looks like it needs more work. We start on the left.

$$\frac{\cot^2 u}{1 + \csc u} = \frac{\csc^2 u - 1}{1 + \csc u} \qquad \text{Pythagorean Identity}$$

$$= \frac{(\csc u + 1)(\csc u - 1)}{\csc u + 1} \qquad \text{Factor}$$

$$= \csc u - 1$$

At this point it is not clear how we can get from this expression to the one on the right-hand side of our identity. However, we now have reason to believe that the right-hand side must simplify to $\csc u - 1$, so we try simplifying the right-hand side.

$$(\cot u)(\sec u - \tan u) = \left(\frac{\cos u}{\sin u}\right)\left(\frac{1}{\cos u} - \frac{\sin u}{\cos u}\right)$$

$$= \frac{1}{\sin u} - 1$$

$$= \csc u - 1$$

Now we can reconstruct the proof by going through $\csc u - 1$ as an intermediate step.

$$\frac{\cot^2 u}{1 + \csc u} = \frac{\csc^2 u - 1}{1 + \csc u}$$

$$= \frac{(\csc u + 1)(\csc u - 1)}{\csc u + 1}$$

$$= \csc u - 1$$

$$= \frac{1}{\sin u} - 1$$

$$= \left(\frac{\cos u}{\sin u}\right)\left(\frac{1}{\cos u} - \frac{\sin u}{\cos u}\right)$$

$$= (\cot u)(\sec u - \tan u)$$

## Disproving Non-Identities

Obviously, not every equation involving trigonometric expressions is an identity. How can we spot a non-identity before embarking on a futile attempt at a proof? Try the following exploration.

---

**Exploration 2** **Confirming a Non-Identity**

Prove or disprove that this is an identity: $\cos 2x = 2 \cos x$.

1. Graph $y = \cos 2x$ and $y = 2 \cos x$ in the same window. Interpret the graphs to make a conclusion about whether or not the equation is an identity.

2. With the help of the graph, find a value of $x$ for which $\cos 2x \neq 2 \cos x$.

3. Does the existence of the $x$ value in part (2) *prove* that the equation is *not* an identity? Yes

4. Graph $y = \cos 2x$ and $y = \cos^2 x - \sin^2 x$ in the same window. Interpret the graphs to make a conclusion about whether or not $\cos 2x = \cos^2 x - \sin^2 x$ is an identity.

5. Do the graphs in part (4) *prove* that $\cos 2x = \cos^2 x - \sin^2 x$ is an identity? Explain your answer.

---

Exploration 2 suggests that we can use graphers to help confirm a *non-identity*, since we only have to produce a single value of $x$ for which the two compared expressions are defined but unequal. On the other hand, we cannot use graphers to prove that an equation *is* an identity, since, for example, the graphers can never prove that two irrational numbers are equal. Also, graphers cannot show behavior over infinite domains.

## Identities in Calculus

In most calculus problems where identities play a role, the object is to make a complicated expression simpler for the sake of computational ease. Occasionally it is actually necessary to make a *simple* expression *more complicated* for the sake of computational ease. Each of the following identities (just a sampling of many) represents a useful substitution in calculus wherein the expression on the right is simpler to deal with (even though it does not look that way). We prove one of these identities as Example 6 and leave the rest for the exercises or for future sections.

1. $\cos^3 x = (1 - \sin^2 x)(\cos x)$

2. $\sec^4 x = (1 + \tan^2 x)(\sec^2 x)$

**3.** $\sin^2 x = \dfrac{1}{2} - \dfrac{1}{2}\cos 2x$

**4.** $\cos^2 x = \dfrac{1}{2} + \dfrac{1}{2}\cos 2x$

**5.** $\sin^5 x = (1 - 2\cos^2 x + \cos^4 x)(\sin x)$

**6.** $\sin^2 x \cos^5 x = (\sin^2 x - 2\sin^4 x + \sin^6 x)(\cos x)$

### Example 6   PROVING AN IDENTITY USEFUL IN CALCULUS

Prove the following identity:
$\sin^2 \cos^5 x = (\sin^2 x - 2\sin^4 x + \sin^6 x)(\cos x).$

**Solution**   We begin with the expression on the left.

$$\sin^2 x \cos^5 x = \sin^2 x \cos^4 x \cos x$$
$$= (\sin^2 x)(\cos^2 x)^2(\cos x)$$
$$= (\sin^2 x)(1 - \sin^2 x)^2(\cos x)$$
$$= (\sin^2 x)(1 - 2\sin^2 x + \sin^4 x)(\cos x)$$
$$= (\sin^2 x - 2\sin^4 x + \sin^6 x)(\cos x)$$

**Assignment Guide**

Day 1: Ex. 3–36, multiples of 3
Day 2: Ex. 39–66, multiples of 3

**Cooperative Learning**

Group Activity: Ex. 62

**Notes on Exercises**

Ex. 7–47 should be done using algebraic skills; however, feel free to incorporate the use of a grapher to support answers or for further exploration. Ex. 69 is related to the derivative of the sine function.

**Ongoing Assessment**

Self-Assessment: Ex. 5, 11, 17, 23, 35, 43, 56
Embedded Assessment: Ex. 24, 57

## Quick Review 5.2

In Exercises 1–6, write the expression in terms of sines and cosines only. Express your answer as a single fraction.

**1.** $\csc x + \sec x \quad \dfrac{\sin x + \cos x}{\sin x \cos x}$    **2.** $\tan x + \cot x$

**3.** $\cos x \csc x + \sin x \sec x$    **4.** $\sin \theta \cot \theta - \cos \theta \tan \theta$

**5.** $\dfrac{\sin x}{\csc x} + \dfrac{\cos x}{\sec x} \quad 1$    **6.** $\dfrac{\sec \alpha}{\cos \alpha} - \dfrac{\sin \alpha}{\csc \alpha \cos^2 \alpha}$

In Exercises 7–12, determine whether or not the equation is an identity. If not, find a single value of $x$ for which the two expressions are not equal.

**7.** $\sqrt{x^2} = x$    **8.** $\sqrt[3]{x^3} = x \quad$ yes

**9.** $\sqrt{1 - \cos^2 x} = \sin x$    **10.** $\sqrt{\sec^2 x - 1} = \tan x$

**11.** $\ln \dfrac{1}{x} = -\ln x \quad$ yes    **12.** $\ln x^2 = 2 \ln x \quad$ yes

## Section 5.2 Exercises

In Exercises 1–6, tell whether or not $f(x) = \sin x$ is an identity.

**1.** $f(x) = \dfrac{\sin^2 x + \cos^2 x}{\csc x} \quad$ yes   **2.** $f(x) = \dfrac{\tan x}{\sec x} \quad$ yes

**3.** $f(x) = \cos x \cdot \cot x \quad$ no       **4.** $f(x) = \cos(x - \pi/2) \quad$ yes

**5.** $f(x) = (\sin^3 x)(1 + \cot^2 x)$    **6.** $f(x) = \dfrac{\sin 2x}{2} \quad$ no

In Exercises 7–47, prove the identity.

**7.** $(\cos x)(\tan x + \sin x \cot x) = \sin x + \cos^2 x$

**8.** $(\sin x)(\cot x + \cos x \tan x) = \cos x + \sin^2 x$

**9.** $(1 - \tan x)^2 = \sec^2 x - 2\tan x$

**10.** $(\cos x - \sin x)^2 = 1 - 2\sin x \cos x$

**11.** $\dfrac{(1 - \cos u)(1 + \cos u)}{\cos^2 u} = \tan^2 u$

**12.** $\tan x + \sec x = \dfrac{\cos x}{1 - \sin x}$

**13.** $\dfrac{\cos^2 x - 1}{\cos x} = -\tan x \sin x$

**14.** $\dfrac{\sec^2 \theta - 1}{\sin \theta} = \dfrac{\sin \theta}{1 - \sin^2 \theta}$

**15.** $(1 - \sin \beta)(1 + \csc \beta) = 1 - \sin \beta + \csc \beta - \sin \beta \csc \beta$

**16.** $\dfrac{1}{1 - \cos x} + \dfrac{1}{1 + \cos x} = 2\csc^2 x$

**17.** $(\cos t - \sin t)^2 + (\cos t + \sin t)^2 = 2$

**18.** $\sin^2 \alpha - \cos^2 \alpha = 1 - 2\cos^2 \alpha$

**19.** $\dfrac{1 + \tan^2 x}{\sin^2 x + \cos^2 x} = \sec^2 x$

**20.** $\dfrac{1}{\tan \beta} + \tan \beta = \sec \beta \csc \beta$

**21.** $\dfrac{\cos \beta}{1 + \sin \beta} = \dfrac{1 - \sin \beta}{\cos \beta}$

**22.** $\dfrac{\sec x + 1}{\tan x} = \dfrac{\tan x}{\sec x - 1}$

**23.** $\dfrac{\tan^2 x}{\sec x + 1} = \dfrac{1 - \cos x}{\cos x}$

**24.** $\dfrac{\cot v - 1}{\cot v + 1} = \dfrac{1 - \tan v}{1 + \tan v}$

**25.** $\cot^2 x - \cos^2 x = \cos^2 x \cot^2 x$

**26.** $\tan^2 \theta - \sin^2 \theta = \tan^2 \theta \sin^2 \theta$

**27.** $\cos^4 x - \sin^4 x = \cos^2 x - \sin^2 x$

**28.** $\tan^4 t + \tan^2 t = \sec^4 t - \sec^2 t$

**29.** $(x \sin \alpha + y \cos \alpha)^2 + (x \cos \alpha - y \sin \alpha)^2 = x^2 + y^2$

**30.** $\dfrac{1 - \cos \theta}{\sin \theta} = \dfrac{\sin \theta}{1 + \cos \theta}$

**31.** $\dfrac{\tan x}{\sec x - 1} = \dfrac{\sec x + 1}{\tan x}$

**32.** $\dfrac{\sin t}{1 + \cos t} + \dfrac{1 + \cos t}{\sin t} = 2 \csc t$

**33.** $\dfrac{\sin x - \cos x}{\sin x + \cos x} = \dfrac{2 \sin^2 x - 1}{1 + 2 \sin x \cos x}$

**34.** $\dfrac{1 + \cos x}{1 - \cos x} = \dfrac{\sec x + 1}{\sec x - 1}$

**35.** $\dfrac{\sin t}{1 - \cos t} + \dfrac{1 + \cos t}{\sin t} = \dfrac{2(1 + \cos t)}{\sin t}$

**36.** $\dfrac{\sin A \cos B + \cos A \sin B}{\cos A \cos B - \sin A \sin B} = \dfrac{\tan A + \tan B}{1 - \tan A \tan B}$

**37.** $\sin^2 x \cos^3 x = (\sin^2 x - \sin^4 x)(\cos x)$

**38.** $\sin^5 x \cos^2 x = (\cos^2 x - 2 \cos^4 x + \cos^6 x)(\sin x)$

**39.** $\cos^5 x = (1 - 2 \sin^2 x + \sin^4 x)(\cos x)$

**40.** $\sin^3 x \cos^3 x = (\sin^3 x - \sin^5 x)(\cos x)$

**41.** $\dfrac{\tan x}{1 - \cot x} + \dfrac{\cot x}{1 - \tan x} = 1 + \sec x \csc x$

**42.** $\dfrac{\cos x}{1 + \sin x} + \dfrac{\cos x}{1 - \sin x} = 2 \sec x$

**43.** $\dfrac{2 \tan x}{1 - \tan^2 x} + \dfrac{1}{2 \cos^2 x - 1} = \dfrac{\cos x + \sin x}{\cos x - \sin x}$

**44.** $\dfrac{1 - 3 \cos x - 4 \cos^2 x}{\sin^2 x} = \dfrac{1 - 4 \cos x}{1 - \cos x}$

**45.** $\cos^3 x = (1 - \sin^2 x)(\cos x)$

**46.** $\sec^4 x = (1 + \tan^2 x)(\sec^2 x)$

**47.** $\sin^5 x = (1 - 2 \cos^2 x + \cos^4 x)(\sin x)$

In Exercises 48–53, match the function with one of the following. Then confirm the match with a proof. (The matching is not one-to-one.)

**(a)** $\sec^2 x \csc^2 x$    **(b)** $\sec x + \tan x$    **(c)** $2 \sec^2 x$

**(d)** $\tan x \sin x$    **(e)** $\sin x \cos x$

**48.** $\dfrac{1 + \sin x}{\cos x}$ (b)

**49.** $(1 + \sec x)(1 - \cos x)$ (d)

**50.** $\sec^2 x + \csc^2 x$ (a)

**51.** $\dfrac{1}{1 + \sin x} + \dfrac{1}{1 - \sin x}$ (c)

**52.** $\dfrac{1}{\tan x + \cot x}$ (e)

**53.** $\dfrac{1}{\sec x - \tan x}$ (b)

## Explorations

In Exercises 54–59, identify a simple function that has the same graph. Then confirm your choice with a proof.

**54.** $\sin x \cot x$   cos $x$

**55.** $\cos x \tan x$   sin $x$

**56.** $\dfrac{\sin x}{\csc x} + \dfrac{\cos x}{\sec x}$   1

**57.** $\dfrac{\csc x}{\sin x} - \dfrac{\cot x \csc x}{\sec x}$   1

**58.** $\dfrac{\sin x}{\tan x}$   cos $x$

**59.** $(\sec^2 x)(1 - \sin^2 x)$   1   ■

**60. Writing to Learn** Let $\theta$ be any number that is in the domain of all six trig functions. Explain why the natural logarithms of all six basic trig functions of $\theta$ sum to 0.

**61.** If $A$ and $B$ are complementary angles, prove that $\sin^2 A + \sin^2 B = 1$.

**62. Group Activity** If your class contains $2n$ students, write the two expression from $n$ different identities on separate pieces of paper. (You can use identities from Exercises 1–47 of this section or from other textbooks, but be sure to write them all in the same variable.) Mix up the slips of paper and give one to each student. Then see how long it takes the class, without looking at the book, to pair themselves off as identities. (This activity takes on an added degree of difficulty if graphers are not allowed.)

    If the class contains an odd number of students, the teacher can join in to make it even.

## Extending the Ideas

In Exercises 62–67, confirm the identity.

**63.** $\sqrt{\dfrac{1 - \sin t}{1 + \sin t}} = \dfrac{1 - \sin t}{|\cos t|}$

**64.** $\sqrt{\dfrac{1 + \cos t}{1 - \cos t}} = \dfrac{1 + \cos t}{|\sin t|}$

**65.** $\sin^6 x + \cos^6 x = 1 - 3 \sin^2 x \cos^2 x$

**66.** $\cos^6 x - \sin^6 x = (\cos^2 x - \sin^2 x)(1 - \cos^2 x \sin^2 x)$

**67.** $\ln |\tan x| = \ln |\sin x| - \ln |\cos x|$

**68.** $\ln |\sec \theta + \tan \theta| + \ln |\sec \theta - \tan \theta| = 0$

**69. Writing to Learn** Let
$y_1 = [\sin (x + 0.001) - \sin x]/0.001$ and $y_2 = \cos x$.

**(a)** Use graphs and tables to decide whether $y_1 = y_2$.

**(b)** Find a value for $h$ so that the graph of $y_3 = y_1 - y_2$ in $[-2\pi, 2\pi]$ by $[-h, h]$ appears to be a sinusoid. Give a convincing argument that $y_3$ is a sinusoid.

**70. Hyperbolic Functions** The *hyperbolic trigonometric functions* are defined as follows:

$$\sinh x = \dfrac{e^x - e^{-x}}{2} \qquad \cosh x = \dfrac{e^x + e^{-x}}{2} \qquad \tanh x = \dfrac{\sinh x}{\cosh x}$$

$$\operatorname{csch} x = \dfrac{1}{\sinh x} \qquad \operatorname{sech} x = \dfrac{1}{\cosh x} \qquad \coth x = \dfrac{1}{\tanh x}$$

Confirm the identity.

**(a)** $\cosh^2 x - \sinh^2 x = 1$

**(b)** $1 - \tanh^2 x = \operatorname{sech}^2 x$

**(c)** $\coth^2 x - 1 = \operatorname{csch}^2 x$

**71. Writing to Learn**  Write a paragraph to explain why

$$\cos x = \cos x + \sin (10\pi x)$$

appears to be an identity when the two sides are graphed in a decimal window. Give a convincing argument that it is not an identity.

## Sum and Difference Identities

Cosine of a Difference  •  Cosine of a Sum  •  Sine of a Difference or Sum  •  Tangent of a Difference or Sum  •  Verifying a Sinusoid Algebraically

### Cosine of a Difference

There is a powerful instinct in all of us to believe that all functions obey the following law of additivity:

$$f(u + v) = f(u) + f(v).$$

In fact, very few do. If there were a Hall of Fame for algebraic blunders, the following would probably be the first two inductees:

$$(u + v)^2 = u^2 + v^2$$

$$\sqrt{u + v} = \sqrt{u} + \sqrt{v}$$

So, before we derive the true sum formulas for sine and cosine, let us clear the air with the following exploration.

---

**Exploration 1**   **Getting Past the Obvious but Incorrect Formulas**

1. Let $u = \pi$ and $v = \pi/2$.
   Find sin $(u + v)$. Find sin $(u)$ + sin $(v)$.
   Does sin $(u + v)$ = sin $(u)$ + sin $(v)$?  No

2. Let $u = 0$ and $v = 2\pi$.
   Find cos $(u + v)$. Find cos $(u)$ + cos $(v)$.
   Does cos $(u + v)$ = cos $(u)$ + cos $(v)$?  No

3. Find your own values of *u* and *v* that will confirm that
   tan $(u + v)$ ≠ tan $(u)$ + tan $(v)$.

---

We could also show easily that

$$\cos (u - v) \neq \cos (u) - \cos (v) \quad \text{and} \quad \sin (u - v) \neq \sin (u) - \sin (v).$$

As you might expect, there *are* formulas for sin $(u \pm v)$, cos $(u \pm v)$, and tan $(u \pm v)$, but Exploration 1 shows that they are not the ones our instincts would suggest. In a sense, that makes them all the more interesting. We will derive them all, beginning with the formula for cos $(u - v)$.

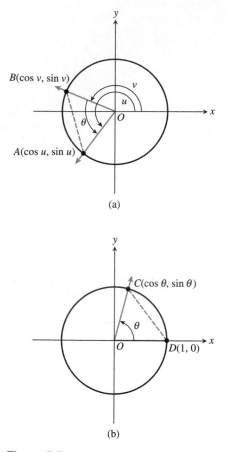

**Figure 5.7** Angles $u$ and $v$ are in standard position in (a), while angle $\theta = u - v$ is in standard position in (b). The chords shown in the two circles are equal in length.

Figure 5.7a shows angles $u$ and $v$ in standard position on the unit circle, determining points $A$ and $B$ with coordinates $(\cos u, \sin u)$ and $(\cos v, \sin v)$, respectively. Figure 5.7b shows the triangle $ABO$ rotated so that the angle $\theta = u - v$ is in standard position. The angle $\theta$ determines point $C$ with coordinates $(\cos \theta, \sin \theta)$.

The chord opposite angle $\theta$ has the same length in both circles, even though the coordinatization of the endpoints is different. Use the distance formula to find the length in each case, and set the formulas equal to each other:

$$AB = CD$$

$$\sqrt{(\cos v - \cos u)^2 + (\sin v - \sin u)^2} = \sqrt{(\cos \theta - 1)^2 + (\sin \theta - 0)^2}$$

Square both sides to eliminate the radical and expand the binomials to get

$$\cos^2 u - 2 \cos u \cos v + \cos^2 v + \sin^2 u - 2 \sin u \sin v + \sin^2 v$$
$$= \cos^2 \theta - 2 \cos \theta + 1 + \sin^2 \theta$$

$$(\cos^2 u + \sin^2 u) + (\cos^2 v + \sin^2 v) - 2 \cos u \cos v - 2 \sin u \sin v$$
$$= (\cos^2 \theta + \sin^2 \theta) + 1 - 2 \cos \theta$$

$$2 - 2 \cos u \cos v - 2 \sin u \sin v = 2 - 2 \cos \theta$$

$$\cos u \cos v + \sin u \sin v = \cos \theta$$

Finally, since $\theta = u - v$, we can write

$$\cos (u - v) = \cos u \cos v + \sin u \sin v.$$

### Example 1    USING THE COSINE DIFFERENCE IDENTITY

Find the exact value of $\cos 15°$ without using a calculator.

**Solution**  The trick is to write $\cos 15°$ as $\cos (45° - 30°)$; then we can use our knowledge of the special angles.

$$\cos 15° = \cos (45° - 30°)$$
$$= \cos 45° \cos 30° + \sin 45° \sin 30° \qquad \text{Cosine difference identity}$$
$$= \left(\frac{\sqrt{2}}{2}\right)\left(\frac{\sqrt{3}}{2}\right) + \left(\frac{\sqrt{2}}{2}\right)\left(\frac{1}{2}\right)$$
$$= \frac{\sqrt{6} + \sqrt{2}}{4}$$

## Cosine of a Sum

Now that we have the formula for the cosine of a difference, we can get the formula for the cosine of a sum almost for free by using the odd–even identities.

$$\cos (u + v) = \cos (u - (-v))$$
$$= \cos u \cos (-v) + \sin u \sin (-v) \qquad \text{Cosine difference identity}$$
$$= \cos u \cos v + \sin u (-\sin v) \qquad \text{Odd–even identities}$$
$$= \cos u \cos v - \sin u \sin v$$

We can combine the sum and difference formulas for cosine as follows:

> ### Cosine of a Sum or Difference
>
> $$\cos (u \pm v) = \cos u \cos v \mp \sin u \sin v$$
>
> (Note the sign switch in either case.)

We pointed out in Section 5.1 that the Cofunction Identities would be easier to prove with the results of Section 5.3. Here is what we mean.

### Example 2 CONFIRMING COFUNCTION IDENTITIES

Prove the identities **(a)** $\cos ((\pi/2) - x) = \sin x$ and **(b)** $\sin ((\pi/2) - x) = \cos x$.

**Solution**

**(a)** $\cos \left( \dfrac{\pi}{2} - x \right) = \cos \left( \dfrac{\pi}{2} \right) \cos x + \sin \left( \dfrac{\pi}{2} \right) \sin x$      Cosine sum identity

$$= 0 \cdot \cos x + 1 \cdot \sin x$$

$$= \sin x$$

**(b)** $\sin \left( \dfrac{\pi}{2} - x \right) = \cos \left( \dfrac{\pi}{2} - \left( \dfrac{\pi}{2} - x \right) \right)$      Because $\sin \theta = \cos ((\pi/2) - \theta)$ by previous proof

$$= \cos (0 + x)$$

$$= \cos x$$

## Sine of a Difference or Sum

We can use the cofunction identities in Example 2 to get the formula for the sine of a sum from the formula for the cosine of a difference.

$\sin (u + v) = \cos \left( \dfrac{\pi}{2} - (u + v) \right)$      Cofunction identity

$$= \cos \left( \left( \dfrac{\pi}{2} - u \right) - v \right)$$

$= \cos \left( \dfrac{\pi}{2} - u \right) \cos v + \sin \left( \dfrac{\pi}{2} - u \right) \sin v$      Cosine difference identity

$= \sin u \cos v + \cos u \sin v$      Cofunction identities

Then we can use the odd–even identities to get the formula for the sine of a difference from the formula for the sine of a sum.

$\sin (u - v) = \sin (u + ( -v))$

$= \sin u \cos (-v) + \cos u \sin (-v)$      Sine sum identity

$= \sin u \cos v + \cos u (-\sin v)$      Odd–even identities

$= \sin u \cos v - \cos u \sin v$

We can combine the sum and difference formulas for sine as follows:

## Alert

Students often make sign errors when applying the sum and difference identities.

> **Sine of a Sum of Difference**
>
> $$\sin (u \pm v) = \sin u \cos v \pm \cos u \sin v$$
>
> (Note that the sign does *not* switch in either case.)

If one of the angles in a sum or difference is a quadrantal angle (that is, a multiple of 90° or of $\pi/2$ radians), then the sum-difference identities yield single-termed expressions. Since the effect is to reduce the complexity, the resulting identity is called a **reduction formula**.

**Example 3** PROVING REDUCTION FORMULAS

Prove the reduction formulas:

**(a)** $\sin (x + \pi) = -\sin x$

**(b)** $\cos \left(x + \dfrac{3\pi}{2}\right) = \sin x$

**Solution**

**(a)** $\sin (x + \pi) = \sin x \cos \pi + \cos x \sin \pi$

$$= \sin x \cdot (-1) + \cos x \cdot 0$$

$$= -\sin x$$

**(b)** $\cos \left(x + \dfrac{3\pi}{2}\right) = \cos x \cos \dfrac{3\pi}{2} - \sin x \sin \dfrac{3\pi}{2}$

$$= \cos x \cdot 0 - \sin x \cdot (-1)$$

$$= \sin x$$

## Tangent of a Difference or Sum

We can derive a formula for $\tan (u \pm v)$ directly from the corresponding formulas for sine and cosine, as follows:

$$\tan (u \pm v) = \frac{\sin (u \pm v)}{\cos (u \pm v)} = \frac{\sin u \cos v \pm \cos u \sin v}{\cos u \cos v \mp \sin u \sin v}.$$

There is also a formula for $\tan (u \pm v)$ that is written entirely in terms of tangent functions:

$$\tan (u \pm v) = \frac{\tan u \pm \tan v}{1 \mp \tan u \tan v}$$

We will leave the proof of the all-tangent formula to the Exercises.

**Example 4** PROVING A TANGENT REDUCTION FORMULA

Prove the reduction formula: $\tan (\theta - (3\pi/2)) = -\cot \theta$.

**Solution** We can't use the all-tangent formula (Do you see why?), so we convert to sines and cosines.

$$\tan \left(\theta - \frac{3\pi}{2}\right) = \frac{\sin (\theta - (3\pi/2))}{\cos (\theta - (3\pi/2))}$$

$$= \frac{\sin \theta \cos (3\pi/2) - \cos \theta \sin (3\pi/2)}{\cos \theta \cos (3\pi/2) + \sin \theta \sin (3\pi/2)}$$

$$= \frac{\sin \theta \cdot 0 - \cos \theta \cdot (-1)}{\cos \theta \cdot 0 + \sin \theta \cdot (-1)}$$

$$= -\cot \theta$$

## Verifying a Sinusoid Algebraically

Example 1 of Section 4.6 asked us to verify that the function $f(x) = 2 \sin x + 5 \cos x$ is a sinusoid. We solved graphically, concluding that $f(x) \approx 5.39 \sin (x + 1.19)$. We now have a way of solving this kind of problem algebraically, with exact values for the amplitude and phase shift. Example 5 illustrates the technique.

### Example 5  EXPRESSING A SUM OF SINUSOIDS AS A SINUSOID

Express $f(x) = 2 \sin x + 5 \cos x$ as a sinusoid in the form $f(x) = a \sin (bx + c)$.

**Solution**  Since $a \sin (bx + c) = a (\sin bx \cos c + \cos bx \sin c)$, we have

$$2 \sin x + 5 \cos x = a (\sin bx \cos c + \cos bx \sin c)$$

$$= (a \cos c) \sin bx + (a \sin c) \cos bx.$$

Comparing coefficients, we see that $b = 1$ and that $a \cos c = 2$ and $a \sin c = 5$.

We can solve for $a$ as follows:

$$(a \cos c)^2 + (a \sin c)^2 = 2^2 + 5^2$$

$$a^2 \cos^2 c + a^2 \sin^2 c = 29$$

$$a^2(\cos^2 c + \sin^2 c) = 29$$

$$a^2 = 29$$

$$a = \pm\sqrt{29}$$

If we choose $a$ to be positive, then $\cos c = 2/\sqrt{29}$ and $\sin c = 5/\sqrt{29}$. We can identify an acute angle $c$ with those specifications as either $\cos^{-1} (2/\sqrt{29})$ or $\sin^{-1} (5/\sqrt{29})$, which are equal. So, an exact sinusoid for $f$ is

$$f(x) = 2 \sin x + 5 \cos x$$

$$= a \sin (bx + c)$$

$$= \sqrt{29} \sin (x + \cos^{-1} (2/\sqrt{29})) \text{ or } \sqrt{29} \sin (x + \sin^{-1} (5/\sqrt{29}))$$

# Quick Review 5.3

In Exercises 1–6, express the angle as a sum or difference of special angles (multiples of 30°, 45°, $\pi/6$, or $\pi/4$).

**1.** 15°  45° − 30°

**2.** 75°  45° + 30°

**3.** 165°  210° − 45°

**4.** $\pi/12$  $\pi/3 − \pi/4$

**5.** $5\pi/12$  $2\pi/3 − \pi/4$

**6.** $7\pi/12$  $5\pi/6 − \pi/4$

In Exercises 7–10, tell whether or not the identity $f(x + y) = f(x) + f(y)$ holds for the function $f$.

**7.** $f(x) = \ln x$  no

**8.** $f(x) = e^x$  no

**9.** $f(x) = 32x$  yes

**10.** $f(x) = x + 10$  no

# Section 5.3 Exercises

In Exercises 1–10, use a sum or difference identity to find an exact value.

**1.** $\sin 15°$  $(\sqrt{6} − \sqrt{2})/4$

**2.** $\tan 15°$  $2 − \sqrt{3}$

**3.** $\sin 75°$  $(\sqrt{6} + \sqrt{2})/4$

**4.** $\cos 75°$  $(\sqrt{6} − \sqrt{2})/4$

**5.** $\cos \dfrac{\pi}{12}$  $(\sqrt{2} + \sqrt{6})/4$

**6.** $\sin \dfrac{7\pi}{12}$  $(\sqrt{6} + \sqrt{2})/4$

**7.** $\tan \dfrac{5\pi}{12}$  $2 + \sqrt{3}$

**8.** $\tan \dfrac{11\pi}{12}$  $\sqrt{3} − 2$

**9.** $\cos \dfrac{7\pi}{12}$  $(\sqrt{2} − \sqrt{6})/4$

**10.** $\sin \dfrac{-\pi}{12}$  $(\sqrt{2} − \sqrt{6})/4$

In Exercises 11–22, write the expression as the sine, cosine, or tangent of an angle.

**11.** $\sin 42° \cos 17° − \cos 42° \sin 17°$  $\sin 25°$

**12.** $\cos 94° \cos 18° + \sin 94° \sin 18°$  $\cos 76°$

**13.** $\sin \dfrac{\pi}{5} \cos \dfrac{\pi}{2} + \sin \dfrac{\pi}{2} \cos \dfrac{\pi}{5}$  $\sin 7\pi/10$

**14.** $\sin \dfrac{\pi}{3} \cos \dfrac{\pi}{7} − \sin \dfrac{\pi}{7} \cos \dfrac{\pi}{3}$  $\sin 4\pi/21$

**15.** $\dfrac{\tan 19° + \tan 47°}{1 − \tan 19° \tan 47°}$

**16.** $\dfrac{\tan (\pi/5) − \tan (\pi/3)}{1 + \tan (\pi/5) \tan (\pi/3)}$

**17.** $\cos \dfrac{\pi}{7} \cos x + \sin \dfrac{\pi}{7} \sin x$

**18.** $\cos x \cos \dfrac{\pi}{7} − \sin x \sin \dfrac{\pi}{7}$

**19.** $\sin 3x \cos x − \cos 3x \sin x$  $\sin 2x$

**20.** $\cos 7y \cos 3y − \sin 7y \sin 3y$  $\cos 10y$

**21.** $\dfrac{\tan 2y + \tan 3x}{1 − \tan 2y \tan 3x}$

**22.** $\dfrac{\tan 3\alpha − \tan 2\beta}{1 + \tan 3\alpha \tan 2\beta}$

In Exercises 23–30, prove the identity.

**23.** $\sin \left( x − \dfrac{\pi}{2} \right) = −\cos x$

**24.** $\tan \left( x − \dfrac{\pi}{2} \right) = −\cot x$

**25.** $\cos \left( x − \dfrac{\pi}{2} \right) = \sin x$

**26.** $\cos \left[ \left( \dfrac{\pi}{2} − x \right) − y \right] = \sin (x + y)$

**27.** $\sin \left( x + \dfrac{\pi}{6} \right) = \dfrac{\sqrt{3}}{2} \sin x + \dfrac{1}{2} \cos x$

**28.** $\cos \left( x − \dfrac{\pi}{4} \right) = \dfrac{\sqrt{2}}{2} (\cos x + \sin x)$

**29.** $\tan \left( \theta + \dfrac{\pi}{4} \right) = \dfrac{1 + \tan \theta}{1 − \tan \theta}$

**30.** $\cos \left( \theta + \dfrac{\pi}{2} \right) = −\sin \theta$

In Exercises 31–34, match each graph with a pair of the following equations. Use your knowledge of identities and transformations, not your grapher.

**(a)** $y = \cos (3 − 2x)$

**(b)** $y = \sin x \cos 1 + \cos x \sin 1$

**(c)** $y = \cos (x − 3)$

**(d)** $y = \sin (2x − 5)$

**(e)** $y = \cos x \cos 3 + \sin x \sin 3$

**(f)** $y = \sin (x + 1)$

**(g)** $y = \cos 3 \cos 2x + \sin 3 \sin 2x$

**(h)** $y = \sin 2x \cos 5 − \cos 2x \sin 5$

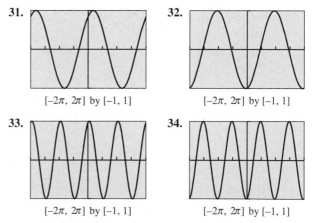

**31.**

**32.**

$[−2\pi, 2\pi]$ by $[−1, 1]$

$[−2\pi, 2\pi]$ by $[−1, 1]$

**33.**

**34.**

$[−2\pi, 2\pi]$ by $[−1, 1]$

$[−2\pi, 2\pi]$ by $[−1, 1]$

In Exercises 35 and 36, use sum or difference identities (and not your grapher) to solve the equation exactly.

**35.** $\sin 2x \cos x = \cos 2x \sin x$

**36.** $\cos 3x \cos x = \sin 3x \sin x$

In Exercises 37–42, prove the reduction formula.

**37.** $\sin \left( \dfrac{\pi}{2} − u \right) = \cos u$

**38.** $\tan \left( \dfrac{\pi}{2} − u \right) = \cot u$

**39.** $\cot\left(\dfrac{\pi}{2} - u\right) = \tan u$  **40.** $\sec\left(\dfrac{\pi}{2} - u\right) = \csc u$

**41.** $\csc\left(\dfrac{\pi}{2} - u\right) = \sec u$  **42.** $\cos\left(x + \dfrac{\pi}{2}\right) = -\sin x$

In Exercises 43–46, express the function as a sinusoid in the form $y = a \sin(bx + c)$.

**43.** $y = 3 \sin x + 4 \cos x$  **44.** $y = 5 \sin x - 12 \cos x$

**45.** $y = \cos 3x + 2 \sin 3x$  **46.** $y = 3 \cos 2x - 2 \sin 2x$

In Exercises 47–55, prove the identity.

**47.** $\sin(x - y) + \sin(x + y) = 2 \sin x \cos y$

**48.** $\cos(x - y) + \cos(x + y) = 2 \cos x \cos y$

**49.** $\cos 3x = \cos^3 x - 3 \sin^2 x \cos x$

**50.** $\sin 3u = 3 \cos^2 u \sin u - \sin^3 u$

**51.** $\cos 3x + \cos x = 2 \cos 2x \cos x$

**52.** $\sin 4x + \sin 2x = 2 \sin 3x \cos x$

**53.** $\tan(x + y) \tan(x - y) = \dfrac{\tan^2 x - \tan^2 y}{1 - \tan^2 x \tan^2 y}$

**54.** $\tan 5u \tan 3u = \dfrac{\tan^2 4u - \tan^2 u}{1 - \tan^2 4u \tan^2 u}$

**55.** $\dfrac{\sin(x + y)}{\sin(x - y)} = \dfrac{(\tan x + \tan y)}{(\tan x - \tan y)}$

## Explorations

**56.** Prove the identity $\tan(u + v) = \dfrac{\tan u + \tan v}{1 - \tan u \tan v}$.

**57.** Prove the identity $\tan(u - v) = \dfrac{\tan u - \tan v}{1 + \tan u \tan v}$.

**58. Writing to Learn** Explain why the identity in Exercise 56 cannot be used to prove the reduction formula $\tan(x + \pi/2) = -\cot x$. Then prove the reduction formula.

**59. Writing to Learn** Explain why the identity in Exercise 57 cannot be used to prove the reduction formula $\tan(x - 3\pi/2) = -\cot x$. Then prove the reduction formula.

**60. An Identity for Calculus** Prove the following identity, which is used in calculus to prove an important differentiation formula.

$$\frac{\sin(x + h) - \sin x}{h} = \sin x\left(\frac{\cos h - 1}{h}\right) + \cos x\,\frac{\sin h}{h}.$$

**61. An Identity for Calculus** Prove the following identity, which is used in calculus to prove another important differentiation formula.

$$\frac{\cos(x + h) - \cos x}{h} = \cos x\left(\frac{\cos h - 1}{h}\right) - \sin x\,\frac{\sin h}{h}.$$

**62. Group Activity** Place 24 points evenly spaced around the unit circle, starting with the point $(1, 0)$. Using only your knowledge of the special angles and the sum and difference identities, work with your group to find the exact coordinates of all 24 points.

## Extending the Ideas

In Exercises 63–66, assume that $A$, $B$, and $C$ are the three angles of some $\triangle ABC$. (Note, then, that $A + B + C = \pi$.) Prove the following identities.

**63.** $\sin(A + B) = \sin C$

**64.** $\cos C = \sin A \sin B - \cos A \cos B$

**65.** $\tan A + \tan B + \tan C = \tan A \tan B \tan C$

**66.** $\cos A \cos B \cos C - \sin A \sin B \cos C - \sin A \cos B \sin C$
$\quad - \cos A \sin B \sin C = -1$

**67. Writing to Learn** The figure shows graphs of $y_1 = \cos 5x \cos 4x$ and $y_2 = -\sin 5x \sin 4x$ in one viewing window. Discuss the question, "How many solutions are there to the equation $\cos 5x \cos 4x = -\sin 5x \sin 4x$ in the interval $[-2\pi, 2\pi]$?" Give an algebraic argument that answers the question more convincingly than the graph does. Then support your argument with an *appropriate* graph.

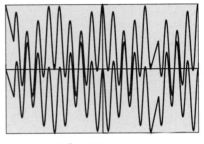

$[-2\pi, 2\pi]$ by $[-1, 1]$

**68. Harmonic Motion** Alternating electric current, an oscillating spring, or any other harmonic oscillator can be modeled by the equation

$$x = a \cos\left(\frac{2\pi}{T}t + \delta\right),$$

where $T$ is the time for one period and $\delta$ is the phase constant. Show that this motion can also be modeled by the following sum of cosine and sine, each with zero phase constant:

$$a_1 \cos\left(\frac{2\pi}{T}\right)t + a_2 \sin\left(\frac{2\pi}{T}\right)t,$$

where $a_1 = a \cos \delta$ and $a_2 = -a \sin \delta$.

**69. Magnetic Fields** A magnetic field $B$ can sometimes be modeled as the sum of an incident and a reflective field as

$$B = B_{\text{in}} + B_{\text{ref}},$$

where  $\quad B_{\text{in}} = \dfrac{E_0}{c} \cos\left(\omega t - \dfrac{\omega x}{c}\right),\quad$ and

$$B_{\text{ref}} = \frac{E_0}{c} \cos\left(\omega t + \frac{\omega x}{c}\right).$$

Show that  $\quad B = 2\dfrac{E_0}{c} \cos \omega t \cos \dfrac{\omega x}{c}.$

# **5.4 Multiple-Angle Identities**

Double-Angle Identities • Power-Reducing Identities • Half-Angle
Identities • Solving Trigonometric Equations

## **Objective**

Students will be able to apply the double-angle identities, power-reducing
identities, and half-angle identities.

## **Motivate**

Have students apply the cosine-of-a-
sum identity in order to evaluate
$\cos (x + x)$. $(\cos^2 x - \sin^2 x)$.

## **Lesson Guide**

Day 1: Skills
Day 2: Applications

## **Double-Angle Identities**

The formulas that result from letting $u = v$ in the angle sum identities are
called the *double-angle identities*. We will state them all and prove one, leaving the rest of the proofs as exercises. (See Exercises 1–4.)

---

**Double-Angle Identities**

$$\sin 2u = 2 \sin u \cos u$$

$$\cos 2u = \begin{cases} \cos^2 u - \sin^2 u \\ 2 \cos^2 u - 1 \\ 1 - 2 \sin^2 u \end{cases}$$

$$\tan 2u = \frac{2 \tan u}{1 - \tan^2 u}$$

---

There are three identities for $\cos 2u$. This is not unusual; indeed, there
are plenty of other identities one could supply for $\sin 2u$ as well, such as
$2 \sin u \sin (\pi/2 - u)$. We list the three identities for $\cos 2u$ because they are all
*useful* in various contexts and therefore worth memorizing.

**Example 1** PROVING A DOUBLE-ANGLE IDENTITY

Prove the identity: $\sin 2u = 2 \sin u \cos u$.

**Solution**

$$\sin 2u = \sin (u + u)$$

$$= \sin u \cos u + \cos u \sin u \qquad \text{Sine of a sum } (v = u)$$

$$= 2 \sin u \cos u$$

## **Power-Reducing Identities**

One immediate use for two of the three formulas for $\cos 2u$ is to derive the
*Power-Reducing Identities*. Some simple-looking functions like $y = \sin^2 u$
would be quite difficult to handle in certain calculus contexts if it were not for
the existence of these identities.

---

**Power-Reducing Identities**

$$\sin^2 u = \frac{1 - \cos 2u}{2}$$

$$\cos^2 u = \frac{1 + \cos 2u}{2}$$

$$\tan^2 u = \frac{1 - \cos 2u}{1 + \cos 2u}$$

---

We will also leave the proofs of these identities as Exercises. (See Exercises 37 and 38.)

### Example 2  PROVING AN IDENTITY

Prove the identity: $\cos^4 \theta - \sin^4 \theta = \cos 2\theta$.

**Solution**

$$
\begin{aligned}
\cos^4 \theta - \sin^4 \theta &= (\cos^2 \theta + \sin^2 \theta)(\cos^2 \theta - \sin^2 \theta) \\
&= 1 \cdot (\cos^2 \theta - \sin^2 \theta) \qquad \text{Pythagorean identity} \\
&= \cos 2\theta \qquad\qquad\qquad \text{Double angle identity}
\end{aligned}
$$

### Example 3  REDUCING A POWER OF 4

Rewrite $\cos^4 x$ in terms of trigonometric functions with no power greater than 1.

**Solution**

$$
\begin{aligned}
\cos^4 x &= (\cos^2 x)^2 \\
&= \left( \frac{1 + \cos 2x}{2} \right)^2 \qquad\qquad\qquad \text{Power-reducing identity} \\
&= \left( \frac{1 + 2\cos 2x + \cos^2 2x}{4} \right) \\
&= \frac{1}{4} + \frac{1}{2}\cos 2x + \frac{1}{4}\left( \frac{1 + \cos 4x}{2} \right) \qquad \text{Power-reducing identity} \\
&= \frac{1}{4} + \frac{1}{2}\cos 2x + \frac{1}{8} + \frac{1}{8}\cos 4x \\
&= \frac{1}{8}(3 + 4\cos 2x + \cos 4x)
\end{aligned}
$$

## Half-Angle Identities

The power-reducing identities can be used to extend our stock of "special" angles whose trigonometric ratios can be found without a calculator. As usual, we are not suggesting that this algebraic procedure is any more practical than using a calculator, but we are suggesting that this sort of exercise helps you to understand how the functions behave. In Exploration 1, for example, we use a power-reduction formula to find the exact values of $\sin (\pi/8)$ and $\sin (9\pi/8)$ without a calculator.

> **Exploration 1  Finding the Sine of Half an Angle**
>
> Recall the power-reducing formula $\sin^2 u = (1 - \cos 2u)/2$.
>
> 1. Use the power-reducing formula to show that
>    $\sin^2 (\pi/8) = (2 - \sqrt{2})/4$.
>
> 2. Solve for $\sin (\pi/8)$. Do you take the positive or negative square root? Why?
>
> 3. Use the power-reducing formula to show that
>    $\sin^2 (9\pi/8) = (2 - \sqrt{2})/4$.
>
> 4. Solve for $\sin (9\pi/8)$. Do you take the positive or negative square root? Why?

A little alteration of the power-reducing identities results in the *Half-Angle Identities*, which can be used directly to find trigonometric functions of $u/2$ in terms of trigonometric functions of $u$. As Exploration 1 suggests, there is an unavoidable ambiguity of sign involved with the square root that must be resolved in particular cases by checking the quadrant in which $u/2$ lies.

**Did We Miss Two ± Signs?**

You might have noticed that all of the half-angle identities have unresolved ± signs except for the last two. The fact that we can omit them on the last two identities for $\tan u/2$ is a fortunate consequence of two facts: (1) $\sin u$ and $\tan u/2$ always have the same sign (easily observed from the graphs of the two functions in Figure 5.8), and (2) $1 \pm \cos u$ is never negative.

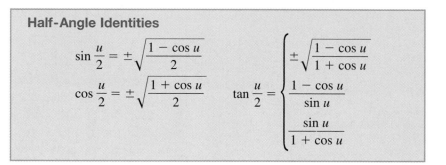

**Half-Angle Identities**

$$\sin \frac{u}{2} = \pm\sqrt{\frac{1 - \cos u}{2}}$$

$$\cos \frac{u}{2} = \pm\sqrt{\frac{1 + \cos u}{2}} \qquad \tan \frac{u}{2} = \begin{cases} \pm\sqrt{\dfrac{1 - \cos u}{1 + \cos u}} \\[2mm] \dfrac{1 - \cos u}{\sin u} \\[2mm] \dfrac{\sin u}{1 + \cos u} \end{cases}$$

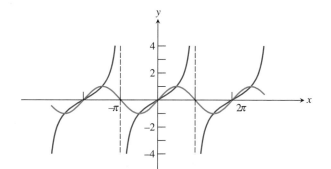

**Figure 5.8** The functions $\sin u$ and $\tan u/2$ always have the same sign.

## Solving Trigonometric Equations

New identities always provide new tools for solving trigonometric equations algebraically. Under the right conditions, they even lead to exact solutions. We assert again that we are not presenting these algebraic solutions for their practical value (as the calculator solutions are certainly sufficient for most applications and unquestionably much quicker to obtain), but rather as ways to observe the behavior of the trigonometric functions and their interwoven tapestry of identities.

### Example 4 USING A DOUBLE-ANGLE IDENTITY

Solve algebraically in the interval $[0, 2\pi)$: $\sin 2x = \cos x$.

**Solution**

$$\sin 2x = \cos x$$
$$2 \sin x \cos x = \cos x$$
$$2 \sin x \cos x - \cos x = 0$$
$$\cos x\, (2 \sin x - 1) = 0$$
$$\cos x = 0 \quad \text{or} \quad 2 \sin x - 1 = 0$$
$$\cos x = 0 \quad \text{or} \quad \sin x = \frac{1}{2}$$

The two solutions of $\cos x = 0$ are $x = \pi/2$ and $x = 3\pi/2$. The two solutions of $\sin x = 1/2$ are $x = \pi/6$ and $x = 5\pi/6$. Therefore, the solutions of $\sin 2x = \cos x$ are

$$\frac{\pi}{6}, \ \frac{\pi}{2}, \ \frac{5\pi}{6}, \ \frac{3\pi}{2}.$$

We can **support** this result **graphically** by verifying the four $x$-intercepts of the function $y = \sin 2x - \cos x$ in the interval $[0, 2\pi)$ (Figure 5.9).

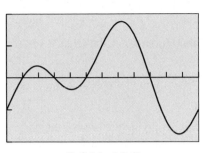

$[0, 2\pi]$ by $[-2, 2]$

**Figure 5.9** The function $y = \sin 2x - \cos x$ for $0 \le x \le 2\pi$. The scale on the $x$-axis shows intervals of length $\pi/6$. This graph supports the solution found algebraically in Example 4.

### Example 5 USING HALF-ANGLE IDENTITIES

Solve $\sin^2 x = 2 \sin^2 (x/2)$.

**Solution** The graph of $y = \sin^2 x - 2 \sin^2 x/2$ in Figure 5.10 suggests that this function is periodic with period $2\pi$ and that the equation $\sin^2 x - 2 \sin^2 (x/2)$ has three solutions in $[0, 2\pi)$.

**Solve Algebraically**

$$\sin^2 x = 2 \sin^2 \frac{x}{2}$$
$$\sin^2 x = 2\left(\frac{1 - \cos x}{2}\right) \qquad \text{Half-angle identity}$$
$$1 - \cos^2 x = 1 - \cos x \qquad \text{Convert to all cosines}$$
$$\cos x - \cos^2 x = 0$$
$$\cos x\, (1 - \cos x) = 0$$
$$\cos x = 0 \quad \text{or} \quad \cos x = 1$$
$$x = \frac{\pi}{2} \quad \text{or} \quad \frac{3\pi}{2} \quad \text{or} \quad 0$$

The rest of the solutions are obtained by periodicity:

$$x = 2n\pi, \quad x = \frac{\pi}{2} + 2n\pi, \quad x = \frac{3\pi}{2} + 2n\pi, \quad n = 0, \pm 1, \pm 2, \dots$$

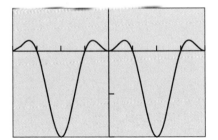

$[-2\pi, 2\pi]$ by $[-2, 1]$

**Figure 5.10** The graph of $y = \sin^2 x - 2 \sin^2 (x/2)$ suggests that $\sin^2 x = 2 \sin^2 (x/2)$ has three solutions in $[0, 2\pi)$. (Example 5)

**Follow-up**

Have students discuss how to decide whether to use a + or − sign when applying half-angle identities.

**Assignment Guide**

Day 1: Ex. 3-36, multiples of 3
Day 2: Ex. 37, 39, 42, 43, 46, 47, 51, 54

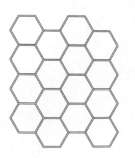

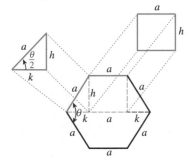

**Figure 5.11** Hexagon-shaped wire mesh. (Example 6)

**Figure 5.12** The Mishmash Mesh hexagon subdivides into two rectangles of area *ah* and four triangles of area 1/2 *kh* (Example 6).

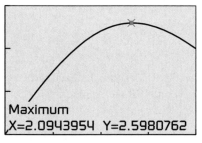

Maximum
X=2.0943954 Y=2.5980762

$[0, \pi]$ by $[0, 3]$

**Figure 5.13** The maximum value of $y = 2 \sin \theta/2 + \sin \theta$ occurs at about 2.0944 radians. (Example 6)

**Example 6   SOLVING AN OPTIMAL DESIGN PROBLEM**

The Mishmash Mesh Company manufactures a wire mesh used in the construction industry. The individual cells in this mesh are hexagons with a horizontal line of symmetry and edges of all the same length, as shown in Figure 5.11. Company design engineers want to find the measure of the "side" angle $\theta$ of the hexagon that will maximize the area of the hexagon.

**(a) Model** Express the area of the hexagon in terms of the angle $\theta$ and the length $a$ of a side.

**(b)** Find the value of $\theta$ that will maximize the area of the hexagon.

**Solution**

**(a)** In light of the symmetry condition, we can divide the area of the hexagon into two rectangles, each of area $ah$, and four triangles, each of area 1/2 $kh$ (Figure 5.12).

The area of the entire hexagon (two rectangles and four triangles) is therefore

$$A = 2ah + 4\left(\frac{1}{2} kh\right) = 2ah + 2kh.$$

**Solve Algebraically**

By right triangle trigonometry, $h = a \sin \theta/2$ and $k = a \cos \theta/2$.

So

$$A = 2a\left(a \sin \frac{\theta}{2}\right) + 2\left(a \cos \frac{\theta}{2}\right)\left(a \sin \frac{\theta}{2}\right)$$

$$= a^2\left(2 \sin \frac{\theta}{2} + 2 \sin \frac{\theta}{2} \cos \frac{\theta}{2}\right)$$

$$= a^2\left(2 \sin \frac{\theta}{2} + \sin \theta\right) \qquad \text{Double-angle identity}$$

**(b)**

**Solve Graphically**

Regardless of the value of $a$, the area function
$A = a^2(2 \sin \theta/2 + \sin \theta)$ will be maximized when the factor
$2 \sin \theta/2 + \sin \theta$ is maximized. Graphing $y = 2 \sin \theta/2 + \sin \theta$, we find that the maximum occurs when $\theta \approx 2.0944$ radians (Figure 5.13).

If you convert this answer to degrees, you will notice that it is 120°. This is the angle that yields a "regular" hexagon with equal sides and equal angles all around, familiar to us from the hexagonal patterns found in bathroom tiles, honeycombs, and (yes) wire mesh.

# Quick Review 5.4

In Exercises 1–8, find the general solution of the equation.

**1.** $\tan x - 1 = 0$         **2.** $\tan x + 1 = 0$

**3.** $(\cos x)(1 - \sin x) = 0$    **4.** $(\sin x)(1 + \cos x) = 0$

**5.** $\sin x + \cos x = 0$       **6.** $\sin x - \cos x = 0$

**7.** $(2 \sin x - 1)(2 \cos x + 1) = 0$

**8.** $(\sin x + 1)(2 \cos x - \sqrt{2}) = 0$

**9.** Find the area of the trapezoid.

**10.** Find the height of the isosceles triangle.  $2\sqrt{2}$

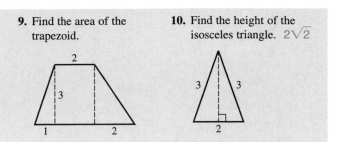

# Section 5.4 Exercises

In Exercises 1–4, use the appropriate sum or difference identity to prove the double-angle identity.

**1.** $\cos 2u = \cos^2 u - \sin^2 u$    **2.** $\cos 2u = 2 \cos^2 u - 1$

**3.** $\cos 2u = 1 - 2 \sin^2 u$    **4.** $\tan 2u = \dfrac{2 \tan u}{1 - \tan^2 u}$

In Exercises 5–10, find all solutions to the equation in the interval $[0, 2\pi)$.

**5.** $\sin 2x = 2 \sin x$  $0, \pi$    **6.** $\sin 2x = \sin x$

**7.** $\cos 2x = \sin x$       **8.** $\cos 2x = \cos x$

**9.** $\sin 2x - \tan x = 0$    **10.** $\cos^2 x + \cos x = \cos 2x$

In Exercises 11–14, write the expression as one involving only $\sin \theta$ and $\cos \theta$.

**11.** $\sin 2\theta + \cos \theta$    **12.** $\sin 2\theta + \cos 2\theta$

**13.** $\sin 2\theta + \cos 3\theta$    **14.** $\sin 3\theta + \cos 2\theta$

In Exercises 15–22, prove the identity.

**15.** $\sin 4x = 2 \sin 2x \cos 2x$    **16.** $\cos 6x = 2 \cos^2 3x - 1$

**17.** $2 \csc 2x = \csc^2 x \tan x$    **18.** $2 \cot 2x = \cot x - \tan x$

**19.** $\sin 3x = (\sin x)(4 \cos^2 x - 1)$

**20.** $\sin 3x = (\sin x)(3 - 4 \sin^2 x)$

**21.** $\cos 4x = 1 - 8 \sin^2 x \cos^2 x$

**22.** $\sin 4x = (4 \sin x \cos x)(2 \cos^2 x - 1)$

In Exercises 23–30, solve algebraically for exact solutions in the interval $[0, 2\pi)$. Use your grapher only to support your algebraic work.

**23.** $\cos 2x + \cos x = 0$    **24.** $\cos 2x + 2 \sin x = 0$

**25.** $\cos x + \cos 3x = 0$    **26.** $\sin x + \sin 3x = 0$

**27.** $\sin 2x + \sin 4x = 0$    **28.** $\cos 2x + \cos 4x = 0$

**29.** $\sin 2x - \cos 3x = 0$    **30.** $\sin 3x + \cos 2x = 0$

In Exercises 31–36, use half-angle identities to find an exact value without a calculator.

**31.** $\sin 15°$  $(1/2)\sqrt{2 - \sqrt{3}}$    **32.** $\tan 195°$  $2 - \sqrt{3}$

**33.** $\cos 75°$  $(1/2)\sqrt{2 - \sqrt{3}}$    **34.** $\sin 5\pi/12$  $(1/2)\sqrt{2 + \sqrt{3}}$

**35.** $\tan 7\pi/12$  $-2 - \sqrt{3}$    **36.** $\cos \pi/8$  $(1/2)\sqrt{2 + \sqrt{2}}$

**37.** Prove the power-reducing identities:

**(a)** $\sin^2 u = \dfrac{1 - \cos 2u}{2}$    **(b)** $\cos^2 u = \dfrac{1 + \cos 2u}{2}$

**38. (a)** Use the identities in Exercise 37 to prove the power-reducing identity $\tan^2 u = \dfrac{1 - \cos 2u}{1 + \cos 2u}$.

**(b) Writing to Learn** Explain why the identity in (a) does not imply that $\tan u = \sqrt{\dfrac{1 - \cos 2u}{1 + \cos 2u}}$.

In Exercises 39–42, use the power-reducing identities to prove the identity.

**39.** $\sin^4 x = \dfrac{1}{8}(3 - 4 \cos 2x + \cos 4x)$

**40.** $\cos^3 x = \left(\dfrac{1}{2} \cos x\right)(1 + \cos 2x)$

**41.** $\sin^3 2x = \left(\dfrac{1}{2} \sin 2x\right)(1 - \cos 4x)$

**42.** $\sin^5 x = \left(\dfrac{1}{8} \sin x\right)(3 - 4 \cos 2x + \cos 4x)$

In Exercises 43–46, use the half-angle identities to find all solutions in the interval $[0, 2\pi)$. Then find the general solution.

**43.** $\cos^2 x = \sin^2\left(\dfrac{x}{2}\right)$    **44.** $\sin^2 x = \cos^2\left(\dfrac{x}{2}\right)$

**45.** $\tan\left(\dfrac{x}{2}\right) = \dfrac{1 - \cos x}{1 + \cos x}$    **46.** $\sin^2\left(\dfrac{x}{2}\right) = 2 \cos^2 x - 1$

# Explorations

**47. Connecting Trigonometry and Geometry** In a regular polygon all sides are the same length and all angles are equal in measure.

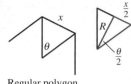

Regular polygon
with $n$ sides

**(a)** If the perpendicular distance from the center of the polygon with $n$ sides to the midpoint of a side is $R$, and if the length of the side of the polygon is $x$, show that

$$x = 2R \tan \frac{\theta}{2}$$

where $\theta = 2\pi/n$ is the central angle subtended by one side.

**(b)** If the length of one side of a regular 11-sided polygon is approximately 5.87 and $R$ is a whole number, what is the value of $R$?

**48. Connecting Trigonometry and Geometry** A rhombus is a quadrilateral with equal sides. The diagonals of a rhombus bisect the angles of the rhombus and are perpendicular bisectors of each other. Let $\angle ABC = \theta$, $d_1$ = length of $AC$, and $d_2$ = length of $BD$.

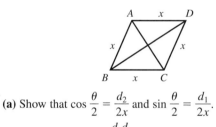

**(a)** Show that $\cos \frac{\theta}{2} = \frac{d_2}{2x}$ and $\sin \frac{\theta}{2} = \frac{d_1}{2x}$.

**(b)** Show that $\sin \theta = \frac{d_1 d_2}{2x^2}$. ■

**49. Group Activity Maximizing Volume** The ends of a 10-foot-long water trough are isosceles trapezoids as shown in the figure. Find the value of $\theta$ that maximizes the volume of the trough and the maximum volume.

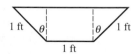

**50. Group Activity Tunnel Problem** A rectangular tunnel is cut through a mountain to make a road. The upper vertices of the rectangle are on the circle $x^2 + y^2 = 400$, as illustrated in the figure.

**(a)** Show that the cross-sectional area of the end of the tunnel is $400 \sin 2\theta$.

**(b)** Find the dimensions of the rectangular end of the tunnel that maximizes its cross-sectional area.

# Extending the Ideas

In Exercises 51–55, prove the double-angle formulas.

**51.** $\csc 2u = \frac{1}{2} \csc u \sec u$

**52.** $\cot 2u = \frac{\cot^2 u - 1}{2 \cot u}$

**53.** $\sec 2u = \frac{\csc^2 u}{\csc^2 u - 2}$

**54.** $\sec 2u = \frac{\sec^2 u}{2 - \sec^2 u}$

**55.** $\sec 2u = \frac{\sec^2 u \csc^2 u}{\csc^2 u - \sec^2 u}$

**56. Writing to Learn** Explain why

$$\sqrt{\frac{1 - \cos 2x}{2}} = |\sin x|$$

is an identity but

$$\sqrt{\frac{1 - \cos 2x}{2}} = \sin x$$

is not an identity.

**57. Sahara Sunset** Table 5.2 gives the time of day that astronomical twilight began in northeastern Mali on the first day of each month of 1999.

| Table 5.2 Astronomical Twilight | | | |
|---|---|---|---|
| Date | Day | Time | 6:00 + |
| Jan 1 | 1 | 6:51 | 51 |
| Feb 1 | 32 | 7:08 | 68 |
| Mar 1 | 60 | 7:18 | 78 |
| Apr 1 | 91 | 7:28 | 88 |
| May 1 | 121 | 7:42 | 102 |
| Jun 1 | 152 | 7:59 | 119 |
| Jul 1 | 182 | 8:07 | 127 |
| Aug 1 | 213 | 7:56 | 116 |
| Sep 1 | 244 | 7:30 | 90 |
| Oct 1 | 274 | 7:01 | 61 |
| Nov 1 | 305 | 6:40 | 40 |
| Dec 1 | 335 | 6:37 | 37 |

*Source: The World Almanac 1999.*

The second column gives the date as the day of the year, and the fourth column gives the time as number of minutes past 6:00.

**(a)** Enter the numbers in column 2 (day) in list L1 and the numbers in column 4 (minutes past 6:00) in list L2. Make a scatter plot with *x*-coordinates from L1 and *y*-coordinates from L2.

**(b)** Using sine regression, find the regression curve through the points and store its equation in Y1. Superimpose the graph of the curve on the scatter plot. Is it a good fit?

**(c)** Make a new column showing the *residuals* (the difference between the *y*-value predicted by the regression curve and the actual *y*-value at each data point) and store

them in list L3. Your calculator might have a list called RESID, in which case the command RESID → L3 will perform this operation. You could also enter Y1(L1) − L2 → L3.

**(d)** Make a scatter plot with *x*-coordinates from L1 and *y*-coordinates from L3. Find the sine regression curve through *these* points and superimpose it on the scatter plot. Is it a good fit?

**(e) Writing to Learn** Interpret what the two regressions seem to indicate about the periodic behavior of astronomical twilight as a function of time. This is not an unusual phenomenon in astronomical data, and it kept ancient astronomers baffled for centuries.

---

## 5.5  The Law of Sines

Deriving the Law of Sines  •  Solving Triangles (AAS, ASA)
•  The Ambiguous Case (SSA)  •  Applications

### Deriving the Law of Sines

Recall from geometry that a triangle has six parts (three sides (S), three angles (A)), but that its size and shape can be completely determined by fixing only three of those parts, provided they are the right three. These threesomes that determine *triangle congruence* are known by their acronyms: AAS, ASA, SAS, and SSS. The other two acronyms represent match-ups that don't quite work: AAA determines similarity only, while SSA does not even determine similarity.

With trigonometry we have the tools to find the other parts of the triangle once congruence is established. For example, given three sides of a triangle we can determine the three angles. We can even come up with possibilities for the SSA case, which is at least predictably ambiguous. The tools we need are the Law of Sines and the Law of Cosines, the subjects of our last two trigonometric sections.

The **Law of Sines** states that the ratio of the sine of an angle to the length of its opposite side is the same for all three angles of any triangle.

> **Law of Sines**
>
> In any $\triangle ABC$ with angles $A$, $B$, and $C$ opposite sides $a$, $b$, and $c$ respectively, the following equation is true:
>
> $$\frac{\sin A}{a} = \frac{\sin B}{b} = \frac{\sin C}{c}$$

The derivation of the Law of Sines refers to the two triangles in Figure 5.14, in each of which we have drawn an altitude to side $c$. Right triangle trigonometry applied to either of the triangles in Figure 5.11 tells us that

$$\sin A = \frac{h}{b}.$$

In the acute triangle on the top,

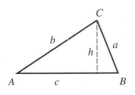

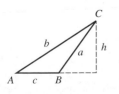

**Figure 5.14** The Law of Sines.

$$\sin B = \frac{h}{a},$$

while in the obtuse triangle on the bottom,

$$\sin (\pi - B) = \frac{h}{a}.$$

But $\sin (\pi - B) = \sin B$, so in either case

$$\sin B = \frac{h}{a}.$$

Solving for $h$ in both equations yields $h = b \sin A = a \sin B$. The equation $b \sin A = a \sin B$ is equivalent to

$$\frac{\sin A}{a} = \frac{\sin B}{b}.$$

If we were to draw an altitude to side $a$ and repeat the same steps as above, we would reach the conclusion that

$$\frac{\sin B}{b} = \frac{\sin C}{c}.$$

Putting the results together,

$$\frac{\sin A}{a} = \frac{\sin B}{b} = \frac{\sin C}{c}.$$

## Solving Triangles (AAS, ASA)

Two angles and a side of a triangle, in any order, determine the size and shape of a triangle completely. Of course, two angles of a triangle determine the third, so we really get one of the missing three parts for free. We solve for the remaining two parts (the unknown sides) with the Law of Sines.

### Example 1 SOLVING A TRIANGLE GIVEN TWO ANGLES AND A SIDE

Solve $\triangle ABC$ given that $\angle A = 36°$, $\angle B = 48°$, and $a = 8$. (See Figure 5.15.)

**Solution** First, we note that $\angle C = 180° - 36° - 48° = 96°$.

We then apply the Law of Sines:

$$\frac{\sin A}{a} = \frac{\sin B}{b} \qquad \text{and} \qquad \frac{\sin A}{a} = \frac{\sin C}{c}$$

$$\frac{\sin 36°}{8} = \frac{\sin 48°}{b} \qquad\qquad \frac{\sin 36°}{8} = \frac{\sin 96°}{c}$$

$$b = \frac{8 \sin 48°}{\sin 36°} \qquad\qquad c = \frac{8 \sin 96°}{\sin 36°}$$

$$b \approx 10.115 \qquad\qquad c \approx 13.536$$

The six parts of the triangle are:

$$\angle A = 36° \qquad a = 8$$

$$\angle B = 48° \qquad b \approx 10.115$$

$$\angle C = 96° \qquad c \approx 13.536$$

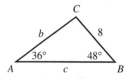

**Figure 5.15** A triangle determined by AAS. (Example 1)

# The Ambiguous Case (SSA)

While two angles and a side of a triangle are always sufficient to determine its size and shape, the same can not be said for two sides and an angle. Perhaps unexpectedly, it depends on where that angle is. If the angle is included between the two sides (the SAS case), then the triangle is uniquely determined up to congruence. If the angle is opposite one of the sides (the SSA case), then there might be one, two, or zero triangles determined.

Solving a triangle in the SAS case involves the Law of Cosines and will be handled in the next section. Solving a triangle in the SSA case is done with the Law of Sines, but with an eye toward the possibilities, as seen in the following Exploration.

**Figure 5.16** The diagram for part 1. (Exploration 1)

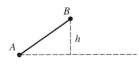

**Figure 5.17** The diagram for part 2. (Exploration 1)

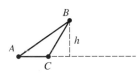

**Figure 5.18** The diagram for parts 3–5. (Exploration 1)

---

**Exploration Extensions**

See Exercise 48.

---

**Exploration 1**   Determining the Number of Triangles

We wish to construct △*ABC* given angle *A*, side *AB*, and side *BC*.

1. Suppose ∠*A* is obtuse and that side *AB* is as shown in Figure 5.16. To complete the triangle, side *BC* must determine a point on the dotted horizontal line (which extends infinitely to the left). Explain from the picture why a *unique* triangle △*ABC* is determined if *BC* > *AB*, but *no* triangle is determined if *BC* ≤ *AB*.

2. Suppose ∠*A* is acute and that side *AB* is as shown in Figure 5.17. To complete the triangle, side *BC* must determine a point on the dotted horizontal line (which extends infinitely to the right). Explain from the picture why a *unique* triangle △*ABC* is determined if *BC* = *h*, but *no* triangle is determined if *BC* < *h*.

3. Suppose ∠*A* is acute and that side *AB* is as shown in Figure 5.18. If *AB* > *BC* > *h*, then we can form a triangle as shown. Find a *second* point *C* on the dotted horizontal line that gives a side *BC* of the same length, but determines a different triangle. (This is the "ambiguous case.")

4. Explain why sin *C* is the same in both triangles in the ambiguous case. (This is why the Law of Sines is also ambiguous in this case.)

5. Explain from Figure 5.18 why a *unique* triangle is determined if *BC* ≥ *AB*.

---

Now that we know what can happen, let us try the algebra.

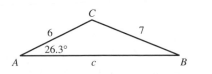

**Figure 5.19** A triangle determined by SSA. (Example 2)

## Example 2   SOLVING A TRIANGLE GIVEN TWO SIDES AND AN ANGLE

Solve △*ABC* given that $a = 7$, $b = 6$, and ∠*A* = 26.3°. (See Figure 5.19.)

**Solution** By drawing a reasonable sketch (Figure 5.19), we can assure ourselves that this is not the ambiguous case. (In fact, this is the case described in step 5 of Exploration 1.)

Begin by solving for the *acute* angle *B*, using the Law of Sines:

$$\frac{\sin A}{a} = \frac{\sin B}{b} \qquad \text{Law of Sines}$$

$$\frac{\sin 26.3°}{7} = \frac{\sin B}{6}$$

$$\sin B = \frac{6 \sin 26.3°}{7}$$

$$B = \sin^{-1}\left(\frac{6 \sin 26.3°}{7}\right)$$

$$B = 22.3° \qquad \text{Round to match accuracy of given angle}$$

Then, find the obtuse angle $C$ by subtraction:

$$C = 180° - 26.3° - 22.3°$$

$$= 131.4°$$

Finally, find side $c$:

$$\frac{\sin A}{a} = \frac{\sin C}{c}$$

$$\frac{\sin 26.3°}{7} = \frac{\sin 131.4°}{c}$$

$$c = \frac{7 \sin 131.4°}{\sin 26.3°}$$

$$c \approx 11.9$$

The six parts of the triangle are:

$$\angle A = 26.3° \qquad a = 7$$

$$\angle B = 22.3° \qquad b = 6$$

$$\angle C = 131.4° \qquad c \approx 11.9$$

### Example 3   HANDLING THE AMBIGUOUS CASE

Solve $\triangle ABC$ given that $a = 6$, $b = 7$, and $\angle A = 30°$.

**Solution**  By drawing a reasonable sketch (Figure 5.20), we see that two triangles are possible with the given information. We keep this in mind as we proceed.

We would like to begin by solving for an *acute* angle as we did in Example 2, but the essence of the ambiguous case is that we cannot guarantee that *either* of the unknown angles is acute! We begin arbitrarily by finding angle $B$.

$$\frac{\sin A}{a} = \frac{\sin B}{b} \qquad \text{Law of Sines}$$

$$\frac{\sin 30°}{6} = \frac{\sin B}{7}$$

$$\sin B = \frac{7 \sin 30°}{6}$$

$$B = \sin^{-1}\left(\frac{7 \sin 30°}{6}\right)$$

$$B = 35.7° \qquad \text{Round to match accuracy of given angle}$$

Notice that the calculator gave us one value for $B$, not two. That is because we used the *function* $\sin^{-1}$, which can not give two output values for the

**Notes on Examples**

To stimulate discussion about the ambiguous case, you may wish to have students investigate numerically what would happen in Example 3 if the length of side *a* were changed to 2 or 3.5, instead of 6.

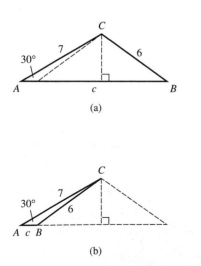

**Figure 5.20** Two triangles determined by the same SSA values. (Example 3)

same input value. Indeed, the function $\sin^{-1}$ will *never give an obtuse angle*, which is why we chose to start with the acute angle in Example 2. For the moment we assume that $B = 35.7°$ and proceed.

Find the obtuse angle $C$ by subtraction:

$$C = 180° - 30.0° - 35.7°$$
$$= 114.3°$$

Finally, find side $c$:

$$\frac{\sin A}{a} = \frac{\sin C}{c}$$
$$\frac{\sin 30.0°}{6} = \frac{\sin 114.3°}{c}$$
$$c = \frac{6 \sin 114.3°}{\sin 30°}$$
$$c \approx 10.9$$

So, under the assumption that angle $B$ is *acute*, the six parts of the triangle are:

$$\angle A = 30.0° \qquad a = 6$$
$$\angle B = 35.7° \qquad b = 7$$
$$\angle C = 114.3° \qquad c \approx 10.9$$

If angle $B$ is *obtuse*, then we can see from Figure 5.20 that it has measure $180° - 35.7° = 144.3°$.

By subtraction, the acute angle $C = 180° - 30.0° - 144.3° = 5.7°$. We then re-compute $c$:

$$c = \frac{6 \sin 5.7°}{\sin 30°} \qquad \text{Substitute 5.7° for 114.3° in earlier computation}$$
$$\approx 1.2$$

So, under the assumption that angle $B$ is *obtuse*, the six parts of the triangle are:

$$\angle A = 30.0° \qquad a = 6$$
$$\angle B = 144.3° \qquad b = 7$$
$$\angle C = 5.7° \qquad c \approx 1.2$$

## Applications

Many problems involving angles and distances can be solved by superimposing a triangle onto the situation and solving the triangle.

### Example 4 LOCATING A FIRE

Forest Ranger Chris Johnson at ranger station $A$ sights a fire in the direction 32° east of north. Ranger Rick Thorpe at ranger station $B$, 10 miles due east of $A$, sights the same fire on a line 48° west of north. Find the distance from each ranger station to the fire.

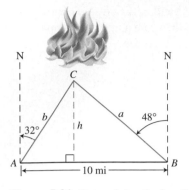

**Figure 5.21** Determining the location of a fire. (Example 4)

**Solution** Let $C$ represent the location of the fire. A sketch (Figure 5.21) shows the superimposed triangle, $\triangle ABC$, in which angles $A$ and $B$ and their included side ($AB$) are known. This is a setup for the Law of Sines.

Note that $\angle A = 90° - 32° = 58°$ and $\angle B = 90° - 48° = 42°$. By subtraction, we find that $\angle C = 180° - 58° - 42° = 80°$.

$$\frac{\sin A}{a} = \frac{\sin C}{c} \quad \text{and} \quad \frac{\sin B}{b} = \frac{\sin C}{c} \qquad \text{Law of Sines}$$

$$\frac{\sin 58°}{a} = \frac{\sin 80°}{10} \qquad\qquad \frac{\sin 42°}{b} = \frac{\sin 80°}{10}$$

$$a = \frac{10 \sin 58°}{\sin 80°} \qquad\qquad b = \frac{10 \sin 42°}{\sin 80°}$$

$$a \approx 8.6 \qquad\qquad b \approx 6.8$$

The fire is about 6.8 miles from ranger station $A$ and about 8.6 miles from ranger station $B$.

### Example 5  FINDING THE HEIGHT OF A POLE

A road slopes $10°$ above the horizontal, and a vertical telephone pole stands beside the road. The angle of elevation of the sun is $62°$, and the pole casts a 14.5-foot shadow downhill along the road. Find the height of the telephone pole.

**Solution** This is an interesting variation on a typical application of right-triangle trigonometry. The slope of the road eliminates the convenient right angle, but we can still solve the problem by solving a triangle.

Figure 5.22 shows the superimposed triangle, $\triangle ABC$. A little preliminary geometry is required to find the measure of angles $A$ and $C$. Due to the slope of the road, angle $A$ is $10°$ less than the angle of elevation of the sun and angle $B$ is $10°$ more than a right angle. That is,

$$\angle A = 62° - 10° = 52°$$

$$\angle B = 90° + 10° = 100°$$

$$\angle C = 180° - 52° - 100° = 28°$$

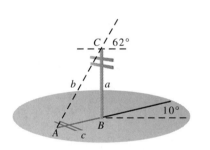

**Figure 5.22** A telephone pole on a slope. (Example 5)

Therefore

$$\frac{\sin A}{a} = \frac{\sin C}{c} \qquad \text{Law of Sines}$$

$$\frac{\sin 52°}{a} = \frac{\sin 28°}{14.5}$$

$$a = \frac{14.5 \sin 52°}{\sin 28°}$$

$$a \approx 24.3 \qquad \text{Round to match accuracy of input}$$

The pole is approximately 24.3 feet high.

# Quick Review 5.5

In Exercises 1–4, solve the equation $a/b = c/d$ for the given variable.

**1.** $a$  $bc/d$    **2.** $b$  $ad/c$

**3.** $c$  $ad/b$    **4.** $d$  $bc/a$

In Exercises 5 and 6, evaluate the expression.

**5.** $\dfrac{7 \sin 48°}{\sin 23°}$  13.314    **6.** $\dfrac{9 \sin 121°}{\sin 14°}$  31.888

In Exercises 7–10, solve for the angle $x$.

**7.** $\sin x = 0.3,\quad 0° < x < 90°$  17.458°

**8.** $\sin x = 0.3,\quad 90° < x < 180°$  162.542°

**9.** $\sin x = -0.7,\quad 180° < x < 270°$

**10.** $\sin x = -0.7,\quad 270° < x < 360°$

# Section 5.5 Exercises

In Exercises 1–4, solve the triangle.

**1.**

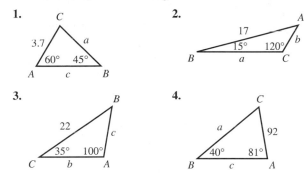

**2.**

**3.**

**4.**

In Exercises 5–8, solve the triangle.

**5.** $A = 40°,\quad B = 30°,\quad b = 10$

**6.** $A = 50°,\quad B = 62°,\quad a = 4$

**7.** $A = 33°,\quad B = 70°,\quad b = 7$

**8.** $B = 16°,\quad C = 103°,\quad c = 12$

In Exercises 9–12, solve the triangle.

**9.** $A = 32°,\quad a = 17,\quad b = 11$

**10.** $A = 49°,\quad a = 32,\quad b = 28$

**11.** $B = 70°,\quad b = 14,\quad c = 9$

**12.** $C = 103°,\quad b = 46,\quad c = 61$

In Exercises 13–18, state whether the given measurements determine zero, one, or two triangles.

**13.** $A = 36°,\quad a = 2,\quad b = 7$  zero

**14.** $B = 82°,\quad b = 17,\quad c = 15$  one

**15.** $C = 36°,\quad a = 17,\quad c = 16$  two

**16.** $A = 73°,\quad a = 24,\quad b = 28$  zero

**17.** $C = 30°,\quad a = 18,\quad c = 9$  two

**18.** $B = 88°,\quad b = 14,\quad c = 62$  zero

In Exercises 19–22, two triangles can be formed using the given measurements. Solve both triangles.

**19.** $A = 64°,\quad a = 16,\quad b = 17$

**20.** $B = 38°,\quad b = 21,\quad c = 25$

**21.** $C = 68°,\quad a = 19,\quad c = 18$

**22.** $B = 57°,\quad a = 11,\quad b = 10$

**23.** Determine the values of $b$ that will produce the given number of triangles if $a = 10$ and $B = 42°$.

 **(a)** two triangles    **(b)** one triangle    **(c)** zero triangles

**24.** Determine the values of $c$ that will produce the given number of triangles if $b = 12$ and $C = 53°$.

 **(a)** two triangles    **(b)** one triangle    **(c)** zero triangles

In Exercises 25 and 26, decide whether the triangle can be solved using the Law of Sines. If so, solve it. If not, explain why not.

**25.**

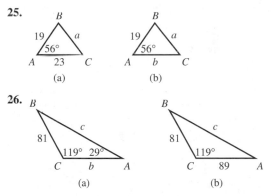

**26.**

In Exercises 27–36, respond in one of the following ways:

**(a)** State, "Cannot be solved with the Law of Sines."

**(b)** State, "No triangle is formed."

**(c)** Solve the triangle.

**27.** $A = 61°,\quad a = 8,\quad b = 21$  no triangle is formed

**28.** $B = 47°,\quad a = 8,\quad b = 21$

**29.** $A = 136°,\quad a = 15,\quad b = 28$  no triangle is formed

**30.** $C = 115°,\quad b = 12,\quad c = 7$  no triangle is formed

**31.** $B = 42°,\quad c = 18,\quad C = 39°$

**32.** $A = 19°$, $b = 22$, $B = 47°$

**33.** $C = 75°$, $b = 49$, $c = 48$

**34.** $A = 54°$, $a = 13$, $b = 15$

**35.** $B = 31°$, $a = 8$, $c = 11$

**36.** $C = 65°$, $a = 19$, $b = 22$

**37. Surveying a Canyon** Two markers $A$ and $B$ on the same side of a canyon rim are 56 ft apart. A third marker $C$, located across the rim, is positioned so that $\angle BAC = 72°$ and $\angle ABC = 53°$.

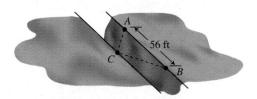

**(a)** Find the distance between $C$ and $A$.  54.6 ft

**(b)** Find the distance between the two canyon rims. (Assume they are parallel.)  51.9 ft

**38. Weather Forecasting** Two meteorologists are 25 mi apart located on an east-west road. The meteorologist at point $A$ sights a tornado 38° east of north. The meteorologist at point $B$ sights the same tornado at 53° west of north. Find the distance from each meteorologist to the tornado. Also find the distance between the tornado and the road.

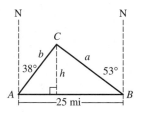

**39. Engineering Design** A vertical flagpole stands beside a road that slopes at an angle of 15° with the horizontal. When the angle of elevation of the sun is 62°, the flagpole casts a 16-ft shadow downhill along the road. Find the height of the flagpole.  ≈ 24.9 ft

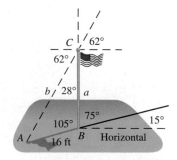

**40. Altitude** Observers 2.32 mi apart see a hot-air balloon directly between them but at the angles of elevation shown in the figure. Find the altitude of the balloon.  ≈ 0.7 mi

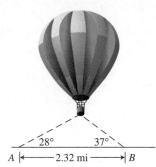

**41. Reducing Air Resistance** A 4-ft airfoil attached to the cab of a truck reduces wind resistance. If the angle between the airfoil and the cab top is 18° and angle $B$ is 10°, find the length of a vertical brace positioned as shown in the figure.

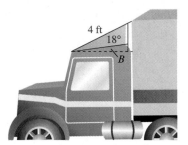

**42. Group Activity Ferris Wheel Design** A Ferris wheel has 16 evenly spaced cars. The distance between adjacent chairs is 15.5 ft. Find the radius of the wheel (to the nearest 0.1 ft.)  39.7 ft

**43. Finding Height** Two observers are 600 ft apart on opposite sides of a flagpole. The angles of elevation from the observers to the top of the pole are 19° and 21°. Find the height of the flagpole.  ≈ 108.9 ft

**44. Finding Height** Two observers are 400 ft apart on opposite sides of a tree. The angles of elevation from the observers to the top of the tree are 15° and 20°. Find the height of the tree.  ≈ 61.7 ft

**45. Finding Distance** Two lighthouses $A$ and $B$ are known to be exactly 20 mi apart on a north-south line. A ship's captain at $S$ measures $\angle ASB$ to be 33°. A radio operator at B measures $\angle ABS$ to be 52°. Find the distance from the ship to each lighthouse.

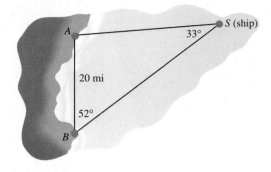

## Explorations

**46. Writing to Learn**

**(a)** Show that there are infinitely many triangles with AAA given if the sum of the three positive angles is 180°.

**(b)** Give three examples of triangles where $A = 30°$, $B = 60°$, and $C = 90°$.

**(c)** Give three examples where $A = B = C = 60°$.

**47.** Use the Law of Sines and the cofunction identities to derive the following formulas from right triangle trigonometry:

**(a)** $\sin A = \dfrac{opp}{hyp}$    **(b)** $\cos A = \dfrac{adj}{hyp}$    **(c)** $\tan A = \dfrac{opp}{adj}$

**48. Wrapping up Exploration 1**  Refer to Figures 5.17 and 5.18 in Exploration 1 of this section.

**(a)** Express $h$ in terms of angle $A$ and length $AB$.

**(b)** In terms of the given angle $A$ and the given length $AB$, state the conditions on length $BC$ that will result in no triangle being formed.  $BC < AB \sin A$

**(c)** In terms of the given angle $A$ and the given length $AB$, state the conditions on length $BC$ that will result in a unique triangle being formed.  $BC \geq AB$ or $BC = AB \sin A$

**(d)** In terms of the given angle $A$ and the given length $AB$, state the conditions on length $BC$ that will result in two possible triangles being formed.  $AB \sin A < BC < AB$   ■

## Extending the Ideas

**49.** Solve this triangle assuming that $\angle B$ is obtuse. (*Hint*: Draw a perpendicular from $A$ to the line through $B$ and $C$.)

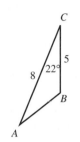

**50. Pilot Calculations**  Towers $A$ and $B$ are known to be 4.1 mi apart on level ground. A pilot measures the angles of depression to the towers to be 36.5° and 25°, respectively, as shown in the figure. Find distances $AC$ and $BC$ and the height of the airplane.

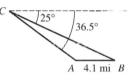

---

| **5.6** | # The Law of Cosines |
|---|---|

Deriving the Law of Cosines • Solving Triangles (SAS, SSS)
• Triangle Area and Heron's Formula • Applications

## Deriving the Law of Cosines

**Objective**

Students will be able to apply the Law of Cosines to solve acute and obtuse triangles and to determine the area of a triangle in terms of the measures of the sides and angles.

**Motivate**

For $\triangle ABC$, ask students what can be said about angle $C$ if $c^2 > a^2 + b^2$. **(It is obtuse.)**

Having seen the Law of Sines, you will probably not be surprised to learn that there is a Law of Cosines. There are many such parallels in mathematics. What you might find surprising is that the Law of Cosines has absolutely no resemblance to the Law of Sines. Instead, it resembles the Pythagorean Theorem. In fact, the Law of Cosines is often called the "generalized Pythagorean Theorem" because it contains that classic theorem as a special case.

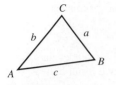

**Figure 5.23** A triangle with the usual labeling (angles *A*, *B*, *C*; opposite sides *a*, *b*, *c*).

**Lesson Guide**

Day 1: Skills
Day 2: Applications

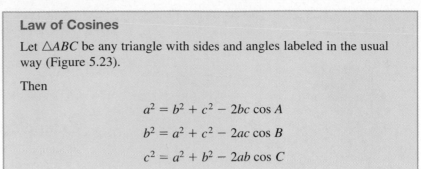

**Law of Cosines**

Let $\triangle ABC$ be any triangle with sides and angles labeled in the usual way (Figure 5.23).

Then

$$a^2 = b^2 + c^2 - 2bc \cos A$$

$$b^2 = a^2 + c^2 - 2ac \cos B$$

$$c^2 = a^2 + b^2 - 2ab \cos C$$

We derive only the first of the three equations, since the other two are derived in exactly the same way. Set the triangle in a coordinate plane so that the angle that appears in the formula (in this case, *A*) is at the origin in standard position, with side *c* along the positive *x*-axis. Depending on whether angle *A* is right (Figure 5.24a), acute (Figure 5.24b), or obtuse (Figure 5.24c), the point *C* will be on the *y*-axis, in QI, or in QII.

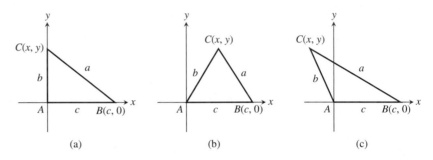

(a)  (b)  (c)

**Figure 5.24** Three cases for proving the Law of Cosines.

In each of these three cases, *C* is a point on the terminal side of angle *A* in standard position, at distance *b* from the origin. Denote the coordinates of *C* by $(x, y)$. By our definitions for trigonometric functions of any angle (Section 4.3), we can conclude that

$$\frac{x}{b} = \cos A \quad \text{and} \quad \frac{y}{b} = \sin A,$$

and therefore

$$x = b \cos A \quad \text{and} \quad y = b \sin A.$$

Now set *a* equal to the distance from *C* to *B* using the distance formula:

$$a = \sqrt{(x - c)^2 + (y - 0)^2} \qquad \text{Distance formula}$$

$$a^2 = (x - c)^2 + y^2 \qquad \text{Square both sides}$$

$$= (b \cos A - c)^2 + (b \sin A)^2 \qquad \text{Substitution}$$

$$= b^2 \cos^2 A - 2bc \cos A + c^2 + b^2 \sin^2 A$$

$$= b^2(\cos^2 A + \sin^2 A) + c^2 - 2bc \cos A$$

$$= b^2 + c^2 - 2bc \cos A \qquad \text{Pythagorean identity}$$

## Solving Triangles (SAS, SSS)

While the Law of Sines is the tool we use to solve triangles in the AAS and ASA cases, the Law of Cosines is the required tool for SAS and SSS. (Both methods can be used in the SSA case, but remember that there might be 0, 1, or 2 triangles.)

### Example 1   SOLVING A TRIANGLE (SAS)

Solve $\triangle ABC$ given that $a = 11$, $b = 5$, and $C = 20°$. (See Figure 5.25.)

**Solution**

$$c^2 = a^2 + b^2 - 2ab \cos C$$
$$= 11^2 + 5^2 - 2(11)(5) \cos 20°$$
$$= 42.6338 \ldots$$
$$c = \sqrt{42.6338 \ldots} \approx 6.5$$

We could now use either the Law of Cosines or the Law of Sines to find one of the two unknown angles. As a general rule, it is better to use the Law of Cosines to find angles, since the arccosine function will distinguish obtuse angles from acute angles.

$$a^2 = b^2 + c^2 - 2bc \cos A$$
$$11^2 = 5^2 + (6.529\ldots)^2 - 2(5)(6.529\ldots) \cos A$$
$$\cos A = \frac{5^2 + (6.529\ldots)^2 - 11^2}{2(5)(6.529\ldots)}$$
$$A = \cos^{-1}\left(\frac{5^2 + (6.529\ldots)^2 - 11^2}{2(5)(6.529\ldots)}\right)$$
$$\approx 144.8°$$
$$B = 180° - 144.8° - 20°$$
$$= 15.2°$$

So the six parts of the triangle are:

$$A = 144.8° \qquad a = 11$$
$$B = 15.2° \qquad b = 5$$
$$C = 20° \qquad c \approx 6.5$$

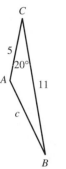

**Figure 5.25**  A triangle with two sides and an included angle known. (Example 1)

### Notes on Examples

Remind studens that $\cos^{-1}$ has range $[0, \pi]$ but $\sin^{-1}$ has range $[-\pi/2, \pi/2]$

### Example 2   SOLVING A TRIANGLE (SSS)

Solve $\triangle ABC$ if $a = 9$, $b = 7$, and $c = 5$. (See Figure 5.26.)

**Solution**  We use the Law of Cosines to find two of the angles. The third angle can be found by subtraction from 180°.

$$a^2 = b^2 + c^2 - 2bc \cos A \qquad\qquad b^2 = a^2 + c^2 - 2ac \cos B$$
$$9^2 = 7^2 + 5^2 - 2(7)(5) \cos A \qquad\qquad 7^2 = 9^2 + 5^2 - 2(9)(5) \cos B$$
$$70 \cos A = -7 \qquad\qquad\qquad 90 \cos B = 57$$
$$A = \cos^{-1}(-0.1) \qquad\qquad\qquad B = \cos^{-1}(57/90)$$
$$\approx 95.7° \qquad\qquad\qquad\qquad \approx 50.7°$$

Then $C = 180° - 95.7° - 50.7° = 33.6°$.

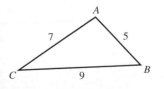

**Figure 5.26**  A triangle with three sides known. (Example 2)

## Triangle Area and Heron's Formula

The same parts that determine a triangle also determine its area. If the parts happen to be two sides and an included angle (SAS), we get a simple area formula in terms of those three parts that does not require finding an altitude.

Observe in Figure 5.24 (used in explaining the Law of Cosines) that each triangle has base $c$ and altitude $y = b \sin A$. Applying the standard area formula, we have

$$\triangle \text{Area} = \frac{1}{2}(base)(height) = \frac{1}{2}(c)(b \sin A) = \frac{1}{2}bc \sin A.$$

This is actually three formulas in one, as it does not matter which side we use as the base.

**Teaching Note**

Note that the formula for the area of a triangle holds for obtuse triangles as well as acute and right triangles.

---

**Area of a Triangle**

$$\triangle \text{Area} = \frac{1}{2}bc \sin A = \frac{1}{2}ac \sin B = \frac{1}{2}ab \sin C$$

---

**Example 3** FINDING THE AREA OF A REGULAR POLYGON

Find the area of a regular octagon (8 equal sides, 8 equal angles) inscribed inside a circle of radius 9 inches.

**Solution** Figure 5.27 shows that we can split the octagon into 8 congruent triangles. Each triangle has two 9-inch sides with an included angle of $\theta = 360°/8 = 45°$. The area of each triangle is

$$\triangle \text{Area} = (1/2)(9)(9) \sin 45° = (81/2) \sin 45° = 81\sqrt{2}/4.$$

Therefore, the area of the octagon is

$$\triangle \text{Area} = 8 \triangle \text{Area} = 162\sqrt{2} \approx 229 \text{ square inches.}$$

**Figure 5.27** A regular octagon inscribed inside a circle of radius 9 inches. (Example 3)

**Heron's Formula**

The formula is named after Heron of Alexandria, whose proof of the formula is the oldest on record, but ancient Arabic scholars claimed to have known it from the works of Archimedes of Syracuse centuries earlier. Archimedes (c. 287–212 BC) is considered to be the greatest mathematician of all antiquity.

There is also an area formula that can be used when the three sides of the triangle are known.

Although Heron proved this theorem using only classical geometric methods, we prove it as most people do today, by using the tools of trigonometry.

---

**Theorem** Heron's Formula

Let $a$, $b$, and $c$ be the sides of $\triangle ABC$, and let $s$ denote the **semiperimeter** $(a + b + c)/2$. Then the area of $\triangle ABC$ is given by
Area $= \sqrt{s(s - a)(s - b)(s - c)}$.

---

**Alert**

Students often make arithmetic errors when applying Heron's Formula. You may wish to encourage students to write a graphing calculator program to apply this formula.

**Proof**

$$\text{Area} = \frac{1}{2}ab \sin C$$

$$4(\text{Area}) = 2ab \sin C$$

$$16(\text{Area})^2 = 4a^2b^2 \sin^2 C$$

$$= 4a^2b^2(1 - \cos^2 C) \qquad \text{Pythagorean identity}$$

$$= 4a^2b^2 - 4a^2b^2 \cos^2 C$$

$$= 4a^2b^2 - (2ab \cos C)^2$$

$$= 4a^2b^2 - (a^2 + b^2 - c^2)^2 \qquad \text{Law of Cosines}$$

$$= (2ab - (a^2 + b^2 - c^2))(2ab + (a^2 + b^2 - c^2)) \qquad \text{Difference of squares}$$

$$= (c^2 - (a^2 - 2ab + b^2))((a^2 + 2ab + b^2) - c^2)$$

$$= (c^2 - (a - b)^2)((a + b)^2 - c^2)$$

$$= (c - (a - b))(c + (a - b))((a + b) - c)((a + b) + c) \qquad \text{Difference of squares}$$

$$= (c - a + b)(c + a - b)(a + b - c)(a + b + c)$$

$$= (2s - 2a)(2s - 2b)(2s - 2c)(2s) \qquad 2s = a + b + c$$

$$16(\text{Area})^2 = 16s(s - a)(s - b)(s - c)$$

$$(\text{Area})^2 = s(s - a)(s - b)(s - c)$$

$$\text{Area} = \sqrt{s(s - a)(s - b)(s - c)}$$

### Example 4  USING HERON'S FORMULA

Find the area of a triangle with sides 13, 15, 18.

**Solution**  First we compute the semiperimeter: $s = (13 + 15 + 18)/2 = 23$.

Then we use Heron's Formula

$$\text{Area} = \sqrt{23\,(23 - 13)(23 - 15)(23 - 18)}$$

$$= \sqrt{23 \cdot 10 \cdot 8 \cdot 5} = \sqrt{9200} = 20\sqrt{23}.$$

The approximate area is 96 square units.

## Applications

We end this section with a few applications.

### Example 5  MEASURING A DIHEDRAL ANGLE (SOLID GEOMETRY)

A regular tetrahedron is a solid with four faces, each of which is an equilateral triangle. Find the measure of the *dihedral angle* formed along the common edge of two intersecting faces of a regular tetrahedron with edges of length 2.

**Solution**  Figure 5.28 shows the tetrahedron. Point $B$ is the midpoint of edge $DE$, and $A$ and $C$ are the endpoints of the opposite edge. The measure of $\angle ABC$ is the same as the measure of the dihedral angle formed along edge $DE$, so we will find the measure of $\angle ABC$.

Because both $\triangle ADB$ and $\triangle CDB$ are 30°–60°–90° triangles, $AB$ and $BC$ both have length $\sqrt{3}$. If we apply the Law of Cosines to $\triangle ABC$, we obtain

$$2^2 = (\sqrt{3})^2 + (\sqrt{3})^2 - 2\sqrt{3}\sqrt{3} \cos (\angle ABC)$$

$$\cos (\angle ABC) = \frac{1}{3}$$

$$\angle ABC = \cos^{-1}\left(\frac{1}{3}\right) \approx 70.53°$$

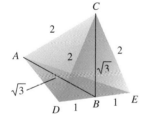

**Figure 5.28** The measure of $\angle ABC$ is the same as the measure of any dihedral angle formed by two of the tetrahedron's faces. (Example 5)

### Platonic Solids

The regular tetrahedron in Example 5 is one of only 5 *regular* solids (solids with faces that are congruent polygons having equal angles and equal sides). The others are the cube (6 square faces), the octahedron (8 triangular faces), the dodecahedron (12 pentagonal faces), and the icosahedron (20 triangular faces). Plato did not discover them, but they are featured in his cosmology as being the stuff of which everything in the universe is made. The Platonic universe itself is a dodecaderon, a favorite symbol of the Pythagoreans.

The dihedral angle has the same measure as $\angle ABC$, approximately $70.53°$. (We chose sides of length 2 for computational convenience, but in fact this is the measure of a dihedral angle in a regular tetrahedron of any size.)

### Example 6  MEASURING A BASEBALL DIAMOND

The bases on a baseball diamond are 90 feet apart, and the front edge of the pitcher's rubber is 60.5 feet from the back corner of home plate. Find the distance from the center of the front edge of the pitcher's rubber to the far corner of first base.

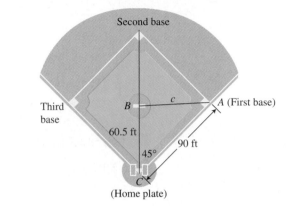

**Figure 5.29** The diamond-shaped part of a baseball diamond. (Example 6)

**Solution**  Figure 5.29 shows first base as $A$, the pitcher's rubber as $B$, and home plate as $C$. The distance we seek is side $c$ in $\triangle ABC$.

By the Law of Cosines,

$$c^2 = 60.5^2 + 90^2 - 2(60.5)(90) \cos 45°$$
$$c = \sqrt{60.5^2 + 90^2 - 2(60.5)(90) \cos 45°}$$
$$\approx 63.7$$

The distance from first base to the pitcher's rubber is about 63.7 feet.

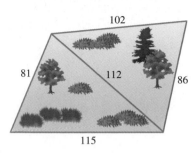

**Figure 5.30** Dimensions (in paces) of an irregular plot of land. (Exploration 1)

**Exploration Extensions**

Explain why Jim and Barbara needed to find the diagonal measurement (112 paces) in order to find the area of the lot.

---

**Exploration 1** **Estimating Acreage of a Plot of Land**

Jim and Barbara are house-hunting and need to estimate the size of an irregular adjacent lot that is described by the owner as "a little more than an acre." With Barbara stationed at a corner of the plot, Jim starts at another corner and walks a straight line toward her, counting his paces. They then shift corners and Jim paces again, until they have recorded the dimensions of the lot (in paces) as in Figure 5.30. They later measure Jim's pace as 2.2 feet. What is the approximate acreage of the lot?

1. Use Heron's formula to find the area in square paces.

2. Convert the area to square feet, using the measure of Jim's pace.

3. There are 5280 feet in a mile. Convert the area to square miles.

4. There are 640 square acres in a square mile. Convert the area to acres.

5. Is there good reason to doubt the owner's estimate of the acreage of the lot?

6. Would Jim and Barbara be able to modify their system to estimate the area of an irregular lot with five straight sides?

---

**Problem**

If a triangular region with sides of 3 kilometers, 4 kilometers, and 6 kilometers has a population of 50 deer, how close is the population on this land to the average national park population?

**Solution**

We can find the area of the land region

By using Heron's formula with

$$s = (3 + 4 + 6)/2 = 13/2$$

and

$$\text{Area} = \sqrt{s(s - a)(s - b)(s - c)}$$
$$= \sqrt{\frac{13}{2}\left(\frac{13}{2} - 3\right)\left(\frac{13}{2} - 4\right)\left(\frac{13}{2} - 6\right)}$$
$$= \sqrt{\frac{13}{2}\left(\frac{7}{2}\right)\left(\frac{5}{2}\right)\left(\frac{1}{2}\right)} \approx 5.3$$

so the area of the land region is 5.3 km².

If this land were to support 14 deer/km², it would have (5.3... km²)(14 deer/km²) = 74.7 ≈ 75 deer. Thus, the land supports 25 deer less than the average.

# Quick Review 5.6

In Exercises 1–4, find an angle between 0° and 180° that is a solution to the equation.

**1.** $\cos A = 3/5$      **2.** $\cos C = -0.23$

**3.** $\cos A = -0.68$      **4.** $3 \cos C = 1.92$

In Exercises 5 and 6, solve the equation (in terms of $x$ and $y$) for
**(a)** $\cos A$ and **(b)** $A$, $0 \le A \le 180°$.

**5.** $9^2 = x^2 + y^2 - 2xy \cos A$      **6.** $y^2 = x^2 + 25 - 10 \cos A$

In Exercises 7–10, find a quadratic polynomial with real coefficients that satisfies the given condition.

**7.** Has two positive zeros   One answer: $(x - 1)(x - 2)$

**8.** Has one positive and one negative zero

**9.** Has no real zeros   One answer $(x - i)(x + i) = x^2 + 1$

**10.** Has exactly one positive zero   One answer $(x - 1)^2$

# Section 5.6 Exercises

In Exercises 1–4, solve the triangle.

**1.**

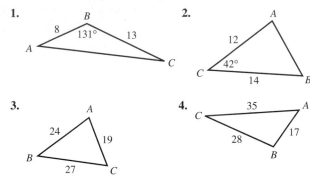

**2.**

**3.**

**4.**

In Exercises 5–16, solve the triangle.

**5.** $A = 55°$,   $b = 12$,   $c = 7$

**6.** $B = 35°$,   $a = 43$,   $c = 19$

**7.** $a = 12$,   $b = 21$,   $C = 95°$

**8.** $b = 22$,   $c = 31$,   $A = 82°$

**9.** $a = 1$,   $b = 5$,   $c = 4$   No triangles possible

**10.** $a = 1$,   $b = 5$,   $c = 8$   No triangles possible

**11.** $a = 3.2$,   $b = 7.6$,   $c = 6.4$

**12.** $a = 9.8$,   $b = 12$,   $c = 23$   No triangles possible

**13.** $A = 42°$,   $a = 7$,   $b = 10$

**14.** $A = 57°$,   $a = 11$,   $b = 10$

**15.** $A = 63°$,   $a = 8.6$,   $b = 11.1$   no triangle

**16.** $A = 71°$,   $a = 9.3$,   $b = 8.5$

In Exercises 17–20, find the area of the triangle.

**17.** $A = 47°$,   $b = 32$ ft,   $c = 19$ ft   $\approx 222.33$ ft$^2$

**18.** $A = 52°$,   $b = 14$ m,   $c = 21$ m   $\approx 115.84$ m$^2$

**19.** $B = 101°$,   $a = 10$ cm,   $c = 22$ cm   $\approx 107.98$ cm$^2$

**20.** $C = 112°$,   $a = 1.8$ in.,   $b = 5.1$ in.   $\approx 4.26$ in.$^2$

In Exercises 21–28, decide whether a triangle can be formed with the given side lengths. If so, use Heron's formula to find the area of the triangle.

**21.** $a = 4$,   $b = 5$,   $c = 8$   $\approx 8.18$

**22.** $a = 5$,   $b = 9$,   $c = 7$   $\approx 17.41$

**23.** $a = 3$,   $b = 5$,   $c = 8$   no triangle is formed

**24.** $a = 23$,   $b = 19$,   $c = 12$   $\approx 113.84$

**25.** $a = 19.3$,   $b = 22.5$,   $c = 31$   $\approx 216.15$

**26.** $a = 8.2$,   $b = 12.5$,   $c = 28$   no triangle is formed

**27.** $a = 33.4$,   $b = 28.5$,   $c = 22.3$   $\approx 314.05$

**28.** $a = 18.2$,   $b = 17.1$,   $c = 12.3$   $\approx 101.34$

**29.** Find the radian measure of the largest angle in the triangle with sides of 4, 5, and 6.   $\approx 1.445$ radians

**30.** A parallelogram has sides of 18 and 26 ft, and an angle of 39°. Find the shorter diagonal.   $\approx 16.5$ ft

**31.** **Measuring Distance Indirectly** Juan wants to find the distance between two points $A$ and $B$ on opposite sides of a building. He locates a point $C$ that is 110 ft from $A$ and 160 ft from $B$, as illustrated in the figure. If the angle at $C$ is 54°, find distance $AB$.   $\approx 130.42$ ft

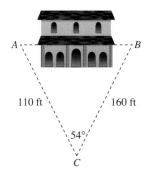

**32.** **Designing a Baseball Field**

**(a)** Find the distance from the center of the front edge of the pitcher's rubber to the far corner of second base. How does this distance compare with the distance from the pitcher's rubber to first base? (See Example 6.)

**(b)** Find $\angle B$ in $\triangle ABC$. 92.8°

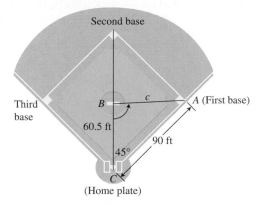

Second base

Third base

*B*    *c*    *A* (First base)

60.5 ft

90 ft

45°

*C*

(Home plate)

**33. Designing a Softball Field** In softball, adjacent bases are 60 ft apart. The distance from the center of the front edge of the pitcher's rubber to the far corner of home plate is 40 ft.

**(a)** Find the distance from the center of the pitcher's rubber to the far corner of first base. ≈ 42.5 ft

**(b)** Find the distance from the center of the pitcher's rubber to the far corner of second base.

**(c)** Find $\angle B$ in $\triangle ABC$. 93.3°

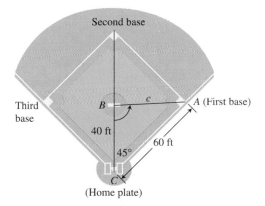

Second base

Third base

*B*    *c*    *A* (First base)

40 ft

60 ft

45°

*C*

(Home plate)

**34. Surveyor's Calculations** Tony must find the distance from *A* to *B* on opposite sides of a lake. He locates a point C that is 860 ft from *A* and 175 ft from *B*. He measures the angle at *C* to be 78°. Find distance *AB*. 841.2 ft

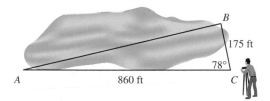

*B*

175 ft

78°

*A*    860 ft    *C*

**35. Construction Engineering** A manufacturer is designing the roof truss that is modeled in the figure shown.

**(a)** Find the measure of $\angle CAE$. $\tan^{-1}(1/3) \approx 18.4°$

**(b)** If $AF = 12$ ft, find the length *DF*. ≈ 4.5 ft

**(c)** Find the length *EF*. ≈ 7.6 ft

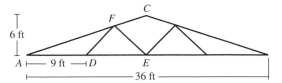

*C*

*F*

6 ft

*A* ⊢— 9 ft —⊣*D*    *E*

36 ft

**36. Navigation** Two airplanes flying together in formation take off in different directions. One flies due east at 350 mph, and the other flies east-northeast at 380 mph. How far apart are the two airplanes 2 h after they separate, assuming that they fly at the same altitude? ≈ 290.8 mi

**37. Football Kick** The player waiting to receive a kickoff stands at the 5 yard line (point *A*) as the ball is being kicked 65 yd up the field from the opponent's 30 yard line. The kicked ball travels 73 yd at an angle of 8° to the right of the receiver, as shown in the figure (point *B*). Find the distance the receiver runs to catch the ball. 12.5 yd

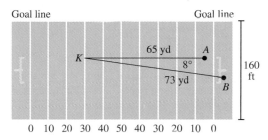

Goal line            Goal line

65 yd    *A*

*K*      8°

73 yd    *B*

160 ft

0   10   20   30   40   50   40   30   20   10   0

**38. Architectural Design** Building Inspector Julie Wang checks a building in the shape of a regular octagon, each side 20 ft long. She checks that the contractor has located the corners of the foundation correctly by measuring several of the diagonals. Calculate what the lengths of diagonals *HB*, *HC*, and *HD* should be.

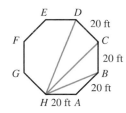

*E*    *D*

*F*          20 ft    *C*

*G*          20 ft    *B*

20 ft

*H* 20 ft *A*

**39. Group Activity Connecting Trigonometry and Geometry** $\angle CAB$ is inscribed in a rectangular box whose sides are 1, 2 and 3 ft long as shown. Find the measure of $\angle CAB$. ≈ 37.9°

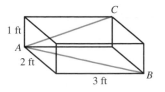

*C*

1 ft

*A*

2 ft

3 ft

*B*

**40. Group Activity Connecting Trigonometry and Geometry** A cube has edges of length 2 ft. Point *A* is the midpoint of an edge. Find the measure of $\angle ABC$.

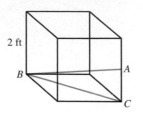

2 ft

**45.** A **segment** of a circle is the region enclosed between a chord of a circle and the arc intercepted by the chord. Find the area of a segment intercepted by a 7-inch chord in a circle of radius 5 inches.

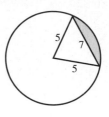

## Explorations

**41.** Find the area of a regular polygon with *n* sides inscribed inside a circle of radius *r*. (Express your answer in terms of *n* and *r*.) Area $= (nr^2/2) \sin (360°/n)$

**42. (a)** Prove the identity: $\dfrac{\cos A}{a} = \dfrac{b^2 + c^2 - a^2}{2abc}$.

**(b)** Prove the (tougher) identity:

$$\frac{\cos A}{a} + \frac{\cos B}{b} + \frac{\cos C}{c} = \frac{a^2 + b^2 + c^2}{2abc}.$$

[*Hint*: use the identity in (a), along with its other variations.]

**43. Navigation** Two ships leave a common port at 8:00 A.M. and travel at a constant rate of speed. Each ship keeps a log showing its distance from port and its distance from the other ship. Portions of the logs from later that morning for both ships are shown below.

| Time | Naut mi from port | Naut mi from ship B | Time | Naut mi from port | Naut mi from ship A |
|------|------|------|------|------|------|
| 9:00 | 15.1 | 8.7 | 9:00 | 12.4 | 8.7 |
| 10:00 | 30.2 | 17.3 | 11:00 | 37.2 | 26.0 |

**(a)** How fast is each ship traveling? (Express your answer in knots, which are nautical miles per hour.)

**(b)** What is the angle of intersection of the courses of the two ships? 35.18°

**(c)** How far apart are the ships at 12:00 noon if they maintain the same courses and speeds? 34.8 nautical mi. ■

## Extending the Ideas

**44.** Prove that the area of a triangle can be found with the formula

$$\triangle \text{ Area} = \frac{a^2 \sin B \sin C}{2 \sin A}.$$

## Math at Work

I got into medicine because I always liked a good challenge. Medicine can be like solving a puzzle, which I enjoy. I chose anesthesiology because it's even more challenging than other fields of medicine. Any surgical specialty tends to offer more difficult problems than, say, treating the sniffles.

What I enjoy most about my job is trying to gain people's confidence in a 5 minute interview. I can tell that some people will accept my judgments after our first interview, and others will question everything I do.

One good example of how we use math in medicine is when a patient goes into shock. Typically, the patient's blood pressure bottoms out, and his natural mechanisms are unable to raise it. We give the patient dopamine to raise his blood pressure, but we need to make sure there is a consistent level of dopamine entering the bloodstream. For instance, if we have 400 mg of dopamine mixed into 250 cc of saline, we need to calculate how fast the intravenous drip should be so that there is 5 mg of dopamine in the bloodstream per kilo of body weight. Since all patients have different body weights, all patients require different rates of intravenous drip.

Ernest Newkirk, M.D.,

---

| Chapter 5 | **Key Ideas** |

### Concepts

### Properties, Theorems, and Formulas

#### Reciprocal Identities

$$\csc \theta = \frac{1}{\sin \theta} \qquad \sec \theta = \frac{1}{\cos \theta} \qquad \cot \theta = \frac{1}{\tan \theta}$$

$$\sin \theta = \frac{1}{\csc \theta} \qquad \cos \theta = \frac{1}{\sec \theta} \qquad \tan \theta = \frac{1}{\cot \theta}$$

#### Quotient Identities

$$\tan \theta = \frac{\sin \theta}{\cos \theta} \qquad \cot \theta = \frac{\cos \theta}{\sin \theta}$$

#### Pythagorean Identities

$$\cos^2 \theta + \sin^2 \theta = 1$$

$$1 + \tan^2 \theta = \sec^2 \theta$$

$$\cot^2 \theta + 1 = \csc^2 \theta$$

## Cofunction Identities

$$\sin\left(\frac{\pi}{2} - \theta\right) = \cos\theta \qquad \cos\left(\frac{\pi}{2} - \theta\right) = \sin\theta$$

$$\tan\left(\frac{\pi}{2} - \theta\right) = \cot\theta \qquad \cot\left(\frac{\pi}{2} - \theta\right) = \tan\theta$$

$$\sec\left(\frac{\pi}{2} - \theta\right) = \csc\theta \qquad \csc\left(\frac{\pi}{2} - \theta\right) = \sec\theta$$

## Odd-Even Identities

$$\sin(-x) = -\sin x \qquad \cos(-x) = \cos x \qquad \tan(-x) = -\tan x$$

$$\csc(-x) = -\csc x \qquad \sec(-x) = \sec x \qquad \cot(-x) = -\cot x$$

## Sum/Difference Identities

$$\sin(u \pm v) = \sin u \cos v \pm \cos u \sin v$$

$$\cos(u \pm v) = \cos u \cos v \mp \sin u \sin v$$

$$\tan(u \pm v) = \frac{\sin u \cos v \pm \cos u \sin v}{\cos u \cos v \mp \sin u \sin v} = \frac{\tan u \pm \tan v}{1 \mp \tan u \tan v}$$

## Double-Angle Identities

$$\sin 2u = 2\sin u \cos u$$

$$\cos 2u = \begin{cases} \cos^2 u - \sin^2 u \\ 2\cos^2 u - 1 \\ 1 - 2\sin^2 u \end{cases}$$

$$\tan 2u = \frac{2\tan u}{1 - \tan^2 u}$$

## Power-Reducing Identities

$$\sin^2 u = \frac{1 - \cos 2u}{2}$$

$$\cos^2 u = \frac{1 + \cos 2u}{2}$$

$$\tan^2 u = \frac{1 - \cos 2u}{1 + \cos 2u}$$

## Half-Angle Identities

$$\sin\frac{u}{2} = \pm\sqrt{\frac{1 - \cos u}{2}}$$

$$\cos\frac{u}{2} = \pm\sqrt{\frac{1 + \cos u}{2}}$$

$$\tan\frac{u}{2} = \begin{cases} \pm\sqrt{\dfrac{1 - \cos u}{1 + \cos u}} \\[2ex] \dfrac{1 - \cos u}{\sin u} \\[2ex] \dfrac{\sin u}{1 + \cos u} \end{cases}$$

## Law of Sines

In any $\triangle ABC$ with angles $A$, $B$, and $C$ opposite sides $a$, $b$, and $c$ respectively, the following equation is true:

$$\frac{\sin A}{a} = \frac{\sin B}{b} = \frac{\sin C}{c}$$

## Law of Cosines

In any $\triangle ABC$ with angles $A$, $B$, and $C$ opposite sides $a$, $b$, and $c$ respectively, the following equations are true:

$$a^2 = b^2 + c^2 - 2bc \cos A$$

$$b^2 = a^2 + c^2 - 2ac \cos B$$

$$c^2 = a^2 + b^2 - 2ab \cos C$$

## Triangle Area

The area of any $\triangle ABC$ with angles $A$, $B$, and $C$ opposite sides $a$, $b$, and $c$ respectively, is given by any of the following formulas:

$$\text{Area} = \frac{1}{2}bc \sin A = \frac{1}{2}ac \sin B = \frac{1}{2}ab \sin C.$$

## Heron's Formula

Let $a$, $b$, and $c$ be the sides of $\triangle ABC$, and let $s$ denote the semiperimeter $\dfrac{a+b+c}{2}$.

Then the area of $\triangle ABC$ is given by $A = \sqrt{s(s-a)(s-b)(s-c)}$.

# Procedures

## Strategies for Proving an Identity

1. Begin with the more complicated expression and work toward the less complicated expression.

2. If no other move suggests itself, convert the entire expression to one involving sines and cosines.

3. Combine fractions by writing them over a common denominator.

4. Use the algebraic identity $(a+b)(a-b) = a^2 - b^2$ to set up applications of the Pythagorean identities.

5. Always be mindful of the "target" expression, and favor manipulations that bring you closer to your goal.

## Chapter 5    Review Exercises

The collection of exercises marked in red could be used as a chapter test.

In Exercises 1 and 2, write the expression as the sine, cosine, or tangent of an angle.

**1.** $2 \sin 100° \cos 100°$

**2.** $\dfrac{2 \tan 40°}{1 - \tan^2 40°}$ $\quad \tan 80°$

In Exercises 3 and 4, simplify the expression to a single term. Support your answer graphically.

**3.** $(1 - 2 \sin^2 \theta)^2 + 4 \sin^2 \theta \cos^2 \theta$   1

**4.** $1 - 4 \sin^2 x \cos^2 x$   $\cos^2 2x$

In Exercises 5–22, prove the identity.

**5.** $\cos 3x = 4 \cos^3 x - 3 \cos x$

**6.** $\cos^2 2x - \cos^2 x = \sin^2 x - \sin^2 2x$

**7.** $\tan^2 x - \sin^2 x = \sin^2 x \tan^2 x$

**8.** $2 \sin \theta \cos^3 \theta + 2 \sin^3 \theta \cos \theta = \sin 2\theta$

**9.** $\csc x - \cos x \cot x = \sin x$

**10.** $\dfrac{\tan \theta + \sin \theta}{2 \tan \theta} = \cos^2 \left( \dfrac{\theta}{2} \right)$

**11.** $\dfrac{1 + \tan \theta}{1 - \tan \theta} + \dfrac{1 + \cot \theta}{1 - \cot \theta} = 0$

**12.** $\sin 3\theta = 3 \cos^2 \theta \sin \theta - \sin^3 \theta$

**13.** $\cos^2 \left( \dfrac{t}{2} \right) = \dfrac{1 + \sec t}{2 \sec t}$

**14.** $\dfrac{\tan^3 \gamma - \cot^3 \gamma}{\tan^2 \gamma + \csc^2 \gamma} = \tan \gamma - \cot \gamma$

**15.** $\dfrac{\cos \phi}{1 - \tan \phi} + \dfrac{\sin \phi}{1 - \cot \phi} = \cos \phi + \sin \phi$

**16.** $\dfrac{\cos (-z)}{\sec (-z) + \tan (-z)} = 1 + \sin z$

**17.** $\sqrt{\dfrac{1 - \cos y}{1 + \cos y}} = \dfrac{1 - \cos y}{|\sin y|}$

**18.** $\sqrt{\dfrac{1 - \sin \gamma}{1 + \sin \gamma}} = \dfrac{|\cos \gamma|}{1 + \sin \gamma}$

**19.** $\tan\left(u + \dfrac{3\pi}{4}\right) = \dfrac{\tan u - 1}{1 + \tan u}$

**20.** $\dfrac{1}{4}\sin 4\gamma = \sin \gamma \cos^3 \gamma - \cos \gamma \sin^3 \gamma$

**21.** $\tan \dfrac{1}{2}\beta = \csc \beta - \cot \beta$

**22.** $\arctan t = \dfrac{1}{2}\arctan \dfrac{2t}{1 - t^2}, \quad -1 < t < 1$

In Exercises 23 and 24, use a grapher to conjecture whether the equation is likely to be an identity. Confirm your conjecture.

**23.** $\sec x - \sin x \tan x = \cos x$

**24.** $(\sin^2 \alpha - \cos^2 \alpha)(\tan^2 \alpha + 1) = \tan^2 \alpha - 1$

In Exercises 25–28, write the expression in terms of $\sin x$ and $\cos x$ only.

**25.** $\sin 3x + \cos 3x$       **26.** $\sin 2x + \cos 3x$

**27.** $\cos^2 2x - \sin 2x$       **28.** $\sin 3x - 3 \sin 2x$

In Exercises 29–34, find the general solution without using a calculator. Give exact answers.

**29.** $\sin 2x = 0.5$      **30.** $\cos x = \dfrac{\sqrt{3}}{2}$   $\pm\dfrac{\pi}{6} + 2n\pi$

**31.** $\tan x = -1$   $-\dfrac{\pi}{4} + n\pi$      **32.** $2 \sin^{-1} x = \sqrt{2}$   $\sin\dfrac{\sqrt{2}}{2}$

**33.** $\tan^{-1} x = 1$   $\tan 1$      **34.** $2 \cos 2x = 1$

In Exercises 35–38, solve the equation graphically.

**35.** $\sin^2 x - 3 \cos x = -0.5$   $x \approx 1.12$

**36.** $\cos^3 x - 2 \sin x - 0.7 = 0$   $x \approx 0.14$ or $x \approx 3.79$

**37.** $\sin^4 x + x^2 = 2$   $x \approx 1.15$

**38.** $\sin 2x = x^3 - 5x^2 + 5x + 1$   $x \approx 1.85$ or $x \approx 3.59$

In Exercises 39–44, find all solutions in the interval $[0, 2\pi)$ without using a calculator. Give exact answers.

**39.** $2 \cos x = 1$   $\pi/3, 5\pi/3$      **40.** $\sin 3x = \sin x$

**41.** $\sin^2 x - 2 \sin x - 3 = 0$      **42.** $\cos 2t = \cos t$

**43.** $\sin (\cos x) = 1$      **44.** $\cos 2x + 5 \cos x = 2$

In Exercises 45–48, solve the inequality. Use any method, but give exact answers.

**45.** $2 \cos 2x > 1$ for $0 \le x < 2\pi$

**46.** $\sin 2x > 2 \cos x$ for $0 < x \le 2\pi$   $(\pi/2, 3\pi/2)$

**47.** $2 \cos x < 1$ for $0 \le x < 2\pi$   $(\pi/3, 5\pi/3)$

**48.** $\tan x < \sin x$ for $-\dfrac{\pi}{2} < x < \dfrac{\pi}{2}$   $(-\pi/2, 0)$

In Exercises 49 and 50, find an equivalent equation of the form $y = a \sin (bx + c)$. Support your work graphically.

**49.** $y = 3 \sin 3x + 4 \cos 3x$      **50.** $y = 5 \sin 2x - 12 \cos 2x$

In Exercises 51–58, solve $\triangle ABC$.

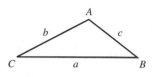

**51.** $A = 79°$,   $B = 33°$,   $a = 7$   $C = 68°, b \approx 3.9, c \approx 6.6$

**52.** $a = 5$,   $b = 8$,   $B = 110°$

**53.** $a = 8$,   $b = 3$,   $B = 30°$   no triangle is formed

**54.** $a = 14.7$,   $A = 29.3°$,   $C = 33°$

**55.** $A = 34°$,   $B = 74°$,   $c = 5$

**56.** $c = 41$,   $A = 22.9°$,   $C = 55.1°$

**57.** $a = 5$,   $b = 7$,   $c = 6$   $A = 44.4°, B = 78.5°, C = 57.1°$

**58.** $A = 85°$,   $a = 6$,   $b = 4$   $B = 41.6°, C = 53.4°, c \approx 4.8$

In Exercises 59 and 60, find the area of $\triangle ABC$.

**59.** $a = 3$,   $b = 5$,   $c = 6$      **60.** $a = 10$,   $b = 6$,   $C = 50°$

**61.** If $a = 12$ and $B = 28°$, determine the values of $b$ that will produce the indicated number of triangles

    **(a)** Two          **(b)** One          **(c)** Zero

**62. Surveying a Canyon** Two markers $A$ and $B$ on the same side of a canyon rim are 80 ft apart, as shown in the figure. A hiker is located across the rim at point $C$. A surveyor determines that $\angle BAC = 70°$ and $\angle ABC = 65°$.

    **(a)** What is the distance between the hiker and point $A$?

    **(b)** What is the distance between the two canyon rims? (Assume they are parallel.)   $\approx 96.4$ ft

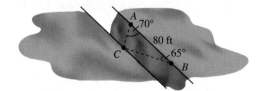

**63. Altitude** A hot-air balloon is seen over Tucson, Arizona, simultaneously by two observers at points $A$ and $B$ that are 1.75 mi apart on level ground and in line with the balloon. The angles of elevation are as shown here. How high above ground is the balloon?   $\approx 0.6$ mi

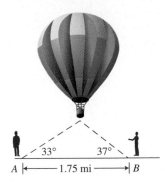

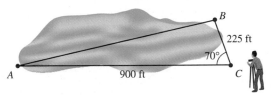

**64. Finding Distance** In order to determine the distance between two points *A* and *B* on opposite sides of a lake, a surveyor chooses a point *C* that is 900 ft from *A* and 225 ft from *B*, as shown in the figure. If the measure of the angle at *C* is 70°, find the distance between *A* and *B*. ≈ 849.77 ft

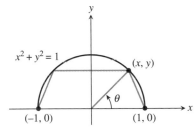

**65. Finding Radian Measure** Find the radian measure of the largest angle of the triangle whose sides have lengths 8, 9, and 10. 1.25 rad

**66. Finding a Parallelogram** A parallelogram has sides of 15 and 24 ft, and an angle of 40°. Find the diagonals.

**67. Maximizing Area** A trapezoid is inscribed in the upper half of a unit circle, as shown in the figure.

(a) Write the area of the trapezoid as a function of *θ*.

(b) Find the value of *θ* that maximizes the area of the trapezoid and the maximum area.

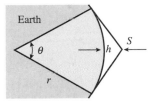

**68. Beehive Cells** A single cell in a beehive is a regular hexagonal prism open at the front with a trihedral cut at the back. Trihedral refers to a vertex formed by three faces of a polyhedron. It can be shown that the surface area of a cell is given by

$$S(\theta) = 6ab + \frac{3}{2}b^2\left(-\cot\theta + \frac{\sqrt{3}}{\sin\theta}\right),$$

where *θ* is the angle between the axis of the prism and one of the back faces, *a* is the depth of the prism, and *b* is the length of the hexagonal front. Assume *a* = 1.75 in. and *b* = 0.65 in.

(a) Graph the function *y* = *S*(*θ*).

(b) What value of *θ* gives the minimum surface area? (*Note:* This answer is quite close to the observed angle in nature.)

(c) What is the minimum surface area? ≈ 7.72 in.²

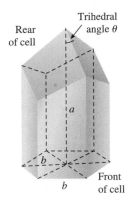

**69. Cable Television Coverage** A cable broadcast satellite *S* orbits a planet at a height *h* (in miles) above the earth's surface, as shown in the figure. The two lines from *S* are tangent to the earth's surface. The part of the earth's surface that is in the broadcast area of the satellite is determined by the central angle *θ* indicated in the figure.

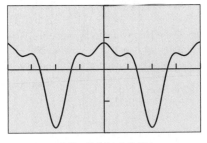

(a) Assuming that the earth is spherical with a radius of 4000 mi, write *h* as a function of *θ*.

(b) Approximate *θ* for a satellite 200 mi above the surface of the earth. ≈ 35.51°

**70. Finding Extremum Values** The graph of

$$y = \cos x - \frac{1}{2}\cos 2x + \frac{1}{3}\cos 3x$$

is shown in the figure. The *x*-values that correspond to local maximum and minimum points are solutions of the equation sin *x* − sin 2*x* + sin 3*x* = 0. Solve this equation algebraically, and support your solution using the graph of *y*.

[−2π, 2π] by [−2, 2]

**71. Using Trigonometry in Geometry** A regular hexagon whose sides are 16 cm is inscribed in a circle. Find the area inside the circle and outside the hexagon.

**72. Using Trigonometry in Geometry** A circle is inscribed in a regular pentagon whose sides are 12 cm. Find the area inside the pentagon and outside the circle.

**73. Using Trigonometry in Geometry** A wheel of cheese in the shape of a right circular cylinder is 18 cm in diameter and 5 cm thick. If a wedge of cheese with a central angle of 15° is cut from the wheel, find the volume of the cheese wedge. $405\pi/24 \approx 53.01$ cm³

**74. Product-to-Sum Formulas** Prove the following identities, which are called the **product-to-sum formulas**.

**(a)** $\sin u \sin v = \dfrac{1}{2}(\cos (u - v) - \cos (u + v))$

**(b)** $\cos u \cos v = \dfrac{1}{2}(\cos (u - v) + \cos (u + v))$

**(c)** $\sin u \cos v = \dfrac{1}{2}(\sin (u + v) + \sin (u - v))$

**75. Sum-to-Product Formulas** Use the product-to-sum formulas in Exercise 74 to prove the following identities, which are called the **sum-to-product formulas.**

**(a)** $\sin u + \sin v = 2 \sin \dfrac{u + v}{2} \cos \dfrac{u - v}{2}$

**(b)** $\sin u - \sin v = 2 \sin \dfrac{u - v}{2} \cos \dfrac{u + v}{2}$

**(c)** $\cos u + \cos v = 2 \cos \dfrac{u + v}{2} \cos \dfrac{u - v}{2}$

**(d)** $\cos u - \cos v = -2 \sin \dfrac{u + v}{2} \sin \dfrac{u - v}{2}$

**76. Catching Students Faking Data** Carmen and Pat both need to make up a missed physics lab. They are to measure the total distance $(2x)$ traveled by a beam of light from point $A$ to point $B$ and record it in 20° increments of $\theta$ as they adjust the mirror $(C)$ upward vertically.

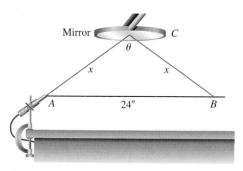

They report the following measurements. However, only one of the students actually did the lab; the other skipped it and faked the data. Who faked the data, and how can you tell?

| CARMEN | | PAT | |
|---|---|---|---|
| $\theta$ | $2x$ | $\theta$ | $2x$ |
| 160° | 24.4″ | 160° | 24.5″ |
| 140° | 25.6″ | 140° | 25.2″ |
| 120° | 28.0″ | 120° | 26.4″ |
| 100° | 31.2″ | 100° | 30.4″ |
| 80° | 37.6″ | 80° | 35.2″ |
| 60° | 48.0″ | 60° | 48.0″ |
| 40° | 70.4″ | 40° | 84.0″ |
| 20° | 138.4″ | 20° | 138.4″ |

**77. An Interesting Fact about (sin A)/a** The ratio $(\sin A)/a$ that shows up in the Law of Sines shows up another way in the geometry of $\triangle ABC$: It is the reciprocal of the radius of the circumscribed circle.

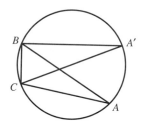

**(a)** Let $\triangle ABC$ be circumscribed as shown in the diagram, and construct diameter $CA'$. Explain why $\angle A'BC$ is a right angle.

**(b)** Explain why $\angle A'$ and $\angle A$ are congruent.

**(c)** If $a$, $b$, and $c$ are the sides opposite angles $A$, $B$, and $C$ as usual, explain why $\sin A' = a/d$, where $d$ is the diameter of the circle.

**(d)** Finally, explain why $(\sin A)/a = 1/d$.

**(e)** Do $(\sin B)/b$ and $(\sin C)/c$ also equal $1/d$? Why?

# Chapter 5 Project

## Modeling the Illumination of the Moon

From the earth, the Moon appears to be a circular disk in the sky that is illuminated to varying degrees by direct sunlight. During each lunar orbit the Moon varies from a status of being a New Moon with no visible illumination to that of a Full Moon which is fully illuminated by direct sunlight. The United States Naval Observatory has developed a mathematical model to find the fraction of the Moon's visible disk that is illuminated by the Sun. The data in the table below (obtained from the U.S. Naval Observatory web site, http://www.usno.navy.mil/, Astronomical Applications Department) shows the fraction of the Moon illuminated at midnight for each day in January 2000.

### Fraction of the Moon Illuminated, January 2000

| Day # | Fraction illuminated | Day # | Fraction illuminated | Day # | Fraction illuminated | Day # | Fraction illuminated |
|-------|---------------------|-------|---------------------|-------|---------------------|-------|---------------------|
| 1 | 0.25 | 9 | 0.05 | 17 | 0.78 | 25 | 0.80 |
| 2 | 0.18 | 10 | 0.11 | 18 | 0.87 | 26 | 0.71 |
| 3 | 0.11 | 11 | 0.18 | 19 | 0.94 | 27 | 0.61 |
| 4 | 0.06 | 12 | 0.26 | 20 | 0.99 | 28 | 0.51 |
| 5 | 0.02 | 13 | 0.36 | 21 | 1.00 | 29 | 0.42 |
| 6 | 0.00 | 14 | 0.46 | 22 | 0.99 | 3 | 0.32 |
| 7 | 0.00 | 15 | 0.57 | 23 | 0.94 | 31 | 0.24 |
| 8 | 0.02 | 16 | 0.68 | 24 | 0.88 | | |

## Explorations

1. Enter the data in the table above into your grapher or computer. Create a scatter plot of the data.

2. Find values for $a$, $b$, $c$, and $d$ so the equation $y = a \cos (b(x - c)) + d$ models the data in the data plot.

3. Verify graphically the cofunction identity $\sin (\pi/2 - \theta) = \cos \theta$ by substituting $(\pi/2 - \theta)$ for $\theta$ in the model above and using sine instead of cosine. (Note $\theta = b(x - c)$.) Observe how well this new model fits the data.

4. Verify graphically the odd-even identity $\cos (\theta) = \cos (-\theta)$ for the model in #2 by substituting $-\theta$ for $\theta$ and observing how well the graph fits the data.

5. Find values for $a$, $b$, $c$, and $d$ so the equation $y = a \sin (b(x - c)) + d$ fits the data in the table.

6. Verify graphically the cofunction identity $\cos (\pi/2 - \theta) = \sin \theta$ by substituting $(\pi/2 - \theta)$ for $\theta$ in the model above and using cosine rather than sine. (Note $\theta = b(x - c)$.) Observe the fit of this model to the data.

7. Verify graphically the odd-even identity $\sin (-\theta) = -\sin (\theta)$ for the model in #5 by substituting $-\theta$ for $\theta$ and graphing $-a \sin (-\theta) + d$. How does this model compare to the original one?

**Bibliography**

**For students:** *The Sky's the Limit in Math-Related Careers (Handbook).* Women's Educational Equity Publishing Center, 55 Chapel Street, Suite 200, Newton, MA 02160.

**For teachers:** *Handbook of Research on Mathematics Teaching and Learning,* Douglas A. Grouws (Ed.) Macmillan, 1992.

*Space Mathematics: A Resource for Secondary School Teachers,* Bernice Kastner. USPGO, 1986. Available through Dale Seymour Publications.

**Teaching Note**

It is worth pointing out that although the exact definition of a vector may vary between texts, the general concept is the same—a vector is defined by its direction and magnitude but *not* by its location.

**Teaching Note**

Attention to what constitutes equal vectors will contribute to success later on. Failure to understand the definition of equal vectors results in a surprising number of errors in students' thinking.

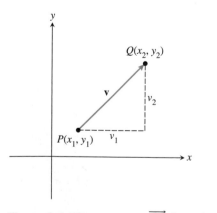
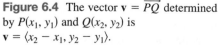

**Figure 6.4** The vector $\mathbf{v} = \overrightarrow{PQ}$ determined by $P(x_1, y_1)$ and $Q(x_2, y_2)$ is $\mathbf{v} = \langle x_2 - x_1, y_2 - y_1 \rangle$.

represented by $\overrightarrow{PQ}$. We denote vectors by lowercase boldface letters such as $\mathbf{u}$, $\mathbf{v}$, $\mathbf{w}$, and so forth. The length and direction of $\mathbf{v} = \overrightarrow{PQ}$ is the same as the length and direction of $\overrightarrow{PQ}$.

Two vectors are **equal** if their corresponding directed line segments are equivalent.

**Example 1   SHOWING VECTORS ARE EQUAL**

Let $\mathbf{u}$ be the vector represented by the directed line segment from $R = (-4, 2)$ to $S = (-1, 6)$, and $\mathbf{v}$ the vector represented by the directed line segment from $O = (0, 0)$ to $P = (3, 4)$. Prove that $\mathbf{u} = \mathbf{v}$ (Figure 6.3).

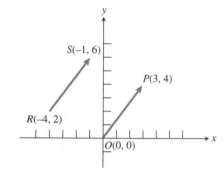

**Figure 6.3** The vectors $\mathbf{u} = \overrightarrow{RS}$ and $\mathbf{v} = \overrightarrow{OP}$. (Example 1)

**Solution**  We must prove that $\overrightarrow{RS}$ and $\overrightarrow{OP}$ are equivalent, that is, $\overrightarrow{RS}$ and $\overrightarrow{OP}$ have the same length and direction. Using the distance formula, we obtain

$$|\overrightarrow{RS}| = \sqrt{(-1 - (-4))^2 + (6 - 2)^2} = \sqrt{3^2 + 4^2} = 5, \qquad \text{Length of } \overrightarrow{RS}$$

$$|\overrightarrow{OP}| = \sqrt{(3 - 0)^2 + (4 - 0)^2} = \sqrt{3^2 + 4^2} = 5. \qquad \text{Length of } \overrightarrow{OP}$$

Thus, $|\overrightarrow{RS}| = |\overrightarrow{OP}|$.

The direction of $\overrightarrow{RS}$ and $\overrightarrow{OP}$ are to the upper right on lines having slope 4/3. So $\overrightarrow{RS}$ and $\overrightarrow{OP}$ have the same length and direction, and $\mathbf{u} = \mathbf{v}$.

## Component Form of a Vector

If $P$ and $Q$ are two points in the coordinate plane, there is one directed line segment equivalent to $\overrightarrow{PQ}$ whose initial point is the origin (see Figure 6.4). It is the representative of the vector $\mathbf{v} = \overrightarrow{PQ}$ in **standard position**, and is the one typically used to represent $\mathbf{v}$.

**Definition  Component Form of a Vector**

If $\mathbf{v}$ is a vector in the plane equal to the vector with initial $(0, 0)$ and terminal point $(v_1, v_2)$, then the **component form** of $\mathbf{v}$ is

$$\mathbf{v} = \langle v_1, v_2 \rangle.$$

Thus a vector in the plane is also an ordered pair $\langle v_1, v_2 \rangle$ of real numbers. The numbers $v_1$ and $v_2$ are the **components** of $\mathbf{v}$. The vector $\langle v_1, v_2 \rangle$ is called the **position vector** of the point $(v_1, v_2)$.

If $\mathbf{v} = \langle v_1, v_2 \rangle$ is represented by the directed line segment $\overrightarrow{PQ}$, when the **initial point** is $P(x_1, y_1)$ and the **terminal point** is $Q(x_2, y_2)$, then $v_1 = x_2 - x_1$ and $v_2 = y_2 - y_1$ are the components of $\overrightarrow{PQ}$ (Figure 6.4). Thus

$$\mathbf{v} = \langle x_2 - x_1, y_2 - y_1 \rangle$$

is the component form of $\mathbf{v}$.

The **magnitude** or **length** of the vector $\mathbf{v} = \overrightarrow{PQ}$ is the length of any of its equivalent directed line segment representatives. In particular, if $\mathbf{v} = \langle x_2 - x_1, y_2 - y_1 \rangle$ is the position vector for $\overrightarrow{PQ}$ (Figure 6.4), then the distance formula gives the magnitude or length of $\mathbf{v}$.

---

**Definition** **Magnitude or Length**

The **magnitude** or **length** of the vector $\mathbf{v} = \overrightarrow{PQ}$ determined by $P(x_1, y_1)$ and $Q(x_2, y_2)$ is

$$|\mathbf{v}| = \sqrt{v_1{}^2 + v_2{}^2} = \sqrt{(x_2 - x_1)^2 + (y_2 - y_1)^2}.$$

---

Two vectors $\langle a, b \rangle$ and $\langle c, d \rangle$ are equal if and only if $a = c$ and $b = d$. The vector $\langle 0, 0 \rangle$ with length 0 and no direction is the **zero vector** and is denoted $\mathbf{0}$.

### Example 2  FINDING THE COMPONENT FORM OF A VECTOR

Find the component form, magnitude, and direction of the vector $\mathbf{v} = \overrightarrow{PQ}$, where $P = (-3, 4)$ and $Q = (-5, 2)$. (See Figure 6.5.)

**Solution**  Because $P = (x_1, y_1) = (-3, 4)$ and $Q = (x_2, y_2) = (-5, 2)$, the component form of $\mathbf{v}$ is

$$\mathbf{v} = \langle x_2 - x_1, y_2 - y_1 \rangle = \langle -5 - (-3), 2 - 4 \rangle = \langle -2, -2 \rangle.$$

The magnitude of $\mathbf{v}$ is

$$|\mathbf{v}| = \sqrt{(-2)^2 + (-2)^2} = 2\sqrt{2}.$$

The direction of $\mathbf{v}$ is lower left on the line with slope $-2/\,-2 = 1$.

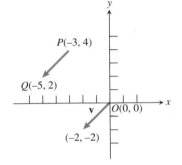

**Figure 6.5** The vector $\mathbf{v}$ of Example 2.

## Vector Operations

Two basic vector operations are *vector addition* and *scalar multiplication* (multiplying a vector by a real number (scalar)). Velocities of objects can be represented by vectors. We can calculate the effect of wind velocity on an airplane by adding the vectors that represent the velocity of the wind and the velocity of the airplane. These operations satisfy many properties similar to those for real numbers. (See Exercise 54.)

**Definition**  Vector Addition and Scalar Multiplication

Let $\mathbf{u} = \langle u_1, u_2 \rangle$ and $\mathbf{v} = \langle v_1, v_2 \rangle$ be vectors and $k$ a real number (scalar). Then the **sum of vectors u and v** is the vector

$$\mathbf{u} + \mathbf{v} = \langle u_1, u_2 \rangle + \langle v_1, v_2 \rangle = \langle u_1 + v_1, u_2 + v_2 \rangle.$$

The **product of the scalar $k$ and the vector u** is

$$k\mathbf{u} = k\langle u_1, u_2 \rangle = \langle ku_1, ku_2 \rangle.$$

The sum of the vectors $\mathbf{u}$ and $\mathbf{v}$ can be obtained geometrically by joining the initial point of the second vector $\mathbf{v}$ to the terminal point of the first vector $\mathbf{u}$, as shown in Figure 6.6a. The sum $\mathbf{u} + \mathbf{v}$ is represented by the directed line segment from the initial point of $\mathbf{u}$ to the terminal point of $\mathbf{v}$, and is a diagonal of the parallelogram formed by vectors $\mathbf{u}$ and $\mathbf{v}$. This procedure is the **parallelogram law** for vector addition.

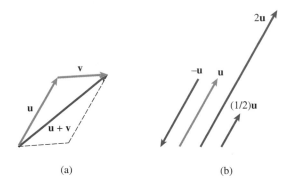

(a)  (b)

**Figure 6.6**  The (a) sum of $\mathbf{u}$ and $\mathbf{v}$, and (b) several scalar multiples of $\mathbf{u}$.

The product $k \cdot \mathbf{u}$ of the scalar $k$ and the vector $\mathbf{u}$ can be obtained geometrically by a stretch or shrink of $\mathbf{u}$ by the factor $k$ when $k > 0$. (See Figure 6.6b.) If $k < 0$, then there is a stretch or shrink by the factor $|k|$, and the direction of $k \cdot \mathbf{u}$ is opposite that of $\mathbf{u}$.

### Example 3  PERFORMING VECTOR OPERATIONS

Let $\mathbf{u} = \langle -1, 3 \rangle$ and $\mathbf{v} = \langle 4, 7 \rangle$. Find the component form of these vectors:

**(a)** $\mathbf{u} + \mathbf{v}$  **(b)** $3\mathbf{u}$  **(c)** $2\mathbf{u} + (-1)\mathbf{v}$

**Solution**

**(a)** Using the component form definition of sum of vectors, we obtain

$$\mathbf{u} + \mathbf{v} = \langle -1, 3 \rangle + \langle 4, 7 \rangle$$
$$= \langle -1 + 4, 3 + 7 \rangle.$$
$$= \langle 3, 10 \rangle$$

Using the parallelogram law: Start with the terminal point $(-1, 3)$ of $\mathbf{u}$. Move right 4 units and up 7 units (the components of $\mathbf{v}$) to arrive at the point $(-1 + 4, 3 + 7) = (3, 10)$, which is the terminal point of $\mathbf{u} + \mathbf{v}$. (See Figure 6.7.)

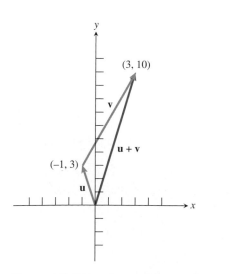

**Figure 6.7**  Using the parallelogram law to find $\mathbf{u} + \mathbf{v}$. (Example 3a)

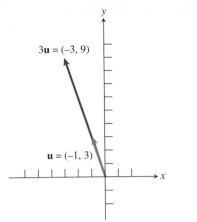

**Figure 6.8** Multiplication by the scalar 3 stretches the vector **u** by the factor 3 both horizontally and vertically. (Example 3b)

**(b)** Using the component form definition of scalar multiplication, we obtain

$$3\mathbf{u} = 3\langle -1, 3\rangle = \langle -3, 9\rangle.$$

Figure 6.8 shows that $3\mathbf{u}$ is a stretch of **u** by the factor 3.

**(c)** Using the component form definitions, we obtain

$$2\mathbf{u} + (-1)\mathbf{v} = 2\langle -1, 3\rangle + (-1)\langle 4, 7\rangle$$
$$= \langle -2, 6\rangle + \langle -4, -7\rangle$$
$$= \langle -6, -1\rangle$$

## Unit Vectors

A vector **u** with length $|\mathbf{u}| = 1$ is a **unit vector**. If **v** is not the zero vector $\langle 0, 0\rangle$, then the vector

$$\mathbf{u} = \frac{\mathbf{v}}{|\mathbf{v}|} = \frac{1}{|\mathbf{v}|}\mathbf{v}$$

is a **unit vector in the direction of v**. Unit vectors provide a way to represent the direction of any nonzero vector. Any vector in the direction of **v**, or the opposite direction, is a scalar multiple of this unit vector **u**.

**Example 4   FINDING A UNIT VECTOR**

If $\mathbf{v} = \langle -3, 2\rangle$, verify that $\mathbf{v}/|\mathbf{v}|$ is a unit vector in the direction of **v**.

**Solution**

$$|\mathbf{v}| = |\langle -3, 2\rangle| = \sqrt{(-3)^2 + (2)^2} = \sqrt{13}, \text{ so}$$

$$\frac{\mathbf{v}}{|\mathbf{v}|} = \frac{1}{\sqrt{13}}\langle -3, 2\rangle$$

$$= \left\langle \frac{-3}{\sqrt{13}}, \frac{2}{\sqrt{13}}\right\rangle$$

The magnitude of this vector is

$$\left|\left\langle \frac{-3}{\sqrt{13}}, \frac{2}{\sqrt{13}}\right\rangle\right| = \sqrt{\left(\frac{-3}{\sqrt{13}}\right)^2 + \left(\frac{2}{\sqrt{13}}\right)^2}$$

$$= \sqrt{\frac{9}{13} + \frac{4}{13}}$$

$$= \sqrt{\frac{13}{13}} = 1$$

Thus, the magnitude of $\mathbf{v}/|\mathbf{v}|$ is 1. Its direction is the same as **v** because it is a positive scalar multiple of **v**.

The two unit vectors $\mathbf{i} = \langle 1, 0\rangle$ and $\mathbf{j} = \langle 0, 1\rangle$ are the **standard unit vectors**. Any vector **v** can be written as an expression in terms of the standard unit vectors:

$$\mathbf{v} = \langle a, b\rangle$$
$$= \langle a, 0\rangle + \langle 0, b\rangle$$
$$= a\langle 1, 0\rangle + b\langle 0, 1\rangle$$
$$= a\mathbf{i} + b\mathbf{j}$$

**Teaching Note**

In Figure 6.9 note that *a* and *b* are simply the Cartesian coordinates of the terminal point of the vector's representative in standard position.

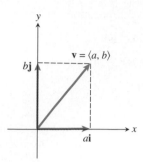

**Figure 6.9** The vector **v** is equal to $a\mathbf{i} + b\mathbf{j}$.

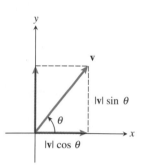

**Figure 6.10** The direction angle $\theta$ of the vector **v**.

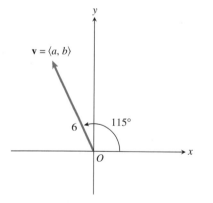

**Figure 6.11** The direction angle of **v** is 115°. (Example 5)

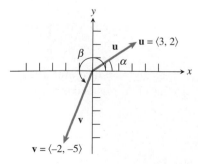

**Figure 6.12** The two vectors of Example 6.

Here the vector $\mathbf{v} = \langle a, b \rangle$ is expressed as the **linear combination** $a\mathbf{i} + b\mathbf{j}$ of the vectors **i** and **j**. The scalars $a$ and $b$ are the **horizontal** and **vertical** components, respectively, of the vector **v**. (See Figure 6.9.)

## Direction Angles

A precise way to specify the direction of a vector **v** is to state its **direction angle**, the angle $\theta$ that **v** makes with the positive $x$-axis (Figure 6.10). Using trigonometry, we see that the horizontal component of **v** is $|\mathbf{v}| \cos \theta$ and the vertical component is $|\mathbf{v}| \sin \theta$. Thus,

$$\mathbf{v} = (|\mathbf{v}| \cos \theta)\mathbf{i} + (|\mathbf{v}| \sin \theta)\mathbf{j}.$$

The unit vector in the direction of **v** is

$$\mathbf{u} = \frac{\mathbf{v}}{|\mathbf{v}|} = (\cos \theta)\mathbf{i} + (\sin \theta)\mathbf{j}.$$

### Example 5 FINDING THE COMPONENTS OF A VECTOR

Find the components of the vector **v** with direction angle 115° (Figure 6.11).

**Solution** If $a$ and $b$ are the horizontal and vertical components, respectively, of **v**, then

$$\mathbf{v} = \langle a, b \rangle = \langle 6 \cos 115°, 6 \sin 115° \rangle.$$

So, $a = 6 \cos 115° \approx -2.54$ and $b = 6 \sin 115° \approx 5.44$.

### Example 6 FINDING THE DIRECTION ANGLE OF A VECTOR

Find the direction angle of each vector:

**(a)** $\mathbf{u} = \langle 3, 2 \rangle$      **(b)** $\mathbf{v} = \langle -2, -5 \rangle$

**Solution** See Figure 6.12.

**(a)** If $\alpha$ is the direction angle of **u**, then $\mathbf{u} = \langle 3, 2 \rangle = \langle |\mathbf{u}| \cos \alpha, |\mathbf{u}| \sin \alpha \rangle$.

$$3 = |\mathbf{u}| \cos \alpha \qquad \text{Horizontal component of } \mathbf{u}$$
$$3 = \sqrt{3^2 + 2^2} \cos \alpha \qquad |\mathbf{u}| = \sqrt{3^2 + 2^2}$$
$$3 = \sqrt{13} \cos \alpha$$
$$\alpha = \cos^{-1}\left(\frac{3}{\sqrt{13}}\right) \approx 33.69° \qquad \alpha \text{ is acute}$$

**(b)** If $\beta$ is the direction angle of **v**, then $\mathbf{v} = \langle -2, -5 \rangle = \langle |\mathbf{v}| \cos \beta, |\mathbf{v}| \sin \beta \rangle$.

$$-2 = |\mathbf{v}| \cos \beta \qquad \text{Horizontal component of } \mathbf{v}$$
$$-2 = \sqrt{(-2)^2 + (-5)^2} \cos \beta \qquad |\mathbf{v}| = \sqrt{(-2)^2 + (-5)^2}$$
$$-2 = \sqrt{29} \cos \beta$$
$$\beta = 360° - \cos^{-1}\left(\frac{-2}{\sqrt{29}}\right) \approx 248.2° \qquad 180° < \beta < 270°$$

## Applications

The **velocity** of a moving object is a vector because velocity has both magnitude and direction. The magnitude of velocity is **speed**.

### Example 7  WRITING VELOCITY AS A VECTOR

A DC-10 jet aircraft is flying on a bearing of 65° at 500 mph. Find the component form of the velocity of the airplane. Recall that the bearing is the angle that the line of travel makes with due north, measured clockwise (see Section 4.1, Figure 4.2).

**Solution**  Let **v** be the velocity of the airplane. A bearing of 65° is equivalent to a direction angle of 25°. The plane's speed, 500 mph, is the magnitude of vector **v**; that is, $|\mathbf{v}| = 500$. (See Figure 6.13.)

The horizontal component of **v** is 500 cos 25° and the vertical component is 500 sin 25°, so

$$\mathbf{v} = (500 \cos 25°)\mathbf{i} + (500 \sin 25°)\mathbf{j}$$
$$= \langle 500 \cos 25°, 500 \sin 25° \rangle \approx \langle 453.15, 211.31 \rangle$$

The components of the velocity give the eastward and northward speeds. That is, the airplane travels about 453.15 mph eastward and about 211.31 mph northward as it travels at 500 mph on a bearing of 65°.

A typical problem for a navigator involves calculating the effect of wind on the direction and speed of the airplane, as illustrated in Example 8.

### Example 8  CALCULATING THE EFFECT OF WIND VELOCITY

Pilot Megan McCarty's flight plan has her leaving San Francisco International Airport and flying a Boeing 727 due east. There is a 65-mph wind with the bearing 60°. Find the compass heading McCarty should follow, and determine what the airplane's ground speed will be (assuming that its speed with no wind is 450 mph).

**Solution**  See Figure 6.14. Vector $\overrightarrow{AB}$ represents the velocity produced by the airplane alone, $\overrightarrow{AC}$ represents the velocity of the wind, and $\theta$ is the angle *DAB*. Vector $\mathbf{v} = \overrightarrow{AD}$ represents the resulting velocity, so

$$\mathbf{v} = \overrightarrow{AD} = \overrightarrow{AC} + \overrightarrow{AB}.$$

We must find the bearing of $\overrightarrow{AB}$ and $|\mathbf{v}|$.

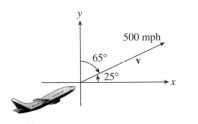

**Figure 6.13**  The airplane's path (bearing) in Example 7.

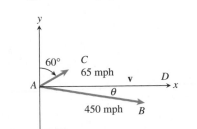

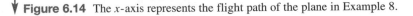

**Figure 6.14**  The *x*-axis represents the flight path of the plane in Example 8.

Using component forms, we obtain

$$\overrightarrow{AC} = \langle 65 \cos 30°, 65 \sin 30° \rangle$$

$$\overrightarrow{AB} = \langle 450 \cos \theta, 450 \sin \theta \rangle$$

$$\overrightarrow{AD} = \overrightarrow{AC} + \overrightarrow{AB}$$

$$= \langle 65 \cos 30° + 450 \cos \theta, 65 \sin 30° + 450 \sin \theta \rangle$$

Because the plane is traveling due east, the second component of $\overrightarrow{AD}$ must be zero.

$$65 \sin 30° + 450 \sin \theta = 0$$

$$\theta = \sin^{-1} \left( \frac{-65 \sin 30°}{450} \right)$$

$$\approx -4.14° \qquad \theta < 0$$

Thus, the compass heading McCarty should follow is

$$90° + |\theta| \approx 94.14°. \qquad \text{Bearing} > 90°$$

The ground speed of the airplane is

$$|\mathbf{v}| = |\overrightarrow{AD}| = \sqrt{(65 \cos 30° + 450 \cos \theta)^2 + 0^2}$$

$$= |65 \cos 30° + 450 \cos \theta|$$

$$\approx 505.12 \qquad \text{Using the unrounded value of } \theta.$$

McCarty should use a bearing of approximately 94.14°. The airplane will travel due east at approximately 505.12 mph.

### Example 9 FINDING THE EFFECT OF GRAVITY

A force of 30 pounds just keeps the box in Figure 6.15 from sliding down the ramp inclined at 20°. Find the weight of the box.

**Solution** We are given that $|\overrightarrow{AD}| = 30$. Let $|\overrightarrow{AB}| = w$, then

$$\sin 20° = \frac{|\overrightarrow{CB}|}{w} = \frac{30}{w}.$$

Thus,

$$w = \frac{30}{\sin 20°} \approx 87.71.$$

The weight of the box is about 87.71 pounds.

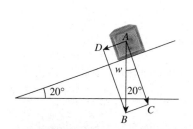

**Figure 6.15** The force of gravity $\overrightarrow{AB}$ has a component $\overrightarrow{AC}$ that holds the box against the surface of the ramp, and a component $\overrightarrow{AD} = \overrightarrow{CB}$ that tends to push the box down the ramp. (Example 9)

**Problem**

During one part of the migration, a salmon is swimming at 6 mph, and the current is flowing downstream at 3 mph at an angle of 7 degrees. How fast is the salmon moving upstream?

### Solution

Assume the salmon is swimming in a plane parallel to the surface of the water.

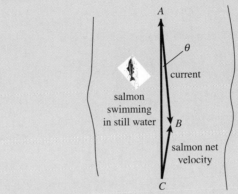

In the figure, vector $\overrightarrow{AB}$ represents the current of 3 mph, $\theta$ is the angle $CAB$, which is 7 degrees, the vector $\overrightarrow{CA}$ represents the velocity of the salmon of 6 mph, and the vector $\overrightarrow{CB}$ is the net velocity at which the fish is moving upstream.

So we have

$$\overrightarrow{AB} = \langle 3 \cos(-90°), 3 \sin(-90°) \rangle \approx \langle 0.3, -3 \rangle$$
$$\overrightarrow{CA} = \langle 6 \cos 83°, 6 \sin 83° \rangle \approx \langle 0.73, 5.96 \rangle$$

Thus $\overrightarrow{CB} = \overrightarrow{CA} + \overrightarrow{AB} = \langle 0.73, 5.96 \rangle + \langle 0, -3 \rangle$
$$= \langle 0.73, 2.96 \rangle$$

The speed of the salmon is then $|\overrightarrow{CB}| = \sqrt{0.73^2 + 2.96^2} \approx 3.04$ mph upstream

## Quick Review 6.1

In Exercises 1–4, find the values of $x$ and $y$.

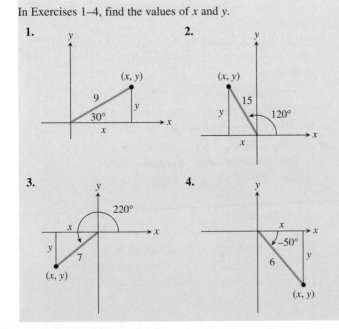

In Exercises 5 and 6, solve for $\theta$ in degrees.

**5.** $\theta = \sin^{-1}\left(\dfrac{3}{\sqrt{29}}\right)$    **6.** $\theta = \cos^{-1}\left(\dfrac{-1}{\sqrt{15}}\right)$

In Exercises 7–9, the point $P$ is on the terminal side of the angle $\theta$. Find the measure of $\theta$ if $0° < \theta < 360°$.

**7.** $P(5, 9)$    **8.** $P(5, -7)$

**9.** $P(-2, -5)$  $180° + \tan^{-1}(5/2) \approx 248.199°$

**10.** A naval ship leaves Port Norfolk and averages 42 knots (nautical mph) traveling for 3 h on a bearing of 40° and then 5 h on a course of 125°. What is the boat's bearing and distance from Port Norfolk after 8 h?

# Section 6.1 Exercises

In Exercises 1–4, show that the directed line segments $\overrightarrow{RS}$ and $\overrightarrow{OP}$ are equivalent.

1. $R = (-4, 7)$, $S = (-1, 5)$, $O = (0, 0)$, and $P = (3, -2)$
2. $R = (7, -3)$, $S = (4, -5)$, $O = (0, 0)$, and $P = (-3, -2)$
3. $R = (2, 1)$, $S = (0, -1)$, $O = (1, 4)$, and $P = (-1, 2)$
4. $R = (-2, -1)$, $S = (2, 4)$, $O = (-3, -1)$, and $P = (1, 4)$

In Exercises 5–12, let $P = (-2, 2)$, $Q = (3, 4)$, $R = (-2, 5)$, and $S = (2, -8)$. Find the component form and magnitude of the vector.

5. $\overrightarrow{PQ}$  $\langle 5, 2 \rangle$; $\sqrt{29}$
6. $\overrightarrow{RS}$  $\langle 4, -13 \rangle$; $\sqrt{185}$
7. $\overrightarrow{QR}$  $\langle -5, 1 \rangle$; $\sqrt{26}$
8. $\overrightarrow{PS}$  $\langle 4, -10 \rangle$, $2\sqrt{29}$
9. $2\overrightarrow{QS}$  $\langle -2, -24 \rangle$, $2\sqrt{145}$
10. $(\sqrt{2})\overrightarrow{PR}$
11. $3\overrightarrow{QR} + \overrightarrow{PS}$
12. $\overrightarrow{PS} - 3\overrightarrow{PQ}$

In Exercises 13–20, let $\mathbf{u} = \langle -1, 3 \rangle$, $\mathbf{v} = \langle 2, 4 \rangle$, and $\mathbf{w} = \langle 2, -5 \rangle$. Find the component form of the vector.

13. $\mathbf{u} + \mathbf{v}$  $\langle 1, 7 \rangle$
14. $\mathbf{u} + (-1)\mathbf{v}$  $\langle -3, -1 \rangle$
15. $\mathbf{u} - \mathbf{w}$  $\langle -3, 8 \rangle$
16. $3\mathbf{v}$  $\langle 6, 12 \rangle$
17. $2\mathbf{u} + 3\mathbf{w}$  $\langle 4, -9 \rangle$
18. $2\mathbf{u} - 4\mathbf{v}$  $\langle -10, -10 \rangle$
19. $-2\mathbf{u} - 3\mathbf{v}$  $\langle -4, -18 \rangle$
20. $-\mathbf{u} - \mathbf{v}$  $\langle -1, -7 \rangle$

In Exercises 21–24, find the unit vector in the direction of the given vector. Write your answer in **(a)** component form and **(b)** as a linear combination of the standard unit vectors **i** and **j**.

21. $\mathbf{u} = \langle 2, 1 \rangle$
22. $\mathbf{u} = \langle -3, 2 \rangle$
23. $\mathbf{u} = \langle -4, -5 \rangle$
24. $\mathbf{u} = \langle 3, -4 \rangle$

In Exercises 25–28, find the component form of the vector **v**.

25.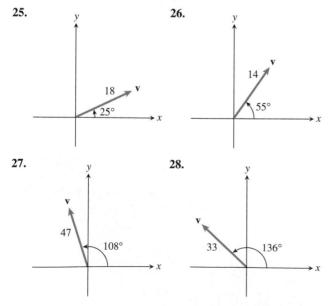

26.

27.

28.

In Exercises 29–34, find the magnitude and direction angle of the vector.

29. $\langle 3, 4 \rangle$
30. $\langle -1, 2 \rangle$
31. $3\mathbf{i} - 4\mathbf{j}$
32. $-3\mathbf{i} - 5\mathbf{j}$
33. $7(\cos 135° \, \mathbf{i} + \sin 135° \, \mathbf{j})$
34. $2(\cos 60° \, \mathbf{i} + \sin 60°)$

In Exercises 35–38, find a unit vector in the direction of the given vector.

35. $\mathbf{u} = \langle -2, 4 \rangle$
36. $\mathbf{v} = \langle 1, -1 \rangle$
37. $\mathbf{w} = -\mathbf{i} - 2\mathbf{j}$
38. $\mathbf{w} = 5\mathbf{i} + 5\mathbf{j}$

In Exercises 39 and 40, find the vector **v** with the given magnitude and the same direction as **u**.

39. $|\mathbf{v}| = 2$, $\mathbf{u} = \langle 3, -3 \rangle$
40. $|\mathbf{v}| = 5$, $\mathbf{u} = \langle -5, 7 \rangle$

41. **Navigation** An airplane is flying on a bearing of 335° at 530 mph. Find the component form of the velocity of the airplane. $\approx \langle -223.99, 480.34 \rangle$

42. **Navigation** An airplane is flying on a bearing of 170° at 460 mph. Find the component form of the velocity of the airplane. $\approx \langle 79.88, -453.01 \rangle$

43. **Flight Engineering** An airplane is flying on a compass heading (bearing) of 340° at 325 mph. A wind is blowing with the bearing 320° at 40 mph.

    **(a)** Find the component form of the velocity of the airplane.

    **(b)** Find the actual ground speed and direction of the plane.

44. **Flight Engineering** An airplane is flying on a compass heading (bearing) of 170° at 460 mph. A wind is blowing with the bearing 200° at 80 mph.

    **(a)** Find the component form of the velocity of the airplane.

    **(b)** Find the actual ground speed and direction of the airplane.

45. **Shooting a Basketball** A basketball is shot at a 70° angle with the horizontal direction with an initial speed of 10 m/sec.

    **(a)** Find the component form of the initial velocity.

    **(b) Writing to Learn** Give an interpretation of the horizontal and vertical components of the velocity.

46. **Moving a Heavy Object** In a warehouse a box is being pushed up a 15° inclined plane with a force of 2.5 lb, as shown in the figure.

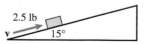

    **(a)** Find the component form of the force.

    **(b) Writing to Learn** Give an interpretation of the horizontal and vertical components of the force.

**47. Moving a Heavy Object** Suppose the box described in Exercise 46 is being towed up the inclined plane, as shown in the figure below. Find the force **w** needed in order for the component of the force parallel to the inclined plane to be 2.5 lb. Give the answer in component form. ≈ ⟨2.20, 1.43⟩

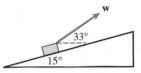

**48. Combining Forces** Juana and Diego Gonzales, ages six and four respectively, own a strong and stubborn puppy named Corporal. It is so hard to take Corporal for a walk that they devise a scheme to use two leashes. If Juana and Diego pull with forces of 23 lb and 18 lb at the angles shown in the figure, how hard is Corporal pulling if the puppy holds the children at a standstill? about 39.26 lb

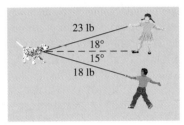

In Exercises 49 and 50, find the direction and magnitude of the resultant force.

**49. Combining Forces** A force of 50 lb acts on an object at an angle of 45°. A second force of 75 lb acts on the object at an angle of −30°. |**F**| ≈ 100.33 lb and θ ≈ −1.22°

**50. Combining Forces** Three forces with magnitudes 100, 50, and 80 lb, act on an object at angles of 50°, 160°, and −20°, respectively. |**F**| ≈ 113.81 lb and θ = 35.66°

**51. Navigation** A ship is heading due north at 12 mph. The current is flowing southwest at 4 mph. Find the actual bearing and speed of the ship. ≈ 342.86°; ≈ 9.6 mph

**52. Navigation** A motor boat capable of 20 mph keeps the bow of the boat pointed straight across a mile wide river. The current is flowing left to right at 8 mph. Find where the boat meets the opposite shore. 0.4 mi downstream

**53. Group Activity** A ship heads due south with the current flowing northwest. Two hours later the ship is 20 miles in the direction 30° west of south from the original starting point. Find the speed with no current of the ship and the rate of the current. ≈ 13.66 mph; ≈ 7.04 mph

**54. Group Activity** Use component form and prove the following properties of vectors.

**(a)** $\mathbf{u} + \mathbf{v} = \mathbf{v} + \mathbf{u}$

**(b)** $(\mathbf{u} + \mathbf{v}) + \mathbf{w} = \mathbf{u} + (\mathbf{v} + \mathbf{w})$

**(c)** $\mathbf{u} + \mathbf{0} = \mathbf{u}$, where $\mathbf{0} = \langle 0, 0 \rangle$

**(d)** $\mathbf{u} + (-\mathbf{u}) = \mathbf{0}$, where $-\langle a, b \rangle = \langle -a, -b \rangle$

**(e)** $a(\mathbf{u} + \mathbf{v}) = a\mathbf{u} + a\mathbf{v}$

**(f)** $(a + b)\mathbf{u} = a\mathbf{u} + b\mathbf{u}$

**(g)** $(ab)\mathbf{u} = a(b\mathbf{u})$

**(h)** $a\mathbf{0} = \mathbf{0}, 0\mathbf{u} = \mathbf{0}$

**(i)** $(1)\mathbf{u} = \mathbf{u}, (-1)\mathbf{u} = -\mathbf{u}$

**(j)** $|a\mathbf{u}| = |a| \, |\mathbf{u}|$

## Explorations

**55. Dividing a Line Segment in a Given Ratio** Let $A$ and $B$ be two points in the plane, as shown in the figure.

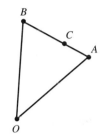

**(a)** Prove that $\overrightarrow{BA} = \overrightarrow{OA} - \overrightarrow{OB}$, where $O$ is the origin.

**(b)** Let $C$ be a point on the line segment $BA$ which divides the segment in the ratio $x : y$ where $x + y = 1$. That is,

$$\frac{|\overrightarrow{BC}|}{|\overrightarrow{CA}|} = \frac{x}{y}.$$

Show that $\overrightarrow{OC} = x\overrightarrow{OA} + y\overrightarrow{OB}$.

**56. Medians of a Triangle** Perform the following steps to use vectors to prove that the medians of a triangle meet at a point $O$ which divides each median in the ratio 1 : 2. $M_1$, $M_2$, and $M_3$ are midpoints of the sides of the triangle shown in the figure.

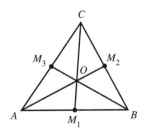

**(a)** Use Exercise 55 to prove that

$$\overrightarrow{OM_1} = \frac{1}{2}\overrightarrow{OA} + \frac{1}{2}\overrightarrow{OB}$$

$$\overrightarrow{OM_2} = \frac{1}{2}\overrightarrow{OC} + \frac{1}{2}\overrightarrow{OB}.$$

$$\overrightarrow{OM_3} = \frac{1}{2}\overrightarrow{OA} + \frac{1}{2}\overrightarrow{OC}$$

**(b)** Prove that each of $2\overrightarrow{OM_1} + \overrightarrow{OC}$, $2\overrightarrow{OM_2} + \overrightarrow{OA}$, $2\overrightarrow{OM_3} + \overrightarrow{OB}$ is equal to $\overrightarrow{OA} + \overrightarrow{OB} + \overrightarrow{OC}$.

**(c) Writing to Learn** Explain why (b) establishes the desired result. ∎

## Extending the Ideas

**57. Vector Equation of a Line** Let $L$ be the line through the two points $A$ and $B$. Prove that $C = (x, y)$ is on the line $L$ if and only if $\overrightarrow{OC} = t\overrightarrow{OA} + (1 - t)\overrightarrow{OB}$, where $t$ is a real number and $O$ is the origin.

**58. Connecting Vectors and Geometry** Prove that the lines which join one vertex of a parallelogram to the midpoints of the opposite sides trisect the diagonal.

---

| **6.2** | **Dot Product of Vectors** |

The Dot Product • Angle Between Vectors • Projecting One Vector Onto Another • Work

### The Dot Product

The *dot product* of two vectors has many geometric and physical applications. For example, we can use the dot product to calculate the angle between two vectors. The dot product of a force vector $\mathbf{F}$ with a vector $\mathbf{v}$ gives the magnitude of $\mathbf{F}$ in the direction of $\mathbf{v}$ and the amount of work done by $\mathbf{F}$ in moving an object from the initial point of $\mathbf{v}$ to its terminal point.

It may surprise you that the dot product of two vectors produces a scalar, unlike the addition and subtraction of vectors or the multiplication of vectors by scalars, which produce vectors.

---

**Definition  Dot Product**

The **dot product** of $\mathbf{u} = \langle u_1, u_2 \rangle$ and $\mathbf{v} = \langle v_1, v_2 \rangle$ is

$$\mathbf{u} \cdot \mathbf{v} = u_1v_1 + u_2v_2.$$

---

Dot products have many important properties that we make use of in this section. We prove the first two and leave the rest for the exercises.

---

**Properties of the Dot Product**

Let $\mathbf{u}$, $\mathbf{v}$, and $\mathbf{w}$ be vectors and let $c$ be a scalar.

1. $\mathbf{u} \cdot \mathbf{v} = \mathbf{v} \cdot \mathbf{u}$
2. $\mathbf{u} \cdot \mathbf{u} = |\mathbf{u}|^2$
3. $\mathbf{0} \cdot \mathbf{u} = 0$
4. $\mathbf{u} \cdot (\mathbf{v} + \mathbf{w}) = \mathbf{u} \cdot \mathbf{v} + \mathbf{u} \cdot \mathbf{w}$
   $(\mathbf{u} + \mathbf{v}) \cdot \mathbf{w} = \mathbf{u} \cdot \mathbf{w} + \mathbf{v} \cdot \mathbf{w}$
5. $(c\mathbf{u}) \cdot \mathbf{v} = \mathbf{u} \cdot (c\mathbf{v}) = c(\mathbf{u} \cdot \mathbf{v})$

---

**Proof**

Let $\mathbf{u} = \langle u_1, u_2 \rangle$ and $\mathbf{v} = \langle v_1, v_2 \rangle$.

**Property 1**

$\mathbf{u} \cdot \mathbf{v} = u_1v_1 + u_2v_2$    Definition of $\mathbf{u} \cdot \mathbf{v}$.

$\qquad = v_1u_1 + v_2u_2$    Commutative property of real numbers.

$\qquad = \mathbf{v} \cdot \mathbf{u}$    Definition of $\mathbf{v} \cdot \mathbf{u}$.

**Property 2**

$$\mathbf{u} \cdot \mathbf{u} = u_1^2 + u_2^2 \qquad \text{Definition of } \mathbf{u} \cdot \mathbf{u}.$$
$$= (\sqrt{u_1^2 + u_2^2})^2$$
$$= |\mathbf{u}|^2 \qquad \text{Definition of } |\mathbf{u}|.$$

**Example 1   FINDING DOT PRODUCTS**

Find each dot product.

**(a)** $\langle 3, 4 \rangle \cdot \langle 5, 2 \rangle$

**(b)** $\langle 1, -2 \rangle \cdot \langle -4, 3 \rangle$

**(c)** $(2\mathbf{i} - \mathbf{j}) \cdot (3\mathbf{i} - 5\mathbf{j})$

**Solution**

**(a)** $\langle 3, 4 \rangle \cdot \langle 5, 2 \rangle = (3)(5) + (4)(2) = 23$

**(b)** $\langle 1, -2 \rangle \cdot \langle -4, 3 \rangle = (1)(-4) + (-2)(3) = -10$

**(c)** $(2\mathbf{i} - \mathbf{j}) \cdot (3\mathbf{i} - 5\mathbf{j}) = (2)(3) + (-1)(-5) = 11$

Property 2 of the dot product gives us another way to find the length of a vector, as illustrated in Example 2.

**Example 2   USING DOT PRODUCT TO FIND LENGTH**

Use the dot product to find the length of the vector $\mathbf{u} = \langle 4, -3 \rangle$.

**Solution**   It follows from Property 2 that $|\mathbf{u}| = \sqrt{\mathbf{u} \cdot \mathbf{u}}$. Thus,

$$|\langle 4, -3 \rangle| = \sqrt{\langle 4, -3 \rangle \cdot \langle 4, -3 \rangle} = \sqrt{(4)(4) + (-3)(-3)} = \sqrt{25} = 5.$$

## Angle Between Vectors

Let $\mathbf{u}$ and $\mathbf{v}$ be two nonzero vectors in standard position as shown in Figure 6.16. The **angle between u and v** is the angle $\theta$, $0 \le \theta \le \pi$ or $0° \le \theta \le 180°$. The angle between any two nonzero vectors is the corresponding angle between their respective standard position representatives.

We can use the dot product to find the angle between nonzero vectors as we prove in the next theorem.

**Figure 6.16**  The angle $\theta$ between nonzero vectors $\mathbf{u}$ and $\mathbf{v}$.

---

**Theorem  Angle Between Two Vectors**

If $\theta$ is the angle between the nonzero vectors $\mathbf{u}$ and $\mathbf{v}$, then

$$\cos \theta = \frac{\mathbf{u} \cdot \mathbf{v}}{|\mathbf{u}||\mathbf{v}|}$$

and $\theta = \cos^{-1}\left(\dfrac{\mathbf{u} \cdot \mathbf{v}}{|\mathbf{u}||\mathbf{v}|}\right)$

### Proof

We apply the Law of Cosines to the triangle determined by **u**, **v**, and **v** − **u** in Figure 6.16, and use the properties of the dot product.

$$|\mathbf{v} - \mathbf{u}|^2 = |\mathbf{u}|^2 + |\mathbf{v}|^2 - 2|\mathbf{u}||\mathbf{v}|\cos\theta$$

$$(\mathbf{v} - \mathbf{u}) \cdot (\mathbf{v} - \mathbf{u}) = |\mathbf{u}|^2 + |\mathbf{v}|^2 - 2|\mathbf{u}||\mathbf{v}|\cos\theta$$

$$\mathbf{v} \cdot \mathbf{v} - \mathbf{u} \cdot \mathbf{v} - \mathbf{v} \cdot \mathbf{u} + \mathbf{u} \cdot \mathbf{u} = |\mathbf{u}|^2 + |\mathbf{v}|^2 - 2|\mathbf{u}||\mathbf{v}|\cos\theta$$

$$|\mathbf{v}|^2 - 2\mathbf{u} \cdot \mathbf{v} + |\mathbf{u}|^2 = |\mathbf{u}|^2 + |\mathbf{v}|^2 - 2|\mathbf{u}||\mathbf{v}|\cos\theta$$

$$-2\mathbf{u} \cdot \mathbf{v} = -2|\mathbf{u}||\mathbf{v}|\cos\theta$$

$$\cos\theta = \frac{\mathbf{u} \cdot \mathbf{v}}{|\mathbf{u}||\mathbf{v}|}$$

$$\theta = \cos^{-1}\left(\frac{\mathbf{u} \cdot \mathbf{v}}{|\mathbf{u}||\mathbf{v}|}\right)$$

### Example 3   FINDING THE ANGLE BETWEEN VECTORS

Find the angle between the vectors **u** and **v**.

**(a)** $\mathbf{u} = \langle 2, 3 \rangle$, $\mathbf{v} = \langle -2, 5 \rangle$      **(b)** $\mathbf{u} = \langle 2, 1 \rangle$, $\mathbf{v} = \langle -1, -3 \rangle$

### Solution

**(a)** See Figure 6.17a. Using the Angle Between Two Vectors Theorem, we have

$$\cos\theta = \frac{\mathbf{u} \cdot \mathbf{v}}{|\mathbf{u}||\mathbf{v}|} = \frac{\langle 2, 3 \rangle \cdot \langle -2, 5 \rangle}{|\langle 2, 3 \rangle||\langle -2, 5 \rangle|} = \frac{11}{\sqrt{13}\,\sqrt{29}}.$$

So,

$$\theta = \cos^{-1}\left(\frac{11}{\sqrt{13}\,\sqrt{29}}\right) \approx 55.5°.$$

**(b)** See Figure 6.17b. Again using the Angle Between Two Vectors Theorem, we have

$$\cos\theta = \frac{\mathbf{u} \cdot \mathbf{v}}{|\mathbf{u}||\mathbf{v}|} = \frac{\langle 2, 1 \rangle \cdot \langle -1, -3 \rangle}{|\langle 2, 1 \rangle||\langle -1, -3 \rangle|} = \frac{-5}{\sqrt{5}\,\sqrt{10}} = \frac{-1}{\sqrt{2}}.$$

So,

$$\theta = \cos^{-1}\left(\frac{-1}{\sqrt{2}}\right) = 135°.$$

If vectors **u** and **v** are perpendicular, that is, if the angle between them is 90°, then

$$\mathbf{u} \cdot \mathbf{v} = |\mathbf{u}||\mathbf{v}|\cos 90° = 0$$

because $\cos 90° = 0$.

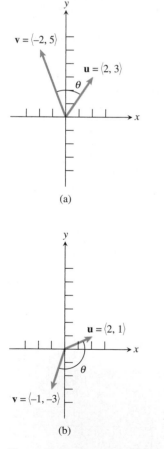

$\mathbf{v} = \langle -2, 5 \rangle$   $\mathbf{u} = \langle 2, 3 \rangle$   $\theta$

(a)

$\mathbf{u} = \langle 2, 1 \rangle$   $\theta$   $\mathbf{v} = \langle -1, -3 \rangle$

(b)

**Figure 6.17** The vectors in (a) Example 3a and (b) Example 3b.

---

**Definition   Orthogonal Vectors**

The vectors **u** and **v** are **orthogonal** if and only if $\mathbf{u} \cdot \mathbf{v} = 0$.

The terms "perpendicular" and "orthogonal" almost mean the same thing. The zero vector has no direction angle, so technically speaking, the zero vector is not perpendicular to any vector. However, the zero vector is orthogonal to every vector. Except for this special case, orthogonal and perpendicular are the same.

### Example 4  PROVING VECTORS ARE ORTHOGONAL

Prove that the vectors $\mathbf{u} = \langle 2, 3 \rangle$ and $\mathbf{v} = \langle -6, 4 \rangle$ are orthogonal.

**Solution**  We must prove that their dot product is zero.

$$\mathbf{u} \cdot \mathbf{v} = \langle 2, 3 \rangle \cdot \langle -6, 4 \rangle = -12 + 12 = 0$$

The two vectors are orthogonal.

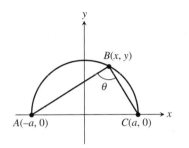

**Figure 6.18**  The angle $\angle ABC$ inscribed in the upper half of the circle $x^2 + y^2 = a^2$. (Exploration 1)

---

#### Exploration 1  Angles Inscribed in Semicircles

Figure 6.18 shows $\angle ABC$ inscribed in the upper half of the circle $x^2 + y^2 = a^2$.

1. For $a = 2$, find the component form of the vectors $\mathbf{u} = \overrightarrow{BA}$ and $\mathbf{v} = \overrightarrow{BC}$.  $\langle -2 - x, -y \rangle, \langle 2 - x, -y \rangle$

2. Find $\mathbf{u} \cdot \mathbf{v}$. What can you conclude about the angle $\theta$ between these two vectors?  $\theta = 90°$

3. Repeat parts 1 and 2 for arbitrary $a$.  Answers will vary

---

## Projecting One Vector onto Another

The **vector projection** of $\mathbf{u} = \overrightarrow{PQ}$ onto a nonzero vector $\mathbf{v} = \overrightarrow{PS}$ is the vector $\overrightarrow{PR}$ determined by dropping a perpendicular from $Q$ to the line $PS$ (Figure 6.19). We have resolved $\mathbf{u}$ into components $\overrightarrow{PR}$ and $\overrightarrow{RQ}$

$$\mathbf{u} = \overrightarrow{PR} + \overrightarrow{RQ}$$

with $\overrightarrow{PR}$ and $\overrightarrow{RQ}$ perpendicular.

The standard notation for $\overrightarrow{PR}$, the vector projection of $\mathbf{u}$ onto $\mathbf{v}$, is $\overrightarrow{PR} = \text{proj}_\mathbf{v}\mathbf{u}$. With this notation, $\overrightarrow{RQ} = \mathbf{u} - \text{proj}_\mathbf{v}\mathbf{u}$. We ask you to establish the following formula in the exercises (see Exercise 56).

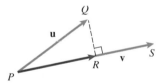

**Figure 6.19**  The vectors $\mathbf{u} = \overrightarrow{PQ}$, $\mathbf{v} = \overrightarrow{PS}$, and the vector projection of $\mathbf{u}$ onto $\mathbf{v}$, $\overrightarrow{PR} = \text{proj}_\mathbf{v}\mathbf{u}$.

> **Projection of u onto v**
>
> If $\mathbf{u}$ and $\mathbf{v}$ are nonzero vectors, the projection of $\mathbf{u}$ onto $\mathbf{v}$ is
>
> $$\text{proj}_\mathbf{v}\mathbf{u} = \left( \frac{\mathbf{u} \cdot \mathbf{v}}{|\mathbf{v}|^2} \right)\mathbf{v}.$$

### Example 5  DECOMPOSING A VECTOR INTO PERPENDICULAR COMPONENTS

Find the vector projection of $\mathbf{u} = \langle 6, 2 \rangle$ onto $\mathbf{v} = \langle 5, -5 \rangle$. Then write $\mathbf{u}$ as the sum of two orthogonal vectors, one of which is $\text{proj}_\mathbf{v}\mathbf{u}$.

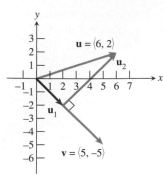

**Figure 6.20** The vectors $\mathbf{u} = \langle 6, 2 \rangle$, $\mathbf{v} = \langle 5, -5 \rangle$, $\mathbf{u}_1 = \text{proj}_{\mathbf{v}}\mathbf{u}$, and $\mathbf{u}_2 = \mathbf{u} - \mathbf{u}_1$. (Example 5)

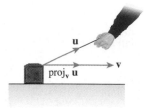

**Figure 6.21** If we pull on a box with force **u**, the effective force in the direction of **v** is $\text{proj}_{\mathbf{v}}\mathbf{u}$, the vector projection of **u** onto **v**.

**Figure 6.22** The sled in Example 6.

**Notes on Exercises**

Ex. 19–20 can be completed by using dot products or by using common sense. Encourage students to try both methods.

Ex. 49–54 involve work done by a force that is not parallel to the direction of motion.

**Ongoing Assessment**

Self-Assessment: Ex. 5, 17, 31, 47
Embedded Assessment: Ex. 59, 60

**Solution** We write $\mathbf{u} = \mathbf{u}_1 + \mathbf{u}_2$ where $\mathbf{u}_1 = \text{proj}_{\mathbf{v}}\mathbf{u}$ and $\mathbf{u}_2 = \mathbf{u} - \mathbf{u}_1$ (Figure 6.20).

$$\mathbf{u}_1 = \text{proj}_{\mathbf{v}}\mathbf{u} = \left( \frac{\mathbf{u} \cdot \mathbf{v}}{|\mathbf{v}|^2} \right)\mathbf{v} = \frac{20}{50} \langle 5, -5 \rangle = \langle 2, -2 \rangle$$

$$\mathbf{u}_2 = \mathbf{u} - \mathbf{u}_1 = \langle 6, 2 \rangle - \langle 2, -2 \rangle = \langle 4, 4 \rangle$$

Thus, $\mathbf{u}_1 + \mathbf{u}_2 = \langle 2, -2 \rangle + \langle 4, 4 \rangle = \langle 6, 2 \rangle = \mathbf{u}$.

If **u** is a force, then $\text{proj}_{\mathbf{v}}\mathbf{u}$ represents the effective force in the direction of **v** (Figure 6.21).

We can use vector projections to determine the amount of force required in problem situations like Example 6.

**Example 6** FINDING A FORCE

Juan is sitting on a sled on the side of a hill inclined at 45°. The combined weight of Juan and the sled is 140 pounds. What force is required for Rafaela to keep the sled from sliding down the hill? (See Figure 6.22.)

**Solution** We can represent the force due to gravity as $\mathbf{F} = -140\mathbf{j}$ because gravity acts vertically downward. We can represent the side of the hill with the vector

$$\mathbf{v} = (\cos 45°)\mathbf{i} + (\sin 45°)\mathbf{j} = \frac{\sqrt{2}}{2}\mathbf{i} + \frac{\sqrt{2}}{2}\mathbf{j}.$$

The force required to keep the sled from sliding down the hill is

$$\mathbf{F}_1 = \text{proj}_{\mathbf{v}}\mathbf{F} = \left( \frac{\mathbf{F} \cdot \mathbf{v}}{|\mathbf{v}|^2} \right)\mathbf{v} = (\mathbf{F} \cdot \mathbf{v})\mathbf{v}$$

because $|\mathbf{v}| = 1$. So,

$$\mathbf{F}_1 = (\mathbf{F} \cdot \mathbf{v})\mathbf{v} = (-140)\left( \frac{\sqrt{2}}{2} \right)\mathbf{v} = -70(\mathbf{i} + \mathbf{j}).$$

The magnitude of the force that Rafaela must exert to keep the sled from sliding down the hill is $70\sqrt{2} \approx 99$ pounds.

## Work

If **F** is a constant force whose direction is the same as the direction of $\overrightarrow{AB}$, then the **work** $W$ done by **F** in moving an object from $A$ to $B$ is

$$W = |\mathbf{F}|\,|\overrightarrow{AB}|.$$

If **F** is a constant force in any direction, then the **work** $W$ done by **F** in moving an object from $A$ to $B$ is

$$W = |\mathbf{F}| \cdot \overrightarrow{AB}$$
$$= |\mathbf{F}||\overrightarrow{AB}| \cos \theta$$

where $\theta$ is the angle between **F** and $\overrightarrow{AB}$. Except for the sign, the work is the magnitude of the effective force in the direction of $\overrightarrow{AB}$ times $\overrightarrow{AB}$.

## Example 7 FINDING WORK

Find the work done by a 10 pound force acting in the direction $\langle 1, 2 \rangle$ in moving an object 3 feet from $(0, 0)$ to $(3, 0)$.

**Solution** The force **F** has magnitude 10 and acts in the direction $\langle 1, 2 \rangle$, so

$$\mathbf{F} = 10 \frac{\langle 1, 2 \rangle}{|\langle 1, 2 \rangle|} = \frac{10}{\sqrt{5}} \langle 1, 2 \rangle.$$

The direction of motion is from $A = (0, 0)$ to $B = (3, 0)$, so $\overrightarrow{AB} = \langle 3, 0 \rangle$. Thus, the work done by the force is

$$\mathbf{F} \cdot \overrightarrow{AB} = \frac{10}{\sqrt{5}} \langle 1, 2 \rangle \cdot \langle 3, 0 \rangle = \frac{30}{\sqrt{5}} \approx 13.42 \text{ foot-pounds.}$$

### Units for Work

Work is usually measured in foot-pounds or Newton-meters. One Newton-meter is commonly referred to as one Joule.

# Quick Review 6.2

In Exercises 1–4, find $|\mathbf{u}|$.

**1.** $\mathbf{u} = \langle 2, -3 \rangle$ $\sqrt{13}$    **2.** $\mathbf{u} = -3\mathbf{i} - 4\mathbf{j}$ 5

**3.** $\mathbf{u} = \cos 35° \, \mathbf{i} + \sin 35° \, \mathbf{j}$ 1  **4.** $\mathbf{u} = 2(\cos 75° \mathbf{i} + \sin 75° \mathbf{j})$

In Exercises 5–8, the points $A$ and $B$ lie on the circle $x^2 + y^2 = 4$. Find the component form of the vector $\overrightarrow{AB}$.

**5.** $A = (-2, 0), B = (1, \sqrt{3})$    **6.** $A = (2, 0), B = (1, \sqrt{3})$

**7.** $A = (2, 0), B = (1, -\sqrt{3})$    **8.** $A = (-2, 0), B = (1, -\sqrt{3})$

In Exercises 9 and 10, find a vector $\mathbf{u}$ with the given magnitude in the direction of $\mathbf{v}$.

**9.** $|\mathbf{u}| = 2, \mathbf{v} = \langle 2, 3 \rangle$    **10.** $|\mathbf{u}| = 3, \mathbf{v} = -4\mathbf{i} + 3\mathbf{j}$

$\left\langle \dfrac{4}{\sqrt{13}}, \dfrac{6}{\sqrt{13}} \right\rangle$    $\left\langle -\dfrac{12}{5}, \dfrac{9}{5} \right\rangle$

# Section 6.2 Exercises

In Exercises 1–8, find the dot product of $\mathbf{u}$ and $\mathbf{v}$.

**1.** $\mathbf{u} = \langle 5, 3 \rangle, \mathbf{v} = \langle 12, 4 \rangle$ 72

**2.** $\mathbf{u} = \langle -5, 2 \rangle, \mathbf{v} = \langle 8, 13 \rangle$ −14

**3.** $\mathbf{u} = \langle 4, 5 \rangle, \mathbf{v} = \langle -3, -7 \rangle$ −47

**4.** $\mathbf{u} = \langle -2, 7 \rangle, \mathbf{v} = \langle -5, -8 \rangle$ −46

**5.** $\mathbf{u} = -4\mathbf{i} - 9\mathbf{j}, \mathbf{v} = -3\mathbf{i} - 2\mathbf{j}$ 30

**6.** $\mathbf{u} = 2\mathbf{i} - 4\mathbf{j}, \mathbf{v} = -8\mathbf{i} + 7\mathbf{j}$ −44

**7.** $\mathbf{u} = 7\mathbf{i}, \mathbf{v} = -2\mathbf{i} + 5\mathbf{j}$ −14

**8.** $\mathbf{u} = 4\mathbf{i} - 11\mathbf{j}, \mathbf{v} = -3\mathbf{j}$ 33

In Exercises 9–12, use the dot product to find $|\mathbf{u}|$.

**9.** $\mathbf{u} = \langle 5, -12 \rangle$ 13    **10.** $\mathbf{u} = \langle -8, 15 \rangle$ 17

**11.** $\mathbf{u} = -4\mathbf{i}$ 4    **12.** $\mathbf{u} = 3\mathbf{j}$ 3

In Exercises 13–22, find the angle $\theta$ between the vectors.

**13.** $\mathbf{u} = \langle -4, -3 \rangle, \mathbf{v} = \langle -1, 5 \rangle$ 115.6°

**14.** $\mathbf{u} = \langle 2, -2 \rangle, \mathbf{v} = \langle -3, -3 \rangle$ 90°

**15.** $\mathbf{u} = \langle 2, 3 \rangle, \mathbf{v} = \langle -3, 5 \rangle$    **16.** $\mathbf{u} = \langle 5, 2 \rangle, \mathbf{v} = \langle -6, -1 \rangle$

**17.** $\mathbf{u} = 3\mathbf{i} - 3\mathbf{j}, \mathbf{v} = -2\mathbf{i} + 2\sqrt{3}\mathbf{j}$ 165°

**18.** $\mathbf{u} = -2\mathbf{i}, \mathbf{v} = 5\mathbf{j}$ 90°

**19.** $\mathbf{u} = \left( 2 \cos \dfrac{\pi}{4} \right)\mathbf{i} + \left( 2 \sin \dfrac{\pi}{4} \right)\mathbf{j}, \mathbf{v} = \left( \cos \dfrac{3\pi}{2} \right)\mathbf{i} + \left( \sin \dfrac{3\pi}{2} \right)\mathbf{j}$

**20.** $\mathbf{u} = \left( \cos \dfrac{\pi}{3} \right)\mathbf{i} + \left( \sin \dfrac{\pi}{3} \right)\mathbf{j}, \mathbf{v} = \left( 3 \cos \dfrac{5\pi}{6} \right)\mathbf{i} + \left( 3 \sin \dfrac{5\pi}{6} \right)\mathbf{j}$

**21.**     94.86°

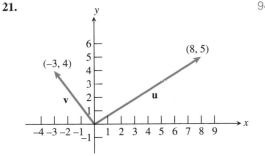

**22.**

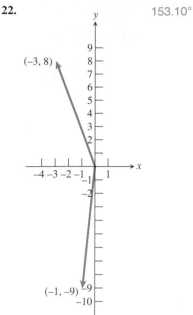

153.10°

In Exercises 23–26, find the vector projection of **u** onto **v**. Then write **u** as a sum of two orthogonal vectors, one of which is proj$_\mathbf{v}$**u**.

**23.** $\mathbf{u} = \langle -8, 3 \rangle$, $\mathbf{v} = \langle -6, -2 \rangle$

**24.** $\mathbf{u} = \langle 3, -7 \rangle$, $\mathbf{v} = \langle -2, -6 \rangle$

**25.** $\mathbf{u} = \langle 8, 5 \rangle$, $\mathbf{v} = \langle -9, -2 \rangle$

**26.** $\mathbf{u} = \langle -2, 8 \rangle$, $\mathbf{v} = \langle 9, -3 \rangle$

In Exercises 27 and 28, find the interior angles of the triangle with given vertices.

**27.** $(-4, 5)$, $(1, 10)$, $(3, 1)$     **28.** $(-4, 1)$, $(1, -6)$, $(5, -1)$

In Exercises 29 and 30, find $\mathbf{u} \cdot \mathbf{v}$ satisfying the given conditions where $\theta$ is the angle between **u** and **v**.

**29.** $\theta = 150°$, $|\mathbf{u}| = 3$, $|\mathbf{v}| = 8$  **30.** $\theta = \dfrac{\pi}{3}$, $|\mathbf{u}| = 12$, $|\mathbf{v}| = 40$

In Exercises 31–36, determine whether the vectors **u** and **v** are parallel, orthogonal, or neither.

**31.** $\mathbf{u} = \langle 5, 3 \rangle$, $\mathbf{v} = \left\langle -\dfrac{10}{4}, -\dfrac{3}{2} \right\rangle$ parallel

**32.** $\mathbf{u} = \langle 2, 5 \rangle$, $\mathbf{v} = \left\langle \dfrac{10}{3}, \dfrac{4}{3} \right\rangle$ neither

**33.** $\mathbf{u} = \langle 15, -12 \rangle$, $\mathbf{v} = \langle -4, 5 \rangle$ neither

**34.** $\mathbf{u} = \langle 5, -6 \rangle$, $\mathbf{v} = \langle -12, -10 \rangle$ orthogonal

**35.** $\mathbf{u} = \langle -3, 4 \rangle$, $\mathbf{v} = \langle 20, 15 \rangle$  **36.** $\mathbf{u} = \langle 2, -7 \rangle$, $\mathbf{v} = \langle -4, 14 \rangle$

In Exercises 37–40, find

**(a)** the $x$-intercept $A$ and $y$-intercept $B$ of the line.

**(b)** the coordinates of the point $P$ so that $\overrightarrow{AP}$ is perpendicular to the line and $|\overrightarrow{AP}| = 1$. (There are two answers.)

**37.** $3x - 4y = 12$          **38.** $-2x + 5y = 10$

**39.** $3x - 7y = 21$          **40.** $x + 2y = 6$

In Exercises 41 and 42, find the vector(s) **v** satisfying the given conditions.

**41.** $\mathbf{u} = \langle 2, 3 \rangle$, $\mathbf{u} \cdot \mathbf{v} = 10$, $|\mathbf{v}|^2 = 17$

**42.** $\mathbf{u} = \langle -2, 5 \rangle$, $\mathbf{u} \cdot \mathbf{v} = -11$, $|\mathbf{v}|^2 = 10$

**43. Sliding Down a Hill** Ojemba is sitting on a sled on the side of a hill inclined at 60°. The combined weight of Ojemba and the sled is 160 pounds. What is the magnitude of the force required for Mandisa to keep the sled from sliding down the hill?

**44. Revisiting Example 6** Suppose Juan and Rafaela switch positions. The combined weight of Rafaela and the sled is 125 pounds. What is the magnitude of the force required for Juan to keep the sled from sliding down the hill?  ≈ 88.39 pounds

**45. Braking Force** A 2000 pound car is parked on a street that makes an angle of 12° with the horizontal (see figure).

**(a)** Find the magnitude of the force required to keep the car from rolling down the hill.  ≈ 415.82 pounds

**(b)** Find the force perpendicular to the street.

**46. Effective Force** A 60 pound force **F** that makes an angle of 25° with an inclined plane is pulling a box up the plane. The inclined plane makes an 18° angle with the horizontal (see figure). What is the magnitude of the effective force pulling the box up the plane?  54.38 pounds

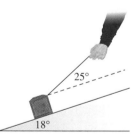

**47. Work** Find the work done lifting a 2600 pound car 5.5 feet.

**48. Work** Find the work done lifting a 100 pound bag of potatoes 3 feet.  300 foot-pounds

**49. Work** Find the work done by a force **F** of 12 pounds acting in the direction $\langle 1, 2 \rangle$ in moving an object 4 feet from $(0, 0)$ to $(4, 0)$.  ≈ 21.47 foot-pounds

**50. Work** Find the work done by a force **F** of 24 pounds acting in the direction $\langle 4, 5 \rangle$ in moving an object 5 feet from $(0, 0)$ to $(5, 0)$.  ≈ 74.96 foot-pounds

**51. Work** Find the work done by a force **F** of 30 pounds acting in the direction $\langle 2, 2 \rangle$ in moving an object 3 feet from $(0, 0)$ to a point in the first quadrant along the line $y = (1/2)x$.

**52. Work** Find the work done by a force **F** of 50 pounds acting in the direction $\langle 2, 3 \rangle$ in moving an object 5 feet from $(0, 0)$ to a point in the first quadrant along the line $y = x$.

**53. Work** The angle between a 200 pound force **F** and $\overrightarrow{AB} = 2\mathbf{i} + 3\mathbf{j}$ is 30°. Find the work done by **F** in moving an object from $A$ to $B$.   $100\sqrt{39} \approx 624.5$ foot-pounds

**54. Work** The angle between a 75 pound force **F** and $\overrightarrow{AB}$ is 60°, where $A = (-1, 1)$ and $B = (4, 3)$. Find the work done by **F** in moving an object from $A$ to $B$.

**55. Properties of the Dot Product** Let **u**, **v**, and **w** be vectors and let $c$ be a scalar. Use the component form of vectors to prove the following properties.

**(a)** $\mathbf{0} \cdot \mathbf{u} = 0$

**(b)** $\mathbf{u} \cdot (\mathbf{v} + \mathbf{w}) = \mathbf{u} \cdot \mathbf{v} + \mathbf{u} \cdot \mathbf{w}$

**(c)** $(\mathbf{u} + \mathbf{v}) \cdot \mathbf{w} = \mathbf{u} \cdot \mathbf{w} + \mathbf{v} \cdot \mathbf{w}$

**(d)** $(c\mathbf{u}) \cdot \mathbf{v} = \mathbf{u} \cdot (c\mathbf{v}) = c(\mathbf{u} \cdot \mathbf{v})$

**56. Group Activity Projection of a Vector** Let **u** and **v** be nonzero vectors. Prove that

**(a)** $\text{proj}_{\mathbf{v}}\mathbf{u} = \left( \dfrac{\mathbf{u} \cdot \mathbf{v}}{|\mathbf{v}|^2} \right)\mathbf{v}$

**(b)** $(\mathbf{u} - \text{proj}_{\mathbf{v}}\mathbf{u}) \cdot (\text{proj}_{\mathbf{v}}\mathbf{u}) = 0$

**57. Group Activity Connecting Geometry and Vectors** Prove that the sum of the squares of the diagonals of a parallelogram is equal to the sum of the squares of its sides.

**58.** If **u** is any vector, prove that we can write **u** as

$$\mathbf{u} = (\mathbf{u} \cdot \mathbf{i})\mathbf{i} + (\mathbf{u} \cdot \mathbf{j})\mathbf{j}.$$

## Explorations

**59. Distance from a Point to a Line** Consider the line $L$ with equation $2x + 5y = 10$ and the point $P = (3, 7)$.

**(a)** Verify that $A = (0, 2)$ and $B = (5, 0)$ are the $y$- and $x$-intercepts of $L$.

**(b)** Find $\mathbf{w}_1 = \text{proj}_{\overrightarrow{AB}} \overrightarrow{AP}$ and $\mathbf{w}_2 = \overrightarrow{AP} - \text{proj}_{\overrightarrow{AB}} \overrightarrow{AP}$.

**(c) Writing to Learn** Explain why $|\mathbf{w}_2|$ is the distance from $P$ to $L$. What is this distance?

**(d)** Find a formula for the distance of any point $P = (x_0, y_0)$ to $L$.

**(e)** Find a formula for the distance of any point $P = (x_0, y_0)$ to the line $ax + by = c$. ∎

## Extending the Ideas

**60. Writing to Learn** Let $\mathbf{w} = (\cos t)\,\mathbf{u} + (\sin t)\,\mathbf{v}$ where **u** and **v** are not parallel.

**(a)** Can the vector **w** be parallel to the vector **u**? Explain.

**(b)** Can the vector **w** be parallel to the vector **v**? Explain.

**(c)** Can the vector **w** be parallel to the vector $\mathbf{u} + \mathbf{v}$? Explain.

**61.** If the vectors **u** and **v** are not parallel, prove that
$$a\mathbf{u} + b\mathbf{v} = c\mathbf{u} + d\mathbf{v} \Rightarrow a = c, b = d.$$

---

## 6.3 Parametric Equations and Motion

Parametric Equations • Parametric Curves • Eliminating the Parameter • Lines and Line Segments • Simulating Motion with a Grapher

### Parametric Equations

Imagine that a rock is dropped from a 420-ft tower. The rock's height $y$ in feet above the ground $t$ seconds later (ignoring air resistance) is modeled by $y = -16t^2 + 420$ as we saw in Section 2.1. Figure 6.23 shows a coordinate system imposed on the scene so that the line of the rock's fall is on the vertical line $x = 2.5$.

The rock's original position and its position after each of the first 5 seconds are the points

$$(2.5, 420),\ (2.5, 404),\ (2.5, 356),\ (2.5, 276),\ (2.5, 164),\ (2.5, 20),$$

which are described by the pair of equations

$$x = 2.5, \qquad y = -16t^2 + 420,$$

when $t = 0, 1, 2, 3, 4, 5$. These two equations are an example of *parametric equations* with *parameter t*. As is often the case, the parameter $t$ represents time.

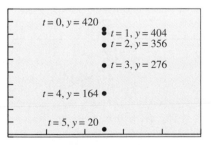

$t = 0, y = 420$

$t = 1, y = 404$
$t = 2, y = 356$

$t = 3, y = 276$

$t = 4, y = 164$

$t = 5, y = 20$

[0, 5] by [−10, 500]

**Figure 6.23** The position of the rock at 0, 1, 2, 3, 4, and 5 seconds.

# Parametric Curves

So far we have studied the graphs of several families of functions in this textbook. In this section we study the graphs of *parametric equations* and investigate motion of objects that can be modeled with parametric equations.

---

**Definition** Parametric Curve, Parametric Equations

The graph of the ordered pairs $(x, y)$ where

$$x = f(t), \quad y = g(t)$$

are functions defined on an interval $I$ of $t$-values is a **parametric curve**. The equations are **parametric equations** for the curve, the variable $t$ is a **parameter**, and $I$ is the **parameter interval**.

---

When we give parametric equations and a parameter interval for a curve, we have **parametrized** the curve. A **parametrization** of a curve consists of the parametric equations and the interval of $t$-values. Time is often the parameter in a problem situation which is why we normally use $t$ for the parameter. Sometimes parametric equations are used by companies in their design plans. It is then easier for the company to make larger and smaller objects efficiently by just changing the parameter $t$.

Graphs of parametric equations can be obtained using parametric mode on a grapher.

### Example 1 GRAPHING PARAMETRIC EQUATIONS

For the given parameter interval, graph the parametric equations

$$x = t^2 - 2, \quad y = 3t.$$

**(a)** $-3 \le t \le 1$ **(b)** $-2 \le t \le 3$ **(c)** $-3 \le t \le 3$

**Solution** In each case, set Tmin equal to the left endpoint of the interval and Tmax equal to the right endpoint of the interval. Figure 6.24 shows a graph of the parametric equations for each parameter interval. The corresponding relations are different because the parameter intervals are different.

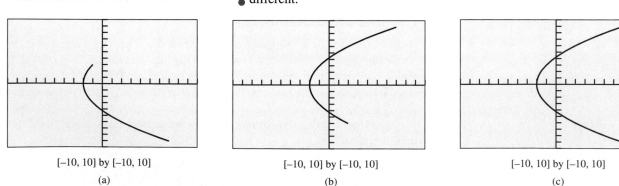

$[-10, 10]$ by $[-10, 10]$     $[-10, 10]$ by $[-10, 10]$     $[-10, 10]$ by $[-10, 10]$

(a)          (b)          (c)

**Figure 6.24** Three different relations defined parametrically. (Example 1)

**Teaching Note**

If students are not familiar with parametric graphing, it might be helpful to show them the graph of the linear function $f(x) = 3x - 2$ and compare it to one defined parametrically as $x = t$ and $y = 3t - 2$, using a trace key to show how $t$, $x$, and $y$ are related.

## Eliminating the Parameter

When a curve is defined parametrically it is sometimes possible to *eliminate the parameter* and obtain a rectangular equation in $x$ and $y$ that represents the curve. This often helps us identify the graph of the parametric curve as illustrated in Example 2.

### Example 2  ELIMINATING THE PARAMETER

Eliminate the parameter and identify the graph of the parametric curve

$$x = 1 - 2t, \quad y = 2 - t, \quad -\infty < t < \infty.$$

**Solution**  We solve the first equation for $t$:

$$x = 1 - 2t$$
$$2t = 1 - x$$
$$t = \frac{1}{2}(1 - x)$$

Then we substitute this expression for $t$ into the second equation:

$$y = 2 - t$$
$$y = 2 - \frac{1}{2}(1 - x)$$
$$y = 0.5x + 1.5$$

The graph of the equation $y = 0.5x + 1.5$ is a line with slope 0.5 and $y$-intercept 1.5 (Figure 6.25).

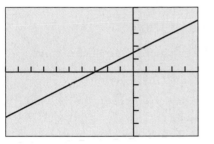

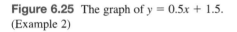

[–10, 5] by [–5, 5]

**Figure 6.25**  The graph of $y = 0.5x + 1.5$. (Example 2)

**Exploration Extensions**

Determine the smallest possible range of $t$-values that produces the graph shown in Figure 6.25, using the given parametric equations.

---

> ### Exploration 1   Graphing the Curve of Example 2 Parametrically
>
> 1. Use the parametric mode of your grapher to reproduce the graph in Figure 6.25. Use $-2$ for Tmin and 5.5 for Tmax.
> 2. Prove that the point $(17, 10)$ is on the graph of $y = 0.5x + 1.5$. Find the corresponding value of $t$ that produces this point.  $t = -8$
> 3. Repeat part 2 for the point $(-23, -10)$.  $t = 12$
> 4. Assume that $(a, b)$ is on the graph of $y = 0.5x + 1.5$. Find the corresponding value of $t$ that produces this point.  $t = 1/2 - a/2$
> 5. How do you have to choose Tmin and Tmax so that the graph in Figure 6.25 fills the window?

**Alert**

Many students will confuse range values of $t$ with range values on the function grapher. Point out that while the scale factor does not affect the way a graph is drawn, the Tstep does affect the way the graph is displayed.

If we do not specify a parameter interval for the parametric equations $x = f(t)$, $y = g(t)$, it is understood that the parameter $t$ can take on all values which produce real numbers for $x$ and $y$. We use this agreement in Example 3.

### Example 3  ELIMINATING THE PARAMETER

Eliminate the parameter and identify the graph of the parametric curve

$$x = t^2 - 2, \quad y = 3t.$$

**Solution** Here $t$ can be any real number. We solve the second equation for $t$ obtaining $t = y/3$ and substitute this value for $y$ into the first equation.

$$x = t^2 - 2$$

$$x = \left(\frac{y}{3}\right)^2 - 2$$

$$x = \frac{y^2}{9} - 2$$

$$y^2 = 9(x + 2)$$

Figure 6.24c shows what the graph of these parametric equations looks like. In Chapter 8 we will call this a parabola that opens to the right. Interchanging $x$ and $y$ we can identify this graph as the inverse of the graph of the parabola $x^2 = 9(y + 2)$.

### Parabolas

The inverse of a parabola that opens up or down is a parabola that opens left or right. We will investigate these curves in more detail in Chapter 8.

### Example 4    ELIMINATING THE PARAMETER

Eliminate the parameter and identify the graph of the parametric curve

$$x = 2 \cos t, \quad y = 2 \sin t, \quad 0 \le t \le 2\pi.$$

**Solution** The graph of the parametric equations in the square viewing window of Figure 6.26 suggests that the graph is a circle of radius 2 centered at the origin. We confirm this result algebraically.

$$\begin{aligned} x^2 + y^2 &= 4 \cos^2 t + 4 \sin^2 t \\ &= 4(\cos^2 t + \sin^2 t) \\ &= 4(1) \qquad \qquad \cos^2 t + \sin^2 t = 1 \\ &= 4 \end{aligned}$$

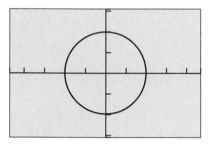

[–4.7, 4.7] by [–3.1, 3.1]

**Figure 6.26** The graph of the circle of Example 4.

The graph of $x^2 + y^2 = 4$ is a circle of radius 2 centered at the origin. Increasing the length of the interval $0 \le t \le 2\pi$ will cause the grapher to trace all or part of the circle more than once. Decreasing the length of the interval will cause the grapher to only draw a portion of the complete circle. Try it!

In Exercise 53, you will find parametric equations for any circle in the plane.

## Lines and Line Segments

We can use vectors to help us find parametric equations for a line as illustrated in Example 5.

### Example 5    FINDING PARAMETRIC EQUATIONS FOR A LINE

Find a parametrization of the line through the points $A = (-2, 3)$ and $B = (3, 6)$.

**Solution** Let $P(x, y)$ be a point on the line through $A$ and $B$ and let $O$ be the origin. Then the vector $\overrightarrow{OP} - \overrightarrow{OA} = \overrightarrow{AP}$ must be a scalar multiple of the vector $\overrightarrow{OB} - \overrightarrow{OA} = \overrightarrow{AB}$ (Figure 6.27). The component forms of $\overrightarrow{OA}$, $\overrightarrow{OB}$, and $\overrightarrow{OP}$ are

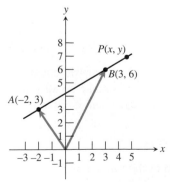

**Figure 6.27** Example 5 uses vectors to produce a parametrization of the line through $A$ and $B$.

$$\overrightarrow{OA} = \langle -2, 3 \rangle, \quad \overrightarrow{OB} = \langle 3, 6 \rangle, \quad \overrightarrow{OP} = \langle x, y \rangle.$$

If we let the scalar be $t$, then we have

$$\overrightarrow{OP} - \overrightarrow{OA} = t(\overrightarrow{OB} - \overrightarrow{OA})$$

$$\langle x + 2, y - 3 \rangle = t\langle 5, 3 \rangle$$

It follows that

$$\begin{matrix} x + 2 = 5t \\ y - 3 = 3t \end{matrix} \quad \text{or} \quad \begin{matrix} x = -2 + 5t \\ y = 3 + 3t \end{matrix}.$$

The parametric equations $x = -2 + 5t$, $y = 3 + 3t$ together with the parameter interval $(-\infty, \infty)$ give a parametrization of the line.

We can confirm our work numerically as follows: If $t = 0$ then $x = -2$ and $y = 3$. This is the point $A$. Similarly, if $t = 1$ then $x = 3$ and $y = 6$. This is the point $B$.

In Example 5, we can also write

$$\overrightarrow{OP} = \overrightarrow{OA} + t\overrightarrow{AB}.$$

If $t = 0$ then $P = A$, and if $t = 1$ then $P = B$. We use this fact in Example 6.

**Example 6  FINDING PARAMETRIC EQUATIONS FOR A LINE SEGMENT**

Find a parametrization of the line segment with endpoints $A = (-2, 3)$ and $B = (3, 6)$.

**Solution**  In Example 5 we found parametric equations for the line through $A$ and $B$:

$$x = -2 + 5t, \quad y = 3 + 3t.$$

We also saw in Example 5 that $t = 0$ produces the point $A$ and $t - 1$ produces the point $B$. A parametrization of the line segment is given by

$$x = -2 + 5t, \quad y = 3 + 3t, \quad 0 \le t \le 1.$$

As $t$ varies between 0 and 1 we pick up every point on the line segment between $A$ and $B$.

## Simulating Motion with a Grapher

Example 7 illustrates several ways to simulate motion along a horizontal line using parametric equations. We use the variable t for the parameter to represent time.

**Example 7  SIMULATING HORIZONTAL MOTION**

Gary walks along a horizontal line (think of it as a number line) with the coordinate of his position (in meters) given by

$$s = -0.1(t^3 - 20t^2 + 110t - 85)$$

where $0 \le t \le 12$. Use parametric equations and a grapher to simulate his motion. Estimate the times when Gary changes direction.

**Solution** We arbitrarily choose the horizontal line $y = 5$ to display this motion. The graph $C_1$ of the parametric equations,

$$C_1: x_1 = -0.1(t^3 - 20t^2 + 110t - 85), y_1 = 5, 0 \le t \le 12,$$

simulates the motion. His position at any time $t$ is given by the point $(x_1(t), 5)$.

Using trace in Figure 6.28 we see that when $t = 0$, Gary is 8.5 m to the right of the $y$-axis at the point $(8.5, 5)$, and that he initially moves left. Five seconds later he is 9 m to to the left of the $y$-axis at the point $(-9, 5)$. And after 8 seconds he is only 2.7 m to the left of the $y$-axis. Gary must have changed direction during the walk. The motion of the trace cursor simulates Gary's motion.

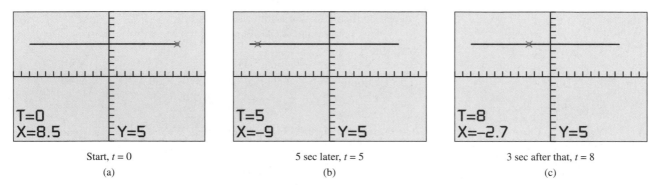

Start, $t = 0$      5 sec later, $t = 5$      3 sec after that, $t = 8$

(a)      (b)      (c)

**Figure 6.28** Three views of the graph $C_1: x_1 = -0.1(t^3 - 20t^2 + 110t - 85), y_1 = 5, 0 \le t \le 12$ in the $[-12, 12]$ by $[-10, 10]$ viewing window. (Example 7)

A variation in $y(t)$,

$$C_2: x_2 = -0.1(t^3 - 20t^2 + 110t - 85), y_2 = -t, 0 \le t \le 12,$$

can be used to help visualize where Gary changes direction. The graph $C_2$ shown in Figure 6.29 suggests that Gary reverses his direction at 3.9 seconds and again at 9.5 seconds after beginning his walk.

**Grapher Note**

The equation $y_2 = t$ is typically used in the parametric equations for the graph $C_2$ in Figure 6.29. We have chosen $y_2 = -t$ to get two curves in Figure 6.29 that do not overlap. Also notice that the $y$-coordinates of $C_1$ are constant ($y_1 = 5$), and that the $y$-coordinates of $C_2$ vary with time $t$ ($y_2 = -t$).

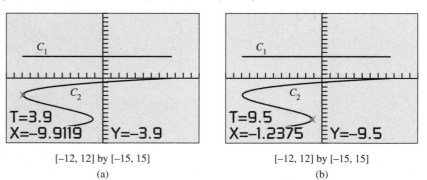

$[-12, 12]$ by $[-15, 15]$      $[-12, 12]$ by $[-15, 15]$

(a)      (b)

**Figure 6.29** Two views of the graph $C_1: x_1 = -0.1(t^3 - 20t^2 + 110t - 85), y_1 = 5,$ $0 \le t \le 12$ and the graph $C_2: x_2 = -0.1(t^3 - 20t^2 + 110t - 85), y_2 = -t,$ $0 \le t \le 12$ in the $[-12, 12]$ by $[-15, 15]$ viewing window. (Example 7)

Example 8 solves a projectile-motion problem. Parametric equations are used in two ways: to find a graph of the modeling equation and to simulate the motion of the projectile.

### Example 8  SIMULATING PROJECTILE MOTION

A distress flare is shot straight up from a ship's bridge 75 ft above the water with an initial velocity of 76 ft/sec. Graph the flare's height against time, give the height of the flare above water at each time, and simulate the flare's motion for each length of time.

**(a)** 1 sec      **(b)** 2 sec      **(c)** 4 sec      **(d)** 5 sec

**Solution**  An equation that models the flare's height above the water $t$ seconds after launch is

$$y = -16t^2 + 76t + 75.$$

A graph of the flare's height against time can be found using the parametric equations

$$x_1 = t, \quad y_1 = -16t^2 + 76t + 75.$$

To simulate the flare's flight straight up and its fall to the water, use the parametric equations

$$x_2 = 5.5, \quad y_2 = -16t^2 + 76t + 75.$$

(We chose $x_2 = 5.5$ so that the two graphs would not intersect.)

Figure 6.30 shows the two graphs in simultaneous graphing mode for (a) $0 \le t \le 1$, (b) $0 \le t \le 2$, (c) $0 \le t \le 4$, and (d) $0 \le t \le 5$. We can read that the height of the flare above the water after 1 sec is 135 ft, after 2 sec is 163 ft, after 4 sec is 123 ft, and after 5 sec is 55 ft.

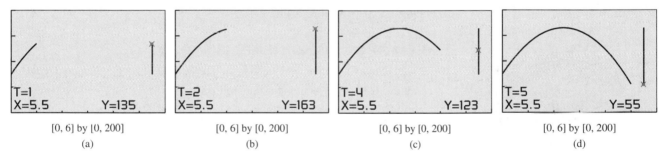

| T=1<br>X=5.5                Y=135 | T=2<br>X=5.5                Y=163 | T=4<br>X=5.5                Y=123 | T=5<br>X=5.5                Y=55 |
| :---: | :---: | :---: | :---: |
| [0, 6] by [0, 200] | [0, 6] by [0, 200] | [0, 6] by [0, 200] | [0, 6] by [0, 200] |
| (a) | (b) | (c) | (d) |

**Figure 6.30**  Simultaneous graphing of $x_1 = t$, $y_1 = -16t^2 + 76t + 75$ (height against time) and $x_2 = 5.5$, $y_2 = -16t^2 + 76t + 75$ (the actual path of the flare). (Example 8)

In Example 8 we modeled the motion of a projectile that was launched straight up. Now we investigate the motion of objects, ignoring air friction, that are launched at angles other than 90° with the horizontal.

Suppose that a baseball is thrown from a point $y_0$ feet above ground level with an initial speed of $v_0$ ft/sec at an angle $\theta$ with the horizontal (Figure 6.31). The initial velocity can be represented by the vector

$$\mathbf{v} = \langle v_0 \cos \theta, v_0 \sin \theta \rangle.$$

The path of the object is modeled by the parametric equations

$$x = (v_0 \cos \theta)t, \quad y = -16t^2 + (v_0 \sin \theta)t + y_0.$$

The $x$-component is simply

$$\text{distance} = (x\text{-component of initial velocity}) \times \text{time}.$$

The $y$-component is the familiar vertical projectile-motion equation using the $y$-component of initial velocity.

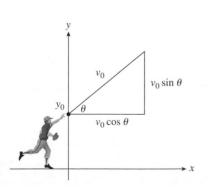

**Figure 6.31**  Throwing a baseball.

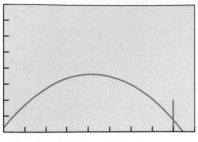

[0, 450] by [0, 80]

**Figure 6.32** The fence and path of the baseball in Example 9.

### Example 9 HITTING A BASEBALL

Kevin hits a baseball at 3 ft above the ground with an initial speed of 150 ft/sec at an angle of 18° with the horizontal. Will the ball clear a 20-ft wall that is 400 ft away?

**Solution** The path of the ball is modeled by the parametric equations

$$x = (150 \cos 18°)t, \quad y = -16t^2 + (150 \sin 18°)t + 3.$$

A little experimentation will show that the ball will reach the fence in less than 3 sec. Figure 6.32 shows a graph of the path of the ball using the parameter interval $0 \le t \le 3$ and the 20-ft wall. The ball does not clear the wall.

---

**Exploration 2** **Extending Example 9**

1. If your grapher has a line segment feature, draw the fence in Example 9.
2. Describe the graph of the parametric equations
$$x = 400, \quad y = 20(t/3), \quad 0 \le t \le 3.$$
3. Repeat Example 9 for the angles 19°, 20°, 21°, and 22°.

---

**Exploration Extensions**

Using trial and error, determine the minimum angle, to the nearest 0.05°, such that the ball clears the fence.

In Example 10 we see how to write parametric equations for position on a moving Ferris wheel using time $t$ as the parameter.

### Example 10 RIDING ON A FERRIS WHEEL

Jane is riding on a Ferris wheel with a radius of 30 ft. As we view it in Figure 6.33, the wheel is turning counterclockwise at the rate of one revolution every 10 sec. Assume the lowest point of the Ferris wheel (6 o'clock) is 10 ft above the ground, and that Jane is at the point marked $A$ (3 o'clock) at time $t = 0$. Find parametric equations to model Jane's path and use them to find Jane's position 22 sec into the ride.

**Solution** Figure 6.34 shows a circle with center (0, 40) and radius 30 that models the Ferris wheel. The parametric equations for this circle in terms of the parameter $\theta$, the central angle of the circle determined by the arc $AP$, are

$$x = 30 \cos \theta, \quad y = 40 + 30 \sin \theta, \quad 0 \le \theta \le 2\pi.$$

To take into account the rate at which the wheel is turning we must describe $\theta$ as a function of time $t$ in seconds. The wheel is turning at the rate of $2\pi$ radians every 10 sec, or $2\pi/10 = \pi/5$ rad/sec. So, $\theta = (\pi/5)t$. Thus, parametric equations that model Jane's path are given by

$$x = 30 \cos\left(\frac{\pi}{5}t\right), \quad y = 40 + 30 \sin\left(\frac{\pi}{5}t\right), \quad t \ge 0.$$

We substitute $t = 22$ into the parametric equations to find Jane's position at that time:

$$x = 30 \cos\left(\frac{\pi}{5} \cdot 22\right) \quad y = 40 + 30 \sin\left(\frac{\pi}{5} \cdot 22\right)$$

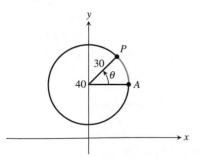

**Figure 6.33** The Ferris wheel of Example 10.

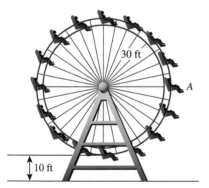

**Figure 6.34** A model for the Ferris wheel of Example 10.

$$x \approx 9.27 \qquad\qquad y \approx 68.53$$

After riding for 22 sec, Jane is approximately 68.5 ft above the ground and approximately 9.3 ft to the right of the $y$-axis using the coordinate system of Figure 6.34.

---

### Exploration 3  Throwing a Ball at a Ferris Wheel

A 20-ft Ferris wheel turns counterclockwise one revolution every 12 sec. (See Figure 6.35.) Eric stands at point $D$, 75 ft from the base of the wheel. At the instant Jane is at point $A$, Eric throws a ball at the Ferris wheel, releasing it from the same height as the bottom of the wheel. If the ball's initial speed is 60 ft/sec and it is released at an angle of 120° with the horizontal, does Jane have a chance to catch the ball?

1. Assign a coordinate system so that the bottom car of the Ferris wheel is at $(0, 0)$ and the center of the wheel is at $(0, 20)$. Then Eric releases the ball at the point $(75, 0)$. Explain why parametric equations for Jane's path are:

$$x_1 = 20 \cos\left(\frac{\pi}{6} t\right), \quad y_1 = 20 + 20 \sin\left(\frac{\pi}{6} t\right), \quad t \geq 0.$$

2. Explain why parametric equations for the path of the ball are:

$$x_2 = -30t + 75, \quad y_2 = -16t^2 + (30\sqrt{3})t, \quad t \geq 0.$$

3. Graph the two paths simultaneously and determine if Jane and the ball arrive at the point of intersection of the two paths at the same time.

4. Find a formula for the distance $d(t)$ between Jane and the ball at any time $t$.

5. Use the graph of the parametric equations

$$x_3 = t, \quad y_3 = d(t)$$

to estimate the minimum distance between Jane and the ball and when it occurs. Do you think Jane has a chance to catch the ball?

---

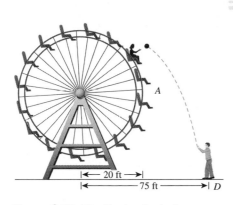

**Figure 6.35** The Ferris wheel of Exploration 3.

### Example 11  MODELING THE PATH OF A STONE

A stone is stuck in the tread of a tire. Assume that the radius of the wheel is 1 ft and find a graph of the stone's path as the wheel rolls along a flat road.

**Solution**  We assign a rectangular coordinate system to agree with the instant the tire picks up the stone. We let $(0, 0)$ be the initial position of the stone and $(0, 1)$ be the center of the wheel, and we assume that the wheel is rolling in the positive $x$ direction. Figure 6.36 shows the wheel after it has rolled a distance $t$.

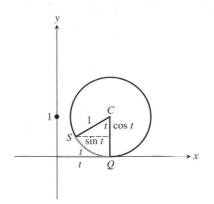

**Figure 6.36** Notice that $t$ is (a) the length of the arc $SQ$, (b) the length of the line segment from the origin to $Q$, and (c) the radian measure of the central angle of the circle with radius 1. (Example 11)

$$t = \text{distance wheel rolls after picking up stone } S$$

$$C = (t, 1)$$

$$S = (t - \sin t, 1 - \cos t)$$

By a careful analysis of Figure 6.36 we see that the path of the stone $S$ can be modeled by the parametric equations

$$x = t - \sin t, \quad y = 1 - \cos t, \quad t \geq 0.$$

Figure 6.37 shows a graph of the parametric equations that model the path of the stone. Whenever $t$ is a multiple of $2\pi$, $\cos t = 1$ and the $y$-coordinate is zero. This is consistent with the fact that the stone in the tire tread touches the road once every revolution.

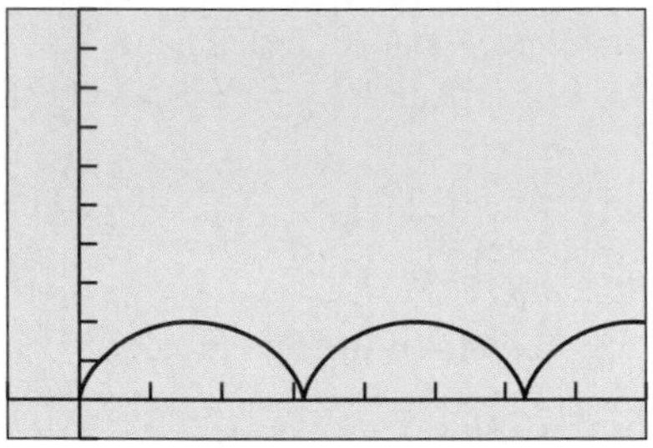

[–2, 16] by [–1, 10]

**Figure 6.37** The path of the stone in Example 10 is a cycloid. (See Exercise 56.)

## Quick Review 6.3

In Exercises 1 and 2, find the component form of the vectors **(a)** $\overrightarrow{OA}$, **(b)** $\overrightarrow{OB}$, and **(c)** $\overrightarrow{AB}$ where $O$ is the origin.

**1.** $A = (-3, -2), B = (4, 6)$     **2.** $A = (-1, 3), B = (4, -3)$

In Exercises 3 and 4, write an equation in point-slope form for the line through the two points.

**3.** $(-3, -2), (4, 6)$            **4.** $(-1, 3), (4, -3)$

In Exercises 5 and 6, graph the equation.

**5.** $y^2 = 8x$                    **6.** $y^2 = -5x$

In Exercises 7 and 8, write an equation for the circle with given center and radius.

**7.** $(0, 0), 2$   $x^2 + y^2 = 4$        **8.** $(-2, 5), 3$

In Exercises 9 and 10, a wheel with radius $r$ spins at the given rate. Find the angular velocity in radians per second.

**9.** $r = 13$ in., 600 rpm        **10.** $r = 12$ in., 700 rpm

# Section 6.3 Exercises

In Exercises 1–4, match the parametric equations with their graph. Identify the viewing window that seems to have been used.

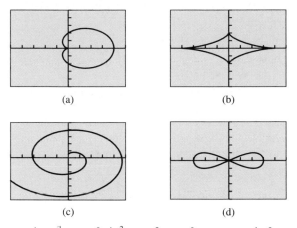

(a)  (b)

(c)  (d)

**1.** $x = 4\cos^3 t$, $y = 2\sin^3 t$    **2.** $x = 3\cos t$, $y = \sin 2t$

**3.** $x = 2\cos t + 2\cos^2 t$, $y = 2\sin t + \sin 2t$

**4.** $x = \sin t - t\cos t$, $y = \cos t + t\sin t$

In Exercises 5 and 6, **(a)** complete the table for the parametric equations and **(b)** plot the corresponding points.

**5.** $x = t + 2$, $y = 1 + 3/t$

| $t$ | $-2$ | $-1$ | $0$ | $1$ | $2$ |
|---|---|---|---|---|---|
| $x$ | | | | | |
| $y$ | | | | | |

**6.** $x = \cos t$, $y = \sin t$

| $t$ | $0$ | $\pi/2$ | $\pi$ | $3\pi/2$ | $2\pi$ |
|---|---|---|---|---|---|
| $x$ | | | | | |
| $y$ | | | | | |

In Exercises 7–22, eliminate the parameter and identify the graph of the parametric curve.

**7.** $x = 1 + t$, $y = t$  $y = x - 1$   **8.** $x = 2 - 3t$, $y = 5 + t$

**9.** $x = 2t - 3$, $y = 9 - 4t$, $3 \le t \le 5$

**10.** $x = 5 - 3t$, $y = 2 + t$, $-1 \le t \le 3$

**11.** $x = t^2$, $y = t + 1$ [*Hint:* Eliminate $t$ and solve for $x$ in terms of $y$.] $x = (y - 1)^2$

**12.** $x = t$, $y = t^2 - 3$        **13.** $x = t$, $y = t^3 - 2t + 3$

**14.** $x = 2t^2 - 1$, $y = t$ [*Hint:* Eliminate $t$ and solve for $x$ in terms of $y$.] $x = 2y^2 - 1$

**15.** $x = 4 - t^2$, $y = t$ [*Hint:* Eliminate $t$ and solve for $x$ in terms of $y$.] $x = 4 - y^2$

**16.** $x = 0.5t$, $y = 2t^3 - 3$, $-2 \le t \le 2$

**17.** $x = t - 3$, $y = 2/t$, $-5 \le t \le 5$

**18.** $x = t + 2$, $y = 4/t$, $t \ge 2$

**19.** $x = 5\cos t$, $y = 5\sin t$        **20.** $x = 4\cos t$, $y = 4\sin t$

**21.** $x = 2\sin t$, $y = 2\cos t$, $0 \le t \le 3\pi/2$

**22.** $x = 3\cos t$, $y = 3\sin t$, $0 \le t \le \pi$

In Exercises 23–28, find a parametrization for the curve.

**23.** The line through the points $(-2, 5)$ and $(4, 2)$.

**24.** The line through the points $(-3, -3)$ and $(5, 1)$.

**25.** The line segment with endpoints $(3, 4)$ and $(6, -3)$.

**26.** The line segment with endpoints $(5, 2)$ and $(-2, -4)$.

**27.** The circle with center $(5, 2)$ and radius 3.

**28.** The circle with center $(-2, -4)$ and radius 2.

Exercises 29–32, refer to the graph of the parametric equations

$$x = 2 - |t|, \quad y = t - 0.5, \quad -3 \le t \le 3$$

given below. Find the values of the parameter $t$ that produces the graph in the indicated quadrant.

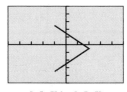

[−5, 5] by [−5, 5]

**29.** Quadrant I  $0.5 < t < 2$   **30.** Quadrant II  $2 < t \le 3$

**31.** Quadrant III  $-3 \le t < -2$  **32.** Quadrant IV  $-2 < t < 0.5$

**33. Famine Relief Air Drop** A relief agency drops food containers from an airplane on a war-torn famine area. The drop was made from an altitude of 1000 ft above ground level.

**(a)** Use an equation to model the height of the containers (during free fall) as a function of time $t$.

**(b)** Use parametric mode to simulate the drop during the first 6 sec.

**(c)** After 4 sec of free fall, parachutes open. How many feet above the ground are the food containers when the parachutes open?  744 ft

**34. Height of a Pop-up** A baseball is hit straight up from a height of 5 ft with an initial velocity of 80 ft/sec.

**(a)** Write an equation that models the height of the ball as a function of time $t$. $y = -16t^2 + 80t + 5$

**(b)** Use parametric mode to simulate the pop-up.

**(c)** Use parametric mode to graph height against time. (*Hint:* Let $x(t) = t$.)

**(d)** How high is the ball after 4 sec?  69 ft

**(e)** What is the maximum height of the ball? How many seconds does it take to reach its maximum height?

**35. Simulating a Foot Race** Ben can sprint at the rate of 24 ft/sec. Jerry sprints at 20 ft/sec. Ben gives Jerry a 10-ft head start. The parametric equations can be used to model a race.

$$x_1 = 20t, \qquad y_1 = 3$$
$$x_2 = 24t - 10, \quad y_2 = 5$$

**(a)** Find a viewing window to simulate a 100-yd dash. Graph simultaneously with $t$ starting at $t = 0$ and $T$step $= 0.05$.

**(b)** Who is ahead after 3 sec and by how much?

**36. Capture the Flag** Two opposing players in "Capture the Flag" are 100 ft apart. On a signal, they run to capture a flag that is on the ground midway between them. The faster runner, however, hesitates for 0.1 sec. The following parametric equations model the race to the flag:

$$x_1 = 10(t - 0.1), \quad y_1 = 3$$
$$x_2 = 100 - 9t, \qquad y_2 = 3$$

**(a)** Simulate the game in a $[0, 100]$ by $[-1, 10]$ viewing window with $t$ starting at 0. Graph simultaneously.

**(b)** Who captures the flag and by how many feet?

50 ft          50 ft

**37.** The complete graph of the parametric equations $x = 2 \cos t$, $y = 2 \sin t$ is the circle of radius 2 centered at the origin. Find an interval of values for $t$ so that the graph is the given portion of the circle.

**(a)** The portion in the first quadrant  $0 < t < \pi/2$

**(b)** The portion above the $x$-axis  $0 < t < \pi$

**(c)** The portion to the left of the $y$-axis  $\pi/2 < t < 3\pi/2$

**38. Writing to Learn** Consider the two pairs of parametric equations $x = 3 \cos t$, $y = 3 \sin t$ and $x = 3 \sin t$, $y = 3 \cos t$ for $0 \le t \le 2\pi$.

**(a)** Give a convincing argument that the graphs of the pairs of parametric equations are the same.

**(b)** Explain how the parametrizations are different.

**39. Hitting a Baseball** Consider Kevin's hit discussed in Example 9.

**(a)** Approximately how many seconds after the ball is hit does it hit the wall?  about 2.80 sec

**(b)** How high up the wall does the ball hit?  $\approx 7.18$ ft

**(c) Writing to Learn** Explain why Kevin's hit might be caught by an outfielder. Then explain why his hit would likely not be caught by an outfielder if the ball had been hit at a 20° angle with the horizontal.

**40. Hitting a Baseball** Kirby hits a ball when it is 4 ft above the ground with an initial velocity of 120 ft/sec. The ball leaves the bat at a 30° angle with the horizontal and heads toward a 30-ft fence 350 ft from home plate.

**(a)** Does the ball clear the fence?  no

**(b)** If so, by how much does it clear the fence? If not, could the ball be caught?  not catchable

**41. Hitting a Baseball** Suppose that the moment Kirby hits the ball in Exercise 40 there is a 5-ft/sec split-second wind gust. Assume the wind acts in the horizontal direction out with the ball.

**(a)** Does the ball clear the fence?  yes

**(b)** If so, by how much does it clear the fence? If not, could the ball be caught?  1.59 ft

**42. Two-Softball Toss** Chris and Linda warm up in the outfield by tossing softballs to each other. Suppose both tossed a ball at the same time from the same height, as illustrated in the figure. Find the minimum distance between the two balls and when this minimum distance occurs.

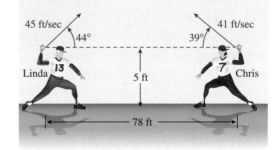

**43. Yard Darts** Tony and Sue are launching yard darts 20 ft from the front edge of a circular target of radius 18 in. on the ground. If Tony throws the dart directly at the target, and releases it 3 ft above the ground with an initial velocity of 30 ft/sec at a 70° angle, will the dart hit the target?  no

**44. Yard Darts** In the game of darts described in Exercise 43 Sue releases the dart 4 ft above the ground with an initial velocity of 25 ft/sec at a 55° angle. Will the dart hit the target?  yes

**45. Hitting a Baseball** Orlando hits a ball when it is 4 ft above ground level with an initial velocity of 160 ft/sec. The ball leaves the bat at a 20° angle with the horizontal and heads toward a 30-ft fence 400 ft from home plate. How strong must a split-second wind gust be (in feet per second) that acts directly with or against the ball in order for the ball to hit within a few inches of the top of the wall? Estimate the answer graphically and solve algebraically.

**46. Hitting Golf Balls** Nancy hits golf balls off the practice tee with an initial velocity of 180 ft/sec with four different clubs. How far down the fairway does the ball hit the ground if it comes off the club making the specified angle with the horizontal?

**(a)** 15°     **(b)** 20°     **(c)** 25°     **(d)** 30°

**47. Analysis of a Ferris Wheel** Ron is on a Ferris wheel of radius 35 ft that turns counterclockwise at the rate of one revolution every 12 sec. The lowest point of the Ferris wheel

(6 o'clock) is 15 ft above ground level at the point (0, 15) on a rectangular coordinate system. Find parametric equations for the position of Ron as a function of time $t$ (in seconds) if the Ferris wheel starts ($t = 0$) with Ron at the point (35, 50).

**48. Throwing a Ball at a Ferris Wheel** A 71-ft-radius Ferris wheel turns counterclockwise one revolution every 20 sec. Tony stands at a point 90 ft to the right of the base of the wheel. At the instant Matthew is at point $A$ (3 o'clock), Tony throws a ball toward the Ferris wheel with an initial velocity of 88 ft/sec at an angle with the horizontal of 100°. Find the minimum distance between the ball and Matthew.

In Exercises 49–52, a particle moves along a horizontal line so that its position at any time $t$ is given by $s(t)$. Write a description of the motion. (*Hint:* See Example 7.)

**49. Group Activity** $s(t) = -t^2 + 3t, \quad -2 \le t \le 4$

**50. Group Activity** $s(t) = -t^2 + 4t, \quad -1 \le t \le 5$

**51. Group Activity** $s(t) = 0.5(t^3 - 7t^2 + 2t), \quad -1 \le t \le 7$

**52. Group Activity** $s(t) = t^3 - 5t^2 + 4t, \quad -1 \le t \le 5$

# Explorations

**53. Parametrizing Circles** Consider the parametric equations

$$x = a \cos t, \quad y = a \sin t, \quad 0 \le t \le 2\pi.$$

**(a)** Graph the parametric equations for $a = 1, 2, 3, 4$ in the same square viewing window.

**(b)** Eliminate the parameter $t$ in the parametric equations to verify that they are all circles. What is the radius?

Now consider the parametric equations

$$x = h + a \cos t, \quad y = k + a \sin t, \quad 0 \le t \le 2\pi.$$

**(c)** Graph the equations for $a = 1$ using the following pairs of values for $h$ and $k$:

| $h$ | 2 | −2 | −4 | 3 |
|---|---|---|---|---|
| $k$ | 3 | 3 | −2 | −3 |

**(d)** Eliminate the parameter $t$ in the parametric equations and identify the graph.

**(e)** Write a parametrization for the circle with center $(-1, 4)$ and radius 3. $x = 3 \cos t - 1; y = 3 \sin t + 4$

**54. Group Activity Parametrization of Lines** Consider the parametrization

$$x = at + b, \quad y = ct + d,$$

where $a$ and $c$ are not both zero.

**(a)** Graph the curve for $a = 2, b = 3, c = -1$, and $d = 2$.

**(b)** Graph the curve for $a = 3, b = 4, c = 1$, and $d = 3$.

**(c) Writing to Learn** Eliminate the parameter $t$ and write an equation in $x$ and $y$ for the curve. Explain why its graph is a line.

**(d) Writing to Learn** Find the slope, $y$-intercept, and $x$-intercept of the line if they exist? If not, explain why not.

**(e)** Under what conditions will the line be horizontal? Vertical? $c = 0; a = 0$ ■

**55. Revisiting Example 5** Eliminate the parameter $t$ from the parametric equations of Example 5 to find an equation in $x$ and $y$ for the line. Verify that the line passes through the points $A$ and $B$ of the example. $(3/5)x + 21/5$

**56. Cycloid** The graph of the parametric equations $x = t - \sin t, y = 1 - \cos t$ of Example 11 is a *cycloid*.

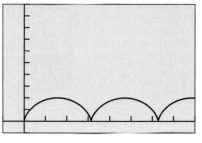

[−2, 16] by [−1, 10]

**(a)** What is the maximum value of $y = 1 - \cos t$? How is that value related to the graph?

**(b)** What is the distance between neighboring $x$-intercepts?

**57. Hypocycloid** The graph of the parametric equations $x = 2 \cos t + \cos 2t, y = 2 \sin t - \sin 2t$ is a *hypocycloid*. The graph is the path of a point $P$ on a circle of radius 1 rolling along the inside of a circle of radius 3, as illustrated in the figure.

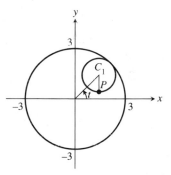

**(a)** Graph simultaneously this hypocycloid and the circle of radius 3.

**(b)** Suppose the large circle had a radius of 4. Experiment! How do you think the equations in (a) should be changed to obtain defining equations? What do you think the hypocycloid would look like in this case? Check your guesses. all 2's should be changed to 3's

# Extending the Ideas

**58. Two Ferris Wheel Problem** Chang is on a Ferris wheel of center (0, 20) and radius 20 ft turning counterclockwise at the rate of one revolution every 12 sec. Kuan is on a Ferris wheel of center (15, 15) and radius 15 turning counterclockwise at the rate of one revolution every 8 sec. Find the minimum distance between Chang and Kuan if both start out ($t = 0$) at 3 o'clock. about 4.11 ft

**59. Two Ferris Wheel Problem** Chang and Kuan are riding the Ferris wheels described in Exercise 58. Find the minimum distance between Chang and Kuan if Chang starts out ($t = 0$) at 3 o'clock and Kuan at 6 o'clock.

Exercises 60–62 refer to the graph $C$ of the parametric equations

$$x = tc + (1 - t)a, \quad y = td + (1 - t)b$$

where $P_1(a, b)$ and $P_2(c, d)$ are two fixed points.

**60. Using Parametric Equations in Geometry** Show that the point $P(x, y)$ on $C$ is equal to

**(a)** $P_1(a, b)$ when $t = 0$.     **(b)** $P_2(c, d)$ when $t = 1$.

**61. Using Parametric Equations in Geometry** Show that if $t = 0.5$ the corresponding point $(x, y)$ on $C$ is the midpoint of the line segment with endpoints $(a, b)$ and $(c, d)$.

**62.** What values of $t$ will find two points that divide the line segment $P_1P_2$ into three equal pieces? Four equal pieces?

---

## 6.4 Polar Coordinates

Polar Coordinate System • Coordinate Conversion • Equation Conversion • Finding Distance Using Polar Coordinates

### Polar Coordinate System

A **polar coordinate system** is a plane with a point $O$, the **pole**, and a ray from $O$, the **polar axis**, as shown in Figure 6.38. Each point $P$ in the plane is assigned **polar coordinates** as follows: $r$ is the **directed distance** from $O$ to $P$, and $\theta$ is the **directed angle** whose initial side is on the polar axis and whose terminal side is on the line $OP$.

As in trigonometry, we measure $\theta$ as positive when moving counterclockwise and negative when moving clockwise. If $r > 0$, then $P$ is on the terminal side of $\theta$. If $r < 0$, then $P$ is on the terminal side of $\theta + \pi$. We can use radian or degree measure for the angle $\theta$ as illustrated in Example 1.

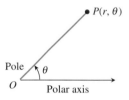

**Figure 6.38** The polar coordinate system.

**Example 1 PLOTTING POINTS IN THE POLAR COORDINATE SYSTEM**

Plot the points with the given polar coordinates.

**(a)** $P(2, \pi/3)$     **(b)** $Q(-1, 3\pi/4)$     **(c)** $R(3, -45°)$

**Solution** Figure 6.39 shows the three points.

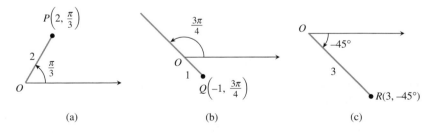

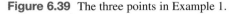

**Figure 6.39** The three points in Example 1.

Each polar coordinate pair determines a unique point. However, the polar coordinates of a point $P$ in the plane are not unique.

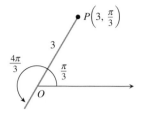

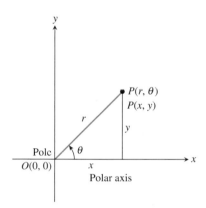

**Figure 6.40** The point *P* in Example 2.

**Figure 6.41** Polar and rectangular coordinates for *P*.

## Example 2 FINDING SEVERAL POLAR COORDINATES FOR A POINT

If the point *P* has polar coordinates $(3, \pi/3)$, find two additional pairs of polar coordinates for *P*.

**Solution** Point *P* is shown in Figure 6.40. Two additional pairs of polar coordinates for *P* are

$$\left(3, \frac{\pi}{3} + 2\pi\right) = \left(3, \frac{7\pi}{3}\right) \quad \text{and} \quad \left(-3, \frac{\pi}{3} + \pi\right) = \left(-3, \frac{4\pi}{3}\right).$$

The coordinates $(r, \theta)$, $(r, \theta + 2\pi)$, and $(-r, \theta + \pi)$ all name the same point. In general, the point with polar coordinates $(r, \theta)$ also has polar coordinates

$$(r, \theta + 2n\pi) \quad \text{or} \quad (-r, \theta + (2n + 1)\pi),$$

where *n* is any integer. In particular, the pole has polar coordinates $(0, \theta)$, where $\theta$ is any angle.

## Coordinate Conversion

When we use both polar coordinates and Cartesian coordinates, the pole is the origin and the polar axis is the positive *x*-axis as shown in Figure 6.41. By applying trigonometry we can find equations that relate the polar coordinates $(r, \theta)$ and the rectangular coordinates $(x, y)$ of a point *P*.

> **Coordinate Conversion Equations**
>
> Let the point *P* have polar coordinates $(r, \theta)$ and rectangular coordinates $(x, y)$. Then
>
> $$x = r \cos \theta, \qquad r^2 = x^2 + y^2,$$
>
> $$y = r \sin \theta, \qquad \tan \theta = \frac{y}{x}.$$

These relationships allow us to convert from one coordinate system to the other.

## Example 3 CONVERTING FROM POLAR TO RECTANGULAR COORDINATES

Find the rectangular coordinates of the points with the given polar coordinates.

**(a)** $P(3, 5\pi/6)$ **(b)** $Q(2, -200°)$

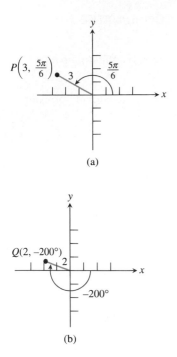

(a)

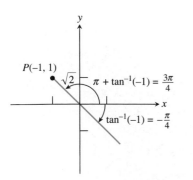

(b)

**Figure 6.42** The points $P$ and $Q$ in Example 3.

---

**Notes on Examples**

Example 4 provides an opportunity to monitor students' use of the inverse keys on their graphers. You may need to assist some students in the correct choice of the quadrant for this example.

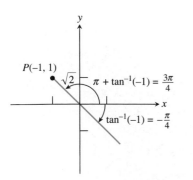

**Figure 6.43** The point $P$ in Example 4a.

---

**Solution**

**(a)** For $P(3, 5\pi/6)$, $r = 3$ and $\theta = 5\pi/6$:

$$x = r\cos\theta \qquad\qquad y = r\sin\theta$$

$$x = 3\cos\frac{5\pi}{6} \qquad \text{and} \quad y = 3\sin\frac{5\pi}{6}$$

$$x = 3\left(-\frac{\sqrt{3}}{2}\right) \approx -2.60 \qquad y = 3\left(\frac{1}{2}\right) = 1.5$$

The rectangular coordinates for $P$ are $(-3\sqrt{3}/2, 1.5) \approx (-2.60, 1.5)$ (Figure 6.42a)

**(b)** For $Q(2, -200°)$, $r = 2$ and $\theta = -200°$:

$$x = r\cos\theta \qquad\qquad y = r\sin\theta$$
$$\qquad\qquad\qquad \text{and}$$
$$x = 2\cos(-200°) \approx -1.88 \qquad y = 2\sin(-200°) \approx 0.68$$

The rectangular coordinates for $Q$ are approximately $(-1.88, 0.68)$ (Figure 6.42b).

When converting rectangular coordinates to polar coordinates, we must remember that there are infinitely many possible polar coordinate pairs. In Example 4 we report two of the possibilities.

**Example 4** CONVERTING FROM RECTANGULAR TO POLAR COORDINATES

Find two polar coordinate pairs for the points with given rectangular coordinates.

**(a)** $P(-1, 1)$ \qquad\qquad **(b)** $Q(-3, 0)$

**Solution**

**(a)** For $P(-1, 1)$, $x = -1$ and $y = 1$:

$$r^2 = x^2 + y^2 \qquad\qquad \tan\theta = \frac{y}{x}$$

$$r^2 = (-1)^2 + (1)^2 \qquad \tan\theta = \frac{-1}{1} = -1$$

$$r = \pm\sqrt{2} \qquad\qquad \theta = \tan^{-1}(-1) + n\pi = -\frac{\pi}{4} + n\pi$$

We use the angles $-\pi/4$ and $-\pi/4 + \pi = 3\pi/4$. Because $P$ is on the ray opposite the terminal side of $-\pi/4$, the value of $r$ corresponding to this angle is negative (Figure 6.43). Because $P$ is on the terminal side of $3\pi/4$, the value of $r$ corresponding to this angle is positive. So two polar coordinates pairs of point $P$ are

$$\left(-\sqrt{2}, -\frac{\pi}{4}\right) \quad \text{and} \quad \left(\sqrt{2}, \frac{3\pi}{4}\right).$$

**(b)** For $Q(-3, 0)$, $x = -3$ and $y = 0$. Thus, $r = \pm 3$ and $\theta = n\pi$. We use the angles $0$ and $\pi$. So two polar coordinates pairs for point $Q$ are

$$(-3, 0) \quad \text{and} \quad (3, \pi).$$

---

### Exploration 1  Using a Grapher to Convert Coordinates

Most graphers have the capability to convert polar coordinates to rectangular and vice-versa. Usually they give just one possible polar coordinate pair for a given rectangular coordinate pair.

1. Use your grapher to check the conversions in Examples 3 and 4.
2. Use your grapher to convert the polar coordinate pairs $(2, \pi/3)$, $(-1, \pi/2)$, $(2, \pi)$, $(-5, 3\pi/2)$, $(3, 2\pi)$, to rectangular coordinate pairs. $(1, \sqrt{3})$, $(0, -1)$, $(-2, 0)$, $(0, 5)$, $(3, 0)$
3. Use your grapher to convert the rectangular coordinate pairs $(-1, -\sqrt{3})$, $(0, 2)$, $(3, 0)$, $(-1, 0)$, $(0, -4)$ to polar coordinate pairs. $(-2, \pi/3)$, $(2, \pi/2)$, $(3, 0)$, $(1, \pi)$, $(4, 3\pi/2)$

---

## Equation Conversion

We can use the Coordinate Conversion Equations to convert polar form to rectangular form and vice-versa. For example, the polar equation $r = 4 \cos \theta$ can be converted to rectangular form as follows:

$$r = 4 \cos \theta$$
$$r^2 = 4r \cos \theta$$
$$x^2 + y^2 = 4x \qquad r^2 = x^2 + y^2, r \cos \theta = x$$
$$x^2 - 4x + 4 + y^2 = 4$$
$$(x - 2)^2 + y^2 = 4$$

Thus the graph of $r = 4 \cos \theta$ is all or part of the circle with center $(2, 0)$ and radius 2.

Figure 6.44 shows the graph of $r = 4 \cos \theta$ for $0 \le \theta \le 2\pi$ obtained using the polar graphing mode of our grapher. So, the graph of $r = 4 \cos \theta$ is the entire circle.

Just as with parametric equations, the domain of a polar equation in $r$ and $\theta$ is understood to be all values of $\theta$ for which the corresponding values of $r$ are real numbers. You must also select a value for $\theta$ min and $\theta$ max to graph in polar mode.

You may be surprised by the polar form for a vertical line in Example 5.

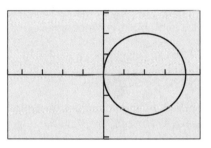

[−4.7, 4.7] by [−3.1, 3.1]

**Figure 6.44**  The graph of the polar equation $r = 4 \cos \theta$ in $0 \le \theta \le 2\pi$.

### Example 5  CONVERTING FROM POLAR FORM TO RECTANGULAR FORM

Convert $r = 4 \sec \theta$ to rectangular form and identify the graph. Support your answer with a polar graphing utility.

**Solution**

$$r = 4 \sec \theta$$
$$\frac{r}{\sec \theta} = 4$$
$$r \cos \theta = 4$$
$$x = 4 \qquad r \cos \theta = x$$

The graph is the vertical line $x = 4$ (Figure 6.45).

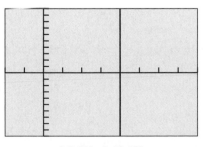

[−2, 8] by [−10, 10]

**Figure 6.45**  The graph of the vertical line $r = 4 \sec \theta$ ($x = 4$). (Example 5)

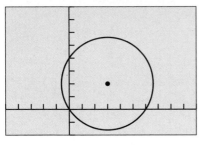

[−5, 10] by [−2, 8]

**Figure 6.46** The graph of the circle
$r = 6 \cos \theta + 4 \sin \theta$. (Example 6)

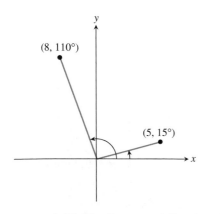

**Figure 6.47** The distance and direction of
two airplanes from a radar source.
(Example 7)

### Example 6  CONVERTING FROM RECTANGULAR FORM TO POLAR FORM

Convert $(x - 3)^2 + (y - 2)^2 = 13$ to polar form.

**Solution**

$$(x - 3)^2 + (y - 2)^2 = 13$$
$$x^2 - 6x + 9 + y^2 - 4y + 4 = 13$$
$$x^2 + y^2 - 6x - 4y = 0$$

Substituting $r^2$ for $x^2 + y^2$, $r \cos \theta$ for $x$, and $r \sin \theta$ for $y$ gives the following:

$$r^2 - 6r \cos \theta - 4r \sin \theta = 0$$
$$r(r - 6 \cos \theta - 4 \sin \theta) = 0$$
$$r = 0 \quad \text{or} \quad r - 6 \cos \theta - 4 \sin \theta = 0$$

The graph of $r = 0$ consists of a single point, the origin, which is also on the graph of $r - 6 \cos \theta - 4 \sin \theta = 0$. Thus, the polar form is

$$r = 6 \cos \theta + 4 \sin \theta.$$

The graph of $r = 6 \cos \theta + 4 \sin \theta$ for $0 \le \theta \le 2\pi$ is shown in Figure 6.46 and appears to be a circle with center $(3, 2)$ and radius $\sqrt{13}$ as expected.

## Finding Distance Using Polar Coordinates

A radar tracking system sends out high-frequency radio waves and receives their reflection from an object. The distance and direction of the object from the radar is often given in polar coordinates.

### Example 7  USING A RADAR TRACKING SYSTEM

Radar detects two airplanes at the same altitude. Their polar coordinates are (8 mi, 110°) and (5 mi, 15°). (See Figure 6.47.) How far apart are the airplanes?

**Solution**  By the Law of Cosines (Section 5.6),

$$d^2 = 8^2 + 5^2 - 2 \cdot 8 \cdot 5 \cos (110° - 15°)$$
$$d = \sqrt{8^2 + 5^2 - 2 \cdot 8 \cdot 5 \cos 95°}$$
$$d \approx 9.80$$

The airplanes are about 9.80 mi apart.

We can also use the Law of Cosines to derive a formula for the distance between points in the polar coordinate system. See Exercise 55.

# Quick Review 6.4

In Exercises 1 and 2, determine the quadrants containing the terminal side of the angles.

**1. (a)** $5\pi/6$   II     **(b)** $-3\pi/4$   III

**2. (a)** $-300°$   I     **(b)** $210°$   III

In Exercises 3–6, find a positive and a negative angle coterminal with the given angle.

**3.** $-\pi/4$   $7\pi/4, -9\pi/4$     **4.** $\pi/3$   $7\pi/3, -5\pi/3$

**5.** $160°$   $520°, -200°$     **6.** $-120°$   $240°, -480°$

In Exercises 7 and 8, write a standard form equation for the circle.

**7.** Center $(3, 0)$ and radius 2     **8.** Center $(0, -4)$ and radius 3

In Exercises 9 and 10, use The Law of Cosines to find the measure of the third side of the given triangle.

**9.**                                               **10.**

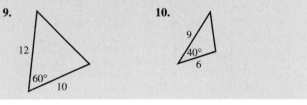

# Section 6.4 Exercises

In Exercises 1–4, the polar coordinates of a point are given. Find its rectangular coordinates.

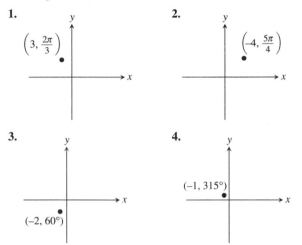

**11.** $(2, 30°)$               **12.** $(3, 210°)$

**13.** $(-2, 120°)$         **14.** $(-3, 135°)$

In Exercises 15–22, find the rectangular coordinates of the point with given polar coordinates.

**15.** $(1.5, 7\pi/3)$         **16.** $(2.5, 17\pi/4)$

**17.** $(-3, -29\pi/7)$     **18.** $(-2, -14\pi/5)$

**19.** $(-2, \pi)$   $(2, 0)$      **20.** $(1, \pi/2)$   $(0, 1)$

**21.** $(2, 270°)$   $(0, -2)$     **22.** $(-3, 360°)$   $(-3, 0)$

In Exercises 23–26, polar coordinates of point $P$ are given. Find all of its polar coordinates.

**23.** $P = (2, \pi/6)$         **24.** $P = (1, -\pi/4)$

**25.** $P = (1.5, -20°)$     **26.** $P = (-2.5, 50°)$

In Exercises 27–30, rectangular coordinates of point $P$ are given. Find all polar coordinates of $P$ that satisfy

**(a)** $0 \le \theta \le 2\pi$     **(b)** $-\pi \le \theta \le \pi$     **(c)** $0 \le \theta \le 4\pi$

**27.** $P = (1, 1)$            **28.** $P = (1, 3)$

**29.** $P = (-2, 5)$         **30.** $P = (-1, -2)$

In Exercises 5 and 6, **(a)** complete the table for the polar equation and **(b)** plot the corresponding points.

**5.** $r = 3 \sin \theta$

| $\theta$ | $\pi/4$ | $\pi/2$ | $5\pi/6$ | $\pi$ | $4\pi/3$ | $2\pi$ |
|---|---|---|---|---|---|---|
| $r$ | | | | | | |

**6.** $r = 2 \csc \theta$

| $\theta$ | $\pi/4$ | $\pi/2$ | $5\pi/6$ | $\pi$ | $4\pi/3$ | $2\pi$ |
|---|---|---|---|---|---|---|
| $r$ | | | | | | |

In Exercises 7–14, plot the point with the given polar coordinates.

**7.** $(3, 4\pi/3)$           **8.** $(2, 5\pi/6)$

**9.** $(-1, 2\pi/5)$        **10.** $(-3, 17\pi/10)$

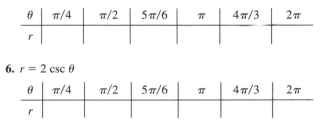

In Exercises 31–34, use your grapher to match the polar equation with its graph.

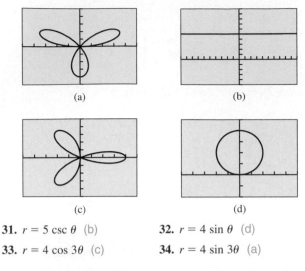

(a)

(b)

(c)

(d)

**31.** $r = 5 \csc \theta$  (b)

**32.** $r = 4 \sin \theta$  (d)

**33.** $r = 4 \cos 3\theta$  (c)

**34.** $r = 4 \sin 3\theta$  (a)

In Exercises 35–42, convert the polar equation to rectangular form and identify the graph. Support your answer by graphing the polar equation.

**35.** $r = 3 \sec \theta$

**36.** $r = -2 \csc \theta$

**37.** $r = -3 \sin \theta$

**38.** $r = -4 \cos \theta$

**39.** $r \csc \theta = 1$

**40.** $r \sec \theta = 3$

**41.** $r = 2 \sin \theta - 4 \cos \theta$

**42.** $r = 4 \cos \theta - 4 \sin \theta$

In Exercises 43–50, convert the rectangular equation to polar form. Graph the polar equation.

**43.** $x = 2$

**44.** $x = 5$

**45.** $2x - 3y = 5$

**46.** $3x + 4y = 2$

**47.** $(x - 3)^2 + y^2 = 9$

**48.** $x^2 + (y - 1)^2 = 1$

**49.** $(x + 3)^2 + (y + 3)^2 = 18$

**50.** $(x - 1)^2 + (y + 4)^2 = 17$

**51. Tracking Airplanes** The location, given in polar coordinates, of two planes approaching the Vicksburg airport are (4 mi, 12°) and (2 mi, 72°). Find the distance between the airplanes. $2\sqrt{3} \approx 3.46$ mi

**52. Tracking Ships** The location of two ships from Mays Landing Lighthouse, given in polar coordinates, are (3 mi, 170°) and (5 mi, 150°). Find the distance between the ships. $\approx 2.410$ mi

**53. Using Polar Coordinates in Geometry** A square with sides of length $a$ and center at the origin has two sides parallel to the $x$-axis. Find polar coordinates of the vertices.

**54. Using Polar Coordinates in Geometry** A regular pentagon whose center is at the origin has one vertex on the positive $x$-axis at a distance $a$ from the center. Find polar coordinates of the vertices.

## Explorations

**55. Polar Distance Formula** Let $P_1$ and $P_2$ have polar coordinates $(r_1, \theta_1)$ and $(r_2, \theta_2)$, respectively.

**(a)** If $\theta_1 - \theta_2$ is a multiple of $\pi$, write a formula for the distance between $P_1$ and $P_2$.

**(b)** Use the Law of Cosines to prove that the distance between $P_1$ and $P_2$ is given by

$$d = \sqrt{r_1^2 + r_2^2 - 2r_1r_2 \cos (\theta_1 - \theta_2)}.$$

**(c) Writing to Learn** Does the formula in (b) agree with the formula(s) you found in (a)? Explain.

**56. Watching Your $\theta$-Step** Consider the polar curve $r = 4 \sin \theta$. Describe the graph for each of the following.

**(a)** $0 \le \theta \le \pi/2$

**(b)** $0 \le \theta \le 3\pi/4$

**(c)** $0 \le \theta \le 3\pi/2$

**(d)** $0 \le \theta \le 4\pi$

In Exercises 57–60, use the results of Exercise 55 to find the distance between the points with given polar coordinates.

**57.** (2, 10°), (5, 130°)  $\approx 6.25$

**58.** (4, 20°), (6, 65°)  $\approx 4.25$

**59.** (−3, 25°), (−5, 160°)

**60.** (6, −35°), (8, −65°)  ∎

## Extending the Ideas

**61. Graphing Polar Equations Parametrically** Find parametric equations for the polar curve $r = f(\theta)$.

**Group Activity** In Exercises 62–65, use what you learned in Exercise 61 to write parametric equations for the given polar equation. Support your answers graphically.

**62.** $r = 2 \cos \theta$

**63.** $r = 5 \sin \theta$

**64.** $r = 2 \sec \theta$

**65.** $r = 4 \csc \theta$

# Graphs of Polar Equations

Symmetry • Analyzing Polar Graphs • Rose Curves • Limaçon Curves • Other Polar Curves

## Symmetry

You learned algebraic tests for symmetry for equations in rectangular form For example, if replacing *x* by $-x$ in an equation in *x* and *y* produces an equivalent equation, then the graph of the original equation is symmetric with respect to the *y*-axis. Similar tests were given for symmetry with respect to the *x*-axis and the origin in Section 1.2. Algebraic tests also exist for polar form.

Figure 6.48 shows a rectangular coordinate system superimposed on a polar coordinate system, with the origin and the pole coinciding and the positive *x*-axis and the polar axis coinciding.

The three types of symmetry figures to be considered will have are:

1. The *x*-axis (polar axis) as a line of symmetry (Figure 6.48a).

2. The *y*-axis (the line $\theta = \pi/2$) as a line of symmetry (Figure 6.48b).

3. The origin (the pole) as a point of symmetry (Figure 6.48c).

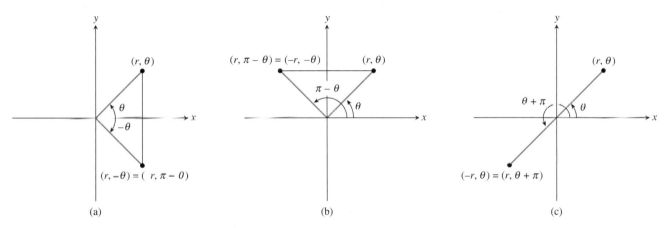

(a)                                    (b)                                    (c)

**Figure 6.48**  Symmetry with respect to (a) the *x*-axis (polar axis), (b) the *y*-axis (the line $\theta = \pi/2$), and (c) the origin (the pole).

All three algebraic tests for symmetry in polar forms require replacing the pair $(r, \theta)$, which satisfies the polar equation, with another coordinate pair and determining whether it also satisfies the polar equation.

### Symmetry Tests for Polar Graphs

The graph of a polar equation has the indicated symmetry if either replacement produces an equivalent polar equation.

| To Test for Symmetry | Replace | By |
|---|---|---|
| **1.** about the *x*-axis, | $(r, \theta)$ | $(r, -\theta)$ or $(-r, \pi - \theta)$. |
| **2.** about the *y*-axis, | $(r, \theta)$ | $(-r, -\theta)$ or $(r, \pi - \theta)$. |
| **3.** about the origin, | $(r, \theta)$ | $(-r, \theta)$ or $(r, \theta + \pi)$. |

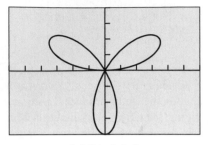

[−6,6] by [−4, 4]

**Figure 6.49** The graph of $r = 4 \sin 3\theta$ is symmetric about the $y$-axis. (Example 1)

---

**Teaching Note**

It is customary to draw graphs of polar curves in radian mode.

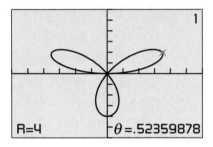

[−6,6] by [−6, 6]
(a)

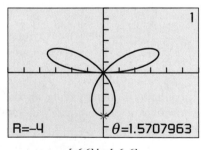

[−6,6] by [−6, 6]
(b)

**Figure 6.50** The values of $r$ in $r = 4 \sin 3\theta$ vary from (a) 4 to (b) −4.

## Example 1   TESTING FOR SYMMETRY

Use the symmetry tests to prove that the graph of $r = 4 \sin 3\theta$ is symmetric about the $y$-axis.

**Solution** Figure 6.49 suggests that the graph of $r = 4 \sin 3\theta$ is symmetric about the $y$-axis and not symmetric about the $x$-axis or origin.

$$r = 4 \sin 3\theta$$

$$-r = 4 \sin 3(-\theta) \qquad \text{Replace } (r, \theta) \text{ by } (-r, -\theta).$$

$$-r = 4 \sin (-3\theta)$$

$$-r = -4 \sin 3\theta \qquad \sin \theta \text{ is an odd function of } \theta.$$

$$r = 4 \sin 3\theta \qquad \text{(Same as original.)}$$

Because the equations $-r = 4 \sin 3(-\theta)$ and $r = 4 \sin 3\theta$ are equivalent, there is symmetry about the $y$-axis.

## Analyzing Polar Graphs

We analyze graphs of polar equations in much the same way that we analyze the graphs of rectangular equations. For example, the function $r$ of Example 1 is a continuous function of $\theta$. Also $r = 0$ when $\theta = 0$ and when $\theta$ is any integer multiple of $\pi/3$. The domain of this function is the set of all real numbers.

Trace can be used to help determine the range of this polar function (Figure 6.50). It can be shown that $-4 \le r \le 4$.

Usually, we are more interested in the maximum value of $|r|$ rather than the range of $r$ in polar equations. In this case, $|r| \le 4$ so we can conclude that the graph is bounded.

A maximum value for $|r|$ is a **maximum $r$-value** for a polar equation. A maximum $r$-value occurs at a point on the curve that is the maximum distance from the pole. In Figure 6.50, a maximum $r$-value occurs at $(4, \pi/6)$ and $(-4, \pi/2)$. In fact, we get a maximum $r$-value at every $(r, \theta)$ which represents the tip of one of the three petals.

To find maximum $r$-values we must find maximum values of $|r|$ as opposed to the directed distance $r$. Example 2 shows one way to find maximum $r$-values graphically.

## Example 2   FINDING MAXIMUM $r$-VALUES

Find the maximum $r$-value of $r = 2 + 2 \cos \theta$.

**Solution** Figure 6.51a shows the graph of $r = 2 + 2 \cos \theta$ for $0 \le \theta \le 2\pi$. Because we are only interested in the values of $r$, we use the graph of the rectangular equation $y = 2 + 2 \cos x$ (Figure 6.51b). From this graph we can see that the maximum value of $r$, or $y$, is 4. It occurs when $\theta$ is any multiple of $2\pi$.

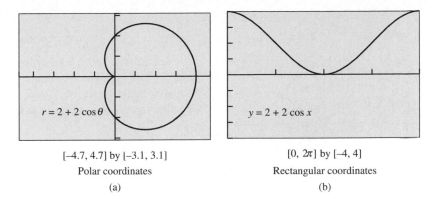

[−4.7, 4.7] by [−3.1, 3.1]
Polar coordinates
(a)

[0, 2π] by [−4, 4]
Rectangular coordinates
(b)

**Figure 6.51** With $\theta = x$, the *y*-values in (b) are the same as the directed distance from the pole to $(r, \theta)$ in (a).

### Example 3  FINDING MAXIMUM *r*-VALUES

Identify the points on the graph of $r = 3 \cos 2\theta$ for $0 \le \theta \le 2\pi$ that give maximum *r*-values.

**Solution**  Using trace in Figure 6.52 we can show that there are four points on the graph of $r = 3 \cos 2\theta$ in $0 \le \theta < 2\pi$ at maximum distance of 3 from the pole:

$$(3, 0), \quad (-3, \pi/2), \quad (3, \pi), \quad \text{and} \quad (-3, 3\pi/2).$$

Figure 6.53a shows the directed distances *r* as the *y*-values of $y_1 = 3 \cos 2x$, and Figure 6.53b shows the distances $|r|$ as the *y*-values of $y_2 = |3 \cos 2x|$. There are four maximum values of $y_2$ (i.e., $|r|$) in (b) corresponding to the four extreme values of $y_1$ (i.e., *r*) in (a).

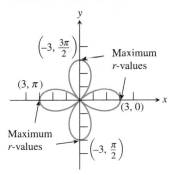

**Figure 6.52** The graph of $r = 3 \cos 2\theta$. (Example 3)

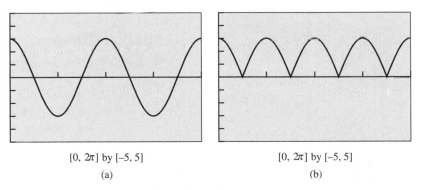

[0, 2π] by [−5, 5]
(a)

[0, 2π] by [−5, 5]
(b)

**Figure 6.53** The graph of (a) $y_1 = 3 \cos 2x$ and (b) $y_2 = |3 \cos 2x|$. (Example 3)

## Rose Curves

The curve in Example 1 is a 3-petal rose curve and the curve in Example 3 is a 4-petal rose curve. The graphs of the polar equations $r = a \cos n\theta$ and $r = a \sin n\theta$, where *n* is an integer greater than 1, are **rose curves**. If *n* is odd there are *n* petals, and if *n* is even there are 2*n* petals.

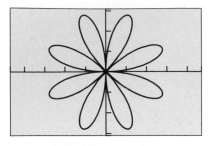

[−4.7, 4.7] by [−3.1, 3.1]

**Figure 6.54** The graph of 8-petal rose curve $r = 3 \sin 4\theta$. (Example 4)

### Example 4 ANALYZING A ROSE CURVE

Analyze the graph of the rose curve $r = 3 \sin 4\theta$.

**Solution** Figure 6.54 shows the graph of the 8-petal rose curve $r = 3 \sin 4\theta$. The maximum $r$-value is 3. The graph appears to be symmetric about the $x$-axis, $y$-axis, and the origin. For example, to prove that the graph is symmetric about the $x$-axis we replace $(r, \theta)$ by $(-r, \pi - \theta)$:

$$r = 3 \sin 4\theta$$
$$-r = 3 \sin 4(\pi - \theta)$$
$$-r = 3 \sin (4\pi - 4\theta)$$
$$-r = 3[\sin 4\pi \cos 4\theta - \cos 4\pi \sin 4\theta]$$
$$-r = 3[(0) \cos 4\theta - (1) \sin 4\theta]$$
$$-r = -3 \sin 4\theta$$
$$r = 3 \sin 4\theta$$

Because the new polar equation is the same as the original equation, the graph is symmetric about the $x$-axis. In a similar way, you can prove that the graph is symmetric about the $y$-axis and the origin. (See Exercise 58.)

Domain: All reals.
Range: $[-3, 3]$
Symmetric about the $x$-axis, the $y$-axis, and the origin.
Continuous
Bounded
Maximum $r$-value: 3
No asymptotes.

## Limaçon Curves

The **limaçon curves** are graphs of polar equations of the form

$$r = a \pm b \sin \theta \quad \text{and} \quad r = a \pm b \cos \theta,$$

where $a > 0$ and $b > 0$. *Limaçon*, pronounced "leemasahn," is Old French for "snail." There are four different shapes of limaçons, as illustrated in Figure 6.55.

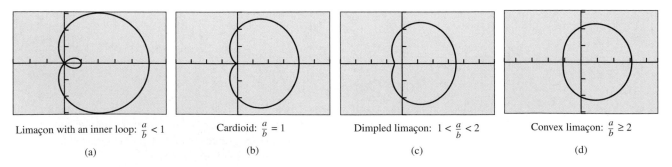

Limaçon with an inner loop: $\frac{a}{b} < 1$

(a)

Cardioid: $\frac{a}{b} = 1$

(b)

Dimpled limaçon: $1 < \frac{a}{b} < 2$

(c)

Convex limaçon: $\frac{a}{b} \geq 2$

(d)

**Figure 6.55** The four types of limaçons.

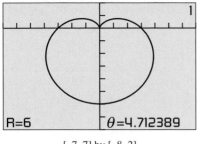

[−7, 7] by [−8, 2]

**Figure 6.56**  The graph of the cardiod of Example 5.

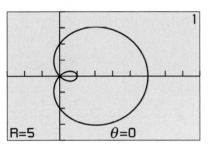

[−3, 8] by [−4, 4]

**Figure 6.57**  The graph of a limaçon with an inner loop. (Example 6)

### Example 5  ANALYZING A LIMAÇON CURVE

Analyze the graph of $r = 3 - 3 \sin \theta$.

**Solution**  We can see from Figure 6.56 that the curve is a cardioid with maximum *r*-value 6. The graph is symmetric only about the *y*-axis.

Domain:  All reals.
Range: $[0, 6]$
Symmetric about the *y*-axis.
Continuous
Bounded
Maximum *r*-value: 6
No asymptotes.

### Example 6  ANALYZING A LIMAÇON CURVE

Analyze the graph of $r = 2 + 3 \cos \theta$.

**Solution**  We can see from Figure 6.57 that the curve is a limaçon with an inner loop and maximum *r*-value 5. The graph is symmetric only about the *x*-axis.

Domain:  All reals.
Range: $[-1, 5]$
Symmetric about the *x*-axis.
Continuous
Bounded
Maximum *r*-value: 5
No asymptotes.

## Other Polar Curves

All the polar curves we have graphed so far have been bounded. The spiral in Example 7 is unbounded.

### Example 7  ANALYZING THE SPIRAL OF ARCHIMEDES

Analyze the graph of $r = \theta$.

**Solution**  We can see from Figure 6.58 that the curve has no maximum *r*-value and is symmetric about the *y*-axis.

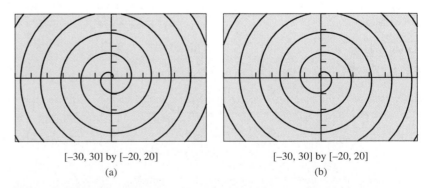

[−30, 30] by [−20, 20]          [−30, 30] by [−20, 20]
(a)                              (b)

**Figure 6.58**  The graph of $r = \theta$ for (a) $\theta \geq 0$ (set $\theta$min = 0, $\theta$max = 45, $\theta$step = 0.1) and (b) $\theta \leq 0$ (set $\theta$min = −45, $\theta$max = 0, $\theta$step = 0.1). (Example 7)

Domain: All reals.
Range: All reals.
Symmetric about the *y*-axis.
Continuous
Unbounded
No maximum *r*-value.
No asymptotes.

The **lemniscate curves** are graphs of polar equations of the form

$$r^2 = a^2 \sin 2\theta \quad \text{and} \quad r^2 = a^2 \cos 2\theta.$$

### Example 8 ANALYZING A LEMNISCATE CURVE

Analyze the graph of $r^2 = 4 \cos 2\theta$ for $[0, 2\pi]$.

**Solution** It turns out that you can get the complete graph using $r = 2\sqrt{\cos 2\theta}$. You also need to choose a very small $\theta$ step to produce the graph in Figure 6.59.

Domain: $[0, \pi/4] \cup [3\pi/4, 5\pi/4] \cup [7\pi/4, 2\pi]$
Range: $[-2, 2]$
Symmetric about the *x*-axis, the *y*-axis, and the origin.
Continuous
Bounded
Maximum *r*-value: 2
No asymptotes.

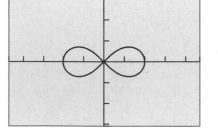

[−4.7, 4.7] by [−3.1, 3.1]

**Figure 6.59** The graph of the lemniscate $r^2 = 4 \cos 2\theta$. (Example 8)

**Exploration Extensions**

Graph $r^2 = 4 \sin 2\theta$. How is this graph related to the graph of $r^2 = 4 \cos 2\theta$?

---

**Exploration 1** **Revisiting Example 8**

1. Prove that $\theta$-values in the intervals $(\pi/4, 3\pi/4)$ and $(5\pi/4, 7\pi/4)$ are not in the domain of the polar equation $r^2 = 4 \cos 2\theta$.

2. Explain why $r = -2\sqrt{\cos 2\theta}$ produces the same graph as $r = 2\sqrt{\cos 2\theta}$ in the interval $[0, 2\pi]$.

3. Use the symmetry tests to show that the graph of $r^2 = 4 \cos 2\theta$ is symmetric about the *x*-axis.

4. Use the symmetry tests to show that the graph of $r^2 = 4 \cos 2\theta$ is symmetric about the *y*-axis.

5. Use the symmetry tests to show that the graph of $r^2 = 4 \cos 2\theta$ is symmetric about the origin.

---

# Quick Review 6.5

In Exercises 1–4, find the absolute maximum value and absolute minimum value in $[0, 2\pi]$ and where they occur.

**1.** $y = 3 \cos 2x$      **2.** $y = 2 + 3 \cos x$

**3.** $y = 2\sqrt{\cos 2x}$      **4.** $y = 3 - 3 \sin x$

In Exercises 5 and 6, determine if the graph of the function is symmetric about the **(a)** *x*-axis, **(b)** *y*-axis, **(c)** origin.

**5.** $y = \sin 2x$   no; no; yes      **6.** $y = \cos 4x$   no; yes; no

In Exercises 7–10, use trig identities to simplify the expression.

**7.** $\sin(\pi - \theta)$   $\sin \theta$      **8.** $\cos(\pi - \theta)$   $-\cos \theta$

**9.** $\cos 2(\pi + \theta)$      **10.** $\sin 2(\pi + \theta)$

# Section 6.5 Exercises

In Exercises 1 and 2, **(a)** complete the table for the polar equation and **(b)** plot the corresponding points.

**1.** $r = 3 \cos 2\theta$

| $\theta$ | 0 | $\pi/4$ | $\pi/2$ | $3\pi/4$ | $\pi$ | $5\pi/4$ | $3\pi/2$ | $7\pi/4$ |
|---|---|---|---|---|---|---|---|---|
| $r$ | | | | | | | | |

**2.** $r = 2 \sin 3\theta$

| $\theta$ | 0 | $\pi/6$ | $\pi/3$ | $\pi/2$ | $2\pi/3$ | $5\pi/6$ | $\pi$ |
|---|---|---|---|---|---|---|---|
| $r$ | | | | | | | |

In Exercises 3–6, draw a graph of the rose curve. State the smallest $\theta$-interval $(0 \le \theta \le k)$ that will produce a complete graph.

**3.** $r = 3 \sin 3\theta$

**4.** $r = -3 \cos 2\theta$

**5.** $r = 3 \cos 2\theta$

**6.** $r = 3 \sin 5\theta$

Exercises 7 and 8 refer to the curves in the given figure.

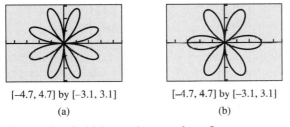

|   |   |
|---|---|
| [–4.7, 4.7] by [–3.1, 3.1] | [–4.7, 4.7] by [–3.1, 3.1] |
| (a) | (b) |

**7.** The graphs of which equations are shown?

$$r_1 = 3 \cos 6\theta \quad r_2 = 3 \sin 8\theta \quad r_3 = 3|\cos 3\theta|$$

**8.** Use trigonometric identities to explain which of these curves is the graph of $r = 6 \cos 2\theta \sin 2\theta$. (a)

In Exercises 9–12, match the equation with its graph without using your graphing calculator.

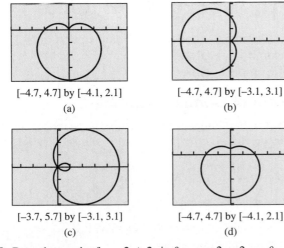

|   |   |
|---|---|
| [–4.7, 4.7] by [–4.1, 2.1] | [–4.7, 4.7] by [–3.1, 3.1] |
| (a) | (b) |
| [–3.7, 5.7] by [–3.1, 3.1] | [–4.7, 4.7] by [–4.1, 2.1] |
| (c) | (d) |

**9.** Does the graph of $r = 2 + 2 \sin \theta$ or $r = 2 - 2 \cos \theta$ appear in the figure? Explain.

**10.** Does the graph of $r = 2 + 3 \cos \theta$ or $r = 2 - 3 \cos \theta$ appear in the figure? Explain.

**11.** Is the graph in (a) the graph of $r = 2 - 2 \sin \theta$ or $r = 2 + 2 \cos \theta$? Explain.

**12.** Is the graph in (d) the graph of $r = 2 + 1.5 \cos \theta$ or $r = 2 - 1.5 \sin \theta$? Explain.

In Exercises 13–16, identify the points for $0 \le \theta \le 2\pi$ where maximum $r$-values occur on the graph of the polar equation.

**13.** $r = 3 \cos 3\theta$

**14.** $r = 4 \sin 2\theta$

**15.** $r = 2 + 3 \cos \theta$

**16.** $r = -3 + 2 \sin \theta$

In Exercises 17–24, use the polar symmetry tests to determine if the graph is symmetric about the $x$-axis, the $y$-axis, or the origin.

**17.** $r = 3 + 3 \sin \theta$

**18.** $r = 1 + 2 \cos \theta$

**19.** $r = 4 - 3 \cos \theta$

**20.** $r = 1 - 3 \sin \theta$

**21.** $r = 5 \cos 2\theta$

**22.** $r = 7 \sin 3\theta$

**23.** $r = \dfrac{3}{1 + \sin \theta}$

**24.** $r = \dfrac{2}{1 - \cos \theta}$

In Exercises 25–44, analyze the graph of the polar curve.

**25.** $r = 3$

**26.** $r = -2$

**27.** $\theta = \pi/3$

**28.** $\theta = -\pi/4$

**29.** $r = 2 \sin 3\theta$

**30.** $r = -3 \cos 4\theta$

**31.** $r = 5 + 4 \sin \theta$

**32.** $r = 6 - 5 \cos \theta$

**33.** $r = 4 + 4 \cos \theta$

**34.** $r = 5 - 5 \sin \theta$

**35.** $r = 5 + 2 \cos \theta$

**36.** $r = 3 - \sin \theta$

**37.** $r = 2 + 3 \cos \theta$

**38.** $r = 3 - 4 \sin \theta$

**39.** $r = 1 - \cos \theta$

**40.** $r = 2 + \sin \theta$

**41.** $r = 2\theta$

**42.** $r = \theta/4$

**43.** $r^2 = \sin 2\theta, 0 \le \theta \le 2\pi$

**44.** $r^2 = 9 \cos 2\theta, 0 \le \theta \le 2\pi$

In Exercises 45–48, find the length of each petal of the polar curve.

**45.** $r = 2 + 4 \sin 2\theta$

**46.** $r = 3 - 5 \cos 2\theta$

**47.** $r = 1 - 4 \cos 5\theta$

**48.** $r = 3 + 4 \sin 5\theta$

In Exercises 49–52, select the two equations whose graphs are the same curve. Then, even though the graphs of the equations are identical, describe how the two paths are different as $\theta$ increases from 0 to $2\pi$.

**49.** $r_1 = 1 + 3 \sin \theta, \quad r_2 = -1 + 3 \sin \theta, \quad r_3 = 1 - 3 \sin \theta$

**50.** $r_1 = 1 + 2 \cos \theta, \quad r_2 = -1 - 2 \cos \theta, \quad r_3 = -1 + 2 \cos \theta$

**51.** $r_1 = 1 + 2 \cos \theta, \quad r_2 = 1 - 2 \cos \theta, \quad r_3 = -1 - 2 \cos \theta$

**52.** $r_1 = 2 + 2 \sin \theta, \quad r_2 = -2 + 2 \sin \theta, \quad r_3 = 2 - 2 \sin \theta$

In Exercises 53–56, **(a)** describe the graph of the polar equation, **(b)** state any symmetry that the graph possesses, and **(c)** state its maximum $r$-value if it exists.

**53.** $r = 2 \sin^2 2\theta + \sin 2\theta$     **54.** $r = 3 \cos 2\theta - \sin 3\theta$

**55.** $r = 1 - 3 \cos 3\theta$     **56.** $r = 1 + 3 \sin 3\theta$

**57. Group Activity** Analyze the graphs of the polar equations $r = a \cos n\theta$ and $r = a \sin n\theta$ when $n$ is an even integer.

**58. Revisiting Example 4** Use the polar symmetry tests to prove that the graph of the curve $r = 3 \sin 4\theta$ is symmetric about the $y$-axis and the origin.

**59. Writing to Learn Revisiting Example 5** Confirm the range stated for the polar function $r = 3 - 3 \sin \theta$ of Example 5 by graphing $y = 3 - 3 \sin x$ for $0 \le x \le 2\pi$. Explain why this works.

**60. Writing to Learn Revisiting Example 6** Confirm the range stated for the polar function $r = 2 + 3 \cos \theta$ of Example 6 by graphing $y = 2 + 3 \cos x$ for $0 \le x \le 2\pi$. Explain why this works.

## Explorations

**61. Analyzing Rose Curves** Consider the polar equation $r = a \cos n\theta$ for $n$ an odd integer.

**(a)** Prove that the graph is symmetric about the $x$-axis.

**(b)** Prove that the graph is not symmetric about the $y$-axis.

**(c)** Prove that the graph is not symmetric about the origin.

**(d)** Prove that the maximum $r$-value is $|a|$.

**(e)** Analyze the graph of this curve.

**62. Analyzing Rose Curves** Consider the polar equation $r = a \sin n\theta$ for $n$ an odd integer.

**(a)** Prove that the graph is symmetric about the $y$-axis.

**(b)** Prove that the graph is not symmetric about the $x$-axis.

**(c)** Prove that the graph is not symmetric about the origin.

**(d)** Prove that the maximum $r$-value is $|a|$.

**(e)** Analyze the graph of this curve.

## Extending the Ideas

In Exercises 63–65, graph each polar equation. Describe how they are related to each other.

**63. (a)** $r_1 = 3 \sin 3\theta$     **(b)** $r_2 = 3 \sin 3\left(\theta + \dfrac{\pi}{12}\right)$

**(c)** $r_3 = 3 \sin 3\left(\theta + \dfrac{\pi}{4}\right)$

**64. (a)** $r_1 = 2 \sec \theta$     **(b)** $r_2 = 2 \sec \left(\theta - \dfrac{\pi}{4}\right)$

**(c)** $r_3 = 2 \sec \left(\theta - \dfrac{\pi}{3}\right)$

**65. (a)** $r_1 = 2 - 2 \cos \theta$     **(b)** $r_2 = r_1\left(\theta + \dfrac{\pi}{4}\right)$

**(c)** $r_3 = r_1\left(\theta + \dfrac{\pi}{3}\right)$

**66. Writing to Learn** Describe how the graphs of $r = f(\theta)$, $r = f(\theta + \alpha)$, and $r = f(\theta - \alpha)$ are related. Explain why you think this generalization is true.

**67. Extended Rose Curves** The graphs of $r_1 = 3 \sin ((5/2)\theta)$ and $r_2 = 3 \sin ((7/2)\theta)$ may be called rose curves.

**(a)** Determine the smallest $\theta$-interval that will produce a complete graph of $r_1$; of $r_2$.

**(b)** How many petals does each graph have?

---

| 6.6 | # De Moivre's Theorem and *n*th Roots |

Trigonometric Form of Complex Numbers • Multiplication and Division of Complex Numbers • Powers of Complex Numbers • Roots of Complex Numbers

## Trigonometric Form of Complex Numbers

The trigonometric form for the complex number $a + bi$ is related to the component form of the vector $a\mathbf{i} + b\mathbf{j}$ (Figure 6.60). In Section 6.1, we showed that

$$\mathbf{v} = (|\mathbf{v}| \cos \theta)\mathbf{i} + (|\mathbf{v}| \sin \theta)\mathbf{j}$$
$$= |\mathbf{v}| (\cos \theta \mathbf{i} + \sin \theta \mathbf{j})$$

We might call this the *trigonometric form* of the vector $\mathbf{v}$. It is not surprising that there is a similar form for complex numbers.

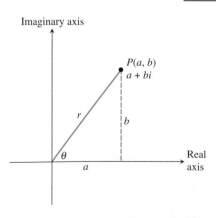

**Figure 6.60** The angle $\theta$ is the direction angle of the vector $\mathbf{v} = \langle a, b \rangle$.

**Objective**

Students will be able to represent complex numbers in trigonometric form and perform operations on them.

**Motivate**

Have students find all solutions of the equation $z^4 = 1$, where z is a complex number. (**z = ±1, z = ±i**)

**Polar Form**

The trigonometric form of a complex number z is also the **polar form** of z. In physics and engineering, the notation

$$\text{cis } \theta = \cos \theta + i \sin \theta$$

is often used.

**Lesson Guide**

Day 1: Trigonometric Form of Complex Numbers; Multiplication and Division of Complex Numbers; Powers of Complex Numbers
Day 2: Roots of Complex Numbers

**Teaching Note**

It may be useful to review complex numbers as introduced in Section 2.5.

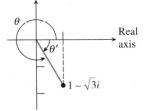

Imaginary axis

Real axis

$\theta$

$\theta'$

$1 - \sqrt{3}i$

**Figure 6.61** The complex number for Example 1a.

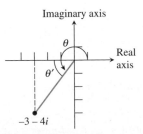

Imaginary axis

Real axis

$\theta$

$\theta'$

$-3 - 4i$

**Figure 6.62** The complex number for Example 1b.

In Figure 6.60, $\theta$ is the angle measured clockwise from the positive *x*-axis to the line segment from the origin to *P*. Because

$$a = r \cos \theta \quad \text{and} \quad b = r \sin \theta$$

we have

$$a + bi = (r \cos \theta)\mathbf{i} + (r \sin \theta)\mathbf{j}$$

where $r = |\mathbf{v}| = \sqrt{a^2 + b^2}$ and $\tan \theta = b/a$.

---

**Definition Trigonometric Form of a Complex Number**

The **trigonometric form** of the complex number $z = a + bi$ is

$$z = r(\cos \theta + i \sin \theta)$$

where $a = r \cos \theta$, $b = r \sin \theta$, $r = \sqrt{a^2 + b^2}$, and $\tan \theta = b/a$. The number *r* is the *absolute value* or *modulus* of z, and $\theta$ is an **argument** of z.

---

An angle $\theta$ for the trigonometric form of z can always be chosen so that $0 \le \theta \le 2\pi$, although any angle coterminal with $\theta$ could be used. Consequently, the *angle* $\theta$ and *argument* of a complex number z are not unique. It follows that the trigonometric form of a complex number z is not unique.

**Example 1 FINDING TRIGONOMETRIC FORMS**

Find the trigonometric form with $0 \le \theta < 2\pi$ for the complex number.

**(a)** $1 - \sqrt{3}i$ **(b)** $-3 - 4i$

**Solution**

**(a)** For $1 - \sqrt{3}i$,

$$r = |1 - \sqrt{3}i| = \sqrt{(1)^2 + (\sqrt{3})^2} = 2.$$

Because the reference angle $\theta'$ for $\theta$ is $-\pi/3$ (Figure 6.61),

$$\theta = 2\pi + \left(-\frac{\pi}{3}\right) = \frac{5\pi}{3}.$$

Thus,

$$1 - \sqrt{3}i = 2 \cos \frac{5\pi}{3} + 2i \sin \frac{5\pi}{3}.$$

**(b)** For $-3 - 4i$,

$$|-3 - 4i| = \sqrt{(-3)^2 + (-4)^2} = 5.$$

The reference angle $\theta'$ for $\theta$ (Figure 6.62) satisfies the equation

$$\tan \theta' = \frac{4}{3}, \quad \text{so}$$

$$\theta' = \tan^{-1} \frac{4}{3} = 0.927\ldots.$$

Because the terminal side of $\theta$ is in the third quadrant, we conclude that

$$\theta = \pi + \theta' \approx 4.07.$$

Therefore,

$$-3 - 4i \approx 5(\cos 4.07 + i \sin 4.07).$$

**Alert**

Some students may confuse the standard unit vector **i** with the imaginary unit *i*. In particular, it is important to recognize that **i** is a horizontal vector, while the imaginary axis is vertical.

**Teaching Note**

The proofs of the product and quotient formulas are good applications of the sum and difference identities studied in Section 5.3

**Teaching Note**

Many of the calculations discussed in this section can be performed using a grapher's built-in functions for converting between rectangular and polar coordinates.

# Multiplication and Division of Complex Numbers

The trigonometric form for complex numbers is particularly convenient for multiplying and dividing complex numbers. The product involves the product of the moduli and the sum of the arguments. (*Moduli* is the plural of *modulus*.) The quotient involves the quotient of the moduli and the difference of the arguments.

---

**Product and Quotient of Complex Numbers**

Let $z_1 = r_1(\cos \theta_1 + i \sin \theta_1)$ and $z_2 = r_2(\cos \theta_2 + i \sin \theta_2)$. Then

1. $z_1 \cdot z_2 = r_1 r_2 [\cos(\theta_1 + \theta_2) + i \sin(\theta_1 + \theta_2)]$.

2. $\dfrac{z_1}{z_2} = \dfrac{r_1}{r_2}[\cos(\theta_1 - \theta_2) + i \sin(\theta_1 - \theta_2)], \quad r_2 \neq 0$.

---

**Proof of the Product Formula**

$$z_1 \cdot z_2 = r_1(\cos \theta_1 + i \sin \theta_1) \cdot r_2(\cos \theta_2 + i \sin \theta_2)$$
$$= r_1 r_2 [(\cos \theta_1 \cos \theta_2 - \sin \theta_1 \sin \theta_2)$$
$$+ i(\sin \theta_1 \cos \theta_2 + \cos \theta_1 \sin \theta_2)]$$
$$= r_1 r_2 [\cos(\theta_1 + \theta_2) + i \sin(\theta_1 + \theta_2)]$$

You will be asked to prove the quotient formula in Exercise 63.

**Example 2   MULTIPLYING COMPLEX NUMBERS**

Express the product of $z_1$ and $z_2$ in standard form:

$$z_1 = 25\sqrt{2}\left(\cos \frac{-\pi}{4} + i \sin \frac{-\pi}{4}\right), \quad z_2 = 14\left(\cos \frac{\pi}{3} + i \sin \frac{\pi}{3}\right).$$

**Solution**

$$z_1 \cdot z_2 = 25\sqrt{2}\left(\cos \frac{-\pi}{4} + i \sin \frac{-\pi}{4}\right) \cdot 14\left(\cos \frac{\pi}{3} + i \sin \frac{\pi}{3}\right)$$
$$= 25 \cdot 14\sqrt{2}\left[\cos\left(\frac{-\pi}{4} + \frac{\pi}{3}\right) + i \sin\left(\frac{-\pi}{4} + \frac{\pi}{3}\right)\right]$$
$$= 350\sqrt{2}\left(\cos \frac{\pi}{12} + i \sin \frac{\pi}{12}\right)$$
$$\approx 478.11 + 128.11i$$

**Example 3   DIVIDING COMPLEX NUMBERS**

Express the quotient $z_1/z_2$ in standard form:

$$z_1 = 2\sqrt{2}(\cos 135° + i \sin 135°), \quad z_2 = 6(\cos 300° + i \sin 300°).$$

**Solution**

$$\frac{z_1}{z_2} = \frac{2\sqrt{2}\,(\cos 135° + i \sin 135°)}{6(\cos 300° + i \sin 300°)}$$

$$= \frac{\sqrt{2}}{3}[\cos (135° - 300°) + i \sin (135° - 300°)]$$

$$= \frac{\sqrt{2}}{3}[\cos (-165°) + i \sin (-165°)]$$

$$\approx -0.46 - 0.12i$$

## Powers of Complex Numbers

We can use the product formula to raise a complex number to a power. For example, let $z = r (\cos \theta + i \sin \theta)$. Then

$$z^2 = z \cdot z$$

$$= r(\cos \theta + i \sin \theta) \cdot r(\cos \theta + i \sin \theta)$$

$$= r^2[\cos (\theta + \theta) + i \sin (\theta + \theta)]$$

$$= r^2(\cos 2\theta + i \sin 2\theta)$$

Figure 6.63 gives a geometric interpretation of squaring a complex number: its argument is doubled and its distance from the origin is multiplied by a factor of $r$, increased if $r > 1$ or decreased if $r < 1$.

We can find $z^3$ by multiplying $z$ by $z^2$:

$$z^3 = z \cdot z^2$$

$$= r(\cos \theta + i \sin \theta) \cdot r^2(\cos 2\theta + i \sin 2\theta)$$

$$= r^3[\cos (\theta + 2\theta) + i \sin (\theta + 2\theta)]$$

$$= r^3(\cos 3\theta + i \sin 3\theta)$$

Similarly,

$$z^4 = r^4(\cos 4\theta + i \sin 4\theta)$$

$$z^5 = r^5(\cos 5\theta + i \sin 5\theta)$$

$$\vdots$$

This pattern can be generalized to the following theorem, named after the mathematician Abraham De Moivre (1667–1754), who also made major contributions to the field of probability.

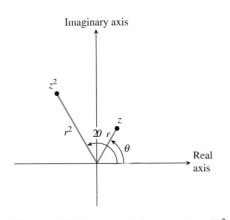

**Figure 6.63**  A geometric interpretation of $z^2$.

**Teaching Note**

This section provides a nice opportunity to bring geometry and algebra together. Providing geometric motivations to numerical work helps students connect different mathematical ideas.

---

**De Moivre's Theorem**

Let $z = r(\cos \theta + i \sin \theta)$ and let $n$ be a positive integer. Then

$$z^n = [r(\cos \theta + i \sin \theta)]^n = r^n(\cos n\theta + i \sin n\theta).$$

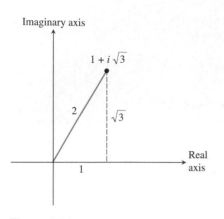

**Figure 6.64** The complex number in Example 4.

### Example 4 USING DE MOIVRE'S THEOREM

Find $(1 + i\sqrt{3})^3$ using De Moivre's theorem.

#### Solution

#### Solve Algebraically

See Figure 6.64. The argument of $z = 1 + i\sqrt{3}$ is $\theta = \pi/3$, and its modulus is $|1 + i\sqrt{3}| = \sqrt{1 + 3} = 2$. Therefore,

$$z = 2\left(\cos\frac{\pi}{3} + i\sin\frac{\pi}{3}\right)$$

$$z^3 = 2^3\left[\cos\left(3\cdot\frac{\pi}{3}\right) + i\sin\left(3\cdot\frac{\pi}{3}\right)\right]$$

$$= 8(\cos\pi + i\sin\pi)$$

$$= 8(-1 + 0i) = -8$$

#### Support Numerically

Figure 6.65a sets the graphing calculator we use in complex number mode. Figure 6.65b supports the result obtained algebraically.

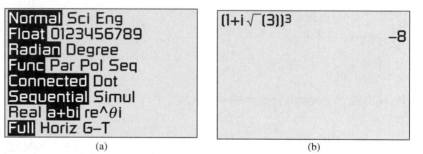

(a)                                    (b)

**Figure 6.65** (a) Setting a graphing calculator in complex number mode. (b) Computing $(1 + i\sqrt{3})^3$ with a graphing calculator.

### Example 5 USING DE MOIVRE'S THEOREM

Find $[(-\sqrt{2}/2) + i(\sqrt{2}/2)]^8$ using De Moivre's theorem.

**Solution** The argument of $z = (-\sqrt{2}/2) + i(\sqrt{2}/2)$ is $\theta = 3\pi/4$, and its modulus is

$$\left|\frac{-\sqrt{2}}{2} + i\frac{\sqrt{2}}{2}\right| = \sqrt{\frac{1}{2} + \frac{1}{2}} = 1.$$

Therefore,

$$z = \cos\frac{3\pi}{4} + i\sin\frac{3\pi}{4}$$

$$z^8 = \cos\left(8\cdot\frac{3\pi}{4}\right) + i\sin\left(8\cdot\frac{3\pi}{4}\right)$$

$$= \cos 6\pi + i\sin 6\pi$$

$$= 1 + i\cdot 0 = 1$$

## Roots of Complex Numbers

The complex number $1 + i\sqrt{3}$ in Example 4 is a solution of $z^3 = -8$ and the complex number $(-\sqrt{2}/2) + i(\sqrt{2}/2)$ in Example 5 is a solution of $z^8 = 1$. The complex number $1 + i\sqrt{3}$ is a third root of $-8$ and $(-\sqrt{2}/2) + i(\sqrt{2}/2)$ is an eighth root of 1.

---

**nth Root**

A complex number $v = a + bi$ is an **nth root of** $z$ if

$$v^n = z.$$

If $z = 1$, then $v$ is an **nth root of unity**.

---

We use De Moivre's theorem to develop a general formula for finding the nth roots of a nonzero complex number.

Suppose that $v = s(\cos \alpha + i \sin \alpha)$ is an nth root of $z = r(\cos \theta + i \sin \theta)$. Then

$$v^n = z$$

$$[s(\cos \alpha + i \sin \alpha)]^n = r(\cos \theta + i \sin \theta)$$

$$s^n(\cos n\alpha + i \sin n\alpha) = r(\cos \theta + i \sin \theta) \qquad (1)$$

Next, we take the absolute value of both sides:

$$|s^n(\cos n\alpha + i \sin na)| = |r(\cos \theta + i \sin \theta)|$$

$$\sqrt{s^{2n}(\cos^2 n\alpha + \sin^2 n\alpha)} = \sqrt{r^2(\cos^2 \theta + \sin^2 \theta)}$$

$$\sqrt{s^{2n}} = \sqrt{r^2}$$

$$s^n = r \qquad\qquad\qquad s > 0, r > 0$$

$$s = \sqrt[n]{r}$$

Substituting $s^n = r$ into Equation (1), we obtain

$$\cos n\alpha + i \sin n\alpha = \cos \theta + i \sin \theta.$$

Therefore, $n\alpha$ can be any angle coterminal with $\theta$. Consequently, for any integer $k$, $v$ is an nth root of $z$ if $s = \sqrt[n]{r}$ and

$$n\alpha = \theta + 2\pi k$$

$$\alpha = \frac{\theta + 2\pi k}{n}.$$

The expression for $v$ takes on $n$ different values for $k = 0, 1, \ldots, n - 1$, and the values start to repeat for $k = n, n + 1, \ldots$.

We summarize this result.

**nth Roots of a Complex Number**

If $z = r(\cos \theta + i \sin \theta)$, then the $n$ distinct complex numbers

$$\sqrt[n]{r}\left(\cos \frac{\theta + 2\pi k}{n} + i \sin \frac{\theta + 2\pi k}{n}\right),$$

where $k = 0, 1, 2, \ldots, n - 1$, are the $n$th roots of the complex number $z$.

**Example 6** FINDING FOURTH ROOTS

Find the fourth roots of $z = 5(\cos(\pi/3) + i \sin(\pi/3))$.

**Solution** The fourth roots of $z$ are the complex numbers

$$\sqrt[4]{5}\left(\cos \frac{\pi/3 + 2\pi k}{4} + i \sin \frac{\pi/3 + 2\pi k}{4}\right)$$

for $k = 0, 1, 2, 3$. Taking into account that $(\pi/3 + 2\pi k)/4 = \pi/12 + \pi k/2$, the list becomes

$$z_1 = \sqrt[4]{5}\left[\cos\left(\frac{\pi}{12} + \frac{0}{2}\right) + i \sin\left(\frac{\pi}{12} + \frac{0}{2}\right)\right]$$

$$= \sqrt[4]{5}\left[\cos \frac{\pi}{12} + i \sin \frac{\pi}{12}\right]$$

$$z_2 = \sqrt[4]{5}\left[\cos\left(\frac{\pi}{12} + \frac{\pi}{2}\right) + i \sin\left(\frac{\pi}{12} + \frac{\pi}{2}\right)\right]$$

$$= \sqrt[4]{5}\left[\cos \frac{7\pi}{12} + i \sin \frac{7\pi}{12}\right]$$

$$z_3 = \sqrt[4]{5}\left[\cos\left(\frac{\pi}{12} + \frac{2\pi}{2}\right) + i \sin\left(\frac{\pi}{12} + \frac{2\pi}{2}\right)\right]$$

$$= \sqrt[4]{5}\left[\cos \frac{13\pi}{12} + i \sin \frac{13\pi}{12}\right]$$

$$z_4 = \sqrt[4]{5}\left[\cos\left(\frac{\pi}{12} + \frac{3\pi}{2}\right) + i \sin\left(\frac{\pi}{12} + \frac{3\pi}{2}\right)\right]$$

$$= \sqrt[4]{5}\left[\cos \frac{19\pi}{12} + i \sin \frac{19\pi}{12}\right]$$

**Example 7** FINDING CUBE ROOTS

Find the cube roots of $-1$ and plot them.

**Solution** First we write the complex number $z = -1$ in trigonometric form

$$z = -1 + 0i = \cos \pi + i \sin \pi.$$

The third roots of $z = -1 = \cos \pi + i \sin \pi$ are the complex numbers

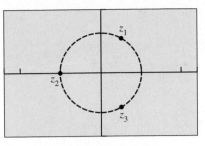

[−2.4, 2.4] by [−1.6, 1.6]

**Figure 6.66** The three cube roots $z_1$, $z_2$, and $z_3$ of $-1$ displayed on the unit circle (dashed). (Example 7)

$$\cos \frac{\pi + 2\pi k}{3} + i \sin \frac{\pi + 2\pi k}{3},$$

for $k = 0, 1, 2$. The three complex numbers are

$$z_1 = \cos \frac{\pi}{3} + i \sin \frac{\pi}{3} \qquad = \frac{1}{2} + \frac{\sqrt{3}}{2}i,$$

$$z_2 = \cos \frac{\pi + 2\pi}{3} + i \sin \frac{\pi + 2\pi}{3} = -1 + 0i,$$

$$z_3 = \cos \frac{\pi + 4\pi}{3} + i \sin \frac{\pi + 4\pi}{3} = \frac{1}{2} - \frac{\sqrt{3}}{2}i.$$

Figure 6.66 shows the graph of the three cube roots $z_1$, $z_2$, and $z_3$. They are evenly spaced (with distance of $2\pi/3$ radians) around the unit circle.

**Example 8  FINDING ROOTS OF UNITY**

Find the eight eighth roots of unity.

**Solution**  First we write the complex number $z = 1$ in trigonometric form

$$z = 1 + 0i = \cos 0 + i \sin 0.$$

The eighth roots of $z = 1 + 0i = \cos 0 + i \sin 0$ are the complex numbers

$$\cos \frac{0 + 2\pi k}{8} + i \sin \frac{0 + 2\pi k}{8},$$

for $k = 0, 1, 2, \ldots, 7$.

$$z_1 = \cos 0 + i \sin 0 \qquad = 1 + 0i$$

$$z_2 = \cos \frac{\pi}{4} + i \sin \frac{\pi}{4} \qquad = \frac{\sqrt{2}}{2} + \frac{\sqrt{2}}{2}i$$

$$z_3 = \cos \frac{\pi}{2} + i \sin \frac{\pi}{2} \qquad = 0 + i$$

$$z_4 = \cos \frac{3\pi}{4} + i \sin \frac{3\pi}{4} = -\frac{\sqrt{2}}{2} + \frac{\sqrt{2}}{2}i$$

$$z_5 = \cos \pi + i \sin \pi \qquad = -1 + 0i$$

$$z_6 = \cos \frac{5\pi}{4} + i \sin \frac{5\pi}{4} = -\frac{\sqrt{2}}{2} - \frac{\sqrt{2}}{2}i$$

$$z_7 = \cos \frac{3\pi}{2} + i \sin \frac{3\pi}{2} = 0 - i$$

$$z_8 = \cos \frac{7\pi}{4} + i \sin \frac{7\pi}{4} = \frac{\sqrt{2}}{2} - \frac{\sqrt{2}}{2}i$$

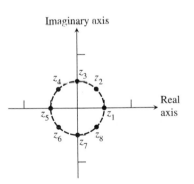

**Figure 6.67** The eight eighth roots of unity are evenly spaced on a unit circle. (Example 8)

Figure 6.67 shows the eight points. They are spaced $2\pi/8 = \pi/4$ radians apart.

# Quick Review 6.6

In Exercises 1 and 2, plot the complex number in the complex plane and find its absolute value.

**1. (a)** $2 + 7i$ **(b)** $-4 + i$ **2. (a)** $6 - 5i$ **(b)** $-1 - 7i$

In Exercises 3 and 4, write the complex number in standard form $a + bi$.

**3.** $(1 + i)^5$ $-4 - 4i$ **4.** $(1 - i)^4$ $-4 + 0i$

In Exercises 5–8, find an angle $\theta$ in $0 \le \theta < 2\pi$ which satisfies both equations.

**5.** $\sin \theta = \dfrac{1}{2}$ and $\cos \theta = -\dfrac{\sqrt{3}}{2}$ $\theta = \dfrac{5\pi}{6}$

**6.** $\sin \theta = -\dfrac{\sqrt{2}}{2}$ and $\cos \theta = \dfrac{\sqrt{2}}{2}$ $\dfrac{7\pi}{4}$

**7.** $\sin \theta = -\dfrac{\sqrt{3}}{2}$ and $\cos \theta = -\dfrac{1}{2}$ $\dfrac{4\pi}{3}$

**8.** $\sin \theta = -\dfrac{\sqrt{2}}{2}$ and $\cos \theta = -\dfrac{\sqrt{2}}{2}$ $\dfrac{5\pi}{4}$

In Exercises 9 and 10, find all real solutions.

**9.** $x^3 - 1 = 0$ $1$ **10.** $x^4 - 1 = 0$ $\pm 1$

# Section 6.6 Exercises

In Exercises 1–12, find the trigonometric form of the complex number where $0 \le \theta < 2\pi$.

**1.** $3i$ **2.** $-2i$

**3.** $2 + 2i$ **4.** $\sqrt{3} + i$

**5.** $-2 + 2i\sqrt{3}$ **6.** $3 - 3i$

**7.** $\dfrac{1}{2} - \dfrac{\sqrt{3}}{2}i$ **8.** $-\dfrac{\sqrt{3}}{2} - \dfrac{1}{2}i$

**9.** $3 + 2i$ **10.** $4 - 7i$

**11.**

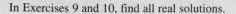

**12.**

In Exercises 13–18, write the complex number in standard form $a + bi$.

**13.** $3(\cos 30° - i \sin 30°)$ **14.** $8(\cos 210° + i \sin 210°)$

**15.** $5[\cos(-60°) + i \sin(-60°)]$ $5/2 - (5/2)\sqrt{3}i$

**16.** $5\left(\cos \dfrac{\pi}{4} + i \sin \dfrac{\pi}{4}\right)$ $\dfrac{5}{2}\sqrt{2} + \dfrac{5}{2}\sqrt{2}i$

**17.** $\sqrt{2}\left(\cos \dfrac{7\pi}{6} + i \sin \dfrac{7\pi}{6}\right)$ **18.** $\sqrt{7}\left(\cos \dfrac{\pi}{12} + i \sin \dfrac{\pi}{12}\right)$

In Exercises 19–22, find the product of $z_1$ and $z_2$. Leave the answer in trigonometric form.

**19.** $z_1 = 7(\cos 25° + i \sin 25°)$ $z_2 = 2(\cos 130° + i \sin 130°)$

**20.** $z_1 = \sqrt{2}(\cos 118° + i \sin 118°)$
$z_2 = 0.5[\cos(-19°) + i \sin(-19°)]$

**21.** $z_1 = 5\left(\cos \dfrac{\pi}{4} + i \sin \dfrac{\pi}{4}\right)$ $z_2 = 3\left(\cos \dfrac{5\pi}{3} + i \sin \dfrac{5\pi}{3}\right)$

**22.** $z_1 = \sqrt{3}\left(\cos \dfrac{3\pi}{4} + i \sin \dfrac{3\pi}{4}\right)$ $z_2 = \dfrac{1}{3}\left(\cos \dfrac{\pi}{6} + i \sin \dfrac{\pi}{6}\right)$

In Exercises 23–26, find the trigonometric form of the quotient.

**23.** $\dfrac{2(\cos 30° + i \sin 30°)}{3(\cos 60° + i \sin 60°)}$ **24.** $\dfrac{5(\cos 220° + i \sin 220°)}{2(\cos 115° + i \sin 115°)}$

**25.** $\dfrac{6(\cos 5\pi + i \sin 5\pi)}{3(\cos 2\pi + i \sin 2\pi)}$ **26.** $\dfrac{\cos(\pi/2) + i \sin(\pi/2)}{\cos(\pi/4) + i \sin(\pi/4)}$

In Exercises 27–30, find the product $z_1 \cdot z_2$ and quotient $z_1/z_2$ in two ways.

**(a)** Using the trigonometric form for $z_1$ and $z_2$.

**(b)** Using the standard form for $z_1$ and $z_2$.

**27.** $z_1 = 3 - 2i$ and $z_2 = 1 + i$

**28.** $z_1 = 1 - i$ and $z_2 = \sqrt{3} + i$

**29.** $z_1 = 3 + i$ and $z_2 = 5 - 3i$

**30.** $z_1 = 2 - 3i$ and $z_2 = 1 - \sqrt{3}i$

In Exercises 31–38, use De Moivre's theorem to find the indicated power of the complex number. Write your answer in standard form $a + bi$.

**31.** $\left(\cos \dfrac{\pi}{4} + i \sin \dfrac{\pi}{4}\right)^3$ **32.** $\left[3\left(\cos \dfrac{3\pi}{2} + i \sin \dfrac{3\pi}{2}\right)\right]^5$

**33.** $\left[2\left(\cos \dfrac{3\pi}{4} + i \sin \dfrac{3\pi}{4}\right)\right]^3$ **34.** $\left[6\left(\cos \dfrac{5\pi}{6} + i \sin \dfrac{5\pi}{6}\right)\right]^4$

**35.** $(1 + i)^5$ $-4 - 4i$ **36.** $(3 + 4i)^{20}$

**37.** $(1 - \sqrt{3}i)^3$ $-8$ **38.** $\left(\dfrac{1}{2} + i\dfrac{\sqrt{3}}{2}\right)^3$ $-1$

In Exercises 39–44, find the cube roots of the complex number.

**39.** $2(\cos 2\pi + i \sin 2\pi)$     **40.** $2\left(\cos \dfrac{\pi}{4} + i \sin \dfrac{\pi}{4}\right)$

**41.** $3\left(\cos \dfrac{4\pi}{3} + i \sin \dfrac{4\pi}{3}\right)$     **42.** $27\left(\cos \dfrac{11\pi}{6} + i \sin \dfrac{11\pi}{6}\right)$

**43.** $3 - 4i$     **44.** $-2 + 2i$

In Exercises 45–50, find the fifth roots of the complex number.

**45.** $(\cos \pi + i \sin \pi)$     **46.** $32\left(\cos \dfrac{\pi}{2} + i \sin \dfrac{\pi}{2}\right)$

**47.** $2\left(\cos \dfrac{\pi}{6} + i \sin \dfrac{\pi}{6}\right)$     **48.** $2\left(\cos \dfrac{\pi}{4} + i \sin \dfrac{\pi}{4}\right)$

**49.** $2i$     **50.** $1 + \sqrt{3}i$

In Exercises 51–56, find the *n*th roots of the complex number for the specified value of *n*.

**51.** $1 + i, \quad n = 4$     **52.** $1 - i, \quad n = 6$

**53.** $2 + 2i, \quad n = 3$     **54.** $-2 + 2i, \quad n = 4$

**55.** $-2i, \quad n = 6$     **56.** $32, \quad n = 5$

In Exercises 57–60, express the roots of unity in standard form $a + bi$. Graph each root in the complex plane.

**57.** Cube roots of unity     **58.** Fourth roots of unity

**59.** Sixth roots of unity     **60.** Square roots of unity

**61.** Determine $z$ and the three cube roots of $z$ if one cube root of $z$ is $1 + \sqrt{3}i$.  $-8$; $-2$ and $1 \pm \sqrt{3}i$

**62.** Determine $z$ and the four fourth roots of $z$ if one fourth root of $z$ is $-2 - 2i$.  $-64$; $2 \pm 2i$ and $-2 \pm 2i$

**63. Quotient Formula** Let $z_1 = r_1(\cos \theta_1 + i \sin \theta_1)$ and $z_2 = r_2(\cos \theta_2 + i \sin \theta_2)$, $r_2 \neq 0$. Verify that $z_1/z_2 = r_1/r_2[\cos (\theta_1 - \theta_2) + i \sin (\theta_1 - \theta_2)]$.

**64. Group Activity** *n*th Roots Show that the *n*th roots of the complex number $r(\cos \theta + i \sin \theta)$ are spaced $2\pi/n$ radians apart on a circle with radius $\sqrt[n]{r}$.

## Explorations

**65. Complex Conjugates** The complex conjugate of $z = a + bi$ is $\bar{z} = a - bi$. Let $z = r(\cos \theta + i \sin \theta)$.

(a) Prove that $\bar{z} = r[\cos (-\theta) + i \sin (-\theta)]$.

(b) Use the trigonometric form to find $z \cdot \bar{z}$.  $r^2$

(c) Use the trigonometric form to find $z/\bar{z}$, if $\bar{z} \neq 0$.

(d) Prove that $-z = r[\cos (\theta + \pi) + i \sin (\theta + \pi)]$.

**66. Modulus of Complex Numbers** Let $z = r(\cos \theta + i \sin \theta)$.

(a) Prove that $|z| = |r|$.

(b) Use the trigonometric form for the complex numbers $z_1$ and $z_2$ to prove that $|z_1 \cdot z_2| = |z_1| \cdot |z_2|$. ■

## Extending the Ideas

**67. Using Polar Form on a Graphing Calculator** The complex number $r(\cos \theta + i \sin \theta)$ can be entered in polar form on some graphing calculators as $re^{i\theta}$.

(a) Support the result of Example 2 by entering the complex numbers $z_1$ and $z_2$ in polar form on your graphing calculator and computing the product with your graphing calculator.

(b) Support the result of Example 3 by entering the complex numbers $z_1$ and $z_2$ in polar form on your graphing calculator and computing the quotient with your graphing calculator.

(c) Support the result of Example 5 by entering the complex number in polar form on your graphing calculator and computing the power with your graphing calculator.

**68. Visualizing Roots of Unity** Set your graphing calculator in parametric mode with $0 \leq T \leq 8$, Tstep = 1, Xmin = $-2.4$, Xmax = 2.4, Ymin = $-1.6$, and Ymax = 1.6.

(a) Let $x = \cos ((2\pi/8)t)$ and $y = \sin ((2\pi/8)t)$. Use trace to visualize the eight eighth roots of unity. We say that $2\pi/8$ *generates* the eighth roots of unity. (Try both dot mode and connected mode.)

(b) Replace $2\pi/8$ in part a by the arguments of other eighth roots of unity. Do any others *generate* the eighth roots of unity? Yes. $6\pi/8$, $10\pi/8$, $14\pi/8$

(c) Repeat (a) and (b) for the fifth, sixth, and seventh roots of unity, using appropriate functions for $x$ and $y$.

(d) What would you conjecture about an *n*th root of unity that generates all the *n*th roots of unity in the sense of (a)?

**69. Parametric Graphing** Write parametric equations that represent $(\sqrt{2} + i)^n$ for $n = t$. Draw and label an *accurate* spiral representing $(\sqrt{2} + i)^n$ for $n = 0, 1, 2, 3, 4$.

**70. Parametric Graphing** Write parametric equations that represent $(-1 + i)^n$ for $n = t$. Draw and label an *accurate* spiral representing $(-1 + i)^n$ for $n = 0, 1, 2, 3, 4$.

**71.** Explain why the triangles formed by 0, 1, and $z_1$ and by 0, $z_2$ and $z_1 z_2$ shown in the figure are similar triangles.

**72. Compass and Straightedge Construction** Using only a compass and straightedge, construct the location of $z_1 z_2$ given the location of 0, 1, $z_1$, and $z_2$.

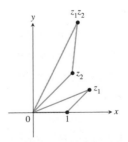

In Exercises 73–78, find all solutions of the equation (real and complex).

**73.** $x^3 - 1 = 0$

**74.** $x^4 - 1 = 0$

**75.** $x^3 + 1 = 0$

**76.** $x^4 + 1 = 0$

**77.** $x^5 + 1 = 0$

**78.** $x^5 - 1 = 0$

## Chapter 6    Key Ideas

### Concepts

angle between vectors (p. 491)
argument of a complex number (p. 525)
cardioid (p. 520)
component form of a vector (p. 480)
components of a vector (p. 480)
DeMoivre's Theorem (p. 527)
directed angle (p. 510)
directed distance (p. 510)
directed line segment (p. 479)
direction angle (p. 484)
dot product (p. 490)
equal vectors (p. 480)
equivalent directed line segments (p. 479)
horizontal component (p. 479)
initial point of a directed line segment
    (p. 479)
lemniscate curve (p. 522)
length (magnitude) of a directed line
    segment (p. 479)
length (magnitude) of a vector (p. 481)

limaçon curve (p. 520)
linear combination of vectors (p. 484)
maximum $r$-value (p. 518)
$n$th root of a complex number
    (p. 529–530)
$n$th root of unity (p. 529)
orthogonal vectors (p. 492)
parallelogram law (p. 482)
parameter (p. 498)
parameter interval (p. 498)
parametric curve (p. 498)
parametric equations (p. 498)
parametrization of a curve (p. 498)
polar axis (p. 510)
polar coordinate system (p. 510)
polar coordinates (p. 510)
polar form of a complex number (p. 525)
pole (p. 510)
position vector (p. 480)
projection of **u** onto **v** (p. 493)

rose curve (p. 519)
scalar multiple of a vector (p. 482)
speed (p. 485)
spiral of Archimedes (p. 521)
standard position of a vector (p. 480)
standard unit vectors (p. 483)
sum of vectors (p. 482)
symmetry about the origin (p. 517)
symmetry about $x$-axis (p. 517)
symmetry about $y$-axis (p. 517)
terminal point of a directed line segment
    (p. 479)
trigonometric form of a complex number
    (p. 525)
unit vector (p. 483)
vector (p. 479)
velocity (p. 485)
vertical component (p. 484)
work (p. 494)
zero vector (p. 481)

### Properties, Theorems, and Formulas

#### Component Form of a Vector

The component form of the vector $\overrightarrow{PQ}$ with initial point $P(x_1, y_1)$ and terminal point $Q(x_2, y_2)$ is
$\mathbf{v} = \langle x_2 - x_1, y_2 - y_1 \rangle$.

#### The Magnitude or Length of a Vector

The magnitude or length of the vector $\mathbf{v} = \overrightarrow{PQ}$ determined by $P(x_1, y_1)$ and $Q(x_2, y_2)$ is
$|\mathbf{v}| = \sqrt{(x_2 - x_1)^2 + (y_2 - y_1)^2}$.

#### Vector Addition and Scalar Multiplication

The sum of vectors $\mathbf{u} = \langle u_1, u_2 \rangle$ and $\mathbf{v} = \langle v_1, v_2 \rangle$ is $\mathbf{u} + \mathbf{v} = \langle u_1 + v_1, u_2 + v_2 \rangle$.

The scalar product of the real number $k$ and the vector $\mathbf{u} = \langle u_1, u_2 \rangle$ is $k\mathbf{u} = k\langle u_1, u_2 \rangle = \langle ku_1, ku_2 \rangle$.

#### Unit Vector in the Direction of the Vector v

The unit vector in the direction of $\mathbf{v} = (|\mathbf{v}| \cos \theta)\mathbf{i} + (|\mathbf{v}| \sin \theta)\mathbf{j}$ is $\mathbf{u} = \dfrac{\mathbf{v}}{|\mathbf{v}|} = (\cos \theta)\mathbf{i} + (\sin \theta)\mathbf{j}$.

#### Dot Product of Two Vectors

The dot product of $\mathbf{u} = \langle u_1, u_2 \rangle$ and $\mathbf{v} = \langle v_1, v_2 \rangle$ is $\mathbf{u} \cdot \mathbf{v} = u_1 v_1 + u_2 v_2$.

#### Properties of the Dot Product

Let $\mathbf{u}$, $\mathbf{v}$, and $\mathbf{w}$ be vectors and let $c$ be a scalar.

**1.** $\mathbf{u} \cdot \mathbf{v} = \mathbf{v} \cdot \mathbf{u}$

**2.** $\mathbf{u} \cdot \mathbf{u} = |\mathbf{u}|^2$

**3.** $\mathbf{0} \cdot \mathbf{u} = 0$

**4.** $\mathbf{u} \cdot (\mathbf{v} + \mathbf{w}) = \mathbf{u} \cdot \mathbf{v} + \mathbf{u} \cdot \mathbf{w}$
   $(\mathbf{u} + \mathbf{v}) \cdot \mathbf{w} = \mathbf{u} \cdot \mathbf{w} + \mathbf{v} \cdot \mathbf{w}$

**5.** $(c\mathbf{u}) \cdot \mathbf{v} = \mathbf{u} \cdot (c\mathbf{v}) = c(\mathbf{u} \cdot \mathbf{v})$

### Theorem Angle Between Two Vectors

If $\theta$ is the angle between the nonzero vectors $\mathbf{u}$ and $\mathbf{v}$, then

$$\cos \theta = \frac{\mathbf{u} \cdot \mathbf{v}}{|\mathbf{u}||\mathbf{v}|} \quad \text{and} \quad \theta = \cos^{-1}\left(\frac{\mathbf{u} \cdot \mathbf{v}}{|\mathbf{u}||\mathbf{v}|}\right).$$

### Projection of the Vector u onto the Vector v

The projection of $\mathbf{u}$ onto $\mathbf{v}$ is $\text{proj}_{\mathbf{v}}\, \mathbf{u} = \left(\frac{\mathbf{u} \cdot \mathbf{v}}{|\mathbf{v}|^2}\right)\mathbf{v}$.

### Work

The work done by a force $\mathbf{F}$ in moving an object from $A$ to $B$ is $\mathbf{F} \cdot \overrightarrow{AB}$.

### Coordinate Conversion Equations

Let the point $P$ have polar coordinates $(r, \theta)$ and rectangular coordinates $(x, y)$. Then

$$x = r \cos \theta\,, y = r \sin \theta,\, r^2 = x^2 + y^2, \text{ and } \tan \theta = \frac{y}{x}.$$

### Symmetry Tests for Polar Graphs

The graph of a polar equation has the indicated symmetry if either replacement produces an equivalent polar equation.

| To Test for Symmetry | Replace | By |
|---|---|---|
| **1.** about the $x$-axis, | $(r, \theta)$ | $(r, -\theta)$ or $(-r, \pi - \theta)$. |
| **2.** about the $y$-axis, | $(r, \theta)$ | $(-r, -\theta)$ or $(r, \pi - \theta)$. |
| **3.** about the origin, | $(r, \theta)$ | $(-r, \theta)$ or $(r, \theta + \pi)$. |

### Trigonometric Form of a Complex Number

The trigonometric form of the complex number $z = a + bi$ is $z = r(\cos \theta + i \sin \theta) = r \operatorname{cis} \theta$, where

$$a = r \cos \theta, b = r \sin \theta, r = \sqrt{a^2 + b^2}, \text{ and } \tan \theta = \frac{b}{a}.$$

### DeMoivre's Theorem

Let $z = r(\cos \theta + i \sin u)$ be the trigonometric (polar) form of the complex number $z$ and let $n$ be a positive integer. Then $z^n = [r(\cos \theta + i \sin \theta)]^n = r^n(\cos n\theta + i \sin n\theta)$.

## Procedures

### Product and Quotient of Complex Numbers

Let $z_1 = r_1(\cos \theta_1 + i \sin \theta_1)$ and $z_2 = r_2(\cos \theta_2 + i \sin \theta_2)$ be the trigonometric (polar) form of the complex numbers $z_1$ and $z_2$. Then

1. $z_1 \cdot z_2 = r_1 r_2[\cos(\theta_1 + \theta_2) + i \sin(\theta_1 + \theta_2)]$

2. $\dfrac{z_1}{z_2} = \dfrac{r_1}{r_2}[\cos(\theta_1 - \theta_2) + i \sin(\theta_1 - \theta_2)], r_2 \neq 0.$

### *n*th Roots of a Complex Number

If $z = r(\cos \theta + i \sin \theta)$, then the $n$ distinct complex numbers

$$\sqrt[n]{r}\left(\cos \frac{\theta + 2\pi k}{n} + i \sin \frac{\theta + 2\pi k}{n}\right),$$

where $k = 0, 1, 2, \ldots, n - 1$, are the $n$th roots of the complex number $z$.

### Gallery of Functions

**Rose Curves:** $r = a \cos n\theta$  and  $r = a \sin n\theta$

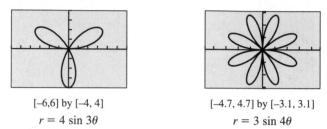

[–6,6] by [–4, 4]     [–4.7, 4.7] by [–3.1, 3.1]
$r = 4 \sin 3\theta$        $r = 3 \sin 4\theta$

**Limaçon Curves:** $r = a \pm b \sin \theta$  and  $r = a \pm b \cos \theta$ with $a > 0$ and $b > 0$

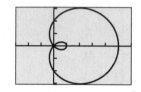

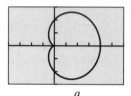

Limaçon with an inner loop: $\dfrac{a}{b} < 1$     Cardiod: $\dfrac{a}{b} = 1$

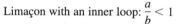

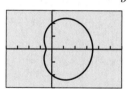

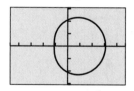

Dimpled limaçon: $1 < \dfrac{a}{b} < 2$     Convex limaçon: $\dfrac{a}{b} \geq 2$

**Spiral of Archimedes:**

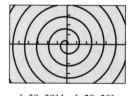

[–30, 30] by [–20, 20]
$r = \theta, 0 \leq \theta \leq 45$

**Lemniscate Curves:** $r^2 = a^2 \sin 2\theta$  and  $r^2 = a^2 \cos 2\theta$

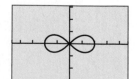

[–4.7, 4.7] by [–3.1, 3.1]
$r^2 = 4 \cos 2\theta$

## Chapter 6 | Review Exercises

The collection of exercises marked in red could be used as a chapter test.

In Exercises 1–6, let $\mathbf{u} = \langle 2, -1 \rangle$, $\mathbf{v} = \langle 4, 2 \rangle$, and $\mathbf{w} = \langle 1, -3 \rangle$ be vectors. Find the indicated expression.

**1.** $\mathbf{u} - \mathbf{v}$  $\langle -2, -3 \rangle$

**2.** $2\mathbf{u} - 3\mathbf{w}$  $\langle 1, 7 \rangle$

**3.** $|\mathbf{u} + \mathbf{v}|$  $\sqrt{37}$

**4.** $|\mathbf{w} - 2\mathbf{u}|$  $\sqrt{10}$

**5.** $\mathbf{u} \cdot \mathbf{v}$  6

**6.** $\mathbf{u} \cdot \mathbf{w}$  5

In Exercises 7–10, let $A = (2, -1)$, $B = (3, 1)$, $C = (-4, 2)$, and $D = (1, -5)$. Find the component form and magnitude of the vector.

**7.** $3\overrightarrow{AB}$  $\langle 3, 6 \rangle$; $3\sqrt{5}$

**8.** $\overrightarrow{AB} + \overrightarrow{CD}$  $\langle 6, -5 \rangle$; $\sqrt{61}$

**9.** $\overrightarrow{AC} + \overrightarrow{BD}$  $\langle -8, -3 \rangle$; $\sqrt{73}$

**10.** $\overrightarrow{CD} - \overrightarrow{AB}$  $\langle 4, -9 \rangle$; $\sqrt{97}$

In Exercises 11 and 12, find **(a)** a unit vector in the direction of $\overrightarrow{AB}$ and **(b)** a vector of magnitude 3 in the opposite direction.

**11.** $A = (4, 0)$, $B = (2, 1)$

**12.** $A = (3, 1)$, $B = (5, 1)$

In Exercises 13 and 14, find **(a)** the direction angles of $\mathbf{u}$ and $\mathbf{v}$ and **(b)** the angle between $\mathbf{u}$ and $\mathbf{v}$.

**13.** $\mathbf{u} = \langle 4, 3 \rangle$, $\mathbf{v} = \langle 2, 5 \rangle$

**14.** $\mathbf{u} = \langle -2, 4 \rangle$, $\mathbf{v} = \langle 6, 4 \rangle$

In Exercises 15–18, convert the polar coordinates to rectangular coordinates.

**15.** $(-2.5, 25°)$

**16.** $(-3.1, 135°)$

**17.** $(2, -\pi/4)$

**18.** $(3.6, 3\pi/4)$

In Exercises 19 and 20, polar coordinates of point $P$ are given. Find all of its polar coordinates.

**19.** $P = (-1, -2\pi/3)$

**20.** $P = (-2, 5\pi/6)$

In Exercises 21–24, rectangular coordinates of point $P$ are given. Find polar coordinates of $P$ that satisfy these conditions:

**(a)** $0 \le \theta \le 2\pi$   **(b)** $-\pi \le \theta \le \pi$   **(c)** $0 \le \theta \le 4\pi$

**21.** $P = (2, -3)$

**22.** $P = (-10, 0)$

**23.** $P = (5, 0)$

**24.** $P = (0, -2)$

In Exercises 25–30, eliminate the parameter $t$ and identify the graph.

**25.** $x = 3 - 5t$, $y = 4 + 3t$

**26.** $x = 4 + t$, $y = -8 - 5t$, $-3 \le t \le 5$

**27.** $x = 2t^2 + 3$, $y = t - 1$

**28.** $x = 3 \cos t$, $y = 3 \sin t$

**29.** $x = e^{2t} - 1$, $y = e^t$

**30.** $x = t^3$, $y = \ln t$, $t > 0$

In Exercises 31 and 32, find a parametrization for the curve.

**31.** The line through the points $(-1, -2)$ and $(3, 4)$.

**32.** The line segment with endpoints $(-2, 3)$ and $(5, 1)$.

Exercises 33 and 34 refer to the complex number $z_1$ shown in the figure.

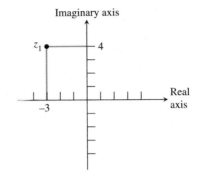

**33.** If $z_1 = a + bi$, find $a$, $b$, and $|z_1|$.

**34.** Find the trigonometric form of $z_1$.

In Exercises 35–38, write the complex number in standard form.

**35.** $6(\cos 30° + i \sin 30°)$

**36.** $3(\cos 150° + i \sin 150°)$

**37.** $2.5 \left( \cos \dfrac{4\pi}{3} + i \sin \dfrac{4\pi}{3} \right)$

**38.** $4(\cos 2.5 + i \sin 2.5)$

In Exercises 39–42, write the complex number in trigonometric form where $0 \le \theta \le 2\pi$. Then write three other possible trigonometric forms for the number.

**39.** $3 - 3i$

**40.** $-1 + i\sqrt{2}$

**41.** $3 - 5i$

**42.** $-2 - 2i$

In Exercises 43 and 44, write the complex numbers $z_1 \cdot z_2$ and $z_1/z_2$ in trigonometric form.

**43.** $z_1 = 3(\cos 30° + i \sin 30°)$ and $z_2 = 4(\cos 60° + i \sin 60°)$

**44.** $z_1 = 5(\cos 20° + i \sin 20°)$ and $z_2 = -2(\cos 45° + i \sin 45°)$

In Exercises 45–48, use De Moivre's theorem to find the indicated power of the complex number. Write your answer in **(a)** trigonometric form and **(b)** standard form.

**45.** $\left[ 3 \left( \cos \dfrac{\pi}{4} + i \sin \dfrac{\pi}{4} \right) \right]^5$

**46.** $\left[ 2 \left( \cos \dfrac{\pi}{12} + i \sin \dfrac{\pi}{12} \right) \right]^8$

**47.** $\left[ 5 \left( \cos \dfrac{5\pi}{3} + i \sin \dfrac{5\pi}{3} \right) \right]^3$

**48.** $\left[ 7 \left( \cos \dfrac{\pi}{24} + i \sin \dfrac{\pi}{24} \right) \right]^6$

In Exercises 49–52, find and graph the $n$th roots of the complex number for the specified value of $n$.

**49.** $3 + 3i$, $n = 4$

**50.** $8$, $n = 3$

**51.** $1$, $n = 5$

**52.** $-1$, $n = 6$

In Exercises 53–60, decide whether the graph of the given polar equation appears among the four graphs shown.

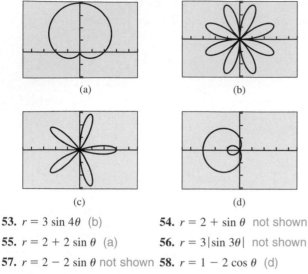

(a)                    (b)

(c)                    (d)

**53.** $r = 3 \sin 4\theta$ (b)         **54.** $r = 2 + \sin \theta$ not shown

**55.** $r = 2 + 2 \sin \theta$ (a)      **56.** $r = 3|\sin 3\theta|$ not shown

**57.** $r = 2 - 2 \sin \theta$ not shown **58.** $r = 1 - 2 \cos \theta$ (d)

**59.** $r = 3 \cos 5\theta$ (c)         **60.** $r = 3 - 2 \tan \theta$ not shown

In Exercises 61–64, convert the polar equation to rectangular form and identify the graph.

**61.** $r = -2$                         **62.** $r = -2 \sin \theta$

**63.** $r = -3 \cos \theta - 2 \sin \theta$  **64.** $r = 3 \sec \theta$

In Exercises 65–68, convert the rectangular equation to polar form. Graph the polar equation.

**65.** $y = -4$                         **66.** $x = 5$

**67.** $(x - 3)^2 + (y + 1)^2 = 10$     **68.** $2x - 3y = 4$

In Exercises 69–72, analyze the graph of the polar curve.

**69.** $r = 2 - 5 \sin \theta$          **70.** $r = 4 - 4 \cos \theta$

**71.** $r = 2 \sin 3\theta$             **72.** $r^2 = 2 \sin 2\theta, 0 \le \theta \le 2\pi$

**73. Graphing Lines Using Polar Equations**

(a) Explain why $r = a \sec \theta$ is a polar form for the line $x = a$.

(b) Explain why $r = b \csc \theta$ is a polar form for the line $y = b$.

(c) Let $y = mx + b$. Prove that

$$r = \frac{b}{\sin \theta - m \cos \theta}$$

is a polar form for the line. What is the domain of $r$?

(d) Illustrate the result in (c) by graphing the line $y = 2x + 3$ using the polar form from (c).

**74. Flight Engineering** An airplane is flying on a bearing of 80° at 540 mph. A wind is blowing with the bearing 100° at 55 mph.

(a) Find the component form of the velocity of the airplane.

(b) Find the actual speed and direction of the airplane.

**75. Flight Engineering** An airplane is flying on a bearing of 285° at 480 mph. A wind is blowing with the bearing 265° at 30 mph.

(a) Find the component form of the velocity of the airplane.

(b) Find the actual speed and direction of the airplane.

**76. Combining Forces** A force of 120 lb acts on an object at an angle of 20°. A second force of 300 lb acts on the object at an angle of −5°. Find the direction and magnitude of the resultant force. $\approx$ 411.89 lb; $\approx$ 2.07°

**77. Braking Force** A 3000 pound car is parked on a street that makes an angle of 16° with the horizontal (see figure).

(a) Find the force required to keep the car from rolling down the hill. $\approx$ 826.91 pounds

(b) Find the component of the force perpendicular to the street.

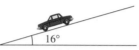

**78. Work** Find the work done by a force **F** of 36 pounds acting in the direction given by the vector $\langle 3, 5 \rangle$ in moving an object 10 feet from $(0, 0)$ to $(10, 0)$. $\approx$ 185.22 foot-pounds

**79. Height of an Arrow** Stewart shoots an arrow straight up from the top of a building with initial velocity of 245 ft/sec. The arrow leaves from a point 200 ft above level ground.

(a) Write an equation that models the height of the arrow as a function of time $t$. $h = -16t^2 + 245t + 200$

(b) Use parametric equations to simulate the height of the arrow.

(c) Use parametric equations to graph height against time.

(d) How high is the arrow after 4 sec? 924 ft

(e) What is the maximum height of the arrow? When does it reach its maximum height? $\approx$ 1138 ft; $t \approx$ 7.66

(f) How long will it be before the arrow hits the ground?

**80. Ferris Wheel Problem** Lucinda is on a Ferris wheel of radius 35 ft that turns at the rate of one revolution every 20 sec. The lowest point of the Ferris wheel (6 o'clock) is 15 ft above ground level at the point $(0, 15)$ of a rectangular coordinate system. Find parametric equations for the position of Lucinda as a function of time $t$ in seconds if Lucinda starts ($t = 0$) at the point $(35, 50)$.

**81. Ferris Wheel Problem** The lowest point of a Ferris wheel (6 o'clock) of radius 40 ft is 10 ft above the ground, and the center is on the $y$-axis. Find parametric equations for Henry's position as a function of time $t$ in seconds if his starting position ($t = 0$) is the point $(0, 10)$ and the wheel turns at the rate of one revolution every 15 sec.

**82. Ferris Wheel Problem** Sarah rides the Ferris wheel described in Exercise 81. Find parametric equations for Sarah's position as a function of time $t$ in seconds if her starting position ($t = 0$) is the point $(0, 90)$ and the wheel turns at the rate of one revolution every 18 sec.

**83. Epicycloid** The graph of the parametric equations

$$x = 4 \cos t - \cos 4t, \quad y = 4 \sin t - \sin 4t$$

is an *epicycloid*. The graph is the path of a point $P$ on a circle of radius 1 rolling along the outside of a circle of radius 3, as suggested in the figure.

**(a)** Graph simultaneously this epicycloid and the circle of radius 3.

**(b)** Suppose the large circle has a radius of 4. Experiment! How do you think the equations in part a should be changed to obtain defining equations? What do you think the epicycloid would look like in this case? Check your guesses.

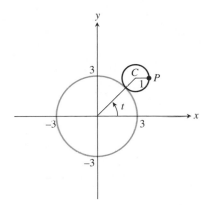

**84. Throwing a Baseball** Sharon releases a baseball 4 ft above the ground with an initial velocity of 66 ft/sec at an angle of 5° with the horizontal. How many seconds after the ball is thrown will it hit the ground? How far from Sharon will the ball be when it hits the ground?

**85. Throwing a Baseball** Diego releases a baseball 3.5 ft above the ground with an initial velocity of 66 ft/sec at an angle of 12° with the horizontal. How many seconds after the ball is thrown will it hit the ground? How far from Diego will the ball be when it hits the ground?

**86. Field Goal Kicking** Spencer practices kicking field goals 40 yd from a goal post with a crossbar 10 ft high. If he kicks the ball with an initial velocity of 70 ft/sec at a 45° angle with the horizontal (see figure), will Spencer make the field goal if the kick sails "true"? it clears the crossbar

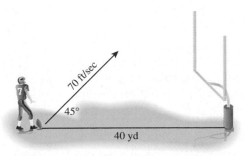

**87. Hang Time** An NFL punter kicks a football downfield with an initial velocity of 85 ft/sec. The ball leaves his foot at the 15 yard line at an angle of 56° with the horizontal. Determine the following:

**(a)** The ball's maximum height above the field.

**(b)** The "hang time" (the total time the football is in the air).

**88. Baseball Hitting** Brian hits a baseball straight toward a 15-ft-high fence that is 400 ft from home plate. The ball is hit when it is 2.5 ft above the ground and leaves the bat at an angle of 30° with the horizontal. Find the initial velocity needed for the ball to clear the fence. just over 125 ft/sec

**89. Throwing a Ball at a Ferris Wheel** A 60-ft-radius Ferris wheel turns counterclockwise one revolution every 12 sec. Sam stands at a point 80 ft to the left of the bottom (6 o'clock) of the wheel. At the instant Kathy is at point $A$ (3 o'clock), Sam throws a ball with an initial velocity of 100 ft/sec and an angle with the horizontal of 70°. He releases the ball from the same height as the bottom of the Ferris wheel. Find the minimum distance between the ball and Kathy. ≈ 17.65 ft

**90. Yard Darts** Gretta and Lois are launching yard darts 20 ft from the front edge of a circular target of radius 18 in. If Gretta releases the dart 5 ft above the ground with an initial velocity of 20 ft/sec and at a 50° angle with the horizontal, will the dart hit the target? no

# Chapter 6 Project

## Parametrizing Ellipses

As you discovered in the Chapter 4 Data Project, it is possible to model the displacement of a swinging pendulum using a sinusoidal equation of the form:

$$x = a \sin (b(t - c)) + d$$

where *x* represents the pendulum's distance from a fixed point and *t* represents total elapsed time. In fact, a pendulum's velocity behaves sinusoidally as well:

$$y = ab \cos (b(t - c))$$

where *y* represents the pendulum's velocity and *a*, *b*, and *c* are constants common to both the displacement and velocity equations.

In this project, you will use a motion detection device to collect distance, velocity, and time data for a swinging pendulum, then determine how a resulting plot of velocity versus displacement (called a phase-space plot) can be modeled using parametric equations.

## Explorations

To start, construct a simple pendulum by fastening about 1 meter of string to the end of a ball. Setup the Calculator Based Laboratory (CBL) system with a motion detector or a Calculator Based Ranger (CBR) system to collect time, distance, and velocity readings for between 2 and 4 seconds (enough time to capture at least one complete swing of the pendulum). See the CBL/CBR guidebook for specific setup instruction. Start the pendulum swinging in front of the detector, then activate the system. The data table below shows a sample set of data collected as a pendulum swung back and forth in front of a CBR where *t* is total elapsed time in seconds, *d* = distance from the CBR in meters, *v* = velocity in meters/second.

| *t* | *d* | *v* | *t* | *d* | *v* | *t* | *d* | *v* |
|---|---|---|---|---|---|---|---|---|
| 0 | 1.021 | 0.325 | 0.7 | 0.621 | −0.869 | 1.4 | 0.687 | 0.966 |
| 0.1 | 1.038 | 0.013 | 0.8 | 0.544 | −0.654 | 1.5 | 0.785 | 1.013 |
| 0.2 | 1.023 | −0.309 | 0.9 | 0.493 | −0.359 | 1.6 | 0.880 | 0.826 |
| 0.3 | 0.977 | −0.598 | 1.0 | 0.473 | −0.044 | 1.7 | 0.954 | 0.678 |
| 0.4 | 0.903 | −0.819 | 1.1 | 0.484 | 0.263 | 1.8 | 1.008 | 0.378 |
| 0.5 | 0.815 | −0.996 | 1.2 | 0.526 | 0.573 | 1.9 | 1.030 | 0.049 |
| 0.6 | 0.715 | −0.979 | 1.3 | 0.596 | 0.822 | 2.0 | 1.020 | −0.260 |

## Explorations

1. If you collected motion data using a CBL or CBR, make a scatter plot of distance versus time on your graphing calculator or computer. If you don't have access to a CBL/CBR, enter the data in the table above into your graphing calculator/computer. Create a scatter plot for the data.

2. With your calculator/computer in function mode, find values for *a*, *b*, *c*, and *d* so that the equation $y = a \sin (b(x - c)) + d$ (where *y* is distance and *x* is time) fits the distance versus time data plot. Refer to the information box on page 374 in Chapter 4 to review sinusoidal graph characteristics.

3. Make a scatter plot of velocity versus time. Using the same *a*, *b*, and *c* values you found in (2), verify that the equation $y = ab \cos (b(x - c))$ (where *y* is velocity and *x* is time) fits the velocity versus time data plot.

4. What do you think a plot of velocity versus distance (with velocity on the vertical axis and distance on the horizontal axis) would look like? Make a rough sketch of your prediction, then create a scatter plot of velocity versus distance. How well did your predicted graph match the actual data plot?

5. You may think it would be difficult to find the equation of the ellipse shown in the velocity versus time scatter plot you just made. But believe it or not, you've already found an equation for this ellipse! With your calculator/computer in parametric mode, graph the parametric curve $x = a \sin (b(t - c)) + d$, $y = ab \cos (b(t - c))$, $0 \le t \le 2$ where $x$ represents distance, $y$ represents velocity, and $t$ is the time parameter. How well does this curve match the scatter plot of velocity versus time?

6. Eliminate the parameter for the parametric curve given in (5) to show that the equation is an ellipse centered at $(d, 0)$ of the form

$$\left(\frac{x - d}{a}\right)^2 + \left(\frac{y}{ab}\right)^2 = 1$$

## Solution

### Solve Algebraically

Solving the first equation for $y$ yields $y = 2x - 10$. Then substitute the expression for $y$ into the second equation.

$$3x + 2y = 1 \qquad \text{Second equation}$$

$$3x + 2(2x - 10) = 1 \qquad \text{Replace } y \text{ by } 2x - 10.$$

$$3x + 4x - 20 = 1$$

$$7x = 21$$

$$x = 3$$

$$y = -4 \qquad \text{Use } y = 2x - 10.$$

### Support Graphically

The graph of each equation is a line. Figure 7.1 shows that the two lines intersect in the single point $(3, -4)$.

### Interpret

The solution of the system is $x = 3$, $y = -4$, or the ordered pair $(3, -4)$.

The method of substitution can sometimes be applied when the equations in the system are not linear, as illustrated in Example 2.

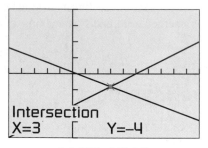

[-5, 10] by [-20, 20]

**Figure 7.1** The two lines $y = 2x - 10$ and $y = -1.5x + 0.5$ intersect in the point $(3, -4)$. (Example 1)

### Example 2   SOLVING A NONLINEAR SYSTEM BY SUBSTITUTION

Find the dimensions of a rectangular garden that has perimeter 100 ft and area 300 ft$^2$.

## Solution

### Model

Let $x$ and $y$ be the lengths of adjacent sides of the garden (Figure 7.2). Then

$$2x + 2y = 100 \qquad \text{Perimeter is 100.}$$

$$xy = 300 \qquad \text{Area is 300.}$$

### Solve Algebraically

Solving the first equation for $y$ yields $y = 50 - x$. Then substitute the expression for $y$ into the second equation.

$$xy = 300 \qquad \text{Second equation}$$

$$x(50 - x) = 300 \qquad \text{Replace } y \text{ by } 50 - x.$$

$$50x - x^2 = 300$$

$$x^2 - 50x + 300 = 0$$

$$x = \frac{50 \pm \sqrt{(-50)^2 - 4(300)}}{2} \qquad \text{Quadratic formula}$$

$$x = 6.972\ldots \quad \text{or} \quad x = 43.027\ldots \qquad \text{Evaluate.}$$

$$y = 43.027\ldots \quad \text{or} \quad y = 6.972\ldots \qquad \text{Use } y = 50 - x.$$

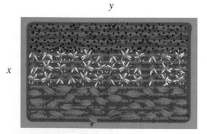

$y$

$x$

**Figure 7.2** The rectangular garden in Example 2.

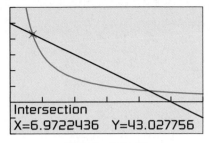

[0, 60] by [−20, 60]

**Figure 7.3**  We can assume $x \geq 0$ and $y \geq 0$ because $x$ and $y$ are lengths. (Example 2)

---

### Rounding at the End

In Example 2, we did *not* round the values found for $x$ until we computed the values for $y$. For the sake of accuracy, do not round *intermediate results*. Carry all decimals on your calculator computations and then round the final answer(s).

---

**Support Graphically**

Figure 7.3 shows that the graphs of $y = 50 - x$ and $y = 300/x$ have two points of intersections.

**Interpret**

The two ordered pairs $(6.972\ldots, 43.027\ldots)$ and $(43.027\ldots, 6.972\ldots)$ produce the same rectangle whose dimensions are approximately 7 ft by 43 ft.

**Example 3**  SOLVING A NONLINEAR SYSTEM BY FACTORING

Solve the system

$$y = x^3 - 6x$$
$$y = 3x \qquad .$$

**Solution**  The graphs of the two equations in Figure 7.4 suggests that this system has three solutions.

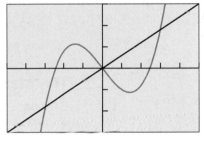

[−5, 5] by [−15, 15]

**Figure 7.4**  The graphs of $y = x^3 - 6x$ and $y = 3x$ have three points of intersection. (Example 3)

Substituting the value of $y$ from the first equation into the second equation yields

$$x^3 - 6x = 3x$$
$$x^3 - 9x = 0$$
$$x(x - 3)(x + 3) = 0$$
$$x = 0,\ x = 3,\ x = -3 \qquad \text{Zero factor property}$$
$$y = 0,\ y = 9,\ y = -9 \qquad \text{Use } y = 3x.$$

The system of equations has three solutions: $(-3, -9)$, $(0, 0)$, and $(3, 9)$.

## Solving Systems Graphically

Sometimes the method of substitution leads to an equation in one variable that we are not able to solve using the standard algebraic techniques we have studied in this text. In these cases we can solve the system graphically by finding intersections as illustrated in Exploration 1.

**Exploration 1** **Solving a System Graphically**

Consider the system:

$$y = \ln x$$
$$y = x^2 - 4x + 2$$

1. Draw the graphs of the two equations in the [0, 10] by [−5, 5] viewing window.
2. Use the graph in part 1 to find the coordinates of the points of intersection shown in the viewing window.
3. Use your knowledge about the graphs of logarithmic and quadratic functions to explain why this system has exactly two solutions.

Substituting the expression for $y$ of the first equation of Exploration 1 into the second equation yields

$$\ln x = x^2 - 4x + 2.$$

We have no standard algebraic technique to solve this equation.

## The Method of Elimination

Consider a system of two linear equations in $x$ and $y$. To **solve by elimination**, we rewrite the two equations as two equivalent equations so that one of the variables has opposite coefficients. Then we add the two equations to eliminate that variable.

**Example 4** USING THE ELIMINATION METHOD

Solve the system

$$2x + 3y = 5$$
$$-3x + 5y = 21.$$

**Solution**

**Solve Algebraically**

Multiply the first equation by 3 and the second equation by 2 to obtain

$$6x + 9y = 15$$
$$-6x + 10y = 42.$$

Then add the two equations to eliminate the variable $x$.

$$19y = 57$$

Next divide by 19 to solve for $y$.

$$y = 3$$

Finally, substitute $y = 3$ into either of the two original equations to determine that $x = -2$.

The solution of the original system is $(-2, 3)$.

### Example 5 FINDING NO SOLUTION

Solve the system

$$x - 3y = -2$$
$$2x - 6y = 4$$

**Solution** We use the elimination method.

**Solve Algebraically**

$$-2x + 6y = 4 \qquad \text{Multiply first equation by } -2.$$

$$2x - 6y = 4 \qquad \text{Second equation}$$

$$0 = 8 \qquad \text{Add.}$$

The last equation is true for *no* values of $x$ and $y$. The system has no solution.

### Support Graphically

Figure 7.5 suggests that the two lines that are the graphs of the two equations in the system are parallel. Solving for $y$ in each equation yields

$$y = \frac{1}{3}x + \frac{2}{3}$$

$$y = \frac{1}{3}x - \frac{2}{3}$$

The two lines have the same slope of 1/3 and are therefore parallel.

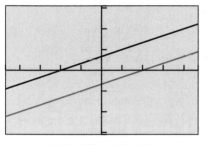

[–4.7, 4.7] by [–3.1, 3.1]

**Figure 7.5** The graph of the two lines in Example 5 in this square viewing window appear to be parallel.

An easy way to determine the *number of solutions* of a system of two linear equations in two variables is to look at the graphs of the two lines. There are three possibilities. The two lines can intersect in a single point, producing exactly *one* solution as in Examples 1 and 4. The two lines can be parallel, producing *no* solution as in Example 5. The two lines can be the same, producing infinitely many solutions as illustrated in Example 6.

### Example 6 FINDING INFINITELY MANY SOLUTIONS

Solve the system

$$4x - 5y = 2$$
$$-12x + 15y = -6$$

**Solution**

$$12x - 15y = 6 \qquad \text{Multiply first equation by 3.}$$

$$-12x + 15y = -6 \qquad \text{Second equation}$$

$$0 = 0 \qquad \text{Add.}$$

The last equation is true for all values of $x$ and $y$. Thus, every ordered pair that satisfies one equation satisfies the other equation. The system has infinitely many solutions.

Another way to see that there are infinitely many solutions is to solve each equation for $y$. Both equations yield

$$y = \frac{4}{5}x - \frac{2}{5}.$$

The two lines are the same.

## Applications

Table 7.1 shows the amount (in billions of dollars) of riding and non-riding lawnmower sales for several years. The amounts for 1999 and 2000 were estimates.

| | Table 7.1 Lawnmower Sales | |
|---|---|---|
| Year | Riding (billions) | Non-riding (billions) |
| 1992 | $1.4 | $0.9 |
| 1997 | $1.8 | $1.2 |
| 1998 | $1.9 | $1.2 |
| 1999 | $2.1 | $1.2 |
| 2000 | $2.4 | $1.3 |

Source: The Freedonia Group as reported in the USA TODAY on Tuesday, October 12, 1999.

### Example 7 ESTIMATING SALES WITH LINEAR MODELS

**(a)** Find linear regression equations for the riding and non-riding lawnmower data in Table 7.1. Superimpose their graphs on a scatter plot of the data.

**(b)** Use the models in (a) to estimate when the sales of riding and non-riding mowers were the same and the amount of sales.

### Solution

**(a)** Let $x = 0$ stand for 1990, $x = 1$ for 1991, and so forth. We use a graphing calculator to find linear regression equations for the riding data $y_R$ and the non-riding data $y_{NR}$:

$$y_R \approx 0.113x + 1.107$$
$$y_{NR} \approx 0.047x + 0.819$$

Figure 7.6 shows the two regression equations together with a scatter plot of the two sets of data.

**(b)** Figure 7.6 shows that the graphs of $y_R$ and $y_{NR}$ intersect at approximately $(-4.41, 0.61)$. $x = -4$ stands for 1986 and $x = -5$ stands for 1985. The sales of riding and non-riding mowers were both about 0.61 billion sometime during 1986.

Suppliers will usually increase production, $x$, if they can get higher prices, $p$, for their products. So, as one variable increases, the other also increases. Normal mathematical practice would be to use $p$ as the independent variable and $x$ as the dependent variable. However, most economists put $x$ on the horizontal axis and $p$ on the vertical axis. In keeping with this practice, we write $p = f(x)$ for a **supply curve**. On one hand, as the price increases (vertical axis) so does the willingness for suppliers to increase production $x$ (horizontal axis).

On the other hand, the demand, $x$, for a product by consumers will decrease as the price, $p$, goes up. So, as one variable increases, the other decreases. Again economists put $x$ (demand) on the horizontal axis and $p$ (price) on the vertical

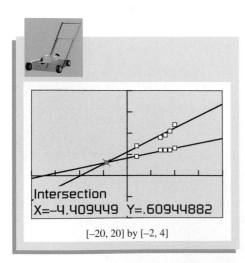

Intersection
X=-4.409449  Y=.60944882

[–20, 20] by [–2, 4]

**Figure 7.6** The scatter plots and regression equations for the data in Table 7.1. (Example 7)

axis, even though it seems like $p$ should be the dependent variable. In keeping with this practice, we write $p = g(x)$ for a **demand curve**.

Finally, a point where the supply curve and demand curve intersect is an **equilibrium point**. The corresponding price is the **equilibrium price**.

### Example 8   DETERMINING THE EQUILIBRIUM PRICE

Nibok Manufacturing has determined that production and price of a new tennis shoe should be geared to the equilibrium point for this system of equations.

$$p = 160 - 5x \qquad \text{Demand curve}$$

$$p = 35 + 20x \qquad \text{Supply curve}$$

The price, $p$, is in dollars and the number of shoes, $x$, is in millions of pairs. Find the equilibrium point.

### Solution

We use substitution to solve the system.

$$160 - 5x = 35 + 20x$$

$$25x = 125$$

$$x = 5$$

Substitute this value of $x$ into the demand curve and solve for $p$.

$$p = 160 - 5x$$

$$p = 160 - 5(5) = 135$$

The equilibrium point is $(5, 135)$. The equilibrium price is \$135, the price for which supply and demand will be equal at 5 million pairs of tennis shoes.

# Quick Review 7.1

In Exercises 1 and 2, solve for $y$ in terms of $x$.

**1.** $2x + 3y = 5$          **2.** $xy + x = 4$   $y = 4/x - 1$

In Exercises 3–6, solve the equation algebraically.

**3.** $3x^2 - x - 2 = 0$        **4.** $2x^2 + 5x - 10 = 0$

**5.** $x^3 = 4x$              **6.** $x^3 + x^2 = 6x$

**7.** Write an equation for the line through the point $(-1, 2)$ and parallel to the line $4x + 5y = 2$.   $y = (-4x + 6)/5$

**8.** Write an equation for the line through the point $(-1, 2)$ and perpendicular to the line $4x + 5y = 2$.   $y = (5x + 13)/4$

**9.** Write an equation equivalent to $2x + 3y = 5$ with coefficient of $x$ equal to $-4$.   $-4x - 6y = -10$

**10.** Find the points of intersection of the graphs of $y = 3x$ and $y = x^3 - 6x$ graphically.

# Section 7.1 Exercises

In Exercises 1 and 2, determine whether the ordered pair is a solution of the system.

**1.** $5x - 2y = 8$
    $2x - 3y = 1$

   **(a)** $(0, 4)$   No      **(b)** $(2, 1)$   Yes      **(c)** $(-2, -9)$   No

**2.** $y = x^2 - 6x + 5$
    $y = 2x - 7$

   **(a)** $(2, -3)$   Yes      **(b)** $(1, -5)$   No      **(c)** $(6, 5)$   Yes

In Exercises 3–12, solve the system by substitution.

**3.** $x + 2y = 5$
   $y = -2$  (9, -2)

**4.**   $x = 3$
   $x - y = 20$  (3, -17)

**5.**   $3x + y = 20$
   $x - 2y = 10$  (50/7, -10/7)

**6.** $2x - 3y = -23$
   $x + y = 0$

**7.** $2x - 3y = -7$
   $4x + 5y = 8$  (-1/2, 2)

**8.** $3x + 2y = -5$
   $2x - 5y = -16$  (-3, 2)

**9.**   $x - 3y = 6$
   $-2x + 6y = 4$  No solution

**10.**   $3x - y = -2$
   $-9x + 3y = 6$

**11.**   $y = x^2$
   $y - 9 = 0$  (±3, 9)

**12.**   $x = y + 3$  (0, -3)
   $x - y^2 = 3y$  and (4, 1)

In Exercises 13–18, solve the system by elimination.

**13.** $x - y = 10$
   $x + y = 6$  (8, -2)

**14.** $2x + y = 10$
   $x - 2y = -5$  (3, 4)

**15.** $3x - 2y = 8$
   $5x + 4y = 28$  (4, 2)

**16.** $4x - 5y = -23$
   $3x + 4y = 6$  (-2, 3)

**17.** $2x - 4y = 8$
   $-x + 2y = -4$

**18.**   $2x - y = 3$
   $-4x + 2y = 5$  No solution

In Exercises 19–22, use the graph to estimate any solutions of the system.

**19.** $y = 1 + 2x - x^2$  (0, 1)
   $y = 1 - x$  and (3, -2)

**20.** $6x - 2y = 7$  (1.5, 1)
   $2x + y = 4$

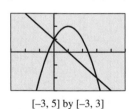
[-3, 5] by [-3, 3]

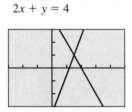
[-3, 5] by [-3, 3]

**21.**   $x + 2y = 0$
   $0.5x + y = 2$  No solution

**22.** $x^2 + y^2 = 16$  (0, -4) and
   $y + 4 = x^2$ ≈ (±2.65, 3)

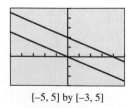
[-5, 5] by [-3, 5]

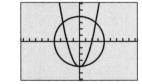
[-9.4, 9.4] by [-6.2, 6.2]

In Exercises 23–26, use graphs to determine the number of solutions the system has.

**23.** $3x + 5y = 7$  One
   $4x - 2y = -3$  solution

**24.** $3x - 9y = 6$  No solution
   $2x - 6y = 1$

**25.** $2x - 4y = 6$  Infinitely
   $3x - 6y = 9$  solutions

**26.** $x - 7y = 9$  One
   $3x + 4y = 1$  solution

In Exercises 27–34, solve the system graphically. Support your answer numerically.

**27.** $y = \ln x$  ≈ (0.69, -0.37)
   $1 = 2x + y$

**28.** $y = 3 \cos x$ ≈ (1.13, 1.27)
   $1 = 2x - y$

**29.** $y = x^3 - 4x$
   $4 = x - 2y$

**30.** $y = x^2 - 3x - 5$
   $1 = 2x - y$

**31.** $x^2 + y^2 = 4$  (-1.2, 1.6)
   $x + 2y = 2$  and (2, 0)

**32.** $x^2 + y^2 = 4$
   $x - 2y = 2$

**33.** $x^2 + y^2 = 9$
   $y = x^2 - 2$

**34.** $x^2 + y^2 = 9$
   $y = 2 - x^2$

In Exercises 35–40, solve the system algebraically. Support your answer graphically.

**35.**   $y = 6x^2$ (-3/2, 27/2)
   $7x + y = 3$  and (1/3, 2/3)

**36.**   $y = 2x^2 + x$ (-4, 28)
   $2x + y = 20$ and (5/2, 15)

**37.** $y = x^3 - x^2$  (0, 0)
   $y = 2x^2$  and (3, 18)

**38.** $y = x^3 + x^2$  (0, 0) and
   $y = -x^2$   (-2, -4)

**39.** $x^2 + y^2 = 9$
   $x - 3y = -1$

**40.**   $x^2 + y^2 = 16$
   $4x + 7y = 13$

In Exercises 41 and 42, find the equilibrium point for the given demand and supply curve.

**41.** $p = 200 - 15x$   Demand curve
   $p = 50 + 25x$   Supply curve   (3.75, 143.75)

**42.** $p = 15 - \dfrac{7}{100}x$   Demand curve

   $p = 2 + \dfrac{3}{100}x$   Supply curve   (130, 5.9)

# Explorations

**43. An Ellipse and a Line**  Consider the system of equations

$$\frac{x^2}{4} + \frac{y^2}{9} = 1.$$
$$x + y = 1$$

**(a)** Solve the equation $x^2/4 + y^2/9 = 1$ for $y$ in terms of $x$ to determine the two implicit functions determined by the equation.  $y = (3/2)\sqrt{4 - x^2}, y = -(3/2)\sqrt{4 - x^2}$

**(b)** Solve the system of equations graphically.

**(c)** Use substitution to confirm the solutions found in (b).

**44. A Hyperbola and a Line**  Consider the system of equations

$$\frac{x^2}{4} - \frac{y^2}{9} = 1.$$
$$x - y = 0$$

**(a)** Solve the equation $x^2/4 - y^2/9 = 1$ for $y$ in terms of $x$ to determine the two implicit functions determined by the equation.  $y = (3/2)\sqrt{x^2 - 4}, y = -(3/2)\sqrt{x^2 - 4}$

**(b)** Solve the system of equations graphically.

**(c)** Use substitution to confirm the solutions found in (b).  ∎

**45. Medicare Expenditure**  Table 7.2 shows expenditures (in billions of dollars) for benefits and administrative cost from federal hospital and medical insurance trust funds for several years. Let $x = 0$ stand for 1970, $x = 1$ for 1971, and so forth.

**(a)** Find the quadratic regression equation and superimpose its graph on a scatter plot of the data.

**(b)** Find the natural logarithmic regression equation and superimpose its graph on a scatter plot of the data.

**(c)** Find when the models in (a) and (b) predict the same expenditure amounts.  1982 and 1993

**(d) Writing to Learn** Which model might an optimist use to predict future medicare expenditures? Explain. Which model might a pessimist use to predict future medicare expenditures? Explain.

### Table 7.2  Medicare National Health Expenditures

| Year | Expenditures (billions) |
|------|-------------------------|
| 1980 | 37.5 |
| 1985 | 72.1 |
| 1990 | 112.1 |
| 1991 | 124.4 |
| 1992 | 141.4 |
| 1993 | 153.0 |
| 1994 | 169.8 |
| 1995 | 187.9 |
| 1996 | 203.1 |

*Source: U.S. Health Care Financing Administration, Health Care Financing Review, fall 1997, in Statistical Abstract of the U.S., 1998.*

**46. Personal Income**  Table 7.3 gives the total personal income (in billions of dollars) for residents of the states of Colorado and Connecticut for several years. Let $x = 0$ stand for 1990, $x = 1$ for 1991, and so forth.

**(a)** Find the linear regression equation for the Colorado data and superimpose its graph on a scatter plot of the Colorado data.

**(b)** Find the linear regression equation for the Connecticut data and superimpose its graph on a scatter plot of the Connecticut data.

**(c)** Using the models in (a) and (b), when will the personal income of the two states be the same?  2009

### Table 7.3  Total Personal Income

| Year | Colorado (billions) | Connecticut (billions) |
|------|---------------------|------------------------|
| 1990 | 63.8 | 87.2 |
| 1995 | 91.7 | 106.5 |
| 1996 | 98.2 | 117.5 |
| 1997 | 105.3 | 118.6 |

*Source: U.S. Bureau of Economic Analysis, Survey of Current Business, May 1998, in Statistical Abstract of the U.S., 1998.*

**47. Population**  Table 7.4 gives the population (in thousands) of the states of Arizona and Maine for several years. Let $x = 0$ stand for 1970, $x = 1$ for 1971, and so forth.

**(a)** Find the linear regression equation for Arizona's data and superimpose its graph on a scatter plot of Arizona's data.

**(b)** Find the linear regression equation for Maine's data and superimpose its graph on a scatter plot of Maine's data.

**(c)** Using the models in (a) and (b), when will the population of the two states be the same?  2116

### Table 7.4  Population

| Year | Arizona (in thousands) | Maine (in thousands) |
|------|------------------------|----------------------|
| 1970 | 1775 | 5689 |
| 1980 | 2718 | 5737 |
| 1985 | 3184 | 5881 |
| 1990 | 3665 | 6016 |
| 1991 | 3763 | 5996 |
| 1992 | 3868 | 5991 |
| 1993 | 3994 | 6008 |
| 1994 | 4149 | 6029 |
| 1995 | 4308 | 6061 |
| 1996 | 4434 | 6085 |
| 1997 | 4556 | 6118 |

*Source: U.S. Bureau of the Census, in Statistical Abstract of the U.S., 1998.*

**48. Group Activity**  Describe all possibilities for the number of solutions to a system of two equations in two variables if the graphs of the two equations are **(a)** a line and a circle, and **(b)** a circle and a parabola.

**49. Garden Problem**  Find the dimensions of a rectangle with a perimeter of 200 m and an area of 500 m$^2$.

**50. Cornfield Dimensions**  Find the dimensions of a rectangular cornfield with a perimeter of 220 yd and an area of 3000 yd$^2$.  50 yd × 60 yd

**51. Rowing Speed**  Hank can row a boat 1 mi upstream (against the current) in 24 min. He can row the same distance downstream in 13 min. If both the rowing speed and current speed are constant, find Hank's rowing speed and the speed of the current.

**52. Airplane Speed**  An airplane flying with the wind from Los Angeles to New York City takes 3.75 hr. Flying against the wind, the airplane takes 4.4 hr for the return trip. If the air distance between Los Angeles and New York is 2500 mi and the airplane speed and wind speed are constant, find the airplane speed and the wind speed.

**53. Food Prices**  At Philip's convenience store the total cost of one medium and one large soda is $1.74. The large soda costs $0.16 more than the medium soda. Find the cost of each soda.  medium: $0.79; large: $0.95

**54. Nut Mixture** A 5-lb nut mixture is worth $2.80 per pound. The mixture contains peanuts worth $1.70 per pound and cashews worth $4.55 per pound. How many pounds of each type of nut are in the mixture?

**55. Connecting Algebra and Functions** Determine $a$ and $b$ so that the graph of $y = ax + b$ contains the two points $(-1, 4)$ and $(2, 6)$. $a = 2/3$ and $b = 14/3$

**56. Connecting Algebra and Functions** Determine $a$ and $b$ so that the graph of $ax + by = 8$ contains the two points $(2, -1)$ and $(-4, -6)$. $a = 5/2$ and $b = -3$

**57. Rental Van** Pedro has two plans to choose from to rent a van.

Company A: a flat fee of $40 plus 10 cents a mile.

Company B: a flat fee of $25 plus 15 cents a mile.

**(a)** How many miles can Pedro drive in order to be charged the same amount by the two companies? 300 miles

**(b) Writing to Learn** Give reasons why Pedro might choose one plan over the other. Explain.

**58. Salary Package** Stephanie is offered two different salary options to sell major household appliances.

Plan A: a $300 weekly salary plus 5% of her sales.

Plan B: a $600 weekly salary plus 1% of her sales.

**(a)** What must Stephanie's sales be to earn the same amount on the two plans? $7500

**(b) Writing to Learn** Give reasons why Stephanie might choose one plan over the other. Explain.

## Extending the Ideas

In Exercises 59 and 60, use the elimination method to solve the system of equations.

**59.** $x^2 - 2y = -6$
     $x^2 + y = 4$

**60.** $x^2 + y^2 = 1$
     $x^2 - y^2 = 1$  $(\pm 1, 0)$

In Exercises 61 and 62, $p(x)$ is the demand curve. The total revenue if $x$ units are sold is $R = px$. Find the number of units sold that gives the maximum revenue.

**61.** $p = 100 - 4x$  12.5 units   **62.** $p = 80 - x^2$  $\approx 5.16$ units

---

## 7.2 Matrix Algebra

Matrices • Matrix Addition and Subtraction • Matrix Multiplication • Identity and Inverse Matrices • Determinant of a Square Matrix • Applications

### Matrices

A *matrix* is a rectangular array of numbers. Matrices provide an efficient way to solve systems of linear equations and to record data. The tables of data presented in this textbook are examples of matrices.

---

**Definition  Matrix**

Let $m$ and $n$ be positive integers. An **$m \times n$ matrix** (read "$m$ by $n$ matrix") is a rectangular array of $m$ rows and $n$ columns of real numbers.

$$\begin{bmatrix} a_{11} & a_{12} & \cdots & a_{1n} \\ a_{21} & a_{22} & \cdots & a_{2n} \\ \vdots & \vdots & & \vdots \\ a_{m1} & a_{m2} & \cdots & a_{mn} \end{bmatrix}$$

We also use the shorthand notation $[a_{ij}]$ for this matrix.

---

Each **element**, or **entry**, $a_{ij}$, of the matrix uses *double subscript* notation. The **row subscript** is the first subscript $i$, and the **column subscript** is $j$. The element $a_{ij}$ is in the $i$th row and $j$th column. In general, the **order of an $m \times n$ matrix** is $m \times n$. If $m = n$, the matrix is a **square matrix**. Two matrices are

**equal matrices** if they have the same order and their corresponding elements are equal.

**Example 1**  DETERMINING THE ORDER OF A MATRIX

(a) The matrix $\begin{bmatrix} 1 & -2 & 3 \\ 2 & 0 & 4 \end{bmatrix}$ has order $2 \times 3$.

(b) The matrix $\begin{bmatrix} 1 & -1 \\ 0 & 4 \\ 2 & -1 \\ 3 & 2 \end{bmatrix}$ has order $4 \times 2$.

(c) The matrix $\begin{bmatrix} 1 & 2 & 3 \\ 4 & 5 & 6 \\ 7 & 8 & 9 \end{bmatrix}$ has order $3 \times 3$ and is a square matrix.

## Matrix Addition and Subtraction

We add or subtract two matrices of the same order by adding or subtracting their corresponding entries. Matrices of different orders can *not* be added or subtracted.

---

**Definition**  **Matrix Addition and Matrix Subtraction**

Let $A = [a_{ij}]$ and $B = [b_{ij}]$ be matrices of order $m \times n$.

1. The **sum $A + B$** is the $m \times n$ matrix

$$A + B = [a_{ij} + b_{ij}].$$

2. The **difference $A - B$** is the $m \times n$ matrix

$$A - B = [a_{ij} - b_{ij}].$$

---

**Example 2**  USING MATRIX ADDITION

Matrix $A$ gives the mean SAT verbal scores for the six New England states over the time period from 1995 to 1998. *(Source: The College Board, 1999 World Almanac and Book of Facts.)* Matrix $B$ gives the mean SAT mathematics scores for the same 4-year period. Express the mean combined scores for the New England states from 1995 to 1998 as a single matrix.

$$A = \begin{array}{c} \\ CT \\ ME \\ MA \\ NH \\ RI \\ VT \end{array} \begin{array}{cccc} 95 & 96 & 97 & 98 \\ \begin{bmatrix} 507 & 507 & 509 & 510 \\ 504 & 504 & 507 & 504 \\ 505 & 507 & 508 & 508 \\ 520 & 520 & 521 & 523 \\ 502 & 501 & 499 & 501 \\ 506 & 506 & 508 & 508 \end{bmatrix} \end{array} \quad B = \begin{array}{c} \\ CT \\ ME \\ MA \\ NH \\ RI \\ VT \end{array} \begin{array}{cccc} 95 & 96 & 97 & 98 \\ \begin{bmatrix} 502 & 504 & 507 & 509 \\ 497 & 498 & 504 & 501 \\ 502 & 504 & 508 & 508 \\ 515 & 514 & 518 & 520 \\ 490 & 491 & 493 & 495 \\ 499 & 500 & 502 & 504 \end{bmatrix} \end{array}$$

**Solution** The combined scores can be obtained by adding the two matrices:

$$
A + B = \begin{array}{c}
\phantom{CT} \\
CT \\
ME \\
MA \\
NH \\
RI \\
VT
\end{array}
\begin{array}{cccc}
95 & 96 & 97 & 98 \\
\left[\begin{array}{cccc}
1009 & 1011 & 1016 & 1019 \\
1001 & 1002 & 1011 & 1005 \\
1007 & 1011 & 1016 & 1016 \\
1035 & 1034 & 1039 & 1043 \\
992 & 992 & 992 & 996 \\
1005 & 1006 & 1010 & 1012
\end{array}\right]
\end{array}
$$

This is a fairly simple result, but it is significant that we found (essentially) 24 pieces of information with a single mathematical operation. That is the power of matrix algebra.

When we work with matrices, real numbers are **scalars**. The product of the real number $k$ and the $m \times n$ matrix $A = [a_{ij}]$ is the $m \times n$ matrix

$$kA = [ka_{ij}].$$

The matrix $kA = [ka_{ij}]$ is a **scalar multiple of $A$**.

### Example 3  USING SCALAR MULTIPLICATION

A consumer advocacy group has computed the mean retail prices for brand name products and generic products at three different stores in a major city. The prices are shown in the $3 \times 2$ matrix below.

$$
\begin{array}{c}
\phantom{Store A} \\
Store\ A \\
Store\ B \\
Store\ C
\end{array}
\begin{array}{cc}
Brand & Generic \\
\left[\begin{array}{cc}
3.97 & 3.64 \\
3.78 & 3.69 \\
3.75 & 3.67
\end{array}\right]
\end{array}
$$

The city has a combined sales tax of 7.25%. Construct a matrix showing the comparative prices with sales tax included.

**Solution** Multiply the original matrix by the scalar 1.0725 to add the sales tax to every price.

$$
1.0725 \times \begin{bmatrix}
3.97 & 3.64 \\
3.78 & 3.69 \\
3.75 & 3.67
\end{bmatrix}
\approx
\begin{array}{c}
Store\ A \\
Store\ B \\
Store\ C
\end{array}
\begin{array}{cc}
Brand & Generic \\
\begin{bmatrix}
4.26 & 3.90 \\
4.05 & 3.96 \\
4.02 & 3.94
\end{bmatrix}
\end{array}
$$

Matrices inherit many properties possessed by the real numbers. Let $A = [a_{ij}]$ be any $m \times n$ matrix. The $m \times n$ matrix $O = [0]$ consisting entirely of zeros is the **zero matrix** because $A + O = A$. In other words, $O$ is the **additive identity** for the set of all $m \times n$ matrices. The $m \times n$ matrix $B = [-a_{ij}]$ consisting of the *additive inverses* of the entries of $A$ is the **additive inverse of $A$** because $A + B = O$. We also write $B = -A$. Just as with real numbers,

$$A - B = [a_{ij} - b_{ij}] = [a_{ij} + (-b_{ij})] = [a_{ij}] + [-b_{ij}] = A + (-B).$$

Thus, subtracting $B$ from $A$ is the same as adding the additive inverse of $B$ to $A$.

Let $A = [a_{ij}]$ and $B = [b_{ij}]$ be $2 \times 2$ matrices with $a_{ij} = 3i - j$ and $b_{ij} = i^2 + j^2 - 3$ for $i = 1, 2$ and $j = 1, 2$.

**1.** Determine $A$ and $B$.

**2.** Determine the additive inverse $-A$ of $A$ and verify that $A + (-A) = [0]$. What is the order of $[0]$?

**3.** Determine $3A - 2B$.

**Exploration Extensions**

Let $C = [c_{ij}]$ be a $2 \times 2$ matrix with $c_{ij} = a_{i1}b_{1j} + a_{i2}b_{2j}$; that is,

$$C = \begin{bmatrix} a_{11}b_{11} + a_{12}b_{21} & a_{11}b_{12} + a_{12}b_{22} \\ a_{21}b_{11} + a_{22}b_{21} & a_{21}b_{12} + a_{22}b_{22} \end{bmatrix}$$

Determine $C$.

## Matrix Multiplication

**Teaching Note**

Help students remember how to do matrix multiplication by emphasizing that in most contexts with matrices, rows come before columns. For example, an $m \times n$ matrix has $m$ rows and $n$ columns; $a_{ij}$ is the $i$th row and $j$th column; the rows of the first matrix are multiplied by the columns of the second matrix. Point out that the product $AB$ is not defined unless the number of entries in a row of $A$ is the same as the number of entries in a column of $B$.

To form the *product AB* of two matrices, the number of columns of the matrix $A$ on the left must be equal to the number of rows of the matrix $B$ on the right. In this case, any row of $A$ has the same number of entries as any column of $B$. Each entry of the product is obtained by summing the products of the entries of a row of $A$ by the corresponding entries of a column of $B$.

**Definition** **Matrix Multiplication**

Let $A = [a_{ij}]$ be an $m \times r$ matrix and $B = [b_{ij}]$ an $r \times n$ matrix. The **product** $AB = [c_{ij}]$ is the $m \times n$ matrix where $c_{ij} = a_{i1}b_{1j} + a_{i2}b_{2j} + \cdots + a_{ir}b_{rj}$.

The key to understanding how to form the product of any two matrices is to first consider the product of a $1 \times r$ matrix $A = [a_{1j}]$ with an $r \times 1$ matrix $B = [b_{j1}]$. According to the definition, $AB = [c_{11}]$ is the $1 \times 1$ matrix where $c_{11} = a_{11}b_{11} + a_{12}b_{21} + \cdots + a_{1r}b_{r1}$. For example, the product $AB$ of the $1 \times 3$ matrix $A$ and the $3 \times 1$ matrix $B$, where

$$A = [1 \quad 2 \quad 3] \quad \text{and} \quad B = \begin{bmatrix} 4 \\ 5 \\ 6 \end{bmatrix} \quad \text{is}$$

**Alert**

Point out that matrix multiplication is not commutative.

$$A \cdot B = [1 \quad 2 \quad 3] \cdot \begin{bmatrix} 4 \\ 5 \\ 6 \end{bmatrix} = [1 \cdot 4 + 2 \cdot 5 + 3 \cdot 6] = [32].$$

Then, the *ij*-entry of the product $AB$ of an $m \times r$ matrix with an $r \times n$ matrix is the product of the *i*th row of $A$ considered as a $1 \times r$ matrix with the *j*th column of $B$ considered as a $r \times 1$ matrix as illustrated in Example 4.

### Example 4  FINDING THE PRODUCT OF TWO MATRICES

Find the product $AB$ if possible, where

**(a)** $A = \begin{bmatrix} 2 & 1 & -3 \\ 0 & 1 & 2 \end{bmatrix}$  and  $B = \begin{bmatrix} 1 & -4 \\ 0 & 2 \\ 1 & 0 \end{bmatrix}$.

**(b)** $A = \begin{bmatrix} 2 & 1 & -3 \\ 0 & 1 & 2 \end{bmatrix}$  and  $B = \begin{bmatrix} 3 & -4 \\ 2 & 1 \end{bmatrix}$.

**Solution**

**(a)** The number of columns of $A$ is 3 and the number of rows of $B$ is 3, so the product $AB$ is defined. The product $AB = [c_{ij}]$ is a $2 \times 2$ matrix where

$$c_{11} = \begin{bmatrix} 2 & 1 & -3 \end{bmatrix} \begin{bmatrix} 1 \\ 0 \\ 1 \end{bmatrix} = 2 \cdot 1 + 1 \cdot 0 + (-3) \cdot 1 = -1,$$

$$c_{12} = \begin{bmatrix} 2 & 1 & -3 \end{bmatrix} \begin{bmatrix} -4 \\ 2 \\ 0 \end{bmatrix} = 2 \cdot (-4) + 1 \cdot 2 + (-3) \cdot 0 = -6,$$

$$c_{21} = \begin{bmatrix} 0 & 1 & 2 \end{bmatrix} \begin{bmatrix} 1 \\ 0 \\ 1 \end{bmatrix} = 0 \cdot 1 + 1 \cdot 0 + 2 \cdot 1 = 2,$$

$$c_{22} = \begin{bmatrix} 0 & 1 & 2 \end{bmatrix} \begin{bmatrix} -4 \\ 2 \\ 0 \end{bmatrix} = 0 \cdot (-4) + 1 \cdot 2 + 2 \cdot 0 = 2.$$

Thus, $AB = \begin{bmatrix} -1 & -6 \\ 2 & 2 \end{bmatrix}$. Figure 7.7 supports this computation.

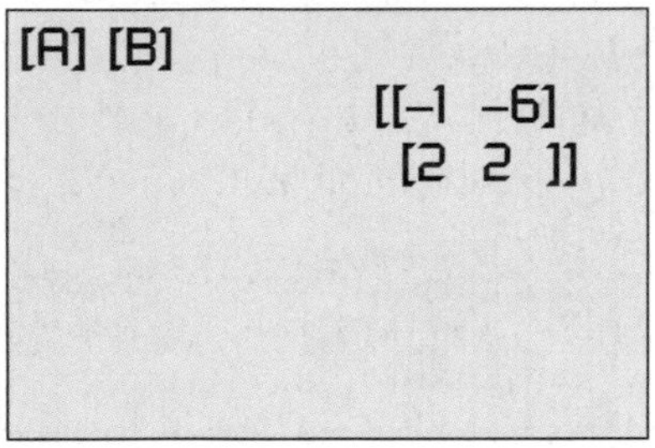

**(b)** The number of columns of $A$ is 3 and the number of rows of $B$ is 2, so the product $AB$ is *not* defined.

```
[A] [B]
        [[-1 -6]
         [2  2 ]]
```

**Figure 7.7** The matrix product $AB$ of Example 4. Notice that the grapher displays the rows of the product as $1 \times 2$ matrices.

### Example 5  USING MATRIX MULTIPLICATION

A florist makes three different cut flower arrangements for Mother's Day (I, II, and III), each involving roses, carnations, and lilies. Matrix $A$ shows the number of each type of flower used in each arrangement.

$$
\begin{array}{c}
\phantom{A = \text{Carnations}} \quad \text{I} \quad \text{II} \quad \text{III} \\
A = \begin{array}{c} \text{Roses} \\ \text{Carnations} \\ \text{Lilies} \end{array} \begin{bmatrix} 5 & 8 & 7 \\ 6 & 6 & 7 \\ 4 & 3 & 3 \end{bmatrix}
\end{array}
$$

The florist can buy his flowers from two different wholesalers (W1 and W2), but wants to give all his business to one or the other. The cost of the three flower types from the two wholesalers is shown in matrix $B$.

$$
\begin{array}{c}
\phantom{B = \text{Carnations}} \quad \text{W1} \quad \text{W2} \\
B = \begin{array}{c} \text{Roses} \\ \text{Carnations} \\ \text{Lilies} \end{array} \begin{bmatrix} 1.50 & 1.35 \\ 0.95 & 1.00 \\ 1.30 & 1.35 \end{bmatrix}
\end{array}
$$

Constuct a matrix showing the cost of making each of the three flower arrangements from flowers supplied by the two different wholesalers.

**Solution**  We can use the labeling of the matricies to help us. We want the columns of $A$ to match up with the rows of $B$ (since that's how the matrix multiplication works). We therefore switch the rows and columns of $A$ to get the flowers along the columns. (The new matrix is called the **transpose** of $A$, denoted by $A^T$.) We then find the product $A^TB$:

$$
\begin{array}{c}
 & \text{Rose Carn Lily} \\
\begin{array}{c}\text{I}\\\text{II}\\\text{III}\end{array} &
\begin{bmatrix} 5 & 6 & 4 \\ 8 & 6 & 3 \\ 7 & 7 & 3 \end{bmatrix}
\end{array}
\times
\begin{array}{c}
 & \text{W1} & \text{W2} \\
\begin{array}{c}\text{Rose}\\\text{Carn}\\\text{Lilly}\end{array} &
\begin{bmatrix} 1.50 & 1.35 \\ 0.95 & 1.00 \\ 1.30 & 1.35 \end{bmatrix}
\end{array}
=
\begin{array}{c}
 & \text{W1} & \text{W2} \\
\begin{array}{c}\text{I}\\\text{II}\\\text{III}\end{array} &
\begin{bmatrix} 18.40 & 18.15 \\ 21.60 & 20.85 \\ 21.05 & 20.50 \end{bmatrix}
\end{array}
$$

Figure 7.8 shows the product $A^TB$ and supports our computation.

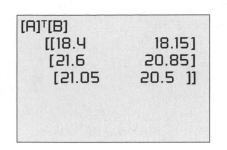

**Figure 7.8** The product $A^TB$ for the matricies $A$ and $B$ of Example 5.

## Identity and Inverse Matrices

We already know that the $n \times n$ matrix consisting entirely of 0's is the *additive identity* for the set of $n \times n$ matrices. The $n \times n$ matrix $I_n$ with 1's on the main diagonal (upper left to lower right) and 0's elsewhere is the **identity matrix of order $n \times n$**

$$
I_n = \begin{bmatrix}
1 & 0 & 0 & \cdots & 0 \\
0 & 1 & 0 & \cdots & 0 \\
0 & 0 & 1 & \cdots & 0 \\
\vdots & \vdots & \vdots & & \vdots \\
0 & 0 & 0 & \cdots & 1
\end{bmatrix}.
$$

For example,

$$
I_2 = \begin{bmatrix} 1 & 0 \\ 0 & 1 \end{bmatrix}, \quad
I_3 = \begin{bmatrix} 1 & 0 & 0 \\ 0 & 1 & 0 \\ 0 & 0 & 1 \end{bmatrix}, \quad \text{and} \quad
I_4 = \begin{bmatrix} 1 & 0 & 0 & 0 \\ 0 & 1 & 0 & 0 \\ 0 & 0 & 1 & 0 \\ 0 & 0 & 0 & 1 \end{bmatrix}.
$$

If $A = [a_{ij}]$ is any $n \times n$ matrix, we can prove (see Exercise 59) that

$$AI_n = I_nA = A,$$

that is, $I_n$ is the **multiplicative identity** for the set of $n \times n$ matrices.

If $a$ is a nonzero real number, then $a^{-1} = 1/a$ is the multiplicative inverse of $a$, that is, $aa^{-1} = a(1/a) = 1$. The definition of the *multiplicative inverse* of a square matrix is similar.

**Definition  Inverse of a Square Matrix**

Let $A = [a_{ij}]$ be an $n \times n$ matrix. If there is a matrix $B$ such that

$$AB = BA = I_n,$$

then $B$ is the **inverse** of $A$. We write $B = A^{-1}$ (read "$A$ inverse").

We will see that not every square matrix (Example 7) has an inverse. If a square matrix $A$ has an inverse, then $A$ is **nonsingular**. If $A$ has no inverse, then $A$ is **singular**.

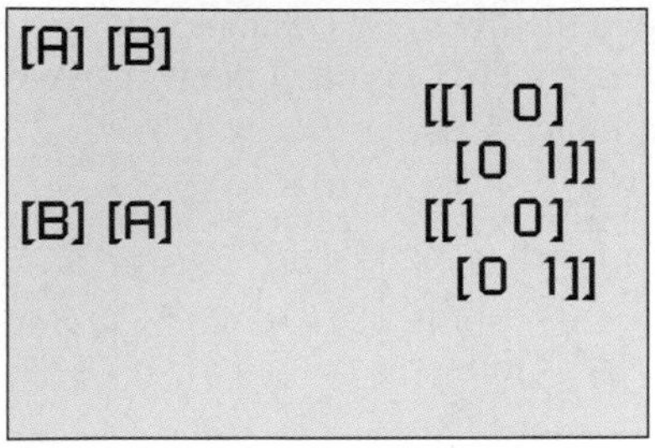

### Example 6  VERIFYING AN INVERSE MATRIX

Prove that

$$A = \begin{bmatrix} 3 & -2 \\ -1 & 1 \end{bmatrix} \quad \text{and} \quad B = \begin{bmatrix} 1 & 2 \\ 1 & 3 \end{bmatrix}$$

are inverse matrices.

**Solution**  Figure 7.9 shows that $AB = BA = I_2$. Thus, $B = A^{-1}$ and $A = B^{-1}$.

### Example 7  SHOWING A MATRIX HAS NO INVERSE

Prove that the matrix $B = \begin{bmatrix} 6 & 3 \\ 2 & 1 \end{bmatrix}$ is singular, that is, $A$ has no inverse.

**Solution**  Suppose $A$ has an inverse $B = \begin{bmatrix} x & y \\ z & w \end{bmatrix}$. Then, $AB = I_2$.

$$AB = \begin{bmatrix} 6 & 3 \\ 2 & 1 \end{bmatrix} \begin{bmatrix} x & y \\ z & w \end{bmatrix} = \begin{bmatrix} 1 & 0 \\ 0 & 1 \end{bmatrix}$$

$$= \begin{bmatrix} 6x + 3z & 6y + 3w \\ 2x + z & 2y + w \end{bmatrix} = \begin{bmatrix} 1 & 0 \\ 0 & 1 \end{bmatrix}$$

Using equality of matrices we obtain:

$$6x + 3z = 1 \qquad 6y + 3w = 0$$

$$2x + z = 0 \qquad 2y + w = 1$$

Multiplying both sides of the equation $2x + z = 0$ by 3 yields $6x + 3z = 0$. There are no values for $x$ and $z$ for which the value of $6x + 3z$ is both 0 and 1. Thus, $A$ does not have an inverse.

## Determinant of a Square Matrix

There is a simple test that determines if a $2 \times 2$ matrix has an inverse.

> **Inverse of a 2 × 2 Matrix**
>
> If $ad - bc \neq 0$, then
> $$\begin{bmatrix} a & b \\ c & d \end{bmatrix}^{-1} = \frac{1}{ad - bc} \begin{bmatrix} d & -b \\ -c & a \end{bmatrix}.$$

The number $ad - bc$ is the **determinant** of the $2 \times 2$ matrix $A = \begin{bmatrix} a & b \\ c & d \end{bmatrix}$ and is denoted

$$\det A = \begin{vmatrix} a & b \\ c & d \end{vmatrix} = ad - bc.$$

To define the determinant of a higher order square matrix we need to introduce the *minors* and *cofactors* associated with the entries of a square matrix.

**Figure 7.9** Showing $A$ and $B$ are inverse matrices. (Example 6)

[A] [B]
        [[1 0]
         [0 1]]
[B] [A]
        [[1 0]
         [0 1]]

**Teaching Note**

On many graphers, the inverse of a square matrix can be calculated using the $x^{-1}$ key. Graphers can also be used to calculate determinants.

Let $A = [a_{ij}]$ be an $n \times n$ matrix. The **minor** (short for "minor determinant") $M_{ij}$ corresponding to the element $a_{ij}$ is the determinant of the $(n - 1) \times (n - 1)$ matrix obtained by deleting the row and column containing $a_{ij}$. The **cofactor** corresponding to $a_{ij}$ is $A_{ij} = (-1)^{i+j}M_{ij}$.

---

**Definition** **Determinant of a Square Matrix**

Let $A = [a_{ij}]$ be a matrix of order $n \times n$ $(n > 2)$. The determinant of $A$, denoted by $\det A$ or $|A|$, is the sum of the entries in any row or any column multiplied by their respective cofactors. For example, expanding by the $i$th row gives

$$\det A = |A| = a_{i1}A_{i1} + a_{i2}A_{i2} + \cdots + a_{in}A_{in}.$$

---

If $A = [a_{ij}]$ is a $3 \times 3$ matrix, then, using the definition of determinant applied to the second row, we obtain

$$\begin{vmatrix} a_{11} & a_{12} & a_{13} \\ a_{21} & a_{22} & a_{23} \\ a_{31} & a_{32} & a_{33} \end{vmatrix} = a_{21}A_{21} + a_{22}A_{22} + a_{23}A_{23}$$

$$= a_{21}(-1)^3\begin{vmatrix} a_{12} & a_{13} \\ a_{32} & a_{33} \end{vmatrix} + a_{22}(-1)^4\begin{vmatrix} a_{11} & a_{13} \\ a_{31} & a_{33} \end{vmatrix}$$

$$+ a_{23}(-1)^5\begin{vmatrix} a_{11} & a_{12} \\ a_{31} & a_{32} \end{vmatrix}$$

$$= -a_{21}(a_{12}a_{33} - a_{13}a_{32}) + a_{22}(a_{11}a_{33} - a_{13}a_{31})$$
$$- a_{23}(a_{11}a_{33} - a_{12}a_{31})$$

The determinant of a $3 \times 3$ matrix involves three determinants of $2 \times 2$ matrices, the determinant of a $4 \times 4$ matrix involves four determinants of $3 \times 3$ matrices, and so forth. This is a tedious definition to apply. Most of the time we use a grapher to evaluate determinants in this textbook.

**Exploration 2** **Investigating the Definition of Determinant**

1. Complete the expansion of the determinant of the $3 \times 3$ matrix $A = [a_{ij}]$ started on the previous page. Explain why each term in the expansion contains an element from each row and each column.

2. Use the first row of the $3 \times 3$ matrix to expand the determinant and compare to the expression in 1.

3. Prove that the determinant of a square matrix with a zero row or a zero column is zero.

4. Prove that the determinant changes sign if two rows or two columns are interchanged. Start with a $3 \times 3$ matrix and compare the expansion by expanding by the same row (or column) before and after the interchange. (*Hint*: Compare without expanding the minors.) How can you generalize from the $3 \times 3$ case?

5. Prove that the determinant of a square matrix with two identical rows or two identical columns is zero.

6. Prove that if a scalar multiple of a row (or column) is added to another row (or column) the value of the determinant of a square matrix is unchanged. (*Hint*: Expand by the row (or column) being added to.)

We can now state the condition under which square matrices have inverses.

**Theorem** **Inverses of $n \times n$ Matrices**

An $n \times n$ matrix $A$ has an inverse if and only if det $A \neq 0$.

There are complicated formulas for finding the inverses of nonsingular matrices of order $3 \times 3$ or higher. We will use a grapher instead of these formulas to find inverses of square matrices.

### Example 8  FINDING INVERSE MATRICES

Determine whether the matrix has an inverse. If so, find its inverse matrix.

**(a)** $A = \begin{bmatrix} 3 & 1 \\ 4 & 2 \end{bmatrix}$   **(b)** $B = \begin{bmatrix} 1 & 2 & -1 \\ 2 & -1 & 3 \\ -1 & 0 & 1 \end{bmatrix}$

**Solution**

**(a)** Since det $A = ad - bc = 3 \cdot 2 - 1 \cdot 4 = 2 \neq 0$, we conclude that $A$ has an inverse. Using the formula for the inverse of a $2 \times 2$ matrix, we obtain

$$A^{-1} = \frac{1}{ad - bc}\begin{bmatrix} d & -b \\ -c & a \end{bmatrix} = \frac{1}{2}\begin{bmatrix} 2 & -1 \\ -4 & 3 \end{bmatrix}$$

$$= \begin{bmatrix} 1 & -0.5 \\ -2 & 1.5 \end{bmatrix}$$

You can check that $A^{-1}A = A^{-1}A = I_2$.

**(b)** Figure 7.10 shows that det $B = -10 \neq 0$ and

$$B^{-1} = \begin{bmatrix} 0.1 & 0.2 & -0.5 \\ 0.5 & 0 & 0.5 \\ 0.1 & 0.2 & 0.5 \end{bmatrix}.$$

You can use your grapher to check that $B^{-1}B = BB^{-1} = I_3$.

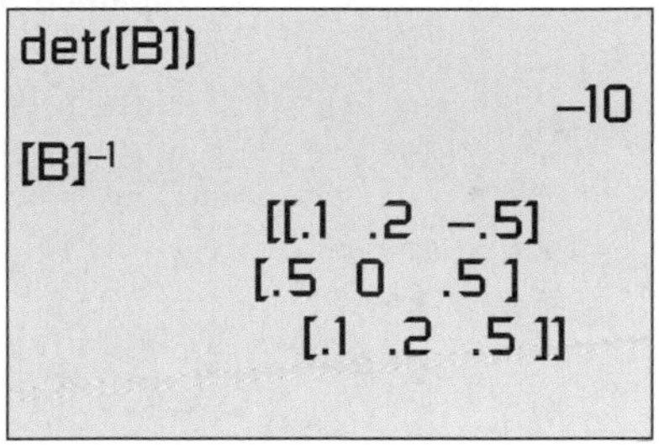

det([B])
                                    −10
[B]⁻¹
                    [[.1 .2 −.5]
                    [.5 0 .5 ]
                    [.1 .2 .5 ]]

**Figure 7.10**  The matrix $B$ is nonsingular and so has an inverse. (Example 8b)

We list some of the important properties of matrices, some of which you will be asked to prove in the exercises.

---

**Properties of Matrices**

Let $A$, $B$, and $C$ be matrices whose orders are such that the following sums, differences, and products are defined.

**1. Commutative property**
Addition:
$A + B = B + A$
Multiplication:
(Does not hold in general)

**2. Associative property**
Addition:
$(A + B) + C = A + (B + C)$
Multiplication:
$(AB)C = A(BC)$

**3. Identity property**
Addition: $A + O = A$
Multiplication: order$(A) = n \times n$
$A \cdot I_n = I_n \cdot A = A$

**4. Inverse property**
Addition: $A + (-A) = O$
Multiplication: order$(A) = n \times n$
$AA^{-1} = A^{-1}A = I_n, \quad |A| \neq 0$

**5. Distributive property**

| **Multiplication over addition** | **Multiplication over subtraction** |
|---|---|
| $A(B + C) = AB + AC$ | $A(B - C) = AB - AC$ |
| $(A + B)C = AC + BC$ | $(A - B)C = AC - BC$ |

---

## Applications

Points in the Cartesian coordinate plane can be represented by $1 \times 2$ matrices. For example, the point $(2, -3)$ can be represented by the $1 \times 2$ matrix $\begin{bmatrix} 2 & -3 \end{bmatrix}$. We can calculate the images of points acted upon by some of the transformations studied in Section 1.5 using matrix multiplication as illustrated in Example 9.

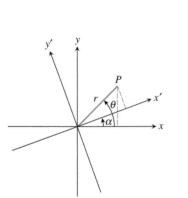

**Figure 7.11** Rotating the $xy$-coordinate system through the angle $\alpha$ to obtain the $x'y'$-coordinate system. (Example 10)

## Example 9  REFLECTING WITH RESPECT TO THE $x$-AXIS AS MATRIX MULTIPLICATION

Prove that the image of a point under a reflection across the $x$-axis can be obtained by multiplying by $\begin{bmatrix} 1 & 0 \\ 0 & -1 \end{bmatrix}$.

**Solution**  The image of the point $(x, y)$ under a reflection across the $x$-axis is $(x, -y)$. The product

$$\begin{bmatrix} x & y \end{bmatrix}\begin{bmatrix} 1 & 0 \\ 0 & -1 \end{bmatrix} = \begin{bmatrix} x & -y \end{bmatrix}$$

shows that the point $(x, y)$ (in matrix form $\begin{bmatrix} x & y \end{bmatrix}$) is moved to the point $(x, -y)$ (in matrix form $\begin{bmatrix} x & -y \end{bmatrix}$).

Figure 7.11 shows the $xy$-coordinate system rotated through the angle $\alpha$ to obtain the $x'y'$-coordinate system. In Example 10, we see that the coordinates of a point in the $x'y'$-coordinate system can be obtained by multiplying the coordinates of the point in the $xy$-coordinate system by an appropriate $2 \times 2$ matrix. In Exercise 51, you will see that the reverse is also true.

## Example 10  ROTATING A COORDINATE SYSTEM

Prove that the $(x', y')$ coordinates of $P$ in Figure 7.11 are related to the $(x, y)$ coordinates of $P$ by the equations

$$\begin{aligned} x' &= x \cos \alpha + y \sin \alpha \\ y' &= -x \sin \alpha + y \cos \alpha \end{aligned}.$$

Then, prove that the coordinates $(x', y')$ can be obtained from the $(x, y)$ coordinates by matrix multiplication. We use this result in Section 8.4 when we study conic sections.

**Solution**  Using the right triangle formed by $P$ and the $x'y'$-coordinate system, we obtain

$$x' = r \cos (\theta - \alpha) \quad \text{and} \quad y' = r \sin (\theta - \alpha).$$

Expanding the above expressions for $x'$ and $y'$ using trigonometric identities for $\cos (\theta - \alpha)$ and $\sin (\theta - \alpha)$ yields

$$x' = r \cos \theta \cos \alpha + r \sin \theta \sin \alpha, \quad \text{and}$$

$$y' = r \sin \theta \cos \alpha - r \cos \theta \sin \alpha.$$

It follows from the right triangle formed by $P$ and the $xy$-coordinate system that $x = r \cos \theta$ and $y = r \sin \theta$. Substituting these values for $x$ and $y$ into the above pair of equations yields

$$x' = x \cos \alpha + y \sin \alpha \quad \text{and} \quad y' = y \cos \alpha - x \sin \alpha$$

$$= -x \sin \alpha + y \cos \alpha,$$

which is what we were asked to prove. Finally, matrix multiplication shows that

$$\begin{bmatrix} x' & y' \end{bmatrix} = \begin{bmatrix} x & y \end{bmatrix}\begin{bmatrix} \cos \alpha & -\sin \alpha \\ \sin \alpha & \cos \alpha \end{bmatrix}.$$

**Problem**

If we have a triangle with vertices at (0, 0), (1, 1), and (2, 0), and we want to double the lengths of the sides of the triangle, where would the vertices of the enlarged triangle be?

**Solution**

Given a triangle with vertices at (0, 0), (1, 1), and (2, 0), as in Figure 7.12, we can find the vertices of a new triangle whose sides are twice as long by multiplying by the scale matrix.

$$\begin{bmatrix} 2 & 0 \\ 0 & 2 \end{bmatrix}$$

For the point (0, 0), we have

$$[x' \quad y'] = [0 \quad 0]\begin{bmatrix} 2 & 0 \\ 0 & 2 \end{bmatrix} = [0 \quad 0]$$

For the point (1, 1), we have

$$[x' \quad y'] = [1 \quad 1]\begin{bmatrix} 2 & 0 \\ 0 & 2 \end{bmatrix} = [2 \quad 2]$$

And for the point (2, 0), we have

$$[x' \quad y'] = [2 \quad 0]\begin{bmatrix} 2 & 0 \\ 0 & 2 \end{bmatrix} = [4 \quad 0]$$

So the new triangle has vertices (0, 0), (2, 2), and (4, 0), as Figure 7.13 shows.

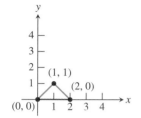

**Figure 7.12**

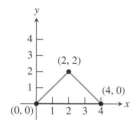

**Figure 7.13**

# Quick Review 7.2

In Exercises 1–4, the points **(a)** (3, −2) and **(b)** (x, y) are reflected across the given line. Find the coordinates of the reflected points.

**1.** The x-axis   (3, 2); (x, −y)   **2.** The y-axis

**3.** The line y = x (−2, 3); (y, x) **4.** The line y = −x

In Exercises 5 and 6, express the coordinates of P in terms of θ.

In Exercises 7–10, expand the expression.

**7.** sin (α + β)                **8.** sin (α − β)

**9.** cos (α + β)                **10.** cos (α − β)

**5.**

**6.**

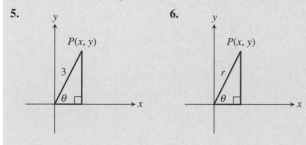

# Section 7.2 Exercises

In Exercises 1–6, determine the order of the matrix. Indicate whether the matrix is square.

**1.** $\begin{bmatrix} 2 & 3 & -1 \\ 1 & 0 & 5 \end{bmatrix}$  **2.** $\begin{bmatrix} 1 & 3 \\ -1 & 2 \end{bmatrix}$  **3.** $\begin{bmatrix} 5 & 6 \\ -1 & 2 \\ 0 & 0 \end{bmatrix}$

**4.** $\begin{bmatrix} -1 & 0 & 6 \end{bmatrix}$  **5.** $\begin{bmatrix} 2 \\ -1 \\ 0 \end{bmatrix}$  **6.** $\begin{bmatrix} 0 \end{bmatrix}$

In Exercises 7–10, identify the element specified for the following matrix.

$$\begin{bmatrix} -2 & 0 & 3 & 4 \\ 3 & 1 & 5 & -1 \\ 1 & 4 & -1 & 3 \end{bmatrix}$$

**7.** $a_{13}$  3  
**8.** $a_{24}$  $-1$  
**9.** $a_{32}$  4  
**10.** $a_{33}$  $-1$

In Exercises 11–16, find **(a)** $A + B$, **(b)** $A - B$, **(c)** $3A$, and **(d)** $2A - 3B$.

**11.** $A = \begin{bmatrix} 2 & 3 \\ -1 & 5 \end{bmatrix}$, $B = \begin{bmatrix} 1 & -3 \\ -2 & -4 \end{bmatrix}$

**12.** $A = \begin{bmatrix} -1 & 0 & 2 \\ 4 & 1 & -1 \\ 2 & 0 & 1 \end{bmatrix}$, $B = \begin{bmatrix} 2 & 1 & 0 \\ -1 & 0 & 2 \\ 4 & -3 & -1 \end{bmatrix}$

**13.** $A = \begin{bmatrix} -3 & 1 \\ 0 & -1 \\ 2 & 1 \end{bmatrix}$, $B = \begin{bmatrix} 4 & 0 \\ -2 & 1 \\ -3 & -1 \end{bmatrix}$

**14.** $A = \begin{bmatrix} 5 & -2 & 3 & 1 \\ -1 & 0 & 2 & 2 \end{bmatrix}$, $B = \begin{bmatrix} -2 & 3 & 1 & 0 \\ 4 & 0 & -1 & -2 \end{bmatrix}$

**15.** $A = \begin{bmatrix} -2 \\ 1 \\ 0 \end{bmatrix}$, $B = \begin{bmatrix} -1 \\ 0 \\ 4 \end{bmatrix}$

**16.** $A = \begin{bmatrix} -1 & -2 & 0 & 3 \end{bmatrix}$, and $B = \begin{bmatrix} 1 & 2 & -2 & 0 \end{bmatrix}$

In Exercises 17–22, use the definition of matrix multiplication to find **(a)** $AB$ and **(b)** $BA$. Support your answer with the matrix feature of your grapher.

**17.** $A = \begin{bmatrix} 2 & 3 \\ -1 & 5 \end{bmatrix}$, $B = \begin{bmatrix} 1 & -3 \\ -2 & -4 \end{bmatrix}$

**18.** $A = \begin{bmatrix} 1 & -4 \\ 2 & 6 \end{bmatrix}$, $B = \begin{bmatrix} 5 & 1 \\ -2 & -3 \end{bmatrix}$

**19.** $A = \begin{bmatrix} 2 & 0 & 1 \\ 1 & 4 & -3 \end{bmatrix}$, $B = \begin{bmatrix} 1 & 2 \\ -3 & 1 \\ 0 & -2 \end{bmatrix}$

**20.** $A = \begin{bmatrix} 1 & 0 & -2 & 3 \\ 2 & 1 & 4 & -1 \end{bmatrix}$, $B = \begin{bmatrix} 5 & -1 \\ 0 & 2 \\ -1 & 3 \\ 4 & 2 \end{bmatrix}$

**21.** $A = \begin{bmatrix} -1 & 0 & 2 \\ 4 & 1 & -1 \\ 2 & 0 & 1 \end{bmatrix}$, $B = \begin{bmatrix} 2 & 1 & 0 \\ -1 & 0 & 2 \\ 4 & -3 & -1 \end{bmatrix}$

**22.** $A = \begin{bmatrix} -2 & 3 & 0 \\ 1 & -2 & 4 \\ 3 & 2 & 1 \end{bmatrix}$, $B = \begin{bmatrix} 4 & -1 & 2 \\ 0 & 2 & 3 \\ -1 & 3 & -1 \end{bmatrix}$

In Exercises 23–28, find **(a)** $AB$ and **(b)** $BA$, or state that the product is not defined.

**23.** $A = \begin{bmatrix} 2 & -1 & 3 \end{bmatrix}$, $B = \begin{bmatrix} -5 \\ 4 \\ 2 \end{bmatrix}$

**24.** $A = \begin{bmatrix} -2 \\ 3 \\ -4 \end{bmatrix}$, $B = \begin{bmatrix} -1 & 2 & 4 \end{bmatrix}$

**25.** $A = \begin{bmatrix} -1 & 2 \\ 3 & 4 \end{bmatrix}$, $B = \begin{bmatrix} -3 & 5 \end{bmatrix}$  not possible; [18  14]

**26.** $A = \begin{bmatrix} -1 & 3 \\ 0 & 1 \\ 1 & 0 \\ -3 & -1 \end{bmatrix}$, $B = \begin{bmatrix} 5 & -6 \\ 2 & 3 \end{bmatrix}$

**27.** $A = \begin{bmatrix} 0 & 0 & 1 \\ 0 & 1 & 0 \\ 1 & 0 & 0 \end{bmatrix}$, $B = \begin{bmatrix} 1 & 2 & 1 \\ 2 & 0 & 1 \\ -1 & 3 & 4 \end{bmatrix}$

**28.** $A = \begin{bmatrix} 0 & 0 & 1 & 0 \\ 0 & 1 & 0 & 0 \\ 1 & 0 & 0 & 0 \\ 0 & 0 & 0 & 1 \end{bmatrix}$, $B = \begin{bmatrix} -1 & 2 & 3 & -4 \\ 2 & 1 & 0 & -1 \\ -3 & 2 & 1 & 3 \\ 4 & 0 & 2 & -1 \end{bmatrix}$

In Exercises 29–32, solve for $a$ and $b$.

**29.** $\begin{bmatrix} a & -3 \\ 4 & 2 \end{bmatrix} = \begin{bmatrix} 5 & -3 \\ 4 & b \end{bmatrix}$  $a = 5, b = 2$

**30.** $\begin{bmatrix} 1 & -1 & 0 \\ a & -2 & 1 \end{bmatrix} = \begin{bmatrix} 1 & b & 0 \\ 3 & -2 & 1 \end{bmatrix}$  $a = 3, b = -1$

**31.** $\begin{bmatrix} 2 & a - 1 \\ 2 & 3 \\ -1 & 2 \end{bmatrix} = \begin{bmatrix} 2 & -3 \\ b + 2 & 3 \\ -1 & 2 \end{bmatrix}$  $a = -2, b = 0$

**32.** $\begin{bmatrix} a + 3 & 2 \\ 0 & 5 \end{bmatrix} = \begin{bmatrix} 4 & 2 \\ 0 & b - 1 \end{bmatrix}$  $a = 1, b = 6$

In Exercises 33 and 34, verify that the matrices are inverses of each other.

**33.** $A = \begin{bmatrix} 2 & 1 \\ 3 & 4 \end{bmatrix}$, $B = \begin{bmatrix} 0.8 & -0.2 \\ -0.6 & 0.4 \end{bmatrix}$

**34.** $A = \begin{bmatrix} -2 & 1 & 3 \\ 1 & 2 & -2 \\ 0 & 1 & -1 \end{bmatrix}$, $B = \begin{bmatrix} 0 & 1 & -2 \\ 0.25 & 0.5 & -0.25 \\ 0.25 & 0.5 & -1.25 \end{bmatrix}$

In Exercises 35–38, find the inverse of the matrix if it has one, or state that the inverse does not exist.

**35.** $\begin{bmatrix} 2 & 3 \\ 2 & 2 \end{bmatrix}$ $\begin{bmatrix} -1 & 1.5 \\ 1 & -1 \end{bmatrix}$

**36.** $\begin{bmatrix} 6 & 3 \\ 10 & 5 \end{bmatrix}$ no inverse

**37.** $\begin{bmatrix} 1 & 2 & -1 \\ 2 & -1 & 3 \\ 3 & 1 & 2 \end{bmatrix}$ no inverse

**38.** $\begin{bmatrix} 2 & 3 & -1 \\ -1 & 0 & 4 \\ 0 & 1 & 1 \end{bmatrix}$

**39.** $A = [a_{ij}]$, $a_{ij} = (-1)^{i+j}$, $1 \le i \le 4$, $1 \le j \le 4$  no inverse

**40.** $B = [b_{ij}]$, $b_{ij} = |i - j|$, $1 \le i \le 3$, $1 \le j \le 3$

In Exercises 41 and 42, use the definition to evaluate the determinant of the matrix.

**41.** $\begin{bmatrix} 2 & 1 & 1 \\ -1 & 0 & 2 \\ 1 & 3 & -1 \end{bmatrix}$ $-14$

**42.** $\begin{bmatrix} 1 & 0 & 2 & 0 \\ 0 & 1 & 2 & 3 \\ 1 & -1 & 0 & 2 \\ 1 & 0 & 0 & 3 \end{bmatrix}$ $10$

In Exercises 43 and 44, solve for X.

**43.** $3X + A = B$, where $A = \begin{bmatrix} 1 \\ 3 \end{bmatrix}$ and $B = \begin{bmatrix} 4 \\ 2 \end{bmatrix} \cdot \begin{bmatrix} 1 \\ -1/3 \end{bmatrix}$

**44.** $2X + A = B$, where $A = \begin{bmatrix} -1 & 2 \\ 0 & 3 \end{bmatrix}$ and $B = \begin{bmatrix} 1 & 4 \\ 1 & -1 \end{bmatrix}$.

**45. Symmetric Matrix** The matrix below gives the road mileage between Atlanta (A), Baltimore (B), Cleveland (C), and Denver (D). (Source: *AAA* Road Atlas)

**(a) Writing to Learn** Explain why the entry in the *i*th row and *j*th column is the same as the entry in the *j*th row and *i*th column. A matrix with this property is **symmetric**.

**(b) Writing to Learn** Why are the entries along the diagonal all 0's?

|   | A | B | C | D |
|---|---|---|---|---|
| A | 0 | 689 | 774 | 1406 |
| B | 689 | 0 | 371 | 1685 |
| C | 774 | 371 | 0 | 1340 |
| D | 1406 | 1685 | 1340 | 0 |

**46. Production** Jordan Manufacturing has two factories, each of which manufactures three products. The number of units of product *i* produced at factory *j* in one week is represented by $a_{ij}$ in the matrix

$$A = \begin{bmatrix} 120 & 70 \\ 150 & 110 \\ 80 & 160 \end{bmatrix}.$$

If production levels are increased by 10%, write the new production levels as a matrix $B$. How is $B$ related to $A$?

**47. Egg Production** Happy Valley Farms produces three types of eggs: 1 (large), 2 (X-large), 3 (jumbo). The number of dozens of type *i* eggs sold to grocery store *j* is represented by $a_{ij}$ in the matrix.

$$A = \begin{bmatrix} 100 & 60 \\ 120 & 70 \\ 200 & 120 \end{bmatrix}.$$

The per dozen price Happy Valley Farms charges for egg type *i* is represented by $b_{i1}$ in the matrix

$$B = \begin{bmatrix} \$0.80 \\ \$0.85 \\ \$1.00 \end{bmatrix}.$$

**(a)** Find the product $B^T A$.

**(b) Writing to Learn** What does the matrix $B^T A$ represent?

**48. Inventory** A company sells four models of one name brand "all in one fax, printer, copier, and scanner machine" at three retail stores. The inventory at store *i* of model *j* is represented by $s_{ij}$ in the matrix.

$$S = \begin{bmatrix} 16 & 10 & 8 & 12 \\ 12 & 0 & 10 & 4 \\ 4 & 12 & 0 & 8 \end{bmatrix}.$$

The wholesale and retail prices of model *i* are represented by $p_{i1}$ and $p_{i2}$, respectively, in the matrix

$$P = \begin{bmatrix} \$180 & \$269.99 \\ \$275 & \$399.99 \\ \$355 & \$499.99 \\ \$590 & \$799.99 \end{bmatrix}.$$

**(a)** Determine the product $SP$.

**(b) Writing to Learn** What does the matrix $SP$ represent?

**49. Profit** A discount furniture store sells four types of 5-piece bedroom sets. The price charged for a bedroom set of type *j* is represented by $a_{1j}$ in the matrix

$$A = [\$398 \quad \$598 \quad \$798 \quad \$998].$$

The number of sets of type *j* sold in one period is represented by $b_{1j}$ in the matrix

$$B = [35 \quad 25 \quad 20 \quad 10].$$

The cost to the furniture store for a bedroom set of type *j* is given by $c_{1j}$ in the matrix

$$C = [\$199 \quad \$268 \quad \$500 \quad \$670].$$

**(a)** Write a matrix product that gives the total revenue made from the sale of the bedroom sets in the one period.

**(b)** Write an expression using matrices that gives the profit produced by the sale of the bedroom sets in the one period.

# Explorations

**50. Continuation of Exploration 2** Let $A = [a_{ij}]$ be an $n \times n$ matrix.

**(a)** Prove that if every element of a row or column of a matrix is multiplied by the real number $c$, then the determinant of the matrix is multiplied by $c$.

**(b)** Prove that if all the entries above the main diagonal (or all below it) of a matrix are zero, the determinant is the product of the elements on the main diagonal.

**51. Continuation of Example 10** The $xy$-coordinate system is rotated through the angle $\alpha$ to obtain the $x'y'$-coordinate system (see Figure 7.11).

**(a)** Show that the inverse of the matrix

$$A = \begin{bmatrix} \cos \alpha & -\sin \alpha \\ \sin \alpha & \cos \alpha \end{bmatrix}$$

of Example 10 is

$$A^{-1} = \begin{bmatrix} \cos \alpha & \sin \alpha \\ -\sin \alpha & \cos \alpha \end{bmatrix}.$$

**(b)** Prove that the $(x, y)$ coordinates of $P$ in Figure 7.11 are related to the $(x', y')$ coordinates of $P$ by the equations

$$x = x' \cos \alpha - y' \sin \alpha$$
$$y = x' \sin \alpha + y' \cos \alpha$$

**(c)** Prove that the coordinates $(x, y)$ can be obtained from the $(x', y')$ coordinates by matrix multiplication. How is this matrix related to $A$?

**52. Writing Equations for Lines Using Determinants** Consider the equation

$$\begin{vmatrix} 1 & x & y \\ 1 & x_1 & y_1 \\ 1 & x_2 & y_2 \end{vmatrix} = 0.$$

**(a)** Verify that the equation is linear in $x$ and $y$.

**(b)** Verify that the two points $(x_1, y_1)$ and $(x_2, y_2)$ lie on the line in (a).

**(c)** Use a determinant to state that the point $(x_3, y_3)$ lies on the line in (a).

**(d)** Use a determinant to state that the point $(x_3, y_3)$ does not lie on the line in (a).

**53. Construction** A building contractor has agreed to build six ranch-style houses, seven Cape Cod-style houses, and 14 colonial-style houses. The number of units of raw materials that go into each type of house are shown in the matrix

|  | Steel | Wood | Glass | Paint | Labor |
|---|---|---|---|---|---|
| Ranch | 5 | 22 | 14 | 7 | 17 |
| $R =$ Cape Cod | 7 | 20 | 10 | 9 | 21 |
| Colonial | 6 | 27 | 8 | 5 | 13 |

Assume that steel costs $1600 a unit, wood $900 a unit, glass $500 a unit, paint $100 a unit, and labor $1000 a unit.

**(a)** Write a $1 \times 3$ matrix $B$ that represents the number of each type of house to be built. [6 7 14]

**(b)** Write a matrix product that gives the number of units of each raw material needed to build the houses. *BR*

**(c)** Write a $5 \times 1$ matrix $C$ that represents the per unit cost of each type of raw material.

**(d)** Write a matrix product that gives the cost of each house.

**(e) Writing to Learn** Compute the product *BRC*. what does this matrix represent? ■

**54. Rotating Coordinate Systems** The $xy$-coordinate system is rotated through the angle 30° to obtain the $x'y'$-coordinate system.

**(a)** If the coordinates of a point in the $xy$-coordinate system are (1, 1), what are the coordinates of the rotated point in the $x'y'$-coordinate system? $\approx$ (1.37 0.37)

**(b)** If the coordinates of a point in the $x'y'$-coordinate system are (1, 1), what are the coordinates of the point in the $xy$-coordinate system that was rotated to it? $\approx$ (0.37 1.37)

**55. Group Activity** Let $A$, $B$, and $C$ be matrices whose orders are such that the following expressions are defined. Prove that the following properties are true.

**(a)** $A + B = B + A$

**(b)** $(A + B) + C = A + (B + C)$

**(c)** $A(B + C) = AB + AC$

**(d)** $(A - B)C = AC - BC$

**56. Group Activity** Let $A$ and $B$ be $m \times n$ matrices and $c$ and $d$ scalars. Prove that the following properties are true.

**(a)** $c(A + B) = cA + cB$

**(b)** $(c + d)A = cA + dA$

**(c)** $c(dA) = (cd)A$

**(d)** $1 \cdot A = A$

**57. Writing to Learn** Explain why the definition given for the determinant of a square matrix agrees with the definition given for the determinant of a $2 \times 2$ matrix. (Assume that the determinant of a $1 \times 1$ matrix is the entry.)

**58. Inverse of a 2 × 2 Matrix** Prove that the inverse of the matrix

$$A = \begin{bmatrix} a & b \\ c & d \end{bmatrix} \quad \text{is} \quad A^{-1} = \frac{1}{ad - bc}\begin{bmatrix} d & -b \\ -c & a \end{bmatrix}$$

provided $ad - bc \neq 0$.

**59. Identity Matrix** Let $A = [a_{ij}]$ be an $n \times n$ matrix. Prove that $AI_n = I_nA = A$.

In Exercises 60–64, prove that the image of a point under the given transformation of the plane can be obtained by matrix multiplication.

**60.** A reflection across the $y$-axis

**61.** A reflection across the line $y = x$

**62.** A reflection across the line $y = -x$

**63.** A vertical stretch or shrink by a factor of $a$

**64.** A horizontal stretch or shrink by a factor of $c$

## Extending the Ideas

**65. Characteristic Polynomial**  Let $A = [a_{ij}]$ be a $2 \times 2$ matrix and define $f(x) = \det(xI_2 - A)$.

**(a)** Expand the determinant to show that $f(x)$ is a polynomial of degree 2. (The **characteristic polynomial** of $A$.)

**(b)** How is the constant term of $f(x)$ related to det $A$?

**(c)** How is the coefficient of $x$ related to $A$?

**(d)** Prove that $f(A) = 0$.

**66. Characteristic Polynomial**  Let $A = [a_{ij}]$ be a $3 \times 3$ matrix and define $f(x) = \det(xI_3 - A)$.

**(a)** Expand the determinant to show that $f(x)$ is a polynomial of degree 3. (The characteristic polynomial of $A$.)

**(b)** How is the constant term of $f(x)$ related to det $A$?

**(c)** How is the coefficient of $x^2$ related to $A$?

**(d)** Prove that $f(A) = 0$.

---

# 7.3 Multivariate Linear Systems and Row Operations

Triangular Form for Linear Systems • Gaussian Elimination • Elementary Row Operations and Row Echelon Form • Reduced Row Echelon Form • Solving Systems with Inverse Matrices • Applications

## Triangular Form for Linear Systems

**Objective**

Students will be able to solve systems of linear equations using Gaussian elimination, the reduced row echelon form of a matrix, or an inverse matrix.

**Motivate**

Ask...

If $4x - 3y = 7$ is a linear equation in two variables, what would be an example of a linear equation in four variables.

**Lesson Guide**

Day 1: Triangular Form; Gaussian Elimination; Elementary Row Operations and Row Echelon Form; Reduced Row Echelon Form
Day 2: Solving Systems with Inverse Matrices; Applications

The method of elimination used in Section 7.1 can be extended to systems of linear equations in more than two variables. The goal of the elimination method is to rewrite the system as an *equivalent system* of equations whose solution is obvious. Two systems of equations are **equivalent** if they have the same solution.

A *triangular form* of a system is an equivalent form from which the solution is easy to read. Here is an example of a system in triangular form.

$$x - 2y + z = 7$$
$$y - 2z = -7$$
$$z = 3$$

This convenient triangular form allows us to solve the system using substitution as illustrated in Example 1.

**Example 1  SOLVING BY SUBSTITUTION**

Solve the system

$$x - 2y + z = 7$$
$$y - 2z = -7.$$
$$z = 3$$

**Solution**  The third equation determines $z$, namely $z = 3$. Substitute the value of $z$ into the second equation to determine $y$.

$$y - 2z = -7 \qquad \text{Second equation}$$
$$y - 2(3) = -7 \qquad \text{Substitute } z = 3.$$
$$y = -1$$

Finally, substitute the values for $y$ and $z$ into the first equation to determine $x$.

$$x - 2y + z = 7 \qquad \text{First equation}$$

$$x - 2(-1) + 3 = 7 \qquad \text{Substitute } y = -1, z = 3.$$

$$x = 2$$

The solution of the system is $x = 2$, $y = -1$, $z = 3$, or the ordered triple $(2, -1, 3)$.

## Gaussian Elimination

Transforming a system to triangular form is **Gaussian elimination**, named after the famous German mathematician Carl Friedrich Gauss (1777–1855). Here are the operations needed to transform a system of linear equations into triangular form.

---

**Equivalent Systems of Linear Equations**

The following operations produce an equivalent system of linear equations.

**1.** Interchange any two equations of the system.

**2.** Multiply (or divide) one of the equations by any nonzero real number.

**3.** Add a multiple of one equation to any other equation in the system.

---

Watch how we use property 3 to bring the system in Example 2 to triangular form.

**Example 2** USING GAUSSIAN ELIMINATION

Solve the system

$$x - 2y + z = 7$$

$$3x - 5y + z = 14.$$

$$2x - 2y - z = 3$$

**Solution** Each step in the following process leads to a system of equations equivalent to the original system.

Multiply the first equation by $-3$ and add the result to the second equation, replacing the second equation. (Leave the first and third equations unchanged.)

$$x - 2y + z = 7$$

$$y - 2z = -7$$

$$2x - 2y - z = 3$$

Multiply the first equation by $-2$ and add the result to the third equation, replacing the third equation.

**Alert**

It is easy to make mistakes when performing Gaussian elimination, so it is particularly important to encourage students to check their answers by substituting them back into the original equations.

$$x - 2y + z = 7$$
$$y - 2z = -7$$
$$2y - 3z = -11$$

Multiply the second equation by $-2$ and add the result to the third equation, replacing the third equation.

$$x - 2y + z = 7$$
$$y - 2z = -7$$
$$z = 3$$

This is the same system of Example 1 and is a triangular form of the original system. We know from Example 1 that the solution is $(2, -1, 3)$.

For a system of equations which has exactly one solution, the final system in Example 2 is in **triangular form**. In this case, the leading term of each equation has coefficient 1, the third equation has one variable ($z$), the second equation has at most two variables including one not in the third equation ($y$), and the first one has the remaining variable, $x$ in this case.

### Example 3  FINDING NO SOLUTION

Solve the system

$$x - 3y + z = 4$$
$$-x + 2y - 5z = 3.$$
$$5x - 13y + 13z = 8$$

**Solution**  Use Gaussian elimination.

$$x - 3y + z = 4$$
$$-y - 4z = 7 \qquad \text{Add 1st equation to 2nd equation.}$$
$$5x - 13y + 13z = 8$$

$$x - 3y + z = 4$$
$$-y - 4z = 7$$
$$2y + 8z = -12 \qquad \text{Multiply 1st equation by } -5 \text{ and add to 3rd equation.}$$

$$x - 3y + z = 4$$
$$-y - 4z = 7$$
$$0 = 2 \qquad \text{Multiply 2nd equation by 2 and add to 3rd equation.}$$

Since $0 = 2$ is never true, we conclude that this system has no solution.

# Elementary Row Operations and Row Echelon Form

When we solve a system of linear equations using Gaussian elimination, all the action is really on the coefficients of the variables. Matrices can be used to record the coefficients as we go through the steps of the Gaussian elimination process. We illustrate with the system of Example 2.

$$x - 2y + z = 7$$

$$3x - 5y + z = 14$$

$$2x - 2y - z = 3$$

The **augmented matrix** of this system of equations is

$$\begin{bmatrix} 1 & -2 & 1 & 7 \\ 3 & -5 & 1 & 14 \\ 2 & -2 & -1 & 3 \end{bmatrix}.$$

**Teaching Note**

Make sure students understand that the entries in each column (except the last) are the coefficients of the *same* variable.

The entries in the last column are the numbers on the right hand side of the equations. For the record, the **coefficient matrix** of this system is

$$\begin{bmatrix} 1 & -2 & 1 \\ 3 & -5 & 1 \\ 2 & -2 & -1 \end{bmatrix}.$$

Here the entries are the coefficients of the variables. We use this matrix to solve certain linear systems later in this section.

We repeat the Gaussian elimination process used in Example 2 and record the corresponding action on the augmented matrix.

Multiply the first equation (row) by $-3$ and add the result to the second equation (row) replacing the second equation (row). (Leave the first and third equations (rows) unchanged.)

<table>
<tr><th>System of Equations</th><th>Augmented Matrix</th></tr>
</table>

$$\begin{aligned} x - 2y + z &= 7 \\ y - 2z &= -7 \\ 2x - 2y - z &= 3 \end{aligned} \qquad \begin{bmatrix} 1 & -2 & 1 & 7 \\ 0 & 1 & -2 & -7 \\ 2 & -2 & -1 & 3 \end{bmatrix}$$

We set $a_{21} = 0$ in the second row of the new augmented matrix because the coefficient of $x$ is zero.

Multiply the first equation (row) by $-2$ and add the result to the third equation (row) replacing the third equation (row).

$$\begin{aligned} x - 2y + z &= 7 \\ y - 2z &= -7 \\ 2y - 3z &= -11 \end{aligned} \qquad \begin{bmatrix} 1 & -2 & 1 & 7 \\ 0 & 1 & -2 & -7 \\ 0 & 2 & -3 & -11 \end{bmatrix}$$

Multiply the second equation (row) by $-2$ and add the result to the third equation (row) replacing the third equation (row).

$$\begin{aligned} x - 2y + z &= 7 \\ y - 2z &= -7 \\ z &= 3 \end{aligned} \qquad \begin{bmatrix} 1 & -2 & 1 & 7 \\ 0 & 1 & -2 & -7 \\ 0 & 0 & 1 & 3 \end{bmatrix}$$

The augmented matrix above, corresponding to the triangular form of the original system of equations, is a *row echelon form* of the augmented matrix of the original system of equations. In general, the last few rows of a row echelon form of a matrix *can* consist of all 0's. We will see examples like this in a moment.

---

**Definition** **Row Echelon Form of a Matrix**

A matrix is in **row echelon form** if the following conditions are satisfied.

1. Rows consisting entirely of 0's (if there are any) occur at the bottom of the matrix.
2. The first entry in any row with nonzero entries is 1.
3. The column subscript of the leading 1 entries increases as the row subscript increases.

---

Another way to phrase parts 2 and 3 of the above definition is to say that the leading 1's move to the right as we move down the rows.

Our goal is to take a system of equations, write the corresponding augmented matrix, and transform it to row echelon form without carrying along the equations. From there we can read off the solutions to the system fairly easily.

The operations that we use to transform a linear system to equivalent triangular form correspond to elementary row operations of the corresponding augmented matrix of the linear system.

---

**Elementary Row Operations on a Matrix**

A combination of the following operations will transform a matrix to row echelon form.

1. Interchange any two rows.
2. Multiply all elements of a row by a nonzero real number.
3. Add a multiple of one row to any other row.

---

Example 4 illustrates how we can transform the augmented matrix to row echelon form to solve a system of linear equations.

### Example 4 FINDING A ROW ECHELON FORM

Solve the system

$$x - y + 2z = -3$$
$$2x + y - z = 0.$$
$$-x + 2y - 3z = 7$$

**Solution** We apply elementary row operations to find a row echelon form of the augmented matrix. The elementary row operations used are

**Row Echelon Form**

A word of caution! You can use your grapher to find a row echelon form of a matrix. However, row echelon form is *not* unique. Your grapher may produce a row echelon form different from the one you obtained by paper-and-pencil. Fortunately, all row echelon forms produce the same solution to the system of equations. (Correspondingly, a triangular form of a linear system is also not unique.)

recorded above the arrows using the notation in the margin.

$$\begin{bmatrix} 1 & -1 & 2 & -3 \\ 2 & 1 & -1 & 0 \\ -1 & 2 & -3 & 7 \end{bmatrix} \xrightarrow{(-2)R_1+R_2} \begin{bmatrix} 1 & -1 & 2 & -3 \\ 0 & 3 & -5 & 6 \\ -1 & 2 & -3 & 7 \end{bmatrix} \xrightarrow{(1)R_1+R_3}$$

$$\begin{bmatrix} 1 & -1 & 2 & -3 \\ 0 & 3 & -5 & 6 \\ 0 & 1 & -1 & 4 \end{bmatrix} \xrightarrow{R_{23}} \begin{bmatrix} 1 & -1 & 2 & -3 \\ 0 & 1 & -1 & 4 \\ 0 & 3 & -5 & 6 \end{bmatrix} \xrightarrow{(-3)R_2+R_3}$$

$$\begin{bmatrix} 1 & -1 & 2 & -3 \\ 0 & 1 & -1 & 4 \\ 0 & 0 & -2 & -6 \end{bmatrix} \xrightarrow{(-1/2)R_3} \begin{bmatrix} 1 & -1 & 2 & -3 \\ 0 & 1 & -1 & 4 \\ 0 & 0 & 1 & 3 \end{bmatrix}$$

The last matrix is in row echelon form. Then we convert each row into equation form and complete the solution by substitution.

$$y - z = 4 \qquad x - y + 2z = -3$$
$$z = 3 \qquad y - 3 = 4 \qquad x - 7 + 2(3) = -3$$
$$y = 7 \qquad x = -2$$

The solution of the original system of equations is $(-2, 7, 3)$.

## Reduced Row Echelon Form

If we continue to apply elementary row operations to a row echelon form of a matrix, we can obtain a matrix in which every column that has a leading 1 has 0's elsewhere. This is a **reduced row echelon form** of the matrix. It is usually easier to read the solution from a reduced row echelon form.

We apply elementary row operations to the row echelon form found in Example 4 until we find a reduced row echelon form.

$$\begin{bmatrix} 1 & -1 & 2 & -3 \\ 0 & 1 & -1 & 4 \\ 0 & 0 & 1 & 3 \end{bmatrix} \xrightarrow{(1)R_2+R_1} \begin{bmatrix} 1 & 0 & 1 & 1 \\ 0 & 1 & -1 & 4 \\ 0 & 0 & 1 & 3 \end{bmatrix} \xrightarrow{(-1)R_3+R_1}$$

$$\begin{bmatrix} 1 & 0 & 0 & -2 \\ 0 & 1 & -1 & 4 \\ 0 & 0 & 1 & 3 \end{bmatrix} \xrightarrow{(1)R_3+R_2} \begin{bmatrix} 1 & 0 & 0 & -2 \\ 0 & 1 & 0 & 7 \\ 0 & 0 & 1 & 3 \end{bmatrix}$$

From this reduced row echelon form, we can immediately read the solution to the system of Example 4: $x = -2$, $y = 7$, $z = 3$. Figure 7.14 shows that the above final matrix is a reduced row echelon form of the augmented matrix of Example 4.

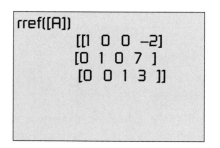

```
rref([A])
         [[1  0  0  -2]
          [0  1  0  7 ]
          [0  0  1  3 ]]
```

**Figure 7.14** $A$ is the augmented matrix of the system of linear equations in Example 4. "rref" stands for the grapher-produced reduced row echelon form of $A$.

### Example 5  FINDING INFINITELY MANY SOLUTIONS

Solve the system

$$x + y + z = 3$$
$$2x + y + 4z = 8.$$
$$x + 2y - z = 1$$

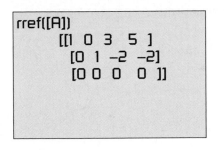

**Figure 7.15** A reduced row echelon form for the augmented matrix of Example 5.

---

**Notes on Examples**

Examples 5 and 6 are important because they show not only how to determine that an equation has infinitely many solutions, but also how to describe these solutions in terms of one or two variables. Although this representation is not unique, the authors consistently solve for the "first" variables in terms of the "last" variables. Encourage students to do the same.

---

**Solution**  Figure 7.15 shows a reduced row echelon form for the augmented matrix of the system. So, the following system of equations is equivalent to the original system.

$$x + 3z = 5$$
$$y - 2z = -2$$
$$0 = 0$$

Solving the first two equations for $x$ and $y$ in terms of $z$ yields:

$$x = -3z + 5$$
$$y = 2z - 2$$

This system has infinitely many solutions because for every value of $z$ we can use these two equations to find a corresponding value for $x$ and $y$.

**Interpret**

The solution is the set of all ordered triples of the form $(-3z + 5, 2z - 2, z)$ where $z$ is any real number.

We can also solve linear systems with more than three variables, or more than three equations, or both, by finding a row (or reduced row) echelon form. The solution set may become more complicated as illustrated in Example 6.

**Example 6  FINDING INFINITELY MANY SOLUTIONS**

Solve the system

$$x + 2y - 3z \qquad = -1$$
$$2x + 3y - 4z + w = -1.$$
$$3x + 5y - 7z + w = -2$$

**Solution**  The $3 \times 5$ augmented matrix is

$$\begin{bmatrix} 1 & 2 & -3 & 0 & -1 \\ 2 & 3 & -4 & 1 & -1 \\ 3 & 5 & -7 & 1 & -2 \end{bmatrix}.$$

Figure 7.16 shows a reduced row echelon form from which we can read that

$$x = -z - 2w + 1$$
$$y = 2z + w - 1.$$

This system has infinitely many solutions because for every pair of values for $z$ and $w$ we can use these two equations to find corresponding values for $x$ and $y$.

**Interpret**

The solution is the set of all ordered 4-tuples of the form $(-z - 2w + 1, 2z + w - 1, z, w)$ where $z$ and $w$ are any real numbers.

## Solving Systems with Inverse Matrices

If a linear system consists of the same number of equations as variables, then the coefficient matrix is square. If this matrix is also nonsingular, then we can solve the system using the technique illustrated in Example 7.

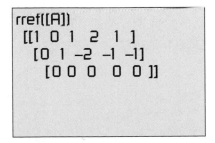

**Figure 7.16** A reduced row echelon form for the augmented matrix of Example 6.

## Linear Equations

If $a$ and $b$ are real numbers with $a \neq 0$, the linear equation $ax = b$ has a unique solution $x = a^{-1}b$. A similar statement holds for the linear matrix equation $AX = B$ when $A$ is a nonsingular square matrix. (See Theorem 7.2.)

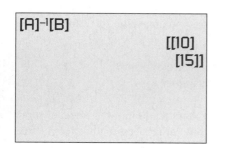

**Figure 7.17** The solution of the matrix equation of Example 7.

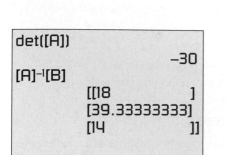

**Figure 7.18** The solution of the system in Example 8.

### Example 7 SOLVING A SYSTEM USING INVERSE MATRICES

Solve the system

$$3x - 2y = 0$$
$$-x + y = 5.$$

**Solution** First we write the system as a matrix equation. Let

$$A = \begin{bmatrix} 3 & -2 \\ -1 & 1 \end{bmatrix}, \qquad X = \begin{bmatrix} x \\ y \end{bmatrix}, \qquad \text{and} \qquad B = \begin{bmatrix} 0 \\ 5 \end{bmatrix}.$$

Then

$$A \cdot X = \begin{bmatrix} 3 & -2 \\ -1 & 1 \end{bmatrix} \cdot \begin{bmatrix} x \\ y \end{bmatrix} = \begin{bmatrix} 3x - 2y \\ -x + y \end{bmatrix}$$

so that

$$AX = B,$$

where $A$ is the coefficient matrix of the system. You can easily check that $\det A = 1$, so $A^{-1}$ exists. From Figure 7.17, we obtain

$$X = A^{-1}B = \begin{bmatrix} 10 \\ 15 \end{bmatrix}.$$

The solution of the system is $x = 10$, $y = 15$, or $(10, 15)$.

Examples 7 and 8 are two instances of the following Theorem.

---

**Theorem  Invertible Square Linear Systems**

Let $A$ be the coefficient matrix of a system of $n$ linear equations in $n$ variables given by $AX = B$, where $X$ is the $n \times 1$ matrix of variables and $B$ is the $n \times 1$ matrix of numbers on the right-hand side of the equations. If $A^{-1}$ exists, then the system of equations has the unique solution

$$X = A^{-1}B.$$

---

### Example 8 SOLVING A SYSTEM USING INVERSE MATRICES

Solve the system

$$3x - 3y + 6z = 20$$
$$x - 3y + 10z = 40.$$
$$-x + 3y - 5z = 30$$

**Solution** Let

$$A = \begin{bmatrix} 3 & -3 & 6 \\ 1 & -3 & 10 \\ -1 & 3 & -5 \end{bmatrix}, \qquad X = \begin{bmatrix} x \\ y \\ z \end{bmatrix}, \qquad \text{and} \qquad B = \begin{bmatrix} 20 \\ 40 \\ 30 \end{bmatrix}.$$

The system of equations can be written as

$$A \cdot X = B.$$

Figure 7.18 shows that $\det A = -30 \neq 0$, so $A^{-1}$ exists and

$$X = A^{-1}B = \begin{bmatrix} 18 \\ 39.\overline{3} \\ 14 \end{bmatrix}.$$

### Interpret

The solution of the system of equations is $x = 18$, $y = 39\frac{1}{3}$, and $z = 14$, or $(18, 39\frac{1}{3}, 14)$.

## Applications

Any three noncollinear points with distinct $x$-coordinates determine exactly one second-degree polynomial as illustrated in Example 9. The graph of a second-degree polynomial is a parabola.

### Example 9  FITTING A PARABOLA TO THREE POINTS

Determine $a$, $b$, and $c$ so that the points $(-1, 5)$, $(2, -1)$, and $(3, 13)$ are on the graph of $f(x) = ax^2 + bx + c$.

### Solution

#### Model

We must have $f(-1) = 5$, $f(2) = -1$, and $f(3) = 13$:

$$f(-1) = a - b + c = 5$$
$$f(2) = 4a + 2b + c = -1$$
$$f(3) = 9a + 3b + c = 13$$

The above system of three linear equations in the three variables $a$, $b$, and $c$ can be written in matrix form $AX = B$, where

$$A = \begin{bmatrix} 1 & -1 & 1 \\ 4 & 2 & 1 \\ 9 & 3 & 1 \end{bmatrix}, \qquad X = \begin{bmatrix} a \\ b \\ c \end{bmatrix}, \qquad \text{and} \qquad B = \begin{bmatrix} 5 \\ -1 \\ 13 \end{bmatrix}.$$

#### Solve Numerically

Figure 7.19a shows that

$$X = A^{-1}B = \begin{bmatrix} 4 \\ -6 \\ -5 \end{bmatrix}.$$

Thus, $a = 4$, $b = -6$, and $c = -5$. The second-degree polynomial $f(x) = 4x^2 - 6x - 5$ contains the points $(-1, 5)$, $(2, -1)$, and $(3, 13)$ (Figure 7.19b).

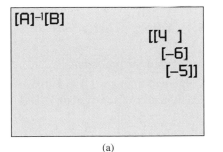

(a)

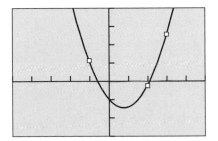

[−5, 5] by [−15, 20]

(b)

**Figure 7.19** (a) The solution of the matrix equation of Example 9. (b) A graph of $f(x) = 4x^2 - 6x - 5$ superimposed on a scatter plot of the three points $(-1, 5)$, $(2, -1)$, and $(3, 13)$.

**Exploration 1** **Mixing Solutions**

Aileen's Drugstore needs to prepare a 60-L mixture that is 40% acid using three concentrations of acid. The first concentration is 15% acid, the second is 35% acid, and the third is 55% acid. Because of the amounts of acid solution on hand, they need to use twice as much of the 35% solution as the 55% solution. How much of each solution should they use?

Let $x$ = the number of liters of 15% solution used, $y$ = the number of liters of 35% solution used, and $z$ = the number of liters of 55% solution used.

1. Explain how the equation $x + y + z = 60$ is related to the problem.

2. Exlain how the equation $0.15x + 0.35y + 0.55z = 24$ is related to the problem.

3. Explain how the equation $y = 2z$ is related to the problem.

4. Write the system of three equations obtained from parts 1–3 in the form $AX = B$, where $A$ is the coefficient matrix of the system. What are $A$, $B$, and $X$?

5. Solve the matrix equation in part 4.

6. Interpret the solution in part 5 in terms of the problem situation.

# Quick Review 7.3

In Exercises 1 and 2, find the amount of pure acid in the solution.

1. 40 L of a 32% acid solution  12.8 liters

2. 60 mL of a 14% acid solution  8.4 milliliters

In Exercises 3 and 4, find the amount of water in the solution.

3. 50 L of a 24% acid solution  38 liters

4. 80 mL of a 70% acid solution  24 milliliters

In Exercises 5 and 6, determine which points are on the graph of the function.

5. $f(x) = 2x^2 - 3x + 1$  $(-1, 6)$

   (a) $(-1, 6)$　　　　　　(b) $(2, 1)$

6. $f(x) = x^3 - 4x - 1$  $(0, -1)$

   (a) $(0, -1)$　　　　　　(b) $(-2, -17)$

In Exercises 7 and 8, solve for $x$ or $y$ in terms of the other variables.

7. $y + z - w = 1$　　　　　　8. $x - 2z + w = 3$

In Exercises 9 and 10, find the inverse of the matrix.

9. $\begin{bmatrix} 1 & 3 \\ -2 & -2 \end{bmatrix}$ $\begin{bmatrix} -0.5 & -0.75 \\ 0.5 & 0.25 \end{bmatrix}$ 10. $\begin{bmatrix} 0 & 0 & 2 \\ -2 & 1 & 3 \\ 0 & 2 & -2 \end{bmatrix}$

# Section 7.3 Exercises

In Exercises 1–4, perform the indicated elementary row operation on the matrix

$$\begin{bmatrix} 2 & -6 & 4 \\ 1 & 2 & -3 \\ -3 & 1 & -2 \end{bmatrix}.$$

1. $(3/2)R_1 + R_3$　　　　　　2. $(1/2)R_1$

3. $(-2)R_2 + R_1$　　　　　　4. $(1)R_1 + R_2$

In Exercises 5–8, what elementary row operations applied to

$$\begin{bmatrix} -2 & 1 & -1 & 2 \\ 1 & -2 & 3 & 0 \\ 3 & 1 & -1 & 2 \end{bmatrix}$$

will yield the given matrix?

**5.** $\begin{bmatrix} 1 & -2 & 3 & 0 \\ -2 & 1 & -1 & 2 \\ 3 & 1 & -1 & 2 \end{bmatrix}$ $R_{12}$   **6.** $\begin{bmatrix} 0 & -3 & 5 & 2 \\ 1 & -2 & 3 & 0 \\ 3 & 1 & -1 & 2 \end{bmatrix}$

**7.** $\begin{bmatrix} -2 & 1 & -1 & 2 \\ 1 & -2 & 3 & 0 \\ 0 & 7 & -10 & 2 \end{bmatrix}$ $(-3)R_2 + R_3$

**8.** $\begin{bmatrix} -2 & 1 & -1 & 2 \\ 1 & -2 & 3 & 0 \\ 0.75 & 0.25 & -0.25 & 0.5 \end{bmatrix}$ $(1/4)R_3$

In Exercises 9–12, find a row echelon form for the matrix.

**9.** $\begin{bmatrix} 1 & 3 & -1 \\ 2 & 1 & 4 \\ -3 & 0 & 1 \end{bmatrix}$   **10.** $\begin{bmatrix} 1 & 2 & -3 \\ -3 & -6 & 10 \\ -2 & -4 & 7 \end{bmatrix}$

**11.** $\begin{bmatrix} 1 & 2 & 3 & -4 \\ -2 & 6 & -6 & 2 \\ 3 & 12 & 6 & 12 \end{bmatrix}$   **12.** $\begin{bmatrix} 3 & 6 & 9 & -6 \\ 2 & 5 & 5 & -3 \end{bmatrix}$

In Exercises 13–16, find a reduced row echelon form for the matrix.

**13.** $\begin{bmatrix} 1 & 0 & 2 & 1 \\ 3 & 2 & 4 & 7 \\ 2 & 1 & 3 & 4 \end{bmatrix}$ $\begin{bmatrix} 1 & 0 & 2 & 1 \\ 0 & 1 & -1 & 2 \\ 0 & 0 & 0 & 0 \end{bmatrix}$

**14.** $\begin{bmatrix} 1 & -2 & 2 & 1 & 1 \\ 3 & -5 & 6 & 3 & -1 \\ -2 & 4 & -3 & -2 & 5 \\ 3 & -5 & 6 & 4 & -3 \end{bmatrix}$ $\begin{bmatrix} 1 & 0 & 0 & 0 & -19 \\ 0 & 1 & 0 & 0 & -4 \\ 0 & 0 & 1 & 0 & 7 \\ 0 & 0 & 0 & 1 & -2 \end{bmatrix}$

**15.** $\begin{bmatrix} 1 & 2 & 3 & 1 \\ -3 & -5 & -7 & -4 \end{bmatrix}$   **16.** $\begin{bmatrix} 3 & -6 & 3 & -3 \\ 2 & -4 & 2 & -2 \\ -3 & 6 & -3 & 3 \end{bmatrix}$

In Exercises 17–20, write the augmented matrix corresponding to the system of equations.

**17.** $\begin{aligned} 2x - 3y + z &= 1 \\ -x + y - 4z &= -3 \\ 3x - z &= 2 \end{aligned}$   **18.** $\begin{aligned} 3x - 4y + z - w &= 1 \\ x + z - 2w &= 4 \end{aligned}$

**19.** $\begin{aligned} 2x - 5y + z - w &= -3 \\ x - 2z + w &= 4 \\ 2y - 3z - w &= 5 \end{aligned}$   **20.** $\begin{aligned} 3x - 2y &= 5 \\ -x + 5y &= 7 \end{aligned}$

In Exercises 21–24, write the system of equations corresponding to the augmented matrix.

**21.** $\begin{bmatrix} 3 & 2 & -1 \\ -4 & 5 & 2 \end{bmatrix}$   **22.** $\begin{bmatrix} 1 & 0 & -1 & 2 & -3 \\ 2 & 1 & 0 & -1 & 4 \\ -1 & 1 & 2 & 0 & 0 \end{bmatrix}$

**23.** $\begin{bmatrix} 2 & 0 & 1 & 3 \\ -1 & 1 & 0 & 2 \\ 0 & 2 & -3 & -1 \end{bmatrix}$   **24.** $\begin{bmatrix} 2 & 1 & -2 & 4 \\ -3 & 0 & 2 & -1 \end{bmatrix}$

In Exercises 25–28, use Gaussian elimination to solve the system of equations. Support your answer numerically.

**25.** $\begin{aligned} x - y + z &= 0 \\ 2x - 3z &= -1 \\ -x - y + 2z &= -1 \end{aligned}$   **26.** $\begin{aligned} 2x - y &= 0 \\ x + 3y - z &= -3 \\ 3y + z &= 8 \end{aligned}$

**27.** $\begin{aligned} x + y - z &= 4 \\ y + w &= -4 \\ x - y &= 1 \\ x + z + w &= 1 \end{aligned}$   **28.** $\begin{aligned} \tfrac{1}{2}x - y + z - w &= 1 \\ -x + y + z + 2w &= -3 \\ x - z &= 2 \\ y + w &= 0 \end{aligned}$

In Exercises 29–36, solve the system of equations by finding the reduced row echelon form for the augmented matrix.

**29.** $\begin{aligned} x + 2y - z &= 3 \\ 3x + 7y - 3z &= 12 \\ -2x - 4y + 3z &= -5 \end{aligned}$   **30.** $\begin{aligned} x - 2y + z &= -2 \\ 2x - 3y + 2z &= 2 \\ 4x - 8y + 5z &= -5 \end{aligned}$

**31.** $\begin{aligned} x + y + 3z &= 2 \\ 3x + 4y + 10z &= 5 \\ x + 2y + 4z &= 3 \end{aligned}$   **32.** $\begin{aligned} x - z &= 2 \\ -2x + y + 3z &= -5 \\ 2x + y - z &= 3 \end{aligned}$

**33.** $\begin{aligned} x + z &= 2 \\ 2x + y + z &= 5 \end{aligned}$   **34.** $\begin{aligned} x + 2y - 3z &= 1 \\ -3x - 5y + 8z &= -29 \end{aligned}$

**35.** $\begin{aligned} x + 2y &= 4 \\ 3x + 4y &= 5 \\ 2x + 3y &= 4 \end{aligned}$ no solution   **36.** $\begin{aligned} x + y &= 3 \\ 2x + 3y &= 8 \\ 2x + 2y &= 6 \end{aligned}$ $(1, 2)$

In Exercises 37 and 38, write the system of equations as a matrix equation $AX = B$, with $A$ as the coefficient matrix of the system.

**37.** $\begin{aligned} 2x + 5y &= -3 \\ x - 2y &= 1 \end{aligned}$   **38.** $\begin{aligned} 5x - 7y + z &= 2 \\ 2x - 3y - z &= 3 \\ x + y + z &= -3 \end{aligned}$

In Exercises 39 and 40, write the matrix equation as a system of equations.

**39.** $\begin{bmatrix} 3 & -1 \\ 2 & 4 \end{bmatrix} \begin{bmatrix} x \\ y \end{bmatrix} = \begin{bmatrix} -1 \\ 3 \end{bmatrix}$

**40.** $\begin{bmatrix} 1 & 0 & -3 \\ 2 & -1 & 3 \\ -2 & 3 & -4 \end{bmatrix} \begin{bmatrix} x \\ y \\ z \end{bmatrix} = \begin{bmatrix} 3 \\ -1 \\ 2 \end{bmatrix}$

In Exercises 41–46, solve the system of equations by using an inverse matrix.

**41.** $\begin{aligned} 2x - 3y &= -13 \\ 4x + y &= -5 \end{aligned}$ $(-2, 3)$   **42.** $\begin{aligned} x + 2y &= -2 \\ 3x - 4y &= 9 \end{aligned}$ $(1, -1.5)$

**43.** $\begin{aligned} 2x - y + z &= -6 \\ x + 2y - 3z &= 9 \\ 3x - 2y + z &= -3 \end{aligned}$   **44.** $\begin{aligned} x + 4y - 2z &= 0 \\ 2x + y + z &= 6 \\ -3x + 3y - 5z &= -13 \end{aligned}$

**45.** $\begin{aligned} 2x - y + z + w &= -3 \\ x + 2y - 3z + w &= 12 \\ 3x - y - z + 2w &= 3 \\ -2x + 3y + z - 3w &= -3 \end{aligned}$ $(-1, 2, -2, 3)$

**46.**
$$2x + y + 2z = 8$$
$$3x + 2y - z - w = 10$$
$$-2x + y - 3w = -1$$
$$4x - 3y + 2z - 5w = 39 \quad (4, -2, 1, -3)$$

In Exercises 47–58, use a method of your choice to solve the system of equations.

**47.** $2x - y = 10$
$\quad x - z = -1$
$\quad y + z = -9 \quad (0, -10, 1)$

**48.** $1.25x + z = -2$
$\quad y - 5.5z = -2.75$
$\quad 3x - 1.5y = -6$

**49.** $x + 2y + 2z + w = 5$
$\quad 2x + y + 2z = 5$
$\quad 3x + 3y + 3z + 2w = 12$
$\quad x + z + w = 1 \quad (3, 3, -2, 0)$

**50.** $x - y + w = -4$
$\quad -2x + y + z = 8$
$\quad 2x - 2y - z = -10$
$\quad -2x + z + w = 5 \quad (-1, 2, 4, -1)$

**51.** $x - y + z = 6$
$\quad x + y + 2z = -2$

**52.** $x - 2y + z = 3$
$\quad 2x + y - z = -4$

**53.** $2x + y + z + 4w = -1$
$\quad x + 2y + z + w = 1$
$\quad x + y + z + 2w = 0$

**54.** $2x + 3y + 3z + 7w = 0$
$\quad x + 2y + 2z + 5w = 0$
$\quad x + y + 2z + 3w = -1$

**55.** $2x + y + z + 2w = -3.5$
$\quad x + y + z + w = -1.5$

**56.** $2x + y + 4w = 6$
$\quad x + y + z + w = 5$

**57.** $x + y - z + 2w = 0$
$\quad y - z + 2w = -1$
$\quad x + y + 3w = 3$
$\quad 2x + 2y - z + 5w = 4$

**58.** $x + y + w = 2$
$\quad x + 4y + z - 2w = 3$
$\quad x + 3y + z - 3w = 2$
$\quad x + y + w = 2$

In Exercises 59–62, determine $f$ so that its graph contains the given points.

**59. Curve Fitting** $f(x) = ax^2 + bx + c$

$(-1, 3), (1, -3), (2, 0) \quad f(x) = 2x^2 - 3x - 2$

**60. Curve Fitting** $f(x) = ax^3 + bx^2 + cx + d.$

$(-2, -37), (-1, -11), (0, -5), (2, 19)$

**61. Family of Curves** $f(x) = ax^2 + bx + c$

$(-1, -4), (1, -2)$

**62. Family of Curves** $f(x) = ax^3 + bx^2 + cx + d$

$(-1, -6), (0, -1), (1, 2)$

**63. Personal Income Data** Table 7.5 gives the personal income data (in billions) in constant 1992 dollars for the states of Mississippi and Utah. Use $x = 0$ for 1990, $x = 1$ for 1991, and so forth.

| Year | Mississippi (billions) | Utah (billions) |
|------|------------------------|-----------------|
| 1990 | 35.3 | 26.5 |
| 1995 | 41.8 | 33.5 |
| 1996 | 43.1 | 35.4 |
| 1997 | 44.3 | 37.3 |

Table 7.5 Personal Income

*Source: U.S. Bureau of Economic Analysis, Survey of Current Business, May 1998, Statistical Abstracts, 118th Edition, 1998.*

**(a)** Find the linear regression equation for the Mississippi data and superimpose its graph on a scatter plot of the data.

**(b)** Find the linear regression equation for the Utah data and superimpose its graph on a scatter plot of the data.

**(c)** Estimate when the personal income for the two states will be the same.

**64. Personal Income Data** Table 7.6 gives the personal income data (in billions) in constant 1992 dollars for the states of Virginia and Washington. Use $x = 0$ for 1990, $x = 1$ for 1991, and so forth.

| Year | Virginia (billions) | Washington (billions) |
|------|---------------------|------------------------|
| 1990 | 134.1 | 103.6 |
| 1995 | 148.6 | 120.8 |
| 1996 | 152.4 | 126.3 |
| 1997 | 158.0 | 133.0 |

Table 7.6 Personal Income

*Source: U.S. Bureau of Economic Analysis, Survey of Current Business, May 1998, Statistical Abstracts, 118th Edition, 1998.*

**(a)** Find the linear regression equation for the Virginia data and superimpose its graph on a scatter plot of the data.

**(b)** Find the linear regression equation for the Washington data and superimpose its graph on a scatter plot of the data.

**(c)** Estimate when the personal income for the two states will be the same. 2031

**65. Train Tickets** At the Pittsburgh zoo, children ride a train for 25 cents, adults pay $1.00, and senior citizens 75 cents. On a given day, 1400 passengers paid a total of $740 for the rides. There were 250 more children riders than all other riders. Find the number of children, adult, and senior riders.

**66. Manufacturing** Stewart's Metals has three silver alloys on hand. One is 22% silver, another is 30% silver, and the third is 42% silver. How many grams of each alloy is required to produce 80 grams of a new alloy that is 34% silver if the amount of 30% alloy used is twice the amount of 22% alloy used.

**67. Investment** Monica receives an $80,000 inheritance. She invests part of it in CDs (certificates of deposit) earning 6.7% APY (annual percentage yield), part in bonds earning

9.3% APY, and the remainder in a growth fund earning 15.6% APY. She invests three times as much in the growth fund as in the other two combined. How much does she have in each investment if she receives $10,843 interest the first year?

**68. Investments** Oscar invests $20,000 in three investments earning 6% APY, 8% APY, and 10% APY. He invests $9000 more in the 10% investment than in the 6% investment. How much does he have invested at each rate if he receives $1780 interest the first year?

**69. Investments** Morgan has $50,000 to invest and wants to receive $5000 interest the first year. He puts part in CDs earning 5.75% APY, part in bonds earning 8.7% APY, and the rest in a growth fund earning 14.6% APY. How much should he invest at each rate if he puts the least amount possible in the growth fund?

**70. Mixing Acid Solutions** Simpson's Drugstore needs to prepare a 40-L mixture that is 32% acid from three solutions: a 10% acid solution, a 25% acid solution, and a 50% acid solution. How much of each solution should be used if Simpson's wants to use as little of the 50% solution as possible?

**71. Loose Change** Matthew has 74 coins consisting of nickels, dimes, and quarters in his coin box. The total value of the coins is $8.85. If the number of nickels and quarters in four more than the number of dimes, find how many of each coin Matthew has in his coin box.

**72. Vacation Money** Heather has saved $177 to take with her on the family vacation. She has 51 bills consisting of $1, $5, and $10 bills. If the number of $5 bills is three times the number of $10 bills, find how many of each bill she has.

In Exercises 73 and 74, use inverse matrices to find the equilibrium point for the demand and supply curves.

**73.** $p = 100 - 5x$    Demand curve   (16/3, 220/3)
     $p = 20 + 10x$    Supply curve

**74.** $p = 150 - 12x$    Demand curve   (10/3, 110)
     $p = 30 + 24x$    Supply curve

## Explorations

**75. Group Activity Investigating the Solution of a System of 3 Linear Equations in 3 Variables** Assume that the graph of a linear equation in three variables is a plane in 3-dimensional space. (You will study these in Chapter 8.)

**(a)** Explain geometrically how such a system can have a unique solution.

**(b)** Explain geometrically how such a system can have no solution. Describe several possibilities.

**(c)** Explain geometrically how such a system can have infinitely many solutions. Describe several possibilities. Construct physical models if you find that helpful. ■

**76. Writing to Learn** Explain why adding one row to another row in a matrix is an elementary row operation.

**77. Writing to Learn** Explain why subtracting one row from another row in a matrix is an elementary row operation.

## Extending the Ideas

**78. Writing to Learn** Explain why a row echelon form of a matrix is not unique. That is, show that a matrix can have two unequal row echelon forms. Give an example.

The roots of the characteristic polynomial $C(x) = \det(xI_n - A)$ of the $n \times n$ matrix $A$ are the **eigenvalues** of $A$ (see Section 7.2, Exercises 65 and 66). Use this information in Exercises 79 and 80.

**79.** Let $A = \begin{bmatrix} 3 & 2 \\ 1 & 5 \end{bmatrix}$.

**(a)** Find the characteristic polynomial $C(x)$ of $A$.

**(b)** Find the graph of $y = C(x)$.

**(c)** Find the eigenvalues of $A$.

**(d)** Compare $\det A$ with the $y$-intercept of the graph of $y = C(x)$.

**(e)** Compare the sum of the main diagonal elements of $A$ ($a_{11} + a_{22}$) with the sum of the eigenvalues.

**80.** Let $A = \begin{bmatrix} 2 & -1 \\ -5 & 2 \end{bmatrix}$.

**(a)** Find the characteristic polynomial $C(x)$ of $A$.

**(b)** Find the graph of $y = C(x)$.

**(c)** Find the eigenvalues of $A$.

**(d)** Compare $\det A$ with the $y$-intercept of the graph of $y = C(x)$.

**(e)** Compare the sum of the main diagonal elements of $A$ ($a_{11} + a_{22}$) with the sum of the eigenvalues.

# 7.4 Partial Fractions

Partial Fraction Decomposition • Denominators with Linear Factors • Denominators with Irreducible Quadratic factors • Applications

## Partial Fraction Decomposition

In Section 2.6 we saw that a polynomial with real coefficients could be factored into a product of factors with real coefficients, where each factor was either a linear factor or an irreducible quadratic factor. In this section we show that a rational function can be expressed as a sum of rational functions where each denominator is a power of a linear factor or a power of an irreducible quadratic factor.

For example,

$$\frac{3x - 4}{x^2 - 2x} = \frac{2}{x} + \frac{1}{x - 2}.$$

Each fraction in the sum is a **partial fraction**, and the sum is a **partial fraction decomposition** of the original rational function.

---

**Partial Fraction Decomposition of $f(x)/d(x)$**

1. Degree of $f \geq$ degree of $d$: Use the division algorithm to divide $f$ by $d$ to obtain the quotient $q$ and remainder $r$ and write

$$\frac{f(x)}{d(x)} = q(x) + \frac{r(x)}{d(x)}.$$

2. Factor $d(x)$ into a product of factors of the form $(mx + n)^u$ or $(ax^2 + bx + c)^v$, where $ax^2 + bx + c$ is irreducible.

3. For each factor $(mx + n)^u$: The partial fraction decomposition of $r(x)/d(x)$ must include the sum

$$\frac{A_1}{mx + n} + \frac{A_2}{(mx + n)^2} + \cdots + \frac{A_u}{(mx + n)^u},$$

where $A_1, A_2, \ldots, A_u$ are real numbers.

4. For each factor $(ax^2 + bx + c)^v$: The partial fraction decomposition of $r(x)/d(x)$ must include the sum

$$\frac{B_1 x + C_1}{ax^2 + bx + c} + \frac{B_2 x + C_2}{(ax^2 + bx + c)^2} + \cdots + \frac{B_v x + C_v}{(ax^2 + bx + c)^v},$$

where $B_1, B_2, \ldots, B_v$ and $C_1, C_2, \ldots, C_v$ are real numbers.

The partial fraction decomposition of the original rational function is the sum of $q(x)$ and the fractions in parts 3 and 4.

---

## Denominators with Linear Factors

Examples 1 and 2 illustrate how the constants $A_i$ in part 3 of the partial fraction decomposition procedure can be found.

**Notes on Examples**

Examples 1–4 use the method of comparing coefficients to determine the constants. Note that two polynomials of the same degree in $x$ are equal if and only if the corresponding coefficients are equal.

**Notes on Examples**

Note that an alternative method of completing Examples 1 and 2 is given in Exploration 1.

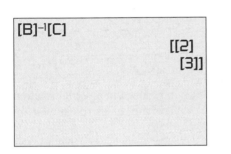

**Figure 7.20** The solution of the system of equations in Example 1.

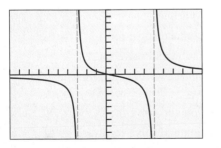

[–10, 10] by [–10, 10]

**Figure 7.21** The graphs of
$y = (5x - 1)/(x^2 - 2x - 15)$ and
$y = 2/(x + 3) + 3/(x - 5)$ appear to be the same. (Example 1)

### Example 1  DECOMPOSING A FRACTION WITH DISTINCT LINEAR FACTORS

Find the partial fraction decomposition of

$$\frac{5x - 1}{x^2 - 2x - 15}.$$

**Solution**  The denominator factors into $(x + 3)(x - 5)$. We write

$$\frac{5x - 1}{x^2 - 2x - 15} = \frac{A_1}{x + 3} + \frac{A_2}{x - 5}$$

and then "clear fractions" by multiplying both sides of the above equation by $x^2 - 2x - 15$ to obtain

$$5x - 1 = A_1(x - 5) + A_2(x + 3)$$
$$5x - 1 = (A_2 + A_2)x + (-5A_1 + 3A_2).$$

Comparing coefficients on the left and right side of the above equation, we obtain the following system of two equations in the two variables $A_1$ and $A_2$:

$$A_1 + A_2 = 5$$
$$-5A_1 + 3A_2 = -1.$$

We can write the above system in matrix form as $BX = C$ where

$$B = \begin{bmatrix} 1 & 1 \\ -5 & 3 \end{bmatrix}, \quad X = \begin{bmatrix} A_1 \\ A_2 \end{bmatrix}, \quad \text{and} \quad C = \begin{bmatrix} 5 \\ -1 \end{bmatrix},$$

and read from Figure 7.20 that

$$X = \begin{bmatrix} 2 \\ 3 \end{bmatrix}.$$

Thus, $A_1 = 2$, $A_2 = 3$, and

$$\frac{5x - 1}{x^2 - 2x - 15} = \frac{2}{x + 3} + \frac{3}{x - 5}.$$

**Support Graphically**

Figure 7.21 suggests that the following two functions are the same:

$$y = \frac{5x - 1}{x^2 - 2x - 15} \quad \text{and} \quad y = \frac{2}{x + 3} + \frac{3}{x - 5}.$$

### Example 2  DECOMPOSING A FRACTION WITH A REPEATED LINEAR FACTOR

Find the partial fraction decomposition of

$$\frac{-x^2 + 2x + 4}{x^3 - 4x^2 + 4x}.$$

**Solution**  The denominator factors into $x(x - 2)^2$. Because the factor $x - 2$ is squared, it contributes two terms to the decomposition:

$$\frac{-x^2 + 2x + 4}{x^3 - 4x^2 + 4x} = \frac{A_1}{x} + \frac{A_2}{x - 2} + \frac{A_3}{(x - 2)^2}.$$

We clear fractions by multiplying both sides of the above equation by $x^3 - 4x^2 + 4x$.

$$-x^2 + 2x + 4 = A_1(x - 2)^2 + A_2x(x - 2) + A_3x$$

Expanding and combining like terms in the above equation we obtain:

$$-x^2 + 2x + 4 = (A_1 + A_2)x^2 + (-4A_1 - 2A_2 + A_3)x + 4A_1.$$

Comparing coefficients of powers of $x$ on the left and right side of the above equation, we obtain the following system of equations:

$$A_1 + A_2 = -1$$
$$-4A_1 - 2A_2 + A_3 = 2.$$
$$4A_1 = 4$$

The reduced row echelon form of the augmented matrix

$$\begin{bmatrix} 1 & 1 & 0 & -1 \\ -4 & -2 & 1 & 2 \\ 4 & 0 & 0 & 4 \end{bmatrix}$$

of the above system of equations is

$$\begin{bmatrix} 1 & 0 & 0 & 1 \\ 0 & 1 & 0 & -2 \\ 0 & 0 & 1 & 2 \end{bmatrix}.$$

Thus $A_1 = 1$, $A_2 = -2$, $A_3 = 2$, and

$$\frac{-x^2 + 2x + 4}{x^3 - 4x^2 + 4x} = \frac{1}{x} + \frac{-2}{x - 2} + \frac{2}{(x - 2)^2}.$$

Sometimes we can solve for the variables introduced in a partial fraction decomposition by substituting strategic values for $x$, as illustrated in Exploration 1.

---

**Exploration 1  Revisiting Examples 1 and 2**

1. When we cleared fractions in Example 1 we obtained the equation
   $5x - 1 = A_1(x - 5) + A_2(x + 3)$.
   
   (a) Substitute $x = 5$ into this equation and solve for $A_2$.  $3 = A_2$
   
   (b) Substitute $x = -3$ into this equation and solve for $A_1$.  $2 = A_1$

2. When we cleared fractions in Example 2 we obtained the equation
   $-x^2 + 2x + 4 = A_1(x - 2)^2 + A_2x(x - 2) + A_3x$.
   
   (a) Substitute $x = 2$ into this equation and solve for $A_3$.  $2 = A_3$
   
   (b) Substitute $x = 0$ into this equation and solve for $A_1$.  $1 = A_1$
   
   (c) Substitute any other value for $x$ and use the values found for $A_1$ and $A_3$ to solve for $A_2$.  $-2 = A_3$

---

## Denominators with Irreducible Quadratic Factors

Example 3 shows how to find the partial fraction decomposition for a rational function whose denominator has an irreducible quadratic factor.

**Example 3** DECOMPOSING A FRACTION WITH AN IRREDUCIBLE QUADRATIC FACTOR

Find the partial fraction decomposition of

$$\frac{x^2 + 4x + 1}{x^3 - x^2 + x - 1}.$$

**Solution** We factor the denominator by grouping terms:

$$x^3 - x^2 + x - 1 = x^2(x - 1) + (x - 1)$$
$$= (x - 1)(x^2 + 1)$$

Each factor occurs once, so each one leads to one term in the decomposition:

$$\frac{x^2 + 4x + 1}{x^3 - x^2 + x - 1} = \frac{A}{x - 1} + \frac{Bx + C}{x^2 + 1}.$$

We clear fractions by multiplying both sides of the above equation by $x^3 - x^2 + x - 1$:

$$x^2 + 4x + 1 = A(x^2 + 1) + (Bx + C)(x - 1).$$

Expanding and combining like terms in the above equation we obtain:

$$x^2 + 4x + 1 = (A + B)x^2 + (-B + C)x + (A - C).$$

Comparing coefficients of powers of $x$ on the left and right side of the above equation, we obtain the following system of equations:

$$A + B = 1$$
$$-B + C - 4.$$
$$A - C = 1$$

Using any of the techniques of Section 7.3, we find $A = 3$, $B = -2$, and $C = 2$. Thus,

$$\frac{x^2 + 4x + 1}{x^3 - x^2 + x - 1} = \frac{3}{x - 1} + \frac{-2x + 2}{x^2 + 1}.$$

**Example 4** DECOMPOSING A FRACTION WITH A REPEATED IRREDUCIBLE QUADRATIC FACTOR

Find the partial fraction decomposition of

$$\frac{2x^3 - x^2 + 5x}{(x^2 + 1)^2}.$$

**Solution** The factor $(x^2 + 1)^2$ in the denominator leads to two terms in the partial fraction decomposition:

$$\frac{2x^3 - x^2 + 5x}{(x^2 + 1)^2} = \frac{B_1 x + C_1}{x^2 + 1} + \frac{B_2 x + C_2}{(x^2 + 1)^2}.$$

We clear fractions by multiplying both sides of the above equation by $(x^2 + 1)^2$:

$$2x^3 - x^2 + 5x = (B_1 x + C_1)(x^2 + 1) + B_2 x + C_2$$
$$= B_1 x^3 + C_1 x^2 + (B_1 + B_2)x + (C_1 + C_2)$$

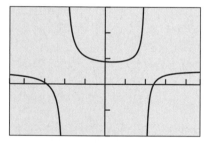

[–4.7, 4.7] by [–8, 12]

**Figure 7.22** The graph of $f(x) = (2x^2 + x - 14)/(x^2 - 4)$. (Example 5)

Comparing coefficients of powers of $x$ on the left and right side of the above equation, we see that $B_1 = 2$, $C_1 = -1$, $B_1 + B_2 = 5$, and $C_1 + C_2 = 0$. It follows that $B_2 = 3$ and $C_2 = 1$. Thus,

$$\frac{2x^3 - x^2 + 5x}{(x^2 + 1)^2} = \frac{2x - 1}{x^2 + 1} + \frac{3x + 1}{(x^2 + 1)^2}.$$

## Applications

Each part of the partial fraction decomposition of a rational function plays a central role in the analysis of its graph. One summand can be used to describe the end behavior of the graph. The other parts can be used to describe the behavior of the graph at one of its vertical asymptotes. Example 5 illustrates.

**Example 5** INVESTIGATING THE GRAPH OF A RATIONAL FUNCTION

Compare the graph of the rational function

$$f(x) = \frac{2x^2 + x - 14}{x^2 - 4}$$

with the graphs of the terms in its partial fraction decomposition.

**Solution** We use division to rewrite $f(x)$ in the form

$$f(x) = 2 + \frac{x - 6}{x^2 - 4}.$$

Then we use the techniques of this section to find the partial fraction decomposition of $(x - 6)/(x^2 - 4)$, and, in turn, that of $f$:

$$f(x) = 2 + \frac{x - 6}{x^2 - 4} = 2 + \frac{2}{x + 2} - \frac{1}{x - 2}.$$

Figure 7.22 shows the graph of $f$. You can see the relation of this graph to the graph of the end behavior asymptote $y = 2$, one of the terms of $f$. The graph of the term $y = 2/(x + 2)$ is very similar to the graph of $f$ near $x = -2$ (Figure 7.23a). The graph of the term $y = -1/(x - 2)$ is very similar to the graph of $f$ near $x = 2$ (Figure 7.23b).

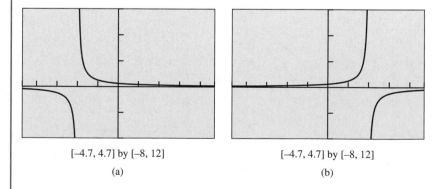

[–4.7, 4.7] by [–8, 12]   (a)       [–4.7, 4.7] by [–8, 12]   (b)

**Figure 7.23** The graphs of (a) $y = 2/(x + 2)$ and (b) $y = -1/(x - 2)$. (Example 5)

## Quick Review 7.4

In Exercises 1–4, perform the indicated operations and write your answer as a single reduced fraction.

**1.** $\dfrac{1}{x-1} + \dfrac{2}{x-3}$

**2.** $\dfrac{5}{x+4} - \dfrac{2}{x+1}$

**3.** $\dfrac{1}{x} + \dfrac{3}{x+1} + \dfrac{1}{(x+1)^2}$

**4.** $\dfrac{3}{x^2+1} - \dfrac{x+1}{(x^2+1)^2}$

In Exercises 5 and 6, divide $f(x)$ by $d(x)$ to obtain a quotient $q(x)$ and remainder $r(x)$. Write a summary statement in fraction form: $q(x) + r(x)/d(x)$.

**5.** $f(x) = 3x^3 - 6x^2 - 2x + 7, \quad d(x) = x - 2$

**6.** $f(x) = 2x^3 + 3x^2 - 14x - 8, \quad d(x) = x^2 + x - 6$

In Exercises 7 and 8, write the polynomial as a product of linear and irreducible quadratic factors with real coefficients.

**7.** $x^4 - 2x^3 + x^2 - 8x - 12$   **8.** $x^4 - x^3 - 15x^2 - 23x - 10$

In Exercises 9 and 10, assume that $f(x) = g(x)$. What can you conclude about $A$, $B$, $C$, and $D$?

**9.** $f(x) = Ax^2 + Bx + C + 1$

$g(x) = 3x^2 - x + 2$

**10.** $f(x) = (A+1)x^3 + Bx^2 + Cx + D$

$g(x) = -x^3 + 2x^2 - x - 5$

## Section 7.4 Exercises

In Exercises 1–4, write the terms for the partial fraction decomposition of the rational function. Do not solve for the constants.

**1.** $\dfrac{x^2 - 7}{x(x^2 - 4)}$

**2.** $\dfrac{x^4 + 3x^2 - 1}{(x^2 + x + 1)^2(x^2 - x + 1)}$

**3.** $\dfrac{x^5 - 2x^4 + x - 1}{x^3(x - 1)^2(x^2 + 9)}$

**4.** $\dfrac{x^2 + 3x + 2}{(x^3 - 1)^3}$

Exercises 5–8, use inverse matrices to find the partial fraction decomposition.

**5.** $\dfrac{x + 22}{(x + 4)(x - 2)} = \dfrac{A}{x + 4} + \dfrac{B}{x - 2} \quad \dfrac{-3}{x+4} + \dfrac{4}{x-2}$

**6.** $\dfrac{x - 3}{x(x + 3)} = \dfrac{A}{x + 3} + \dfrac{B}{x} \quad \dfrac{2}{x+3} - \dfrac{1}{x}$

**7.** $\dfrac{3x^2 + 2x + 2}{(x^2 + 1)^2} = \dfrac{Ax + B}{x^2 + 1} + \dfrac{Cx + D}{(x^2 + 1)^2}$

**8.** $\dfrac{4x + 4}{x^2(x + 2)} = \dfrac{A}{x} + \dfrac{B}{x^2} + \dfrac{C}{x + 2} \quad \dfrac{1}{x} + \dfrac{2}{x^2} - \dfrac{1}{x+2}$

In Exercises 9–12, use a reduced row echelon form for the augmented matrix to find the partial fraction decomposition.

**9.** $\dfrac{x^2 - 2x + 1}{(x - 2)^3} = \dfrac{A}{x - 2} + \dfrac{B}{(x - 2)^2} + \dfrac{C}{(x - 2)^3}$

**10.** $\dfrac{5x^3 - 10x^2 + 5x - 5}{(x^2 + 4)(x^2 + 9)} = \dfrac{Ax + B}{x^2 + 4} + \dfrac{Cx + D}{x^2 + 9}$

**11.** $\dfrac{5x^5 + 22x^4 + 36x^3 + 53x^2 + 71x + 20}{(x + 3)^2(x^2 + 2)^2}$

$= \dfrac{A}{x + 3} + \dfrac{B}{(x + 3)^2} + \dfrac{Cx + D}{x^2 + 2} + \dfrac{Ex + F}{(x^2 + 2)^2}$

**12.** $\dfrac{-x^3 - 6x^2 - 5x + 87}{(x - 1)^2(x + 4)^2}$

$= \dfrac{A}{x - 1} + \dfrac{B}{(x - 1)^2} + \dfrac{C}{x + 4} + \dfrac{D}{(x + 4)^2}$

In Exercises 13–16, find the partial fraction decomposition. Confirm your answer algebraically by combining the partial fractions.

**13.** $\dfrac{2}{(x - 5)(x - 3)}$

**14.** $\dfrac{4}{(x + 3)(x + 7)}$

**15.** $\dfrac{4}{x^2 - 1}$

**16.** $\dfrac{6}{x^2 - 9}$

In Exercises 17–20, find the partial fraction decomposition. Support your answer graphically.

**17.** $\dfrac{1}{x^2 + 2x} \quad \dfrac{1}{2x} - \dfrac{1/2}{x + 2}$

**18.** $\dfrac{-6}{x^2 - 3x} \quad \dfrac{-2}{x - 3} + \dfrac{2}{x}$

**19.** $\dfrac{-x + 10}{x^2 + x - 12}$

**20.** $\dfrac{7x - 7}{x^2 - 3x - 10}$

In Exercises 21–32, find the partial fraction decomposition.

**21.** $\dfrac{x + 17}{2x^2 + 5x - 3}$

**22.** $\dfrac{4x - 11}{2x^2 - x - 3}$

**23.** $\dfrac{2x^2 + 5}{(x^2 + 1)^2}$

**24.** $\dfrac{3x^2 + 4}{(x^2 + 1)^2}$

**25.** $\dfrac{x^2 - x + 2}{x^3 - 2x^2 + x}$

**26.** $\dfrac{-6x + 25}{x^3 - 6x^2 + 9x}$

**27.** $\dfrac{3x^2 - 4x + 3}{x^3 - 3x^2}$

**28.** $\dfrac{5x^2 + 7x - 4}{x^3 + 4x^2}$

**29.** $\dfrac{2x^3 + 4x - 1}{(x^2 + 2)^2}$

**30.** $\dfrac{3x^3 + 6x - 1}{(x^2 + 2)^2}$

**31.** $\dfrac{x^2 + 3x + 2}{x^3 - 1}$

**32.** $\dfrac{2x^2 - 4x + 3}{x^3 + 1}$

In Exercises 33–36, use division to write the rational function in the form $q(x) + r(x)/h(x)$, where the degree of $r(x)$ is less than the degree of $h(x)$. Then find the partial fraction decomposition of $r(x)/h(x)$.

**33.** $\dfrac{2x^2 + x + 3}{x^2 - 1}$

**34.** $\dfrac{3x^2 + 2x}{x^2 - 4}$

**35.** $\dfrac{x^3 - 2}{x^2 + x}$

**36.** $\dfrac{x^3 + 2}{x^2 - x}$

In Exercises 37–42, match the function with its graph. Do this without using your grapher.

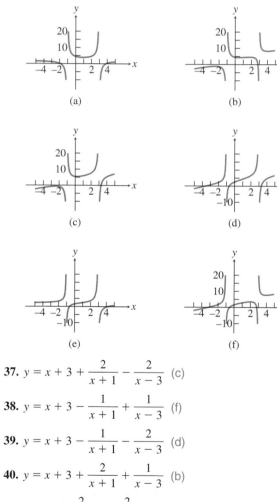

(a)

(b)

(c)

(d)

(e)

(f)

**37.** $y = x + 3 + \dfrac{2}{x + 1} - \dfrac{2}{x - 3}$  (c)

**38.** $y = x + 3 - \dfrac{1}{x + 1} + \dfrac{1}{x - 3}$  (f)

**39.** $y = x + 3 - \dfrac{1}{x + 1} - \dfrac{2}{x - 3}$  (d)

**40.** $y = x + 3 + \dfrac{2}{x + 1} + \dfrac{1}{x - 3}$  (b)

**41.** $y = 2 + \dfrac{2}{x + 1} - \dfrac{2}{x - 3}$  (a)

**42.** $y = 2 - \dfrac{1}{x + 1} - \dfrac{2}{x - 3}$  (e)

**43. Group Activity**  Find the partial fraction decomposition of
$$\dfrac{1}{x(x - a)} \cdot \dfrac{-1}{ax} + \dfrac{1}{a(x - a)}$$

**44. Group Activity**  Find the partial fraction decomposition of
$$\dfrac{-1}{(x - 2)(x - b)} \cdot \dfrac{1}{(b - 2)(x - 2)} - \dfrac{1}{(b - 2)(x - b)}$$

**45. Group Activity**  Find the partial fraction decomposition of
$$\dfrac{3}{(x - a)(x - b)} \cdot \dfrac{-3}{(b - a)(x - a)} + \dfrac{3}{(b - a)(x - b)}$$

**46. Group Activity**  Find the partial fraction decomposition of
$$\dfrac{2}{x^2 - a^2} \cdot \dfrac{-1}{a(x + a)} + \dfrac{1}{a(x - a)}$$

## Explorations

**47. Revisiting Example 3**  When we cleared fractions in Example 3 we obtained the equation
$$x^2 + 4x + 1 = A(x^2 + 1) + (Bx + C)(x - 1).$$

**(a)**  Substitute $x = 1$ into this equation and solve for $A$.

**(b)**  Substitute $x = i$ and $x = -i$ into this equation to find a system of 2 equations to solve for $B$ and $C$. $B = -2, C = 2$

**48. Writing to Learn**  Explain why it is valid in this section to obtain the systems of equations by equating coefficients of powers of $x$. ∎

## Extending the Ideas

**49. Group Activity**  Examine the graph of
$$f(x) = \dfrac{a}{x - 1} + \dfrac{b}{(x - 1)^2} \quad \text{for}$$

**(i)** $a = b = 1$       **(ii)** $a = 1, b = -1$

**(iii)** $a = -1, b = 1$       **(iv)** $a = -1, b = -1$

Based on this examination, which of the two functions $y = a/(x - 1)$ or $y = b/(x - 1)^2$ has the greater effect on the graph of $f(x)$ near $x = 1$? $b/(x - 1)^2$

**50. Writing to Learn**  Use partial fraction decomposition to explain why the graphs of
$$f(x) = \dfrac{2x - 3}{(x - 1)^2} \quad \text{and} \quad g(x) = \dfrac{2x + 3}{(x - 1)^2}$$

are so different near $x = 1$.

**7.5**

# Systems of Inequalities in Two Variables

Graph of an Inequality • Systems of Inequalities • Linear Programming

## Graph of an Inequality

An ordered pair $(a, b)$ of real numbers is a **solution of an inequality** in $x$ and $y$ if the substitution $x = a$ and $y = b$ satisfies the inequality. For example, the ordered pair $(2, 5)$ is a solution of $y < 2x + 3$ because

$$5 < 2(2) + 3 = 7.$$

However, the ordered pair $(2, 8)$ is *not* a solution because

$$8 \not< 2(2) + 3 = 7.$$

When we have found all the solutions we have **solved the inequality**.

The **graph of an inequality** in $x$ and $y$ consists of all pairs $(x, y)$ that are solutions of the inequality. The graph of an inequality involving two variables typically is a region of the coordinate plane.

The point $(2, 7)$ is on the graph of the line $y = 2x + 3$ but is not a solution of $y < 2x + 3$. A point $(2, y)$ below the line $y = 2x + 3$ is on the graph of $y < 2x + 3$, and those above it are not. The graph of $y < 2x + 3$ is the set of all points below the line $y = 2x + 3$ (Figure 7.24).

We can summarize our observations about the graph of an inequality in two variables with the following procedure.

---

**Steps for Drawing the Graph of an Inequality in Two Variables**

1. Draw the graph of the equation obtained by replacing the inequality sign by an equal sign. Use a dashed line if the inequality is $<$ or $>$. Use a solid line if the inequality is $\leq$ or $\geq$.

2. Check a point in each of the two regions of the plane determined by the graph of the equation. If the point satisfies the inequality, then shade the region containing the point.

---

### Example 1  GRAPHING A LINEAR INEQUALITY

Draw the graph of $y \geq 2x + 3$.

**Solution** Because of "$\geq$," the graph of the line $y = 2x + 3$ is part of the graph of the inequality and should be drawn as a solid line. The point $(0, 4)$ is above the line and satisfies the inequality because

$$4 \geq 2(0) + 3 = 3.$$

Thus, the graph of $y \geq 2x + 3$ consists of all points on or above the line $y = 2x + 3$ (Figure 7.25).

The graph of the linear inequality $y \geq ax + b$, $y > ax + b$, $y \leq ax + b$, or $y < ax + b$ is a **half-plane**. The graph of the line $y = ax + b$ is the **boundary** of the region.

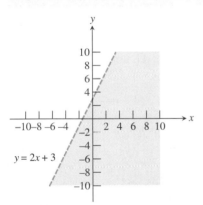

**Figure 7.24** A graph of $y = 2x + 3$ (dashed line) and $y < 2x + 3$ (shaded area). The line is dashed to indicate it is *not* part of the solution of $y < 2x + 3$.

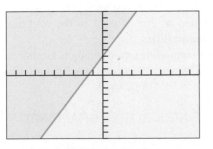

[–10, 10] by [–10, 10]

**Figure 7.25** The graph of $y \geq 2x + 3$. (Example 1)

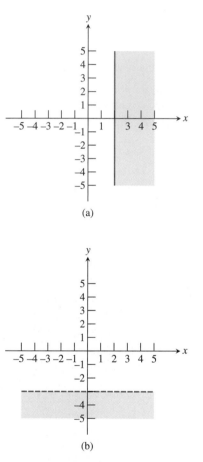

(a)

(b)

**Figure 7.26** The graphs of (a) $x \geq 2$ and (b) $y < -3$. (Example 2)

**Shading Graphs**

Most graphing utilities are capable of shading solutions to inequalities. Check the owner's manual for your grapher.

## Example 2 GRAPHING LINEAR INEQUALITIES

Draw the graph of the inequality. State the boundary of the region.

**(a)** $x \geq 2$                            **(b)** $y < -3$

**Solution**

**(a)** Replacing "$\geq$" by "$=$" we obtain the equation $x = 2$ whose graph is a vertical line. The graph of $x \geq 2$ is the set of all points on and to the right of the vertical line $x = 2$ (Figure 7.26a). The line $x = 2$ is the boundary of the region.

**(b)** Replacing "$<$" by "$=$" we obtain the equation $y = -3$ whose graph is a horizontal line. The graph of $y < -3$ is the set of all points below the horizontal line $y = -3$ (Figure 7.26b). The line $y = -3$ is the boundary of the region.

## Example 3 GRAPHING A QUADRATIC INEQUALITY

Draw the graph of $y \geq x^2 - 3$. State the boundary of the region.

**Solution** Replacing "$\geq$" by "$=$" we obtain the equation $y = x^2 - 3$ whose graph is a parabola. The pair $(0, 1)$ is a solution of the inequality because

$$1 \geq (0)^2 - 3 = -3.$$

Thus, the graph of $y \geq x^2 - 3$ is the parabola together with the region inside the parabola (Figure 7.27). The parabola is the boundary of the region.

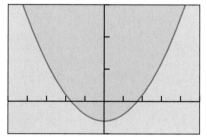

[–5, 5] by [–5, 15]

**Figure 7.27** The graph of $y \geq x^2 - 3$. (Example 3)

## Systems of Inequalities

A **solution** of a system of inequalities in $x$ and $y$ is an ordered pair $(x, y)$ that satisfies each inequality in the system. When we have found all the common solutions we have **solved the system of inequalities**.

The technique for solving a system of inequalities graphically is similar to that for solving a system of equations graphically. We graph each inequality and determine the points common to the individual graphs.

## Example 4 SOLVING A SYSTEM OF INEQUALITIES GRAPHICALLY

Solve the system.

$$y > x^2$$

$$2x + 3y < 4$$

**Solution** The graph of $y > x^2$ is shaded in Figure 7.28a. It does not include its boundary $y = x^2$. The graph of $2x + 3y < 4$ is shaded in Figure 7.28b. It does not include its boundary $2x + 3y = 4$. The solution to the system is the intersection of these two graphs, as shaded in Figure 7.28c.

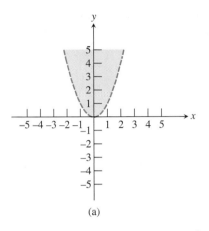

(a)                              (b)                              (c)

**Figure 7.28** The graphs of (a) $y > x^2$, (b) $2x + 3y < 4$, and (c) the system of Example 4.

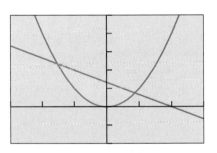

[–3, 3] by [–2, 5]

**Figure 7.29** The solution of the system in Example 4. Most graphers cannot distinguish between dashed and solid boundaries.

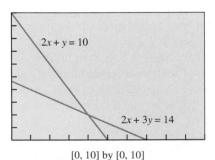

[0, 10] by [0, 10]

**Figure 7.30** The solution (shaded) of the system in Example 5. The boundary points are included.

### Support with a Grapher

Figure 7.29 shows what our grapher produces when we shade above the curve $y = x^2$ and below the curve $2x + 3y = 4$. The shaded portion appears to be identical to the shaded portion in Figure 7.28c.

**Example 5** SOLVING A SYSTEM OF INEQUALITIES

Solve the system.

$$2x + y \leq 10$$
$$2x + 3y \leq 14$$
$$x \geq 0$$
$$y \geq 0$$

**Solution** The solution is in the first quadrant because $x \geq 0$ and $y \geq 0$, it lies below each of the two lines $2x + y = 10$ and $2x + 3y = 14$, and includes all of its boundary points (Figure 7.30).

## Linear Programming

Sometimes decision-making in management science requires that we find a minimum or a maximum of a linear function

$$f = a_1 x_1 + a_2 x_2 + \cdots + a_n x_n,$$

called an **objective function**, over a set of points. Such a problem is a **linear programming problem**. In two dimensions, the function $f$ takes the form $f = ax + by$ and the set of points is the solution of a system of inequalities, called **constraints**, such as the one in Figure 7.30. The solution of the system of inequalities is the set of **feasible** $xy$ points for the optimization problem.

It can be shown that if a linear programming problem has a solution it occurs at one of the **vertex points**, or **corner points**, along the boundary of the region. We use this information in Examples 6 and 7.

### Example 6 SOLVING A LINEAR PROGRAMMING PROBLEM

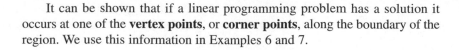

Find the maximum and minimum values of the objective function $f = 5x + 8y$, subject to the constraints given by the system of inequalities.

$$2x + y \leq 10$$

$$2x + 3y \leq 14$$

$$x \geq 0$$

$$y \geq 0$$

**Solution** The feasible $xy$ points are those graphed in Figure 7.30. Figure 7.31 shows that the two lines $2x + 3y = 14$ and $2x + y = 10$ intersect at $(4, 2)$. The corner points are:

$(0, 0)$,

$(0, 14/3)$, the $y$-intercept of $2x + 3y = 14$,

$(5, 0)$, the $x$-intercept of $2x + y = 10$, and

$(4, 2)$, the point of intersection of $2x + 3y = 14$ and $2x + y = 10$.

The following table evaluates $f$ at the corner points of the region in Figure 7.28.

| $(x, y)$ | $(0, 0)$ | $(0, 14/3)$ | $(4, 2)$ | $(5, 0)$ |
|----------|----------|-------------|----------|----------|
| $f$ | 0 | 112/3 | 36 | 25 |

The maximum value of $f$ is 112/3 at $(0, 14/3)$. The minimum value is 0 at $(0, 0)$.

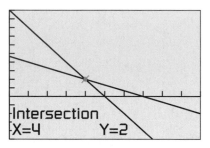

[0, 10] by [−5, 10]

**Figure 7.31** The lines $2x + 3y = 14$ and $2x + y = 10$ intersect at $(4, 2)$. (Example 6)

Here is one way to analyze the linear programming problem in Example 6. By assigning positive values to $f$ in $f = 5x + 8y$, we obtain a family of parallel lines whose distance from the origin increase as $f$ increases. (See Exercise 41.) This family of lines sweep across the region of feasible solutions. Geometrically, we can see that there is a minimum and maximum value for $f$ if the line $f = 5x + 8y$ is to intersect the region of feasible solutions.

### Example 7 PURCHASING FERTILIZER

Johnson's Produce is purchasing fertilizer with two nutrients: N (nitrogen) and P (phosphorous). They need at least 180 units of N and 90 units of P. Their supplier has two brands of fertilizer for them to buy. Brand A costs $10 a bag and has 4 units of N and 1 unit of P. Brand B costs $5 a bag and has 1 unit of each nutrient. Johnson's Produce can pay at most $800 for the fertilizer. How many bags of each brand should be purchased to minimize cost?

**Solution**

**Model**

Let $x$ = number of bags of Brand A

Let $y$ = number of bags of Brand B

Then $C$ = the total cost = $10x + 5y$ is the objective function to be minimized. The constraints are:

$$4x + y \geq 180 \qquad \text{Amount of N is at least 180.}$$

$$x + y \geq 90 \qquad \text{Amount of P is at least 90.}$$

$$10x + 5y \leq 800 \qquad \text{Total cost to be at most \$800.}$$

$$x \geq 0, y \geq 0$$

### Solve Graphically

The region of feasible points is the intersection of the graphs of $4x + y \geq 180$, $x + y \geq 90$, and $10x + 5y \leq 800$ in the first quadrant (Figure 7.32).

The region has three corner points at the points of intersections of the three lines $4x + y = 180$, $x + y = 90$, and $10x + 5y = 800$: (10, 140), (70, 20), and (30, 60). The values of the objective function $C$ at the corner points are:

$$C(10, 140) = 10(10) + 5(140) = 800$$

$$C(70, 20) = 10(70) + 5(20) \;\; = 800$$

$$C(30, 60) = 10(30) + 5(60) \;\; = 600$$

### Interpret

The minimum cost for the fertilizer is $600 when 30 bags of Brand A and 60 bags of Brand B are purchased. For this purchase, Johnson's Produce gets exactly 180 units of nutrient N and 90 units of nutrient P.

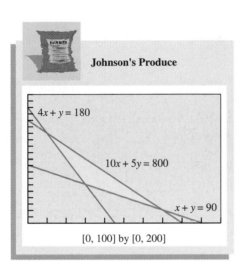

**Johnson's Produce**

$4x + y = 180$

$10x + 5y = 800$

$x + y = 90$

[0, 100] by [0, 200]

**Figure 7.32** The feasible region in Example 7.

The region in Example 8 is unbounded. Using the discussion following Example 6, we can see geometrically that the linear programming problem in Example 8 does not have a maximum value but, fortunately, does have a minimum value.

### Example 8  MINIMIZING OPERATING COST

Gonza Manufacturing has two factories that produce three grades of paper: low grade, medium grade, and high grade. It needs to supply 24 tons of low grade, 6 tons of medium grade, and 30 tons of high grade paper. Factory A produces 8 tons of low grade, 1 ton of medium grade, 2 tons of high grade paper daily, and costs $2000 per day to operate. Factory B produces 2 tons of low grade, 1 ton of medium grade, 8 tons of high grade paper daily, and costs $4000 per day to operate. How many days should each factory operate to fill the orders at minimum cost?

### Solution
### Model

Let $x$ = the number of days Factory A operates.

Let $y$ = the number of days Factory B operates.

Then $C$ = total operating cost = $2000x + 4000y$ is the objective function to be minimized. The constraints are:

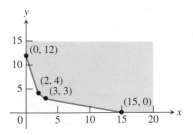

**Figure 7.33** The graph of the feasible points in Example 8.

$$8x + 2y \geq 24 \qquad \text{Amount of low grade is at least 24.}$$

$$x + y \geq 6 \qquad \text{Amount of medium grade is at least 6.}$$

$$2x + 8y \geq 30 \qquad \text{Amount of high grade is at least 30.}$$

$$x \geq 0, y \geq 0$$

**Solve Graphically**

The region of feasible points is the intersection of the graphs of $8x + 2y \geq 24$, $x + y \geq 6$, and $2x + 8y \geq 30$ in the first quadrant (Figure 7.33).

The region has four corner points:

(0, 12), the $y$-intercept of $8x + 2y = 24$,

(2, 4), the point of intersection of $8x + 2y = 24$ and $x + y = 6$,

(3, 3), the point of intersection of $x + y = 6$ and $2x + 8y = 30$,

(15, 0), the $x$-intercept of $2x + 8y = 30$.

The values of the objective function $C$ at the corner points are:

$$C(0, 12) = 2000(0) + 4000(12) = 48,000$$

$$C(2, 4) = 2000(2) + 4000(4) = 20,000$$

$$C(3, 3) = 2000(3) + 4000(3) = 18,000$$

$$C(15, 0) = 2000(15) + 4000(0) = 30,000$$

**Interpret**

The minimum operational cost is $18,000 when the two factories are operated for 3 days each. The two factories will produce 30 tons of low grade, 6 tons of medium grade, and 30 tons of high grade. They will have a surplus of 6 tons of low grade paper.

# Quick Review 7.5

In Exercises 1–4, find the $x$- and $y$-intercepts of the line and draw its graph.

**1.** $2x - 3y = 6$  (3, 0); (0, −2) **2.** $5x + 10y = 30$

**3.** $\dfrac{x}{20} + \dfrac{y}{50} = 1$ **4.** $\dfrac{x}{30} - \dfrac{y}{20} = 1$

In Exercises 5–9, find the point of intersection of the two lines. (We will use these values in Examples 7 and 8.)

**5.** $4x + y = 180$ and $x + y = 90$

**6.** $x + y = 90$ and $10x + 5y = 800$

**7.** $4x + y = 180$ and $10x + 5y = 800$

**8.** $8x + 2y = 24$ and $x + y = 6$

**9.** $x + y = 6$ and $2x + 8y = 30$

**10.** Solve the system of equations:

$$y = x^2$$

$$2x + 3y = 4$$

$(x, y) \approx \{(-1.54, 2.36), (0.87, 0.75)\}$

# Section 7.5 Exercises

In Exercises 1–6, match the inequality with its graph. Indicate whether the boundary is included in or excluded from the graph. All graphs are drawn in $[-4.7, 4.7]$ by $[-3.1, 3.1]$.

**1.** $x \le 3$

**2.** $y > 2$

**3.** $2x - 5y \ge 2$

**4.** $y > (1/2)x^2 - 1$

**5.** $y \ge 2 - x^2$

**6.** $x^2 + y^2 < 4$

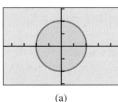

(a)

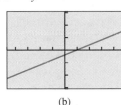

(b)

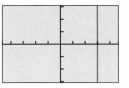

(c)

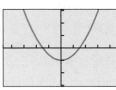

(d)

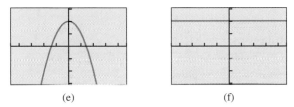

(e)                                  (f)

In Exercises 7–16, graph the inequality. State the boundary of the region.

**7.** $x \le 4$

**8.** $y > -3$

**9.** $2x + 5y \le 7$

**10.** $3x - y > 4$

**11.** $y < x^2 + 1$

**12.** $y \ge x^2 - 3$

**13.** $x^2 + y^2 < 9$

**14.** $x^2 + y^2 \ge 4$

**15.** $y \ge \dfrac{e^x + e^{-x}}{2}$

**16.** $y < \sin x$

In Exercises 17–22, solve the system of inequalities. Give the coordinates of any corner points.

**17.** $5x - 3y > 1$
      $3x + 4y \le 18$

**18.** $4x + 3y \le -6$
      $2x - y \le -8$

**19.** $y \le 2x + 3$
      $y \ge x^2 - 2$

**20.** $x - 3y - 6 < 0$
      $y > -x^2 - 2x + 2$

**21.**   $y \ge x^2$
      $x^2 + y^2 \le 4$

**22.** $x^2 + y^2 \le 9$
      $y \ge |x|$

In Exercises 23–26, solve the system of inequalities. Give the coordinates of any corner points.

**23.** $2x + y \le 80$
      $x + 2y \le 80$
      $x \ge 0$
      $y \ge 0$

**24.** $3x + 8y \ge 240$
      $9x + 4y \ge 360$
      $x \ge 6$
      $y \ge 0$

**25.** $5x + 2y \le 20$
      $2x + 3y \le 18$
      $x + y \ge 2$
      $x \ge 0$
      $y \ge 0$

**26.** $7x + 3y \le 210$
      $3x + 7y \le 210$
      $x + y \ge 30$

In Exercises 27–30, write a system of inequalities whose solution is the region shaded in the given figure. All boundaries are to be included.

**27. Group Activity**

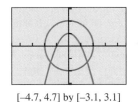

$[-4.7, 4.7]$ by $[-3.1, 3.1]$

**28. Group Activity**

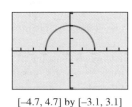

$[-4.7, 4.7]$ by $[-3.1, 3.1]$

**29. Group Activity**

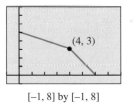

$[-1, 8]$ by $[-1, 8]$

**30. Group Activity**

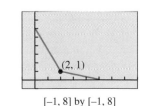

$[-1, 8]$ by $[-1, 8]$

In Exercises 31–36, find the minimum and maximum, if they exist, of the objective function $f$, subject to the constraints.

**31.** Objective function: $f = 4x + 3y$
      Constraints:
$$x + y \le 80$$
$$x - 2y \le 0$$
$$x \ge 0, y \ge 0$$

**32.** Objective function: $f = 10x + 11y$
      Constraints:
$$x + y \le 90$$
$$3x - y \ge 0$$
$$x \ge 0, y \ge 0$$

**33.** Objective function: $f = 7x + 4y$
      Constraints:
$$5x + y \ge 60$$
$$x + 6y \ge 60$$
$$4x + 6y \ge 204$$
$$x \ge 0, y \ge 0$$

**34.** Objective function: $f = 15x + 25y$
Constraints:
$$3x + 4y \geq 60$$
$$x + 8y \geq 40$$
$$11x + 28y \leq 380$$
$$x \geq 0, y \geq 0$$

**35.** Objective function: $f = 5x + 2y$
Constraints:
$$2x + y \geq 12$$
$$4x + 3y \geq 30$$
$$x + 2y \geq 10$$
$$x \geq 0, y \geq 0$$

**36.** Objective function: $f = 3x + 5y$
Constraints:
$$3x + 2y \geq 20$$
$$5x + 6y \geq 52$$
$$2x + 7y \geq 30$$
$$x \geq 0, y \geq 0$$

**37. Mining Ore** Pearson's Metals mines two ores: R and S. The company extracts minerals A and B from each type of ore. It costs $50 per ton to extract 80 lb of A and 160 lb of B from ore R. It costs $60 per ton to extract 140 lb of A and 50 lb of B from ore S. Pearson's must produce at least 4000 lb of A and 3200 lb of B. How much of each ore should be processed to minimize cost? What is the minimum cost?

**38. Planning a Diet** Paul's diet is to contain at least 24 units of carbohydrates and 16 units of protein. Food substance A costs $1.40 per unit and each unit contains 3 units of carbohydrates and 4 units of protein. Food substance B costs $0.90 per unit and each unit contains 2 units of carbohydrates and 1 unit of protein. How many units of each food substance should be purchased in order to minimize cost? What is the minimum cost?

**39. Producing Gasoline** Two oil refineries produce three grades of gasoline: A, B, and C. At each refinery, the three grades of gasoline are produced in a single operation in the following proportions: Refinery 1 produces 1 unit of A, 2 units of B, and 1 unit of C; Refinery 2 produces 1 unit of A, 4 units of B, and 4 units of C. For the production of one operation, Refinery 1 charges $300 and Refinery 2 charges $600. A customer needs 100 units of A, 320 units of B, and 200 units of C. How should the orders be placed if the customer is to minimize his cost?

**40. Maximizing Profit** A manufacturer wants to maximize the profit for two products. Product A yields a profit of $2.25 per unit, and product B yields a profit of $2.00 per unit. Demand information requires that the total number of units produced be no more than 3000 units, and that the number of units of product B produced be greater than or equal to half the number of units of product A produced. How many of each unit should be produced to maximize profit?

## Explorations

**41. Revisiting Example 6** Consider the objective function $f = 5x + 8y$ of Example 6.

(a) Prove that for any two real number values for $f$, the two lines are parallel.

(b) **Writing to Learn** For $f > 0$, give reasons why the line moves further away from the origin as the value of $f$ increases.

(c) **Writing to Learn** Give a geometric explanation of why the region of Example 6 must contain a minimum and a maximum value for $f$.

**42. Writing to Learn** Describe all the possible ways that two distinct parabolas of the form $y = f(x)$ can intersect . Give examples.  ■

## Extending the Ideas

**43. Implicit Functions** The equation
$$\frac{x^2}{9} + \frac{y^2}{4} = 1$$
defines $y$ as two implicit functions of $x$. Solve for $y$ to find the two functions and draw the graph of the equation.

**44. Implicit Functions** The equation
$$x^2 - y^2 = 4$$
defines $y$ as two implicit functions of $x$. Solve for $y$ to find the two functions and draw the graph of the equation.

**45.** Solve the system of inequalities:
$$\frac{x^2}{9} + \frac{y^2}{4} \leq 1$$
$$y \geq x^2 - 1$$

(*Hint*: See Exercise 43.)

**46.** Graph the inequality $x^2 - y^2 \leq 4$. (*Hint*: See Exercise 44.)

## Math at Work

I got into electrical engineering because I enjoy working on computers. Also, it is a skill one can use to get a good job.

One of the ways I use mathematics in my field is when sending a picture between computers. A picture on a computer is made of pixels, which are tiny little dots of color. There are many pixels in a picture, and therefore, storing the picture as the sum total of its pixels makes for a very large file. This large file would take a very long time to send to another computer. Therefore, a mathematical model is used to represent the picture. The mathematical model can use one symbol to represent many pixels, so that only that symbol needs to be sent, instead of many pixels. Then, at the other computer, the picture can be translated back into the pixels.

Another way we represent a picture is by textures. Different mathematical models can be used to represent different textures, and then the computer can differentiate between different textures in one picture.

Ngao Mayuma

## Chapter 7    Key Ideas

### Concepts

additive identity matrix (p. 554)
additive inverse of a matrix (p. 554)
augmented matrix (p. 570)
boundary of a region (p. 587)
characteristic polynomial (p. 567)
coefficient matrix (p. 570)
cofactor (p. 559)
column subscript (p. 552)
constraints (p. 589)
corner (vertex) points (p. 590)
demand curve (p. 549)
determinant of a matrix (p. 558–559)
eigenvalues (p. 579)
element (entry) of a matrix (p. 552)
elementary row operation (p. 571)
elimination method (p. 546)
equal matrices (p. 553)
equilibrium point (p. 549)

equilibrium price (p. 549)
equivalent systems of equations (p. 567)
feasible points (p. 589)
Gaussian elimination (p. 568)
graph of an inequality in $x$ and $y$ (p. 587)
half-plane (p. 587)
linear programming problem (p. 589)
matrix addition (p. 553)
matrix multiplication (p. 555)
matrix subtraction (p. 553)
matrix (p. 552)
minor (p. 559)
multiplicative identity matrix (p. 557)
multiplicative inverse of a matrix
    (p. 557, 560)
nonsingular matrix (p. 558)
objective function (p. 589)
order of a matrix (p. 552)

partial fraction decomposition (p. 580)
partial fraction (p. 580)
reduced row echelon form (p. 572)
row echelon form (p. 571)
row subscript (p. 552)
scalar multiple of a matrix (p. 554)
singular matrix (p. 558)
solution of a system of equations (p. 543)
solution of a system of inequalities
    (p. 588)
solution of an inequality in $x$ and $y$
    (p. 587)
square matrix (p. 552)
supply curve (p. 548)
symmetric matrix (p. 565)
transpose of a matrix (p. 557)
triangular form (p. 569)
zero matrix (p. 554)

## Properties, Theorems, and Formulas

### Matrix Operations

Let $A = [a_{ij}]$ and $B = [b_{ij}]$ be matrices of order $m \times n$, $C = [c_{ij}]$ a matrix of order $n \times r$, and $k$ a real number.

1. **Sum:** $A + B = [a_{ij} + b_{ij}]$

2. **Difference:** $A - B = [a_{ij} - b_{ij}]$

3. **Scalar Multiple:** $kA = [ka_{ij}]$

4. **Multiplication:** $AC = [d_{ij}]$, where $d_{ij} = a_{i1}c_{1j} + a_{i2}c_{2j} + \cdots + a_{in}c_{nr}$.

### Theorem Inverses of $n \times n$ Matrices

An $n \times n$ matrix $A$ has an inverse if and only if $\det A \neq 0$.

### Properties of Matrices

Let $A$, $B$, and $C$ be matrices whose orders are such that the following sums, differenes, and products are defined.

1. **Commutative property**
   Addition:
   $A + B = B + A$
   Multiplication
   (Does not hold in general)

2. **Associative property**
   Addition:
   $(A + B) + C = A + (B + C)$
   Multiplication:
   $(AB)C = A(BC)$

3. **Identity property**
   Addition: $A + O = A$
   Multiplication: order$(A) = n \times n$
   $I_n A = AI_n = A$

4. **Inverse property**
   Addition: $A + (-A) = O$
   Multiplication: order$(A) = n \times n$
   $AA^{-1} = A^{-1}A = I_n$, $\det A \neq 0$

5. **Distributive property**
   **Multiplication over addition**
   $A(B + C) = AB + AC$
   $(A + B)C = AC + BC$

   **Multiplication over subtraction**
   $A(B - C) = AB - AC$
   $(A - B)C = AC - BC$

### Theorem Invertible Square Linear Systems

Let $A$ be the coefficient matrix of a system of $n$ linear equations in $n$ variables given by $AX = B$, where $X$ is the $n \times 1$ matrix of variables and $B$ is the $n \times 1$ matrix of numbers on the right-hand side of the equations. If $A^{-1}$ exists, then the system of equations has the unique solution $X = A^{-1}B$.

## Procedures

### Solving Systems of Equations Algebraically

1. Substitution

2. Elimination

3. Gaussian Elimination

4. Inverse Matrices

### Partial Fraction Decomposition of $\dfrac{f(x)}{d(x)}$

1. degree of $f \geq$ degree of $d$: Use the division algorithm to divide $f$ by $d$ to obtain the quotient $q$ and remainder $r$ and write

$$\frac{f(x)}{d(x)} = q(x) + \frac{r(x)}{d(x)}.$$

**2.** Factor $d(x)$ into a product of factors of the form $(mx + n)^u$ or $(ax^2 + bx + c)^v$, where $ax^2 + bx + c$ is irreducible.

**3.** For each factor $(mx + n)^u$: The partial fraction decomposition of $r(x)/d(x)$ must include the sum

$$\frac{A_1}{mx + n} + \frac{A_2}{(mx + n)^2} + \cdots + \frac{A_u}{(mx + n)^u},$$

where $A_1, A_2, \ldots, A_u$ are real numbers.

**4.** For each factor $(ax^2 + bx + c)^v$: The partial fraction decomposition of $r(x)/d(x)$ must include the sum

$$\frac{B_1x + C_1}{ax^2 + bx + c} + \frac{B_2x + C_2}{(ax^2 + bx + c)^2} + \cdots + \frac{B_vx + C_v}{(ax^2 + bx + c)^v},$$

where $A_1, A_2, \ldots, A_v$ and $B_1, B_2, \ldots, B_v$ are real numbers.

The partial fraction decomposition of the original rational function is the sum of $q(x)$ and the fractions in parts 3 and 4.

# Chapter 7   Review Exercises

The collection of exercises marked in red could be used as a chapter test.

In Exercises 1 and 2, find **(a)** $A + B$, **(b)** $A - B$, **(c)** $-2A$, and **(d)** $3A - 2B$.

**1.** $A = \begin{bmatrix} -1 & 3 \\ 4 & 0 \end{bmatrix}$, $B = \begin{bmatrix} 2 & -1 \\ 4 & 3 \end{bmatrix}$

**2.** $A = \begin{bmatrix} 2 & 3 & -1 & 2 \\ 1 & 4 & -2 & -3 \\ 0 & -3 & 2 & 1 \end{bmatrix}$, $B = \begin{bmatrix} -1 & 2 & 0 & 4 \\ 2 & -1 & 3 & 3 \\ -2 & 4 & 1 & 3 \end{bmatrix}$

In Exercises 3–8, find the products $AB$ and $BA$, or state that a given product is not possible.

**3.** $A = \begin{bmatrix} -1 & 4 \\ 0 & 6 \end{bmatrix}$, $B = \begin{bmatrix} 3 & -1 & 5 \\ 0 & -2 & 4 \end{bmatrix}$

**4.** $A = \begin{bmatrix} -1 & 2 \\ 3 & -1 \\ 4 & 3 \end{bmatrix}$, $B = \begin{bmatrix} -2 & 3 & 1 \\ 2 & 1 & 0 \\ -1 & 2 & -3 \end{bmatrix}$

**5.** $A = [-1 \ \ 4]$, $B = \begin{bmatrix} 5 & -3 \\ 2 & 1 \end{bmatrix}$ [3 7]; not possible

**6.** $A = \begin{bmatrix} -1 & 1 \\ 0 & 1 \end{bmatrix}$, $B = \begin{bmatrix} 3 & -4 \\ 1 & 2 \\ 3 & 1 \\ 1 & 1 \end{bmatrix}$

**7.** $A = \begin{bmatrix} 0 & 1 & 0 \\ 1 & 0 & 0 \\ 0 & 0 & 1 \end{bmatrix}$, $B = \begin{bmatrix} 2 & -3 & 4 \\ 1 & 2 & -3 \\ -2 & 1 & -1 \end{bmatrix}$

**8.** $A = \begin{bmatrix} 0 & 1 & 0 & 0 \\ 1 & 0 & 0 & 0 \\ 0 & 0 & 0 & 1 \\ 0 & 0 & 1 & 0 \end{bmatrix}$, $B = \begin{bmatrix} -2 & 1 & 0 & 1 \\ 3 & 0 & 2 & 1 \\ -1 & 1 & 2 & -1 \\ 3 & -2 & 1 & 0 \end{bmatrix}$

In Exercises 9 and 10, use multiplication to verify that the matrices are inverses.

**9.** $A = \begin{bmatrix} 1 & -2 & 1 & 1 \\ 1 & -1 & 0 & 3 \\ 1 & -1 & 2 & 2 \\ 2 & -4 & 2 & 3 \end{bmatrix}$, $B = \begin{bmatrix} 8 & 1.5 & 0.5 & -4.5 \\ 2 & 0.5 & 0.5 & -1.5 \\ -1 & -0.5 & 0.5 & 0.5 \\ -2 & 0 & 0 & 1 \end{bmatrix}$

**10.** $A = \begin{bmatrix} -1 & 1 & 1 \\ 2 & 1 & 0 \\ -1 & 0 & 2 \end{bmatrix}$, $B = \begin{bmatrix} -0.4 & 0.4 & 0.2 \\ 0.8 & 0.2 & -0.4 \\ -0.2 & 0.2 & 0.6 \end{bmatrix}$

In Exercises 11 and 12, find the inverse of the matrix if it has one. If it does, use multiplication to support your result.

**11.** $\begin{bmatrix} 1 & 2 & 0 & -1 \\ 2 & -1 & 1 & 2 \\ 2 & 0 & 1 & 2 \\ -1 & 1 & 1 & 4 \end{bmatrix}$    **12.** $\begin{bmatrix} -1 & 0 & 1 \\ 2 & -1 & 1 \\ 1 & 1 & 1 \end{bmatrix}$

In Exercises 13 and 14, evaluate the determinant of the matrix.

**13.** $\begin{bmatrix} 1 & -3 & 2 \\ 2 & 4 & -1 \\ -2 & 0 & 1 \end{bmatrix}$ 20    **14.** $\begin{bmatrix} -2 & 3 & 0 & 1 \\ 3 & 0 & 2 & 0 \\ 5 & 2 & -3 & 4 \\ 1 & -1 & 2 & 3 \end{bmatrix}$ 270

In Exercises 15–18, find a reduced row echelon form of the matrix.

**15.** $\begin{bmatrix} 1 & 0 & 2 \\ 3 & 1 & 5 \\ 1 & -1 & 3 \end{bmatrix}$    **16.** $\begin{bmatrix} 2 & 1 & 1 & 1 \\ -3 & -1 & -2 & 1 \\ 5 & 2 & 2 & 3 \end{bmatrix}$

**17.** $\begin{bmatrix} 1 & 2 & 3 & 1 \\ 2 & 3 & 3 & -2 \\ 1 & 2 & 4 & 6 \end{bmatrix}$    **18.** $\begin{bmatrix} 1 & -2 & 0 & 4 \\ -2 & 5 & 3 & -6 \\ 2 & -4 & 1 & 9 \end{bmatrix}$

In Exercises 19–22, state whether the system of equations has a solution. If it does, solve the system.

**19.** $3x - y = 1$  (1, 2)
$x + 2y = 5$

**20.** $x - 2y = -1$  (−3, −1)
$-2x + y = 5$

**21.** $x + 2y = 1$  no solution
$4y - 4 = -2x$

**22.** $x - 2y = 9$  no solution
$3y - \dfrac{3}{2}x = -9$

In Exercises 23–28, use Gaussian elimination to solve the system of equations.

**23.** $x + z + w = 2$
$x + y + z = 3$
$3x + 2y + 3z + w = 8$

**24.** $x + w = -2$
$x + y + z + 2w = -2$
$-x - 2y - 2z - 3w = 2$

**25.** $x + y - 2z = 2$
$3x - y + z = 4$
$-2x - 2y + 4z = 6$

**26.** $x + y - 2z = 2$
$3x - y + z = 1$
$-2x - 2y + 4z = -4$

**27.** $-x - 6y + 4z - 5w = -13$
$2x + y + 3z - w = 4$
$2x + 2y + 2z = 6$
$-x - 3y + z - 2w = -7$

**28.** $-x + 2y + 2z - w = -4$  (−w + 2, −z − 1, z, w)
$y + z = -1$
$-2x + 2y + 2z - 2w = -6$
$-x + 3y + 3z - w = -5$

In Exercises 29–32, solve the system of equations by using inverse matrices.

**29.** $x + 2y + z = -1$
$x - 3y + 2z = 1$
$2x - 3y + z = 5$

**30.** $x + 2y - z = -2$
$2x - y + z = 1$
$x + y - 2z = 3$

**31.** $2x + y + z - w = 1$  no solution
$2x - y + z - w = -2$
$-x + y - z + w = -3$
$x - 2y + z - w = 1$

**32.** $x - 2y + z - w = 2$  (13/3, −8/3, −1/3, 22/3)
$2x + y - z - w = -1$
$x - y + 2z - w = -1$
$x + 3y - z + w = 4$

In Exercises 33–36, solve the system of equations by finding the reduced row echelon form of the augmented matrix.

**33.** $x + 2y - 2z + w = 8$  (2 − w, z + 3, z, w)
$2x + 3y - 3z + 2w = 13$

**34.** $x + 2y - 2z + w = 8$  (2 − w, z + 3, z, w)
$2x + 7y - 7z + 2w = 25$
$x + 3y - 3z + w = 11$

**35.** $x + 2y + 4z + 6w = 6$  (−2, 1, 3, −1)
$3x + 4y + 8z + 11w = 11$
$2x + 4y + 7z + 11w = 10$
$3x + 5y + 10z + 14w = 15$

**36.** $x + 2z - 2w = 5$  (1, −w − 3, w + 2, w)
$2x + y + 4z - 3w = 7$
$4x + y + 7z - 6w = 15$
$2x + y + 5z - 4w = 9$

In Exercises 37 and 38, find the equilibrium point for the demand and supply curves.

**37.** $p = 100 - x^2$  Demand curve  ≈ (7.57, 42.71)
$p = 20 + 3x$  Supply curve

**38.** $p = 80 - \dfrac{1}{10}x^2$  Demand curve  ≈ (13.91, 60.65)
$p = 5 + 4x$  Supply curve

In Exercises 39–44, solve the system of equations graphically.

**39.** $3x - 2y = 5$
$2x + y = -2$

**40.** $y = x - 1.5$
$y = 0.5x^2 - 3$

**41.** $y = -0.5x^2 + 3$
$y = 0.5x^2 - 1$

**42.** $x^2 + y^2 = 4$
$y = 2x^2 - 3$

**43.** $y = 2 \sin x$
$y = 2x - 3$

**44.** $y = \ln 2x$
$y = 2x^2 - 12x + 15$

In Exercises 45 and 46, find the coefficients of the function so that its graph goes through the given points.

**45. Curve Fitting** $f(x) = ax^3 + bx^2 + cx + d$
(2, 8), (4, 5), (6, 3), (9, 4)

**46. Curve Fitting** $f(x) = ax^4 + bx^3 + cx^2 + dx + e$
(−2, −4), (1, 2), (3, 6), (4, −2), (7, 8)

In Exercises 47–52, find the partial fraction decomposition of the rational function.

**47.** $\dfrac{3x - 2}{x^2 - 3x - 4}$

**48.** $\dfrac{x - 16}{x^2 + x - 2}$

**49.** $\dfrac{3x + 5}{x^3 + 4x^2 + 5x + 2}$

**50.** $\dfrac{3(3 + 2x + x^2)}{x^3 + 3x^2 - 4}$

**51.** $\dfrac{5x^2 - x - 2}{x^3 + x^2 + x + 1}$

**52.** $\dfrac{-x^2 - 5x + 2}{x^3 + 2x^2 + 4x + 8}$

In Exercises 53–56, match the function with its graph. Do this without using your grapher.

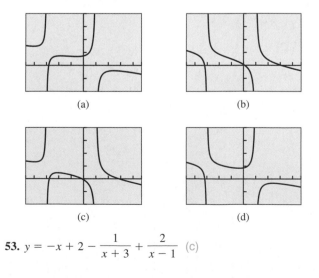

(a)  (b)

(c)  (d)

**53.** $y = -x + 2 - \dfrac{1}{x + 3} + \dfrac{2}{x - 1}$  (c)

**54.** $y = -x + 2 + \dfrac{1}{x+3} - \dfrac{2}{x-1}$ (d)

**55.** $y = -x + 2 + \dfrac{1}{x+3} + \dfrac{2}{x-1}$ (b)

**56.** $y = -x + 2 - \dfrac{1}{x+3} - \dfrac{2}{x-1}$ (a)

In Exercises 57 and 58, graph the inequality.

**57.** $2x - y \leq 1$  **58.** $x + 3y < 2$

In Exercises 59–64, solve the system of inequalities. Give the coordinates of any corner points.

**59.** $4x + 9y \geq 360$
$9x + 4y \geq 360$
$x + y \leq 90$

**60.** $7x + 10y \leq 70$
$2x + y \leq 10$
$x + y \geq 3$
$x \geq 0$
$y \geq 0$

**61.** $x - 3y + 6 < 0$
$y > x^2 - 6x + 7$

**62.** $x + 2y \geq 4$
$y \leq 9 - x^2$

**63.** $x^2 + y^2 \leq 4$
$y \geq x^2$

**64.** $y \leq x^2 + 4$
$x^2 + y^2 \geq 4$

In Exercises 65–68, find the minimum and maximum, if they exist, of the objective function $f$, subject to the constraints.

**65.** Objective function: $f = 7x + 6y$
Constraints:
$$7x + 5y \geq 100$$
$$2x + 5y \geq 50$$
$$x \geq 0, y \geq 0$$

**66.** Objective function: $f = 11x + 5y$
Constraints:
$$5x + 2y \geq 60$$
$$5x + 8y \geq 120$$
$$x \geq 0, y \geq 0$$

**67.** Objective function: $f = 3x + 7y$
Constraints:
$$5x + 2y \geq 100$$
$$x + 4y \geq 110$$
$$5x + 11y \leq 460$$
$$x \geq 0, y \geq 0$$

**68.** Objective function: $f = 9x + 14y$
Constraints:
$$x + y \leq 120$$
$$9x + 2y \geq 240$$
$$3x + 10y \geq 360$$

**69. Rotating Coordinate Systems** The $xy$-coordinate system is rotated through the angle 45° to obtain the $x'y'$-coordinate system.

**(a)** If the coordinates of a point in the $xy$-coordinate system are (1, 2), what are the coordinates of the rotated point in the $x'y'$-coordinate system?

**(b)** If the coordinates of a point in the $x'y'$-coordinate system are (1, 2), what are the coordinates of the point in the $xy$-coordinate system that was rotated to it?

**70. National Health Expenditures** Table 7.7 shows hospital and medical expenditures (in billions of dollars) for the Defense Department for several years. Let $x = 0$ stand for 1970, $x = 1$ for 1971, and so forth.

**(a)** Find a linear regression model and superimpose its graph on a scatter plot of the data.

**(b)** Find a logistic regression model and superimpose its graph on a scatter plot of the data.

**(c)** Find when the models in (a) and (b) predict the same expenditure amounts. 1973, 1984, and 1993

**(d) Writing to Learn** Which model appears to be a better fit for the data? Explain. Which model would you choose to make predictions for 2000 and beyond? Explain.

| Table 7.7 Defense Department Hospital and Medical Expenditures | |
| --- | --- |
| Year | Expenditures (billions) |
| 1980 | 4.4 |
| 1985 | 7.5 |
| 1990 | 11.6 |
| 1991 | 12.8 |
| 1992 | 13.0 |
| 1993 | 13.3 |
| 1994 | 13.2 |
| 1995 | 13.4 |
| 1996 | 13.4 |

*Source: U.S. Health Care Financing Administration, Health Care Financing Review, fall 1997, in Statistical Abstract of the U.S., 1998.*

**71. Population** Table 7.8 gives the population (in thousands) of the states of Hawaii and Idaho for several years. Let $x = 0$ stand for 1970, $x = 1$ for 1971, and so forth.

**(a)** Find a linear regression model for Hawaii's data and superimpose its graph on a scatter plot of Hawaii's data.

**(b)** Find a linear regression model for Idaho's data and superimpose its graph on a scatter plot of Idaho's data.

**(c)** Using the models in (a) and (b), when will the population of the two states be the same? 2026

**Table 7.8 Population**

| Year | Hawaii (in thousands) | Idaho (in thousands) |
|------|------------------------|----------------------|
| 1970 | 770  | 713  |
| 1980 | 965  | 944  |
| 1985 | 1040 | 994  |
| 1990 | 1108 | 1007 |
| 1991 | 1131 | 1039 |
| 1992 | 1150 | 1066 |
| 1993 | 1160 | 1101 |
| 1994 | 1173 | 1135 |
| 1995 | 1179 | 1165 |
| 1996 | 1183 | 1188 |
| 1997 | 1187 | 1210 |

*Source: U.S. Bureau of the Census, in Statistical Abstract of the U.S., 1998.*

**72. (a)** The 1990 population data for three states is listed below. Use the data in the first table on the following page to create a $3 \times 2$ matrix that estimates the number of males and females in each state.

| State | Population (millions) |
|-------|------------------------|
| California | 29.8 |
| Florida | 12.9 |
| Rhode Island | 1.0 |

**(b)** Write the data from the 1990 Census table below in the form of a $3 \times 2$ matrix.

| State | % Pop under 18 years | % Pop 65 years or older |
|-------|-----------------------|--------------------------|
| California | 26.2 | 10.5 |
| Florida | 22.3 | 18.2 |
| Rhode Island | 22.6 | 15.3 |

**(c)** Multply your $3 \times 2$ matrix by the scalar 0.01 to change the values from percentages to decimals.

**(d)** Use matrix multiplication to multiply the transpose of the matrix from (c) by the matrix with numbers of males and females in each state. What information does the resulting matrix provide?

**(e)** How many males under age 18 lived in these three states in 1999? How many females age 65 or older lived in these three states? 5.3 million; 2.9 million

**73. Using Matrices** A stockbroker sold a customer 200 shares of stock A, 400 shares of stock B, 600 shares of stock C, and 250 shares of stock D. The price per share of A, B, C, and D are $80, $120, $200, and $300, respectively.

**(a)** Write a $1 \times 4$ matrix $N$ representing the number of shares of each stock the customer bought.

**(b)** Write a $1 \times 4$ matrix $P$ representing the price per share of each stock.

**(c)** Write a matrix product that gives the total cost of the stocks that the customer bought.

**74. Basketball Attendance** At Whetstone High School 452 tickets were sold for the first basketball game. There were two ticket prices: $0.75 for students and $2.00 for nonstudents. How many tickets of each type were sold if the total revenue from the sale of tickets was $429?

**75. Truck Deliveries** Brock's Discount TV has three types of television sets on sale: a 13-in. portable, a 27-in. remote, and a 50-in. console. They have three types of vehicles to use for delivery: vans, small trucks, and large trucks. The vans can carry 8 portable, 3 remote, and 2 console TVs; the small trucks, 15 portable, 10 remote, and 6 console TVs; and the large trucks, 22 portable, 20 remote, and 5 console TVs. On a given day of the sale they have 115 portable, 85 remote, and 35 console TVs to deliver. How many vehicles of each type are needed to deliver the TVs?

**76. Investments** Jessica invests $38,000, part at 7.5% simple interest and the remainder at 6% simple interest. If her annual interest income is $2600, how much does she have invested at each rate?

**77. Business Loans** Thompson's Furniture Store borrowed $650,000 to expand its facilities and extend its product line. Some of the money was borrowed at 4%, some at 6.5%, and the rest at 9%. How much was borrowed at each rate if the annual interest was $46,250 and the amount borrowed at 9% was twice the amount borrowed at 4%?

**78. Home Remodeling** Sanchez Remodeling has three painters: Sue, Esther, and Murphy. Working together they can paint a large room in 4 hours. Sue and Murphy can paint the same size room in 6 hours. Esther and Murphy can paint the same size room in 7 hours. How long would it take each of them to paint the room alone?

**79. Swimming Pool** Three pipes, A, B, and C, are connected to a swimming pool. When all three pipes are running, the pool can be filled in 3 hr. When only A and B are running, the pool can be filled in 4 hr. When only B and C are running, the pool can be filled in 3.75 hr. How long would it take each pipe running alone to fill the pool?

**80. Writing to Learn** If the products $AB$ and $BA$ are defined for the $n \times n$ matrix $A$, what can you conclude about the order of matrix $B$? Explain.

**81. Writing to Learn** If $A$ is an $m \times n$ matrix and $B$ is a $p \times q$ matrix, and if $AB$ is defined, what can you conclude about their orders? Explain.

# Chapter 7 Project

## Analyzing Census Data

The data below was gathered from the U.S. census Bureau (www.census.gov). Examine the male and female population data from 1990 to 1998.

| Population (in millions) | Male | Female |
|---|---|---|
| 1990 | 121.3 | 127.5 |
| 1995 | 128.3 | 134.5 |
| 1996 | 129.5 | 135.7 |
| 1997 | 130.8 | 137.0 |
| 1998 | 132.0 | 138.3 |

1. Plot the data using 1990 as the year zero.

2. What do the slope and *y*-intercept mean in each equation?

3. What conclusions can you draw? According to these models, will the male population ever become greater than the female population? Was the male population ever greater than the female population? Is this enough data to create a model for a hundred years or more? Explain your answers.

4. Notice that the data above only gave information for a span of eight years. This is often not enough information to accurately answer the questions asked above. Data over a small period often appears to be linear and can be modeled with a linear equation that works well over that limited domain. The chart below gives more data. Now use this data to plot the number of males versus time and females versus time using 1890 as year zero.

| Population (in millions) | Male | Female | Population (in millions) | Male | Female |
|---|---|---|---|---|---|
| 1890 | 32.2 | 30.7 | 1950 | 75.2 | 76.1 |
| 1900 | 38.8 | 37.2 | 1960 | 88.3 | 91.0 |
| 1910 | 47.3 | 44.6 | 1970 | 98.9 | 104.3 |
| 1920 | 53.9 | 51.8 | 1980 | 110.1 | 116.5 |
| 1930 | 62.1 | 60.6 | 1990 | 121.3 | 127.5 |
| 1940 | 66.0 | 65.6 | 1998 | 132.0 | 138.3 |

5. Notice that this data does not seem to be linear. Often times, you may remember from chapter 3, a logistic model is used to model population growth. Find the logistic regression for each data plot. What is the intersection of the curves and what does it represent? Would any of your responses in question number 3 above change? If so, how? Why?

6. Go to the census website (www.census.gov). How does your model predict the populations for the current year?

7. Use the census data for 1990. What percentage of the population is male and what percentage is female?

8. Go to the U.S. Census Bureau website (www.census.gov) and use the most recent data along with the concepts from this chapter to collect and analyze other data.

will prove that a circle is a special case of an ellipse. Because it is atypical and lacks some of the features usually associated with an ellipse, a circle is considered to be a degenerate ellipse.

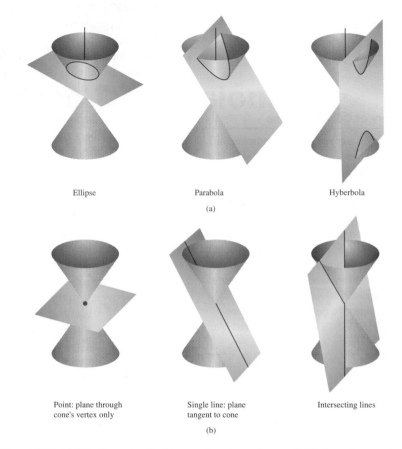

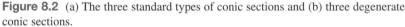

**Figure 8.2** (a) The three standard types of conic sections and (b) three degenerate conic sections.

The conic sections can be defined algebraically in the Cartesian plane as the graphs of **second-degree equations in two variables**, that is, equations of the form

$$Ax^2 + Bxy + Cy^2 + Dx + Ey + F = 0,$$

where $A$, $B$, and $C$ are not all zero.

## Geometry of a Parabola

In Section 2.1 we learned that the graph of a quadratic function is an upward or downward opening parabola. We have seen the role of the parabola in free-fall and projectile motion. We now investigate the geometric properties of parabolas.

## A Degenerate Parabola

If the focus *F* lies on the directrix *l*, the parabola "degenerates" to the line through *F* perpendicular to *l*. Henceforth, we will assume *F* does not lie on *l*.

> **Definition   Parabola**
>
> A **parabola** is the set of all points in a plane equidistant from a particular line (the **directrix**) and a particular point (the **focus**) in the plane.

The line passing through the focus and perpendicular to the directrix is the focal **axis** of the parabola. The axis is the line of symmetry for the parabola. The point where the parabola intersects its axis is the **vertex** of the parabola. The vertex is located midway between the focus and the directrix and is the point of the parabola that is closest to both the focus and the directrix. See Figure 8.3.

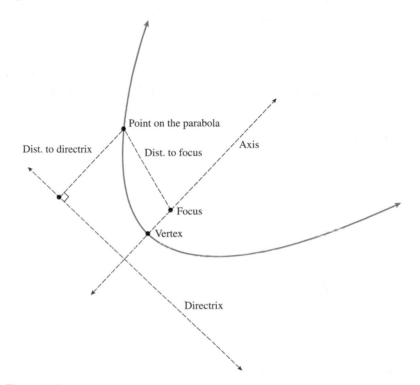

**Figure 8.3** *Structure of a Parabola.* The distance from each point on the parabola to both focus and directrix is the same.

**Exploration Extensions**

Find the *y*-coordinate for a point on the parabola that is a distance of 25 units from the focus.

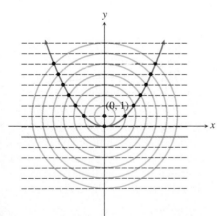

**Figure 8.4** The geometry of a parabola.

---

**Exploration 1   Understanding the Definition of Parabola**

1. Prove that the vertex of the parabola with focus (0, 1) and directrix $y = -1$ is (0, 0). (See Figure 8.4.)
2. Find an equation for the parabola shown in Figure 8.4.   $x^2 = 4y$
3. Find the coordinates of the points of the parabola that are highlighted in Figure 8.4.

Example 1 generalizes the situation in Exploration 1. It shows a connection between the geometric definition of a parabola and its equation.

**Example 1** CONNECTING ALGEBRA AND GEOMETRY

Prove that an equation for the parabola with focus $(0, p)$ and directrix $y = -p$ is $x^2 = 4py$. (See Figure 8.5.)

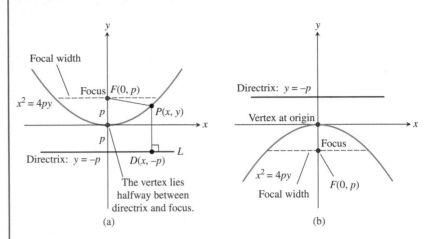

**Figure 8.5** (a) $p > 0$ and (b) $p < 0$. (Example 1)

**Solution** We must show first that a point $P(x, y)$ that is equidistant from $F(0, p)$ and the line $y = -p$ satisfies the equation $x^2 = 4py$, and then that a point satisfying the equation $x^2 = 4py$ is equidistant from $F(0, p)$ and the line $y = -p$:

- Let $P(x, y)$ be equidistant from $F(0, p)$ and the line $y = -p$. Then

$$\sqrt{(x - 0)^2 + (y - p)^2} = \text{distance from } P(x, y) \text{ to } F(0, p).$$
$$\sqrt{(x - x)^2 + (y - (-p))^2} = \text{distance from } P(x, y) \text{ to } y = -p,$$

Equating these distances and squaring yields:

$$(x - 0)^2 + (y - p)^2 = (x - x)^2 + (y - (-p))^2$$
$$x^2 + (y - p)^2 = 0 + (y + p)^2$$
$$x^2 + y^2 - 2py + p^2 = y^2 + 2py + p^2$$
$$x^2 = 4py$$

- By reversing the above steps, we see that a solution $(x, y)$ of $x^2 = 4py$ is equidistant from $(0, p)$ and the line $y = -p$.

The equation $x^2 = 4py$ is the **standard form** of the equation of an upward or downward opening parabola with vertex at the origin. If $p > 0$, the parabola opens upward; if $p < 0$, it opens downward. An alternative algebraic form for such a parabola is $y = ax^2$, where $a = 1/(4p)$. So the graph of $x^2 = 4py$ is also the graph of the quadratic function $f(x) = ax^2$.

When the equation of an upward or downward opening parabola is written as $x^2 = 4py$, the value $p$ is interpreted as the **focal length** of the parabola—the *directed* distance from the vertex to the focus of the parabola. A line segment with endpoints on a parabola is a **chord** of the parabola. The value $|4p|$ is the **focal width** of the parabola—the length of the chord through the focus and perpendicular to the axis.

Parabolas that open to the right or to the left are *inverse relations* of upward or downward opening parabolas. So equations of parabolas with vertex

**Alert**

Note that the standard form of the equation of an upward or downward opening parabola is not the same as the standard form of a quadratic function.

$(0, 0)$ that open to the right or to the left have the standard form $y^2 = 4px$. If $p > 0$, the parabola opens to the right, and if $p < 0$, to the left. (See Figure 8.6.)

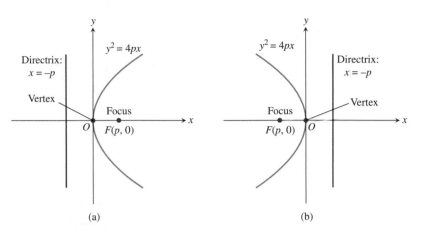

**Figure 8.6** Graph of $y^2 = 4px$ with (a) $p > 0$ and (b) $p < 0$.

**Parabolas with Vertex (0, 0)**

|  |  |  |
|---|---|---|
| • **Standard equation** | $x^2 = 4py$ | $y^2 = 4px$ |
| • **Opens** | Upward or downward | To the right or to the left |
| • **Focus** | $(0, p)$ | $(p, 0)$ |
| • **Directrix** | $y = -p$ | $x = -p$ |
| • **Axis** | $y$-axis | $x$-axis |
| • **Focal length** | $p$ | $p$ |
| • **Focal width** | $|4p|$ | $|4p|$ |

See Figures 8.5 and 8.6.

**Example 2   FINDING THE FOCUS, DIRECTRIX, AND FOCAL WIDTH**

Find the focus, the directrix, and the focal width of the parabola $y = -(1/2)x^2$.

**Solution**  Multiplying both sides of the equation by $-2$ yields the standard form $x^2 = -2y$. The coefficient of $y$ is $4p = -2$, and $p = -1/2$. So the focus is $(0, p) = (0, -1/2)$. Because $-p = -(-1/2) = 1/2$, the directrix is the line $y = 1/2$. The focal width is $|4p| = |-2| = 2$.

**Example 3   FINDING AN EQUATION OF A PARABOLA**

Find an equation in standard form for the parabola whose directrix is the line $x = 2$ and focus is the point $(-2, 0)$.

**Solution**  Because the directrix is $x = 2$ and the focus is $(-2, 0)$, the focal length is $p = -2$ and the parabola opens to the left. The equation of the parabola in standard form is $y^2 = 4px$, or more specifically, $y^2 = -8x$.

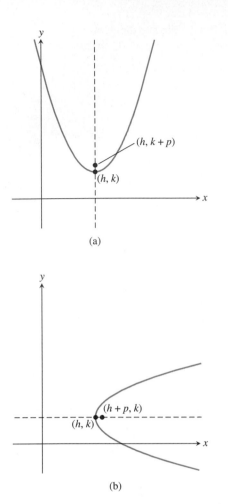

(a)

(b)

**Figure 8.7** Parabolas with vertex $(h, k)$ and focus on (a) $x = h$ and (b) $y = k$.

---

**Teaching Note**

Students should be able to relate the transformations of parabolas (and other conic sections) to the transformations that were studied in Section 1.5.

## Translations of Parabolas

When a parabola with the equation $x^2 = 4py$ or $y^2 = 4px$ is translated horizontally by $h$ units and vertically by $k$ units, the vertex of the parabola moves from $(0, 0)$ to $(h, k)$. (See Figure 8.7.) Such a translation does not change the focal length, the focal width, or the direction the parabola opens.

---

**Parabolas with Vertex ($h$, $k$)**

| | | |
|---|---|---|
| • **Standard equation** | $(x - h)^2 = 4p(y - k)$ | $(y - k)^2 = 4p(x - h)$ |
| • **Opens** | Upward or downward | To the right or to the left |
| • **Focus** | $(h, k + p)$ | $(h + p, k)$ |
| • **Directrix** | $y = k - p$ | $x = h - p$ |
| • **Axis** | $x = h$ | $y = k$ |
| • **Focal length** | $p$ | $p$ |
| • **Focal width** | $|4p|$ | $|4p|$ |

See Figure 8.7.

---

### Example 4 FINDING AN EQUATION OF A PARABOLA

Find the standard form of the equation for the parabola with vertex $(3, 4)$ and focus $(5, 4)$.

**Solution** The axis of the parabola is the line passing through the vertex $(3, 4)$ and the focus $(5, 4)$. This is the line $y = 4$. So the equation has the form

$$(y - k)^2 = 4p(x - h).$$

Because the vertex $(h, k) = (3, 4)$, $h = 3$ and $k = 4$. The directed distance from the vertex $(3, 4)$ to the focus $(5, 4)$ is $p = 5 - 3 = 2$, so $4p = 8$. Thus the equation we seek is

$$(y - 4)^2 = 8(x - 3).$$

When solving a problem like Example 4, it is a good idea to sketch the vertex, the focus, and other features of the parabola as we solve the problem. This makes it easy to see whether the axis of the parabola is horizontal or vertical and the relative positions of its features. Exploration 2 "walks us through" this process.

---

**Exploration 2**  **Building a Parabola**

Carry out the following steps using a sheet of rectangular graph paper.

1. Let the focus $F$ of a parabola be $(2, -2)$ and its directrix be $y = 4$. Draw the $x$- and $y$-axes on the graph paper. Then sketch and label the focus and directrix of the parabola.

2. Locate, sketch, and label the axis of the parabola. What is its equation?

3. Locate and plot the vertex $V$ of the parabola. Label it by name and coordinates.

4. What are the focal length and width of the parabola?

5. Use the focal width to locate, plot, and label the endpoints of a chord of the parabola that parallels the directrix.

6. Sketch the parabola.

7. Which direction does it open?  downward

8. What is its equation in standard form?  $(x - 2)^2 = -12(y - 1)$

---

**Exploration Extensions**

Check your work by using a grapher to graph the equation you found in step 8.

Sometimes it is best to sketch a parabola by hand, as in Exploration 2; this helps us see the structure and relationships of the parabola and its features. At other times, we may want or need an accurate, active graph. If we wish to graph a parabola using a function grapher, we need to solve the equation of the parabola for $y$, as illustrated in Example 5.

### Example 5   GRAPHING A PARABOLA

Use a graphing utility to graph the parabola $(y - 4)^2 = 8(x - 3)$ of Example 4.

### Solution

$$(y - 4)^2 = 8(x - 3)$$
$$y - 4 = \pm\sqrt{8(x - 3)} \qquad \text{Extract square roots.}$$
$$y = 4 \pm \sqrt{8(x - 3)} \qquad \text{Add 4.}$$

Let $y_1 = 4 + \sqrt{8(x - 3)}$ and $y_2 = 4 - \sqrt{8(x - 3)}$, and graph the two equations in a window centered at the vertex, as shown in Figure 8.8.

**Closing the Gap**

In Figure 8.8, we centered the graphing window at the vertex $(3, 4)$ of the parabola to ensure that this point would be plotted. This avoids the common grapher error of a gap between the two functional branches of the conic section being plotted.

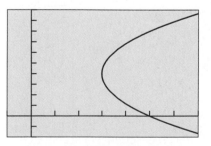

[–1, 7] by [–2, 10]

**Figure 8.8**  The graphs of $y_1 = 4 + \sqrt{8(x - 3)}$ and $y_2 = 4 - \sqrt{8(x - 3)}$ together form the graph of $(y - 4)^2 = 8(x - 3)$. (Example 5)

### Example 6 USING STANDARD FORMS WITH A PARABOLA

Prove that the graph of $y^2 - 6x + 2y + 13 = 0$ is a parabola, and find its vertex, focus, and directrix.

**Solution** Because this equation is quadratic in the variable $y$, we complete the square with respect to $y$ to obtain a standard form.

$$y^2 - 6x + 2y + 13 = 0$$
$$y^2 + 2y = 6x - 13 \qquad \text{Isolate the } y\text{-terms.}$$
$$y^2 + 2y + 1 = 6x - 13 + 1 \qquad \text{Complete the square.}$$
$$(y + 1)^2 = 6x - 12$$
$$(y + 1)^2 = 6(x - 2)$$

This equation is in the standard form $(y - k)^2 = 4p(x - h)$, where $h = 2$, $k = -1$, and $p = 6/4 = 3/2 = 1.5$. It follows that

- the vertex $(h, k)$ is $(2, -1)$;
- the focus $(h + p, k)$ is $(3.5, -1)$, or $(7/2, -1)$;
- the directrix $x = h - p$ is $x = 0.5$, or $x = 1/2$.

## Reflective Property of a Parabola

The main applications of parabolas involve their use as reflectors of sound, light, radio waves, and other electromagnetic waves. If we rotate a parabola in three-dimensional space about its axis, the parabola sweeps out a **paraboloid of revolution**. If we place a signal source at the focus of a reflective paraboloid, the signal reflects off the surface in lines parallel to the axis of symmetry, as illustrated in Figure 8.9a. This property is used by flashlights, headlights, searchlights, microwave relays, and satellite up-links.

The principle works for signals traveling in the reverse direction as well; signals arriving parallel to a parabolic reflector's axis are directed toward the reflector's focus. This property is used to intensify signals picked up by radio telescopes and television satellite dishes, to focus arriving light in reflecting telescopes, to concentrate heat in solar ovens, and to magnify sound for side-line microphones at football games. See Figure 8.9b.

### Example 7 STUDYING A PARABOLIC MICROPHONE

On the sidelines of each of its televised football games, the FBTV network uses a parabolic reflector with a microphone at the reflector's focus to capture the conversations among players on the field. If the parabolic reflector is 3 ft across and 1 ft deep, where should the microphone be placed?

**Solution**

**Model**

We draw a cross section of the reflector as an upward opening parabola in the Cartesian plane, placing its vertex $V$ at the origin (see Figure 8.10). We let the focus $F$ have coordinates $(0, p)$ to yield the equation

$$x^2 = 4py.$$

Because the reflector is 3 ft across and 1 ft deep, the points $(\pm 1.5, 1)$ must lie on the parabola.

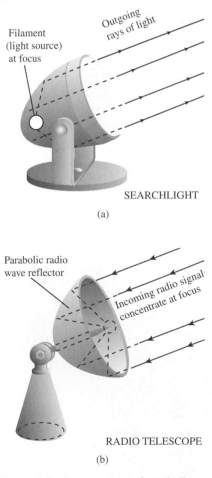

Filament (light source) at focus

Outgoing rays of light

SEARCHLIGHT

(a)

Parabolic radio wave reflector

Incoming radio signal concentrate at focus

RADIO TELESCOPE

(b)

**Figure 8.9** Cross sections of parabolic reflectors.

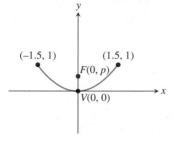

**Figure 8.10** Cross section of parabolic reflector in Example 7.

**Solve Algebraically**

The microphone should be placed at the focus, so we need to find the value of $p$:

$$(\pm1.5)^2 = 4p(1)$$

$$2.25 = 4p$$

$$p = 2.25/4 = 0.5625$$

**Interpret**

Because $p = 0.5625$ ft, or 6.75 inches, the microphone should be placed inside the reflector along its axis and 6.75 inches from its vertex.

## Quick Review 8.1

In Exercises 1 and 2, find the distance between the given points.

**1.** $(-1, 3)$ and $(2, 5)$  $\sqrt{13}$     **2.** $(2, -3)$ and $(a, b)$

In Exercises 3 and 4, solve for $y$ in terms of $x$.

**3.** $2y^2 = 8x$  $y = \pm2\sqrt{x}$     **4.** $3y^2 = 15x$  $y = \pm\sqrt{5x}$

In Exercises 5 and 6, complete the square to rewrite the equation in vertex form.

**5.** $f(x) = -x^2 + 2x - 7$     **6.** $f(x) = 2x^2 + 6x - 5$

In Exercises 7 and 8, find the vertex and axis of the graph of $f$. Describe how the graph of $f$ can be obtained from the graph of $g(x) = x^2$, and graph $f$.

**7.** $f(x) = 3(x - 1)^2 + 5$     **8.** $f(x) = -2x^2 + 12x + 1$

In Exercises 9 and 10, write an equation for the quadratic function whose graph contains the given vertex and point.

**9.** Vertex $(-1, 3)$, point $(0, 1)$   **10.** Vertex $(2, -5)$, point $(5, 13)$

## Section 8.1 Exercises

In Exercises 1–6, find the vertex, focus, directrix, and focal width of the parabola.

**1.** $x^2 = 6y$     **2.** $y^2 = -8x$

**3.** $(y - 2)^2 = 4(x + 3)$     **4.** $(x + 4)^2 =$   $6(y + 1)$

**5.** $3x^2 = -4y$     **6.** $5y^2 = 16x$

In Exercises 7–10, match the graph with its equation.

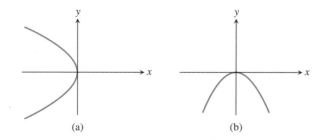

(a)                    (b)

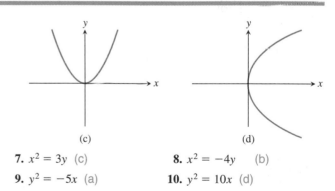

(c)                    (d)

**7.** $x^2 = 3y$  (c)     **8.** $x^2 = -4y$   (b)

**9.** $y^2 = -5x$  (a)     **10.** $y^2 = 10x$  (d)

In Exercises 11–30, find an equation in standard form for the parabola that satisfies the given conditions.

**11.** Vertex $(0, 0)$, focus $(-3, 0)$  $y^2 = -12x$

**12.** Vertex $(0, 0)$, focus $(0, 2)$  $x^2 = 8y$

**13.** Vertex $(0, 0)$, directrix $y = 4$  $x^2 = -16y$

**14.** Vertex $(0, 0)$, directrix $x = -2$  $y^2 = 8x$

**15.** Focus $(0, 5)$, directrix $y = -5$  $x^2 = 20y$

**16.** Focus $(-4, 0)$, directrix $x = 4$  $y^2 = -16x$

**17.** Vertex $(0, 0)$, opens to the right, focal width $= 8$.  $y^2 = 8x$

**18.** Vertex $(0, 0)$, opens to the left, focal width = 12. $y^2 = -12x$

**19.** Vertex $(0, 0)$, opens downward, focal width = 6.

**20.** Vertex $(0, 0)$, opens upward, focal width = 3. $x^2 = 3y$

**21.** Focus $(-2, -4)$ and vertex $(-4, -4)$ $(y + 4)^2 = 8(x + 4)$

**22.** Focus $(-5, 3)$ and vertex $(-5, 6)$ $(x + 5)^2 = -12(y - 6)$

**23.** Focus $(3, 4)$ and directrix $y = 1$ $(x - 3)^2 = 6(y - (5/2))$

**24.** Focus $(2, -3)$ and directrix $x = 5$ $(y + 3)^2 = -6(x - (7/2))$

**25.** Vertex $(4, 3)$ and directrix $x = 6$ $(y - 3)^2 = -8(x - 4)$

**26.** Vertex $(3, 5)$, and directrix $y = 7$ $(x - 3)^2 = -8(y - 5)$

**27.** Vertex $(2, -1)$, opens upward, focal width = 16.

**28.** Vertex $(-3, 3)$, opens downward, focal width = 20.

**29.** Vertex $(-1, -4)$, opens to the left, focal width = 10.

**30.** Vertex $(2, 3)$, opens to the right, focal width = 5.

In Exercises 31–36, sketch the graph of the parabola by hand.

**31.** $y^2 = -4x$      **32.** $x^2 = 8y$

**33.** $(x + 4)^2 = -12(y + 1)$      **34.** $(y + 2)^2 = -16(x - 3)$

**35.** $(y - 1)^2 = 8(x + 3)$      **36.** $(x - 5)^2 = 20(y + 2)$

In Exercises 37–44, graph the parabola using a function grapher.

**37.** $y = 4x^2$      **38.** $y = -\dfrac{1}{6}x^2$

**39.** $x = -8y^2$      **40.** $x = 2y^2$

**41.** $12(y + 1) = (x - 3)^2$      **42.** $6(y - 3) = (x + 1)^2$

**43.** $2 - y = 16(x - 3)^2$      **44.** $(x + 4)^2 = -6(y - 1)$

In Exercises 45–48, prove that the graph of the equation is a parabola, and find its vertex, focus, and directrix.

**45.** $x^2 + 2x - y + 3 = 0$      **46.** $3x^2 - 6x - 6y + 10 = 0$

**47.** $y^2 - 4y - 8x + 20 = 0$      **48.** $y^2 - 2y + 4x - 12 = 0$

In Exercises 49 and 50, write an equation for the parabola.

**49.**

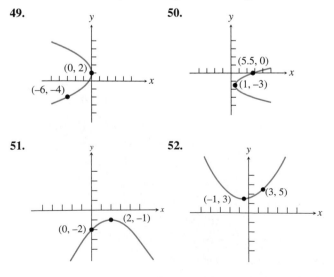

**50.**

**51.**

**52.**

**53. Writing to Learn** Explain why the proof in Example 1 is valid regardless of whether $p > 0$ or $p < 0$.

**54. Writing to Learn** Prove that an equation for the parabola with focus $(p, 0)$ and directrix $x = -p$ is $y^2 = 4px$. [*Hint*: Refer to Example 1.]

**55. Designing a Flashlight Mirror** The mirror of a flashlight is a paraboloid of revolution. Its diameter is 6 cm and its depth is 2 cm. How far from the vertex should the filament of the light bulb be placed for the flashlight to have its beam run parallel to the axis of its mirror?

**56. Designing a Satellite Dish** The reflector of a television satellite dish is a paraboloid of revolution with diameter 5 ft and a depth of 2 ft. How far from the vertex should the receiving antenna be placed?

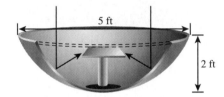

**57. Parabolic Microphones** Sports Channel uses a parabolic microphone to capture all the sounds from the basketball players and coaches during a regular season game. If one of its microphones has a parabolic surface generated by the parabola $10y = x^2$, locate the focus (the electronic receiver) of the parabola.

**58. Parabolic Headlights** Stein Glass, Inc., makes parabolic headlights for a variety of automobiles. If one of its headlights has a parabolic surface generated by the parabola $x^2 = 12y$, where should its light bulb be placed?

**59. Group Activity  Designing a Suspension Bridge** The main cables of a suspension bridge uniformly distribute the weight of the bridge when in the form of a parabola. The main cables of a particular bridge are attached to towers that are 600 ft apart. The cables are attached to the towers at a height of 110 ft above the roadway and are 10 ft above the roadway at their lowest points. If vertical support cables are at 50-ft intervals along the level roadway, what are the lengths of these vertical cables?

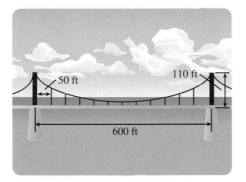

**60. Designing a Bridge Arch** Parabolic arches are known to have greater strength than other arches. A bridge with a supporting parabolic arch spans 60 ft with a 30-ft wide road passing underneath the bridge. In order to have a minimum clearance of 16 ft, what is the maximum clearance?

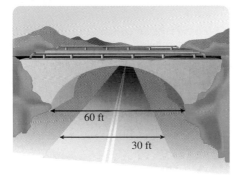

# Explorations

**61. Dynamically Constructing a Parabola** Use a geometry software package, such as *Cabri Geometry II, The Geometer's Sketchpad*, or the TI-92 geometry application, to construct a parabola geometrically from its definition:

**(a)** Start by placing a line *l* (directrix) and a point *F* (focus) not on the line in the construction window.

**(b)** Construct a point *A* on the directrix, and then the segment *AF*.

**(c)** Construct a point *P* where the perpendicular bisector of *AF* meets the line perpendicular to *l* through *A*.

**(d)** What curve does *P* trace out as *A* moves?  a parabola

**(e)** Prove your answer to (d) is correct.

**62. Constructing Points of a Parabola** Use a geometry software package, such as *Cabri Geometry II™, The Geometer's Sketchpad™*, or the TI-92 geometry application, to construct Figure 8.4, associated with Exploration 1.

**(a)** Start by placing the coordinate axes in the construction window.

**(b)** Construct the line $y = -1$ as the directrix and the point $(0, 1)$ as the focus.

**(c)** Construct the horizontal lines and concentric circles shown in Figure 8.4.

**(d)** Construct the points where these horizontal lines and concentric circles meet.

**(e)** Prove these points lie on the parabola with directrix $y = -1$ and focus $(0, 1)$.

**63. Degenerate Cones and Degenerate Conics** Degenerate cones occur when the generator and axis of the cone are parallel or perpendicular.

**(a)** Draw a sketch and describe the "cone" obtained when the generator and axis of the cone are parallel.

**(b)** Draw sketches and name the types of degenerate conics obtained by intersecting the degenerate cone in (a) with a plane.

**(c)** Draw a sketch and describe the "cone" obtained when the generator and axis of the cone are perpendicular.

**(d)** Draw sketches and name the types of degenerate conics obtained by intersecting the degenerate cone in (c) with a plane.

# Extending the Ideas

**64. Tangent Lines** A **tangent line** of a parabola is a line that intersects but does not cross the parabola. Prove that a line tangent to the parabola $x^2 = 4py$ at the point $(a, b)$ crosses the *y* axis at $(0, -b)$.

**65. Focal Chords** A **focal chord** of a parabola is a chord of the parabola that passes through the focus.

**(a)** Prove that the *x*-coordinates of the endpoints of a focal chord of $x^2 = 4py$ are $x = 2p(m \pm \sqrt{m^2 + 1})$, where *m* is the slope of the focal chord.

**(b)** Using (a), prove the minimum length of a focal chord is the focal width $|4p|$.

**66. Latus Rectum** The focal chord of a parabola perpendicular to the axis of the parabola is the **latus rectum**, Latin for "right chord." Using the results from Exercises 64 and 65, prove:

**(a)** that for a parabola the two endpoints of the latus rectum and the point of intersection of the axis and directrix are the vertices of an isosceles right triangle,

**(b)** that the sides of the right angle of this triangle are tangent to the parabola.

## Ellipses

Geometry of an Ellipse • Translations of Ellipses • Orbits and Eccentricity • Reflective Property of an Ellipse

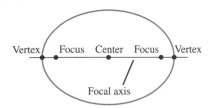

**Figure 8.11** Key points on the focal axis of an ellipse.

### Geometry of an Ellipse

When a plane intersects one nappe of a right circular cylinder and forms a simple closed curve, the curve is an ellipse.

---

**Definition Ellipse**

An **ellipse** is the set of all points in a plane whose distances from two fixed points in the plane have a constant sum. The fixed points are the **foci** (plural of focus) of the ellipse. The line through the foci is the **focal axis**. The point on the focal axis midway between the foci is the **center**. The points where the ellipse intersects its axis are the **vertices** of the ellipse. (See Figure 8.11.)

---

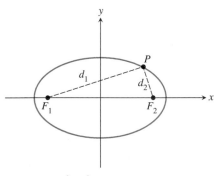

$$d_1 + d_2 = \text{constant}$$

**Figure 8.12** *Structure of an Ellipse.* The sum of the distances from the foci to each point on the ellipse is a constant.

Figure 8.12 shows a point $P$ of an ellipse. Each of the fixed points $F_1$ and $F_2$ is a focus of the ellipse, and the distances whose sum is constant are $d_1$ and $d_2$. We can construct an ellipse using a pencil, a loop of string, and two pushpins. Put the loop around the two pins placed at $F_1$ and $F_2$, pull the string taut with a pencil point $P$, and move the pencil around to trace out the ellipse (Figure 8.13).

We now use the definition to derive an equation for an ellipse. For some constants $a$ and $c$ with $a > c \geq 0$, let $F_1(-c, 0)$ and $F_2(c, 0)$ be the foci (Figure 8.14). Then an ellipse is defined by the set of points $P(x, y)$ such that

$$PF_1 + PF_2 = 2a.$$

Using the distance formula, the equation becomes

$$\sqrt{(x + c)^2 + (y - 0)^2} + \sqrt{(x - c)^2 + (y - 0)^2} = 2a$$

$$\sqrt{(x - c)^2 + y^2} = 2a - \sqrt{(x + c)^2 + y^2}$$

$$x^2 - 2cx + c^2 + y^2 = 4a^2 - 4a\sqrt{(x + c)^2 + y^2} + x^2 + 2cx + c^2 + y^2 \qquad \text{Square}$$

$$a\sqrt{(x + c)^2 + y^2} = a^2 + cx \qquad \text{Simplify.}$$

$$a^2(x^2 + 2cx + c^2 + y^2) = a^4 + 2a^2cx + c^2x^2 \qquad \text{Square.}$$

$$(a^2 - c^2)x^2 + a^2y^2 = a^2(a^2 - c^2) \qquad \text{Simplify.}$$

Letting $b^2 = a^2 - c^2$, we have

$$b^2x^2 + a^2y^2 = a^2b^2$$

which is usually written as

$$\frac{x^2}{a^2} + \frac{y^2}{b^2} = 1.$$

Because these steps can be reversed, a point $P(x, y)$ satisfies this last equation if and only if the point lies on the ellipse defined by $PF_1 + PF_2 = 2a$, provided that $a > c \geq 0$ and $b^2 = a^2 - c^2$. The *Pythagorean relation*

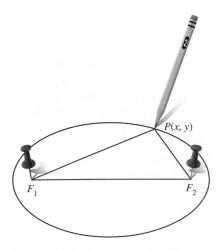

**Figure 8.13** How to draw an ellipse.

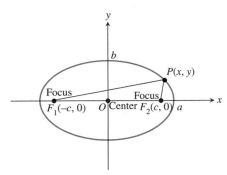

**Figure 8.14** The ellipse defined by $PF_1 + PF_2 = 2a$ is the graph of the equation $x^2/a^2 + y^2/b^2 = 1$, where $b^2 = a^2 - c^2$.

**Motivate**

Ask students how the graph of $x^2 + (y/2)^2 = 1$ is related to the graph of $x^2 + y^2 = 1$.

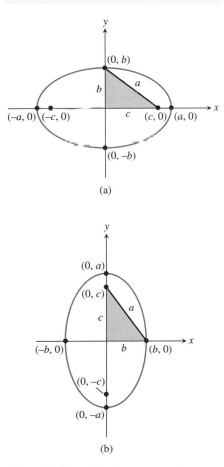

(a)

(b)

**Figure 8.15** Ellipses centered at the origin with foci on (a) the $x$-axis and (b) the $y$-axis. In each case, a right triangle illustrating the Pythagorean relation is shown.

$b^2 = a^2 - c^2$ can be written many ways, including $c^2 = a^2 - b^2$ and $a^2 = b^2 + c^2$.

The equation $x^2/a^2 + y^2/b^2 = 1$ is the **standard form** of the equation of an ellipse centered at the origin with the $x$-axis as its focal axis. An ellipse centered at the origin with the $y$-axis as its focal axis is the *inverse* of $x^2/a^2 + y^2/b^2 = 1$, and thus has an equation of the form

$$\frac{y^2}{a^2} + \frac{x^2}{b^2} = 1.$$

As with circles and parabolas, a line segment with endpoints on an ellipse is a **chord** of the ellipse. The chord lying on the focal axis is the **major axis** of the ellipse. The chord through the center perpendicular to the focal axis is the **minor axis** of the ellipse. The length of the major axis is $2a$, and of the minor axis is $2b$. The number $a$ is the **semimajor axis**, and $b$ is the **semiminor axis**.

**Ellipses with Center (0, 0)**

| | | |
|---|---|---|
| • **Standard equation** | $\dfrac{x^2}{a^2} + \dfrac{y^2}{b^2} = 1$ | $\dfrac{y^2}{a^2} + \dfrac{x^2}{b^2} = 1$ |
| • **Focal axis** | $x$-axis | $y$-axis |
| • **Foci** | $(\pm c, 0)$ | $(0, \pm c)$ |
| • **Vertices** | $(\pm a, 0)$ | $(0, \pm a)$ |
| • **Semimajor axis** | $a$ | $a$ |
| • **Semiminor axis** | $b$ | $b$ |
| • **Pythagorean relation** | $a^2 = b^2 + c^2$ | $a^2 = b^2 + c^2$ |

See Figure 8.15.

**Example 1   FINDING THE VERTICES AND FOCI OF AN ELLIPSE**

Find the vertices and the foci of the ellipse $4x^2 + 9y^2 = 36$.

**Solution**  Dividing both sides of the equation by 36 yields the standard form $x^2/9 + y^2/4 = 1$. Because the larger number is the denominator of $x^2$, the focal axis is the $x$-axis. So $a^2 = 9$, $b^2 = 4$, and $c^2 = a^2 - b^2 = 9 - 4 = 5$. Thus the vertices are $(\pm 3, 0)$, and the foci are $(\pm\sqrt{5}, 0)$.

An ellipse centered at the origin with its focal axis on a coordinate axis is symmetric with respect to the origin and both coordinate axes. Such an ellipse can be sketched by first drawing a rectangle centered at the origin with sides parallel to the coordinate axes and then sketching the ellipse inside the rectangle, as shown in the Drawing Lesson on the next page.

**Drawing Lesson**

**How to Sketch the Ellipse $x^2/a^2 + y^2/b^2 = 1$**

1. Sketch line segments at $x = \pm a$ and $y = \pm b$ and complete the rectangle they determine.

2. Inscribe an ellipse that is tangent to the rectangle at $(\pm a, 0)$ and $(0, \pm b)$.

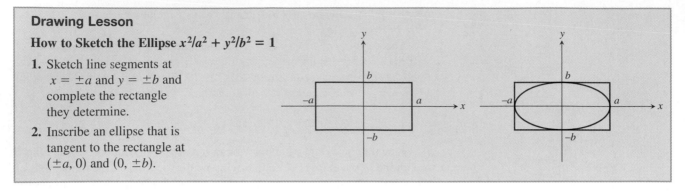

**Lesson Guide**

Day 1: Geometry of an Ellipse;
Translations of Ellipses
Day 2: Orbits and Eccentricity;
Reflective Property of an Ellipse

If we wish to graph an ellipse using a function grapher, we need to solve the equation of the ellipse for $y$, as illustrated in Example 2.

**Example 2** FINDING AN EQUATION AND GRAPHING AN ELLIPSE

Find an equation of the ellipse with foci $(0, -3)$ and $(0, 3)$ whose minor axis has length 4. Sketch the ellipse and support your sketch with a grapher.

**Solution** The center is $(0, 0)$. The foci are on the $y$-axis with $c = 3$. The semiminor axis is $b = 4/2 = 2$. Using $a^2 = b^2 + c^2$, we have $a^2 = 2^2 + 3^2 = 13$. So the standard form of the equation for the ellipse is

$$\frac{y^2}{13} + \frac{x^2}{4} = 1.$$

Using $a = \sqrt{13} \approx 3.61$ and $b = 2$, we can sketch a guiding rectangle and then the ellipse itself. (Try doing this.) To graph the ellipse using a function grapher, we solve for $y$ in terms of $x$.

$$\frac{y^2}{13} = 1 - \frac{x^2}{4}$$
$$y^2 = 13(1 - x^2/4)$$
$$y = \pm\sqrt{13(1 - x^2/4)}$$

Figure 8.16 shows three views of the graphs of

$$y_1 = \sqrt{13(1 - x^2/4)} \quad \text{and} \quad y_2 = -\sqrt{13(1 - x^2/4)}.$$

We must select the viewing window carefully to avoid grapher failure.

**Notes on Examples**

Notice that we chose square viewing windows in Figure 8.16. A nonsquare window would give a distorted view of an ellipse.

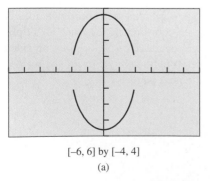

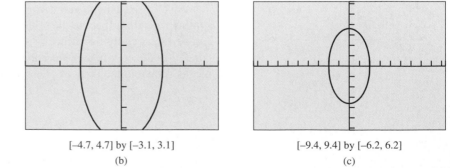

| [–6, 6] by [–4, 4] | [–4.7, 4.7] by [–3.1, 3.1] | [–9.4, 9.4] by [–6.2, 6.2] |
| (a) | (b) | (c) |

**Figure 8.16** Three views of the ellipse $y^2/13 + x^2/4 = 1$. All of the windows are square or approximately square viewing windows so we can see the true shape. Notice that the gaps between the two function branches do not show when the grapher window includes columns of pixels whose $x$-coordinates are $\pm 2$ as in (b) and (c). (Example 2)

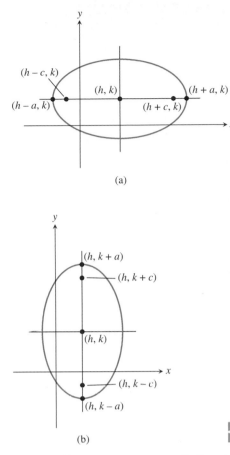

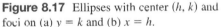

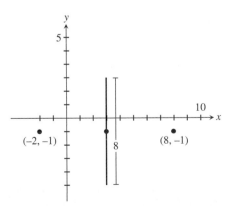

**Figure 8.17** Ellipses with center $(h, k)$ and foci on (a) $y = k$ and (b) $x = h$.

## Translations of Ellipses

When an ellipse with center $(0, 0)$ is translated horizontally by $h$ units and vertically by $k$ units, the center of the ellipse moves from $(0, 0)$ to $(h, k)$, as shown in Figure 8.17. Such a translation does not change the length of the major or minor axis or the Pythagorean relation.

### Ellipses with Center ($h$, $k$)

| | | |
|---|---|---|
| • **Standard equation** | $\dfrac{(x - h)^2}{a^2} + \dfrac{(y - k)^2}{b^2} = 1$ | $\dfrac{(y - k)^2}{a^2} + \dfrac{(x - h)^2}{b^2} = 1$ |
| • **Focal axis** | $y = k$ | $x = h$ |
| • **Foci** | $(h \pm c, k)$ | $(h, k \pm c)$ |
| • **Vertices** | $(h \pm a, k)$ | $(h, k \pm a)$ |
| • **Semimajor axis** | $a$ | $a$ |
| • **Semiminor axis** | $b$ | $b$ |
| • **Pythagorean relation** | $a^2 = b^2 + c^2$ | $a^2 = b^2 + c^2$ |

See Figure 8.17.

### Example 3   FINDING AN EQUATION OF AN ELLIPSE

Find the standard form of the equation for the ellipse whose major axis has endpoints $(-2, -1)$ and $(8, -1)$, and whose minor axis has length 8.

**Solution**   Figure 8.18 shows the major-axis endpoints, the minor axis, and the center of the ellipse. The standard equation of this ellipse has the form

$$\frac{(x - h)^2}{a^2} + \frac{(y - k)^2}{b^2} = 1,$$

where the center $(h, k)$ is the midpoint $(3, -1)$ of the major axis. The semimajor axis and semiminor axis are

$$a = \frac{8 - (-2)}{2} = 5 \quad \text{and} \quad b = \frac{8}{2} = 4.$$

So the equation we seek is

$$\frac{(x - 3)^2}{5^2} + \frac{(y - (-1))^2}{4^2} = 1,$$

$$\frac{(x - 3)^2}{25} + \frac{(y + 1)^2}{16} = 1.$$

**Figure 8.18**  Given information for Example 3.

**Example 4  LOCATING KEY POINTS OF AN ELLIPSE**

Find the center, vertices, and foci of the ellipse

$$\frac{(x+2)^2}{9} + \frac{(y-5)^2}{49} = 1.$$

**Solution**  The standard equation of this ellipse has the form

$$\frac{(y-5)^2}{49} + \frac{(x+2)^2}{9} = 1.$$

The center $(h, k) = (-2, 5)$. Because the semimajor axis $a = \sqrt{49} = 7$, the vertices $(h, k \pm a)$ are

$$(h, k + a) = (-2, 5 + 7) = (-2, 12) \quad \text{and}$$

$$(h, k - a) = (-2, 5 - 7) = (-2, -2).$$

Because $c = \sqrt{a^2 - b^2} = \sqrt{49 - 9} = \sqrt{40}$, the foci $(h, k \pm c)$ are $(-2, 5 \pm \sqrt{40})$, or approximately $(-2, 11.32)$ and $(-2, -1.32)$.

With the information found about the ellipse in Example 4 and knowing that its semiminor axis $b = \sqrt{9} = 3$, we could easily sketch the ellipse. Obtaining an accurate graph of the ellipse using a function grapher is another matter. Generally, the best way to graph an ellipse using a graphing utility is to use parametric equations.

**Exploration Extensions**

Give an equation in standard form for the ellipse defined by the parametric equations $x = -3 + \cos t$, $y = 4 + 5 \sin t$, $0 \leq t < 2\pi$

**Exploration 1  Graphing an Ellipse Using Its Parametric Equations**

1. Use the Pythagorean trigonometry identity $\cos^2 t + \sin^2 t = 1$ to prove that the parameterization $x = -2 + 3 \cos t$, $y = 5 + 7 \sin t$, $0 \leq t \leq 2\pi$ will produce a graph of the ellipse $(x + 2)^2/9 + (y - 5)^2/49 = 1$.

2. Graph $x = -2 + 3 \cos t$, $y = 5 + 7 \sin t$, $0 \leq t \leq 2\pi$ in a square viewing window to support part 1 graphically.

3. Create parameterizations for the ellipses in Examples 1, 2, and 3.

4. Graph each of your parameterizations in part 3 and check the features of the obtained graph to see whether they match the expected geometric features of the ellipse. Revise your parameterization and regraph until all features match.

5. Prove that each of your parameterizations is valid.

## Orbits and Eccentricity

Kepler's first law of planetary motion, published in 1609, states that the path of a planet's orbit is an ellipse with the Sun at one of the foci. Asteroids and many comets and other bodies that orbit the Sun follow elliptical paths. The closest point to the Sun in such an orbit is the *perihelion*, and the farthest point is the *aphelion* (Figure 8.19). Some elliptical orbits are nearly circular, like the orbit of Venus; others are elongated, like the orbit of Halley's comet. The shape of an ellipse is measured by its eccentricity.

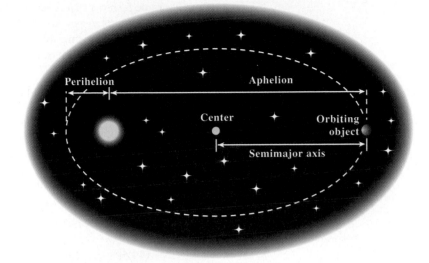

**Figure 8.19**  Many celestial objects have elliptical orbits around the Sun.

**A new *e***

Try not to confuse the eccentricity *e* with the natural base *e* used in exponential and logarithmic functions. The context should clarify which meaning is intended.

---

**Definition  Eccentricity of an Ellipse**

The **eccentricity** of an ellipse is

$$e = \frac{c}{a} = \frac{\sqrt{a^2 - b^2}}{a},$$

where *a* is the semimajor axis and *b* is the semiminor axis.

---

In Section 8.5 we will generalize the concept of eccentricity to all types of conics. For an ellipse, because $a > c \geq 0$, $0 \leq e < 1$. Elongated ellipses have eccentricities close to 1; for example, the orbit of Halley's comet has eccentricity $e \approx 0.97$. Nearly circular ellipses have eccentricities close to 0; for instance, Venus's orbit has eccentricity 0.0068. What happens when $e = 0$?

**Example 5   PROVING A CIRCLE IS AN ELLIPSE**

Prove that the ellipse $(x - h)^2/a^2 + (y - k)^2/b^2 = 1$ is a circle if and only if $e = 0$.

**Solution**  Assume $e = 0$. Because $a > 0$, $e = 0$ implies that $c = 0$ and $a = b$. When $a = b$, it is customary to denote this common value as $r$. So the equation of the ellipse becomes

$$\frac{(x - h)^2}{r^2} + \frac{(y - k)^2}{r^2} = 1, \quad \text{or} \quad (x - h)^2 + (y - k)^2 = r^2,$$

the equation of a circle with center $(h, k)$ and radius $r$.

Any circle has an equation of the form $(x - h)^2 + (y - k)^2 = r^2$. This equation can be written in the form

$$\frac{(x - h)^2}{r^2} + \frac{(y - k)^2}{r^2} = 1,$$

which is an equation of an ellipse with $a^2 = b^2 = r^2$. Thus,

$$e = \frac{\sqrt{a^2 - b^2}}{a} = 0.$$

Example 5 shows that, when eccentricity of an ellipse is 0, the semimajor and semiminor axis *degenerate* to a common value, which is the radius of the resulting circle. Surprising things happen when an ellipse is nearly but not quite a circle, as in the orbit of the planet Earth.

**Example 6   ANALYZING THE EARTH'S ORBIT**

The Earth's orbit has a semimajor axis $a \approx 149.598$ Gm (gigameters) and an eccentricity of $e \approx 0.0167$. Calculate and interpret $b$ and $c$.

**Solution**   Because $e = c/a$, $c = ea \approx 0.0167 \times 149.598 = 2.4982866$ and

$$b = \sqrt{a^2 - c^2} \approx \sqrt{149.598^2 - 2.4982866^2} \approx 149.577.$$

The semiminor axis $b \approx 149.577$ Gm is only 0.014% shorter than the semimajor axis $a \approx 149.598$ Gm. The aphelion is $a + c \approx 149.598 + 2.498 = 152.096$ Gm, and the perihelion is $a - c \approx 149.598 - 2.498 = 147.100$ Gm.

Thus Earth's orbit is nearly a perfect circle, but the distance between the center of the Sun at one focus and the center of Earth's orbit is $c \approx 2.498$ Gm, more than 2 orders of magnitude greater than $a - b$. The eccentricity as a percentage is 1.67%; this measures how far off-center the Sun is.

---

**Exploration 2   Constructing Ellipses to Understand Eccentricity**

You will need a pencil, paper, a 20-cm long loop of string, two pushpins, and a foam board or other appropriate backing material.

1. First use one pushpin and the loop to construct a circle of radius 10 cm.

2. Then using two pushpins placed 2 cm apart, construct an ellipse. Record the values of $a$, $b$, $c$, $e$, and the ratio $b/a$ on the paper right by the ellipse.

3. Repeat step 2, placing the pushpins 4, 6, and 8 cm apart.

4. Write your observations about the ratio $b/a$ as the eccentricity $e$ increases. Plot the ordered pairs $(e, b/a)$ and determine a formula for the ratio $b/a$ as a function of the eccentricity $e$.

---

## Reflective Property of an Ellipse

Because of its geometry, an ellipse can be used to make a reflector of sound, light, and other waves. If we rotate an ellipse in three-dimensional space about its focal axis, the ellipse sweeps out an **ellipsoid of revolution**. If we place a signal source at one focus of a reflective ellipsoid, the signal reflects off the elliptical surface to the other focus, as illustrated in Figure 8.20. This property is used to make mirrors for optical equipment and to study aircraft noise in wind tunnels.

In architecture, ceilings in the shape of an ellipsoid are used to create *whispering galleries*. A person whispering at one focus can be heard across the room by a person at the other focus. Statuary Hall in the U.S. capitol in Washington, D.C. and the Mormon Tabernacle in Salt Lake City, Utah are

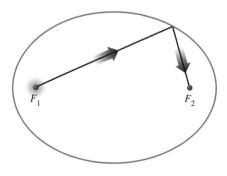

**Figure 8.20** The reflective property of an ellipse.

whispering galleries. An ellipsoid is part of the design of the Texas state capitol; a hand clap made in the center of the main vestibule (at one focus of the ellipsoid) bounces off the inner elliptical dome, passes through the other focus, bounces off the dome a second time, and returns to the person as a distinct echo.

Ellipsoids are used in medicine to avoid surgery in the treatment of kidney stones. An elliptical *lithotripter* emits underwater ultrahigh-frequency (UHF) shock waves from one focus, with the patient's kidney carefully positioned at the other focus (Figure 8.21).

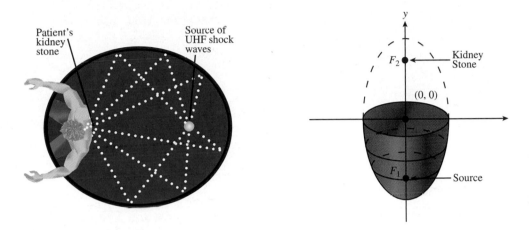

**Figure 8.21**  How a lithotripter breaks up kidney stones.

### Example 7  FOCUSING A LITHOTRIPTER

The ellipse used to generate the ellipsoid of a lithotripter has a major axis of 12 ft and a minor axis of 5 ft. How far from the center are the foci?

**Solution**  From the given information, we know $a = 12/2 = 6$ and $b = 5/2 = 2.5$. So

$$c = \sqrt{a^2 - b^2} \approx \sqrt{6^2 - 2.5^2} \approx 5.4544.$$

So the foci are about 5 ft 5.5 inches from the center of the lithotripter.

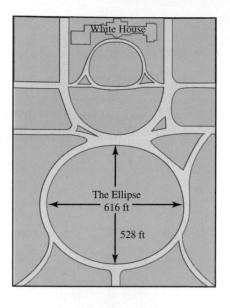

The Ellipse
616 ft

528 ft

**Problem**

If we know the Ellipse behind the White House is 616 feet across at its longest point and 528 feet across at its widest point, what is an equation that models it?

**Solution**

For simplicity's sake, we model the Ellipse as centered at (0, 0) with the $x$-axis as its focal axis. Because the Ellipse is 616 feet long, $a = 616/2 = 308$, and because the Ellipse is 528 feet wide, $b = 528/2 = 264$. Using $x^2/a^2 + y^2 b^2 = 1$, we obtain

$$\frac{x^2}{308^2} + \frac{y^2}{264^2} = 1,$$

$$\frac{x^2}{94{,}864} + \frac{y^2}{69{,}696} = 1.$$

Other models are possible.

## Quick Review 8.2

In Exercises 1 and 2, find the distance between the given points.

**1.** $(-3, -2)$ and $(2, 4)$  $\sqrt{61}$   **2.** $(-3, -4)$ and $(a, b)$

In Exercises 3 and 4, solve for $y$ in terms of $x$.

**3.** $\dfrac{y^2}{9} + \dfrac{x^2}{4} = 1$

**4.** $\dfrac{x^2}{36} + \dfrac{y^2}{25} = 1$

In Exercises 5–8, solve for $x$ algebraically.

**5.** $\sqrt{3x + 12} + \sqrt{3x - 8} = 10$  $x = 8$

**6.** $\sqrt{6x + 12} - \sqrt{4x + 9} = 1$  $x = 4$

**7.** $\sqrt{6x^2 + 12} + \sqrt{6x^2 + 1} = 11$  $x = 2, x = -2$

**8.** $\sqrt{2x^2 + 8} + \sqrt{3x^2 + 4} = 8$  $x = 2, x = -2$

In Exercises 9 and 10, find exact solutions by completing the square.

**9.** $2x^2 - 6x - 3 = 0$

**10.** $2x^2 + 4x - 5 = 0$

## Section 8.2 Exercises

In Exercises 1–6, find the vertices and foci of the ellipse.

**1.** $\dfrac{x^2}{16} + \dfrac{y^2}{7} = 1$

**2.** $\dfrac{y^2}{25} + \dfrac{x^2}{21} = 1$

**3.** $\dfrac{(y - 1)^2}{25} + \dfrac{(x + 2)^2}{16} = 1$

**4.** $\dfrac{(x - 3)^2}{11} + \dfrac{(y - 5)^2}{7} = 1$

**5.** $3x^2 + 4y^2 = 12$

**6.** $9x^2 + 4y^2 = 36$

In Exercises 7–10, match the graph with its equation, given that the ticks on all axes are 1 unit apart.

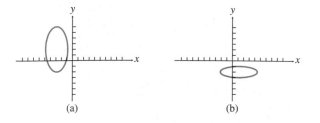

(a)

(b)

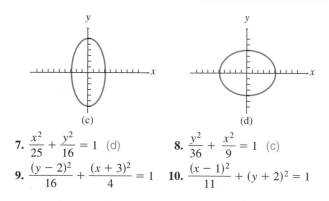

(c)

(d)

**7.** $\dfrac{x^2}{25} + \dfrac{y^2}{16} = 1$  (d)

**8.** $\dfrac{y^2}{36} + \dfrac{x^2}{9} = 1$  (c)

**9.** $\dfrac{(y - 2)^2}{16} + \dfrac{(x + 3)^2}{4} = 1$

**10.** $\dfrac{(x - 1)^2}{11} + (y + 2)^2 = 1$

In Exercises 11–26, find an equation in standard form for the ellipse that satisfies the given conditions.

**11.** Major axis length 6 on $y$-axis, minor axis length 4

**12.** Major axis length 14 on $x$-axis, minor axis length 10

**13.** Foci $(\pm 2, 0)$, major axis length 10  $x^2/25 + y^2/21 = 1$

**14.** Foci $(0, \pm 3)$, major axis length 10  $y^2/25 + x^2/16 = 1$

**15.** Endpoints of axes are $(\pm 4, 0)$ and $(0, \pm 5)$

**16.** Endpoints of axes are $(\pm 7, 0)$ and $(0, \pm 4)$

**17.** Major axis endpoints $(0, \pm 6)$, minor axis length 8

**18.** Major axis endpoints $(\pm 5, 0)$, minor axis length 4

**19.** Minor axis endpoints $(0, \pm 4)$, major axis length 10

**20.** Minor axis endpoints $(\pm 12, 0)$, major axis length 26

**21.** The endpoints of one axis are $(-3, 2)$ and $(5, 2)$, and of the other are $(1, -4)$ and $(1, 8)$.  $(x - 1)^2/16 + (y - 2)^2/36 = 1$

**22.** The endpoints of one axis are $(-2, -3)$ and $(-2, 7)$, and of the other are $(-4, 2)$ and $(0, 2)$.

**23.** The foci are $(1, -4)$ and $(5, -4)$; the major axis endpoints are $(0, -4)$ and $(6, -4)$.  $(x - 3)^2/9 + (y + 4)^2/5 = 1$

**24.** The foci are $(-2, 1)$ and $(-2, 5)$; the major axis endpoints are $(-2, -1)$ and $(-2, 7)$.  $(x + 2)^2/12 + (y - 3)^2/16 = 1$

**25.** The major axis endpoints are $(3, -7)$ and $(3, 3)$; the minor axis length is 6.  $(x - 3)^2/9 + (y + 2)^2/25 = 1$

**26.** The major axis endpoints are $(-5, 2)$ and $(3, 2)$; the minor axis length is 6.  $(x + 1)^2/16 + (y - 2)^2/9 = 1$

In Exercises 27–32, sketch the graph of the ellipse by hand.

**27.** $\dfrac{x^2}{64} + \dfrac{y^2}{36} = 1$     **28.** $\dfrac{x^2}{81} + \dfrac{y^2}{25} = 1$

**29.** $\dfrac{y^2}{9} + \dfrac{x^2}{4} = 1$     **30.** $\dfrac{y^2}{49} + \dfrac{x^2}{25} = 1$

**31.** $\dfrac{(x + 3)^2}{16} + \dfrac{(y - 1)^2}{4} = 1$     **32.** $\dfrac{(x - 1)^2}{2} + \dfrac{(y + 3)^2}{4} = 1$

In Exercises 33–36, graph the ellipse using a function grapher.

**33.** $\dfrac{x^2}{36} + \dfrac{y^2}{16} = 1$     **34.** $\dfrac{y^2}{64} + \dfrac{x^2}{16} = 1$

**35.** $\dfrac{(x + 2)^2}{5} + 2(y - 1)^2 = 1$     **36.** $\dfrac{(x - 4)^2}{16} + 16(y + 4)^2 = 8$

In Exercises 37–40, graph the ellipse using a parametric grapher.

**37.** $\dfrac{y^2}{25} + \dfrac{x^2}{4} = 1$     **38.** $\dfrac{x^2}{30} + \dfrac{y^2}{20} = 1$

**39.** $\dfrac{(x + 3)^2}{12} + \dfrac{(y - 6)^2}{5} = 1$     **40.** $\dfrac{(y + 1)^2}{15} + \dfrac{(x - 2)^2}{6} = 1$

In Exercises 41–44, prove that the graph of the equation is an ellipse, and find its vertices, foci, and eccentricity.

**41.** $9x^2 + 4y^2 - 18x + 8y - 23 = 0$

**42.** $3x^2 + 5y^2 - 12x + 30y + 42 = 0$

**43.** $9x^2 + 16y^2 + 54x - 32y - 47 = 0$

**44.** $4x^2 + y^2 - 32x + 16y + 124 = 0$

In Exercises 45 and 46, write an equation for the ellipse.

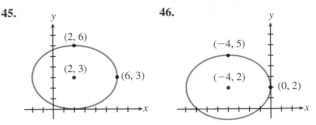

**45.** (2, 6)  (2, 3)  (6, 3)

**46.** (-4, 5)  (-4, 2)  (0, 2)

**47. Writing to Learn**  Prove that an equation for the ellipse with center $(0, 0)$, foci $(0, \pm c)$, and semimajor axis $a > c \geq 0$ is $y^2/a^2 + x^2/b^2 = 1$, where $b^2 = a^2 - c^2$. [*Hint*: Refer to derivation at the beginning of the section.]

**48. The Moon's Orbit**  The moon's apogee (farthest distance from the earth) is 252,710 miles, and perigee (closest distance to the earth) is 221,463 miles. Assuming the moon's orbit of the earth is elliptical with the earth at one focus, calculate and interpret $a$, $b$, $c$, and $e$.

**49. Dancing Planets  Writing to Learn**  Using the data in Table 8.1, prove that the planet with the most eccentric orbit sometimes is closer to the sun than the planet with the least eccentric orbit.

**Table 8.1  Semimajor Axes and Eccentricities of the Planets**

| Planet | Semimajor Axis (Gm) | Eccentricity |
|--------|---------------------|--------------|
| Mercury | 57.9 | 0.2056 |
| Venus | 108.2 | 0.0068 |
| Earth | 149.6 | 0.0167 |
| Mars | 227.9 | 0.0934 |
| Jupiter | 778.3 | 0.0485 |
| Saturn | 1427 | 0.0560 |
| Uranus | 2869 | 0.0461 |
| Neptune | 4497 | 0.0050 |
| Pluto | 5900 | 0.2484 |

*Source: Shupe, et al., National Geographic Atlas of the World (rev. 6th ed.). Washington, DC: National Geographic Society, 1992, plate 116, and other sources*

**50. Hot Mercury**  Given that the diameter of the sun is about 1.392 Gm, how close does Mercury get to the sun's surface?

**51. Saturn**  Find the perihelion and aphelion of Saturn.

**52. Venus and Mars**  Write equations for the orbits of Venus and Mars in the form $x^2/a^2 + y^2/b^2 = 1$.

**53. Sungrazers**  One comet group, known as the *sungrazers*, passes within a sun's diameter (1.392 Gm) of the solar surface. What can you conclude about $a - c$ for orbits of the sungrazers?

**54. Halley's Comet**  The orbit of Halley's comet is 36.18 AU long and 9.12 AU wide. What is its eccentricity?

**55. Group Activity  Lithotripter** For an ellipse that generates the ellipsoid of a lithotripter, the major axis has endpoints $(-8, 0)$ and $(8, 0)$. One endpoint of the minor axis is $(0, 3.5)$. Find the coordinates of the foci.

**56. Group Activity  Lithotripter (Refer to Figure 8.21.)** A lithotripter's shape is formed by rotating the portion of an ellipse below its minor axis about its major axis. If the length of the major axis is 26 in., and the length of the minor axis is 10 in., where should the shock-wave source and the patient be placed for maximum effect?

In Exercises 57 and 58, solve the system of equations algebraically and support your answer graphically.

**57.** $\dfrac{x^2}{4} + \dfrac{y^2}{9} = 1 \; (-2, 0), (2, 0)$  **58.** $\dfrac{x^2}{9} + y^2 = 1 \; (-3, 0), (0, 1)$

$x^2 + y^2 = 4$  $x - 3y = -3$

**59.** Consider the system of equations

$x^2 + 4y^2 = 4$
$y = 2x^2 - 3$

**(a)** Solve the system graphically.

**(b)** If you have access to a grapher that also does symbolic algebra, use it to find the exact solutions to the system.

## Explorations

**60. Area and Perimeter** The area of an ellipse is $A = \pi ab$, but the perimeter cannot be expressed so simply:

$$P \approx \pi(a + b)\left(3 - \frac{\sqrt{(3a + b)(a + 3b)}}{a + b}\right)$$

**(a)** Prove that, when $a = b = r$, these become the familiar formulas for the area and perimeter (circumference) of a circle.

**(b)** Find a pair of ellipses such that the one with greater area has smaller perimeter.  answers will vary

**61. Writing to Learn  Kepler's Laws** We have encountered Kepler's First and Third Laws (p. 181). Using a library or the Internet,

**(a)** Read about Kepler's life, and write in your own words how he came to discovery his three laws of planetary motion.

**(b)** What is Kepler's Second Law? Explain it with both pictures and words.  ■

## Extending the Ideas

**62.** Prove that a nondegenerate graph of the equation

$$Ax^2 + Cy^2 + Dx + Ey + F = 0$$

is an ellipse if $AC > 0$.

**63. Writing to Learn** The graph of the equation

$$\frac{(x - h)^2}{a^2} + \frac{(y - k)^2}{b^2} = 0$$

is considered to be a degenerate ellipse. Describe the graph. How is it like a full-fledged ellipse, and how is it different?

# 8.3  Hyperbolas

Geometry of a Hyperbola • Translations of Hyperbolas • Eccentricity and Orbits • Reflective Property of a Hyperbola • Long-Range Navigation

## Geometry of a Hyperbola

When a plane intersects both nappes of a right circular cylinder, the intersection is a hyperbola. The definition, features, and derivation for a hyperbola closely resemble those for an ellipse. As you read on, you may find it helpful to compare the nature of the hyperbola with the nature of the ellipse.

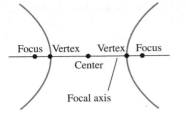

**Figure 8.22** Key points on the focal axis of a hyperbola.

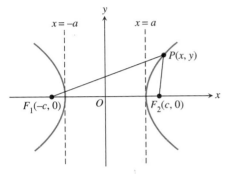

**Figure 8.23** *Structure of a Hyperbola*. The difference of the distances from the foci to each point on the hyperbola is a constant.

> **Definition** Hyperbola
>
> A hyperbola is the set of all points in a plane whose distances from two fixed points in the plane have a constant *difference*. The fixed points are the **foci** of the hyperbola. The line through the foci is the **focal axis**. The point on the focal axis midway between the foci is the **center**. The points where the hyperbola intersects its focal axis are the **vertices** of the hyperbola. (See Figure 8.22.)

Figure 8.23 shows a point $P$ of a hyperbola. Each of the fixed points $F_1$ and $F_2$ is a focus on the hyperbola, and the distances whose difference is constant are $PF_1$ and $PF_2$. Notice the hyperbola has *two branches*. For points on the right-hand branch shown in Figure 8.23, $PF_1 - PF_2 = 2a$. For points on the left-hand branch, $PF_2 - PF_1 = 2a$.

We use the definition to derive an equation for a hyperbola. For some constants $a$ and $c$ with $c > a \geq 0$, let $F_1(-c, 0)$ and $F_2(c, 0)$ be the foci (Figure 8.23). Then a hyperbola is defined by the set of points $P(x, y)$ such that

$$PF_1 - PF_2 = \pm 2a.$$

Using the distance formula, the equation becomes

$$\sqrt{(x + c)^2 + (y - 0)^2} - \sqrt{(x - c)^2 + (y - 0)^2} = \pm 2a.$$

$$\sqrt{(x - c)^2 + y^2} = \pm 2a + \sqrt{(x + c)^2 + y^2}$$

$$x^2 - 2cx + c^2 + y^2 = 4a^2 \pm 4a\sqrt{(x + c)^2 + y^2} + x^2 + 2cx + c^2 + y^2$$
$$\text{Square.}$$

$$\mp a\sqrt{(x + c)^2 + y^2} = a^2 + cx \qquad \text{Simplify.}$$

$$a^2(x^2 + 2cx + c^2 + y^2) = a^4 + 2a^2cx + c^2x^2 \qquad \text{Square.}$$

$$(c^2 - a^2)x^2 - a^2y^2 = a^2(c^2 - a^2) \qquad \text{Simplify.}$$

Letting $b^2 = c^2 - a^2$, we have

$$b^2x^2 - a^2y^2 = a^2b^2,$$

which is usually written as

$$\frac{x^2}{a^2} - \frac{y^2}{b^2} = 1.$$

Because these steps can be reversed, a point $P(x, y)$ satisfies this last equation if and only if the point lies on the hyperbola defined by $PF_1 - PF_2 = \pm 2a$, provided that $c > a \geq 0$ and $b^2 = c^2 - a^2$. The *Pythagorean relation* $b^2 = c^2 - a^2$ can be written many ways, including $a^2 = c^2 - b^2$ and $c^2 = a^2 + b^2$.

The equation $x^2/a^2 - y^2/b^2 = 1$ is the **standard form** of the equation of a hyperbola centered at the origin with the $x$-axis as its focal axis. A hyperbola centered at the origin with the $y$-axis as its focal axis is the *inverse relation* of $x^2/a^2 - y^2/b^2 = 1$, and thus has an equation of the form

$$\frac{y^2}{a^2} - \frac{x^2}{b^2} = 1.$$

As with other conics, a line segment with endpoints on a hyperbola is a **chord** of the hyperbola. The chord lying on the focal axis connecting the vertices is the **transverse axis** of the hyperbola. The length of the transverse axis is $2a$. The line segment of length $2b$ that is perpendicular to the focal axis and that has the center of the hyperbola as its midpoint is the **conjugate axis** of the hyperbola. The number $a$ is the **semitransverse axis**, and $b$ is the **semiconjugate axis**.

The hyperbola

$$\frac{x^2}{a^2} - \frac{y^2}{b^2} = 1$$

has two *asymptotes*. These asymptotes are slant lines that can be found by replacing the 1 on the right-hand side of the hyperbola's equation by a 0:

$$\underbrace{\frac{x^2}{a^2} - \frac{y^2}{b^2} = 1}_{\text{hyperbola}} \to \underbrace{\frac{x^2}{a^2} - \frac{y^2}{b^2} = 0}_{\text{replace 1 by 0}} \Rightarrow \underbrace{y = \pm\frac{b}{a}x}_{\text{asymptotes}}$$

### Naming axes

The word "transverse" comes from the Latin *trans vertere:* to go across. The transverse axis "goes across" from one vertex to the other. The conjugate axis is the transverse axis for the *conjugate hyperbola*, defined in Exercise 60.

A hyperbola centered at the origin with its focal axis a coordinate axis is symmetric with respect to the origin and both coordinate axes. Such a hyperbola can be sketched by drawing a rectangle centered at the origin with sides parallel to the coordinate axes, followed by drawing the asymptotes through opposite corners of the rectangle, and finally sketching the hyperbola using the central rectangle and asymptotes as guides, as shown in the Drawing Lesson.

---

**Drawing Lesson**

**How to Sketch the Hyperbola $x^2/a^2 - y^2/b^2 = 1$**

1. Sketch line segments at $x = \pm a$ and $y = \pm b$ and complete the rectangle they determine.
2. Sketch the asymptotes by extending the rectangle's diagonals.
3. Use the rectangle and asymptotes to guide your drawing.

---

**Hyperbolas with Center (0, 0)**

| | | |
|---|---|---|
| • **Standard equation** | $\dfrac{x^2}{a^2} - \dfrac{y^2}{b^2} = 1$ | $\dfrac{y^2}{a^2} - \dfrac{x^2}{b^2} = 1$ |
| • **Focal axis** | $x$-axis | $y$-axis |
| • **Foci** | $(\pm c, 0)$ | $(0, \pm c)$ |
| • **Vertices** | $(\pm a, 0)$ | $(0, \pm a)$ |
| • **Semitransverse axis** | $a$ | $a$ |
| • **Semiconjugate axis** | $b$ | $b$ |
| • **Pythagorean relation** | $c^2 = a^2 + b^2$ | $c^2 = a^2 + b^2$ |
| • **Asymptotes** | $y = \pm\dfrac{b}{a}x$ | $y = \pm\dfrac{a}{b}x$ |

See Figure 8.24.

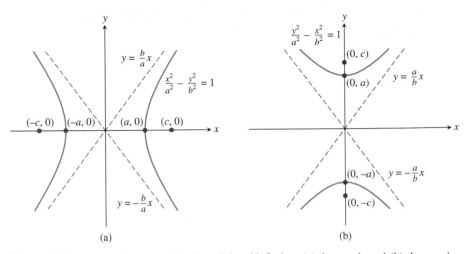

**Figure 8.24** Hyperbolas centered at the origin with foci on (a) the $x$-axis and (b) the $y$-axis.

### Example 1 FINDING THE VERTICES AND FOCI OF A HYPERBOLA

Find the vertices and the foci of the hyperbola $4x^2 - 9y^2 = 36$.

**Solution** Dividing both sides of the equation by 36 yields the standard form $x^2/9 - y^2/4 = 1$. So $a^2 = 9$, $b^2 = 4$, and $c^2 = a^2 + b^2 = 9 + 4 = 13$. Thus the vertices are $(\pm 3, 0)$, and the foci are $(\pm\sqrt{13}, 0)$.

If we wish to graph a hyperbola using a function grapher, we need to solve the equation of the hyperbola for $y$, as illustrated in Example 2.

### Example 2 FINDING AN EQUATION AND GRAPHING A HYPERBOLA

Find an equation of the hyperbola with foci $(0, -3)$ and $(0, 3)$ whose conjugate axis has length 4. Sketch the hyperbola and its asymptotes, and support your sketch with a grapher.

**Solution** The center is $(0, 0)$. The foci are on the $y$-axis with $c = 3$. The semiconjugate axis is $b = 4/2 = 2$. Using $a^2 = c^2 - b^2$, we have $a^2 = 3^2 - 2^2 = 5$. So the standard form of the equation for the hyperbola is

$$\frac{y^2}{5} - \frac{x^2}{4} = 1.$$

Using $a = \sqrt{5} \approx 2.24$ and $b = 2$, we can sketch the central rectangle, the asymptotes, and the hyperbola itself. Try doing this. To graph the hyperbola using a function grapher, we solve for $y$ in terms of $x$.

$$\frac{y^2}{5} = 1 + \frac{x^2}{4}$$

$$y^2 = 5(1 + x^2/4)$$

$$y = \pm\sqrt{5(1 + x^2/4)}$$

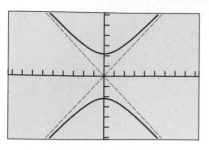

[–9.4, 9.4] by [–6.2, 6.2]

**Figure 8.25** The hyperbola $y^2/5 - x^2/4 = 1$, shown with its asymptotes. (Example 2)

Figure 8.25 shows the graphs of

$$y_1 = \sqrt{5(1 + x^2/4)} \quad \text{and} \quad y_2 = -\sqrt{5(1 + x^2/4)},$$

together with the asymptotes of the hyperbola

$$y_3 = \frac{\sqrt{5}}{2}x \quad \text{and} \quad y_4 = -\frac{\sqrt{5}}{2}x.$$

In Example 2, because the hyperbola had a vertical focal axis, selecting a viewing rectangle was easy. When a hyperbola has a horizontal focal axis, we try to select a viewing window to include the two vertices in the plot and thus avoid gaps in the graph of the hyperbola.

## Translations of Hyperbolas

When a hyperbola with center $(0, 0)$ is translated horizontally by $h$ units and vertically by $k$ units, the center of the hyperbola moves from $(0, 0)$ to $(h, k)$, as shown in Figure 8.26. Such a translation does not change the length of the transverse or conjugate axis or the Pythagorean relation.

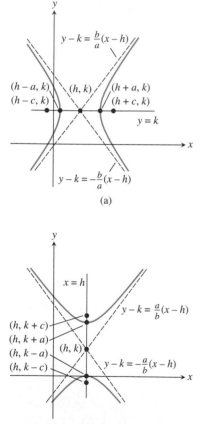

(a)

(b)

**Figure 8.26** Hyperbolas with center $(h, k)$ and foci on (a) $y = k$ and (b) $x = h$.

| **Hyperbolas with Center ($h$, $k$)** | | |
|---|---|---|
| • **Standard equation** | $\dfrac{(x-h)^2}{a^2} - \dfrac{(y-k)^2}{b^2} = 1$ | $\dfrac{(y-k)^2}{a^2} - \dfrac{(x-h)^2}{b^2} = 1$ |
| • **Focal axis** | $y = k$ | $x = h$ |
| • **Foci** | $(h \pm c, k)$ | $(h, k \pm c)$ |
| • **Vertices** | $(h \pm a, k)$ | $(h, k \pm a)$ |
| • **Semitransverse axis** | $a$ | $a$ |
| • **Semiconjugate axis** | $b$ | $b$ |
| • **Pythagorean relation** | $c^2 = a^2 + b^2$ | $c^2 = a^2 + b^2$ |
| • **Asymptotes** | $y = \pm\dfrac{b}{a}(x - h) + k$ | $y = \pm\dfrac{a}{b}(x - h) + k$ |

See Figure 8.26.

### Example 3   FINDING AN EQUATION OF A HYPERBOLA

Find the standard form of the equation for the hyperbola whose transverse axis has endpoints $(-2, -1)$ and $(8, -1)$, and whose conjugate axis has length 8.

**Solution**   Figure 8.27 shows the transverse-axis endpoints, the conjugate axis, and the center of the hyperbola. The standard equation of this hyperbola has the form

$$\frac{(x - h)^2}{a^2} - \frac{(y - k)^2}{b^2} = 1,$$

where the center $(h, k)$ is the midpoint $(3, -1)$ of the transverse axis. The semitransverse axis and semiconjugate axis are

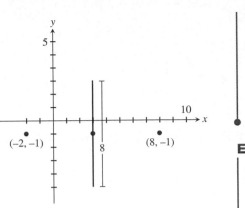

**Figure 8.27** Given information for Example 3.

**Exploration Extensions**

Create a parametrization for the hyperbola
$$\frac{(x-h)^2}{a^2} - \frac{(y-k)^2}{b^2} = 1.$$

$$a = \frac{8-(-2)}{2} = 5 \quad \text{and} \quad b = \frac{8}{2} = 4.$$

So the equation we seek is
$$\frac{(x-3)^2}{5^2} - \frac{(y-(-1))^2}{4^2} = 1,$$
$$\frac{(x-3)^2}{25} - \frac{(y+1)^2}{16} = 1.$$

### Example 4  LOCATING KEY POINTS OF A HYPERBOLA

Find the center, vertices, and foci of the hyperbola
$$\frac{(x+2)^2}{9} - \frac{(y-5)^2}{49} = 1.$$

**Solution**  The center $(h, k) = (-2, 5)$. Because the semitransverse axis $a = \sqrt{9} = 3$, the vertices are
$$(h+a, k) = (-2+3, 5) = (1, 5) \quad \text{and}$$
$$(h-a, k) = (-2-3, 5) = (-5, 5).$$
Because $c = \sqrt{a^2 + b^2} = \sqrt{9+49} = \sqrt{58}$, the foci $(h \pm c, k)$ are $(-2 \pm \sqrt{58}, 5)$, or approximately $(5.62, 5)$ and $(-9.62, 5)$.

With the information found about the hyperbola in Example 4 and knowing that its semiconjugate axis $b = \sqrt{49} = 7$, we could easily sketch the hyperbola. Obtaining an accurate graph of the hyperbola using a function grapher is another matter. Often, the best way to graph a hyperbola using a graphing utility is to use parametric equations.

---

**Exploration 1**  **Graphing a Hyperbola Using Its Parametric Equations**

1. Use the Pythagorean trigonometry identity $\sec^2 t - \tan^2 t - 1$ to prove that the parameterization $x = -1 + 3/\cos t$, $y = 1 + 2 \tan t$ $(0 \le t \le 2\pi)$ will produce a graph of the hyperbola $(x+1)^2/9 - (y-1)^2/4 = 1$.

2. Using Dot graphing mode, graph $x = -1 + 3/\cos t$, $y = 1 + 2 \tan t$ $(0 \le t \le 2\pi)$ in a square viewing window to support part 1 graphically. Switch to Connected graphing mode, and regraph the equation. What do you observe? Explain.

3. Create parameterizations for the hyperbolas in Examples 1, 2, 3, and 4.

4. Graph each of your parameterizations in part 3 and check the features of the obtained graph to see whether they match the expected geometric features of the hyperbola. If necessary, revise your parameterization and regraph until all features match.

5. Prove that each of your parameterizations is valid.

# Eccentricity and Orbits

> **Definition Eccentricity of a Hyperbola**
>
> The **eccentricity** of a hyperbola is
>
> $$e = \frac{c}{a} = \frac{\sqrt{a^2 + b^2}}{a},$$
>
> where $a$ is the semitransverse axis and $b$ is the semiconjugate axis.

For a hyperbola, because $c > a$, the eccentricity $e > 1$. In Section 8.2 we learned that the eccentricity of an ellipse satisfies the inequality $0 \leq e < 1$ and that, for $e = 0$, the ellipse is a circle. In Section 8.5 we will generalize the concept of eccentricity to all types of conics and learn that the eccentricity of a parabola is $e = 1$.

Kepler's first law of planetary motion says that a planet's orbit is elliptical with the Sun at one focus. Since 1609, astronomers have generalized Kepler's law; the current theory states: A celestial body that travels within the gravitational field of a much more massive body follows a path that closely approximates a conic section that has the more massive body as a focus. Two bodies that do not differ greatly in mass (such the Earth and the Moon, or Pluto and its moon Charon) actually revolve around their balance point, or *barycenter*. In theory, a comet can approach the Sun from interstellar space, make a partial loop about the sun, and then leave the solar system returning to deep space; such a comet follows a path that is one branch of a hyperbola.

### Example 5  ANALYZING A COMET'S ORBIT

A comet following a hyperbolic path about the Sun has a perihelion of 90 Gm. When the line from the comet to the Sun is perpendicular to the focal axis of the orbit, the comet is 281.25 Gm from the Sun. Calculate $a$, $b$, $c$, and $e$. What are the coordinates of the center of the Sun if we coordinatize space so that the hyperbola is given by

$$\frac{x^2}{a^2} - \frac{y^2}{b^2} = 1?$$

**Solution**  The perihelion is $c - a = 90$. When $x = c$, $y = \pm b^2/a$ (see Exercise 61). So $b^2/a = 281.25$, or $b^2 = 281.25a$. Because $b^2 = c^2 - a^2$, we have the system

$$c - a = 90 \quad \text{and} \quad c^2 - a^2 = 281.25a,$$

which yields the equation:

$$(a + 90)^2 - a^2 = 281.25a$$
$$a^2 + 180a + 8100 - a^2 = 281.25a$$
$$8100 = 101.25a$$
$$a = 80$$

So $a = 80$ Gm, $b = 150$ Gm, $c = 170$ Gm, and $e = 17/8 = 2.125$. If the comet's path is taken to be the branch of the hyperbola with positive $x$-coordinates, then the Sun is centered at the focus $(c, 0) = (170, 0)$. See Figure 8.28.

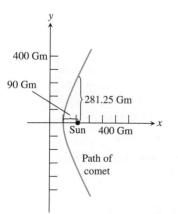

**Figure 8.28** The graph of one branch of $x^2/6400 - y^2/22{,}500 = 1$. (Example 5)

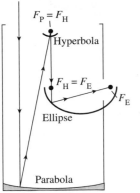

**Figure 8.29** Cross section of a reflecting telescope.

## Reflective Property of a Hyperbola

Like other conics, a hyperbola can be used to make a reflector of sound, light, and other waves. If we rotate a hyperbola in three-dimensional space about its focal axis, the hyperbola sweeps out a **hyperboloid of revolution**. If a signal is directed toward a focus of a reflective hyperboloid, the signal reflects off the hyperbolic surface to the other focus. In Figure 8.29 light reflects off a primary parabolic mirror toward the mirror's focus $F_P = F_H$, which is also the focus of a small hyperbolic mirror. The light is then reflected off the hyperbolic mirror, toward the hyperboloid's other focus $F_H = F_E$, which is also the focus of an elliptical mirror. Finally the light is reflect into the observer's eye, which is at the second focus of the ellipsoid $F_E$.

Reflecting telescopes date back to the 1600s when Isaac Newton used a primary parabolic mirror in combination with a flat secondary mirror, slanted to reflect the light out the side to the eyepiece. French optician G. Cassegrain was the first to use a hyperbolic secondary mirror, which directed the light through a hole at the vertex of the primary mirror (see Exercise 57). Today, reflecting telescopes such as the Hubble Telescope have become quite sophisticated and must have nearly perfect mirrors to focus properly.

## Long-Range Navigation

Hyperbolas and radio signals are the basis of the LORAN (long-range navigation) system. Example 6 illustrates this system using the definition of hyperbola and the fact that radio signals travel 980 ft per microsecond (1 microsecond = 1 $\mu$sec = $10^{-6}$ sec).

### Example 6   USING THE LORAN SYSTEM

Radio signals are sent simultaneously from transmitters located at points $O$, $Q$, and $R$ (Figure 8.30). $R$ is 100 mi due north of $O$, and $Q$ is 80 mi due east of $O$. The LORAN receiver on sloop *Gloria* receives the signal from $O$ 323.27 $\mu$sec after the signal from $R$, and 258.61 $\mu$sec after the signal from $Q$. What is the sloop's bearing and distance from $O$?

**Solution**

**Model**

The *Gloria* is at a point of intersection between two hyperbolas: one with foci $O$ and $R$, the other with foci $O$ and $Q$.

The hyperbola with foci $O(0, 0)$ and $R(0, 100)$ has center $(0, 50)$ and transverse axis

$$2a = (323.27 \ \mu\text{sec})(980 \ \text{ft}/\mu\text{sec})(1 \ \text{mi}/5280 \ \text{ft}) \approx 60 \ \text{mi}.$$

Thus $a \approx 30$ and $b = \sqrt{c^2 - a^2} \approx \sqrt{50^2 - 30^2} = 40$, yielding the equation

$$\frac{(y - 50)^2}{30^2} - \frac{x^2}{40^2} = 1.$$

The hyperbola with foci $O(0, 0)$ and $Q(80, 0)$ has center $(40, 0)$ and transverse axis

$$2a = (258.61 \ \mu\text{sec})(980 \ \text{ft}/\mu\text{sec})(1 \ \text{mi}/5280 \ \text{ft}) \approx 48 \ \text{mi}.$$

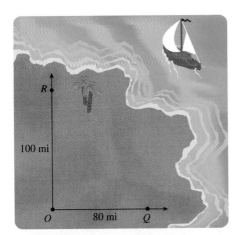

**Figure 8.30** Strategically located LORAN transmitters $O$, $Q$, and $R$. (Example 6)

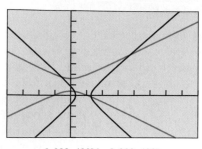

[−200, 400] by [−200, 400]

**Figure 8.31** Graphs for Example 6.

Thus $a \approx 24$ and $b = \sqrt{c^2 - a^2} \approx \sqrt{40^2 - 24^2} = 32$, yielding the equation

$$\frac{(x - 40)^2}{24^2} - \frac{y^2}{32^2} = 1.$$

**Solve Graphically**

The *Gloria* is at point $P$ where upper and right-hand branches of the hyperbolas meet (see Figure 8.31). Using a grapher we find that $P \approx (187.09, 193.49)$. So the bearing from point $O$ is

$$\theta \approx 90° - \tan^{-1}\left(\frac{193.49}{187.09}\right) \approx 44.04°,$$

and the distance from point $O$ is

$$d \approx \sqrt{187.09^2 + 193.49^2} \approx 269.15.$$

**Interpret**

The *Gloria* is about 187.1 mi east and 193.5 mi north of point $O$ on a bearing of roughly 44°. The sloop is about 269 mi from point $O$.

# Quick Review 8.3

In Exercises 1 and 2, find the distance between the given points.

**1.** $(4, -3)$ and $(-7, -8)$ $\sqrt{146}$ **2.** $(a, -3)$ and $(b, c)$

In Exercises 3 and 4, solve for $y$ in terms of $x$.

**3.** $\dfrac{y^2}{16} - \dfrac{x^2}{9} = 1$ **4.** $\dfrac{x^2}{36} - \dfrac{y^2}{4} = 1$

In Exercises 5–8, solve for $x$ algebraically.

**5.** $\sqrt{3x + 12} - \sqrt{3x - 8} = 10$ no solution

**6.** $\sqrt{4x + 12} - \sqrt{x + 8} = 1$ $x = 1$

**7.** $\sqrt{6x^2 + 12} - \sqrt{6x^2 + 1} = 1$ $x = 2, x = -2$

**8.** $\sqrt{2x^2 + 12} - \sqrt{3x^2 + 4} = -8$ $x \approx 25.55, x \approx -25.55$

In Exercises 9 and 10, solve the system of equations.

**9.** $c - a = 2$ and $c^2 - a^2 = 16a/3$ $a = 3, c = 5$

**10.** $c - a = 1$ and $c^2 - a^2 = 25a/12$ $a = 12, c = 13$

# Section 8.3 Exercises

In Exercises 1–6, find the vertices and foci of the hyperbola.

**1.** $\dfrac{x^2}{16} - \dfrac{y^2}{7} = 1$ **2.** $\dfrac{y^2}{25} - \dfrac{x^2}{21} = 1$

**3.** $\dfrac{(y - 1)^2}{25} - \dfrac{(x + 2)^2}{16} = 1$ **4.** $\dfrac{(x - 3)^2}{11} - \dfrac{(y - 5)^2}{7} = 1$

**5.** $3x^2 - 4y^2 = 12$ **6.** $9x^2 - 4y^2 = 36$

In Exercises 7–10, match the graph with its equation.

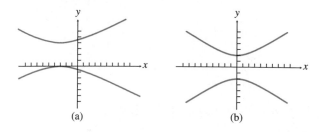

(a)  (b)

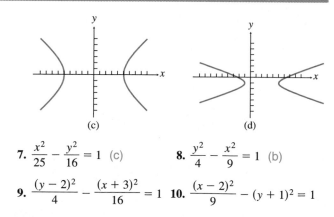

(c)  (d)

**7.** $\dfrac{x^2}{25} - \dfrac{y^2}{16} = 1$ (c) **8.** $\dfrac{y^2}{4} - \dfrac{x^2}{9} = 1$ (b)

**9.** $\dfrac{(y - 2)^2}{4} - \dfrac{(x + 3)^2}{16} = 1$ **10.** $\dfrac{(x - 2)^2}{9} - (y + 1)^2 = 1$

In Exercises 11–24, find an equation in standard form for the hyperbola that satisfies the given conditions.

**11.** Foci $(\pm 3, 0)$, transverse axis length 4 $x^2/4 - y^2/5 = 1$

**12.** Foci $(0, \pm 3)$, transverse axis length 4 $y^2/4 - x^2/5 = 1$

13. Foci $(0, \pm 15)$, transverse axis length 8 $y^2/16 - x^2/209 = 1$

14. Foci $(\pm 5, 0)$, transverse axis length 3

15. Center at $(0, 0)$, $a = 5$, $e = 2$, horizontal focal axis

16. Center at $(0, 0)$, $a = 4$, $e = 3/2$, vertical focal axis

17. Center at $(0, 0)$, $b = 5$, $e = 13/12$, vertical focal axis

18. Center at $(0, 0)$, $c = 6$, $e = 2$, horizontal focal axis

19. The transverse axis endpoints are $(-1, 3)$ and $(5, 3)$, and the slope of one asymptote is 4/3.

20. The transverse axis endpoints are $(-2, -2)$ and $(-2, 7)$, and the slope of one asymptote is 4/3.

21. The foci are $(-4, 2)$ and $(2, 2)$, and the transverse axis endpoints are $(-3, 2)$ and $(1, 2)$.

22. The foci are $(-3, -11)$ and $(-3, 0)$, and the transverse axis endpoints are $(-3, -9)$ and $(-3, -2)$.

23. Center at $(-3, 6)$, $a = 5$, $e = 2$, vertical focal axis

24. Center at $(1, -4)$, $c = 6$, $e = 2$, horizontal focal axis

In Exercises 25–30, sketch the graph of the hyperbola by hand.

25. $\dfrac{x^2}{49} - \dfrac{y^2}{25} = 1$

26. $\dfrac{y^2}{64} - \dfrac{x^2}{25} = 1$

27. $\dfrac{y^2}{25} - \dfrac{x^2}{16} = 1$

28. $\dfrac{x^2}{169} - \dfrac{y^2}{144} = 1$

29. $\dfrac{(x + 3)^2}{16} - \dfrac{(y - 1)^2}{4} = 1$

30. $\dfrac{(x - 1)^2}{2} - \dfrac{(y + 3)^2}{4} = 1$

In Exercises 31–36, graph the hyperbola using a function grapher.

31. $\dfrac{x^2}{36} - \dfrac{y^2}{16} = 1$

32. $\dfrac{y^2}{64} - \dfrac{x^2}{16} = 1$

33. $\dfrac{x^2}{4} - \dfrac{y^2}{9} = 1$

34. $\dfrac{y^2}{16} - \dfrac{x^2}{9} = 1$

35. $\dfrac{x^2}{4} - \dfrac{(y - 3)^2}{5} = 1$

36. $\dfrac{(y - 3)^2}{9} - \dfrac{(x + 2)^2}{4} = 1$

In Exercises 37–40, graph the hyperbola using a parametric grapher.

37. $\dfrac{y^2}{25} - \dfrac{x^2}{4} = 1$

38. $\dfrac{x^2}{30} - \dfrac{y^2}{20} = 1$

39. $\dfrac{(x + 3)^2}{12} - \dfrac{(y - 6)^2}{5} = 1$

40. $\dfrac{(y + 1)^2}{15} - \dfrac{(x - 2)^2}{6} = 1$

In Exercises 41–44, graph the hyperbola, and find its vertices, foci, and eccentricity.

41. $4(y - 1)^2 - 9(x - 3)^2 = 36$

42. $4(x - 2)^2 - 9(y + 4)^2 = 1$

43. $9x^2 - 4y^2 - 36x + 8y - 4 = 0$

44. $25y^2 - 9x^2 - 50y - 54x - 281 = 0$

In Exercises 45 and 46, write an equation for the hyperbola.

45.

46.

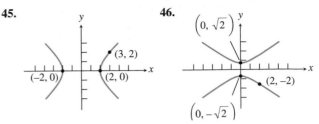

$(0, \sqrt{2})$

$(3, 2)$

$(-2, 0)$ $(2, 0)$

$(2, -2)$

$(0, -\sqrt{2})$

47. **Writing to Learn** Prove that an equation for the hyperbola with center $(0, 0)$, foci $(0, \pm c)$, and semitransverse axis $a$ is $y^2/a^2 - x^2/b^2 = 1$, where $c > a \geq 0$ and $b^2 = a^2 - c^2$. [*Hint*: Refer to derivation at the beginning of the section.]

48. **Degenerate Hyperbolas** Graph the degenerate hyperbola.

(a) $\dfrac{x^2}{4} - y^2 = 0$

(b) $\dfrac{y^2}{9} - \dfrac{x^2}{16} = 0$

49. **Group Activity Long-Range Navigation** Three LORAN radio transmitters are positioned as shown in the figure, with $R$ due north of $O$. The cruise ship *Princess Ann* receives simultaneous signals from the three transmitters. The signal from $O$ arrives 323.27 $\mu$ sec after the signal from $R$, and 646.53 $\mu$ sec after the signal from $Q$. Determine the ship's bearing and distance from point $O$.

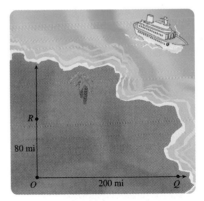

$R$

80 mi

$O$   200 mi   $Q$

**50. Gun Location** Observers are located at positions $A$, $B$, and $C$ with $A$ due north of $B$. A cannon is located somewhere in the first quadrant as illustrated in the figure. $A$ hears the sound of the cannon 2 sec before $B$, and $C$ hears the sound 4 sec before $B$. Determine the bearing and distance of the cannon from point $B$. (Assume that sound travels at 1100 ft/sec.)

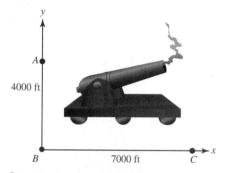

**51. Rogue Comet** A comet following a hyperbolic path about the Sun has a perihelion of 120 Gm. When the line from the comet to the Sun is perpendicular to the focal axis of the orbit, the comet is 250 Gm from the Sun. Calculate $a$, $b$, $c$, and $e$. What are the coordinates of the center of the Sun if the center of the hyperbolic orbit is $(0, 0)$ and the Sun lies on the positive $x$-axis?

**52. Rogue Comet** A comet following a hyperbolic path about the Sun has a perihelion of 140 Gm. When the line from the comet to the Sun is perpendicular to the focal axis of the orbit, the comet is 405 Gm from the Sun. Calculate $a$, $b$, $c$, and $e$. What are the coordinates of the center of the Sun if the center of the hyperbolic orbit is $(0, 0)$ and the Sun lies on the positive $x$-axis?

In Exercises 53 and 54, solve the system of equations algebraically and support your answer graphically.

**53.** $\dfrac{x^2}{4} - \dfrac{y^2}{9} = 1$ $(-2, 0)$, **54.** $\dfrac{x^2}{4} - y^2 = 1$

$x - \dfrac{2\sqrt{3}}{3}y = -2$ $(4, 3\sqrt{3})$ $x^2 + y^2 = 9$

**55.** Consider the system of equations

$$\frac{x^2}{4} - \frac{y^2}{25} = 1$$
$$\frac{x^2}{25} + \frac{y^2}{4} = 1$$

**(a)** Solve the system graphically.

**(b)** If you have access to a grapher that also does symbolic algebra, use it to find the the exact solutions to the system.

## Explorations

**56. Constructing Points of a Hyperbola** Use a geometry software package, such as *Cabri Geometry II, The Geometer's Sketchpad*, or the TI-92 geometry application, to carry out the following construction.

**(a)** Start by placing the coordinate axes in the construction window.

**(b)** Construct two points on the $x$-axis at $(\pm 5, 0)$ as the foci.

**(c)** Construct concentric circles of radii $r = 1, 2, 3, \ldots, 12$.

**(d)** Construct the points where these concentric circles meet and have a difference of radii of $2a = 6$.

**(e)** Find the equation whose graph includes all of these points. $x^2/9 - y^2/16 = 1$

**57. Cassegrain Telescope** A Cassegrain telescope as described in the section has the dimensions shown in the figure. Find the standard form for the equation of the hyperbola centered at the origin with the focal axis the $x$-axis.

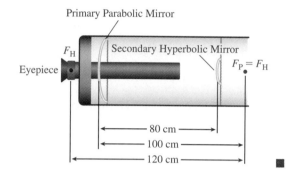

## Extending the Ideas

**58.** Prove that a nondegenerate graph of the equation

$$Ax^2 + Cy^2 + Dx + Ey + F = 0$$

is a hyperbola if $AC < 0$.

**59. Writing to Learn** The graph of the equation

$$\frac{(x - h)^2}{a^2} - \frac{(y - k)^2}{b^2} = 0$$

is considered to be a degenerate hyperbola. Describe the graph. How is it like a full-fledged hyperbola, and how is it different?

**60. Conjugate Hyperbolas** The hyperbolas

$$\frac{(x - h)^2}{a^2} - \frac{(y - k)^2}{b^2} = 1 \text{ and } \frac{(y - k)^2}{b^2} - \frac{(x - h)^2}{a^2} = 1$$

obtained by switching the order of subtraction in their standard equations are **conjugate hyperbolas**. Prove that these hyperbolas have the same asymptotes and that the conjugate axis of each of these hyperbolas is the transverse axis of the other hyperbola.

**61. Focal Width of a Hyperbola** Prove that, for the hyperbola

$$\frac{x^2}{a^2} - \frac{y^2}{b^2} = 1,$$

if $x = c$, then $y = \pm b^2/a$. Why is it reasonable to define the **focal width** of such hyperbolas to be $2b^2/a$?

**62. Writing to Learn** Show how the standard form equations for the conics are related to

$$Ax^2 + Bxy + Cy^2 + Dx + Ey + F = 0.$$

##  Translation and Rotation of Axes

Second-Degree Equations in Two Variables   •   Translating Axes versus Translating Graphs   •   Rotation of Axes   •   Discriminant Test

## Second-Degree Equations in Two Variables

In Section 8.1 we began with a unified approach to conic sections, learning that parabolas, ellipses, and hyperbolas are all cross sections of a right circular cone. In Sections 8.1–8.3, we gave separate plane-geometry definitions for parabolas, ellipses, and hyperbolas that led to separate kinds of equations for each type of curve. In this section and the next, we once again consider parabolas, ellipses, and hyperbolas as a unified family of interrelated curves.

In Section 8.1 we claimed that the conic sections can be defined algebraically in the Cartesian plane as the graphs of *second-degree equations in two variables*, that is, equations of the form

$$Ax^2 + Bxy + Cy^2 + Dx + Ey + F = 0,$$

where $A$, $B$, and $C$ are not all zero. In this section we investigate equations of this type, which are really just *quadratic equations in x and y*. Because they are quadratic equations, we can adapt familiar methods to this unfamiliar setting. That is exactly what we do in Examples 1–3.

### Example 1   GRAPHING A SECOND-DEGREE EQUATION

Use a function graphing utility to graph the conic

$$9x^2 + 16y^2 - 18x + 64y - 71 = 0.$$

**Solution**   Because we wish to use a function grapher, we need to solve the equation for $y$ in terms of $x$. Rearranging terms yields the equation:

$$16y^2 + 64y + (9x^2 - 18x - 71) = 0$$

The quadratic formula gives us

$$y = \frac{-64 \pm \sqrt{64^2 - 4(16)(9x^2 - 18x - 71)}}{2(16)}$$

$$= \frac{-8 \pm 3\sqrt{-x^2 + 2x + 15}}{4}$$

$$= -2 \pm \frac{3}{4}\sqrt{-x^2 + 2x + 15}$$

Let

$$y_1 = -2 + 0.75\sqrt{-x^2 + 2x + 15} \text{ and } y_2 = -2 - 0.75\sqrt{-x^2 + 2x + 15},$$

and graph the two equations in the same viewing window, as shown in Figure 8.32. The combined figure appears to be an ellipse.

In the equation in Example 1, there was no $Bxy$ term. None of the examples in Sections 8.1–8.3 included such a *cross-product* term. A cross-product term in the equation causes the graph to tilt relative to the coordinate axes, as illustrated in Examples 2 and 3.

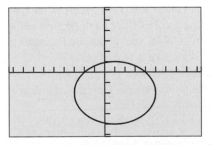

[−9.4, 9.4] by [−6.2, 6.2]

**Figure 8.32**  The graph of $9x^2 + 16y^2 - 18x + 64y - 71 = 0$. (Example 1)

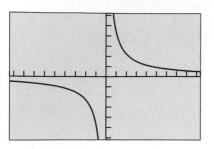

[−9.4, 9.4] by [−6.2, 6.2]

**Figure 8.33** The graph of $2xy − 9 = 0$.
(Example 2)

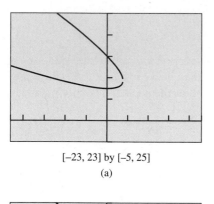

[−23, 23] by [−5, 25]
(a)

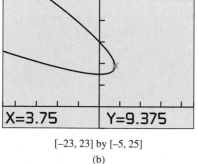

X=3.75        Y=9.375

[−23, 23] by [−5, 25]
(b)

**Figure 8.34** The graph of
$x^2 + 4xy + 4y^2 − 30x − 90y + 450 = 0$
(a) with a gap and (b) with the trace feature
activated at the connecting point.
(Example 3)

**Example 2   GRAPHING A SECOND-DEGREE EQUATION**

Use a function graphing utility to graph the conic

$$2xy − 9 = 0.$$

**Solution**   This equation can be rewritten as $2xy = 9$ or as $y = 9/(2x)$. The graph of this equation is shown in Figure 8.33. It appears to be a hyperbola with a slant focal axis.

**Example 3   GRAPHING A SECOND-DEGREE EQUATION**

Use a function graphing utility to graph the conic

$$x^2 + 4xy + 4y^2 − 30x − 90y + 450 = 0.$$

**Solution**   We rearrange the terms as a quadratic equation in $y$:

$$4y^2 + (4x − 90)y + (x^2 − 30x + 450) = 0$$

The quadratic formula gives us

$$y = \frac{-(4x − 90) \pm \sqrt{(4x − 90)^2 − 4(4)(x^2 − 30x + 450)}}{2(4)}$$

$$= \frac{45 − 2x \pm \sqrt{225 − 60x}}{4}$$

Let

$$y_1 = \frac{45 − 2x + \sqrt{225 − 60x}}{4} \text{ and } y_2 = \frac{45 − 2x − \sqrt{225 − 60x}}{4},$$

and graph the two equations in the same viewing window, as shown in Figure 8.34a. The combined figure appears to be a parabola, with a slight gap due to grapher failure. The combined graph should connect at a point for which the radicand $225 − 60x = 0$, that is, when $x = 225/60 = 15/4 = 3.75$. Figure 8.34b supports this analysis.

The graphs obtained in Examples 1–3 all *appear* to be conic sections, but at this point we have no proof that they are. If they are conic sections, then we have most likely classified Examples 1 and 2 correctly. The classification of Example 3 as a parabola is less certain. Could the graph in Example 3 (Figure 8.34) be part of an ellipse or one branch of a hyperbola? We now set out to answer this question, to confirm our conclusions in Examples 1–3, and to develop general methods for sketching and classifying second-degree equations in two variables.

## Translating Axes versus Translating Graphs

The coordinate axes are often viewed as a permanent fixture of the plane, but this just isn't so. Geometry was done in the plane for thousands of years without coordinate axes, and the location of the axes relative to other figures in the plane is totally arbitrary. So we can shift the position of axes just as we have been shifting the position of graphs since Chapter 1. Such a **translation of axes** produces a new set of axes parallel to the original axes, as shown in Figure 8.35.

Figure 8.35 shows a plane containing a point $P$ that is named in two ways: using the ordered pair $(x, y)$ and using the ordered pair $(x', y')$. The name

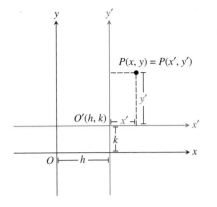

**Figure 8.35** A translation of Cartesian coordinate axes. (Example 2)

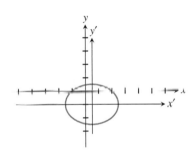

**Figure 8.36** The graph of $(x')^2/16 + (y')^2/9 = 1$. (Example 4)

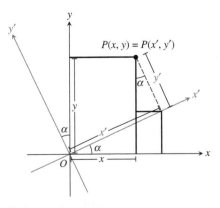

**Figure 8.37** A rotation of Cartesian coordinate axes.

$P(x, y)$ is based on the original $x$- and $y$- axes and the original origin $O$; the name $P(x', y')$ is based on the translated $x'$- and $y'$-axes and the corresponding origin $O'$. From the geometry of Figure 8.35 we obtain the following **translation formulas**:

$$x = x' + h \quad \text{and} \quad y = y' + k$$

The translation formulas can also be written in the form

$$x' = x - h \quad \text{and} \quad y' = y - k.$$

(See Exercise 44.) We use this latter form in Example 4.

### Example 4 REVISITING EXAMPLE 1

Prove that $9x^2 + 16y^2 - 18x + 64y - 71 = 0$ is the equation of an ellipse. Translate the coordinate axes so that the origin is at the center of this ellipse.

**Solution** We complete the square of both $x$ and $y$:

$$9x^2 - 18x + 16y^2 + 64y = 71$$

$$9(x^2 - 2x + 1) + 16(y^2 + 4y + 4) = 71 + 9(1) + 16(4)$$

$$9(x - 1)^2 + 16(y + 2)^2 = 144$$

$$\frac{(x - 1)^2}{16} + \frac{(y + 2)^2}{9} = 1$$

This is a standard equation of an ellipse. If we let $x' = x - 1$ and $y' = y + 2$, then the equation of the ellipse becomes

$$\frac{(x')^2}{16} + \frac{(y')^2}{9} = 1.$$

Figure 8.36 shows the graph of this final equation in the new $x'y'$ coordinate system, with the original $xy$-axes overlaid. Compare Figures 8.32 and 8.36.

## Rotation of Axes

To show that the equation in Example 2 or 3 is the equation of a conic section, we need to rotate the coordinate axes so that one of the coordinate axes aligns with the (focal) axis of the conic. In such a **rotation of axes** the origin stays fixed, and we rotate the $x$- and $y$-axes through an angle of rotation $\alpha$ such that $0 < \alpha < \pi/2$, as shown in Figure 8.37.

Figure 8.37 shows a plane containing a point $P$ named in two ways: as $(x, y)$ and as $(x', y')$. The name $P(x, y)$ is based on the original $x$- and $y$-axes; the name $P(x', y')$ is based on the rotated $x'$- and $y'$-axes. In Example 10 of Section 7.2 we proved the validity of the following **rotation formulas**:

$$x' = x \cos \alpha + y \sin \alpha \quad \text{and} \quad y' = -x \sin \alpha + y \cos \alpha,$$

which can be expressed in the form

$$x = x' \cos \alpha - y' \sin \alpha \quad \text{and} \quad y = x' \sin \alpha + y' \cos \alpha.$$

(See Exercises 45 and 46.) We use this latter form in Example 5.

### Example 5  REVISITING EXAMPLE 2

Prove that $2xy - 9 = 0$ is the equation of a hyperbola by rotating the coordinate axes through an angle $\alpha = \pi/4$.

**Solution**  Because $\cos(\pi/4) = \sin(\pi/4) = 1/\sqrt{2}$, the rotation equations become

$$x = \frac{x' - y'}{\sqrt{2}} \quad \text{and} \quad y = \frac{x' + y'}{\sqrt{2}}.$$

So by rotating the axes, the equation $2xy - 9 = 0$ becomes

$$2\left(\frac{x' - y'}{\sqrt{2}}\right)\left(\frac{x' + y'}{\sqrt{2}}\right) - 9 = 0$$

$$(x')^2 - (y')^2 - 9 = 0.$$

To see that this is the equation of a hyperbola, we put it in standard form:

$$(x')^2 - (y')^2 = 9$$

$$\frac{(x')^2}{9} - \frac{(y')^2}{9} = 1$$

If we apply the rotation formulas to the general second-degree equation in $x$ and $y$, we obtain a second-degree equation in $x'$ and $y'$ of the form

$$A'x'^2 + B'x'y' + C'y'^2 + D'x' + E'y' + F' = 0,$$

where the coefficients are

$$A' = A\cos^2\alpha + B\cos\alpha\sin\alpha + C\sin^2\alpha$$

$$B' = B\cos 2\alpha + (C - A)\sin 2\alpha$$

$$C' = C\cos^2\alpha - B\cos\alpha\sin\alpha + A\sin^2\alpha$$

$$D' = D\cos\alpha + E\sin\alpha$$

$$E' = E\cos\alpha - D\sin\alpha$$

$$F' = F$$

In order to eliminate the cross-product term and thus align the coordinate axes with the focal axis of the conic, we rotate the coordinate axes through an angle $\alpha$ that causes $B'$ to equal 0. Setting $B' = B\cos 2\alpha + (C - A)\sin 2\alpha = 0$ yields

$$\cot 2\alpha = \frac{A - C}{B}.$$

### Example 6  REVISITING EXAMPLE 3

Prove that $x^2 + 4xy + 4y^2 - 30x - 90y + 450 = 0$ is the equation of a parabola by rotating the coordinate axes through a suitable angle $\alpha$.

**Solution**  The angle of rotation $\alpha$ must satisfy the equation

$$\cot 2\alpha = \frac{A - C}{B} = \frac{1 - 4}{4} = -\frac{3}{4}.$$

So

$$\cos 2\alpha = -\frac{3}{5},$$

and thus

$$\cos \alpha = \sqrt{\frac{1 + \cos 2\alpha}{2}} = \sqrt{\frac{1 + (-3/5)}{2}} = \frac{1}{\sqrt{5}},$$

$$\sin \alpha = \sqrt{\frac{1 - \cos 2\alpha}{2}} = \sqrt{\frac{1 - (-3/5)}{2}} = \frac{2}{\sqrt{5}}.$$

Therefore the coefficients of the transformed equation are

$$A' = 1 \cdot \frac{1}{5} + 4 \cdot \frac{2}{5} + 4 \cdot \frac{4}{5} = \frac{25}{5} = 5$$

$$B' = 0$$

$$C' = 4 \cdot \frac{1}{5} - 4 \cdot \frac{2}{5} + 1 \cdot \frac{4}{5} = 0$$

$$D' = -30 \cdot \frac{1}{\sqrt{5}} - 90 \cdot \frac{2}{\sqrt{5}} = -\frac{210}{\sqrt{5}} = -42\sqrt{5}$$

$$E' = -90 \cdot \frac{1}{\sqrt{5}} + 30 \cdot \frac{2}{\sqrt{5}} = -\frac{30}{\sqrt{5}} = -6\sqrt{5}$$

$$F' = 450$$

So the equation $x^2 + 4xy + 4y^2 - 30x - 90y + 450 = 0$ becomes

$$5x'^2 - 42\sqrt{5}x' - 6\sqrt{5}y' + 450 = 0.$$

If we translate using $h = 21/\sqrt{5}$ and $k = 3\sqrt{5}/10$, then the equation becomes

$$(x'')^2 = \frac{6}{\sqrt{5}}(y''),$$

a standard equation of a parabola.

Figure 8.38 shows the graph of the original equation in the original $xy$ coordinate system with the $x''y''$-axes overlaid.

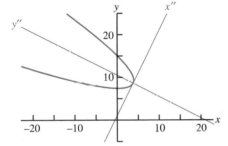

**Figure 8.38** The graph of $x^2 + 4xy + 4y^2 - 30x - 90y + 450 = 0$. (Example 6)

## Discriminant Test

Example 6 demonstrates that the algebra of rotation can get ugly. Fortunately, we can determine which type of conic a second-degree equation represents by looking at the sign of the **discriminant** $B^2 - 4AC$.

---

**Discriminant Test**

The second-degree equation $Ax^2 + Bxy + Cy^2 + Dx + Ey + F = 0$ graphs as

- a hyperbola if $B^2 - 4AC > 0$,
- a parabola if $B^2 - 4AC = 0$,
- an ellipse if $B^2 - 4AC < 0$,

except for degenerate cases.

---

**Assignment Guide**

Day 1: Ex. 3, 4, 9, 12, 25–30
Day 2: Ex. 6, 15, 18, 21, 33, 36, 38, 39, 42, 45, 47, 50

**Cooperative Learning**

Group Activity: Ex. 48

**Notes on Exercises**

Ex. 13–22 and 33–40 are basic problems involving rotation of axes.
Ex. 45–46 require students to prove the rotation formulas.

**Ongoing Assessment**

Self-Assessment: Ex. 7, 17, 23, 37
Embedded Assessment: Ex. 43

This test hinges on the fact that the discriminant $B^2 - 4AC$ is **invariant under rotation**; in other words, even though $A$, $B$, and $C$ do change when we rotate the coordinate axes, the combination $B^2 - 4AC$ maintains its value.

### Example 7 REVISITING EXAMPLES 5 AND 6

**(a)** In Example 5, before the rotation $B^2 - 4AC = (2)^2 - 4(0)(0) = 4$, and after the rotation $B'^2 - 4A'C' = (0)^2 - 4(1)(-1) = 4$. The positive discriminant tells us the conic is a hyperbola.

**(b)** In Example 6, before the rotation $B^2 - 4AC = (4)^2 - 4(1)(4) = 0$, and after the rotation $B'^2 - 4A'C' = (0)^2 - 4(5)(0) = 0$. The zero discriminant tells us the conic is a parabola.

Not only is the discriminant $B^2 - 4AC$ invariant under rotation, but also its *sign* is invariant under translation and under algebraic manipulations that preserve the equivalence of the equation, such as multiplying both sides of the equation by a nonzero constant.

The discriminant test can be applied to degenerate conics. Table 8.2 displays the three basic types of conic sections grouped with their associated degenerate conics. Each conic or degenerate conic is shown with a sample equation and the sign of its discriminant.

**Follow-up**

Ask students to explain why it was okay to assme $\cos \alpha \geq 0$ and $\sin \alpha \geq 0$ in Example 6.

### Table 8.2 Conics and the Equation $Ax^2 + Bxy + Cy^2 + Dx + Ey + F = 0$

| Conic | Sample Equation | $A$ | $B$ | $C$ | $D$ | $E$ | $F$ | Sign of Discriminant |
|---|---|---|---|---|---|---|---|---|
| **Hyperbola** | $x^2 - 2y^2 = 1$ | 1 | | $-2$ | | | $-1$ | Positive |
| Intersecting lines | $x^2 + xy = 0$ | 1 | 1 | | | | | Positive |
| **Parabola** | $x^2 = 2y$ | 1 | | | | $-2$ | | Zero |
| Parallel lines | $x^2 = 4$ | 1 | | | | | $-4$ | Zero |
| One line | $y^2 = 0$ | | | 1 | | | | Zero |
| No graph | $x^2 = -1$ | 1 | | | | | 1 | Zero |
| **Ellipse** | $x^2 + 2y^2 = 1$ | 1 | | 2 | | | $-1$ | Negative |
| Circle | $x^2 + y^2 = 9$ | 1 | | 1 | | | $-9$ | Negative |
| Point | $x^2 + y^2 = 0$ | 1 | | 1 | | | | Negative |
| No graph | $x^2 + y^2 = -1$ | 1 | | 1 | | | 1 | Negative |

## Quick Review 8.4

In Exercises 1–10, assume $0 \leq \alpha < \pi/2$.

**1.** Given that $\cot 2\alpha = 5/12$, find $\cos 2\alpha$. $\cos 2\alpha = 5/13$

**2.** Given that $\cot 2\alpha = 8/15$, find $\cos 2\alpha$. $\cos 2\alpha = 8/17$

**3.** Given that $\cot 2\alpha = 1/\sqrt{3}$, find $\cos 2\alpha$. $\cos 2\alpha = 1/2$

**4.** Given that $\cot 2\alpha = 2/\sqrt{5}$, find $\cos 2\alpha$. $\cos 2\alpha = 2/3$

**5.** Given that $\cot 2\alpha = 0$, find $\alpha$. $\alpha = \pi/4$

**6.** Given that $\cot 2\alpha = \sqrt{3}$, find $\alpha$. $\alpha = \pi/12$

**7.** Given that $\cot 2\alpha = 3/4$, find $\cos \alpha$. $\cos \alpha = 2/\sqrt{5}$

**8.** Given that $\cot 2\alpha = 3/\sqrt{7}$, find $\cos \alpha$.

**9.** Given that $\cot 2\alpha = \sqrt{11}/5$, find $\sin \alpha$. $\sin \alpha \approx 0.47$

**10.** Given that $\cot 2\alpha = 45/28$, find $\sin \alpha$.

# Section 8.4 Exercises

In Exercises 1–4, using the point $P(x, y)$ and the translation information, find the coordinates of $P$ in the translated $x'y'$ coordinate system.

1. $P(x, y) = (2, 3), h = -2, k = 4$  $(x', y') = (4, -1)$

2. $P(x, y) = (-2, 5), h = -4, k = -7$  $(x', y') = (2, 12)$

3. $P(x, y) = (6, -3), h = 1, k = \sqrt{5}$  $(x', y') = (5, -3 - \sqrt{5})$

4. $P(x, y) = (-5, -4), h = \sqrt{2}, k = -3$

In Exercises 5–8, using the point $P(x, y)$ and the rotation information, find the coordinates of $P$ in the rotated $x'y'$ coordinate system.

5. $P(x, y) = (-2, 5), \alpha = \pi/4$  $(3\sqrt{2}/2, 7\sqrt{2}/2)$

6. $P(x, y) = (6, -3), \alpha = \pi/3$

7. $P(x, y) = (-5, -4), \cot 2\alpha = -3/5 \approx (-5.94, 2.38)$

8. $P(x, y) = (2, 3), \cot 2\alpha = 0$  $(5\sqrt{2}/2, \sqrt{2}/2)$

In Exercises 9–12, write an equation in standard form for the conic shown.

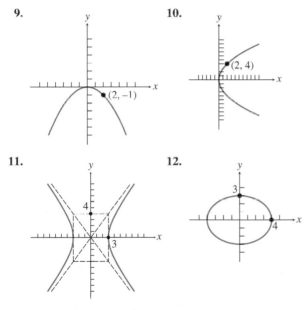

9.

10.

$(2, 4)$

$(2, -1)$

11.

12.

$4$

$3$

$3$

$4$

In Exercises 13–22, use the discriminant $B^2 - 4AC$ to decide whether the equation represents a parabola, an ellipses, or a hyperbola.

13. $x^2 - 4xy + 10y^2 + 2y - 5 = 0$  $-24 < 0$; ellipse

14. $x^2 - 4xy + 3x + 25y - 6 = 0$  $16 > 0$; hyperbola

15. $2x^2 - xy + 9y^2 - 7x - 3y = 0$  $-71 < 0$; ellipse

16. $-xy + 3y^2 - 4x + 2y + 8 = 0$  $1 > 0$; hyperbola

17. $8x^2 - 4xy + 2y^2 + 6 = 0$  $-48 < 0$; ellipse

18. $3x^2 - 12xy + 4y^2 + x - 5y - 4 = 0$  $96 > 0$; hyperbola

19. $x^2 - 3y^2 - y - 22 = 0$  $12 > 0$; hyperbola

20. $5x^2 + 4xy + 3y^2 + 2x + y = 0$  $-44 < 0$; ellipse

21. $4x^2 - 2xy + y^2 - 5x + 18 = 0$  $-12 < 0$; ellipse

22. $6x^2 - 4xy + 9y^2 - 40x + 20y - 56 = 0$ $-200 < 0$; ellipse

In Exercises 23–32, identify the type of conic, write the equation in standard form, translate the conic to the origin, and sketch it in the translated coordinate system.

23. $4y^2 - 9x^2 - 18x - 8y - 41 = 0$

24. $2x^2 + 3y^2 + 12x - 24y + 60 = 0$

25. $x^2 + 2x - y + 3 = 0$  26. $3x^2 - 6x - 6y + 10 = 0$

27. $9x^2 + 4y^2 - 18x + 16y - 11 = 0$

28. $16x^2 - y^2 - 32x - 6y - 57 = 0$

29. $y^2 - 4y - 8x + 20 = 0$  30. $2x^2 - 4x + y^2 - 6y = 9$

31. $2x^2 - y^2 + 4x + 6 = 0$  32. $y^2 - 2y + 4x - 12 = 0$

In Exercises 33–36, identify the type of conic, solve for $y$, and graph the conic. Approximate to two decimal places the angle of rotation needed to eliminate the $xy$ term.

33. $2x^2 - xy + 3y^2 - 3x + 4y - 6 = 0$

34. $-x^2 + 3xy + 4y^2 - 5x - 10y - 20 = 0$

35. $2x^2 - 4xy + 8y^2 - 10x + 4y - 13 = 0$

36. $2x^2 - 4xy + 2y^2 - 5x + 6y - 15 = 0$

In Exercises 37–40, identify the type of conic, and rotate the coordinate axes to eliminate the $xy$ term. Graph the transformed equation.

37. $xy = 8$

38. $2x^2 + \sqrt{3}xy + y^2 - 10 = 0$

39. $3x^2 + 2\sqrt{3}xy + y^2 - 14 = 0$

40. $16x^2 - 24xy + 9y^2 - 40 = 0$

41. **Revisiting Example 5**  Use the results of Example 5, find the center, vertices, and foci of the hyperbola $2xy - 9 = 0$ in the original coordinate system.

42. **Revisiting Examples 3 and 6**  Use information from Examples 3 and 6,

   (a) to prove the point $P(x, y) = (3.75, 9.375)$ where the graphs of $y_1 = (45 - 2x + \sqrt{225 - 60x})/4$ and $y_2 = (45 - 2x - \sqrt{225 - 60x})/4$ meet is not the vertex of the parabola,

   (b) to prove the point $V(x, y) = (3.6, 8.7)$ is the vertex of the parabola.

43. **Writing to Learn  Translation Formulas**  Use the geometric relationships illustrated in Figure 8.35 to explain the translation formulas $x = x' + h$ and $y = y' + k$.

44. **Translation Formulas**  Prove that if $x = x' + h$ and $y = y' + k$, then $x' = x - h$ and $y' = y - k$.

**45. Rotation Formulas** Prove $x = x' \cos \alpha - y' \sin \alpha$ and $y = x' \sin \alpha + y' \cos \alpha$ using the geometric relationships illustrated in Figure 8.37.

**46. Rotation Formulas** Prove that if $x' = x \cos \alpha + y \sin \alpha$ and $y' = -x \sin \alpha + y \cos \alpha$, then $x = x' \cos \alpha - y' \sin \alpha$ and $y = x' \sin \alpha + y' \cos \alpha$.

## Explorations

**47. The Discriminant** Determine what happens to the sign of $B^2 - 4AC$ within the equation $Ax^2 + Bxy + Cy^2 + Dx + Ey + F = 0$ when

**(a)** the axes are translated $h$ units horizontally and $k$ units vertically, no sign change

**(b)** a constant $k$ is added to both sides of the equation,

**(c)** both sides of the equation are multiplied by the same nonzero constant $k$. no sign change ∎

## Extending the Ideas

**48. Group Activity** Working together prove that the formulas for $A'$, $B'$, $C'$, $D'$, $E'$, and $F'$ given between Examples 5 and 6 are correct.

**49. Identifying a Conic** Develop a way to decide whether $Ax^2 + Cy^2 + Dx + Ey + F = 0$, with $A$ and $C$ not both 0, represents a parabola, an ellipse, or a hyperbola. Write an example to illustrate each of the three cases.

**50. Rotational Invariant** Prove that $B'^2 - 4A'C' = B^2 - 4AC$ when the $xy$ coordinate system is rotated through an angle $\alpha$.

**51. Other Rotational Invariants** Prove each of the following are invariants under rotation:

**(a)** $F$, **(b)** $A + C$, **(c)** $D^2 + E^2$.

**52. Degenerate Conics** Graph all of the degenerate conics given in Table 8.2. Recall that *degenerate cones* occur when the generator and axis of the cone are parallel or perpendicular. Explain the occurrence of all of the degenerate conics in Table 8.2 on the basis of cross sections of typical or degenerate right circular cones.

---

## 8.5 Polar Equations of Conics

Eccentricity Revisited • Writing Polar Equations for Conics • Analyzing Polar Equations of Conics • Orbits Revisited

### Eccentricity Revisited

Eccentricity and polar coordinates provide ways to see once again that parabolas, ellipses, and hyperbolas are a unified family of interrelated curves. We can define these three curves simultaneously by generalizing the focus-directrix definition of parabola given in Section 8.1.

> **Focus-Directrix Definition** Conic Section
>
> A **conic section** is the set of all points in a plane whose distances from a particular point (the **focus**) and a particular line (the **directrix**) in the plane have a constant ratio. (We assume that the focus does not lie on the directrix.)

The line passing through the focus and perpendicular to the directrix is the **(focal) axis** of the conic section. The axis is a line of symmetry for the conic. The point where the conic intersects its axis is a **vertex** of the conic. If $P$ is a point of the conic, $F$ is the focus, and $D$ is the point of the directrix closest to $P$, then the constant ratio $PF/PD$ is the **eccentricity $e$** of the conic (see Figure 8.39). A parabola has one focus and one directrix. Ellipses and hyperbolas have two focus-directrix pairs, and either focus-directrix pair can be used with the eccentricity to generate the entire conic section.

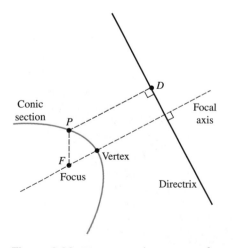

**Figure 8.39** The geometric structure of a conic section.

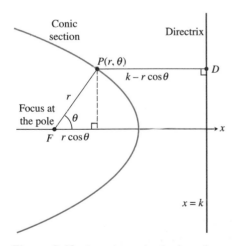

**Figure 8.40** A conic section in the polar plane.

### Focus-Directrix-Eccentricity Relationship

If $P$ is a point of a conic section, $F$ is the conic's focus, and $D$ is the point of the directrix closest to $F$, then

$$e = \frac{PF}{PD} \quad \text{and} \quad PF = e \cdot PD,$$

where the constant $e$ is the eccentricity of the conic. Moreover, the conic is

- a hyperbola if $e > 1$,
- a parabola if $e = 1$,
- an ellipse if $e < 1$.

In this approach to conic sections, the eccentricity $e$ is a strictly positive constant, and there are no circles or other degenerate conics.

## Writing Polar Equations for Conics

Our focus-directrix definition of conics works best in combination with polar coordinates. Recall in polar coordinates that the origin is the *pole* and the $x$-axis is the *polar axis*. To obtain a polar equation for a conic section, we position the pole at the conic's focus and the polar axis along the focal axis with the directrix to the right of the pole (Figure 8.40). If the distance from the focus to the directrix is $k$, the Cartesian equation of the directrix is $x = k$. From Figure 8.40, we see that

$$PF = r \quad \text{and} \quad PD = k - r \cos \theta.$$

So the equation $PF = e \cdot PD$ becomes

$$r = e(k - r \cos \theta),$$

which when solved for $r$ is

$$r = \frac{ke}{1 + e \cos \theta}$$

In Exercise 46 you are asked to show that this equation is still valid if $r < 0$ or $r \cos \theta > k$. This one equation can produce any nondegenerate conic section. In Exploration 1 you will discover how changing the value of $e$ affects the graph of $r = ke/(1 + e \cos \theta)$.

**Exploration 1**    **Graphing Polar Equations of Conics**

Set your grapher to Polar and Dot graphing modes, and to Radian mode. Using $k = 3$, an $xy$ window of $[-12, 24]$ by $[-12, 12]$, $\theta$ min $= 0$, $\theta$ max $= 2\pi$, and $\theta$ step $= \pi/48$, graph

$$r = \frac{ke}{1 + e \cos \theta}$$

for $e = 0.7, 0.8, 1, 1.5, 3$. Identify the type of conic section obtained for each $e$ value. Overlay the five graphs, and explain how changing the value of $e$ affects the graph of $r = ke/(1 + e \cos \theta)$. Explain how the five graphs are similar and how they are different.

**Exploration Extensions**

How would the graphs change if you used the equation

$$r = \frac{ke}{1 + e \cos (\alpha - \pi/2)}?$$

Simplify this equation.

### Example 1    WRITING POLAR EQUATIONS FOR CONICS

**(a)** Sample ellipse: Letting $e = 1/2$ yields $r = k/(2 + \cos \theta)$.

**(b)** Typical parabola: Because $e = 1$, we have $r = k/(1 + \cos \theta)$.

**(c)** Sample hyperbola: Letting $e = 2$ yields $r = 2k/(1 + 2 \cos \theta)$.

Figure 8.41 shows the graphs of these three types of conic sections.

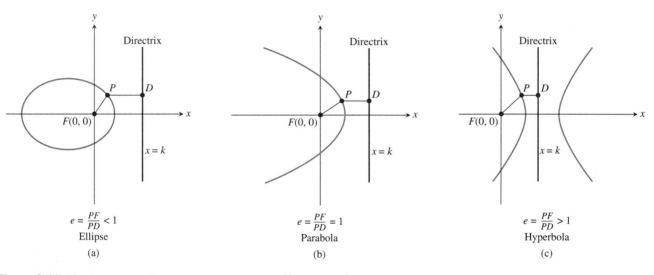

$e = \dfrac{PF}{PD} < 1$
Ellipse
(a)

$e = \dfrac{PF}{PD} = 1$
Parabola
(b)

$e = \dfrac{PF}{PD} > 1$
Hyperbola
(c)

**Figure 8.41**  The three types of conics possible for $r = ke/(1 + e \cos \theta)$. (Example 1)

Variations of the equation $r = ke/(1 + e \cos \theta)$ can be obtained by placing the directrix in different positions. See Figure 8.42 on page 645.

**Notes on Examples**

If possible, work with students through all of the examples in this section, allowing time for exploration. Consider having some students present a few of the examples in class.

**Teaching Note**

This section provides a good opportunity to observe students' selection of viewing window settings in the polar mode on the grapher. Students should have a good notion of how the graphs of conics should appear and can now be in a position to refine their ideas about range.

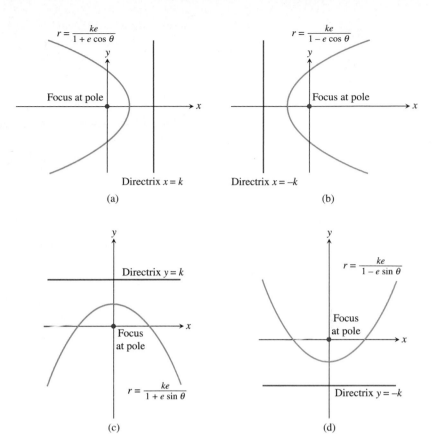

**Figure 8.42** The four standard orientations of a conic in the polar plane.

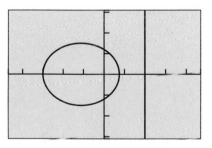

[−4.7, 4.7] by [−3.1, 3.1]

(a)

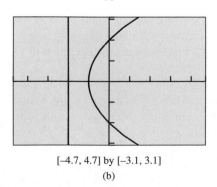

[−4.7, 4.7] by [−3.1, 3.1]

(b)

**Figure 8.43a and b** Graphs for Example 2.

### Example 2 WRITING AND GRAPHING POLAR EQUATIONS OF CONICS

Given that the focus is at the pole, write a polar equation for the specified conic, and graph it.

**(a)** Eccentricity $e = 3/5$, directrix $x = 2$.

**(b)** Eccentricity $e = 1$, directrix $x = -2$.

**(c)** Eccentricity $e = 3/2$, directrix $y = 4$.

**Solution**

**(a)** Setting $e = 3/5$ and $k = 2$ in $r = \dfrac{ke}{1 + e \cos \theta}$ yields

$$r = \frac{2(3/5)}{1 + (3/5) \cos \theta}$$

$$= \frac{6}{5 + 3 \cos \theta}.$$

Figure 8.43a shows this ellipse and the given directrix.

**(b)** Setting $e = 1$ and $k = 2$ in $r = \dfrac{ke}{1 - e \cos \theta}$ yields

$$r = \frac{2}{1 - \cos \theta}.$$

Figure 8.43b shows this parabola and its directrix.

**(c)** Setting $e = 3/2$ and $k = 4$ in $r = \dfrac{ke}{1 + e \sin \theta}$ yields

$$r = \frac{4(3/2)}{1 + (3/2) \sin \theta}$$

$$= \frac{12}{2 + 3 \sin \theta}.$$

Figure 8.43c shows this hyperbola and the given directrix.

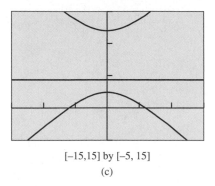

[−15,15] by [−5, 15]

(c)

**Figure 8.43c** Graph for Example 2.

## Analyzing Polar Equations of Conics

The first step in analyzing the polar equations of a conic section is to identify which type of conic the equation represents and to determine the equation of its directrix.

### Example 3 IDENTIFYING CONICS FROM THEIR POLAR EQUATIONS

Identify the type of conic and the directrix for the equation.

**(a)** $r = \dfrac{6}{2 + 3 \cos \theta}.$  **(b)** $r = \dfrac{6}{4 - 3 \sin \theta}.$

**Solution**

**(a)** Dividing numerator and denominator by 2 yields $r = 3/(1 + 1.5 \cos \theta)$. So the eccentricity $e = 1.5$, and thus the conic is a hyperbola. The numerator $ke = 3$, so $k = 2$, and thus the equation of the directrix is $x = 2$.

**(b)** Dividing numerator and denominator by 4 yields $r = 1.5/(1 - 0.75 \sin \theta)$. So the eccentricity $e = 0.75$, and thus the conic is an ellipse. The numerator $ke = 1.5$, so $k = 2$, and thus the equation of the directrix is $y = -2$.

All of the geometric properties and features of parabolas, ellipses, and hyperbolas developed in Sections 8.1–8.3 still apply in the polar coordinate setting. In Example 4 we use this prior knowledge.

### Example 4 ANALYZING A CONIC

Analyze the conic section given by the equation $r = 16/(5 - 3 \cos \theta)$. Include in the analysis the values of $e$, $a$, $b$, and $c$.

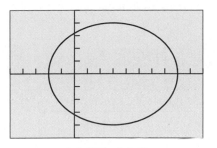

[–5, 10] by [–5, 5]

**Figure 8.44** Graph of the ellipse $r = 16/(5 - 3 \cos \theta)$. (Example 4)

**Solution** Dividing numerator and denominator by 5 yields

$$r = \frac{3.2}{1 - 0.6 \cos \theta}.$$

So the eccentricity $e = 0.6$, and thus the conic is an ellipse. Figure 8.44 shows this ellipse. The vertices (endpoints of the major axis) have polar coordinates $(8, 0)$ and $(2, \pi)$. So $2a = 8 + 2 = 10$, and thus $a = 5$.

The vertex $(2, \pi)$ is 2 units to the left of the pole, and the pole is a focus of the ellipse. So $a - c = 2$, and thus $c = 3$. An alternative way to find $c$ is to use the fact that the eccentricity of an ellipse is $e = c/a$, and thus $c = ae = 5 \cdot 0.6 = 3$.

To find $b$ we use the Pythagorean relation of an ellipse:

$$b = \sqrt{a^2 - c^2} = \sqrt{25 - 9} = 4$$

With all of this information, we can write the Cartesian equation of the ellipse:

$$\frac{(x - 3)^2}{25} + \frac{y^2}{16} = 1$$

## Orbits Revisited

The polar equations developed for conics in this section are used extensively in celestial mechanics, the branch of astronomy born out of Kepler's work on planetary motion that studies the motion of celestial bodies. The polar equations of conic sections are well suited to the *two-body problem* of celestial mechanics for several reasons. First, the same equations are used for ellipses, parabolas, and hyperbolas—the paths of one body traveling about another. Second, a focus of the conic is always at the pole. This situation has two immediate advantages in celestial mechanics:

- The pole can be thought of as the center of the larger body, such as the sun, with the smaller body, such as the earth, following a polar conic about the larger body.

- The polar coordinates given by the equation give the distance $r$ between the two bodies and the direction $\theta$ from the larger body to the smaller body relative to the axis of the conic path of motion.

For these reasons, polar coordinates are preferred over Cartesian coordinates for studying orbital motion.

To use the data in Table 8.3 to create polar equations for the elliptical orbits of the planets, we need to express the equation $r = ke/(1 + e \cos \theta)$ in terms of $a$ and $e$. We apply the formula $PF = e \cdot PD$ to the ellipse shown in Figure 8.45:

$$e \cdot PD = PF$$

$$e(c + k - a) = a - c$$

$$e(ae + k - a) = a - ae \qquad \text{Use } e = c/a.$$

$$ae^2 + ke - ae = a - ae$$

$$ae^2 + ke = a$$

$$ke = a - ae^2$$

$$ke = a(1 - e^2)$$

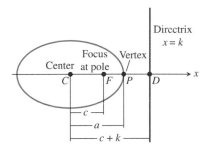

**Figure 8.45** Geometric relationships within an ellipse.

So the equation $r = ke/(1 + e \cos \theta)$ can be rewritten as

$$r = \frac{a(1 - e^2)}{1 + e \cos \theta}.$$

In this form of the equation, when $e = 0$, the equation reduces to $r = a$, the equation of a circle with radius $a$.

### Table 8.3 Semimajor Axes and Eccentricities of the Planets

| Planet | Semimajor Axis (Gm) | Eccentricity |
|--------|--------------------|--------------|
| Mercury | 57.9 | 0.2056 |
| Venus | 108.2 | 0.0068 |
| Earth | 149.6 | 0.0167 |
| Mars | 227.9 | 0.0934 |
| Jupiter | 778.3 | 0.0485 |
| Saturn | 1427 | 0.0560 |
| Uranus | 2869 | 0.0461 |
| Neptune | 4497 | 0.0050 |
| Pluto | 5900 | 0.2484 |

*Source: Shupe, et al., National Geographic Atlas of the World (rev. 6th ed.).*
*Washington, DC: National Geographic Society, 1992, plate 116, and other sources*

**Follow-up**

Ask students how to determine what type of conic is represented by an equation of the form

$$v = \frac{a}{b + c \cos \theta}.$$

**Assignment Guide**

Day 1: Ex. 1–12, 23–25, 28–30
Day 2: Ex. 15, 17, 18, 33–40, 42, 43, 47

**Cooperative Learning**

Group Activity: Ex. 43–44

**Notes on Exercises**

The exercises in this section require that students bring together their knowledge of polar-coordinate graphing and conic sections. This is a challenging set of exercises for most students.
Ex. 43–44 deal with circular orbits.
Ex. 50–51 require students to work with the directrix of an ellipse and hyperbola in rectangular coordinates.

**Ongoing Assessment**

Self-Assessment: Ex. 13, 19, 27, 31
Embedded Assessment: Ex. 46–51

### Example 5   ANALYZING A PLANETARY ORBIT

Find a polar equation for the orbit of Mercury, and use it to approximate its aphelion (farthest distance from the sun) and perihelion (closest distance to the sun).

**Solution**  Setting $e = 0.2056$ and $a = 57.9$ in

$$r = \frac{a(1 - e^2)}{1 + e \cos \theta} \quad \text{yields} \quad r = \frac{57.9(1 - 0.2056^2)}{1 + 0.2056 \cos \theta}.$$

Mercury's aphelion is

$$r = \frac{57.9(1 - 0.2056^2)}{1 - 0.2056} \approx 69.8 \text{ Gm.}$$

Mercury's perihelion is

$$r = \frac{57.9(1 - 0.2056^2)}{1 + 0.2056} \approx 46.0 \text{ Gm.}$$

## Quick Review 8.5

In Exercises 1 and 2, solve for $r$.

**1.** $(3, \theta) = (r, \theta + \pi)$  $r = -3$

**2.** $(-2, \theta) = (r, \theta + \pi)$  $r = 2$

In Exercises 3 and 4, solve for $\theta$.

**3.** $(1.5, \pi/6) = (-1.5, \theta)$, $-2\pi \leq \theta \leq 2\pi$

**4.** $(-3, 4\pi/3) = (3, \theta)$, $-2\pi \leq \theta \leq 2\pi$  $\theta = \pi/3, \theta = -5\pi/3$

In Exercises 5 and 6, find the focus and directrix of the parabola.

**5.** $x^2 = 16y$

**6.** $y^2 = -12x$

In Exercises 7–10, find the foci and vertices of the conic.

**7.** $\dfrac{x^2}{9} + \dfrac{y^2}{4} = 1$

**8.** $\dfrac{x^2}{9} + \dfrac{y^2}{25} = 1$

**9.** $\dfrac{x^2}{16} - \dfrac{y^2}{9} = 1$

**10.** $\dfrac{y^2}{36} - \dfrac{x^2}{4} = 1$

# Section 8.5 Exercises

In Exercises 1–6, match the polar equation with its graph, and identify the viewing window.

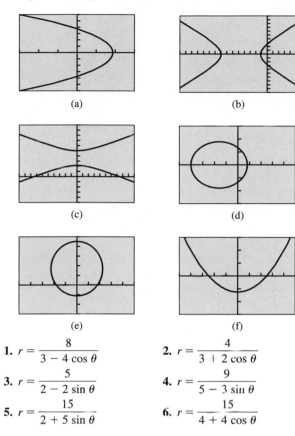

(a)

(b)

(c)

(d)

(e)

(f)

**1.** $r = \dfrac{8}{3 - 4\cos\theta}$

**2.** $r = \dfrac{4}{3 + 2\cos\theta}$

**3.** $r = \dfrac{5}{2 - 2\sin\theta}$

**4.** $r = \dfrac{9}{5 - 3\sin\theta}$

**5.** $r = \dfrac{15}{2 + 5\sin\theta}$

**6.** $r = \dfrac{15}{4 + 4\cos\theta}$

In Exercises 7–12, find a polar equation for the conic with a focus at the pole and the given eccentricity and directrix. Identify the conic, and graph it.

**7.** $e = 1,\ x = -2$

**8.** $e = 5/4,\ x = 4$

**9.** $e = 3/5,\ y = 4$

**10.** $e = 1,\ y = 2$

**11.** $e = 7/3,\ y = -1$

**12.** $e = 2/3,\ x = -5$

In Exercises 13–20, identify the conic without graphing it. State the eccentricity and directrix.

**13.** $r = \dfrac{2}{1 + \cos\theta}$

**14.** $r = \dfrac{6}{1 + 2\cos\theta}$

**15.** $r = \dfrac{5}{2 - 2\sin\theta}$

**16.** $r = \dfrac{2}{4 - \cos\theta}$

**17.** $r = \dfrac{20}{6 + 5\sin\theta}$

**18.** $r = \dfrac{42}{2 - 7\sin\theta}$

**19.** $r = \dfrac{6}{5 + 2\cos\theta}$

**20.** $r = \dfrac{20}{2 + 5\sin\theta}$

In Exercises 21–24, find a polar equation for the ellipse with a focus at the pole and the given polar coordinates as the endpoints of its major axis.

**21.** $(1.5, 0)$ and $(6, \pi)$

**22.** $(1.5, 0)$ and $(1, \pi)$

**23.** $(1, \pi/2)$ and $(3, 3\pi/2)$

**24.** $(3, \pi/2)$ and $(0.75, -\pi/2)$

In Exercises 25–28, find a polar equation for the hyperbola with a focus at the pole and the given polar coordinates as the endpoints of its transverse axis.

**25.** $(3, 0)$ and $(-15, \pi)$

**26.** $(-3, 0)$ and $(1.5, \pi)$

**27.** $\left(2.4, \dfrac{\pi}{2}\right)$ and $\left(-12, \dfrac{3\pi}{2}\right)$

**28.** $\left(-6, \dfrac{\pi}{2}\right)$ and $\left(2, \dfrac{3\pi}{2}\right)$

In Exercises 29 and 30, find a polar equation for the conic with a focus at the pole.

**29.**

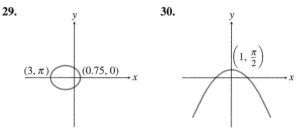

**30.**

In Exercises 31–36, graph the conic, and find the values of $e$, $a$, $b$, and $c$.

**31.** $r = \dfrac{21}{5 - 2\cos\theta}$

**32.** $r = \dfrac{11}{6 - 5\sin\theta}$

**33.** $r = \dfrac{24}{4 + 2\sin\theta}$

**34.** $r = \dfrac{16}{5 + 3\cos\theta}$

**35.** $r = \dfrac{16}{3 + 5\cos\theta}$

**36.** $r = \dfrac{12}{1 - 5\sin\theta}$

In Exercises 37 and 38, determine a Cartesian equation for the given polar equation.

**37.** $r = \dfrac{4}{2 - \sin\theta}$

**38.** $r = \dfrac{6}{1 + 2\cos\theta}$

In Exercises 39 and 40, use the fact that $k = 2p$ is twice the focal length and half the focal width, to determine a Cartesian equation of the parabola whose polar equation is given.

**39.** $r = \dfrac{4}{2 - 2\cos\theta}$

**40.** $r = \dfrac{12}{3 + 3\cos\theta}$

**41. Halley's Comet** The orbit of Halley's comet has a semimajor axis of 18.09 AU and an orbital eccentricity of 0.97. Compute its perihelion and aphelion.

**42. Uranus** The orbit of the planet Uranus has a semimajor axis of 19.18 AU and an orbital eccentricity of 0.0461. Compute its perihelion and aphelion.

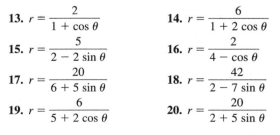

In Exercises 43 and 44, the velocity of an object traveling in a circular orbit of radius $r$ (distance from center of planet in meters) around a planet is given by

$$v = \sqrt{\frac{3.99 \times 10^{14} k}{r}} \text{ m/sec,}$$

where $k$ is a constant related to the masses of the planet and the orbiting object.

**43. Group Activity Lunar Module** A lunar excursion module is in a circular orbit 250 km above the surface of the Moon. Assume that the Moon's radius is 1740 km and that $k = 0.012$. Find the following.

**(a)** The velocity of the lunar module

**(b)** The length of time required for the lunar module to circle the moon once

**44. Group Activity Mars Satellite** A satellite is in a circular orbit 1000 mi above Mars. Assume that the radius of Mars is 2100 mi and that $k = 0.11$. Find the velocity of the satellite.

## Explorations

**45. Planetary Orbits** Use the equation $r = a(1 - e^2)/(1 + e \cos \theta)$ to prove the perihelion of any planet is $a(1 - e)$ and the aphelion is $a(1 + e)$.

**(a)** Confirm that $a(1 - e) = a - c$ and $a(1 + e) = a + c$.

**(b)** Use the formulas for $a(1 - e)$ and $a(1 + e)$ to compute the perihelion and the aphelion of each planet listed in Table 8.4.

**(c)** For which of these planets is the difference between the perihelion and the aphelion the greatest? What does this mean astronomically?

**Table 8.4 Semimajor Axes and Eccentricities of the Six Innermost Planets**

| Planet | Semimajor Axis (AU) | Eccentricity |
|---|---|---|
| Mercury | 0.3871 | 0.206 |
| Venus | 0.7233 | 0.007 |
| Earth | 1.0000 | 0.017 |
| Mars | 1.5237 | 0.093 |
| Jupiter | 5.2026 | 0.048 |
| Saturn | 9.5547 | 0.056 |

*Source: Encrenaz & Bibring. The Solar System (2nd ed.). New York: Springer, p. 5.*

## Extending the Ideas

**46. Revisiting Figure 8.40** In Figure 8.40, if $r < 0$ or $r \cos \theta > k$, then we must use $PD = |k - r \cos \theta|$ and $PF = |r|$. Prove that, even in these cases, the resulting equation is still $r = ke/(1 + e \cos \theta)$.

**47. Deriving Other Forms of the Polar Equation for Conics** Using Figure 8.40 as a guide draw an appropriate diagram for and derive the equation.

**(a)** $r = \dfrac{ke}{1 - e \cos \theta}$

**(b)** $r = \dfrac{ke}{1 + e \sin \theta}$

**(c)** $r = \dfrac{ke}{1 - e \sin \theta}$

**48. Focal Widths** Using polar equations, derive formulas for the focal width of an ellipse and the focal width of a hyperbola. Begin by defining focal width for these conics in a manner analogous to the definition of the focal width of a parabola given in Section 8.1.

**49.** Prove that for a hyperbola the formula $r = ke/(1 - e \cos \theta)$ is equivalent to $r = a(e^2 - 1)/(1 - e \cos \theta)$, where $a$ is the semitransverse axis of the hyperbola.

**50. Connecting Polar to Rectangular** Consider the ellipse

$$\frac{x^2}{a^2} + \frac{y^2}{b^2} = 1,$$

where half the length of the major axis is $a$, and the foci are $(\pm c, 0)$ such that $c^2 = a^2 - b^2$. Let $L$ be the vertical line $x = a^2/c$.

**(a)** Prove that $L$ is a directrix for the ellipse. (*Hint*: Prove that $PF/PD$ is the constant $c/a$, where $P$ is a point on the ellipse, and $D$ is the point on $L$ such that $PD$ is perpendicular to $L$.)

**(b)** Prove that the eccentricity is $e = c/a$.

**(c)** Prove that the distance from $F$ to $L$ is $a/e - ea$.

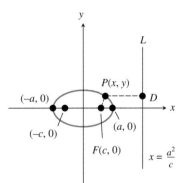

**51. Connecting Polar to Rectangular** Consider the hyperbola

$$\frac{x^2}{a^2} - \frac{y^2}{b^2} = 1,$$

where half the length of the transverse axis is $a$, and the foci are $(\pm c, 0)$ such that $c^2 = a^2 + b^2$. Let $L$ be the vertical line $x = a^2/c$.

**(a)** Prove that $L$ is a directrix for the hyperbola. (*Hint*: Prove that $PF/PD$ is the constant $c/a$, where $P$ is a point on the hyperbola, and $D$ is the point on $L$ such that $PD$ is perpendicular to $L$.)

**(b)** Prove that the eccentricity is $e = c/a$.

**(c)** Prove that the distance from $F$ to $L$ is $ea - \dfrac{a}{e}$.

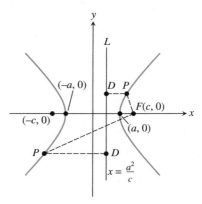

---

<table>
<tr><td>**8.6**</td><td></td></tr>
</table>

# Three-Dimensional Cartesian Coordinate System

Three-Dimensional Cartesian Coordinates • Distance and Midpoint Formulas • Equation of a Sphere • Planes and Other Surfaces • Vectors in Space • Lines in Space

## Three-Dimensional Cartesian Coordinates

In Sections P.2 and P.4 we studied Cartesian coordinates and the associated basic formulas and equations for the two-dimensional plane; we now extend these ideas to three-dimensional space. In the plane, we used two axes and ordered pairs to name points; in space, we use three mutually perpendicular axes and ordered triples of real numbers to name points. See Figure 8.46.

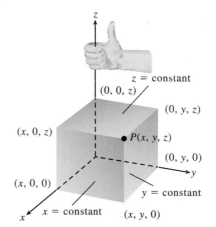

**Figure 8.46** The point $P(x, y, z)$ in Cartesian space.

Notice that Figure 8.46 exhibits several important features of the three-dimensional Cartesian coordinate system:

- The axes are labeled $x$, $y$, and $z$, and these three **coordinate axes** form a **right-handed coordinate frame**: When you hold your right hand with fingers curving from the positive $x$-axis toward the positive $y$-axis, your thumb points in the direction of the positive $z$-axis.

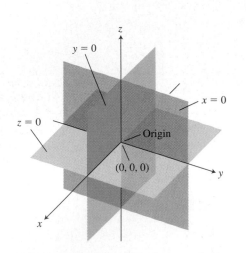

**Figure 8.47** The coordinate planes divide space into eight octants.

- A point $P$ in space is uniquely paired with an order triple $(x, y, z)$ of real numbers. The numbers $x$, $y$, and $z$ are the **Cartesian coordinates of $P$**.

- Points on the coordinate axes have the form $(x, 0, 0)$, $(0, y, 0)$, or $(0, 0, z)$, with $(x, 0, 0)$ on the $x$-axis, $(0, y, 0)$ on the $y$-axis, and $(0, 0, z)$ on the $z$-axis.

In Figure 8.47, the coordinate axes are paired to determine the **coordinate planes**:

- The coordinate planes are the **$xy$-plane**, the **$xz$-plane**, and the **$yz$-plane**, and have equations $z = 0$, $y = 0$, and $x = 0$, respectively.

- Points on the coordinate planes have the form $(x, y, 0)$, $(y, 0, z)$, or $(0, y, z)$, with $(x, y, 0)$ on the $xy$-plane, $(x, 0, z)$ on the $xz$-plane, and $(0, y, z)$ on the $yz$-plane.

- The coordinate planes meet at the **origin,** $(0, 0, 0)$.

- The coordinate planes divide space into eight regions called **octants**, with the **first octant** containing all points in space with three nonnegative coordinates.

### Example 1    LOCATING A POINT IN CARTESIAN SPACE

Draw a sketch that shows the point $(2, 3, 5)$.

**Solution**  To locate the point $(2, 3, 5)$, we first sketch a right-handed three-dimensional coordinate frame. We then draw the planes $x = 2$, $y = 3$, and $z = 5$, which parallel the coordinate planes $x = 0$, $y = 0$, and $z = 0$, respectively. The point $(2, 3, 5)$ lies at the intersection of the planes $x = 2$, $y = 3$, and $z = 5$, as shown in Figure 8.48.

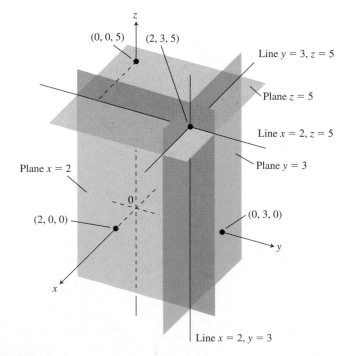

**Figure 8.48** The planes $x = 2$, $y = 3$, and $z = 5$ determine the point $(2, 3, 5)$. (Example 1)

## Distance and Midpoint Formulas

The distance and midpoint formulas for space are natural generalizations of the corresponding formulas for the plane.

---

**Distance Formula (Cartesian Space)**

The distance $d(P, Q)$ between the points $P(x_1, y_1, z_1)$ and $Q(x_2, y_2, z_2)$ in space is

$$d(P, Q) = \sqrt{(x_1 - x_2)^2 + (y_1 - y_2)^2 + (z_1 - z_2)^2}.$$

---

Just as in the plane, the coordinates of the midpoint of a line segment are the averages for the coordinates of the endpoints of the segment.

---

**Midpoint Formula (Cartesian Space)**

The midpoint $M$ of the line segment $PQ$ with endpoints $P(x_1, y_1, z_1)$ and $Q(x_2, y_2, z_2)$ is

$$M = \left( \frac{x_1 + x_2}{2}, \frac{y_1 + y_2}{2}, \frac{z_1 + z_2}{2} \right).$$

---

**Example 2**  CALCULATING A DISTANCE AND FINDING A MIDPOINT

Find the distance between the points $P(-2, 3, 1)$ and $Q(4, -1, 5)$, and find the midpoint of line segment $PQ$.

**Solution**  The distance is given by

$$d(P, Q) = \sqrt{(-2 - 4)^2 + (3 + 1)^2 + (1 - 5)^2}$$
$$= \sqrt{36 + 16 + 16}$$
$$= \sqrt{68} \approx 8.25$$

The midpoint is

$$M = \left( \frac{-2 + 4}{2}, \frac{3 - 1}{2}, \frac{1 + 5}{2} \right) = (1, 1, 3).$$

## Equation of a Sphere

A *sphere* is the three-dimensional analogue of a circle: In space, the set of points that lie a fixed distance from a fixed point is a **sphere**. The fixed distance is the **radius**, and the fixed point is the **center** of the sphere. The point $P(x, y, z)$ is a point of the sphere with center $(h, k, l)$ and radius $r$ if and only if

$$\sqrt{(x - h)^2 + (y - k)^2 + (z - l)^2} = r.$$

Squaring both sides gives the standard equation shown on page 655.

**Teaching Note**

Many of the results discussed in this chapter are fairly intuitive extensions of previously studied two-dimensional results. You may wish to present a comparison between the two-dimensional results and their three-dimensional counterparts.

**Drawing Lesson**

## How to Draw Three-Dimensional Objects to Look Three-Dimensional

1. Make the angle between the positive *x*-axis and the positive *y*-axis large enough.

2. Break lines. When one line passes behind another, break it to show that it doesn't touch and that part of it is hidden.

3. Dash or omit hidden portions of lines. Don't let the line touch the boundary of the parallelogram that represents the plane, unless the line lies in the plane.

4. Spheres: Draw the sphere first (outline and equator); draw axes, if any, later. Use line breaks and dashed lines.

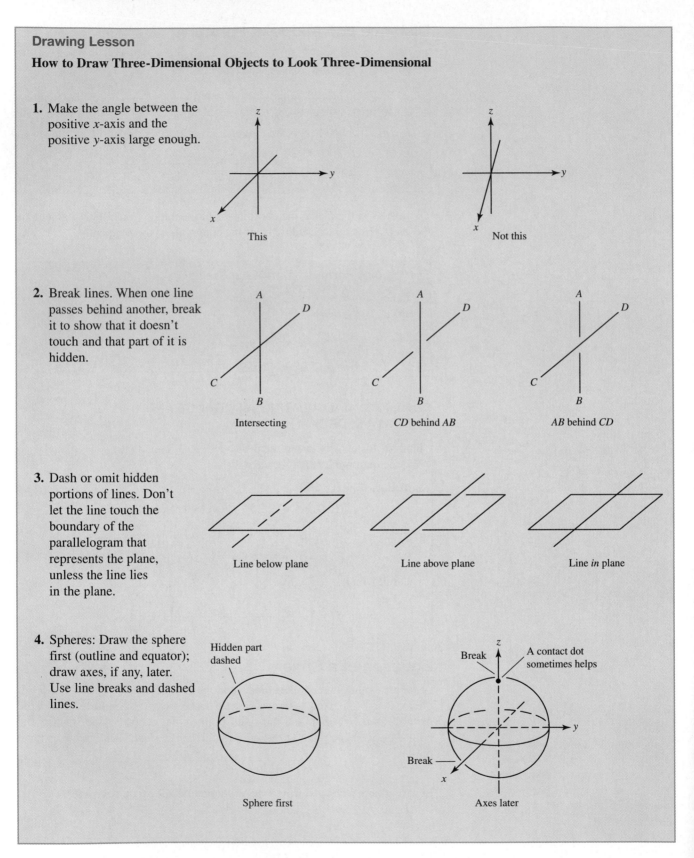

> ### Standard Equation of a Sphere
>
> A point $P(x, y, z)$ is on the sphere with center $(h, k, l)$ and radius $r$ if and only if
>
> $$(x - h)^2 + (y - k)^2 + (z - l)^2 = r^2.$$

**Example 3  FINDING THE STANDARD EQUATION OF A SPHERE**

The standard equation of the sphere with center $(2, 0, -3)$ and radius 7 is

$$(x - 2)^2 + y^2 + (z + 3)^2 = 49.$$

## Planes and Other Surfaces

In Section P.4 we learned that every line in the Cartesian plane can be written as a linear (first-degree) equation in two variables; that is, every line can be written as

$$Ax + By + C = 0,$$

where $A$ and $B$ are not both zero. Conversely, every linear equation in two variables represents a line in the Cartesian plane.

In an analogous way, every **plane** in Cartesian space can be written as a **linear (first-degree) equation in three variables**; that is, every plane can be written as

$$Ax + By + Cz + D = 0,$$

where $A$, $B$, and $C$ are not all zero. Conversely, every linear equation in three variables represents a plane in Cartesian space.

**Example 4  SKETCHING A PLANE IN SPACE**

Sketch the graph of $12x + 15y + 20z = 60$.

**Solution**  Because this is a linear equation, its graph is a plane. Three points determine a plane. To find three points, we first divide both sides of $12x + 15y + 20z = 60$ by 60:

$$\frac{x}{5} + \frac{y}{4} + \frac{z}{3} = 1$$

In this form, it is easy to see that the points $(5, 0, 0)$, $(0, 4, 0)$, and $(0, 0, 3)$ satisfy the equation. These are the points where the graph crosses a coordinate axis. Figure 8.49 shows the completed sketch.

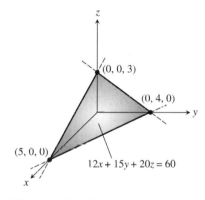

**Figure 8.49** The intercepts $(5, 0, 0)$, $(0, 4, 0)$, and $(0, 0, 3)$ determine the plane $12x + 15y + 20z = 60$. (Example 4)

Equations in the three variables $x$, $y$, and $z$ generally graph as surfaces in three-dimensional space. Just as in the plane, second-degree equations are of particular interest. Recall that second-degree equations in two variables yield conic sections in the Cartesian plane. In space, second-degree equations in *three* variables yield **quadric surfaces**: The paraboloids, ellipsoids, and hyperboloids of revolution that have special reflective properties are all quadric surfaces, as are such exotic-sounding surfaces as hyperbolic paraboloids and elliptic hyperboloids.

Other surfaces of interest include graphs of **functions of two variables**, whose equations have the form $z = f(x, y)$. Some examples are $z = x \ln y$, $z = \sin(xy)$, and $z = \sqrt{1 - x^2 - y^2}$. The last equation graphs as a hemisphere (see Exercise 53). Equations of the form $z = f(x, y)$ can be graphed using some graphing calculators and most computer algebra software. Quadric surfaces and functions of two variables are studied in most university-level calculus course sequences.

## Vectors in Space

In space, just as in the plane, the sets of equivalent directed line segments are called *vectors*. They are used to represent forces, displacements, and velocities in three dimensions. In space, we use ordered triples to denote vectors:

$$\mathbf{v} = \langle v_1, v_2, v_3 \rangle.$$

The **zero vector** is $\mathbf{0} = \langle 0, 0, 0 \rangle$, and the **standard unit vectors** are $\mathbf{i} = \langle 1, 0, 0 \rangle$, $\mathbf{j} = \langle 0, 1, 0 \rangle$, and $\mathbf{k} = \langle 0, 0, 1 \rangle$. As shown in Figure 8.50, the vector $\mathbf{v}$ can be expressed in terms of these standard unit vectors:

$$\mathbf{v} = \langle v_1, v_2, v_3 \rangle = v_1\mathbf{i} + v_2\mathbf{j} + v_3\mathbf{k}.$$

The vector $\mathbf{v}$ that is represented by the directed line segment $\overrightarrow{PQ}$ from $P(a, b, c)$ to $Q(x, y, z)$ is

$$\mathbf{v} = \overrightarrow{PQ} = \langle x - a, y - b, z - c \rangle = (x - a)\mathbf{i} + (y - b)\mathbf{j} + (z - c)\mathbf{k}.$$

A vector $\mathbf{v} = \langle v_1, v_2, v_3 \rangle$ can be multiplied by a scalar (real number) $c$ as follows:

$$c\mathbf{v} = c\langle v_1, v_2, v_3 \rangle = \langle cv_1, cv_2, cv_3 \rangle.$$

Many other properties of vectors generalize in a natural way when we move from two to three dimensions:

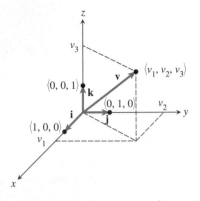

**Figure 8.50** The vector $\mathbf{v} = \langle v_1, v_2, v_3 \rangle$.

---

### Vector Relationships in Space

For vectors $\mathbf{v} = \langle v_1, v_2, v_3 \rangle$ and $\mathbf{w} = \langle w_1, w_2, w_3 \rangle$,

- **Equality:** $\mathbf{v} = \mathbf{w}$ if and only if $v_1 = w_1$, $v_2 = w_2$, and $v_3 = w_3$.
- **Addition:** $\mathbf{v} + \mathbf{w} = \langle v_1 + w_1, v_2 + w_2, v_3 + w_3 \rangle$
- **Subtraction:** $\mathbf{v} - \mathbf{w} = \langle v_1 - w_1, v_2 - w_2, v_3 - w_3 \rangle$
- **Magnitude:** $|\mathbf{v}| = \sqrt{v_1^2 + v_2^2 + v_3^2}$
- **Dot product:** $\mathbf{v} \cdot \mathbf{w} = v_1w_1 + v_2w_2 + v_3w_3$
- **Unit vector:** $\mathbf{u} = \mathbf{v}/|\mathbf{v}|$, $\mathbf{v} \neq \mathbf{0}$, is the unit vector in the direction of $\mathbf{v}$.

---

### Example 5   COMPUTING WITH VECTORS

**(a)** $3\langle -2, 1, 4 \rangle = \langle 3 \cdot -2, 3 \cdot 1, 3 \cdot 4 \rangle = \langle -6, 3, 12 \rangle$.

**(b)** $\langle 0, 6, -7 \rangle + \langle -5, 5, 8 \rangle = \langle 0 - 5, 6 + 5, -7 + 8 \rangle = \langle -5, 11, 1 \rangle$.

**(c)** $\langle 1, -3, 4 \rangle - \langle -2, -4, 5 \rangle = \langle 1 + 2, -3 + 4, 4 - 5 \rangle = \langle 3, 1, -1 \rangle$.

**(d)** $|\langle 2, 0, -6 \rangle| = \sqrt{2^2 + 0^2 + 6^2} = \sqrt{40} \approx 6.32$.

**(e)** $\langle 5, 3, -1 \rangle \cdot \langle -6, 2, 3 \rangle = 5 \cdot (-6) + 3 \cdot 2 + (-1) \cdot 3$
$$= -30 + 6 - 3 = -27.$$

### Example 6 USING VECTORS IN SPACE

A jet airplane just after takeoff is pointed due east. Its air velocity vector makes an angle of 30° with flat ground with an airspeed of 250 mph. If the wind is out of the southeast at 32 mph, calculate a vector that represents the plane's velocity relative to the point of takeoff.

**Solution** Let **i** point east, **j** point north, and **k** point up. The plane's air velocity is

$$\mathbf{a} = \langle 250 \cos 30°, 0, 250 \sin 30° \rangle \approx \langle 216.506, 0, 125 \rangle,$$

and the wind velocity, which is pointing northwest, is

$$\mathbf{w} = \langle 32 \cos 135°, 32 \sin 135°, 0 \rangle \approx \langle -22.627, 22.627, 0 \rangle.$$

The velocity relative to the ground is $\mathbf{v} = \mathbf{a} + \mathbf{w}$, so

$$\mathbf{v} \approx \langle 216.506, 0, 125 \rangle + \langle -22.627, 22.627, 0 \rangle$$

$$\approx \langle 193.88, 22.63, 125 \rangle$$

$$= 193.88\mathbf{i} + 22.63\mathbf{j} + 125\mathbf{k}$$

In Exercise 54 you will be asked to interpret the meaning of the velocity vector obtained in Example 6.

## Lines in Space

We have seen that linear equations in three variables graph as planes in space. So how do we get lines? The answer is, several ways. First notice that to specify the $x$-axis, which is a line, we could use the two linear equations $y = 0$ and $z = 0$. As alternatives to using a pair of linear Cartesian equations, we can specify any line in space using

- one vector equation, or

- a set of three parametric equations.

Suppose $\ell$ is a line through the point $P_0(x_0, y_0, z_0)$ and in the direction of a nonzero vector $\mathbf{v} = \langle a, b, c \rangle$ (Figure 8.51). Then for any point $P(x, y, z)$ on $\ell$,

$$\overrightarrow{P_0P} = t\mathbf{v}$$

for some real number $t$. The vector **v** is a **direction vector** for line $\ell$. If $\mathbf{r} = \overrightarrow{OP} = \langle x, y, z \rangle$ and $\mathbf{r}_0 = \overrightarrow{OP_0} = \langle x_0, y_0, z_0 \rangle$, then $\mathbf{r} - \mathbf{r}_0 = t\mathbf{v}$. So the vector equation of the line $\ell$ is

$$\mathbf{r} = \mathbf{r}_0 + t\mathbf{v}.$$

Writing this in component form yields

$$\langle x, y, z \rangle = \langle x_0, y_0, z_0 \rangle + t\langle a, b, c \rangle$$

$$= \langle x_0 + at, y_0 + bt, z_0 + ct \rangle,$$

which can be expressed as the parametric equations

$$x = x_0 + at, \quad y = y_0 + bt, \quad \text{and} \quad z = z_0 + ct.$$

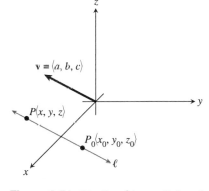

**Figure 8.51** The line $\ell$ is parallel to the direction vector $\mathbf{v} = \langle a, b, c \rangle$.

> ### Equations for a Line in Space
>
> If $\ell$ is a line through the point $P_0(x_0, y_0, z_0)$ in the direction of a nonzero vector $\mathbf{v} = \langle a, b, c \rangle$, then a point $P(x, y, z)$ is on $\ell$ if and only if
>
> - **Vector form:** $\quad \mathbf{r} = \mathbf{r}_0 + t\mathbf{v}$, where $\mathbf{r} = \langle x, y, z \rangle$ and $\mathbf{r}_0 = \langle x_0, y_0, z_0 \rangle$; or
> - **Parametric form:** $\quad x = x_0 + at$, $y = y_0 + bt$, and $z = z_0 + ct$, where $t$ is a real number.

### Example 7 FINDING EQUATIONS FOR A LINE

The line through $P_0(4, 3, -1)$ with direction vector $\mathbf{v} = \langle -2, 2, 7 \rangle$ can be written

- in vector form as $\quad \mathbf{r} = \langle 4, 3, -1 \rangle + t\langle -2, 2, 7 \rangle$; or
- in parametric form as $\quad x = 4 - 2t$, $y = 3 + 2t$, and $z = -1 + 7t$.

### Example 8 FINDING EQUATIONS FOR A LINE

Using the standard unit vectors $\mathbf{i}$, $\mathbf{j}$, and $\mathbf{k}$, write a vector equation for the line containing the points $A(3, 0, -2)$ and $B(-1, 2, -5)$, and compare it to the parametric equations for the line.

**Solution** The line is in the direction of

$$\mathbf{v} = \overrightarrow{AB} = \langle -1 - 3, 2 - 0, -5 + 2 \rangle = \langle -4, 2, -3 \rangle.$$

So using $\mathbf{r}_0 = \overrightarrow{OA}$, the vector equation of the line becomes:

$$\mathbf{r} = \mathbf{r}_0 + t\mathbf{v}$$
$$\langle x, y, z \rangle = \langle 3, 0, -2 \rangle + t\langle -4, 2, -3 \rangle$$
$$\langle x, y, z \rangle = \langle 3 - 4t, 2t, -2 - 3t \rangle$$
$$x\mathbf{i} + y\mathbf{j} + z\mathbf{k} = (3 - 4t)\mathbf{i} + 2t\mathbf{j} + (-2 - 3t)\mathbf{k}$$

The parametric equations are the three component equations

$$x = 3 - 4t, \quad y = 2t, \quad \text{and} \quad z = -2 - 3t.$$

**Follow-up**

Ask students to give alternate vector and parametric forms for the line in Example 7.

**Assignment Guide**

Day 1: Ex. 1–11 odds
Day 2: Ex. 13–21 odds
Day 3: Ex. 33–54, multiples of 3

**Cooperative Learning**

Group Activity: Ex. 53

**Notes on Exercises**

Ex. 49, 51, and 52 are proof exercises.
Ex. 55–58 introduce the cross product of two vectors.

**Ongoing Assessment**

Self-Assessment: Ex. 5, 11, 13, 17, 25, 37, 47
Embedded Assessment: Ex. 43–48, 53

## Quick Review 8.6

In Exercises 1–3, let $P(x, y)$ and $Q(2, -3)$ be points in the $xy$-plane.

**1.** Compute the distance between $P$ and $Q$.

**2.** Find the midpoint of the line segment $PQ$.

**3.** If $P$ is 5 units from $Q$, describe the position of $P$.

In Exercises 4–6, let $\mathbf{v} = \langle -4, 5 \rangle = -4\mathbf{i} + 5\mathbf{j}$ be a vector in the $xy$-plane.

**4.** Find the magnitude of $\mathbf{v}$. $\sqrt{41}$

**5.** Find a unit vector in the direction of $\mathbf{v}$.

**6.** Find a vector 7 units long in the direction of $-\mathbf{v}$.

**7.** Give a geometric description of the graph of $(x + 1)^2 + (y - 5)^2 = 25$ in the $xy$-plane.

**8.** Give a geometric description of the graph of $x = 2 - t$, $y = -4 + 2t$ in the $xy$-plane.

**9.** Find the center and radius of the circle $x^2 + y^2 + 2x - 6y + 6 = 0$ in the $xy$-plane.

**10.** Find a vector from $P(2, 5)$ to $Q(-1, -4)$ in the $xy$-plane.

# Section 8.6 Exercises

In Exercises 1–4, draw a sketch that shows the point.

**1.** $(3, 4, 2)$

**2.** $(2, -3, 6)$

**3.** $(1, -2, -4)$

**4.** $(-2, 3, -5)$

In Exercises 5–8, compute the distance between the points.

**5.** $(-1, 2, 5), (3, -4, 6)$   $\sqrt{53}$

**6.** $(2, -1, -8), (6, -3, 4)$   $2\sqrt{41}$

**7.** $(a, b, c), (1, -3, 2)$   $\sqrt{(a-1)^2 + (b+3)^2 + (c-2)^2}$

**8.** $(x, y, z), (p, q, r)$   $\sqrt{(x-p)^2 + (y-q)^2 + (z-r)^2}$

In Exercises 9–12, find the midpoint of the segment $PQ$.

**9.** $P(-1, 2, 5), Q(3, -4, 6)$   $(1, -1, 11/2)$

**10.** $P(2, -1, -8), Q(6, -3, 4)$   $(4, -2, -2)$

**11.** $P(2x, 2y, 2z), Q(-2, 8, 6)$   $(x - 1, y + 4, z + 3)$

**12.** $P(-a, -b, -c), Q(3a, 3b, 3c)$   $(a, b, c)$

In Exercises 13–16, write an equation for the sphere with the given point as its center and the given number as its radius.

**13.** $(5, -1, -2), 8$

**14.** $(-1, 5, 8), \sqrt{5}$

**15.** $(1, -3, 2), \sqrt{a}, a > 0$

**16.** $(p, q, r), 6$

In Exercises 17–22, sketch a graph of the equation. Label all intercepts.

**17.** $x + y + 3z = 9$

**18.** $x + y - 2z = 8$

**19.** $x + z = 3$

**20.** $2y + z = 6$

**21.** $x - 3y = 6$

**22.** $x = 3$

In Exercises 23–32, evaluate the expression using $\mathbf{r} = \langle 1, 0, -3 \rangle$, $\mathbf{v} = \langle -3, 4, -5 \rangle$, and $\mathbf{w} = \langle 4, -3, 12 \rangle$.

**23.** $\mathbf{r} + \mathbf{v}$   $\langle -2, 4, -8 \rangle$

**24.** $\mathbf{r} - \mathbf{w}$   $\langle -3, 3, -15 \rangle$

**25.** $\mathbf{v} \cdot \mathbf{w}$   $-84$

**26.** $|\mathbf{w}|$   $13$

**27.** $\mathbf{r} \cdot (\mathbf{v} + \mathbf{w})$   $-20$

**28.** $(\mathbf{r} \cdot \mathbf{v}) + (\mathbf{r} \cdot \mathbf{w})$   $-20$

**29.** $\mathbf{w}/|\mathbf{w}|$

**30.** $\mathbf{i} \cdot \mathbf{r}$   $1$

**31.** $\langle \mathbf{i} \cdot \mathbf{v}, \mathbf{j} \cdot \mathbf{v}, \mathbf{k} \cdot \mathbf{v} \rangle$

**32.** $(\mathbf{r} \cdot \mathbf{v})\mathbf{w}$   $\langle 48, -36, 144 \rangle$

In Exercises 33–42, use the points $A(-1, 2, 4), B(0, 6, -3)$, and $C(2, -4, 1)$.

**33.** Find the distance from $A$ to the midpoint of $BC$.   $\sqrt{30}$

**34.** Find the vector from $A$ to the midpoint of $BC$.

**35.** Write a vector equation of the line through $A$ and $B$.

**36.** Write a vector equation of the line through $A$ and the midpoint of $BC$.   $r = \langle -1, 2, 4 \rangle + t\langle 2, -1, -5 \rangle$

**37.** Write parametric equations for the line through $A$ and $C$.

**38.** Write parametric equations for the line through $B$ and $C$.

**39.** Write parametric equations for the line through $B$ and the midpoint of $AC$.

**40.** Write parametric equations for the line through $C$ and the midpoint of $AB$.

**41.** Is $\triangle ABC$ equilateral, isosceles, or scalene?   scalene

**42.** If $M$ is the midpoint of $BC$, what is the midpoint of $AM$?

In Exercises 43–48, **(a)** sketch the line defined by the pair of equations, and **(b) Writing to Learn** give a geometric description of the line, including its direction and its position relative to the coordinate frame.

**43.** $x = 0, y = 0$

**44.** $x = 0, z = 2$

**45.** $x = -3, y = 0$

**46.** $y = 1, z = 3$

**47.** $y = 2x + 1, z = 3$

**48.** $x = 0, z = 3y$

**49.** Write a vector equation for the line through the distinct points $P(x_1, y_1, z_1)$ and $Q(x_2, y_2, z_2)$.

**50.** Write parametric equations for the line through the distinct points $P(x_1, y_1, z_1)$ and $Q(x_2, y_2, z_2)$.

**51.** Draw a sketch of the points $P(x_1, y_1, z_1)$ and $Q(x_2, y_2, z_2)$, and use the distance formula for points in the plane to prove $d(P, Q) = \sqrt{(x_1 - x_2)^2 + (y_1 - y_2)^2 + (z_1 - z_2)^2}$.

**52.** Referring to the theorem in Section 6.2 (page 491), prove the corresponding statement for the angle $\theta$ between nonzero three-dimensional vectors $\mathbf{u}$ and $\mathbf{v}$; that is, prove

$$\cos \theta = \frac{\mathbf{u} \cdot \mathbf{v}}{|\mathbf{u}||\mathbf{v}|}.$$

## Exploration

**53. Group Activity  Writing to Learn**  The figure shows a graph of the ellipsoid $x^2/9 + y^2/4 + z^2/16 = 1$ drawn in a *box* using *Mathematica* computer software.

**(a)** Describe its cross sections in each of the three coordinate planes, that is, for $z = 0$, $y = 0$, and $x = 0$. In your description, include the name of each cross section and its position relative to the coordinate frame.

**(b)** Explain algebraically why the graph of $z = \sqrt{1 - x^2 - y^2}$ is half of a sphere. What is the equation of the related whole sphere?

**(c)** By hand, sketch the graph of the hemisphere $z = \sqrt{1 - x^2 - y^2}$. Check your sketch using a 3D grapher if you have access to one.

**(d)** Explain how the graph of an ellipsoid is related to the graph of a sphere and why a sphere is a *degenerate* ellipsoid.

## Extending the Ideas

The **cross product u $\times$ v** of the vectors $\mathbf{u} = u_1\mathbf{i} + u_2\mathbf{j} + u_3\mathbf{k}$ and $\mathbf{v} = v_1\mathbf{i} + v_2\mathbf{j} + v_3\mathbf{k}$ is

$$\mathbf{u} \times \mathbf{v} = \begin{vmatrix} \mathbf{i} & \mathbf{j} & \mathbf{k} \\ u_1 & u_2 & u_3 \\ v_1 & v_2 & v_3 \end{vmatrix}$$

$$= (u_2v_3 - u_3v_2)\mathbf{i} + (u_3v_1 - u_1v_3)\mathbf{j} + (u_1v_2 - u_2v_1)\mathbf{k}.$$

Use this definition in Exercises 55–58.

**55.** $\langle 1, -2, 3 \rangle \times \langle -2, 1, -1 \rangle$  $\langle -1, -5, -3 \rangle$

**56.** $\langle 4, -1, 2 \rangle \times \langle 1, -3, 2 \rangle$  $\langle 4, -6, -11 \rangle$

**57.** Prove that $\mathbf{i} \times \mathbf{j} = \mathbf{k}$.

**58.** Use the result of Exercise 52 to prove that $\mathbf{u} \times \mathbf{v}$ is perpendicular to both $\mathbf{u}$ and $\mathbf{v}$ if they are nonzero.

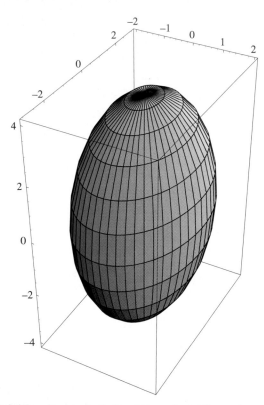

**54. Revisiting Example 6**  Read Example 6. Then using $\mathbf{v} \approx 193.879\mathbf{i} + 22.627\mathbf{j} + 125\mathbf{k}$, establish the following:

**(a)** The plane's compass bearing is 83.34°.

**(b)** The plane is climbing at an angle of 32.63°.

**(c)** Its speed downrange (that is, ignoring the vertical component) is 195.2 mph.

**(d)** The plane's overall speed is 231.8 mph.  ■

## Chapter 8   Key Ideas

### Concepts

Cartesian coordinates of a point (p. 652)
center of a hyperbola (p. 625)
center of a sphere (p. 653)
center of an ellipse (p. 614)
chord of a conic (p. 606, 615, 625)
conic (p. 603)
conic section (p. 603, 642)
conjugate axis of a hyperbola (p. 626)
coordinate axes (p. 651)
coordinate planes (p. 652)
cross product terms (p. 635)
degenerate conic sections (p. 603)
direction vector (p. 657)
directrix of a conic (p. 642)
discriminant of a second-degree equation
    (p. 638)
double-napped right circular cone
    (p. 603)
eccentricity of a conic (p. 642)
ellipse (p. 614, 643)
ellipsoid of revolution (p. 620)
first octant (p. 652)

focal axis of a conic (p. 642)
focal chord of a parabola (p. 613)
focal length of a parabola (p. 606)
focal width of a parabola (p. 606)
focus (foci) of a conic (p. 642)
functions of two variables (p. 655)
hyperbola (p. 625, 643)
hyperboloid of revolution (p. 631)
invariant under rotation (p. 640)
latus rectum of a parabola (p. 613)
linear equation in three variables (p. 615)
major axis of an ellipse (p. 615)
minor axis of an ellipse (p. 652)
nappes of a cone (p. 603)
octants (p. 652)
origin (p. 652)
parabola (p. 605, 643)
paraboloid of revolution (p. 610)
plane in Cartesian space (p. 655)
quadric surfaces (p. 655)
radius of a sphere (p. 653)
right-handed coordinate frame (p. 651)

rotation of axes (p. 637)
second-degree equation in two variables
    (p. 604)
semiconjugate axis of a hyperbola
    (p. 626)
semimajor axis of an ellipse (p. 615)
semiminor axis of an ellipse (p. 615)
semitransverse axis of a hyperbola
    (p. 626)
sphere (p. 653)
standard unit vectors (p. 656)
tangent line of a parabola (p. 613)
translation of axes (p. 636)
transverse axis of a hyperbola (p. 625)
vector in space (p. 656)
vertex (vertices) of a conic (p. 605, 614,
    625)
vertex of a cone (p. 603)
$xy$-plane (p. 652)
$xz$-plane (p. 652)
$yz$-plane (p. 652)
zero vector (p. 656)

### Properties, Theorems, and Formulas

#### Parabolas with Vortex ($h, k$)

| | | |
|---|---|---|
| • **Standard equation** | $(x - h)^2 = 4p(y - k)$ | $(y - k)^2 = 4p(x - h)$ |
| • **Opens** | Upward or downward | To the right or to the left |
| • **Focus** | $(h, k + p)$ | $(h + p, k)$ |
| • **Directrix** | $y = k - p$ | $x = h - p$ |
| • **Axis** | $x = h$ | $y = k$ |
| • **Focal length** | $p$ | $p$ |
| • **Focal width** | $|4p|$ | $|4p|$ |

#### Ellipses with Center ($h, k$)

| | | |
|---|---|---|
| • **Standard equation** | $\dfrac{(x - h)^2}{a^2} + \dfrac{(y - k)^2}{b^2} = 1$ | $\dfrac{(y - k)^2}{a^2} + \dfrac{(x - h)^2}{b^2} = 1$ |
| • **Focal axis** | $y = k$ | $x = h$ |
| • **Foci** | $(h \pm c, k)$ | $(h, k \pm c)$ |
| • **Vertices** | $(h \pm a, k)$ | $(h, k \pm a)$ |
| • **Semimajor axis** | $a$ | $a$ |
| • **Semiminor axis** | $b$ | $b$ |
| • **Pythagorean relation** | $a^2 = b^2 + c^2$ | $a^2 = b^2 + c^2$ |

### Hyperbolas with Center ($h$, $k$)

| | | |
|---|---|---|
| • **Standard equation** | $\dfrac{(x-h)^2}{a^2} - \dfrac{(y-k)^2}{b^2} = 1$ | $\dfrac{(y-k)^2}{a^2} - \dfrac{(x-h)^2}{b^2} = 1$ |
| • **Focal axis** | $y = k$ | $x = h$ |
| • **Foci** | $(h \pm c, k)$ | $(h, k \pm c)$ |
| • **Vertices** | $(h \pm a, k)$ | $(h, k \pm a)$ |
| • **Semitransverse axis** | $a$ | $a$ |
| • **Semiconjugate axis** | $b$ | $b$ |
| • **Pythagorean relation** | $c^2 = a^2 + b^2$ | $c^2 = a^2 + b^2$ |
| • **Asymptotes** | $y = \pm\dfrac{b}{a}(x-h) + k$ | $y = \pm\dfrac{a}{b}(x-h) + k$ |

### Translation Formulas

$$x = x' + h \quad \text{and} \quad y = y' + k,$$
$$x' = x - h \quad \text{and} \quad y' = y - k.$$

### Rotation Formulas

$$x' = x \cos \alpha + y \sin \alpha \quad \text{and} \quad y' = -x \sin \alpha + y \cos \alpha,$$
$$x = x' \cos \alpha - y' \sin \alpha \quad \text{and} \quad y = x' \sin \alpha + y' \cos \alpha.$$

### Discriminant Test

The second-degree equation $Ax^2 + Bxy + Cy^2 + Dx + Ey + F = 0$ graphs as

• a hyperbola if $B^2 - 4AC > 0$,

• a parabola if $B^2 - 4AC = 0$,

• an ellipse if $B^2 - 4AC < 0$,

except for degenerate cases.

### Focus–Directrix–Eccentricity Relationship

If $P$ is a point of a conic section, $F$ is the conic's focus, and $D$ is the point of the directrix closest to $F$, then

$$e = \frac{PF}{PD} \quad \text{and} \quad PF = e \cdot PD,$$

where the constant $e$ is the eccentricity of the conic. Moreover, the conic is

• a hyperbola if $e > 1$,

• a parabola if $e = 1$,

• an ellipse if $e < 1$.

### Distance Formula (Cartesian Space)

The distance $d(P, Q)$ between the points $P(x_1, y_1, z_1)$ and $Q(x_2, y_2, z_2)$ in space is

$$d(P, Q) = \sqrt{(x_1 - x_2)^2 + (y_1 - y_2)^2 + (z_1 - z_2)^2}.$$

### Midpoint Formula (Cartesian Space)

Just as in the plane, the coordinates of the midpoint of a line segment are the averages of the coordinates of the endpoints of the segment.

The midpoint $M$ of the line segment $PQ$ with endpoints $P(x_1, y_1, z_1)$ and $Q(x_2, y_2, z_2)$ is

$$M = \left( \frac{x_1 + x_2}{2}, \frac{y_1 + y_2}{2}, \frac{z_1 + z_2}{2} \right).$$

### Standard Equation of a Sphere

A point $P(x, y, z)$ is on the sphere with center $(h, k, l)$ and radius $r$ if and only if

$$(x - h)^2 + (y - k)^2 + (z - l)^2 = r^2.$$

### Vector Relationships in Space

For vectors $\mathbf{v} = \langle v_1, v_2, v_3 \rangle$ and $\mathbf{w} = \langle w_1, w_2, w_3 \rangle$,

- **Equality:**          $\mathbf{v} = \mathbf{w}$ if and only if $v_1 = w_1$, $v_2 = w_2$ and $v_3 = w_3$.

- **Addition:**          $\mathbf{v} + \mathbf{w} = \langle v_1 + w_1, v_2 + w_2, v_3 + w_3 \rangle$.

- **Subtraction:**       $\mathbf{v} - \mathbf{w} = \langle v_1 - w_1, v_2 - w_2, v_3 - w_3 \rangle$.

- **Magnitude:**        $|\mathbf{v}| = \sqrt{v_1{}^2 + v_2{}^2 + v_3{}^2}$.

- **Dot product:**       $\mathbf{v} \cdot \mathbf{w} = v_1 w_1 + v_2 w_2 + v_3 w_3$.

- **Unit vector:**        $\mathbf{u} = \mathbf{v}/|\mathbf{v}|$, $\mathbf{v} \neq 0$ is in the direction of $\mathbf{v}$.

### Equations for a Line in Space

If $\ell$ is a line through the point $P(x_0, y_0, z_0)$ in the direction of a nonzero vector $\mathbf{v} = \langle a, b, c \rangle$, then a point $P(x, y, z)$ is on $\ell$ if and only if

**1. Vector form:**      $\mathbf{r} = \mathbf{r}_0 + t\mathbf{v}$, where $\mathbf{r} = \langle x, y, z \rangle$ and $\mathbf{r}_0 = \langle x_0, y_0, z_0 \rangle$, or

**2. Parametric form:**    $x = x_0 + at$, $y = y_0 + bt$, and $z = z_0 + ct$, where $t$ is a real number.

## Procedures

### How to Sketch the Ellipse $\dfrac{x^2}{a^2} + \dfrac{y^2}{b^2} = 1$

**1.** Sketch line segments at $x = \pm a$ and $y = \pm b$ and complete the rectangle they determine.

**2.** Inscribe an ellipse that is tangent to the rectangle at $(\pm a, 0)$ and $(0, \pm b)$.

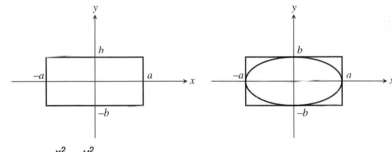

### How to Sketch the Hyperbola $\dfrac{x^2}{a^2} - \dfrac{y^2}{b^2} = 1$

**1.** Sketch line segments at $x = \pm a$ and $y = \pm b$ and complete the rectangle they determine.

**2.** Sketch the asymptotes by extending the rectangle's diagonals.

**3.** Use the rectangle and asymptotes to guide your drawing.

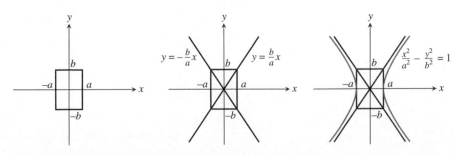

## Chapter 8 Review Exercises

The collection of exercises marked in red could be used as a chapter test.

In Exercises 1–4, find the vertex, focus, directrix, and focal width of the parabola, and sketch the graph.

**1.** $y^2 = 12x$

**2.** $x^2 = -8y$

**3.** $(x + 2)^2 = -4(y - 1)$

**4.** $(y + 2)^2 = 16x$

In Exercises 5–12, find the center, vertices, and foci of the conic. Identify the conic, and sketch its graph.

**5.** $\dfrac{x^2}{5} + \dfrac{y^2}{8} = 1$

**6.** $\dfrac{y^2}{16} - \dfrac{x^2}{49} = 1$

**7.** $\dfrac{x^2}{25} - \dfrac{y^2}{36} = 1$

**8.** $\dfrac{x^2}{49} - \dfrac{y^2}{9} = 1$

**9.** $\dfrac{(x + 3)^2}{18} - \dfrac{(y - 5)^2}{28} = 1$

**10.** $\dfrac{(y - 3)^2}{9} - \dfrac{(x - 7)^2}{12} = 1$

**11.** $\dfrac{(x - 2)^2}{16} + \dfrac{(y + 1)^2}{7} = 1$

**12.** $\dfrac{y^2}{36} + \dfrac{(x + 6)^2}{20} = 1$

In Exercises 13–20, match the equation with its graph.

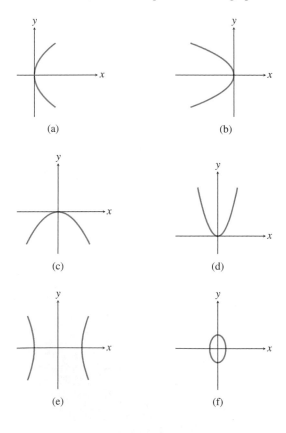

(a)

(b)

(c)

(d)

(e)

(f)

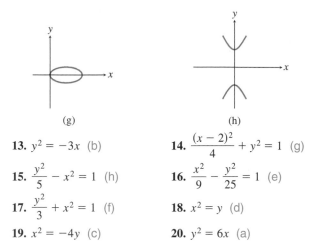

(g)

(h)

**13.** $y^2 = -3x$   (b)

**14.** $\dfrac{(x - 2)^2}{4} + y^2 = 1$   (g)

**15.** $\dfrac{y^2}{5} - x^2 = 1$   (h)

**16.** $\dfrac{x^2}{9} - \dfrac{y^2}{25} = 1$   (e)

**17.** $\dfrac{y^2}{3} + x^2 = 1$   (f)

**18.** $x^2 = y$   (d)

**19.** $x^2 = -4y$   (c)

**20.** $y^2 = 6x$   (a)

In Exercises 21–28, identify the conic. Then complete the square to write the conic in standard form, and sketch the graph.

**21.** $x^2 - 6x - y - 3 = 0$

**22.** $x^2 + 4x + 3y^2 - 5 = 0$

**23.** $x^2 - y^2 - 2x + 4y - 6 = 0$

**24.** $x^2 + 2x + 4y - 7 = 0$

**25.** $y^2 - 6x - 4y - 13 = 0$

**26.** $3x^2 - 6x - 4y - 9 = 0$

**27.** $2x^2 - 3y^2 - 12x - 24y + 60 = 0$

**28.** $12x^2 - 4y^2 - 72x - 16y + 44 = 0$

**29.** Prove that the parabola with focus $(0, p)$ and directrix $y = -p$ has the equation $x^2 = 4py$.

**30.** Prove that the parabola with equation $x = ay^2$ has focus $(1/(4a), 0)$ and directrix $x = -1/(4a)$.

In Exercises 31–36, identify the conic. Solve the equation for $y$ and graph it.

**31.** $3x^2 - 8xy + 6y^2 - 5x - 5y + 20 = 0$

**32.** $10x^2 - 8xy + 6y^2 + 8x - 5y - 30 = 0$

**33.** $3x^2 - 2xy - 5x + 6y - 10 = 0$

**34.** $5xy - 6y^2 + 10x - 17y + 20 = 0$

**35.** $-3x^2 + 7xy - 2y^2 - x + 20y - 15 = 0$

**36.** $-3x^2 + 7xy - 2y^2 - 2x + 3y - 10 = 0$

In Exercises 37–40, find the equation for the conic in standard form.

**37.** Parabola: vertex $(0, 0)$, focus $(2, 0)$.   $y^2 = 8x$

**38.** Parabola: vertex $(0, 0)$, opens downward, focal width = 12.

**39.** Parabola: vertex $(-3, 3)$, directrix $y = 0$.

**40.** Parabola: vertex $(1, -2)$, opens to the left, focal length = 2.

**41.** Ellipse: center $(0, 0)$, foci $(\pm12, 0)$, vertices $(\pm13, 0)$.

**42.** Ellipse: center $(0, 0)$, foci $(0, \pm2)$, vertices $(0, \pm6)$.

**43.** Ellipse: center $(0, 2)$, semimajor axis $= 3$, one focus is $(2, 2)$. $x^2/9 + (y - 2)^2/5 = 1$

**44.** Ellipse: center $(-3, -4)$, semimajor axis $= 4$, one focus is $(0, -4)$.

**45.** Hyperbola: center $(0, 0)$, foci $(0, \pm6)$, vertices $(0, \pm5)$

**46.** Hyperbola: center $(0, 0)$, vertices $(\pm2, 0)$, asymptotes $y = \pm2x$ $x^2/4 - y^2/16 = 1$

**47.** Hyperbola: center $(2, 1)$, vertices $(2 \pm 3, 1)$, one asymptote is $y = (4/3)(x - 2) + 1$

**48.** Hyperbola: center $(-5, 0)$, one focus is $(-5, 3)$, one vertex is $(-5, 2)$. $y^2/4 - (x + 5)^2/5 = 1$

In Exercises 49–54, find the equation for the conic in standard form.

**49.** $x = 5 \cos t$, $y = 2 \sin t$, $0 \le t \le 2\pi$

**50.** $x = 4 \sin t$, $y = 6 \cos t$, $0 \le t \le 4\pi$

**51.** $x = -2 + \cos t$, $y = 4 + \sin t$, $2\pi \le t \le 4\pi$

**52.** $x = 5 + 3 \cos t$, $y = -3 + 3 \sin t$, $-2\pi \le t \le 0$

**53.** $x = 3 \sec t$, $y = 5 \tan t$, $0 \le t \le 2\pi$

**54.** $x = 4 \csc t$, $y = 3 \cot t$, $0 \le t \le 2\pi$

In Exercises 55–62, identify and graph the conic, and rewrite it in Cartesian coordinates.

**55.** $r = \dfrac{4}{1 + \cos \theta}$

**56.** $r = \dfrac{5}{1 - \sin \theta}$

**57.** $r = \dfrac{4}{3 - \cos \theta}$

**58.** $r = \dfrac{3}{4 + \sin \theta}$

**59.** $r = \dfrac{35}{2 - 7 \sin \theta}$

**60.** $r = \dfrac{15}{2 + 5 \cos \theta}$

**61.** $r = \dfrac{2}{1 + \cos \theta}$

**62.** $r = \dfrac{4}{4 - 4 \cos \theta}$

In Exercises 63–74, use the points $P(-1, 0, 3)$, $Q(3, -2, -4)$ and the vectors $\mathbf{v} = \langle-3, 1, -2\rangle$, and $\mathbf{w} = \langle3, -4, 0\rangle$.

**63.** Compute the distance from $P$ to $Q$. $\sqrt{69}$

**64.** Find the midpoint of segment $PQ$. $(1, -1, -1/2)$

**65.** Compute $\mathbf{v} + \mathbf{w}$. $\langle0, -3, -2\rangle$

**66.** Compute $\mathbf{v} - \mathbf{w}$. $\langle-6, 5, -2\rangle$

**67.** Compute $\mathbf{v} \cdot \mathbf{w}$. $-13$

**68.** Compute the magnitude of $\mathbf{v}$. $\sqrt{14}$

**69.** Write the unit vector in the direction of $\mathbf{w}$. $\langle3/5, -4/5, 0\rangle$

**70.** Compute $(\mathbf{v} \cdot \mathbf{w})(\mathbf{v} + \mathbf{w})$. $\langle0, 39, 26\rangle$

**71.** Write an equation for the sphere centered at $P$ with radius 4.

**72.** Write parametric equations for the line through $P$ and $Q$.

**73.** Write a vector equation for the line through $P$ in the direction of $\mathbf{v}$. $\langle-1, 0, 3\rangle + t\langle-3, 1, -2\rangle$

**74.** Write parametric equations for the line in the direction of $\mathbf{w}$ through the midpoint of $PQ$.

**75. Parabolic Microphones** Sports Channel uses a parabolic microphone to capture all the sounds from the basketball players and coaches during each regular season game. If one of its microphones has a parabolic surface generated by the parabola $18y = x^2$, locate the focus (the electronic receiver) of the parabola. $(0, 4.5)$

**76. Parabolic Headlights** Stein Glass, Inc., makes parabolic headlights for a variety of automobiles. If one of its headlights has a parabolic surface generated by the parabola $y^2 = 15x$ (see figure), where should its lightbulb be placed?

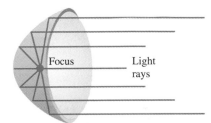

Focus — Light rays

**77. Elliptical Pool Table** Elliptical pool tables have been constructed with a single pocket at one focus and a spot marking the other focus. Suppose such a table has a major axis of 6 ft and minor axis of 4 ft.

**(a)** Explain how a "pool shark" who knows conic geometry has a great advantage in a game of pool on this table over a "mark" who knows no conic geometry.

**(b)** How should the ball be hit so that it bounces off the cushion directly into the pocket? Give specific measurements.

**78. Weather Satellite** The Nimbus weather satellite travels in a north-south circular orbit 500 meters above the Earth. Find the following. (Assume the earth's radius is 6380 km.)

**(a)** The velocity of the satellite using the formula for velocity $v$ given for Exercises 43 and 44 in Section 8.5 with $k = 1$. $7.908$ km/sec

**(b)** The time required for Nimbus to circle the earth once.

**79. Elliptical Orbits** The velocity of a body in an elliptical earth orbit at a distance $r$ (in meters) from the focus (center of earth) is

$$v = \sqrt{3.99 \times 10^{14}\left(\frac{2}{r} - \frac{1}{a}\right)} \text{ m/sec},$$

where $a$ is the length of the semimajor axis of the ellipse. An earth satellite has an *apogee* (maximum distance from earth) of 18,000 km and a *perigee* (minimum distance from earth) of 170 km. Assuming the earth's radius is 6380 km, find the velocity of the satellite at its apogee and perigee.

**80. Icarus** The asteroid Icarus is about 1 mi wide. It revolves around the sun once every 409 Earth days and has an orbital eccentricity of 0.83. Use Kepler's first and third laws to determine Icarus's semimajor axis, perihelion, and aphelion.

**81. Group Activity** Let $c^2 = b^2 - a^2$ with $a < b$ and $d_1 + d_2 = 2b$. Show that the standard form equation for the ellipse with foci $(0, \pm c)$ is

$$\frac{x^2}{a^2} + \frac{y^2}{b^2} = 1$$

**82. Group Activity**
Let $c^2 = a^2 + b^2$ and $d_1 - d_2 = \pm 2b$. Show that the standard form equation for the hyperbola with foci $(0, \pm c)$ is

$$\frac{y^2}{b^2} - \frac{x^2}{a^2} = 1.$$

## Chapter 8 Project

### Parametrizing Ellipses

As a simple pendulum swings back and forth, a plot of its velocity with respect to position is elliptical in nature and can be modeled using the standard form of the equation of an ellipse:

$$\frac{(x - h)^2}{a^2} + \frac{(y - k)^2}{b^2} = 1 \text{ or } \frac{(y - k)^2}{a^2} + \frac{(x - h)^2}{b^2} = 1$$

where $x$ represents the pendulum's distance from a fixed point and $y$ represents the velocity of the pendulum. In this project, you will use a motion detection device to collect distance and velocity data for a swinging pendulum, then find a mathematical model that describes the pendulum's velocity with respect to position.

### Collecting the Data

Construct a simple pendulum by fastening about 0.5 meter of string to the end of a ball. Setup the Calculator Based Laboratory (CBL) system with a motion detector or a Calculator Based Ranger (CBR) system to collect distance and velocity readings for 4 seconds (enough time to capture at least one complete swing of the pendulum). See the CBL/CBR guidebook for specific setup instruction. Start the pendulum swinging in front of the detector, then activate the system. The data table below shows a sample set of data collected as a pendulum swung back and forth in front of a CBR.

| Time (sec) | Distance from the CBR (meters) | Velocity (mps) | Time (sec) | Distance from the CBR (meters) | Velocity (mps) |
|---|---|---|---|---|---|
| 0 | 0.682 | −0.3 | 0.706 | 0.476 | 0.429 |
| 0.59 | 0.659 | −0.445 | 0.765 | 0.505 | 0.544 |
| 0.118 | 0.629 | −0.555 | 0.824 | 0.54 | 0.616 |
| 0.176 | 0.594 | −0.621 | −0.882 | −0.576 | −0.639 |
| 0.235 | 0.557 | −0.638 | 0.941 | 0.612 | 0.612 |
| 0.294 | 0.521 | −0.605 | 1 | 0.645 | 0.536 |
| 0.353 | 0.489 | −0.523 | 1.059 | 0.672 | 0.418 |
| 0.412 | 0.463 | −0.4 | 1.118 | 0.69 | 0.266 |
| 0.471 | 0.446 | −0.246 | 1.176 | 0.699 | 0.094 |
| 0.529 | 0.438 | −0.071 | 1.235 | 0.698 | −0.086 |
| 0.647 | 0.454 | 0.279 | | | |

### Explorations

1. If you collected data using a CBL or CBR, a plot of distance versus time may be shown on your graphing calculator or computer screen. Go to the plot setup screen and create a scatter plot of velocity versus distance. If you do not have access to a CBL/CBR, enter the distance and velocity data from the table above into your graphing calculator/computer and create a scatter plot.

2. Find values for $a$, $b$, $h$, and $k$ so that the equation $\dfrac{(x-h)^2}{a^2} + \dfrac{(y-k)^2}{b^2} = 1$ or $\dfrac{(y-k)^2}{a^2} + \dfrac{(x-h)^2}{b^2} = 1$ fits the velocity versus position data plot. (Which equation you use depends on where the focal axis is located.) To graph this model you will have to solve the appropriate equation for $y$ and enter it into the calculator in two parts as shown below.

```
 Plot1    Plot2   Plot3
\Y1=B√ (1−(X-H)²/
A²)+K
\Y2=−B√ (1−(X-H)²
/A²)+K
\Y3=
\Y4=
\Y5=
```

```
 Plot1    Plot2   Plot3
\Y1=A√ (1−(X-H)²/
B²)+K
\Y2=−A√ (1−(X-H)²
/B²)+K
\Y3=
\Y4=
\Y5=
```

3. With respect to the ellipse modeled above, what do the variables $a$, $b$, $h$, and $k$ represent?

4. Find the coordinates of the foci of the ellipse modeled above.

5. What is the eccentricity of the ellipse modeled above?

6. Set up plots of distance versus time and velocity versus time. Find models for both of these plots and use them to graph the plot of the ellipse using parametric equations.

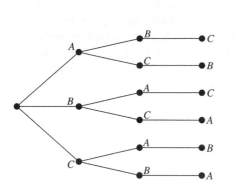
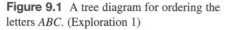

**Figure 9.1** A tree diagram for ordering the letters *ABC*. (Exploration 1)

---

**Exporation Extensions**

A car manufacturing company asked 1000 people to view their latest two-door coupe shown in four colors (Alabaster White, Black, Cherry Red, and Dusk Blue). The viewers were asked to list their color preferences in order from favorite. How many ways could the colors be ordered? **(24)** Of the 1000 viewers, how many would you expect to choose the preference order Alabaster White, Black, Cherry Red,and Dusk Blue? **(41 2/3 ≈ 42 people assuming all preferencs are equally likely to be chosen.)**

---

| **Exploration 1** | **Questionable Product Claims** |

A salesman for a copying machine company is trying to convince a client to buy his $2000 machine instead of his competitor's $5000 machine. To make his point, he lines up an original document, a copy made by his machine, and a copy made by the more expensive machine on a table and asks 60 office workers to identify which is which. To everyone's surprise, not a single worker identifies all three correctly. The salesman states triumphantly that this proves that all three documents look the same to the naked eye and that therefore the client should buy his company's less expensive machine.

What do you think?

1. Call the three documents *A*, *B*, and *C*. Each worker is essentially being asked to put the three letters in the correct order. How many ways can the three letters be ordered?

2. Suppose all three documents really *do* look alike. What fraction of the workers would you expect to put them into the correct order by chance alone?

3. If zero people out of 60 put the documents in the correct order, should we conclude that "all three documents look the same to the naked eye"?

4. Can you suggest a more likely conclusion that we might draw from the results of the salesman's experiment?

You probably had no problem coming up with all the orderings of *A*, *B*, and *C* in the exploration above, but would you have been able to count the orderings of *ABCDE*? It helps to have a way of visualizing the choices, and that's where a *tree diagram* can help. Figure 9.1 shows how we initially have three choices for the first letter. Then, branching off each of those three choices are two choices for the second letter. Finally, branching off each of the $3 \times 2 = 6$ branches formed so far is one choice for the third letter. By beginning at the "root" of the tree, we can proceed to the left along any of $3 \times 2 \times 1 = 6$ branches. We conclude that there are six ways to order the letters *ABC*.

You would not want to draw the tree diagram for *ABCDE*, but you should be able to see in your mind that it would have $5 \times 4 \times 3 \times 2 \times 1 = 120$ branches. A tree diagram is a geometric visualization of a fundamental counting principle known as the *Multiplication Principle*.

---

**Multiplication Principle of Counting**

If a procedure *P* has a sequence of stages $S_1, S_2, \ldots, S_n$ and if

$$S_1 \text{ can occur in } r_1 \text{ ways,}$$

$$S_2 \text{ can occur in } r_2 \text{ ways,}$$

$$\vdots$$

$$S_n \text{ can occur in } r_n \text{ ways,}$$

then the number of ways that the procedure *P* can occur is the product

$$r_1 r_2 \cdots r_n.$$

It is important to be mindful of how the choices at each stage are affected by the choices at preceding stages. For example, when choosing an order for the letters *ABC* we have 3 choices for the first letter, but only 2 choices for the second and 1 for the third.

### Example 1   USING THE MULTIPLICATION PRINCIPLE

Regular motor vehicle license plates in Tennessee in 1999 consisted of three letters of the alphabet followed by three numerical digits (0 through 9). Find the number of different license plates that could be formed

**(a)** if there is no restriction on the letters or digits that can be used;

**(b)** if no two letters and no two digits can be repeated.

**Solution**   Consider each license plate as having six blanks to be filled in: three letters followed by three numerical digits.

**(a)** If there are no restrictions on letters or digits, then we can fill in the first blank 26 ways, the second blank 26 ways, the third blank 26 ways, the fourth blank 10 ways, the fifth blank 10 ways, and the sixth blank 10 ways. By the Multiplication Principle, we can fill in all six blanks in $26 \times 26 \times 26 \times 10 \times 10 \times 10 = 17{,}576{,}000$ ways. There are 17,576,000 possible license plates with no restrictions on letters or digits.

**(b)** If no letter or digit can be repeated, then we can fill in the first blank 26 ways, the second blank 25 ways, the third blank 24 ways, the fourth blank 10 ways, the fifth blank 9 ways, and the sixth blank 8 ways. By the Multiplication Principle, we can fill in all six blanks in $26 \times 25 \times 24 \times 10 \times 9 \times 8 = 11{,}232{,}000$ ways. There are 11,232,000 possible license plates with no letters or digits repeated.

## Permutations

One important application of the Multiplication Principle is to count the number of ways that a set of $n$ objects (called an **$n$-set**) can be arranged in order. Each such ordering is called a **permutation** of the set. Exploration 1 was about permutations of the letters *ABC*.

### Example 2   PERMUTATIONS OF AN *n*-SET

How many different permutations of an $n$-set can be formed?

**Solution**   Line up $n$ blanks to be filled in order. The first blank can be filled in $n$ ways, the second in $n - 1$ ways (since one object has already been used to fill the first blank), the third in $n - 2$ ways, and so on until there is one object left to fill in the final blank. By the Multiplication Principle, we can fill in all $n$ blanks in $n(n - 1)(n - 2) \cdots 2 \cdot 1 = n!$ ways. Therefore there are $n!$ permutations of an $n$-set.

In many counting problems we are interested in using $n$ objects to fill $r$ blanks in order, where $r < n$. These are called **permutations of $n$ objects taken $r$ at a time**. The procedure for counting them is the same; only this time we run out of blanks before we run out of objects.

The first blank can be filled in $n$ ways, the second in $n - 1$ ways, and so on until we come to the $r$th blank, which can be filled in $n - (r - 1)$ ways. By

---

**Teaching Note**

Encourage students to ask the following questions prior to actually doing any counting. (1) What is the process that is being completed? Does the order mater (either in terms of completing the problem correctly or simplifying calculations)? (2) What is the first stage? How many ways can it be completed? (3) What is the second stage? How many wys can it be completed? And so on.

**License Plate Restrictions**

Although prohibiting repeated letters and digits as in Example 1 would make no practical sense (why rule out more than 6 million possible plates for no good reason?), states do impose some restrictions on license plates. They rule out certain letter progressions that could be considered obscene or offensive.

**Factorials**

If $n$ is a positive integer, the symbol $n!$ (read "$n$ factorial") represents the product $n(n - 1)(n - 2)(n - 3) \cdots 2 \cdot 1$. We also define $0! = 1$.

the Multiplication Principle, we can fill in all $r$ blanks in $n(n-1)(n-2) \cdots (n-r+1)$ ways. This expression can be written in a more compact (but less easily computed) way as $n!/(n-r)!$.

### Permutations on a Calculator

Most modern calculators have an $_nP_r$ selection built in. They also compute factorials, but remember that factorials get very large. If you want to count the number of permutations of 90 objects taken 5 at a time, be sure to use the $_nP_r$ function. The expression 90!/85! is likely to lead to an overflow error.

> ### Permutation Counting Formula
>
> The number of permutations of $n$ objects taken $r$ at a time is denoted $_nP_r$ and is given by
>
> $$_nP_r = \frac{n!}{(n-r)!} \quad \text{for} \quad 0 \le r \le n.$$
>
> If $r > n$, then $_nP_r = 0$.

Notice that $_nP_n = n!/(n-n)! = n!/0! = n!/1 = n!$, which we have already seen is the number of permutations of a complete set of $n$ objects. This is why we define $0! = 1$.

### Notes on Examples

Example 3 shows some paper-and-pencil methods for calculating permutations. It is important that students have the algebraic skills to perform these operations since the numbers in some counting problems may exceed the capacity of a calculator.

### Example 3   COUNTING PERMUTATIONS

Evaluate each expression without a calculator.

**(a)** $_6P_4$        **(b)** $_{11}P_3$        **(c)** $_nP_3$

### Solution

**(a)** By the formula, $_6P_4 = 6!/(6-4)! = 6!/2! = (6 \cdot 5 \cdot 4 \cdot 3 \cdot 2!)/2! = 6 \cdot 5 \cdot 4 \cdot 3 = 360$.

**(b)** Although you could use the formula again, you might prefer to apply the Multiplication Principle directly. We have 11 objects and 3 blanks to fill:

$$_{11}P_3 = 11 \cdot 10 \cdot 9 = 990.$$

**(c)** This time it is definitely easier to use the Multiplication Principle. We have $n$ objects and 3 blanks to fill:

$$_nP_3 = n(n-1)(n-2).$$

### Example 4   APPLYING PERMUTATIONS

Sixteen actors answer a casting call to try out for roles as dwarfs in a production of *Snow White and the Seven Dwarfs*. In how many different ways can the director cast the seven roles?

**Solution**   The 7 different roles can be thought of as 7 blanks to be filled, and we have 16 actors with which to fill them. The director can cast the roles in $_{16}P_7 = 57,657,600$ ways.

## Combinations

When we count permutations of $n$ objects taken $r$ at a time, we consider different orderings of the same $r$ selected objects as being different permutations. In many applications we are only interested in the ways to *select* the $r$ objects, regardless of the order in which we arrange them. These unordered selections are called **combinations of $n$ objects taken $r$ at a time**.

> **Combination Counting Formula**
>
> The number of combinations of $n$ objects taken $r$ at a time is denoted $_nC_r$ and is given by
>
> $$_nC_r = \frac{n!}{r!(n-r)!} \quad \text{for} \quad 0 \le r \le n.$$
>
> If $r > n$, then $_nC_r = 0$.

We can verify the $_nC_r$ formula with the Multiplication Principle. Since every permutation can be thought of as an *unordered* selection of $r$ objects *followed* by a particular *ordering* of the objects selected, the Multiplication Principle gives $_nP_r = {}_nC_r \cdot r!$.

Therefore

$$_nC_r = \frac{_nP_r}{r!} = \frac{1}{r!} \cdot \frac{n!}{(n-r)!} = \frac{n!}{r!(n-r)!}.$$

## A Word on Notation

Some textbooks use $P(n, r)$ instead of $_nP_r$ and $C(n, r)$ instead of $_nC_r$. Much more common is the notation $\binom{n}{r}$ for $_nC_r$. Both $\binom{n}{r}$ and $_nC_r$ are often read "$n$ choose $r$."

### Example 5  PICKING MISS AMERICA FINALISTS

In the Miss America pageant, 51 contestants must be narrowed down to 10 finalists who will compete on national television. In how many possible ways can the ten finalists be selected?

**Solution** Notice that the *order* of the finalists does not matter at this phase; all that matters is which women are selected. So we count combinations rather than permutations.

$$_{51}C_{10} = \frac{51!}{10!41!} = 12{,}777{,}711{,}870.$$

The 10 finalists can be chosen in 12,777,711,870 ways.

### Example 6  PICKING LOTTERY NUMBERS

The Georgia Lotto requires winners to pick 6 integers between 1 and 46. The order in which you select them does not matter; indeed, the lottery tickets are always printed with the numbers in ascending order. How many different lottery tickets are possible?

**Solution** There are $_{46}C_6 = 9{,}366{,}819$ possible lottery tickets of this type. The slim chance of picking correctly does not keep many people from trying.

## Subsets of an *n*-Set

As a final application of the counting principle, consider the pizza topping problem.

### Example 7  SELECTING PIZZA TOPPINGS

Armando's Pizzeria offers patrons any combination of up to 10 different toppings: pepperoni, mushroom, sausage, onion, green pepper, bacon, prosciutto, black olive, green olive, and anchovies. How many different pizzas can be ordered

**(a)** if we can choose any three toppings?

**(b)** If we can choose any number of toppings (0 through 10)?

**Solution**

**(a)** Order does not matter (for example, the sausage-pepperoni-mushroom pizza is the same as the pepperoni-mushroom-sausage pizza), so the number of possible pizzas is $_{10}C_3 = 120$.

**(b)** We could add up all the numbers of the form $_{10}C_r$ for $r = 0, 1, \ldots, 10$, but there is an easier way to count the possibilities. Consider the ten options to be lined up as in the statement of the problem. In considering each option, we have two choices: yes or no. (For example, the pepperoni-mushroom-sausage pizza would correspond to the sequence YYYNNNNNNN.) By the multiplication Principle, the number of such sequences is $2 \cdot 2 \cdot 2 \cdot 2 \cdot 2 \cdot 2 \cdot 2 \cdot 2 \cdot 2 \cdot 2 = 1024$, which is the number of possible pizzas.

The solution to Example 7b suggests a general rule that will be our last counting formula of the section.

> **Formula for Counting Subsets of an *n*-Set**
>
> There are $2^n$ subsets of a set with $n$ objects (including the empty set and the entire set).

### Example 8  ANALYZING AN ADVERTISED CLAIM

A national hamburger chain used to advertise that it fixed its hamburgers "256 ways," since patrons could order whatever toppings they wanted. How many toppings must have been available to choose from?

**Solution**  We need to solve the equation $2^n = 256$ for $n$. We could solve this easily enough by trial and error, but we will solve it with logarithms just to keep the method fresh in our minds.

$$2^n = 256$$
$$\log 2^n = \log 256$$
$$n \log 2 = \log 256$$
$$n = \frac{\log 256}{\log 2}$$
$$n = 8$$

There must have been 8 toppings from which to choose.

**Problem**

In February of 2000, 230 area codes were in use in the U.S. How many additional three-digit area codes are possible? In a given area code, how many unique phone numbers are possible?

**Solution**

Each area code has three unique digits, represented as

$$d_1 \, d_2 \, d_3$$

Each digit, $d_i$, where $i = 1, 2, 3$, has 10 possible values: 0, 1, ..., 9. So there are

$$10 \cdot 10 \cdot 10 = 10^3 = 1000 \text{ unique area codes possible.}$$

As of February, 2000, there were 230 area codes in use in the U.S., so there are

$$1000 - 230 = 770 \text{ additional area codes possible.}$$

In a given area code, each phone number has seven unique digits, represented as

$$d_1 \, d_2 \, d_3 \quad d_4 \, d_5 \, d_6 \, d_7$$

Each digit, $d_i$ where $i = 1, 2, ..., 7$, has 10 possible values: 0, 1, ..., 9. So there are

$$10 \cdot 10 \cdot 10 \cdot 10 \cdot 10 \cdot 10 \cdot 10 = 10^7 = 10,000,000 \text{ unique phone}$$
numbers in a given area code.

# Quick Review 9.1

In Exercises 1–10, give the number of objects described. In some cases you might have to do a little research or ask a friend.

1. The number of cards in a standard deck  52

2. The number of cards of each suit in a standard deck  13

3. The number of faces on a cubical die  6

4. The number of possible totals when two dice are rolled  11

5. The number of vertices of a decagon  10

6. The number of musicians in string quartet  4

7. The number of players on a soccer team  11

8. The number of prime numbers between 1 and 10 inclusive  4

9. The number of squares on a chessboard  64

10. The number of cards in a contract bridge hand  13

# Section 9.1 Exercises

In Exercises 1–6, evaluate each expression without a calculator. Then check with your calculator to see if your answer is correct.

1. $4!$  24

2. $(3!)(0!)$  6

3. $_6P_2$  30

4. $_9P_2$  72

5. $_{10}C_7$  120

6. $_{10}C_3$  120

In Exercises 7–10, tell whether permutations (ordered) or combinations (unordered) are being described.

7. 13 cards are selected from a deck of 52 to form a bridge hand  combinations

8. 7 digits are selected (with possible repetition) to form a telephone number  permutations

9. 4 students are selected from the senior class to form a committee to advise the cafeteria director about food

10. 4 actors are chosen to play the Beatles in a film biography.

11. **Permuting Letters** How many 9-letter "words" (not necessarily in any dictionary) can be formed from the letters of the word LOGARITHM? (Curiously, one such arrangement spells another word related to mathematics. Can you name it?) 362,880 (ALGORITHM)

12. **Three-Letter Crossword Entries** Excluding J, Q, X, and Z, how many 3-letter crossword puzzle entries can be formed that contain no repeat letters? (It has been conjectured that all of them have appeared in puzzles over the years, sometimes with painfully contrived definitions.)

13. **Possible Routes** There are three roads from town *A* to town *B* and four roads from town *B* to town *C*. How many different routes are there from *A* to *C* by way of *B*? 12

14. **Possible Routes** Using the information in Exercise 13, how many different routes are there from *A* to *C* and back to *A*, passing through *B* in both directions? 144

15. **Homecoming King and Queen** There are four candidates for homecoming queen and three candidates for king. How many king-queen pairs are possible? 12

16. **Airline Tickets** When ordering an airline ticket you can request first class, business class, or coach. You can also choose a window, aisle, or middle seat, except that there are no middle seats in first class. How many different ways can you order a ticket? 8

17. **Phone Numbers** How many seven-digit telephone numbers are possible? (A number may not begin with a 0 or a 1. Why?) 8,000,000

18. **Social Security Numbers** How many nine-digit social security numbers are there? $10^9 = 1,000,000,000$

19. **License Plates** How many different license plates begin with two digits, followed by two letters and then three digits if no letters or digits repeat? 19,656,000

20. **License Plates** How many different license plates consist of five symbols, either digits or letters? 60,466,176

21. **Tumbling Dice** Suppose that two dice, one red and one green, are rolled. How many different outcomes are possible for the pair of dice? 36

22. **Coin Toss** How many different sequences of heads and tails are there if a coin is tossed 10 times? 1024

23. **Forming Committees** A 3-woman committee is to be elected from a 25-member sorority. How many different committees can be elected? 2300

24. **Straight Poker** In the original version of poker known as "straight" poker, a five-card hand is dealt from a standard deck of 52. How many different straight poker hands are possible? 2,598,960

25. **Buying Discs** Juan has money to buy only three of the 48 compact discs available. How many different sets of discs can he purchase? 17,296

26. **Coin Toss** A coin is tossed 20 times and the heads and tails sequence is recorded. From among all the possible sequences of heads and tails, how many have exactly seven heads? 77,520

27. **Drawing Cards** How many different 13-card hands include the ace and king of spades? 37,353,738,800

28. The head of the personnel department interviews eight people for three identical openings. How many different groups of three can be employed? 56

29. **Scholarshop Nominations** Six seniors at Rydell High school meet the qualifications for a competitive honor scholarshop at a major university. The university allows the school to nominate up to three candidates, and the school always nominates at least one. How many different choices could the nominating committee make? 41

30. **Pu-pu Platters** A Chinese restaurant will make a Pu-pu platter "to order" containing any one, two, or three selections from its appetizer menu. If the menu offers five different appetizers, how many different platters could be made? 25

31. **Yahtzee** In the game of Yahtzee, five dice are tossed simultaneously. How many outcomes can be distinguished if all the dice are different colors? 7776

32. **Indiana Jones and the Final Exam** Professor Indiana Jones gives his class 20 study questions, from which he will select 8 to be answered on the final exam. How many ways can he select the questions? 125,970

33. **Salad Bar** Mary's lunch always consists of a full plate of salad from Ernestine's salad bar. She always takes equal amounts of each salad she chooses, but she likes to vary her selections. If she can choose from among 9 different salads, how many essentially different lunches can she create? 511

34. **Buying a New Car** A new car customer has to choose from among 3 models, each of which comes in 4 exterior colors, 3 interior colors, and with any combination of up to 6 optional accessories. How many essentially different ways can the customer order the car? 2304

35. **True-False Tests** How many different answer keys are possible for a 10-question True-False test? 1024

36. **Multiple-Choice Tests** How many different answer keys are possible for a 10-question multiple-choice test in which each question leads to choice *a*, *b*, *c*, *d*, or *e*? 9,765,625

## Explorations

37. **Group Activity** For each of the following numbers, make up a counting problem that has it as the answer.

   (a) $_{52}C_3$      (b) $_{12}C_3$      (c) $_{25}P_{11}$

   (d) $2^5$      (e) $3 \cdot 2^{10}$

38. **Writing to Learn** You have a fresh carton containing a dozen eggs and you need to choose two for breakfast. Give a counting argument based on this scenario to explain why $_{12}C_2 = {}_{12}C_{10}$.

**39. Permutations with Repeated Letters**

(a) Explain why there are 6! different "words" that can be made using all 6 letters of ROMANS, but only 6!/2 different "words" that can be made using all 6 letters of GREEKS.

(b) Explain why there are only 6!/3! possible words that can be made using all 6 letters of STRESS.

(c) How many different 11-letter words can be formed using the 11 letters of MISSISSIPPI? Is this number more or less than the number possible with CHATTANOOGA?

**40. Group Activity  Diagonals of a Regular Polygon**  In Exploration 1 of Section 1.7 you reasoned from data points and quadratic regression that the number of diagonals of a regular polygon with $n$ vertices was $(n^2 - 3n)/2$.

(a) Explain why the number of segments connecting all pairs of vertices is $_nC_2$.

(b) Use the result from (a) to prove that the number of diagonals in $(n^2 - 3n)/2$. ■

## Extending the Ideas

**41. Writing to Learn**  Suppose that a chain letter (illegal if money is involved) is sent to five people the first week of the year. Each of these five people sends a copy of the letter to five more people during the second week of the year. Assume that everyone who receives a letter participates. Explain how you know with certainty that someone will receive a second copy of this letter later in the year.

**42. A Round Table**  How many different seating arrangements are possible for 4 people sitting around a round table?  6

**43. Colored Beads**  Four beads—red, blue, yellow, and green—are arranged on a string to make a simple necklace as shown in the figure. How many arrangements are possible?  3

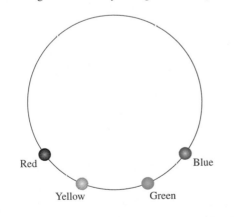

Red  Blue
Yellow  Green

**44. Casting a Play**  A director is casting a play with two female leads and wants to have a chance to audition the actresses two at a time to get a feeling for how well they would work together. His casting director and his administrative assistant both prepare charts to show the amount of time that would be required, depending on the number of actresses who came to the audition. Which time chart is more reasonable, and why?

| Number who audition | Time required (minutes) | Number who audition | Time required (minutes) |
|---|---|---|---|
| 3 | 10 | 3 | 10 |
| 6 | 45 | 6 | 30 |
| 9 | 110 | 9 | 60 |
| 12 | 200 | 12 | 100 |
| 15 | 320 | 15 | 150 |

**45. Bridge Around the World**  Suppose that a contract bridge hand is dealt somewhere in the world every second. What is the fewest number of years required for every possible bridge hand to be dealt? (See Quick Review Exercise 10.)  ≈ 20,123 years

**46. Basketball Lineups**  Each NBA basketball team has 12 players on its roster. If each coach choses 5 starters without regard to position, how many different sets of 10 players can start when two given teams play a game?  627,264

# The Binomial Theorem

Powers of Binomials • Pascal's Triangle • The Binomial Theorem • Applications

## Powers of Binomials

**Objective**

Students will be able to expand a power of a binomial using the binomial theorem or Pascal's triangle. They will also find the coefficient of a given term of a binomial expansion.

**Motivate**

Have students expand $(2x + y)^4$ using the distributive property.
$(16x^4 + 32x^3y + 24x^2y^2 + 8xy^3 + y^4)$
Explain that this section will give them easier ways to do this kind of calculation.

**Lesson Guide**

Day 1: Powers of Binomials; Pascal's Triangle; The Binomial Theorem
Day 2: Applications

Many important mathematical discoveries have begun with the study of patterns. In this chapter we want to introduce an important polynomial theorem called the Binomial Theorem, for which we will set the stage by observing some patterns.

If you expand $(a + b)^n$ for $n = 0, 1, 2, 3, 4,$ and 5, here is what you get:

$$(a + b)^0 = 1$$
$$(a + b)^1 = 1a^1b^0 + 1a^0b^1$$
$$(a + b)^2 = 1a^2b^0 + 2a^1b^1 + 1a^0b^2$$
$$(a + b)^3 = 1a^3b^0 + 3a^2b^1 + 3a^1b^2 + 1a^0b^3$$
$$(a + b)^4 = 1a^4b^0 + 4a^3b^1 + 6a^2b^2 + 4a^1b^3 + 1a^0b^4$$
$$(a + b)^5 = 1a^5b^0 + 5a^4b^1 + 10a^3b^2 + 10a^2b^3 + 5a^1b^4 + 1a^0b^5$$

Can you observe the patterns and predict what the expansion of $(a + b)^6$ will look like? You can probably predict the following:

1. The powers of $a$ will decrease from 6 to 0 by 1's.

2. The powers of $b$ will increase from 0 to 6 by 1's.

3. The first two coefficients will be 1 and 6.

4. The last two coefficients will be 6 and 1.

You at first might not see the pattern that would enable you to find the other so-called *binomial coefficients*, but you should see it after the following Exploration.

**Exploration Extensions**

Have the students find the coefficient of $ab^2$ in $(a + b)^3$ by using the distributive property or using exponents and without combining like terms. The relevant terms in the expansion will be *abb*, *bab*, and *bba*. There are three of these terms, just as there are 3 ways to choose 2 positions out of 3 places, so the coefficient of $ab^2$ is $_3C_2 = 3$. Now have the students list all possible ways to have $a^3b^2$ then state the coefficient of this term in the expansion of $(a + b)^5$.
(*aaabb, aabba, abba, …; 10*)

**Exploration 1** **Exploring the Binomial Coefficients**

1. Compute $_3C_0$, $_3C_1$, $_3C_2$, and $_3C_3$. Where can you find these numbers in the binomial expansions above?

2. The expression $4_nC_r \{0, 1, 2, 3, 4\}$ tells the calculator to compute $_4C_r$ for each of the numbers $r = 0, 1, 2, 3, 4$ and display them as a list. Where can you find these numbers in the binomial expansions above?

3. Compute $5_nC_r \{0, 1, 2, 3, 4, 5\}$. Where can you find these numbers in the binomial expansions above?

4. Find the binomial coefficients that would appear in the expansions of $(a + b)^6$ and $(a + b)^7$ and use them to write out the actual expansions.

By now you are probably ready to conclude that the binomial coefficients in the expansion of $(a + b)^n$ are just the values of $_nC_r$ for $r = 0, 1, 2, 3, 4, \ldots, n$. We also hope you are wondering *why* this is true.

The expansion of

$$(a + b)^n = \underbrace{(a + b)(a + b)(a + b) \cdots (a + b)}_{n \text{ factors}}$$

consists of all possible products that can be formed by taking one letter (either *a* or *b*) from each factor $(a + b)$. The number of ways to form the product $a^r b^{n-r}$ is exactly the same as the number of ways to choose *r* factors to contribute an *a*, since the rest of the factors will obviously contribute a *b*. The number of ways to choose *r* factors from *n* factors is $_nC_r$.

---

**Definition**  **Binomial Coefficient**

The binomial coefficients that appear in the expansion of $(a + b)^n$ are the values of $_nC_r$ for $r = 0, 1, 2, 3, \ldots, n$.

A classical notation for $_nC_r$, especially in the context of binomial coefficients, is $\binom{n}{r}$. Both notations are read "*n* choose *r*."

## Pascal's Triangle

If we eliminate the plus signs and the powers of the variables *a* and *b* in the "triangular" array of binomial coefficients with which we began this section, we get:

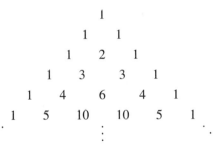

This is called **Pascal's triangle** in honor of Blaise Pascal (1623–1662), who used it in his work but certainly did not invent it. It appeared in 1303 in a Chinese text, the *Precious Mirror*, by Chu Shih-chieh, who referred to it even then as a "diagram of the old method for finding eighth and lower powers."

For convenience, we refer to the top "1" in Pascal's triangle as row 0. That allows us to associate the numbers along row *n* with the expansion of $(a + b)^n$.

Pascal's triangle is so rich in patterns that people still write about them today. One of the simplest patterns is the one that we use for getting from one row to the next, as in the following example.

### Example 1 EXTENDING PASCAL'S TRIANGLE

Show how row 5 of Pascal's triangle can be used to obtain row 6, and use the information to write the expansion of $(x + y)^6$.

**Solution** The two outer numbers of every row are 1's. Each number between them is the sum of the two numbers immediately above it. So row 6 can be found from row 5 as follows:

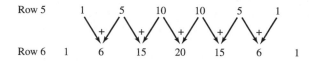

These are the binomial coefficients for $(x + y)^6$, so

$$(x + y)^6 = x^6 + 6x^5y + 15x^4y^2 + 20x^3y^3 + 15x^2y^4 + 6xy^5 + y^6.$$

### Example 2 EXPLAINING THE PATTERN BY COUNTING

Give a counting argument to explain why each interior number in Pascal's triangle is the sum of the two numbers above it.

**Solution** If you count along the row, you will find that the two numbers directly above $_nC_r$ in Pascal's triangle are the numbers $_{n-1}C_{r-1}$ and $_{n-1}C_r$.

Suppose we are choosing $r$ objects from $n$ objects. As we have seen, this can be done in $_nC_r$ ways.

Now identify one of the $n$ objects with a special tag. How many ways can we choose $r$ objects from the $n$ objects if the tagged object is among them? Well, we have $r - 1$ objects yet to be chosen from among $n - 1$ objects available, so $_{n-1}C_{r-1}$. How many ways can we choose $r$ objects from $n$ objects if the tagged object is *not* among them? This time we must choose all $r$ objects from among $n - 1$ available, so $_{n-1}C_r$. Since our selection of $r$ objects must either contain the tagged object or not contain it, $_{n-1}C_{r-1} + {}_{n-1}C_r$ counts all the possibilities with no overlap. Therefore, $_{n-1}C_{r-1} + {}_{n-1}C_r = {}_nC_r$.

It is not necessary to construct Pascal's triangle to find specific binomial coefficients, since we already have a formula for computing them: $_nC_r = n!/(r!(n - r)!)$. This formula can be used to construct an algebraic proof of the pattern cited in Example 2, but we will leave this as an exercise for the end of the section.

### Example 3 COMPUTING BINOMIAL COEFFICIENTS

Find the coefficient of $x^{10}$ in the expansion of $(x + 2)^{15}$.

**Solution** The only term in the expansion that we need to deal with is $_{15}C_{10}x^{10}2^5$. This is

$$\frac{15!}{10!5!} \cdot 2^5 \cdot x^{10} = 3003 \cdot 32 \cdot x^{10} = 96{,}096 \, x^{10}.$$

The coefficient of $x^{10}$ is 96,096.

# The Binomial Theorem

We now state formally the theorem about expanding powers of binomials, known as the Binomial Theorem. For tradition's sake, we will use the symbol $\binom{n}{r}$ instead of $_nC_r$.

## The Binomial Theorem in $\Sigma$ Notation

For those who are already familiar with it, here is how the Binomial Theorem looks in summation notation:

$$(a+b)^n = \sum_{r=0}^{n} \binom{n}{r} a^{n-r} b^r.$$

Those who are not familiar with this notation will learn more about it in Section 9.4.

---

### The Binomial Theorem

For any positive integer $n$,

$$(a+b)^n = \binom{n}{0}a^n + \binom{n}{1}a^{n-1}b + \cdots + \binom{n}{r}a^{n-r}b^r + \cdots + \binom{n}{n}b^n,$$

where

$$\binom{n}{r} = {}_nC_r = \frac{n!}{r!(n-r)!}.$$

### Example 4   EXPANDING A BINOMIAL

Expand $(2x - y^2)^4$.

**Solution**  We use the Binomial theorem to expand $(a+b)^4$ where $a = 2x$ and $b = -y^2$.

$$(a+b)^4 = a^4 + 4a^3b + 6a^2b^2 + 4ab^3 + b^4$$
$$(2x-y^2)^4 = (2x)^4 + 4(2x)^3(-y^2) + 6(2x)^2(-y^2)^2$$
$$+ 4(2x)(-y^2)^3 + (-y^2)^4$$
$$= 16x^4 - 32x^3y^2 + 24x^2y^4 - 8xy^6 + y^8$$

## Applications

The most basic application of the Binomial Theorem is to expand binomials, but it can also be used in proofs, as in the following example.

### Example 5   USING THE BINOMIAL THEOREM IN A PROOF

Prove that $\binom{n}{0} + \binom{n}{1} + \binom{n}{2} + \cdots + \binom{n}{n} = 2^n$ for any positive integer $n$.

**Solution**

$$2^n = (1+1)^n = \binom{n}{0}1^n + \binom{n}{1}1^{n-1}1^1 + \cdots + \binom{n}{n}1^n$$
$$= \binom{n}{0} + \binom{n}{1} + \cdots + \binom{n}{n}$$

Example 5 verifies another well-known pattern in Pascal's triangle: Every row adds up to a power of 2.

## Quick Review 9.2

In Exercises 1–10, use the distributive property to expand the binomial.

**1.** $(x + y)^2$ $x^2 + 2xy + y^2$ **2.** $(a + b)^2$ $a^2 + 2ab + b^2$

**3.** $(5x - y)^2$ $25x^2 - 10xy + y^2$ **4.** $(a - 3b)^2$ $a^2 - 6ab + 9b^2$

**5.** $(3s + 2t)^2$

**6.** $(3p - 4q)^2$

**7.** $(u + v)^3$

**8.** $(b - c)^3$

**9.** $(2x - 3y)^3$

**10.** $(4m + 3n)^3$

## Section 9.2 Exercises

In Exercises 1 and 2, find the indicated row of Pascal's triangle.

**1.** Row 8

**2.** Row 10

In Exercises 3–8, expand the binomial using Pascal's triangle to find the coefficients.

**3.** $(a + b)^5$

**4.** $(x + y)^6$

**5.** $(x - y)^7$

**6.** $(a - b)^6$

**7.** $(3x - y)^4$

**8.** $(a - 2b^2)^4$

In Exercises 9–12, evaluate the expression by hand (using the formula) before checking your answer on a grapher.

**9.** $\binom{9}{2}$ 36

**10.** $\binom{15}{11}$ 1365

**11.** $\binom{166}{166}$ 1

**12.** $\binom{166}{0}$ 1

In Exercises 13–16, use the Binomial Theorem to find a polynomial expansion for the function.

**13.** $f(x) = (x - 2)^5$

**14.** $g(x) = (x + 3)^6$

**15.** $h(x) = (2x - 1)^7$

**16.** $f(x) = (3x + 4)^5$

In Exercises 17–22, use the Binomial Theorem to expand each binomial.

**17.** $(2x + y)^4$

**18.** $(2y - 3x)^5$

**19.** $(\sqrt{x} - \sqrt{y})^6$

**20.** $(\sqrt{x} + \sqrt{3})^4$

**21.** $(x^{-2} + 3)^5$

**22.** $(a - b^{-3})^7$

In Exercises 23–26, find the coefficient of the given term in the binomial expansion.

**23.** $x^{11}y^3$ term, $(x + y)^{14}$ 364 **24.** $x^5y^8$ term, $(x + y)^{13}$ 1287

**25.** $x^4$ term, $(x - 2)^{12}$ 126,720 **26.** $x^7$ term, $(x - 3)^{11}$ 26,730

**27.** Determine the largest integer $n$ for which your calculator will compute $n!$.

**28.** Determine the largest integer $n$ for which your calculator will compute $\binom{n}{100}$.

**29.** Prove that $\binom{n}{1} = \binom{n}{n-1} = n$ for all integers $n \geq 1$.

**30.** Prove that $\binom{n}{r} = \binom{n}{n-r}$ for all integers $n \geq r \geq 0$.

**31.** Use the formula $\binom{n}{r} = \dfrac{n!}{r!(n-r)!}$ to prove that

$\binom{n}{r} = \binom{n-1}{r-1} + \binom{n-1}{r}$. (This is the pattern in Pascal's

triangle that appears in Example 2.)

**32.** Find a counterexample to show that each statement is *false*.

(a) $(n + m)! = n! + m!$ (b) $(nm)! = n!m!$

**33.** Prove that $\binom{n}{2} + \binom{n+1}{2} = n^2$ for all integers $n \geq 2$.

**34.** Prove that $\binom{n}{n-2} + \binom{n+1}{n-1} = n^2$ for all integers $n \geq 2$.

## Explorations

**35. Triangular Numbers** Numbers of the form $1 + 2 + \cdots + n$ are called **triangular numbers** because they count numbers in triangular arrays, as shown below:

(a) Compute the first 10 triangular numbers.

(b) Where do the triangular numbers appear in Pascal's triangle?

(c) **Writing to Learn** Explain why the diagram below shows that the $n$th triangular number can be written as $n(n + 1)/2$.

(d) Write the formula in (c) as a binomial coefficient. (This is why the triangular numbers appear as they do in Pascal's triangle.)

**36. Group Activity  Exploring Pascal's Triangle** Break into groups of two or three. Just by looking at patterns in Pascal's triangle, guess the answers to the following questions. (It is easier to make a conjecture from a pattern than it is to construct a proof!)

**(a)** What positive integer appears the least number of times?

**(b)** What number appears the most number of times?  1

**(c)** Is there any positive integer that does *not* appear in Pascal's triangle?  no

**(d)** If you go along any row alternately adding and subtracting the numbers, what is the result?  0

**(e)** If $p$ is a prime number, what do all the interior numbers along the $p$th row have in common?  All are divisible by $p$.

**(f)** Which rows have all even interior numbers?

**(g)** Which rows have all odd numbers?

**(h)** What other patterns can you find? Share your discoveries with the other groups.  ■

## Extending the Ideas

**37.** Use the Binomial Theorem to prove that the alternating sum along any row of Pascal's triangle is zero. That is,

$$\binom{n}{0} - \binom{n}{1} + \binom{n}{2} - \cdots + (-1)^n \binom{n}{n} = 0.$$

**38.** Use the Binomial Theorem to prove that

$$\binom{n}{0} + 2\binom{n}{1} + 4\binom{n}{2} + \cdots + 2^n\binom{n}{n} = 3^n.$$

---

## 9.3 Probability

Sample Spaces and Probability Functions • Determining Probabilities • Venn Diagrams and Tree Diagrams • Conditional Probability • Binomial Distributions

### Sample Spaces and Probability Functions

They say that knowledge only gets you in trouble if you know a little bit about something, and everybody knows a little bit about probability. It is that little bit of intuitive knowledge that makes us vulnerable to scam artists, casinos, and advertisers. Rather than lament that fact, we want to build on your intuitive understanding of probability and show you at least some of the mathematics behind it.

**Is Probability Just For Games?**

Probability theory got its start in letters between Blaise Pascal (1623–1662) and Pierre de Fermat (1601–1665) concerning games of chance, but it has come along way since then. Modern mathematicians like David Blackwell (1919), the first African-American to receive a fellowship to the Institute for Advanced Study at Princeton, have greatly extended both the theorey and the applications of probability, especially in the areas of statistics, quantum physics, and information theory. Moreover, the work of John Von Neumann (1903–1957) has led to a separate branch of modern discrete mathematics that really is about games, called Game Theory.

### Example 1  TESTING YOUR INTUITION ABOUT PROBABILITY

Find the probability of each of the following events.

**(a)** Tossing a head on one toss of a fair coin.

**(b)** Tossing two heads in a row on two tosses of a fair coin.

**(c)** Drawing a queen from a standard deck of 52 cards.

**(d)** Rolling a total of 4 on a single roll of two fair dice.

**(e)** Guessing all 6 numbers in a state lottery that requires you to pick 6 numbers from 1 to 46.

**Solution**

**(a)** There are two equally likely outcomes: {T, H}. The probability is 1/2.

**(b)** There are four equally likely outcomes: {TT, TH, HT, HH}. The probability is 1/4.

**(c)** There are 52 equally likely outcomes, 4 of which are queens. The probability is 4/52, or 1/13.

**Figure 9.2** A sum of 4 on a roll of two dice. (Example 1d)

**(d)** By the Multiplication Principle of Counting (Section 9.1), there are $6 \times 6 = 36$ equally likely outcomes. Of these, three $\{(1, 3), (3, 1), (2, 2)\}$ give a total of 4 (Figure 9.2). The probability is 3/36, or 1/12.

**(e)** There are $_{46}C_6 = 9{,}366{,}819$ equally likely ways that 6 numbers can be chosen from 46 numbers without regard to order. Only one of these choices wins the lottery. The probability is $1/9{,}366{,}819 \approx 0.00000010676$.

Notice in each of these cases that we first counted the number of possible outcomes of the experiment in question. The set of all possible outcomes of an experiment is a **sample space**. An **event** is a subset of a sample space. Each of our sample spaces consisted of a finite number of **equally likely outcomes**, which enabled us to find the probability of an event by counting.

---

**Probability of an Event (Equally Likely Outcomes)**

If $E$ is an event in a finite, nonempty sample space $S$ of equally likely outcomes, then the **probability** of the event $E$ is

$$P(E) = \frac{\text{the number of outcomes in } E}{\text{the number of outcomes in } S}.$$

---

The hypothesis of equally likely outcomes is critical here. Many people guess wrongly on the probability in Example 1d because they figure that there are 11 possible outcomes for the total on two fair dice: $\{2, 3, 4, 5, 6, 7, 8, 9, 10, 11, 12\}$, and that 4 is one of them. (That reasoning is correct so far.) The reason that 1/11 is not the probability of rolling a 4 is that all those totals are *not equally likely*, as anyone who has ever played a board game with dice will surely realize.

On the other hand, we can *assign* probabilities to the 11 outcomes in this smaller sample space in a way that is consistent with the number of ways each total can occur. The table below shows a **probability distribution**, in which each outcome is assigned a unique probability by a *probability function*.

| Outcome | Probability |
|---------|-------------|
| 2 | 1/36 |
| 3 | 2/36 |
| 4 | 3/36 |
| 5 | 4/36 |
| 6 | 5/36 |
| 7 | 6/36 |
| 8 | 5/36 |
| 9 | 4/36 |
| 10 | 3/36 |
| 11 | 2/36 |
| 12 | 1/36 |

We see that the outcomes are *not* equally likely, but we can find the probabilities of events by adding up the probabilities of the outcomes in the event, as in the following example.

### Example 2 ROLLING THE DICE

Find the probability of rolling a total divisible by 3 on a single roll of two fair dice.

**Solution** The event $E$ consists of the outcomes $\{3, 6, 9, 12\}$. To get the probability of $E$ we add up the probabilities of the outcomes in $E$ (see the table of the probability distribution):

$$P(E) = \frac{2}{36} + \frac{5}{36} + \frac{4}{36} + \frac{1}{36} = \frac{12}{36} = \frac{1}{3}.$$

Notice that this method would also have worked just fine with our 36-outcome sample space, in which every outcome has probability 1/36. In general, it is easier to work with sample spaces of equally likely events because it is not necessary to write out the probability distribution. When outcomes do have unequal probabilities, we need to know what probabilities to assign to the outcomes.

Not every function that assigns numbers to outcomes can qualify as a probability function.

---

**Empty Set**

A set with no elements is the *empty set*, denoted by $\varnothing$.

---

> ### Definition Probability Function
>
> A **probability function** is a function $P$ that assigns to each outcome in a sample space a unique real number, subject to the following conditions:
>
> **1.** $0 \le P(O) \le 1$ for every outcome $O$;
> **2.** the sum of the probabilities of all outcomes in $S$ is 1;
> **3.** $P(\varnothing) = 0$.

The probability of any event can then be defined in terms of the probability function.

> ### Probability of an Event (Outcomes not Equally Likely)
>
> Let $S$ be a finite, nonempty sample space in which every outcome has a probability assigned by a probability function $P$. If $E$ is any event in $S$, the **probability** of the event $E$ is the sum of the probabilities of all the outcomes contained in $E$.

### Example 3 TESTING A PROBABILITY FUNCTION

Is it possible to weight a standard 6-sided die in such a way that the probability of rolling each number $n$ is exactly $1/(n^2 + 1)$?

**Solution** The probability distribution would look like this:

| Outcome | Probability |
|---------|-------------|
| 1 | 1/2 |
| 2 | 1/5 |
| 3 | 1/10 |
| 4 | 1/17 |
| 5 | 1/26 |
| 6 | 1/37 |

This is not a valid probability function, because $1/2 + 1/5 + 1/10 + 1/17 + 1/26 + 1/37 \neq 1$.

## Determining Probabilities

It is not always easy to determine probabilities, but the arithmetic involved is fairly simple. It usually comes down to multiplication, addition, and (most importantly) counting. Here is the strategy we will follow:

---

**Strategy for Determining Probabilities**

1. Determine the sample space of all possible outcomes. When possible, choose outcomes that are equally likely.

2. If the sample space has equally likely outcomes, the probability of an event $E$ is determined by counting:

$$P(E) = \frac{\text{the number of outcomes in } E}{\text{the number of outcomes in } S}.$$

3. If the sample space does not have equally likely outcomes, determine the probability function. (This is not always easy to do.) Check to be sure that the conditions of a probability function are satisfied. Then the probability of an event $E$ is determined by adding up the probabilities of all the outcomes contained in $E$.

---

### Example 4   CHOOSING CHOCOLATES, SAMPLE SPACE I

Sal opens a fresh box of a dozen chocolate cremes and generously offers two of them to Val. Val likes vanilla cremes the best, but all the chocolates look alike on the outside. If four of the twelve cremes happen to be vanilla, what is the probability that both of Val's picks turn out by chance to be vanilla?

**Solution** The experiment in question is the selection of two chocolates, without regard to order, from a box of 12. There are $_{12}C_2 = 66$ outcomes of this experiment, and all of them are equally likely. We can therefore determine the probability by counting.

The event $E$ consists of all possible pairs of vanilla cremes that can be chosen, without regard to order, from 4 vanilla cremes available. There are $_4C_2 = 6$ ways to form such pairs.

Therefore, $P(E) = 6/66 = 1/11$.

Many probability problems require that we think of events happening in succession, often with the occurrence of one event affecting the probability of the occurrence of another event. In these cases we use a law of probability called the Multiplication Principle of Probability.

---

### Multiplication Principle of Probability

Suppose an event $A$ has probability $p_1$ and an event $B$ has probability $p_2$ under the assumption that $A$ occurs. Then the probability that both $A$ and $B$ occur is $p_1 p_2$.

---

If the events $A$ and $B$ are **independent**, we can omit the phrase "under the assumption that $A$ occurs," since that assumption would not matter.

As an example of this principle at work, we will solve the *same problem* as that posed in Example 4, this time using a sample space that appears at first to be simpler, but which consists of events that are not equally likely.

### Example 5  CHOOSING CHOCOLATES, SAMPLE SPACE II

Sal opens a fresh box of a dozen chocolate cremes and generously offers two of them to Val. Val likes vanilla cremes the best, but all the chocolates look alike on the outside. If four of the twelve cremes happen to be vanilla, what is the probability that both of Val's picks turn out by chance to be vanilla?

**Ordered or Unordered?**

Notice that in Example 4 we had a sample space in which order was disregarded, whereas in Example 5 we have a sample space in which order matters. (For example, *UV* and *VU* are distinct outcomes.) The order matters in Example 5 because we are considering the probabilities of two events (first draw, second draw), one of which affects the other. In Example 4 we are simply counting unordered combinations.

**Solution**  As far as Val is concerned, there are two kinds of chocolate cremes: vanilla ($V$) and unvanilla ($U$). When choosing two chocolates, there are four possible outcomes: *VV*, *VU*, *UV*, and *UU*. We need to determine the probability of the outcome *VV*.

Notice that these four outcomes are *not equally likely*! There are twice as many $U$ chocolates as $V$ chocolates. So we need to consider the distribution of probabilities, and we may as well begin with $P(VV)$, as that is the probability we seek.

The probability of picking a vanilla creme on the first draw is 4/12. The probability of picking a vanilla creme on the second draw, *under the assumption that a vanilla creme was drawn on the first*, is 3/11. By the multiplication Principle, the probability of drawing a vanilla creme on both draws is

$$\frac{4}{12} \cdot \frac{3}{11} = \frac{1}{11}.$$

Since this is the probability we are looking for, we do not need to compute the probabilities of the other outcomes. However, you should verify that the other probabilities would be:

$$P(VU) = \frac{4}{12} \cdot \frac{8}{11} = \frac{8}{33}$$

$$P(UV) = \frac{8}{12} \cdot \frac{4}{11} = \frac{8}{33}$$

$$P(UU) = \frac{8}{12} \cdot \frac{7}{11} = \frac{14}{33}$$

Notice that $P(VV) + P(VU) + P(UV) + P(UU) = (1/11) + (8/33) + (8/33) + (14/33) = 1$, so the probability function checks out.

## Venn Diagrams and Tree Diagrams

We have seen many instances in which geometric models help us to understand algebraic models more easily, and probability theory is yet another setting in which this is true. **Venn diagrams**, associated mainly with the world of set theory, are good for visualizing relationships among events within sample spaces. **Tree diagrams**, which we first met in Section 9.1 as a way to visualize the Multiplication Principle of Counting, are good for visualizing the Multiplication Principle of Probability.

### Example 6 USING A VENN DIAGRAM

In a large high school, 54% of the students are girls and 62% of the students play sports. Half of the girls at the school play sports.

**(a)** What percentage of the students who play sports are boys?

**(b)** If a student is chosen at random, what is the probability it is a boy who does not play sports?

**Solution** There are several correct ways to approach this problem, but in any approach it helps to organize all the categories (boy, girl, sports, no sports). As a visualization, we draw a large rectangle to represent the sample space (all students at the school). Inside the rectangle we draw two regions to represent "girls" and "sports." Since there are girls who play sports, we show the two regions overlapping (Figure 9.3).

We fill in the regions with the percentages (or probabilities), beginning with the overlap, which represents girls who play sports. The number that belongs in this region is half of 0.54, or 0.27. That leaves 0.27 for the "girls and not sports" part of the left-hand circle and 0.35 for the "sports and not girls" part of the right-hand circle (Figure 9.4). Since all probabilities must add up to 1, we can also fill in the region outside the two circles with 0.11.

We can now answer the questions by reading the Venn diagram.

**(a)** We see from the diagram that the ratio of *boys* who play sports to *all students* who play sports is 0.35/(0.35 + 0.27) = 35/62. Therefore, $35/62 \approx 56.45\%$ of the students who play sports at the school are boys.

**(b)** We see directly from the diagram that the probability of "boy and not sports" is 0.11.

**John Venn**

John Venn (1834–1923) was an English logician and clergyman, just like his contemporary, Charles L. Dodgson. Although both men used overlapping circles to illustrate their logical syllogisms, it is Venn whose name lives on in connection with these diagrams. Dodgson's name barely lives on at all, and yet he is far the more famous of the two: under the pen name of Lewis Carroll, he wrote *Alice's Adventures in Wonderland* and *Through the Looking Glass*.

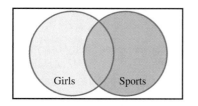

**Figure 9.3** A Venn diagram for Example 6. The overlapping region common to both circles represents "girls who play sports." The region outside both circles (but inside the rectangle) represents "boys who do not play sports."

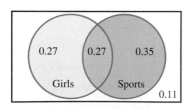

**Figure 9.4** A Venn diagram for Example 6 with the probabilities filled in.

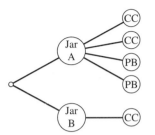

**Figure 9.5** A tree diagram for Example 7.

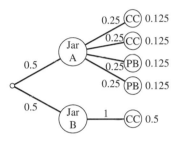

**Figure 9.6** The tree diagram for Example 7 with the probabilities filled in. Notice that the five cookies are not equally likely to be drawn. Notice also that the probabilities of the five cookies do add up to 1.

### Example 7   USING A TREE DIAGRAM

Two identical cookie jars are on a counter. Jar *A* contains 2 chocolate chip and 2 peanut butter cookies, while jar *B* contains 1 chocolate chip cookie. We select a cookie at random. What is the probability that it is a chocolate chip?

**Solution**  It is tempting to say 3/5, since there are 5 cookies in all, 3 of which are chocolate chip. Indeed, this would be the answer if the cookies were all in the same jar. However, the fact that they are in different jars means that the 5 cookies are *not equally likely outcomes*. That lone chocolate chip in jar *B* has a much better chance of being chosen than any of the four cookies in jar *A*.

We need to think of this as a two-step experiment: first pick a jar, then pick a cookie. A tree diagram allows us to "see" the choices (Figure 9.5).

The probability of picking either jar *A* or jar *B* is 1/2. We fill in those probabilities along the branches connecting the root of the tree to the two jar choices. Inside jar *A*, each of the four cookies has probability 1/4. We fill in those probabilities along the branches connecting jar *A* to the four cookies. Inside jar *B*, the probability of picking the lone chocolate chip is 1. We fill in that probability along that branch.

Now we travel from the root to the end of each branch, *multiplying probabilities* as we go, by the Multiplication Principle. We write each probability at the end of the branch (Figure 9.6). *Those* probabilities are the true probabilities of the cookies being chosen, and (as predicted) they are not equally likely events.

The event "chocolate chip" is a set containing three outcomes. We add the probabilities of the three outcomes to get the probability of the event:

$$P(\text{chocolate chip}) = 0.125 + 0.125 + 0.5 = 0.75.$$

## Conditional Probability

The probability of drawing a chocolate chip cookie in Example 7 is an example of **conditional probability**, since the "cookie" probability is **dependent** on the "jar" outcome. A convenient symbol to use with conditional probability is $P(A|B)$, pronounced "*P* of *A* given *B*," meaning "the probability of the event *A*, given that event *B* occurs." In the cookie jars of Example 7,

$$P(\text{chocolate chip}|\text{jar } A) = \frac{2}{4} \quad \text{and} \quad P(\text{chocolate chip}|\text{jar } B) = 1.$$

(In the tree diagram, these are the probabilities along the *branches* that come out of the two jars, not the probabilities at the *ends* of the branches.)

The Multiplication Principle of Probability can be stated succinctly with this notation as follows:

$$P(A \text{ and } B) = P(A) \cdot P(B|A).$$

This is how we found the numbers at the ends of the branches in Example 7.

As our final example of a probability problem, we will show how to use this formula in a different but equivalent form, sometimes called the **conditional probability formula**:

$$P(B|A) = \frac{P(A \text{ and } B)}{P(A)}.$$

**Example 8   USING THE CONDITIONAL PROBABILITY FORMULA**

Suppose we have drawn a cookie at random from one of the jars described in Example 7. Given that it is a chocolate chip, what is the probability that it came from jar *A*?

**Solution**   By the formula,

$$P(\text{jar } A \,|\, \text{chocolate chip}) = \frac{P(\text{jar } A \text{ and chocolate chip})}{P(\text{chocolate chip})}$$

$$= \frac{(1/2)(2/4)}{0.75} = \frac{0.25}{0.75} = \frac{1}{3}$$

**Exploration 1**   **Testing Positive for HIV**

As of December 1998, the probability of an adult (age 15–49) in North America having HIV/AIDS was 0.0056 (Source: *UNAIDS/WHO*). The ELISA test is used to detect the virus antibody in blood. If the antibody is present, the test reports positive with probability 0.997 and negative with probability 0.003. If the antibody is not present, the test reports positive with probability 0.015 and negative with probability 0.985.

1. Draw a tree diagram with branches to nodes "antibody present" and "antibody absent" branching from the root. Fill in the probabilities for North American adults (age 15–49) along the branches. (Note that these two probabilities must add up to 1.)

2. From the node at the end of each of the two branches, draw branches to "positive" and "negative." Fill in the probabilities along the branches.

3. Use the Multiplication Principle to fill in the probabilities at the ends of the four branches. Check to see that they add up to 1.

4. Find the probability of a positive test result. (Note that this event consists of two outcomes).

5. Use the conditional probability formula to find the probability that a person with a positive test result actually *has* the antibody, i.e., $P(\text{antibody present} \,|\, \text{positive})$.

You might be surprised that the answer to part 5 is so low, but it should be compared with the probability of the antibody being present *before* seeing the positive test result, which was 0.0056. Nonetheless, that is why a positive ELISA test is followed by further testing before a diagnosis of HIV/AIDS is made. This is the case with many diagnostic tests.

## Binomial Distributions

We noted in our "Strategy for Determining Probabilities" that it is not always easy to determine a probability distribution for a sample space with unequal probabilities. An interesting exception for those who have studied the Binomial Theorem (Section 9.2) is the binomial distribution.

### Example 9  REPEATING A SIMPLE EXPERIMENT

We roll a fair die four times. What is the probability that we roll:

**(a)** all 3's.          **(b)** no 3's          **(c)** exactly two 3's.

### Solution

**(a)** We have a probability 1/6 of rolling a 3 each time. By the Multiplication Principle, the probability of rolling a 3 all four times is $(1/6)^4 \approx 0.00077$.

**(b)** There is a probability 5/6 of rolling something other than 3 each time. By the Multiplication Principle, the probability of rolling a non-3 all four times is $(5/6)^4 \approx 0.48225$.

**(c)** The probability of rolling two 3's followed by two non-3's (again by the Multiplication Principle) is $(1/6)^2(5/6)^2 \approx 0.01929$. However, that is not the only outcome we must consider. In fact, the two 3's could occur *anywhere* among the four rolls, in exactly $\binom{4}{2} = 6$ ways. That gives us 6 outcomes, each with probability $(1/6)^2(5/6)^2$. The probability of the event "exactly two 3's" is therefore $\binom{4}{2}(1/6)^2(5/6)^2 \approx 0.11574$.

Did the form of those answers look a little familiar? Watch what they look like when we let $p = 1/6$ and $q = 5/6$:

$$P(\text{four 3's}) = p^4$$

$$P(\text{no 3's}) = q^4$$

$$P(\text{two 3's}) = \binom{4}{2} p^2 q^2$$

You can probably recognize these as three of the terms in the expansion of $(p + q)^4$. This is no coincidence. In fact, the terms in the expansion

$$(p + q)^4 = p^4 + 4p^3q^1 + 6p^2q^2 + 4p^1q^3 + q^4$$

give the exact probabilities of 4, 3, 2, 1, and 0 three's (respectively) when we toss a fair die four times! That is why this is called a *binomial probability distribution*. The general theorem follows.

## Binomial Probabilities on a Calculator

Your calculator might be programmed to find values for the binomial probability distribution function (binompdf). The solution to Example 10 in one calculator syntax, for example, could be obtained by:

**(a)** binompdf(20, .9, 20) (20 repetitions, 0.9 probability, 20 successes)

**(b)** binompdf(20, .9, 18) (20 repetitions, 0.9 probability, 18 successes)

**(c)** $1 -$ binomcdf(20, .9, 17) (1 minus the *cumulative* probability of 17 or fewer successes)

Check your *Owner's Manual* for more information.

### Follow-up

Assume that the probability that a newborn child is a particular sex is 50%. In a family of four children, what is the probability of the given event.

a. All the children are girls. **(1/16)**

b. None of the children are the same sex. **(0)**

c. All the children are boys or girls. **(1)**

d. A least two of the children are boys. **(11/16)**

### Assignment Guide

Day 1: Ex. 3–39, multiples of 3
Day 2: Ex. 41, 42, 45, 46, 48, 49, 50, 51

### Cooperative Learing

Group Activity: Ex. 33, 34

### Notes on Exercises

Ex. 1–34 are typical probability practice problems.
Ex. 37–40, 42–45, 47–50 are application problems requiring students to decide which probability equation to use in each situation.
Ex. 51 and 52 are expected value problems.

### Ongoing Assessment

Self-Assessment: Ex. 7, 11, 19, 31, 43
Embedded Assessment: Ex. 45, 46, 50, 51

### Theorem  Bionomial Distribution

Suppose an experiment consists of $n$ independent repetitions of an experiment with two outcomes, called "success" and "failure." Let $P(\text{success}) = p$ and $P(\text{failure}) = q$. (Note that $q = 1 - p$.)

Then the terms in the binomial expansion of $(p + q)^n$ give the respective probabilities of exactly $n, n - 1, \ldots, 2, 1, 0$ successes. The distribution is shown below:

| Number of successes out of $n$ independent repetitions | Probability |
|---|---|
| $n$ | $p^n$ |
| $n - 1$ | $\binom{n}{n-1}p^{n-1}q$ |
| $\vdots$ | $\vdots$ |
| $r$ | $\binom{n}{r}p^r q^{n-r}$ |
| $\vdots$ | $\vdots$ |
| $1$ | $\binom{n}{1}pq^{n-1}$ |
| $0$ | $q^n$ |

### Example 10  SHOOTING FREE THROWS

Suppose Michael is a 90% free throw shooter. If he shoots 20 free throws, and if his chance of making each one is 90%, independent of the other shots (an assumption you might question in a game situation), what is the probability that he makes

**(a)** all 20?

**(b)** exactly 18?

**(c)** at least 18?

**Solution** We could get the probabilities of all possible outcomes by expanding $(0.9 + 0.1)^{20}$, but that is not necessary in order to answer these three questions. We just need to compute three specific terms.

**(a)** $P(20 \text{ successes}) = (0.9)^{20} \approx 0.12158$.

**(b)** $P(18 \text{ successes}) = \binom{20}{18}(0.9)^{18}(0.1)^2 \approx 0.28518$.

**(c)** $P(\text{at least 18 successes}) = P(18) + P(19) + P(20)$

$$= \binom{20}{18}(0.9)^{18}(0.1)^2 + \binom{20}{19}(0.9)^{19}(0.1) + (0.9)^{20}$$

$$\approx 0.6769$$

# Quick Review 9.3

In Exercises 1–8, tell how many outcomes are possible for the experiment.

1. A single coin is tossed.  2

2. A single 6-sided die is rolled.  6

3. Three different coins are tossed.  8

4. Three different 6-sided dice are rolled.  216

5. Five different cards are drawn from a standard deck of 52.

6. Two chips are drawn simultaneously without replacement from a jar containing 10 chips.  45

7. Five people are lined up for a photograph.  120

8. Three digit numbers are formed from the numbers $\{1, 2, 3, 4, 5\}$ without repetition.  60

In Exercises 9 and 10, evaluate the expression by pencil and paper. Verify your answer with a calculator.

9. $\dfrac{_5C_3}{_{10}C_3}$  $\dfrac{1}{12}$

10. $\dfrac{_5C_2}{_{10}C_2}$  $\dfrac{2}{9}$

# Section 9.3 Exercises

In Exercises 1–8, a red die and a green die have been rolled. What is the probability of the event?

1. The sum is 9.  1/9

2. The sum is even.  1/2

3. The number on the red die is greater than the number on the green die.  5/12

4. The sum is less than 10.  5/6

5. Both dice are odd.  1/4

6. Both dice are even.  1/4

7. The sum is prime.  5/12

8. The sum is 7 or 11.  2/9

The maker of a popular chocolate candy that is covered in a thin colored shell has released information about the overall color proportions in its production of the candy, summarized in the following table.

| Color | Brown | Red | Yellow | Green | Orange | Tan |
|---|---|---|---|---|---|---|
| Proportion | 0.3 | 0.2 | 0.2 | 0.1 | 0.1 | 0.1 |

In Exercises 9–14, a single candy of this type is selected at random from a newly-opened bag. What is the probability that the candy has the given color(s)?

9. Brown or tan  0.4

10. Red, green, or orange  0.4

11. Red  0.2

12. Not red  0.8

13. Neither orange nor yellow

14. Neither brown nor tan  0.6

A peanut version of the same candy has all the same colors except tan. The proportions of the peanut version are given in the following table.

| Color | Brown | Red | Yellow | Green | Orange |
|---|---|---|---|---|---|
| Proportion | 0.3 | 0.2 | 0.2 | 0.2 | 0.1 |

In Exercises 15–20, a candy of this type is selected at random from *each of two* newly-opened bags. What is the probability that the two candies have the given color(s)?

15. Both are brown.  0.09

16. Both are orange.  0.01

17. One is red and one is green.  0.08

18. The first is brown and the second is yellow.  0.06

19. Neither is yellow.  0.64

20. The first is not red and the second is not orange.  0.72

Exercises 21–24 concern a version of the card game "bid Euchre" that uses a pack of 24 cards, consisting of ace, king, queen, jack, 10, and 9 in each of the four suits (spades, hearts, diamonds, and clubs). In bid Euchre a hand consists of 6 cards. Find the probability of each event.

21. **Euchre**  A hand is all spades.  1/34,569

22. **Euchre**  All six cards are from the same suit.  1/33,649

23. **Euchre**  A hand includes all four aces.  5/3542

24. **Euchre**  A hand includes two jacks of the same color (called the right and left bower).  190/1771

In Exercises 25–32, ten dimes dated 1990 through 1999 are tossed. Find the probability of each event.

25. **Tossing Ten Dimes**  Heads on the 1990 dime only.

26. **Tossing Ten Dimes**  Heads on the 1991 and 1996 dimes only.  1/1024

27. **Tossing Ten Dimes**  Heads on all 10 dimes.  1/1024

28. **Tossing Ten Dimes**  Heads on all but one dime.  5/512

29. **Tossing Ten Dimes**  Exactly two heads.  45/1024

30. **Tossing Ten Dimes**  Exactly three heads.  15/128

31. **Tossing Ten Dimes**  At least one head.  1023/1024

32. **Tossing Ten Dimes**  At least two heads.  1013/1024

33. **Group Activity  Using Venn Diagrams**  $A$ and $B$ are events in a sample space $S$ such that $P(A) = 0.6$, $P(B) = 0.5$, and $P(A \text{ and } B) = 0.3$.

   **(a)** Draw a Venn diagram showing the overlapping sets $A$ and $B$ and fill in the probabilities of the four regions formed.

**(b)** Find the probability that $A$ occurs but $B$ does not.  0.3

**(c)** Find the probability that $B$ occurs but $A$ does not.  0.2

**(d)** Find the probability that neither $A$ nor $B$ occurs.  0.2

**(e)** Are events $A$ and $B$ independent? (That is, does $P(A|B) = P(A)$?)  yes

**34. Group Activity  Using Venn Diagrams** $A$ and $B$ are events in a sample space $S$ such that $P(A) = 0.7$, $P(B) = 0.4$, and $P(A \text{ and } B) = 0.2$.

**(a)** Draw a Venn diagram showing the overlapping sets $A$ and $B$ and fill in the probabilities of the four regions formed.

**(b)** Find the probability that $A$ occurs but $B$ does not.  0.5

**(c)** Find the probability that $B$ occurs but $A$ does not.  0.2

**(d)** Find the probability that neither $A$ nor $B$ occurs.  0.1

**(e)** Are events $A$ and $B$ independent? (That is, does $P(A|B) = P(A)$?)  no

In Exercises 35 and 36, it will help to draw a tree diagram.

**35. Piano Lessons** If it rains tomorrow, the probability is 0.8 that John will practice his piano lesson. If it does not rain tomorrow, there is only a 0.4 chance that John will practice. Suppose that the chance of rain tomorrow is 60%. What is the probability that John will practice his piano lesson? 0.64

**36. Predicting Cafeteria Food** If the school cafeteria serves meat loaf, there is a 70% chance that they will serve peas. If they do not serve meat loaf, there is a 30% chance that they will serve peas anyway. The students know that meat loaf will be served exactly once during the 5-day week, but they do not know which day. If tomorrow is Monday, what is the probability that

**(a)** the cafeteria serves meat loaf?  0.20

**(b)** the cafeteria serves meat loaf and peas?  0.14

**(c)** the cafeteria serves peas?  0.38

**37. Renting Cars** Floppy Jalopy Rent-a-Car has 25 cars available for rental—20 big bombs and 5 midsize cars. If two cars are selected at random, what is the probability that both are big bombs?  19/30

**38. Defective Calculators** Dull Calculators, Inc., knows that a unit coming off an assembly line has a probability of 0.037 of being defective. If four units are selected at random during the course of a workday, what is the probability that none of the units are defective?  $\approx 0.860$

**39. Causes of Death** The government designates a single cause for each death in the United States. The resulting data indicate that 45% of deaths are due to heart and other cardiovascular disease and 22% are due to cancer.

**(a)** What is the probability that the death of a randomly selected person will be due to cardiovascular disease or cancer?  0.67

**(b)** What is the probability that the death will be due to some other cause?  0.33

**40. Yahtzee** In the game of *Yahtzee*, on the first roll five dice are tossed simultaneously. What is the probability of rolling five of a kind (which is Yahtzee!) on the first roll?  1/1296

**41. Writing to Learn** Explain why the following statement cannot be true. The probabilities that a computer salesperson will sell zero, one, two, or three computers in any one day are 0.12, 0.45, 0.38, and 0.15, respectively.

**42. HIV Testing** A particular test for HIV, the virus that causes AIDS, is 0.7% likely to produce a false positive result—a result indicating that the human subject has HIV when in fact the person is not carrying the virus. If 60 individuals who are HIV-negative are tested, what is the probability of obtaining at least one false result?  $\approx 34.4\%$

**43. Graduate School Survey** The Earmuff Junction College Alumni Office surveys selected members of the class of 1996. Of the 254 who graduated that year, 172 were women, 124 of whom went on to graduate school. Of the male graduates, 58 went on to graduate school. What is the probability of the given event?

**(a)** The graduate is a woman.  86/127

**(b)** The graduate went on to graduate school.  91/127

**(c)** The graduate was a woman who went on to graduate school.  62/127

**44. Indiana Jones and the Final Exam** Professor Indiana Jones gives his class a list of 20 study questions, from which he will select 8 to be answered on the final exam. If a given student knows how to answer 14 of the questions, what is the probability that the student will be able to answer correctly the given number of questions?

**(a)** All 8 questions.  77/3230  **(b)** Exactly 5 questions.

**(c)** At least 6 questions.  176/323

**45. Graduation Requirement** To complete the kinesiology requirement at Palpitation Tech you must pass two classes chosen from aerobics, aquatics, defense arts, gymnastics, racket sports, recreational activities, rhythmic activities, soccer, and volleyball. If you decide to choose your two classes at random by drawing two class names from a box, what is the probability that you will take racket sports and rhythmic activities?  1/36

**46. Writing to Learn** During July in Gunnison, Colorado, the probability of at least 1 hour a day of sunshine is 0.78, the probability of at least 30 minutes of rain is 0.44, and the probability that it will be cloudy all day is 0.22. Write a paragraph explaining whether this statement could be true.

## Explorations

**47. Empirical Probability** In real applications it is often necessary to approximate the probabilities of the various outcomes of an experiment by performing the experiment a large number of times and recording the results. Barney's Bread Basket offers five different kinds of bagels. Barney records the sales of the first 500 bagels in a given week in the table shown below:

| Type of Bagel | No. Sold |
|---|---|
| Plain | 185 |
| Onion | 60 |
| Rye | 55 |
| Cinnamon Raisin | 125 |
| Sourdough | 75 |

**(a)** Use the observed sales number to approximate the probability that a random customer buys a plain bagel. Do the same for each other bagel type and make a table showing the approximate probability distribution.

**(b)** Assuming independence of the events, find the probability that three customers in a row all order plain bagels.  0.051

**(c) Writing to Learn** Do you think it is reasonable to assume that the orders of three consecutive customers actually are independent? Explain.

**48. Straight Poker** In the original version of poker known as "straight" poker, a 5-card hand is dealt from a standard deck of 52. What is the probability of the given event?

**(a)** A hand will contain at least one king.  18,472/54,145

**(b)** A hand will be a "full house" (any three of one kind and a pair of another kind).  6/4165

**49. Married Students** Suppose that 23% of all college students are married. Answer the following questions for a random sample of eight college students.

**(a)** How many would you expect to be married?  ≈ 2

**(b)** Would you regard it as unusual if the sample contained five married students?  yes

**(c)** What is the probability that five or more of the eight students are married?  ≈ 1.913%

**50. Investigating an Athletic Program** A university widely known for its track and field program claims that 75% of its track athletes get degrees. A journalist investigates what happened to the 32 athletes who began the program over a 6-year period that ended 7 years ago. Of these athletes, 17 have graduated and the remaining 15 are no longer attending any college. If the university's claim is true, the number of athletes who graduate among the 32 examined should have been governed by binomial probability with $p = 0.75$.

**(a)** What is the probability that exactly 17 athletes should have graduated?  ≈ 0.396%

**(b)** What is the probability that 17 or fewer athletes should have graduated?  ≈ 0.596%

**(c)** If you were the journalist, what would you say in your story on the investigation?

## Extending the Ideas

**51. Expected Value** If the outcomes of an experiment are given numerical values (such as the total on a roll of two dice, or the payoff on a lottery ticket), we define the **expected value** to be the sum of all the numerical values times their respective probabilities.

For example, suppose we roll a fair die. If we roll a multiple of 3, we win $3; otherwise we lose $1. The probabilities of the two possible payoffs are shown in the table below:

| Value | Probability |
|---|---|
| +3 | 2/6 |
| −1 | 4/6 |

The expected value is $3 \times (2/6) + (-1) \times (4/6) = (6/6) - (4/6) = 1/3$.

We interpret this to mean that we would win an average of 1/3 dollar per game in the long run.

**(a)** A game is called *fair* if the expected value of the payoff is zero. Assuming that we still win $3 for a multiple of 3, what should we pay for any other outcome, in order to make the game fair?  $1.50

**(b)** Suppose we roll *two* fair dice and look at the total under the original rules. That is, we win $3 for rolling a multiple of 3 and lose $1 otherwise. What is the expected value of this game?  1/3

**52. Expected Value** (Continuation of Exercise 51)  Gladys has a personal rule never to enter the lottery (picking 6 numbers from 1 to 46) until the payoff reaches 4 million dollars. When it does reach 4 million, she always buys ten different $1 tickets.

**(a)** Assume that the payoff for a winning ticket is 4 million dollars. What is the probability that Glady holds a winning ticket? (Refer to Example 1 of this section for the probability of any ticket winning.)

**(b)** Fill in the probability distribution for Gladys's possible payoffs in the table below. (Note that we subtract $10 from the $4 million, since Gladys has to pay for her tickets even if she wins.)

| Value | Probability |
|---|---|
| −10 | |
| +3,999,990 | |

**(c)** Find the expected value of the game for Gladys.

**(d) Writing to Learn** In terms of the answer in part (b), explain to Gladys the long-term implications of her strategy.

**Sequences and Series**

Infinite Sequences • Arithmetic and Geometric Sequences • Sequences and Graphing Calculators • Summation Notation • Sums of Arithmetic and Geometric Sequences • Infinite Series • Convergence of Geometric Series

## Infinite Sequences

**Objective**

Students will be able to express arithmetic and geometric sequences explicitly and recursively; they will also be able to use sigma notation and basic summation formulas to find the sum of a finite series or a converging infinite geometric series.

**Motivate**

Have the students speculate what the sum of 1/2 + 1/4 + 1/8 + ⋯ might be. **(1)** (this will be examined more thoroughly under the topic of "Convergence of Geometric Series," Example 13d.)

**Lesson Guide**

Day 1: Infinite Sequences; Arithmetic and Geometric Sequences; Sequences and Graphing Calculators; Summation Notation; Sums of Arithmetic and Geometric Sequences
Day 2: Infinite Series; Convergence of Geometric Series

One of the most natural ways to study patterns in mathematics is to look at an ordered progression of numbers, called a **sequence**. Here are some examples of sequences:

**1.** 5, 10, 15, 20, 25

**2.** 2, 4, 8, 16, 32, . . . , $2^k$, . . .

**3.** $\left\{ \dfrac{1}{k} : k = 1, 2, 3, \ldots \right\}$

**4.** $\{a_1, a_2, a_3, \ldots, a_k, \ldots\}$, which is sometimes abbreviated $\{a_k\}$

The first of these is a **finite sequence**, while the other three are **infinite sequences**. Notice that in (2) and (3) we were able to define a rule that gives the $k$th number in the sequence (called the **$k$th term**) as a function of $k$. In (4) we do not have a rule, but notice how we can use subscript notation ($a_k$) to identify the $k$th term of a "general" infinite sequence. In this sense, an infinite sequence can be thought of as a *function* that assigns a unique number ($a_k$) to each natural number $k$.

We will not concern ourselves much with finite sequences, so a "sequence" in this book will mean an infinite sequence unless otherwise specified. We will look at two different ways to define a sequence: explicitly and recursively.

### Example 1 DEFINING A SEQUENCE EXPLICITLY

Find the first 6 terms and the 100th term of the sequence $\{a_k\}$ in which $a_k = k^2 - 1$.

**Solution** Since we know the $k$th term *explicitly* as a function of $k$, we need only to evaluate the function to find the required terms:

$$a_1 = 1^2 - 1 = 0, \ \ a_2 = 3, \ \ a_3 = 8, \ \ a_4 = 15, \ \ a_5 = 24, \ \ a_6 = 35, \quad \text{and}$$

$$a_{100} = 100^2 - 1 = 9999.$$

Explicit formulas are the easiest to work with, but there are other ways to define sequences. For example, we can specify the value of the first term, $a_1$, and then give a rule for obtaining each succeeding term **recursively** from the one preceding it.

## Example 2  DEFINING A SEQUENCE RECURSIVELY

Find the first 6 terms and the 100th term for the sequence defined recursively by the conditions:

$$b_1 = 3$$

$$b_n = b_{n-1} + 2 \quad \text{for} \quad \text{all} \quad n > 1$$

**Solution**  We proceed a term at a time, starting with $b_1 = 3$ and getting each succeeding term by adding 2 to the term just before it:

$$b_1 = 3$$

$$b_2 = b_1 + 2 = 5$$

$$b_3 = b_2 + 2 = 7$$

etc.

Eventually it becomes apparent that we are building the sequence of odd natural numbers, beginning with 3:

$$\{3, 5, 7, 9, \ldots\}.$$

The 100th term is 99 terms beyond the first, which means that we can get there quickly by adding 99 2's to the number 3:

$$b_{100} = 3 + 99 \times 2 = 201.$$

## Arithmetic and Geometric Sequences

There are all kinds of rules by which we can construct sequences, but two particular types of sequences dominate in mathematical applications: those in which pairs of successive terms all have a common *difference* (**arithmetic** sequences), and those in which pairs of successive terms all have a common quotient, or *ratio* (**geometric** sequences). We will take a closer look at those in this section.

> **Definition  Arithmetic Sequence**
>
> A sequence $\{a_n\}$ is an **arithmetic sequence** if it can be written in the form
>
> $$\{a, a + d, a + 2d, \ldots, a + (n - 1)d, \ldots\} \text{ for some constant } d.$$
>
> The number $d$ is called the **common difference**.
>
> Each term in an arithmetic sequence can be obtained recursively from its preceding term by adding $d$:
>
> $$a_n = a_{n-1} + d \text{ (for all } n \geq 2).$$

## Example 3  IDENTIFYING ARITHMETIC SEQUENCES

Determine whether the sequence could be arithmetic. If so, find the 100th term.

**(a)** $-6, -2, 2, 6, 10, \ldots$     **(b)** $48, 24, 12, 6, 3, \ldots$

**(c)** $\ln 3, \ln 6, \ln 12, \ln 24, \ldots$

**Solution**

**(a)** Note that the difference between each pair of successive terms is 4. If the pattern continues, this sequence is arithmetic with constant difference $d = 4$.

Therefore, $a_{100} = -6 + 99 \times 4 = 390$.

**(b)** This sequence is not arithmetic, since

$$a_2 - a_1 = 24 - 48 = -24 \quad \text{and} \quad a_3 - a_2 = 12 - 24 = -12.$$

**(c)** This sequence might not look arithmetic at first, but remember that $\ln u - \ln v = \ln (u/v)$.

$$a_2 - a_1 = \ln 6 - \ln 3 = \ln \frac{6}{3} = \ln 2$$

$$a_3 - a_2 = \ln 12 - \ln 6 = \ln \frac{12}{6} = \ln 2$$

$$a_4 - a_3 = \ln 24 - \ln 12 = \ln \frac{24}{12} = \ln 2$$

If the pattern continues, this sequence is arithmetic with constant difference $d = \ln 2$.

Therefore, $a_{100} = \ln 3 + 99 \ln 2 = \ln 3 + \ln 2^{99} = \ln (3 \cdot 2^{99}) \approx 69.720$.

---

**Definition** **Geometric Sequence**

A sequence $\{a_n\}$ is a **geometric sequence** if it can be written in the form

$$\{a, a \cdot r, a \cdot r^2, \ldots, a \cdot r^{n-1}, \ldots\} \text{ for some non-zero constant } r.$$

The number $r$ is called the **common ratio**.

Each term in a geometric sequence can be obtained recursively from its preceding term by multiplying by $r$:

$$a_n = a_{n-1} \cdot r \text{ (for all } n \geq 2\text{)}.$$

---

**Example 4** IDENTIFYING GEOMETRIC SEQUENCES

Determine whether the sequence could be geometric. If so, find the 20th term.

**(a)** $2, \dfrac{2}{3}, \dfrac{2}{9}, \dfrac{2}{15}, \dfrac{2}{21}, \ldots$

**(b)** $3, 6, 12, 24, 48, \ldots$

**(c)** $10^{-3}, 10^{-1}, 10^1, 10^3, 10^5, \ldots$

## Solution

**(a)** We compute the ratio of each term to the preceding term:

$$\frac{a_2}{a_1} = \frac{2/3}{2} = \frac{1}{3}$$

$$\frac{a_3}{a_2} = \frac{2/9}{2/3} = \frac{1}{3}$$

$$\frac{a_4}{a_3} = \frac{2/15}{2/9} = \frac{3}{5}$$

The sequence is not geometric because the ratios are not constant.

**(b)** Again, we compute the ratios, but this time the ratios are all 2:

$$\frac{6}{3} = \frac{12}{6} = \frac{24}{12} = \frac{48}{24} = 2.$$

If this pattern continues, the sequence is geometric with common ratio $r = 2$.

Therefore, $a_{20} = 3 \cdot 2^{19} = 1{,}572{,}864$.

**(c)** The ratios this time involve some work with the laws of exponents, but they are all $10^2$:

$$\frac{10^{-1}}{10^{-3}} = \frac{10^1}{10^{-1}} = \frac{10^3}{10^1} = \frac{10^5}{10^3} = 10^2.$$

If this pattern continues, the sequence is geometric with common ratio $r = 10^2$.

Therefore, the 20th term is $a_{20} = 10^{-3} \cdot (10^2)^{19} = 10^{-3+38} = 10^{35}$.

**Example 5  DETERMINING ARITHMETIC AND GEOMETRIC SEQUENCES**

A sequence begins 12, 4, . . . . Find a formula for the 10th term if the sequence is

**(a)** arithmetic          **(b)** geometric.

## Solution

**(a)** Assume that the sequence is arithmetic. Then $12 + d = 4$, and so $d = -8$. It follows that $a_{10} = 12 + (9)(-8) = -60$.

**(b)** Assume that the sequence is geometric. Then $12 \cdot r = 4$, and $r = 1/3$. It follows that $a_{10} = 12 \cdot (1/3)^9 = 4/6561$.

## Sequences and Graphing Calculators

**Sequence Graphing**

Most graphers enable you to graph in "sequence mode." Check your *Owner's Manual* to see how to use this mode.

As with other kinds of functions, it helps to be able to represent a sequence geometrically with a graph. There are at least two ways to obtain a sequence graph on a graphing calculator. One way to graph explicitly-defined sequences is as scatter plots of points of the form $(k, a_k)$. A second way is to use the sequence graphing mode on a graphing calculator.

### Example 6  GRAPHING A SEQUENCE DEFINED EXPLICITLY

Produce on a graphing calculator a graph of the sequence $\{a_k\}$ in which $a_k = k^2 - 1$.

### Solution

#### Method 1

The command seq(K, K, 1, 10) → $L_1$ puts the first 10 natural numbers in list $L_1$. (You could change the 10 if you wanted to graph more or fewer points.)

The command $L_1^2 - 1 \to L_2$ puts the corresponding terms of the sequence in list $L_2$. A scatter plot of $L_1$, $L_2$ produces the graph in Figure 9.7a.

#### Method 2

With your calculator in Sequence mode, enter the sequence $a_k = k^2 - 1$ in the Y = list as $u(n) = n^2 - 1$ with $n$Min = 1, $n$Max = 10, and $u(n$Min) = 0. (You could change the 10 if you wanted to graph more or fewer points.) Figure 9.7b shows the graph in the same window as Figure 9.7a.

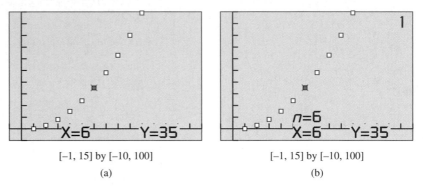

[−1, 15] by [−10, 100]
(a)

[−1, 15] by [−10, 100]
(b)

**Figure 9.7** The sequence $a_k = k^2 - 1$ graphed (a) as a scatter plot and (b) using the sequence graphing mode. Tracing along the points gives values of $a_k$ for $k = 1$, 2, 3, . . . . (Example 6)

### Example 7  GENERATING A SEQUENCE WITH A CALCULATOR

Using a graphing calculator, generate the specific terms of the following sequences:

**(a)** (Explicit)     $a_k = 3k - 5$   for   $k = 1, 2, 3, \ldots$ .

**(b)** (Recursive)   $a_1 = -2$   and   $a_n = a_{n-1} + 3$   for   $n = 2, 3, 4, \ldots$ .

### Solution

**(a)** On the home screen, type the two commands shown in Figure 9.8. The calculator will then generate the terms of the sequence as you push the ENTER key repeatedly.

**(b)** On the home screen, type the two commands shown in Figure 9.9. The first command gives the value of $a_1$. The calculator will generate the remaining terms of the sequence as you push the ENTER key repeatedly.

Notice that these two definitions generate the very same sequence!

A recursive definition of $a_n$ can actually be made in terms of any combination of preceding terms, just as long as those preceding terms have already

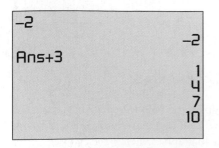

**Figure 9.8** Typing these two commands (on the left of the viewing screen) will generate the terms of the explicitly-defined sequence $a_k = 3k - 5$. (Example 7a)

**Figure 9.9** Typing these two commands (on the left of the viewing screen) will generate the terms of the recursively-defined sequence with $a_1 = -2$ and $a_n = a_{n-1} + 3$ (Example 7b).

been determined. A famous example is the **Fibonacci sequence**, named for Leonardo of Pisa (ca. 1170–1250), who wrote under the name Fibonacci. You can generate it with the two commands shown in Figure 9.10.

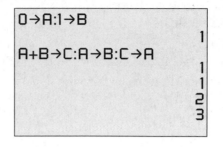

**Figure 9.10** The two commands on the left will generate the Fibonacci sequence as the ENTER key is pressed repeatedly.

### Fibonacci Numbers

The numbers in the Fibonacci sequence have fascinated professional and amateur mathematicians alike since the thirteenth century. Not only is the sequence, like Pascal's triangle, a rich source of curious internal patterns, but the Fibonacci numbers seem to appear everywhere in nature. If you count the leaflets on a leaf, the leaves on a stem, the whorls on a pine cone, the rows on an ear of corn, the spirals in a sunflower, or the branches from a trunk of a tree, they tend to be Fibonacci numbers. (Check **phyllotaxy** in a biology book.)

**Example 8**  DEFINING THE FIBONACCI SEQUENCE

Give a recursive definition of the Fibonacci sequence $\{1, 1, 2, 3, 5, 8, \ldots\}$ by specifying $a_1$, $a_2$, and a rule for obtaining $a_n$ from $a_{n-2}$ and $a_{n-1}$.

**Solution**  We see directly that $a_1 = 1$ and $a_2 = 1$. From that point, each succeeding term is obtained by adding the two terms immediately preceding it. Thus, $a_3 = a_1 + a_2$, $a_4 = a_2 + a_3$, and in general, $a_n = a_{n-2} + a_{n-1}$. The complete recursive definition consists of three equations:

$$a_1 = 1$$
$$a_2 = 1$$
$$a_n = a_{n-2} + a_{n-1}$$

## Summation Notation

We want to look at the formulas for summing the terms of arithmetic or geometric sequences, but first we need a notation for writing the sum of an indefinite number of terms. The capital Greek letter sigma ($\Sigma$) provides our shorthand notation for a "summation."

---

**Definition**  Summation Notation

In **summation notation**, the sum of the terms of the sequence $\{a_1, a_2, \ldots, a_n\}$ is denoted

$$\sum_{k=1}^{n} a_k$$

which is read "the sum of $a_k$ from $k = 1$ to $n$."

The variable $k$ is called the **index of summation**.

## Summations on a Calculator

If you think of summations as summing sequence values, it is not hard to translate sigma notation into calculator syntax. Here, in calculator syntax, are the first three summations in Exploration 1. (Don't try these on your calculator until you have figured out the answers with pencil and paper.)

1. sum (seq (3K, K, 1, 5))

2. sum (seq (K^2, K, 5, 8))

3. sum (seq (cos (N$\pi$), N, 0, 12))

---

**Summing with Sigma**

Sigma notation is actually even more versatile than the definition above suggests. See if you can determine the number represented by each of the following expressions.

1. $\displaystyle\sum_{k=1}^{5} 3k$  2. $\displaystyle\sum_{k=5}^{8} k^2$  3. $\displaystyle\sum_{n=0}^{12} \cos(n\pi)$  4. $\displaystyle\sum_{n=1}^{\infty} \sin(n\pi)$  5. $\displaystyle\sum_{k=1}^{\infty} \frac{3}{10^k}$

(If you're having trouble with number 5, here's a hint: Write the sum as a decimal!)

Although you probably computed them correctly, there is more going on in number 4 and number 5 in the above exploration than first meets the eye. We will have more to say about these "infinite" summations toward the end of this section.

## Sums of Arithmetic and Geometric Sequences

One of the most famous legends in the lore of mathematics concerns the German mathematician Karl Friedrich Gauss (1777–1855), whose mathematical talent was apparent at a very early age. One version of the story has Gauss, at age ten, being in a class that was challenged by the teacher to add up all the numbers from 1 to 100. While his classmates were still writing down the problem, Gauss walked to the front of the room to present his slate to the teacher. The teacher, certain that Gauss could only be guessing, refused to look at his answer. Gauss simply placed it face down on the teacher's desk, declared "There it is," and returned to his seat. Later, after all the slates had been collected, the teacher looked at Gauss' work, which consisted of a single number: the correct answer. No other student (the legend goes) got it right.

The important feature of this legend for mathematicians is *how* the young Gauss got the answer so quickly. We'll let you reproduce his technique in Exploration 2.

**Gauss' Insight**

Your challenge is to find the sum of the natural numbers from 1 to 100 without a calculator.

1. On a wide piece of paper, write the sum
   "1 +  2 +  3 + ⋯ + 98 + 99 + 100."

2. Underneath this sum, write the sum
   "100 + 99 + 98 + ⋯ +  3 +  2 +  1."

3. Add the numbers two-by-two in *vertical* columns and notice that you get the same identical sum 100 times. What is it?

4. What is the sum of the 100 identical numbers referred to in part 3?

5. Explain why half the answer in part 4 is the answer to the challenge. Can you find it without a calculator?

If this story is true, then the youthful Gauss had discovered a fact that his elders knew about arithmetic sequences. If you write an arithmetic sequence forward on one line and backward on the line below it, then all the pairs stacked vertically sum to the same number. Multiplying this number by $n$ and dividing by 2 gives us a shortcut to the sum of the $n$ terms. We state this result as a theorem.

---

**Theorem**   **Sum of a Finite Arithmetic Sequence**

Let $\{a_1, a_2, a_3, \ldots, a_n\}$ be a finite arithmetic sequence with common difference $d$. Then the **sum** of the terms of the sequence is

$$\sum_{k=1}^{n} a_k = a_1 + a_2 + \cdots + a_n$$

$$= n\left(\frac{a_1 + a_n}{2}\right)$$

$$= \frac{n}{2}(2a_1 + (n-1)d)$$

---

**Proof**

We can construct the sequence forward by starting with $a_1$ and *adding d* each time, or we can construct the sequence backward by starting at $a_n$ and *subtracting d* each time. We thus get two expressions for the sum we are looking for:

$$\sum_{k=1}^{n} a_k = a_1 + (a_1 + d) + (a_1 + 2d) + \cdots + (a_1 + (n-1)d)$$

$$\sum_{k=1}^{n} a_k = a_n + (a_n - d) + (a_n - 2d) + \cdots + (a_n - (n-1)d)$$

Summing the pairs vertically, we get

$$2\sum_{k=1}^{n} a_k = (a_1 + a_n) + (a_1 + a_n) + \cdots + (a_1 + a_n)$$

$$2\sum_{k=1}^{n} a_k = n(a_1 + a_n)$$

$$\sum_{k=1}^{n} a_k = n\left(\frac{a_1 + a_n}{2}\right)$$

If we substitute $a_1 + (n-1)d$ for $a_n$, we get the alternate formula

$$\sum_{k=1}^{n} a_k = \frac{n}{2}(2a_1 + (n-1)d).$$

**Example 9**   SUMMING THE TERMS OF AN ARITHMETIC SEQUENCE

A corner section of a stadium has 8 seats along the front row. Each successive row has two more seats than the row preceding it. If the top row has 24 seats, how many seats are in the entire section?

**Solution**   The numbers of seats in each row form an arithmetic sequence with

$$a_1 = 8, \quad a_n = 24, \quad \text{and} \quad d = 2.$$

Solving $a_n = a_1 + (n - 1)d$, we find that

$$24 = 8 + (n - 1)(2)$$
$$16 = (n - 1)(2)$$
$$8 = n - 1$$
$$n = 9$$

Applying the Sum of a Finite Arithmetic Sequence Theorem, the total number of seats in the section is $9(8 + 24)/2 = 144$.

We can support this answer numerically by computing the sum on a calculator:

$$\text{sum}(\text{seq}(8 + (N - 1)2, N, 1, 9) = 144.$$

As you might expect, there is also a convenient formula for summing the terms of a geometric sequence.

---

**Theorem** **Sum of a Finite Geometric Sequence**

Let $\{a_1, a_2, a_3, \ldots, a_n\}$ be a finite geometric sequence with common ratio $r \neq 1$.

Then the **sum** of the terms of the sequence is

$$\sum_{k=1}^{n} a_k = a_1 + a_2 + \cdots + a_n$$
$$= \frac{a_1(1 - r^n)}{1 - r}$$

---

**Proof**

Because the sequence is geometric, we have

$$\sum_{k=1}^{n} a_k = a_1 + a_1 \cdot r + a_1 \cdot r^2 + \cdots + a_1 \cdot r^{n-1}$$

Therefore,

$$r \cdot \sum_{k=1}^{n} a_k = \qquad a_1 \cdot r + a_1 \cdot r^2 + \cdots + a_1 \cdot r^{n-1} + a_1 \cdot r^n$$

If we now *subtract* the lower summation from the one above it, we have (after eliminating a lot of zeros):

$$\left( \sum_{k=1}^{n} a_k \right) - r \cdot \left( \sum_{k=1}^{n} a_k \right) = a_1 - a_1 \cdot r^n$$

$$\left( \sum_{k=1}^{n} a_k \right)(1 - r) = a_1(1 - r^n)$$

$$\sum_{k=1}^{n} a_k = \frac{a_1(1 - r^n)}{1 - r}$$

**Example 10** SUMMING THE TERMS OF A GEOMETRIC SEQUENCE

Find the sum of the geometric sequence
$4 - (4/3) + (4/9) - (4/27) + \cdots + 4(-(1/3))^{10}$.

**Solution** We can see that $a_1 = 4$ and $r = -1/3$. The $n$th term is $4(-1/3)^{10}$, which means that $n = 11$. (Remember that the exponent on the $n$th term is $n - 1$, not $n$.) Applying the Sum of a Finite Geometric Sequence Theorem, we find that

$$\sum_{n=1}^{11} 4\left(-\frac{1}{3}\right)^{n-1} = \frac{4(1 - (-1/3)^{11})}{1 - (-1/3)} \approx 3.000016935.$$

We can support this answer by having the calculator do the actual summing:

$$\text{sum(seq}(4(-1/3)\wedge(N - 1), N, 1, 11) = 3.000016935.$$

As one practical application of the Sum of a Finite Geometric Sequence Theorem, we will tie up a loose end from Section 3.6, wherein you learned that the future value *FV* of an ordinary annuity consisting of $n$ equal periodic payments of $R$ dollars at an interest rate $i$ per compounding period (payment interval) is

$$FV = R\frac{(1 + i)^n - 1}{i}.$$

We can now consider the mathematics behind this formula.

### Example 11   FUTURE VALUE OF AN ORDINARY ANNUITY

Use the Sum of a Finite Geometric Sequence Theorem to derive the future value formula for an ordinary annuity.

**Solution** The $n$ payments remain in the account for different lengths of time and so earn different amounts of interest. The total value of the annuity after $n$ payment periods (see Example 8 in Section 3.6) is

$$FV = R + R(1 + i) + R(1 + i)^2 + \cdots + R(1 + i)^{n-1}.$$

These numbers form a geometric sequence with first term $R$ and common ratio $(1 + i)$. Applying Theorem 2, the sum of the $n$ terms is

$$FV = \frac{R(1 - (1 + i)^n)}{1 - (1 + i)}$$

$$= R\frac{1 - (1 + i)^n}{-i}$$

$$= R\frac{(1 + i)^n - 1}{i}$$

## Infinite Series

If you change the "11" in the calculator sum in Example 10 to higher and higher numbers, you will find that the sum approaches a value of 3. This is no coincidence, either. In the language of limits,

$$\lim_{n \to \infty} \sum_{k=1}^{n} 4\left(-\frac{1}{3}\right)^{k-1} = \lim_{n \to \infty} \frac{4(1 - (-1/3)^n)}{1 - (-1/3)}$$

$$= \frac{4(1 - 0)}{4/3} \qquad \text{Since } \lim_{n \to \infty}(-1/3)^n = 0.$$

$$= 3$$

This gives us the opportunity to extend the usual meaning of the word "sum," which always applies to a *finite* number of terms being added together. By using limits, we can make sense of expressions in which an *infinite* number of terms are added together. Such expressions are called **infinite series**.

---

**Definition** Infinite Series

An **infinite series** is an expression of the form

$$\sum_{n=1}^{\infty} a_n = a_1 + a_2 + \cdots + a_n + \cdots.$$

---

The first thing to understand about an infinite series is that it is not a true sum. There are properties of real number addition that allow us to extend the definition of $a + b$ to sums like $a + b + c + d + e + f$, but not to "infinite sums." For example, we can add any finite number of 2's together and get a real number, but if we add an *infinite* number of 2's together we do not get a real number at all. Sums do not behave that way.

What makes series so interesting is that sometimes (as in Example 10) the sequence of **partial sums**, all of which are true sums, approaches a finite limit $S$:

$$\lim_{n \to \infty} \sum_{k=1}^{n} a_k = \lim_{n \to \infty} (a_1 + a_2 + \cdots + a_n) = S.$$

In this case we say that the series **converges** to $S$, and it makes sense to define $S$ as the **sum** of the series. In sigma notation,

$$\sum_{k=1}^{\infty} a_k = \lim_{n \to \infty} \sum_{k=1}^{n} a_k = S.$$

If the limit of partial sums does not exist, then the series **diverges** and has no sum.

### Example 12   LOOKING AT LIMITS OF PARTIAL SUMS

For each of the following series, find the first five terms in the sequence of partial sums. Which of the series appear to converge?

**(a)** $0.1 + 0.01 + 0.001 + 0.0001 + \cdots$

**(b)** $10 + 20 + 30 + 40 + \cdots$

**(c)** $1 - 1 + 1 - 1 + \cdots$

**Solution**

**(a)** The first five partial sums are $\{0.1, 0.11, 0.111, 0.1111, 0.11111\}$. These appear to be approaching a limit of $0.\overline{1} = 1/9$, which would suggest that the series converges to a sum of $1/9$.

**(b)** The first five partial sums are $\{10, 30, 60, 100, 150\}$. These numbers increase without bound and do not approach a limit. The series diverges and has no sum.

**(c)** The first five partial sums are $\{1, 0, 1, 0, 1\}$. These numbers oscillate and do not approach a limit. The series diverges and has no sum.

**Alert**

Some students will expect a seres with $r = -1$ to converge because the partial sums do not go to infinity. Explain that the partial sums alternate between two values, so the series does not converge.

You might have been tempted to "pair off" the terms in Example 12c to get an infinite summation of 0's (and hence a sum of 0), but you would be applying a rule (namely the *associative property of addition*) that works on *finite* sums but not, in general, on infinite series. The sequence of partial sums does not have a limit, so any manipulation of the series in Example 12c that appears to result in a sum is actually meaningless.

## Convergence of Geometric Series

Determining the convergence or divergence of infinite series is an important part of a calculus course, in which series are used to represent functions. Most of the convergence tests are well beyond the scope of this course, but we are in a position to settle the issue completely for geometric series.

> **Theorem  Sum of an Infinite Geometric Series**
>
> The geometric series $\sum_{k=1}^{\infty} a \cdot r^{k-1}$ converges if and only if $|r| < 1$. If it does converge, the sum is $a/(1 - r)$.

**Teaching Note**

The grapher allows students to visualize the convergence of an infinite geometric series. The ease of visual understanding was not available before recent technological advances.

**Proof**

If $r = 1$, the series is $a + a + a + \cdots$, which is unbounded and hence diverges.

If $r = -1$, the series is $a - a + a - a + \cdots$, which diverges. (See Example 12c.)

If $r \neq 1$, then by the Sum of a Finite Geometric Sequence Theorem, the $n$th partial sum of the series is $\sum_{k=1}^{n} a \cdot r^{k-1} = a(1 - r^n)/(1 - r)$. The limit of the $n$th partial sum is $\lim_{n \to \infty}[a(1 - r^n)]/(1 - r)$, which converges if and only if $\lim_{n \to \infty} r^n$ exists. But $\lim_{n \to \infty} r^n$ is 0 when $|r| < 1$ and unbounded when $|r| > 1$. Therefore, the sequence of partial sums converges if and only if $|r| < 1$, in which case the sum of the series is

$$\lim_{n \to \infty}[a(1 - r^n)]/(1 - r) = a(1 - 0)/(1 - r) = a/(1 - r).$$

### Example 13  SUMMING INFINITE GEOMETRIC SERIES

**Notes on Examples**

Example 13d can be easily demonstrated with a blank sheet of paper. Tear the paper in half and label one half "1/2." Tear the remaining half in half and label one part "1/4." Tear the remaining fourth in half and label one part "1/8." Continue this process until the students are able to visualize that by putting the fractional parts back together again they will very nearly make one "whole" piece of paper with an "infinitely small" portion missing.

Determine whether the series converges. If it converges, give the sum.

(a) $\sum_{k=1}^{\infty} 3(0.75)^{k-1}$    (b) $\sum_{n=0}^{\infty} \left(-\frac{4}{5}\right)^n$

(c) $\sum_{n=1}^{\infty} \left(\frac{\pi}{2}\right)^n$    (d) $1 + \frac{1}{2} + \frac{1}{4} + \frac{1}{8} + \cdots$

**Solution**

(a) Since $|r| = |0.75| < 1$, the series converges. The first term is $3(0.75)^0 = 3$, and so the sum is $a/(1 - r) = 3/(1 - 0.75) = 12$.

(b) Since $|r| = |-4/5| < 1$, the series converges. The first term is $(-4/5)^0 = 1$, and so the sum is $a/(1 - r) = 1/(1 - (-4/5)) = 5/9$.

(c) Since $|r| = |\pi/2| > 1$, the series diverges.

(d) Since $|r| = |1/2| < 1$, the series converges. The first term is 1, and so the sum is $a/(1 - r) = 1/(1 - 1/2) = 2$.

**Example 14 CONVERTING A REPEATING DECIMAL TO FRACTION FORM**

Express $0.\overline{234} = 0.234234234\ldots$ in fraction form.

**Solution** We can write this number as a sum: $0.234 + 0.000234 + 0.000000234 + \cdots$.

This is a convergent infinite geometric series in which $a = 0.234$ and $r = 0.001$. The sum is

$$\frac{a}{1-r} = \frac{0.234}{1 - 0.001} = \frac{0.234}{0.999} = \frac{234}{999} = \frac{26}{111}.$$

# Quick Review 9.4

In Exercises 1 and 2, evaluate each expression when $a = 3$, $d = 4$, and $n = 5$.

**1.** $a + (n - 1)d$   19

**2.** $\dfrac{n}{2}(2a + (n - 1)d)$   55

In Exercises 3 and 4, evaluate each expression when $a = 5$, $r = 4$, and $n = 3$.

**3.** $a \cdot r^{n-1}$   80

**4.** $\dfrac{a(1 - r^n)}{1 - r}$   105

In Exercises 5–10, find $a_{10}$.

**5.** $a_k = \dfrac{k}{k + 1}$   10/11

**6.** $a_k = 5 + (k - 1)3$   32

**7.** $a_k = 5 \cdot 2^{k-1}$   2560

**8.** $a_k = \dfrac{4}{3}\left(\dfrac{1}{2}\right)^{k-1}$   1/384

**9.** $a_k = 32 - a_{k-1}$ and $a_9 = 17$   15

**10.** $a_k = \dfrac{k^2}{2^k}$   25/256

# Section 9.4 Exercises

In Exercises 1–4, find the first 6 terms and the 100th term of the explicitly-defined sequence.

**1.** $u_n = \dfrac{n + 1}{n}$

**2.** $v_n = \dfrac{4}{n + 2}$

**3.** $c_n = n^3 - n$

**4.** $d_n = n^2 - 5n$

In Exercises 5–10, find the first 4 terms and the eighth term of the recursively-defined sequence.

**5.** $a_1 = 8$ and $a_n = a_{n-1} - 4$, for $n \geq 2$   8, 4, 0, −4; −20

**6.** $u_1 = -3$ and $u_{k+1} = u_k + 10$, for $k \geq 1$   −3, 7, 17, 27; 67

**7.** $b_1 = 2$ and $b_{k+1} = 3b_k$, for $k \geq 1$   2, 6, 18, 54; 4374

**8.** $v_1 = 0.75$ and $v_n = (-2)v_{n-1}$, for $n \geq 2$

**9.** $c_1 = 2$, $c_2 = -1$, and $c_{k+2} = c_k + c_{k+1}$, for $k \geq 1$

**10.** $c_1 = -2$, $c_2 = 3$, and $c_k = c_{k-2} + c_{k-1}$, for $k \geq 3$

In Exercises 11–14, the sequences are arithmetic. Find

**(a)** the common difference,

**(b)** the tenth term,

**(c)** a recursive rule for the $n$th term, and

**(d)** an explicit rule for the $n$th term.

**11.** 6, 10, 14, 18, . . .

**12.** −4, 1, 6, 11, . . .

**13.** −5, −2, 1, 4, . . .

**14.** −7, 4, 15, 26, . . .

In Exercises 15–18, the sequences are geometric. Find

**(a)** the common ratio,

**(b)** the eighth term,

**(c)** a recursive rule for the $n$th term, and

**(d)** an explicit rule for the $n$th term.

**15.** 2, 6, 18, 54, . . .

**16.** 3, 6, 12, 24, . . .

**17.** 1, −2, 4, −8, 16, . . .

**18.** −2, 2, −2, 2, . . .

**19.** The fourth and seventh terms of an arithmetic sequence are −8 and 4, respectively. Find the first term and a recursive rule for the $n$th term.   $a_1 = -20$; $a_n = a_{n-1} + 4$

**20.** The fifth and ninth terms of an arithmetic sequence are −5 and −17, respectively. Find the first term and a recursive rule for the $n$th term.   $a_1 = 7$; $a_n = a_{n-1} - 3$.

**21.** The second and eighth terms of a geometric sequence are 3 and 192, respectively. Find the first term, common ratio, and an explicit rule for the $n$th term.

**22.** The third and sixth terms of a geometric sequence are −75 and −9375, respectively. Find the first term, common ratio, and an explicit rule for the $n$th term

In Exercises 23–26, graph the sequence.

**23.** $a_n = 2 - \dfrac{1}{n}$

**24.** $b_n = \sqrt{n} - 3$

**25.** $c_n = n^2 - 5$    **26.** $d_n = 3 + 2n$

In Exercises 27–32, write each series using summation notation, assuming the suggested pattern continues.

**27.** $-7 - 1 + 5 + 11 + \cdots + 53$

**28.** $2 + 5 + 8 + 11 + \cdots + 29$

**29.** $1 + 4 + 9 + \cdots + (n + 1)^2$

**30.** $1 + 8 + 27 + \cdots + (n + 1)^3$

**31.** $6 - 12 + 24 - 48 + \cdots$

**32.** $5 - 15 + 45 - 135 + \cdots$

In Exercises 33–38, find the sum of the arithmetic series.

**33.** $-7 - 3 + 1 + 5 + 9 + 13$  18

**34.** $-8 - 1 + 6 + 13 + 20 + 27$  57

**35.** $1 + 2 + 3 + 4 + \cdots + 80$  3240

**36.** $2 + 4 + 6 + 8 + \cdots + 70$  1260

**37.** $117 + 110 + 103 + \cdots + 33$  975

**38.** $111 + 108 + 105 + \cdots + 27$  2001

In Exercises 39–42, find the sum of the geometric series.

**39.** $3 + 6 + 12 + \cdots + 12{,}288$  24,573

**40.** $5 + 15 + 45 + \cdots + 98{,}415$  147,620

**41.** $42 + 7 + \dfrac{7}{6} + \cdots + 42\left(\dfrac{1}{6}\right)^8$  $50.4(1 - 6^{-9}) \approx 50.4$

**42.** $42 - 7 + \dfrac{7}{6} - \cdots + 42\left(-\dfrac{1}{6}\right)^9$  $36 - 6^{-8} \approx 36$

In Exercises 43–48, find the sum of the first $n$ terms of the sequence. The sequence is either arithmetic or geometric.

**43.** $2, 5, 8, \ldots; n = 10$  155    **44.** $14, 8, 2, \ldots; n = 9$  −90

**45.** $4, -2, 1, -\dfrac{1}{2}, \ldots; n = 12$  $\dfrac{8}{3} \cdot (1 - 2^{-12}) \approx 2.666$

**46.** $6, -3, \dfrac{3}{2}, -\dfrac{3}{4}, \ldots; n = 11$  $4 \cdot (1 + 2^{-11}) \approx 4.002$

**47.** $-1, 11, -121, \ldots; n = 9$  −196,495,641

**48.** $-2, 24, -288, \ldots; n = 8$  66,151,030

In Exercises 49–54, determine whether the infinite geometric series converges. If it does, find its sum.

**49.** $6 + 3 + \dfrac{3}{2} + \dfrac{3}{4} + \cdots$    **50.** $4 + \dfrac{4}{3} + \dfrac{4}{9} + \dfrac{4}{27} + \cdots$

**51.** $\dfrac{1}{64} + \dfrac{1}{32} + \dfrac{1}{16} + \dfrac{1}{8} + \cdots$  diverges

**52.** $\dfrac{1}{48} + \dfrac{1}{16} + \dfrac{3}{16} + \dfrac{9}{16} + \cdots$  diverges

**53.** $\displaystyle\sum_{j=1}^{\infty} 3\left(\dfrac{1}{4}\right)^j$  converges; 1    **54.** $\displaystyle\sum_{n=1}^{\infty} 5\left(\dfrac{2}{3}\right)^n$  converges; 10

In Exercises 55–58, express the rational number as a fraction of integers.

**55.** $7.14141414\ldots$  707/99    **56.** $5.93939393\ldots$  196/33

**57.** $-17.268268268\ldots$    **58.** $-12.876876876\ldots$

**59. Rain Forest Growth**  The bungy-bungy tree in the Amazon rain forest grows an average 2.3 cm per week. Write a sequence that represents the weekly height of a bungy-bungy over the course of 1 year if it is 7 meters tall today. Display the first four terms and the last two terms.

**60. Half-Life**  (See Section 3.2) Thorium-232 has a half-life of 14 billion years. Make a table showing the half-life decay of a sample of Thorium-232 from 16 grams to 1 gram; list the time (in years, starting with $t = 0$) in the first column and the mass (in grams) in the second column. Which type of sequence is each column of the table?

**61. Population—Conroe vs. Huntsville**  The population of Huntsville, Texas, was 25,854 in 1985 and has been increasing by 1.55% each year. The population of Conroe, Texas, was 22,314 in 1985 and has been increasing by 4.35% each year. *Source:* The Chambers of Commerce for the two cities.

**(a)** Make a table, based on the given data, that predicts the population of the two cities for 10 years after 1985.

**(b)** In which year should Conroe's population first exceed Huntsville's?

**62. Rainy Days**  Table 9.1 lists the average number of days with rain each month for four major U.S. cities. Answer each question by summing the appropriate series.

**(a)** Which of the four cities has the most rainy days? The fewest?  Seattle; Los Angeles

**(b)** Which month is the rainiest for the four cities? The least rainy?  January; August

**Table 9.1  Number of Rainy Days**

| Month | Chicago | Los Angeles | New York | Seattle |
|---|---|---|---|---|
| January | 11 | 6 | 12 | 18 |
| February | 10 | 6 | 10 | 16 |
| March | 12 | 6 | 12 | 16 |
| April | 11 | 4 | 11 | 13 |
| May | 12 | 2 | 11 | 12 |
| June | 11 | 1 | 10 | 9 |
| July | 9 | 0 | 12 | 4 |
| August | 9 | 0 | 10 | 5 |
| September | 9 | 1 | 9 | 8 |
| October | 9 | 2 | 9 | 13 |
| November | 10 | 3 | 9 | 17 |
| December | 11 | 6 | 10 | 19 |

*Source: National Geographic Atlas of the World (rev. 6th ed., 1992)*

**63. Group Activity  Follow the Bouncing Ball**  When "superballs" sprang upon the scene in the 1960s, kids across the United States were amazed that these hard rubber balls could bounce to 90% of the height from which they dropped. If a superball is dropped from a height of

2 m, how far does it travel by the time it hits the ground for the tenth time? (*Hint:* The ball goes down to the first bounce, then up *and* down thereafter.) ≈ 22.5 m

64. **Group Activity  End Behavior** Does

$$f(x) = 2\left(\frac{1 - 1.05^x}{1 - 1.05}\right)$$

have a horizontal asymptote? How does this relate to the convergence or divergence of the series

$$2 + 2.1 + 2.205 + 2.31525 + \cdots ?$$

## Explorations

65. **Rabbit Populations** Assume that 2 months after birth, each male-female pair of rabbits begins producing one new male-female pair of rabbits each month. Further assume that the rabbit colony begins with one newborn male-female pair of rabbits and no rabbits die for 12 months. Let $a_n$ represent the number of *pairs* of rabbits in the colony after $n - 1$ months.

    (a) **Writing to Learn** Explain why $a_1 = 1$, $a_2 = 1$, and $a_3 = 2$.

    (b) Find $a_4, a_5, a_6, \ldots, a_{13}$.

    (c) **Writing to Learn** Explain why the sequence $\{a_n\}$, $1 \le n \le 13$, is a model for the size of the rabbit colony for a 1-year period.

66. **Fibonacci Sequence** Compute the first seven terms of the sequence whose $n$th term is

$$a_n = \frac{1}{\sqrt{5}}\left(\frac{1 + \sqrt{5}}{2}\right)^n - \frac{1}{\sqrt{5}}\left(\frac{1 - \sqrt{5}}{2}\right)^n.$$

How do these seven terms compare with the first seven terms of the Fibonacci sequence?

67. **Connecting Geometry and Sequences** In the following sequence of diagrams, regular polygons are inscribed in unit circles with at least one side of each polygon perpendicular to the positive $x$-axis.

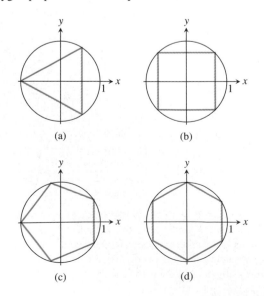

(a) Prove that the perimeter of each polygon in the sequence is given by $a_n = 2n \sin{(\pi/n)}$, where $n$ is the number of sides in the polygon.

(b) Investigate the value of $a_n$ for $n = 10, 100, 1000,$ and 10,000. What conclusion can you draw? ■

68. **Recursive Sequence** The population of Centerville was 525,000 in 1992 and is growing annually at the rate of 1.75%. Write a recursive sequence $\{P_n\}$ for the population. State the first term $P_1$ for your sequence.

69. **Writing to Learn** If $\{a_n\}$ is a geometric sequence with all positive terms, explain why $\{\log{(a_n)}\}$ must be arithmetic.

70. **Writing to Learn** If $\{b_n\}$ is an arithmetic sequence, explain why $\{10^{b_n}\}$ must be geometric.

71. **Population Density** The *National Geographic Picture Atlas of Our Fifty States* (1991) groups the states into 10 regions. The two largest groupings are the Heartland (Table 9.2) and the Southeast (Table 9.3). Population and area data for the two regions are given in the tables. The populations are official 1990 U.S. Census figures.

    (a) What is the total population of each region?

    (b) What is the total area of each region?

    (c) What is the population density (in persons per square mile) of each region?

    (d) **Writing to Learn** For the two regions, compute the population density of each state. What is the average of the seven state population densities for each region? Explain why these answers differ from those found in part (c).

**Table 9.2  The Heartland**

| State | Population | Area (mi²) |
|---|---|---|
| Iowa | 2,776,755 | 56,275 |
| Kansas | 2,477,574 | 82,277 |
| Minnesota | 4,375,099 | 84,402 |
| Missouri | 5,117,073 | 69,697 |
| Nebraska | 1,578,385 | 77,355 |
| North Dakota | 638,800 | 70,703 |
| South Dakota | 696,004 | 77,116 |

**Table 9.3 The Southeast**

| State | Population | Area (mi²) |
|---|---|---|
| Alabama | 4,040,587 | 51,705 |
| Arkansas | 2,350,725 | 53,187 |
| Florida | 12,937,926 | 58,644 |
| Georgia | 6,478,216 | 58,910 |
| Louisiana | 4,219,973 | 47,751 |
| Mississippi | 2,573,216 | 47,689 |
| S. Carolina | 3,486,703 | 31,113 |

72. **Finding a Pattern** Write the finite series $-1 + 2 + 7 + 14 + 23 + \cdots + 62$ in summation notation.

# Extending the Ideas

**73. A Sequence of Matrices** Write out the first seven terms of the "geometric sequence" with the first term the matrix $\begin{bmatrix} 1 & 1 \end{bmatrix}$ and the common ratio the matrix $\begin{bmatrix} 0 & 1 \\ 1 & 1 \end{bmatrix}$. How is this sequence of matrices related to the Fibonacci sequence?

**74.** Prove that the sum of the first $n$ odd positive integers is $n^2$.

**75. Fibonacci Sequence and Series** Complete the following table, where $F_n$ is the $n$th term of the Fibonacci sequence and $S_n$ is the $n$th partial sum of the Fibonacci series. Make a conjecture based on the numerical evidence in the table.

$$S_n = \sum_{k=1}^{n} F_k.$$

| $n$ | $F_n$ | $S_n$ | $F_{n+2} - 1$ |
|-----|-------|-------|---------------|
| 1 | 1 | 1 | 1 |
| 2 | 1 | 2 | 2 |
| 3 | 2 | 4 | 4 |
| 4 | 3 | 7 | 7 |
| 5 | 5 | 12 | 12 |
| 6 | 8 | 20 | 20 |
| 7 | 13 | 33 | 33 |
| 8 | 21 | 54 | 54 |
| 9 | 34 | 88 | 88 |

Conjecture: $S_n = F_{n+2} - 1$

**76. Triangular Numbers Revisited** Exercise 35 in Section 9.2 introduced **triangular numbers** as numbers that count objects arranged in triangular arrays:

1    3    6    10    15

In that exercise you gave a geometric argument that the $n$th triangular number was $n(n + 1)/2$. Prove that formula algebraically using the Sum of a Finite Arithmetic Sequence Theorem.

**77. Square Numbers and Triangular Numbers** Prove that the sum of consecutive triangular numbers is a square number; that is, prove

$$T_{n-1} + T_n = n^2$$

for all positive integers $n \geq 2$. Use both a geometric and an algebraic approach.

**78. Harmonic Series** Graph the sequence of partial sums of the *harmonic series:*

$$1 + \frac{1}{2} + \frac{1}{3} + \frac{1}{4} + \cdots + \frac{1}{n} + \cdots.$$

Overlay on it the graph of $f(x) = \ln x$. The resulting picture should support the claim that

$$1 + \frac{1}{2} + \frac{1}{3} + \frac{1}{4} + \cdots + \frac{1}{n} \geq \ln n,$$

for all positive integers $n$. Make a table of values to further support this claim. Explain why the claim implies that the harmonic series must diverge.

---

## 9.5 Mathematical Induction

The Tower of Hanoi Problem • Principle of Mathematical Induction

### The Tower of Hanoi Problem

You might be familiar with a game that is played with a stack of round washers of different diameters and a stand with three vertical pegs (Figure 9.11). At the start of the game, the washers are stacked on the leftmost peg so that they decrease in diameter from bottom to top. The object of the game is to move the entire stack of washers to the rightmost peg, moving only one washer at a time and never placing a larger washer on top of a smaller washer. The game is not difficult to win once you get the hang of it, but it takes a while to move all the washers even when you know what you are doing. A mathematician presented with this game, of course wants to figure out the *minimum* number of moves required to win the game—not because of impatience, but because it is an interesting mathematical problem.

In case mathematics is not sufficient motivation to look into the problem, there is a legend attached to the game that provides a sense of urgency. The

**Figure 9.11** The Tower of Hanoi Game. The object is to move the entire stack of washers to another peg, one washer at a time, never placing a larger washer on top of a smaller washer.

## Tower of Hanoi History

The legend notwithstanding, the Tower of Hanoi dates back to 1883, when Édouard Lucas marketed the game as "La Tour de Hanoï," brought back from the Orient by "Professor N. Claus de Siam"—an anagram of "Professor Lucas d'Amiens." The legend appeared shortly thereafter. The game has been a favorite among computer programmers, so a web search on "Tower of Hanoi" will bring up multiple sites that allow you to play it on your home computer.

## Objective

Students will be able to us the principle of mathematical induction to prove mathematical generalizations.

## Motivate

Ask…

Suppose a sequence is defined recursively by $a_1 = 2$ and $a_{n+1} = a_n{}^2 + 3a_n$. How can you be sure that $a_{364}$ is defined?

legend has it that a game of this sort with 64 golden washers was created at the beginning of time. A special order of far eastern monks has been moving the washers at a move per second ever since, always using the minimum number of moves required to win the game. When the final washer is moved, that will be the end of time. The Tower of Hanoi Problem is simply to figure out how much time we have left.

We will answer the question by proving a general theorem that gives the minimum number of moves for any number of washers. The technique of proof we use is called the principle of mathematical induction, the topic of this section.

---

**Theorem** The Tower of Hanoi Solution

The minimum number of moves required to move a stack of $n$ washers in a Tower of Hanoi game is $2^n - 1$.

---

### Proof

**(The Anchor)** First, we note that the assertion is true when $n = 1$. We can certainly move the one washer to the right peg in (minimally) one move, and $2^1 - 1 = 1$.

**(The Inductive Hypothesis)** Now let us assume that the assertion holds for $n = k$; that is, the minimum number of moves required to move $k$ washers is $2^k - 1$. (So far the only $k$ we are sure of is 1, but keep reading.)

**(The Inductive Step)** We next consider the case when $n = k + 1$ washers. To get at the bottom washer, we must first move the entire stack of $k$ washers sitting on top of it. *By the assumption we just made*, this will take a minimum of $2^k - 1$ moves. We can then move the bottom washer to the free peg (1 move). Finally, we must move the stack of $k$ washers back onto the bottom washer— again, *by our assumption*, a minimum of $2^k - 1$ moves. Altogether, moving $k + 1$ washers requires

$$(2^k - 1) + 1 + (2^k - 1) = 2 \cdot 2^k - 1 = 2^{k+1} - 1$$

moves. Since that agrees with the formula in the statement of the proof, we have shown the assertion to be true for $n = k + 1$ washers—under the assumption that it is true for $n = k$.

Remarkably, we are done. Recall that we *did* prove the theorem to be true for $n = 1$. Therefore, by the Inductive Step, it must also be true for $n = 2$. By the Inductive Step again, it must be true for $n = 3$. And so on, for all positive integers $n$.

If we apply the Tower of Hanoi Solution to the legendary Tower of Hanoi Problem, the monks will need $2^{64} - 1$ seconds to move the 64 golden washers. The largest current conjecture for the age of the universe is something on the order of 20 billion years. If you convert $2^{64} - 1$ seconds to years, you will find that the end of time (at least according to this particular legend) is not exactly imminent. In fact, you might be surprised at how much time is left!

**Exploration Extensions**

A shop owner has 500 cans of tomatoes that are near their expiration date. She decides to display them in a triangular pyramid and lower the price to try to sell them quickly. How many cans will need to be set up on the bottom layer, and will there be any cans left over?

> **Exploration 1** **Winning the Game**
>
> One thing that the Tower of Hanoi Solution does not settle is how to get the stack to finish on the rightmost peg rather than the middle peg. Predictably, it depends on where you move the first washer, but it also depends on the height of the stack. Using a website game, or coins of different sizes, or even the real game if you have one, play the game with 1 washer, then 2, then 3, then 4, and so on, keeping track of what your first move must be in order to have the stack wind up on the rightmost peg in $2^n - 1$ moves. What is the general rule for a stack of $n$ washers?

## Principle of Mathematical Induction

The proof of the Tower of Hanoi Solution used a general technique known as the principle of mathematical induction. It is a powerful tool for proving all kinds of theorems about positive integers. We *anchor* the proof by establishing the truth of the theorem for 1, then we show the *inductive hypothesis* that "true for $k$" implies "true for $k + 1$."

> ### Principle of Mathematical Induction
>
> Let $P_n$ be a statement about the integer $n$. Then $P_n$ is true for all positive integers $n$ provided the following conditions are satisfied:
>
> **1.** (the anchor) $P_1$ is true;
> **2.** (the inductive step) if $P_k$ is true, then $P_{k+1}$ is true.

A good way to visualize how the principle works is to imagine an infinite sequence of dominoes stacked upright, each one close enough to its neighbor so that any $k$th domino, if it falls, will knock over the $(k + 1)$st domino (Figure 9.12). Given that fact (the inductive step), the toppling of domino 1 guarantees the toppling of the entire infinite sequence of dominoes.

Let us use the principle to prove a fact that we already know.

**Figure 9.12** The Principle of Mathematical Induction visualized by dominoes. If the dominoes are close enough together so that any $k$th domino will knock over the $(k + 1)$st domino, then the toppling of domino #1 guarantees the toppling of domino $n$ for all positive integers $n$.

### Example 1  USING MATHEMATICAL INDUCTION

Prove that $1 + 3 + 5 + \cdots + (2n - 1) = n^2$ is true for all positive integers $n$.

**Solution**  Call the statement $P_n$. We could verify $P_n$ by using the formula for the sum of an arithmetic sequence, but here is how we prove it by mathematical induction.

**(The anchor)**  For $n = 1$, the equation reduces to $P_1$: $1 = 1^2$, which is true.

**(The inductive hypothesis)**  Assume that the equation is true for $n = k$. That is, assume

$$P_k: 1 + 3 + \cdots + (2k - 1) = k^2 \text{ is true.}$$

**(The inductive step)** The next term on the left-hand side would be $2(k + 1) - 1$. We add this to both sides of $P_k$ and get

$$1 + 3 + \cdots + (2k - 1) + (2(k + 1) - 1) = k^2 + (2(k + 1) - 1)$$
$$= k^2 + 2k + 1$$
$$= (k + 1)^2$$

This is exactly the statement $P_{k+1}$, so the equation is true for $n = k + 1$. Therefore, $P_n$ is true for all positive integers, by mathematical induction.

Notice that we did *not* plug in $k + 1$ on both sides of the equation $P_n$ in order to verify the inductive step; if we had done that, there would have been nothing to verify. If you find yourself verifying the inductive step without using the inductive hypothesis, you can assume that you have gone astray.

**Example 2  USING MATHEMATICAL INDUCTION**

Prove that $1^2 + 2^2 + 3^2 + \cdots + n^2 = [n(n + 1)(2n + 1)]/6$ is true for all positive integers $n$.

**Solution** Let $P_n$ be the statement
$1^2 + 2^2 + 3^2 + \cdots + n^2 = [n(n + 1)(2n + 1)]/6$.

**(The anchor)** $P_1$ is true because $1^2 = [1(2)(3)]/6$.

**(The inductive hypothesis)** Assume that $P_k$ is true, so that

$$1^2 + 2^2 + \cdots + k^2 = \frac{k(k + 1)(2k + 1)}{6}.$$

**(The inductive step)** The next term on the left-hand side would be $(k + 1)^2$. We add this to both sides of $P_k$ and get

$$1^2 + 2^2 + \cdots + k^2 + (k + 1)^2 = \frac{k(k + 1)(2k + 1)}{6} + (k + 1)^2$$
$$= \frac{k(k + 1)(2k + 1) + 6(k + 1)^2}{6}$$
$$= \frac{(k + 1)(2k^2 + k + 6k + 6)}{6}$$
$$= \frac{(k + 1)(k + 2)(2k + 3)}{6}$$
$$= \frac{(k + 1)((k + 1) + 1)(2(k + 1) + 1)}{6}$$

This is exactly the statement $P_{k+1}$, so the equation is true for $n = k + 1$. Therefore, $P_n$ is true for all positive integers, by mathematical induction.

Applications of mathematical induction can be quite different from the first two examples. Here is one involving divisibility.

**Example 3  PROVING DIVISIBILITY**

Prove that $4^n - 1$ is evenly divisible by 3 for all positive integers $n$.

**Solution** Let $P_n$ be the statement that $4^n - 1$ is evenly divisible by 3 for all positive integers $n$.

**(The anchor)** $P_1$ is true because $4^1 - 1 = 3$ is divisible by 3.

**Ongoing Assessment**

Self-Assessment: Ex. 3, 13, 27, 33, 35
Embedded Assessment: Ex. 8, 16, 20,
22, 28, 36, 40

**(The inductive hypothesis)** Assume that $P_k$ is true, so that $4^k - 1$ is divisible by 3.

**(The inductive step)** We need to prove that $4^{k+1} - 1$ is divisible by 3. Using a little algebra, we see that $4^{k+1} - 1 = 4 \cdot 4^k - 1 = 4(4^k - 1) + 3$.

By the inductive hypothesis, $4^k - 1$ is divisible by 3. Of course, so is 3. Therefore, $4(4^k - 1) + 3$ is a sum of multiples of 3, and hence divisible by 3. This is exactly the statement $P_{k+1}$, so $P_{k+1}$ is true. Therefore, $P_n$ is true for all positive integers, by mathematical induction.

# Quick Review 9.5

In Exercises 1–3, expand the product.

**1.** $n(n + 5)$  $n^2 + 5n$      **2.** $(n + 2)(n - 3)$ $n^2 - n - 6$

**3.** $k(k + 1)(k + 2)$ $k^3 + 3k^2 + 2k$

In Exercises 4–6, factor the polynomial.

**4.** $n^2 + 2n - 3$ $(n + 3)(n - 1)$ **5.** $k^3 + 3k^2 + 3k + 1$ $(k + 1)^3$

**6.** $n^3 - 3n^2 + 3n - 1$ $(n - 1)^3$

In Exercises 7–10, evaluate the function at the given domain values or variable expressions.

**7.** $f(x) = x + 4$;   $f(1), f(t), f(t + 1)$   $5; t + 4; t + 5$

**8.** $f(n) = \dfrac{n}{n + 1}$;   $f(1), f(k), f(k + 1)$   $\dfrac{1}{2}; \dfrac{k}{k + 1}; \dfrac{k + 1}{k + 2}$

**9.** $P(n) = \dfrac{2n}{3n + 1}$;   $P(1), P(k), P(k + 1)$   $\dfrac{1}{2}; \dfrac{2k}{3k + 1}; \dfrac{2k + 2}{3k + 4}$

**10.** $P(n) = 2n^2 - n - 3$;   $P(1), P(k), P(k + 1)$

# Section 9.5 Exercises

In Exercises 1–4, use mathematical induction to prove that the statement holds for all positive integers.

**1.** $2 + 4 + 6 + \cdots + 2n = n^2 + n$

**2.** $8 + 10 + 12 + \cdots + (2n + 6) = n^2 + 7n$

**3.** $6 + 10 + 14 + \cdots + (4n + 2) = n(2n + 4)$

**4.** $14 + 18 + 22 + \cdots + (4n + 10) = 2n(n + 6)$

In Exercises 5–8, state an explicit rule for the $n$th term of the recursively defined sequence. Then use mathematical induction to prove the rule.

**5.** $a_n = a_{n-1} + 5, a_1 = 3$      **6.** $a_n = a_{n-1} + 2, a_1 = 7$

**7.** $a_n = 3a_{n-1}, a_1 = 2$      **8.** $a_n = 5a_{n-1}, a_1 = 3$

In Exercises 9–12, write the statements $P_1$, $P_k$, and $P_{k+1}$. (Do not write a proof.)

**9.** $P_n$: $1 + 2 + \cdots + n = \dfrac{n(n + 1)}{2}$

**10.** $P_n$: $1^2 + 3^2 + 5^2 + \cdots + (2n - 1)^2 = \dfrac{n(2n - 1)(2n + 1)}{3}$

**11.** $P_n$: $\dfrac{1}{1 \cdot 2} + \dfrac{1}{2 \cdot 3} + \cdots + \dfrac{1}{n \cdot (n + 1)} = \dfrac{n}{n + 1}$

**12.** $P_n$: $\displaystyle\sum_{k=1}^{n} k^4 = \dfrac{n(n + 1)(2n + 1)(3n^2 + 3n - 1)}{30}$

In Exercises 13–20, use mathematical induction to prove that the statement holds for all positive integers.

**13.** $1 + 5 + 9 + \cdots + (4n - 3) = n(2n - 1)$

**14.** $1 + 2 + 2^2 + \cdots + 2^{n-1} = 2^n - 1$

**15.** $\dfrac{1}{1 \cdot 2} + \dfrac{1}{2 \cdot 3} + \dfrac{1}{3 \cdot 4} + \cdots + \dfrac{1}{n(n + 1)} = \dfrac{n}{n + 1}$

**16.** $\dfrac{1}{1 \cdot 3} + \dfrac{1}{3 \cdot 5} + \cdots + \dfrac{1}{(2n - 1)(2n + 1)} = \dfrac{n}{2n + 1}$

**17.** $2^n \geq 2n$      **18.** $3^n \geq 3n$

**19.** 3 is a factor of $n^3 + 2n$.    **20.** 6 is a factor of $7^n - 1$.

In Exercises 21 and 22, use *mathematical induction* to prove that the statement holds for all positive integers. (We have already seen each proved in another way.)

**21.** The sum of the first $n$ terms of a geometric sequence with first term $a_1$ and common ratio $r \neq 1$ is $a_1(1 - r^n)/(1 - r)$.

**22.** The sum of the first $n$ terms of an arithmetic sequence with its first term $a_1$ and common difference $d$ is

$$S_n = \frac{n}{2}[2a_1 + (n - 1)d].$$

In Exercises 23 and 24, use mathematical induction to prove that the formula holds for all positive integers.

**23. Triangular Numbers** $\displaystyle\sum_{k=1}^{n} k = \dfrac{n(n + 1)}{2}$

**24. Sum of the First *n* Cubes** $\displaystyle\sum_{k=1}^{n} k^3 = \frac{n^2(n+1)^2}{4}$

[Note that if you put the results from Exercises 23 and 24 together, you obtain the pleasantly surprising equation

$$1^3 + 2^3 + 3^3 + \cdots + n^3 = (1 + 2 + 3 + \cdots + n)^2.]$$

In Exercises 25–30, use the results of Exercises 21–24 and Example 2 to find the sums.

**25.** $1 + 2 + 3 + \cdots + 500$     **26.** $1^2 + 2^2 + \cdots + 250^2$

**27.** $4 + 5 + 6 + \cdots + n$     **28.** $1^3 + 2^3 + 3^3 + \cdots + 75^3$

**29.** $1 + 2 + 4 + 8 + \cdots + 2^{34}$   **30.** $1 + 8 + 27 + \cdots + 3375$

In Exercises 31–34, use the results of Exercises 21–24 and Example 2 to find the sum in terms of *n*.

**31.** $\displaystyle\sum_{k=1}^{n} (k^2 - 3k + 4)$     **32.** $\displaystyle\sum_{k=1}^{n} (2k^2 + 5k - 2)$

**33.** $\displaystyle\sum_{k=1}^{n} (k^3 - 1)$     **34.** $\displaystyle\sum_{k=1}^{n} (k^3 + 4k - 5)$

**35. Group Activity** Here is a proof by mathematical induction that any gathering of *n* people must all have the same blood type.

**(Anchor)** If there is 1 person in the gathering, everyone in the gathering obviously has the same blood type.

**(Inductive hypothesis)** Assume that any gathering of *k* people must all have the same blood type.

**(Inductive step)** Suppose $k + 1$ people are gathered. Send one of them out of the room. The remaining *k* people must all have the same blood type (by the inductive hypothesis). Now bring the first person back and send someone else out of the room. You get another gathering of *k* people, all of whom must have the same blood type. Therefore all $k + 1$ people must have the same blood type, and we are done by mathematical induction.

This result is obviously false, so there must be something wrong with the proof. Explain where the proof goes wrong.

**36. Writing to Learn** Kitty is having trouble understanding mathematical induction proofs because she does not understand the inductive hypothesis. If we can assume it is true for *k*, she asks, why can't we assume it is true for *n* and be done with it? After all, a variable is a variable! Write a response to Kitty to clear up her confusion.

## Explorations

**37.** Use mathematical induction to prove that 2 is a factor of $(n + 1)(n + 2)$ for all positive integers *n*.

**38.** Use mathematical induction to prove that 6 is a factor of $n(n + 1)(n + 2)$ for all positive integers *n*. (You may assume the assertion in Exercise 37 to be true.)

**39.** Give an alternate proof of the assertion in Exercise 37 based on the fact that $(n + 1)(n + 2)$ is a product of two consecutive integers.

**40.** Give an alternate proof of the assertion in Exercise 38 based on the fact that $n(n + 1)(n + 2)$ is a product of three consecutive integers. ∎

## Extending the Ideas

In Exercises 41 and 42, use mathematical induction to prove that the statement holds for all positive integers.

**41. Fibonacci Sequence and Series** $F_{n+2} - 1 = \displaystyle\sum_{k=1}^{n} F_k$, where $\{F_n\}$ is the Fibonacci sequence.

**42.** If $\{a_n\}$ is the sequence $\sqrt{2}, \sqrt{2 + \sqrt{2}}, \sqrt{2 + \sqrt{2 + \sqrt{2}}}, \ldots$, then $a_n < 2$.

**43.** Let *a* be any integer greater than 1. Use mathematical induction to prove that $a - 1$ divides $a^n - 1$ evenly for all positive integers *n*.

**44.** Give an alternate proof of the assertion in Exercise 43 based on the Factor Theorem of Section 2.4.

It is not necessary to anchor a mathematical induction proof at $n = 1$; we might only be interested in the integers greater than or equal to some integer *c*. In this case, we simply modify the anchor and inductive step as follows:

Anchor: $P_c$ is true

Inductive Step: If $P_k$ is true for some $k \geq c$, then $P_{k+1}$ is true.

(This is sometimes called the **Extended Principle of Mathematical Induction**.) Use this principle to prove the statements in Exercises 45 and 46.

**45.** $3n - 4 \geq n$, for all $n \geq 2$    **46.** $2^n \geq n^2$, for all $n \geq 4$

**47. Proving the interior angle formula** Use extended mathematical induction to prove $P_n$ for $n \geq 3$.
$P_n$: The sum of the interior angles of an *n*-sided polygon is $180°(n - 2)$.

**9.6** # Statistics and Data (Graphical)

Statistics • Displaying Categorical Data • Stemplots
• Frequency Tables • Histograms • Time Plots

## Statistics

**Objective**

Students will be able to distinguish between categorical and quantitative variables and use various kinds of graphs to display data.

**Motivate**

Discuss ways to display a set of data graphically.

**Lesson Guide**

Day 1: Statistics; Displaying Categorical Data; Stemplots; Frequency Tables; Histograms; Time Plots
Day 2: Applications

Statistics is a branch of science that draws from both discrete mathematics and the mathematics of the continuum. The aim of statistics is to draw meaning from data and communicate it to others.

The objects described by a set of data are **individuals**, which may be people, animals, or things. The characteristic of the individuals being identified or measured is a **variable**. Variables are either *categorical* or *quantitative*. If the variable identifies each individual as belonging to a distinct class, such as male or female, then the variable is a **categorical variable**. If the variable takes on numerical values for the characteristic being measured, then the variable is a **quantitative variable**.

Examples of quantitative variables are heights of people, weights of lobsters, and lengths of rhododendron leaves. You have already seen many tables of quantitative data in this book; indeed, most of the data-based exercises you have seen in this book have been solved using techniques that are basic tools for statisticians. So far, however, most of our attention has been restricted to finding models that relate quantitative variables to each other, since that fit in well with our study of functions. In the last two sections of this chapter we will look at some of the other graphical and algebraic tools that can be used to draw meaning from data and communicate it to others.

## Displaying Categorical Data

The National Center for Health Statistics reported that the leading causes of death in 1998 were heart disease, cancer, and stroke. Table 9.4 gives more detailed information.

**Table 9.4  Leading Causes of Death in the United States in 1998**

| Cause of Death | Number of Deaths | Percentage |
|---|---|---|
| Heart Disease | 742,269 | 31.0 |
| Cancer | 538,947 | 23.0 |
| Stroke | 158,060 | 6.8 |
| Other | 916,799 | 39.2 |

*Source: National Vital Statistics Reports, Volume 47 Number 25, National Center for Health Statistics*

**Teaching Note**

Note that the *individuals* in this example are individual people who died in 1998.

Because the causes of death are categories, not numbers, "cause of death" is a categorical variable. The numbers of deaths and percentages, while certainly numerical, are not values of a *variable*, because they do not describe the *individuals*. Nonetheless, the numbers can communicate information about the categorical variables by telling us the relative size of the categories in the 1998 population.

We can get that information directly from the numbers, but it is very helpful to see the comparative sizes visually. This is why you will often see categorical data displayed graphically, as a **bar chart** (Figure 9.13a), a **pie chart** (Figure 9.13b), or a **circle graph** (Figure 9.13c). For variety, the popular press also makes use of **picture graphs** suited to the categories being displayed. For example, the bars in Figure 9.13a could be made to look like tombstones of different sizes to emphasize that these are causes of death. In each case, the graph provides a visualization of the relative sizes of the categories, with the pie chart and circle graph providing the added visualization of how the categories are parts of a whole population.

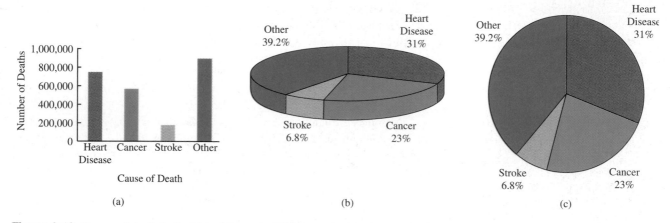

**Figure 9.13** Causes of death in the United States in 1998 shown in (a) a bar graph, (b) a 3-D pie chart, and (c) a circle graph.

In bar charts of categorical data the *y*-axis has a numerical scale, but the *x*-axis is labeled by category. The rectangular bars are separated by spaces to show that no continuous numerical scale is implied. (In this respect a bar graph differs from a *histogram*, to be described later in this section.) A circle graph or a pie chart consists of shaded or colored sections of a circle or "pie." The central angles for the sectors are found by multiplying the percentages by 360°. For example, the angle for the sector representing cancer victims in Figure 9.13c is

$$23\% \cdot 360° = 82.8°.$$

It used to require time, skill, and mathematical savvy to draw data displays that were both visually appealing and geometrically accurate, but modern spreadsheet programs have made it possible for anyone with a computer to produce high-quality graphs from tabular data with the click of a button.

## Stemplots

A quick way to organize and display a small set of quantitative data is with a **stemplot**, also called a **stem-and-leaf plot**. Each number in the data set is split into a **stem**, consisting of its initial digit or digits, and a **leaf**, which is its final digit.

**Example 1**  MAKING A STEMPLOT

Table 9.5 gives the percentage of each state's population in 1998 that was 65 or older. Make a stemplot for the data.

**Table 9.5  Percentages of State Residents in 1998 That Were 65 or Older**

| | | | | | | | | | |
|---|---|---|---|---|---|---|---|---|---|
| AL | 13.1 | HI | 13.3 | MA | 14.0 | NM | 11.4 | SD | 14.3 |
| AK | 5.5 | ID | 11.3 | MI | 12.5 | NY | 13.3 | TN | 12.5 |
| AZ | 13.2 | IL | 12.4 | MN | 12.3 | NC | 12.5 | TX | 10.1 |
| AR | 14.3 | IN | 12.5 | MS | 12.2 | ND | 14.4 | UT | 8.8 |
| CA | 11.1 | IO | 15.1 | MO | 13.7 | OH | 13.4 | VT | 12.3 |
| CO | 10.1 | KS | 13.5 | MT | 13.3 | OK | 13.4 | VA | 11.3 |
| CT | 14.3 | KY | 12.5 | NE | 13.8 | OR | 13.2 | WA | 11.5 |
| DE | 13.0 | LA | 11.5 | NV | 11.5 | PA | 15.9 | WV | 15.2 |
| FL | 18.3 | ME | 14.1 | NH | 12.0 | RI | 15.6 | WI | 13.2 |
| GA | 9.9 | MD | 11.5 | NJ | 13.6 | SC | 12.2 | WY | 11.5 |

*Source: U.S. Census Bureau, 1999*

**Solution**  To form the stem-and-leaf plot, we use the first two digits of each number as the stem and the last digit as the leaf. We write the stems in order down the first column and, for each number, write the leaf in the appropriate stem row. Then we arrange the leaves in each stem row in ascending order. The final plot looks like this:

| Stem | Leaf |
|---|---|
| 5 | 5 |
| 6 | |
| 7 | |
| 8 | 8 |
| 9 | 9 |
| 10 | 1 1 |
| 11 | 1 3 3 4 5 5 5 5 5 |
| 12 | 0 2 2 3 3 4 5 5 5 5 5 |
| 13 | 0 1 2 2 2 3 3 3 4 4 5 6 7 8 |
| 14 | 0 1 3 3 3 4 |
| 15 | 1 2 6 9 |
| 16 | |
| 17 | |
| 18 | 3 |

Notice that we include the "leafless" stems (6, 7, 16, 17) in our plot, as those empty gaps are significant features of the visualization. For the same reason, be sure that each "leaf" takes up the same space along the stem. A branch with twice as many leaves should appear to be twice as long.

**Exploration 1** **Using Information from a Stemplot**

By looking at both the stemplot and the table, answer the following questions about the distribution of senior citizens among the 50 states.

1. Judging from the stemplot, what was the approximate *average* national percentage of residents who were 65 or older?

2. In how many states were more than 15% of the residents 65 or older? 5

3. Which states were in the bottom tenth of all states in this statistic?

4. The numbers 5.5 and 18.3 are so far above or below the other numbers in this stemplot that statisticians would call them *outliers*. Quite often there is some special circumstance that sets outliers apart from the other individuals under study and explains the unusual data. What could explain the two outliers in this stemplot?

Sometimes the data are so tightly clustered that a stemplot has too few stems to give a meaningful visualization of the data. In such cases we can spread the data out by splitting the stems, as in Example 2.

**Example 2**  MAKING A SPLIT-STEM STEMPLOT

The 1997 per capita incomes for the forty most populous metropolitan areas in the United States are given in Table 9.6. (The names are omitted, as the definitions of some "metropolitan areas" are rather involved. For example, the one we might think of as "Philadelphia" involves four different states.) Make a stemplot that provides a good visualization of the data.

**Table 9.6** The 1997 Per Capita Incomes (in dollars) for the 40 Most Populous Metropolitan Areas in the United States in 1997

| | | | | | | | |
|---|---|---|---|---|---|---|---|
| 26,193.7 | 37,775.0 | 31,158.5 | 32,218.6 | 29,485.7 | 34,482.1 | 28,260.0 | 28,495.0 |
| 27,931.2 | 30,319.8 | 20,331.2 | 24,259.7 | 30,353.6 | 24,865.6 | 29,966.9 | 33,772.4 |
| 27,488.8 | 27,785.0 | 26,275.3 | 30,798.4 | 33,379.6 | 24,862.8 | 27,873.3 | 23,931.6 |
| 34,938.2 | 29,956.0 | 27,389.5 | 27,072.4 | 40,776.7 | 39,452.2 | 36,145.1 | 26,497.0 |
| 24,569.8 | 23,129.0 | 22,573.3 | 26,760.5 | 25,389.9 | 28,204.8 | 23,037.7 | 26,418.3 |

*Source: Regional Financial Associates as quoted by The Wall Street Journal Almanac, 1999*

**Solution**  First, the units (dollars) are inappropriate for any stemplot, so we round the data to units of $1000. (The precision we lose by rounding the data is not necessary for the visualization.) This results in Table 9.7:

**Table 9.7**  The Data of Table 9.6 Rounded to the Nearest Thousand Dollars and Shown in Units of $1000

| | | | | | | | |
|---|---|---|---|---|---|---|---|
| 26 | 38 | 31 | 32 | 29 | 34 | 28 | 28 |
| 28 | 30 | 20 | 24 | 30 | 25 | 30 | 34 |
| 27 | 28 | 26 | 31 | 33 | 25 | 28 | 24 |
| 35 | 30 | 27 | 27 | 41 | 39 | 36 | 26 |
| 25 | 23 | 23 | 27 | 25 | 28 | 23 | 26 |

If we form the stem plot as in Example 1, there will only be three stems—not enough to give a very good visualization of the spread of the data. So we *split* each stem, putting leaves 0–4 on the lower stem and leaves 5–9 on the upper stem:

| Stem | Leaf |
|---|---|
| 2 | 0 3 3 3 4 4 |
| 2 | 5 5 5 5 6 6 6 6 7 7 7 7 8 8 8 8 8 8 9 |
| 3 | 0 0 0 0 1 1 2 3 4 4 |
| 3 | 5 6 8 9 |
| 4 | 1 |
| 4 | |

Sometimes it is easier to compare two sets of data if we have a visualization that allows us to view both stemplots simultaneously. **Back-to-back stemplots** use the same stems, but leaves from one set of data are added to the left, while leaves from another set are added to the right.

### Example 3  MAKING BACK-TO-BACK STEMPLOTS

From 1986 to 1999, Mark McGwire averaged 21.75 home runs per year. Sammy Sosa, who entered the major leagues three years later, averaged 30.55 home runs per year from 1989 to 1999. Their annual home run totals (HR) and times-at-bat (AB) are given in Table 9.8. Compare their annual home run totals with a back-to-back stemplot. Can you tell which player has been more consistent as a home run hitter?

| | Table 9.8 Major League Home Run Totals for Mark McGwire and Sammy Sosa Through 1999 | | | | |
|---|---|---|---|---|---|
| | Mark McGwire | | | Sammy Sosa | |
| Year | AB | HR | AB | HR | |
| 1986 | 53 | 3 | | | |
| 1987 | 557 | 49 | | | |
| 1988 | 550 | 32 | | | |
| 1989 | 490 | 33 | 183 | 4 | |
| 1990 | 523 | 39 | 532 | 15 | |
| 1991 | 483 | 22 | 316 | 10 | |
| 1992 | 467 | 42 | 262 | 8 | |
| 1993 | 84 | 9 | 598 | 33 | |
| 1994 | 135 | 9 | 426 | 25 | |
| 1995 | 317 | 39 | 564 | 36 | |
| 1996 | 423 | 52 | 498 | 40 | |
| 1997 | 540 | 58 | 642 | 36 | |
| 1998 | 509 | 70 | 643 | 66 | |
| 1999 | 521 | 65 | 625 | 63 | |

*Source: Major League Baseball Enterprises, 1999*

**Solution** We form a back-to-back stemplot using the numbers in the HR column for both players, with McGwire's totals branching off the stem to the left and Sosa's to the right.

| Mark McGwire | | Sammy Sosa |
|---|---|---|
| 9 9 3 | 0 | 4 8 |
| | 1 | 0 5 |
| 2 | 2 | 5 |
| 9 9 3 2 | 3 | 3 6 6 |
| 9 2 | 4 | 0 |
| 8 2 | 5 | |
| 5 | 6 | 3 6 |
| 0 | 7 | |

The single-digit home run years for both players can be explained by fewer times at bat, either because of late entry into the major leagues (McGwire 1986, Sosa 1989) or injuries (McGwire 1993, 1994, heel injury; Sosa 1992, hand injury). If those years are ignored, the tighter distribution of McGwire's numbers seems to indicate more consistency.

## Frequency Tables

The visual impact of a stemplot comes from the lengths of the various rows of leaves, which is just a way of seeing *how many* leaves branch off each stem. The number of leaves for a particular stem is the **frequency** of observations

**Teaching Note**

In this context the frequency of an observation is simply the number of times the data value (or any data value in a particular range of values) occurs.

within each stem interval. Frequencies are often recorded in a **frequency table**. Table 9.9 shows a frequency table for Mark McGwire's yearly home run totals from 1986 to 1999 (see Example 3). The table shows the **frequency distribution**—literally the way that the total frequency of 14 is "distributed" among the various home run intervals. This same information is conveyed visually in a stemplot, but notice that the stemplot has the added numerical advantage of displaying what the numbers in each interval actually are.

**Table 9.9** Frequency Table for Mark McGwire's Yearly Home Run Totals, 1986–1999 (Higher frequencies in a table correspond to longer leaf rows in a stemplot. Unlike a stemplot, a frequency table does not display what the numbers in each interval actually are.)

| Home Runs | Frequency |
|-----------|-----------|
| 0–9 | 3 |
| 10–19 | 0 |
| 20–29 | 1 |
| 30–39 | 4 |
| 40–49 | 2 |
| 50–59 | 2 |
| 60–69 | 1 |
| 70–79 | 1 |
| Total | 14 |

## Histograms

A **histogram**, closely related to a stemplot, displays the information of a frequency table. Visually, a histogram is to quantitative data what a bar chart is to categorical data. Unlike a bar chart, however, both axes of a histogram have numerical scales, and the rectangular bars on adjacent intervals have no gaps between them.

Figure 9.14 shows a histogram of the information in Table 9.9, where each bar corresponds to an interval in the table and the height of each bar represents the frequency of observations within the interval.

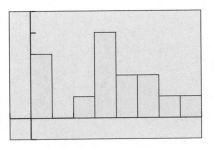

[−10, 80] by [−1, 5]

**Figure 9.14** A histogram showing the distribution of Mark McGwire's annual home run totals from 1986 to 1999. This is a visualization of the information in Table 9.9.

### Example 4   GRAPHING A HISTOGRAM ON A CALCULATOR

Make a histogram of Hank Aaron's annual home run totals given in Table 9.10, using intervals of width 5.

**Table 9.10  Regular Season Home Run Statistics for Hank Aaron**

| Year | Home Runs | Year | Home Runs | Year | Home Runs |
|------|-----------|------|-----------|------|-----------|
| 1954 | 13 | 1962 | 45 | 1970 | 38 |
| 1955 | 27 | 1963 | 44 | 1971 | 47 |
| 1956 | 26 | 1964 | 24 | 1972 | 34 |
| 1957 | 44 | 1965 | 32 | 1973 | 40 |
| 1958 | 30 | 1966 | 44 | 1974 | 20 |
| 1959 | 39 | 1967 | 39 | 1975 | 12 |
| 1960 | 40 | 1968 | 29 | 1976 | 10 |
| 1961 | 34 | 1969 | 44 |  |  |

*Source: The Baseball Encyclopedia (7th ed., 1988, New York: MacMillan) p. 695*

**Solution**  We first make a frequency table for the data, using intervals of width 5. (This is not needed for the calculator to produce the histogram, but we will compare this with the result.)

| Home Runs | Frequency |
|-----------|-----------|
| 10–14 | 3 |
| 15–19 | 0 |
| 20–24 | 2 |
| 25–29 | 3 |
| 30–34 | 4 |
| 35–39 | 3 |
| 40–44 | 6 |
| 45–49 | 2 |
| Total | 23 |

To scale the $x$-axis to be consistent with the intervals of the table, let Xmin = 0, Xmax = 50, and Xscl = 5. Notice that the maximum frequency is 6 years (40–44 home runs), so the $y$-axis needs to extend at least to 6. Enter the data from Table 9.10 into list L1 and plot a histogram in the window [0, 50] by [−2, 6]. (See Figure 9.15a.) Tracing along the histogram should reveal the same frequencies as in the frequency table we made. (See Figure 9.15b.)

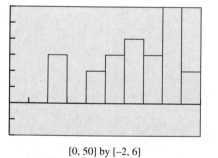

[0, 50] by [−2, 6]

(a)

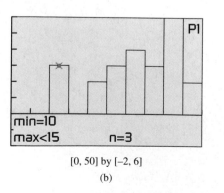

min=10
max<15          n=3

[0, 50] by [−2, 6]

(b)

**Figure 9.15** Calculator histograms of Hank Aaron's yearly home run totals. (Example 4)

## Time Plots

We have seen in this book many examples of functions in which the input variable is time. It is also quite common to consider quantitative data as a function of time. If we make a scatter plot of the data ($y$) against the time ($x$) that it was

**Teaching Note**

When time is one of the varying factors, a time plot is usually the proper way to represent the data.

measured, we can analyze the patterns as the variable changes over time. To help with the visualization, the discrete points from left to right are connected by line segments, just as a grapher would do in connect mode. The resulting **line graph** is called a **time plot**.

Time plots reveal trends in data over time. These plots frequently appear in magazines and newspapers and on the Internet, a typical example being the graph of the Dow Jones Industrial Average (DJIA) in Figure 9.16.

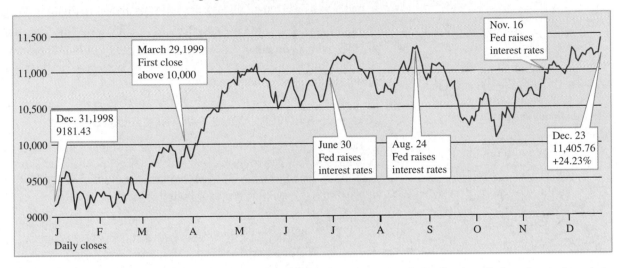

**Figure 9.16** Time plot of the Dow Jones Industrial Average. Investors get a good visualization of where the stock market has been, although the trick is to figure out where it is going. *(Source: Quote.com, as reported by the Associated Press in the Chattanooga Times/Free Press)*

## Example 5 DRAWING A TIME PLOT

Table 9.11 gives the number of compact disc (CD) units shipped to retailers each year in the ten-year period from 1988 to 1997. (The numbers are shown in millions, net after returns.) Display the data in a time plot and analyze the ten-year trend.

**Table 9.11 The Number of CDs Shipped to Retailers Each Year from 1988 to 1997, in Millions of Units**

| Year | CDs Shipped (in millions) |
| --- | --- |
| 1988 | 149.7 |
| 1989 | 207.2 |
| 1990 | 286.5 |
| 1991 | 333.3 |
| 1992 | 407.5 |
| 1993 | 495.4 |
| 1994 | 662.1 |
| 1995 | 722.9 |
| 1996 | 778.9 |
| 1997 | 753.1 |

*Source: Recording Industry Association of America as reported in the Wall Street Journal Almanac, 1999*

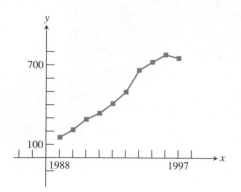

**Figure 9.17** A time plot of CD volume over the years from 1988 to 1997, as reflected in units shipped to retailers. (Example 5)

**Solution** The horizontal axis represents time (in years) from 1988 to 1997. The vertical axis represents the number of CD's shipped that year, in millions of units. Since the visualization is enhanced by showing both axes in the viewing window, the vertical axis of a time plot is usually translated to cross the $x$-axis at or near the beginning of the time interval in the data. You can create this effect on your grapher by entering the years as $\{1, 2, 3, \ldots, 10\}$ rather than $\{1988, 1989, 1990, \ldots, 1997\}$. The labeling on the $x$-axis can then display the years, as shown in Figure 9.17.

The time plot shows that CD shipments increased steadily from 1988 to 1996, with a particularly sharp jump from 1993 to 1994. There was then a slight drop from 1996 to 1997, attributed by some analysts to a sudden age shift in the popular music culture.

### Example 6   OVERLAYING TWO TIME PLOTS

Table 9.12 gives the number of cassette units shipped to retailers each year in the ten-year period from 1988 to 1997. (The numbers are shown in millions, net after returns.) Compare the cassette trend with the CD trend by overlaying the time plots for both products.

**Table 9.12  The Number of Cassettes Shipped to Retailers Each Year from 1988 to 1997, in Millions of Units**

| Year | Cassettes Shipped (in millions) |
| --- | --- |
| 1988 | 450.1 |
| 1989 | 446.2 |
| 1990 | 442.2 |
| 1991 | 360.1 |
| 1992 | 366.4 |
| 1993 | 339.5 |
| 1994 | 345.4 |
| 1995 | 272.6 |
| 1996 | 225.3 |
| 1997 | 172.6 |

*Source: Recording Industry Association of America as reported in the Wall Street Journal Almanac, 1999*

**Solution** The two time plots are shown in Figure 9.18. The popularity of cassette tapes declined as more and more consumers switched to CD technology. The CD spurt in 1994 foreshadowed a steeper decline in cassette shipments thereafter.

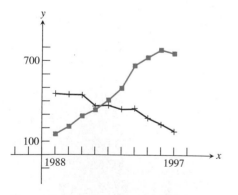

**Figure 9.18** A time plot comparing cassette shipments and CD shipments for the time period from 1988 to 1997. (Example 6)

Time plots are often used to seek, exhibit, or clarify trends. Example 7 illustrates how a time plot can show a trend even when the time intervals for the data are not uniform.

## Example 7 OVERLAYING TWO TIME PLOTS

Use the data in Table 9.13 to overlay time plots of the life expectancy for males and females in the United States from 1920 to 1997. What can you conclude about life expectancy in the United States, especially as related to gender?

| Table 9.13 Life Expectancy in the United States | | |
|---|---|---|
| Year | Male | Female |
| 1920 | 53.6 | 54.6 |
| 1930 | 58.1 | 61.6 |
| 1940 | 60.8 | 65.2 |
| 1950 | 65.6 | 71.1 |
| 1960 | 66.6 | 73.1 |
| 1965 | 66.8 | 73.7 |
| 1970 | 67.1 | 74.7 |
| 1980 | 70.0 | 77.5 |
| 1985 | 71.2 | 78.2 |
| 1990 | 71.8 | 78.8 |
| 1993 | 72.1 | 78.9 |
| 1995 | 72.5 | 78.9 |
| 1997 | 73.6 | 79.2 |

*Source: National Center for Health Statistics as reported in The World Almanac and Book of Facts, 1999*

**Solution** To get both axes in the visualization, we subtract 1910 from each of the years and 50 from each of the life expectancies when entering the data. We then label the picture according to the original data. The result is shown in Figure 9.19.

Figure 9.19 shows several trends about life expectancy in the U.S. from 1920 to 1997:

- The life expectancies of both males and females increased.

- In general, women lived longer than men.

- The gap in life expectancy between men and women widened from 1920 to about 1970, then stabilized briefly. The last part of the plot suggests that the gender gap in life expectancy was starting to close in the 1990s.

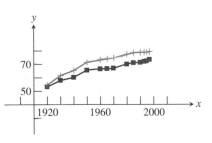

**Figure 9.19** Time plots of male and female life expectancies in the United States from 1920 to 1997. (Example 7)

# Quick Review 9.6

In Exercises 1–6, solve for the required value.

1. 457 is what percent of 2953?  ≈ 15.48%

2. 827 is what percent of 3950?  ≈ 20.94%

3. 52° is what percent of 360°?  ≈ 14.44%

4. 98° is what percent of 360°?  ≈ 27.22%

5. 734 is 42.6% of what number?  ≈ 1723

6. 5106 is 55.5% of what number?  9200

In Exercises 7–10, round the given value to the nearest whole number in the specified units.

7. $234,598.43 (thousands of dollars)  $235 thousand

8. 237,834,289 (millions)  238 million

9. 848.36 thousands (millions)  1 million

10. 1432 millions (billions)  1 billion

# Section 9.6 Exercises

In Exercises 1–4, construct the indicated stemplot from the data in Table 9.14. Then write a brief interpretation of the stemplot.

**Table 9.14** Life Expectancy by Gender for the Nations of South America

| Nation | Male | Female |
|--------|------|--------|
| Argentina | 70.9 | 78.3 |
| Bolivia | 58.0 | 63.9 |
| Brazil | 59.4 | 69.6 |
| Chile | 72.0 | 78.5 |
| Colombia | 66.2 | 74.1 |
| Ecuador | 69.2 | 74.5 |
| Guyana | 59.5 | 65.3 |
| Paraguay | 70.3 | 74.3 |
| Peru | 67.8 | 72.3 |
| Suriname | 68.1 | 73.3 |
| Uruguay | 72.4 | 78.8 |
| Venezuela | 69.7 | 75.9 |

*Source: The World Almanac and Book of Facts, 1999*

1. A stemplot showing life expectancies of males in the nations of South America. (Round to nearest year and use split stems.)

2. A stemplot showing life expectancies of females in the nations of South America. (Round to nearest year and use split stems.)

3. A back-to-back stemplot for life expectancies of males and females in the nations of South America. (Round to nearest year and use split stems.)

4. A stemplot showing the difference between female and male life expectancies in the nations of South America. (Use unrounded data and do not split stems.)

In Exercises 5 and 6, use the data in Table 9.14 to construct the indicated frequency table, using intervals 55.0–59.9, 60.0–64.9, 65.0–69.9, etc.

5. Life expectancies of males in the nations of South America.

6. Life expectancies of females in the nations of South America.

Table 9.15 shows the regular-season home run statistics for Roger Maris during his major league career.

**Table 9.15** Regular Season Home Run Statistics for Roger Maris

| Year | Home Runs |
|------|-----------|
| 1957 | 14 |
| 1958 | 28 |
| 1959 | 16 |
| 1960 | 39 |
| 1961 | 61 |
| 1962 | 33 |
| 1963 | 23 |
| 1964 | 26 |
| 1965 | 8 |
| 1966 | 13 |
| 1967 | 9 |
| 1968 | 5 |

*Source: The Baseball Encyclopedia, 7th ed., 1988*

7. Make a stemplot of the data in Table 9.15. Are there any outliers?

8. Make a back-to-back stemplot comparing the annual home run production of Roger Maris (Table 9.15) with that of Hank Aaron (Table 9.10 on page 724). Write a brief interpretation of the stemplot.

In Exercises 9–12, draw a histogram for the given table.

9. The frequency table in Exercise 5.

10. The frequency table in Exercise 6.

11. Table 9.16 of Willie Mays's annual home run totals, using intervals 1–5, 6–10, 11–15, etc.

12. Table 9.16 of Mickey Mantle's annual home run totals, using intervals 1–5, 6–10, 11–15, etc.

**Table 9.16** Regular Season Home Run Statistics for Willie Mays and Mickey Mantle

| Year | Mays | Mantle | Year | Mays | Mantle |
|------|------|--------|------|------|--------|
| 1951 | 20 | 13 | 1962 | 38 | 30 |
| 1952 | 4 | 23 | 1963 | 47 | 15 |
| 1953 | 41 | 21 | 1964 | 52 | 35 |
| 1954 | 51 | 27 | 1965 | 37 | 19 |
| 1955 | 36 | 37 | 1966 | 22 | 23 |
| 1956 | 35 | 52 | 1967 | 23 | 22 |
| 1957 | 29 | 34 | 1968 | 13 | 18 |
| 1958 | 34 | 42 | 1969 | 28 | |
| 1959 | 29 | 31 | 1970 | 18 | |
| 1960 | 40 | 40 | 1971 | 8 | |
| 1961 | 49 | 54 | 1972 | 6 | |

*Source: The Baseball Encyclopedia, 7th ed., 1988*

In Exercises 13–16, make a time plot for the indicated data.

**13.** Willie Mays's annual home run totals given in Table 9.16.

**14.** Mickey Mantle's annual home run totals given in Table 9.16.

**15.** Mark McGwire's home run totals given in Table 9.8 on page 722.

**16.** Hank Aaron's home run totals given in Table 9.10 on page 724.

Table 9.17 shows the total amount of money won (in units of $1000, rounded to the nearest whole number) by the leading money winners in women's (LPGA) and men's (PGA) professional golf for selected years between 1955 and 1997.

**Table 9.17** Yearly Earnings (in Thousands of Dollars) of the Top Money Winners in Men's and Women's Golf for Selected Years from 1955 to 1997 (Rounded to the Nearest $1000)

| Year | Men (PGA) | Women (LPGA) |
|------|-----------|--------------|
| 1955 | 65 | 16 |
| 1960 | 75 | 17 |
| 1965 | 141 | 29 |
| 1970 | 157 | 30 |
| 1975 | 323 | 95 |
| 1980 | 531 | 231 |
| 1985 | 542 | 416 |
| 1989 | 1395 | 654 |
| 1992 | 1344 | 693 |
| 1995 | 1655 | 667 |
| 1996 | 1780 | 1002 |
| 1997 | 2067 | 1237 |

*Source: The World Almanac and Book of Facts, 1999*

**17.** Make a time plot for the men's winnings in Table 9.17. Write a brief interpretation of the time plot.

**18.** Make a time plot for the women's winnings in Table 9.17. Write a brief interpretation of the time plot.

**19.** Compare the trends in Table 9.17 by overlaying the time plots. Write a brief interpretation.

**20.** **Writing to Learn** (Continuation of Exercise 19) The data in Table 9.17 show that the earnings for the top PGA player rose by a modest 68% in the decade from 1975 to 1985, while the earnings for the top LPGA player rose by a whopping 338%. Although this was, in fact, a strong growth period for women's sports, statisticians would be unlikely to draw any conclusions from a comparison of these two numbers. Use the visualization from the comparative time plot in Exercise 19 to explain why.

In Exercises 21 and 22, compare performances by overlaying time plots.

**21.** The time plots from Exercises 13 and 14 to compare the performances of Mays and Mantle.

**22.** The time plots from Exercises 15 and 16 to compare the performances of McGwire and Aaron.

In Exercises 23 and 24, analyze the data as indicated.

**23.** The salaries of the workers in one department of the Garcia Brothers Company (given in thousands of dollars) are as follows:

33.5, 35.3, 33.8, 29.3, 36.7, 32.8, 31.7, 36.3, 33.5, 28.2, 34.8, 33.5, 35.3, 29.7, 38.5, 32.7, 34.8, 34.2, 31.6, 35.4

(a) Complete a stemplot for this data set.

(b) Create a frequency table for the data.

(c) Draw a histogram for the data. What viewing window did you use?

(d) Why does a time plot not work well for the data?

**24.** The average wind speeds for one year at 44 climatic data centers around the United States are as follows:

9.0, 6.9, 9.1, 9.2, 10.2, 12.5, 12.0, 11.2, 12.9, 10.3, 10.6, 10.9, 8.7, 10.3, 11.0, 7.7, 11.4, 7.9, 9.6, 8.0, 10.7, 9.3, 7.9, 6.2, 8.3, 8.9, 9.3, 11.6, 10.6, 9.0, 8.2, 9.4, 10.6, 9.5, 6.3, 9.1, 7.9, 9.7, 8.8, 6.9, 8.7, 9.0, 8.9, 9.3

(a) Complete a stemplot for this data set.

(b) Create a frequency table for the data.

(c) Draw a histogram for the data. What viewing window did you use?

(d) Why does a circle graph not work well for the data?

In Exercises 25 and 26, compare by overlaying time plots for the data in Table 9.18.

**Table 9.18** Population (in millions) of California, Florida, Illinois, New York, Pennsylvania, and Texas

| Year | CA | FL | IL | NY | PA | TX |
|------|------|------|------|------|------|------|
| 1900 | 1.5 | 0.5 | 4.8 | 7.3 | 6.3 | 3.0 |
| 1910 | 2.4 | 0.8 | 5.6 | 9.1 | 7.7 | 3.9 |
| 1920 | 3.4 | 1.0 | 6.5 | 10.4 | 8.7 | 4.7 |
| 1930 | 5.7 | 1.5 | 7.6 | 12.6 | 9.6 | 5.8 |
| 1940 | 6.9 | 1.9 | 7.9 | 13.5 | 9.9 | 6.4 |
| 1950 | 10.6 | 2.7 | 8.7 | 14.8 | 10.5 | 7.7 |
| 1960 | 15.7 | 5.0 | 10.0 | 16.8 | 11.3 | 9.6 |
| 1970 | 20.0 | 6.8 | 11.1 | 18.2 | 11.8 | 11.2 |
| 1980 | 23.7 | 9.7 | 11.4 | 17.6 | 11.9 | 14.2 |
| 1990 | 29.8 | 12.9 | 11.4 | 18.0 | 11.9 | 17.0 |

*Source: U.S. Census Bureau, as reported in The World Almanac and Books of Facts (1995, Mahwah, N.J.: Funk & Wagnalls), p. 377*

**25.** The populations of California, New York, and Texas from 1900 through 1990.

**26.** The populations of Florida, Illinois, and Pennsylvania from 1900 through 1990.

## Explorations

**27. Group Activity** Measure the resting pulse rates (beats per minute) of the members of your class. Make a stemplot for the data. Are there any outliers? Can they be explained?

**28. Group Activity** Measure the heights (in inches) of the members of your class. Make a back-to-back stemplot to compare the distributions of the male heights and the female heights. Write a brief interpretation of the stemplot. ∎

## Extending the Ideas

**29. Time Plot of Periodic Data** Some data are periodic functions of time. If data vary in an annual cycle, the period is 1 year. Use the information in Table 9.19 and overlay the time plots for the average daily high and low temperatures for Beijing, China.

**Table 9.19** Average Daily High and Low Temperature in °C for Beijing, China

| Month | High | Low |
|-----------|------|-----|
| January | 2 | −9 |
| February | 5 | −7 |
| March | 12 | −1 |
| April | 20 | 7 |
| May | 27 | 13 |
| June | 31 | 18 |
| July | 32 | 22 |
| August | 31 | 21 |
| September | 27 | 14 |
| October | 21 | 7 |
| November | 10 | −1 |
| December | 3 | −7 |

*Source: National Geographic Atlas of the World (rev. 6th ed., 1992, Washington, D.C.), plate 132*

**30.** Find a sinusoidal function that models each time plot in Exercise 29. (See Sections 4.4 and 4.8.)

---

## 9.7 Statistics and Data (Algebraic)

**Objective**

Students will be able to use measures of center, the five-number summary, a boxplot, standard deviation, and normal distribution to describe quantitative data.

**Motivate**

Discuss ways to use a number to describe a set of quantitative data.

**Lesson Guide**

Day 1: Parameters and Statistics; Mean, Median, and Mode; The Five-Number Summary; Boxplots
Day 2: Variance and Standard Deviation; Normal Distributions

Parameters and Statistics • Mean, Median, and Mode • The Five-Number Summary • Boxplots • Variance and Standard Deviation • Normal Distributions

### Parameters and Statistics

The various numbers that are associated with a data set are called **statistics**. They serve to describe the individuals from which the data came, so the gathering and processing of such numerical information is often called **descriptive statistics**. You saw many examples of descriptive statistics in Section 9.6.

The *science* of statistics comes in when we use descriptive statistics (like the results of a study of 1500 smokers) to make judgments, called *inferences*, about entire *populations* (like all smokers). Statisticians are really interested in the numbers called **parameters** that are associated with entire populations. Since it is usually either impractical or impossible to measure entire populations, statisticians gather statistics from carefully chosen **samples**, then use the science of **inferential statistics** to make inferences about the parameters.

### Example 1  DISTINGUISHING A PARAMETER FROM A STATISTIC

A 1996 study called *Kids These Days: What Americans Really Think About the Next Generation* reported that 33% of adolescents say there is no adult at home when they return from school. The report was based on a survey of 600 randomly selected people aged 12 to 17 years old and had a margin of error of ±4% (Source: *Public Agenda*). Did the survey measure a parameter or a statistic, and what does that "margin of error" mean?

**Solution**  The survey did not measure all adolescents in the population, so it did not measure a parameter. They *sampled* 600 adolescents and found a statistic. On the other hand, the statement "33% of adolescents say there is no adult at home" is making an inference about *all* American adolescents. We should interpret that statement in terms of the margin of error, as meaning "between 29% and 37% of all American adolescents would say that there is no adult at home when they return from school." In other words, the statisticians are confident that the parameter is within ±4% of their sample statistic, even though they only sampled 600 adolescents—a tiny fraction of the adolescent population! The mathematics that gives them that confidence is based on the laws of probability and is scientifically reliable, but we will not go into it here.

## Mean, Median, and Mode

If you wanted to study the effect of chicken feed additives on the thickness of egg shells, you would need to sample many eggs from different hens under various feeding conditions. Suppose you were to gather data from 50 eggs from hens eating feed A and 50 eggs from hens eating feed B. How would you compare the two? The simplest way would be to find the *average* egg shell thickness for each feed and compare those two numbers.

The word "average," however, can have several different meanings, all of them somehow *measures of center*.

- If we say, "The average on last week's test was 83.4," we are referring to the **mean**. (This is what most people usually think of as "average.")

- If we say, "The average test score puts you right in the middle of the class," we are referring to the **median**.

- If we say, "The average American student starts college at age 18," we are referring to the **mode**.

We will look at each of these measures separately.

---

**Definition  Mean**

The **mean** of a list of $n$ numbers $\{x_1, x_2, \ldots, x_n\}$ is

$$\bar{x} = \frac{x_1 + x_2 + \cdots + x_n}{n} = \frac{1}{n}\sum_{i=1}^{n} x_i.$$

---

The mean is also called the *arithmetic mean*, *arithmetic average*, or *average value*.

### Example 2   COMPUTING A MEAN

Find the mean regular-season home run total for Roger Maris's major league career, 1957–1968 (Table 9.15 on page 728).

**Solution**   According to Table 9.15, we are looking for the mean of the following list of 12 numbers: {14, 28, 16, 39, 61, 33, 23, 26, 8, 13, 9, 5}.

$$\bar{x} = \frac{14 + 28 + 16 + \cdots + 9 + 5}{12} = \frac{275}{12} \approx 22.9.$$

As common as it is to use the mean as a measure of center, sometimes it can be misleading. For example, if you were to find the mean annual salary of a Geography major from the University of North Carolina working in Chicago in 1997, it would probably be a number in the millions of dollars. This is because the group being measured, which is not very large, includes an outlier named Michael Jordan. The mean can be strongly affected by outliers.

We call a statistic **resistant** if it is not strongly affected by outliers (See Exploration 1, Section 9.6). While the mean is not a resistant measure of center, the *median* is.

---

**Definition   Median**

The **median** of a list of $n$ numbers $\{x_1, x_2, \ldots, x_n\}$ arranged in order (either ascending or descending) is

- the middle number if $n$ is odd, and
- the mean of the two middle numbers if $n$ is even.

---

### Example 3   FINDING A MEDIAN

Find the median of Roger Maris's annual home run totals. (See Example 2.)

**Solution**   First, we arrange the list in ascending order: {5, 8, 9, 13, 14, 16, 23, 26, 28, 33, 39, 61}. Since there are 12 numbers, the median is the mean of the 6th and 7th number:

$$\frac{16 + 23}{2} = 19.5.$$

Notice that this number is quite a bit smaller than the mean (22.9). The mean is strongly affected by the outlier representing Maris's record-breaking season, while the median is not. The median would still have been 19.5 if he had hit only 41 home runs that season, or, indeed, if he had hit 81.

Both the mean and the median are important measures of center. A somewhat less important measure of center (but a measure of possible significance statistically) is the *mode*.

**À La Mode**

The mode can also be used for categorical variables.

---

**Definition   Mode**

The **mode** of a list of numbers is the number that appears most frequently in the list.

---

### Example 4 FINDING A MODE

Find the mode for the annual home run totals of Hank Aaron (Table 9.10 on page 724). Is this number of any significance?

**Solution** It helps to arrange the list in ascending order: {10, 12, 13, 20, 24, 26, 27, 29, 30, 32, 34, 34, 38, 39, 39, 40, 40, 44, 44, 44, 44, 45, 47}.

Most numbers in the list appear only once. Three numbers appear twice, and the number 44 appears four times. The mode is 44.

It is rather unusual to have that many repeats of a number in a list of this sort. (By comparison, the Maris list has no repeats—and hence no mode— while the lists for Mays, Mantle, McGwire, and Sosa contain no number more than twice.) The mode in this case has special significance only to baseball trivia buffs, who recognize 44 as Aaron's uniform number!

**Notes on Examples**

Example 4 mentions a situation in which the mode does not exist.

Example 4 demonstrates why the mode is less useful as a measure of center. The mode (44) is a long way from the median (34) and the mean (32.83), either of which does a much better job of representing Aaron's annual home run output over the course of his career.

### Example 5 USING A FREQUENCY TABLE

A teacher gives a 10-point quiz and records the scores in a frequency table (Table 9.20) as shown below. Find the mode, median, and mean of the data.

**Table 9.20 Quiz Scores for Example 5**

| Score | 10 | 9 | 8 | 7 | 6 | 5 | 4 | 3 | 2 | 1 | 0 |
|-------|----|----|----|----|----|----|----|----|----|----|----|
| Frequency | 2 | 2 | 3 | 8 | 4 | 3 | 3 | 2 | 1 | 1 | 1 |

**Solution** The total of the frequencies is 30, so there are 30 scores.

The *mode* is 7, since that is the score with the highest frequency.

The *median* of 30 numbers will be the mean of the 15th and 16th numbers. The table is already arranged in descending order, so we count the frequencies from left to right until we come to 15. We see that the 15th number is a 7 and the 16th number is a 6. The median, therefore, is 6.5.

To find the *mean*, we multiply each number by its frequency, add the products, and divide the total by 30:

$$\bar{x} = \frac{\begin{aligned}[10(2) + 9(2) + 8(3) + 7(8) + 6(4) + 5(3) \\ + 4(3) + 3(2) + 2(1) + 1(1) + 0(1)]\end{aligned}}{30}$$

$$= 5.9\overline{3}$$

The formula for finding the mean of a list of numbers $\{x_1, x_2, \ldots, x_n\}$ with frequencies $\{f_1, f_2, \ldots, f_n\}$ is

$$\bar{x} = \frac{x_1 f_1 + x_2 f_2 + \cdots + x_n f_n}{f_1 + f_2 + \cdots + f_n} = \frac{\sum x_i f_i}{\sum f_i}.$$

This same formula can be used to find a **weighted mean**, in which numbers $\{x_1, x_2, \ldots, x_n\}$ are given **weights** before the mean is computed. The weights act the same way as frequencies.

### Example 6  WORKING WITH A WEIGHTED MEAN

At Marty's school it is an administrative policy that the final exam must count 25% of the semester grade. If Marty has an 88.5 average going into the final exam, what is the minimum exam score needed to earn a 90 for the semester?

**Solution**  The preliminary average (88.5) is given a weight of 0.75 and the final exam ($x$) is given a weight of 0.25. We will assume that a semester average of 89.5 will be rounded to a 90 on the transcript. Therefore,

$$\frac{88.5(0.75) + x(0.25)}{0.75 + 0.25} = 89.5$$

$$0.25x = 89.5(1) - 88.5(0.75)$$

$$x = 92.5$$

Interpreting the answer, we conclude that Marty needs to make a 93 on the final exam.

**Teaching Note**

Note that the weights used can be the values 25 and 75 or 0.25 and 0.75. Either method will give the same result for the average value.

## The Five-Number Summary

Measures of center tell only part of the story of a data set. They do *not* indicate how widely distributed or highly variable the data are. *Measures of spread* do. The simplest and crudest measure of spread is the **range**, which is the difference between the maximum and minimum values in the data set:

$$\text{Range} = \text{maximum} - \text{minimum}.$$

For example, the range of numbers in Roger Maris's annual home run production is $61 - 5 = 56$. Like the mean, the range is a statistic that is strongly influenced by outliers, so it can be misleading. A more resistant (and therefore more useful) measure is the *interquartile range*, which is the range of the middle half of the data.

Just as the median separates the data into halves, the **quartiles** separate the data into fourths. The **first quartile** $Q_1$ is the median of the lower half of the data, the **second quartile** is the median, and the **third quartile** $Q_3$ is the median of the upper half of the data. The **interquartile range** (*IQR*) measures the spread between the first and third quartiles, comprising the middle half of the data:

$$IQR = Q_3 - Q_1.$$

Taken together, the maximum, the minimum, and the three quartiles give a fairly complete picture of both the center and the spread of a data set.

**Alert**

Stress the difference between the range of a function (which is a set of numbers) and the range of a data set (which is a number).

**Teaching Note**

Boxplots provide one of the simplest and visually clearest ways to compare data. Ask students to collect real data and, working in cooperative learning groups, use boxplots to compare the data.

> **Definition  Five-Number Summary**
>
> The **five-number summary** of a data set is the collection
>
> $$\{\text{minimum}, Q_1, \text{median}, Q_3, \text{maximum}\}.$$

### Example 7   FIVE-NUMBER SUMMARY AND SPREAD

Find the five-number summaries for the male and female life expectancies in South American nations (Table 9.14 on page 728) and compare the spreads.

**Solution**   Here are the lists in ascending order.

Males:
{58.0, 59.4, 59.5,   66.2, 67.8, 68.1,   69.2, 69.7, 70.3,   70.9, 72.0, 72.4}

Females:
{63.9, 65.3, 69.6,   72.3, 73.3, 74.1,   74.3, 74.5, 75.9,   78.3, 78.5, 78.8}

We have spaced the lists to show where the quartiles appear. The median of the 12 values is midway between the 6th and 7th values. The first quartile is the median of the lower 6 values (i.e., midway between the 3rd and 4th) and the third quartile is the median of the upper 6 values (i.e., midway between the 9th and 10th).

The five-number summaries are shown below.

Males:     {58.0, 62.85, 68.65, 70.6, 72.4}
Females:   {63.9, 70.95, 74.2, 77.1, 78.8}

The males have a range of $72.4 - 58.0 = 14.4$ and an *IQR* of $70.6 - 62.85 = 7.75$.

The females have a range of $78.8 - 63.9 = 14.9$ and an *IQR* of $77.1 - 70.95 = 6.15$.

Not only do the women live longer, but there is less variability in their life expectancies (as measured by the *IQR*). Male life expectancy is more strongly affected by different political conditions within countries (war, civil strife, crime, etc.).

You can learn a lot about the data by considering the *shape* of the distribution, as visualized in a histogram. Try answering the questions in Exploration 1.

**Computing Statistics on a Calculator**

Modern calculators will usually process lists of data and give statistics like the mean, median, and quartiles with a push of a button. Consult your *Owner's Manual*.

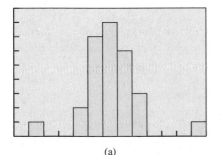

(a)

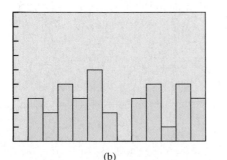

(b)

**Figure 9.20**  Which data set has more variability? (Exploration 1)

**Exploration Extensions**

Have the students discuss which might be the better statistic to use (mean or median) for interpreting the data in the graphs of Figure 9.21.

**Exploration 1**   **Interpreting Histogram**

1. Of the two histograms shown in Figure 9.20, which displays a data set with more variability?
2. Of the three histograms in Figure 9.21, which has a median less than its mean? Which has a median greater than its mean? Which has a median approximately equal to its mean?

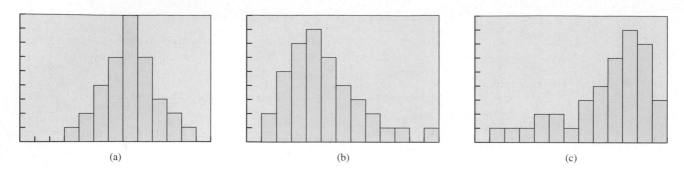

(a)  (b)  (c)

**Figure 9.21** Which graph shows a data set in which the mean is less than the median? Greater than the median? Approximately equal to the median? (Exploration 1)

> **Alert**
>
> Some students may interpret the box-plot in Figure 9.24 to mean that there are more data items in the fourth quartile than in the other quartiles. Stress that each quartile has the same number of data items.

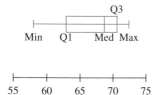

**Figure 9.22** A boxplot for the 5-number summary of the male life expectancies in Example 7. (The features on the box are labeled here for illustrative purposes; it is not necessary in general to label the min, the quartiles, or the max.)

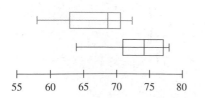

**Figure 9.23** A single graph showing boxplots for male and female life expectancies in the nations of South America gives a good visualization of the differences in the two data sets. (Example 8)

The distribution in Figure 9.21a is **symmetric** because it looks approximately the same when reflected about a vertical line through the median. The distribution in Figure 9.21b is **skewed right** because it has a longer "tail" to the right. The values in the tail will pull non-resistant measures (like the mean) to the *right*, leaving resistant measures (like the median) behind. The distribution in Figure 9.21c is **skewed left**, because non-resistant measures are pulled to the *left*.

You can also see symmetry and skewedness in stemplots, which have the same shapes vertically as histograms have horizontally.

## Boxplots

A **boxplot** (sometimes called a **box-and-whisker plot**) is a graph that depicts the five-number summary of a data set. The plot consists of a central rectangle (box) that extends from the first quartile to the third quartile, with a vertical segment marking the median. Line segments (whiskers) extend at either end of the box to the minimum and maximum values. For example, the 5-number summary for the male life expectancies in South American nations (Example 7) was {58.0, 62.85, 68.65, 70.6, 72.4}. The boxplot for the data is shown in Figure 9.22. Notice that the box and the whisker extend further to the left of the median than to the right, suggesting a distribution that is skewed left. (The histogram obtained in Exercise 9 of Section 9.6 confirms that this is the case.)

### Example 8 COMPARING BOXPLOTS

Draw boxplots for the male and female data in Example 7 and describe briefly the information displayed in the visualization.

**Solution** The 5-number summaries are:

$$\text{Men:} \quad \{58.0, 62.85, 68.65, 70.6, 72.4\}$$
$$\text{Women:} \quad \{63.9, 70.95, 74.2, 77.1, 78.8\}$$

The boxplots can be graphed simultaneously (Figure 9.23).

From this graph we can see that the middle half of the female life expectancies are all greater than the third quartile of the male life expectancies. The smaller "box" for the female plot shows the smaller variability. The median life expectancy for the women among South American nations is greater than the maximum for the men.

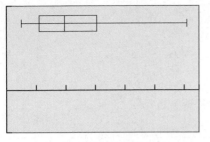

[0, 65] by [–5, 10]

**Figure 9.24** A boxplot of Roger Maris's annual home run production (Table 9.15 on page 728). The outlier (61) results in a long whisker on the right because the maximum is much larger than $Q_3$.

### Defining Outliers

It must be pointed out that the rule of thumb given here for identifying outliers is not a universal definition. The only sure way to characterize an outlier is that it lies outside the expected range of the data, and that "expected range" can be a judgment call.

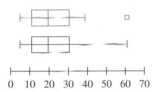

**Figure 9.25** A modified boxplot and a regular boxplot of Roger Maris's annual home run totals.

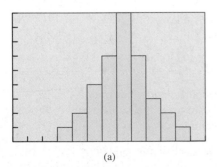

(a)

**Figure 9.26** Histograms based on data gathered from real-world sources tend to be symmetric and higher in the middle, without outliers. Frequency distributions with graphs of this shape are called *normal.*

If we look at the boxplot for Roger Maris's annual home run totals {14, 28, 16, 39, 61, 33, 23, 26, 8, 13, 9, 5}, we see that the whisker on the right is very long (Figure 9.24). This is a visualization of the effect of the outlier (61), which is much larger than the third quartile (30.5).

In fact, a boxplot gives us a convenient way to think of an outlier: a number that makes one of the whiskers noticeably longer than the box. The usual rule of thumb for "noticeably longer" in this case is 1.5 times as long. Since the length of the box is the *IQR*, that leads us to the following numerical check.

> A number in a data set can be considered an **outlier** if it is more than $1.5 \times IQR$ below the first quartile or above the third quartile.

### Example 9 IDENTIFYING AN OUTLIER

Is 61 an outlier in Roger Maris's home run data according to the $1.5 \times IQR$ criterion?

**Solution** Maris's totals, in order: {5, 8, 9, 13, 14, 16, 23, 26, 28, 33, 39, 61}.

His 5-number summary: {5, 11, 19.5, 30.5, 61}

His *IQR*: $30.5 - 11 = 19.5$.

So,

$$Q_3 + 1.5 \times IQR = 30.5 + 1.5 \times 19.5 = 59.75.$$

Since $61 > 59.75$, the rule of thumb identifies it as an outlier.

By their very nature, outliers can distort the overall picture we get of the data. For that reason, statisticians will frequently look for reasons to omit them from their statistical displays and calculations. (This can, of course, be risky. You want to omit a strange laboratory reading if you suspect equipment error, but you do not want to ignore a potential scientific discovery.) A **modified boxplot** is a compromise visualization that separates outliers off as separate points, extending the whiskers only to the farthest non-outliers. Figure 9.25 shows a modified boxplot of Roger Maris's home run data, as compared to a regular boxplot.

## Variance and Standard Deviation

You might be surprised that the 5-number summary and its boxplot graph do not even make reference to the mean, which is a more familiar measure than a median or a quartile. This is because the mean, being a non-resistant measure, is less reliable in the presence of outliers or skewed data.

On the other hand, the mean is an excellent measure of center when outliers and skewedness are not present, which is quite often the case. Indeed, histograms of data from all kinds of real-world sources tend to look something like Figure 9.26, in which frequencies are higher close to the mean and lower as you move away from the mean in either direction. Statisticians call these *normal* distributions. (We will make that term more precise shortly.)

For normally-distributed data, the mean is the preferred measure of center. There is also a measure of variability for normal data that is better than the *IQR*, called the *standard deviation*. Like the mean, the standard deviation is strongly affected by outliers and can be misleading if outliers are present.

---

**Definition** Standard Deviation

The **standard deviation** of the numbers $\{x_1, x_2, \ldots, x_n\}$ is

$$\sigma = \sqrt{\frac{1}{n} \sum_{i=1}^{n} (x_i - \bar{x})^2},$$

where $\bar{x}$ denotes the mean. The **variance** is $\sigma^2$, the square of the standard deviation.

---

If we define the "deviation" of a data value to be how much it differs from the mean, then the variance is just the mean of the squared deviations. The standard deviation is the square *root* of the *mean* of the *squared deviations*, which is why it is sometimes called the *root mean square deviation*. The symbol "$\sigma$" is a lower-case Greek letter sigma.

Calculating a standard deviation by hand can be tedious, but with modern calculators it is usually only necessary to enter the list of data and push a button. In fact, most calculators give you a choice of two standard deviations, one slightly larger than the other. The larger one (usually called $s$) is based on the formula

$$s = \sqrt{\frac{1}{n-1} \sum_{i=1}^{n} (x_i - \bar{x})^2}.$$

The difference is that the $\sigma$ formula is for finding the true parameter, which means that it only applies when $\{x_1, x_2, \ldots, x_n\}$ is the whole **population**. If $\{x_1, x_2, \ldots, x_n\}$ is a *sample* from the population, then the $s$ formula actually gives a better estimate of the parameter than the $\sigma$ formula does. So use the larger standard deviation when your data come from a sample (which is almost always the case).

**Example 10** FINDING STANDARD DEVIATION WITH A CALCULATOR

A researcher measured 30 newly-hatched loon chicks and recorded their weights in grams as shown in Table 9.21.

**Table 9.21 Weights in Grams of 30 Loon Chicks**

| | | | | | | | | | |
|---|---|---|---|---|---|---|---|---|---|
| 79.5 | 87.5 | 88.5 | 89.2 | 91.6 | 84.5 | 82.1 | 82.3 | 85.7 | 89.8 |
| 84.0 | 84.8 | 88.2 | 88.2 | 82.9 | 89.8 | 89.2 | 94.1 | 88.0 | 91.1 |
| 91.8 | 87.0 | 87.7 | 88.0 | 85.4 | 94.4 | 91.3 | 86.4 | 85.7 | 86.0 |

Based on the sample, estimate the mean and standard deviation for the weights of newly-hatched loon chicks. Are these measures useful in this case, or should we use the 5-number summary?

**Solution** We enter the list of data into a calculator and choose the command that will produce statistics of a single variable. The output from one such calculator is shown in Figure 9.27.

```
1-Var Stats
  x̄=87.49
  Σx=2624.7
  Σx²=229992.25
  Sx=3.509823652
  σ=3.450830818
↓n=30
```

**Figure 9.27** Single-variable statistics in a typical calculator display. (Example 10)

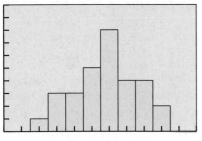

[75, 100] by [0, 10]

**Figure 9.28** The weights of the loon chicks in Example 10 appear to be normal, with no outliers or strong skewedness. We conclude that the mean and standard deviation are appropriate measures of center and variability, respectively.

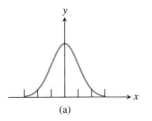

(a)

**Figure 9.29** The graph of $y = e^{-x^2/2}$. This is a Gaussian (or normal) curve.

The mean is $\overline{x} = 87.49$ grams. For standard deviation we choose $Sx = 3.51$ grams because the calculations are based on a sample of loon chicks, not the entire population of loon chicks.

A histogram (Xscl = 2) in the window [75, 100] by [0, 10] shows that the distribution is normal (as we would expect from nature), containing no outliers or strong skewedness. Therefore, the mean and standard deviation are appropriate measures (Figure 9.28).

## Normal Distributions

Although we use the word *normal* in many contexts to suggest typical behavior, in the context of statistics and data distributions it is really a technical term. If you graph the function

$$y = e^{-x^2/2}$$

in the window [−3, 3] by [0, 1], you will see what **normal** means mathematically (Figure 9.29).

The shape corresponds to the kind of distribution we have been calling "normal." In fact, this curve, called a **Gaussian curve** or **normal curve** is a *precise mathematical model* for normal behavior. That is where mean and standard deviation come in.

The standard deviation of the curve in Figure 9.29 is 1. Using calculus, we can find that about 68% of the total area under this curve lies between −1 and 1 (Figure 9.30a). Since any normal distribution has this shape, that implies that *about 68% of the data in any normal distribution lie within 1 standard deviation of the mean.*

Similarly, we can find that about 95% of the total area under the Gaussian curve lies between −2 and 2 (Figure 9.30b), implying that *about 95% of the data in any normal distribution lie within 2 standard deviations of the mean.*

Similarly we can find that about 99.7% (nearly all) of the total area under the Gaussian curve lies between −3 and 3 (Figure 9.30c), implying that *about 99.7% of the data in any normal distribution lie within 3 standard deviations of the mean.*

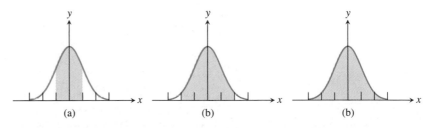

(a)                (b)                (b)

**Figure 9.30** (a) About 68% of the area under a Gaussian curve lies within 1 unit of the mean. (b) About 95% of the area lies within 2 units of the mean. (c) About 99.7% of the area lies within 3 units of the mean. If we think of the units as standard deviations, this gives us a model for *any* normal distribution.

**Assignment Guide**

Day 1: Ex. 3–27, multiples of 3
Day 2: Ex. 30, 33, 37, 39, 41, 42, 43, 44, 46

**Cooperative Learning**

Group Activity: Ex. 39, 40

**Notes on Exercises**

Ex. 1–16 deal with measures of center.
Ex. 17–28 deal with five-number summaries and boxplots.
Ex. 29–38 deal with standard deviation and variance and normal distribution.
Ex. 39 demonstrates that there are no generalizations concerning the order of the mean, median, and mode for a set of data.

**Ongoing Assessment**

Self-Assessment: Ex. 5, 11, 13, 19, 25, 29, 45
Embedded Assessment: Ex. 28, 36, 43

---

**The 68–95–99.7 Rule**

If the data for a population are normally distributed with mean $\mu$ and standard deviation $\sigma$, then

- approximately 68% of the data lie between $\mu - 1\sigma$ and $\mu + 1\sigma$;
- approximately 95% of the data lie between $\mu - 2\sigma$ and $\mu + 2\sigma$;
- approximately 99.7% of the data lie between $\mu - 3\sigma$ and $\mu + 3\sigma$.

---

What makes this rule so useful is that normal distributions are common in a wide variety of statistical applications. We close the chapter with a simple application.

**Example 11**   USING THE 68–95–99.7 RULE

Based on the research data presented in Example 10, would a loon chick weighing 95 grams be in the top 2.5% of all newly-hatched loon chicks?

**Solution**   We assume that the weights of newly-hatched loon chicks are normally distributed in the whole population. Since we do not know the mean and standard deviation for the whole population (the parameters $\mu$ and $\sigma$), we use $\overline{x} = 87.49$ and $Sx = 3.51$ as estimates.

Look at Figure 9.30b. The shaded region contains 95% of the area, so the two identical white regions at either end must each contain 2.5% of the area. That is, to be in the top 2.5%, a loon chick will have to weigh at least 2 standard deviations above the mean:

$$\overline{x} + 2Sx = 87.49 + 2(3.51) = 94.51 \text{ grams.}$$

Since $95 > 94.51$, a 95-gram loon chick is indeed in the top 2.5%.

If you study statistics more deeply someday you will learn that there is more going on in Example 11 than meets the eye. For starters, we need to know that the chicks are really a random sample of all loon chicks (not, for example, from the same geographical area). Also, we lose some accuracy by using a statistic to estimate the true standard deviation, and statisticians have ways of taking that into account.

This section has only offered a brief glimpse into how statisticians use mathematics. If you are interested in learning more, we urge you to find a good statistics textbook and pursue it!

# Quick Review 9.7

In Exercises 1–6, write the sum in expanded form.

**1.** $\displaystyle\sum_{i=1}^{7} x_i$

**2.** $\displaystyle\sum_{i=1}^{5} (x_i - \overline{x})$

**3.** $\dfrac{1}{7} \displaystyle\sum_{i=1}^{7} x_i$

**4.** $\dfrac{1}{5} \displaystyle\sum_{i=1}^{5} (x_i - \overline{x})$

**5.** $\dfrac{1}{5} \displaystyle\sum_{i=1}^{5} (x_i - \overline{x})^2$

**6.** $\sqrt{\dfrac{1}{5} \displaystyle\sum_{i=1}^{5} (x_i - \overline{x})^2}$

In Exercises 7–10, write the sum in sigma notation.

**7.** $x_1 f_1 + x_2 f_2 + x_3 f_3 + \cdots + x_8 f_8$

**8.** $(x_1 - \overline{x})^2 + (x_2 - \overline{x})^2 + \cdots + (x_{10} - \overline{x})^2$

**9.** $\dfrac{1}{50}[(x_1 - \overline{x})^2 + (x_2 - \overline{x})^2 + \cdots + (x_{50} - \overline{x})^2]$

**10.** $\sqrt{\dfrac{1}{7}[(x_1 - \overline{x})^2 + (x_2 - \overline{x})^2 + \cdots + (x_7 - \overline{x})^2]}$

# Section 9.7 Exercises

In Exercises 1–4, find the mean of the data set.

**1.** {12, 23, 15, 48, 36}  26.8    **2.** {4, 8, 11, 6, 21, 7}  9.5

**3.** {32.4, 48.1, 85.3, 67.2, 72.4, 55.3}  ≈ 60.12

**4.** {27.4, 3.1, 9.7, 32.3, 12.8, 39.4, 73.7}  ≈ 28.3

In Exercises 5 and 6, find the mean population of the six states listed in Table 9.18 (page 730) for the indicated year.

**5.** 1900  3.9 million          **6.** 1990  ≈ 16.83 million

In Exercises 7 and 8, find the mean of the indicated data.

**7.** The number of satellites (moons), from the data in Table 9.22.  ≈ 6.8 satellites

### Table 9.22  Planetary Satellites

| Planet | Number of Satellites |
|---|---|
| Mercury | 0 |
| Venus | 0 |
| Earth | 1 |
| Mars | 2 |
| Jupiter | 16 |
| Saturn | 18 |
| Uranus | 15 |
| Neptune | 8 |
| Pluto | 1 |

*Source: National Geographic Atlas of the World
(rev. 6th ed., 1992)*

**8.** The area of the continents, from the data in Table 9.23.

### Table 9.23  Size of Continent

| Continent | Area (km$^2$) |
|---|---|
| Africa | 30,269,680 |
| Antarctica | 13,209,000 |
| Asia | 44,485,900 |
| Australia | 7,682,300 |
| Europe | 10,530,750 |
| North America | 24,235,280 |
| South America | 17,820,770 |

*Source: National Geographic Atlas of the World (rev. 6th ed., 1992)*

**9. Home Run Production** Determine the average annual home run production for Willie Mays and for Mickey Mantle for their career totals of 660 over 22 years and 536 over 18 years, respectively. Who had the greater production rate?  30 runs/yr; ≈ 29.8 runs/yr; Mays

**10. Painting House** A painting crew in State College, Pennsylvania, painted 12 houses in 5 days, and a crew in College Station, Texas, painted 15 houses in 7 days.

Determine the average number of houses per day each crew painted. Which crew had the greater rate?

**11. Skirt Production** The Hip-Hop House produced 1147 scooter skirts in 4 weeks, and What-Next Fashion produced 1516 scooter skirts in 4 weeks. Which company had the greater production rate?  What-Next Fashion

**12. Per Capita Income** Per capita income (PCI) is an average found by dividing a nation's gross national product (GNP) by its population. India has 882,575,000 people and a GNP of 311 billion dollars, and Mexico has 87,715,000 people and a GNP of 218 billion dollars. Determine the PCI for India and for Mexico. Which nation has the greater income per person?  ≈ $352; ≈ $2485; Mexico

**13.** Find the median and mode of the numbers in Table 9.7 on page 721 (rounded per capita income in thousands of dollars).

**14.** Find the median and mode of the numbers in Table 9.21 on page 738 (weights in grams of newly hatched loon chicks).

In Exercises 15 and 16, use the data in Table 9.19 (page 730).

**(a)** Find the average of the indicated temperatures for Beijing.

**(b)** Find the weighted average using the number of days in the month as the weight. (Assume no leap year.)

**(c)** Compare your results in (a) and (b). Do the weights have an effect on the average? Why or why not? Which average is the better indicator for these temperatures?

**15.** The monthly high temperatures.

**16.** The monthly low temperatures.

In Exercises 17–20, determine the five-number summary, the range, and the interquartile range for the sets of data specified. Identify any outliers.

**17.** The annual home run production data for Mark McGwire and Sammy Sosa in Table 9.8 (page 722).

**18.** The annual home run production data for Willie Mays and Mickey Mantle in Table 9.16 (page 729).

**19.** The following average annual wind speeds at 44 climatic data centers around the United States:

9.0, 6.9, 9.1, 9.2, 10.2, 12.5, 12.0 ,11.2, 12.9, 10.3, 10.6, 10.9, 8.7, 10.3, 11.0, 7.7, 11.4, 7.9, 9.6, 8.0, 10.7, 9.3, 7.9, 6.2, 8.3, 8.9, 9.3, 11.6, 10.6, 9.0, 8.2, 9.4, 10.6, 9.5, 6.3, 9.1, 7.9, 9.7, 8.8, 6.9, 8.7, 9.0, 8.9, 9.3

**20.** The following salaries for employees in one department of the Garcia Brothers Company (in thousands of dollars):

33.5, 35.3, 33.8, 29.3, 36.7, 32.8, 31.7, 37.3, 33.5, 28.2, 34.8, 33.5, 29.7, 38.5, 32.7, 34.8, 34.2, 31.6, 35.4

In Exercises 21–24, make **(a)** a boxplot and **(b)** a modified boxplot for the data.

**21.** The annual home run production data for Mark McGwire in Table 9.8 (page 722).

**22.** The annual home run production data for Sammy Sosa in Table 9.8 (page 722).

**23.** The rounded per capita incomes (in thousands of dollars) in Table 9.7 (page 721).

**24.** The annual CD shipments in Table 9.11 (page 725).

In Exercises 25 and 26, refer to the wind speed data analyzed in Exercise 19.

**25.** Some wind turbine generators, to be efficient generators of power, require average wind speeds of at least 10.5 mph. Approximately what fraction of the climatic centers are suited for these wind turbine generators?

**26.** If technology improves the efficiency of the wind turbines so that they are efficient in winds that average at least 7.5 mph, approximately what fraction of the climatic centers are suited for these improved wind generators?

In Exercises 27 and 28, use simultaneous boxplots of the annual home run production data for Willie Mays and Mickey Mantle in Table 9.16 (page 729) to answer the question.

**27. (a)** Which data set has the greater range?  Mays

**(b)** Which data set has the greater interquartile range?  Mays

**28. Writing to Learn** Write a paragraph explaining the difference in home run production between Willie Mays and Mickey Mantle.

In Exercises 29–34, find the standard deviation and variance of the data set. (Since the data set is the population under consideration, use $\sigma$ in each case, rather than $s$.)

**29.** $\{23, 45, 29, 34, 39, 41, 19, 22\}$  $\sigma \approx 9.08; \sigma^2 = 82.5$

**30.** $\{28, 84, 67, 71, 92, 37, 45, 32, 74, 96\}$

**31.** The CD shipment data in Table 9.11 (page 725).

**32.** The cassette shipment data in Table 9.12 (page 726).

**33.** The data in Exercise 19.  $\sigma \approx 1.53, \sigma^2 \approx 2.34$

**34.** The data in Exercise 20.  $\sigma \approx 2.60, \sigma^2 \approx 6.77$

**35.** Mark McGwire's annual home run production (Table 9.8, page 722) ranges from a low of 3 to a high of 70. Is either of these numbers an outlier by the $1.5 \times IQR$ criterion?  no

**36. Writing to Learn** Is it possible for the standard deviation of a set to be negative? To be zero? Explain your answer in both cases.

**37. SAT Scores** In 1995, scores on the Scholastic Aptitude Tests were re-scaled to their original mean (500) with an approximate standard deviation of 100. SAT scores in the general population have a normal distribution.

**(a)** Approximately what percentage of the 1995 scores fell between 400 and 600?  68%

**(b)** Approximately what percentage of the 1995 scores fell below 300?  2.5%

**(c)** By 1998 the national average on the SAT Math section had risen to 512. Is this number a parameter or a statistic?

**38. ACT Scores** In 1995, the national mean ACT Math score was 20.2, with an approximate standard deviation of 6. ACT scores in the general population have a normal distribution.

**(a)** Approximately what percentage of the 1995 scores were higher than 26.2?  16%

**(b)** Approximately what ACT Math score would one need to make in 1995 to be ranked among the top 2.5% of all who took the test?  32.2

**(c) Writing to Learn** Mean ACT scores are published state by state. If we add up the 50 state means and divide by 50, will the result be a good estimate for the national mean score? Explain your answer.

## Explorations

**39. Group Activity** List a set of data for which the inequality holds.

**(a)** Mode < median < mean

**(b)** Median < mean < mode

**(c)** Mean < mode < median

**40. Group Activity** List a set of data for which the inequality holds.

**(a)** Standard deviation < interquartile range

**(b)** Interquartile range < standard deviation

**(c)** Range = interquartile range

**41.** Is it possible for the standard deviation of a data set to be greater than the range? Explain.

**42.** Why can we find the mode of categorical data but not the mean or median?

**43.** Draw a boxplot for which the inequality holds.

**(a)** Median < mean

**(b)** $2 \times$ interquartile range < range

**(c)** Range < $2 \times$ interquartile range

**44.** Construct a set of data with median 5, mode 6, and mean 7.  Possible answer: $\{2, 3, 4, 6, 6, 21\}$  ∎

## Extending the Ideas

**Weighting Data by Population** The average life expectancies for males and females in South American nations were given in Table 9.14 (page 728). To find an overall average life expectancy for males or females in *all* of these nations, we would need to weight the national data according to the various national populations. Table 9.24 is an extension of Table 9.14, showing the populations (in millions). Assume that males and females appear in roughly equal numbers in every nation.

**T**

**Table 9.24** Life Expectancy by Gender and Population (in millions) for the Nations of South America

| Nation | Male | Female | Population |
|--------|------|--------|-----------|
| Argentina | 70.9 | 78.3 | 36.3 |
| Bolivia | 58.0 | 63.9 | 7.8 |
| Brazil | 59.4 | 69.6 | 169.8 |
| Chile | 72.0 | 78.5 | 14.8 |
| Colombia | 66.2 | 74.1 | 38.6 |
| Ecuador | 69.2 | 74.5 | 12.3 |
| Guyana | 59.5 | 65.3 | 0.7 |
| Paraguay | 70.3 | 74.3 | 5.3 |
| Peru | 67.8 | 72.3 | 26.1 |
| Suriname | 68.1 | 73.3 | 0.4 |
| Uruguay | 72.4 | 78.8 | 3.3 |
| Venezuela | 69.7 | 75.9 | 22.8 |

*Source: The World Almanac and Book of Facts, 1999*

In Exercises 45 and 46, use the data in Table 9.24 to find the mean life expectancy for each group.

**45.** Women living in South American nations. ≈ 72.28 years

**46.** Men living in South American nations. ≈ 63.94 years

**47. Quality Control** A plant manufactures ball bearings to the purchaser's specifications, rejecting any output with a diameter that deviates more than 0.1 mm from the specified value. If the ball bearings are produced with the specified mean and a standard deviation of 0.05 mm, what percentage of the output will be rejected? 5%

**48. Quality Control** A machine fills 12-ounce cola cans with a mean of 12.08 ounces of cola and a standard deviation of 0.04 ounces. Approximately what percentage of the cans will actually contain less than the advertised 12 ounces of cola? 2.5%

## Math at Work

In the course of getting my bachelor's degree in communications, I took a class in math communications research, and I enjoyed it immensely. After I graduated from college, I interviewed in print journalism, but I wasn't a good editor. Since I really enjoyed research and analysis, I got started in research management.

What I like best about my job is that I get to act as a detective, using math to find out about people. Basically, my job is to bring the voice of the customers into a product development plan, so we can figure out how to make a product for them. There are two ways to get to know your customers: mathematics and psychology. Using psychology, the people developing the product have a tendency to use their gut feelings, which are sometimes entirely incorrect, to surmise what the customers want. That's why we use mathematics.

We use a set of questions, called a "Customer Satisfaction Survey" to find out what values are important to the customers. Then we can analyze the raw information and find out such things as the differences between frequent and infrequent users, and how males and females differ in their uses of the product. In this way, I learn all about people, and we, as a company, know how to tailor a product to our customers.

*Veronica Guerrerro*

## Chapter 9   Key Ideas

### Concepts

Anchor (p. 713)
Arithmetic sequence (p. 697)
Back-to-back stemplot (p. 721)
Bar chart (p. 718)
Binomial coefficient (p. 679)
Binomial distribution (p. 692)
Binomial theorem (p. 681)
Boxplot (p. 736)
Box-and-whisker plot (p. 736)
Categorical variable (p. 717)
Circle graph (p. 718)
Combination (p. 672)
Conditional probability (p. 689)
Continuous (p. 669)
Continuum (p. 669)
Convergent series (p. 706)
Dependent events (p. 687, 689)
Descriptive statistics (p. 730)
Discrete (p. 669)
Divergent series (p. 706)
Event (p. 684)
Explanatory variable (p. 669)
Explicitly-defined sequence (p. 696)

Fibonacci sequence (p. 701)
Five-number summary (p. 734)
Frequency distribution (p. 723)
Frequency table (p. 723)
Gaussian curve (p. 739)
Geometric sequence (p. 697–698)
Histogram (p. 723)
Independent events (p. 687)
Inductive hypothesis (p. 713)
Inductive step (p. 713)
Inferential statistics (p. 730)
Interquartile range (p. 734)
Line graph (p. 725)
Mathematical induction (p. 713)
Mean (p. 731)
Median (p. 731–732)
Mode (p. 731–732)
Normal curve (p. 739)
Normal distribution (p. 739)
$n$-set (p. 671)
Outcome (p. 684)
Outlier (p. 737)
Parameter (p. 730)

Partial sums (p. 706)
Pascal's triangle (p. 679)
Permutation (p. 671)
Picture graph (p. 718)
Pie chart (p. 718)
Population (p. 730, 738)
Probability distribution (p. 684)
Quantitative variable (p. 717)
Quartile (p. 734)
Range (p. 734)
Recursively-defined sequence (p. 696)
Response variable (p. 669)
Sample (p. 730)
Sample space (p. 684)
Sequence (p. 696)
Series (p. 705–706)
Skewed left (p. 736)
Skewed right (p. 736)
Split stemplot (p. 720)
Standard deviation (p. 738)
Statistic (p. 730)
Stem-and-leaf plot (p. 718)
Stemplot (p. 718)

## Properties, Theorems, and Formulas

### Multiplication Principle of Counting

If a procedure $P$ has a sequence of stages $S_1, S_2, \ldots, S_n$ and if

$S_1$ can occur in $r_1$ ways

$S_2$ can occur in $r_2$ ways,

$\vdots$

$S_n$ can occur in $r_n$ ways,

then the number of ways that the procedure $P$ can occur is the product

$$r_1 r_2 \ldots r_n.$$

### Permutation Counting Formula

The number of permutations of $n$ objects taken $r$ at a time is denoted $_nP_r$ and is given by

$$_nP_r = \frac{r!}{(n-r)!} \qquad \text{for } 0 \leq r \leq n.$$

If $r > n$, then $_nP_r = 0$.

### Combination Counting Formula

The number of combinations of $n$ objects taken $r$ at a time is denoted $_nC_r$ and is given by

$$_nC_r = \frac{n!}{r!(n-r)!} \qquad \text{for } 0 \leq r \leq n.$$

If $r > n$, then $_nC_r = 0$.

### Formula for Counting Subsets of an *n*-Set

There are $2^n$ subsets of a set with $n$ objects (including the empty set and the entire set).

### The Binomial Theorem

For any positive integer $n$,

$$(a+b)^n = \binom{n}{0}a^n + \binom{n}{1}a^{n-1}b + \cdots + \binom{n}{r}a^{n-r}b^r + \cdots + \binom{n}{n}b^n,$$

where $\binom{n}{r} = {}_nC_r = \dfrac{r!}{r!(n-r)!}.$

### Probability of an Event (Equally Likely Outcomes)

If $E$ is an event in a finite, nonempty sample space $S$ of equally likely outcomes, then the **probability** of the event $E$ is

$$P(E) = \frac{\text{the number of outcomes in } E}{\text{the number of outcomes in } S}.$$

### Probability of an Event (Outcomes not Equally Likely)

Let $S$ be a finite, nonempty sample space in which every outcome has a probability assigned by a probability function $P$. If $E$ is any event in $S$, the **probability** of the event $E$ is the sum of the probabilities of all the outcomes contained in $E$.

### Multiplication Principle of Probability

Suppose an event $A$ has probability $p_1$ and an event $B$ has probability $p_2$ under the assumption that $A$ occurs. Then the probabity that both $A$ and $B$ occur is $p_1 p_2$.

### Conditional Probability Formula

$$P(B|A) = \frac{P(A \text{ and } B)}{P(A)}.$$

### Theorem  Sum of a Finite Arithmetic Progression

Let $\{a_1, a_2, a_3, \ldots, a_n\}$ be a finite arithmetic sequence with common difference $d$.

Then the *sum* of the terms of the sequence is

$$\sum_{k=1}^{\infty} = a_1 + a_2 + \ldots + a_n$$

$$= n\left(\frac{a_1 + a_n}{2}\right)$$

$$= \frac{n}{2}\left(2a_1 + (n-1)d\right)$$

### Theorem  Sum of a Finite Geometric Progression

Let $\{a_1, a_2, a_3, \ldots, a_n\}$ be a finite geometric sequence with common ratio $r \neq 1$.

Then the **sum** of the terms of the sequence is

$$\sum_{k=1}^{n} a_k = a_1 + a_2 + \ldots + a_n$$

$$= \frac{a_1(1 - r^n)}{1 - r}$$

### Theorem  Sum of an Infinite Geometric Series

The geometric series $\sum_{k=1}^{\infty} a \cdot r^{k-1}$ converges if and only if $|r| < 1$. If it does converge, the sum is $\dfrac{a}{1 - r}$.

### Principle of Mathematical Induction

Let $P_n$ be a statement about the integer $n$. Then $P_n$ is true for all positive integers $n$ provided the following conditions are satisfied:

1. (the anchor) $P_1$ is true;

2. (the inductive step) if $P_k$ is true, then $P_{k+1}$ is true.

### The 68-95-99.7 Rule

If the data for a population are normally distributed with mean $\mu$ and standard deviation $\sigma$, then

1. approximately 68% of the data lie between $\mu - 1\sigma$ and $\mu + 1\sigma$;

2. approximately 95% of the data lie between $\mu - 2\sigma$ and $\mu + 2\sigma$;

3. approximately 99.7% of the data lie between $\mu - 3\sigma$ and $\mu + 3\sigma$.

## Procedures

### Strategy for Determining Probabilities

1. Determine the sample space of all possible outcomes. When possible, choose outcomes that are equally likely.

2. If the sample space has equally likely outcomes, the probability of an event E is determined by counting:

$$P(E) = \frac{\text{the number of outcomes in } E}{\text{the number of outcomes in } S}.$$

3. If the sample space does not have equally likely outcomes, determine the probability function. (This is not always easy to do.) Check to be sure that the conditions of a probability function are satisfied. Then the probability of an event $E$ is determined by adding up the probabilities of all the outcomes contained in $E$.

---

| Chapter 9 | **Review Exercises** |
|---|---|

The collection of exercises marked in red could be used as a chapter test.

In Exercises 1–6, evaluate the expression by hand, then check your result with a calculator.

**1.** $\binom{12}{5}$   792

**2.** $\binom{789}{787}$   310,866

**3.** $_{18}C_{12}$   18,564

**4.** $_{35}C_{28}$   6,724,520

**5.** $_{12}P_7$   3,991,680

**6.** $_{15}P_8$   259,459,200

**7. Code Words** How many five-character code words are there if the first character is always a letter and the other characters are letters and/or digits?   43,670,016

**8. Scheduling Trips** A travel agent is trying to schedule a client's trip from city $A$ to city $B$. There are three direct flights, three flights from $A$ to a connecting city $C$, and four flights from this connecting city $C$ to city $B$. How many trips are possible?   15

**9. License Plates** How many license plates begin with two letters followed by four digits or begin with three digits followed by three letters? Assume that no letters or digits are repeated.   14,508,000

**10. Forming Committees** A club has 45 members, and its membership committee has three members. How many different membership committees are possible?   14,190

**11. Bridge Hands** How many 13-card bridge hands include the ace, king, and queen of spades?   8,217,822,536

**12. Bridge Hands** How many 13-card bridge hands include all four aces and exactly one king?   708,930,508

**13. Coin Toss** Suppose that a coin is tossed five times. How many different outcomes include at least two heads?   26

**14. Forming Committees** A certain small business has 35 employees, 21 women and 14 men. How many different employee representative committees are there if the committee must consist of two women and two men?

**15. Code Words** How many code words of any length can be spelled out using game tiles of five different letters (including single-letter code words)?   325

**16. A Pocket of Coins** Sean tells Moira that he has less than 50 cents in American coins in his pocket and no two coins of the same denomination. How many possible total amounts could be in Sean's pocket?   16

In Exercises 17–22, expand the binomial.

**17.** $(2x + y)^5$

**18.** $(4a - 3b)^7$

**19.** $(3x^2 + y^3)^5$

**20.** $\left(1 + \dfrac{1}{x}\right)^6$

**21.** $(2a^3 - b^2)^9$

**22.** $(x^{-2} + y^{-1})^4$

**23.** Find the coefficient of $x^8$ in the expansion of $(x - 2)^{11}$.

**24.** Find the coefficient of $x^2y^6$ in the expansion of $(2x + y)^8$.

In Exercises 25–28, list the elements of the sample space.

**25. Spinners** A game spinner on a circular region divided into 6 equal sectors is spun.   {1, 2, 3, 4, 5, 6}

**26. Rolling Dice** A red die and a green die are rolled.

**27. Code Words** A two-digit code is selected from the digits {1, 3, 6}, where no digits are to be repeated.

**28. Production Line** A product is inspected as it comes off the production line and is classified as either defective or nondefective.   {defective, nondefective}

In Exercises 29–32, a penny, a nickel, and a dime are tossed.

**29.** List all possible outcomes.

**30.** List all outcomes in the event "two heads or two tails."

**31.** List all outcomes in the complement of the event in Exercise 30 (i.e., the event "neither two heads nor two tails.")   {HHH, TTT}

**32.** Find the probability of tossing at least one head. 7/8

**33. Coin Toss** A fair coin is tossed six times. Find the probability of the event "*HHTHTT*." 1/64

**34. Coin Toss** A fair coin is tossed five times. Find the probability of obtaining two heads and three tails. 5/16

**35. Coin Toss** A fair coin is tossed four times. Find the probability of obtaining one head and three tails. 1/4

**36. Assembly Line** In a random check on an assembly line, the probability of finding a defective item is 0.003. Find the probability of a nondefective item occurring 10 times in a row. ≈ 0.97

**37. Success or Failure** An experiment has only two possible outcomes—success (S) or failure (F)—and repetitions are independent events. If $P(S) = 0.5$, find the probability of obtaining three successes and one failure in four repetitions. 0.25

**38. Success or Failure** For the experiment in Exercise 37, explain why the probability of one success and three failures is equal to the probability of three successes and one failure.

In Exercises 39–42, an experiment has only two possible outcomes—success (S) or failure (F)—and repetitions are independent events. The probability of success is 0.4.

**39.** Find the probability of SF on two repetitions. 0.24

**40.** Find the probability of SFS on three repetitions. 0.096

**41.** Find the probability of at least one success on two repetitions. 0.64

**42.** Explain why the probability of one success and three failures is not equal to the probability of three successes and one failure.

**43. Mixed Nuts** Two cans of mixed nuts of different brands are open on a table. Brand *A* consists of 30% cashews, while brand *B* consists of 40% cashews. A can is chosen at random, and a nut is chosen at random from the can. Find the probability that the nut is

**(a)** from the brand *A* can 0.5

**(b)** a brand *A* cashew 0.15

**(c)** a cashew 0.35

**(d)** from the brand *A* can, given that it is a cashew. ≈ 0.43

**44. Horse Racing** If the track is wet, Mudder Earth has a 70% chance of winning the fifth race at Keeneland. If the track is dry, she only has a 40% chance of winning. Weather forecasts predict an 80% chance that the track will be wet. Find the probability that

**(a)** the track is wet and Mudder Earth wins 0.56

**(b)** the track is dry and Mudder earth wins 0.08

**(c)** Mudder Earth wins 0.64

**(d)** (in retrospect) the track was wet, given that Mudder Earth wins. 0.875

In Exercises 45 and 46, find the first 6 terms and the 40th term of the sequence.

**45.** $a_n = \dfrac{n^2 - 1}{n + 1}$  **46.** $b_k = \dfrac{(-2)^k}{k + 1}$

In Exercises 47–52, find the first 6 terms and the 12th term of the sequence.

**47.** $a_1 = -1$ and $a_n = a_{n-1} + 3$, for $n \geq 2$

**48.** $b_1 = 5$ and $b_k = 2b_{k-1}$, for $k \geq 2$

**49.** Arithmetic sequence, with $a_1 = -5$ and $d = 1.5$

**50.** Geometric sequence, with $a_1 = 3$ and $r = 1/3$

**51.** $v_1 = -3$, $v_2 = 1$ and $v_k = v_{k-2} + v_{k-1}$, for $k \geq 3$

**52.** $w_1 = -3$, $w_2 = 2$ and $w_k = w_{k-2} + w_{k-1}$, for $k \geq 3$

In Exercises 53–60, the sequences are arithmetic or geometric. Find an explicit formula for the *n*th term. State the common difference or ratio.

**53.** 12, 9.5, 7, 4.5, . . .  **54.** −5, −1, 3, 7, . . .

**55.** 10, 12, 14.4, 17.28, . . .  **56.** $\dfrac{1}{8}, -\dfrac{1}{4}, \dfrac{1}{2}, -1, . . .$

**57.** $a_1 = -11$ and $a_n = a_{n-1} + 4.5$ for $n \geq 2$

**58.** $b_1 = 7$ and $b_n = (1/4) b_{n-1}$ for $n \geq 2$

**59.** The fourth and ninth terms of a geometric sequence are −192 and 196,608, respectively. $a_n = 3(-4)^{n-1}$; $r = -4$

**60.** The third and eighth terms of an arithmetic sequence are 14 and −3.5, respectively. $a_n = 24.5 - 3.5n$; $d = -3.5$

In Exercises 61–64, find the sum of the terms of the arithmetic sequence.

**61.** $-11 - 8 - 5 - 2 + 1 + 4 + 7 + 10$  −4

**62.** $13 + 9 + 5 + 1 - 3 - 7 - 11$  7

**63.** $2.5 - 0.5 - 3.5 - \cdots - 75.5$  −985.5

**64.** $-5 - 3 - 1 + 1 + \cdots + 55$  775

In Exercises 65–68, find the sum of the terms of the geometric sequence.

**65.** $4 - 2 + 1 - \dfrac{1}{2} + \dfrac{1}{4} - \dfrac{1}{8}$  21/8

**66.** $-3 - 1 - \dfrac{1}{3} - \dfrac{1}{9} - \dfrac{1}{27}$  −121/27

**67.** $2 + 6 + 18 + \cdots + 39,366$  59,048

**68.** $1 - 2 + 4 - 8 + \cdots - 8192$  −5461

In Exercises 69 and 70, find the sum of the first 10 terms of the arithmetic or geometric sequence.

**69.** 2187, 729, 243, . . .  **70.** 94, 91, 88, . . .  805

In Exercises 71 and 72, graph the sequence.

**71.** $a_n = 1 + \dfrac{(-1)^n}{n}$  **72.** $a_n = 2n^2 - 1$

In Exercises 73–78, determine whether the geometric series converges. If it does, find its sum.

**73.** $\sum_{j=1}^{\infty} 2\left(\frac{3}{4}\right)^j$ converges; 6  **74.** $\sum_{k=1}^{\infty} 2\left(-\frac{1}{3}\right)^k$ converges; $-\frac{1}{2}$

**75.** $\sum_{j=1}^{\infty} 4\left(-\frac{4}{3}\right)^j$ diverges  **76.** $\sum_{k=1}^{\infty} 5\left(\frac{6}{5}\right)^k$ diverges

**77.** $\sum_{k=1}^{\infty} 3(0.5)^k$ converges; 3  **78.** $\sum_{k=1}^{\infty} (1.2)^k$ diverges

In Exercises 79–82, write the sum in sigma notation.

**79.** $-8 - 3 + 2 + \cdots + 92$

**80.** $4 - 8 + 16 - 32 + \cdots - 2048$

**81.** $1^2 + 3^2 + 5^2 + \cdots$

**82.** $1 + \frac{1}{2} + \frac{1}{2^2} + \frac{1}{2^3} + \cdots$  $\sum_{k=0}^{\infty} \left(\frac{1}{2}\right)^k$ or $\sum_{k=1}^{\infty} \left(\frac{1}{2}\right)^{k-1}$

In Exercises 83–86, use summation formulas to evaluate the expression.

**83.** $\sum_{k=1}^{n} (3k+1)$  $\frac{n(3n+5)}{2}$  **84.** $\sum_{k=1}^{n} 3k^2$  $\frac{3n(n+1)(2n+1)}{6}$

**85.** $\sum_{k=1}^{25} (k^2 - 3k + 4)$  4650  **86.** $\sum_{k=1}^{175} (3k^2 - 5k + 1)$

In Exercises 87–90, use mathematical induction to prove that the statement is true for all positive integers $n$.

**87.** $1 + 3 + 6 + \cdots + \frac{n(n+1)}{2} = \frac{n(n+1)(n+2)}{6}$

**88.** $1 \cdot 2 + 2 \cdot 3 + 3 \cdot 4 + \cdots + n(n+1) = \frac{n(n+1)(n+2)}{3}$

**89.** $2^{n-1} \le n!$

**90.** $n^3 + 2n$ is divisible by 3.

In Exercises 91–94, construct (a) a stemplot, (b) a frequency table, and (c) a histogram for the indicated data.

**91. Real Estate Prices**  Use intervals of $10,000. The median sales prices (in units of $10,000) for homes in 30 randomly selected metropolitan areas in 1997 were as follows:

10.8, 19.6, 8.2, 12.4, 11.1, 11.2, 9.7, 14.1, 12.0, 9.4, 11.7, 9.1, 10.4, 8.6, 9.7, 11.8, 17.8, 20.5, 17.7, 16.4, 7.7, 9.5, 11.4, 8.7, 15.3, 8.7, 12.9, 17.1, 16.6, 13.3

(Source: *National Association of Realtors,* as reported in *The 1999 Wall Street Journal Almanac*)

**92. Music Albums**  Use intervals of 500,000. The number of albums sold during the year (in millions) for the top 20 music albums of 1997 were as follows:

5.3, 4.3, 3.4, 3.4, 3.2, 3.2, 3.1, 3.0, 3.0, 3.0, 2.8, 2.7, 2.7, 2.6, 2.5, 2.2, 2.2, 2.1, 2.1, 2.0

(Source: *Soundscan, Inc.,* as reported in *The 1999 Wall Street Journal Almanac*)

**93. Beatle Songs**  Use intervals of length 10. The lengths (in seconds) of 24 randomly-selected Beatle songs that appeared on singles are as follows, in order of release date:

143, 120, 120, 139, 124, 144, 131, 132, 148, 163, 140, 177, 136, 124, 179, 131, 180, 137, 156, 202, 191, 197, 230, 190

(Source: Personal collection)

**94. Passing Yardage**  In 1995, Warren Moon of the Minnesota Vikings became the first pro quarterback to pass for 60,000 total yards. Use intervals of 1000 yards for Moon's regular season passing yards given in Table 9.25.

**Table 9.25  Regular Season Passing Yardage Statistics for Warren Moon**

| Year | Yards | Year | Yards |
|---|---|---|---|
| 1978 | 1112 | 1987 | 2806 |
| 1979 | 2382 | 1988 | 2327 |
| 1980 | 3127 | 1989 | 3631 |
| 1981 | 3959 | 1990 | 4689 |
| 1982 | 5000 | 1991 | 4690 |
| 1983 | 5648 | 1992 | 2521 |
| 1984 | 3338 | 1993 | 3485 |
| 1985 | 2709 | 1994 | 4264 |
| 1986 | 3489 | 1995 | 4228 |

*Source: The Minnesota Vikings, as reported by Julie Stacey in USA Today on September 25, 1995 and www.nfl.com*

In Exercises 95–98, find the five-number summary, the range, the interquartile range, the standard deviation, and the variance for the specified data (use $\sigma$ and $\sigma^2$). Identify any outliers.

**95.** The data in Exercise 91.  **96.** The data in Exercise 92.

**97.** The data in Exercise 93.  **98.** The data in Exercise 94.

In Exercises 99–102, construct (a) a boxplot and (b) a modified boxplot for the specified data.

**99.** The data in Exercise 91.  **100.** The data in Exercise 92.

**101.** The data in Exercise 93.  **102.** The data in Exercise 94.

**103.** Make a back-to-back stemplot of the data in Exercise 93, showing the earlier 12 songs in one plot and the later 12 songs in the other. Write a sentence interpreting the stemplot.

**104.** Make simultaneous boxplots of the data in Exercise 93, showing the earlier 12 songs in one boxplot and the later 12 songs in the other.

(a) Which set of data has the greater range?

(b) Which set of data has the greater interquartile range?

**105. Time Plots**  Make a time plot for the data in Exercise 93, assuming equal time intervals between songs. Interpret the trend revealed in the time plot.

**106. Damped Time Plots**  Statisticians sometimes use a technique called *damping* to smooth out random fluctuations in a time plot. Find the mean of the first four numbers in Exercise 93, the mean of the next four numbers, and so on. Then graph the six means as a function of time. Is there a clear trend?

**107.** Find row 9 of Pascal's triangle.

**108.** Show algebraically that

$$_nP_k \times {}_{n-k}P_j = {}_nP_{k+j}.$$

**109. Baseball Bats** Suppose that the probability of producing a defective baseball bat is 0.02. Four bats are selected at random. What is the probability that the lot of four bats contains the following?

**(a)** No defective bats. $\approx 0.922$

**(b)** One defective bat. $\approx 0.075$

**110. Light Bulbs** Suppose that the probability of producing a defective light bulb is 0.0004. Ten light bulbs are selected at random. What is the probability that the lot of 10 contains the following?

**(a)** No defective light bulbs. $\approx 0.996$

**(b)** Two defective light bulbs. $\approx 7.18 \times 10^{-6}$

# Chapter 9 Project

## Analyzing Height Data

The set of data below was gathered from a class of 30 precalculus students. Use this set of data or collect the data for your own class and use it for analysis.

| Heights of students in inches | | | | |
|---|---|---|---|---|
| 66 | 69 | 72 | 64 | 68 |
| 70 | 71 | 66 | 65 | 63 |
| 72 | 59 | 64 | 63 | 66 |
| 68 | 63 | 64 | 71 | 71 |
| 69 | 62 | 61 | 67 | 69 |
| 64 | 73 | 75 | 61 | 70 |

1.  Create a stem and leaf plot of the data using split stems. From this data, what is the approximate average height of a student in the class?

2.  Create a frequency table for the data using an interval of 2. What information does this give?

3.  Create a histogram for the data using an interval of 2. What conclusions can you draw from this representation of the data? Can you estimate the average height for males and the average height for females?

4.  Compute the mean, median and mode for the data set. Discuss whether each is a good measure of the average height of a student in the class. Is each a good predictor for average height of students in other precalculus classes?

5.  What can you say about the data if the mean and median values are close?

6.  Find the five-number summary for the class heights.

7.  Create a boxplot and explain what information it gives about the data set.

8.  A new student is now added to the class. He is a 7'2" star basketball player. Add his height to the data set. Recalculate the mean, median and five-number summary. Create a new boxplot and use your calculator to plot it underneath the boxplot for the original class. How does this new student affect the statistics?

9.  Explain why this new student would be considered an outlier and the importance of identifying outliers when calculating statistics and making predictions from them.

10. Suppose now that three additional basketball players transferred into the class. They are 7'0", 6'11" and 6'10". Recalculate the statistics from number 9 and discuss the implications of using these statistics to make predictions for other precalculus classes.

## Instantaneous Velocity

Galileo experimented with gravity by rolling a ball down an inclined plane and recording its approximate velocity as a function of elapsed time. Here is how he might have answered his own question when he began his experiments:

### Example 2   FINDING INSTANTANEOUS VELOCITY

A ball rolls a distance of two vertical feet in 4 seconds. If I freeze the ball at a moment of time 3 seconds after it starts to fall, what is its instantaneous velocity at that moment?

**A Proposed Solution**  The ball has zero velocity, because it is frozen at a moment of time! (You might well ask: Is this a trick question?)

Example 2 might seem a little foolish, but it is actually quite profound—and, far from being a trick question, it is exactly the question that Gaileo (among many others) was trying to answer. Notice how easy it is to find the *average* velocity:

$$v_{ave} = \frac{\triangle s}{\triangle t} = \frac{2 \text{ feet}}{4 \text{ seconds}} = 0.5 \text{ feet per second.}$$

Now, notice how inadequate our algebra becomes when we try to apply the same formula to *instantaneous* velocity:

$$v_{ave} = \frac{\triangle s}{\triangle t} = \frac{0 \text{ feet}}{0 \text{ seconds}},$$

which involves division by 0—and is therefore undefined!

So Galileo did the best he could by making $\triangle t$ as small as experimentally possible, measuring the small values of $\triangle s$, and then finding the quotients. It only *approximated* the instantaneous velocity, but finding the exact value appeared to be algebraically out of the question, since division by zero was impossible.

## Limits Revisited

Newton invented "fluxions" and Leibniz invented "differentials" to explain instantaneous rates of change without resorting to zero denominators. Both involved mysterious quantities that could be infinitesimally small without really being zero. (Their 17th-century colleague Bishop Berkeley called them "ghosts of departed quantities" and dismissed them as nonsense.) Though not well understood by many, the strange quantities modeled the behavior of moving bodies so effectively that most scientists were willing to accept them on faith until a better explanation could be developed. That development, which took about a hundred years, led to our modern understanding of limits.

Since you are already familiar with limit notation, we can show you how this works with a simple example.

### Example 3   USING LIMITS TO AVOID ZERO DIVISION

A ball slides down a ramp so that its distance $s$ from the top of the ramp after $t$ seconds is exactly $t^2$ feet. What is its instantaneous velocity after 3 seconds?

**A Disclaimer**

Readers who know a little calculus will recognize that the ball in Example 2 really does have a non-zero instantaneous velocity after 3 seconds (as we will eventually show). The point of the example is to show how difficult it is to demonstate that fact at a single instant, since both time and the position of the ball appear not to change.

**Solution** We might try to answer this question by computing average velocity over smaller and smaller time intervals.

On the interval $[3, 3.1]$:

$$\frac{\triangle s}{\triangle t} = \frac{(3.1)^2 - 3^2}{3.1 - 3} = \frac{0.61}{0.1} = 6.1 \text{ feet per second.}$$

On the interval $[3, 3.05]$:

$$\frac{\triangle s}{\triangle t} = \frac{(3.05)^2 - 3^2}{3.05 - 3} = \frac{0.3025}{0.05} = 6.05 \text{ feet per second.}$$

Continuing this process we would would eventually conclude that the **instantaneous velocity** must be 6.

However, we can see *directly* what is happening to the quotient by treating it as a *limit* of the average velocity on the interval $[3, t]$ as $t$ approaches 3:

$$\lim_{t \to 3} \frac{\triangle s}{\triangle t} = \lim_{t \to 3} \frac{t^2 - 3^2}{t - 3}$$

$$= \lim_{t \to 3} \frac{(t + 3)(t - 3)}{t - 3} \qquad \text{Factor the numerator.}$$

$$= \lim_{t \to 3} (t + 3) \cdot \frac{t - 3}{t - 3}$$

$$= \lim_{t \to 3} (t + 3) \qquad \text{Since } t \neq 3, \frac{t - 3}{t - 3} = 1$$

$$= 6$$

Notice that $t$ is *not equal to* 3 but is *approaching* 3 as a limit, which allows us to make the crucial cancellation in the second to last line of Example 3. If $t$ were actually equal to 3, the algebra above would lead to the incorrect conclusion that $0/0 = 6$. The difference between equaling 3 and approaching 3 as a limit is a subtle one, but it makes all the difference algebraically.

It is not easy to formulate a rigorous algebraic definition of a limit (which is why Newton and Leibniz never really did). We have used an intuitive approach to limits so far in this book and will continue to do so, deferring the rigorous definitions to your calculus course. For now, we will use the following *informal* definition.

**Arbitrarily Close**

This definition is useless for mathematical proofs until one defines "arbitrarily close," but if you have a sense of how it applies to the solution in Example 3 above, then you are ready to use limits to study motion problems.

---

**Definition (informal)** Limit at *a*

When we write "$\lim_{x \to a} f(x) = L$," we mean that $f(x)$ gets arbitrarily close to $L$ as $x$ gets arbitrarily close (but not equal) to $a$.

---

## The Connection to Tangent Lines

What Galileo discovered by rolling balls down ramps was that the distance traveled was proportional to the *square* of the elapsed time. For simplicity, let us suppose that the ramp was tilted just enough so that the relation between $s$, the distance from the top of the ramp, and $t$, the elapsed time, was given (as in Example 3) by

$$s = t^2.$$

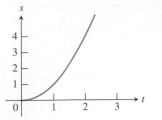

**Figure 10.1** The graph of $s = t^2$ shows the distance ($s$) traveled by a ball rolling down a ramp as a function of the elapsed time $t$.

**Exploration Extensions**

Repeat steps 1–3 using the points (0, 0) and (2, 4).

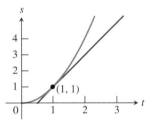

**Figure 10.2** A line tangent to the graph of $s = t^2$ at the point (1, 1). The slope of this line appears to be the instantaneous velocity at $t = 1$, even though $\triangle s/\triangle t = 0/0$. The geometry succeeds where the algebra fails!

Graphing $s$ as a function of $t \geq 0$ gives the right half of a parabola (Figure 10.1).

**Exploration 1**  **Seeing Average Velocity**

Copy Figure 10.1 on a piece of paper and connect the points (1, 1) and (2, 4) with a straight line. (This is called a *secant line* because it connects two points on the curve.)

**1.** Find the slope of the line.  3

**2.** Find the average velocity of the ball over the time interval $[1, 2]$.

**3.** What is the relationship between the numbers that answer questions 1 and 2?  They are the same.

**4.** In general, how could you represent the average velocity of the ball over the time interval $[a, b]$ geometrically?

Exploration 1 suggests an important general fact: If $(a, s(a))$ and $(b, s(b))$ are two points on a distance-time graph, then the *average velocity* over the time interval $[a, b]$ can be thought of as the *slope* of the line connecting the two points. In fact, we designate both quantities with the same symbol: $\triangle s/\triangle t$.

Galileo knew this. He also knew that he wanted to find instantaneous velocity by letting the two points become one—resulting in $\triangle s/\triangle t = 0/0$, an algebraic impossibility. The picture, however, told a different story *geometrically*. If, for example, we were to connect pairs of points closer and closer to (1, 1), our secant lines would look more and more like a line that is tangent to the curve at (1, 1) (Figure 10.2).

It seemed obvious to Galileo and the other scientists of his time that the slope of the tangent line was the long-sought-after answer to the quest for instantaneous velocity. They could *see* it, but how could they *compute* it without dividing by zero? That was the "tangent line problem," eventually solved for general functions by Newton and Leibniz in slightly different ways. We will solve it with limits as illustrated in Example 4.

**Example 4**  FINDING THE SLOPE OF A TANGENT LINE

Use limits to find the slope of the tangent line to the graph of $s = t^2$ at the point (1, 1) (Figure 10.2).

**Solution**  This will look a lot like the solution to Example 3.

$$\lim_{t \to 1} \frac{\triangle s}{\triangle t} = \lim_{t \to 1} \frac{t^2 - 1^2}{t - 1}$$

$$= \lim_{t \to 1} \frac{(t + 1)(t - 1)}{t - 1} \qquad \text{Factor the numerator.}$$

$$= \lim_{t \to 1} (t + 1) \cdot \frac{t - 1}{t - 1}$$

$$= \lim_{t \to 1} (t + 1) \qquad \text{Since } t \neq 1, \frac{t - 1}{t - 1} = 1$$

$$= 2$$

## The Tangent Line Problem

Although we have focused on Galileo's work with motion problems in order to follow a coherent story, it was Pierre de Fermat (1601–1665) who first developed a "method of tangents" for general curves, recognizing its usefulness for finding relative maxima and minima. Fermat is best remembered for his work in number theory, particularly for Fermat's Last Theorem, which states that there are no positive integers $x$, $y$, and $z$ that satisfy the equation $x^n + y^n = z^n$ if $n$ is an integer greater than 2. Fermat wrote in the margin of a textbook, "I have a truly marvelous proof that this margin is too narrow to contain," but if he had one, he apparently never wrote it down. Although mathematicians tried for over 330 years to prove (or disprove) Fermat's Last Theorem, nobody succeeded until Andrew Wiles of Princeton University finally proved it in 1994.

## Differentiability

We say a function is "differentiable" at $a$ if $f'(a)$ exists, because we can find the limit of the "quotient of differences."

If you compare Example 4 to Example 3 it should be apparent that a method for solving the tangent line problem can be used to solve the instantaneous velocity problem, and vice-versa. They are geometric and algebraic versions of the same problem!

## The Derivative

Velocity, the rate of change of position with respect to time, is only one application of the general concept of "rate of change." If $y = f(x)$ is *any* function, we can speak of how $y$ changes as $x$ changes.

**Definition  Average Rate of Change**

If $y = f(x)$, then the **average rate of change** of $y$ with respect to $x$ on the interval $[a, b]$ is

$$\frac{\triangle y}{\triangle x} = \frac{f(b) - f(a)}{b - a}.$$

Geometrically, this is the slope of the **secant line** through $(a, f(a))$ and $(b, f(b))$.

Using limits, we can proceed to a definition of the *instantaneous* rate of change of $y$ with respect to $x$ at the point where $x = a$. This instantaneous rate of change is called the *derivative*.

**Definition  Derivative at a Point**

The **derivative of the function $f$ at $x = a$**, denoted by $f'(a)$ and read "$f$ prime of $a$" is

$$f'(a) = \lim_{x \to a} \frac{f(x) - f(a)}{x - a},$$

provided the limit exists.

Geometrically, this is the slope of the **tangent line** through $(a, f(a))$.

A more computationally useful formula for the derivative is obtained by letting $x = a + h$ and looking at the limit as $h$ approaches 0 (equivalent to letting $x$ approach $a$).

**Definition  Derivative at a Point (easier for computing)**

The **derivative of the function $f$ at $x = a$**, denoted by $f'(a)$ and read "$f$ prime of $a$" is

$$f'(a) = \lim_{h \to 0} \frac{f(a + h) - f(a)}{h},$$

provided the limit exists.

The fact that the derivative of a function at a point can be viewed geometrically as the slope of the line tangent to the curve $y = f(x)$ at that point provides us with some insight as to how a derivative might fail to exist. Unless a function has a well-defined "slope" when you zoom in on it at $a$, the derivative at $a$ will not exist. For example, Figure 10.3 shows three cases for which $f(0)$ exists but $f'(0)$ does not.

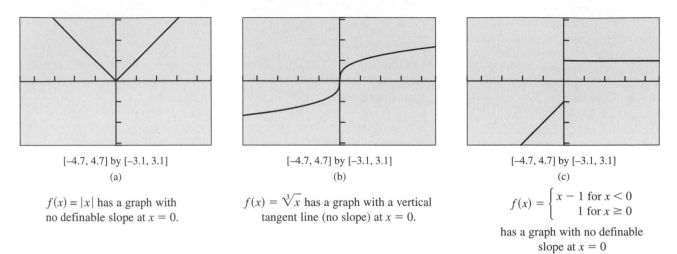

| [–4.7, 4.7] by [–3.1, 3.1] | [–4.7, 4.7] by [–3.1, 3.1] | [–4.7, 4.7] by [–3.1, 3.1] |
| (a) | (b) | (c) |

$f(x) = |x|$ has a graph with no definable slope at $x = 0$.

$f(x) = \sqrt[3]{x}$ has a graph with a vertical tangent line (no slope) at $x = 0$.

$f(x) = \begin{cases} x - 1 \text{ for } x < 0 \\ 1 \text{ for } x \geq 0 \end{cases}$

has a graph with no definable slope at $x = 0$

**Figure 10.3** Three examples of functions defined at $x = 0$ but not differentiable at $x = 0$.

## Example 5  FINDING A DERIVATIVE AT A POINT

Find $f'(4)$ if $f(x) = 2x^2 - 3$.

**Solution**

$$f'(4) = \lim_{h \to 0} \frac{f(4 + h) - f(4)}{h}$$

$$= \lim_{h \to 0} \frac{2(4 + h)^2 - 3 - (2 \cdot 4^2 - 3)}{h}$$

$$= \lim_{h \to 0} \frac{2(16 + 8h + h^2) - 32}{h}$$

$$= \lim_{h \to 0} \frac{16h + 2h^2}{h}$$

$$= \lim_{h \to 0} (16 + 2h) \qquad \frac{h}{h} = 1, \text{ since } h \neq 0.$$

$$= 16$$

The derivative can also be thought of as a function of $x$. Its domain consists of all values in the domain of $f$ for which $f$ is differentiable. The function $f'$ can be defined by adapting the second definition above.

**Definition  Derivative**

If $y = f(x)$, then the **derivative of the function $f$ with respect to $x$** is the function $f'$ whose value at $x$ is

$$f'(x) = \lim_{h \to 0} \frac{f(x + h) - f(x)}{h},$$

for all values of $x$ where the limit exists.

To emphasize the connection with slope $\triangle y / \triangle x$, Leibniz used the notation $dy/dx$ for the derivative. (The $dy$ and $dx$ were his "ghosts of departed quantities.") This **Leibniz notation** has several advantages over the "prime" notation, as you will learn when you study calculus. We will use both notations in our examples and exercises.

### Example 6  FINDING THE DERIVATIVE OF A FUNCTION

**(a)** Find $f'(x)$ if $f(x) = x^2$.

**(b)** Find $\dfrac{dy}{dx}$ if $y = \dfrac{1}{x}$.

**Solution**

**(a)**
$$f'(x) = \lim_{h \to 0} \frac{f(x+h) - f(x)}{h}$$
$$= \lim_{h \to 0} \frac{(x+h)^2 - x^2}{h}$$
$$= \lim_{h \to 0} \frac{x^2 + 2xh + h^2 - x^2}{h}$$
$$= \lim_{h \to 0} \frac{2xh + h^2}{h}$$
$$= \lim_{h \to 0} (2x + h) \qquad \text{Since } h \neq 0, \frac{h}{h} = 1$$
$$= 2x$$

So $f'(x) = 2x$.

**(b)**
$$\frac{dy}{dx} = \lim_{h \to 0} \frac{f(x+h) - f(x)}{h}$$
$$= \lim_{h \to 0} \frac{\dfrac{1}{x+h} - \dfrac{1}{x}}{h}$$
$$= \lim_{h \to 0} \frac{\dfrac{x - (x+h)}{x(x+h)}}{h}$$
$$= \lim_{h \to 0} \frac{-h}{x(x+h)} \cdot \frac{1}{h}$$
$$= \lim_{h \to 0} \frac{-1}{x(x+h)}$$
$$= -\frac{1}{x^2}$$

So $\dfrac{dy}{dx} = -\dfrac{1}{x^2}$.

**Problem**

If we have a windmill with propeller radius 5 meters, what is the rate of change in power generated when the wind velocity is 7 meters/sec (about 15 miles/hour)?

**Solution**

Power generated in watts is given by the equation

$$P = kr^2v^3$$

Since the radius is 5 meters, the equation is

$$P = kr^2v^3 = 5^2kv^3 = 25kv^3$$

Or $P(v) = 25kv^3$.

The rate of change in power generated when the wind velocity is 7 meters per second is then

$$P'(7) = \lim_{h \to 0} \frac{P(7 + h) - P(7)}{h}$$

$$= \lim_{h \to 0} \frac{25k(7 + h)^3 - 25k(7)^3}{h}$$

$$= \lim_{h \to 0} \frac{25k(7^3 + 147h + 21h^2 + h^3) - 25k(7)^3}{h}$$

$$= \lim_{h \to 0} \frac{25k(7^3) + 25k(147h) + 25k(21h^2) + 25k(h^3) - 25k(7)^3}{h}$$

$$= \lim_{h \to 0} \frac{25k(147h) + 25k(21h^2) + 25k(h^3)}{h}$$

$$= \lim_{h \to 0} \left[ 25k(147) + 25k(21h) + 25k(h^2) \right]$$

$$= 25k(147)$$

$$= 3675k$$

The rate of change in power generated is $3675k$ watts per meter/sec.

# Quick Review 10.1

In Exercises 1 and 2, find the slope of the line determined by the points.

**1.** $(-2, 3), (5, -1)$  $-4/7$     **2.** $(-3, -1), (3, 3)$  2/3

In Exercises 3–6, write an equation for the specified line.

**3.** Through $(-2, 3)$ with slope $= 3/2$  $y = (3/2)x + 6$

**4.** Through $(1, 6)$ and $(4, -1)$  $y - 6 = (-7/3)(x - 1)$

**5.** Through $(1, 4)$ and parallel to $y = (3/4)x + 2$

**6.** Through $(1, 4)$ and perpendicular to $y = (3/4)x + 2$

In Exercises 7–10, simplify the expression assuming $h \neq 0$.

**7.** $\dfrac{(2 + h)^2 - 4}{h}$  $h + 4$

**8.** $\dfrac{(3 + h)^2 + 3 + h - 12}{h}$  $h + 7$

**9.** $\dfrac{1/(2 + h) - 1/2}{h}$  $-\dfrac{1}{2(h + 2)}$

**10.** $\dfrac{1/(x + h) - 1/x}{h}$  $-\dfrac{1}{x(x + h)}$

# Section 10.1 Exercises

In Exercises 1–4, use the graph to estimate the slope of the tangent line, if it exists, to the graph at the given point.

**1.** $x = 0$  1

**2.** $x = 1$  $-1$

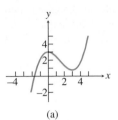

(a)

**3.** $x = 2$  no tangent

**4.** $x = 4$  no tangent

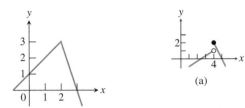

(a)

In Exercises 5–8, graph the function in a square viewing window and, without doing any calculations, estimate the derivative of the function at the given point by interpreting it as the tangent line slope, if it exists at the point.

**5.** $f(x) = x^2 - 2x + 5$  at  $x = 3$  4

**6.** $f(x) = \frac{1}{2}x^2 + 2x - 5$  at  $x = 2$  4

**7.** $f(x) = x^3 - 6x^2 + 12x - 9$  at  $x = 0$  12

**8.** $f(x) = 2 \sin x$  at  $x = \pi$  $-2$

In Exercises 9–14, find the derivative, if it exists, of the function at the specified point.

**9.** $f(x) = 1 - x^2$  at  $x = 2$  $-4$

**10.** $f(x) = 2x + \frac{1}{2}x^2$  at  $x = 2$  4

**11.** $f(x) = 3x^2 + 2$  at  $x = -2$  $-12$

**12.** $f(x) = x^2 - 3x + 1$  at  $x = 1$  $-1$

**13.** $f(x) = |x + 2|$  at  $x = -2$  does not exist

**14.** $f(x) = \frac{1}{x + 2}$  at  $x = -1$  $-1$

In Exercises 15–18, use the limit definition to find

**(a)** the slope of the graph of the function at the indicated point,

**(b)** an equation of the tangent line at the point.

**(c)** Sketch a graph of the curve near the point without using your graphing calculator.

**15.** $f(x) = 2x^2$  at  $x = -1$

**16.** $f(x) = 2x - x^2$  at  $x = 2$

**17.** $f(x) = 2x^2 - 7x + 3$  at  $x = 2$

**18.** $f(x) = \frac{1}{x + 2}$  at  $x = 1$

In Exercises 19 and 20, estimate the slope of the tangent line to the graph of the function, if it exists, at the indicated points.

**19.** $f(x) = |x|$  at  $x = -2, 2,$ and 0.  $-1; 1;$ none

**20.** $f(x) = \tan^{-1}(x + 1)$  at  $x = -2, 2,$ and 0.  0.5; 0.1; 0.5

In Exercises 21–24, find the derivative of $f$.

**21.** $f(x) = 2 - 3x$  $-3$

**22.** $f(x) = 2 - 3x^2$  $-6x$

**23.** $f(x) = 3x^2 + 2x - 1$

**24.** $f(x) = \frac{1}{x - 2}$  $-\frac{1}{(x - 2)^2}$

In Exercises 25–28, complete the following.

**(a)** Draw a graph of the function.

**(b)** Find the derivative of the function at the given point if it exists.

**(c) Writing to Learn**  If the derivative does not exist at the point, explain why not.

**25.** $f(x) = \begin{cases} 4 - x & x \le 2 \\ x + 3 & x > 2 \end{cases}$  at  $x = 2$

**26.** $f(x) = \begin{cases} 1 + (x - 2)^2 & x \le 2 \\ 1 - (x - 2)^2 & x > 2 \end{cases}$  at  $x = 2$

**27.** $f(x) = \begin{cases} \dfrac{|x - 2|}{x - 2} & x \ne 2 \\ 1 & x = 2 \end{cases}$  at  $x = 2.$

**28.** $f(x) = \begin{cases} \dfrac{\sin x}{x} & x \ne 0 \\ 1 & x = 0 \end{cases}$  at  $x = 0$

**29. A Rock Toss**  A rock is thrown straight up from level ground. The distance (in ft) the ball is above the ground (the position function) is $f(t) = 3 + 48t - 16t^2$ at any time $t$ (in sec). Find

**(a)** $f'(0)$.  48

**(b)** the initial velocity of the rock.  48 ft/sec

**30. Rocket Launch**  A toy rocket is launched straight up in the air from level ground. The distance (in ft) the rocket is above the ground (the position function) is $f(t) = 170t - 16t^2$ at any time $t$ (in sec). Find

**(a)** $f'(0)$.  170

**(b)** the initial velocity of the rocket.  170 ft/sec

**31. Average speed** A lead ball is held at water level and dropped from a boat into a lake. The distance the ball falls at 0.1 sec time intervals is given in Table 10.1.

**Table 10.1 Distance Data of the Lead Ball**

| Time (sec) | Distance (ft) |
|------------|---------------|
| 0 | 0 |
| 0.1 | 0.1 |
| 0.2 | 0.4 |
| 0.3 | 0.8 |
| 0.4 | 1.5 |
| 0.5 | 2.3 |
| 0.6 | 3.2 |
| 0.7 | 4.4 |
| 0.8 | 5.8 |
| 0.9 | 7.3 |

**(a)** Compute the average speed from 0.5 to 0.6 seconds and from 0.8 to 0.9 seconds. 9 ft/sec; 15 ft/sec

**(b)** Find a quadratic regression model for the distance data and overlay its graph on a scatter plot of the data.

**(c)** Use the model in (b) to estimate the depth of the lake if the ball hits the bottom after 2 seconds. $\approx 35.9$ ft

**32. Finding Derivatives from Data** A ball is dropped from the roof of a two story building. The distance in feet above ground of the falling ball is given in Table 10.2 where $t$ is in seconds.

**Table 10.2**

| Time (sec) | Distance (ft) |
|------------|---------------|
| 0.2 | 30.00 |
| 0.4 | 28.36 |
| 0.6 | 25.44 |
| 0.8 | 21.24 |
| 1.0 | 15.76 |
| 1.2 | 9.02 |
| 1.4 | 0.95 |

**(a)** Use the data to estimate the average velocity of the ball in the interval $0.8 \leq t \leq 1$. $-27.4$ ft/sec

**(b)** Find a quadratic regression model $s$ for the data in Table 10.2 and overlay its graph on a scatter plot of the data.

**(c)** Find the derivative of the regression equation and use it to estimate the velocity of the ball at time $t = 1$.

In Exercises 33–36, the position of an object at time $t$ is given by $s(t)$. Find the instantaneous velocity at the indicated value of $t$.

**33.** $s(t) = 3t - 5$ at $t = 4$ 3

**34.** $s(t) = \dfrac{2}{t + 1}$ at $t = 2$ $-2/9$

**35.** $s(t) = at^2 + 5$ at $t = 2$ $4a$

**36.** $s(t) = \sqrt{t + 1}$ at $t = 1$ (*Hint:* "rationalize the numerator.") $1/(2\sqrt{2})$

**37. Writing to Learn** Explain why you can find the derivative of $f(x) = ax + b$ without doing any computations. What is $f'(x)$? The slope of the line is $a$; $f'(x) = a$

**38. Writing to Learn** Use the *first* definition of derivative at a point to express the derivative of $f(x) = |x|$ at $x = 0$ as a limit. Then explain why the limit does not exist. (A graph of the quotient for $x$ values near 0 might help.)

In Exercises 39–42, sketch a possible graph for a function that has the stated properties.

**39.** The domain of $f$ is $[0, 5]$ and the derivative at $x = 2$ is 3.

**40.** The domain of $f$ is $[0, 5]$ and the derivative is 0 at both $x = 2$ and $x = 4$.

**41.** The domain of $f$ is $[0, 5]$ and the derivative at $x = 2$ is undefined.

**42.** The domain of $f$ is $[0, 5]$, $f$ is non-decreasing on $[0, 5]$, and the derivative at $x = 2$ is 0.

## Explorations

Graph each function in Exercises 43–46 and then answer the following questions.

**(a) Writing to Learn** Does the function have a derivative at $x = 0$? Explain.

**(b)** Does the function appear to have a tangent line at $x = 0$? If so, what is an equation of the tangent line?

**43.** $f(x) = |x|$      **44.** $f(x) = |x^{1/3}|$

**45.** $f(x) = x^{1/3}$      **46.** $f(x) = \tan^{-1} x$

**47. Free Fall** A water balloon dropped from a window will fall a distance of $s = 16t^2$ feet during the first $t$ seconds. Find the balloon's **(a)** average velocity during the first 3 seconds of falling and **(b)** instantaneous velocity at $t = 3$.

**48. Free Fall on Another Planet** It can be established by experimentation that heavy objects dropped from rest free fall near the surface of another planet according to the formula $y = gt^2$, where $y$ is the distance in meters the object falls in $t$ seconds after being dropped. An object falls from the top of a 125 m spaceship which landed on the surface. It hits the surface in 5 seconds.

**(a)** Find the value of $g$. 5 m/sec²

**(b)** Find the average speed for the fall of the object.

**(c)** With what speed did the object hit the surface? ■

## Extending the Ideas

**49. Graphing the derivative** The graph of $f(x) = x^2 e^{-x}$ is shown below. Use your knowledge of the geometric interpretation of the derivative to sketch a rough graph of the derivative $y = f'(x)$.

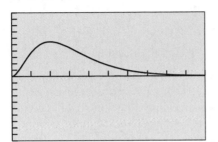

[0, 10] by [−1, 1]

**50. Group Activity** The graph of $y = f'(x)$ is shown below. Determine a possible graph for the function $y = f(x)$.

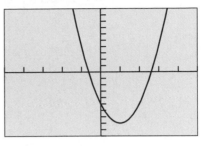

[−5, 5] by [−10, 10]

---

<table>
<tr><td>**10.2**</td><td>## Limits and Motion: The Area Problem</td></tr>
</table>

Distance from a Constant Velocity  •  Distance from a Changing Velocity  •  Limits at Infinity  •  The Connection to Areas  •  The Definite Integral

### Distance from a Constant Velocity

"Distance equals rate times time" is one of the earliest problem-solving formulas that we learn in school mathematics. Given a velocity and a time period, we can use the formula to compute distance traveled—as in the following standard example.

**Objective**

Students will be able to calculate definite integrals using areas.

**Motivate**

Ask students to guess the meaning of the following statement: $\lim_{x \to \infty} f(x) = 5$.

**Lesson Guide**

Day 1: Distance from a Constant Velocity; Distance from a Changing Velocity; Limits at Infinity; The Connection to Areas
Day 2: The Definite Integral

**Example 1  COMPUTING DISTANCE TRAVELED**

An automobile travels at a constant rate of 48 miles per hour for 2 hours and 30 minutes. How far does the automobile travel?

**Solution** We apply the formula $D = RT$:

$$D = (48 \text{ mi/hr})(2.5 \text{ hr}) = 120 \text{ miles}.$$

The similarity to Example 1 in Section 10.1 is intentional. In fact, if we represent distance traveled (i.e., the change in position) by $\triangle s$ and the time interval by $\triangle t$, the formula becomes

$$\triangle s = (48 \text{ mph}) \triangle t,$$

which is equivalent to

$$\frac{\triangle s}{\triangle t} = 48 \text{ mph}.$$

So the two Example 1s are nearly identical—except that Example 1 of Section 10.1 did not make an assumption about constant velocity. What we computed in that instance was the *average* velocity over the 2.5-hour interval. This suggests that we could actually have solved the following, slightly different, problem to open this section.

**Example 2** COMPUTING DISTANCE TRAVELED

An automobile travels at an *average* rate of 48 miles per hour for 2 hours and 30 minutes. How far does the automobile travel?

**Solution** The distance traveled is $\triangle s$, the time interval has length $\triangle t$, and $\triangle s / \triangle t$ is the average velocity.

Therefore,

$$\triangle s = \frac{\triangle s}{\triangle t} \cdot \triangle t = (48 \text{ mph})(2.5 \text{ hr}) = 120 \text{ miles.}$$

So, given average velocity over a time interval, we can easily find distance traveled. But suppose we have a velocity function $v(t)$ that gives instantaneous velocity as a changing function of time. How can we use the instantaneous velocity function to find distance traveled over a time interval? This was the other intriguing problem about instantaneous velocity that puzzled the 17th century scientists—and once again, algebra was inadequate for solving it, as we shall see.

## Distance from a Changing Velocity

Here's how Galileo might have answered his own question about finding distance from a changing velocity when he began his experiments.

**Example 3** DISTANCE

Suppose a ball rolls down a ramp so that its velocity after $t$ seconds is always $2t$ feet per second. How far does it fall during the first 3 seconds?

**A Proposed Solution** Velocity times $\triangle t$ gives $\triangle s$. But instantaneous velocity occurs at an instant of time, so $\triangle t = 0$. That means $\triangle s = 0$. So, at any given instant of time, the ball doesn't move. Since any time interval consists of instants of time, the ball never moves at all! (You might well ask: Is this another trick question?)

As was the case with Example 2 in Section 10.1, this foolish-looking example conceals a very subtle algebraic dilemma—and, far from being a trick question, it is exactly the question that needed to be answered in order to compute the distance traveled by an object whose velocity varied as a function of time. The scientists who were working on the tangent line problem realized that the distance-traveled problem must be related to it, but, surprisingly, their geometry led them in another direction. The distance traveled problem led them not to tangent lines, but to areas.

## Limits at Infinity

Before we see the connection to areas, let us revisit another limit concept that will make instantaneous velocity easier to handle, just as in the last section. We will again be content with an informal definition.

**Zeno's Paradoxes**

The Greek philosopher Zeno of Elea (490–425 B.C.) was noted for presenting paradoxes similar to Example 3. One of the most famous concerns the race between Achilles and a slow-but-sure tortoise. Achilles sportingly gives the tortoise a head start, then sets off to catch him. He must first get halfway to the tortoise, but by the time he runs halfway to where the tortoise was when he started, the tortoise has moved ahead. Now Achilles must close half of that distance, but by the time he does, the tortoise has moved ahead again. Continuing this argument forever, we see that Achilles can never even catch the tortoise, let alone pass him, so the tortoise must win the race.

> **Definition (Informal)  Limit at Infinity**
>
> When we write "$\lim\limits_{x \to \infty} f(x) = L$," we mean that $f(x)$ gets arbitrarily close to $L$ as $x$ gets arbitrarily large.

---

**Exploration 1**  **An Infinite imit**

A gallon of water is divided equally and poured into teacups. Find the amount in each teacup and the *total amount* in *all* the teacups if there are

1. 10 teacups  0.1 gal; 1 gal
2. 100 teacups  0.01 gal; 1 gal
3. 1 billion teacups  0.000000001 gal; 1 gal
4. an infinite number of teacups  0 gal;  1 gal

---

**Exploration Extensions**

Find the number of teacups needed if the amount of water in each teacup is (1) 1 cup, (2) 1 tablespoon, (3) an infinitely small amount.
(Hint: 1 gallon = 16 cups; 1 cup = 16 tablespoons.)

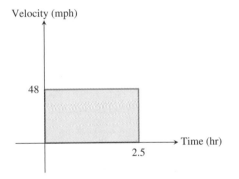

Velocity (mph)

48

2.5  Time (hr)

**Figure 10.4**  For constant velocity, the area of the rectangle is the same as the distance traveled, since it represents the product of the same two quantities:
(48 mph)(2.5 hr) = 120 miles.

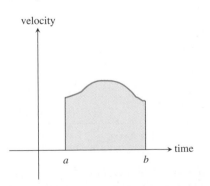

velocity

$a$      $b$      time

**Figure 10.5**  If the velocity varies over the time interval $[a, b]$, does the shaded region give the distance traveled?

The preceding exploration probably went pretty smoothly until you came to the infinite number of teacups. At that point you were probably pretty comfortable in saying what the *total amount* would be, and probably a little uncomfortable in saying how much would be in each teacup. (Theoretically it would be zero, which is just one reason why the actual experiment cannot be performed.) In the language of limits, the total amount of water in the infinite number of teacups would look like this:

$$\lim_{x \to \infty}\left(n \cdot \frac{1}{n}\right) = \lim_{x \to \infty} \frac{n}{n} = 1 \text{ gallon}$$

while the total amount in each teacup would look like this:

$$\lim_{x \to \infty} \frac{1}{n} = 0 \text{ gallons.}$$

Summing up an infinite number of nothings to get something is mysterious enough when we use limits; *without* limits it seems to be an algebraically impossibility. That is the dilemma that faced the 17th-century scientists who were trying to work with instantaneous velocity. Once again, it was geometry that showed the way when the algebra failed.

## The Connection to Areas

If we graph the constant velocity $v = 48$ in Example 1 as a function of time $t$, we notice that the area of the shaded rectangle is the same as the distance traveled (Figure 10.4). This is no mere coincidence, either, as the area of the rectangle and the distance traveled over the time interval are both computed by multiplying the same two quantities:

$$(48 \text{ mph})(2.5 \text{ hr}) = 120 \text{ miles.}$$

Now suppose we graph a velocity function that varies continuously as a function of time (Figure 10.5). Would the area of this irregularly-shaped region still give the total distance traveled over the time interval $[a, b]$?

Newton and Leibniz (and, actually, many others who had considered this question) were convinced that it obviously would, and that is why they were interested in a calculus for finding areas under curves. They imagined the time interval being partitioned into many tiny subintervals, each one so small that the velocity over it would essentially be constant. Geometrically, this was equivalent to slicing the area into narrow strips, each one of which would be nearly indistinguishable from a narrow rectangle (Figure 10.6).

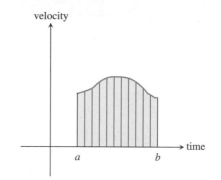

**Figure 10.6**   The region is partitioned into vertical strips. If the strips are narrow enough, they are almost indistinguishable from rectangles. The sum of the areas of these "rectangles" will give the total area and can be interpreted as distance traveled.

The idea of partitioning irregularly-shaped areas into approximating rectangles was not new. Indeed, Archimedes had used that very method to approximate the area of a circle with remarkable accuracy. However, it was an exercise in patience and perseverance, as Example 4 will show.

### Example 4   APPROXIMATING AN AREA WITH RECTANGLES

Use the six rectangles in Figure 10.7 to approximate the area of the region below the graph of $f(x) = x^2$ over the interval [0, 3].

**Solution**   The base of each approximating rectangle is 1/2. The height is determined by the function value at the right hand endpoint of each subinterval. The areas of the six rectangles and the total area are computed in the table below:

| Subinterval | Base of rectangle | Height of rectangle | Area of rectangle |
|---|---|---|---|
| [0, 1/2] | 1/2 | $f(1/2) = (1/2)^2 = 1/4$ | $(1/2)(1/4) = 0.125$ |
| [1/2, 1] | 1/2 | $f(1) = (1)^2 = 1$ | $(1/2)(1) = 0.500$ |
| [1, 3/2] | 1/2 | $f(3/2) = (3/2)^2 = 9/4$ | $(1/2)(9/4) = 1.125$ |
| [3/2, 2] | 1/2 | $f(2) = (2)^2 = 4$ | $(1/2)(4) = 2.000$ |
| [2, 5/2] | 1/2 | $f(5/2) = (5/2)^2 = 25/4$ | $(1/2)(25/4) = 3.125$ |
| [5/2, 3] | 1/2 | $f(3) = (3)^2 = 9$ | $(1/2)(9) = 4.500$ |
| | | **Total Area:** | 11.375 |

The six rectangles give a (rather crude) approximation of 11.375 square units for the area under the curve from 0 to 3.

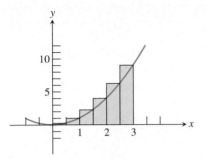

**Figure 10.7** The area under the graph of $f(x) = x^2$ is approximated by six rectangles, each with base 1/2. The height of each rectangle is the function value at the right-hand endpoint of the subinterval (Example 4).

Figure 10.7 shows that the *right rectangular approximation method* (RRAM) in Example 4 overestimates the true area. If we were to use the function values at the left-hand endpoints of the subintervals (LRAM), we would obtain a rectangular approximation (6.875 square units) that underestimates the true area (Figure 10.8). The average of the two approximations is 9.125 square units, which is actually a pretty good estimate of the true area of 9 square units. If we were to repeat the process with 20 rectangles, the average would be 9.01125. This method of converging toward an unknown area by refining approximations is tedious, but it works—Archimedes used a variation of it 2200 years ago to estimate the area of a circle, and in the process demonstrated that the ratio of the circumference to the diameter was between 3.140845 and 3.142857.

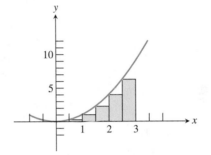

**Figure 10.8** If we change the rectangles in Figure 10.7 so that their heights are determined by function values at the *left-hand* endpoints, we get an area approximation (6.875 square units) that underestimates the true area.

The calculus step is to move from a finite number of rectangles (yielding an approximate area) to an infinite number of rectangles (yielding an exact area). This brings us to the definite integral.

## The Definite Integral

In general, begin with a continuous function $y = f(x)$ over an interval $[a, b]$. Divide $[a, b]$ into $n$ subintervals of length $\triangle x = (b - a)/n$. Choose any value $x_1$ in the first subinterval, $x_2$ in the second, and so on. Compute $f(x_1)$, $f(x_2)$, $f(x_3)$, ..., $f(x_n)$, multiply each value by $\triangle x$, and sum up the products. In sigma notation, the sum of the products is

$$\sum_{i=1}^{n} f(x_i)\triangle x.$$

The *limit* of this sum as $n$ approaches infinity is the solution to the area problem, and hence the solution to the problem of distance traveled. Indeed, it solves a variety of other problems as well, as you will learn when you study calculus. The limit, if it exists, is called a *definite integral*.

### Riemann Sums

A sum of the form $\sum_{i=1}^{n} f(x_i)\triangle x$ in which $x_1$ is in the first subinterval, $x_2$ is in the second, and so on, is called a **Riemann sum**, in honor of Georg Riemann (1826–1866), who determined the functions for which such sums had limits as $n \to \infty$.

## Definite Integral Notation

Notice that the notation for the definite integral (another legacy of Leibniz) parallels the sigma notation of the sum for which it is a limit. The "$\Sigma$" in the limit becomes a stylized "S," for "sum." The "$\triangle x$" becomes "$dx$" (as it did in the derivative), and the "$f(x_i)$" becomes simply "$f(x)$" because we are effectively summing up *all* the $f(x)$ values along the interval (times an arbitrarily small change in $x$), rendering the subscripts unnecessary.

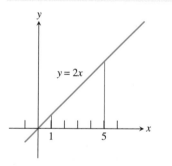

**Figure 10.9** The area of the trapezoid equals $\int_1^5 2x\, dx$. (Example 5)

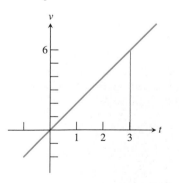

**Figure 10.10** The area under the velocity graph $v(t) = 2t$, over the interval $[0, 3]$ is the distance traveled by the ball in Example 6 during the first 3 seconds.

---

> **Definition** Definite Integral
>
> Let $f$ be a function defined on $[a, b]$ and let $\sum\limits_{i=1}^{n} f(x_i)\triangle x$ be defined as on page 767. The definite integral of $f$ over $[a, b]$, denoted $\int_a^b f(x)\, dx$, is given by
>
> $$\int_a^b f(x)\,dx = \lim_{n \to \infty} \sum_{i=1}^{n} f(x_i)\triangle x,$$
>
> provided the limit exists. If the limit exists, we say $f$ is **integrable** on $[a, b]$

The solution to Example 4 shows that it can be tedious to approximate a definite integral by working out the sum for a large value of $n$. One of the crowning achievements of calculus was to demonstrate how the exact value of a definite integral could be obtained without summing up any products at all. You will have to wait until calculus to see how that is done; meanwhile, you will learn in Section 10.4 how to use a calculator to take the tedium out of finding definite integrals by summing.

You can also use the area connection to your advantage, as shown in these next two examples.

**Example 5  COMPUTING AN INTEGRAL**

Find $\int_1^5 2x\, dx$.

**Solution**  This will be the area under the line $y = 2x$ over the interval $[1, 5]$. The graph in Figure 10.9 shows that this is the area of a trapezoid.

Using the formula $A = h\left(\dfrac{b_1 + b_2}{2}\right)$, we find that

$$\int_1^5 2x\, dx = 4\left(\frac{2(1) + 2(5)}{2}\right) = 24$$

We end this section by revisiting Example 3 and giving its actual solution.

**Example 6  COMPUTING AN INTEGRAL**

Suppose a ball rolls down a ramp so that its velocity after $t$ seconds is always $2t$ feet per second. How far does it fall during the first 3 seconds?

**Solution**  The distance traveled will be the same as the area under the velocity graph, $v(t) = 2t$, over the interval $[0, 3]$. The graph is shown in Figure 10.10. Since the region is triangular, we can find its area: $A = (1/2)(3)(6) = 9$. The distance traveled in the first 3 seconds, therefore, is $\triangle s = (1/2)(3 \text{ sec})(6 \text{ feet/sec}) = 9$ feet.

# Quick Review 10.2

In Exercises 1 and 2, list the elements of the sequence.

**1.** $a_k = \dfrac{1}{2}\left(\dfrac{1}{2}k\right)^2$  for  $k = 1, 2, 3, 4, \ldots, 9, 10$

**2.** $a_k = \dfrac{1}{4}\left(2 + \dfrac{1}{4}k\right)^2$  for  $k = 1, 2, 3, 4, \ldots, 9, 10$

In Exercises 3–6, find the sum.

**3.** $\displaystyle\sum_{k=1}^{10} \dfrac{1}{2}(k + 1)$   $\dfrac{65}{2}$

**4.** $\displaystyle\sum_{k=1}^{n} (k + 1)$   $\dfrac{n(n + 3)}{2}$

**5.** $\displaystyle\sum_{k=1}^{10} \dfrac{1}{2}(k + 1)^2$   $\dfrac{505}{2}$

**6.** $\displaystyle\sum_{k=1}^{n} \dfrac{1}{2}k^2$   $\dfrac{n(n + 1)(2n + 1)}{12}$

**7.** A truck travels at an average speed of 57 mph for 4 hours. How far does it travel?  228 miles

**8.** A pump working at 5 gal/min pumps for 2 hours. How many gallons are pumped?  600 gal

**9.** Water flows over a spillway at a steady rate of 200 cubic feet per second. How many cubic feet of water pass over the spillway in 6 hours?  4,320,000 ft³

**10.** A county has a population density of 560 people per square mile in an area of 35,000 square miles. What is the population of the county?  19,600,000 people

# Section 10.2 Exercises

In Exercises 1–4, estimate the area of the region above the *x*-axis and under the graph of the function from $x = 0$ to $x = 5$.

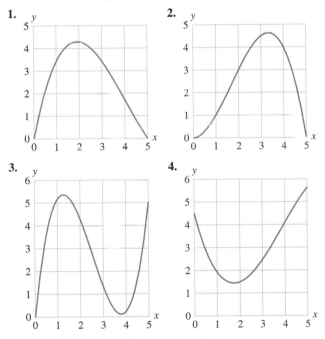

**1.**

**2.**

**3.**

**4.**

In Exercises 5–8, partition the given interval into the indicated number of subintervals.

**5.** [0, 2];  4

**6.** [0, 2];  8

**7.** [1, 4];  6

**8.** [1, 5];  8

In Exercises 9–12, complete the following.

**(a)** Draw the graph of the function for *x* in the specified interval. Verify the function is non-negative in that interval.

**(b)** On the graph in (a), draw and shade the approximating rectangles for the RRAM using the specified partition. Compute the RRAM area estimate without using a calculator.

**(c)** Repeat (b) using the LRAM.

**(d)** Average the RRAM and LRAM approximations from (b) and (c) to find an average estimate of the area.

**9.** $f(x) = x^2$;  [0, 4];  4 subintervals

**10.** $f(x) = x^2 + 2$;  [0, 6];  6 subintervals

**11.** $f(x) = 4x - x^2$;  [0, 4];  4 subintervals

**12.** $f(x) = x^3$;  [0, 3];  3 subintervals

In Exercises 13–20, find the definite integral by computing an area. (It may help to look at a graph of the function.)

**13.** $\displaystyle\int_{3}^{7} 5\,dx$  20

**14.** $\displaystyle\int_{-1}^{4} 6\,dx$  30

**15.** $\displaystyle\int_{0}^{5} 3x\,dx$  37.5

**16.** $\displaystyle\int_{1}^{7} 0.5x\,dx$  12

**17.** $\displaystyle\int_{1}^{4} (x + 3)\,dx$  16.5

**18.** $\displaystyle\int_{1}^{4} (3x - 2)\,dx$  16.5

**19.** $\displaystyle\int_{-2}^{2} \sqrt{4 - x^2}\,dx$  $2\pi$

**20.** $\displaystyle\int_{0}^{6} \sqrt{36 - x^2}\,dx$  $9\pi$

It can be shown that the area enclosed between the *x*-axis and one arch of the sine curve is 2. Use this fact in Exercises 21–30 to compute the definite integral. (It may help to look at a graph of the function.)

**21.** $\displaystyle\int_{0}^{\pi} \sin x\,dx$  2

**22.** $\displaystyle\int_{0}^{\pi} (\sin x + 2)\,dx$  $2 + 2\pi$

**23.** $\displaystyle\int_{2}^{\pi+2} \sin (x - 2)\,dx$  2

**24.** $\displaystyle\int_{-\pi/2}^{\pi/2} \cos x\,dx$  2

**25.** $\displaystyle\int_{0}^{\pi/2} \sin x\,dx$  1

**26.** $\displaystyle\int_{0}^{\pi/2} \cos x\,dx$  1

**27.** $\int_0^\pi 2 \sin x \, dx$ (Hint: All the rectangles are twice as tall.)  4

**28.** $\int_0^{2\pi} \sin\left(\dfrac{x}{2}\right) dx$ (Hint: All the rectangles are twice as wide.)

**29.** $\int_0^{2\pi} |\sin x| \, dx$  4     **30.** $\int_{-\pi}^{3\pi/2} |\cos x| \, dx$  5

In Exercises 31–34, explain how to represent the problem situation as an area question and then solve the problem.

**31.** A train travels at 65 mph for 3 hours. How far does it travel?

**32.** A pump working at 15 gal/min pumps for one-half hour. How many gallons are pumped?  450 gal

**33.** Water flows over a spillway at a steady rate of 150 cubic feet per second. How many cubic feet of water pass over the spillway in one hour?  540,000 ft³

**34.** A city has a population density of 650 people per square mile in an area of 20 square miles. What is the population of the city?  13,000 people

In Exercises 35–38, find the integral assuming that k is a number between 0 and 4.

**35.** $\int_0^4 (kx + 3) \, dx$  8k + 12     **36.** $\int_0^k (4x + 3) \, dx$  2k² + 3k

**37.** $\int_0^4 (3x + k) \, dx$  24 + 4k     **38.** $\int_k^4 (4x + 3) \, dx$

**39. Writing to Learn** Let $g(x) = -f(x)$ where $f$ has non-negative function values on an interval $[a, b]$. Explain why the area above the graph of $g$ is the same as the area under the graph of $f$ in the same interval.

**40. Writing to Learn** Explain how you can find the area under the graph of $f(x) = \sqrt{16 - x^2}$ from $x = 0$ to $x = 4$ by mental computation only.

**41. Rock Toss** A rock is thrown straight up from level ground. The velocity of the rock at any time $t$ (sec) is $v(t) = 48 - 32t$ ft/sec.

(a) Graph the velocity function.

(b) At what time does the rock reach its maximum height?

(c) Find how far the rock has traveled at its maximum height.

**42. Rocket Launch** A toy rocket is launched straight up from level ground. Its velocity function is $f(t) = 170 - 32t$ feet per second, where $t$ is the number of seconds after launch.

(a) Graph the velocity function.

(b) At what time does the rocket reach its maximum height?

(c) Find how far the rocket has traveled at its maximum height.  ≈ 451.6 ft

**43. Finding Distance Traveled as Area** A ball is pushed off the roof of a three-story building. Table 10.3 gives the velocity (in feet per second) of the falling ball at 0.2-second intervals until it hits the ground 1.4 seconds later.

| Table 10.3  Velocity Data of the Ball | |
|---|---|
| Time | Velocity |
| 0.2 | −5.05 |
| 0.4 | −11.43 |
| 0.6 | −17.46 |
| 0.8 | −24.21 |
| 1.0 | −30.62 |
| 1.2 | −37.06 |
| 1.4 | −43.47 |

(a) Draw a scatter plot of the data.

(b) Find the approximate building height using RRAM areas as in Example 4. Use the fact that if the velocity function is always negative the distance traveled will be the same as if the absolute value of the velocity values were used.  33.86 ft

**44. Work** Work is defined as force times distance. A full water barrel weighing 1250 pounds has a significant leak and must be lifted 35 feet. Table 10.4 displays the weight of the barrel measured after each 5 feet of movement. Find the approximate work in foot-pounds done in lifting the barrel 35 feet.  31,500 ft-pounds

| Table 10.4  Weight of a Leaking Water Barrel | |
|---|---|
| Distance (ft) | Weight (lb) |
| 0 | 1250 |
| 5 | 1150 |
| 10 | 1050 |
| 15 | 950 |
| 20 | 850 |
| 25 | 750 |
| 30 | 650 |

## Explorations

**45. Group Exploration** You may have erroneously assumed that the function $f$ had to be positive in the definition of the definite integral. It is a fact that $\int_0^{2\pi} \sin x \, dx = 0$. Use the definition of the definite integral to explain why this is so. What does this imply about $\int_0^1 (x - 1) \, dx$?

**46. Area Under a Discontinuous Function** Let

$$f(x) = \begin{cases} 1 & x < 2 \\ x & x > 2 \end{cases}.$$

(a) Draw a graph of $f$. Determine its domain and range.

(b) **Writing to Learn** How would you define the area under $f$ from $x = 0$ to $x = 4$? Does it make a difference if the function has no value at $x = 2$?

■

## Extending the Ideas

**Group Activity** From what you know about definite integrals, decide whether each of the following statements is true or false for integrable functions (in general). Work with your classmates to justify your answers.

**47.** $\int_a^b f(x)\,dx + \int_a^b g(x)\,dx = \int_a^b (f(x) + g(x))\,dx$  true

**48.** $\int_a^b 8 \cdot f(x)\,dx = 8 \cdot \int_a^b f(x)\,dx$  true

**49.** $\int_a^b f(x) \cdot g(x)\,dx = \int_a^b f(x)\,dx \cdot \int_a^b g(x)\,dx$  false

**50.** $\int_a^c f(x)\,dx + \int_c^b f(x)\,dx = \int_a^b f(x)\,dx$ for $a < c < b$  true

**51.** $\int_a^b f(x) = \int_b^a f(x)$  false

**52.** $\int_a^a f(x)\,dx = 0$  true

---

## 10.3 More on Limits

A Little History • Defining a Limit Informally • Properties of Limits • Limits of Continuous Functions • One-Sided and Two-Sided Limits • Limits Involving Infinity

### A Little History

Progress in mathematics occurs much as it does in other walks of life, which means gradually and without much fanfare in the early stages. The fanfare occurs much later, after discoveries and innovations have been put into perspective and, quite often, cleaned up a bit from their earlier forms. Calculus is certainly a case in point. Most of the ideas in this chapter pre-dated Newton and Leibniz; indeed, Newton had heard all about them from Isaac Barrow (1630–1677), his predecessor as Lucasian Professor of Mathematics at Cambridge. Fermat, Descartes, John Wallis (1616–1703), James Gregory (1638–1675), Galileo, Kepler, Christiaan Huygens (1629–1695)—even Archimedes of Syracuse (287–212 B.C.)—were all solving calculus problems before calculus was officially "discovered." What Newton and Leibniz did was to develop the rules of the game, so that derivatives and integrals could be computed algebraically and, most importantly, so that the connection between the "tangent line problem" and the "area problem" could be explained. (That connection has come to be called the Fundamental Theorem of Calculus.)

But recall that the methods of both Newton and Leibniz had to depend on "infinitesimal" quantities that were small enough to vanish and yet were not zero. Since the existence of these strange quantities was questioned from the very beginning, there was no shortage of mathematicians who tried to liberate calculus from that dependence. Jean Le Rond d'Alembert (1717–1783) was a strong proponent of replacing infinitesimals with limits (the strategy that would eventually work), but his definition of limit was unfortunately no more precise than our informal definition in Section 10.1. It took another century of tinkering before Karl Weierstrass (1815–1897) and his student Heinrich Eduard Heine (1821–1881) introduced the formal, unassailable definition that is used in our higher mathematics courses today. By that time, although calculus was certainly alive and well, Newton and Leibniz had been dead for over 150 years.

### Defining a Limit Informally

There is nothing difficult about the following limit statements:

- $\lim_{x \to 3} (2x - 1) = 5$

---

**Objective**

Students will be able to use the properties of limits and evaluate one-sided limits, two-sided limits, and limits involving infinity.

**Motivate**

Discuss whether $\lim_{x \to 2} \text{int } x$ exists, where int $x$ represents the greatest integer that is less than or equal to $x$.

**Lesson Guide**

Day 1: A Little History; Defining a Limit Informally; Properties of Limits; Limits of Continuous Functions; One-Sided and Two-Sided Limits
Day 2: Limits Involving Infinity

- $\displaystyle\lim_{x\to\infty} (x^2 + 3) = \infty$

- $\displaystyle\lim_{n\to\infty} \frac{1}{n} = 0$

That is why we have used limit notation throughout this book. Particularly when electronic graphers are available, analyzing the limiting behavior of functions algebraically, numerically, and graphically can tell us much of what we need to know about the functions.

What *is* difficult is to come up with an air-tight definition of what a limit really is. If it had been easy, it would not have taken 150 years. The subtleties of the "epsilon-delta" definiton of Weierstrass and Heine are as beautiful as they are profound, but they are not the stuff of a precalculus course. Therefore, even as we look more closely at limits and their properties in this section, we will continue to refer to our "informal" definiton of limit (essentially that of d'Alembert). We repeat it here for ready reference:

---

**Definition (informal)** Limit at *a*

When we write "$\displaystyle\lim_{x\to a} f(x) = L$," we mean that $f(x)$ gets arbitrarily close to $L$ as $x$ gets arbitrarily close (but not equal) to $a$.

---

**Teaching Note**

For reference, the formal "epsilon-delta" definition of a limit is given below. Note that "$f(x)$ gets arbitrarily close to $L$" means $f(x)$ is within $\epsilon$ units of $L$, and "$x$ gets arbitrarily close to $a$" means that $x$ is within $\delta$ units of $a$.

$f$ has a limit $L$ as $x$ approaches $a$ (that is, $\displaystyle\lim_{x\to a} f(x) = L$) if and only if for any $\epsilon > 0$ there exists a number $\delta > 0$ such that, whenever $|x - a| < \delta$, we have $|f(x) - L| < \epsilon$.

**Exploration Extensions**

Discuss the following limit statement and decide whether it is true:
$\displaystyle\lim_{x\to 0} x \sin (1/x) = 0.$

---

**Exploration 1**  **What's the Limit?**

As a class, discuss the following two limit statements until you really understand why they are true. Look at them every way you can. Use your calculators. Do you see how the above definition verifies that they are true? In particular, can you defend your position against the challenges that follow the statements? (This exploration is intended to be free-wheeling and philosophical. You can't *prove* these statements without a stronger definition.)

1. $\displaystyle\lim_{x\to 2} 7x \neq 14.000000000000000001$

   Challenges:

   - Isn't $7x$ getting "arbitrarily close" to that number as $x$ approaches 2?

   - How can you tell that 14 is the limit and 14.000000000000000001 is not?

2. $\displaystyle\lim_{x\to 0} \frac{x^2 + 2x}{x} = 2$

   Challenges:

   - How can the limit be 2 when the quotient isn't even defined at 0?

   - Won't there be an asymptote at $x = 0$? The denominator equals 0 there.

   - How can you *tell* that 2 is the limit and 1.99999999999999999999 is not?

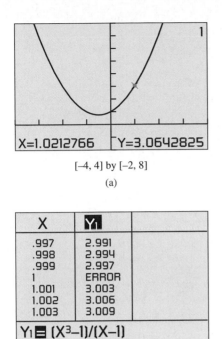

[−4, 4] by [−2, 8]

(a)

| X | Y₁ |
|---|---|
| .997 | 2.991 |
| .998 | 2.994 |
| .999 | 2.997 |
| 1 | ERROR |
| 1.001 | 3.003 |
| 1.002 | 3.006 |
| 1.003 | 3.009 |

Y₁ ▤ (X³−1)/(X−1)

(b)

**Figure 10.11** A graph and table of values for $f(x) = (x^3 - 1)/(x - 1)$. (Example 1)

### Example 1   FINDING A LIMIT

Find $\lim\limits_{x \to 1} \dfrac{x^3 - 1}{x - 1}$.

**Solution**   The graph of $f(x) = (x^3 - 1)/(x - 1)$ in Figure 10.11a suggests that the limit exists and is about 3. The table in Figure 10.11b also gives compelling evidence that the limit is 3. As convincing as the graphical and numerical evidence is, however, the best evidence is algebraic:

$$\lim_{x \to 1} \frac{x^3 - 1}{x - 1} = \lim_{x \to 1} \frac{(x - 1)(x^2 + x + 1)}{x - 1} \qquad \text{Factor}$$

$$= \lim_{x \to 1} (x^2 + x + 1) \qquad \text{Since } x \neq 1, \frac{x - 1}{x - 1} = 1$$

$$= 1 + 1 + 1$$

$$= 3$$

## Properties of Limits

When limits exist, there is nothing unusual about the way they interact algebraically with each other. You could easily predict that the following properties would hold. These are all theorems that one could prove with a rigorous definition of limit, but we must state them without proof here.

---

**Properties of Limits**

If $\lim\limits_{x \to c} f(x)$ and $\lim\limits_{x \to c} g(x)$ both exist, then

1. *Sum Rule:*   $\lim\limits_{x \to c} (f(x) + g(x)) = \lim\limits_{x \to c} f(x) + \lim\limits_{x \to c} g(x)$

2. *Difference Rule:*   $\lim\limits_{x \to c} (f(x) - g(x)) = \lim\limits_{x \to c} f(x) - \lim\limits_{x \to c} g(x)$

3. *Product Rule:*   $\lim\limits_{x \to c} (f(x) \cdot g(x)) = \lim\limits_{x \to c} f(x) \cdot \lim\limits_{x \to c} g(x)$

4. *Constant Multiple Rule:*   $\lim\limits_{x \to c} (k \cdot g(x)) = k \cdot \lim\limits_{x \to c} g(x)$

5. *Quotient Rule:*   $\lim\limits_{x \to c} \dfrac{f(x)}{g(x)} = \dfrac{\lim\limits_{x \to c} f(x)}{\lim\limits_{x \to c} g(x)},$

   provided $\lim\limits_{x \to c} g(x) \neq 0$

6. *Power Rule:*   $\lim\limits_{x \to c} (f(x))^n = (\lim\limits_{x \to c} f(x))^n$ for $n$
   a positive integer

7. *Root Rule:*   $\lim\limits_{x \to c} \sqrt[n]{f(x)} = \sqrt[n]{\lim\limits_{x \to c} f(x)}$ for $n \geq 2$
   a positive integer, provided $\sqrt[n]{\lim\limits_{x \to c} f(x)}$ is
   a real number.

---

### Example 2   USING THE LIMIT PROPERTIES

You will learn in Example 11 that $\lim\limits_{x \to 0} \dfrac{\sin x}{x} = 1$. Use this fact, along with the limit properties, to find the following limits:

**(a)** $\lim\limits_{x \to 0} \dfrac{x + \sin x}{x}$  **(b)** $\lim\limits_{x \to 0} \dfrac{1 - \cos^2 x}{x^2}$  **(c)** $\lim\limits_{x \to 0} \dfrac{\sqrt[3]{\sin x}}{\sqrt[3]{x}}$

**Solution**

**(a)** $\lim\limits_{x \to 0} \dfrac{x + \sin x}{x} = \lim\limits_{x \to 0} \left( \dfrac{x}{x} + \dfrac{\sin x}{x} \right)$

$\qquad\qquad = \lim\limits_{x \to 0} \dfrac{x}{x} + \lim\limits_{x \to 0} \dfrac{\sin x}{x}$  Sum Rule

$\qquad\qquad = 1 + 1$

$\qquad\qquad = 2$

**(b)** $\lim\limits_{x \to 0} \dfrac{1 - \cos^2 x}{x^2} = \lim\limits_{x \to 0} \dfrac{\sin^2 x}{x^2}$  Pythagorean identity

$\qquad\qquad = \lim\limits_{x \to 0} \left( \dfrac{\sin x}{x} \right)\left( \dfrac{\sin x}{x} \right)$

$\qquad\qquad = \lim\limits_{x \to 0} \left( \dfrac{\sin x}{x} \right) \cdot \lim\limits_{x \to 0} \left( \dfrac{\sin x}{x} \right)$  Product Rule

$\qquad\qquad = 1 \cdot 1$

$\qquad\qquad = 1$

**(c)** $\lim\limits_{x \to 0} \dfrac{\sqrt[3]{\sin x}}{\sqrt[3]{x}} = \lim\limits_{x \to 0} \sqrt[3]{\dfrac{\sin x}{x}}$

$\qquad\qquad = \sqrt[3]{\lim\limits_{x \to 0} \dfrac{\sin x}{x}}$  Root Rule

$\qquad\qquad = \sqrt[3]{1}$

$\qquad\qquad = 1$

## Limits of Continuous Functions

Recall from Section 1.2 that a function is continuous at $a$ if $\lim\limits_{x \to a} f(x) = f(a)$.

This means that the limit (at $a$) of a function can be found by "plugging in $a$" provided the function is continuous at $a$. (The condition of continuity is essential when employing this strategy. For example, plugging in 0 does not work on any of the limits in Example 2.)

**Example 3**  FINDING LIMITS BY SUBSTITUTION

Find the limits.

**(a)** $\lim\limits_{x \to 0} \dfrac{e^x - \tan x}{\cos^2 x}$  **(b)** $\lim\limits_{n \to 16} \dfrac{\sqrt{n}}{\log_2 n}$

**Solution**

You might not recognize these functions as being continuous, but you can use the limit properties to write the limits in terms of limits of basic functions.

**(a)** $\lim\limits_{x \to 0} \dfrac{e^x - \tan x}{\cos^2 x} = \dfrac{\lim\limits_{x \to 0} (e^x - \tan x)}{\lim\limits_{x \to 0} (\cos^2 x)}$     Quotient Rule

$$= \dfrac{\lim\limits_{x \to 0} e^x - \lim\limits_{x \to 0} \tan x}{(\lim\limits_{x \to 0} \cos x)^2}$$     Difference and Power Rules

$$= \dfrac{e^0 - \tan 0}{(\cos 0)^2}$$     Limits of continuous functions

$$= \dfrac{1 - 0}{1}$$

$$= 1$$

**(b)** $\lim\limits_{n \to 16} \dfrac{\sqrt{n}}{\log_2 n} = \dfrac{\lim\limits_{n \to 16} \sqrt{n}}{\lim\limits_{n \to 16} \log_2 n}$     Quotient Rule

$$= \dfrac{\sqrt{16}}{\log_2 16}$$     Limits of continuous functions

$$= \dfrac{4}{4}$$

$$= 1$$

**Notes on Examples**

The functions in Example 3 are in fact continuous at the limit points, so we could have found the limits using direct substitution instead of using the method shown.

Example 3 hints at some important properties of continuous functions that follow from the properties of limits. If $f$ and $g$ are both continuous at $x = a$, then so are $f + g, f - g, fg,$ and $f/g$ (with the assumption that $g(a)$ does not create a zero denominator in the quotient). Also, the $n$th power and $n$th root of a function that is continuous at $a$ will also be continuous at $a$ (with the assumption that $\sqrt{f(a)}$ is real).

## One-sided and Two-sided Limits

We can see that the limit of the function in Figure 10.11 is 3 whether $x$ approaches 1 from the left or right. Sometimes the values of a function $f$ can approach different values as $x$ approaches a number $c$ from opposite sides. When this happens, the limit of $f$ as $x$ approaches $c$ from the left is the **left-hand limit** of $f$ at $c$ and the limit of $f$ as $x$ approaches $c$ from the right is the **right-hand limit** of $f$ at $c$. Here is the notation we use:

left-hand:     $\lim\limits_{x \to c^-} f(x)$     *The limit of f as x approaches c from the left.*

right-hand:     $\lim\limits_{x \to c^+} f(x)$     *The limit of f as x approaches c from the right.*

**Example 4**   FINDING LEFT- AND RIGHT-HAND LIMITS

Find $\lim\limits_{x \to 2} f(x)$ where $f(x) = \begin{cases} -x^2 + 4x - 1 & x \le 2 \\ 2x - 3 & x > 2 \end{cases}$.

**Solution**   Figure 10.12 suggests that the left- and right-hand limits of $f$ exist but are not equal. Using algebra we find:

$$\lim\limits_{x \to 2^-} f(x) = \lim\limits_{x \to 2^-} (-x^2 + 4x - 1)$$     Definition of f.

$$= -2^2 + 4 \cdot 2 - 1$$

$$= 3$$

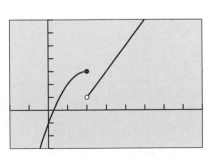

[−2, 8] by [−3, 7]

**Figure 10.12** A graph of the piecewise-defined function.

$f(x) = \begin{cases} -x^2 + 4x - 1 & x \le 2 \\ 2x - 3 & x > 2 \end{cases}$.

(Example 4)

$$\lim_{x\to2^+} f(x) = \lim_{x\to2^+} (2x - 3) \qquad \text{Definition of } f.$$
$$= 2 \cdot 2 - 3$$
$$= 1$$

You can use trace or tables to support the above results.

The limit $\lim_{x\to c} f(x)$ is sometimes called the **two-sided limit** of $f$ at $c$ to distinguish it from the *one-sided* left-hand and right-hand limits of $f$ at $c$. The following theorem indicates how these limits are related.

---

**Theorem   One-sided and Two-sided Limits**

A function $f(x)$ has a limit as $x$ approaches $c$ if and only if the left-hand and right-hand limits at $c$ exist and are equal. That is,

$$\lim_{x\to c} f(x) = L \iff \lim_{x\to c^-} f(x) = L \text{ and } \lim_{x\to c^+} f(x) = L.$$

---

The limit of the function $f$ of Example 4 as $x$ approaches 2 does not exist, so $f$ is discontinuous at $x = 2$. However, discontinuous functions can have a limit at a point of discontinuity. The function $f$ of Example 1 is discontinuous at $x = 1$ because $f(1)$ does not exist, but it has the limit 3 as $x$ approaches 1. Example 5 illustrates another way a function can have a limit and still be discontinuous.

### Example 5   FINDING A LIMIT AT A POINT OF DISCONTINUITY

Let

$$f(x) = \begin{cases} \dfrac{x^2 - 9}{x - 3} & x \neq 3 \\ 2 & x = 3 \end{cases}.$$

Find $\lim_{x\to3} f(x)$ and prove that $f$ is discontinuous at $x = 3$.

**Solution**   Figure 10.13 suggests that the limit of $f$ as $x$ approaches 3 exists. Using algebra we find

$$\lim_{x\to3} \frac{x^2 - 9}{x - 3} = \lim_{x\to3} \frac{(x - 3)(x + 3)}{x - 3}$$
$$= \lim_{x\to3} (x + 3) \qquad \text{We can assume } x \neq 3.$$
$$= 6.$$

Because $f(3) = 2 \neq \lim_{x\to3} f(x)$, $f$ is discontinuous at $x = 3$.

### Example 6   FINDING ONE-SIDED AND TWO-SIDED LIMITS

Let $f(x) = \text{int}(x)$ (the greatest integer function). Find:

**(a)** $\lim_{x\to2^-} \text{int}(x)$ **(b)** $\lim_{x\to2^+} \text{int}(x)$ **(c)** $\lim_{x\to2} \text{int}(x)$

**Solution**   Recall that int $(x)$ is equal to *the greatest integer less than or equal to x*. For example, int$(2) = 2$. From the definition of $f$ and its graph in Figure 10.14 we can see that

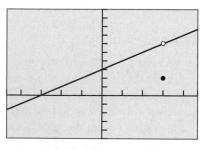

[–4.7, 4.7] by [–5, 10]

**Figure 10.13**  A graph of the function in Example 5.

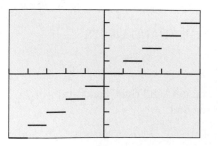

[–5, 5] by [–5, 5]

**Figure 10.14**  The graph of $f(x) = \text{int}(x)$. (Example 6)

**(a)** $\lim\limits_{x \to 2^-} \text{int}(x) = 1$

**(b)** $\lim\limits_{x \to 2^+} \text{int}(x) = 2$

**(c)** $\lim\limits_{x \to 2} \text{int}(x)$ does not exist.

## Limits Involving Infinity

The informal definition that we have for a limit refers to $\lim\limits_{x \to a} f(x) = L$ where both $a$ and $L$ are real numbers. In Section 10.2 we adapted the definition to apply to limits of the form $\lim\limits_{x \to \infty} f(x) = L$ so that we could use this notation in describing definite integrals. This is one type of "limit at infinity." Notice that the limit itself ($L$) is a finite real number, assuming the limit exists, but that the values of $x$ are approaching infinity.

> ### Definition  Limits at Infinity
>
> When we write "$\lim\limits_{x \to \infty} f(x) = L$," we mean that $f(x)$ gets arbitrarily close to $L$ as $x$ gets arbitrarily large. We say that $f$ **has a limit $L$ as $x$ approaches $\infty$**.
>
> When we write "$\lim\limits_{x \to -\infty} f(x) = L$," we mean that $f(x)$ gets arbitrarily close to $L$ as $-x$ gets arbitrarily large. We say that $f$ **has a limit $L$ as $x$ approaches $-\infty$**.

Notice that limits, whether at $a$ or at infinity, are always finite real numbers; otherwise, the limits do not exist. For example, it is correct to write

$$\lim\limits_{x \to 0} \frac{1}{x^2} \text{ does not exist,}$$

since it approaches no real number $L$. In this case, however, it is also convenient to write

$$\lim\limits_{x \to 0} \frac{1}{x^2} = \infty,$$

which gives us a little more information about *why* the limit fails to exist. (It increases without bound.) Similarly, it is convenient to write

$$\lim\limits_{x \to 0^+} \ln x = -\infty,$$

since $\ln x$ decreases without bound as $x$ approaches $0$ from the right. In this context, the symbols "$\infty$" and "$-\infty$" are sometimes called **infinite limits**.

### Example 7  INVESTIGATING LIMITS AS $x \to \pm \infty$

Let $f(x) = (\sin x)/x$. Find $\lim\limits_{x \to \infty} f(x)$ and $\lim\limits_{x \to -\infty} f(x)$.

**Solution**  The graph of $f$ in Figure 10.15 suggests that

$$\lim\limits_{x \to \infty} \frac{\sin x}{x} = \lim\limits_{x \to -\infty} \frac{\sin x}{x} = 0.$$

In Section 1.3, we used limits to describe the *unbounded behavior* of the function $f(x) = x^3$ as $x \to \pm \infty$:

$$\lim\limits_{x \to \infty} x^3 = \infty \quad \text{and} \quad \lim\limits_{x \to -\infty} x^3 = -\infty$$

---

### Notes on Examples

Examples 5 and 6 demonstrate that it is not necessarily true that $\lim\limits_{x \to c} f(x) = f(c)$. Make certain that students understand the role of continuity in evaluating limits.

### Infinite Limits are not Limits

It is important to realize that *an infinite limit is not a limit*, despite what the name might imply. It describes a special case of a limit that does not exist. Recall that a saw horse is not a horse and a badminton bird is not a bird.

### Archimedes (287–212 B.C.)

The Greek mathematician Archimedes found the area of a circle using a method involving infinite limits. See Exercise 79 for a modern version of his method.

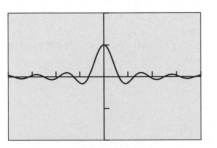

[–20, 20] by [–2, 2]

**Figure 10.15**  The graph of $f(x) = (\sin x)/x$. (Example 7)

The behavior of the function $g(x) = e^x$ as $x \to \pm\infty$ can be described by the following two limits:

$$\lim_{x \to \infty} e^x = \infty \quad \text{and} \quad \lim_{x \to -\infty} e^x = 0$$

The function $g(x) = e^x$ has unbounded behavior as $x \to \infty$ and has a finite limit as $x \to -\infty$.

### Example 8   USING TABLES TO INVESTIGATE LIMITS AS $x \to \pm\infty$

Let $f(x) = xe^{-x}$. Find $\lim\limits_{x \to \infty} f(x)$ and $\lim\limits_{x \to -\infty} f(x)$.

**Solution**   The tables in Figure 10.16 suggest that

$$\lim_{x \to \infty} xe^{-x} = 0 \quad \text{and} \quad \lim_{x \to -\infty} xe^{-x} = -\infty.$$

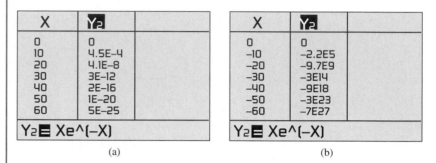

(a)                                        (b)

**Figure 10.16**   The table in (a) suggests that the values of $f(x) = xe^{-x}$ approach 0 as $x \to \infty$ and the table in (b) suggests that the values of $f(x) = xe^{-x}$ approach $-\infty$ as $x \to -\infty$. (Example 8)

The graph of $f$ in Figure 10.17 supports these results.

### Example 9   INVESTIGATING LIMITS AS $x \to \pm\infty$

Find $\lim\limits_{x \to \infty} f(x)$ and $\lim\limits_{x \to -\infty} f(x)$.

**(a)** $f(x) = \tan^{-1} x$                    **(b)** $f(x) = \sin x$

**Solution**

**(a)** We know from our work in trigonometry that

$$\lim_{x \to \infty} \tan^{-1} x = \frac{\pi}{2} \quad \text{and} \quad \lim_{x \to -\infty} \tan^{-1} x = -\frac{\pi}{2}.$$

Figure 10.18 supports these conclusions.

**(b)** The values of $f(x) = \sin x$ vary between $-1$ and 1 in every interval of length $2\pi$. Thus, the two limits do not exist.

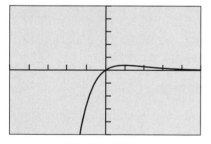

[–5, 5] by [–5, 5]

(a)

**Figure 10.17**   The graph of the function $f(x) = xe^{-x}$. (Example 8)

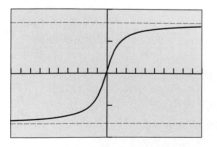

[–10, 10] by [–2, 2]

**Figure 10.18**   The graph of $f(x) = \tan^{-1} x$ with its two horizontal asymptotes overlaid. (Example 9a)

---

**Exploration 2** **Investigating a Logistic Function**

Let $f(x) = \dfrac{50}{1 + 2^{3-x}}$.

1. Use tables and graphs to find $\lim\limits_{x \to \infty} f(x)$ and $\lim\limits_{x \to -\infty} f(x)$.  50; 0
2. Identify any horizontal asymptotes.  $y = 50$; $y = 0$
3. How is the numerator of the fraction for *f* related to part 2?

---

In Section 2.7 we used the graph of $f(x) = 2/(x + 3)$ to state that

$$\lim_{x \to -3^+} \frac{2}{x + 3} = \infty \quad \text{and} \quad \lim_{x \to -3^-} \frac{2}{x + 3} = -\infty.$$

Either one of these unbounded limits allows us to conclude that the vertical line $x = -3$ is a vertical asymptote of the graph of *f* (Figure 10.19).

### Example 10 INVESTIGATING UNBOUNDED LIMITS

Find $\lim\limits_{x \to 2} 1/(x - 2)^2$.

**Solution** The graph of $f(x) = 1/(x - 2)^2$ in Figure 10.20 suggests that

$$\lim_{x \to 2^-} \frac{1}{(x - 2)^2} = \infty \quad \text{and} \quad \lim_{x \to 2^+} \frac{1}{(x - 2)^2} = \infty.$$

This means that the limit of *f* as *x* approaches 2 does not exist. The graph of *f* has a vertical asymptote at $x = 2$.

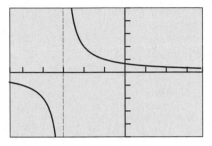

[–5.7, 3.7] by [–5, 5]

**Figure 10.19** The graph of $f(x) = 2/(x + 3)$ with its vertical asymptote overlaid.

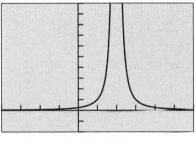

[–4, 6] by [–2, 10]

**Figure 10.20** The graph of $f(x) = 1/(x - 2)^2$ in Example 10.

Not all zeros of denominators correspond to vertical asymptotes as illustrated in Examples 5–7.

### Example 11 INVESTIGATING A LIMIT AT *x* = 0

Find $\lim\limits_{x \to 0} (\sin x)/x$.

**Solution** The graph of $f(x) = (\sin x)/x$ in Figure 10.15 suggests this limit exists. The table of values in Figure 10.21 suggest that

$$\lim_{x \to 0} \frac{\sin x}{x} = 1.$$

| X | Y₁ | |
|------|--------|---|
| –.03 | .99985 | |
| –.02 | .99993 | |
| –.01 | .99998 | |
| 0 | ERROR | |
| .01 | .99998 | |
| .02 | .99993 | |
| .03 | .99985 | |

Y₁ ◘ sin(X)/X

**Figure 10.21** A table of values for $f(x) = (\sin x)/x$. (Example 11)

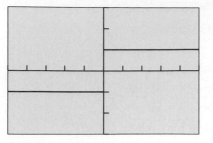

[–5, 5] by [–3, 3]

**Figure 10.22** The graph of the function $f(x) = |x|/x$. (Example 12)

**Notes on Examples**

Encourage students to view graphs of the function in Example 13 using smaller and smaller windows, changing xmin and xmax only. This should help convince students that the limit does not exist.

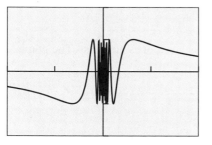

[–2, 2] by [–2, 2]

**Figure 10.23** The graph of the function $f(x) = \sin(1/x)$. (Example 13)

### Example 12   INVESTIGATING A LIMIT AT $x = 0$

Find $\lim\limits_{x \to 0} (|x|/x)$.

**Solution**   The graph of $f(x) = |x|/x$ in Figure 10.22 suggests that

$$\lim_{x \to 0^-} \frac{|x|}{x} = -1 \quad \text{and} \quad \lim_{x \to 0^+} \frac{|x|}{x} = 1.$$

We can use the fact that

$$|x| = \begin{cases} -x & x < 0 \\ x & x > 0 \end{cases},$$

to confirm these results:

$$\lim_{x \to 0^-} \frac{|x|}{x} = \frac{-x}{x} = -1 \quad \text{and} \quad \lim_{x \to 0^+} \frac{|x|}{x} = \frac{x}{x} = 1.$$

Thus, $\lim\limits_{x \to 0} (|x|/x)$ does not exist.

The function in Example 13 has surprising behavior near $x = 0$.

### Example 13   INVESTIGATING A LIMIT AT $x = 0$

Find $\lim\limits_{x \to 0} \sin(1/x)$.

**Solution**   The graph of $f(x) = \sin(1/x)$ in Figure 10.23 suggests that this limit does not exist. In fact, it can be shown that in every interval centered at $x = 0$ the function $f$ takes on all values between $-1$ and $1$. For example, if

$$x = \frac{2}{\pi}, \frac{2}{3\pi}, \frac{2}{5\pi}, \frac{2}{7\pi}, \ldots, \text{then}$$

$$\sin\left(\frac{1}{x}\right) = 1, -1, 1, -1, \ldots; \text{ and if}$$

$$x = -\frac{2}{\pi}, -\frac{2}{3\pi}, -\frac{2}{5\pi}, -\frac{2}{7\pi}, \ldots, \text{then}$$

$$\sin\left(\frac{1}{x}\right) = -1, 1, -1, 1, \ldots.$$

This is more than enough to show that the limit does not exist.

## Quick Review 10.3

In Exercises 1 and 2, find **(a)** $f(-2)$, **(b)** $f(0)$, and **(c)** $f(2)$.

**1.** $f(x) = \dfrac{2x + 1}{(2x - 4)^2}$       **2.** $f(x) = \dfrac{\sin x}{x}$

In Exercises 3 and 4, find the **(a)** vertical asymptotes and **(b)** horizontal asymptotes of the graph of $f$, if any.

**3.** $f(x) = \dfrac{2x^2 + 3}{x^2 - 4}$       **4.** $f(x) = \dfrac{x^3 + 1}{2 - x - x^2}$

In Exercises 5 and 6, the end behavior asymptote of the function $f$ is one of the following. Which one is it?

**(a)** $y = 2x^2$    **(b)** $y = -2x^2$    **(c)** $y = x^3$    **(d)** $y = -x^3$

**5.** $f(x) = \dfrac{2x^3 - 3x^2 + 1}{3 - x}$  (b)   **6.** $f(x) = \dfrac{x^4 + 2x^2 + x + 1}{x - 3}$

In Exercises 7 and 8, find **(a)** the points of continuity and **(b)** the points of discontinuity of the function.

**7.** $f(x) = \sqrt{x + 2}$       **8.** $g(x) = \dfrac{2x + 1}{x^2 - 4}$

Exercises 9 and 10 refer to the piecewise-defined function

$$f(x) = \begin{cases} 3x + 1 & x \le 1 \\ 4 - x^2 & x > 1 \end{cases}.$$

**9.** Draw the graph of $f$.

**10.** Find the points of continuity and the points of discontinuity of $f$.

# Section 10.3 Exercises

In Exercises 1–10, find the limit by direct substitution if it exists.

**1.** $\lim\limits_{x \to -1} x\,(x-1)^2$  $-4$

**2.** $\lim\limits_{x \to 3} (x-1)^{12}$  4096

**3.** $\lim\limits_{x \to 2} (x^3 - 2x + 3)$  7

**4.** $\lim\limits_{x \to -2} (x^3 - x + 5)$  $-1$

**5.** $\lim\limits_{x \to 2} \sqrt{x+5}$  $\sqrt{7}$

**6.** $\lim\limits_{x \to -2} (x-4)^{2/3}$  $\approx 3.30$

**7.** $\lim\limits_{x \to 0} (e^x \sin x)$  0

**8.** $\lim\limits_{x \to \pi} \ln\left(\sin \dfrac{x}{2}\right)$  0

**9.** $\lim\limits_{x \to a} (x^2 - 2)$  $a^2 - 2$

**10.** $\lim\limits_{x \to a} \dfrac{x^2 - 1}{x^2 + 1}$  $\dfrac{a^2 - 1}{a^2 + 1}$

In Exercises 11–18, **(a)** explain why you cannot use substitution to find the limit and **(b)** find the limit algebraically if it exists.

**11.** $\lim\limits_{x \to -3} \dfrac{x^2 + 7x + 12}{x^2 - 9}$

**12.** $\lim\limits_{x \to 3} \dfrac{x^2 - 9}{x^2 + 2x - 15}$

**13.** $\lim\limits_{x \to -1} \dfrac{x^3 + 1}{x + 1}$

**14.** $\lim\limits_{x \to 2} \dfrac{x^3 - 2x^2 + x - 2}{x - 2}$

**15.** $\lim\limits_{x \to -2} \dfrac{x^2 - 4}{x + 2}$

**16.** $\lim\limits_{x \to -2} \dfrac{|x^2 - 4|}{x + 2}$

**17.** $\lim\limits_{x \to 0} \sqrt{x} - 3$

**18.** $\lim\limits_{x \to 0} \dfrac{x - 2}{x^2}$

In Exercises 19–22, use the given graph to find the limits or to explain why the limits do not exist.

**19. (a)** $\lim\limits_{x \to 2^-} f(x)$  3  **(b)** $\lim\limits_{x \to 2^+} f(x)$  1  **(c)** $\lim\limits_{x \to 2} f(x)$  none

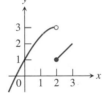

**20. (a)** $\lim\limits_{x \to 3^-} f(x)$  2  **(b)** $\lim\limits_{x \to 3^+} f(x)$  4  **(c)** $\lim\limits_{x \to 3} f(x)$  none

**21. (a)** $\lim\limits_{x \to 3^-} f(x)$  4  **(b)** $\lim\limits_{x \to 3^+} f(x)$  4  **(c)** $\lim\limits_{x \to 3} f(x)$  4

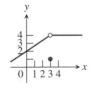

**22. (a)** $\lim\limits_{x \to 1^-} f(x)$  1  **(b)** $\lim\limits_{x \to 1^+} f(x)$  3  **(c)** $\lim\limits_{x \to 1} f(x)$  none

In Exercises 23 and 24, the graph of a function $y = f(x)$ is given. Which of the statements about the function are true and which are false?

**23. (a)** $\lim\limits_{x \to -1^+} f(x) = 1$  true  **(b)** $\lim\limits_{x \to 0^-} f(x) = 0$  true

**(c)** $\lim\limits_{x \to 0^-} f(x) = 1$  false  **(d)** $\lim\limits_{x \to 0^-} f(x) = \lim\limits_{x \to 0^+} f(x)$

**(e)** $\lim\limits_{x \to 0} f(x)$ exists  false  **(f)** $\lim\limits_{x \to 0} f(x) = 0$  false

**(g)** $\lim\limits_{x \to 0} f(x) = 1$  false  **(h)** $\lim\limits_{x \to 1} f(x) = 1$  true

**(i)** $\lim\limits_{x \to 1} f(x) = 0$  false  **(j)** $\lim\limits_{x \to 2^-} f(x) = 2$  true

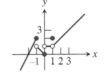

**24. (a)** $\lim\limits_{x \to -1^+} f(x) = 1$  true  **(b)** $\lim\limits_{x \to 2} f(x)$ does not exist

**(c)** $\lim\limits_{x \to 2} f(x) = 2$  false  **(d)** $\lim\limits_{x \to 1^-} f(x) = 2$  true

**(e)** $\lim\limits_{x \to 1^+} f(x) = 1$  **(f)** $\lim\limits_{x \to 1} f(x)$ does not exist

**(g)** $\lim\limits_{x \to 0^+} f(x) = \lim\limits_{x \to 0^-} f(x)$  false

**(h)** $\lim\limits_{x \to c} f(x)$ exists for every $c$ in $(-1, 1)$.  false (not $x = 0$)

**(i)** $\lim\limits_{x \to c} f(x)$ exists for every $c$ in $(1, 3)$.  true

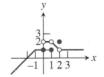

In Exercises 25 and 26, use a graph of $f$ to find **(a)** $\lim\limits_{x \to 0^-} f(x)$, **(b)** $\lim\limits_{x \to 0^+} f(x)$, and **(c)** $\lim\limits_{x \to 0} f(x)$ if they exist.

**25.** $f(x) = (1 + x)^{1/x}$

**26.** $f(x) = (1 + x)^{1/(2x)}$

In Exercises 27–32, find the limit.

**27.** $\lim\limits_{x \to 2^+} \text{int } x$  2

**28.** $\lim\limits_{x \to 2^-} \text{int } x$  1

**29.** $\lim\limits_{x \to 0.0001} \text{int } x$  0

**30.** $\lim\limits_{x \to 5/2^-} \text{int } (2x)$  4

**31.** $\lim\limits_{x \to -3^+} \dfrac{x + 3}{|x + 3|}$  1

**32.** $\lim\limits_{x \to 0^-} \dfrac{5x}{|2x|}$  $-\dfrac{5}{2}$

**33. Group Activity** Assume that $\lim_{x \to 4} f(x) = -1$ and $\lim_{x \to 4} g(x) = 4$. Find the limit.

(a) $\lim_{x \to 4} (g(x) + 2)$   6     (b) $\lim_{x \to 4} xf(x)$   $-4$

(c) $\lim_{x \to 4} g^2(x)$   16     (d) $\lim_{x \to 4} \dfrac{g(x)}{f(x) - 1}$   $-2$

**34. Group Activity** Assume that $\lim_{x \to a} f(x) = 2$ and $\lim_{x \to a} g(x) = -3$. Find the limit.

(a) $\lim_{x \to a} (f(x) + g(x))$   $-1$     (b) $\lim_{x \to a} (f(x) \cdot g(x))$   $-6$

(c) $\lim_{x \to a} (3g(x) + 1)$   $-8$     (d) $\lim_{x \to a} \dfrac{f(x)}{g(x)}$   $-\dfrac{2}{3}$

In Exercises 35–38, complete the following for the given piecewise-defined function $f$.

(a) Draw the graph of $f$.

(b) Determine $\lim_{x \to a^+} f(x)$ and $\lim_{x \to a^-} f(x)$.

(c) **Writing to Learn** Does $\lim_{x \to a} f(x)$ exist? If it does, give its value. If it does not exist, give an explanation.

**35.** $a = 1, f(x) = \begin{cases} 2 - x & x < 1 \\ x + 1 & x \geq 1 \end{cases}$

**36.** $a = 2, f(x) = \begin{cases} 2 - x & x < 2 \\ 1 & x = 2 \\ x^2 - 4 & x > 2 \end{cases}$

**37.** $a = 0, f(x) = \begin{cases} |x - 3| & x < 0 \\ x^2 - 2x & x \geq 0 \end{cases}$

**38.** $a = -3, f(x) = \begin{cases} 1 - x^2 & x \geq -3 \\ 8 - x & x < -3 \end{cases}$

In Exercises 39–44, use graphs and tables to find the limit and identify any vertical asymptotes.

**39.** $\lim_{x \to 3^-} \dfrac{1}{x - 3}$   $-\infty; x = 3$     **40.** $\lim_{x \to 3^+} \dfrac{x}{x - 3}$   $\infty; x = 3$

**41.** $\lim_{x \to -2^+} \dfrac{1}{x + 2}$   $\infty; x = -2$     **42.** $\lim_{x \to -2^-} \dfrac{x}{x + 2}$   $\infty; x = -2$

**43.** $\lim_{x \to 2} \dfrac{1}{(x - 2)^2}$   $\infty; x = 2$     **44.** $\lim_{x \to 2} \dfrac{1}{x^2 - 4}$

In Exercises 45–52, determine the limit algebraically if possible. Support your answer graphically.

**45.** $\lim_{x \to 0} \dfrac{(1 + x)^3 - 1}{x}$   3     **46.** $\lim_{x \to 0} \dfrac{1/(3 + x) - 1/3}{x}$   $-\dfrac{1}{9}$

**47.** $\lim_{x \to 0} \dfrac{\sin x}{2x^2 - x}$   $-1$     **48.** $\lim_{x \to 0} \dfrac{\sin 3x}{x}$   3

**49.** $\lim_{x \to 0} \dfrac{\sin^2 x}{x}$   0     **50.** $\lim_{x \to 0} \dfrac{x + \sin x}{2x}$   1

**51.** $\lim_{x \to 0} \dfrac{\tan x}{x}$   1     **52.** $\lim_{x \to 2} \dfrac{x - 4}{x^2 - 4}$   none

In Exercises 53–56, match the function $y = f(x)$ with the table.

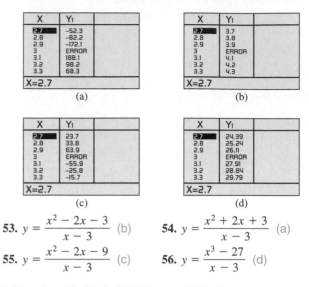

| X | Y1 |
|---|---|
| **2.7** | -52.3 |
| 2.8 | -82.2 |
| 2.9 | -172.1 |
| 3 | ERROR |
| 3.1 | 188.1 |
| 3.2 | 98.2 |
| 3.3 | 68.3 |

X=2.7

(a)

| X | Y1 |
|---|---|
| **2.7** | 3.7 |
| 2.8 | 3.8 |
| 2.9 | 3.9 |
| 3 | ERROR |
| 3.1 | 4.1 |
| 3.2 | 4.2 |
| 3.3 | 4.3 |

X=2.7

(b)

| X | Y1 |
|---|---|
| **2.7** | 23.7 |
| 2.8 | 33.8 |
| 2.9 | 63.9 |
| 3 | ERROR |
| 3.1 | -55.9 |
| 3.2 | -25.8 |
| 3.3 | -15.7 |

X=2.7

(c)

| X | Y1 |
|---|---|
| **2.7** | 24.39 |
| 2.8 | 25.24 |
| 2.9 | 26.11 |
| 3 | ERROR |
| 3.1 | 27.91 |
| 3.2 | 28.84 |
| 3.3 | 29.79 |

X=2.7

(d)

**53.** $y = \dfrac{x^2 - 2x - 3}{x - 3}$   (b)     **54.** $y = \dfrac{x^2 + 2x + 3}{x - 3}$   (a)

**55.** $y = \dfrac{x^2 - 2x - 9}{x - 3}$   (c)     **56.** $y = \dfrac{x^3 - 27}{x - 3}$   (d)

In Exercises 57–60, find (a) $\lim_{x \to \infty} y$ and (b) $\lim_{x \to -\infty} y$.

**57.** $y = \dfrac{\cos x}{1 + x}$   0; 0     **58.** $y = \dfrac{x + \sin x}{x}$   1; 1

**59.** $y = 1 + 2^x$   $\infty$; 1     **60.** $y = \dfrac{x}{1 + 2^x}$   0; $-\infty$

In Exercises 61–64, find the limit.

**61.** $\lim_{x \to 1} \dfrac{x^2 + 1}{x - 1}$   undefined     **62.** $\lim_{x \to \infty} \dfrac{\ln x^2}{\ln x}$   2

**63.** $\lim_{x \to \infty} \dfrac{\ln x}{\ln x^2}$   $\dfrac{1}{2}$     **64.** $\lim_{x \to \infty} 3^{-x}$   0

## Explorations

In Exercises 65–68, complete the following for the given piecewise-defined function $f$.

(a) Draw the graph of $f$.

(b) At what points $c$ in the domain of $f$ does $\lim_{x \to c} f(x)$ exist?

(c) At what points $c$ does only the left-hand limit exist?

(d) At what points $c$ does only the right-hand limit exist?

**65.** $f(x) = \begin{cases} \cos x & -\pi \leq x < 0 \\ -\cos x & 0 \leq x \leq \pi \end{cases}$

**66.** $f(x) = \begin{cases} \sin x & -\pi \leq x < 0 \\ \csc x & 0 \leq x \leq \pi \end{cases}$

**67.** $f(x) = \begin{cases} \sqrt{1 - x^2} & -1 \leq x < 0 \\ x & 0 \leq x < 1 \\ 2 & x = 1 \end{cases}$

**68.** $f(x) = \begin{cases} x^2 & -2 \leq x < 0 \text{ or } 0 < x \leq 2 \\ 1 & x = 0 \\ 2x & x < -2 \text{ or } x > 2 \end{cases}$

**69. Rabbit Population** The population of rabbits over a 2 year period in a certain county is given in Table 10.4

**Table 10.5** Rabbit Population

| Beginning of Month | Number (in thousands) |
|---|---|
| 0 | 10 |
| 2 | 12 |
| 4 | 14 |
| 6 | 16 |
| 8 | 22 |
| 10 | 30 |
| 12 | 35 |
| 14 | 39 |
| 16 | 44 |
| 18 | 48 |
| 20 | 50 |
| 22 | 51 |

**(a)** Draw a scatter plot of the data in Table 10.5.

**(b)** Find a logistic regression model for the data. Find the limit of that model as time approaches infinity.

**(c)** What can you conclude about the limit of the rabbit population growth in the county?

**(d)** Provide a reasonable explanation for the population growth limit.

In Exercises 70–73, **(a)** make a table of function values for $x = 1, 0.1, 0.01, 0.001, 0.0001, 0.00001$ and $0.000001$, **(b)** use the information in (a) to estimate $\lim_{x \to 0} f(x)$, if it exists, and **(c)** draw a graph to support your limit estimate in (b).

**70.** $f(x) = \dfrac{2^x - 1}{2^x}$

**71.** $f(x) = x \sin \dfrac{1}{x}$

**72.** $f(x) = \dfrac{1}{x} \sin \dfrac{1}{x}$

**73.** $f(x) = x^2 \sin (\ln |x|)$ ■

In Exercises 74–77, sketch a graph of a function $y = f(x)$ that satisfies the stated conditions. Include any asymptotes.

**74. Group Activity** $\lim_{x \to 0} f(x) = \infty, \lim_{x \to \infty} f(x) = \infty,$ $\lim_{x \to -\infty} f(x) = 2$

**75. Group Activity** $\lim_{x \to 4} f(x) = -\infty, \lim_{x \to \infty} f(x) = -\infty,$ $\lim_{x \to -\infty} f(x) = 2$

**76. Group Activity** $\lim_{x \to \infty} f(x) = 2, \lim_{x \to -2^+} f(x) = -\infty,$ $\lim_{x \to -2^-} f(x) = -\infty, \lim_{x \to -\infty} f(x) = \infty$

**77. Group Activity** $\lim_{x \to 1} f(x) = \infty, \lim_{x \to 2^+} f(x) = -\infty,$ $\lim_{x \to 2^-} f(x) = -\infty, \lim_{x \to -\infty} f(x) = \infty$

## Extending the Ideas

**78. Properties of Limits** Find the limits of $f$, $g$, and $fg$ as $x$ approaches $c$.

**(a)** $f(x) = \dfrac{2}{x^2}, \quad g(x) = x^2, \quad c = 0$   $\infty$; 0; 2

**(b)** $f(x) = \left| \dfrac{1}{x} \right|, \quad g(x) = \sqrt[3]{x}, \quad c = 0$   $\infty$; 0; none

**(c)** $f(x) = \left| \dfrac{3}{x - 1} \right|, \quad g(x) = (x - 1)^2, \quad c = 1$   $\infty$; 0; 0

**(d)** $f(x) = \dfrac{1}{(x - 1)^4}, \quad g(x) = (x - 1)^2, \quad c = 1$   $\infty$; 0; $\infty$

**(e) Writing to Learn** Suppose that $\lim_{x \to c} f(x) = \infty$ and $\lim_{x \to c} g(x) = 0$. Based on your results in parts (a)–(d), what can you say about $\lim_{x \to a} (f(x) \cdot g(x))$? nothing

**79. Limits and the area of a circle** Consider an $n$-sided regular polygon made up of $n$ congruent isosceles triangles, each with height $h$ and base $b$. The figure shows an 8-sided regular polygon.

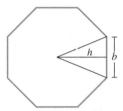

**(a)** Show that the area of an 8-sided regular polygon is $A = 4hb$ and the area of the $n$-sided regular polygon is $A = (1/2)nhb$.

**(b)** Show that the base $b$ of the $n$-sided regular polygon is $b = 2h \tan (180/n)°$.

**(c)** Show that the Area $A$ of the $n$-sided regular polygon is $A = nh^2 \tan (180/n)°$.

**(d)** Let $h = 1$. Construct a table of values for $n$ and $A$ for $n = 4, 8, 16, 100, 500, 1000, 5000, 10000, 100000$. Does $A$ have a limit as $n \to \infty$?

**(e)** Repeat part (d) with $h = 3$.

**(f)** Give a convincing argument that $\lim_{n \to \infty} A = \pi h^2$, the area of a circle of radius $h$.

**80. Continuous Extension of a Function** Let

$$f(x) = \begin{cases} x^2 - 3x + 3 & x \neq 2 \\ a & x = 2 \end{cases}.$$

**(a)** Sketch several possible graphs for $f$.

**(b)** Find a value for $a$ so that the function is continuous at $x = 2$.   1

In Exercises 81–83, **(a)** graph the function, **(b)** verify that the function has one removable discontinuity, and **(c)** give a formula for a continuous extension of the function. (*Hint*: See Exercise 80.)

**81.** $y = \dfrac{2x + 4}{x + 2}$

**82.** $y = \dfrac{x - 5}{5 - x}$

**83.** $y = \dfrac{x^3 - 1}{x - 1}$

## Numerical Derivatives and Integrals

Derivatives on a Calculator • Definite Integrals on a Calculator • Computing a Derivative from Data • Computing a Definite Integral from Data

**Objective**

Students will be able to estimate derivatives and integrals using numerical techniques.

**Motivate**

Discuss the commands used to calculate numerical derivatives and integrals on a graphing calculator.

**Lesson Guide**

Day 1: Derivatives on a Calculator; Definite Integrals on a Calculator
Day 2: Computing a Derivative from Data; Computing a Definite Integral from Data

### Derivatives on a Calculator

As computers and sophisticated calculators have become indispensable tools for modern engineers and mathematicians (and, ultimately, for modern students of mathematics), numerical techniques of differentiation and integration have re-emerged as primary methods of problem-solving. This is no small irony, as it was precisely to avoid the tedious computations inherent in such methods that calculus was invented in the first place. Although nothing can diminish the magnitude of calculus as a significant human achievement, and although nobody can get far in mathematics or science without it, the modern fact is that applying the old-fashioned methods of limiting approximations—with the help of a calculator—is often the most efficient way to solve a calculus problem.

Most graphing calculators have built-in algorithms that will approximate derivatives of functions numerically with good accuracy at most points of their domains. We will use the notation NDER $f(a)$ to denote such a calculator derivative approximation to $f'(a)$.

For small values of $h$, the regular difference quotient

$$\frac{f(a + h) - f(a)}{h}$$

is often a good approximation of $f'(a)$. However, the same value of $h$ will usually produce a better approximation of $f'(a)$ if we use the **symmetric difference quotient**

$$\frac{f(a + h) - f(a - h)}{2h},$$

as illustrated in Figure 10.24.

Many graphing utilities use the symmetric difference quotient with a default value of $h = 0.001$ for computing NDER $f(a)$. When we refer to the numerical derivative in this book, we will assume that it is the symmetric difference quotient with $h = 0.001$.

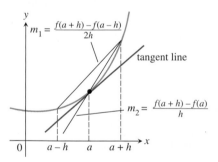

**Figure 10.24** The symmetric difference quotient (slope $m_1$) usually gives a better approximation of the derivative for a given value of $h$ than does the regular difference quotient (slope $m_2$).

---

**Definition  Numerical Derivative**

In this book, we define the **numerical derivative of $f$ at $a$** to be

$$\text{NDER } f(a) = \frac{f(a + 0.001) - f(a - 0.001)}{0.002}.$$

Similarly, we define the **numerical derivative of $f$** to be the function

$$\text{NDER } f(x) = \frac{f(x + 0.001) - f(x - 0.001)}{0.002}.$$

**Example 1**   COMPUTING A NUMERICAL DERIVATIVE

Let $f(x) = x^3$. Compute NDER $f(2)$ by calculating the symmetric difference quotient with $h = 0.001$. Compare it to the actual value of $f'(x)$.

**Solution**

$$\text{NDER } f(x) = \frac{f(2 + 0.001) - f(2 - 0.001)}{0.002}$$

$$= \frac{f(2.001) - f(1.999)}{0.002}.$$

$$= 12.000001$$

The actual value is

$$f'(2) = \lim_{h \to 0} \frac{(2 + h)^3 - 2^3}{h}$$

$$= \lim_{h \to 0} \frac{8 + 12h + 6h^2 + h^3 - 8}{h}$$

$$= \lim_{h \to 0} (12 + 6h + h^2)$$

$$= 12$$

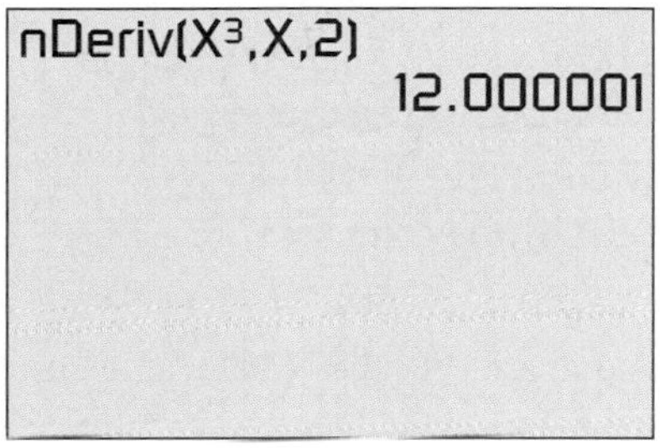

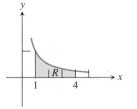

nDeriv(X³,X,2)

                12.000001

**Figure 10.25**  Applying the numerical derivative command on a graphing calculator. (Example 1)

The numerical derivative in this case is obviously quite accurate. In practice, it is not necessary to key in the symmetric difference quotient, as it is done by the calculator with its built-in algorithm. Figure 10.25 shows the command that would be used on one such calculator to find the numerical derivative in Example 1.

If $f'(a)$ exists, then NDER $f(a)$ usually gives a good approximation to the actual value. On the other hand, the algorithm will sometimes return a value for NDER $f(a)$ when $f'(a)$ does not exist. (See Exercise 39.)

## Definite Integrals on a Calculator

Recall from the history of the area problem (Section 10.2) that the strategy of summing up thin rectangles to approximate areas is ancient. The thinner the rectangles, the better the approximation—and, of course, the more tedious the computation. Today, thanks to technology, we can employ the ancient strategy without the tedium.

Many graphing calculators have built-in algorithms to compute definite integrals with great accuracy. We use the notation NINT $(f(x), x, a, b)$ to denote such a calculator approximation to $\int_a^b f(x)\, dx$. Unlike NDER, which uses a fixed value of $\triangle x$, NINT will vary the value of $\triangle x$ until the numerical integral gets close to a limiting value, often resulting in an exact answer (at least to the number of digits in the calculator display). Because the algorithm for NINT finds the definite integral by Riemann sum approximation rather than by calculus, we call it a **numerical integral**.

**Example 2**   FINDING A NUMERICAL INTEGRAL

Use NINT to find the area of the region $R$ enclosed between the $x$-axis and the graph of $y = 1/x$ from $x = 1$ to $x = 4$.

**Solution**  The region is shown in Figure 10.26.

**Figure 10.26**  The graph of $f(x) = 1/x$ with the area under the curve between $x = 1$ and $x = 4$ shaded. (Example 2)

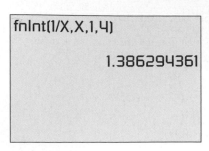

**Figure 10.27** A numerical integral approximation for $\int_1^4 \frac{1}{x}\, dx$. (Example 2)

The area can be written as the definite integral $\int_1^4 \frac{1}{x}\, dx$, which we find on a graphing calculator: NINT $(1/x, x, 1, 4) = 1.386294361$. The exact answer (as you will learn in a calculus course) is ln 4, which agrees in every displayed digit with the NINT value! Figure 10.27 shows the syntax for numerical integration on one type of calculator.

---

**Exploration 1    A Do-It-Yourself Numerical Integrator**

Recall that a definite integral is the limit at infinity of a Riemann sum—that is, a sum of the form $\sum_{k=1}^{n} f(x_k)\triangle x$. You can use your calculator to evaluate sums of sequences using LIST commands. (It is not as accurate as NINT, and certainly not as easy, but at least you can see the summing that takes place.)

1. The integral in Example 2 can be computed using the command
   $$\text{sum(seq}(1/(1 + K \cdot 3/50) \cdot 3/50, K, 1, 50)).$$
   This uses 50 RRAM rectangles, each with width $\triangle x = 3/50$. Find the sum on your calculator and compare it to the NINT value.

2. Study the command until you see how it works. Adapt the command to find the RRAM approximation for 100 rectangles and compute it on your calculator. Does the approximation get better?

3. What definite integral is approximated by the command
   $$\text{sum(seq}(\sin(0 + K \cdot \pi/50) \cdot \pi/50, K, 1, 50))?$$
   Compute it on your calculator and compare it to the NINT value for the same integral.

4. Write a command that uses 50 RRAM rectangles to approximate $\int_4^9 \sqrt{x}\, dx$. Compute it on your calculator and compare it to the NINT value for the same integral.

---

**Exploration Extensions**

Which of the following would you expect typically to give the most accurate estimate of the area under a curve using $n$ rectangles: the LRAM approximation, the RRAM approximation, or the average of these two?

Remember that we were originally motivated to find areas because of their connection to the problem of distance traveled. To show just one of the many applications of integration, we use the numerical integral to solve a distance problem in Example 3.

**Who's driving???**

We will candidly admit that the conditions in Example 3 would be virtually impossible to replicate in a real setting, even if one could imagine a reason for doing so. Mathematics textbooks are filled with such unreal real-world problems, but they do serve a purpose when students are being exposed to new material. Real real-world problems are often either too easy to illustrate the concept or too hard for beginners to solve.

**Example 3    FINDING DISTANCE TRAVELED**

An automobile is driven at a variable rate along a test track for 2 hours so that its velocity at any time $t$ $(0 \le t \le 2)$ is given by $v(t) = 30 + 10 \sin 6t$ miles per hour. How far does the automobile travel during the 2-hour test?

**Solution**  According to the analysis found in Section 10.2, the distance traveled is given by $\int_0^2 (30 + 10 \sin (6t))\, dt$. We use a calculator to find the numerical integral:

$$\text{NINT } (30 + 10 \sin (6t), t, 0, 2) \approx 60.26.$$

Interpreting the answer, we conclude that the automobile travels 60.26 miles.

## Computing a Derivative From Data

Sometimes all we are given about a problem situation is a scatter plot obtained from a set of data—a numerical model of the problem. There are two ways to get information about the derivative of the model.

1. To approximate the derivative at a point: Remember that the average rate of change over a small interval, $\triangle y / \triangle x$, approximates the derivative at points in that interval. (Generally, the approximation is better near the middle of the interval than it is near the endpoints.) The average rate of change on an interval between two data points can be computed directly from the data.

2. To approximate the derivative function: Regression techniques can be used to fit a curve to the data, and then NDER can be applied to the regression model to approximate the derivative. Alternatively, the values of $\wedge y / \triangle x$ can be plotted, and then regression techniques can be used to fit a derivative approximation through those points.

**Table 10.6 Falling Ball**

| Time (sec) | Position (ft) |
| --- | --- |
| 0.04 | 6.80 |
| 0.08 | 6.40 |
| 0.12 | 5.95 |
| 0.16 | 5.45 |
| 0.20 | 4.90 |
| 0.24 | 4.30 |
| 0.28 | 3.60 |
| 0.32 | 2.90 |
| 0.36 | 2.15 |
| 0.40 | 1.30 |
| 0.44 | 0.40 |

**Table 10.6** The data for the falling ball in Example 4.

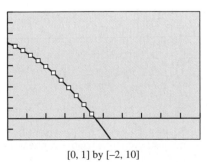

[0, 1] by [–2, 10]

(a)

**Figure 10.28** A scatter plot of the data in Table 10.6 together with its quadratic regression model. (Example 4)

**Example 4** FINDING DERIVATIVES FROM DATA

Table 10.6 shows the height (in feet) of a falling ball above ground level as measured by a motion detector at time intervals of 0.04 seconds.

(a) Estimate the instantaneous speed of the ball at $t = 0.2$ seconds.

(b) Draw a scatter plot of the data and use quadratic regression to model the height $s$ of the ball above the ground as a function of $t$.

(c) Use NDER to approximate $s'(0.2)$ and compare it to the value found in part (a).

**Solution**

(a) Since 0.2 is the midpoint of the time interval [0.16, 0.24], the average rate of change $\wedge s / \triangle t$ on the interval [0.16, 0.24] should give a good approximation to $s'(0.2)$.

$$s'(0.2) = \frac{\triangle s}{\triangle t} = \frac{4.30 - 5.45}{0.24 - 0.16} = -14.375.$$

The speed is about 14.375 feet per second at $t = 0.2$.

(b) The scatter plot is shown in Figure 10.28, along with the quadratic regression curve. A graphing calculator gives $s(t) \approx -17.12t^2 - 7.74t + 7.13$ as the equation of the quadratic regression model.

(c) The calculator computes NDER $s(0.2)$ to be $-14.588$, which agrees quite well with the approximation in part (a). In fact, the difference is only 0.213 feet per second, less than 1.5% of the speed of the ball.

| Table 10.7 Change over Intervals from Table 10.6 | |
|---|---|
| Midpoint | $\triangle s/\triangle t$ |
| 0.06 | −10.00 |
| 0.10 | −11.25 |
| 0.14 | −12.50 |
| 0.18 | −13.75 |
| 0.22 | −15.00 |
| 0.26 | −17.50 |
| 0.30 | −17.50 |
| 0.34 | −18.75 |
| 0.38 | −21.25 |
| 0.42 | −22.50 |

**Table 10.7** The midpoints of the time intervals in Table 10.6 and the average velocities over those time intervals. (Example 5)

**Example 5** FINDING DERIVATIVES FROM DATA

This example also uses the falling ball data in Table 10.6.

**(a)** Compute the average velocity, $\triangle s/\triangle t$, on each subinterval of length 0.4. Make a table showing the midpoints of the subintervals in one column and the values of $\triangle s/\triangle t$ in the second column.

**(b)** Make a scatter plot showing the numbers in the second column as a function of the numbers in the first column and find a linear regression model to model the data.

**(c)** Use the linear regression model in part (b) to approximate the velocity of the ball at $t = 0.2$ and compare it to the values found in Example 4.

**Solution**

**(a)** The first subinterval, which begins at 0.04 and ends at 0.08, has a midpoint of $(0.04 + 0.08)/2 = 0.06$. On that interval, $\triangle s/\triangle t = (6.40 - 6.80)/(0.08 - 0.04)$. The rest of the midpoints and values of $\triangle s/\triangle t$ are computed similarly and are shown in Table 10.7.

**(b)** The scatter plot and the regression line are shown in Figure 10.29. A graphing calculator gives $v(t) = -34.470t - 7.727$ as the regression line.

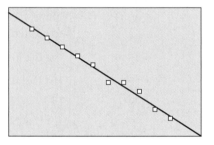

[0, 0.5] by [−25, −7]

**Figure 10.29** A scatter plot of the data in Table 10.7 together with its linear regression model. (Example 5)

**(c)** The linear regression model gives $v(0.2) \approx -14.62$, which is close to the values found in Example 4.

## Computing a Definite Integral from Data

If we are given a set of data points, the $x$-coordinates of the points define subintervals between the smallest and largest $x$-values in the data. We can form a Riemann sum

$$\sum_{k=1}^{n} f(x_k) \triangle x$$

using the lengths of the subintervals for $\triangle x$ and either the left or right endpoints of the intervals as the $x_k$'s. The Riemann sum then approximates the definite integral of the function over the interval.

Example 6 illustrates how this is done.

| Table 10.8  Velocity of the Moving Body | |
|---|---|
| Time (sec) | Velocity (m/sec) |
| 0.00 | 0.00 |
| 0.25 | 0.28 |
| 0.50 | 0.53 |
| 0.75 | 0.73 |
| 1.00 | 0.90 |
| 1.25 | 1.01 |
| 1.50 | 1.11 |
| 1.75 | 1.18 |
| 2.00 | 1.21 |
| 2.25 | 1.17 |
| 2.50 | 1.13 |
| 2.75 | 1.05 |
| 3.00 | 0.91 |
| 3.25 | 0.72 |
| 3.50 | 0.55 |
| 3.75 | 0.26 |

**Table 10.8**  Data for the moving body in Example 6.

### Example 6  FINDING A DEFINITE INTEGRAL USING DATA

Table 10.8 shows the velocity of a moving body (in meters per second) measured at regular quarter-second intervals. Estimate the distance traveled by the body from $t = 0$ to $t = 3.75$.

**Solution**  Figure 10.30 gives a scatter plot of the velocity data.

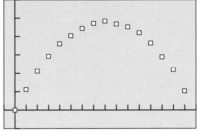

[−0.25, 4] by [−0.25, 1.5]

**Figure 10.30**  A scatter plot of the velocity data in Table 10.8. (Example 6)

The distance traveled is $\int_0^{3.75} v(t)\, dx$, which we approximate with a

Riemann sum constructed directly from the data. We sum up 15 products of the form $v(t_k)\, \triangle t$, using the right endpoint for $t_k$ each time. (This is the RRAM approximation in the notation of Section 10.2.) Note that $\triangle t = 0.25$ for every subinterval.

$$\int_0^{3.75} v(t)\, dx = \sum_{k=1}^{15} v(t_k)\, \triangle t$$

$$= 0.25(0.28 + 0.53 + 0.73 + 0.90 + 1.01 + 1.11 + 1.18 + 1.21 + 1.17 + 1.13 + 1.05 + 0.91 + 0.72 + 0.55 + 0.26)$$

$$= 3.185$$

So the distance traveled by the body is about 3.2 meters.

## Quick Review 10.4

In Exercises 1–6, find $\triangle y / \triangle x$ on the interval $[1, 4]$ under the given conditions.

**1.** $y = x^2$  5

**2.** $y = \sqrt{x}$  1/3

**3.** $y = \log_2 x$  2/3

**4.** $y = 3^x$  26

**5.** $y = 2$ when $x = 1$ and $y = 11$ when $x = 4$.  3

**6.** The graph of $y = f(x)$ passes through points $(4, 10)$ and $(1, -2)$.  4

In Exercises 7–10, compute the quotient $(f(1 + h) - f(1 - h))/2h$ with the given $f$ and $h$.

**7.** $f(x) = \sin x$, $h = 0.01$

**8.** $f(x) = x^4$, $h = 0.001$

**9.** $f(x) = \ln x$, $h = 0.001$

**10.** $f(x) = e^x$, $h = 0.0001$

# Section 10.4 Exercises

In Exercises 1–10, use NDER on a calculator to find the numerical derivative of the function at the specified point.

**1.** $f(x) = 1 - x^2$ at $x = 2$ $-4$

**2.** $f(x) = 2x + \dfrac{1}{2}x^2$ at $x = 2$ $4$

**3.** $f(x) = 3x^2 + 2$ at $x = -2$ $-12$

**4.** $f(x) = x^2 - 3x + 1$ at $x = 1$ $-1$

**5.** $f(x) = |x + 2|$ at $x = -2$ $0$

**6.** $f(x) = \dfrac{1}{x + 2}$ at $x = -1$ $-1$

**7.** $f(x) = \ln 2x$ at $x = 1$ **8.** $f(x) = 2 \ln x$ at $x = 1$

**9.** $f(x) = 3 \sin x$ at $x = \pi$ **10.** $f(x) = \sin 3x$ at $x = \pi$

In Exercises 11–20, use NINT on a calculator to find the numerical integral of the function over the specified interval.

**11.** $f(x) = x^2$, $[0, 4]$ $64/3$ **12.** $f(x) = x^2$, $[-4, 0]$ $64/3$

**13.** $f(x) = \sin x$, $[0, \pi]$ $2$ **14.** $f(x) = \sin x$, $[\pi, 2\pi]$ $-2$

**15.** $f(x) = \cos x$, $[0, \pi]$ $\approx 0$ **16.** $f(x) = |\cos x|$, $[0, \pi]$ $2$

**17.** $f(x) = 1/x$, $[1, e]$ $1$ **18.** $f(x) = 1/x$, $[e, 2e]$

**19.** $f(x) = \dfrac{2}{1 + x^2}$, $[0, 10^8]$ $\approx 3.1416$

**20.** $f(x) = \sec^2 x - \tan^2 x$, $[0, 10]$ $10$

**21. Finding Derivatives from Data** A ball is dropped from the roof of a 30-story building. The height in feet above the ground of the falling ball is measured at 1/2-second intervals and recorded in the table below.

### Height Data of the Falling Ball

| Time (sec) | Height (ft) |
|:---:|:---:|
| 0 | 500 |
| 0.5 | 495 |
| 1.0 | 485 |
| 1.5 | 465 |
| 2.0 | 435 |
| 2.5 | 400 |
| 3.0 | 355 |
| 3.5 | 305 |
| 4.0 | 245 |
| 4.5 | 175 |
| 5.0 | 100 |
| 5.5 | 15 |

**(a)** Use the average velocity on the interval $[1, 2]$ to estimate the velocity of the ball at $t = 1.5$ seconds. $-50$ ft/sec

**(b)** Draw a scatter plot of the data.

**(c)** Find a quadratic regression model for the data.

**(d)** Use NDER of the model in (c) to estimate the velocity of the ball at $t = 1.5$ seconds. $\approx -47.88$ ft/sec

**(e)** Use the model to estimate how fast the ball is going when it hits the ground. $\approx 179.28$ ft/sec

**22. Estimating Average Rate of Change from Data** Table 10.9 gives the U.S. Gross Domestic Product Data in billions of dollars for the years 1989–1998.

**(a)** Find the average rate of change of the gross domestic product from 1990 to 1991 and then from 1996 to 1997.

**(b)** Find a quadratic regression model for the data in Table 10.9 and overlay its graph on a scatter plot of the data. Let $x = 0$ stand for 1980, $x = 1$ for 1981, and so forth.

**(c)** Use the model in (b), and a calculator NDER computation, to estimate the rate of change of the gross domestic product in 1993 and in 1996.

**(d) Writing to Learn** Use the model in (b) to predict the gross domestic product in 2002. Is this reasonable? Why or why not?

### Table 10.9 U.S. Gross Domestic Product Data

| Year | Amount (billions) |
|:---:|:---:|
| 1989 | 6062 |
| 1990 | 6136 |
| 1991 | 6079 |
| 1992 | 6244 |
| 1993 | 6390 |
| 1994 | 6611 |
| 1995 | 6742 |
| 1996 | 6995 |
| 1997 | 7270 |
| 1998 | 7552 |

*Source: Geospatial and Statistical Data Center, http://fisher.lib.virginia.edu/ or http://viva.lib.virginia.edu/socsci/nipa/nipa.html*

**23. Writing to Learn** Analyze the following program, which produces an LRAM approximation for the function entered in Y1 in the calculator. Then write a short paragraph explaining how it works.

```
PROGRAM:LRAM
:Input "A",A
:Input "B",B
:Input "N",N
:sum(seq(((B-A)/
N)Y₁(A+K((B-A)/N
)),K,0,N-1))→C
:Disp "AREA =",C
```

**24. Writing to Learn** Analyze the following program, which produces an RRAM approximation for the function entered in Y1 in the calculator. Then write a short paragraph explaining how it works.

```
PROGRAM:RRAM
:Input "A",A
:Input "B",B
:Input "N",N
:sum(seq(((B−A)/
N)Y₁(A+K((B−A)/N
)),K,1,N))→C
:Disp "AREA =",C
```

In Exercises 25–36, complete the following for the indicated interval $[a, b]$.

**(a)** Verify the given function is non-negative.

**(b)** Use a calculator to find the LRAM, RRAM, and average approximations for the area under the graph of the function from $x = a$ to $x = b$ with 10, 20, 50, and 100 approximating rectangles. (You may want to use the programs in Exercises 23 and 24.)

**(c)** Compare the average area estimate in (b) using 100 approximating rectangles with the calculator NINT area estimate, if your calculator has this feature.

**25.** $f(x) = x^2 - x + 1$; $[0, 4]$

**26.** $f(x) = 2x^2 - 2x + 1$; $[0, 6]$

**27.** $f(x) = x^2 - 2x + 1$; $[0, 4]$ **28.** $f(x) = x^2 + x + 5$; $[0, 6]$

**29.** $f(x) = x^2 + x + 5$; $[2, 6]$ **30.** $f(x) = x^3 + 1$; $[1, 5]$

**31.** $f(x) = \sqrt{x + 2}$; $[0, 4]$ **32.** $f(x) = \sqrt{x - 2}$; $[3, 6]$

**33.** $f(x) = \cos x$; $[0, \pi/2]$ **34.** $f(x) = x \cos x$; $[0, \pi/2]$

**35.** $f(x) = xe^{-x}$; $[0, 2]$ **36.** $f(x) = \dfrac{1}{x - 2}$; $[3, 5]$

## Explorations

**37.** Let $f(x) = 2x^2 + 3x + 1$ and $g(x) = x^3 + 1$.

**(a)** Compute the derivative of $f$. $4x + 3$

**(b)** Compute the derivative of $g$. $3x^2$

**(c)** Using $x = 2$ and $h = 0.001$, compute the standard difference quotient 11.002

$$\frac{f(x + h) - f(x)}{h}$$

and the symmetric difference quotient 11

$$\frac{f(x + h) - f(x - h)}{2h}.$$

**(d)** Using $x = 2$ and $h = 0.001$, compare the approximations to $f'(2)$ in (c). Which is the better approximation?

**(e)** Repeat (c) and (d) for $g$.

**38. When are Derivatives and Areas Equal?** Let $f(x) = 1 + e^x$.

**(a)** Draw a graph of $f$ for $0 \le x \le 1$.

**(b)** Use NDER on your calculator to compute the derivative of $f$ at 1. $\approx 2.72$

**(c)** Use NINT on your calculator to compute the area under $f$ from $x = 0$ to $x = 1$ and compare it to the answer in (b).

**(d) Group Activity** What do you think the exact answers to (b) and (c) are? ∎

**39. Calculator Failure** Many calculators report that NDER of $f(x) = |x|$ evaluated at $x = 0$ is equal to 0. Explain why this is incorrect. Explain why this error occurs.

**40. Grapher Failure** Graph the function $f(x) = |x|/x$ in the window $[-5, 5]$ by $[-3, 3]$ and explain why $f'(0)$ does not exist. Find the value of NDER $f(0)$ on the calculator and explain why it gives an incorrect answer.

## Extending the Ideas

**41. Group Activity Finding Total Area** The **total area** bounded by the graph of the function $y = f(x)$ and the $x$-axis from $x = a$ to $x = b$ is the area below the graph of $y = |f(x)|$ from $x = a$ to $x = b$.

**(a)** Find the total area bounded by the graph of $f(x) = \sin x$ and the $x$-axis from $x = 0$ to $x = 2\pi$. 4

**(b)** Find the total area bounded by the graph of $f(x) = x^2 - 2x - 3$ and the $x$-axis from $x = 0$ to $x = 5$.

**42. Writing to Learn** If a function is *unbounded* in an interval $[a, b]$ it may have *finite* area. Use your knowledge of limits at infinity to explain why this might be the case.

**43. Writing to Learn** Let $f$ and $g$ be two continuous functions with $f(x) \ge g(x)$ on an interval $[a, b]$. Devise a limit of sums definition of the area of the region between the two curves. Explain how to compute the area if the area under both curves is already known.

**44. Area as a function** Consider the function $f(t) = 2t$.

**(a)** Use NINT on a calculator to compute $A(x)$ where $A$ is the area under the graph of $f$ from $t = 0$ to $t = x$ for $x = 0.25, 0.5, 1, 1.5, 2, 2.5$, and 3.

**(b)** Make a table of pairs $(x, A(x))$ for the values of $x$ given in (a) and plot them using graph paper. Connect the plotted points with a smooth curve.

**(c)** Use a quadratic regression equation to model the data in (b) and overlay its graph on a scatter plot of the data.

**(d)** Make a conjecture about the exact value of $A(x)$ for any $x$ greater than zero.

**(e)** Find the derivative of the $A(x)$ found in (d). Record any observations.

**45. Area as a Function** Consider the function $f(t) = 3t^2$.

**(a)** Use NINT on a calculator to compute $A(x)$ where $A$ is the area under the graph of $f$ from $t = 0$ to $t = x$ for $x = 0.25, 0.5, 1, 1.5, 2, 2.5$, and 3.

**(b)** Make a table of pairs $(x, A(x))$ for the values of $x$ given in (a) and plot them using graph paper. Connect the plotted points with a smooth curve.

**(c)** Use a cubic regression equation to model the data in (b) and overlay its graph on a scatter plot of the data.

**(d)** Make a conjecture about the exact value of $A(x)$ for any $x$ greater than zero.

**(e)** Find the derivative of the $A(x)$ found in (d). Record any observations.

**46. Group Activity** Based on Exercises 44 and 45, discuss how derivatives (slope functions) and integrals (area functions) may be connected.

---

## Chapter 10 Key Ideas

### Concepts

| | | |
|---|---|---|
| Area problem (p. 766) | Instantaneous velocity (p. 755) | Numerical integral (p. 785) |
| Average rate of change (p. 757) | Integrable function (p. 768) | Riemann sum (p. 767) |
| Average velocity (p. 753) | Left-hand limit (p. 775) | Right-hand limit (p. 775) |
| Definite integral (p. 768) | Limit at a point (p. 755) | RRAM (p. 767) |
| Derivative at a point (p. 757) | Limit at infinity (p. 765, 777) | Secant line (p. 757) |
| Derivative of a function (p. 758) | LRAM (p. 767) | Symmetric difference quotient (p. 784) |
| Differentiable (p. 757) | NDER $f(a)$ (p. 784) | Tangent line (p. 757) |
| $dy/dx$ (p. 758) | NINT ($f(x)$, $x$, $a$, $b$) (p. 785) | Tangent line problem (p. 756) |
| Infinite limit (p. 777) | Numerical derivative (p. 784) | Two-sided limits (p. 776) |
| Instantaneous rate of change (p. 759) | Numerical derivative of $f$ at $a$ (p. 784) | |

### Properties, Theorems, and Formulas

#### Limit at *a* (informal definition)

When we write "$\lim_{x \to a} f(x) = L$," we mean that $f(x)$ gets arbitrarily close to $L$ as $x$ gets arbitrarily close (but not equal) to $a$.

#### Average Rate of Change

If $y = f(x)$, then the **average rate of change** of $y$ with respect to $x$ on the interval $[a, b]$ is

$$\frac{\triangle y}{\triangle x} = \frac{f(a) - f(b)}{b - a}.$$

#### Derivative at a Point

The **derivative of the function $f$ at $x = a$**, denoted by $f'(a)$ and read "$f$ prime of $a$" is

$$f'(a) = \lim_{x \to a} \frac{f(x) - f(a)}{x - a},$$

provided the limit exists.

#### Derivative at a Point (easier for computing)

The **derivative of the function $f$ at $x = a$,** denoted by $f'(a)$ and read "$f$ prime of $a$" is

$$f'(a) = \lim_{h \to 0} \frac{f(a + h) - f(a)}{h},$$

provided the limit exists.

### Derivative

If $y = f(x)$, then the **derivative of the function $f$ with respect to $x$** is the function $f'$ whose value at $x$ is

$$f'(x) = \lim_{h \to 0} \frac{f(x + h) - f(a)}{h},$$

for all values of $x$ where the limit exists.

### Limit at Infinity (informal definition)

When we write "$\lim_{x \to a} f(x) = L$," we mean that $f(x)$ gets arbitrarily close to $L$ as $x$ gets arbitrarily large.

### Definite Integral

Let $f$ be a function defined on $[a, b]$. The definite integral of $f$ over $[a, b]$, denoted $\int_a^b f(x)\,dx$, is given by

$$\int_a^b f(x)\,dx = \lim_{n \to \infty} \sum_{i=1}^{n} f(x_i)\,\triangle x,$$

provided the limit exists. If the limit exists, we say $f$ is **integrable** on $[a, b]$.

### Properties of Limits

If $f(x)$ and $g(x)$ both exist, then

**1.** *Sum Rule:* $\qquad\qquad\qquad\quad \lim_{x \to c} (f(x) + g(x)) = \lim_{x \to c} f(x) + \lim_{x \to c} g(x)$

**2.** *Difference Rule:* $\qquad\qquad\quad \lim_{x \to c} (f(x) - g(x)) = \lim_{x \to c} f(x) - \lim_{x \to c} g(x)$

**3.** *Product Rule:* $\qquad\qquad\qquad \lim_{x \to c} (f(x) \cdot g(x)) = \lim_{x \to c} f(x) \cdot \lim_{x \to c} g(x)$

**4.** *Constant Multiple Rule:* $\qquad \lim_{x \to c} (k \cdot f(x)) = k \cdot \lim_{x \to c} f(x)$

**5.** *Quotient Rule:* $\qquad\qquad\quad \lim_{x \to c} \dfrac{f(x)}{g(x)} = \dfrac{\lim_{x \to c} f(x)}{\lim_{x \to c} g(x)}$, provided $\lim_{x \to c} g(x) \neq 0$

**6.** *Power Rule:* $\qquad\qquad\qquad \lim_{x \to c} (f(x))^n = (\lim_{x \to c} f(x))^n$ for $n$ a positive integer

**7.** *Root Rule:* $\qquad\qquad\qquad\quad \lim_{x \to c} \sqrt[n]{f(x)} = \sqrt[n]{\lim_{x \to c} f(x)}$ for $n \geq 2$ a positive integer,

$\qquad\qquad\qquad\qquad\qquad\qquad$ provided $\sqrt[n]{\lim_{x \to c} f(x)}$ is a real number.

### Limits at Infinity (informal definition)

When we write "$\lim_{x \to \infty} f(x) = L$," we mean that $f(x)$ gets arbitrarily close to $L$ as $x$ gets arbitrarily large. We say that $f$ **has a limit $L$ as $x$ approaches** $\infty$.

When we write "$\lim_{x \to -\infty} f(x) = L$," we mean that $f(x)$ gets arbitrarily close to $L$ as $-x$ gets arbitrarily large. We say that $f$ **has a limit $L$ as $x$ approaches** $-\infty$.

### Symmetric Difference Quotient

$$\frac{f(a + h) - f(a - h)}{}$$

### Numerical Derivative of $f$ at $a$

$$\text{NDER}\, f(a) = \frac{f(a + 0.001) - f(a - 0.001)}{}$$

### Numerical Derivative of f

$$\text{NDER } f(x) = \frac{f(x + 0.001) - f(x - 0.001)}{0.002}$$

### Numerical Integral

An approximating Riemann sum: $\sum_{k=1}^{n} f(x_k)\, \triangle x$

## Procedures

### Computing a Derivative From Data

**1.** To approximate the derivative at a point: Remember that the average rate of change over a small interval, $\triangle y / \triangle x$, approximates the derivative at points in that interval. (Generally, the approximation is better near the middle of the interval than it is near the endpoints.) The average rate of change on an interval between two data points can be computed directly from the data.

**2.** To approximate the derivative function: Regression techniques can be used to fit a curve to the data, and then NDER can be applied to the regression model to approximate the derivative. Alternatively, the values of $\triangle y / \triangle x$ can be plotted, and then regression techniques can be used to fit a derivative approximation through those points.

## Chapter 10   Review Exercises

The collection of exercises marked in red could be used as a chapter test.

In Exercises 1–4, use the graph of the function $y = f(x)$ to find **(a)** $\lim\limits_{x \to 1^-} f(x)$ and **(b)** $\lim\limits_{x \to 1} f(x)$.

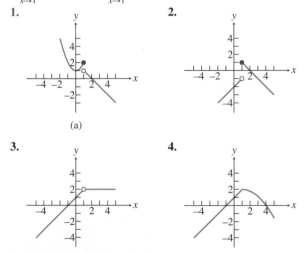

**1.**

**2.**

(a)

**3.**

**4.**

In Exercises 5–10, find the limit at the indicated point, if it exists. Support your answer graphically.

**5.** $f(x) = \dfrac{x - 1}{x^2 + 1}, x = -1$   $-1$

**6.** $f(x) = \dfrac{\sin 5x}{x}, x = 0$   $5$

**7.** $f(x) = \dfrac{x^2 - 3x - 10}{x + 2}, x = -2$   $-7$

**8.** $f(x) = |x - 1|, x = 1$   $0$

**9.** $f(x) = 2 \tan^{-1} x, x = 0$   $0$

**10.** $f(x) = \dfrac{2}{1 - 2^x}, x = 0$   none

In Exercises 11–14, find the limit. Support your answer with an appropriate table.

**11.** $\lim\limits_{x \to -\infty} \dfrac{-1}{(x + 2)^2}$   $0$

**12.** $\lim\limits_{x \to \infty} \dfrac{x + 5}{x - 3}$   $1$

**13.** $\lim\limits_{x \to \infty} \dfrac{2 - x^2}{x}$   $-\infty$

**14.** $\lim\limits_{x \to -\infty} \dfrac{x^2}{x - 2}$   $-\infty$

In Exercises 15–18, find the limit.

**15.** $\lim\limits_{x \to 2^+} \dfrac{1}{x - 2}$   $\infty$

**16.** $\lim\limits_{x \to 2^-} \dfrac{1}{x^2 - 4}$   $-\infty$

**17.** $\lim\limits_{x \to 0} \dfrac{1/(2 + x) - 1/2}{x}$   $-\dfrac{1}{4}$

**18.** $\lim\limits_{x \to 0} \dfrac{(2 + x)^3 - 8}{x}$   $12$

In Exercises 19–20, find the vertical and horizontal asymptotes, if any.

**19.** $f(x) = \dfrac{x - 5}{x^2 + 6x + 5}$

**20.** $f(x) = \dfrac{x^2 + 1}{2x - 4}$

In Exercises 21–26, find the limit algebraically.

**21.** $\lim\limits_{x \to 3} \dfrac{x^2 + 2x - 15}{3 - x}$   $-8$

**22.** $\lim\limits_{x \to 1} \dfrac{x^2 - 4x + 3}{x - 1}$   $-2$

**23.** $\lim\limits_{x \to 0} \dfrac{1/(-3 + x) + 1/3}{x}$   $-\dfrac{1}{9}$

**24.** $\lim\limits_{x \to 2} \dfrac{\tan (3x - 6)}{x - 2}$   $3$

**25.** $\lim\limits_{x \to 2} \dfrac{x^2 - 5x + 6}{x^2 - 3x + 2}$   $-1$

**26.** $\lim\limits_{x \to 3} \dfrac{(x - 3)^2}{x - 3}$   $0$

In Exercises 27 and 28, state a formula for the continuous extension of the function. (See Exercise 80, Section 10.3.)

**27.** $f(x) = \dfrac{x^3 - 1}{x - 1}$

**28.** $f(x) = \dfrac{x^2 - 6x + 5}{x - 5}$

In Exercises 29 and 30, use the limit definition to find the derivative of the function at the specified point, if it exists. Support your answer numerically with an NDER calculator estimate.

**29.** $f(x) = 1 - x - 2x^2$ at $x = 2$ $-9$

**30.** $f(x) = (x + 3)^2$ at $x = 2$ $10$

In Exercises 31 and 32, find **(a)** the average rate of change of the function over the interval $[3, 3.01]$ and **(b)** the instantaneous rate of change at $x = 3$.

**31.** $f(x) = x^2 + 2x - 3$ $8.01; 8$ **32.** $f(x) = \dfrac{3}{x + 2}$

In Exercises 33 and 34, **(a)** find the slope of and **(b)** an equation of the line tangent to the graph of the function at the indicated point

**33.** $f(x) = x^3 - 2x + 1$ at $x = 1$ $1; y = x - 1$

**34.** $f(x) = \sqrt{x - 4}$ at $x = 7$

In Exercises 35 and 36, find the derivative of $f$.

**35.** $f(x) = 5x^2 + 7x - 1$ **36.** $f(x) = 2 - 8x + 3x^2$

In Exercises 37 and 38, complete the following for the indicated interval $[a, b]$.

**(a)** Verify the given function is non-negative.

**(b)** Use a calculator to find the LRAM, RRAM, and average approximations for the area under the graph of the function from $x = a$ to $x = b$ with 50 approximating rectangles.

**37.** $f(x) = (x - 5)^2$; $[0, 4]$

**38.** $f(x) = 2x^2 - 3x + 1$; $[1, 5]$

**39. Take a Wild Ride on the DOW** The data given in the Table 10.10 represents the closing Dow Jones Industrial Averages at the end of each month in 1999.

**Table 10.10 Monthly Dow Jones Averages in 1999**

| Month | Average |
|---|---|
| January | 9400 |
| February | 9350 |
| March | 9800 |
| April | 10,760 |
| May | 10,550 |
| June | 10,975 |
| July | 10,800 |
| August | 10,050 |
| September | 10,400 |
| October | 10,650 |
| November | 10,900 |
| December | 11,490 |

**(a)** Draw a scatter plot of the data in Table 10.10. Use $x = 1$ for January, $x = 2$ for February, and so forth.

**(b)** Find the average rate of change from April to May and from June to July. $-\$210/\text{month}; -\$175/\text{month}$

**(c)** From what month to the next consecutive month does the average rate of change exhibit the greatest increase.

**(d)** From what month to the next consecutive month does the average rate of change exhibit the greatest decrease?

**(e)** Find a linear regression model for the data and overlay its graph on a scatter plot of the data.

**(f) Group Activity** Find a cubic regression model for the data and overlay its graph on a scatter plot of the data. Discuss pro and con arguments that this is a good model for the DOW data. Compare the cubic model with the linear model. Which one does your group think is best? Why?

**(g) Writing to Learn** Use the cubic regression model found in (f) and NDER to find the instantaneous rate of change at the end of March, the end of June, the end of August, and the end of December. How could the cubic model give some misleading information?

**(h) Writing to Learn** Use the cubic regression model found in (f) to predict the DOW average at the end of December in the year 2000. Do you think this is a reasonable estimate? Give reasons. Find the actual DOW average on the last day of December of 2000 (use the Internet) and compare with your answer.

**40. An Interesting Connection**
Let $A(x) = \text{NINT}(\cos t, t, 0, x)$

**(a)** Draw a scatter plot of the pairs $(x, A(x))$ for $x = 0, 0.4, 0.8, 0.12, \ldots, 6.0, 6.4$.

**(b)** Find a function that seems to model the data in (a) and overlay its graph on a scatter plot of the data.

**(c)** Assuming that the function found in (b) agrees with $A(x)$ for all vales of $x$, what is the derivative of $A(x)$?

**(d) Writing to Learn** Describe what seems to be true about the derivative of $\text{NINT}(f(t), t, 0, x)$.

## Chapter 10 Project

### Estimating Population Growth Rates

Las Vegas, Nevada and its surrounding cities represent one of the fastest-growing areas in the United States. Clark County, the county in which Las Vegas is located, has grown by over one million people in the past three decades. The data in the table below (obtained from the web site www.unlv.edu/Research_Centers/cber/pop.html) summarizes the growth of Clark County from 1970 through 1998.

| Year | Population |
|------|-----------|
| 1970 | 277,230 |
| 1980 | 463,087 |
| 1990 | 770,280 |
| 1995 | 1,036,290 |
| 1996 | 1,115,940 |
| 1997 | 1,192,200 |
| 1998 | 1,255,200 |

### Explorations

1. Enter the data in the table above into your graphing calculator or computer. (Let $t = 0$ represent 1970.) Draw a scatter plot of the data.

2. Find the average population growth rates for Clark County from 1970 to 1998, from 1980 to 1998, from 1990 to 1998, and from 1996 to 1998.

3. Use your calculator or computer to find an exponential regression equation to model the population data set (see your grapher's guidebook for instructions on how to do this).

4. Use the exponential model you just found in question (3) and your calculator/computer NDER feature to estimate the instantaneous population growth rate in 1998. Which of the average growth rates you found in question (2) most closely matches this instantaneous growth rate? Explain why this makes sense.

5. Use the exponential regression model you found in question (3) to predict the population of Clark County in the years 2000, 2010, 2020, and 2030. Then go to the web site www.co.clark.nv.us/COMPPLAN/ADVPLAN/Rpteam /Demo/Popbroc2.htm which gives the Ten Year Projections, and record the predicted populations for 2000, 2010, 2020, and 2030. Are these predictions reasonable? Explain why or why not.

**Appendixes Overview**

This section contains a review of some basic algebraic skills. (You should read Section P.1 before reading this appendix.) Radical and rational expressions are introduced and radical expressions are simplified algebraically. We add, subtract, and multiply polynomials and factor simple polynomials by a variety of techniques. These factoring techniques are used to add, subtract, multiply, and divide fractional expressions.

## A.1 Radicals and Rational Exponents

Radicals • Simplifying Radical Expressions • Rationalizing the Denominator • Rational Exponents

### Radicals

If $b^2 = a$, then $b$ is a **square root** of $a$. For example, both 2 and $-2$ are square roots of 4 because $(2)^2 = (-2)^2 = 4$. Similarly, $b$ is a **cube root** of $a$ if $b^3 = a$. For example, 2 is a cube root of 8 because $2^3 = 8$.

Every real number has exactly one real *nth root* whenever $n$ is odd. For example, 2 is the only real cube root of 8. When $n$ is even, positive real numbers have two real *nth roots* and negative real numbers have no real *n*th roots. For example, the real fourth roots of 16 are $\pm 2$, and $-16$ has no real fourth roots. The *principal* fourth root of 16 is 2.

---

**Definition  Real *n*th Root of a Real Number**

Let $n$ be an integer greater than 1 and $a$ and $b$ real numbers.

**1.** If $b^n = a$, then $b$ is an ***n*th root** of $a$.

**2.** If $a$ has an $n$th root, the **principal *n*th root** of $a$ is the $n$th root having the same sign as $a$.

The principal $n$th root of $a$ is denoted by the **radical expression** $\sqrt[n]{a}$. The positive integer $n$ is the **index** of the radical and $a$ is the **radicand**.

---

When $n = 2$, special notation is used for roots. We omit the index and write $\sqrt{a}$ instead of $\sqrt[2]{a}$. If $a$ is a positive real number and $n$ a positive even integer, its two $n$th roots are denoted by $\sqrt[n]{a}$ and $-\sqrt[n]{a}$.

**Example 1  FINDING PRINCIPAL *n*TH ROOTS**

**(a)** $\sqrt{36} = 6$ because $6^2 = 36$.

**(b)** $\sqrt[3]{\dfrac{27}{8}} = \dfrac{3}{2}$ because $\left(\dfrac{3}{2}\right)^3 = \dfrac{27}{8}$.

**(c)** $\sqrt[3]{-\dfrac{27}{8}} = -\dfrac{3}{2}$ because $\left(-\dfrac{3}{2}\right)^3 = -\dfrac{27}{8}$.

**(d)** $\sqrt[4]{-625}$ is not a real number because the index 4 is even and the radicand $-625$ is negative (there is *no* real number whose fourth power is negative).

### Principal *n*th Roots and Calculators

Most calculators have a key for the principal *n*th root. Use this feature of your calculator to check the computations in Example 1.

### Caution

Without the restriction that preceded the list, Property 5 would need special attention. For example,

$$\sqrt{(-3)^4} \neq (\sqrt{-3})^4$$

because $\sqrt{-3}$ on the right is not a real number.

Here are some properties of radicals together with examples that help illustrate their meaning.

---

**Properties of Radicals**

Let $u$ and $v$ be real numbers, variables, or algebraic expressions, and $m$ and $n$ be positive integers greater than 1. We assume that all of the roots are real numbers and all of the denominators are not zero.

| Property | Example |
|---|---|
| **1.** $\sqrt[n]{uv} = \sqrt[n]{u} \cdot \sqrt[n]{v}$ | $\sqrt{75} = \sqrt{25 \cdot 3}$ |
| | $\qquad = \sqrt{25} \cdot \sqrt{3} = 5\sqrt{3}$ |
| **2.** $\sqrt[n]{\dfrac{u}{v}} = \dfrac{\sqrt[n]{u}}{\sqrt[n]{v}}$ | $\dfrac{\sqrt[4]{96}}{\sqrt[4]{6}} = \sqrt[4]{\dfrac{96}{6}} = \sqrt[4]{16} = 2$ |
| **3.** $\sqrt[m]{\sqrt[n]{u}} = \sqrt[m \cdot n]{u}$ | $\sqrt{\sqrt[3]{7}} = \sqrt[2 \cdot 3]{7} = \sqrt[6]{7}$ |
| **4.** $(\sqrt[n]{u})^n = u$ | $(\sqrt[4]{5})^4 = 5$ |
| **5.** $\sqrt[n]{u^m} = (\sqrt[n]{u})^m$ | $\sqrt[3]{27^2} = (\sqrt[3]{27})^2 = 3^2 = 9$ |
| **6.** $\sqrt[n]{u^n} = \begin{cases} |u| & n \text{ even} \\ u & n \text{ odd.} \end{cases}$ | $\sqrt{(-6)^2} = |-6| = 6$ |
| | $\sqrt[3]{(-6)^3} = -6$ |

---

## Simplifying Radical Expressions

Many simplifying techniques for roots of real numbers have been rendered obsolete because of calculators. For example, when determining the decimal form of $1/\sqrt{2}$, it was once very common first to change the fraction so that the radical was in the numerator:

$$\frac{1}{\sqrt{2}} = \frac{1}{\sqrt{2}} \cdot \frac{\sqrt{2}}{\sqrt{2}} = \frac{\sqrt{2}}{2}$$

Using paper and pencil, it was then easier to divide a decimal for $\sqrt{2}$ by 2 than to divide the decimal into 1. Now either form is quickly computed with a calculator. However, these techniques are still valid for radicals involving algebraic expressions and for numerical computations when you need exact answers. Example 2 illustrate the technique of *removing factors from radicands*.

### Example 2   REMOVING FACTORS FROM RADICANDS

**(a)** $\sqrt[4]{80} = \sqrt[4]{16 \cdot 5}$        Find greatest fourth-power factor.

$\qquad = \sqrt[4]{2^4 \cdot 5}$

$\qquad = \sqrt[4]{2^4} \cdot \sqrt[4]{5}$        Property 1

$\qquad = 2\sqrt[4]{5}$        Property 6

**(b)** $\sqrt{18x^5} = \sqrt{9x^4 \cdot 2x}$        Finding greatest square factor.

$\qquad = \sqrt{(3x^2)^2 \cdot 2x}$

$\qquad = 3x^2\sqrt{2x}$        Properties 1 and 6

**(c)** $\sqrt[4]{x^4 y^4} = \sqrt[4]{(xy)^4}$        Finding greatest fourth-power factor.

$\qquad = |xy|$        Property 6

**(d)** $\sqrt[3]{-24y^6} = \sqrt[3]{(-2y^2)^3 \cdot 3}$  Finding greatest cube factor.

$= -2y^2\sqrt[3]{3}$  Properties 1 and 6

## Rationalizing the Denominator

The process of rewriting fractions containing radicals so that the denominator is free of radicals is **rationalizing the denominator**. When the denominator has the form $\sqrt[n]{u^k}$, multiplying numerator and denominator by $\sqrt[n]{u^{n-k}}$ and using Property 6 will eliminate the radical from the denominator because

$$\sqrt[n]{u^k} \cdot \sqrt[n]{u^{n-k}} = \sqrt[n]{u^{k+n-k}} = \sqrt[n]{u^n}.$$

Example 3 illustrates the process.

**Example 3**  RATIONALIZING THE DENOMINATOR

**(a)** $\sqrt{\dfrac{2}{3}} = \dfrac{\sqrt{2}}{\sqrt{3}} = \dfrac{\sqrt{2}}{\sqrt{3}} \cdot \dfrac{\sqrt{3}}{\sqrt{3}} = \dfrac{\sqrt{6}}{3}$

**(b)** $\dfrac{1}{\sqrt[4]{x}} = \dfrac{1}{\sqrt[4]{x}} \cdot \dfrac{\sqrt[4]{x^3}}{\sqrt[4]{x^3}} = \dfrac{\sqrt[4]{x^3}}{\sqrt[4]{x^4}} = \dfrac{\sqrt[4]{x^3}}{|x|}$

**(c)** $\sqrt[5]{\dfrac{x^2}{y^3}} = \dfrac{\sqrt[5]{x^2}}{\sqrt[5]{y^3}} = \dfrac{\sqrt[5]{x^2}}{\sqrt[5]{y^3}} \cdot \dfrac{\sqrt[5]{y^2}}{\sqrt[5]{y^2}} = \dfrac{\sqrt[5]{x^2y^2}}{\sqrt[5]{y^5}} = \dfrac{\sqrt[5]{x^2y^2}}{y}$

## Rational Exponents

We know how to handle exponential expressions with integer exponents. For example, $x^3 \cdot x^4 = x^7$, $(x^3)^2 = x^6$, $x^5/x^2 = x^3$, $x^{-2} = 1/x^2$, and so forth. But exponents can also be rational numbers. How should we define $x^{1/2}$? If we assume that the same rules that apply for integer exponents also apply for rational exponents we get a clue. For example, we want

$$x^{1/2} \cdot x^{1/2} = x^1.$$

This equation suggests that $x^{1/2} = \sqrt{x}$. In general, we have the following definition.

**Definition**  **Rational Exponents**

Let $u$ be a real number, variable, or algebraic expression, and $n$ an integer greater than 1. Then

$$u^{1/n} = \sqrt[n]{u}.$$

If $m$ is a positive integer, $m/n$ is in reduced form, and all roots are real numbers, then

$$u^{m/n} = (u^{1/n})^m = (\sqrt[n]{u})^m \quad \text{and} \quad u^{m/n} = (u^m)^{1/n} = \sqrt[n]{u^m}.$$

The numerator of a rational exponent is the *power* to which the base is raised, and the denominator is the *root* to be taken. The fraction $m/n$ needs to be in reduced form because, for example,

$$u^{2/3} = (\sqrt[3]{u})^2$$

is defined for all real numbers $u$ (every real number has a cube root), but

$$u^{4/6} = (\sqrt[6]{u})^4$$

is defined only for $u \geq 0$ (only nonnegative real numbers have sixth roots).

### Example 4   CONVERTING RADICALS TO EXPONENTIALS AND VICE VERSA

**(a)** $\sqrt{(x + y)^3} = (x + y)^{3/2}$   **(b)** $3x\sqrt[5]{x^2} = 3x \cdot x^{2/5} = 3x^{7/5}$

**(c)** $x^{2/3}y^{1/3} = (x^2y)^{1/3} = \sqrt[3]{x^2y}$   **(d)** $z^{-3/2} = \dfrac{1}{z^{3/2}} = \dfrac{1}{\sqrt{z^3}}$

**Simplifying Radicals**

If you also want the radical form in Example 4d to be simplified, then continue as follows:

$$\frac{1}{\sqrt{z^3}} = \frac{1}{\sqrt{z^3}} \cdot \frac{\sqrt{z}}{\sqrt{z}} = \frac{\sqrt{z}}{z^2}$$

An expression involving powers is *simplified* if each factor appears only once and all exponents are positive. Example 5 illustrates.

### Example 5   SIMPLIFYING EXPONENTIAL EXPRESSIONS

**(a)** $(x^2y^9)^{1/3}(xy^2) = (x^{2/3}y^3)(xy^2) = x^{5/3}y^5$

**(b)** $\left(\dfrac{3x^{2/3}}{y^{1/2}}\right)\left(\dfrac{2x^{-1/2}}{y^{2/5}}\right) = \dfrac{6x^{1/6}}{y^{9/10}}$

Example 6 suggests how to simplify a sum or difference of radicals.

### Example 6   COMBINING RADICALS

**(a)** $2\sqrt{80} - \sqrt{125} = 2\sqrt{16 \cdot 5} - \sqrt{25 \cdot 5}$   Find greatest square factors.

$\qquad\qquad\qquad\quad = 8\sqrt{5} - 5\sqrt{5}$   Remove factors from radicands.

$\qquad\qquad\qquad\quad = 3\sqrt{5}$   Distributive property.

**(b)** $\sqrt{4x^2y} - \sqrt{y^3} = \sqrt{(2x)^2y} - \sqrt{y^2y}$   Find greatest square factors.

$\qquad\qquad\qquad\quad = 2|x|\sqrt{y} - |y|\sqrt{y}$   Remove factors from radicands.

$\qquad\qquad\qquad\quad = (2|x| - |y|)\sqrt{y}$   Distributive property.

Here is a summary of the procedures we use to **simplify expressions** involving radicals.

1. Remove factors from the radicand (see Example 2).

2. Eliminate radicals from denominators and denominators from radicands (see Example 3).

3. Combine sums and differences of radicals, if possible (see Example 6).

# Appendix A.1 Exercises

In Exercises 1–6, find the indicated real roots.

**1.** Square roots of 81   9 or $-9$   **2.** Fourth roots of 81

**3.** Cube roots of 64   4   **4.** Fifth roots of 243   3

**5.** Square roots of 16/9   **6.** Cube roots of $-27/8$

In Exercises 7–12, evaluate the expression without using a calculator.

**7.** $\sqrt{144}$   12   **8.** $\sqrt{-16}$   **9.** $\sqrt[3]{-216}$   $-6$

**10.** $\sqrt[3]{216}$   6   **11.** $\sqrt[3]{-\dfrac{64}{27}}$   $-\dfrac{4}{3}$   **12.** $\sqrt{\dfrac{64}{25}}$   $\dfrac{8}{5}$

In Exercises 13–22, use a calculator to evaluate the expression.

**13.** $\sqrt[4]{256}$  4

**14.** $\sqrt[5]{3125}$  5

**15.** $\sqrt[3]{15.625}$  2.5

**16.** $\sqrt{12.25}$  3.5

**17.** $81^{3/2}$  729

**18.** $16^{5/4}$  32

**19.** $32^{-2/5}$  0.25

**20.** $27^{-4/3}$

**21.** $\left(-\dfrac{1}{8}\right)^{-1/3}$  $-2$

**22.** $\left(-\dfrac{125}{64}\right)^{-1/3}$  $-0.8$

In Exercises 23–26, use the information from the grapher screens below to evaluate the expresssion.

| 1.5^3 | |
|---|---|
| | 3.375 |
| 4.41^2 | |
| | 19.4481 |

| 1.3^2 | |
|---|---|
| | 1.69 |
| 2.1^4 | |
| | 19.4481 |

**23.** $\sqrt{1.69}$  1.3

**24.** $\sqrt{19.4481}$  4.41

**25.** $\sqrt[4]{19.4481}$  2.1

**26.** $\sqrt[3]{3.375}$  1.5

In Exercises 27–36, simplify by removing factors from the radicand.

**27.** $\sqrt{288}$  $12\sqrt{2}$

**28.** $\sqrt[3]{500}$  $5\sqrt[3]{4}$

**29.** $\sqrt[3]{-250}$  $-5\sqrt[3]{2}$

**30.** $\sqrt[4]{192}$  $2\sqrt[4]{12}$

**31.** $\sqrt{2x^3y^4}$  $|x|y^2\sqrt{2x}$

**32.** $\sqrt[3]{-27x^3y^6}$  $-3xy^2$

**33.** $\sqrt[4]{3x^8y^6}$  $x^2|y|\sqrt[4]{3y^2}$

**34.** $\sqrt[3]{8x^6y^4}$  $2x^2y\sqrt[3]{y}$

**35.** $\sqrt[5]{96x^{10}}$  $2x^2\sqrt[5]{3}$

**36.** $\sqrt{108x^4y^9}$  $6x^2y^4\sqrt{3y}$

In Exercises 37–42, rationalize the denominator.

**37.** $\dfrac{4}{\sqrt[3]{2}}$  $2\sqrt[3]{4}$

**38.** $\dfrac{1}{\sqrt{5}}$  $\dfrac{\sqrt{5}}{5}$

**39.** $\dfrac{1}{\sqrt[5]{x^2}}$  $\dfrac{\sqrt[5]{x^3}}{x}$

**40.** $\dfrac{2}{\sqrt[4]{y}}$  $\dfrac{2\sqrt[4]{y^3}}{y}$

**41.** $\sqrt[3]{\dfrac{x^2}{y}}$  $\dfrac{\sqrt[3]{x^2y^2}}{y}$

**42.** $\sqrt[5]{\dfrac{a^3}{b^2}}$  $\dfrac{\sqrt[5]{a^3b^3}}{b}$

In Exercises 43–46, convert to exponential form.

**43.** $\sqrt[3]{(a+2b)^2}$  $(a+2b)^{2/3}$

**44.** $\sqrt[5]{x^2y^3}$  $x^{2/5}y^{3/5}$

**45.** $2x\sqrt[3]{x^2y}$  $2x^{5/3}y^{1/3}$

**46.** $xy\sqrt[4]{xy^3}$  $x^{5/4}y^{7/4}$

In Exercises 47–50, convert to radical form.

**47.** $a^{3/4}b^{1/4}$  $\sqrt[4]{a^3b}$

**48.** $x^{2/3}y^{1/3}$  $\sqrt[3]{x^2y}$

**49.** $x^{-5/3}$  $\sqrt[3]{x^{-5}}=1/\sqrt[3]{x^5}$

**50.** $(xy)^{-3/4}$

In Exercises 51–56, write using a single radical.

**51.** $\sqrt{\sqrt{2x}}$  $\sqrt[4]{2x}$

**52.** $\sqrt{\sqrt[3]{3x^2}}$  $\sqrt[6]{3x^2}$

**53.** $\sqrt[4]{\sqrt{xy}}$  $\sqrt[8]{xy}$

**54.** $\sqrt[3]{\sqrt{ab}}$  $\sqrt[6]{ab}$

**55.** $\dfrac{\sqrt[5]{a^2}}{\sqrt[3]{a}}$  $\sqrt[15]{a}$

**56.** $\sqrt{a}\sqrt[3]{a^2}$  $\sqrt[6]{a^7}$

In Exercises 57–64, simplify the exponential expression.

**57.** $\dfrac{a^{3/5}a^{1/3}}{a^{3/2}}$  $a^{-17/30}$

**58.** $(x^2y^4)^{1/2}$  $|x|y^2$

**59.** $(a^{5/3}b^{3/4})(3a^{1/3}b^{5/4})$

**60.** $\left(\dfrac{x^{1/2}}{y^{2/3}}\right)^6$

**61.** $\left(\dfrac{-8x^6}{y^{-3}}\right)^{2/3}$  $4x^4y^2$

**62.** $\dfrac{(p^2q^4)^{1/2}}{(27q^3p^6)^{1/3}}$  $\dfrac{q}{3|p|}$

**63.** $\dfrac{(x^9y^6)^{-1/3}}{(x^6y^2)^{-1/2}}$  $\dfrac{|x|}{x|y|}$

**64.** $\left(\dfrac{2x^{1/2}}{y^{2/3}}\right)\left(\dfrac{3x^{-2/3}}{y^{1/2}}\right)$

In Exercises 65–74, simplify the radical expression.

**65.** $\sqrt{9x^{-6}y^4}$  $3y^2/|x^3|$

**66.** $\sqrt{16y^8z^{-2}}$  $4y^4/|z|$

**67.** $\sqrt[4]{\dfrac{3x^8y^2}{8x^2}}$

**68.** $\sqrt[5]{\dfrac{4x^6y}{9x^3}}$

**69.** $\sqrt[3]{\dfrac{4x^2}{y^2}}\cdot\sqrt[3]{\dfrac{2x^2}{y}}$  $\dfrac{2x\sqrt[3]{x}}{y}$

**70.** $\sqrt[5]{9ab^6}\cdot\sqrt[5]{27a^2b^{-1}}$

**71.** $3\sqrt{48}-2\sqrt{108}$  0

**72.** $2\sqrt{175}-4\sqrt{28}$  $2\sqrt{7}$

**73.** $\sqrt{x^3}-\sqrt{4xy^2}$

**74.** $\sqrt{18x^2y}+\sqrt{2y^3}$

In Exercises 75–82, replace $\bigcirc$ with $<$, $=$, or $>$ to make a true statement.

**75.** $\sqrt{2+6}\bigcirc\sqrt{2}+\sqrt{6}$

**76.** $\sqrt{4}+\sqrt{9}\bigcirc\sqrt{4+9}$

**77.** $(3^{-2})^{-1/2}\bigcirc 3$  $(3^{-2})^{-1/2}=3$

**78.** $(2^{-3})^{1/3}\bigcirc 2$  $(2^{-3})^{1/3}<2$

**79.** $\sqrt[4]{(-2)^4}\bigcirc-2$

**80.** $\sqrt[3]{(-2)^3}\bigcirc-2$

**81.** $2^{2/3}\bigcirc 3^{3/4}$  $2^{2/3}<3^{3/4}$

**82.** $4^{-2/3}\bigcirc 3^{-3/4}$

**83.** The time $t$ (in seconds) that it takes for a pendulum to complete one period is aproximately $t=1.1\sqrt{L}$, where $L$ is the length (in feet) of the pendulum. How long is the period of a pendulum of length 10 ft?  $\approx 3.48$ sec

**84.** The time $t$ (in seconds) that it takes for a rock to fall a distance $d$ (in meters) is approximately $t=0.45\sqrt{d}$. How long does it take for the rock to fall a distance of 200 m?

**85. Writing to Learn** Explain why $\sqrt[n]{a}$ and a real $n$th root of $a$ need not have the same value.

## Polynomials and Factoring

Adding, Subtracting, and Multiplying Polynomials  •  Special Products  •  Factoring Polynomials Using Special Products  •  Factoring Trinomials  •  Factoring by Grouping

### Adding, Subtracting, and Multiplying Polynomials

A **polynomial in $x$** is any expression that can be written in the form

$$a_n x^n + a_{n-1} x^{n-1} + \cdots + a_1 x + a_0,$$

where $n$ is a nonnegative integer, and $a_n \neq 0$. $a_{n-1}, \ldots, a_1, a_0$ are real numbers called **coefficients**. The **degree of the polynomial** is $n$ and the **leading coefficient** is $a_n$. Polynomials with one, two, or three terms are **monomials, binomials**, or **trinomials**, respectively. A polynomial written with powers of $x$ in *descending order* is in **standard form**.

To add or subtract polynomials we add or subtract *like terms* using the distributive property. Terms of polynomials that have the same variable each raised to the same power are **like terms**.

**Example 1   ADDING AND SUBTRACTING POLYNOMIALS**

(a) $(2x^3 - 3x^2 + 4x - 1) + (x^3 + 2x^2 - 5x + 3)$

(b) $(4x^2 + 3x - 4) - (2x^3 + x^2 - x + 2)$

**Solution**

(a) We group like terms and then combine them as follows:

$$(2x^3 + x^3) + (-3x^2 + 2x^2) + (4x + (-5x)) + (-1 + 3)$$
$$= 3x^3 - x^2 - x + 2$$

(b) We group like terms and then combine them as follows:

$$(0 - 2x^3) + (4x^2 - x^2) + (3x - (-x)) + (-4 - 2)$$
$$= -2x^3 + 3x^2 + 4x - 6$$

To **expand the product** of two polynomials we use the distributive property. Here is what the procedure looks like when we multiply the binomials $3x + 2$ and $4x - 5$.

$(3x + 2)(4x - 5)$

$$= 3x(4x - 5) + 2(4x - 5) \qquad\qquad \text{Distributive property}$$
$$= (3x)(4x) - (3x)(5) + (2)(4x) - (2)(5) \qquad \text{Distributive property}$$
$$= \underbrace{12x^2}_{\substack{\text{Product of} \\ \text{First terms}}} - \underbrace{15x}_{\substack{\text{Product of} \\ \text{Outer terms}}} + \underbrace{8x}_{\substack{\text{Product of} \\ \text{Inner terms}}} - \underbrace{10}_{\substack{\text{Product of} \\ \text{Last terms}}}$$

In the above **FOIL method** for products of binomials, the outer ($O$) and inner ($I$) terms are like terms and can be added to give

$$(3x + 2)(4x - 5) = 12x^2 - 7x - 10.$$

Multiplying two polynomials requires multiplying each term of one polynomial by every term of the other polynomial. A convenient way to compute a product is to arrange the polynomials in standard form one on top of another so their terms align vertically, as illustrated in Example 2.

**Example 2  MULTIPLYING POLYNOMIALS IN VERTICAL FORM**

Write $(x^2 - 4x + 3)(x^2 + 4x + 5)$ in standard form.

**Solution**

$$
\begin{array}{r}
x^2 - 4x + 3 \\
x^2 + 4x + 5 \\
\hline
x^4 - 4x^3 + \phantom{0}3x^2 \\
4x^3 - 16x^2 + 12x \\
5x^2 - 20x + 15 \\
\hline
x^4 + 0x^3 - \phantom{0}8x^2 - \phantom{0}8x + 15
\end{array}
$$

$= x^2(x^2 - 4x + 3)$

$= 4x(x^2 - 4x + 3)$

$= 5(x^2 - 4x + 3)$

Add.

Thus,

$$(x^2 - 4x + 3)(x^2 + 4x + 5) = x^4 - 8x^2 - 8x + 15.$$

## Special Products

Certain products provide patterns that will be useful when we factor polynomials. Here is a list of some special products for binomials.

---

**Special Binomial Products**

Let $u$ and $v$ be real numbers, variables, or algebraic expressions.

1. **Product of a sum and a difference:**  $(u + v)(u - v) = u^2 - v^2$
2. **Square of a sum:**  $(u + v)^2 = u^2 + 2uv + v^2$
3. **Square of a difference:**  $(u - v)^2 = u^2 - 2uv + v^2$
4. **Cube of a sum:**  $(u + v)^3 = u^3 + 3u^2v + 3uv^2 + v^3$
5. **Cube of a difference.**  $(u - v)^3 = u^3 - 3u^2v + 3uv^2 - v^3$

---

Identifying a product as one of the special types simplifies the multiplication process as illustrated in Example 3.

**Example 3  USING SPECIAL PRODUCTS**

Expand the products.

**(a)** $(3x + 8)(3x - 8) = (3x)^2 - 8^2$      Product of a sum and difference

$\phantom{(3x + 8)(3x - 8)} = 9x^2 - 64$      Simplify.

**(b)** $(5y - 4)^2 = (5y)^2 - 2(5y)(4) + 4^2$      Square of a difference

$\phantom{(5y - 4)^2} = 25y^2 - 40y + 16$      Simplify.

**(c)** $(2x - 3y)^3 = (2x)^3 - 3(2x)^2(3y)$

$\phantom{(2x - 3y)^3 =} + 3(2x)(3y)^2 - (3y)^3$      Cube of a difference

$\phantom{(2x - 3y)^3} = 8x^3 - 36x^2y + 54xy^2 - 27y^3$      Simplify.

## Factoring Polynomials Using Special Products

When we write a polynomial as a product of two or more **polynomial factors** we are **factoring a polynomial**. Unless specified otherwise, we factor polynomials into factors of lesser degree and with integer coefficients in this appendix. A polynomial that cannot be factored using integer coefficients is a **prime polynomial**.

A polynomial is **completely factored** if it is written as a product of its prime factors. For example,

$$2x^2 + 7x - 4 = (2x - 1)(x + 4)$$

and

$$x^3 + x^2 + x + 1 = (x + 1)(x^2 + 1)$$

are completely factored (it can be shown that $x^2 + 1$ is prime). However,

$$x^3 - 9x = x(x^2 - 9)$$

is not completely factored because $(x^2 - 9)$ is *not* prime. In fact, $x^2 - 9 = (x - 3)(x + 3)$ and

$$x^3 - 9x = x(x - 3)(x + 3)$$

is completely factored.

The first step in factoring a polynomial is to remove common factors from its terms using the distributive property as illustrated by Example 4.

### Example 4  REMOVING COMMON FACTORS

**(a)** $2x^3 + 2x^2 - 6x = 2x(x^2 + x - 3)$     $2x$ is the common factor.

**(b)** $u^3v + uv^3 = uv(u^2 + v^2)$     $uv$ is the common factor.

Recognizing the expanded form of the five special binomial products will help us factor them. The special form that is easiest to identify is the difference of two squares. The two binomial factors have opposite signs:

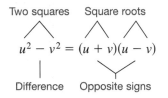

### Example 5  FACTORING THE DIFFERENCE OF TWO SQUARES

**(a)** $25x^2 - 36 = (5x)^2 - 6^2$     Difference of two squares

  $\qquad\qquad = (5x + 6)(5x - 6)$     Factors are prime.

**(b)** $4x^2 - (y + 3)^2 = (2x)^2 - (y + 3)^2$     Difference of two squares

  $\qquad\qquad = [2x + (y + 3)][2x - (y + 3)]$     Factors are prime.

  $\qquad\qquad = (2x + y + 3)(2x - y - 3)$

A perfect square trinomial is the square of a binomial and has one of the two forms shown here. The first and last terms are squares of $u$ and $v$, and the middle term is twice the product of $u$ and $v$. The operation signs before the middle term and in the binomial factor are the same.

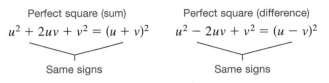

Perfect square (sum)

$$u^2 + 2uv + v^2 = (u + v)^2$$

Same signs

Perfect square (difference)

$$u^2 - 2uv + v^2 = (u - v)^2$$

Same signs

**Example 6** FACTORING PERFECT SQUARE TRINOMIALS

(a) $9x^2 + 6x + 1 = (3x)^2 + 2(3x)(1) + 1^2$      $u = 3x, v = 1$

               $= (3x + 1)^2$

(b) $4x^2 - 12xy + 9y^2 = (2x)^2 - 2(2x)(3y) + (3y)^2$      $u = 2x, v = 3y$

                    $= (2x - 3y)^2$

In the sum and difference of two cubes, notice the pattern of the signs.

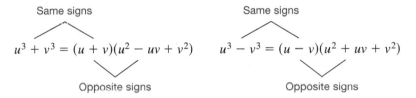

Same signs

$$u^3 + v^3 = (u + v)(u^2 - uv + v^2)$$

Opposite signs

Same signs

$$u^3 - v^3 = (u - v)(u^2 + uv + v^2)$$

Opposite signs

**Example 7** FACTORING THE SUM AND DIFFERENCE OF TWO CUBES

(a) $x^3 - 64 = x^3 - 4^3$          Difference of two cubes

         $= (x - 4)(x^2 + 4x + 16)$      Factors are prime.

(b) $8x^3 + 27 = (2x)^3 + 3^3$        Sum of two cubes

         $= (2x + 3)(4x^2 - 6x + 9)$      Factors are prime.

## Factoring Trinomials

Factoring the trinomial $ax^2 + bx + c$ into a product of binomials with integer coefficients requires factoring the integers $a$ and $c$.

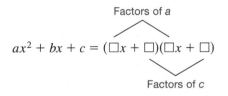

Factors of $a$

$$ax^2 + bx + c = (\Box x + \Box)(\Box x + \Box)$$

Factors of $c$

Because the number of integer factors of $a$ and $c$ is finite, we can list all possible binomial factors. Then we begin checking each possibility until we find a pair that works. (If no pair works, then the trinomial is prime.) Example 8 illustrates.

**Example 8** FACTORING A TRINOMIAL WITH LEADING
COEFFICIENT = 1

Factor $x^2 + 5x - 14$.

**Solution** The only factor pair of the leading coefficient is 1 and 1. The factor pairs of 14 are 1 and 14, and 2 and 7. Here are the four possible factorizations of the trinomial:

$$(x + 1)(x - 14) \qquad (x - 1)(x + 14)$$
$$(x + 2)(x - 7) \qquad (x - 2)(x + 7)$$

If you check the middle term from each factorization you will find that

$$x^2 + 5x - 14 = (x - 2)(x + 7).$$

With practice you will find that it usually is not necessary to list all possible binomial factors. Often you can test the possibilities mentally.

**Example 9** FACTORING A TRINOMIAL WITH LEADING
COEFFICIENT ≠ 1

Factor $35x^2 - x - 12$.

**Solution** The factor pairs of the leading coefficient are 1 and 35, and 5 and 7. The factor pairs of 12 are 1 and 12, 2 and 6, and 3 and 4. The possible factorizations must be of the form

$$(x - *)(35x + ?), \qquad (x + *)(35x - ?),$$
$$(5x - *)(7x + ?), \qquad (5x + *)(7x - ?),$$

where * and ? are one of the factor pairs of 12. Because the two binomial factors have opposite signs, there are 6 possibilities for each of the four forms—a total of 24 possibilities in all. If you try them, mentally and systematically, you should find that

$$35x^2 - x - 12 = (5x - 3)(7x + 4).$$

We can extend the technique of Examples 8 and 9 to trinomials in two variables as illustrated in Example 10.

**Example 10** FACTORING TRINOMIALS IN $x$ AND $y$

Factor $3x^2 - 7xy + 2y^2$.

**Solution** The only way to get $-7xy$ as the middle term is with $3x^2 - 7xy + 2y^2 = (3x - ?y)(x - ?y)$.

The signs in the binomials must be negative because the coefficient of $y^2$ is positive *and* the coefficient of the middle term is negative. Checking the two possibilities, $(3x - y)(x - 2y)$ and $(3x - 2y)(x - y)$, shows that

$$3x^2 - 7xy + 2y^2 = (3x - y)(x - 2y).$$

## Factoring by Grouping

Notice that $(a + b)(c + d) = ac + ad + bc + bd$. If a polynomial with four terms is the product of two binomials, we can group terms to factor. There are only three ways to group the terms and two of them work. So, if two of the possibilities fail, then it is not factorable.

**Example 11   FACTORING BY GROUPING**

(a) $3x^3 + x^2 - 6x - 2$

$\qquad = (3x^3 + x^2) - (6x + 2)$       Group terms.

$\qquad = x^2(3x + 1) - 2(3x + 1)$       Factor each group.

$\qquad = (3x + 1)(x^2 - 2)$       Distributive property

(b) $2ac - 2ad + bc - bd$

$\qquad = (2ac - 2ad) + (bc - bd)$       Group terms.

$\qquad = 2a(c - d) + b(c - d)$       Factor each group.

$\qquad = (c - d)(2a + b)$       Distributive property

Here is a checklist for factoring polynomials.

---

**Factoring Polynomials**

**1.** Look for common factors.

**2.** Look for special polynomial forms.

**3.** Use factor pairs.

**4.** If there are four terms, try grouping.

---

# Appendix A.2 Exercises

In Exercises 1–4, write the polynomial in standard form and state its degree.

**1.** $2x - 1 + 3x^2$

**2.** $x^2 - 2x - 2x^3 + 1$

**3.** $1 - x^7 - x^7 + 1$; degree 7

**4.** $x^2 - x^4 + x - 3$

In Exercises 5–8, state whether the expression is a polynomial.

**5.** $x^3 - 2x^2 + x^{-1}$  no

**6.** $\dfrac{2x - 4}{x}$  no

**7.** $(x^2 + x + 1)^2$  yes

**8.** $1 - 3x + x^4$  yes

In Exercises 9–18, simplify the expression. Write your answer in standard form.

**9.** $(x^2 - 3x + 7) + (3x^2 + 5x - 3)$  $4x^2 + 2x + 4$

**10.** $(-3x^2 - 5) - (x^2 + 7x + 12)$  $-4x^2 - 7x - 17$

**11.** $(4x^3 - x^2 + 3x) - (x^3 + 12x - 3)$  $3x^3 - x^2 - 9x + 3$

**12.** $-(y^2 + 2y - 3) + (5y^2 + 3y + 4)$  $4y^2 + y + 7$

**13.** $2x(x^2 - x + 3)$

**14.** $y^2(2y^2 + 3y - 4)$

**15.** $-3u(4u - 1) - 12u^2 + 3u$

**16.** $-4v(2 - 3v^3)$

**17.** $(2 - x - 3x^2)(5x)$

**18.** $(1 - x^2 + x^4)(2x)$

In Exercises 19–40, expand the product.

**19.** $(x - 2)(x + 5)$  $x^2 + 3x - 10$

**20.** $(2x + 3)(4x + 1)$

**21.** $(3x - 5)(x + 2)$

**22.** $(2x - 3)(2x + 3)$

**23.** $(3x - y)(3x + y)$  $9x^2 - y^2$

**24.** $(3 - 5x)^2$

**25.** $(3x + 4y)^2$

**26.** $(x - 1)^3$

**27.** $(2u - v)^3$

**28.** $(u + 3v)^3$

**29.** $(2x^3 - 3y)(2x^3 + 3y)$

**30.** $(5x^3 - 1)^2$

**31.** $(x^2 - 2x + 3)(x + 4)$

**32.** $(x^2 + 3x - 2)(x - 3)$

**33.** $(x^2 + x - 3)(x^2 + x + 1)$  $x^4 + 2x^3 - x^2 - 2x - 3$

**34.** $(2x^2 - 3x + 1)(x^2 - x + 2)$  $2x^4 - 5x^3 + 8x^2 - 7x + 2$

**35.** $(x - \sqrt{2})(x + \sqrt{2})$   $x^2 - 2$   **36.** $(x^{1/2} - y^{1/2})(x^{1/2} + y^{1/2})$

**37.** $(\sqrt{u} + \sqrt{v})(\sqrt{u} - \sqrt{v})$   **38.** $(x^2 - \sqrt{3})(x^2 + \sqrt{3})$

**39.** $(x - 2)(x^2 + 2x + 4)$   $x^3 - 8$ **40.** $(x + 1)(x^2 - x + 1)$

In Exercises 41–44, factor out the common factor.

**41.** $5x - 15$   $5(x - 3)$   **42.** $5x^3 - 20x$   $5x(x^2 - 4)$

**43.** $yz^3 - 3yz^2 + 2yz$   **44.** $2x(x + 3) - 5(x + 3)$

In Exercises 45–48, factor the difference of two squares.

**45.** $z^2 - 49$   $(z + 7)(z - 7)$   **46.** $9y^2 - 16$

**47.** $64 - 25y^2$   $(8 + 5y)(8 - 5y)$ **48.** $16 - (x + 2)^2$

In Exercises 49–52, factor the perfect square trinomial.

**49.** $y^2 + 8y + 16$   $(y + 4)^2$   **50.** $36y^2 + 12y + 1$

**51.** $4z^2 - 4z + 1$   $(2z - 1)^2$   **52.** $9z^2 - 24z + 16$ $(3z - 4)^2$

In Exercises 53–58, factor the sum or difference of two cubes.

**53.** $y^3 - 8$   $(y - 2)(y^2 + 2y + 4)$ **54.** $z^3 + 64$

**55.** $27y^3 - 8$   **56.** $64z^3 + 27$

**57.** $1 - x^3$   **58.** $27 - y^3$

In Exercises 59–68, factor the trinomial.

**59.** $x^2 + 9x + 14$   $(x + 2)(x + 7)$ **60.** $y^2 - 11y + 30$

**61.** $z^2 - 5z - 24$   $(z - 8)(z + 3)$ **62.** $6t^2 + 5t + 1$

**63.** $14u^2 - 33u - 5$   **64.** $10v^2 + 23v + 12$

**65.** $12x^2 + 11x - 15$   **66.** $2x^2 - 3xy + y^2$

**67.** $6x^2 + 11xy - 10y^2$   **68.** $15x^2 + 29xy - 14y^2$

In Exercises 69–74, factor by grouping.

**69.** $x^3 - 4x^2 + 5x - 20$   **70.** $2x^3 - 3x^2 + 2x - 3$

**71.** $x^6 - 3x^4 + x^2 - 3$   **72.** $x^6 + 2x^4 + x^2 + 2$

**73.** $2ac + 6ad - bc - 3bd$   $(c + 3d)(2a - b)$

**74.** $3uw + 12uz - 2vw - 8vz$   $(w + 4z)(3u - 2v)$

In Exercises 75–90, factor completely.

**75.** $x^3 + x$   $x(x^2 + 1)$   **76.** $4y^3 - 20y^2 + 25y$

**77.** $18y^3 + 48y^2 + 32y$   **78.** $2x^3 - 16x^2 + 14x$

**79.** $16y - y^3$   $y(4 + y)(4 - y)$ **80.** $3x^4 + 24x$

**81.** $5y + 3y^2 - 2y^3$   **82.** $z - 8z^4$

**83.** $2(5x + 1)^2 - 18$   **84.** $5(2x - 3)^2 - 20$

**85.** $12x^2 + 22x - 20$   **86.** $3x^2 + 13xy - 10y^2$

**87.** $2ac - 2bd + 4ad - bc$   **88.** $6ac - 2bd + 4bc - 3ad$

**89.** $x^3 - 3x^2 - 4x + 12$   **90.** $x^4 - 4x^3 - x^2 + 4x$

**91.** Show that the grouping

$$(2ac + bc) - (2ad + bd)$$

leads to the same factorization as in Example 11b.

---

## A.3 Fractional Expressions

Domain of an Algebraic Expression • Reducing Rational Expressions • Operations with Rational Expressions • Compound Rational Expressions

### Domain of an Algebraic Expression

A quotient of two algebraic expressions, besides being another algebraic expression, is a **fractional expression**, or simply a fraction. If the quotient can be written as the ratio of two polynomials, the fractional expression is a **rational expression**. Here are examples of each.

$$\frac{x^2 - 5x + 2}{\sqrt{x^2 + 1}} \qquad \frac{2x^3 - x^2 + 1}{5x^2 - x - 3}$$

The one on the left is a fractional expression but not a rational expression. The other is both a fractional expression and a rational expression.

Unlike polynomials, which are defined for all real numbers, some algebraic expressions are not defined for some real numbers. The set of real numbers for which an algebraic expression is defined is the **domain of the algebraic expression**.

### Example 1  FINDING DOMAINS OF ALGEBRAIC EXPRESSIONS

Find the domain of the algebraic expression.

(a) $3x^2 - x + 5$  (b) $\sqrt{x - 1}$  (c) $\dfrac{x}{x - 2}$

**Solution**

(a) The domain of $3x^2 - x + 5$, like that of any polynomial, is the set of all real numbers.

(b) Because only nonnegative numbers have square roots, $x - 1 \geq 0$, or $x \geq 1$. In interval notation the domain is $[1, \infty)$.

(c) Because division by zero is undefined, $x - 2 \neq 0$, or $x \neq 2$. The domain is the set of all real numbers except 2.

## Reducing Rational Expressions

Let $u$, $v$, and $z$ be real numbers, variables, or algebraic expressions. We can write rational expressions in simpler form using.

$$\frac{uz}{vz} = \frac{u}{v}$$

provided $z \neq 0$. This requires that we first factor the numerator and denominator into prime factors. When all factors common to numerator and denominator have been removed, the rational expression (or rational number) is in **reduced form**.

### Example 2  REDUCING RATIONAL EXPRESSIONS

Write $(x^2 - 3x)/(x^2 - 9)$ in reduced form.

**Solution**

$$\frac{x^2 - 3x}{x^2 - 9} = \frac{x(x - 3)}{(x + 3)(x - 3)} \qquad \text{Factor completely.}$$

$$= \frac{x}{x + 3}, \quad x \neq 3 \qquad \text{Remove common factors.}$$

We include $x \neq 3$ as part of the reduced form because 3 is not in the domain of the original rational expression and thus should not be in the domain of the final rational expression.

Two rational expressions are **equivalent** if they have the same domain and have the same value for all numbers in the domain. The reduced form of a rational expression must have the same domain as the original rational expression. This is why we attached the restriction $x \neq 3$ to the reduced form in Example 2.

## Operations with Rational Expressions

Two fractions are **equal**, $u/v = z/w$, if and only if $uw = vz$. Here is how we operate with fractions.

---

### Operations with Fractions

Let $u$, $v$, $w$, and $z$ be real numbers, variables, or algebraic expressions. All of the denominators are assumed to be different from zero.

| Operation | Example |
|---|---|
| 1. $\dfrac{u}{v} + \dfrac{w}{v} = \dfrac{u+w}{v}$ | $\dfrac{2}{3} + \dfrac{5}{3} = \dfrac{2+5}{3} = \dfrac{7}{3}$ |
| 2. $\dfrac{u}{v} + \dfrac{w}{z} = \dfrac{uz+vw}{vz}$ | $\dfrac{2}{3} + \dfrac{4}{5} = \dfrac{2 \cdot 5 + 3 \cdot 4}{3 \cdot 5} = \dfrac{22}{15}$ |
| 3. $\dfrac{u}{v} \cdot \dfrac{w}{z} = \dfrac{uw}{vz}$ | $\dfrac{2}{3} \cdot \dfrac{4}{5} = \dfrac{2 \cdot 4}{3 \cdot 5} = \dfrac{8}{15}$ |
| 4. $\dfrac{u}{v} \div \dfrac{w}{z} = \dfrac{u}{v} \cdot \dfrac{z}{w} = \dfrac{uz}{vw}$ | $\dfrac{2}{3} \div \dfrac{4}{5} = \dfrac{2}{3} \cdot \dfrac{5}{4} = \dfrac{10}{12} = \dfrac{5}{6}$ |

5. For subtraction, replace "$+$" by "$-$" in 1 and 2.

---

The division step shown above in 4 is often referred to as *invert and multiply*.

### Example 3   MULTIPLYING AND DIVIDING RATIONAL EXPRESSIONS

(a) $\dfrac{2x^2 + 11x - 21}{x^3 + 2x^2 + 4x} \cdot \dfrac{x^3 - 8}{x^2 + 5x - 14}$

$= \dfrac{(2x-3)\cancel{(x+7)}}{x\cancel{(x^2+2x+4)}} \cdot \dfrac{\cancel{(x-2)}\cancel{(x^2+2x+4)}}{\cancel{(x-2)}\cancel{(x+7)}}$    Factor completely.

$= \dfrac{2x-3}{x}, \quad x \neq 2, \quad x \neq -7$    Remove common factors.

(b) $\dfrac{x^3 + 1}{x^2 - x - 2} \div \dfrac{x^2 - x + 1}{x^2 - 4x + 4}$

$= \dfrac{(x^3+1)(x^2-4x+4)}{(x^2-x-2)(x^2-x+1)}$    Invert and multiply.

$= \dfrac{\cancel{(x+1)}\cancel{(x^2-x+1)}(x-2)^{\cancel{2}1}}{\cancel{(x+1)}\cancel{(x-2)}\cancel{(x^2-x+1)}}$    Factor completely.

$= x - 2, \quad x \neq -1, \quad x \neq 2$    Remove common factors.

### Example 4   ADDING RATIONAL EXPRESSIONS

$\dfrac{x}{3x-2} + \dfrac{3}{x-5} = \dfrac{x(x-5) + 3(3x-2)}{(3x-2)(x-5)}$    Definition of addition.

$= \dfrac{x^2 - 5x + 9x - 6}{(3x-2)(x-5)}$    Distributive property.

$= \dfrac{x^2 + 4x - 6}{(3x-2)(x-5)}$    Combine like terms.

**Note**

---

The numerator, $x^2 + 4x - 6$, of the final expression in Example 4 is a prime polynomial. Thus there are no common factors.

If the denominators of fractions have common factors, then it is often more efficient to find the LCD before adding or subtracting the fractions. The **LCD (least common denominator)** is the product of all the prime factors in the denominators, where each factor is raised to the greatest power found in any one denominator for that factor.

### Example 5  USING THE LCD

Write the following expression as a fraction in reduced form.

$$\frac{2}{x^2 - 2x} + \frac{1}{x} - \frac{3}{x^2 - 4}$$

**Solution**  The factored denominators are $x(x - 2)$, $x$, and $(x - 2)(x + 2)$, respectively. The LCD is $x(x - 2)(x + 2)$.

$$\frac{2}{x^2 - 2x} + \frac{1}{x} - \frac{3}{x^2 - 4}$$

$$= \frac{2}{x(x - 2)} + \frac{1}{x} - \frac{3}{(x - 2)(x + 2)} \qquad \text{Factor.}$$

$$= \frac{2(x + 2)}{x(x - 2)(x + 2)} + \frac{(x - 2)(x + 2)}{x(x - 2)(x + 2)} - \frac{3x}{x(x - 2)(x + 2)} \qquad \begin{array}{l}\text{Equivalent}\\ \text{fractions.}\end{array}$$

$$= \frac{2(x + 2) + (x - 2)(x + 2) - 3x}{x(x - 2)(x + 2)} \qquad \begin{array}{l}\text{Combine}\\ \text{numerators.}\end{array}$$

$$= \frac{2x + 4 + x^2 - 4 - 3x}{x(x - 2)(x + 2)} \qquad \begin{array}{l}\text{Expand}\\ \text{terms.}\end{array}$$

$$= \frac{x^2 - x}{x(x - 2)(x + 2)} \qquad \text{Simplify.}$$

$$= \frac{x(x - 1)}{x(x - 2)(x + 2)} \qquad \text{Factor.}$$

$$= \frac{x - 1}{(x - 2)(x + 2)}, \quad x \neq 0 \qquad \text{Reduce.}$$

## Compound Rational Expressions

Sometimes a complicated algebraic expression needs to be changed to a more familiar form before we can work on it. A **compound fraction** (sometimes called a **complex fraction**), in which the numerators and denominators may themselves contain fractions, is such an example. One way to simplify a compund fraction is to write both the numerator and denominator as single fractions and then invert and multiply. If the fraction then takes the form of a rational expression, then we write the expression in reduced or simplest form.

**Example 6**  SIMPLIFYING A COMPOUND FRACTION

$$\frac{3 - \dfrac{7}{x + 2}}{1 - \dfrac{1}{x - 3}} = \frac{\dfrac{3(x + 2) - 7}{x + 2}}{\dfrac{(x - 3) - 1}{x - 3}} \qquad \text{Combine fractions.}$$

$$= \frac{\dfrac{3x - 1}{x + 2}}{\dfrac{x - 4}{x - 3}} \qquad \text{Simplify.}$$

$$= \frac{(3x - 1)(x - 3)}{(x + 2)(x - 4)}, \quad x \neq 3 \qquad \text{Invert and multiply.}$$

A second way to simplify a compound fraction is to multiply the numerator and denominator by the LCD of all fractions in the numerator and denominator as illustrated in Example 7.

**Example 7**  SIMPLIFYING ANOTHER COMPOUND FRACTION

Use the LCD to simplify the compound fraction

$$\frac{\dfrac{1}{a^2} - \dfrac{1}{b^2}}{\dfrac{1}{a} - \dfrac{1}{b}}$$

**Solution**  The LCD of the four fractions in the numerator and denominator is $a^2b^2$.

$$\frac{\dfrac{1}{a^2} - \dfrac{1}{b^2}}{\dfrac{1}{a} - \dfrac{1}{b}} = \frac{\left(\dfrac{1}{a^2} - \dfrac{1}{b^2}\right)a^2b^2}{\left(\dfrac{1}{a} - \dfrac{1}{b}\right)a^2b^2} \qquad \begin{array}{l}\text{Multiply numerator and}\\\text{denominator by LCD.}\end{array}$$

$$= \frac{b^2 - a^2}{ab^2 - a^2b} \qquad \text{Simplify.}$$

$$= \frac{(b + a)(b - a)}{ab(b - a)} \qquad \text{Factor.}$$

$$= \frac{b + a}{ab}, \quad a \neq b \qquad \text{Reduce.}$$

# Appendix A.3 Exercises

In Exercises 1–8, rewrite as a single fraction.

**1.** $\dfrac{5}{9} + \dfrac{10}{9}$  $\dfrac{5}{3}$

**2.** $\dfrac{17}{32} - \dfrac{9}{32}$  $\dfrac{1}{4}$

**3.** $\dfrac{20}{21} \cdot \dfrac{9}{22}$  $\dfrac{30}{77}$

**4.** $\dfrac{33}{25} \cdot \dfrac{20}{77}$  $\dfrac{12}{35}$

**5.** $\dfrac{2}{3} \div \dfrac{4}{5}$  $\dfrac{5}{6}$

**6.** $\dfrac{9}{4} \div \dfrac{15}{10}$  $\dfrac{3}{2}$

**7.** $\dfrac{1}{14} + \dfrac{4}{15} - \dfrac{5}{21}$  $\dfrac{1}{10}$

**8.** $\dfrac{1}{6} + \dfrac{6}{35} - \dfrac{4}{15}$  $\dfrac{1}{14}$

In Exercises 9–18, find the domain of the algebraic expression.

**9.** $5x^2 - 3x - 7$

**10.** $2x - 5$

**11.** $\sqrt{x - 4}$  $x \geq 4$ or $[4, \infty)$

**12.** $\dfrac{2}{\sqrt{x + 3}}$

**13.** $\dfrac{2x + 1}{x^2 + 3x}$  $x \neq 0$ and $x \neq -3$

**14.** $\dfrac{x^2 - 2}{x^2 - 4}$

**15.** $\dfrac{x}{x - 1}$,  $x \neq 2$

**16.** $\dfrac{3x - 1}{x - 2}$,  $x \neq 0$

**17.** $x^2 + x^{-1}$   $x \neq 0$

**18.** $x(x + 1)^{-2}$   $x \neq -1$

In Exercises 19–26, find the missing numerator or denominator so that the two rational expressions are equal.

**19.** $\dfrac{2}{3x} = \dfrac{?}{12x^3}$   $8x^2$

**20.** $\dfrac{5}{2y} = \dfrac{15y}{?}$   $6y^2$

**21.** $\dfrac{x - 4}{2x} = \dfrac{x^2 - 4x}{?}$   $x^2$  $x$

**22.** $\dfrac{x}{x + 2} = \dfrac{?}{x^2 - 4}$   $x^2 -$

**23.** $\dfrac{x + 3}{x - 2} = \dfrac{?}{x^2 + 2x - 8}$

**24.** $\dfrac{x - 4}{x + 5} = \dfrac{x^2 - x - 12}{?}$

**25.** $\dfrac{x^2 - 3x}{?} = \dfrac{x - 3}{x^2 + 2x}$   $x^3 + 2x^2$   **26.** $\dfrac{?}{x^2 - 9} = \dfrac{x^2 + x - 6}{x - 3}$

In Exercises 27–32, consider the original fraction and its reduced form from the specified example. Explain why the given restriction is needed on the reduced form.

**27.** Example 3a, $x \neq 2$, $x \neq -7$

**28.** Example 3b, $x \neq -1$, $x \neq 2$

**29.** Example 4, none

**30.** Example 5, $x \neq 0$

**31.** Example 6, $x \neq 3$

**32.** Example 7, $a \neq b$

In Exercises 33–44, write the expression in reduced form.

**33.** $\dfrac{18x^3}{15x}$   $\dfrac{6x^2}{5}$, $x \neq 0$

**34.** $\dfrac{75y^2}{9y^4}$   $\dfrac{25}{3y^2}$

**35.** $\dfrac{x^3}{x^2 - 2x}$   $\dfrac{x^2}{x - 2}$, $x \neq 0$

**36.** $\dfrac{2y^2 + 6y}{4y + 12}$   $\dfrac{y}{2}$, $y \neq -3$

**37.** $\dfrac{z^2 - 3z}{9 - z^2}$   $-\dfrac{z}{z + 3}$, $z \neq 3$

**38.** $\dfrac{x^2 + 6x + 9}{x^2 - x - 12}$

**39.** $\dfrac{y^2 - y - 30}{y^2 - 3y - 18}$   $\dfrac{y + 5}{y + 3}$, $y \neq 6$   **40.** $\dfrac{y^3 + 4y^2 - 21y}{y^2 - 49}$

**41.** $\dfrac{8z^3 - 1}{2z^2 + 5z - 3}$

**42.** $\dfrac{2z^3 + 6z^2 + 18z}{z^3 - 27}$

**43.** $\dfrac{x^3 + 2x^2 - 3x - 6}{x^3 + 2x^2}$

**44.** $\dfrac{y^2 + 3y}{y^3 + 3y^2 - 5y - 15}$

In Exercises 45–62, simplify.

**45.** $\dfrac{3}{x - 1} \cdot \dfrac{x^2 - 1}{9}$   $\dfrac{x + 1}{3}$, $x \neq 1$   **46.** $\dfrac{x + 3}{7} \cdot \dfrac{14}{2x + 6}$

**47.** $\dfrac{x + 3}{x - 1} \cdot \dfrac{1 - x}{x^2 - 9}$

**48.** $\dfrac{18x^2 - 3x}{3xy} \cdot \dfrac{12y^2}{6x - 1}$

**49.** $\dfrac{x^3 - 1}{2x^2} \cdot \dfrac{4x}{x^2 + x + 1}$

**50.** $\dfrac{y^3 + 2y^2 + 4y}{y^3 + 2y^2} \cdot \dfrac{y^2 - 4}{y^3 - 8}$

**51.** $\dfrac{2y^2 + 9y - 5}{y^2 - 25} \cdot \dfrac{y - 5}{2y^2 - y}$

**52.** $\dfrac{y^2 + 8y + 16}{3y^2 - y - 2} \cdot \dfrac{3y^2 + 2y}{y + 4}$

**53.** $\dfrac{1}{2x} \div \dfrac{1}{4}$   $\dfrac{2}{x}$

**54.** $\dfrac{4x}{y} \div \dfrac{8y}{x}$   $\dfrac{x^2}{2y^2}$, $x \neq 0$

**55.** $\dfrac{x^2 - 3x}{14y} \div \dfrac{2xy}{3y^2}$

**56.** $\dfrac{7x - 7y}{4y} \div \dfrac{14x - 14y}{3y}$

**57.** $\dfrac{\dfrac{2x^2y}{(x - 3)^2}}{\dfrac{8xy}{x - 3}}$

**58.** $\dfrac{\dfrac{x^2 - y^2}{2xy}}{\dfrac{y^2 - x^2}{4x^2y}}$

**59.** $\dfrac{2x + 1}{x + 5} - \dfrac{3}{x + 5}$   $\dfrac{2x - 2}{x + 5}$   **60.** $\dfrac{3}{x - 2} + \dfrac{x + 1}{x - 2}$   $\dfrac{x + 4}{x - 2}$

**61.** $\dfrac{3}{x^2 + 3x} - \dfrac{1}{x} - \dfrac{6}{x^2 - 9}$

**62.** $\dfrac{5}{x^2 + x - 6} - \dfrac{2}{x - 2} + \dfrac{4}{x^2 - 4}$

In Exercises 63–70, simplify the compound fraction.

**63.** $\dfrac{\dfrac{x}{y^2} - \dfrac{y}{x^2}}{\dfrac{1}{y^2} - \dfrac{1}{x^2}}$

**64.** $\dfrac{\dfrac{1}{x} + \dfrac{1}{y}}{\dfrac{1}{x^2} - \dfrac{1}{y^2}}$

**65.** $\dfrac{2x + \dfrac{13x - 3}{x - 4}}{2x + \dfrac{x + 3}{x - 4}}$

**66.** $\dfrac{2 - \dfrac{13}{x + 5}}{2 + \dfrac{3}{x - 3}}$

**67.** $\dfrac{\dfrac{1}{(x + h)^2} - \dfrac{1}{x^2}}{h}$

**68.** $\dfrac{\dfrac{x + h}{x + h + 2} - \dfrac{x}{x + 2}}{h}$

**69.** $\dfrac{\dfrac{b}{a} - \dfrac{a}{b}}{\dfrac{1}{a} - \dfrac{1}{b}}$

**70.** $\dfrac{\dfrac{1}{a} + \dfrac{1}{b}}{\dfrac{b}{a} - \dfrac{a}{b}}$

In Exercises 71–74, write with positive exponents and simplify.

**71.** $\left(\dfrac{1}{x} + \dfrac{1}{y}\right)(x + y)^{-1}$

**72.** $\dfrac{(x + y)^{-1}}{(x - y)^{-1}}$

**73.** $x^{-1} + y^{-1}$   $\dfrac{x + y}{xy}$

**74.** $(x^{-1} + y^{-1})^{-1}$

## B.1 Formulas from Algebra

### Exponents

If all bases are nonzero:

$$u^m u^n = u^{m+n}$$

$$\frac{u^m}{u^n} = u^{m-n}$$

$$u^0 = 1$$

$$u^{-n} = \frac{1}{u^n}$$

$$(uv)^m = u^m v^m$$

$$(u^m)^n = u^{mn}$$

$$\left(\frac{u}{v}\right)^m = \frac{u^m}{v^m}$$

### Radicals and Rational Exponents

If all roots are real numbers:

$$\sqrt[n]{uv} = \sqrt[n]{u} \cdot \sqrt[n]{v}$$

$$\sqrt[n]{\frac{u}{v}} = \frac{\sqrt[n]{u}}{\sqrt[n]{v}} \quad (v \neq 0)$$

$$\sqrt[m]{\sqrt[n]{u}} = \sqrt[mn]{u}$$

$$(\sqrt[n]{u})^n = u$$

$$\sqrt[n]{u^m} = (\sqrt[n]{u})^m$$

$$\sqrt[n]{u^n} = \begin{cases} |u| & n \text{ even} \\ u & n \text{ odd} \end{cases}$$

$$u^{1/n} = \sqrt[n]{u}$$

$$u^{m/n} = (u^{1/n})^m = (\sqrt[n]{u})^m$$

$$u^{m/n} = (u^m)^{1/n} = \sqrt[n]{u^m}$$

### Special Products

$$(u + v)(u - v) = u^2 - v^2$$

$$(u + v)^2 = u^2 + 2uv + v^2$$

$$(u - v)^2 = u^2 - 2uv + v^2$$

$$(u + v)^3 = u^3 + 3u^2v + 3uv^2 + v^3$$

$$(u - v)^3 = u^3 - 3u^2v + 3uv^2 - v^3$$

### Factoring Polynomials

$$u^2 - v^2 = (u + v)(u - v)$$

$$u^2 + 2uv + v^2 = (u + v)^2$$

$$u^2 - 2uv + v^2 = (u - v)^2$$

$$u^3 + v^3 = (u + v)(u^2 - uv + v^2)$$

$$u^3 - v^3 = (u - v)(u^2 + uv + v^2)$$

### Inequalities

If $u < v$ and $v < w$, then $u < w$.

If $u < v$, then $u + w < v + w$.

If $u < v$ and $c > 0$, then $uc < vc$.

If $u < v$ and $c < 0$, then $uc > vc$.

If $c > 0$, $|u| < c$ is equivalent to $-c < u < c$.

If $c > 0$, $|u| > c$ is equivalent to $u < -c$ or $u > c$.

### Quadratic Formula

If $a \neq 0$, the solutions of the equation $ax^2 + bx + c = 0$ are given by

$$x = \frac{-b \pm \sqrt{b^2 - 4ac}}{2a}.$$

### Logarithms

If $0 < b \neq 1, 0 < a \neq 1, x, R, S, > 0$

$y = \log_b x$ if and only if $b^y = x$

$$\log_b 1 = 0 \qquad \log_b b = 1$$

$$\log_b b^y = y \qquad b^{\log_b x} = x$$

$$\log_b RS = \log_b R + \log_b S \qquad \log_b \frac{R}{S} = \log_b R - \log_b S$$

$$\log_b R^c = c \log_b R \qquad \log_b x = \frac{\log_a x}{\log_a b}$$

### Determinants

$$\begin{vmatrix} a & b \\ c & d \end{vmatrix} = ad - bc$$

### Arithmetic Sequences and Series

$$a_n = a_1 + (n - 1)d$$

$$S_n = n\left(\frac{a_1 + a_n}{2}\right) \text{ or } S_n = \frac{n}{2}[2a_1 + (n - 1)d]$$

### Geometric Sequences and Series

$$a_n = a_1 \cdot r^{n-1}$$

$$S_n = \frac{a_1(1 - r^n)}{1 - r} \quad (r \neq 1)$$

$$S = \frac{a_1}{1 - r} \quad (|r| < 1) \text{ infinite geometric series}$$

### Factorial

$$n! = n \cdot (n - 1) \cdot (n - 2) \cdot \ldots \cdot 3 \cdot 2 \cdot 1$$

$$n \cdot (n - 1)! = n!, 0! = 1$$

### Binomial Coefficient

$$\binom{n}{r} = \frac{n!}{r!(n - r)!} \text{ (integer } n \text{ and } r, n \geq r \geq 0)$$

### Binomial Theorem

If $n$ is a positive integer

$$(a + b)^n = \binom{n}{0} a^n + \binom{n}{1} a^{n-1} b$$

$$+ \cdots + \binom{n}{r} a^{n-r} b^r + \cdots + \binom{n}{n} b^n$$

## B.2 Formulas from Geometry

### Triangle

$h = a \sin \theta$

$\text{Area} = \dfrac{1}{2}bh$

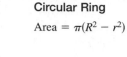

### Trapezoid

$\text{Area} = \dfrac{h}{2}(a + b)$

### Circle

$\text{Area} = \pi r^2$

$\text{Circumference} = 2\pi r$

### Sector of Circle

$\text{Area} = \dfrac{\theta r^2}{2}$ ($\theta$ in radians)

$s = r\theta$ ($\theta$ in radians)

### Right Circular Cone

$\text{Volume} = \dfrac{\pi r^2 h}{3}$

$\text{Lateral surface area} = \pi r \sqrt{r^2 + h^2}$

### Right Circular Cylinder

$\text{Volume} = \pi r^2 h$

$\text{Lateral surface area} = 2\pi rh$

### Right Triangle

Pythagorean Theorem:

$c^2 = a^2 + b^2$

### Parallelogram

$\text{Area} = bh$

### Circular Ring

$\text{Area} = \pi(R^2 - r^2)$

### Ellipse

$\text{Area} = \pi ab$

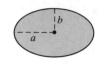

### Cone

$\text{Volume} = \dfrac{Ah}{3}$ ($A = \text{Area of base}$)

### Sphere

$\text{Volume} = \dfrac{4}{3}\pi r^3$

$\text{Surface area} = 4\pi r^2$

## B.3 Formulas from Trigonometry

### Angular Measure

$\pi$ radians $= 180°$

So, 1 radian $= \dfrac{180}{\pi}$ degrees,

and 1 degree $= \dfrac{\pi}{180}$ radians.

### Reciprocal Identities

$\sin x = \dfrac{1}{\csc x}$ $\qquad$ $\csc x = \dfrac{1}{\sin x}$

$\cos x = \dfrac{1}{\sec x}$ $\qquad$ $\sec x = \dfrac{1}{\cos x}$

$\tan x = \dfrac{1}{\cot x}$ $\qquad$ $\cot x = \dfrac{1}{\tan x}$

### Quotient Identities

$\tan x = \dfrac{\sin x}{\cos x}$ $\qquad$ $\cot x = \dfrac{\cos x}{\sin x}$

### Pythagorean Identities

$\sin^2 x + \cos^2 x = 1$

$\tan^2 x + 1 = \sec^2 x$

$1 + \cot^2 x = \csc^2 x$

## Odd-Even Identities

$\sin(-x) = -\sin x$ $\qquad$ $\csc(-x) = -\csc x$

$\cos(-x) = \cos x$ $\qquad$ $\sec(-x) = \sec x$

$\tan(-x) = -\tan x$ $\qquad$ $\cot(-x) = -\cot x$

## Sum and Difference Identities

$\sin(u + v) = \sin u \cos v + \cos u \sin v$

$\sin(u - v) = \sin u \cos v - \cos u \sin v$

$\cos(u + v) = \cos u \cos v - \sin u \sin v$

$\cos(u - v) = \cos u \cos v + \sin u \sin v$

$\tan(u + v) = \dfrac{\tan u + \tan v}{1 - \tan u \tan v}$

$\tan(u - v) = \dfrac{\tan u - \tan v}{1 + \tan u \tan v}$

## Cofunction Identities

$\cos\left(\dfrac{\pi}{2} - u\right) = \sin u$

$\sin\left(\dfrac{\pi}{2} - u\right) = \cos u$

$\tan\left(\dfrac{\pi}{2} - u\right) = \cot u$

$\cot\left(\dfrac{\pi}{2} - u\right) = \tan u$

$\sec\left(\dfrac{\pi}{2} - u\right) = \csc u$

$\csc\left(\dfrac{\pi}{2} - u\right) = \sec u$

## Double-Angle Identities

$\sin 2u = 2 \sin u \cos u$

$\cos 2u = \cos^2 u - \sin^2 u$
$\qquad = 2 \cos^2 u - 1$
$\qquad = 1 - 2 \sin^2 u$

$\tan 2u = \dfrac{2 \tan u}{1 - \tan^2 u}$

## Power-Reducing Identities

$\sin^2 u = \dfrac{1 - \cos 2u}{2}$

$\cos^2 u = \dfrac{1 + \cos 2u}{2}$

$\tan^2 u = \dfrac{1 - \cos 2u}{1 + \cos 2u}$

## Half-Angle Identities

$\sin\dfrac{u}{2} = \pm\sqrt{\dfrac{1 - \cos u}{2}}$

$\cos\dfrac{u}{2} = \pm\sqrt{\dfrac{1 + \cos u}{2}}$

$\tan\dfrac{u}{2} = \pm\sqrt{\dfrac{1 - \cos u}{1 + \cos u}}$

$\qquad = \dfrac{1 - \cos u}{\sin u} = \dfrac{\sin u}{1 + \cos u}$

## Triangles

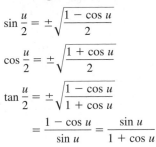

Law of sines:

$\dfrac{\sin A}{a} = \dfrac{\sin B}{b} = \dfrac{\sin C}{c}$

Law of cosines:

$a^2 = b^2 + c^2 - 2bc \cos A$

$b^2 = a^2 + c^2 - 2ac \cos B$

$c^2 = a^2 + b^2 - 2ab \cos C$

Area:

$\text{Area} = \dfrac{1}{2} bc \sin A$

$\qquad = \dfrac{1}{2} ac \sin B = \dfrac{1}{2} ab \sin C$

$\text{Area} = \sqrt{s(s - a)(s - b)(s - c)}$

$\qquad$ where $s = \dfrac{1}{2}(a + b + c)$

## Trigonometric Form of a Complex Number

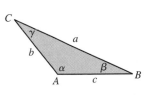

$z = a + bi = (r \cos\theta) + (r \sin\theta)i$
$\qquad = r(\cos\theta + i \sin\theta)$

## De Moivre's Theorem

$$z^n = [r(\cos\theta + i\sin\theta)]^n$$
$$= r^n(\cos n\theta + i\sin n\theta)$$

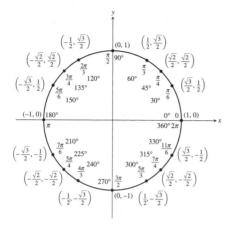

## B.4 Formulas from Analytic Geometry

### Basic Formulas

Distance $d$ between points $P(x_1, y_1)$ and $Q(x_2, y_2)$:

$$d(P, Q) = \sqrt{(x_1 - x_2)^2 + (y_1 - y_2)^2}$$

Midpoint: $\left(\dfrac{x_1 + x_2}{2}, \dfrac{y_1 + y_2}{2}\right)$

Slope of a line: $m = \dfrac{y_2 - y_1}{x_2 - x_1}$

Condition for parallel lines: $m_1 = m_2$

Condition for perpendicular lines: $m_2 = \dfrac{-1}{m_1}$

### Equations of a Line

The point-slope form, slope $m$ and through $(x_1, y_1)$:

$$y - y_1 = m(x - x_1)$$

The slope-intercept form, slope $m$ and $y$-intercept $b$: $y = mx + b$

### Equation of a Circle

The circle with center $(h, k)$ and radius $r$: $(x - h)^2 + (y - k)^2 = r^2$

### Parabolas with Vertex ($h, k$)

**Standard equation** $(x - h)^2 = 4p(y - k)$   $(y - k)^2 = 4p(x - h)$

**Opens**    Upward or downward   To the right or to the left

**Focus**    $(h, k + p)$    $(h + p, k)$

**Directrix**   $y = k - p$    $x = h - p$

**Axis**    $x = h$    $y = k$

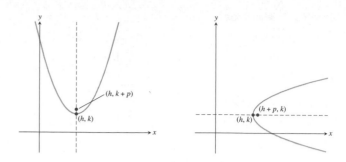

### Ellipses with Center ($h, k$)

| | | |
|---|---|---|
| **Standard equation** | $\dfrac{(x - h)^2}{a^2} + \dfrac{(y - k)^2}{b^2} = 1$ | $\dfrac{(y - k)^2}{a^2} + \dfrac{(x - h)^2}{b^2} = 1$ |
| **Focal axis** | $y = k$ | $x = h$ |
| **Foci** | $(h \pm c, k)$ | $(h, k \pm c)$ |
| **Vertices** | $(h \pm a, k)$ | $(h, k \pm a)$ |
| **Pythagorean relation** | $a^2 = b^2 + c^2$ | $a^2 = b^2 + c^2$ |

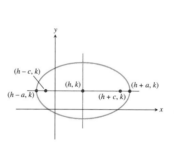

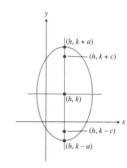

### Hyperbolas with Center ($h, k$)

| | | |
|---|---|---|
| **Standard equation** | $\dfrac{(x - h)^2}{a^2} - \dfrac{(y - k)^2}{b^2} = 1$ | $\dfrac{(y - k)^2}{a^2} - \dfrac{(x - h)^2}{b^2} = 1$ |
| **Focal axis** | $y = k$ | $x = h$ |
| **Foci** | $(h \pm c, k)$ | $(h, k \pm c)$ |
| **Vertices** | $(h \pm a, k)$ | $(h, k \pm a)$ |
| **Pythagorean relation** | $c^2 = a^2 + b^2$ | $c^2 = a^2 + b^2$ |
| **Asymptotes** | $y = \pm\dfrac{b}{a}(x - h) + k$ | $y = \pm\dfrac{a}{b}(x - h) + k$ |

## B.5 Gallery of Basic Functions

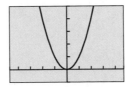

[−4.7, 4.7] by [−3.1, 3.1]

Identity Function
$f(x) = x$
Domain $= (-\infty, \infty)$
Range $= (-\infty, \infty)$

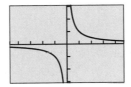

[−4.7, 4.7] by [−1, 5]

Squaring Function
$f(x) = x^2$
Domain $= (-\infty, \infty)$
Range $= [0, \infty)$

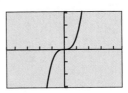

[−4.7, 4.7] by [−3.1, 3.1]

Cubing Function
$f(x) = x^3$
Domain $= (-\infty, \infty)$
Range $= (-\infty, \infty)$

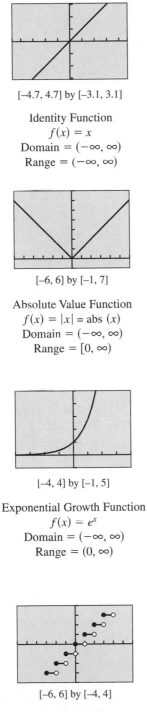

[−6, 6] by [−1, 7]

Absolute Value Function
$f(x) = |x| = \text{abs}(x)$
Domain $= (-\infty, \infty)$
Range $= [0, \infty)$

[−4.7, 4.7] by [−3.1, 3.1]

Reciprocal Function
$f(x) = \dfrac{1}{x}$
Domain $= (-\infty, 0) \cup (0, \infty)$
Range $= (-\infty, \infty) \cup (0, \infty)$

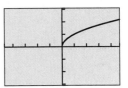

[−4.7, 4.7] by [−3.1, 3.1]

Square Root Function
$f(x) = \sqrt{x}$
Domain $= [0, \infty)$
Range $= [0, \infty)$

[−4, 4] by [−1, 5]

Exponential Growth Function
$f(x) = e^x$
Domain $= (-\infty, \infty)$
Range $= (0, \infty)$

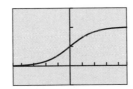

[−4.7, 4.7] by [−0.5, 1.5]

Basic Logistic Function
$f(x) = \dfrac{1}{1 + e^{-x}}$
Domain $= (-\infty, \infty)$
Range $= (0, 1)$

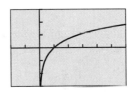

[−2, 6] by [−3, 3]

Natural Logarithmic Function
$f(x) = \ln x$
Domain $= (0, \infty)$
Range $= (-\infty, \infty)$

[−6, 6] by [−4, 4]

Greatest Integer Function
$f(x) = \text{int}(x)$
Domain $= (-\infty, \infty)$
Range $=$ all integers

[−2π, 2π] by [−4, 4]

Sine Function
$f(x) = \sin(x)$
Domain $= (-\infty, \infty)$
Range $= [-1, 1]$

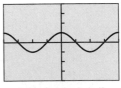

[−2π, 2π] by [−4, 4]

Cosine Function
$f(x) = \cos(x)$
Domain $= (-\infty, \infty)$
Range $= [-1, 1]$

# GLOSSARY

**Absolute maximum** A value $f(c)$ is an absolute maximum value of $f$ if $f(c) \geq f(c)$ for all $x$ in the domain of $f$, p. 89.

**Absolute minimum** A value $f(c)$ is an absolute minimum value of $f$ if $f(c) \leq f(c)$ for all $x$ in the domain of $f$, p. 89.

**Absolute value of a complex number** The absolute value of the complex number $z = a + bi$ is given by $\sqrt{a^2 + b^2}$; also, the length of the segment from the origin to $z$ in the complex plane, p. 217.

**Absolute value of a real number** Denoted by $|a|$, represents the number $a$ if , or the positive number $-a$ if $a < 0$, p. 12.

**Acute angle** An angle whose measure is between 0° and 90°, p. 346.

**Acute triangle** A triangle in which all angles measure less than 90°, p. 453.

**Addition property of equality** If $u = v$ and $w = z$, then $u + w = v + z$, p. 19.

**Addition property of inequality** If $u < v$, then $u + w < v + w$, p. 22.

**Additive identity for the complex numbers** $0 + 0i$ is the complex number zero, p. 213.

**Additive inverse of a real number** The opposite of $b$, or $-b$, p. 5.

**Additive inverse of a complex number** The opposite of $a + bi$, or $-a - bi$, p. 213.

**Algebraic expression** A collection of variables and constants that are combined by addition, subtraction, multiplication, division, radicals and rational exponents, p. 5.

**Ambiguous case** A triangle in which two sides and a nonincluded angle are known, p. 455.

**Amplitude** See *sinusoid*.

**Anchor** See *mathematical induction*.

**Angle** Union of two rays with a common endpoint (the vertex). The beginning ray (the initial side) can be rotated about its endpoint to obtain the final position (the terminal side), p. 355.

**Angle of depression** The acute angle formed by the line of sight (downward) and the horizontal, p. 404.

**Angle of elevation** The acute angle formed by the line of sight (upward) and the horizontal, p. 404.

**Angular speed** Speed of rotation, typically measured in radians or revolutions per unit time, p. 341.

**Annual percentage rate (APR)** The annual interest rate, p. 323.

**Annual percentage yield (APY)** The rate that would give the same return if interest were computed just once a year, p. 320.

**Annuity** A sequence of equal periodic payments, p. 321.

**Aphelion** The farthest point from the Sun in a planet's orbit, p. 618.

**Arc length formula** The length of an arc in a circle of radius $r$ intercepted by a central angle of $\theta$ is $s = r\theta$, p. 340–341.

**Arccosine function** See *inverse cosine function*.

**Arcsine function** See *inverse sine function*.

**Arctangent function** See *inverse tangent function*.

**Argument of a complex number** The argument of $a + bi$ is the direction angle of the vector $\langle a, b \rangle$, p. 525.

**Arithmetic sequence** A sequence $\{a_n\}$ in which $a_n = a_{n-1} + d$ for every integer $n \geq 2$. The number $d$ is the common difference, p. 697.

**Associative properties** $a + (b + c) = (a + b) + c$, $a(bc) = (ab)c$, p. 6.

**Augmented matrix** A matrix that represents a system of equations, p. 570.

**Average rate of change of $f$ over $(a, b)$** The number $\dfrac{f(b) - f(a)}{b - a}$, provided $a \neq b$, p. 157.

**Average velocity** The change in position divided by the change in time, p. 753.

**Axis of symmetry** See *line of symmetry*.

**Back-to-back stemplot** A stemplot with leaves on either side used to compare two distributions, p. 721.

**Bar chart** A rectangular graphical display of categorical data, p. 718.

**Base** See logarithmic function, $n$th power of $a$.

**Basic logistic function** The function $f(x) = \dfrac{1}{1 + e^{-x}}$, p. 267.

**Bearing** Measure of the clockwise angle that the line of travel makes with due north, p. 338.

**Binomial** A polynomial with exactly two terms, p. 802.

**Binomial coefficients** The numbers in Pascal's triangle:

$$_nC_r = \binom{n}{r} = \frac{n!}{r!(n-r)!}, \text{ p. 679.}$$

**Binomial probability** In an experiment with two possible outcomes, the probability of one outcome occurring $k$ times in $n$ independent trials is $P(E) = \dfrac{n!}{k!(n-k)!}p^k(1-p)^{n-k}$, where $P$ is the probability of the outcome occurring once, p. 692.

**Binomial theorem** A theorem that gives an expansion formula for $(a + b)^n$, p. 681.

**Boundary** The set of points on the "edge" of a region, p. 587.

**Bounded** A function $f$ is bounded if there are numbers $b$ and $B$ such that $b \leq f(x) \leq B$ for all $x$ in the domain of $f$, p. 88.

**Bounded above** A function $f$ is bounded above if there is a number $B$ such that $f(x) \leq B$ for all $x$ in the domain of $f$, p. 88.

**Bounded below** A function $f$ is bounded below if there is a number $b$ such that $b \leq f(x)$ for all $x$ in the domain of $f$, p. 88.

**Bounded interval** An interval that has finite length (does not extend to $\infty$ or $-\infty$), p. 4.

**Boxplot (or box-and-whisker plot)** A graph that displays a five-number summary, p. 736.

**Branches** The two separate curves that make up a hyperbola, p. 625.

**Cardioid** A limaçon whose polar equation is $r = a \pm b \sin \theta$, or $r = a \pm b \cos \theta$, where $a > 0$ and $b > 0$, p. 520.

**Cartesian coordinate system** An association between the points in a plane and ordered pairs of real numbers; or an association between the points in three dimensional space and ordered triples of real numbers, pp. 11, 652.

**Categorical variable** A variable (in statistics) that identifies each individual as belonging to a distinct class, p. 717.

**Center** The central point in a circle, ellipse, hyperbola, or sphere, pp.15, 614, 625, 653.

**Central angle** An angle whose vertex is the center of a circle, p. 338.

**Characteristic polynomial of a square matrix $A$** $\det(xI_n - A)$, where $A$ is an $n \times n$ matrix, p. 567.

**Chord of a conic** A line segment with endpoints on the conic, pp. 606, 615, 625.

**Circle** A set of points in a plane equally distant from a fixed point called the center, p. 15.

**Circle graph** A circular graphical display of categorical data, p. 718.

**Circular functions** Trigonometric functions when applied to real numbers are circular functions, p. 363.

**Closed interval** An interval that includes its endpoints, p. 4.

**Coefficient** The real number multiplied by the variable(s) in a term, p. 187.

**Coefficient of determination** The number $r^2$ or $R^2$ that measures how well a regression curve fits the data, p. 139.

**Coefficient matrix** A matrix whose elements are the coefficients in a system of linear equations, p. 570.

**Cofunction identity** An identity that relates the sine, secant, or tangent to the cosine, cosecant, or cotangent, respectively, p. 426.

**Combination** An arrangement of elements of a set, in which order is not important, p. 672.

**Combinations of $n$ objects taken $r$ at a time** $_nC_r = \dfrac{n!}{r!(n - r)!}$, p. 672.

**Combinatorics** A branch of mathematics related to determining the number of elements of a set or the number of ways objects can be arranged or combined, p. 669.

**Common difference** See *Arithmetic sequence.*

**Common logarithm** A logarithm with base 10, p. 284.

**Common ratio** See *Geometric sequence.*

**Commutative properties** $a + b = b + a$, $ab = ba$, p. 6.

**Complements or complementary angles** Two angles of positive measure whose sum is 90°, p. 425.

**Completely factored polynomial** A polynomial written in factored form with all prime factors, p. 804.

**Completing the square** A method of adding a constant to an expression in order to form a perfect square, p. 42.

**Complex conjugates** Complex numbers $a + bi$ and $a - bi$, p. 215.

**Complex fraction** See *Compound fraction.*

**Complex number** An expression $a + bi$, where $a$ (the real part) and $b$ (the imaginary part) are real numbers, p. 213.

**Complex plane** A coordinate plane used to represent the complex numbers. The $x$-axis of the complex plane is called the real axis and the $y$-axis is the imaginary axis, p. 216.

**Component form of a vector** If a vector's representative in standard position has a terminal point $(a, b)$ (or $(a, b, c)$), then ($\langle a, b \rangle$ or $\langle a, b, c \rangle$) is the component form of the vector, and $a$ and $b$ are the horizontal and vertical components of the vector (or $a$, $b$, and $c$ are the $x$-, $y$-, and $z$-components of the vector, respectively), pp. 480, 656.

**Components of a vector** See *Component form of a vector.*

**Composition of functions** $(f \circ g)(x) = f(g(x))$, p. 108.

**Compound fraction** A fractional expression in which the numerator or denominator may contain fractions, p. 811.

**Compound interest** Interest that becomes part of the investment, p. 316.

**Compounded annually** See *Compounded k times per year.*

**Compounded continuously** Interest compounded using the formula $A = Pe^{rt}$, p. 319.

**Compounded $k$ times per year** Interest compounded using the formula $A = P\left(1 + \dfrac{r}{k}\right)^{kt}$ where $k = 1$ is compounded annually, $k = 4$ is compounded quarterly, $k = 12$ is compounded monthly, etc., p. 316.

**Compounded monthly** See *Compounded k times per year.*

**Conditional probability** The probability of an event $A$ given that an event $B$ has already occured ($P(A|B)$), p. 689.

**Cone** See *Double-napped right circular cone.*

**Conic section (or conic)** A curve obtained by intersecting a double-napped right circular cone with a plane, p. 603.

**Conjugate axis of a hyperbola** The line segment of length $2b$ that is perpendicular to the focal axis and has the center of the hyperbola as its midpoint, p. 626.

**Constant** A letter or symbol that stands for a specific number, p. 5.

**Constant function (on an interval)** $f(x_1) = f(x_2)$ for any $x_1$ and $x_2$ (in the interval), p. 86.

**Constant term** See *Polynomial function.*

**Constant of variation** See *Power function.*

**Constraints** See *Linear programming problem.*

**Continuous function** A function that is continuous on its entire domain, p. 102.

**Continuous at $x = a$** $\lim_{x \to a} f(x) = f(a)$, p. 84.

**Converge** An infinite series $\sum_{k=1}^{\infty} a_k$ converges to a number $S$, or has sum $S$, if $\lim_{n \to \infty} \sum_{k=1}^{n} a_k = S$, p. 706.

**Conversion factor** $\Lambda$ ratio equal to 1, p. 142.

**Coordinate(s) of a point** The number associated with a point on a number line, or the ordered pair associated with a point in the Cartesian coordinate plane, or the ordered triple associated with a point in the Cartesian three-dimensional space, pp. 2, 11, 652.

**Coordinate plane** See *Cartesian coordinate system.*

**Correlation coefficient** A measure of the strength of the linear relationship between two variables, pp. 139, 160.

**Cosecant** The function $y = \csc x$, p. 382.

**Cosine** The function $y = \cos x$, p. 369.

**Cotangent** The function $y = \cot x$, p. 381.

**Coterminal angles** Two angles having the same initial side and the same terminal side, p. 356.

**Course** See *Bearing.*

**Cube root** $n$th root, where $n = 3$ (see *Principal nth root*), p. 797.

**Cubic** A degree 3 polynomial function, p. 186.

**Cycloid** The graph of the parametric equations $x = t - \sin t$, $y = 1 - \cos t$, pp. 506, 509.

**Damping factor** The factor $Ae^{-at}$ in an equation such as $y = Ae^{-at} \cos bt$, p. 391.

**Data** Facts collected for statistical purposes (singular form is *datum*), pp. 12, 717, 730.

**De Moivre's theorem**

$$(r(\cos \theta + i \sin \theta))^n = r^n(\cos n\theta + i \sin n\theta), \text{ p. 527.}$$

**Decreasing on an interval** A function $f$ is decreasing on an interval $I$ if, for any two points in $I$, a positive change in $x$ results in a negative change in $f(x)$, p. 86.

**Definite integral** The definite integral of the function $f$ over $[a, b]$ is $f(x)\, dx = \lim_{n \to \infty} \sum_{i=1}^{n} f(x_i)\, \triangle x$ provided the limit of the Riemann sums exists, p. 768.

**Degree** Unit of measurement (represented by the symbol °) for angles or arcs, equal to 1/360 of a complete revolution, p. 338.

**Degree of a polynomial (function)** The largest exponent on the variable in any of the terms of the polynomial (function), pp. 155, 802.

**Demand curve** $p = g(x)$, where $x$ represents demand and $p$ represents price, p. 549.

**Denominator** The expression below the line in a fraction, p. 809.

**Dependent event** An event whose probability depends on another event already occuring, pp. 687, 689.

**Dependent variable** Variable representing the range value of a function (usually $y$), p. 78.

**Derivative of $f$** The function $f'$ defined by $f'(a) = \dfrac{f(x + h) - f(x)}{h}$ for all of $x$ where the limit exists, p. 758.

**Derivative of $f$ at $x = a$** $f'(a) = \lim_{x \to a} \dfrac{f(x) - f(a)}{x - a}$ provided the limit exists, p. 757.

**Descartes' rule of signs** A rule for determining the possible number of positive and negative zeros of a polynomial function, p. 212.

**Descriptive statistics** The gathering and processing of numerical information, p. 730.

**Determinant** A number that is associated with a square matrix, pp. 558-559.

**Difference identity** An identity involving a trigonometric function of $u - v$, pp. 441-442.

**Difference of complex numbers**

$$(a + bi) - (c + di) = (a - c) + (b - d)i, \text{ p. 213}$$

**Difference of functions** $(f - g)(x) = f(x) - g(x)$, p. 107.

**Difference of two vectors** $\langle u_1, u_2 \rangle - \langle v_1, v_2 \rangle$ $= \langle u_1 - v_1, u_2 - v_2 \rangle$ or $\langle u_1, u_2, u_3 \rangle - \langle v_1, v_2, v_3 \rangle$ $= \langle u_1 - v_1, u_2 - v_2, u_3 - v_3 \rangle$, pp. 482, 656.

**Differentiable at $x = a$** $f'(a)$ exists, p. 757.

**Directed angle** See *Polar coordinates.*

**Directed distance** See *Polar coordinates.*

**Directed line segment** The notation $\overrightarrow{PQ}$ denoting the directed line segment with initial point $P$ and terminal point $Q$, p. 479.

**Direction angle of a vector** The angle that the vector makes with the positive $x$-axis, p. 484.

**Direction vector for a line** A vector in the direction of a line in three-dimensional space, p. 657.

**Direction of a directed line segment** The angle the line segment makes with the horizontal, p. 479.

**Directrix of a parabola, ellipse, or hyperbola** A line used to determine the conic, pp. 605, 643.

**Discriminant** For the equation $ax^2 + bx + c$, the expression $b^2 - 4ac$; for the equation $Ax^2 + Bxy + Cy^2 + Dx + Ey + F = 0$, the expression $B^2 - 4AC$, pp. 47, 215, 639.

**Distance (in a coordinate plane)** The distance $d(P, Q)$ between $P(x, y)$ and $Q(x, y)$, $d(P, Q) = \sqrt{(x_1 - x_2)^2 + (y_1 - y_2)^2}$, p. 14.

**Distance (on a number line)** The distance between real numbers $a$ and $b$, or $|a - b|$, p. 13.

**Distance (in Cartesian space)** The distance $d(P, Q)$ between and $P(x, y, z)$ and $Q(x, y, z)$, or $d(P, Q) \sqrt{(x_1 - x_2)^2 + (y_1 - y_2)^2 + (z_1 - z_2)^2}$, p. 653.

**Distributive property** $a(b + c) = ab + ac$ and related properties, p. 6.

**Diverge** An infinite series diverges if it does not converge, p. 706.

**Division** $\dfrac{a}{b} = a\left(\dfrac{1}{b}\right)$, $b \neq 0$, p. 5.

**Division algorithm for polynomials** Given $f(x)$, $d(x) \neq 0$ there are unique polynomials $q(x)$ (quotient) and $r(x)$ (remainder) with $f(x) = d(x)q(x) + r(x)$ with either $r(x) = 0$ or degree of $r(x) <$ degree of $d(x)$, p. 200.

**Divisor of a polynomial** See *Division algorithm for polynomials*.

**DMS measure** The measure of an angle in degrees, minutes, and seconds, p. 338.

**Domain of a function** The set of all input values for a function, p. 78.

**Domain of an expression** The set of numbers for which an expression is defined, p. 809.

**Dot product** The number found when the corresponding components of two vectors are multiplied and then summed, pp. 490, 656.

**Double-angle identity** An identity involving a trigonometric function of $2u$, p. 446.

**Double inequality** A statement that describes a bounded interval, such as $3 \leq x < 5$, p. 23.

**Double-napped right circular cone** A surface generated by rotating a line about an axis that intersects the line (at the vertex of the cone), maintaining a constant angle between the rotating line and the axis, p. 603.

**Double root** Root of multiplicity 2 (see *Multiplicity*), p. 192.

**Eccentricity** A positive number that specifies the shape of a conic, p. 642.

**Elementary row operations** The following three row operations: Multiply all elements of a row by a nonzero constant; interchange two rows; and add a multiple of one row to another row, p. 571.

**Elements of a matrix** See *Matrix element*.

**Elimination method** A method of solving a system of linear equations, p. 546.

**Ellipse** The set of all points in the plane such that the sum of the distances from a pair of fixed points (the foci), is a constant, p. 614.

**Ellipsoid of revolution** A surface generated by rotating an ellipse about its major axis, p. 620.

**Empty set** A set with no elements, p. 685.

**End behavior** The behavior of a graph of a function as $|x| \to \infty$, p. 94.

**End behavior asymptote of a rational function** A polynomial that the function approaches as $|x| \to \infty$, p. 230.

**Endpoint of an interval** A real number that represents one "end" of an interval, p. 4.

**Equal complex numbers** Complex numbers whose real parts are equal and whose imaginary parts are equal, p. 213.

**Equal fractions** $\dfrac{u}{v} = \dfrac{z}{w}$ if and only if $uw = vz$ ($v \neq 0$, $w \neq 0$), p. 810.

**Equal matrices** Matrices that have the same order and equal corresponding elements, p. 553.

**Equal vectors** Vectors with the same magnitude and direction, pp. 480, 656.

**Equally likely outcomes** Outcomes of an experiment that have the same probability of occurring, p. 684.

**Equation** A statement of equality between two expressions, p. 19.

**Equilibrium point** A point where the supply curve and demand curve intersect. The corresponding price is the equilibrium price, p. 549.

**Equilibrium price** See *Equilibrium point*.

**Equivalent (algebraic) expressions** Expressions that have the same domain and the same value for each number in the domain, p. 809.

**Equivalent directed line segments** Directed line segments that have the same length and direction, pp. 479, 656.

**Equivalent equations (inequalities)** Equations (inequalities) that have the same solutions, pp. 20, 22.

**Equivalent systems of equations** Systems of equations that have the same solution, p. 567.

**Even function** A function whose graph is symmetric about the $y$-axis ($f(-x) = f(x)$ for all $x$ in the domain of $f$), p. 91.

**Event** A subset of a sample space, p. 684.

**Expanded form** The right side of $u(v + w) = uv + uw$, p. 6.

**Expanded form of a series** A series written explicitly as a sum of terms (not in summation notation), p. 706.

**Experiment** Any process in which the individual results are uncertain, p. 691.

**Explicitly-defined sequence** A sequence in which the $k$ term is given as a function of $k$, p. 696

**Exponent** See *nth power of a*, p. 7.

**Exponential decay function** Decay modeled by $f(x) = a \cdot b^x$, $a > 0$ with $0 < b < 1$, p. 262.

**Exponential form** An equation written with exponents instead of logarithms, pp. 302, 304.

**Exponential function** A function of the form $f(x) = a \cdot b^x$, where $a \neq 0$, $b > 0$, $b \neq 1$, p. 260.

**Exponential growth function** Growth modeled by $f(x) = a \cdot b^x$, $a > 0$, $b > 1$, p. 262.

**Exponential regression** A procedure for fitting an exponential function to a set of data, p. 138.

**Extracting square roots** A method for solving equations in the form $x^2 = k$, p. 41.

**Extraneous solution** Any solution of the resulting equation that is not a solution of the original equation, pp. 44, 231.

**Factor Theorem** $x - c$ is a factor of a polynomial if and only if $c$ is a zero of the polynomial, p. 202.

**Factored form** The left side of $u(v + w) = uv + uw$, p. 6.

**Factoring (a polynomial)** Writing a polynomial as a product of two or more polynomial factors, p. 804.

**Feasible points** Points which satisfy the constraints in a linear programming problem, p. 589.

**Fibonacci numbers** The terms of the Fibonacci sequence, p. 701.

**Fibonacci sequence** The sequence 1, 1, 2, 3, 5, 8, 13, …, p. 701.

**Finite sequence** A function whose domain is the first $n$ positive integers for some fixed integer $n$, p. 696.

**Finite series** Sum of a finite number of terms, pp. 703–704.

**First octant** The points $(x, y, z)$ in space with $x \geq 0$, $y \geq 0$, and $z \geq 0$, p. 652.

**First quartile** See *Quartile.*

**Fitting a line or curve to data** Finding a line or curve that comes close to passing through all the points in a scatter plot, p. 137.

**Five-number summary** The minimum, first quartile, median, third quartile, and maximum of a data set, p. 734.

**Focal axis** The line through the focus and perpendicular to the directrix of a conic, p. 642.

**Focal chord of a parabola** A chord of a parabola that passes through the focus, p. 613.

**Focal length of a parabola** The *directed distance* from the vertex to the focus, p. 606.

**Focal width of a parabola** The length of the chord through the focus and perpendicular to the axis, p. 606.

**Foci, focus** See *Ellipse, Hyperbola, Parabola.*

**Fractional expression** Quotient of two algebraic expressions, p. 808.

**Frequency** Reciprocal of the period, pp.372, 406.

**Frequency (in statistics)** The number of individuals or observations with a certain characteristic, p. 723.

**Frequency distribution** See *Frequency table.*

**Frequency table (in statistics)** A table showing frequencies, p. 723.

**Function** A relation that associates each value in the domain with exactly one value in the range, pp. 78, 655.

**Fundamental Theorem of Algebra** A polynomial function of degree $n > 0$ has $n$ complex zeros (counting multiplicity), p. 220.

**Future value of an annuity** The net amount of money returned from an annuity, p. 322.

**Gaussian curve** See *Normal curve.*

**Gaussian elimination** A method of solving a system of $n$ linear equations in n unknowns, p. 568.

**General form (of a line)** $Ax + By + C = 0$, where $A$ and $B$ are not both zero, p. 28.

**Geometric sequence** A sequence $\{a_n\}$ in which $a_n = a_{n-1} \cdot r$ for every positive integer $n \geq 2$. The nonzero number $r$ is called the common ratio, pp. 697–698.

**Geometric series** A series whose terms form a geometric sequence, pp. 704–707.

**Graph of a function** $f$ The set of all points in the coordinate plane corresponding to the pairs $(x, f(x))$ for $x$ in the domain of $f$, p. 79.

**Graph of a polar equation** The set of all points in the polar coordinate system corresponding to the ordered pairs $(r, \theta)$ that are solutions of the polar equation, p. 518.

**Graph of a relation** The set of all points in the coordinate plane corresponding to the ordered pairs of the relation, p. 112.

**Graph of an equation in $x$ and $y$** The set of all points in the coordinate plane corresponding to the pairs $(x, y)$ that are solutions of the equation, p. 29.

**Graph of an inequality in $x$ and $y$** The set of all points in the coordinate plane corresponding to the solutions $(x, y)$ of the inequality, p. 587.

**Graph of parametric equations** The set of all points in the coordinate plane corresponding to the ordered pairs determined by the parametric equations, p. 498.

**Graph of a second degree equation in *x* and *y*** The set of all points in Cartesian space corresponding to the solutions $(x, y, z)$ of the equation, p. 636.

**Grapher or graphing utility** Graphing calculator or a computer with graphing software, p. 29.

**Half-angle identity** identity involving a trigonometric function of $u/2$, p. 448.

**Half-life** The amount of time required for half of a radioactive substance to decay, p. 274.

**Half-plane** The graph of the linear inequality $y \geq ax + b$, $y > ax + b$, $y \leq ax + b$, or $y < ax + b$, p. 587.

**Heron's formula** The area of $\triangle ABC$ with semiperimeter $s$ is given by $\sqrt{s(s - a)(s - b)(s - c)}$, p. 464.

**Higher-degree polynomial function** A polynomial function whose degree is $\geq 3$, p. 155.

**Histogram** A graph with rectangles of equal width that visually represents the information in a frequency table, p. 723.

**Horizontal asymptote** The line $y = b$ is a horizontal asymptote of the graph of a function $f$ if $\lim_{x \to -\infty} f(x) = b$ or $\lim_{x \to \infty} f(x) = b$, p. 94.

**Horizontal component** *See Component form of a vector.*

**Horizontal line** $y = b$, p. 28.

**Horizontal Line Test** A test for determining whether the inverse of a relation is a function, p. 118.

**Horizontal shrink or stretch** See *Shrink, stretch.*

**Horizontal translation** A shift of a graph to the left or right, pp. 125–126.

**Hyperbola** A set of points in a plane, the absolute value of the difference of whose distances from two fixed points (the foci) is a constant, p. 625

**Hypotenuse** Side opposite the right angle in a right triangle, p. 346.

**Identity** An equation that is always true throughout its domain, p. 423.

**Identity function** The function $f(x) = x$, p. 99.

**Identity matrix** A square matrix with 1's in the main diagonal and 0's elsewhere, p. 557.

**Identity properties** $a + 0 = a$, $a \cdot 1 = a$, p. 6.

**Imaginary axis** See *Complex plane.*

**Imaginary part of a complex number** See *Complex number.*

**Imaginary unit** The complex number $i = \sqrt{-1}$, p. 212.

**Increasing on an interval** A function $f$ is increasing on an interval $I$ if, for any two points in $I$, a positive change in $x$ results in a positive change in $f(x)$, p. 86.

**Independent events** The occurrence of one event has no effect on the probability of the occurrence of the other event, p. 687.

**Independent variable** Variable representing the domain value of a function (usually $x$), p. 78.

**Index** See *Radical.*

**Index of summation** See *Summation notation.*

**Individuals** The objects described by a set of data, p. 717.

**Inductive step** See *Mathematical induction.*

**Inequality** A statement that compares two quantities using an inequality symbol, p. 3.

**Inequality symbol** $<$, $>$, $\leq$, or $\geq$, p. 3.

**Inferential statistics** Using the science of statistics to make inferences about the parameters in a population from a sample, p. 730.

**Infinite discontinuity at $x = a$** $\lim_{x \to a^+} f(x) = \pm\infty$ or $\lim_{x \to a^-} f(x) = \pm\infty$, p. 84.

**Infinite limit** A special case of a limit that does not exist, p. 777.

**Infinite sequence** A function whose domain is the set of all natural numbers, p. 696.

**Initial point** See *Directed line segment.*

**Initial side of an angle** See *Angle.*

**Initial value of a function** $f(0)$, p. 159.

**Instantaneous rate of change** See *Derivative at $x = a$.*

**Instantaneous velocity** The instantaneous rate of change of a position function with respect to time, p. 755.

**Integers** The numbers $\ldots, -3, -2, -1, 0, 1, 2, \ldots$, p. 1.

**Integrable over $[a, b]$** $\int_a^b f(x)\,dx$ exists, p. 768.

**Intercept** Point where a curve crosses the $x$-, $y$-, or $z$-axis in a graph, pp. 28, 68, 655.

**Intercepted arc** Arc of a circle between the initial side and terminal side of a central angle, p. 340.

**Intermediate Value Theorem** If $f$ is a polynomial function and $a < b$, then $f$ assumes every value between $f(a)$ and $f(b)$, p. 193.

**Interquartile range** The difference between the third quartile and the first quartile, p. 734.

**Interval** Connected subset of the real number line with at least two points, p. 4.

**Interval notation** Notation used to specify intervals, p. 4.

**Inverse cosine function** The function $y = \cos^{-1} x$, p. 397.

**Inverse function** The inverse relation of a one-to-one function, p. 118.

**Inverse of a matrix** The inverse of matrix $A$, if it exists, is a matrix $B$, such that $AB = BA = I$, where $I$ is an identity matrix, p. 557.

**Inverse properties** $a + (-a) = 0$, $a \cdot \dfrac{1}{a} = 1 (a \neq 0)$, p. 6.

**Inverse relation (of the relation $R$)**  A relation that consists of all ordered pairs $(b, a)$ for which $(a, b)$ belongs to $R$, p. 117.

**Inverse rule of logarithms**  $b^{\log_b x} = x$ and $\log_b b^y = y$,  p. 283.

**Inverse sine function**  The function $y = \sin^{-1} x$, p. 395.

**Inverse tangent function**  The function $y = \tan^{-1} x$, pp. 397–398.

**Invertible linear system**  A system of $n$ linear equations in $n$ variables whose coefficient matrix has nonzero determinant, p. 574.

**Irrational numbers**  Real numbers that are not rational, p. 2.

**Irrational zeros**  Zeros of a function that are irrational numbers, p. 204.

**Irreducible quadratic over the reals**  A quadratic polynomial that cannot be factored using real coefficients, p. 224.

**Jump discontinuity at $x = a$**  $\lim\limits_{x \to a^-} f(x)$ and $\lim\limits_{x \to a^+} f(x)$ exist but are not equal, p. 84.

**$k$th term of a sequence**  The $k$th expression in the sequence, p. 696.

**Law of cosines**  $a^2 = b^2 + c^2 - 2bc \cos A$, $b^2 = a^2 + c^2 - 2ac \cos B$, $c^2 = a^2 + b^2 - 2ab \cos C$, p. 462.

**Law of sines**  $\dfrac{\sin A}{a}$, $\dfrac{\sin B}{b}$, $\dfrac{\sin C}{c}$, p. 453.

**Leading coefficient**  See *Polynomial function in x*.

**Leading term**  See *Polynomial function in x*.

**Leaf**  The final digit of a number in a stemplot, p. 718.

**Least common denominator (LCD)**  The least common multiple of the denominators of fractions, p. 811.

**Least squares line**  See *Linear regression line*.

**Leibniz notation**  The notation $dy/dx$ for the derivative of $f$, p. 768.

**Left-hand limit of $f$ at $x = a$**  The limit of $f$ as $x$ approaches $a$ from the left, p. 775.

**Lemniscate**  A graph of a polar equation of the form $r^2 = a^2 \sin 2\theta$ or $r^2 = a^2 \cos 2\theta$, p. 522.

**Length of a directed line segment**  The length or magnitude $|\overline{PQ}|$ of the directed line segment is the distance between $P$ and $Q$, pp. 479, 653.

**Length of a vector**  The length of one of a vector's representative directed line segments, pp. 481, 656.

**Like terms**  Terms in a polynomial that have the same variable part, p. 802.

**Limaçon**  A graph of a polar equation $r = a \pm b \sin \theta$ or $r = a \pm b \cos \theta$ with $a > 0$, $b > 0$, p. 520.

**Limit**  $\lim\limits_{x \to a} f(x) = L$ means that $f(x)$ gets arbitrarily close to $L$ as $x$ gets arbitrarily close (but not equal) to $a$, pp. 84, 755.

**Limit to growth**  See *Logistic growth function*.

**Limit at infinity**  $\lim\limits_{x \to \infty} f(x) = L$ means that $f(x)$ gets arbitrarily close to $L$ as $x$ gets arbitrarily large; $\lim\limits_{x \to -\infty} f(x)$ means that $f(x)$ gets arbitrarily close to $L$ as $-x$ gets arbitrarily large, pp. 765, 777.

**Line graph**  A graph of data in which consecutive data points are connected by line segments, p. 725.

**Line of symmetry**  A line over which a graph is the mirror image of itself, p. 163.

**Line of travel**  The path along which an object travels, p. 338.

**Linear combination of vectors u and v**  An expression $a\mathbf{u} + b\mathbf{v}$, where $a$ and $b$ are real numbers, p. 484.

**Linear correlation**  A scatter plot with points clustered along a line. Correlation is positive if the slope is positive and negative if the slope is negative, pp. 139, 159–160.

**Linear equation in $x$**  An equation that can be written in the form $ax + b = 0$, where $a$ and $b$ are real numbers and $a \neq 0$, p. 20.

**Linear equation in $x$, $y$, and $z$**  An equation that can be written in the form $Ax + By + Cz + D = 0$, p. 655.

**Linear equation in $x_1 x_2, ..., x_n$**  An equation that can be written in the form $a_1 x_1 + a_2 x_2 + \cdots + a_n x_n = b$, where $a_1, a_2, ..., a_n$ and $b$ are real numbers, p. 567.

**Linear Factorization Theorem**  A polynomial $f(x)$ of degree $n > 0$ has the factorization $f(x) = a(x - z_1)(x - z_2) \cdots (x - z_n)$ where the $z_i$ are the zeros of $f$, p. 220.

**Linear function**  A function that can be written in the form $f(x) = mx + b$, where $m \neq 0$ and $b$ are real numbers, p. 156.

**Linear inequality in two variables $x$ and $y$**  An inequality that can be written in one of the following forms: $y < mx + b$, $y \leq mx + b$, $y > mx + b$, or $y \geq mx + b$ with $m \neq 0$, p. 587.

**Linear inequality in $x$**  An inequality that can be written in the form $ax + b < 0$, $ax + b \leq 0$, $ax + b > 0$, or $ax + b \geq 0$, where $a$ and $b$ are real numbers and $a \neq 0$, p. 22.

**Linear programming problem**  A method of solving certain problems involving maximizing or minimizing a function of two variables (called an objective function) subject to restrictions (called constraints), p. 589.

**Linear regression**  A procedure for finding the straight line that is the best fit for the data, p. 137.

**Linear regression equation**  Equation of a linear regression line, p. 137.

**Linear regression line**  The line for which the sum of the squares of the residuals is the smallest possible, p. 173.

**Linear system**  A system of linear equations, p. 567.

**Local extremum**  A local maximum or a local minimum, p. 89.

**Local maximum**  A value $f(c)$ is a local maximum of $f$ if there is an open interval $I$ containing $c$ such that $f(x) \leq f(c)$ for all values of $x$ in $I$, p. 89.

**Local minimum** A value $f(c)$ is a local minimum of $f$ if there is an open interval $I$ containing $c$ such that $f(x) \geq f(c)$ for all values of $x$ in $I$, p. 89.

**Logarithm** An expression of the form $\log_b x$ (see *Logarithmic function*), p. 282.

**Logarithmic form** An equation written with logarithms instead of exponents, pp. 302–303.

**Logarithmic function with base $b$** The inverse of the exponential function $y = b_x$, denoted by $y = \log_b x$, p. 282.

**Logarithmic re-expression of data** Transformation of a data set involving the natural logarithm: exponential regression, natural logarithmic regression, power regression, p. 310.

**Logarithmic regression** See *Natural logarithmic regression*.

**Logistic curve** The graph of the logistic growth function, p. 267.

**Logistic growth function** A model of population growth:
$$f(x) = \frac{c}{1 + a \cdot b^x} \text{ or } f(x) = \frac{c}{1 + ae^{-kx}}, \text{ or } ,$$
where $a$, $b$, $c$, and $k$ are positive with $b < 1$. $c$ is the limit to growth, p. 267.

**Logistic regression** A procedure for fitting a logistic curve to a set of data, p. 138.

**Lower bound of $f$** Any number $b$ for which $b \leq f(x)$ for all $x$ in the domain of $f$, p. 88.

**Lower bound for real zeros** A number $c$ is a lower bound for the set of real zeros of $f$ if $f(x) \neq 0$ whenever $x < c$, p. 206.

**Lower bound test for real zeros** A test for finding a lower bound for the real zeros of a polynomial, p. 206.

**LRAM** A Riemann sum approximation of the area under a curve $f(x)$ from $x = a$ to $x = b$ using $x_i$ as the left-hand endpoint of each subinterval, p. 767.

**Magnitude of a directed line segment** See *Length of a directed line segment*.

**Magnitude of a real number** See *Absolute value of a real number*.

**Magnitude of a vector** See *length of a vector*.

**Main diagonal** The diagonal from the top left to the bottom right of a square matrix, p. 557.

**Major axis** The line segment through the foci of an ellipse with endpoints on the ellipse, p. 615.

**Mapping** A function viewed as a mapping of the elements of the domain onto the elements of the range, p. 78.

**Mathematical induction** A process for proving that a statement is true for all natural numbers $n$ by showing that it is true for $n = 1$ (the anchor) and that, if it is true for $n = k$, then it must be true for $n = k + 1$ (the inductive step), p. 713.

**Matrix, $m \times n$** A rectangular array of $m$ rows and $n$ columns of real numbers, p. 552.

**Matrix element** Any of the real numbers in a matrix, p.552.

**Maximum $r$-value** The value of $|r|$ at the point on the graph of a polar equation that has the maximum distance from the pole, p. 518.

**Mean (of a set of data)** The sum of all the data divided by the total number of items, p. 731.

**Measure of an angle** The number of degrees or radians in an angle, p. 355.

**Measure of center** A measure of the typical, middle, or average value for a data set, p. 734.

**Measure of spread** A measure that tells how widely distributed data are, p. 734.

**Median (of a data set)** The middle number (or the mean of the two middle numbers) if the data are listed in order, p. 732.

**Midpoint (in a coordinate plane)** For the line segment with endpoints $(a, b)$ and $(c, d)$, $\left(\dfrac{a + c}{2}, \dfrac{b + d}{2}\right)$, p. 15.

**Midpoint (on a number line)** For the line segment with endpoints $a$ and $b$, $\dfrac{a + b}{2}$, p. 14.

**Midpoint (in Cartesian space)** For the line segment with endpoints $(x_1, y_1, z_1)$ and $(x_2, y_2, z_2)$, $\left(\dfrac{x_1 + x_2}{2}, \dfrac{y_1 + y_2}{2}, \dfrac{z_1 + z_2}{2}\right)$, p. 653.

**Minor axis** The perpendicular bisector of the major axis of a ellipse with endpoints on the ellipse, p. 615.

**Minute** Angle measure equal to 1/60 of a degree, p. 338.

**Mode of a data set** The category or number that occurs most frequently in the set, p. 732.

**Modified boxplot** A boxplot with the outliers removed, p. 737.

**Modulus** See *Absolute value of a complex number*.

**Monomial function** A polynomial with exactly one term, p. 177.

**Multiplication principle of counting** A principle used to find the number of ways an event can occur, p. 670.

**Multiplication property of equality** If $u = v$ and $w = z$, then $uw = vz$, p. 19.

**Multiplication property of inequality** If $u < v$ and $c > 0$, then $uc < vc$. If $u < v$ and $c < 0$, then $uc > vc$, p. 22.

**Multiplicative identity for matrices** See *Identity matrix*.

**Multiplicative inverse of a complex number** The reciprocal of $a + bi$, or $\dfrac{1}{a + bi} = \dfrac{a}{a^2 + b^2} - \dfrac{b}{a^2 + b^2} i$, p. 215.

**Multiplicative inverse of a matrix** See *Inverse of a matrix*.

**Multiplicative inverse of a real number** The reciprocal of $b$, or $1/b$, $b \neq 0$, p. 5.

**Multiplicity** The multiplicity of a zero $c$ of a polynomial $f(x)$ of degree $n > 0$ is the number of times the factor $(x - c)$ occurs in the linear factorization $f(x) = a(x - z_i)(x - z_2) \cdots (x - z_n)$, p. 192.

**Natural exponential function** The function $f(x) = e^x$, p. 264.

**Natural logarithm** A logarithm with base $e$, p. 286.

**Natural logarithmic function** The inverse of the exponential function $y = e^x$, denoted by $y = \ln x$, p. 287.

**Natural logarithmic regression** A procedure for fitting a curve to a set of data, p. 138.

**Natural numbers** The numbers 1, 2, 3, …, p. 1.

**Nautical mile** Length of 1 minute of arc along the earth's equator, p. 341.

**NDER $f(a)$** See *Numerical derivative of $f$ at $x = a$.*

**Negative angle** Angle generated by clockwise rotation, p. 355.

**Negative linear correlation** See *Linear correlation.*

**Negative numbers** Real numbers shown to the left of the origin on a number line, p. 2.

**Newton's law of cooling** $T(t) = T_m + (T_0 - T_m)e^{-kt}$, p. 307.

**$n$ factorial** For any positive integer $n$, $n$ factorial is $n! = n \cdot (n - 1) \cdot (n - 2) \cdot \cdots \cdot 3 \cdot 2 \cdot 1$; zero factorial is $0! = 1$, p. 671.

**NINT($f(x), x, a, b$)** A calculator approximation to $\int_a^b f(x)\,dx$, p. 785.

**Nonsingular matrix** A square matrix with nonzero determinant, p. 558.

**Normal curve** The graph of $f(x) = e^{-x^2/2}$, p. 739.

**Normal distribution** A distribution of data shaped like the *normal curve*, p. 739.

**$n$-set** A set of $n$ objects, p. 671.

**$n$th power of $a$** The number $a^n = a \cdot a \cdot \cdots \cdot a$ (with $n$ factors of $a$), where $n$ is the exponent and $a$ is the base, p. 7.

**$n$th root** See *Principal $n$th root.*

**$n$th root of a complex number $z$** A complex number $v$ such that $v^n = z$, p. 529.

**$n$th root of unity** A complex number $v$ such that $v^n = 1$, p. 529.

**Number line graph of a linear inequality** The graph of the solutions of a linear inequality (in $x$) on a number line, p. 23.

**Numerator** The expression above the line in a fraction, p. 809.

**Numerical derivative of $f$ at $a$**

NDER $f(a) = \dfrac{f(a + 0.001) - f(a - 0.001)}{}$, p. 784.

**Objective function** See *Linear programming problem.*

**Obtuse angle** An angle whose measure is between 90° and 180°, p. 453.

**Obtuse triangle** A triangle in which one angle is greater than 90°, p. 454.

**Octants** The eight regions of space determined by the coordinate planes, p. 652.

**Odd-even identity** An identity involving a trigonometric function of $-x$, p. 426.

**Odd function** A function whose graph is symmetric about the origin ($f(-x) = -f(x)$ for all $x$ in the domain of $f$), p. 91.

**One-to-one function** A function in which each element of the range corresponds to exactly one element in the domain, p. 118.

**One-to-one rule of exponents** $x = y$ if and only if $b^x = b^y$, p. 302.

**One-to-one rule of logarithms** $x = y$ if and only if $\log_b x = \log_b y$, p. 302.

**Open interval** An interval that does not include its endpoints, p. 4.

**Opens upward or downward** A parabola $y = ax^2 + bx + c$ opens upward if $a > 0$ and opens downward if $a < 0$, p. 163.

**Opposite** See *Additive inverse of a real number and Additive inverse of a complex number.*

**Order of magnitude (of $n$)** $\log n$, p. 305.

**Order of an $m \times n$ matrix** The order of an $m \times n$ matrix is $m \times n$, p. 552.

**Ordered pair** A pair of real numbers $(x, y)$, p. 11.

**Ordered set** A set is ordered if it is possible to compare any two elements and say that one element is "less than" or "greater than" the other, p. 2.

**Ordinary annuity** An annuity in which deposits are made at the same time interest is posted, p. 321.

**Origin** The number zero on a number line, or the point where the $x$- and $y$-axes cross in the Cartesian coordinate system, or the point where the $x$-, $y$-, and $z$-axes cross in Cartesian three-dimensional space, pp. 2, 11, 652.

**Orthogonal vectors** Two vectors $\mathbf{u}$ and $\mathbf{v}$ with $\mathbf{u} \cdot \mathbf{v} = 0$, p. 492.

**Outcomes** The various possible results of an experiment, p. 684.

**Outliers** Data items more than 1.5 times the IQR below the first quartile or above the third quartile, p. 737.

**Parabola** The graph of a quadratic function, or the set of points in a plane that are equidistant from a fixed point (the focus) and a fixed line (the directrix), pp. 99, 605.

**Paraboloid of revolution** A surface generated by rotating a parabola about its line of symmetry, p. 610.

**Parallel lines** Two lines that are both vertical or have equal slopes, p. 30.

**Parallelogram law** Rule for adding vectors geometrically, p. 482.

**Parameter** See *Parametric equations.*

**Parameter interval** See *Parametric equations.*

**Parametric curve** The graph of parametric equations, p. 498.

**Parametric equations** The graph of the ordered pairs $(x, y)$ where $x = f(t)$ and $y = g(t)$ are functions defined on an interval $I$. $t$ is the parameter and $I$ the parameter interval, p. 498.

**Parametric equations for a line in space** The line through $P_0(x_0, y_0, z_0)$ in the direction of the nonzero vector $\mathbf{v} = \langle a, b, c \rangle$ has parametric equations $x = x_0 + at, y = y_0 + bt, z = z_0 + ct$, p. 658.

**Parametrization** A set of parametric equations for a curve, p. 498.

**Partial fraction decomposition** See *Partial fractions*.

**Partial fractions** The process of expanding a fraction into a sum of fractions. The sum is called the partial fraction decomposition of the original fraction, p. 580.

**Partial sums** See *Sequence of partial sums*.

**Pascal's triangle** A number pattern in which row $n$ (beginning with $n = 0$) consists of the coefficients of the expanded form of $(a + b)^n$, p. 679.

**Perihelion** The closest point to the Sun in a planet's orbit, p. 618.

**Period** See *Periodic function*.

**Periodic function** A function $f$ for which there is a positive number $c$ such that $f(t + c) = f(t)$ for every value $t$ in the domain of $f$. The smallest such number $c$ is the period of the function, p. 363.

**Permutation** An arrangement of elements of a set, in which order is important, p. 671.

**Permutations of $n$ objects taken $r$ at a time** $_nP_r = \dfrac{n!}{(n - r)!}$, p. 671.

**Perpendicular lines** Two lines that are at right angles to each other, p. 30.

**pH** The measure of acidity, p. 307.

**Phase shift** See *Sinusoid*.

**Piecewise-defined function** A function whose domain is divided into several parts and a different function rule is applied to each part, p. 103.

**Pie chart** See *Circle graph*.

**Plane in Cartesian space** The graph of $Ax + By + Cz + D = 0$, where $A$, $B$, and $C$ are not all zero, p. 655.

**Point-slope form (of a line)** $y - y_1 = m(x - x_1)$, p. 27.

**Polar axis** See *Polar coordinate system*.

**Polar coordinate system** A coordinate system whose ordered pair is based on the directed distance from a central point (the pole) and the angle measured from a ray from the pole (the polar axis), p. 510.

**Polar coordinates** The numbers $(r, \theta)$ that determine a point's location in a polar coordinate system. The number $r$ is the directed distance and $\theta$ is the directed angle, p. 510.

**Polar distance formula** The distance between the points with polar coordinates $(r_1, \theta_1)$ and $(r_2, \theta_2)$

$= \sqrt{r_1^2 + r_2^2 - 2r_1r_2 \cos (\theta_2 - \theta_2)}$, p. 516.

**Polar equation** An equation in $r$ and $\theta$, p. 513.

**Polar form of a complex number** See *Trigonometric form of a complex number.*

**Pole** See *Polar coordinate system*.

**Polynomial factor of $p(x)$** A polynomial that can be multiplied by another polynomial to obtain $p(x)$, p. 804.

**Polynomial function** A function in which $f(x)$ is a polynomial in $x$, p. 155.

**Polynomial in $x$** An expression that can be written in the form $a_n x^n + a_{n-1}x^{n-1} + \cdots + a_1 x + a_0$, where $n$ is a nonnegative integer, the coefficients are real numbers, and $a_n \neq 0$. The degree of the polynomial is $n$, the leading coefficient is $a_n$, the leading term is $a_n x^n$, and the constant term is $a_0$. (The number 0 is the zero polynomial), p. 802.

**Polynomial interpolation** The process of fitting a polynomial of degree $n$ to $(n + 1)$ points, p. 194.

**Position vector of the point $(a, b)$** The vector $\langle a, b \rangle$, p. 480.

**Positive angle** Angle generated by a counterclockwise rotation, p. 355.

**Positive linear correlation** See *Linear correlation*.

**Positive numbers** Real numbers shown to the right of the origin on a number line, p. 2.

**Power function** A function of the form $f(x) = k \cdot x^a$, where $k$ and $a$ are nonzero constants. $k$ is the constant of variation and $a$ is the power, p. 174.

**Power-reducing identity** An identity involving the square of a trigonometric function, p. 446.

**Power regression** A procedure for fitting a curve $y = a \cdot x^b$ to a set of data, p. 138.

**Power rule of logarithms** $\log_b R^c = c \log_b R, R > 0$, p. 293.

**Present value of an annuity** The net amount of your money put into an annuity, p. 323.

**Prime polynomial** A polynomial whose only polynomial factors (with integer coefficients) are itself, its opposite, and $\pm 1$, p. 804.

**Principal $n$th root** If $b^n = a$, then $b$ is an $n$th root of $a$. If $b^n = a$ and $a$ and $b$ have the same sign, $b$ is the principal $n$th root of $a$ (see *Radical*), p. 797.

**Principle of mathematical induction** A principle related to mathematical induction, p. 713.

**Probability distribution** The collection of probabilities of each event assigned by a probability function, p. 684.

**Probability of an event** The number of elements in the event divided by the number of elements in the sample space, p. 684.

**Probability function** A function $P$ that assigns a real number to each outcome $O$ in a sample space satisfying: $0 \leq P(\text{O}) \leq 1$, $P(\varnothing) = 0$, and the sum of the probabilities of all outcomes is 1, p. 685.

**Product of complex numbers** $(a + bi)(c + di) = (ac - bd) + (ad + bc)i$, pp. 214, 526.

**Product of a scalar and a vector** The product of scalar $k$ and vector $\mathbf{u} = \langle u_1, u_2 \rangle$ (or $\mathbf{u} = \langle u_1, u_2, u_3 \rangle$) is $k \cdot \mathbf{u} = \langle ku_1, ku_2 \rangle$ (or $k \cdot \mathbf{u} = \langle ku_1, ku_2, ku_3 \rangle$), p. 482, 656.

**Product of functions** $(fg)(x) = f(x)g(x)$, p. 107.

**Product of matrices $A$ and $B$** The matrix in which each entry is obtained by multiplying the entries of a row of $A$ by the corresponding entries of a column of $B$ and then adding, p. 555.

**Product rule of logarithms** $\log_b (RS) = \log_b R + \log_b S$, $R > 0$, $S > 0$, p. 293.

**Projectile motion** The movement of an object that is subject only to the force of gravity, p. 52.

**Projection of u onto v** The vector $\text{proj}_v \mathbf{u} = \left( \dfrac{\mathbf{u} \cdot \mathbf{v}}{|\mathbf{v}|} \right)^2 \mathbf{v}$, p. 493.

**Pythagorean identities** $\sin^2 \theta + \cos^2 \theta = 1$, $1 + \tan^2 \theta = \sec^2 \theta$, and $1 + \cot^2 \theta = \csc^2 \theta$, p. 425.

**Pythagorean theorem** In a right triangle with sides $a$ and $b$ and hypotenuse $c$, $c^2 = a^2 + b^2$, p. 13.

**Quadrant** Any one of the four parts into which a plane is divided by the perpendicular coordinate axes, pp. 12, 356.

**Quadrantal angle** An angle in standard position whose terminal side lies on an axis, p. 359.

**Quadratic equation in $x$** An equation that can be written in the form $ax^2 + bx + c = 0$ $(a \neq 0)$, p. 41.

**Quadratic formula** The formula $x = \dfrac{-b \pm \sqrt{b^2 - 4ac}}{2a}$ used to solve $ax^2 + bx + c = 0$, p. 42.

**Quadratic function** A function that can be written in the form $f(x) = ax^2 + bx + c$, where $a$, $b$, and $c$ are real numbers, and $a \neq 0$, p. 162.

**Quadratic regression** A procedure for fitting a quadratic function to a set of data, p. 138.

**Quadric surface** The graph in three-dimensions of a second degree equation in three variables, p. 655.

**Quantitative variable** A variable (in statistics) that takes on numerical values for a characteristic being measured, p. 717.

**Quartic function** A degree 4 polynomial function, p. 186.

**Quartic regression** A procedure for fitting a quartic function to a set of data, p. 138.

**Quartile** The first quartile is the median of the lower half of a set of data, the second quartile is the median, and the third quartile is the median of the upper half of the data, p. 734.

**Quotient identities** $\tan \theta = \dfrac{\sin \theta}{\cos \theta}$ and $\cot \theta = \dfrac{\cos \theta}{\sin \theta}$, p. 424.

**Quotient of complex numbers** $\dfrac{a + bi}{c + di} = \dfrac{ac + bd}{c^2 + d^2} + \dfrac{bc - ad}{c^2 + d^2}i$, pp. 215, 526.

**Quotient of functions** $\left( \dfrac{f}{g} \right)(x) = \dfrac{f(x)}{g(x)}$, $g(x) \neq 0$, p. 107.

**Quotient rule of logarithms** $\log_b \left( \dfrac{R}{S} \right) = \log_b R - \log_b S$, $R > 0$, $S > 0$, p. 293.

**Quotient polynomial** See *Division algorithm for polynomials.*

**Radian** The measure of a central angle whose intercepted arc has a length equal to the circle's radius, p. 339.

**Radian measure** The measure of an angle in radians, or, for a central angle, the ratio of the length of the intercepted arc to the radius of the circle, p. 340.

**Radical** Any expression of the form $\sqrt[n]{a}$ denoting the principal $n$th root of $a$ $(n > 1)$. The integer $n$ is the index of the radical, and $a$ is the radicand, p. 797.

**Radicand** See *Radical.*

**Radius** The distance from a point on a circle to the center of the circle, pp. 15, 653.

**Range of a function** The set of all output values corresponding to elements in the domain, p. 78.

**Range (in statistics)** The difference between the greatest and least values in a data set, p. 734.

**Range screen** See *Viewing window.*

**Rational expression** An expression that can be written as a ratio of two polynomials, p. 808.

**Rational function** Function of the form $\dfrac{f(x)}{g(x)}$, where $f(x)$ and $g(x)$ are polynomials and $g(x)$ is not the zero polynomial, p. 227.

**Rational numbers** Numbers that can be written as $a/b$, where $a$ and $b$ are integers, and $b \neq 0$, p. 1.

**Rational zeros** Zeros of a function that are rational numbers, p. 204.

**Rational Zeros Theorem** A procedure for finding the possible rational zeros of a polynomial, p. 205.

**Rationalizing the denominator** Rewriting fractions so that the denominator is free of radicals, p. 799.

**Real axis** See *Complex plane.*

**Real number** Any number that can be written as a decimal, p. 1.

**Real number line** A horizontal line that represents the set of real numbers, p. 2.

**Real part of a complex number** See *Complex number.*

**Real zeros** Zeros of a function that are real numbers, p. 204.

**Reciprocal function** The function $f(x) = \dfrac{1}{x}$, p. 228.

**Reciprocal identity** An identity that equates a trigonometric function with the reciprocal of another trigonometric function, p. 424.

**Reciprocal of a real number** See *Multiplicative inverse of a real number.*

**Rectangular coordinate system** See *Cartesian coordinate syatem.*

**Recursively defined sequence** A sequence defined by giving the first term (or the first few terms) along with a procedure for finding the subsequent terms, p. 696.

**Reduced form of a fraction** The numerator and denominator have no common prime factors, p. 809.

**Reduced row echelon form** A matrix in row echelon form with every column that has a leading 1 having 0's in all other positions, p. 572.

**Re-expression of data** A transformation of a data set, p. 297.

**Reference angle** See *Reference triangle.*

**Reference triangle** For an angle $\theta$ in standard position, the reference triangle is the triangle formed by the terminal side of angle $\theta$, the $x$-axis, and a perpendicular dropped from a point on the terminal side to the $x$-axis. The angle in the reference triangle at the origin is the reference angle, p. 358.

**Reflection** Two points that are symmetric with respect to a line or a point, p. 127.

**Reflection across the $x$-axis** $(x, y)$ and $(-x, -y)$ are reflections of each other across the $x$-axis, p. 127.

**Reflection across the $y$-axis** $(x, y)$ and $(-x, -y)$ are reflections of each other across the $y$-axis, p. 127.

**Reflexive property of equality** $a = a$, p. 19.

**Regression model** An equation found by regression and which can be used to predict unknown values, p. 138.

**Relation** A set of ordered pairs of real numbers, p. 111.

**Remainder polynomial** See *Division algorithm for polynomials.*

**Remainder Theorem** If a polynomial $f(x)$ is divided by $x - c$, the remainder is $f(c)$, p. 201.

**Removable discontinuity at $x = a$** $\lim_{x \to a^-} f(x) = \lim_{x \to a^+} f(x)$ but either the common limit is not equal to $f(a)$ or $f(a)$ is not defined, p. 83.

**Repeated zeros** Zeros of multiplicity $\geq 2$ (see *Multiplicity*), p. 192.

**Residual** The difference $y_1 - (ax_1 + b)$, where $(x_1, y_1)$ is a point in a scatter plot and $y = ax + b$ is a line that fits the set of data, p. 173.

**Resistant measure** A statistical measure that does not change much in response to outliers, p. 732.

**Richter scale** A logarithmic scale used in measuring the intensity of an earthquake, p. 300.

**Riemann sum** A sum $\sum_{i=1}^{n} f(x_i) \, \triangle x$ where the interval $[a, b]$ is divided into $n$ subintervals of equal length $\triangle x$ and $x_i$ is in the $i$th subinterval, p. 767.

**Right angle** A 90° angle, p. 346.

**Right-hand limit of $f$ at $x = a$** The limit of $f$ as $x$ approaches $a$ from the right, p. 775.

**Right triangle** A triangle with a 90° angle, p. 13.

**Rigid transformation** A transformation that leaves the basic shape of a graph unchanged, p. 125.

**Root of a number** See *Principal nth root.*

**Root of an equation** A solution, p. 68.

**Rose curve** A graph of a polar equation $r = a \cos n\theta$ or $r = a \sin n\theta$, p. 519.

**Row echelon form** A matrix in which rows consisting of all 0's occur only at the bottom of the matrix, the first nonzero entry in any row with nonzero entries is 1, and the leading 1's move to the right as we move down the rows, p. 571.

**Row operations** See *Elementary row operations.*

**RRAM** A Riemann sum approximation of the area under a curve $f(x)$ from $x = a$ to $x = b$ using $x_i$ as the right-hand endpoint of each subinterval, p. 767.

**Sample space** Set of all possible outcomes of an experiment, p. 684.

**Sample standard deviation** The *standard deviation* computed using only a sample of the entire population, p. 738.

**Scalar** A real number, pp. 481, 554.

**Scatter plot** A plot of all the ordered pairs of a two-variable data set on a coordinate plane, p. 12.

**Scientific notation** A positive number written as $c \times 10^m$, where $1 \leq c < 10$ and $m$ is an integer, p. 8.

**Secant** The function $y = \sec x$, p. 381.

**Secant line of $f$** A line joining two points of the graph of $f$, p. 756.

**Second** Angle measure equal to 1/60 of a minute, p. 338.

**Second quartile** See *Quartile.*

**Second-degree equation** $Ax^2 + Bxy + Cy^2 + Dx + Ey + F = 0$, where $A$, $B$, and $C$ are not all zero, p. 604.

**Semimajor axis** Line segment with endpoints at the center of and on an ellipse, containing one of the foci, p. 615.

**Semiminor axis** Line segment with endpoints at the center of and on an ellipse, and perpendicular to the major axis, p. 615.

**Semiperimeter of a triangle** One-half of the sum of the lengths of the sides of a triangle, p. 464.

**Sequence** See *Finite sequence, Infinite sequence.*

**Sequence of partial sums** The sequence $\{S_n\}$, where $S_n$ is the $n$th partial sum of the series, that is, the sum of the first $n$ terms of the series, p. 706.

**Series** A finite or infinite sum of terms, p. 705.

**Shrink of factor $c$** A transformation of a graph obtained by multiplying all the $x$-coordinates (horizontal shrink) by the constant $1/c$ or all of the $y$-coordinates (vertical shrink) by the constant $c$, $0 < c < 1$, p. 130.

**Similar figures** Figures that have the same shape but not necessarily the same size, p. 345.

**Simple harmonic motion** Motion described by $d = a \sin \omega t$ or $d = a \cos \omega t$, p. 406.

**Sine** The function $y = \sin x$, p. 368.

**Singular matrix** A square matrix with zero determinant, p. 558.

**Sinusoid** A function that can be written in the form $f(x) = a \sin (b(x - h)) + k$ or $f(x) = a \cos (b(x - h)) + k$. The number $a$ is the amplitude, and the number $h$ is the phase shift, p. 374.

**Sinusoidal regression** A procedure for fitting a curve $y = a \sin (bx + c) + d$ to a set of data, p. 138.

**Slant asymptote** An end behavior asymptote that is a line and is not horizontal or vertical, p. 230.

**Slope** Ratio $\dfrac{\text{change in } y}{\text{change in } x}$, p. 26.

**Slope-intercept form (of a line)** $y = mx + b$, p. 28.

**Solution set of an inequality** The set of all solutions of an inequality, p. 22.

**Solution of a system in two variables** An ordered pair of real numbers that satisfies all of the equations or inequalities in the system, pp. 543, 588.

**Solution of an equation or inequality** A value of the variable (or values of the variables) for which the equation or inequality is true, pp. 19, 22, 68, 587.

**Solve an equation or inequality** To find all solutions of the equation or inequality, pp. 19, 22, 68, 587.

**Solve a triangle** To find one or more unknown sides or angles of a triangle, p. 350.

**Solve a system** To find all solutions of a system, pp. 543, 588.

**Solve by elimination or substitution** Methods for solving systems of linear equations, pp. 543, 546.

**Speed** The magnitude of the velocity vector, given by distance/time, p. 485.

**Sphere** A set of points in Cartesian space equally distant from a fixed point called the center, p. 653.

**Spiral of Archimedes** The graph of the polar curve $r = \theta$, p. 521.

**Square matrix** A matrix whose number of rows equals the number of columns, p. 552.

**Square root** $n$th root, where $n = 2$ (see *Principal $n$th root*), p. 797.

**Standard deviation** A measure of how a data set is spread, p. 738.

**Standard form:**

    **equation of a circle** $(x - h)^2 + (y - k)^2 = r^2$, p. 16.

**equation of an ellipse** $\dfrac{(x - h)^2}{a^2} + \dfrac{(y - k)^2}{b^2} = 1$ or $\dfrac{(y - k)^2}{a^2} + \dfrac{(x - h)^2}{b^2} = 1$, p. 617.

**equation of a hyperbola** $\dfrac{(x - h)^2}{a^2} + \dfrac{(y - k)^2}{b^2} = 1$ or $\dfrac{(y - k)^2}{a^2} + \dfrac{(x - h)^2}{b^2} = 1$, p. 628.

**equation of a parabola** $(x - h)^2 = 4p(y - k)$ or $(y - k)^2 = 4p(x - h)$, p. 608.

**of a polynomial** $a_n x^n + a_{n-1} x^{n-1} + \cdots + a_1 x + a_0$, p. 187, 802.

**equation of a quadratic function** $f(x) = ax^2 + bx + c$ $(a \neq 0)$, p. 163.

**Standard form of a complex number** $a + bi$, where $a$ and $b$ are real numbers, p. 213.

**Standard form polar equation of a conic** $r = \dfrac{ke}{1 \pm e \cos \theta}$ or $r = \dfrac{ke}{1 \pm e \cos \theta}$, p. 645.

**Standard position (angle)** An angle positioned on a rectangular coordinate system with its vertex at the origin and its initial side on the positive $x$-axis, pp. 346, 355.

**Standard position of a vector** A representative directed line segment with its initial point at the origin, pp. 480, 656.

**Standard unit vectors** In the plane $\mathbf{i} = \langle 1, 0 \rangle$ and $\mathbf{j} = \langle 0, 1 \rangle$; in space $\mathbf{i} = \langle 1, 0, 0 \rangle$, $\mathbf{j} = \langle 0, 1, 0 \rangle$, and $\mathbf{k} = \langle 0, 0, 1 \rangle$, pp. 480, 656.

**Statistic** A number that describes some characteristic of a data set, p. 730.

**Statute mile** 5280 feet, p. 342.

**Stem** The initial digit or digits of a number in a stemplot, p. 718.

**Stemplot (or stem-and-leaf plot)** An arrangement of a numerical data set into a specific tabular format, p. 718.

**Stretch of factor $c$** A transformation of a graph obtained by multiplying all the $x$-coordinates (horizontal stretch) by the constant $1/c$, or all of the $y$-coordinates (vertical stretch) of the points by a constant $c$, $c > 1$, p. 130.

**Subtraction** $a - b = a + (-b)$, p. 5.

**Sum identity** An identity involving a trigonometric function of $u + v$, pp. 441–442.

**Sum of a finite arithmetic series** $S_n = n\left(\dfrac{a_1 + a_2}{2}\right)$ $= \dfrac{n}{2}[2a_1 + 9(n - 1)d]$, p. 703.

**Sum of a finite geometric series** $S_n = \dfrac{a_1(1 - r^n)}{1 - r}$, p. 704.

**Sum of an infinite geometric series** $S_n = \dfrac{a}{1 - r}$, $|r| < 1$, p. 707.

**Sum of an infinite series** See *Converge*.

**Sum of complex numbers** $(a + bi) + (c + di) = (a + c) + (b + d)i$, p. 213.

**Sum of functions** $(f + g)(x) = f(x) + g(x)$ p. 107.

**Sum of two vectors** $\langle u_1, u_2 \rangle + \langle v_1, v_2 \rangle = \langle u_1 + v_1, u_2 + v_2 \rangle$ or $\langle u_1, u_2, u_3 \rangle + \langle v_1, v_2, v_3 \rangle = \langle u_1 + v_1, u_2 + v_2, u_3 + v_3 \rangle$, pp. 482, 656.

**Summation notation** The series $\sum_{k=1}^{b} a_k$, where $n$ is a natural number (or $\infty$) is in summation notation and is read "the sum of $a_k$ from $k = 1$ to $n$ (or infinity)." $k$ is the index of summation and is the $k$th term of the series, p. 701.

**Supply curve** $p = f(x)$, where $x$ represents production and $p$ represents price, p. 548.

**Symmetric difference quotient of $f$ at $a$** $\dfrac{f(x + h) - f(x - h)}{2h}$, p. 784.

**Symmetric matrix** A matrix $A = [a_{ij}]$ with the property $a_{ij} = a_{ji}$ for all $i$ and $j$, p. 565.

**Symmetric about the origin** A graph in which $(-x, -y)$ is on the graph whenever $(x, y)$ is; or a graph in which $(-r, \theta)$ or $(r, \theta + \pi)$ is on the graph whenever $(r, \theta)$ is, pp. 91, 517.

**Symmetric about the $x$-axis** A graph in which $(x, -y)$ is on the graph whenever is; or a graph in which $(r, -\theta)$ or $(-r, \pi - \theta)$ is on the graph whenever $(r, \theta)$ is, pp. 91, 517.

**Symmetric about the $y$-axis** A graph in which $(-x, y)$ is on the graph whenever $(x, y)$ is; or a graph in which $(-r, -\theta)$ or $(r, \pi - \theta)$ is on the graph whenever $(r, \theta)$ is, pp. 91, 517.

**Symmetric property of equality** If $a = b$, then $b = a$, p. 19.

**Synthetic division** A procedure used to divide a polynomial by a linear factor, $x - a$, p. 202.

**System** A set of equations or inequalities, pp. 543, 588.

**Tangent** The function $y = \tan x$, p. 379.

**Tangent line of $f$ at $x = a$** The line through $(a, f(a))$ with slope $f'(a)$ provided $f'(a)$ exists, p. 757.

**Terminal point** See *Directed line segment.*

**Terminal side of an angle** See *Angle.*

**Term of a polynomial (function)** An expression of the form $a_n x^n$ in a polynomial (function), pp. 187, 802.

**Terms** The parts of an expression separated by addition, pp. 187, 802.

**Terms of a sequence** The range elements of a sequence, p. 696.

**Third quartile** See *Quartile.*

**Time plot** The points from left to right in a scatter plot are connected with straight line segments, p. 725.

**Transformation** A function that maps real numbers to real numbers, p. 125.

**Transitive property** If $a = b$ and $b = c$, then $a = c$. Similar properties hold for the inequality symbols $<, \leq, >, \geq$, pp. 19, 22.

**Translation** See *Horizontal translation, Vertical translation.*

**Transpose of a matrix** The matrix $A^T$ obtained by interchanging the rows and columns of $A$, p. 557.

**Transverse axis** The line segment whose endpoints are the vertices of a hyperbola, p. 625.

**Tree diagram** A visualization of the *Multiplication Principle of Probability*, p. 688.

**Triangular form** A special form for a system of linear equations that facilitates finding the solution, p. 569.

**Triangular number** A number that is a sum of the arithmetic series $1 + 2 + 3 + \cdots + n$ for some natural number $n$, p. 682.

**Trichotomy property** For real numbers $a$ and $b$, exactly one of the following is true: $a < b$, $a = b$, or $a > b$, p. 3.

**Trigonometric form of a complex number** $r(\cos \theta + i \sin \theta)$, p. 525.

**Trinomial** A polynomial with exactly three terms, p. 802.

**Unbounded interval** An interval that extends to $-\infty$ or $\infty$ (or both), p. 4.

**Union of two sets $A$ and $B$** The set of all elements that belong to $A$ or $B$ or both, p. 49.

**Unit circle** A circle with radius 1 centered at the origin, p. 362.

**Unit ratio** See *Conversion factor.*

**Unit vector** Vector of length 1, pp. 483, 656.

**Unit vector in the direction of a vector** A unit vector that has the same direction as the given vector, pp. 483, 656.

**Upper bound for $f$** Any number $B$ for which $f(x) \leq B$ for all $x$ in the domain of $f$, p. 88.

**Upper bound for real zeros** A number $d$ is an upper bound for the set of real zeros of $f$ if $f(x) \neq 0$ whenever $x > d$, p. 206.

**Upper bound test for real zeros** A test for finding an upper bound for the real zeros of a polynomial, p. 206.

**Value of an annuity** $FV = \dfrac{R(1 + i)^n - 1}{i}$, p. 322.

**Value of an investment** $A = P\left(1 + \dfrac{r}{k}\right)^{kt}$ or $A = Pe^{rt}$, p. 319.

**Variable** A letter that represents an unspecified number, p. 5.

**Variable (in statistics)** A characteristic of individuals that is being identified or measured, p. 717.

**Variance** The square of the standard deviation, p. 738.

**Vector (determined by a directed line segment)** The set of all directed line segments equivalent to a particular line segment, pp. 479, 656.

**Vector equation for a line in space** The line through $P_0(x_0, y_0, z_0)$ in the direction of the nonzero vector $\mathbf{v} = \langle a, b, c \rangle$ has vector equation $\mathbf{r} = \mathbf{r}_0 + t\mathbf{v}$, where $\mathbf{r} = \langle x, y, z \rangle$, p. 658.

**Velocity** A vector that specifies the motion of an object in terms of its speed and direction, p. 485.

**Venn diagram** A visualization of the relationships among events within a sample space, p. 688.

**Vertex of a cone** See *Double-napped right circular cone*.

**Vertex of a parabola** The point of intersection of a parabola and its line of symmetry, pp. 163, 605.

**Vertex of an angle** See *Angle*.

**Vertex form for a quadratic function** $f(x) = a(x - h)^2 + k$, p. 164.

**Vertical asymptote** The line $x = a$ is a vertical asymptote of the graph of the function $f$ if $\lim_{x \to a^+} f(x) = \pm\infty$ or $\lim_{x \to a^-} f(x) = \pm\infty$, pp. 94, 230.

**Vertical component** See *Component form of a vector*.

**Vertical line** $x = a$, p. 28.

**Vertical Line Test** A test for determining whether a graph is a function, p. 80.

**Vertical stretch or shrink** See *Stretch, Shrink*.

**vertical translation** A shift of a graph up or down, pp. 125–126.

**Vertices of an ellipse** The points where the ellipse intersects its focal axis, p. 614.

**Vertices of a hyperbola** The points where a hyperbola intersects the line containing its foci, p. 625.

**Viewing window** The rectangular portion of the coordinate plane specified by the dimensions [Xmin, Xmax] by [Ymin, Ymax], p. 29.

**Weighted mean** A mean calculated in such a way that some elements of the data set have higher weights (that is, are counted more strongly in determining the mean) than others, p. 734.

**Weights** See *Weighted mean*.

**Whole numbers** The numbers 0, 1, 2, 3, …, p. 1.

**Window dimensions** The restrictions on $x$ and $y$ that specify a viewing window. See *Viewing window*.

**$x$-axis** Usually the horizontal coordinate line in a Cartesian coordinate system with positive direction to the right, pp. 11, 651.

**$x$-coordinate** The directed distance from the $y$-axis ($yz$-plane) to a point in a plane (space), or the first number in an ordered pair (triple), pp. 11, 652.

**$x$-intercept** A point that lies on both the graph and the $x$-axis, pp. 29, 68, 230.

**Xmax** The $x$-value of the right side of the viewing window, p. 29.

**Xmin** The $x$-value of the left side of the viewing window, p. 29.

**Xscl** The scale of the tick marks on the $x$-axis in a viewing window, p. 29.

**$xy$-plane** The points $(x, y, 0)$ in Cartesian space, p. 652.

**$xz$-plane** The points $(x, 0, z)$ in Cartesian space, p. 652.

**$y$-axis** Usually the vertical coordinate line in a Cartesian coordinate system with positive direction up, p. 11, 651.

**$y$-coordinate** The directed distance from the $x$-axis to a point in a plane ($xz$-plane), or the second number in an ordered pair (triple), pp. 11, 651.

**$y$-intercept** A point that lies on both the graph and the $y$-axis, pp. 29, 230.

**Ymax** The $y$-value of the top of the viewing window, p. 29.

**Ymin** The $y$-value of the bottom of the viewing window, p. 29.

**Yscl** The scale of the tick marks on the $y$-axis in a viewing window, p. 29.

**$yz$-plane** The points $(0, y, z)$ in Cartesian space, p. 652.

**$z$-axis** Usually the third dimension in Cartesian space, p. 651.

**$z$-coordinate** The directed distance from the $xy$-plane to a point in space, or the third number in an ordered triple, p. 651.

**Zero factor property** If $ab = 0$, then either $a = 0$ or $b = 0$, p. 67.

**Zero factorial** See *n factorial*.

**Zero of a function** A value in the domain that makes the function value zero, p. 68.

**Zero matrix** A matrix consisting entirely of zeros, p. 554.

**Zero polynomial** The constant 0, p. 155.

**Zero vector** The vector $\langle 0, 0 \rangle$ or $\langle 0, 0, 0 \rangle$, pp. 481, 656.

**Zoom out** A procedure of a graphing utility used to view more of the coordinate plane (used, for example, to find the end behavior of a function), p. 190.

# ADDITIONAL
# ANSWERS

## Chapter P

### Section P.1 (pp. 1–11)

#### Quick Review

**1.** $\{1, 2, 3, 4, 5, 6\}$
**2.** $\{-2, -1, 0, 1, 2, 3, 4, 5, 6\}$
**5.** **(a)** $1,187.75$      **(b)** $-4.72$
**6.** **(a)** $20.65$      **(b)** $0.10$

#### Exercises

**1.** $-4.625$ (terminating)
**5.**

**6.**

**7.**

**8.**

**9.**

**10.**

**12.** $-\infty < x \leq 4$, or $x \leq 4$
**13.** $-\infty < x < 5$, or $x < 5$
**14.** $-2 \leq x < 2$
**23.** The real numbers greater than 4 and less than or equal to 9.
**24.** The real numbers greater than or equal to $-1$, or the real numbers which are at least $-1$.
**25.** The real numbers greater than or equal to $-3$, or the real numbers which are at least $-3$.
**26.** The real numbers between $-5$ and 7, or the real numbers greater than $-5$ and less than 7.
**27.** The real numbers greater than $-1$.
**28.** The real numbers between $-3$ and 0 (inclusive), or . . . greater than or equal to $-3$ and less than or equal to 0.
**29.** Endpoints $-3$ and 4; bounded; half-open
**30.** Endpoints $-3$ and $-1$; bounded; open
**31.** Endpoint 5; unbounded; open
**32.** Endpoint $-6$; unbounded; closed
**33.** $x \geq 29$ or $[29, \infty)$; $x =$ Bill's age
**34.** $0 \leq x \leq 2$ or $[0, 2]$; $x =$ cost of an item

**35.** $1.099 \leq x \leq 1.399$ or $[1.099, 1.399]$; $x = \$$ per gallon of gasoline
**36.** $0.02 < x < 0.065$ or $(0.02, 0.065)$; $x =$ average percent of all salary raises
**43.** **(a)** Associative property of multiplication
     **(b)** Commutative property of multiplication
     **(c)** Addition inverse property
     **(d)** Addition identity property
     **(e)** Distributive property of multiplication over addition
**44.** **(a)** Multiplication inverse property
     **(b)** Multiplication identity property – or distributive property of multiplication over addition, followed by the multiplication identity property. Note that we also use the multiplicative commutative property to say that $1 \cdot u = u \cdot 1 = u$.
     **(c)** Distributive property of multiplication over subtraction
     **(d)** Definition of subtraction; associative property of addition; definition of subtraction
     **(e)** Associative property of multiplication; multiplicative inverse; multiplicative identity
**45.** **(a)** Because $a^m \neq 0$, $a^m a^0 = a^{m+0} = a^m$ implies that $a^0 = 1$.
     **(b)** Because $a^m \neq 0$, $a^m a^{-m} - a^{m-m} = a^0 = 1$ implies that $a^{-m} = \dfrac{1}{a^m}$.

**46.**

| Step | Quotient | Remainder |
|------|----------|-----------|
| 1 | 0 | 1 |
| 2 | 0 | 10 |
| 3 | 5 | 15 |
| 4 | 8 | 14 |
| 5 | 8 | 4 |
| 6 | 2 | 6 |
| 7 | 3 | 9 |
| 8 | 5 | 5 |
| 9 | 2 | 16 |
| 10 | 9 | 7 |
| 11 | 4 | 2 |
| 12 | 1 | 3 |
| 13 | 1 | 13 |
| 14 | 7 | 11 |
| 15 | 6 | 8 |
| 16 | 4 | 12 |
| 17 | 7 | 1 |

**(b)** When the remainder is repeated, the quotients generated in the long division process will also repeat.

**(c)** When any remainder is first repeated, the next quotient will be the same number as the quotient resulting after the first occurrence of the remainder, since the decimal representation does not terminate.

**55.** $1.3 \times 10^8$      **56.** $2.86 \times 10^8$

**59.** 0.000 000 033 3      **60.** 673,000,000,000

**62.** 0.000 000 000 000 000 000 000 001 674 7

**63.** $2.6028 \times 10^{-8}$      **64.** $6.364 \times 10^{-8}$

**65.** 0, 1, 2, 3, 4, 5, 6      **66.** 1, 2, 3, 4, 5, 6

**67.** $-6, -5, -4, -3, -2, -1, 0, 1, 2, 3, 4, 5, 6$

## Section P.2 (pp. 11–19)

### Quick Review

**1.**

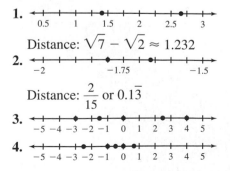

Distance: $\sqrt{7} - \sqrt{2} \approx 1.232$

**2.**

Distance: $\dfrac{2}{15}$ or $0.1\overline{3}$

**3.**

**4.**

**5.**

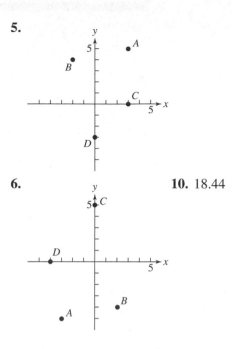

**6.**                   **10.** 18.44

### Exercises

**1.** $A(1, 0)$; $B(2, 4)$; $C(-3, -2)$; $D(0, -2)$

**2.** $A(0, 3)$; $B(-3, 1)$; $C(-2, 0)$; $D(4, -1)$

**3.** **(a)** First quadrant.
     **(b)** On the *y*-axis, between quadrants I and II.
     **(c)** Second quadrant.
     **(d)** Third quadrant.

**4.** **(a)** First quadrant.
     **(b)** On the *x*-axis, between quadrants II and III.
     **(c)** Third quadrant.
     **(d)** Third quadrant.

**10.** $\dfrac{5}{2} - \sqrt{5}$ or $2.5 - \sqrt{5}$

**14.** $\sqrt{41} \approx 6.403$      **16.** $\sqrt{34} \approx 5.830$

**19.** Perimeter $= 2\sqrt{41} + \sqrt{82} \approx 21.86$;
     Area $= \dfrac{1}{2}(\sqrt{41})(\sqrt{41}) = 20.5$

**20.** Perimeter $= 16$; Area $= 16$

**21.** Perimeter $= 2\sqrt{20} + 16 \approx 24.94$;
     Area $= 8 \cdot 4 = 32$

**29.**

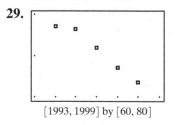

[1993, 1999] by [60, 80]

**30.**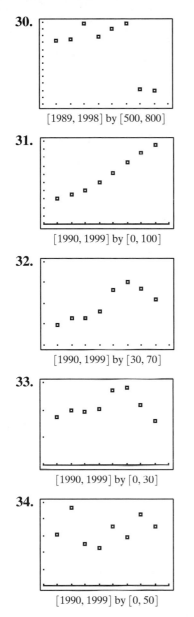

[1989, 1998] by [500, 800]

**31.**

[1990, 1999] by [0, 100]

**32.**

[1990, 1999] by [30, 70]

**33.**

[1990, 1999] by [0, 30]

**34.**

[1990, 1999] by [0, 50]

**37.** The three sides have lengths 5, 5, $5\sqrt{2}$. Since two sides have the same length, the triangle is isosceles.

**38.** (a) Both diagonals have midpoint $(-2, -1)$.
(b) Both diagonals have midpoint $(2, 2)$.

**39.** (a) 8; 5; $\sqrt{89}$

**40.** (a) $\sqrt{32}$; $\sqrt{50}$; $\sqrt{18}$
(b) $(\sqrt{32})^2 + (\sqrt{18})^2 = (\sqrt{50})^2$

**41.** $(x - 1)^2 + (y - 2)^2 = 25$

**42.** $(x + 3)^2 + (y - 2)^2 = 1$

**43.** $(x + 1)^2 + (y + 4)^2 = 9$

**44.** $x^2 + y^2 = 3$     **47.** $(0, 0)$; $\sqrt{5}$

**48.** $(2, -6)$; 5

**50.** $|y - (-2)| \geq 4$, or $|y + 2| \geq 4$

**51.** $|x - c| < d$     **52.** $|y - c| > d$

**53.** (a) 4; 6

(c) $\dfrac{2a + b}{3}$; $\dfrac{a + 2b}{3}$

(e) $\left(\dfrac{2a + c}{3}, \dfrac{2b + d}{3}\right)$; $\left(\dfrac{a + 2c}{3}, \dfrac{b + 2d}{3}\right)$

**55.** Show that two sides have length $2\sqrt{5}$ and the third side has length $2\sqrt{10}$.

**56.** Midpoint is $\left(\dfrac{5}{2}, \dfrac{7}{2}\right)$. Distances from this point to vertices are equal $\sqrt{18.5}$.

**59.** If the legs have lengths $a$ and $b$, and the hypotenuse is $c$ units long, then without loss of generality, we can assume the vertices are $(0, 0)$, $(a, 0)$, and $(0, b)$. Then the midpoint of the hypotenuse is $\left(\dfrac{a + 0}{2}, \dfrac{b + 0}{2}\right) = \left(\dfrac{a}{2}, \dfrac{b}{2}\right)$. The distance to the other vertices is $\sqrt{\left(\dfrac{a}{2}\right)^2 + \left(\dfrac{b}{2}\right)^2}$ $= \sqrt{\dfrac{a^2}{4} + \dfrac{b^2}{4}} = \dfrac{c}{2} = \dfrac{1}{2}c$.

**63.** $Q(-a, -b)$

**64.** Let the points on the number line be $(a, 0)$ and $(b, 0)$. The distance between them is $\sqrt{(a - b)^2 + (0 - 0)^2} = \sqrt{(a - b)^2} = |a - b|$

## Section P.3 (pp. 19–25)

### Quick Review

**6.** $\dfrac{4y - 5}{(y - 1)(y - 2)}$     **9.** $\dfrac{11x + 18}{10}$

### Exercises

**17.** $y = -\dfrac{4}{5} = -0.8$     **22.** $x = \dfrac{9}{4} = 2.25$

**25.** $x = \dfrac{17}{10} = 1.7$     **26.** $x = \dfrac{7}{2} = 3.5$

**27.** $t = \dfrac{31}{9}$     **28.** $t = -\dfrac{5}{7}$

**29.** (a) The figure shows that $x = -2$ is a solution of the equation $2x^2 + x - 6 = 0$
(b) The figure shows that $x = \dfrac{3}{2}$ is a solution of the equation $2x^2 + x - 6 = 0$

**30.** (a) The figure shows that $x = 2$ is a solution of the equation $7x + 5 = 4x - 7$.
(b) The figure shows that $x = -4$ is a solution of the equation $7x + 5 = 4x - 7$.

**35.**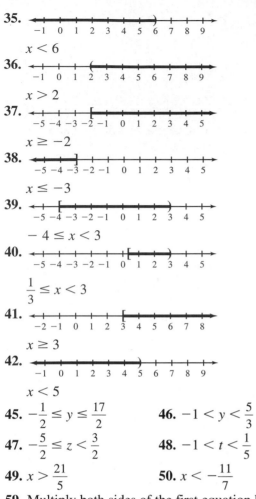
$x < 6$

**36.**
$x > 2$

**37.**
$x \geq -2$

**38.**
$x \leq -3$

**39.**
$-4 \leq x < 3$

**40.**
$\frac{1}{3} \leq x < 3$

**41.**
$x \geq 3$

**42.**
$x < 5$

**45.** $-\frac{1}{2} \leq y \leq \frac{17}{2}$   **46.** $-1 < y < \frac{5}{3}$

**47.** $-\frac{5}{2} \leq z < \frac{3}{2}$   **48.** $-1 < t < \frac{1}{5}$

**49.** $x > \frac{21}{5}$   **50.** $x < -\frac{11}{7}$

**59.** Multiply both sides of the first equation by 2.

**60.** Divide both sides of the first equation by 2.

**63. (e)** If your calculator returns 0 when you enter $2x + 1 < 4$, you can conclude that the value stored in $x$ is not a solution of the inequality $2x + 1 < 4$.

**64.** $W = \frac{1}{2}P - L = \frac{P - 2L}{2}$

**65.** $b_1 = \frac{2A}{h} - b_2$

## Section P.4 (pp. 26–38)

**Exploration 1**

**1.**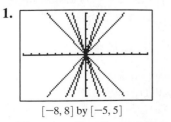
$[-8, 8]$ by $[-5, 5]$

These graphs all pass through the origin. They have different slopes.

**2.** If $m > 0$, then the graphs of $y = mx$ and $y = -mx$ have the same steepness, but one increases from left to right, and the other decreases left to right.

**3.**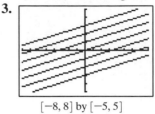
$[-8, 8]$ by $[-5, 5]$

These graphs have the same slope, but different $y$-intercepts.

**4.** The graphs of $y = mx + b$ and $y = mx + c$ have the same slope but different $y$-intercepts.

**5.**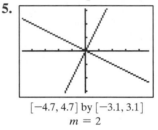
$[-4.7, 4.7]$ by $[-3.1, 3.1]$
$m = 2$

The angle between the two lines appears to be 90°.

**6.**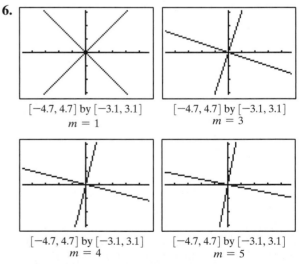

$[-4.7, 4.7]$ by $[-3.1, 3.1]$
$m = 1$

$[-4.7, 4.7]$ by $[-3.1, 3.1]$
$m = 3$

$[-4.7, 4.7]$ by $[-3.1, 3.1]$
$m = 4$

$[-4.7, 4.7]$ by $[-3.1, 3.1]$
$m = 5$

In each case, the two lines appear to be at right angles to one another.

**Quick Review**

**3.** $x = 12$   **4.** $x = \frac{3}{29}$

**5.** $y = \frac{2}{5}x - \frac{21}{5}$   **6.** $y = -\frac{4}{3}x + 8$

**7.** $y = \dfrac{17}{5}$

**8.** $y = x - \dfrac{1}{3}x^2$

## Exercises

**1.** $-2$

**2.** $\dfrac{2}{3}$

**3.**

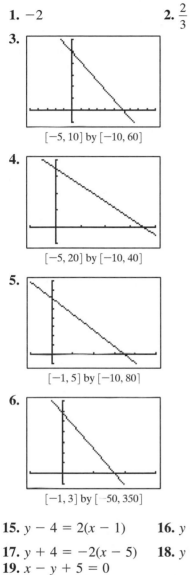

$[-5, 10]$ by $[-10, 60]$

**4.**

$[-5, 20]$ by $[-10, 40]$

**5.**

$[-1, 5]$ by $[-10, 80]$

**6.**

$[-1, 3]$ by $[-50, 350]$

**15.** $y - 4 = 2(x - 1)$  **16.** $y - 3 = -\dfrac{2}{3}(x + 4)$

**17.** $y + 4 = -2(x - 5)$  **18.** $y - 4 = 3(x + 3)$

**19.** $x - y + 5 = 0$

**20.** $x - y - 5 = 0$

**21.** $y + 3 = 0$

**22.** $x + y + 6 = 0$

**23.** $x - y + 3 = 0$

**24.** $x = 4$

**27.** $y = -\dfrac{1}{4}x + 4$

**28.** $y = \dfrac{1}{7}x + \dfrac{10}{7}$

**31.** (a): the slope is 1.5, compared to 1 in (b).

**32.** (b): the slopes are $\dfrac{7}{4}$ and 4, respectively.

**33.** $x = 4$; $y = 21$

**34.** $x = 2$; $y = -18$

**35.** $x = -10$; $y = -7$

**36.** $x = 14$; $y = 20$

**37.** Ymin $= -30$, Ymax $= 30$, Yscl $= 3$

**38.** Ymin $= -50$, Ymax $= 50$, Yscl $= 5$

**39.** Ymin $= -20/3$, Ymax $= 20/3$, Yscl $= 2/3$

**40.** Ymin $= -12.5$, Ymax $= 12.5$, Yscl $= 1.25$

**41.** **(a)** $y = 3x - 1$  **(b)** $y = -\dfrac{1}{3}x + \dfrac{7}{3}$

**42.** **(a)** $y = -2x - 1$  **(b)** $y = \dfrac{1}{2}x + 4$

**43.** **(a)** $y = -\dfrac{2}{3}x + 3$  **(b)** $y = \dfrac{3}{2}x - \dfrac{7}{2}$

**44.** **(a)** $y = \dfrac{3}{5}x - \dfrac{13}{5}$  **(b)** $y = -\dfrac{5}{3}x + 11$

**45.** **(a)** 3,187.5; 42,000
   **(b)** 9.57 years
   **(c)** $3,187.5t + 42,000 = 74,000$; $t = 10.04$
   **(d)** $t = 12$ years

**46.** **(a)** $0 \le x \le 18,000$
   **(b)** $I = 0.05x + 0.08(18,000 - x)$
   **(c)** $x = 14,000$ dollars
   **(d)** $x = 8,500$ dollars

**48.** **(c)** 2,217.6 ft

**49.** $m = \dfrac{3}{8} > \dfrac{4}{12}$, so asphalt shingles are acceptable.

**50.** American's income in the months August 1998 through July 1999 was, respectively, 7.16, 7.19, 7.22, 7.25, 7.28, 7.32, 7.35, 7.38, 7.41, 7.44, 7.47, and 7.5 trillion dollars.

**51.** **(a)** $y = 0.031\overline{6}x + 5.82$
   **(b)** Americans' spending in the months August 1998 through July 1999 was, respectively, 5.85, 5.88, 5.92, 5.95, 5.98, 6.01, 6.04, 6.07, 6.11, 6.14, 6.17, and 6.2 trillion dollars.

**52.** **(b)**

$[1990, 1999]$ by $[0, 100]$

**53.** **(a)**

$[0, 10]$ by $[5500, 6000]$

**(b)** $y = 79.2(x - 3) + 5{,}523 = 79.2x + 5{,}285.4$

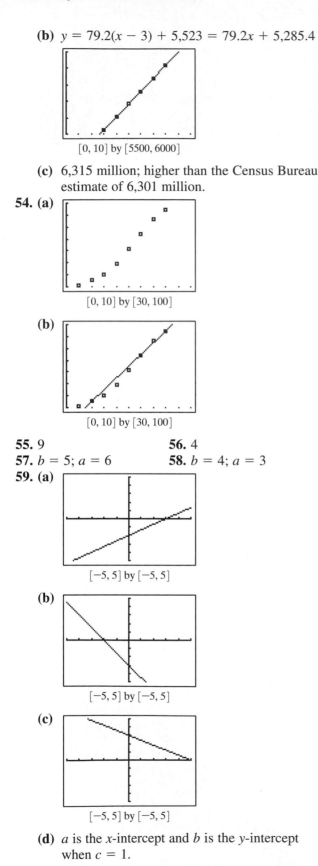

$[0, 10]$ by $[5500, 6000]$

**(c)** 6,315 million; higher than the Census Bureau estimate of 6,301 million.

**54. (a)**

$[0, 10]$ by $[30, 100]$

**(b)**

$[0, 10]$ by $[30, 100]$

**55.** 9               **56.** 4

**57.** $b = 5$; $a = 6$       **58.** $b = 4$; $a = 3$

**59. (a)**

$[-5, 5]$ by $[-5, 5]$

**(b)**

$[-5, 5]$ by $[-5, 5]$

**(c)**

$[-5, 5]$ by $[-5, 5]$

**(d)** $a$ is the $x$-intercept and $b$ is the $y$-intercept when $c = 1$.

**(e)**

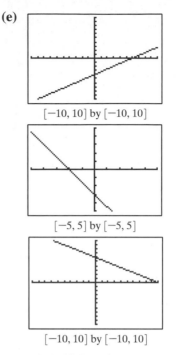

$[-10, 10]$ by $[-10, 10]$

$[-5, 5]$ by $[-5, 5]$

$[-10, 10]$ by $[-10, 10]$

$a$ is half the $x$-intercept and $b$ is half the $y$-intercept when $c = 2$.

**(f)** when $c = -1$, $a$ is the opposite of the $x$-intercept and $b$ is opposite of the $y$-intercept.

**61. (a)** If $b = 0$, both lines are vertical; otherwise, both have slope $m = -\dfrac{a}{b}$. If $c = d$, the lines are coincident.

**(b)** If either $a$ or $b$ equals 0, then one line is horizontal and the other is vertical. Otherwise, their slopes are $-\dfrac{a}{b}$ and $\dfrac{b}{a}$, respectively. In either case, they are perpendicular.

**62.** As in the diagram, we can choose one point to be the origin, and another to be on the $x$-axis. The midpoints of the sides, starting from the origin and working around counterclockwise in the diagram, are then $A\left(\dfrac{a}{2}, 0\right)$, $B\left(\dfrac{a + b}{2}, \dfrac{c}{2}\right)$, $C\left(\dfrac{b + d}{2}, \dfrac{c + e}{2}\right)$, and $D\left(\dfrac{d}{2}, \dfrac{e}{2}\right)$.

The opposite sides are therefore parallel, since the slopes of the four lines connecting those points are: $m_{AB} = \dfrac{c}{b}$; $m_{BC} = \dfrac{e}{d - a}$; $m_{CD} = \dfrac{c}{b}$; $m_{DA} = \dfrac{e}{d - a}$.

**63.** $y - 4 = -\dfrac{3}{4}(x - 3)$.

**64.** $A$ has coordinates $\left(\dfrac{b}{2}, \dfrac{c}{2}\right)$, while $B$ is $\left(\dfrac{a+b}{2}, \dfrac{c}{2}\right)$,

so the line containing $A$ and $B$ is the horizontal

line $y = \dfrac{c}{2}$, and the distance from $A$ to $B$ is

$\left|\dfrac{a+b}{2} - \dfrac{b}{2}\right| = \dfrac{a}{2}$.

## Section P.5 (pp. 38–47)

### Exploration 1

**1.**

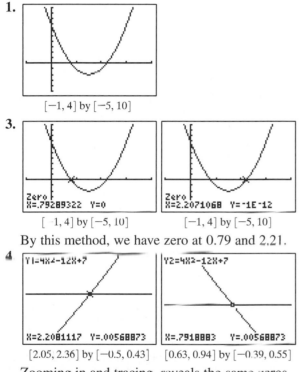

$[-1, 4]$ by $[-5, 10]$

**3.**

$[-1, 4]$ by $[-5, 10]$ 

$[-1, 4]$ by $[-5, 10]$

Zero
X=.79289322  Y=0

Zero
X=2.2071068  Y=-1E-12

By this method, we have zero at 0.79 and 2.21.

**4.**

Y1=4X²-12X+7

Y2=4X²-12X+7

X=2.2081117  Y=.00568873

X=.7918883  Y=.00568873

$[2.05, 2.36]$ by $[-0.5, 0.43]$   $[0.63, 0.94]$ by $[-0.39, 0.55]$

Zooming in and tracing, reveals the same zeros, correct to 2 decimal places.

**5.** The answers in parts 2, 3, and 4 are the same.

**6.** On a calculator, evaluating $4x^2 - 12x + 7$ when $x = 0.79$ gives $y = 0.0164$ and when $x = 2.21$ gives $y = 0.0164$, so the numbers 0.79 and 2.21 are approximate zeros.

### Quick Review

**3.** $6x^2 - 7x - 5$          **4.** $15y^2 + 7y - 4$

**6.** $x(5x - 4)(3x - 2)$      **7.** $(3x + 1)(x^2 - 5)$

**8.** $(y - 2)(y + 2)(y - 3)(y + 3)$

**9.** $\dfrac{(x - 2)(x + 1)}{(2x + 1)(x - 3)}$      **10.** $\dfrac{-2(x + 4)}{(x - 2)(x + 2)}$ if $x \neq 3$

### Exercises

**2.** $x = 5 \pm \sqrt{8.5}$          **3.** $x = -4 \pm \sqrt{\dfrac{8}{3}}$

**4.** $u = -1 \pm \sqrt{4.5}$       **5.** $y = \pm\sqrt{\dfrac{7}{2}}$

**6.** $x = -8$ or $x = 5$         **7.** $x = -7$ or $x = 1$

**8.** $x = -2.5 - \sqrt{15.25} = -6.405$ or

$x = -2.5 + \sqrt{15.25} = 1.405$

**9.** $x = \dfrac{7}{2} - \sqrt{11} \approx 0.183$ or $x = \dfrac{7}{2} + \sqrt{11}$

$\approx 6.817$

**10.** $x = -3 - \sqrt{13} \approx -6.606$ or

$x = -3 + \sqrt{13} \approx 0.606$

**12.** $x = \dfrac{4}{3} - \dfrac{1}{3}\sqrt{46} \approx -0.927$ or

$x = \dfrac{4}{3} + \dfrac{1}{3}\sqrt{46} \approx 3.594$

**13.** $x = -4 \pm 3\sqrt{2} \approx -8.243$ or $\approx 0.243$

**14.** $x = \dfrac{1}{2}$ or $x = 1$

**15.** $x = -1$ or $x = 4$

**16.** $x = \dfrac{1}{2}\sqrt{3} \pm \dfrac{1}{2}\sqrt{23} \approx -1.532$ or $3.264$

**19.** $x = -4; x = 5$          **20.** $x = -3; x = 0.5$

**21.** $x = 0.5$ or $x = 1.5$    **22.** $x = 3; x = 5$

**23.** $x = -\dfrac{2}{3}$ or $x = 3$     **24.** $x = \dfrac{4}{3}$ or $x = -5$

**25.** $x$-intercept: 3; $y$-intercept: 2

**26.** $x$-intercept: 1, 3; $y$-intercept 3

**27.** $x$-intercepts: $-2$, 0, 2; $y$-intercept 0

**28.** no $x$-intercepts; no $y$-intercepts

**29.**

Zero
X=.61803399  Y=0

Zero
X=-1.618034  Y=0

$[-5, 5]$ by $[-5, 5]$     $[-5, 5]$ by $[-5, 5]$

**30.**

Zero
X=-1.792893  Y=0

Zero
X=-3.207107  Y=0

$[-5, 5]$ by $[-5, 5]$     $[-5, 5]$ by $[-5, 5]$

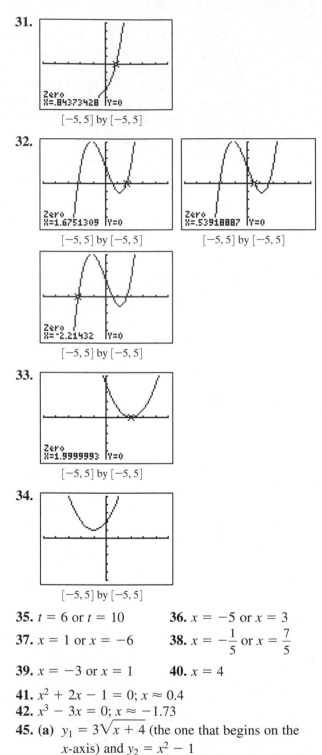

**31.**

[−5, 5] by [−5, 5]

**32.**

[−5, 5] by [−5, 5]          [−5, 5] by [−5, 5]

[−5, 5] by [−5, 5]

**33.**

[−5, 5] by [−5, 5]

**34.**

[−5, 5] by [−5, 5]

**35.** $t = 6$ or $t = 10$          **36.** $x = -5$ or $x = 3$

**37.** $x = 1$ or $x = -6$          **38.** $x = -\dfrac{1}{5}$ or $x = \dfrac{7}{5}$

**39.** $x = -3$ or $x = 1$          **40.** $x = 4$

**41.** $x^2 + 2x - 1 = 0$; $x \approx 0.4$

**42.** $x^3 - 3x = 0$; $x \approx -1.73$

**45. (a)** $y_1 = 3\sqrt{x+4}$ (the one that begins on the $x$-axis) and $y_2 = x^2 - 1$

   **(b)** $y = 3\sqrt{x+4} - x^2 + 1$

   **(c)** The $x$-coordinates of the intersections in the first picture are the same as the $x$-coordinates where the second graph crosses the $x$-axis.

**46.** Any number between 1.324 and 1.325 must have the digit 4 in its thousandths position. Such a number would round to 1.32.

**47.** $x = -2$ or $x = 1$          **48.** $x \approx -4.24$, $x \approx 4.24$

**49.** $x = 3$ or $x = -2$          **50.** $x = \sqrt{8} \approx 2.83$

**51.** $x \approx -4.56$, $x \approx -0.44$, $x = 1$

**52.** $x = -3.5 = -\dfrac{7}{2}$          **53.** $x = 2$ or $x = 5$

**54.** $x = -5$ or $x = 3$          **59.** $x = -2 \pm 2\sqrt{3}$

**61.** $x \approx -2.41$ or $x \approx 2.91$

**62.** $x \approx -1.64$ or $x \approx 1.45$

**63. (a)**
$$ax^2 + bx + c = 0$$
$$ax^2 + bx = -c$$
$$x^2 + \frac{b}{a}x = -\frac{c}{a}$$
$$x^2 + \frac{b}{a}x + \left(\frac{1}{2}\cdot\frac{b}{a}\right)^2 = -\frac{c}{a} + \left(\frac{1}{2}\cdot\frac{b}{a}\right)^2$$
$$x^2 + \frac{b}{a}x + \left(\frac{b}{2a}\right)^2 = -\frac{c}{a} + \frac{b^2}{4a^2}$$
$$\left(x + \frac{b}{2a}\right)\left(x + \frac{b}{2a}\right) = -\frac{4ac}{4a^2} + \frac{b^2}{4a^2}$$
$$\left(x + \frac{b}{2a}\right)^2 = \frac{b^2 - 4ac}{4a^2}$$
$$x + \frac{b}{2a} = \pm\sqrt{\frac{b^2 - 4ac}{4a^2}}$$
$$x + \frac{b}{2a} = \pm\frac{\sqrt{b^2 - 4ac}}{2a}$$
$$x = -\frac{b}{2a} \pm \frac{\sqrt{b^2 - 4ac}}{2a}$$
$$x = \frac{-b \pm \sqrt{b^2 - 4ac}}{2a}$$

**64. (a)** There must be 2 distinct real zeros, because $b^2 - 4ac > 0$ implies that $\pm\sqrt{b^2 - 4ac}$ are 2 distinct real numbers.

   **(b)** There must be 1 real zero, because $b^2 - 4ac = 0$ implies that $\pm\sqrt{b^2 - 4ac} = 0$, so the root must $x = -\dfrac{b}{a}$

   **(c)** There must be no real zeros, because $b^2 - 4ac < 0$ implies that $\pm\sqrt{b^2 - 4ac}$ are not real numbers.

**65. (a)** Answers may vary. $x^2 + 2x - 3$

   **(b)** Answers may vary. $x^2 + 2x + 1$

   **(c)** Answers may vary. $x^2 + 2x + 2$

**68.** $\approx 11.98$ ft

**69. (e)** There are no other possible number of solutions of this equation. For any $c$, the solution involves solving two quadratic equations, each of which can have 0, 1, or 2 solutions.

**70. (a)** Let $D = b^2 - 4ac$. The two solutions are
$\dfrac{-b \pm \sqrt{D}}{2a}$; adding them gives
$$\frac{-b + \sqrt{D}}{2a} + \frac{-b - \sqrt{D}}{2a} = \frac{-2b + \sqrt{D} - \sqrt{D}}{2a}$$
$$= -\frac{b}{a}$$

**(b)** Let $D = b^2 - 4ac$. The two solutions are
$\dfrac{-b \pm \sqrt{D}}{2a}$; multiplying them gives
$$\frac{-b + \sqrt{D}}{2a} \cdot \frac{-b - \sqrt{D}}{2a} = \frac{(-b)^2 - (\sqrt{D})^2}{4a^2}$$
$$= \frac{b^2 - (b^2 - 4ac)}{4a^2} = \frac{c}{a}$$

## Section P.6 (pp. 48–54)

### Quick Review

**1.** $-2 < x < 5$      **2.** $x \le -3$

**9.** $\dfrac{4x^2 - 4x - 1}{(x - 1)(3x - 4)}$     **10.** if $x \ne 2$   $\dfrac{(3x + 1)}{(x + 1)(x - 1)}$

### Exercises

**1.** $(-\infty, -9] \cup [1, \infty)$

**2.** $(-\infty, -1.3) \cup (2.3, \infty)$

**3.** $(1, 5)$

**4.** $[-8, 2]$:   $-8 \le x \le 2$

**5.** $\left(-\dfrac{2}{3}, \dfrac{10}{3}\right)$

**6.** $(-\infty, 0) \cup (3, \infty)$

**7.** $(-\infty, -11) \cup (7, \infty)$

**8.** $[-19, 29]$

**9.** $\left[-7, -\dfrac{3}{2}\right]$     **10.** $\left(-\infty, \dfrac{2}{3}\right] \cup \left[\dfrac{3}{2}, \infty\right)$

**11.** $\left(-\infty, -5\right) \cup \left(\dfrac{3}{2}, \infty\right)$    **12.** $\left(\dfrac{1}{4}, 2\right)$

**13.** $(-\infty, -2) \cup \left(\dfrac{1}{3}, \infty\right)$    **15.** $[-1, 0] \cup \left[1, \infty\right)$

**16.** $(-\infty, -5] \cup [0, 6]$    **18.** $\left(-\infty, \dfrac{3}{4}\right] \cup \left[\dfrac{4}{3}, \infty\right)$

**19.** $(-2, 0) \cup (2, \infty)$     **20.** $[-2, 0] \cup [2, \infty)$

**21.** $\left(-\infty, -\dfrac{1}{2}\right) \cup \left(\dfrac{4}{3}, \infty\right)$

**23.** $(-\infty, -1.41] \cup [0.08, \infty)$

**24.** $(0.79, 2.21)$      **25.** $\left(-\infty, \dfrac{1}{2}\right) \cup \left(\dfrac{1}{2}, \infty\right)$

**27.** no solution      **28.** $(-\infty, \infty)$

**29.** $(-3, 0) \cup (3, \infty)$    **30.** $\left[-\dfrac{2}{3}, 0\right] \cup \left[\dfrac{2}{3}, \infty\right)$

**33.** $(-\infty, -5) \cup (-3, \infty)$   **34.** $(-\infty, 2] \cup (9, \infty)$

**36.** $(-\infty, -1] \cup \left(-\dfrac{2}{3}, \infty\right)$

**37.** Answers may vary.
  **(a)** $x^2 + 1 > 0$      **(b)** $x^2 + 1 < 0$
  **(c)** $x^2 \le 0$
  **(d)** $(x + 2)(x - 5) \le 0$
  **(e)** $(x + 1)(x - 4) > 0$
  **(f)** $x(x - 4) \ge 0$

**39. (a)** Either $x \approx 0.94$ in. or $x \approx 3.78$ in.
  **(c)** $(0, 0.94) \cup (3.78, 6)$

**40. (c)** when $t$ is in the interval $(0, 4]$ or $[12, 16)$

**41. (c)** when $t$ is in the interval $(0, 5]$ or $[12, 17)$

**42.** reveals the boundaries of the solution set

**44. (b)** when $x$ is in the interval $(1, 25]$

**47.** $(-5.69, -4.11) \cup (0.61, 2.19)$

**48.** $(-\infty, -4.69] \cup [-3.11, 1.61] \cup [3.19, \infty)$

## Chapter P Review (pp. 57–59)

**1.** Endpoints 0 and 5; bounded

**2.** Endpoint 2; unbounded

**10.** 0.000 000 000 000 000 000 000 000 910 66
   (27 zeros between the decimal point and the 9)

**11. (a)** $3.35 \times 10^8$

**14. (a)** $\approx 9.849$      **(b)** $\left(\dfrac{1}{2}, 1\right)$

**17.** $(x - 0)^2 + (y - 0)^2 = 2^2$, or $x^2 + y^2 = 4$

**18.** $(x - 5)^2 + [y - (-3)]^2 = 4^2$, or
   $(x - 5)^2 + (y + 3)^2 = 16$

**19.** the center is $(-5, -4)$ and the radius is 3

**20.** the center is $(0, 0)$ and the radius is 1.

**21. (a)** $\sqrt{20} \approx 4.47$, $\sqrt{80} \approx 8.94$, 10
   **(b)** $(\sqrt{20})^2 + (\sqrt{80})^2 = 20 + 80 = 100 = 10^2$

**22.** $|z - (-3)| \le 1$, or $|z + 3| \le 1$

**26.** $9x + 7y + 17 = 0$

**27.** $y = \frac{4}{5}x - 4.4$          **28.** $y = \frac{3}{2}x - \frac{5}{2}$

**33. (a)**

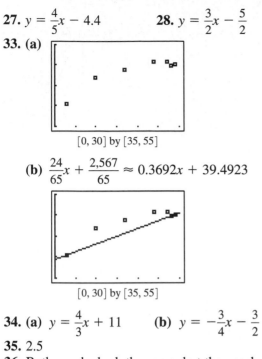

[0, 30] by [35, 55]

**(b)** $\frac{24}{65}x + \frac{2{,}567}{65} \approx 0.3692x + 39.4923$

[0, 30] by [35, 55]

**34. (a)** $y = \frac{4}{3}x + 11$      **(b)** $y = -\frac{3}{4}x - \frac{3}{2}$

**35.** 2.5

**36.** Both graphs look the same, but the graph on the left has slope $\frac{2}{3}$ – less than the slope of the one on the right, which is $\frac{12}{15} = \frac{4}{5}$. The different horizontal and vertical scales for the two windows make it difficult to judge by looking at the graphs.

**38.** $x = -\frac{9}{5}$

**40.** $x = \frac{1}{3} - \frac{\sqrt{7}}{3} \approx -0.55;\ x\ x = \frac{1}{3} + \frac{\sqrt{7}}{3} \approx 1.22$

**41.** $x = 2 - \sqrt{7} \approx -0.65;\ x = 2 + \sqrt{7} \approx 4.65$

**42.** $x = \frac{3}{4} - \frac{\sqrt{2}}{4} \approx 0.40;\ x = \frac{3}{4} + \frac{\sqrt{2}}{4} \approx 1.10$

**43.** $x = \frac{1}{3}$ or $x = -\frac{3}{2}$      **45.** $x = \frac{7}{2}$ or $x = -4$

**46.** $x = \frac{1}{2}$ or $x = -1$      **48.** $x = \frac{2}{3}$

**50.** $x = 1$ or $x = -3$

**51.** $x = \frac{3}{4} - \frac{\sqrt{17}}{4} \approx -0.28$ or $x = \frac{3}{4} + \frac{\sqrt{17}}{4} \approx 1.78$

**52.** $x = -\frac{2}{3} - \frac{\sqrt{7}}{3} \approx -1.55$ or

$x = -\frac{2}{3} + \frac{\sqrt{7}}{3} \approx 0.22$

**53.** $x = 0$ or $x = -\frac{2}{3}$ or $x = 7$

**54.** $x = -2$ or $x = 2$      **55.** $x \approx 2.36$

**56.** $x = -1$ or $x \approx 1.45$

**57.** $(-6, 3]$

**58.** $\left[ -\frac{5}{3}, \infty \right)$

**61.** $(-\infty, -2] \cup \left[ -\frac{2}{3}, \infty \right)$   **62.** $(-\infty, -2) \cup \left( \frac{5}{4}, \infty \right)$

**63.** $(-\infty, -3] \cup [0, 3]$      **64.** $\left( -\frac{3}{2}, 0 \right) \cup \left( \frac{3}{2}, \infty \right)$

**65.** $(-\infty, 2) \cup (7, \infty)$      **66.** $(-\infty, -8) \cup (2, \infty)$

**67.** $(-\infty, -17) \cup (3, \infty)$   **68.** $\left( -5, \frac{7}{2} \right)$

**69.** $(-\infty, \infty)$               **70.** no solution

**71. (b)** when $t$ is in the interval $(0, 8]$ or $[12, 20)$ (approximately)

**(c)** when $t$ is in the interval $[8, 12]$ (approximately)

**73. (a)** when $x$ is in the interval $(0, 18.5]$

**(b)** when $x$ is in the interval $(22.19, \infty)$ (approximately)

# Chapter 1

## Section 1.1 (pp. 61–78)

### Exploration 2

**3.** A statistician might look for adverse economic factors in 1989, especially those that would affect people near or below the poverty line.

**4.** Yes. Table 1 shows that the minimum wage worker had less purchasing power in 1989 than in any other year since 1950.

### Quick Review 1.1

**2.** $(x + 5)(x + 5)$

**4.** $3x(x - 2)(x - 3)$

**5.** $(4h^2 + 9)(2h + 3)(2h - 3)$

**6.** $(x + h)(x + h)$

**9.** $(2x - 1)(x - 5)$

**10.** $(x^2 + 5)(x + 2)(x - 2)$

### Exercises 1.1

**11.** increasing

**13.**

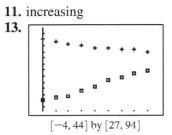

$[-4, 44]$ by $[27, 94]$

**15.** Men: $y = -0.26x + 85.4$
Women: $y = 0.58x + 35.7$

**18.** Percentages below 0% and above 100% are impossible.

**19.** The lower line shows the minimum salaries, since they are lower than the average salaries.

**20.** 11th from the left on each line. Both graphs show increases that year.

**21.** third from the right on both graphs. There is a clear drop in the average salary right after the 1994 strike.

**22.** Answers may vary.

**33.**

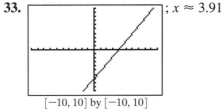

$; x \approx 3.91$

$[-10, 10]$ by $[-10, 10]$

**34.** 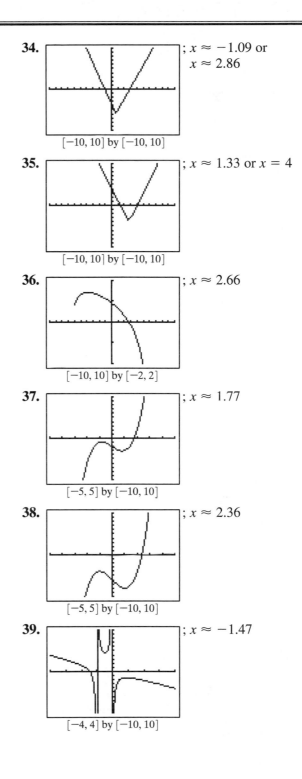 $; x \approx -1.09$ or $x \approx 2.86$

$[-10, 10]$ by $[-10, 10]$

**35.** $; x \approx 1.33$ or $x = 4$

$[-10, 10]$ by $[-10, 10]$

**36.** $; x \approx 2.66$

$[-10, 10]$ by $[-2, 2]$

**37.** $; x \approx 1.77$

$[-5, 5]$ by $[-10, 10]$

**38.** $; x \approx 2.36$

$[-5, 5]$ by $[-10, 10]$

**39.** $; x \approx -1.47$

$[-4, 4]$ by $[-10, 10]$

**40.**

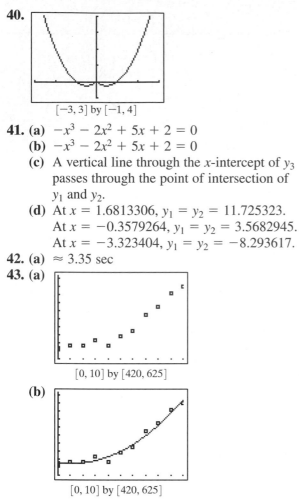

[−3, 3] by [−1, 4]

**41. (a)** $-x^3 - 2x^2 + 5x + 2 = 0$
**(b)** $-x^3 - 2x^2 + 5x + 2 = 0$
**(c)** A vertical line through the $x$-intercept of $y_3$ passes through the point of intersection of $y_1$ and $y_2$.
**(d)** At $x = 1.6813306$, $y_1 = y_2 = 11.725323$.
At $x = -0.3579264$, $y_1 = y_2 = 3.5682945$.
At $x = -3.323404$, $y_1 = y_2 = -8.293617$.

**42. (a)** $\approx 3.35$ sec

**43. (a)**

[0, 10] by [420, 625]

**(b)**

[0, 10] by [420, 625]

**(c)** Thirteen years after 1987, which is 2000.
**(d)** It is unlikely that the number of passengers will continue to grow according to this quadratic model, as factors like airport size, available routes, and numbers of aircraft limit the potential for growth. Indeed, the scatterplot already appears to be bending in a different direction from the parabola.

**44.** The length of each side of the square is $x + b$, so the area of the whole square is $(x + b)^2$. The square is made up of one square with area $x \cdot x = x^2$, one square with area $b \cdot b = b^2$, and two rectangles, each with area $b \cdot x = bx$. Using these four figures, the area of the square is $x^2 + 2bx + b^2$.

**45. (b)** $\dfrac{b}{2} \cdot \dfrac{b}{2} = \left(\dfrac{b}{2}\right)^2$

**(c)** $x^2 + bx + \left(\dfrac{b}{2}\right)^2 = \left(x + \dfrac{b}{2}\right)^2$ is the algebraic formula for completing the square, just as the area $\left(\dfrac{b}{2}\right)^2$ completes the area $x^2 + bx$ to form the area $\left(x + \dfrac{b}{2}\right)^2$.

**47. (a)** $y = (x^{200})^{1/200} = x^{\frac{200}{200}} = x^1 = x$ for all $x \geq 0$.
**(b)**

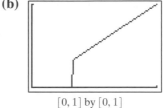

[0, 1] by [0, 1]

**(d)** For values of $x$ close to 0, $x^{200}$ is so small that the calculator is unable to distinguish it from zero. It returns a value of $0^{1/200} = 0$ rather than $x$.

**48.** One possible story: The jogger travels at an approximately constant speed throughout her workout. She jogs to the far end of the course, turns around and returns to her starting point, then goes out again for a second trip.

**49. (b)** $120
**(c)** June, after three months of poor performance.
**(e)** After reaching a low in June, the stock climbed back to a price near $140 by December. Tracy's shares had gained $2,000 by that point.

**50. (a)**

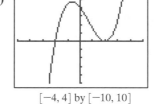

[−4, 4] by [−10, 10]

**(b)** Factoring, we find $y = (x + 2)(x - 2)(x - 2)$. There is a double zero at $x = 2$, a zero at $x = -2$, and no other zeros (since it is cubic).
**(c)** Same visually as graph in (a).
**(d)** Since the discriminant of the quadratic $x^2 - 4x + 4.01$ is negative, the only real zero of the product $y = (x + 2)(x^2 - 4x + 4.01)$ is at $x = -2$
**(e)** Same visually as graph in (a).
**(f)** Since the discriminant of the quadratic $x^2 - 4x + 3.99$ is positive, there are two real zeros of the quadratic and three real zeros of the product $y = (x + 2)(x^2 - 4x + 3.99)$.

**51. (a)** L3 = 3.90     **(b)** L3 = 3.90
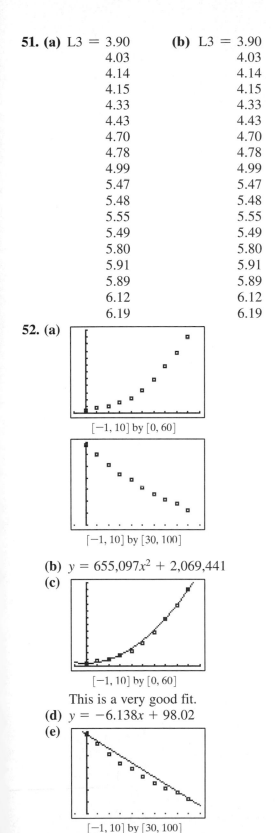

       4.03              4.03
       4.14              4.14
       4.15              4.15
       4.33              4.33
       4.43              4.43
       4.70              4.70
       4.78              4.78
       4.99              4.99
       5.47              5.47
       5.48              5.48
       5.55              5.55
       5.49              5.49
       5.80              5.80
       5.91              5.91
       5.89              5.89
       6.12              6.12
       6.19              6.19

**52. (a)**

[−1, 10] by [0, 60]

[−1, 10] by [30, 100]

**(b)** $y = 655{,}097x^2 + 2{,}069{,}441$

**(c)**

[−1, 10] by [0, 60]

This is a very good fit.

**(d)** $y = -6.138x + 98.02$

**(e)**

[−1, 10] by [30, 100]

This is a fairly good fit.

**(f)** Yes. The points seem to be curved beneath the line and might be better approximated by a parabola.

**(g)** $y = 0.3x^2 - 8.7x + 98.02$

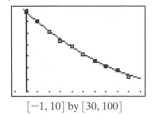

[−1, 10] by [30, 100]

This fits better than the line.

**53.** One possible answer: The number of cell phone users is increasing dramatically (as the quadratic model shows), and the average monthly bill is going down as more people share the industry cost. The model shows that the number of users will continue to rise, although the quadratic model cannot hold up indefinitely. The decline in the average monthly bill is tapering off as the graph approaches a vertex, but the model will fail before the parabola begins to climb.

**54.** Let $n$ be any integer.

$n^2 + 2n = n(n + 2)$, which is either the product of two odd integers or the product of two even integers.

The product of two odd integers is odd.

The product of two even integers is a multiple of 4, since each even integer in the product contributes a factor of 2 to the product,

Therefore, $n^2 + 2n$ is either odd or a multiple of 4.

## Section 1.2 (pp. 78–98)

### Exploration 1

**1.** From left to right, the tables are (c) constant, (b) decreasing, and (a) increasing.

**2.**

| X moves from | ΔX | ΔY1 | | X moves from | ΔX | ΔY2 |
|---|---|---|---|---|---|---|
| −2 to −1 | 1 | 0 | | −2 to −1 | 1 | −2 |
| −1 to 0 | 1 | 0 | | −1 to 0 | 1 | −1 |
| 0 to 1 | 1 | 0 | | 0 to 1 | 1 | −2 |
| 1 to 3 | 2 | 0 | | 1 to 3 | 2 | −4 |
| 3 to 7 | 4 | 0 | | 3 to 7 | 4 | −6 |

| X moves from | ΔX | ΔY3 |
|:---:|:---:|:---:|
| −2 to −1 | 1 | 2 |
| −1 to 0 | 1 | 2 |
| 0 to 1 | 1 | 2 |
| 1 to 3 | 2 | 3 |
| 3 to 7 | 4 | 6 |

**3.** positive; negative; 0

**4.** For lines, $\Delta Y/\Delta X$ is the slope. Lines with positive slope are increasing, lines with negative slope are decreasing, and lines with 0 slope are constant.

### Exercises 1.2

**1.**

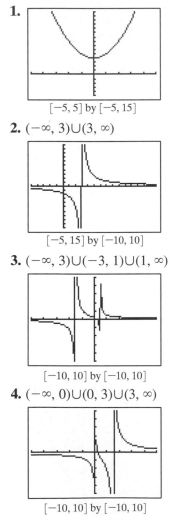

$[-5, 5]$ by $[-5, 15]$

**2.** $(-\infty, 3) \cup (3, \infty)$

$[-5, 15]$ by $[-10, 10]$

**3.** $(-\infty, 3) \cup (-3, 1) \cup (1, \infty)$

$[-10, 10]$ by $[-10, 10]$

**4.** $(-\infty, 0) \cup (0, 3) \cup (3, \infty)$

$[-10, 10]$ by $[-10, 10]$

**5.** $(-\infty, 5) \cup (5, \infty)$

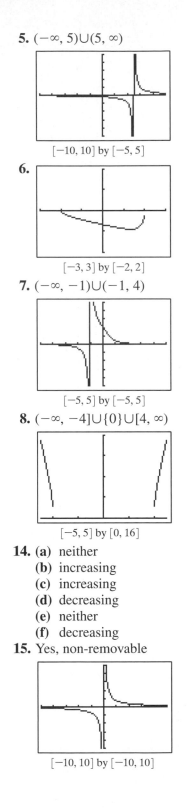

$[-10, 10]$ by $[-5, 5]$

**6.**

$[-3, 3]$ by $[-2, 2]$

**7.** $(-\infty, -1) \cup (-1, 4)$

$[-5, 5]$ by $[-5, 5]$

**8.** $(-\infty, -4] \cup \{0\} \cup [4, \infty)$

$[-5, 5]$ by $[0, 16]$

**14. (a)** neither
 **(b)** increasing
 **(c)** increasing
 **(d)** decreasing
 **(e)** neither
 **(f)** decreasing

**15.** Yes, non-removable

$[-10, 10]$ by $[-10, 10]$

**16.** Yes, removable

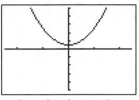

$[-5, 5]$ by $[-10, 10]$

**17.** Yes, non-removable

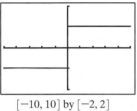

$[-10, 10]$ by $[-2, 2]$

**18.** Yes, non-removable

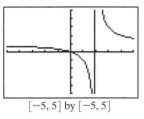

$[-5, 5]$ by $[-5, 5]$

**19.** Local maxima at $(-1, 4)$ and $(5, 5)$, local minimum at $(2, 2)$. The function increases on $(-\infty, -1]$ decreases on $[-1, 2]$ increases on $[2, 5]$, and decreases on $[5, \infty)$.

**20.** Local minimum at $(1, 2)$, $(3, 3)$ is neither, and $(5, 7)$ is a local maximum. The function decreases on $(-\infty, 1]$, increases on $[1, 5]$, and decreases on $[5, \infty)$

**21.** $(-1, 3)$ and $(3, 3)$ are neither. $(1, 5)$ is a local maximum, and $(5, 1)$ is a local minimum. The function increases on $(-\infty, 1]$ decreases on $[1, 5]$ increases on $[5, \infty)$.

**22.** $(-1, 1)$ and $(3, 1)$ are local minima, while $(1, 6)$ and $(5, 4)$ are local maxima. The function decreases on $(-\infty, -1]$ increases on $[-1, 1]$, decreases on $(1, 3]$, increases on $[3, 5]$, and decreases on $[5, \infty)$.

**23.** $f$ has a local minimum when $x = 0.5$, where $y = 3.75$. It has no maximum.

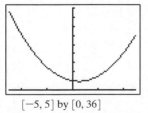

$[-5, 5]$ by $[0, 36]$

**24.** Local maximum: $y \approx 4.08$ at $x \approx -1.15$. Local minimum: $y \approx -2.08$ at $x \approx 1.15$.

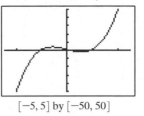

$[-5, 5]$ by $[-50, 50]$

**25.** Local minimum: $y \approx -4.09$ at $x \approx -0.82$. Local maximum: $y \approx -1.91$ at $x \approx 0.82$.

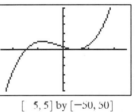

$[-5, 5]$ by $[-50, 50]$

**26.** Local minimum: $y \approx 9.48$ at $x \approx -1.67$. Local maximum: $y = 0$ when $x = 1$.

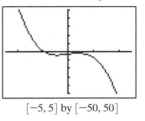

$[-5, 5]$ by $[-50, 50]$

**27.** Local maximum: $y \approx 9.16$ at $x \approx -3.20$. Local minimum: $y = 0$ at $x = 0$ and $y = 0$ at $x = -4$.

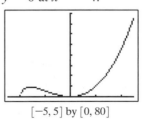

$[-5, 5]$ by $[0, 80]$

**28.** Local maximum: $y = 0$ at $x = -2.5$. Local minimum: $y \approx -3.13$ at $x = -1.25$.

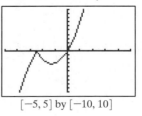

$[-5, 5]$ by $[-10, 10]$

**29.** Decreasing on $(-\infty, -2]$; increasing on $[-2, \infty)$.

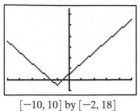

[−10, 10] by [−2, 18]

**29.** Decreasing on $(-\infty, -2)$; increasing on $(-2, \infty)$.

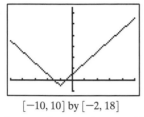

[−10, 10] by [−2, 18]

**30.** Decreasing on $(-\infty, -1]$; constant on $[-1, 1]$; increasing on $[1, \infty)$.

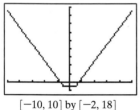

[−10, 10] by [−2, 18]

**31.** Decreasing on $(-\infty, -2]$; constant on $[-2, 1]$; increasing on $[1, \infty)$.

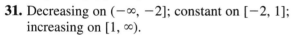

[−10, 10] by [0, 20]

**32.** Decreasing on $(-\infty, -2]$; increasing on $[-2, \infty)$.

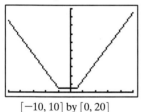

[−7, 3] by [−2, 13]

**33.** Increasing on $(-\infty, 1]$; decreasing on $[1, \infty)$.

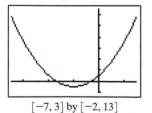

[−4, 6] by [−25, 25]

**34.** Increasing on $(-\infty, -0.5]$; decreasing on $[-0.5, 1.2]$; increasing on $[1.2, \infty)$. The middle values are approximate–they are actually at about $-0.549$ and $1.215$. The values given are what might be observed on the decimal window.

**35.** Even

**36.** Odd

**37.** Even

**38.** Even

**39.** Neither

**40.** Neither

**41.** Odd

**42.** Odd

**43.**

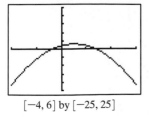

[−10, 10] by [−10, 10]

**44.**

[−10, 10] by [−10, 10]

**45.**　　　　　　　　　　　　　　　; $x = 3$; $y = -1$

[−8, 12] by [−10, 10]

**46.**

[−10, 10] by [−10, 10]

**47.**　　　　　　　　　　　　　; $x = 1$; $x = -1$; $y = 1$

[−10, 10] by [−10, 10]

**48.**

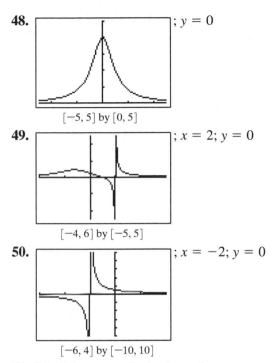

$; y = 0$

$[-5, 5]$ by $[0, 5]$

**49.** $; x = 2; y = 0$

$[-4, 6]$ by $[-5, 5]$

**50.** $; x = -2; y = 0$

$[-6, 4]$ by $[-10, 10]$

**51. (a)** $f(x)$ crosses the horizontal asymptote at $(0, 0)$.

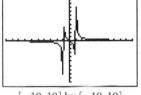

$[-10, 10]$ by $[-10, 10]$

**(b)** $g(x)$ crosses the horizontal asymptote at $(0, 0)$.

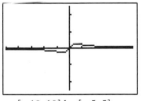

$[-10, 10]$ by $[-5, 5]$

**(c)** $h(x)$ crosses the horizontal asymptote at $(0, 0)$.

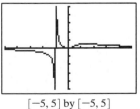

$[-5, 5]$ by $[-5, 5]$

**53. (a)** The vertical asymptote is $x = 0$, and this function is undefined at $x = 0$ (because a denominator can't be zero).

**(b)**

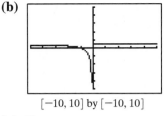

$[-10, 10]$ by $[-10, 10]$

**(c)** Yes

**54.** The horizontal asymptotes are determined by the two limits $\lim\limits_{x \to -\infty} f(x)$ and $\lim\limits_{x \to +\infty} f(x)$. These are at most two different numbers.

**56.** bounded above

**57.** bounded below

**58.** bounded below

**59.** $y$ is bounded

**60.** Not bounded above or below

**61.** $(-\infty, -1) \cup [0, \infty)$

**62.** $(-\infty, -1) \cup [0.75, \infty)$

**63. (a)**

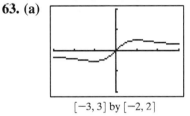

$[-3, 3]$ by $[-2, 2]$

**(b)** $\dfrac{x}{1 + x^2} < 1 \Leftrightarrow x < 1 + x^2 \Leftrightarrow$
$x^2 - x + 1 > 0$
But the discriminant of $x^2 - x + 1$ is negative ($-3$ to be exact), so the graph never crosses the $x$-axis on the interval $(0, \infty)$.

**(d)** $\dfrac{x}{1 + x^2} > -1 \Leftrightarrow x > -1 - x^2 \Leftrightarrow$
$x^2 + x + 1 > 0$
But the discriminant of $x^2 - x + 1$ is negative ($-3$ again), so the graph never crosses the $x$-axis on the interval $(-\infty, 0)$

**64. (a)** increasing

**(b)** **(c)**

| $\Delta y$ | $\Delta\Delta y$ |
|---|---|
| 1 | 0.05 |
| 1.05 | $-0.53$ |
| 0.52 | $-0.09$ |
| 0.43 | $-0.07$ |
| 0.36 | $-0.03$ |
| 0.33 | $-0.02$ |
| 0.31 | $-0.03$ |
| 0.28 | |

$\Delta y$ is none of these since it first increases

from 1 to 1.05 and then decreases.

(d) The graph rises, but bends downward as it rises.

70. As the graph moves continuously from the point $(-1, 5)$ down to the point $(1, -5)$, it must cross the $x$-axis somewhere along the way. That $x$-intercept will be a zero of the function in the interval $[-1, 1]$.

71. Since $f$ is odd, $f(-x) = -f(x)$ for all $x$. In particular, $f(-0) = -f(0)$. This is equivalent to saying that $f(0) = -f(0)$, and the only number which equals its opposite is 0. Therefore $f(0) = 0$, which means the graph must pass through the origin.

72. $-1 \le \dfrac{3x^2 - 1}{2x^2 + 1} \le \dfrac{3}{2}$

$0 \le 1 + \dfrac{3x^2 - 1}{2x^2 + 1} \le \dfrac{5}{2}$

$0 \le 2x^2 + 1 + 3x^2 - 1 \le 5x^2 + \dfrac{5}{2}$

$0 \le 5x^2 \le 5x^2 + \dfrac{5}{2}$

True for all $x$.

## Section 1.3 (pp. 99–107)

### Exercises 1.3

1. $y = x^3 + 1$     2. $y = |x| - 2$
3. $y = -\sqrt{x}$
4. $y = -\sin x$ or $y = \sin(-x)$
5. $y = -x$     6. $y = (x - 1)^2$
7. $y = -1/x$     8. $y = (x + 2)^3$
9. $y = e^x - 2$     10. $y = \cos x + 1$
11. $y = x,\ y = x^3,\ y = 1/x,\ y = x,\ y = \sin x$
14. $y = \sin x,\ y = \cos x$
16. $y = x,\ y = x^3,\ y = \ln x$
18. $y = x,\ y = x^3$
19. $y = x,\ y = x^3,\ y = 1/x,\ y = \sin x$
21. Domain: all reals; Range: $[-5, \infty)$
22. Domain: all reals; Range: $[0, \infty)$
23. Domain: $(-6, \infty)$; Range: all reals
24. Domain: $(-\infty, 0) \cup (0, \infty)$; Range: reals $\ne 3$
25. Domain: all reals; Range: $[0, \infty)$
26. Domain: $[10, \infty)$; Range: $[0, \infty)$
27. Domain: all reals; Range: $[4, 6]$
28. Domain: all reals; Range: $(2, \infty)$
29. Domain: all reals; Range: $[-10, \infty)$
30. Domain: all reals; Range: $[-4, 4]$

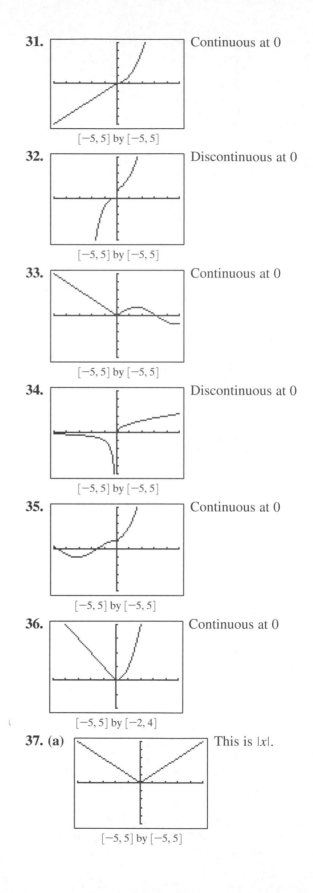

31. Continuous at 0
$[-5, 5]$ by $[-5, 5]$

32. Discontinuous at 0
$[-5, 5]$ by $[-5, 5]$

33. Continuous at 0
$[-5, 5]$ by $[-5, 5]$

34. Discontinuous at 0
$[-5, 5]$ by $[-5, 5]$

35. Continuous at 0
$[-5, 5]$ by $[-5, 5]$

36. Continuous at 0
$[-5, 5]$ by $[-2, 4]$

37. (a) This is $|x|$.
$[-5, 5]$ by $[-5, 5]$

**(b)** Squaring $x$ and taking the (positive) square root has the same effect as the absolute value function.

**38. (a)**

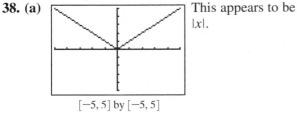

[−5, 5] by [−5, 5]

This appears to be $|x|$.

**(b)** For example, $g(2) = 1.990025$, whereas $|2| = 2$

**39. (a)**

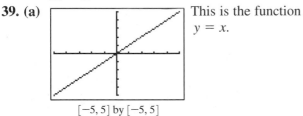

[−5, 5] by [−5, 5]

This is the function $y = x$.

**(b)** The fact that $\ln(e^x) = x$ shows that the natural logarithm function takes on arbitrarily large values. In particular, it takes on the value $L$ when $x = e^L$.

**40. (a)** Answers will vary.
   **(b)** In this window, it appears that $\sqrt{x} < x < x^2$

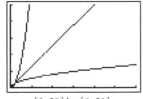

[0, 30] by [0, 20]

   **(c)**

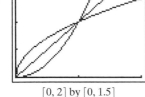

[0, 2] by [0, 1.5]

   **(d)** On the interval $(0, 1)$, $x^2 < x < \sqrt{x}$
   On the interval $(1, \infty)$, $\sqrt{x} < x < x^2$
   All three functions equal 1 when $x = 1$.

**41. (a)** A product of two odd functions is even.
   **(b)** A product of two even functions is even.
   **(c)** A product of an odd function and an even function is odd.

**42.** Answers vary.

**43. (a)** Pepperoni count ought to be proportional to the area of the pizza, which is proportional to the square of the radius.
   **(c)** Yes, very well.

**(d)** The fact that the pepperoni count fits the expected quadratic model so perfectly suggests that the pizzeria uses such a chart. If repeated observations produced the same results, there would be little doubt.

**44. (c)** With domain $[0, \infty)$, $x^2$ becomes the inverse of $\sqrt{x}$.

**45. (e)** The odd functions: $x$, $x^3$, $1/x$, $\sin x$

## Section 1.4 (pp. 107–125)

### Exploration 1

**1.** T starts at $-6$, at the point $(-5, 24)$. It stops at $T = 4$, at the point $(5, 24)$. 101 points are computed.

**2.** The graph is smoother because the plotted points are closer together.

**3.** The graph is less smooth because the plotted points are further apart. In CONNECT mode, they are connected by straight lines.

**4.** The smaller the Tstep, the slower the graphing proceeds. This is because the calculator has to compute more X and Y values.

**5.** The grapher skips directly from the point $(-1, 0)$ to the point $(1, 0)$, corresponding to the T-values $T = -2$ and $T = 0$. The two points are connected by a straight line, hidden by the X-axis.

**6.** With Tmin set at $-1$, the grapher begins at the point $(0, -1)$, missing the left-hand side of the curve entirely.

**7.** Leave everything else the same, but change Tmax to $-1$.

### Quick Review 1.4

**9.** $y = x^2 - 3$, $y \geq -3$ and $x \geq 0$
**10.** $y = x^2 + 2$, $y \geq 2$ and $x \geq 0$

### Exercises 1.4

**1.** $(f + g)(x) = 2x - 1 + x^2$;
   $(f - g)(x) = 2x - 1 - x^2$;
   $(fg)(x) = (2x - 1)(x^2) = 2x^3 - x^2$.
   There are no restrictions on any of the domains, so all three domains are $(-\infty, \infty)$.

**2.** $(f + g)(x) = x^2 - 3x + 4$;
   $(f - g)(x) = x^2 - x - 2$;
   $(fg)(x) = (x - 1)^2(3 - x) = -x^3 + 5x^2 - 7x + 3$.
   There are no restrictions on any of the domains, so all three domains are $(-\infty, \infty)$.

**3.** $(f + g)(x) = \sqrt{x} + \sin x;$
$(f - g)(x) = \sqrt{x} - \sin x;$
$(fg)(x) = \sqrt{x} \sin x.$
Domain in each case is $[0, \infty)$. For $\sqrt{x}$, $x \geq 0$.
For $\sin x$, $-\infty \leq x \leq \infty$.

**4.** $(f + g)(x) = \sqrt{x + 5} + |x + 3|;$
$(f - g)(x) = \sqrt{x + 5} - |x + 3|;$
$(fg)(x) = |x + 3|\sqrt{x + 5}.$
All three expressions contain $\sqrt{x + 5}$, so
$x + 5 \geq 0$ and $x \geq -5$; all three domains are
$[-5, \infty)$. For $|x + 3|$, $-\infty \leq x \leq \infty$.

**5.** $(f/g)(x) = \dfrac{\sqrt{x + 3}}{x^2};$ $x + 3 \geq 0$ and $x \neq 0$, so the
domain is $[-3, 0) \cup (0, \infty)$.

$(g/f)(x) = \dfrac{x^2}{\sqrt{x + 3}};$ $x + 3 > 0$, so the domain is
$(-3, \infty)$.

**6.** $(f/g)(x) = \dfrac{\sqrt{x - 2}}{\sqrt{x + 4}} = \sqrt{\dfrac{x - 2}{x + 4}};$ $x - 2 \geq 0$ and
$\quad x + 4 > 0$, so $x \geq 2$ and $x$
$> -4$; the domain is $[2, \infty)$.

$(g/f)(x) = \dfrac{\sqrt{x + 4}}{\sqrt{x - 2}} = \sqrt{\dfrac{x + 4}{x - 2}};$ $x - 2 > 0$ and
$x + 4 \geq 0$, so $x > 2$ and $x \geq -4$; the domain is
$(2, \infty)$.

**7.**

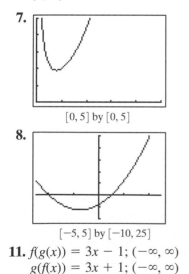

$[0, 5]$ by $[0, 5]$

**8.**

$[-5, 5]$ by $[-10, 25]$

**11.** $f(g(x)) = 3x - 1;$ $(-\infty, \infty)$
$g(f(x)) = 3x + 1;$ $(-\infty, \infty)$

**12.** $f(g(x)) = \left(\dfrac{1}{x - 1}\right)^2 - 1 = \dfrac{1}{(x - 1)^2} - 1;$
$(-\infty, 1) \cup (1, \infty)$

$g(f(x)) = \dfrac{1}{(x^2 - 1) - 1} = \dfrac{1}{x^2 - 2};$
$(-\infty, -\sqrt{2}) \cup (-\sqrt{2}, \sqrt{2}) \cup (\sqrt{2}, \infty)$
$f(g(x)) = x - 1;$ $[-1, \infty)$
$g(f(x)) = \sqrt{x^2 - 1};$ $(-\infty, 1] \cup [1, \infty)$

**13.** $f(g(x)) = x - 1;$ $[-1, \infty)$
$g(f(x)) = \sqrt{x^2 - 1};$ $(-\infty, 1] \cup [1, \infty)$

**14.** $f(g(x)) = \dfrac{1}{\sqrt{x} - 1};$ $[0, 1) \cup (1, \infty)$

$g(f(x)) = \dfrac{1}{\sqrt{x - 1}};$ $(1, \infty)$

**15.** One possibility: $f(x) = \sqrt{x}$ and $g(x) = x^2 - 5x$
**16.** One possibility: $f(x) = (x + 1)^2$ and $g(x) = x^3$
**17.** One possibility: $f(x) = |x|$ and $g(x) = 3x - 2$
**18.** One possibility: $f(x) = 1/x$ and
$\quad g(x) = x^3 - 5x + 3$
**19.** One possibility: $f(x) = x^5 - 2$ and $g(x) = x - 3$
**20.** One possibility: $f(x) = e^x$ and $g(x) = \sin x$
**25.** $1.5x - 1;$ It is a function.
**26.** $x^2 - 4x + 3;$ It is a function.
**27.** $x = (y + 2)^2;$ It is not a function.
**28.** $y = 2x^2 - 5;$ It is a function.
**29.** $y = \sqrt{25 - x^2}$ and $-\sqrt{25 - x^2}$
**30.** $y = \sqrt{25 - x}$ and $-\sqrt{25 - x}$
**31.** $y = \sqrt{x^2 - 25}$ and $-\sqrt{x^2 - 25}$
**32.** $y = \sqrt{3x^2 - 25}$ and $y = -\sqrt{3x^2 - 25}$
**33.** $f(g(x)) = 3\left[\dfrac{1}{3}(x + 2)\right] - 2 = x + 2 - 2 = x;$
$g(f(x)) = \dfrac{1}{3}[(3x - 2) + 2] = \dfrac{1}{3}(3x) = x.$
**34.** $f(g(x)) = \dfrac{1}{4}[(4x - 3) + 3] = \dfrac{1}{4}(4x) = x;$

$g(f(x)) = 4\left[\dfrac{1}{4}(x + 3)\right] - 3 = x + 3 - 3 = x.$
**35.** $f(g(x)) = [(x - 1)^{1/3}]^3 + 1 = (x - 1)^1 + 1$
$\quad = x - 1 + 1 = x;$
$g(f(x)) = [(x^3 + 1) - 1]^{1/3} = (x^3)^{1/3} = x^1 = x.$
**36.** $f(g(x)) = \dfrac{7}{\frac{7}{x}} = \dfrac{7}{1} \cdot \dfrac{x}{7} = x;$ $g(f(x)) = \dfrac{7}{\frac{7}{x}} = \dfrac{7}{1} \cdot \dfrac{x}{7}$

$= x$

**37.** $f(g(x)) = \dfrac{\dfrac{1}{x-1}+1}{\dfrac{1}{x-1}} = (x-1)\left(\dfrac{1}{x-1}+1\right)$

$= 1 + x - 1 = x;$

$g(f(x)) = \dfrac{1}{\dfrac{x+1}{x}-1} = \left(\dfrac{1}{\dfrac{x+1}{x}-1}\right)\cdot\dfrac{x}{x}$

$= \dfrac{x}{x+1-x} = \dfrac{x}{1} = x.$

**38.** $f(g(x)) = \dfrac{\dfrac{2x+3}{x-1}+3}{\dfrac{2x+3}{x-1}-2} = \left(\dfrac{\dfrac{2x+3}{x-1}+3}{\dfrac{2x+3}{x-1}-2}\right)\cdot\left(\dfrac{x-1}{x-1}\right)$

$= \dfrac{2x+3+3(x-1)}{2x+3-2(x-1)} = \dfrac{5x}{5} = x;$

$g(f(x)) = \dfrac{2\left(\dfrac{x+3}{x-2}\right)+3}{\dfrac{x+3}{x-2}-1} = \left[\dfrac{2\left(\dfrac{x+3}{x-2}\right)+3}{\dfrac{x+3}{x-2}-1}\right]\cdot\left(\dfrac{x-2}{x-2}\right)$

$= \dfrac{2(x+3)+3(x-2)}{x+3-(x-2)} = \dfrac{5x}{5} = x.$

**39.** one-to-one

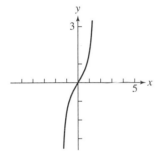

**40.** not one-to-one

**41.** one-to-one

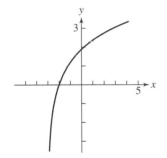

**42.** not one-to-one

**43.** $f^{-1}(x) = \dfrac{1}{3}x + 2,\ (-\infty, \infty)$

**44.** $f^{-1}(x) = \dfrac{1}{2}x - \dfrac{5}{2},\ (-\infty, \infty)$

**45.** $f^{-1}(x) = \dfrac{x+3}{2-x},\ (-\infty, 2)\cup(2, \infty)$

**46.** $f^{-1}(x) = \dfrac{2x+3}{x-1},\ (-\infty, 1)\cup(1, \infty)$

**47.** $f^{-1}(x) = x^2 - 3,\ x \geq 0$
**48.** $f^{-1}(x) = x^2 - 2,\ x \geq 0$
**49.** $f^{-1}(x) = \sqrt[3]{x},\ (-\infty, \infty)$
**50.** $f^{-1}(x) = \sqrt[3]{x} - 5,\ (-\infty, \infty)$
**51.** $f^{-1}(x) = x^3 - 5,\ (-\infty, \infty)$
**52.** $f^{-1}(x) = x^3 + 2,\ (-\infty, \infty)$
**53.** $V = \dfrac{4}{3}\pi r^3 = \dfrac{4}{3}\pi(48 + 0.03t)^3;\ 775{,}734.6\ \text{in.}^3$

**54.** 3.6 in.
**55.** $t \approx 3.63$ sec
**57.** (a) 631.4 francs
    (b) $0.1584y$. This converts francs to dollars.
    (c) \$19.80
**58.** (a) $c^{-1}(x) = \dfrac{9}{5}x + 32$. This converts Celsius
    temperature to Fahrenheit temperature.
    (b) $\dfrac{5}{9}x + 255.38$. This converts Fahrenheit
    temperature to Kelvin temperature.
**59.** (Answers may vary.)
    (a) If the graph of $f$ is unbroken, its reflection in
    the line $y = x$ will be also.
    (b) Both $f$ and its inverse must be one-to-one in
    order to be inverse functions.
    (c) Since $f$ is odd, $(-x, -y)$ is on the graph
    whenever $(x, y)$ is. This implies that $(-y, -x)$
    is on the graph of $f^{-1}$ whenever $(x, y)$ is. That
    implies that $f^{-1}$ is odd.
    (d) Let $y = f(x)$. Since the ratio of $\Delta y$ to $\Delta x$ is
    positive, the ratio of $\Delta x$ to $\Delta y$ is positive. Any
    ratio of $\Delta y$ to $\Delta x$ on the graph of $f^{-1}$ is the
    same as some ratio of $\Delta x$ to $\Delta y$ on the graph
    of $f$, hence positive. This implies that $f^{-1}$ is
    increasing.
**60.** (Answers may vary.)
    (a) $f(x) = e^x$ has a horizontal asymptote;
    $f^{-1}(x) = \ln x$ does not.
    (b) $f(x) = e^x$ has domain all real numbers;
    $f^{-1}(x) = \ln x$ does not.
    (c) $f(x) = e^x$ has a graph that is bounded below;
    $f^{-1}(x) = \ln x$ does not.
    (d) $f(x) = \dfrac{x^2 - 25}{x - 5}$ has a removeable
    discontinuity at $x = 5$ because its graph is the
    line $y = x + 5$ with the point $(5, 10)$
    removed. The inverse function is the line
    $y = x - 5$ with the point $(10, 5)$ removed.
    This function has a removable discontinuity,
    but not at $x = 5$.
**61.** (b) $y = \dfrac{4}{3}(x - 31)$. It converts scaled scores to
    raw scores.

**62.** The function must be increasing so that the *order* of the students' grades, top to bottom, will remain the same after scaling as it is before scaling. A student with a raw score of 136 gets dropped to 133, but that will still be higher than the scaled score for a student with 134.

**63.** (a) No
   (b) No
   (c) 45°. Yes

**64.** (a) $y = \dfrac{30}{3^{1.7}}(x - 1)^{1.7} + 65$. This can be used to convert GPA's to percentage grades.
   (b) Yes; $x$ is restricted to the domain [1, 4.28].
   (c) Graph the composition function and verify that it is the line $y = x$ on the restricted domain (either [65, 100] or [100, 4.28], depending on the order of composition).

**65.** When $k = 1$, the scaling function is linear. Opinions will vary as to which is the best value of $k$.

**66.** $y = \dfrac{-x^2 \pm \sqrt{x^4 + 20}}{2}$

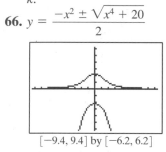
$[-9.4, 9.4]$ by $[-6.2, 6.2]$

# Section 1.5 (pp. 125–135)

## Exploration 1

   **1.** They raise or lower the parabola along the $y$-axis
   **2.** They move the parabola left and right along the $x$-axis.
   **3.** Yes

## Exploration 2

   **1.**

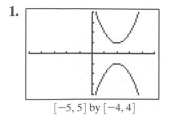
$[-5, 5]$ by $[-4, 4]$

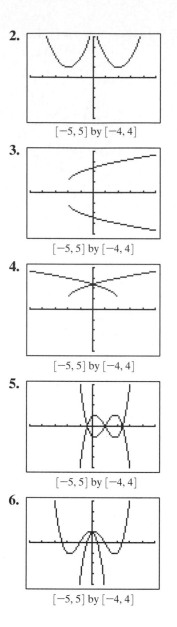

**2.**
$[-5, 5]$ by $[-4, 4]$

**3.**
$[-5, 5]$ by $[-4, 4]$

**4.**
$[-5, 5]$ by $[-4, 4]$

**5.**
$[-5, 5]$ by $[-4, 4]$

**6.**
$[-5, 5]$ by $[-4, 4]$

## Exploration 3

   **1.**

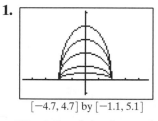
$[-4.7, 4.7]$ by $[-1.1, 5.1]$

The 1.5 and the 2 stretch the graph vertically; the 0.5 and the 0.25 shrink the graph vertically.

**2.**

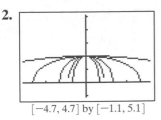

$[-4.7, 4.7]$ by $[-1.1, 5.1]$

The 1.5 and the 2 shrink the graph horizontally; the 0.5 and the 0.25 stretch the graph horizontally.

**Quick Review 1.5**

**7.** $x^2 - x + 2$

**8.** $2x^2 + 7x + 1$

**Exercises 1.5**

**1.** vertical translation down 3 units

**2.** vertical translation up 5.2 units

**3.** horizontal translation left 4 units

**4.** horizontal translation right 3 units

**5.** horizontal translation to the right 100 units

**6.** vertical translation downward 100 units

**7.** horizontal translation to the right 1 unit, and vertical translation up 3 units

**8.** Horizontal translation to the left 50 units and vertical translation downward 279 units

**9.** reflection across $x$-axis

**10.** horizontal translation right 5 units

**11.** reflection across $y$-axis

**12.** This can be written as $y = \sqrt{-(x-3)}$ or $y = \sqrt{-x + 3}$. The first of these can be interpreted as reflection across the $y$-axis followed by a horizontal translation to the right 3 units. The second may be viewed as a horizontal translation left 3 units followed by a reflection across the $y$-axis.

**13.** vertically stretch by 2

**14.** horizontally shrink by $\frac{1}{2}$, or vertically stretch by $2^3 = 8$

**15.** horizontally stretch by $\frac{1}{0.2} = 5$, or vertically shrink by $0.2^3 = 0.008$

**16.** vertically shrink by 0.3

**17.** translate right 6 units to get $g$

**18.** translate left 4 units, and reflect across the $x$-axis to get $g$

**19.** translate left 4 units, and reflect across the $x$-axis to get $g$

**20.** vertically stretch by 2 to get $g$

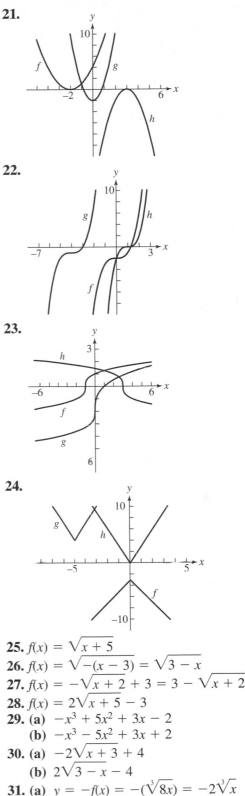

**21.**

**22.**

**23.**

**24.**

**25.** $f(x) = \sqrt{x + 5}$

**26.** $f(x) = \sqrt{-(x - 3)} = \sqrt{3 - x}$

**27.** $f(x) = -\sqrt{x + 2} + 3 = 3 - \sqrt{x + 2}$

**28.** $f(x) = 2\sqrt{x + 5} - 3$

**29. (a)** $-x^3 + 5x^2 + 3x - 2$

   **(b)** $-x^3 - 5x^2 + 3x + 2$

**30. (a)** $-2\sqrt{x + 3} + 4$

   **(b)** $2\sqrt{3 - x} - 4$

**31. (a)** $y = -f(x) = -(\sqrt[3]{8x}) = -2\sqrt[3]{x}$

   **(b)** $y = f(-x) = \sqrt[3]{8(-x)} = -2\sqrt[3]{x}$

**32. (a)** $-3|x + 5|$
  **(b)** $3|5 - x|$

**33.** vertical stretch by 9, or a horizontal shrink by $\frac{1}{3}$.

**34.** vertical shrink by $\frac{1}{4}$, or horizontal stretch by 2.

**35.** vertical stretch by 16, or a horizontal shrink by $\frac{1}{4}$.

**36.** vertical stretch by 5, or horizontal shrink by $\frac{1}{\sqrt{5}}$.

**37. (a)** $2x^3 - 8x$
  **(b)** $27x^3 - 12x$

**38. (a)** $2|x + 2|$
  **(b)** $|3x + 2|$

**39. (a)** $2x^2 + 2x - 4$

  **(b)** $9x^2 + 3x - 2$

**40. (a)** $\dfrac{2}{x + 2}$

  **(b)** $\dfrac{1}{(3x + 2)}$

**41.** Starting with $y = x^2$, translate right 3 units, vertically stretch by 2, and translate down 4 units.

**42.** Starting with $y = \sqrt{x}$, translate left 1 unit, vertically stretch by 3, and reflect across $x$-axis.

**43.** Starting with $y = x^2$, horizontally shrink by $\frac{1}{3}$ and translate down 4 units.

**44.** Starting with $y = |x|$, translate left 4 units, vertically stretch by 2, and reflect across $x$-axis, and translate up 1 unit.

**48.** $y = |2x + 2| - 4$

**49.**

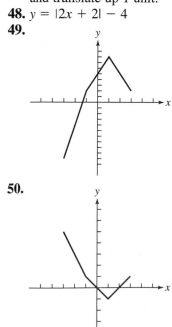

**50.**

**51.**

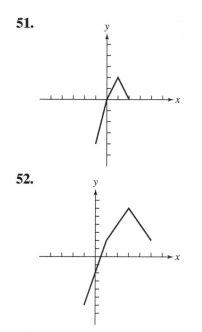

**52.**

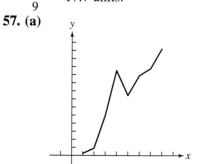

**53.** Reflections have more effect on points that are farther away from the line of reflection. Translations affect the distance of points from the axes, and hence change the effect of the reflections.

**54.** The $x$-intercepts are the values at which the funciton equals zero. The stretching (or shrinking) factors have no effect on the number zero, so those $y$-coordinates do not change.

**55.** First vertically stretch by $\frac{9}{5}$, then translate up 32 units.

**56.** First vertically shrink by $\frac{5}{9}$, then translate down $\frac{160}{9} = 17.\overline{7}$ units.

**57. (a)**

**(b)** Change the $y$-value according to the conversion rate from dollars to yen. This results in a vertical stretch by a number that changes according to international market conditions.

**58.** Apply the same transformation to the Ymin, Ymax, and Yscl as you apply to transform the function.

**59. (a)** The original graph is on the top; the graph of $y = |f(x)|$ is on the bottom.

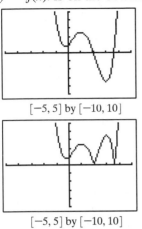

[−5, 5] by [−10, 10]

[−5, 5] by [−10, 10]

**(b)** The original graph is on the top; the graph of $y = f(|x|)$ is on the bottom.

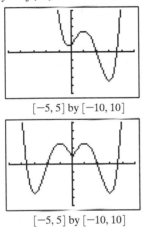

[−5, 5] by [−10, 10]

[−5, 5] by [−10, 10]

**(c)** The original graph is on the top; the graph of $y = |g(x)|$ is on the bottom.

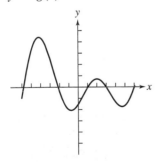

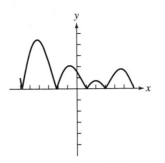

**(d)** The original graph is on the top; the graph of $y = g(|x|)$ is on the bottom.

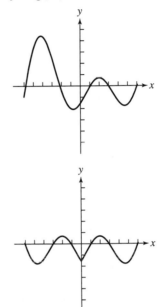

**60. (a)** You should get a graph that looks like this:

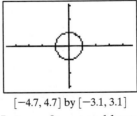

[−4.7, 4.7] by [−3.1, 3.1]

**(b)** Let $x = 2\cos t$ and leave $y$ unchanged.

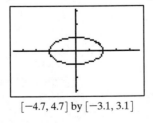

[−4.7, 4.7] by [−3.1, 3.1]

**(c)** Let $x = 3\cos t$ and $y = 3\sin t$.

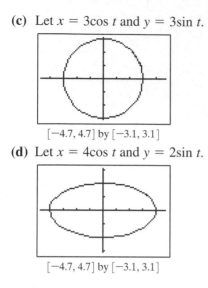

$[-4.7, 4.7]$ by $[-3.1, 3.1]$

**(d)** Let $x = 4\cos t$ and $y = 2\sin t$.

$[-4.7, 4.7]$ by $[-3.1, 3.1]$

## Section 1.6 (pp. 135–147)

### Exploration 1

**1.**

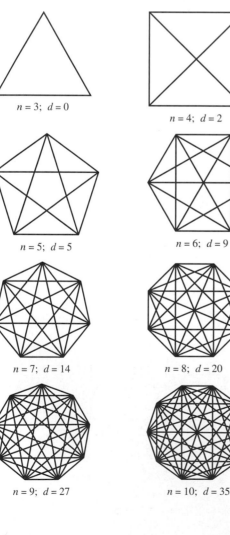

$n = 3$; $d = 0$

$n = 4$; $d = 2$

$n = 5$; $d = 5$

$n = 6$; $d = 9$

$n = 7$; $d = 14$

$n = 8$; $d = 20$

$n = 9$; $d = 27$

$n = 10$; $d = 35$

**2.**

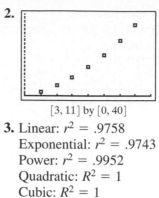

$[3, 11]$ by $[0, 40]$

**3.** Linear: $r^2 = .9758$
Exponential: $r^2 = .9743$
Power: $r^2 = .9952$
Quadratic: $R^2 = 1$
Cubic: $R^2 = 1$
Quartic $R^2 = 1$

**5.** Since the quadratic curve fits the points perfectly, there is nothing to be gained by adding a cubic term or a quartic term. The coefficients of these terms in the regressions are zero.

### Quick Review 1.6

**2.** $h = \dfrac{2A}{(b_1 + b_2)}$

**5.** $r = \sqrt[3]{\dfrac{3V}{4\pi}}$

**6.** $r = \sqrt{\dfrac{A}{4\pi}}$

**7.** $h = \dfrac{A - 2\pi r^2}{2\pi r} = \dfrac{A}{2\pi r} - r$

**9.** $P = \dfrac{A}{(1 + r/n)^{nt}} = A(1 + r/n)^{-nt}$

### Exercises 1.6

**11.** Let $C$ be the total cost and $n$ be the number of items produced; $C = 34{,}500 + 5.75n$.

**12.** Let $C$ be the total cost and $n$ be the number of items produced; $C = (1.09)28{,}000 + 19.85n$.

**13.** Let $R$ be the revenue and $n$ be the number of items sold: $R = 3.75n$.

**14.** Let $P$ be the profit, and $s$ be the amount of sales; then $P = 200{,}000 + 0.12s$.

**15.** $2\pi r^3$

**21.** $x + 4x = 620$; $x = 124$; $4x = 496$

**22.** $x + 2x + 3x = 714$, so $x = 119$; the second and third numbers are 238 and 357.

**24.** $1.017x = 161.3$

**25.** $182 = 52yt$, so $t = 3.5$ hr

**26.** $560 = 45t + 55(t + 2)$, so $t = 4.5$ hours on local highways.

**27.** The \$33 shirt is a better bargain, because the sale price is cheaper.

**31. (a)** $0.10x + 0.45(100 - x) = 0.25(100)$
**(b)** Use $x \approx 57.14$ gallons of the 10% solution and about 42.86 gal of the 45% solution.

**32.** $0.20x + 0.35(25 - x) = 0.26(25)$. Use $x = 15$ liters of the 20% solution and 10 liters of the 35% solution.

**33.** $2x + 2(x + 16) = 136$. Two pieces that are $x = 26$ ft long are needed, along with two 42 ft pieces.

**35.** $900 = 0.07x + 0.085(12,000 - x)$; 8,000 dollars was invested at 7%; the other $4,000 was invested at 8.5%.

**36.** $1,571 = 0.055x + 0.083(25,000 - x)$; $x = 18,000$ dollars was invested at 7%; the other $7,000 was invested at 8.3%.

**37. (d)** The graphs cross when $x = 5000$ pairs of shoes

**39. (e)**

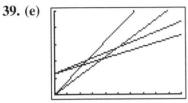

[0, 10000] by [0, 500000]

**(f)** You should recommend stringing the rackets; fewer strund rackets need to be sold to begin making a profit (since the intersection of $y_2$ and $y_4$ occurs for smaller $x$ than the intersection of $y_1$ and $y_3$).

**40. (a)**

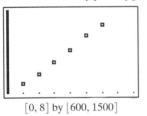

[0, 8] by [600, 1500]

**(d)** $y = 4.75x^2 + 88.31x + 503.74$ (same units as b)

**(f)** The linear function predicts 2,387,310, while the quadratic function predicts 2,897,140. These differ by 509,830 students!

**(g)** Both the regression functions model the data well enough to give us great confidence in their accuracy; nonetheless, the estimations that they give 8 years beyond the last data point are quite different. Both estimates cannot be right, so at least one of the excellent-fitting curves will lead to a bad estimate.

**41. (a)** $y = 118.07(0.951)^x$

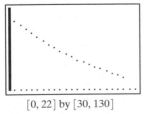

[0, 22] by [30, 130]

**(b)** List L3 = {112.3, 106.5, 101.5, 96.6, 92.0, 87.2, 83.1, 79.8, 75.0, 71.7, 68, 64.1, 61.5, 58.5, 55.9, 53.0, 50.8, 47.9, 45.2, 43.2}

**(c)** $y = 118.07*0.951^x$. It fits the data extremely well.

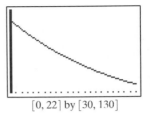

[0, 22] by [30, 130]

**42.** Answers will vary in (a)-(e)

**(f)** Some possible answers: the thickness of the liquid, the darkness of the liquid, the type of cup it is in, the amount of surface exposed to the air, the specific heat of the substance (a technical term that may have been learned in physics), etc.

## Chapter 1 Review (pp. 148–152)

| | |
|---|---|
| **1.** d | **2.** f |
| **3.** i | **4.** h |
| **5.** b | **6.** j |
| **7.** g | **8.** c |
| **9.** a | **10.** e |
| **11. (a)** all reals | **(b)** all reals |
| **12. (a)** all reals | **(b)** all reals |
| **13. (a)** all reals | **(b)** $[0, \infty)$ |
| **14. (a)** all reals | **(b)** $[5, \infty)$ |
| **15. (a)** all reals | **(b)** $[8, \infty)$ |
| **16. (a)** $[-2, 2]$ | **(b)** $[-2, \infty)$ |
| **17. (a)** all reals except 2 | **(b)** all reals except 0 |
| **18. (a)** $(-3, 3)$ | **(b)** $\left[\frac{1}{3}, \infty\right)$ |

**19.** continuous    **20.** continuous

**21. (a)** vertical asymptotes at $x = 0$ and $x = 5$
   **(b)** none

| | |
|---|---|
| **23. (a)** none | **(b)** $y = 7$ and $y = -7$ |
| **24. (a)** $x = -1$ | **(b)** $y = 1$ and $y = -1$ |

**27.** $(-\infty, -1), (-1, 1), (1, \infty)$

**28.** $(-\infty, -2), (-2, 0]$

**29.** not bounded    **31.** bounded above

**32.** not bounded above or below

| | |
|---|---|
| **33. (a)** none | **(b)** $-7$, at $x = -1$ |
| **34. (a)** 2, at $x = -1$ | **(b)** $-2$, at $x = 1$ |
| **35. (a)** $-1$, at $x = 0$ | **(b)** none |
| **36. (a)** 1, at $x = 2$ | **(b)** $-1$, at $x = -2$ |

**45.**

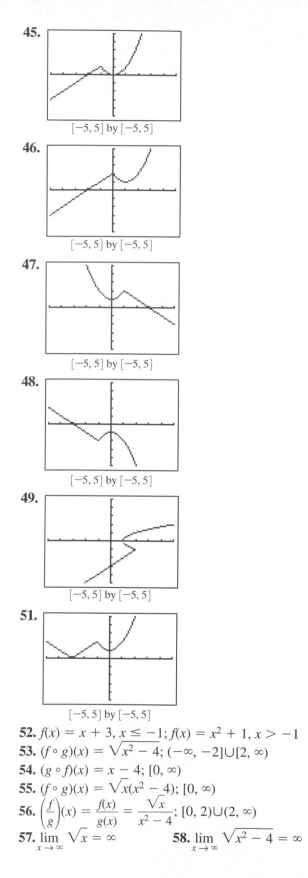

$[-5, 5]$ by $[-5, 5]$

**46.**

$[-5, 5]$ by $[-5, 5]$

**47.**

$[-5, 5]$ by $[-5, 5]$

**48.**

$[-5, 5]$ by $[-5, 5]$

**49.**

$[-5, 5]$ by $[-5, 5]$

**51.**

$[-5, 5]$ by $[-5, 5]$

**52.** $f(x) = x + 3, x \le -1; f(x) = x^2 + 1, x > -1$

**53.** $(f \circ g)(x) = \sqrt{x^2 - 4}; (-\infty, -2] \cup [2, \infty)$

**54.** $(g \circ f)(x) = x - 4; [0, \infty)$

**55.** $(f \circ g)(x) = \sqrt{x}(x^2 - 4); [0, \infty)$

**56.** $\left(\dfrac{f}{g}\right)(x) = \dfrac{f(x)}{g(x)} = \dfrac{\sqrt{x}}{x^2 - 4}; [0, 2) \cup (2, \infty)$

**57.** $\lim\limits_{x \to \infty} \sqrt{x} = \infty$      **58.** $\lim\limits_{x \to \infty} \sqrt{x^2 - 4} = \infty$

**61.** $100\pi h$

**65. (a)**

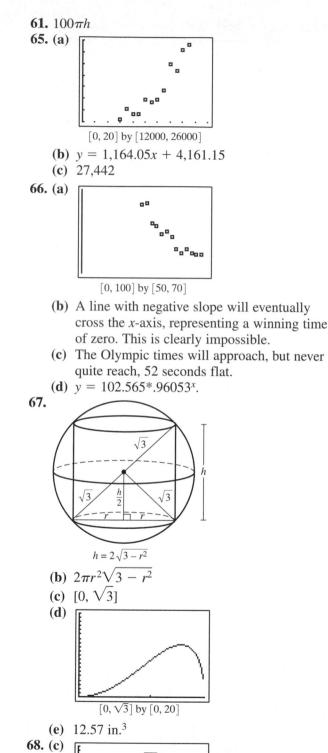

$[0, 20]$ by $[12000, 26000]$

  **(b)** $y = 1,164.05x + 4,161.15$

  **(c)** 27,442

**66. (a)**

$[0, 100]$ by $[50, 70]$

  **(b)** A line with negative slope will eventually cross the *x*-axis, representing a winning time of zero. This is clearly impossible.

  **(c)** The Olympic times will approach, but never quite reach, 52 seconds flat.

  **(d)** $y = 102.565 * .96053^x$.

**67.**

$h = 2\sqrt{3 - r^2}$

  **(b)** $2\pi r^2 \sqrt{3 - r^2}$

  **(c)** $[0, \sqrt{3}]$

  **(d)**

$[0, \sqrt{3}]$ by $[0, 20]$

  **(e)** 12.57 in.$^3$

**68. (c)**

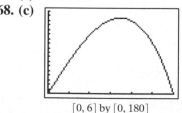

$[0, 6]$ by $[0, 180]$

## Chapter 1 Project

**1.**
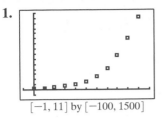
$[-1, 11]$ by $[-100, 1500]$

**2.** $y \approx 20.575 \cdot 1.539^x$

**3.** predicted: 2360, 3632; actual: 1886, 2200

**4.** $y \approx \dfrac{4009.711}{1 + 264.197e^{-0.497x}}$

**5.** predicted: 1899, 2392, 4010

# Chapter 2

## Section 2.1 (pp. 155–174)

### Exploration 1

**1.** $-\$2000$ per year

**2.** $v(t) = -2000t + 50,000$

### Quick Review 2.1

**1.** $y = 8x + 3.6$    **2.** $y = -1.8x - 2$

**3.** $y = -0.6x + 2.8$

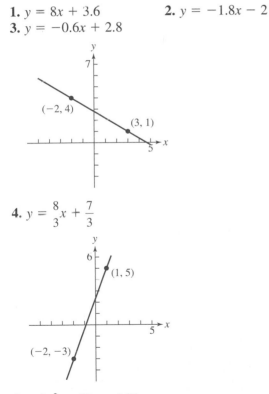

**4.** $y = \dfrac{8}{3}x + \dfrac{7}{3}$

**8.** $-3x^2 - 42x - 147$

**3.** Polynomial of degree 5 with leading coefficient 2.

**4.** Polynomial of degree 0 with leading coefficient 13.

**5.** Not a polynomial function because of the radical.

**6.** Polynomial of degree 2 with leading coefficient $-5$.

**7.** $f(x) = \dfrac{5}{7}x + \dfrac{18}{7}$

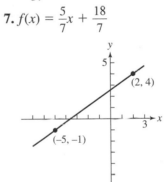

**8.** $f(x) = -\dfrac{7}{9}x + \dfrac{8}{3}$

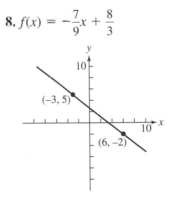

### Section 2.1 Exercises

**1.** Not a polynomial function because of the exponent $-5$.

**2.** Polynomial of degree 1 with leading coefficient 2.

**9.** $f(x) = -\dfrac{4}{3}x + \dfrac{2}{3}$

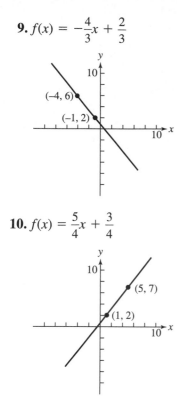

**10.** $f(x) = \dfrac{5}{4}x + \dfrac{3}{4}$

**11.** $f(x) = -x + 3$

**12.** $f(x) = \dfrac{1}{2}x + 2$

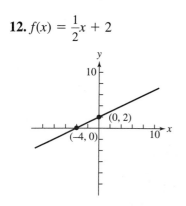

**19.** Vertex: $(1, 5)$; axis: $x = 1$
**20.** Vertex: $(-2, -1)$; axis: $x = -2$
**21.** Vertex: $(1, -7)$; axis: $x = 1$

**22.** Vertex: $(\sqrt{3}, 4)$; axis: $x = \sqrt{3}$
**23.** Vertex: $\left(-\dfrac{5}{6}, -\dfrac{73}{12}\right)$; axis: $x = -\dfrac{5}{6}$
**24.** Vertex: $(1.75, 3.125)$; axis: $x = 1.75$
**25.** Vertex: $(4, 19)$; axis: $x = 4$
**26.** Vertex: $(0.25, 5.75)$; axis: $x = 0.25$
**27.** Vertex: $(0.6, 2.2)$; axis: $x = 0.6$
**28.** Vertex: $(-1.75, 2.125)$; axis: $x = 1.75$
**29.** $f(x) = (x - 2)^2 + 2$.
Vertex: $(2, 2)$; axis: $x = 2$; opens upward; does not intersect $x$-axis.

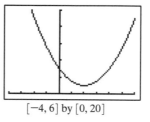

$[-4, 6]$ by $[0, 20]$

**30.** Vertex: $(3, 3)$; axis: $x = 3$; opens upward; does not intersect $x$-axis.

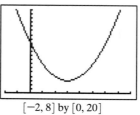

$[-2, 8]$ by $[0, 20]$

**31.** Vertex: $(-8, 74)$; axis: $x = -8$; opens downward; intersects $x$-axis at about $-16.602$ and $0.602$, or $(-8 \pm \sqrt{74})$.

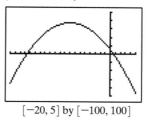

$[-20, 5]$ by $[-100, 100]$

**32.** Vertex: $(1, 9)$; axis: $x = 1$; opens downward; intersects $x$-axis at $-2$ and $4$.

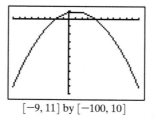

$[-9, 11]$ by $[-100, 10]$

**33.** Vertex: $\left(-\dfrac{3}{2}, \dfrac{5}{2}\right)$; axis of: $x = -\dfrac{3}{2}$; opens upward; does not intersect $x$-axis; vertically stretched by 2.

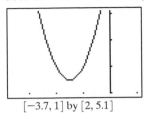

$[-3.7, 1]$ by $[2, 5.1]$

**34.** Vertex: $(2.5, -19.25)$; axis: $x = 2.5$; opens upward; intersects $x$–axis at about 0.538 and 4.462, or $\left(2.5 \pm \dfrac{1}{10}\sqrt{385}\right)$; vertically stretched by 5.

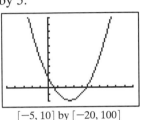

$[-5, 10]$ by $[-20, 100]$

**35.** $y = 2(x + 1)^2 - 3$  **36.** $y = 3(x - 2)^2 - 7$

**37.** $y = -2(x - 1)^2 + 11$  **38.** $y = -2(x + 1)^2 + 5$

**39.** $y = 2(x - 1)^2 + 3$  **40.** $y = -\dfrac{11}{2}(x + 2)^2 - 5$

**41.** Strong positive  **42.** Strong negative

**43.** Weak positive  **44.** No correlation

**45. (a)**

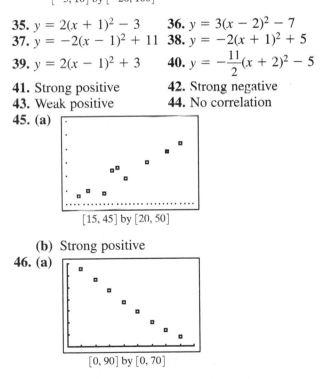

$[15, 45]$ by $[20, 50]$

**(b)** Strong positive

**46. (a)**

$[0, 90]$ by $[0, 70]$

**(b)** Strong negative

**49. (a)** $y \approx 782.5x + 33{,}778$; $m \approx 782.5$. This represents the increase in $y$ for an increase by 1 in the $x$ value. That is, it is the typical annual increase in salary.

**50. (a)** $[0, 100]$ by $[0, 1000]$ is one possibility.

**(b)** either 107,335 units or 372,665 units.

**52.** 11 ft by 14 ft

**54. (a)** $R(x) = (800 + 20x)(300 - 5x)$.

**(b)** $[0, 25]$ by $[200000, 260000]$ is one possibility (shown).

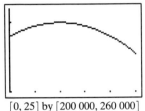

$[0, 25]$ by $[200\,000, 260\,000]$

**55. (a)** $R(x) = (26000 - 1000x)(0.50 + 0.05x)$.

**(b)** Many choices of Xmax and Ymin are reasonable. Shown is $[0, 15]$ by $[10000, 17000]$.

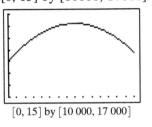

$[0, 15]$ by $[10\,000, 17\,000]$

**(c)** 90 cents per can; $16,200

**57. (b)** 39.5 ft, 1.5 sec

**58. (a)** $h = -16t^2 + 80t - 10$

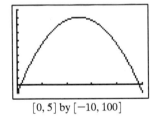

$[0, 5]$ by $[-10, 100]$

**59.** $32\sqrt{3}$, or about 55.426 ft/sec

**61. (c)** 2217.6 ft

**62. (a)**

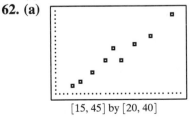

$[15, 45]$ by $[20, 40]$

**(c)** On average, female children gain 0.68 pounds per month.

**(d)**

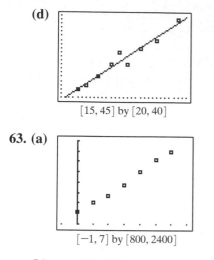

$[15, 45]$ by $[20, 40]$

**63. (a)**

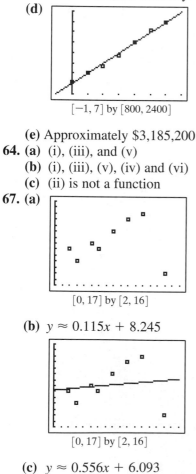

$[-1, 7]$ by $[800, 2400]$

**(b)** $y \approx 198.887x + 997.438$, where $x =$ number of years after 1990 in which the season started.

**(c)** Players' average salaries are increasing by about \$198,887 annually.

**(d)**

$[-1, 7]$ by $[800, 2400]$

**(e)** Approximately \$3,185,200

**64. (a)** (i), (iii), and (v)

**(b)** (i), (iii), (v), (iv) and (vi)

**(c)** (ii) is not a function

**67. (a)**

$[0, 17]$ by $[2, 16]$

**(b)** $y \approx 0.115x + 8.245$

$[0, 17]$ by $[2, 16]$

**(c)** $y \approx 0.556x + 6.093$

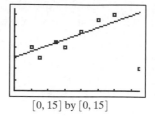

$[0, 15]$ by $[0, 15]$

**(d)** The median–median line appears to be the better fit, because it approximates more of the data values more closely.

**70.** $\left( \dfrac{a + b}{2}, \ -\dfrac{(a - b)^2}{4} \right)$

**71.** $x_1$ and $x_2$ are given by the quadratic formula $\dfrac{-b \pm \sqrt{b^2 - 4ac}}{2a}$; then $x_1 + x_2 = -\dfrac{b}{a}$, and the line of symmetry is $x = -\dfrac{b}{2a}$, which is exactly equal to $\dfrac{x_1 + x_2}{2}$.

## Section 2.2 (pp. 174–186)

### Exploration 1

**1.**

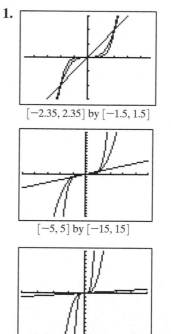

$[-2.35, 2.35]$ by $[-1.5, 1.5]$

$[-5, 5]$ by $[-15, 15]$

$[-20, 20]$ by $[-200, 200]$

The pairs $(0, 0)$, $(1, 1)$ and $(-1, -1)$ are common to all three graphs.

**2.**

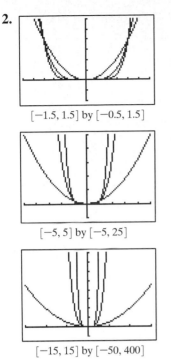

$[-1.5, 1.5]$ by $[-0.5, 1.5]$

$[-5, 5]$ by $[-5, 25]$

$[-15, 15]$ by $[-50, 400]$

The pairs $(0, 0)$, $(1, 1)$ and $(-1, 1)$ are common to all three graphs.

**Exploration 2**

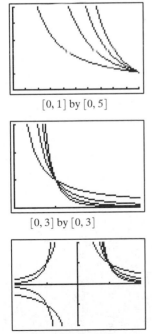

$[0, 1]$ by $[0, 5]$

$[0, 3]$ by $[0, 3]$

$[-2, 2]$ by $[-2, 2]$

The pair $(1, 1)$ is common to all four functions.

**Exploration 3**

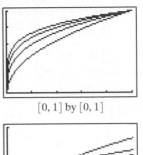

$[0, 1]$ by $[0, 1]$

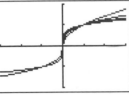

$[0, 3]$ by $[0, 2]$

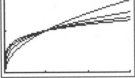

$[-3, 3]$ by $[-2, 2]$

The pairs $(0, 0)$, $(1, 1)$ are common to all 4 functions.

**Exercises 2.2**

1. power $= 5$, constant $= -\dfrac{1}{2}$

2. power $= \dfrac{5}{3}$, constant $= 9$

3. not a power function

4. power $= 0$, constant $= 13$

5. power $= 1$, constant $= c^2$

6. power $= 5$, constant $= \dfrac{k}{2}$

7. power $= 2$, constant $= \dfrac{g}{2}$, indep. variable $= t$

8. power $= 3$, constant $= \dfrac{4\pi}{3}$, indep. variable $= r$

9. power $= -2$, constant $= k$, indep. variable $= d$

10. power $= 1$, constant $= m$, indep. variable $= a$

11. degree $= 0$, coefficient $= -4$

12. not a monomial function (degree $= -5$)

13. degree $= 7$, coefficient $= 6$

14. not a monomial function (it is exponential!)

15. degree $= 2$, coefficient $= 4\pi$, variable $= r$

16. degree $= 1$, coefficient $= l$, variable $= w$

23. The weight $w$ of an object varies directly with its mass $m$, with the constant of variation $g$.

24. The circumference $C$ of a circle is proportional to its diameter $D$, with the constant of variation $\pi$.

**25.** The refractive index $n$ of a medium is inversely proportional to $v$, the velocity of light in the medium, with constant of variation $c$, the constant velocity of light in free space.

**26.** The distance $d$ traveled by a free-falling object dropped from rest varies directly with the square of its speed $p$,

with the constant of variation $\dfrac{1}{2g}$.

**27.** $y = \dfrac{8}{x^2}$, power $= -2$, constant $= 8$

**28.** $y = -2\sqrt{x}$, power $= \dfrac{1}{2}$, constant $= -2$

**35.** shrink vertically by $\dfrac{2}{3}$; $f$ is even.

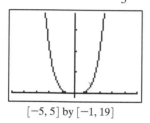

$[-5, 5]$ by $[-1, 19]$

**36.** stretch vertically by 5; $f$ is odd.

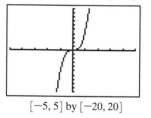

$[-5, 5]$ by $[-20, 20]$

**37.** stretch vertically by 1.5 and rotate across the $x$-axis; $f$ is odd.

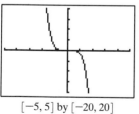

$[-5, 5]$ by $[-20, 20]$

**38.** stretch vertically by 2 and rotate over the $x$-axis; $f$ is even.

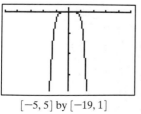

$[-5, 5]$ by $[-19, 1]$

**39.** shrink vertically by $\dfrac{1}{4}$; $f$ is even.

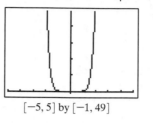

$[-5, 5]$ by $[-1, 49]$

**40.** shrink vertically by $\dfrac{1}{8}$; $f$ is odd.

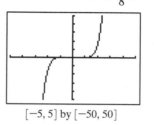

$[-5, 5]$ by $[-50, 50]$

**41.** power $= 4$, constant $= 2$
Domain: $(-\infty, \infty)$
Range: $(0, \infty)$
Continuous
Decreasing on $(-\infty, 0)$. Increasing on $(0, \infty)$.
Even. Symmetric with respect to $y$-axis.
Bounded below, but not above
Local minimum at $x = 0$.
End Behavior: $\displaystyle\lim_{x \to -\infty} 2x^4 = \infty$, $\displaystyle\lim_{x \to \infty} 2x^4 = \infty$

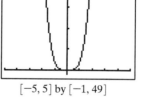

$[-5, 5]$ by $[-1, 49]$

**42.** power $= 3$, constant $= -3$
Domain: $(-\infty, \infty)$
Range: $(-\infty, \infty)$
Continuous
Decreasing for all $x$
Odd. Symmetric with respect to origin
Not bounded above or below
No local extrema
End Behavior: $\lim\limits_{x \to -\infty} -3x^3 = \infty$,
$\lim\limits_{x \to \infty} -3x^3 = -\infty$

[−5, 5] by [−20, 20]

**43.** power $= \dfrac{1}{4}$, constant $= \dfrac{1}{2}$

Domain: $[0, \infty)$
Range: $[0, \infty)$
Continuous
Increasing for all $x$
Bounded below
Neither even nor odd
Local minimum at $(0, 0)$

End Behavior: $\lim\limits_{x \to \infty} \dfrac{1}{2}\sqrt[4]{x} = \infty$

[−1, 99] by [−1, 4]

**44.** power $= -3$, constant $= -2$
Domain: $(-\infty, 0) \cup (0, \infty)$
Range: $(-\infty, 0) \cup (0, \infty)$
Discontinuous at $x = 0$
Increasing on $(-\infty, 0)$. Increasing on $(0, \infty)$.
Odd. Symmetric with respect to origin
Not bounded above or below
No local extrema
Asymptote at $x = 0$.
End Behavior: $\lim\limits_{x \to -\infty} -2x^{-3} = 0$,
$\lim\limits_{x \to \infty} -2x^{-3} \to 0$.

[−5, 5] by [−5, 5]

**45.** $k = 3$, $a = \dfrac{1}{4}$. $f$ is undefined for $x < 0$.

**46.** $k = -4$, $a = \dfrac{2}{3}$. $f$ is even.

**47.** $k = -2$, $a = \dfrac{4}{3}$. $f$ is even.

**48.** $k = \dfrac{2}{5}$, $a = \dfrac{5}{2}$. $f$ is undefined for $x < 0$.

**49.** $k = \dfrac{1}{2}$, $a = -3$. $f$ is odd.

**50.** $k = -1$, $a = -4$. $f$ is even.

**52.** 3.87 L

**54.**

| Wind Speed (mph) | Power (W) |
|---|---|
| 10 | 15 |
| 20 | 120 |
| 40 | 960 |
| 80 | 7680 |

**55. (a)**

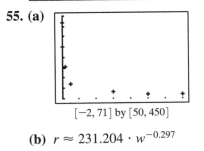

[−2, 71] by [50, 450]

**(b)** $r \approx 231.204 \cdot w^{-0.297}$

**(c)**

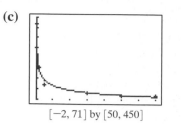

[−2, 71] by [50, 450]

**(d)** Approximately 37.67 beats/min, which is very close to Clark's observed value.

**56. (a)**

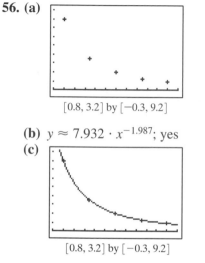

[0.8, 3.2] by [−0.3, 9.2]

**(b)** $y \approx 7.932 \cdot x^{-1.987}$; yes

**(c)**

[0.8, 3.2] by [−0.3, 9.2]

**(d)** Approximately $2.76 \dfrac{W}{m^2}$ and $0.697 \dfrac{W}{m^2}$, respectively.

**57.** *n* even: *f* is undefined for $x < 0$.
*m* even, *n* odd: *f* is even.
*m* odd, *n* odd: *f* is odd.

**63. (a)** The force *F* acting on an object varies jointly as the mass *m* of the object and the acceleration *a* of the object.

**(b)** The kinetic energy *KE* of an object varies jointly as the mass *m* of the object and square of the velocity *v* of the object.

**(c)** The force of gravity *F* acting on two objects varies jointly as the product $m_1m_2$ of the objects' masses and the inverse of the distance *r* between their centers, with the constant of variation *G*, the universal gravitational constant.

# Section 2.3 (pp. 186–199)

## Exploration 1

**1. (a)** $\infty$; $-\infty$
**(b)** $-\infty$; $\infty$
**(c)** $\infty$; $-\infty$

**(d)** $-\infty$; $\infty$
**2. (a)** $-\infty$; $-\infty$
**(b)** $\infty$; $\infty$
**(c)** $\infty$; $\infty$
**(d)** $-\infty$; $-\infty$
**3. (a)** $-\infty$; $\infty$
**(b)** $-\infty$; $-\infty$
**(c)** $\infty$; $\infty$
**(d)** $\infty$; $-\infty$

## Exploration 2

**1.** $y \approx 0.0061x^3 + 0.0177x^2 - 0.5007x + 0.9769$
**2.** $y \approx -0.375x^4 + 6.917x^3 - 44.125x^2 + 116.583x - 111$

## Quick Review 2.3

**2.** $(x - 7)(x - 4)$
**3.** $(3x - 2)(x - 3)$
**4.** $(2x - 1)(3x - 1)$
**5.** $x(3x - 2)(x - 1)$
**6.** $2x(3x - 2)(x - 3)$
**8.** $x = 0, x = -2, x = 5$

## Section 2.3 Exercises

**1.** Shift to the right by 3 units, stretch vertically by
**2.** *y*-intercept: $(0, -54)$

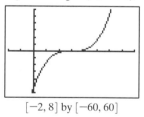

[−2, 8] by [−60, 60]

**2.** Shift to the left by 5 units and then rotate over the *x*-axis. *y*-intercept: (0, −125)

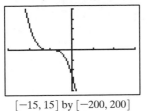

$[-15, 15]$ by $[-200, 200]$

**3.** Shift to the left by 1 unit, vertically shrink by $\frac{1}{2}$, rotate over the *x*-axis and then vertically shift up 2 units. *y*-intercept: $\left(0, \frac{3}{2}\right)$

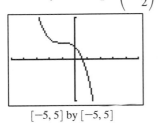

$[-5, 5]$ by $[-5, 5]$

**4.** shift to the right by 3 units, vertically shrink by $\frac{2}{3}$, and vertically shift up 1 unit. *y*-intercept: (0, −17)

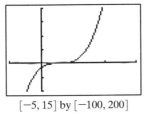

$[-5, 15]$ by $[-100, 200]$

**5.** Shift to the left 2 units, vertically stretch by 2, rotate over the *x*-axis, and vertically shift down 3 units. *y*-intercept: (0, −35)

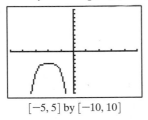

$[-5, 5]$ by $[-10, 10]$

**6.** Shift to the right 1 unit, vertically stretch by 3, and vertically shift down 2 units. *y*-intercept: (0, 1)

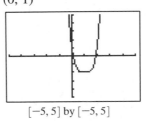

$[-5, 5]$ by $[-5, 5]$

**7.** degree: 3, zeros: *x* = 0 (one, crosses *x*-axis), *x* = 3 (two, does not cross *x*-axis)

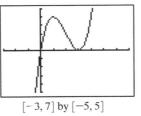

$[-3, 7]$ by $[-5, 5]$

**8.** degree: 4, zeros: *x* = 0 (3, crosses *x*-axis), *x* = 2 (one, crosses *x*-axis)

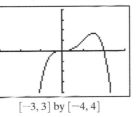

$[-3, 3]$ by $[-4, 4]$

**9.** degree: 5, zeros: *x* = 2 (3, crosses *x*-axis), *x* = −1 (two, docs not cross *x*-axis)

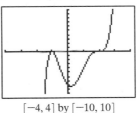

$[-4, 4]$ by $[-10, 10]$

**10.** degree: 5, zeros: *x* = 3 (3, crosses *x*-axis), *x* = −1 (two, does not cross *x*-axis)

**11.** local maximum: ≈ (0.79, 1.19), zeros: *x* = 0 and *x* ≈ 1.26.

**12.** local max at (0, 0) and local minima at (1.12, −3.13) and (−1.12, −3.13). zeros: *x* = 0, *x* ≈ 1.58, *x* ≈ −1.58.

**17.** One possibility:

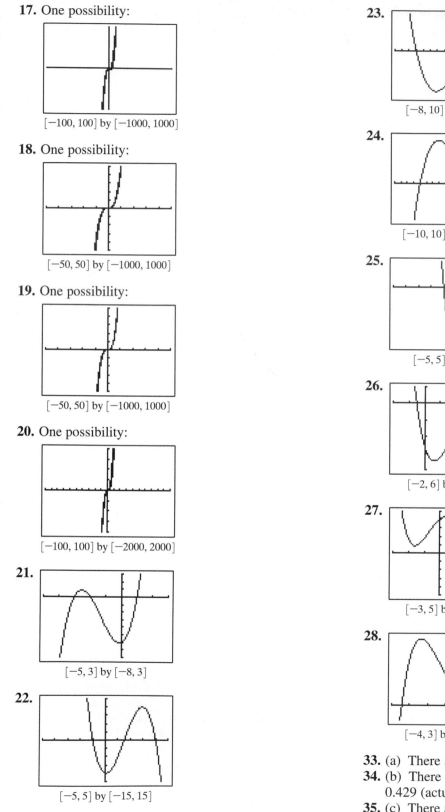

$[-100, 100]$ by $[-1000, 1000]$

**18.** One possibility:

$[-50, 50]$ by $[-1000, 1000]$

**19.** One possibility:

$[-50, 50]$ by $[-1000, 1000]$

**20.** One possibility:

$[-100, 100]$ by $[-2000, 2000]$

**21.**

$[-5, 3]$ by $[-8, 3]$

**22.**

$[-5, 5]$ by $[-15, 15]$

**23.**

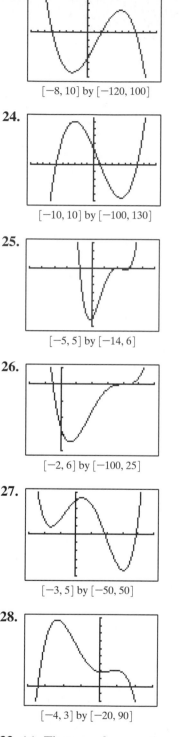

$[-8, 10]$ by $[-120, 100]$

**24.**

$[-10, 10]$ by $[-100, 130]$

**25.**

$[-5, 5]$ by $[-14, 6]$

**26.**

$[-2, 6]$ by $[-100, 25]$

**27.**

$[-3, 5]$ by $[-50, 50]$

**28.**

$[-4, 3]$ by $[-20, 90]$

**33.** (a)  There are 3 zeros: they are $-2.5$, 1, and 1.1.

**34.** (b)  There are 3 zeros: they are 0.4, approximately 0.429 (actually 3/7), and 3.

**35.** (c)  There are 3 zeros: approximately $-0.273$ (actually $-3/11$), $-0.25$, and 1.

**36.** (d)  There are 3 zeros: $-2$, 0.5, and 3.

**37.** −4 and 2      **38.** −2 and 2/3
**39.** 2/3 and −1/3      **40.** 0, −5, and 5
**41.** 0, −2/3, and 1      **42.** 0, −1, and 2
**43.** −2.43, −0.74, 1.67      **44.** −1.73, 0.26, 4.47
**45.** −2.47, −1.46, 1.94      **46.** −4.53, 2
**47.** −4.90, −0.45, 1, 1.35
**48.** −1.98, −0.16, 1.25, 2.77, 3.62
**53.** $x^3 - 5x^2 - 18x + 72$    **54.** $x^3 + 4x^2 - 11x - 30$
**55.** $x^3 - 4x^2 - 3x + 12$    **56.** $x^3 - 3x^2 + x + 1$
**57.** $y \approx 0.25x^3 - 1.25x^2 - 6.75x + 19.75$
**58.** $y \approx 0.074x^3 - 0.167x^2 + 0.611x + 4.48$
**59.** $y \approx -2.21x^4 + 45.75x^3 - 339.79x^2 + 1075.25x - 1231$
**60.** $y \approx -0.017x^4 + 0.226x^3 + 0.289x^2 - 3.202x - 21$
**61.** **(a)** between 30 and 541 customers
     **(b)** either 201 or 429 customers
**62.** **(a)**

[0, 60] by [−10, 210]

     **(b)** $y \approx 0.051x^2 + 0.97x + 0.26$
     **(c)**

[0, 60] by [−10, 210]

     **(d)** $y(25) \approx 56.39$ ft
     **(e)** $x = 67.74$ mph
**63.** **(a)**

[0, 0.8] by [0, 1.20]

     **(b)** 0.3391 cm
**64.** **(a)** The height of the box will be $x$, the width will be $15 - 2x$, and the length $60 - 2x$.
     **(b)** Any value of $x$ between approximately 0.550 and 6.786 inches.
**65.** $0 < x \le 0.929$ or $3.644 \le x < 5$
**66.** $0 < x < 21.5$

**69.** The exact behavior near $x = 1$ is hard to see. A zoomed–in view around the point $(1, 0)$ suggests that the graph just touches the $x$-axis at 0 without actually crossing it — that is, $(1, 0)$ is a local maximum. One possible window is $[0.9999, 1.0001]$ by $[-1 \times 10^{-7}, 1 \times 10^{-7}]$.
**70.** This also has a maximum near $x = 1$ — but this time a window such as $[0.6, 1.4]$ by $[-0.1, 0.1]$ reveals that the graph actually rises above the $x$-axis and has a maximum at $(0.999, 0.025)$.
**71.** A maximum and minimum are not visible in the standard window, but can be seen on the window $[0.2, 0.4]$ by $[5.29, 5.3]$.
**72.** A maximum and minimum are not visible in the standard window, but can be seen on the window $[0.95, 1.05]$ by $[-6.0005, -5.9995]$.
**73.** The graph of $y = 3(x^3 - x)$ increases, then decreases, then increases; the graph of $y = x^3$ only increases. Therefore, this graph can not be obtained from the graph of $y = x^3$ by the transformations studied in Chapter 2 (translations, reflections, and stretching/shrinking). Since the right side includes only these transformations, there can be no solution.
**74.** The graph of $y = x^4$ has a "flat bottom," while the graph of $y = x^4 + 3x^3 - 2x - 3$ is "bumpy." Therefore, this graph cannot be obtained from the graph of $y = x^4$ through the transformations of Chapter 2. Since the right side includes only these transformations, there can be no solution.
**75.** **(a)** Substituting $x = 2$, $y = 7$, we find that $7 = 5(2 - 2) + 7$, so $Q$ is on line $L$, and also $f(2) = -8 + 8 + 18 - 11 = 7$, so $Q$ is on the graph of $f(x)$.
     **(b)**

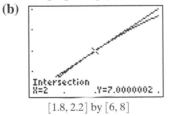

[1.8, 2.2] by [6, 8]

     **(c)** The line $L$ also crosses the graph of $f(x)$ at $(-2, -13)$.
**76.** **(a)** Note that $f(a) = a^n$ and $f(-a) = -a^n$.
$$m = \frac{y_2 - y_1}{x_2 - x_1} = \frac{-a^n - a^n}{-a - a} = \frac{-2a^n}{-2a} = a^{n-1}.$$
     **(b)** $y - a^{n/(n-1)} = a^{n-1}(x - a^{1/(n-1)})$.

**(c)** $y = 9x - 6\sqrt{3}$, $y = x^3$

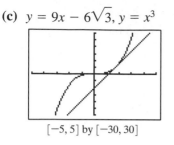

$[-5, 5]$ by $[-30, 30]$

## Section 2.4 (pp. 199–212)

### Quick Review 2.4

**2.** $x^2 - \dfrac{5}{2}x - 3$

**3.** $7x^3 + x^2 - 3$

**4.** $2x^2 - \dfrac{2}{3}x + \dfrac{7}{3}$

**7.** $4(x + 5)(x - 3)$

**8.** $x(3x - 2)(5x - 4)$

**9.** $(x + 2)(x + 1)(x - 1)$

**10.** $x(x + 1)(x + 3)(x - 3)$

### Section 2.4 Exercises

**2.** $f(x) = (x^2 - x + 1)(x + 1) - 2$

**3.** $f(x) = (x^2 + x + 4)(x + 3) - 21$

**4.** $f(x) = \left(2x^2 - 5x + \dfrac{7}{2}\right)(2x + 1) - \dfrac{9}{2}$

**5.** $f(x) = (x^2 - 4x + 12)(x^2 + 2x - 1) - 32x + 18$

**6.** $f(x) = (x^2 - 3x + 5)(x^2 + 1)$

**7.** $x^2 - 6x + 9 + \dfrac{-11}{x + 1}$

**8.** $2x^3 + x^2 + 10x + 27 + \dfrac{82}{x - 3}$

**9.** $9x^2 + 97x + 967 + \dfrac{9670}{x - 10}$

**10.** $3x^3 - 14x^2 + 66x - 321 + \dfrac{1602}{x + 5}$

**11.** $-5x^3 - 20x^2 - 80x - 317 + \dfrac{1269}{4 - x}$

**12.** $x^7 - 2x^6 + 4x^5 - 8x^4 + 16x^3 - 32x^2 + 64x - 128 + \dfrac{255}{x + 2}$

**19.** Yes

**20.** Yes

**21.** No

**22.** Yes

**23.** Yes

**24.** No

**25.** $f(x) = (x + 3)(x - 1)(5x - 17)$

**26.** $f(x) = (x + 2)(x - 3)(5x - 7)$

**27.** $2x^3 - 6x^2 - 12x + 16$

**28.** $2x^3 + 6x^2 - 26x - 30$

**29.** $2x^3 - 8x^2 + \dfrac{19}{2}x - 3$

**30.** $2x^4 + 3x^3 - 14x^2 - 15x$

**33.** $\dfrac{\pm 1}{\pm 1, \pm 2, \pm 3, \pm 6}; 1$

**34.** $\dfrac{\pm 1, \pm 2, \pm 7, \pm 14}{\pm 1, \pm 3}; \dfrac{7}{3}$

**35.** $\dfrac{\pm 1, \pm 3, \pm 9}{\pm 1, \pm 2}; \dfrac{3}{2}$

**36.** $\dfrac{\pm 1, \pm 2, \pm 3, \pm 4, \pm 6, \pm 12}{\pm 1, \pm 2, \pm 3, \pm 6}; -\dfrac{4}{3}$ and $\dfrac{3}{2}$

**45.** No zeros outside window.

**46.** No zeros outside window.

**47.** There *are* zeros not shown (approx. $-11.002$ and $12.003$)

**48.** There *are* zeros not shown (approx. $-8.036$ and $9.038$)

**49.** Rational zero: $\dfrac{3}{2}$; irrational zeros: $\pm\sqrt{2}$

**50.** Rational zero: $-3$; irrational zeros: $\pm\sqrt{3}$

**51.** Rational: $-3$; irrational: $1 \pm \sqrt{3}$

**52.** Rational: $4$; irrational: $1 \pm \sqrt{2}$

**53.** Rational: $-1$ and $4$; irrational: $\pm\sqrt{2}$

**54.** Rational: $-1$ and $2$; irrational: $\pm\sqrt{5}$

**55.** Rational: $-\dfrac{1}{2}$ and $4$; irrational: none

**56.** Rational: $\dfrac{2}{3}$; irrational: about $-0.6823$

**57.** \$36.27; 53.7

**58.** \$106.99; 1010.15

**61. (c)** $(x - 2)(x^3 + 4x^2 - 3x - 19)$

**(d)** an irrational zero of $x$ is $x \approx 2.04$

**(e)** $f(x) \approx (x - 2)(x - 2.04)(x^2 + 6.04x + 9.3116)$

**62. (a)** $y_1 \approx -0.0097x^4 + 0.4492x^3 - 7.3508x^2 + 49.6201x - 97.6144$

$[0, 20]$ by $[12, 22]$

**(b)** $y_2 \approx 0.0668x^2 - 1.3335x + 21.8505$

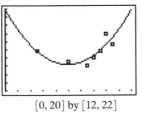

[0, 20] by [12, 22]

**(c)** 20.6 billion (quadratic); 7.8 billion (quartic)

**63. (d)** $x \approx 0.6527$ m

**65. (a)** Shown is one possible view, on the window [0, 600] by [0, 500].

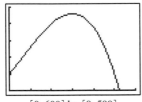

[0, 600] by [0, 500]

**(b)** The maximum population, after 300 days, is 460 turkeys.

**(c)** $P = 0$ when $t \approx 523.22$ — about 523 days after release.

**71. (b)** Zeros: $-\dfrac{7}{3}, \dfrac{1}{2},$ and 3.

**(c)** There are no rational zeros.

**73. (a)** Approximate zeros: $-3.126, -1.075, 0.910, 2.291$.

**(b)** $f(x) \approx g(x) = (x + 3.126)(x + 1.075)(x - 0.910)(x - 2.291)$.

**(c)** Graphically: graph the original function and the approximate factorization on a variety of windows and observe their similarity. Numerically: Compute $f(c)$ and $g(c)$ for several values of $c$.

**74. (a)** 0 or 2 positive zeros, 1 negative zero

**(b)** no positive zeros, 1 or 3 negative zeros

**(c)** 1 positive zero, no negative zeros

**(d)** 1 positive zero, 1 negative zero

# Section 2.5 (pp. 212–220)

## Exploration 1

**1.** $\left\{1, -\dfrac{1}{2} + \dfrac{i\sqrt{3}}{2}, -\dfrac{1}{2} - \dfrac{i\sqrt{3}}{2}\right\}$ exactly one unit

**2.** $\left\{-1, \dfrac{1}{2} + \dfrac{i\sqrt{3}}{2}, \dfrac{1}{2} - \dfrac{i\sqrt{3}}{2}\right\}$ exactly one unit

## Quick Review 2.5

**2.** $x + 2y$

**3.** $a + 2d$

**5.** $x^2 - x - 6$

**6.** $2x^2 + 5x - 3$

## Exercises 2.5

**2.** $5 - 7i$

**14.** $15 + 18i$

**16.** $-(10 + 5\sqrt{2}) + (12 + 6\sqrt{2})i$

**21.** $x = 2, y = 3$

**22.** $x = 3, y = -7$

**38.** $\dfrac{26}{29} + \dfrac{7}{29}i$

**40.** $\dfrac{2\sqrt{2} - 1}{3} - \dfrac{2\sqrt{2} + 1}{3}i$

**45.** $x = -1 \pm 2i$

**46.** $x = -\dfrac{1}{6} \pm \dfrac{\sqrt{23}}{6}i$

**47.** $x = \dfrac{7}{8} \pm \dfrac{\sqrt{15}}{8}i$

**48.** $x = 2 \pm \sqrt{15}i$

**49.** length = 8, midpoint: (2, 0)

**50.** length = 14, midpoint: (−3, 0)

**51.** length = $\sqrt{13}$, midpoint: $-\dfrac{1}{2} - 5i$

**52.** length = $\sqrt{73}$, midpoint: $-1 + \dfrac{5}{2}i$

**56. (a)**

Imaginary axis

• (2 + 3i)

Real axis

**57. (a)** real = 8, imaginary = −8; real = 16, imaginary = 0.

**(b)**

Imaginary axis

(0, 2)
(−2, 2)• •(1, 1)
(−4, 0)• •(1, 0)  Real axis
(−4, −4)•
(0, −8)•

**(c)** Answers will vary.

**(d)** $(1 + i)^7 = 8 - 8i$, $(1 + i)^8 = 16$

**61.** (d), (h), (e), (c), (f), (b), (g), (a)
**63.** **(a)** Not always true.
  **(b)** Always true.
  **(c)** Not always true.
**64.** **(a)** Always true.
  **(b)** Not true.
  **(c)** Not true.

## Section 2.6 (pp. 220–227)

### Quick Review 2.6

**2.** $(3x + 1)(2x - 5)$
**3.** $(5x - 4)(2x + 1)$
**4.** $(3x + 2)(3x - 4)$
**5.** $\dfrac{5}{2} \pm \dfrac{\sqrt{19}}{2}i$
**6.** $-\dfrac{3}{4} \pm \dfrac{\sqrt{47}}{4}i$
**8.** $\pm 1, \pm 3, \pm\dfrac{1}{2}, \pm\dfrac{3}{2}, \pm\dfrac{1}{4}, \pm\dfrac{3}{4}$
**9.** $\pm 1, \pm 2, \pm 4, \pm\dfrac{1}{5}, \pm\dfrac{2}{5}, \pm\dfrac{4}{5}$
**10.** $\pm 1, \pm\dfrac{1}{2}, \pm\dfrac{1}{3}, \pm\dfrac{1}{4}, \pm\dfrac{1}{6}, \pm\dfrac{1}{12}$

### Exercises 2.6

**2.** $x^2 - 2x + 5$
**3.** $x^3 - x^2 + 9x - 9$
**4.** $x^3 + 2x^2 - 6x + 8$
**5.** $x^4 - 5x^3 + 7x^2 - 5x + 6$
**6.** $x^4 - 3x^3 + 2x^2 + 2x - 4$
**7.** $x^3 - 11x^2 + 43x - 65$
**8.** $x^3 + x + 10$
**9.** $x^5 + 4x^4 + x^3 - 10x^2 - 4x + 8$
**11.** $x^4 - 10x^3 + 38x^2 - 64x + 40$
**12.** $x^4 + 6x^3 + 14x^2 + 14x + 5$
**23.** Zeros: $x = 1, x = -\dfrac{1}{2} \pm \dfrac{\sqrt{19}}{2}i$

$f(x) = \dfrac{1}{4}(x - 1)(2x + 1 + \sqrt{19}i)$

$(2x + 1 - \sqrt{19}i)$

**24.** Zeros: $x = 3, x = \dfrac{7}{2} \pm \dfrac{\sqrt{43}}{2}i$

$f(x) = \dfrac{1}{4}(x - 3)(2x - 7 + \sqrt{43}i)$

$(2x - 7 - \sqrt{43}i)$

**25.** Zeros: $x = \pm 1, x = -\dfrac{1}{2} \pm \dfrac{\sqrt{23}}{2}i$

$f(x) = \dfrac{1}{4}(x - 1)(x + 1)(2x + 1 + \sqrt{23}i)$

$(2x + 1 - \sqrt{23}i)$

**26.** Zeros: $x = -2, x = \dfrac{1}{3}, x = -\dfrac{1}{2} \pm \dfrac{\sqrt{3}}{2}i$

$f(x) = \dfrac{1}{4}(x + 2)(3x - 1)(2x + 1 + \sqrt{3}i)$

$(2x + 1 - \sqrt{3}i)$

**27.** Zeros: $x = -\dfrac{7}{3}, x = \dfrac{3}{2}, x = 1 \pm 2i$

$f(x) = (3x + 7)(2x - 3)(x - 1 + 2i)(x - 1 - 2i)$

**28.** Zeros: $x = -\dfrac{3}{5}, x = 5, x = 1.5 \pm i$

$f(x) = (5x + 3)(x - 5)(2x - 3 + 2i)$
$(2x - 3 - 2i)$

**29.** Zeros: $x = \pm\sqrt{3}, x = 1 \pm i$
$f(x) = (x - \sqrt{3})(x + \sqrt{3})(x - 1 + i)$
$(x - 1 - i)$

**30.** Zeros: $x = \pm\sqrt{3}, x = \pm 4i$
$f(x) = (x - \sqrt{3})(x + \sqrt{3})(x - 4i)(x + 4i)$

**31.** Zeros: $x = \pm 2, x = 3 \pm 2i$
$f(x) = (x - \sqrt{2})(x + \sqrt{2})(x - 3 + 2i)$
$(x - 3 - 2i)$

**32.** Zeros: $x = \pm\sqrt{5}, x = 1 \pm 3i$
$f(x) = (x - \sqrt{5})(x + \sqrt{5})(x - 1 + 3i)$
$(x - 1 - 3i)$

**45.** $f(x) = -2x^4 + 12x^3 - 20x^2 - 4x + 30$
**46.** $f(x) = 2x^4 - 8x^3 + 22x^2 - 28x + 20$
**47.** **(a)** $d \approx 0.0669t^3 - 0.7420t^2 + 2.1759t + 0.8250$

$[-1, 8.25]$ by $[0, 5]$

  **(b)** 0.825 m
  **(c)** $t \approx 2.02$ sec ($d \approx 2.74$ m) and
    $t \approx 5.38$ sec ($d \approx 1.47$ m)
**48.** **(a)** $d \approx -0.0820t^3 + 0.9162t^2 - 2.5126t$
    $+ 3.3779$

$[-1, 9]$ by $[0, 5]$

  **(b)** Sally walks toward the detector, turns and
    walks away (or walks backwards), then walks
    toward the detector again.
  **(c)** $t \approx 1.81$ sec ($d \approx 1.35$ m) and
    $t \approx 5.64$ sec ($d \approx 3.65$ m).

**49. (a)** $d \approx 0.2434t^2 - 1.7159t + 4.4241$

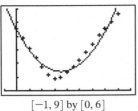

$[-1, 9]$ by $[0, 6]$

**(b)** Jacob walks toward the detector, then turns and walks away (or walks backwards).

**(c)** The model "changes direction" at $t \approx 3.52$ (when $d \approx 1.40$m).

**50. (a)** $a^2 + 2abi - b^2$

**(c)** $\left\{ \left( \sqrt{\dfrac{1}{2}}, \sqrt{\dfrac{1}{2}} \right), \left( -\sqrt{\dfrac{1}{2}}, -\sqrt{\dfrac{1}{2}} \right) \right\}$

**(e)** $\left( \sqrt{\dfrac{1}{2}} + \sqrt{\dfrac{1}{2}}i \right)$ and $\left( -\sqrt{\dfrac{1}{2}} - \sqrt{\dfrac{1}{2}}i \right)$

**51.** $f(i) = i^3 - i(i)^2 + 2i(i) + 2 = -i + i - 2 + 2$
$= 0$

**52.** $f(-2i)$
$= (-2i)^3 - (2 - i)(-2i)^2 + (2 - 2i)(-2i) - 4$
$= 8i + (2 - i)(4) - (2 - 2i)(2i) - 4$
$= 8i + 8 - 4i - 4i - 4 - 4 = 0$

**53.** Synthetic division shows that $f(i) = 0$ (the remainder), and at the same time gives
$f(x) \div (x - i) = x^2 + 3x - i = h(x)$, so
$f(x) = (x - i)(x^2 + 3x - i)$.

**54.** Synthetic division shows that $f(1 + i) = 0$ (the remainder), and at the same time gives
$f(x) \div (x - 1 - i) = x^2 + 1 = h(x)$, so
$f(x) = (x - 1 - i)(x^2 + 1)$.

**55.** None

**56.** $x = \{2, -1 + \sqrt{3}i, -1 - \sqrt{3}i\}$

**57.** $x = \{-4, 2 + 2\sqrt{3}i, 2 - 2\sqrt{3}i\}$

## Section 2.7 (pp. 227–238)

### Exploration 1

**1.** $g(x) = \dfrac{1}{x - 2}$

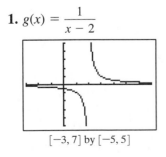

$[-3, 7]$ by $[-5, 5]$

**2.** $h(x) = -\dfrac{1}{x - 5}$

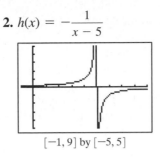

$[-1, 9]$ by $[-5, 5]$

**3.** $\dfrac{3}{x + 4} - 2$

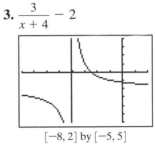

$[-8, 2]$ by $[-5, 5]$

### Quick Review 2.7

**1.** $x = -3$ or $x = \dfrac{1}{2}$

**2.** $x = -\dfrac{4}{3}$ or $x = 2$

**6.** no real zeros

### Exercises 2.7

**1.** Yes; $(-\infty, -0.682) \cup (-0.682, \infty)$

**2.** No; because of the square root expression in denominator.

**3.** No; because of the square root expression in numerator.

**4.** Yes; $\left( -\infty, -\dfrac{4}{3} \right) \cup \left( -\dfrac{4}{3}, \infty \right)$

**5.** Yes; $(-\infty, 0) \cup (0, \infty)$

**6.** Yes; $(-\infty, 0) \cup (0, \infty)$

**7.** Yes; $(-\infty, -1) \cup (-1, \infty)$

**8.** Yes; $(-\infty, -3) \cup (-3, \infty)$

**9.** Translate right 3 units. Asymptotes: $x = 3$, $y = 0$.

**10.** Translate left 3 units. Asymptotes: $x = -3$, $y = 0$.

**11.** Translate left 5 units, reflect across $x$-axis, vertically stretch by 2. Asymptotes: $x = -5$, $y = 0$.

**12.** Translate left 2 units, vertically stretch by 2. Asymptotes: $x = -2$, $y = 0$.

**13.** Translate left 3 units, reflect across $x$-axis, vertically stretch by 7, translate up 2 units. Asymptotes: $x = -3$, $y = 2$.

**14.** Translate right 1 unit, translate up 3 units. Asymptotes: $x = 1$, $y = 3$.

**15.** Translate left 4 units, vertically stretch by 13, translate down 2 units. Asymptotes: $x = -4$, $y = -2$.

**16.** Translate right 5 units, reflect across $x$-axis, vertically stretch by 11, translate down 3 units. Asymptotes: $x = 5$, $y = -3$.

**25.** Intercept: $\left(0, -\dfrac{2}{3}\right)$. Asymptotes: $x = 3$ and $y = 0$.

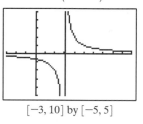

$[-3, 10]$ by $[-5, 5]$

**26.** Intercept: $\left(0, -\dfrac{3}{2}\right)$. Asymptotes: $x = -2$ and $y = 0$.

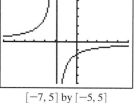

$[-7, 5]$ by $[-5, 5]$

**27.** Intercepts: $\left(0, \dfrac{2}{3}\right)$ and $(2, 0)$. Asymptotes: $x = -1$, $x = 3$ and $y = 0$.

$[-4, 6]$ by $[-5, 5]$

**28.** Intercepts: $\left(0, -\dfrac{2}{3}\right)$ and $(-2, 0)$. Asymptotes: $x = -3$, $x = 1$ and $y = 0$.

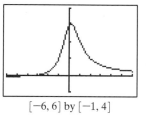

$[-6, 4]$ by $[-5, 5]$

**29.** No intercepts. Asymptotes: $x = -1$, $x = 0$, $x = 1$ and $y = 0$.

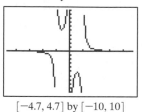

$[-4.7, 4.7]$ by $[-10, 10]$

**30.** No intercepts. Asymptotes: $x = -2$, $x = 0$, $x = 2$ and $y = 0$.

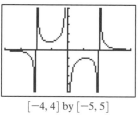

$[-4, 4]$ by $[-5, 5]$

**31.** Intercepts: $\left(0, -\dfrac{1}{3}\right)$ and $(1, 0)$. Asymptote: $y = 0$.

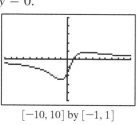

$[-10, 10]$ by $[-1, 1]$

**32.** Intercepts: $(0, 3)$ and $(-3, 0)$. Asymptote: $y = 0$.

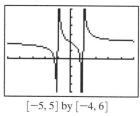

$[-6, 6]$ by $[-1, 4]$

**33.** Intercepts: $(0, 2)$, $(-1.28, 0)$ and $(0.78, 0)$. Asymptotes: $x = 1$, $x = -1$, and $y = 2$.

$[-5, 5]$ by $[-4, 6]$

**34.** Intercepts: $(0, -3)$, $(-1.84, 0)$ and $(2.17, 0)$. Asymptotes: $x = -2$, $x = 2$, and $y = -3$.

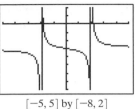

[−5, 5] by [−8, 2]

**35.** Intercepts: $\left(0, \dfrac{3}{2}\right)$. Asymptotes: $x = -2$, $y = x - 4$.

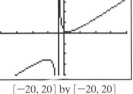

[−20, 20] by [−20, 20]

**36.** Intercepts: $\left(0, -\dfrac{7}{3}\right)$, $(-1.54, 0)$ and $(4.54, 0)$. Asymptotes: $x = -3$, $y = x - 6$.

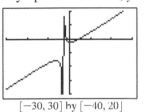

[−30, 30] by [−40, 20]

**37.** (d); Xmin $= -2$, Xmax $= 8$, Xscl $= 1$, and Ymin $= -3$, Ymax $= 3$, Yscl $= 1$.

**38.** (b); Xmin $= -6$, Xmax $= 2$, Xscl $= 1$, and Ymin $= -3$, Ymax $= 3$, Yscl $= 1$.

**39.** (a); Xmin $= -3$, Xmax $= 5$, Xscl $= 1$, and Ymin $= -5$, Ymax $= 10$, Yscl $= 1$.

**40.** (f); Xmin $= -6$, Xmax $= 2$, Xscl $= 1$, and Ymin $= -5$, Ymax $= 5$, Yscl $= 1$.

**41.** (e); Xmin $= -2$, Xmax $= 8$, Xscl $= 1$, and Ymin $= -3$, Ymax $= 3$, Yscl $= 1$.

**42.** (c); Xmin $= -3$, Xmax $= 5$, Xscl $= 1$, and Ymin $= -3$, Ymax $= 8$, Yscl $= 1$.

**43.** $x = -\dfrac{1}{3}$ or $x = 2$; the latter is extraneous.

**44.** $x = -\dfrac{3}{4}$ or $x = 1$; the latter is extraneous.

**45.** $x = 5$ or $x = 0$; the latter is extraneous.
**46.** $x = 3$ or $x = 0$; the latter is extraneous.
**47.** $x = -2$ or $x = 0$; both of these are extraneous (there are no real solutions).
**48.** $x = -1$ or $x = -3$; the latter is extraneous.

**49. (a)** The total amount of solution is $(125 + x)$ mL; of this, the amount of acid is $x$ plus 60% of the original amount, or $x + 0.6(125)$.

**(c)** $C(x) = \dfrac{x + 75}{x + 125} \geq 0.83$. $x \geq 169.12$ mL (approximately).

**50. (a)** $C(x) = \dfrac{x + 0.35(100)}{x + 100} = \dfrac{x + 35}{x + 100}$

**(b)** For $x < 160$, $C(x) < 0.75$

**(c)** $x < 160$ mL

**51. (a)** $C(x) = \dfrac{3000 + 2.12x}{x}$

**(b)** 4762 hats per week

**(c)** 6350 hats

**52. (a)** 200, 350, 425

**(b)** Yes; $y = 500$

**(c)** 500

**53. (a)** No: the domain of $f$ is $(-\infty, 3) \cup (3, \infty)$; the domain of $g$ is all real numbers.

**(b)** No: while it is not defined at 3, it does not tend toward $\pm\infty$ on either side.

**(c)** Most grapher viewing windows do not reveal that $f$ is undefined at 3.

**(d)** Almost—but not quite, they are equal for all $x \neq 3$.

**54. (a)** The functions are identical at all points except $x = 1$

**(b)** The functions are identical except at $x = -1$.

**(c)** The functions are identical except at $x = -1$.

**(d)** The functions are identical except at $x = 1$.

**55. (b)** If $f(x) = kx^a$, where $a$ is a negative integer, then the power function $f$ is also a rational function.

**56. (a)**

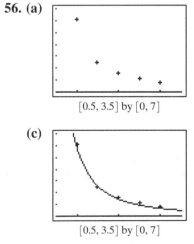

[0.5, 3.5] by [0, 7]

**(c)**

[0.5, 3.5] by [0, 7]

**57.** End behavior asymptote: $y = \frac{1}{2}x^2 - \frac{3}{4}x - \frac{1}{8}$;

vertical asymptote: $x = \frac{1}{2}$; $x$-intercept: approx.

1.75; $y$-intercept: $(0, 1)$. Intermediate behavior:

$f(x) \to \infty$ as $x \to \frac{1}{2}^-$; $f(x) \to -\infty$ as $x \to \frac{1}{2}^+$.

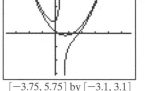
$[-3.75, 5.75]$ by $[-3.1, 3.1]$

**58.** End behavior asymptote: $y = 2x^2 + 2x + 3$;
vertical asymptote: $x = 2$; $x$-intercept: approx.
$-1.19$; $y$-intercept: $(0, -2.5)$. Intermediate
behavior: $g(x) \to -\infty$ as $x \to 2^-$; $g(x) \to \infty$ as
$x \to 2^+$.

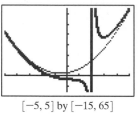
$[-5, 5]$ by $[-15, 65]$

**59.** End behavior asymptote: $y = x^3 + 2x^2 - 3x + 1$;

vertical asymptote: $x = \frac{5}{2}$; $x$-intercepts: approx.

$-3.09$ and $2.52$; $y$-intercept: $(0, 1.2)$. Intermediate

behavior: $f(x) \to \infty$ as $x \to \frac{5}{2}^-$;

$f(x) \to -\infty$ as $x \to \frac{5}{2}^+$.

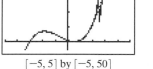
$[-5, 5]$ by $[-5, 50]$

Note: Try plotting the two functions separately to
observe the behavior along the asymptote.

**60.** End behavior asymptote:
$y = 2x^4 + 2x^3 - x^2 - x + 1$; vertical asymptote:
$x = 1$; $x$-intercept: approx. 1.33; $y$-intercept:
$(0, 4)$. Intermediate behavior: $g(x) \to \infty$ as
$x \to 1^-$; $g(x) \to -\infty$ as $x \to 1^+$.

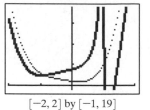
$[-2, 2]$ by $[-1, 19]$

**61.** Horizontal asymptotes: $y = -2$ and $y = 2$.

Intercepts: $\left(0, -\frac{3}{2}\right), \left(\frac{3}{2}, 0\right)$

$$h(x) = \begin{cases} \dfrac{2x - 3}{x + 2} & x \geq 0 \\[2mm] \dfrac{2x - 3}{-x + 2} & x < 0 \end{cases}$$

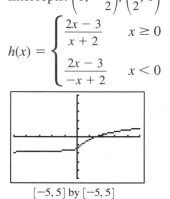
$[-5, 5]$ by $[-5, 5]$

**62.** Horizontal asymptotes: $y = \pm 3$.

Intercepts: $\left(0, \frac{5}{3}\right), \left(-\frac{5}{3}, 0\right)$

$$h(x) = \begin{cases} \dfrac{3x + 5}{x + 3} & x \geq 0 \\[2mm] \dfrac{3x + 5}{-x + 3} & x < 0 \end{cases}$$

$[-5, 5]$ by $[-5, 5]$

**63.** Horizontal asymptotes: $y = \pm 3$.
Intercepts: $\left(0, \dfrac{5}{4}\right), \left(\dfrac{5}{3}, 0\right)$

$$h(x) = \begin{cases} \dfrac{5 - 3x}{x + 4} & x \geq 0 \\[2ex] \dfrac{5 - 3x}{-x + 4} & x < 0 \end{cases}$$

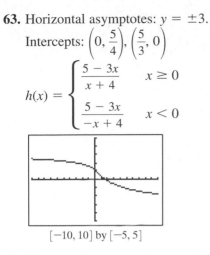

$[-10, 10]$ by $[-5, 5]$

**64.** Horizontal asymptotes: $y = \pm 2$.
Intercepts: $(0, 2), (1, 0)$

$$h(x) = \begin{cases} \dfrac{2 - 2x}{x + 1} & x \geq 0 \\[2ex] 2 & x < 0 \end{cases}$$

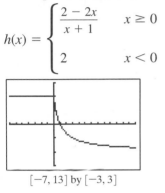

$[-7, 13]$ by $[-3, 3]$

# Section 2.8 (pp. 239–248)

## Exploration 1

**1. (a)**

$$\underset{\substack{\\ \text{Negative}}}{(+)(-)(+)} \underset{\substack{-3}}{\Big|} \underset{\substack{\\ \text{Negative}}}{(+)(-)(+)} \underset{\substack{2}}{\Big|} \underset{\substack{\\ \text{Positive} \quad x}}{(+)(+)(+)}$$

**(b)**

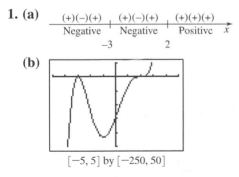

$[-5, 5]$ by $[-250, 50]$

**2. (a)**

$$\underset{\substack{\\ \text{Positive}}}{(-)(+)(-)(+)} \underset{\substack{-2}}{\Big|} \underset{\substack{\\ \text{Positive}}}{(-)(+)(-)(+)} \underset{\substack{-1}}{\Big|} \underset{\substack{\\ \text{Negative} \quad x}}{(-)(+)(+)(+)}$$

**(b)**

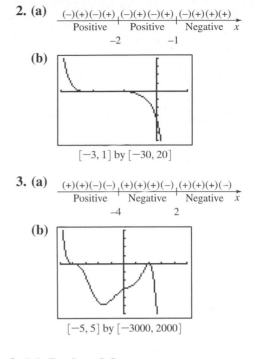

$[-3, 1]$ by $[-30, 20]$

**3. (a)**

$$\underset{\substack{\\ \text{Positive}}}{(+)(+)(-)(-)} \underset{\substack{-4}}{\Big|} \underset{\substack{\\ \text{Negative}}}{(+)(+)(+)(-)} \underset{\substack{2}}{\Big|} \underset{\substack{\\ \text{Negative} \quad x}}{(+)(+)(+)(-)}$$

**(b)**

$[-5, 5]$ by $[-3000, 2000]$

## Quick Review 2.8

**2.** $\lim\limits_{x \to \infty} f(x) = -\infty, \quad \lim\limits_{x \to -\infty} f(x) = -\infty$

**3.** $\lim\limits_{x \to \infty} g(x) = \infty, \quad \lim\limits_{x \to -\infty} g(x) = \infty$

**4.** $\lim\limits_{x \to \infty} g(x) = \infty, \quad \lim\limits_{x \to -\infty} g(x) = -\infty$

**7.** $\dfrac{x^2 - 7x - 2}{2x^2 - 5x - 3}$

**8.** $\dfrac{4x^2 - 4x - 1}{3x^2 - 7x + 4}$

**9. (a)** $\pm 1, \pm\dfrac{1}{2}, \pm 3, \pm\dfrac{3}{2}$

**(b)** $(x + 1)(2x - 3)(x + 1)$

**10. (a)** $\pm 1, \pm\dfrac{1}{3}, \pm 2, \pm\dfrac{2}{3}, \pm 4, \pm\dfrac{4}{3}, \pm 8, \pm\dfrac{8}{3}$

**(b)** $(x + 2)(x - 1)(3x - 4)$

## Exercises 2.8

**1. (a)** $x = -2, -1, 5$

**(b)** $-2 < x < -1$ or $x > 5$

**(c)** $x < -2$ or $-1 < x < 5$

**2. (a)** $x = 7, -\dfrac{1}{3}, -4$

**(b)** $-4 < x < -\dfrac{1}{3}$ or $x > 7$

**(c)** $x < -4$ or $-\dfrac{1}{3} < x < 7$

**3. (a)** $x = -7, -4, 6$

**(b)** $x < -7$ or $-4 < x < 6$ or $x > 6$

**(c)** $-7 < x < -4$

**4. (a)** $x = -\dfrac{3}{5}, 1$

**(b)** $x < -\dfrac{3}{5}$ or $x > 1$

**(c)** $-\dfrac{3}{5} < x < 1$

**5. (a)** $x = 8, -1$
**(b)** $-1 < x < 8$ or $x > 8$
**(c)** $x < -1$
**6. (a)** $x = -2, 9$
**(b)** $-2 < x < 9$ or $x > 9$
**(c)** $x < -2$
**9.** $(-\infty, -1) \cup (1, 2)$
**10.** $\left(\dfrac{7}{2}, \infty\right)$
**11.** $\left[-2, \dfrac{1}{2}\right] \cup [3, \infty)$
**12.** $(-\infty, -1] \cup [2, 3]$
**13.** $[-1, 0] \cup [2, \infty)$
**14.** $(-\infty, 0) \cup \left(1, \dfrac{3}{2}\right)$
**15.** $\left(-1, \dfrac{3}{2}\right) \cup (2, \infty)$
**16.** $(-\infty, -1] \cup [1, 4]$
**17.** $[-1.15, \infty)$
**18.** $(0.39, \infty)$
**20.** $\left(-\infty, -\dfrac{4}{3}\right] \cup [-1, \infty)$
**|21. (a)** $x = 1$

**(b)** $x = -\dfrac{3}{2}, 4$

**(c)** $-\dfrac{3}{2} < x < 1$ or $x > 4$

**(d)** $x < -\dfrac{3}{2}$, or $1 < x < 4$

**22. (a)** $x = \dfrac{7}{2}, -1$

**(b)** $x = -5$

**(c)** $-5 < x < -1$ or $x > \dfrac{7}{2}$

**(d)** $x < -5$ or $-1 < x < \dfrac{7}{2}$

**23. (a)** $x = 0, -3$
**(b)** $x < -3$
**(c)** $x > 0$
**(d)** $-3 < x < 0$
**24. (a)** $x = 0, -\dfrac{9}{2}$

**(b)** None.

**(c)** $x \neq -\dfrac{9}{2}, 0$

**(d)** None.

**25. (a)** $x = -5$

**(b)** $x = -\dfrac{1}{2}, x = 1, x < -5$

**(c)** $-5 < x < -\dfrac{1}{2}$ or $x > 1$

**(d)** $-\dfrac{1}{2} < x < 1$

**26. (a)** $x = 1$
**(b)** $x = 4, x \leq -2$
**(c)** $-2 < x < 1$ or $x > 4$
**(d)** $1 < x < 4$
**27. (a)** $x = 3$
**(b)** $x = 4, x < 3$
**(c)** $3 < x < 4$ or $x > 4$
**(d)** $f(x)$ is never negative
**28. (a)** None
**(b)** $x \leq 5$
**(c)** $5 < x < \infty$
**(d)** None
**29.** $(-\infty, -2) \cup (1, 2)$
**30.** $(-\infty, -3) \cup (-2, 3)$
**32.** $(-\infty, -2) \cup (2, \infty)$
**33.** $(-\infty, -4) \cup (3, \infty)$
**35.** $[-1, 0] \cup [1, \infty)$
**36.** $(-\infty, -2] \cup [0, 2]$
**37.** $(0, 2) \cup (2, \infty)$
**38.** $(-\infty, -2) \cup (-2, 3)$
**39.** $\left(-4, \dfrac{1}{2}\right)$
**40.** $\left[\dfrac{4}{3}, \infty\right)$
**42.** $(-\infty, -3) \cup (0, 5]$
**43.** $(-\infty, 0) \cup (\sqrt[3]{2}, \infty)$
**44.** $(-\infty, -\sqrt[3]{4}] \cup (0, \infty)$
**45.** $(-\infty, -1) \cup [1, 3)$
**46.** $(-\infty, -5) \cup (-2, 1)$
**47.** $[-3, \infty)$
**48.** no solution
**49.** $[5, \infty)$
**50.** $(-1, 0) \cup (0, 4)$
**52.** no more than 4 hours 7.5 minutes (which rounds to 4 hours)
**53.** 1 in. $< x < 34$ in.
**55.** 0 in. $\leq x \leq 0.69$ in. or 4.20 in. $\leq x \leq 6$ in.
**56.** 1.68 in. $\leq x \leq 9.10$ in.
**57. (a)** $R = \dfrac{2.3x}{x + 2.3}$

**(b)** at least 6.5 ohms
**58. (b)** 1.12 cm $\leq x \leq 11.37$ cm
1.23 cm $\leq h \leq 126.88$ cm

**(c)** about 348.73 cm$^2$

**59. (a)** $y \approx 895.9483x + 19113.4828$

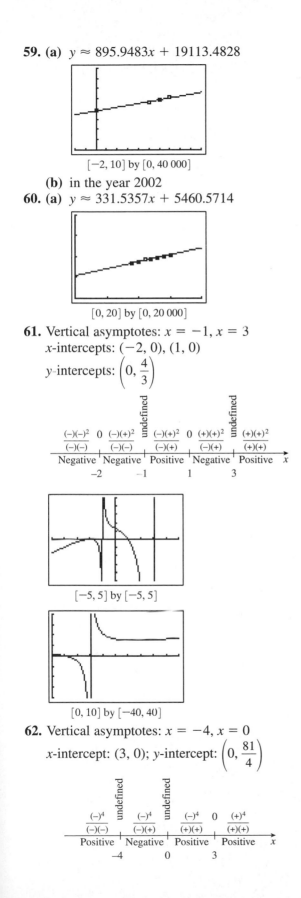

$[-2, 10]$ by $[0, 40\,000]$

**(b)** in the year 2002

**60. (a)** $y \approx 331.5357x + 5460.5714$

$[0, 20]$ by $[0, 20\,000]$

**61.** Vertical asymptotes: $x = -1$, $x = 3$

$x$-intercepts: $(-2, 0)$, $(1, 0)$

$y$-intercepts: $\left(0, \dfrac{4}{3}\right)$

$[-5, 5]$ by $[-5, 5]$

$[0, 10]$ by $[-40, 40]$

**62.** Vertical asymptotes: $x = -4$, $x = 0$

$x$-intercept: $(3, 0)$; $y$-intercept: $\left(0, \dfrac{81}{4}\right)$

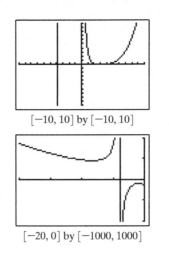

$[-10, 10]$ by $[-10, 10]$

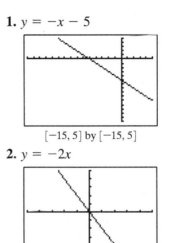

$[-20, 0]$ by $[-1000, 1000]$

# Chapter 2 Review (pp. 253–256)

**1.** $y = -x - 5$

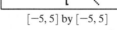

$[-15, 5]$ by $[-15, 5]$

**2.** $y = -2x$

$[-5, 5]$ by $[-5, 5]$

**3.** Starting from $y = x^2$, translate right 2 units and vertically stretch by 3 (either order), then translate up 4 units.

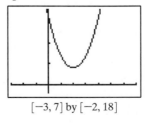

$[-3, 7]$ by $[-2, 18]$

**4.** Starting from $y = x^2$, translate left 3 units and reflect across $x$-axis (either order), then translate up 1 unit.

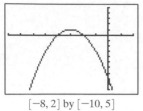

$[-8, 2]$ by $[-10, 5]$

**5.** Vertex: $(-3, 5)$; axis: $x = -3$
**6.** Vertex: $(5, -7)$; axis: $x = 5$
**7.** Vertex: $(-4, 1)$; axis: $x = -4$
**8.** Vertex: $(1, -1)$; axis: $x = 1$
**11.** $y = \frac{1}{2}(x - 3)^2 - 2$
**12.** $y = -\frac{1}{2}(x + 4)^{-2} + 5$

**13.**

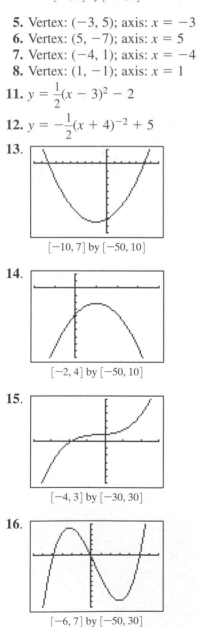

$[-10, 7]$ by $[-50, 10]$

**14.**

$[-2, 4]$ by $[-50, 10]$

**15.**

$[-4, 3]$ by $[-30, 30]$

**16.**

$[-6, 7]$ by $[-50, 30]$

**19.** The force $F$ needed varies directly with the distance $x$ from its resting position, with constant of variation $k$.
**20.** The area of a circle $A$ varies directly with the square of its radius.
**21.** $k = 4$, $a = \frac{1}{3}$, $f$ is odd.
**22.** $k = -2$, $a = \frac{3}{4}$, $f$ is not defined for $x < 0$.
**23.** $k = -2$, $a = -3$, $f$ is odd.
**24.** $k = \frac{2}{3}$, $a = -4$, $f$ is even.
**25.** $2x^2 - x + 1 - \dfrac{2}{x - 3}$
**26.** $x^3 + x^2 - x - 1 + \dfrac{5}{x + 2}$
**27.** $2x^2 - 3x + 1 + \dfrac{-2x + 3}{x^2 + 4}$
**28.** $x^3 - 2x^2 + 1 + \dfrac{-7}{3x + 1}$
**29.** $-39$
**30.** $-2$
**31.** Yes
**32.** No
**37.** $\pm 1$, $\pm 2$, $\pm 3$, $\pm 6$, $\pm\frac{1}{2}$, $\pm\frac{3}{2}$; $-\frac{3}{2}$ and $2$ are zeros.
**38.** $\pm 1$, $\pm 7$, $\pm\frac{1}{2}$, $\pm\frac{7}{2}$, $\pm\frac{1}{3}$, $\pm\frac{7}{3}$, $\pm\frac{1}{6}$, $\pm\frac{7}{6}$; $\frac{7}{3}$ is a zero.
**39.** $1 + 3i$
**40.** $2 - 5i$
**48.** $1 \pm \sqrt{3}i$
**53.** Rational: $0$. Irrational: $5 \pm \sqrt{2}$. No nonreal zeros.
**54.** Rational: $\pm 2$. Irrational: $\pm\sqrt{3}$. No nonreal zeros.
**55.** Rational: none. Irrational: approximately $-2.34$, $0.57$, $3.77$. No nonreal zeros.
**56.** Rational: none. Irrational: approximately $-3.97$, $-0.19$. Two nonreal zeros.
**57.** Rational: $-\frac{3}{2}$. Nonreal complex: $3 + i$
$f(x) = (2x + 3)(x - 3 + i)(x - 3 - i)$
**58.** Rational: $\frac{4}{5}$. Irrational: $2 \pm \sqrt{7}$
$f(x) = (5x - 4)(x - 2 - \sqrt{7})(x - 2 + \sqrt{7})$
**59.** Rational: $1$, $-1$, $\frac{2}{3}$, and $-\frac{5}{2}$.
$f(x) = (3x - 2)(2x + 5)(x - 1)(x + 1)$
**60.** Nonreal complex: $3 \pm i$, $1 \pm 2i$
$f(x) = (x - 1 - 2i)(x - 1 + 2i)(x - 3 - i)$
$(x - 3 + i)$
**61.** $f(x) = (x - 2)(x^2 + x + 1)$
**62.** $f(x) = (x + 1)(9x^2 - 12x - 1)$
**63.** $f(x) = (2x - 3)(x - 1)(x^2 - 2x + 5)$

**64.** $f(x) = (3x + 2)(x + 1)(x^2 - 4x + 5)$

**67.** $6x^4 - 5x^3 - 38x^2 - 5x + 6$

**69.** $x^4 - 4x^3 - 12x^2 + 32x + 64$

**70.** $2x^3 - 6x^2 + 2x + 10$

**71.** Translate right 5 units and vertically stretch by 2 (either order), then translate down 1 unit. Horizontal asymptote: $y = -1$; vertical asymptote: $x = 5$.

**72.** Translate left 2 units and reflect across $x$-axis (either order), then translate up 3 units. Horizontal asymptote: $y = 3$; vertical asymptote: $x = -2$.

**73.** Asymptotes: $y = 1$, $x = -1$, and $x = 1$. Intercept: $(0, -1)$.

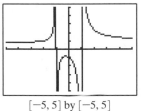

$[-5, 5]$ by $[-5, 5]$

**74.** Asymptotes: $y = 2$, $x = -3$, and $x = 2$. Intercept: $\left(0, -\dfrac{7}{6}\right)$.

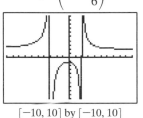

$[-10, 10]$ by $[-10, 10]$

**75.** End behavior asymptote: $y = x - 7$. Vertical asymptote: $x = -3$. Intercept: $\left(0, \dfrac{5}{3}\right)$.

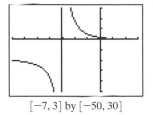

$[-7, 3]$ by $[-50, 30]$

**76.** End behavior asymptote: $y = x - 6$. Vertical asymptote: $x = -3$. Intercepts: approx. $(-1.54, 0)$, $(4.54, 0)$ and $\left(0, -\dfrac{7}{3}\right)$.

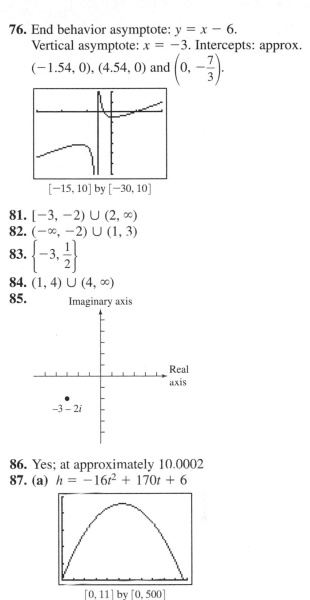

$[-15, 10]$ by $[-30, 10]$

**81.** $[-3, -2) \cup (2, \infty)$

**82.** $(-\infty, -2) \cup (1, 3)$

**83.** $\left\{-3, \dfrac{1}{2}\right\}$

**84.** $(1, 4) \cup (4, \infty)$

**85.**

Imaginary axis

Real axis

$-3 - 2i$

**86.** Yes; at approximately $10.0002$

**87.** **(a)** $h = -16t^2 + 170t + 6$

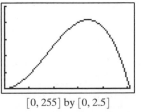

$[0, 11]$ by $[0, 500]$

**(b)** When $t \approx 5.3125$, $h \approx 457.5625$

**(c)** about $10.66$ sec.

**88.** **(a)** $V = x(30 - 2x)(70 - 2x)$ in.$^3$

**(b)** Either $x \approx 4.57$ or $x \approx 8.63$ in.

**89.** **(a)** & **(b)**

$[0, 255]$ by $[0, 2.5]$

**(c)** When $d \approx 170$ ft, $s \approx 2.088$ ft.

**90.** **(a)** $V = \dfrac{4}{3}\pi x^3 + \pi x^2(140 - 2x)$

**(b)**

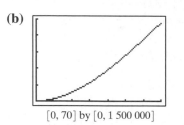

[0, 70] by [0, 1 500 000]

**(c)** The largest volume occurs when $x = 70$ (so it is actually a sphere). This volume is $\frac{4}{3}\pi 70^3 \approx 1,436,755$ ft$^3$.

**91. (a)** $y \approx 2.0167x^3 - 48.4429x^2 + 374.4690x - 918.7171$

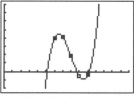

[0, 15] by [−10, 40]

**(b)** $y \approx -4.8833x^3 + 116.8857x^2 - 919.5881x + 2392.0543$

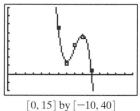

[0, 15] by [−10, 40]

**(c)** As $x \to \infty$, the Dog stock model goes to $\infty$, the Star stock model goes to $-\infty$.

**92. (a)** $y \approx 1.0055x + 6.5194$

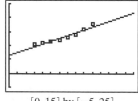

[0, 15] by [−5, 25]

**(b)** $y \approx 0.1550x^2 - 1.0095x + 12.2544$

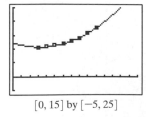

[0, 15] by [−5, 25]

**(c) Linear:** in the year 2003

**Quadratic:** in the year 2001

**93. (a)** $C = 4.32 + \dfrac{4000}{x}$

**(b)** Solve $x\left(5.25 - 4.32 - \dfrac{4000}{x}\right) = 8000$:

$0.93x = 12000$, so $x \approx 12\,903.23$ — round up to 12,904.

**94. (a)** $P(15) = 325$, $P(70) = 600$, $P(100) = 648$.

**(b)** $y = \dfrac{640}{0.8} = 800$

**(c)** The deer population approaches (but never equals) 800.

**95. (a)** $R_2 = \dfrac{1.2x}{x - 1.2}$    **(b)** 2 ohms

**96. (a)** $C(x) = \dfrac{50}{50 + x}$

**(b)** about 33.33 ounces of distilled water

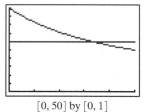

[0, 50] by [0, 1]

**(c)** $x = \dfrac{100}{3} \approx 33.3$

**97. (a)** $S = 2\pi x^2 + \dfrac{2000}{x}$

**(b)** Either $x \approx 2.31$ cm and $h \approx 59.75$ cm, or $x \approx 10.65$ and $h \approx 2.81$.

**(c)** Approximately $2.31 < x < 10.65$ (graphically) and $2.81 < h < 59.75$.

**98. (a)** $S = x^2 + \dfrac{4000}{x}$

**(b)** 20 ft by 20 ft by 2.5 ft. or $x \approx 7.32$, giving approximate dimensions 7.32 by 7.32 by 18.66.

**(c)** $7.32 < x < 20$ (lower bound approximate), so $y$ must be between 2.5 and about 18.66.

**Chapter 2 Project**

Answers are based on the sample data shown in the table.

**1.**

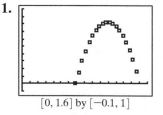

[0, 1.6] by [−0.1, 1]

**2.** $y = -4.914(x - 1.097)^2 + 0.829$

**3.** The sign of *a* affects the direction the parabola opens. The magnitude of *a* affects the vertical stretch of the graph. Changes to *h* cause horizontal shifts to the graph while changes to *k*

cause vertical shifts.

**4.** $y = -4.914x^2 + 10.781x - 5.085$

**5.** $y \approx -4.968x^2 + 10.913x - 5.160$

**6.** $y \approx -4.968(x - 1.098)^2 - 0.833$

# Chapter 3

## Section 3.1 (pp. 259–272)

### Exploration 1

**1.** (0, 1)

Domain: $(-\infty, \infty)$

Range: $(0, \infty)$

Continuous

Intercepts: (0, 1)

Always increasing

Not symmetric

No local extrema

Bounded below by $y = 0$, which is also the only asymptote

$\lim\limits_{x \to \infty} f(x) \to \infty$. $\lim\limits_{x \to -\infty} f(x) \to 0$

Concave up

**2.** (0, 1)

Domain: $(-\infty, \infty)$

Range: $(0, \infty)$

Continuous

Intercepts: (0, 1)

Always decreasing

Not symmetric

No local extrema

Bounded below by $y = 0$, which is also the only asymptote

$\lim\limits_{x \to \infty} g(x) \to 0$, $\lim\limits_{x \to -\infty} g(x) \to \infty$

Concave up

### Exploration 2

**1.**

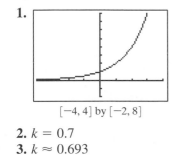

$[-4, 4]$ by $[-2, 8]$

**2.** $k = 0.7$

**3.** $k \approx 0.693$

### Exercises 3.1

**1.** Not an exponential function

**2.** Exponential function, initial value of 1 and base of 3.

**3.** Exponential function, initial value of 1 and base of 5.

**4.** Not an exponential function

**5.** Not an exponential function

**6.** Not an exponential function

**15.** Translate $f(x) = 2^x$ by 3 units to the right.

**16.** Translate $f(x) = 3^x$ by 4 units to the left.

**17.** Reflect $f(x) = 4^x$ over the *y*-axis.

**18.** Reflect $f(x) = 2^x$ over the *y*-axis and then shift by 5 units to the right.

**19.** Vertically stretch $f(x) = 0.5^x$ by a factor of 3 and then shift 4 units up.

**20.** Vertically stretch $f(x) = 0.6^x$ by a factor of 2 and then horizontally shrink by a factor of 3.

**21.** Reflect $f(x) = e^x$ across the *y*-axis and horizontally shrink by a factor of 2.

**22.** Reflect $f(x) = e^x$ across the *x*-axis and *y*-axis. Then, horizontally shrink by a factor of 3.

**23.** Reflect $f(x) = e^x$ across the *y*-axis, horizontally shrink by a factor of 3, translate 3 units to the right and vertically stretch by a factor of 2.

**24.** Horizontally shrink $f(x) = e^x$ by a factor of 2, vertically stretch by a factor of 3 and shift down one unit.

**25.** Graph (a) is the only graph "shaped like the graph of $y = 2^x$" and having the correct intercept.

**26.** Graph (d) is the only graph "shaped like the graph of $y = \left(\dfrac{1}{2}\right)^x = 2^{-x}$" and having the correct intercept.

**27.** Graph (c) is the reflection of $y = 2^x$ across the *x*-axis.

**28.** Graph (e) is the reflection of $y = 0.5^x$ across the *x*-axis.

**29.** Graph (b) is the graph of $y = 3^{-x}$ (shaped like the graph of $y = 2^{-x}$) translated down 2 units.

**30.** Graph (*f*) is the graph of $y = 1.5^x$ (shaped like the graph of $y = 2^x$) translated down 2 units.

**31.** Exponential decay; $\lim\limits_{x \to \infty} f(x) \to 0$, $\lim\limits_{x \to -\infty} f(x) \to \infty$

**32.** Exponential decay; $\lim\limits_{x \to \infty} f(x) \to 0$, $\lim\limits_{x \to -\infty} f(x) \to \infty$

**33.** Exponential decay; $\lim\limits_{x \to \infty} f(x) \to 0$, $\lim\limits_{x \to -\infty} f(x) \to \infty$

**34.** Exponential growth; $\lim\limits_{x \to \infty} f(x) \to \infty$, $\lim\limits_{x \to -\infty} f(x) \to \infty$

**39.** $y_1 = y_3$ since $3^{2x + 4} = 3^{2(x + 2)} = (3^2)^{x + 2}$ $= 9^{x + 2}$.

**40.** $y_2 = y_3$ since $2 \cdot 2^{3x - 2} = 2^1 \cdot 2^{3x - 2}$ $= 2^{1 + 3x - 2} = 2^{3x - 1}$.

**41.** $y$-intercept: $(0, 4)$. Horizontal asymptote: $y = 12$.

**42.** $y$-intercept: $(0, 3)$. Horizontal asymptote: $y = 18$.

**43.** $y$-intercept: $(0, 4)$. Horizontal asymptote: $y = 16$.

**44.** $y$-intercept: $(0, 3)$. Horizontal asymptote: $y = 9$.

**45.**

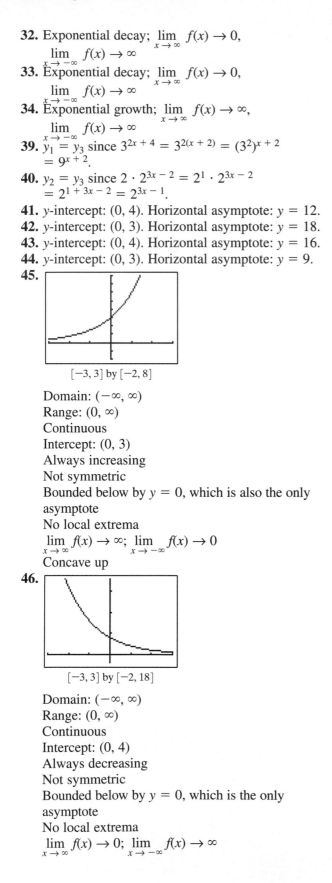

$[-3, 3]$ by $[-2, 8]$

Domain: $(-\infty, \infty)$
Range: $(0, \infty)$
Continuous
Intercept: $(0, 3)$
Always increasing
Not symmetric
Bounded below by $y = 0$, which is also the only asymptote
No local extrema
$\lim\limits_{x \to \infty} f(x) \to \infty$; $\lim\limits_{x \to -\infty} f(x) \to 0$
Concave up

**46.**

$[-3, 3]$ by $[-2, 18]$

Domain: $(-\infty, \infty)$
Range: $(0, \infty)$
Continuous
Intercept: $(0, 4)$
Always decreasing
Not symmetric
Bounded below by $y = 0$, which is the only asymptote
No local extrema
$\lim\limits_{x \to \infty} f(x) \to 0$; $\lim\limits_{x \to -\infty} f(x) \to \infty$
Concave up

**47.**

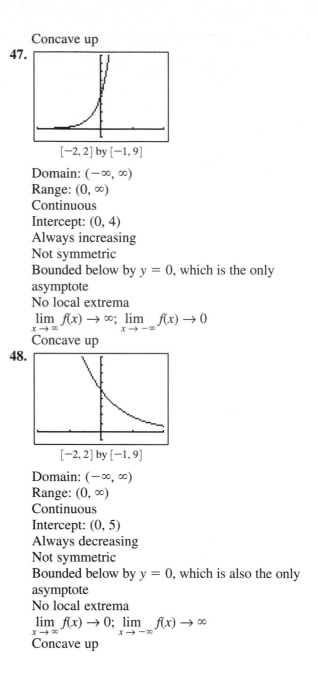

$[-2, 2]$ by $[-1, 9]$

Domain: $(-\infty, \infty)$
Range: $(0, \infty)$
Continuous
Intercept: $(0, 4)$
Always increasing
Not symmetric
Bounded below by $y = 0$, which is the only asymptote
No local extrema
$\lim\limits_{x \to \infty} f(x) \to \infty$; $\lim\limits_{x \to -\infty} f(x) \to 0$
Concave up

**48.**

$[-2, 2]$ by $[-1, 9]$

Domain: $(-\infty, \infty)$
Range: $(0, \infty)$
Continuous
Intercept: $(0, 5)$
Always decreasing
Not symmetric
Bounded below by $y = 0$, which is also the only asymptote
No local extrema
$\lim\limits_{x \to \infty} f(x) \to 0$; $\lim\limits_{x \to -\infty} f(x) \to \infty$
Concave up

**49.**

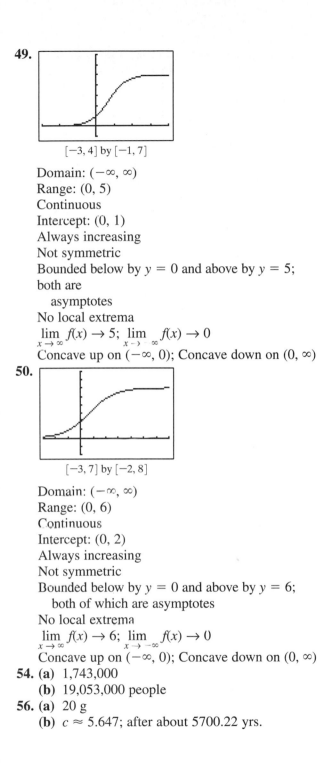

$[-3, 4]$ by $[-1, 7]$

Domain: $(-\infty, \infty)$
Range: $(0, 5)$
Continuous
Intercept: $(0, 1)$
Always increasing
Not symmetric
Bounded below by $y = 0$ and above by $y = 5$; both are
　asymptotes
No local extrema
$\lim\limits_{x \to \infty} f(x) \to 5$; $\lim\limits_{x \to -\infty} f(x) \to 0$
Concave up on $(-\infty, 0)$; Concave down on $(0, \infty)$

**50.**

Domain: $(-\infty, \infty)$
Range: $(0, 6)$
Continuous
Intercept: $(0, 2)$
Always increasing
Not symmetric
Bounded below by $y = 0$ and above by $y = 6$;
　both of which are asymptotes
No local extrema
$\lim\limits_{x \to \infty} f(x) \to 6$; $\lim\limits_{x \to -\infty} f(x) \to 0$
Concave up on $(-\infty, 0)$; Concave down on $(0, \infty)$

**54. (a)** 1,743,000
**(b)** 19,053,000 people
**56. (a)** 20 g
**(b)** $c \approx 5.647$; after about 5700.22 yrs.

**57. (a)**

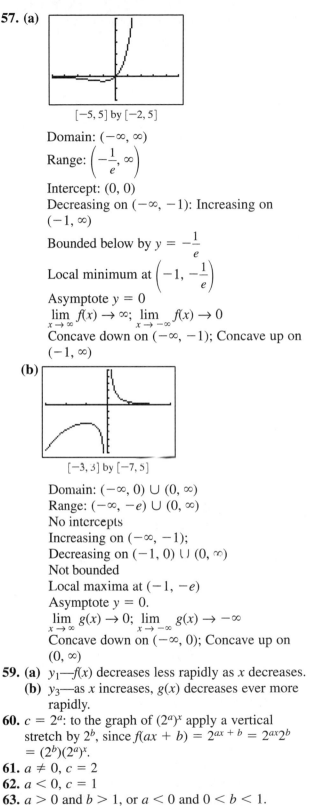

$[-5, 5]$ by $[-2, 5]$

Domain: $(-\infty, \infty)$
Range: $\left(-\dfrac{1}{e}, \infty\right)$
Intercept: $(0, 0)$
Decreasing on $(-\infty, -1)$: Increasing on $(-1, \infty)$

Bounded below by $y = -\dfrac{1}{e}$

Local minimum at $\left(-1, -\dfrac{1}{e}\right)$
Asymptote $y = 0$
$\lim\limits_{x \to \infty} f(x) \to \infty$; $\lim\limits_{x \to -\infty} f(x) \to 0$
Concave down on $(-\infty, -1)$; Concave up on $(-1, \infty)$

**(b)**

$[-3, 3]$ by $[-7, 5]$

Domain: $(-\infty, 0) \cup (0, \infty)$
Range: $(-\infty, -e) \cup (0, \infty)$
No intercepts
Increasing on $(-\infty, -1)$;
Decreasing on $(-1, 0) \cup (0, \infty)$
Not bounded
Local maxima at $(-1, -e)$
Asymptote $y = 0$.
$\lim\limits_{x \to \infty} g(x) \to 0$; $\lim\limits_{x \to -\infty} g(x) \to -\infty$
Concave down on $(-\infty, 0)$; Concave up on $(0, \infty)$

**59. (a)** $y_1$—$f(x)$ decreases less rapidly as $x$ decreases.
**(b)** $y_3$—as $x$ increases, $g(x)$ decreases ever more rapidly.
**60.** $c = 2^a$: to the graph of $(2^a)^x$ apply a vertical stretch by $2^b$, since $f(ax + b) = 2^{ax+b} = 2^{ax}2^b = (2^b)(2^a)^x$.
**61.** $a \neq 0$, $c = 2$
**62.** $a < 0$, $c = 1$
**63.** $a > 0$ and $b > 1$, or $a < 0$ and $0 < b < 1$.
**64.** $a > 0$ and $0 < b < 1$, or $a < 0$ and $b > 1$.

**65.** As $x \to -\infty$, $b^x \to 0$, so $1 + ab^x \to 1$;
as $x \to \infty$, $b^x \to \infty$, so $1 + ab^x \to \infty$.

## Section 3.2 (pp. 272–281)

### Exercises 3.2

1. exponential growth, 9%
2. exponential growth, 1.8%
3. exponential decay, 3.2%
4. exponential decay, 0.32%
5. exponential growth, 100%
6. exponential decay, 95%
7. $5 \cdot 1.17^x$
8. $52 \cdot 1.023^x$
9. $16 \cdot 0.5^x$
10. $5 \cdot 0.9941^x$
13. $18 \cdot 1.052^x$
14. $15 \cdot 0.954^x$
16. $250 \cdot 2^{2x/15}$
17. $592 \cdot 2^{-x/6}$
18. $17 \cdot 2^{-x/32}$
21. $4 \cdot 2.72^x$
22. $3 \cdot 0.64^x$
27. $\dfrac{20}{1 + 3 \cdot 0.58^x}$
28. $\dfrac{60}{1 + 3 \cdot 0.87^x}$
29. 20.222 years, or about 20 years and 2.7 months
30. 32.824 years, or about 32 years and 9.9 months
31. **(a)** 12,315; 24,265
    **(b)** 1966 or 1967
32. **(a)** 6554; 9151
    **(b)** 1980
33. **(a)** $y = 6.6\left(\dfrac{1}{2}\right)^{t/14}$, where $t$ is time in days.
    **(b)** After 38.11 days.
34. **(a)** $y = 3.5\left(\dfrac{1}{2}\right)^{t/65}$, where $t$ is time in days.
    **(b)** After 117.48 days.
35. One possible answer: Exponential and linear functions are similar in that they are always increasing or always decreasing. However, the two functions vary in how *quickly* they increase or decrease. While a linear function will increase or decrease at a steady rate over a given interval, the rate at which exponential functions increase or decrease over a given interval will vary.
36. One possible answer: Exponential functions and logistic functions are similar in the sense that they are always increasing or always decreasing. They differ, however, in the sense that logistic functions

have both an upper and lower limit to their growth (or decay), while exponential functions generally have only a lower limit. (Exponential functions just keep growing.)

37. One possible answer: The amount of time it takes for a population to double is dependent on its rate of growth; the higher the rate of growth, the faster the population increases, which reduces the amount of time it takes for a population to increase by 100%, or double.
38. One possible answer: The amount of time it takes for the radio activity of a substance to decrease to half of its original radioactivity is dependent on how *quickly* the substance decays, i.e., the rate of decay. If a substance decays slowly (i.e., the rate of decay is small), it will take more time to reach half of its original value of radioactivity. If a substance decays quickly (i.e., the rate of decay is large), it will take less time for the radioactivity level to reach half of its initial value.
42. about 9.20 miles above sea level
43. **(a)** 16
    **(b)** about 14 days
44. **(b)** 24 or 25 years
46. the same model
47. the same model
48. will not surpass
49. **(a)** 272,410,000
50. **(a)** Mexico will eventually have a larger population. 2136
    **(c)** According to the logistics growth models, the maximum sustainable populations are: Mexico—322.56 million people. U.S.—732.48 million people
    **(d)** Answers will vary.

## Section 3.3 (pp. 282–293)

### Exploration 1

1.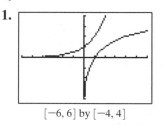
$[-6, 6]$ by $[-4, 4]$

### Exercises 3.3

32. $\approx 0.7076$
35. $\approx 0.8339$
36. $\approx 0.3213$
39. $\approx 169,824,365$

**40.** ≈ 177,828

**45.** Starting from $y = \ln x$: reflect across the $y$-axis and translate up 3 units.

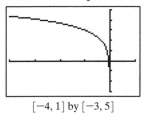

[−4, 1] by [−3, 5]

**46.** Starting from $y = \ln x$: reflect across the $y$-axis and translate down 2 units.

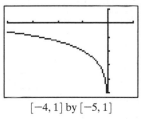

[−4, 1] by [−5, 1]

**47.** Starting from $y = \ln x$: reflect across the $y$-axis and translate right 2 units.

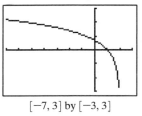

[−7, 3] by [−3, 3]

**48.** Starting from $y = \ln x$: reflect across the $y$-axis and translate right 5 units.

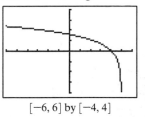

[−6, 6] by [−4, 4]

**49.** Starting from $y = \log x$: reflect across both axes and vertically stretch by 2.

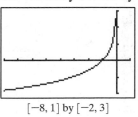

[−8, 1] by [−2, 3]

**50.** Starting from $y = \log x$: reflect across both axes and vertically stretch by 3.

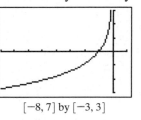

[−8, 7] by [−3, 3]

**51.** Starting from $y = \log x$: reflect across the $y$-axis, translate right 3 units, vertically stretch by 2, translate down 1 unit.

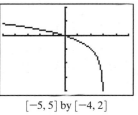

[−5, 5] by [−4, 2]

**52.** Starting from $y = \log x$: reflect across both axes, translate right 1 units, vertically stretch by 3, translate up 1 unit.

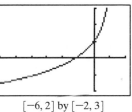

[−6, 2] by [−2, 3]

**53.**

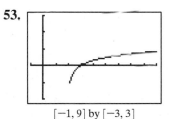

[−1, 9] by [−3, 3]

Domain: $(2, \infty)$
Range: $(-\infty, \infty)$
Continuous
Intercept: $(3, 0)$
Always increasing
Not symmetric
Not bounded
No local extrema
Asymptote at $x = 2$
$\lim\limits_{x \to \infty} f(x) \to \infty$; $\lim\limits_{x \to 2^+} f(x) \to -\infty$
Concave down

**54.**

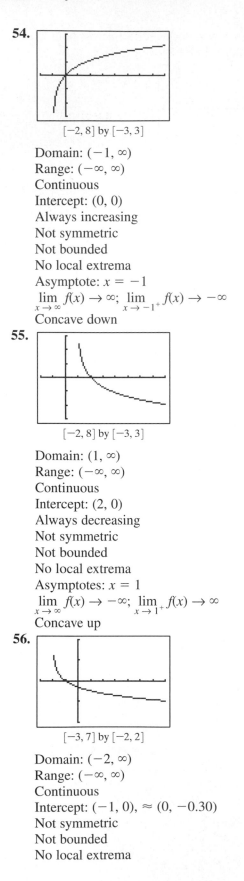

$[-2, 8]$ by $[-3, 3]$

Domain: $(-1, \infty)$
Range: $(-\infty, \infty)$
Continuous
Intercept: $(0, 0)$
Always increasing
Not symmetric
Not bounded
No local extrema
Asymptote: $x = -1$
$\lim_{x \to \infty} f(x) \to \infty$; $\lim_{x \to -1^+} f(x) \to -\infty$
Concave down

**55.**

$[-2, 8]$ by $[-3, 3]$

Domain: $(1, \infty)$
Range: $(-\infty, \infty)$
Continuous
Intercept: $(2, 0)$
Always decreasing
Not symmetric
Not bounded
No local extrema
Asymptotes: $x = 1$
$\lim_{x \to \infty} f(x) \to -\infty$; $\lim_{x \to 1^+} f(x) \to \infty$
Concave up

**56.**

$[-3, 7]$ by $[-2, 2]$

Domain: $(-2, \infty)$
Range: $(-\infty, \infty)$
Continuous
Intercept: $(-1, 0)$, $\approx (0, -0.30)$
Not symmetric
Not bounded
No local extrema

Asymptotes: $x = -2$
$\lim_{x \to \infty} f(x) \to -\infty$; $\lim_{x \to -2^+} f(x) \to \infty$

**57. (b)** $\approx 54.7319$
**(c)**

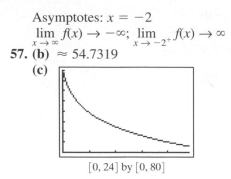

$[0, 24]$ by $[0, 80]$

**58.** $\approx 10.2019$ lumens
**61.** $y = -66.296 + 34.654 \ln x$; when $x = 30$,
$y \approx 51.57$ metric tons.
**62.** $y = -76.721 + 46.715 \ln x$; when $x = 35$,
$y \approx 89.37$ metric tons.

**63.**

| $f(x)$ | $3^x$ | $\log_3 x$ |
|---|---|---|
| Domain | $(-\infty, \infty)$ | $(0, \infty)$ |
| Range | $(0, \infty)$ | $(-\infty, \infty)$ |
| Intercepts | $(0, 1)$ | $(1, 0)$ |
| Asymptotes | $y = 0$ | $x = 0$ |

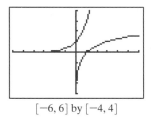

$[-6, 6]$ by $[-4, 4]$

**64.**

| $f(x)$ | $5^x$ | $\log_5 x$ |
|---|---|---|
| Domain | $(-\infty, \infty)$ | $(0, \infty)$ |
| Range | $(0, \infty)$ | $(-\infty, \infty)$ |
| Intercepts | $(0, 1)$ | $(1, 0)$ |
| Asymptotes | $y = 0$ | $x = 0$ |

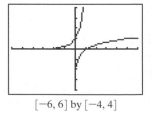

$[-6, 6]$ by $[-4, 4]$

**66.** 0 is not in the domain of the logarithm functions because 0 is not in the range of exponential functions; that is, $a^x$ is never equal to 0.

## Section 3.4 (pp. 293–301)

### Exercises 3.4

**2.** $\approx 1.8295$

**3.** $\approx 2.4837$

**4.** $\approx 2.2362$

**5.** $\approx -3.5850$

**6.** $\approx -2.0922$

**13.** $-\dfrac{\log(x + y)}{\log 2}$

**14.** $-\dfrac{\log(x - y)}{\log 3}$

**15.** Domain: $(0, \infty)$, range: $(-\infty, \infty)$

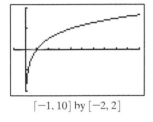

$[-1, 10]$ by $[-2, 2]$

**16.** Domain: $(0, \infty)$, range: $(-\infty, \infty)$

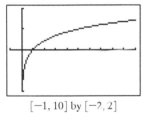

$[-1, 10]$ by $[-2, 2]$

**17.** Domain: $(2, \infty)$, range: $(-\infty, \infty)$

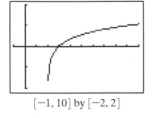

$[-1, 10]$ by $[-2, 2]$

**18.** Domain: $(-\infty, 2)$, range: $(-\infty, \infty)$

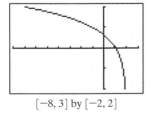

$[-8, 3]$ by $[-2, 2]$

**19.** (b): $[-5, 5] \times [-3, 3]$, with Xscl $= 1$ and

Yscl $= 1$

**20.** (c): $[-2, 8] \times [-3, 3]$, with Xscl $= 1$ and Yscl $= 1$

**21.** (d): $[-2, 8] \times [-3, 3]$, with Xscl $= 1$ and Yscl $= 1$

**22.** (a): $[-8, 4] \times [-8, 8]$, with Xscl $= 1$ and Yscl $= 1$

**29.** $3 \log x + 2 \log y$

**30.** $\log x + 3 \log y$

**43.** $\log\left(\dfrac{x^4}{z^3}\right)$

**44.** $\ln(x^9 y^5 z^4)$

**45.**

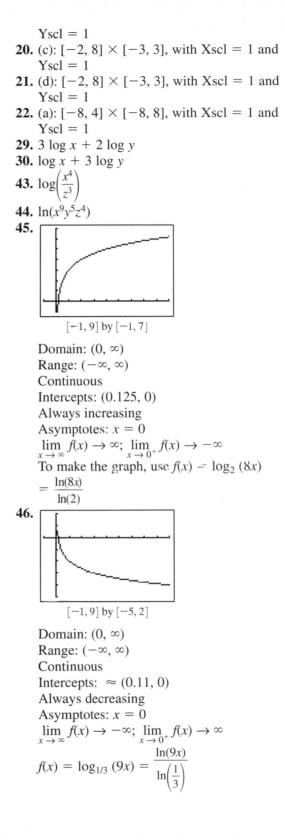

$[-1, 9]$ by $[-1, 7]$

Domain: $(0, \infty)$
Range: $(-\infty, \infty)$
Continuous
Intercepts: $(0.125, 0)$
Always increasing
Asymptotes: $x = 0$
$\lim\limits_{x \to \infty} f(x) \to \infty$; $\lim\limits_{x \to 0^+} f(x) \to -\infty$
To make the graph, use $f(x) = \log_2(8x)$
$= \dfrac{\ln(8x)}{\ln(2)}$

**46.**

$[-1, 9]$ by $[-5, 2]$

Domain: $(0, \infty)$
Range: $(-\infty, \infty)$
Continuous
Intercepts: $\approx (0.11, 0)$
Always decreasing
Asymptotes: $x = 0$
$\lim\limits_{x \to \infty} f(x) \to -\infty$; $\lim\limits_{x \to 0^+} f(x) \to \infty$
$f(x) = \log_{1/3}(9x) = \dfrac{\ln(9x)}{\ln\left(\dfrac{1}{3}\right)}$

**47.**

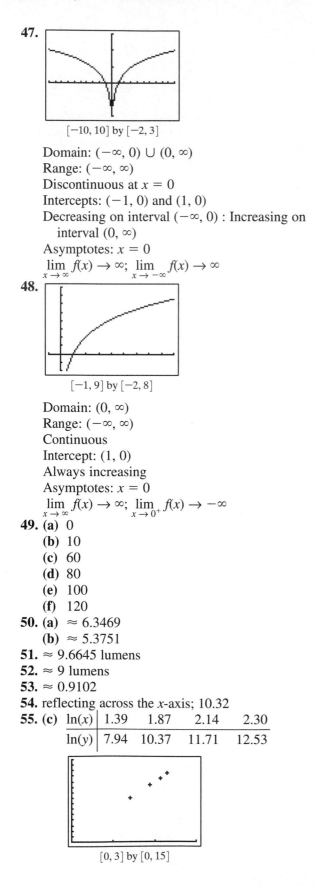

[−10, 10] by [−2, 3]

Domain: $(-\infty, 0) \cup (0, \infty)$
Range: $(-\infty, \infty)$
Discontinuous at $x = 0$
Intercepts: $(-1, 0)$ and $(1, 0)$
Decreasing on interval $(-\infty, 0)$ : Increasing on
   interval $(0, \infty)$
Asymptotes: $x = 0$
$\lim\limits_{x \to \infty} f(x) \to \infty$; $\lim\limits_{x \to -\infty} f(x) \to \infty$

**48.**

[−1, 9] by [−2, 8]

Domain: $(0, \infty)$
Range: $(-\infty, \infty)$
Continuous
Intercept: $(1, 0)$
Always increasing
Asymptotes: $x = 0$
$\lim\limits_{x \to \infty} f(x) \to \infty$; $\lim\limits_{x \to 0^+} f(x) \to -\infty$

**49. (a)** 0
  **(b)** 10
  **(c)** 60
  **(d)** 80
  **(e)** 100
  **(f)** 120

**50. (a)** $\approx 6.3469$
  **(b)** $\approx 5.3751$

**51.** $\approx 9.6645$ lumens

**52.** $\approx 9$ lumens

**53.** $\approx 0.9102$

**54.** reflecting across the $x$-axis; 10.32

**55. (c)**

| $\ln(x)$ | 1.39 | 1.87 | 2.14 | 2.30 |
|---|---|---|---|---|
| $\ln(y)$ | 7.94 | 10.37 | 11.71 | 12.53 |

[0, 3] by [0, 15]

**(e)** $a \approx 5$, $b \approx 1$ so $f(x) = e^1 x^5 = ex^5 \approx 2.72x^5$.
   The two equations are the same.

**56. (c)**

| $\ln(x)$ | 0.69 | 1.10 | 1.57 | 2.04 |
|---|---|---|---|---|
| $\ln(y)$ | 2.01 | 1.97 | 1.92 | 1.86 |

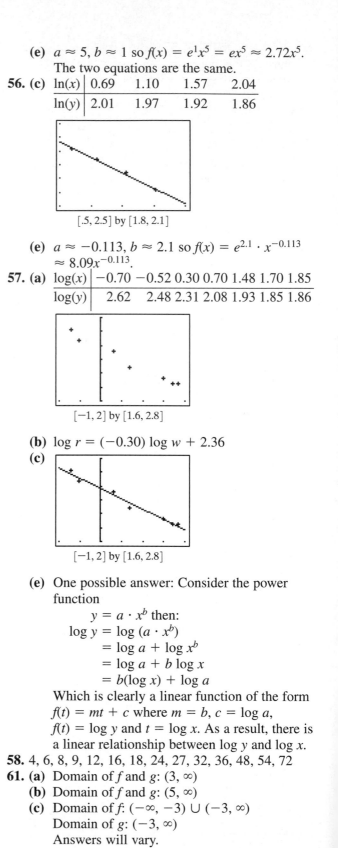

[.5, 2.5] by [1.8, 2.1]

**(e)** $a \approx -0.113$, $b \approx 2.1$ so $f(x) = e^{2.1} \cdot x^{-0.113}$
   $\approx 8.09x^{-0.113}$.

**57. (a)**

| $\log(x)$ | −0.70 | −0.52 | 0.30 | 0.70 | 1.48 | 1.70 | 1.85 |
|---|---|---|---|---|---|---|---|
| $\log(y)$ | 2.62 | 2.48 | 2.31 | 2.08 | 1.93 | 1.85 | 1.86 |

[−1, 2] by [1.6, 2.8]

**(b)** $\log r = (-0.30) \log w + 2.36$

**(c)**

[−1, 2] by [1.6, 2.8]

**(e)** One possible answer: Consider the power
   function
$$y = a \cdot x^b \text{ then:}$$
$$\log y = \log (a \cdot x^b)$$
$$= \log a + \log x^b$$
$$= \log a + b \log x$$
$$= b(\log x) + \log a$$
   Which is clearly a linear function of the form
   $f(t) = mt + c$ where $m = b$, $c = \log a$,
   $f(t) = \log y$ and $t = \log x$. As a result, there is
   a linear relationship between $\log y$ and $\log x$.

**58.** 4, 6, 8, 9, 12, 16, 18, 24, 27, 32, 36, 48, 54, 72

**61. (a)** Domain of $f$ and $g$: $(3, \infty)$
  **(b)** Domain of $f$ and $g$: $(5, \infty)$
  **(c)** Domain of $f$: $(-\infty, -3) \cup (-3, \infty)$
     Domain of $g$: $(-3, \infty)$
     Answers will vary.

**65.**

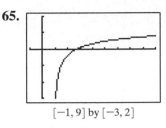

$[-1, 9]$ by $[-3, 2]$

Domain: $(1, \infty)$
Range: $(-\infty, \infty)$
Continuous
Intercepts: $(e, 0) \approx (2.72, 0)$
Not Symmetric
$\lim\limits_{x \to \infty} f(x) \to \infty$; $\lim\limits_{x \to 1^+} f(x) \to -\infty$
Invertible $(f^{-1}(x) = e^{xe})$
Concave down

## Section 3.5 (pp. 302–315)

### Exploration 1

1.60206, 2.60206, 3.60206, 4.60206, 5.60206,
6.60206, 7.60206, 8.60206, 9.60206, 10.60206
The integers increase by 1. The decimal parts are
exactly equal. One order of magnitude greater.

### Quick Review 3.5

1. $f(g(x)) = e^{2 \ln(x^{1/2})} = e^{\ln x} = x$ and
   $g(f(x)) = \ln(e^{2x})^{1/2} = \ln(e^x) = x$.
2. $f(g(x)) = 10^{(\log x^2)/2} = 10^{\log x} = x$ and
   $g(f(x)) = \log(10^{x/2})^2 = \log(10^x) = x$.
3. $f(g(x)) = \frac{1}{3} \ln(e^{3x}) = \frac{1}{3}(3x) = x$ and
   $g(f(x)) = e^{3(\frac{1}{3} \ln x)} = e^{\ln x} = x$.
4. $f(g(x)) = 3 \log(10^{x/6})^2 = 6 \log(10^{x/6}) = 6(x/6) =$
   $x$ and $g(f(x)) = 10^{(3 \log x^2)/6} = 10^{(6 \log x)/6} = 10^{\log x}$
   $= x$.
5. $7.783 \times 10^8$ km
6. $1 \times 10^{-15}$ m
7. 602,000,000,000,000,000,000,000
8. 0.000 000 000 000 000 000 000 000 001 66
   (26 zeros between the decimal point and the 1)
9. $5.766 \times 10^{12}$

### Exercises 3.5

1. $\approx 24.2151$
2. $\approx -23.2644$
5. $\approx 39.6084$
6. $\approx 24.4136$
9. $\approx 5.2877$
11. $\approx -0.4055$
12. $\approx -0.5108$

15. $\approx 4.3956$
19. 1.5
21. Domain: $(-\infty, -1) \cup (0, \infty)$; graph (e).
22. Domain: $(0, \infty)$; graph (f).
23. Domain: $(-\infty, -1) \cup (0, \infty)$; graph (d).
24. Domain: $(0, \infty)$; graph (c).
25. Domain: $(0, \infty)$; graph (a).
26. Domain: $(-\infty, 0) \cup (0, \infty)$; graph (b).

For #27–34, algebraic solutions are shown (and are
generally the only way to get *exact* answers). In many
cases solving graphically would be faster; graphical
support is also useful.

27. $x \approx 3.5949$
28. $x \approx \pm 2.5431$
29. $x \approx \pm 2.0634$
30. $x \approx -0.6931$.
31. $x \approx -9.3780$
32. $x \approx 6.7384$
33. $x \approx 2.3028$
34. $x \approx 103.8516$
36. $x \approx 3.1098$
37. 20 times greater
38. 4 times greater
39. (a) $1.26 \times 10^{-4}$; $1.26 \times 10^{-12}$
    (b) $10^8$
40. (a) $1 \times 10^{-2}$; $3.98 \times 10^{-8}$
    (b) $2.51 \times 10^5$

The equations in #41–42 can be solved either
algebraically or graphically; the latter approach is
generally faster.

42. 29.59 minutes.
43. (a)

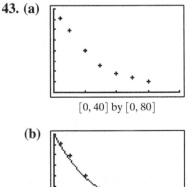

$[0, 40]$ by $[0, 80]$

(b)

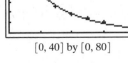

$[0, 40]$ by $[0, 80]$

$T(x) \approx 79.47 \cdot 0.93^x$

**44. (a)**

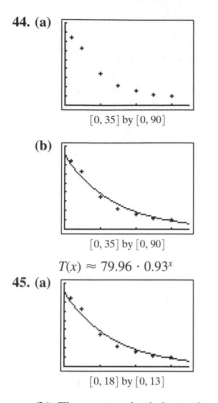

[0, 35] by [0, 90]

**(b)**

[0, 35] by [0, 90]

$T(x) \approx 79.96 \cdot 0.93^x$

**45. (a)**

[0, 18] by [0, 13]

**(b)** The scatter plot is better because it accurately represents the times between the measurements. The equal spacing on the bar graph suggests that the measurements were taken at equally spaced intervals, which distorts our perceptions of how the consumption has changed over time.

**46.** Answers will vary.

**47.** Logarithmic seems best — the scatterplot of $(x, y)$ looks most logarithmic. (The data can be modeled by $y = 3 + 2 \ln x$.)

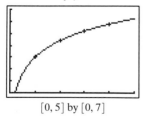

[0, 5] by [0, 7]

**48.** Exponential — the scatterplot of $(x, y)$ is *exactly* exponential. (The data can be modeled by $y = 2 \cdot 3^x$.)

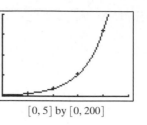

[0, 5] by [0, 200]

**49.** Exponential — the scatterplot of $(x, y)$ is *exactly* exponential. (The data can be modeled by $y = \dfrac{3}{2} \cdot 2^x$.)

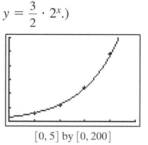

[0, 5] by [0, 200]

**50.** Linear — the scatterplot of $(x, y)$ is *exactly* linear $(y = 2x + 3.)$

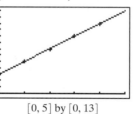

[0, 5] by [0, 13]

**53. (a)** As $k$ increases, the height of the bell curve increases and the slope of the curve seems to steepen.

**(b)** As $c$ increases, the slope of the bell curve seems to steepen, increasing more rapidly to (0, 1) and decreasing more rapidly from (0, 1).

**54.** $\log u - \log v = n(1) = n$
$u$ and $v$ vary by an order of magnitude $n$.

**55. (a)** $r$ cannot be negative since it is a distance.
**(b)** $[0, 10] \times [-5, 3]; 2.3807$

**56.** 0.0750

**57.** $y = a \ln x + b$, a logarithmic regression.

**58.** $cx^a$, where $c = e^b$, exactly the power regression

**59.** $x \approx 1.3066$

**60.** $x \approx 0.4073$ or $x \approx 0.9333$

**61.** $0 < x < 1.7115$ (approx.)

**62.** $x \leq -20.0855$ (approx.)

**63.** $x > 9$

**64.** $-1 < x < 5$

## Section 3.6 (pp. 315–326)

### Exploration 1

**1.**

| $k$ | $A$ |
|-----|--------|
| 10  | 1104.6 |
| 20  | 1104.9 |
| 30  | 1105   |
| 40  | 1105   |
| 50  | 1105.1 |
| 60  | 1105.1 |
| 70  | 1105.1 |
| 80  | 1105.1 |
| 90  | 1105.1 |
| 100 | 1105.1 |

$A$ approaches 1105.1 and doesn't exceed this limit.

**2.** 1105.171

### Exercises 3.6

**19.** 6.63 years — round to 6 years 9 months
**20.** 7.73 years — round to 7 years 9 months
**21.** 13.71 years — round to 13 years 9 months
**22.** 11.57 years — round to 11 years 9 months
**25.** 7.07%
**27.** 12.14 — round to 12 years 3 months.
**28.** 17.62 — round to 17 years 8 months
**29.** 7.7016 years; $48,217.82
**30.** 8.6643 years; $106,243.74
**31.** 17.33%; $127,816.26
**32.** 11.55%; $95,035.15
**33.** 17.42 — round to 17 years 6 months
**34.** 8.751 — round to 9 years
**35.** 10.24 — round to 11 years
**36.** 9.99— round to 10 years
**37.** 9.93 — round to 10 years
**38.** 9.90 years
**46.** $80,367.73
**48.** $158.03
**51.** $676.57
**53.** **(a)** 172 months (14 years, 4 months)
**55.** One possible answer: The APY is the percentage increase from the initial balance $S(0)$ to the end-of-year balance $S(1)$; specifically, it is $S(1)/S(0) - 1$. Multiplying the initial balance by $P$ results in the end-of-year balance being multiplied by the same amount, so that the ratio remains unchanged. Whether we start with a $1 investment, or a $1000 investment,

$$APY = \left(1 + \frac{r}{k}\right)^k - 1.$$

**56.** One possible answer: The APR will be lower than the APY (except under annual compounding), so the bank's offer looks more attractive when the APR is given. Assuming monthly compounding, the APY is about 4.594%; quarterly and daily compounding give approximately 4.577% and 4.602%, respectively.

**57.** One possible answer: Some of these situations involve counting things (e.g., populations), so that they can only take on whole number values — exponential models which predict, e.g., 439.72 fish, have to be interpreted in light of this fact.

Technically, bacterial growth, radioactive decay, and compounding interest also are "counting problems" — for example, we cannot have fractional bacteria, or fractional molecules of radioactive material, or fractions of pennies. However, because these are generally very large numbers, it is easier to ignore the fractional parts. (This might also apply when one is talking about, e.g., the population of the whole world.)

Another distinction: while we often use an exponential model for all these situations, it generally fits better (over long periods of time) for radioactive decay than for most of the others. Rates of growth in populations (esp. human populations) tend to fluctuate more than exponential models suggest. Of course, an exponential model also fits well in compound interest situations where the interest rate is held constant, but there are many cases where interest rates change over time.

**58.** **(a)** Steve's balance will always remain $1000, since interest is not added to it. Every year he receives 6% of that $1000 in interest: 6% in the first year, then another 6% in the second year (for a total of $2 \cdot 6\% = 12\%$), then another 6% (totaling $3 \cdot 6\% = 18\%$), etc. After $t$ years, he has earned $6t\%$ of the $1000 investment, meaning that altogether he has $1000 + 1000 \cdot 0.06t = 1000(1 + 0.06t)$.

**(b)** The table is shown below; the second column gives values of $1000(1.06)^t$. The effects of annual compounding show up beginning in year 2.

| Years | Not Compounded | Compounded |
|---|---|---|
| 0 | 1000.00 | 1000.00 |
| 1 | 1060.00 | 1060.00 |
| 2 | 1120.00 | 1123.60 |
| 3 | 1180.00 | 1191.02 |
| 4 | 1240.00 | 1262.48 |
| 5 | 1300.00 | 1338.23 |
| 6 | 1360.00 | 1418.52 |
| 7 | 1420.00 | 1503.63 |
| 8 | 1480.00 | 1593.85 |
| 9 | 1540.00 | 1689.48 |
| 10 | 1600.00 | 1790.85 |

**60.** One possible answer:

(c) Most accurately represents the loan balance, as the loan balance is reduced linearly, dependent only on the rate of interest charged. A mortgage loan at twice the interest rate and spread over 30 years would have the same shape because its balance would also be linearly decreasing; the slope, however, would probably be different.

## Chapter 3 Review (pp. 331–334)

**3.** $3 \cdot 2^{x/2}$

**4.** $2^{-x+1}$

**5.** $f(x) = 2^{-2x} + 3$ — starting from $2^x$, horizontally shrink by $\frac{1}{2}$, reflect across $y$-axis, and translate up 3 units.

**6.** $f(x) = 2^{-2x}$ — starting from $2^x$, horizontally shrink by $\frac{1}{2}$, reflect across the $y$-axis, reflect across $x$-axis.

**7.** $f(x) = -2^{-3x} - 3$ — starting from $2^x$, horizontally shrink by $\frac{1}{3}$, reflect across the $y$-axis, reflect across $x$-axis, translate down 3 units.

**8.** $f(x) = 2^{-3x} + 3$ — starting from $2^x$, horizontally shrink by $\frac{1}{3}$, reflect across the $y$-axis, translate up 3 units.

**9.** Starting from $e^x$, horizontally shrink by $\frac{1}{2}$ then translate right $\frac{3}{2}$ units — or translate right 3 units, then horizontally shrink by $\frac{1}{2}$.

**10.** Starting from $e^x$, horizontally shrink by $\frac{1}{3}$, then translate right $\frac{4}{3}$ units — or translate right 4 units, then horizontally shrink by $\frac{1}{3}$.

**11.** $y$-intercept: $(0, 12.5)$ Asymptotes: $y = 0$ and $y = 20$

**12.** $y$-intercept: $\left(0, \frac{50}{3}\right) \approx (0, 16.67)$
Asymptotes: $y = 0$, $y = 10$ and $x \approx -22.91$

**13.** exponential decay; 2; $\infty$

**14.** exponential growth; $\infty$; 1

**15.** Domain: $(-\infty, \infty)$
Range: $(1, \infty)$
Continuous
Intercepts: $(0, \approx 21.09)$
Always decreasing
Not symmetric
Bounded below by $y = 1$, which is also the only asymptote
No local extrema
$\lim\limits_{x \to \infty} f(x) \to 1$; $\lim\limits_{x \to -\infty} f(x) \to \infty$
Concave up

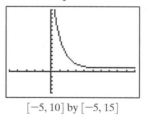

$[-5, 10]$ by $[-5, 15]$

**16.** Domain: $(-\infty, \infty)$
Range: $(-2, \infty)$
Continuous
Intercepts: $(0, 10)$ and approx $(-1.29, 0)$
Always increasing
Not symmetric
Bounded below by $y = -2$, which is also the only asymptote
No local extrema
$\lim\limits_{x \to \infty} f(x) \to \infty$; $\lim\limits_{x \to -\infty} f(x) \to -2$
Concave up

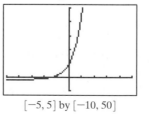

$[-5, 5]$ by $[-10, 50]$

**17.** Domain: $(-\infty, \infty)$
Range: $(0, 25)$
Continuous
Intercepts: $\left(0, \dfrac{100}{6}\right) \approx (0, 16.67)$
Always increasing
Not symmetric
Bounded above by $y = 25$, and below by $y = 0$,
   the two asymptotes
No local extrema
$\lim\limits_{x \to \infty} f(x) \to 25$; $\lim\limits_{x \to -\infty} f(x) \to 0$
Concave up on $(-\infty, 0)$; Concave down $(0, \infty)$

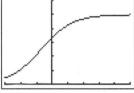

$[-300, 500]$ by $[0, 30]$

**18.** Domain: $(-\infty, \approx 91.63) \cup (\approx 91.63, \infty)$
Range: $(-\infty, 0) \cup (12, \infty)$
Discontinuous at $x \approx 91.63$
Intercepts: $(0, -8)$
Decreasing on $(-\infty, \approx 91.63)$;
Increasing on $(\approx 91.63, \infty)$
Not symmetric
Not bounded
No local extrema
$\lim\limits_{x \to \infty} f(x) \to 12$; $\lim\limits_{x \to -\infty} f(x) \to 0$
Asymptotes: $y = 12$ and $y = 0$
Concave down on $(-\infty, \approx 91.63)$;
Concave up on $(\approx 91.63, \infty)$

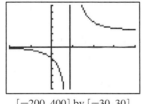

$[-200, 400]$ by $[-30, 30]$

**19.** $f(x) = 24 \cdot (1.053)^x$
**21.** $f(x) = 18 \cdot (2)^{x/21}$
**22.** $f(x) = 117 \cdot 2^{-x/262}$

**25.** $f(x) = \dfrac{20}{1 + 3e^{-0.37x}}$

**26.** $f(x) = \dfrac{44}{1 + 3e^{-0.22x}}$

**35.** Translate left 4 units.
**36.** Reflect across $y$-axis and translate right 4 units —
   or translate left 4 units, then reflect across $y$-axis.
**37.** Translate right 1 unit, reflect across $x$-axis, and

translate up 2 units.
**38.** Translate left 1 unit, reflect across $x$-axis, and
   translate up 4 units.
**39.** Domain: $(0, \infty)$
Range: $\left(-\dfrac{1}{e}, \infty\right) \approx (-0.37, \infty)$
Continuous
Intercepts: $(1, 0)$
Decreasing on $(0, \approx 0.37)$; increasing on
   $(\approx 0.37, \infty)$
Not symmetric
Bounded below by $y = \dfrac{1}{e}$
Local minimum at $\left(\dfrac{1}{e}, -\dfrac{1}{e}\right)$
$\lim\limits_{x \to \infty} f(x) \to \infty$; $\lim\limits_{x \to 0^+} f(x) \to 0$
Concave up

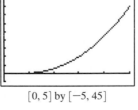

$[0, 10]$ by $[0, 25]$

**40.** Domain: $(0, \infty)$
Range: $(\approx -0.18, \infty)$
Continuous
Intercepts: $(1, 0)$
Decreasing on $(0, \approx 0.61)$; increasing on
   $(\approx 0.61, \infty)$
Not symmetric
Bounded below by $y \approx -0.18$
Local minimum at $(\approx 0.61, -0.18)$
No asymptotes
$\lim\limits_{x \to \infty} f(x) \to \infty$; $\lim\limits_{x \to 0^+} f(x) \to 0$
Concave up

$[0, 5]$ by $[-5, 45]$

**41.** Domain: $(-\infty, 0) \cup (0, \infty)$
Range: $(\approx -0.18, \infty)$
Discontinuous at $x = 0$
Intercepts: $(1, 0)$ and $(-1, 0)$
Decreasing on $(-\infty, \approx -0.61) \cup (0, \approx 0.61)$;
Increasing on $(\approx -0.61, 0) \cup (\approx 0.61, \infty)$
Symmetric across $y$-axis
Bounded below by $y \approx -0.18$
Local minimums at $(\approx -0.61, \approx -0.18)$
    and $(\approx 0.61, -0.18)$
No asymptotes
$\lim\limits_{x \to \infty} f(x) \to \infty$; $\lim\limits_{x \to -\infty} f(x) \to \infty$
Concave up

$[-5, 5]$ by $[-5, 25]$

**42.** Domain: $(0, \infty)$
Range: $\left(-\infty, \dfrac{1}{e}\right) \approx (-\infty, 0.37)$
Continuous
Intercepts: $(1, 0)$
Increasing on $(0, e) \approx (0, 2.72)$,
Decreasing $(e, \infty) \approx (2.72, \infty)$
Not symmetric

Bounded above by $y = \dfrac{1}{e} \approx 0.37$

Local maximum at $\left(e, \dfrac{1}{e}\right) \approx (2.72, 0.37)$

Asymptote: $y = 0$ and $x = 0$
$\lim\limits_{x \to \infty} f(x) \to 0$; $\lim\limits_{x \to 0^+} f(x) \to -\infty$
Concave down

$[0, 15]$ by $[-4, 1]$

**44.** $x \approx -1.3863$          **45.** $x \approx 22.5171$
**47.** $x = 0.0000001$          **48.** $x \approx 4.4650$
**49.** $x = 4$                              **50.** $x \approx 46.7654$
**51.** $x \approx 2.1049$.          **52.** $x \approx -0.3031$.
**53.** $x \approx 99.5112$          **54.** $x \approx -0.4915$
**56.** $-1 - 1.12 \ln(x)$          **58.** $-2 - 9.97 \log x$

**66.** $FV = R \cdot \dfrac{\left(1 + \dfrac{r}{k}\right)^{kt} - 1}{\left(\dfrac{r}{k}\right)}$

**69.** $-0.3054$                    **70.** $-0.4055$
**71.** $P(x) = 2.16 \cdot 1.01^x$, $P(105) \approx 6.92$ million
**72.** $P(x) = \dfrac{14.03}{(1 + 1.95^{-0.026x})}$, $P(110) \approx 12.2$ million

**73. (c)**

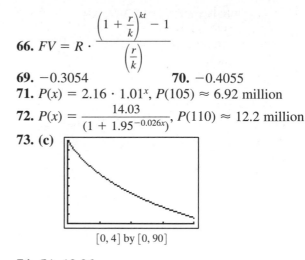

$[0, 4]$ by $[0, 90]$

**74. (b)** 12.86 years
**75. (b)** 31.74 years
**76. (b)** 80.6824 — 80 or 81 students.
    **(c)** 3.3069 — sometime on the fourth day.
**77. (a)** $P(t) = 20 \cdot 2^t$
    **(b)** 81,920; $2.3058 \times 10^{19}$
**78. (a)** $P(t) = 2^{t + 2}$
**79. (c)** 1,099,500 metric tons.
**80. (c)** 16,777.216 kg
**82. (a)** $2.51 \times 10^{-8}$; $3.16 \times 10^{-11}$
    **(b)** 794.33
**85.** 137.7940 — about 11 years 6 months
**90.** $e^{-1} < b < e$; $0 < b < e^{-1}$ or $b > e$.
**91.** $\dfrac{1}{10} < b < 10$; $0 < b < \dfrac{1}{10}$ or $b > 10$
**92.** $g(x) = \ln[a \cdot b^x] = \ln a + \ln b^x = \ln a + x \ln b$
    This has a slope $\ln b$ and $y$-intercept $\ln a$.
**93. (c)** 8.7413 — about 8 or 9 days.
**99. (a)** Grace's balance will always remain $1000, since interest is not added to it. Every year she receives 5% of that $1000 in interest; after $t$ years, she has been paid $5t\%$ of the $1000 investment, meaning that altogether she has $1000 + 1000 \cdot 0.05t$ $= 1000(1 + 0.05t)$.

**(b)** The table is shown below; the second column gives values of $1000e^{0.05t}$. The effects of compounding continuously show up immediately.

| Years | Not Compounded | Compounded |
|---|---|---|
| 0 | 1000.00 | 1000.00 |
| 1 | 1050.00 | 1051.27 |
| 2 | 1100.00 | 1105.17 |
| 3 | 1150.00 | 1161.83 |
| 4 | 1200.00 | 1221.40 |
| 5 | 1250.00 | 1284.03 |
| 6 | 1300.00 | 1349.86 |
| 7 | 1350.00 | 1419.07 |
| 8 | 1400.00 | 1491.82 |
| 9 | 1450.00 | 1568.31 |
| 10 | 1500.00 | 1648.72 |

### Chapter 3 Project

Answers are based on the sample data shown in the table.

**2.** 79%, 77%, 76%, 78%, 79%

**3.**

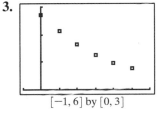

$[-1, 6]$ by $[0, 3]$

**4.** Each successive height will be predicted by multiplying the previous height by the same percentage of rebound. The rebound height can therefore be predicted by the equation $Y = HP^x$ where $x$ is the bounce number.

**5.** $y \approx 2.7188 \cdot 0.78^x$

**6.** $y \approx 2.733 \cdot 0.776^x$

**7.** A different ball would change the rebound percentage, $P$.

**8.** $H$ would be changed by varying the height and $P$ would be changed by using a different type of ball.

**9.** $Y = He^{\ln(P)x}$ so $y = 2.7188e^{-0.248x}$

**10.** Linear

**11.** The linear regression is $Y = -0.253X + 1.005$. Since $\ln Y = (\ln P)x + \ln H$, the slope is $\ln P$ and the $y$-intercept is $\ln H$.

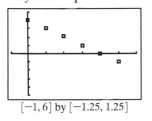

$[-1, 6]$ by $[-1.25, 1.25]$

# Chapter 4

## Section 4.1 (pp. 337–345)

### Exploration 1

**1.** $2\pi r$

**2.** $2\pi$ radians ($2\pi$ lengths of thread)

**3.** No, not quite, since the distance $\pi r$ would require a piece of thread $\pi$ times as long, and $\pi > 3$.

### Exercises 4.1

**17.** $\theta = \dfrac{9}{11}$ rad and $s_2 = 36$ cm

**18.** $\theta = 4.5$ rad and $r_2 = 16$ km

**39.** $\approx 4.23$ statute miles

**40.** $\approx 52.36$ mph

**43.**

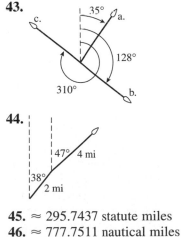

**44.**

**45.** $\approx 295.7437$ statute miles

**46.** $\approx 777.7511$ nautical miles

**47. (a)** $\pi \approx 3.142$

 **(b)** $\approx 15.708$ m

**48. (a)** $16\pi \approx 50.265$ in.
  **(b)** $2\pi \approx 6.283$ ft

**49.** $\dfrac{13}{45}\pi \approx 0.908$ ft

**52.** $38°03'$
**53.** $43°12'$
**54.** $5°38'$
**55.** $42°09'$
**60.** The whole circle's area is $\pi r^2$; the sector with central angle $\theta$ makes up $\dfrac{\theta}{2\pi}$ of that area, or

$$\dfrac{\theta}{2\pi} \cdot \pi r^2 = \dfrac{1}{2}\theta r^2.$$

**61. (a)** $= 3.481\pi \approx 10.936$ ft$^2$
  **(b)** $4.736$ km$^2$

**62.**

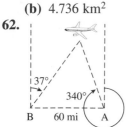

**63.** wheel: $\approx 56.5714$ rad/sec
  wheel sprocket: $\approx 56.5714$ rad/sec
  pedal sprocket: $\approx 18.8571$ rad/sec

## Section 4.2 (pp. 345–354)

### Exploration 1

**1.** sin and csc, cos and sec, and tan and cot.
**2.** $\tan \theta$ **3.** $\sec \theta$
**5.** $\sin \theta$ and $\cos \theta$

### Exploration 2

**1.** Let $\theta = 60°$. Then

$\sin \theta = \dfrac{\sqrt{3}}{2} \approx 0.866$ $\qquad \csc \theta = \dfrac{2}{\sqrt{3}} \approx 1.155$

$\cos \theta = \dfrac{1}{2}$ $\qquad \sec \theta = 2$

$\tan \theta = \sqrt{3} \approx 1.732$ $\qquad \cot \theta = \dfrac{1}{\sqrt{3}} \approx 0.577$

**2.** The values are the same, but for different functions. For example, sin 30° is the same as cos 60°, cot 30° is the same as tan 60°, etc.
**3.** The value of a trig function at $\theta$ is the same as the value of its co-function at $90° - \theta$.

### Quick Review 4.2

**2.** $4\sqrt{13}$ **6.** $\approx 0.17803$ mi

---

**7.** 7.9152 km **8.** $\approx 13.895$ ft
**9.** $\approx 1.0101$ (no units) **10.** $\approx 4.18995$ (no units)

### Exercises 4.2

**1.** $\sin \theta = \dfrac{4}{5}$, $\cos \theta = \dfrac{3}{5}$, $\tan \theta = \dfrac{4}{3}$, $\csc \theta = \dfrac{5}{4}$,
  $\sec \theta = \dfrac{5}{3}$, $\cot \theta = \dfrac{3}{4}$.

**2.** $\sin \theta = \dfrac{8}{\sqrt{113}}$, $\cos \theta = \dfrac{7}{\sqrt{113}}$, $\tan \theta = \dfrac{8}{7}$;
  $\csc \theta = \dfrac{\sqrt{113}}{8}$, $\sec \theta = \dfrac{\sqrt{113}}{7}$, $\cot \theta = \dfrac{7}{8}$.

**3.** $\sin \theta = \dfrac{12}{13}$, $\cos \theta = \dfrac{5}{13}$, $\tan \theta = \dfrac{12}{5}$; $\csc \theta = \dfrac{13}{12}$,
  $\sec \theta = \dfrac{13}{5}$, $\cot \theta = \dfrac{5}{12}$.

**4.** $\sin \theta = \dfrac{8}{17}$, $\cos \theta = \dfrac{15}{17}$, $\tan \theta = \dfrac{8}{15}$; $\csc \theta = \dfrac{17}{8}$,
  $\sec \theta = \dfrac{17}{15}$, $\cot \theta = \dfrac{15}{8}$.

**5.** $\sin \theta = \dfrac{7}{\sqrt{170}}$, $\cos \theta = \dfrac{11}{\sqrt{170}}$, $\tan \theta = \dfrac{7}{11}$;
  $\csc \theta = \dfrac{\sqrt{170}}{7}$, $\sec \theta = \dfrac{\sqrt{170}}{11}$, $\cot \theta = \dfrac{11}{7}$.

**6.** $\sin \theta = \dfrac{3}{4}$, $\cos \theta = \dfrac{\sqrt{7}}{2}$, $\tan \theta = \dfrac{3}{\sqrt{7}}$; $\csc \theta = \dfrac{4}{3}$,
  $\sec \theta = \dfrac{2}{\sqrt{7}}$, $\cot \theta = \dfrac{\sqrt{7}}{3}$.

**7.** $\sin \theta = \dfrac{\sqrt{57}}{11}$, $\cos \theta = \dfrac{8}{11}$, $\tan \theta = \dfrac{\sqrt{57}}{8}$;
  $\csc \theta = \dfrac{11}{\sqrt{57}}$, $\sec \theta = \dfrac{11}{8}$, $\cot \theta = \dfrac{8}{\sqrt{57}}$.

**8.** $\sin \theta = \dfrac{9}{13}$, $\cos \theta = \dfrac{2\sqrt{22}}{13}$, $\tan \theta = \dfrac{9}{2\sqrt{22}}$;
  $\csc \theta = \dfrac{13}{9}$, $\sec \theta = \dfrac{13}{2\sqrt{22}}$, $\cot \theta = \dfrac{2\sqrt{22}}{9}$.

**9.** $\sin \theta = \dfrac{3}{7}$, $\cos \theta = \dfrac{2\sqrt{10}}{7}$, $\tan \theta = \dfrac{3}{2\sqrt{10}}$;
  $\csc \theta = \dfrac{7}{3}$, $\sec \theta = \dfrac{7}{2\sqrt{10}}$, $\cot \theta = \dfrac{2\sqrt{10}}{3}$.

**10.** $\sin \theta = \dfrac{2}{3}$, $\cos \theta = \dfrac{\sqrt{5}}{3}$, $\tan \theta = \dfrac{2}{\sqrt{5}}$; $\csc \theta = \dfrac{3}{2}$,
  $\sec \theta = \dfrac{3}{\sqrt{5}}$, $\cot \theta = \dfrac{\sqrt{5}}{2}$.

**11.** $\sin \theta = \dfrac{4\sqrt{6}}{11}$, $\cos \theta = \dfrac{5}{11}$, $\tan \theta = \dfrac{4\sqrt{6}}{5}$;
  $\csc \theta = \dfrac{11}{4\sqrt{6}}$, $\sec \theta = \dfrac{11}{5}$, $\cot \theta = \dfrac{5}{4\sqrt{6}}$.

**12.** $\sin \theta = \dfrac{\sqrt{39}}{8}$, $\cos \theta = \dfrac{5}{8}$, $\tan \theta = \dfrac{\sqrt{39}}{5}$;

$\csc \theta = \dfrac{8}{\sqrt{39}}$, $\sec \theta = \dfrac{8}{5}$, $\cot \theta = \dfrac{5}{\sqrt{39}}$.

**13.** $\sin \theta = \dfrac{5}{\sqrt{106}}$, $\cos \theta = \dfrac{9}{\sqrt{106}}$, $\tan \theta = \dfrac{5}{9}$;

$\csc \theta = \dfrac{\sqrt{106}}{5}$, $\sec \theta = \dfrac{\sqrt{106}}{9}$, $\cot \theta = \dfrac{9}{5}$.

**14.** $\sin \theta = \dfrac{12}{\sqrt{313}}$, $\cos \theta = \dfrac{13}{\sqrt{313}}$, $\tan \theta = \dfrac{12}{13}$;

$\csc \theta = \dfrac{\sqrt{313}}{12}$, $\sec \theta = \dfrac{\sqrt{313}}{13}$, $\cot \theta = \dfrac{13}{12}$.

**15.** $\sin \theta = \dfrac{3}{\sqrt{130}}$, $\cos \theta = \dfrac{11}{\sqrt{130}}$, $\tan \theta = \dfrac{3}{11}$;

$\csc \theta = \dfrac{\sqrt{130}}{3}$, $\sec \theta = \dfrac{\sqrt{130}}{11}$, $\cot \theta = \dfrac{11}{3}$.

**16.** $\sin \theta = \dfrac{5}{12}$, $\cos \theta = \dfrac{\sqrt{119}}{12}$, $\tan \theta = \dfrac{5}{\sqrt{119}}$;

$\csc \theta = \dfrac{12}{5}$, $\sec \theta = \dfrac{12}{\sqrt{119}}$, $\cot \theta = \dfrac{\sqrt{119}}{5}$.

**17.** $\sin \theta = \dfrac{9}{23}$, $\cos \theta = \dfrac{8\sqrt{7}}{23}$, $\tan \theta = \dfrac{9}{8\sqrt{7}}$;

$\csc \theta = \dfrac{23}{9}$, $\sec \theta = \dfrac{23}{8\sqrt{7}}$, $\cot \theta = \dfrac{8\sqrt{7}}{9}$.

**18.** $\sin \theta = \dfrac{2\sqrt{66}}{17}$, $\cos \theta = \dfrac{5}{17}$, $\tan \theta = \dfrac{2\sqrt{66}}{5}$;

$\csc \theta = \dfrac{17}{2\sqrt{66}}$, $\sec \theta = \dfrac{17}{5}$, $\cot \theta = \dfrac{5}{2\sqrt{66}}$.

**45.** $\dfrac{15}{\sin 34°} \approx 26.82$ **46.** $\dfrac{23}{\cos 39°} \approx 29.60$

**47.** $\dfrac{32}{\tan 57°} \approx 20.78$ **48.** $14 \sin 43° \approx 9.55$

**49.** $\dfrac{6}{\sin 35°} \approx 10.46$ **50.** $50 \cos 66° \approx 20.34$

**51.** $b \approx 33.79$, $c \approx 35.96$, $\beta = 70°$
**52.** $a \approx 6.56$, $b \approx 7.55$, $\beta = 49°$
**53.** $b \approx 22.25$, $c \approx 27.16$, $\alpha = 35°$
**54.** $b \approx 8.32$, $c \approx 9.71$, $\alpha = 31°$
**55.** As $\theta$ gets smaller and smaller, the side opposite $\theta$ gets smaller and smaller, so its ratio to the hypotenuse approaches 0 as a limit.
**56.** As $\theta$ gets smaller and smaller, the side adjacent to $\theta$ approaches the hypotenuse in length, so its ratio to the hypotenuse approaches 1 as a limit.
**57.** $\approx 205.26$ ft
**62.** Strip B width $= 3.244$ ft, Strip C width $= 2.168$ ft
**63.** The distance $d_A$ from $A$ to the mirror is $5 \cos 30°$; the distance from $B$ to the mirror is $d_B = d_A - 2$. Then

$PB = \dfrac{d_B}{\cos \beta} = \dfrac{d_A - 2}{\cos 30°} = 5 - \dfrac{2}{\cos 30°}$

$= 5 - \dfrac{4}{\sqrt{3}} \approx 2.69$ m

**64.** Aim 18 in to the right of $C$ (or 12 in to the left of $D$).
**65.** One possible proof:

$(\sin \theta)^2 + (\cos \theta)^2 = \left(\dfrac{a}{c}\right)^2 + \left(\dfrac{b}{c}\right)^2$

$= \dfrac{a^2}{c^2} + \dfrac{b^2}{c^2}$

$= \dfrac{a^2 + b^2}{c^2}$

$= \dfrac{c^2}{c^2}$

$= 1$  (Pythagorean theorem: $a^2 + b^2 = c^2$.)

**66.** Let $h$ be the length of the altitude to base $b$ and denote the area of the triangle by $A$. Then

$$\dfrac{h}{a} = \sin \theta$$
$$\therefore h = a \sin \theta$$

Since $A = \dfrac{1}{2}bh$, we can substitute $h = a \sin \theta$ to

get $A = \dfrac{1}{2}ab \sin \theta$.

## Section 4.3 (pp. 355–367)

### Exploration 1

**1.** The side opposite $\theta$ in the triangle has length $y$ and the hypotenuse has length $r$. Therefore

$\sin \theta = \dfrac{\text{opp}}{\text{hyp}} = \dfrac{y}{r}$.

**4.** $\cot \theta = \dfrac{x}{y}$; $\sec \theta = \dfrac{r}{x}$; $\csc \theta = \dfrac{r}{y}$.

### Exploration 2

**1.** The $x$-coordinates on the unit circle lie between $-1$ and 1, and $\cos t$ is always an $x$-coordinate on the unit circle.
**2.** The $y$-coordinates on the unit circle lie between $-1$ and 1, and $\sin t$ is always a $y$-coordinate on the unit circle.
**3.** The points corresponding to $t$ and $-t$ on the number line are wrapped to points above and below the $x$-axis with the same $x$-coordinates. Therefore $\cos t$ and $\cos (-t)$ are equal.
**4.** The points corresponding to $t$ and $-t$ on the number line are wrapped to points above and below the $x$-axis with exactly opposite $y$-coordinates. Therefore $\sin t$ and $\sin (-t)$ are opposites.

**5.** Since $2\pi$ is the distance around the unit circle, both $t$ and $t + 2\pi$ get wrapped to the same point.

**6.** The points corresponding to $t$ and $t + \pi$ get wrapped to points on either end of a diameter on the unit circle. These points are symmetric with respect to the origin and therefore have coordinates $(x, y)$ and $(-x, -y)$. Therefore $\sin t$ and $\sin (t + \pi)$ are opposites, as are $\cos t$ and $\cos (t + \pi)$.

**7.** By the observation in (6), $\tan t$ and $\tan(t + \pi)$ are ratios of the form $\dfrac{y}{x}$ and $\dfrac{-y}{-x}$, which are either equal to each other or both undefined.

**8.** The sum is always of the form $x^2 + y^2$ for some $(x, y)$ on the unit circle. Since the equation of the unit circle is $x^2 + y^2 = 1$, the sum is always 1.

**9.** Answers will vary. For example, there are similar statements that can be made about the functions cot, sec, and csc.

## Quick Review 4.3

**9.** $\sin \theta = \dfrac{5}{13}$, $\cos \theta = \dfrac{12}{13}$, $\tan \theta = \dfrac{5}{12}$; $\csc \theta = \dfrac{13}{5}$, $\sec \theta = \dfrac{13}{12}$, $\cot \theta = \dfrac{12}{5}$.

**10.** $\sin \theta = \dfrac{8}{17}$, $\cos \theta = \dfrac{15}{17}$, $\tan \theta = \dfrac{8}{15}$; $\csc \theta = \dfrac{17}{8}$, $\sec \theta = \dfrac{17}{15}$, $\cot \theta = \dfrac{15}{8}$.

## Exercises 4.3

**1.** $\sin \theta = \dfrac{2}{\sqrt{5}}$, $\cos \theta = -\dfrac{1}{\sqrt{5}}$, $\tan \theta = -2$; $\csc \theta = \dfrac{\sqrt{5}}{2}$, $\sec \theta = -\sqrt{5}$, $\cot \theta = -\dfrac{1}{2}$.

**2.** $\sin \theta = -\dfrac{3}{5}$, $\cos \theta = \dfrac{4}{5}$, $\tan \theta = -\dfrac{3}{4}$; $\csc \theta = -\dfrac{5}{3}$, $\sec \theta = \dfrac{5}{4}$, $\cot \theta = -\dfrac{4}{3}$.

**3.** $\sin \theta = -\dfrac{1}{\sqrt{2}}$, $\cos \theta = -\dfrac{1}{\sqrt{2}}$, $\tan \theta = 1$; $\csc \theta = -\sqrt{2}$, $\sec \theta = -\sqrt{2}$, $\cot \theta = 1$.

**4.** $\sin \theta = -\dfrac{5}{\sqrt{34}}$, $\cos \theta = \dfrac{3}{\sqrt{34}}$, $\tan \theta = -\dfrac{5}{3}$; $\csc \theta = -\dfrac{\sqrt{34}}{5}$, $\sec \theta = \dfrac{\sqrt{34}}{3}$, $\cot \theta = -\dfrac{3}{5}$.

**5.** $\sin \theta = \dfrac{4}{5}$, $\cos \theta = \dfrac{3}{5}$, $\tan \theta = \dfrac{4}{3}$; $\csc \theta = \dfrac{5}{4}$, $\sec \theta = \dfrac{5}{3}$, $\cot \theta = \dfrac{3}{4}$.

**6.** $\sin \theta = 0$, $\cos \theta = -1$, $\tan \theta = 0$; $\csc \theta$ undefined, $\sec \theta = -1$, $\cot \theta$ undefined.

**7.** $\sin \theta = 1$, $\cos \theta = 0$, $\tan \theta$ undefined; $\csc \theta = 1$, $\sec \theta$ undefined, $\cot \theta = 0$.

**8.** $\sin \theta = -\dfrac{3}{\sqrt{13}}$, $\cos \theta = -\dfrac{2}{\sqrt{13}}$, $\tan \theta = \dfrac{3}{2}$; $\csc \theta = -\dfrac{\sqrt{13}}{3}$, $\sec \theta = -\dfrac{\sqrt{13}}{2}$, $\cot \theta = \dfrac{2}{3}$.

**9.** $\sin \theta = -\dfrac{2}{\sqrt{29}}$, $\cos \theta = \dfrac{5}{\sqrt{29}}$, $\tan \theta = -\dfrac{2}{5}$; $\csc \theta = -\dfrac{\sqrt{29}}{2}$, $\sec \theta = \dfrac{\sqrt{29}}{5}$, $\cot \theta = -\dfrac{5}{2}$.

**10.** $\sin \theta = -\dfrac{1}{\sqrt{2}}$, $\cos \theta = \dfrac{1}{\sqrt{2}}$, $\tan \theta = -1$; $\csc \theta = -\sqrt{2}$, $\sec \theta = \sqrt{2}$, $\cot \theta = -1$.

**25. (a)** $-1$. **(b)** 0. **(c)** undefined.
**26. (a)** 1. **(b)** 0. **(c)** undefined.
**27. (a)** 0. **(b)** $-1$. **(c)** 0.
**28. (a)** $-1$. **(b)** 0. **(c)** undefined.
**29. (a)** 1. **(b)** 0. **(c)** undefined.
**30. (a)** 0. **(b)** 1. **(c)** 0.

**43.** $\sin \theta = \dfrac{\sqrt{5}}{3}$; $\tan \theta = \dfrac{\sqrt{5}}{2}$.

**44.** $\cos \theta = -\dfrac{\sqrt{15}}{4}$; $\cot \theta = -\sqrt{15}$.

**45.** $\tan \theta = -\dfrac{2}{\sqrt{21}}$; $\sec \theta = \dfrac{5}{\sqrt{21}}$.

**46.** $\sin \theta = -\dfrac{7}{\sqrt{58}}$; $\cos \theta = -\dfrac{3}{\sqrt{58}}$.

**47.** $\sec \theta = -\dfrac{5}{4}$; $\csc \theta = \dfrac{5}{3}$.

**48.** $\csc \theta = \dfrac{5}{4}$; $\cot \theta = -\dfrac{3}{4}$.

**52.** $\tan \left( \dfrac{3\pi - 70000\pi}{2} \right) = \tan \left( \dfrac{3\pi}{2} \right) =$ undefined.

**53.** The calculator's value of the irrational number $\pi$ is necessarily an approximation. When multiplied by a very large number, the slight error of the original approximation is magnified sufficiently to throw the trigonometric functions off.

**54.** $\sin t$ is the $y$-coordinate of the point on the unit circle after measuring counterclockwise $t$ units from $(1, 0)$. This will repeat every $2\pi$ units (and not before), since the distance around the circle is $2\pi$.

**55.** $\mu = \dfrac{\sin 83°}{\sin 36°} \approx 1.69$

**56.** $\approx 26.12°$.

**58.** When $t = 0$, $\theta = .25$ rad. When $t = 2.5$, $y \approx -0.2003$.

**59.** The difference in the elevations is 600 ft, so $d = 600/\sin \theta$. Then:

(a)  ≈ 848.53 ft
(b)  = 600 ft
(c)  ≈ 933.43 ft

**60.** January = 103.25.

April ($t = 4$): $72.4 + 61.7 \sin \dfrac{2\pi}{3} \approx 125.83$.

June ($t = 6$): $72.4 + 61.7 \sin \pi = 72.4$.

October ($t = 10$): $72.4 + 61.7 \sin \dfrac{5\pi}{3} \approx 18.97$.

December ($t = 12$): $72.4 + 61.7 \sin 2\pi = 72.4$.
June and December are the same; perhaps by June most people have suits for the summer, and by December they are beginning to purchase them for next summer (or as Christmas presents, or for mid-winter vacations).

**65.** The two triangles are congruent: both have hypotenuse 1, and the corresponding angles are congruent—the smaller acute angle has measure $t$ in both triangles, and the two acute angles in a right triangle add up to $\pi/2$.

**66.** These coordinates give the lengths of the legs of the triangles from #65, and these triangles are congruent. For example, the length of the horizontal leg of the triangle with vertex $P$ is given by the (absolute value of the) $x$ coordinate of $P$; this must be the same as the (absolute value of the) $y$ coordinate of $Q$.

**67.** One possible answer: Starting from the point $(a, b)$ on the unit circle—at an angle of $t$, so that $\cos t = a$—then measuring a quarter of the way around the circle (which corresponds to adding $\pi/2$ to the angle), we end at $(-b, a)$, so that $\sin(t + \pi/2) = a$. For $(a, b)$ in quadrant I, this is shown in the figure; similar illustrations can be drawn for the other quadrants.

**68.** One possible answer: Starting from the point $(a, b)$ on the unit circle—at an angle of $t$, so that $\sin t = b$—then measuring a quarter of the way around the circle (which corresponds to adding $\pi/2$ to the angle), we end at $(-b, a)$, so that $\cos(t + \pi/2) = -b = -\sin t$. For $(a, b)$ in quadrant I, this is shown in the figure; similar illustrations can be drawn for the other quadrants.

**69.** Starting from the point $(a, b)$ on the unit circle— at an angle of $t$, so that $\cos t = a$—then measuring a quarter of the way around the circle (which corresponds to adding $\pi/2$ to the angle), we end at $(-b, a)$, so that $\sin(t + \pi/2) = a$. This holds true when $(a, b)$ is in quadrant II, just as it did for quadrant I.

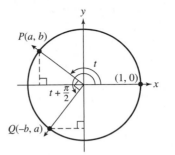

**70.** **(a)** Both triangles are right triangles with hypotenuse 1, and the angles at the origin are both $t$ (for the triangle on the left, the angle is the supplement of $\pi - t$). Therefore the vertical legs are also congruent; their lengths correspond to the sines of $t$ and $\pi - t$.

   **(b)** The points $P$ and $Q$ are reflections of each other across the $y$-axis, so they are the same distance (but opposite directions) from the $y$-axis. Alternatively, use the congruent triangles argument from part (a).

**71.** $|\theta| < 0.2441$ (approximately).

**72.** Let $(x, y)$ be the coordinates of the point that corresponds to $t$ under the wrapping. Then

$$1 + (\tan t)^2 = 1 + \left(\dfrac{y}{x}\right)^2 = \dfrac{x^2 + y^2}{x^2} = (\sec t)^2.$$

(Note that $x^2 + y^2 = 1$ because $(x, y)$ is on the unit circle.)

**73.** This Taylor polynomial is generally a very good approximation for $\sin \theta$—in fact, the relative error (see #71) is less than 1% for $|\theta| < 1$ (approx.). It is better for $\theta$ close to 0; it is slightly larger than $\sin \theta$ when $\theta < 0$ and slightly smaller when $\theta > 0$.

**74.** This Taylor polynomial is generally a very good approximation for $\cos \theta$ —in fact, the relative error (see #71) is less than 1% for $|\theta| < 1.2$ (approx.). It is better for $\theta$ close to 0; it is slightly larger than $\cos \theta$ when $\theta \neq 0$.

## Section 4.4 (pp. 368–379)

### Exploration 1

1. $\pi/2$  (at the point $(0, 1)$)
2. $3\pi/2$  (at the point $(0, -1)$)
3. Both graphs cross the $x$-axis when the $y$-coordinate on the unit circle is 0.
4. (Calculator exploration)
5. The sine function tracks the $y$-coordinate of the point as it moves around the unit circle. After the point has gone completely around the unit circle

(a distance of $2\pi$), the same pattern of $y$-coordinates starts over again.

6. Leave all the settings as they are shown at the start of the Exploration, except change Y2(T) to cos(T).

### Quick Review 4.4

7. vertically stretch by 3
8. reflect across $y$-axis
9. vertically shrink by 0.5
10. translate down 2 units

### Exercises 4.4

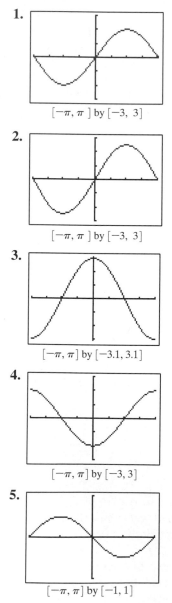

1.

$[-\pi, \pi]$ by $[-3, 3]$

2.

$[-\pi, \pi]$ by $[-3, 3]$

3.

$[-\pi, \pi]$ by $[-3.1, 3.1]$

4.

$[-\pi, \pi]$ by $[-3, 3]$

5.

$[-\pi, \pi]$ by $[-1, 1]$

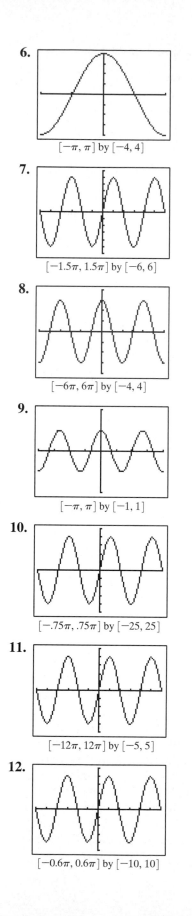

6.

$[-\pi, \pi]$ by $[-4, 4]$

7.

$[-1.5\pi, 1.5\pi]$ by $[-6, 6]$

8.

$[-6\pi, 6\pi]$ by $[-4, 4]$

9.

$[-\pi, \pi]$ by $[-1, 1]$

10.

$[-.75\pi, .75\pi]$ by $[-25, 25]$

11.

$[-12\pi, 12\pi]$ by $[-5, 5]$

12.

$[-0.6\pi, 0.6\pi]$ by $[-10, 10]$

13. Period $\pi$; amplitude 1.5; $[-2\pi, 2\pi] \times [-2, 2]$
14. Period $2\pi/3$; amplitude 2; $\left[-\dfrac{2\pi}{3}, \dfrac{2\pi}{3}\right] \times [-4, 4]$
15. Period $\pi$; amplitude 3; $[-2\pi, 2\pi] \times [-4, 4]$
16. Period $4\pi$; amplitude 5; $[-4\pi, 4\pi] \times [-10, 10]$
17. Period 6; amplitude 4; $[-3, 3] \times [-5, 5]$
18. Period 2; amplitude 3; $[-4, 4] \times [-5, 5]$
19. Maximum: $2 \left(\text{at } -\dfrac{3\pi}{2} \text{ and } \dfrac{\pi}{2}\right)$;

    minimum: $-2 \left(\text{at } -\dfrac{\pi}{2} \text{ and } \dfrac{3\pi}{2}\right)$;

    Zeros: $0, \pm\pi, \pm 2\pi$.
20. Maximum: 3 (at 0); minimum: $-3$ (at $\pm 2\pi$);
    Zeroes: $\pm\pi$.
21. Maximum: 1 (at $0, \pm\pi, \pm 2\pi$); minimum:

    $-1 \left(\text{at } \pm\dfrac{\pi}{2} \text{ and } \pm\dfrac{3\pi}{2}\right)$; Zeroes: $\pm\dfrac{\pi}{4}, \pm\dfrac{3\pi}{4}, \pm\dfrac{5\pi}{4},$

    $\pm\dfrac{7\pi}{4}$
22. Maximum: $\dfrac{1}{2} \left(\text{at } -\dfrac{3\pi}{2} \text{ and } \dfrac{\pi}{2}\right)$; minimum:

    $-\dfrac{1}{2} \left(\text{at } -\dfrac{\pi}{2} \text{ and } \dfrac{3\pi}{2}\right)$; Zeroes: $0, \pm\pi, \pm 2\pi$
23. Maximum: $1 \left(\text{at } \pm\dfrac{\pi}{2} \text{ and } \pm\dfrac{3\pi}{2}\right)$; minimum: $-1$

    (at $0, \pm\pi, \pm 2\pi$);

    Zeroes: $\pm\dfrac{\pi}{4}, \pm\dfrac{3\pi}{4}, \pm\dfrac{5\pi}{4}, \pm\dfrac{7\pi}{4}$
24. Maximum: $2 \left(\text{at } -\dfrac{\pi}{2} \text{ and } \dfrac{3\pi}{2}\right)$;

    minimum: $-2 \left(\text{at } -\dfrac{3\pi}{2} \text{ and } \dfrac{\pi}{2}\right)$;

    Zeroes: $0, \pm\pi, \pm 2\pi$

35. Starting from $y = \sin x$, horizontally shrink by $\dfrac{1}{3}$ and vertically shrink by 0.5.
36. Starting from $y = \cos x$, horizontally shrink by $\dfrac{1}{4}$ and vertically stretch by 1.5.
37. Starting from $y = \cos x$, horizontally stretch by 3, vertically shrink by $\dfrac{2}{3}$, reflect across $x$-axis.
38. Starting from $y = \sin x$, horizontally stretch by 5 and vertically shrink by $\dfrac{3}{4}$.
39. Starting from $y = \cos x$, horizontally shrink by $\dfrac{3}{2\pi}$ and vertically stretch by 3.
40. Starting from $y = \sin x$, horizontally stretch by $\dfrac{4}{\pi}$, vertically stretch by 2, and reflect across $x$-axis.

41. Starting with $y_1$, vertically stretch by $\dfrac{5}{3}$.
42. Starting with $y_1$, translate right $\dfrac{\pi}{12}$ units and vertically shrink by $\dfrac{1}{2}$.
43. Starting with $y_1$, horizontally shrink by $\dfrac{1}{2}$.
44. Starting with $y_1$, horizontally stretch by 2 and vertically shrink by $\dfrac{2}{3}$.
49. Amplitude 2, period $2\pi$, phase shift $\dfrac{\pi}{4}$, vertical translation 1 unit up.
50. Amplitude 3.5, period $\pi$, phase shift $\dfrac{\pi}{4}$, vertical translation 1 unit down.
51. Amplitude 5, period $\dfrac{2\pi}{3}$, phase shift $\dfrac{\pi}{18}$, vertical translation $\dfrac{1}{2}$ units up.
52. Amplitude 3, period $2\pi$, phase shift $-3$, vertical translation 2 units down.
53. Amplitude 2, period 1, phase shift 0, vertical translation 1 unit up.
54. Amplitude 4, period $\dfrac{2}{3}$, phase shift 0, vertical translation 2 units down.
55. Amplitude $\dfrac{7}{3}$, period $2\pi$, phase shift $-\dfrac{5}{2}$, vertical translation 1 unit down.
56. Amplitude $\dfrac{2}{3}$, period $8\pi$, phase shift 3, vertical translation 1 unit up.
57. $y = 2 \sin 2x$ $(a = 2, b = 2, h = 0, k = 0)$.
58. $y = 3 \sin[2(x + 0.5)]$ $(a = 3, b = 2, h = 0.5, k = 0)$.
59. (a) two
    (b) $(0, 1)$ and $(2\pi, 1.3^{-2\pi}) \approx (6.28, 0.19)$
62. 972,000 ft, or about 184 miles
63. (c)

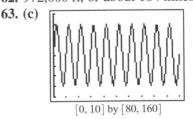

$[0, 10]$ by $[80, 160]$

**65. (a)**

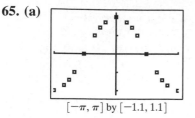

$[-\pi, \pi]$ by $[-1.1, 1.1]$

**(b)** $0.0246x^4 + 0x^3 - 0.4410x^2 + 0x + 0.9703$
**(c)** the coefficients are fairly similar.

**66. (a)**

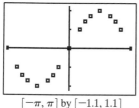

$[-\pi, \pi]$ by $[-1.1, 1.1]$

**(b)** $-0.0872x^3 + 0x^2 + 0.8263x + 0$

**(c)** the coefficients are somewhat similar.

**67. (b)** $f = 262\dfrac{1}{\sec}$ ("cycles per sec"), or 262 Hertz

**(c)**

$[0, 0.025]$ by $[-2, 2]$

**68.** Since the cursor moves at a constant rate, its distance from the center must be made up of linear pieces as shown (the slope of the line is the rate of motion). A graph of a sinusoid is included for comparison.
**69. (a)** $a - b$ must equal 1.
  **(b)** $a - b$ must equal 2.
  **(c)** $a - b$ must equal $k$.
**70. (a)** $a - b$ must equal 1.
  **(b)** $a - b$ must equal 2.
  **(c)** $a - b$ must equal $k$.
**71.** $B = (0, 3)$, $C = \left(\dfrac{3\pi}{4}, 0\right)$

**72.** $B = \left(\dfrac{3\pi}{4}, 4.5\right)$, $C = \left(\dfrac{9\pi}{4}, 0\right)$

**73.** $B = \left(\dfrac{\pi}{4}, 2\right)$, $C = \left(\dfrac{3\pi}{4}, 0\right)$

**74.** $A = \left(\dfrac{\pi}{2}, 0\right)$, $B = \left(\dfrac{3\pi}{4}, 3\right)$, $C = \left(\dfrac{3\pi}{2}, 0\right)$

## Section 4.5 (pp. 379–387)

### Exploration 1

**2.** Set the expressions equal and solve for *x:*
$$-k \cos x = \sec x$$
$$-k \cos x = 1/\cos x$$
$$-k(\cos x)^2 = 1$$
$$(\cos x)^2 = -1/k$$

Since $k > 0$, this requires that the square of $\cos x$ be negative, which is impossible. This proves that there is no value of *x* for which the two functions are equal, so the graphs do not intersect.

### Quick Review 4.5

**5.** Zero: 3. Asymptote: $x = -4$
**6.** Zero: $-5$. Asymptote: $x = 1$
**7.** Zero: $-1$. Asymptotes: $x = 2$ and $x = -2$
**8.** Zero: $-2$. Asymptotes: $x = 0$ and $x = 3$

### Exercises 4.5

**1.** The graph of $y = 2 \csc x$ must be vertically stretched by 2 compared to $y = \csc x$, so $y_1 = 2 \csc x$ and $y_2 = \csc x$.
**2.** The graph of $y = 5 \tan x$ must be vertically stretched by 10 compared to $y = 0.5 \tan x$, so $y_1 = 5 \tan x$ and $y_2 = 0.5 \tan x$.
**3.** $y_1 = 3 \csc 2x$, $y_2 = \csc x$
**4.** $y_1 = \cot(x - 0.5) + 3$, $y_2 = \cot x$
**5.**

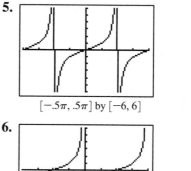

$[-.5\pi, .5\pi]$ by $[-6, 6]$

**6.**

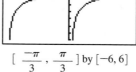

$\left[\dfrac{-\pi}{3}, \dfrac{\pi}{3}\right]$ by $[-6, 6]$

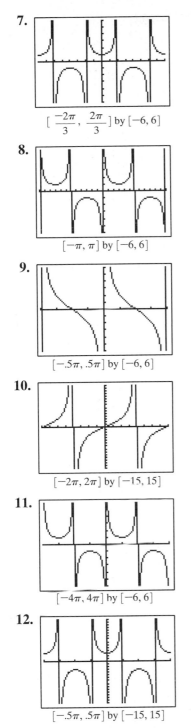

**7.**
$$\left[ \frac{-2\pi}{3}, \frac{2\pi}{3} \right] \text{ by } [-6, 6]$$

**8.**
$$[-\pi, \pi] \text{ by } [-6, 6]$$

**9.**
$$[-.5\pi, .5\pi] \text{ by } [-6, 6]$$

**10.**
$$[-2\pi, 2\pi] \text{ by } [-15, 15]$$

**11.**
$$[-4\pi, 4\pi] \text{ by } [-6, 6]$$

**12.**
$$[-.5\pi, .5\pi] \text{ by } [-15, 15]$$

**13.** Graph (a); Xmin $= -\pi$ and Xmax $= \pi$
**14.** Graph (d); Xmin $= -\pi$ and Xmax $= \pi$
**15.** Graph (c); Xmin $= -\pi$ and Xmax $= \pi$
**16.** Graph (b); Xmin $= -\pi$ and Xmax $= \pi$
**17.** Starting with $y = \tan x$, vertically stretch by 3.
**18.** Starting with $y = \tan x$, reflect across $x$-axis.
**19.** Starting with $y = \csc x$, vertically stretch by 3.

**20.** Starting with $y = \tan x$, vertically stretch by 2.
**21.** Starting with $y = \cot x$, horizontally stretch by 2, vertically stretch by 3, and reflect across $x$-axis.
**22.** Starting with $y = \sec x$, horizontally stretch by 2, vertically stretch by 2, and reflect across $x$-axis.
**23.** Starting with $y = \tan x$, horizontally shrink by $\frac{2}{\pi}$ and reflect across $x$-axis and shift up by 2 units.
**24.** Starting with $y = \tan x$, horizontally shrink by $\frac{1}{\pi}$ and vertically stretch by 2 and shift down by 2 units.
**41.** (a) $d = 350 \sec x$
**42.** (c) $9°$
**44.** $\approx 0.666$ or $\approx 2.475$
**45.** $\approx \pm 1.107$ or $\approx \pm 2.034$
**46.** $\approx 1.082$ or $\approx 2.060$
**47.** about $(-0.44, 0) \cup (0.44, \pi)$

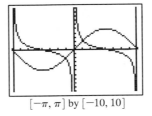
$$[-\pi, \pi] \text{ by } [-10, 10]$$

**48.** about $(-\pi, -2.24) \cup \left( -\frac{\pi}{2}, 0 \right) \cup \left( \frac{\pi}{2}, 2.24 \right)$

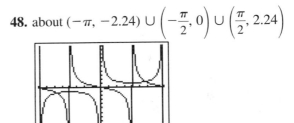
$$[-\pi, \pi] \text{ by } [-10, 10]$$

**49.** $\cot x$ is not defined at 0; the definition of "increasing on $(a, b)$" requires that the function be defined everywhere in $(a, b)$. Also, choosing $a = -\pi/4$ and $b = \pi/4$, we have $a < b$ but $f(a) = 1 > f(b) = -1$.

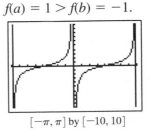
$$[-\pi, \pi] \text{ by } [-10, 10]$$

**50.** They look similar on this window, but they are noticeably different at the edges (near 0 and $\pi$). Also, if $f$ were equal to $g$, then it would follow that $\dfrac{1}{f} = -\cos x = \dfrac{1}{g} = x - \dfrac{\pi}{2}$ on this interval, which we know to be false.

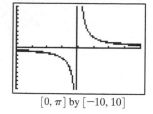

$[0, \pi]$ by $[-10, 10]$

**51.** $\csc x = \sec\!\left(x - \dfrac{\pi}{2}\right)$

**53.** $d = 30 \sec x = \dfrac{30}{\cos x}$

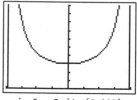

$[-.5\pi, .5\pi]$ by $[0, 100]$

**56. (a)** $\dfrac{1}{y} = \dfrac{1}{a\,\sec(bx)} = \dfrac{1}{a} \cdot \dfrac{1}{\sec(bx)} = \dfrac{1}{a}\cos(bx) = \dfrac{1}{a}\sin(bx + \pi/2)$

**(b)** $y = 0.2\sin\!\left(\dfrac{1}{6}x + \dfrac{\pi}{2}\right)$

**(c)** $a = 1/0.2 = 5$ and $b = 1/6$

**(d)** $y = 5\sec\!\left(\dfrac{x}{6}\right)$ The scatter plot is shown below,

and the fit is very good—so good that you should realize that we made the data up!

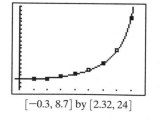

$[-0.3, 8.7]$ by $[2.32, 24]$

## Section 4.6 (pp. 387–394)

### Exploration 1

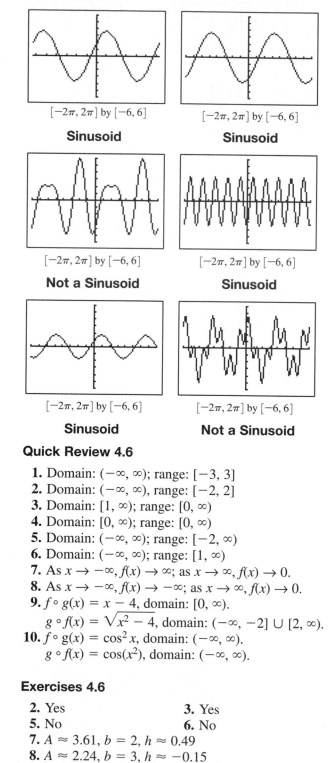

$[-2\pi, 2\pi]$ by $[-6, 6]$
**Sinusoid**

$[-2\pi, 2\pi]$ by $[-6, 6]$
**Sinusoid**

$[-2\pi, 2\pi]$ by $[-6, 6]$
**Not a Sinusoid**

$[-2\pi, 2\pi]$ by $[-6, 6]$
**Sinusoid**

$[-2\pi, 2\pi]$ by $[-6, 6]$
**Sinusoid**

$[-2\pi, 2\pi]$ by $[-6, 6]$
**Not a Sinusoid**

### Quick Review 4.6

**1.** Domain: $(-\infty, \infty)$; range: $[-3, 3]$
**2.** Domain: $(-\infty, \infty)$, range: $[-2, 2]$
**3.** Domain: $[1, \infty)$; range: $[0, \infty)$
**4.** Domain: $[0, \infty)$; range: $[0, \infty)$
**5.** Domain: $(-\infty, \infty)$; range: $[-2, \infty)$
**6.** Domain: $(-\infty, \infty)$; range: $[1, \infty)$
**7.** As $x \to -\infty,\, f(x) \to \infty$; as $x \to \infty,\, f(x) \to 0$.
**8.** As $x \to -\infty,\, f(x) \to -\infty$; as $x \to \infty,\, f(x) \to 0$.
**9.** $f \circ g(x) = x - 4$, domain: $[0, \infty)$.
$g \circ f(x) = \sqrt{x^2 - 4}$, domain: $(-\infty, -2] \cup [2, \infty)$.
**10.** $f \circ g(x) = \cos^2 x$, domain: $(-\infty, \infty)$.
$g \circ f(x) = \cos(x^2)$, domain: $(-\infty, \infty)$.

### Exercises 4.6

**2.** Yes  **3.** Yes
**5.** No  **6.** No
**7.** $A \approx 3.61,\ b = 2,\ h \approx 0.49$
**8.** $A \approx 2.24,\ b = 3,\ h \approx -0.15$
**9.** $A \approx 2.24,\ b = \pi,\ h \approx 0.35$

**10.** $A \approx 3.16$, $b = 2\pi$, $h \approx -0.05$
**11.** $A \approx 2.24$, $b = 1$, $h \approx -1.11$
**12.** $A \approx 3.16$, $b = 2$, $h \approx 0.16$

**13.**

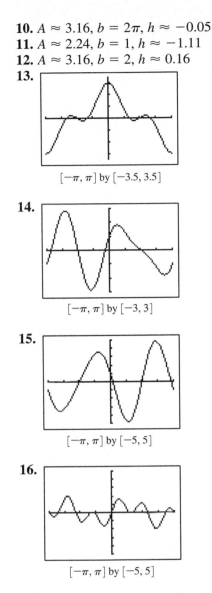

$[-\pi, \pi]$ by $[-3.5, 3.5]$

**14.**

$[-\pi, \pi]$ by $[-3, 3]$

**15.**

$[-\pi, \pi]$ by $[-5, 5]$

**16.**

$[-\pi, \pi]$ by $[-5, 5]$

**21.** Period $\pi$ — the cosine function has the property that $\cos(x + \pi) = -\cos x$, so that $[\cos(x + \pi)]^2 = (-\cos x)^2 = \cos^2 x$.

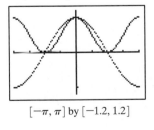

$[-\pi, \pi]$ by $[-1.2, 1.2]$

**22.** Period $2\pi$, the same as the cosine function: $[\cos(x + 2\pi)]^3 = [\cos x]^3 = \cos^3 x$.

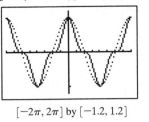

$[-2\pi, 2\pi]$ by $[-1.2, 1.2]$

**23.** Period $\pi$ — the cosine function has the property that $\cos(x + \pi) = -\cos x$, and this function is the same as $|\cos x|$.

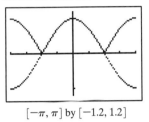

$[-\pi, \pi]$ by $[-1.2, 1.2]$

**24.** Period $\pi$ — the cosine function has the property that $\cos(x + \pi) = -\cos x$, so that $|[\cos(x + \pi)]^3| = |(-\cos x)^3| = |\cos^3 x|$.

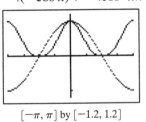

$[-\pi, \pi]$ by $[-1.2, 1.2]$

**25.** Domain: $(-\infty, \infty)$. Range: $[0, 1]$. Possible window: $[-2\pi, 2\pi] \times [-0.25, 1.25]$.

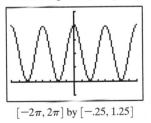

$[-2\pi, 2\pi]$ by $[-.25, 1.25]$

**26.** Domain: $(-\infty, \infty)$. Range: $[0, 1]$.

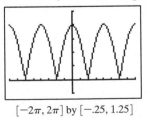

$[-2\pi, 2\pi]$ by $[-.25, 1.25]$

**27.** Domain: all $x \neq n\pi$, $n$ an integer. Range: $[0, \infty)$.

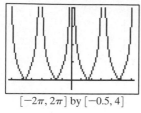

$[-2\pi, 2\pi]$ by $[-0.5, 4]$

**28.** Domain: $(-\infty, \infty)$. Range: $[-1, 1]$.

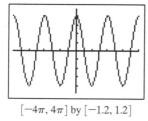

$[-4\pi, 4\pi]$ by $[-1.2, 1.2]$

**29.** Domain: all $x \neq \dfrac{\pi}{2} + n\pi$, $n$ an integer. Range: $(-\infty, 0]$.

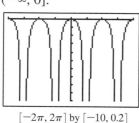

$[-2\pi, 2\pi]$ by $[-10, 0.2]$

**30.** Domain: $(-\infty, \infty)$. Range: $[-1, 0]$.

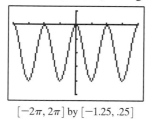

$[-2\pi, 2\pi]$ by $[-1.25, .25]$

**31.** $2x - 1 \leq y \leq 2x + 1$.

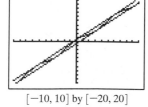

$[-10, 10]$ by $[-20, 20]$

**32.** $-0.5x \leq y \leq 2 - 0.5x$

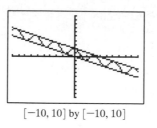

$[-10, 10]$ by $[-10, 10]$

**33.** $1 - 0.3x \leq y \leq 3 - 0.3x$.

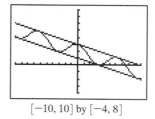

$[-10, 10]$ by $[-4, 8]$

**34.** $x \leq y \leq x + 2$.

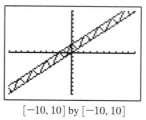

$[-10, 10]$ by $[-10, 10]$

**35.** The graph of $f(x)$ lies between $y = x$ and $y = -x$ (in the vertical direction).

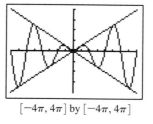

$[-4\pi, 4\pi]$ by $[-4\pi, 4\pi]$

**36.** The graph of $f(x)$ lies (vertically) between $y = -x$ and the $x$-axis.

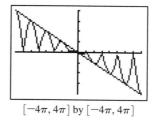

$[-4\pi, 4\pi]$ by $[-4\pi, 4\pi]$

**37.** The graph of $f(x)$ lies above the $x$-axis and below the graphs of $y = -x$ (on the left) and $y = x$ (on the right)—that is, it lies below the graph of $y = |x|$.

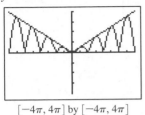

$$[-4\pi, 4\pi] \text{ by } [-4\pi, 4\pi]$$

**38.** The graph of $f(x)$ lies between $y = x$ and $y = -x$ (in the vertical direction).

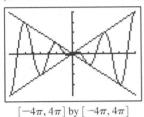

$$[-4\pi, 4\pi] \text{ by } [-4\pi, 4\pi]$$

**39.** $f$ oscillates up and down between $1.2^{-x}$ and $-1.2^{-x}$. As $x \to \infty$, $f(x) \to 0$.

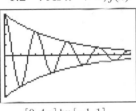

$$[0, 4\pi] \text{ by } [-1, 1]$$

**40.** $f$ oscillates up and down between $2^{-x}$ and $-2^{-x}$. As $x \to \infty$, $f(x) \to 0$.

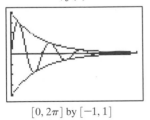

$$[0, 2\pi] \text{ by } [-1, 1]$$

**41.** $f$ oscillates up and down between $\dfrac{1}{x}$ and $-\dfrac{1}{x}$. As $x \to \infty$, $f(x) \to 0$.

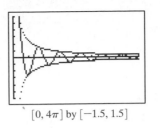

$$[0, 4\pi] \text{ by } [-1.5, 1.5]$$

**42.** Window: $[0, 1.5\pi] \times [-1, 1]$. $f$ oscillates up and down between $e^{-x}$ and $-e^{-x}$. As $x \to \infty$, $f(x) \to 0$.

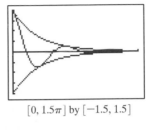

$$[0, 1.5\pi] \text{ by } [-1.5, 1.5]$$

**43.**

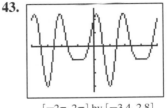

$$[-2\pi, 2\pi] \text{ by } [-3.4, 2.8]$$

**44.**

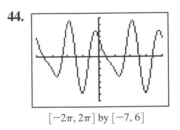

$$[-2\pi, 2\pi] \text{ by } [-7, 6]$$

**45.**

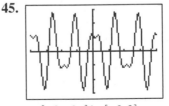

$$[-2\pi, 2\pi] \text{ by } [-3, 3]$$

**46.**

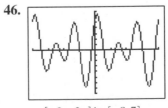

$$[-2\pi, 2\pi] \text{ by } [-8, 7]$$

**47.** Period $2\pi$

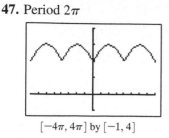

$[-4\pi, 4\pi]$ by $[-1, 4]$

**48.** Not periodic

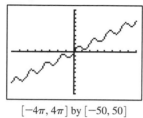

$[-4\pi, 4\pi]$ by $[-50, 50]$

**49.** Not periodic

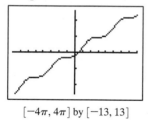

$[-4\pi, 4\pi]$ by $[-13, 13]$

**50.** Not periodic

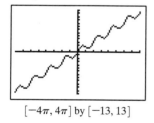

$[-4\pi, 4\pi]$ by $[-13, 13]$

**51.** Not periodic

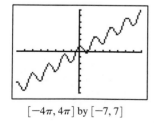

$[-4\pi, 4\pi]$ by $[-7, 7]$

**52.** Not periodic

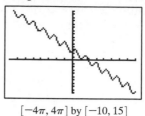

$[-4\pi, 4\pi]$ by $[-10, 15]$

**53.** domain is $(-\infty, \infty)$. Range: $(-\infty, \infty)$.
**54.** domain is $(-\infty, \infty)$. Range: $(-\infty, \infty)$.
**55.** domain is $(-\infty, \infty)$. Range: $[1, \infty)$.
**56.** domain is $(-\infty, \infty)$. Range: $(-\infty, \infty)$.
**57.** domain is $\ldots \cup [-2\pi, -\pi] \cup [0, \pi] \cup [2\pi, 3\pi]$
$\cup \ldots$; that is, all $x$ with $2n\pi \le x \le (2n + 1)\pi$, $n$
an integer. Range: $[0, 1]$.
**58.** domain is $(-\infty, \infty)$. Range: $[-1, 1]$.
**59.** domain is $(-\infty, \infty)$. Range: $[0, 1]$.
**60.** domain is $\ldots \cup \left[-\dfrac{5\pi}{2}, -\dfrac{3\pi}{2}\right] \cup \left[-\dfrac{\pi}{2}, \dfrac{\pi}{2}\right] \cup$

$\left[\dfrac{3\pi}{2}, \dfrac{5\pi}{2}\right] \cup \ldots$;

that is, all $x$ with $\dfrac{(4n - 1)\pi}{2} \le x \le \dfrac{(4n + 1)\pi}{2}$, $n$
an integer. Range: $[0, 1]$.

**61. (a)**

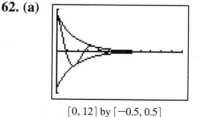

$[0, 15]$ by $[70, 140]$

**(b)** 2 years
**(c)** $\approx 94.90$ million dollars.
**62. (a)**

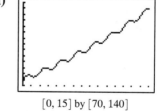

$[0, 12]$ by $[-0.5, 0.5]$

**(b)** For $t > 0.51$ (approximately).
**67.** Graph (d), shown on $[-2\pi, 2\pi] \times [-4, 4]$
**68.** Graph (a), shown on $[-2\pi, 2\pi] \times [-4, 4]$
**69.** Graph (b), shown on $[-2\pi, 2\pi] \times [-4, 4]$
**70.** Graph (c), shown on $[-2\pi, 2\pi] \times [-4, 4]$
**71. (a)** Answers will vary — for example,

on a TI-81: $\dfrac{\pi}{47.5} = 0.0661\ldots \approx 0.07$;

on a TI-82: $\dfrac{\pi}{47} = 0.0668\ldots \approx 0.07$;

on a TI-85: $\dfrac{\pi}{63} = 0.0498\ldots \approx 0.05$;

on a TI-92: $\dfrac{\pi}{119} = 0.0263\ldots \approx 0.03$.

**(b)** Period: $p = \pi/125 = 0.0251\ldots$. For any of the TI graphers, there are from 1 to 3 cycles between each pair of pixels; the graphs produced are therefore inaccurate, since so much detail is lost.

**72.** $186 \sin\!\left[\dfrac{2\pi}{365}(x - 80)\right] + 731$

**73.** Domain: $(-\infty, \infty)$. Range: $[-1, 1]$. Horizontal asymptote: $y = 1$. Zeros at $\ln\!\left(\dfrac{\pi}{2} + n\pi\right)$, $n$ a non-negative integer.

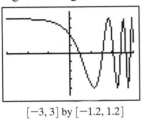

$[-3, 3]$ by $[-1.2, 1.2]$

**74.** Period: $\pi$. Domain: $x \neq n\pi + \dfrac{\pi}{2}$. Range: $(0, \infty)$. Vertical asympotes at missing points of domain: $x = n\pi + \dfrac{\pi}{2}$.

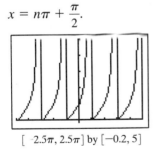

$[-2.5\pi, 2.5\pi]$ by $[-0.2, 5]$

**75.** Domain: $[0, \infty)$. Range: $(-\infty, \infty)$. Zeroes at $n\pi$, $n$ a nonnegative integer.

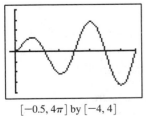

$[-0.5, 4\pi]$ by $[-4, 4]$

**76.** Domain: $[-2, 2]$. Range: $[0, 2.94]$ (approximately). Zeros at $-2$ and $2$.

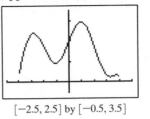

$[-2.5, 2.5]$ by $[-0.5, 3.5]$

**77.** Domain: $(-\infty, 0) \cup (0, \infty)$. Range: approximately $[-0.22, 1)$. Horizontal asymptote: $y = 0$. Zeros at $n\pi$, $n$ a non-zero integer.

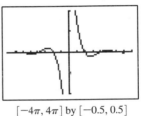

$[-5\pi, 5\pi]$ by $[-0.5, 1.2]$

**78.** Domain: $(-\infty, 0) \cup (0, \infty)$. Range: $(-\infty, \infty)$. Horizontal asymptote: $y = 0$. Vertical asymptote: $x = 0$. Zeros at $n\pi$, $n$ a non-zero integer.

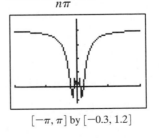

$[-4\pi, 4\pi]$ by $[-0.5, 0.5]$

**79.** Domain: $(-\infty, 0) \cup (0, \infty)$. Range: approximately $[-0.22, 1)$. Horizontal asymptote: $y = 1$. Zeros at $\dfrac{1}{n\pi}$, $n$ a non-zero integer.

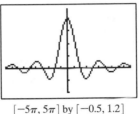

$[-\pi, \pi]$ by $[-0.3, 1.2]$

**80.** Domain: $(-\infty, 0) \cup (0, \infty)$. Range: $(-\infty, \infty)$. Zeros at $\dfrac{1}{n\pi}$, $n$ a non-zero integer. Note: the graph also suggests the end-behavior asymptote $y = x$.

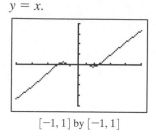

[−1, 1] by [−1, 1]

## Section 4.7 (pp. 394–403)

### Exploration 1

**4.** $\dfrac{x}{\sqrt{1 + x^2}}$

**5.** $\sqrt{1 + x^2}$

**6.** The hypotenuse is positive in either quadrant. The ratios in the six basic trig functions are the same in every quadrant, so the functions are still valid regardless of the sign of $x$. (Also, the sign of the answer in (4) is negative, as it should be, and the sign of the answer in (5) is negative, as it should be.)

### Quick Review 4.7

**1.** sin $x$: positive; cos $x$: positive; tan $x$: positive
**2.** sin $x$: positive; cos $x$: negative; tan $x$: negative
**3.** sin $x$: negative; cos $x$: negative; tan $x$: positive
**4.** sin $x$: negative; cos $x$: positive; tan $x$: negative

### Exercises 4.7

**33.** Domain: $\left[-\dfrac{1}{2}, \dfrac{1}{2}\right]$. Range: $\left[-\dfrac{\pi}{2}, \dfrac{\pi}{2}\right]$. Starting from $y = \sin^{-1} x$, horizontally shrink by $\dfrac{1}{2}$.

**34.** Domain: $\left[-\dfrac{1}{2}, \dfrac{1}{2}\right]$. Range: $[0, 3\pi]$. Starting from $y = \cos^{-1} x$, horizontally shrink by $\dfrac{1}{2}$ and vertically stretch by 3 (either order).

**35.** Domain: $(-\infty, \infty)$. Range: $\left(-\dfrac{5\pi}{2}, \dfrac{5\pi}{2}\right)$. Starting from $y = \tan^{-1} x$, horizontally stretch by 2 and vertically stretch by 5 (either order).

**36.** Domain: $[-2, 2]$. Range: $[0, 3\pi]$. Starting from $y = \arccos x = \cos^{-1} x$, horizontally stretch by 2 and vertically stretch by 3 (either order).

**38.** $1 + 2n\pi$ or $-1 + 2n\pi$, for all integers $n$.

**39.** $\sin \dfrac{1}{2} \approx 0.479$

**40.** $\tan(-1) \approx -1.557$

**41.** $\dfrac{1}{3}$

**42.** $\dfrac{\pi}{10}$

**47.** $\dfrac{1}{\sqrt{1 + 4x^2}}$

**49. (b)**

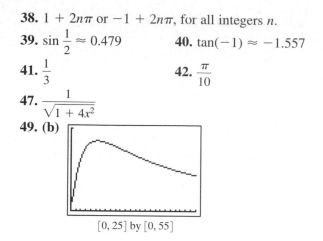

[0, 25] by [0, 55]

**(c)** $\approx 1.83$ or $\approx 15.31$

**50. (b)** Graph is shown (using DEGREE mode). Negative values of $x$ correspond to the point $Q$ being "upshore" from $P$ ("into" the picture) instead of downshore (as shown in the illustration). Positive angles are angles that point downshore; negative angles point upshore.

**51. (a)** $\theta = \tan^{-1} \dfrac{s}{500}$.

**(b)** As $s$ changes from 10 to 20 ft, $\theta$ changes from about $1.1458°$ to $2.2906°$—it almost exactly doubles (a 99.92% increase). As $s$ changes from 200 to 210 ft, $\theta$ changes from about $21.80°$ to $22.78°$—an increase of less than $1°$, and a very small relative change (only about 4.25%).

**(c)** The $x$-axis represents the height and the $y$-axis represents the angle: the angle cannot grow past $90°$ (in fact, it *approaches* but never exactly equals $90°$).

**52. (a)** $[-1, 1]$      **(b)** $[-1, 1]$
**(c)** $(-\infty, \infty)$      **(d)** $[-1, 1]$
**(e)** $(-\infty, \infty)$

**53.** The cotangent function restricted to the interval $(0, \pi)$ is one-to-one and has an inverse. The unique angle $y$ between 0 and $\pi$ (non-inclusive) such that $\cot y = x$ is called the inverse cotangent (or arccotangent) of $x$, denoted $\cot^{-1} x$ or arccot $x$. The domain of $y = \cot^{-1} x$ is $(-\infty, \infty)$ and the range is $(0, \pi)$.

**54.** In the triangle below, $A = \sin^{-1} x$ and $B = \cos^{-1} x$. Since $A$ and $B$ are complementary angles, $A + B = \pi/2$. The left-hand side of the equation is only defined for $-1 \le x \le 1$.

**55. (a)** Domain all reals, range $[-\pi/2, \pi/2]$, period $2\pi$.

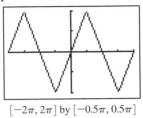

$[-2\pi, 2\pi]$ by $[-0.5\pi, 0.5\pi]$

**(b)** Domain all reals, range $[0, \pi]$, period $2\pi$.

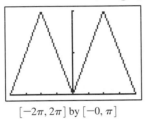

$[-2\pi, 2\pi]$ by $[-0, \pi]$

**(c)** Domain all reals except $\pi/2 + n\pi$, range $(-\pi/2, \pi/2)$, period $\pi$. Discontinuity is not removable.

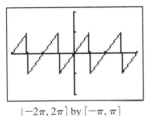

$[-2\pi, 2\pi]$ by $[-\pi, \pi]$

**57.** $y = \dfrac{\pi}{2} - \tan^{-1} x$.

**58. (a)** $\cos(\sin^{-1} x)$ or $\sin(\cos^{-1} x)$

**(b)** $\sin(\tan^{-1} x)$ or $\cos(\cot^{-1} x)$

**(c)** $\tan(\sin^{-1} x)$ or $\cot(\cos^{-1} x)$

**59.** $\dfrac{18}{\pi} \tan^{-1} x + 33$

**60. (a)** As in Example 5, $\theta$ can only be in QI or QIV, so the horizontal side of the triangle can only be positive.

**(b)** $\tan\left(\sin^{-1}\left(\dfrac{1}{x}\right)\right) = \dfrac{x}{s} = \dfrac{x}{\sqrt{x^4 - x^2}}$

**(c)** $\sin\left(\cos^{-1}\left(\dfrac{1}{x}\right)\right) = \dfrac{s}{x^2} = \dfrac{\sqrt{x^4 - x^2}}{x^2}$

(See figure below.)

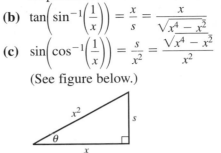

## Section 4.8 (pp. 404–415)

### Exploration 1

**2.** The grapher is actually graphing the unit circle, but the $y$-window is so large that the point never seems to get above or below the $x$-axis. It is flattened vertically.

**3.** Since the grapher is plotting points along the unit circle, it covers the circle at a constant speed. Toward the extremes its motion is mostly vertical, so not much horizontal progress (which is all that we see) occurs. Toward the middle, the motion is mostly horizontal, so it moves faster.

**4.** The directed distance of the point from the origin at any T is exactly cos T, and $d = \cos t$ models simple harmonic motion.

### Quick Review 4.8

**1.** $b = 15 \cot 31° \approx 24.964$,
$c = 15 \csc 31° \approx 29.124$

**2.** $a = 25 \cos 68° \approx 9.365$,
$b = 25 \sin 68° \approx 23.179$

**3.** $b = 28 \cot 28° - 28 \cot 44° \approx 23.665$,
$c = 28 \csc 28° \approx 59.642$,
$a = 28 \csc 44° \approx 40.308$

**4.** $b = 21 \cot 31° - 21 \cot 48° \approx 16.041$,
$c = 21 \csc 31° \approx 40.774$,
$a = 21 \csc 48° \approx 28.258$

**5.** complement: 58°, supplement: 148°

**6.** complement: 17°, supplement: 107°

**8.** 202.5°

**9.** Amplitude: 3; period: $\pi$

**10.** Amplitude: 4; period: $\pi/2$

### Exercises 4.8

**4.** $90 \cot 14° \approx 360.970$ ft

**5.** wire length 5 sec 80° $\approx 28.794$ ft;
tower height $= 5 \tan 80° \approx 28.356$ ft

**6.** $\ell = 16 \sec 62° \approx 34.081$ ft;
$h = 16 \tan 62° \approx 30.092$ ft

**7.** $185 \tan 80°1'12'' \approx 1051.333$ ft

**8.** $1580 \tan 38° \approx 1234.431$ ft.

**9.** $100 \tan 83°12' \approx 838.625$ ft.

**11.** $10 \tan 55° \approx 14.281$

**14.** $\dfrac{1000}{\cot 30° - \cot 35°} \approx 3290.5$ ft

**15.** $200(\tan 40° - \tan 30°) \approx 52.350$ ft

**16.** $100(\cot 15° - \cot 33°) \approx 219.219$ ft

**17.** distance: $60\sqrt{2} \approx 84.853$ naut mi; bearing is $140°$

**18.** $d = 80\sqrt{5} \approx 178.885$ naut mi; bearing is $65° + \tan^{-1} 2 \approx 128.435°$

**19.** $1097 \cot 19° \approx 3185.919$ ft

**21.** $325 \tan 63° \approx 637.848$ ft

**22.** $\tan 17° = \dfrac{h}{12 \text{ mi}}$, so $h = 12 \tan 17° \approx 3.669$ mi

**24.** $760 \cot 5.25° \approx 8271.020$ ft

**25.** $\dfrac{550}{\cot 70° - \cot 80°} \approx 2931.094$ ft

**27.** **(b)** $d = 6 \cos 16\pi t$ inches.
**(c)** about 4.1 in. left of the starting position

**28.** **(b)** $d = 18 \cos \pi t$ cm
**(c)** 30 cycles/min

**32.** **(b)**

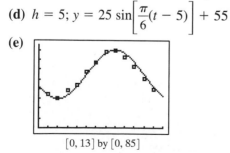

[0, 30] by [−1, 20]

**(c)** $h(4) \approx 6.528$ m; $h(10) = 17$ m.

**33.** **(b)** The amplitude is half this difference (the maximum minus the minimum): $a = 25°$.

**(d)** $h = 5$; $y = 25 \sin\left[\dfrac{\pi}{6}(t - 5)\right] + 55$

**(e)**

[0, 13] by [0, 85]

**(f)** $y = 70$ when $t \approx 6.23$ (about July 7) or $t \approx 9.77$ (about October 24).

**34.** **(d)** All have the correct period, but the others are incorrect in various ways. Equation (a) oscillates between −25 and +25, while

equation (b) oscillates between −17 and +33. Equation (c) is the closest among the incorrect formulas: it has the right maximum and minimum values, but it does not have the property that $h(0) = 8$. This is accomplished by the horizontal shift in (d).

**35.** **(a)**

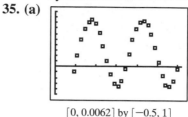

[0, 0.0062] by [−0.5, 1]

**(c)** About $\dfrac{2464}{2\pi} = \dfrac{1232}{\pi} \approx 392$ oscillations/sec

**36.** **(a)** Newborn: about 6 hours. Four-year-old: about 18 hours. Adult: about 24 hours.

**(b)** The adult sleep cycle is perhaps most like a sinusoid, though one might also pick the newborn cycle. At least one can perhaps say that the four-year-old sleep cycles is *least* like a sinusoid.

**37.** $2.5 \cot \dfrac{\pi}{7} \approx 5.191$ cm

**39.** $AC \approx 33.609$ in.; $BD \approx 12.901$ in.

**40.** **(c)** $AE + ED = 20 \sec 50° + 45 \sec 20° \approx 79.002$ ft, so the total distance across the top of the roof is about 158 ft.

**41.** $\tan^{-1} 0.06 \approx 3.434$

**43.** **(a)**

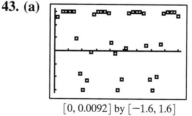

[0, 0.0092] by [−1.6, 1.6]

**(b)** One pretty good match is
$y = 1.51971 \sin[2467(t - 0.0002)]$ (that is, $a = 1.51971$, $b = 2467$, $h = 0.0002$). Answers will vary but should be close to these values.

**(c)** Frequency: about $\dfrac{2467}{2\pi} \approx \dfrac{1}{0.0025472} \approx 392$ or 393 Hz. It appears to be a G.

## Chapter 4 Review (pp. 418–420)

**9.** $270°$ or $\dfrac{3\pi}{2}$ radians

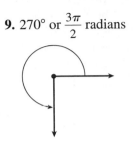

**10.** $900°$ or $5\pi$ radians

**15.** $360° + \tan^{-1} -2 \approx 296.565 \approx 5.176$ radians

**16.** $\tan^{-1} 2 \approx 63.435° \approx 1.107$ radians

**29.** $\sin\left(-\dfrac{\pi}{6}\right) = -\dfrac{1}{2}, \cos\left(-\dfrac{\pi}{6}\right) = \dfrac{\sqrt{3}}{2}$,

$\tan\left(-\dfrac{\pi}{6}\right) = -\dfrac{1}{\sqrt{3}}; \csc\left(-\dfrac{\pi}{6}\right) = -2$,

$\sec\left(-\dfrac{\pi}{6}\right) = \dfrac{2}{\sqrt{3}}; \cot\left(-\dfrac{\pi}{6}\right) = -\sqrt{3}$.

**30.** $\sin\dfrac{19\pi}{4} = \dfrac{1}{\sqrt{2}}, \cos\dfrac{19\pi}{4} = -\dfrac{1}{\sqrt{2}}, \tan\dfrac{19\pi}{4} = -1$;

$\csc\dfrac{19\pi}{4} = \sqrt{2}, \sec\dfrac{19\pi}{4} = -\sqrt{2}; \cot\dfrac{19\pi}{4} = -1$.

**31.** $\sin(-135°) = -\dfrac{1}{\sqrt{2}}, \cos(-135°) = -\dfrac{1}{\sqrt{2}}$,

$\tan(-135°) = 1; \csc(-135°) = -\sqrt{2}$,

$\sec(-135°) = -\sqrt{2}, \cot(-135°) = 1$.

**32.** $\sin 420° = \dfrac{\sqrt{3}}{2}, \cos 420° = \dfrac{1}{2}, \tan 420° = \sqrt{3}$;

$\csc 420° = \dfrac{2}{\sqrt{3}}, \sec 420° = 2, \cot 420° = \dfrac{1}{\sqrt{3}}$.

**33.** $\cos\alpha = \dfrac{12}{13}, \tan\alpha = \dfrac{5}{12}, \csc\alpha = \dfrac{13}{5}, \sec\alpha = \dfrac{13}{12}$,

$\cot\alpha = \dfrac{12}{5}$.

**34.** $\sin\theta = \dfrac{2\sqrt{6}}{7}, \cos\theta = \dfrac{5}{7}, \tan\theta = \dfrac{2\sqrt{6}}{5}$;

$\csc\theta = \dfrac{7}{2\sqrt{6}}, \sec\theta = \dfrac{7}{5}, \cot\theta = \dfrac{5}{2\sqrt{6}}$.

**35.** $\sin\theta = \dfrac{15}{17}, \cos\theta = \dfrac{8}{17}, \tan\theta = \dfrac{15}{8}; \csc\theta = \dfrac{17}{15}$,

$\sec\theta = \dfrac{17}{8}, \cot\theta = \dfrac{8}{15}$.

**38.** $\approx 0.220$ or $\approx 2.922$ radians

**39.** $a = 15\sin 35° \approx 8.604$,

$b = 15\cos 35° \approx 12.287, \beta = 55°$

**40.** $a = 6, \alpha \approx 36.87°, \beta \approx 53.13°$

**41.** $b = 7\tan 48° \approx 7.774, c = \dfrac{7}{\cos 48°} \approx 10.461$,

$\alpha = 42°$

**42.** $a = 8\sin 28° \approx 3.756, b = 8\cos 28° \approx 7.064$,

$\beta = 62°$

**43.** $a = 2\sqrt{6} \approx 4.90, \alpha \approx 44.42°, \beta \approx 45.58°$

**44.** $c = \sqrt{59.54} \approx 7.716, \alpha \approx 18.90°, \beta \approx 71.10°$.

**49.** $\sin\theta = \dfrac{2}{\sqrt{5}}, \cos\theta = -\dfrac{1}{\sqrt{5}}, \tan\theta = -2$;

$\csc\theta = \dfrac{\sqrt{5}}{2}, \sec\theta = -\sqrt{5}, \cot\theta = -\dfrac{1}{2}$.

**50.** $\sin\theta = \dfrac{7}{\sqrt{193}}, \cos\theta = \dfrac{12}{\sqrt{193}}, \tan\theta = \dfrac{7}{12}$;

$\csc\theta = \dfrac{\sqrt{193}}{7}, \sec\theta = \dfrac{\sqrt{193}}{12}, \cot\theta = \dfrac{12}{7}$.

**51.** $\sin\theta = -\dfrac{3}{\sqrt{34}}, \cos\theta = -\dfrac{5}{\sqrt{34}}, \tan\theta = \dfrac{3}{5}$;

$\csc\theta = -\dfrac{\sqrt{34}}{3}, \sec\theta = -\dfrac{\sqrt{34}}{5}, \cot\theta = \dfrac{5}{3}$.

**52.** $\sin\theta = \dfrac{9}{\sqrt{97}}, \cos\theta = \dfrac{4}{\sqrt{97}}, \tan\theta = \dfrac{9}{4}$;

$\csc\theta = \dfrac{\sqrt{97}}{9}, \sec\theta = \dfrac{\sqrt{97}}{4}, \cot\theta = \dfrac{4}{9}$.

**53.** Starting from $y = \sin x$, translate left $\pi$ units.

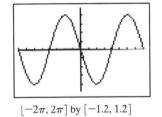

$[-2\pi, 2\pi]$ by $[-1.2, 1.2]$

**54.** Starting from $y = \cos x$, vertically stretch by 2 then translate up 3 units.

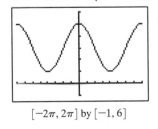

$[-2\pi, 2\pi]$ by $[-1, 6]$

**55.** Starting from $y = \cos x$, translate left $\frac{\pi}{2}$ units, reflect across $x$-axis, and translate up 4 units.

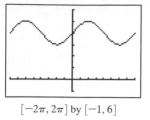

$[-2\pi, 2\pi]$ by $[-1, 6]$

**56.** Starting from $y = \sin x$, translate right $\pi$ units, vertically stretch by 3, reflect across $x$-axis, and translate down 2 units.

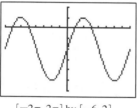

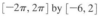

$[-2\pi, 2\pi]$ by $[-6, 2]$

**57.** Starting from $y = \tan x$, horizontally shrink by $\frac{1}{2}$.

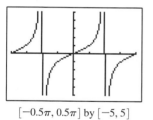

$[-0.5\pi, 0.5\pi]$ by $[-5, 5]$

**58.** Starting from $y = \cot x$, horizontally shrink by $\frac{1}{3}$, vertically stretch by 2, and reflect across $x$-axis (in any order).

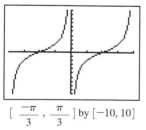

$\left[ \frac{-\pi}{3}, \frac{\pi}{3} \right]$ by $[-10, 10]$

**59.** Starting from $y = \sec x$, horizontally stretch by 2, vertically stretch by 2, and reflect across $x$-axis (in any order).

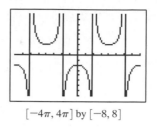

$[-4\pi, 4\pi]$ by $[-8, 8]$

**60.** Starting from $y = \csc \pi x$, horizontally shrink by $\frac{1}{\pi}$.

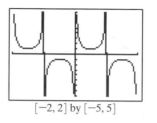

$[-2, 2]$ by $[-5, 5]$

**61.** Amplitude: 2; period: $\frac{2\pi}{3}$; phase shift: 0; domain: $(-\infty, \infty)$; range: $[-2, 2]$.

**62.** Amplitude: 3; period: $\frac{\pi}{2}$; phase shift: 0; domain: $(-\infty, \infty)$; range: $[-3, 3]$.

**63.** Amplitude: 1.5; period: $\pi$; phase shift: $\frac{\pi}{8}$; domain: $(-\infty, \infty)$; range: $[-1.5, 1.5]$.

**64.** Amplitude: 2; period: $\frac{2\pi}{3}$; phase shift: $\frac{\pi}{9}$; domain: $(-\infty, \infty)$; range: $[-2, 2]$.

**65.** Amplitude: 4; period: $\pi$; phase shift: $\frac{1}{2}$; domain: $(-\infty, \infty)$; range: $[-4, 4]$.

**66.** Amplitude: 2; period: $\frac{2\pi}{3}$; phase shift: $-\frac{1}{3}$; domain: $(-\infty, \infty)$; range: $[-2, 2]$.

**68.** $a \approx 3.61$, $b = 2$, and $h \approx -1.08$

**69.** $\approx 49.996° \approx 0.873$ radians

**70.** $\approx 61.380° \approx 1.071$ radians

**72.** $60° = \frac{\pi}{3}$ radians

**73.** Starting from $y = \sin^{-1} x$, horizontally shrink by $\frac{1}{3}$. Domain: $\left[ -\frac{1}{3}, \frac{1}{3} \right]$. Range: $\left[ -\frac{\pi}{2}, \frac{\pi}{2} \right]$.

**74.** Starting from $y = \tan^{-1} x$, horizontally shrink by $\frac{1}{2}$. Domain: $(-\infty, \infty)$. Range: $\left( -\frac{\pi}{2}, \frac{\pi}{2} \right)$.

**75.** Starting from $y = \sin^{-1} x$, translate right 1 unit, horizontally shrink by $\frac{1}{3}$, translate up 2 units.

Domain: $\left[ 0, \frac{2}{3} \right]$. Range: $\left[ 2 - \frac{\pi}{2}, 2 + \frac{\pi}{2} \right]$.

**76.** Starting from $y = \cos^{-1} x$, translate left 1 unit, horizontally shrink by $\frac{1}{2}$, translate down 3 units. Domain: $[-1, 0]$. Range: $[-3, \pi - 3]$.

**83.** As $|x| \to \infty$, $\frac{\sin x}{x^2} \to 0$.

**84.** As $x \to \infty$, $\frac{3}{5}e^{-x/12} \sin(2x - 3) \to 0$; as $x \to -\infty$, the function oscillates from positive to negative, and tends to $\infty$ in absolute value.

**89.** Periodic; period $\pi$. Domain $x \neq \frac{\pi}{2} + n\pi$, $n$ an integer. Range: $[1, \infty)$.

**90.** Not periodic. Domain: $(-\infty, \infty)$. Range: $[-1, 1]$.

**91.** Not periodic. Domain: $x \neq \frac{\pi}{2} + n\pi$, $n$ an integer. Range: $(-\infty, \infty)$.

**92.** Periodic; period $2\pi$. Domain: $(-\infty, \infty)$. Range: approximately $[-5, 4.65]$.

**95.** $100 \tan 78° \approx 470.463$ m

**97.** $= 150 \cot 18° - 150 \cot 42° \approx 295.061$ ft

**99.**

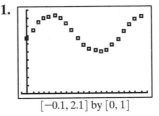

**100.** $\dfrac{855}{\tan 33° - \tan 25°} \approx 4669.581$ ft

**102.(a)** $75 \sin 22° \approx 28.095$ ft
 **(b)** $75 \sin 27° \approx 34.049$ ft

**104.** $368.614$ in.$^2$

**105.** $x = 112{,}270$ which corresponds to April 22 and Sept 27.

**106.** $\approx 72.73$ ft

## Chapter 4 Project

Answers are based on the sample data shown in the table.

**1.**

$[-0.1, 2.1]$ by $[0, 1]$

**2.** $y \approx 0.22 \cos (3.87(x - 0.45)) + 0.71$

**3.** The constant $a$ represents the distance the pendulum bob swings as it moves from its highest point to its lowest point; $d$ represents the distance from the detector to the pendulum bob when it is in the lowest part of its swing.

**4.** Only $c$; $c \approx 0.05$

**5.** $y \approx 0.22 \sin (3.87x - 0.16) + 0.71$
Most calculator/ computer regression models are expressed in the form $y = a \sin(px + q) + d$, where $p = b$ and $-q/p = c$ in the equation $y = a \sin(b(x - c)) + d$

# Chapter 5

## Section 5.1 (pp. 423–431)

### Exploration 1

**1.** $\cos \theta = \dfrac{1}{\sec \theta}$, $\sec \theta = \dfrac{1}{\cos \theta}$, and $\tan \theta = \dfrac{\sin \theta}{\cos \theta}$

**2.** $\sin \theta = \dfrac{1}{\csc \theta}$ and $\tan \theta = \dfrac{1}{\cot \theta}$

**3.** $\csc \theta = \dfrac{1}{\sin \theta}$, $\cot \theta = \dfrac{1}{\tan \theta}$, and $\cot \theta = \dfrac{\cos \theta}{\sin \theta}$

### Quick Review 5.1

**1.** $1.1760$ rad $= 67.380°$    **2.** $0.9273$ rad $= 53.130°$
**3.** $2.4981$ rad $= 143.130°$
**4.** $-0.3948$ rad $= -22.620°$
**7.** $(2x + y)(x - 2y)$         **8.** $(2v + 1)(v - 3)$

### Section 5.1 Exercises

**8.** $\sec u$
**26.** $2 \sec^2 x$
**30.** $2 \csc x$
**32.** $(1 - \sin x)^2$
**34.** $(\sin x - 1)(\sin x + 2)$
**35.** $(2\cos x - 1)(\cos x + 1)$
**36.** $(\sin x + 1)^2$

**25.** $2 \sin \theta$
**28.** $2 \cot^2 x$
**31.** $(\cos x + 1)^2$
**33.** $(1 - \sin x)^2$

**46.** $\left\{0, \dfrac{\pi}{4}, \dfrac{3\pi}{4}, \pi, \dfrac{5\pi}{4}, \dfrac{7\pi}{4}\right\}$

**47.** $\left\{\dfrac{\pi}{3}, \dfrac{2\pi}{3}, \dfrac{4\pi}{3}, \dfrac{5\pi}{3}\right\}$    **48.** $\left\{\dfrac{\pi}{4}, \dfrac{3\pi}{4}, \dfrac{5\pi}{4}, \dfrac{7\pi}{4}\right\}$

**49.** $\left\{\pm\dfrac{\pi}{3} + 2n\pi \mid n = 0,\ \pm 1, \pm 2, \ldots\right\}$

**50.** $\left\{-\dfrac{\pi}{6} + 2n\pi, -\dfrac{5\pi}{6} + 2n\pi, -\dfrac{\pi}{2} + 2n\pi \mid n = 0, \right.$

$\left. \pm 1, \pm 2, \ldots \right\}$

**51.** $\{n\pi \mid n = 0, \pm 1, \pm 2, \ldots \}$

**52.** $\left\{\dfrac{\pi}{6} + 2n\pi, \dfrac{5\pi}{6} + 2n\pi \mid n = 0, \pm 1, \pm 2, \ldots \right\}$

**53.** $\{n\pi \mid n = 0, \pm 1, \pm 2, \ldots \}$

**54.** $\left\{\dfrac{\pi}{6} + 2n\pi, \dfrac{5\pi}{6} + 2n\pi \mid n = 0, \pm 1, \pm 2, \ldots \right\}$

**55.** $|\sin \theta|$      **56.** $|\sec \theta|$

**57.** $3|\tan \theta|$      **58.** $6|\cos \theta|$

**61.** $\sin x$, $\cos x = \pm\sqrt{1 - \sin^2 x}$,

$\tan x = \pm\dfrac{\sin x}{\sqrt{1 - \sin^2 x}}$, $\csc x = \dfrac{1}{\sin x}$,

$\sec x = \pm\dfrac{1}{\sqrt{1 - \sin^2 x}}$, $\cot x = \pm\dfrac{\sqrt{1 - \sin^2 x}}{}$

**62.** $\cos x$, $\sin x = \pm\sqrt{1 - \cos^2 x}$

$\tan x = \pm\dfrac{\sqrt{1 - \cos^2 x}}{}$, $\csc x = \pm\dfrac{\cos x}{\sqrt{1 - \cos^2 x}}$

$\sec x = \dfrac{1}{\cos x}$, $\cot x = \pm\dfrac{\cos x}{\sqrt{1 - \cos^2 x}}$

**63.** The two functions are parallel to each other, separated by 1 unit for every $x$. At any $x$, the distance between the two graphs is
$\sin^2 x - (-\cos^2 x) = \sin^2 x + \cos^2 x = 1$.

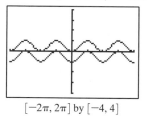

$[-2\pi, 2\pi]$ by $[-4, 4]$

**64.** The two functions are parallel to each other, separated by 1 unit for every $x$. At any $x$, the distance between the two graphs is $\sec^2 x - \tan^2 x = 1$.

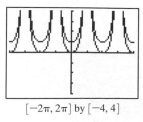

$[-2\pi, 2\pi]$ by $[-4, 4]$

**65. (a)**

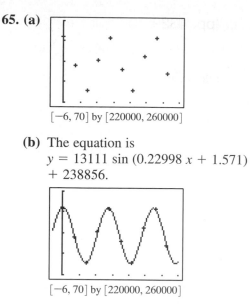

$[-6, 70]$ by $[220000, 260000]$

**(b)** The equation is
$y = 13111 \sin (0.22998 x + 1.571) + 238856$.

$[-6, 70]$ by $[220000, 260000]$

**(c)** $(2\pi)/0.22998 \approx 27.32$ days. This is the number of days that it takes the moon to make one complete orbit of the earth (known as the moon's sidereal period).

**(d)** 225745 miles

**(e)** $y = 13111 \cos(-0.22998x) + 238856$, or
$y = 13111 \cos(0.22998x) + 238856$.

**67.** Factor the left-hand side:
$\sin^4 \theta - \cos^4 \theta$
$= \sin^2 \theta - \cos^2 \theta)(\sin^2 \theta + \cos^2 \theta)$
$= (\sin^2 \theta - \cos^2 \theta) \cdot 1$
$= \sin^2 \theta - \cos^2 \theta$

**68.** Any $k$ satisfying $k \geq 2$ or $k \leq -2$.

**69.** Use the hint:
$\sin(\pi - x) = \sin(\pi/2 - (x - \pi/2))$
$= \cos(x - \pi/2)$    Cofunction identity
$= \cos(\pi/2 - x)$    Since cos is even
$= \sin x$    Cofunction identity

**70.** Use the hint:
$\cos(\pi - x) = \cos(\pi/2 - (x - \pi/2))$
$= \sin(x - \pi/2)$    Cofunction identity
$= -\sin(\pi/2 - x)$    Since sin is odd
$= -\cos x$    Cofunction identity

**71.** Since $A$, $B$, and $C$ are angles of a triangle,
$A + B = \pi - C$. So: $\sin(A + B) = \sin(\pi - C)$
$= \sin C$

**72.** Using the identities from Exercises 69 and 70, we
have: $\tan(\pi - x) = \dfrac{\sin(\pi - x)}{\cos(\pi - x)} = \dfrac{\sin x}{-\cos x}$
$= -\tan x$

## Section 5.2 (pp. 432–439)

### Exploration 1

Here's one possible path:
LOVE
MOVE
MOTE
MATE
MATH

### Exploration 2

**1.** The graphs lead us to conclude that this is not an identity.

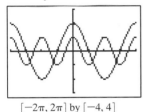

$[-2\pi, 2\pi]$ by $[-4, 4]$

**2.** For example, $\cos(2 \cdot 0) = 1$, whereas $2\cos(0) = 2$.

**4.** The graphs lead us to conclude that this is an identity.

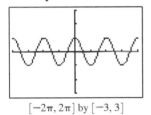

$[-2\pi, 2\pi]$ by $[-3, 3]$

**5.** No. The graph window can not show the full graphs, so they could differ outside the viewing window. Also, the function values could be so close that the graphs *appear* to coincide.

### Quick Review 5.2

**2.** $\dfrac{1}{\sin x \cos x}$

**3.** $\dfrac{1}{\sin x \cos x}$

**4.** $\cos \theta - \sin \theta$

**6.** $1$

**7.** No. (Any negative $x$.)

**9.** No. (Any $x$ for which $\sin x < 0$, e.g. $x = -\pi/2$.)

**10.** No. (Any $x$ for which $\tan x < 0$, e.g. $x = -\pi/4$.)

### Section 5.2 Exercises

**5.** $(\sin^3 x)(1 + \cot^2 x) = (\sin^3 x)(\csc^2 x)$
$= \dfrac{\sin^3 x}{\sin^2 x} = \sin x$. Yes

**7.** $(\cos x)(\tan x + \sin x \cot x)$
$= \cos x \cdot \dfrac{\sin x}{\cos x} + \cos x \sin x \cdot \dfrac{\cos x}{\sin x}$
$= \sin x + \cos^2 x$

**8.** $(\sin x)(\cot x + \cos x \tan x)$
$= \sin x \cdot \dfrac{\cos x}{\sin x} + \sin x \cos x \cdot \dfrac{\sin x}{\cos x}$
$= \cos x + \sin^2 x$

**9.** $(1 - \tan x)^2 = 1 - 2\tan x + \tan^2 x$
$= (1 + \tan^2 x) - 2\tan x = \sec^2 x - 2\tan x$

**10.** $(\cos x - \sin x)^2 = \cos^2 x - 2\sin x \cos x + \sin^2 x$
$= (\cos^2 x + \sin^2 x) - 2\sin x \cos x$
$= 1 - 2\sin x \cos x$

**11.** One possible proof:
$\dfrac{(1 - \cos u)(1 + \cos u)}{\cos^2 u} = \dfrac{1 - \cos^2 u}{\cos^2 u} = \dfrac{\sin^2 u}{\cos^2 u}$
$= \tan^2 u$

**12.** $\tan x + \sec x = \dfrac{\sin x}{\cos x} + \dfrac{1}{\cos x} = \dfrac{\sin x + 1}{\cos x}$
$= \dfrac{\cos x(\sin x + 1)}{\cos^2 x} = \dfrac{\cos x(\sin x + 1)}{1 - \sin^2 x} = \dfrac{\cos x}{1 - \sin x}$

**13.** $\dfrac{\cos^2 x - 1}{\cos x} = \dfrac{-\sin^2 x}{\cos x} = -\dfrac{\sin x}{\cos x} \cdot \sin x$
$= -\tan x \sin x$

**14.** $\dfrac{\sec^2 \theta - 1}{\sin \theta} = \dfrac{\tan^2 \theta}{\sin \theta} = \dfrac{1}{\sin \theta} \cdot \left(\dfrac{\sin \theta}{\cos \theta}\right)^2 = \dfrac{\sin \theta}{\cos^2 \theta}$
$= \dfrac{\sin \theta}{1 - \sin^2 \theta}$

**15.** Multiply out the expression on the left side.

**16.** $\dfrac{1}{1 - \cos x} + \dfrac{1}{1 + \cos x} = \dfrac{(1 + \cos x) + (1 - \cos x)}{(1 - \cos x)(1 + \cos x)}$
$= \dfrac{2}{1 - \cos^2 x} = \dfrac{2}{\sin^2 x} = 2\csc^2 x$

**17.** $(\cos t - \sin t)^2 + (\cos t + \sin t)^2$
$= \cos^2 t - 2\cos t \sin t + \sin^2 t + \cos^2 t$
$\quad + 2\cos t \sin t + \sin^2 t = 2\cos^2 t + 2\sin^2 t$
$= 2$

**18.** $\sin^2 \alpha - \cos^2 \alpha = (1 - \cos^2 \alpha) - \cos^2 \alpha$
$= 1 - 2\cos^2 \alpha$

**19.** $\dfrac{1 + \tan^2 x}{\sin^2 x + \cos^2 x} = \dfrac{\sec^2 x}{1} = \sec^2 x$

**20.** $\dfrac{1}{\tan \beta} + \tan \beta = \dfrac{\cos \beta}{\sin \beta} + \dfrac{\sin \beta}{\cos \beta} = \dfrac{\cos^2 \beta + \sin^2 \beta}{\cos \beta \sin \beta}$
$= \dfrac{1}{\cos \beta \sin \beta} = \sec \beta \csc \beta$

**21.** $\dfrac{\cos \beta}{1 + \sin \beta} = \dfrac{\cos^2 \beta}{\cos \beta(1 + \sin \beta)} = \dfrac{1 - \sin^2 \beta}{\cos \beta(1 + \sin \beta)}$
$= \dfrac{(1 - \sin \beta)(1 + \sin \beta)}{\cos \beta(1 + \sin \beta)} = \dfrac{1 - \sin \beta}{\cos \beta}$

**22.** $\dfrac{\sec x + 1}{\tan x} = \dfrac{\tan x(1 + \sec x)}{\tan^2 x} = \dfrac{\tan x(1 + \sec x)}{\sec^2 x - 1}$

$= \dfrac{\tan x(1 + \sec x)}{(\sec x - 1)(\sec x + 1)} = \dfrac{\tan x}{\sec x - 1}$

**23.** $\dfrac{\tan^2 x}{\sec x + 1} = \dfrac{\sec^2 x - 1}{\sec x + 1} = \sec x - 1 = \dfrac{1}{\cos x} - 1$

$= \dfrac{1 - \cos x}{\cos x}$

**24.** $\dfrac{\cot v - 1}{\cot v + 1} = \dfrac{\cot v - 1}{\cot v + 1} \cdot \dfrac{\tan v}{\tan v} = \dfrac{\cot v \tan v - \tan v}{\cot v \tan v + \tan v}$

$= \dfrac{1 - \tan v}{1 + \tan v}$ since $\cot v \tan v = \dfrac{\cos v}{\sin v} \cdot \dfrac{\sin v}{\cos v} = 1$

**25.** $\cot^2 x - \cos^2 x = \left(\dfrac{\cos x}{\sin x}\right)^2 - \cos^2 x$

$= \dfrac{\cos^2 x(1 - \sin^2 x)}{\sin^2 x} = \dfrac{\cos^4 x}{\sin^2 x} = \cos^2 x \cdot \dfrac{\cos^2 x}{\sin^2 x}$

$= \cos^2 x \cot^2 x$

**26.** $\tan^2\theta - \sin^2\theta = \left(\dfrac{\sin \theta}{\cos \theta}\right)^2 - \sin^2\theta$

$= \dfrac{\sin^2 \theta(1 - \cos^2 \theta)}{\cos^2 \theta} = \dfrac{\sin^4 \theta}{\cos^2 \theta} = \sin^2 \theta \cdot \dfrac{\sin^2 \theta}{\cos^2 \theta}$

$= \sin^2 \theta \tan^2 \theta$

**27.** $\cos^4 x - \sin^4 x$

$= (\cos^2 x + \sin^2 x)(\cos^2 x - \sin^2 x)$

$= 1(\cos^2 x - \sin^2 x) = \cos^2 x - \sin^2 x$

**28.** $\tan^4 t + \tan^2 t = \tan^2 t(\tan^2 t + 1)$

$= (\sec^2 t - 1)(\sec^2 t) = \sec^4 t - \sec^2 t$

**29.** $(x \sin \alpha + y \cos \alpha)^2 + (x \cos \alpha - y \sin \alpha)^2$

$= (x^2 \sin^2 \alpha + 2xy \sin \alpha \cos \alpha + y^2 \cos^2 \alpha)$

$+ (x^2 \cos^2 \alpha - 2xy \cos \alpha \sin \alpha + y^2 \sin^2 \alpha)$

$= x^2 \sin^2 \alpha + y^2 \cos^2 \alpha + x^2 \cos^2 \alpha + y^2 \sin^2 \alpha$

$= (x^2 + y^2)(\sin^2 \alpha + \cos^2 \alpha) = x^2 + y^2$

**30.** $\dfrac{1 - \cos \theta}{\sin \theta} = \dfrac{1 - \cos^2 \theta}{\sin \theta(1 + \cos \theta)} = \dfrac{\sin^2 \theta}{\sin \theta(1 + \cos \theta)}$

$= \dfrac{\sin \theta}{1 + \cos \theta}$

**31.** $\dfrac{\tan x}{\sec x - 1} = \dfrac{\tan x(\sec x + 1)}{\sec^2 x + 1} = \dfrac{\tan x(\sec x + 1)}{\tan^2 x}$

$= \dfrac{\sec x + 1}{\tan x}$. See also #22.

**32.** $\dfrac{\sin t}{1 + \cos t} + \dfrac{1 + \cos t}{\sin t} = \dfrac{\sin^2 t + (1 + \cos t)^2}{}$

$= \sin^2 t + 1 + 2 \cos t + \cos^2 t = \dfrac{2 + 2 \cos t}{(\sin t)(1 + \cos t)}$

$= \dfrac{2}{\sin t} = 2 \csc t$

**33.** $\dfrac{\sin x - \cos x}{\sin x + \cos x} = \dfrac{(\sin x - \cos x)(\sin x + \cos x)}{(\sin x + \cos x)^2}$

$= \dfrac{\sin^2 x - \cos^2 x}{\sin^2 x + 2 \sin x \cos x + \cos^2 x}$

$= \dfrac{\sin^2 x - (1 - \sin^2 x)}{1 + 2 \sin x \cos x} = \dfrac{2 \sin^2 x - 1}{1 + 2 \sin x \cos x}$

**34.** $\dfrac{1 + \cos x}{1 - \cos x} = \dfrac{1 + \cos x}{1 - \cos x} \cdot \dfrac{\sec x}{\sec x} = \dfrac{\sec x + \cos x \sec x}{\sec x - \cos x \sec x}$

$= \dfrac{\sec x + 1}{\sec x - 1}$ since $\cos x \sec x = \cos x \cdot \dfrac{1}{\cos x} = 1$.

**35.** $\dfrac{\sin t}{1 - \cos t} + \dfrac{1 + \cos t}{\sin t}$

$= \dfrac{\sin^2 t + (1 + \cos t)(1 - \cos t)}{(\sin t)(1 - \cos t)}$

$= \dfrac{\sin^2 t + 1 - \cos^2 t}{(\sin t)(1 - \cos t)} = \dfrac{2(1 - \cos^2 t)}{(\sin t)(1 - \cos t)}$

$= \dfrac{2(1 + \cos t)}{\sin t}$

**36.** $\dfrac{\sin A \cos B + \cos A \sin B}{\cos A \cos B - \sin A \sin B}$

$= \left(\dfrac{\dfrac{1}{\cos A \cos B}}{\dfrac{1}{\cos A \cos B}}\right) \cdot \dfrac{\sin A \cos B + \cos A \sin B}{\cos A \cos B - \sin A \sin B}$

$= \dfrac{\dfrac{\sin A}{\cos A} + \dfrac{\sin B}{\cos B}}{1 - \dfrac{\sin A \sin B}{\cos A \cos B}} = \dfrac{\tan A + \tan B}{1 - \tan A \tan B}$

**37.** $\sin^2 x \cos^3 x = \sin^2 x \cos^2 x \cos x$

$= \sin^2 x(1 - \sin^2 x)\cos x = (\sin^2 x - \sin^4 x)\cos x$

**38.** $\sin^5 x \cos^2 x = \sin^4 x \cos^2 x \sin x$

$= (\sin^2 x)^2 \cos^2 x \sin x$

$= (1 - \cos^2 x)^2 \cos^2 x \sin x$

$= (1 - 2 \cos^2 x + \cos^4 x)\cos^2 x \sin x$

$= (\cos^2 x - 2 \cos^4 x + \cos^6 x)\sin x$

**39.** $\cos^5 x = \cos^4 x \cos x = (\cos^2 x)^2 \cos x$

$= (1 - \sin^2 x)^2 \cos x$

$= (1 - 2 \sin^2 x + \sin^4 x) \cos x$

**40.** $\sin^3 x \cos^3 x = \sin^3 x \cos^2 x \cos x$

$= \sin^3 x (1 - \sin^2 x)\cos x$

$= (\sin^3 x - \sin^5 x)\cos x$

**41.** $\dfrac{\tan x}{1 - \cot x} + \dfrac{\cot x}{1 - \tan x}$

$= \dfrac{\tan x}{1 - \cot x} \cdot \dfrac{\sin x}{\sin x} + \dfrac{\cot x}{1 - \tan x} \cdot \dfrac{\cos x}{\cos x}$

$= \left(\dfrac{\sin^2 x/\cos x}{\sin x - \cos x} + \dfrac{\cos^2 x/\sin x}{\cos x - \sin x}\right)\dfrac{\sin x \cos x}{\sin x \cos x}$

$= \dfrac{\sin^3 x - \cos^3 x}{\sin x \cos x(\sin x - \cos x)}$

$= \dfrac{\sin^2 x + \sin x \cos x + \cos^2 x}{\sin x \cos x} = \dfrac{1 + \sin x \cos x}{\sin x \cos x}$

$= \dfrac{1}{\sin x \cos x} + 1 = \csc x \sec x + 1$.

This involves rewriting $a^3 - b^3$ as

$(a - b)(a^2 + ab + b^2)$, where $a = \sin x$ and

$b = \cos x$.

**42.** $\dfrac{\cos x}{1 + \sin x} + \dfrac{\cos x}{1 - \sin x}$

$= \dfrac{(\cos x)[(1 - \sin x) + (1 + \sin x)]}{(1 + \sin x)(1 - \sin x)} = \dfrac{2 \cos x}{1 - \sin^2 x}$

$= \dfrac{2 \cos x}{\cos^2 x} = 2 \sec x$

**43.** $\dfrac{2 \tan x}{1 - \tan^2 x} + \dfrac{1}{2 \cos^2 x - 1}$

$= \dfrac{2 \tan x}{1 - \tan^2 x} \cdot \dfrac{\cos^2 x}{\cos^2 x} + \dfrac{1}{\cos^2 x - \sin^2 x}$

$= \dfrac{2 \sin x \cos x}{\cos^2 x - \sin^2 x} + \dfrac{\cos^2 x + \sin^2 x}{\cos^2 x - \sin^2 x}$

$= \dfrac{2 \sin x \cos x + \cos^2 x + \sin^2 x}{(\cos x - \sin x)(\cos x + \sin x)}$

$= \dfrac{(\cos x + \sin x)^2}{(\cos x - \sin x)(\cos x + \sin x)} = \dfrac{\cos x + \sin x}{\cos x - \sin x}$

**44.** $1 - 3 \cos x - 4 \cos^2 x = \dfrac{(1 + \cos x)(1 - 4 \cos x)}{1 - \cos^2 x}$

$= \dfrac{(1 + \cos x)(1 - 4 \cos x)}{(1 + \cos x)(1 - \cos x)} = \dfrac{1 - 4 \cos x}{1 - \cos x}$

**45.** $\cos^3 x = (\cos^2 x)(\cos x) = (1 - \sin^2 x)(\cos x)$

**46.** $\sec^4 x = (\sec^2 x)(\sec^2 x) = (1 + \tan^2 x)(\sec^2 x)$

**47.** $\sin^5 x = (\sin^4 x)(\sin x) = (\sin^2 x)^2 (\sin x)$

$= (1 - \cos^2 x)^2 (\sin x)$

$= (1 - 2 \cos^2 x + \cos^4 x)(\sin x)$

**60.** Since the sum of the logarithms is the logarithm of the product, and since the product of the absolute values of all six basic trig functions is 1, the logarithms sum to ln 1, which is 0.

**61.** If $A$ and $B$ are complimentary angles, then $\sin^2 A + \sin^2 B = \sin^2 A + \sin^2(\pi/2 - A)$

$= \sin^2 A + \cos^2 A = 1$

**63.** Multiply and divide by $1 - \sin t$ under the radical:

$\sqrt{\dfrac{1 - \sin t}{1 + \sin t} \cdot \dfrac{1 - \sin t}{1 - \sin t}} = \sqrt{\dfrac{(1 - \sin t)^2}{1 - \sin^2 t}}$

$= \sqrt{\dfrac{(1 - \sin t)^2}{\cos^2 t}} = \dfrac{|1 - \sin t|}{|\cos t|}$ since $\sqrt{a^2} = |a|$.

Now, since $1 - \sin t \geq 0$, we can dispense with the absolute value in the numerator, but it must stay in the denominator.

**64.** Multiply and divide by $1 + \cos t$ under the radical:

$\sqrt{\dfrac{1 + \cos t}{1 - \cos t} \cdot \dfrac{1 + \cos t}{1 + \cos t}} = \sqrt{\dfrac{(1 + \cos t)^2}{1 - \cos^2 t}}$

$= \sqrt{\dfrac{(1 + \cos t)^2}{\sin^2 t}} = \dfrac{|1 + \cos t|}{|\sin t|}$ since $\sqrt{a^2} = |a|$.

Now, since $1 + \cos t \geq 0$, we can dispense with the absolute value in the numerator, but it must stay in the denominator.

**65.** $\sin^6 x + \cos^6 x = (\sin^2 x)^3 + \cos^6 x$

$= (1 - \cos^2 x)^3 + \cos^6 x$

$= (1 - 3 \cos^2 x + 3 \cos^4 x - \cos^6 x) + \cos^6 x$

$= 1 - 3 \cos^2 x(1 - \cos^2 x)$

$= 1 - 3 \cos^2 x \sin^2 x.$

**66.** Note that $a^3 - b^3 = (a - b)(a^2 + ab + b^2)$. Also note that $a^2 + ab + b^2 = a^2 + 2ab + b^2 - ab$

$= (a + b)^2 - ab$. Taking $a = \cos^2 x$ and $b = \sin^2 x$, we have $\cos^6 x - \sin^6 x$

$= (\cos^2 x - \sin^2 x)(\cos^4 x + \cos^2 x \sin^2 x + \sin^4 x)$

$= (\cos^2 x - \sin^2 x)[(\cos^2 x + \sin^2 x)^2 - \cos^2 x \sin^2 x]$

$= (\cos^2 x - \sin^2 x)(1 - \cos^2 x \sin^2 x).$

**67.** One possible proof: $\ln|\tan x| = \ln\dfrac{|\sin x|}{|\cos x|}$

$= \ln|\sin x| - \ln|\cos x|.$

**68.** One possible proof:

$\ln|\sec \theta + \tan \theta| + \ln|\sec \theta - \tan \theta|$

$= \ln|\sec^2 \theta - \tan^2 \theta| = \ln 1 = 0$

**69. (a)** They are not equal. Shown is the window $[-2\pi, 2\pi] \times [-2, 2]$; graphing on nearly any viewing window does not show any apparent difference — but using TRACE, one finds that the $y$ coordinates are not identical. Likewise, a table of values will show slight differences; for example, when $x = 1$, $y_1 = 0.53988$ while $y_2 = 0.54030$

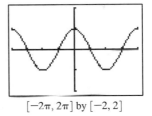

$[-2\pi, 2\pi]$ by $[-2, 2]$

**(b)** One choice for $h$ is 0.001 (shown). The function $y_3$ is a combination of three sinusoidal functions ($1000 \sin(x + 0.001)$, $1000 \sin x$, and $\cos x$), all with period $2\pi$.

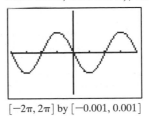

$[-2\pi, 2\pi]$ by $[-0.001, 0.001]$

**70. (a)** $\cosh^2 x - \sinh^2 x$

$= \frac{1}{4}(e^x + e^{-x})^2 - \frac{1}{4}(e^x - e^{-x})^2$

$= \frac{1}{4}[e^{2x} + 2 + e^{-2x} - (e^{2x} - 2 + e^{-2x})]$

$= \frac{1}{4}(4) = 1.$

**(b)** $1 - \tanh^2 x = 1 - \dfrac{\sinh^2 x}{\cosh^2 x} = \dfrac{\cosh^2 x - \sinh^2 x}{\cosh^2 x}$

$= \dfrac{1}{\cosh^2 x}$, using the result from (a). This equals $\operatorname{sech}^2 x.$

**(c)** $\coth^2 x - 1 = \dfrac{\cosh^2 x}{\sinh^2 x} - 1 = \dfrac{\cosh^2 x - \sinh^2 x}{\sinh^2 x}$

$= \dfrac{1}{\sinh^2 x}$, using the result from (a). This equals $\operatorname{csch}^2 x.$

**71.** In the decimal window, the $x$ coordinates used to plot the graph on the calculator are (e.g.) 0, 0.1, 0.2, 0.3, etc. — that is, $x = n/10$, where $n$ is an integer. Then $10\pi x = \pi n$, and the sine of integer multiples of $\pi$ is 0; therefore, $\cos x + \sin 10\pi x = \cos x + \sin \pi n = \cos x + 0 = \cos x$. However, for other choices of $x$, such as $x = \dfrac{1}{\pi}$, we have $\cos x + \sin 10\pi x = \cos x + \sin 10 \neq \cos x.$

## Section 5.3 (pp. 439–445)

### Exploration 1

**3.** $\tan\left(\dfrac{\pi}{3} + \dfrac{\pi}{3}\right) = -\sqrt{3}, \tan\dfrac{\pi}{3} + \tan\dfrac{\pi}{3} = 2\sqrt{3}.$
(Many other answers are possible.)

### Exercises 5.3

**15.** $\tan 66°$

**16.** $\tan\left(-\dfrac{2\pi}{15}\right)$

**17.** $\cos\left(x - \dfrac{\pi}{7}\right)$

**18.** $\cos\left(x + \dfrac{\pi}{7}\right)$

**21.** $\tan(2y + 3x)$

**22.** $\tan(3\alpha - 2\beta)$

**23.** $\sin\left(x - \dfrac{\pi}{2}\right) = \sin x \cos\dfrac{\pi}{2} - \cos x \sin\dfrac{\pi}{2}$

$= \sin x \cdot 0 - \cos x \cdot 1 = -\cos x$

**24.** Using the difference identity for the tangent function, we encounter $\tan\dfrac{\pi}{2}$, which is undefined.

However, we can compute $\tan\left(x - \dfrac{\pi}{2}\right)$

$= \dfrac{\sin(x - \pi/2)}{\cos(x - \pi/2)}$. From #23, $\sin\left(x - \dfrac{\pi}{2}\right) = -\cos x$.

Since the cosine function is even,

$\cos\left(x - \dfrac{\pi}{2}\right) = \cos\left(\dfrac{\pi}{2} - x\right) = \sin x$ (see Example 4, or #25). Therefore this simplifies to

$-\dfrac{\cos x}{\sin x} = -\cot x.$

**25.** $\cos\left(x - \dfrac{\pi}{2}\right) = \cos x \cos\dfrac{\pi}{2} + \sin x \sin\dfrac{\pi}{2}$

$= \cos x \cdot 0 + \sin x \cdot 1 = \sin x$

**26.** The simplest way is to note that

$\left(\dfrac{\pi}{2} - x\right) - y = \dfrac{\pi}{2} - x - y = \dfrac{\pi}{2} - (x + y)$, so

that $\cos\left[\left(\dfrac{\pi}{2} - x\right) - y\right] = \cos\left[\dfrac{\pi}{2} - (x + y)\right].$

Now use Example 2 to conclude that

$\cos\left[\dfrac{\pi}{2} - (x + y)\right] = \sin(x + y).$

**27.** $\sin\left(x + \dfrac{\pi}{6}\right) = \sin x \cos\dfrac{\pi}{6} + \cos x \sin\dfrac{\pi}{6}$

$= \sin x \cdot \dfrac{\sqrt{3}}{2} + \cos x \cdot \dfrac{1}{2}$

**28.** $\cos\left(x - \dfrac{\pi}{4}\right) = \cos x \cos\dfrac{\pi}{4} + \sin x \sin\dfrac{\pi}{4}$

$= \cos x \cdot \dfrac{\sqrt{2}}{2} + \sin x \cdot \dfrac{\sqrt{2}}{2} = \dfrac{\sqrt{2}}{2}(\cos x + \sin x)$

**29.** $\tan\left(\theta + \dfrac{\pi}{4}\right) = \dfrac{\tan\theta + \tan(\pi/4)}{1 - \tan\theta \tan(\pi/4)}$

$= \dfrac{\tan\theta + 1}{1 - \tan\theta \cdot 1} = \dfrac{1 + \tan\theta}{1 - \tan\theta}$

**30.** $\cos\left(\theta + \dfrac{\pi}{2}\right) = \cos\theta \cos\dfrac{\pi}{2} - \sin\theta \sin\dfrac{\pi}{2}$

$= \cos\theta \cdot 0 - \sin\theta \cdot 1 = -\sin\theta$

**31.** Equations (b) and (f).

**32.** Equations (c) and (e).

**33.** Equations (d) and (h).

**34.** Equations (a) and (g).

**35.** $x = n\pi$, $n$ an integer.

**36.** $\dfrac{\pi}{8} + n\dfrac{\pi}{4}$, $n$ an integer.

**37.** $\sin\left(\dfrac{\pi}{2} - u\right) = \sin\dfrac{\pi}{2}\cos u - \cos\dfrac{\pi}{2}\sin u$

$= 1 \cdot \cos u - 0 \cdot \sin u = \cos u.$

**38.** Using the difference identity for the tangent function, we encounter $\tan \dfrac{\pi}{2}$, which is undefined. However, we can compute $\tan\left(\dfrac{\pi}{2} - u\right)$
$$= \frac{\sin(\pi/2 - u)}{\cos(\pi/2 - u)};$$ using the first two confunction identities, we find that the numerator and denominator are $\cos u$ and $\sin u$, respectively, so that this is $\cot u$. Or, use #24, and the fact that the tangent function is odd.

**39.** $\cot\left(\dfrac{\pi}{2} - u\right) = \dfrac{\cos(\pi/2 - u)}{\sin(\pi/2 - u)} = \dfrac{\sin u}{\cos u} = \tan u$ using the first two cofunction identities.

**40.** $\sec\left(\dfrac{\pi}{2} - u\right) = \dfrac{1}{\cos(\pi/2 - u)} = \dfrac{1}{\sin u} = \csc u$ using the first cofunction identity.

**41.** $\csc\left(\dfrac{\pi}{2} - u\right) = \dfrac{1}{\sin(\pi/2 - u)} = \dfrac{1}{\cos u} = \sec u$ using the second cofunction identity.

**42.** $\cos\left(x + \dfrac{\pi}{2}\right) = \cos x \cos\left(\dfrac{\pi}{2}\right) - \sin x \sin\left(\dfrac{\pi}{2}\right)$
$$= \cos x \cdot 0 - \sin x \cdot 1 = -\sin x$$

**43.** $y \approx 5\cos(x + 0.9273)$

**44.** $y \approx 13\sin(x - 1.176)$

**45.** $y \approx 2.236\sin(x + 0.4636)$

**46.** $y \approx -3.606\sin(2x - 0.9828)$

**47.** $\sin(x - y) + \sin(x + y)$
$$= (\sin x \cos y - \cos x \sin y)$$
$$+ (\sin x \cos y + \cos x \sin y) = 2\sin x \cos y$$

**48.** $\cos(x - y) + \cos(x + y)$
$$= (\cos x \cos y + \sin x \sin y)$$
$$+ (\cos x \cos y - \sin x \sin y) = 2\cos x \cos y$$

**49.** $\cos 3x = \cos[(x + x) + x]$
$$= \cos(x + x)\cos x - \sin(x + x)\sin x$$
$$= (\cos x \cos x - \sin x \sin x)\cos x$$
$$- (\sin x \cos x + \cos x \sin x)\sin x$$
$$= \cos^3 x - \sin^2 x \cos x - 2\cos x \sin^2 x$$
$$= \cos^3 x - 3\sin^2 x \cos x$$

**50.** $\sin 3u = \sin[(u + u) + u]$
$$= \sin(u + u)\cos u + \cos(u + u)\sin u$$
$$= (\sin u \cos u + \cos u \sin u)\cos u$$
$$+ (\cos u \cos u - \sin u \sin u)\sin u$$
$$= 2\cos^2 u \sin u + \cos^2 u \sin u - \sin^3 u$$
$$= 3\cos^2 u \sin u - \sin^3 u$$

**51.** $\cos 3x + \cos x = \cos(2x + x) + \cos(2x - x)$; use #48 with $x$ replaced with $2x$ and $y$ replaced with $x$.

**52.** $\sin 4x + \sin 2x = \sin(3x + x) + \sin(3x - x)$; use #47 with $x$ replaced with $3x$ and $y$ replaced with $x$.

**53.** $\tan(x + y)\tan(x - y) = \left(\dfrac{\tan x + \tan y}{1 - \tan x \tan y}\right)$
$$\left(\frac{\tan x - \tan y}{1 + \tan x \tan y}\right) = \frac{\tan^2 x - \tan^2 y}{1 - \tan^2 x \tan^2 y}$$ since both the numerator and denominator are factored forms for differences of squares.

**54.** $\tan 5u \tan 3u = \tan(4u + u)\tan(4u - u)$; use #53 with $x = 4u$ and $y = u$.

**55.** Prove the identity:
$$\frac{\sin(x + y)}{\sin(x - y)} = \frac{(\tan x + \tan y)}{(\tan x - \tan y)}.$$
Solution:
$$\frac{\sin(x + y)}{\sin(x - y)}$$
$$= \frac{\sin x \cos y + \cos x \sin y}{\sin x \cos y - \cos x \sin y}$$
$$= \frac{\sin x \cos y + \cos x \sin y}{\sin x \cos y - \cos x \sin y} \cdot \frac{1/(\cos x \cos y)}{1/(\cos x \cos y)}$$
$$= \frac{(\sin x \cos y)/(\cos x \cos y) + (\cos x \sin y)/(\cos x \cos y)}{(\sin x \cos y)/(\cos x \cos y) - (\cos x \sin y)/(\cos x \cos y)}$$
$$= \frac{(\sin x/\cos x) + (\sin y/\cos y)}{(\sin x/\cos x) - (\sin y/\cos y)}$$
$$= \frac{\tan x + \tan y}{\tan x - \tan y}$$

**56.** $\tan(u + v) = \dfrac{\sin(u + v)}{\cos(u + v)}$
$$= \frac{\sin u \cos v + \cos u \sin v}{\cos u \cos v - \sin u \sin v}$$
$$= \frac{\dfrac{\sin u \cos v}{\cos u \cos v} + \dfrac{\cos u \sin v}{\cos u \cos v}}{\dfrac{\cos u \cos v}{\cos u \cos v} - \dfrac{\sin u \sin v}{\cos u \cos v}}$$
$$= \frac{\dfrac{\sin u}{\cos u} + \dfrac{\sin v}{\cos v}}{1 - \dfrac{\sin u \sin v}{\cos u \cos v}}$$
$$= \frac{\tan u + \tan v}{1 - \tan u \tan v}$$

**57.** $\tan(u - v) = \dfrac{\sin(u - v)}{\cos(u - v)}$
$$= \frac{\sin u \cos v - \cos u \sin v}{\cos u \cos v + \sin u \sin v}$$
$$= \frac{\dfrac{\sin u \cos v}{\cos u \cos v} - \dfrac{\cos u \sin v}{\cos u \cos v}}{\dfrac{\cos u \cos v}{\cos u \cos v} + \dfrac{\sin u \sin v}{\cos u \cos v}}$$
$$= \frac{\dfrac{\sin u}{\cos u} - \dfrac{\sin v}{\cos v}}{1 + \dfrac{\sin u \sin v}{\cos u \cos v}}$$
$$= \frac{\tan u - \tan v}{1 + \tan u \tan v}$$

**58.** The identity would involve $\tan\left(\dfrac{\pi}{2}\right)$ which does not exist.

$$\tan\left(x + \frac{\pi}{2}\right) = \frac{\sin\left(x + \dfrac{\pi}{2}\right)}{\cos\left(x + \dfrac{\pi}{2}\right)}$$

$$= \frac{\sin x \cos \dfrac{\pi}{2} + \cos x \sin \dfrac{\pi}{2}}{\cos x \cos \dfrac{\pi}{2} - \sin x \sin \dfrac{\pi}{2}}$$

$$= \frac{\sin x \cdot 0 + \cos x \cdot 1}{\cos x \cdot 0 - \sin x \cdot 1} = -\cot x$$

**59.** The identity would involve $\tan\left(\dfrac{3\pi}{2}\right)$, which does not exit.

$$\tan\left(x - \frac{3\pi}{2}\right) = \frac{\sin\left(x - \dfrac{3\pi}{2}\right)}{\cos\left(x - \dfrac{3\pi}{2}\right)}$$

$$= \frac{\sin x \cos \dfrac{3\pi}{2} - \cos x \sin \dfrac{3\pi}{2}}{\cos x \cos \dfrac{3\pi}{2} + \sin x \sin \dfrac{3\pi}{2}}$$

$$= \frac{\sin x \cdot 0 - \cos x \cdot (-1)}{\cos x \cdot 0 + \sin x \cdot (-1)}$$

$$= -\cot x$$

**60.** $\dfrac{\sin(x + h) - \sin x}{h} = \dfrac{\sin x \cos h + \cos x \sin h - \sin x}{h}$

$$= \frac{\sin x(\cos h - 1) + \cos x \sin h}{h}$$

$$= \sin x\left(\frac{\cos h - 1}{h}\right) + \cos x \frac{\sin h}{h}$$

**61.** $\dfrac{\cos(x + h) - \cos x}{h}$

$$= \frac{\cos x \cos h - \sin x \sin h - \cos x}{h}$$

$$= \frac{\cos x(\cos h - 1) - \sin x \sin h}{h}$$

$$= \cos x\left(\frac{\cos h - 1}{h}\right) - \sin x \frac{\sin h}{h}$$

**62.** Coordinates in the first quadrant are $(1, 0)$, $\left(\dfrac{\sqrt{2} + \sqrt{6}}{4}, \dfrac{\sqrt{6} - \sqrt{2}}{4}\right)$, $\left(\dfrac{\sqrt{3}}{2}, \dfrac{1}{2}\right)$, $\left(\dfrac{\sqrt{2}}{2}, \dfrac{\sqrt{2}}{2}\right)$, $\left(\dfrac{1}{2}, \dfrac{\sqrt{3}}{2}\right)$, $\left(\dfrac{\sqrt{6} - \sqrt{2}}{4}, \dfrac{\sqrt{2} + \sqrt{6}}{4}\right)$, $(0, 1)$

**63.** $\sin(A + B) = \sin(\pi - C)$
$$= \sin \pi \cos C - \cos \pi \sin C$$
$$= 0 \cdot \cos C - (-1)\sin C$$
$$= \sin C$$

**64.** $\cos C = \cos(\pi - (A + B))$
$$= \cos \pi \cos(A + B) + \sin \pi \sin(A + B)$$
$$= (-1)(\cos A \cos B - \sin A \sin B)$$
$$\quad + 0 \cdot \sin(A + B)$$
$$= \sin A \sin B - \cos A \cos B$$

**65.** $\tan A + \tan B + \tan C = \dfrac{\sin A}{\cos A} + \dfrac{\sin B}{\cos B} + \dfrac{\sin C}{\cos C}$

$$= \frac{\sin A(\cos B \cos C) + \sin B(\cos A \cos C)}{\cos A \cos B \cos C}$$

$$+ \frac{\sin C(\cos A \cos B)}{\cos A \cos B \cos C}$$

$$= \frac{\cos C(\sin A \cos B + \cos A \sin B) + \sin C(\cos A \cos B)}{\cos A \cos B \cos C}$$

$$= \frac{\cos C \sin(A + B) + \sin C(\cos(A + B) + \sin A \sin B)}{\cos A \cos B \cos C}$$

$$= \frac{\cos C \sin(\pi - C) + \sin C(\cos(\pi - C) + \sin A \sin B)}{\cos A \cos B \cos C}$$

$$= \frac{\cos C \sin C + \sin C(-\cos C) + \sin C \sin A \sin B}{\cos A \cos B \cos C}$$

$$= \frac{\sin A \sin B \sin C}{\cos A \cos B \cos C}$$

$$= \tan A \tan B \tan C$$

**66.** $\cos A \cos B \cos C - \sin A \sin B \cos C$
$$\quad - \sin A \cos B \sin C - \cos A \sin B \sin C$$
$$= \cos A(\cos B \cos C - \sin B \sin C)$$
$$\quad - \sin A(\sin B \cos C + \cos B \sin C)$$
$$= \cos A \cos(B + C) - \sin A \sin(B + C)$$
$$= \cos(A + B + C)$$
$$= \cos \pi$$
$$= -1$$

**67.** This equation is easier to deal with after rewriting it as $\cos 5x \cos 4x + \sin 5x \sin 4x = 0$. The left side of this equation is the expanded form of $\cos(5x - 4x)$, which of course equals $\cos x$; the graph shown is simply $y = \cos x$. The equation $\cos x = 0$ is easily solved on the interval $[-2\pi, 2\pi]$: $x = \pm\dfrac{\pi}{2}$ or $x = \pm\dfrac{3\pi}{2}$. The original graph is so crowded that one cannot see where crossings occur. The window shown is $[-2\pi, 2\pi] \times [-1.1, 1.1]$.

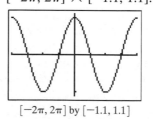

$[-2\pi, 2\pi]$ by $[-1.1, 1.1]$

**68.** $x = a \cos\left(\dfrac{2\pi t}{T} + \delta\right)$

$= a\left[\cos\left(\dfrac{2\pi t}{T}\right)\cos\delta - \sin\left(\dfrac{2\pi t}{T}\right)\sin\delta\right]$

$= (a\cos\delta)\cos\left(\dfrac{2\pi t}{T}\right) + (-a\sin\delta)\sin\left(\dfrac{2\pi t}{T}\right)$

**69.** $B = B_{\text{in}} + B_{\text{ref}}$

$= \dfrac{E_0}{c}\cos\left(\omega t - \dfrac{\omega x}{c}\right) + \dfrac{E_0}{c}\cos\left(\omega t + \dfrac{\omega x}{c}\right)$

$= \dfrac{E_0}{c}\left(\cos\omega t\cos\dfrac{\omega x}{c} + \sin\omega t\sin\dfrac{\omega x}{c}\right.$

$\left. + \cos\omega t\cos\dfrac{\omega x}{c} - \sin\omega t\sin\dfrac{\omega x}{c}\right)$

$= \dfrac{E_0}{c}\left(2\cos\omega t\cos\dfrac{\omega x}{c}\right) = 2\dfrac{E_0}{c}\cos\omega t\cos\dfrac{\omega x}{c}$

## Section 5.4 (pp. 446–453)

### Exploration 1

**1.** $\sin^2\dfrac{\pi}{8} = \dfrac{1 - \cos(\pi/4)}{2}$

$= \dfrac{1 - (\sqrt{2}/2)}{2} \cdot \dfrac{2}{2}$

$= \dfrac{2 - \sqrt{2}}{4}$

**2.** $\sin\dfrac{\pi}{8} = \dfrac{\sqrt{2 - \sqrt{2}}}{2}$. We take the

positive square root because $\dfrac{\pi}{8}$ is a first-quadrant

angle.

**3.** $\sin^2\dfrac{9\pi}{8} = \dfrac{1 - \cos(9\pi/4)}{2}$

$= \dfrac{1 - (\sqrt{2}/2)}{2} \cdot \dfrac{2}{2}$

$= \dfrac{2 - \sqrt{2}}{4}$

**4.** $\sin\dfrac{9\pi}{8} = -\dfrac{\sqrt{2 - \sqrt{2}}}{2}$

We take the negative square root because $\dfrac{\pi}{8}$ is a

third-quadrant angle.

### Quick Review 5.4

**1.** $x = \dfrac{\pi}{4} + n\pi$, $n$ an integer

**2.** $x = -\dfrac{\pi}{4} + n\pi$, $n$ an integer

**3.** $x = \dfrac{\pi}{2} + n\pi$, $n$ an integer.

**4.** $x = n\pi$, $n$ an integer.

**5.** $x = -\dfrac{\pi}{4} + n\pi$, $n$ an integer

**6.** $x = \dfrac{\pi}{4} + n\pi$, $n$ an integer

**7.** $x = \dfrac{\pi}{6} + 2n\pi$ or

$x = \dfrac{5\pi}{6} + 2n\pi$ or $x = \pm\dfrac{2\pi}{3} + 2n\pi$, $n$ an integer.

**8.** $x = \dfrac{3\pi}{2} + 2n\pi$ or $x = \pm\dfrac{\pi}{4} + 2n\pi$, $n$ an integer.

**9.** $A = (2)(3) + \dfrac{1}{2}(1)(3) + \dfrac{1}{2}(2)(3)$

$= 10.5$ square units.

### Exercises 5.4

**1.** $\cos 2u = \cos(u + u) = \cos u\cos u - \sin u\sin u$
$= \cos^2 u - \sin^2 u$

**2.** Starting with the result of #1: $\cos 2u$
$= \cos^2 u - \sin^2 u = \cos^2 u - (1 - \cos^2 u)$
$= 2\cos^2 u - 1$

**3.** Starting with the result of #1: $\cos 2u$
$= \cos^2 u - \sin^2 u = (1 - \sin^2 u) - \sin^2 u$
$= 1 - 2\sin^2 u$

**4.** $\tan 2u = \tan(u + u) = \dfrac{\tan u + \tan u}{1 - \tan u\tan u}$

$= \dfrac{2\tan u}{1 - \tan^2 u}$

**6.** $\left\{0, \pi, \dfrac{\pi}{3}, \dfrac{5\pi}{3}\right\}$

**7.** $\left\{\dfrac{\pi}{6}, \dfrac{5\pi}{6}, \dfrac{3\pi}{2}\right\}$

**8.** $\left\{0, \dfrac{2\pi}{3}, \dfrac{4\pi}{3}\right\}$

**9.** $\left\{0, \dfrac{\pi}{4}, \dfrac{3\pi}{4}, \pi, \dfrac{5\pi}{4}, \dfrac{7\pi}{4}\right\}$

**10.** $\{2.2370, 4.0461\}$

**11.** $(\cos\theta)(2\sin\theta + 1)$

**12.** $2\sin\theta\cos\theta + 1 - 2\sin^2\theta$
or $2\sin\theta\cos\theta + 2\cos^2\theta - 1$

**13.** $2\sin\theta\cos\theta + 4\cos^3\theta - 3\cos\theta$
or $2\sin\theta\cos\theta + \cos^3\theta - 3\sin^2\theta\cos\theta$

**14.** $3\sin\theta\cos^2\theta - \sin^3\theta + \cos^2\theta - \sin^2\theta$

**15.** $\sin 4x = \sin 2(2x) = 2\sin 2x\cos 2x$

**16.** $\cos 6x = \cos 2(3x) = 2\cos^2 3x - 1$

**17.** $2\csc 2x = \dfrac{2}{\sin 2x} = \dfrac{2}{2\sin x\cos x}$

$= \dfrac{1}{\sin^2 x} \cdot \dfrac{\sin x}{\cos x} = \csc^2 x\tan x$

**18.** $2 \cot 2x = \dfrac{2}{\tan 2x} = \dfrac{2(1 - \tan^2 x)}{2 \tan x} = \dfrac{1}{\tan x} - \tan x$
$= \cot x - \tan x$

**19.** $\sin 3x = \sin 2x \cos x + \cos 2x \sin x$
$= 2 \sin x \cos^2 x + (2 \cos^2 x - 1) \sin x$
$= (\sin x)(4 \cos^2 x - 1)$

**20.** $\sin 3x = \sin 2x \cos x + \cos 2x \sin x$
$= 2 \sin x \cos^2 x + (1 - 2 \sin^2 x) \sin x$
$= (\sin x)(2 \cos^2 x + 1 - 2 \sin^2 x)$
$= (\sin x)(3 - 4 \sin^2 x)$

**21.** $\cos 4x = \cos 2(2x) = 1 - 2 \sin^2 2x$
$= 1 - 2(2 \sin x \cos x)^2 = 1 - 8 \sin^2 x \cos^2 x$

**22.** $\sin 4x = \sin 2(2x) = 2 \sin 2x \cos 2x$
$= 2(2 \sin x \cos x)(2 \cos^2 x - 1)$
$= (4 \sin x \cos x)(2 \cos^2 x - 1)$

**23.** $\left\{\dfrac{\pi}{3}, \pi, \dfrac{5\pi}{3}\right\}$

**24.** $\{3.5163, \text{ or } 5.9085\}$

**25.** $\left\{\dfrac{\pi}{4}, \dfrac{\pi}{2}, \dfrac{3\pi}{4}, \dfrac{5\pi}{4}, \dfrac{3\pi}{2}, \dfrac{7\pi}{4}\right\}$

**26.** $\left\{0, \dfrac{\pi}{2}, \pi, \dfrac{3\pi}{2}\right\}$

**27.** $\left\{0, \dfrac{\pi}{3}, \dfrac{\pi}{2}, \dfrac{2\pi}{3}, \pi, \dfrac{4\pi}{3}, \dfrac{3\pi}{2}, \dfrac{5\pi}{3}\right\}$

**28.** $\left\{\dfrac{\pi}{6}, \dfrac{\pi}{2}, \dfrac{5\pi}{6}, \dfrac{7\pi}{6}, \dfrac{3\pi}{2}, \dfrac{11\pi}{6}\right\}$

**29.** $\left\{\dfrac{\pi}{2}, \dfrac{3\pi}{2}, 0.1\pi, 0.9\pi, 1.3\pi, 1.7\pi\right\}$

**30.** $\left\{\dfrac{3\pi}{2}, 0.3\pi, 0.7\pi, 1.1\pi, 1.9\pi\right\}$

**37. (a)** Starting from the right side: $\dfrac{1}{2}(1 - \cos 2u)$

$= \dfrac{1}{2}[1 - (1 - 2 \sin^2 u)] = \dfrac{1}{2}(2 \sin^2 u)$
$= \sin^2 u.$

**(b)** Starting from the right side: $\dfrac{1}{2}(1 + \cos 2u)$

$= \dfrac{1}{2}[1 + (2 \cos^2 u - 1)] = \dfrac{1}{2}(2 \cos^2 u)$
$= \cos^2 u.$

**38. (a)** $\tan^2 u = \dfrac{\sin^2 u}{\cos^2 u} = \dfrac{(1 - \cos 2u)/2}{(1 + \cos 2u)/2} = \dfrac{1 - \cos 2u}{1 + \cos 2u}$

**(b)** The equation is false when $\tan u$ is a negative number. It would be an identity if it were written as $|\tan u| = \sqrt{\dfrac{1 - \cos u}{1 + \cos u}}$.

**39.** $\sin^4 x = (\sin^2 x)^2 = \left[\dfrac{1}{2}(1 - \cos 2x)\right]^2$

$= \dfrac{1}{4}(1 - 2 \cos 2x + \cos^2 2x)$

$= \dfrac{1}{4}\left[1 - 2 \cos 2x + \dfrac{1}{2}(1 + \cos 4x)\right]$

$= \dfrac{1}{8}(2 - 4 \cos 2x + 1 + \cos 4x)$

$= \dfrac{1}{8}(3 - 4 \cos 2x + \cos 4x)$

**40.** $\cos^3 x = \cos x \cos^2 x = \cos x \cdot \dfrac{1}{2}(1 + \cos 2x)$

$= \dfrac{1}{2}(\cos x)(1 + \cos 2x)$

**41.** $\sin^3 2x = \sin 2x \sin^2 2x = \sin 2x \cdot \dfrac{1}{2}(1 - \cos 4x)$

$= \dfrac{1}{2}(\sin 2x)(1 - \cos 4x)$

**42.** $\sin^5 x = (\sin x)(\sin^2 x)^2 = (\sin x)\left[\dfrac{1}{2}(1 - \cos 2x)\right]^2$

$= \dfrac{1}{4}(\sin x)(1 - 2 \cos 2x + \cos^2 2x)$

$= \dfrac{1}{4}(\sin x)\left[1 - 2 \cos 2x + \dfrac{1}{2}(1 + \cos 4x)\right]$

$= \dfrac{1}{8}(\sin x)(2 - 4 \cos 2x + 1 + \cos 4x)$

$= \dfrac{1}{8}(\sin x)(3 - 4 \cos 2x + \cos 4x).$

Alternatively, take $\sin^5 x = \sin x \sin^4 x$ and apply the result of #39.

**43.** $\left\{\dfrac{\pi}{3}, \pi, \dfrac{5\pi}{3}\right\}$ General solution: $\pm\dfrac{\pi}{3} + 2n\pi$ or $\pi + 2n\pi$, $n$ an integer.

**44.** $\left\{\dfrac{\pi}{3}, \pi, \dfrac{5\pi}{3}\right\}$ General solution: $\pm\dfrac{\pi}{3} + 2n\pi$ or $\pi + 2n\pi$, $n$ an integer.

**45.** $\left\{0, \dfrac{\pi}{2}\right\}$ The general solution is $2n\pi$ or $\dfrac{\pi}{2} + 2n\pi$, $n$ an integer.

**46.** Let $\alpha \approx .7227$, $\{\alpha, \pi, 2\pi - \alpha\}$. General solution: $\pm\alpha + 2n\pi$ or $\pi + 2n\pi$, $n$ an integer.

**47. (a)** In the figure, the triangle with side lengths $x/2$ and $R$ is a right triangle, since $R$ is given as the perpendicular distance. Then the tangent of the angle $\theta/2$ is the ratio "opposite over adjacent": $\tan \dfrac{\theta}{2} = \dfrac{x/2}{R}$ Solving for $x$ gives the equation desired. The central angle $\theta$ is $2\pi/n$ since one full revolution of $2\pi$ radians is divided evenly into $n$ sections.

**(b)** $5.87 \approx 2R \tan \dfrac{\theta}{2}$, where $\theta = \dfrac{2\pi}{11}$, so

$$R \approx 5.87 \Big/ \left(2 \tan \frac{\pi}{11}\right) \approx 9.9957, \; R = 10.$$

**48. (a)** Call the center of the rhombus $E$. Consider right $\triangle ABE$, with lengths $d_2/2$ and $d_1/2$, and hypotenuse length $x$. $\angle ABE$ has measure $\theta/2$, and using "sine equals $\dfrac{\text{opp}}{\text{hyp}}$," and "cosine equals $\dfrac{\text{adj}}{\text{hyp}}$," we have

$$\cos \frac{\theta}{2} = \frac{d_2/2}{x} = \frac{d_2}{2x} \text{ and } \sin \frac{\theta}{2} = \frac{d_1/2}{x} = \frac{d_1}{2x}.$$

**(b)** Use the double angle formula for the sine function:

$$\sin \theta = \sin 2\left(\frac{\theta}{2}\right) = 2 \sin \frac{\theta}{2} \cos \frac{\theta}{2} = 2\frac{d_1}{2x} \cdot \frac{d_2}{2x}$$

$$= \frac{d_1 d_2}{2x^2}$$

**49.** $\theta = \dfrac{\pi}{6}$ The maximum value is about $12.99 \text{ ft}^3$.

**50. (a)** The height of the tunnel is $y$, and the width is $2x$, so the area is $2xy$. The $x$ and $y$ coordinates of the vertex are $20 \cos \theta$ and $20 \sin \theta$, so the area is $2(20 \cos \theta)(20 \sin \theta) = 400(2 \cos \theta \sin \theta) = 400 \sin 2\theta$.

**(b)** width of about $28.28$, and a height of $y \approx 14.14$

**51.** $\csc 2u = \dfrac{1}{\sin 2u} = \dfrac{1}{2 \sin u \cos u} = \dfrac{1}{2} \cdot \dfrac{1}{\sin u} \cdot \dfrac{1}{\cos u}$

$$= \frac{1}{2} \csc u \sec u$$

**52.** $\cot 2u = \dfrac{1}{\tan 2u} = \dfrac{1 - \tan^2 u}{2 \tan u}$

$$= \left(\frac{1 - \tan^2 u}{2 \tan u}\right)\left(\frac{\cot^2 u}{\cot^2 u}\right) = \frac{\cot^2 u - 1}{2 \cot u}$$

**53.** $\sec 2u = \dfrac{1}{\cos 2u} = \dfrac{1}{1 - 2 \sin^2 u}$

$$= \left(\frac{1}{1 - 2 \sin^2 u}\right)\left(\frac{\csc^2 u}{\csc^2 u}\right) = \frac{\csc^2 u}{\csc^2 u - 2}$$

**54.** $\sec 2u = \dfrac{1}{\cos 2u} = \dfrac{1}{2 \cos^2 u - 1}$

$$= \left(\frac{1}{2 \cos^2 u - 1}\right)\left(\frac{\sec^2 u}{\sec^2 u}\right) = \frac{\sec^2 u}{2 - \sec^2 u}$$

**55.** $\sec 2u = \dfrac{1}{\cos 2u} = \dfrac{1}{\cos^2 u - \sin^2 u}$

$$= \left(\frac{1}{\cos^2 u - \sin^2 u}\right)\left(\frac{\sec^2 u \csc^2 u}{\sec^2 u \csc^2 u}\right)$$

$$= \frac{\sec^2 u \csc^2 u}{\csc^2 u - \sec^2 u}$$

**56.** The second equation cannot work for any values of $x$ for which $\sin x < 0$, since the square root cannot be negative. The first is correct since a double angle identity for the cosine gives $\cos 2x = 1 - 2 \sin^2 x$; solving for $\sin x$ gives $\sin^2 x = \dfrac{1}{2}(1 - \cos 2x)$, so that

$\sin x = \pm \sqrt{\dfrac{1}{2}(1 - \cos 2x)}$. The absolute value of both sides removes the "$\pm$."

**57. (a)**

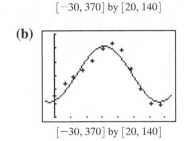

$[-30, 370]$ by $[20, 140]$

**(b)**

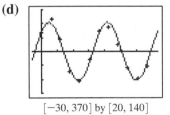

$[-30, 370]$ by $[20, 140]$

$y = 40.184 \sin(0.018x - 1.259) + 82.154$
This is a fairly good fit, but not really as good as one might expect from data generated by a sinusoidal physical model.

**(c)** The residual list:
$\{6.87, 11.69, 4.16, -7.33, -10.58, -2.76,$
$7.18, 8.82, 2.64, -5.69, -9.76, -5.20\}$

**(d)**

![Graph of residuals with sinusoidal curve, window [-30, 370] by [20, 140]]

$[-30, 370]$ by $[20, 140]$

$y = 10.401 (\sin 0.034x + 0.749) + 0.009$
This is another fairly good fit, which indicates that the residuals are not due to chance. There is a periodic variation that is most probably due to physical causes.

**(e)** The first regression indicates that the data are periodic and nearly sinusoidal. The second regression indicates that the *variation* of the data around the predicted values is also periodic and nearly sinusoidal. Periodic variation around periodic models is a predictable consequence of bodies orbiting bodies, but ancient astronomers had a difficult time reconciling the data with their simpler models of the universe.

## Section 5.5 (pp. 453–461)

### Exploration 1

1. If $BC \leq AB$, the segment will not reach from point $B$ to the dotted line. On the other hand, if $BC > AB$, then a circle of radius $BC$ will intersect the dotted line in a unique point. (Note that the line only extends to the left of point $A$.)
2. A circle of radius $BC$ will be tangent to the dotted line at $C$ if $BC = h$, thus determining a unique triangle. It will miss the dotted line entirely if $BC < h$, thus determining zero triangles.
3. The second point $(C_2)$ is the reflection of the first point $(C_1)$ on the other side of the altitude, as shown below.
4. $\sin C_2 = \sin (\pi - C_1)$
   $= \sin \pi \cos C_1 - \cos \pi \sin C_1 = \sin C_1$.
5. If $BC \geq AB$, then $BC$ can only extend to the right of the altitude, thus determining a unique triangle.

### Quick Review 5.5

9. 224.427
10. 315.573

### Exercises 5.5

1. $C = 75°$; $a \approx 4.5$; $c \approx 5.1$
2. $A = 45°$; $a \approx 13.9$; $b \approx 5.1$
3. $B = 45°$; $b \approx 15.8$; $c \approx 12.8$
4. $C = 59°$; $a \approx 141.4$; $c \approx 122.7$
5. $C = 110°$; $a \approx 12.9$; $c \approx 18.8$
6. $C = 68°$; $b \approx 4.6$; $c \approx 4.8$
7. $C = 77°$; $a \approx 4.1$; $c \approx 7.3$
8. $A = 61°$; $a \approx 10.8$; $b \approx 3.4$
9. $B \approx 20.1°$; $C \approx 127.9°$; $c \approx 25.3$
10. $B \approx 41.3°$; $C \approx 89.7°$; $c \approx 42.4$
11. $C \approx 37.2°$; $A \approx 72.8°$; $a \approx 14.2$
12. $B \approx 47.3°$; $A \approx 29.7°$; $a \approx 31.0$
19. $B_1 \approx 72.7°$; $C_1 \approx 43.3°$; $c_1 \approx 12.2$
    $B_2 \approx 107.3°$; $C_2 \approx 8.7°$; $c_2 \approx 2.7$
20. $C_1 \approx 47.1°$; $A_1 \approx 94.9°$; $a_1 \approx 34.0$

$C_2 \approx 132.9°$; $A_2 \approx 9.1°$; $a_2 \approx 5.4$
21. $A_1 \approx 78.2°$; $B_1 \approx 33.8°$; $b_1 \approx 10.8$
    $A_2 \approx 101.8°$; $B_2 \approx 10.2°$; $b_2 \approx 3.4$
22. $A_1 \approx 67.3°$; $C_1 \approx 55.7°$; $c_1 \approx 9.9$
    $A_2 \approx 112.7°$; $C_2 \approx 10.3°$; $c_2 \approx 2.1$
23. **(a)** $6.691 < b < 10$.
    **(b)** $b \approx 6.691$ or $b \geq 10$.
    **(c)** $b < 6.691$
24. **(a)** $9.584 < c < 12$.
    **(b)** $c \approx 9.584$ or $c \geq 12$.
    **(c)** $c < 9.584$
25. **(a)** No: this is an SAS case
    **(b)** No: only two pieces of information given.
26. **(a)** Yes: this is an AAS case.
    $B = 32°$; $b \approx 88.5$; $c \approx 146.1$
    **(b)** No: this is an SAS case.
28. $A \approx 16.2°$; $C \approx 116.8°$; $c \approx 25.6$
31. $A = 99°$; $a \approx 28.3$; $b \approx 19.1$
32. $C = 114°$; $a \approx 9.8$; $c \approx 27.5$
33. $A_1 \approx 24.6°$; $B_1 \approx 80.4°$; $a_1 \approx 20.7$
    $A_2 \approx 5.4°$; $B_2 \approx 99.6°$; $a_2 \approx 4.7$
34. $B_1 \approx 69.0°$; $C_1 \approx 57.0°$; $c_1 \approx 13.5$
    $B_2 \approx 111.0°$; $C_2 \approx 15.0°$; $c_2 \approx 4.2$
35. Cannot be solved by law of sines (an SAS case).
36. Cannot be solved by law of sines (an SAS case).
38. $a \approx 19.7$ mi; $b \approx 15.0$ mi; $h \approx 11.9$ mi
41. 1.3 ft
45. $a \approx 36.6$ mi; $b \approx 28.9$ mi
46. **(b)** Possible answers: $a = 1$, $b = \sqrt{3}$, $c = 2$ (or any set of three numbers proportional to these).
    **(c)** Any set of three identical numbers.
47. In each proof, assume that sides $a$, $b$, and $c$ are opposite angles $A$, $B$, and $C$, and that $c$ is the hypotenuse.
    **(a)** $\dfrac{\sin A}{a} = \dfrac{\sin 90°}{c}$
    $\dfrac{\sin A}{a} = \dfrac{1}{c}$
    $\sin A = \dfrac{a}{c} = \dfrac{\text{opp}}{\text{hyp}}$
    **(b)** $\dfrac{\sin B}{b} = \dfrac{\sin 90°}{c}$
    $\dfrac{\cos(\pi/2 - B)}{b} = \dfrac{1}{c}$
    $\cos A = \dfrac{b}{c} = \dfrac{\text{adj}}{\text{hyp}}$

**(c)** $\dfrac{\sin A}{a} = \dfrac{\sin B}{b}$

$\dfrac{\sin A}{\sin B} = \dfrac{a}{b}$

$\dfrac{\sin A}{\cos A} = \dfrac{a}{b}$

$\tan A = \dfrac{a}{b} = \dfrac{\text{opp}}{\text{adj}}$

**48. (a)** $h = AB \sin A$

**49.** $c \approx 3.9; A \approx 29.1°; B \approx 128.9°$

**50.** $AC \approx 8.7$ mi; $BC \approx 12.2$ mi; $h \approx 5.2$ mi

## Section 5.6 (pp. 461–470)

### Exploration 1

1. 8475.742818 paces$^2$
2. 41022.59524 square feet
3. 0.0014714831 square miles
4. 0.94175 acres
5. The estimate of "a little over an acre" seems questionable, but the roughness of their measurement system does not provide firm evidence that it is incorrect. If Jim and Barbara wish to make an issue of it with the owner, they would be well-advised to get some more reliable data.
6. Yes. In fact, any polygonal region can be subdivided into triangles.

### Quick Review 5.6

1. $A \approx 53.130°$
2. $C \approx 103.297°$
3. $A \approx 132.844°$
4. $C \approx 50.208°$
5. **(a)** $\cos A = \dfrac{x^2 + y^2 - 81}{2xy}$

   **(b)** $A = \cos^{-1}\!\left(\dfrac{x^2 + y^2 - 81}{2xy}\right)$
6. **(a)** $\cos A = \dfrac{x^2 - y^2 + 25}{10}$

   **(b)** $A = \cos^{-1}\!\left(\dfrac{x^2 - y^2 + 25}{10}\right)$
8. One answer: $(x - 1)(x + 1) = x^2 - 1$.

### Section 5.6 Exercises

1. $A \approx 30.7°; C \approx 18.3°, b \approx 19.2$
2. $A \approx 80.3°; B \approx 57.7°, c \approx 9.5$
3. $A \approx 76.8°; B \approx 43.2°, C \approx 60°$
4. $A \approx 52.2°; B \approx 99.2°, C \approx 28.6°$
5. $B \approx 89.3°; C \approx 35.7°, a \approx 9.8$

6. $A \approx 123.3°; C \approx 21.7°, b \approx 29.5$
7. $A \approx 28.5°; B \approx 56.5°, c \approx 25.1$
8. $B \approx 37.9°; C \approx 60.1°, a \approx 35.4$
11. $A \approx 24.6°; B \approx 99.2°, C \approx 56.2$
13. $B_1 \approx 72.9°; C_1 \approx 65.1°, c \approx 9.487$
    $B_2 \approx 107.1°; C_2 \approx 30.9°, c_2 \approx 5.376$
14. $B \approx 49.7°; C \approx 73.3°; c = 12.564$
16. $c \approx 7.447; B \approx 59.8°; C \approx 49.2°$
32. **(a)** 66.8 ft; This is a bit more than 63.7 ft
33. **(b)** The home-to-second segment is the hypotenuse of a right triangle, so the distance from the pitcher's rubber to second base is $60\sqrt{2} - 40 \approx 44.9$ ft
38. $HB \approx 37.0$ ft, $HC \approx 48.3$ ft, $HD \approx 52.3$ ft
40. $19.5°$
42. **(a)** $\dfrac{b^2 + c^2 - a^2}{2abc} = \dfrac{b^2 + c^2 - (b^2 + c^2 - 2bc \cos A)}{2abc}$

   Law of Cosines

   $= \dfrac{2bc \cos A}{2abc} = \dfrac{\cos A}{a}$

   **(b)** The identity in (a) has two other equivalent forms:

   $\dfrac{\cos B}{b} = \dfrac{a^2 + c^2 - b^2}{2abc}$

   $\dfrac{\cos C}{c} = \dfrac{a^2 + b^2 - c^2}{2abc}$

   We use them all in the proof:

   $\dfrac{\cos A}{a} + \dfrac{\cos B}{b} + \dfrac{\cos C}{c}$

   $= \dfrac{b^2 + c^2 - a^2}{2abc} + \dfrac{a^2 + c^2 - b^2}{2abc}$

   $\quad + \dfrac{a^2 + b^2 - c^2}{2abc}$

   $= \dfrac{b^2 + c^2 - a^2 + a^2 + c^2 - b^2 + a^2 + b^2 - c^2}{2abc}$

   $= \dfrac{a^2 + b^2 + c^2}{2abc}$

43. **(a)** Ship A: $\dfrac{30.2 - 15.1}{1 \text{ hr}} = 15.1$ knots;

   Ship B: $\dfrac{37.2 - 12.4}{2 \text{ hrs}} = 12.4$ knots

44. Use the area formula and the Law of Sines:

   $A_\triangle = \dfrac{1}{2}ab \sin C$

   $= \dfrac{1}{2}a\!\left(\dfrac{a \sin B}{\sin A}\right)\sin C \ \left(\text{Law of Sines} \Rightarrow b = \dfrac{a \sin B}{\sin A}\right)$

   $= \dfrac{a^2 \sin B \sin C}{2 \sin a}$

45. 6.9 in.$^2$

## Chapter 5 Review (pp. 473–476)

1. $\sin 200°$

**5.** $\cos 3x = \cos(2x + x)$
$= \cos 2x \cos x - \sin 2x \sin x$
$= (\cos^2 x - \sin^2 x) \cos x - (2 \sin x \cos x) \sin x$
$= \cos^3 x - 3 \sin^2 x \cos x$
$= \cos^3 x - 3(1 - \cos^2 x)\cos x$
$= \cos^3 x - 3 \cos x + 3 \cos^3 x$
$= 4 \cos^3 x - 3 \cos x$

**6.** $\cos^2 2x - \cos^2 x = (1 - \sin^2 2x) - (1 - \sin^2 x)$
$= \sin^2 x - \sin^2 2x$

**7.** $\tan^2 x - \sin^2 x = \sin^2 x \left( \dfrac{1 - \cos^2 x}{\cos^2 x} \right)$

$= \sin^2 x \cdot \dfrac{\sin^2 x}{\cos^2 x} = \sin^2 x \tan^2 x$

**8.** $2 \sin \theta \cos^3 \theta + 2 \sin^3 \theta \cos \theta$
$= (2 \sin \theta \cos \theta)(\cos^2 \theta + \sin^2 \theta)$
$= (2 \sin \theta \cos \theta)(1) = \sin 2\theta.$

**9.** $\csc x - \cos x \cot x = \dfrac{1}{\sin x} - \cos x \cdot \dfrac{\cos x}{\sin x}$

$= \dfrac{1 - \cos^2 x}{\sin x} = \dfrac{\sin^2 x}{\sin x} = \sin x$

**10.** $\dfrac{\tan \theta + \sin \theta}{2 \tan \theta} = \dfrac{1 + \cos \theta}{2} = \left( \pm \sqrt{\dfrac{1 + \cos \theta}{2}} \right)^2$

$= \left( \dfrac{\cos \theta}{2} \right)^2$

**11.** Recall that $\tan \theta \cot \theta = 1$. $\dfrac{1 + \tan \theta}{1 - \tan \theta} + \dfrac{1 + \cot \theta}{1 - \cot \theta}$

$= \dfrac{(1 + \tan \theta)(1 - \cot \theta) + (1 + \cot \theta)(1 - \tan \theta)}{(1 - \tan \theta)(1 - \cot \theta)}$

$= \dfrac{(1 + \tan \theta - \cot \theta - 1) + (1 + \cot \theta - \tan \theta - 1)}{(1 - \tan \theta)(1 - \cot \theta)}$

$= \dfrac{0}{(1 + \tan \theta)(1 - \cot \theta)} = 0$

**12.** $\sin 3\theta = \sin(2\theta + \theta) = \sin 2\theta \cos \theta + \cos 2\theta \sin \theta$
$= 2 \sin \theta \cos^2 \theta + (\cos^2 \theta - \sin^2 \theta) \sin \theta$
$= 3 \sin \theta \cos^2 \theta - \sin^3 \theta$

**13.** $\cos^2 \dfrac{t}{2} = \left[ \pm \sqrt{\dfrac{1}{2}(1 + \cos t)} \right]^2 = \dfrac{1}{2}(1 + \cos t)$

$= \left( \dfrac{1 + \cos t}{2} \right) \left( \dfrac{\sec t}{\sec t} \right) = \dfrac{\sec t + 1}{2 \sec t}$

**14.** $\dfrac{\tan^3 \gamma - \cot^3 \gamma}{\tan^2 \gamma + \csc^2 \gamma}$

$= (\tan \gamma - \cot \gamma)(\tan^2 \gamma + \tan \gamma \cot \gamma + \cot^2 \gamma)$

$= (\tan \gamma - \cot \gamma)(\tan^2 \gamma + 1 + \cot^2 \gamma)$

$= (\tan \gamma - \cot \gamma)(\tan^2 \gamma + \csc^2 \gamma) = \tan \gamma - \cot \gamma$

**15.** $\dfrac{\cos \phi}{1 - \tan \phi} + \dfrac{\sin \phi}{1 - \cot \phi}$

$= \left( \dfrac{\cos \phi}{1 - \tan \phi} \right) \left( \dfrac{\cos \phi}{\cos \phi} \right) + \left( \dfrac{\sin \phi}{1 - \cot \phi} \right) \left( \dfrac{\sin \phi}{\sin \phi} \right)$

$= \dfrac{\cos^2 \phi}{\cos \phi - \sin \phi} + \dfrac{\sin^2 \phi}{\sin \phi - \cos \phi}$

$= \dfrac{\cos^2 \phi - \sin^2 \phi}{\cos \phi - \sin \phi} = \cos \phi + \sin \phi$

**16.** $\dfrac{\cos(-z)}{\sec(-z) + \tan(-z)} = \dfrac{\cos(-z)}{[1 + \sin(-z)]/\cos(-z)}$

$= \dfrac{\cos^2(-z)}{1 + \sin(-z)} = \dfrac{1 - \sin^2 z}{1 - \sin z} = 1 + \sin z$

**17.** $\sqrt{\dfrac{1 - \cos y}{1 + \cos y}} = \sqrt{\dfrac{(1 - \cos y)^2}{(1 + \cos y)(1 - \cos y)}}$

$= \sqrt{\dfrac{(1 - \cos y)^2}{1 - \cos^2 y}} = \sqrt{\dfrac{(1 - \cos y)^2}{\sin^2 y}}$

$= \dfrac{|1 - \cos y|}{|\sin y|} = \dfrac{1 - \cos y}{|\sin y|}$ —since

$1 - \cos y \geq 0$, we can drop that absolute value.

**18.** $\sqrt{\dfrac{1 - \sin \gamma}{1 + \sin \gamma}} = \sqrt{\dfrac{(1 - \sin \gamma)(1 + \sin \gamma)}{(1 + \sin \gamma)^2}}$

$= \dfrac{|\cos \gamma|}{|1 + \sin \gamma|} = \dfrac{|\cos \gamma|}{1 + \sin \gamma}$ —since $1 + \sin \gamma \geq 0$,

we can drop that absolute value.

**19.** $\tan \left( u + \dfrac{3\pi}{4} \right) = \dfrac{\tan u + \tan (3\pi/4)}{1 - \tan u \tan (3\pi/4)}$

$= \dfrac{\tan u + (-1)}{1 - \tan u \, (-1)} = \dfrac{\tan u - 1}{1 + \tan u}$

**20.** $\dfrac{1}{4} \sin 4\gamma = \dfrac{1}{4} \sin 2(2\gamma) = \dfrac{1}{4}(2 \sin 2\gamma \cos 2\gamma)$

$= \dfrac{1}{2}(2 \sin \gamma \cos \gamma)(\cos^2 \gamma - \sin^2 \gamma)$

$= \sin \gamma \cos^3 \gamma - \cos \gamma \sin^3 \gamma$

**21.** $\tan \dfrac{1}{2}\beta = \dfrac{1 - \cos \beta}{\sin \beta} = \dfrac{1}{\sin \beta} - \dfrac{\cos \beta}{\sin \beta}$

$= \csc \beta - \cot \beta$

**22.** Let $\theta = \arctan t$, so that $\tan \theta = t$. Then

$\tan 2\theta = \dfrac{2 \tan \theta}{1 - \tan^2 \theta} = \dfrac{2t}{1 - t^2}$. Note also that since

$-1 < t < 1$, $-\dfrac{\pi}{4} < \theta < \dfrac{\pi}{4}$, and therefore $-\dfrac{\pi}{2} < \theta < \dfrac{\pi}{2}$.

That means that $2\theta$ is in the range of the arctan

function, and so $2\theta = \arctan \dfrac{2t}{1 - t^2}$, or

equivalently $\theta = \dfrac{1}{2} \arctan \dfrac{2t}{1 - t^2}$ — and of course,

$\theta = \arctan t$.

**23.** Yes: $\sec x - \sin x \tan x = \dfrac{1}{\cos x} - \dfrac{\sin^2 x}{\cos x}$

$= \dfrac{1 - \sin^2 x}{\cos x} = \dfrac{\cos^2 x}{\cos x} = \cos x$

**24.** Yes: $(\sin^2 \alpha - \cos^2 \alpha)(\tan^2 \alpha + 1)$

$$= (\sin^2 \alpha - \cos^2 \alpha)(\sec^2 \alpha) = \frac{\sin^2 \alpha - \cos^2 \alpha}{\cos^2 \alpha}$$

$$= \frac{\sin^2 \alpha}{\cos^2 \alpha} - 1 = \tan^2 \alpha - 1$$

**25.** Many answers are possible, for example,
$(\cos x - \sin x)(1 + 4 \sin x \cos x)$.

**26.** Many answers are possible, for example,
$\cos x(2 \sin x + 1 - 4 \sin^2 x)$.

**27.** Many answers are possible, for example,
$1 - 4 \sin^2 x \cos^2 x - 2 \sin x \cos x$.

**28.** Many answers are possible, for example
$\sin x(4 \cos^2 x - 1 - 6 \cos x)$.

**29.** $\frac{\pi}{12} + n\pi$ or $\frac{5\pi}{12} + n\pi$

**34.** $\frac{\pi}{6} + 2n\pi$ or $\frac{5\pi}{6} + 2n\pi$

**40.** $\left\{0, \frac{\pi}{4}, \frac{3\pi}{4}, \pi, \frac{5\pi}{4}, \frac{7\pi}{4}\right\}$

**41.** $\left\{\frac{3\pi}{2}\right\}$

**42.** $\left\{0, \frac{2\pi}{3}, \frac{4\pi}{3}\right\}$

**43.** no solutions

**44.** $\left\{\frac{\pi}{3}, \frac{5\pi}{3}\right\}$

**45.** $\left[0, \frac{\pi}{6}\right) \cup \left(\frac{5\pi}{6}, \frac{7\pi}{6}\right) \cup \left(\frac{11\pi}{6}, 2\pi\right)$

**49.** $y \approx 5 \sin(3x + 0.9273)$

**50.** $y \approx 13 \sin(2x - 1.176)$

**52.** $A \approx 36.0°$; $C \approx 34.0°$, $c \approx 4.8$

**54.** $B = 117.7°$; $b \approx 26.6$, $c \approx 16.4$

**55.** $C = 72°$; $a \approx 2.9$, $b \approx 5.1$

**56.** $B = 102°$; $a \approx 19.5°$, $b \approx 48.9$

**59.** 7.5

**60.** 23.0

**61.** (a) $\approx 5.6 < b < 12$.
  (b) $b \approx 5.6$ or $b \geq 12$.
  (c) $b < 5.6$.

**62.** (a) 102.5 ft

**66.** 15.794 ft, 36.777 ft

**67.** (a) $\sin \theta + \frac{1}{2} \sin 2\theta$.
  (b) $\theta = 60°$; 1.30 square units.

**68.** (a)

$[0, 9.4]$ by $[-6.2, 6.2]$

  (b) $\theta \approx 54.74°$

**69.** (a) $h = 4000 \sec \frac{\theta}{2} - 4000$ miles.

**70.** $n\frac{\pi}{2}$ or $\pm\frac{\pi}{3} + 2n\pi$

**71.** 139.140 cm$^2$ = area outside hexagon
area inside circle = $256\pi$ = 804.248 cm$^2$

**74.** (a) $\frac{1}{2}(\cos(u - v) - \cos(u + v))$

$$= \frac{1}{2}(\cos u \cos v + \sin u \sin v$$
$$- (\cos u \cos v - \sin u \sin v))$$
$$= \frac{1}{2}(2 \sin u \sin v) = \sin u \sin v$$

  (b) $\frac{1}{2}(\cos(u - v) + \cos(u + v))$

$$= \frac{1}{2}(\cos u \cos v + \sin u \sin v + \cos u \cos v$$
$$- \sin u \sin v)$$
$$= \frac{1}{2}(2 \cos u \cos v)$$
$$= \cos u \cos v$$

  (c) $\frac{1}{2}(\sin(u + v) + \sin(u - v))$

$$= \frac{1}{2}(\sin u \cos v + \cos u \sin v + \sin u \cos v$$
$$- \cos u \sin v)$$
$$= \frac{1}{2}(2 \sin u \cos v)$$
$$= \sin u \cos v$$

**75.** (a) By the product-to-sum formula in 74 (c),

$$2 \sin \frac{u + v}{2} \cos \frac{u - v}{2}$$
$$= 2 \cdot \frac{1}{2}\left(\sin \frac{u + v + u - v}{2}\right.$$
$$\left. + \sin \frac{u + v - (u - v)}{2}\right)$$
$$= \sin u + \sin v$$

  (b) By the product-to-sum formula in 74 (c),

$$2 \sin \frac{u - v}{2} \cos \frac{u + v}{2}$$
$$= 2 \cdot \frac{1}{2}\left(\sin \frac{u - v + u + v}{2}\right.$$
$$\left. + \sin \frac{u - v - (u + v)}{2}\right)$$
$$= \sin u + \sin(-v)$$
$$= \sin u - \sin v$$

  (c) By the product-to-sum formula in 74 (b),

$$2 \cos \frac{u + v}{2} \cos \frac{u - v}{2}$$
$$= 2 \cdot \frac{1}{2}\left(\cos \frac{u + v - (u - v)}{2}\right.$$
$$\left. + \cos \frac{u + v + u - v}{2}\right)$$

$$= \cos v + \cos u$$
$$= \cos u + \cos v$$

**(d)** By the product-to-sum formula in 74 (a),

$$-2 \sin \frac{u+v}{2} \sin \frac{u-v}{2}$$

$$= -2 \cdot \frac{1}{2} \left( \cos \frac{u+v-(u-v)}{2} \right.$$

$$\left. - \cos \frac{u+v+u-v}{2} \right)$$

$$= -(\cos v - \cos u)$$
$$= \cos u - \cos v$$

**76.** Pat faked the data. The Law of Cosines can be solved to show that $x = 12\sqrt{\dfrac{2}{1 - \cos \theta}}$. Only Carmen's values are consistent with the formula.

**77. (a)** Any inscribed angle that intercepts an arc of $180°$ is a right angle.

**(b)** Two inscribed angles that intercept the same arc are congruent.

**(c)** In right $\triangle A'BC$, $\sin A' = \dfrac{\text{opp}}{\text{hyp}} = \dfrac{a}{d}$.

**(d)** Because $\angle A'$ and $\angle A$ are congruent,

$$\frac{\sin A}{a} = \frac{\sin A'}{a} = \frac{a/d}{a} = \frac{1}{d}.$$

**(e)** Of course. They both equal $\dfrac{\sin A}{a}$ by the Law of Sines.

## Chapter 5 Project

**1.**

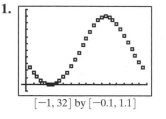

$[-1, 32]$ by $[-0.1, 1.1]$

**2.** One possible model is

$$y = 0.5 \cos \left( \frac{2\pi}{29}(x - 21) \right) + 0.5$$

**5.** One possible model is

$$y = 0.5 \cos \left( \frac{2\pi}{29}(x - 13.75) \right) + 0.5$$

# Chapter 6

## Section 6.1 (pp. 479–490)

### Exploration 1

**1.** $[-1, 3]$; $[4, 7]$
**2.** $[3, 10]$
**3.** $[-3, 9]$; $[-6, 10]$

### Quick Review 6.1

**1.** $\dfrac{9\sqrt{3}}{2}$, $4.5$

**2.** $-7.5$, $\dfrac{15\sqrt{3}}{2}$

**3.** $-5.362$, $-4.499$
**4.** $3.856$, $-4.596$
**5.** $33.854°$
**6.** $104.963°$
**7.** $60.945°$
**8.** $305.537°$
**9.** $248.199°$
**10.** Distance: $254.14$ naut mi. Bearing: $95.40°$

### Section 6.1 Exercises

**10.** $\langle 0, 3\sqrt{2} \rangle$, $3\sqrt{2}$
**11.** $\langle -11, -7 \rangle$, $\sqrt{170}$
**12.** $\langle -12, -16 \rangle$, $20$

**21. (a)** $\left\langle \dfrac{2}{\sqrt{5}}, \dfrac{1}{\sqrt{5}} \right\rangle$

**(b)** $\dfrac{2}{\sqrt{5}}\mathbf{i} + \dfrac{1}{\sqrt{5}}\mathbf{j}$

**22. (a)** $\left\langle -\dfrac{3}{\sqrt{13}}, \dfrac{2}{\sqrt{13}} \right\rangle$

**(b)** $-\dfrac{3}{\sqrt{13}}\mathbf{i} + \dfrac{2}{\sqrt{13}}\mathbf{j}$

**23. (a)** $\left\langle -\dfrac{4}{\sqrt{41}}, -\dfrac{5}{\sqrt{41}} \right\rangle$

**(b)** $-\dfrac{4}{\sqrt{41}}\mathbf{i} - \dfrac{5}{\sqrt{41}}\mathbf{j}$

**24. (a)** $\left\langle \dfrac{3}{5}, -\dfrac{4}{5} \right\rangle$

**(b)** $\dfrac{3}{5}\mathbf{i} - \dfrac{4}{5}\mathbf{j}$

**25.** $\approx \langle 16.314, 7.607 \rangle$
**26.** $\approx \langle 8.030, 11.468 \rangle$
**27.** $\approx \langle -14.524, 44.700 \rangle$
**28.** $\approx \langle -23.738, 22.924 \rangle$
**29.** $5; \approx 53.13°$
**30.** $\sqrt{5}; \approx 116.57°$
**31.** $5; \approx 306.87°$
**32.** $\sqrt{34}; \approx 239.04°$
**33.** $7; 135°$
**34.** $2; 60°$
**35.** $\approx -0.45\mathbf{i} + 0.89\mathbf{j}$
**36.** $\approx 0.71\mathbf{i} - 0.71\mathbf{j}$
**37.** $\approx -0.45\mathbf{i} - 0.89\mathbf{j}$
**38.** $\approx 0.71\mathbf{i} - 0.71\mathbf{j}$
**39.** $\langle \sqrt{2}, -\sqrt{2} \rangle$
**40.** $\langle -2.91, 4.07 \rangle$
**43.** **(a)** $\approx \langle -111.16, 305.40 \rangle$
  **(b)** $362.84$ mph; $337.84°$
**44.** **(a)** $\approx \langle 79.88, -453.01 \rangle$
  **(b)** $530.79$ mph; $174.32°$
**45.** **(a)** $\approx \langle 3.42, 9.40 \rangle$
**46.** **(a)** $\approx \langle 2.41, 0.65 \rangle$

## Section 6.2 (pp. 490–497)

### Quick Review 6.2

**4.** $2$
**5.** $\langle 3, \sqrt{13} \rangle$
**6.** $\langle -1, \sqrt{13} \rangle$
**7.** $\langle -1, -\sqrt{13} \rangle$
**8.** $\langle 3, -\sqrt{13} \rangle$
**9.** $\left\langle \dfrac{4}{\sqrt{13}}, \dfrac{6}{\sqrt{13}} \right\rangle$
**10.** $\left\langle -\dfrac{12}{5}, \dfrac{9}{5} \right\rangle$

### Section 6.2 Exercises

**15.** $\approx 64.65°$
**16.** $\approx 167.66°$
**19.** $135°$
**20.** $90°$
**23.** $-\dfrac{21}{10}\langle 3, 1 \rangle; -\dfrac{21}{10}\langle 3, 1 \rangle + \dfrac{17}{10}\langle -1, 3 \rangle$
**24.** $-\dfrac{9}{5}\langle 1, 3 \rangle; -\dfrac{9}{5}\langle 1, 3 \rangle + \dfrac{8}{5}\langle 3, -1 \rangle$
**25.** $\dfrac{82}{85}\langle 9, 2 \rangle; \dfrac{82}{85}\langle 9, 2 \rangle + \dfrac{29}{85}\langle -2, 9 \rangle$
**26.** $\dfrac{7}{5}\langle -3, 1 \rangle; \dfrac{7}{5}\langle -3, 1 \rangle + \dfrac{1}{5}\langle 11, 33 \rangle$
**27.** $47.73°, 74.74°, 57.53°$

**28.** $63.87°, 41.93°$
**29.** $-20.78$
**30.** $240$
**35.** Orthogonal
**36.** Parallel
**37.** **(a)** $(4, 0)$ and $(0, -3)$
  **(b)** $(4.6, -0.8)$ or $(3.4, 0.8)$
**38.** **(a)** $(-5, 0)$ and $(0, 2)$
  **(b)** $\approx (-5.37, 0.93)$ or $(-4.63, -0.93)$
**39.** **(a)** $(7, 0)$ and $(0, -3)$
  **(b)** $\approx (7.39, -0.92)$ or $(6.61, 0.92)$
**40.** **(a)** $(6, 0)$ and $(0, 3)$
  **(b)** $\approx (6.45, 0.89)$ or $(5.55, -0.89)$
**41.** $\langle -1, 4 \rangle$ or $\left\langle \dfrac{5}{13}, \dfrac{8}{13} \right\rangle$
**42.** $\langle 3, -1 \rangle$ or $\left\langle -\dfrac{43}{29}, -\dfrac{81}{29} \right\rangle$
**43.** $138.56$ pounds
**45.** **(b)** $1956.30$ pounds
**47.** $14{,}300$ foot-pounds
**51.** $85.38$ foot-pounds
**52.** $245.15$ foot-pounds
**54.** $201.9$ foot-pounds
**59.** **(a)** $2 \cdot 0 + 5 \cdot 2 = 10$ and $2 \cdot 5 + 5 \cdot 0 = 10$
  **(b)** $\dfrac{5}{29}\langle 5, -2 \rangle; \dfrac{1}{29}\langle 62, 155 \rangle$
  **(c)** $|\mathbf{w}_2| = \dfrac{31\sqrt{29}}{29}$
  **(d)** $\dfrac{1}{29}\sqrt{(4x_0 + 10y_0 - 20)^2 + (10x_0 + 25y_0 - 50)^2}$
  **(e)** $d = \dfrac{ax_0 + by_0 - c}{\sqrt{a^2 + b^2}}$
**60.** **(a)** Yes, if $\mathbf{v} = \langle 0, 0 \rangle$ or $t = n\pi$, $n = $ any integer
  **(b)** Yes, if $\mathbf{u} = \langle 0, 0 \rangle$ or $t = \dfrac{n\pi}{2}$, $n = $ odd integer
  **(c)** Generally, no, because $\sin t \neq \cos t$ for most $t$. Exceptions, however, would occur when $t = \dfrac{\pi}{4} + n\pi$, $n = $ any integer, or if $\mathbf{u} = \langle 0, 0 \rangle$ and/or $\mathbf{v} = \langle 0, 0 \rangle$.

## Section 6.3 (pp. 497–510)

### Exploration 1

**1.**

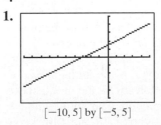

$[-10, 5]$ by $[-5, 5]$

**5.** Tmin $\leq -2$ and Tmax $\geq 5.5$

## Exploration 2

**2.** It looks like the line in figure 6.28.

**3.**

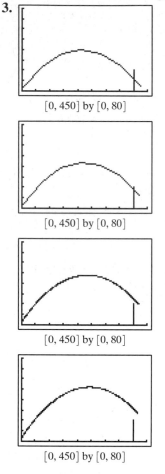

[0, 450] by [0, 80]

[0, 450] by [0, 80]

[0, 450] by [0, 80]

[0, 450] by [0, 80]

## Exploration 3

**1.** Jane is travelling in a circle of radius 20 feet and origin (0, 20), which yields $x_1 = 20 \cos (nt)$ and $y_1 = 20 + 20 \sin (nt)$. Since the ferris wheel is making one revolution ($2\pi$) every 12 seconds, $2\pi = 12n$, so $n = \dfrac{2\pi}{12} = \dfrac{\pi}{6}$. Thus,

$$x_1 = 20 \cos \left(\frac{\pi}{6}t\right) \text{ and } y_1 = 20 + 20 \sin \left(\frac{\pi}{6}t\right),$$

in radian mode.

**2.** Since the ball was released as 75 ft in the positive *x*-direction and gravity acts in the negative *y*-direction at 16 ft/s², we have $x_2 = at + 75$ and $y_2 = -16t^2 + bt$, where *a* is the initial speed of the ball in the *x*-direction and is the initial speed of the ball in the *y*-direction. The initial velocity vector of the ball is 60 $\langle \cos 120°, \sin 120° \rangle$

$= \langle -30, 30\sqrt{3} \rangle$, so $a = -30$ and $b = 30\sqrt{3}$. As a result $x_2 = -30t + 75$ and $y_2 = -16t^3 + 30\sqrt{3}t$ are the parametric equations for the ball.

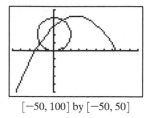

[−50, 100] by [−50, 50]

**3.** Jane and the ball will be close to each other but not at the exact same point at $t = 2.1$ seconds.

**4.** $d(t) = \sqrt{\left(20 \cos \left(\dfrac{\pi}{6}t\right) + 30t - 75\right)^2 + \left(20 + 20 \sin \left(\dfrac{\pi}{6}t\right) + 16t^2 - (30\sqrt{3}t)\right)^2}$

**5.** The minimum distance occurs at $t = 2.2$, when $d(t) = 1.64$ feet.

## Quick Review 6.3

**1. (a)** $\langle -3, -2 \rangle$
   **(b)** $\langle 4, 6 \rangle$
   **(c)** $\langle 7, 8 \rangle$
**2. (a)** $\langle -1, 3 \rangle$
   **(b)** $\langle 4, -3 \rangle$
   **(c)** $\langle 5, -6 \rangle$
**3.** $y + 2 = \dfrac{8}{7}(x + 3)$ or $y - 6 = \dfrac{8}{7}(x + 4)$
**4.** $y - 3 = -\dfrac{6}{5}(x + 1)$ or $y + 3 = -\dfrac{6}{5}(x - 4)$
**5.**

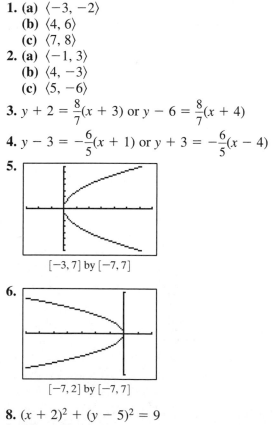

[−3, 7] by [−7, 7]

**6.**

[−7, 2] by [−7, 7]

**8.** $(x + 2)^2 + (y - 5)^2 = 9$
**9.** $20\pi$ rad/sec

**10.** $\dfrac{70}{3}\pi$ rad/sec

## Section 6.3 Exercises

**1. (b)** $[-5, 5]$ by $[-5, 5]$
**2. (d)** $[-5, 5]$ by $[-5, 5]$
**3. (a)** $[-5, 5]$ by $[-5, 5]$
**4. (c)** $[-10, 10]$ by $[-12, 10]$

**5. (b)**

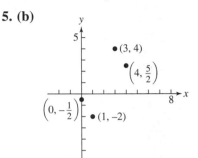

**6. (b)**

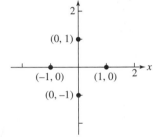

**7.** $y = x - 1$: line through $(0, -1)$ and $(1, 0)$
**8.** $y = -\dfrac{1}{3}x + \dfrac{17}{3}$: line through $\left(0, \dfrac{17}{3}\right)$ and $\left(\dfrac{17}{3}, 0\right)$
**9.** $y = -2x + 3$, $3 \le x \le 7$: line segment with endpoints $(3, -3)$ and $(7, -11)$
**10.** $y = -\dfrac{1}{3}x + \dfrac{11}{3}$, $-4 \le x \le 8$: line segment with endpoints $(8, 1)$ and $(-4, 5)$
**11.** $x = (y - 1)^2$: parabola that opens to right with vertex at $(0, 1)$
**12.** $y = x^2 - 3$: parabola that opens upward with vertex at $(0, -3)$
**13.** $y = x^3 - 2x + 3$: cubic polynomial
**16.** $t = 2x$, so $y = 16x^3 - 3$: cubic, $-1 \le x \le 1$
**17.** $t = x + 3$, so $y = \dfrac{2}{x + 3}$, on domain: $[8, -3) \cup (-3, 2]$
**18.** $t = x - 2$, so: $y = \dfrac{4}{x - 2}$, $x \ge 4$
**19.** $x^2 + y^2 = 25$, circle of radius 5 centered at $(0, 0)$
**20.** $x^2 + y^2 = 16$, circle of radius 4 centered at $(0, 0)$
**21.** $x^2 + y^2 = 4$, circle of radius 2 centered at $(0, 0)$ (not in Quadrant II)

**22.** $x^2 + y^2 = 9$, semicircle of radius 3, $y \ge 0$ only.
**23.** $x = 6t - 2$; $y = -3t + 5$
**24.** $x = 8t - 3$; $y = 4t - 3$

For #25–28, many answers are possible; one of the simplest is given.
**25.** $x = 3t + 3$, $y = 4 - 7t$, $0 \le t \le 1$
**26.** $x = 5 - 7t$, $y = 2 - 6t$, $0 \le t \le 1$
**27.** $x = 5 + 3 \cos t$, $y = 2 + 3 \sin t$, $0 \le t \le 2\pi$
**28.** $x = -2 + 2 \cos t$, $y = -4 + 2 \sin t$, $0 \le t \le 2\pi$
**33. (a)** $y = -16t^2 + 1000$
**34. (c)** Graph $x = t$ and $y = -16t^2 + 80t + 5$ with $0 \le t \le 5.1$ (upper limit may vary).

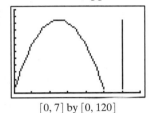

$[0, 7]$ by $[0, 120]$

**(e)** 105 ft; 2.5 sec
**35. (b)** Ben is ahead by 2 ft.
**36. (b)** The left runner by 4.1 ft
**42.** 6.60 ft; 1.206 sec
**45.** $v \approx -10.0015$ ft/sec
**46. (a)** 506.25 ft.
**(b)** 650.82 ft.
**(c)** 775.62 ft.
**(d)** 876.85 ft.

**47.** $x = 35 \cos\left(\dfrac{\pi}{6}t\right)$ and $y = 50 + 35 \sin\left(\dfrac{\pi}{6}t\right)$
**48.** about 3.38 ft.
**53. (a)**

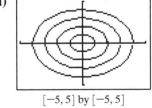

$[-5, 5]$ by $[-5, 5]$

**(b)** $a$
**(c)**

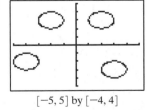

$[-5, 5]$ by $[-4, 4]$

**(d)** circle of radius $a$ centered at $(h, k)$.
**(e)** $x = 3 \cos t - 1$; $y = 3 \sin t + 4$

**54. (a)**

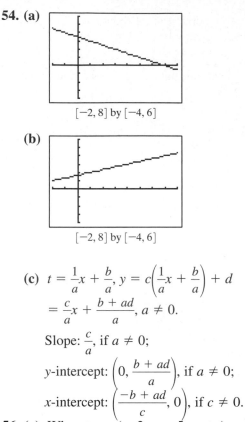

[−2, 8] by [−4, 6]

**(b)**

[−2, 8] by [−4, 6]

**(c)** $t = \dfrac{1}{a}x + \dfrac{b}{a}$, $y = c\left(\dfrac{1}{a}x + \dfrac{b}{a}\right) + d$

$= \dfrac{c}{a}x + \dfrac{b + ad}{a}$, $a \neq 0$.

Slope: $\dfrac{c}{a}$, if $a \neq 0$;

$y$-intercept: $\left(0, \dfrac{b + ad}{a}\right)$, if $a \neq 0$;

$x$-intercept: $\left(\dfrac{-b + ad}{c}, 0\right)$, if $c \neq 0$.

**56. (a)** When $t = \pi$ (or $3\pi$, or $5\pi$, etc.), $y = 2$. This corresponds to the highest points on the graph.

**(b)** $2\pi$ units

**57. (a)**

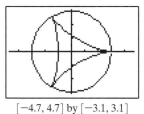

[−4.7, 4.7] by [−3.1, 3.1]

**59.** about 10.48 ft

**62.** $t = \dfrac{1}{3}$; $t = \dfrac{1}{4}$

## Section 6.4 (pp. 510–516)

### Quick Review 6.4

**7.** $(x - 3)^2 + y^2 = 4$
**8.** $x^2 + (y + 4)^2 = 9$
**9.** $\approx 11.14$
**10.** $\approx 5.85$

### Exercises 6.4

**1.** $\left(-\dfrac{3}{2}, \dfrac{3\sqrt{3}}{2}\right)$
**2.** $(2\sqrt{2}, 2\sqrt{2})$
**3.** $(-1, -\sqrt{3})$
**4.** $\left(-\dfrac{\sqrt{2}}{2}, \dfrac{\sqrt{2}}{2}\right)$
**5. (a)**

| $\theta$ | $\dfrac{\pi}{4}$ | $\dfrac{\pi}{2}$ | $\dfrac{5\pi}{6}$ | $\pi$ | $\dfrac{4\pi}{3}$ | $2\pi$ |
|---|---|---|---|---|---|---|
| $r$ | $\dfrac{3\sqrt{2}}{2}$ | 3 | $\dfrac{3}{2}$ | 0 | $\dfrac{-3\sqrt{3}}{2}$ | 0 |

**(b)**

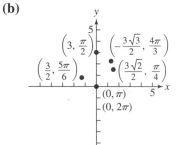

**6. (a)**

| $\theta$ | $\dfrac{\pi}{4}$ | $\dfrac{\pi}{2}$ | $\dfrac{5\pi}{6}$ | $\pi$ | $\dfrac{4\pi}{3}$ | $2\pi$ |
|---|---|---|---|---|---|---|
| $r$ | $2\sqrt{2}$ | 2 | 4 | error | $\dfrac{-4\sqrt{3}}{3}$ | error |

**(b)**

**7.**

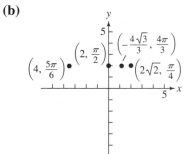

$\left(3, \dfrac{4\pi}{3}\right)$

**8.** $\left(2, \dfrac{5\pi}{6}\right)$

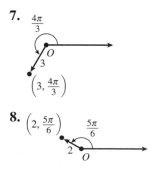

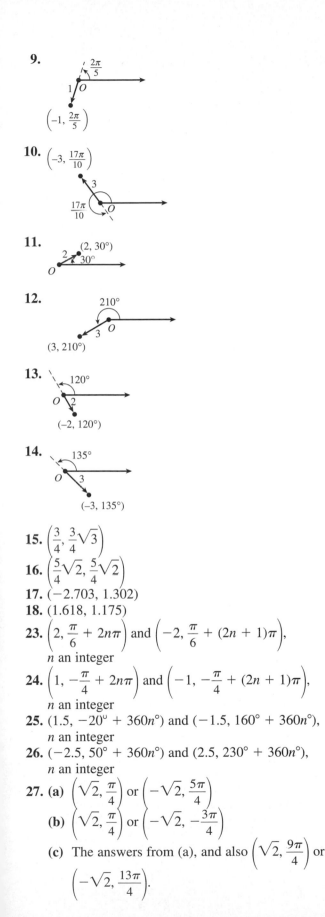

**9.**

$\left(-1, \frac{2\pi}{5}\right)$

**10.** $\left(-3, \frac{17\pi}{10}\right)$

**11.** $(2, 30°)$

**12.** $(3, 210°)$

**13.** $(-2, 120°)$

**14.** $(-3, 135°)$

**15.** $\left(\frac{3}{4}, \frac{3}{4}\sqrt{3}\right)$

**16.** $\left(\frac{5}{4}\sqrt{2}, \frac{5}{4}\sqrt{2}\right)$

**17.** $(-2.703, 1.302)$

**18.** $(1.618, 1.175)$

**23.** $\left(2, \frac{\pi}{6} + 2n\pi\right)$ and $\left(-2, \frac{\pi}{6} + (2n + 1)\pi\right)$, $n$ an integer

**24.** $\left(1, -\frac{\pi}{4} + 2n\pi\right)$ and $\left(-1, -\frac{\pi}{4} + (2n + 1)\pi\right)$, $n$ an integer

**25.** $(1.5, -20° + 360n°)$ and $(-1.5, 160° + 360n°)$, $n$ an integer

**26.** $(-2.5, 50° + 360n°)$ and $(2.5, 230° + 360n°)$, $n$ an integer

**27. (a)** $\left(\sqrt{2}, \frac{\pi}{4}\right)$ or $\left(-\sqrt{2}, \frac{5\pi}{4}\right)$

   **(b)** $\left(\sqrt{2}, \frac{\pi}{4}\right)$ or $\left(-\sqrt{2}, -\frac{3\pi}{4}\right)$

   **(c)** The answers from (a), and also $\left(\sqrt{2}, \frac{9\pi}{4}\right)$ or $\left(-\sqrt{2}, \frac{13\pi}{4}\right)$.

**28. (a)** $(\sqrt{10}, 1.249)$ or $(-\sqrt{10}, 4.391)$

   **(b)** $(\sqrt{10}, 1.249)$ or $(-\sqrt{10}, -1.893)$

   **(c)** The answers from (a), and also $(\sqrt{10}, 7.532)$ or $(-\sqrt{10}, 10.674)$

**29. (a)** $(\sqrt{29}, 1.951)$ or $(-\sqrt{29}, 5.093)$

   **(b)** $(-\sqrt{29}, -1.190)$ or $(\sqrt{29}, 1.951)$

   **(c)** The answers from (a), plus $(\sqrt{29}, 8.234)$ or $(-\sqrt{29}, 11.376)$

**30. (a)** $(-\sqrt{5}, 1.107)$ or $(\sqrt{5}, 4.248)$

   **(b)** $(-\sqrt{5}, 1.107)$ or $(\sqrt{5}, -2.034)$

   **(c)** The answers from (a), plus $(-\sqrt{5}, 7.390)$ or $(\sqrt{5}, 10.532)$

**35.** $x = 3$ — a vertical line

**36.** $y = -2$ — a horizontal line

**37.** $x^2 + \left(y + \frac{3}{2}\right)^2 = \frac{9}{4}$ — a circle centered at $\left(0, -\frac{3}{2}\right)$ with radius $\frac{3}{2}$.

**38.** $(x + 2)^2 + y^2 = 4$ — a circle centered at $(-2, 0)$ with radius 2

**39.** $x^2 + \left(y - \frac{1}{2}\right)^2 = \frac{1}{4}$ — a circle centered at $\left(0, \frac{1}{2}\right)$ with radius $\frac{1}{2}$.

**40.** $\left(x - \frac{3}{2}\right)^2 + y^2 = \frac{9}{4}$ — a circle centered at $\left(\frac{3}{2}, 0\right)$ with radius $\frac{3}{2}$.

**41.** $(x + 2)^2 + (y - 1)^2 = 5$ — a circle centered at $(-2, 1)$ with radius $\sqrt{5}$.

**42.** $(x - 2)^2 + (y + 2)^2 = 8$ — a circle centered at $(2, -2)$ with radius $2\sqrt{2}$.

**43.** $r = 2/\cos\theta = 2\sec\theta$

$[-5, 5]$ by $[-5, 5]$

**44.** $r = 5/\cos\theta = 5\sec\theta$

$[0, 10]$ by $[-5, 5]$

**45.** $r = \dfrac{5}{2 \cos \theta - 3 \sin \theta}$

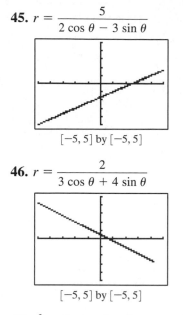

[−5, 5] by [−5, 5]

**46.** $r = \dfrac{2}{3 \cos \theta + 4 \sin \theta}$

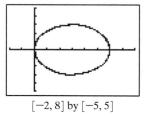

[−5, 5] by [−5, 5]

**47.** $r^2 - 6r \cos \theta = 0$, so $r = 6 \cos \theta$

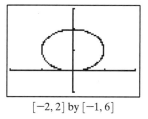

[−2, 8] by [−5, 5]

**48.** $r^2 - 2r \sin \theta = 0$, so $r = 2 \sin \theta$

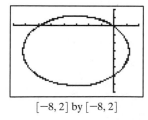

[−2, 2] by [−1, 6]

**49.** $r^2 + 6r \cos \theta + 6r \sin \theta = 0$, so
$r = -6 \cos \theta - 6 \sin \theta$

[−8, 2] by [−8, 2]

**50.** $r^2 - 2r \cos \theta + 8r \sin \theta = 0$, so
$r = 2 \cos \theta - 8 \sin \theta$

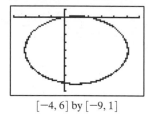

[−4, 6] by [−9, 1]

**53.** $\left( \dfrac{a}{\sqrt{2}}, \dfrac{\pi}{4} \right), \left( \dfrac{a}{\sqrt{2}}, \dfrac{3\pi}{4} \right), \left( \dfrac{a}{\sqrt{2}}, \dfrac{5\pi}{4} \right),$ and $\left( \dfrac{a}{\sqrt{2}}, \dfrac{7\pi}{4} \right)$

**54.** $(a, 0), \left( a, \dfrac{2\pi}{5} \right), \left( a, \dfrac{4\pi}{5} \right), \left( a, \dfrac{6\pi}{5} \right),$ and $\left( a, \dfrac{8\pi}{5} \right)$

**55. (a)** If $\theta_1 - \theta_2$ is an odd integer multiple of $\pi$, then the distance is $|r_1 + r_2|$. If $\theta_1 - \theta_2$ is an even integer multiple of $\pi$, then the distance is $|r_1 - r_2|$.

**56. (a)** The right half of a circle centered at (0, 2) of radius 2.
**(b)** Three quarters of the same circle, starting at (0, 0) and moving counter-clockwise.
**(c)** The full circle (plus another half circle found through the trace function).
**(d)** 4 counter-clockwise rotations of the same circle.

**59.** $\approx 7.43$
**60.** $\approx 4.11$
**61.** $x = f(\theta) \cos (\theta), y = f(\theta) \sin (\theta)$
**62.** $x = 2 \cos^2 \theta$
    $y = 2(\cos \theta)(\sin \theta)$
**63.** $x = 5(\cos \theta)(\sin \theta)$
    $y = 5 \sin^2 \theta$
**64.** $x = 2$
    $y = 2 \tan \theta$
**65.** $x = 4 \cot \theta$
    $y = 4$

## Section 6.5 (pp. 517–524)

### Quick Review 6.5

**1.** Minimum: $-3$ at $x \left\{ \dfrac{\pi}{2}, \dfrac{3\pi}{2} \right\}$;
    Maximum: 3 at $x = \{0, \pi, 2\pi\}$
**2.** Minimum: $-1$ at $x = \pi$;
    Maximum: 5 at $x = \{0, 2\pi\}$
**3.** Minimum: approaches 0 but undefined at
    $x = \left\{ \dfrac{\pi}{4}, \dfrac{3\pi}{4}, \dfrac{5\pi}{4}, \dfrac{7\pi}{4} \right\}$; Maximum: 2 at
    $x = \{0, \pi, 2\pi\}$

**4.** Minimum: 0 at $x = \dfrac{\pi}{2}$; Maximum: 6 at $x = \dfrac{3\pi}{2}$

**9.** $\cos^2 \theta - \sin^2 \theta$

**10.** $2 \sin \theta \cos \theta$

## Section 6.5 Exercises

**1. (a)**

| $\theta$ | 0 | $\dfrac{\pi}{4}$ | $\dfrac{\pi}{2}$ | $\dfrac{3\pi}{4}$ | $\pi$ | $\dfrac{5\pi}{4}$ | $\dfrac{3\pi}{2}$ | $\dfrac{7\pi}{4}$ |
|---|---|---|---|---|---|---|---|---|
| $r$ | 3 | 0 | $-3$ | 0 | 3 | 0 | $-3$ | 0 |

**(b)**

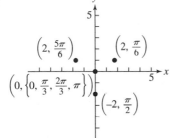

$\left(0, \left\{\dfrac{\pi}{4}, \dfrac{3\pi}{4}, \dfrac{5\pi}{4}, \dfrac{7\pi}{4}\right\}\right)$

$\left(-3, \dfrac{3\pi}{2}\right)$

$(3, 0)$

$(3, \pi)$

$\left(-3, \dfrac{\pi}{2}\right)$

**2. (a)**

| $\theta$ | 0 | $\dfrac{\pi}{6}$ | $\dfrac{\pi}{3}$ | $\dfrac{\pi}{2}$ | $\dfrac{2\pi}{3}$ | $\dfrac{5\pi}{6}$ | $\pi$ |
|---|---|---|---|---|---|---|---|
| $r$ | 0 | 2 | 0 | $-2$ | 0 | 2 | 0 |

**(b)**

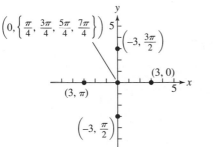

$\left(2, \dfrac{5\pi}{6}\right)$

$\left(2, \dfrac{\pi}{6}\right)$

$\left(0, \left\{0, \dfrac{\pi}{3}, \dfrac{2\pi}{3}, \pi\right\}\right)$

$\left(-2, \dfrac{\pi}{2}\right)$

**3.** $k = \pi$

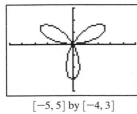

$[-5, 5]$ by $[-4, 3]$

**4.** $k = 2\pi$

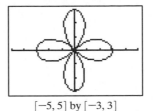

$[-5, 5]$ by $[-3, 3]$

**5.** $k = 2\pi$

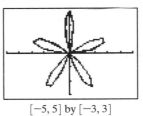

$[-5, 5]$ by $[-3, 3]$

**6.** $k = \pi$

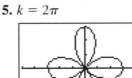

$[-5, 5]$ by $[-3, 3]$

**7.** $r_3$ is graph (b)

**9.** Graph (b) is $r = 2 - 2 \cos \theta$

**10.** Graph (c) is $r = 2 + 3 \cos \theta$

**11.** Graph (a) is $r = 2 - 2 \sin \theta$

**12.** Graph (d) is $r = 2 - 1.5 \sin \theta$

**13.** Maximum $r$ is 3 (along with $-3$)—when $\theta = 2n\pi/3$ for any integer $n$.

**14.** Maximum $r$ is 4 (along with $-4$)—when $\theta = n\pi/4$ for any odd integer $n$.

**15.** Maximum $r$ is 5 — when $\theta = 2n\pi$ for any integer $n$.

**16.** Maximum $r$ is 5 (actually, $-5$) — when $\theta = \dfrac{3\pi}{2} + 2n\pi$ for any integer $n$.

**17.** Symmetric about the $y$-axis.

**18.** Symmetric about the $x$-axis.

**19.** Symmetric about the $x$-axis.

**20.** Symmetric about the $y$-axis.

**21.** All three symmetries.

**22.** Symmetric about the $y$-axis.

**23.** Symmetric about the $y$-axis.

**24.** Symmetric about the $x$-axis.

**25.** Domain: All reals
Range: $[-3, 3]$
Symmetric about the $x$-axis, $y$-axis, and origin

Continuous
Bounded
Maximum *r*-value: 3
No asymptotes

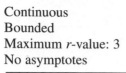

[−4, 4] by [−4, 4]

**26.** Domain: All reals
Range: [−2, 2]
Symmetric about the *x*-axis, *y*-axis, and origin
Continuous
Bounded
Maximum *r*-value: 2
No asymptotes

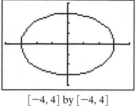

[−3, 3] by [−3, 3]

**27.** Domain: $\theta = \pi/3 + n\pi$, *n* is an integer
Range: $(-\infty, \infty)$
Symmetric about the origin
Continuous
Unbounded
Maximum *r*-value: none
No asymptotes

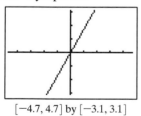

[−4.7, 4.7] by [−3.1, 3.1]

**28.** Domain: $\theta = -\pi/4 + n\pi$, *n* is an integer
Range: $(-\infty, \infty)$
Symmetric about the origin
Continuous
Unbounded
Maximum *r*-value: none

No asymptotes

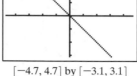

[−4.7, 4.7] by [−3.1, 3.1]

**29.** Domain: All reals
Range: [−2, 2]
Symmetric about the *y*-axis
Continuous
Bounded
Maximum *r*-value: 2
No asymptotes

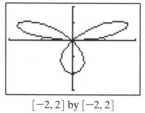

[−2, 2] by [−2, 2]

**30.** Domain: All reals
Range: [−3, 3]
Symmetric about the *x*-axis, *y*-axis, and origin
Continuous
Bounded
Maximum *r*-value: 3
No asymptotes

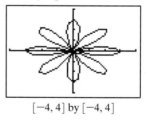

[−4, 4] by [−4, 4]

**31.** Domain: All reals
Range: [1, 9]
Symmetric about the *y*-axis
Continuous
Bounded
Maximum *r*-value: 9
No asymptotes

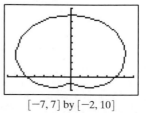

[−7, 7] by [−2, 10]

**32.** Domain: All reals
Range: [1, 11]
Symmetric about the *x*-axis
Continuous
Bounded
Maximum *r*-value: 11
No asymptotes

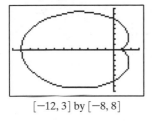

[−12, 3] by [−8, 8]

**33.** Domain: All reals
Range: [0, 8]
Symmetric about the *x*-axis
Continuous
Bounded
Maximum *r*-value: 8
No asymptotes

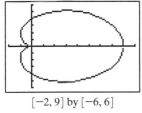

[−2, 9] by [−6, 6]

**34.** Domain: All reals
Range: [0, 10]
Symmetric about the *y*-axis
Continuous
Bounded
Maximum *r*-value: 10
No asymptotes

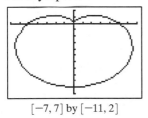

[−7, 7] by [−11, 2]

**35.** Domain: All reals
Range: [3, 7]
Symmetric about the polar *x*-axis
Continuous
Bounded
Maximum *r*-value: 7
No asymptotes

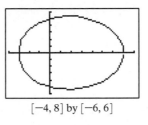

[−4, 8] by [−6, 6]

**36.** Domain: All reals
Range: [2, 4]
Symmetric about the *y*-axis
Continuous
Bounded
Maximum *r*-value: 4
No asymptotes

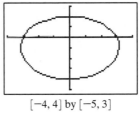

[−4, 4] by [−5, 3]

**37.** Domain: All reals
Range: [−1, 5]
Symmetric about the polar *x*-axis
Continuous
Bounded
Maximum *r*-value: 5
No asymptotes

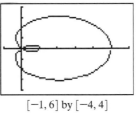

[−1, 6] by [−4, 4]

**38.** Domain: All reals
Range: [−1, 7]
Symmetric about the *y*-axis
Continuous
Bounded
Maximum *r*-value: 7
No asymptotes

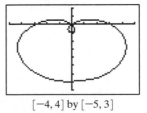

[−4, 4] by [−5, 3]

**39.** Domain: All reals
Range: [0, 2]
Symmetric about the *x*-axis
Continuous
Bounded
Maximum *r*-value: 2
No asymptotes

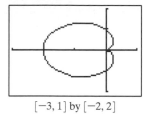

[−3, 1] by [−2, 2]

**40.** Domain: All reals
Range: [1, 3]
Symmetric about the *y*-axis
Continuous
Bounded
Maximum *r*-value: 3
No asymptotes

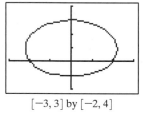

[−3, 3] by [−2, 4]

**41.** Domain: All reals
Range: [0, ∞]
Continuous
Unbounded
Maximum *r*-value: none
No asymptotes

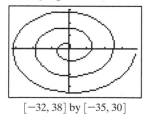

[−32, 38] by [−35, 30]

**42.** Domain: All reals
Range: [0, ∞]
Continuous
Unbounded
Maximum *r*-value: none
No asymptotes

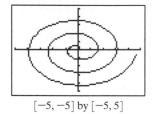

[−5, −5] by [−5, 5]

**43.** Domain: $\left[0, \dfrac{\pi}{2}\right] \cup \left[\dfrac{3\pi}{2}, 2\pi\right]$
Range: [0, 1]
Symmetric about the origin
Continuous on domain
Bounded
Maximum *r*-value: 1
No asymptotes

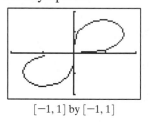

[−1, 1] by [−1, 1]

**44.** Domain: $\left[0, \dfrac{\pi}{4}\right] \cup \left[\dfrac{3\pi}{4}, \dfrac{5\pi}{4}\right] \cup \left[\dfrac{7\pi}{4}, 2\pi\right]$
Range: (0, 3)
Symmetric about the polar *x*-axis, *y*-axis, and origin
Continuous on domain
Bounded
Maximum *r*-value: 3
No asymptotes

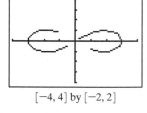

[−4, 4] by [−2, 2]

**45.** {6, 2, 6, 2}
**46.** {2, 8, 2, 8}
**47.** {3, 5, 3, 5, 3, 5, 3, 5, 3, 5}
**48.** {7, 1, 7, 1, 7, 1, 7, 1, 7, 1}
**49.** $r_1$ and $r_2$
**50.** $r_1$ and $r_3$

**51.** $r_2$ and $r_3$

**52.** $r_1$ and $r_2$

**53. (a)** A 4-petal rose curve with 2 short petals of length 1 and 2 long petals of length 3.

   **(b)** Symmetric about the origin.

   **(c)** Maximum $r$-value: 3.

**54. (a)** A 4-petal rose curve with petals of about length 1, 3.3, and 4 units.

   **(b)** Symmetric about the $y$-axis.

   **(c)** Maximum $r$-value: 4.

**55. (a)** A 6-petal rose curve with 3 short petals of length 2 and 3 long petals of length 4.

   **(b)** Symmetric about the $x$-axis.

   **(c)** Maximum $r$-value: 4.

**56. (a)** A 6-petal rose curve with 3 short petals of length 2 and 3 long petals of length 4.

   **(b)** Symmetric about the $y$-axis.

   **(c)** Maximum $r$-value: 4.

**61. (e)** Domain: All reals

   Range: $[-|a|, |a|]$

   Symmetric about the polar $x$-axis

   Continuous

   Bounded

   Maximum $r$-value: 2

   No asymptotes

**62. (e)** Domain: All reals

   Range: $[-|a|, |a|]$

   Symmetric about $y$-axis

   Continuous

   Bounded

   Maximum $r$-value: $|a|$

   No asymptotes

**63.** Starting with the graph of $r_1$, if we rotate clockwise (centered at the origin) by $\pi/12$ radians (15°), we get the graph of $r_2$; rotating $r_1$ clockwise by $\pi/4$ radians (45°) gives the graph of $r_3$.

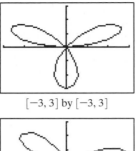

$[-3, 3]$ by $[-3, 3]$

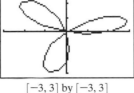

$[-3, 3]$ by $[-3, 3]$

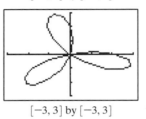

$[-3, 3]$ by $[-3, 3]$

**64.** Starting with the graph of $r_1$, if we rotate counterclockwise (centered at the origin) by $\pi/4$ radians (45°), we get the graph of $r_2$; rotating $r_1$ counterclockwise by $\pi/3$ radians (60°) gives the graph of $r_3$.

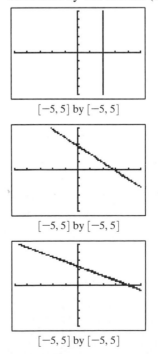

$[-5, 5]$ by $[-5, 5]$

$[-5, 5]$ by $[-5, 5]$

$[-5, 5]$ by $[-5, 5]$

**65.** Starting with the graph of $r_1$, if we rotate clockwise (centered at the origin) by $\pi/4$ radians (45°), we get the graph of $r_2$; rotating $r_1$ clockwise by $\pi/3$ radians (60°) gives the graph of $r_3$.

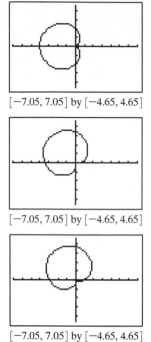

$[-7.05, 7.05]$ by $[-4.65, 4.65]$

$[-7.05, 7.05]$ by $[-4.65, 4.65]$

$[-7.05, 7.05]$ by $[-4.65, 4.65]$

**67. (a)** For $r_1$: $0 \le \theta \le 4\pi$ (or any interval that is $4\pi$ units long). For $r_2$: same answer.
  **(b)** $r_1$: 10 (overlapping) petals. $r_2$: 14 (overlapping) petals.

## Section 6.6 (pp. 524–534)

### Quick Review 6.6

**1. (a)**

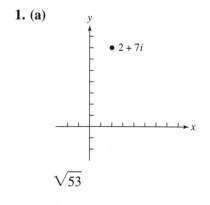

$\sqrt{53}$

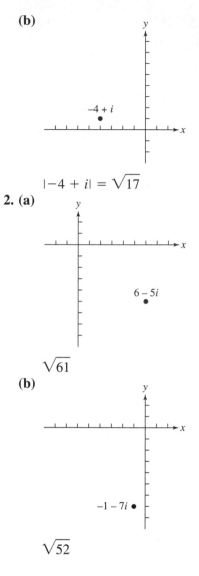

**(b)**

$-4 + i$

$|-4 + i| = \sqrt{17}$

**2. (a)**

$6 - 5i$

$\sqrt{61}$

**(b)**

$-1 - 7i$

$\sqrt{52}$

### Section 6.6 Exercises

**1.** $3\left(\cos \dfrac{\pi}{2} + i \sin \dfrac{\pi}{2}\right)$

**2.** $2\left(\cos \dfrac{3\pi}{2} + i \sin \dfrac{3\pi}{2}\right)$

**3.** $2\sqrt{2}\left(\cos \dfrac{\pi}{4} + i \sin \dfrac{\pi}{4}\right)$

**4.** $2\left(\cos \dfrac{\pi}{6} + i \sin \dfrac{\pi}{6}\right)$

**5.** $4\left(\cos \dfrac{2\pi}{3} + i \sin \dfrac{2\pi}{3}\right)$

**6.** $3\sqrt{2}\left(\cos \dfrac{7\pi}{4} + i \sin \dfrac{7\pi}{4}\right)$

**7.** $\cos \dfrac{5\pi}{3} + i \sin \dfrac{5\pi}{3}$

**8.** $\cos \dfrac{7\pi}{6} + i \sin \dfrac{7\pi}{6}$

**9.** $\approx \sqrt{13}(\cos 0.588 + i \sin 0.588)$

**10.** $\approx \sqrt{65} \ (\cos 5.232 + i \sin 5.232)$

**11.** $3\left(\cos \dfrac{\pi}{6} + i \sin \dfrac{\pi}{6}\right)$

**12.** $4\left(\cos \dfrac{5\pi}{4} + i \sin \dfrac{5\pi}{4}\right)$

**13.** $\dfrac{3}{2}\sqrt{3} - \dfrac{3}{2}i$

**14.** $-4\sqrt{3} - 4i$

**17.** $\dfrac{\sqrt{6}}{2} - \dfrac{\sqrt{2}}{2}i$

**18.** $\approx 2.56 + 0.68i$

**19.** $14 \ (\cos 155° + i \sin 155°)$

**20.** $\dfrac{\sqrt{2}}{2}(\cos 99° + i \sin 99°)$

**21.** $15\left(\cos \dfrac{23\pi}{12} + i \sin \dfrac{23\pi}{12}\right)$

**22.** $\dfrac{1}{\sqrt{3}}\left(\cos \dfrac{11\pi}{12} + i \sin \dfrac{11\pi}{12}\right)$

**23.** $\dfrac{2}{3}(\cos 30° - i \sin 30°)$

**24.** $\dfrac{5}{2}(\cos 105° + i \sin 105°)$

**25.** $2(\cos \pi + i \sin \pi)$

**26.** $\cos \dfrac{\pi}{4} + i \sin \dfrac{\pi}{4}$

**27. (a)** $5 + i; \dfrac{1}{2} - \dfrac{5}{2}i$
  **(b)** Same as (a).

**28. (a)** $\approx 2.732 - 0.732i; \approx 0.183 - 0.683i$
  **(b)** Same as (a).

**29. (a)** $18 - 4i; \approx 0.353 + 0.412i$
  **(b)** Same as (a).

**30. (a)** $-3.196 - 6.464i; 1.799 + 0.116i$
  **(b)** Same as (a).

**31.** $-\dfrac{\sqrt{2}}{2} + i\dfrac{\sqrt{2}}{2}$

**32.** $243i$

**33.** $4\sqrt{2} + 4\sqrt{2}\,i$

**34.** $-648 - 648\sqrt{3}\,i$

**36.** $\approx 5^{20}(0.954 - 0.299)$

**39.** $\dfrac{-1 + \sqrt{3}i}{\sqrt[3]{4}}; \dfrac{-1 - \sqrt{3}i}{\sqrt[3]{4}}; \sqrt[3]{2}$

**40.** $\sqrt[3]{2}\left(\cos \dfrac{\pi}{12} + i \sin \dfrac{\pi}{12}\right), \dfrac{-1 + i}{\sqrt[6]{2}},$

$\sqrt[3]{2}\left(\cos \dfrac{17\pi}{12} + i \sin \dfrac{17\pi}{12}\right)$

**41.** $\sqrt[3]{3}\left(\cos \dfrac{4\pi}{9} + i \sin \dfrac{4\pi}{9}\right),$

$\sqrt[3]{3}\left(\cos \dfrac{10\pi}{9} + i \sin \dfrac{10\pi}{9}\right),$

$\sqrt[3]{3}\left(\cos \dfrac{16\pi}{9} + i \sin \dfrac{16\pi}{9}\right)$

**42.** $3\left(\cos \dfrac{11\pi}{18} + i \sin \dfrac{11\pi}{18}\right), 3\left(\cos \dfrac{23\pi}{18} + i \sin \dfrac{23\pi}{18}\right),$

$3\left(\cos \dfrac{35\pi}{18} + i \sin \dfrac{35\pi}{18}\right)$

**43.** $\approx \sqrt[3]{5}(\cos 1.785 + i \sin 1.785),$
  $\approx \sqrt[3]{5}(\cos 3.879 + i \sin 3.879),$
  $\approx \sqrt[3]{5}(\cos 5.974 + i \sin 5.974)$

**44.** $\sqrt{2}\left(\cos \dfrac{\pi}{4} + i \sin \dfrac{\pi}{4}\right), \sqrt{2}\left(\cos \dfrac{11\pi}{12} + i \sin \dfrac{11\pi}{12}\right),$

$\sqrt{2}\left(\cos \dfrac{19\pi}{12} + i \dfrac{19\pi}{12}\right)$

**45.** $\cos \dfrac{\pi}{5} + i \sin \dfrac{\pi}{5}, \cos \dfrac{3\pi}{5} + i \sin \dfrac{3\pi}{5},$

$\cos \pi + i \sin \pi, \cos \dfrac{7\pi}{5} + i \sin \dfrac{7\pi}{5},$

$\cos \dfrac{9\pi}{5} + i \sin \dfrac{9\pi}{5}$

**46.** $2\left(\cos \dfrac{\pi}{10} + i \sin \dfrac{\pi}{10}\right), 2\left(\cos \dfrac{\pi}{2} + i \sin \dfrac{\pi}{2}\right),$

$2\left(\cos \dfrac{9\pi}{10} + i \sin \dfrac{9\pi}{10}\right), 2\left(\cos \dfrac{13\pi}{10} + i \sin \dfrac{13\pi}{10}\right),$

$2\left(\cos \dfrac{17\pi}{10} + i \sin \dfrac{17\pi}{10}\right)$

**47.** $\sqrt[5]{2}\left(\cos \dfrac{\pi}{30} + i \sin \dfrac{\pi}{30}\right),$

$\sqrt[5]{2}\left(\cos \dfrac{13\pi}{30} + i \sin \dfrac{13\pi}{30}\right),$

$\sqrt[5]{2}\left(\cos \dfrac{5\pi}{6} + i \sin \dfrac{5\pi}{6}\right),$

$\sqrt[5]{2}\left(\cos \dfrac{37\pi}{30} + i \sin \dfrac{37\pi}{30}\right),$

$\sqrt[5]{2}\left(\cos \dfrac{49\pi}{30} + i \sin \dfrac{49\pi}{30}\right)$

**48.** $\sqrt[5]{2}\left(\cos \dfrac{\pi}{20} + i \sin \dfrac{\pi}{20}\right), \sqrt[5]{2}\left(\cos \dfrac{9\pi}{20} + i \sin \dfrac{9\pi}{20}\right),$

$\sqrt[5]{2}\left(\cos \dfrac{17\pi}{20} + i \sin \dfrac{17\pi}{20}\right),$

$\sqrt[5]{2}\left(\cos \dfrac{5\pi}{4} + i \sin \dfrac{5\pi}{4}\right),$

$\sqrt[5]{2}\left(\cos \dfrac{33\pi}{20} + i \sin \dfrac{33\pi}{20}\right)$

**49.** $\sqrt[5]{2}\left(\cos \dfrac{\pi}{10} + i \sin \dfrac{\pi}{10}\right), \sqrt[5]{2}\left(\cos \dfrac{\pi}{2} + i \sin \dfrac{\pi}{2}\right),$

$\sqrt[5]{2}\left(\cos \dfrac{9\pi}{10} + i \sin \dfrac{9\pi}{10}\right),$

$\sqrt[5]{2}\left(\cos \dfrac{13\pi}{10} + i \sin \dfrac{13\pi}{10}\right),$

$\sqrt[5]{2}\left(\cos \dfrac{17\pi}{10} + i \sin \dfrac{17\pi}{10}\right)$

**50.** $\sqrt[5]{2}\left(\cos\dfrac{\pi}{15} + i\sin\dfrac{\pi}{15}\right),$ $\sqrt[5]{2}\left(\cos\dfrac{7\pi}{15} + i\sin\dfrac{7\pi}{15}\right),$
$\sqrt[5]{2}\left(\cos\dfrac{13\pi}{15} + i\sin\dfrac{13\pi}{15}\right),$
$\sqrt[5]{2}\left(\cos\dfrac{19\pi}{15} + i\sin\dfrac{19\pi}{15}\right),$
$\sqrt[5]{2}\left(\cos\dfrac{5\pi}{3} + i\sin\dfrac{5\pi}{3}\right)$

**51.** $\sqrt[8]{2}\left(\cos\dfrac{\pi}{16} + i\sin\dfrac{\pi}{16}\right),$ $\sqrt[8]{2}\left(\cos\dfrac{9\pi}{16} + i\sin\dfrac{9\pi}{16}\right),$
$\sqrt[8]{2}\left(\cos\dfrac{17\pi}{16} + i\sin\dfrac{17\pi}{16}\right),$
$\sqrt[8]{2}\left(\cos\dfrac{25\pi}{16} + i\sin\dfrac{25\pi}{16}\right)$

**52.** $\sqrt[12]{2}\left(\cos\dfrac{7\pi}{24} + i\sin\dfrac{7\pi}{24}\right),$ $\sqrt[12]{2}\left(\cos\dfrac{5\pi}{8} + i\sin\dfrac{5\pi}{8}\right),$
$\sqrt[12]{2}\left(\cos\dfrac{23\pi}{24} + i\sin\dfrac{23\pi}{24}\right),$
$\sqrt[12]{2}\left(\cos\dfrac{31\pi}{24} + i\sin\dfrac{31\pi}{24}\right),$
$\sqrt[12]{2}\left(\cos\dfrac{13\pi}{8} + i\sin\dfrac{13\pi}{8}\right),$
$\sqrt[12]{2}\left(\cos\dfrac{47\pi}{24} + i\sin\dfrac{47\pi}{24}\right)$

**53.** $\sqrt{2}\left(\cos\dfrac{\pi}{12} + i\sin\dfrac{\pi}{12}\right),$ $\sqrt{2}\left(\cos\dfrac{3\pi}{4} + i\sin\dfrac{3\pi}{4}\right),$
$\sqrt{2}\left(\cos\dfrac{17\pi}{12} + i\sin\dfrac{17\pi}{12}\right)$

**54.** $\sqrt[8]{8}\left(\cos\dfrac{3\pi}{16} + i\sin\dfrac{3\pi}{16}\right),$
$\sqrt[8]{8}\left(\cos\dfrac{11\pi}{16} + i\sin\dfrac{11\pi}{16}\right),$
$\sqrt[8]{8}\left(\cos\dfrac{19\pi}{16} + i\sin\dfrac{19\pi}{16}\right),$
$\sqrt[8]{8}\left(\cos\dfrac{27\pi}{16} + i\sin\dfrac{27\pi}{16}\right)$

**55.** $\sqrt[6]{2}\left(\cos\dfrac{\pi}{4} + i\sin\dfrac{\pi}{4}\right),$ $\sqrt[6]{2}\left(\cos\dfrac{7\pi}{12} + i\sin\dfrac{7\pi}{12}\right),$
$\sqrt[6]{2}\left(\cos\dfrac{11\pi}{12} + i\sin\dfrac{11\pi}{12}\right),$
$\sqrt[6]{2}\left(\cos\dfrac{5\pi}{4} + i\sin\dfrac{5\pi}{4}\right),$
$\sqrt[6]{2}\left(\cos\dfrac{19\pi}{12} + i\sin\dfrac{19\pi}{12}\right),$
$\sqrt[6]{2}\left(\cos\dfrac{23\pi}{12} + i\sin\dfrac{23\pi}{12}\right)$

**56.** $2(\cos 0 + i\sin 0) = 2,$ $2\left(\cos\dfrac{2\pi}{5} + i\sin\dfrac{2\pi}{5}\right),$
$2\left(\cos\dfrac{4\pi}{5} + i\sin\dfrac{4\pi}{5}\right),$ $2\left(\cos\dfrac{6\pi}{5} + i\sin\dfrac{6\pi}{5}\right),$
$2\left(\cos\dfrac{8\pi}{5} + i\sin\dfrac{8\pi}{5}\right)$

**57.** $1,\ -\dfrac{1}{2} \pm \dfrac{\sqrt{3}}{2}i$

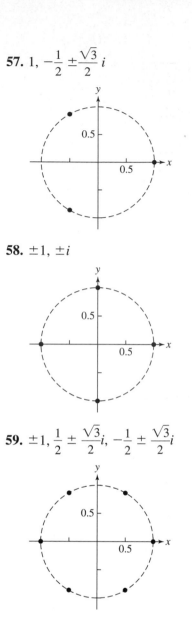

**58.** $\pm 1,\ \pm i$

**59.** $\pm 1,\ \dfrac{1}{2} \pm \dfrac{\sqrt{3}}{2}i,\ -\dfrac{1}{2} \pm \dfrac{\sqrt{3}}{2}i$

**60.** $1,\ -1$

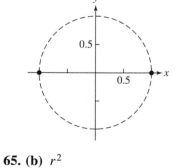

**65. (b)** $r^2$
**(c)** $\cos(2\theta) + i\sin(2\theta)$

**67. (a)**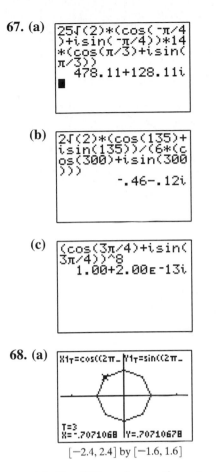

**(b)**

**(c)**

**68. (a)**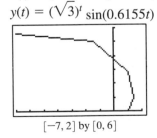

$[-2.4, 2.4]$ by $[-1.6, 1.6]$

**(c)** For fifth and seventh roots of unity, all of the roots except the complex number 1 generate the corresponding roots of unity. For the sixth roots of unity, only $2\pi/6$ and $10\pi/6$ generate the sixth roots of unity.

**(d)** $2\pi k/n$ generates the $n$th roots of unity if and only if $k$ and $n$ have no common factors other than 1.

**69.** $x(t) = (\sqrt{3})^t \cos(0.6155t)$
$y(t) = (\sqrt{3})^t \sin(0.6155t)$

$[-7, 2]$ by $[0, 6]$

**70.** $x(t) = (\sqrt{2})^t \cos(0.75\pi t)$
$y(t) = (\sqrt{2})^t \sin(0.75\pi t)$

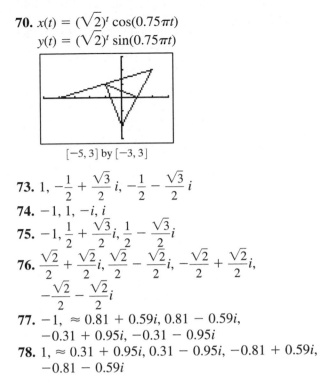

$[-5, 3]$ by $[-3, 3]$

**73.** $1, -\dfrac{1}{2} + \dfrac{\sqrt{3}}{2}i, -\dfrac{1}{2} - \dfrac{\sqrt{3}}{2}i$

**74.** $-1, 1, -i, i$

**75.** $-1, \dfrac{1}{2} + \dfrac{\sqrt{3}}{2}i, \dfrac{1}{2} - \dfrac{\sqrt{3}}{2}i$

**76.** $\dfrac{\sqrt{2}}{2} + \dfrac{\sqrt{2}}{2}i, \dfrac{\sqrt{2}}{2} - \dfrac{\sqrt{2}}{2}i, -\dfrac{\sqrt{2}}{2} + \dfrac{\sqrt{2}}{2}i,$
$-\dfrac{\sqrt{2}}{2} - \dfrac{\sqrt{2}}{2}i$

**77.** $-1, \approx 0.81 + 0.59i, 0.81 - 0.59i,$
$-0.31 + 0.95i, -0.31 - 0.95i$

**78.** $1, \approx 0.31 + 0.95i, 0.31 - 0.95i, -0.81 + 0.59i,$
$-0.81 - 0.59i$

## Chapter 6 Review (pp. 537–539)

**11. (a)** $\left\langle \dfrac{-2}{\sqrt{5}}, \dfrac{1}{\sqrt{5}} \right\rangle$
**(b)** $\left\langle \dfrac{6}{\sqrt{5}}, \dfrac{-3}{\sqrt{5}} \right\rangle$

**12. (a)** $\langle 1, 0 \rangle$
**(b)** $\langle -3, 0 \rangle$

**13. (a)** $\tan^{-1}\left(\dfrac{3}{4}\right) \approx 0.643, \tan^{-1}\left(\dfrac{5}{2}\right) \approx 1.190$
**(b)** $\approx 0.547$

**14. (a)** $\cos^{-1}\left(-\dfrac{1}{\sqrt{5}}\right) \approx 2.034, \tan^{-1}\left(\dfrac{2}{3}\right) \approx 0.588$
**(b)** $\approx 1.446$

**15.** $\approx (-2.266, -1.057)$

**16.** $(1.55\sqrt{2}, 1.55\sqrt{2})$

**17.** $(\sqrt{2}, -\sqrt{2})$

**18.** $(-1.8\sqrt{2}, 1.8\sqrt{2})$

**19.** $\left(1, -\dfrac{2\pi}{3} + (2n + 1)\pi\right)$ and $\left(-1, -\dfrac{2\pi}{3} + 2n\pi\right)$, $n$ an integer

**20.** $\left(2, \dfrac{5\pi}{6} + (2n + 1)\pi\right)$ and $\left(-2, \dfrac{5\pi}{6} + 2n\pi\right)$, $n$ an integer

**21. (a)** $\left(-\sqrt{13}, \pi + \tan^{-1}\left(-\frac{3}{2}\right)\right) \approx (-\sqrt{13}, 2.159)$

or $\left(\sqrt{13}, 2\pi + \tan^{-1}\left(-\frac{3}{2}\right)\right) \approx (\sqrt{13}, 5.300)$

**(b)** $\left(\sqrt{13}, \tan^{-1}\left(-\frac{3}{2}\right)\right) \approx (\sqrt{13}, -0.983)$ or

$\left(-\sqrt{13}, \pi + \tan^{-1}\left(-\frac{3}{2}\right)\right) \approx (-\sqrt{13}, 2.158)$

**(c)** The answers from (a), and also

$\left(-\sqrt{13}, 3\pi + \tan^{-1}\left(-\frac{3}{2}\right)\right)$

$\approx (-\sqrt{13}, 8.441)$ or

$\left(\sqrt{13}, 4\pi + \tan^{-1}\left(-\frac{3}{2}\right)\right)$

$\approx (\sqrt{13}, 11.583)$

**22. (a)** $(-10, 0)$ or $(10, \pi)$ or $(-10, 2\pi)$
**(b)** $(10, -\pi)$ or $(-10, 0)$ or $(10, \pi)$
**(c)** The answers from (a), and also $(10, 3\pi)$ or $(-10, 4\pi)$.

**23. (a)** $(5, 0)$ or $(-5, \pi)$ or $(5, 2\pi)$
**(b)** $(-5, -\pi)$ or $(5, 0)$ or $(-5, \pi)$
**(c)** The answers from (a), and also $(-5, 3\pi)$ or $(5, 4\pi)$.

**24. (a)** $\left(-2, \frac{\pi}{2}\right)$ or $\left(2, \frac{3\pi}{2}\right)$
**(b)** $\left(2, -\frac{\pi}{2}\right)$ or $\left(-2, \frac{\pi}{2}\right)$.

**(c)** The answers from (a), and also $\left(-2, \frac{5\pi}{2}\right)$

or $\left(2, \frac{7\pi}{2}\right)$

**25.** $-\frac{3}{5}x + \frac{29}{5}$: line through $\left(0, \frac{29}{5}\right)$ with slope $m = -\frac{3}{5}$.

**26.** $-5x + 12$, $1 \le x \le 9$: segment from $(1, 7)$ to $(9, -33)$.

**27.** $2(y + 1)^2 + 3$: Parabola that opens to right with vertex at $(3, 1)$.

**28.** $x^2 + y^2 = 9$: Circle of radius 3 centered at $(0, 0)$.

**29.** $\sqrt{x + 1}$: square root function starting at $(-1, 0)$

**30.** $\frac{1}{3}\ln x$: logarithmic function

**31.** $x = 2t + 3$, $y = 3t + 4$
**32.** $x = 7t + 5$, $y = -2t + 1$, $-1 \le t \le 0$
**33.** $a = -3$, $b = 4$, $|z_1| = 5$

**34.** $z_1 = 5\left\{\cos\left[\cos^{-1}\left(-\frac{3}{5}\right)\right] + i \sin\left[\cos^{-1}\left(-\frac{3}{5}\right)\right]\right\}$
$\approx 5[\cos (2.214) + i \sin (2.214)]$

**35.** $3\sqrt{3} + 3i$
**36.** $-1.5\sqrt{3} + 1.5i$
**37.** $-1.25 - 1.25\sqrt{3}i$

**38.** $\approx -3.205 + 2.394i$
**39.** $32\left(\cos \frac{7\pi}{4} + i \sin \frac{7\pi}{4}\right)$. Other representations would use angles $\frac{7\pi}{4} + 2n\pi$, $n$ an integer.

**40.** $\approx \sqrt{3}[\cos (2.186) + i \sin (2.186)]$. Other representations would use angles $2.186 + 2n\pi$, $n$ an integer.

**41.** $\approx \sqrt{34}[\cos(-1.030) + i \sin(-1.030)]$
$\approx \sqrt{34}[\cos (5.253) + i \sin (5.253)]$
Other representations would use angles $\approx 5.253 + 2n\pi$, $n$ an integer.

**42.** $2\sqrt{2}\left(\cos \frac{5\pi}{4} + i \sin \frac{5\pi}{4}\right)$. Other representations would use angles $\frac{5\pi}{4} + 2n\pi$, $n$ an integer.

**43.** $12(\cos 90° + i \sin 90°)$; $\frac{3}{4}(\cos 330° + i \sin 330°)$

**44.** $10(\cos 245° + i \sin 245°)$;
$2.5(\cos 155° + i \sin 155°)$

**45. (a)** $243\left(\cos \frac{5\pi}{4} + i \sin \frac{5\pi}{4}\right)$
**(b)** $-\frac{243\sqrt{2}}{2} - \frac{243\sqrt{2}}{2}i$

**46. (a)** $256\left(\cos \frac{2\pi}{3} + i \sin \frac{2\pi}{3}\right)$
**(b)** $-128 + 128\sqrt{3}i$

**47. (a)** $125(\cos \pi + i \sin \pi)$
**(b)** $-125$

**48. (a)** $117{,}649\left(\cos \frac{\pi}{4} + i \sin \frac{\pi}{4}\right)$
**(b)** $\frac{117{,}649\sqrt{2}}{2} + \frac{117{,}649\sqrt{2}}{2}i$

**49.** $\sqrt[8]{18}\left(\cos \frac{\pi}{16} + i \sin \frac{\pi}{16}\right)$,
$\sqrt[8]{18}\left(\cos \frac{9\pi}{16} + i \sin \frac{9\pi}{16}\right)$,
$\sqrt[8]{18}\left(\cos \frac{17\pi}{16} + i \sin \frac{17\pi}{16}\right)$,
$\sqrt[8]{18}\left(\cos \frac{25\pi}{16} + i \sin \frac{25\pi}{16}\right)$

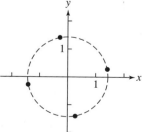

**50.** $2(\cos 0 + i \sin 0) = 2, 2\left(\cos \dfrac{2\pi}{3} + i \sin \dfrac{2\pi}{3}\right),$

$2\left(\cos \dfrac{4\pi}{3} + i \sin \dfrac{4\pi}{3}\right)$

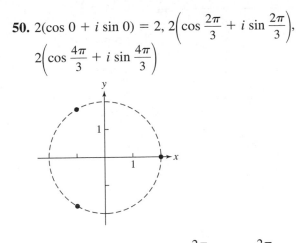

**51.** $2(\cos 0 + i \sin 0) = 1, \cos \dfrac{2\pi}{5} + i \sin \dfrac{2\pi}{5},$

$\cos \dfrac{4\pi}{5} + i \sin \dfrac{4\pi}{5}, \cos \dfrac{6\pi}{5} + i \sin \dfrac{6\pi}{5},$

$\cos \dfrac{8\pi}{5} + i \sin \dfrac{8\pi}{5}$

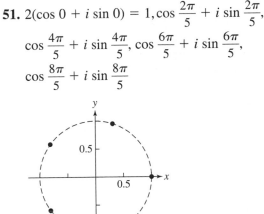

**52.** $\cos \dfrac{\pi}{6} + i \sin \dfrac{\pi}{6}, \cos \dfrac{\pi}{2} + i \sin \dfrac{\pi}{2},$

$\cos \dfrac{5\pi}{6} + i \sin \dfrac{5\pi}{6}, \cos \dfrac{7\pi}{6} + i \sin \dfrac{7\pi}{6},$

$\cos \dfrac{3\pi}{2} + i \sin \dfrac{3\pi}{2}, \cos \dfrac{11\pi}{6} + i \sin \dfrac{11\pi}{6}$

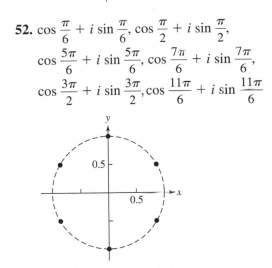

**61.** $x^2 + y^2 = 4$ — a circle with center $(0, 0)$ and radius 2

**62.** $x^2 + (y^2 + 1)^2 = 1$ — a circle of radius 1 with center $(0, -1)$.

**63.** $\left(x + \dfrac{3}{2}\right)^2 + (y + 1)^2 = \dfrac{13}{4}$

— a circle of radius $\dfrac{\sqrt{13}}{2}$ with center $\left(-\dfrac{3}{2}, -1\right)$.

**64.** $x = 3$ — a vertical line through $(3, 0)$.

**65.** $r = -4 \csc \theta$

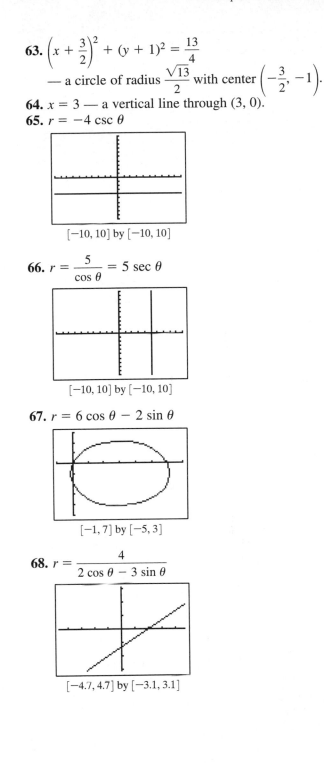

$[-10, 10]$ by $[-10, 10]$

**66.** $r = \dfrac{5}{\cos \theta} = 5 \sec \theta$

$[-10, 10]$ by $[-10, 10]$

**67.** $r = 6 \cos \theta - 2 \sin \theta$

$[-1, 7]$ by $[-5, 3]$

**68.** $r = \dfrac{4}{2 \cos \theta - 3 \sin \theta}$

$[-4.7, 4.7]$ by $[-3.1, 3.1]$

**69.**

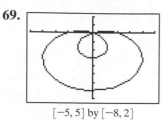

[-5, 5] by [-8, 2]

Domain: All reals
Range: [-3, 7]
Symmetric about the *y*-axis
Continuous
Bounded
Maximum *r*-value: 7
No asymptotes

**70.**

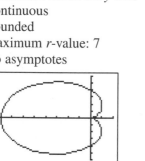

[-9, 2] by [-6, 6]

Domain: All reals
Range: [0, 8]
Symmetric about the *x*-axis
Continuous
Bounded
Maximum *r*-value: 8
No asymptotes

**71.**

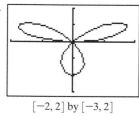

[-2, 2] by [-3, 2]

Domain: All reals
Range: [-2, 2]
Symmetric about the *y*-axis
Continuous
Bounded
Maximum *r*-value: 2
No asymptotes

**72.**

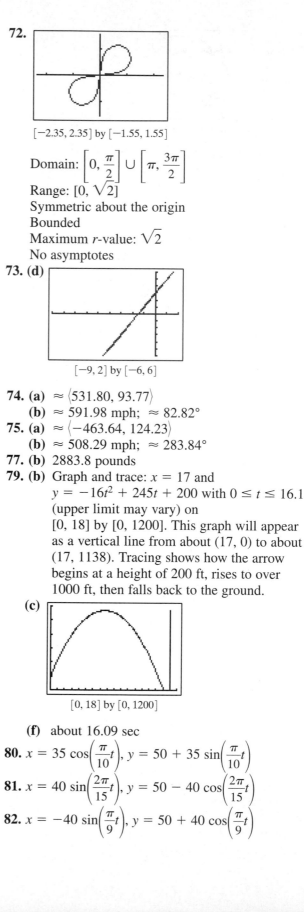

[-2.35, 2.35] by [-1.55, 1.55]

Domain: $\left[0, \dfrac{\pi}{2}\right] \cup \left[\pi, \dfrac{3\pi}{2}\right]$
Range: $[0, \sqrt{2}]$
Symmetric about the origin
Bounded
Maximum *r*-value: $\sqrt{2}$
No asymptotes

**73. (d)**

[-9, 2] by [-6, 6]

**74. (a)** $\approx \langle 531.80, 93.77 \rangle$
   **(b)** $\approx 591.98$ mph; $\approx 82.82°$
**75. (a)** $\approx \langle -463.64, 124.23 \rangle$
   **(b)** $\approx 508.29$ mph; $\approx 283.84°$
**77. (b)** 2883.8 pounds
**79. (b)** Graph and trace: $x = 17$ and
   $y = -16t^2 + 245t + 200$ with $0 \le t \le 16.1$
   (upper limit may vary) on
   [0, 18] by [0, 1200]. This graph will appear
   as a vertical line from about (17, 0) to about
   (17, 1138). Tracing shows how the arrow
   begins at a height of 200 ft, rises to over
   1000 ft, then falls back to the ground.

**(c)**

[0, 18] by [0, 1200]

**(f)** about 16.09 sec
**80.** $x = 35 \cos\left(\dfrac{\pi}{10}t\right)$, $y = 50 + 35 \sin\left(\dfrac{\pi}{10}t\right)$
**81.** $x = 40 \sin\left(\dfrac{2\pi}{15}t\right)$, $y = 50 - 40 \cos\left(\dfrac{2\pi}{15}t\right)$
**82.** $x = -40 \sin\left(\dfrac{\pi}{9}t\right)$, $y = 50 + 40 \cos\left(\dfrac{\pi}{9}t\right)$

**83. (a)**

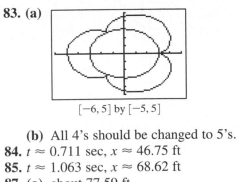

$[-6, 5]$ by $[-5, 5]$

**(b)** All 4's should be changed to 5's.
**84.** $t \approx 0.711$ sec, $x \approx 46.75$ ft
**85.** $t \approx 1.063$ sec, $x \approx 68.62$ ft
**87. (a)** about 77.59 ft
**(b)** $\approx 4.404$ sec
**90.** No

## Chapter 6 Project

Answers are based on the sample data shown in the table.

**1.**

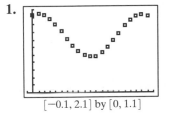

$[-0.1, 2.1]$ by $[0, 1.1]$

**2.** $y \approx 0.28 \sin(3.46(x - 1.46)) + 0.75$
**3.**

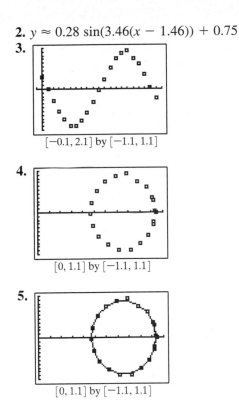

$[-0.1, 2.1]$ by $[-1.1, 1.1]$

**4.**

$[0, 1.1]$ by $[-1.1, 1.1]$

**5.**

$[0, 1.1]$ by $[-1.1, 1.1]$

# Chapter 7

## Section 7.1 (pp. 543–552)

### Exploration 1

**1.**

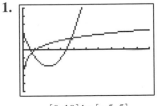

$[0, 10]$ by $[-5, 5]$

**2.** About $(0.71, -0.34)$ and $(3.83, 1.34)$

### Quick Review

**1.** $y = \dfrac{5}{3} - \dfrac{2}{3}x$

**3.** $x = -\dfrac{2}{3}, x = 1$

**4.** $x = \dfrac{-5 \pm \sqrt{105}}{4}$

**5.** $0, 2, -2$      **6.** $0, -3, 2$
**10.** $(0, 0), (3, 9), (-3, -9)$

### Exercises 7.1

**6.** $\left(-\dfrac{23}{5}, \dfrac{23}{5}\right)$

**10.** Any pair $(x, 3x + 2)$

**17.** Any pair $\left(x, \dfrac{1}{2}x - 2\right)$

**29.** $\approx(-2.32, -3.16), (0.47, -1.77)$ or $(-1.85, -1.08)$
**30.** $\approx(-0.70, -2.40)$ or $(5.70, 10.40)$
**32.** $(-1.2, -1.6)$ or $(2, 0)$
**33.** $\approx(2.05, 2.19)$ or $(-2.05, 2.19)$
**34.** $\approx(2.05, -2.19)$ or $(-2.05, -2.19)$
**39.** $(-0.1 - 3\sqrt{0.89}, 0.3 - \sqrt{0.89})$
$(-0.1 + 3\sqrt{0.89}, 0.3 + \sqrt{0.89})$
**40.** $\left(\dfrac{91 + 4\sqrt{871}}{65}, \dfrac{91 - 4\sqrt{871}}{65}\right)$ or
$\left(\dfrac{91 - 4\sqrt{871}}{65}, \dfrac{91 + 4\sqrt{871}}{65}\right)$
**43. (b)** $\approx(-1.29, 2.29)$ or $(1.91, -0.91)$

**44. (b)** $\approx(2.68, 2.68)$ or $(-2.68, -2.68)$
**45. (a)** $y = 0.44x^2 - 5.71x + 52.55$
  **(b)** $y = -364.98 + 166.57 \ln x$
  **(d)** 1982 and 1993
**46. (a)** $y = 5.84x + 63.48$
  **(b)** $y = 4.61x + 86.72$
**47. (a)** $y = 101.78x + 1703.15$
  **(b)** $y = 16.56x + 5643.45$
**48. (a)** None, one, or two
  **(b)** None, one, two, three or four
**49.** $5.28$ m $\times$ $94.72$ m
**51.** current speed $\approx 1.06$ mph;
  rowing speed $\approx 3.56$ mph
**52.** wind speed $= 49.24$ mph;
  airplane speed $= 617.42$ mph
**54.** Cashews: 1.93 lb, peanuts: 3.07 lb
**59.** $\left(\pm\sqrt{\dfrac{2}{3}}, \dfrac{10}{3}\right)$

## Section 7.2 (pp. 552–567)

### Exploration 1

**1.** $A = \begin{bmatrix} 2 & 1 \\ 5 & 4 \end{bmatrix}$; $B = \begin{bmatrix} -1 & 2 \\ 2 & 5 \end{bmatrix}$

**2.** $-A = \begin{bmatrix} -2 & -1 \\ -5 & -4 \end{bmatrix}$

The order of $[0]$ is $2 \times 2$.

**3.** $\begin{bmatrix} 8 & -1 \\ 11 & 2 \end{bmatrix}$

### Exploration 2

**1.** $\det A = -a_{12}a_{21}a_{33} + a_{13}a_{21}a_{32} + a_{11}a_{22}a_{33}$
  $- a_{13}a_{22}a_{31} - a_{11}a_{23}a_{32} + a_{12}a_{23}a_{31}$
**2.** $\det A = -a_{12}a_{21}a_{33} + a_{13}a_{21}a_{32} + a_{11}a_{22}a_{33}$
  $- a_{13}a_{22}a_{31} - a_{11}a_{23}a_{32} + a_{12}a_{23}a_{31}$

### Quick Review

**2. (a)** $(-3, 2)$ **(b)** $(-x, y)$
**4. (a)** $(2, -3)$ **(b)** $(-y, -x)$
**5.** $(3 \cos \theta, 3 \sin \theta)$ **6.** $(r \cos \theta, r \sin \theta)$
**7.** $\sin \alpha \cos \beta + \cos \alpha \sin \beta$
**8.** $\sin \alpha \cos \beta - \cos \alpha \sin \beta$
**9.** $\cos \alpha \cos \beta - \sin \alpha \sin \beta$
**10.** $\cos \alpha \cos \beta + \sin \alpha \sin \beta$

### Exercises 7.2

**1.** $2 \times 3$; not square **2.** $2 \times 2$; square

**3.** $3 \times 2$; not square **4.** $1 \times 3$; not square
**5.** $3 \times 1$; not square **6.** $1 \times 1$; square

**11. (a)** $\begin{bmatrix} 3 & 0 \\ -3 & 1 \end{bmatrix}$ **(b)** $\begin{bmatrix} 1 & 6 \\ 1 & 9 \end{bmatrix}$

  **(c)** $\begin{bmatrix} 6 & 9 \\ -3 & 15 \end{bmatrix}$ **(d)** $\begin{bmatrix} 1 & 15 \\ 4 & 22 \end{bmatrix}$

**12. (a)** $\begin{bmatrix} 1 & 1 & 2 \\ 3 & 1 & 1 \\ 6 & -3 & 0 \end{bmatrix}$ **(b)** $\begin{bmatrix} -3 & -1 & 2 \\ 5 & 1 & -3 \\ -2 & 3 & 2 \end{bmatrix}$

  **(c)** $\begin{bmatrix} -3 & 0 & 6 \\ 12 & 3 & -3 \\ 6 & 0 & 2 \end{bmatrix}$ **(d)** $\begin{bmatrix} -8 & -3 & 4 \\ 11 & 2 & -8 \\ -8 & 9 & 5 \end{bmatrix}$

**13. (a)** $\begin{bmatrix} 1 & 1 \\ -2 & 0 \\ -1 & 0 \end{bmatrix}$ **(b)** $\begin{bmatrix} -7 & 1 \\ 2 & -2 \\ 5 & 2 \end{bmatrix}$

  **(c)** $\begin{bmatrix} -9 & 3 \\ 0 & -3 \\ 6 & 3 \end{bmatrix}$ **(d)** $\begin{bmatrix} -18 & 2 \\ 6 & -5 \\ 13 & 5 \end{bmatrix}$

**14. (a)** $\begin{bmatrix} 3 & 1 & 4 & 1 \\ 3 & 0 & 1 & 0 \end{bmatrix}$ **(b)** $\begin{bmatrix} 7 & -5 & 2 & 1 \\ -5 & 0 & 3 & 4 \end{bmatrix}$

  **(c)** $\begin{bmatrix} 15 & -6 & 9 & 3 \\ -3 & 0 & 6 & 6 \end{bmatrix}$ **(d)** $\begin{bmatrix} 16 & -13 & 3 & 2 \\ -14 & 0 & 7 & 10 \end{bmatrix}$

**15. (a)** $\begin{bmatrix} -3 \\ 1 \\ 4 \end{bmatrix}$ **(b)** $\begin{bmatrix} -1 \\ 1 \\ -4 \end{bmatrix}$

  **(c)** $\begin{bmatrix} -6 \\ 3 \\ 0 \end{bmatrix}$ **(d)** $\begin{bmatrix} -1 \\ 2 \\ -12 \end{bmatrix}$

**16. (a)** $[0 \quad 0 \quad -2 \quad 3]$ **(b)** $[-2 \quad -4 \quad 2 \quad 3]$
  **(c)** $[-3 \quad -6 \quad 0 \quad 9]$ **(d)** $[-5 \quad -10 \quad 6 \quad 6]$

**17. (a)** $\begin{bmatrix} -4 & -18 \\ -11 & -17 \end{bmatrix}$ **(b)** $\begin{bmatrix} 5 & -12 \\ 0 & -26 \end{bmatrix}$

**18. (a)** $\begin{bmatrix} 13 & 13 \\ -2 & -16 \end{bmatrix}$ **(b)** $\begin{bmatrix} 7 & -14 \\ -8 & -10 \end{bmatrix}$

**19. (a)** $\begin{bmatrix} 2 & 2 \\ -11 & 12 \end{bmatrix}$ **(b)** $\begin{bmatrix} 4 & 8 & -5 \\ -5 & 4 & -6 \\ -2 & -8 & 6 \end{bmatrix}$

**20. (a)** $\begin{bmatrix} & \\ & \end{bmatrix}$

**(b)** $\begin{bmatrix} 3 & -1 & -14 & 16 \\ 4 & 2 & 8 & -2 \\ 5 & 3 & 14 & -6 \\ 8 & 2 & 0 & 10 \end{bmatrix}$

**21. (a)** $\begin{bmatrix} 6 & -7 & -2 \\ 3 & 7 & 3 \\ 8 & -1 \end{bmatrix}$ **(b)** $\begin{bmatrix} 2 & 1 & 3 \\ 5 & 0 & 0 \\ -18 & -3 & 10 \end{bmatrix}$

**22. (a)** $\begin{bmatrix} -8 & 8 & 5 \\ 0 & 7 & -8 \\ 11 & 4 & 11 \end{bmatrix}$ **(b)** $\begin{bmatrix} -3 & 18 & -2 \\ 11 & 2 & 11 \\ 2 & -11 & \end{bmatrix}$

**23. (a)** $[-8]$ **(b)** $\begin{bmatrix} -10 & 5 & -15 \\ 8 & -4 & 12 \\ 4 & -2 & 6 \end{bmatrix}$

**24. (a)** $\begin{bmatrix} 2 & -4 & -8 \\ -3 & 6 & 12 \\ 4 & -8 & -16 \end{bmatrix}$

**(b)** $[-8]$

**26. (a)** $\begin{bmatrix} 1 & 15 \\ 2 & 3 \\ 5 & -6 \\ -17 & 15 \end{bmatrix}$ **(b)** not possible

**27. (a)** $\begin{bmatrix} -1 & 3 & 4 \\ 2 & 0 & 1 \\ 1 & 2 & 1 \end{bmatrix}$ **(b)** $\begin{bmatrix} 1 & 2 & 1 \\ 1 & 0 & 2 \\ 4 & 3 & -1 \end{bmatrix}$

**28. (a)** $\begin{bmatrix} -3 & 2 & 1 & 3 \\ 2 & 1 & 0 & -1 \\ -1 & 2 & 3 & -4 \\ 4 & 0 & 2 & -1 \end{bmatrix}$

**(b)** $\begin{bmatrix} 3 & 2 & -1 & -4 \\ 0 & 1 & 2 & -1 \\ 1 & 2 & -3 & 3 \\ 2 & 0 & 4 & -1 \end{bmatrix}$

**33.** $AB = BA = I_2$ **34.** $AB = BA = I_3$

**38.** $\begin{bmatrix} 1 & 1 & -3 \\ -0.25 & -0.5 & 1.75 \\ 0.25 & 0.5 & -0.75 \end{bmatrix}$

**40.** $\begin{bmatrix} -0.25 & 0.5 & 0.25 \\ 0.5 & -1.0 & 0.5 \\ 0.25 & 0.5 & -0.25 \end{bmatrix}$

**44.** $\begin{bmatrix} 1 & 1 \\ \frac{1}{2} & -2 \end{bmatrix}$

**46.** $\begin{bmatrix} 132 & 77 \\ 165 & 121 \\ 88 & 176 \end{bmatrix}$; $B = 1.1A$

**47. (a)** $[382 \quad 227.50]$

**48. (a)** $\begin{bmatrix} \$15{,}550 & \$21{,}919.54 \\ \$8070 & \$11{,}439.74 \\ \$8740 & \$12{,}279.76 \end{bmatrix}$

**49. (a)** $AB$ or $BA$ **(b)** $(A - C)B^T$

**51. (a)** $\begin{bmatrix} 1 & 0 \\ 0 & 1 \end{bmatrix}$

**(c)** It is the inverse of $A$

**52. (c)** $\begin{vmatrix} 1 & x_3 & y_3 \\ 1 & x_1 & y_1 \\ 1 & x_2 & y_2 \end{vmatrix} = 0$

**(d)** $\begin{vmatrix} 1 & x_3 & y_3 \\ 1 & x_1 & y_1 \\ 1 & x_2 & y_2 \end{vmatrix} \neq 0$

**53. c)** $C = \begin{bmatrix} \$1600 \\ \$900 \\ \$500 \\ \$100 \\ \$1000 \end{bmatrix}$ **(d)** $RC$

**(e)** $[\$1{,}427{,}300]$

**58.** $A \cdot A^{-1} = I_2$

**60.** $[x \quad y]\begin{bmatrix} -1 & 0 \\ 0 & 1 \end{bmatrix}$ **61.** $[x \quad y]\begin{bmatrix} 0 & 1 \\ 1 & 0 \end{bmatrix}$

**62.** $[x \quad y]\begin{bmatrix} 0 & -1 \\ -1 & 0 \end{bmatrix}$ **63.** $[x \quad y]\begin{bmatrix} 1 & 0 \\ 0 & a \end{bmatrix}$

**64.** $[x \quad y]\begin{bmatrix} c & 0 \\ 0 & 1 \end{bmatrix}$

**65. (a)** $\det(xI_2 - A) = x^2 + (-a_{22} - a_{11})x + (a_{11}a_{22} - a_{12}a_{21})$

**(b)** They are equal.
**(c)** The coefficient of $x$ is the opposite of the sum of the elements of the main diagonal in $A$.

**66. (b)** The constant term equals $-\det A$.
**(c)** The coefficient of $x^2$ is the opposite of the sum of the elements of the main diagonal in $A$.

## Section 7.3 (pp. 567–576)

### Exploration 1

**1.** $x + y + z$ must equal 60 L.

**2.** $0.15x + 0.35y + 0.55z$ must equal $0.40(60) = 24$ L.

**3.** The number of liters of 35% solution must equal twice the number of liters of 55% solution.

**4.** $A = \begin{bmatrix} 1 & 1 & 1 \\ 0.15 & 0.35 & 0.55 \\ 0 & 1 & -2 \end{bmatrix}$; $X = \begin{bmatrix} x \\ y \\ z \end{bmatrix}$; $B = \begin{bmatrix} 60 \\ 24 \\ 0 \end{bmatrix}$

**5.** $\begin{bmatrix} 3.75 \\ 37.5 \\ 18.75 \end{bmatrix}$

**6.** 3.75 L of 15% acid, 37.5 L of 35% acid, and 18.75 L of 55% acid are required to make 60 L of a 40% acid solution.

### Quick Review

**7.** $y = w - z + 1$ 

**8.** $x = 2z - w + 3$

**10.** $\begin{bmatrix} 1 & -0.5 & 0.25 \\ 0.5 & 0 & 0.5 \\ 0.5 & 0 & 0 \end{bmatrix}$

### Exercises 7.3

**1.** $\begin{bmatrix} 2 & -6 & 4 \\ 1 & 2 & -3 \\ 0 & -8 & 4 \end{bmatrix}$

**2.** $\begin{bmatrix} 1 & -3 & 2 \\ 1 & 2 & -3 \\ -3 & 1 & -2 \end{bmatrix}$

**3.** $\begin{bmatrix} 0 & -10 & 10 \\ 1 & 2 & -3 \\ -3 & 1 & -2 \end{bmatrix}$

**4.** $\begin{bmatrix} 2 & -6 & 4 \\ 3 & 4 & 1 \\ -3 & 1 & -2 \end{bmatrix}$

**6.** $(2)R_2 + R_1$

For 9–12, possible answers are given.

**9.** $\begin{bmatrix} 1 & 3 & -1 \\ 0 & 1 & -1.2 \\ 0 & 0 & 1 \end{bmatrix}$

**10.** $\begin{bmatrix} 1 & 2 & -3 \\ 0 & 0 & 1 \\ 0 & 0 & 0 \end{bmatrix}$

**11.** $\begin{bmatrix} 1 & 2 & 3 & -4 \\ 0 & 1 & 0 & -0.6 \\ 0 & 0 & 1 & -9.2 \end{bmatrix}$

**12.** $\begin{bmatrix} 1 & 2 & 3 & -2 \\ 0 & 1 & -1 & 1 \end{bmatrix}$

**15.** $\begin{bmatrix} 1 & 0 & -1 & 3 \\ 0 & 1 & 2 & -1 \end{bmatrix}$

**16.** $\begin{bmatrix} 1 & -2 & 1 & -1 \\ 0 & 0 & 0 & 0 \\ 0 & 0 & 0 & 0 \end{bmatrix}$

**17.** $\begin{bmatrix} 2 & -3 & 1 & 1 \\ -1 & 1 & -4 & -3 \\ 3 & 0 & -1 & 2 \end{bmatrix}$

**18.** $\begin{bmatrix} 3 & -4 & 1 & -1 & 1 \\ 1 & 0 & 1 & -2 & 4 \end{bmatrix}$

**19.** $\begin{bmatrix} 2 & -5 & 1 & -1 & -3 \\ 1 & 0 & -2 & 1 & 4 \\ 0 & 2 & -3 & -1 & 5 \end{bmatrix}$

**20.** $\begin{bmatrix} 3 & -2 & 5 \\ -1 & 5 & 7 \end{bmatrix}$

In #21–24, the variable names are arbitrary.

**21.** $\begin{aligned} 3x + 2y &= -1 \\ -4x + 5y &= 2 \end{aligned}$

**22.** $\begin{aligned} x - z + 2w &= -3 \\ 2x + y - w &= 4 \\ -x + y + 2z &= 0 \end{aligned}$

**23.** $\begin{aligned} 2x + z &= 3 \\ -x + y &= 2 \\ 2y - 3z &= -1 \end{aligned}$

**24.** $\begin{aligned} 2x + y - 2z &= 4 \\ -3x + 2z &= -1 \end{aligned}$

**25.** $(1, 2, 1)$

**26.** $\left( \dfrac{5}{13}, \dfrac{10}{13}, \dfrac{74}{13} \right)$

**27.** $\left( \dfrac{9}{2}, \dfrac{7}{2}, 4, -\dfrac{15}{2} \right)$

**28.** $(2, 1, 0, -1)$

**29.** $(-2, 3, 1)$

**30.** $(7, 6, 3)$

**31.** no solution

**32.** $(z + 2, -z - 1, z)$

**33.** $(2 - z, 1 + z, z)$

**34.** $(z + 53, z - 26, z)$

**37.** $\begin{bmatrix} 2 & 5 \\ 1 & -2 \end{bmatrix} \begin{bmatrix} x \\ y \end{bmatrix} = \begin{bmatrix} -3 \\ 1 \end{bmatrix}$

**38.** $\begin{bmatrix} 5 & -7 & 1 \\ 2 & -3 & -1 \\ 1 & 1 & 1 \end{bmatrix} \begin{bmatrix} x \\ y \\ z \end{bmatrix} = \begin{bmatrix} 2 \\ 3 \\ -3 \end{bmatrix}$

**39.** $\begin{aligned} 3x - y &= -1 \\ 2x + 4y &= 3 \end{aligned}$

**40.** $\begin{aligned} x - 3z &= 3 \\ 2x - y + 3z &= -1 \\ -2x + 3y - 4z &= 2 \end{aligned}$

**43.** $(-2, -5, -7)$

**44.** $(3, -0.5, 0.5)$

**48.** $(-2, 0, 0.5)$

**51.** $\left( 2 - \dfrac{3}{2}z, -\dfrac{1}{2}z - 4, z \right)$ **52.** $\left( \dfrac{1}{5}z - 1, \dfrac{3}{5}z - 2, z \right)$

**53.** $(-1 - 2w, w + 1, -w, w)$

**54.** $(w, 1 - 2w, -w - 1, w)$

**55.** $(-w - 2, 0.5 - z, z, w)$

**56.** $(z - 3w + 1, 2w - 2z + 4, z, w)$

**57.** no solution

**58.** $(1, 1 - w, 6w - 2, w)$

**60.** $f(x) = 3x^3 - x^2 + 2x - 5$

**61.** $f(x) = (-c - 3)x^2 + x + c$, for any $c$

**62.** $f(x) = (4 - c)x^3 - x^2 + cx - 1$, for any $c$

**63.** **(a)** $y = 1.29x + 35.31$

**(b)** $y = 1.51x + 26.37$

**(c)** 2031

**64. (a)** $y = 3.26x + 133.62$

**(b)** $y = 4.01x + 102.90$

**65.** 825 children, 410 adults, 165 senior citizens

**66.** about 14.55 g 22% alloy, 29.09 g 30% alloy, 36.36 g 42% alloy

**67.** \$14,500 CDs, \$5,500 bonds, \$60,000 growth funds

**68.** \$z at 10%, \$(z − 9000) at 6%, \$(29,000 − 2z) at 8%

**69.** \$0 CDs, \$38,983.05 bonds, \$11,016.95 growth fund

**70.** 0 L of 10% solution, 28.8 L of 25% solution, 11.2 L of 50% solution

**71.** 22 nickels, 35 dimes, and 17 quarters

**72.** 27 one-dollar bills, 18 fives, and 6 tens

**79. (a)** $C(x) = x^2 - 8x + 13$

**(b)**

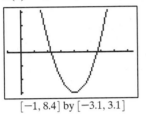

$[-1, 8.4]$ by $[-3.1, 3.1]$

**(c)** $4 \pm \sqrt{13}$

**(d)** $\det A = C(0) = 13$

**(e)** $a_{11} + a_{22} = (4 - \sqrt{13}) + (4 + \sqrt{13}) = 8$

**80. (a)** $C(x) = x^2 - 4x - 1$

**(b)**

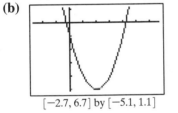

$[-2.7, 6.7]$ by $[-5.1, 1.1]$

**(c)** $2 \pm \sqrt{5}$

**(d)** $\det A = C(0) = -1$

**(e)** $a_{11} + a_{22} = (2 - \sqrt{5}) + (2 + \sqrt{5}) = 4,$

## Section 7.4 (pp. 580–584)

### Quick Review

**1.** $\dfrac{3x - 5}{x^2 - 4x + 3}$

**2.** $\dfrac{3(x - 1)}{x^2 + 5x + 4}$

**3.** $\dfrac{4x^2 + 6x + 1}{x^3 + 2x^2 + x}$

**4.** $\dfrac{3x^2 - x + 2}{(x^2 + 1)^2}$

**5.** $3x^2 - 2 + \dfrac{3}{x - 2}$

**6.** $2x + 1 - \dfrac{3x + 2}{x^2 + x - 6}$

**7.** $(x + 1)(x - 3)(x^2 + 4)$   **8.** $(x + 1)^2(x + 2)(x - 5)$

**9.** $A = 3, B = -1, C = 1$

**10.** $A = -2, B = 2, C = -1, D = -5$

### Exercises 7.4

**1.** $\dfrac{A_1}{x} + \dfrac{A_2}{x - 2} + \dfrac{A_3}{x + 2}$

**2.** $\dfrac{B_1x + C_1}{x^2 + x + 1} + \dfrac{B_2x + C_2}{(x^2 + x + 1)^2} + \dfrac{B_3x + C_2}{(x^2 - x + 1)}$

**3.** $\dfrac{A_1}{x} + \dfrac{A_2}{x^2} + \dfrac{A_3}{x^3} + \dfrac{A_4}{x - 1} + \dfrac{A_5}{(x - 1)^2} + \dfrac{B_1x + C_1}{x^2 + 9}$

**4.** $\dfrac{A_1}{x - 1} + \dfrac{A_2}{(x - 1)^2} + \dfrac{A_3}{(x - 1)^3} + \dfrac{B_1x + C_1}{x^2 + x + 1}$
$+ \dfrac{B_2x + C_2}{(x^2 + x + 1)^2} + \dfrac{B_3x + C_3}{(x^2 + x + 1)^3}$

**7.** $\dfrac{3}{x^2 + 1} + \dfrac{2x - 1}{(x^2 + 1)^2}$

**9.** $\dfrac{1}{x - 2} + \dfrac{2}{(x - 2)^2} + \dfrac{1}{(x - 2)^3}$

**10.** $\dfrac{-3x + 7}{x^2 + 4} + \dfrac{8x - 17}{x^2 + 9}$

**11.** $\dfrac{2}{x + 3} - \dfrac{1}{(x + 3)^2} + \dfrac{3x - 1}{x^2 + 2} + \dfrac{x + 2}{(x^2 + 2)^2}$

**12.** $\dfrac{-2}{x - 1} + \dfrac{3}{(x - 1)^2} + \dfrac{1}{x + 4} + \dfrac{3}{(x + 4)^2}$

**13.** $\dfrac{1}{x - 5} - \dfrac{1}{x - 3}$

**14.** $\dfrac{1}{x + 3} - \dfrac{1}{x + 7}$

**15.** $\dfrac{2}{x - 1} - \dfrac{2}{x + 1}$

**16.** $\dfrac{1}{x - 3} - \dfrac{1}{x + 3}$

**19.** $\dfrac{1}{x - 3} - \dfrac{2}{x + 4}$

**20.** $\dfrac{4}{x - 5} + \dfrac{3}{x + 2}$

**21.** $-\dfrac{2}{x + 3} + \dfrac{5}{2x - 1}$

**22.** $\dfrac{3}{x + 1} - \dfrac{2}{2x - 3}$

**23.** $\dfrac{2}{x^2 + 1} + \dfrac{3}{(x^2 + 1)^2}$

**24.** $\dfrac{3}{x^2 + 1} + \dfrac{1}{(x^2 + 1)^2}$

**25.** $\dfrac{2}{x} - \dfrac{1}{x - 1} + \dfrac{2}{(x - 1)^2}$

**26.** $\dfrac{7/3}{(x - 3)^2} - \dfrac{25/9}{x - 3} + \dfrac{25/9}{x}$

**27.** $\dfrac{1}{x} - \dfrac{1}{x^2} + \dfrac{2}{x - 3}$

**28.** $\dfrac{2}{x} - \dfrac{1}{x^2} + \dfrac{3}{x + 4}$

**29.** $\dfrac{2x}{x^2 + 2} - \dfrac{1}{(x^2 + 2)^2}$

**30.** $\dfrac{3x}{x^2 + 2} - \dfrac{1}{(x^2 + 2)^2}$

**31.** $\dfrac{2}{x - 1} - \dfrac{x}{x^2 + x + 1}$

**32.** $\dfrac{3}{x + 1} - \dfrac{x}{x^2 - x + 1}$

**33.** $2 + \dfrac{x + 5}{x^2 - 1}; \dfrac{3}{x - 1} - \dfrac{2}{x + 1}$

**34.** $3 + \dfrac{2x + 12}{x^2 - 4}; \dfrac{4}{x - 2} - \dfrac{2}{x + 2}$

**35.** $x - 1 + \dfrac{x - 2}{x^2 + x}; \dfrac{3}{x + 1} - \dfrac{2}{x}$

**36.** $x + 1 + \dfrac{x + 2}{x^2 - x}; \dfrac{3}{x - 1} - \dfrac{2}{x}$

**47.** $A = 3$

# Section 7.5 (pp. 587–592)

## Quick Review

**2.** (6, 0); (0, 3)
**3.** (20, 0); (0, 50)
**4.** (30, 0); (0, −20)

**5.** $\begin{bmatrix} 30 \\ 60 \end{bmatrix}$     **6.** $\begin{bmatrix} 70 \\ 20 \end{bmatrix}$     **7.** $\begin{bmatrix} 10 \\ 140 \end{bmatrix}$

**8.** $\begin{bmatrix} 2 \\ 4 \end{bmatrix}$     **9.** $\begin{bmatrix} 3 \\ 3 \end{bmatrix}$

## Exercises 7.5

**1.** Graph (c); boundary included
**2.** Graph (f); boundary excluded
**3.** Graph (b); boundary included
**4.** Graph (d); boundary excluded
**5.** Graph (e); boundary included
**6.** Graph (a); boundary excluded

**7.**

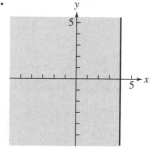

boundary included

**8.**

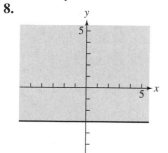

boundary included

**9.**

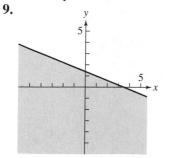

boundary included

**10.**

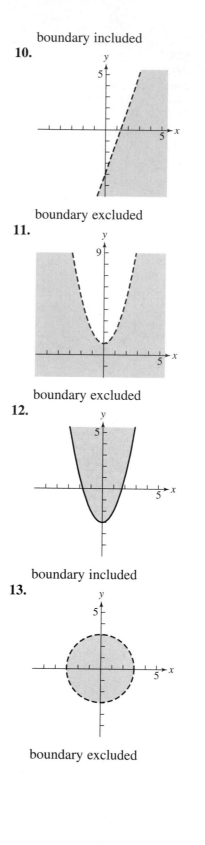

boundary included

boundary excluded

**11.**

boundary excluded

**12.**

boundary included

**13.**

boundary excluded

**14.**

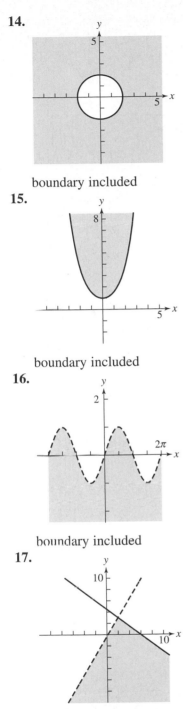

boundary included

**15.**

boundary included

**16.**

boundary included

**17.**

Corner at (2, 3)

**18.**

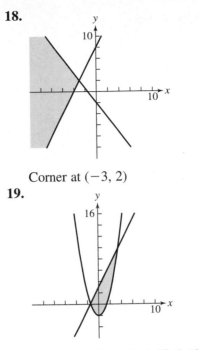

Corner at (−3, 2)

**19.**

Corners at about (−1.45, 0.10) and (3.45, 9.90)

**20.**

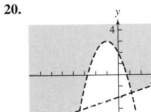

Corners at about (−3.48, −3.16) and (1.15, −1.62)

**21.**

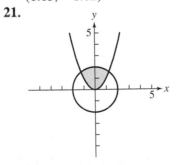

Corners at about (±1.25, 1.56)

**22.**

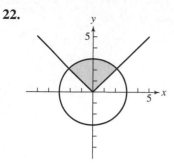

Corners at about $(\pm 2.12, 2.12)$

**23.**

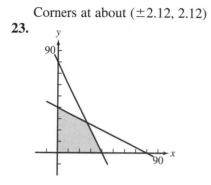

Corners at $(0, 40)$, $(26.7, 26.7)$, $(0, 0)$, and $(40, 0)$

**24.**

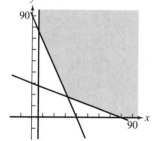

Corners at $(0, 90)$, $(32, 8)$, and $(80, 0)$

**25.**

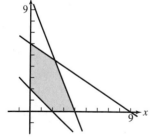

Corners at $(0, 2)$, $(0, 6)$, $(2.18, 4.55)$, $(4, 0)$, and $(2, 0)$

**26.**

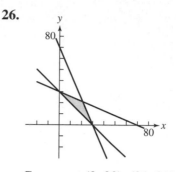

Corners at $(0, 30)$, $(21, 21)$, and $(30, 0)$

**27.** $x^2 + y^2 \geq 4$
   $y \geq -x^2 + 1$

**28.** $x^2 + y^2 \geq 4$
   $y \geq 0$

**29.** $y \leq -\frac{1}{2}x + 5$

   $y \leq -\frac{3}{2}x + 9$

   $x \geq 0$

   $y \geq 0$

**30.** $y \geq -\frac{5}{2}x + 6$

   $y \geq -\frac{1}{3}x + \frac{5}{3}$

   $y \geq 0$

   $x \geq 0$

**31.** $f_{\min} = 0; f_{\max} \approx 293.33$
**32.** $f_{\min} = 0; f_{\max} = 990$
**33.** $f_{\min} = 162; f_{\max} = $ none
**34.** $f_{\min} = 360; f_{\max} = 505$
**35.** $f_{\min} = 24; f_{\max} = $ none
**36.** $f_{\min} = 34; f_{\max} = $ none
**37.** $C_{\min} = \$1,926.09$   **38.** $C_{\min} = \$10.88$
**39.** $C_{\min} = \$48,000$   **40.** $P_{\max} = \$6,500$

**43.** $y_1 = 2\sqrt{1 - \frac{x^2}{9}}; y_2 = -2\sqrt{1 - \frac{x^2}{9}}$

**44.** $y_1 = \sqrt{x^2 - 4}; y_2 = -\sqrt{x^2 - 4}$

**45.**

## Chapter Review (pp. 597–600)

**1. (a)** $\begin{bmatrix} 1 & 2 \\ 8 & 3 \end{bmatrix}$   **(b)** $\begin{bmatrix} -3 & 4 \\ 0 & -3 \end{bmatrix}$

**(c)** $\begin{bmatrix} 2 & -6 \\ -8 & 0 \end{bmatrix}$   **(d)** $\begin{bmatrix} -7 & 11 \\ 4 & -6 \end{bmatrix}$

**2. (a)** $\begin{bmatrix} 1 & 5 & -1 & 6 \\ 3 & 3 & 1 & 0 \\ -2 & 1 & 3 & 4 \end{bmatrix}$

**(b)** $\begin{bmatrix} 3 & 1 & -1 & -2 \\ -1 & 5 & -5 & -6 \\ 2 & -7 & 1 & -2 \end{bmatrix}$

**(c)** $\begin{bmatrix} 4 & -6 & 2 & -4 \\ -2 & -8 & 4 & 6 \\ 0 & 6 & -4 & 2 \end{bmatrix}$

**(d)** $\begin{bmatrix} 8 & 5 & -3 & -2 \\ -1 & 14 & -12 & -15 \\ 4 & -17 & 4 & -3 \end{bmatrix}$

**3.** $\begin{bmatrix} -3 & -7 & 11 \\ 0 & -12 & 12 \end{bmatrix}$; not possible

**4.** not possible; $\begin{bmatrix} 15 & -4 \\ 1 & 3 \\ -5 & -13 \end{bmatrix}$

**6.** not possible; $\begin{bmatrix} -3 & -1 \\ -1 & 3 \\ -3 & 4 \\ -1 & 2 \end{bmatrix}$

**7.** $\begin{bmatrix} 1 & 2 & -3 \\ 2 & -3 & 4 \\ -2 & 1 & -1 \end{bmatrix}$ $\begin{bmatrix} -3 & 2 & 4 \\ 2 & 1 & -3 \\ 1 & -2 & -1 \end{bmatrix}$

**8.** $\begin{bmatrix} 3 & 0 & 2 & 1 \\ -2 & 1 & 0 & 1 \\ 3 & -2 & 1 & 0 \\ -1 & 1 & 2 & -1 \end{bmatrix}$; $\begin{bmatrix} 1 & -2 & 1 & 0 \\ 0 & 3 & 1 & 2 \\ 1 & -1 & -1 & 2 \\ -2 & 3 & 0 & 1 \end{bmatrix}$

**9.** $AB = BA = I_4$ **10.** $AB = BA = I_3$

**11.** $\begin{bmatrix} -2 & -5 & 6 & -1 \\ 0 & -1 & 1 & 0 \\ 10 & 24 & -27 & 4 \\ -3 & -7 & 8 & -1 \end{bmatrix}$

**12.** $\begin{bmatrix} -0.4 & 0.2 & 0.2 \\ -0.2 & -0.4 & 0.6 \\ 0.6 & 0.2 & 0.2 \end{bmatrix}$

**15.** $\begin{bmatrix} 1 & 0 & 2 \\ 0 & 1 & -1 \\ 0 & 0 & 0 \end{bmatrix}$ **16.** $\begin{bmatrix} 1 & 0 & 0 & 1 \\ 0 & 1 & 0 & 2 \\ 0 & 0 & 1 & -3 \end{bmatrix}$

**17.** $\begin{bmatrix} 1 & 0 & 0 & 8 \\ 0 & 1 & 0 & -11 \\ 0 & 0 & 1 & 5 \end{bmatrix}$ **18.** $\begin{bmatrix} 1 & 0 & 0 & 2 \\ 0 & 1 & 0 & -1 \\ 0 & 0 & 1 & 1 \end{bmatrix}$

**23.** $(2 - z - w, w + 1, z, w)$
**24.** $(-w - 2, -z - w, z, w)$
**25.** no solution
**26.** $\left(\frac{1}{4}z + \frac{3}{4}, \frac{7}{4}z + \frac{5}{4}, z\right)$
**27.** $(1 - 2z + w, 2 + z - w, z, w)$
**29.** $\left(\frac{9}{4}, -\frac{3}{4}, -\frac{7}{4}\right)$ **30.** $\left(\frac{1}{2}, -\frac{5}{2}, -\frac{5}{2}\right)$
**39.** $x \approx 0.14$ and $y \approx -2.29$
**40.** $(x, y) = (-1, 2.5)$ or $(x, y) = (3, 1.5)$
**41.** $(x, y) = (-2, 1)$ or $(x, y) = (2, 1)$
**42.** $(x, y) \approx (-1.47, 1.35)$ or $(x, y) \approx (1.47, 1.35)$ or $(x, y) \approx (0.76, -1.85)$ or $(x, y) \approx (-0.76, -1.85)$
**43.** $(x, y) \approx (2.27, 1.53)$
**44.** $(x, y) \approx (4.62, 2.22)$ or $(x, y) \approx (1.56, 1.14)$
**45.** $(a, b, c, d) = \left(\frac{17}{840}, -\frac{33}{280}, -\frac{571}{420}, \frac{386}{35}\right)$
**46.** $(a, b, c, d, e) = \left(\frac{19}{108}, -\frac{29}{18}, \frac{59}{36}, \frac{505}{54}, -\frac{68}{9}\right)$
**47.** $\frac{1}{x + 1} + \frac{2}{x - 4}$ **48.** $\frac{6}{x + 2} - \frac{5}{x - 1}$
**49.** $-\frac{1}{x + 2} + \frac{1}{x + 1} + \frac{2}{(x + 1)^2}$
**50.** $\frac{2}{x - 1} + \frac{1}{x + 2} - \frac{3}{(x + 2)^2}$
**51.** $\frac{2}{x + 1} + \frac{3x - 4}{x^2 + 1}$ **52.** $\frac{1}{x + 2} - \frac{2x + 1}{x^2 + 4}$

**57.**

**58.**

**59.**

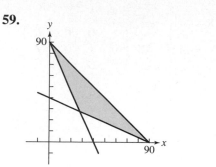

Corners at $(0, 90)$, $(90, 0)$, $\left(\dfrac{360}{13}, \dfrac{360}{13}\right)$.
Boundaries included.

**60.**

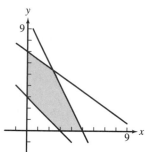

Corners at $(0, 3)$, $(0, 7)$, $\left(\dfrac{30}{13}, \dfrac{70}{13}\right)$, $(3, 0)$, $(5, 0)$.
Boundaries included.

**61.**

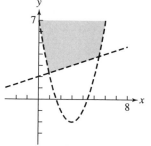

Corners at $(0.92, 2.31)$ and $(5.41, 3.80)$.
Boundaries excluded.

**62.**

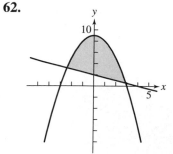

Corners at $(-2.41, 3.20)$ and $(2.91, 0.55)$.
Boundaries included.

**63.**

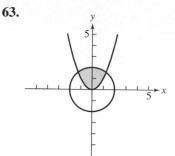

Corners at $(-1.25, 1.56)$ and $(1.25, 1.56)$.
Boundaries included.

**64.**

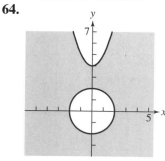

No corner points. Boundaries included

**65.** $f_{\min} = 106$; $f_{\max} = $ none
**66.** $f_{\min} = 138$; $f_{\max} = $ none
**67.** $f_{\min} = 205$; $f_{\max} = 292$
**68.** $f_{\min} = 600$; $f_{\max} = 1680$

**69. (a)** $\begin{bmatrix} 2.12 \\ 0.71 \end{bmatrix}$      **(b)** $\begin{bmatrix} -0.71 \\ 2.12 \end{bmatrix}$

**70. (a)** $y = 0.61x - 1.18$

$[0, 30]$ by $[0, 15]$

**(b)** $y = \dfrac{14.57}{1 + 28.54e^{-0.239x}}$

$[0, 30]$ by $[0, 15]$

**71. (a)** $y = 15.62x + 792.5$

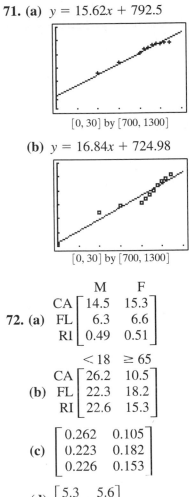

$[0, 30]$ by $[700, 1300]$

**(b)** $y = 16.84x + 724.98$

$[0, 30]$ by $[700, 1300]$

**72. (a)**

$$\begin{array}{c} & \text{M} & \text{F} \\ \text{CA} & \\ \text{FL} & \\ \text{RI} & \end{array} \begin{bmatrix} 14.5 & 15.3 \\ 6.3 & 6.6 \\ 0.49 & 0.51 \end{bmatrix}$$

**(b)**

$$\begin{array}{c} & < 18 & \geq 65 \\ \text{CA} & \\ \text{FL} & \\ \text{RI} & \end{array} \begin{bmatrix} 26.2 & 10.5 \\ 22.3 & 18.2 \\ 22.6 & 15.3 \end{bmatrix}$$

**(c)**
$$\begin{bmatrix} 0.262 & 0.105 \\ 0.223 & 0.182 \\ 0.226 & 0.153 \end{bmatrix}$$

**(d)**
$$\begin{bmatrix} 5.3 & 5.6 \\ 2.7 & 2.9 \end{bmatrix}$$

This matrix provides total numbers of males and females who are under 18 or are 65 or older in all three states.

**73. (a)** $N = [200 \quad 400 \quad 600 \quad 250]$
   **(b)** $P = [\$80 \quad \$120 \quad \$200 \quad \$300]$
   **(c)** $NP^T = \$259,000$

**74.** 380 student, 72 non student

**75.** 2 vans, 4 small trucks, 3 large trucks

**76.** about $21,333.33 at 7.5% and $16,666.67 at 6%

**77.** $160,000 at 4%, $170,000 at 6.5%, $320,000 at 9%

**78.** Sue: 9.3 hours; Esther: 12 hours, Murphy: 16.8 hours

**79.** Pipe A: 15 hours. Pipe B: $\dfrac{60}{11} = 5.45$ hours.

Pipe C: 12 hours

**Chapter 7 Project**

**1.**

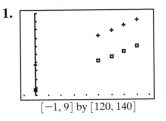

$[-1, 9]$ by $[120, 140]$

males $y \approx 1.346x + 121.4$;
females $y \approx 1.353x + 127.6$

**2.** Slope is the rate of change of people (in millions) per year. The $y$-intercept is the number of people (either males or females) in 1990.

**3.** no; yes; no

**4.**

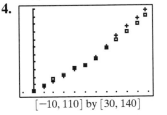

$[-10, 110]$ by $[30, 140]$

**5.** males: $y = \dfrac{312.98}{1 + 8.18e^{-0.017x}}$;

females: $y = \dfrac{284.79}{1 + 8.11e^{-0.019x}}$;

$(46, 65.2)$; This represents the time when the female population became greater than the male population.

**7.** 48.8% male and 51.2% female

# Chapter 8

## Section 8.1 (pp. 603–613)

### Exploration 1

**3.** $\{(-2\sqrt{6}, 6), (-2\sqrt{5}, 5), (-4, 4), (-2\sqrt{3}, 3),$
$(-2\sqrt{2}, 2), (-2, 1), (0, 0), (2, 1), (2\sqrt{2}, 2),$
$(2\sqrt{3}, 3), (4, 4), (2\sqrt{5}, 5), (2\sqrt{6}, 6)\}$

### Exploration 2

**1.**

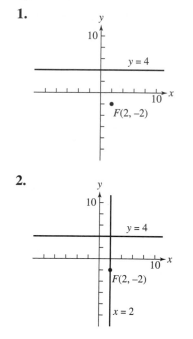

**2.**

The equation of the axis is $x = 2$.

**3.**

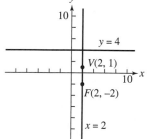

**4.** The focal length $p$ is $-3$ and the focal width $|4p|$ is 12.

**5.**

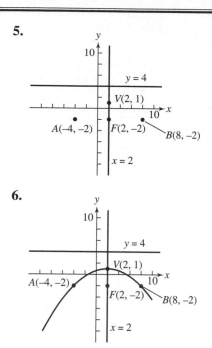

**6.**

### Quick Review 1.1

**2.** $\sqrt{(a - 2)^2 + (b + 3)^2}$

**5.** $f(x) + 6 = -(x - 1)^2$

**6.** $f(x) + \dfrac{19}{2} = 2\left(x + \dfrac{3}{2}\right)^2$

**7.** Vertex: $(1, 5)$. $f(x)$ can be obtained from $g(x)$ by stretching $x^2$ by 3, shifting up 5 units, and shifting right 1 unit.

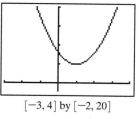

$[-3, 4]$ by $[-2, 20]$

**8.** Vertex: $(3, 17)$. $f(x) = -2(x - 3)^2 + 17$. $f(x)$ can be obtained from $g(x)$ by stretching $x^2$ by 2, rotating over the $x$-axis, shifting up 19 units and shifting right 3 units.

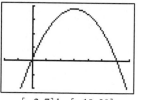

$[-2, 7]$ by $[-10, 20]$

**9.** $y = -2(x + 1)^2 + 3$

**10.** $y = 2(x - 2)^2 - 5$

## Exercises 8.1

**1.** Vertex: $(0, 0)$, Focus: $\left(0, \frac{3}{2}\right)$, Directrix: $y = -\frac{3}{2}$,

Focal width: $|4p| = \left|4 \cdot \frac{3}{2}\right| = 6$

**2.** Vertex: $(0, 0)$, Focus: $(-2, 0)$, Directrix: $x = 2$,
Focal width: $|4p| = |4(-2)| = 8$

**3.** Vertex: $(-3, 2)$, Focus: $(-2, 2)$, Directrix:
$x = -4$, Focal width: $|4p| = |4(1)| = 4$

**4.** Vertex: $(-4, -1)$, Focus: $\left(-4, -\frac{5}{2}\right)$, Directrix:

$y = \frac{1}{2}$, Focal width: $|4p| = \left|4\left(\frac{-3}{2}\right)\right| = 6$

**5.** Vertex: $(0, 0)$, Focus: $\left(0, \frac{-1}{3}\right)$, Directrix: $y = \frac{1}{3}$,

Focal width: $|4p| = \left|4\left(\frac{-1}{3}\right)\right| = \frac{4}{3}$

**6.** Vertex: $(0, 0)$, Focus: $\left(\frac{4}{5}, 0\right)$, Directrix: $x = \frac{-4}{5}$,

Focal width: $|4p| = \left|4\left(\frac{4}{5}\right)\right| = \frac{16}{5}$

**19.** $x^2 = -6y$

**27.** $(x - 2)^2 = 16(y + 1)$

**28.** $(x + 3)^2 = -20(y - 3)$

**29.** $(y + 4)^2 = 10(x + 1)$

**30.** $(y - 3)^2 = 5(x - 2)$

**31.**

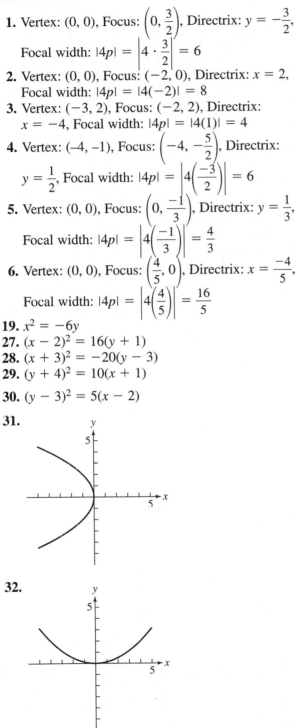

**32.**

**33.**

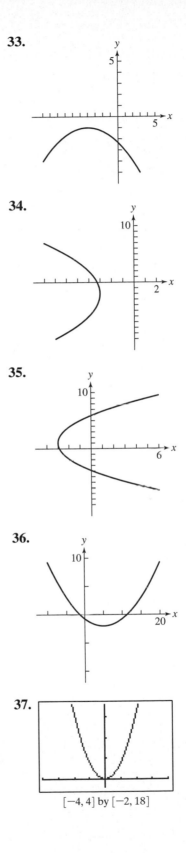

**34.**

**35.**

**36.**

**37.**

$[-4, 4]$ by $[-2, 18]$

**38.**

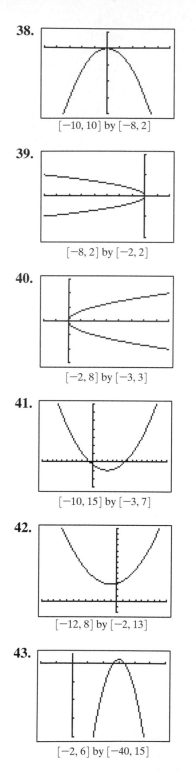

$[-10, 10]$ by $[-8, 2]$

**39.**

$[-8, 2]$ by $[-2, 2]$

**40.**

$[-2, 8]$ by $[-3, 3]$

**41.**

$[-10, 15]$ by $[-3, 7]$

**42.**

$[-12, 8]$ by $[-2, 13]$

**43.**

$[-2, 6]$ by $[-40, 15]$

**44.**

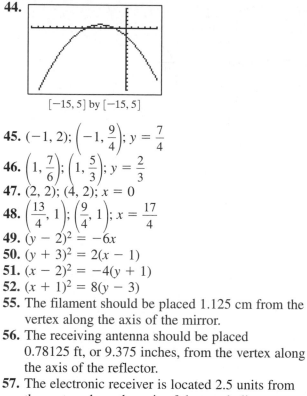

$[-15, 5]$ by $[-15, 5]$

**45.** $(-1, 2); \left(-1, \dfrac{9}{4}\right); y = \dfrac{7}{4}$

**46.** $\left(1, \dfrac{7}{6}\right); \left(1, \dfrac{5}{3}\right); y = \dfrac{2}{3}$

**47.** $(2, 2); (4, 2); x = 0$

**48.** $\left(\dfrac{13}{4}, 1\right); \left(\dfrac{9}{4}, 1\right); x = \dfrac{17}{4}$

**49.** $(y - 2)^2 = -6x$

**50.** $(y + 3)^2 = 2(x - 1)$

**51.** $(x - 2)^2 = -4(y + 1)$

**52.** $(x + 1)^2 = 8(y - 3)$

**55.** The filament should be placed 1.125 cm from the vertex along the axis of the mirror.

**56.** The receiving antenna should be placed 0.78125 ft, or 9.375 inches, from the vertex along the axis of the reflector.

**57.** The electronic receiver is located 2.5 units from the vertex along the axis of the parabolic microphone.

**58.** The light bulb should be placed 3 units from the vertex along the axis of the headlight.

**59.** Starting at the leftmost tower, the lengths of the cables are: $\approx$ {79.44, 54.44, 35, 21.11, 12.78, 10, 12.78, 21.11, 35, 54.44, 79.44}

**60.** The maximum clearance must be at least 21.33 feet.

**61. (a)–(c)**

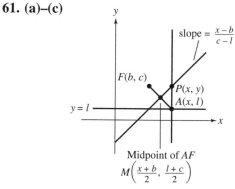

**(d)** parabola

**62. (a)–(d)**

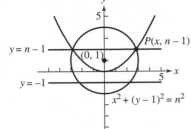

**63. (a)**

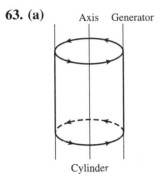

Cylinder

**(b)**

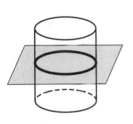

Circle          Single line

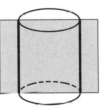

Two parallel lines

**(c)**

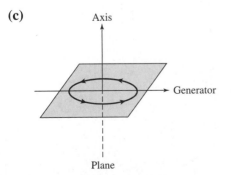

Plane

**(d)**

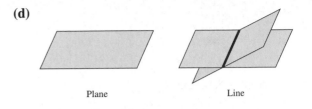

Plane          Line

## Section 8.2 (pp. 614–624)

### Exploration 1

**2.**

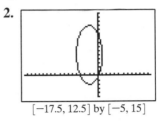

$[-17.5, 12.5]$ by $[-5, 15]$

**3.** Example 1: $x = 3 \cos(t)$ and $y = 2 \sin(t)$.
Example 2: $y = \sqrt{13} \sin(t)$ and $x = 2 \cos(t)$.
Example 3: $x = 5 \cos(t) + 3$ and $y = 4 \sin(t) - 1$.

**4.**

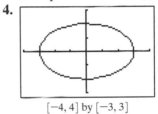

$[-4, 4]$ by $[-3, 3]$

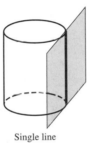

$[-6, 6]$ by $[-4, 4]$

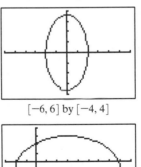

$[-3, 9]$ by $[-6, 4]$

### Quick Review 8.2

**2.** $\sqrt{(a + 3)^2 + (b + 4)^2}$

**3.** $y = \pm \dfrac{3}{2}\sqrt{4 - x^2}$

**4.** $y = \pm\sqrt{\dfrac{900 - 25x^2}{36}} = \pm\dfrac{5}{6}\sqrt{36 - x^2}$

**9.** $x = \dfrac{3 \pm \sqrt{15}}{2}$

**10.** $x = -1 \pm \sqrt{\dfrac{7}{2}}$

### Section 8.2 Exercises

**1.** Vertices: $(4, 0)$, $(-4, 0)$; Foci: $(3, 0)$, $(-3, 0)$

**2.** Vertices: $(0, 5)$, $(0, -5)$; Foci: $(0, 2)$, $(0, -2)$

**3.** Vertices: $(-2, 6)$, $(-2, -4)$; Foci: $(-2, 4)$, $(-2, -2)$

**4.** Vertices: $(3 + \sqrt{11}, 5)$, $(3 - \sqrt{11}, 5)$; Foci: $(5, 5)$, $(1, 5)$

**5.** Vertices: $(2, 0)$, $(-2, 0)$; Foci: $(1, 0)$, $(-1, 0)$

**6.** Vertices: $(0, 3)$, $(0, -3)$; Foci: $(0, \sqrt{5})$, $(0, -\sqrt{5})$

**9.** (a)

**10.** (b)

**11.** $\dfrac{x^2}{4} + \dfrac{y^2}{9} = 1$

**12.** $\dfrac{x^2}{49} + \dfrac{y^2}{25} = 1$

**15.** $\dfrac{x^2}{16} + \dfrac{y^2}{25} = 1$

**16.** $\dfrac{x^2}{49} + \dfrac{y^2}{16} = 1$

**17.** $\dfrac{x^2}{16} + \dfrac{y^2}{36} = 1$

**18.** $\dfrac{x^2}{25} + \dfrac{y^2}{4} = 1$

**19.** $\dfrac{x^2}{25} + \dfrac{y^2}{16} = 1$

**20.** $\dfrac{x^2}{144} + \dfrac{y^2}{169} = 1$

**22.** $\dfrac{(x + 2)^2}{4} + \dfrac{(y - 2)^2}{25} = 1$

**27.**

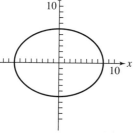

**28.**

**29.**

**30.**

**31.**

**32.**

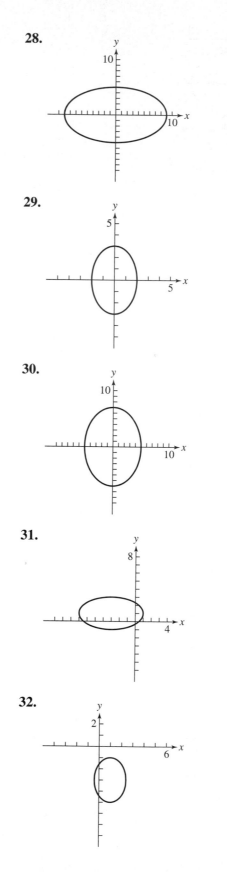

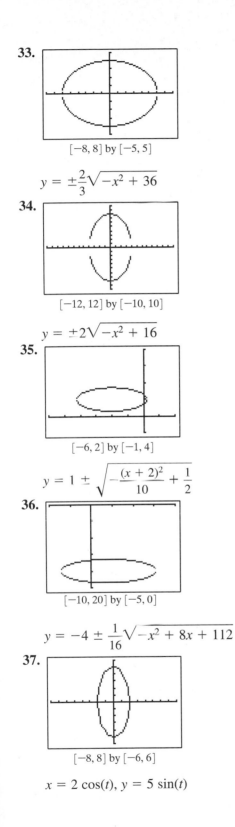

**33.**

$[-8, 8]$ by $[-5, 5]$

$$y = \pm\frac{2}{3}\sqrt{-x^2 + 36}$$

**34.**

$[-12, 12]$ by $[-10, 10]$

$$y = \pm 2\sqrt{-x^2 + 16}$$

**35.**

$[-6, 2]$ by $[-1, 4]$

$$y = 1 \pm \sqrt{-\frac{(x + 2)^2}{10} + \frac{1}{2}}$$

**36.**

$[-10, 20]$ by $[-5, 0]$

$$y = -4 \pm \frac{1}{16}\sqrt{-x^2 + 8x + 112}$$

**37.**

$[-8, 8]$ by $[-6, 6]$

$$x = 2 \cos(t), \quad y = 5 \sin(t)$$

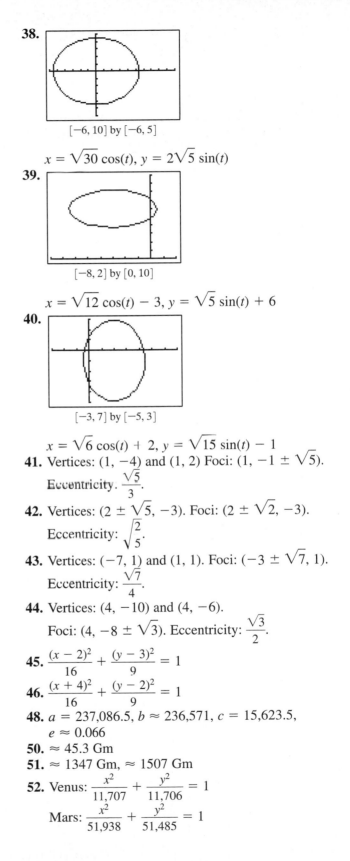

**38.**

$[-6, 10]$ by $[-6, 5]$

$$x = \sqrt{30} \cos(t), \quad y = 2\sqrt{5} \sin(t)$$

**39.**

$[-8, 2]$ by $[0, 10]$

$$x = \sqrt{12} \cos(t) - 3, \quad y = \sqrt{5} \sin(t) + 6$$

**40.**

$[-3, 7]$ by $[-5, 3]$

$$x = \sqrt{6} \cos(t) + 2, \quad y = \sqrt{15} \sin(t) - 1$$

**41.** Vertices: $(1, -4)$ and $(1, 2)$ Foci: $(1, -1 \pm \sqrt{5})$. Eccentricity. $\dfrac{\sqrt{5}}{3}$.

**42.** Vertices: $(2 \pm \sqrt{5}, -3)$. Foci: $(2 \pm \sqrt{2}, -3)$. Eccentricity: $\sqrt{\dfrac{2}{5}}$.

**43.** Vertices: $(-7, 1)$ and $(1, 1)$. Foci: $(-3 \pm \sqrt{7}, 1)$. Eccentricity: $\dfrac{\sqrt{7}}{4}$.

**44.** Vertices: $(4, -10)$ and $(4, -6)$. Foci: $(4, -8 \pm \sqrt{3})$. Eccentricity: $\dfrac{\sqrt{3}}{2}$.

**45.** $\dfrac{(x - 2)^2}{16} + \dfrac{(y - 3)^2}{9} = 1$

**46.** $\dfrac{(x + 4)^2}{16} + \dfrac{(y - 2)^2}{9} = 1$

**48.** $a = 237{,}086.5$, $b \approx 236{,}571$, $c = 15{,}623.5$, $e \approx 0.066$

**50.** $\approx 45.3$ Gm

**51.** $\approx 1347$ Gm, $\approx 1507$ Gm

**52.** Venus: $\dfrac{x^2}{11{,}707} + \dfrac{y^2}{11{,}706} = 1$

Mars: $\dfrac{x^2}{51{,}938} + \dfrac{y^2}{51{,}485} = 1$

**53.** For sungrazers, $a - c < 2(1.392) = 2.784$. The eccentricity of their ellipses would be very close to 1.

**54.** $e \approx 0.97$

**55.** Foci at $(\pm\sqrt{51.75}, 0) \approx (\pm 7.19, 0)$.

**56.** Place the source and the patient at opposite foci — 12 inches from the center along the major axis.

**59. (a)** Approximate solutions:
$(\pm 1.04, -0.86), (\pm 1.37, 0.73)$

**(b)** $\left( \pm \dfrac{\sqrt{94 - 2\sqrt{161}}}{8}, -\dfrac{1 + \sqrt{161}}{16} \right),$

$\left( \pm \dfrac{\sqrt{94 + 2\sqrt{161}}}{8}, \dfrac{-1 + \sqrt{161}}{16} \right)$

## Section 8.3 (pp. 624–634)

### Exploration 1

**2.**

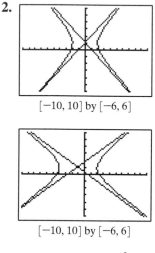

$[-10, 10]$ by $[-6, 6]$

$[-10, 10]$ by $[-6, 6]$

**3.** Example 1: $x = \dfrac{3}{\cos (t)}, y = 2 \tan(t)$

Example 2: $x = 2 \tan(t), y = \sqrt{5}/\cos (t)$

Example 3: $x = 3 + 5/\cos(t), y = -1 + 4 \tan(t)$

Example 4: $x = -2 + 3/\cos(t), y = 5 + 7 \tan(t)$

**4.**

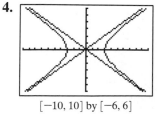

$[-10, 10]$ by $[-6, 6]$

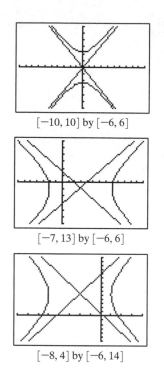

$[-10, 10]$ by $[-6, 6]$

$[-7, 13]$ by $[-6, 6]$

$[-8, 4]$ by $[-6, 14]$

### Quick Review 8.3

**2.** $\sqrt{(b - a)^2 + (c + 3)^2}$

**3.** $y = \pm\dfrac{4}{3}\sqrt{9 + x^2}$

**4.** $y = \pm\dfrac{1}{3}\sqrt{x^2 - 36}$

### Section 8.3 Exercises

**1.** Vertices: $(\pm 4, 0)$; Foci: $(\pm\sqrt{23}, 0)$

**2.** Vertices: $(0, \pm 5)$; Foci: $(0, \pm\sqrt{46})$

**3.** Vertices: $(-2, 6), (-2, -4)$; Foci: $(-2, 1 + \sqrt{41})$

**4.** Vertices: $(3 \pm \sqrt{11}, 5)$; Foci: $(3 \pm \sqrt{18}, 5)$

**5.** Vertices: $(\pm 2, 0)$; Foci: $(\pm\sqrt{7}, 0)$

**6.** Vertices: $(\pm 2, 0)$; Foci: $(\pm\sqrt{13}, 0)$

**9.** (a)

**10.** (d)

**14.** $\dfrac{x^2}{9/4} - \dfrac{y^2}{91/4} = 1$

**15.** $\dfrac{x^2}{25} - \dfrac{y^2}{75} = 1$

**16.** $\dfrac{x^2}{16} - \dfrac{y^2}{20} = 1$

**17.** $\dfrac{x^2}{144} - \dfrac{y^2}{25} = 1$

**18.** $\dfrac{x^2}{9} - \dfrac{y^2}{27} = 1$

**19.** $\dfrac{(x - 2)^2}{9} - \dfrac{(y - 3)^2}{16} = 1$

**20.** $\dfrac{(x + 2)^2}{729/64} - \dfrac{(y - 5/2)^2}{81/4} = 1$

**21.** $\dfrac{(x + 1)^2}{4} - \dfrac{(y - 2)^2}{5} = 1$

**22.** $\dfrac{(y + 5.5)^2}{3.5^2} - \dfrac{(x + 3)^2}{18} = 1$

**23.** $\dfrac{(y - 6)^2}{25} - \dfrac{(x + 3)^2}{75} = 1$

**24.** $\dfrac{(x - 1)^2}{9} - \dfrac{(y + 4)^2}{27} = 1$

**25.**

**26.**

**27.**

**28.**

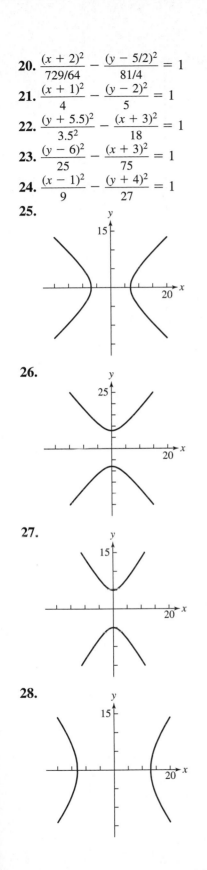

**29.**

**30.**

**31.**

$[-15, 15]$ by $[-9, 9]$

**32.**

$[-10, 10]$ by $[-20, 20]$

**33.**

$[-10, 10]$ by $[-15, 15]$

**34.**

$[-10, 10]$ by $[-20, 20]$

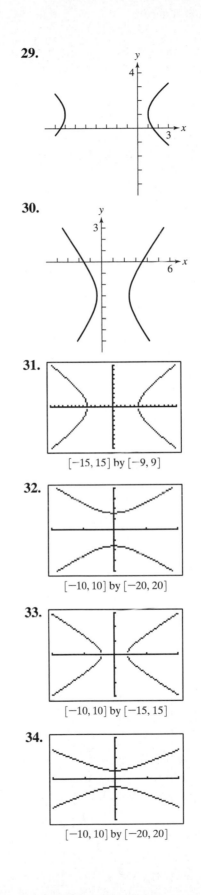

**35.**

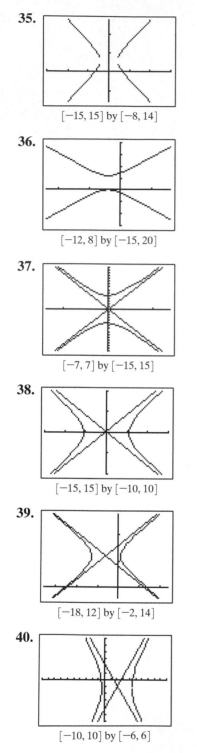

$[-15, 15]$ by $[-8, 14]$

**36.**

$[-12, 8]$ by $[-15, 20]$

**37.**

$[-7, 7]$ by $[-15, 15]$

**38.**

$[-15, 15]$ by $[-10, 10]$

**39.**

$[-18, 12]$ by $[-2, 14]$

**40.**

$[-10, 10]$ by $[-6, 6]$

**41.** Vertices: $(3, -2)$ and $(3, 4)$, Foci: $(3, 1 \pm \sqrt{13})$,
$e = \dfrac{\sqrt{13}}{3}$.

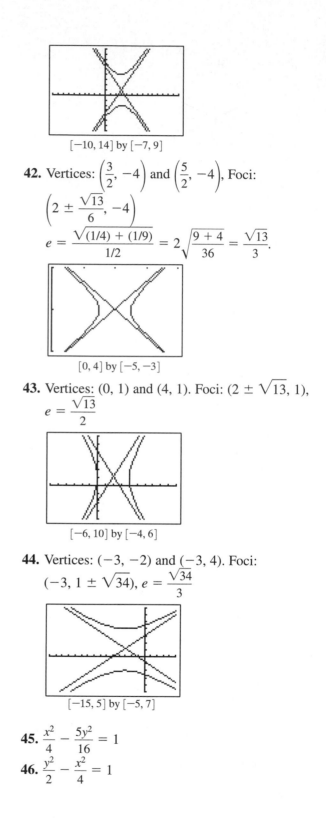

$[-10, 14]$ by $[-7, 9]$

**42.** Vertices: $\left(\dfrac{3}{2}, -4\right)$ and $\left(\dfrac{5}{2}, -4\right)$, Foci:
$\left(2 \pm \dfrac{\sqrt{13}}{6}, -4\right)$
$e = \dfrac{\sqrt{(1/4) + (1/9)}}{1/2} = 2\sqrt{\dfrac{9 + 4}{36}} = \dfrac{\sqrt{13}}{3}.$

$[0, 4]$ by $[-5, -3]$

**43.** Vertices: $(0, 1)$ and $(4, 1)$. Foci: $(2 \pm \sqrt{13}, 1)$,
$e = \dfrac{\sqrt{13}}{2}$

$[-6, 10]$ by $[-4, 6]$

**44.** Vertices: $(-3, -2)$ and $(-3, 4)$. Foci:
$(-3, 1 \pm \sqrt{34})$, $e = \dfrac{\sqrt{34}}{3}$

$[-15, 5]$ by $[-5, 7]$

**45.** $\dfrac{x^2}{4} - \dfrac{5y^2}{16} = 1$

**46.** $\dfrac{y^2}{2} - \dfrac{x^2}{4} = 1$

**48. (a)**

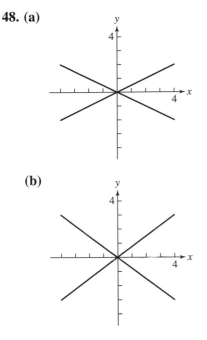

**(b)**

y

**49.** A bearing and distance of about 40.29° and 1371.11 miles, respectively.

**50.** A bearing and distance of about 50.11° and 15628.2 ft (2.89 mi), respectively.

**51.** The Sun is centered at focus $(c, 0) = (1,560, 0)$.

**52.** The Sun is centered at focus $(c, 0) = (297, 0)$.

**54.** four solutions: $(\pm 2\sqrt{2}, \pm 1)$

**55. (a)**

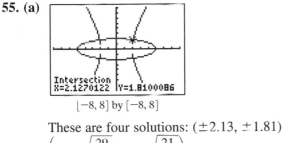

$[-8, 8]$ by $[-8, 8]$

These are four solutions: $(\pm 2.13, \pm 1.81)$

**(b)** $\left(\pm 10\sqrt{\dfrac{29}{641}}, \pm 10\sqrt{\dfrac{21}{641}}\right)$

**56. (e)** $\dfrac{x^2}{9} - \dfrac{y^2}{16} = 1$

**57.** $\dfrac{x^2}{1600} - \dfrac{y^2}{2000} = 1$

## Section 8.4 (pp. 635–642)

### Quick Review 8.4

**8.** $\cos \alpha = \sqrt{\dfrac{7}{8}}$

**10.** $\sin \alpha = \dfrac{2}{\sqrt{53}}$

### Exercises 8.4

**4.** $(x', y') = (-5 - \sqrt{2}, -1)$

**6.** $\left(\dfrac{6 - 3\sqrt{3}}{2}, \dfrac{-6\sqrt{3} - 3}{2}\right) \approx (0.40, -6.70)$

**9.** $-4y = x^2$

**10.** $8x = y^2$

**11.** $\dfrac{x^2}{9} - \dfrac{y^2}{16} = 1$

**12.** $\dfrac{x^2}{16} + \dfrac{y^2}{9} = 1$

**23.** $\dfrac{(y - 1)^2}{9} - \dfrac{(x + 1)^2}{4} = 1$. This is a hyperbola, with $a = 3$, $b = 2$, and $c = \sqrt{13}$.

Foci: $(-1, 1 \pm \sqrt{13})$. Center $(-1, 1)$, so $\dfrac{(y')^2}{9} - \dfrac{(x')^2}{4} = 1$

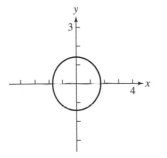

**24.** $\dfrac{(x + 3)^2}{3} + \dfrac{(y - 4)^2}{2} = 1$. This is an ellipse with $a = \sqrt{3}$, $b = \sqrt{2}$, and $c = 1$.

y

**25.** $y - 2 = (x + 1)^2$, a parabola. The vertex is $(h, k) = (-1, 2)$, so $y' = (x')^2$.

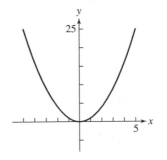

**26.** $2\left(y - \dfrac{7}{6}\right) = (x - 1)^2$, a parabola. The vertex is
$(h, k) = \left(1, \dfrac{7}{6}\right)$, so $2y' = (x')^2$

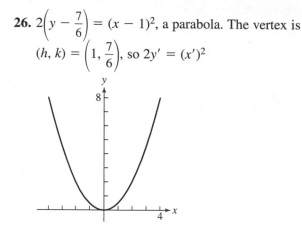

**27.** $\dfrac{(x - 1)^2}{4} + \dfrac{(y + 2)^2}{9} = 1$.

This is an elipse, with $a = 2$, $b = 3$, and $c = \sqrt{5}$.
Foci: $(1, -2 \pm \sqrt{5})$. Center $(1, -2)$, so
$\dfrac{(x')^2}{4} + \dfrac{(y')^2}{9} = 1$.

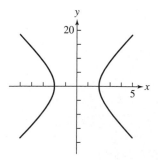

**28.** $\dfrac{(x - 1)^2}{4} - \dfrac{(y + 3)^2}{64} = 1$. This is a hyperbola, with
$a = 2$, $b = 8$, and $c = \sqrt{68} = 2\sqrt{17}$. Foci:
$(1 \pm 2\sqrt{17}, -3)$. Center $(1, -3)$, so $\dfrac{(x')^2}{4} - \dfrac{(y')^2}{64} = 1$.

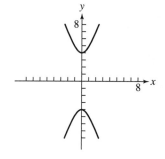

**29.** $8(x - 2) - (y - 2)^2$, a parabola. The vertex is
$(h, k) = (2, 2)$, so $8x' = (y')^2$.

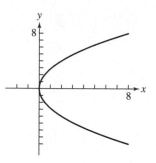

**30.** $\dfrac{(x - 1)^2}{10} + \dfrac{(y - 3)^2}{20} = 1$. This is an ellipse, with
$a = \sqrt{10}$, $b = \sqrt{20} = 2\sqrt{5}$, and $c = \sqrt{10}$.
Foci: $(1, 3 \pm \sqrt{10})$. Center $(1, 3)$, so $\dfrac{(x')^2}{10} + \dfrac{(y')^2}{20} = 1$.

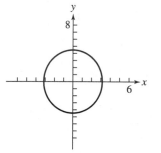

**31.** $\dfrac{y^2}{4} - \dfrac{(x + 1)^2}{2} = 1$. This is a hyperbola, with
$a = \sqrt{2}$, $b = 2$, and $c = \sqrt{6}$. Foci: $(-1, \pm\sqrt{6})$.
Center $(-1, 0)$, so $\dfrac{(y')^2}{4} - \dfrac{(x')^2}{2} = 1$.

**32.** $-4(x - 3.25) = (y - 1)^2$, a parabola. The vertex
is $(h, k) = (3.25, 1)$, so $-4x' = (y')^2$

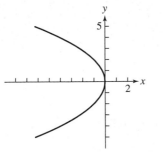

**33.** $y = \frac{1}{6}[x - 4 \pm \sqrt{-23x^2 + 28x + 88}]$ — an ellipse. The angle of rotation is

$\alpha = \frac{1}{2} \cot^{-1}\left(\frac{2-3}{-1}\right) = \frac{1}{2} \cot^{-1} 1 = \frac{\pi}{8}.$

$[-1.7, 7.7]$ by $[-3.1, 3.1]$

**34.** $y = \frac{1}{8}[10 - 3x \pm \sqrt{25x^2 + 20x + 420}]$
— a hyperbola. The angle of rotation is
$\alpha = \frac{1}{2} \cot^{-1}\left(\frac{-1-4}{3}\right) = \frac{1}{2} \cot^{-1}\left(-\frac{5}{3}\right)$
$= \frac{1}{2} \tan^{-1}\left(-\frac{3}{5}\right)$
$\approx -0.270$ radians (another possibility is
$\frac{\pi}{2} + \frac{1}{2} \tan^{-1}\left(-\frac{3}{5}\right) \approx 1.301$ radians).

$[-10, 10]$ by $[-8, 8]$

**35.** $y = \frac{1}{4}[x - 1 \pm \sqrt{3(-x^2 + 6x + 9)}]$ — an ellipse. The angle of rotation is

$\alpha = \frac{1}{2} \cot^{-1}\left(\frac{2-8}{-4}\right) = \frac{1}{2} \cot^{-1} \frac{3}{2} = \frac{1}{2} \tan^{-1} \frac{2}{3}$
$\approx 0.294$ radians.

$[-2, 8]$ by $[-3, 3]$

**36.** $y = x - \frac{3}{2} \pm \frac{1}{2} \sqrt{39 - 2x}$ — a parabola. The

angle of rotation is $\alpha = \frac{1}{2} \cot^{-1}\left(\frac{2-2}{-4}\right)$

$= \frac{1}{2} \cot^{-1} 0 = \frac{\pi}{4}.$

$[-10, 30]$ by $[-5, 20]$

**37.** hyperbola; $\alpha = \frac{\pi}{4}, \frac{(x')^2}{16} - \frac{(y')^2}{16} = 1,$

$[-10, 10]$ by $[-8, 8]$

**38.** ellipse; $\alpha = \frac{\pi}{6}, \frac{(x')^2}{4} + \frac{(y')^2}{20} = 1$

$[-10, 10]$ by $[-6, 6]$

**39.** parabola; $\alpha = \frac{\pi}{6}, x' = \pm\frac{\sqrt{14}}{2}$. This is a degenerate form consisting of only two points on a number line.

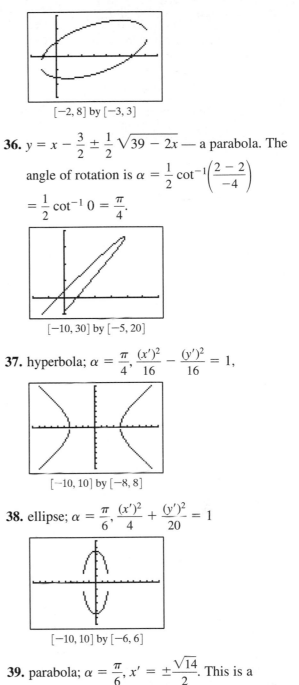

**40.** ellipse; $\alpha \approx 0.927$; $y' = \pm \dfrac{2\sqrt{10}}{5}$.

This is a degenerate form consisting of only two points on a number line.

**41.** Under the "old" coordinate system, the center $(x, y) = (0, 0)$, the vertices occured at $\left(\dfrac{3\sqrt{2}}{2}, \dfrac{3\sqrt{2}}{2}\right)$ and $\left(-\dfrac{3\sqrt{2}}{2}, -\dfrac{3\sqrt{2}}{2}\right)$, and the foci are located at $(3, 3)$ and $(-3, -3)$.

**47. (b)** no sign change

**52.**

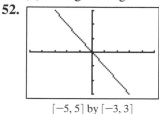

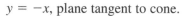

$y = -x$, plane tangent to cone.

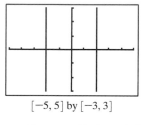

$x = 2$, $x = -2$, focus of parabola lies on directrix.

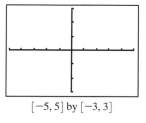

$y = 0$, plane tangent to cone rotated through same angle $\alpha$.

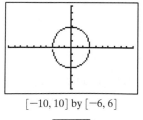

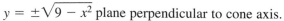

$y = \pm\sqrt{9 - x^2}$ plane perpendicular to cone axis.

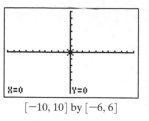

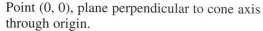

Point $(0, 0)$, plane perpendicular to cone axis through origin.

## Section 8.5 (pp. 642–651)

### Exploration 1

**1.**

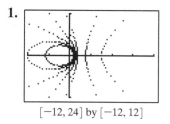

One possible answer: The graphs are similar in that they are all conic sections with a common vertex at $(1.24, 0)$ they are different, however, in that values of $e < 1$ produce a closed figure (an ellipse), $e = 1$ produces a parabola and $e > 1$ produce hyperbolas.

### Quick Review 8.5

**3.** $\theta = \dfrac{7\pi}{6}$ or $-\dfrac{5\pi}{6}$

**5.** The focus is $(0, 4)$ and the directrix is $y = -4$.

**6.** The focus is $(-3, 0)$ and the directrix is $x = 3$.

**7.** $(\pm\sqrt{5}, 0)$; $(\pm 3, 0)$

**8.** $(0, \pm 4)$; $(0, \pm 5)$

**9.** $(\pm 5, 0)$; $(\pm 4, 0)$

**10.** $(0, \pm 4\sqrt{2})$; $(0, \pm 6)$

### Section 8.5 Exercises

**1. (b)** $[-15, 5]$ by $[-10, 10]$

**2. (d)** $[-5, 5]$ by $[-3, 3]$

**3. (f)** $[-5, 5]$ by $[-3, 3]$

**4. (e)** $[-5, 5]$ by $[-3, 5]$

**5. (c)** $[-10, 10]$ by $[-5, 10]$

**6. (a)** $[-3, 3]$ by $[-6, 6]$

**7.** $r = \dfrac{2}{1 - \cos\theta}$ — a parabola.

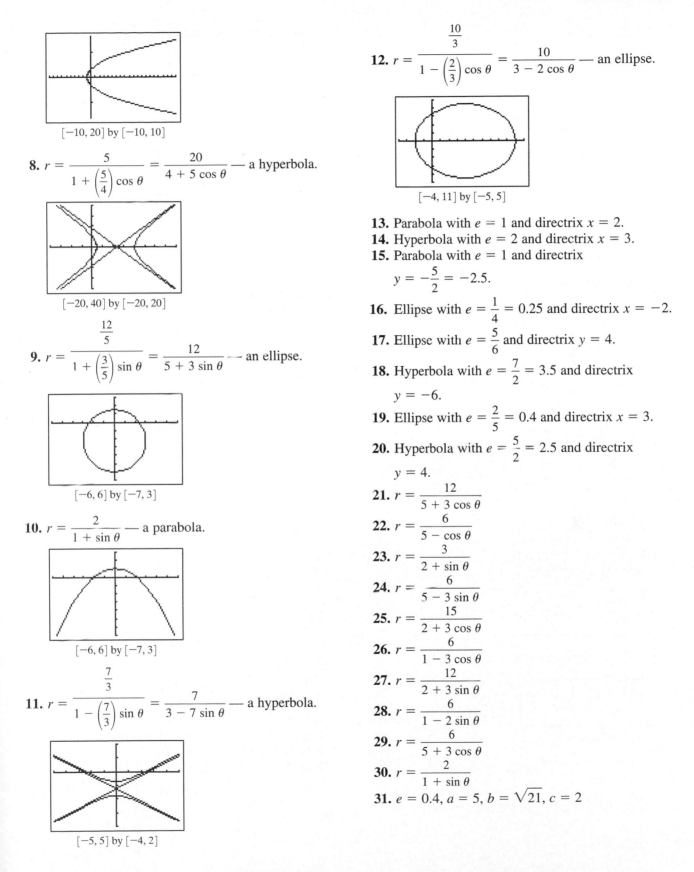

[−10, 20] by [−10, 10]

**8.** $r = \dfrac{5}{1 + \left(\dfrac{5}{4}\right)\cos\theta} = \dfrac{20}{4 + 5\cos\theta}$ — a hyperbola.

[−20, 40] by [−20, 20]

**9.** $r = \dfrac{\dfrac{12}{5}}{1 + \left(\dfrac{3}{5}\right)\sin\theta} = \dfrac{12}{5 + 3\sin\theta}$ — an ellipse.

[−6, 6] by [−7, 3]

**10.** $r = \dfrac{2}{1 + \sin\theta}$ — a parabola.

[−6, 6] by [−7, 3]

**11.** $r = \dfrac{\dfrac{7}{3}}{1 - \left(\dfrac{7}{3}\right)\sin\theta} = \dfrac{7}{3 - 7\sin\theta}$ — a hyperbola.

[−5, 5] by [−4, 2]

**12.** $r = \dfrac{\dfrac{10}{3}}{1 - \left(\dfrac{2}{3}\right)\cos\theta} = \dfrac{10}{3 - 2\cos\theta}$ — an ellipse.

[−4, 11] by [−5, 5]

**13.** Parabola with $e = 1$ and directrix $x = 2$.
**14.** Hyperbola with $e = 2$ and directrix $x = 3$.
**15.** Parabola with $e = 1$ and directrix
$$y = -\dfrac{5}{2} = -2.5.$$

**16.** Ellipse with $e = \dfrac{1}{4} = 0.25$ and directrix $x = -2$.

**17.** Ellipse with $e = \dfrac{5}{6}$ and directrix $y = 4$.

**18.** Hyperbola with $e = \dfrac{7}{2} = 3.5$ and directrix
$$y = -6.$$

**19.** Ellipse with $e = \dfrac{2}{5} = 0.4$ and directrix $x = 3$.

**20.** Hyperbola with $e = \dfrac{5}{2} = 2.5$ and directrix
$$y = 4.$$

**21.** $r = \dfrac{12}{5 + 3\cos\theta}$

**22.** $r = \dfrac{6}{5 - \cos\theta}$

**23.** $r = \dfrac{3}{2 + \sin\theta}$

**24.** $r = \dfrac{6}{5 - 3\sin\theta}$

**25.** $r = \dfrac{15}{2 + 3\cos\theta}$

**26.** $r = \dfrac{6}{1 - 3\cos\theta}$

**27.** $r = \dfrac{12}{2 + 3\sin\theta}$

**28.** $r = \dfrac{6}{1 - 2\sin\theta}$

**29.** $r = \dfrac{6}{5 + 3\cos\theta}$

**30.** $r = \dfrac{2}{1 + \sin\theta}$

**31.** $e = 0.4$, $a = 5$, $b = \sqrt{21}$, $c = 2$

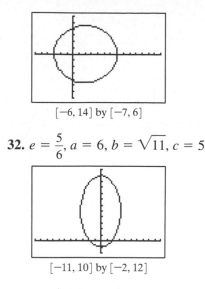

$[-6, 14]$ by $[-7, 6]$

**32.** $e = \dfrac{5}{6}$, $a = 6$, $b = \sqrt{11}$, $c = 5$

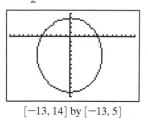

$[-11, 10]$ by $[-2, 12]$

**33.** $e = \dfrac{1}{2}$, $a = 8$, $b = 4\sqrt{3}$, $c = 4$

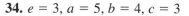

$[-13, 14]$ by $[-13, 5]$

**34.** $e = 3$, $a = 5$, $b = 4$, $c = 3$

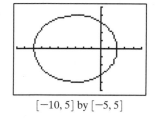

$[-10, 5]$ by $[-5, 5]$

**35.** $e = \dfrac{5}{3}$, $a = 3$, $b = 4$, $c = 5$

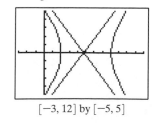

$[-3, 12]$ by $[-5, 5]$

**36.** $e = 5$, $a = \dfrac{1}{2}$, $b = \sqrt{6}$, $c = \dfrac{5}{2}$

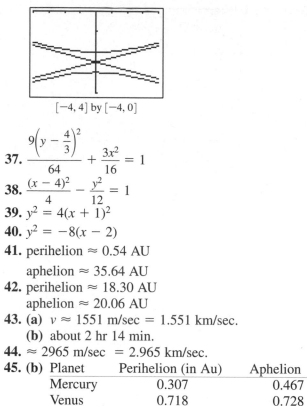

$[-4, 4]$ by $[-4, 0]$

**37.** $\dfrac{9\left(y - \dfrac{4}{3}\right)^2}{64} + \dfrac{3x^2}{16} = 1$

**38.** $\dfrac{(x - 4)^2}{4} - \dfrac{y^2}{12} = 1$

**39.** $y^2 = 4(x + 1)^2$

**40.** $y^2 = -8(x - 2)$

**41.** perihelion $\approx 0.54$ AU

aphelion $\approx 35.64$ AU

**42.** perihelion $\approx 18.30$ AU

aphelion $\approx 20.06$ AU

**43.** **(a)** $v \approx 1551$ m/sec $= 1.551$ km/sec.

**(b)** about 2 hr 14 min.

**44.** $\approx 2965$ m/sec $= 2.965$ km/sec.

**45.** **(b)**

| Planet | Perihelion (in Au) | Aphelion |
|---|---|---|
| Mercury | 0.307 | 0.467 |
| Venus | 0.718 | 0.728 |
| Earth | 0.983 | 1.017 |
| Mars | 1.382 | 1.665 |
| Jupiter | 4.953 | 5.452 |
| Saturn | 9.020 | 10.090 |

**(c)** The difference is greatest for Saturn.

## Section 8.6 (pp. 651–660)

### Quick Review 8.6

**1.** $\sqrt{(x - 2)^2 + (y + 3)^2}$

**2.** $\left(\dfrac{x + 2}{2}, \dfrac{y - 3}{2}\right)$

**3.** $P$ lies on the circle of radius 5 centered at $(2, -3)$

**5.** $\left\langle \dfrac{-4}{\sqrt{41}}, \dfrac{5}{\sqrt{41}} \right\rangle$

**6.** $\left\langle \dfrac{28}{\sqrt{41}}, \dfrac{-35}{\sqrt{41}} \right\rangle$

**7.** Circle of radius 5 centered at $(-1, 5)$

**8.** A line of slope $-2$, passing through $(2, -4)$

**9.** Center: $(-1, 3)$, radius: 2

**10.** $\langle 3, 9 \rangle$

## Section 8.6 Exercises

**1.**

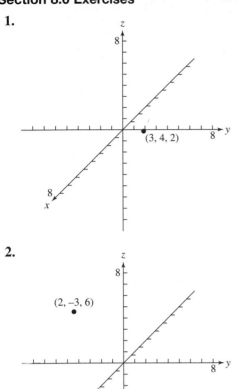

**2.**

**3.**

**4.**

**13.** $(x - 5)^2 + (y + 1)^2 + (z + 2)^2 = 64$
**14.** $(x + 1)^2 + (y - 5)^2 + (z - 8)^2 = 5$
**15.** $(x - 1)^2 + (y + 3)^2 + (z - 2)^2 = a$
**16.** $(x - p)^2 + (y - q)^2 + (z - r)^2 = 36$
**17.**

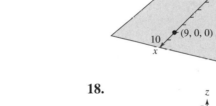

**18.**

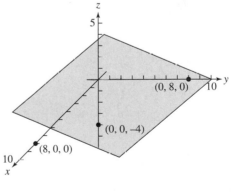

**19.**

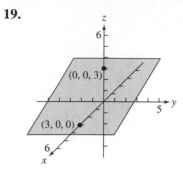

**20.**

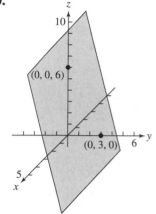

**21.**

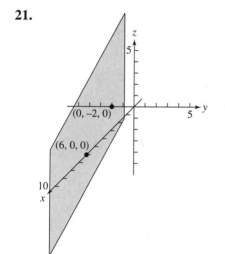

**22.**

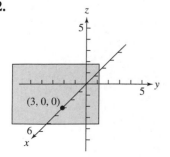

**29.** $\left\langle \dfrac{4}{13}, -\dfrac{3}{13}, \dfrac{12}{13} \right\rangle$

**31.** $\langle -3, 4, -5 \rangle$

**34.** $\langle 2, -1, -5 \rangle$

**35.** $\mathbf{r} = \langle -1, 2, 4 \rangle + t\langle 1, 4, -7 \rangle$

**37.** $x = -1 + 3t,\ y = 2 - 6t,\ z = 4 - 3t$

**38.** $x = 2t,\ y = 6 - 10t,\ z = -3 + 4t$

**39.** $x = \dfrac{1}{2}t,\ y = 6 - 7t,\ z = -3 + \dfrac{11}{2}t$

**40.** $x = 2 - \dfrac{5}{2}t,\ y = -4 + 8t,\ z = 1 - \dfrac{1}{2}t$

**42.** $\left(0, \dfrac{3}{2}, \dfrac{3}{2}\right)$

**43. (a)**

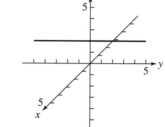

**(b)** the $z$-axis

**44. (a)**

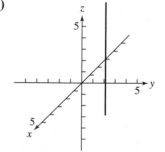

**(b)** the intersection of the $yz$ plane (at $x = 0$) and $xy$ plane (at $z = 2$); a line parallel to the $y$-axis through $(0, 0, 2)$

**45. (a)**

**(b)** the intersection of the $xz$ plane (at $y = 0$) and $yz$ plane (at $x = -3$); a line parallel to the $z$-axis through $(-3, 0, 0)$

**46. (a)**

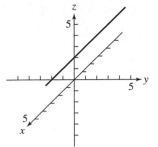

**(b)** the intersection of the $xz$ plane (at $y = 1$) and $xy$ plane (at $z = 3$); a line through $(0, 1, 3)$ parallel to the $x$-axis.

**47. (a)**

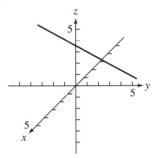

**(b)** the intersection of the $xy$ plane (at $z = 3$) and the plane $2x - y + 1 = 0$.

**48. (a)**

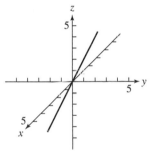

**(b)** the intersection of the $yz$ plane (at $x = 0$) and the plane $3y - z = 0$.

**49.** $\mathbf{r} = \langle x_1 + (x_2 - x_1)t, y_1 + (y_2 - y_1)t, z_1 + (z_2 - z_1)t \rangle$

**50.** $x = x_1 + (x_2 - x_1)t, y = y_1 + (y_2 - y_1)t, z = z_1 + (z_2 - z_1)t$

**54. (a)** $83.34°$

## Chapter 8 Review (pp. 664–666)

**1.** Vertex: $(0, 0)$, focus: $(3, 0)$, directrix: $x = -3$, focal width: 12

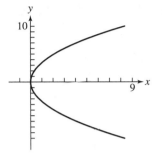

**2.** Vertex: $(0, 0)$, focus: $(0, -2)$, directrix: $y = 2$, focal width: 8

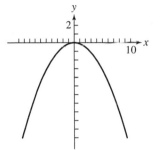

**3.** Vertex: $(-2, 1)$, focus: $(-2, 0)$, directrix: $y = 2$, focal width: 4

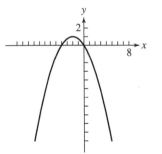

**4.** Vertex: $(0, -2)$, focus: $(4, -2)$, directrix: $x = -4$, focal width: 16

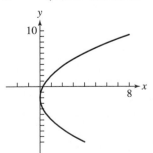

**5.** Ellipse. Center $(0, 0)$. Foci: $(0, \pm\sqrt{3})$ Vertices: $(0, \pm2\sqrt{2})$.

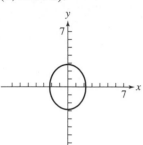

**6.** Hyperbola. Center: $(0, 0)$. Foci: $(0, \pm\sqrt{65})$ Vertices: $(0, \pm4)$.

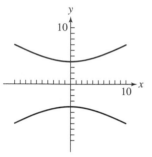

**7.** Hyperbola. Center: $(0, 0)$. Vertices: $(\pm5, 0)$. Foci: $(\pm\sqrt{61}, 0)$

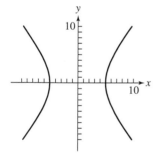

**8.** Hyperbola. Center: $(0, 0)$. Vertices: $(\pm7, 0)$. Foci: $(\pm\sqrt{58}, 0)$

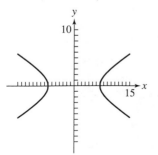

**9.** Hyperbola. Center: $(-3, 5)$. Vertices:

$(-3 \pm 3\sqrt{2}, 5)$. Foci: $(-3 \pm \sqrt{46}, 5)$

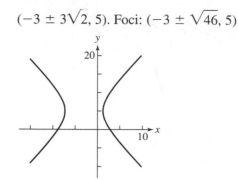

**10.** Hyperbola. Center: $(7, 3)$. Vertices: $(7, 3 \pm 3)$ $= (7, 0)$ and $(7, 6)$. Foci: $(7, 3 \pm \sqrt{21})$

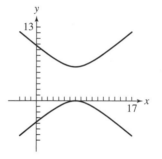

**11.** Ellipse. Center: $(2, -1)$. Vertices: $(2 \pm 4, -1)$ $= (6, -1)$ and $(-2, -1)$. Foci: $(2 \pm 3, -1)$ $= (5, -1)$ and $(-1, -1)$

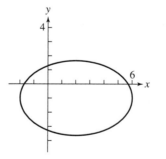

**12.** Ellipse. Center: $(-6, 0)$. Vertices: $(-6, \pm 6)$. Foci: $(-6, \pm4)$

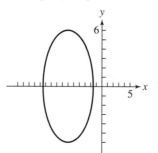

**21.** parabola, $(x - 3)^2 = y + 12$

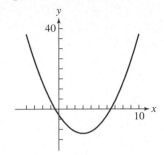

**22.** ellipse $\dfrac{(x + 2)^2}{9} + \dfrac{y^2}{3} = 1$

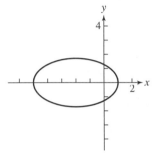

**23.** hyperbola $\dfrac{(x - 1)^2}{3} - \dfrac{(y - 2)^2}{3} = 1$

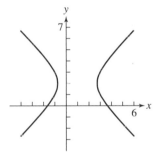

**24.** parabola $(x + 1)^2 = -4(y - 2)$

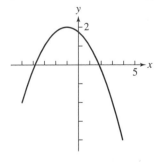

**25.** parabola $(y - 2)^2 = 6\left(x + \dfrac{17}{6}\right)$

**26.** parabola $(x - 1)^2 = \dfrac{4}{3}(y + 3)$

**27.** hyperbola $\dfrac{(y + 4)^2}{30} - \dfrac{(x - 3)^2}{45} = 1$

$[-10, 15]$ by $[-15, 10]$

**28.** hyperbola $\dfrac{(x - 3)^2}{4} - \dfrac{(y + 2)^2}{12} = 1$

$[-1.7, 7.7]$ by $[-5.1, 1.1]$

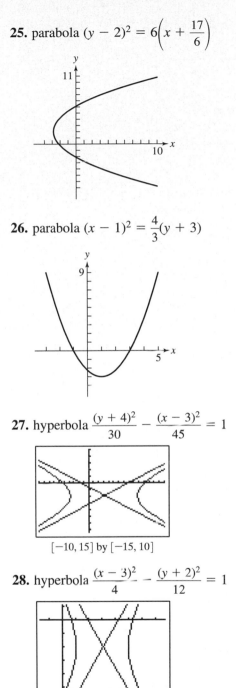

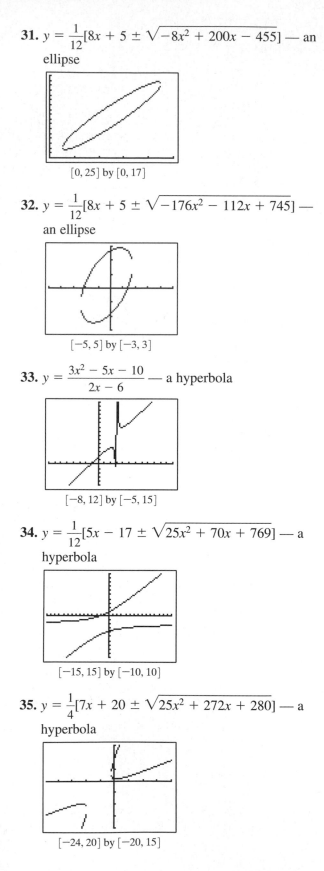

**31.** $y = \frac{1}{12}[8x + 5 \pm \sqrt{-8x^2 + 200x - 455}]$ — an ellipse

$[0, 25]$ by $[0, 17]$

**32.** $y = \frac{1}{12}[8x + 5 \pm \sqrt{-176x^2 - 112x + 745}]$ — an ellipse

$[-5, 5]$ by $[-3, 3]$

**33.** $y = \frac{3x^2 - 5x - 10}{2x - 6}$ — a hyperbola

$[-8, 12]$ by $[-5, 15]$

**34.** $y = \frac{1}{12}[5x - 17 \pm \sqrt{25x^2 + 70x + 769}]$ — a hyperbola

$[-15, 15]$ by $[-10, 10]$

**35.** $y = \frac{1}{4}[7x + 20 \pm \sqrt{25x^2 + 272x + 280}]$ — a hyperbola

$[-24, 20]$ by $[-20, 15]$

**36.** $y = \frac{1}{4}[7x + 3 \pm \sqrt{25x^2 + 26x - 71}]$ — a hyperbola

$[-15, 15]$ by $[-15, 15]$

**38.** $x^2 = -12y$      **39.** $(x + 3)^2 = 12(y - 3)$

**40.** $(y + 2)^2 = -8(x - 1)$   **41.** $\dfrac{x^2}{169} + \dfrac{y^2}{25} = 1$

**42.** $\dfrac{y^2}{36} + \dfrac{x^2}{32} = 1$

**44.** $\dfrac{(x + 3)^2}{16} + \dfrac{(y + 4)^2}{7} = 1$

**45.** $\dfrac{y^2}{25} - \dfrac{x^2}{11} = 1$

**47.** $\dfrac{(x - 2)^2}{9} - \dfrac{(y - 1)^2}{16} = 1$

**49.** $\dfrac{x^2}{25} + \dfrac{y^2}{4} = 1$ — an ellipse

**50.** $\dfrac{x^2}{16} + \dfrac{y^2}{36} = 1$ — an ellipse

**51.** $(x + 2)^2 + (y - 4)^2 = 1$ — an ellipse (a circle).

**52.** $\dfrac{(x - 5)^2}{9} + \dfrac{(y + 3)^2}{9} = 1$, or $(x - 5)^2 + (y + 3)^2 = 9$ — an ellipse (a circle)

**53.** $\dfrac{x^2}{9} - \dfrac{y^2}{25} = 1$ — a hyperbola

**54.** $\dfrac{x^2}{16} - \dfrac{y^2}{9} = 1$ — a hyperbola

**55.**

$[-8, 3]$ by $[-10, 10]$

parabola, $y^2 = -8(x - 2)$

**56.**

$[-10, 10]$ by $[-4, 10]$

parabola, $x^2 = -10\left(y - \dfrac{5}{2}\right)$

**57.**

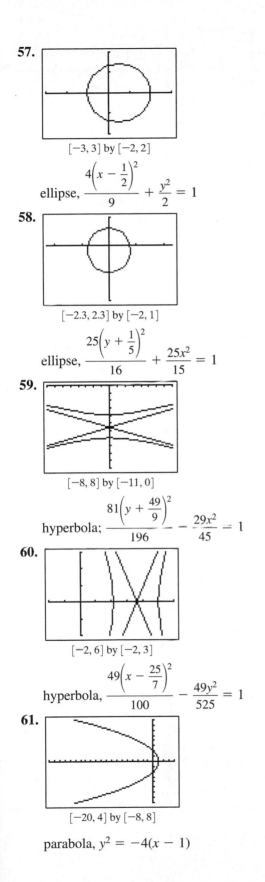

$[-3, 3]$ by $[-2, 2]$

ellipse, $\dfrac{4\left(x - \dfrac{1}{2}\right)^2}{9} + \dfrac{y^2}{2} = 1$

**58.**

$[-2.3, 2.3]$ by $[-2, 1]$

ellipse, $\dfrac{25\left(y + \dfrac{1}{5}\right)^2}{16} + \dfrac{25x^2}{15} = 1$

**59.**

$[-8, 8]$ by $[-11, 0]$

hyperbola; $\dfrac{81\left(y + \dfrac{49}{9}\right)^2}{196} - \dfrac{29x^2}{45} = 1$

**60.**

$[-2, 6]$ by $[-2, 3]$

hyperbola, $\dfrac{49\left(x - \dfrac{25}{7}\right)^2}{100} - \dfrac{49y^2}{525} = 1$

**61.**

$[-20, 4]$ by $[-8, 8]$

parabola, $y^2 = -4(x - 1)$

**62.**

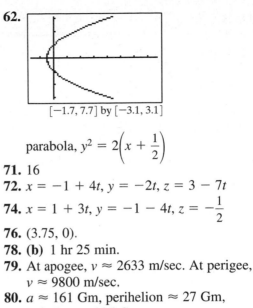

$[-1.7, 7.7]$ by $[-3.1, 3.1]$

parabola, $y^2 = 2\left(x + \dfrac{1}{2}\right)$

**71.** 16

**72.** $x = -1 + 4t,\ y = -2t,\ z = 3 - 7t$

**74.** $x = 1 + 3t,\ y = -1 - 4t,\ z = -\dfrac{1}{2}$

**76.** $(3.75, 0)$.

**78. (b)** 1 hr 25 min.

**79.** At apogee, $v \approx 2633$ m/sec. At perigee, $v \approx 9800$ m/sec.

**80.** $a \approx 161$ Gm, perihelion $\approx 27$ Gm, aphelion $\approx 295$ Gm

## Chapter 8 Project

Answers are based on the sample data shown in the table.

**1.**

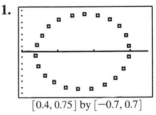

$[0.4, 0.75]$ by $[-0.7, 0.7]$

**2.** $\dfrac{(y - 0)^2}{(0.64)^2} + \dfrac{(x - 0.569)^2}{(0.134)^2} = 1$; $a = 0.64$; $b = 0.134$; $h = 0.569$; $k = 0$

**3.** With respect to the graph of the ellipse, the point $(h, k)$ represents the center of the ellipse. The value $a$ is the length of the semimajor axis and $b$ is the length of the semiminor axis.

**4.** $(0.569, 0.626)$; $(0.569, -0.626)$

**5.** 0.978

**6.** The parametric equations for the sample data set are $x_{1T} = 0.131 \sin(4.82(T + 0.433)) + 0.569$
$y_{1T} = 0.641 \sin(4.82(T - 0.553)) + 0.001$

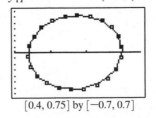

$[0.4, 0.75]$ by $[-0.7, 0.7]$

# Chapter 9

## Section 9.1 (pp. 669–677)

### Exploration 1

**1.** Six
**2.** Approximately 1 person out of 6
**3.** No
**4.** It is likely that the salesman rigged the test.

### Section 9.1 Exercises

**9.** combinations
**10.** permutations
**12.** 9240
**39. (a)** In GREEKS, exchanging the E's does not change the word.
   **(b)** Since rearranging the S's does not change the word, the number 6! counts every word 3! times.
   **(c)** MISSISSIPPI: 34,650
   CHATTANOOGA: 1,663,200
**40. (a)** Each combination of the $n$ vertices taken 2 at a time determines a segment that is either an edge or a diagonal. There are $_nC_2$ such combinations.
**44.** The chart on the left is more reasonable. Each pair of actresses will require about the same amount of time to interview.

## Section 9.2 (pp. 678–683)

### Exploration 1

**1.** 1, 3, 3, 1. These are (in order) the coefficients in the expansion of $(a + b)^3$.
**2.** {1 4 6 4 1}. These are (in order) the coefficients in the expansion of $(a + b)^4$.
**3.** {1 5 10 10 5 1}. These are (in order) the coefficients in the expansion of $(a + b)^5$.
**4.** $(a + b)^6 = a^6 + 6a^5b + 15a^4b^2 + 20a^3b^3 + 15a^2b^4 + 6ab^5 + b^6$
$(a + b)^7 = a^7 + 7a^6b + 21a^5b^2 + 35a^4b^3 + 35a^3b^4 + 21a^2b^5 + 7ab^6 + b^7$

### Quick Review 9.2

**5.** $9s^2 - 12st + 4t^2$
**6.** $9p^2 - 24pq + 16q^2$
**7.** $u^3 + 3u^2v + 3uv^2 + v^3$
**8.** $b^3 - 3b^2c + 3bc^2 - c^3$
**9.** $8x^3 - 36x^2y + 54xy^2 - 27y^3$
**10.** $64m^3 + 144m^2n + 108mn^2 + 27n^3$

### Exercises 9.2

**1.** 1  8  28  56 70  56  28  8  1
**2.** 1  10  45  120 210  252  210  120  45  10  1
**3.** $a^5 + 5a^4b + 10a^3b^2 + 10a^2b^3 + 5ab^4 + b^5$
**4.** $x^6 + 6x^5y + 15x^4y^2 + 20x^3y^3 + 15x^2y^4 + 6xy^5 + y^6$
**5.** $x^7 - 7x^6y + 21x^5y^2 - 35x^4y^3 + 35x^3y^4 - 21x^2y^5 + 7xy^6 - y^7$
**6.** $a^6 - 6a^5b + 15a^4b^2 - 20a^3b^3 + 15a^2b^4 - 6ab^5 + b^6$
**7.** $81x^4 - 108x^3y + 54x^2y^2 - 12xy^3 + y^4$
**8.** $a^4 - 8a^3b^2 + 24a^2b^4 - 32ab^6 + 16b^8$
**13.** $x^5 - 10x^4 + 40x^3 - 80x^2 + 80x - 32$
**14.** $x^6 + 18x^5 + 135x^4 + 540x^3 + 1215x^2 + 1458x + 729$
**15.** $128x^7 - 448x^6 + 672x^5 - 560x^4 + 280x^3 - 84x^2 + 14x - 1$
**16.** $243x^5 + 1620x^4 + 4320x^3 + 5760x^2 + 3840x + 1024$
**17.** $16x^4 + 32x^3y + 24x^2y^2 + 8xy^3 + y^4$
**18.** $32y^5 - 240y^4x + 720y^3x^2 - 1080y^2x^3 + 810yx^4 - 243x^5$
**19.** $x^3 - 6x^{5/2}y^{1/2} + 15x^2y - 20x^{3/2}y^{3/2} + 15xy^2 - 6x^{1/2}y^{5/2} + y^3$
**20.** $= x^2 + 4x\sqrt{3x} + 18x + 12\sqrt{3x} + 9$
**21.** $= x^{-10} + 15x^{-8} + 90x^{-6} + 270x^{-4} + 405x^{-2} + 243$
**22.** $a^7 - 7a^6b^{-3} + 21a^5b^{-6} - 35a^4b^{-9} + 35a^3b^{-12} - 21a^2b^{-15} + 7ab^{-18} - b^{-21}$
**35. (a)** 1, 3, 6, 10, 15, 21, 28, 36, 45, 55
   **(b)** They appear diagonally down the triangle, starting with either of the 1's in row 2.
**36. (a)** 2
   **(f)** Rows that are powers of 2: 2, 4, 8, 16, etc.
   **(g)** Rows that are 1 less than a power of 2: 1, 3, 7, 15, etc.

## Section 9.3 (pp. 683–695)

### Exploration 1

**1.**

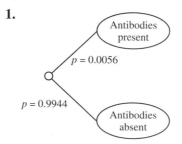

**2.**

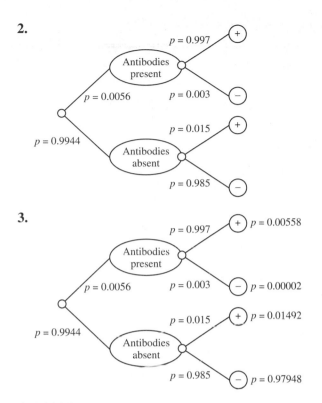

**3.**

**4.** 0.0205
**5.** ≈ 0.272

## Quick Review 9.3

**5.** 2,598,960

## Section 9.3 Exercises

**13.** 0.7
**25.** $\dfrac{1}{1024}$
**33. (a)**

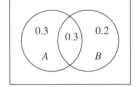

**34. (a)**

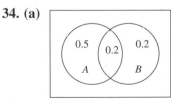

**44. (b)** $\dfrac{308}{969}$

**47. (a)**

| Type of Bagel | Probability |
|---|---|
| Plain | 0.37 |
| Onion | 0.12 |
| Rye | 0.11 |
| Cinnamon Raisin | 0.25 |
| Sourdough | 0.15 |

**50. (c)** The university's graduation rate seems to be exaggerated.

**52. (a)** $\dfrac{10}{9,366,819} \approx 0.0000010676$

**(b)**

| Value | Probability |
|---|---|
| $-10$ | $\dfrac{9,366,809}{9,366,819}$ |
| $+3,999,990$ | $\dfrac{10}{9,366,819}$ |

**(c)** $-5.73$

# Section 9.4 (pp. 696–711)

## Exploration 1

**1.** 45
**2.** 174
**3.** 1
**4.** 0
**5.** $\dfrac{1}{3}$

## Exploration 2

**1.** $1 + \quad 2 + \quad 3 + ... + \quad 99 + 100$
**2.** $100 + \quad 99 + \quad 98 + ... + \quad 2 + \quad 1$
**3.** $101 + 101 + 101 + ... + 101 + 101$
**4.** $100(101) = 10,100$
**5.** The sum in 4 involves two copies of the same progression, so it doubles the sum of the progression. The answer is 5,050.

## Section 9.4 Exercises

**1.** $2, \dfrac{3}{2}, \dfrac{4}{3}, \dfrac{5}{4}, \dfrac{6}{5}, \dfrac{7}{6}; \dfrac{101}{100}$
**2.** $\dfrac{4}{3}, 1, \dfrac{4}{5}, \dfrac{2}{3}, \dfrac{4}{7}, \dfrac{1}{2}; \dfrac{2}{51}$
**3.** 0, 6, 24, 60, 120, 210; 999,900
**4.** $-4, -6, -6, -4, 0, 6$; 9500
**8.** $0.75, -1.5, 3, -6; -96$
**9.** $2, -1, 1, 0; 3$
**10.** $-2, 3, 1, 4; 23$

**11. (a)** 4
   **(b)** 42
   **(c)** $a_n = a_{n-1} + 4$
   **(d)** $a_n = 6 + 4(n-1)$
**12. (a)** 5
   **(b)** 41
   **(c)** $a_n = a_{n-1} + 5$
   **(d)** $a_n = -4 + 5(n-1)$
**13. (a)** 3
   **(b)** 22
   **(c)** $a_n = a_{n-1} + 3$
   **(d)** $a_n = -5 + 3(n-1)$
**14. (a)** 11
   **(b)** 92
   **(c)** $a_n = a_{n-1} + 11$
   **(d)** $a_n = -7 + 11(n-1)$
**15. (a)** 3
   **(b)** 4374
   **(c)** $a_1 = 2; a_n = 3a_{n-1}$
   **(d)** $a_n = 2 \cdot 3^{n-1}$
**16. (a)** 2
   **(b)** 384
   **(c)** $a_1 = 3; a_n = 2a_{n-1}$
   **(d)** $a_n = 3 \cdot 2^{n-1}$
**17. (a)** $-2$
   **(b)** $-128$
   **(c)** $a_1 = 1; a_n = -2a_{n-1}$
   **(d)** $a_n = (-2)^{n-1}$
**18. (a)** $r = -1$
   **(b)** 2
   **(c)** $a_1 = -2; a_n = -a_{n-1}$
   **(d)** $a_n = 2 \cdot (-1)^n$

**21.** $a_1 = \pm\dfrac{3}{2}$, $r = \pm 2$, and $a_n = 3(\pm 2)^{n-2}$

**22.** $a_1 = -3$, $r = 5$, and $a_n = -3 \cdot 5^{n-1}$

**23.**

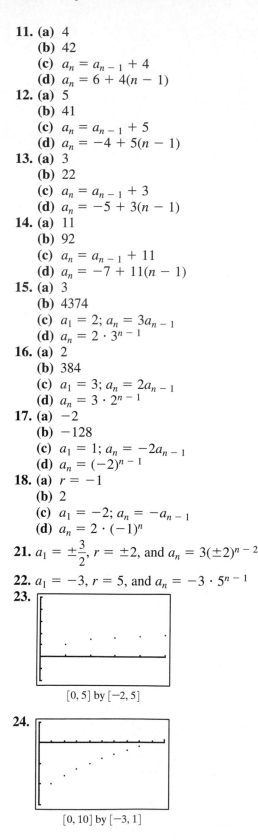

[0, 5] by [−2, 5]

**24.**

[0, 10] by [−3, 1]

**25.**

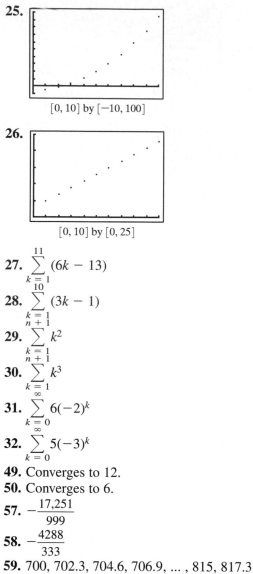

[0, 10] by [−10, 100]

**26.**

[0, 10] by [0, 25]

**27.** $\displaystyle\sum_{k=1}^{11} (6k - 13)$

**28.** $\displaystyle\sum_{k=1}^{10} (3k - 1)$

**29.** $\displaystyle\sum_{k=1}^{n+1} k^2$

**30.** $\displaystyle\sum_{k=1}^{n+1} k^3$

**31.** $\displaystyle\sum_{k=0}^{\infty} 6(-2)^k$

**32.** $\displaystyle\sum_{k=0}^{\infty} 5(-3)^k$

**49.** Converges to 12.
**50.** Converges to 6.
**57.** $-\dfrac{17{,}251}{999}$

**58.** $-\dfrac{4288}{333}$

**59.** 700, 702.3, 704.6, 706.9, ... , 815, 817.3
**60.** The first column is an arithmetic sequence with common difference $d = 14$. The second column is a geometric sequence with common ratio $r = \dfrac{1}{2}$.

| Time (billions of years) | Mass (g) |
|---|---|
| 0 | 16 |
| 14 | 8 |
| 28 | 4 |
| 42 | 2 |
| 56 | 1 |

**61. (a)**

| Year | Huntsville | Conroe |
|------|-----------|--------|
| 1985 | 25,854 | 22,314 |
| 1986 | 26,255 | 23,285 |
| 1987 | 26,662 | 24,298 |
| 1988 | 27,075 | 25,354 |
| 1989 | 27,495 | 26,457 |
| 1990 | 27,921 | 27,608 |
| 1991 | 28,354 | 28,809 |
| 1992 | 28,793 | 30,062 |
| 1993 | 29,239 | 31,370 |
| 1994 | 29,693 | 32,735 |
| 1995 | 30,153 | 34,159 |

  **(b)** Around 1991

**64.** Yes.

**65. (b)** 3, 5, 8, 13, 21, 34, 55, 89, 144, 233

**66.** 1, 1, 2, 3, 5, 8, 13. These are the first seven terms of the Fibonacci sequence.

**67. (b)** $a_n \to 2\pi$ as $n \to \infty$

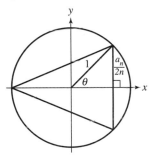

**68.** $P_1 = 525{,}000$; $P_n = 1.0175 P_{n-1}$, $n \geq 2$

**71. (a)** Heartland: 17,659,690 people
      Southeast: 36,087,346 people

  **(b)** Heartland: 517,825 mi$^2$
      Southeast: 348,999 mi$^2$

  **(c)** Heartland: $\approx$ 34.10 people/mi$^2$
      Southeast: $\approx$ 103.40 people/mi$^2$

**72.** $\displaystyle\sum_{k=1}^{8} (k^2 - 2)$

**73.** $a_1 = [1\ 1]$, $a_2 = [1\ 2]$, $a_3 = [2\ 3]$, $a_4 = [3\ 5]$, $a_5 = [5\ 8]$, $a_6 = [8\ 13]$, $a_7 = [13\ 21]$. The entries in the terms of this sequence are successive pairs of terms from the Fibonacci sequence.

**75.** $S_n = F_{n+2} - 1$

# Section 9.5 (pp. 711–716)

## Exploration 1

**1.** Start with the rightmost peg if $n$ is odd and the middle peg if $n$ is even.

## Quick Review 9.5

**10.** $-2$; $2k^2 - k - 3$; $2k^2 + 3k - 2$

## Exercises 9.5

**1.** $P_n$: $2 + 4 + 6 + \cdots + 2n = n^2 + n$.
$P_1$ is true: $2(1) = 1^2 + 1$.
Now assume $P_k$ is true: $2 + 4 + 6 + \cdots + 2k = k^2 + k$. Add $2(k+1)$ to both sides:
$2 + 4 + 6 + \cdots + 2k + 2(k+1)$
$= k^2 + k + 2(k+1) = k^2 + 3k + 2$
$= k^2 + 2k + 1 + k + 1 = (k+1)^2 + (k+1)$,
so $P_{k+1}$ is true. Therefore $P_n$ is true for all $n \geq 1$.

**2.** $P_n$: $8 + 10 + 12 + \cdots + (2n + 6) = n^2 + 7n$.
$P_1$ is true: $2(1) + 6 = 1^2 + 7 \cdot 1$.
Now assume $P_k$ is true:
$8 + 10 + 12 + \cdots + (2k + 6) = k^2 + 7k$.
Add $2(k+1) + 6 = 2k + 8$ to both sides:
$8 + 10 + 12 + \cdots + (2k+6) + [2(k+1)+6]$
$= k^2 + 7k + 2k + 8 = (k^2 + 2k + 1) + 7k + 7$
$= (k+1)^2 + 7(k+1)$, so $P_{k+1}$ is true.
Therefore $P_n$ is true for all $n \geq 1$.

**3.** $P_n$: $6 + 10 + 14 + \cdots + (4n + 2) = n(2n + 4)$.
$P_1$ is true: $4(1) + 2 = 1(2 + 4)$.
Now assume $P_k$ is true:
$6 + 10 + 14 + \cdots + (4k + 2) = k(2k + 4)$.
Add $4(k+1) + 2 = 4k + 6$ to both sides:
$6 + 10 + 14 + \cdots + (4k+2) + [4(k+1)+2]$
$= k(2k+4) + 4k + 6 = 2k^2 + 8k + 6$
$= (k+1)(2k+6) = (k+1)[2(k+1)+4]$, so
$P_{k+1}$ is true. Therefore, $P_n$ is true for all $n \geq 1$.

**4.** $P_n$: $14 + 18 + 22 + \cdots + (4n + 10)$
$= 2n(n + 6)$.
$P_1$ is true: $4(1) + 10 = 2 \cdot 1(1 + 6)$. Now assume $P_k$ is true:
$14 + 18 + 22 + \cdots + (4k + 10) = 2k(k + 6)$.
Add $4(k+1) + 10 = 4k + 14$ to both sides:
$14 + 18 + 22 + \cdots + (4k+10) + [4(k+1) + 10] = 2k(k+6) + (4k+14)$
$= 2(k^2 + 8k + 7) = 2(k+1)(k+7)$
$= 2(k+1)(k+1+6)$, so $P_{k+1}$ is true.
Therefore, $P_n$ is true for all $n \geq 1$.

**5.** $P_n$: $5n - 2$. $P_1$ is true: $a_1 = 5 \cdot 1 - 2 = 3$.
Now assume $P_k$ is true: $a_k = 5k - 2$.
To get $a_{k+1}$, add 5 to $a_k$; that is,
$a_{k+1} = (5k - 2) + 5 = 5(k+1) - 2$. This
shows that $P_{k+1}$ is true. Therefore, $P_n$ is true for all $n \geq 1$.

**6.** $P_n$: $a_n = 2n + 5$. $P_1$ is true: $a_1 = 2 \cdot 1 + 5 = 7$.
Now assume $P_k$ is true: $a_k = 2k + 5$.
To get $a_{k+1}$, add 2 to $a_k$; that is,
$a_{k+1} = (2k + 5) + 2 = 2(k + 1) + 5$. This
shows that $P_{k+1}$ is true. Therefore, $P_n$ is true for
all $n \geq 1$.

**7.** $P_n$: $a_n = 2 \cdot 3^{n-1}$.
$P_1$ is true: $a_1 = 2 \cdot 3^{1-1} = 2 \cdot 3^0 = 2$.
Now assume $P_k$ is true: $a_k = 2 \cdot 3^{k-1}$.
To get $a_{k+1}$, multiply $a_k$ by 3; that is,
$a_{k+1} = 3 \cdot 2 \cdot 3^{k-1} = 2 \cdot 3k = 2 \cdot 3^{k+1-1}$.
This shows that $P_{k+1}$ is true. Therefore, $P_n$ is
true for all $n \geq 1$.

**8.** $P_n$: $a_n = 3 \cdot 5^{n-1}$.
$P_1$ is true: $a_1 = 3 \cdot 5^{1-1} = 3 \cdot 5^0 = 3$.
Now assume $P_k$ is true: $a_k = 3 \cdot 5^{k-1}$.
To get $a_{k+1}$, multiply $a_k$ by 5; that is,
$a_{k+1} = 5 \cdot 3 \cdot 5^{k-1} = 3 \cdot 5^k = 3 \cdot 5^{k+1-1}$.
This shows that $P_{k+1}$ is true. Therefore, $P_n$ is
true for all $n \geq 1$.

**9.** $P_1$: $1 = \dfrac{1(1 + 1)}{2}$.

$P_k$: $1 + 2 + \cdots + k = \dfrac{k(k + 1)}{2}$.

$P_{k+1}$: $1 + 2 + \cdots + k + (k + 1)$
$= \dfrac{(k + 1)(k + 2)}{2}$.

**10.** $P_1$: $(2(1) - 1)^2 = \dfrac{1(2 - 1)(2 + 1)}{3}$.

$P_k$: $1^2 + 3^2 + \cdots + (2k - 1)^2 = \dfrac{k(2k - 1)(2k + 1)}{3}$.

$P_{k+1}$: $1^2 + 3^2 + \cdots + (2k - 1)^2 + (2k + 1)^2$
$= \dfrac{(k + 1)(2k + 1)(2k + 3)}{3}$.

**11.** $P$: $\dfrac{1}{1 \cdot 2} = \dfrac{1}{1 + 1}$.

$P_k$: $\dfrac{1}{1 \cdot 2} + \dfrac{1}{2 \cdot 3} + \cdots + \dfrac{1}{k(k + 1)} = \dfrac{k}{k + 1}$.

$P_{k+1}$: $\dfrac{1}{1 \cdot 2} + \dfrac{1}{2 \cdot 3} + \cdots + \dfrac{1}{k(k + 1)}$
$+ \dfrac{1}{(k + 1)(k + 2)} = \dfrac{k + 1}{k + 2}$.

**12.** $P_1$: $1^4 = \dfrac{1(1 + 1)(2 + 1)(3 + 3 - 1)}{30}$.

$P_k$: $1^4 + 2^4 + \cdots + k^4$
$= \dfrac{k(k + 1)(2k + 1)(3k^2 + 3k - 1)}{30}$.

$P_{k+1}$: $1^4 + 2^4 + \cdots + k^4 + (k + 1)^4$
$= \dfrac{(k + 1)(k + 2)(2k + 3)(3k^2 + 9k + 5)}{30}$.

**13.** $P_n$: $1 + 5 + 9 + \cdots + (4n - 3) = n(2n - 1)$.
$P_1$ is true: $4(1) - 3 = 1 \cdot (2 \cdot 1 - 1)$.
Now assume $P_k$ is true:
$1 + 5 + 9 + \cdots + (4k - 3) = k(2k - 1)$.
Add $4(k + 1) - 3 = 4k + 1$ to both sides:
$1 + 5 + 9 + \cdots + (4k - 3) + [4(k + 1) - 3]$
$= k(2k - 1) + 4k + 1 = 2k^2 + 3k + 1$
$= (k + 1)(2k + 1) = (k + 1)[2(k + 1) - 1]$,
so $P_{k+1}$ is true.
Therefore $P_n$ is true for all $n \geq 1$.

**14.** $P_n$: $1 + 2 + 2^2 + \cdots + 2^{n-1} = 2^n - 1$.
$P_1$ is true: $2^{1-1} = 2^1 - 1$.
Now assume $P_k$ is true:
$1 + 2 + 2^2 + \cdots + 2^{k-1} = 2^k - 1$.
Add $2^k$ to both sides:
$1 + 2 + 2^2 + \cdots + 2^{k-1} + 2^k$
$= 2^k - 1 + 2^k = 2 \cdot 2^k - 1 = 2^{k+1} - 1$, so
$P_{k+1}$ is true.
Therefore $P_n$ is true for all $n \geq 1$.

**15.** $P_n$: $\dfrac{1}{1 \cdot 2} + \dfrac{1}{2 \cdot 3} + \cdots + \dfrac{1}{n(n + 1)} = \dfrac{n}{n + 1}$.

$P_1$ is true: $\dfrac{1}{1 \cdot 2} = \dfrac{1}{1 + 1}$.
Now assume $P_k$ is true:
$\dfrac{1}{1 \cdot 2} + \dfrac{1}{2 \cdot 3} + \cdots + \dfrac{1}{k(k + 1)} = \dfrac{k}{k + 1}$.

Add $\dfrac{1}{(k + 1)(k + 2)}$ to both sides:
$\dfrac{1}{1 \cdot 2} + \dfrac{1}{2 \cdot 3} + \cdots + \dfrac{1}{k(k + 1)} + \dfrac{1}{(k + 1)(k + 2)}$
$= \dfrac{k}{k + 1} + \dfrac{1}{(k + 1)(k + 2)} = \dfrac{k(k + 2) + 1}{(k + 1)(k + 2)}$
$= \dfrac{(k + 1)(k + 1)}{(k + 1)(k + 2)} = \dfrac{k + 1}{k + 2} = \dfrac{k + 1}{k + 1 + 1}$,
so $P_{k+1}$ is true.
Therefore $P_n$ is true for all $n \geq 1$.

**16.** $P_n$: $\dfrac{1}{1 \cdot 3} + \dfrac{1}{3 \cdot 5} + \cdots + \dfrac{1}{(2n - 1)(2n + 1)}$
$= \dfrac{n}{2n + 1}$.

$P_1$ is true: $\dfrac{1}{1 \cdot 3} = \dfrac{1}{(2 - 1)(2 + 1)}$.
Now assume $P_k$ is true:
$\dfrac{1}{1 \cdot 3} + \dfrac{1}{3 \cdot 5} + \cdots + \dfrac{1}{(2k - 1)(2k + 1)} = \dfrac{k}{2k + 1}$.
Add $\dfrac{1}{[2(k + 1) - 1][2(k + 1) + 1]}$
$= \dfrac{1}{(2k + 1)(2k + 3)}$ to both sides, and we have
$\dfrac{1}{1 \cdot 3} + \dfrac{1}{3 \cdot 5} + \cdots + \dfrac{1}{(2k - 1)(2k + 1)}$
$+ \dfrac{1}{[2(k + 1) - 1][2(k + 1) + 1]}$

$$= \frac{k}{2k+1} + \frac{1}{(2k+1)(2k+3)} = \frac{k(2k+3)+1}{(2k+1)(2k+3)}$$

$$= \frac{(2k+1)(k+1)}{(2k+1)(2k+3)} = \frac{k+1}{2(k+1)+1},$$

so $P_{k+1}$ is true.
Therefore $P_n$ is true for all $n \geq 1$.

**17.** $P_n$: $2^n \geq 2n$.
$P_1$ is true: $2^1 \geq 2 \cdot 1$ (in fact, they are equal).
Now assume $P_k$ is true: $2^k \geq 2k$.
Then $2^{k+1} = 2 \cdot 2^k \geq 2 \cdot 2k$
$= 2 \cdot (k + k) \geq 2(k+1)$, so $P_{k+1}$ is true.
Therefore $P_n$ is true for all $n \geq 1$.

**18.** $P_n$: $3^n \geq 3n$.
$P_1$ is true: $3^1 \geq 3 \cdot 1$ (in fact, they are equal).
Now assume $P_k$ is true: $3^k \geq 3k$.
Then $3^{k+1} = 3 \cdot 3^k \geq 3 \cdot 3k$
$= 3 \cdot (k + 2k) \geq 3(k+1)$, so $P_{k+1}$ is true.
Therefore $P_n$ is true for all $n \geq 1$.

**19.** $P_n$: 3 is a factor of $n^3 + 2n$.
$P_1$ is true: 3 is a factor of $1^3 + 2 \cdot 1 = 3$.
Now assume $P_k$ is true: 3 is a factor of $k^3 + 2k$.
Then $(k+1)^3 + 2(k+1)$
$= (k^3 + 3k^2 + 3k + 1) + (2k + 2)$
$= (k^3 + 2k) + 3(k^2 + k + 1)$.
Since 3 is a factor of both terms, it is a factor of the sum, so $P_{k+1}$ is true. Therefore $P_n$ is true for all $n \geq 1$.

**20.** $P_n$: 6 is a factor of $7^n - 1$.
$P_1$ is true: 6 is a factor of $7^1 - 1 = 6$.
Now assume $P_k$ is true, so that 6 is a factor of $7^k - 1 = 6$.
Then $7^{k+1} - 1 = 7 \cdot 7^k - 1 = 7(7^k - 1) + 6$.
Since 6 is a factor of both terms of this sum, it is a factor of the sum, so $P_{k+1}$ is true.
Therefore, $P_n$ is true for all positive integers $n$.

**21.** $P_n$: The sum of the first $n$ terms of a geometric sequence with first term $a_1$ and common ratio $r \neq 1$ is $\dfrac{a_1(1 - r^n)}{1 - r}$.
$P_1$ is true: $a_1 = \dfrac{a_1(1 - r^1)}{1 - r}$.
Now assume $P_k$ is true so that
$$a_1 + a_1 r + \cdots + a_1 r^{k-1} = \frac{a_1(1 - r^k)}{(1 - r)}.$$
Add $a_1 r^k$ to both sides: $a_1 + a_1 r + \ldots + a_1 r^{k-1}$
$$+ a_1 r^k = \frac{a_1(1 - r^k)}{(1 - r)} + a_1 r^k$$
$$= \frac{a_1(1 - r^k) + a_1 r^k(1 - r)}{1 - r}$$
$$= \frac{a_1 - a_1 r^k + a_1 r^k - a_1 r^{k+1}}{1 - r} = \frac{a_1 - a_1 r^{k+1}}{1 - r},$$
so $P_{k+1}$ is true. Therefore, $P_n$ is true for all

positive integers $n$.

**22.** $P_n$: $S_n = \dfrac{n}{2}[2a_1 + (n-1)d]$.
First note that $a_n = a_1 + (n-1)d$. $P_1$ is true:
$$S_1 = \frac{1}{2}[2a_1 + (1-1)d] = \frac{1}{2}(2a_1) = a_1.$$
Now assume $P_k$ is true: $S_k = \dfrac{k}{2}[2a_1 + (k-1)d]$.
Add $a_{k+1} = a_1 + kd$ to both sides, and observe that $S_k + a_{k+1} = S_{k+1}$.
Then we have
$$S_{k+1} = \frac{k}{2}[2a_1 + (k-1)d] + a_1 + kd$$
$$= ka_1 + \frac{1}{2}k(k-1)d + a_1 + kd$$
$$= (k+1)a_1 + \frac{1}{2}k(k+1)d$$
$$= \frac{k+1}{2}[2a_1 + (k+1-1)d].$$
Therefore $P_{k+1}$ is true, and $P_n$ is true for all $n \geq 1$.

**23.** $P_n$: $\displaystyle\sum_{k=1}^{n} = \dfrac{n(n+1)}{2}$.
$P_1$ is true: $\displaystyle\sum_{k=1}^{1} k = 1 = \dfrac{1 \cdot 2}{2}$.
Now assume $P_k$ is true: $\displaystyle\sum_{i=1}^{k} i = \dfrac{k(k+1)}{2}$.
Add $(k+1)$ to both sides, and we have
$$\sum_{i=1}^{k+1} i = \frac{k(k+1)}{2} + (k+1)$$
$$= \frac{k(k+1)}{2} + \frac{2(k+1)}{2} = \frac{(k+1)(k+2)}{2}$$
$$= \frac{(k+1)(k+1+1)}{2}, \text{ so } P_{k+1} \text{ is true.}$$
Therefore, $P_n$ is true for all $n \geq 1$.

**24.** $P_n$: $\displaystyle\sum_{k=1}^{n} k^3 = \dfrac{n^2(n+1)^2}{4}$. $P_1$ is true: $1^3 = \dfrac{1^2 \cdot 2^2}{4}$.
Now assume $P_k$ is true so that
$1^3 + 2^3 + \ldots + k^3 = \dfrac{k^2(k+1)^2}{4}$. Add $(k+1)^3$ to both sides and we have
$1^3 + 2^3 + \ldots + k^3 + (k+1)^3$
$$= \frac{k^2(k+1)^2}{4} + (k+1)^3 = \frac{k^2(k+1)^2 + 4(k+1)^3}{4}$$
$$= \frac{(k+1)^2(k^2 + 4k + 4)}{4} = \frac{(k+1)^2((k+1)+1)^2}{4}$$
so $P_{k+1}$ is true. Therefore, $P_n$ is true for all positive integers.

**25.** 125,250

**26.** 5,239,625

27. $\dfrac{(n-3)(n+4)}{2}$

28. 8,122,500

29. $\approx 3.44 \times 10^{10}$

30. 14,400

31. $\dfrac{n(n^2-3n+8)}{3}$

32. $\dfrac{n(n+5)(4n+1)}{6}$

33. $\dfrac{n(n-1)(n^2+3n+4)}{4}$

34. $\dfrac{n(n-1)(n^2+3n+12)}{4}$

35. The inductive step does not work for 2 people. Sending them alternately out of the room leaves 1 person (and one blood type) each time, but we cannot conclude that their blood types will match *each other.*

36. The number $k$ is a *fixed* number for which the statement $P_k$ is known to be true. Once the anchor is established, we can assume that such a number $k$ exists. We can not assume that $P_n$ is true, because $n$ is not fixed.

37. $P_n$: 2 is a factor of $(n+1)(n+2)$. $P_1$ is true because 2 is a factor of $(2)(3)$. Now assume $P_k$ is true so that 2 is a factor of $(k+1)(k+2)$. Then
$[(k+1)+1][(k+2)+2]$
$= (k+2)(k+3) = k^2+5k+6$
$= k^2+3k+2+2k+4$
$= (k+1)(k+2)+2(k+2)$. Since 2 is a factor of both terms of this sum, it is a factor of the sum, and so $P_{k+1}$ is true. Therefore, $P_n$ is true for all positive integers $n$.

38. $P_n$: 6 is a factor of $n(n+1)(n+2)$. $P_1$ is true because 6 is a factor of $(1)(2)(3)$. Now assume $P_k$ is true so that 6 is a factor of $(k+1)(k+2)$. Then
$(k+1)[(k+1)+1][(k+1)+2]$
$= k(k+1)(k+2)+3(k+1)(k+2)$. Since 2 is a factor of $(k+1)(k+2)$, 6 is a factor of both terms of the sum and thus of the sum itself, and so $P_{k+1}$ is true.

39. Given any two consecutive integers, one of them must be even. Therefore, their product is even. Since $n+1$ and $n+2$ are consecutive integers, their product is even. Therefore, 2 is a factor of $(n+1)(n+2)$.

40. Given any three consecutive integers, one of them must be a multiple of 3, and at least one of them must be even. Therefore, their product is a multiple of 6. Since $n$, $n+1$ and $n+2$ are three consecutive integers, 6 is a

multiple of $n(n+1)(n+2)$.

41. $P_n$: $F_{n+2}-1 = \displaystyle\sum_{k=1}^{n} F_k$. $P_1$ is true since
$F_{1+2}-1 = F_3-1 = 2-1 = 1$, which equals
$\displaystyle\sum_{k=1}^{1} F_k = 1$. Now assume that $P_k$ is true:
$F_{k+2}-1 = \displaystyle\sum_{i=1}^{k} F_i$. Then $F_{(k+1)+2}-1$
$= F_{k+3}-1 = F_{k+1}+F_{k+2}-1$
$= (F_{k+2}-1)+F_{k+1} = \left(\displaystyle\sum_{i=1}^{k} F_i\right)+F_{k+1}$
$= \displaystyle\sum_{i=1}^{k+1} F_i$, so $P_{k+1}$ is true. Therefore $P_n$ is true for all $n \geq 1$.

42. $P_n$: $a_n < 2$. $P_1$ is easy: $a_1 = \sqrt{2} < 2$. Now assume that $P_k$ is true: $a_k < 2$. Note that $a_{k+1} = \sqrt{2+a_k}$, so that
$a_{k+1}^2 = 2+a_k < 2+2 = 4$; therefore
$a_{k+1} < 2$, so $P_{k+1}$ is true. Therefore $P_n$ is true for all $n \geq 1$.

43. $P_n$: $a-1$ is a factor of $a^n-1$. $P_1$ is true because $a-1$ is a factor of $a-1$. Now assume $P_k$ is true so that $a-1$ is a factor of $a^k-1$. Then
$a^{k+1}-1 = a \cdot a^k-1 = a(a^k-1)+(a-1)$.
Since $a-1$ is a factor of both terms in the sum, it is a factor of the sum, and so $P_{k+1}$ is true. Therefore, $P_n$ is true for all positive integers $n$.

44. Let $P(a) = a^n-1$. Since $P(1) = 1^n-1 = 0$, the Factor Theorem for polynomials allows us to conclude that $a-1$ is a factor of $P$.

45. $P_n$: $3n-4 \geq n$ for $n \geq 2$. $P_2$ is true since $3 \cdot 2-4 \geq 2$. Now assume that $P_k$ is true: $3k-4 \geq k$. Then $3(k+1)-4 = 3k+3-4$
$= (3k-4)+3 \geq k+3 \geq k+1$, so $P_{k+1}$ is true. Therefore $P_n$ is true for all $n \geq 2$.

46. $P_n$: $2^n \geq n^2$ for $n \geq 4$. $P_4$ is true since $2^4 \geq 4^2$. Now assume that $P_k$ is true: $2^k \geq k^2$. Then
$2^{k+1} = 2 \cdot 2^k \geq 2 \cdot k^2 \geq 2 \cdot k^2 \geq k^2+2k+1$
$= (k+1)^2$. (The inequality $2k^2 \geq k^2+2k+1$, or equivalently, $k^2 \geq 2k+1$, is true for all $k \geq 4$ because $k^2 = k \cdot k \geq 4k = 2k+2k > 2k+1$.) Thus $P_{k+1}$ is true, so $P_n$ is true for all $n \geq 4$.

47. Use $P_3$ as the anchor and obtain the inductive step by representing any $n$-gon as the union of a triangle and an $(n-1)$-gon.

## Section 9.6 (pp. 717–730)

### Exploration 1

1. **(a)** The average is a bit below 13.
   **(c)** Alaska, Colorado, Georgia, Texas, and Utah
   **(d)** The low outlier is Alaska, where older people would be less willing or able to cope with the harsh winter conditions. The high outlier is Florida, where the mild weather and abundant retirement communities attract older residents.

### Exercises 9.6

1. 5 | 8 9
   6 | 0
   6 | 6 8 8 9
   7 | 0 0 1 2 2
   This stemplot shows most of the life expectancies to be clustered near 70, with three lower values clustered near 60.

2. 6 | 4
   6 | 5
   7 | 0 2 3 4 4
   7 | 5 6 8 9 9
   This stemplot shows most of the life expectancies to be spread through the 70's, with two lower values near 65.

3.     Male    Female
       9 8 | 5 |
         0 | 6 | 4
     9 8 8 6 | 6 | 5
   2 2 1 0 0 | 7 | 0 2 3 4 4
         | 7 | 5 6 8 9 9
   This stemplot shows two distributions of similar shapes, but the women's life expectancies are uniformly about 5 years higher than the men's.

4. 4 | 0 5
   5 | 2 3 8 9
   6 | 2 4 5
   7 | 4 9
   8 |
   9 |
   10 | 2
   Women in South American nations have higher life expectancies than the men, by about 5 years in most cases.

5.

| Life expectancy (years) | Frequency |
|---|---|
| 55.0 — 59.9 | 3 |
| 60.0 — 64.9 | 0 |
| 65.0 — 69.9 | 5 |
| 70.0 — 74.9 | 4 |

6.

| Life expectancy (years) | Frequency |
|---|---|
| 60.0 — 64.9 | 1 |
| 65.0 — 69.9 | 2 |
| 70.0 — 74.9 | 5 |
| 75.0 — 79.9 | 4 |

7. 0 | 5 8 9
   1 | 3 4 6
   2 | 3 6 8
   3 | 3 9
   4 |
   5 |
   6 | 1
   The 61 is an outlier.

8.     Maris    Aaron
       9 8 5 | 0 |
     6 4 3 | 1 | 0 2 3
     8 6 3 | 2 | 0 4 6 7 9
     9 3 | 3 | 0 2 4 4 8 9 9
       | 4 | 0 0 4 4 4 4 5 7
       | 5 |
       1 | 6 |
   Except for Maris's one record-breaking year, his home run output falls well short of Aaron's.

9.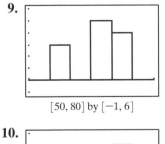
   [50, 80] by [−1, 6]

10.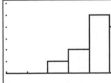
    [50, 80] by [−1, 6]

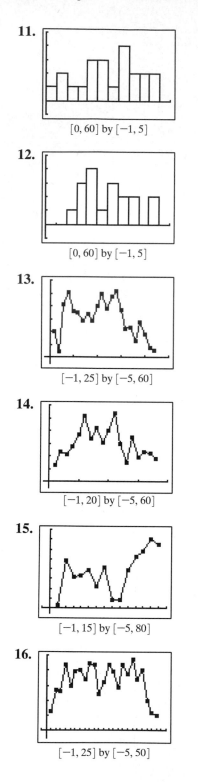

**11.**

$[0, 60]$ by $[-1, 5]$

**12.**

$[0, 60]$ by $[-1, 5]$

**13.**

$[-1, 25]$ by $[-5, 60]$

**14.**

$[-1, 20]$ by $[-5, 60]$

**15.**

$[-1, 15]$ by $[-5, 80]$

**16.**

$[-1, 25]$ by $[-5, 50]$

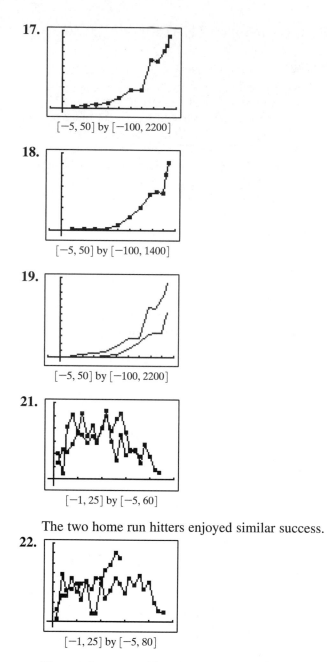

**17.**

$[-5, 50]$ by $[-100, 2200]$

**18.**

$[-5, 50]$ by $[-100, 1400]$

**19.**

$[-5, 50]$ by $[-100, 2200]$

**21.**

$[-1, 25]$ by $[-5, 60]$

The two home run hitters enjoyed similar success.

**22.**

$[-1, 25]$ by $[-5, 80]$

The two home run hitters were comparable for the first seven years of their careers.

**23. (a)**

| Stem | Leaf |
|------|------|
| 28 | 2 |
| 29 | 3 7 |
| 30 | |
| 31 | 6 7 |
| 32 | 7 8 |
| 33 | 5 5 5 8 |
| 34 | 2 8 8 |
| 35 | 3 3 4 |
| 36 | 3 7 |

```
37 |
38 | 5
```

**(b)**

| Interval | Frequency |
|----------|-----------|
| 25–29 | 3 |
| 30–34 | 11 |
| 35–39 | 6 |

**(c)**

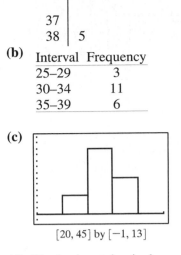

[20, 45] by [−1, 13]

**(d)** The horizontal axis does not represent time.

**24. (a)**

| Stem | Leaf |
|------|------|
| 6 | 2 3 9 9 |
| 7 | 7 9 9 9 |
| 8 | 0 2 3 7 7 8 9 9 |
| 9 | 0 0 0 1 1 2 3 3 3 4 5 6 7 |
| 10 | 2 3 3 6 6 6 7 9 |
| 11 | 0 2 4 6 |
| 12 | 0 5 9 |

**(b)**

| Interval | Frequency |
|----------|-----------|
| 6.0–6.9 | 4 |
| 7.0–7.9 | 4 |
| 8.0–8.9 | 8 |
| 9.0–9.9 | 13 |
| 10.0–10.9 | 8 |
| 11.0–11.9 | 4 |
| 12.0–12.9 | 3 |

**(c)**

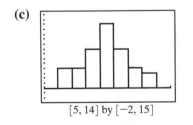

[5, 14] by [−2, 15]

**(d)** The data are not categorical.

**25.**

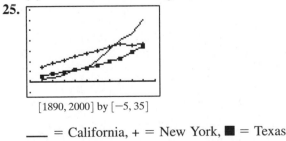

[1890, 2000] by [−5, 35]

_____ = California, + = New York, ■ = Texas

**26.**

[1890, 2000] by [−2, 14]

_____ = Pennsylvania, + = Illinois, ■ = Texas

**29.**

[0, 13] by [−15, 40]

**30.** High temperature:
$$f(t) \approx 17 + 15 \sin\left[\frac{\pi}{6}(t-4)\right]$$
Low temperature:
$$g(t) \approx 6.5 + 15.5 \sin\left[\frac{\pi}{6}(t-4)\right]$$

# Section 9.7 (pp. 730–743)

**Exploration 1**

1. Figure (b)
2. Figure (b); Figure (c); Figure (a)

**Quick Review 9.7**

1. $x_1 + x_2 + x_3 + x_4 + x_5 + x_6 + x_7$
2. $x_1 + x_2 + x_3 + x_4 + x_5 - 5\bar{x}$

   Note that, since $x = \frac{1}{5}\sum_{i=1}^{5} x_i$, this simplifies to 0.

3. $\frac{1}{7}(x_1 + x_2 + x_3 + x_4 + x_5 + x_6 + x_7)$

4. $\frac{1}{5}(x_1 + x_2 + x_3 + x_4 + x_5) - \bar{x}$.

5. $\frac{1}{5}[(x_1 - \bar{x})^2 + (x_2 - \bar{x})^2 + \ldots + (x_5 - \bar{x})^2]$

6. $\sqrt{\frac{1}{5}[(x_1 - \bar{x})^2 + (x_2 - \bar{x})^2 + \ldots + (x_5 - \bar{x})^2]}$

7. $\sum_{i=1}^{8} x_i f_i$

8. $\sum_{i=1}^{10} (x_i - \bar{x})^2$

9. $\frac{1}{50}\sum_{i=1}^{50} (x_i - \bar{x})^2$

**10.** $\sqrt{\dfrac{1}{7}\displaystyle\sum_{i=1}^{7}(x_i - \overline{x})^2}$

## Section 9.7 Exercises

**8.** 21,176,240 km$^2$

**10.** State College: 2.4 houses/day. College Station: $\approx$ 2.14 houses/day. The State College workers were faster.

**13.** median: 28; mode: 28

**14.** median: 87.85; mode: None

**15. (a)** $\approx$ 18.42°C

    **(b)** $\approx$ 18.49°C

    **(c)** The weighted average is the better indicator.

**16. (a)** $\approx$ 6.42°C

    **(b)** $\approx$ 6.49°C

    **(c)** The weighted average is the better indicator.

**17.** Mark McGwire:

5-number summary: {3, 22, 39, 52, 70}

Range: 67

IQR: 30

No outliers

Sammy Sosa:

5-number summary: {4, 10, 33, 40, 66}

Range: 62

IQR: 30

No outliers

**18.** Willie Mays:

5-number summary: {4, 20, 31.5, 40, 52}

Range: 48

IQR: 20

No outliers

Mickey Mantle:

5-number summary: {13, 21, 28.5, 37, 54}

Range: 41

IQR: 16

No outliers

**19.** {6.2, 8.5, 9.25, 10.6, 12.9}; 6.7; 2.1; No outliers

**20.** {28.2, 31.7, 33.5, 35.3, 38.5}; 10.3; 3.6

**21. (a)**

$[-4, 80]$ by $[-1, 3]$

**(b)**

$[-4, 80]$ by $[-1, 3]$

**22.(a)**

$[-3, 70]$ by $[-1, 2]$

**(b)**

$[-3, 70]$ by $[-1, 2]$

**23. (a)**

$[15, 45]$ by $[-1, 1]$

**(b)**

$[15, 45]$ by $[-1, 1]$

**24. (a)**

$[80, 850]$ by $[-1, 1]$

**(b)**

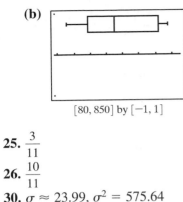

[80, 850] by [−1, 1]

**25.** $\dfrac{3}{11}$

**26.** $\dfrac{10}{11}$

**30.** $\sigma \approx 23.99$, $\sigma^2 = 575.64$

**31.** $\sigma \approx 224.5$; $\sigma^2 \approx 50{,}405.9$

**32.** $\sigma \approx 89.8$; $\sigma^2 \approx 8065.6$

**37. (c)** A parameter

**38. (c)** No. The mean would have to be *weighted* according to the number of people in each state who took the ACT.

**39.** There are many possible answers; examples are given.
  **(a)** {2, 2, 2, 3, 6, 8, 20}
  **(b)** {1, 2, 3, 4, 6, 48, 48}
  **(c)** {−20, 1, 1, 1, 2, 3, 4, 5, 6}

**40.** There are many possible answers; examples are given.
  **(a)** {2, 4, 6, 8}
  **(b)** {1, 5, 5, 6, 6, 9}
  **(c)** {3, 3, 3, 9, 9, 9}

**41.** No

## Chapter 9 Review (pp. 747–750)

**17.** $32x^5 + 80x^4y + 80x^3y^2 + 40x^2y^3 + 10xy^4 + y^5$

**18.** $16{,}384a^7 - 86{,}016a^6b + 193{,}536a^5b^2$
  $- 241{,}920a^4b^3 + 181{,}440a^3b^4 - 81{,}648a^2b^5$
  $+ 20{,}412ab^6 - 21{,}87b^7$

**19.** $243x^{10} + 405x^8y^3 + 270x^6y^6 + 90x^4y^9 + 15x^2y^{12}$
  $+ y^{15}$

**20.** $1 + 6x^{-1} + 15x^{-2} + 20x^{-3} + 15x^{-4} + 6x^{-5}$
  $+ x^{-6}$

**21.** $512a^{27} - 2304a^{24}b^2 + 4608a^{21}b^4 - 5376a^{18}b^6$
  $+ 4032a^{15}b^8 - 2016a^{12}b^{10} + 672a^9b^{12}$
  $- 144a^6b^{14} + 18a^3b^{16} - b^{18}$

**22.** $x^{-8} + 4x^{-6}y^{-1} + 6x^{-4}y^{-2} + 4x^{-2}y^{-3} + y^{-4}$

**23.** $-1320$

**24.** 112

**26.** {(1, 1), (1, 2), (1, 3), (1, 4), (1, 5), (1, 6), (2, 1), (2, 2), … , (6, 6)}

**27.** {13, 16, 31, 36, 61, 63}

**29.** {*HHH, HHT, HTH, HTT, THH, THT, TTH, TTT*}

**30.** {*HHT, HTH, THH, TTH, THT, HTT*}

**38.** The experiment is symmetrical, because $P(\text{S}) = P(\text{F})$.

**42.** Successes are less likely than failures, so the two are not interchangeable.

**45.** 0, 1, 2, 3, 4, 5; 39

**46.** $-1, \dfrac{4}{3}, -2, \dfrac{16}{5}, -\dfrac{16}{3}, \dfrac{64}{7}; \approx 2.68 \times 10^{10}$

**47.** −1, 2, 5, 8, 11, 14; 32

**48.** 5, 10, 20, 40, 80, 160; 10, 240

**49.** −5, −3.5, −2, −0.5, 1, 2.5; 11.5

**50.** $3, 1, \dfrac{1}{3}, \dfrac{1}{9}, \dfrac{1}{27}, \dfrac{1}{81}; 3^{-10} = \dfrac{1}{59{,}049}$

**51.** −3, 1, −2, −1, −3, −4; −76

**52.** −3, 2, −1, 1, 0, 1; 13

**53.** Arithmetic with $d = -2.5$; $a_n = 14.5 - 2.5n$

**54.** Arithmetic with $d = 4$; $a_n = 4n - 9$

**55.** Geometric with $r = 1.2$; $a_n = 10 \cdot (1.2)^{n-1}$

**56.** Geometric with $r = -2$; $a_n = -\dfrac{1}{16}(-2)^n$

**57.** Arithmetic with $d = 4.5$; $a_n = 4.5n - 15.5$

**58.** Geometric with $r = \dfrac{1}{4}$;
  $b_n = 28 \cdot \left(\dfrac{1}{4}\right)^n$

**69.** $3280.\overline{4}$

**71.**

[0, 15] by [0, 2]

**72.**

[0, 16] by [−10, 460]

**79.** $\displaystyle\sum_{k=1}^{21} (5k - 13)$

**80.** $\displaystyle\sum_{k=1}^{10} (-2)^{k+1}$

**81.** $\displaystyle\sum_{k=0}^{\infty} (2k + 1)^2$ or $\displaystyle\sum_{k=1}^{\infty} (2k - 1)^2$

**86.** 5,328,575

**87.** $P_n$: $1 + 3 + 6 + \cdots + \dfrac{n(n+1)}{2} = \dfrac{n(n+1)(n+2)}{6}$.

$P_1$ is true: it says that $1(1+1) = \dfrac{1(1+1)(1+2)}{3}$.

Now assume $P_k$ is true:

$1 \cdot 2 + 2 \cdot 3 + 3 \cdot 4 + \cdots + k(k+1)$

$= \dfrac{k(k+1)(k+2)}{3}$. Add $(k+1)(k+2)$ to both

sides, and we have

$1 \cdot 2 + 2 \cdot 3 + \cdots + k(k+1) + (k+1)(k+2)$

$= \dfrac{k(k+1)(k+2)}{3} + (k+1)(k+2)$

$= (k+1)(k+2)\left(\dfrac{1}{3}k + 1\right)$

$= (k+1)(k+2)\left(\dfrac{k+3}{3}\right)$

$= \dfrac{(k+1)(k+1+1)(k+1+2)}{3}$

**89.** $P_n$: $2^{n-1} \le n!$. $P_1$ is true: its says that $2^{1-1} \le 1!$
(they are equal). Now assume $P_k$ is true:
$2^{k-1} \le k!$. Then $2^{k+1-1}$
$= 2 \cdot 2^{k-1} \le 2 \cdot k! \le (k+1)k! = (k+1)!$, so
$P_{k+1}$ is true. Therefore $P_n$ is true for all $n \ge 1$.

**90.** $P_n$: $n^3 + 2n$ is divisible by 3. $P_1$ is true because
$1^3 + 2 \cdot 1 = 3$ is divisible by 3. Now assume $P_k$
is true: $k^3 + 2k$ is divisible by 3. Then note that
$(k+1)^3 + 2(k+1)$
$= (k^3 + 3k^2 + 3k + 1) + (2k + 2)$
$= (k^3 + 2k) + 3(k^2 + k + 1)$.

Since both terms are divisible by 3, so is the sum,
so $P_{k+1}$ is true. Therefore $P_n$ is true for all $n \ge 1$.

**91. (a)**    7|7
         8|2 6 7 7
         9|1 4 5 7 7
        10|4 8
        11|1 2 4 7 8
        12|0 4 9
        13|3
        14|1
        15|3
        16|4 6
        17|1 7 8
        18|
        19|6
        20|5

**(b)**    70,000—79,999|1
        80,000—89,999|4
        90,000—99,999|5
       100,000—109,999|2
       110,000—119,999|5
       120,000—129,999|3
       130,000—139,999|1

       140,000—149,999|1
       150,000—159,999|1
       160,000—169,999|2
       170,000—179,999|3
       180,000—189,999|0
       190,000—199,999|1
       200,000—209,999|1

**(c)**

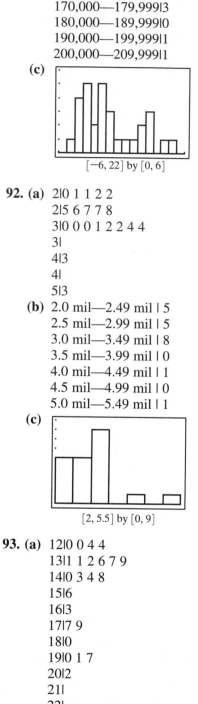

$[-6, 22]$ by $[0, 6]$

**92. (a)** 2|0 1 1 2 2
         2|5 6 7 7 8
         3|0 0 0 1 2 2 4 4
         3|
         4|3
         4|
         5|3

**(b)** 2.0 mil—2.49 mil | 5
        2.5 mil—2.99 mil | 5
        3.0 mil—3.49 mil | 8
        3.5 mil—3.99 mil | 0
        4.0 mil—4.49 mil | 1
        4.5 mil—4.99 mil | 0
        5.0 mil—5.49 mil | 1

**(c)**

$[2, 5.5]$ by $[0, 9]$

**93. (a)** 12|0 0 4 4
         13|1 1 2 6 7 9
         14|0 3 4 8
         15|6
         16|3
         17|7 9
         18|0
         19|0 1 7
         20|2
         21|1
         22|1
         23|0

**(b)** 120—129|4
130—139|6
140—149|4
150—159|1
160—169|1
170—179|2
180—189|1
190—199|3
200—209|1
210—219|0
220—229|0
230—239|1

**(c)**

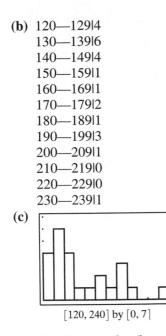

[120, 240] by [0, 7]

**94. (a)** For the stemplot (but not for the other frequency table or histogram), round to nearest hundred yards first:

Stem  Leaf
1 | 1
2 | 3 4 5 7 8
3 | 1 3 5 5 6
4 | 0 3 7 7
5 | 0 6

**(b)**

| Interval | Frequency |
|---|---|
| 1000–1999 | 1 |
| 2000–2999 | 5 |
| 3000–3999 | 6 |
| 4000–4999 | 3 |
| 5000–5999 | 2 |

**(c)**

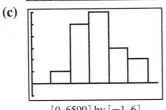

[0, 6500] by [−1, 6]

**95.** 5-number summary: {7.7, 9.5, 11.55, 15.3, 20.5}
Range: 12.8 ($128,000)
IQR: 5.8 ($58,000)
$\sigma \approx 3.55$, $\sigma^2 \approx 12.63$

**96.** 5-number summary: {2.0, 2.35, 2.9, 3.2, 5.3}
Range: 3.3 (3.3 million)
IQR: 0.85 (0.85 million)
$\sigma \approx 0.77$, $\sigma^2 \approx 0.592$

**97.** 5-number summary: {120, 131.5, 143.5, 179.5, 230}
Range: 110
IQR: 48
$\sigma = 29.9$, $\sigma^2 = 891.4$

**98.** 5-number summary: {1112, 2709, 3487, 4264, 5648}
Range: 4536
IQR: 1555
$\sigma \approx 1095$, $\sigma^2 \approx 1,119,223$

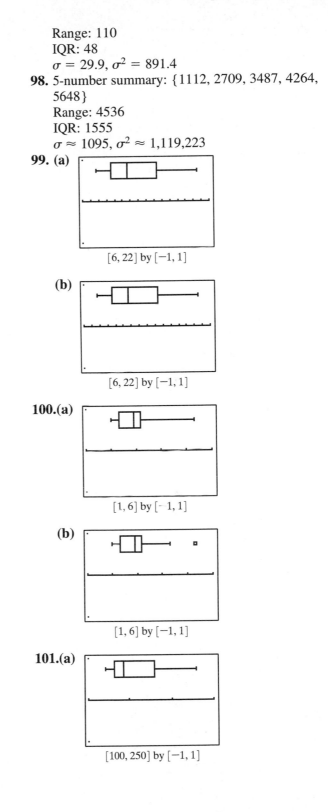

**99. (a)**

[6, 22] by [−1, 1]

**(b)**

[6, 22] by [−1, 1]

**100.(a)**

[1, 6] by [−1, 1]

**(b)**

[1, 6] by [−1, 1]

**101.(a)**

[100, 250] by [−1, 1]

**(b)**

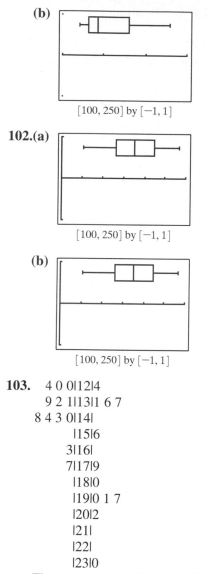

$[100, 250]$ by $[-1, 1]$

**102.(a)**

$[100, 250]$ by $[-1, 1]$

**(b)**

$[100, 250]$ by $[-1, 1]$

**103.**   4 0 0|12|4
     9 2 1|13|1 6 7
     8 4 3 0|14|
          |15|6
          3|16|
          7|17|9
          |18|0
          |19|0 1 7
          |20|2
          |21|
          |22|
          |23|0

The songs released in the earlier years tended to be shorter.

**104.** The range and interquartile range are both greater in the lower graph, which shows the times for later years.

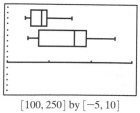

$[100, 250]$ by $[-5, 10]$

**105.**

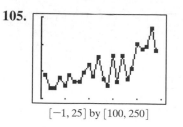

$[-1, 25]$ by $[100, 250]$

Again, the data demonstrates that songs appearing later tended to be longer.

**106.**

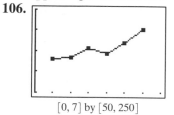

$[0, 7]$ by $[50, 250]$

The trend is clearly increasing overall.

**107.** 1 9 36 84 126 84 36 9 1

**108.** $_nP_k \times _{n-k}P_j = \dfrac{n!}{(n-k)!} \cdot \dfrac{(n-k)!}{[(n-k)-j]!}$

$= \dfrac{n!}{(n-k-j)!} = \dfrac{n!}{[n-(k+j)]!} = \, _nP_{k+j}$

## Chapter 9 Project

Answers are based on the sample data shown in the table.

**1.** Step | Leaf
     5|
     5|9
     6|1 1 2 3 3 3 4 4 4 4
     6|5 6 6 6 7 8 8 9 9 9
     7|0 0 1 1 1 2 2 3
     7|5

67 or 68 inches

**2.** A large number of students are between 63-64 inches and also 69-72 inches.

| Height | Frequency |
|--------|-----------|
| 59–60  | 1 |
| 61–62  | 3 |
| 63–64  | 7 |
| 65–66  | 4 |
| 67–68  | 3 |
| 69–70  | 5 |
| 71–72  | 5 |
| 73–74  | 1 |
| 75–76  | 1 |

**3.**

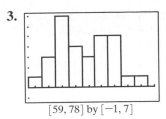

$[59, 78]$ by $[-1, 7]$

**4.** 66.9 in.; 66.5 in.; 64 in.

**5.** The data set is well distributed and probably does not have outliers.

**6.** $\{59, 64, 66.5, 70, 75\}$

**7.**

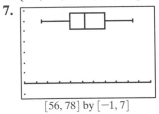

$[56, 78]$ by $[-1, 7]$

**8.** 67.5; 67; $\{59, 64, 67, 71, 86\}$;

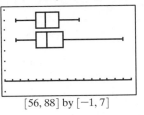

$[56, 88]$ by $[-1, 7]$

67.5; 67; $\{59, 64, 67, 71, 86\}$

**10.** 68.9; 68; $\{59, 64, 68, 71, 86\}$

# Chapter 10

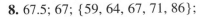

## Section 10.1 (pp. 753–762)

### Exploration 1

**2.** 3 ft/sec

**4.** As the slope of the line joining $(a, (s(a))$ and $(b, s(b))$.

### Quick Review 10.1

**5.** $(y - 4) = \dfrac{3}{4}(x - 1)$     **6.** $(y - 4) = -\dfrac{4}{3}(x - 1)$

### Exercises 10.1

**5.**

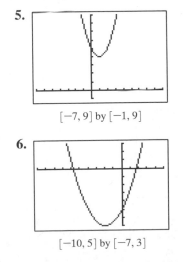

$[-7, 9]$ by $[-1, 9]$

**6.**

$[-10, 5]$ by $[-7, 3]$

**7.**

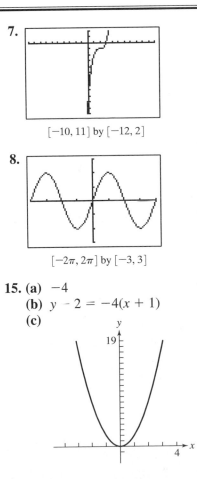

$[-10, 11]$ by $[-12, 2]$

**8.**

$[-2\pi, 2\pi]$ by $[-3, 3]$

**15. (a)** $-4$

**(b)** $y - 2 = -4(x + 1)$

**(c)**

**16. (a)** $-2$

**(b)** $y = -2(x - 2)$

**(c)**

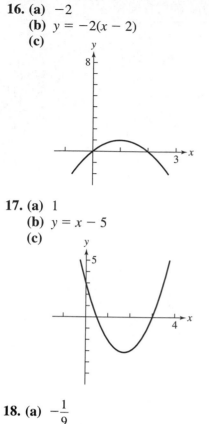

**17. (a)** 1

**(b)** $y = x - 5$

**(c)**

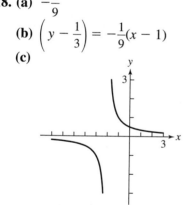

**18. (a)** $-\dfrac{1}{9}$

**(b)** $\left(y - \dfrac{1}{3}\right) = -\dfrac{1}{9}(x - 1)$

**(c)**

**23.** $6x + 2$

**25. (a)**

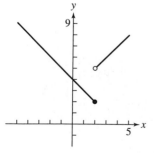

**(b)** Since the graph of the function does not have a definable slope at $x = 2$, the derivative of $f$ does not exist at $x = 2$.

**(c)** Derivatives do not exist at points where functions have discontinuities.

**26. (a)**

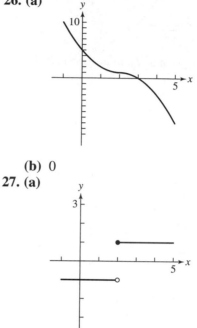

**(b)** 0

**27. (a)**

**(b)** Since the graph of the function does not have a definable slope at $x = 2$, the derivative of $f$ does not exist at $x = 2$.

**(c)** Derivatives do not exist at points where functions have discontinuities.

**28. (a)**

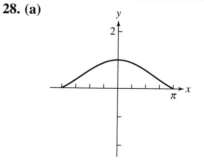

**(b)** 0

**31. (b)** $f(x) = 8.94x^2 + 0.05x + 0.01$, $x$ = time in seconds

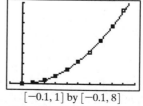

$[-0.1, 1]$ by $[-0.1, 8]$

**32. (b)** $s(t) = -16.015t^2 + 1.43t + 30.35$

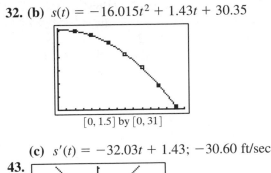

[0, 1.5] by [0, 31]

**(c)** $s'(t) = -32.03t + 1.43$; $-30.60$ ft/sec

**43.**

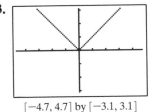

[−4.7, 4.7] by [−3.1, 3.1]

**(a)** No, there is no derivative because the graph has a corner at $x = 0$.

**(b)** No

**44.**

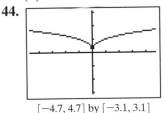

[−4.7, 4.7] by [−3.1, 3.1]

**(a)** No, there is no derivative because the graph has a cusp ("spike") at $x = 0$.

**(b)** Yes, $x = 0$

**45.**

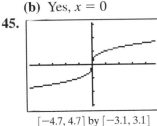

[−4.7, 4.7] by [−3.1, 3.1]

**(a)** No, there is no derivative because the graph has a vertical tangent (no slope) at $x = 0$.

**(b)** Yes, $x = 0$

**46.**

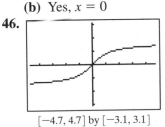

[−4.7, 4.7] by [−3.1, 3.1]

**(a)** Yes, there is a derivative because the graph has a nonvertical tangent line at $x = 0$.

**(b)** Yes, $y = x$

**47. (a)** 48 ft/sec

**(b)** 96 ft/sec

**48. (b)** 25 m/sec

**(c)** 50 m/sec

**49.**

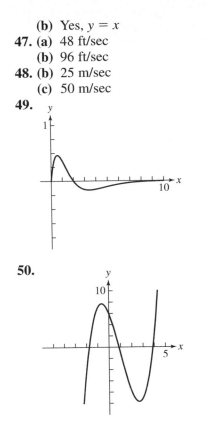

**50.**

## Section 10.2 (pp. 763–771)

### Quick Review 10.2

**1.** $\dfrac{1}{8}, \dfrac{1}{2}, \dfrac{9}{8}, 2, \dfrac{25}{8}, \dfrac{9}{2}, \dfrac{49}{8}, 8, \dfrac{81}{8}, \dfrac{25}{2}$

**2.** $\dfrac{81}{64}, \dfrac{25}{16}, \dfrac{121}{64}, \dfrac{9}{4}, \dfrac{169}{64}, \dfrac{49}{16}, \dfrac{225}{64}, 4, \dfrac{289}{64}, \dfrac{81}{16}$

### Exercises 10.2

**1.** 13; Answers will vary.

**2.** $12\dfrac{1}{2}$; Answers will vary.

**3.** 13; Answers will vary.

**4.** $17\dfrac{1}{2}$; Answers will vary.

**5.** $\left[0, \dfrac{1}{2}\right], \left[\dfrac{1}{2}, 1\right], \left[1, \dfrac{3}{2}\right], \left[\dfrac{3}{2}, 2\right]$

**6.** $\left[0, \dfrac{1}{4}\right], \left[\dfrac{1}{4}, \dfrac{1}{2}\right], \left[\dfrac{1}{2}, \dfrac{3}{4}\right], \left[\dfrac{3}{4}, 1\right], \left[1, \dfrac{5}{4}\right], \left[\dfrac{5}{4}, \dfrac{3}{2}\right],$ $\left[\dfrac{3}{2}, \dfrac{7}{8}\right], \left[\dfrac{7}{8}, 1\right]$

**7.** $\left[1, \dfrac{3}{2}\right], \left[\dfrac{3}{2}, 2\right], \left[2, \dfrac{5}{2}\right], \left[\dfrac{5}{2}, 3\right], \left[3, \dfrac{7}{2}\right], \left[\dfrac{7}{2}, 4\right]$

**8.** $\left[1, \frac{3}{2}\right], \left[\frac{3}{2}, 2\right], \left[2, \frac{5}{2}\right], \left[\frac{5}{2}, 3\right], \left[3, \frac{7}{2}\right], \left[\frac{7}{2}, 4\right],$
$\left[4, \frac{9}{2}\right], \left[\frac{9}{2}, 5\right]$

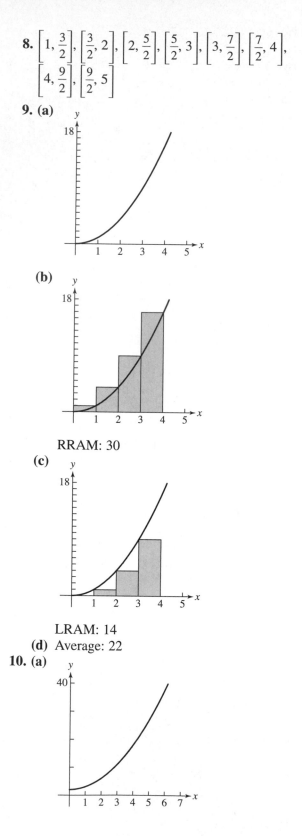

**9. (a)**

**(b)**

RRAM: 30

**(c)**

LRAM: 14

**(d)** Average: 22

**10. (a)**

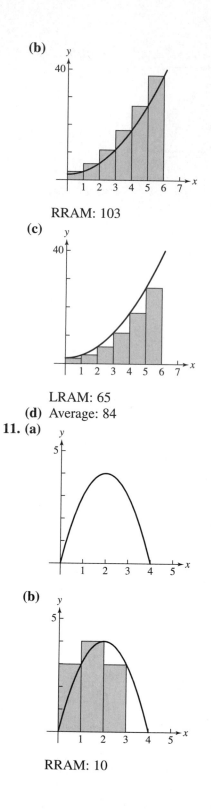

**(b)**

RRAM: 103

**(c)**

LRAM: 65

**(d)** Average: 84

**11. (a)**

**(b)**

RRAM: 10

**(c)**

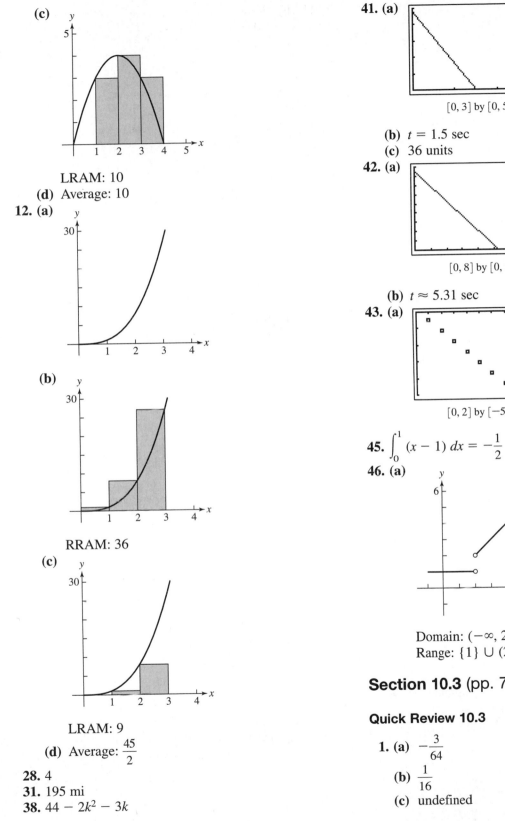

LRAM: 10

**(d)** Average: 10

**12. (a)**

**(b)**

RRAM: 36

**(c)**

LRAM: 9

**(d)** Average: $\dfrac{45}{2}$

**28.** 4

**31.** 195 mi

**38.** $44 - 2k^2 - 3k$

**41. (a)**

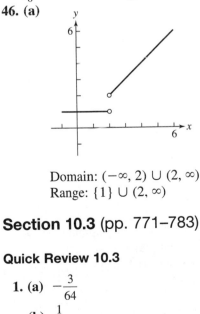

$[0, 3]$ by $[0, 50]$

**(b)** $t = 1.5$ sec
**(c)** 36 units

**42. (a)**

$[0, 8]$ by $[0, 180]$

**(b)** $t \approx 5.31$ sec

**43. (a)**

$[0, 2]$ by $[-50, 0]$

**45.** $\displaystyle\int_0^1 (x - 1)\, dx = -\dfrac{1}{2}$

**46. (a)**

Domain: $(-\infty, 2) \cup (2, \infty)$
Range: $\{1\} \cup (2, \infty)$

## Section 10.3 (pp. 771–783)

### Quick Review 10.3

**1. (a)** $-\dfrac{3}{64}$

 **(b)** $\dfrac{1}{16}$

 **(c)** undefined

**2. (a)** $\dfrac{\sin 2}{2} \approx 0.45$

 **(b)** undefined

 **(c)** $\dfrac{\sin 2}{2} \approx 0.45$

**3. (a)** $x = -2$ and $x = 2$

 **(b)** $y = 2$

**4. (a)** $x = -2$ and $x = 1$

 **(b)** no horizontal asymptotes

**6.** (c)

**7. (a)** $[-2, \infty)$

 **(b)** None

**8. (a)** $(-\infty, -2) \cup (-2, 2) \cup (2, \infty)$

 **(b)** $x = -2, x = 2$

**9.**

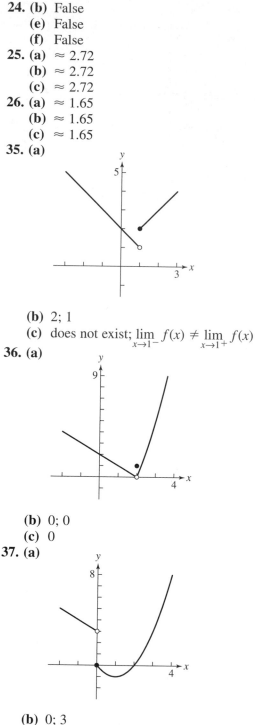

**10.** continuous on $(-\infty, 1) \cup (1, \infty)$; discontinuous at $x = 1$

**Exercises 10.3**

**11. (a)** division by zero

 **(b)** $-\dfrac{1}{6}$

**12. (a)** division by zero

 **(b)** $\dfrac{3}{4}$

**13. (a)** division by zero

 **(b)** 3

**14. (a)** division by zero

 **(b)** 5

**15. (a)** division by zero

 **(b)** $-4$

**16. (a)** division by zero

 **(b)** 4

**17. (a)** The square root of negative numbers is not defined in the real plane.

 **(b)** The limit does not exist.

**18. (a)** division by zero

 **(b)** The limit does not exist.

**19. (c)** not defined; $\lim\limits_{x \to 2^-} f(x) \neq \lim\limits_{x \to 2^+} f(x)$

**20. (c)** not defined; $\lim\limits_{x \to 3^-} f(x) \neq \lim\limits_{x \to 3^+} f(x)$

**22. (c)** not defined; $\lim\limits_{x \to 1^-} f(x) \neq \lim\limits_{x \to 1^+} f(x)$

**23. (d)** False

**24. (b)** False

 **(e)** False

 **(f)** False

**25. (a)** $\approx 2.72$

 **(b)** $\approx 2.72$

 **(c)** $\approx 2.72$

**26. (a)** $\approx 1.65$

 **(b)** $\approx 1.65$

 **(c)** $\approx 1.65$

**35. (a)**

 **(b)** 2; 1

 **(c)** does not exist; $\lim\limits_{x \to 1^-} f(x) \neq \lim\limits_{x \to 1^+} f(x)$

**36. (a)**

 **(b)** 0; 0

 **(c)** 0

**37. (a)**

 **(b)** 0; 3

 **(c)** does not exist; $\lim\limits_{x \to 0^-} f(x) \neq \lim\limits_{x \to 0^+} f(x)$

**38. (a)**

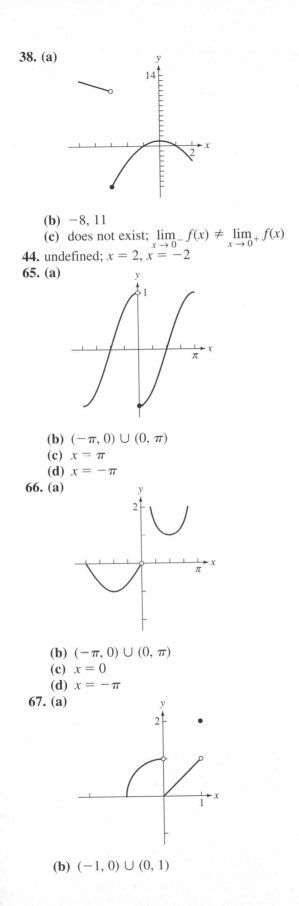

**(b)** −8, 11

**(c)** does not exist; $\lim\limits_{x \to 0^-} f(x) \neq \lim\limits_{x \to 0^+} f(x)$

**44.** undefined; $x = 2, x = -2$

**65. (a)**

**(b)** $(-\pi, 0) \cup (0, \pi)$

**(c)** $x = \pi$

**(d)** $x = -\pi$

**66. (a)**

**(b)** $(-\pi, 0) \cup (0, \pi)$

**(c)** $x = 0$

**(d)** $x = -\pi$

**67. (a)**

**(b)** $(-1, 0) \cup (0, 1)$

**(c)** $x = 1$

**(d)** $x = -1$

**68. (a)**

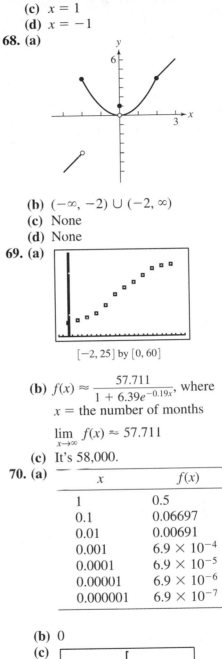

**(b)** $(-\infty, -2) \cup (-2, \infty)$

**(c)** None

**(d)** None

**69. (a)**

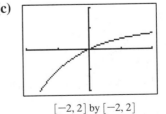

$[-2, 25]$ by $[0, 60]$

**(b)** $f(x) \approx \dfrac{57.711}{1 + 6.39e^{-0.19x}}$, where

$x$ = the number of months

$\lim\limits_{x \to \infty} f(x) \approx 57.711$

**(c)** It's 58,000.

**70. (a)**

| $x$ | $f(x)$ |
|---|---|
| 1 | 0.5 |
| 0.1 | 0.06697 |
| 0.01 | 0.00691 |
| 0.001 | $6.9 \times 10^{-4}$ |
| 0.0001 | $6.9 \times 10^{-5}$ |
| 0.00001 | $6.9 \times 10^{-6}$ |
| 0.000001 | $6.9 \times 10^{-7}$ |

**(b)** 0

**(c)**

$[-2, 2]$ by $[-2, 2]$

**71. (a)**

| $x$ | $f(x)$ |
|---|---|
| 1 | 0.84 |
| 0.1 | $-0.054$ |
| 0.01 | $-0.0051$ |
| 0.001 | $8.3 \times 10^{-4}$ |
| 0.0001 | $-3 \times 10^{-5}$ |
| 0.00001 | $3.6 \times 10^{-7}$ |
| 0.000001 | $-3.5 \times 10^{-7}$ |

**(b)** 0

**(c)**

$[-1, 1]$ by $[-1, 1]$

**72. (a)**

| $x$ | $f(x)$ |
|---|---|
| 1 | 0.84147 |
| 0.1 | $-5.44$ |
| 0.01 | $-50.64$ |
| 0.001 | 826.88 |
| 0.0001 | $-3056$ |
| 0.00001 | 3574.9 |
| 0.000001 | $-3.5 \times 10^5$ |

**(b)** undefined

**(c)**

$\left[-\dfrac{\pi}{2}, \dfrac{\pi}{2}\right]$ by $[-3, 3]$

**73. (a)**

| $x$ | $f(x)$ |
|---|---|
| 1 | 0 |
| 0.1 | $-0.00744$ |
| 0.01 | $9.9 \times 10^{-5}$ |
| 0.001 | $-5.8 \times 10^{-7}$ |
| 0.0001 | $-2.1 \times 10^{-9}$ |
| 0.00001 | $8.7 \times 10^{-11}$ |
| 0.000001 | $-9.5 \times 10^{-13}$ |

**(b)** 0

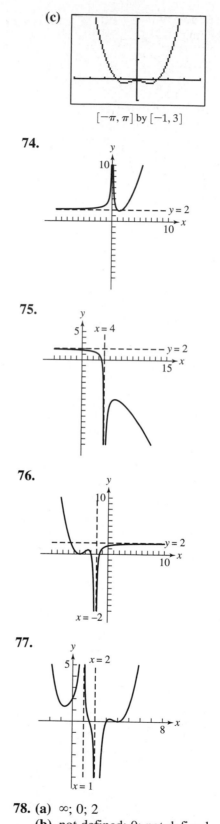

**(c)**

$[-\pi, \pi]$ by $[-1, 3]$

**74.**

**75.**

**76.**

**77.**

**78. (a)** $\infty$; 0; 2

**(b)** not defined; 0; not defined

**(c)** not defined; 0; 0

**(d)** not defined; 0; not defined

**(e)** It can be anything.

**79. (d)**

| n | A |
|---|---|
| 4 | 4 |
| 8 | 3.3137 |
| 16 | 3.1826 |
| 100 | 3.1426 |
| 500 | 3.1416 |
| 1000 | 3.1416 |
| 5000 | 3.1416 |
| 10,000 | 3.1416 |
| 100,000 | 3.1416 |

Yes, A $\to \pi$, $n \to \infty$

**(e)**

| n | A |
|---|---|
| 4 | 36 |
| 8 | 29.823 |
| 16 | 28.643 |
| 100 | 28.284 |
| 500 | 28.275 |
| 1,000 | 28.274 |
| 5,000 | 28.274 |
| 10,000 | 28.274 |
| 100,000 | 28.274 |

As $n \to \infty$, $A \to 9\pi$

**(f)** One possible answer:

$$\lim_{n\to\infty} A = \lim_{n\to\infty} nh^2 \tan\left(\frac{360°}{2n}\right)$$
$$= h^2 \lim_{n\to\infty} n \tan\left(\frac{360°}{n}\right)$$
$$= h^2\pi = \pi h^2$$

As the number of sides of the polygon increases, the distance between $h$ and the edge of the circle becomes progressively smaller. As $n \to \infty$ $h \to$ radius of the circle.

**80. (a)**

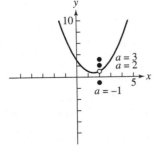

**(b)** 1

**81. (a)**

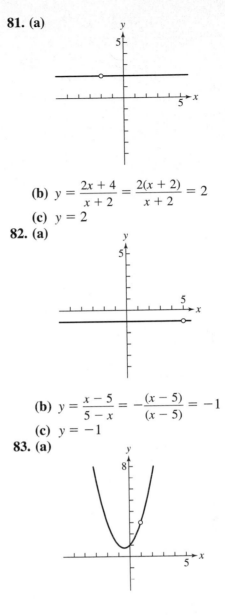

**(b)** $y = \dfrac{2x + 4}{x + 2} = \dfrac{2(x + 2)}{x + 2} = 2$

**(c)** $y = 2$

**82. (a)**

**(b)** $y = \dfrac{x - 5}{5 - x} = -\dfrac{(x - 5)}{(x - 5)} = -1$

**(c)** $y = -1$

**83. (a)**

**(b)** $y = \dfrac{x^3 - 1}{x - 1} = \dfrac{(x - 1)(x^2 + x + 1)}{x - 1}$
$= x^2 + x + 1$

**(c)** $y = x^2 + x + 1$

## Section 10.4 (pp. 784–792)

### Exploration 1

**1.** 1.364075504

**2.** 1.375114667; yes

**3.** $\displaystyle\int_0^\pi \sin x \, dx$; sum(seq(sin(0 + K*$\pi$/50)*$\pi$/50, K, 1, 50)) = 1.999341983; fnInt(sin(X), X, 0, $\pi$) = 2

**4.** sum(seq($\sqrt{\ }$ (4 + K*5/50)*50/50, K, 1, 50))
= 12.71659722; fnInt($\sqrt{\ }$ (X), X, 4, 9)
= 12.66666667

## Quick Review 10.4

**7.** $\approx 0.5403$
**8.** $\approx 4.000$
**9.** $\approx 1.000$
**10.** $\approx 2.7183$

## Exercises 10.4

**7.** $\approx 1.0000$
**8.** $\approx 2.0000$
**9.** $\approx -3.0000$
**10.** $\approx -3.0000$
**18.** $\approx 0.69315$
**21. (b)**

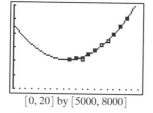

[−1, 6] by [0, 550]

**(c)** $s(t) = -16.08t^2 + 0.36t + 499.77$
**22. (a)** 1990–1991: −57 billion dollars per year;
1996–1997: 275 billion dollars per year
**(b)** $y \approx 16.894x^2 - 288.603x + 7285.945$

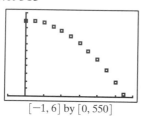

[0, 20] by [5000, 8000]

**(c)** 1993: 150.64 billion dollars per year; 1996:
252.00 billion dollars per year
**(d)** 9113 billion dollars
**25. (b)**

| N | LRAM | RRAM | Average |
|---|------|------|---------|
| 10 | 15.04 | 19.84 | 17.44 |
| 20 | 16.16 | 18.56 | 17.36 |
| 50 | 16.86 | 17.82 | 17.34 |
| 100 | 17.09 | 17.57 | 17.33 |

**(c)** fn Int gives 17.33, at $N_{100}$, the average is
17.3344

**26. (b)**

| N | LRAM | RRAM | Average |
|---|------|------|---------|
| 10 | 96.72 | 132.72 | 114.72 |
| 20 | 105.18 | 123.18 | 114.18 |
| 50 | 110.43 | 117.63 | 114.03 |
| 100 | 112.21 | 115.81 | 114.01 |

**(c)** fn Int gives 114, $N_{100}$ is very close at 114.01
**27. (b)**

| N | LRAM | RRAM | Average |
|---|------|------|---------|
| 10 | 7.84 | 11.04 | 9.44 |
| 20 | 8.56 | 10.16 | 9.36 |
| 50 | 9.02 | 9.66 | 9.34 |
| 100 | 9.17 | 9.49 | 9.33 |

**(c)** fn Int gives 9.33, $N_{100}$ has the same result
**28. (b)**

| N | LRAM | RRAM | Average |
|---|------|------|---------|
| 10 | 107.76 | 132.96 | 120.36 |
| 20 | 113.79 | 126.39 | 120.09 |
| 50 | 117.49 | 122.53 | 120.01 |
| 100 | 118.74 | 121.26 | 120 |

**(c)** fn Int gives 120, $N_{100}$ has the same result
**29. (b)**

| N | LRAM | RRAM | Average |
|---|------|------|---------|
| 10 | 98.24 | 112.64 | 105.44 |
| 20 | 101.76 | 108.96 | 105.36 |
| 50 | 103.90 | 106.78 | 105.34 |
| 100 | 104.61 | 106.05 | 105.33 |

**(c)** fn Int gives 105.33, $N_{100}$ has the same result
**30. (b)**

| N | LRAM | RRAM | Average |
|---|------|------|---------|
| 10 | 136.16 | 185.76 | 160.96 |
| 20 | 147.84 | 172.64 | 160.24 |
| 50 | 155.08 | 165.00 | 160.04 |
| 100 | 157.53 | 162.49 | 160.01 |

**(c)** fn Int gives 160, very close to $N_{100}$ of 160.01
**31. (b)**

| N | LRAM | RRAM | Average |
|---|------|------|---------|
| 10 | 7.70 | 8.12 | 7.91 |
| 20 | 7.81 | 8.02 | 7.91 |
| 50 | 7.87 | 7.95 | 7.91 |
| 100 | 7.89 | 7.93 | 7.91 |

**(c)** fn Int gives 7.91, the same result as $N_{100}$
**32. (b)**

| N | LRAM | RRAM | Average |
|---|------|------|---------|
| 10 | 4.51 | 4.81 | 4.66 |
| 20 | 4.59 | 4.74 | 4.67 |
| 50 | 4.64 | 4.70 | 4.67 |
| 100 | 4.65 | 4.68 | 4.67 |

**(c)** fn Int gives 4.67, the same result as $N_{100}$

**33. (b)**

| N | LRAM | RRAM | Average |
|---|------|------|---------|
| 10 | 1.08 | 0.92 | 1 |
| 20 | 1.04 | 0.96 | 1 |
| 50 | 1.02 | 0.98 | 1 |
| 100 | 1.01 | 0.99 | 1 |

**(c)** fn Int gives 1, the same result as $N_{100}$

**34. (b)**

| N | LRAM | RRAM | Average |
|---|------|------|---------|
| 10 | 0.57 | 0.57 | 0.57 |
| 20 | 0.57 | 0.57 | 0.57 |
| 50 | 0.57 | 0.57 | 0.57 |
| 100 | 0.57 | 0.57 | 0.57 |

**(c)** fn Int = 0.57, the same result as $N_{100}$

**35. (b)**

| N | LRAM | RRAM | Average |
|---|------|------|---------|
| 10 | 0.56 | 0.62 | 0.59 |
| 20 | 0.58 | 0.61 | 0.593 |
| 50 | 0.59 | 0.60 | 0.594 |
| 100 | 0.59 | 0.60 | 0.594 |

**(c)** fn Int = 0.594, which is the same as the $N_{100}$ result.

**36. (b)**

| N | LRAM | RRAM | Average |
|---|------|------|---------|
| 10 | 1.17 | 1.03 | 1.10 |
| 20 | 1.13 | 1.07 | 1.10 |
| 50 | 1.11 | 1.09 | 1.10 |
| 100 | 1.11 | 1.09 | 1.10 |

**(c)** fn Int = 1.10, which is the same result as $N_{100}$

**37. (d)** The symmetric method provides a closer approximation to $f'(2) = 11$.

**(e)** $\approx 12.006001$; $12.000001$; symmetric

**38. (a)**

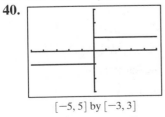

$[0, 1]$ by $[0, 4]$

**(c)** $\approx 2.72$

**(d)** $e \approx 2.72$.

**39.** The value of $f(0 + h)$ and $f(0 - h)$ are the same.

**40.**

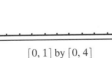

$[-5, 5]$ by $[-3, 3]$

$f'(0)$ does not exist because $f(x)$ is discontinuous at $x = 0$. The calculator gives an incorrect answer because

**41. (b)** $\approx 19.67$

**44. (b)**

| x | A(x) |
|---|------|
| 0.25 | 0.0625 |
| 0.5 | 0.25 |
| 1.1 | 1 |
| 1.5 | 2.25 |
| 2 | 4 |
| 2.5 | 6.25 |
| 3 | 9 |

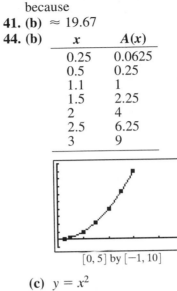

$[0, 5]$ by $[-1, 10]$

**(c)** $y = x^2$

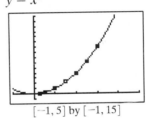

$[-1, 5]$ by $[-1, 15]$

**(d)** The data seems to support a curve of $A(x) = x^2$

**(e)** $A'(x) = 2x$

**45. (b)**

| x | A(x) |
|---|------|
| 0.25 | 0.0156 |
| 0.5 | 0.125 |
| 1 | 1 |
| 1.5 | 3.375 |
| 2 | 8 |
| 2.5 | 15.625 |
| 3 | 27 |

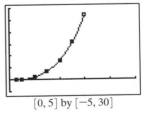

$[0, 5]$ by $[-5, 30]$

**(c)** $f(x) \approx x^3$

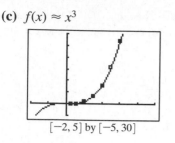

$[-2, 5]$ by $[-5, 30]$

**(d)** The exact value of $A(x)$ for any $x$ greater than zero appears to be $x^3$.

**(e)** $A'(x) = 3x^2$

## Chapter 10 Review (pp. 794–795)

**1. (a)** 2
   **(b)** Does not exist.
**2. (a)** $-1$
   **(b)** Does not exist.
**3. (a)** 2
   **(b)** 2
**4. (a)** 2
   **(b)** 2
**19.** $f$ has vertical asymptotes at $x = -1$ and $x = -5$; $f$ has a horizontal asymptote at $y = 0$.
**20.** $f$ has a vertical asymptote at $x = 2$; $f$ has no horizontal asymptotes.

**27.** $y = \begin{cases} \dfrac{x^3 - 1}{x - 1} & x \neq 1 \\ 3 & x = 1 \end{cases}$

**28.** $y = \begin{cases} \dfrac{x^2 - 6x + 5}{x - 5} & x \neq 5 \\ 4 & x = 5 \end{cases}$

**32. (a)** $\approx -0.12$
   **(b)** $-\dfrac{3}{25}$

**34. (a)** $\dfrac{\sqrt{3}}{6}$
   **(b)** $y = \dfrac{\sqrt{3}}{6}x - \dfrac{\sqrt{3}}{6}$

**35.** $10x + 7$
**36.** $6x - 8$
**37.** LRAM: 42.2976; RRAM: 40.3776; 41.3376
**38.** LRAM: 49.2352; RRAM: 52.1152; 50.6752
**39. (a)**

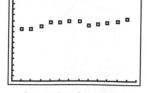

$[0, -13]$ by $[0, 15\,000]$

**(b)** April to May: -210; June to July: -175
**(c)** March to April
**(d)** July to August
**(e)** $y \approx 137.8147x + 9531.2879$

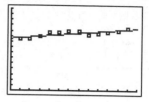

$[0, -13]$ by $[0, 15\,000]$

**(f)** $y \approx 8.5723x^3 - 178.0470x^2 + 1183.7304x + 8030.9091$

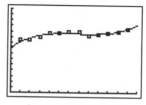

$[0, 13]$ by $[0, 15\,000]$

**(g)** March: 346.9; June: $-27.0$; August: $-19.1$; December: 613.8
**(h)** 52,388; Does not seem reasonable.

**40 (a)**

| $x$ | $A(x)$ |
|---|---|
| 0 | 0 |
| 0.4 | 0.38942 |
| 0.8 | 0.71736 |
| 1.2 | 0.93204 |
| 1.6 | 0.99957 |
| 2.0 | 0.90930 |
| 2.4 | 0.67546 |
| 2.8 | 0.33499 |
| 3.2 | -0.05837 |
| 3.6 | -0.44252 |
| 4.0 | -0.75680 |
| 4.4 | -0.95160 |
| 4.8 | -0.99616 |
| 5.2 | -0.88345 |
| 5.6 | -0.63127 |
| 6.0 | -0.27942 |
| 6.4 | 0.11655 |

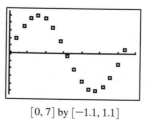

$[0, 7]$ by $[-1.1, 1.1]$

**(b)** $f(x) = \sin x$

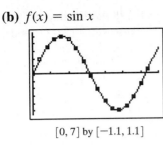

$[0, 7]$ by $[-1.1, 1.1]$

**(c)** $f'(x) = \cos x$, the function being integrated.

## Chapter 10 Project

**1.**

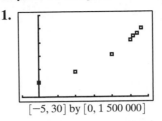

$[-5, 30]$ by $[0, 1\,500\,000]$

**2.** 34,743 people/year; 43,571 people/year; 57,677 people/year; 63,193 people/year

**3.** $y \approx 270,565 \cdot 1.05566^x$

**4.** 63,263 people/year; matches 1996–1998 growth rate closely

**5.** model predictions: 1,374,000; 2,362,000; 4,060,000; 6,978,000; web site predictions: 1,355,368; 1,874,431; 2,178,006; 2,408,197

# Appendix A

## Appendix A.1 (pp. 797–801)

**2.** 3 or $-3$

**5.** $\dfrac{4}{3}$ or $-\dfrac{4}{3}$

**6.** $-\dfrac{3}{2}$

**8.** no real number

**20.** $\dfrac{1}{81}$

**50.** $\dfrac{1}{\sqrt[4]{x^3y^3}}$

**59.** $3a^2b^2$ $(b \geq 0)$

**60.** $\dfrac{x^3}{y^4}$ $(x \geq 0)$

**64.** $\dfrac{6}{x^{1/6}y^{7/6}}$

**67.** $\sqrt[4]{\dfrac{3}{8}}\,\sqrt{|x^3y|}$

**68.** $\sqrt[5]{\dfrac{4}{9}}\,x^{3/5}y^{1/5}$

**70.** $3b\sqrt[5]{a^3}$

**73.** $(x - 2|y|)\sqrt{x}$

**74.** $(3|x| + y)\sqrt{2y}$

**75.** $<$

**76.** $>$

**79.** $>$

**80.** $=$

**82.** $<$

**84.** $4.5\sqrt{2} \approx 6.36$ sec

**85.** If $n$ is even, then there are two real $n$th roots of $a$ (when $a > 0$): $\sqrt[n]{a}$ and $-\sqrt[n]{a}$.

## Appendix A.2 (pp. 802–808)

**1.** $3x^2 + 2x - 1$; degree 2

**2.** $-2x^3 + x^2 - 2x + 1$; degree 3

**4.** $-x^4 + x^2 + x - 3$; degree 4

**13.** $2x^3 - 2x^2 + 6x$

**14.** $2y^4 + 3y^3 - 4y^2$

**16.** $12v^4 - 8v$

**17.** $-15x^3 - 5x^2 + 10x$

**18.** $2x^5 - 2x^3 + 2x$

**20.** $8x^2 + 14x + 3$

**21.** $3x^2 + x - 10$

**22.** $4x^2 - 9$

**24.** $25x^2 - 30x + 9$

**25.** $9x^2 + 24xy + 16y^2$

**26.** $x^3 - 3x^2 + 3x - 1$

**27.** $8u^3 - 12u^2v + 6uv^2 - v^3$

**28.** $u^3 + 9u^2v + 27uv^2 + 27v^3$

**29.** $4x^6 - 9y^2$

**30.** $25x^6 - 10x^3 + 1$

**31.** $x^3 + 2x^2 - 5x + 12$

**32.** $x^3 - 11x + 6$

**36.** $x - y$, $x \geq 0$ and $y \geq 0$

**37.** $u - v$, $u \geq 0$ and $v \geq 0$

**38.** $x^4 - 3$

**40.** $x^3 + 1$

**43.** $yz(z^2 - 3z + 2)$

**44.** $(x + 3)(2x - 5)$

**46.** $(3y + 4)(3y - 4)$

**48.** $(6 + x)(2 - x)$

**50.** $(6y + 1)^2$

**54.** $(z + 4)(z^2 - 4z + 16)$

**55.** $(3y - 2)(9y^2 + 6y + 4)$

**56.** $(4z + 3)(16z^2 - 12z + 9)$

**57.** $(1 - x)(x^2 + x + 1)$

**58.** $(3 - y)(y^2 + 3y + 9)$

**60.** $(y - 5)(y - 6)$

**62.** $(2t + 1)(3t + 1)$

**63.** $(2u - 5)(7u + 1)$

**64.** $(2v + 3)(5v + 4)$

**65.** $(3x + 5)(4x - 3)$

**66.** $(x - y)(2x - 7)$

**67.** $(2x + 5y)(3x - 2y)$

**68.** $(3x + 7y)(5x - 2y)$

**69.** $(x - 4)(x^2 + 5)$

**70.** $(2x - 3)(x^2 + 1)$

**71.** $(x^2 - 3)(x^4 + 1)$

**72.** $(x^2 + 2)(x^4 + 1)$

**76.** $y(2y - 5)^2$

**77.** $2y(3y + 4)^2$

**78.** $2x(x - 1)(x - 7)$

**80.** $3x(x + 2)(x^2 - 2x + 4)$

**81.** $y(y + 1)(5 - 2y)$

**82.** $z(1 - 2z)(4z^2 + 2z + 1)$

**83.** $2(5x + 4)(5x - 2)$

**84.** $5(2x - 1)(2x - 5)$

**85.** $2(2x + 5)(3x - 2)$

**86.** $(x + 5y)(3x - 2y)$

**87.** $(2a - b)(c + 2d)$

**88.** $(3a + 2b)(2c - d)$

**89.** $(x - 3)(x + 2)(x - 2)$

**90.** $x(x - 4)(x + 1)(x - 1)$

**91.** $(2ac + bc) - (2ad + bd) = c(2a + b)$
$- d(2a + b) = (2a + b)(c - d)$

## Appendix A.3 (pp. 808–813)

**9.** all real numbers

**10.** all real numbers

**12.** $(-3, \infty)$

**14.** $x \neq -2$ and $x \neq 2$

**15.** $x \neq 2$ and $x \neq 1$

**16.** $x \neq 2$ and $x \neq 0$

**23.** $x^2 + 7x + 12$

**24.** $x^2 + 8x + 15$

**26.** $x^3 + 4x^2 - 3x - 18$

**27.** $(x - 2)(x + 7)$ cancels out during simplification; the restriction indicates that the values 2 and $-7$ were not valid in the original expression.

**28.** $(x + 1)(x - 2)$ cancels out during simplification; the restriction indicates that the values $-1$ and 2 were not valid in the original expression.

**29.** No factors were removed from the expresson; we can see by inspection that $\frac{2}{3}$ and 5 are not valid.

**30.** $x$ cancels out during simplification; the restriction indicates that 0 was not valid in the original expression.

**31.** $(x - 3)$ ends up in the numerator of the simplified expression; the restriction reminds us that it began in the denominator, so that 3 is not allowed.

**32.** When $a = b$ in the original, we get division by 0; this is not apparent in the simplified expression because we canceled a factor of $b - a$.

**38.** $\dfrac{x + 3}{x - 4}, x \neq -3$

**40.** $\dfrac{y(y - 3)}{(y - 7)}, y \neq -7$

**41.** $\dfrac{4z^2 + 2z + 1}{z + 3}, z \neq \dfrac{1}{2}$

**42.** $\dfrac{2z}{z - 3}$

**43.** $\dfrac{x^2 - 3}{x^2}, x \neq -2$

**44.** $\dfrac{y}{y^2 - 5}, y \neq -3$

**46.** $1, x \neq -3$

**47.** $-\dfrac{1}{x - 3}, x \neq 1$ and $x \neq -3$

**48.** $12y, x \neq 0, y \neq 0$, and $x \neq \dfrac{1}{6}$

**49.** $\dfrac{2(x - 1)}{x}$

**50.** $\dfrac{1}{y}, y \neq -2$ and $y \neq 2$

**51.** $\dfrac{1}{y}, y \neq 5, y \neq -5$, and $y \neq \dfrac{1}{2}$

**52.** $\dfrac{y(y + 4)}{y - 1}, y \neq -4$ and $y \neq -\dfrac{2}{3}$

**55.** $\dfrac{3(x - 3)}{28}, x \neq 0$ and $y \neq 0$

**56.** $\dfrac{3}{8}, x \neq y$ and $y \neq 0$

**57.** $\dfrac{x}{4(x - 3)}, x \neq 0$ and $y \neq 0$

**58.** $-2x, x \neq 0, y \neq 0, x \neq y, x \neq -y$

**61.** $\dfrac{1}{3 - x}, x \neq 0$ and $x \neq -3$

**62.** $-\dfrac{2x + 5}{x^2 + 5x + 6}, x \neq 2$

**63.** $\dfrac{x^2 + xy + y^2}{x + y}$, $x \neq y$, $x \neq 0$, and $y \neq 0$

**64.** $\dfrac{xy}{y - x}$, $x \neq 0$, $y \neq 0$, and $x \neq -y$

**65.** $\dfrac{x + 3}{x - 3}$, $x \neq 4$ and $x \neq \dfrac{1}{2}$

**66.** $\dfrac{x - 3}{x + 5}$, $x \neq 3$ and $x \neq \dfrac{3}{2}$

**67.** $-\dfrac{2x + h}{x^2(x + h)^2}$, $h \neq 0$

**68.** $\dfrac{2}{(x + h + 2)(x + 2)}$, $h \neq 0$

**69.** $a + b$, $a \neq 0$, $b \neq 0$, and $a \neq b$

**70.** $\dfrac{1}{b - a}$, $a \neq 0$, $b \neq 0$, and $a \neq -b$

**71.** $\dfrac{1}{xy}$, $x \neq -y$

**72.** $\dfrac{x - y}{x + y}$, $x \neq y$

**74.** $\dfrac{xy}{y + x}$, $x \neq 0$ and $y \neq 0$

# APPLICATIONS INDEX

# INDEX